MAMMAL
SPECIES
OF THE
WORLD

MAMMAL SPECIES OF THE WORLD

A Taxonomic and Geographic Reference

THIRD EDITION
Volume 2

Edited by Don E. Wilson
and DeeAnn M. Reeder

The Johns Hopkins University Press
Baltimore

The editors assume full responsibility for the contents and form
of volumes 1 and 2.

© 2005 The Johns Hopkins University Press
All rights reserved. Published 2005
Printed in the United States of America on acid-free paper
9 8 7 6 5 4 3 2 1

The Johns Hopkins University Press
2715 North Charles Street
Baltimore, Maryland 21218-4363
www.press.jhu.edu

Library of Congress Cataloging-in-Publication Data

Mammal species of the world : a taxonomic and geographic
reference / edited by Don E. Wilson and DeeAnn M. Reeder. —
3rd ed.
    p. cm.
  Includes bibliographical references and index.
  ISBN 0-8018-8221-4 (hardcover set (vols. 1 and 2) : alk. paper)
   1. Mammals—Classification. 2. Mammals—Geographical
distribution. I. Wilson, Don E. II. Reeder, DeeAnn M.
QL708.M35 2005
599′.012—dc22   2005001870

A catalog record for this book is available from the British Library.

In memory of
Peter Cannell, Charles O. Handley, Jr., and Karl F. Koopman

Contents

Volume 1

Volume 2

Museum Abbreviations

AM	Australian Museum, Sydney
AMNH	American Museum of Natural History, New York
ANSP	Philadelphia Academy of Natural Sciences, Philadelphia
BBM	Bernice P. Bishop Museum, Hawaii
BMNH	The Natural History Museum, London
CIB	Centro de Investigaciones Biológicas del Noroeste, S. C., Mexico
CM	Carnegie Museum of Natural History, Pittsburgh
CMNH	Cincinnati Museum of Natural History
FMNH	Field Museum of Natural History, Chicago
HUNHM	Hokkaido University Natural History Museum, Hokkaido
IEBR	Institute of Ecology and Biological Resources, Hanoi
IRSNB	Institute de Royale de Sciences Naturelles Belgique, Brussels
IZAC	Institute of Zoology, Academia Sinica (Chinese Academy of Sciences), Beijing
KNMB	Koninklijk Belgisch Instituut voor Natuurwetenschappen, Brussels
MCZ	Museum of Comparative Zoology, Harvard University, Cambridge
MNHN	Museum National d'Histoire Naturelle, Paris
MRAC	Musée royal de l'Afrique centrale, Tervuren
MVZ	Museum of Vertebrate Zoology, University of California, Berkeley
MZB	Museum Zoologicum Bogoriense, Bogor (now Balai penelitian Zoologi-Pusat Penelitian Biologi-LIPI)
NMW	Naturhistorisches Museum Wien, Wien
OMNH	Oklahoma Museum of Natural History, University of Oklahoma
RMBR	Raffles Museum of Biodiversity Research, Department of Biological Sciences, National University of Singapore, Singapore
RMNH	Rijksmuseum van Natuurlijke Historie, Leiden (now National Natuurhistorisch Museum, Leiden)
ROM	Royal Ontario Museum, Toronto
TCWC	Texas Cooperative Wildlife Collection, Texas A&M University
UCA	Universitair Centrum, Antwerpen
UMMZ	University of Michigan Museum of Zoology, Ann Arbor
USNM	United States National Museum of Natural History, Smithsonian Institution
ZFMK	Zoologisches Forschungsinstitut und Museum Alexander Koenig, Bonn
ZMA	Zoological Museum of the University of Amsterdam, Amsterdam
ZMB	Zoologisches Museum und Institut für Spezielle Zoologie, Museum für Naturkunde der Humboldt-Universität zu Berlin, Berlin
ZMO	Zoological Museum, Oslo
ZMVNU	Zoological Museum, Vietnam National University, Hanoi
ZSI	Zoological Survey of India Collections, Calcutta

CHECKLIST OF MAMMAL SPECIES OF THE WORLD
(CLASS MAMMALIA)

ORDER RODENTIA

by Michael D. Carleton and Guy G. Musser

Order Rodentia Bowdich, 1821:7.

COMMENTS: Rodentia is the largest order of living Mammalia, encompassing 2277 species as recognized herein, or approximately 42% of worldwide mammalian biodiversity. Following the mid-1900s era of uncritical application of the biological species concept and consequent obscuration of species richness, views on the size of the Order continue to appreciate substantially (1591 species—Corbet and Hill, 1980; 1719—Honacki et al., 1982; 1738—Corbet and Hill, 1986; 2015—Wilson and Reeder, 1993; 2277—this volume). Thus, "a checklist of species" considered valid is an appropriate taxonomic focus in the current work, as emphasized by its title and that of its predecessors (Honacki et al., 1982; Wilson and Reeder, 1993). In contrast to the period covered by the second edition (Wilson and Reeder, 1993), however, systematic research on Rodentia since 1993 has been equally as prolific, multifaceted, and informative at taxonomic levels above the species and requires introductory comment. Issues of monophyly, phyletic relationship, and corresponding classification at the genus- and family-group ranks are addressed to varying depths in the following chapters. This brief foreword offers remarks on the Order as a whole, particularly evidence that bears on its monophyly and on suprafamilial relationships. Although 20 years old, the 1984 Paris symposium on rodent evolutionary relationships, edited by Luckett and Hartenberger (1985a), today remains an excellent and invaluable primer to these same issues.

Rodent monophyly and ordinal relationships.—The evolution of living rodents from some Paleocene common ancestor and their acceptance as a monophyletic order of Mammalia have not been seriously challenged until recently. The provocative inquiry "Is the guinea-pig a rodent?" (Graur et al., 1991) leapt precipitously to the conclusions that "The guinea-pig is not a rodent" (D'Erchia et al., 1996), that, by extension, hystricognaths represent a separate mammalian order, and that Rodentia as conventionally viewed is polyphyletic (also Frye and Hedges, 1995; Graur et al., 1992; Li et al., 1992). Attention in molecular investigations just as swiftly turned to, foremost, the adequacy of taxon sampling persuant to the taxonomic question posed, to the appropriateness of models that account among-site rate heterogeneity, and to a preference for nuclear over mitchondrial genes (and for multiple over single) in illuminating deeper divisions of mammalian phylogenesis (e.g., see Adkins et al., 2001; Cao et al., 1997; Corneli, 2002; Y.-H. Lin et al., 2002; Luckett and Hartenberger, 1993; Philippe, 1997; Sullivan and Swofford, 1997). The subsequent wave of sleeves-up research, characterized by wider sampling of taxa and genes, has made the case for rodent monophyly vastly more secure (Adkins et al., 2001, 2003; Amrine-Madsen et al., 2003; Debry, 2003; Debry and Sagel, 2001; Huchon et al., 1999, 2002; Y.-H. Lin et al., 2002; Nedbal et al., 1996; Waddell and Shelley, 2003). And in those molecular studies that have focused on hystricognaths in general or cavioids in particular (Huchon and Douzery, 2001; Kramerov and Vassetzky, 2001; Lara et al., 1996; Leite and Patton, 2002; Marivaux et al., 2002; Mouchaty et al., 2001; Rowe and Honeycutt, 2002; and see references under Hystricognathi), the guinea pig nests unremarkably within the kinship web so long predicated by so many past studies and classifications.

An implicit but faulty premise in these early molecular studies perhaps engendered the over-eager acceptance of rodent polyphyly: that is, the evolution of large, chisel-like incisors in other mammalian groups (e.g., the primate *Daubentonia* and diprotodont marsupials) underscores the likelihood for the independent acquisition of enlarged incisors in other distantly related lineages, including and especially those that had been mistakenly lumped under Rodentia based on this single, homoplasious trait. The point is that the hypertrophied incisors in Rodentia represent retained deciduous second incisors (dI2/2), whereas the gliriform teeth in those other orders consist of permanent first incisors (I1/1). The embryogenesis of rodent incisors from the dI2 was appreciated by the late 1800s and constitutes a persuasive synapomorphy for the Order (see Landry, 1999, Luckett, 1985, and references therein). To demonstrate that the adult incisors in one or more other rodent groups develop from an anlage other than the dI2 would more severely cripple the case for monophyly of Rodentia. The morphological character information historically used to argue the common ancestry of rodents is wonderfully richer than this single feature, however, as anyone who peruses the plates of Tullberg's (1899) monograph will readily appreciate. See Carleton (1984) for general morphological characterization of the Order; Hartenberger (1985), Landry (1999), and Luckett and Hartenberger (1993) provide phylogenetic interpretation of derived characters and integrated character suites that support monophyly of Rodentia. In retrospect, we are fortunate that Tullberg based his phylogenetic study on more than a single species each of *Cavia*, *Mus*, and *Rattus*.

Rodentia and Lagomorpha have long been advocated as sister taxa, with the two groups initially placed in the same order, Rodentia or Glires, usually as the suborders Simplicidentata and Duplicidentata (Alston, 1876; Brandt, 1855; Gregory, 1910; Thomas, 1896; Tullberg, 1899). After elevation to separate orders (Gidley, 1912; Miller and Gidley, 1918), indication of cognate affinity was still conveyed, e.g., as the Cohort or Superorder Glires (Landry, 1974, 1999; Luckett and Hartenberger, 1985*b*; Simpson, 1945). For some time, a paleontological foundation for their supposed shared ancestry proved disconcertingly elusive (Gregory, 1910; Simpson, 1945, 1961), but such hard fossil evidence has proliferated rapidly in the later 1900s (Bryant and McKenna, 1995; Dashzeveg et al., 1998; Li and Ting, 1985, 1993; C.-K. Li et al., 1987). The current picture of earliest gliriform radiation and classification is complex, and the cladistic position of various groups, notably eurymylids in the broad sense, as basal to or subtended by the crown-clade Rodentia has been variously interpreted (see Meng and Wyss, 2001, and Meng et al., 2003, for phylogenetic analysis, review, and discussion). Sister-group stature of Lagomorpha and Rodentia is decisively sustained in modern phylogenetic studies, both those drawing on morphological (Ade, 1999; Frahnert, 1999; Giere et al., 1999; Landry, 1999; Luckett and Hartenberger, 1993; Martin, 1999; Meng and Wyss, 2001; Meng et al., 2003; Mess, 1999; Novacek, 1985, 1986; Novacek and Wyss, 1986; Novacek et al., 1988; Shoshani and McKenna, 1998) and molecular data sources (Amrine-Madsen et al., 2003; Delsuc et al., 2002; Eizirik et al. 2001; Huchon et al., 2002; Y.-H. Lin et al., 2002; Murphy et al., 2001*a*, *b*; Waddell and Shelley, 2003). Other orders besides Lagomorpha have been identified as closely related to Rodentia, in particular Primates (Patterson and Wood, 1982; Wood, 1962). Such proposals merit some vindication, albeit phyletically more remote, in the emerging recognition of the superorder Euarchontoglires (Dermoptera-Scandentia-Primates + Lagomorpha-Rodentia—Archibald, 2003; Delsuc et al., 2002; Helgen, 2003*a*; Huchon et al., 2002; Y.-H. Lin et al., 2002; Murphy et al., 2001*a*, *b*; Scally et al., 2001; Waddell and Shelley, 2003). Zalambdalestids from the late Cretaceous of Asia have also been claimed as sister-group to Rodentia-Lagomorpha (Archibald, 2003; Archibald et al., 2001), but their close link is unsupported by analyses employing a greater range of relevant taxa and characters (Ji et al., 2002; Meng et al., 2003; Wible et al., 2004).

If Eurymylidae, including *Heomys*, is considered a family of Rodentia (e.g., Li and Ting, 1993), then the geological range of the Order extends from the early-middle Paleocene. If not (e.g., Meng and Wyss, 2001; Wyss and Meng, 1996), then several ischyromyid genera from the late Paleocene appear to be the earliest true rodents yet recorded (Hartenberger, 1998; McKenna and Bell, 1997). Explosive diversification into principal lineages (suborders) transpired in the early Eocene, and examples of most modern families are encountered by the middle to late Oligocene (Hartenberger, 1996, 1998; Wood, 1959). Based on the earliest fossils so far discovered and their cladistically basal stature within Glires or Rodentia, "Asia" (a vast and diverse region to be sure) is presently accepted as the area of origin for the Order (Beard, 1998; Bryant and McKenna, 1995; Hartenberger, 1996; Meng and Wyss, 2001).

Rodent suborders?—The now classical and familiar rodent suborders—Sciuromorpha, Myomorpha, and Hystricomorpha—issue from Brandt (1855), who eponomously based his names on Waterhouse's (1839) characterizations of sciuromorphous, myomorphous, and hystricomorphous zygomasseteric morphologies. Thereafter, the question of rodent suborders has become a century and a half long work-in-progress. The most important conceptual landmark since Brandt is Tullberg's (1899) *Eine Phylogenetische Studie*, in which he integrated mandibular conformation (sciurognathy versus hystricognathy) and Brandtian zygomasseteric criteria to delineate two major rodent groups (each called a Tribus), the Sciurognathi (Sciuromorphi and Myomorphi) and Hystricognathi (Bathyergomorphi and Hystricomorphi). Subsequently, most mammalian classifiers and rodent systematists have adopted either the Brantian tri-subordinal (Alston, 1876; Lavocat, 1978; Miller and Kellogg, 1955; Simpson, 1945; Wilson, 1949; Wood, 1965) or Tullbergian dual subordinal themes (Chaline and Mein, 1979; Ellerman, 1940; Landry, 1999; Lavocat, 1973; Patterson and Wood, 1982; Woods, 1972), some in more or less their orthodox forms and others with due amendment of lesser ranks and reallocation of taxa among them. Other variations have recognized more primary divisions within Rodentia, whether or not called a suborder per se, anywhere from five to 16 (Hartenberger, 1985, 1998; McKenna and Bell, 1997; Miller and Gidley, 1918; Thaler, 1966; Thomas, 1896; Weber, 1904; Wood, 1955, 1959). Within each of these classifications, problematic and-or poorly understood groups have been regularly acknowledged through the time-honored qualifiers of '*incertae sedis*' or 'suborder indeterminate,' some to an extent that little sense of higher-level relationship within the Order is conveyed (e.g., Carleton, 1984). See Carleton (1984), Ellerman (1940), Hartenberger (1985), Landry (1999), Simpson (1945), Wilson (1949), and Wood (1955, 1959, 1965) for additional review and commentary on rodent classifications, in particular the historical treatment of rodent suborders.

Compared with the vacillation and disagreement over suborders, the number of families of living rodents considered valid has remained fairly stable over the past half-century, around 30 to 35 (Anderson and Jones, 1967, 1984; Corbet and Hill, 1986; Grassé and De-keyser, 1955; Hartenberger, 1985; Honacki et al., 1982; Simpson, 1945; Wilson and Reeder, 1993). Variation among these, and other, works pivots on the authors' decisions concerning the rank of certain groups as subfamily versus family. Thirty-three families are acknowledged herein (Table 1), and recent molecular investigations have generally sustained the monophyletic status of most of these, where the taxonomic sampling is suitably robust and critically focused (see various family-group Comments). Moreover, certain kinship hypotheses linking these families recur in the proliferation of gene-sequencing investigations over the past decade. The stability of these genetic clades, superposed upon the morphological evidence mustered and classifications promulgated over the past 150 years, lends some empirical confidence that agreement about rodent suborders will be eventually realized, perhaps in less than a decade. Five family-group phyletic associations are briefly reviewed below as a basis for the provisional subordinal groupings observed here.

(1). Aplodontidae-Sciuridae + Gliridae: Aplodontids and sciurids form the core of Sciuromorpha, as reflected in classifications generated from the late 1800s to the present (e.g., Alston, 1876; Hartenberg, 1998; Landry, 1999; McKenna and Bell, 1997; Simpson, 1945; Thomas, 1896; Tullberg, 1899). Their common evolutionary origin has been historically postulated (Wilson, 1949), and this phyletic link has been reinforced by a suite of morphological traits (Landry, 1999; Luckett and Hartenberger, 1985*b*; Tullberg, 1896, 1899), in particular craniodental features (Vianey-Liaud, 1985), cranial foramina (Wahlert, 1985), middle ear anatomy (Lavocat and Parent, 1985; Meng, 1990), and fetal membrane development (Luckett, 1985). That Aplodontiidae and Sciuridae are each others closest relative (among living rodents) is also overwhelmingly indicated by albumin immunology (Sarich, 1985) and by phylogenetic comparisons of several mitochondrial and nuclear genes (Adkins et al., 2001, 2003; DeBry, 2003; DeBry and Sagel, 2001; Huchon et al., 1999, 2000, 2002; Montgelard et al., 2002; Nedbal et al., 1996; Waddell and Shelley, 2003).

Association of Gliridae with aplodontids and sciurids is not novel, but their union in an expanded clade (Sciuroidea or Sciuromorpha) has recently garnered additional confidence. Most classifications and phylogenetic reconstructions have allied dormice with Myomorpha because the zygomasseteric structure of extant glirids (excluding graphiurines) was characterized as myomorphous (Alston, 1876; Chaline and Mein, 1979; Ellerman, 1940; Landry, 1999; Simpson; 1945; Thomas, 1896; Wahlert et al., 1993; Wilson, 1949; Wood, 1965). However, the "myomorphy" common to most living dormice is now recognized as convergent to the zygomasseteric configuration in true Myomorpha (Dipodidae-Muroidea): i.e., "pseudomyomorphy" as per Maier et al. (2002) and Vianey-Liaud (1985) or "pseudosciuromorphy" as per Landry (1999).

Exclusion of dormice from Myomorpha is further bolstered by certain morphological traits, especially middle ear anatomy (Lavocat and Parent, 1985; Meng, 1990) and the cephalic arterial pattern (Bugge, 1971a, 1985), and notably by the rich fossil record that traces glirids back to the early Eocene (Hartenberger, 1971, 1994, 1998; Vianey-Liaud, 1974, 1985, 1989, 1994; Vianey-Liaud et al., 1994; Vianey-Liaud and Jaeger, 1996). These inquiries identify glirids as sciuromorph rodents, not myomorphs. Sequence analyses based on both mitochondrial and nuclear genes similarly represent dormice as sister-group to Sciuridae or to a sciurid-aplodontiid clade (Adkins et al., 2001, 2003; Bentz and Montgelard, 1999; De-Bry and Sagel, 2001; Huchon et al., 1999, 2002; Kramerov, 1999; Kramerov and Vassetzky, 2001; Kramerov et al., 1999; Y.-H. Lin et al., 2002; Montgelard et al., 2001, 2002; Murphy et al., 2001a; Nedbal et al., 1996; Reyes et al., 1998). See Gliridae for expansion.

(2). Castoridae + Heteromyidae-Geomyidae: Castorids have been traditionally allied with groups placed in Sciuromorpha or Sciurognathi, whether in a Brantian or Tullberg-ian scheme (Alston, 1876; Chaline and Mein, 1979; Ellerman, 1940; Hartenberger, 1985, 1998; Landry, 1999; McKenna and Bell, 1997; Miller and Gidley, 1918; Miller and Kellogg, 1955; Simpson, 1945; Thomas, 1896; Wilson, 1949). However, their sciuromorphous zygomasseteric anatomy, sciurognathus mandible, primitive cranial morphology but highly derived dentition, and deep and rich fossil history beginning in the late Eocene of North America (Korth, 2001) have spawned other views that conflict with the traditional arrangement. Some have divorced beavers from any close phyletic link with sciurids (Schaub, 1953; Wood, 1955), or frustratingly viewed their affinities as intractable (Bugge, 1985:347—"one of the most isolated groups, near to no other Recent family"; Lavocat and Parent, 1985; Wood, 1959, 1965). Meng (1990:27) also emphasized the uncertain kinship of castorids and instead speculated that they are "derived from an ancestral stock giving rise to muroids than [sic] to sciurids."

Origin of Geomyoidea (Heteromyidae-Geomyidae) is documented by fossils from the North American Eocene (McKenna and Bell, 1997), and the subsequent evolutionary diversification of the group has unfolded in the New World. Although Thaler (1966) isolated geomyoids as a separate suborder, the Geomorpha, their sciuromorphous and sciurognathus crania have influenced many researchers to retain them within Sciuromorpha (e. g., Ellerman, 1940; Hartenberger, 1998; Miller and Gidley, 1918; Simpson, 1945), and others to suggest, based on different morphological traits, a closer phylogenetic link with myomorph rodents (Alston, 1876; Hill, 1937; Landry, 1999; McKenna and Bell, 1997; Wahlert, 1985; Wilson, 1949; Wood, 1955). Nevertheless, affiliation with myomorphs, cautioned Fahlbusch (1985:627), is not strongly supported by morphology, and as "long as zygomasseteric structure is not replaced by a more reliable feature in classification, there is scarcely any reason to transfer the Geomyoidea from Sciuromorpha to Myomorpha." Recent phylogenetic analyses of mitochondrial and nuclear gene sequences regularly disclose a castorid-geomyoid clade (Adkins et al., 2001, 2003; DeBry, 2003; Huchon et al., 2002; Montgelard et al., 2002; Murphy et al., 2001a; Waddell and Shelley, 2003). These results, while surprising in view of the usual subordinal separation of castorids (Sciuromorpha) and geomyoids (Myomorpha), mirror the evolutionary tree developed by Tullberg (1899:481), who more than a century ago depicted the common ancestry of "Castoroidei" (fossil and extant beavers) and "Geomyoidei" (Geomys, Heteromys, Perognathus, Dipodomys) within his Sciuromorphi.

Molecular data generally support inclusion of the castorid-geomyoid clade in a larger group that variously contains muroids, dipodoids, anomalurids, and Pedetes (e. g., Adkins et al., 2003; DeBry, 2003; Eizirik et al., 2001; Huchon et al., 2002; Montgelard et al., 2002). Integrity of such an assemblage requires further testing, as urged especially by Adkins et al. (2003). They considered this large clade to be poorly resolved based on their sampled genes and suggested that "the base of the rodent phylogeny may be resolved by a much larger amount of sequence data or a unique class of molecular characters." Further paleontological study may also clarify the reputed alliance between the castorid-geomyoid and dipodoid-muroid (Myodonta) clusters. For example, North American Eocene Sciuravidae has been suggested as ancestral to both geomyoids (Vianey-Liaud, 1985; Wilson, 1949) and myodonts (see Dipodidae for discussion and references), and Walton (1993) identified a basic occlusal pattern shared by some Eocene Eutypomyidae (extinct sister-group to Castoridae—Korth, 2001; Wahlert, 1977), Eocene Eomyidae (extinct sister-family to geomyoids—Falbusch, 1985; Wahlert, 1985), and early sciuravids.

(3). Dipodidae + Muroidea (Platacanthomyidae through Muridae): Close phylogenetic affinity between Dipodoidea and Muroidea has long been recognized, notably as developed by the insightful arguments of Tullberg (1899), Wilson (1949), and Klingener (1964). Schaub (1958) formalized their cognate relationship as Myodonta, used as a suborder or infraorder, and a very substantial body of recent paleontological, morphological, and molecular results also ratifies this phyletic union (see Dipodidae and Muroidea for references).

The clade Dipodidae-Muroidea corresponds to the nucleus of Myomorpha in the usually accepted sense, and their joint kinship with Geomyoidea (Heteromyidae-Geomyidae), and sometimes with Gliridae, has been commonly affirmed by classifications advanced in the middle to late 1900s (e.g., Carleton, 1984; Chaline and Mein, 1979; McKenna and Bell, 1997; Simpson, 1945; Wood, 1955, 1965). The purported relationship to Gliridae finds no support in recent phylogenetic studies, but the evolutionary connection to Geomyoidea receives moderate corroboration, particularly in certain molecular results (see above commentary under 1 and 2).

(4). Anomaluridae + Pedetidae: Along with ctenodactylids, the hystricomorphous and sciurognathus anomalurids and pedetids have been often treated as *incertae sedis*, whether either one or both families, in past classifications (e.g., Carleton, 1984; Chaline and Mein, 1979; Hartenberger, 1998; Simpson, 1945; Wood, 1965). Although recent molecular studies have credited Simpson (1945) for perception of their cognate affinity (e.g., Montgelard et al., 2002), the idea properly dates at least to Winge (1887) and Tullberg (1899), both of whom provided characters and arranged the two groups in the same family or superfamily. This phyletic association has earned recent support based on arterial patterns (Bugge, 1974, 1985; George, 1981), middle ear anatomy (Lavocat and Parent, 1985; Meng, 1990), and mitochondrial genes (Montgelard et al., 2001, 2002); McKenna and Bell (1997) acknowledged their possible sister-group affinity as the Suborder Anomaluromorpha. Not all investigators agree and some specifically dispute any special relationship between Anomaluridae and Pedetidae (e.g., Jaeger, 1988b; Landry, 1999), especially given fundamental microstructural differences in their incisor enamel (T. Martin, 1993, 1995).

Evidence for close relationship to any of the other four suprafamilial groups is similarly mixed. In molecular studies that have sampled only *Anomalurus* or only *Pedetes*, the represented family usually emerges basally in a large clade encompassing Castoridae-Geomyoidea and-or Dipodidae-Muroidea (Adkins et al., 2001, 2003; Debry, 2003; Huchon et al., 2002; Nedbal et al., 1996); in those that have sampled both, using two mitochondrial genes, Anomaluridae-Pedetidae is depicted as sister group to Sciuromorpha-Hystricognathi (Montgelard et al., 2001, 2002). The possible link with Dipodidae-Muroidea recalls the older and broader notion of "Myomorphi" (Ellerman, 1940; Tullberg, 1899). On the other hand, some morphological interpretations have implicated early genealogical ties with Ctenodactylidae-Hystricognathi (Bryant and McKenna, 1995; Meng, 1990); George (1985), Landry (1999), T. Martin (1993, 1995), and Thomas (1896) cautiously endorsed this kinship hypothesis for Pedetidae but not Anomaluridae. This alternative association recalls the formative views of "Hystricomorpha" before the character weighting accorded to mandibular morphology, a feature emphasized by Tullberg (1899).

Although researchers are again probing into the relationships of Anomaluridae and Pedetidae, resolution of their phylogenetic placement has improved little since the summary of Luckett and Hartenberger (1985b). Two interrelated and systematically critical questions remain at issue: do anomalurids and pedetids constitute a natural clade and where does this putative sister-group cladistically fit among the other major rodent associations listed here, in particular Myomorpha versus Hystricomorpha?

(5). Ctenodactylidae + Hystricognathi (Bathyergidae through Heptaxodontidae): The African gundis, Ctenodactylidae, represent another group whose phylogenetic placement has perplexed systematists: early placed in Hystricomorpha (Thomas, 1896); arranged with Sciurognathi after Tullberg (1899) stressed the cardinal importance of sciurognathy and hystricognathy (Chaline et al., 1977; Ellerman, 1940); isolated (among living families) in its own suborder (Hartenberger, 1998; McKenna and Bell, 1997); or left as infraorder or suborder *incertae sedis* (Carleton, 1984; Simpson, 1945; Wood, 1955, 1965). Of these prior interpretations of relationship, or concessions to indeterminate relationship, a substantive and jointly impressive body of information supplies support for Thomas' association of ctenodactylids with Hystricomorpha *sensu lato* (He actually arranged gundis as a subfamily of Octodontidae, a family assignment contraindicated by this same literature). Such corroborative studies broadly draw upon morphology, both individual character systems and multicharacter cladistic analyses (Bryant and McKenna, 1995; Bugge, 1971*b*, 1985; Flynn et al., 1986; George, 1985; Jaeger, 1988*b*; Landry, 1957, 1999; Lavocat and Parent, 1985; Luckett, 1985; Luckett and Hartenberger, 1985*b*; T. Martin, 1993, 1994*a*; Meng, 1990), paleontology (Flynn et al., 1986; Hussain et al., 1978; Marivaux et al., 2002; Marivaux and Welcomme, 2003), and mitochondrial and nuclear DNA sequences (Adkins et al., 2001, 2003; Huchon and Douzery, 2001; Huchon et al., 2000, 2002; Montgelard et al., 2002). Certain of these analyses posit southern Asia as the continental area of origin of the common ancestor of Ctenodactylidae-Hystricognathi (Huchon and Douzery, 2001; Marivaux et al., 2002). Landry (1999) introduced the name Entodacrya (reflecting the internal course of the nasolacrimal duct) to identify the Ctenodactylidae-Hystricognathi clade, and Huchon et al. (2000) soon after coined Ctenohystrica, a contraction whose taxonomic meaning is self evident.

The story of the hystricognathus and hystricomorphous rodents is a suspenseful and exciting read in systematic mammalogy. Ever since Tullberg (1899) defined the Hystricognathi, the explicit union of New and Old World groups, respectively neo- and paleotropical in principal distribution, has sparked lively debate over issues of monophyly and polyphyly, character homology and homoplasy, biogeography and areas of origin, evolutionary and phylogenetic criteria for classification. It would be pretentious to encapsulate the details and nuanced conclusions of these exchanges in a brief foreword (see summaries in Huchon and Douzery, 2001, Jaeger, 1988*b*, Landry, 1999, and Martin, 1993, as well as remarks and references under Hystricognathi). Important bookmarks in the long dialogue include Wood (1950, 1965, 1974, 1980), Landry (1957), Wood and Patterson (1959), Lavocat (1969, 1971, 1974, 1980), Hoffstetter and Lavocat (1970; Hoffstetter, 1975), Woods (1972, 1982), Patterson and Wood (1982), and many thoughtful papers in Luckett and Hartenberger (1985*a*). As summarized by Luckett and Hartenberger (1985*b*:691), a wealth and heterogeneity of data provide "overwhelming support for hystricognath monophyly, and all indicate that it is virtually impossible to distinguish between Old and New World hystricognaths as separate taxa." As fitting denouement to their assessment, ensuing hypothesis-testing using a variety of mitochondrial and nuclear genes has only reinforced the phylogenetic perspective for monophyly of Hystricognathi *sensu* Tullberg (Adkins et al., 2001, 2003; Catzeflis et al., 1995; Huchon and Douzery, 2001; Huchon et al., 1999, 2000, 2002; Murphy et al., 2001*a*; Nedbal et al., 1994, 1996).

Provisional subordinal classification.—Although this systematic compendium is aimed at the species, a higher level classification is an obvious, and natural, means to collate and convey such alpha-level information. With regard to suborders, we provisionally use the names Sciuromorpha, Castorimorpha, Myomorpha, Anomaluromorpha, and Hystricomorpha, respectively, to designate suborders for the five clades reviewed above (Table 1). Based on our impressionistic consensus, that research indicates that Sciuromorpha, Myomopha, and Hystricomorpha, as arranged, receive good to excellent support as monophyletic taxa. The evidence for Castorimorpha and Anomaluromorpha in this regard is overall less persuasive. Future phylogenetic investigation of these groups must address both the question of their naturalness and the possibility that they belong within one of the other three suborders. Ranks between the suborder and family were avoided, *except where* already assigned authorship at certain higher levels (namely the Superfamily Muroidea and Infraorder Hystricognathi) required coordinate groupings for consistency and balance of classificatory arrangement. Proper exposition of the nomenclatural and taxonomic background and of phylogenetic and biogeographic issues that bear on these intermediate ranks (superfamily to infraorder) is beyond the scope of a species checklist (and would certainly overtax the patience of our editors, whose kind indulgence to add this chapter we much appreciate). See McKenna and Bell (1997) for one classificatory view of those ranks, their past usage, and allocation of synonyms.

We retain Brandt's (1855) names for three of the five suborders (Sciuromorpha, Myomorpha, and Hystricomorpha), whereas some foremost rodent authorities have point-edly advised against continued employment of these Brantian epithets, especially at the subordinal rank (e.g., Landry, 1999:283; Wood, 1985). They have argued that the descrip-tive meaning of the terms does not strictly concord with the morphologies of included members, and that this contradiction will only continue to engender confusion. That the logical connotation of a taxon's name should uniformly correspond to its taxonomic in-tention strikes us as puzzling. Not all members of the Superorder Afrotheria live in Africa, nor are all species of the Order Carnivora carnivorous. The species- to family-group ranks are rife with valid names whose logical meaning, as intended by the original descriptor, is partially inconsistent or actually misleading in terms of the currently accepted contents of the taxon, whether in strict morphological accuracy, indication of distribution, or im-plication of phylogenetic alliance. In reviewing 150 years of rodent classifications, we are impressed that few of our predecessors applied the features of infraorbital configuration and jaw shape in an overridingly typological touchstone of all rodent classification. Most made the distinction between, e.g., Myomorpha in its taxic sense and myomorphy as a morphological condition and routinely consulted additional traits. Thus, the hystrico-morphous Dipodidae have been nearly always included within Myomorpha (in retro-spect, correctly it would seem), and the sciuromorphous Geomyoidea have been consid-ered by many to also fit within Myomorpha (in retrospect, perhaps incorrectly). And while Brandt's taxonomic names qua morphological descriptors may not perfectly corre-spond to all members, they nonetheless do conform very well. The core family members of Sciuromorpha, Myomorpha, and Hystricomorpha have remained largely the same, and most, not all, do exhibit those fundamental zygomasseteric structures. More impor-tantly, these core assemblages have survived recent scrutiny using ever more explicit prin-ciples for systematic classification and sophisticated methods for inferring phylogenetic relationship and assessing descent from a common ancestor.

Brandt's names, intended as a suprafamilial rank, do not merit any special consider-ation as far as the International Code of Zoological Nomenclature (Int. Comm. Zool. Nom., 1999) is concerned. Aside from requirements of formal publication and avail-ability, the Code does not prescribe standard suffices to denote relative rank or stipulate strict adherence to priority for taxa recognized above the family-group. Such novel names for recently apprehended clades are beginning to appear within Rodentia (e.g., Bryant and McKenna, 1995; Huchon et al., 2000; Landry, 1999; McKenna and Bell, 1997). Crite-ria such as familiarity, historical usage, and nomenclatural priority, however, persuaded us to retain Brandt's terms, particularly when recent evidence for monophyly has not sub-stantively altered their earlier taxonomic meaning. Perhaps an argument could be devel-oped for a new subordinal name to contain the apparent clade Castoridae-Geomyoidea; however, the case for common ancestry is still tenuous, and in the meanwhile, we con-ventionally applied nomenclatural priority in adopting Castorimorpha Wood, 1955, over Geomorpha Thaler, 1966. For other viewpoints on naming, or the need to rename, order-level taxa, see Archibald (2003:356-357), Bryant and McKenna (1995:32-34), and McKenna and Bell (1997:513-514).

Table 1

Provisional suprafamilial classification of Rodentia.

Order Rodentia Bowdich, 1821
(33 families, 481 genera, 2277 species)

Suborder Sciuromorpha Brandt, 1855 (61 genera, 307 species)
    Family Aplodontiidae Brandt, 1855 (1 genus, 1 species)
    Family Sciuridae Fischer, 1817 (51 genera, 278 species)
    Family Gliridae Muirhead, 1819 (9 genera, 28 species)

Suborder Castorimorpha A. E. Wood, 1955 (13 genera, 102 species)
    Family Castoridae Hemprich, 1820 (1 genus, 2 species)
    Family Heteromyidae Gray, 1868 (6 genera, 60 species)
    Family Geomyidae Bonaparte, 1845 (6 genera, 40 species)

Suborder Myomorpha Brandt, 1855 (326 genera, 1569 species)
  Superfamily Dipodoidea Fischer, 1817
    Family Dipodidae Fischer, 1817 (16 genera, 51 species)
  Superfamily Muroidea Illiger, 1811
    Family Platacanthomyidae Alston, 1876 (2 genera, 2 species)
    Family Spalacidae Gray, 1821 (6 genera, 36 species)
    Family Calomyscidae Vorontsov and Potapova, 1979 (1 genus, 8 species)
    Family Nesomyidae Major, 1897 (21 genera, 61 species)
    Family Cricetidae Fischer, 1817 (130 genera, 681 species)
    Family Muridae Illiger, 1811 (150 genera, 730 species)

Suborder Anomaluromorpha Bugge, 1974 (4 genera, 9 species)
    Family Anomaluridae Gervais, 1849 (3 genera, 7 species)
    Family Pedetidae Gray, 1825 (1 genus, 2 species)

Suborder Hystricomorpha Brandt, 1855 (77 genera, 290 species)
  Infraorder Ctenodactylomorphi Chaline and Mein, 1979
    Family Ctenodactylidae Gervais, 1853 (4 genera, 5 species)
  Infraorder Hystricognathi Tullberg, 1899
    Family Bathyergidae Waterhouse, 1841 (5 genera, 17 species)
    Family Hystricidae Fischer, 1817 (3 genera, 11 species)
    Family Petromuridae Tullberg, 1899 (1 genus, 1 species)
    Family Thryonomyidae Fitzinger, 1867 (1 genus, 2 species)
    Family Erethizontidae Bonaparte, 1845 (5 genera, 16 species)
    Family Chinchillidae Bennett, 1833 (3 genera, 7 species)
    Family Dinomyidae Peters, 1873 (1 genus, 1 species)
    Family Caviidae Fischer, 1817 (6 genera, 18 species)
    Family Dasyproctidae Bonaparte, 1838 (2 genera, 13 species)
    Family Cuniculidae Miller and Gidley, 1918 (1 genus, 2 species)
    Family Ctenomyidae Lesson, 1842 (1 genus, 60 species)
    Family Octodontidae Waterhouse, 1840 (8 genera, 13 species)
    Family Abrocomidae Miller and Gidley, 1918 (2 genus, 10 species)
    Family Echimyidae Gray, 1825 (21 genera, 87 species)
    Family Myocastoridae Ameghino, 1904 (1 genus, 1 species)
    Family Capromyidae Smith, 1842 (8 genera, 20 species)
    Family Heptaxodontidae Anthony, 1917 (4 genera, 4 species)

SUBORDER SCIUROMORPHA Brandt, 1855.

FAMILY APLODONTIIDAE
by Kristofer M. Helgen

Family Aplodontiidae Brandt, 1855. Mem. Acad. Imp. Sci. St. Petersbourg, ser. 6, VII; Sci. Nat., p. 151.
> SYNONYMS: Aplodontidae Trouessart, 1897; Haplodontidae Alston, 1876; Haploodontidae Lilljeborg, 1866; Haploodontini Brandt, 1855 [original spelling].
> COMMENTS: The correct spelling of the family name is Aplodontiidae (Thomas, 1896). Traditionally classified as the sole extant representative of a unique suborder or infraorder (Protrogomorpha), the monotypic family Aplodontiidae may constitute the sister group to the sciurids (Landry, 1999; Montgelard et al., 2002). Consult McKenna and Bell (1997) for additional fossil synonyms.

Aplodontia Richardson, 1829. Zool. J., 4:334.
> TYPE SPECIES: *Aplodontia leporina* Richardson, 1829 (= *Anisonyx rufa* Rafinesque, 1817).
> SYNONYMS: *Apludontia* Fischer, 1830; *Apluodontia* Richardson, 1837; *Haplodon* Wagler, 1830; *Haplodus* Coues, 1877; *Haploodon* Brandt, 1855; *Haploodus* Coues, 1877; *Haploudon* Coues, 1877; *Haploudontia* Coues, 1889; *Haploudus* Coues, 1877; *Hapludon* Brandt 1855.

Aplodontia rufa (Rafinesque, 1817). Am. Mon. Mag., 2:45.
> COMMON NAME: Sewellel.
> TYPE LOCALITY: USA, Oregon, neighborhood of Columbia River. Specimens from Marmot, Clackamas County, Oregon are considered typical (Taylor, 1918).
> DISTRIBUTION: W coast of North America from SW British Columbia (Canada) to N California (USA), isolated populations in N and C California, extending into W Nevada. See Hall (1981:335).
> STATUS: U. S. ESA – Endangered as *A. r. nigra*; IUCN – Vulnerable as *A. r. nigra* and *A. r. phaea*; otherwise Lower Risk (nt).
> SYNONYMS: *chryseola* L. Kellogg, 1914; *grisea* Taylor, 1916; *leporina* Richardson, 1829; *olympica* Merriam, 1899; **californica** (Peters, 1864); *major* Merriam, 1886; **humboldtiana** Taylor, 1916; **nigra** Taylor, 1914; **pacifica** Merriam, 1899; **phaea** Merriam, 1899; **rainieri** Merriam, 1899; *columbiana* Taylor, 1916.
> COMMENTS: Reviewed by Carraway and Verts (1993; Mammalian Species, 431).

FAMILY SCIURIDAE
by Richard W. Thorington, Jr., and Robert S. Hoffmann

Family Sciuridae Fischer de Waldheim, 1817. Adversaria zoologica, p. 408.

COMMENTS: Ellerman (1940) reviewed the history of sciurid classification. The first modern classification (Pocock, 1923) recognized six subfamilies: Sciurinae, Tamiasciurinae, Funambulinae, Callosciurinae, Xerinae, and Marmotinae. We do not accept the use of Nannosciurinae (Forsyth Major, 1895) as a senior synonym for Callosciurinae (Pocock, 1923), as proposed by McKenna and Bell (1997). Simpson (1945) recognized the same taxa, but all at the tribal level. Ellerman (1940) avoided formal designation, but recognized seven "sections" in the *Sciurus* "group," which often do not conform with the above. Moore (1959) recognized Simpson's six tribes and, in addition, Ratufini and Protoxerini for certain genera that had previously been included in Funambulini. Black (1963) elevated Tamiini to tribal level (previously in Marmotini). Gromov et al. (1965) elevated the ground squirrels to subfamily rank, Marmotinae, which included the tribes Tamiini Black, Otospermophilini Gromov, Citellini Gromov, Marmotini Simpson (part), and Cynomyini Gromov. He also recognized the subfamilies Xerinae and Sciurinae, but did not treat any other groups. Heaney (1985) recommended that the subtribe Hyosciurina of Moore (1959) be discarded because it lacks defining characters, and that the component genera (*Hyosciurus, Prosciurillus, Rubisciurus,* and *Exilisciurus*) be retained in tribe Callosciurini Simpson, 1945. Emry and Thorington (1984) submerged Tamiasciurini in Sciurini and Tamiini in Marmotini. They also (1982) described the oldest known sciurid, *Douglassciurus* (formerly *Protosciurus*), which they interpreted as an arboreal form similar to *Sciurus* from which both flying squirrels and ground squirrels have been derived. These authors treated flying squirrels as a monophyletic sister group to other squirrels, either as a family (Pocock, 1923) or subfamily (other authors). The monophyly of flying squirrels has been both supported (Thorington, 1984) and questioned (Hight et al., 1974). A catalogue of Indian sciurids was presented by Agrawal and Chakraborty (1979). Molecular studies (Mercer and Roth, 2003; Steppan et al., 2004) force a revision of the higher level systematic arrangement of the Sciuridae, which is adopted here.

Table 1

Classification of Sciuridae (51 genera, 278 species) developed here.

Family Sciuridae Fischer de Waldheim, 1817

Subfamily Ratufinae Moore, 1959 (1 genus, 4 species)
Ratufa Gray, 1867 (4)

Subfamily Sciurillinae Moore, 1959 (1 genus, 1 species)
Sciurillus Thomas, 1914 (1)

Subfamily Sciurinae Fischer de Waldheim, 1817 (20 genera, 81 species)

Tribe Sciurini Fischer de Waldheim, 1817 (5 genera, 37 species)
Microsciurus J. A. Allen, 1895 (4)
Rheithrosciurus Gray, 1867 (1)
Sciurus Linnaeus, 1758 (28)
Syntheosciurus Bangs, 1902 (1)
Tamiasciurus Trouessart, 1880 (3)

Tribe Pteromyini Brandt, 1855 (15 genera, 44 species)
Aeretes G. M. Allen, 1940 (1)

Aeromys Robinson and Kloss, 1915 (2)
Belomys Thomas, 1908 (1)
Biswamoyopterus Saha, 1981 (1)
Eoglaucomys A. H. Howell, 1915 (1)
Eupetaurus Thomas, 1888 (1)
Glaucomys Thomas, 1908 (2)
Hylopetes Thomas, 1908 (9)
Iomys Thomas, 1908 (2)
Petaurillus Thomas, 1908 (3)
Petaurista Link, 1795 (8)
Petinomys Thomas, 1908 (9)
Pteromys G. Cuvier, 1800 (2)
Pteromyscus Thomas, 1908 (1)
Trogopterus Heude, 1898 (1)

Subfamily Callosciurinae Pocock 1923 (14 genera, 64 species)
Callosciurus Gray, 1867 (15)
Dremomys Heude, 1898 (6)
Exilisciurus Moore, 1958 (3)
Funambulus Lesson, 1835 (5)
Glyphotes Thomas, 1898 (1)
Hyosciurus Archbold and Tate, 1935 (2)
Lariscus Thomas and Wroughton, 1909 (4)
Menetes Thomas, 1908 (1)
Nannosciurus Trouessart, 1880 (1)
Prosciurillus Ellerman, 1947 (5)
Rhinosciurus Blyth, 1856 (1)
Rubrisciurus Ellerman, 1954 (1)
Sundasciurus Moore, 1958 (15)
Tamiops J. A. Allen, 1906 (4)

Subfamily Xerinae Osborn 1910 (15 genera, 128 species)

Tribe Xerini Osborn 1910 (3 genera, 6 species)
Atlantoxerus Forsyth Major, 1893 (1)
Spermophilopsis Blasius, 1884 (1)
Xerus Hemprich and Ehrenberg, 1833 (4)

Tribe Protoxerini Moore 1959 (6 genera, 30 species)
Epixerus Thomas, 1909 (1)
Funisciurus Trouessart, 1880 (9)
Heliosciurus Trouessart, 1880 (6)
Myosciurus Thomas, 1909 (1)
Paraxerus Forsyth Major, 1893 (11)
Protoxerus Forsyth Major, 1893 (2)

Tribe Marmotini Pocock, 1923 (6 genera, 92 species)
Ammospermophilus Merriam, 1892 (5)
Cynomys Rafinesque, 1817 (5)
Marmota Blumenbach, 1779 (14)
Sciurotamias Miller, 1901 (2)
Spermophilus F. Cuvier, 1825 (41)
Tamias Illiger, 1811 (25)

Subfamily Ratufinae Moore, 1959. Bull. Am. Mus. Nat. Hist. 118, p. 167.
> COMMENTS: Moore (1959) recognized the tribe Ratufini. Treated as subfamily Ratufinae by Steppan et al. (2004) on the basis of their molecular evidence and that of Mercer and Roth (2003).

Ratufa Gray, 1867. Ann. Mag. Nat. Hist., ser. 3, 20:273.
> TYPE SPECIES: *Sciurus indicus* Erxleben, 1777.
> SYNONYMS: *Eosciurus* Trouessart, 1880; *Rukaia* Gray, 1867.
> COMMENTS: Tribe Ratufini (Moore, 1959). Reviewed in part by Moore and Tate (1965).

Ratufa affinis (Raffles, 1821). Trans. Linn. Soc. Lond., 13:259.
> COMMON NAME: Pale Giant Squirrel.
> TYPE LOCALITY: "Singapore Island".
> DISTRIBUTION: Malay Peninsula, Sumatra, Borneo, and adjacent small islands.
> STATUS: CITES – Appendix II; IUCN – Lower Risk (lc).
> SYNONYMS: *albiceps* (Jentink, 1897); *aureiventer* (I. Geoffroy, 1831); *frontalis* Kloss, 1932; *interposita* Kloss, 1932; *johorensis* Robinson and Kloss, 1911; *klossi* J. E. Hill, 1960; *pyrsonota* Miller, 1900; **bancana** Lyon, 1906; **baramensis** Bonhote, 1900; *banguei* Chasen and Kloss, 1932; *dulitensis* Lönnberg and Mjöberg, 1925; *lumholtzi* Lönnberg and Mjoberg, 1925; *sandakanensis* Bonhote, 1900; **bunguranensis** (Thomas and Hartert, 1894); *confinis* Miller, 1906; *nanogigas* (Thomas and Hartert, 1895); *notabilis* Miller, 1902; *sirhassenensis* Bonhote, 1900; **cothurnata** Lyon, 1911; *griseicollis* Lyon, 1911; **ephippium** (Müller, 1838); *vittata* Lyon, 1911; *vittatula* Lyon, 1911; **hypoleucos** (Horsfield, 1823); *arusinus* Lyon, 1907; *balae* Miller, 1903; *catemana* Lyon, 1907; *femoralis* Miller, 1903; *masae* Miller, 1903; *nigrescens* Miller, 1903; *piniensis* Miller, 1903; **insignis** Miller, 1903; *bulana* Lyon, 1909; *carimonensis* Miller, 1906; *condurensis* Miller, 1906; *conspicua* Miller, 1903; **polia** Lyon, 1906.
> COMMENTS: Reviewed by Corbet and Hill (1992); they doubted the validity of a record from Vietnam (Cao, 1984).

Ratufa bicolor (Sparrman, 1778). Samhelle Hand. (Wet. Afd.), 1:70.
> COMMON NAME: Black Giant Squirrel.
> TYPE LOCALITY: Indonesia, W Java, Anjer.
> DISTRIBUTION: E Nepal; SE Tibet to S Yunnan and Hainan (China); Assam (India), Burma, Thailand, Laos, Cambodia, and Vietnam, south through the Malay Peninsula to Java and Bali.
> STATUS: CITES – Appendix II; IUCN – Lower Risk (lc).
> SYNONYMS: *albiceps* (Desmarest, 1817); *baliensis* Thomas, 1913; *humeralis* (Coulon, 1836); *javensis* (Zimmerman, 1780); *leschnaultii* (Desmarest, 1822); *major* Miller, 1911; *sondaica* (Müller and Schlegel, 1844); **condorensis** Kloss, 1920; **felli** Thomas and Wroughton, 1916; **gigantea** (McClelland, 1839); *lutrina* Thomas and Wroughton, 1916; *macruroides* (Hodgson, 1849); **hainana** J. A. Allen, 1906; *stigmosa* Thomas, 1923; **leucogenys** Kloss, 1916; *sinus* Kloss, 1916; **melanopepla** Miller, 1900; *anambae* Miller, 1900; *angusticeps* Miller, 1901; *dicolorata* Robinson and Kloss, 1914; *fretensis* Thomas and Wroughton, 1909; *penangensis* Robinson and Kloss, 1911; *peninsulae* (Miller, 1913); *tiomanensis* Miller, 1900; **palliata** Miller, 1902; *batuana* Lyon, 1916; *laenata* Miller, 1903; **phaeopepla** Miller, 1913; *celaenopepla* Miller, 1913; *marana* Thomas and Wroughton, 1916; **smithi** Robinson and Kloss, 1922. **Unassigned**: *tennentii* Layard, in Blyth, 1849.
> COMMENTS: Reviewed by Corbet and Hill (1992).

Ratufa indica (Erxleben, 1777). Syst. Regn. Anim., 1:420.
> COMMON NAME: Indian Giant Squirrel.
> TYPE LOCALITY: "in India orientali" [Bombay, India]. Based on "Pennant's Bombay squirrel" (Moore and Tate, 1965:43).
> DISTRIBUTION: C and S India, excluding central lowlands.
> STATUS: CITES – Appendix II; IUCN – Extinct as *R. i. dealbata*, Vulnerable as *R. i. centralis*, *R. i. indica* and *R. i. maxima*.
> SYNONYMS: *bombaya* (Boddaert, 1785); *elphinstoni* (Sykes, 1831); *purpureus* (Zimmermann,

1777); *superans* Ryley, 1913; **centralis** Ryley, 1913; **dealbata** (Blanford, 1897); **maxima** (Schreber, 1784); *bengalensis* (Blanford, 1897); *malabarica* (Scopoli, 1786).
COMMENTS: Pennant (1771:281) described, but did not name this species, and said it "Inhabits Bombay."

Ratufa macroura (Pennant, 1769). Indian Zool., 1, pl. 1.
COMMON NAME: Sri Lankan Giant Squirrel.
TYPE LOCALITY: "Ceylon and Malabar. . . . Malacca . . . Goa and Amboina." Restricted by Phillips (1933) to highlands of Central and Uva Provs., Sri Lanka.
DISTRIBUTION: Sri Lanka and S India.
STATUS: CITES – Appendix II. IUCN – Vulnerable
SYNONYMS: *albipes* (Blyth, 1859); *ceilonensis* (Boddaert, 1785); *ceylonica* (Erxleben, 1777); *macrura* Blanford, 1891; *montana* (Kalaart, 1852); *tennentii* (Blyth, 1849); **dandolena** Thomas and Wroughton, 1915; *sinhala* Phillips, 1931; **melanochra** Thomas and Wroughton, 1915. **Unassigned:** *zeylanicus* (Ray, 1693).
COMMENTS: Reviewed by Corbet and Hill (1992). Joshua (1996) reported apparent hybridization with *R. indica*, but had no cytogenetic data.

Subfamily Sciurillinae Moore, 1959. Bull. Am. Mus. Nat. Hist. 118:180.
COMMENTS: Moore (1959) recognized the subtribe Sciurillina, noting that "*Sciurillus* is one of the most distinctive genera of squirrels in the world . . . In this measure of divergence . . . they are approached only by the Ratufini." Supporting this prescient observation, recent molecular work places *Sciurillus* as a basal member of the Sciuridae (Mercer and Roth, 2003; Steppan et al., 2004), leading to its recognition as a distinct subfamily, Sciurillinae (Steppan et al., 2004).

Sciurillus Thomas, 1914. Proc. Zool. Soc. Lond., 1914:416.
TYPE SPECIES: *Sciurus pusillus* Desmarest, 1817, as designated by Thomas (= *Sciurus pusillus* E. Geoffroy, 1803).

Sciurillus pusillus (E. Geoffroy, 1803). *In* Cat. Mamm. Mus. Hist. Nat., Paris, p. 178.
COMMON NAME: Neotropical Pygmy Squirrel.
TYPE LOCALITY: French Guiana, Cayenne.
DISTRIBUTION: Brazil, French Guiana, Peru, Suriname.
STATUS: IUCN – Lower Risk (lc).
SYNONYMS: **glaucinus** Thomas, 1914; *hoehnei* Miranda Ribeiro, 1941; **kuhlii** (Gray, 1867).
COMMENTS: Husson (1978) designated a lectotype. Allelic data first indicated that this species falls outside the clade that contains other New World tree squirrels (Hafner et al., 1994*a*).

Subfamily Sciurinae Fischer de Waldheim, 1817. Adversaria zoologica, p. 408.
COMMENTS: Subfamily modified by Steppan et al. (2004) to include the tribes Sciurini and Pteromyini.

Tribe Sciurini Fischer de Waldheim, 1817. Adversaria zoologica, p. 408.

Microsciurus J. A. Allen, 1895. Bull. Am. Mus. Nat. Hist., 7:332.
TYPE SPECIES: *Sciurus alfari* J. A. Allen, 1895.
COMMENTS: Tribe Microsciurini, according to Moore (1959). Evidence suggests that this genus is polyphyletic (Mercer and Roth, 2003). See Emmons and Feer (1990).

Microsciurus alfari (J. A. Allen, 1895). Bull. Am. Mus. Nat. Hist., 7:333.
COMMON NAME: Central American Dwarf Squirrel.
TYPE LOCALITY: Costa Rica, Jimenez.
DISTRIBUTION: Colombia, Costa Rica, Nicaragua, Panama.
STATUS: IUCN – Lower Risk (lc).
SYNONYMS: **alticola** Goodwin, 1943; **browni** (Bangs, 1902); **fusculus** (Thomas, 1910); **septentrionalis** Anthony, 1920; **venustulus** Goldman, 1912.
COMMENTS: See Hall (1981:439-440).

Microsciurus flaviventer (Gray, 1867). Ann. Mag. Nat. Hist., ser. 3, 20:432.
    COMMON NAME: Amazon Dwarf Squirrel.
    TYPE LOCALITY: Brazil. Cabrera (1961) suggested the type locality could be restricted to
        Pebas, based on Thomas (1928b).
    DISTRIBUTION: Amazon basin of Colombia, Ecuador, Peru, and Brazil west of the Rios Negro
        and Jurua.
    STATUS: IUCN – Lower Risk (lc).
    SYNONYMS: *manarius* Thomas, 1920; **napi** (Thomas, 1900); *avunculus* Thomas, 1914;
        florenciae J. A. Allen, 1914; **otinus** (Thomas, 1901); **peruanus** (J. A. Allen, 1897);
        rubrirostris J. A. Allen, 1914; *rubicollis* Thomas, 1914 [*lapsus* for *rubrirostris*];
        sabanillae Anthony, 1922; **similis** (Nelson, 1899); **simonsi** (Thomas, 1900).

Microsciurus mimulus (Thomas, 1898). Ann. Mag. Nat. Hist., ser. 7, 2:266.
    COMMON NAME: Western Dwarf Squirrel.
    TYPE LOCALITY: Ecuador, Esmeraldas, Cachavi, 665 ft. (203 m).
    DISTRIBUTION: NW Ecuador, N Colombia, and Panama.
    STATUS: IUCN – Lower Risk (lc).
    SYNONYMS: **boquetensis** (Nelson, 1903); **isthmius** (Nelson, 1899); *palmeri* (Thomas, 1909);
        vivatus Goldman, 1912.
    COMMENTS: See Hall (1981:440) and Handley (1966a).

Microsciurus santanderensis (Hernandez-Camacho, 1957). *In* Borerro and Hernandez-Camacho,
    Anal. Soc. Biol. Bogota, 7:219.
    COMMON NAME: Santander Dwarf Squirrel.
    TYPE LOCALITY: Colombia, Santander Dept., Meseta de los Caballeros, NE of La Albania.
    DISTRIBUTION: Colombia, between the Magdalena River and the Cordillera Oriental.
    STATUS: IUCN – Lower Risk (lc).
    COMMENTS: See Eisenberg (1989) and Hernandez-Camacho (1960).

Rheithrosciurus Gray, 1867. Ann. Mag. Nat. Hist., ser. 3, 20:273.
    TYPE SPECIES: *Sciurus macrotis* Gray, 1857.
    SYNONYMS: *Rhithrosciurus* Hose, 1893.
    COMMENTS: Tribe Sciurini according to Moore (1959:177), or *incertae sedis* (Simpson, 1945:78).
        Recent molecular studies (Mercer and Roth, 2003) support the placement of
        Rheithrosciurus in the tribe Sciurini.

Rheithrosciurus macrotis (Gray, 1857). Proc. Zool. Soc. Lond., 1856:341 [1857].
    COMMON NAME: Tufted Ground Squirrel.
    TYPE LOCALITY: "Sarawak," [Malaysia].
    DISTRIBUTION: Borneo.
    STATUS: IUCN – Lower Risk (lc).

Sciurus Linnaeus, 1758. Syst. Nat., 10th ed., 1:63.
    TYPE SPECIES: *Sciurus vulgaris* Linnaeus, 1758.
    SYNONYMS: *Aphrontis* Schultze, 1893; *Araeosciurus* Nelson, 1899; *Baiosciurus* Nelson, 1899;
        Echinosciurus Trouessart, 1880; *Guerlinguetus* Gray, 1821; *Hadrosciurus* J. A. Allen, 1915;
        Hesperosciurus Nelson, 1899; *Histriosciurus* J. A. Allen, 1915; *Leptosciurus* J. A. Allen, 1915;
        Macroxus F. Cuvier, 1823; *Mesosciurus* J. A. Allen, 1915; *Neosciurus* Trouessart, 1880;
        Oreosciurus Ognev, 1935; *Otosciurus* Nelson, 1899; *Parasciurus* Trouessart, 1880;
        Simosciurus J. A. Allen, 1915; *Tenes* Thomas, 1909; *Urosciurus* J. A. Allen, 1915.
    COMMENTS: Tribe Sciurini (Moore, 1959:177); includes *Guerlinguetus, Hesperosciurus, Otosciurus,*
        Sciurus (Hall, 1981:417-436), *Tenes* (Corbet, 1978c:76), *Hadrosciurus* (Cabrera, 1961:374),
        and *Urosciurus* (Moore, 1959:198) as subgenera. Moore (1959) considered *Guerlinguetus* a
        distinct genus (*Hadrosciurus, Urosciurus* as subgenera) and included *Otosciurus* and
        Hesperosciurus in subgenus *Sciurus*.

Sciurus aberti Woodhouse, 1853. Proc. Acad. Nat. Sci. Philadelphia, 1852, 6:110, 220 [1853].
    COMMON NAME: Abert's Squirrel.
    TYPE LOCALITY: ". . . in the San Francisco Mountains, New Mexico" [= Coconino Co., Arizona, USA].

DISTRIBUTION: SE Utah, S and W Colorado, extreme SE Wyoming, W and C New Mexico, and
Arizona (USA); Chihuahua, Durango, and Sonora (NW Mexico).

STATUS: IUCN – Lower Risk (lc).

SYNONYMS: *castanonotus* Baird, 1858; *castanotus* Baird, 1855; *dorsalis* Woodhouse, 1853; *navajo*
Durrant and Kelson, 1947; **barberi** J. A. Allen, 1904; **chuscensis** Goldman, 1931; **durangi**
Thomas, 1893; *phaeurus* J. A. Allen, 1904; **ferreus** True, 1900; *concolor* True, 1894; *mimus*
Merriam, 1904; **kaibabensis** (Merriam, 1904).

COMMENTS: Subgenus *Otosciurus* (Hall, 1981:434). Reviewed by Nash and Seaman (1977,
Mammalian Species, 80) and by Hoffmeister and Diersing, (1978), whose subspecies
allocations are followed here. Post-Pleistocene dispersal analyzed by Davis and Brown
(1989). Phylogeography was studied by Lamb et al. (1997), who found two major
assemblages.

Sciurus aestuans Linnaeus, 1766. Syst. Nat., 12th ed., 1:88.

COMMON NAME: Guianan Squirrel.

TYPE LOCALITY: Surinam.

DISTRIBUTION: Brazil, French Guiana, Guyana, Suriname, and Venezuela.

STATUS: IUCN – Lower Risk (lc).

SYNONYMS: *bancrofti* Kerr, 1792; *guajanensis* Kerr, 1792; *guerlingus* (Shaw, 1801); *guianensis*
Peters, 1863; *olivascens* Illiger, 1815; **alphonsei** Thomas, 1903; *roberti* Thomas, 1903;
garbei (Pinto, 1931); **georgihernandezi** Barriga-Bonilla, 1966; **henseli** Miranda Ribeiro,
1941; **ingrami** Thomas, 1901; **macconnelli** Thomas, 1901; **poaiae** (Moojen, 1942);
quelchii Thomas, 1901; **venustus** (J. A. Allen, 1940).

COMMENTS: Subgenus *Guerlinguetus*; see Cabrera (1961:359). Formerly included *gilvigularis*;
see Avila-Pires (1964). Subspecies recognized based on Cabrera (1961).

Sciurus alleni Nelson, 1898. Proc. Biol. Soc. Wash., 12:147.

COMMON NAME: Allen's Squirrel.

TYPE LOCALITY: "Monterey, Tamaulipas [=Nuevo Leon]," [Mexico].

DISTRIBUTION: SE Coahuila through C Nuevo Leon, south through W Tamaulipas to
extreme N San Luis Potosi (Mexico).

STATUS: IUCN – Lower Risk (lc).

COMMENTS: Subgenus *Sciurus* (Hall, 1981:430). Reviewed by Best (1995*a*, Mammalian
Species No. 501).

Sciurus anomalus Gmelin, 1778. Syst. Nat., 13th ed., 1:148.

COMMON NAME: Caucasian Squirrel.

TYPE LOCALITY: "without exact indication of locality" (Ognev, 1966:368). Restricted by
Güldenstadt (1785, see Ognev, 1940:423) to Sabeka, 25 km SW of Kutais, Georgia.

DISTRIBUTION: Turkey, Transcaucasia (Armenia, Azerbaijan, Georgia), N and W Iran, Syria,
Lebanon, Israel, Palestine, Jordan, and Iraq. Newly reported from Lesbos Isl, Greece
(Hecht-Markou, 1995).

STATUS: IUCN – Lower Risk (nt).

SYNONYMS: *caucasicus* Pallas, 1811; *fulvus* Blanford, 1875 ; *russatus* Wagner, 1842; **pallescens**
(Gray, 1867); **syriacus** Ehrenberg, 1828; *historicus* Gray, 1867; *persicus auctorum ignotus*
[author unknown; the name *persicus* was assigned by Erxleben, 1777, to S. G. Gmelin's
description (1774, p. 379, fig. 43) of *Sciurus persicus,* which was, however, based on a
dormouse, *Myoxus glis*. It is still used for *S. anomalus* (cf. Pavlinov et al., 1995)].

COMMENTS: Subgenus *Tenes* (Corbet, 1978*c*:76). Reviewed by Harrison and Bates (1991).

Sciurus arizonensis Coues, 1867. Am. Nat., 1:357.

COMMON NAME: Arizona Gray Squirrel.

TYPE LOCALITY: "Fort Whipple," [Yavapai Co., Arizona, USA].

DISTRIBUTION: C and SE Arizona and WC New Mexico (USA); NE Sonora (Mexico).

STATUS: IUCN – Lower Risk (nt).

SYNONYMS: **catalinae** Doutt, 1931; **huachuca** J. A. Allen, 1894.

COMMENTS: Subgenus *Sciurus* (Hall, 1981:432). Reviewed by Best and Riedel (1995,
Mammalian Species No. 496).

Sciurus aureogaster F. Cuvier, 1829. *In* E. Geoffroy and F. Cuvier, Hist. Nat. Mammifères, pt. 3, 6(59):1-2 "Ecureuil de la Californie".
COMMON NAME: Red-bellied Squirrel.
TYPE LOCALITY: "California" Restricted by Nelson (1899*b*:38) to Alta Mira, Tamaulipas, Mexico.
DISTRIBUTION: SW and C Guatemala to Guanajuato to Nayarit and Nuevo Leon (Mexico).
STATUS: IUCN – Lower Risk (lc).
SYNONYMS: *chrysogaster* Giebel, 1855; *ferruginiventris* Audubon and Bachman, 1841; *hypopyrrhus* Wagler, 1831; *hypoxanthus* I. Geoffroy, 1855; *leucogaster* F. Cuvier, 1831; *maurus* Gray, 1867; *morio* Gray, 1867; *mustelinus* Audubon and Bachman, 1841; *raviventer* Lichtenstein, 1830; *rufiventris* Rovirosa, 1887; **nigrescens** Bennett, 1833; *affinis* Alston, 1878; *albipes* Wagner, 1837; *cervicalis* J. A. Allen, 1890; *chiapensis* Nelson, 1899; *cocos* Nelson, 1898; *colimensis* Nelson, 1898; *effugius* Nelson, 1898; *frumentor* Nelson, 1898; *griseoflavus* Gray, 1867; *hernandezi* Nelson, 1898; *hirtus* Nelson, 1898; *leucops* Gray, 1867; *littoralis* Nelson, 1907; *nelsoni* Merriam, 1893; *nemoralis* Nelson, 1898; *perigrinator* Nelson, 1904; *poliopus* Fitzinger, 1867; *quercinus* Nelson, 1898; *rufipes* Fitzinger, 1867; *senex* Nelson, 1904; *socialis* Wagner, 1837; *tepicanus* J. A. Allen, 1906; *varius* Wagner, 1843; *wagneri* J. A. Allen, 1898.
COMMENTS: Subgenus *Sciurus* (Hall, 1981:418). Revised by Musser (1968), who also designated a lectotype and confirmed the type locality restriction (Musser, 1970*c*). Includes *griseoflavus*, *nelsoni*, *poliopus*, and *socialis* (Musser, 1968). Introduced to Elliot Key, Dade County, Florida (USA) (Brown and McGuire, 1975).

Sciurus carolinensis Gmelin, 1788. *In* Linnaeus, Syst. Nat., 13th ed., 1:148.
COMMON NAME: Eastern Gray Squirrel.
TYPE LOCALITY: "Carolina."
DISTRIBUTION: E Texas (USA) to Saskatchewan (Canada) and east to Atlantic Coast. Introduced into Britain, Scotland, Ireland, Italy, Australia, South Africa, and various localities in W North America.
SYNONYMS: **extimus** Bangs, 1896; *matecumbei* Bailey, 1937; *minutus* Bailey, 1937; **fuliginosus** Bachman, 1839; **hypophaeus** Merriam, 1886; **pennsylvanicus** Ord, 1815; *hiemalis* Ord, 1815; *leucotis* Gapper, 1830; *migratorius* Audubon and Bachman, 1849.
COMMENTS: Subgenus *Sciurus* (Hall, 1981:417). Reviewed by Koprowski (1994*b*, Mammalian Species No. 480).

Sciurus colliaei Richardson, 1839. Zool. Capt. Beechey's Voy., p. 8.
COMMON NAME: Collie's Squirrel.
TYPE LOCALITY: "San Blas, Tepic, [Nayarit,] Mexico."
DISTRIBUTION: Mexico: WC coast including Sonora, Chihuahua, Sinaloa, Durango, Nayarit, Jalisco, and Colima.
STATUS: IUCN – Lower Risk (lc).
SYNONYMS: **nuchalis** Nelson, 1899; **sinaloensis** Nelson, 1899; **truei** Nelson, 1899.
COMMENTS: Subgenus *Sciurus* (Hall, 1981:421). Includes *sinaloensis* and *truei* (Anderson, 1962). Reviewed by Best (1995*b*, Mammalian Species No. 497).

Sciurus deppei Peters, 1863. Monatsb. K. Preuss. Akad. Wiss. Berlin, 1863:654.
COMMON NAME: Deppe's Squirrel.
TYPE LOCALITY: "Papantla, Veracruz, Mexico." (Nelson, 1899*b*:101).
DISTRIBUTION: Tamaulipas (Mexico) to Costa Rica.
STATUS: CITES – Appendix III (Costa Rica); IUCN – Lower Risk (lc).
SYNONYMS: *taeniurus* Gray, 1867; *tephrogaster* Gray, 1867; **matagalpae** J. A. Allen, 1908; **miravallensis** Harris, 1931; **negligens** Nelson, 1898; *vivax* Nelson, 1901.
COMMENTS: Subgenus *Sciurus* (Hall, 1981:426). Reviewed by Best (1995*c*, Mammalian Species No. 505).

Sciurus flammifer Thomas, 1904. Ann. Mag. Nat. Hist., ser. 7, 14:33.
COMMON NAME: Fiery Squirrel.
TYPE LOCALITY: Venezuela, Bolivar, Caura Valley, La Union.
DISTRIBUTION: Venezuela south of Orinoco River from the Colombian border to Cuidad Bolivar.

STATUS: IUCN – Lower Risk (lc).
COMMENTS: Subgenus *Hadrosciurus*; see Cabrera (1961:374).

Sciurus gilvigularis Wagner, 1842. Arch. Naturgesch., 2:43.
   COMMON NAME: Yellow-throated Squirrel.
   TYPE LOCALITY: Brazil, Borba, Rio Madeira.
   DISTRIBUTION: N Brazil, Guyana, Venezuela.
   STATUS: IUCN – Lower Risk (lc).
   SYNONYMS: *gilviventris* Pelzeln, 1883; ***paraensis*** Goeldi and Hagmann, 1904.
   COMMENTS: Subgenus *Guerlinguetus*; see Avila-Pires (1964). Cabrera (1961:359) included
      gilvigularis and *paraensis* in *aestuans*.

Sciurus granatensis Humboldt, 1811. Rec. Observ. Zool., 1(1805):8.
   COMMON NAME: Red-tailed Squirrel.
   TYPE LOCALITY: Colombia, Dept. Bolivar, Cartagena.
   DISTRIBUTION: Costa Rica, Colombia, Ecuador, Margarita Isl, Panama, Trinidad, Tobago,
      Venezuela.
   STATUS: IUCN – Lower Risk (lc).
   SYNONYMS: ***agricolae*** Hershkovitz, 1947; ***bondae*** J. A. Allen, 1899; ***candelensis*** (J. A. Allen,
      1914); ***carchensis*** Harris and Hershkovitz, 1938; ***chapmani*** J. A. Allen, 1899;
      quebradensis J. A. Allen, 1899; *tobagensis* Ogsood, 1910; ***chiriquensis*** Bangs, 1902;
      chrysuros Pucheran, 1845; *hyporrhodus* Gray, 1867; *rufoniger* Pucheran, 1845; ***ferminae***
      (Cabrera, 1917); ***gerrardi*** Gray, 1861; ***griseimembra*** (J. A. Allen, 1914); ***griseogena***
      (Gray, 1867); ***klagesi*** Thomas, 1914; ***hoffmanni*** Peters, 1863; *xanthotus* (Gray, 1867);
      imbaburae Harris and Hershkovitz, 1938; ***llanensis*** Mondolfi and Boher, 1984;
      manavi (J. A. Allen, 1914); ***maracaibensis*** Hershkovitz, 1947; ***meridensis*** Thomas,
      1901; *tamae* Osgood, 1912; ***morulus*** Bangs, 1900; *baudensis* J. A. Allen, 1915; *choco*
      Goldman, 1915; *salaquensis* J. A. Allen, 1914; ***nesaeus*** G. M. Allen, 1902; ***norosiensis***
      Hershkovitz, 1947; ***perijae*** Hershkovitz, 1947; ***quindianus*** (J. A. Allen, 1914);
      saltuensis Bangs, 1898; ***soederstroemi*** Stone, 1914; *rhoadsi* (J. A. Allen, 1914);
      splendidus Gray, 1842; *magdalenae* J. A. Allen, 1914; ***sumaco*** (Cabrera, 1917); ***tarrae***
      Hershkovitz, 1947; ***valdiviae*** (J. A. Allen, 1915); ***variabilis*** I. Geoffroy St. Hilaire, 1832;
      versicolor Thomas, 1900; *inconstans* Osgood, 1921; *leonis* Lawrence, 1933; *milleri*
      J. A. Allen, 1912; ***zuliae*** Osgood, 1910; *cucutae* J. A. Allen, 1914.
   COMMENTS: Subgenus *Guerlinguetus*; see Hall (1981:436) and Nitikman (1985, Mammalian
      Species, 246). Subspecies recognized according to Cabrera (1961).

Sciurus griseus Ord, 1818. J. Phys. Chim. Hist. Nat. Arts Paris, 87:152.
   COMMON NAME: Western Gray Squirrel.
   TYPE LOCALITY: "The Dalles of the Columbia" [River, Wasco Co., Oregon, USA].
   DISTRIBUTION: C Washington, W Oregon, and California (USA) to Baja California Norte
      (Mexico).
   STATUS: IUCN – Lower Risk (lc).
   SYNONYMS: *fossor* Peale, 1848; *heermanni* Le Conte, 1852; *leporinus* Audubon and Bachman,
      1841; ***anthonyi*** Mearns, 1897; ***nigripes*** Bryant, 1889.
   COMMENTS: Subgenus *Hesperosciurus* (Hall, 1981:433). Reviewed by Carraway and Verts
      (1994, Mammalian Species No. 474).

Sciurus ignitus (Gray, 1867). Ann. Mag. Nat. Hist., ser. 3, 20:429.
   COMMON NAME: Bolivian Squirrel.
   TYPE LOCALITY: Bolivia, near Yungas, upper Rio Beni.
   DISTRIBUTION: Argentina, Bolivia, Brazil, and Peru.
   STATUS: IUCN – Lower Risk (lc).
   SYNONYMS: *cuscinus* Thomas, 1899; *ochrescens* Thomas, 1914; ***argentinius*** Thomas, 1921;
      boliviensis Osgood, 1921; *leucogaster* (J. A. Allen, 1915); ***cabrerai*** Moojen, 1958;
      irroratus (Gray, 1867).
   COMMENTS: Subgenus *Guerlinguetus*; subspecies according to Cabrera (1961:370-371).

Sciurus igniventris Wagner, 1842. Arch. Naturg., 1:360.
COMMON NAME: Northern Amazon Red Squirrel.
TYPE LOCALITY: Brazil, Amazonas, north of the Rio Negro, Marabitanos.
DISTRIBUTION: Brazil, Colombia, Ecuador, Peru, Venezuela.
STATUS: IUCN – Lower Risk (lc).
SYNONYMS: *taedifer* Thomas, 1900; *cocalis* Thomas, 1900; *zamorae* J. A. Allen, 1914.
Unassigned: *duida* J. A. Allen, 1914; *fulminatus* Thomas, 1926; *manhanensis* (Moojen, 1942).
COMMENTS: Subgenus *Urosciurus*; see Patton (1984). Lawrence (1988:1) restricted *duida* to this species although it has also been referred to *S. spadiceus*; the holotype is a composite.

Sciurus lis Temminck, 1844. Fauna Japonica, 1(Mamm.), p. 45.
COMMON NAME: Japanese Squirrel.
TYPE LOCALITY: "Japan". Restricted by Corbet (1978c:78) to Honshu [Japan].
DISTRIBUTION: Honshu, Shikoku, and Kyushu (Japan).
STATUS: IUCN – Lower Risk (lc).
COMMENTS: Subgenus *Sciurus* (Corbet, 1978c:77). Oshida et al. (1996a), and Oshida and Yoshida (1997) described chromosomes.

Sciurus nayaritensis J. A. Allen, 1890. Bull. Am. Mus. Nat. Hist., 2:7, footnote.
COMMON NAME: Mexican Fox Squirrel.
TYPE LOCALITY: "Sierra Valparaiso, Zacatecas," [Mexico.]
DISTRIBUTION: Jalisco (Mexico) north to SE Arizona (USA).
STATUS: IUCN – Lower Risk (lc).
SYNONYMS: *alstoni* J. A. Allen, 1889; **apache** J. A. Allen, 1893; **chiricahuae** Goldman, 1933.
COMMENTS: Subgenus *Sciurus* (Hall, 1981:431). Includes *chiricahuae* and *apache* (Hall, 1981:431; Lee and Hoffmeister, 1963). Reviewed by Best (1995d, Mammalian Species No. 492).

Sciurus niger Linnaeus, 1758. Syst. Nat., 10th ed., 1:64.
COMMON NAME: Eastern Fox Squirrel.
TYPE LOCALITY: "in America septentrionalis." Restricted by Thomas (1911a:149) to S South Carolina (USA).
DISTRIBUTION: Texas (USA) and adjacent Mexico, north to Manitoba (Canada) east to the Atlantic Coast.
STATUS: U.S. ESA – Endangered as *S. n. cinereus*; nonessential experimental population in Sussex Co., Delaware (USA); IUCN – Data Deficient as *S. n. vulpinus*, Lower Risk (conservation dependent) as *S. n. avicinnia* and *S. n. cinereus,* Lower Risk (nt) as *S. n. shermani*, otherwise Lower Risk (lc) as *S. niger.*
SYNONYMS: *capistratus* Bosc, 1802; **avicinnia** A H. Howell, 1919; **bachmani** Lowery and Davis, 1942; **cinereus** Linnaeus, 1758; *bryanti* Bailey, 1920; *neglectus* (Gray, 1867); **limitis** Baird, 1855; **ludovicianus** Custis, 1806; *texianus* Bachman, 1839; **rufiventer** E. Geoffroy, 1803; *macroura* Say, 1823; *magnificaudatus* Harlan, 1825; *ruber* Rafinesque, 1820; *rubicaudatus* Audubon and Bachman, 1851; *sayii* Audubon and Bachman, 1851; **shermani** Moore, 1956; **subauratus** Bachman, 1839; *auduboni* Bachman, 1839; **vulpinus** Gmelin, 1788; *vicinus* Bangs, 1896.
COMMENTS: Subgenus *Sciurus* (Hall, 1981:427). Reviewed by Koprowski (1994a, Mammalian Species No. 479).

Sciurus oculatus Peters, 1863. Monatsb. K. Preuss. Akad. Wiss. Berlin, 1863:653.
COMMON NAME: Peters's Squirrel.
TYPE LOCALITY: "Eastern Mexico." Restricted by Nelson (1899b:88) to "near Las Vigas [Veracruz] Mexico."
DISTRIBUTION: Mexico: San Luis Potosi, Hidalgo, Veracruz, Puebla, Mexico, Queretaro and Guanajuato.
STATUS: IUCN – Lower Risk (lc).
SYNONYMS: *capistratus* Lichtenstein, 1830; *melanonotus* Thomas, 1890; **shawi** Dalquest, 1950; **tolucae** Nelson, 1898.

COMMENTS: Subgenus *Sciurus* (Hall, 1981:430). Reviewed by Best (1995*e*, Mammalian Species No. 498).

Sciurus pucheranii (Fitzinger, 1867). Sitzb. Math. Naturw. Cl., 55, Abth., 1:487.
    COMMON NAME: Andean Squirrel.
    TYPE LOCALITY: Colombia, near Bogota.
    DISTRIBUTION: Colombian Andes.
    STATUS: IUCN – Lower Risk (lc).
    SYNONYMS: *minor* (Alston, 1878); **caucensis** Nelson, 1899; **medellinensis** (Gray, 1867); *salentensis* (J. A. Allen, 1914).
    COMMENTS: Subgenus *Guerlinguetus*; see Cabrera (1961:372) and Eisenberg (1989). Moore (1959) placed *pucheranii* in the genus *Microsciurus*. Subspecies according to Cabrera (1961).

Sciurus pyrrhinus Thomas, 1898. Ann. Mag. Nat. Hist., ser. 7, 2:265.
    COMMON NAME: Junín Red Squirrel.
    TYPE LOCALITY: Peru, Junin Dept., Vitoc, Garita del Sol.
    DISTRIBUTION: E slopes of the Andes of Peru.
    STATUS: IUCN – Lower Risk (lc).
    SYNONYMS: *variabilis* Tschudi, 1844.
    COMMENTS: Subgenus *Hadrosciurus*; see Cabrera (1961:378). J. A. Allen (1915*b*) placed *pyrrhinus* in the genus *Mesosciurus*.

Sciurus richmondi Nelson, 1898. Proc. Biol. Soc. Wash., 12:146.
    COMMON NAME: Richmond's Squirrel.
    TYPE LOCALITY: "Escondido River (50 mi. [80 km] above Bluefields) Nicaragua" (Nelson, 1899*b*:100).
    DISTRIBUTION: Nicaragua.
    STATUS: IUCN – Lower Risk (nt).
    COMMENTS: Subgenus *Guerlinguetus* (Hall, 1981:436). Reviewed by Jones and Genoways (1975*b*, Mammalian Species No. 53).

Sciurus sanborni Osgood, 1944. Field Mus. Nat. Hist. Publ., Zool. Ser., 29:191.
    COMMON NAME: Sanborn's Squirrel.
    TYPE LOCALITY: Peru, Madre de Dios Dept., La Pampa, between the Rio Inambari and Rio Tambopata, 33 km N of Santo Domingo, 570 m.
    DISTRIBUTION: Madre de Dios Dept., Peru.
    STATUS: IUCN – Lower Risk (nt).
    COMMENTS: Subgenus *Guerlinguetus*; see Cabrera (1961:373).

Sciurus spadiceus Olfers, 1818. *In* Eschwege, J. Brasilien Neue Bibliothek. Reisenb., 15(2):208.
    COMMON NAME: Southern Amazon Red Squirrel.
    TYPE LOCALITY: Brazil, restricted by Hershkovitz (1959*a*) to Cuyabá, Matto Grosso.
    DISTRIBUTION: Bolivia, Brazil, Colombia, Ecuador, Peru.
    STATUS: IUCN – Lower Risk (lc).
    SYNONYMS: *langsdorffi* Brandt, 1835; **steinbachi** J. A. Allen, 1914; **tricolor** Tschudi, 1844.
        Unassigned: *brunneo-niger* (Gray, 1867); *castus* Thomas, 1903; *fumigatus* (Gray, 1867); *juralis* Thomas, 1926; *morio* Wagner, 1848; *nigratus* (Pinto, 1931); *purusianus* (Moojen, 1942); *pyrrhonotus* Wagner, 1842; *rondoniae* (Moojen, 1942); *taparius* Thomas, 1926; *urucumus* J. A. Allen, 1914.
    COMMENTS: Subgenus *Urosciurus*; for subspecies see Patton (1984).

Sciurus stramineus Eydoux and Souleyet, 1841. *In* Vaillant, Voy. autour du monde . . . la Bonite, Zool., 1:73.
    COMMON NAME: Guayaquil Squirrel.
    TYPE LOCALITY: Peru, Piura Dept., Omatope.
    DISTRIBUTION: Extreme NW Peru and SW Ecuador in the area surrounding the Gulf of Guayaquil.
    STATUS: IUCN – Lower Risk (lc).

SYNONYMS: *fraseri* (Gray, 1867); *guayanus* Thomas, 1900; *nebouxii* I. Geoffroy St. Hilaire, 1855; *zarumae* J. A. Allen, 1914.

COMMENTS: Subgenus *Guerlinguetus*; see Cabrera (1961:373).

Sciurus variegatoides Ogilby, 1839. Proc. Zool. Soc. Lond., 1839:117.

COMMON NAME: Variegated Squirrel.

TYPE LOCALITY: " . . . west coast of South America." Restricted by Nelson (1899b:80) to "[El] Salvador."

DISTRIBUTION: S Chiapas (Mexico), through Central America to Panama.

STATUS: IUCN – Lower Risk (lc).

SYNONYMS: *griseocaudatus* Gray, 1843; *pyladei* Lesson, 1842; *adolphei* Lesson, 1842; *annalium* Thomas, 1905; *atrirufus* Harris, 1930; *bangsi* Dickey, 1928; *belti* Nelson, 1899; *boothiae* Gray, 1843; *fuscovariegatus* Schinz, 1845; *richardsoni* Gray, 1842; *dorsalis* Gray, 1849; *goldmani* Nelson, 1898; *helveolus* Goldman, 1912; *loweryi* McPherson, 1972; *managuensis* Nelson, 1898; *melania* (Gray, 1867); *rigidus* Peters, 1863; *austini* Harris, 1933; *intermedius* Gray, 1867; *nicoyana* Gray, 1867; *thomasi* Nelson, 1899; *underwoodi* Goldman, 1932.

COMMENTS: Subgenus *Sciurus*; includes *goldmani* (Hall, 1981:424). Revised by Harris (1937). Reviewed by Best (1995f, Mammalian Species No. 500).

Sciurus vulgaris Linnaeus, 1758. Syst. Nat., 10th ed., 1:63.

COMMON NAME: Eurasian Red Squirrel.

TYPE LOCALITY: "in Europae arboribus." Restricted by Thomas (1911a:148) to Uppsala, Sweden.

DISTRIBUTION: Forested regions of Palearctic, from Iberia and Great Britain east to Kamchatka Peninsula and Sakhalin Isl (Russia); south to Mediterranean and Black Seas, N Mongolia, W and NE China.

STATUS: IUCN – Near Threatened.

SYNONYMS: *albonotatus* Billberg, 1827; *albus* Billberg, 1827; *carpathicus* Pietruski, 1853; *europaeus* Gray, 1843; *niger* Billberg, 1827; *rufus* Kerr, 1792; *typicus* Barret-Hamilton, 1899; *alpinus* Desmarest, 1822; *baeticus* Cabrera, 1905; *hoffmanni* Valverde, 1967; *infuscatus* Cabrera, 1905; *italicus* Bonaparte, 1838; *meridionalis* Lucifero, 1907; *numantius* Miller, 1907; *segurae* Miller, 1909; *silanus* Hecht, 1931; *altaicus* Serebrennikov, 1928; *anadyrensis* Ognev, 1929; *arcticus* Trouessart, 1906; *jacutensis* Ognev, 1929; *balcanicus* Heinrich, 1936; *istrandjae* Heinrich, 1936; *rhodopensis* Heinrich, 1936; *chiliensis* Sowerby, 1921; *cinerea* Hermann, 1804; *dulkeiti* Ognev, 1929; *exalbidus* Pallas, 1778; *argenteus* Kerr, 1792; *kalbinensis* Selevin, 1924; *fedjushini* Ognev, 1935; *formosovi* Ognev, 1935; *fuscoater* Altum, 1876; *brunnea* Altum, 1876; *gotthardi* Fatio, 1905; *graeca* Altum, 1876; *nigrescens* Altum, 1876; *russus* Miller, 1907; *rutilans* Miller, 1907; *fusconigricans* Dwigubski, 1804; *leucourus* Kerr, 1792; *lilaeus* Miller, 1907; *ameliae* Cabrera, 1924; *croaticus* Wettstein, 1927; *mantchuricus* Thomas, 1909; *coreae* Sowerby, 1921; *coreanus* Kishida, 1924; *martensi* Matschie, 1901; *jenissejensis* Ognev, 1935; *ognevi* Migulin, 1928; *bashkiricus* Ognev, 1935; *golzmajeri* Smirnov, 1960; *uralensis* Ognev, 1935; *orientis* Thomas, 1906; *rupestris* Thomas, 1907; *ukrainicus* Migulin, 1928; *kessleri* Migulin, 1928; *varius* Gmelin, 1789. **Unassigned**: *fuscorubens* Dwigubski, 1804; *nadymensis* Serebrennikov, 1928; *subalpinus* Burg, 1920; *talahutky* Brass, 1911.

COMMENTS: Subgenus *Sciurus*. For discussion of taxonomy, see Sidorowicz (1971), Corbet (1978c), Wiltafsky (1978), and Gromov and Erbajeva (1995). Oshida et al. (1993) and Oshida and Yoshida (1997) described chromosomes of *S. v. orientis*.

Sciurus yucatanensis J. A. Allen, 1877. *In* Coues and Allen, Mongr. N Am. Rodentia (U.S. Geol. Geograph. Survey Terr., Rep., 11:705).

COMMON NAME: Yucatan Squirrel.

TYPE LOCALITY: "Merida, Yucatan," [Mexico].

DISTRIBUTION: Yucatan Peninsula (Mexico); N and SW Belize; N Guatemala.

STATUS: IUCN – Lower Risk (lc).

SYNONYMS: *baliolus* Nelson, 1901; *phaeopus* Goodwin, 1932.

COMMENTS: Subgenus *Sciurus* (Hall, 1981:422). Reviewed by Best et al. (1995, Mammalian Species No. 506).

Syntheosciurus Bangs, 1902. Bull. Mus. Comp. Zool., 39:25.
TYPE SPECIES: *Syntheosciurus brochus* Bangs, 1902.
COMMENTS: Tribe Microsciurini according to Moore (1959). Goodwin (1946) suggested that *Syntheosciurus* might be a subgenus of *Sciurus*; but see Heaney and Hoffmann (1978), Enders (1980), and Hall (1981:438). Moore (1959:179) included *Mesosciurus* (*granatensis, pyrrhinus*) as a subgenus, but see *Sciurus*.

Syntheosciurus brochus Bangs, 1902. Bull. Mus. Comp. Zool., 39:25.
COMMON NAME: Bangs's Mountain Squirrel.
TYPE LOCALITY: "Boquete, 7,000 ft." [2134 m, Chiriqui, Panama].
DISTRIBUTION: Costa Rica to N Panama.
STATUS: IUCN – Lower Risk (nt).
SYNONYMS: *poasensis* (Goodwin, 1942).
COMMENTS: Reviewed by Enders (1953:509), and Wells and Giacalone (1985, Mammalian Species, 249). Includes *poasensis* (Hall, 1981:438; Heaney and Hoffmann, 1978; Enders, 1980).

Tamiasciurus Trouessart, 1880. Le Naturaliste, 2(37):292.
TYPE SPECIES: [*Sciurus vulgaris*] *hudsonicus* Erxleben, 1777.
COMMENTS: Tribe Tamiasciurini (*Tamiasciurus* + *Sciurotamias*) according to Moore (1959), but Black (1963) and Thorington et al. (1998) argued that the genus is closely related to *Sciurus*, which has been supported by molecular data (Mercer and Roth, 2003; Steppan et al., 2004).

Tamiasciurus douglasii (Bachman, 1839). Proc. Zool. Soc. Lond., 1838:99 [1839].
COMMON NAME: Douglas's Squirrel.
TYPE LOCALITY: "Shores of Columbia River" Restricted by J. A. Allen (1898:284) to mouth of Columbia River, [Clatsop Co., Oregon, USA].
DISTRIBUTION: Coast and Cascade ranges and Sierra Nevada of SW British Columbia (not Vancouver Isl) (Canada) to S California (USA).
STATUS: IUCN – Lower Risk (lc).
SYNONYMS: *belcheri* (Gray, 1842); *suckleyi* (Baird, 1855); **mollipilosus** (Audubon and Bachman, 1841); *cascadensis* (J. A. Allen, 1898); *orarius* (Bangs, 1897).
COMMENTS: Formerly included *mearnsi*; see Lindsay (1981). Hall (1981:466) suggested that *douglasii* might be conspecific with *hudsonicus*, but Lindsay (1982) showed that apparent hybrids were probably due to character convergence. Reviewed by Steele (1999, Mammalian Species No. 630). Arbogast et al. (2001) found that mitochondrial DNA differed only slightly among the three species of *Tamiasciurus*.

Tamiasciurus hudsonicus (Erxleben, 1777). Syst. Regn. Anim., 1:416.
COMMON NAME: Red Squirrel.
TYPE LOCALITY: "ad fretum Hudsonis" Restricted by A. H. Howell (1936:134) to the mouth of Severn River, Hudson Bay, Ontario, Canada.
DISTRIBUTION: Alaska (USA), throughout Canada (south of tundra); including Vancouver Isl, W USA in mountain states; NE USA, south to NW South Carolina.
STATUS: U.S. ESA – Endangered as *T. h. grahamensis*; IUCN – Critically Endangered as *T. h. grahamensis*; otherwise Lower Risk (lc).
SYNONYMS: *rubrolineatus* (Desmarest, 1822); **abieticola** (A. H. Howell, 1929); **baileyi** (J. A. Allen, 1898); *columbiensis* A. H. Howell, 1936; **dakotensis** (J. A. Allen, 1894); **dixiensis** Hardy, 1942; **fremonti** (Audubon and Bachman, 1853); *wasatchensis* Hardy, 1950; **grahamensis** (J. A. Allen, 1894); *gymnicus* (Bangs, 1899); **kenaiensis** A. H. Howell, 1936; **lanuginosus** (Bachman, 1839); *vancouverensis* (J. A. Allen, 1890); **laurentianus** Anderson, 1942; **loquax** (Bangs, 1896); **lychnuchus** (Stone and Rehn, 1903); **minnesota** (J. A. Allen, 1899); *murii* A. H. Howell, 1943; **mogollonensis** (Mearns, 1890); *neomexicanus* (J. A. Allen, 1898); **pallescens** A. H. Howell, 1942; **petulans** (Osgood, 1900); **picatus** (Swarth, 1921); **preblei** A. H. Howell, 1936; **regalis**

A. H. Howell, 1936; *richardsoni* (Bachman, 1839); *streatori* (J. A. Allen, 1898); *ungavensis* Anderson, 1942; *ventorum* (J. A. Allen, 1898).
COMMENTS: Includes *fremonti*; see Hardy (1950). See also *douglasii*. Reviewed by Steele (1998, Mammalian Species No. 586). Arbogast et al. (2001) found that mitochondrial DNA differed only slightly among the three species of *Tamiasciurus*.

Tamiasciurus mearnsi (Townsend, 1897). Proc. Biol. Soc. Wash., 11:146.
COMMON NAME: Mearns's Squirrel.
TYPE LOCALITY: "San Pedro Martir Mountains, Lower California (altitude about 7,000 feet)" [Baja California Norte, Mexico, 2134 m].
DISTRIBUTION: Sierra San Pedro Martir Mtns (Baja California Norte, Mexico).
STATUS: IUCN – Lower Risk (lc).
COMMENTS: Formerly included in *douglasii*; see Lindsay (1981). Arbogast et al. (2001) found that mitochondrial DNA differed only slightly among the three species of *Tamiasciurus*.

Tribe Pteromyini Brandt, 1855. Beitrage zur nahern Kenntniss der Saugethiere Russland's. Kaiserlich. Akad. Wiss. Méth. Math, Phys. Nat., 7:157.
SYNONYMS: Petauristinae Miller, 1912.
COMMENTS: Flying squirrels are considered a monophyletic group (Thorington 1984) contrary to Black (1963) and Hight et al. (1974). Paleontological evidence (Mein, 1970; Bruijn and Uenay, 1989) has suggested they are not members of the Sciuridae, but molecular studies by Mercer and Roth (2003) and by Steppan et al. (2004) strongly support the view that they are the sister group of the Sciurini. Both morphological studies (Thorington et al., 2002) and molecular studies (Mercer and Roth, 2003) support division into two subtribes. Bibliographies by Lin et al. (1985) and Robins (1998).

Aeretes G. M. Allen, 1940. Nat. Hist. Cent. Asia, II, 2:745.
TYPE SPECIES: *Pteromys melanopterus* Milne-Edwards, 1867.
COMMENTS: Subtribe Pteromyina. *Aeretes* is often dated as G. M. Allen, "p.vii, September 2, 1938," but this is a *nomen nudum*.

Aeretes melanopterus (Milne-Edwards, 1867). Ann. Sci. Nat. Zool., 8:375.
COMMON NAME: Northern Chinese Flying Squirrel.
TYPE LOCALITY: "Les forêts qui couvrent la chaine montagneuse du Tscheli" [= Chihli; old name for Hebei Prov., China].
DISTRIBUTION: Hebei and Sichuan (China).
STATUS: IUCN – Lower Risk (nt).
SYNONYMS: *sulcatus* (A. B. Howell, 1927); *szechuanensis* Wang, Tu, and Wang, 1966.
COMMENTS: Known only from two widely separated areas.

Aeromys Robinson and Kloss, 1915. J. Fed. Malay St. Mus., 6:23.
TYPE SPECIES: *Pteromys tephromelas* Günther, 1873.
COMMENTS: Subtribe Pteromyina.

Aeromys tephromelas (Günther, 1873). Proc. Zool. Soc. Lond., 1873:413.
COMMON NAME: Black Flying Squirrel.
TYPE LOCALITY: "Pinang" [Wellesley, Penang Isl, Malaysia].
DISTRIBUTION: Malay Peninsula, Sumatra, and Borneo.
STATUS: IUCN – Lower Risk (lc).
SYNONYMS: *bartelsi* (Sody, 1936); *phaeomelas* (Günther, 1873).
COMMENTS: Includes *phaeomelas*; see Medway (1977:101).

Aeromys thomasi (Hose, 1900). Ann. Mag. Nat. Hist., ser. 7, 5:215.
COMMON NAME: Thomas's Flying Squirrel.
TYPE LOCALITY: "Silat River, about 70 miles [113 km] south of Claudetown, Eastern Sarawak" [Baram, Sarawak, Malaysia].
DISTRIBUTION: Borneo, except SE.
STATUS: IUCN – Lower Risk (lc).

Belomys Thomas, 1908. Ann. Mag. Nat. Hist., ser. 8, 1:2.
    TYPE SPECIES: *Sciuropterus pearsonii* Gray, 1842.
    COMMENTS: Subtribe Pteromyina.

    Belomys pearsonii (Gray, 1842). Ann. Mag. Nat. Hist., [ser. 1], 10:263.
        COMMON NAME: Hairy-footed Flying Squirrel.
        TYPE LOCALITY: "India, [Assam,] Dargellan" (= Darjeeling).
        DISTRIBUTION: Sikkim and Assam (India) to Hunan, Sichuan, Yunnan, Guizhou, Guangxi, Hainan, Taiwan (China); Bhutan; Indochina, and N Burma (see Agrawal and Chakraborty, 1979).
        STATUS: IUCN – Lower Risk (nt).
        SYNONYMS: *kaleensis* (Swinhoe, 1863); *trichotis* Thomas, 1908; *villosus* (Blyth, 1847); ***blandus*** Osgood, 1932.
        COMMENTS: Subspecies tentatively recognized by Corbet and Hill (1992), who synonymized this monotypic genus with *Trogopterus*.

Biswamoyopterus Saha, 1981. Bull. Zool. Surv. India, 4:331.
    TYPE SPECIES: *Biswamoyopterus biswasi* Saha, 1981.
    COMMENTS: Subtribe Pteromyina.

    Biswamoyopterus biswasi Saha, 1981. Bull. Zool. Surv. India, 4:333.
        COMMON NAME: Namdapha Flying Squirrel.
        TYPE LOCALITY: "Namdapha, Tirap District, Arunachal Pradesh, India."
        DISTRIBUTION: Known only from type locality, western slope, Patkai Range.
        STATUS: IUCN – Critically Endangered.
        COMMENTS: "Close to *Aeromys*" (Saha, 1981:333).

Eoglaucomys A. H. Howell, 1915. Proc. Bio. Soc. Wash., 28:109.
    TYPE SPECIES: *Sciuropterus fimbriatus* Gray, 1837.
    COMMENTS: Subtribe Glaucomyina (new name; type genus: *Glaucomys*). Characterized by origin of tibiocarpalis muscle on distal tibial tuberosity, pisiform bone with elevated scapalunate tuberosity but no triquetral process; teeth lacking crosslophs, hypocone usually absent (Thorington et al., 2002).

    Eoglaucomys fimbriatus (Gray, 1837). Ann. Mag. Nat. Hist., ser. 1:584.
        COMMON NAME: Kashmir Flying Squirrel.
        TYPE LOCALITY: "India." Restricted by Robinson and Kloss (1918a:184) to "Western Himalayas," and by Ellerman and Morrison-Scott (1955:468) to Simla, Punjab.
        DISTRIBUTION: Kashmir and Punjab (India) east to Raniket, Uttar Pradesh (Pasha and Suhail, 1997) from 1800 to 3600 m, mountains of EC and NW Afghanistan, between 1600 and 3500 m.
        STATUS: IUCN – Lower Risk (nt) as *E. fimbriatus* and as *Hylopetes baberi*.
        SYNONYMS: ***baberi*** (Blyth, 1847).
        COMMENTS: Genus included in *Hylopetes* by Ellerman (1947a) but see Thorington et al. (1996). Subspecies *baberi* elevated to specific status by Chakraborty (1981), but see Thorington et al. (1996).

Eupetaurus Thomas, 1888. J. Asiat. Soc. Bengal, 57:256.
    TYPE SPECIES: *Eupetaurus cinereus* Thomas, 1888.
    COMMENTS: Subtribe Pteromyina.

    Eupetaurus cinereus Thomas, 1888. J. Asiat. Soc. Bengal, 57:258.
        COMMON NAME: Woolly Flying Squirrel.
        TYPE LOCALITY: "Gilgit [Valley], . . . about 6000 feet" [Pakistan, 1829 m].
        DISTRIBUTION: High elevations from N Pakistan and Kashmir to Sikkim (India) to Tibet and possibly Yunnan (China; Agrawal and Chakraborty, 1970).
        STATUS: IUCN – Endangered.
        COMMENTS: Reviewed by McKenna (1962). In 1994, the first specimen to be found in " . . . over a century . . . " was obtained in the Gilgit district (Zahler, 1996).

Glaucomys Thomas, 1908. Ann. Mag. Nat. Hist., ser. 8, 1:5.
  TYPE SPECIES: *Mus volans* Linnaeus, 1758.
  COMMENTS: Subtribe Glaucomyina (see also *Eoglaucomys*). Revised by A. H. Howell (1918); karyotypes and evolution evaluated by Rausch and Rausch (1982).

Glaucomys sabrinus (Shaw, 1801). Gen. Zool., 2:157.
  COMMON NAME: Northern Flying Squirrel.
  TYPE LOCALITY: Not specified. Restricted by A. H. Howell (1918:33) to the mouth of Severn River, Ontario, Canada.
  DISTRIBUTION: Alaska and Canada, NW USA to S California and W South Dakota (Black Hills), NE USA to S Appalachian Mtns.
  STATUS: U.S. ESA – Endangered as *G. s. coloratus* and *G. s. fuscus*; IUCN – Endangered as *G. s. griseifrons*, Data Deficient as *G. s. californicus*, Vulnerable as *G. s. coloratus* and *G. s. fuscus*, otherwise Lower Risk (lc) as *G. sabrinus*.
  SYNONYMS: *canadensis* (E. Geoffroy, 1803); *hudsonius* Gmelin, 1788; **alpinus** (Richardson, 1828); **bangsi** (Rhoads, 1897); *bullatus* A. H. Howell, 1915; **californicus** (Rhoads, 1897); **canescens** A. H. Howell, 1915; **coloratus** Handley, 1953; **columbiensis** A. H. Howell, 1915; **flaviventris** A. H. Howell, 1915; **fuliginosus** (Rhoads, 1897); **fuscus** Miller, 1936; **goodwini** Anderson, 1943; **gouldi** Anderson, 1943; **griseifrons** A. H. Howell, 1934; **klamathensis** (Merriam, 1897); **lascivus** (Bangs, 1899); **latipes** A. H. Howell, 1915; **lucifugus** Hall, 1934; **macrotis** (Mearns, 1898); **makkovikensis** (Sornborger, 1900); **murinauralis** Musser, 1961; **oregonensis** (Bachman, 1839); *olympicus* (Elliot, 1899); **reductus** Cowan, 1937; **stephensi** (Merriam, 1900); **yukonensis** (Osgood, 1900); **zaphaeus** (Osgood, 1905).
  COMMENTS: Reviewed by Wells-Gosling and Heaney (1984, Mammalian Species No. 229). Subspecies according to Hall (1981); also see Arbogast (1999) who presented evidence that a western clade consisting of *oregonensis* and *californicus* may be of full species rank.

Glaucomys volans (Linnaeus, 1758). Syst. Nat., 10th ed., 1:63.
  COMMON NAME: Southern Flying Squirrel.
  TYPE LOCALITY: "Virginia, Mexico"; restricted by Elliot (1901:109) to "Virginia (U.S.A.)".
  DISTRIBUTION: Texas, Kansas, and Minnesota (USA) to Nova Scotia (Canada) and E USA; montane populations scattered from NW Mexico to Honduras.
  STATUS: IUCN – Lower Risk (lc) as *G. volans*, Not Evaluated as *G. v. goldmani*, *G. v. guerreroensis*, and *G. v. oaxacensis*.
  SYNONYMS: *americana* (Oken, 1816); *cucullatus* (Fischer, 1829); *nebrascensis* (Swenk, 1915); *silus* (Bangs, 1896); *virginianus* (Tiedemann, 1808); *volucella* (Pallas, 1778); **chontali** Goodwin, 1961; **goldmani** (Nelson, 1904); **guerreroensis** Diersing, 1980; **herreranus** Goldman, 1936; **madrensis** Goldman, 1936; **oaxacensis** Goodwin, 1961; **querceti** (Bangs, 1896); **saturatus** A. H. Howell, 1915; **texensis** A. H. Howell, 1915; **underwoodi** Goodwin, 1936.
  COMMENTS: Reviewed by Dolan and Carter (1973, Mammalian Species, 78). For systematics and distribution of Mesoamerican populations, see Diersing (1980*a*) and Braun (1988). Subspecies according to Hall (1981). A phylogeographic study using mitochondrial DNA showed little divergence between two subspecies (Arbogast, 1999).

Hylopetes Thomas, 1908. Ann. Mag. Nat. Hist., ser. 8, 1:6.
  TYPE SPECIES: *Sciuropterus everetti* Thomas, 1908 (= *Sciuropterus spadiceus* Blyth, 1847).
  COMMENTS: Subtribe Glaucomyina (see also *Eoglaucomys*). Ellerman (1947*a*) included *Eoglaucomys* in *Hylopetes*, but see McKenna (1962) and Thorington et al. (1996).

Hylopetes alboniger (Hodgson, 1836). J. Asiat. Soc. Bengal, 5:231.
  COMMON NAME: Particolored Flying Squirrel.
  TYPE LOCALITY: "central and Northern regions of Nipál" [Nepal].
  DISTRIBUTION: Nepal and Assam (India) to Sichuan, Yunnan, and Hainan (China) and Indochina.
  STATUS: IUCN – Endangered.
  SYNONYMS: *leachii* (Gray, 1837); *turnbulli* (Gray, 1838); **chianfengensis** Wang and Lu, 1966; *leonardi* (Thomas, 1921); **orinus** (G. M. Allen, 1940).

Hylopetes bartelsi (Chasen, 1939). Treubia, 17:185.
> COMMON NAME: Bartel's Flying Squirrel.
> TYPE LOCALITY: "Tjilondong, Mt. Pangrango, West Java, about 900 metres" [Indonesia].
> DISTRIBUTION: Java.
> STATUS: IUCN – Lower Risk (nt).
> COMMENTS: Formerly included in *Petinomys*; but see Corbet and Hill (1992).

Hylopetes lepidus (Horsfield, 1822). Zool. Res. Java, 5: pl. and 2 unno. pp.
> COMMON NAME: Gray-cheeked Flying Squirrel.
> TYPE LOCALITY: "..only found in the closest forests of Java," [Indonesia].
> DISTRIBUTION: Java and Borneo.
> STATUS: IUCN – Lower Risk (lc).
> COMMENTS: Formerly called *sagitta* Linnaeus, 1766; see Medway (1977:104). Formerly
> included *platyurus* (Corbet and Hill, 1992; Hill, 1961c; Medway, 1977:104), but see
> Thorington et al. (1996).

Hylopetes nigripes (Thomas, 1893). Ann. Mag. Nat. Hist., ser. 6, 12:30.
> COMMON NAME: Palawan Flying Squirrel.
> TYPE LOCALITY: "Puerta Princesa, Palawan" [Philippines].
> DISTRIBUTION: Palawan and Bancalan Isls (Philippines).
> STATUS: IUCN – Lower Risk (nt).
> SYNONYMS: *elassodontus* (Osgood, 1918).

Hylopetes phayrei (Blyth, 1859). J. Asiat. Soc. Bengal, 28:278.
> COMMON NAME: Indochinese Flying Squirrel.
> TYPE LOCALITY: "Rangoon, Merqui" [Burma].
> DISTRIBUTION: Burma; Thailand; Laos; S Vietnam; Fukien and Hainan (China).
> STATUS: IUCN – Lower Risk (lc).
> SYNONYMS: *anchises* G. M. Allen and Coolidge, 1940; *laotum* (Thomas, 1914); *probus*
> (Thomas, 1914); *electilis* (G. M. Allen, 1925).

Hylopetes platyurus (Jentink, 1890). Notes Leyden Museum 12:147.
> COMMON NAME: Jentink's Flying Squirrel.
> TYPE LOCALITY: Deli, NE Sumatra.
> DISTRIBUTION: Sumatra, Malaysia.
> COMMENTS: Formerly included in *lepidus* (Hill, 1962c; Pavlinov et al., 1995a), but see
> Thorington et al. (1996). Corbet and Hill (1992) listed *aurantiacus* as a synonym, here
> placed in *H. spadiceus*.

Hylopetes sipora Chasen, 1940. Bull. Raffles Mus., 15:117.
> COMMON NAME: Sipora Flying Squirrel.
> TYPE LOCALITY: "Sipora Island, Mentawi Islands, West Sumatra" [Indonesia].
> DISTRIBUTION: Sipora Isl (Indonesia).
> STATUS: IUCN – Endangered.
> COMMENTS: Formerly included in *spadiceus*; see Hill (1962c) who noted that an adult
> specimen is needed to clarify the status of this taxon. Part of the rodent fauna endemic
> to the Mentawai Archipelago (see comments under *Leopoldamys siporanus*).

Hylopetes spadiceus (Blyth, 1847). J. Asiat. Soc. Bengal, 16:867.
> COMMON NAME: Red-cheeked Flying Squirrel.
> TYPE LOCALITY: "Arracan" [Arakan, Burma].
> DISTRIBUTION: Burma, Thailand, S Vietnam, Sumatra, and Malaysia.
> STATUS: IUCN – Lower Risk (lc).
> SYNONYMS: *amoenus* (Miller, 1906); *aurantiacus* (Wagner, 1841); *belone* (Thomas, 1908);
> *everetti* (Thomas, 1895); *harrisoni* (Stone, 1900); *sumatrae* Sody, 1949; *caroli*
> Gyldenstolpe, 1920.
> COMMENTS: Includes *harrisoni*; see Medway (1977:105). Formerly included *sipora*; see Hill
> (1961c). Corbet and Hill (1980:137) listed *spadiceus* in *lepidus*, without comment.

Hylopetes winstoni (Sody, 1949). Treubia Buitenzorg, 20:75.
> COMMON NAME: Sumatran Flying Squirrel.

TYPE LOCALITY: "Baleq, E. Atjeh, N. Sumatra, 1200 m," [Indonesia].
DISTRIBUTION: Known only from the type locality.
STATUS: IUCN – Critically Endangered.
COMMENTS: Originally described as a species of *Iomys*, but placed in *Hylopetes* by Corbet and Hill (1992).

Iomys Thomas, 1908. Ann. Mag. Nat. Hist., ser. 8, 1:1.
TYPE SPECIES: *Pteromys horsfieldii* Waterhouse, 1838.
COMMENTS: Subtribe Glaucomyina (see also *Eoglaucomys*). Formerly included *winstoni*, here placed in *Hylopetes*.

Iomys horsfieldii (Waterhouse, 1838) Proc. Zool. Soc. Lond., 1837:87 [1838].
COMMON NAME: Javanese Flying Squirrel.
TYPE LOCALITY: "Either from Java or Sumatra." Restricted by Chasen (1940:114) to Sumatra.
DISTRIBUTION: Malay Peninsula to Java; Borneo.
STATUS: IUCN – Lower Risk (lc).
SYNONYMS: **davisoni** (Thomas, 1886); *everetti* (G. M. Allen and Coolidge, 1940) [not Thomas, 1895]; **penangensis** Chasen, 1940; **thomsoni** Thomas, 1900; *lepidus* Lyon, 1911 [not Horsfield, 1824].
COMMENTS: Formerly included *Iomys sipora* and *Hylopetes winstoni* (Corbet and Hill, 1980, 1986).

Iomys sipora Chasen and Kloss, 1928. Proc. Zool. Soc. Lond., 1927(4):819 [1928].
COMMON NAME: Mentawai Flying Squirrel.
TYPE LOCALITY: "Sipora Island, West Sumatra." [Indonesia].
DISTRIBUTION: Mentawai Isls (Indonesia).
STATUS: IUCN – Vulnerable.
COMMENTS: Part of the rodent fauna endemic to the Mentawai Archipelago (see comments under *Leopoldamys siporanus*).

Petaurillus Thomas, 1908. Ann. Mag. Nat. Hist., ser. 8, 1:3.
TYPE SPECIES: *Sciuropterus hosei* Thomas, 1900.
COMMENTS: Subtribe Glaucomyina (see also *Eoglaucomys*).

Petaurillus emiliae Thomas, 1908. Ann. Mag. Nat. Hist., ser. 8, 1:8.
COMMON NAME: Lesser Pygmy Flying Squirrel.
TYPE LOCALITY: "Baram, E. Sarawak," [Malaysia].
DISTRIBUTION: Sarawak.
STATUS: IUCN – Lower Risk (lc).

Petaurillus hosei (Thomas, 1900). Ann. Mag. Nat. Hist., ser. 7, 5:275.
COMMON NAME: Hose's Pygmy Flying Squirrel.
TYPE LOCALITY: "Baram District, Eastern Sarawak . . . Toyut River" [Malaysia].
DISTRIBUTION: Sarawak.
STATUS: IUCN – Lower Risk (lc).

Petaurillus kinlochii (Robinson and Kloss, 1911). J. Fed. Malay St. Mus., 4:171.
COMMON NAME: Selangor Pygmy Flying Squirrel.
TYPE LOCALITY: Kapar, Selangor, Malaysia.
DISTRIBUTION: Selangor (Malay Peninsula).
STATUS: IUCN – Lower Risk (lc).
COMMENTS: Corbet and Hill (1992) included this form in *hosei*.

Petaurista Link, 1795. Zool. Beytr., 1(2):52, 78.
TYPE SPECIES: *Sciurus petaurista* Pallas, 1766.
SYNONYMS: *Galeolemur* Lesson, 1840.
COMMENTS: Subtribe Pteromyina.

Petaurista alborufus (Milne-Edwards, 1870). C. R. Acad. Sci. (Paris), 70:342.
COMMON NAME: Red and White Giant Flying Squirrel.

TYPE LOCALITY: Moupin [= Baoxing, Sichuan, China].

DISTRIBUTION: Taiwan, S and C China.

STATUS: IUCN – Lower Risk (lc).

SYNONYMS: *alborusus* (Hilzheimer, 1906); ***castaneus*** Thomas, 1923; ***lena*** Thomas, 1907; *pectoralis* (Swinhoe, 1871); ***leucocephalus*** (Hilzheimer, 1905); ***ochraspis*** Thomas, 1923.

COMMENTS: Includes *lena*, which was treated as a separate species by Kuntz and Ming (1970); see Jones (1975). Provisionally includes *pectoralis* (Corbet and Hill, 1992). Reviewed from the literature by Day (1988). Subspecies provisionally recognized by Corbet and Hill (1992). Chromosomes described by Oshida et al. (1993).

Petaurista elegans (Müller, 1840). *In* Temminck, Verhandl. Nat. Gesch. Nederland. Overz. Bezitt., Zool., Zoogd. Indisch. Archipel, pp. 35, 56 [1840], see comments.

COMMON NAME: Spotted Giant Flying Squirrel.

TYPE LOCALITY: "Java" [Indonesia].

DISTRIBUTION: Nepal, Sikkim (India), Sichuan and Yunnan (China), N and W Burma, Laos, Tonkin (Vietnam), Malay Peninsula, Sumatra, Java (Indonesia), Borneo.

STATUS: IUCN – Lower Risk (lc).

SYNONYMS: *slamatensis* Sody, 1949; ***banksi*** Chasen, 1933; ***caniceps*** (Gray, 1842); *clarkei* Thomas, 1922; *gorkhali* (Lindsay, 1929); *senex* (Hodgson, 1844); ***marica*** Thomas, 1912; ***punctatus*** (Gray, 1846); ***sumatrana*** Kloss, 1921; ***sybilla*** Thomas and Wroughton, 1916.

COMMENTS: Includes *clarkei* and *marica*; see Ellerman and Morrison-Scott (1966:460-461). This species was described in greater detail by Schlegel and Müller, *in* Temminck, Verh. Nat. Gesch. Nederland. Overz. Bezitt., Zool., Mammalia, pp. 107, 112 [1845]. Corbet and Hill (1992) considered *caniceps* from Nepal, Sikkim, N Burma, and W China, and *sybilla* from a few localities in Burma and W China, as distinct species, sympatric with *elegans* in W Yunnan (China).

Petaurista leucogenys (Temminck, 1827). Monogr. Mamm., 1:27.

COMMON NAME: Japanese Giant Flying Squirrel.

TYPE LOCALITY: " . . . de forêts dans les provinces de Figo." Restricted by Kuroda (1938:50) to "Higo, Kiusiu" [Kyushu, Japan].

DISTRIBUTION: Japan, except Hokkaido.

STATUS: IUCN – Lower Risk (lc).

SYNONYMS: *tosae* Thomas, 1905; ***hintoni*** Mori, 1923; *thomasi* Kuroda and Mori, 1923 [not Hose, 1900]; *watasei* Mori, 1927; ***nikkonis*** Thomas, 1905; *osiui* Kuroda, 1938; ***oreas*** Thomas, 1905.

COMMENTS: Formerly included *xanthotis*, see Corbet (1978c:86). McKenna (1962) and Corbet and Hill (1991:145) considered it distinct. Subspecies follow Corbet (1978c). Chromosomal variation studied by Oshida and Obara (1991, 1993). Specimens reported from Korea and Manchuria were purchased in markets (Jones and Johnson, 1965).

Petaurista magnificus (Hodgson, 1836). J. Asiat. Soc. Bengal, 5:231.

COMMON NAME: Hodgson's Giant Flying Squirrel.

TYPE LOCALITY: "central and Northern regions of Nipál" [Nepal].

DISTRIBUTION: Tibet (China), Nepal, Bhutan, Sikkim (India).

STATUS: IUCN – Lower Risk (nt).

SYNONYMS: *hodgsoni* Ghose and Saha, 1981.

COMMENTS: Formerly included *nobilis* (Ellerman and Morrison-Scott, 1966:464); but also see Ghose and Saha (1981:95).

Petaurista nobilis (Gray, 1842). Ann. Mag. Nat. Hist., ser. 1, 10:263.

COMMON NAME: Bhutan Giant Flying Squirrel.

TYPE LOCALITY: "India, Dargellan" [Darjeeling].

DISTRIBUTION: C Nepal, Sikkim (India), Bhutan.

STATUS: IUCN – Lower Risk (nt).

SYNONYMS: *chrysotrix* (Hodgson, 1844); ***singhei*** Saha, 1977.

COMMENTS: Formerly included in *magnificus*; elevated to specific rank by Ghose and Saha (1981:95) and Corbet and Hill (1992).

Petaurista petaurista (Pallas, 1766). Misc. Zool., p. 54.
  COMMON NAME: Red Giant Flying Squirrel.
  TYPE LOCALITY: Not stated. Restricted by Robinson and Kloss (1918*a*:172, 221) to Preanger Regencies, Western Java, Indonesia.
  DISTRIBUTION: E Afghanistan, Kashmir and Punjab east to Assam (India); Yunnan, Sichuan, Fukien (China); Burma; Thailand; Indochina; Malaysia; Sumatra, Java (Indonesia); Borneo.
  STATUS: IUCN – Lower Risk (lc).
  SYNONYMS: *nitida* (Desmarest, 1818); *taguan* (Link, 1795); **albiventer** (Gray, 1834); *barroni* Kloss, 1916; *birrelli* Wroughton, 1911; *fulvinus* Wroughton, 1911; *inornatus* (Geoffroy, 1844); **batuana** Miller, 1903; **candidula** Wroughton, 1911; **cicur** Robinson and Kloss, 1914; **interceptio** Sody, 1949; **lumholtzi** Gyldenstolpe, 1920; **marchio** Thomas, 1908; *mimicus* Miller, 1913; **melanotus** (Gray, 1837); **nigrescens** Medway, 1965; **nigricaudatus** Robinson and Kloss, 1918; **nitidula** Thomas, 1900; **penangensis** Robinson and Kloss, 1918; **rajah** Thomas, 1908; **rufipes** Sody, 1949; *sodyi* Harris, 1951; **stellaris** Chasen, 1940; **taylori** Thomas, 1914; **terutaus** Lyon, 1907.
  COMMENTS: Reviewed by Corbet and Hill (1992), who recognized *philippensis* as distinct and allocated many forms to it that were formerly assigned to *petaurista*; the two species are widely sympatric. Also reviewed from the literature by Day (1988); karyotypic variation reported by Yong and Dhaliwal (1976). Taxonomic revision and comprehensive review of geographic variation needed (Corbet and Hill, 1992). Ghose and Bhattacharya (1995) proposed full species status for *P. fulvinus*, based on limited specimen comparison.

Petaurista philippensis (Elliot, 1839). Madras J. Litt. Sci., 10:217.
  COMMON NAME: Indian Giant Flying Squirrel.
  TYPE LOCALITY: Near Madras [India].
  DISTRIBUTION: Sri Lanka; India, north to Bombay and Rajastan, S Bihar; Burma, Thailand; S China, including Hainan and Taiwan.
  STATUS: IUCN – Lower Risk (lc).
  SYNONYMS: *cinderella* Wroughton, 1911; *griseiventer* (Gray, 1843); *lanka* Wroughton, 1911; *oral* (Tickell, 1842); **annamensis** Thomas, 1914; **badiatus** Thomas, 1925; *miloni* Bourret, 1942; **cineraceus** (Blyth, 1847); **grandis** (Swinhoe, 1863); **lylei** Bonhote, 1900; *stockleyi* Carter, 1933; *venningi* Thomas, 1914; **mergulus** Thomas, 1922; *primrosei* Thomas, 1926; *reguli* Thomas, 1926; *yunanensis* (Anderson, 1875); *hainana* G. M. Allen, 1925; *nigra* Wang, 1981; *rubicundus* A. B Howell, 1927; *rufipes* G. M. Allen, 1925.
  COMMENTS: Formerly included in *petaurista*; but see Corbet and Hill (1992), who reviewed the species. Subspecies listed are those recognized by Corbet and Hill (1992), who note that further work is needed on the geographic variation of this species.

Petaurista xanthotis (Milne-Edwards, 1872). Rech. Hist. Nat. Mammifères, p. 301.
  COMMON NAME: Chinese Giant Flying Squirrel.
  TYPE LOCALITY: Moupin [= Baoxing, Sichuan, China].
  DISTRIBUTION: Mountains of W China (Sichuan, Yunnan, E Tibet, Gansu).
  STATUS: IUCN – Lower Risk (lc).
  SYNONYMS: *buechneri* (Matschie, 1907); *filchnerinae* (Matschie, 1907).
  COMMENTS: Formerly included in *leucogenys*; but see Corbet and Hill (1992).

Petinomys Thomas, 1908. Ann. Mag. Nat. Hist., ser. 8, 1:6.
  TYPE SPECIES: *Sciuropterus lugens* Thomas, 1895.
  SYNONYMS: *Olisthomys* Carter, 1942.
  COMMENTS: Subtribe Glaucomyina (see also *Eoglaucomys*). Formerly included *bartelsi* and *electilis*, here included in *Hylopetes*; see McKenna (1962:35) and Corbet and Hill (1992).

Petinomys crinitus (Hollister, 1911). Proc. Biol. Soc. Wash., 24:185.
 COMMON NAME: Basilan Flying Squirrel.
 TYPE LOCALITY: "Basilan Island, Philippines."
 DISTRIBUTION: Basilan Island, Philippines.
 STATUS: IUCN – Lower Risk (lc).
 COMMENTS: Smaller than *Petinomys mindanensis*, with flattened tail. Paratype of *mindanensis*
  is probably not *P. crinitus* as previously thought, but a distinct species of *Petinomys;* see
  below.

Petinomys fuscocapillus (Jerdon, 1847). J. Asiat. Soc. Bengal, 16:867.
 COMMON NAME: Travancore Flying Squirrel.
 TYPE LOCALITY: "S. India" Travancore.
 DISTRIBUTION: S India, Sri Lanka.
 STATUS: IUCN – Vulnerable as *P. f. fuscocapillus*; otherwise Lower Risk (lc).
 SYNONYMS: *layardi* (Kelaart, 1850).

Petinomys genibarbis (Horsfield, 1822). Zool. Res. Java, 4:pl., 6 unno. pp.
 COMMON NAME: Whiskered Flying Squirrel.
 TYPE LOCALITY: " . . . the forests of Pugar . . . Eastern portion of Java." [Indonesia].
 DISTRIBUTION: Malaya to Sumatra, Java, and Borneo.
 STATUS: IUCN – Lower Risk (lc).
 SYNONYMS: *borneoensis* (Thomas, 1908); *malaccanus* (Thomas, 1908).
 COMMENTS: Perhaps conspecific with *sagitta*; see comment therein, and Medway (1977:102).

Petinomys hageni (Jentink, 1888). Notes Leyden Mus., 11:26.
 COMMON NAME: Hagen's Flying Squirrel.
 TYPE LOCALITY: "Deli" [North East Sumatra, Indonesia].
 DISTRIBUTION: Borneo, Sumatra.
 STATUS: IUCN – Lower Risk (lc).
 SYNONYMS: *ouwensi* Sody, 1949.
 COMMENTS: Formerly included *lugens* (Chasen, 1940:119); see comments therein. Medway
  (1977:112) considered the status of *ouwensi* doubtful.

Petinomys lugens (Thomas, 1895). Ann. Mus. Civ. Storia Nat. Geneva, Ser. 2a, 14:666.
 COMMON NAME: Sipora Flying Squirrel.
 TYPE LOCALITY: Indonesia, Mentawai Isls, Sipora Isl, "Si Oban".
 DISTRIBUTION: Sipora and N Pagai Islands (Sumatra, Indonesia).
 STATUS: IUCN – Lower Risk (nt).
 SYNONYMS: *maerens* (Miller, 1903).
 COMMENTS: Formerly included in *hageni*; but see Chasen and Kloss (1927:819) and Jenkins
  and Hill (1982). Part of the rodent fauna endemic to the Mentawai Archipelago (see
  comments under *Leopoldamys siporanus*).

Petinomys mindanensis (Rabor, 1939). Philippine Jour. Sci., 69:390.
 COMMON NAME: Mindanao Flying Squirrel.
 TYPE LOCALITY: Badiangon, Gingoog, Oriental Misamis Province, northern coast of
  Mindanao, Philippines.
 DISTRIBUTION: Dinagat, Siargao, and Mindanao Isls (Philippines; Heaney and Rabor, 1982).
 SYNONYMS: *nigricaudus* Sanborn, 1954.
 COMMENTS: Holotype of *Hylopetes mindanensis* destroyed during World War II; paratype in
  National Museum of Natural History is larger than *P. crinitus* with a round, not a
  flattened tail.

Petinomys sagitta (Linnaeus, 1766). Syst. Nat., 12th ed., 1:88.
 COMMON NAME: Arrow Flying Squirrel.
 TYPE LOCALITY: Indonesia, Java.
 DISTRIBUTION: Java.
 STATUS: IUCN – Lower Risk (lc).
 COMMENTS: Formerly included in *Hylopetes*; see Medway (1977:102). May be conspecific with
  genibarbis; see Ellerman and Morrison-Scott (1955:31). Medway (1977:102) considered

it inadvisable to synonymize *sagitta* and *genibarbis* until the relationship has been certainly established. Corbet and Hill (1992) considered *sagitta* as *incertae sedis*.

Petinomys setosus (Temminck, 1844). Fauna Japonica, 1(Mamm.), p. 49.
COMMON NAME: Temminck's Flying Squirrel.
TYPE LOCALITY: " . . . Padang, êle de Sumatra" [Indonesia].
DISTRIBUTION: Burma, Thailand, Malaysia, Sumatra, Borneo [Indonesia].
STATUS: IUCN – Lower Risk (lc).
SYNONYMS: *morrisi* (Carter, 1942).
COMMENTS: Includes *morrisi*; see Muul and Thonglongya (1971) and Corbet and Hill (1980:137). McKenna (1962) considered *morrisi* assignable to a distinct genus, *Olisthomys*. Oshida and Yoshida (1998) described chromosomes.

Petinomys vordermanni (Jentink, 1890). Notes Leyden Mus., 12:150.
COMMON NAME: Vordermann's Flying Squirrel.
TYPE LOCALITY: "Billiton" [Belitung Isl, Indonesia].
DISTRIBUTION: S Burma, Thailand, Malaya, Borneo.
STATUS: IUCN – Lower Risk (lc).
SYNONYMS: *phipsoni* (Thomas, 1916).
COMMENTS: McKenna (1962) considered *vordermanni* representative of an undescribed genus, but Muul and Thonglongya (1971) and Hill (1961c) retained it in *Petinomys*.

Pteromys G. Cuvier, 1800. Lecon's Anat. Comp., I, tab. 1.
TYPE SPECIES: *Sciurus volans* Linnaeus, 1758.
SYNONYMS: *Sciuropterus* F. Cuvier, 1825.
COMMENTS: Subtribe Pteromyina. *Sciuropterus* was previously employed for this genus by Simpson (1945:80); he believed *Pteromys* to be a synonym of *Petaurista*, but Ellerman and Morrison-Scott (1951:466) presented evidence for the validity of *Pteromys*. However, *Pteromys* may be the sister group to *Petaurista* (Oshida et al., 2000b).

Pteromys momonga Temminck, 1844. Fauna Japonica, 1(Mamm.), p. 47.
COMMON NAME: Japanese Flying Squirrel.
TYPE LOCALITY: " . . . les forêts de l'interieur," Restricted by Kishida (in Kuroda, 1938:51) to "Kiusiu" [Kyushu, Japan].
DISTRIBUTION: Kyushu and Honshu (Japan).
STATUS: IUCN – Lower Risk (lc).
SYNONYMS: *amygdali* (Thomas, 1906); *interventus* (Kuroda, 1941); *momoga* Temminck, 1845.
COMMENTS: The sister species to *P. momonga* is *P. volans* (Oshida et al., 2000b). Oshida et al. (1996b, 2000a) described chromosomes.

Pteromys volans (Linnaeus, 1758). Syst. Nat., 10th ed., 1:64.
COMMON NAME: Siberian Flying Squirrel.
TYPE LOCALITY: "in borealibus Europae, Asiae, et Americae." Restricted by Thomas (1911a:149) to "Finland." Ognev (1966:268) proposed restriction to "central Sweden," but the species does not occur there (Sulkava, 1978:76).
DISTRIBUTION: Palearctic taiga, from N Finland east to Chukotka (Russia); south to E Baltic shore; S Ural Mtns, Altai Mtns (Russia), Mongolia; N China; Korea; Sakhalin Isl (Russia), and Hokkaido (Japan); perhaps mtns of W China.
STATUS: IUCN – Lower Risk (nt).
SYNONYMS: *aluco* (Thomas, 1907); *anadyrensis* Ognev, 1940; *arsenjevi* Ognev, 1935; *betulinus* Serebrennikov, 1930; *gubari* Ognev, 1935; *incanus* Miller, 1918; *ognevi* Stroganov, 1936; *russicus* Tiedemann, 1808; *sibiricus* Desmarest, 1922; *turovi* Ognev, 1919; *vulgaris* Wagner, 1842; *wulungshanensis* (Mori, 1939); **athene** (Thomas, 1907); **buechneri** Satunin, 1903; **orii** (Kuroda, 1921).
COMMENTS: Chromosomes described by Rausch and Rausch (1982), and Oshida et al. (2000a). Subspecies follow Corbet (1978c).

Pteromyscus Thomas, 1908. Ann. Mag. Nat. Hist., ser. 8, 1:3.
TYPE SPECIES: *Sciuropterus pulverulentus* Günther, 1873.

COMMENTS: Subtribe Pteromyina.

Pteromyscus pulverulentus (Günther, 1873). Proc. Zool. Soc. Lond., 1873:413.
COMMON NAME: Smoky Flying Squirrel.
TYPE LOCALITY: "Pinang" [Penang, Malaysia].
DISTRIBUTION: S Thailand to Sumatra; Borneo.
STATUS: IUCN – Lower Risk (nt).
SYNONYMS: ***borneanus*** Thomas, 1908.

Trogopterus Heude, 1898. Mem. Hist. Nat. Emp. Chin., 4(1): 46-47.
TYPE SPECIES: *Pteromys xanthipes* Milne-Edwards, 1867.
COMMENTS: Subtribe Pteromyina. *Belomys* included in *Trogopterus* by Corbet and Hill (1992).

Trogopterus xanthipes (Milne-Edwards, 1867). Ann. Sci. Nat. Zool., 8: 376.
COMMON NAME: Complex-toothed Flying Squirrel.
TYPE LOCALITY: "Les forêts Qui couvrent la chaine montagneuse du Tscheli" [= Chihli, old name for Hebei Prov., China].
DISTRIBUTION: Montane forests, from Yunnan to C and E China.
STATUS: IUCN – Endangered.
SYNONYMS: *edithae* Thomas, 1923; *himalaicus* Thomas, 1914; *minax* Thomas, 1923; *mordax* Thomas, 1914.
COMMENTS: No subspecies recognized (Corbet and Hill, 1992).

Subfamily Callosciurinae Pocock 1923. Proc. Zool. Soc. Lond., 1923: 239.

Callosciurus Gray, 1867. Ann. Mag. Nat. Hist., ser. 3, 20:277.
TYPE SPECIES: *Sciurus rafflesii* Vigors and Horsfield, 1828 (= *Sciurus prevostii* Desmarest, 1822).
SYNONYMS: *Baginia* Gray, 1867; *Erythrosciurus* Gray, 1867; *Hessonoglyphotes* Moore, 1959; *Heterosciurus* Trouessart, 1880; *Tomeutes* Thomas, 1915.
COMMENTS: Tribe Callosciurini according to Moore (1959). Formerly included *Sundasciurus* and *Prosciurillus* (in part; see Moore, 1958) and *Tamiops* (Moore and Tate, 1965). See also Corbet and Hill (1992). Reviewed in part by Chakraborty (1985).

Callosciurus adamsi (Kloss, 1921). J. Str. Br. Roy. Asiat. Soc., 83:151.
COMMON NAME: Ear-spot Squirrel.
TYPE LOCALITY: "Long Mujan, 150 mi. (241 km) up Baram River, Baram, Sarawak,700-900 ft." (213-274 m) (Malaysia). See Corbet and Hill (1992:292).
DISTRIBUTION: Lowlands and hills in Sabah and Sarawak, Borneo, below the altitudinal range of *orestes*.
STATUS: IUCN – Lower Risk (lc).
COMMENTS: Closely related to *orestes*. Distinct from *albescens*; see Medway (1977:93). Chromosomes of "*C. albescens*" from Sabah by Harada and Kobayashi (1980) refer to either this species or *C. orestes*.

Callosciurus albescens (Bonhote, 1901). Ann. Mag. Nat. Hist., ser. 7, 7:446.
COMMON NAME: Kloss's Squirrel.
TYPE LOCALITY: Indonesia, Sumatra, Acheen (= Atjeh).
DISTRIBUTION: Sumatra. Restricted to N Sumatra.
STATUS: IUCN – Lower Risk (lc).
SYNONYMS: *albiculus* G. Miller, 1942.
COMMENTS: Considered a distinct species by Medway (1977:93). Corbet and Hill (1992:291) considered *albescens* and *albiculus* assignable to *C. notatus*.

Callosciurus baluensis (Bonhote, 1901). Ann. Mag. Nat. Hist., ser. 7, 7:174.
COMMON NAME: Kinabalu Squirrel.
TYPE LOCALITY: Malaysia, Sabah, Mount Kinabalu, Borneo, 1000 ft. (305 m).
DISTRIBUTION: Above 300 m elevation in Sabah and Sarawak, Malaysia.
STATUS: IUCN – Lower Risk (lc).
SYNONYMS: *baramensis* (Chasen, 1940); *medialis* G. M. Allen and Coolidge, 1940.
COMMENTS: Sometimes considered a subspecies of *prevostii* (see Medway, 1977:86), but often

sympatric with that species (Payne et al., 1985). Corbet and Hill (1992:290-91) placed *prevostii erythromelas* (Temminck, 1853) in this species, with *schlegeli* (Gray, 1867) a synonym.

Callosciurus caniceps (Gray, 1842). Ann. Mag. Nat. Hist., [ser. 1], 10:263.
    COMMON NAME: Gray-bellied Squirrel.
    TYPE LOCALITY: "Bhotan." Restricted by Robinson and Kloss (1918*a*:206) to N Tenasserim, Burma.
    DISTRIBUTION: Thailand, peninsular Burma, peninsular Malaysia, and adjacent islands.
    STATUS: IUCN – Lower Risk (lc).
    SYNONYMS: *epomophorus* (Bonhote, 1901); **adangensis** (Miller, 1903); **bimaculatus** (Temminck, 1853); *davisoni* (Bonhote, 1901); *milleri* (Robinson and Wroughton, 1911); **casensis** (Miller, 1903); **concolor** (Blyth, 1856); *erubescens* Cabrera, 1917; **domelicus** (Miller, 1903). **Unassigned**: *altinsularis* (Miller, 1903); *bentincanus* (Miller, 1903); *chrysonotus* (Blyth, 1847); *fallax* (Robinson and Kloss, 1914); *fluminalis* (Robinson and Wroughton, 1911); *hastilis* Thomas, 1923; *helgei* (Gyldenstolpe, 1917); *helvus* (Shamel, 1930); *inexpectatus* (Kloss, 1916); *lancavensis* (Miller, 1903); *lucas* (Miller, 1903); *mapravis* Thomas and Robinson, 1921; *matthaeus* (Miller, 1903); *moheius* Thomas and Robinson, 1921; *mohillius* Thomas and Robinson, 1921; *nakanus* Thomas and Robinson, 1921; *panjioli* Thomas and Robinson, 1921; *panjius* Thomas and Robinson, 1921; *pipidonis* Thomas and Robinson, 1921; *samuiensis* (Robinson and Kloss, 1914); *sullivanus* (Miller, 1903); *tabaudius* Thomas, 1922; *tacopius* Thomas and Robinson, 1921; *telibius* Thomas and Robinson, 1921; *terutavensis* (Thomas and Wroughton, 1909).
    COMMENTS: Revised by Moore and Tate (1965), and subsequently by Corbet and Hill (1992).

Callosciurus erythraeus (Pallas, 1779). Nova Spec. Quad. Glir. Ord., p. 377.
    COMMON NAME: Pallas's Squirrel.
    TYPE LOCALITY: Not known; restricted to Assam, India by Bonhote (1901*a*); further restricted to the Garo Hills of Assam by Moore and Tate (1965).
    DISTRIBUTION: West of Irrawaddy River in India, Burma, and SE China. East of Irrawaddy River in Burma, Thailand, Peninsular Malaysia, Indochina, S China, and Taiwan. Introduced into Argentina (Aprila and Chicco, 1999).
    STATUS: IUCN – Lower Risk (lc).
    SYNONYMS: *wellsi* Wroughton, 1921; **atrodorsalis** (Gray, 1942); **bartoni** (Thomas 1914); *careyi* Thomas and Wroughton, 1916; *fryanus* Thomas and Wroughton, 1916; *millardi* Thomas and Wroughton, 1916; *shortridgei* Thomas and Wroughton, 1916; **bhutanensis** (Bonhote, 1901); *crumpi* Wroughton, 1916; **bonhotei** (Robinson and Wroughton, 1911); **castaneoventris** (Gray, 1842); *insularis* (J. A. Allen, 1906); **erythrogaster** (Blyth, 1842); *crotalius* Thomas and Wroughton, 1916; *kinneari* Thomas and Wroughton, 1916; *nagarum* Thomas and Wroughton, 1916; *punctatissimus* (Gray, 1867); **flavimanus** (G. St. Hilaire, 1831); *bolovensis* Osgood, 1932; *contumax* Thomas, 1927; *dactylinus* Thomas, 1927; *pirata* Thomas, 1929; *quantulus* Thomas, 1927; **gloveri** (Thomas, 1921); **gordoni** (Anderson, 1871); **griseimanus** (Milne-Edwards, 1867); *fumigatus* Bonhote, 1907; *leucopus* Gray, 1867; *phanrangis* Robinson and Kloss, 1922; *vassali* Bonhote, 1907; **harringtoni** (Thomas, 1905); *solutus* Thomas, 1914; **hendeei** Osgood, 1932; **hyperythrus** (Blyth, 1855); **intermedius** (Anderson, 1879); *aquilo* Wroughton, 1921; **michianus** (Robinson and Wroughton, 1911); *haemobaphes* (G. M. Allen, 1912); **ningpoensis** (Bonhote, 1901); *tsingtanensis* (Hilzheimer, 1905); **pranis** (Kloss, 1916); **rubeculus** (Miller, 1903); *youngi* (Robinson and Kloss, 1914); **shanicus** (Ryley, 1914); **siamensis** (Gray 1860); *tachin* (Kloss, 1916); **sladeni** (Anderson, 1871); *kemmisi* (Wroughton, 1908); *midas* (Thomas, 1914); *rubex* (Thomas, 1914); *vernayi* Carter, 1942; **styani** (Thomas, 1894); *albifer* (Hilzheimer, 1906); *canigenus* A. B Howell, 1927; *griseopectus* (Milne-Edwards, 1874); *leucurus* (Hilzheimer, 1905); *woodi* Harris, 1931; **thai** (Kloss, 1917); **thaiwanensis** (Bonhote, 1901); *centralis* (Bonhote, 1901); *nigridorsalis* Kuroda, 1935; *roberti* (Bonhote, 1901); **zimmeensis** Robinson and Wroughton, 1916; *primus* G. M. Allen and Coolidge, 1940. **Unassigned**:

cinnamomeiventris (Swinhoe, 1862); *cucphuongis* Dao Van Tien, 1965; *dabshanensis* Xu and Chen, 1989; *gongshanensis* Wang, 1981; *griseopectus* (Blyth, 1847); *quinlingensis* Xu and Chen, 1989; *tsingtauensis* (Hilzheimer, 1906); *wuliangshanensis* Li and Wang, 1981.

COMMENTS: Includes *sladeni*; see Moore and Tate (1965). Includes *flavimanus* G. St. Hillaire (1831) as a junior synonym (Corbet and Hill, 1992:286); both were previously considered distinct (Moore and Tate, 1965). Subspecies allocations follow Moore and Tate (1965). Chromosomes described (as *flavimanus*) by Nadler et al. (1975*b*). Indian populations reviewed by Agrawal and Chakraborty (1979) and by Oshida et al. (1996*a*, *b*), and Vietnamese populations by Dao (1965).

Callosciurus finlaysonii (Horsfield, 1823). Zool. Res. Java, 7:151.
COMMON NAME: Finlayson's Squirrel.
TYPE LOCALITY: "the Islands called Sichang, in the Gulf of Siam", Koh Si Chang (Gulf of Thailand).
DISTRIBUTION: SC Burma, Thailand, Cambodia, Laos, Vietnam. Introduced into Italy (Bertolino et al., 2000).
STATUS: IUCN – Lower Risk (lc).
SYNONYMS: Insular: **albivexilli** (Kloss, 1916); **folletti** (Kloss, 1915); **frandseni** (Kloss, 1916); **germaini** (Milne-Edwards, 1867); **harmandi** (Milne-Edwards, 1877); *pierrei* Robinson and Kloss, 1922; **trotteri** (Kloss, 1916); Mainland: **annellatus** Thomas, 1929; **bocourti** (Milne-Edwards, 1867); *cockerelli* Thomas, 1928; *floweri* (Bonhote, 1901); *leucocephalus* (Bonhote, 1901); *leucogaster* (Milne-Edwards, 1867); *prachin* Kloss, 1920; *rajasima* (Kloss, 1920); *tachardi* Robinson, 1916; **boonsongi** Moore and Tate, 1965; **cinnamomeus** (Temminck, 1853); *herberti* Robinson and Kloss, 1922; *splendens* (Gray, 1861); **ferrugineus** (F. Cuvier, 1829); **menamicus** Thomas, 1929; **nox** (Wroughton, 1908; **sinistralis** (Wroughton, 1908); *dextralis* (Wroughton, 1908); *grutei* (Gyldenstolpe, 1917); *lylei* (Wroughton, 1908); **williamsoni** Robinson and Kloss, 1922. **Unassigned**: *hyperythrus* Blyth, 1856; *keraudrenii* (Lesson, 1830); *portus* (Kloss, 1915).
COMMENTS: Revised by Moore and Tate (1965). Subsequently revised and expanded to include *ferrugineus* by Corbet and Hill (1992:286-87). Chromosomes described by Nadler et al. (1975*b*).

Callosciurus inornatus (Gray, 1867). Ann. Mag. Nat. Hist., ser. 3, 20:282.
COMMON NAME: Inornate Squirrel.
TYPE LOCALITY: "Loo Mountains." Restricted by Moore and Tate (1965:209) to "Mountains in Laos."
DISTRIBUTION: Laos, N Vietnam, S Yunnan (China).
STATUS: IUCN – Lower Risk (lc).
SYNONYMS: *imitator* Thomas, 1925.
COMMENTS: Revised by Moore and Tate (1965); distribution allopatric, separated from *C. canipes* by the Mekong River; sympatric with the larger *C. erythraeus hendeei* (Corbet and Hill, 1992:289).

Callosciurus melanogaster (Thomas, 1895). Ann. Mus. Civ. Stor. Nat. Genova, 14:668.
COMMON NAME: Mentawai Squirrel.
TYPE LOCALITY: Indonesia, Mentawai Isls, Sipora Isl, "Si Oban".
DISTRIBUTION: Mentawai Isls (Indonesia), Siberut, Sipora, and N Pagai.
STATUS: IUCN – Lower Risk (lc).
SYNONYMS: **atratus** (Miller, 1903); **mentawi** (Chasen and Kloss, 1927).
COMMENTS: Reviewed by Corbet and Hill (1992:292). Part of the rodent fauna endemic to the Mentawai Archipelago (see comments under *Leopoldamys siporanus*).

Callosciurus nigrovittatus (Horsfield, 1823). Zool. Res. Java, 7:149.
COMMON NAME: Black-striped Squirrel.
TYPE LOCALITY: Indonesia, "The island of Java." Restricted to W Java by Kloss (1921).
DISTRIBUTION: Vietnam, Malaysia, peninsular Thailand, Sumatra, Borneo, Java, adjacent small islands.
STATUS: IUCN – Lower Risk (lc).

SYNONYMS: *bantamensis* Sody, 1949; *besuki* (Kloss, 1921); *grisei-venter* (Geoffroy, 1831); *madsoedi* Sody, 1929; *phoenicurus* Sody, 1949; *salakensis* Sody, 1949; *tenggerensis* Sody, 1949; **bilimitatus** (Miller, 1903); *johorensis* (Robinson and Wroughton, 1911); *microrhynchus* (Kloss, 1908); **bocki** (Robinson and Wroughton, 1911); *acraeus* (Miller, 1942); **klossi** (Miller, 1900).

COMMENTS: Reviewed by Corbet and Hill (1992:292).

Callosciurus notatus (Boddaert, 1785). Elench. Anim., p. 119.

COMMON NAME: Plantain Squirrel.

TYPE LOCALITY: Indonesia, West Java.

DISTRIBUTION: From peninsular Malaysia and Thailand to Java, Bali, and Borneo; Lombok Isl; Salayer Isl (south of Sulawesi; Musser, 1987*a* considered this population an introduction, possibly from Java); widespread on smaller islands.

STATUS: IUCN – Lower Risk (lc).

SYNONYMS: The following taxa are from Java: *badjing* (Kerr, 1792); *balstoni* (Robinson and Wroughton, 1911); *bilineatus* (Geoffroy St. Hilaire, 1817); *dschinschicus* (Gmelin, 1788); *ginginianus* (Shaw, 1801); *plantani* (Ljungh, 1801); *tamansari* (Kloss, 1921); **diardii** (Jentink, 1879); *andrewsii* (Bonhote, 1901); *madurae* (Thomas, 1910); *magnificus* Sody, 1949; *prinsulae* Sody, 1949; *vanheurni* Sody, 1929; *verbeeki* Sody, 1929. The following taxa are from Sumatra: **vittatus** (Raffles, 1821); *albescens* (Bonhote, 1901); *albiculus* Miller, 1942; *bivittatus* (Desmarest, 1822); *ictericus* (Miller, 1903); *kalianda* (Sody, 1949); *nicotianicae* (Sody, 1936); *percommodus* (Chasen, 1940); *pretiosus* (Miller, 1903); *rupatius* (Lyon, 1908); *saturatus* (Miller, 1903); *tapanulius* (Lyon, 1907); *tarussanus* (Lyon, 1907); *tedongus* (Lyon, 1906); *toupai* (Lesson, 1827); *ubericolor* (Miller, 1903). The following is from Borneo: **suffusus** (Bonhote, 1901). The following taxa are from mainland Malaya, S Thailand, and adjacent islands: **miniatus** (Miller, 1900); *aoris* (Miller, 1903); *famulus* (Robinson, 1912); *guillemardi* (Kloss, 1926); *lighti* (Chasen and Kloss, 1924); *pemangiliensis* (Miller, 1903); *peninsularis* (Miller, 1903); *plasticus* (Kloss, 1911); *scottii* (Kloss, 1911); *singapurensis* (Robinson, 1916); *subluteus* (Thomas and Wroughton, 1909); *tenuirostris* (Miller, 1900); *tinggius* Hill, 1960; *watsoni* (Kloss, 1911). Remaining names (most from small islands), not allocated to subspecies: *abbottii* (Miller, 1900); *anambensis* (Miller, 1900); *arendsis* (Lyon, 1911); *atristriatus* (Miller, 1913); *billitonus* (Lyon, 1906); *conipus* (Lyon, 1911); *datus* (Lyon, 1911); *dilutus* (Miller, 1913); *director* (Lyon, 1909); *dulitensis* (Bonhote, 1901); *lamucotanus* (Lyon, 1911); *lautensis* (Miller, 1901); *lunaris* (Chasen and Kloss, 1924); *lutescens* (Miller, 1901); *malawali* (Chasen and Kloss, 1932); *maporensis* (Robinson, 1916); *marinsularis* (Lyon, 1911); *microtis* (Jentink, 1879); *nesiotes* (Thomas and Wroughton, 1909); *pannovianus* (Miller, 1903); *perhentiani* (Kloss, 1911); *poliopus* (Lyon, 1911); *proteus* (Kloss, 1911); *raptor* Hill, 1960; *rubidiventris* (Miller, 1901); *rutiliventris* (Miller, 1901); *seraiae* (Miller, 1901); *serutus* (Miller, 1906); *siriensis* (Lyon, 1911); *stellaris* (Chasen and Kloss, 1924); *stresemanni* (Thomas, 1913); *vinocastaneus* Sody, 1949.

COMMENTS: Corbet and Hill (1992:291) commented on the enormous number of named forms, in contrast to which is the slight degree of geographic variation compared to other widespread species of *Callosciurus*. Chromosomes described by Nadler et al. (1975*b*).

Callosciurus orestes (Thomas, 1895). Ann. Mag. Nat. Hist., ser. 6, 15:529.

COMMON NAME: Borneo Black-banded Squirrel.

TYPE LOCALITY: Malaysia, Sarawak, Mount Dulit, 4000 ft (1219 m)

DISTRIBUTION: Sabah and Sarawak (Borneo), Malaysia, at middle elevations (Payne et al., 1985).

STATUS: IUCN – Lower Risk (lc).

SYNONYMS: *canalvus* (Moore, 1959); *venetus* (Chasen, 1940).

COMMENTS: Formerly considered a subspecies of *C. nigrovittatus* (e.g., Medway, 1977:92), but clearly distinct (Payne et al., 1985). Includes *Glyphotes canalvus*, which Moore (1959) placed in subgenus *Hessonoglyphotes*; these names are now synonyms of *Callosciurus*. Closely related to *adamsi* (see comments therein).

Callosciurus phayrei (Blyth, 1856). J. Asiat. Soc. Bengal, 24:472.
COMMON NAME: Phayre's Squirrel.
TYPE LOCALITY: Burma, Martaban, "Sent from Moulmein" (Robinson and Kloss, 1918a:225).
DISTRIBUTION: Upper Irrawaddy River and Sittang River eastward to Salween River, S Burma.
STATUS: IUCN – Lower Risk (lc).
SYNONYMS: *blanfordii* (Blyth, 1862); *heinrichi* Tate, 1954.
COMMENTS: Considered a distinct species by Moore and Tate (1965) instead of a subspecies of *flavimanus*. Closely allied to *caniceps* and *pygerythrus* (Corbet and Hill, 1992:289).

Callosciurus prevostii (Desmarest, 1822). Mammalogie, *in* Encyclop. Méth., p. 335.
COMMON NAME: Prevost's Squirrel.
TYPE LOCALITY: Malaysia, Malacca Prov., "Settlement of Malacca."
DISTRIBUTION: Peninsular Thailand to Sumatra, Borneo, and adjacent small islands; N Sulawesi. Musser (1987a) considered the population in Sulawesi as introduced.
STATUS: IUCN – Lower Risk (lc).
SYNONYMS: The following taxa are from mainland Malaysia: *humei* (Bonhote, 1901); *indica* (Muller and Schlegel, 1842); *wrayi* (Kloss, 1910). The following taxa are from S Sumatra: **rafflesii** (Vigors and Horsfield, 1828); *harrisoni* (Stone and Rehn, 1902); *sumatranus* (Schlegel, 1863). The following are from E Sumatra: **melanops** (Miller, 1902); *nyx* (Lyon, 1908); *penialius* (Lyon, 1908). The following are from N Sumatra: **piceus** (Peters, 1866); *erebus* (Miller, 1903); *pluto* (Gray, 1867). The following are from Borneo, including Sarawak: **atricapillus** (Schlegel, 1863); *armalis* (Lyon, 1911); *carimatae* (Miller, 1906); *pelapius* (Lyon, 1911); *proserpinae* (Lyon, 1906); *sanggaus* (Lyon, 1907); **sarawakensis** (Gray, 1867); *caroli* (Bonhote, 1901). **Unassigned**: *atrox* (Miller, 1913); *bangkanus* (Schlegel, 1863); *banksi* (Chasen, 1933); *borneoensis* (Müller and Schlegel, 1842); *caedis* (Chasen and Kloss, 1932); *carimonensis* (Miller, 1906); *condurensis* (Miller, 1906); *coomansi* Sody, 1949; *erythromelas* (Temminck, 1853); *griseicauda* (Bonhote, 1901); *indica* (Müller and Schlegel, 1842) [not *indicus* Erxleben, 1777]; *kuchingensis* (Bonhote, 1901); *mendanauus* (Lyon, 1906); *mimellus* (Miller, 1900); *mimiculus* (Miller, 1900); *navigator* (Bonhote, 1901); *palustris* (Lyon, 1907); *redimitus* (Boon Mesch, 1829); *rufogularis* (Gray, 1842); *rufoniger* (Motley and Dillwyn, 1855); *rufonigra* (Gray, 1842); *schlegeli* (Gray, 1867); *suffusus* (Bonhote, 1901); *waringensis* Sody, 1949.
COMMENTS: Reviewed by Corbet and Hill (1992:289-90). Formerly included *baluensis*; but the two are often sympatric (Payne et al., 1985). Heaney (1978) discussed body size variation in this species. Oshida et al. (1996b) described chromosomes.

Callosciurus pygerythrus (I. Geoffroy Saint Hilaire, 1833). Mag. Zool. Paris, vol. 2: p. 5, pl. 4-6.
COMMON NAME: Irrawaddy Squirrel.
TYPE LOCALITY: "from forest of Syriam, near Pegu, Burma" (Moore and Tate, 1965:217).
DISTRIBUTION: Nepal and NE India to Burma, N Vietnam, and Yunnan (China).
STATUS: IUCN – Vulnerable.
SYNONYMS: **blythii** (Tytler, 1854); **janetta** (Thomas, 1914); **lokroides** (Hodgson, 1836); *assamensis* (Gray, 1843); *similis* (Gray, 1867); **mearsi** (Bonhote, 1906); *bellona* (Thomas and Wroughton, 1916); *virgo* (Thomas and Wroughton, 1916); **owensi** (Thomas and Wroughton, 1916); **stevensi** (Thomas, 1908).
COMMENTS: Revised by Moore and Tate (1965:209); reviewed by Corbet and Hill (1992:289).

Callosciurus quinquestriatus (Anderson, 1871). Proc. Zool. Soc. Lond., 1871:142.
COMMON NAME: Anderson's Squirrel.
TYPE LOCALITY: "common at Ponsee, on the Kakhyen range of hills, east of Bhamo, at an elevation of from 2000 to 3000 ft" [Burma].
DISTRIBUTION: NE Burma, Yunnan (China).
STATUS: IUCN – Vulnerable.
SYNONYMS: *beebei* (J. A. Allen, 1911); *sylvester* Thomas, 1926; **imarius** Thomas, 1926.
COMMENTS: Included in *C. erythraeus* by Moore and Tate (1965:209); but this distinctly marked species with a very limited range is sympatric with the more widely distributed *C. erythraeus* in W Yunnan (Corbet and Hill, 1992:286).

Dremomys Heude, 1898. Mem. Hist. Nat. Emp. Chin., 4(2):54.
TYPE SPECIES: *Sciurus pernyi* Milne-Edwards, 1867.
SYNONYMS: *Zetis* Thomas, 1908 [Type species *S. rufigenis* Blanford].
COMMENTS: Tribe Callosciurini according to Moore (1959). Reviewed by Moore and Tate (1965), and Corbet and Hill (1992).

Dremomys everetti (Thomas, 1890). Ann. Mag. Nat. Hist., ser. 6, 6:171.
COMMON NAME: Bornean Mountain Ground Squirrel.
TYPE LOCALITY: "Mount Penrisen, West Sarawak," [Malaysia].
DISTRIBUTION: Mountains of N and W Borneo (Kalimantan, Sarawak, Sabah), above 3200 ft (975 m).
STATUS: IUCN – Lower Risk (lc).

Dremomys gularis Osgood, 1932. Field Mus. Nat. Hist., Publ. Zool., 18:284.
COMMON NAME: Red-throated Squirrel.
TYPE LOCALITY: Mt. Fan Si Pan, near Chapa, Tonkin, Vietnam.
DISTRIBUTION: Red River Valley from N Vietnam to SC Yunnan (China) (Zhang et al., 1997).
COMMENTS: This species lives in geographic sympatry with *D. rufigenis,* but at higher altitudes. It resembles *D. pyrrhomerus* in large size of rostrum.

Dremomys lokriah (Hodgson, 1836). J. Asiat. Soc. Bengal, 5:232.
COMMON NAME: Orange-bellied Himalayan Squirrel.
TYPE LOCALITY: "central and Northern regions of Nipál" [Nepal].
DISTRIBUTION: C Nepal east to Salween River; Xizang (China); N Burma; mountains in E India; Bhutan.
STATUS: IUCN – Lower Risk (lc).
SYNONYMS: *bhotia* Thomas and Wrougton, 1916; **garonum** Thomas, 1922; *subflaviventris* Thomas, 1922; **macmillani** Thomas and Wroughton, 1916; **motuoensis** Cai and Zhang, 1980; **pagus** Moore, 1956.
COMMENTS: Revised by Moore and Tate (1965); see also Agrawal and Chakraborty (1979).

Dremomys pernyi (Milne-Edwards, 1867). Rev. Mag. Zool. (Paris), ser. 2, 19:19.
COMMON NAME: Perny's Long-nosed Squirrel.
TYPE LOCALITY: "les montagnes de la principaute de Moupin [Muping]" [= Baoxing, Sichuan, China].
DISTRIBUTION: NE India; N Burma; N Vietnam; Xizang; Sichuan, Yunnan, Guizhou, Hunan, Hubei, Jiangxi, Fujian, Anhui, Taiwan (China).
STATUS: IUCN – Lower Risk (lc).
SYNONYMS: *griselda* Thomas, 1916; *lentus* A. B. Howell, 1927; **flavior** G. M. Allen, 1912; *lichiensis* Thomas, 1922; **howelli** Thomas, 1922; *calidior* Thomas, 1916; *chintalis* Thomas, 1916; *modestus* Thomas, 1916; **imus** Thomas, 1922; *mentosus* Thomas, 1922; **owstoni** (Thomas, 1908); **senex** G. M. Allen, 1912.
COMMENTS: Reviewed by Corbet and Hill (1992:298-99); see also G. M. Allen (1940:647-54).

Dremomys pyrrhomerus (Thomas, 1895). Ann. Mag. Nat. Hist., ser. 6, 16:242.
COMMON NAME: Red-hipped Squirrel.
TYPE LOCALITY: "Ichang, Yang-tse-kiang [river] [Hupei, China]."
DISTRIBUTION: C and S China, extreme N Vietnam, Hainan Isl (China).
STATUS: IUCN – Lower Risk (lc).
SYNONYMS: *melli* Matschie, 1922; **riudonensis** J. A. Allen, 1906.
COMMENTS: Moore and Tate (1965) treated *pyrrhomerus* as a species separate from *rufigenis* based on evidence of geographic parapatry, and relative rostral length. Corbet and Hill (1992) listed *gularis* as a separate species, but considered other forms of *pyrrhomerus* to be conspecific with *rufigenis,* whereas Zhang et al. (1997) followed Moore and Tate (op. cit.), providing additional Chinese locality records showing sympatry of the two taxa at 10 localities in Yunnan, 3 in Guangxi, and one each in Hunan and Anhui provinces.

Dremomys rufigenis (Blanford, 1878). J. Asiat. Soc. Bengal, 47(2):156.
COMMON NAME: Asian Red-cheeked Squirrel.

TYPE LOCALITY: Burma, Tenasserim, Mt. Mooleyit.

DISTRIBUTION: NE India, N and C Burma; Anhui, Hunan, Guangxi and Yunnan (China), Laos, south through Vietnam, Thailand, peninsular Malaysia.

STATUS: IUCN – Lower Risk (lc).

SYNONYMS: *laomache* Thomas, 1921; **adamsoni** Thomas, 1914; *opimus* Thomas and Wroughton, 1916; **belfieldi** (Bonhote, 1908); **fuscus** (Bonhote, 1907); **ornatus** Thomas, 1914.

COMMENTS: Formerly included *pyrrhomerus*; but see Moore and Tate (1965), and comments under *D. pyrrhomerus*. Chromosomes described by Nadler and Hoffmann (1970).

Exilisciurus Moore, 1958. Am. Mus. Novit., 1914:4.

TYPE SPECIES: *Sciurus exilis* Müller, 1838.

COMMENTS: Tribe Callosciurini according to Moore (1959). The species included in *Exilisciurus* were formerly included in *Nannosciurus*; see Moore (1958, 1959) and Heaney (1985).

Exilisciurus concinnus (Thomas, 1888). Ann. Mag. Nat. Hist., ser. 6, 2:407.

COMMON NAME: Philippine Pygmy Squirrel.

TYPE LOCALITY: Philippines, Zamboanga Prov., Basilan Isl, Isabela.

DISTRIBUTION: Mindanao faunal region (Heaney et al., 1987) including Mindanao, Basilan, Biliran, Bohol, Dinagat, Leyte, Samar, and Siargao Isls (Heaney, 1985).

STATUS: IUCN – Lower Risk (lc).

SYNONYMS: *luncefordi* (Taylor, 1934); *samaricus* (Thomas, 1897); *surrutilus* (Hollister, 1913).

COMMENTS: Revised by Heaney (1985).

Exilisciurus exilis (Müller, 1838). Tijdschr. Nat. Gesch. Physiol., 5:138.

COMMON NAME: Least Pygmy Squirrel.

TYPE LOCALITY: Indonesia, Kalimantan, Kapuas River Basin, Tanah Laut (see Medway, 1977, and Heaney, 1985).

DISTRIBUTION: Borneo and Banggi Isl.

STATUS: IUCN – Lower Risk (lc).

SYNONYMS: *retectus* (Thomas, 1910); *sordidus* (Chasen and Kloss, 1928).

Exilisciurus whiteheadi (Thomas, 1887). Ann. Mag. Nat. Hist., ser. 5, 20:127.

COMMON NAME: Tufted Pygmy Squirrel.

TYPE LOCALITY: Malaysia, Sabah, Mt Kinabalu.

DISTRIBUTION: Mountains of Sabah and Sarawak (Malaysia), above 900 m, and adjacent parts of West Kalimantan, Indonesia (Medway, 1977).

STATUS: IUCN – Lower Risk (lc).

Funambulus Lesson, 1835. Illustr. Zool., pl. 43.

TYPE SPECIES: *Sciurus indicus* Lesson, 1835 (= *Sciurus palmarum* Linnaeus, 1776).

SYNONYMS: *Palmista* Gray, 1867; *Prasadsciurus* Moore and Tate, 1965; *Tamiodes* Pocock, 1923.

COMMENTS: Tribe Funambulini according to Moore (1959:170). Reviewed by Moore (1960) and Moore and Tate (1965). Includes *Funambulus* and *Prasadsciurus* as subgenera. Prasad (1957) proposed a separate subfamily for the genus, based on the supposedly unique anatomy of the male reproductive tract. Trends in evolution of karyotypes analyzed by Aswathanarayana (1987). Grouped with African squirrels by Moore (1959). However, recent molecular studies suggest that *Funambulus* is a basal member of the Callosciurinae (Mercer and Roth, 2003), as suggested earlier on the basis of bacular morphology (Pocock, 1923: 239).

Funambulus layardi (Blyth, 1849). J. Asiat. Soc. Bengal, 18:602.

COMMON NAME: Layard's Palm Squirrel.

TYPE LOCALITY: "Upland districts" [Sri Lanka]. Restricted by Thomas (1924a:241) to "Ambigamoa Hills . . . Central Province (7° N., 80° 3′E.). . . . "

DISTRIBUTION: S and C Sri Lanka, and mountains of S India.

STATUS: IUCN – Lower Risk (lc).

SYNONYMS: *signatus* Thomas, 1924; **dravidianus** Robinson, 1917.

COMMENTS: Subgenus *Funambulus*. The presence of this species in S India is based on very few specimens.

Funambulus palmarum (Linnaeus, 1766). Syst. Nat., 12th ed., 1:86.
> COMMON NAME: Indian Palm Squirrel.
> TYPE LOCALITY: "in America, Asia, Africa." Restricted by Wroughton (1905*a*:409) to E coast of Madras, India.
> DISTRIBUTION: C and S India, Sri Lanka.
> STATUS: IUCN – Lower Risk (lc).
> SYNONYMS: *bellaricus* Wroughton, 1916]; *comorinus* Wroughton, 1905; *gossei* Wroughton and Davidson, 1919; *indicus* (Lesson, 1835) [not Erxleben, 1777]; *penicillatus* (Leach, 1814); **brodiei** (Blyth, 1849); *favonicus* Thomas and Wroughton, 1915; *kelaarti* (Layard, in Blyth, 1851); *matugamensis* Lindsay, 1926; *olympius* Thomas and Wroughton, 1915; **robertsoni** Wrougton, 1916; *bengalensis* Wroughton, 1916.
> COMMENTS: Subgenus *Funambulus*. Distribution largely allopatric with that of *F. pennantii*, but limited area of sympatry in Central Provinces and along W coast of India, south of 16° N lat. (Corbet and Hill, 1992; Moore, 1960).

Funambulus pennantii Wroughton, 1905. J. Bombay Nat. Hist. Soc., 16:411.
> COMMON NAME: Northern Palm Squirrel.
> TYPE LOCALITY: "Mandvi Taluka of Surat District," Guzerath (= Gudjerat), India.
> DISTRIBUTION: SE Iran through Pakistan to Nepal and N and C India. Perhaps adjacent Afghanistan. Introduced to Andaman Isls and vicinity of Perth in Western Australia.
> STATUS: IUCN – Lower Risk (lc).
> SYNONYMS: *lutescens* Wroughton, 1916; **argentescens** Wroughton, 1905.
> COMMENTS: Subgenus *Prasadsciurus* (Moore and Tate, 1965:71). Reviewed by Agrawal and Chakraborty (1979), and Corbet and Hill (1992).

Funambulus sublineatus (Waterhouse, 1838). Proc. Zool. Soc. Lond., 1838:19.
> COMMON NAME: Dusky Palm Squirrel.
> TYPE LOCALITY: India, Madras, Nilgiri Hills.
> DISTRIBUTION: SW India, C Sri Lanka.
> STATUS: IUCN – Lower Risk (lc).
> SYNONYMS: *delesserti* (Gervais, 1841); *trilineatus* (Kelaart, 1852); **obscurus** (Pelzeln and Kohl, 1886); *kathleenae* Thomas and Wroughton, 1915.
> COMMENTS: Subgenus *Funambulus*.

Funambulus tristriatus (Waterhouse, 1837). Mag. Nat. Hist. [Charlesworth's], 1:499.
> COMMON NAME: Jungle Palm Squirrel.
> TYPE LOCALITY: " . . . the more southern parts of Hindostan." Restricted by Wroughton (1905*a*:411) to Travancore; further restricted by Moore and Tate (1965:89) to Western Ghats, south of 12 deg. N. lat. (India).
> DISTRIBUTION: West coast of India, from below 20° N to southern tip.
> STATUS: IUCN – Lower Risk (nt).
> SYNONYMS: *annandalei* Robinson, 1917; *dussumieri* (Milne-Edwards, 1867); *wroughtoni* Ryley, 1913; **numarius** Wroughton, 1916; *thomasi* Wroughton and Davidson, 1919.
> COMMENTS: Subgenus *Funambulus*.

Glyphotes Thomas, 1898. Ann. Mag. Nat. Hist., ser. 7, 2:251.
> TYPE SPECIES: *Glyphotes simus* Thomas, 1898.
> COMMENTS: Tribe Callosciurini according to Moore (1959). Corbet and Hill (1992) placed this genus in *Callosciurus*. Both Payne et al. (1985) and Corbet and Hill (1992) considered *Hessonoglyphotes* Moore (1959), originally described as a subgenus of *Glyphotes*, to be a junior synonym of *Callosciurus*.

Glyphotes simus Thomas, 1898. Ann. Mag. Nat. Hist., ser. 7, 2:251.
> COMMON NAME: Sculptor Squirrel.
> TYPE LOCALITY: "Mount Kina Balu, N. Bórneo" [Sabah, Malaysia].
> DISTRIBUTION: Mountains of Borneo; Sabah and Sarawak (Malaysia), 1000 to 1700 m.
> STATUS: IUCN – Lower Risk (lc).
> COMMENTS: Reviewed by Hill (1959).

Hyosciurus Archbold and Tate, 1935. Am. Mus. Novit., 801:2.
 TYPE SPECIES: *Hyosciurus heinrichi* Tate and Archbold, 1935.
 COMMENTS: Tribe Callosciurini according to Moore (1959:174).

Hyosciurus heinrichi Archbold and Tate, 1935. Am. Mus. Novit., 801:2.
 COMMON NAME: Montane Long-nosed Squirrel.
 TYPE LOCALITY: "Latimodjong Mtns., central Celebes, 2300 meters" [C Sulawesi, Indonesia].
 DISTRIBUTION: Restricted to mountains of C Sulawesi (Musser and Dagosto, 1987:44).
 STATUS: IUCN – Vulnerable.
 COMMENTS: Formerly included *ileile* (Tate and Archbold, 1936:1); see comments therein.

Hyosciurus ileile Tate and Archbold, 1936. Am. Mus. Novit., 846:1.
 COMMON NAME: Lowland Long-nosed Squirrel.
 TYPE LOCALITY: "Ile-ile, North Celebes, 1700 meters" [N Sulawesi, Indonesia].
 DISTRIBUTION: Mountains of Sulawesi.
 STATUS: IUCN – Vulnerable.
 COMMENTS: Formerly considered a subspecies of *heinrichi* (Tate and Archbold, 1936), but see Musser (1987*a*).

Lariscus Thomas and Wroughton, 1909. Proc. Zool. Soc. Lond., 1909:389.
 TYPE SPECIES: *Sciurus insignis* F. Cuvier, 1821.
 SYNONYMS: *Laria* Gray, 1867 [not Scopoli, 1763]; *Paralariscus* Ellerman, 1947.
 COMMENTS: Tribe Callosciurini according to Moore (1959:173). Includes *Paralariscus*, treated as a genus by Ellerman (1947*b*:259) and Moore (1959:173).

Lariscus hosei (Thomas, 1892). Ann. Mag. Nat. Hist., ser. 6, 10:215.
 COMMON NAME: Four-striped Ground Squirrel.
 TYPE LOCALITY: "Batu Sang Mount [Mt. Batu Song], Baram River, N. Borneo (5000 feet)" [Baram Dist., Sarawak, Malaysia, 1524 m].
 DISTRIBUTION: Mountains of Sarawak and Sabah (Malaysia).
 STATUS: IUCN – Vulnerable.
 COMMENTS: Sometimes put in the genus *Paralariscus*; see Ellerman (1947*b*:259); but also see Medway (1977:97).

Lariscus insignis (F. Cuvier, 1821). *In* E. Geoffroy St. Hilaire and F. Cuvier, Hist. Nat. Mammifères, Pt. 2, 4(34):2 pp.
 COMMON NAME: Three-striped Ground Squirrel.
 TYPE LOCALITY: "Sumatra," [Indonesia]. Restricted by Sody (1949*a*:113) to Lampongs, S Sumatra.
 DISTRIBUTION: Peninsular Malaysia and Thailand, Sumatra, Java, Borneo, adjacent isls.
 STATUS: IUCN – Lower Risk (lc).
 SYNONYMS: **diversus** (Thomas, 1898); **javanus** (Thomas and Wroughton, 1909); *murianus* Sody, 1937; **peninsulae** (Miller, 1903); *fornicatus* Robinson, 1917; *meridionalis* Robinson and Kloss, 1911; **rostratus** (Miller, 1903); *atchinensis* Sody, 1949; *diversoides* Sody, 1949; *jalorensis* (Bonhote, 1903). **Unassigned**: *castaneus* (Miller, 1900); *saturatus* Chasen, 1935.
 COMMENTS: Formerly included *niobe*, and *obscurus*; see Chasen (1940:145-146); but see also Corbet and Hill (1992:300); see comments under *niobe*.

Lariscus niobe (Thomas, 1898) Ann. Mag. Nat. Hist., ser. 7, 2:249.
 COMMON NAME: Niobe Ground Squirrel.
 TYPE LOCALITY: "Pajo, Sumatra" [Sumatra, Indonesia]. Restricted by Robinson and Kloss (1918*b*:37) to Pajokombo, Padang Highlands, West Sumatra.
 DISTRIBUTION: Mountains of Sumatra, and Java (Idjen Mtns).
 STATUS: IUCN – Lower Risk (nt).
 SYNONYMS: **vulcanus** Kloss, 1921.
 COMMENTS: Given full specific status by Moore (1959); but see Chasen (1940:145). The subspecies *vulcanus* is restricted to Java.

Lariscus obscurus (Miller, 1903). Smithson. Misc. Coll., 45:23.
COMMON NAME: Mentawai Three-striped Squirrel.
TYPE LOCALITY: Indonesia, Mentawai Isls, "South Pagi Island, Sumatra".
DISTRIBUTION: Siberut, Sipora, N and S Pagai Island (Sumatra, Indonesia).
STATUS: IUCN – Lower Risk (nt).
SYNONYMS: *auroreus* Sody, 1949; *siberu* Chasen and Kloss, 1928.
COMMENTS: Formerly included in *insignis* (Chasen, 1940) or *niobe* (Chasen and Kloss (1927).
Part of the rodent fauna endemic to the Mentawai Archipelago (see comments under
Leopoldamys siporanus).

Menetes Thomas, 1908. J. Bombay Nat. Hist. Soc., 18:244.
TYPE SPECIES: *Sciurus berdmorei* Blyth, 1849.
COMMENTS: Tribe Callosciurini according to Moore (1959:173). Reviewed by Moore and Tate
(1965).

Menetes berdmorei (Blyth, 1849). J. Asiat. Soc. Bengal, 18(30):603.
COMMON NAME: Indochinese Ground Squirrel.
TYPE LOCALITY: "Thougyeen district" [Tenasserim, Burma].
DISTRIBUTION: S Vietnam, Cambodia, S Laos, Thailand, S Yunnan (China) to C Burma.
STATUS: IUCN – Lower Risk (lc).
SYNONYMS: *amotus* (Miller, 1913); *consularis* Thomas, 1914; *koratensis* (Gyldenstolpe, 1917);
decoratus Thomas, 1914; *moerescens* Thomas, 1914; *mouhotei* (Gray, 1861);
peninsularis Kloss, 1919; *rufescens* Kloss, 1916; *umbrosus* Kloss, 1916; *pyrrocephalus*
(Milne-Edwards, 1867).
COMMENTS: For taxonomic history, see Moore and Tate (1965:294). Chromosomes described
by Nadler and Hoffmann (1970). Corbet and Hill (1992) opined "Although there is
considerable geographic variation . . . it is unlikely that discrete subspecies can be
recognized."

Nannosciurus Trouessart, 1880. Le Naturaliste, p. 292.
TYPE SPECIES: *Sciurus melanotis* Müller, 1840.
COMMENTS: Tribe Callosciurini according to Moore (1959). Formerly included the species now
placed in *Exilisciurus*; see Moore (1958) and Heaney (1985).

Nannosciurus melanotis (Müller, 1840). *In* Temminck, Verh. Nat. Gesch. Nederland. Overz. Bezitt.,
Zool., Zoogd. Indisch. Archipel., p. 35 [1840], see comments.
COMMON NAME: Black-eared Squirrel.
TYPE LOCALITY: "Borneo". Restricted to Java by Lyon (1906) and by lectotype designation of
Heaney (1985).
DISTRIBUTION: Sumatra, Java, Borneo, adjacent small islands.
STATUS: IUCN – Lower Risk (lc).
SYNONYMS: *soricinus* (Waterhouse, 1838) [*nomen nudum*]; *bancanus* Lyon, 1906; *borneanus*
Lyon, 1906; *pallidus* Chasen and Kloss, 1928; *pulcher* Miller, 1902; *sumatranus* Lyon,
1906.
COMMENTS: This species was further described by Müller and Schlegel, *in* Temminck, Verh.
Nat. Gesch. Nederland. Overz. Bezitt., Zool., Mammalia, pp. 87, 88, 98 [1844], pl. 14,
fig 4, 5 [1841] (from which this species is often cited). Reviewed by Heaney (1985).

Prosciurillus Ellerman, 1947. Proc. Zool. Soc. Lond., 117:259.
TYPE SPECIES: *Sciurus murinus* Müller and Schlegel, 1844.
COMMENTS: Tribe Callosciurini according to Moore (1959); closely related to *Sundasciurus* (see
Heaney, 1985). Reviewed by Moore (1958), who transferred *leucomus* from *Callosciurus* to
this genus.

Prosciurillus abstrusus Moore, 1958. Am. Mus. Novit., 1890:3.
COMMON NAME: Secretive Dwarf Squirrel.
TYPE LOCALITY: " . . . 1500 meters elevation, Gunong Tanke Salokko, 'Mengkoka Geb.'
[Mekongqa Gebirgte], in the southeastern peninsula of Celebes, latitude 3° 40' S.,
longitude 121° 13' E.," [Sulawesi, Indonesia].

DISTRIBUTION: Known only from the type locality in SE Sulawesi.

STATUS: IUCN – Vulnerable.

SYNONYMS: *obscurus* Moore, 1959.

COMMENTS: Includes *obscurus*, probably a *nomen nudum*; see Moore (1959:203).

Prosciurillus leucomus (Müller and Schlegel, 1844). *In* Temminck, Verh. Nat. Gesch. Nederland. Overz. Bezitt., Zool., Mammalia, p. 87 [1844].

COMMON NAME: Whitish Dwarf Squirrel.

TYPE LOCALITY: "Celebes." Restricted by Meyer (1898) to Minahassa, NE Celebes [Sulawesi, Indonesia].

DISTRIBUTION: Sulawesi; and neighboring islands, Buton and E Kabaena.

STATUS: IUCN – Lower Risk (lc).

SYNONYMS: **hirsutus** (Hayman, 1946); *sarasinorum* (Meyer, 1898); **occidentalis** (Meyer, 1898); **tonkeanus** (Meyer, 1896); *elbertae* (Schwarz, 1911); *mowewensis* (Roux, 1910); *topapuensis* (Roux, 1910).

COMMENTS: Formerly included in *Callosciurus*; see Moore (1958).

Prosciurillus murinus (Müller and Schlegel, 1844). *In* Temminck, Verh. Nat. Gesch. Nederland. Overz. Bezitt., Zool., Mammalia, p. 87 [1844].

COMMON NAME: Celebes Dwarf Squirrel.

TYPE LOCALITY: "Celebes." Restricted by Sody (1949*a*) to NE Celebes [Sulawesi, Indonesia].

DISTRIBUTION: NE and C Sulawesi.

STATUS: IUCN – Lower Risk (lc).

SYNONYMS: *evidens* (Miller and Hollister, 1921); **griseus** (Sody, 1949); **necopinus** (Miller and Hollister, 1921).

COMMENTS: For authority and date, see comments under *Rubrisciurus rubriventer*.

Prosciurillus rosenbergii (Jentink 1879). Notes Leyden Mus. 1:37.

COMMON NAME: Sanghir Squirrel

TYPE LOCALITY: "Sangi-islands, Siao".

DISTRIBUTION: Sanghir Island

SYNONYMS: *tingahi* (Meyer, 1896)

COMMENTS: Full species status accorded by Feiler (1990) and follwed by Flannery (1995*b*).

Prosciurillus weberi (Jentink, 1890). Weber's Zool. Ergebn. 1:115.

COMMON NAME: Weber's Dwarf Squirrel.

TYPE LOCALITY: Indonesia, C Celebes, Luwu (near Palopo).

DISTRIBUTION: C Sulawesi (Indonesia).

STATUS: IUCN – Lower Risk (nt).

COMMENTS: Elevated to specific status by Corbet and Hill (1992).

Rhinosciurus Blyth, 1856. J. Asiat. Soc. Bengal, 14:477.

TYPE SPECIES: *Sciurus laticaudatus* Müller, 1840. *Rhinosciurus tupaioides* Blyth (=*Sciurus laticaudatus* Müller, 1840).

COMMENTS: Tribe Callosciurini according to Moore (1959:173).

Rhinosciurus laticaudatus (Müller, 1840). *In* Temminck, Verh. Nat. Gesch. Nederland. Overz. Bezitt., Zool., Zoogd. Indisch. Archipel., p. 34 [1840], see comments.

COMMON NAME: Shrew-faced Squirrel.

TYPE LOCALITY: "Pontianak" [W Kalimantan, Indonesia].

DISTRIBUTION: Malay Peninsula, Sumatra, Borneo, and small adjacent islands.

STATUS: IUCN – Lower Risk (lc).

SYNONYMS: *tupaioides* Gray, 1843 [*nomen nudum*]; *tupaioides* Blyth, 1856; **alacris** (Thomas, 1908); *leo* Thomas and Wroughton, 1909; *peracer* Thomas and Wroughton, 1909; *robinsoni* Thomas, 1908; **saturatus** Robinson and Kloss, 1919. **Unassigned**: *incultus* Lyon, 1916; *rhionis* Thomas and Wroughton, 1909.

COMMENTS: This species was further described by Müller and Schlegel, *in* Temminck, Verh. Nat. Gesch. Nederland. Overz. Bezitt., Zool., Mammalia, pp. 87, 100 [1844], pl. 15 [1841] (from which this species is often cited).

Rubrisciurus Ellerman, 1954. *In* Laurie and Hill, Land Mammals of New Guinea, Celebes, and Adjacent Islands, p. 94.
> TYPE SPECIES: *Sciurus rubriventer* Müller and Schlegel, 1844.
> COMMENTS: Tribe Callosciurini according to Moore (1959:174). Sometimes considered a subgenus of *Callosciurus*; see Laurie and Hill (1954:93) and Corbet and Hill (1992); but see also Moore (1959) and McLaughlin (1984:273), who considered it a distinct genus.

Rubrisciurus rubriventer (Müller and Schlegel, 1844). *In* Temminck, Verh. Nat. Gesch. Nederland. Overz. Bezitt., Zool., Mammalia, p. 86 [1844].
> COMMON NAME: Sulawesi Giant Squirrel.
> TYPE LOCALITY: "Celebes" [Minahassa, NE Sulawesi, Indonesia].
> DISTRIBUTION: N, C, and SE Sulawesi.
> STATUS: IUCN – Lower Risk (lc).
> COMMENTS: Ellerman (1940:375) dated the species description as 1839, and ascribed it to Forsten, who was listed by Müller and Schlegel as the author, but Forsten was the collector, not the author.

Sundasciurus Moore, 1958. Am. Mus. Novit., 1914:2.
> TYPE SPECIES: *Sciurus lowii* Thomas, 1892.
> SYNONYMS: *Aletesciurus* Moore, 1958.
> COMMENTS: Tribe Callosciurini according to Moore (1959:173). Reviewed in part by Heaney (1979). Includes *Aletesciurus* and *Sundasciurus* as subgenera (Moore, 1958:3). Ellerman (1940) placed *Sundasciurus* in *Callosciurus*, and Chasen (1940), in *Sciurus*. Corbet and Hill (1992) recognized seven species.

Sundasciurus brookei (Thomas, 1892). Ann. Mag. Nat. Hist., ser. 6, 9:253.
> COMMON NAME: Brooke's Squirrel.
> TYPE LOCALITY: "Mount Dulit, N. Borneo" [Sarawak, Malaysia].
> DISTRIBUTION: Mountains of Borneo, from 600 to 1500 m.
> STATUS: IUCN – Lower Risk (nt).
> COMMENTS: Subgenus *Sundasciurus*. Status discussed by Corbet and Hill (1992).

Sundasciurus davensis (Sanborn, 1952). Fieldiana Zool., 33:117.
> COMMON NAME: Davao Squirrel.
> TYPE LOCALITY: "Madaum, 25 feet [8 m] altitude, Tagum Municipality, Davao Province, Mindanao Island, Philippines Islands."
> DISTRIBUTION: Known only from the type locality.
> STATUS: IUCN – Lower Risk (lc).
> COMMENTS: Subgenus *Aletesciurus*. May be conspecific with *mindanensis*, *philippinensis*, and *samarensis* (Heaney et al., 1987; Corbet and Hill, 1992).

Sundasciurus fraterculus (Thomas, 1895). Ann. Mus. Civ. Storia Nat. Genova, Ser. 2a, 14:669.
> COMMON NAME: Fraternal Squirrel.
> TYPE LOCALITY: Indonesia, Mentawai Isls, Sipora Isl, "Sereinu".
> DISTRIBUTION: Sipora Isl, Siberut Isl, and S Pagai Isl (Sumatra, Indonesia).
> STATUS: IUCN – Lower Risk (lc).
> SYNONYMS: *pumilus* (Miller, 1903); *siberu* (Chasen and Kloss, 1928).
> COMMENTS: Subgenus *Sundasciurus* formerly included in *lowii* (Chasen, 1940:144); but see Moore (1959). Geographic variation reviewed by Jenkins and Hill (1982). Part of the rodent fauna endemic to the Mentawai Archipelago (see comments under *Leopoldamys siporanus*).

Sundasciurus hippurus (I. Geoffroy, 1831). *In* Bélanger (ed.), Voy. Indes Orient., Mamm., 3(Zoologie):149.
> COMMON NAME: Horse-tailed Squirrel.
> TYPE LOCALITY: "Java." Restricted by Robinson and Kloss (1918a:226) to Malacca, Malaysia.
> DISTRIBUTION: S Vietnam, S to Malay Peninsula, Sumatra, and Borneo.
> STATUS: IUCN – Lower Risk (lc).
> SYNONYMS: *rufogaster* (Gray, 1842); **borneensis** (Gray, 1867); *grayi* (Bonhote, 1901); *hippurellus* (Lyon, 1907); **pryeri** (Thomas, 1892); *inquinatus* (Thomas, 1908); **hippurosus** (Lyon, 1907); **ornatus** Dao and Cao, 1990.

COMMENTS: Subgenus *Aletesciurus*. Corbet and Hill (1992: 297) recognized four subspecies on Borneo, including *hippurellus* and *inquinatus*.

Sundasciurus hoogstraali (Sanborn, 1952). Fieldiana Zool., 33:115.
COMMON NAME: Busuanga Squirrel.
TYPE LOCALITY: "Dimaniang, Busuanga Island, Calamianes group, Philippine Islands."
DISTRIBUTION: Busuanga Isl (Philippines).
STATUS: IUCN – Lower Risk (lc).
COMMENTS: Subgenus *Aletesciurus*. Member of the *steerii* group; see comment under *steerii*. Placed in *moellendorffi* by Corbet and Hill (1991:140), without comment.

Sundasciurus jentinki (Thomas, 1887). Ann. Mag. Nat. Hist., ser. 5, 20:128.
COMMON NAME: Jentinck's Squirrel.
TYPE LOCALITY: "Mount Kina-Balu," [Sabah, Malaysia].
DISTRIBUTION: Mountains of N Borneo.
STATUS: IUCN – Vulnerable.
SYNONYMS: *subsignanus* (Chasen, 1937).
COMMENTS: Subgenus *Sundasciurus*. Chromosomes described by Harada and Kobayashi (1980).

Sundasciurus juvencus (Thomas, 1908). Ann. Mag. Nat. Hist., ser. 8, 2:498.
COMMON NAME: Northern Palawan Tree Squirrel.
TYPE LOCALITY: "the islands of Palawan. . . . Puerto Princesa [Philippines]".
DISTRIBUTION: N Palawan Isl.
STATUS: IUCN – Endangered.
COMMENTS: Subgenus *Aletesciurus*. Member of the *steerii* group; see comment under *steerii*.

Sundasciurus lowii (Thomas, 1892). Ann. Mag. Nat. Hist., ser. 6, 2:253.
COMMON NAME: Low's squirrel.
TYPE LOCALITY: "Lumbidan, on the mainland opposite Labuan" [Sarawak, Malaysia].
DISTRIBUTION: Malay Peninsula, Sumatra, Borneo, and adjacent small islands.
STATUS: IUCN – Lower Risk (lc).
SYNONYMS: **balae** (Miller, 1903); *piniensis* (Miller, 1903); **bangueyae** (Thomas, 1910); **humilis** (Miller, 1913); *vanakeni* (Robinson and Kloss, 1916); **natunensis** (Thomas, 1895); *lingungensis* (Miller, 1901); **robinsoni** (Bonhote, 1903); *alacris* (Thomas, 1908); **seimundi** (Thomas and Wroughton, 1909).
COMMENTS: Subgenus *Sundasciurus*. Formerly included *fraterculus* and *pumilus* (Chasen, 1940:144).

Sundasciurus mindanensis (Steere, 1890). List of the Birds and Mammals collected by the Steere Expedition to the Philippines. Ann Arbor, Mich., p. 29.
COMMON NAME: Mindanao Squirrel.
TYPE LOCALITY: "Mindanao," [Philippines].
DISTRIBUTION: Mindanao and adjacent small islands (Philippines).
STATUS: IUCN – Lower Risk (lc).
COMMENTS: Subgenus *Aletesciurus*. May be conspecific with *davensis*, *philippinensis*, and *samarensis* (Heaney et al., 1987; Corbet and Hill, 1992).

Sundasciurus moellendorffi (Matschie, 1898). Sitzb. Gesell. Naturf. Fr., Berlin, 5:41.
COMMON NAME: Culion Tree Squirrel.
TYPE LOCALITY: "Calamianes. . . . von Culion" [Calamian Isls, Philippines].
DISTRIBUTION: Calamian Isls other than Busuanga.
STATUS: IUCN – Lower Risk (nt).
SYNONYMS: *albicauda* (Matschie, 1898).
COMMENTS: Subgenus *Aletesciurus*. The form *albicauda* was listed as a distinct species by Corbet and Hill (1980:132) without comment, but included in *mollendorffi* by Corbet and Hill (1986:149), as was *hoogstraali* (in Corbet and Hill, 1991:140) without comment. Spelling of name changed in accordance with Art. 32.5.2.1 of the International Code of Zoological Nomenclature (International Commission on Zoological Nomenclature, 1999).

Sundasciurus philippinensis (Waterhouse, 1839). Proc. Zool. Soc. Lond., 1839:117.
>COMMON NAME: Philippine Tree Squirrel.
>TYPE LOCALITY: "Mindanado" [Mindanao Isl, Philippines].
>DISTRIBUTION: S and W Mindanao, and Basilan (Philippines).
>STATUS: IUCN – Lower Risk (lc).
>COMMENTS: Subgenus *Aletesciurus*. May be conspecific with *davensis*, *mindanensis*, and
>>*samarensis* (Heaney et al., 1987; Corbet and Hill, 1992).

Sundasciurus rabori Heaney, 1979. Proc. Biol. Soc. Wash., 92:281.
>COMMON NAME: Palawan Montane Squirrel.
>TYPE LOCALITY: "Magtaguimbong, Mt. Mantalingajan, Palawan Isl, Republic of the
>>Philippines, Island between 3600 and 4350 ft [1097 and 1326 m] elevation . . .
>>approximately 8° 48′N, 117° 40′E."
>DISTRIBUTION: Above 800 m in mountains on Palawan (Philippines).
>STATUS: IUCN – Vulnerable.
>SYNONYMS: *cagsi* (Meyer, 1890).
>COMMENTS: Subgenus *Aletesciurus* (see Heaney, 1979).

Sundasciurus samarensis (Steere, 1890). List of the Birds and Mammals Collected by the Steere
>Expedition to the Philippines. Ann Arbor, Mich., p. 30.
>COMMON NAME: Samar Squirrel.
>TYPE LOCALITY: "Samar and Leyte," [Philippines].
>DISTRIBUTION: Samar and Leyte Isls (Philippines).
>STATUS: IUCN – Vulnerable.
>COMMENTS: Subgenus *Aletesciurus*. May be conspecific with *davensis*, *mindanensis*, and
>>*philippinensis* (Corbet and Hill, 1992; Heaney et al., 1987).

Sundasciurus steerii (Günther, 1877). Proc. Zool. Soc. Lond., 1876:735 [1877].
>COMMON NAME: Southern Palawan Tree Squirrel.
>TYPE LOCALITY: "Balabac" [Isl, Philippines].
>DISTRIBUTION: Balabac and S Palawan Isls in lowlands (Philippines).
>STATUS: IUCN – Lower Risk (nt).
>COMMENTS: Subgenus *Aletesciurus*. Member of the *steerii* group, which contains 1-3 species;
>>see Heaney (1979).

Sundasciurus tenuis (Horsfield, 1824). Zool. Res. Java, 1824:153.
>COMMON NAME: Slender Squirrel.
>TYPE LOCALITY: Singapore.
>DISTRIBUTION: Malay Peninsula, Sumatra, Borneo and adjacent small islands.
>STATUS: IUCN – Lower Risk (lc).
>SYNONYMS: *gunong* (Robinson and Kloss, 1914); *sordidus* (Kloss, 1911); *surdus* (Miller, 1900);
>>*tahan* (Bonhote, 1908); **bancarus** (Miller, 1903); **modestus** (Müller, 1840); *altitudinus*
>>(Robinson and Kloss, 1916); **parvus** (Miller, 1901); **procerus** (Miller, 1901).
>>**Unassigned**: *batus* (Lyon, 1916); *mansalaris* (Miller, 1903); *sianticus* (Chasen and Kloss,
>>1928); *tiomanicus* (Robinson, 1917).
>COMMENTS: Subgenus *Sundasciurus*. The name *pumilus* (see synonym in *fraterculus*) was
>>previously assigned to *tenuis* (Robinson and Kloss, 1918a:229); but see Chasen
>>(1940:144).

Tamiops J. A. Allen, 1906. Bull. Am. Mus. Nat. Hist., 22: 475.
>TYPE SPECIES: *Tamiops macclellandi hainanus* J. A. Allen, 1906 (= *T. maritimus hainanus*).
>COMMENTS: Tribe Callosciurini according to Moore (1959:173). Revised by Moore and Tate
>>(1965).

Tamiops mcclellandii (Horsfield, 1840). Proc. Zool. Soc. Lond., 1839:152 [1840].
>COMMON NAME: Himalayan Striped Squirrel.
>TYPE LOCALITY: "Bengal as well as Assam" [India]. Restricted by Ellerman (1940:354) to Assam.
>DISTRIBUTION: E Nepal through Assam and Mizoram (India); N and C Burma, and Yunnan
>>(China), and south through Thailand, Vietnam, Laos, and Cambodia to the S Malay
>>Peninsula.

STATUS: IUCN – Lower Risk (lc).

SYNONYMS: *manipurensis* (Bonhote, 1900); *pembertoni* (Blyth, 1842); **barbei** (Blyth, 1847); **collinus** Moore, 1958; **inconstans** Thomas, 1920; **kongensis** (Bonhote, 1901); **leucotis** (Temminck, 1853); *novemlineatus* Miller, 1903.

Tamiops maritimus (Bonhote, 1900). Ann. Mag. Nat. Hist., ser. 7, 5:51.
COMMON NAME: Maritime Striped Squirrel.
TYPE LOCALITY: "Foochow, province of Fokien" [Fukien, China].
DISTRIBUTION: Hubei, Anhui and Zhejiang, south through Guangxi and Guangdong (China), to S Vietnam and Laos; Hainan and Taiwan (China).
STATUS: IUCN – Lower Risk (lc).
SYNONYMS: *formosanus* (Bonhote, 1900); *sauteri* J. A. Allen, 1911; **hainanus** J. A. Allen, 1906; *laotum* Robinson and Kloss, 1922; *riudoni* J. A. Allen, 1906; **moi** Robinson and Kloss, 1922; **monticolus** (Bonhote, 1900).

Tamiops rodolphii (Milne-Edwards, 1867). Rev. Mag. Zool. Paris, ser. 2, 19:227.
COMMON NAME: Cambodian Striped Squirrel.
TYPE LOCALITY: "Cochin China near Saigon," [Vietnam].
DISTRIBUTION: E Thailand, Cambodia, S Laos, S Vietnam.
STATUS: IUCN – Lower Risk (lc).
SYNONYMS: *dolphoides* Kloss, 1921; *holti* (Ellerman, 1940) [to replace *lylei*, preoccupied]; *liantis* Kloss, 1919; *lylei* Thomas, 1920; **elbeli** Moore, 1958.

Tamiops swinhoei (Milne-Edwards, 1874). Rech. Hist. Nat. Mammifères, 1:308.
COMMON NAME: Swinhoe's Striped Squirrel.
TYPE LOCALITY: "Moupin [Muping]," [= Baoxing, Sichuan, China]. Restricted by G. M. Allen (1940:673) to "Hongchantin . . . about 6000 feet [1829 m]."
DISTRIBUTION: Extreme SW Gansu south through Tibet, Sichuan and Yunnan (China) to N Burma and N Vietnam; isolated population in Hebei (China).
STATUS: IUCN – Lower Risk (lc).
SYNONYMS: *clarkei* Thomas, 1920; *forresti* Thomas, 1920; **olivaceus** Osgood, 1932; **spencei** Thomas, 1921; *chingpingensis* Lu and Qyan, 1965; *russeolus* Jacobi, 1923; **vestitus** Miller, 1915.
COMMENTS: The form *forresti* was originally named as a subspecies of *T. maritimus*, but assigned to *swinhoei* by Moore and Tate (1965:248).

Subfamily Xerinae Osborn, 1910. The Age of Mammals: 535.

Tribe Xerini Osborn, 1910. The Age of Mammals: 535.

Atlantoxerus Forsyth Major, 1893. Proc. Zool. Soc. Lond., 1893:189.
TYPE SPECIES: *Sciurus getulus* Linnaeus, 1758.
COMMENTS: Placed in Subfamily Xerinae by Gromov et al. (1965:70), but in tribe Xerini by Moore (1959). *Atlantoxerus* was originally a subgenus of *Xerus*, and later raised to full generic rank by Thomas (1909a). Closely related to *Xerus* (Corbet, 1978c:79).

Atlantoxerus getulus (Linnaeus, 1758). Syst. Nat., 10th ed., 1:64.
COMMON NAME: Barbary Ground Squirrel.
TYPE LOCALITY: "in Africa." Restricted by Thomas (1911a:149) to "Barbary;" and by Cabrera (1932:217) to Agadir, Morocco.
DISTRIBUTION: Grand and Middle Atlas south to Agadir and N edge of Sahara (Morocco), NW Algeria.
STATUS: IUCN – Lower Risk (lc).
SYNONYMS: *praetextus* Wagner, 1842; *trivittatus* (Gray, 1842); see Corbet (1978c:79).

Spermophilopsis Blasius, 1884. Tageblatt. Versamml. Deutsch. Naturf. Magdeburg, 57:325.
TYPE SPECIES: *Arctomys leptodactylus* Lichtenstein, 1823.
COMMENTS: Placed in Subfamily Xerinae by Gromov et al. (1965:70), but in tribe Xerini by Moore (1959).

Spermophilopsis leptodactylus (Lichtenstein, 1823). Naturh. Abh. Eversmann's Reise, p. 119.
COMMON NAME: Long-clawed Ground Squirrel.
TYPE LOCALITY: "Vicinity of Kara Ata, 140 km northwest of the Old Town of Bukhara" [Uzbekistan] (Ognev, 1966:394).
DISTRIBUTION: SE Kazakhstan, Turkmenistan, Uzbekistan, W Tajikistan, NE Iran, NW Afghanistan.
STATUS: IUCN – Lower Risk (lc).
SYNONYMS: *schumakovi* (Satunin, 1908); *turcomanus* (Eichwald, 1834); **bactrianus** (Scully, 1888); **heptopotamicus** Heptner and Ismagilov, 1952.
COMMENTS: Related to African xerine squirrels (*Atlantoxerus* and *Xerus*) (Nadler and Hoffmann, 1974; Nadler et al., 1969).

Xerus Hemprich and Ehrenberg, 1833. Symb. Phys. Mamm., vol. 1, sig. Ee, pl. 9.
TYPE SPECIES: *Sciurus* (*Xerus*) *brachyotus* Hemprich and Ehrenberg, 1832 (= *Sciurus rutilus* Cretzchmar, 1828).
SYNONYMS: *Euxerus* Thomas, 1909; *Geosciurus* Smith, 1834.
COMMENTS: Tribe Xerini according to Simpson (1945:79). Includes *Euxerus*, *Geosciurus*, and *Xerus* as subgenera (Amtmann, 1975; Ellerman, 1940; Moore, 1959).

Xerus erythropus (E. Geoffroy, 1803). *In* Cat. Mamm. Mus. Hist. Nat., Paris, p. 178.
COMMON NAME: Striped Ground Squirrel.
TYPE LOCALITY: Senegal (neotype). Origin of original type unknown. "Inconnue."
DISTRIBUTION: SE Morocco, S Mauritania, Senegal, Gambia, Guinea Bissau, Guinea, Sierra Leone, Côte d'Ivoire, S Mali, Burkina Faso, Ghana, Togo, Benin, SE Niger, Nigeria, Cameroon, NE Republic of Congo, SE Chad, NE Central African Republic, Sudan, Dem. Rep. Congo, NW Uganda, Rwanda, W Ethiopia, W Kenya, N Tanzania.
STATUS: IUCN – Lower Risk (lc).
SYNONYMS: *agadius* (Thomas and Hinton, 1921); *albovittatus* (Desmarest, 1817); *lessonii* (Fitzinger, 1867) [new name for *marabutus* Lesson]; *maestus* (Thomas, 1910); *marabutus* (Lesson, 1838) [preoccupied]; *prestigiator* (Lesson, 1838); *chadensis* (Thomas, 1905); **lacustris** (Thomas, 1905); **leucoumbrinus** (Rüppell, 1835); **limitaneus** (Thomas and Hinton, 1923); **microdon** Thomas, 1905; *fulvior* (Thomas, 1905).
COMMENTS: Placed in *Euxerus* which is considered a subgenus of *Xerus* by Ellerman (1940), Moore (1959), and Amtmann (1975). The spelling *erythopus* was used by E. Geoffroy, 1803; Shinz, 1845, used the spelling *erythropus*. Opinion 945 of the International Commission on Zoological Nomenclature (1971*b*) ruled that *erythopus* Geoffroy, 1803 be changed to *erythropus* as an incorrect original spelling; the proper latin root is "erythro", and in the last 100 years nearly all authors have used the spelling *erythropus*. In the interest of orthographic stability we advocate that the specific name be spelled *erythropus*. It is not desirable to perpetuate the lapsus in spelling by early workers.

Xerus inauris (Zimmermann, 1780). Geogr. Gesch. Mensch. Vierf. Thiere, 2:344.
COMMON NAME: South African Ground Squirrel.
TYPE LOCALITY: South Africa, Kaffirland, 100 mi. (160 km) N of Cape of Good Hope. "Bewohnt die Cafferen, über hundert Meilen nordwärts des Vorgebürges der guten Hofnung . . . ".
DISTRIBUTION: S Angola, Namibia, Botswana, W Zimbabwe, South Africa.
STATUS: IUCN – Lower Risk (lc).
SYNONYMS: *africanus* (Shaw, 1801); *capensis* (Kerr, 1792); *dschinshicus* (Gmelin, 1788); *ginginianus* (Shaw, 1801); *levaillantii* (Kuhl, 1820); *namaquensis* (Lichtenstein, 1793); *setosus* (Smuts, 1832).
COMMENTS: Subgenus *Geosciurus*. Distinction from *X. princeps* supported by minor chromosomal differences reported by Robinson et al. (1986).

Xerus princeps (Thomas, 1929). Proc. Zool. Soc. Lond., 1929:106.
COMMON NAME: Damara Ground Squirrel.
TYPE LOCALITY: N Namibia, C Koakoveld, Otjitundua.
DISTRIBUTION: W Namibia, S Angola, restricted to the Kaokoland escarpment of Namibia and Angola as far north as 14°10'S 16°0'E.

STATUS: IUCN – Lower Risk (lc).

COMMENTS: Subgenus *Geosciurus*. Distinction from *X. inauris* supported by minor chromosomal differences reported by Robinson et al. (1986).

Xerus rutilus (Cretzschmar, 1828). *In* Rüppell, Atlas Reise Nordl. Afr., Zool. Säugeth., p. 59.
COMMON NAME: Unstriped Ground Squirrel.
TYPE LOCALITY: Ethiopia, eastern slope of Abyssynia. "Der östliche Abhang Abyssiniens, wo es häufig vorkommmt." Probably Massawa, according to Mertens (1925:26) " . . . wahrsheinlich Massaua".
DISTRIBUTION: SE Sudan, E and S Ethiopia, Djibouti, Eritrea, Somalia, Kenya, NE Uganda, NE Tanzania.
STATUS: IUCN – Lower Risk (lc).
SYNONYMS: *abessinicus* (Gmelin, 1788); *brachyotis* (Hemprich and Ehrenberg, 1833); *fuscus* (Huet, 1880); *dabagala* Heuglin, 1861; *dorsalis* Dollman, 1911; *intensus* Thomas, 1904; *massaicus* Toschi, 1945; *rufifrons* Dollman, 1911; *saturatus* (Neumann, 1900); *stephanicus* Thomas, 1906.
COMMENTS: Subgenus *Xerus*. Chromosomes described by Nadler and Hoffmann (1974). Reviewed by O'Shea (1991, Mammalian Species, 370).

Tribe Protoxerini Moore, 1959. Bull. Am. Mus. Nat. Hist. 118, p. 168.

Epixerus Thomas, 1909. Ann. Mag. Nat. Hist., ser. 8, 3:472.
TYPE SPECIES: *Sciurus wilsoni* Du Chaillu, 1860 (=*Sciurus ebii* Temminck, 1853).
COMMENTS: Tribe Protoxerini (Moore, 1959).

Epixerus ebii (Temminck, 1853). Esquisses Zool. sur la Côte de Guine, p. 129.
COMMON NAME: Palm Squirrel.
TYPE LOCALITY: Ghana. " . . . les grandes forêts de la Guiné, et se trouve dans les mêmes localités que l'espèce précédente, [. . . les confins du pays des Fantes . . .] mais parait être moins abondante dans les parties boisées de Dabocrom."
DISTRIBUTION: Sierra Leone, Liberia, Côte d'Ivoire, Ghana, Cameroon, Rio Muni, Gabon.
STATUS: CITES – Appendix III (Ghana); IUCN – Lower Risk (nt) as *E. ebii* and as *E. wilsoni*.
SYNONYMS: *jonesi* Hayman, 1954; *wilsoni* (Du Chaillu, 1860); *mayumbicus* Verheyen, 1959.
COMMENTS: *E. ebii* and *E. wilsoni* were formerly considered distinct species by Perret and Aellen (1956), Verheyen (1959), and Rosevear (1969), but Kuhn (1964) revised the genus and concluded that they are conspecific.

Funisciurus Trouessart, 1880. Le Naturaliste, 2(37):293.
TYPE SPECIES: *Sciurus isabella* Gray, 1862.
COMMENTS: Tribe Funambulini according to Moore (1959). Raised to full generic rank by Thomas (1897b). This treatment follows Amtmann (1975), and Kingdon (1997). Recent molecular work suggests that *Funisciurus* is not closely related to *Funambulus* (Mercer and Roth, 2003). Instead, it should be considered a member of the Tribe Protoxerini, according to Steppan et al. (2004).

Funisciurus anerythrus (Thomas, 1890). Proc. Zool. Soc. Lond., 1890:447.
COMMON NAME: Thomas's Rope Squirrel.
TYPE LOCALITY: Uganda, "Buguera", S of Lake Albert.
DISTRIBUTION: Nigeria, Cameroon, Gabon, Central African Republic, NE Dem. Rep. Congo, Uganda; SW Dem. Rep. Congo and N Shaba Prov. (Dem. Rep. Congo).
STATUS: IUCN – Lower Risk (lc).
SYNONYMS: *niapu* J. A. Allen, 1922; *ochrogaster* Cabrera and Ruxton, 1926; *bandarum* Thomas, 1915; *mystax* de Winton, 1898; *raptorum* Thomas, 1903.

Funisciurus bayonii (Bocage, 1890). J. Sci. Math. Phys. Nat. Lisboa, ser. 2, 2:3.
COMMON NAME: Lunda Rope Squirrel.
TYPE LOCALITY: N Angola, " . . . du Duque de Bragança . . . "
DISTRIBUTION: NE Angola, SW Dem. Rep. Congo.
STATUS: IUCN – Lower Risk (lc).

Funisciurus carruthersi Thomas, 1906. Ann. Mag. Nat. Hist., ser. 7, 18:140.
   COMMON NAME: Carruther's Mountain Squirrel.
   TYPE LOCALITY: Uganda, "Ruwenzori East, 6500' (1900 m)."
   DISTRIBUTION: Ruwenzori (S Uganda), Rwanda, Burundi.
   STATUS: IUCN – Vulnerable.
   SYNONYMS: *birungensis* Gyldenstolpe, 1927; *chrysippus* Thomas, 1923; *tanganyikae*
      Thomas, 1909.

Funisciurus congicus (Kuhl, 1820). Beitr. Zool. Vergl. Anat. Abt., 2:66.
   COMMON NAME: Congo Rope Squirrel.
   TYPE LOCALITY: "Congo". No specific locale given, probably Angola (Hill and Carter,
      1941:71).
   DISTRIBUTION: Dem. Rep. Congo, Angola, Namibia.
   STATUS: IUCN – Lower Risk (lc).
   SYNONYMS: *damarensis* (Roberts, 1938); *flavinus* Thomas, 1904; *interior* Thomas, 1916; *oenone*
      Thomas, 1926; *olivellus* Thomas, 1904; *poolii* (Jentink, 1906); *praetextus* Wagner, 1843.

Funisciurus isabella (Gray, 1862). Proc. Zool. Soc. Lond., 1862:180.
   COMMON NAME: Lady Burton's Rope Squirrel.
   TYPE LOCALITY: Cameroon, " . . . from the Camaroon Mountains, 7000 feet (2100 m) above
      the level of the sea . . . "
   DISTRIBUTION: Cameroon, Equatorial Guinea, Gabon, Central African Republic, Republic of
      Congo.
   STATUS: IUCN – Lower Risk (nt).
   SYNONYMS: *duchaillui* Sanborn, 1953; *dubosti* Eisentraut, 1969.

Funisciurus lemniscatus (Le Conte, 1857). Proc. Acad. Nat. Sci. Philadelphia, p. 11.
   COMMON NAME: Ribboned Rope Squirrel.
   TYPE LOCALITY: Equatorial Guinea, Rio Muni. " . . . from Western Africa."
   DISTRIBUTION: S of Sanaga River (Cameroon), Equatorial Guinea, Gabon, Central African
      Republic, Dem. Rep. Congo.
   STATUS: IUCN – Lower Risk (lc).
   SYNONYMS: *sharpei* (Gray, 1873); *mayumbicus* Kershaw, 1923.

Funisciurus leucogenys (Waterhouse, 1842). Ann. Mag. Nat. Hist., [ser. 1], 10: 202.
   COMMON NAME: Red-cheeked Rope Squirrel.
   TYPE LOCALITY: Equatorial Guinea, Bioko " . . . brought from Fernando Po".
   DISTRIBUTION: Ghana, Togo, Benin, Nigeria, Cameroon, Central African Republic, Rio Muni,
      Bioko.
   STATUS: IUCN – Lower Risk (lc).
   SYNONYMS: *erythrogenys* (Waterhouse, 1843); *auriculatus* (Matschie, 1891); *beatus* (Thomas,
      1910); *boydi* (Thomas, 1910); *oliviae* (Dollman, 1911).
   COMMENTS: Waterhouse (1842[1843]) renamed this species *erythrogenys*, "red-cheeked" in an
      attempt to replace the inappropriate name *leucogenys*, "white-cheeked". This is an
      unjustified emendation.

Funisciurus pyrropus (F. Cuvier, 1833). *In* E. Geoffroy and F. Cuvier, Hist. Nat. Mammifères, vii,
      No. 66, "Ecureuil aux pieds roux," 2 unno. pp and pl. [1833]; Tab 4:240 [1842].
   COMMON NAME: Fire-footed Rope Squirrel.
   TYPE LOCALITY: Gabon. " . . . et elle venoit de l'île Fernandopô, dans le gulfe de Guinée . . . "
   DISTRIBUTION: Gambia, S Senegal, Guinea Bissau, W Guinea, Sierra Leone, Liberia, S Côte
      d'Ivoire, SW Ghana, Nigeria, W Cameroon, Rio Muni (Equatorial Guinea), W Republic
      of Congo, Uganda, Rwanda, Burundi, Dem. Rep. Congo, NW Angola.
   STATUS: IUCN – Lower Risk (lc).
   SYNONYMS: *erythrops* (Gray, 1867); *rubripes* (Du Chaillu, 1860); *akka* de Winton, 1895; *emini*
      (de Winton, 1895) [not Stuhlman, 1894]; *victoriae* G. M. Allen and Loveridge, 1942;
      wintoni Neumann, 1900; *leonis* Thomas, 1905; *leucostigma* (Temminck, 1853);
      mandingo Thomas, 1903; *nigrensis* Thomas, 1909; *niveatus* Thomas, 1923;
      pembertoni Thomas, 1904; *talboti* Thomas, 1909.
   COMMENTS: Cuvier wrote, "Je donnerai à cet Écuriel le nom de *Pyrropus*, á cause de la couleur

rousse de ses pieds." Schinz (1845) spelled the species name *pyrrhopus* with no mention of the previous spelling. This constitutes an unjustified emendation. Not found on Bioko, Equatorial Guinea (= "Fernandopô", see type locality), and since the animal was a pet, it probably was captured on the mainland (Thomas, 1890:447).

Funisciurus substriatus de Winton, 1899. Ann. Mag. Nat. Hist., ser. 7, 4:357.
    COMMON NAME: Kintampo Rope Squirrel.
    TYPE LOCALITY: Ghana. " . . . near Kintampo, Gold Coast hinterland, 800 feet (240 m)."
    DISTRIBUTION: S Ghana, Togo, Benin, Burkina.
    STATUS: IUCN – Lower Risk (lc).

Heliosciurus Trouessart, 1880. Le Naturaliste, 2nd year, 1:292.
    TYPE SPECIES: *Sciurus gambianus* Ogilby, 1835 as designated by Opinion 464 of the International Commission on Zoological Nomenclature (1957*d*) in which *Sciurus annulatus* Desmarest, 1822 was suppressed.
    COMMENTS: Tribe Protoxerini (Moore, 1959). Reviewed in part by Grubb (1982*b*).

Heliosciurus gambianus (Ogilby, 1835). Proc. Zool. Soc. Lond., 1835:103.
    COMMON NAME: Gambian Sun Squirrel.
    TYPE LOCALITY: Gambia, possibly near Ft. St. Mary. " . . . brought from the Gambia . . . " "Through . . . Mr. Rendall, who has lately arrived from the Gambia, where his brother is lieutenent-governor of Fort St. Mary and the other British possessions in that neighbourhood . . . "
    DISTRIBUTION: Senegal, Gambia, Guinea Bissau, Guinea, Sierra Leone, Liberia, Côte d'Ivoire, Burkina Faso, Ghana, Togo, Benin, Nigeria, Chad, Central African Republic, Sudan, Ethiopia, Uganda, Kenya, Burundi, Tanzania, Dem. Rep. Congo, Angola, Zimbabwe, Zambia.
    STATUS: IUCN – Lower Risk (lc).
    SYNONYMS: *albina* (Gray, 1867); *annularis* (Schinz, 1845); *annulatus* (Desmarest, 1822); *simplex* (Lesson, 1838); *abassensis* (Neumann, 1902); *bongensis* Heuglin, 1877; *canaster* Thomas and Hinton, 1923; *dysoni* (St. Leger, 1937); *elegans* Thomas, 1909; *hoogstraali* Setzer, 1954; *kaffensis* (Neumann, 1902); *lateris* (Thomas, 1909); *limbatus* Schwarz, 1915; *loandicus* Thomas, 1923; *madogae* (Heller, 1911); *multicolor* (Rüppell 1835); *omensis* (Thomas, 1904); *rhodesiae* (Wroughton, 1907); *senescens* Thomas, 1909.

Heliosciurus mutabilis (Peters, 1852) Bericht Verhandl. K. Preuss Akad. Wiss., Berlin, 17:273.
    COMMON NAME: Mutable Sun Squirrel.
    TYPE LOCALITY: Mozambique, Boror, 19 km NW of Quelimane, 17°S . "Africa orientalis, Boror, 17° Lat. Austr."
    DISTRIBUTION: Malawi; Zambia; S and SW highlands, Tanzania; NW of the Zambezi River near Beira (Mozambique); Chirinda Forest, Melsetter Dist., Sabi/Lundi River confluence, Vumba, Umtali (SE Zimbabwe).
    STATUS: IUCN – Lower Risk (lc).
    SYNONYMS: *beirae* Roberts, 1913; *chirindensis* Roberts, 1913; *shirensis* (Gray, 1867); *smithersi* (Lundholm, 1955); *vumbae* Roberts, 1937.
    COMMENTS: G. M. Allen (1939) considered *mutabilis* a distinct species. Ellerman (1940) treated it as a subspecies of *H. gambianus* following Ingoldby (1927). Rosevear (1963), followed by Amtmann (1975), treated it as a subspecies of *H. rufobrachium*. Grubb (1982*b*) again treated *H. mutabilis* as a distinct species and recognized the five subspecies above.

Heliosciurus punctatus (Temminck, 1853) Esquisses Zool. sur la Côte de Guiné, p. 138.
    COMMON NAME: Small Sun Squirrel.
    TYPE LOCALITY: Guinea coast, no exact locality given. " . . . dans toutes les forêts de la Guiné . . . " Given by Ingoldby (1927) as Ghana: "Secondi and Bibiani, Gold Coast."
    DISTRIBUTION: Sierra Leone, E Liberia, S Côte d'Ivoire, S Ghana (E to Lake Volta).
    STATUS: IUCN – Lower Risk (lc).
    SYNONYMS: *savannius* Thomas, 1923.

COMMENTS: Treated as a subspecies of *H. gambianus* by Ingoldby (1927), Ellerman (1940), Rosevear (1969), and Amtmann (1975). Considered a distinct species by G. M. Allen (1939), and Roth and Thorington (1982).

Heliosciurus rufobrachium (Waterhouse, 1842). Ann. Mag. Nat. Hist., [ser. 1], 10:202.
  COMMON NAME: Red-legged Sun Squirrel.
  TYPE LOCALITY: Equatorial Guinea, Bioko " . . . brought from Fernando Po . . . ".
  DISTRIBUTION: Senegal, W Gambia, W Guinea Bissau, W Guinea, Sierra Leone, Liberia, S Côte d'Ivoire, S Ghana, S Togo, Benin, S Nigeria, Cameroon, Bioko and Rio Muni (Equatorial Guinea), SW Central African Republic, SE Sudan, Uganda, Rwanda, Burundi, SW and SE Kenya, Dem. Rep. Congo.
  STATUS: IUCN – Lower Risk (lc).
  SYNONYMS: *acticola* Thomas, 1923; *rufo-brachiatus* (Waterhouse, 1843); *arrhenii* (Lönnberg, 1917); *aubryi* (Milne-Edwards, 1867); *benga* Cabrera, 1917; *brauni* St. Leger, 1935; *caurinus* Thomas, 1923; *coenosus* (Thomas, 1909); *emissus* Thomas, 1923; *hardyi* Thomas, 1923; *isabellinus* (Gray, 1867); *keniae* (Neumann, 1902); *leakyi* Toschi, 1946; *leonensis* Thomas, 1923; *lualabae* Thomas, 1923; *maculatus* (Temminck, 1853); *aschantiensis* (Neumann, 1902); *libericus* (Miller, 1900); *waterhousii* (Gray, 1867); *medjianus* J. A. Allen, 1922; *nyansae* (Neumann, 1902); *obfuscatus* Thomas, 1923; *occidentalis* (Monard, 1941); *pasha* (Schwann, 1904); *rubricatus* J. A. Allen, 1922; *semlikii* Thomas, 1907.
  COMMENTS: Inclusion of *aubryi* and *emissus* questioned (Amtmann, 1975; Grubb, 1978:158).

Heliosciurus ruwenzorii (Schwann, 1904). Ann. Mag. Nat. Hist., ser. 7, 13:71.
  COMMON NAME: Ruwenzori Sun Squirrel.
  TYPE LOCALITY: E Dem. Rep. Congo, Ruwenzori, Wimi Valley.
  DISTRIBUTION: Ruwenzori Mtns in E Dem. Rep. Congo; Rwanda; Burundi; SW Uganda.
  STATUS: IUCN – Lower Risk (lc).
  SYNONYMS: *ituriensis* (Prigogone, 1954); *schoutedeni* (Prigogine, 1954); *vulcanius* Thomas, 1909.
  COMMENTS: Formerly included in *Aethosciurus*, which is here included in *Paraxerus* following Moore (1959).

Heliosciurus undulatus (True, 1892) Proc. U.S. Natl. Mus., 15:465.
  COMMON NAME: Zanj Sun Squirrel.
  TYPE LOCALITY: Tanzania, Mt. Kilimanjaro. "Male. Mount Kilima-Njaro, June 12, 1888. 6,000 feet (1800 m). Female. Kahé, south of Mount Kilima-Njaro, September 6, 1888."
  DISTRIBUTION: SE Kenya; NE Tanzania, including Mafia and Zanzibar Isls.
  STATUS: IUCN – Lower Risk (lc).
  SYNONYMS: *daucinus* Thomas, 1909; *dolosus* Thomas, 1909; *marwitzi* Müller, 1911; *shindi* (Heller, 1914).
  COMMENTS: Considered a subspecies of *H. rufobrachium* by Amtmann (1975), but treated as a separate monotypic species by Grubb (1982*b*).

Myosciurus Thomas, 1909. Ann. Mag. Nat. Hist., ser. 8, 3:474.
  TYPE SPECIES: *Sciurus minutus* Du Chaillu, 1860 (= *Sciurus pumilio* Le Conte, 1857).
  COMMENTS: Tribe Funambulini according to Moore (1959). The recent molecular study of Mercer and Roth (2003) places *Myosciurius* with the other African tree squirrels, to the exclusion of *Funambulus*.

Myosciurus pumilio (Le Conte, 1857). Proc. Acad. Nat. Sci. Philadelphia, 9:11.
  COMMON NAME: African Pygmy Squirrel.
  TYPE LOCALITY: Gabon. " . . . the head waters of the Ovenga River . . . ".
  DISTRIBUTION: SE Nigeria, Cameroon, Gabon, Rio Muni (Equatorial Guinea).
  STATUS: IUCN – Vulnerable.
  SYNONYMS: *minutulus* Hollister, 1921; *minutus* (Du Chaillu, 1860).
  COMMENTS: Reviewed by Gharaibeh and Jones (1996, Mammalian Species No. 523).

Paraxerus Forsyth Major, 1893. Proc. Zool. Soc. Lond., 1893:189.
TYPE SPECIES: *Sciurus cepapi* A. Smith, 1836.
SYNONYMS: *Aethosciurus* Thomas, 1916; *Montisciurus* Eisentraut, 1976; *Tamiscus* Thomas, 1918.
COMMENTS: Tribe Funambulini according to Moore (1959). Originally a subgenus of *Xerus*; includes *Aethosciurus* (with the exception of *ruwenzorii*, here included in *Heliosciurus*), *Montisciurus*, and *Tamiscus*. Recent molecular work of Mercer and Roth (2003) suggests that *Paraxerus* is not closely related to *Funambulus*, but to the rest of the African squirrels, in which case, it should be placed in the tribe Protoxerini.

Paraxerus alexandri (Thomas and Wroughton, 1907). Ann. Mag. Nat. Hist., ser. 7, 19:376.
COMMON NAME: Alexander's Bush Squirrel.
TYPE LOCALITY: Dem. Rep. Congo, Gudima, River Iri, Upper Welle.
DISTRIBUTION: NE Dem. Rep. Congo, Uganda.
STATUS: IUCN – Lower Risk (nt).

Paraxerus boehmi (Reichenow, 1886). Zool. Anz., 9:315.
COMMON NAME: Boehm's Bush Squirrel.
TYPE LOCALITY: S Dem. Rep. Congo, Marungu. "Marungu (Inner-Afrika)."
DISTRIBUTION: S Sudan, N and E Dem. Rep. Congo, Uganda, W Kenya, NW Tanzania, N Zambia.
STATUS: IUCN – Lower Risk (lc).
SYNONYMS: *antoniae* Thomas and Wroughton, 1907; *emini* (Stuhlman, 1894) [not de Winton, 1895]; *lunaris* (Thomas, 1918); *tanganyikae* (Thomas, 1918); *ugandae* (Neumann, 1902); *vulcanorum* (Thomas, 1918); *gazellae* (Thomas, 1918).

Paraxerus cepapi (A. Smith, 1836). Rept. Exped. Exploring Central Africa, p. 43.
COMMON NAME: Smith's Bush Squirrel.
TYPE LOCALITY: South Africa, North West Prov., Rustenberg Dist., Marico River.
DISTRIBUTION: S Angola, Zambia, SE Dem. Rep. Congo, Malawi, SW Tanzania, Mozambique, N Namibia, N Botswana, Zimbabwe, NE South Africa.
STATUS: IUCN – Lower Risk (lc).
SYNONYMS: *cepate* [*lapsus* for *cepapi*; Gray, 1843]; *bororensis* Roberts, 1946; *carpi* Lundholm, 1955; *cepapoides* Roberts, 1946; *chobiensis* Roberts, 1932; *kalaharicus* Roberts, 1932; *maunensis* Roberts, 1932; *phalaena* Thomas, 1926; *quotus* Wroughton, 1909; *sindi* Thomas and Wroughton, 1908; *soccatus* Wroughton, 1909; *yulei* (Thomas, 1902).

Paraxerus cooperi Hayman, 1950. Ann. Mag. Nat. Hist., ser. 12, 3:262.
COMMON NAME: Cooper's Mountain Squirrel.
TYPE LOCALITY: Cameroon, Kumba Div., Rumpi Hills, 5°N, 9°15′E.
DISTRIBUTION: Cameroon.
STATUS: IUCN – Vulnerable.
COMMENTS: Eisentraut (1976) put *cooperi* in a separate genus, *Montisciurus*.

Paraxerus flavovittis (Peters, 1852). Bericht Verhandl. K. Preuss. Akad. Wiss. Berlin, 17:274.
COMMON NAME: Striped Bush Squirrel.
TYPE LOCALITY: NE Mozambique, Mocímboa, 11°S on the coast. "Africa orientalis, Mossimboa, Quitangonha, a 11° ad 15° Lat. Austr."
DISTRIBUTION: S Kenya, Tanzania, N Mozambique.
STATUS: IUCN – Lower Risk (lc).
SYNONYMS: *exgeanus* (Hinton, 1920); *ibeanus* (Hinton, 1920); *mossambicus* (Thomas, 1919).
COMMENTS: Commonly spelled *flavivittis* but this was an unjustified emendation by Peters (1852).

Paraxerus lucifer (Thomas, 1897). Proc. Zool. Soc. Lond., 1897:430.
COMMON NAME: Black and Red Bush Squirrel.
TYPE LOCALITY: Malawi, Kombe Forest, Misuku Mtns, 9°43′S, 33°31′E.
DISTRIBUTION: N Malawi, SW Tanzania.
STATUS: IUCN – Lower Risk (lc).

Paraxerus ochraceus (Huet, 1880). Nouv. Arch. Mus. Hist. Nat. Paris, ser. 2, 3:154.
 COMMON NAME: Ochre Bush Squirrel.
 TYPE LOCALITY: Tanzania, Bagamoyo, (6°25′S, 38°54′E). "Cette petite espèce provient de
  Bagamoyo, station de nos missionnaires, sur la côte de Zanguebar, . . . "
 DISTRIBUTION: S Ethiopia, Kenya, Tanzania.
 STATUS: IUCN – Lower Risk (lc).
 SYNONYMS: *salutans* Thomas, 1909; *affinis* (Trouessart, 1897); *percivali* Dollman, 1911;
  animosus Dollman, 1911; *aruscensis* (Pagenstecher, 1885); *augustus* Dollman, 1911;
  pauli Matschie, 1894; *electus* Thomas, 1909; *ganana* (Rhoads, 1896); *jacksoni*
  (de Winton, 1897); *capitis* (Thomas, 1909); *kahari* Heller, 1911.

Paraxerus palliatus (Peters, 1852). Bericht Verhandl. K. Preuss. Akad. Wiss., Berlin, 17:273.
 COMMON NAME: Red Bush Squirrel.
 TYPE LOCALITY: Mozambique, mainland near Mocambique Isl "Africa orientalis,
  Quintangonha, 15° Lat. Austr."
 DISTRIBUTION: S Somalia, E Kenya, E Tanzania, Malawi, Mozambique, Zimbabwe, KwaZulu-
  Natal (South Africa).
 STATUS: IUCN – Vulnerable.
 SYNONYMS: *suahelicus* (Neumann, 1902); *bridgemani* Dollman, 1914; *auriventris* Roberts,
  1926; *tongensis* Roberts, 1931; *frerei* (Gray, 1873); *lastii* Thomas, 1906; *ornatus* (Gray,
  1864); *sponsus* Thomas and Wroughton, 1907; *swynnertoni* Wroughton, 1908; *tanae*
  (Neumann, 1902); *barawensis* (Neumann, 1902).
 COMMENTS: Kingdon (1974b) recognized three subspecies in the northern part of the range,
  Viljoen (1989) recognized four in the southern part.

Paraxerus poensis (A. Smith, 1830). S. Afr. Quart. J., 2:128.
 COMMON NAME: Green Bush Squirrel.
 TYPE LOCALITY: Equatorial Guinea, Bioko "Fernando Póo."
 DISTRIBUTION: Sierra Leone, SE Guinea, Liberia, Côte d'Ivoire, Ghana, Cameroon, Equatorial
  Guinea, Gabon, Republic of Congo, W Dem. Rep. Congo.
 STATUS: IUCN – Lower Risk (lc).
 SYNONYMS: *affinis* (Rhoads, 1896); *musculinus* (Temminck, 1853); *olivaceus* (Milne-Edwards,
  1867); *subviridescens* (Le Conte, 1857).
 COMMENTS: Subgenus *Aethosciurus* according to Moore (1959).

Paraxerus vexillarius (Kershaw, 1923). Ann. Mag. Nat. Hist., ser. 9, 11:591.
 COMMON NAME: Swynnerton's Bush Squirrel.
 TYPE LOCALITY: Tanzania, Usambara, Lushoto, Wilhelmsthal.
 DISTRIBUTION: C and E Tanzania.
 STATUS: IUCN – Vulnerable.
 SYNONYMS: *byatti* (Kershaw, 1923); *laetus* (G. M. Allen and Loveridge, 1933).
 COMMENTS: It is possible that *vexillarius* and *byatti* are separate species, as treated by
  G. M. Allen (1939) and Ellerman (1940). Amtmann (1975) combined them but stated
  that they may be distinct species.

Paraxerus vincenti Hayman, 1950. Ann. Mag. Nat. Hist., ser. 12, 3:263.
 COMMON NAME: Vincent's Bush Squirrel.
 TYPE LOCALITY: Mozambique, Namuli Mtn, north of the Zambezi River, (15°21′S, 37°4′E).
  " . . . collected at Namuli Mountain, Portuguese East Africa (15°21′S, 37°4′E), at 5000 ft
  (1500 m) . . . "
 DISTRIBUTION: N Mozambique.
 STATUS: IUCN – Vulnerable.

Protoxerus Forsyth Major, 1893. Proc. Zool. Soc. Lond., 1893:189.
 TYPE SPECIES: *Sciurus stangeri* Waterhouse, 1842.
 SYNONYMS: *Allosciurus* Conisbee, 1953.
 COMMENTS: Tribe Protoxerini (Moore, 1959). Includes *Allosciurus* (replaced *Myrsilas* Thomas,
  1909, which was preoccupied by *Myrsilas* Stål, 1865 [Hemiptera]) and *Protoxerus* as
  subgenera. Originally a subgenus of *Xerus*, raised to full generic rank by Thomas (1897b).

Protoxerus aubinnii (Gray, 1873). Ann. Mag. Nat. Hist., ser. 4, 12:65.
  COMMON NAME: Slender-tailed Squirrel.
  TYPE LOCALITY: Ghana, Ashanti Prov., Fanti. "Fantee".
  DISTRIBUTION: Liberia, Côte d'Ivoire, Ghana.
  STATUS: IUCN – Lower Risk (lc).
  SYNONYMS: *salae* (Jentink, 1881).
  COMMENTS: Subgenus *Allosciurus*.

Protoxerus stangeri (Waterhouse, 1842). Ann. Mag. Nat. Hist., [ser. 1], 10:202 (footnote).
  COMMON NAME: Forest Giant Squirrel.
  TYPE LOCALITY: Equatorial Guinea, Bioko. " . . . brought from Fernando Po . . . ".
  DISTRIBUTION: Sierra Leone, Liberia, Côte d'Ivoire, W Ghana, Togo, Nigeria, Cameroon, Rio
      Muni, Bioko (Equatorial Guinea), Gabon, E Republic of Congo, N Angola, S Central
      African Republic, Dem. Rep. Congo, Uganda, Rwanda, Burundi, W Kenya, N Tanzania.
  STATUS: IUCN – Lower Risk (lc).
  SYNONYMS: *nordhoffi* (Du Chaillu, 1860); *subalbidus* (Du Chaillu, 1860); *bea* Heller, 1912;
      centricola (Thomas, 1906); *moerens* Thomas, 1923; *notabilis* Thomas, 1923; *torrentium*
      Thomas, 1923; *cooperi* Kingdon, 1971; *eborivorus* (Du Chaillu, 1860); *calliurus* (Peters,
      1874); *dissonus* Thomas, 1923; *kabobo* Verheyen, 1960; *kwango* Verheyen, 1960;
      loandae (Thomas, 1906); *nigeriae* (Thomas, 1906); *personatus* Kershaw, 1923;
      signatus Thomas, 1910; *temminckii* (Anderson, 1879); *caniceps* (Temminck, 1853).
  COMMENTS: Subgenus *Protoxerus*.

Tribe Marmotini Pocock, 1923. Proc. Zool. Soc. Lond., 1923:240.

Ammospermophilus Merriam, 1892. Proc. Biol. Soc. Wash., 7:27
  TYPE SPECIES: *Spermophilus leucurus* Merriam, 1892.
  COMMENTS: Tribe Otospermophilini (Gromov et al., 1965), Marmotini, or
      Ammospermophilini. Formerly included in *Spermophilus* (Hershkovitz, 1949*b*); Bryant
      (1945) considered *Ammospermophilus* a distinct genus. Recent sequence, biochemical and
      chromosomal data support Bryant's recognition of generic status and basal position
      among ground squirrels (Harrison et al., 2003).

Ammospermophilus harrisii (Audubon and Bachman, 1854). Viviparous Quadrupeds of North
  America, 3:267.
  COMMON NAME: Harris's Antelope Squirrel.
  TYPE LOCALITY: Unknown. Restricted by Mearns (1896:444) to Santa Cruz Valley at the
      Mexican boundary, Santa Cruz Co., Arizona [USA].
  DISTRIBUTION: Arizona to SW New Mexico (USA) and adjoining Sonora, Mexico.
  STATUS: IUCN – Lower Risk (lc).
  SYNONYMS: *saxicolus* (Mearns, 1896); *kinoensis* Huey, 1937.
  COMMENTS: See *A. insularis*; reviewed by Best et al. (1990*c*, Mammalian Species No. 366).

Ammospermophilus insularis Nelson and Goldman, 1909. Proc. Biol. Soc. Wash., 22:24.
  COMMON NAME: Espiritu Santo Island Antelope Squirrel.
  TYPE LOCALITY: "Espiritu Santo Island, Lower California [=Baja California Sur], Mexico."
  DISTRIBUTION: Known only from the type locality.
  STATUS: IUCN – Lower Risk (lc).
  COMMENTS: Considered a distinct species (Hall, 1981:381); may be most closely related to
      A. harrisii (Mascarello and Bolles, 1980). Reviewed by Best et al. (1990*a*, Mammalian
      Species No. 364).

Ammospermophilus interpres (Merriam, 1890). N. Am. Fauna, 4:21.
  COMMON NAME: Texas Antelope Squirrel.
  TYPE LOCALITY: "El Paso, [El Paso Co.], Texas [USA]."
  DISTRIBUTION: New Mexico and W Texas (USA) to Coahuila, Chihuahua, and Durango
      (Mexico).
  STATUS: IUCN – Lower Risk (lc).

COMMENTS: Most divergent species of the genus (Bolles, 1981), and probable primitive sister-species to remainder (Hafner, 1981). Reviewed by Best et al. (1990b, Mammalian Species No. 365).

Ammospermophilus leucurus (Merriam, 1889). N. Am. Fauna, 2:19.
   COMMON NAME: White-tailed Antelope Squirrel.
   TYPE LOCALITY: "San Gorgonio Pass, [Riverside Co.], California [USA]."
   DISTRIBUTION: E California and SE Oregon to Colorado and New Mexico (USA), south to Baja California Sur (Mexico).
   STATUS: IUCN – Lower Risk (lc).
   SYNONYMS: *vinnulus* (Elliot, 1904); *canfieldiae* Huey, 1929; *cinamomeus* (Merriam, 1890); *escalante* (Hansen, 1955); *extimus* Nelson and Goldman, 1929; *notom* (Hansen, 1955); *peninsulae* (J. A. Allen, 1893); *pennipes* A. H. Howell, 1931; *tersus* Goldman, 1929.
   COMMENTS: Reviewed by Belk and Smith (1990, Mammalian Species No. 368).

Ammospermophilus nelsoni (Merriam, 1893). Proc. Biol. Soc. Wash., 8:129.
   COMMON NAME: Nelson's Antelope Squirrel.
   TYPE LOCALITY: "Tipton, San Joachin Valley [Tulare Co.], California [USA]."
   DISTRIBUTION: San Joaquin Valley (S California, USA).
   STATUS: May now be restricted to southern half of its former range (Hafner, 1981). IUCN – Endangered.
   SYNONYMS: *amplus* Taylor, 1916.
   COMMENTS: Most closely related to *interpres* (Hafner, 1981). Reviewed by Best et al. (1990d, Mammalian Species No. 367).

Cynomys Rafinesque, 1817. Am. Mon. Mag., 2:43.
   TYPE SPECIES: *Cynomys socialis* Rafinesque, 1817 (= *Arctomys ludoviciana* Ord, 1815).
   SYNONYMS: *Arctomys* Ord, 1815; *Cynomomus* Osborn, 1894; *Leucocrossuromys* Hollister, 1916; *Mamcynomiscus* Herrera, 1899; *Monax* Warden, 1819.
   COMMENTS: Tribe Cynomyini (Gromov et al., 1965), or Marmotini (McKenna and Bell, 1997). Revised by Pizzimenti (1975). Clark et al. (1971) published a key to the genus. Includes *Cynomys* and *Leucocrossuromys* as subgenera. Relationships of *Cynomys* to other ground squirrels are in flux. While generally regarded as monophyletic, and a sister-group to North American *Spermophilus*, recent evidence suggests that *Cynomys* may be most closely related to the subgenera *Ictidomys* and *Xerospermophilus*, making genus *Spermophilus* paraphyletic (Hafner, 1984:17; Harrison et al., 2003). Goodwin (1995) reviewed the biogeographic history of the genus.

Cynomys gunnisoni (Baird, 1855). Proc. Acad. Nat. Sci. Philadelphia, 7:334.
   COMMON NAME: Gunnison's Prairie Dog.
   TYPE LOCALITY: "Cochitope [Cochetopa] Pass of Rocky Mountains." (Saguache Co., Colorado, USA).
   DISTRIBUTION: SE Utah, SW Colorado, NE Arizona, and NW New Mexico (USA).
   STATUS: IUCN – Lower Risk (lc).
   SYNONYMS: *zuniensis* Hollister, 1916.
   COMMENTS: Subgenus *Leucocrossuromys* (Hall, 1981). Reviewed by Pizzimenti and Hoffmann (1973, Mammalian Species No. 25) and Pizzimenti (1976).

Cynomys leucurus Merriam, 1890. N. Am. Fauna, 3:59.
   COMMON NAME: White-tailed Prairie Dog.
   TYPE LOCALITY: "Fort Bridger, [Uinta Co.] Wyoming" (Merriam, 1890:33).
   DISTRIBUTION: SC Montana, W and C Wyoming, NE Utah, and NW Colorado (USA).
   STATUS: IUCN – Lower Risk (lc).
   COMMENTS: Subgenus *Leucocrossuromys* (Hall, 1981). Reviewed by Clark et al. (1971, Mammalian Species No. 7) and Pizzimenti (1976).

Cynomys ludovicianus (Ord, 1815). *In* Guthrie, New Geogr., Hist. Coml. Grammar, Philadelphia, 2nd ed., 2:292.
   COMMON NAME: Black-tailed Prairie Dog.

TYPE LOCALITY: "vicinity of the Missouri." Restricted by Hollister (1916a:14) to "Upper Missouri River" (USA).

DISTRIBUTION: Saskatchewan (Canada); Montana to E Nebraska, W Texas, New Mexico, and SE Arizona (USA); NE Sonora, and N Chihuahua (Mexico).

STATUS: U.S. ESA – Candidate taxon; IUCN – Lower Risk (nt).

SYNONYMS: *cinereus* Richardson, 1829; *grisea* Rafinesque, 1817; *latrans* (Harlan, 1825); *missouriensis* (Warden, 1819); *pyrrotrichus* Elliot, 1905; *socialis* Rafinesque, 1817; **arizonensis** Mearns, 1890.

COMMENTS: Subgenus *Cynomys*. Intraspecific variation reviewed by Chesser (1983). Reviewed by Hoogland (1996, Mammalian Species No. 535).

Cynomys mexicanus Merriam, 1892. Proc. Biol. Soc. Wash., 7:157.

COMMON NAME: Mexican Prairie Dog.

TYPE LOCALITY: "La Ventura, Coahuila, Mexico."

DISTRIBUTION: Coahuila, and San Luis Potosi; perhaps Nuevo Leon, and Zacatecas (NC Mexico).

STATUS: CITES – Appendix I; U.S. ESA – Endangered; IUCN – Endangered.

COMMENTS: Subgenus *Cynomys* (Hall, 1981:412); reviewed by Ceballos-G. and Wilson (1985, Mammalian Species No. 248) and Treviño-V. (1991).

Cynomys parvidens J. A. Allen, 1905. Mus. Brooklyn Inst. Arts and Sci., Sci. Bull., 1:119.

COMMON NAME: Utah Prairie Dog.

TYPE LOCALITY: USA, "Buckskin Valley, Iron County, Utah."

DISTRIBUTION: SC Utah (USA).

STATUS: U.S. ESA – Threatened; IUCN – Lower Risk (conservation dependent).

COMMENTS: Subgenus *Leucocrossuromys* (Hall, 1981). Reviewed by Pizzimenti and Collier (1975, Mammalian Species No. 52) and Pizzimenti and Nadler (1972).

Marmota Blumenbach, 1779. Hand. Hilfsb. Nat., 1:79.

TYPE SPECIES: *Mus marmota* Linnaeus, 1758.

SYNONYMS: *Arctomys* Schreber, 1780; *Glis* Erxleben, 1777; *Lagomys* Storr, 1780; *Lipura* Storr, 1780; *Marmotops* Pocock, 1922; *Petromarmota* Steppan et al., 1999.

COMMENTS: Tribe Marmotini (Moore, 1959). North American species reviewed by A. H. Howell (1915); Eurasian species revised by Gromov et al. (1965); amphiberingian species reviewed by Hoffmann et al. (1979); phylogeny tested by Steppan et al. (1999). Frase and Hoffmann (1980) provided a key to North American species.

Marmota baibacina Kastschenko, 1899. Rezul't. Altaisk. Zool. Exp. 1898, p. 62.

COMMON NAME: Gray Marmot.

TYPE LOCALITY: " . . . Multa River, near Nizhne-Uimon in the Altai Mountains" [Altaisk. Krai, Russia] (Ognev, 1963a:252). Alternatively, Aktol' River near Cherga, Gorno-Altaisk. A.O. (Kuznetsov, *in* Ellerman and Morrison-Scott, 1951:514).

DISTRIBUTION: Altai and Tien Shan Mtns, SW Siberia (Russia), SE Kazakhstan, Kyrgyzstan; Mongolia; Xinjiang (China). Introduced into Caucasus Mtns (Dagestan, Russia; Gromov et al., 1965:360).

STATUS: IUCN – Lower Risk (lc).

SYNONYMS: *aphanasievi* Kuznetsov, 1965; *lewisi* (Audubon and Bachman, 1854) [*nomen oblitum*]; *ognevi* Scalon, 1950; **centralis** (Thomas, 1909); **kastschenkoi** Stroganov and Yudin, 1956.

COMMENTS: Subgenus *Marmota* (Steppan et al., 1999). Placed by Ellerman and Morrison-Scott (1951:514) in *marmota*, and by Corbet (1978c:81) in *bobak*; Kapitonov (1966) analyzed purported hybridization between *baibacina* and *bobak*, while Nikol'skii (1974) and Nikol'skii et al. (1983) found species-specific vocalizations. Most Russian authors retain both as distinct species (Gromov et al., 1965:337-387; Zholnerovskaya et al., 1990; Zimina, 1978) and include *centralis* in this species. Steppan et al.(1999) found that *bobac* and *baibacina* are sister species. Kapitonov (1966) indicated that the population called *aphanasievi* is included in this species; but also see Corbet (1978c:81). Includes *lewisi*, a *nomen oblitum* (Hoffmann, 1977); *baibacina* (Brandt, 1843) is a *nomen nudum*. See also *bobak*, *sibirica*.

Marmota bobak (Müller, 1776). Linné's Vollstand. Natursyst. Suppl., p. 40.
COMMON NAME: Bobak Marmot.
TYPE LOCALITY: "Poland." Restricted by Ognev (1963*a*:221) to "right [W] bank of the Dnepr [River]", Ukraine.
DISTRIBUTION: Steppes of E Europe, east through Belarus, Ukraine, and Russia to N and C Kazakhstan.
STATUS: Regaining parts of former range (Bibikov, 1991). IUCN – Lower Risk (conservation dependent).
SYNONYMS: *bobac* (Schreber, 1780); *arctomys* (Pallas, 1779); *baibac* (Pallas, 1811); **tschaganensis** Bazhanov, 1930; *kozlovi* Fokanov, 1966.
COMMENTS: Subgenus *Marmota* (Steppan et al., 1999). See comments under *baibacina*, *himalayana*, and *sibirica*. Includes *kozlovi* (Fokanov, 1966) and *tschaganensis* (Gromov et al., 1965).

Marmota broweri Hall and Gilmore, 1934. Canadian Field Nat., 48:57.
COMMON NAME: Alaska Marmot.
TYPE LOCALITY: USA, "Point Lay, Arctic Coast of Alaska" Restricted by Rausch (1953:117) to head of Kukpowruk River, Alaska.
DISTRIBUTION: Brooks Range of N Alaska (USA) from near coast of Chukchi Sea to Alaska-Yukon border; perhaps also N Yukon (Canada).
STATUS: IUCN – Lower Risk (lc).
COMMENTS: Subgenus *Marmota* (Steppan et al., 1999). Regarded as a synonym of *caligata* (Hall, 1981) but Rausch and Rausch (1965, 1971) and Hoffmann et al. (1979) considered *broweri* a distinct species. Steppan et al. (1999) found that *broweri* is more closely related to the Old World *caudata* than to Nearctic marmots.

Marmota caligata (Eschscholtz, 1829). Zool. Atlas, Part 2, p. 1, pl. 6.
COMMON NAME: Hoary Marmot.
TYPE LOCALITY: Not specified (?). Restricted by J. A. Allen (1877:927) to near Bristol Bay, Alaska, USA.
DISTRIBUTION: C Alaska (USA), Yukon and Northwest Territories (Canada) south to W and NE Washington, C Idaho, and W Montana (USA).
STATUS: IUCN – Data Deficient as *M. c. sheldoni* and *M. c. vigilis*; otherwise – Lower Risk (lc).
SYNONYMS: *sheldoni* A. H. Howell, 1914; *vigilis* Heller, 1909; **cascadensis** A. H. Howell, 1914; *raceyi* Anderson, 1932; **okanagana** (King, 1836); *nivaria* A. H. Howell, 1914; *oxytona* Hollister, 1914; *sibila* Hollister, 1912.
COMMENTS: Subgenus *Petromarmota* (Steppan et al., 1999).

Marmota camtschatica (Pallas, 1811). Zoogr. Rosso-Asiat., p. 156.
COMMON NAME: Black-capped Marmot.
TYPE LOCALITY: "Kamchatka" [Kamchatsk. Obl., Russia].
DISTRIBUTION: E Siberia from Transbaikalia to Chukotka and Kamchatka (Russia), in several geographically isolated populations (Nikol'skii et al., 1991).
STATUS: IUCN – Lower Risk (lc).
SYNONYMS: **bungei** (Kastschenko, 1901); *cliftoni* (Thomas, 1902); **doppelmayri** Birula, 1922.
COMMENTS: Subgenus *Marmota* (Steppan et al., 1999). Regarded as a synonym of *marmota* (Ellerman and Morrison-Scott, 1951; Rausch, 1953). Hoffmann et al. (1979) reviewed this and related species, and affirmed its specific status. Kapitonov (1978) concluded that morphological differences justified independent specific status for *doppelmayeri*, but Nikol'skii et al. (1991) showed similarity of vocalization between it and *bungei*, while the nominate form differed, and recommended that *doppelmayeri* be retained provisionally in this species. Boyeskorov et al. (1999) then showed that *camtschatica* and *doppelmayeri* were most divergent morphologically and immunologically, but the geographically intermediate Yakutian subspecies was also intermediate in these characters, and referred to the group as a whole as a superspecies. Steppan et al. (1999) found *himalayana* and *sibirica* to be sister species, and in the same clade as *camtschatica*. Lyapunova et al.(1992) regarded *camtschatica* as closely related to Nearctic marmots based on pattern of chromosome evolution, as had Hoffmann et al. (1979) based on morphology, but molecular data do not support this interpretation.

Marmota caudata (Geoffroy, 1844). *In* Jacquemont, Voy. dans l'Inde, 4, Zool., p. 66.
    COMMON NAME: Long-tailed Marmot.
    TYPE LOCALITY: "Hombur [Ghombur] area, upper reaches of the Indus in Kashmir [India]"
        (Ognev, 1963*a*:284).
    DISTRIBUTION: W Tien Shan through the Pamirs (Kyrgyzstan, Tajikistan) to Hindu Kush
        (Afghanistan), Pakistan, Kashmir (India), and mountains of extreme W Xinjiang and
        Xizang (China).
    STATUS: CITES – Appendix III (India); IUCN – Lower Risk (nt).
    SYNONYMS: *aurea* (Blanford, 1875); *flavina* (Thomas, 1909); *littledalei* (Thomas, 1909);
        dichrous (Anderson, 1875); *stirlingi* Thomas, 1916.
    COMMENTS: Subgenus *Marmota* (Steppan et al., 1999). Includes *dichrous* (Corbet, 1978*c*:82);
        but also see Gromov et al. (1965:440) who listed it as a distinct species.

Marmota flaviventris (Audubon and Bachman, 1841). Proc. Acad. Nat. Sci. Philadelphia, 1:99.
    COMMON NAME: Yellow-bellied Marmot.
    TYPE LOCALITY: "Mountains between Texas and California" Restricted by A. H. Howell
        (1915) to Mt. Hood (Oregon, USA).
    DISTRIBUTION: SC British Columbia and S Alberta (Canada) south to N New Mexico, S Utah,
        Nevada, and California (USA).
    STATUS: IUCN – Lower Risk (lc).
    SYNONYMS: *fortirostris* Grinnell, 1921; *sierrae* A. H. Howell, 1915; *avara* (Bangs, 1899);
        parvula A. H. Howell, 1915; *dacota* (Merriam, 1889); *luteola* A. H. Howell, 1914;
        campioni Figgins, 1915; *warreni* A. H. Howell, 1914; *nosophora* A. H. Howell, 1914;
        engelhardti J. A. Allen, 1905; *notioros* Warren, 1934; *obscura* A. H. Howell, 1914.
    COMMENTS: Subgenus *Petromarmota* (Steppan et al., 1999). Reviewed by Frase and Hoffmann
        (1980, Mammalian Species No. 135).

Marmota himalayana (Hodgson, 1841). J. Asiat. Soc. Bengal, 10:777.
    COMMON NAME: Himalayan Marmot.
    TYPE LOCALITY: "Himalaya . . . and sandy plains of Tibet"; "potius Tibetensis" (Hodgson,
        1843). Restricted by Blanford (1875*a*) to "the Kachar of Nepal."
    DISTRIBUTION: Montane regions of W China, Nepal, and N India to Ladak.
    STATUS: CITES – Appendix III (India); IUCN – Lower Risk (lc).
    SYNONYMS: *hemachalana* (Hodgson, 1843); *hodgsoni* (Blanford, 1879); *tataricus* (Jameson,
        1847); *tibetanus* (Gray, 1847); *robusta* (Milne-Edwards, 1872).
    COMMENTS: Subgenus *Marmota* (Steppan et al., 1999). Placed in *bobak* (Ellerman and
        Morrison-Scott, 1951:515; Corbet, 1978*c*:81), but geographically separated from that
        species; evidence for specific status in Gromov et al. (1965); sister species of *sibirica*; see
        also comment under *baibacina* and *sibirica*.

Marmota marmota (Linnaeus, 1758). Syst. Nat., 10th ed., 1:60.
    COMMON NAME: Alpine Marmot.
    TYPE LOCALITY: "in alpibus Helveticis" Restricted by Thomas (1911*a*:147) to Swiss Alps.
    DISTRIBUTION: Swiss, Italian, and French Alps; W Austria; S Germany; Carpathian (Romania)
        and Tatra Mtns (Czech Republic, Poland); introduced into French Pyrenees, E Austria
        and N Serbia and Montenegro.
    STATUS: IUCN – Lower Risk (lc).
    SYNONYMS: *alba* (Bechstein, 1801); *alpina* Blumenbach, 1779; *marmotta* Trouessart, 1904;
        nigra (Bechstein, 1801); *tigrina* (Bechstein, 1801); *latirostris* Kratochvil, 1961.
    COMMENTS: Subgenus *Marmota* (Steppan et al., 1999). Formerly included *baibacina*, *broweri*,
        caligata, *camtschatica*, and *menzbieri* (Ellerman and Morrison-Scott, 1951; Rausch,
        1953).

Marmota menzbieri (Kashkarov, 1925). Trans. Turk. Sci. Soc., 2:47.
    COMMON NAME: Menzbier's Marmot.
    TYPE LOCALITY: "Chigyr-Tash, in the headwaters of the Ugam River, Talass Ala Tau" [Yuzhno-
        Kazakhstansk. Obl., Kazakhstan].
    DISTRIBUTION: W Tien Shan Mtns, in S Kazakhstan and NW Kirgizia.
    STATUS: IUCN – Vulnerable.

SYNONYMS: *zachidovi* Petrov, 1963.

COMMENTS: Subgenus *Marmota* (Steppan et al., 1999). Regarded by Ellerman and Morrison-Scott (1951:514) as a probable synonym of *marmota*; but see Corbet (1978c:81). Steppan et al. (1999) found *menzbieri* to be the sister species to *caudata*.

Marmota monax (Linnaeus, 1758). Syst. Nat., 10th ed., 1:60.

COMMON NAME: Woodchuck.

TYPE LOCALITY: "in America septentrionalis." Restricted by Thomas (1911a:147) to Maryland (USA).

DISTRIBUTION: Alaska (USA) through S Canada to S Labrador to NE and SC USA; south in Rocky Mtns, possibly to N Idaho.

STATUS: IUCN – Lower Risk (lc).

SYNONYMS: *bunkeri* Black, 1935; **canadensis** (Erxleben, 1777); *empetra* (Pallas, 1778); *johnsoni* Anderson, 1943; *melanopus* (Kuhl, 1820); *ochracea* Swarth, 1911; *petrensis* A. H. Howell, 1915; *sibila* (Wolf, 1808); **ignava** (Bangs, 1899); **rufescens** A. H. Howell, 1914; *preblorum* A. H. Howell, 1914.

COMMENTS: Subgenus *Marmota* (Steppan et al., 1999). Reviewed by Kwiecinski (1998, Mammalian Species No. 591).

Marmota olympus (Merriam, 1898). Proc. Acad. Nat. Sci. Philadelphia, 50:352.

COMMON NAME: Olympic Marmot.

TYPE LOCALITY: "From Timberline at head of Soleduc River, Olympic Mountains, [Olympic Nat. Park] Washington [USA]."

DISTRIBUTION: Olympic Mtns of W Washington (USA).

STATUS: IUCN – Lower Risk (lc).

COMMENTS: Subgenus *Petromarmota*; basal to the *caligata-vancouverensis* clade (Steppan et al., 1999). Considered a subspecies of *marmota* by Rausch (1953), but reviewed by Hoffmann et al. (1979) who confirmed its specific status. Also see Edelman (2003, Mammalian Species No. 736).

Marmota sibirica (Radde, 1862). Reise in den Suden von Ost-Sibierien, p. 159.

COMMON NAME: Tarbagan Marmot.

TYPE LOCALITY: "Kulusutai, near Lake Torei-Nor, southeast Transbaikal" [Chitinsk Obl., Russia].

DISTRIBUTION: SW Siberia, Tuva, Transbaikalia (Russia); N and W Mongolia; Heilungjiang and Inner Mongolia (China).

STATUS: IUCN – Lower Risk (lc).

SYNONYMS: *dahurica* (Dybowski, 1922); **caliginosus** Bannikov and Skalon, 1949.

COMMENTS: Subgenus *Marmota* (Steppan et al., 1999). Placed (with *baibacina*) in *bobak* (Ellerman and Morrison-Scott, 1951:515; Corbet, 1978c:81). Gromov et al. (1965) and Zimina (1978) provided evidence of specific distinctness and included *caliginosus* in this species; see comment under *baibacina* and *sibirica*. Steppan et al. (1999) found *sibirica* and *himalayana* to be sister species. Nikol'skii (1974) and Smirin et al. (1985) analyzed contact between *baibacina* and *sibirica* in Tuva and the Mongolian Altai; Sokolov and Orlov (1980:329) also indicated sympatry in NW Mongolia; limited hybridization is possible.

Marmota vancouverensis Swarth, 1911. Univ. California Publ. Zool., 7:201.

COMMON NAME: Vancouver Island Marmot.

TYPE LOCALITY: "Mt. Douglas (altitude 4,200 feet [1280 m]), twenty miles [32 km] south of Alberni, Vancouver Island, British Columbia" [Canada].

DISTRIBUTION: Mountains of Vancouver Isl (British Columbia, Canada).

STATUS: U.S. ESA – Endangered; IUCN – Endangered.

COMMENTS: Subgenus *Petromarmota*; sister species to *M. caligata* (Steppan et al., 1999). Considered a subspecies of *marmota* by Rausch (1953). Reviewed by Hoffmann et al. (1979) who confirmed its specific status, and by Nagorsen (1987, Mammalian Species No. 270).

Sciurotamias Miller, 1901. Proc. Biol. Soc. Wash., 14:23.
    TYPE SPECIES: *Sciurus davidianus* Milne-Edwards, 1867.
    SYNONYMS: *Rupestes* Thomas, 1922.
    COMMENTS: Includes *Rupestes* and *Sciurotamias* as subgenera; reviewed by Moore and Tate
        (1965). *S. davidianus* has a penile duct and Cowper's glands, and its glans and baculum are
        similar to those of *Ratufa*; therefore, Callahan and Davis (1982) removed this taxon from
        the Tamiasciurini (Moore, 1959:182) or Tamiini (Gromov et al., 1965:124) and tentatively
        referred it to Ratufini. Molecular studies now support its placement in Tamiini (Steppan
        et al., 2004), as does morphology (Thorington et al., 1998).

Sciurotamias davidianus (Milne-Edwards, 1867). Rev. Mag. Zool. Paris, ser. 2, 19:196.
    COMMON NAME: Père David's Rock Squirrel.
    TYPE LOCALITY: "Mountains of Peking," Hebei Prov., China.
    DISTRIBUTION: S Gansu to Hebei and Shandong to S Liaoning, Shanxi, Shaanxi, Ningxia,
        Henan, Hubei, Guizhou, Guangxi, Anhui and SW Sichuan (China) (Zhang et al.,
        1997).
    STATUS: IUCN – Lower Risk (lc).
    SYNONYMS: *latro* Heude, 1898; *owstoni* J. A. Allen, 1909; *saltitans* Heude, 1898; **consobrinus**
        (Milne-Edwards, 1868-1874); *thayeri* G. M. Allen, 1912. **Unassigned**: *collaris* Heude,
        1898.
    COMMENTS: Subgenus *Sciurotamias*. Distribution shown by Corbet and Hill (1992) is more
        restricted than that of Zhang et al. (1997).

Sciurotamias forresti (Thomas, 1922). Ann. Mag. Nat. Hist., ser. 9, 10:399.
    COMMON NAME: Forrest's Rock Squirrel.
    TYPE LOCALITY: "Mekong-Yangtze Divide on 27° 20′ N 7000-9000′." [=Hengduan Shan]
        Yunnan Prov., China.
    DISTRIBUTION: Yunnan Province,(China). See Zhang et al. (1997).
    STATUS: IUCN – Vulnerable.
    COMMENTS: Subgenus *Rupestes* (Moore and Tate, 1965). Apparently parapatric with
        davidianus ESE of Kunming.

Spermophilus F. Cuvier, 1825. Dentes des Mammiferes, p. 255.
    TYPE SPECIES: *Mus citellus* Linnaeus, 1766.
    SYNONYMS: *Anisonyx* Rafinesque, 1817; *Arctomys* Schreber, 1780; *Callospermophilus* Merriam,
        1897; *Citellus* Oken, 1816; *Citillus* Lichtenstein, 1830; *Colobates* Milne-Edwards, 1874;
        Colobotis Brandt, 1844; *Ictidomoides* Mearns, 1907; *Ictidomys* J. A. Allen, 1877; *Notocitellus*
        A. H. Howell, 1938; *Otocolobus* Brandt, 1844; *Otospermophilus* Brandt, 1844; *Poliocitellus*
        A. H. Howell, 1938; *Spermatophilus* Wagler, 1830; *Spermophilis* Richardson, 1839;
        Urocitellus Obolenskij, 1927; *Xerospermophilus* Merriam, 1892.
    COMMENTS: Tribe Marmotini (Moore, 1959). *Citellus* Oken, 1816 has been widely used, but is
        invalid (Corbet, 1978c:82; Hershkovitz, 1949b). Includes *Otospermophilus*,
        Xerospermophilus, *Ictidomys*, *Poliocitellus*, *Callospermophilus*, and *Spermophilus* as subgenera
        (Hall, 1981:382); Gromov et al. (1965) gave the first three taxa generic rank, and
        considered *Poliocitellus* a subgenus of *Ictidomys*, while placing *Callospermophilus* and
        Otospermophilus as sister subgenera under genus *Otospermophilus*. North American species
        revised by A. H. Howell (1938). Eurasian species revised by Gromov et al. (1965) who also
        recognized *Colobotis* and *Urocitellus* as subgenera; but see Hall (1981:382), who included
        them in subgenus *Spermophilus*. Holarctic species reviewed by Nadler et al. (1982, 1984). A
        key to the genus was given by Rickart and Yensen (1991).

Spermophilus adocetus (Merriam, 1903). Proc. Biol. Soc. Wash., 16:79.
    COMMON NAME: Tropical Ground Squirrel.
    TYPE LOCALITY: "La Salada, 40 miles [64 km] south of Uruapan, Michoacan, Mexico."
    DISTRIBUTION: E Jalisco, Michoacan, and N Guerrero (WC Mexico).
    STATUS: IUCN – Lower Risk (lc).
    SYNONYMS: *arceliae* Villa-R., 1942; **infernatus** Alvarez and Ramírez-P., 1968.
    COMMENTS: Subgenus *Notocitellus* (A. H. Howell, 1938) or *Otospermophilus* (Hall, 1981:399);
        but see also Birney and Genoways (1973) who suggested it is closer to subgenus

Ictidomys. Molecular sequence data indicate that *adocetus* and *annulatus* are sister species (Harrison et al., 2003), and basal to all other spermophiline ground squirrels, except *Ammospermophilus*. Reviewed by Best (1995*g*, Mammalian Species No. 504).

Spermophilus alashanicus Büchner, 1888. Wiss. Res. Przewalski Cent. Asien Zool. I:(Säugeth.):11.
COMMON NAME: Alashan Ground Squirrel.
TYPE LOCALITY: "Southern Ala Shan" [Desert, China] (Ognev, 1963*a*:150).
DISTRIBUTION: SC Mongolia; Ala Shan and E Nan Shan (N China).
STATUS: IUCN – Lower Risk (lc).
SYNONYMS: *dilutus* (Formozov, 1929); *obscurus* Büchner, 1888; *siccus* (G. M. Allen, 1925).
COMMENTS: Subgenus *Spermophilus* (Gromov et al. 1965:208). Placed by Corbet (1978*c*) in *dauricus*; Orlov and Davaa (1975) provided evidence of specific distinctness. Molecular sequence data (Harrison et al., 2003) suggest a sister species relationship with *pallidicauda*.

Spermophilus annulatus Audubon and Bachman, 1842. J. Acad. Nat. Sci. Philadelphia, 8:319.
COMMON NAME: Ring-tailed Ground Squirrel.
TYPE LOCALITY: "Western prairies." Restricted by A. H. Howell (1938:163) to Manzanillo, Colima, Mexico.
DISTRIBUTION: Nayarit to N Guerrero (Mexico).
STATUS: IUCN – Lower Risk (lc).
SYNONYMS: *goldmani* Merriam, 1902.
COMMENTS: Subgenus *Notocitellus* (A. H. Howell, 1938:162) or *Otospermophilus* (Hall, 1981:403). See *adocetes*, above. Reviewed by Best (1995*h*, Mammalian Species No. 508).

Spermophilus armatus Kennicott, 1863. Proc. Acad. Nat. Sci. Philadelphia, 15:158.
COMMON NAME: Uinta Ground Squirrel.
TYPE LOCALITY: "In the foothills of the Uinta Mountains, near Fort Bridger, [Uinta Co.] Wyo[ming]." [USA].
DISTRIBUTION: SC Utah to S Montana, SE Idaho to W Wyoming (USA).
STATUS: IUCN – Lower Risk (lc).
COMMENTS: Subgenus *Spermophilus* (Hall, 1981:386). Largely allopatric with its sister species, *beldingi* (Harrison et al., 2003). Reviewed by Eshelman and Sonnemann (2000; Mammalian Species No. 637).

Spermophilus atricapillus W. Bryant, 1889. Proc. California Acad. Sci., ser. 2, 2:26.
COMMON NAME: Baja California Rock Squirrel.
TYPE LOCALITY: "Comondu, Lower California" [Baja California Sur, Mexico].
DISTRIBUTION: Baja California (Mexico).
STATUS: IUCN – Lower Risk (lc).
COMMENTS: Subgenus *Otospermophilus* (Hall, 1981:402). Sister species to *beecheyi* (Harrison et al., 2003). Reviewed by Alvarez-Casteñeda et al. (1996, Mammalian Species No. 521).

Spermophilus beecheyi (Richardson, 1829). Fauna Boreali-Americana, 1:170.
COMMON NAME: California Ground Squirrel.
TYPE LOCALITY: "neighborhood of San Francisco and Monterey, in Calfornia." Restricted by Grinnell (1933) to Monterey, Monterey Co., California, USA.
DISTRIBUTION: W Washington (USA) to Baja California Norte (Mexico).
STATUS: IUCN – Lower Risk (lc).
SYNONYMS: *douglasii* (Richardson, 1829); *fisheri* Merriam, 1893; *nesioticus* (Elliot, 1904); *nudipes* (Huey, 1931); *parvulus* (A. H. Howell, 1931); *rupinarum* (Huey, 1931); *sierrae* (A. H. Howell, 1938).
COMMENTS: Subgenus *Otospermophilus* (Hall, 1981:401).

Spermophilus beldingi Merriam, 1888. Ann. N.Y. Acad. Sci., 4:317.
COMMON NAME: Belding's Ground Squirrel.
TYPE LOCALITY: "Donner, [Placer Co.,] California [USA]."
DISTRIBUTION: E Oregon, SW Idaho, NE California, N Nevada, and NW Utah (USA).
SYNONYMS: *creber* (Hall, 1940); *oregonus* Merriam, 1898.

COMMENTS: Subgenus *Spermophilus* (Hall, 1981:387). Reviewed by Jenkins and Eshelman (1984, Mammalian Species No. 221).

Spermophilus brevicauda Brandt, 1843. Bull. Acad. Sci., St. Petersbourg, pp. 364.
COMMON NAME: Brandt's Ground Squirrel.
TYPE LOCALITY: "Habitat, ut videtur, in provincis Altaicis australiorbis versus lacum Balchasch" (Ognev, 1947:83). Ellerman and Morrison-Scott (1966:508) however, cite "Zaisan basin", after Kuznetsov (1944).
DISTRIBUTION: Zaisan depression south and westward along the Tien Shan mountains to the vicinity of Almaty, on both sides of the Kazakh-Chinese border (see Ma et al., 1987).
SYNONYMS: *carruthersi* (Thomas, 1912); *intermedius* (Brandt, 1844); *ilensis* (Belyaev, 1945); *saryarka* (Selevin, 1937); *selevini* (Argyropolu, 1941).
COMMENTS: Subgenus *Colobotis* according to Gromov et al. (1965:315), but see Hall (1981:381) who included *Colobotis* in subgenus *Spermophilus*. See comments in *erythrogenys*, of which *brevicauda* was long considered a subspecies. A phylogeny based on molecular sequence data separates *brevicauda* from *erythrogenys*, *pallidicauda*, and *alashanicus* (Harrison et al., 2003).

Spermophilus brunneus (A. H. Howell, 1928). Proc. Biol. Soc. Wash., 41:211.
COMMON NAME: Idaho Ground Squirrel.
TYPE LOCALITY: "New Meadows, Adams County, Idaho [USA]."
DISTRIBUTION: WC Idaho (USA), in three isolated areas; north of Payette River to Hitt and Cuddy Mtns; between Cuddy and Seven Devils Mtns, and east of West Mtns.
STATUS: " . . . limited ranges and small breeding populations . . . vulnerable . . . " (Yensen, 1991:597). U.S. ESA – Candidate taxon as *S. b. endemicus*, Threatened as *S. b. brunneus*; IUCN - Critically Endangered as *S. b. brunneus*, Vulnerable as *S. b. endemicus*; otherwise Endangered as *S. brunneus*.
SYNONYMS: *endemicus* Yensen, 1991.
COMMENTS: Subgenus *Spermophilus* (Hall, 1981:385). Includes *endemicus*, which "may be reaching species-level separation" (Yensen, 1991:597). Included in clade which also contains *canus* and *mollis*. Population genetics analyzed by Gavin et al., 1999. Reviewed by Yensen and Sherman (1997, Mammalian Species No. 560).

Spermophilus canus Merriam, 1898. Proc. Biol. Soc. Wash., 12:70.
COMMON NAME: Merriam's Ground Squirrel.
TYPE LOCALITY: "Antelope, Wasco County, Oregon." [USA].
DISTRIBUTION: USA: E Oregon, except NE and SE corners; extreme NW Nevada; W side of Snake River in WC Idaho.
STATUS: IUCN – Lower Risk (lc).
SYNONYMS: *vigilis* (Merriam, 1913).
COMMENTS: Subgenus *Spermophilus* (Hall, 1981:383). Includes *vigilis* (see Vorontsov and Lyapunova, 1970; Nadler et al., 1984, wherein that junior synonym was employed as the species name). Formerly considered a subspecies of *townsendii* (Hall, 1981:383-384), but differs in diploid chromosome number (Nadler et al., 1984). No hybridization between *canus* (2n=46) and adjacent *mollis* (2n=38) or *townsendii* (2n=36) has been reported (Rickart et al., 1985). Reviewed (in part) by Rickart (1987), as *S. townsendii*.

Spermophilus citellus (Linnaeus, 1766). Syst. Nat., 12th ed., 1:80.
COMMON NAME: European Ground Squirrel.
TYPE LOCALITY: "Austria"; restricted by Martino and Martino (1940) to "Wagram, Niederosterrich" (Bauer, 1960:254).
DISTRIBUTION: SE Germany, Czech Republic, SW Poland through SE Europe to European Turkey, Moldovia and W Ukraine.
STATUS: IUCN – Vulnerable.
SYNONYMS: *citillus* (Pallas, 1779); **gradojevici** (Martino, 1929); *karamani* (Martino, 1940); **istricus** (Calinescu, 1934); *laskarevi* (Martino, 1940); **martinoi** (Peshchev, 1955); *balcanicus* (Markov, 1957); *thracius* (Mursaloglu, 1964). Not allocated to subspecies: *macedonicus* Fraguedakis-Tsolis and Ondrias, 1977.

COMMENTS: Subgenus *Spermophilus*. Formerly included *dauricus* and *xanthoprymnus* as in Ellerman and Morrison-Scott (1951:506); but see Gromov et al. (1965:208, 237), Vorontsov and Lyapunova (1970), and Orlov and Davaa (1975). Cytogenetics described by Belcheva and Peshev (1985) and Soldatovic et al. (1984); *xanthoprymnus* is now considered to be more closely aligned with *suslicus* and *dauricus* (Harrison et al., 2003).

Spermophilus columbianus (Ord, 1815). *In* Guthrie, New Geogr., Hist., Coml. Grammar, Philadelphia, 2nd ed., 2:292.
COMMON NAME: Columbian Ground Squirrel.
TYPE LOCALITY: "Between the forks of the Clearwater and Kooskooskie rivers," [Idaho Co., Idaho, USA].
DISTRIBUTION: SE British Columbia and W Alberta (Canada) to NE Oregon, C Idaho, and C Montana (USA).
STATUS: IUCN – Lower Risk (lc).
SYNONYMS: *albertae* (J. A. Allen, 1903); *brachiura* (Rafinesque, 1817); *erythrogluteia* (Richardson, 1829); **ruficaudus** (A. H. Howell, 1928).
COMMENTS: Subgenus *Urocitellus* according to Gromov et al. (1965:196), but Hall (1981:381) included *Urocitellus* in subgenus *Spermophilus*. Chromosomes described by Nadler et al. (1975a). Reviewed by Elliot and Flinders (1991, Mammalian Species No. 372).

Spermophilus dauricus Brandt, 1843. Bull. Phys. Math. Acad. Sci. St. Petersbourg, 2:379.
COMMON NAME: Daurian Ground Squirrel.
TYPE LOCALITY: " . . . circa Torei lacum exiccatum Dauuriae et ad Onon Bursa rivum." Torei-Nor (Lake), Chitinsk. Obl., Russia.
DISTRIBUTION: Transbaikalia (Russia), Mongolia, N China.
STATUS: IUCN – Lower Risk (lc).
SYNONYMS: *mongolicus* (Milne-Edwards, 1867); *ramosus* (Thomas, 1909); *umbratus* (Thomas, 1908); *yamashinae* (Kuroda, 1939).
COMMENTS: Subgenus *Spermophilus* (Gromov et al., 1965:244). Corbet (1978c:83) tentatively included *alashanicus* in this species, but see Orlov and Davaa (1975) who provided evidence of specific distinctness. See comment under *alashanicus*. Ellerman and Morrison-Scott (1951:506) included *dauricus* in *citellus*; but see Gromov et al. (1965:244) who considered *dauricus* a distinct species. Molecular sequence data suggest a sister species relationship with *xanthoprymnus* (Harrison et al., 2003).

Spermophilus elegans Kennicott, 1863. Acad. Nat. Sci. Philadelphia, p. 158.
COMMON NAME: Wyoming Ground Squirrel.
TYPE LOCALITY: "Fort Bridger," [Uinta Co., Wyoming, USA].
DISTRIBUTION: NE Nevada, SE Oregon, S Idaho, and SW Montana to C Colorado and W Nebraska (USA).
STATUS: IUCN – Data Deficient as *S. e. nevadensis*; otherwise Lower Risk (lc).
SYNONYMS: **aureus** (Davis, 1939); **nevadensis** (A. H. Howell, 1928).
COMMENTS: Subgenus *Spermophilus* (Hall, 1981:385). Regarded by A. H. Howell (1938) and Hall (1981:385) as a subspecies of *richardsonii*; but Nadler et al. (1971a), Robinson and Hoffmann (1975), Koeppl et al. (1978) and Fagerstone (1982) provided evidence of specific distinctness and included *aureus* and *nevadensis* in *elegans*. Sequence data (Harrison et al., 2003) support this. Reviewed by Zegers (1984, Mammalian Species No. 214).

Spermophilus erythrogenys Brandt, 1841. Bull. Acad. Sci. St. Petersbourg, p. 43.
COMMON NAME: Red-cheeked Ground Squirrel.
TYPE LOCALITY: " . . . vicinity of Barnaul" [Altaisk Krai, Russia] (Ognev, 1963a:60).
DISTRIBUTION: E Kazakhstan, SW Siberia (Russia). Formerly included an isolated population (*pallidicauda*) in Mongolia and Inner Mongolia (China), which is here given full species status along with *brevicauda*.
STATUS: IUCN – Lower Risk (lc).
SYNONYMS: *brunnescens* (Belyaev, 1943); *heptneri* (Vasil'eva, 1964); *ungae* (Martino, 1923).
COMMENTS: Subgenus *Colobotis* according to Gromov et al. (1965:315), but see Hall

(1981:381) who included *Colobotis* in subgenus *Spermophilus*. Ellerman and Morrison-Scott (1951:508, 511) regarded *brevicauda*, *intermedius* and *carruthersi* as synonyms of *pygmaeus*, and *pallidicauda* as a full species. Sludskii et al. (1969) considered *intermedius* (= *brevicauda*) a full species; while Corbet (1978c:84) provisionally included these taxa in *major*; but see Gromov et al. (1965:315), Vorontsov and Lyapunova (1970), and Nikol'skii (1984) for evidence of specific distinctness from *major*. The names *ungae* (Martino, 1923), *ilensis* (Belyaev, 1945), and *heptneri* (Vasil'eva, 1964) are particularly unstable. Ognev (1947) placed *ungae* in *erythrogenys*, and omitted *ilensis*, as did Ellerman and Morrison-Scott (1955). Gromov et al. (1965) placed *ungae*, and the newly described *heptneri*, in *major*, but *heptneri* was considered a junior synonym of *ungae*; *ilensis* was placed in *brevicauda*, but considered invalid because it is infrasubspecific. Pavlinov and Rossolimo (1987) placed all three, as well as *brevicauda*, in *erythrogenys*, without comment. Finally, Hoffmann et al. (1993) placed *ilensis* in *erythrogenys*, and *ungae* plus *heptneri* in *major*. The present treatment was reached after plotting the type localities of all taxa on distribution maps of the three species (*brevicauda*, *erythrogenys*, and *major*); *ungae* and *heptneri* fall within the range of *erythrogenys*, while *ilensis* is within *brevicauda*.

Spermophilus franklinii (Sabine, 1822). Trans. Linn. Soc. Lond., 13:587.
COMMON NAME: Franklin's Ground Squirrel.
TYPE LOCALITY: None specified. Restricted by Preble (1908:165) to Carlton House, Saskatchewan, Canada.
DISTRIBUTION: N Great Plains; Alberta, Saskatchewan, and Manitoba (Canada), south to Kansas, Illinois, and Indiana (USA).
STATUS: IUCN – Vulnerable.
COMMENTS: Subgenus *Poliocitellus* (Hall, 1981:397). Molecular sequence data (Harrison et al., 2003) group this species with *Xerospermophilus*, *Ictidomys* and *Cynomys*. Reviewed by Ostroff and Finck (2003, Mammalian Species No. 724).

Spermophilus fulvus (Lichtenstein, 1823). Naturh. Abh. Eversmann's Reise, p. 119.
COMMON NAME: Yellow Ground Squirrel.
TYPE LOCALITY: "near the Kuvandzhur River, east of Mugodzhary Mountains, north of Aral Sea" [Kazakhstan] (Ognev, 1963a:29).
DISTRIBUTION: Kazakhstan, from the Caspian Sea and the Volga River to Lake Balkash; south through Uzbekistan, W Tajikistan and Turkmenistan to NE Iran, and N Afghanistan; W Xinjiang (China).
STATUS: IUCN – Lower Risk (lc).
SYNONYMS: *concolor* (Fischer, 1829); *concolor* I. Geoffroy, 1831; *giganteus* (Fischer, 1829); *maximus* (Pallas, 1778); *nanus* (Fischer, 1829); **hypoleucos** (Satunin, 1909); *parthianus* (Thomas, 1915); **oxianus** (Thomas, 1915); *nigrimontanus* (Antipin, 1942).
COMMENTS: Subgenus *Colobotis* according to Gromov et al. (1965:276), but see Hall (1981:381) who included *Colobotis* in subgenus *Spermophilus*. See also *major*, which may be the sister species to *fulvus* (Harrison et al., 2003); *fulvus* occurs sympatrically with *major* in the southern third of the latter's range, where sporadic hybridization occurs south of Saratov (Ermakov, 1996).

Spermophilus lateralis (Say, 1823). *In* Long, Account Exped. Pittsburgh to Rocky Mtns, 2:46.
COMMON NAME: Golden-mantled Ground Squirrel.
TYPE LOCALITY: "near Cañon City." Restricted by Merriam (1905:163) to Arkansas River, about 26 mi. [42 km] below Canyon City, Fremont Co., Colorado (USA).
DISTRIBUTION: Montane W North America, from C British Columbia to S New Mexico in the Rocky Mtns, and the Columbia River south to S California and Nevada.
STATUS: IUCN – Data Deficient as *S. l. wortmani*; otherwise Lower Risk (lc).
SYNONYMS: **arizonensis** (V. Bailey, 1913); **bernardinus** Merriam, 1898; *brevicauda* Merriam, 1893 [not Brandt, 1843]; **castanurus** (Merriam, 1890); *caryi* (A. H. Howell, 1917); **certus** (Goldman, 1921); **chrysodeirus** (Merriam, 1890); **cinerascens** (Merriam, 1890); **connectens** (A. H. Howell, 1931); **mitratus** (A. H. Howell, 1931); **tescorum** (Hollister, 1911); **trepidus** (Taylor, 1910); **trinitatus** (Merriam, 1901); **wortmani** (J. A. Allen, 1895).

COMMENTS: Subgenus *Callospermophilus* (Hall, 1981:382-406). Gromov et al. (1965:70;150), in contrast to Hall (1981) considered *Callospermophilus* a subgenus of the genus *Otospermophilus*. Reviewed by Bartels and Thompson (1993, Mammalian Species No. 440).

Spermophilus madrensis (Merriam, 1901). Proc. Washington Acad. Sci., 3:563.
COMMON NAME: Sierra Madre Ground Squirrel.
TYPE LOCALITY: "from Sierra Madre, near Guadalupe y Calvo, Chihuahua, Mexico (7,000 feet [2134 m] altitude)."
DISTRIBUTION: SW Chihuahua (Mexico).
STATUS: IUCN – Lower Risk (nt).
COMMENTS: Subgenus *Callospermophilus* (Hall, 1981:410). Gromov et al. (1965:150) considered *Callospermophilus* a subgenus of the genus *Otospermophilus*. *S. madrensis* and *S. lateralis* are sister species (Harrison et al., 2003). Reviewed by Best and Thomas (1991*b*, Mammalian Species No. 378).

Spermophilus major (Pallas, 1778). Nova Spec. Quad. Glir. Ord., p. 125.
COMMON NAME: Russet Ground Squirrel.
TYPE LOCALITY: "Steppe near Samara," [Kuibyshev, Kuibyshevsk. Obl., Russia] (Ognev, 1963*a*:34).
DISTRIBUTION: Steppe between Volga and Irtysh rivers (Russia; N Kazakhstan). Formerly, steppe between Don and Volga rivers (Russia; Gromov et al., 1965:291). Reported from Xinjiang (Ma et al., 1987); but probably a misidentified *brevicauda*.
STATUS: IUCN – Lower Risk (nt).
SYNONYMS: *argyropuloi* (Bazhanov, 1947); *rufescens* (Keyserling and Blasius, 1840); *selevini* (Argyropulo, 1941).
COMMENTS: Subgenus *Colobotis* according to Gromov et al. (1965:290), but see Hall (1981:381) who included *Colobotis* in subgenus *Spermophilus*. Occasionally hybridizes with *brevicauda* and *fulvus* (Denisov, 1963; Nikol'skii and Starikov, 1997; Ognev, 1947), and more widely with *pygmaeus* and *suslicus* (Ermakov, 1996). Corbet (1978*c*:84) provisionally included *erythrogenys* and *brevicauda* in this species, but Gromov et al. (1965:290) and Vorontsov and Lyapunova (1970) considered *erythrogenys* a distinct species, and Gromov et al. (1965:315) included *brevicauda* in *erythrogenys*; see comment under those species. *S. major* is geographically cohesive, and allopatrically distributed with respect to *brevicauda*, and allopatric (Bobrinskii et al., 1965:61) or narrowly sympatric to *erythrogenys* (Sludskii et al., 1969:162). In contrast, it it genetically unstable, grouping with either *brevicauda* or *pygmaeus* in sequence data (Harrison et al., 2003), probably as a result of hybridization (see specimens examined). Formerly included *relictus* (Ellerman and Morrison-Scott, 1955).

Spermophilus mexicanus (Erxleben, 1777). Syst. Regn. Anim., 1:428.
COMMON NAME: Mexican Ground Squirrel.
TYPE LOCALITY: "in nova Hispania?" Restricted by Mearns (1896:443) to "Toluca, [Mexico,] Mexico."
DISTRIBUTION: S New Mexico and W Texas (USA) to Jalisco and S Puebla (C Mexico).
STATUS: IUCN – Lower Risk (lc).
SYNONYMS: **parvidens** Mearns, 1896.
COMMENTS: Subgenus *Ictidomys* (Hall, 1981:394). Known to hybridize at several localities with *tridecemlineatus* (Cothran and Honeycutt, 1984; Cothran et al., 1977). Reviewed by Young and Jones (1982, Mammalian Species No. 164).

Spermophilus mohavensis Merriam, 1889. N. Am. Fauna, 2:15.
COMMON NAME: Mohave Ground Squirrel.
TYPE LOCALITY: "Mohave River, California [USA]." Restricted by Grinnell and Dixon (1918) to near Rabbit Springs, about 15 mi. (24 km) E Hesperia, San Bernardino Co.
DISTRIBUTION: NW Mohave Desert and Owens Valley (S California, USA).
STATUS: IUCN – Vulnerable.
COMMENTS: Subgenus *Xerospermophilus*; (Hall, 1981:405). Most closely related to *S. tereticaudus* (Hafner and Yates, 1983); hybridizes at three localities, but the hybrid

zone is narrow and stable (Hafner, 1992). Reviewed by Best (1995*i*, Mammalian Species No. 509).

Spermophilus mollis Kennicott, 1863. Proc. Acad. Nat. Sci. Philadelphia, 15:157.

COMMON NAME: Piute Ground Squirrel.

TYPE LOCALITY: "Camp Floyd, near Fairfield, [Utah Co.,] Utah [USA]."

DISTRIBUTION: Two disjunct populations: in Washington, N of Yakima, and W of Columbia, rivers; and SE corner of Oregon, Snake River valley (Idaho) southward through Nevada (except extreme S), extreme EC California, and W Utah.

STATUS: IUCN – Lower Risk (lc).

SYNONYMS: *leurodon* (Merriam, 1913); *stephensi* Merriam, 1898; *washoensis* (Merriam, 1913); **artemesiae** (Merriam, 1913); *pessimus* (Merriam, 1913); **idahoensis** (Merriam, 1913).

COMMENTS: Subgenus *Spermophilus* (Hall, 1981:383). Formerly considered a subspecies of *townsendii* (Hall, 1981:383), but differs chromosomally (Nadler et al., 1984). No hybridization between *mollis* (2n=38) and adjacent *canus* (incl. *vigilis*) (2n=46) has been reported (Rickart et al., 1985:97-98). Reviewed (in part) by Rickart (1987) as *S. townsendii*; see that account below.

Spermophilus musicus Ménétries, 1832. Cat. Raisonne des Objets de Zoologie, St. Petersbourg, p. 21.

COMMON NAME: Caucasian Mountain Ground Squirrel.

TYPE LOCALITY: "Il habite le Caucase sur les montagnes le plus élevées et pas loin des nieges éternalles." Restricted by Sviridenko (1927, see Ognev, 1963*a*:114) to "Ush-Kulan [Georgia]".

DISTRIBUTION: N Caucasus Mtns (Georgia).

STATUS: IUCN – Lower Risk (lc).

SYNONYMS: *boehmii* (Krassovskii, 1932); *magisteri* (Heptner, 1948); *saturatus* (Ognev, 1947) [not Rhoads, 1895]; *typicus* (Satunin, 1908).

COMMENTS: Subgenus *Spermophilus* (Gromov et al., 1965:249). Regarded by Ellerman and Morrison-Scott (1951:508) and Corbet (1978*c*:83) as a subspecies of *pygmaeus*. Gromov et al. (1965:249) and Vorontsov and Lyapunova (1970) provided evidence of specific distinctness. Molecular sequence data place it close to *pygmaeus* (Harrison et al., 2003).

Spermophilus pallidicauda (Satunin, 1903). Ezhegodnik Zoologicheskii Muzeya, Akademii Nauk, St. Petersburg, 7:5-6.

COMMON NAME: Pallid Ground Squirrel.

TYPE LOCALITY: "vicinity of Lake Khulu-Nur", Ullyn Bulyk, Baidarak river, Mongolian Atlai. Mongolia, Gobi Altai.

DISTRIBUTION: This monotypic species is endemic to Mongolia and the adjacent Nei Mongol Autonomous Region (Inner Mongolia).

COMMENTS: Subgenus *Spermophilus* (Gromov et al., 1965). In addition to constant morphological differences, it also differs in chromosome number and molecular sequence from *major* and *erythrogenys*, in which it was previously placed (Corbet, 1978*c*; Sokolov and Orlov, 1980), having a sister species relationship with *alashanicus* (Harrison et al., 2003).

Spermophilus parryii (Richardson, 1825). *In* Parry, Voy. discovery Northwest Passage, Vol. 6- app. second voy., p. 316.

COMMON NAME: Arctic Ground Squirrel.

TYPE LOCALITY: Restricted by Preble (1902:46) to "Five Hawser Bay, Lyon Inlet, Melville Peninsula, [Hudson Bay, Keewatin District, Northwest Territories], Canada."

DISTRIBUTION: NW Canada; Alaska (USA); NE Yakutia, Anadyrsk. Krai, and Chukotka (Russia).

STATUS: IUCN – Date Deficient as *S. p. kodiacensis, S. p. lyratus,* and *S. p. nebulicola*; otherwise Lower Risk (lc).

SYNONYMS: *phaeognatha* (Richardson, 1829); **ablusus** (Osgood, 1903); *stonei* (J. A. Allen, 1903); **kennicottii** (Ross, 1861); *barrowensis* Merriam, 1900; *beringensis* Merriam, 1900; **kodiacensis** J. A. Allen, 1874; **leucostictus** Brandt, 1844; *buxtoni* J. A. Allen, 1903; *tschuktschorum* (Chernyavskii, 1972); **lyratus** (Hall and Gilmore, 1932); **nebulicola** (Osgood, 1903); **osgoodi** Merriam, 1900; **plesius** Osgood, 1900; **stejnegeri** (J. A. Allen,

1903); *brunniceps* Kittlitz, 1858 [*nomen nudum*]; *coriakorum* (Portenko, 1963); *janensis* (Ognev, 1937).

COMMENTS: Subgenus *Urocitellus* according to Gromov et al. (1965:184), but see Hall (1981:381) who included *Urocitellus* in subgenus *Spermophilus*. Regarded by Ellerman and Morrison-Scott (1951:511) and Hall and Kelson (1959:343) as a synonym of *undulatus*. Gromov et al. (1965:184) and Nadler et al. (1974) provided evidence of specific distinctness. Reviewed by Chernyavskii (1972), Nadler and Hoffmann (1977), Serdyuk (1979) (Palearctic), and Pearson (1981) (Nearctic). Nikol'skii and Wallschläger (1982) noted differences in alarm calls between Siberian and Alaskan populations.

Spermophilus perotensis Merriam, 1893. Proc. Biol. Soc. Wash., 8:131.
COMMON NAME: Perote Ground Squirrel.
TYPE LOCALITY: "Perote, Veracruz, Mexico."
DISTRIBUTION: Veracruz and Puebla (EC Mexico).
STATUS: IUCN – Lower Risk (nt).
COMMENTS: Subgenus *Ictidomys* (Hall, 1981:397). Sister species to *spilosoma* (Harrison et al., 2003). Chromosomes described by Uribe-Alcocer et al. (1979). Reviewed by Best and Ceballos (1995, Mammalian Species No. 507).

Spermophilus pygmaeus (Pallas, 1778). Nova Spec. Quad. Glir. Ord., p. 122.
COMMON NAME: Little Ground Squirrel.
TYPE LOCALITY: "Maximos et paene dixerim monstrosos Citillós passim ad inferiorum laikum in campis squalidis." Restricted by Ognev (1963a:102) to "lower reaches of the Ural River" "Indersk [Kazakhstan]".
DISTRIBUTION: SW Ukraine; S Ural Mtns to Crimea (Russia); Kazakhstan; NW Uzbekistan; Dagestan (Georgia).
STATUS: IUCN – Lower Risk (lc).
SYNONYMS: *arenicola* (Rall, 1935); *binominatus* (Ellerman, 1940); *ellermani* (Harris, 1944); *flavescens* (Pallas, 1779); *orlovi* (Ellerman, 1940); *pallidus* (Orlov and Fenyuk, 1927) [not J. A. Allen, 1874]; *planicola* (Satunin, 1909); *ralli* (Heptner, 1948) [for *arenicola*, Rall; unavailable]; *satunini* (Sveridenko, 1922); *saturatus* (Ognev, 1947) [preoccupied, not Rhoads, 1895]; **brauneri** (Martino, 1914); *kalabuchovi* (Ognev, 1937); **herbicolus** (Martino, 1914); *atricapilla* (Orlov, 1927) [preoccupied; not Bryant, 1889]; *herbidus* (Martino, 1915) [*nomen nudum*]; *septentrionalis* (Obolenskii, 1927); **mugosaricus** (Lichtenstein, 1823); *kazakstanicus* (Goodwin, 1935); *nikolskii* (Heptner, 1934).
COMMENTS: Subgenus *Spermophilus*, see Gromov et al. (1965:257) and comment under *musicus*. Hybridizes rarely with *erythrogenys* and *fulvus*, and more frequently with *major* and *suslicus* (Bazhanov 1944; Denisov, 1964); in zone of contact SW of Saratov (Russia) (Denisov and Smirnova, 1976; Ermakov, 1996).

Spermophilus ralli (Kuznetsov, 1948). Zveri Kirgizii [Animals of Kirgiziya, Moscow Soc. of Naturalists] pp. 39.
COMMON NAME: Tien Shan Ground Squirrel.
TYPE LOCALITY: "Tip iz kotlovinny Oz. Issyk-Kul' " [Type from basin of lake Issyk-Kul'] (Kuznetsov, 1965:264).
DISTRIBUTION: Mountains and valleys surrounding the eastern end of the Issyk-Kul' valley, from the Terskii-Alatau in the southeast to the Ketmen' Alatau in the northeast. Its range is separated from that of *S. relictus*, which occupies the western Tien Shan.
COMMENTS: Subgenus *Spermophilus*. This species is usually included in *S. relictus* as an allopatric subspecies, but molecular sequence data indicate that the two taxa are separate (Harrison et al., 2003). Kuznetsov's name is regarded as having priority over *ralli* (Heptner, 1948a; a replacement name for *pygmaeus arenicola*; and thus preoccupied), but Pavlinov and Rossolimo (1987) indicated that Kuznetsov (1948) might not have been published in May, but later, in July-August, while *pygmaeus ralli* was published in May, 1948. If so, Kuznetsov, 1948 is preoccupied.

Spermophilus relictus (Kashkarov, 1923). Trans. Turk. Sci. Soc., 1:185.
COMMON NAME: Relict Ground Squirrel.

TYPE LOCALITY: "Kara-Bura Gorge and Kumysh-Tagh Gorge in the Talus Ala Tau" [Talassk. Obl., Kyrgyzstan] (Ognev, 1963a:70).

DISTRIBUTION: Tien Shan Mtns in Kyrgyzstan and SE Kazakhstan.

STATUS: IUCN – Lower Risk (lc).

COMMENTS: Subgenus *Spermophilus* (Gromov et al., 1965:198).

Spermophilus richardsonii (Sabine, 1822). Trans. Linn. Soc. Lond., 13:589.
COMMON NAME: Richardson's Ground Squirrel.
TYPE LOCALITY: "Carleton-House," [Saskatchewan, Canada].
DISTRIBUTION: N Great Plains in S Alberta, S Saskatchewan, S Manitoba (Canada), Montana (see Swenson, 1981), North Dakota, NE South Dakota, W Minnesota, and NW Iowa (USA).
STATUS: IUCN – Lower Risk (lc).
COMMENTS: Subgenus *Spermophilus* (Hall, 1981:385). Formerly included *elegans*; see comment under that species. Reviewed by Michener and Koeppl (1985, Mammalian Species No. 243).

Spermophilus saturatus (Rhoads, 1895). Proc. Acad. Nat. Sci. Philadelphia, 47:43.
COMMON NAME: Cascade Golden-mantled Ground Squirrel.
TYPE LOCALITY: "Lake Kichelos [= Keechelus], Kittitas Co., Wash[ingto]n, (elevation 8,000 feet [2438 m])."
DISTRIBUTION: Cascade Mtns of W Washington (USA) and SW British Columbia (Canada).
STATUS: IUCN – Lower Risk (lc).
COMMENTS: Subgenus *Callospermophilus* (Hall, 1981:409). Gromov et al. (1965:150) considered *Callospermophilus* a subgenus of genus *Otospermophilus*. Reviewed by Trombulak (1988, Mammalian Species No. 322). Molecular sequence data indicate that this species is basal to other *Callospermophilus* (Harrison et al., 2003).

Spermophilus spilosoma Bennett, 1833. Proc. Zool. Soc. Lond., 1833:40.
COMMON NAME: Spotted Ground Squirrel.
TYPE LOCALITY: "that part of California which adjoins to Mexico." Restricted by A. H. Howell (1938:122) to Durango [City], Durango, Mexico.
DISTRIBUTION: C Mexico to S Texas, SW South Dakota, and NW Arizona (USA).
STATUS: IUCN – Lower Risk (lc).
SYNONYMS: *altiplanensis* Anderson, 1972; *ammophilus* Hoffmeister, 1959; *annectens* Merriam, 1893; *bavicorensis* Anderson, 1972; *cabrerai* (Dalquest, 1951); *canescens* Merriam, 1890; *arens* V. Bailey, 1902; *macrospilotus* Merriam, 1890; *microspilotus* Elliot, 1901; *cryptospilotus* Merriam, 1890; *marginatus* V. Bailey, 1902; *major* Merriam, 1890 [not Pallas, 1778]; *obsoletus* Kennicott, 1863; *oricolus* Alvarez, 1962; *pallescens* (A. H. Howell, 1928); *pratensis* Merriam, 1890; *obsidianus* Merriam, 1890.
COMMENTS: Subgenus *Ictidomys* (Hall, 1981:395). Reviewed by Streubel and Fitzgerald (1978a, Mammalian Species No. 101). See comment under *perotensis*. Chromosomes of the race *cabrerai* described by Uribe-Alcocer et al. (1978).

Spermophilus suslicus (Güldenstaedt, 1770). Nova Comm. Acad. Sci. Petropoli, 14:389.
COMMON NAME: Speckled Ground Squirrel.
TYPE LOCALITY: " . . . in campis vastissimus tanaicensibus precipue urbes et Tambov" [Voronezh area, Voronezhsk. Obl., Russia].
DISTRIBUTION: Steppes of E and S Europe, including Poland, E Romania, Ukraine north to Oka River and east to the Volga River (Russia).
STATUS: IUCN – Vulnerable.
SYNONYMS: *averini* (Migulin, 1927); *meridioccidentalis* (Migulin, 1927); *odessana* Nordmann, 1842; *ognevi* (Reshetnik, 1946); *volhynensis* (Reshetnik, 1946); *boristhenicus* (Pusanov, 1958); *guttatus* (Pallas, 1770); *guttulatus* Schinz, 1845; *leucopictus* (Dondorff, 1792).
COMMENTS: Subgenus *Spermophilus* Gromov et al. (1965:212). Reviewed by Corbet (1978c) and Gromov and Erbajeva (1995), who listed only four subspecies. Hybridizes with *major* west of Kazan' (Ermakov, 1996). See also comment under *pygmaeus*.

Spermophilus tereticaudus Baird, 1858. Mammalia, *in* Repts. U.S. Expl. Surv., 8(1):315.
COMMON NAME: Round-tailed Ground Squirrel.

TYPE LOCALITY: [Old] "Fort Yuma" [Imperial Co., California, USA].

DISTRIBUTION: Deserts of SE California, S Nevada, W Arizona (USA), NE Baja California and Sonora (Mexico).

STATUS: U.S. ESA – Candidate taxon as *S. t. chlorus*; IUCN – Data Deficient as *S. t. chlorus*, otherwise Lower Risk (lc).

SYNONYMS: *eremonomus* (Elliot, 1904); *vociferans* (Huey, 1926); ***apricus*** Ç(Huey, 1927); ***chlorus*** (Elliot, 1904); ***neglectus*** Merriam, 1889; *arizonae* (Grinnell, 1918); *sonoriensis* Ward, 1891.

COMMENTS: Subgenus *Xerospermophilus* (Hall, 1981:405). Reviewed by Ernest and Mares (1987, Mammalian Species No. 274). See also *mohavensis*.

Spermophilus townsendii Bachman, 1839. J. Acad. Nat. Sci. Philadelphia, 8:61.

COMMON NAME: Townsend's Ground Squirrel.

TYPE LOCALITY: "On the Columbia River, about 300 miles [483 km] above its mouth." Restricted by A. H. Howell (1938:60, 62) to west bank of Walla Walla River near confluence with Columbia River [near Wallula, Walla Walla Co., Washington, USA].

DISTRIBUTION: SE Washington (USA), S of Yakima River and W and N of Columbia River.

STATUS: IUCN – Data Deficient.

SYNONYMS: *yakimensis* Merriam, 1898; ***nancyae*** Nadler, 1968.

COMMENTS: Subgenus *Spermophilus* (Hall, 1981:382). Formerly included two cytotypes (*mollis, canus*) now considered distinct species (Nadler et al., 1984; Vorontsov and Lyapunova, 1970). Reviewed by Rickart (1987, Mammalian Species No. 268).

Spermophilus tridecemlineatus (Mitchill, 1821). Med. Repos. (NY), (n.s.), 6(21):248.

COMMON NAME: Thirteen-lined Ground Squirrel.

TYPE LOCALITY: " . . . region bordering the sources of the river Mississippi . . . "; restricted by J. A. Allen (1895*b*:338) to C Minnesota [USA].

DISTRIBUTION: Great Plains, from C Texas to E Utah, Ohio (USA) and SC Canada.

STATUS: IUCN – Data Deficient as *S. t. alleni*, otherwise Lower Risk (lc).

SYNONYMS: *hoodii* (Sabine, 1822); ***alleni*** Merriam, 1898; ***arenicola*** (A. H. Howell, 1928); ***blanca*** Armstrong, 1971; ***hollisteri*** (V. Bailey, 1913); ***monticola*** (A. H. Howell, 1928); ***olivaceous*** J. A. Allen, 1895; ***pallidus*** J. A. Allen, 1874; ***parvus*** J. A. Allen, 1895; ***texensis*** Merriam, 1898; *badius* Bangs, 1899.

COMMENTS: Subgenus *Ictidomys* (Hall, 1981:391). Reviewed by Streubel and Fitzgerald (1978*b*, Mammalian Species No. 103). See also comment under *mexicanus*.

Spermophilus undulatus (Pallas, 1778). Nova Spec. Quad. Glir. Ord., p. 122.

COMMON NAME: Long-tailed Ground Squirrel.

TYPE LOCALITY: "Selenga River valley," [Buryat ASSR, Russia].

DISTRIBUTION: E Kazakhstan; S Siberia, Transbaikalia (Russia); N Mongolia; Heilungjiang and Xinjiang (China).

STATUS: IUCN – Lower Risk (lc).

SYNONYMS: *undulatum* (Pallas, 1779); ***altaica*** (Brandt, 1841); *eversmanni* (Brandt, 1841); ***jacutensis*** (Brandt, 1844); ***menzbieri*** (Ognev, 1937); ***stramineus*** (Obolenskii, 1927); ***transbaikalicus*** (Obolenskii, 1927); *intercedens* (Ognev, 1937).

COMMENTS: Subgenus *Urocitellus* according to Gromov et al. (1965:162), but see Hall (1981:381) who included *Urocitellus* in subgenus *Spermophilus*. Formerly included *parryii* (Hall and Kelson, 1959:343); but see Nadler et al. (1974) and comments under *parryii*. Chromosomes described by Nadler et al. (1975*a*).

Spermophilus variegatus (Erxleben, 1777). Syst. Regn. Anim., 1:421.

COMMON NAME: Rock Squirrel.

TYPE LOCALITY: "in Mexico." Restricted by Nelson (1898:898) to "Valley of Mexico, near City of Mexico," [Distrito Federal, Mexico].

DISTRIBUTION: S Nevada to SW Texas and Utah (USA) to Puebla (C Mexico).

STATUS: IUCN – Lower Risk (lc).

SYNONYMS: *buccatus* (Lichtenstein, 1830); *macrourus* Bennett, 1833; ***buckleyi*** Slack, 1861; ***couchii*** Baird, 1855; ***grammurus*** (Say, 1823); *juglans* (V. Bailey, 1913); *tiburonensis* Jones and Manning, 1989; ***robustus*** (Durrant and Hansen, 1954); ***rupestris*** (J. A. Allen, 1903); ***tularosae*** (Benson, 1932); ***utah*** (Merriam, 1903).

COMMENTS: Subgenus *Otospermophilus* (Hall, 1981:399); belongs to *variegatus* group, which also includes *beecheyi* and *atricapillus* (Harrison et al., 2003). Reviewed by Oaks et al. (1987, Mammalian Species No. 272).

Spermophilus washingtoni (A. H. Howell, 1938). N. Am. Fauna, 56:69.
COMMON NAME: Washington Ground Squirrel.
TYPE LOCALITY: "Touchet, Walla Walla Co., Wash[ington]." [USA].
DISTRIBUTION: SE Washington, NE Oregon (USA).
STATUS: U.S. ESA – Candidate taxon; IUCN – Vulnerable.
SYNONYMS: *loringi* A. H. Howell, 1938.
COMMENTS: Subgenus *Spermophilus* (Hall, 1981:384). Current status assessed by Betts (1999). Reviewed by Rickart and Yensen (1991, Mammalian Species No. 371).

Spermophilus xanthoprymnus (Bennett, 1835). Proc. Zool. Soc. Lond., 1835:90.
COMMON NAME: Asia Minor Ground Squirrel.
TYPE LOCALITY: "Erzurum" [Turkey].
DISTRIBUTION: Transcaucasia (Armenia, possibly Azerbaijan), Turkey, Syria, and Israel.
STATUS: IUCN – Lower Risk (lc).
SYNONYMS: *schmidti* (Satunin, 1908).
COMMENTS: Subgenus *Spermophilus* (Gromov et al. 1965:237). Regarded by Ellerman and Morrison-Scott (1951:506) and Corbet (1978c:83) as a synonym of *citellus*; but see Vasil'eva (1961), Gromov et al. (1965:237), and Vorontsov and Lyapunova (1970), who provided evidence of specific distinctness. Molecular sequence data suggest that this species may form a clade together with *dauricus* and *suslicus* (Harrison et al., 2003). Karyotypes analyzed by Doğramaci et al. (1994a).

Tamias Illiger, 1811. Prodr. Syst. Mamm. Avium., p. 83.
TYPE SPECIES: *Sciurus striatus* Linnaeus, 1758.
SYNONYMS: *Eutamias* Trouessart, 1880; *Neotamias* A. H. Howell, 1929.
COMMENTS: Tribe Marmotini (Moore, 1959). Nearctic forms revised by A. H. Howell (1929). Sutton (1992) provided a key to the species. Includes *Eutamias* (*sibiricus*), *Tamias* (*striatus*), and *Neotamias* as subgenera (Corbet, 1978c:85; Ellerman, 1940:428; Levenson et al., 1985; Nadler et al., 1977). Disagreement exists regarding the status of *Eutamias* and *Neotamias*; see White (1953), Ellis and Maxson (1979), Hall (1981:337), Patterson and Heaney (1987), and Jameson (1999). Levenson et al. (1985) found that *T.* (*T.*) *striatus* and *T.* (*E.*) *sibiricus* together formed the primitive sister group to *Neotamias* species. Piaggio and Spicer (2001) supported this hypothesis. Thus, a single genus, *Tamias*, may be employed for all chipmunks (Levenson et al., 1985), but two genera (*Tamias* and *Neotamias*) could be recognized, or all three could be recognized as genera.

Tamias alpinus Merriam, 1893. Proc. Biol. Soc. Wash., 8:137.
COMMON NAME: Alpine Chipmunk.
TYPE LOCALITY: "Big Cottonwood Meadows, . . . just south of Mount Whitney, altitude 3,050 meters or 10,000 feet" [Tulare Co., California, USA].
DISTRIBUTION: Alpine zone in Sierra Nevada, from Tuolumne to Tulare Counties (EC California, USA).
STATUS: IUCN – Lower Risk (lc).
COMMENTS: Subgenus *Neotamias*. Reviewed by Clawson et al. (1994a, Mammalian Species No. 461).

Tamias amoenus J. A. Allen, 1890. Bull. Am. Mus. Nat. Hist., 3:90.
COMMON NAME: Yellow-pine Chipmunk.
TYPE LOCALITY: "Fort Klamath, [Klamath Co.,] Oregon." [USA].
DISTRIBUTION: C British Columbia (Canada) south to C California east to C Montana and W Wyoming (USA).
STATUS: IUCN – Lower Risk (lc).
SYNONYMS: *propinquus* (Anthony, 1913); **affinis** J. A. Allen, 1890; **albiventris** (Booth, 1947); **canicaudus** (Merriam, 1903); **caurinus** (Merriam, 1898); **celeris** (Hall and Johnson, 1940); **cratericus** (Blossom, 1937); **felix** Rhoads, 1895; **ludibundus** (Hollister, 1911);

luteiventris J. A. Allen, 1890; ***monoensis*** (Grinnell and Storer, 1916); ***ochraceus*** (A. H. Howell, 1925); ***septentrionalis*** (Cowan, 1946); ***vallicola*** (A. H. Howell, 1922).
COMMENTS: Subgenus *Neotamias*. Reviewed by Sutton (1992, Mammalian Species No. 390). DNA variation and introgressive hybridization between *T. amoenus* and *T. ruficaudus* reported by Good et al. (2003) and Demboski and Sullivan (2003).

Tamias bulleri J. A. Allen, 1889. Bull. Am. Mus. Nat. Hist., 2:173.
COMMON NAME: Buller's Chipmunk.
TYPE LOCALITY: "Sierra de Valparaiso, Zacatecas," [Mexico].
DISTRIBUTION: Sierra Madre, in S Durango, W Zacatecas, and N Jalisco (Mexico).
STATUS: IUCN – Lower Risk (lc).
COMMENTS: Subgenus *Neotamias*. Formerly included *durangae* and *solivagus*, which were considered *incertae sedis* by Callahan (1980); see *durangae*. Reviewed by Bartig et al. (1993, Mammalian Species No. 438).

Tamias canipes (V. Bailey, 1902). Proc. Biol. Soc. Wash., 15:117.
COMMON NAME: Gray-footed Chipmunk.
TYPE LOCALITY: "Guadalupe Mts" [Culberson Co., Texas, USA]. Restricted by A. H. Howell (1929:101) to head of Dog Canyon, 7000 ft [2130 m].
DISTRIBUTION: Mountains of SE New Mexico and W Texas (USA).
STATUS: IUCN – Lower Risk (nt).
SYNONYMS: ***sacramentoensis*** Flaherty, 1960.
COMMENTS: Subgenus *Neotamias*. Elevated from subspecies of *cinereicollis* by Fleharty (1960). Reviewed by Findley et al. (1975:103-112), and by Best et al. (1992, Mammalian Species No. 411).

Tamias cinereicollis J. A. Allen, 1890. Bull. Am. Mus. Nat. Hist., 3:94.
COMMON NAME: Gray-collared Chipmunk.
TYPE LOCALITY: "San Francisco Mountain, [Coconino Co.,] Arizona." [USA].
DISTRIBUTION: Mountains of C and E Arizona and C and SW New Mexico (USA).
STATUS: IUCN – Lower Risk (lc).
SYNONYMS: ***cinereus*** (V. Bailey, 1911).
COMMENTS: Subgenus *Neotamias*. Reviewed by Hilton and Best (1993, Mammalian Species No. 436).

Tamias dorsalis Baird, 1855. Proc. Acad. Nat. Sci. Philadelphia, 7:332.
COMMON NAME: Cliff Chipmunk.
TYPE LOCALITY: "Fort Webster, Coppermines of the Mimbres" Restricted by A. H. Howell (1929:131) to near present site of Santa Rita, Grant Co., New Mexico.
DISTRIBUTION: E Nevada, S Idaho, Utah, SW Wyoming, and NW Colorado south through Arizona and W New Mexico (USA) to NW Durango, W Coahuila, and coastal Sonora (Mexico).
STATUS: IUCN – Lower Risk (lc).
SYNONYMS: *canescens* (J. A. Allen, 1904); ***carminis*** (Goldman, 1938); ***grinnelli*** (Burt, 1931); ***nidoensis*** (Lidicker, 1960); ***sonoriensis*** (Callahan and Davis, 1977); ***utahensis*** (Merriam, 1897).
COMMENTS: Subgenus *Neotamias*. Reviewed by Callahan and Davis (1977) who included *sonoriensis* in this species, and by Hart (1992, Mammalian Species No. 399).

Tamias durangae (J. A. Allen, 1903). Bull. Am. Mus. Nat. Hist., 19:594.
COMMON NAME: Durango Chipmunk.
TYPE LOCALITY: "Arroyo de Bucy, Sierra de Candella, . . . about 7,500 feet [2134 m], northwestern Durango, Mexico."
DISTRIBUTION: SW Chihuahua to WC Durango; SE Coahuila (Mexico).
STATUS: IUCN – Lower Risk (lc).
SYNONYMS: *nexus* Elliot, 1905; *solivagus* (A. H. Howell, 1922).
COMMENTS: Subgenus *Neotamias*. Formerly included in *bulleri*; includes *solivagus*; perhaps conspecific with *canipes*; see Callahan (1980), who considered both *durangae* and *solivagus*, *incertae sedis*. Reviewed by Best et al. (1993, Mammalian Species No. 437).

Tamias merriami J. A. Allen, 1889. Bull. Am. Mus. Nat. Hist., 2:176.
> COMMON NAME: Merriam's Chipmunk.
> TYPE LOCALITY: "San Bernardino Mts" [N of San Bernardino, 4500 feet (1372 m),
> San Bernardino Co., California, USA].
> DISTRIBUTION: San Francisco Bay southward in the Coast Range, and south of Columbia
> (California, USA) in the Sierra Nevada, to extreme N Baja California (Mexico).
> STATUS: IUCN – Lower Risk (lc).
> SYNONYMS: *mariposae* (Grinnell and Storer, 1916); ***kernensis*** (Grinnell and Storer, 1916);
> ***pricei*** J. A. Allen, 1895.
> COMMENTS: Subgenus *Neotamias*. Formerly included *meridionalis* and *obscurus*; see Callahan
> (1977). See also comment under *obscurus*. Reviewed by Best and Granai (1994*a*,
> Mammalian Species No. 476), and Larson (1981-1987).

Tamias minimus Bachman, 1839. J. Acad. Nat. Sci. Philadelphia, 8:71.
> COMMON NAME: Least Chipmunk.
> TYPE LOCALITY: "Green River, near mouth of Big Sandy Creek, [Sweetwater Co.,]
> Wyo[ming]." [USA].
> DISTRIBUTION: C Yukon (Canada) south through Sierra Nevada and S New Mexico, east to
> Michigan (USA) and W Quebec (Canada).
> STATUS: IUCN – Critically Endangered as *T. m. atristriatus,* Vulnerable as *T. m. selkirki*,
> otherwise Lower Risk (lc).
> SYNONYMS: ***atristriatus*** (V. Bailey, 1913); ***borealis*** J. A. Allen, 1877; ***cacodemus*** (Cary, 1906);
> ***caniceps*** (Osgood, 1900); ***caryi*** (Merriam, 1908); ***confinis*** (A. H. Howell, 1925);
> ***consobrinus*** J. A. Allen, 1890; *clarus* (V. Bailey, 1918); *lectus* (J. A. Allen, 1905);
> *grisescens* (A. H. Howell, 1925); ***hudsonius*** (Anderson and Rand, 1944); ***neglectus***
> J. A. Allen, 1890; *jacksoni* (A. H. Howell, 1925); *operarius* (Merriam, 1905); *arizonensis*
> (A. H. Howell, 1922); ***oreocetes*** (Merriam, 1897); ***pallidus*** J. A. Allen, 1874; ***pictus***
> J. A. Allen, 1890; *melanurus* Merriam, 1890; ***scrutator*** (Hall and Hatfield, 1934);
> ***selkirki*** (Cowan, 1946); ***silvaticus*** (White, 1952).
> COMMENTS: Subgenus *Neotamias*. Southern Rocky Mtn populations revised by Sullivan
> (1985). Reviewed by Verts and Carraway (2001, Mammalian Species No. 653).

Tamias obscurus J. A. Allen, 1890. Bull. Am. Mus. Nat. Hist., 3:70.
> COMMON NAME: California Chipmunk.
> TYPE LOCALITY: "San Pedro [Martir] Mountains" [near Vallecitos, Baja California Norte,
> Mexico].
> DISTRIBUTION: S California (San Bernardino Co., USA) to C Baja California (Mexico).
> STATUS: IUCN – Lower Risk (lc).
> SYNONYMS: ***davisi*** (Callahan, 1977); ***meridionalis*** (Nelson and Goldman, 1909).
> COMMENTS: Subgenus *Neotamias*. Regarded by A. H. Howell (1929) as a synonym of
> *merriami*; but see Callahan (1977) who provided evidence of specific distinctness and
> included *davisi* and *meridionalis* in this species. Reviewed by Best and Granai (1994*b*,
> Mammalian Species No. 472).

Tamias ochrogenys (Merriam, 1897). Proc. Biol. Soc. Wash., 11:195, 206.
> COMMON NAME: Yellow-cheeked Chipmunk.
> TYPE LOCALITY: "Mendocino, [Mendocino Co.,] California." [USA].
> DISTRIBUTION: Coast of N California from Van Duzen River south to S Sonoma Co. (USA).
> STATUS: IUCN – Lower Risk (lc).
> COMMENTS: Subgenus *Neotamias*. Elevated from subspecies of *townsendii* by Sutton and
> Nadler (1974); supported by Kain (1985), Sutton (1987), and Gannon and Lawlor
> (1989). Reviewed by Gannon et al. (1993, Mammalian Species No. 445).

Tamias palmeri (Merriam, 1897). Proc. Biol. Soc. Wash., 11:208.
> COMMON NAME: Palmer's Chipmunk.
> TYPE LOCALITY: "Charleston Peak, [Clark Co.,] Nevada [USA] (altitude about 2450 meters or
> 8000 feet)."
> DISTRIBUTION: Charleston Mtns (S Nevada, USA).
> STATUS: IUCN – Vulnerable.

COMMENTS: Subgenus *Neotamias*. Reviewed by Best (1993*a*, Mammalian Species No. 443).

Tamias panamintinus Merriam, 1893. Proc. Biol. Soc. Wash., 8:134.
COMMON NAME: Panamint Chipmunk.
TYPE LOCALITY: "Johnson Cañon . . . Panamint Mountains, [Inyo Co.,] California." [USA] Restricted by Grinnell (1933:128) to vicinity of Hungry Bill's Ranch, about 5000 ft [1524 m].
DISTRIBUTION: Mountains of SE California and SW Nevada (USA).
STATUS: IUCN – Lower Risk (lc).
SYNONYMS: *juniperus* (Burt, 1931); ***acrus*** (Johnson, 1943).
COMMENTS: Subgenus *Neotamias*. Reviewed by Best et al. (1994*b*, Mammalian Species No. 468).

Tamias quadrimaculatus Gray, 1867. Ann. Mag. Nat. Hist., ser. 3, 20:435.
COMMON NAME: Long-eared Chipmunk.
TYPE LOCALITY: "California, Michigan Bluff (Gruber)" [Placer Co., USA].
DISTRIBUTION: Sierra Nevada of EC California (Plumas to Mariposa cos.); C and adjacent WC Nevada (USA).
STATUS: IUCN – Lower Risk (lc).
SYNONYMS: *macrorhabdotes* Merriam, 1886.
COMMENTS: Subgenus *Neotamias*. Reviewed by Clawson et al. (1994*b*, Mammalian Species No. 469).

Tamias quadrivittatus (Say, 1823). *In* James, Account Exped. Pittsburgh to Rocky Mtns, 2:45.
COMMON NAME: Colorado Chipmunk.
TYPE LOCALITY: "[Arkansas River] . . . the place where the river leaves the mountains," Restricted by Merriam (1905:163) to about 26 mi. [42 km] below Cañon City, Fremont Co., Colorado [USA].
DISTRIBUTION: Mountains of Colorado and E Utah south to NE Arizona and S New Mexico (USA).
STATUS: IUCN – Vulnerable as *T. q. australis*, otherwise Lower Risk (lc).
SYNONYMS: *animosus* (Warren, 1909); *australis* (Patterson, 1980); *gracilis* J. A. Allen, 1890.
COMMENTS: Subgenus *Neotamias*. Formerly included *hopiensis* (Merriam, 1905), but see Patterson (1984:452), who regarded it a *nomen dubium*, and *rufus*, now considered distinct (Patterson, 1984); but see Hoffmeister and Ellis (1979). See also *umbrinus*. Reviewed by Best et al. (1994*a*, Mammalian Species No. 466).

Tamias ruficaudus (A. H. Howell, 1920). Proc. Biol. Soc. Wash., 33:91.
COMMON NAME: Red-tailed Chipmunk.
TYPE LOCALITY: "Upper St. Mary's Lake, [Glacier Co.,] Montana." [USA].
DISTRIBUTION: NE Washington to W Montana (USA), and SE British Columbia (Canada).
STATUS: IUCN – Lower Risk (lc).
SYNONYMS: ***simulans*** (A. H. Howell, 1922).
COMMENTS: Subgenus *Neotamias*. Patterson and Heaney (1987) presented both cranial and bacular data indicating that *simulans* may be specifically distinct from *ruficaudus*, but the nature of contact between the two forms is not known. Reviewed by Best (1993*b*, Mammalian Species No. 452). Introgressive hybridization with *T. amoenus* reported in Good et al. (2003).

Tamias rufus (Hoffmeister and Ellis, 1979). Southwest. Nat., 24:656.
COMMON NAME: Hopi Chipmunk.
TYPE LOCALITY: USA, "10 mi SW Page, Coconino County, Arizona".
DISTRIBUTION: E and S Utah, extreme W Colorado, and NE Arizona, USA.
STATUS: IUCN – Lower Risk (lc).
COMMENTS: Subgenus *Neotamias*. Formerly included in *quadrivittatus*, but see Patterson (1984). Reviewed by Burt and Best (1994, Mammalian Species No. 460).

Tamias senex J. A. Allen, 1890. Bull. Am. Mus. Nat. Hist., 3:83.
COMMON NAME: Shadow Chipmunk.
TYPE LOCALITY: "Summit of Donner Pass, Placer Co., Cal[ifornia, USA]."

DISTRIBUTION: Sierra Nevada of EC California and WC Nevada to N coast of California, and NC Oregon (USA).

STATUS: IUCN – Lower Risk (lc).

SYNONYMS: *pacifica* Sutton and Patterson, 2000.

COMMENTS: Subgenus *Neotamias*. Elevated from a subspecies of *townsendii* by Sutton and Nadler (1974); distinction also suggested by Kain (1985), Sutton (1987), and Gannon and Lawlor (1989). Reviewed by Gannon and Forbes (1995, Mammalian Species No. 502).

Tamias sibiricus (Laxmann, 1769). Sibirische Briefe, Gottingen, p. 69.

COMMON NAME: Siberian Chipmunk.

TYPE LOCALITY: "Vicinity of Barnaul" [Altaisk Krai, Russia] (Chaworth-Musters, 1937).

DISTRIBUTION: N European and Siberian Russia to Sakhalin; S Kurile Isls (Russia); extreme E Kazakhstan to N Mongolia, China, Korea, and Hokkaido (Japan). Introduced into Austria, France, Germany, Holland and Italy (Niethammer and Krapp, 1978).

STATUS: IUCN – Lower Risk (lc).

SYNONYMS: *altaicus* (Hollister, 1912); *asiaticus* (Gmelin, 1788); *uthensis* (Pallas, 1811); *jacutensis* Ognev, 1935; *lineatus* (Siebold, 1824); *okadae* (Kuroda, 1932); *ordinalis* (Thomas, 1908); *albogularis* (J. A. Allen, 1909); *orientalis* Bonhote, 1899; *barberi* (Johnson and Jones, 1955); *pallasi* Baird, 1856; *striatus* (Pallas, 1778) [not Linnaeus, 1758]; *senescens* (Miller, 1898); *intercessor* (Thomas, 1908); *umbrosus* (A. H. Howell, 1927).

COMMENTS: Subgenus *Eutamias*; see Gromov et al. (1965:125). Koh (1994) compared Korean and Manchurian forms. Oshida and Yoshida (1994) described the chromosomes of *T. s. lineatus*.

Tamias siskiyou (A. H. Howell, 1922). J. Mammal., 3:180.

COMMON NAME: Siskiyou Chipmunk.

TYPE LOCALITY: "Near summit of White Mountain, Siskiyou Mountains, altitude 6,000 feet [1829 m], [Siskiyou Co.,] California" [USA].

DISTRIBUTION: Siskiyou Mtns and coast of N California to C Oregon (USA).

STATUS: IUCN – Lower Risk (lc).

SYNONYMS: *humboldti* Sutton and Patterson, 2000.

COMMENTS: Subgenus *Neotamias*. Elevated from a subspecies of *townsendii* by Sutton and Nadler (1974); distinction supported by Kain (1985), Sutton (1987), and Gannon and Lawlor (1989).

Tamias sonomae (Grinnell, 1915). Univ. California Publ. Zool., 12:321.

COMMON NAME: Sonoma Chipmunk.

TYPE LOCALITY: "One mile [1.6 km] west of Guerneville, Sonoma County, California." [USA].

DISTRIBUTION: NW California, from San Francisco Bay north to Siskiyou Co. (USA).

STATUS: IUCN – Lower Risk (lc).

SYNONYMS: *alleni* (A. H. Howell, 1922).

COMMENTS: Subgenus *Neotamias*. Reviewed by Best (1993c, Mammalian Species No. 444).

Tamias speciosus Merriam, 1890. *In* J. A. Allen, Bull. Am. Mus. Nat. Hist., 3:86.

COMMON NAME: Lodgepole Chipmunk.

TYPE LOCALITY: "San Bernardino Mts., [San Bernardino Co.,] Cal[ifornia, USA]." Restricted by A. H. Howell (1929:89) to Whitewater Creek, 7500 ft [2255 m].

DISTRIBUTION: USA: Sierra Nevada from Mt. Lassen to San Bernardino Mtns (California); W Nevada.

STATUS: IUCN – Lower Risk (lc).

SYNONYMS: *callipeplus* Merriam, 1893; *frater* J. A. Allen, 1890; *sequoiensis* (A. H. Howell, 1922).

COMMENTS: Subgenus *Neotamias*. Reviewed by Best et al. (1994c, Mammalian Species No. 478).

Tamias striatus (Linnaeus, 1758). Syst. Nat., 10th ed., 1:64.

COMMON NAME: Eastern Chipmunk.

TYPE LOCALITY: Restricted by A. H. Howell (1929:14) to "upper Savannah River, S[outh] C[arolina]." [USA].

DISTRIBUTION: S Manitoba and Nova Scotia (Canada) to Louisiana, Alabama, and Georgia, east to Atlantic Coast (USA).

STATUS: IUCN – Lower Risk (lc).

SYNONYMS: *americanus* Gmelin, 1788; **doorsiensis** Long, 1971; *fisheri* A. H. Howell, 1925; **griseus** Mearns, 1891; **lysteri** (Richardson, 1829); **ohioensis** Bole and Moulthrop, 1942; **peninsulae** Hooper, 1942; **pipilans** Lowery, 1943; **quebecensis** Cameron, 1950; **rufescens** Bole and Moulthrop, 1942; **venustus** Bangs, 1896.

COMMENTS: Subgenus *Tamias*; see Gromov et al. (1965:134). Reviewed by Snyder (1982, Mammalian Species No. 168).

Tamias townsendii Bachman, 1839. J. Acad. Nat. Sci. Philadelphia, 8(1):68.

COMMON NAME: Townsend's Chipmunk.

TYPE LOCALITY: "lower Columbia River, near lower mouth of Willamette River, [Multnomah Co.,] Oreg[on]." [USA].

DISTRIBUTION: SW British Columbia (Canada), W Washington and Oregon to the Rogue River (USA).

STATUS: IUCN – Lower Risk (lc).

SYNONYMS: *hindei* Gray, 1842; *littoralis* Elliot, 1903; **cooperi** Baird, 1855.

COMMENTS: Subgenus *Neotamias*. Formerly included *ochrogenys*, *siskiyou* and *senex*; see comments under those species. Reviewed by Sutton (1993, Mammalian Species No. 435).

Tamias umbrinus J. A. Allen, 1890. Bull. Am. Mus. Nat. Hist., 3:96.

COMMON NAME: Uinta Chipmunk.

TYPE LOCALITY: "Uintah Mountains, south of Ft. Bridger" Restricted by A. H. Howell (1929:94) to Blacks Fork, about 8000 ft [2438 m], Summit Co., Utah [USA].

DISTRIBUTION: E California and N Arizona to N Colorado, SE and NW Wyoming, and extreme SW Montana (USA).

STATUS: IUCN – Critically Endangered as *T. u. nevadensis*, Data Deficient as *T. u. sedulus*, otherwise Lower Risk (lc).

SYNONYMS: **adsitus** (J. A. Allen, 1905); **fremonti** (White, 1953); **inyoensis** (Merriam, 1897); **montanus** (White, 1953); **nevadensis** (Burt, 1931); **sedulus** (White, 1953).

COMMENTS: Subgenus *Neotamias*. Bergstrom and Hoffmann (1991) found species-specific vocalizations, habitats, and bacular characters in sympatric *umbrinus* and *quadrivittatus*, but convergence in electromorphs.

FAMILY GLIRIDAE
by Mary Ellen Holden

Family Gliridae Muirhead, 1819:433 (see McKenna and Bell, 1997). Mazology [sic]. Pp. 393-480, pls. 353-358, *in* Edinburgh Encyclopedia, Vol. 13 (D. Brewster, ed).

SYNONYMS: Glirini Muirhead, 1819 (Gliridae Thomas, 1896; Gliroidea Simpson, 1945); Leithiidae Lydekker, 1895; Muscardinidae Palmer, 1899; Myoxidae Zimmerman, 1780 (Myosidae Gray, 1821; Myoxina Gray, 1825; Myoxidae Waterhouse, 1839; Myoxini Giebel, 1855; Myoxida Haekel, 1866; Myoxoidea Gill, 1872); Seleviniidae Bashanov and Belosludov, 1939 (see McKenna and Bell, 1997).

COMMENTS: Simpson (1945) deemed Myoxidae Gray, 1821 invalid due to the apparent synonymy of the type genus *Myoxus* Zimmermann, 1790, with *Glis* Brisson, 1762, and used Gliridae Thomas, 1896. Hopwood (1947) argued that Brisson's names are invalid because they are not Linnaean or binomial, and noted that *Glis* is valid in Erxleben (1777) for marmots, ground-squirrels, voles, and lemmings, rendering *Glis* Storr, 1780 (which included pedetids, dormice, and other rodents) invalid. Thus the oldest available name to replace *Glis* Brisson is *Myoxus* Zimmerman, valid in Linnaeus (1788) for dormice, and, but for the ruling discussed below, the correct family name for dormice would be Myoxidae. Despite this clear historical evidence given in support of recognizing Myoxidae as the valid family name for dormice (see also Wahlert et al., 1993), and despite the willingness of many dormouse experts to employ Myoxidae as the valid family name (Hutterer, 1996; and usage of Myoxidae in Filippucci, 1995), the unfortunate ruling by the International Commission on Zoological Nomenclature (1998) to conserve *Glis* Brisson requires that the valid family name for dormice is henceforth Gliridae.

Glirids are one of the oldest extant rodent families. They first appear as fossils in early Eocene deposits (Daams, 1999; Daams and De Bruijn, 1995; Uhlig, 2001), suggesting a late Paleocene-early Eocene origin (Hartenberger 1994, 1998), which is concordant with the most recent molecular dating estimate based upon combined markers (Adkins et al., 2003; Huchon et al., 2002). Gnaw marks made by the extinct Eocene glirid *Glamys* were discovered on fossilized Eocene seeds of a freshwater aquatic floating plant, providing evidence of early glirid behavior (Collinson and Hooker, 2000). Early middle Eocene sediments in Germany have yielded a perfectly preserved *Eogliravus wildi* (Storch et al., 2000; Storch and Seiffert, 2002) represented by a completely articulated skeleton, soft body outline of pelage over body and tail, gut contents, and baculum. Apparently, *E. wildi* was agile, arboreal, and fed predominantly on seeds, fruits and buds (Storch et al., 2000; Storch and Seiffert, 2002). Most extant glirid genera were clearly differentiated and exhibited their greatest species diversity by the early to middle Miocene (Daams, 1999; Daams and De Bruijn, 1995; Hartenberger, 1994). The Graphiurinae are an exception; definite examples of *Graphiurus* are known only as far back as Pliocene (Hendey, 1981; Pocock, 1976), although Denys (1990a), Senut et al. (1992), and Mein et al. (2000a) recorded late Miocene "graphiurines" from South Africa and Namibia. The living Palaearctic genera are but relics of a rich adaptive radiation of up to 15 genera, and it is hypothesized that Miocene European glirids were ecologically equivalent to certain recent murid assemblages (Hartenberger, 1994).

The origin of glirids, and the evolutionary relationship between glirids and other rodent families are subjects of much morphological and molecular research with contradictory results. Meng (1990) suggested ancestry rooted in reithroparamyids (in Infraorder Sciurida, which also contains Sciuridae; see McKenna and Bell, 1997). Hartenberger (1971, 1994, 1998), Vianey-Liaud (1994) and Vianey-Liaud and Jaeger (1996) arranged glirids as sciuromorphs rooted in the subfamily Microparamyinae (Ischyromyoidea). Analyses of middle ear anatomy (Lavocat and Parent, 1985) and cephalic arterial supply (Bugge, 1971a, 1985) support a strong phylogenetic affinity between glirids and sciurids. Landry (1999), however, placed dormice in the Phaneraulata of Sciuromorpha, which excludes squirrels but contains dipodoids, muroids, geomyoids, and theridomyids. He regarded the theridomyid pseudosciurines as possibly ancestral to Phaneraulata, with one lineage leading to murids and geomyoids, another to glirids: "It would not be too far off the mark to think of *Graphiurus* as a surviving pseudosciurine" (Landry, 1999:313). Modern glirids (excluding graphiurines) have been characterized as myomorphous, and several authors included Gliridae within suborder Myomorpha (Chaline and Mein, 1979; McKenna and Bell, 1997; Simpson, 1945; Wahlert, 1978, 1983, 1985; Wahlert et al., 1993; Wood, 1965). Wahlert et al. (1993) placed them within Myomorpha based on derived cranial characters; their phylogenetic reconstruction indicated that the hystricomorphous *Graphiurus* is the most primitive extant glirid, and the myomorphy exhibited by all other extant glirids is convergent to that of true myomorphs. Vianey-Liaud (1985) employed "pseudomyomorphy" to distinguish the zygomasseteric muscular arrangement exhibited by glirids from murid myomorphy, and this term was endorsed by Maier et al. (2002) to describe the configuration derived from their ontological study of the medial masseter muscle. Landry (1999) agreed that glirids do not exhibit myomorphy, but stated that "pseudosciuromorphy" would be more appropriate for the muscular arrangement, based on his view that the hystricomorphous graphiurines represent the primitive dormouse condition.

The inclusion of glirids in Myomorpha has not been supported by recent morphological and molecular research (Adkins et al., 2001, 2003; Bentz and Montgelard, 1999; Bugge, 1971*a*, 1985; Corneli, 2002; DeBry and Sagel, 2001; Eizirik et al., 2001; Hartenberger 1971, 1994, 1998; Huchon et al., 1999, 2002; Kramerov, 1999; Kramerov and Vassetzky, 2001; Kramerov et al., 1999; Lavocat and Parent, 1985; Lin et al., 2002; Meng, 1990; Montgelard et al., 2001, 2002; Murphy et al., 2001*a*; Nedbal et al., 1996; Nikaido et al., 2003; Reyes et al., 1998; Vianey-Liaud 1974, 1985, 1989, 1994; Waddell and Shelley, 2003; Yachontov and Potapova, 1991). Brandt's (1855) subordinal divisions (Sciuromorpha, Myomorpha and Hystricomorpha), which were based on differences in zygomasseteric musculature arrangement, have likewise been contested (Landry, 1999; Luckett and Hartenberger, 1985*b*; Nedbal et al., 1996). Most authors now associate glirids with sciurids (or with Hystricognathi, when sciurids were not included in analyses); they agree with Wahlert et al. (1993) that the myomorphous condition exhibited by most modern glirids is convergent with muroids, though most advocate glirid myomorphy as being derived from a protrogomorphous ancestor. Moreover, recent molecular analyses support the grouping of sciurids and aplodontids (Suborder Sciuromorpha of McKenna and Bell, 1997) with glirids, and suggest that Gliridae is the sister clade to that containing Sciuridae and Aplodontidae (Adkins et al., 2001, 2003; Huchon et al., 1999, 2002; Montgelard et al., 2002). If protrogomorphy is the primitive state for glirids, the hystricomorphy exhibited by graphiurines was either 1) derived from the primitive protrogomorphous morphology, thus representing a lineage independent of all other dormice, which would be concordant with several morphological studies (e.g., Yachontov and Potapova, 1991; Wahlert et al., 1993); or 2) derived from the myomorphous state (Vianey-Liaud's, 1985, "pseudomyomorphy") shared by other modern glirids, which would be consistent with other morphological studies (e.g., Daams and De Bruijn, 1995; Koenigswald, 1993, 1995).

Vianey-Liaud and Jaeger (1996) proposed Gliridae to be paraphyletic, hypothesizing that graphiurines (which they advocated placing in a separate family, Graphiuridae) are most closely related to anomalurids. They further suggested that anomalurids and graphiurines should possibly be placed in the same family, based on their view that both groups are descended from zegdoumyids. All other recently published molecular and morphological studies support the monophyly of Gliridae (Catzeflis et al., 1995; Daams and de Bruijn, 1995; Debry and Sagel, 2001; Hanni et al., 1995; Hartenberger, 1994, 1998; Huchon et al., 1999, 2002; Meng, 1990; Montgelard et al., 2001, 2002, 2003; Robinson et al., 1997; Storch 1995*b*; Suzuki et al., 1997; Wahlert et al., 1993), and results of molecular analyses by Bentz and Montgelard (1999) and Montgelard et al. (2001, 2002) explicitly refute the hypothesis of paraphyly and phylogenetic alliance with anomalurids.

Wahlert et al. (1993) incorporated dental characters plus forty-three osteological traits in a phylogenetic analysis of extant glirid genera. Except for the inclusion of *Muscardinus* in Leithiinae (Montgelard et al., 2003) instead of Glirinae, Wahlert et al.'s classification is followed here because it reflects the most comprehensive phylogenetic analysis of glirids thus far. Several recent reviews and analyses address intrafamilial relationships using a variety of characters: molecular data (Bentz and Montgelard, 1999; Filippucci and Kotsakis, 1995; Montgelard et al., 2003; Suzuki et al. 1997); karyotypic information (Zima et al., 1995); dental morphology (Daams and De Bruijn, 1995; Pavlinov, 2001b); incisor enamel microstructure (Koenigswald, 1993, 1995); cranial and dental morphology (Vianey-Liaud and Jaeger, 1996); cranial, dental and genital morphology (Storch, 1995b); cranial, masticatory and tongue structure (Yachontov and Potapova, 1991). Each study has provided valuable insight, data and hypotheses regarding the relationships among dormice. Some of their conclusions support the hypotheses of Wahlert et al. (1993), others are contradictory. It is not possible to follow any one of these alternate classifications here because usually the methodology used to deduce relationships among glirid genera was not given. Dendrograms depicting hypothesized relationships were provided, as were descriptions of characters, but essential information regarding character states and polarities, numbers and origins of specimens examined, and methodology was either absent or inconsistent. In the few studies that clearly documented methodology and character states, the results were either inconclusive (Bentz and Montgelard, 1999; Zima et al., 1995) or sampled too few taxa to provide a comprehensive hypothesis of intrafamilial relationships (Filippucci and Kotsakis, 1995; Suzuki et al., 1997; Zima et al., 1995). The phylogenetic analyses of DNA nuclear fragments and mitochondrial gene sequences by Montgelard et al. (2003: 1953) is an exception: ". . . except for the position of *Muscardinus* . . ., the identification of three major glirid clades (*Graphiurus*, *Muscardinus* + Leithiinae and *Glis* + *Glirulus*) on molecular grounds confirms and extends the classification proposed by Wahlert et al. (1993) on the basis of morphological data."

Rossolimo et al. (2001) provided a comprehensive review of recent and fossil glirids, including descriptions and illustrations of cranial characters, zygomasseteric myology, internal bullar morphology and review of phylogeny. Each detailed species account contains illustrations of live animals, measurements, karyological data, ecological information, vocalization data, ectoparasite records and a distribution map. Review and comparison of vocalization data among glirid genera reported by Hutterer and Peters (2001) and Nowakowski and Rachwald (2000). The Dormouse Hollow (www.glirarium .de/dormouse), edited by Werner Haberl, provides a variety of information about dormice, including past and current research by glirid specialists.

Subfamily Graphiurinae Winge, 1887. E Museu Lundii, 1:109, 123.

> SYNONYMS: Graphiurinae Palmer, 1899 (Graphiuridae Miller and Gidley, 1918); see McKenna and Bell, 1997.
>
> COMMENTS: Results of Wahlert et al. (1993) identified the Graphiurinae as the earliest or most primitive branch of glirids, a position also supported by Meng (1990) and Yachontov and Potapova (1991). Hartenberger (1994) considered many aspects of graphiurine morphology to be primitive, and postulated that graphiurines may have been in Africa since the Miocene; late Miocene graphiurines are recorded from Namibia (Mein et al., 2000a; Senut et al., 1992) and South Africa (Denys, 1990a). Vianey-Liaud and Jaeger (1996) placed *Graphiurus* in its own family; they considered Gliridae to be paraphyletic, and hypothesized that *Graphiurus* is closely related to anomalurids (a position not supported by other published works, see discussion under Gliridae). Daams and de Bruijn (1995) postulated that *Graphiurus* is a descendant of *Eliomys*, and included both within their Dryomyinae (along with *Dryomys*, *Glirulus* and *Chaetocauda*). Based on the longitudinal orientation of the Hunter-Shreger bands of incisor enamel, Koenigswald (1993, 1995) arranged *Graphiurus* within his derived "Group III" that also contained *Muscardinus*, *Myomimus* and *Selevinia* (he did not employ subfamily names to describe his groups). In a phylogenetic analysis based on analysis of mitochondrial genes, Bentz and Montgelard (1999) were unable to resolve the position of *Graphiurus* within Gliridae, but phylogenetic analyses of DNA nuclear fragments and mitochondrial 12S rRNA sequences reinforced the conclusions of Wahlert et al. (1993) in which *Graphiurus* constitutes one of the three primary evolutionary lineages in Gliridae and is basal to the lineages forming Glirinae and Leithiinae (Montgelard et al., 2003). Using molecular-clock estimates, Montgelard et al. (2003) estimated that the evolutionary radiation of species in *Graphiurus* began 8-10 million years ago, which predates the oldest fossil *Graphiurus* (about 5 million years old), but

is consistent with occurrence of the late Miocene (10-11 million years old) Namibian graphiurine *Otaviglis* (Mein et al., 2000*a*). This age is roughly equivalent to the lower Vallesian of Europe at which time low sea levels may have facilitated extensive faunal interchange between Europe and Africa. Subsequent speciation in *Graphiurus* represents "a burst of evolution with rapid divergence from a single ancestral form . . . driven by novel ecological opportunities. . .," which is reflected in the strikingly greater number of species in *Graphiurus* compared with extant Eurasiatic glirid genera, and the short internal branches forming unresolved nodes among the species in the molecular analyses (Montgelard et al., 2003:1954).

Graphiurus Smuts, 1832. Enumer. Mamm. Capensium, pp. 32-33.

TYPE SPECIES: *Sciurus ocularis* Smith, 1829.

SYNONYMS: *Aethoglis* G. M. Allen, 1936; *Claviglis* Jentink, 1888; *Gliriscus* Thomas and Hinton, 1925; *Graphidurus* Brandt, 1855 [see Ellerman, 1940; Ellerman et al., 1953; McKenna and Bell, 1997; Rosevear, 1969].

COMMENTS: *Graphiurus* has been divided into as many as four separate genera (e.g., G. M. Allen, 1939; Holden, 1996*b*; Pavlinov and Potapova, 2003): *Aethoglis*, containing the largest African dormouse, *G. nagtglasii* (sometimes erroneously including *G. monardi*); *Graphiurus*, comprised of *G. ocularis*, with its reduced, simple premolar; *Gliriscus*, consisting of the rupicolous *G. platyops* and *G. rupicola*, with their flattened skulls; and *Claviglis*, the so-called "tree dormice", to which the remaining species of *Graphiurus* were assigned. Two of these, *Graphiurus* and *Claviglis*, have often been retained as subgenera (e.g., Ellerman et al., 1953; Rosevear, 1969). No published studies based upon a broad sample of species have addressed the validity of these subgeneric boundaries as used by past authorities or presented hypotheses of relationships among species. However, a phylogenetic study based on cranial and middle ear morphology, supports the recognition of three traditionally recognized subgenera of *Graphiurus*: *Aethoglis* containing *G. nagtglasii*, *Claviglis* comprised of *G. crassicaudatus*, and *Graphiurus* which would include all other graphiurines that were sampled (Pavlinov and Potapova, 2003). Their analyses showed that first *nagtglasii*, then *crassicaudatus*, diverged early in the evolution of African Dormice, and that the rest of the graphiurines sampled (*angolensis*, *christyi*, *kelleni*, *lorraineus*, *murinus*, *ocularis*, and *surdus*) form a monophyletic group. Based on Pavlinov and Potapova's results, three subgenera are recognized below.

The revision of *Graphiurus* by Genest-Villard (1978), based mostly on size grades, underestimated species diversity, particularly in the *G. murinus* group. Subsequently, species limits were defined in reports covering different African regions (e.g., Ansell and Dowsett, 1988; Holden, 1996*b*; Robbins and Schlitter, 1981) The species recognized below reflect information in the literature, as well as my examination of museum specimens and preliminary, mostly unpublished multivariate analyses of cranial and dental measurements.

Graphiurus angolensis de Winton, 1897. Ann. Mag. Nat. Hist, ser 6, 20:320.

COMMON NAME: Angolan African Dormouse.

TYPE LOCALITY: Angola, Caconda.

DISTRIBUTION: Angola (Hill and Carter, 1941) and NW Zambia (Ansell, 1978).

SYNONYMS: *dasilvai* (Roberts, 1938); *parvulus* Monard, 1933.

COMMENTS: Subgenus *Graphiurus*. Following the arrangement of Ellerman et al. (1953), Holden (1993) provisionally assigned *dasilvai* and *parvulus* as synonyms of *G. platyops*. Ansell (1974, 1978) recognized that the NW Zambian population (identified by him as *G. platyops parvulus*) is morphologically and ecologically different from *G. platyops*. Based on my study of type specimens and large series of Angolan and Zambian specimens, the Angolan and NW Zambia populations exhibit a distinctive skull morphology that is consistently separable from that of *G. platyops* and *G. rupicola*. Ansell (1974, 1978) correctly surmised that these populations are probably phylogenetically aligned with *G. microtis*.

Graphiurus angolensis de Winton, 1897, precedes *Graphiurus parvulus* Monard, 1932, and is the correct name for these populations. Assigning *parvulus* as a junior synonym is problematic. In describing *parvulus*, Monard (1933) did not designate a holotype, based his description upon several specimens, and listed three localities where he caught this dormouse. He later provided a more detailed description of *parvulus*, identified nine specimens as representing the species, along with a table of measurements (Monard, 1935). I examined several of Monard's specimens; most are either young or poorly preserved, and except for their paler dorsal pelage color, the material appears inseparable from samples of *G. angolensis*.

Graphiurus ocularis, which occurs in South Africa, was reported from Parc National de l'Upemba, Dem. Rep. Congo (Verschuren, 1987). The specimen upon which the *G. ocularis* record appears to be based (examined via loan from IRSNB) consists of only a skin that closely resembles *G. angolensis*, although it could possibly represent a large or overstuffed individual of *G. microtis*; the skin is not an example of *G. ocularis*. The relationship between populations representing *G. angolensis* and *G. microtis* needs to be studied and documented. Reviewed by Holden (In Press).

Graphiurus christyi Dollman, 1914. Rev. Zool. Afr., 4(1):80.
  COMMON NAME: Christy's African Dormouse.
  TYPE LOCALITY: Dem. Rep. Congo, Mambaka.
  DISTRIBUTION: N Dem. Rep. Congo (Hatt, 1940*a*; Holden, 1996*b*; Schlitter et al., 1985), S Cameroon (Holden 1996*b*; Robbins and Schlitter, 1981).
  STATUS: IUCN – Lower Risk (lc).
  COMMENTS: Subgenus *Graphiurus*. Morphological data and comparison with morphologically similar species given by Holden (1996*b*), Robbins and Schlitter (1981) and Schlitter et al. (1985). Reviewed by Rossolimo et al. (2001) and Holden (In Press). A much needed systematic revision of the *G. murinus* species group should include careful comparisons with *G. christyi*, as some E African *G. murinus* populations (particularly montane) closely resemble that species. See comments under *G. murinus*.

Graphiurus crassicaudatus (Jentink, 1888). Notes Leyden Mus., 10:41.
  COMMON NAME: Thick-tailed African Dormouse.
  TYPE LOCALITY: Liberia, Du (Du Queah) River, Hill Town.
  DISTRIBUTION: West Africa: Liberia (Kuhn, 1965), Guinea, Côte d'Ivoire (in Mt Nimba reserve; see Heim de Balsac and Lamotte, 1958), Ghana, Togo, Nigeria, Cameroon (Robbins and Schlitter, 1981; Schlitter et al., 1985), and Bioko (Eisentraut, 1973). See Rosevear (1969) and Grubb et al. (1998).
  STATUS: IUCN – Lower Risk (lc).
  SYNONYMS: *dorotheae* Dollman, 1912.
  COMMENTS: Subgenus *Claviglis*. Some researchers hypothesized that the morphological similarity between *G. crassicaudatus* and *G. nagtglasii* indicates a close phylogenetic relationship between the two species (Holden, 1996*b*; Rosevear, 1969). Results of Pavlinov and E. G. Potapova's (2003) cladistic analysis of African dormice based on cranial and middle ear characters does not support this conclusion; they showed that characters shared by *G. crassicaudatus* and *G. nagtglasii* may be primitive for the genus, and that *G. crassicaudatus* shares several derived characters with other species of *Graphiurus*. Additionally, *G. nagtglasii* and *G. crassicaudatus* each exhibit a suite of autapomorphies. Reviewed by Rossolimo et al. (2001) and Holden (In Press).

Graphiurus johnstoni (Thomas, 1898). Proc. Zool. Soc. Lond., 1897:934 [1898].
  COMMON NAME: Johnston's African Dormouse.
  TYPE LOCALITY: Malawi, Zomba.
  DISTRIBUTION: S Malawi (Happold and Happold, 1997*a* [as *G. lorraineus*]; Thomas, 1898); limits unknown.
  COMMENTS: Subgenus *Graphiurus*. Ansell and Dowsett (1988), Ansell (1989*b*), and Holden (1993) synonymized *johnstoni* under *G. kelleni*. My recent reexamination and comparisons of museum specimens (including all holotypes), and preliminary unpublished multivariate analyses indicate that *G. johnstoni* represents a species distinct from, but morphologically closely related to, *G. lorraineus*, not *G. kelleni* (before my analyses were completed, I identified a specimen of *G. johnstoni* as *G. lorraineus* in Happold and Happold, 1997*a*, 1998). Few specimens exist that can be attributed to *G. johnstoni*. If future research using larger samples of *G. johnstoni* indicates that *G. johnstoni* and *G. lorraineus* are conspecific, the latter would be a junior

synonym of *G. johnstoni* (see account of *G. lorraineus*). Holden (In press) reviewed *G. johnstoni* as a species pending additional study. See comments under *G. kelleni* and *G. murinus*.

Graphiurus kelleni (Reuvens, 1890). Notes Leyden Mus., 13:74.
    COMMON NAME: Kellen's African Dormouse.
    TYPE LOCALITY: Angola, Mossamedes district, "Damara-land" (see Hill and Carter, 1941).
    DISTRIBUTION: Savannahs of SubSaharan Africa, excluding Sierra Leone, Liberia, Namibia, Botswana, and South Africa, but including: Senegal (Hubert et al., 1973, as *G. murinus*), The Gambia, Guinea-Bissau, Côte d'Ivoire, Ghana, Togo, Benin, ? Burkina Faso (Gautun et al., 1985, as *G. murinus*), Nigeria, Mali, Niger (Dobigny et al., 2002*b*), Sudan, N Dem. Rep. Congo (Verschuren, 1987–Parc National de la Garamba, as *G. lorraineus*), N Uganda (Heller, 1911, as *G. personatus*; Hollister, 1919, as *G. personatus*), Kenya (Hollister, 1919), Ethiopia (Yalden et al., 1996), Somalia, Tanzania (Stanley et al., 2002; Swynnerton and Hayman, 1951), Malawi (Ansell and Dowsett, 1988), Mozambique, Angola (Hayman, 1963*b*; Hill and Carter, 1941), Zambia (Ansell, 1978, as *G. johnstoni*) and Zimbabwe. In W Africa see Grubb et al. (1998) and Rosevear (1969).
    STATUS: IUCN – Lower Risk (lc) as *G. kelleni, G. olga,* and *G. parvus*.
    SYNONYMS: *ansorgei* Dollman, 1912; *brockmani* Dollman, 1910; *cuanzensis* (Hill and Carter, 1937); *dollmani* Osgood, 1910; *foxi* Dollman, 1914; *internus* Dollman, 1912; *nanus* (de Winton, 1896); *olga* (Thomas, 1925); *parvus* (True, 1893); *personatus* Heller, 1911; *tasmani* (Roberts, 1929).
    COMMENTS: Subgenus *Graphiurus*. Schlitter et al. (1985) discussed taxonomic problems associated with the small dormice occurring in savannah woodlands of West, East, and Southern Africa. Holden (1993) provisionally recognized three species: *G. kelleni, G. olga* and *G. parvus*. My recent comparisons of museum specimens, including holotypes of all the synonyms listed, and results of preliminary multivariate analyses indicate that only one species of small-bodied African savannah dormouse can be diagnosed solely on the basis of cranial morphology: Analyses showed that *olga* is morphologically indistinguishable from West and East African *parvus*, and Angolan and Zambian *kelleni* cannot be discriminated on the basis of morphometric characters from samples of populations that include *parvus* and *olga*.
        Gautun et al. (1985) reported *G. murinus* from the vicinity of Ouagadougou, Burkina Faso, but based on the distributions and habitat preferences of *G. murinus, G. microtis* and *G. kelleni*, the specimen likely represents *G. kelleni*. Description and comparison of the vocal repertoire of *G. kelleni* was reported by Hutterer and Peters (2001). Karyotype of Niger sample with 2n = 70 (Dobigny et al., 2002*b*; reported as *parvus*). Reviewed by Rossolimo et al. (2001) and Holden (In Press).

Graphiurus lorraineus Dollman, 1910. Ann. Mag. Nat. Hist., ser. 8, 5:285.
    COMMON NAME: Lorraine's African Dormouse.
    TYPE LOCALITY: Dem. Rep. Congo, Welle (Uele) River, Molegbwe, south of Setema Rapids.
    DISTRIBUTION: W and C Africa: Sierra Leone, Guinea, Liberia (Kuhn, 1965), Côte d'Ivoire (Aellen, 1965; Mt Nimba Reserve see Heim de Balsac and Lamotte, 1958), Ghana (Decher and Bahian, 1999), Nigeria (Happold, 1987), Cameroon (Eisentraut, 1973; Robbins and Schlitter, 1981; Schlitter et al., 1985), Central African Republic, Bioko (Eisentraut, 1973), Equatorial Guinea, Gabon, Republic of Congo (Dowsett and Granjon, 1991), N Angola (Hayman, 1963*b*), Dem. Rep. Congo (Hatt, 1940*a*; Petter, 1967*a*; Verheyen and Verschuren, 1966), and N Zambia (Ansell, 1978). See Grubb et al. (1998) and Rosevear (1969) for W Africa reviews.
    STATUS: IUCN – Lower Risk (lc).
    SYNONYMS: *haedulus* Dollman, 1912; *spurrelli* Dollman, 1912.
    COMMENTS: Subgenus *Graphiurus*. Schlitter et al. (1985) discussed The Gambia specimens (identified by them as *G. lorraineus*) and their possible synonymy with *G. coupeii* (Cuvier, 1822) described from Senegal. Following Schlitter et al. (1985), Holden (1993) tentatively included "perhaps Gambia" in the distribution of *G. lorraineus*, but the Gambia specimens have been reidentified as *G. kelleni* (*cf. parvus*) (Grubb et al., 1998). I examined the holotype of *coupeii* (in MNHN), a stuffed skin mounted in live pose, and

it appears too large for *G. kelleni*, unless the skin was overstuffed. The specimen resembles West African populations of *G. lorraineus*, but *G. crassicaudatus* cannot be ruled out because distinguishing these two species is difficult using only skin characters. Accurate identification to species is elusive using only the skin, but samples taken the holotype for molecular analyses might allow identification.

Hubert et al. (1973) reported *G. " murinus"* (presumably cf. *spurrelli*) from Senegal, caught in wooded savanna. They did not describe the specimen and provided no measurements, so its identification is uncertain. Other specimens from Senegal in collections of natural history museums in Bonn, Brussels, and Paris that I examined appear to represent *G. kelleni*, but these examples consist of a tattered adult skin, juveniles, and skull fragments, and are difficult to allocate with certainty. The western distributional limit for *G. lorraineus* is therefore Sierra Leone, and a series of specimens from Seredou, Guinea (in MNHN), also appear to represent this species. Morphological data and comparison with similar species given by Holden (1996*b*) and Robbins and Schlitter (1981). Reviewed by Rossolimo et al. (2001) and Holden (In Press). See comments under *G. kelleni* and *G. murinus*.

Graphiurus microtis (Noack, 1887). Zool. Jahrb., 2:248.
  COMMON NAME: Large Savanna African Dormouse.
  TYPE LOCALITY: Dem. Rep. Congo, Marungu, Qua Mpala.
  DISTRIBUTION: SubSaharan Africa excluding West Africa: Chad, Sudan (Setzer, 1956), Eritrea, Ethiopia (Corbet and Yalden, 1972; Yalden et al., 1996), Uganda (Delany, 1975), Rwanda (Geider and Kock, 1991; Misonne, 1965*b*; Monfort, 1992; Verschuren, 1987–Parc National de l'Akagera, as *G. murinus*), Kenya, Tanzania (Hatt, 1940*b*; Swynnerton and Hayman, 1951), Mozambique (Smithers and Lobão Tello, 1976), Malawi (Ansell and Dowsett, 1988), E and S Dem. Rep. Congo (Verschuren, 1987–Parc National de l'Upemba, as *G. murinus*), Zambia (Ansell, 1978), Botswana (Smithers, 1971), Namibia, Zimbabwe (Smithers and Wilson, 1979) and South Africa (Lynch, 1983, 1989, 1994; Rautenbach, 1982; Roberts, 1951; P. J. Taylor, 1998; P. J. Taylor et al., 1994*a*, as *G.* cf. *murinus*) (for Southern Africa see de Graaff, 1981; Smithers, 1983). In some of these regional works, *G. microtis* is included in *G. murinus*; thus the mapped localities and natural history data are a composite for both species.
  STATUS: IUCN – Lower Risk (lc).
  SYNONYMS: *albolineata* (Frechkop, 1947); *butleri* Dollman, 1912; *etoschae* (Roberts, 1938); *griselda* Schwann, 1906; *littoralis* (Roberts, 1929); *marrensis* Setzer, 1956; ? *orobinus* (Wagner, 1845); *pretoriae* Roberts, 1913; *schneideri* (Roberts, 1938); *smithii* (Thomas, 1893); *streeteri* Roberts, 1913; *sudanensis* Setzer, 1953; *tzaneenensis* Roberts, 1913; *vandami* (Roberts, 1929); *woosnami* Dollman, 1910.
  COMMENTS: Subgenus *Graphiurus*. In concordance with Ansell (1989*b*) and Ansell and Dowsett (1988), *G. microtis* is recognized as a species distinct from *G. murinus*. Holden (1993) did not attempt to separate the synonyms associated with *G. murinus* and *G. microtis*. My improved provisional arrangement here results from further examination of specimens, data and preliminary multivariate analyses, but is no substitute for a much needed careful revision of the *G. microtis* and *G. murinus* species groups. *Graphiurus microtis* is a grade of savannah African dormouse; significant geographic variation exists, and it is likely that at least two species are contained within *G. microtis* as understood here.

  I examined the holotype of *orobinus* and concluded that it could conceivably be *parvus*, but most likely represents *G. microtis*. If so, *orobinus* would have priority over *microtis* and be the correct name for the species, but the young age and poor condition of the holotype makes positive identification difficult. Once an adequate molecular sampling of species of *Graphiurus* has been obtained, a sample from the holotype for molecular analyses might allow definitive allocation of *orobinus*. The holotype of *G. microtis* is lost. Dorsal, lateral and ventral views of the skull were figured in Noack (1887). Those drawings, and the dimensions given in the accompanying description, make it clear that Noack was describing a Large Savannah African Dormouse.

  Photograph of live animal provided in Geider and Kock (1991). Reviewed by Rossolimo et al. (2001), though some populations of *G. microtis* are included under their account of *G. murinus*, and by Holden (In Press). Late Quaternary fossils of *G. microtis* from C Zambia recorded by Avery (1996). Specimens reported as *G. cf. murinus* from Swaziland and NE South Africa by P. J. Taylor et al. (1994a) probably represent a population of *G. microtis*, based on their descriptions of cranial and skin characters. *Graphiurus microtis* is sympatric with *G. kelleni* over much of its range. See comments under *G. angolensis* regarding its relationship to *G. microtis*.

Graphiurus monardi (St. Leger, 1936). Ann. Mag. Nat. Hist., ser. 10, 17:465.
    COMMON NAME: Monard's African Dormouse.
    TYPE LOCALITY: Angola, Tyihumbwe (Chiumbe) River, 15 km above Dala, Kioko, 1250 m (see Hill and Carter, 1941).
    DISTRIBUTION: NE Angola (Hayman, 1963b), NW Zambia (Ansell, 1978), and S Dem. Rep. Congo; limits unknown.
    STATUS: IUCN – Lower Risk (lc).
    SYNONYMS: *schoutedeni* Frechkop, 1947 [see Ansell, 1989].
    COMMENTS: Subgenus *Graphiurus*. G. M. Allen (1939) listed *monardi* as a subspecies of *G. hueti* (=*nagtglasii*; see that account), an arrangement followed by Genest-Villard (1978). Ellerman et al. (1953) correctly stated that *monardi* has no close affinity with *G. hueti* and provided a detailed description and comparison of cranial and external characters, which I was able to verify, that clearly distinguish the two. Schouteden (1948) listed a specimen of *G. ocularis* from Dilolo, Dem. Rep. Congo. I examined the Dilolo specimens (in MRAC). Although certainly not representative of *G. ocularis*, these individuals are difficult to identify. Their pelage is similar to that of *G. rupicola*, yet the skulls resemble *G. monardi*. Photographs of live animal given by Hayman (1963b). Reviewed by Rossolimo et al. (2001) and Holden (In Press).

Graphiurus murinus (Desmarest, 1822). Mammalogie, *in* Encyclop. Méth., 2(Suppl.):542.
    COMMON NAME: Forest African Dormouse.
    TYPE LOCALITY: South Africa, Cape of Good Hope.
    DISTRIBUTION: C, E and Southern Africa: E Dem. Rep. Congo (Rahm and Christiaensen, 1963; Verschuren, 1987–Parc National des Virunga), Uganda (Clausnitzer and Kityo, 2001; Delany, 1975), Rwanda (Monfort, 1992), Burundi, Ethiopia (Corbet and Yalden, 1972; Yalden et al., 1996), Kenya (Hollister, 1919), Tanzania (Grimshaw et al., 1995; Stanley et al., 1998, 2000; Swynnerton and Hayman, 1951), Malawi (Ansell and Dowsett, 1988), Mozambique (Smithers and Lobão Tello, 1976), Zambia (Ansell, 1974, as *Graphiurus* sp.; 1978), Zimbabwe (Smithers and Wilson, 1979), South Africa (Kryštufek et al., 2004a; Lynch 1989; Rautenbach, 1982; Roberts, 1951; P. J. Taylor, 1998) (for Southern Africa see de Graaff, 1981; Smithers, 1983). In some of these regional works, *G. microtis* is included in *G. murinus*; thus the mapped localities and natural history data are a composite for both species.
    STATUS: IUCN – Lower Risk (lc).
    SYNONYMS: *alticola* (Roberts, 1929); *cineraceus* (Rûppell, 1842); *cinerascens* (Schinz, 1845); *collaris* (Allen and Loveridge, 1933); *erythrobronchus* (Smith, 1829); *griseus* G. M. Allen, 1912; *isolatus* Heller, 1912; *lalandianus* (Schinz, 1825); *johnstoni* Heller, 1912 [not Thomas, 1898]; *raptor* Dollman, 1910; *saturatus* Dollman, 1910; *selindensis* (Roberts, 1937); *soleatus* Thomas and Wroughton, 1910; ? *subrufus* (Neumann, 1900); *vulcanicus* Lönnberg and Gyldenstolpe, 1925; *zuluensis* (Roberts, 1931).
    COMMENTS: Subgenus *Graphiurus*. The synonyms included here under *G. murinus* represent populations inhabiting forests (predominantly on plateaus and mountains) in C, E and Southern Africa. As with the *G. microtis* group, significant variation in pelage color and skull morphology exists among populations of the *G. murinus* group, and it is likely that more than one species comprises this group. The *selindensis* and *collaris* populations are distinctive, as are populations from Rwanda and Burundi, and several populations from South Africa. Three different karyotypes were found in the *G. murinus* species group in Southern Africa (Dippenaar et al., 1983). Furthermore, Kryštufek et al. (2004a) reported that two South African samples, one from lowland riverine forest and the other from Afromontane forest, had the same karyotypes but could be clearly separated by discriminant function analyses of cranial and dental

measurements. These studies provide examples reflecting the complexity of relationships among populations now allocated to *G. murinus*.

In Dem. Rep. Congo, *G. christyi* (which is probably a close relative of *G. murinus*) and *G. lorraineus* occur sympatrically, and the two are easily distinguished (Holden, 1996*b*). *Graphiurus murinus* and *G. lorraineus* are not sympatric; in SC Africa, some plateau and montane populations of *G. murinus* (e.g., *collaris*, *selindensis*, and southern populations of cf. *saturatus*) approach *G. lorraineus* in size and pelage color. The skull of *G. lorraineus* is generally smaller, has a wider interorbit and blunt rostrum as compared with *G. murinus*; *G. johnstoni* exhibits these characters and appears closely related to *G. lorraineus*. Careful systematic revision and comparisons of *G. murinus* and *G. lorraineus* is required; the *G. murinus* revision should also include comprehensive comparisons with *G. microtis*.

A specimen from the Zambian Nyika plateau (BMNH 66.875) identified by Ingles (1965) as *G. johnstoni*, by Ansell (1974) as *G. sp.*, and by Ansell and Dowsett (1988) as *G. murinus* subsp., morphologically most closely resembles the Ruwenzori population cf. *soleatus*. Specimens reported as *G. cf. murinus* from Swaziland and NE South Africa by P. J. Taylor et al. (1994*b*) probably represent a population of *G. microtis*, based on their descriptions of cranial and skin characters. The holotype of *subrufus* is lost, but another specimen I examined (in ZMB) matching its description was also collected from Tanga, NE Tanzania in 1898, and could be designated as a neotype for *subrufus*. This additional specimen represents the *G. murinus* species group. Reviewed by Rossolimo et al. (2001), although that species account is a composite that includes data for both *G. murinus* and *G. microtis* (see synonyms listed therein). Reviewed by Holden (In Press). See comments under *G. microtis*.

Graphiurus nagtglasii Jentink, 1888. Notes Leyden Mus., 10:38-39.
    COMMON NAME: Nagtglas's African Dormouse.
    TYPE LOCALITY: Liberia, Du (DuQueah) River, Hill Town (see Holden, In Press).
    DISTRIBUTION: West Africa: Sierra Leone, Liberia (Coe, 1975; Kuhn, 1965), SE Guinea, Côte d'Ivoire (Aellen, 1965; Heim de Balsac and Lamotte, 1958), Ghana (Jeffrey, 1973), Togo, Nigeria (Happold, 1987), Cameroon (Robbins and Schlitter, 1981; Schlitter et al., 1985), Central African Republic, and Gabon. For West Africa reviews see Rosevear (1969) and Grubb et al. (1998).
    STATUS: IUCN – Lower Risk (lc) as *G. hueti*.
    SYNONYMS: *argenteus* (G. M. Allen, 1936); *hueti* de Rochebrune, 1883 [see G. M. Allen, 1939, and Grubb and Ansell, 1996].
    COMMENTS: Subgenus *Aethoglis*. Does not include *monardi* (see that account). Grubb and Ansell (1996) recommended applying the name *G. nagtglasii* to the large West African dormouse traditionally known as *G. hueti* de Rochebrune. Their argument is supported by the dubious nature of the type locality of *G. hueti* given by de Rochebrune (1883), the lack of an available or likely holotype for *G. hueti* and existence of a holotype for *nagtglasii*, and because the animal used by de Rochebrune as a model to figure *G. hueti* is probably from Gabon, not Senegal (the type locality of *G. hueti*). Despite de Rochebrune's (1883) claims, the occurrence of Nagtglas' Dormouse in Senegal and The Gambia has never been substantiated (Grubb and Ansell, 1996). Karyotype of Côte d'Ivoire specimen given by Tranier and Dosso (1979). Reviewed by Rossolimo et al. (2001) and Holden (In Press). See comments under *G. crassicaudatus*.

Graphiurus ocularis (Smith, 1829). Zool. J., 4:439.
    COMMON NAME: Spectacled African Dormouse.
    TYPE LOCALITY: South Africa, Western Cape Prov., near Plettenberg Bay.
    DISTRIBUTION: South Africa, Eastern, Northern and Western Cape Provinces (Channing, 1984; de Graaff, 1991; Roberts, 1951; Smithers, 1983).
    STATUS: IUCN – Lower Risk (lc).
    SYNONYMS: *capensis* (Cuvier, 1829); *elegans* Ogilby, 1838; *typicus* Smith, 1834.
    COMMENTS: Subgenus *Graphiurus*. Other historical distributional records for *G. ocularis* are questionable. Lorenz (1894) recorded a specimen from Linokana (Dinokana), northwest of Zeerust, North West Province, although no other specimen has ever been

taken from this region (see Rautenbach, 1982). The Linokana specimen is missing
from the Naturhistorisches Museum Wien, where most of the Lorenz collection is
housed (B. Herzig, pers.comm.), so unfortunately the historical occurrence of the
Spectacled African Dormouse as far north as the Zeerust region cannot be verified;
possibly the specimen represented either *G. platyops* or *G. microtis* and was misidentified as *G. ocularis*. An example of *G. ocularis* recorded from Damaraland, Namibia (type
specimen of *G. elegans*) may have come from Namaqualand, South Africa; the
occurrence of *G. ocularis* north of the Orange River is doubtful (see Thomas, 1926*e* for
discussion). Though Roberts (1951) and Ellerman et al. (1953) included [Orange] Free
State in the distribution of this species, Lynch (1983) stated that it has not been
collected in the province. It is doubtful that *G. ocularis* occurs in KwaZulu-Natal,
though its range may have historically included the province (P. J. Taylor 1998). See
comments under *G. monardi* regarding a specimen reported from Dilolo, Dem. Rep.
Congo, and discussion under *G. angolensis* covering a specimen reported from Parc
National de l'Upemba, Dem. Rep. Congo. Comparative ecological data reported by
Channing (1984, 1987, 1997) and Van Hensbergen and Channing (1989). Reviewed by
Rossolimo et al. (2001) and Holden (In Press).

Graphiurus platyops Thomas, 1897. Ann. Mag. Nat. Hist., ser. 6, 19:388.
    COMMON NAME: Flat-headed African Dormouse.
    TYPE LOCALITY: S Zimbabwe, Mashonaland, Enkeldorn.
    DISTRIBUTION: Eastern and E Southern Africa: S Malawi (Ansell and Dowsett, 1988), E and
        S Zambia (Ansell, 1978), Zimbabwe (Smithers and Wilson, 1979), E Botswana
        (Smithers, 1971), Mozambique (Smithers and Lobão Tello, 1976) and NE South Africa
        (de Graaff, 1981; Rautenbach, 1982; Roberts, 1951; Smithers, 1983).
    STATUS: IUCN – Lower Risk (lc).
    SYNONYMS: *eastwoodae* Roberts, 1913; *jordani* Roberts, 1929.
    COMMENTS: Subgenus *Graphiurus*. Does not include *angolensis*, *dasilvai* or *parvulus* (see
        comments under *G. angolensis* and *G. rupicola*). Previously thought to occur in
        C Botswana (De Graaff 1981), but my measurements of that specimen are well outside
        the range of *G. platyops*, and are concordant with those of Southern African
        populations of *G. microtis*. A specimen in the Amathole (formerly Kaffrarian) Museum,
        King William's Town, is labeled as the holotype of "*Gliriscus angolensis albicaudatus*
        Roberts". Apparently the name *albicaudatus* was never formally published (F. Kigozi,
        pers. comm.). Possibly A. Roberts labeled the specimen with the intention of
        describing it (the specimen was originally part of the Albany Museum collection). The
        specimen represents *G. platyops*, and is from Isoka, Zambia, the type locality of
        G. platyops jordani Roberts. Reviewed by Rossolimo et al. (2001) and Holden (In Press).

Graphiurus rupicola (Thomas and Hinton, 1925). Proc. Zool. Soc. Lond., 1925:232.
    COMMON NAME: Rupicolous African Dormouse.
    TYPE LOCALITY: Namibia, Karibib, 3842 ft (1171 m).
    DISTRIBUTION: Angola (M. E. Holden, unpubl.), Namibia (Thomas and Hinton, 1925) and
        NW South Africa (Shortridge and Carter, 1938), in a narrow strip from Mt Soque,
        Angola, south to Port Nolloth and Eenriet in Little Namaqualand, South Africa
        (see also Roberts, 1951).
    STATUS: IUCN – Lower Risk (lc).
    SYNONYMS: *australis* (Shortridge and Carter, 1938); *kaokoensis* (Roberts, 1938); *montosus*
        (Thomas and Hinton, 1925).
    COMMENTS: Subgenus *Graphiurus*. Ellerman et al. (1953) and Genest-Villard (1978) listed
        rupicola as a subspecies of *G. platyops*, but Roberts (1951) recognized it as a distinct
        species, a position followed here based on my study of museum specimens, including
        holotypes. The northern distributional limit for *G. rupicola* was previously thought to
        be Kamanjab, Namibia, but I found two specimens from Mt Soque, Angola (in FMNH)
        that represent *G. rupicola*. Reviewed by Rossolimo et al. (2001) and Holden (In Press).
        See comments under *G. monardi* regarding enigmatic specimens from Dilolo, Dem.
        Rep. Congo.

Graphiurus surdus Dollman, 1912. Ann. Mag. Nat. Hist., ser. 8, 9:314.

COMMON NAME: Short-eared African Dormouse.

TYPE LOCALITY: Equatorial Guinea, Rio Muni, Benito River.

DISTRIBUTION: WC and C Africa: Southern Cameroon south to Equatorial Guinea and Gabon, and NE and SC Dem. Rep. Congo (Holden, 1996*b*).

STATUS: IUCN – Data Deficient.

SYNONYMS: *schwabi* G. M. Allen, 1912 [see Holden, 1996*b*].

COMMENTS: Subgenus *Graphiurus*. Morphological revision and comparisons with similar species contributed by Holden (1996*b*). Reviewed by Rossolimo et al. (2001) and Holden (In Press).

Subfamily Leithiinae Lydekker, 1896. Proc. Zool. Soc. Lond., 1895:862 [1896].

SYNONYMS: Dryomyinae de Bruijn, 1967; Leithiinae Trouessart, 1897; Muscardininae Palmer, 1904; Myomiminae Daams, 1981; Seleviniidae Bashanov and Belosludov, 1939 (Seleviniinae Ognev, 1947); see McKenna and Bell (1997).

COMMENTS: Lydekker (1895) proposed Leithiidae to separate the giant Pleistocene dormouse of Malta from other glirids, and *Leithia* for the type genus. Major (1899) argued that *Leithia* was in fact a glirid, and Leithiidae a junior synonym of Gliridae. De Bruijn (1967) proposed Dryomyinae, which included *Leithia*, *Dryomys*, *Eliomys*, and other genera. The International Commission on Zoological Nomenclature (1999:45) mandates that when a nominal taxon is lowered in rank in the family group, its type genus remains the same. Because Dryomyinae de Bruijn contains *Leithia*, the correct name for the subfamily is Leithiinae. The results of the phylogenetic analysis by Wahlert et al. (1993) indicated that the genera listed below [excluding *Chaetocauda*, which was not available for study] form a monophyletic group composed of two tribes: Leithiini (*Dryomys* and *Eliomys*) and Seleviniini (*Myomimus* and *Selevinia*).

Daams and de Bruijn (1995) argued that it is imprudent to revive the subfamily name Leithiinae because *Leithia* does not exhibit all the character states outlined for the subfamily by Wahlert et al. (1993). Additionally, they do not advocate use of Leithiini and Seleviini, because they believe that too little published information is available for *Selevinia* to serve as type genus for the tribe. Rossolimo et al. (2001) employed both Leithiinae and Seleviniinae. Daams and de Bruijn (1995) arranged *Dryomys*, *Eliomys*, *Graphiurus*, *Glirulus* and *Chaetocauda* under the Dryomyinae de Bruijn, 1967, and placed *Myomimus* in its own subfamily.

Yachontov and Potapova (1991) considerd *Selevinia* to be closely related to Glirinae genera *Glis* and *Muscardinus*. Wahlert et al. (1993) concluded that *Myomimus* and *Selevinia* are related, and placed them both in Seleviniini, uniting them with Leithiini (containing *Eliomys* and *Dryomys)* in Leithiinae. Storch (1995*b*) grouped *Dryomys* and *Eliomys* into Dryomyinae, and placed *Selevinia*, *Chaetocauda*, and *Myomimus* in a separate subfamily, Seleviniinae. Koenigswald (1993, 1995) placed *Dryomys* and *Eliomys* into his derived "Group II", and *Myomimus* and *Selevinia* (along with *Muscardinus* and *Graphiurus*) in his derived "Group III". Simson et al. (1995) found that *Myomimus* clustered with *Dryomys*, and that *Glis* clustered with *Eliomys*, based on phallic and bacular characters. Bentz and Mongelard's (1999) mitochondrial analyses suggest that *Eliomys* and *Dryomys* (the only two genera sampled from Leithiinae) form a monophyletic group; however, in some of their analyses, *Eliomys* also appeared in the basal position within Gliridae. Rossolimo (2001) included *Dryomys*, *Eliomys*, and "?*Chaetocauda*" in Leithiinae, and recognized Seleviniinae containing Myomimini (comprised of *Myomimus*), and Seleviniini (comprised of *Selevinia*). Potapova's (2001) analysis of middle ear morphology supports a close relationship of *Dryomys* and *Eliomys*, and of *Selevinia* with *Myomimus*. Phylogenetic analyses of nuclear DNA fragments and mitochondrial 12S rRNA sequences identified *Eliomys*, *Dryomys*, *Myomimus*, and *Muscardinus* as comprising a major evolutionary lineage, with *Muscardinus* basal to the other three genera (Mongelard et al., 2003). This cladistic position reflects that of Rossolimo et al. (2001), who did not allocate *Muscardinus* to subfamily, but suggested it was related to *Dryomys* and *Eliomys*; Potapova (2001), who hypothesized that *Muscardinus* represents an independent group that is more closely related to *Dryomys* and *Eliomys* than to *Glis*; and Pavlinov's (2001*b*) conclusion, based upon upper molar crown pattern, that *Glis* and *Muscardinus* are "most dissimilar both from each other and from the remainder of the glirids studied."

Chaetocauda Wang, 1985. Acta Theriol. Sinica, 5(1):67.
> TYPE SPECIES: *Chaetocauda sichuanensis* Wang, 1985.
> COMMENTS: Holden (1993) formerly arranged *C. sichuanensis* as a species of *Dryomys*, based on shape and other morphological characters. Although *Chaetocauda* resembles *Myomimus* in some features of the skull, other cranial characters resemble those found in *Dryomys* (particularly *D. niethammeri*); additionally, the continuous endoloph on the upper molars supports its inclusion in Leithiinae. Because there is so little information regarding this animal, the genus should be recognized until its relationship with other dormouse genera, particularly leithiines, can be analyzed. Wang (1985a) considered *Chaetocauda* a close relative of *Myomimus*, and placed it within Myomiminae. Based on the presence of a continuous endoloph on the upper molars, Rossolimo (2001) tentatively placed *Chaetocauda* in Leithiinae, along with *Dryomys* and *Eliomys*. The same character served as the basis for Daams and de Bruijn's (1995) arrangement, in which they provisionally placed *Chaetocauda* in their Dryomyinae (which also contains *Dryomys*, *Eliomys*, *Graphiurus* and *Glirulus*). They hypothesized that the "general appearance of the ridge pattern is reminiscent of a simplified *Eliomys*."

Chaetocauda sichuanensis Wang, 1985. Acta Theriol. Sinica, 5(1):67.
> COMMON NAME: Sichuan Dormouse.
> TYPE LOCALITY: China, N Sichuan Sheng, Pinwu county, Wang-lang Natural Reserve.
> DISTRIBUTION: Known only from the type locality, a subalpine deciduous and coniferous forest in the Sichuan highlands; limits unknown.
> STATUS: IUCN – Endangered as *Dryomys sichuanensis*.
> COMMENTS: Reviewed by Rossolimo et al. (2001); see also Vorontsov (1986) and Wang (2003).

Dryomys Thomas, 1906. Proc. Zool. Soc. Lond., 1905(2):348 [1906].
> TYPE SPECIES: *Mus nitedula* Pallas, 1778.
> SYNONYMS: *Afrodryomys* Jaeger, 1975 [see Daams and de Bruijn, 1995; Rossolimo et al., 2001]; *Dyromys* Thomas, 1907 [see Ellerman and Morrison-Scott, 1951; McKenna and Bell, 1997; and footnote in Simpson, 1945]; *Elius* Schulze, 1900 [excluding *glis*].
> COMMENTS: Evolutionary patterns of dental morphology during Pliocene and Pleistocene discussed by Nadachowski and Daoud (1995). Comparative vocalization data reviewed by Hutterer and Peters (2001). It is generally (but not universally) accepted that *Eliomys* is the closest living relative of *Dryomys* (see comments under Subfamily Leithiinae and references therein). However, phylogenetic analyses of nuclear DNA fragments and mitochondrial 12S rRNA sequences clearly identified *Dryomys* as sister to *Eliomys* (Montgelard et al., 2003).

Dryomys laniger Felten and Storch, 1968. Senckenberg. Biol., 49(6):429.
> COMMON NAME: Woolly Forest Dormouse.
> TYPE LOCALITY: Turkey, Antalya Prov., 20 km SSE Elmali, Bey Mtns, Ciglikara, 2000 m.
> DISTRIBUTION: S Turkey, Toros Daglari (Taurus Mtns) and E Anatolia (Kivanç et al., 1997a; Krystufek and Vohralík, 2001; Obuch, 2001; Spitzenberger, 1976).
> STATUS: IUCN – Lower Risk (nt).
> COMMENTS: Morphology, ecology and distribution reviewed by Rossolimo et al. (2001). Morphology and histology of plantar surface studied by Spitzenberger and Eberl-Rothe (1974). Photographs provided by Felten and Storch (1968). Karyology and phallic morphology given by Kivanç et al. (1997a). Distribution mapped in Kivanç et al. (1997a).

Dryomys niethammeri Holden, 1996. Bonn. Zool. Bieträge 46(1-4):116.
> COMMON NAME: Niethammer's Forest Dormouse.
> TYPE LOCALITY: Pakistan, Baluchistan Prov., 1 mi. (1.6 km) E of Ziarat.
> DISTRIBUTION: Pakistan, NE Baluchistan; limits unknown.
> STATUS: IUCN – Vulnerable.

COMMENTS: Known only from three specimens, although an additional five specimens may exist in the collections of the Pakistan Agricultural Research Council (T. J. Roberts, A. Mian, pers. comm.). Though *D. niethammeri* is larger than *D. laniger*, and differs in other proportions, both species have relatively inflated auditory bullae, a character that may be convergent (Holden, 1996a). Reviewed by Rossolimo et al. (2001). A much needed critical revision of *D. nitedula* should include phylogenetic comparisons with *D. laniger* and *D. niethammeri* to elucidate intrageneric relationships (see Holden, 1996a).

Dryomys nitedula (Pallas, 1778). Nova Spec. Quad. Glir. Ord., p. 88.
  COMMON NAME: Forest Dormouse.
  TYPE LOCALITY: Russia, lower Volga River.
  DISTRIBUTION: Europe, the Middle East and C Asia: SE Germany (Faltin, 1988), Switzerland (Catzeflis, 1995b), Austria (Niethammer, 1960; Spitzenberger, 1983; Spitzenberger et al., 1995), Czech Republic and Slovakia (Andera, 1987, 1995; Kratochvíl, 1967; Obuch, 1998), Poland (Daoud, 1989; Jurczyszyn and Wolk, 1998; Kosior, 1996; Nowakowski, 2000; Nowakowski and Boratynski, 2001; Pucek, 1983b); Ukraine (Bezrodny, 1991) and Belorus (Serñanin, 1961) north to Lithuania (Balciauskas, 1996; Juskaitis, 1995a) and Latvia (Pilāts, 1995), Russia east to Kazan region and upper Volga River, south to lower Dnepr River, mouth of Volga River, and Caucasus Mtns (Gromov and Erbajeva, 1995; Lichaev, 1972; Ognev, 1947; Vereshchagin, 1959; also Kuznetsov, 1965; Pavlinov and Rossolimo, 1987), Italy (Amori et al., 1995, 1999; Filippucci, 1986; Paolucci et al., 1987), Hungary (Bakó et al., 1998; Kryštufek and Vohralík, 1994), Slovenia (Kryštufek, 1991; Kryštufek and Vohralík, 1994), Croatia (Tvrtković et al., 1995), Bosnia and Herzegovina, Serbia and Montenegro (Gazaryan, 1985; Krystufek, 1985a; Kryštufek and Vohralík; 1994; Petrov, 1992), Romania (Istrate, 1998; Kryštufek and Vohralík, 1994), Moldavia (Lozan, 1970), Albania (Prigioni, 1996), Macedonia (Petrov, 1992), Bulgaria (Mitev et al., 1994; Peshev, 1996; Peshev and Mitev, 1979; Vohralík, 1985), Greece (Ondrias, 1966; Sofianidou and Vohralík, 1991; Vohralík and Sofianidou, 1987, 1992a), Turkey (Kryštufek and Vohralík, 2001; Mursaloğlu, 1973b; Obuch, 2001; Pamukoglu and Albayrak, 1996), Arabia (Harrison and Bates, 1991), W and E Syria (Obuch, 2001; von Lehmann, 1965), N Israel (Atallah, 1978; Mendelssohn and Yom-Tov, 1999; Nevo and Amir, 1961a, 1964; Qumsiyeh, 1996), N Iraq (Jawdat, 1977), Iran (Lay, 1967; Obuch, 2001), Afghanistan (Hassinger, 1973), N Pakistan (Roberts, 1977, 1997); Tajikistan (Allobergenov, 1986; Davydov, 1984), Uzbekistan, Turkmenistan, Kyrgyzstan, and C Kazakhstan north to the S Altai Gobi (see Gromov and Erbajeva, 1995; Kuznetsov, 1965; Ognev, 1947; Pavlinov and Rossolimo, 1987), the Tarbagaty Mtns east to the eastern limit of the Tien Shan Mtns in Xinjiang, China (Ma et al., 1981, 1987; Wang, 2003; Wang and Yang, 1983); perhaps Lebanon (Lewis et al., 1967). In Europe see also Kryštufek (1999b), Kryštufek and Vohralík (1994) and Storch (1978a).
  STATUS: IUCN – Lower Risk (nt).
  SYNONYMS: *angelus* (Thomas, 1906); *aspromontis* Lehmann, 1963; *bilkjewiczi* Ognev and Heptner, 1928; *carpathicus* Brohmer, 1927; *caucasicus* Ognev and Turov, 1935; *daghestanicus* Ognev and Turov, 1935; *diamesus* Lehmann, 1959; *dryas* (Schreber, 1782); *intermedius* (Nehring, 1902); *kurdistanicus* Ognev and Turov, 1935; *milleri* Thomas, 1912; *obolenskii* Ognev and Worobiev, 1923; *ognevi* Heptner and Formozov, 1928; *pallidus* Ognev and Turov, 1935; *phrygius* Thomas, 1907; *pictus* (Blanford, 1875); *ravijojla* Paspalev et al., 1952; *robustus* Miller, 1910; *saxatilis* Rosanov, 1935; *tanaiticus* Ognev and Turov, 1935; *tichomirowi* Satunin, 1920; *wingei* (Nehring, 1902); see Ellerman and Morrison-Scott (1951) and Corbet (1978c).
  COMMENTS: There has been no critical revision of this species throughout its range, and *D. nitedula* may actually contain two or more species. An allozymic and biometric study by Filippucci et al. (1995) of southern populations indicates that the Israeli population probably represents a separate species, a conclusion supported by ecological data (Nevo and Amir, 1961b) and phallic and bacular morphology (Simson et al., 1995). It would have been treated as a species here, but unfortunately no name has been given to this population. Historically, the Israeli population has been

included in *D. n. phrygius* (described from Murat Dagi, W Turkey), but Filippucci et al. (1995) showed that the two populations are morphologically different, and topotypes of *D. n. phrygius* clustered with the European population from Turkish Thrace. Mursaloğlu (1973*b*) suggested that the E Anatolia (Turkey) population represents a separate species, *D. pictus*, originally described from Kohrud, south of the Caspian Sea, Iran (Krystufek and Vohralík, 2001). Documented comparisons with representative geographic samples of *D. nitedula* are critical, as are comparisons with the NC Iranian population from the type locality of *pictis* to determine if that name applies.

Rossolimo et al. (1990) reviewed *D. nitedula*, including illustrations of live animal, distributional, morphological, ecological, behavioral and other characteristics. Taxonomy of subspecies occuring in the independent republics of the former USSR studied by Rossolimo (1971). Taxonomic study and key to subspecies in Europe provided by Roesler and Witte (1969), in Serbia and Montenegro by Kryštufek (1985*a*), in Greece and SE Europe by Ondrias (1966), and in the independant republics of the former USSR by Ognev (1947). Morphometric study of Bulgarian populations given by Mitev et al. (1994), Peshev and Delov (1995*b*), and Markov (2001*b*). Markov found significant sexual dimorphism within Bulgarian populations, an uncommon finding within the family, whereas Mitev et al. (1994) found no significant sexual dimorphism. Comparison of dental pattern with fossil *Dryomys* from Poland given by Daoud (1993). Chromosomal data, and review of past cytogenetic studies provided by Zima et al. (1995), additional karyotypic data from Bulgaria reported by Peshev and Delov (1995*a*), Markov et al. (1997); from Macedonia by Zima et al. (1997*b*); from Turkey by Doğramaci and Kefelioğlu (1990); from Italy, Israel and Turkey by Civitelli et al. (1995*b*); and from Kazakhstan, Russia, and Turkmenistan by Graphodatsky and Fokin (1993). Allozyme variation analyzed by Filippucci and Kotsakis (1995). Phallic and bacular structure and variation studied by Hrabe (1969) and Simson et al. (1995). Illustrations and taxonomic implications of the os and glans penis, and stomach anatomy provided by Kratochvil (1973). Patterns of daily activity, winter activity, and importance of air temperature studied by Nowakowski (1998, 2001*b*). Detailed maps of European distribution, and discussion and review of distribution and ecology for each European country reported by Kryštufek and Vohralík (1994). Documented in faunal studies made in N Italy (Locatelli and Paolucci, 1996*a*; Paolucci et al., 1993) and S Italy (Cagnin and Aloise, 1995). Status in the E Baltic region reviewed by Timm et al. (1998) and discussed in a review of ecological strategies of Baltic rodents by Miljutin (1998). Recorded from Pleistocene sediments in Europe (Horacek, 1987; Kowalski, 2001).

Eliomys Wagner, 1840. Abh. math-phys. Cl. Kgl. Bayer Akad. Wiss., 3(1):176.
    TYPE SPECIES: *Eliomys melanurus* (Wagner, 1840).
    SYNONYMS: *Bifa* Lataste, 1885 [see Corbet, 1978*c* and Ellerman and Morrison-Scott, 1951].
    COMMENTS: Vocalization data reviewed by Hutterer and Peters (2001). Evolutionary patterns of dental morphology during Pliocene and Pleistocene discussed by Nadachowski and Daoud (1995). Dobson (1998) speculated that *Eliomys* probably reached Africa at least two times, once during the Messinian from Iberia, and later during the late Pleistocene from the eastern Mediterranean. Multiple colonizations may account for the lack of mor- phometric cohesion among what are presumed to be conspecific North African popula- tions of *E. munbyanus* (the Tunisian population, for example). It is generally (but not uni- versally) accepted that *Dryomys* is the closest living relative of *Eliomys* (see comments under Subfamily Leithiinae and references therein). However, phylogenetic analyses of nuclear DNA fragments and 12S rRNA sequences clearly identified *Eliomys* and *Dryomys* as sister-genera within a major evolutionary lineage that includes *Myomimus* and *Muscar- dinus* (Montgelard et al., 2003).

Eliomys melanurus (Wagner, 1840). Gelehrte Anz. I. K. Bayer. Akad. Wiss., München, 8(37):299.
    COMMON NAME: Large-eared Garden Dormouse (see comments).
    TYPE LOCALITY: Sinai (restricted to vicinity of Mt Sinai by Nader et al., 1983).
    DISTRIBUTION: E North Africa, the Middle East and S Turkey: Libya east from Barqah

(Cyrenaica) (Ranck, 1968); Egypt (Osborn and Helmy, 1980), the Sinai Peninsula (Haim and Tchernov, 1974; Kahmann, 1981; Osborn and Helmy, 1980; Wassif and Hoogstraal, 1954), Saudi Arabia (Harrison and Bates, 1991; Kahmann, 1981; Nader et al., 1981; Vesey-Fitzgerald, 1953), Israel (Bodenheimer, 1958; Ilani and Shalmon, 1983; Kahmann, 1981; Mendelssohn and Yom-Tov, 1999; Obuch, 2001; Qumsiyeh, 1996), Jordan (Atallah, 1978; Bodenheimer, 1958; Kahmann, 1981; Qumsiyeh, 1996; Tristram, 1877), Lebanon (G. M. Allen, 1915; Lewis et al., 1967; Qumsiyeh, 1996), Syria (Kahmann, 1981; Obuch, 2001; Qumsiyeh, 1996), Iraq (Kahmann, 1981; Nadachowski et al., 1978) and S Turkey (Misonne, 1957). In North Africa see Kahmann and Thoms (1981) and Niethammer (1959, 1987c).

STATUS: IUCN – Lower Risk (nt).

SYNONYMS: *cyrenaicus* Festa 1922 [see Ellerman and Morrison-Scott, 1951; Corbet, 1978c; Kryštufek and Kraft, 1997].

COMMENTS: See Kryštufek and Kraft (1997) for clarification of the publication date for *Eliomys melanurus* (Wagner (1839 vs. 1840)). Allozymic and karyological data analyzed by Filippucci et al. (1988a, b), Filippucci and Kotsakis (1995), and Filippucci and Capanna (1996) indicated that North African and Middle Eastern populations are genetically differentiated from European *E. quercinus*. They recommended the recognition of *E. melanurus* (containing all North African, Middle Eastern and Turkish populations) as a distinct species of *Eliomys*. Their conclusion was supported by a study utilizing phallic and bacular characters (Simson et al., 1995). Based on multivariate analyses of cranial morphology, Kryštufek and Kraft (1997) agreed that *E. melanurus* and *E. quercinus* represent two distinct species, but concluded that North African populations eastward to Tripolitania represent *E. quercinus*, and restricted the distribution of *E. melanurus* to E North Africa (east from Cyrenaica), the Middle East, and S Turkey.

An alternate interpretation of the available, seemingly contadictory morphological data versus karyological and allozymic data is that three species are represented: *E. quercinus* (Europe), *E. munbyanus* (W North Africa), and *E. melanurus* (E North Africa, Middle East and S Turkey), the latter two species seemingly closely related. The recognition of *E. munbyanus* as a distinct species was suggested by Delibes et al. (1980) and Filippucci and Capanna (1996) advocated recognizing *munbyanus* as a subspecies of *E. melanurus*. Utilizing this interpretation, the morphometric, karyological and allozymic data could be viewed as being concordant. *Eliomys munbyanus* is clearly not as morphologically differentiated from *E. quercinus* as is *E. melanurus*, but (aside from the Tunisia population) they form a discreet cluster in multivariate analyses (see figures in Kryštufek and Kraft, 1997). Moreover, *E. munbyanus* shows an extensive amount of karyological and genetic divergence from *E. quercinus*, as well as substantial genetic divergence and moderate karyological differences from *E. melanurus* (see Delibes et al., 1980; Filippucci and Capanna, 1996; and Filippucci and Kotsakis, 1995; Filippucci et al., 1988b). Based on these three data sets, both *E. melanurus* and *E. munbyanus* are recognized here as species distinct from *E. quercinus*.

The geographic distributions (and assignment of synonyms) given here for *E. munbyanus* and *E. melanurus* are based primarily on the results of Kryštufek and Kraft's (1997) morphological study. Additional genetic and karyological sampling of *E. munbyanus* and North African *E. melanurus* populations is needed, as only the Moroccan population of *E. munbyanus* has been sampled for allozyme variation, and karyological data has only been reported from Moroccan and Tunisian populations of that species. The hypothesized distribution of *E. munbyanus* is concordant with that of other Maghreb (Barbarian) mammalian endemics (see Carleton and Van der Straeten, 1997, and references therein). Reviewed by Rossolimo et al. (2001). Comparisons of North African *E. melanurus* population with *E. munbyanus* populations, including descriptions of holotypes, karyotypes, color plates of skins, and notes on biology, reported by Kahmann and Thoms (1981); phallic and bacular structure and variation reported by Simson et al. (1995); chromosomal studies and karyotypic variation within *E. melanurus* reviewed and discussed by Zima et al. (1995). Additional karyotypic data analyzed by Filippucci and Capanna (1996). The frequently used common name "Black-tailed Garden Dormouse" has been replaced here to avoid confusion, as the tails of individuals in certain populations of *E. quercinus*, *E. melanurus* and *E. munbyanus* have black undersides (Kryštufek and Kraft, 1997).

Eliomys munbyanus (Pomel, 1856). C. R. Hebdomadaires des Séances de L'Academie des Sciences, 42:653.

COMMON NAME: Maghreb Garden Dormouse.

TYPE LOCALITY: Algeria, Region d'Oran (Province of Oran).

DISTRIBUTION: W North Africa: Western Sahara (Kahmann and Thoms, 1981); Morocco (Aulagnier and Thevenot, 1986; Moreno and Delibes, 1981); Algeria (Khidas, 1993; Kowalski and Rzebik-Kowalska, 1991); Tunisia (Kahmann and Thoms, 1987); east to Tarbulus (Tripolitania) and Fazzan regions of E Libya (Ranck, 1968). See also Kahmann and Thoms (1981), Kock (1985) and Niethammer (1959).

SYNONYMS: *denticulatus* Ranck, 1968; *lerotina* (Lataste, 1885); *occidentalis* Thomas, 1903; *tunetae* Thomas, 1903.

COMMENTS: See comments under *E. melanurus* regarding data supporting recognition of *E. munbyanus*, and for discussion of geographic distributions of the two species. Morphological taxonomic study of circum-Mediterranean populations reported by Kryštufek and Kraft (1997). Comparative study of African populations by Kahmann and Thoms (1973b). Comparisons of North African *E. melanurus* population with North African *E. munbyanus* populations, including descriptions of type specimens, karyotypes, color plates of skins, and notes on biology, reported by Kahmann and Thoms (1981). Biometric study of Tunesian populations provided by Kahmann and Thoms (1987). Systematics of Moroccan population examined by Moreno (1989) and Moreno and Delibes (1981). Phallic and bacular structure and variation reported by Simson et al. (1995). Chromosomal studies reviewed by Zima et al. (1995); additional karyotypic data analyzed by Filippucci and Capanna (1996). Comprehensive analyses of allozyme data and genetic relationships reported by Filippucci and Kotsakis (1995) and Filippucci and Capanna (1996). Recorded from Pleistocene in Tunisia (Mein and Pickford, 1992).

Eliomys quercinus (Linnaeus, 1766). Syst. Nat., 12th ed., 1:84.

COMMON NAME: Garden Dormouse.

TYPE LOCALITY: Germany.

DISTRIBUTION: Primarily Europe, also Russia: Portugal (Santos-Reis and Mathias, 1996), Spain (Abad, 1987; Castién and Gosálbez, 1992, Moreno, 1989; Torre et al., 1996), Balearic Isls (Alcover, 1986; Alcover and Gosalbez, 1988; Kahmann and Alcover, 1974; Kahmann and Thoms, 1973a), Andorra (Gosalbez-Noguera et al., 1989), France (Geissert and Merkel, 1994), Corsica (Orsini and Cheylan, 1988), Belgium (Libois, 1996; Luyts, 1986), Germany (Bitz, 1991; Faltin, 1988; Feustel, 1984; Gorner and Henkel, 1988; Mockel, 1986; Rehage, 1984; Schoppe, 1986), Netherlands (Foppen and Bergers, 1992; Foppen et al., 1989; Thissen and Hollander, 1996), Czech Republic and Slovakia (Andera, 1986, 1995; Andera and Cerveny, 1994; Hůrka, 1990; Kratochvíl, 1967; Obuch, 1998; Smaha, 1996), Poland (Daoud, 1989; Jurczyszyn and Wolk, 1998; Pucek, 1983a), Belarus, Lithuania (Balciauskas, 1996; Juskaitis, 1995a), Latvia (Pilāts, 1995), Estonia including Suur Tutarsaar Isl (Ernits, 1991a; b; Masing and Timm, 1988), Finland, Ukraine (Bezrodny, 1991; Zagorodnyuk, 1998), European Russia to S Urals (Gromov and Erbajeva, 1995; Kuznetsov, 1965; Lichaev, 1972; Ognev, 1947), Switzerland (Catzeflis, 1995a; Maurizio, 1994), Austria (Spitzenberger, 1983, 1996; Spitzenberger et al., 1995), Italy, including Sardegna, Eolian archipelago and Sicilia (Amori et al., 1995, 1999, 2002a; Cagnin and Aloise, 1995; Cristaldi and Amori, 1988; Locatelli and Paolucci, 1996a; Sarà and Casamento, 1995a; Scaravelli et al., 1995), Slovenia (Kryštufek, 2003), Croatia (Petrov, 1992; Tvrtković et al., 1995; Vujošević et al., 1993), Adriatic Isls (Petrov, 1992; Tvrtković et al., 1995; Vujošević et al., 1993), Bosnia and Herzegovina (Petrov, 1992; Tvrtković et al., 1995), possibly Albania (Prigioni, 1996), possibly Bulgaria, and perhaps Romania (Istrate, 1998). See also Filippucci (1999), Kryštufek and Vohralík (1994) and Storch (1978a).

STATUS: IUCN – Vulnerable.

SYNONYMS: *amori* (Graells, 1897); *cincticauda* Miller, 1901; *dalmaticus* Dulic and Felten, 1962; *dichrurus* (Rafinesque, 1814); *gotthardus* Burg, 1920; *gymnesicus* Thomas, 1903; *hamiltoni* Cabrera, 1907; *hortualis* Cabrera, 1904; *jurrasicus* Burg, 1920; *liparensis* Kahmann, 1960; *lusitanicus* Reuvens, 1890; *nitela* (Schreber, 1782); *ophiusae* Thomas, 1925; *pallidus* Barrett-Hamilton, 1899; *räticus* Burg, 1920; *sardus* Barrett-Hamilton, 1901; *superans* Ognev and Stroganov, 1936; *valverdei* Palacios, Castroviejo, and Garzon, 1974; see Ellerman and Morrison-Scott (1951) and Corbet (1978*c*).

COMMENTS: Roman record in England (probably introduced) reported by O'Conner (1986). It has been proposed (e.g., Petrov, 1992) that the E Adriatic population was introduced from S Italy, but analyses of chromosomal and allozyme data contradict this hypothesis and indicate that the E Adriatic and Italian populations exhibit many allelic differences and are characterized by chromosomal rearrangements and a relatively high Nei's genetic distance (Filippucci and Capanna, 1996; Vujošević et al., 1993). These studies also indicate that the Dalmatian population was one of several that recolonized C Europe after the cold glacial periods of the late Pleistocene.

Rossolimo et al. (2001) reviewed *E. quercinus*, including illustrations of live animal, distributional, morphological, ecological, behavioral and other characteristics. Kryštufek and Kraft (1997) provided morphological taxonomic study of circum-Mediterranean populations. Moreno (1989) did a taxonomic study of Spanish populations, and Moreno et al. (1986) provided a systematic study of C European and Spanish populations. Grulich and Jurik (1994) studied morphometrics of a Slovakian population. Status discussed in a review of E Baltic mammals (Timm et al., 1998) and included in a review of the ecological strategies of Baltic rodents (Miljutin, 1998). Phallic and bacular structure and variation reported by Simson et al. (1995). Os penis figured by Didier (1953). Review of chromosomal studies and discussion of karyotypic variation within *E. quercinus* provided by Zima et al. (1995). Additional karyotypic data and analysis from European Russia given by Graphodatsky and Fokin (1993), from Czech Republic by Zima et al. (1997*b*); from Formentera Isl by Ramalhinho and Libois (2001), from circum-Mediterranean populations and Europe by Filippucci and Capanna (1996), and from Serbia and Montenegro by Vujošević et al. (1993). Comprehensive analyses of allozyme data and genetic relationships reported by Filippucci and Kotsakis (1995), Filippucci and Capanna (1996) and Vujošević et al. (1993). Coronary arterial anatomy described by Sans-Coma et al. (1995). Social organization, home range and movement of NW Italy population studied by Bertolino et al. (1997). Ecology in Alpine habitat studied by Bertolino and Currado (2001). Human influence on geographical distribution discussed by Carpaneto and Cristaldi (1995). See comments under *E. melanurus* for allocation of North African populations. Recorded from Pleistocene in Europe (Capasso Barbato and Gliozzi, 2001; Chaline, 1980; Cuenca Bescos, et al., 2001; Horacek, 1986; Kowalski, 2001; Storch, 1978*a*).

Muscardinus Kaup, 1829. Skizz. Entwikel.-Gesch. Nat. Syst. Europ. Thierwelt, 1:139.

TYPE SPECIES: *Mus avellanarius* Linnaeus, 1758.

SYNONYMS: *Eomuscardinus* Hartenberger, 1966; *Pentaglis* Kretzoi, 1943; see Daams and de Bruijn (1995); Rossolimo et al. (2001).

COMMENTS: Kratochvil (1973) proposed to place *Muscardinus* in its own subfamily, Muscardininae, based on the uniquely derived stomach anatomy, molar morphology, and phallic features. De Bruijn (1967) and Daams (1981) both included *Muscardinus* in the Glirinae based on dental morphology. This arrangement was supported by the results of Wahlert et al. (1993). However, phylogenetic analyses of nuclear DNA fragments and mitochondrial 12S rRNA sequences clearly identified *Muscardinus* as a member of Leithiinae and basal to *Myomimus*, and the clade comprised of *Dromys*, and *Eliomys*. See also comments under Leithiinae. Comparative vocalization data reviewed by Hutterer and Peters (2001). Evolutionary patterns of dental morphology during Pliocene and Pleistocene discussed by Nadachowski and Daoud (1995).

Muscardinus avellanarius (Linnaeus, 1758). Syst. Nat., 10th ed., 1:62.

COMMON NAME: Hazel Dormouse.

TYPE LOCALITY: Sweden.

DISTRIBUTION: Europe and N Turkey: Cumbria and S England (Bright and Morris, 1992; Bright et al., 1996; Coult, 2001), France (Jourde, 2001; Papillon et al., 2000), Switzerland (Catzeflis, 1995d; Maurizio, 1994), Italy, including Sicilia (Amori et al., 1995, 1999; Sarà and Casamento, 1995a; Sarà et al., 2002), Austria (Spitzenberger, 1983; Spitzenberger et al., 1995), Germany (Bitz, 1991; Faltin, 1988; Gorner and Henkel, 1988; Harsch, 1993; Mockel, 1988; Rehage and Steinborn, 1984; Schoppe, 1986; Schulze, 1986, 1987), Belgium (Christiaens, 1995; Libois, 1996), Netherlands (van Laar, 1984, 1992; Thissen and Hollander, 1996), Denmark (Jensen, 1980), S and C Sweden (Berg, 1996, 1997), Poland (Daoud, 1989; Jurczyszyn and Wolk, 1998; Kaluza, 1987; Pucek, 1983d; Wilk, 1987), Ukraine (Bezrodny, 1991), Belorus, Lithuania (Juškaitis, 1995a), Latvia (Pilāts, 1995), and Estonia (Ernits, 1991b) east to Russia (Gromov and Erbajeva, 1995; Kuznetsov, 1965; Lichaev, 1972; Ognev, 1947); Czech Republic and Slovakia (Anděra, 1987, 1995; Anděra and Cerveny, 1994; Danko, 1994; Hůrka, 1990; Obuch, 1998; Smaha, 1996; Stanko and Mosansky, 2000), Hungary (Bakó et al., 1998), Slovenia (Kryštufek, 1991), Croatia (Tvrtković et al., 1995), Bosnia and Herzegovina, Serbia and Montenegro (Petrov, 1992), Romania (Istrate, 1998), Albania (Prigioni, 1996), Macedonia, Bulgaria (Belcheva et al., 1989; Peshev, 1996), Greece (Ondrias, 1966), Corfu, N Turkey (Doğramaci and Kefelioğlu, 1992; Kivanc, 1983; Obuch, 2002). In Europe see also Morris (1999) and Storch (1978a).

STATUS: IUCN – Lower Risk (nt).

SYNONYMS: *abanticus* Kivanc, 1983; *anglicus* Barrett-Hamilton, 1900; *corilinum* (Fatio, 1869); *kroecki* Niethammer and Bohmann, 1950; *muscardinus* (Schreber, 1782); *niveus* Altobello, 1920; *pulcher* Barrett-Hamilton, 1898; *speciosus* (Dehne, 1855); *trapezius* Miller, 1908; *zeus* Chaworth-Musters, 1932; see Ellerman and Morrison-Scott (1951) and Corbet (1978c, 1984).

COMMENTS: Comprehensive review of *M. avellanarius*, including illustrations of live animal, distributional, morphological, ecological, behavioral and other characteristics, contributed by Rossolimo et al. (2001). Systematic study of W and C European subspecies provided by Witte (1962) and Roesler and Witte (1969), and of Turkish subspecies by Kivanc (1983). Morphometric study of Bulgarian populations reported by Peshev et al. (1990a) and Peshev and Delov (1995b). Comparison of dental pattern with fossil *Muscardinus* from Poland given by Daoud (1993). Review of chromosomal studies given by Zima et al. (1995), additional karyotypic data from Bulgaria reported by Peshev and Delov (1995a), from Turkey by Doğramaci and Kefelioğlu (1992), and from Russia by Graphodatsky and Fokin (1993). Allozyme variation and genetic relationships analyzed by Filippucci and Kotsakis (1995). Phallic and bacular structure and variation reported by Hrabe (1969) and Simson et al. (1995). Illustrations and taxonomic implications of os and glans penis, and stomach anatomy, provided by Kratochvil (1973). Tendon-locking mechanism as an adaptation for climbing by grasping documented by Haffner (1996). Hibernation in nature studied by Vogel and Frey (1995). Range, movement, nest-box interactions and annual weight fluctuations in Lithuania reported by Juškaitis (1995b, 1997, 2001). Occurrence of inividuals with white (albino) tail tips in Lithuania analyzed by Juškaitis (1995c). Status of populations in the E Baltic region summarized by Timm et al. (1998), and included in a review of ecological strategies of Baltic rodents by Miljutin (1998). Documented in regional faunal studies in N Italy (Cantini, 1991; Cresti et al., 1994; Paolucci et al., 1993; Scaravelli et al., 1995), C Italy (Amori et al., 2002a; Cerone and Aloise, 1994), S Italy (Cagnin and Aloise, 1995; Cagnin et al., 1996) and Sicilia (Sara et al., 2001). Human influence on geographical distribution discussed by Carpaneto and Cristaldi (1995). Recorded from Pleistocene in Europe (Capasso Barbato and Gliozzi, 2001; Horacek, 1987; Kowalski, 2001; Viriot et al., 1991).

Myomimus Ognev, 1924. Priroda Okhota Ukraine [Nat. and Hunting in Ukraine], Kharkov, 1-2:115.

TYPE SPECIES: *Myomimus personatus* Ognev, 1924.

SYNONYMS: *Philistomys* Bate, 1937 [see Kowalski, 1963; Rossolimo et al., 2001].

COMMENTS: This genus requires systematic revision. A new, undescribed species of *Myomimus* has been found in the C Zagros Mtns, Bakhtaran Prov., Iran, 1300 m. Known only from owl pellets, the mandible of this species is larger than that of *M. setzeri*, but smaller than *M. personatus* (Obuch, 2001; J. Obuch, pers. comm.); Obuch (2001) included photographs of dentaries of the new species compared with those of *M. personatus* and *M. setzeri*. Daams and de Bruijn (1995) arranged *Myomimus* in subfamily Myominae; according to their hypothesis myomimines and glirines originated from gliravines in the late Oligocene. There seems to be a general (but not universal, see Yachontov and Potapova, 1991, and Simson et al., 1995) consensus that *Myomimus* and *Selevinia* are related (Koenigswald, 1993, 1995; Potapova, 2001; Rossolimo et al., 2001; Storch, 1995*b*; Wahlert et al., 1993). Phylogenetic analyses of nuclear DNA fragments and mitochondrial 12S rRNA sequences identified *Myomimus* as sister to a clade containing *Dryomys* and *Eliomys* within a larger evolutionary lineage corresponding to the Leithiinae of Wahlert et al. (1993). See comments under Subfamily Leithiinae for further taxonomic discussion. Evolutionary patterns of dental morphology during Pliocene and Pleistocene discussed by Nadachowski and Daoud (1995). Recorded from Pleistocene in E and SE Europe (Kowalski, 2001).

Myomimus personatus Ognev, 1924. Priroda Okhota Ukraine [Nat. and Hunting in Ukraine], Kharkov, 1-2:115.

COMMON NAME: Masked Mouse-tailed Dormouse.

TYPE LOCALITY: Turkmenistan, Koppeh-Dagh Mtns, near Kaine-Kassyr on the Sumbar River, Turkmenistan-Iran border.

DISTRIBUTION: Bulgaria (Peshev et al., 1964); W Turkey (Mursaloğlu, 1973*b*); Koppet Dag Mtns, NE Iran (Obuch, 2001); Koppet Dag and Malyy Balkhan Mtns, Turkmenistan (Csorba, 1993; Kurbanov et al., 1990; Kuznetsov, 1965; Marinina et al., 1987; Ognev, 1947; Shcherbina et al., 1988); Iskander, Uzbekistan (Zykov, 1987); limits unknown.

STATUS: IUCN – Vulnerable.

COMMENTS: Review of morphology, ecology and distribution, and illustration of live animal, contributed by Rossolimo et al. (2001). External morphology of the glans penis and its taxonomic significance discussed by Rossolimo and Pavlinov (1985). Mensural data, and figures of skin, feet, cranium and teeth given by Peshev et al. (1964). Skull, teeth, os penis, feet, and ear ossicles figured by Ognev (1947). Karyotypic data and analysis provided by Graphodatsky and Fokin (1993). Russian range, taxonomy and other characters reviewed by Gromov and Erbajeva (1995). Comparison with *M. roachi* provided by Rossolimo (1976*b*).

Myomimus roachi (Bate, 1937). Ann. Mag. Nat. Hist., ser. 10, 20:399.

COMMON NAME: Roach's Mouse-tailed Dormouse.

TYPE LOCALITY: Israel, Mt Carmel, Tabun Cave, upper Pleistocene layers.

DISTRIBUTION: Extant in SE Bulgaria (Filippucci and Peshev, 1999; Peshev, 1996), Turkish Thrace (Filippucci and Peshev, 1999; Kurtonur and Özkan, 1991), and W Turkey (Mursaloğlu, 1973*a*); limits unknown. Early to late Pleistocene fossils are known from Israel (Tchernov, 1992, 1994), Greece (Macedonia, Athens, Chios Isl and Kalimnos Isl) (Kuss and Storch, 1978; Storch, 1975, 1978*b*), and from S Anatolia, Turkey (Corbet and Morris, 1967; Storch, 1988).

STATUS: IUCN – Vulnerable.

SYNONYMS: *bulgaricus* Rossolimo, 1976 [see Storch, 1978*b*].

COMMENTS: Morphology, ecology and distribution reviewed by Rossolimo et al. (2001). Distribution and review provided by Storch (1978*b*). Detailed description and figures of SE Bulgaria population given by Pechev et al. (1964). Taxonomic study of Bulgarian samples and comparison with *M. personatus* provided by Rossolimo (1976*b*). Distribution, measurements, and ecological data of Turkish Thrace population recorded by Kurtonur and Özkan (1991). Cranial abnormality in wild Turkish population documented by Kryštufek et al. (2004*b*). Phallic and bacular structure and variation reported by Simson et al. (1995). Chromosomal data and analysis given by Civitelli et al. (1995*b*). Allozyme variation analyzed by Filippucci and Kotsakis (1995). Comparative behavior and hibernation of wild-caught captives reported by Buruldağ and Kurtonur (2001).

Myomimus setzeri Rossolimo, 1976. Vestn. Zool., 4:51.
>COMMON NAME: Setzer's Mouse-tailed Dormouse.
>TYPE LOCALITY: Iran, Kordestan Prov., 4 km W of Bane.
>DISTRIBUTION: From 1500-2800 m in E Turkey (Obuch, 1994, 2001), and Azarbaijan-e Gharbi and Kordestan Prov., NW Iran (Obuch, 1994, 2001; Rossolimo, 1976a); limits unknown.
>STATUS: IUCN – Endangered.
>COMMENTS: Reviewed by Rossolimo et al. (2001). Skull and teeth figured in Rossolimo (1976a); dentaries illustrated and compared with *M. personatus* and *Myomimus sp.* (see comments under *Myomimus*) by Obuch (2001).

Selevinia Belosludov and Bazhanov, 1939. Uchen. Zap. Kaz. Univ. Alma-Ata, 1(1):81.
>TYPE SPECIES: *Selevinia betpakdalaensis* Belosludov and Bazhanov, 1939.
>SYNONYMS: *Salevinia* Argyropulo and Vinogradov, 1939.
>COMMENTS: Belosludov and Bazhanov (1939) recognized *Selevinia* as a new genus of murids, and subsequently as the only member of a separate family of rodents seemingly allied to glirids (Bazhanov and Belosludov, 1941). Based upon a suite of morphological characters, Ognev (1947) hypothesized that *Selevinia* was a highly differentiated dormouse most closely related to *Myomimus*. Because of its distinct dental formula and structure, Ognev (1947) listed *Selevinia* in a separate subfamily, a hypothesis independantly suggested by Ellerman (1949a). Published descriptions and figures of *Selevinia* included derived character states that allowed Wahlert et al. (1993) to tentatively place this genus within Myomimini, supporting Ognev's (1947) hypothesis. Though the nearest extant relative of *Selevinia* is proposed to be *Myomimus*, its closest relative may be *Plioselevinia*, from early Pliocene breccia of Poland (McKenna and Bell, 1997). Reviewed by Gromov and Erbajeva (1995). For further taxonomic discussion see comments under Leithiinae and *Myomimus*, and references therein.

Selevinia betpakdalaensis Belosludov and Bazhanov, 1939. Uchen. Zap. Kaz. Univ. Alma-Ata, 1(1):81.
>COMMON NAME: Desert Dormouse.
>TYPE LOCALITY: S Kazakhstan, N Betpak-Dala Desert, Kyzyl-Ui.
>DISTRIBUTION: S (=SE) and E Kazakhstan, deserts east, north and west of Lake Balkhash (Burdelov and Rossinskaya, 1959; Ilchenko and Volodin, 1992; Ismagilov, 1961; Kuznetsov, 1965).
>STATUS: IUCN – Endangered.
>SYNONYMS: *paradoxa* Argyropulo and Vinogradov, 1939 [see Bazhanov and Belosludov, 1941].
>COMMENTS: Review of morphology, ecology and distribution, and illustrations of live animals, contributed by Rossolimo et al. (2001). Measurements and illustration of cranium, and biological observations provided by Bazhanov and Belosludov (1941). Skull, ear, ear ossicles, and foot figured in Ognev (1947). Photographs of live specimen, distribution, and biological data contributed by Sludskii (1977). Teeth and embryo illustrated by Bazhanov (1951). Mastoid portion of the bullae discussed by Pavlinov (1988). Included in an analysis of the relationship between size of ear pinna and auditory bulla in specialized desert rodents (Pavlinov and Rogovin, 2000).

Subfamily Glirinae Muirhead, 1819:433 (see McKenna and Bell, 1997). Mazology [sic]. Pp. 393-480, pls. 353-358, *in* Edinburgh Encyclopedia, Vol. 13 (D. Brewster, ed.).
>SYNONYMS: Glirinae Thomas, 1897; Glirulinae de Bruijn, 1966; Myosidae Gray, 1821 (Myoxina Gray, 1825; Myoxinae Huxley, 1872).
>COMMENTS: The results of Wahlert et al. (1993) and Storch (1995b) indicated that *Myoxus*, *Muscardinus*, and *Glirulus* form a monophyletic group. Yachontov and Potapova (1991) hypothesized that *Glis* and *Muscardinus* are closely related to *Selevinia*. Daams and de Bruijn (1995) included only *Glis* and *Muscardinus* in Glirinae, and arranged *Glirulus* under their Dryomyinae. Koenigswald (1993, 1995) placed *Glis* and *Glirulus* together in his most primitive cluster, "Group I", and *Muscardinus* in his most derived cluster, "Group III". Bentz and Mongelard's (1999) mitochondrial analyses moderately support a *Glis-Muscardinus* grouping, but there was no support for the inclusion of *Glirulus* in this clade. Suzuki

et al. (1997) asserted that *Glirulus* should be placed in its own subfamily based on its high level of genetic divergence, a position concordant with the arrangement of Rossolimo et al. (2001), and with Potapova's (2001) findings based on its unique middle ear morphology compared with that of other extant dormice. Rossolimo et al. (2001) arranged *Glis* (*Myoxus*) as the sole member of *Glirinae* (*Myoxinae*), and did not allocate *Muscardinus* to subfamily, but suggested it was related to *Dryomys* and *Eliomys*. Potapova (2001) hypothesized that *Muscardinus* represents an independent group that is more closely related to *Dryomys* and *Eliomys* than to *Glis*. Based upon upper molar crown pattern, Pavlinov (2001b) concluded that *Glis* and *Muscardinus* are "most dissimilar both from each other and from the remainder of the glirids studied." Phylogenetic analyses of nuclear DNA fragments and mitochondrial 12S rRNA sequences identified *Glis* and *Glirulus* as members of a major lineage comparable to the Glirinae of Wahlert et al. (1993), divorced *Muscardinus* from that clade and placed it in the Leithiinae as a basal member to *Myomimus*, *Dryomysi* and *Eliomys* (Mongelard et al., 2003). Only *Glis* and *Glirulus* are listed here as members of Glirinae (see discussion under Leithiinae).

Glirulus Thomas, 1906. Proc. Zool. Soc. Lond., 1905(2):347 [1906].
    TYPE SPECIES: *Myoxus javanicus* Schinz, 1845 (= *lapsus* for *japonicus*; see Thomas (1905a) for an explanation of the emendation of *javanicus* to *japonicus*).
    SYNONYMS: *Amphidyromys* Heller, 1936 [see Daams and de Bruijn, 1995; Rossolimo et al., 2001]; *Paraglirulus* Engesser, 1972 [see McKenna and Bell, 1997, and Wu et al., 2000].
    COMMENTS: The phylogenetic study by Wahlert et al. (1993) supports inclusion of *Glirulus* within Glirinae, as did Montgelard et al.'s (2003) phylogenetic analyses of nuclear and mitochondrial gene sequences (see also comment under Glirinae). Daams and de Bruijn (1995) arranged this genus within their Dryomyinae (in which they also include *Dryomys*, *Eliomys*, *Graphiurus* and *Chaetocauda*). Suzuki et al. (1997) presented genetic data indicating that *Glirulus* may belong in its own subfamily, though their study only included *Dryomys* and *Muscardinus* for comparison. Although the geographic range of the sole extant species of *Glirulus* is restricted to Japan, the genus is represented in Europe and Turkey by early Miocene, Pliocene and early Pleistocene fossils (Bednarczyk, 1993; Daams and de Bruijn, 1995; Hugueney and Mein, 1965; Kowalski, 1963, 2001; Nadachowski and Daoud, 1995; Ünay, 1994). Molars identified as *Glirulus* from late Oligocene beds in Xinjiang Prov. of NW China provide the oldest record of not only a dormouse from China but of *Glirulus* itself (Wu et al., 2000). A late Miocene fossil from Ardèche, France, suggests that an extinct species of *Glirulus* was capable of gliding (Mein and Romaggi, 1991). Evolutionary patterns of dental morphology during Pliocene and Pleistocene discussed by Nadachowski and Daoud (1995).

Glirulus japonicus (Schinz, 1845). Syst. Verzeichniss Säugeth., 2:530.
    COMMON NAME: Japanese Dormouse.
    TYPE LOCALITY: Japan.
    DISTRIBUTION: Japan, three main islands of Honshu, Shikoku, and Kyushu, and small island of Dogo in the Oki group north of E Honshu (Iguchi et al., 1996; Kaneko, 1994).
    STATUS: IUCN – Endangered.
    SYNONYMS: As proposed by Smeenk and Kaneko (2000), the International Commission on Zoological Nomenclature (2001a) has ruled (Opinion 1978) that the following synonyms are invalid: *elegans* (Temminck, 1844) [not of Ogilby, 1838], *javanicus* (Schinz, 1845), and *lasiotis* (Thomas, 1880); see also Ellerman and Morrison-Scott (1951) and Corbet (1978c).
    COMMENTS: Based on mitochondrial and nuclear genetic data (Suzuki et al., 1997), there are at least two distinct populations of *G. japonicus* that may represent separate species. Unpublished morphological and behavioral differences were also cited. External genital morphology and its taxonomic significance were examined by Rossolimo and Pavlinov (1985). Chromosomal study provided by Tsuchiya (1979). Analysis of nest materials as indication of foraging behavior reported by Minato and Doei (1995). Included as part of the endemic Japanese fauna by Dobson (1994) who described distribution patterns of Japanese endemic mammals. Comprehensive review, including color photographs of living animals, by Kaneko (1994). Review of morphology, ecology, distribution, and

illustrations of live animals contributed by Rossolimo et al. (2001). Middle and late Pleistocene representatives of *G. japonicus* on Japan were discussed by Kawamura (1989, 1991, 1994) and Kowalski and Hasegawa (1976).

Glis Brisson, 1762. Regnum animale in classes IX distributum, sive synopsis methodica, 2nd ed.:113
TYPE SPECIES: *Sciurus glis* (Linnaeus, 1766).
SYNONYMS: *Elius* Schulze, 1900 [excluding *dryas*]; *Myorus* Reichenbach, 1835; *Myoxus* Zimmerman, 1780 [see Opinion 1894 of the International Commission on Zoological Nomenclature, 1998; McKenna and Bell, 1997].
COMMENTS: See comments under Gliridae and discussion in Wahlert et al. (1993) for explanation behind validity of *Myoxus* versus *Glis*. Comparative vocalization data reviewed by Hutterer and Peters (2001). Evolutionary patterns of dental morphology during Pliocene and Pleistocene discussed by Nadachowski and Daoud (1995). Morphological (Wahlert et al., 1993) and molecular studies (Montgelard et al., 2003) identify *Glis* and *Glirulus* as the only extant members of Glirinae (see subfamily account).

Glis glis (Linnaeus, 1766). Syst. Nat., 12th ed., 1(1):87.
COMMON NAME: Fat Dormouse.
TYPE LOCALITY: Slovenia, Carniola (see comments below).
DISTRIBUTION: Europe, N Turkey, the Caucasus, N Iran and SW Turkmenistan: N Spain (Castien and Gosalbez, 1992), France (Gautherin, 1988; Geissert and Merkel, 1994), Switzerland (Catzeflis, 1995*c*; Maurizio, 1994), Belgium (Christiaens, 1995; Libois, 1996), Netherlands, Germany (Bitz, 1991; Faltin, 1988; Feustel, 1984; Gorner and Henkel, 1988; Grunwald, 1992; Harsch, 1993; Labes et al., 1987; Nachtigall, 1996; Pankow, 1989; Rehage and Preywisch, 1984; Schoppe, 1986; Schulze, 1986; von Vietinghoff-Riesch, 1960), Poland (Bielecka, 1986; Daoud, 1989; Indyk and Pawlowska-Indyk, 1994; Jurczyszyn and Wolk, 1998; Jurczyszyn et al., 2001; Kazmierczak and Kaliczewski, 1989; Nowakowski and Terlecki, 1991; Profus, 2000; Pucek, 1983*c*; Wuczynski and Garbowski, 2000); Ukraine (Bezrodny, 1991) north to Belorus, Lithuania (Balciauskas, 1996; Juskaitis, 1995*a*) and Latvia (Pilāts, 1995), east to Volga River, south to Saratov and Voronezh (Gromov and Erbajeva, 1995); Caucasus Mtns (Gromov and Erbajeva, 1995) south to N Iran (Lay, 1967; Obuch, 2001) and SW Turkmenistan (see Bezrodny, 1991; Kuznetsov, 1965; Lichaev, 1972; Ognev, 1947; Ruprecht and Szwagrzak, 1986; Vereshchagin, 1959); the Mediterranean (except S and C Iberia, Balearic Isls), Corsica, Sardinia, Sicily, Elba, Italy (Amori et al., 1995, 1999), including Sicilia (Sarà and Casamento, 1995*a*), Eolia (Cristaldi and Amori, 1988) and N Adriatic Isls (Petrov, 1992; Tvrtković et al., 1995), Austria (Spitzenberger, 1983; Spitzenberger et al., 1995), Czech Republic and Slovakia (Anděra, 1986, 1995; Anděra and Cerveny, 1994; Danko, 1994; Hůrka, 1990; Obuch, 1998; Smaha, 1996; Stanko and Mosansky, 2000), Hungary (Bakó et al., 1998; Becsy, 1982), Slovenia (Kryštufek, 1991; Kryštufek and Haberl, 2001; Polak, 1997); Croatia (Tvrtković et al., 1995), Bosnia and Herzegovina, Serbia and Montenegro (Petrov, 1992), Romania (Istrate, 1998; Vasiliu, 1961), Albania (Prigioni, 1996), Macedonia (Petrov, 1992), Bulgaria (Peshev, 1996; Peshev et al., 1990*b*), Greece (Ondrias, 1966; in Andros Isl see Dimaki, 1999; in Macedonia see Vohralík and Sofianidou, 1987; in Thrace see Vohralík and Sofianidou, 1992*a*), Crete, Corfu, Cephalonia, Turkish Thrace (Kurtonur, 1992), N Turkey (Doğramaci and Tez, 1991; Kock, 1990; Obuch, 2001). In Europe see also Kryštufek (1999*a*) and Storch (1978*a*). Introduced to England (Morris, 1997*a*, *b*).
STATUS: IUCN – Lower Risk (nt).
SYNONYMS: *abruttii* Altobello, 1924; *argenteus* Zimmermann, 1953; *avellanus* (Owen, 1840); *caspicus* Satunin, 1906; *caspius* (Satunin, 1905); *esculentus* Blumenbach, 1779; *germanicus* Violani and Zava, 1995; *giglis* (Cuvier, 1832); *insularis* Barrett-Hamilton, 1899; *intermedius* Altobello, 1920; *italicus* Barrett-Hamilton, 1898; *martinoi* Miri, 1960; *melonii* Thomas, 1907; *minutus* Martino, 1930; *orientalis* (Nehring, 1903); *persicus* (Erxleben, 1777); *petruccii* Goodwin, 1939; *pindicus* Ondrias, 1966; *postus* Montagu, 1923; *pyrenaicus* Cabrera, 1908; *spoliatus* Thomas, 1906; *subalpinus* Burg, 1920; *tschetshenicus* Satunin, 1920; *vagneri* Martino and Martino, 1941; *vulgaris* Oken, 1816. See Ellerman (1941), Thomas (1906), Morrison-Scott (1951), and Corbet (1978*c*).

COMMENTS: Violani and Zava (1995) restricted the type locality of *G. glis* to Southern Carniola, Slovenia, based on a letter written by Scopoli describing the Fat Dormouse to Linnaeus, who was unacquainted with the animal. Scopoli gave the following locality information, *"Habitat in Carniola, in primis inferiore"*, which Violani and Zava (1995) translated to mean "Southern Carniola". An alternate translation of *"Habitat in Carniola, in primis inferiore."* is "dwells in Carniola, principally Southern." Krystufek is of the opinion that Scopoli likely became acquainted with the Fat Dormouse in the vicinity of Idrija, central Slovenia, where Scopoli worked as a physician, and where Fat Dormice are still common (B. Kryštufek, pers. comm.). Unless additional locality information is uncovered regarding the Scopoli specimen, it seems more accurate to list the type locality as simply Carniola.

The significance of geographical variation within *G. glis*, in the context of subspecific or specific level differentiation among populations, has not been investigated throughout its range. Comprehensive review of *G. glis*, including illustrations of live animal, distributional, morphological, ecological, behavioral and other characteristics, contributed by Rossolimo et al. (2001). Biometry and taxonomy of Asiago Plateau population by Franco (1988); morphometric study of Bulgarian populations by Peshev and Delov (1995b); variability of non-metric characters of Bulgarian populations analyzed by Markov (2001a); morphometric and chromosomal study of Turkish populations by Doğramaci and Tez (1991); diagnosis and distribution of subspecies in SE Europe by Ondrias (1966). Comparison of dental pattern with fossil *Glis* from Poland given by Daoud (1993). Review of chromosomal studies produced by Zima et al. (1995); additional karyotypic data from Bulgaria reported by Peshev and Delov (1995a), from Turkey by Doğramaci and Tez (1991), from Italy and Turkey by Civitelli et al. (1995b), and from Russia by Graphodatsky and Fokin (1993); complete mitochondrial DNA sequence analyzed by Reyes et al. (1998); allozyme variation and genetic relationships analyzed by Filippucci and Kotsakis (1995); phallic and bacular structure and variation reported by Hrabe (1969) and Simson et al. (1995); illustrations and taxonomic significance of the os and glans penis, and stomach anatomy, provided by Kratochvil (1973). Functional morphology and histology of the feet reported by Krattli and Haffner (1995); morphology and functional significance of vibrissae studied by Kulikov (1988); variation in number of teats in Slovenian samples and its possible significance elucidated by Kryštufek (2004). Ecology, seasonal variation in body mass, variation in mass of individual organs, and other aspects of population biology of Slovenian population studied by Kryštufek (2001c), Kryštufek and M. Zavodnik (2003), and Kryštufek et al. (2003). Daily torpor in wild population reported by Nowakowski (2001a). Use of tibial length for age determination analyzed by Schlund (1997). Status in E Baltic region summarized by Timm et al. (1998), and treated in a review of ecological stategies of Baltic rodents (Miljutin, 1998). Documented in faunal studies made in N Italy (Cantini, 1991; Paolucci et al., 1993; Cresti et al., 1994; Locatelli and Paolucci, 1996a; Scaravelli et al., 1995), C Italy (Amori et al., 2002a; Cerone and Aloise, 1994), and S Italy (Cagnin and Aloise, 1995; Cagnin et al., 1996). Human influence on geographical distribution discussed by Carpaneto and Cristaldi (1995). Recorded from middle and late Pleistocene in Europe (Capasso Barbato and Gliozzi, 2001; Horacek, 1986; Kowalski, 2001; Storch, 1978a; Viriot et al., 1991).

SUBORDER CASTORIMORPHA A. E. Wood, 1955.

FAMILY CASTORIDAE
by Kristofer M. Helgen

Family Castoridae Hemprich, 1820. Grundriss Naturgesch., p. 33.
  SYNONYMS: Castorini Giebel, 1855; Castorida Haeckel, 1866; Castoroidea Gill, 1872.
  COMMENTS: Generally placed as the sole extant family of the sciurognath infraorder
    Castorimorpha (Carleton, 1984; McKenna and Bell, 1997), beavers are probably most
    closely allied to geomyoid rodents (Geomyidae + Heteromyidae; see Murphy et al., 2001*a*;
    Montgelard et al., 2002). There are a number of additional fossil synonyms that do not
    refer to crown-group beavers (McKenna and Bell, 1997).

Castor Linnaeus, 1758. Syst. Nat., 10th ed, 1:58.
  TYPE SPECIES: *Castor fiber* Linnaeus, 1758.
  SYNONYMS: *Fiber* Dumeril, 1806 [not of G. Cuvier, 1800]; *Mamcastorus* Herrera, 1899.
  COMMENTS: The two species differ in cranial features and chromosome number (Heidecke,
    1986).

Castor canadensis Kuhl, 1820. Beitr. Zool. Vergl. Anat., 1:64.
  COMMON NAME: American Beaver.
  TYPE LOCALITY: Canada, Hudson Bay.
  DISTRIBUTION: Alaska and Canada south of the Arctic Circle (including Vancouver and
    Newfoundland), most of the continental United States (absent from parts of SW USA
    and from most of Florida), extending into N Mexico. See Hall (1981:604). Introduced
    to Tierra del Fuego (South America) and Eurasia, including Finland, NW Russia,
    Poland, Germany, and Austria.
  STATUS: IUCN – Data Deficient as *C. c. phaeus* (from Admiralty Island in Alaska's Alexander
    Archipelago); Not Evaluated as *C. c. frondator* and *C. c. mexicanus*; otherwise Lower Risk
    (lc).
  SYNONYMS: *acadicus* Bailey and Doutt, 1942; *baileyi* Nelson, 1927; *belugae* Taylor, 1916;
    caecator Bangs, 1913; *carolinensis* Rhoads, 1898; *concisor* Warren and Hall, 1939;
    duchesnei Durrant and Crane, 1948; *frondator* Mearns, 1897; *idoneus* Jewett and Hall,
    1940; *labradorensis* Bailey and Doutt, 1942; *leucodontus* Gray, 1869; *mexicanus* Bailey,
    1913; *michiganensis* Bailey, 1913; *missouriensis* Bailey, 1919; *pacificus* Rhoads, 1898;
    pallidus Durrant and Crane, 1948; *phaeus* Heller, 1909; *repentinus* Goldman, 1932;
    rostralis Durrant and Crane, 1948; *sagittatus* Benson, 1933; *shastensis* Taylor, 1916;
    subauratus Taylor, 1912; *taylori* Davis, 1939; *texensis* Bailey, 1905.
  COMMENTS: Reviewed by Jenkins and Busher (1979, Mammalian Species, 120). With the
    exception of *pacificus* (type locality Cascade Mtns, Washington), which was held in
    synonymy with *leucodontus* (type locality Vancouver Isl, British Columbia), Hall
    (1981) recognized all named forms as valid subspecies. Pending a critical review, no
    subspecies are recognized here.

Castor fiber Linnaeus, 1758. Syst. Nat., 10th ed, 1:58.
  COMMON NAME: Eurasian Beaver.
  TYPE LOCALITY: Sweden.
  DISTRIBUTION: Throughout N Eurasia, including Austria, Belarus, Belgium, China, Croatia,
    Czech Republic, Estonia, Finland, France, Germany, Hungary, Latvia, Liechtenstein,
    Lithuania, Moldova, Mongolia, Netherlands, Norway, Poland, Russia (populations
    throughout), Slovakia, Slovenia, Sweden, Switzerland, Ukraine; formerly extinct but
    reintroduced in many of these countries. See Nolet and Rosell (1998) and Véron
    (1992*a*).
  STATUS: U. S. ESA – Endangered as *C. fiber birulai*; IUCN – Critically Endangered as
    C. f. tuvinicus, Vulnerable as *C. f. birulai* and *C. f. pohlei*; otherwise Near Threatened.
  SYNONYMS: *albicus* Matschie, 1907; *albus* Kerr, 1792; *balticus* Matschie, 1907; *belarusicus*

Lavrov, 1974; *belorussicus* Lavrov, 1981; *bielorussieus* Lavrov, 1983; *birulai* Serebrennikov, 1929; *flavus* Desmarest, 1822; *fulvus* Bechstein, 1801; *galliae* É. Geoffroy, 1803; *gallicus* Fischer, 1829; *introductus* Saveljev, 1997 [*nomen nudum*]; *niger* Desmarest, 1822; *orientoeuropaeus* Lavrov, 1981; *osteuropaeus* Lavrov, 1974; *pohlei* Serebrennikov, 1929; *proprius* Billberg, 1833; *solitarius* Kerr, 1792; *tuvinicus* Lavrov, 1969; *variegatus* Bechstein, 1801; *varius* Desmarest, 1822; *vistulanus* Matschie, 1907.

COMMENTS: Reviewed by Heidecke (1986), who recognized eight subspecies, and Véron (1992*b*), who recognized six; neither supported Lavrov (1983) in treating *C. albicus* as a separate species. Subspecific boundaries and resultant synonymies are unclear and further obscured by historical translocations and reintroductions.

FAMILY HETEROMYIDAE
by James L. Patton

Family Heteromyidae Gray, 1868. Proc. Zool. Soc. Lond., 1868:201.
> COMMENTS: Currently divided into three subfamilies: Dipodomyinae containing the Recent genera *Dipodomys* and *Microdipodops*, Heteromyinae with *Heteromys* and *Liomys*, and Perognathinae comprised of *Chaetodipus* and *Perognathus*. Content defined by Wood (1935), Hafner and Hafner (1983), Wahlert (1985, 1993), and Korth (1994). Considered as a subfamily in the family Geomyidae (along with the extant pocket gophers, Geomyinae, and fossil Entotychinae) by McKenna and Bell (1997), following earlier suggestions of Shotwell (1967*b*) and Lindsay (1972), with the subfamilies recognized here lowered to the rank of tribes. The hierarchical rank of both heteromyids and geomyids is partly a matter of taxonomic philosophy but it is also a decision that stems from the phylogenetic placement of the fossil entotychines. I accept the evidence presented by Wahlert (1988) and Korth (1994) that entotychines are the sister to the modern pocket gophers, and thus follow their arguments for maintaining the traditional separation of the Heteromyidae as a family apart from the Geomyidae in the accounts here. Syntheses of the taxonomy, systematics, morphological diversity, cytogenetics, molecular genetics, and other aspects of the biology of living and fossil heteromyids can be found in Genoways and Brown (1993). Of particular note, Williams et al. (1993) provide a synopsis of the taxonomy of species and subspecies, with keys to all extant genera and species recognized at that time.

Subfamily Dipodomyinae Gervais, 1853. Ann. Sci. Nat., Paris, ser. 3, 20:245.
> COMMENTS: Wood (1935) allocated only the Recent genus *Dipodomys* to the subfamily; Hafner and Hafner (1983; see also Hafner, 1982) included *Microdipodops*, as did Wahlert (1985, 1993), Ryan (1989*a*), and Williams et al. (1993).

Dipodomys Gray, 1841. Ann. Mag. Nat. Hist., [ser. 1], 7:521.
> TYPE SPECIES: *Dipodomys phillipsii* Gray 1841.
> SYNONYMS: *Dipodops* Merriam, 1890; *Macrocolus* Wagner, 1846; *Perodipus* Fitzinger, 1867.
> COMMENTS: Interspecific relationships summarized by Setzer (1949), Lidicker (1960), Johnson and Selander (1971), Stock (1974), Best and Schnell (1974), Schnell et al. (1978), Best (1993*d*), Baumgardner and Kennedy (1994), and Carrasco (2000). Hall (1981:563-564), Best (1991), and Williams et al. (1993) provide keys to Recent species.

Dipodomys agilis Gambel, 1848. Proc. Acad. Nat. Sci. Philadelphia, 4:77.
> COMMON NAME: Agile Kangaroo Rat.
> TYPE LOCALITY: USA, California, Los Angeles Co., Los Angeles (see Grinnell, 1922).
> DISTRIBUTION: SW and SC California (USA).
> STATUS: IUCN – Lower Risk (lc).
> SYNONYMS: *wagneri* LeConte, 1853; ***perplexus*** Merriam, 1907; *fuscus* Boulware, 1943.
> COMMENTS: Specific distinctness of northern populations with 2n=62 (*agilis*) from southern forms with 2n=60 (*simulans*) documented by Best et al. (1986) and Sullivan and Best (1997*a*). Reviewed, in part, by Best (1978, 1983*a*) and Lackey (1967); see also Hall (1981:1179). Sullivan and Best (1997) and Williams et al. (1993) each listed two valid subspecies, with synonyms.

Dipodomys californicus Merriam, 1890. N. Am. Fauna, 4:49.
> COMMON NAME: California Kangaroo Rat.
> TYPE LOCALITY: USA, California, Mendocino Co., Ukiah.
> DISTRIBUTION: SC Oregon, NW Nevada (Stangl et al., 1999), and N California to north of San Francisco Bay (USA).
> STATUS: IUCN – Lower Risk (lc).
> SYNONYMS: *gabrielsoni* Goldman, 1925; *pallidulus* Bangs, 1899; *trinitatus* Kellogg, 1916; ***eximius*** Grinnell, 1919; ***saxatilis*** Grinnell and Linsdale, 1929.
> COMMENTS: Considered distinct from *heermanni* based on chromosomal (Fashing, 1973) and

biochemical data (Patton et al., 1976). Hall (1981:578) listed *californicus* as a subspecies of *heermanni*, without discussion of Patton et al. (1976). Reviewed by Kelt (1988*b*, Mammalian Species No. 324). Williams et al. (1993) regarded *eximius* Grinnell and saxatilis Grinnell and Linsdale as valid subspecies.

Dipodomys compactus True, 1889. Proc. U.S. Natl. Mus., 11:160.
    COMMON NAME: Gulf Coast Kangaroo Rat.
    TYPE LOCALITY: USA, Texas, Cameron Co., Padre Isl.
    DISTRIBUTION: Mainland, Padre and Mustang Isls of S Texas (USA) and barrier islands of
        N Tamaulipas (Mexico).
    STATUS: IUCN – Lower Risk (lc).
    SYNONYMS: *largus* Hall, 1951; *parvabullatus* Hall, 1951; *sennetti* J. A. Allen, 1891.
    COMMENTS: Considered distinct from *ordii* by Johnson and Selander (1971), Schmidly and
        Hendricks (1976), and Baumgardner and Schmidly (1981), who also documented
        sympatry between *compactus* and *ordii*. Hall (1981:565) provisionally retained
        compactus as a subspecies of *ordii*. Reviewed by Baumgardner (1991, Mammalian
        Species No. 369). Williams et al. (1993) regarded *sennetti* J. A. Allen as a valid
        subspecies.

Dipodomys deserti Stephens, 1887. Am. Nat., 21:42.
    COMMON NAME: Desert Kangaroo Rat.
    TYPE LOCALITY: USA, California, San Bernardino, Mohave River (3 to 4 mi [5-7 km] from,
        and opposite, Hesperia; see Hall, 1981:588).
    DISTRIBUTION: Deserts of E California, to S and W Nevada, SW Utah, W and SC Arizona
        (USA), NW Sonora and NE Baja California (Mexico).
    STATUS: IUCN – Lower Risk (lc).
    SYNONYMS: *helleri* Elliot, 1903; *aquilus* Nader, 1965; *arizonae* Huey, 1955; *sonoriensis*
        Goldman, 1923.
    COMMENTS: Revised by Nader (1978). Reviewed by Best et al. (1989, Mammalian Species No.
        339). Valid subspecies follow Williams et al. (1993).

Dipodomys elator Merriam, 1894. Proc. Biol. Soc. Wash., 9:109.
    COMMON NAME: Texas Kangaroo Rat.
    TYPE LOCALITY: USA, Texas, Clay Co., Henrietta.
    DISTRIBUTION: SW Oklahoma and NC Texas (USA).
    STATUS: IUCN – Vulnerable; Texas Parks and Wildlife Department – Threatened.
    COMMENTS: Probably no longer occurs in Oklahoma (Caire et al., 1989). Reviewed by Carter
        et al. (1985, Mammalian Species No. 232) and Williams et al. (1993). Sexual
        dimorphism and morphometric variation reviewed by Best (1987, 1993*d*) and
        molecular systematic relationships by Mantooth et al. (2000).

Dipodomys gravipes Huey, 1925. Proc. Biol. Soc. Wash., 38:83.
    COMMON NAME: San Quintin Kangaroo Rat.
    TYPE LOCALITY: Mexico, Baja California, 2 mi (3 km) W Santo Domingo Mission, 30°45′N,
        115°58′W.
    DISTRIBUTION: NW Baja California (Mexico).
    STATUS: IUCN – Endangered. Mexico – Endangered.
    COMMENTS: Considered distinct from *agilis* by Best (1978). Reviewed by Best (1983*b*), Best
        and Lackey (1985, Mammalian Species No. 236), Williams et al. (1993), and Patton and
        Alvarez-Castañeda (1999).

Dipodomys heermanni Le Conte, 1853. Proc. Acad. Nat. Sci. Philadelphia, 6:224.
    COMMON NAME: Heermann's Kangaroo Rat.
    TYPE LOCALITY: USA, California, Sierra Nevada; restricted to Calaveras River, Calaveras Co.
        by Grinnell (1922:47).
    DISTRIBUTION: Inner coastal ranges and western slopes of Sierra Nevada in C California (USA).
    STATUS: U.S. ESA and California Dept. of Fish and Game – Endangered as *D. h. morroensis*;
        U.S. ESA – Presumed Extinct as *D. h. berkleyensis* [sic]; IUCN – Critically Endangered as
        D. h. morroensis, Vulnerable as *D. h. berkeleyensis*, Lower Risk (nt) as *D. h. dixoni*,
        otherwise Lower Risk (lc).

SYNONYMS: *streatori* (Merriam, 1894); *arenae* Boulware, 1943; *berkeleyensis* Grinnell, 1919; *dixoni* (Grinnell, 1919); *goldmani* (Merriam, 1904); *jolonensis* Grinnell, 1919; *morroensis* (Merriam, 1907); *swarthi* (Grinnell, 1919); *tularensis* (Merriam, 1904).

COMMENTS: Revised by Grinnell (1922). Does not include *californicus*, see Patton et al. (1976), Williams et al. (1993), and comment under that species. Reviewed by Kelt (1988a, Mammalian Species No. 323). Williams et al. (1993) recognized all but *streatori* Merriam as valid subspecies.

Dipodomys ingens (Merriam, 1904). Proc. Biol. Soc. Wash., 17:141.

COMMON NAME: Giant Kangaroo Rat.

TYPE LOCALITY: USA, California, San Luis Obispo Co., Carrizo Plain, Painted Rock, 20 [=25.5] mi (32 km) SE of Simmler.

DISTRIBUTION: Western edge of Joaquin Valley, adjacent Carrizo and Elkhorn plains and upper Cuyama Valley of WC California (USA).

STATUS: U.S. ESA Endangered; IUCN – Critically Endangered; and California Dept. of Fish and Game – Endangered. Extirpated over much of its original range.

COMMENTS: Reviewed by Williams and Kilburn (1991, Mammalian Species No. 377) and Williams et al. (1993).

Dipodomys merriami Mearns, 1890. Bull. Am. Mus. Nat. Hist., 2:290.

COMMON NAME: Merriam's Kangaroo Rat.

TYPE LOCALITY: USA, Arizona, Maricopa Co., New River, between Phoenix and Prescott (see Lidicker, 1960:165).

DISTRIBUTION: NW Nevada and NE California to Texas (USA), south to Baja California Sur, N Sinaloa, and Mexican Plateau to San Luis Potosí (Mexico).

STATUS: U.S. ESA – Endangered as *D. m. parvus*. IUCN – Critically Endangered as *D. margaritae* and *D. insularis*; Data Deficient as *D. m. collinus* and *D. m. parvus*; otherwise Lower Risk (lc).

SYNONYMS: *kernensis* Merriam, 1907; *mortivallis* Elliot, 1904; *nevadensis* Merriam, 1894; *nitratus* Merriam, 1894; *regillus* Goldman, 1937; *similis* Rhoads, 1894; *simiolus* Rhoads, 1894; *ambiguus* Merriam, 1890; *annulus* Huey, 1951; *arenivagus* Elliot, 1904; *atronasus* Merriam, 1894; *brunensis* Huey, 1951; *collinus* Lidicker, 1960; *frenatus* Bole, 1936; *insularis* Merriam, 1907; *margaritae* Merriam, 1907; *mayensis* Goldman, 1928; *melanurus* Merriam, 1893; *llanoensis* Huey, 1951; *mitchelli* Mearns, 1897; *olivaceus* Swarth, 1929; *parvus* Rhoads, 1894; *platycephalus* Merriam, 1907; *semipallidus* Huey, 1927; *quintinensis* Huey, 1951; *trinidadensis* Huey, 1951; *vulcani* Benson, 1934.

COMMENTS: Revised by Lidicker (1960). Includes *insularis* Merriam, viewed as a separate species by Lidicker (1960; see also Huey, 1964, Hall, 1981, and Best and Thomas, 1991a, Mammalian Species No. 374), but as a subspecies of *merriami* by Best and Janecek (1992), Williams et al. (1993), and Patton and Alvarez-Castañeda (2000). Also includes *margaritae* Merriam (see Lidicker, 1960; Williams et al., 1993; Patton and Alvarez-Castañeda, 1999), which has been considered by others to be a distinct species (Best, 1992, Mammalian Species No. 400; Hall, 1981; Huey, 1964). The inclusion of both *insularis* and *margaritae* within *merriami* is supported by mitochondrial DNA sequence data (Riddle et al., 2000b), although these same data may eventually result in separation of populations of *merriami* (including *insularis* and *margaritae*) from the southern half of the Baja California Peninsula as a species separate from the remaining parts of the species' range.

Dipodomys microps (Merriam, 1904). Proc. Biol. Soc. Wash., 17:145.

COMMON NAME: Chisel-toothed Kangaroo Rat.

TYPE LOCALITY: USA, California, Inyo Co., Owens Valley, Lone Pine.

DISTRIBUTION: SE Oregon and SW Idaho, south through NW and SE California, Nevada, and W Utah, to NW Arizona (USA).

STATUS: IUCN – Vulnerable as *D. m. leucotis*, Data Deficient as *D. m. alfredi*, otherwise Lower Risk (lc).

SYNONYMS: *alfredi* Goldman, 1937; *aquilonius* Willett, 1935; *bonnevillei* Goldman, 1937; *celsus* Goldman, 1924; *woodburyi* Hardy, 1942; *centralis* Hall and Dale, 1939;

idahoensis Hall and Dale, 1939; *leucotis* Goldman, 1931; *levipes* (Merriam, 1904); *occidentalis* Hall and Dale, 1939; *preblei* (Goldman, 1921); *russeolus* Goldman, 1939; *subtenuis* Goldman, 1939.

COMMENTS: Revised by Hall and Dale (1939); reviewed by Csuti (1979) and Hayssen (1991, Mammalian Species No. 389). Williams et al. (1993) listed valid subspecies.

Dipodomys nelsoni Merriam, 1907. Proc. Biol. Soc. Wash., 20:75.
COMMON NAME: Nelson's Kangaroo Rat.
TYPE LOCALITY: Mexico, Coahuila, La Ventura.
DISTRIBUTION: Mexican Plateau from N Coahuila and S Chihuahua to N San Luis Potosi and S Nuevo Leon (Mexico).
STATUS: IUCN – Lower Risk (lc).
COMMENTS: Nader (1978) included *nelsoni* in *spectabilis*; Anderson (1972) and Matson (1980) presented evidence of specific distinctness. Williams et al. (1993) summarized the evidence of the status of *nelsoni* and concluded that it is a distinct species. Reviewed by Best (1988b, Mammalian Species No. 326).

Dipodomys nitratoides Merriam, 1894. Proc. Biol. Soc. Wash., 9:112.
COMMON NAME: San Joaquin Valley Kangaroo Rat.
TYPE LOCALITY: USA, California, Tulare Co., San Joaquin Valley, Tipton.
DISTRIBUTION: S San Joaquin Valley, WC California (USA).
STATUS: U.S. ESA – Endangered as *D. n. nitratoides* and *D. n. exilis*. IUCN – Critically Endangered as *D. n. exilis* and *D. n. nitratoides*, Lower Risk (nt) as *D. nitratoides* and as *D. n. brevinasus*. California Dept. of Fish and Game - Endangered as *D. n. exilis*; *D. n. nitratoides* is of Special Concern; and *D. n. brevinasus* is California Fully Protected.
SYNONYMS: *brevinasus* Grinnell, 1920; *exilis* Merriam, 1894.
COMMENTS: Revised by Grinnell (1922) and reviewed by Best (1991, Mammalian Species No. 381) and Williams et al. (1993). Considered closely related to *merriami*, but clearly distinct based on morphological (e.g., baculum, Best and Schnell, 1974), chromosomal (Stock, 1974), and molecular (Johnson and Selander, 1971; Patton et al., 1976) characters.

Dipodomys ordii Woodhouse, 1853. Proc. Acad. Nat. Sci. Philadelphia, 6:224.
COMMON NAME: Ord's Kangaroo Rat.
TYPE LOCALITY: USA, Texas, El Paso Co., El Paso.
DISTRIBUTION: SW Saskatchewan and SE Alberta (Canada) and SE Washington south through Great Plains and intermontane basins of W USA, to Mexican Plateau as far south as Hidalgo (Mexico).
STATUS: IUCN – Lower Risk (lc).
SYNONYMS: *celeripes* Durrant and Hall, 1939; *chapmani* Mearns, 1890; *cinderensis* Hardy, 1944; *cineraceus* Goldman, 1939; *columbianus* (Merriam, 1894); *cupidineus* Goldman, 1924; *durranti* Setzer, 1952 [replacement name for *fuscus* Setzer, 1949, which is preoccupied by *fuscus* Boulware, 1943, a subspecies of *D. agilis*]; *evexus* Goldman, 1933; *extractus* Setzer, 1949; *fetosus* Durrant and Hall, 1939; *fremonti* Durrant and Setzer, 1945; *inaquosus* Hall, 1941; *longipes* (Merriam, 1890); *cleomophila* Goldman, 1933; *luteolus* (Goldman, 1917); *marshalli* Goldman, 1937; *medius* Setzer, 1949; *monoensis* (Grinnell, 1919); *montanus* Baird, 1855; *nexilis* Goldman, 1933; *obscurus* (J. A. Allen, 1903); *attenuatus* Bryant, 1939; *idoneus* Setzer, 1949; *oklahomae* Trowbridge and Whitaker, 1940; *pallidus* Durrant and Setzer, 1945; *palmeri* (J. A. Allen, 1891); *panguitchensis* Hardy, 1942; *priscus* Hoffmeister, 1942; *pullus* Anderson, 1972; *richardsoni* (J. A. Allen, 1891); *sanrafaeli* Durrant and Setzer, 1945; *terrosus* Hoffmeister, 1942; *uintensis* Durrant and Setzer, 1945; *utahensis* (Merriam, 1904).
COMMENTS: Revised by Setzer (1949) and reviewed by Garrison and Best (1990, Mammalian Species No. 353); subspecies follow Williams et al. (1993). Does not include *compactus*, see Schmidly and Hendricks (1976), Baumgardner and Schmidly (1981), and comment under that species. Williams et al. (1993) provide a list of what they consider as valid subspecies.

Dipodomys panamintinus (Merriam, 1894). Proc. Biol. Soc. Wash., 9:114.
  COMMON NAME: Panamint Kangaroo Rat.
  TYPE LOCALITY: USA, California, Inyo Co., Panamint Mtns, head of Willow Creek.
  DISTRIBUTION: Deserts of E California and W Nevada (USA).
  STATUS: IUCN – Lower Risk (lc).
  SYNONYMS: *argusensis* Huey, 1945; *caudatus* Hall, 1946; *leucogenys* (Grinnell, 1919);
    mohavensis (Grinnell, 1918).
  COMMENTS: Reviewed by Intress and Best (1990, Mammalian Species No. 354) and Williams
    et al. (1993).

Dipodomys phillipsii Gray, 1841. Ann. Mag. Nat. Hist., [ser. 1], 7:522.
  COMMON NAME: Phillips's Kangaroo Rat.
  TYPE LOCALITY: Mexico, Hidalgo, Valley of Mexico, near Real del Monte (= Mineral
    de Monte).
  DISTRIBUTION: C Durango south to N Oaxaca (Mexico).
  STATUS: IUCN – Lower Risk (nt).
  SYNONYMS: *oaxacae* Hooper, 1947; *ornatus* Merriam, 1894; *perotensis* Merriam, 1894.
  COMMENTS: Systematics reviewed by Genoways and Jones (1971); biology reviewed by Jones
    and Genoways (1975a, Mammalian Species No. 51).

Dipodomys simulans (Merriam, 1904). Proc. Biol. Soc. Wash., 17:144.
  COMMON NAME: Dulzura Kangaroo Rat.
  TYPE LOCALITY: USA, California, San Diego Co., Dulzura.
  DISTRIBUTION: Coastal southern California (USA) south to southern Baja California Sur
    (Mexico).
  SYNONYMS: *australis* Huey, 1951; *antiquarius* Huey, 1962; *cabezonae* (Merriam, 1904);
    eremoecus Huey, 1951; *latimaxilaris* Huey, 1925; *martirensis* Huey, 1927; *paralius* Huey,
    1951; *pedionomus* Huey, 1951; *plectilis* Huey, 1951; *peninsularis* (Merriam, 1907).
  COMMENTS: Systematics reviewed by Sullivan and Best (1997a), who documented the
    specific distinctness of *simulans* from *D. agilis* (see account of that species). Several
    included taxa (e.g., *peninsularis, antiquarius, paralius*) listed by Hall (1981) as distinct
    species. Two subspecies (*simulans* and *peninsularis*) recognized as valid by Sullivan and
    Best (1997a) and Williams et al. (1993), but these authors disagree in their allocation of
    synonyms to each; the account here follows Sullivan and Best (1997a). Phenotypic
    variation reviewed by Sullivan and Best (1997b).

Dipodomys spectabilis Merriam, 1890. N. Am. Fauna, 4:46.
  COMMON NAME: Banner-tailed Kangaroo Rat.
  TYPE LOCALITY: USA, Arizona, Cochise Co., Dos Cabezos [=Dos Cabezas].
  DISTRIBUTION: SC Arizona, New Mexico, W Texas (USA) south to N Sonora, Chihuahua and
    San Luis Potosí (Mexico).
  STATUS: IUCN – Lower Risk (lc).
  SYNONYMS: *baileyi* Goldman, 1923; *clarencei* Goldman, 1933; *cratodon* Merriam, 1907;
    intermedius Nader, 1965; *perblandus* Goldman, 1933; *zygomaticus* Goldman, 1923.
  COMMENTS: Revised by Nader (1978) who included *nelsoni*; but also see Anderson (1972),
    Matson (1980), Hall (1981:581), and Williams et al. (1993) who presented evidence of
    specific distinctness. Reviewed by Best (1988a, Mammalian Species No. 311).
    Subspecies listed by Best (1988a) and Williams et al. (1993).

Dipodomys stephensi (Merriam, 1907). Proc. Biol. Soc. Wash., 20:78.
  COMMON NAME: Stephens's Kangaroo Rat.
  TYPE LOCALITY: USA, California, Riverside Co., San Jacinto Valley, west of Winchester.
  DISTRIBUTION: Riverside, San Bernardino, and San Diego Cos. of S California (USA).
  STATUS: U.S. ESA – Endangered; IUCN – Lower Risk (conservation dependent); California
    Dept. of Fish and Game – Threatened.
  SYNONYMS: *cascus* Huey, 1962.
  COMMENTS: Relationships to other species of the *heermanni* group studied by Lackey (1967).
    Reviewed by Bleich (1977, Mammalian Species No. 73).

Dipodomys venustus (Merriam, 1904). Proc. Biol. Soc. Wash., 17:142.
  COMMON NAME: Narrow-faced Kangaroo Rat.
  TYPE LOCALITY: USA, California, Santa Cruz Co., Santa Cruz.
  DISTRIBUTION: Outer coast ranges from S San Francisco Bay to Estero Bay and Gabilan Range
      of San Benito and Monterey counties, WC California (USA).
  STATUS: IUCN – Lower Risk (lc).
  SYNONYMS: *elephantinus* (Grinnell, 1919); *sanctiluciae* Grinnell, 1919.
  COMMENTS: Revised by Grinnell (1922), who considered *elephantinus* a separate species.
      Best et al. (1996) concluded that *elephantinus* was only a subspecies of *venustus*, based
      on both molecular and morphological comparisons; also see Stock (1974), Schnell
      et al. (1978), and Hall (1981:574). *Dipodomys venustus,* exclusive of *elephantinus,* was
      reviewed by Best (1992, Mammalian Species No. 403); *elephantinus* as a distinct species
      was reviewed by Best (1986, Mammalian Species No. 255).

Microdipodops Merriam, 1891. N. Am. Fauna, 5:115.
  TYPE SPECIES: *Microdipodops megacephalus* Merriam, 1891.
  COMMENTS: Revised by Hall (1941); also see Hafner et al. (1979). Wood (1935), Hafner (1978),
      and Hall (1981) considered *Microdipodops* a member of the subfamily Perognathinae.
      Hafner (1982), Hafner and Hafner (1983), Wahlert (1985), and Ryan (1989*a*) summarized
      evidence for referring the genus to the subfamily Dipodomyinae. Hall (1981:560) and
      Williams et al. (1993:92) provide keys to species.

Microdipodops megacephalus Merriam, 1891. N. Am. Fauna, 5:116.
  COMMON NAME: Dark Kangaroo Mouse.
  TYPE LOCALITY: USA, Nevada, Elko Co., Halleck.
  DISTRIBUTION: SE Oregon, S Idaho, NE and EC California, N and C Nevada, and WC Utah
      (USA).
  STATUS: IUCN – Vulnerable as *M. m. atrirelictus*, Data Deficient as *M. m. nexus*, otherwise
      Lower Risk (lc).
  SYNONYMS: *albiventer* Hall and Durrant, 1937; *ambiguus* Hall, 1941; *atrirelictus* Hafner,
      1985; *californicus* Merriam, 1901; *leucotis* Hall and Durrant, 1941; *medius* Hall, 1941;
      nasutus Hall, 1941; *nexus* Hall, 1941; *oregonus* Merriam, 1901; *paululus* Hall and
      Durrant, 1941; *polionotus* Grinnell, 1914; *sabulonis* Hall, 1941.
  COMMENTS: Reviewed by O'Farrell and Blaustein (1974*a*, Mammalian Species No. 46).
      Hafner et al. (1979) discounted the suggestion by Hall (1941:380-382) of hybridization
      between *megacephalus* and *pallidus*. Also, Hall's suggestion (1981:560) that *leucotis* may
      warrant specific status is not supported (see Hafner and Hafner, 1983). Subspecies
      listed by Williams et al. (1993).

Microdipodops pallidus Merriam, 1901. Proc. Biol. Soc. Wash., 14:127.
  COMMON NAME: Pale Kangaroo Mouse.
  TYPE LOCALITY: USA, Nevada, Churchill Co., Mountain Well.
  DISTRIBUTION: EC California, W and SC Nevada (USA).
  STATUS: IUCN – Vulnerable as *M. p. restrictus*, otherwise Lower Risk (lc).
  SYNONYMS: *dickeyi* Goldman, 1927; *lucidus* Goldman, 1926; *ammophilus* Hall, 1941; *purus*
      Hall, 1941; *restrictus* Hafner, 1985; *ruficollaris* Hall, 1941.
  COMMENTS: Reviewed by O'Farrell and Blaustein (1974*b*, Mammalian Species No. 47).
      Subspecies follow Williams et al. (1993).

Subfamily Heteromyinae Gray, 1868. Proc. Zool. Soc. Lond., 1868:201.
  COMMENTS: Contains the Recent genera *Heteromys* and *Liomys*, following Wood (1935),
      Wahlert (1985, 1993), and Ryan (1989*a*). However, a review of the generic limits is
      warranted because biochemical data (Rogers, 1990) suggest that *Heteromys* as currently
      defined is paraphyletic relative to *Liomys*. Williams et al. (1993:100) provided a key to
      genera.

Heteromys Desmarest, 1817. Nouv. Dict. Hist. Nat., Nouv. ed., 14:181.
  TYPE SPECIES: *Mus anomalus* Thompson, 1815.

SYNONYMS: *Xylomys* Merriam, 1902.

COMMENTS: Includes *Xylomys* as a subgenus for *H. nelsoni* (see also Hall, 1981, and Williams et al., 1993), although the validity of subgenera is questionable (R. P. Anderson, pers. comm.). Revised by Goldman (1911), with systematics currently under review by R. P. Anderson (pers. comm.). Rogers and Schmidly (1982) partially revised the *desmarestianus* group. Rogers (1989, 1990) discussed phylogenetic relationships among species, and remarked on at least one undescribed species from Costa Rica, as did Handley (1976) from Venezuela. Distribution and systematics of Colombian species reviewed by Anderson (1999 [2000]). Key to subgenera and species given by Williams et al. (1993:101); Schmidt et al. (1989) also provide a key to species.

Heteromys anomalus (Thompson, 1815). Trans. Linn. Soc. Lond., 11:161.

COMMON NAME: Caribbean Spiny Pocket Mouse.

TYPE LOCALITY: Trinidad and Tobago, Trinidad.

DISTRIBUTION: N Colombia, including Magdalena Valley, east to N Venezuela, Trinidad, Tobago, and Margarita Isl. It does not enter Panama (contra Rogers, 1990; Méndez, 1993; Williams et al., 1993; Nowak, 1999; see Anderson, 1999 [2000]).

STATUS: IUCN – Lower Risk (lc).

SYNONYMS: *bicolor* (Gray, 1868); *thompsonii* Lesson, 1827; *melanoleucus* Gray, 1868; **brachialis** Osgood, 1912; **hershkovitzi** Hernández-Camacho, 1956; **jesupi** J. A. Allen, 1899.

COMMENTS: Subgenus *Heteromys*. Subspecies listed by Williams et al. (1993) and distribution mapped by Eisenberg (1989) and Anderson (1999 [2000]). As currently understood, *anomalus* is likely composite. Anderson (2002) recently described *H. oasicus* from NW Venezuela, a taxon that Handley (1976) regarded as distinct from *anomalus*. Populations from the Cordillera de la Costa of Venezuela, above 1000 m, also represent an undescribed species (R. P. Anderson, pers. comm.).

Heteromys australis Thomas, 1901. Ann. Mag. Nat. Hist., ser. 7, 7:194.

COMMON NAME: Southern Spiny Pocket Mouse.

TYPE LOCALITY: Ecuador, Esmeraldas Prov., Cachabi River, San Javier, below Cachabi.

DISTRIBUTION: E Panama south to SW Colombia and NW Ecuador, both slopes of the Cordillera Occidental and Cordillera Central and western slope of the Cordillera Oriental of Colombia. Disjunct population present in Cordillera de Mérida in Venezuela (Anderson and Soriano, 1999; Anderson, 1999 [2000]).

STATUS: IUCN – Lower Risk (lc).

SYNONYMS: *lomitensis* J. A. Allen, 1912; **conscius** Goldman, 1913, **pacificus** Pearson, 1939.

COMMENTS: Subgenus *Heteromys*. Subspecies listed by Williams et al. (1993) and distribution mapped by Anderson (1999 [2000]).

Heteromys desmarestianus Gray, 1868. Proc. Zool. Soc. Lond., 1868:204.

COMMON NAME: Desmarest's Spiny Pocket Mouse.

TYPE LOCALITY: Guatemala, Coban.

DISTRIBUTION: SE Tobasco (Mexico) south to NW Colombia.

STATUS: IUCN – Lower Risk (lc) as *H. desmarestianus*; Lower Risk (nt) as *H. goldmani*.

SYNONYMS: *griseus* Merriam, 1902; *lepturus* Merriam, 1902; *longicaudatus* Gray, 1868; *nigricaudatus* Goodwin, 1956; *psakastus* Dickey, 1928; **chiriquensis** Enders, 1938; **crassirostris** Goldman, 1912; **fuscatus** J. A. Allen, 1908; **goldmani** Merriam, 1902; **panamensis** Goldman, 1912; **planifrons** Goldman, 1937; **repens** Bangs, 1902; **subaffinis** Goldman, 1937; **temporalis** Goldman, 1911; **underwoodi** Goodwin, 1943; **zonalis** Goldman, 1912.

COMMENTS: Subgenus *Heteromys*. Goodwin (1969) and Rogers and Schmidly (1982) provided partial revisions. I follow Rogers (1990) and Williams et al. (1993) and include *goldmani* (contra Patton, 1993b). Subspecies listed by Williams et al. (1993). Given the extent of chromosomal and allozymic diversity (Rogers, 1989, 1990), *desmarestianus* is likely to be composite, with several described taxa to be elevated to species status and additional species to be described (R. P. Anderson, pers. comm.).

Heteromys gaumeri J. A. Allen and Chapman, 1897. Bull. Am. Mus. Nat. Hist., 9:9.
COMMON NAME: Gaumer's Spiny Pocket Mouse.
TYPE LOCALITY: Mexico, Yucatan, Chichén-Itza.
DISTRIBUTION: Endemic to Yucatan Peninsula (Mexico), N Belize, and N Guatemala.
STATUS: IUCN – Lower Risk (lc).
COMMENTS: Subgenus *Heteromys*. Reviewed by Engstrom et al. (1987*a*) and Schmidt et al. (1989, Mammalian Species No. 345). Both Engstrom et al. (1987*a*) and Rogers (1990) suggested that this taxon might deserve separate subgeneric status, as yet unnamed.

Heteromys nelsoni Merriam, 1902. Proc. Biol. Soc. Wash., 15:43.
COMMON NAME: Nelson's Spiny Pocket Mouse.
TYPE LOCALITY: Mexico, Chiapas, Pinabete, 8,200 ft. (2,499 m).
DISTRIBUTION: Known only from S Chiapas (Mexico) and W Guatemala.
STATUS: IUCN – Critically Endangered.
COMMENTS: Type species of subgenus *Xylomys* Merriam, which is of debatable validity (R. P. Anderson, pers. comm.). Reviewed by Rogers and Rogers (1992*b*).

Heteromys oasicus Anderson, 2003. Am. Mus. Novit., 3396:9.
COMMON NAME: Paraguaná Spiny Pocket Mouse.
TYPE LOCALITY: Venezuela, Estado Falcón, 42 km N, 32 km W of Coro, Cerro Santa Ana, 550 m.
DISTRIBUTION: Known only from Cerro Santa Ana and the Fila de Monte Cano on the Penínsulsa de Paraguaná, Estado Falcón, Venezuela.
COMMENTS: Subgenus *Heteromys*. This is the species recognized by Handley (1976) as undescribed at that time; reported as *H. anomalus* by Bisbal-E. (1990).

Heteromys oresterus Harris, 1932. Occas. Pap. Mus. Zool., Univ. Michigan, 248:4.
COMMON NAME: Mountain Spiny Pocket Mouse.
TYPE LOCALITY: Costa Rica, Cordillera de Talamanca, El Copey de Dota, 6,000 ft. (1,829 m).
DISTRIBUTION: Talamanca Range of Costa Rica.
STATUS: IUCN – Lower Risk (nt).
COMMENTS: Hall (1981) placed *oresterus* in the subgenus *Xylomys* but Rogers (1989, 1990) presented evidence that *oresterus* was not closely related to *nelsoni*, the type species of *Xylomys*. Reviewed by Rogers and Rogers (1992*a*, Mammalian Species No. 396).

Heteromys teleus Anderson and Jarrín-V., 2002. Am. Mus. Novit., 3382:6.
COMMON NAME: Ecuadoran Spiny Pocket Mouse.
TYPE LOCALITY: Ecuador, Prov. Guayas, Cerro Manglar Alto, western slope, 1,500 ft. (457 m).
DISTRIBUTION: Pacific lowlands of Ecuador.
COMMENTS: Subgenus *Heteromys*.

Liomys Merriam, 1902. Proc. Biol. Soc. Wash., 15:44.
TYPE SPECIES: *Heteromys alleni* Coues, 1881 (= *L. irroratus alleni*, see Genoways, 1973).
COMMENTS: Revised by Genoways (1973) with current taxonomy reviewed by Williams et al. (1993). Rogers (1989, 1990) discussed phylogenetic relationships among species and relative to *Heteromys*. Genoways (1973:44-45), Dowler and Genoways (1978), and Williams et al. (1993) provide keys to species. Cervantes et al. (1999*b*) summarized karyotypic variability and relationships among Mexican species.

Liomys adspersus (Peters, 1874). Monatsb. K. Preuss. Akad. Wiss., Berlin, p. 357.
COMMON NAME: Panamanian Spiny Pocket Mouse.
TYPE LOCALITY: Panama, City of Panama (restricted by Goldman, 1920).
DISTRIBUTION: C Panama, principally Pacific versant.
STATUS: IUCN – Lower Risk (nt).

Liomys irroratus (Gray, 1868). Proc. Zool. Soc. Lond., 1868:205.
COMMON NAME: Mexican Spiny Pocket Mouse.
TYPE LOCALITY: Mexico, Oaxaca, Oaxaca (restricted by Genoways, 1973:111).
DISTRIBUTION: S Texas (USA), and SC Chihuahua to Oaxaca (Mexico).
STATUS: IUCN – Lower Risk (lc).

SYNONYMS: *albolimbatus* (Gray, 1868); *yautepecus* Goodwin, 1956; **alleni** (Coues, 1881); *acutus* Hall and Villa-R., 1948; *canus* Merriam, 1902; *pullus* Hooper, 1947; **bulleri** (Thomas, 1893); **guerrerensis** (Goldman, 1911); **jaliscensis** (J. A. Allen, 1906); **texensis** Merriam, 1902; *pretiosus* Goldman, 1911; **torridus** Merriam, 1902; *exiguus* (Elliot, 1903); *minor* Merriam, 1902.

COMMENTS: Reviewed by Dowler and Genoways (1978, Mammalian Species No. 82). Currently recognized subspecies delineated by Williams et al. (1993).

Liomys pictus (Thomas, 1893). Ann. Mag. Nat. Hist., ser. 6, 12:233.

COMMON NAME: Painted Spiny Pocket Mouse.

TYPE LOCALITY: Mexico, Jalisco, San Sebastian, 4,300 ft. (1,311 m).

DISTRIBUTION: West coast of Mexico from Sonora to Chiapas, and east coast in Veracruz south to extreme NW Guatemala.

STATUS: IUCN – Lower Risk (lc).

SYNONYMS: *isthmius* Merriam, 1902; *obscurus* Merriam, 1902; *orbitalis* Merriam, 1902; *paralius* (Elliot, 1903); *phaeura* Merriam, 1902; *pinetorum* Goodwin, 1956; *rostratus* Merriam, 1902; *veraecrucis* Merriam, 1902; **annectens** (Merriam, 1902); **hispidus** (J. A. Allen, 1897); *escuinapae* (J. A. Allen, 1906); *sonorana* Merriam, 1902; **plantinarensis** Merriam, 1902; *parviceps* Goldman, 1904.

COMMENTS: Reviewed by McGhee and Genoways (1978, Mammalian Species No. 83). Genic data suggest that *pictus* as currently comprised is paraphyletic; additional species are likely to be recognized with further analyses (see Morales and Engstrom, 1989, and Rogers, 1990). Williams et al. (1993) reviewed subspecific taxonomy.

Liomys salvini (Thomas, 1893). Ann. Mag. Nat. Hist., ser. 6, 11:331.

COMMON NAME: Salvin's Spiny Pocket Mouse.

TYPE LOCALITY: Guatemala, Sacatépequez, Dueñas.

DISTRIBUTION: E Oaxaca (Mexico) south to C Costa Rica.

STATUS: IUCN – Lower Risk (lc).

SYNONYMS: *anthonyi* Goodwin, 1932; *aterrimus* Goodwin, 1938; *heterothrix* Merriam, 1902; *nigrescens* (Thomas, 1893); **crispus** Merriam, 1902; *setosus* Merriam, 1902; **vulcani** (J. A. Allen, 1908).

COMMENTS: Reviewed by Carter and Genoways (1978, Mammalian Species No. 84). Williams et al. (1993) delimited subspecies.

Liomys spectabilis Genoways, 1971. Occas. Papers Mus. Nat. Hist., Univ. Kansas, 5:1.

COMMON NAME: Jaliscan Spiny Pocket Mouse.

TYPE LOCALITY: Mexico, Jalisco, 2.2 mi (3.5 km) NE Contla, 3,850 ft. (1,173 m).

DISTRIBUTION: SE Jalisco (Mexico).

STATUS: IUCN – Lower Risk (nt).

Subfamily Perognathinae Coues, 1875. Proc. Acad. Nat. Sci. Philadelphia, 27:277.

COMMENTS: Subfamily name emended by Wood (1935); originally given by Coues as Perognathidinae. Wood (1935:89), Hafner (1978), and Hall (1981) included *Microdipodops* in the subfamily, a course not followed by Hafner and Hafner (1983), Wahlert (1985, 1988), Ryan (1989a), and Williams et al. (1993). Williams et al. (1993:121) provide a key to the genera.

Chaetodipus Merriam, 1889. N. Am. Fauna, 1:5.

TYPE SPECIES: *Perognathus spinatus* Merriam, 1889.

SYNONYMS: *Burtognathus* Hoffmeister, 1986.

COMMENTS: Species revised by Merriam (1889) and Osgood (1900) under the generic name *Perognathus*; both authors considered *Chaetodipus* as a valid subgenus (see also Hall, 1981). Raised to generic status by Hafner and Hafner (1983), an action followed by most subsequent authors. Includes *Burtognathus* (see Hoffmeister, 1986), defined as a subgenus to contain the single species *C. hispidus*; see also Williams et al. (1993). Chromosomal and biochemical systematics summarized by Patton and Rogers (1993a, b). Best (1993f) and Williams et al. (1993:122-124) provide keys to Recent species.

Chaetodipus arenarius (Merriam, 1894). Proc. California Acad. Sci., ser. 2, 4:461.
COMMON NAME: Little Desert Pocket Mouse.
TYPE LOCALITY: Mexico, Baja California Sur, San Jorge, near Comondú.
DISTRIBUTION: Baja California Peninsula (Mexico).
STATUS: IUCN – Lower Risk (lc).
SYNONYMS: *albescens* (Huey, 1926); *albulus* (Nelson and Goldman, 1923); *ambiguus* (Nelson and Goldman, 1929); *ammophilus* (Osgood, 1907); *helleri* (Elliot, 1903); *mexicalis* (Huey, 1939); *paralios* (Huey, 1964); *sabulosus* (Huey, 1964); *siccus* (Osgood, 1907); *sublucidus* (Nelson and Goldman, 1929).
COMMENTS: Reviewed by Huey (1964) and Lackey (1991*a*, Mammalian Species No. 384). Subspecies reviewed by Williams et al. (1993) and Patton and Alvarez-Castañeda (1999).

Chaetodipus artus (Osgood, 1900). N. Am. Fauna, 18:55.
COMMON NAME: Narrow-skulled Pocket Mouse.
TYPE LOCALITY: Mexico, Chihuahua, Batopilas.
DISTRIBUTION: S Sonora, SW Chihuahua, W Durango, Sinaloa, and N Nayarit (Mexico).
STATUS: IUCN – Lower Risk (lc).
COMMENTS: Revised by Anderson (1964). Taxonomy and distribution reviewed by Williams et al. (1993) and Patton and Alvarez-Castañeda (1999); biology reviewed by Best and Lackey (1992*a*, Mammalian Species No. 418).

Chaetodipus baileyi (Merriam, 1894). Proc. Acad. Nat. Sci. Philadelphia, 46:262.
COMMON NAME: Bailey's Pocket Mouse.
TYPE LOCALITY: Mexico, Sonora, Magdalena.
DISTRIBUTION: S Arizona, SW New Mexico (USA), south to N Sinaloa (Mexico).
STATUS: IUCN – Lower Risk (lc).
SYNONYMS: *domensis* Goldman, 1928; *insularis* Townsend, 1912.
COMMENTS: Reviewed by Paulson (1988*a*, Mammalian Species No. 297). Species now restricted to populations and subspecies from east of the Colorado River in Arizona, SW New Mexico, Sonora, and N Sinaloa; those from west of the Colorado River in California and Baja California are now regarded as *C. rudinoris* (see account below and Riddle et al., 2000*a*).

Chaetodipus californicus (Merriam, 1889). N. Am. Fauna, 1:26.
COMMON NAME: California Pocket Mouse.
TYPE LOCALITY: USA, California, Alameda Co., Berkeley.
DISTRIBUTION: C California (USA) to N Baja California (Mexico).
STATUS: IUCN – Data Deficient as *C. c. femoralis*, otherwise Lower Risk (lc);.
SYNONYMS: *armatus* (Merriam, 1889); *bensoni* (von Bloeker, 1938); *bernardinus* (Benson, 1930); *dispar* (Osgood, 1900); *femoralis* (J. A. Allen, 1891); *marinensis* (von Bloeker, 1938); *mesopolius* (Elliot, 1903); *ochrus* (Osgood, 1904).
COMMENTS: Subspecies listed by Hall (1981) and Williams et al. (1993).

Chaetodipus dalquesti (Roth, 1976). J. Mammal., 57:562.
COMMON NAME: Dalquest's Pocket Mouse.
TYPE LOCALITY: Mexico, Baja California Sur, 4 mi SE Migriño.
DISTRIBUTION: Cape Region of Baja California Sur (Mexico).
STATUS: IUCN – Lower Risk (lc) as included in *C. arenarius*.
COMMENTS: Considered a subspecies of *C. arenarius* by Patton (1993*b*), Williams et al. (1993), and Patton and Alvarez-Castañeda (1999), with which it shares an identical karyotype (Hafner and Hafner, 1983). However, both mitochondrial DNA (Riddle et al., 2000*b*) and morphology (D. J. Hafner, pers. comm.) suggest separate species status, although a complete analysis of the relationships of *dalquesti* to *arenarius* has not as yet been made.

Chaetodipus eremicus (Mearns, 1898). Bull. Amer. Mus. Nat. Hist., 10:300.
COMMON NAME: Chihuahuan Pocket Mouse.
TYPE LOCALITY: USA, Texas, El Paso [now in Hudspeth] Co., Fort Hancock.
DISTRIBUTION: Chihuahuan Desert from S New Mexico and Trans-Pecos Texas (USA)

through Chihuahua, Coahuila, Durango, Zacatecas, Nuevo Leon, and San Luis Potosí (Mexico).

STATUS: IUCN – Lower Risk (lc) as included in *C. penicillatus*.

SYNONYMS: *atrodorsalis* (Dalquest, 1951).

COMMENTS: Hoffmeister and Lee (1967), Hall (1981), Patton (1993*b*), and Williams et al. (1993) considered both *eremicus* and *atrodorsalis* subspecies of *C. penicillatus*. Lee et al. (1996) elevated *eremicus* to species status, with *atrodorsalis* a subspecies, based on mito-chondrial DNA sequence divergence and the previously recognized sharply divergent karyotypes (Patton, 1969*a*) and allozyme differences (J. L. Patton et al., 1981).

Chaetodipus fallax (Merriam, 1889). N. Am. Fauna, 1:19.

COMMON NAME: San Diego Pocket Mouse.

TYPE LOCALITY: USA, California, San Bernardino Co., Reche Canyon (3 mi [5 km] SE Colton), 1,250 ft. (381 m).

DISTRIBUTION: SW California (USA) to W Baja California (Mexico).

STATUS: IUCN – Data Deficient as *C. h. fallax* and *C. h. pallidus*, otherwise Lower Risk (lc).

SYNONYMS: *anthonyi* (Osgood, 1900); *inopinus* (Nelson and Goldman, 1929); *majusculus* (Huey, 1960); *pallidus* (Mearns, 1901); *xerotrophicus* (Huey, 1960).

COMMENTS: Reviewed by Huey (1960; 1964) and Lackey (1996, Mammalian Species No. 517). Includes *anthonyi*, considered a full species by Hall (1981) but a subspecies by Williams et al. (1993). Subspecies reviewed by Williams et al. (1993).

Chaetodipus formosus (Merriam, 1889). N. Am. Fauna, 1:17.

COMMON NAME: Long-tailed Pocket Mouse.

TYPE LOCALITY: USA, Utah, Washington Co., St. George.

DISTRIBUTION: W Utah, Nevada, E California, and NW Arizona (USA), and E coast of Baja California to Bahía Concepcion (Baja California Sur, Mexico).

STATUS: IUCN – Lower Risk (lc).

SYNONYMS: *domisaxensis* (Cockrum, 1956); *melanocaudus* (Cockrum, 1956); *cinerascens* (Nelson and Goldman, 1929); *incolatus* (Hall, 1941); *infolatus* (Huey, 1954); *melanurus* (Hall, 1941); *mesembrinus* (Elliot, 1904); *mohavensis* (Huey, 1938).

COMMENTS: Reviewed by Huey (1964). Included in *Perognathus* by Hall (1981:542) and earlier workers, but allocated to *Chaetodipus* by J. L. Patton et al. (1981) and Hafner and Hafner (1983). Subspecies reviewed by Williams et al. (1993).

Chaetodipus goldmani (Osgood, 1900). N. Am. Fauna, 18:54.

COMMON NAME: Goldman's Pocket Mouse.

TYPE LOCALITY: Mexico, Sinaloa, Sinaloa.

DISTRIBUTION: NE to S Sonora, SW Chihuahua, and N Sinaloa (Mexico); see Straney and Patton (1980).

STATUS: IUCN – Lower Risk (lc).

COMMENTS: Revised by Anderson (1964). Chromosome races described by Patton (1969*b*). Reviewed by Lackey and Best (1992, Mammalian Species No. 419).

Chaetodipus hispidus (Baird, 1858). Mammalia *in* Repts. U.S. Expl. Surv., 8(1):421.

COMMON NAME: Hispid Pocket Mouse.

TYPE LOCALITY: Mexico, Tamaulipas, Charco Escondido.

DISTRIBUTION: Great Plains from S North Dakota to SE Arizona and W Louisiana (USA), south to Tamaulipas and Hidalgo (Mexico).

STATUS: IUCN – Lower Risk (lc).

SYNONYMS: *paradoxus* (Merriam, 1889); *conditi* (J. A. Allen, 1894); *latirostris* (Rhoads, 1894); *spilotus* (Merriam, 1889); *maximus* (Elliot, 1904); *zacatecae* (Osgood, 1900).

COMMENTS: Revised by Glass (1947); subspecies listed by Hall (1981) and Williams et al. (1993). Reviewed by Paulson (1988*b*, Mammalian Species No. 320). Type species of monotypic subgenus *Burtognathus* Hoffmeister.

Chaetodipus intermedius (Merriam, 1889). N. Am. Fauna, 1:18.

COMMON NAME: Rock Pocket Mouse.

TYPE LOCALITY: USA, Arizona, Mohave Co., Mud Spring.

DISTRIBUTION: SC Utah and Arizona to W Texas (USA), south to C Sonora and C Chihuahua (Mexico).

STATUS: IUCN – Lower Risk (lc).

SYNONYMS: *nigrimontis* (Blossom, 1933); *obscurus* (Merriam, 1889); *ater* (Dice, 1929); *beardi* (Weckerly, Gennaro, and Best, 1988); *crinitus* (Benson, 1934); *lithophilus* (Huey, 1937); *minimus* (Burt, 1932); *phasma* (Goldman, 1918); *pinacate* (Blossom, 1933); *rupestris* (Benson, 1932); *umbrosus* (Benson, 1934).

COMMENTS: Subspecies listed by Hoffmeister (1974), Hall (1981), and Williams et al. (1993).

Chaetodipus lineatus (Dalquest, 1951). J. Washington Acad. Sci, 41:362.
COMMON NAME: Lined Pocket Mouse.
TYPE LOCALITY: Mexico, San Luis Potosí, 1 km south of Arriaga.
DISTRIBUTION: San Luis Potosí and SE Zacatecas (Mexico).
COMMENTS: Possibly conspecific with *nelsoni* (Williams et al., 1993). Reviewed by Best (1993*f*, Mammalian Species No. 451).

Chaetodipus nelsoni (Merriam, 1894). Proc. Acad. Nat. Sci. Philadelphia, 46:266.
COMMON NAME: Nelson's Pocket Mouse.
TYPE LOCALITY: Mexico, San Luis Potosí, Hacienda La Parada, about 25 mi (40 km) NW Ciudad San Luis Potosí.
DISTRIBUTION: Chihuahuan desert plateau from SE New Mexico and W Texas (USA) to Jalisco and San Luis Potosí (Mexico).
STATUS: IUCN – Lower Risk (lc).
SYNONYMS: *canescens* (Merriam, 1894); *collis* (Blair, 1938); *popei* (Blair, 1938).
COMMENTS: Subspecies listed by Hall (1981) and Williams et al. (1993); may include *lineatus* (see above). Reviewed by Best (1994*b*, Mammalian Species No. 484). Two karyotypic races known (Patton, 1970; Lee, 1990), which may prove to be distinct species.

Chaetodipus penicillatus (Woodhouse, 1852). Proc. Acad. Nat. Sci. Philadelphia, 6:200.
COMMON NAME: Desert Pocket Mouse.
TYPE LOCALITY: USA, San Francisco Mountains, New Mexico (fixed to Arizona, Yuma Co., 1 mi [1.6 km] SW Parker by Hoffmeister and Lee, 1967).
DISTRIBUTION: SE California and S Nevada to S Arizona and SW New Mexico (USA) to NE Baja California and Sonora (Mexico).
STATUS: IUCN – Lower Risk (lc).
SYNONYMS: *angustirostris* (Osgood, 1900); *pricei* (J. A. Allen, 1894); *seri* (Nelson, 1912) [renaming of *goldmani* Townsend, 1912, which is preoccupied by *goldmani* Osgood, 1900]; *sobrinus* (Goldman, 1939) [renaming of *seorsus* Goldman, 1939, which is preoccupied by *seorsus* Burt, 1932, a subspecies of *C. spinatus*]; *stephensi* (Merriam, 1894).
COMMENTS: Revised by Hoffmeister and Lee (1967). Subspecies reviewed by Williams et al. (1993). Excludes Chihuahuan Desert subspecies *atrodorsalis* and *eremicus*, now regarded as a separate species, *C. eremicus* (see account above and Lee et al., 1996).

Chaetodipus pernix (J. A. Allen, 1898). Bull. Am. Mus. Nat. Hist., 10:149.
COMMON NAME: Sinaloan Pocket Mouse.
TYPE LOCALITY: Mexico, Sinaloa, Rosario.
DISTRIBUTION: Coastal lowlands from S Sonora to N Nayarit (Mexico).
STATUS: IUCN – Lower Risk (lc).
SYNONYMS: *rostratus* (Osgood, 1900).
COMMENTS: Chromosomal and biochemical evidence suggest that the northern subspecies *rostratus* is specifically distinct from *pernix* (J. L. Patton et al., 1981). Reviewed by Best and Lackey (1992*b*, Mammalian Species No. 420).

Chaetodipus rudinoris (Elliot, 1903). Field Columb. Mus. Publ. 74, Zool. Ser. 3(10):167.
COMMON NAME: Baja California Pocket Mouse.
TYPE LOCALITY: Mexico, Baja California, San Quintín.
DISTRIBUTION: SE California (USA) south to Cape Region of Baja California Sur (Mexico).
STATUS: IUCN – Lower Risk (lc) as included in *C. baileyi*.

SYNONYMS: *extimus* (Nelson and Goldman, 1929); *fornicatus* (Burt, 1932); *hueyi* (Nelson and Goldman, 1929); *knekus* (Elliot, 1903); *mesidios* (Huey, 1964).

COMMENTS: Reviewed, in part as *C. baileyi*, by Paulson (1988a, Mammalian Species No. 297). Chromosomal (Patton and Rogers, 1993), allozyme (J. L. Patton et al., 1981), and mitochondrial DNA (Riddle et al., 2000b) support species status for populations from west of the Colorado River in California and from throughout the Baja California Peninsula traditionally allocated to *C. baileyi* (Hall, 1981; Patton, 1993b; Patton and Alvarez-Castañeda, 1999; Williams et al., 1993). Subspecies delineated by Riddle et al. (2000b).

Chaetodipus spinatus (Merriam, 1889). N. Am. Fauna, 1:21.

COMMON NAME: Spiny Pocket Mouse.

TYPE LOCALITY: USA, California, San Bernardino Co., 25 mi (40 km) below The Needles, Colorado River.

DISTRIBUTION: S Nevada, SE California (USA) south to Cape Region of Baja California Peninsula (Mexico).

STATUS: IUCN – Lower Risk (lc).

SYNONYMS: *broccus* (Huey, 1960); *bryanti* (Merriam, 1894); *evermanni* (Nelson and Goldman, 1929); *guardiae* (Burt, 1932); *lambi* (Benson, 1930); *latijugularis* (Burt, 1932); *lorenzi* (Banks, 1967); *magdalenae* (Osgood, 1907); *macrosensis* (Burt, 1932); *margaritae* (Merriam, 1894); *occultus* (Nelson, 1912); *oribates* (Huey, 1960); *peninsulae* (Merriam, 1894); *prietae* (Huey, 1930); *pullus* (Burt, 1932); *rufescens* (Huey, 1930); *seorsus* (Burt, 1932).

COMMENTS: Reviewed by Lackey (1991b, Mammalian Species No. 385). Subspecies reviewed by Williams et al. (1993) and Patton and Alvarez-Castañeda (1999).

Perognathus Wied-Neuwied, 1839. Nova Acta Phys.-Med. Acad. Caes. Leop.-Carol., 19(1):368.

TYPE SPECIES: *Perognathus fasciatus* Wied-Neuwied, 1839.

SYNONYMS: *Abromys* Gray, 1868; *Cricetodipus* Peale, 1848; *Otognosis* Coues, 1875.

COMMENTS: Revised by Merriam (1889) and Osgood (1900), who also included those species listed here for *Chaetodipus*. Chromosomal relationships reviewed by Patton (1967b) and Williams (1978a). Taxonomy reviewed by Williams et al. (1993). Williams et al. (1993) and Best (1994a) provide keys to Recent species.

Perognathus alticolus Rhoads, 1894. Proc. Acad. Nat. Sci. Philadelphia, 45:412.

COMMON NAME: White-eared Pocket Mouse.

TYPE LOCALITY: USA, California, San Bernardino Co., San Bernardino Mtns, Squirrel Inn, Little Bear Valley, 5,500 ft. (1,576 m).

DISTRIBUTION: SC California (USA).

STATUS: IUCN – Critically Endangered as *P. a. alticola* [sic]; Lower Risk (nt) as *P. a. inexpectatus*.

SYNONYMS: *inexpectatus* Huey, 1926.

COMMENTS: Subspecies listed by Hall (1981), Williams et al. (1993); may be only subspecifically distinct from *parvus*. Reviewed by Best (1994a, Mammalian Species No. 463).

Perognathus amplus Osgood, 1900. N. Am. Fauna, 18:32.

COMMON NAME: Arizona Pocket Mouse.

TYPE LOCALITY: USA, Arizona, Yavapai Co., Fort Verde.

DISTRIBUTION: W and C Arizona (USA) to NW Sonora (Mexico).

STATUS: IUCN – Lower Risk (nt).

SYNONYMS: *jacksoni* Goldman, 1933; *rotundus* Goldman, 1932; *cineris* Benson, 1933; *ammodytes* Benson, 1933; *pergracilis* Goldman, 1932; *taylori* Goldman, 1932.

COMMENTS: Subspecies listed in Hall (1981) and reviewed by Williams et al. (1993).

Perognathus fasciatus Wied-Neuwied, 1839. Nova Acta Phys.-Med. Acad. Caes. Leop.-Carol., 19(1):369.

COMMON NAME: Olive-backed Pocket Mouse.

TYPE LOCALITY: USA, North Dakota, Williams Co., upper Missouri River near jct. with the Yellowstone, near Buford.

DISTRIBUTION: Great Plains from SE Alberta, Saskatchewan, and SW Manitoba (Canada) to NE Utah, S Colorado, and E South Dakota (USA).

STATUS: IUCN – Lower Risk (lc).

SYNONYMS: *infraluteus* Thomas 1893; *litus* Cary, 1911; *olivaceogriseus* Swenk, 1940; **callistus** Osgood, 1900.

COMMENTS: Revised by Williams and Genoways (1979). Reviewed by Manning and Jones (1988*a*, Mammalian Species No. 303).

Perognathus flavescens Merriam, 1889. N. Am. Fauna, 1:11.

COMMON NAME: Plains Pocket Mouse.

TYPE LOCALITY: USA, Nebraska, Cherry Co., Kennedy.

DISTRIBUTION: Great Plains and intermountain basins from Minnesota and N Utah (USA) to N Chihuahua (Mexico).

STATUS: IUCN – Lower Risk (lc).

SYNONYMS: **apache** Merriam, 1889; *cleomophila* Goldman, 1918; **caryi** Goldman, 1918; **cockrumi** Hall, 1954; **copei** Rhoads, 1894; **melanotis** Osgood, 1900; *gypsi* Dice, 1929; **perniger** Osgood, 1904; **relictus** Goldman, 1938.

COMMENTS: Reviewed by Williams (1978*b*). Hoffmeister (1986) considered *apache* a distinct species. Subspecies reviewed by Williams et al. (1993).

Perognathus flavus Baird, 1855. Proc. Acad. Nat. Sci. Philadelphia, 7:332.

COMMON NAME: Silky Pocket Mouse.

TYPE LOCALITY: USA, Texas, El Paso Co., El Paso.

DISTRIBUTION: SW Great Plains and intermountain plateaus from South Dakota, E Wyoming, and SE Utah (USA) south to Sonora and Puebla (Mexico).

STATUS: IUCN – Lower Risk (nt) as *P. f. goodpasteri*; otherwise Lower Risk (lc).

SYNONYMS: **bimaculatus** Merriam, 1889; **bunkeri** Cockrum, 1951; **fuliginosus** Merriam, 1890; **fuscus** Anderson, 1972; **goodpasteri** Hoffmeister, 1956; **hopiensis** Goldman, 1932; **medius** Baker, 1954; **mexicanus** Merriam, 1894; **pallescens** Baker, 1954; **parviceps** Baker, 1954; **piperi** Goldman, 1917; **sanluisi** Hill, 1952; **sonoriensis** Nelson and Goldman, 1934.

COMMENTS: Revised by Baker (1954); subspecies listed by Hall (1981) and reviewed by Williams et al. (1993). Wilson (1973) considered *merriami* conspecific, but Anderson (1972) Lee and Engstrom (1991) documented species distinctness of *merriami* from *flavus* (see below). Reviewed by Best and Skupski (1994*a*, Mammalian Species No. 471).

Perognathus inornatus Merriam, 1889. N. Am. Fauna, 1:15.

COMMON NAME: San Joaquin Pocket Mouse.

TYPE LOCALITY: USA, California, Fresno Co., Fresno.

DISTRIBUTION: Sacramento, San Joaquin and Salinas valleys and adjacent foothills, and western Mojave Desert, of California (USA).

STATUS: IUCN – Lower Risk (nt) as *P. i. neglectus* and *P. i. psammophilus*; otherwise Lower Risk (lc).

SYNONYMS: **neglectus** Taylor, 1912; **psammophilus** von Bloeker, 1937; *sillimani* von Bloeker, 1937.

COMMENTS: Revised by Osgood (1918); subspecies listed by Williams et al. (1993) who allocated *psammophilus* (with *sillimani* as a synonym) to this species, not to *longimembris* where this name has been usually placed (Hall, 1981; Patton, 1993*b*). As noted by Williams et al. (1993), at least two and possibly three distinct species are currently included under the name *inornatus*. Reviewed by Best (1993*e*, Mammalian Species No. 450).

Perognathus longimembris (Coues, 1875). Proc. Acad. Nat. Sci. Philadelphia, 27:305.

COMMON NAME: Little Pocket Mouse.

TYPE LOCALITY: USA, California, Kern Co., Tehachapi Mtns, Old Fort Tejon.

DISTRIBUTION: SE Oregon and W Utah (USA) south to N Sonora and Baja California and Baja California Sur (Mexico) (see Alvarez-Castañeda et al., 2001).

STATUS: U.S. ESA – Endangered as *P. l. pacificus*. IUCN – Critically Endangered as *P. l. pacificus*, Vulnerable as *P. l. brevinasus*, Data Deficient as *P. l. bangsi* and *P. l. internationalis*; otherwise Lower Risk (lc).

SYNONYMS: *elibatus* Elliot, 1904; *pericalles* Elliot, 1904; *aestivus* Huey, 1928; *arizonensis* Goldman, 1931; *arcus* Benson, 1935; *virginis* Huey, 1939; *bangsi* Mearns, 1898; *arenicola* Stephens, 1900; *bombycinus* Osgood, 1907; *brevinasus* Osgood, 1900; *gulosus* Hall, 1941; *internationalis* Huey, 1939; *kinoensis* Huey, 1935; *nevadensis* Merriam, 1894; *pacificus* Mearns, 1898; *cantwelli* von Bloeker, 1932; *panamintinus* Merriam, 1894; *pimensis* Huey, 1937; *salinensis* Bole, 1937; *tularensis* Richardson, 1937; *venustus* Huey, 1930.

COMMENTS: Revised by Osgood (1918); subspecies listed by Hall (1981) and Williams et al. (1993), who placed *psammophilus* von Bloeker with *P. inornatus* instead of *longimembris*, where this taxon has usually been assigned (Hall, 1981; Patton, 1993*b*).

Perognathus merriami J. A. Allen, 1892. Bull. Am. Mus. Nat. Hist., 4:45.

COMMON NAME: Merriam's Pocket Mouse.

TYPE LOCALITY: USA, Texas, Cameron Co., Brownsville.

DISTRIBUTION: SE New Mexico east to S Texas (USA), east from N Chihuahua to Tamaulipas (Mexico).

STATUS: IUCN – Lower Risk (lc).

SYNONYMS: *mearnsi* J. A. Allen, 1896; *gilvus* Osgood, 1900.

COMMENTS: Synonymized with *flavus* by Wilson (1973), but Lee and Engstrom (1991) considered *merriami* a separate species based on biochemical genetics, as did Anderson (1972) using morphological criteria. Reviewed by Best and Skupski (1994*b*, Mammalian Species No. 473).

Perognathus parvus (Peale, 1848). Mammalia *in* Repts. U.S. Expl. Surv., 8:53.

COMMON NAME: Great Basin Pocket Mouse.

TYPE LOCALITY: USA, Oregon, Wasco Co., probably near The Dalles.

DISTRIBUTION: Great Basin from S British Columbia (Canada), south to E California and east to SE Wyoming and NW Arizona (USA).

STATUS: IUCN – Lower Risk (lc) as *P. parvus* and *P. xanthanotus*.

SYNONYMS: *monticola* Baird, 1858; *bullatus* Durrant and Lee, 1956; *clarus* Goldman, 1917; *columbianus* Merriam, 1894; *idahoensis* Goldman, 1922; *laingi* Anderson, 1932; *lordi* (Gray, 1868); *mollipilosus* Coues, 1875; *olivaceus* Merriam, 1889; *amoenus* Merriam, 1889; *magruderensis* Osgood, 1900; *plerus* Goldman, 1939; *trumbullensis* Benson, 1937; *xanthanotus* Grinnell, 1912; *yakimensis* Broadbrooks, 1954.

COMMENTS: Reviewed by Verts and Kirkland (1988, Mammalian Species No. 318). Includes *xanthonotus* (Williams et al., 1993), often considered a distinct species (Hall, 1981; Patton, 1993*b*; Verts and Kirkland, 1988, by omission).

FAMILY GEOMYIDAE
by James L. Patton

Family Geomyidae Bonaparte, 1845. Cat. Meth. Mamm. Europe, p. 5.

    COMMENTS: Revised by Russell (1968*a*), who placed all Recent genera in one of two tribes in the subfamily Geomyinae: Geomyini (*Geomys*, *Pappogeomys*, *Orthogeomys*, and *Zygogeomys*) and Thomomyini (*Thomomys*); he included *Macrogeomys* and *Heterogeomys* as subgenera of *Orthogeomys*, an action followed by more recent authors (Hafner, 1982, 1991; Hall, 1981), and *Cratogeomys* as a subgenus of *Pappogeomys* (see also Russell, 1968*b*). The latter action was accepted by Hall (1981) and Patton (1993*a*) but not by J. K. Jones et al. (1986), Lee and Baker (1987), Davidow-Henry et al. (1989), Hollander (1990), and C. Jones et al. (1997) who considered *Cratogeomys* as a separate genus. Hooper (1946) synonymized *Platygeomys* Merriam with *Cratogeomys*. McKenna and Bell (1997) included heteromyids as a subfamily within Geomyidae; these are excluded here (see comments under Family Heteromyidae). Williams (1982*b*) reviewed phallic morphology; Hafner (1982), Honeycutt and Williams (1982), and Hafner et al. (1994*b*) reported on phylogenetic relationships based on biochemical characters; and Wahlert (1985) provided a classification of the superfamily Geomyoidea, including Geomyidae. Russell (1968*a*) and Hall (1981) provided keys to Recent genera. Korth (1994) summarized the fossil record and provided a classification of fossil taxa. Species boundaries within all pocket gopher genera are often difficult to define because of the characteristic allopatric distributions of named taxa (see Patton, 1990) and because of the common discordances among character data sets (see Patton and Smith, 1994). As a result, the history of pocket gopher systematics has been one where early on new taxa were described as separate species followed by a period of aggregation of names as subspecies or as simple synonyms (see, for example, Anderson, 1966, for a history of the allocation of names in the genus *Thomomys*). Applications of chromosomal and, more recently, molecular genetic methods have begun to reverse this historical trend, and will undoubtedly continue to do so. These methodologies, along with a shift in the species concept used by systematists to one emphasizing phyletic principles, will undoubtedly result in the elevation of several of the synonyms listed for some species below to full species status in the future. This trend is already evident, for example, in the species now recognized for *Geomys* relative to those listed in Patton (1993*a*). In the accounts that follow, names listed under "synonyms" that were considered distinct species at the time of their original description are indicated by an asterisk.

Cratogeomys Merriam, 1895. N. Amer. Fauna, 8:150.

    TYPE SPECIES: *Geomys merriami* Thomas, 1893.

    SYNONYMS: *Craterogeomys* J. A. Allen, 1895 [misprint]; *Platygeomys* Merriam, 1895.

    COMMENTS: Revised by Nelson and Goldman (1934*a*) and Russell (1968*a*, *b*), who included *Cratogeomys* as a valid subgenus of *Pappogeomys*, a position followed by Hall (1981) and Patton (1993*a*). However, Honeycutt and Williams (1982), based on a phylogenetic analysis of allozyme data, and DeWalt et al. (1993*a*) and Demastes et al. (2002), using mitochondrial DNA sequences, confirmed Russell's hypothesis of the sister relationship between *Pappogeomys* and *Cratogeomys*, but treated the two as separate genera. While it remains a matter of personal opinion as to the distinction between genera and subgenera, I choose here to follow the current consensus of workers on this group (e.g., Lee and Baker, 1987; Davidow-Henry et al., 1989; Hollander, 1990). Russell (1968*b*) recognized two species groups, the *castanops*-group (with *castanops* and *merriami*) and the *gymnurus*-group (with *fumosus*, *gymnurus*, *neglectus*, *tylorhinus*, and *zinseri*). While these groups have been confirmed by molecular phylogenetic studies (e.g., DeWalt et al., 1993*a*; Demastes et al., 2002), both studies call into question the species boundaries recognized by Russell and earlier workers. Some changes are incorporated into the accounts below, but additional ones are likely following the completion of on-going studies by M. S. Hafner, J. W. Demastes, and co-workers (pers. comm.). Hooper (1946) included *Platygeomys* within *Cratogeomys*. This name, with *gymnurus* as its type species, would be available for Russell's *gymnurus*-group, should that group warrant formal recognition. Key to eight species listed below given in León et al. (2001, Mammalian Species No. 685).

Cratogeomys castanops (Baird, 1852). *In* Stansbury, Expl. Surv. Valley Great Salt Lake, Utah, App. C (Zool.), p. 313.
   COMMON NAME: Yellow-faced Pocket Gopher.
   TYPE LOCALITY: USA, Colorado, Bent Co., along prairie road to Bent's Fort, near present town of Las Animas.
   DISTRIBUTION: SE Colorado and SW Kansas (USA) to E Durango and S Coahuila (Mexico).
   STATUS: IUCN – Lower Risk (lc) as *Pappogeomys castanops*.
   SYNONYMS: *angusticeps* Nelson and Goldman, 1934; *bullatus* Russell and Baker, 1955; **clarkii* (Baird, 1855); *convexus* Nelson and Goldman, 1934; *pratensis* Russell, 1968; *torridus* Russell, 1968; *consitus* Nelson and Goldman, 1934; *dalquesti* Hollander, 1990; *excelsus* Nelson and Goldman, 1934; *hirtus* Nelson and Goldman, 1934; *jucundus* Russell and Baker, 1955; *parviceps* Russell, 1968; *perexiguus* Russell, 1968; *perplanus* Nelson and Goldman, 1934; *lacrimalis* Nelson and Goldman, 1934; *simulans* Russell, 1968; *pratensis* Russell, 1968; *sordidulus* Russell and Baker, 1955; *subsimus* Nelson and Goldman, 1934; *surculus* Russell, 1968; *tamaulipensis* Nelson and Goldman, 1934; *ustulatus* Russell and Baker, 1955.
   COMMENTS: Equalivalent to the 2n=46 cytotype of Berry and Baker (1972) and Lee and Baker (1987), shown by mitochondrial DNA analysis to be phylogenetically separate from the 2n=42 cytotype of these authors (DeWalt et al., 1993*a*), which is herein referred to as *C. goldmani*. Largely coincides to the "*excelsus* subspecies-group" of Russell (1968*b*), but contains some taxa in his "*subnubilis* subspecies-group" (see Davidow-Henry et al., 1989, Mammalian Species No. 338). However, allocations of some taxa belonging to Russell's "*subnubilus*-group" to either *castanops* or *goldmani* need verification by karytoypic and other character analyses. Grouped with *merriami* in the *castanops* species-group by Russell (1968*b*). Subspecies listed above are those of Davidow-Henry et al. (1989), except as revised by Hollander (1990) for those in the USA.

Cratogeomys goldmani Merriam, 1895. N. Amer. Fauna, 8:160.
   COMMON NAME: Goldman's Pocket Gopher.
   TYPE LOCALITY: Mexico, Zacatecas, Cañitas.
   DISTRIBUTION: E Zacatecas, S Nuevo Leon, and San Luis Potosí (Mexico).
   SYNONYMS: *elibatus* Russell, 1968; *maculatus* Alvarez and Alvarez-Castañeda, 1996; *peridoneus* Nelson and Goldman, 1934; *planifrons* Nelson and Goldman, 1934; *rubellus* Nelson and Goldman, 1934; *subnubilus* Nelson and Goldman, 1934.
   COMMENTS: Included within *castanops* by earlier workers (Nelson and Goldman, 1934*a*; Russell, 1968*a*; Hall, 1981), but karyotypically (2n=42 – Berry and Baker, 1972; Lee and Baker, 1987) and phylogenetically distinct by mitochondrial DNA analysis (DeWalt et al., 1993*a*; Demastes et al., 2002); see also Alvarez and Alvarez-Castañeda (1996). Validity of subspecies listed is based on Russell (1968*a*), who included *goldmani* and its component taxa within *Cratogeomys castanops*.

Cratogeomys fumosus (Merriam, 1892). Proc. Biol. Soc. Wash., 7:165.
   COMMON NAME: Smoky Pocket Gopher.
   TYPE LOCALITY: Mexico, Colima, 3 mi. (5 km) W of Colima, 1,700 ft. (518 m).
   DISTRIBUTION: Plain of E Colima (Mexico).
   STATUS: IUCN – Lower Risk (nt) as *Pappogeomys fumosus*.
   COMMENTS: Included in *gymnurus* species-group by Russell (1968*b*). However, *fumosus* is likely a species that includes *gynmurus*, *neglectus*, *tylorhinus*, and *zinseri*; see Demastes et al. (2002).

Cratogeomys gymnurus (Merriam, 1892). Proc. Biol. Soc. Wash., 7:166.
   COMMON NAME: Llano Pocket Gopher.
   TYPE LOCALITY: Mexico, Jalisco, Zapotlán (= Ciudad Guzmán), 4,000 ft. (1,219 m).
   DISTRIBUTION: S and C Jalisco and NE Michoacan (Mexico).
   STATUS: IUCN – Lower Risk (lc) as *Pappogeomys gymnurus*.

SYNONYMS: *inclarus* (Goldman, 1939); *morulus* Russell, 1953; ***imparilis*** (Goldman, 1939); ***russelli*** Genoways and Jones, 1969; ***tellus*** Russell, 1953.

COMMENTS: Polytypic; subspecies reviewed by Russell (1968*b*), but all are probably only geographic units of *fumosus* (Demastes et al., 2002).

Cratogeomys merriami (Thomas, 1893). Ann. Mag. Nat. Hist., ser. 6, 12:271.

COMMON NAME: Merriam's Pocket Gopher.

TYPE LOCALITY: Mexico, "Southern Mexico," probably Valley of Mexico.

DISTRIBUTION: WC Veracruz to Distrito Federal, Morelos, and surrounding areas, including SE Central Plateau and S Sierra Madre Oriental (Mexico).

STATUS: IUCN – Lower Risk (lc) as *Pappogeomys merriami*.

SYNONYMS: *oreocetes Merriam, 1895; *peregrinus Merriam, 1895; ****estor*** Merriam, 1895; ****fulvescens*** Merriam, 1895; ***irolonis*** Nelson and Goldman, 1934; ***peraltus*** Goldman, 1937; *perotensis Merriam, 1895; ***saccharalis*** Nelson and Goldman, 1934.

COMMENTS: Included in the *castanops* species-group by Russell (1968*b*) and close phyletic relationship to *castanops* and *goldmani* confirmed by mitochondrial DNA analyses (DeWalt et al., 1993*a*; Demastes et al., 2002). Polytypic; subspecies reviewed by Russell (1968*b*).

Cratogeomys neglectus (Merriam, 1902). Proc. Biol. Soc. Wash., 15:68.

COMMON NAME: Querétaro Pocket Gopher.

TYPE LOCALITY: Mexico, Querétaro, Cerro de la Calentura, about 8 mi (13 km) NW of Pinal de Amoles, 9,500 ft. (2,896 m).

DISTRIBUTION: Known only from the type locality.

STATUS: IUCN – Critically Endangered as *Pappogeomys neglectus*.

COMMENTS: Probably only a recent geographic isolate of *tylorhinus* (Monterrubio et al., 2000) or, along with *tylorhinus*, part of *fumosus* (Demastes et al., 2002). Reviewed by León et al. (2001, Mammalian Species No. 685).

Cratogeomys tylorhinus (Merriam, 1895). N. Am. Fauna, 8:167.

COMMON NAME: Naked-nosed Pocket Gopher.

TYPE LOCALITY: Mexico, Hidalgo, Tula, 6,800 ft. (2,073 m).

DISTRIBUTION: Distrito Federal and Hidalgo to C Jalisco (Mexico).

STATUS: IUCN – Lower Risk (nt) as *Pappogeomys tylorhinus*.

SYNONYMS: *arvalis* Hooper, 1947; ***angustirostris*** (Merriam, 1903), *varius Goldman, 1939; ***atratus*** Russell, 1953; ***brevirostris*** Russell, 1968; ****planiceps*** (Merriam, 1895); ***zodius*** Russell, 1953.

COMMENTS: Included in the *gymnurus* species-group by Russell (1968*b*) and probably same species as *fumosus* (Demastes et al., 2002). As currently understood, however, *tylorhinus* is polytypic; subspecies reviewed by Russell (1968*b*). Reviewed by Cervantes et al. (1993*a*, Mammalian Species No. 433) as *Pappogeomys tylorhinus*.

Cratogeomys zinseri (Goldman, 1939). J. Mammal., 20:91.

COMMON NAME: Zinser's Pocket Gopher.

TYPE LOCALITY: Mexico, Jalisco, Lagos, 6,150 ft. (1,875 m).

DISTRIBUTION: NE Jalisco (Mexico).

STATUS: IUCN – Lower Risk (nt) as *Pappogeomys zinseri*.

COMMENTS: Another geographic isolate that is part of the *gymnurus* species-group (Russell, 1968*b*) but likely the same species as *fumosus* (Demastes et al., 2002).

Geomys Rafinesque, 1817. Am. Mon. Mag., 2(1):45.

TYPE SPECIES: *Geomys pinetis* Rafinesque, 1817.

SYNONYMS: *Ascomys* Lichtenstein, 1825; *Diplostoma* Rafinesque, 1817; *Mamgeomyscus* Herrera, 1899; *Neterogeomys* Gazin, 1942; *Parageomys* Hibbard, 1944; *Progeomys* Dalquest, 1983; *Pseudostoma* Say, 1823; *Saccophorus* Kuhl, 1820.

COMMENTS: Revised by Merriam (1895*a*). Species boundaries in this genus, as in other pocket gophers, are difficult to define. Complex relationships have been described for several diagnosable geographic units based on a variety of contact zone analyses that have employed morphological, karyological, allozyme, and/or mitochondrial and nuclear

DNA biochemical analyses. These have indicated varying degrees of hybridization between geographically differentiated forms, and authors have varied in their recognition of these entities at the specific or subspecific levels. Hall (1981) recognized five species (*arenarius*, *bursarius*, *personatus*, *pinetis*, and *tropicalis*). While this set of taxa was regarded as an incomplete recognition of likely species in the genus, Patton (1993*a*) followed Hall, waiting until workers of the genus reached consensus. This has now been achieved, with the nine species listed below generally recognized (see Elrod et al., 2000; Jolley et al., 2000). Baker and Williams (1974) and Hall (1981) provide keys to some species of *Geomys*, and Davis and Schmidly (1994) reviewed and mapped the species that occur in Texas. Phylogenetic relationships among species have been examined by comparative karyology (Smolen and Bickham, 1994, 1995), mitochondrial DNA sequences (Jolley et al., 2000), and morphometrics (Mauk et al., 1999). Qumsiyeh et al. (1988*a*) summarized karyotypic diversity in the genus.

Geomys attwateri Merriam, 1895. N. Am. Fauna, 8:135.
> COMMON NAME: Attwater's Pocket Gopher.
> TYPE LOCALITY: USA, Texas, Caldwell Co., Rockport.
> DISTRIBUTION: SC coast of Texas (USA).
> SYNONYMS: **ammophilus** Davis, 1940.
> COMMENTS: Reviewed by Williams and Cameron (1991, Mammalian Species No. 382). Considered a subspecies of *bursarius* by Hall (1981) and Patton (1993*a*) but a distinct species by Tucker and Schmidly (1981) and Block and Zimmerman (1991). Burt and Dowler (1999) reviewed the evidence for species status and Jolley et al. (2000) documented a phylogenetic relationship to *personatus*, not *bursarius*.

Geomys arenarius Merriam, 1895. N. Am. Fauna, 8:139.
> COMMON NAME: Desert Pocket Gopher.
> TYPE LOCALITY: USA, Texas, El Paso Co., El Paso.
> DISTRIBUTION: Extreme W Texas, SW and SC New Mexico (USA); N Chihuahua (Mexico).
> STATUS: IUCN – Lowered Risk (nt).
> SYNONYMS: **brevirostris** Hall, 1932.
> COMMENTS: Reviewed by Williams and Baker (1974, Mammalian Species No. 36) and Williams and Genoways (1978). Considered a subspecies of *bursarius* by Hafner and Geluso (1983); phylogenetic separation of *arenarius* and *bursarius* established by Jolley et al. (2000).

Geomys breviceps Baird, 1855. Proc. Acad. Nat. Sci., Philadelphia 7:335.
> COMMON NAME: Baird's Pocket Gopher.
> TYPE LOCALITY: USA, Louisiana, Morehouse Parish, Mer Rouge.
> DISTRIBUTION: W Louisiana, E Texas, E Oklahoma, and SW Arkansas (USA).
> SYNONYMS: **sagittalis** Merriam, 1895; *brazensis* Davis, 1938; *dutcheri* Davis, 1940; *ludemani* Davis, 1940; *pratincolus* Davis, 1940; *terricolus* Davis, 1940.
> COMMENTS: Reviewed by Sulentich et al. (1991, Mammalian Species No. 383). Considered a subspecies of *bursarius* by Hall (1981) and Patton (1993*a*). Contact zones between *breviceps* and *bursarius* studied by Bohlin and Zimmerman (1982) and Zimmerman and Gayden (1981) and between *breviceps* and *attwateri* by Tucker and Schmidly (1981), Dowler (1989), and Burt and Dowler (1999). Jolley et al. (2000) examined the phylogenetic position of *breviceps* relative to other species of *Geomys*. Two subspecies recognized (Sulentich et al., 1991), but *sagittalis* may be a distinct species based on mitochondrial DNA variation (Demastes, 1994).

Geomys bursarius (Shaw, 1800). Philosophical Magazine, 6:215.
> COMMON NAME: Plains Pocket Gopher.
> TYPE LOCALITY: USA, upper Mississippi Valley (restricted to Minnesota, Sherburne Co., Elk River by Swenk, 1939).
> DISTRIBUTION: SC Manitoba (Canada) to NW Indiana, SC Texas, and NE New Mexico (USA).
> STATUS: IUCN – Lower Risk (lc).
> SYNONYMS: *alba* (Rafinesque, 1817); *canadensis* (Lichtenstein, 1825); *fusca* (Rafinesque, 1817); *saccatus* (Mitchell, 1821); **illinoensis** Komarek and Spencer, 1931; **industrius**

Villa and Hall, 1947; *jugossicularis* Hooper, 1940; *lutescens* Merriam, 1890; *hylaeus* Blossom, 1938; *levisagittalis* Swenk, 1940; *vinaceus* Swenk, 1940; *major* Davis, 1940; *majusculus* Swenk, 1939; *missouriensis* McLaughlin, 1958; *ozarkensis* Elrod, Zimmerman, Sudman, and Heidt, 2000; *wisconsinensis* Jackson, 1957.

COMMENTS: Revised by Merriam (1895*a*) and, in part, by Honeycutt and Schmidly (1979) and by Heaney and Timm (1983*a*), who considered *lutescens* a valid species (see also Heaney and Timm, 1985; Jolley et al., 2000). Burns et al. (1985), Sudman et al (1987), and Elrod et al. (2000) listed *lutescens* as a subspecies of *bursarius*, which is followed here. Elrod et al. (2000) mapped the ranges of all subspecies recognized by Hall (1981).

Geomys knoxjonesi Baker and Genoways, 1975. Occas. Pap. Mus. Texas Tech Univ., 29:1.

COMMON NAME: Knox Jones's Pocket Gopher.

TYPE LOCALITY: USA, Texas, Winkler Co., 4.1 mi. (6.6 km) N and 5.1 mi. (8.2 km) E Kermit.

DISTRIBUTION: W Texas and SE New Mexico (USA).

COMMENTS: Considered a subspecies of *bursarius* by Hall (1981) and Patton (1993*a*) but a distinct species by Bradley et al. (1991) and Block and Zimmerman (1991). Baker and Davis (1989) described a hybrid zone with partial reproductive isolation between *knoxjonesi* and *bursarius major*. Phylogenetic position within *Geomys* established by Jolley et al. (2000).

Geomys personatus True, 1889. Proc. U.S. Natl. Mus., 11:159.

COMMON NAME: Texas Pocket Gopher.

TYPE LOCALITY: USA, Texas, Cameron Co., Padre Isl.

DISTRIBUTION: S Texas, south of San Antonio and Del Rio, including Padre and Mustang Isls (USA); barrier beaches of extreme NE Tamaulipas (Mexico).

STATUS: IUCN – Vulnerable as *G. p. maritimus* and *G. p. streckeri*, Lower Risk (nt) as *G. p. fuscus* and as *G. personatus*.

SYNONYMS: *davisi* Williams and Genoways, 1981; *fallax* Merriam, 1895; *fuscus* Davis, 1940; *maritimus* Davis, 1940; *megapotamus* Davis, 1940; *streckeri* Davis, 1943.

COMMENTS: Reviewed by Davis (1940), Williams and Genoways (1981), and Williams (1982*a*, Mammalian Species No. 170). Subspecies *streckeri* may warrant specific status, as it does not share a monophyletic relationship with other taxa of *personatus* based on mitochondrial DNA analyses (Jolley et al., 2000). Historical distribution and geographic variation examined by Wilkins and Swearingen (1990).

Geomys pinetis Rafinesque, 1817. Amer. Monthly Mag., 2:45.

COMMON NAME: Southeastern Pocket Gopher.

TYPE LOCALITY: USA, Georgia, in the region of the pines (restricted to Screven Co. by Harper, 1952).

DISTRIBUTION: C Florida to S Georgia and S Alabama (USA).

STATUS: IUCN – Extinct as *G. p. goffi*; Vulnerable as *G. p. cumberlandius* and *G. p. frontanelus*; Lower Risk (nt) as *G. p. colonus*, otherwise Lower Risk (lc).

SYNONYMS: *austrinus* Bangs, 1898; **colonus* Bangs, 1898; **cumberlandius* Bangs, 1898; **floridanus* (Audubon and Bachman, 1853); *goffi* Sherman, 1944; *mobilensis* Merriam, 1895; *tuza* (Barton, 1806); **fontanelus* Sherman, 1940.

COMMENTS: *Mus tuza* Barton, 1806, a senior synonym of *pinetis* according to Merriam (1895*a*), was considered of uncertain application and not available by Harper (1952). Hall (1981:505) regarded *colonus, frontanelus*, and *cumberlandius* as distinct species. Laerm (1981) supported the synonymy of *cumberlandius*. Reviewed by Pembleton and Williams (1978, Mammalian Species No. 86), Williams and Genoways (1980*b*), and Wilkins (1987*b*).

Geomys texensis Merriam, 1895. N. Amer. Fauna, 8:137.

COMMON NAME: Central Texas Pocket Gopher.

TYPE LOCALITY: USA, Texas, Mason Co., Mason.

DISTRIBUTION: SC Texas (USA).

STATUS: IUCN – Lowered Risk (nt) as *G. texensis* and as *G. t. bakeri*.

SYNONYMS: *bakeri* Smolen, Pitts, and Bickham, 1993; *llanensis* Bailey, 1905.

COMMENTS: Considered a subspecies of *bursarius* by Hall (1981) and Patton (1993*a*). Block and Zimmerman (1991), Smolen et al. (1993), and Jolley et al. (2000) provided evidence for species status. Reviewed by Cramer and Cameron (2001, Mammalian Species No. 679).

Geomys tropicalis Goldman, 1915. Proc. Biol. Soc. Wash., 28:134.
> COMMON NAME: Tropical Pocket Gopher.
> TYPE LOCALITY: Mexico, Tamaulipas, Altamira.
> DISTRIBUTION: Vicinity of Altamira and Tampico in SE Tamaulipas (Mexico).
> STATUS: IUCN – Vulnerable.
> COMMENTS: Elevated to species status by Alvarez (1963a). Reviewed by Baker and Williams (1974, Mammalian Species No. 35) and Williams and Genoways (1977).

Orthogeomys Merriam, 1895. N. Am. Fauna, 8:23, 26, 172.
> TYPE SPECIES: *Geomys scalops* Thomas, 1894.
> SYNONYMS: *Heterogeomys* Merriam, 1895; *Macrogeomys* Merriam, 1895.
> COMMENTS: Russell (1968a) revised the genus, and included *Heterogeomys* and *Macrogeomys* as valid subgenera. Hafner (1991) examined molecular phylogenetics of species in the subgenus *Macrogeomys* while Sudman and Hafner (1992) examined relationships among species within each of the subgenera. Species limits for those taxa known from single localities or otherwise geogeographically restricted areas are poorly defined.

Orthogeomys cavator (Bangs, 1902). Bull. Mus. Comp. Zool., 39:42.
> COMMON NAME: Chiriqui Pocket Gopher.
> TYPE LOCALITY: Panama, Chiriqui Prov., Boquete, 4,800 ft. (1,463 m).
> DISTRIBUTION: NW Panama to C Costa Rica.
> STATUS: IUCN – Lower Risk (lc).
> SYNONYMS: *nigrescens* (Goodwin, 1943); ***pansa** (Bangs, 1902).
> COMMENTS: Subgenus *Macrogeomys*. Reviewed by Goodwin (1946).

Orthogeomys cherriei (J. A. Allen, 1893). Bull. Am. Mus. Nat. Hist., 5:337.
> COMMON NAME: Cherrie's Pocket Gopher.
> TYPE LOCALITY: Costa Rica, Limón Prov., Santa Clara.
> DISTRIBUTION: NC Costa Rica (see Hafner and Hafner, 1987; Hafner, 1991).
> STATUS: IUCN – Lowered Risk (nt).
> SYNONYMS: *carlosensis* (Goodwin, 1934); ***costaricensis** (Merriam, 1895).
> COMMENTS: Subgenus *Macrogeomys*. Revised by Goodwin (1946). Phylogeography examined by Demastes et al. (1996). May include *matagalpae* (M. S. Hafner, pers. comm.).

Orthogeomys cuniculus Elliot, 1905. Proc. Biol. Soc. Wash., 18:224.
> COMMON NAME: Oaxacan Pocket Gopher.
> TYPE LOCALITY: Mexico, Oaxaca, Zanatepec (corrected from Yautepec, Oaxaca by Elliot, 1907a).
> DISTRIBUTION: Known only from the type locality.
> STATUS: IUCN – Critically Endangered.
> COMMENTS: Subgenus *Orthogeomys*. Reviewed by Nelson and Goldman (1930).

Orthogeomys dariensis (Goldman, 1912). Smithson. Misc. Coll., 60(2):8.
> COMMON NAME: Darien Pocket Gopher.
> TYPE LOCALITY: Panama, Darien Prov., Cana, upper Rio Tuyra, 2,000 ft. (610 m).
> DISTRIBUTION: E Panama.
> STATUS: IUCN – Lower Risk (lc).
> COMMENTS: Subgenus *Macrogeomys*. May include *thaeleri* (see Sudman and Hafner, 1992).

Orthogeomys grandis (Thomas, 1893). Ann. Mag. Nat. Hist., ser. 6, 12:270.
> COMMON NAME: Giant Pocket Gopher.
> TYPE LOCALITY: Guatemala, Sacatépequez Prov., Dueñas.
> DISTRIBUTION: Honduras to Jalisco (Mexico).
> STATUS: IUCN – Lower Risk (lc).
> SYNONYMS: *alleni* Nelson and Goldman, 1930; *alvarezi* Schaldach, 1966; *annexus* Nelson and Goldman, 1933; *carbo* Goodwin, 1956; *engelhardi* Felten, 1957; *felipensis* Nelson

and Goldman, 1930; *guerrerensis* Nelson and Goldman, 1930; *huixtlae* Villa, 1944; **latifrons* Merriam, 1895; **nelsoni* Merriam, 1895; *pluto* Lawrence, 1933; **pygacanthus* Dickey, 1928; **scalops* (Thomas, 1894); *soconuscensis* Villa, 1949; *vulcani* Nelson and Goldman, 1931.

COMMENTS: Subgenus *Orthogeomys*. Burt and Stirton (1961) and Hall (1981) included *pygacanthus* in *grandis*; Russell (1968a) considered it a distinct species.

Orthogeomys heterodus (Peters, 1865). Monatsb. K. Preuss. Akad. Wiss. Berlin, p. 177.
COMMON NAME: Variable Pocket Gopher.
TYPE LOCALITY: Costa Rica, exact locality unknown.
DISTRIBUTION: C Costa Rica.
STATUS: IUCN – Lowered Risk (nt).
SYNONYMS: *cartagoensis* (Goodwin, 1943); **dolichocephalus* (Merriam, 1895).
COMMENTS: Subgenus *Macrogeomys*. Reviewed by Goodwin (1946). Phylogenetic relationships given by Hafner (1991).

Orthogeomys hispidus (Le Conte, 1852). Proc. Acad. Nat. Sci., Philadelphia, 6:158.
COMMON NAME: Hispid Pocket Gopher.
TYPE LOCALITY: Mexico (restricted to Veracruz, near Jalapa, by Merriam, 1895a).
DISTRIBUTION: Yucatan Peninsula, Belize, Guatemala, and NW Honduras, to S Tamaulipas (Mexico).
STATUS: IUCN – Lower Risk (lc).
SYNONYMS: *cayoensis* (Burt, 1937); *chiapensis* (Nelson and Goldman, 1929); *concavus* (Nelson and Goldman, 1929); **hondurensis* (Davis, 1966); *isthmicus* (Nelson and Goldman, 1929); *latirostris* (Hall and Alvarez, 1961); *negatus* (Goodwin, 1953); *teapensis* (Goldman, 1939); *tehuantepecus* (Goldman, 1939); **torridus* (Merriam, 1895); *yucantanensis* (Nelson and Goldman, 1929).
COMMENTS: Subgenus *Heterogeomys*. May include *lanius* (see Hall, 1981:511-512). Revised by Nelson and Goldman (1929).

Orthogeomys lanius (Elliot, 1905). Proc. Biol. Soc. Wash., 18:235.
COMMON NAME: Big Pocket Gopher.
TYPE LOCALITY: Mexico, Veracruz, Xuchil, SE side of Mt. Orizaba.
DISTRIBUTION: Known only from the type locality.
STATUS: IUCN – Lower Risk (lc).
COMMENTS: Subgenus *Heterogeomys*. May be conspecific with *hispidus* (Hall, 1981:512).

Orthogeomys matagalpae (J. A. Allen, 1910). Bull. Amer. Mus. Nat. Hist., 28:97.
COMMON NAME: Nicaraguan Pocket Gopher.
TYPE LOCALITY: Nicaragua, Matagalpa, Peña Blanca.
DISTRIBUTION: NC Nicaragua to SC Honduras.
STATUS: IUCN – Lower Risk (lc).
COMMENTS: Subgenus *Macrogeomys*. May be conspecific with *cherriei* (M. S. Hafner, pers. comm.).

Orthogeomys thaeleri Alberico, 1990. *In* Peters and Hutterer (eds.), Vertebrates in the Tropics, Mus. Alex. Koenig, Bonn, p. 104.
COMMON NAME: Thaeler's Pocket Gopher.
TYPE LOCALITY: *ca.* 7 km S. Bahía Solano, Municipio Bahía Solano, Depto. Chocó, Colombia, 100 m.
DISTRIBUTION: Serranía de Baudó (extreme NW Colombia).
STATUS: IUCN – Lower Risk (lc).
COMMENTS: Subgenus *Macrogeomys*. May be conspecific with *dariensis* (see Sudman and Hafner, 1992).

Orthogeomys underwoodi (Osgood, 1931). Field Mus. Nat. Hist. Publ., Zool. Ser., 295, 18:143.
COMMON NAME: Underwood's Pocket Gopher.
TYPE LOCALITY: Costa Rica, San Jose Prov., Alto de Jabillo Pirris, between San Geronimo and Pozo Azul.
DISTRIBUTION: Central Pacific coast of Costa Rica.

STATUS: IUCN – Lower Risk (lc).

COMMENTS: Subgenus *Macrogeomys*. Molecular phylogeography examined by Demastes et al. (1996).

Pappogeomys Merriam, 1895. N. Am. Fauna, 8:145.

TYPE SPECIES: *Geomys bulleri* Thomas, 1892.

COMMENTS: Revised, in part, by Russell (1968*a, b*), who included *Cratogeomys* as a valid subgenus (see above).

Pappogeomys alcorni Russell, 1957. Univ. Kansas Publ., Mus. Nat. Hist., 9:359.

COMMON NAME: Alcorn's Pocket Gopher.

TYPE LOCALITY: Mexico, Jalisco, 4 mi. (6 km) west of Mazamitla, 6,600 ft. (2,012 m).

DISTRIBUTION: S Jalisco (Mexico), in Sierra del Tigre.

STATUS: IUCN – Vulnerable.

COMMENTS: Probably conspecific with *bulleri*, based on molecular studies (J. W. Demastes and A. Butt, pers. comm.).

Pappogeomys bulleri (Thomas, 1892). Ann. Mag. Nat. Hist., ser. 6, 10:196.

COMMON NAME: Buller's Pocket Gopher.

TYPE LOCALITY: Mexico, Jalisco, Talpa, W slope Sierra de Mascota, 8,500 ft. (2,591 m) (probably about 5000 ft., 1500 m).

DISTRIBUTION: Nayarit, Jalisco, and Colima (Mexico).

STATUS: IUCN – Lower Risk (lc).

SYNONYMS: *flammeus* Goldman, 1939; *lagunensis* Goldman, 1939; **nelsoni* (Merriam, 1895); ***albinasus** Merriam, 1895; *amecensis* Goldman, 1939; **burti** Goldman, 1939; **infuscus** Russell, 1968; **lutulentus** Russell, 1968; **melanurus** Genoways and Jones, 1969; **nayaritensis** Goldman, 1939.

COMMENTS: Reviewed by Genoways and Jones (1969*a*). Polytypic; subspecies reviewed by Russell (1968*b*) and listed in Hall (1981). May include *alcorni* (see above).

Thomomys Wied-Neuwied, 1839. Nova Acta Phys.-Med. Acad. Caes. Leop.-Carol., 19(1):377.

TYPE SPECIES: *Thomomys rufescens* Wied-Neuwied, 1939 (= *T. talpoides rufescens*).

SYNONYMS: *Megascapheus* Elliot, 1903; *Plesiothomomys* Gidley and Gazin, 1933; *Tomomys* Brandt, 1955.

COMMENTS: Recent species allocated to two subgenera (Thaeler, 1980): *Thomomys* and *Megascapheus*. Patton and Smith (1981) and M. F. Smith (1998) examined phylogenetic relationships among members of the subgenus *Megascapheus* based on molecular characters. Species boundaries for some forms are poorly defined, but Patton and Smith (1989, 1994) provided an operational definition that can be applied to all pocket gophers.

Thomomys bottae (Eydoux and Gervais, 1836). Mag. Zool., Paris, 6:23.

COMMON NAME: Botta's Pocket Gopher.

TYPE LOCALITY: USA, coast of California (restricted to vicinity of Monterey, Monterey Co., by Baird, 1857).

DISTRIBUTION: SW and W USA, north to Oregon, east to Colorado, and south to the Cape region of Baja California, Sinaloa and Nuevo Leon (Mexico).

STATUS: IUCN – Lower Risk (lc).

SYNONYMS: **altivallis* Rhoads, 1895; **angularis* Merriam, 1897; **argusensis* Huey, 1931; **diaboli* Grinnell, 1914; **infrapallidus* Grinnell, 1914; *lorenzi* Huey, 1940; **neglectus* Bailey, 1914; *pallescens* Rhoads, 1895; *perpes* Merriam, 1901; *piutensis* Grinnell and Hill, 1936; *sanctidiegi* Huey, 1945; **scapterus* Elliot, 1904; **abbotti** Huey, 1928; **abstrusus** Hall and Davis, 1935; **actuosus** Kelson, 1951; **albatus* Grinnell, 1912; *aderrans* Huey, 1939; *boregoensis* Huey, 1939; *crassus* Chattin, 1941; *flavidus* Goldman, 1931; **harquahalae* Grinnell, 1936; *patulus* Goldman, 1938; **albicaudatus** Hall, 1930; **alexandrae* Goldman, 1933; **alpinus* Merriam, 1897; **alticolus** J. A. Allen, 1899; **analogus** Goldman, 1938; **angustidens** Baker, 1953; **anitae** J. A. Allen, 1898; **apache* Bailey, 1910; **aphrastus* Elliot, 1903; **aureiventris** Hall, 1930; **aureus* J. A. Allen, 1893, **latirostris* Merriam, 1901; **awahnee** Merriam, 1908; **baileyi* Merriam, 1901; **basilicae** Benson and Tillotson, 1940 [a renaming of *occipitalis* Benson and Tillotson, 1939, which is preoccupied by *occipitalis* Dice, 1925, a fossil *Thomomys* from Rancho La Brea, California]; ***birdseyei*** Goldman, 1937; ***bonnevillei*** Durant, 1946; ***borjasensis*** Huey, 1945; ***brazierhowelli*** Huey, 1960; ***brevidens*** Hall, 1932; ***cactophilus*** Huey, 1929; ***camoae*** Burt, 1937; **canus* Bailey, 1910; ***catalinae*** Goldman, 1931; *hueyi* Goldman, 1938; *parvulus* Goldman, 1938; ***catavinensis*** Huey, 1931; ***centralis*** Hall, 1930; **cervinus* J. A. Allen, 1895; **chrysonotus* Grinnell, 1912; ***cinereus*** Hall, 1932; ***collis*** Hooper, 1940; ***concisor*** Hall and Davis, 1935; ***confinalis*** Goldman, 1936; ***connectens*** Hall, 1936; ***contractus*** Durant, 1946; ***convergens*** Nelson and Goldman, 1934; ***convexus*** Durrant, 1939; ***cultellus*** Kelson, 1951; ***cunicularius*** Huey, 1945; ***curtatus*** Hall, 1932; ***depressus*** Hall, 1932; **desertorum* Merriam, 1901; *cedrinus* Huey, 1955; *desitus* Goldman, 1936; *hualpaiensis* Goldman, 1936; **muralis* Goldman, 1936; *suboles* Goldman, 1928; ***detumidus*** Grinnell, 1935; ***dissimilis*** Goldman, 1931; ***divergens*** Nelson and Goldman, 1934; *estanciae* Benson and Tillotson, 1939; **fulvus* Woodhouse, 1852; *mutabilis* Goldman, 1933, *nasutus* Hall, 1932; *operosus* Hatfield, 1942; ***fumosus*** Hall, 1932; ***guadalupensis*** Goldman, 1936; ***homorus*** Huey, 1949; ***howelli*** Goldman, 1936; ***humilis*** Baker, 1953; ***imitabilis*** Goldman, 1939; ***incomptus*** Goldman, 1939; ***internatus*** Goldman, 1936; ***jojobae*** Huey, 1945; ***juarezensis*** Huey, 1945; ***lachuguilla*** Bailey, 1902; ***lacrymalis*** Hall, 1932; **laticeps* Baird, 1855; *minor* Bailey, 1914; *silvifugus* Grinnell, 1935; ***latus*** Hall and Davis, 1935; ***lenis*** Goldman, 1942; **leucodon* Merriam, 1897; ***levidensis*** Goldman, 1942; ***limitaris*** Goldman, 1936; ***limpiae*** Blair, 1939; ***litoris*** Burt, 1940; ***lucidus*** Hall, 1932; ***lucrificus*** Hall and Durham, 1938; **magdalenae* Nelson and Goldman, 1909; ***martirensis*** J. A. Allen, 1898; **mearnsi* Bailey, 1914; *alienus* Goldman, 1938; *caneloensis* Lange, 1959; *carri* Lange, 1959; *chiricahuae* Nelson and Goldman, 1934; *collinus* Goldman, 1931; *extenuatus* Goldman, 1935; *grahamensis* Goldman, 1931; **mewa* Merriam, 1908; ***minimus*** Durrant, 1939; ***modicus*** Goldman, 1931; *proximus* Burt and Campbell, 1934; ***morulus*** Hooper, 1940; ***nanus*** Hall, 1932; ***navus*** Merriam, 1901; *acrirostratus* Grinnell, 1935; *agricolaris* Grinnell, 1935; **neglectus* Bailey, 1914; ***nesophilus*** Durrant, 1936; ***nigricans*** Rhoads, 1895; *affinis* Huey, 1945; **cabezonae* Merriam, 1901; **jacinteus* Grinnell and Swarth, 1914; *puertae* Grinnell, 1914; **operarius* Merriam, 1897; ***optabilis*** Goldman, 1936; ***opulentus*** Goldman, 1935; *osgoodi* Goldman, 1931; ***paguatae*** Hooper, 1940; ***pascalis*** Merriam, 1901; *ingens* Grinnell, 1932; **pectoralis* Goldman, 1936; ***peramplus*** Goldman, 1931; *rufidulus* Hoffmeister, 1955; ***perditus*** Merriam, 1901; ***perpallidus*** Merriam, 1886; *amargosae* Grinnell, 1921; **melanotis* Grinnell, 1918; *mohavensis* Grinnell, 1918; **oreoecus* Burt, 1932; **providentialis* Grinnell, 1931; ***pervagus*** Merriam, 1901; ***pervarius*** Goldman, 1938; **phelleoecus* Burt, 1933; ***pinalensis*** Goldman, 1938; ***planirostris*** Burt, 1931; *absonus* Goldman, 1931; *boreorarius* Durham, 1952; *nicholi* Goldman, 1938; *trumbullensis* Hall and Davis, 1934; *virgineus* Goldman, 1937; ***planorum*** Hooper, 1940; ***powelli*** Durrant, 1955; ***proximarinus*** Huey, 1945; ***pusillus*** Goldman, 1931; *aridicola* Huey, 1937; *comobabiensis* Huey, 1937; *depauperatus* Grinnell and Hill, 1936; *growlerensis* Huey, 1937; *phasma* Goldman, 1933; ***retractus*** Baker, 1953; ***rhizophagus*** Huey, 1949; ***riparius*** Grinnell and Hill, 1936; ***robustus*** Durrant, 1946; ***rubidus*** Youngman, 1958; ***ruidosae*** Hall, 1932; ***rupestris*** Chattin, 1941; ***ruricola*** Huey, 1949; ***russeolus*** Nelson and Goldman, 1909; ***saxatilis*** Grinnell, 1934; ***scotophilus*** Davis, 1940; ***sevieri*** Durrant, 1946; ***siccovallis*** Huey, 1945;**simulus* Nelson and Goldman, 1934; **sinaloae* Merriam, 1901; **solitarius* Grinnell, 1926; ***spatiosus*** Goldman, 1938; ***stansburyi*** Durrant, 1946; **sturgisi* Goldman, 1938; ***subsimilis*** Goldman, 1933; ***texensis*** Bailey, 1902; ***tivius*** Durrant, 1937; **toltecus* J.A. Allen, 1893; ***tularosae*** Hall, 1932; ***vanrosseni*** Huey, 1934; ***varus*** Hall and Long, 1960; ***vescus*** Hall and Davis, 1935; ***villai*** Baker, 1953; ***wahwahensis*** Durrant, 1937; ***winthropi*** Nelson and Goldman, 1934; ***xerophilus*** Huey, 1945.

COMMENTS: Subgenus *Megascapheus*. The taxonomic history of this, and related species has been contentious. Hall and Kelson (1959) included *bottae* within *umbrinus* but considered *baileyi* and *townsendii* as separate species, but Hall (1981) included all four of these taxa within his concept of *umbrinus*. Anderson (1966, 1972), Hoffmeister (1969, 1986), and Patton and co-workers (Patton, 1973; Patton and Dingman, 1968; Patton and Smith, 1981) considered *bottae* (including *baileyi*) separate from *umbrinus*. Thaeler

(1968*b*), Patton et al. (1984), Patton and Smith (1989, 1994), and Rogers (1991*a, b*) supported the specific separation of *townsendii* from Hall's (1981) *umbrinus*. As envisioned here, *bottae* includes a number of taxa considered by earlier workers to be full species but each of which are now relegated to subspecific status or are recognized as synonyms of other subspecies (taxa marked by asterisks in the list of synonyms, above). Hoffmeister (1969), Patton and Dingman (1968), and Patton (1973) reported on hybridization between *bottae* and *umbrinus* in S Arizona; Thaeler (1968*b*), Patton et al. (1984), and Patton and Smith (1993) examined hybridization between *bottae* and *townsendii* in NE California. For subspecies, see Hall (1981), exclusive of those herein allocated to *townsendii* and *umbrinus*, and as modified for Arizona by Hoffmeister (1986) and for California by Patton and Smith (1990).

Thomomys bulbivorus (Richardson, 1829). Quadrupeds, *in* Fauna Boreali-Americana, 1:206.
COMMON NAME: Camas Pocket Gopher.
TYPE LOCALITY: USA, banks of Columbia River (probably Portland, Multnomah Co., Oregon; see V. E. Bailey, 1915).
DISTRIBUTION: Willamette Valley (NW Oregon, USA).
STATUS: IUCN – Lower Risk (lc).
COMMENTS: Subgenus *Megascapheus*. Revised by V. E. Bailey (1915). Reviewed by Verts and Carraway (1987*b*, Mammalian Species No. 273). Genetic variation delineated by Carraway and Kennedy (1994).

Thomomys clusius Coues, 1875. Proc. Acad. Nat. Sci., Philadelphia, 27:138.
COMMON NAME: Wyoming Pocket Gopher.
TYPE LOCALITY: USA, Wyoming, Carbon Co., Bridger Pass, 18 mi (29 km) SW Rawlins.
DISTRIBUTION: Carbon and Sweetwater Cos., SC Wyoming (USA).
STATUS: IUCN – Lower Risk (lc).
COMMENTS: Subgenus *Thomomys*. Included in *talpoides* by Hall (1981:457); revised by Thaeler and Hinesley (1979).

Thomomys idahoensis Merriam, 1901. Proc. Biol. Soc. Wash., 14:114.
COMMON NAME: Idaho Pocket Gopher.
TYPE LOCALITY: USA, Idaho, Clark Co., Birch Creek (10 mi [16 km] S Nicholia [Lemhi Co.], about 6,400 ft. [1940 m]).
DISTRIBUTION: EC Idaho, adjacent Montana and W Wyoming, and N Utah (USA).
STATUS: IUCN – Lowered Risk (nt) as *T. idahoensis* and as *T. i. confinus*.
SYNONYMS: *confinus* Davis, 1937; ***pygmaeus** Merriam, 1901.
COMMENTS: Subgenus *Thomomys*. Thaeler (1972, 1977) revised this species. Formerly included in *talpoides* by Hall and Kelson (1959:441) and Hall (1981:457, 463; but see Hall, 1981:1179).

Thomomys mazama Merriam, 1897. Proc. Biol. Soc. Wash., 11:214.
COMMON NAME: Western Pocket Gopher.
TYPE LOCALITY: USA, Oregon, Klamath Co., Anna Creek near Crater Lake, Mt. Mazama, 6,000 ft. (1,829 m).
DISTRIBUTION: NW Washington through C Oregon to N California (USA).
STATUS: U.S. ESA and Washington Department of Fish and Game – Candidate; IUCN – Extinct as *T. m. tacomensis*; Critically Endangered as *T. m. louiei*; Vulnerable as *T. m. couchi, T. m. glacialis, T. m. pugetensis, T. m. tumuli,* and *T. m. yelmensis;* Lower Risk (nt) as *T. mazama* and as *T. m. helleri* and *T. m. melanops*.
SYNONYMS: *couchi* Goldman, 1939; *glacialis* Dalquest and Scheffer, 1942; ***helleri** Elliot, 1903; ***hesperus** Merriam, 1901; *louiei* Gardner, 1950; ***melanops** Merriam, 1899; ***nasicus** Merriam, 1897; ***niger** Merriam, 1901; *oregonus* Merriam, 1901; *premaxillaris* Grinnell, 1914; *pugetensis* Dalquest and Scheffer, 1942; *tacomensis* Taylor, 1919; *tumuli* Dalquest and Scheffer, 1942; *yelmensis* Merriam, 1899.
COMMENTS: Subgenus *Thomomys*. Revised by Johnson and Benson (1960); considered a subspecies of *monticola* by Bailey (1915) and Hall and Kelson (1959) but geographic sympatry with *monticola* in N California documented by Thaeler (1968*a*). Subspecies

listed in Hall (1981:465-467). Reviewed by Verts and Carraway (2000, Mammalian Species No. 641).

Thomomys monticola J. A. Allen, 1893. Bull. Am. Mus. Nat. Hist., 5:48.
COMMON NAME: Mountain Pocket Gopher.
TYPE LOCALITY: USA, California, El Dorado Co., Mt. Tallac, 7,500 ft. (2,286 m).
DISTRIBUTION: Sierra Nevada Mtns of C and N California and extreme WC Nevada (USA).
STATUS: IUCN – Lower Risk (lc).
SYNONYMS: *pinetorum* Merriam 1899.
COMMENTS: Subgenus *Thomomys*. Revised by Johnson and Benson (1960); Thaeler (1968a) provided details of distribution in N California.

Thomomys talpoides (Richardson, 1828). Zool. J., 3:518.
COMMON NAME: Northern Pocket Gopher.
TYPE LOCALITY: Restricted to Canada, Saskatchewan, North Saskatchewan River, Carlton House near Fort Carlton by V. E. Bailey (1915:97).
DISTRIBUTION: S British Columbia to C Alberta and SW Manitoba (Canada), south to C South Dakota and N New Mexico, N Arizona, N Nevada, and NE California (USA).
STATUS: IUCN – Vulnerable as *T. t. douglasii*, Lower Risk (nt) as *T. t. limosus* and *T. t. segregatus*, otherwise Lower Risk (lc).
SYNONYMS: **borealis* (Richardson, 1837); **unisulcatus* (Gray, 1843); *aequalidens* Dalquest, 1942; *agrestis* Merriam, 1908; *andersoni* Goldman, 1939; *attenuatus* Hall and Montague, 1951; **bridgeri* Merriam, 1901; *bullatus* Bailey, 1914; *caryi* Bailey, 1914; *cheyennensis* Swenk, 1941; *cognatus* Johnstone, 1955; *columbianus* Bailey, 1914; *devexus* Hall and Dalquest, 1939; **douglasii* (Richardson, 1829); *duranti* Kelson, 1949; **falcifer* Grinnell, 1926; *fisheri* Merriam, 1901; **fossor* J. A. Allen, 1893; *fuscus* Merriam, 1891, **myops* Merriam, 1901; *gracilis* Durrant, 1939; *immunis* Hall and Dalquest, 1939; *incensus* Goldman, 1939; *kaibabensis* Goldman, 1938; *kelloggi* Goldman, 1939; *levis* Goldman, 1938; **limosus* Merriam, 1901; *loringi* Bailey, 1914; *macrotis* Miller, 1930; *medius* Goldman, 1939; *meritus* Hall, 1951; *monoensis* Huey, 1934; *moorei* Goldman, 1938; *nebulosus* Bailey, 1914; *ocius* Merriam, 1901; *oquirrhensis* Durrant, 1939; *parowanensis* Goldman, 1938; *pierreicolus* Swenk, 1941; **pryori* Bailey, 1914; **quadratus* Merriam, 1897; *ravus* Durrant, 1946; *relicinus* Goldman, 1939; *retrorsus* Hall, 1951; *rostralis* Hall and Montague, 1951; **rufescens* Wied-Neuwied, 1839; *saturatus* Bailey, 1914; *segregatus* Johnstone, 1955; *shawi* Taylor, 1921; *taylori* Hooper, 1940; *tenellus* Goldman, 1939; *trivialis* Goldman, 1939; **uinta* Merriam, 1901; *wallowa* Hall and Orr, 1933; *wasatchensis* Durrant, 1946; *whitmani* Drake and Booth, 1952; *yakimensis* Hall and Dalquest, 1939; *badius* Goldman, 1939.
COMMENTS: Subgenus *Thomomys*. Formerly included *idahoensis* and *clusius*; see Thaeler (1972) and Thaeler and Hinesley (1979). Partial revision by V. E. Bailey (1915) and Thaeler (1985). The considerable degree of chromosomal differentiation among geographic representatives of this form (Thaeler, 1985) suggests that more than one biological species is currently included under the name *talpoides*. Reviewed by Verts and Carraway (1999, Mammalian Species No. 618) who included *pygmaeus*, here considered a subspecies of *idahoensis* (following Thaeler, 1972).

Thomomys townsendii (Bachman, 1839). J. Acad. Nat. Sci., Philadelphia, 8:105.
COMMON NAME: Townsend's Pocket Gopher.
TYPE LOCALITY: USA, erroneously given as "Columbia River", but restricted to Idaho, Canyon Co., near Nampa, by V. E. Bailey (1915).
DISTRIBUTION: Snake River Valley of Idaho south and west to SE Oregon, NE California, and N Nevada (USA).
STATUS: IUCN – Lower Risk (lc).
SYNONYMS: *atrogriseus* Bailey, 1914; *owyhensis* Davis, 1937; *similis* Davis, 1937; **nevadensis* Merriam, 1897; *bachmani* Davis, 1937; *elkoensis* Davis, 1937; **relictus* Grinnell, 1926.
COMMENTS: Subgenus *Megascapheus*. Revised by Davis (1937) and Rogers (1991a, b), who recognized only two valid subspecies (*townsendii* and *nevadensis*). Considered a distinct species by Thaeler (1968b), Patton et al. (1984), Patton and Smith (1989, 1994), and

Rogers (1991*a*, *b*), despite limited hybridization with *bottae* in NE California. Hall (1981:469, 495) reviewed Thaeler's evidence and included *townsendii* in *umbrinus* (*sensu* Hall).

Thomomys umbrinus (Richardson, 1829). Quadrupeds, *in* Fauna Boreali-Americana, 1:202.
COMMON NAME: Southern Pocket Gopher.
TYPE LOCALITY: Originally given as Cadadaguios, SW Louisiana, but restricted to Mexico, Veracruz (probably Puebla), vicinity of Boca del Monte, by Bailey (1906).
DISTRIBUTION: SC Arizona and SW New Mexico (USA) south to Puebla and Veracruz (Mexico).
STATUS: IUCN – Lowered Risk (nt) as *T. u. emotus*, otherwise Lower Risk (lc).
SYNONYMS: *albigularis* Nelson and Goldman, 1934; *martinensis* Nelson and Goldman, 1934; **orizabae* Merriam, 1893; **peregrinus* Merriam, 1893; *tolucae* Nelson and Goldman, 1934; *vulcanius* Nelson and Goldman, 1934; **arriagensis** Dalquest, 1951; **atrodorsalis** Nelson and Goldman, 1934; **atrovarius* J. A. Allen, 1898; **camargensis** Anderson, 1972; **chihuahuae** Nelson and Goldman, 1934; **crassidens** Nelson and Goldman, 1934; **durangi** Nelson and Goldman, 1934; **emotus** Goldman, 1933; **enixus** Nelson and Goldman, 1934; **eximius** Nelson and Goldman, 1934; **extimus** Nelson and Goldman, 1934; **goldmani* Merriam, 1901; **intermedius** Mearns, 1897; **burti* Huey, 1932; *quercinus* Burt and Campbell, 1934; **juntae** Anderson, 1972; **madrensis** Nelson and Goldman, 1934; *caliginosus* Nelson and Goldman, 1934; **sheldoni* Bailey, 1915; **musculus** Nelson and Goldman, 1934; **nelsoni* Merriam, 1901; *evexus* Nelson and Goldman, 1934; **newmani** Dalquest, 1951; **parviceps** Nelson and Goldman, 1934; **potosinus** Nelson and Goldman, 1934; **pullus** Hall and Villa, 1948; **sonoriensis** Nelson and Goldman, 1934; **supernus** Nelson and Goldman, 1934; **zacatecae** Nelson and Goldman, 1934.
COMMENTS: Subgenus *Megascapheus*. Hybridization with *bottae* reported in Patagonia Mtns of S Arizona by Patton and Dingman (1968), Patton (1973), and Hoffmeister (1969, 1986). Considered a species separate from *bottae* by these authors and by Anderson (1966, 1972). Hall (1981:469) included *baileyi*, *bottae*, and *townsendii* in this species based on his view of the hybridization data in Hoffmeister (1969), Patton (1973), and Thaeler (1968*b*); he reported intergradation between *umbrinus* and *bottae* in Nuevo Leon (Mexico), which has not been verified by genetic analyses. Hafner et al. (1987) examined phylogenetic relationships among geographic units of *umbrinus*. Revised, in part, by Nelson and Goldman (1934*b*), Anderson (1972; for Chihuahua), and Castro-Campillo and Ramírez-Pulido (2000; for C Mexico).

Zygogeomys Merriam, 1895. N. Am. Fauna, 8:195.
TYPE SPECIES: *Zygogeomys trichopus* Merriam, 1895.
SYNONYMS: *Gygogeomys* J. A. Allen 1895 [misprint].

Zygogeomys trichopus Merriam, 1895. N. Am. Fauna, 8:196.
COMMON NAME: Michoacan Pocket Gopher.
TYPE LOCALITY: Mexico, Michoacan, Nahuatzen, 8,000-8,500 ft. (2,438-2,591 m).
DISTRIBUTION: Known only from small areas in the general vicinity of Lago Pátzcuaro, NC Michoacan (Mexico).
STATUS: IUCN – Endangered. Local temporal extirpation documented by Hafner and Barkley (1984).

SUBORDER MYOMORPHA Brandt, 1855.
SUPERFAMILY DIPODOIDEA Fischer, 1817.

FAMILY DIPODIDAE
by Mary Ellen Holden and Guy G. Musser

Family Dipodidae Fischer de Waldheim, 1817. Mem. Soc. Imp. Nat., Moscow, 5:372.

SYNONYMS: Allactagidae Vinogradov, 1925; Dipodes Fischer de Waldheim, 1817; Dipodina, Bonaparte, 1838; Dipodum Fischer de Waldheim, 1817; Dipsidae Gray, 1821; Jaculidae Gill, 1872; Sicistidae Weber, 1928; Sminthidae Brandt, 1855; Zapodidae Coues, 1875.

COMMENTS: The monophyly of dipodids is strongly established (Ellerman, 1940; Klingener, 1964, 1984; Shenbrot, 1992; Shenbrot et al., 1995; Stein, 1990; Vinogradov, 1930). Authors have usually recognized either a single family, Dipodidae (Ellerman, 1940, 1961; Ellerman and Morrison-Scott, 1951; Holden, 1993; Hugueney and Vianey-Liaud, 1980; Klingener, 1964, 1984; Kowalski, 2001; McKenna and Bell, 1997; Ognev, 1948; Qiu and Storch, 2000; Savinov, 1970; Thomas, 1896; Vinogradov, 1930, 1937; Wang and Qiu, 2000; Winge, 1887), or two families, Zapodidae (including sicistines and zapodines), and Dipodidae (Corbet, 1978c; Corbet and Hill, 1992; Daxner-Höck, 1999; Lyon, 1901; R. A. Martin, 1994; Miller and Gidley, 1918; Pocock, 1922b; Schaub, 1958; Simpson, 1945; Vinogradov, 1925; B. Y. Wang, 1985; Wilson, 1949; Wood, 1955). Whether one family or two, modern arrangements of subfamilies stem from Vinogradov's (1925, 1930, 1937) classic research and classification, which provided the foundation for systematic research of dipodids. Phylogenetic studies have supported much of Vinogradov's original classificatory arrangement, though the proposed relationships among some of the taxa have changed (Ellerman, 1940; Pavlinov and Shenbrot, 1983; Shenbrot, 1992; Stein, 1990; Zazhigin and Lopatin, 2000a).

Vinogradov went from accepting two families in 1925 to one in 1930 when he recognized Zapodinae, Euchoreutinae, Cardiocraniinae, Allactaginae, and Dipodinae, which was followed by Ellerman (1940), who also listed Sicistinae. Pavlinov and Shenbrot (1983) surveyed anatomy of male reproductive tracts among dipodines and recognized Euchoreutinae, Allactaginae, Cardiocraniinae, Salpingotinae, and Dipodinae within a Dipodidae. Later, Shenbrot (1992) incorporated male reproductive morphology, coronal structure of molars, and bullar anatomy that resulted in an eclectic classification formed from both cladistic analysis and degree of morphological divergence (phenetic distance). He recognized Allactagidae, Dipodidae (containing Cardiocraniinae, Paradipodinae, and Dipodinae), Sminthidae (with Sminthinae [= Sicistinae] and Euchoreutinae) and Zapodidae within a larger Dipodoidea, an arrangement also used by Pavlinov et al. (1995) and Shenbrot et al. (1995). In a phylogenetic study based on limb myology, Stein (1990) proposed retention of only two families, a primitive Sicistidae and derived Dipodidae (including zapodines, Euchoreutinae, and all other dipodids); in her scheme, Euchoreutinae is a sister group to zapodines and allactagines. Shenbrot's arrangement differed significantly from that of Steins' in that *Sicista* was not shown to be the most primitive dipodoid (Zapodidae and Sicistidae form an unresolved dichotomy), and Euchoreutinae was hypothesized to be most closely related to Sicistinae and united with Sicistinae in Sminthidae. From a paleontological perspective, Zazhigin and Lopatin (2000a) proposed a classification listing Zapodidae (containing Sicistinae and Zapodinae), Allactagidae (with Allactaginae and Euchoreutinae) and Dipodidae (which includes Cardiocraniinae, Dipodinae, and the extinct Lophocricetinae).

Sicistines are generally regarded as representing the most primitive dipodids morphologically (e.g., Dawson and Krishtalka, 1984; R. A. Martin, 1994; Miljutin, 1999; Stein, 1990). This cladistic position is supported by the few published analyses of gene sequences (nuclear LCAT, IRBP, and vWF) where in all phylogenetic reconstructions, *Sicista* is basal to different combinations of *Zapus, Napaeozapus, Dipus, Allactaga*, and *Jaculus* (DeBry, 2003; DeBry and Sagel, 2001; Jansa and Weksler, 2004; Michaux and Catzeflis, 2000; Michaux et al., 2001a), endorsing Stein's (1990:310) results and conclusion "that sicistines are, in fact, the sister group to all other zapodid and dipodid rodents." In his review of early Tertiary North American rodents, Wilson (1949:128) wrote that the Sicistinae "may have been ancestral to the Zapodinae on the one hand, and to the Dipodidae on the other," a suggestion repeated by R. A. Martin (1994:7), who opined that the "sicistines form a recognizable stem group from which all later zapodids and probably dipodids evolved." This possibility was echoed by Miljutin (1999:119): "Although recent Sicistinae have their own trends of adaptive evolution, which are inconsistent or even opposite to those of Dipodoidea as a whole, their morphological characters and relatively low level of specialization make them suitable candidates for the role of ancestral stock for all other dipodoids." Sokolov et al. (1987b) and Vorontsov (1969), however, hypothesized that zapodines possess the most primitive karyotype of dipodids, from which those of sicistines and other dipodids are derived.

Despite the research by Pavlinov and Shenbrot (1983), Shenbrot (1992), Stein (1990), and Zazhigin and Lopatin (2000a), classification of dipodoid genera at either subfamily or family levels remains unresolved, reflecting different interpretations of analyzed morphological systems (teeth, head and postcranial skeletons, male reproductive anatomy, hind limb myology), lack of molecular data from a broader range of genera representing all subfamilies, and no phylogenetic reconstructions based on careful cladistic methodologies that employ character traits derived from a greater variety of morphological systems. Results from multi-trait (morphological and molecular) integrative cladistic analyses, which include appropriate outgroups, would test the hypotheses presented by Vinogradov, Pavlinov and Shenbrot, Shenbrot, Stein, and Zazhigin and Lopatin. Until such information become available, we maintain a modification of Vinogradov's classification and recognize a single family, Dipodidae, with subfamilies Sicistinae, Zapodinae, Cardiocraniinae, Euchoreutinae, Allactaginae, and Dipodinae; hypotheses defended by Pavlinov and Shenbrot (1983), Stein (1990), Shenbrot (1992), and Zazhigin and Lopatin 2000a are discussed at the subfamilial level.

In this context, Klingener's (1984:387) exposition is still relevant. He noted that dipodid genera show a gradation in hind limb osteology and myology (with cardiocraniines retaining many sicistine and zapodine traits); that arranging sicistines and zapodines in a separate family "is unwarranted on the basis of morphology of the hind limb and obscures the close relationship of these rodents to the jerboas"; that cardiocraniines exhibit extreme hypertrophy of the auditory bullae and extreme fusion of cervical vertebrae, which are derivations characteristic of dipodines, but allactagines have relatively small bullae and unfused cervical vertebrae; that "evolution of cranial and cervical structures has been subject to selective forces different from those acting on locomotor structures, and distribution of derived character states in different structural complexes in the spectrum of Recent genera has a mosaic pattern" and that treating "the dipodoid spectrum as a single family Dipodidae . . . is a better reflection of biological reality than dividing it."

Reviews of dipodid research and classification were contributed by Gambaryan et al. (1980), Heptner (1984), Klingener (1964, 1984), Shenbrot (1986, 1992), and Stein (1990); see also references cited in those reports. Cranial and dental characters were investigated by Vinogradov (1930). Comparative myology studied by Klingener (1964); facial myology by Gambaryan et al. (1980) and Meinertz (1941); myology of postcrania by Fokin (1971) and Stein (1990). Review of distribution and habitat of dipodids (excluding sicistines and zapodines) given by Kulik (1980). Chromosome numbers of members of each subfamily provided by Vorontsov (1969). Male genitalia studied by Vinogradov (1925) and Pavlinov and Shenbrot (1983). Comparative behavior and its taxonomic significance studied by Rogovin (1984). Significance of the relation between size of pinna and auditory bulla in allactagine and dipodine species chronicled by Pavlinov and Rogovin (2000). Geographic distributions, ecologies, and comparative data for Mongolian jerboas monographed by Sokolov et al. (1996, 1998). Taxonomy, characteristics, geographic ranges, ecologies, and other topics covering dipodoids occurring in Russia and adjacent regions reviewed in detail by Shenbrot et al. (1995). Panteleyev (1998) provided distribution maps for the Palaearctic species. Distributions of species occuring in the former USSR were verified and much enhanced by personal communication with G. I. Shenbrot, who also highlighted taxonomic problems; range and taxonomy of Chinese forms were illuminated by the efforts of Lin Yong-lie.

That Muroidea (Platacanthomyidae, Spalacidae, Calomyscidae, Nesomyidae, Cricetidae, and Muridae) is the sister group to Dipodidae (Dipodoidea) has been consistently demonstrated by analyses of the following DNA sequences: nuclear LCAT gene (Michaux and Catzeflis, 2000); nuclear vWF (Huchon et al., 1999, 2000); LCAT and vWF combined (Michaux et al., 2001*b*); nuclear IRBP gene (DeBry and Sagel, 2001; Jansa and Weksler, 2004): nuclear CB1, IRBP, and RAG2 (DeBry, 2003); mitochondrial cytochrome *b* and ND4 (Conroy and Cook, 1999); mitochondrial cytochrome *b* and 12S rRNA (Montgelard et al., 2002); mitochondrial 12S rRNA combined with nuclear GHR, BRCA1, and vWF genes (Adkins et al., 2001); mitochondrial 12S rRNA (Nedbal et al., 1996); nuclear GHR and BRCA1 (Adkins et al., 2003); nuclear vWF, IRBP, and A2AB (Huchon et al., 2002); (samples are from *Sicista, Zapus, Napaeozapus, Allactaga, Dipus,* and *Jaculus* in different combinations). The molecular data strongly reinforces a significant body of past research based upon a range of morphological systems derived from extant (Alston, 1876; Forsyth Major, 1896; Hartenberger, 1985; Klingener, 1964; Méhely, 1913; Meinertz, 1941; Thomas, 1896; Tullberg, 1899; Winge, 1887) and fossil (Emry, 1981; Emry et al., 1998; Lindsay, 1977; Schaub, 1934; Walsh, 1997; B. Y. Wang, 1985; R. W. Wilson, 1949) species that indicates dipodids to be a sister group of, or closely related to muroid rodents (see also the reviews by Klingener, 1964, 1984). Schaub (1958) and McKenna and Bell (1997) expressed this relationship in their classifications by arranging Dipodoidea and Muroidea in the infraorder Myodonta. B. Y. Wang (1985) noted the close similarity in molar occlusal patterns between cricetines and early Oligocene "Zapodidae," and similarities in molar patterns between the middle Eocene Asian dipodid *Aksyiromys* and cricetines prompted Emry et al. (1998:222) to suggest that *Aksyiromys* "was close to a common ancestor of Cricetidae and Zapodidae; the Asian cricetine *Phodopus* was used as an outgroup in Stein's (1990) phylogenetic study and Cricetinae in general (as defined by by Carleton and Musser, 1984) was implied as an outgroup by Shenbrot (1991*a*). Alternative alliances between dipodoids and non-muroid rodent groups proposed in the older literature (e.g., Dobson, 1882*b*; Miller and Gidley, 1918; Parsons, 1894; Zittel, 1893) and even more recent contributions (e.g., Hartenberger, 1998) are not supported by morphological, molecular, or paleontological data.

North American and Asian Eocene strata contain evidence for the evolutionary beginnings of dipodoids. Current consensus identifies the North American *Elymys* (middle Eocene), *Simimys* (middle to late Eocene), *Simiacritomys* (late Eocene), and possibly *Nonomys* (late Eocene), along with Asian *Aksyiromys, Banyuesminthus, Blentosomys, Primisminthus,* and *Ulkenulastomys* (all middle Eocene) as the oldest members of Dipodoidea (Emry and Korth, 1989; Kelly, 1992; Walsh, 1997; Wang and Dawson, 1994; Wang and Qiu, 2000). Whether any or all genera are ancestral to dipodids, belong in Sicistinae or Zapodinae, or represent the sister group to Dipodidae within a larger Dipodoidea has yet to be determined (see review of these Paleogene genera and analysis by Walsh, 1997). *Armintomys tullbergi*, from the early part of the middle Eocene of North America and the only recorded member of Armintomyidae (Dawson et al., 1990), is the oldest known rodent with hystricomorphous zygomasseteric structure and incisor enamel transitional from pauciserial to uniserial (Wang and Dawson, 1994). It is apparently a primitive myomorph, and its morphological traits (sciuravid-like molar occlusal patterns, hystricomorphy, and intermediate incisor enamel architecture) indicate that *A. tullbergi* "may be derived from the sciuravids" (Dawson et al., 1990:145) and also "may represent the closest known sister group to a unit including dipodoids and cricetids" (Wang and Dawson, 1994:251). In Walsh's (1997) phylogenetic analysis, *Armintomys,* along with the more derived North American *Pauromys* (early to middle Eocene), formerly considered a sciuravid (McKenna and Bell, 1997), is placed in Myodonta (the infraorder also containing dipodoids and muroids; see McKenna and Bell, 1997) and basal to the branching point of Dipodoidea and Muroidea (see McKenna and Bell, 1997). The basic outline of this phylogenetic pattern was discerned more than 50 years ago by Wilson (1949;127), when in reviewing limited material compared to that now available, he suggested that "the muroids and dipodoids, although arising from a common (mid-Eocene?) ancestor, have been distinct groups since the late Eocene. Moreover, this common ancestor apparently arose from the sciuravines, if any stem is known at present."

Subfamily Allactaginae Vinogradov, 1925. Proc. Zool. Soc. Lond., 1925(1):578.

SYNONYMS: Alactaginae Vinogradov, 1925; Allactaginae Vinogradov, 1930; Allactodipodini Zazhigin and Lopatin, 2000 [*nomen nudum*].

COMMENTS: Dental morphology and evolution studied by Shenbrot (1984). Allactodipodini was proposed for *Allactodipus* (Zazhigin and Lopatin, 2000*a*), but no type genus was explicitly indicated. Modern range of the subfamily is Asian (only *Allactaga tetradactyla* occurs in north Africa) and its evolutionary roots are in that continent. The first allactagine may have evolved from Asian late Oligocene *Gobiosminthus* (Zazhigin and Lopatin, 2000*c*) or *Parasminthus* (Wang and Qiu, 2000; McKenna and Bell, 1997, treat both genera as synonyms of *Plesiosminthus*) into two major lineages (Zazhigin and Lopatin, 2000*c*). One is represented by species of *Protalactaga*, the first true allactagines, from latter portion of Asian early Miocene, and may have been ancestral to *Allactaga*, which first appears in the later half of Asian late Miocene and is represented by an array of species in the Pliocene. *Pygerethmus* dates from the late Pliocene; it and the extinct Asian *Brachyscirtetes* (late Miocene to early Pliocene) may have been derived from *Allactaga*. *Allactodipus bobrinskii*, unrepresented by fossils, is the other lineage, and Zazhigin and Lopatin (2000*c*) speculated a North African or Near East origin for the genus. To taxonomically reflect this evolutionary duality, Zazhigin and Lopatin (2000*a*) separated *Allactodipus* in a tribe separate from the other allactagines. Outside of Asia, Miocene Allactaginae are represented by *Arabosminthus* from Saudia Arabia (late Miocene, usually treated as a zapodid; R. A. Martin, 1994; McKenna and Bell, 1997) and a middle Miocene sample from Morocco originally identified as "*Protalactaga moghrebiensis*" (Jaeger, 1977*b*) that likely represents an undescribed genus (Zazhigin and Lopatin, 2000*c*). *Himalayactaga*, from late Miocene Tibetan strata, is usually regarded as an allactagine (e.g., McKenna and Bell, 1997), but Zazhigin and Lopatin (2000*c*:554) considered it of "uncertain taxonomic position within the Myomorpha." European late Pliocene and Pleistocene records of *Allactaga* and *Pygerethmus* are summarized by Kowalski (2001).

Allactaga F. Cuvier, 1837. Proc. Zool. Soc. Lond., 1836:141 [1837].

TYPE SPECIES: *Mus jaculus* Pallas, 1778 (= *Dipus sibericus major* Kerr, 1792); see comments below.

SYNONYMS: *Alactaga* F. Cuvier, 1838; *Beloprymnus* Gloger, 1841 [*nomen nudum*]; *Cuniculus* Brisson, 1762 [*nomen nudum*]; *Mesoallactaga* Shenbrot, 1974; *Microallactaga* Shenbrot, 1974; *Orientallactaga* Shenbrot, 1984; *Paralactaga* Young, 1927; *Proalactaga* Savinov, 1970; *Scarturus* Gloger, 1841; *Scirteta* Brandt, 1844; *Scirtetes* Wagner, 1841 [not Hartig, 1838]; *Scirtomys* Brandt, 1844; see Ellerman and Morrison-Scott (1951) and Pavlinov and Rossolimo (1987, 1998).

COMMENTS: Does not include *Allactodipus* (see comment under *Allactodipus bobrinskii*). Shenbrot's (1984) subgeneric classification is followed below except where noted. Subspecific revision of species occuring in Belorussia, Ukraine, Kazakhstan, W Uzbekistan, Kyrgyzstan, S Russia, and NW Mongolia provided by Shenbrot (1993). Generic review and composite range map of species presented by Shenbrot et al. (1995). *Mus jaculus* Linnaeus, 1758 is the type species of *Jaculus* (see comments therein). Shenbrot (1984) described several subgenera, which we use here, but Zazhigin and Lopatin (2000*c*) claimed their distinctness is doubtful. Shenbrot et al. (1995) included the extinct *Protalactaga* (Miocene of North Africa and Asia) as a synonym, but Zazhigin and Lopatin (2000*c*) and McKenna and Bell (1997) retained it as a separate genus. McKenna and Bell (1997) recognized *Proalactaga* (Asian late Miocene), and Kowalski (2001) used *Paralactaga* (Pliocene-Pleistocene), but Shenbrot (1984) and Zahigin and Lopatin (2000*b*) explained why the two genera should be synonymized with *Allactaga*.

Allactaga balikunica Hsia and Fang, 1964. Acta Zootaxon. Sin., 6:16.

COMMON NAME: Balikun Jerboa.

TYPE LOCALITY: China, Xinjiang, Balikun.

DISTRIBUTION: Mongolia, from Altai Sumon east to Bordzon-Gobi (Sokolov et al., 1981*a*), and NE Xinjiang, China (Ma et al., 1987).

STATUS: IUCN – Lower Risk (lc).

SYNONYMS: *nataliae* Sokolov, 1981 [see Sokolov and Shenbrot, 1987*a*].

COMMENTS: Subgenus *Orientallactaga*. Closely related to *A. bullata*, with which it is parapatric in S Mongolia where the two species occur together in a narrow strip several kilometers wide (G. Shenbrot, in litt., 2003). Listed as a subspecies of *A. bullata* by Wang (2003).

Allactaga bullata Allen, 1925. Am. Mus. Novit., 161:2.

COMMON NAME: Gobi Jerboa.

TYPE LOCALITY: Mongolia, Altai Gobi, Tsagan-Nur (Tsagaan Nuur).

DISTRIBUTION: Deserts of S and W Mongolia (Bannikov, 1954; Sokolov et al., 1981*a*; adjacent Chinese provinces of Nei Mongolia, E Xinjiang, Ningxia (Ma et al., 1987), Gansu (Chen and Wang, 1985; Zheng and Zhang, 1990) and N Shaanxi (T.-z. Wang, 1990).

STATUS: IUCN – Lower Risk (nt).

COMMENTS: Subgenus *Orientallactaga*. Closely related to *A. balikunica* (see that account).

Allactaga elater (Lichtenstein, 1828). Abh. König. Akad. Wiss., Berlin, 1825:155 [1828].

COMMON NAME: Small Five-toed Jerboa.

TYPE LOCALITY: W Kazakhstan, Kirgiz Steppe. The type locality given by Lichtenstein (1828) is in W Kazakhstan, according to Vinogradov (1937), Kuznetsov (1944, 1965), and Ognev (1963*b*), not E Kazakhstan as reported by Ellerman and Morrison-Scott (1951:529) and Corbet (1978*c*:154).

DISTRIBUTION: SW Pakistan (Roberts, 1977, 1997); Afghanistan (Hassinger, 1973); Iran (Lay, 1967); E Turkey (Kryštufek and Vohralík, 2001); Armenia, Azerbaijan, Georgia, N Caucasus, north along W Caspian Sea to Lower Volga south to Turkmenistan, east through Kazakhstan (see Kuznetsov, 1965; Sludskii, 1977; and Shenbrot, 1993) to NE Xinjiang, Nei Mongol, and N Gansu, China (Ma et al., 1987; Chinese range mapped in Zhang et al., 1997), and western Mongolia (Sokolov and Orlov, 1980), in desert and semi-desert zones. Overall range mapped in Shenbrot et al. (1995).

STATUS: IUCN – Lower Risk (lc).

SYNONYMS: *aralychensis* Satunin, 1901; *bactriana* Blyth, 1863; *caucasicus* Nehring, 1900; *dzungariae* Thomas, 1912; *heptneri* Pavlenko and Denisov, 1976; *indica* Gray, 1842 [not Gray, 1830 or Bechstein, 1800]; *kizljaricus* Satunin, 1907; *strandi* Hepner, 1934; *turkmeni* Goodwin, 1940; *zaisanicus* Shenbrot, 1993 [see Corbet, 1978*c*, 1984; Shenbrot, 1993].

COMMENTS: Subgenus *Allactaga*. Does not include *vinogradovi* (see comment therein). Detailed review provided by Ognev (1963*b*). Karyotype contributed by Vorontsov et al. (1969*c*). Subspecific revision and additional distributional data provided by Shenbrot (1993). Segments in Russia, Turkmenistan, Uzbekistan, and Kazakhstan reviewed by Gromov and Erbajeva (1995) and Shenbrot et al. (1995). Shenbrot (in litt., 2003) is skeptical about distribution of this species in Nei Mongol, and N Gansu, China due to possible misidentification with *Pygeretmus pumilio*.

Allactaga euphratica Thomas, 1881. Ann. Mag. Nat. Hist., ser. 5, 8:15.

COMMON NAME: Euphrates Jerboa.

TYPE LOCALITY: Iraq (no other information).

DISTRIBUTION: Steppe and semi-desert from SE Turkey (Colak et al., 1994) south through Syria and Iraq to Jordan (Qumsiyeh, 1996), N Saudi Arabia and Kuwait (essentially as outlined by Ellerman and Morrison-Scott, 1951; see also Colak et al., 1994, and references therein; Hatt, 1959).

STATUS: IUCN – Lower Risk (nt).

SYNONYMS: *kivanci* Çolak and Yiğit, 1998 [see Corbet, 1978*c*; Kryštufek and Vohralík, 2001].

COMMENTS: Subgenus *Paralactaga*. Many authors incorrectly employ the name *williamsi* for this species, but that name refers to a different species (see account of *A. williamsi*). Following Corbet (1978*c*:155), *A. euphratica* is considered a separate species from *A. hotsoni*. Reviewed by Colak et al. (1994).

Allactaga firouzi Womochel, 1978. Fieldiana Zool., 72(5):65.

COMMON NAME: Iranian Jerboa.

TYPE LOCALITY: Iran, Isfahan Prov., 18 mi (29 km) S Shah Reza (Qomisheh), 2253 m.
DISTRIBUTION: Known only from the type locality, a flat plain with a gravel substrate and sparse, mountain steppe vegetation (Womochel, 1978).
STATUS: IUCN – Critically Endangered.
COMMENTS: Subgenus *Allactaga* (not allocated to subgenus by Shenbrot, 1984). *Allactaga firouzi* appears to be morphologically distinct from *A. euphratica* and *A. hotsoni* (Womochel, 1978), but its relationship with these and other species of allactagines needs further study. Shenbrot (1993) tenatively synonymized *firouzi* with *A. elater turkmeni*, but later (in litt.) examinied the type specimen and considered *firouzi* synonymous with *hotsoni*, which is where Pavlinov et al. (1995) listed it. We recognize the species until published data indicates otherwise.

Allactaga hotsoni Thomas, 1920. J. Bombay Nat. Hist. Soc., 26(4):936.
COMMON NAME: Hotson's Jerboa.
TYPE LOCALITY: Iran, SE Kerman Prov., Sib, 20 mi (32 km) SW Kant (Kont; Lay, 1967, provides coordinates), 3950 ft (1204 m).
DISTRIBUTION: N, C, and SE Iran, SW Pakistan (Brown, 1980; Roberts, 1977, 1997; specimens in USNM), and S Afghanistan (Hassinger, 1973) in gravelly or stony peneplains "where practically no other rodent exists" (Roberts, 1997:352).
STATUS: IUCN – Lower Risk (lc).
COMMENTS: Subgenus *Allactaga* (see Pavlinov and Shenbrot, 1985). Lay (1967) thought the holotype might represent *A. williamsi*, but Corbet (1978c:155) noted that "the very large bullae, short tooth-row and short, wide incisive foramina place it well outside the range of variation otherwise known" in *A. williamsi* (Corbet regarded *williamsi* as a synonym of *A. euphratica*). Morphological distinctions between *A. hotsoni* and *A. elator* in Pakistan are described by Roberts (1997). Available published data support recognition of *A. hotsoni* as separate from *A. elater* and *A. euphratica*; see comment under *A. firouzi*.

Allactaga major (Kerr, 1792). *In* Linnaeus, Anim. Kingdom, p. 274.
COMMON NAME: Great Jerboa.
TYPE LOCALITY: Kazakhstan, between Caspian Sea and Irtysh River.
DISTRIBUTION: Steppes and deserts from Caucasus N to Moscow and Kiev E to Ob River (W Siberia), Kazakhstan, and N Uzbekistan (west of Aral Sea), and W Xinjiang, China (Wang, 2003); range figured by Kuznetsov (1965), Shenbrot et al. (1995), and Sludskii (1977).
STATUS: IUCN – Lower Risk (lc).
SYNONYMS: *aulacotis* (Wagner, 1840); *brachyotis* Brandt, 1844; *chachlovi* Martino, 1921; *decumanus* (Lichtenstein, 1825); *djetysuensis* Shenbrot, 1993; *flavescens* Brandt, 1844; *fuscus* Ognev, 1924; *hochlovi* Martino, 1922; *intermedius* Ognev, 1948; *jaculus* (Pallas, 1779) [not Linneaus, 1758]; *macrotis* Brandt, 1844; *nigricans* Brandt, 1844; *spiculum* (Lichtenstein, 1825); *vexillarius* (Eversmann, 1840); see Ellerman and Morrison-Scott (1951), Corbet (1978c), and Shenbrot (1991d).
COMMENTS: Subgenus *Allactaga*. Includes *jaculus* Pallas (Ognev, 1963b:94; Pavlinov and Rossolimo, 1987:154). Subspecific revision and additional distributional data provided by Shenbrot (1993). Karyotype contributed by Vorontsov et al. (1969c). Reviewed by Ognev (1963b), Gromov and Erbajeva (1995), and Shenbrot et al. (1995). Detailed habitat data provided by Naumov and Lobachev (1975). Although modern western margins of range are near Moscow and Kiev, the species occurred as far west as Germany and Austria during the Pleistocene (Kowalski, 2001).

Allactaga severtzovi Vinogradov, 1925. Proc. Zool. Soc. Lond., 1925:583.
COMMON NAME: Severtzov's Jerboa.
TYPE LOCALITY: SE Kazakhstan, Taldy-Kurgan (Kopal) dist., Tamar-Utkul.
DISTRIBUTION: Kazakhstan, Uzbekistan, N Turkmenistan, and N and SW Tajikistan; range figured by Kuznetsov (1965), Shenbrot et al. (1995), and Sludskii (1977).
STATUS: IUCN – Lower Risk (lc).
SYNONYMS: *chorezmi* Shenbrot, 1993.
COMMENTS: Subgenus *Allactaga*. Taxonomic study provided by Shenbrot (1991d), who

described *chorezmi* as a subspecies of *A. severtzovi*. Reviewed by Ognev (1963*b*) and Shenbrot et al. (1995). Karyotype elaborated by Vorontsov et al. (1969*c*). Detailed habitat data provided by Naumov and Lobachev (1975).

Allactaga sibirica (Forster, 1778). K. Svenska Vet.-Akad. Handl. Stockholm, 39:112.
COMMON NAME: Mongolian Five-toed Jerboa.
TYPE LOCALITY: SE Transbaikalia, Chitinskaya Oblast, near Lake Tarei-Nur.
DISTRIBUTION: From lower Ural River (Kazakhstan) and Caspian Sea east to Chitinskaya Oblast and south to N Turkmenistan (Kuznetsov, 1965; Shenbrot, 1993; Sludskii, 1977); Mongolia (Bannikov, 1954); China: Nei Mongol, Xinjiang, Qinghai, Gansu, Nigxia, Shaanxi, Shanxi, N Hebei, W Liaoning, W Julin, and W Heilongjiang (in China see Ho et al., 1986; Laing and Zhang, 1985; Liu et al., 1990; Ma et al., 1987; Mi et al., 1990; Qin, 1991; Shou, 1962; Zhang and Wang, 1963; Wang, 2003; Zhang et al., 1997; and Zheng and Zhang, 1990); Shenbrot et al. (1995) provided overall distribution map; no valid record from Korea (see Corbet, 1978*c*:154).
STATUS: IUCN – Lower Risk (lc).
SYNONYMS: *alactaga* (Olivier, 1800); *alpinus* Shnitnikov, 1936; *altorum* Ognev, 1946; *annulata* (Milne-Edwards, 1867); *brachyurus* (Blainville, 1817); *bulganensis* Shenbrot, 1993; *dementiewi* Toktosunov, 1958; *grisescens* Hollister, 1912; *halticus* (Illiger, 1825); *longior* Miller, 1911; *media* (Pallas, 1779); *mongolica* (Radde, 1861); *ognevi* Shenbrot, 1991; *ruckbeili* Thomas, 1914; *salicus* Ognev, 1924; *saliens* (Gmelin, 1760); *saliens* (Shaw, 1790); *saltator* (Eversmann, 1848); *semideserta* Bannikov, 1947; *suschkini* Satunin, 1900; see Ellerman (1940), Ellerman and Morrison-Scott (1951), Corbet (1978*c*, 1984), Pavlinov and Rossolimo (1987, 1998), Shenbrot (1993), and Shenbrot et al. (1995).
COMMENTS: Subgenus *Orientallactaga*. Reviewed by Gromov and Erbajeva (1995), Ognev (1963*b*), and Shenbrot et al. (1995). Subspecific revision and additional distributional data provided by Shenbrot (1993). Geographic variation studied by Varshavsky (1991). Karyotype described by Vorontsov et al. (1969*c*). Detailed habitat data provided by Naumov and Lobachev (1975).

Allactaga tetradactyla (Lichtenstein, 1823). Verz. Doublet. Zool. Mus. Univ. Berlin, p. 2.
COMMON NAME: Four-toed Jerboa.
TYPE LOCALITY: Libyan Desert between Siwa and Alexandria.
DISTRIBUTION: Coastal gravel plains of Egypt and E Libya, from near Alexandria to the Gulf of Sirte (see Ranck, 1968, and Osborn and Helmy, 1980).
STATUS: IUCN – Endangered.
SYNONYMS: *brucii* (Lesson, 1827).
COMMENTS: Subgenus *Scarturus*.

Allactaga vinogradovi Argyropulo, 1941. Fauna SSSR, Mlekopitaiushchiy, Opredelitel grizunov, Izdatelstvo Akademii Nauk, SSSR, p. 138.
COMMON NAME: Vinogradov's Jerboa.
TYPE LOCALITY: Kazakhstan, Dzhambul region, Burnoye and Rovnoye.
DISTRIBUTION: S Kazakhstan, E Uzbekistan, and Kyrgyzstan (see Shenbrot, 1993; Shenbrot et al., 1995).
STATUS: IUCN – Lower Risk (lc).
COMMENTS: Subgenus *Allactaga*. Shenbrot (1993) showed that *A. vinogradovi* differs from *A. elater* in toothrow length and phallic morphology, and that the two are sympatric along the Talas River and in SE Betpak-Dala (Kazakhstan). While the species is recognized by most workers (Pavlinov and Rossolimo, 1998; Pavlinov et al., 1995; Shenbrot et al., 1995), Gromov and Erbajeva (1995) retained *vinogradovi* in the synonymy of *A. elater*. Reviewed by Shenbrot et al. (1995). There are no recent records of *A. vinogradovi* from Tajikistan, but the species occurred there during middle Pleistocene (G. Shenbrot, in litt., 2003).

Allactaga williamsi Thomas, 1897. Ann. Mag. Nat. Hist., 20:309.
COMMON NAME: Williams's Jerboa.
TYPE LOCALITY: Turkey, near Van Gölü.
DISTRIBUTION: Anatolian *Artemisia* steppe of Turkey except in SE, Caucasia, Afghanistan,

and Iran (see Colak et al., 1994, and references therein; Kryštufek and Vohralík, 2001; range mapped as *euphratica* by Shenbrot et al., 1995, and recorded from N Turkey as *euphratica* by Pamukoglu and Albayrak, 1996).

SYNONYMS: *caprimulga* Ellerman, 1948; *laticeps* Nehring, 1903; *schmidti* Satunin, 1907.

COMMENTS: Subgenus *Paralactaga*. From the time it was described, *williamsi* was considered a species. Ellerman and Morrison-Scott (1951:530) recognized its specific status but noted that *A. williamsi* "is very close to *euphratica*, possibly merely a further series of larger races of that." Based upon a sample from Syria that seemed to show traits intermediate between *euphratica* and *williamsi*, Atallah and Harrison (1968) reduced the latter to subspecific rank. Colak et al. (1994), however, using reproductive, external, morphological, cranial, chromosomal, and behavioral traits, showed that *euphratica* and *williamsi* are separate species without evidence of hybridization in any geographic region. Gromov and Erbajeva (1995), Pavlinov and Rossolimo (1998), Pavlinov et al. (1995), and Shenbrot et al. (1995) retained *williamsi* and *schmidti* in the synonymy of *A. euphratica*. Reviewed by Colak et al. (1994) and Ognev (1963*b*). Detailed habitat data provided by Naumov and Lobachev (1975). Recorded in owl pellets from E Turkey (Obuch, 1994).

Allactodipus Kolesnikov, 1937. Bull. Sredne-Az. Gos. Univ., 22(29):255.

TYPE SPECIES: *Allactodipus bobrinskii* Kolesnikov, 1937.

COMMENTS: Reviewed by Shenbrot (1974, 1984), who showed that in postcranial and dental characters, particularly the height of the molars and alveolar pattern, *A. bobrinskii* falls outside the range of variation of species in the genus *Allactaga* where it was placed after Kolesnikov's description. He restored *bobrinskii* to *Allactodipus*, which is recognized by Pavlinov and Rossolimo (1987, 1998), Pavlinov et al. (1995), Shenbrot et al. (1995), and Zazhigin and Lopatin (2000*a, b*). A careful phylogenetic study incorporating the traits used by Shenbrot, along with additional morphological differences, between *A. bobrinskii* and other allactagines and dipodines that also includes gene sequences would be welcome.

Allactodipus bobrinskii Kolesnikov, 1937. Bull. Sredne-Az. Gos. Univ., 22(29):255.

COMMON NAME: Bobrinski's Jerboa.

TYPE LOCALITY: Uzbekistan, Kizil-kum (Kyzylkum) Desert, 140 km NW of Bukhara, Khala-Ata.

DISTRIBUTION: W and N Turkmenistan and C and W Uzbekistan, in the Kyzylkum and Karakumy deserts; figured by Kuznetsov (1965), Shenbrot et al. (1995), and Sludskii (1977).

STATUS: IUCN – Lower Risk (lc).

COMMENTS: Karyotype provided by Vorontsov et al. (1969*c*). Reviewed by Gromov and Erbajeva (1995) and Shenbrot et al. (1995).

Pygeretmus Gloger, 1841. Gemeinn. Hand. Hilfsbuch. Nat., 1:106.

TYPE SPECIES: *Dipus platurus* Lichtenstein, 1823 (emended to *D. platyurus* by Lichtenstein, 1828).

SYNONYMS: *Alactagulus* Nehring, 1897; *Platycercomys* Brandt, 1843; *Pliopygerethmus* Topachevskii and Skorik, 1971; *Pseudoalactaga* Topachevskii, 1971; *Pygerethmus* Vinogradov, 1930; *Pygeretmus* Gloger, 1841; see McKenna and Bell (1997) and Zazhigin and Lopatin (2000*c*).

COMMENTS: Vorontsov et al. (1969*b*) found no distinguishing karyological characters between subgenera *Alactagulus* and *Pygeretmus*; in the same study, karyotypes and morphometric comparisons of subgenus *Pygeretmus* were provided. Shenbrot's (1984) subgeneric classification is followed below. Gromov and Erbajeva (1995) recognized *Alactagulus* as a separate genus, and Kowalski (2001) used *Pliopygerethmus*, but both are considered synonyms of *Pygeretmus* (Heptner, 1984; Pavlinov and Rossolimo, 1987; Shenbrot, 1984; Zazhigin and Lopatin, 2000*c*). Generic review and distribution provided by Shenbrot et al. (1995).

Pygeretmus platyurus (Lichtenstein, 1823). Naturh. Abh. Eversmann's Reise, p. 121.

COMMON NAME: Lesser Fat-tailed Jerboa.

TYPE LOCALITY: Kazakhstan, E shore of Aral Sea Kuwan-Darya River.

DISTRIBUTION: W, C, and E Kazakhstan, and NW Turkmenistan (G. Shenbrot, in litt., 2003; see map in Shenbrot et al., 1995).

STATUS: IUCN – Lower Risk (lc).

SYNONYMS: *platurus* (Lichtenstein, 1823); *vinogradovi* Vorontsov, 1958.

COMMENTS: Subgenus *Pygeretmus*. Corbet (1978c) and Gromov and Baranova (1981) synonymized *P. vinogradovi* with *P. platyurus* without comment. Heptner (1984) agreed because he felt the differences between the two forms were not sharp enough to warrant specific recognition, a view followed by Pavlinov and Rossolimo (1987, 1998), Pavlinov et al. (1995), and Shenbrot et al. (1995). Shenbrot's (1988) results indicated that *P. vinogradovi* falls within the range of variaiton of *P. platyurus*, and that *vinogradovi* should not even be recognized at the subspecific level. Reviewed by Gromov and Erbajeva (1995), Shenbrot et al. (1995), Silverstov et al. (1969) and Sludskii (1977); detailed habitat data provided by Naumov and Lobachev (1975).

Pygeretmus pumilio (Kerr, 1792). *In* Linnaeus, Anim. Kingdom, p. 275.

COMMON NAME: Dwarf Fat-tailed Jerboa.

TYPE LOCALITY: Kazakhstan, between Caspian Sea and Irtysh River. Pavlinov and Rossolimo (1987) listed Kirghiz Steppe, "the old Russian name for Central Kazakhstan" (G. Shenbrot, in litt., 2003) as the type locality for this species.

DISTRIBUTION: From the Don River (Russia) through Kazakhstan to the Irtysh River (Kuznetsov, 1965; Sludskii, 1977), south to NE Iran (Lay, 1967); E to S Mongolia (Bannikov, 1954); China: W Nei Mongol (Ma et al., 1987), N Xinjiang (Chen and Wang, 1985; Ma et al., 1987), Gansu, and Ningxia (Wang, 2003; Chinese range mapped in Zhang et al., 1997).

STATUS: IUCN – Lower Risk (lc).

SYNONYMS: *acontion* (Pallas, 1811); *aralensis* (Ognev, 1948); *brachyotis* (Ostrouchov, 1889) [*nomen nudum*]; *dinniki* (Satunin, 1920); *minor* (Pallas, 1779); *minutus* (Blainville, 1817); *pallidus* (Vinogradov, 1933); *potanini* (Vinogradov, 1926); *pumilio* (Kerr, 1792); *pygmaea* (Pallas, 1779); *tanaiticus* (Ognev, 1948); *turcomanus* (Heptner and Samorodov, 1939); see Ellerman and Morrison-Scott (1951), Corbet, (1978c), and Shenbrot et al. (1995).

COMMENTS: Subgenus *Alactagulus* (see Pavlinov and Rossolimo, 1987). The name *pygmaeus* is preoccupied and is an invalid junior synonym of *pumilio*, (Corbet, 1978c; Ellerman and Morrison-Scott, 1951). Reviewed by Ognev (1963b) and Shenbrot et al. (1995). Reviewed also by Gromov and Erbajeva (1995) who retained *Alactagulus* as the genus, not subgenus, for *pumilio*.

Pygeretmus shitkovi (Kuznetsov, 1930). Doklady Acad. Nauk S.S.S.R., Leningrad, 1930A:623.

COMMON NAME: Greater Fat-tailed Jerboa.

TYPE LOCALITY: SE Kazakhstan, Taldy-Kurgan district, on NW shore of Ala-Kul (Alakol) Lake, Rybalnoje.

DISTRIBUTION: E Kazakhstan, in region of Lake Balkhash (Corbet, 1978c; Kuznetsov, 1965; Shenbrot et al., 1995; Sludskii, 1977).

STATUS: IUCN – Lower Risk (nt).

SYNONYMS: *schitkovi* (Vinogradov, 1930); *zhitkovi* (Vinogradov, 1937); see Ellerman and Morrison-Scott (1951) and Pavlinov and Rossolimo (1987, 1998).

COMMENTS: Subgenus *Pygeretmus*. Corbet (1978c) regarded the emendation of *shitkovi* to *zhitkovi* invalid, but other workers disagree; there has been no ruling by the International Commission on Zoological Nomenclature. Reviewed by Ognev (1963b), Gromov and Erbajeva (1995), and Shenbrot et al. (1995).

Subfamily Cardiocraniinae Vinogradov, 1925. Proc. Zool. Zoc. Lond., 1925(1):578.

SYNONYMS: Cardiocraniini Pavlinov, 1980; Salpingotinae Vinogradov, 1925 (Salpingotini Pavlinov, 1980).

COMMENTS: Following Pavlinov (1980b) two tribes, Cardiocraniini and Salpingotini, are recognized. Pavlinov and Shenbrot (1983) recognized Cardiocraniinae and Salpingotinae in a Dipodidae, which was modified by Shenbrot (1992), who recognized only Cardiocraniinae within Dipodidae, an arrangement also employed by Zazhigin and

Lopatin (2000a). Shenbrot et al. (1995) include the extinct Lophocricetinae, but other researchers recognize it as a subfamily in Zapodidae (R. A. Martin, 1994; Qiu, 1985), or tribe in Zapodinae (McKenna and Bell, 1997), or a subfamily in Dipodidae (where Zapodidae, Allactagidae, and Dipodidae are recognized; Zazhigin and Lopatin, 2000a). Characters of subfamily and geographic distribution reviewed by Shenbrot et al. (1995).

Cardiocranius Satunin, 1903. Ann. Mus. Zool. Acad. Imp. Sci. St. Petersbourg, 7:582.
TYPE SPECIES: *Cardiocranius paradoxus* Satunin, 1903.
COMMENTS: Cardiocraniini. Detailed review provided by Ognev (1963b) and Shenbrot et al. (1995).

Cardiocranius paradoxus Satunin, 1903. Ann. Mus. Zool. Acad. Imp. Sci. St. Petersbourg, 7:584.
COMMON NAME: Five-toed Pygmy Jerboa.
TYPE LOCALITY: China, NW Gansu, Nan Shan, Shargol-Dzhin.
DISTRIBUTION: China (N Xinjiang, C Nei Mongol, N Ningxia, and Gansu; see G. Shenbrot, in litt., 2003; Ma et al., 1987; Mi et al., 1990; Qin, 1991; Wang, 2003; Zhang et al., 1997, and Zhou et al., 1985); Mongolia; S Tuviskaya Oblast, and E Kazakhstan (see Ilchenko and Volodin, 1992, and Sokolov and Shenbrot, 1988). Shenbrot et al. (1995) provide overall distribution map.
STATUS: IUCN – Vulnerable.
COMMENTS: A morphometric study by Sokolov and Shenbrot (1988) indicated that the Kazakhstan population falls within the range of intraspecific variation for *C. paradoxus*, and is not a separate species as was suggested by Gromov and Baranova (1981). For detailed habitat data see Naumov and Lobachev (1975). Reviewed by Gromov and Erbajeva (1995) and Shenbrot et al. (1995).

Salpingotulus Pavlinov, 1980. Vest. Zool., 2:50.
TYPE SPECIES: *Salpingotus michaelis* Fitzgibbon, 1966.
COMMENTS: Salpingotini (Pavlinov, 1980b). Proposed by Pavlinov (1980b) based upon genital morphology, tooth characters, and shape of the condylar process on the dentary. Results of a morphometric and qualitative study of *Salpingotus* by Vorontsov and Shenbrot (1984) showed substructure within the genus based on overall morphological similarity (phenetic analyses), with *michaelis* joining the clusters of other *Salpingotus* at a high level of dissimilarity, followed by *S. kozlovi*. Their results indicated that generic separation of *michaelis* was not supported, particularly if *kozlovi* was retained in *Salpingotus*, so they treated *Salpingotulus* as a subgenus. Pavlinov and Rossolimo (1987:162) suggested that *kozlovi* might belong in *Salpingotulus*, and the latter has either been included in *Salpingotus* (e.g., Holden, 1993; Shenbrot et al., 1995; Vorontsov and Shenbrot, 1984) or recognized as a separate genus (Pavlinov and Rossolimo, 1998; Pavlinov et al., 1995; Zazhigin and Lopatin, 2000a). We treat *Salpingotulus* as a separate monophyletic group until analyses prepared within a phylogenetic (cladistic) context demonstrate otherwise; such a study has yet to be done.

Salpingotulus michaelis (Fitzgibbon, 1966). Mammalia, 30(3):431.
COMMON NAME: Baluchistan Pygmy Jerboa.
TYPE LOCALITY: Pakistan, NW Baluchistan, Nushki Plateau, appox. 29°N, 66°E, 3500 ft (1067 m).
DISTRIBUTION: Pakistan, SW Baluchistan (Roberts, 1977, 1997).
STATUS: IUCN – Lower Risk (lc) as *Salpingotus michaelis*.
COMMENTS: For discussion of *Salpingotulus michaelis* versus *Salpingotus thomasi*, and the possible occurance of *Salpingotulus michaelis* in Afghanistan, see Hassinger (1973) and Roberts (1977, 1997).

Salpingotus Vinogradov, 1922. *In* Kozlov, Mongolia and Amdo, p. 540.
TYPE SPECIES: *Salpingotus kozlovi* Vinogradov, 1922.
SYNONYMS: *Anguistodontus* Vorontsov and Shenbrot, 1984; *Prosalpingotus* Vorontsov and Shenbrot, 1984; see Pavlinov and Rossolimo (1987) and Shenbrot et al. (1995).
COMMENTS: Salpingotini. The subgeneric arrangement hypothesized by Vorontsov and

Shenbrot (1984), based on morphometric analyses and skull, dental, and genital morphology, is followed here, except that *Salpingotulus* is recognized at the generic level (see that account). Comparative myology of pelvic girdle studied by Fokin (1971). Distribution of species of *Salpingotus* shown in Shenbrot et al. (1995) and Vorontsov and Shenbrot (1984).

Salpingotus crassicauda Vinogradov, 1924. Zool. Anz., 61:150.
COMMON NAME: Thick-tailed Pygmy Jerboa.
TYPE LOCALITY: China, N Xinjiang, Altai Gobi, near Schara-sumé (Sharasume), approx. 160km S Russia-Mongolian border. Most authors list the type locality as being in W Mongolia, but it is actually in N Xinjiang, China (G. Shenbrot, pers. comm.).
DISTRIBUTION: Steppes and deserts of NW China (Xinjiang, Nei Mongol, and Gansu; Chen and Wang, 1985; Ma et al., 1987; Wang, 2003; Zheng and Zhang, 1990; Chinese range mapped by Zhang et al., 1997), S and SW Mongolia, and adjacent E Kazakhstan in Lake Zaysan basin (see Naumov and Lobachev, 1975; Vorontsov and Shenbrot, 1984; Vorontsov et al., 1969a). General distribution mapped by Shenbrot et al. (1995).
STATUS: IUCN – Vulnerable.
SYNONYMS: *gobicus* Sokolov and Shenbrot, 1988.
COMMENTS: Subgenus *Anguistodontus*. Populations south of Lake Balkhash and north of Aral Sea previously included in *S. crassicauda* now recognized as a distinct species, *S. pallidus*. Reviewed by Ognev (1963b). Taxonomic study of Mongolia and Zaysan Basin populations provided by Sokolov and Shenbrot (1988). Karyotype elaborated by Vorontsov et al. (1969d). Reviewed by Ognev (1963b), Gromov and Erbajeva (1995) and Shenbrot et al. (1995).

Salpingotus heptneri Vorontsov and Smirnov, 1969. *In* Vorontsov (ed.) [The mammals: evolution, karyology, taxonomy, fauna.], Novosibirsk, p. 60.
COMMON NAME: Heptner's Pygmy Jerboa.
TYPE LOCALITY: Uzbekistan, NW Kizil-Kum (Kyzylkum) Desert, 80 km NE Takhta-Kupir, 8 km E Gori Kok-Tobe.
DISTRIBUTION: Uzbekistan and S Kazakhstan, NW and N Kyzylkum desert (see Shenbrot et al., 1995; Vorontsov and Shenbrot, 1984).
STATUS: IUCN – Lower Risk (nt).
COMMENTS: Subgenus *Prosalpingotus*. Reviewed by Gromov and Erbajeva (1995) and Shenbrot et al. (1995).

Salpingotus kozlovi Vinogradov, 1922. *In* Kozlov, Mongolia and Amdo, p. 542.
COMMON NAME: Kozlov's Pygmy Jerboa.
TYPE LOCALITY: Mongolia: Gobi desert, Khara-Khoto.
DISTRIBUTION: Deserts of S and SE Mongolia; and China: Nei Mongol, Xinjiang, Gansu, N Shaanxi, and Ningxia (see Chen and Wang, 1985; Ma et al., 1987; Mi et al., 1990; Qian et al., 1965; Qin, 1991; T.-z. Wang, 1990; Y. Wang, 2003; and Zheng and Zhang, 1990; Chinese range mapped by Zhang et al., 1997).
STATUS: IUCN – Lower Risk (nt).
SYNONYMS: *xiangi* Hou and Jiang, 1994.
COMMENTS: Subgenus *Salpingotus*. See Corbet (1978c) for comment regarding allocation of specimen from Irtysh River on Kazakhstan-Chinese border to *S. crassicauda*. Reviewed by Ognev (1963b). Study of geographic variation in Mongolian samples provided by Sokolov and Shenbrot (1988).

Salpingotus pallidus Vorontsov and Shenbrot, 1984. Zool. Zh., 63(5):740.
COMMON NAME: Pallid Pygmy Jerboa.
TYPE LOCALITY: Kazakhstan, Aktyubinskaya, Chelkarskii, Peski Bol'shiye Barsuki.
DISTRIBUTION: Deserts of N Aral and S Balkhash regions (see Shenbrot et al., 1995).
STATUS: IUCN – Lower Risk (lc).
SYNONYMS: *sludskii* Shenbrot and Mazin, 1989.
COMMENTS: Subgenus *Prosalpingotus*. Shenbrot and Mazin (1989) named *sludskii* as a subspecies of *S. pallidus*, but we are not recognizing subspecies. Reviewed by Shenbrot et al. (1995).

Salpingotus thomasi Vinogradov, 1928. Ann. Mag. Nat. Hist., ser. 10, 1:373.
>   COMMON NAME: Thomas's Pygmy Jerboa.
>   TYPE LOCALITY: Afghanistan or S Tibet (Ognev, 1963*b*; Vinogradov, 1928).
>   DISTRIBUTION: Known only from the type specimen, of which the country of origin,
>   "Afghanistan", is questionable (see discussion in Hassinger, 1973, and Roberts, 1977,
>   1997). Listed as occurring in S Xizang, China by Wang (2003).
>   STATUS: IUCN – Data Deficient.
>   COMMENTS: Subgenus *Prosalpingotus* (see Pavlinov and Shenbrot, 1985). Reviewed by Ognev
>   (1963*b*).

Subfamily Dipodinae Fischer, 1817. Mem. Soc. Imp. Nat. Moscow, 5:372.
>   SYNONYMS: Dipsidae Gray, 1821 (Dipina Gray, 1825); Dipodes Fischer de Waldheim, 1817;
>   (Dipodinae Murray, 1866; Dipodina Bonaparte, 1838; Dipodini Brandt, 1855; Dipodum
>   Fischer de Waldheim, 1817); Gerboidae Waterhouse, 1839 [*nomen nudum*]; Jaculini
>   Brandt, 1855 (Jaculina Haeckel, 1866; Jaculinae Alston, 1876); Paradipodini Pavlinov and
>   Schenbrot, 1983; Stylodipodina Zazhigin and Lopatin, 2000 [*nomen nudum*].
>   COMMENTS: Stylodipodina was proposed for *Stylodipus* by Zazhigin and Lopatin (2000*a*) but no
>   type genus was explicitly indicated. Zazhigin and Lopatin (2001) recorded four dipodine
>   genera as occurring in Asia by late Miocene, the extinct *Scirtodipus* and *Plioscirtopoda*, and
>   extant *Dipus* and *Jaculus*. Those authors also noted that by late Miocene-early Pliocene
>   times, three phylogenetic lineages were already apparent. One consists only of *Dipus*;
>   another contains *Scirtodipus* (late Miocene), *Stylodipus*, and *Plioscirtopoda* (late Miocene to
>   early Pleistocene; see also Kowalski, 2001); and the third is represented by *Jaculus*,
>   *Eremodipus*, and the extinct *Jaculodipus* (early Pliocene). Both extinct and living forms are
>   basically Eurasian in distribution, and only two species of *Jaculus* extend beyond that
>   region to North Africa (see accounts of *J. jaculus* and *J. orientalis*).

Dipus Zimmermann, 1780. Geogr. Gesch. Mensch. Vierf. Thiere, 2:354.
>   TYPE SPECIES: *Mus sagitta* Pallas, 1773.
>   SYNONYMS: *Dipodipus* Trouessart, 1910; *Dipsus* Gray, 1821; *Sminthoides* Schlosser, 1924; see
>   Ellerman and Morrison-Scott (1951), McKenna and Bell (1997), and Zazhigin and Lopatin
>   (2001).
>   COMMENTS: Dipodini. Shenbrot et al. (1995) and Zazhigin and Lopatin (2001) included the
>   extinct *Sminthoides* (late Miocene to late Pliocene of Asia), but McKenna and Bell (1997)
>   retained it as a separate genus in Dipodinae.

Dipus sagitta (Pallas, 1773). Reise Prov. Russ. Reichs., 2:706.
>   COMMON NAME: Northern Three-toed Jerboa.
>   TYPE LOCALITY: N Kazakhstan, Pavlodarskaya Oblast, right bank of Irtysh River near
>   Yamyshevskaya at Podpusknoi (see Ellerman and Morrison-Scott, 1951:535).
>   DISTRIBUTION: Desert, steppe and dry woodland from Don River (Russia), NW coast of
>   Caspian Sea, and N Iran, through Turkmenistan, Uzbekistan, and Kazakhstan to
>   S Tuva, Russia; Mongolia (Bannikov, 1954); China (Nei Mongol, Xinjiang, Qinghai,
>   Gansu, Ningxia, N Shaanxi, N Shanxi, Liaoning, and Jilin; see Chen and Wang, 1985;
>   Liu et al., 1990; Ma et al., 1987; Mi et al., 1990; Qian et al., 1965; Qin, 1991; Shou,
>   1962; T.-z. Wang, 1990; Y. Wang, 2003; Zhang and Wang, 1963; Zheng and Zhang,
>   1990; and Zhou et al., 1985; Zhang et al., 1997, provide map for Chinese range); see
>   Shenbrot et al. (1995).
>   STATUS: IUCN – Lower Risk (lc).
>   SYNONYMS: *aksuensis* Wang, 1964; *austrouralensis* Shenbrot, 1991; *bulganensis* Shenbrot,
>   1991; *deasyi* Barrett-Hamilton, 1900; *fuscocanus* Wang, 1964; *halli* Sowerby, 1920);
>   *innae* (Ognev, 1930); *kalmikensis* Kazantseva, 1940; *lagopus* Lichtenstein, 1823;
>   *megacranius* Shenbrot, 1991; *nogai* Satunin, 1907; *sowerbyi* Thomas, 1908; *turanicus*
>   Shenbrot, 1991; *ubsanensis* Bannikov, 1947; *usuni* Shenbrot, 1991; *zaissanensis* Selevin,
>   1934 (see Corbet, 1984; Ellerman and Morrison-Scott, 1951; Pavlinov and Rossolimo,
>   1987; Shenbrot, 1991*a*, *c*; and Wang, 1964).
>   COMMENTS: Karyotype described by Vorontsov et al. (1969*d*). Taxonomic study, analysis of
>   geographic variation, and distribution throughout most of range provided by

Shenbrot (1991*a*, *c*). Reviewed by Gromov and Erbajeva (1995), Ognev (1963*b*), and Shenbrot et al. (1995).

Eremodipus Vinogradov, 1930. Izv. Akad. Nauk S.S.S.R., Leningrad, Otdel. Phyz.-Math, p. 334.
    TYPE SPECIES: *Scirtopoda lichtensteini* Vinogradov, 1927.
    COMMENTS: Following Vinogradov (1930), Heptner (1975), Gromov and Baranova (1981), Pavlinov and Rossolimo (1987, 1998), Shenbrot (1990*b*), and Shenbrot et al. (1995), *Eremodipus* is recognized as a genus distinct from *Jaculus*. The oldest record of *Eremodipus* comes from the early Pliocene of Kazakhstan (Zazhigin and Lopatin, 2001).

Eremodipus lichtensteini (Vinogradov, 1927). Z. Säugetierk., 2(1):92.
    COMMON NAME: Lichtenstein's Jerboa.
    TYPE LOCALITY: Turkmenistan, vicinity of Merv.
    DISTRIBUTION: Kazakhstan, Turkmenistan, and Uzbekistan, from Caspian Sea to Aral Sea, and south of Lake Balkhash (see map in Shenbrot et al., 1995).
    STATUS: IUCN – Lower Risk (lc).
    SYNONYMS: *balkashensis* Shenbrot, 1990; *jaxartensis* Shenbrot, 1990. See Shenbrot et al. (1995).
    COMMENTS: Reviewed by Gromov and Erbajeva (1995), Ognev (1963*b*) and Shenbrot et al. (1995). Taxonomic study, distribution, and review contributed by Shenbrot (1990*b*), who described *balkashensis* and *jaxartensis* as subspecies (we do not recognize subspecies here). Detailed habitat data provided by Naumov and Lobachev (1975). Karyotype given by Vorontsov et al. (1969*d*).

Jaculus Erxleben, 1777. Syst. Regni Anim., 1:404.
    TYPE SPECIES: *Mus jaculus* Linnaeus, 1758, as fixed by Opinion 730 of the International Commission on Zoological Nomenclature (1965); not *Jaculus orientalis* Erxleben, 1777.
    SYNONYMS: *Haltomys* Brandt, 1844; *Scirtopoda* Brandt, 1844 [see Ellerman and Morrison-Scott, 1951; and Corbet, 1978*c*].
    COMMENTS: Dipodini. Myology, in context of adaptive and phylogenetic significance, studied by Klingener (1964). Generic review and composite distribution map of all species provided by Shenbrot et al. (1995). Evolutionary history extends back to late Miocene of Kazakhstan (Zazhigin and Lopatin, 2001), late Pliocene of Morocco (Jaeger, 1970) and Ethiopia (Wesselman, 1984), and Plio-Pleistocene in Kenya (Black and Krishtalka, 1986); see review by Denys (1999).

Jaculus blanfordi (Murray, 1884). Ann. Mag. Nat. Hist., ser. 5, 14:98.
    COMMON NAME: Blanford's Jerboa.
    TYPE LOCALITY: Iran, Bushire.
    DISTRIBUTION: SE coast of Caspian Sea through Turkmenistan to the Kyzylkum Desert, C Uzbekistan (Kuznetsov, 1965; Shenbrot et al., 1995), E and S Iran (Lay, 1967), S and W Afghanistan (Hassinger, 1973), and SW Pakistan (Roberts, 1977, 1997).
    STATUS: IUCN – Lower Risk (lc).
    SYNONYMS: *margianus* Shenbrot, 1990; *turcmenicus* Vinogradov and Bondar, 1949.
    COMMENTS: The taxon *turcmenicus* has been considered a separate species (Corbet, 1978*c*; Holden, 1993; Shenbrot, 1990*a*), but Heptner (1975) treated it as conspecific with *J. blanfordi*, even though measurements given for the two species differ considerably, no tests of significance were performed, and only two specimens of *J. blanfordi* were included in the study. Holden (1993) recognized *J. turcmenicus*, as in Shenbrot (1990*a*), pending critical revision. Recent research corroborates Heptner's view (Shenbrot et al., 1995) and *turcmenicus* is now included in *J. blanfordi* (Pavlinov and Rossolimo, 1998; Shenbrot et al., 1995). Presence of spines on the bacula of *J. blanfordi* and *J. orientalis* (figured by Didier and Petter, 1960), is shared by *turcmenicus* (figured in Heptner, 1975), reinforcing its status as a geographic segment of *J. blanfordi*, and suggesting that the two species form a monophyletic group that excludes *J. jaculus*; see also comment in Corbet (1978*c*) under *J. blanfordi*. Taxonomic study, distribution, and review of the *turcmenicus* section provided by Gromov and Erbajeva (1995), Shenbrot (1990*a*), and Shenbrot et al. (1995); morphology and habitat discussed by Stalmakova (1957); and karyotype described by Vorontsov et al. (1969*d*).

Jaculus jaculus (Linnaeus, 1758). Syst. Nat., 10th ed., 1:63.

COMMON NAME: Lesser Egyptian Jerboa.

TYPE LOCALITY: Egypt, Giza Pyramids.

DISTRIBUTION: N Africa in Senegal (Bâ et al., 2000; Duplantier and Granjon, 1992), NE Nigeria (Happold, 1987) and Niger, from S Mauritania to Morocco (see range map in Aulagnier and Thévenot. 1986, and Bâ et al., 2001), E through Algeria (Kowalski and Rzebik-Kowalska, 1991), Tunisia (Vesmanis, 1984) and Libya (Ranck, 1968) to Sudan (Setzer, 1956), Ethiopia and Eritrea (Yalden et al., 1996), Egypt (Osborn and Helmy, 1980), and Somalia; throughout Arabia (Harrsion and Bates, 1991; Al-Jumaily, 1998, for Yemen), the Sinai and Israel (Mendelssohn and Yom-Tov, 1999) through Iraq (Hatt, 1959) to SW Iran (Lay, 1967). Granjon et al. (1992:272) stated that this species "has probably only recently reached Senegal from Mauritania where it was previously known to occur."

STATUS: IUCN – Lower Risk (lc).

SYNONYMS: *aegyptius* (Lichtenstein, 1827); *airensis* Thomas and Hinton, 1921; *arenaceous* Ranck, 1968; *butleri* Thomas, 1922; *centralis* Thomas and Hinton, 1921; *collinsi* Ranck, 1968; *cufrensis* Ranck, 1968; *darricarrerei* Lataste, 1883; *deserti* Loche, 1867; *elbaensis* Setzer, 1955; *favillus* Setzer, 1955; *favonicus* Thomas, 1913; *florentiae* Cheesman and Hinton, 1924; *fuscipes* Ranck, 1968; *gordoni* Thomas, 1903; *hirtipes* (Lichtenstein, 1823); *loftusi* Blanford, 1875; *macromystax* (Lichtenstein, 1828); *macrotarsus* (Wagner, 1840); *microtis* Reichenow, 1887; *oralis* Cheesman and Hinton, 1924; *rarus* Ranck, 1968; *schlueteri* Nehring, 1901; *sefrius* Thomas and Hinton, 1921; *syrius* Thomas, 1922; *tripolitanicus* Ranck, 1968; *vastus* Ranck, 1968; *vocator* Thomas, 1921; *vulturnus* Thomas, 1913; *whitchurchi* Ranck, 1968 [see Ellerman and Morrison-Scott, 1951; and Corbet, 1978c].

COMMENTS: Ranck (1968) recognized two species in this complex, *J. jaculus* and *J. deserti*, but Harrison (1978) showed that they are conspecific based on Ranck's criteria (see also discussion in Corbet, 1978c:152). Mendelssohn and Yom-Tov (1999) suggested that *schlueteri* is a species because it occurs adjacent to *J. j. vocator* in Israel without apparently intergrading as was claimed by Harrison and Bates (1991). Karyotype given by Al Saleh and Khan (1984) and Granjon et al. (1992).

Jaculus orientalis Erxleben, 1777. Syst. Regn. Anim., 1:404.

COMMON NAME: Greater Egyptian Jerboa.

TYPE LOCALITY: Egypt, in the "mountains separating Egypt from Arabia" (G. M. Allen, 1939:424).

DISTRIBUTION: Arid or semiarid regions of N Africa and Israel, from Morocco (see the range map in Aulagnier and Thévenot, 1986) E through Algeria (Kowalski and Rzebik-Kowalska, 1991), Tunisia (Vesmanis, 1984), and Libya (Ranck, 1968) to Egypt (Osborn and Helmy, 1980), Sinai and S Israel (Mendelssohn and Yom-Tov, 1999; "a narrow strip in northern Negev," G. Shenbrot, in litt., 2003).

STATUS: IUCN – Lower Risk (nt).

SYNONYMS: *bipes* (Lichtenstein, 1823); *gerboa* (Olivier, 1800); *locusta* (Illiger, 1815); *mauritanicus* (Duvernoy, 1841) [see Ellerman and Morrison-Scott, 1951].

COMMENTS: *Jaculus orientalis* has been identified from the late Pliocene in Ethiopia (Wesselman, 1984) and Plio-Pleistocene in Kenya (Black and Krishtalka, 1986).

Paradipus Vinogradov, 1930. Izv. Acad. Sci. U.S.S.R., p. 333.

TYPE SPECIES: *Scirtopoda ctenodactyla* Vinogradov, 1929.

COMMENTS: Paradipodini. Shenbrot (1992) showed that *Paradipus* is highly differentiated from other genera in Dipodini (in which it has traditionally been placed) and appears to be most closely related to Cardiocraniinae, based on molar and mastoid characters. Study of the male reproductive tract by Pavlinov and Shenbrot (1983) supported molar and mastoid data, and those authors segregated *Paradipus* in its own tribe within Dipodinae, which is where Zazhigin and Lopatin (2000a) placed it in their classification. Shenbrot (1992) and Shenbrot et al. (1995) recognized Paradipodinae within Dipodidae. *Paradipus* was not included in Stein's (1990) study of limb myology. The tribe is represented by living *P. ctenodactylus* and extinct *P. badhysus* from early Pleistocene of Turkmenistan (Shenbrot, 1986; Zazhigin and Lopatin, 2001).

Paradipus ctenodactylus (Vinogradov, 1929). Doklady Akad. Nauk S.S.S.R., Leningrad, 1929:248.
COMMON NAME: Comb-toed Jerboa.
TYPE LOCALITY: E Turkmenistan, near Repetek.
DISTRIBUTION: Sand deserts of Turkmenistan, Uzbekistan, and E Aral region of Kazakhstan; see Kuznetsov (1965), Sludskii (1977), and Shenbrot et al. (1995).
STATUS: IUCN – Lower Risk (lc).
COMMENTS: Karyotype described by Vorontsov et al. (1969*d*). Os penis described and figured by Shenbrot (1992). Range, taxonomy, and other characteristics reviewed by Gromov and Erbajeva (1995), Ognev (1963*b*), and Shenbrot et al. (1995).

Stylodipus G. M. Allen, 1925. Am. Mus. Novit., 161:4.
TYPE SPECIES: *Stylodipus andrewsi* Allen, 1925.
SYNONYMS: *Halticus* Brandt, 1844 [see Ellerman and Morrison-Scott, 1951].
COMMENTS: *Scirtopoda* is often incorrectly used for this genus, but is a junior synonym of *Jaculus* (Corbet, 1978*c*:153; Ellerman and Morrison-Scott, 1951:536). Shenbrot et al. (1995) included the extinct *Scirtodipus* (late Miocene of Kazakhstan) as a synonym, but Zazhigin and Lopatin (2001) and McKenna and Bell (1997) listed it as a separate genus in Dipodinae. Zazhigin and Lopatin (2001) also regard *Scirtodipus* as ancestral to *Stylodipus*, which is represented by Pleistocene samples (*S. telum*) and early Pliocene material from Mongolia (*S. iderensis* and *S. perfectus*). Generic review provided by Shenbrot et al. (1995).

Stylodipus andrewsi Allen, 1925. Am. Mus. Novit., 161:4.
COMMON NAME: Andrew's Three-toed Jerboa.
TYPE LOCALITY: S Mongolia, near Mt. Uskuk (Ussuk), Camp Ondai Sair (Andrews, 1932:101).
DISTRIBUTION: NW, S, and C Mongolia east of Barun Khurai (Baruun Huuray) Valley; and adjacent China: Nei Mongolia, N Hebei, N Shanxi, N Shaanxi, Gansu, and Ningxia (see Ma et al., 1987; Qin, 1991; Wang, 2003; and Zheng and Zhang, 1990).
STATUS: IUCN – Lower Risk (lc).
COMMENTS: Ellerman and Morrison-Scott (1951) and Corbet (1978*c*) included *andrewsi* as a subspecies of *S. telum*, but Sokolov and Orlov (1980) recognized it as a distinct species. Sokolov and Shenbrot (1987*b*) showed that in addition to the retention of a rudimentary P4, *S. andrewsi* is differentiated from *S. sungorus* and *S. telum* in dental and phallic characters, and in greater bullar inflation. The ranges of *S. andrewsi* and *S. sungorus* are adjacent but do not overlap (Sokolov and Shenbrot, 1987*b*). See comments under *S. sungorus* and *S. telum*.

Stylodipus sungorus Sokolov and Shenbrot, 1987. Zool. Zh., 66(4):580.
COMMON NAME: Mongolian Three-toed Jerboa.
TYPE LOCALITY: SW Mongolia, Altai Gobi, north slope Takhin-Shara-Nuru (Tahiyn-Shar-Nuruu) range, 15 km E Tsargin.
DISTRIBUTION: SW Mongolia, possibly Xinjiang, China (see Sokolov and Shenbrot, 1987*b*:585).
STATUS: IUCN – Lower Risk (lc).
COMMENTS: On the basis of the length and breadth of the molar row, size of auditory bullae, and phallic characters (Sokolov and Shenbrot, 1987*b*), *S. sungorus* appears to be distinct from *S. telum*. Relationship and distribution relative to *S. andrewsi* and *S. telum* are unclear and need further study (see Sokolov and Shenbrot, 1987*b*). See comment under *S. andrewsi*.

Stylodipus telum (Lichtenstein, 1823). Naturhist. Anhang (or Eversmann's Reise Orenburg), p. 120.
COMMON NAME: Thick-tailed Three-toed Jerboa.
TYPE LOCALITY: Kazakhstan, steppe along NE shore of Aral Sea (Ognev, 1963*b*:303).
DISTRIBUTION: E Ukraine, N Caucasus, W Turkmenistan, W Uzbekistan and Kazakhstan (Kuznetsov, 1965; Shenbrot, 1991*b*); E to N Xinjiang, China (see Chen and Wang, 1985; Ma et al., 1987; Mi et al., 1990; Qian et al., 1965; Shou, 1962; Wang, 2003; and Zheng and Zhang, 1990; but see comment below; Chinese range mapped by Zhang et al., 1997). See Shenbrot et al. (1995) for overall distribution
STATUS: IUCN – Lower Risk (lc).
SYNONYMS: *amankaragai* (Selewin, 1934); *birulae* (Martino, 1922); *falzfeini* (Brauner, 1913);

halticus (Brandt, 1844); *karelini* (Selewin, 1934); *nastjukovi* Shenbrot, 1991; *proximus* (Fairmaire, 1853); *turovi* (Heptner, 1934); see Ellerman and Morrison-Scott (1951), Shenbrot (1991*b*), Shenbrot et al. (1995).

COMMENTS: Reviewed by Gromov and Erbajeva (1995), Ognev (1963*b*) and Shenbrot et al. (1995); subspecific revision contributed by Shenbrot (1991*b*). Karyotype provided by Vorontsov et al. (1969*d*). Because *S. andrewsi* is considered a synonym of *S. telum* by some workers, some of the earlier published records of *S. telum* from China represent *S. andrewsi*. Wang (2003) records *S. telum* only from N Xinjiang. See comments under *S. andrewsi* and *S. sungorus*. European Pleistocene records reviewed by Kowalski (2001).

Subfamily Euchoreutinae Lyon, 1901. Proc. U. S. Natl. Mus., 23 (1228):666.

COMMENTS: Based on crown structure of the molars and characters of the mastoid region, Shenbrot (1992) proposed that Euchoreutinae is most closely related to Sicistinae in Sminthidae. The results of Stein's (1990) study of limb musculature placed Euchoreutinae as a sister group of Zapodinae and Allactaginae in Dipodidae, Zazhigin and Lopatin (2000*a*, *b*) arranged it as a subfamily of Allactagidae, and Potapova (2000) espoused family rank based on bullar morphology. Lyon (1901) had originally proposed Euchoreutinae as a subfamily in Dipodidae. See comment under Dipodidae.

Euchoreutes Sclater, 1891. Proc. Zool. Soc. Lond., 1890:610 [1891].

TYPE SPECIES: *Euchoreutes naso* Sclater, 1891.

COMMENTS: Detailed review provided by Ognev (1963*b*).

Euchoreutes naso Sclater, 1891. Proc. Zool. Soc. Lond., 1890:610 [1891].

COMMON NAME: Long-eared Jerboa.

TYPE LOCALITY: NW China; W Xinjiang, W of Taklimakan Shamo (Takla-Makan Desert), near Shache (Yarkand).

DISTRIBUTION: S Mongolia; China (Nei Mongolia, Xinjiang, Qinghai, Gansu, and Ningxia; see Chen and Wang, 1985; Liu et al., 1990; Ma et al., 1987; Qian et al., 1965; Shou, 1962; Zhang and Wang, 1963; Wang, 2003; and Zheng and Zhang, 1990; Chinese range mapped by Zhang et al., 1997).

STATUS: IUCN – Endangered.

SYNONYMS: *alashanicus* Howell, 1928; *yiwuensis* Ma and Li, 1979 [see Corbet, 1978*c*, 1984].

Subfamily Sicistinae J. A. Allen, 1901. Proc. Biol. Soc. Wash., 14:185.

SYNONYMS: Sminthi Brandt, 1855; Sminthinae Murray, 1866; Sminthidae Schulze, 1890; Sicistidae Weber, 1928; Lophocricetinae Savinov, 1970.

COMMENTS: Some authors, mostly Russian (Pavlinov and Rossolimo, 1987, 1998; Pavlinov et al., 1995; Shenbrot, 1992; Shenbrot et al., 1995), use the family name Sminthidae for this group, because Brandt's (1855) supergeneric taxon Sminthi predates Sicistinae J. A. Allen, 1901*a*. Holden (1993) noted that the 1985 International Code of Zoological Nomenclature indicated that when two genera are united their respective type species remain the same, and the valid name of the newly formed taxon is that of the component taxon with the oldest valid name (ICZN, 1985*d*, article 66:125-127). *Sicista* Gray, 1827 predates *Sminthus* Nordmann, 1840, and thus *Sicista* is correct. Sicistinae is the valid subfamily (or family or tribe) name according to article 23 of the 1985 ICZN (p. 47). However, Article 40.2 of the 1999 ICZN states if " a family-group name was replaced before 1961 because of the synonymy of the type genus, the substitute name is to be maintained if it is in prevailing usage" (p. 47), a ruling, in our view, supporting use of Sicistinae, which has been employed (as subfamily or tribe) in a variety of major works since 1901 (Corbet, 1978*c*; Ellerman, 1940, 1961; Ellerman and Morrison-Scott, 1951; Klingener, 1984; Kowalski, 2001; R. A. Martin, 1994; McKenna and Bell, 1997; Miller and Gidley, 1918; Simpson, 1945; Stein, 1990; Vinogradov, 1925, 1930, 1937; B. Y. Wang, 1985; Zazhigin and Lopatin, 2000*a*). The separation of Sicistinae from Zapodinae, suggested by Ellerman (1940), is supported by analyses of morphology (Shenbrot, 1986, 1992; Stein, 1990), chromosomes (Sokolov et al., 1987*b*; Vorontsov, 1969), and nuclear gene sequences (DeBry and Sagel, 2001; Jansa and Weksler, 2004; Michaux and Catzeflis, 2000; Michaux et al., 2001*b*).

Extant species of *Sicista* range through the Palaearctic of Eurasia, and the group is the only Recent remnant of an impressive extinct radiation of species and genera already present by the early Oligocene of Asia (*Allosminthus, Heosminthus, Tatalsminthus, Shamosminthus,* and *Sinosminthus*) that extended through the later Oligocene to the late Miocene of Asia (*Plesiosminthus, Parasminthus, Gobiosminthus, Heterosminthus, Lophocricetus, Lophosminthus, Sibirosminthus,* and *Xenosminthus*), and late Oligocene and early Miocene of Europe (*Pleisiosminthus*). Sicistinae are also recorded from late Oligocene to Early Pleistocene deposits in North America (*Pleisiosminthus, Schaubeumys, Macrognathomys, Megasminthus, Miosicista, and Tyrannomys*) (New and Old World records presented in various contexts in Daxner-Höck, 1999, 2001; Daxner-Höck and Wu, 2003; Huang, 1992; Hugueney and Vianey-Liaud, 1980; Korth, 1993; Lopatin and Zazhigin, 2000; R. A. Martin, 1989*a*; 1994; McKenna and Bell, 1997; B. Y. Wang, 1985; Wang and Qiu, 2000). *Heterosminthus, Lophocricetus, Lophosminthus,* and *Sibirosminthus* are often brought together in Lophocricetinae and placed in Zapodidae (R. A. Martin, 1994; Qiu, 1985), Dipodidae (Savinov, 1970; Topachevskii et al., 1984; Zazhigin and Lopatin, 2000*b*, 2002; Zazhigin et al., 2002), or Cardiocraniinae (Shenbrot et al., 1995), or as a tribe (Lophocricetini) in Sicistinae (McKenna and Bell, 1997). *Primisminthus* and *Banyuesminthus*, from middle Eocene deposits of Shanxi and Henan provinces in China, possibly represent even older representatives of Sicistinae; Wang and Qiu (2000) speculated that *Primisminthus* may be the oldest member of an inclusive Dipodidae, and *Banyuesminthus* could be a sister-group to dipodids.

Sicista Gray, 1827. *In* Griffith et al., Anim. Kingdom, 5:228.
TYPE SPECIES: *Mus subtilis* Pallas, 1773.
SYNONYMS: *Clonomys* Thilesius, 1850; *Sminthus* Nordmann, 1840 [see Ellerman and Morrison-Scott, 1951].
COMMENTS: Reviewed by Ognev (1963*b*). Karyological research and systematic problems in the genus reviewed by Sokolov et al. (1987*b*). Several species in the *S. concolor* complex (*S. armenica, S. caucasica, S. caudata, S. kazbegica, S. kluchorica, S. tianshanica*) have been distinguished by Sokolov and colleagues primarily by karyotypic and spermatozoal differences (Baskevich, 1996*a*, provides the most current summary of the morphological, chromosomal, and spermatozoal traits used to distinguish these species). All these species are provisionally recognized here, but need further documentation and corroborative data sets to firmly establish their specific status. Diagnoses, characteristics, distribution, geographical variation, ecology, and economic importance of species in Russia and adjacent regions extensively reviewed by Shenbrot et al. (1995). Myology, in context of adaptive and phylogenetic significance, studied by Klingener (1964) and Stein (1990). Comparative myology of pelvic girdle studied by Fokin (1971). European late Pliocene and Pleistocene records of *Sicista* reviewed by Kowalski (2001).

Sicista armenica Sokolov and Baskevich, 1988. Zool. Zh., 67(2):301.
COMMON NAME: Armenian Birch Mouse.
TYPE LOCALITY: NW Armyanskaya, Malyy Kavkaz, Pazdanskiy Region, Pambakskiy Range, near Ankavan, head of Marmarik River, subalpine zone, 2200 m.
DISTRIBUTION: Known only from the type locality.
STATUS: IUCN – Critically Endangered.
COMMENTS: Sokolov and Baskevich (1988) presented karyological and spermatozoal characters that distinguished *S. armenica* from *S. caucasica, S. kluchorica,* and *S. kazbegica.* See also comment under *Sicista.* Recognized by Pavlinov and Rossolimo (1998) and reviewed by Gromov and Erbajeva (1995) and Shenbrot et al. (1995).

Sicista betulina (Pallas, 1779). Nova Spec. Quad. Glir. Ord., p. 332.
COMMON NAME: Northern Birch Mouse.
TYPE LOCALITY: SW Siberia, birch plain on bank of Ishim River, and Barabinskaya Step.
DISTRIBUTION: Boreal and montane forests from Norway and Denmark, east to Lake Baikal region, north to the Artic Circle at the White Sea and Usa River, south to Austria, Carpathian and Sayan Mtns (Corbet, 1978*c*). See Gromov and Erbajeva (1995), Kuznetsov (1965), Sludskii (1977), and Shenbrot et al. (1995) for range in Russia.

Corbet (1978c) included the Ussuri region of SE Siberia, but Sokolov et al. (1989) considered the Ussuri region records questionable, and Pavlinov (in litt., 1994) and Shenbrot et al. (1995) indicate that the species does not extend east of Lake Baikal.

STATUS: IUCN – Lower Risk (nt).

SYNONYMS: *montana* Méhely, 1913; *norvegica* Chaworth-Musters, 1927; *taigica* Stroganov and Potapkina, 1950; *tatricus* Méhely, 1913 [see Ellerman and Morrison-Scott, 1951; and Corbet, 1978c].

COMMENTS: Sokolov et al. (1982, 1987b) gave karyological and spermatozoal characters that distinguished this species from *S. napaea* and *S. pseudonapaea*. Pallas' type specimen was probably not preserved (Ognev, 1963b:33). Review of taxonomy, characteristics, ecology, and distribution available for Europe (Pucek, 1982; Mitchell-Jones, 1999), Austria (Spitzenberger, et al., 1995), East Baltic region (Timm et al., 1998), E Carpathian Mtns of Slovakia (Danko, 1994), Sumava Mtns of SW Bohemia (Anděra and Červený, 1994), Svjatoj Nos peninsula and isthmus in Lake Baikal (Reiter et al., 1995), and Russia (Gromov and Erbajeva, 1995; Shenbrot et al., 1995). Fragments in owl pellets from Schleswig-Holstein is the first and only record for *S. betulina* since 1950 (Borkenhagen, 1996). Miljutin (1999) provided a comprehensive review of the morphology of *S. betulina* and its phylogenetic and adaptive significance. He (Miljutin, 1997, 1998) also included the species in a treatise on ecomorphology of Baltic rodents and review of ecological strategies in those populations. See comment under *S. strandi*, a taxon that was once included in *S. betulina*.

Sicista caucasica Vinogradov, 1925. Proc. Zool. Soc. Lond., 1925:584.

COMMON NAME: Caucasian Birch Mouse.

TYPE LOCALITY: Russia, N Caucasus, Krasnodarskiy Kray (Kuban Prov.), Maykop (Maikop) District, 7000-9000 ft.

DISTRIBUTION: NE Turkey (E Black Sea Mtns; Kryštufek and Vohralík, 2001) and NW Caucasus (see Sokolov et al., 1987a).

STATUS: IUCN – Lower Risk (lc).

COMMENTS: Sokolov et al. (1981b, 1987b) gave karyological and morphological characters that distinguished this species from *S. kluchorica* and *S. concolor*, *S. armenica* (Sokolov and Baskevich, 1988), and *S. kazbegika* (Sokolov et al., 1986b). Recognized by Pavlinov and Rossolimo (1998) and reviewed by Shenbrot et al. (1995). See also comment under *Sicista*, and in Corbet (1984:25).

Sicista caudata Thomas, 1907. Proc. Zool. Soc. Lond., 1907:413.

COMMON NAME: Long-tailed Birch Mouse.

TYPE LOCALITY: Russia, Sakhalin Oblast, Sakhalin Isl, 17 miles NW Korsakov.

DISTRIBUTION: Ussuri region of NE China (Heilongjiang, Jilin; Wang, 2003) and Primorski Kray, Sikhote-Alin range, N Korea, and Sakhalin Isl, Russia; see map in Shenbrot et al. (1995).

STATUS: IUCN – Endangered.

COMMENTS: Sokolov et al. (1982, 1987b) and Sokolov and Kovalskaya (1990) gave karyological and spermatozoal characters that distinguished this species from *S. tianshanica*, and from *S. concolor* (Sokolov et al., 1980). See also comment under *Sicista* and *S. concolor*. Korean range is discussed by Won and Smith (1999). Populations in Russia reviewed by Gromov and Erbajeva (1995) and Shenbrot et al. (1995).

Sicista concolor (Büchner, 1892). Bull. Sci. Acad. Imp. Sci. St. Petersbourg, 35(3):107.

COMMON NAME: Chinese Birch Mouse.

TYPE LOCALITY: China, Gansu, N slope of the mountains of Xining, Guiduisha.

DISTRIBUTION: China: Heilongjiang, Jilin, Xinjiang, Qinghai, Gansu, Shaanxi, W Sichuan, and Yunnan (Ma et al., 1987; T.-z. Wang, 1990; Y. Wang, 2003; Zhang et al., 1997; and Zheng and Zhang, 1990); W Kashmir and N Pakistan (Corbet and Hill, 1992; Roberts, 1997).

STATUS: IUCN – Lower Risk (lc).

SYNONYMS: *flavus* (True, 1894); *leathemi* (Thomas, 1893); *weigoldi* Jacobi, 1923 [see Ellerman and Morrison-Scott, 1951].

COMMENTS: *S. concolor* has been reported from the Chinese provinces of Heilongjinag and Jilin (Yang et al., 1991); the relationship of these populations to *S. caudata* needs further study. Does not include *S. armenica*, *S. caucasica*, *S. caudata*, *S. kazbegica*, *S. kluchorica*, or *S. tianshanica*; see comments under respective species, under *Sicista*, and in Corbet (1978c:149, 1984:25).

Sicista kazbegica Sokolov, Baskevich, and Kovalskaya, 1986. Zool. Zh., 65(6):949.
    COMMON NAME: Kazbeg Birch Mouse.
    TYPE LOCALITY: Georgia, Kazbegi District, 14 km NW Kobi, Suatisi Gap, upper reaches Terek River, subalpine zone, 2200 m.
    DISTRIBUTION: Greater Caucasus: southern flanks in N Ossetia region of Russia, and the Kazbegi District on the northern flanks in Georgia (Shenbrot et al. 1995; Sokolov and Baskevich, 1992; Sokolov et al., 1986b, 1987a).
    STATUS: IUCN – Data Deficient.
    COMMENTS: Sokolov et al. (1986b) gave karyological and spermatozoal characters that distinguished the Georgian population of this species from *S. caucasica* and *S. kluchorica*, and *S. armenica* (Sokolov and Baskevich, 1988). Sokolov and Baskevich (1992) reported chromosomal, morphological, and spermatozoal data for a population from the North Ossetia region, which shows some chromosomal differences that they interpreted as simply geographic variation within *S. kazbegica*. Recognized by Pavlinov and Rossolimo (1998) and reviewed by Gromov and Erbajeva (1995) and Shenbrot et al. (1995). See also comment under *Sicista*.

Sicista kluchorica Sokolov, Kovalskaya, and Baskevich, 1980. Gryzuny Severnovo Kavkaza., p. 38.
    COMMON NAME: Klochor Birch Mouse.
    TYPE LOCALITY: Russia, N Caucasus, Karachayevo-Cherkess Autonomous Region, upper North Klukhor River at Klukhor Pass, 2100 m.
    DISTRIBUTION: NW Caucasus; see Shenbrot et al. (1995) and Sokolov et al. (1987a).
    STATUS: IUCN – Data Deficient.
    COMMENTS: Sokolov et al. (1981b, 1987b) gave karyological and morphological characters that distinguished this species from *S. caucasica* and *S. concolor*, *S. armenica* (Sokolov and Baskevich, 1988), and *S. kazbegika* (Sokolov et al., 1986b). See also comment under *Sicista* and in Corbet (1984:25). Recognized by Pavlinov and Rossolimo (1998) and reviewed by Shenbrot et al. (1995).

Sicista napaea Hollister, 1912. Smithson. Misc. Coll., 60(14):2.
    COMMON NAME: Altai Birch Mouse.
    TYPE LOCALITY: Russia, Altai Krai, Altai Mtns, Seminsk Ridge, Tapuchii (Tapucha).
    DISTRIBUTION: E Kazakhstan and Russia, Altai Krai, NW Altai Mtns; see Kuznetsov (1965), Sludskii (1977) and Shenbrot et al. (1995).
    STATUS: IUCN – Lower Risk (lc).
    COMMENTS: Sokolov et al. (1982, 1987b) gave karyological, and spermatozoal characters that distinguished this species from *S. pseudonapaea* and *S. betulina*. Listed by Pavlinov and Rossolimo (1998) and reviewed by Gromov and Erbajeva (1995) and Shenbrot et al. (1995).

Sicista pseudonapaea Strautman, 1949. Vestn. Akad. Nauk Kazakh. SSR, 5:109.
    COMMON NAME: Gray Birch Mouse.
    TYPE LOCALITY: E Kazakhstan, Altai Mtns, N slope of Narymskiy Range, Katon-Karagay.
    DISTRIBUTION: E Kazakhstan, Taiga of S Altai Mtns; see Kuznetsov (1965), Sludskii (1977), and Shenbrot et al. (1995).
    STATUS: IUCN – Data Deficient.
    COMMENTS: Sokolov et al. (1982, 1987b) gave karyological, spermatozoal, and phallic characters that distinguished this species from *S. napaea* and *S. betulina*. *Sicista pseudonapaea* is provisionally recognized here, but requires further documentation, and the inclusion of other data sets, to establish its specific status. Recognized by Pavlinov and Rossolimo (1998) and reviewed by Gromov and Erbajeva (1995) and Shenbrot et al. (1995).

Sicista severtzovi Ognev, 1935. Byullet. Nauchno-issled. Inst. Zool., Mosk., 2:54.
    COMMON NAME: Severtzov's Birch Mouse.
    TYPE LOCALITY: Russia, Voronezh Oblast, Bobrov District, Kamennaya Steppe Experimental Station.

DISTRIBUTION: S Russia (S Voronezh Region and N Rostov Region; Koval'skaya et al., 2000) and E Ukraine (Zagorodnyuk and Kondratenko, 2000). See Shenbrot et al. (1995).

STATUS: IUCN – Lower Risk (lc).

SYNONYMS: *cimlanica* Koval'skaya, Tikhonov, Tikhonova, Surov, and Bogomolov, 2000.

COMMENTS: Sokolov et al. (1986*a*) separated this species from *S. subtilis* by its distinctive karyotype. Holden (1993) provisionally recognized *S. severtzovi* but noted the need for further documentation, and the inclusion of other data sets, to bolster its specific status. Recently, Koval'skaya et al. (2000) sampled more geographic samples and amplified the chromosomal and geographic definition of *S. severtzovi* relative to *S. subtilis*, and Zagorodnyuk and Kondratenko (2000) provided more chromosomal and distributional information. Reviewed by Shenbrot et al. (1995).

Sicista strandi (Formozov, 1931). Folia Zool. Hydrob. Riga, 3:79.

COMMON NAME: Strand's Birch Mouse.

TYPE LOCALITY: Russia, Caucasus, Stavropol Krai, Karachayevo-Cherkess Region, Karachayevsk District, Uchkulan (Utschkulak), Igera, 2100 m; shown in Sokolov et al. (1989).

DISTRIBUTION: N Caucasus, north to Kursk District of S Russia; see Shenbrot et al. (1995) and Sokolov et al. (1989).

STATUS: IUCN – Lower Risk (lc).

COMMENTS: Reviewed by Shenbrot et al. (1995). Sokolov et al. (1989) distinguished this species from *S. betulina* primarily by its different karyotype. Pavlinov and Rossolimo (1987, 1998) listed *S. strandi* as a separate species, a hypothesis provisionally followed here. Further documentation and incorporation of other character sets are essential in order to assess whether or not *strandi* should be included in *S. betulina*.

Sicista subtilis (Pallas, 1773). Reise Prov. Russ. Reichs., 1(2):705.

COMMON NAME: Southern Birch Mouse.

TYPE LOCALITY: Russia, Kurgan Oblast, on Tobol River near Kaminskaya Kur'ya (suburb), on road from Zverinogolovskoye to Kurgan (Ognev, 1963*b*:27).

DISTRIBUTION: Steppes from SE Poland, Hungary, E Serbia, Romania and NE Bulgaria (see Mitchell-Jones, 1999, for European range) through S Russia, N Kazakhstan, and SW Siberia to the Altai Range, Lake Balkhash, Lake Baikal, and NW Xinjiang, China (Li and Wang, 1981; Ma et al., 1987; Wang, 2003; see distribution map for Chinese segment in Zhang et al., 1997). See Ilchenko and Volodin (1992), Kuznetsov (1965), Sludskii (1977) and Shenbrot et al. (1995) for range in Russia.

STATUS: IUCN – Lower Risk (nt).

SYNONYMS: *interstriatus* (Petenyi, 1882); *interzonus* (Petenyi, 1882); *lineatus* (Lichtenstein, 1823); *loriger* (Nathusius, 1840); *nordmanni* (Keyserling and Blasius, 1840); *pallida* Kashkarov, 1926); *siberica* Ognev, 1935; *tripartitus* (Petenyi, 1882); *tristriatus* (Petenyi, 1882); *trizona* (Petenyi, 1882); *vagus* (Pallas, 1779); *virgulosus* (Petenyi, 1882) [see Ellerman and Morrison-Scott, 1951; Shenbrot et al., 1995].

COMMENTS: Holotype was probably not preserved (Ognev, 1963*b*:27). Karyology studied by Sokolov et al. (1986*a*). Review and distribution in Europe provided by Pucek (1982; in Mitchell-Jones, 1999), in Serbia by Petrov (1992), and in Russia by Shenbrot et al. (1995). Chromosomal characteristics of samples of *S. subtilis* in S Russia (Volgograd and E Rostov Regions) where that species ranges close to *S. severtzovi* is documented by Koval'skaya et al. (2000). See comment under *S. severtzovi*.

Sicista tianshanica (Salensky, 1903). Ezheg. Zool. Muz. Akad. Nauk, 8:17.

COMMON NAME: Tien Shan Birch Mouse.

TYPE LOCALITY: China, Xinjiang, S slope Tien Shan Mtns, between Kapchagay (Chapzagai-gol) and Tsaima (Zanma) Rivers.

DISTRIBUTION: Tien Shan Mtns of Kazakhstan (see Sludskii, 1977); Tien Shan Mtns and E Tarbagatay Mtns of Xinjiang, China (see Ma et al., 1987; Y. Wang, 2003); see overall distribution map in Shenbrot et al. (1995).

STATUS: IUCN – Lower Risk (lc).

COMMENTS: Sokolov et al. (1982, 1987*b*) and Sokolov and Kovalskaya (1990) gave karyological and spermatozoal characters that distinguish this species from *S. caudata*, and from *S. concolor* (Sokolov et al., 1980). See also comment under *Sicista*. Listed by Pavlinov and Rossolimo (1998) and reviewed by Gromov and Erbajeva (1995) and Shenbrot et al. (1995).

Subfamily Zapodinae Coues, 1875. Bull. U.S. Geol. Geog. Surv. Terr., ser. 2, 5(3):253.

COMMENTS: The recognition of three genera of living zapodines (Ellerman, 1940; Klingener, 1963; Krutzsch, 1954; R. A. Martin, 1994), which we retain here, has been challenged by Corbet (1978c), Corbet and Hill (1992), and Simpson (1945), who included *Eozapus* in *Zapus*. Higher level relationships among zapodines have been addressed by Preble (1899), who described *Eozapus* and *Napaeozapus* as new subgenera of *Zapus*, and by Krutzsch (1954), who supported generic separation based on differences in tooth number and occlusal pattern, bacula, and ear ossicles. Klingener (1964:75) found no consistant differences in the myology of *Zapus* versus *Napaeozapus* (*Eozapus* was not included in his study), but favored generic separation of the two based on dental morphology. The dental differences between *Eozapus* on one hand, and *Zapus* and *Napaeozapus* on the other, as documented in Klingener (1963), R. A. Martin (1994), Preble (1899), van de Weerd (1976), and Krutzsch (1954), are particularly striking and phylogenetically significant (e.g., "dental features of *Eozapus* . . . are completely different from those of . . . *Zapus* and *Napaeozapus*," van de Weerd, 1976:139). Molar occlusal patterns and degree of hypsodonty in *Eozapus*, especially the lowers, are primitive and closely resemble molar structure in the extinct sicistine *Plesiosminthus* (late Oligocene to Miocene in North America and Eurasia); those of *Zapus* and *Napaeozapus* are highly derived (Klingener, 1963; R. A. Martin, 1994). External traits of *Eozapus*, by contrast, closely resemble *Napaeozapus* and *Zapus* in pelage color and pattern and in saltatory adaptations (elongate hind legs and feet, long tail relative to body length) characterizing the other two genera, and possibly "the appearance of the first zapodine dentition in the fossil record also signals the first filling of the jumping mouse adaptive zone" (R. A. Martin, 1994:105).

Evolutionary history of zapodines, as documented by fossils, extends back to early Miocene of Kazakhstan and Mongolia, late Miocene in North America, Europe, and China, and is represented by species in the three recent genera and the extinct *Asiazapus, Javazapus, Pliozapus, Sinozapus,* and *Sminthozapus* (Fahlbusch, 1992; Lopatin and Zazhigin, 2000; Qiu and Storch, 2000; R. A. Martin, 1994). Judged by fossil evidence, zapodines may have originated in Asia with *Eozapus prosimilis* (early Miocene of Mongolia) basal to Eurasian species, and *Asiazapus* (late Miocene of Kazakhstan) closely related to North American *Zapus* and *Napaeozapus* (Lopatin and Zazhigin, 2000).

Eozapus Preble, 1899. N. Am. Fauna, 15:37.

TYPE SPECIES: *Zapus setchuanus* Pousargues, 1896.

SYNONYMS: *Protozapus* Bachmayer and Wilson, 1970.

COMMENTS: Oldest record is from early Miocene of Mongolia represented by *E. prosimilis* (Lopatin and Zazhigin, 2000). *Eozapus similis* from late Miocene sediments in Nei Mongol (Ertemte and Harr Obo, N China) may be ancestral to living *E. setchuanus* (Fahlbusch, 1992), and extinct *E. intermedius* (type species of *Protozapus*) from late Miocene strata in Europe. *Protozapus*, proposed by Bachmayer and Wilson (1970) and documented from the late Miocene of Spain, Austria, and Poland (R. A. Martin, 1994; van de Weerd, 1976), has been considered a junior synonym of either *Eozapus* (van de Weerd, 1976) or *Sminthozapus* (Farjanel and Mein, 1984). Van de Weerd's allocation of *Protozapus* has been endorsed by Fahlbusch (1992), who also noted that molar occlusal patterns of Polish *janossyi*, the type species of *Sminthozapus* (Pliocene), closely resemble those of *Eozapus*, that many late Miocene European samples identified as *Sminthozapus* are examples of *Eozapus*, and that *Sminthozapus* may be another synonym of *Eozapus*. Daxner-Höck (1999) followed Fahlbusch and considered European *Eozapus* (containing *E. intermedius* and *E.* sp.) to be restricted to the late Miocene, but Fahlbusch and Bolliger (1996) extended the time range to early Pliocene. Qiu and Storch (2000:188) regarded the early Pliocene *Sinozapus* from Nei Mongol sediments to be a sister group to "the *Eozapus-Sminthozapus* complex and not a descendent of the late Miocene *E. similis* from the same general region." The significance of these new discoveries and reidentifications is revealed within an evolutionarily and biogeographic context: the extant Chinese *E. setchuanus*, which is relictual in its primitive molar morphology and geographic distribution, is the only living representative of a clade containing other species of *Eozapus* that is rooted in the early Miocene of Central Asia and once ranged from Europe to eastern Asia during later Miocene times.

Eozapus setchuanus (Pousargues, 1896). Bull. Mus. Hist. Nat. Paris, 2:13.
    COMMON NAME: Chinese Jumping Mouse.
    TYPE LOCALITY: China, Sichuan Prov., Tatsienlu (Kangding).
    DISTRIBUTION: China: Qinghai, Gansu, Ningxia, Shaanxi, Sichuan, and NW Yunnan;
        3000-4000 m (Qin, 1991; T.-z. Wang, 1990; Y. Wang, 2003; Zhang and Wang, 1963;
        Zheng and Zhang, 1990; Zhang et al., 1997, provided distribution map).
    STATUS: IUCN – Vulnerable.
    SYNONYMS: *vicinus* (Thomas, 1912).
    COMMENTS: R. A. Martin (1994:5) wrote "Although distinctly more hypsodont and with
        lophs and lophids connecting the primary molar cusps, *Eozapus* retains the
        generalized, underived zapodid pattern of *Plesiosminthus*." The latter is the oldest
        known sicistine found in late Oligocene and Miocene sediments in both Europe and
        North America. Reviewed by Corbet (1978*c*) and Corbet and Hill (1992).

Napaeozapus Preble, 1899. N. Am. Fauna, 15:33.
    TYPE SPECIES: *Zapus insignis* Miller, 1891.
    COMMENTS: For verification of the absence of cheek pouches in *Napaeozapus* see Klingener
        (1971). Documented in fossil record from the middle Pleistocene (R. A. Martin, 1994;
        McKenna and Bell, 1997)

Napaeozapus insignis (Miller, 1891). Am. Nat., 25:742.
    COMMON NAME: Woodland Jumping Mouse.
    TYPE LOCALITY: Canada, New Brunswick, Restigouche River.
    DISTRIBUTION: Canada: SE Manitoba, SW and E Ontario, S and E Quebec north to
        S Labrador. USA: E Minnesota, N and C Wisconsin, upper peninsular and N lower
        peninsular Michigan, E Ohio, Pennsylvania; north and east to NW New Jersey, New
        York, Connecticut, Rhode Island, W Massachusetts (isolated population in Martha's
        Vinyard), Vermont, New Hampshire, and Maine; south to West Virginia, W Virginia,
        E Kentucky, E Tennessee, W North Carolina, NW South Carolina, and NE Georgia.
    STATUS: IUCN – Lower Risk (lc).
    SYNONYMS: *abietorum* (Preble, 1899); *algonquinensis* Prince, 1941; *frutectanus* Jackson, 1919;
        gaspensis Anderson, 1942; *roanensis* (Preble, 1899); *saguenayensis* Anderson, 1942 [see
        Hall, 1981].
    COMMENTS: A specimen of *Napaeozapus* was collected in Park County, Indiana (Lyon, 1942),
        though subsequent trapping failed to yield further examples (Mumford, 1969). The
        identity of the specimen was verified by Klingener (1965:645) and Wrigley (1972:42).
        Systematic revision and biology provided by Wrigley (1972). Myology, in context of
        adaptive and phylogenetic significance, studied by Klingener (1964). Diagnosis, range
        map, and records provided by Hall (1981). Population in E Kentucky discussed by
        Meade (1992). Reviewed by Whitaker and Wrigley (1972, Mammalian Species, 14),
        Whitaker and Hamilton (1998), and Whitaker (1999*a*).

Zapus Coues, 1875. Bull. U.S. Geol. Geog. Surv. Terr., ser. 2, 5(3):253.
    TYPE SPECIES: *Dipus hudsonius* Zimmermann, 1780.
    COMMENTS: Revised by Preble (1899) and Krutzsch (1954). Myology, in context of adaptive and
        phylogenetic significance, studied by Klingener (1964). Dental evolution investigated by
        Klingener (1963). For verification of the absence of cheek pouches in *Zapus* see Klingener
        (1971). Phallic morphology described by Shenbrot (1992). Evolutionary history extends
        back to late Pliocene in North America (Klingener, 1963; R. A. Martin, 1994).

Zapus hudsonius (Zimmermann, 1780). Geogr. Gesch. Mensch. Vierf. Thiere, 2:358.
    COMMON NAME: Meadow Jumping Mouse.
    TYPE LOCALITY: Canada, Ontario, Hudson Bay, Fort Severn (see Anderson, 1942*b*).

DISTRIBUTION: USA and Canada: S Alaska to S Coast Hudson Bay to Labrador, south to E North Carolina and NW South Carolina, southwest to NW Alabama, north to NE Mississippi and Tennessee, west to NE Oklahoma, northwest to SE Montana, northeast to SE Saskatchewan, northwest to C and S British Columbia. Isolated populations in S Wyoming, NC Colorado, N and C New Mexico, and EC Arizona.

STATUS: U. S. ESA – Threatened as *Z. h. preblei*; IUCN – Endangered as *Z. h. preblei*, Vulnerable as *Z. h. campestris*, Lower Risk (nt) as *Z. h luteus*, otherwise Lower Risk (lc).

SYNONYMS: *acadicus* Anderson, 1942; *alascensis* Merriam, 1897; *americanus* (Barton, 1799); *australis* Bailey, 1913; *brevipes* Bole and Moulthrop, 1942; *campestris* Preble, 1899; *canadensis* (Davies, 1798); *hardyi* Batchelder, 1899; *intermedius* Krutzsch, 1954; *labradorius* (Kerr, 1792); *ladas* Bangs, 1899; *luteus* Miller, 1911; *microcephalus* (Harlan, 1839); *ontarioensis* Anderson, 1943; *pallidus* Cockrum and Baker, 1950; *preblei* Krutzsch, 1954; *rafinesquei* Bole and Moulthrop, 1942; *tenellus* Merriam, 1897 [see Hall, 1981].

COMMENTS: Distinguishing *Z. hudsonius* from *Z. princeps* using repeated cranial measurements elaborated by Conner and Shenk (2003). *The* S Rocky Mtn population originally described as *luteus*, formerly assigned to *princeps*, was shown to represent *hudsonius* by Hafner et al. (1981). Diagnosis, records and range map (excluding Mississippi), provided by Hall (1981). Mississippi record given by Kennedy et al. (1982). Results of new survey for the species in New Mexico reported by Morrison (1992), and population in NE Oklahoma reviewed by Kasper et al. (1993). Reviewed by Whitaker (1972, Mammalian Species, 11; 1999*b*), and Whitaker and Hamilton (1998).

Zapus princeps Allen, 1893. Bull. Am. Mus. Nat. Hist., 5:71.

COMMON NAME: Western Jumping Mouse.

TYPE LOCALITY: USA, Colorado, La Plata Co., Florida.

DISTRIBUTION: Canada and USA: S Yukon southeast to NE South Dakota, west to C Montana, southeast to SE Wyoming, S to NC New Mexico; northwest to N and C Utah (isolated population in SE Utah), N and C Nevada, EC California north to SW, C and E Oregon, SE Washington northwest to S Yukon.

STATUS: IUCN – Lower Risk (lc).

SYNONYMS: *alleni* Elliot, 1898; *chrysogenys* Lee and Durrant, 1960; *cinereus* Hall, 1931; *curtatus* Hall, 1931; *idahoensis* Davis, 1934; *kootenayensis* Anderson, 1932; *major* Preble, 1899; *minor* Preble, 1899; *nevadensis* Preble, 1899; *oregonus* Preble, 1899; *pacificus* Merriam, 1897; *palatinus* Hall, 1931; *saltator* Allen, 1899; *utahensis* Hall, 1934 [see Hall, 1981].

COMMENTS: Formerly included *luteus*; see comment under *Z. hudsonius*. Diagnosis, records, and range map provided by Hall (1981). Discrimination from *Z.* hudsonius by repeated cranial measurements documented by Conner and Shenk (2003). Reviewed by Cranford (1999).

Zapus trinotatus Rhoads, 1895. Proc. Acad. Nat. Sci. Philadelphia, 1894, 46:421 [1895].

COMMON NAME: Pacific Jumping Mouse.

TYPE LOCALITY: Canada, British Columbia, mouth of the Frazer River, Lulu Isl.

DISTRIBUTION: Canada and USA: SW British Columbia, W Washington, coastal and WC Oregon, along the N California coast south to the Marin Peninsula.

STATUS: IUCN – Lower Risk (conservation dependent) as *Z. t. orarius*, otherwise Lower Risk (nt).

SYNONYMS: *eureka* A. B. Howell, 1920; *montanus* Merriam, 1897; *orarius* Preble, 1899 [see Hall, 1981].

COMMENTS: Diagnosis, records, and range map provided by Hall (1981). Reviewed by Gannon (1988, Mammalian Species, 315; 1999).

SUPERFAMILY MUROIDEA

by Guy G. Musser and Michael D. Carleton

Superfamily Muroidea Illiger, 1811.

SYNONYMS: Murina Illiger, 1811 (Myoidea Gill, 1872; Muriformes Tullberg, 1899; Muroidae Miller and Gidley, 1918; Muroidea Simpson, 1945); see family or subfamily accounts for other family-group synonyms.

COMMENTS: Although a loose group concept of Muroidea emerged with Gill's 1872 classification (as Myoidea), its contents and definition were most substantively advanced in Tullberg's (1899) seminal monograph (as Muriformes) and its usage popularized in Miller and Gidley's (1918) supergeneric arrangement of Rodentia (as Muroidae). The landmark classification of Simpson (1945) standardized the superfamily's spelling, using the now familiar family-group suffix. With the exception of Gliridae, included by some early authors within Muroidea (e.g., Ellerman, 1941; Miller and Gidley, 1918), monophyly of the superfamily has been widely ratified from paleontological, morphological, and molecular perspectives (Bugge, 1970, 1971a; Emry, 1981; Flynn et al., 1985; Klingener, 1964; Luckett, 1985; Michaux et al., 2001b; Tullberg, 1899; Wilson, 1949; and see character summary and systematic review by Carleton and Musser, 1984:290-300). This same body of literature identifies Dipodoidea as sister group to Muroidea, a coordinate relationship first appreciated by Tullberg (1899), formalized by Schaub (1958) as the infraorder Myodonta (followed by McKenna and Bell, 1997), and sustained by recent molecular surveys of appropriate breadth (Adkins et al., 2001, 2003; DeBry and Sagel, 2001; Huchon et al., 1999, 2000, 2002; Jansa and Weksler, 2004; Martin et al., 2000; Michaux and Catzeflis, 2000; Michaux et al., 2001b; Montgelard et al., 2002; Nedbal et al., 1996; Sarich, 1985); see Hartenberger (1998) for a dissenting view of their close kinship. The geological range of the superfamily covers the middle Eocene to Recent in Asia, early Oligocene to Recent of Europe, early Miocene to Recent in Africa, early Eocene to Recent in North America, and early Pliocene to Recent in South America (McKenna and Bell, 1997).

Research on muroid systematics, from the demic to family-group level, has been dynamic and prodigious over the past decade prodigious, and in some cases disconcertingly inconsistent when faced with integrating and explaining results across those publications in the context of a bare-bones systematic checklist. Below, we offer comments and caveats regarding the bases for taxonomies adopted, format of our accounts, and the family-group classification employed (Table 1).

Alpha-level classification and format.—Changes of taxa recognized at the generic and specific levels since 1993 generally reflect the authors' decisions in recent systematic studies. Molecular contributions, especially genetic sequencing results, are conspicuous in the prolific 10-year output on muroid alpha systematics and have impelled many of the changes so recorded here. Unsurprisingly, the application of late 20th century investigative methodologies to taxonomies forged in the early to middle 1900s, when the biological species concept held sway over the prosecution of systematic revision, has uncovered greater diversity of muroid rodents, often labeled as "cryptic" species in titles and discussions. Our general impression is that systematists who have consulted morphological traits along with nucleotide sequences seldom construe their results as disclosing truly cryptic entities.

Understandably, not all muroid taxa have received equal attention in the last decade. As in 1993, we have emphasized the original descriptive and primary revisionary literature over secondary compilations and checklists in representing those many cases, sometimes as a basis for returning to earlier classifications where recent information is absent, ambiguous, or incorrect. Authorities sometimes disagree, of course, about the rank or application of a name, and we attempted to explain the bases for disagreement and offer a rationale for the course we adopted. Readers should consult the primary literature referenced and judge for themselves the worthiness of the authors' data, analyses, and conclusions, not our interpretations of them. Where time and access to critical specimens allowed, we consulted type material and museum series to resolve, or at least understand, conflicting opinions and to otherwise aid taxonomic decisions. In certain instances, museum abbreviations (see Appendix I) and sometimes registration numbers, are given to substantiate a taxonomic interpretation or unpublished geographic occurrence.

Citations given within Distribution are understood to usually supply maps, sometimes noted to figure or page number, and/or collecting localities that explicitly vouch the distributions outlined. Users are urged to consult such sources rather than rely upon our necessarily cursory sketches of specific ranges. Throughout the text, we use the abbreviations M1-3 or m1-m3 to individually reference the upper (maxillary) and lower (dentary) molars, respectively.

Common names.—Early natural historians were regrettably unimaginative in coining vernacular names for small rodents, especially when compared with the rich and colorful appellations available for bird and butterfly species. The etymological archetypes of "rat" and "mouse" were seared into European vocabularies either as scurrilous purveyors of the Black Death and other pandemic miseries or as unrelenting predators on agricultural fields and food stores, well before the appearance of Linnaeus and the earliest glimmerings of natural history understanding. Western European explorers and naturalists of the latter 2nd millenium indiscriminately extended those commensal vernaculars to similarly appearing indigenous mammals around the world, foremost on the basis of size and to a lesser extent on pelage texture.

As a result, "rat" or "mouse" has become loosely and inconsistently applied to mammals that are non-murine or non-murid, commonly non-muroid or non-myomorph (spiny rats and pocket mice), and even non-rodent (moon rats and marsupial mice). Coupled with "rat" or "mouse" is some descriptive modifier that identifies collector or color, size or form, geographic place or habitat. Notwithstanding such modifiers, redundancies across Muroidea are commonplace (e.g., climbing rat or mouse, spiny or harvest mouse, short- or long-tailed rat or mouse, and so on) and are duplicated for muroid species in different genera and families. We believe that emulating the ornithological convention of adopting the genus as the vulgar name is helpful (e.g., as was done by Osgood [1912] and is already accepted for many Moluccan, New Guinean, and Australian mammals [Flannery, 1995*a*, *b*; Strahan, 1995]); or contracting dual standard names to create a new and unique vernacular (e.g., deermouse a la woodrat); or impressing regional vernaculars from languages other than English (e.g., as suggested by Thomas [1919], a practice which may occasionally impart logical redundancies for speakers of those languages). Even so, we sometimes fell back on Wilson and Cole's (2000) dictionary of common names for expediency.

Common names, like the languages to which they belong, are regionally fluid and continually evolving, and the need to prescribe "standard" names seems an ever elusive and fruitless goal. While at times an entertaining diversion, this exercise pointedly reminded us why a stable system of scientific nomenclature, founded on a language long dead, was conceived in the first place. Thank you Linnaeus.

Synonyms.—The International Code of Zoological Nomenclature (International Commission on Zoological Nomenclature, 1999) is explicit about the coordinate status of family-group taxa and identification of authorship for such names (Article 36), whatever their subsequent usage at other ranks. Although other rodent classifiers have inconsistently adhered to these guidelines (e.g., Simpson, 1945; Reig, 1980, 1981), we observe them here. Family-group synonyms, therefore, are listed in the orthography given by the original describer, that is, a generic root in plural construction which provides the formal availability and authorship of the name; in parentheses following that original form are spelling amendments that effected the standard suffix, homonymous spellings for first subsequent usage at different ranks, and other irregularly formed variants. A full and proper family-group synonymy is beyond the scope of the present work; additional citations for subsequent employment at various ranks may be found in McKenna and Bell (1997). While we include genus-group synonyms of extinct muroids, where confidently known, we did not do so at the family-group level, although many are mentioned in the introductory discussions to families and subfamilies. Readers are again directed to McKenna and Bell (1997) for such family-group synonymies, especially the many extinct groups of Cricetidae.

In specific accounts (with one exception), we alphabetically list all species-group synonyms rather than delineate formally recognized subspecies (in bold-face). Even assuming consensus among mammalogists in the meaning and usage of subspecies, their recognition is impractical for most of the world's muroid species, given the unrefined alpha level comprehension over vast geographic regions and the usual dearth of modern infraspecific studies to objectively delimit races and to vouch their distributions. Even within the exhaustively reported rodent faunas of North America and Europe, the recent upheaval in definitions and distributions of supposedly well known species of *Apodemus*, *Microtus*, *Neotoma*, *Peromyscus*, and *Sigmodon* renders perpetuation of subspecific classifications in such cases as superficial, if not actually misleading taxonomically. Users who wish conventional arrangements of subspecies, mostly for the North American and European faunas, should refer to compendia such as Hall (1981), Wilson and Ruff (1997), and Mitchell-Jones et al. (1999), unless another authority is deliberately referenced under Comments. The sole exception is *Mus musculus*, for which the plethora of old names and their inconsistent usage persuaded us to group synonyms under five "subspecies" as a means to improve understanding of their application.

Family-group classification.—In 1984 (p. 299), we posed the question: "Should rats and mice be assigned to just one family, to the two families Muridae and Cricetidae, or to more than two?" The systematic research over the last two decades has resoundingly answered the last. Still, exactly how many more is less clear, and we have steered a conservative course in recognizing just six: Platacanthomyidae, Spalacidae, Calomyscidae, Nesomyidae, Cricetidae, and Muridae. Such a primary subdivision of muroid evolutionary diversity draws upon numerous studies generated over the past century, but most notably blends elements of certain influential classifications (Chaline et al., 1977; Ellerman, 1941; Lavocat, 1978; Reig, 1981, 1984; Simpson, 1945; Thomas, 1896; Tullberg, 1899) with the phylogenetic perspective lent by recent molecular investigations, particularly those that have incorporated broad taxon representation in addressing questions of suprageneric relationships (e.g., Ducroz et al., 2001; Engel et al., 1998; Jansa and Weksler, 2004; Jansa et al., 1999; Martin et al., 2000; Michaux and Catzeflis, 2000; Michaux et al., 2001*b*; Watts and Baverstock, 1995*b*).

Increased employment of family-group ranks, from subtribe to family, will concomitantly issue from firmer understanding of genealogical hierarchies among some 1500 species (terminal taxa) of Muroidea and the desire to represent those patterns of evolutionary descent in our classification. As such comprehension continues to unfold, we note two areas of systematic investigation that might improve the dialogue concerning muroid intrarelationships. First, explicit diagnoses of newly revealed (or rediscovered) supraspecific associations has lagged, especially at the family-group level but even for genus-group ranks. Taxa such as Acomyinae, Arvicanthini, and Zygodontomyini have appeared in narrative discussion without any intent to differentiate or awareness of older available names. Although older descriptive habits were sufficiently loose to permit such casual anointing of family-group names, the International Code of Zoological Nomenclature (International Commission on Zoological Nomenclature, 1999) has stipulated progressively more rigorous standards for creation of such names (particularly Articles 13 and 16). Perhaps a differential diagnosis in this implicitly morphological sense is meaningless when considering the distribution of molecular traits on maximum parsimony and likelihood trees or consensus renditions thereof. The effort nonetheless seems worthwhile, e.g., as Smith and Patton (1999) attempted for nucleotide changes of cytochrome *b* and amino acid synapomorphies for tribal clades of Sigmodontinae. At the genus-group level, it is well to remember that the subgenus serves as a legitimate rank for expression of interspecific relationships and could be employed more regularly where diverse character data substantially concur. Second, there is comparable need for syntheses of paleontological and neontological results that bear on family-group classification. As investigators continue to probe deep muroid phylogeny, one anticipates that certain Miocene, perhaps even Oligocene, assemblages should be implicated as ancestral or cognate groups, but integrative work by those facile with both paleontological and neontological evidentiary sources is wanting. Certain studies on Arvicolinae (Chaline et al., 1999) and Otomyinae (Chevret et al., 1993*b*) stand apart as notable and salutary exceptions, but more syncretic initiatives of this kind are called for.

Table 1. Classification of Muroidea (310 genera, 1517 species) developed here.

Superfamily Muroidea Illiger, 1811

Family Platacanthomyidae Alston, 1876 (2 genera, 2 species)
  Platacanthomys Blyth, 1859 (1)
  Typhlomys Milne-Edwards, 1877 (1)

Family Spalacidae Gray, 1821 (6 genera, 36 species)

Subfamily Myospalacinae Lilljeborg, 1866 (2 genera, 6 species)
  Eospalax G. M. Allen, 1938 (3)
  Myospalax Laxmann, 1769 (3)

Subfamily Rhizomyinae Wing, 1887 (2 genera, 4 species)
  Cannomys Thomas, 1915 (1)
  Rhizomys Gray, 1831 (3)

Subfamily Spalacinae Gray, 1821 (1 genus, 13 species)
  Spalax Guldenstaedt, 1770 (13)

Subfamily Tachyoryctinae Miller and Gidley, 1918 (1 genus, 13 species)
  Tachyoryctes Rüppell, 1835 (13)

Family Calomyscidae Vorontsov and Potapova, 1979 (1 genus, 8 species)
  Calomyscus Thomas, 1905 (8)

Family Nesomyidae Major, 1897 (21 genera, 61 species)

Subfamily Cricetomyinae Roberts, 1951 (3 genera, 8 species)
Tribe Cricetomyini Roberts, 1951 (2 genera, 6 species)
  Beamys Thomas, 1909 (2)
  Cricetomys Waterhouse, 1840 (4)

Tribe Saccostomurini Roberts, 1951 (1 genus, 2 species)
  Saccostomus Peters, 1846 (2)

Subfamily Delanymyinae, new taxon (1 genus, 1 species)
  Delanymys Hayman, 1962 (1)

Subfamily Dendromurinae G. M. Allen, 1939 (6 genera, 24 species)
  Dendromus Smith, 1829 (12)
  Dendroprionomys F. Petter, 1966 (1)
  Malacothrix Wagner, 1843 (1)
  Megadendromus Dieterlen and Rupp, 1978 (1)
  Prionomys Dollman, 1910 (1)
  Steatomys Peters, 1846 (8)

Subfamily Mystromyinae Vorontsov, 1966 (1 genus, 1 species)
  Mystromys Wagner, 1841 (1)

Subfamily Nesomyinae Major, 1897 (9 genera, 23 species)
  Brachytarsomys Günter, 1875 (2)
  Brachyuromys Major, 1896 (2)
  Eliurus Milne-Edwards, 1885 (10)

Gymnuromys Major, 1896 (1)
Hypogeomys A. Grandidier, 1869 (1)
Macrotarsomys Milne-Edwards and G. Grandidier, 1898 (2)
Monticolomys Carleton and Goodman, 1996 (1)
Nesomys Peters, 1870 (3)
Voalavo Carleton and Goodman, 1998 (1)

Subfamily Petromyscinae Roberts, 1951 (1 genus, 4 species)
Petromyscus Thomas, 1926 (4)

Family Cricetidae Fischer, 1817 (130 genera, 681 species)

Subfamily Arvicolinae Gray, 1821 (28 genera, 151 species)
Tribe Arvicolini Gray, 1821 (10 genera, 82 species)
Arvicola Lacepede, 1799 (3)
Blanfordimys Argyropulo, 1933 (2)
Chionomys Miller, 1908 (3)
Lasiopodomys Lataste, 1887 (3)
Lemmiscus Thomas, 1912 (1)
Microtus Schrank, 1798 (62)
Neodon Horsfield, 1841 (4)
Phaiomys Blyth, 1863 (1)
Proedromys Thomas, 1911 (1)
Volemys Zagorodnyuk, 1990 (2)

Tribe Dicrostonychini Kretzoi, 1955 (1 genus, 8 species)
Dicrostonyx Gloger, 1841(8)

Tribe Ellobiusini Gill, 1872 (1 genus, 5 species)
Ellobius Fischer, 1814 (5)

Tribe Lagurini Kretzoi, 1955 (2 genera, 3 species)
Eolagurus Argyropulo, 1946 (2)
Lagurus Gloger, 1841 (1)

Tribe Lemmini Gray, 1825 (3 genera, 8 species)
Lemmus Link, 1795 (5)
Myopus Miller, 1910 (1)
Synaptomys Baird, 1857 (2)

Tribe Myodini Kretzoi, 1969 (5 genera, 37 species)
Alticola Blanford, 1881 (12)
Caryomys Thomas, 1911 (2)
Eothenomys Miller, 1896 (9)
Hyperacrius Miller, 1896 (2)
Myodes Pallas, 1811 (12)

Tribe Neofibrini Hooper and Hart, 1962 (1 genus, 1 species)
Neofiber True, 1884 (1)

Tribe Ondatrini Gray, 1825 (1 genus, 1 species)
Ondatra Link, 1795 (1)

Tribe Pliomyini Kretzoi, 1955 (1 genus, 1 species)
Dinaromys Kretzoi, 1955 (1)

Tribe Prometheomyini Kretzoi, 1955 (1 genus, 1 species)
  Prometheomys Satunin, 1901 (1)

Arvicolinae *incertae sedis* (2 genera, 5 species)
  Arborimus Taylor, 1915 (3)
  Phenacomys Merriam, 1889 (2)

Subfamily Cricetinae Fischer, 1817 (7 genera, 18 species)
  Allocricetulus Argyropulo, 1932 (2)
  Cansumys G. M. Allen, 1928 (1)
  Cricetulus Milne-Edwards, 1867 (6)
  Cricetus Leske, 1779 (1)
  Mesocricetus Nehring, 1898 (4)
  Phodopus Miller, 1910 (3)
  Tscherskia Ognev, 1914 (1)

Subfamily Lophiomyinae Milne-Edwards, 1867 (1 genus, 1 species)
  Lophiomys Milne-Edwards, 1867 (1)

Subfamily Neotominae Merriam, 1894 (16 genera, 124 species)
Tribe Baiomyini, new taxon (2 genera, 4 species)
  Baiomys True, 1894 (2)
  Scotinomys Thomas, 1913 (2)

Tribe Neotomini Merriam, 1894 (4 genera, 26 species)
  Hodomys Merriam, 1894 (1)
  Nelsonia Merriam, 1897 (2)
  Neotoma Say and Ord, 1825 (22)
  Xenomys Merriam, 1892 (1)

Tribe Ochrotomyini, new taxon (1 genus, 1 species)
  Ochrotomys Osgood, 1909 (1)

Tribe Reithrodontomyini Vorontsov, 1959 (9 genera, 93 species)
  Habromys Hooper and Musser, 1964 (6)
  Isthmomys Hooper and Musser, 1964 (2)
  Megadontomys Merriam, 1898 (3)
  Neotomodon Merriam, 1898 (1)
  Onychomys Baird, 1857 (3)
  Osgoodomys Hooper and Musser, 1964 (1)
  Peromyscus Gloger, 1841 (56)
  Podomys Osgood, 1909 (1)
  Reithrodontomys Giglioli, 1874 (20)

Subfamily Sigmodontinae Wagner, 1843 (74 genera, 377 species)
Tribe Akodontini Vorontsov, 1959 (19 genera, 106 species)
  Abrothrix Waterhouse, 1837 (9)
  Akodon Meyen, 1833 (41)
  Bibimys Massoia, 1979 (3)
  Blarinomys Thomas, 1896 (1)
  Brucepattersonius Hershkovitz, 1998 (8)
  Chelemys Thomas, 1903 (3)
  Deltamys Thomas, 1917 (1)
  Geoxus Thomas, 1919 (1)
  Juscelinomys Moojen, 1965 (3)

Kunsia Hershkovitz, 1966 (2)
Lenoxus Thomas, 1909 (1)
Necromys Ameghino, 1889 (9)
Notiomys Thomas, 1890 (1)
Oxymycterus Waterhouse, 1837 (16)
Pearsonomys Patterson, 1992 (1)
Podoxymys Anthony, 1929 (1)
Scapteromys Waterhouse, 1837 (2)
Thalpomys Thomas, 1916 (2)
Thaptomys Thomas, 1916 (1)

Tribe Ichthyomyini Vorontsov, 1959 (5 genera, 16 species)
Anotomys Thomas, 1906 (1)
Chibchanomys Voss, 1988 (2)
Ichthyomys Thomas, 1893 (4)
Neusticomys Anthony, 1921 (5)
Rheomys Thomas 1906 (4)

Tribe Oryzomyini Vorontsov, 1959 (18 genera, 114 species)
Amphinectomys Malygin, 1994 (1)
Handleyomys Voss, Gómez-Laverde, and Pacheco, 2002 (2)
Holochilus Brandt, 1835 (3)
Lundomys Voss and Carleton, 1993 (1)
Megalomys Trouessart, 1881 (2)
Melanomys Thomas, 1902 (3)
Microakodontomys Hershkovitz, 1993 (1)
Microryzomys Thomas, 1917 (2)
Neacomys Thomas, 1900 (8)
Nectomys Peters, 1861 (5)
Nesoryzomys Heller, 1904 (4)
Noronhomys Carleton and Olson, 1999 (1)
Oecomys Thomas, 1906 (15)
Oligoryzomys Bangs, 1900 (18)
Oryzomys Baird, 1857 (43)
Pseudoryzomys Hershkovitz, 1962 (1)
Sigmodontomys J. A. Allen, 1897 (2)
Zygodontomys J. A. Allen, 1897 (2)

Tribe Phyllotini Vorontsov, 1959 (13 genera, 47 species)
Andalgalomys Williams and Mares, 1978 (3)
Andinomys Thomas, 1902 (1)
Auliscomys Osgood, 1915 (3)
Calomys Waterhouse, 1837 (12)
Chinchillula Thomas, 1898 (1)
Eligmodontia F. Cuvier, 1837 (4)
Galenomys Thomas, 1916 (1)
Graomys Thomas, 1916 (4)
Loxodontomys Osgood, 1947 (2)
Paralomys Thomas, 1926 (1)
Phyllotis Waterhouse, 1837 (13)
Salinomys Braun and Mares, 1995 (1)
Tapecomys Anderson and Yates, 2000 (1)

Tribe Reithrodontini Vorontsov, 1959 (3 genera, 7 species)
Euneomys Coues, 1874 (4)

Neotomys Thomas, 1894 (1)
Reithrodon Waterhouse, 1837 (2)

Tribe Sigmodontini Wagner, 1843 (1 genus, 14 species)
Sigmodon Say and Ord, 1825 (14)

Tribe Thomasomyini Steadman and Ray, 1982 (4 genera, 56 species)
Aepeomys Thomas, 1898 (2)
Chilomys Thomas, 1897 (1)
Rhipidomys Tschudi, 1845 (17)
Thomasomys Coues, 1884 (36)

Tribe Wiedomyini Reig, 1980 (1 genus, 1 species)
Wiedomys Hershkovitz, 1959 (1)

Sigmodontinae *incertae sedis* (10 genera, 16 species)
Abrawayaomys Souza Cunha and Cruz, 1979 (1)
Delomys Thomas, 1917 (3)
Irenomys Thomas, 1919 (1)
Juliomys González, 2000 (2)
Megaoryzomys Lenglet and Coppois, 1979 (1)
Phaenomys Thomas, 1917 (1)
Punomys Osgood, 1943 (2)
Rhagomys Thomas, 1917 (2)
Scolomys Anthony, 1924 (2)
Wilfredomys Avila-Pires, 1960 (1)

Subfamily Tylomyinae Reig, 1984 (4 genera, 10 species)
Tribe Nyctomyini, new taxon (2 genera, 2 species)
Nyctomys Saussure, 1860 (1)
Otonyctomys Anthony, 1932 (1)

Tribe Tylomyini Reig, 1984 (2 genera, 8 species)
Ototylomys Merriam, 1901 (1)
Tylomys Peters, 1866 (7)

Family Muridae Illiger, 1811 (150 genera, 730 species)

Subfamily Leimacomyinae, new taxon (1 genus, 1 species)
Leimacomys Matschie, 1893 (1)

Subfamily Deomyinae Thomas, 1888 (4 genera, 42 species)
Acomys I. Geoffroy, 1838 (19)
Deomys Thomas, 1888 (1)
Lophuromys Peters, 1874 (21)
Uranomys Dollman, 1909 (1)

Subfamily Gerbillinae Gray, 1825 (16 genera, 103 species)
Tribe Ammodillini Pavlinov, 1981 (1 genus, 1 species)
Ammodillus Thomas, 1904 (1)

Tribe Taterillini Chaline, Mein and F. Petter, 1977 (5 genera, 26 species)
Subtribe Gerbillurina Pavlinov, 1982
Desmodillus Thomas and Schwann, 1904 (1)
Gerbillurus Shortridge, 1942 (4)

Subtribe Taterillina Chaline, Mein, and F. Petter, 1977
    Gerbilliscus Thomas, 1897 (11)
    Tatera Lataste, 1882 (1)
    Taterillus Thomas, 1910 (9)

Tribe Gerbillini Gray, 1825 (10 genera, 76 species)
Subtribe Desmodilliscina Pavlinov, 1982
    Desmodilliscus Wettstein, 1916 (1)

Subtribe Gerbillina Gray, 1825
    Dipodillus Lataste, 1881 (13)
    Gerbillus Desmarest, 1804 (38)
    Microdillus Thomas, 1910 (1)

Subtribe Pachyuromyina Pavlinov, 1982
    Pachyuromys Lataste, 1880 (1)

Subtribe Rhombomyina Heptner, 1933
    Brachiones Thomas, 1925 (1)
    Meriones Illiger, 1811 (17)
    Psammomys Cretzschmar, 1828 (2)
    Rhombomys Wagner, 1841 (1)
    Sekeetamys Ellerman, 1947 (1)

Subfamily Murinae Illiger, 1811 (126 genera, 561 species)
Aethomys Division
    Aethomys Thomas, 1915 (11)
    Micalaemys Ellerman, 1941 (2)
Apodemus Division
    Apodemus Kaup, 1829 (20)
    Rhagamys Major, 1905 (1)
    Tokudaia Kuroda, 1943 (2)
Arvicanthis Division
    Arvicanthis Lesson, 1842 (7)
    Desmomys Thomas, 1910 (2)
    Lemniscomys Trouessart, 1881 (11)
    Mylomys Thomas, 1906 (2)
    Pelomys Peters, 1852 (5)
    Rhabdomys Thomas, 1916 (2)
Chrotomys Division
    Apomys Mearns, 1905 (9)
    Archboldomys Musser, 1982 (2)
    Chrotomys Thomas, 1895 (4)
    Rhynchomys Thomas, 1895 (2)
Colomys Division
    Colomys Thomas and Wroughton, 1907 (1)
    Nilopegamys Osgood, 1928 (1)
    Zelotomys Osgood, 1910 (2)
Crunomys Division
    Crunomys Thomas, 1897 (4)
    Sommeromys Musser and Durden, 2002 (1)
Dacnomys Division
    Anonymomys Musser, 1981 (1)
    Chiromyscus Thomas, 1925 (1)
    Dacnomys Thomas, 1916 (1)

Leopoldamys Ellerman, 1947 (6)
Niviventer Marshall, 1976 (17)
Srilankamys Musser, 1981 (1)
Dasymys Division
Dasymys Peters, 1875 (9)
Echiothrix Division
Echiothrix Gray, 1867 (2)
Golunda Division
Golunda Gray, 1837 (1)
Hadromys Division
Hadromys Thomas, 1911 (2)
Hybomys Division
Dephomys Thomas, 1926 (2)
Hybomys Thomas, 1910 (6)
Stochomys Thomas, 1926 (1)
Hydromys Division
Crossomys Thomas, 1907 (1)
Hydromys E. Geoffroy, 1804 (5)
Microhydromys Tate and Archbold, 1941 (2)
Parahydromys Poche, 1906 (1)
Paraleptomys Tate and Archbold, 1941 (2)
Lorentzimys Division
Lorentzimys Jentink, 1911 (1)
Malacomys Division
Malacomys Milne-Edwards, 1877 (3)
Maxomys Division
Maxomys Sody, 1936 (17)
Melasmothrix Division
Melasmothrix Miller and Hollister, 1921 (1)
Tateomys Musser, 1969 (2)
Micromys Division
Chiropodomys Peters, 1869 (6)
Haeromys Thomas, 1911 (3)
Hapalomys Blyth, 1859 (2)
Micromys Dehne, 1841 (1)
Vandeleuria Gray, 1842 (3)
Vernaya Anthony, 1941 (1)
Millardia Division
Cremnomys Wroughton, 1912 (2)
Diomys Thomas, 1917 (1)
Madromys Sody, 1941 (1)
Millardia Thomas, 1911 (4)
Mus Division
Muriculus Thomas, 1903 (1)
Mus Linnaeus, 1758 (38)
Oenomys Division
Grammomys Thomas, 1915 (12)
Lamottemys F. Petter, 1986 (1)
Malpaisomys Hutterer, Lopez-Martinez, and Michaux, 1988 (1)
Oenomys Thomas, 1904 (2)
Thallomys Thomas, 1920 (4)
Thamnomys Thomas, 1907 (3)
Phloeomys Division
Batomys Thomas, 1895 (4)
Carpomys Thomas, 1895 (2)

    Crateromys Thomas, 1895 (4)
    Phloeomys Waterhouse, 1839 (2)
Pithecheir Division
    Eropeplus Miller and Hollister, 1921 (1)
    Lenomys Thomas, 1898 (1)
    Lenothrix Miller, 1903 (1)
    Margaretamys Musser, 1981 (3)
    Pithecheir Lesson, 1840 (2)
    Pithecheirops Emmons, 1993 (1)
Pogonomys Division
    Abeomelomys Menzies, 1990 (1)
    Anisomys Thomas, 1904 (1)
    Chiruromys Thomas, 1888 (3)
    Coccymys Menzies, 1990 (2)
    Coryphomys Schaub, 1937 (1)
    Hyomys Thomas, 1904 (2)
    Macruromys Stein, 1933 (2)
    Mallomys Thomas, 1898 (4)
    Mammelomys Menzies, 1996 (2)
    Pogonomelomys Rümmler, 1936 (2)
    Pogonomys Milne-Edwards, 1877 (5)
    Spelaeomys Hooijer, 1957 (1)
    Xenuromys Tate and Archbold, 1941 (1)
Pseudomys Division
    Conilurus Ogilby, 1838 (2)
    Leggadina Thomas, 1910 (2)
    Leporillus Thomas, 1906 (2)
    Mastacomys Thomas, 1982 (1)
    Mesembriomys Palmer, 1906 (2)
    Notomys Lesson, 1842 (9)
    Pseudomys Gray, 1832 (24)
    Zyzomys Thomas, 1909 (5)
Rattus Division
    Abditomys Musser, 1982 (1)
    Bandicota Gray, 1873 (3)
    Berylmys Ellerman, 1947 (4)
    Bullimus Mearns, 1905 (3)
    Bunomys Thomas, 1910 (6)
    Diplothrix Thomas, 1916 (1)
    Kadarsanomys Musser, 1981 (1)
    Komodomys Musser and Boeadi, 1980 (1)
    Limnomys Mearns, 1905 (2)
    Nesokia Gray, 1842 (2)
    Nesoromys Thomas, 1922 (1)
    Palawanomys Musser and Newcomb, 1983 (1)
    Papagomys Sody, 1941 (2)
    Paruromys Ellerman, 1954 (1)
    Paulamys Musser, 1986 (1)
    Rattus Fischer, 1803 (66)
    Sundamys Musser and Newcomb, 1983 (3)
    Taeromys Sody, 1941 (7)
    Tarsomys Mearns, 1905 (2)
    Tryphomys Miller, 1910 (1)
Stenocephalemys Division
    Heimyscus Misonne, 1969 (1)

Hylomyscus Thomas, 1926 (8)
Mastomys Thomas, 1915 (8)
Myomyscus Shortridge, 1942 (4)
Praomys Thomas, 1915 (16)
Stenocephalemys Frick, 1914 (4)
Uromys Division
Melomys Thomas, 1922 (23)
Paramelomys Rümmler, 1936 (9)
Protochromys Menzies, 1996 (1)
Solomys Thomas, 1922 (5)
Uromys Peters, 1867 (10)
Xeromys Division
Leptomys Thomas, 1897 (3)
Pseudohydromys Rümmler, 1934 (4)
Xeromys Thomas, 1889 (1)

Subfamily Otomyinae Thomas, 1897 (3 genera, 23 species)
Myotomys Thomas, 1918 (2)
Otomys F. Cuvier, 1824 (19)
Parotomys Thomas, 1918 (2)

Family Platacanthomyidae Alston, 1876. Proc. Zool. Soc. Lond., 1876:81.
SYNONYMS: Platacanthomyinae Alston, 1876 (Platacanthomyidae Miller and Gidley, 1918;
Platacanthomyini Ognev, 1947); Typhlomyinae Ognev, 1947.
COMMENTS: Detailed diagnosis, general characteristics, and natural history provided by Carleton and Musser (1984). Since Blyth (1859) described *Platacanthomys* as a genus of Gliridae,
Platacanthomys and *Typhlomys* have been regarded: as a subfamily of dormice (Ellerman,
1940, 1961; Ognev, 1947; Thomas, 1896); a family related to Gliridae (Simpson, 1945); a
family allied with Cricetidae (G. M. Allen, 1940; Miller and Gidley, 1918; Pavlinov et al.,
1995*a*; Qiu, 1989); a subfamily or tribe within Cricetidae (Chaline et al., 1977; Engesser,
1972; Fahlbusch, 1966; Mein and Freudenthal, 1971; Reig, 1980; Schaub and Zapfe,
1953); allied to "the Murine family of Rodents" (Peters, 1865); or as a subfamily of a
broadly defined Muridae (Alston, 1876; Carleton and Musser, 1984; Musser and Carleton,
1993). Separation from dormice and alliance with Muroidea have been supported by paleontologists, our own examinations, and those of others (Fejfar, 1999*b*; Stehlin and
Schaub, 1951; Vorontsov, 1979). Morphological differences between *Platacanthomys* and
Typhlomys are pronounced, to an extent that Ognev (1947) arranged each in its own glirid
subfamily, a treatment that Vorontsov (1979) thought extreme and unnecessary.
    While molar similarities between platacanthomyids and dormice are superficial, those
between platacanthomyids and the Madagascan nesomyine *Gymnuromys* are actually
closer (Ellerman, 1940; Wood, 1955). Stehlin and Schaub (1951) also demonstrated that
occlusal patterns in *Platacanthomys* are not homologous to those of dormice and can be
derived from a "cricetid plan." In naming the Miocene *Neocometes*, Schaub and Zapfe
(1953) arranged it, along with *Platacanthomys* and *Typhlomys*, as a tribe within Cricetinae.
Platacanthomyids have three cheekteeth (all molars) in each quadrant of the jaw (four in
dormice, premolar and three molars), small muroid-type ectotympanic bullae without
septa (bullae globular with conspicuous transbullar septa in dormice), and the dentary is
neither inflected lingually nor perforated (traits characteristic of dormice).
    Extinct Pleistocene and Miocene species of *Platacanthomys* and *Typhlomys* have now
been described from China (Fejfar, 1999*b*; Ni and Qiu, 2002; see generic accounts), and a
"platacanthomyid" has been identified from early middle Miocene (17 million year ago)
strata in the Siwaliks of N Pakistan (Flynn, 2003). Apparently, platacanthomyids originated in Asia and reached Europe in the Miocene, as far as Spain (de Bruijn and Moltzer,
1974). *Platacanthomys* and *Typhlomys*, although characterized by many specialized features, are relicts of a clade recognizable as platacanthomyid by the early Miocene. Because
extinct forms are contemporaneous with early to late Miocene cricetids and their denti-

tions are so strikingly unlike the cricetid genera, platacanthomyids likely originated from some Eocene or Oligocene, probably Asian, muroid stock (Carleton and Musser, 1984). The platacanthomyid molar patterns and their relative sizes (upper and lower second and third molars large relative to first) could have evolved from a morphology resembling that in *Eucricetodon*, which flourished from early Eocene to early Miocene in Asia and early Oligocene to early Miocene in Europe (McKenna and Bell, 1997). Species of *Eucricetodon* were replaced by Miocene cricetids unrelated to *Eucricetodon* and its allies (Hugueney, 1999) or to platacanthomyids.

Closest relatives of *Platacanthomys* and *Typhlomys* are *Neocometes brunonis* and *N. similis* from early to late Miocene of Europe (de Bruijn and Moltzer, 1974; Fahlbusch, 1966; Mein and Freudenthal, 1971; Schaub and Zapfe, 1953; Ziegler, 1995), recorded as far northeast as Poland (Kowalski, 1993); *Neocometes* sp. from early Miocene in S China (Qiu and Li, 2003); and *N. orientalis* from early Miocene in N Thailand (Mein et al., 1990; Mein and Ginsburg, 1997), which is the oldest and regarded as the most primitive of the three species (see review by Fejfar, 1999*b*). Schaub and Zapfe (1953) considered *Neocometes* to be morphologically close to *Typhlomys*, a relationship endorsed by Fejfar (1999*b*), who noted that *Platacanthomys* has more derived dental traits and represents a separate evolutionary branch.

Current common names (e.g., Wilson and Cole, 2000) are Malabar spiny dormouse (*Platacanthomys*) and pygmy dormouse (*Typhlomys*). Either "tree mouse" or incorporation of the genus is preferred; both arrangements avoid perpetuating the incorrect phylogenetic alliance implied by the usual vernaculars.

Platacanthomys Blyth, 1859. J. Asiat. Soc. Bengal, 28:288.
> TYPE SPECIES: *Platacanthomys lasiurus* Blyth, 1859.
> SYNONYMS: *Platyacanthomus* Marschall, 1873 [*lapsus*]; *Platyacanthomys* Coues, 1890 [*lapsus*].
> COMMENTS: Although currently known only from the Indian Peninsula, the genus is represented in the Indomalayan region by the late Miocene *P. dianensis*, described from isolated molars in Yuanmou and Lufeng, Yunnan, China (Ni and Qiu, 2002; Qiu, 1989); the "living *Platacanthomys lasiurus* is closely related to *P. dianensis*, and very probably descended from it" (Qiu, 1989:281).

Platacanthomys lasiurus Blyth, 1859. J. Asiat. Soc. Bengal, 28:289.
> COMMON NAME: Spiny Tree Mouse.
> TYPE LOCALITY: India, Kerala State, Malabar, Alipi.
> DISTRIBUTION: Forests below 3000 ft (914 m) in SW Peninsular India to 14°north latitude (Kerala and Karnataka states; Agrawal, 2000).
> STATUS: IUCN – Lower Risk (lc).
> COMMENTS: Known by few museum specimens but apparently common in the right habitats. Peters (1865) redescribed the species in some detail; Ellerman (1961) also furnished an informative description, excellent cranial illustrations, and measurements of BMNH specimens. Reviewed by Agrawal (2000), who still regarded the species as "a specialised and aberrant arboreal dormouse." Jayson and Christopher (1995) encountered *P. lasiurus* in the Peppara Wildlife Sanctuary, Western Ghats in Kerala State, and provided what little ecological information is known.

Typhlomys Milne-Edwards, 1877. Bull. Sci. Soc. Philom. Paris, ser. 6, 12:9 [1877].
> TYPE SPECIES: *Typhlomys cinereus* Milne-Edwards, 1877.
> COMMENTS: Miocene *Neocometes* is morphologically close to *Typhlomys*. Fossil *Typhlomys* are known from late Miocene in China (Lufeng and Yuanmou, Yunnan), isolated molars described as *T. primitivus* and *T. hipparionum* (Ni and Qiu, 2002; Qiu, 1989). Two additional species, *T. macrourus* and *T. intermedius*, are also represented by molars from the late Pliocene in the Sichuan-Guizhou region, and Pleistocene samples of living *T. cinereus* come from the same area (Zheng, 1993) and Guangxi in S China (Chen et al., 2002). Zheng (1993) postulated two lineages: one leading from the late Miocene *T. primitivus*, through Pleistocene *T. intermedius*, to extant *T. cinereus*; the other from late Miocene *T. hipparionum* to late Pliocene *T. macrourus*. Of the two late Miocene species, occlusal patterns of *T. primitivus* are most like those of *Neocometes*.

Typhlomys cinereus Milne-Edwards, 1877. Bull. Sci. Soc. Philom. Paris, ser. 6, 12:9 [1877].

    COMMON NAME: Soft-furred Tree Mouse.

    TYPE LOCALITY: China, W Fujian.

    DISTRIBUTION: Montane forests in mountains of S China (S Shaanxi, SE Gansu, W Sichuan, SW Hubei, SW Yunnan, Guizhou, Chongqing, Guangxi, W Hunan, Fujian, Jiangxi, W Zhejiang, and S Anhui) and NW Vietnam (northwest of the Red River); see G. M. Allen (1940 and references cited therein), Osgood (1932), Liu et al. (1985), Wang et al. (1996), Wang (2003), Wu and Wang (1984), Zhang et al. (1997).

    STATUS: IUCN – Critically Endangered as *T. chapensis* (but see below), Lower Risk (lc) as *T. cinereus*.

    SYNONYMS: *chapensis* Osgood, 1932; *daloushanensis* Wang and Li, 1996 (in Wang et al., 1996); *guangxiensis* Wang and Chen, 1996 (in Wang et al., 1996); *jingdongensis* Wu and Wang, 1984.

    COMMENTS: Since G. M. Allen's (1940) monograph, which provided a good description of the species and its geographic range as then known, the number of specimens identified as *T. cinereus* and its distribution in S China have increased dramatically (Wang et al., 1996). The new material has substantiated considerable geographic variation in pelage coloration and morphometric traits (Corbet and Hill, 1992; Wang et al., 1996; Wu and Wang, 1984), and led to the description of three new Chinese taxa as subspecies (Wang, 2003; Wang et al., 1996; Wu and Wang, 1984).

        The species also occurs in NW Vietnam, where it was originally described as a distinctive subspecies of *T. cinereus* (*chapensis* Osgood, 1932). Based on study of older museum series, Musser and Carleton (1993) treated *chapensis* as a species because of its larger size (noted by Osgood), dark hind feet, and dark buffy underparts (grayish white in *cinereus*), which together distinguish it from typical *cinereus* in Fujian. With the improved samples and expanded geographic representation now at hand, the Vietnam specimens have been recently interpreted as falling within the variation that defines a single species, *T. cinereus*, with all five species-group taxa retained as subspecies (Wang et al., 1996). According to the morphometric results of Wang et al. (1996), populations of *chapensis* and *guangxiensis* are phenetically most alike, clustering apart from samples representing the other three subspecies. All populations are montane, and those in the southern portion of the range tend to be geographic outliers from the northern core of the species distribution (see Wang et al., 1996:61). Significance of the morphometric and chromatic variation described should now be tested by other data sources to confirm the hypothesis of a single species.

Family Spalacidae Gray, 1821. Lond. Med. Repos., 15:303.

    COMMENTS: This family contains species of fossorial and subterranean muroids arranged in the Myospalacinae (zokors, *Eospalax* and *Myospalax*), Rhizomyinae (bamboo rats, *Cannomys* and *Rhizomys*), Spalacinae (blind mole rats, *Spalax*), and Tachyoryctinae (African mole rats, *Tachyoryctes*). All extant species are characterized by extreme morphological, physiological, and behavioral specializations associated with subterranean life in tubular burrows (Gambaryan, 1960; Gambaryan and Gasc, 1993; Nevo et al., 2001; Tullberg, 1899). Although each subfamily can be readily diagnosed by unique traits (Carleton and Musser, 1984), and although molar occlusal patterns among living species are dissimilar, all share a comparable cranial and postcranial skeletal architecture integrated with a myological system that characterizes highly specialized fossorial rodents. Such a phenotypic resemblance was interpreted as phylogenetic relationship by Tullberg (1899), who placed *Spalax*, *Rhizomys*, *Tachyoryctes*, and *Myospalax* in Spalacidae, an ordering also used by Ognev (1947, 1963a; he did not discuss *Tachyoryctes*).

        Most other classificatory arrangements, implicitly or explicitly, reflect the view that the shared fossorial adaptations indicate evolutionary convergence, not descent from a common ancestor, and have variously combined the genera (also see Topachevskii, 1969, 1976): e.g., *Myospalax* (Myospalacinae) in Muridae, *Spalax* and *Rhizomys* in Spalacidae (Alston, 1876); *Rhizomys* and *Tachyoryctes* (Rhizomyinae) and *Spalax* (Spalacinae) in Spalacidae, *Myospalax* (Myospalacinae) in Muridae (Thomas, 1896); *Rhizomys* (Rhizomyinae) and *Tachyoryctes* (Tachyoryctinae) in Rhizomyidae, *Myospalax* (Myospalacinae) and *Spalax* (Spalacinae) in Spalacidae (Miller and Gidley, 1918); *Spalax* in Spalacidae, *Rhiz-*

omys and *Cannomys* in Rhizomyidae, *Tachyoryctes* (Tachyoryctinae) and *Myospalax* (Myospalacinae) in Muridae (Ellerman, 1940, 1941); *Rhizomys and Cannomys* in Rhizomyidae, Myospalacinae and Spalacinae in Spalacidae (G. M. Allen, 1940); *Spalax* in Spalacidae, *Rhizomys* and *Cannomys* in Rhizomyidae, *Myospalax* as a tribe or subfamily in Cricetidae (Pavlinov et al., 1995a; Simpson, 1945); Myospalacinae and Spalacinae in Cricetidae, Rhizomyinae and Tachyoryctinae in Rhizomyidae (Chaline et al., 1977); Spalacines and Rhizomyines in Spalacidae of Theridomyoidea (Schaub, 1958); Spalacidae and Myospalacidae in Muroidea, Rhizomyidae in Theridomyoidea (Reig, 1980); Spalacinae (*Spalax*), Rhizomyinae (*Cannomys, Rhizomys, Tachyoryctes*), and Myospalacinae (*Myospalax*) in Muridae (Carleton and Musser, 1984; McKenna and Bell, 1997; Musser and Carleton, 1993). These compilations, which recognize three or four separate subfamilies and two or three families, imply independent origin of each, a pattern that echos Ellerman's (1940:638) pronouncement, "I do not think that there is much doubt that the grouping together of *Spalax, Myospalax, Tachyoryctes* and *Rhizomys* is a very unnatural arrangement," and opinions of some fossil rodent specialists: "Spalacidae and Rhizomyidae originate separately from different muroids" (Flynn et al., 1985:608).

Much doubt, nonetheless, is supplied by recent studies in which the cladistic union of *Spalax* and *Rhizomys* is supported by the cephalic arterial pattern (Bugge, 1971a, 1985) and by multiple gene-sequence analyses (nuclear LCAT, Michaux and Catzeflis, 2000, and Robinson et al., 1997; LCAT and vWF, Michaux et al., 2001b; nuclear IRBP, DeBry and Sagel, 2001). Further support for uniting *Spalax, Myospalax, Rhizomys*, and *Tachyoryctes* as a basal clade relative to other muroids stems from other sequence analyses with wider taxon sampling (Jansa and Weksler, 2004; Norris et al., 2004). These molecular data collectively point to zokors, bamboo rats, blind mole rats, and African mole rats as a monophyletic radiation that is sister-group to all other muroid family-group taxa so far investigated, inferring an early divergence, possibly one of the earliest, from a middle or late Oligocene ancestral stock.

Morphology of fossil molars attributed to spalacines and rhizomyines suggests their common origin. In describing late Miocene African species of *Nakalimys* and *Harasibomys*, Mein et al. (2000a:385) allocated them to "Family Rhizomyidae or Spalacidae," and noted that the early Miocene Spalacine *Debruijnia* from Turkey (at about 20 million years old, the earliest record among the four subfamiles) "could represent an ancestral form for the African burrowing rodents." *Nakalimys* was first described from the late Miocene of Kenya and placed in Rhizomyidae (Flynn, 1990; Flynn and Sabatier, 1984), and *Pronakalimys* from middle Miocene deposits in Kenya is regarded "as the most primitive known African Rhizomyidae" (Tong and Jaeger, 1993:59). The resemblance of Myospalacines to some arvicolines, with their simple occlusal patterns and rootless molars (some extinct species are rooted) impressed Hinton (1926a), but he also noted that the head skeleton and jaw musculature were definitely unlike arvicolines. Lawrence (1991:282) reinforced his observation, remarking that the "character complex myospalacines share with arvicolines is a set of parallel adaptations for propalinal chewing of tough, fibrous plant material," and listing those parallel features as well as synapomorphic traits distinctive to myospalacines and indicative of great phylogenetic distance from arvicolines. Although the molars have different coronal outlines, occlusal patterns in myospalacines are basically similar to those in spalacines and also could have been derived from a form similar to *Debruijnia*. In turn, the dentition of the Turkish, African, and some Asian Miocene forms recalls occlusal patterns common to *Eucricetodon* (our observations; Lindsay, 1994, suggested that rhizomyines are closely linked to eucricetodontines), which is known from the late Eocene to early Miocene in Asia, Oligocene in Mediterranean region, and early Oligocene to early Miocene in Europe (Hugueney, 1999; McKenna and Bell, 1997). Hugueney and Mein (1993) wondered if a hypothetical Oligocene ancestor could be unambiguously distinguished from an ancient "cricetid."

A return to Tullberg's (1899) interpretation of phylogenetic affinities among *Myospalax, Spalax, Rhizomys*, and *Tachyoryctes* is the hypothesis best supported by the accumulation of recent morphological and molecular information. This arrangement should be tested by further cladistic analyses of multi-morphological systems and a wider array of genes.

Subfamily Myospalacinae Lilljeborg, 1866. Syst. Öfversigt. Gnag. Däggdjuren, p. 25.

SYNONYMS: Mesosiphneinae Zheng, 1994; Myospalacini Lilljeborg, 1866 (Myospalacinae Miller and Gidley, 1918; Myospalacidae Kretzoi, 1961); Myotalpinae Miller, 1896; Prosiphneinae Leroy, 1940; Siphneinae Gill, 1872 (Siphneidae Zheng, 1994).

COMMENTS: All family-group names listed in the synonymy have been used at one time or another, and recently Zheng (1994) preferred to employ Siphneidae. Regardless of preference, Lilljeborg's (1866) Myospalacinae is the oldest available name, whether used as family or subfamily. Diagnosis, morphological and chromosomal characteristics, distribution, and remarks on habits and habitat are provided by Carleton and Musser (1984). Phylogenetic arrangements of myospalacines were attempted by G. M. Allen (1940) and Leroy (1940) dealing with extant species; by Teilhard de Chardin (1942) studying extant and extinct species; by Kretzoi (1961), who reorganized the group using all named taxa up to 1961; and by Lawrence (1991) and Zheng (1994), who conducted phylogenetic analyses of living and extinct species.

Lawrence (1991) consolidated synapomorphic morphological traits defining Myospalacinae and recognized four species groups: 1) *M. psilurus*; 2) *M. myospalax*, *M. epsilanus*, *M. aspalax*, and the extinct *M. youngi* and *M. pseudarmandi* (which Zheng, 1994, regarded as an immature *youngi*); 3) *M. tingi* and related species, all extinct; 4) *M. fontanieri*, *M. smithi*, and *M. rothschildi*. Extant species in groups 1 and 2 are traditionally included in the subgenus and genus *Myospalax* (G. M. Allen, 1940; Corbet, 1978c; Ellerman, 1941; Kretzoi, 1961, proposed the genus *Episiphneus* for *pseudarmandi* and *youngi*) and have been arranged in subfamily Myospalacinae by Zheng (1994). *Mesosiphneus* (Kretzoi, 1961), *Chardinia*, and *Youngia* (Zheng, 1994; invalid as no type species was identified) were erected to contain the species that Lawrence recognized in group 3, as well some additional extinct species, and these three genera were placed in the subfamily Mesosiphneinae by Zheng. Species in group 4 are traditionally separated as the subgenus *Eospalax* of *Myospalax* (G. M. Allen, 1940; Corbet, 1978c; Ellerman, 1941, proposed *Zokor* for the same assemblage), and Zheng listed these species in the subfamily Prosiphneinae, along with the genera *Prosiphneus*, *Myotalpavus*, *Pliosiphneus*, and *Allosiphneus*.

Zheng's (1994) classification, reflecting his hypothesis of evolution in the family, recognizes three subfamilies and ten genera, including those already named, all those proposed by Kretzoi (1961), and three new ones. While Zheng's subfamilial clusters correspond to three of Lawrence's species groups, and apparently represent different lineages, elevation of each to subfamily level inflates their phylogenetic status relative to other spalacid assemblages. Although we do not employ tribes, we do recognize *Myospalax* for Lawrence's groups 1 and 2, the extinct *Mesosiphneus* for Lawrence's group 3 (with *Chardina* and *Youngia* as synonyms), and *Eospalax* for her group 4. Zheng's (1994) Myospalacinae is thus equivalent to *Myospalax*, his Prosiphneinae to *Eospalax* and *Prosiphneus* (*Myotalpavus* and *Pliosiphneus* as synonyms), and his Mesosiphneinae to the extinct species of *Mesosiphneus*. Whether extant and extinct species are eventually separated into many genera and three subfamilies or some other arrangement, they unquestionably constitute a monophyletic group as indicated by analyses of cranial and dental characters (Lawrence, 1991; Zheng, 1994).

Phylogenetic position of myospalacines has been differently conveyed in classifications and taxonomic compilations by treating them as a separate family in Muroidea (Pavlinov and Rossilimo, 1987; Reig, 1980; Zheng, 1994), subfamily of Muridae (Alston, 1876; Carleton and Musser, 1984; Ellerman, 1941; Musser and Carleton, 1993; Thomas, 1896), subfamily of Cricetidae (Chaline et al., 1977; Corbet, 1978c; Lindsay, 1994; Pavlinov et al., 1995a; Pavlinov and Rossilimo, 1998; Simpson, 1945), closely related to arvicolines (G. M. Allen, 1940; Miller, 1896; Milne-Edwards, 1868-1874), or as a genus or subfamily in Spalacidae (Miller and Gidley, 1918; Ognev, 1947; Tullberg, 1899). Gambaryan and Gasc (1993) documented the musculoskeletal system and burrowing behavior of *Myospalax myospalax*, contrasted it with other burrowing rodents, primarily *Spalax nehringi*, and noted that *Myospalax* was unrelated to these other burrowers and belonged in its own monotypic family. Phylogenetic analyses of the nuclear gene LCAT (Michaux and Catzeflis, 2000) and LACT and nuclear vWF in combination (Michaux et al., 2001b) nested *Myospalax* next to *Phodopus* within a cricetine clade (also *Cricetulus* and *Meso-*

cricetus), prompting those authors to strongly urge its transfer to Cricetinae as the tribe Myspalacini. Such an arrangement would revitalize Winge's (1941) classification of zokors, which he included in the "Cricetini." We unsuccessfully tried to locate their voucher and suspect that it was misidentified. Michaux and Catzeflis (2000:283) listed their tissue sample as *Myospalax* sp. from an unknown locality in Russia, a vast region that could have produced a true hamster or a zokor. The analyses of mitochondrial (Norris et al., 2004) and nuclear (Jansa and Weksler, 2004) sequences tellingly unites zokors, based on accurately identified myospalacines, with *Spalax*, *Rhizomys*, and *Tachyoryctes* in a clade distinct from other muroid groups sampled. Furthermore, scrutiny of morphological diagnoses for myospalacines (Carleton and Musser, 1984; Lawrence, 1991; Ognev, 1947; Tullberg, 1899) and first-hand study of specimens reveal little about extant myospalacines that indicates their close affinity to living hamsters, a conclusion earlier reached by Ognev (1947).

When zokors diverged from the ancestral spalacid stock is so far untraced in the fossil record. Earliest fossils readily identified as myospalacine are about 11 million years old (beginning of late Miocene; Zheng, 1994). Zheng (1993) speculated that zokors derived from middle Miocene *Plesiodipus*, which some researchers place close to the ancestry of arvicolines (see discussion in Lindsay, 1994). Considering the extreme fossorial adaptations of myospalacines, their evolutionary history must be rooted in some ancestral group existing well before the Miocene, probably in late or middle Oligocene times, a view echoed by Lawrence (1991:279): "myospalacines, fossil and Recent, are a closely related group of species derived from a primitive muroid stock."

Eospalax G. M. Allen, 1938. Mammals of China and Mongolia, Nat. Hist. Central Asia, 2, pt. 1:vii.
  TYPE SPECIES: *Siphneus fontanieri* Milne-Edwards, 1867.
  SYNONYMS: *Allosiphneus* Kretzoi, 1961; *Zokor* Ellerman, 1941.
  COMMENTS: Extant and extinct species recorded only from China. Phylogenetic relationships reviewed by Lawrence (1991, as *Myospalax*; see above discussion) and Zheng (1994), both of whom brought together *fontanieri*, *rothschildi*, and *smithi* in a cluster defined by a suite of derived morphological features (convex occipital shield, long incisive foramina bisected by premaxillary-maxillary suture, carotid canal at basioccipital-basisphenoid suture, configuration of pterygoid fossa, hypsodent and rootless molars). All three have always been recognized as species, usually grouped at the species-group or subgeneric level (G. M. Allen, 1940; Corbet, 1978c; Kuzhyakin, 1965; Leroy, 1940). The three were carefully described by G. M. Allen (1940), who also provided distributional and habitat information.

Zheng (1994) recognized *Allosiphneus* for the extinct species *arvicolinus*, which has rootless molars and belongs in this group, but the traits distinguishing this genus define only a different species to us. He also included an assemblage of rooted species sorted into the genera *Prosiphneus*, *Myotalpavus*, and *Pliosiphneus*. We do not appreciate the cladistic support for the latter two and include them in *Prosiphneus*. The distinguishing traits among them are not nearly of the same magnitude as those between *Eospalax*, *Myospalax*, and *Mesosiphneus*; *Prosiphneus* thus contains extinct species with rooted molars. Lawrence (1991:282) did not recognize this dichotomy: "Placing species that have not achieved complete rootless hypsodonty in the separate genus *Prosiphneus* is an artifical division of a monophyletic group." Her conclusion was influenced by study of fossil *youngi*, which at the time was included in *Prosiphneus*, and her analysis placed it in the same group as *Myospalax*, an arrangement also proposed by Zheng (1994). Lawrence otherwise did not have available any of the species placed in *Prosiphneus* by Zheng, and we cannot assess the generic validity from Zheng's report because he offered no diagnosis or definition of it. Until phylogenetic relationships among the extinct species with rooted molars in *Prosiphneus* and those extant and extinct forms in *Eospalax* are illuminated, the two genera should be retained. If shown to be congeneric, *Eospalax* G. M. Allen (1939) would be replaced by *Prosiphneus* Tielhard de Chardin (1926).

Eospalax fontanierii (Milne-Edwards, 1867). Ann. Sci. Nat. (Paris), 7:376.
  COMMON NAME: Chinese Zokor.
  TYPE LOCALITY: China, Kansu.

DISTRIBUTION: Forests, steppes, scrub, and farmland in NC China (N Qinghai, N Sichuan, NGansu, Ningxia, N Shaanxi, Henen, N Shanxi, Shandong, Beijing, Jilin, Hebei, and Nei Mongol; Qin, 1991; Zhang et al., 1997).

STATUS: IUCN – Vulnerable as *Myospalax fontanierii*.

SYNONYMS: *baileyi* Thomas, 1911; *cansus* Lyon, 1907; *fontanus* Thomas, 1912; *kukunoriensis* Lönnberg, 1926; *rufescens* (J. A. Allen, 1909); *shenseius* Thomas, 1911.

COMMENTS: In same monophyletic group as *E. smithii* and *E. rothschildi*, which is comparable to Ellerman's (1941) *fontanierii* and *smithii* groups, and Zheng's (1994) Prosiphneinae. Fan and Shi (1982) recognized *cansus* and *baileyi* as species, but Li and Chen (1987, 1989) argued for their recognition as subspecies of *M. fontanierii* based upon analyses of morphology, karyology, and LDH isozymes. They were included in *E. fontanierii* by G. M. Allen (1940) and also by Zhang et al. (1997); Wang (2003) treated *rufescens* as a species with *baileyi* as a subspecies. Burrowing behavior recorded and quantified by Wang et al. (1994, as *M. baileyi*).

Eospalax rothschildi (Thomas, 1911). Ann. Mag. Nat. Hist., ser. 8, 8:722.
   COMMON NAME: Rothschild's Zokor.
   TYPE LOCALITY: China, Gansu, 40 mi (64 km) SE Tao-chou.
   DISTRIBUTION: Forest, scrub, grassland and farmland in NC China (Henan, Shaanxi, N Gansu, N Sichuan, and Hubei; see Zhang et al., 1997).
   STATUS: IUCN – Lower Risk (nt) as *Myospalax rothschildi*.
   SYNONYMS: *hubeinensis* Li and Chen, 1989; *minor* Lönnberg, 1926.
   COMMENTS: The sister-species of *E. smithii* (Lawrence, 1991). The taxon *hubeinensis* (Li and Chen, 1989) was described as subspecies of *M. rothschildi*.

Eospalax smithii (Thomas, 1911). Ann. Mag. Nat. Hist., ser. 8, 8:720.
   COMMON NAME: Smith's Zokor.
   TYPE LOCALITY: China, Gansu, 30 mi (48 km) SE Tao-chou.
   DISTRIBUTION: Steppe and agricultural fields in NC China; recorded from provinces of N Gansu N Shaanxi, and Ningxia (Qin, 1991; Zhang et al., 1997).
   STATUS: IUCN – Lower Risk (nt) as *Myospalax smithii*.
   COMMENTS: Phylogenetically closely allied to *E. rothschildi* and in same monophyletic group with *E. fontainierii* (Lawrence, 1991).

Myospalax Laxmann, 1769. Sibirische Briefe, Gottingen, p. 75.
   TYPE SPECIES: *Mus myospalax* Laxmann, 1773.
   SYNONYMS: *Aspalomys* Gervais, 1841; *Episiphneus* Kretzoi, 1961; *Myotalpa* Kerr, 1792; *Siphneus* Brants, 1827.
   COMMENTS: Phylogenetic relationships reviewed by Lawrence (1991) and Zheng (1994), both of whose analyses affiliated extant species listed here and extinct Pliocene and Pleistocene species in the same monophyletic group (flat occipital shield, short incisive foramina within premaxillary bone, rooted or rootless molars, and other traits considered primitive for the subfamily). Zheng (1994) revived *Episiphneus* for the extinct *youngi* and *pseudarmandi* (which he called *sinensis*) and placed rootless forms in *Myospalax*. Lawrence (1991) demonstrated the artificiality of this arrangement, and we follow her arrangement in recognizing one genus with the *M. psilurus* and *M. myospalax* species groups, the latter also including the Pliocene and Pleistocene *M. youngi* and *M. pseudarmandi*. *Myospalax* as represented here is equivalent to Zheng's (1994) Myospalacinae. Corbet (1978c), following Kuzhyakin (1965), treated all *Myospalax* (they used subgenus *Myospalax*) as a single highly variable species, a lumping which clashes with Ognev's (1947) recognition of *aspalax*, *myospalax*, and *psilurus* as distinct; his arrangement proves to be a sharper insight into the real species diversity. Morphological, geographic, and other information for species with ranges in Russia and adjacent regions reviewed by Ognev (1947) and Gromov and Erbajeva (1995).

Myospalax aspalax (Pallas, 1776). Reise Prov. Russ. Reichs., 3:692.
   COMMON NAME: Steppe Zokor.
   TYPE LOCALITY: Russia, Transbaikalia, Dauuria ("Doldogo, on Onon River, below Atchinsk," Ellerman and Morrison-Scott, 1951:652).

DISTRIBUTION: Steppes and farmland of Russia on banks of Onon and Ingoda Rivers in the Upper Amur basin, N Mongolia (Sokolov and Orlov, 1980), and NE China (Heilongjiang, Jilin, Liaoning, Nei Mongol, Hebei, and N Shanxi; see Zhang et al., 1997).

STATUS: IUCN – Lower Risk (lc).

SYNONYMS: *armandii* (Milne-Edwards, 1867); *dybowskii* Sherskey, 1873; *hangaicus* (Orlov and Baskevich, 1992); *talpinus* (Pallas, 1811); *zokor* (Desmarest, 1822).

COMMENTS: *M. myospalax* species group. Listed, with a question mark, as a subspecies of *M.* myospalax by Ellerman and Morrison-Scott (1951), and unequivocally as *M. m aspalax* by Ellerman (1941) and Corbet (1978c). Separated as a species by Lawrence (1991) based on morphology, a status supported by karyotypic data (*M. myospalax*, 2n = 44, FN = 80-84; *M. aspalax*, 2n = 62, FN =110-114; *M. psilurus*, 2n = 64, FN =106-108; Martynova, 1975; Nevo, 1999; Vorontsov and Martynova, 1976) and analysis of blood proteins (Martynova et al., 1977). Closest phylogenetic relative is the early Pleistocene *M. pseudarmandi*, and both belong to the same species group containing the modern *M. myospalax* and Pliocene *M. youngi* (Lawrence, 1991).

Myospalax myospalax (Laxmann, 1773). Kongl. Svenska. Vet.-Akad. Handl. Stockholm, 34:134.

COMMON NAME: Altai Zokor.

TYPE LOCALITY: Russia, Altai Krai, 100 km SE of Barnaul, Sommaren, near Paniusheva on Alei River.

DISTRIBUTION: Altai Mtns and upper basin of Ob and Irtysh River drainages in S Russia, and E Kazakhstan (entire N region of Cisaltai plain and foothills as well as the W and C Altai; see Ognev, 1947).

STATUS: IUCN – Lower Risk (lc).

SYNONYMS: *incertus* Ognev, 1936; *komurai* Mori, 1927 (see Kaneko and Maeda, 2002); *laxmanni* Sherskey, 1873; *tarbagataicus* Ognev, 1936.

COMMENTS: *M. myospalax* species group. As part of a study of 25 extant rodent genera, morphology and schmelzmuster of rootless molars in *M. myospalax* and their relation to jaw movement were documented (Koenigswald et al., 1994). Gambaryan and Gasc (1993) presented a cinefluorographiccal, anatomical, and biomechanical analyses of burrowing in *M. myospalax* to reveal the adaptive properties of its musculoskeletal system. See account of *M. aspalax* for chromosomal contrasts.

Myospalax psilurus (Milne-Edwards, 1874). Rech. Hist. Nat. Mammifères, p. 126.

COMMON NAME: North China Zokor.

TYPE LOCALITY: China, Chihli (=Hebei), south of Beijing.

DISTRIBUTION: Agricultural fields and grasslands from NE and C China (Heilongjiang, Jilin, Liaoning, Nei Mongol, Hebei, Beijing, Tianjin, Shandong, Henan, N Shaanxi, Ningxia, N Gansu, and Anhui; see Zhang et al., 1997) to SE Mongolia (Sokolov and Orlov, 1980), and adjacent parts of Russia in the Amur region.

STATUS: IUCN – Lower Risk (nt) as *M. epsilanus*, Lower Risk (lc) as *M. psilurus*.

SYNONYMS: *epsilanus* Thomas, 1912; *spilurus* (Trouessart, 1897).

COMMENTS: *M. psilurus* species group. Sole member of group, which is defined, except for rootless hypsodonty, by retention of the greatest number of primitive traits of any *Myospalax* studied by Lawrence (1991). The status of the taxon *epsilanus* (type locality, N Manchuria, Khingan Mtns, 3400 ft) has been variously interpreted. It was described as a species (Thomas, 1912b) but subsequently included in *M. myospalax* (Corbet, 1978c) or in *M. psilurus* (Ellerman, 1941; Ellerman and Morrison-Scott, 1951; Gromov and Erbajeva, 1995; Ognev, 1947). Lawrence (1991) recognized *epsilanus* as a separate species, the most primitive of the *M. myospalax* species group, which also includes Pliocene and Pleistocene fossil species. Her viewpoint has been incorporated into mammalian checklists (Musser and Carleton, 1993; Pavlinov and Rossolimo, 1998; Pavlinov et al., 1995a; Wang, 2003), but not all Chinese researchers have discriminated the two (Zhang et al., 1997; Zheng, 1994). Both Thomas (1912c) and Ognev (1947) noted morphological traits distinguishing *epsilanus* from *psilurus*, but Ognev did not regard them as persuasive specific differences; furthermore, the characters used by Lawrence (1991) to separate the two are qualitative in nature and may reflect individual or geographic variation. Until a more rigorous revision and documentation of geographic variation are provided to illuminate the significance of the contrasting features recorded by Ognev (1947) and Lawrence (1991), we follow Ognev's arrangement. See account of *M. aspalax* for chromosomal contrasts.

Subfamily Rhizomyinae Winge, 1887. E Museo Lundii, 1:109.

SYNONYMS: Rhizomyini Winge, 1887 (Rhizomyinae Thomas, 1896; Rhizomyidae Miller and Gidley, 1918).

COMMENTS: Diagnosed and reviewed by Ellerman (1961), Carleton and Musser (1984), and also by Flynn (1990), who summarized cladistic relationships among extant and extinct genera. Differences in dentition, infraorbital canal, and zygomatic plate have been used to separate African mole rats as a subfamily from the Asian Rhizomyinae (Chaline et al., 1977; Miller and Gidley, 1918). Ellerman (1940, 1941) interpreted these distinctions as demonstrating *Tachyoryctes* to be phyletically remote from Asian rhizomyines, instead emphasized its dental similarities to the nesomyine *Brachyuromys*, a link recognized earlier by Major (1897), and combined the two genera in Tachyoryctinae of an inclusively defined Muridae. The distant relationship of *Tachyoryctes* to Asian rhizomyines was also supported by Lavocat (1978), who listed Tachyoryctinae as one of six subfamiles of Nesomyidae, all occurring primarily in Africa and presumably descended from afrocricetodontines rather than some ancient Asian stock. Recent analysis of masticatory muscles and skeletal architecture in *Rhizomys*, *Cannomys*, and *Tachyoryctes* demonstrated different masticatory systems, implying functional convergence between African *Tachyoryctes* and Asian *Cannomys* and *Rhizomys* (Endo et al., 2001). Their findings are consistent with the dual subfamily classification favored by Flynn (1990), who recognizes Tachyoryctinae for *Tachyoryctes* and related extinct genera and Rhizomyinae for *Rhizomys* and *Cannomys* and their extinct relatives, both within Rhizomyidae.

Rhizomyines and tachyoryctines were more geographically widespread and taxonomically diverse during the Miocene and Pliocene, a radiation succinctly summarized by Flynn (1990) and reflected in the eleven fossil genera described from Africa (Flynn and Sabatier, 1984; Mein et al., 2000a; Tong and Jaeger, 1993), Thailand (Mein and Ginsburg, 1997), and Pakistan, India, and China (Black, 1972; de Bruijn et al., 1981; Flynn, 1982a, b; Patnaik, 2001). Two lineages have been identified in the Siwalik localities of Pakistan and India, one leading to extant African *Tachyoryctes*, the other to living Asian *Rhizomys* and *Cannomys* (Black, 1972; Flynn, 1982a). *Prokanisamys arifi*, from 19 million-year-old (early Miocene) sediments in Pakistan (de Bruijn et al., 1981), is the earliest Asian "rhizomyid," and the middle Miocene *Pronakalimys* (Tong and Jaeger, 1993) and late Miocene *Nakalimys* and *Harasibomys* (Flynn and Sabatier, 1984; Mein et al., 2000a) represent the earliest African "rhizomyid" examples. Whether these forms can be classified as rhizomyine or tachyoryctine is unclear, although they do resemble middle Miocene Asian taxa (Flynn, 1990) or even early spalacines (Mein et al., 2000a). Flynn (1990) surmised that the group originated in Asia, with an earlier migration to Africa represented by *Nakalimys* (and presumably *Pronakalimys* and *Harasibomys*), possibly before 14 million years ago (Tong and Jaeger, 1993), and a second resulting in extant *Tachyoryctes*. The oldest record of the latter is *T. pliocaenicus*, represented by fragments from late Pliocene sediments in Ethiopia (Sabatier, 1978); *Tachyoryctes* shares more derived features with late Miocene Siwalik taxa than with the older African forms (Flynn, 1990). Although all extant and some extinct rhizomyines and tachyoryctines are fossorial and subterranean, osteological fragments from the early fossil record of both groups indicate that early forms were not subterranean. Specialized fossorial adaptations appear in rhizomyines around 8.5 million years ago and after 7 million years in tachyoryctines (Flynn, 1980).

The subfamilial separation of African mole rats from Asian bamboo rats is a reasonable hypothesis supported to date by morphological data from extant and extinct species and one testable by cladistic analyses that incorporate a wider range of anatomical systems and molecular sources.

Cannomys Thomas, 1915. Ann. Mag. Nat. Hist., ser. 8, 16:57.

TYPE SPECIES: *Rhizomys badius* Hodgson, 1841.

COMMENTS: The sister-genus of *Rhizomys*, a relationship based on cladistic analyses of skeletal and dental traits (Flynn, 1990). No fossil *Cannomys* have been identified.

Cannomys badius (Hodgson, 1841). Calcutta J. Nat. Hist., 2:60.
    COMMON NAME: Lesser Bamboo Rat.
    TYPE LOCALITY: Nepal.
    DISTRIBUTION: E Nepal, through NE India (West Bengal, Assam, Meghalaya, Manipur, Nagaland, and Mizoram; Agrawal, 2000), Bhutan, SE Bangladesh, Burma (Ellerman, 1961), S China (SW Yunnan; Zhang et al., 1997), NW Vietnam (Dang et al., 1994), Thailand (J. T. Marshall, Jr., 1977*a*; Robinson et al., 1995), and Cambodia; see Kock and Posamentier (1983) and Lekagul and McNeely (1977).
    STATUS: IUCN – Lower Risk (lc).
    SYNONYMS: *castaneus* (Blyth, 1843); *lönnbergi* Gyldenstolpe, 1917; *minor* (Gray, 1842); *pater* Thomas, 1915; *plumbescens* Thomas, 1915.
    COMMENTS: Reviewed by Corbet and Hill (1992). Regional faunal treatises provide informative accounts for populations in Thailand (J. T. Marshall, Jr., 1977*a*), Burma (Ellerman, 1961), and China (G. M. Allen, 1940). Agrawal (2000) reviewed Indian populations and could find no significant differences in fur coloration and other traits among the three subspecies (*castaneus*, *pater*, and *plumbescens*) recognized by Ellerman (1961) and treated them as full synonyms of *C. badius*.

Rhizomys Gray, 1831. Proc. Zool. Soc. Lond., 1831:95.
    TYPE SPECIES: *Rhizomys sinensis* Gray, 1831.
    SYNONYMS: *Brachyrhizomys* Teilhard de Chardin, 1942; *Nyctocleptes* Temminck, 1832.
    COMMENTS: Anatomy of gastrointestinal tract and comparisons with other muroids provided by Vorontsov (1979); cranium and dentition figured by Ellerman (1940). Evolutionary history extends back to the very late Miocene (5.7 million years ago) in China (Flynn, 1993). The extinct Chinese *shansius*, the genotype of *Brachyrhizomys*, is regarded as a species of *Rhizomys* with *Brachyrhizomys* conserved as a subgenus (Flynn, 1990, 1993). Other extinct Pakistan and Indian species formally allocated to *Brachyrhizomys* (Flynn, 1982*a, b*) are rhizomyines, not tachyoryctines, that are morphologically more closely related to *Rhizomys* than to *Cannomys* and lack a generic name (Flynn, 1990).

Rhizomys pruinosus Blyth, 1851. J. Asiat. Soc. Bengal, 20:519.
    COMMON NAME: Hoary Bamboo Rat.
    TYPE LOCALITY: India, Cherrapunji, Khasi Hills.
    DISTRIBUTION: S China (Yunnan, Ghizhou, Sichuan, Jiangxi, Hunan, Guangxi, Guangdong, and Fujian; Zhang et al., 1997), NE India (Meghalaya, Nagaland, and Manipur; Agrawal, 2000), E Burma, Thailand, Laos, Cambodia, Vietnam (Dang et al., 1994), south to Perak on Malay Peninsula (Chasen, 1940; Medway, 1969); 1000 to 4000 m in mountains.
    STATUS: IUCN – Lower Risk (lc).
    SYNONYMS: *latouchei* Thomas, 1915; *pannosus* Thomas, 1915; *prusianus* Shih, 1930; *senex* Thomas, 1915; *umbriceps* Thomas, 1916.
    COMMENTS: Reviewed by Corbet and Hill (1992). Good regional reviews exist for India (Agrawal, 2000), Burma (Ellerman, 1961), Thailand (J. T. Marshall, Jr., 1977*a*), and China (G. M. Allen, 1940).

Rhizomys sinensis Gray, 1831. Proc. Zool. Soc. Lond., 1831:95.
    COMMON NAME: Chinese Bamboo Rat.
    TYPE LOCALITY: China, Guangdon, near Canton.
    DISTRIBUTION: S China (Yunnan, Guizhou, Sichuan, north to S Gansu and S Shaanxi, east and south through Hubei, Hunan, Guangxi, Guangdong, Fujian, Anhui, Zhejiang, and Jiangxi; G. M. Allen, 1940; Ellerman and Morrison-Scott, 1951; T.-z. Wang, 1990; Zhang et al., 1997; Zheng and Zhang, 1990), N Burma (Ellerman, 1961), and N Vietnam (Dang et al., 1994).
    STATUS: IUCN – Lower Risk (lc).
    SYNONYMS: *chinensis* Swinhoe, 1870; *davidi* Thomas, 1911; *neowardi* Wang, 2003; *pediculus* Wang, 2003; *reductus* Dao and Cao, 1990; *vestitus* Milne-Edwards, 1871; *wardi* Thomas, 1921.

COMMENTS: Reviewed by Corbet and Hill (1992); G. M. Allen (1940) and Ellerman (1961) offered excellent portrayals concerning morphology, distribution, taxonomy, and ecology. Dao and Cao (1990) described *reductus* as a subspecies of *R. sinensis* from N Vietnam. Corbet and Hill (1992) listed *troglodytes*, which is based on Chinese fossils, in the synonymy of *R. sinensis* following Cao (1985), but Zheng (1993) documented fossils identified as *R. sinensis* and *R. troglodytes* from Pleistocene cave sediments in the Sichuan-Guizhou region. Wang (2003) recognized *R. wardi* as a species, with *neowardi* as subspecies.

Rhizomys sumatrensis (Raffles, 1821). Trans. Linn. Soc. Lond., 13:258.

COMMON NAME: Indomalayan Bamboo Rat.

TYPE LOCALITY: Malaysia, Malacca.

DISTRIBUTION: Sumatra and Malay Peninsula (Chasen, 1940; Medway, 1969), Thailand (J. T. Marshall, Jr., 1977*a*), Laos, Cambodia, Vietnam (Dang et al., 1994), S China (SW Yunnan; Zhang et al., 1997), and Burma (Ellerman, 1961).

STATUS: IUCN – Lower Risk (lc).

SYNONYMS: *cinereus* M'Clelland, 1842; *dekan* (Temminck, 1832); *erythrogenys* Anderson, 1877; *insularis* Thomas, 1915; *javanus* (Cuvier, 1829); *padangensis* Brongersma, 1936.

COMMENTS: Reviewed by Corbet and Hill (1992). Informative regional reports covering distribution, taxonomy, morphology, and natural history exist for Thailand (J. T. Marshall, Jr. 1977*a*), Burma (Ellerman, 1961), and China (G. M. Allen, 1940). Karyotype (2n = 50, FN = 100) documented by Hsu and Johnson (1963).

Subfamily Spalacinae Gray, 1821. Lond. Med. Repos., 15:303.

SYNONYMS: Aspalacidae Gray, 1825 (Aspalacina, Gray, 1825; Aspalacina Bonaparte, 1837); Spalacidae Gray 1821 (Spalasina Reichenbach, 1836; Spalacini Giebel, 1855; Spalacinae Thomas, 1896; Spalacini Fejfar, 1972, un-necessary renaming).

COMMENTS: First comprehensive monograph by Méhely (1909), who arranged all taxa into three subgenera of *Spalax*. A single genus had been recognized in most influential check-lists and faunal accounts (Ellerman, 1940; Ellerman and Morrison-Scott, 1951; Ognev, 1963*a*; Simpson, 1945) until Topachevskii (1969, 1976) monographed the subfamily and defended two genera, *Spalax* and *Microspalax* (= *Nannospalax*; proposed by Palmer, 1903, to replace the preoccupied *Microspalax*). His dual arrangement is accepted by some (de Bruijn, 1984; Gromov and Erbajeva, 1995; Kretzoi, 1970-71; Lyapunova et al., 1974; McKenna and Bell, 1997; Mitchell-Jones et al., 1999; Musser and Carleton, 1993; Pavlinov and Rossolimo, 1987, 1998; Pavlinov et al., 1995*a*; Saviè, 1982*a*; Vorontsov et al., 1977*b*) but not others (Carleton and Musser, 1984; Corbet, 1978*c*; Kowalski, 2001; Nevo et al., 2001; Saviè and Nevo, 1990; Ünay, 1996, 1999). Topachevskii (1969) documented variation in a suite of cranial, postcranial, and dental traits that formed the basis for his taxonomy, and de Bruijn (1984) found appreciable differences in dentition to distinguish the two genera among fragmentary fossils. Species of *Spalax* include high diploid and fundamental numbers, no acrocentric chromosomes, smaller interspecific differences in karyotypic structure, and lack of significant interpopulation variation (Saviè and Nevo, 1990). Those in *Nannospalax* have low diploid and fundamental numbers and exhibit significant interpopulational differences that correlate with biological and ecological distinctions that have prompted investigators to suspect that this group is composed of superspecies, each comprised of many separate biological species (Saviè and Nevo, 1990). Unfortunately, no study integrates variation in morphological and dental traits, karyotypes, and biochemical data within the framework of a cladistic analyses that would test Topachevskii's monophyletic groups. Until such results are available, we recognize only *Spalax*, an expedient endorsed by researchers who focus on living (Nevo et al., 2001; Saviè and Nevo, 1990) or fossil (Ünay, 1996, 1999) spalacines.

In addition to the classical monographs (Méhely, 1913; Topachevskii, 1969), taxonomy and distribution of Spalacinae are reviewed by Corbet (1978*c*, 1984), and illuminating regional reviews are available for species in Europe (Mitchell-Jones et al., 1999; Saviè, 1982*a*), Russia and adjacent regions (Gromov and Erbajeva, 1995; Ognev, 1963*a*), and Egypt (Osborn and Helmy, 1980). Carleton and Musser (1984) provided a diagnosis of the subfamily based upon morphology and summarized general characters and distribution;

and Ünay (1996, 1999) reiterated the diagnostic dental traits. Saviè and Nevo (1990) gave an excellent synopsis of the evolutionary history, speciation, and population biology of blind mole rats, updated by Nevo et al. (2001) in a review of Israeli populations. The answer to why spalacines live underground may be found in the transition during the early Tertiary from a subtropical to generally temperate world where an "underground ecotype provided subterranean mammals with shelter from extreme climatic fluctuations, and predators" (Ünay, 1999:421).

The panoply of adaptations necessary for living underground in tubular burrows has dramatically defined the blind mole rat phenotype, which facilitates their diagnosis but obscures phylogenetic connections to other muroid rodents. *Spalax* has been classified in a family with all burrowing myomorphous and even some hystricomorphous species (Alston, 1876; Murray, 1866), with *Myospalax*, *Rhizomys*, and *Tachyoryctes* (Weber, 1928; Tullberg, 1899), with just *Myospalax* (G. M. Allen, 1940; Miller and Gidley, 1918), with only *Rhizomys* and *Tachyoryctes* (Sen, 1977; Thomas, 1896), or by itself (de Bruijn, 1984; Ellerman, 1940; Ellerman and Morrison-Scott, 1951; Nevo et al., 2001; Pavlinov and Rossolimo, 1987, 1998; Pavlinov et al., 1995*a*; Reig, 1980; Simpson, 1945; Topachevskii, 1969; Ünay, 1996, 1999). Even the fundamental muroid affinity of Spalacidae was questioned by Stehlin and Schaub (1951) and Schaub (1958), who arranged *Spalax* and *Rhizomys* together in a separate suborder Pentalophodonta that also contained Theridomyidae, Oligocene rodents with premolars and molars characterized by complex lophate occlusal patterns. Petter (1961*b*) reassociated blind mole rats with muroid rodents based on their possession of only three molars with simple cricetine occlusal patterns. He homologized their pattern to that of Malagasy *Brachyuromys*, placed in Cricetidae at the time, and suggested that *Spalax* is within the range of dental variation characteristic of Cricetidae. Petter's view was ratified by Chaline et al. (1977), who allocated *Spalax* to a subfamily of Cricetidae, a significant departure from previous classifications. Whether placed in its own family or subfamily, the cladistic affinity of *Spalax* is indisputably with other muroid rodents, a connection established not only by morphology, but also by analyses of the nuclear gene IRBP (DeBry and Sagel, 2001), nuclear genes LCAT and vWF (Huchon et al., 1999; Michaux and Catzeflis, 2000; Michaux et al., 2001*b*; Robinson et al., 1997), and mitochondrial cytochrome *b* and 12S rRNA genes (Montgelard et al., 2002). Sequencing results from alpha crystalline A chain, a lens protein of the eye, and its significance for molecular clock hypothesis and phyletic analysis in *Spalax* were discussed by McKenna (1992).

SE Europe, SW Russia and Ukraine around Black and Caspian Seas, Asia Minor, Caucasia, Middle East adjacent to the Mediterranean, and N Africa encompass the modern geographic distribution of spalacines (see summary map in Nevo et al., 2001), which is generally concordant with range of described fossils (Ünay, 1999). Evolutionary history, based upon those fossils, dates from about 20 million years ago in the early Miocene of Anatolian Turkey (*Debruijnia* Ünay, 1996), somewhat later during the early Miocene of Greece (*Heramys* Hofmeijer and de Bruijn, 1985), and Pleistocene of N Africa (McKenna and Bell, 1997). Those early Miocene genera are replaced by the extinct and specious *Pliospalax*, ranging from early middle Miocene to Pliocene in Turkey, and from middle Miocene to early Pleistocene in Europe (Kowalski, 2001; see review by Ünay, 1999). Living *Spalax* is apparently a relict of a once extremely diverse group, especially during the Miocene, that Ünay (1999:422) speculated originated and evolved in what is now Anatolia because it is there where the oldest and most numerous fossil species have been found: from Anatolia, spalacines "probably spread into the Balkans, Russia, the Near East and Africa at different times." Using DNA-DNA hybridization, Catzeflis et al. (1989) estimated spalacines to have diverged from the other muroids sampled (arvicolines and murines) about 19 million years ago, which is consistent with antiquity claimed for the earliest fossil spalacines, and also for the divergence time estimated from LCAT nuclear gene sequences (Michaux and Catzeflis, 2000; Robinson et al., 1997).

Spalacines were once phylogenetically linked with the extinct Anomalomyidae (*Anomalomys*, *Anomalospalax*, *Prospalax*), ranging from early Miocene to late Pliocene mostly in Europe, because of similar molar patterns; the latter are now viewed as fossorial rodents that are dentally convergent with spalacines and have a separate evolutionary history (de Bruijn, 1984; Fejfar, 1972; see reviews by Bolliger, 1996, 1999, and Hugueney and Mein, 1993) with possible derivation from very early Miocene *Eumaryion* (de Bruijn et al., 1981). Whether the earliest spalacines, *Debruijnia* and *Heramys*, were subterranean dwellers is unknown, but fossorial adaptations characterize the Miocene and Pliocene *Pliospalax* (en, 1977), and spalacines may have evolved into a fossorial and subterranean habitus around 15-14 million years ago, much earlier than either rhizomyines or tachyoryctines (Flynn, 1990).

Spalax Guldenstaedt, 1770. Nova Comm. Acad. Sci. Petropoli, ser. 14, 1:410.
> TYPE SPECIES: *Spalax microphthalmus* Guldenstaedt, 1770.
> SYNONYMS: *Anotis* Rafinesque, 1815; *Aspalax* Desmarest, 1804; *Glis* Erxleben, 1777 [not Brisson, 1762]; *Macrospalax* Méhely, 1909; *Mesospalax* Méheley, 1909 [not Nehring, 1897, a *nomen nudum*]; *Microspalax* Méhely, 1909 [not Nehring, 1898, a *nomen nudum* preoccupied by Megnin and Trouessart, 1885]; *Myospalax* Hermann, 1783 [not Laxmann, 1769, or Blyth, 1846]; *Nannospalax* Palmer, 1903; *Ommatostergus* Nordmann, 1840; *Talpoides* Lacepède, 1799; *Ujhelyiana* Strand, 1922.
> COMMENTS: Vorontsov et al. (1977*b*) distinguished four species groups based upon biochemical data: the monotypic *S. nehringi*, *S. leucodon*, and *S. microphthalmus* groups, and the last containing *S. graecus*, *S. polonicus*, *S. arenarius*, and *S. giganteus*. The species discriminated biochemically are the same as those defined by Ognev (1963*a*) and Topachevskii (1969) using morphological traits and by Lyapunova et al. (1974) using chromosomal evidence. To these seven was added *S. ehrenbergi*. Although Saviè and Nevo (1990:133) acknowledged eight extant species, they concluded that the systematics is unrealistic because "it is based primarily on classical morphology, ignoring the central phenomenon of Spalacid evolution, i.e., chromosomal speciation which suggests that more than 30 living karyotypes, or species have been described and the end is not yet in sight." Based upon new information published since 1993 or reinterpretation of older data, we add five additional species, realizing that 13 likely underestimates actual species diversity in *Spalax*, for Nevo et al. (1995:226) later remarked that the "40-50 karyotypes described in Spalacidae represent presumptive good biological sibling species" and that "the morphological species concept does not hold in *Spalax*."
> Peshev (1989*a*, *b*) illuminated interspecific differences among five species, as well as sexual dimorphism within each, using morphometric techniques. The cephalic arterial system and its phylogenetic significance was described by Bugge (1971*a*, 1985), and aspects of various morphological systems, particularly masseter musculature and gastrointestinal structure, summarized by Vorontsov (1979, 1982). Pasichnyk (1992) contrasted structure and function of the maxillary region in several species of *Spalax*, contrasting them with *Ellobius*, *Rhombomys*, *Rattus*, and *Cricetus*. Zagorodnyuk (1992*b*) discussed taxonomic status of names based on Ukrainian samples. Evolutionary history of *Spalax* extends from the early Pliocene of Europe (Kowalski, 2001; Topachevskii et al., 1998), Pleistocene of North Africa, and middle Pleistocene of SW Asia (McKenna and Bell, 1997; see reviews in Nevo et al., 2001, and Ünay, 1999).

Spalax arenarius Reshetnik, 1939. Reports Zool. Mus. Kiev, 23:11.
> COMMON NAME: Sandy Blind Mole Rat.
> TYPE LOCALITY: Ukraine, Nikolaev Region, Golaya Pristan, NW shore of Black Sea.
> DISTRIBUTION: Small range in S Ukraine (see Vorontsov et al., 1977*b*, and Gromov and Erbajeva, 1995).
> STATUS: IUCN – Vulnerable.
> COMMENTS: Listed as a subspecies of *S. microphthalmus* by Corbet (1978*c*) and included in *S. giganteus* by Pavlinov and Rossolimo (1987), but elevated to species by Topachevskii (1969) and recognized as such by Lyapunova et al. (1974), Pavlinov and Rossolimo (1998), and Pavlinov et al. (1995*a*). Topachevskii (1969) viewed the restricted geographic range of *S. arenarius* as relictual. Chromosomal traits (2n = 62, FN = 124) documented by Lyapunova et al. (1974). Reviewed by Gromov and Erbajeva (1995).

Spalax carmeli Nevo, Ivanitskaya, and Beiles, 2001. Adaptive Radiation of Blind Subterranean Mole Rats, p. 23.

COMMON NAME: Mt Carmel Blind Mole Rat.

TYPE LOCALITY: N Israel, Mt Carmel, Muhraka.

DISTRIBUTION: Recorded from Kabri, Zippori, Mt Carmel, and Afiq in N Israel south of
ranges of *S. galili* and *S. golani* (see Nevo et al., 2001:19).

COMMENTS: Member of the *S. ehrenbergi* superspecies in Israel with 2n = 58 (Nevo et al.,
2001). See account of *S. ehrenbergi*.

Spalax ehrenbergi (Nehring, 1898). Sitzb. Ges. Naturf. Fr. Berlin (for December, 1897), 178, pl. 2
[1898].

COMMON NAME: Middle East Blind Mole Rat.

TYPE LOCALITY: Israel, Jaffa.

DISTRIBUTION: Middle East: SE Turkey (Kryštufek and Vohralík, 2001), N Iraq (Hatt, 1959),
Syria, Lebanon, Jordan, and Jaffa, Israel. North Africa: Mediterranean coastal region
from west of Nile Delta in N Egypt to N Libyan Cyrenaica (Lay and Nadler, 1972;
Osborn and Helmy, 1980; Ranck, 1968).

STATUS: IUCN – Lower Risk (lc) as *Nannospalax ehrenbergi*.

SYNONYMS: *aegyptiacus* Nehring, 1898; *berytensis* Miller, 1903; *fritschi* Nehring, 1902;
intermedius Nehring, 1898; *kirgisorum* Nehring, 1898; *nevoi* Cokun, 1996; *tuncelicus*
Cokun, 1996.

COMMENTS: A cluster of allopatric or parapatric species, each primarily diagnosed by unique
chromosomal traits but lacking strong attendant morphological or other
distinguishing features, expresses the current perception of *S. ehrenbergi*, and the
reason it is referred to as the *S. ehrenbergi* superspecies (see reviews by Nevo, 1991; Nevo
et al., 2001; Saviè and Nevo, 1990). For more than 30 years, Israeli populations of this
"superspecies" have been the subjects of intensive research concerning evolutionary
theory and the processes of speciation and adaptive radiation (Nevo, 1991; Nevo et al.,
2001). These studies have resulted in the exclusion of *ehrenbergi* (as historically
recognized) from Israel and its replacement by four newly described species (see
accounts of *S. carmeli*, *S. galili*, *S. golani*, and *S. judaei*) that Nevo et al. (2001:20) defined
as "genetically cohesive, interbreeding reproductive communities, adapted to four
different climatic-ecological regimes." Each have parapatric ranges separated by
narrow hybrid zones but without introgression (e.g., impaired gametogenesis; Zuccotti
et al., 1995; strong selection against unfit hybrids, restricted dispersal of hybrids into
parental territories, and lowered fitness of hybrids; see Nevo et al., 2001). The four are
defined primarily by chromosomal traits, although each can be also distinguished by
subtle distinctions in anatomy of middle ear ossicles (Burda et al., 1990), baculum
(Simson et al., 1993), mandible (Corti et al., 1996a), incisor enamel (Flynn et al., 1987),
and molars (Butler et al., 1993); morphometrics (Nevo et al., 1988); genetic patterns
(Ben-Shlomo and Nevo, 1993; Ben-Shlomo et al., 1993, 1996; Nevo and Beiles, 1992;
Nevo et al., 1992a, b, 1993, 1994a, 1997, 1999); DNA/DNA hybridization (Catzeflis et
al., 1989), and patterns associated with physiology, population structures, and
behavior (see additional references in the comprehensive review by Nevo et al., 2001).

Confusingly, Nevo et al. (2001:23) did not use *ehrenbergi* for any of the four Israeli
species they recognized, all described as new, because the formal descriptions are based
on chromosomal differences and the holotype of *ehrenbergi* comes from Jaffa, a place
located in a hybrid zone between *S. carmeli* (2n = 58) and *S. judaei* (2n = 60). Because
the karyotype is unretrievable from the holotype, Nevo et al. (2001) reserve *ehrenbergi*
to designate the superspecies but recognize no species per se. No nomenclatural
problem exists for those unwilling to accept Nevo's species definitions (see Nevo et al.,
2001:2), for then *S. ehrenbergi* is simply a species containing great diversity in
chromosomal morphology and other traits. On the other hand, if most populations
now sequestered in the superspecies are divided into species in a manner apropos
those from Israel, the status of the name *ehrenbergi* will have to be formally clarified,
either by resorting to legal petition that renders it unavailable or by identifying the
holotype through genetic analyses based on dry tissue.

Chromosomal, allozymic, and body size data from Jordanian samples of the
S. ehrenbergi complex has also revealed the presence of an additional four "putative"
species with geographic distributions associated with different habitats and climatic

regimes (Ivanitskaya and Nevo, 1998; Nevo et al., 2000). Significant chromosomal, allozymic, ribosomal and mitochondrial DNA sequences, and dental diversity also characterize samples of *S. ehrenbergi* from SE Turkey (Butler et al., 1993; Ivanitskaya et al., 1997; Nevo et al., 1994*b*, 1995; Sözen et al., 1999; Suzuki et al., 1996*a*).

Chromosomal data from an Egyptian sample were provided by Lay and Nadler (1972), who noted its close similarity to central Israeli specimens based on the shared 2n of 60. However, a multidisciplinary approach reveals that the Egyptian population differs from the Israel species by bacular traits (Simson et al., 1993), molar variation (Butler et al., 1993), allozymic diversity (Nevo et al., 1994*a*), cranial morphology (Ellerman and Morrison-Scott, 1951), behavioral traits, and a different karyotypic complement even though the 2n is 60 (Nevo et al., 1992*c*). Nevo et al. (1992*c*:431) proposed that the Egyptian "isolate is a recently speciating derivative of the *Spalax ehrenbergi* superspecies, which possibly evolved, because of the disjunction of the Sinai desert from the Israeli mole rats 10,000-20,000 years ago." If the Egyptian population is treated as a separate species (and the same as the Libyan population), *aegyptiacus* (type locality, Ramleh, near Alexandria; see Ellerman and Morrison-Scott, 1951) is the name that should be applied to this North African segment of the *S. ehrenbergi* complex.

Record of the *S. ehrenbergi* complex in Israel goes back at least 1.4 million years years before present, but notable diversification in the superspecies is not evident until the last 120,000 years before present (Nevo, 1993; see reviews by Tchernov, 1992, 1994, and references cited therein). A Lebanese sample, available from prehistoric Natufian occupation, dates from 10,500-8,000 years before present (Churcher, 1994).

Spalax galili Nevo, Ivanitskaya, and Beiles, 2001. Adaptive Radiation of Blind Subterranean Mole Rats, p. 23.
COMMON NAME: Upper Galilee Mountains Blind Mole Rat.
TYPE LOCALITY: Extreme N Israel, Upper Galilee Mtns, Kerem Ben Zimra.
DISTRIBUTION: Known from Ma'alot, Kerem Ben Zimra, and Qiryat Shemona in Upper Galilee Mtns near border with Lebanon (see Nevo et al., 2001:19).
COMMENTS: Member of *S. ehrenbergi* superspecies in Israel, 2n = 52 (Nevo et al., 2001). See account of *S. ehrenbergi*.

Spalax giganteus Nehring, 1898. Sitzb. Ges. Naturf. Fr. Berlin, p. 169.
COMMON NAME: Giant Blind Mole Rat.
TYPE LOCALITY: Russia, Daghestan, W shore of Caspian Sea, near Makhachkala (see Topachevskii, 1969).
DISTRIBUTION: Steppes north of the Caucuses and west of the Caspian Sea (Puzachenko, 1993). Distribution summarized by other researchers includes the Kazakhstan *S. uralensis* (Gromov and Erbajeva, 1995; Topachevskii, 1969; Vorontsov et al., 1977*b*).
STATUS: IUCN – Vulnerable.
COMMENTS: Once included in *S. microphthalmus* (Ellerman and Morrison-Scott, 1951), but revised by Topachevskii (1969) as a separate species and listed as such by Pavlinov and Rossolimo (1987, 1998) and Pavlinov et al. (1995*a*). Geographic range is parapatric with that of *S. microphthalmus* (Corbet, 1978*c*). The taxon *uralensis* has usually been included in *S. giganteus* (e. g., Gromov and Erbajeva, 1995; Topachevskii, 1969; Vorontsov et al., 1977*b*), but analyses of cranial dimensions and proportions indicated significant differences between populations from west of the Caspian Sea (*giganteus*) and the isolated segment in Kazakhstan (*uralensis*) and prompted the recognition of *uralensis* as a species (Puzachenko, 1993), an arrangement anticipated by Topachevskii (1969) and accepted by Pavlinov et al. (1995*a*) and Pavlinov and Rossolimo (1998). Chromosomal traits (2n = 64, FN = 124) documented by Lyapunova et al. (1974). Reviewed by Gromov and Erbajeva (1995, Russian segment).

Spalax golani Nevo, Ivanitskaya, and Beiles, 2001. Adaptive Radiation of Blind Subterranean Mole Rats, p. 23.
COMMON NAME: Golan Heights Blind Mole Rat.
TYPE LOCALITY: NE Israel, Golan Heights, Quneitra.

DISTRIBUTION: NE Israel, recorded from Mt Hermon, Quneitra, and El-Al on the Golan Heights (see Nevo et al., 2001:19).

COMMENTS: Member of *S. ehrenbergi* superspecies in Israel, 2n = 54 (Nevo et al., 2001). See account of *S. ehrenbergi*.

Spalax graecus Nehring, 1898. Zool. Anz., 21:228.

COMMON NAME: Balkan Blind Mole Rat.

TYPE LOCALITY: Ukraine, Bukovina region, vicinity of Chernovtsy (as restriced by Topachevskii, 1969).

DISTRIBUTION: Romania (Suceava, Craiova, Transylvania, and lower Danube Valley) and SW Ukraine, a European endemic (Gromov and Erbajeva, 1995; Mitchell-Jones et al., 1999; Saviè, 1982d; Topachevskii, 1969).

STATUS: IUCN – Vulnerable.

SYNONYMS: *antiquus* Méhely, 1909; *istricus* Méhely, 1909; *mezöségiensis* Szunyoghy, 1937.

COMMENTS: Included in *S. microphthalmus* by Ellerman and Morrison-Scott (1951) and Corbet (1978c), but arranged as a distinctive species by Topachevskii (1969) and so listed by Pavlinov and Rossolimo (1987, 1998) and Pavlinov et al. (1995a). Saviè (1982d) reviewed in detail the European segment; Gromov and Erbajeva (1995) discussed the portion the Ukraine. Topachevskii (1969) recognized *istricus*, apparently restricted to Romania, as a subspecies, and Murariu and Torcea (1984) separated it as a species based on cranial traits. Their sample, however, was small, and traits believed to discriminate *istricus* from *S. graecus* should be reassessed using larger samples from both Romania and the Ukraine. Chromosomal traits (2n = 62, FN = 124) illuminated by Lyapunova et al. (1974).

Spalax judaei Nevo, Ivanitskaya, and Beiles, 2001. Adaptive Radiation of Blind Subterranean Mole Rats, p. 23.

COMMON NAME: Judean Mountains Blind Mole Rat.

TYPE LOCALITY: C Israel, Judean Mtns, Lahav.

DISTRIBUTION: Recorded from Anza, Jerusalem, Lahav, Sede Boqer, Wadi Fara, Jiftlik, Dimona, and Ramat Hovav in Judean highlands of C Israel (see Nevo et al., 2001:19).

COMMENTS: Member of *S. ehrenbergi* superspecies in Israel, 2n = 60 (Nevo et al., 2001). See account of *S. ehrenbergi*.

Spalax leucodon Nordmann, 1840. Demidoff Voy., 3:34.

COMMON NAME: Lesser Blind Mole Rat.

TYPE LOCALITY: Ukraine, near Odessa.

DISTRIBUTION: From E and S Hungary through Balkan region (see Petrov, 1992), Greece (including Samothraki Isl off coast of SE Thracian Greece; Vohralík and Sofianidou, 1992a), Romania, and Bulgaria to NW Turkey (Thrace and possibly Marmara; see Kryštufek and Vohralík, 2001), to SW Ukraine just east of Dnestr River in Odessa region (see Gromov and Erbajeva, 1995; Mitchell-Jones et al., 1999; Saviè, 1982b; Vorontsov, 1977b).

STATUS: IUCN – Vulnerable as *Nannospalax leucodon*.

SYNONYMS: *bulgaricus* (Saviè and Soldatoviè, 1984) [see Kryštufek, 1997]; *dolbrogeae* Miller, 1903; *ehiki* Petrov, 1991 [*nomen nudum*]; *epiroticus* (Saviè, 1982); *hellenicus* Méhely, 1909; *hercegovinensis* Méhely, 1909; *hungaricus* Nehring, 1898; *ilici* Petrov, 1992 [*nomen nudum*]; *insularis* Thomas, 1917; *intermedius* Petrov, 1992 [*nomen nudum*]; *makedonicus* Soldatoviè, 1977; *martinoi* Petrov, 1971; *montanoserbicus* Soldatoviè, 1977; *montano-syrmiensis* Soldatoviè, 1977; *monticola* Nehring, 1898; *ovchepolensis* Soldatoviè, 1977; *peloponnesiacus* Ondrias, 1966; *petrovi* Petrov, 1992 [*nomen nudum*]; *rhodopiensis* (Saviè and Soldastoviè, 1984) [see Kryštufek, 1997]; *serbicus* Méhely, 1909; *sofiensis* (Saviè and Soldatoviè, 1984) [see Kryštufek, 1997]; *srebarnensis* (Saviè and Soldatoviè, 1984) [see Kryštufek, 1997]; *strumiciensis* Soldatoviè, 1977; *syrmiensis* Méhely, 1909; *thermaicus* Hinton, 1920; *thessalicus* Ondrias, 1966; *thracius* (Saviè, 1982); *tranensis* (Saviè and Soldatoviè, 1984) [see Kryštufek, 1997]; *transsylvanicus* Méhely, 1909; *turcicus* Méhely, 1913.

COMMENTS: Morphologically characterized by Topachevskii (1969) and now viewed as another superspecies based on chromosomal studies (Giagia et al., 1982; Ivanitskaya et al., 1997; Peshev, 1983; Saviè, 1982b; Saviè and Nevo, 1990; Saviè and Soldatoviè, 1977, 1979). More than 20 chromosomal forms have been uncovered (2n ranging from 38 to 62, FN between 74 and 96 (summarized in Mitchell-Jones et al., 1999, and Nevo et al., 2001). Saviè (1982b) delineated six clusters that he interpreted as chromosomal species: *leucodon* group (with *hungaricus*, *montanosyrmiensis*, *monticola*, and *transsylvanicus*), *makedonicus* group, *strumiciensis* group (with *ovchepolensis* and *serbicus*), *epiroticus* (with *hellenicus*), *turcicus* group (with *thracius*), and *montanoserbicus* group (with *hercegovinensis* and *syrmiensis*). Morphological and biometric analyses by Kivanç (see reference in Saviè and Nevo, 1990) indicated that *S. leucodon* occurs in most of Turkey (subspecies *anatolicus*, *armeniacus*, *cilicicus*, *nehringi*, and *turcicus*) and that *S. ehrenbergi* extends into SW Turkey (subspecies *intermedius* and *kirgisorum*). Saviè and Nevo (1990) skeptically viewed these results and urged elucidation using chromosomal evidence. Mikes et al. (1982) analysed the pelvis of *S. leucodon* in the context of assessing sexual dimorphism and taxonomic differences.

Currently, most researchers view the *S. leucodon* superspecies as embracing *S. nehringi*, whether occurring only in N Turkey (Pamukoglu and Albayrak, 1996) or throughout most of the country (Butler et al., 1993; Nevo et al., 1994a, b, 1995, 2001; Saviè and Nevo, 1990; Sözen et al., 1999; Suzuki et al., 1996a). Biochemical and chromosomal data do not support the conspecificity of *S. leucodon* and *S. nehringi* (Vorontsov et al., 1977b, and references therein), although specific limits of each are unresolved (Mitchell-Jones et al., 1999). We retain the two entities as separate, while acknowledging that each may represent a complex of allopatric or parapatric species and eventually lose their current geographic identity. Comprehensive taxonomic revision is needed to unravel relationships among populations of the two purported superspecies and to meaningfully delineate species boundaries. Chromosomal data from Macedonian samples and comparisons with other Macedonian populations documented by Zima et al. (1997a). See account of *S. nehringi*. Pleistocene occurrences of *S. leucodon* in Europe are summarized by Kowalski (2001).

Spalax microphthalmus Guldenstaedt, 1770. Nova Comm. Acad. Sci. Petropoli, 14:1.
> COMMON NAME: Greater Blind Mole Rat.
> TYPE LOCALITY: Russia, Voronezhskaya Oblast, Novokhoper Steppe.
> DISTRIBUTION: Steppes in Ukraine and S Russia between Dnieper and Volga Rivers, north to Orel-Kursk line, and south to Ciscaucasia (see Topachevskii, 1969; Vorontsov et al., 1977b; clearly mapped by Gromov and Erbajeva, 1995).
> STATUS: IUCN – Vulnerable.
> SYNONYMS: *pallasii* Nordmann, 1839 [see Pavlinov et al., 1995a for status]; *typhlus* (Pallus, 1779).
> COMMENTS: Monographed by Topachevskii (1969) and maintained as a species by Pavlinov and Rossolimo (1987, 1998) and Pavlinov et al. (1995a). Zagorodnyuk (1992b) endorsed the synonymy. Chromosomal features (2n = 60, FN = 115-120) recorded by Lyapunova et al. (1974) and summarized by Nevo et al. (2001). Reviewed by Gromov and Erbajeva (1995). Although not currently recorded from Romania, it has been identified at Pleistocene localities in that country (Kowalski, 2001).

Spalax nehringi (Satunin, 1898). Zool. Anz., 21:314.
> COMMON NAME: Nehring's Blind Mole Rat.
> TYPE LOCALITY: Armenia, Kasikoporan.
> DISTRIBUTION: Most of Turkey (except Thrace and SE region; also on isls of Gökçeada and Bozcaada; see Kryštufek and Vohralík, 2001), Armenia, and Georgia (see Gromov and Erbajeva, 1995; Topachevskii, 1969; Vorontsov et al., 1977b).
> STATUS: IUCN – Lower Risk (lc) as *Nannospalax nehringi*.
> SYNONYMS: *anatolicus* Méhely, 1909; *armeniacus* Méhely, 1909; *captorum* Hinton, 1920; *cilicicus* Méhely, 1909; *corybantium* Hinton, 1920; *labaumei* Matschie, 1919; *xanthodon* Nordmann, 1840.
> COMMENTS: Ellerman and Morrison-Scott (1951) and Corbet (1978c) listed *nehringi* as a synonym of *S. leucodon*, but Topachevskii (1969) treated it as a distinct species (under subgenus *Mesospalax*, genus *Microspalax*), a revision followed by Corbet (1984) and

other researchers (Kryštufek and Vohralík, 2001; Musser and Carleton, 1993; Pavlinov et al., 1995*a*; Pavlinov and Rossolimo, 1987, 1998). Topachevskii (1969) included *turcicus* in the synonymy, but Saviè (1982*b*) placed it in *S. leucodon*. Analyses of morphometric, chromosomal, and genetic character sets intimate that *S. nehringi* and its synonyms are geographic variants of *S. leucodon* (Butler et al., 1993; Nevo et al., 1994*a*, *b*, 1995, 2001; Saviè and Nevo, 1990; Sözen et al., 1999; Suzuki et al., 1996*a*). Russian portion reviewed by Gromov and Erbajeva (1995). See account of *S. leucodon*.

Spalax uralensis Tiflov and Usov, 1939. Vestn. Microbiol. Epidemiol. And Parasitol., 17:141.
    COMMON NAME: Kazakhstan Blind Mole Rat.
    TYPE LOCALITY: W Kazakhstan, Chingerlauz region (see Pavlinov and Rossolimo, 1987:164).
    DISTRIBUTION: Steppes in W Kazakhstan between the Ural and Emba Rivers (Puzachenko, 1993).
    COMMENTS: An allopatric, morphologically close relative of *S. giganteus*, in which it was once synonymized (Gromov and Erbajeva, 1995; Pavlinov and Rossolimo, 1987; Topachevskii, 1969; Vorontsov et al., 1977*b*) until shown to be a separate entity by Puzachenko's (1993) morphometric analyses. Distribution as summarized by other researchers includes *S. giganteus* from steppes west of the Caspian Sea (Gromov and Erbajeva, 1995; Topachevskii, 1969; Vorontsov et al., 1977*b*).

Spalax zemni (Erxleben, 1777). Syst. Regni Anim., 1:370-371.
    COMMON NAME: Podolsk Blind Mole Rat.
    TYPE LOCALITY: Ukraine, Ternopolsk region (see Topachevskii, 1969).
    DISTRIBUTION: SE Poland east into Ukraine between Dnestr and Dnepr Rivers and south to margin of Black Sea (see Saviè, 1982*c*; Topachevskii, 1969; Vorontsov et al., 1977*b*).
    STATUS: IUCN – Lower Risk (lc).
    SYNONYMS: *diluvii* Nordmann, 1858; *podolicus* Trouessart, 1897; *polonicus* Méhely, 1909.
    COMMENTS: Included in *S. microphthalmus* by Ellerman and Morrison-Scott (1951) and Corbet (1978*c*), but monographed as a distinct species (as *S. polonicus*) by Topachevskii (1969) and subsequently listed or reviewed as such (Corbet, 1984; Gromov and Erbajeva, 1995; Pavlinov and Rossolimo, 1987, 1998; Pavlinov et al., 1995*a*; Saviè, 1982*c*). This species is often referenced as *S. polonicus* because Topachevskii (1969) characterized *zemni* as a *nomen nudem*, but the name is properly available (Ellerman, 1949*b*) and is the earliest that can be applied (Pavlinov and Rossolimo, 1987; Zagorodnyuk, 1992*b*). The synonyms are documented by Zagorodnyuk (1992*b*). Chromosomal traits (2n = 62, FN = 124) summarized in Nevo et al. (2001, as *S. podolicus*).

Subfamily Tachyoryctinae Miller and Gidley, 1918. J. Wash. Acad. Sci., 8:437.
    SYNONYMS: Tachyoryctinae Miller and Gidley, 1918 (Tachyoryctini McKenna and Bell, 1997).
    COMMENTS: Included in Rhizomyinae (Carleton and Musser, 1984; Musser and Carleton, 1993), recognized as a subfamily of Rhizomyidae (Chaline et al., 1977; Miller and Gidley, 1918; Flynn, 1990, and references therein), or as a subfamily of Muridae (Ellerman, 1941). See also discussion under Rhizomyinae.

Tachyoryctes Rüppell, 1835. Neue wirbelth. Fauna Abyssin. Gehörig., Säugeth., 1:35, footnote.
    TYPE SPECIES: *Bathyergus splendens* Rüppell, 1835.
    SYNONYMS: *Chrysomys* Gray, 1843.
    COMMENTS: Hollister (1919:40) listed eight species of *Tachyorytces*, noted that "all have constant characters of differentiation, and intergradation between any two of them is not indicated by this material," but speculated that the "numerous forms will doubtlessly be connected by complete chains of intergrades and the final monographer of the genus will be obliged to reduce many of the named forms to the rank of subspecies." G. M. Allen (1939) and Ellerman (1941) listed 14 species, but these were reduced to *T. macrocephalus* and *T. splendens* by Misonne (1974), an arrangement followed by Corbet and Hill (1991). Of the twenty forms described, Rahm (1980) considered about 16 of them to be valid subspecies of *T. splendens*.
      The abrupt reduction from 14 to 2 species was not based on reevaluation of

morphological differences that characterize the named forms. Aside from Bekele's (1986) inconclusive univariate analyses of craniometric data from samples of *Tachyoryctes*, no assessment of morphological variation in the genus is available in context of a systematic revision. Until such a study provides documentation and definition of specific limits and geographic distributions, we return to a modified treatment of the arrangement by G. M. Allen (1939) and Ellerman (1941), which reflects the diagnostic information now available from specimen study and provides a working estimate of biological reality. Morphology and morphometry of the lungs of a Kenyan sample of *Tachyoryctes* contrasted with *Heterocephalus* by Maina et al. (1992), stomach morphology described by Rahm (1976), morphogenesis of fetal membranes and placenta by Makori et al. (1991), and lower incisor shape compared with other muroids by Millien-Parra (2000*a*).

Tachyoryctes ankoliae Thomas, 1909. Ann. Mag. Nat. Hist., ser. 8, 4:545.
> COMMON NAME: Ankole African Mole Rat.
> TYPE LOCALITY: SW Uganda, Burumba, Ankole.
> DISTRIBUTION: SW Uganda (Hollister, 1919; Lunde and Sarmiento, 2002) NW Tanzania (Swynnerton and Hayman, 1951); limits unknown.
> STATUS: IUCN – Vulnerable.
> COMMENTS: Recorded as a species by Swynnerton and Hayman (1951) but as a subspecies of *T. splendens* by Delany (1975), who nevertheless noted the cranial traits by which *ankoliae* could be distinguished from the adjacent *ruddi*, which was also treated as a subspecies of *T. splendens* by Delany. Both *T. ankoliae* and *T. ruddi* were caught in the Kalinzu forest preserve in SW Uganda (Lunde and Sarmiento, 2002) and the two species apparently occur in NW Tanzania (see account of *T. ruddi*).

Tachyoryctes annectens (Thomas, 1891). Ann. Mag. Nat. Hist., ser. 6, 7:304.
> COMMON NAME: Mianzini African Mole Rat.
> TYPE LOCALITY: SW Kenya, Mianzini, E of Lake Naivasha (Hollister, 1919).
> DISTRIBUTION: Known only from vicinity of type locality.
> STATUS: IUCN – Endangered.
> COMMENTS: Hollister (1919) noted the distinctively large size of the holotype and commented that none of the large series (at USNM) from S and SW of Lake Naivasha, identified as the small-bodied *T. naivashae*, approached it. Apparently the type locality has never been fixed ("either Masai-land or inland British East Africa" were given in the original description), but assumed to be in the vicinity of Mianzini because other field specimens of mammals accompanying the holotype of *annectens* were labeled with this locality. Hollister mentioned that the only other *Tachyoryctes* of so large a body size is the Mt Kenyan endemic *T. rex*. Locating the geographic source of the holotype of *T. annectens* and ascertaining its relationship to *T. rex* warrant further inquiry.

Tachyoryctes audax Thomas, 1910. Ann. Mag. Nat. Hist., ser. 8, 5:421.
> COMMON NAME: Aberdare Mountains African Mole Rat.
> TYPE LOCALITY: Kenya, summit of Aberdare Range, 10,000 ft (3048 m).
> DISTRIBUTION: Kenya, Aberdare Range, 2775-3200 m (Hollister, 1919).
> STATUS: IUCN – Vulnerable.
> COMMENTS: A smaller-bodied and paler-furred relative of *T. rex*, which is endemic to high altitudes on Mt Kenya (Hollister, 1919).

Tachyoryctes daemon Thomas, 1909. Ann. Mag. Nat. Hist., ser. 8, 4:545.
> COMMON NAME: Demon African Mole Rat.
> TYPE LOCALITY: N Tanzania, Mt Kilimanjaro, 5000 ft (1524 m).
> DISTRIBUTION: N Tanzania in foothills of Mt Kilimanjaro, the Arusha area, upper limits of forest at 4000 m on Mt Meru, and in the west at Banagi in Mara Prov near Lake Victoria; limits of distribution unresolved.
> STATUS: IUCN – Lower Risk (nt).
> COMMENTS: Thomas (1909*b*) remarked that *daemon* was "markedly smaller than in *T. ibeanus*" and could be distinguished from *T. ankoliae* by its linear sagittal crest and from *T. ruddi* by its "more abruptly and widely expanded zygomata." Genetic

relationships between the small-bodied *T. daemon*, *T. naivashae* (north of Kenya-Tanzania border), and *T. ruddi* (SW Kenya, Mt Elgon, SW Uganda, and NW Tanzania) need to be evaluated to determine whether a single species is represented. Distribution and ecology of *T. daemon* on Mt Kilimanjaro reviewed by Grimshaw et al. (1995). Distributional records are from Hollister (1919) and Swynnerton and Hayman (1951), who also listed *daemon* as a species. The latter authors recorded the species from NW Tanzania but those occurrences likely represent *T. ruddi* (see that account).

Tachyoryctes ibeanus Thomas, 1900. Proc. Zool. Soc. Lond., 1900:179.

COMMON NAME: Kenyan African Mole Rat.

TYPE LOCALITY: SC Kenya, Machakos.

DISTRIBUTION: S Kenya, near Nairobi and western margin of the Athi Plains (Hollister, 1919); limits require resolution.

COMMENTS: Musser and Carleton (1993) included *ibeanus* in *T. splendens* because *ibeanus* is geographically close to the range of *T. splendens* and was originally described as its subspecies (G. M. Allen, 1939). Since no empirical data as yet support the inclusion of *ibeanus* in *T. splendens*, we follow Hollister (1919) who identified series from SC Kenya as *T. ibeanus*. The genetic and distributional relationships between *T. ibeanus* and other named forms in S Kenya and N Tanzania require careful taxonomic revision.

Tachyoryctes macrocephalus (Rüppell, 1842). Mus. Senckenberg., 3:97.

COMMON NAME: Ethiopian African Mole Rat.

TYPE LOCALITY: Ethiopia, Shoa.

DISTRIBUTION: Ethiopia; endemic to high southern plateau, 3000-4150 m (Rupp, 1980; Yalden and Largen, 1992; Yalden et al., 1996).

STATUS: IUCN – Lower Risk (lc).

SYNONYMS: *hecki* Neumann and Rümmler, 1928.

COMMENTS: Inhabitant of Afro-alpine moorland and grassland, which the species shares with the other Ethiopian endemics *Stenocephalemys albocaudata*, *Lophuromys melanonyx*, and *Arvicanthis blicki* (Yalden, 1988). The congenor *T. splendens* ranges from below 1000 m to about 3000 m, where it apparently overlaps the lower altitudinal distribution of *T. macrocephalus* (Rupp, 1980; Yalden et al., 1976). Yalden (1975) provided ecological observations for *T. macrocephalus* and morphologically contrasted it with *T. splendens*. Habitat selection and daily activity patterns in relation to the endangered Ethiopian wolf (*Canis simensis*) recorded by Sillero-Zubiri et al. (1995*b*). See Yalden (1985, Mammalian Species, 237).

Tachyoryctes naivashae Thomas, 1909. Ann. Mag. Nat. Hist., ser. 8, 4:547.

COMMON NAME: Navivasha African Mole Rat.

TYPE LOCALITY: S Kenya, Lake Naivasha, 6350 ft (1935 m).

DISTRIBUTION: Kenya; recorded from plains S and SW of Lake Naivasha to near the Tanzanian border (Hollister, 1919); may also occur in N Tanzania.

STATUS: IUCN – Endangered.

COMMENTS: Considered by Thomas (1909*b*:547) to be the smallest in body size of the East African *Tachyoryctes* and occurring "quite close" to the two largest (*T. annectens* and *T. storeyi*). "Its small size and flattened skull will readily separate it from any other form" (Thomas, 1909*b*). See Hollister (1919) for notes on habitat and collection localities.

Tachyoryctes rex Heller, 1910. Smithson. Misc. Coll., 56:4.

COMMON NAME: King African Mole Rat.

TYPE LOCALITY: Kenya, western slopes of Mt Kenya, 10,000 ft (3048 m).

DISTRIBUTION: Kenya, western slopes of Mt Kenya, 2600-3350 m (Hollister, 1919).

COMMENTS: A large bodied and distinctive species that Hollister (1919) believed to be related to *T. audax*, an endemic of the Aberdare Range to the west of Mt Kenya. Hollister provided notes on habitat, habits, and coloration, and described changes in molar morphology associated with age.

Tachyoryctes ruandae Lönnberg and Gyldenstolpe, 1925. Ark. Zool., 17B, no. 5:6.

COMMON NAME: Rwanda African Mole Rat.

TYPE LOCALITY: Rwanda, Mt Muhavura.

DISTRIBUTION: E Dem. Rep. Congo (Kivu), Rwanda, and Burundi.

STATUS: IUCN – Lower Risk (lc).

COMMENTS: See Elbl et al. (1966) and Rahm (1967) for distributional and ecological information. Stomach morphology described by Rahm (1976, as *splendens*) and chromosomal data (2n = 48, FN = 54) reported by Matthey (1967a).

Tachyoryctes ruddi Thomas, 1909. Ann. Mag. Nat. Hist., ser. 8, 4:547.

COMMON NAME: Rudd's African Mole Rat.

TYPE LOCALITY: Kenya, Mt Elgon, Kirui, 6000 ft (1830 m).

DISTRIBUTION: SW Kenya between foothills of Mt Elgon and Port Florence on NE margin of Lake Victoria (Hollister, 1919), Ugandan and Kenyan slopes of Mt Elgon, SW Uganda (Lunde and Sarmiento, 2002), and NW Tanzania (N Kagera Prov.).

STATUS: IUCN – Lower Risk (lc).

SYNONYMS: *badius* Thomas, 1909.

COMMENTS: Thomas (1909b) described *T. ruddi* as being about the same body size as *T. daemon* but differing in a suite of cranial traits; he also described *badius* as a subspecies of *T. ruddi*. This species and *T. ankoliae* have been taken together in the Kalinzu Forest Reserve, SW Uganda (Lunde and Sarmiento, 2002). Localities in NW Tanzania represent *T. daemon* according to Swynnerton and Hayman (1951), but we suspect that they represent *T. ruddi*. *Tachyoryctes ankoliae* has been reported from the same region in NE Tanzania (Bukoba; Swynnerton and Hayman, 1951). Ranges of the small-bodied *T. ruddi*, *T. naivashae*, and *T. daemon* are allopatric in mountains and high plains encircling Lake Victoria; specimens from intervening regions between the ranges and critical systematic revision should reveal the actual genetic relationships among these populations. Hollister (1919) commented on variation in pelage traits. Habitat and altitudinal distribution of *T. ruddi* on Ugandan slopes of Mt Elgon documented by Clausnitzer (2001) and Clausnitzer and Kityo (2001).

Tachyoryctes spalacinus Thomas, 1909. Ann. Mag. Nat. Hist., ser. 8, 4:547.

COMMON NAME: Embi African Mole Rat.

TYPE LOCALITY: Kenya, Embi (=Embu), plains near base of Mt Kenya, 5400 ft (1646 m).

DISTRIBUTION: Kenya, plains and lower slopes of Mt Kenya (Hollister, 1919).

STATUS: IUCN – Lower Risk (nt).

COMMENTS: Body size about like *T. ruddi*, according to Thomas (1909b), but "Readily recognizable by its sloping occipital plane." This species occurs at the base of Mt Kenya and adjacent plains but is replaced at altitudes above 2600 m by the much larger-bodied *T. rex* (Hollister, 1919). Morphological, distributional, and genetic relationships between *T. spalacinus* and *T. ibeanus*, which occurs to the south along the western margin of the Athi Plains, need to be evaluated.

Tachyoryctes splendens (Rüppell, 1835). Neue wirbelt. Fauna Abyssin. Gehörig., Säugeth., 1:36.

COMMON NAME: Northeast African Mole Rat.

TYPE LOCALITY: Ethiopia, Dembea Prov., Gondar.

DISTRIBUTION: Ethiopia (500-3900 m; Rupp, 1980), Somalia, and NW Kenya; limits unresolved.

STATUS: IUCN – Lower Risk (lc).

SYNONYMS: *canicaudus* Osgood, 1936; *cheesmani* Thomas, 1928; *gallarum* Osgood, 1936; *omensis* Neumann and Rümmler, 1928; *pontifex* Neumann and Rümmler, 1928; *somalicus* Osgood, 1910.

COMMENTS: Yalden et al. (1976:57) claimed that the synonyms listed here clearly apply to one species, but conceded that "The possibility of distinct races associated with different mountain blocks, or of a cline in size with altitude, remains to be properly examined." Osgood (1936) noted that two species, *splendens* and *cheesemani*, could be recognized among the Ethiopian samples he examined, and Bekele could distinguish samples of the two by morphometric traits but found that *gallarum* was intermediate. Early records indicated that *T. splendens* reached 3900 m in the Bale Mtns (Yalden et al., 1976), but Yalden et al. (1995) observed that "neither the live animal nor its burrows have ever been observed above 3200" and that predators may have carried the specimens (mainly extracted from owl pellets) from lower elevations. Chromosomal characteristics of an Ethiopian sample (2n = 48, FN = 62) documented and contrasted with other rhizomyine samples by Baskevich et al. (1992, 1993). Distribution in the isolated Harenna Forest of S Ethiopia documented by Lavrenchenko (2000).

Tachyoryctes storeyi Thomas, 1909. Ann. Mag. Nat. Hist., ser.8, 5:276.

COMMON NAME: Storey's African Mole Rat.

TYPE LOCALITY: SW Kenya, Lake Elmenteita.

DISTRIBUTION: Recorded only from the type locality.

COMMENTS: Thomas (1909b) remarked that *T. storeyi* "is readily distinguishable by its comparatively large size, long and well-defined brain-case, and long nasals." Only *T. annectens* exceeds *T. storeyi* in body size, and the relationships and stature between the two and other large-bodied taxa remain to be uncovered.

Family Calomyscidae Vorontsov and Potapova, 1979. Zool. Zh., 58:1393.

SYNONYMS: Calomyscini Vorontsov and Potapova, 1979 (Calomyscinae Musser and Carleton, 1993).

COMMENTS: Earlier associated either with Palearctic hamsters in Cricetinae (e.g., Corbet, 1978c; Corbet and Hill, 1991; Ellerman and Morrison-Scott, 1951) or with the New World neotomine *Peromyscus* (Ellerman, 1941, 1961). Nomenclatural association with Palaearctic hamsters seemed to reflect more the uncertainty and indecision concerning its relationships rather than conviction of its membership in Palearctic Cricetinae (Carleton and Musser, 1984; Corbet, 1978c). On the other hand, Ellerman (1941) did not question the close phyletic grouping of *Calomyscus* and *Peromyscus*, and Osgood (1947:166) even viewed it as being "so similar to the American *Peromyscus* that it probably signifies a late Pleistocene invasion from North America." An alliance with New World species was echoed by Pavlinov (1980c), who retained *Calomyscus* with neotomines because of close similarities between it and *Reithrodontomys* in morphology of the auditory ossicles. Graphodatsky et al. (2000:303) noted that conventionally stained and banded karyotypes of *Calomyscus* actually resemble those of *Peromyscus*, "particularly in the presence and variation of the heterochromatic small arms," but they did not discover clear G-band homologies between the two.

After comprehensive review of morphological systems, Vorontsov and Potapova (1979) recognized *Calomyscus'* distinctive combination of features by setting it apart as a tribe in the subfamily Cricetinae, Cricetidae (elevated to subfamily in an inclusive Muridae by Musser and Carleton, 1993). Agusti (1989:387) regarded *Calomyscus* as the only living member of the Myocricetodontinae, an allocation supported only by this statement: "We also consider the genus *Calomyscus* as belonging to the Myocricetodontinae." His action has been followed by researchers reporting isolated fossil teeth identified as species of Myocricetodontinae (e. g., Agustí and Casanovas-Villar, 2003; Wessels, 1996, 1998, 1999). Wessels (1998) also included the South African *Mystromys* (the sole living member of Mystromyinae; see that account) within the Myocricetodontinae, which he arranged within Gerbillidae. We compared molar occlusal patterns of living species of *Calomyscus* with published illustrations of various myocricetodontines and cannot appreciate the resemblance indicated by Agusti and Wessels. No critical study exists supporting either Agusti's allocation, and the subsequent arrangements based upon it, or Fahlbusch's (1969) earlier suggested derivation of *Calomyscus* from the Miocene *Democricetodon*.

The cladistic isolation of *Calomyscus* within Muroidea has been recently supported by phylogenetic analyses of nuclear DNA sequences (Michaux et al., 2001b). *Calomyscus* emerges as the sole member of a clade removed from the lineage including *Mystromys* (Nesomyidae) and basal to those radiations representing Cricetidae and Muridae. Until further studies refine the preliminary morphological and molecular evidence, family status for *Calomyscus* is an interim nomenclatural treatment that reflects its relationships as so far understood.

Several extinct species of *Calomyscus* have been recovered from late Miocene strata in Turkey, Spain, and France, and early Pliocene beds on Rhodes (see reviews by Agustí and Casanovas-Vilar, 2003; Wessels, 1998, 1999); origin of the genus certainly occurred earlier. Species of *Calomyscus* are generally known as "mouse-like hamsters" (Wilson and

Cole, 2000), which unfortunately perpetuates its inaccurate association with true hamsters in subfamily Cricetinae. The Greek *kalos* means beautiful and the generic name translates as "beautiful mouse," an improvement over "mouse-like hamster" and one implicit in using Calomyscus as part of the common name.

Calomyscus Thomas, 1905. Abstr. Proc. Zool. Soc. Lond., 1905(24):23.

  TYPE SPECIES: *Calomyscus bailwardi* Thomas, 1905.

  COMMENTS: Although several taxa were originally described as species, *Calomyscus* was long
      considered monotypic (Ellerman, 1941, 1961; Ellerman and Morrison-Scott, 1951;
      Peshev, 1991). Character variation in the genus was concisely summarized by Corbet
      (1978c:89): "There is considerable variation in colour, size and proportions but this seems
      to form a mosaic pattern and it seems unlikely that any major, discrete regional groups
      can be recognized." However, Vorontsov et al. (1979) comprehensively revised the genus
      and treated most former subspecies of *C. bailwardi* as separate species. The morphological
      and geographic integrity of some has been subsequently tested with additional
      chromosomal data (Graphodatsky et al., 2000; Malikov et al., 1999; Meyer and Malikov,
      1995, 2000), mitochondrial cytochrome *b* sequences (Morshed and Patton, 2002),
      comparative postnatal growth and development (Meyer and Malikov, 1996), and
      multivariate analyses of cranial and dental measurements (Lebedev et al., 1998). While
      much research has focused on the geographical distribution of different chromosomal
      morphologies and their taxonomic significance (summarized in Graphodatsky et al.,
      2000), the multivariate analysis by Lebedev et al. (1998) demonstrated distinct
      morphological clusters that correspond to the karyotypic differences. More integrative
      studies of this kind are required to better understand species diversity and geographic
      distributions.

      All species occupy well drained, barren, rocky habitats in foothills and mountains
      (Malikov et al., 1999; Morshed and Patton, 2002), as is depicted in the descriptions and
      photographs of collection localities for AMNH and FMNH specimens from Iran and
      Afghanistan (Goodwin, 1938; Hassinger, 1968; Lay, 1967). Populations appear to be
      patchily distributed and some are geographically isolated, "which promotes the effect of
      random genetic drift and may be an important factor of rapid karyotype evolution"
      (Graphodatsky et al., 2000:303). Natural hybridization between populations with very
      different karyotypes seems usual in *Calomyscus* (see account of *C. elburzensis*) and no
      unequivocal evidence suggests that chromosomal change is responsible for speciation in
      the populations studied (Graphodatsky et al., 2000). Modern geographic range of the
      genus (SW Syria, S Caucusus, Iran, S Turkmenistan, Afghanistan, and W Pakistan) is a
      remnant of a broader distribution that in the late Miocene extended as far west as Spain
      (see Agusti, 1989; Wessels, 1998, 1999).

Calomyscus bailwardi Thomas, 1905. Abstr. Proc. Zool. Soc. Lond., 1905(24):23.

  COMMON NAME: Zagros Mountains Calomyscus.

  TYPE LOCALITY: WC Iran, E Khuzistan Prov., Zagros Mtns, 120 km SE Ahwaz (= Ahaz), Mala-
      i-Mir (= Izeh).

  DISTRIBUTION: Zagros Mtns of W Iran in the provinces of Kordistan, Ilam, W Esfahan,
      E Khuzistan, Luristan, Fars, and W Kerman (Vorontsov et al., 1979, and our study of
      specimens in AMNH, FMNH, and USNM); actual range has yet to be defined.

  STATUS: IUCN – Lower Risk (lc).

  COMMENTS: True *C. bailwardi* has been recorded only from the Zagros Mtns in W Iran; the
      range as usually described in the literature (Corbet, 1978c) represents other species as
      well. It is medium in body size (*C. urartensis*, *C. mystax*, *C. elburzensis*, *C. hotsoni*, and
      C. tsolovi smaller-bodied; *C. grandis* larger-bodied) and may be most closely allied to
      the medium-sized *C. baluchi*; nature of the phylogenetic affinites between *C. bailwardi*
      and all other named forms has yet to be resolved by careful morphometric,
      chromosomal, and molecular analyses. Graphodatsky et al. (2000) reported a
      karyotypic variability for samples from the northern Zagros Mtns in Bahtaran Prov.
      (2n = 37, FN = 44), southern Zagros Mtns in Fars Prov. (2n = FN = 50), and mountains
      of southern Kerman Prov. (2n = 52, FN = 56), just east of the southern Zagros Mtns.
      They could not attach a species name to any sample. Whether one or all of the

karyotypes represent *C. bailwardi* or three separate species cannot be determined until the geographic ranges of the karyotypes are determined and, as Graphodatsky et al. (2000) noted, animals from the type locality of *C. bailwardi* are karyotyped. Mitochondrial cytochrome *b* divergences documented among samples identified as *C. bailwardi* from Hormozgan and Kerman Provs., S Iran (Morshed and Patton, 2002).

Calomyscus baluchi Thomas, 1920. J. Bombay Nat. Hist. Soc., 26:939.
COMMON NAME: Pakistan Calomyscus.
TYPE LOCALITY: W Pakistan, NW Balochistan, Kelat Dist.
DISTRIBUTION: W Pakistan west of the Indus River from N Balochistan (Kalat Dist.) in the south northward through Sibi, Quetta, Punjab, and Parachinar districts to Malakand Dist. in the northwest (Roberts, 1977; specimens in AMNH and USNM); and E and NC Afghanistan (specimens in FMNH).
STATUS: IUCN – Lower Risk (lc).
SYNONYMS: *mustersi* Ellerman, 1948.
COMMENTS: Originally described as a species, later included in *C. bailwardi* (Corbet, 1978*c*; Ellerman, 1941, 1961; Ellerman and Morrison-Scott, 1951), but reinstated as a separate species by Vorontsov et al. (1979), an allocation followed by others (Musser and Carleton, 1993; Pavlinov et al., 1995*a*). Body and cranial dimensions are similar to *C. bailwardi* but the level of their genetic relationship has yet to be uncovered. Mitochrondrial DNA differences between *C. baluchi* and samples identified as *C. bailwardi* from SE Iran documented by Moshed and Patton (2002). See Vorontsov et al. (1979) for status of *mustersi*. Distribution based on Vorontsov et al. (1979) and identification of specimens in FMNH and USNM.

Calomyscus elburzensis Goodwin, 1938. Am. Mus. Novit., 1050:1.
COMMON NAME: Goodwin's Calomyscus.
TYPE LOCALITY: NE Iran, N Khorasan Prov., Kurkhud Mtns, Dergermatie ("Degermatie" in original description), 4000 ft (1220 m).
DISTRIBUTION: Mountains of N and NE Iran (from southern foothills of Elburz Mtns in Semnan Prov. near Semnan and adjacent Sang-i-sar eastward through N Khorasan Prov. in NE Iran to the Mashhad region), SW and S Turkmenistan (from Little Balkhan Mtn east through the Kopet Dag Mtns to east of the Tedzhen River), and NW Afghanistan (Herat Prov.). (Range based on identification of AMNH, FMNH, and USNM specimens and those used in the multivariate analysis by Lebedev et al., 1998:723).
SYNONYMS: *firiusaensis* Meyer and Malikov, 2000; *zykovi* Meyer and Malikov, 2000.
COMMENTS: Included in *C. mystax* by Vorontsov et al. (1979; see that account) but recognized as a separate species by Pavlinov and Rossolimo (1998). Samples from C and E Kopet-Dag Mtns in S Turkmenistan and NE Iran (2n = 44, FN = 58; Graphodatsky et al., 2000) were first reported as *Calomyscus* sp. (Meyer and Malikov, 1995) and then described as *C. firiusaensis* (Meyer and Malikov, 2000). Specimens from C and W Kopet-Dag Mtns and Little Balkhan Mtn (2n = 30, FN = 44; Graphodatsky et al., 2000) were reported as *Calomyscus* sp. (Meyer and Malikov, 1995) and subsequently described as *zykovi*, a subspecies of *C. mystax* (Meyer and Malikov, 2000). Although the two karyotypes are very different, natural hybrids were recorded in a contact zone in the central Kopet-Dag Mtns and viable hybrids were obtained in the laboratory (Graphodatsky et al., 2000). Multivariate analysis of cranial and dental measurements clustered the 2n = 44 karyotypes with samples from N Khorasan Prov. that presumably represent Goodwin's *elburzensis*, and sorted the 2n = 30 samples into another group (Lebedev et al., 1998). Because there is natural hybridization between the two cytotypes, we treat all samples as representing a single species, for which *elburzensis* is the oldest name, as implied by Graphodatsky et al. (2000) and Lebedev et al. (1998) and suggested by Malikov (in litt.).

Calomyscus grandis Schlitter and Setzer, 1973. Proc. Biol. Soc. Wash., 68:163.
COMMON NAME: Noble Calomyscus.
TYPE LOCALITY: N Iran, Teheran Prov., foothills of Elburz Mtns, 11 km ENE Fasham.
DISTRIBUTION: N Iran; documented only from S foothills and ridges (8500 ft) of Mt Demavend, C Elburz Mtns (specimens in USNM), at the crest of the C Elburz Mtns

(Doab, specimens in FMNH) and on the N slopes in Mazandaran Prov. at Abass-Abad (3644' N, 5108' E; specimens in USNM) ; limits unresolved.

COMMENTS: The largest, darkest, and longest-tailed species of *Calomyscus*. Originally described as a subspecies of *C. bailwardi*, as recognized by Corbet (1978*c*) and Musser and Carleton (1993), but listed as a separate species by Pavlinov et al. (1995*a*). Morphometric analysis by Lebedev et al. (1998) demonstrated the great phenetic distance of *C. grandis* from *C. mystax* and *C. elburzensis* (S Turkmenistan and NE Iran) and *C. urartensis* (S Caucusus). Schlitter and Setzer (1973) enumerated some of these distinctions and also the contrasts between *grandis* and *C. bailwardi*. The much smaller-bodied *C. elburzensis* has been collected in the southern foothills of the Elburz Mtns at Semnan, about 120 km east of the localities where *grandis* was obtained, an identification we made from specimen examination and independently confirmed by the multivariate analysis of Lebedev et al. (1998). Karyotype of a specimen from Tehran Prov., south of the distribution outlined above, was reported as 2n = 44, FN = 46, which is identical to the karyotype of *C. mystax* (Graphodatsky et al., 2000; Malikov et al. 1999). However, Graphodatsky et al. (2000:303) noted that this karyotypic identity "can be explained either as persistence of the ancestral karyotype in different areas of the range or as a consequence of homoplasy."

Calomyscus hotsoni Thomas, 1920. J. Bombay Nat. Hist. Soc., 26:938.

COMMON NAME: Hotson's Calomyscus.

TYPE LOCALITY: SW Pakistan, W Balochistan, Makran Dist., Gwambuk Kaul, 50 km SW Panjgur (26 30'N, 63 50'E).

DISTRIBUTION: Recorded from vicinity of type locality and Baluchistan Prov. of SE Iran (specimens in USNM); range extent unresolved.

STATUS: IUCN – Endangered.

COMMENTS: A smaller-bodied *Calomyscus* (Vorontsov et al., 1979) that was described as a species but later arranged as a subspecies of *C. bailwardi*, including *mystax* as a synonym (Corbet, 1978*c*; Ellerman, 1941, 1961; Ellerman and Morrison-Scott, 1951). Based on size, *C. hotsoni* is allied with *C. mystax* and *C. elburzensis* of S Turkmenistan, NW Afghanistan, and NE Iraq, not with the larger-bodied *C. baluchi* that occurs to the north in W Pakistan. Affinities of *C. hotsoni* uncertain and karyotype undocumented.

Calomyscus mystax Kashkarov, 1925. Trans. Turkestansk. Nauch. ob-va pri Sredniaziatsk. Univ. (Tashkent), 2:43.

COMMON NAME: Great Balkhan Calomyscus.

TYPE LOCALITY: SW Turkmenistan, Bolshoi (= Greater) Balkan Mtn (Nibit-Dag region), Bashi-Mugur.

DISTRIBUTION: SW Turkmenistan: Ersarybaba (= Ersary Ridge) and Greater Balkan Mtns (Lebedev et al., 1998:723, collection localities 3 and 18).

STATUS: IUCN – Lower Risk (nt).

COMMENTS: Samples from Ersarybaba and the type locality have 2n = 44, FN = 46 (Graphodatsky et al., 2000; Malikov et al., 1999; Meyer and Malikov, 1995, 2000), which led Graphodatsky et al. (2000) to consider of *C. mystax* as endemic to the Great Balkan Mtns. This chromosomal and geographic distinction is concordant with results from multivariate discriminant analysis of cranial and dental measurements (Lebedev et al., 1998). Vorontsov's et al. (1979) broad distributional outline of *C. mystax* extended from the Greater Balkan Mtns eastward through the Lesser Balkan and Kopet Dag Mtns of S Turkmanistan and extreme NE Iran and westward through NE (Khorassan Prov.) and NC (Mazanderan Prov.) Iran, a range that includes *C. elburzensis* (which they treated as a synonym of *C. mystax*) and *C. grandis* (which was not included in their study). Meyer and Malikov (1995, 2000) and Pavlinov and Rossilimo (1998) also included the Lesser Balkans and Kopet Dag Mtns within the distribution of *C. mystax*. Meyer and Malikov (2000) described *zykovi* as a subspecies of *C. mystax*, but we include it in *C. elburzensis* (see above). Reasons for treating *C. mystax* and *C. urartensis* as distinct are provided by Meyer and Malikov (1995).

Calomyscus tsolovi Peshev, 1991. Mammalia, 55:112.

COMMON NAME: Syrian Calomyscus.

TYPE LOCALITY: Syria, Thafas, 18 km NW Derra (= Der'a).

DISTRIBUTION: SW Syria; limits unknown.

STATUS: IUCN – Lower Risk (nt).

COMMENTS: Described as a subspecies of *C. bailwardi*, but its small size and short tail clearly distinguish *C. tsolovi* from any other recognized species; such a possibility was even suggested by Peshev (1991:112). Phylogenetic affinities unstudied.

Calomyscus urartensis Vorontsov and Kartavseva, 1979. Zool. Zh., 58:1218.

COMMON NAME: Urar Calomyscus.

TYPE LOCALITY: Azerbaijan, Nakhichevanskaya., Alindzhachai River, 7 km N Dzhul'fa.

DISTRIBUTION: Extreme S Transcaucasus (Azerbaijan), far NW Iran (NW Azarbaijan Prov., specimens in FMNH).

STATUS: IUCN – Lower Risk (nt).

COMMENTS: *Calomyscus urartensis* (2n = 32, FN = 42; Graphodatsky et al., 2000) is morphologically similar to *C. mystax* (2n = 44, FN = 46) and *C. elburzensis* (Vorontsov et al., 1979, and material in FMNH). Meyer and Malikov (1995) cited hybridization results in addition to chromosomal distinctions to support the status of *C. urartensis* and *C. mystax* as separate species. The morphological integrity of *C. urartensis* and our identification of the species in NW Iran are corroborated by multivariate analysis of cranial and dental measurements (Lebedev et al., 1998).

Family Nesomyidae Major, 1897. Proc. Zool. Soc. Lond., 1897:718.

COMMENTS: Although a Nesomyidae per se had been earlier recognized by Tullberg (1899), Weber (1904), and later Chaline et al. (1977), theirs was essentially a grouping of the indigenous Malagasy rodents as already identified by Major (1897; Chaline et al. also included the otomyines, a clade clearly allied with murids—see those accounts). Although differing in contents, the family composition observed here owes its conceptual roots to Lavocat (1973, 1978), who identified a number of small but morphologically well-defined groups as relicts of a middle Tertiary (late Oligocene-early Miocene?) cricetodontine presence in Africa and expanded the definition of Nesomyidae to embrace their diverse descendants (also see Carleton and Musser, 1984:344). The results of DNA hybridization and mitochondrial and nuclear gene-sequence studies, although not wholly concordant, have supplied some empirical basis for Lavocat's view of Nesomyidae, associating Cricetomyinae, Dendromurinae, Mystromyinae, and Nesomyinae in a monophyletic lineage basal to other muroid taxa representing Cricetidae and Muridae (DuBois et al., 1996; Jansa et al., 1999; Michaux and Catzeflis, 2000; Michaux et al., 2001*b*).

Fossil representatives of Nesomyidae, as construed here, date from the early Miocene of Africa (see individual subfamily comments), with extinct taxa such as *Notocricetodon* and *Protarsomys* (Afrocricetodontinae *sensu* Lavocat, 1973, 1978) variously implicated in the origin of several of the living genera and-or subfamilies. The paleontological argument for such phyletic links remains sketchy and the hard evidence from critical middle Tertiary beds is scanty. Such Tertiary discoveries nonetheless continue to emerge from Africa and will help much to illuminate the validity of a family Nesomyidae upon their critical study. The alternative possibility obviously demands attention: do the subfamilies associated here form a polyphyletic wastebasket for the remnants of early evolutionary branches within other major radiations of Muroidea, by implication Cricetidae or Muridae? This is the considered view of Tong and Jaeger (1993:56) regarding Lavocat's expansive concept of Nesomyidae. Mustering evidence to tease apart these hypotheses will require denser taxonomic sampling in molecular studies and broader appeal to organ systems other than the dentition in morphological investigations.

Subfamily Cricetomyinae Roberts, 1951. Mammals South Africa, p. 434.

SYNONYMS: Cricetomyinae Roberts, 1951 (Cricetomyidae Chaline, Mein, and F. Petter, 1977); Saccostomurinae Roberts, 1951.

COMMENTS: Although initially classified within Murinae (e.g., Ellerman, 1941; Simpson, 1945; Thomas, 1896), systematists have acknowledged a closer affinity of African pouched rats to one another than to other murines, a distinction later formalized by Roberts' (1951) diagnosis of Cricetomyinae. Most subsequent systematic arrangements have followed

F. Petter (1966a) in allying cricetomyines with cricetids (Reig, 1980, 1981; Rosevear, 1969) based on his interpretation of molar cusp homologies. Lavocat (1973, 1978) first suggested phylogenetic association with other archaic African groups in a broadly defined Nesomyidae, an interpretation generally endorsed by Carleton and Musser (1984). Configuration of molar cusps (F. Petter, 1966a), anatomy of internal cheek pouches (Ryan, 1989b), and mitochondrial and nuclear DNA sequence data (Jansa et al., 1999; Michaux et al., 2001b; Michaux and Catzeflis, 2000) support the monophyly of the subfamily, which forms an unresolved polytomy with species of Dendromurinae and *Mystromys* according to phyletic evaluation of nuclear protein-coding genes (Michaux et al., 2001b; Michaux and Catzeflis, 2000).

General and diagnostic morphological traits, taxonomic history, and phylogenetic interpretations summarized by Carleton and Musser (1984). Certain early to middle Miocene taxa of East Africa (e.g., *Leakeymys* and *Notocricetodon*) have been speculatively linked to the autochthonous African origin of Cricetomyinae, but their phyletic association so far lacks persuasive demonstration (see Chaline et al., 1977; Mein et al., 2004; Tong and Jaeger, 1993). Indisputable fossil representatives extend from the early Pliocene to the Recent of eastern and southern Africa (McKenna and Bell, 1997). Roberts (1951) taxonomically segregated *Saccostomus* (Saccostomurinae) from other African pouched rats (Cricetomyinae), a division not recognized by later systematists (F. Petter, 1966a; Ryan, 1989b). Many morphological traits, however, depict *Beamys* and *Cricetomys* as sister species apart from *Saccostomus*, as enumerated by Carleton and Musser (1984), and we employ those family-group taxa as tribes. The closer affinity of *Beamys* and *Cricetomys* is also reflected in gene-sequence analyses (Corti et al., 2004).

Beamys Thomas, 1909. Ann. Mag. Nat. Hist., ser. 8, 4:107.

    TYPE SPECIES: *Beamys hindei* Thomas, 1909.

    COMMENTS: Cricetomyini, new rank. Two species have been generally recognized as originally described (e.g., G. M. Allen, 1939; Ellerman, 1941; Ellerman et al., 1953; Hanney, 1965; Misonne, 1974), until Hubbard (1970), without presentation of data, suggested their synonymy, a classification adopted by some (e.g., Ansell, 1978; Ansell and Ansell, 1973; Corbet and Hill, 1991; Honacki et al., 1982) but not others (Musser and Carleton, 1993). Ansell and Ansell (1973), while following Hubbard's opinion, did present limited measurements that demonstrate the uniformly larger size of *major* compared with *hindei*. The number of species, their morphological discrimination, and vouchered distribution deserve the attention of a full generic revision. Emphasis should be given to those *Beamys* populations that occur in the Eastern Arc Mtns, whether more closely related to *major* proper from uplands in NE Zambia and Malawi or to *hindei* from lower dry coastal forest in SE Kenya and E Tanzania.

Beamys hindei Thomas, 1909. Ann. Mag. Nat. Hist., ser. 8, 4:108.

    COMMON NAME: Hinde's Pouched Rat.

    TYPE LOCALITY: Kenya, Coast Prov., Taveta.

    DISTRIBUTION: SE Kenya and E Tanzania (FitzGibbon et al., 1995:Fig. 1).

    STATUS: IUCN – Vulnerable.

    COMMENTS: Karyotype reported (2n = 52), distribution amplified, and natural history discussed by FitzGibbon et al. (1995); univariate analyses with small sample sizes suggest clinal variation along coastal Tanzania, not the existence of two species, but the matter of synonymy was left as inconclusive.

Beamys major Dollman, 1914. Ann. Mag. Nat. Hist., ser. 8, 14:428.

    COMMON NAME: Greater Pouched Rat.

    TYPE LOCALITY: Malawi, Southern Region, Mulanje.

    DISTRIBUTION: Evergreen forest, 700-2000 m, in NE Zambia (Ansell, 1978) and Malawi (Ansell and Dowsett, 1988); possibly in Zambezia Dist., Mozambique (Smithers and Tello, 1976).

    STATUS: IUCN – Lower Risk (nt).

    COMMENTS: Most information on biology and ecology contained in the report of Hanney and Morris (1962).

Cricetomys Waterhouse, 1840. Proc. Zool. Soc. Lond., 1840:2.

TYPE SPECIES: *Cricetomys gambianus* Waterhouse, 1840.

COMMENTS: Cricetomyini. Six nominal species were recognized (e.g., Allen, 1939) before Ellerman (1941) summarily reduced all to subspecies of *C. gambianus*. Genest-Villard's (1967) revision provided evidence of two kinds, a predominantly savanna-dwelling species (*C. gambianus*) and a lowland forest form (*C. emini*), an arrangement broadly compatible with Hinton's (1919b) perception of harsh-furred and sleek-furred groups. Her character states and univariate ratios for discriminating two species are generally workable in West Africa but break down in eastern and southern Africa; based on series in MCZ and USNM, four kinds of *Cricetomys* can be morphologically sorted. For studies of the gastrointestinal tract interpreted in the context of taxonomic comparisons, trophic niche, and digestive function see Knight and Knight-Eloff (1987) and Perrin and Curtis (1980).

Cricetomys ansorgei Thomas, 1904. Ann. Mag. Nat. Hist., ser. 7, 13:412.

COMMON NAME: Southern Giant Pouched Rat.

TYPE LOCALITY: Angola, Pungo Andongo.

DISTRIBUTION: Kenya and E and S Uganda, southwards exclusive of Congo forest block, to W 2nd S Angola (Crawford-Cabral, 1998:Map 21), S Zambia, E Zimbabwe, and NE South pfrica (Limpopo Province and NE KwaZulu-Natal); including Zanzibar.

SYNONYMS: *adventor* Thomas and Wroughton, 1907; *cosensi* Hinton, 1919; *cunctator* Thomas and Wroughton, 1908; *elgonis* Thomas, 1910; *enguvi* Heller, 1912; *haagneri* Roberts, 1926; *kenyensis* Osgood, 1910; *luteus* Dollman, 1911; *microtis* Lönnberg, 1917; *osgoodi* Heller, 1912; *raineyi* Heller, 1912; *selindensis* Roberts, 1946; *vaughanjonesi* St. Leger, 1937; *viator* Thomas, 1904.

COMMENTS: Cranial proportions simililar to *C. gambianus* but size larger and skull robustly constructed (LM1-3 typically 11-12 mm versus 10-11 mm), dominant tones of dorsal pelage brown compared with grayish of *C. gambianus*; as equally differentiated in multivariate space from *C. gambianus* as from *C. emini*. Other morphological and ecological contrasts to *C. gambianus* enumerated (as *C. g. ansorgei*) by Genest-Villard (1967). Even as arranged, *ansorgei* is likely a species complex. Populations that occur on mountains in East Africa (*elgonis, enguvi, kenyensis, microtis*), each retained as well differentiated but localized subspecies of *C. gambianus* by Genest-Villard (1967), invite detailed investigation; certain populations may plausibly link with *C. kivuensis* from Western Rift Mtns. Also, *cosensi*, described as a species from Zanzibar, was considered by Hinton (1919b) to be related to the *emini* complex.

Cricetomys emini Wroughton, 1910. Ann. Mag. Nat. Hist., ser. 8, 5:269.

COMMON NAME: Forest Giant Pouched Rat.

TYPE LOCALITY: Dem. Rep. Congo, Gadda, Monbuttu.

DISTRIBUTION: Lowland, closed-canopied forest—in West Africa, from W Gambia (Grubb et al., 1998), S Guinea (Barnett et al., 1996), and Sierra Leone to S Nigeria; in Central Africa, from S Cameroon to SW Uganda, through Gabon, Republic of Congo, and much of Dem. Rep. Congo, to Cabinda, Angola (Crawford-Cabral, 1998).

STATUS: IUCN – Lower Risk (lc).

SYNONYMS: *dissimilis* Rochebrune, 1885 [*nomen dubium* as per Genest-Villard, 1967:421]; *dolichops* Osgood, 1910; *liberiae* Osgood, 1910; *luteus* Dollman, 1911; *poensis* Osgood, 1910; *proparator* Wroughton, 1910; *sanctus* Hinton, 1919; *tephrus* (Rochebrune, 1885) [*nomen dubium*?].

COMMENTS: As noted by Genest-Villard (1967), pelage color and cranial morphology are fairly conservative across the range of *C. emini* proper. Genest-Villard declared that Rochebrune's *dissimilis* is a *nomen dubium* in defending Wroughton's *emini* as the oldest available name for populations in the Guinea and Congo forest blocks. In his description of *C. ansorgei*, however, Thomas' (1904c:413) remarks intimate that he may have seen type or topotypic specimens: "The two forms of the group [*dissimilis* and *tephrus*], no doubt synonymous with each other, from Landana described by Rochebrune are both far smaller and have their bellies 'albocinereis.'" G. M. Allen (1939) followed Thomas in treating *tephrus* as full synonym of *dissimilis*, the latter as a subspecies of *C. gambianus* (as erroneously repeated by Musser and Carleton, 1993); Genest-Villard and Crawford-Cabral (1998) did not mention *tephrus*. Further museum sleuthing may disclose whether type material exists and settle the nomenclatural status of Rochebrune's taxa with regard to both *emini* and *ansorgei*.

Cricetomys gambianus Waterhouse, 1840. Proc. Zool. Soc. Lond., 1840:2.
COMMON NAME: Northern Giant Pouched Rat.
TYPE LOCALITY: Gambia, Gambia River.
DISTRIBUTION: Subsaharan savanna belt (Sudan and Guinea) and forest edges, from Gambia and Senegal eastwards to NE Dem. Rep. Congo, S Sudan, and N Uganda.
STATUS: IUCN – Lower Risk (lc).
SYNONYMS: *buchanani* Thomas and Hinton, 1921; *dichrurus* Osgood, 1910; *gambiensis* Temminck, 1853 [spelling *lapsus*]; *goliath* (Ruppell, 1842); *grahami* Hinton, 1919; *langi* Hatt, 1934; *oliviae* Dollman, 1911; *servorum* Hinton, 1919.
COMMENTS: The range outlined here basically corresponds to Hinton's (1919*b*) grouping of relatively small-bodied *Cricetomys* with gray, moderately long and hispid pelage. Morphological, behavioral, and ecological contrasts between *C. gambianus* proper and *C. emini* elucidated by Genest-Villard (1967) and Rosevear (1969).

Cricetomys kivuensis Lönnberg, 1917. Kungl. Svenska Vet-Akad. Handl., Stockholm, 58: art. 2, p. 75.
COMMON NAME: Kivu Giant Pouched Rat.
TYPE LOCALITY: Dem. Rep. Congo, Masisi.
DISTRIBUTION: Mountains of easternmost Dem. Rep. Congo (Kivu), S Uganda, Rwanda, and Burundi; limits unknown.
COMMENTS: Although Genest-Villard (1967) retained *kivuensis* as a subspecies of *C. emini*, she considered its differentiation as marginal. Compared with *C. emini*, examples of *kivuensis* are larger, their skulls fully as robust as the biggest *C. ansorgei*; dorsal color is fuscous gray (not the bright browns of *emini*) and the underparts are dull grayish-white, weakly demarcated from the dorsum (pure white and cleanly delineated in *emini*); and fur is long, its texture soft and luxurient (close-cropped and slightly hispid in *emini*). The union of such disparate morphologies under one species deserves unambiguous demonstration.

Saccostomus Peters, 1846. Bericht Verhandl. K. Preuss. Akad. Wiss., Berlin, 11:258.
TYPE SPECIES: *Saccostomus campestris* Peters, 1846.
SYNONYMS: *Eosaccomys* Palmer, 1903 [unjustified replacement name].
COMMENTS: Saccostomurini, new rank. Ellerman (1941) thought that all nominal forms would prove to be races of the single species *S. campestris*, a speculation widely and uncritically observed in later classifications and faunal reports (e.g., Delany, 1975; Ellerman et al., 1953; Kingdon, 1974*b*; Misonne, 1974). Hubert (1978*a*), however, marshalled karyotypic and morphological evidence that differentiates an eastern African species (*S. mearnsi*) from a southern African one (*S. campestris*), as earlier intimated by G. M. Allen and Lawrence (1936) and later reinforced by Denys (1988) in her morphological study of Tanzanian populations. Relationship of the two living species to the middle Pliocene *S. major* evaluated by Denys (1987*a*, 1988), who placed their divergence some time after 3.7 million years ago. Geological range from middle Pliocene through the Pleistocene of Tanzania (Denys, 1988) and late Miocene to Pleistocene of Namibia (Mein et al., 2004; Senut et al., 1992).

Saccostomus campestris Peters, 1846. Bericht Verhandl. K. Preuss. Akad. Wiss., Berlin, 11:258.
COMMON NAME: Southern African Pouched Mouse.
TYPE LOCALITY: Mozambique, Tete Dist., Zambezi River, Tete.
DISTRIBUTION: Arid to mesic southern savannahs and grasslands: from SW Tanzania across to W Angola (see Crawford-Cabral, 1998: Map 20); south through most of Malawi, Zambia, Zimbabwe, Botswana, and Namibia; to S Mozambique and S South Africa (Western and Eastern Cape provinces).
STATUS: IUCN – Lower Risk (lc).
SYNONYMS: *anderssoni* de Winton, 1898; *angolae* Roberts, 1938; *elegans* Thomas, 1897; *fuscus*

Peters, 1852; *hildae* Schwann, 1906; *lapidarius* Peters, 1852; *limpopoensis* Roberts, 1914; *mashonae* de Winton, 1897; *pagei* Thomas and Hinton, 1923; *streeteri* Roberts, 1914.

COMMENTS: A composite of two or more species as suggested by extraordinary chromosomal, morphological, and genetic variation (Corti et al., 2004; Denys, 1988; Gordon, 1986; Gordon and Rautenbach, 1980); the status of *anderssoni* and *mashonae* especially deserves attention in regard to *campestris* proper. Ellison (1993) demonstrated and discussed geographic variability in torpor response in the context of the species distribution and its karyotypic variants; Ellison et al. (1993) examined correlations between body size and climatic variables in southern Africa.

Saccostomus mearnsi Heller, 1910. Smithson. Misc. Coll., 54:3.

COMMON NAME: East African Pouched Mouse.

TYPE LOCALITY: Kenya, Coast Prov., Changamwe.

DISTRIBUTION: Extreme S Ethiopia and S Somalia, through E Uganda and Kenya, to NE Tanzania.

STATUS: IUCN – Lower Risk (lc).

SYNONYMS: *cricetulus* G. M. Allen and Lawrence, 1936; *isiolae* Heller, 1912; *umbriventer* Miller, 1910.

COMMENTS: Perhaps also a species complex as suggested by newly reported karyotypic, cranial, and genetic variation (Corti et al., 2004; Fadda et al., 2001*b*).

Subfamily Delanymyinae new subfamily.

SYNONYMS: Delanomyinae McKenna and Bell, 1997 [*nomen nudum*, the cited publication (Denys et al., 1995) containing no family-group description].

COMMENTS: Type genus–*Delanymys* Hayman, 1962. Definition–very small-bodied (head and body less than 65 mm in adults, 5-7 grams), arboreal muroid with a thinly haired (caudal hairs less than two tail scales long), very long (twice as long as head and body), and semi-prehensile tail (conformation of body and appendages similar to Sulawesian *Haeromys*; see Musser, 1990); four digits and rudimentary fifth on front foot (fifth very short, middle two long, second longer than fifth), five on hind foot (first short, middle three very long, and fifth two-thirds as long as middle digits and opposable), sharp and moderately curved claws on all digits; hind foot very long (relative to body size) and narrow, with four inter-digital pads, thenar and hypothenar elongate and subequal in length; pelage short, dense and soft (not silky) with short guard hairs; four pairs of teats; cranium small and delicate with very short and narrow rostrum compared with the large, globular, and smooth (faint temporal ridges) braincase; interorbit narrow and without ridges, postorbit broad; zygomatic plate narrow, with slightly projecting anterior margin (very shallow zygomatic notch), indistinct masseteric tubercle; alisphenoid strut absent; large subsquamosal foramen bounded ventrally by prominent hamular process; incisive foramina long but not extending past anterior margins of molars, posterior palatine foramina at level of posterior half of M1; bony palate wide, with posterior inverted V-shaped margin at level of second molar; mesopterygoid fossa wide and open to level of second molar, sphenopalatine vacuities large; parapterygoid fossae long, wide, moderately excavated; bullae small; carotid circulation fully derived (minute stapedial foramen, no squamosal-alisphenoid groove or sphenofrontal foramen; pattern 3 as per Carleton, 1980, and Voss, 1988); soft palate with three premolar and four intermolar ridges; mandible short and robust, pronounced coronoid process, incisor alveolus projects prominently between coronoid and condyloid processes; upper incisors ungrooved, slightly compressed, orthodont or slightly opisthodont; lower incisors narrow; molars brachyodont with three-rooted uppers and two-rooted lowers; primary cusps large, arranged in two rows, connected by short longitudinal mure and transverse enamel crests; anterocone(id) complete on M1 and m1(appears bifurcate on m1 with wear because anteroconid coalesces with anterolophid); accessory lingual cusp large, next to protocone on M1 and M2 and confluent after wear; anteroloph wide and short (style-like), mesoloph a long and prominent ridge reaching cingular margin on both M1 and M2; anterolophid large (comma-shaped) on m1 and m2, mesolophid smaller but connects with style on cingulum; posterior cingular ridge (posteroloph) connects posterior margins of metacone and hypocone on M1 and M2, large posterior cingulum on m1 and m2; M2 subequal to M1, M3 more than half as long as M2 with similar occlusal pattern (based upon study of specimens in AMNH and information in Delany [1975], Dieterlen [1969*b*], Hayman [1962*a*], and W. Verheyen [1965*b*]; see dental terminology in Carleton and Musser [1989]). Contents–*Delanymys* Hayman, 1962; *Stenodontomys* Pocock, 1987.

The discoverer Hayman (1962*a*) regarded *Delanymys* as a dendromurine, morphologically close to *Petromyscus*. According to Lavocat (1964:184), however, "*Delanymys* shows typical Cricetid dental structures, particularly in the presence of the longitudinal crest. Thus, *Petromyscus* and *Delanymys* are perfect structural links between *Mystromys* and the typical Dendromurinae and show us the clear systematic affinities of these forms." W. Verheyen (1965:12) was similarly convinced that "*Delanymys* represents in its tooth structure an almost perfect morphological link between *Mystromys* and the Dendromurinae." F. Petter (1967*b*) treated *Delanymys* and *Petromyscus* as cricetids, but united the two in Petromyscinae, divorcing them from dendromurines because he thought the shared lingual cusp to be nonhomologous; he identified it as the protocone and the mesial cusp as neomorphic, having developed from the longitudinal crest (mure). Jaeger (1977*b*) discounted Petter's hypothesis and reemphasized a cingular origin for the accessory cusp, noting that lingual accessory cusps are found in some Miocene myocricetodontines as well as dendromurines; because occlusal patterns are otherwise so different among myocricetodontines, dendromurines, and petromyscines, he viewed their shared lingual cusps as parallelisms reflecting an origin from a distant common ancestor.

Carleton and Musser (1984:332) enumerated the general characters of *Petromyscus* and *Delanymys*, as Petromyscinae, but pointed out that, aside from the lingual accessory cusps, "no other derived features are shared by the two that would indicate they belong in a natural group separate from other African cricetids. Even the configuration of the lingual cusps and the extent of their attachment to the protocones are different in the two genera." This view draws support from Denys (1994*c*), who described the dental similarities between *Delanymys* and *Stenodontomys* (type species originally named as a *Mystromys* by Lavocat, 1956) from the Plio-Pleistocene of South Africa (Pocock, 1987) and Namibia (Senut et al., 1992); she further contrasted their molar patterns with *Petromyscus* and regarded the Petromyscinae as an unnatural group, the genera originating independently from an early Miocene lineage different from that leading to dendromurines. Accessory lingual cusps are common to cricetomyines, dendromurines, leimacomyines, some Miocene myocricetodontines, deomyines and murines (cusp t4), as well as delanymyines and petromyscines, and apparently were independently acquired in several Old World muroid groups.

Delanymys Hayman, 1962. Rev. Zool. Bot. Afr., 65:1-2.
> TYPE SPECIES: *Delanymys brooksi* Hayman, 1962.

Delanymys brooksi Hayman, 1962. Rev. Zool. Bot. Afr., 65:1-2.
> COMMON NAME: Delany's Swamp Mouse.
> TYPE LOCALITY: SW Uganda, Kigezi, near Kanaba, Echuya (or Muchuya) Swamp, 7500 ft (2286 m) (Hayman, 1962*a*).
> DISTRIBUTION: Sedge swamps in bamboo and montane forest in SW Uganda, Dem. Rep. Congo (Kivu), Rwanda (Volcanos and Nyungwe forest), and Burundi (Kibira National Park; Peterhans Kerbis, in litt., 1992).
> STATUS: IUCN – Lower Risk (lc).
> COMMENTS: Available information on morphology, distribution, and ecology covered in Hayman (1962*a*, *b*, 1963*a*), W. Verheyen (1965*b*), Dieterlen (1969*b*), Delany (1975), and Van der Straeten and Verheyen (1983).

Subfamily Dendromurinae G. M. Allen, 1939. Bull. Mus. Comp. Zool. Harv. Coll., 83:349.
> SYNONYMS: Dendromurinae G. M. Allen, 1939 (Dendromuridae Chaline, Mein, and F. Petter, 1977); Dendromyinae Alston, 1876.
> COMMENTS: Dendromurines had been earlier allied with murines (Miller and Gidley, 1918; Simpson, 1945) but later included with "cricetids" (Lavocat, 1959, 1964; Lindsay, 1988) or treated as a separate family (Chaline et al., 1977). Carleton and Musser (1984) attempted a diagnosis of the subfamily as then understood and cautioned that more research was re-

quired to determine whether dendromurines represent a natural group or polyphyletic assemblage of specialized relicts. Monophyly of Dendromurinae was also questioned by Rosevear (1969). Recent analyses of morphology (Denys et al., 1995), spermatozoa (Breed, 1995*d*), and DNA sequences (Michaux et al., 2001*b*; E. Verheyen et al., 1996) have exposed the polyphyletic structure of Dendromurinae and uniformly sustained a distant phylogenetic affinity between murines and dendromurines. These data collectively support removal of *Deomys* and *Leimacomys* to different subfamilies of Muridae: the former phylogenetically associated with *Acomys*, *Lophuromys*, and *Uranomys* (see account of Deomyinae); the latter as sole member of Leimacomyinae (see that account).

Divorced of *Deomys* and *Leimacomys*, the Dendromurinae contains medium to small-bodied muroids whose range of morphologies reflects varied evolutionary responses to different ecological constraints. *Dendromus* and *Megadendromus*, while active at ground level, are primarily adept climbers of tall grass and shrubs where they forage and construct nests and inhabit places where such vegetation is predominant (marshes, savannas, forest edges, alpine bamboo and heath zones). *Malacothrix* is terrestrial, granivorous, and gerbil-like in its morphology, habits, and habitat. *Steatomys* is terrestrial, dwelling primarily in savanna habitats and accumulating fat to remain inactive during unfavorable environmental conditions. *Dendroprionomys* and *Prionomys* are arboreal and insectivorous inhabitants of tropical lowland evergreen rainforest. Compared with the core genera (*Dendromus*, *Megadendromus*, *Steatomys*), tribal separation of *Prionomys* and *Dendroprionomys* is plausibly indicated by dental traits (Denys et al., 1995), and the highly specialized *Malacothrix* may also warrant tribal segregation.

Fossil antecedants of dendromurines are unresolved, although Tong and Jaeger (1993) suggested derivation from an early Miocene ancestor related to *Notocricetodon* of East Africa or to *Potwarmus* of Pakistan (apparently also the late Miocene in Libya; Savage, 1988). Extant dendromurines live only in Subsaharan Africa, where they are represented by Miocene, Pliocene, and Pleistocene fossils of *Dendromus*, *Steatomys*, and unidentified "Dendromurinae" (Avery, 1977, 1995, 1998, 2000; de Graaff, 1961; Denys, 1987*a*, 1987*b*, 1994*b*; Geraads, 2001; Jaeger, 1979; Jaeger and Wesselman, 1976; Jaeger et al., 1985; Lavocat, 1965, 1978; Mein et al., 2004; Senut et al., 1992; Tong and Jaeger, 1993). During the late Miocene, however, the subfamily's geographic range extended to Algeria (Ameur, 1984), S Spain (Aguilar et al., 1984), and United Arab Emirates (de Bruijn, 1999; de Bruijn and Whybrow, 1994). The earliest true dendromurines are two species of *Ternania* from the middle Miocene (14-13.9 million years ago) of Kenya (Tong and Jaeger, 1993). Winkler (1998) described *Mabokomys* from middle Miocene sediments (slightly older than 14.7 million years ago) of Kenya as the oldest African dendromurine, but to us its molar pattern is not dendromurine. We also agree with Tong and Jaeger (1993) in excluding those extinct genera (*Dakkamys*, *Paradakkamys*, *Potwarmus*) from the middle Miocene of NW Africa, Pakistan and Thailand that Lindsay (1988, 1994) defined as dendromurines (also Mein and Ginsburg, 1997; Winkler, 1998). Jaeger (1977*a*, *b*) considered *Dakkamys* to be a member of Myocricetodontinae, where it properly belongs according to Tong and Jaeger (1993) and Wessels (1996).

Alston's (1876) Dendromyinae is the oldest family-group name and was accepted in checklists and classifications until the 1940s (Ellerman, 1941; Miller and Gidley, 1918; Thomas, 1896); it was abandoned for Dendromurinae (G. M. Allen, 1939), as since employed. Because *Dendro* is the stem from which the subfamily name was derived, either Dendromyinae or Dendromurinae is available (ICZN, 1999, Art. 11.7); however, *Dendromys*, the type genus of Alston's Dendromyinae, is junior synonym of *Dendromus*, and when "a family-group name was replaced before 1961 because of the synonymy of the type genus, the substitute name is to be maintained if it is in prevailing usage" (ICZN, 1999, Art.40.2), a stipulation supporting continued use of Dendromurinae.

Dendromus Smith, 1829. Zool. J. Lond., 4:438.

    TYPE SPECIES: *Dendromus typus* Smith, 1829 (= *Mus mesomelas* Brants, 1827).

    SYNONYMS: *Chortomys* Thomas, 1916; *Dendromys* J. B. Fischer, 1830; *Dendromys* Smuts, 1832; *Poemys* Thomas, 1916.

    COMMENTS: Rosevear (1969) thoroughly reviewed historical usages of generic names given in the synonymy. In his checklist of African mammals, G. M. Allen (1939) listed 29 species of *Dendromus*, soon after reduced to four by Bohmann's (1942) revision. Based on morphological traits and ecological associations, Dieterlen (1971) recognized five species within the central African region. We recognize twelve based on specimen examination from North American and European museums and on literature sources; most have clear morphological and geographic characteristics, a few are heterogeneous entities and require further taxonomic revision. Chromosomal information for several species were reported by Matthey (1967a, 1970). The species have been traditionally segregated into the subgenera *Poemys* (hallux with a nail—*D. melanotis, D. nyikae*) and *Dendromus* (hallux bears a claw—all other species) (see Ansell, 1974b, for utility of the hallucial trait and other characters proposed to define subgenera).

The extant diversity of *Dendromus* in Subsaharan Africa mirrors an equal or greater radiation of extinct species. Recorded earliest from the late Miocene of S Ethiopia (Geraads, 2001) and Namibia (Mein et al., 2004); additional species uncovered in Pliocene-Pleistocene sediments in South Africa (Avery, 1995, 1998; Denys, 1994b) and East Africa (Denys, 1987a, b; Jaeger, 1979). Of the two species Denys (1994b) named from the Pliocene Langebaanweg site, southern Africa, *D. darti* is most closely related to living *D. melanotis* and *D. averyi* to extant *D. mesomelas*. The latter is also morphologically similar to the late Miocene *D. denysae* from Namibia (Mein et al., 2004). Two species of *Dendromus* are also known outside Africa, from late Miocene sediments in the United Arab Emirates (de Bruijn, 1999; de Bruijn and Whybrow, 1994); their molars are strikingly similar to the South African and Namibian species.

Dendromus insignis (Thomas, 1903). Ann. Mag. Nat. Hist., ser. 7, 12:341.
COMMON NAME: Montane African Climbing Mouse.
TYPE LOCALITY: Kenya, Nandi.
DISTRIBUTION: Discontinuous in bamboo, heath, and alpine zones, ca. 3000-4700 m, of East Africa—Ethiopia (Yalden et al., 1976, 1996; AMNH 81105) through W Kenya (Hollister, 1919; AMNH, MCZ, and USNM series) and Mt Elgon on the Kenya-Uganda border (Clausnitzer, 2001; reported as *mesomelas*), to Mt Kilimanjaro (Shore and Garbett, 1991; Grimshaw et al., 1995; FMNH specimens); also Western Rift mountains from W Uganda (Rwenzoris, Kerbis Peterhans et al., 1998) south to W Rwanda and E Dem. Rep. Congo (Kivu region; large samples in AMNH, BMNH, and FMNH); distributional extent unknown.
STATUS: IUCN – Lower Risk (lc).
SYNONYMS: *abyssinicus* Osgood, 1936; *kilimandjari* Bohmann, 1939; *percivali* Heller, 1912.
COMMENTS: Although *insignis* has been included in *D. mesomelas* (Bohmann, 1942; Misonne, 1974), Thomas (1916b) cautioned that the "lumping of *insignis* with the southern *D. mesomelas*" appeared to be "unfounded," an evaluation sustained by specimen study. *Dendromus insignis* is distinguished easily from the South African *D. mesomelas* by larger body size, shorter pelage, darker upperparts with a more prominent stripe, dark gray or grayish buff underparts (white washed with buff or ochraceous in *D. mesomelas*), larger skull and longer molar rows, and interorbital and postorbital shape. Bohmann (1939) described the large-bodied *kilimandjari* as a subspecies of *D. mesomelas*, but the name identifies another montane population of *D. insignis* (study of BMNH paratypes); Thomas (1916b) explained why *percivali* is a synonym; and allocation of *abyssinicus* is based upon our inspection of the FMNH holotype. Prior to 1991, most literature references to *D. mesomelas* in mountains north of the Southern African Subregion actually represent either *D. insignis* or *D. nyasae* (see account), which may co-occur in the Western Rift mountains. In the Rwenzori Mountains National Park, Uganda, *D. insignis* and *D. nyasae* (recorded as *kivu*) occur together in the moorlands above 3000 m (Kerbis Peterhans et al., 1998). On Mt Kilimanjaro, *D. insignis* inhabits heath and alpine zones, 3500-4700 m, and the savannah species, *D. melanotis*, is found near the forest margin, 1500 m (Grimshaw et al., 1995).

Dendromus kahuziensis Dieterlen, 1969. Z. Säugetierk., 34:348-353.
COMMON NAME: Mt Kahuzi African Climbing Mouse.

TYPE LOCALITY: Dem. Rep. Congo, Kivu, Mt Kahuzi, 2100 m.

DISTRIBUTION: Kivu region, E Dem. Rep. Congo.

STATUS: IUCN – Lower Risk (nt).

COMMENTS: A distinctive species with an extremely long tail, broad middorsal stripe, and long gracile rostrum. Still known by few specimens from montane forest-bamboo habitat (Dieterlen, 1969*a*, 1976*c*).

Dendromus leucostomus Monard, 1933. Bull. Soc. Nat. Neuchâtel, Sci. Nat. 57:55.

COMMON NAME: Monard's African Climbing Mouse.

TYPE LOCALITY: Angola, Caluquembe.

DISTRIBUTION: Recorded only from the type locality, EC Angola.

COMMENTS: Described as a species by Monard (1933, 1935) and known only by his specimens. The name has usually been included in *D. melanotis*, but Crawford-Cabral (1998) questioned that allocation because the animal has no dorsal stripe and is brighter than *D. melanotis*. Hill and Carter (1941) treated *leucostomus* as a separate species and summarized Monard's description (upperparts without middoral stripe and less russet than *D. mystacalis*, ventral fur with gray bases and yellowish white tips) and measurements (body size similar to *D. melanotis* and *D. mystacalis*). We follow Hill and Carter until the taxonomic ambiguity associated with *leucostomus* is resolved.

Dendromus lovati (de Winton, 1900). Proc. Zool. Soc. Lond., 1899:986 [1900].

COMMON NAME: Lovat's African Climbing Mouse.

TYPE LOCALITY: Ethiopia, Managasha, near Addis Ababa.

DISTRIBUTION: Ethiopian Plateau, 2500-3550 m (Rupp, 1980).

STATUS: IUCN – Lower Risk (nt).

COMMENTS: The type species of *Chortomys*, reviewed as an Ethiopian endemic by Yalden and Largen (1992) and Yalden et al. (1996). A grassland species differing from other *Dendromus* in its terrestrial habits (Yalden, 1988; Yalden et al., 1976); sample statistics of external measurements provided by Sillero-Zubiri et al. (1995*a*).

Dendromus melanotis (Smith, 1834). S. Afr. Quart. J., 2:158.

COMMON NAME: Gray African Climbing Mouse.

TYPE LOCALITY: South Africa, KwaZulu-Natal Prov., near Port Natal (Durban).

DISTRIBUTION: From South Africa (de Graaff, 1997*dd*; Skinner and Smithers, 1990:306; P. J. Taylor, 1998) northward in the west through Botswana (Smithers, 1971) to C Angola (Crawford-Cabral, 1998; AMNH specimens from Chitau); northward in the east through Zimbabwe (Smithers and Wilson, 1979), Mozambique (Smithers and Lobao Tello, 1976), Zambia (Ansell, 1978), Malawi (Ansell and Dowsett, 1988; Ansell, 1989), and Tanzania (Mt Kilimanjaro; Grimshaw et al., 1995; Swynnerton and Hayman, 1951) to Uganda (Delany, 1975); westward through Nigeria and Ghana (Grubb et al., 1998) to S Guinea (Rosevear, 1969); also from Ethiopia (AMNH 81119 and four other specimens as reported by Duckworth et al., 1993, and Yalden et al., 1976). Range limits unresolved.

STATUS: IUCN – Lower Risk (lc).

SYNONYMS: *arenarius* Roberts, 1924; *basuticus* Roberts, 1927; *capensis* Roberts, 1931; *chiversi* Roberts, 1929; *concinnus* Thomas, 1926; *exoneratus* Thomas, 1918; *insignis* (Shortridge and Carter, 1938) [not of Thomas, 1903]; *nigrifrons* (True, 1892); *pallidus* (Heuglin, 1877); *pecilei* (Milne-Edwards, 1886); *pretoriae* Roberts, 1931; *shortridgei* St. Leger, 1930; *spectabilis* Heller, 1911; *subtilis* (Sundevall, 1846); *thorntoni* Roberts, 1931; *vulturnus* Thomas, 1916.

COMMENTS: Apparently most closely related to *D. nyikae* (see that account). A very widespread species and one requiring taxonomic revision, especially West African populations. Specimens from Mt Nimba, on the Guinea-Côte d'Ivoire border (Heim de Balsac and Lamotte, 1958; Rosevear, 1969), have ochraceous-red upperparts with indistinct stripe and white underparts, chromatic traits that fall outside the fur color and dorsal pattern of southern African *D. melanotis* (dorsum gray with prominent dark stripe, venter grayish), and may represent a separate species (either undescribed or a population of *D. messorius*). Crawford-Cabral (1998) questioned the inclusion of *concinnus*, but Thomas' (1926*e*) description fits within morphological variation currently accepted as *D. melanotis* (see Roberts, 1951:449).

Dendromus mesomelas (Brants, 1827). Het. Geslacht der Muizen, p. 122.

COMMON NAME: Brants's African Climbing Mouse.

TYPE LOCALITY: South Africa, Eastern Cape Prov., Sunday's River, east of Port Elizabeth.

DISTRIBUTION: Discontinuous range in Southern African Subregion—in South Africa across Western and Eastern Cape Provs., north through KwaZulu-Natal (P. J. Taylor, 1998), to Mpumalanga and Limpopo; Okavango swamps in N Botswana and Caprivi region of NE Namibia (Smithers, 1971); NW Zambia (Ansell, 1978, as *D. m. major*); and Gorongoza Mtn in C Mozambique (Smithers and Lobao Tello, 1976); may eventually be found in wet habitats of SE Angola and SW Zambia. Not recorded north of the Zambeze River in Mozambique or north of about 12 latitude elsewhere.

STATUS: IUCN – Lower Risk (lc).

SYNONYMS: *ayresi* Roberts, 1913; *major* St. Leger, 1930; *pumilio* (Wagner, 1841); *typicus* Smith, 1834; *typus* Smith, 1829.

COMMENTS: Reviewed by de Graaff (1997*cc*). The indecisive shifting of *pumilio* between *D. mesomelas* and *D. mystacalis* is reviewed Meester et al. (1986), who documented its proper association with the former. Some names once included in *D. mesomelas* (see Misonne, 1974) are here treated as distinct species (see *D. insignis, D. nyasae, D. oreas*, and *D. vernayi*). Judged by specimens, *D. mesomelas* from the Southern African Subregion is morphologically separable from northern populations in East Africa, Angola, and Cameroon. It is a larger version of, and likely closely related to, *D. mystacalis. Dendromus mesomelas* inhabits grasslands in swamps and other wet habitats; whereas, *D. mystacalis* and *D. melanotis*, which occur in the same regions, occupy dryer environments (Smithers, 1971; Smithers and Lobao Tello, 1976).

Dendromus messorius (Thomas, 1903). Ann. Mag. Nat. Hist., ser. 7, 12:340.

COMMON NAME: Banana African Climbing Mouse.

TYPE LOCALITY: Cameroon, Efulen.

DISTRIBUTION: Tropical lowlands and lower mountain slopes, from Ghana (Grubb et al., 1998) through Benin (USNM 422451), Nigeria (USNM 410307), and Cameroon (series in AMNH and MCZ), to N Dem. Rep. Congo (series in AMNH and USNM) and extreme S Sudan (Torit region; USNM 299833, 299834); southwards to NW (USNM 165270) and W Uganda (Ruwenzori Mtns; series in AMNH and MCZ), W Kenya near Lake Victoria (USNM series) and Mt Elgon (holotype of *ruddi*; MCZ specimens; Clausnitzer and Kityo, 2001); limits unresolved (Dieterlen, 1971).

STATUS: IUCN – Lower Risk (lc).

SYNONYMS: *haymani* Hatt, 1934; *kumasi* Bohmann, 1939; *ruddi* Wroughton, 1910.

COMMENTS: Treated as part of *D. mystacalis* by Misonne (1974), but regarded as a separate species by Hatt (1940) and Dieterlen (1971), who pointed out its sympatric occurrence with *D. mystacalis*. Morphology and pelage coloration similar to *D. mystacalis*, but body size larger, middorsal stripe absent (present but indistinct in some *D. mystacalis*, absent in many), and underparts white (white to buffy gray in *D. mystacalis*). Both species may be phylogenetically close to southern African *D. mesomelas*. Hatt (1940) and Rosevear (1969) provided excellent descriptions of morphology and habits.

Dendromus mystacalis (Heuglin, 1863). Nova Acta Acad. Caes. Leop.-Carol., Halle, 30:2, suppl. 5.

COMMON NAME: Chestnut African Climbing Mouse.

TYPE LOCALITY: Ethiopia, Baschlo region (Ellerman et al., 1953, offered additional comments).

DISTRIBUTION: Much of C and E Africa (G. M. Allen and Loveridge, 1942; Ansell, 1978, 1989; Ansell and Dowsett, 1988; Delany, 1975; Lawrence and Loveridge, 1953; Smithers, 1971; Smithers and Lobao Tello, 1976; Smithers and Wilson, 1979; Stanley et al., 1998*b*), including S Sudan (Setzer, 1956) and Ethiopia (Lavrenchenko, 2000; Yalden et al., 1996), as far south as Angola (Crawford-Cabral, 1966*b*, 1998) and E South Africa (de Graaff, 1997*ee*; Skinner and Smithers, 1990:308; P. J. Taylor, 1998); range based also on specimens in AMNH, BMNH, FMNH, and MCZ.

STATUS: IUCN – Lower Risk (lc).

SYNONYMS: *acraeus* Wroughton, 1909; *ansorgei* Thomas and Wroughton, 1905; *capitis* Heller, 1912; *jamesoni* Wroughton, 1909; *lineatus* Heller, 1911; *nairobae* Osgood, 1910;

ochropus Osgood, 1910; *pallescens* Osgood, 1910; *pongolensis* Roberts, 1931; *uthmoelleri* Bohmann, 1939; *whytei* Wroughton, 1909.

COMMENTS: Upperparts are bright ochraceous-buff without a stripe in most samples, a faint strip in others; underparts range from white (usual) to buffy gray, the coloration typical of N Dem. Rep. Congo and AMNH series from Angola (*ansorgei*, as per Thomas and Wroughton, 1905; Hill and Carter, 1941; and Hayman, 1963b). Those samples without a faint middorsal stripe closely resemble *D. messorius* but are smaller and usually have buffy gray underparts (always white in *messorius*). This species is a morphological miniature of southern African *D. mesomelas*, the two are probably close phylogenetic allies, and both constitute one of the two species-pairs occurring in southern Africa (Avery, 1998). Morphological variation within this widely ranging species should be investigated to determine whether more than one species is represented.

Dendromus nyasae Thomas, 1916. Ann. Mag. Nat. Hist., ser. 8, 18:241.

COMMON NAME: Kivu African Climbing Mouse.

TYPE LOCALITY: N Malawi, Nyika Plateau, 6500 ft (1981 m).

DISTRIBUTION: Western Rift Mtns—Ruwenzoris in E Dem. Rep. Congo and W Uganda (Kerbis Peterhans et al., 1998; Osgood, 1936; AMNH and FMNH), south through the E Dem. Rep. Congo (Kivu region; AMNH, BMNH, and MCZ) to the Marungu Mtns in SE Dem. Rep. Congo (AMNH 55735), Mbizi Mtns in W Tanzania (FMNH), SW Tanzania (Rungwe and Ukinga; AMNH, FMNH, MCZ; listed by G. M. Allen and Loveridge, 1933, as *kivu*), Nyika Plateau in NE Zambia (Ansell, 1978; Ansell and Ansell, 1973) and N Malawi, and highlands in S Malawi (Ansell and Dowsett, 1988; Lawrence and Loveridge, 1953); also through the Eastern Arc Mtns, E Tanzania (Uzungwa Mtns, FMNH and MCZ; Uluguru highlands, holotype of *hintoni*; Ukaguru Mtns, FMNH; Nguru Mtns, FMNH; South Pare Mtns, Stanley et al., 1998a, as *mesomelas*).

STATUS: IUCN – Lower Risk (lc) as *D. kivu*.

SYNONYMS: *hintoni* Bohmann, 1939; *kivu* Thomas, 1916; *lunaris* Osgood, 1936.

COMMENTS: *Dendromus nyasae* and the three names in synonymy (based on study of all holotypes) have been treated as subspecies of the southern African *D. mesomelas* (Bohmann, 1942; Misonne, 1974). The latter is larger in external dimensions; has longer pelage with upperparts brighter and buffier (darker, tawny hue in *D. nyasae*), underparts white usually suffused with pale buff or ochraceous (dark grayish white in *D. nyasae*); its skull is more robust with narrow interorbital and postorbital regions and more flaring zygomatic arches (interorbit constricted but postorbit wider), longer molar rows; and occurs commonly in lowlands (*D. nyasae* is strictly montane). Ansell (1974b, 1978) noted that specimens from NW Zambia identified as *D. mesomelas major* (= *D. mesomelas*) were "quite distinct" from *D. mesomelas nyasae* (= *D. nyasae*) on the Nyika Plateau, NE Zambia. In the Western Rift mtns, *D. nyasae* may occur sympatrically with *D. insignis* (see account). In morphology, *D. nyasae* seems to be most closely related to the small-bodied *D. vernayi* from the Angolan Plateau, *D. oreas* from the E Cameroon mountains, and the larger-bodied *D. insignis* from the mountains of East Africa. The four species may form a monophyletic group restricted to mountains and plateaus north of the Southern African Subregion and south of the Sahara.

Dendromus nyikae Wroughton, 1909. Ann. Mag. Nat. Hist., ser. 8, 3:248.

COMMON NAME: Nyika African Climbing Mouse.

TYPE LOCALITY: Malawi, Nyika Plateau.

DISTRIBUTION: Patchy range in S Subsaharan Africa (about 5 to 24 S latitude)—SC Dem. Rep. Congo (Kananga, AMNH series), N and C Angola (Crawford-Cabral, 1998; Hill and Carter, 1941; FMNH), N South Africa (Tzaneen Dist. of Limpopo; de Graaff, 1997bb), E Zimbabwe (de Graaff, 1997bb), Zambia (Ansell, 1978), Malawi (Ansell, 1989b; Ansell and Dowsett, 1988; Chitaukali et al., 2001), and Eastern Arc Mtns, NE Tanzania (Stanley et al., 1998b; Udzungwa Mtns, FMNH); limits unknown.

STATUS: IUCN – Lower Risk (lc).

SYNONYMS: *angolensis* Roberts, 1929; *bernardi* Lundholm, 1955; *longicaudatus* Roberts, 1913.

COMMENTS: A dark-furred and large-bodied relative of *D. melanotis*, as indicated by synapomorphic possession of a nail on the fifth hind digit (claw in all other *Dendromus*), an elongate front fifth digit (vestigial and scarely evident in others), and burrowing habit to construct underground nests (other species are better climbers and nest above ground); see Ansell, 1974b. The two comprise one of two species-pairs that commonly co-occur in southern Africa (Avery, 1998), in grassland and savanna woodland biomes at low and high altitudes (Chitaukali et al., 2001; Mugo et al., 1995). The report of *D. melanotis nyikae* from Ukerewe Isl in Lake Victoria (MCZ 26613; G. M. Allen and Loveridge, 1933) is an example of *D. mystacalis*.

Dendromus oreas Osgood, 1936. Field Mus. Nat. Hist., Zool. Ser., 20:236.
COMMON NAME: Cameroon African Climbing Mouse.
TYPE LOCALITY: Cameroon, Southwest Prov., southwest side of Mt Cameroon, 9000 ft (2743 m).
DISTRIBUTION: Mt Cameroon, Mts Kupe and Manenguba in the massif about 70 mi (113 km) NE Mt Cameroon (Eisentraut, 1963, 1968; Osgood, 1936; Rosevear, 1969), and possibly in mtns of extreme E Cameroon.
STATUS: IUCN – Vulnerable.
COMMENTS: Although described by Osgood (1936) as a species, *oreas* was later included in *D. mesomelas* (Bohmann, 1942; Misonne, 1974; Rosevear, 1969). Study of the holotype supports Osgood's (1936) view that *oreas* is related to what he called *D. lunaris* (=*D. nyasae*) and is not a geographic outlier of either the montane East African *D. insignis* or southern African *D. mesomelas* (see those accounts). Judged from the wide range in molar length that Rosevear (1969:472) recorded for *oreas*, his Cameroon sample is either extremely variable, which is unusual in the genus, or consists of two species.

Dendromus vernayi Hill and Carter, 1937. Am. Mus. Novit., 913:4.
COMMON NAME: Vernay's African Climbing Mouse.
TYPE LOCALITY: Angola, Angolan Plateau, Chitau, 4930 ft (1502 m).
DISTRIBUTION: Known only from the type locality (Hill and Carter, 1941; Crawford-Cabral, 1986a, 1998), EC Angola.
STATUS: IUCN – Critically Endangered.
COMMENTS: Hill and Carter (1941) described *vernayi* as a subspecies of *D. mesomelas*, an association which has not been questioned (Bohmann, 1942; Crawford-Cabral, 1986a; Misonne, 1974). Re-examination of the holotype and original series (in AMNH) reveals *vernayi* to be distinguished from *mesomelas* by its much smaller body size and strikingly shorter tail, buffy gray venter (white, often with pale buff in *D. mesomelas*), much smaller skull, postorbital region not constricted (as narrow as interorbit in *D. mesomelas*), shorter and wider rostrum, and longer molar rows relative to skull length. No morphological data supports its affiliation with *D. mesomelas* or with the much larger *D. insignis*. Externally, *D. vernayi* resembles *D. mystacalis*, which also occurs at Chitau, but that species is much smaller and recalls *D. mesomelas* in coloration and cranial conformation. While also morphologically distinct from *D. nyasae* and *D. oreas*, *D. vernayi* will likely prove to be phylogenetically allied to those species than to any other *Dendromus*.

Dendroprionomys F. Petter, 1966. Mammalia, 30:129.
TYPE SPECIES: *Dendroprionomys rousseloti* F. Petter, 1966.

Dendroprionomys rousseloti F. Petter, 1966. Mammalia, 30:129.
COMMON NAME: Velvet African Climbing Mouse.
TYPE LOCALITY: SE Republic of the Congo, Brazzaville, Zoological Gardens.
DISTRIBUTION: Republic of the Congo, recorded only from vicinity of Brazzaville.
STATUS: IUCN – Lower Risk (nt).
COMMENTS: A species combining features of *Dendromus* with those of *Prionomys* that is still known only by a few specimens (F. Petter, 1966b).

Malacothrix Wagner, 1843. *In* Schreber, Die Säugethiere, Suppl., 3:496.
TYPE SPECIES: *Otomys typicus* A. Smith, 1834.
SYNONYMS: *Otomys* Smith, 1834 [preoccupied by *Otomys* Cuvier, 1823, a murid].
COMMENTS: Morphologically highly distinctive compared with species of *Dendromus*, *Megadendromus*, and *Steatomys* (Denys et al., 1995). See Matthey (1967*a*) for chromosomal information and karyological contrasts with other dendromurine genera. Known from late Pliocene and early Pleistocene in southern Africa (Avery, 1998; Senut et al., 1992). Generally considered monotypic (Meester et al., 1986), but this view needs affirmation by systematic revision.

Malacothrix typica (A. Smith, 1834). S. Afr. Quart. J., 2:148.
COMMON NAME: Large-eared African Desert Mouse.
TYPE LOCALITY: South Africa, Eastern Cape Prov., Graaff Reinet Dist.
DISTRIBUTION: Eastern portion of Southern African Subregion in semidesert regions (mean annual rainfall = 150-500 mm) where sandy plains, short grassy velds, and karroid shrubs on hard substrates predominate; ranges in C and E South Africa (Eastern, Western and Northern Cape Provs., Free State, Northwest Prov.), S Botswana (Smithers, 1971), most of Namibia, and extreme SW Angola (Crawford-Cabral, 1998).
STATUS: IUCN – Lower Risk (lc).
SYNONYMS: *damarensis* Roberts, 1932; *egeria* Thomas, 1926; *fryi* Roberts, 1917; *harveyi* Roberts, 1951; *kalaharicus* Roberts, 1932; *molopensis* Roberts, 1933.
COMMENTS: Reviewed by Skinner and Smithers (1990) and de Graaff (1997*aa*). The species is now absent from KwaZulu-Natal in E South Africa but was present until about 60,000 years ago (Avery, 1991). Closely related to the extinct *M. makapani*, early Pleistocene Swartkrans cave sediments in South Africa (Avery, 1998).

Megadendromus Dieterlen and Rupp, 1978. Z. Säugetierk., 43:129.
TYPE SPECIES: *Megadendromus nikolausi* Dieterlen and Rupp, 1978.
COMMENTS: A spectacular dendromurine that resembles a giant *Dendromus*, having derived molar traits that influenced Dieterlen and Rupp (1978) to pose close relationship between Dendromurinae and Murinae. Denys et al. (1995), however, showed *Megadendromus* to be part of a clade that includes *Dendromus*, *Steatomys*, and *Malacothrix*, but excludes murines.

Megadendromus nikolausi Dieterlen and Rupp, 1978. Z. Säugetierk., 43:131.
COMMON NAME: Nikolaus's African Climbing Mouse.
TYPE LOCALITY: Ethiopia, S Goba, Bale Mtns.
DISTRIBUTION: Bale Mtns and Mt Chilalo in Ethiopia in ericaceous scrub moorland, 3000-3800 m.
STATUS: IUCN – Vulnerable.
COMMENTS: Known by less than ten specimens (Demeter and Topal, 1982; Dieterlen and Rupp, 1978; Denys et al., 1995; MCZ 57324). Morphology and ecology covered in detail by Dieterlen and Rupp (1978); Demeter and Topal (1982) provided additional habitat information. Reviewed as another Ethiopian endemic by Yalden and Largen (1992) and Yalden et al. (1996).

Prionomys Dollman, 1910. Ann. Mag. Nat. Hist., ser. 8, 6:226.
TYPE SPECIES: *Prionomys batesi* Dollman, 1910.
COMMENTS: Dollman's (1910) generic description is excellent and highlights its many unique traits among dendromurines. The interorbital and postorbital conformation, shape and size of incisive foramina, and molar patterns are similar to *Dendroprionomys*. A cladistic analysis based primarily on dental traits excludes *Prionomys* and *Dendroprionomys* from dendromurines (Denys et al., 1995), but this hypothesis should be tested with other data.

Prionomys batesi Dollman, 1910. Ann. Mag. Nat. Hist., ser. 8, 6:228.
COMMON NAME: Dollman's African Tree Mouse.
TYPE LOCALITY: Cameroon, Ja River, Bitye, 2000 ft (610 m).
DISTRIBUTION: W (55 km NE Obala, AMNH 241344) and S Cameroon, and S Central African Republic; limits unknown.
STATUS: IUCN – Lower Risk (nt).

COMMENTS: F. Petter (1964, 1966b) discussed *P. batesi* from Central African Republic, and contrasted its morphology with that of *Dendroprionomys*.

Steatomys Peters, 1846. Bericht Verhandl. K. Preuss. Akad. Wiss. Berlin, 11:258.
    TYPE SPECIES: *Steatomys pratensis* Peters, 1846.
    COMMENTS: At least 18 species were recognized (G. M. Allen, 1939; Ellerman, 1941) until Coetzee (1977a) provisionally acknowledged only three. Based upon study of relevant literature and museum specimens, we identify eight, but relationship between the three West African forms and the five outside that region has yet to be resolved and the widely distributed *S. pratensis* and *S. parvus* need systematic revision. Rosevear (1969) provided an excellent exposition of the West African species, and Swanepoel and Schlitter (1978) produced a more comprehensive revision of them. Still, their definitions are unsatisfactory and distributional limits unresolved; Coetzee (1977a), e.g., included some West African species recognized by Swanepoel and Schlitter (1978) in either *S. pratensis* or *S. parvus*. Fossil history extends to the Pliocene (Avery, 1998, 2000; Denys, 1987a) and late Miocene (Mein et al., 2004; Senut et al., 1992).

Steatomys bocagei Thomas, 1892. Ann. Mag. Nat. Hist. ser. 6, 10:264.
    COMMON NAME: Bocage's African Fat Mouse.
    TYPE LOCALITY: Angola, Caconda.
    DISTRIBUTION: Moist savannas and woodlands from the Central Plateau of Angola (Crawford-Cabral, 1998; Hill and Carter, 1941) eastward through NE Angola to SC Dem. Rep. Congo at Luluabourg (= Kananga; type series of *kasaicus* in AMNH, see Hatt, 1934); eastern limits unresolved.
    SYNONYMS: *kasaicus* Hatt, 1934.
    COMMENTS: A distinctive species characterized by very large body size and four pairs of teats (pectoral, postaxillary, and two inguinal). Treated as a species by Hill and Carter (1941) and Ellerman et al. (1953), but synonymized with *S. pratensis* by Coetzee (1977a) and listed that way in most contemporary faunal accounts. Crawford-Cabral (1998) recognized the specific integrity of *bocagei*, and explained that it has only four pairs of teats in contrast to *S. pratensis* with five pairs (sometimes more; see Coetzee, 1977a). That distinguishing trait, its very large body size compared to *S. pratensis* from southern and eastern Africa, and study of the large series reported by Hill and Carter (1941) identify a species separate from *S. pratensis* or the large-bodied *S. opimus*.
        Crawford-Cabral identified the large-bodied *Steatomys* from NE Angola as *S. pratensis kasaicus* (teat number not counted); Coetzee (1977a), however, included *kasaicus* in *bocagei*, the correct allocation. Skins in the type series of *kasaicus* (AMNH 85073, 86080) possess only four pairs of teats; this pattern, their large body and cranial dimensions, and geographic continuity with Angolan populations attest to their identification as *S. bocagei*. Coetzee claimed intergradation between *bocagei* and *pratensis* in the Katanga region of E Dem. Rep. Congo, but no published data supports that contention. Furthermore, typical *S. pratensis* has been collected in S Angola and those examples exhibit no trace of morphological intergradation with *S. bocagei*. Relationship with the other large-bodied species of *Steatomys*—*S. jacksoni* from West Africa, number of teats unknown; or *S. opimus* from northern savannas, five or more pairs of teats—awaits resolution. Rosevear (1969) regarded West African *S. jacksoni* to closely resemble *S. bocagei* in body size and pelage coloration.

Steatomys caurinus Thomas, 1912. Ann. Mag. Nat. Hist., ser. 9, 9:271.
    COMMON NAME: Northwestern African Fat Mouse.
    TYPE LOCALITY: Nigeria, Panyam, 4000 ft (1219 m).
    DISTRIBUTION: West Africa, Senegal (Duplantier and Granjon, 1992), through S Mali (Meinig, 2000) and Ghana (Grubb et al., 1998), to C Nigeria; range limits unknown.
    STATUS: IUCN – Lower Risk (lc).
    SYNONYMS: *roseveari* Swanepoel and Schlitter, 1978.
    COMMENTS: Listed as a subspecies of *S. pratensis* by Coetzee (1977a), but documented as species by Rosevear (1969) and Swanepoel and Schlitter (1978). The relationship between *S. caurinus* and *S. pratensis* from southern and East Africa warrants inquiry.

Steatomys cuppedius Thomas and Hinton, 1920. Novit. Zool., 27:318.

COMMON NAME: Dainty African Fat Mouse.

TYPE LOCALITY: Nigeria, Farniso (= Panisau), near Kano, 1700 ft (518 m).

DISTRIBUTION: West Africa, in Senegal (Duplantier and Granjon, 1992), NC Nigeria, and SC Niger; range unresolved.

STATUS: IUCN – Lower Risk (nt).

COMMENTS: Revised by Swanepoel and Schlitter (1978), who regarded *cuppedius* as a distinct species, as did Rosevear (1969), not a member of *S. parvus* as per Coetzee (1977a). Thomas and Hinton (1923a) thought the holotype of *aquilo* from EC Sudan was related to *S. cuppedius*. Whether West African *cuppedius* represents a separate species or segment of the widely ranging *S. parvus* has yet to be resolved. Specimens identified as *S. cuppedius* from SE Ghana by Yeboah (1982-1988) are examples of *Uranomys ruddi* (Grubb et al., 1998:189).

Steatomys jacksoni Hayman, 1936. Proc. Zool. Soc. Lond., 1935:930 [1936].

COMMON NAME: Jackson's African Fat Mouse.

TYPE LOCALITY: Ghana, Ashanti, Wenchi.

DISTRIBUTION: Spottily recorded from the type locality (Grubb et al., 1998), possibly Togo (Rosevear, 1969), and SW Nigeria (Anadu, 1979); limits indeterminate.

STATUS: IUCN – Vulnerable.

COMMENTS: Included in *S. pratensis* by Coetzee (1977a), but provisionally retained as species by Rosevear (1969) and Swanepoel and Schlitter (1978) until significance of the diagnostic character (size and shape of interparietal) is assessed by additional specimens and other information. The relationship between *S. jacksoni* (females not yet collected) and the other large-bodied species, *S. bocagei* and *S. opimus*, needs to be examined.

Steatomys krebsii Peters, 1852. Reise nach Mossambique, Säugethiere, p. 165.

COMMON NAME: Kreb's African Fat Mouse.

TYPE LOCALITY: South Africa, Kaffraria.

DISTRIBUTION: Patchy distribution in Southern African Subregion—C and SW Angola (Crawford-Cabral, 1998), NE Namibia (Caprivi Strip and vicinity; specimens in USNM), N Botswana (Smithers, 1971), W Zambia (Ansell, 1978), and South Africa (Western, Eastern and Northern Cape Provs., E Free State, Northwest Prov., and NW KwaZulu-Natal).

STATUS: IUCN – Lower Risk (lc).

SYNONYMS: *angolensis* Hill and Carter, 1937; *bensoni* Ansell, 1958; *bradleyi* Hill and Carter, 1937; *chiversi* Roberts, 1931; *mariae* Ansell, 1958; *orangiae* Roberts, 1929; *pentonyx* (Sclater, 1899); *tongensis* Roberts, 1931; *transvaalensis* Roberts, 1929.

COMMENTS: A distinctive small-bodied species with a long tail and small hind feet; reviewed by Coetzee (1977a), Meester et al. (1986), and Ansell (1978). The holotypes of *angolensis* and *bradleyi* (in AMNH) are unproblematic examples of *S. krebsii*. Populations in KwaZulu-Natal Prov., E South Africa, reviewed by P. J. Taylor (1998), who transferred *tongensis* from *S. parvus* to *S. krebsii*. De Graaff (1997hh) mapped the spotty distribution but suspected that it reflected lack of focused collection.

Steatomys opimus Pousargues, 1894. Bull. Soc. Zool. De France, 19:131.

COMMON NAME: Pousargues's African Fat Mouse.

TYPE LOCALITY: Central African Republic, near Balao, Dakoas Dist., 05°26'N., 17°40'E.

DISTRIBUTION: Moist savanna and forest margins from Cameroon, through S Central African Republic and N Dem. Rep. Congo (Faradje and Niangara per AMNH specimens), to extreme SW Sudan.

SYNONYMS: *gazellae* Thomas and Hinton, 1923.

COMMENTS: Closely similar to the large-bodied *S. bocagei* in body and cranial dimensions, but with five pairs of teats (abdominal pair in addition to the other four; AMNH specimens from Faradje and Niangara). Included in *S. pratensis* by Coetzee (1977a), but *opimus* is much larger in body and cranial dimensions than samples from eastern and southern Africa. Hatt (1940a) redescribed *S. opimus* as a species based on AMNH series, emphasized traits that distinguish it from other *Steatomys*, and explained that the holotype of *gazellae* is another example of it. Status of *S. opimus* with respect to the large-bodied West African *S. jacksoni* requires resolution.

Steatomys parvus Rhoads, 1896. Proc. Acad. Nat. Sci. Philadelphia, p. 529.

COMMON NAME: Tiny African Fat Mouse.

TYPE LOCALITY: Ethiopia, northern shore Lake Turkana, Rusia (see Yalden et al., 1996).

DISTRIBUTION: East and Southern Africa—S and EC Sudan (holotypes of *aquilo* and *thomasi*; USNM specimens), S Ethiopia, and Somalia; south through Kenya (Hollister, 1919; specimens in AMNH), Uganda (Delany, 1975; AMNH 119143), and Tanzania (Swynnerton and Hayman, 1951); to SW Angola (Coetzee, 1977*a*; Crawford-Cabral, 1998; AMNH 81949), NE Namibia, NW Botswana (Smithers, 1971), W Zambia (Ansell, 1978), and W Zimbabwe (Smithers and Wilson, 1979).

STATUS: IUCN – Lower Risk (lc).

SYNONYMS: *aquilo* Thomas and Hinton, 1923; *athi* Heller, 1910; *kalaharicus* Roberts, 1932; *loveridgei* Thomas, 1919; *minutus* Thomas and Wroughton, 1905; *muanzae* Kershaw, 1923; *swalius* Thomas, 1926; *thomasi* Setzer, 1956; *umbratus* Thomas, 1926.

COMMENTS: Distribution and conservation status discussed as a savanna woodland species by Mugo et al. (1995). The single Ethiopian record was first listed as *S. pratensis* (Yalden et al., 1976) and later reidentified as *S. parvus* (Yalden et al., 1996). Coetzee (1977*a*) omitted *thomasi* from his review; Setzer's (1956) large type series matches the range of variation in what is now regarded as *S. parvus*. Coetzee (1977*a*) included the holotype and only specimen of *aquilo* (EC Sudan, Jebel Marra) in *S. parvus*, although Thomas and Hinton (1923*a*) believed it to be related to West African *S. cuppedius*. Southern African records reviewed and mapped by de Graaff (1997*gg*), who included an isolate in KwaZulu-Natal, but P. J. Taylor (1998) reallocated those specimens to *S. krebsii*. *Steatomys parvus* has not been found south of NE Namibia and NW Botswana, nor have we located any samples between S Sudan and the eastern range limits of West African *S. cuppedius* (see account).

Steatomys pratensis Peters, 1846. Bericht Verhandl. K. Preuss. Akad. Wiss. Berlin, 11:258.

COMMON NAME: Common African Fat Mouse.

TYPE LOCALITY: Mozambique, Zambezi River, Tete.

DISTRIBUTION: Southern and East Africa—S Angola (Carter and Hill, 1941) and N Namibia; eastward through N Botswana (Smithers, 1971), Zimbabwe (Smithers and Wilson, 1979), N South Africa (Limpopo, Mpumalanga, Gauteng, and N KwaZulu-Natal; de Graaff, 1997*ff*; P. J. Taylor, 1998), and Mozambique (Smithers and Lobao Tello, 1976); north through Zambia (Ansell, 1978), Malawi (Ansell and Dowsett, 1988; Chitaukali et al., 2001; Lawrence and Loveridge, 1953), and Tanzania (AMNH material) to EC Ethiopia (Demeter, 1982).

STATUS: IUCN – Lower Risk (lc).

SYNONYMS: *edulis* Peters, 1852; *leucorhynchus* Hill and Carter, 1937; *maunensis* Roberts, 1932; *natalensis* Roberts, 1929; *nyasae* Lawrence and Loveridge, 1953.

COMMENTS: Formerly included *bocagei* and *opimus* (see those accounts). Demeter (1982) identified the lone Ethiopian record (Sabober Plains, Awash National Park) as *S. pratensis* because of their large skull size compared with *S. parvus*; his range of values match those for *S. pratensis* in the AMNH, not the larger *S. opimus* which occurs no closer than extreme SW Sudan. Although *leucorhynchus* is usually included in *S. krebsii* (Coetzee, 1977*a*; Meester et al., 1986), Crawford-Cabral (1998) provisionally recognized it as a species because relative ear length of the holotype (Angola, Capelongo) is more similar to *S. pratensis*; he also suggested that *leucorhynchus* may prove to be another geographic sample of the latter. After studying the holotype of *leucorhynchus* and others from Capelongo (AMNH 85815, 85816, 86976) in S Angola, we agree with that assessment. We have not seen material between Ethiopia and the West African *S. caurinus*; that region is mostly occupied by the much larger-bodied *S. opimus*. Identified from the early Pleistocene of South Africa based on isolated molars (Avery, 1998).

Subfamily Mystromyinae Vorontsov, 1966. Zool. Zh., 45:437.

SYNONYMS: Mystromyini Vorontsov, 1966 (Mystromyinae Lavocat, 1973).

COMMENTS: Phylogenetic allocation of *Mystromys* so puzzled Ellerman (1941:445) that he wrote "I am entirely at a loss to suggest the relationships of this genus, which seems not only isolated from the Palaearctic and Neotropical genera, but to have no marked generic characters." Based on a comprehensive study of morphological characters, Vorontsov (1966) concluded that *M. albicaudatus* is not closely related to Palearctic hamsters and placed it in the monotypic tribe Mystromyini. Carleton and Musser (1984:313) remarked that *Mystromys* is possibly a survivor of an ancient phyletic line, pointing out that Lavocat (1973, 1978) "raised such a novel possibility for *Mystromys* by noting its probable derivation from afrocricetodontine rodents and by recognizing the subfamily Mystromyinae in the Nesomyidae, a family composed of archaic African cricetids that he derived from the afrocricetodontines, the African counterpart to European and Asian cricetodontines." Pocock (1987), Skinner and Smithers (1990), and Denys (1991) retained *Mystromys* in the Cricetinae, family Cricetidae. Wessels (1996), based upon molar patterns, regarded *Mystromys* and *Calomyscus* as members of Myocricetodontinae, which she later arranged within Gerbillidae (Wessels, 1998), an alliance endorsed by Mein et al. (2000*b*).

No other morphological data suggest such a close relationship with either gerbils or *Calomyscus* (Vorontsov, 1966), and molecular data also contradict either allocation. In a phylogenetic analysis of two nuclear protein-coding genes (Michaux et al., 2001*b*), *Mystromys* is arrayed with representatives of Madagascan Nesomyinae and African Cricetomyinae and Dendromurinae, a clade completely separate from that containing *Calomyscus* and another circumscribing Gerbillinae, Deomyinae, and Murinae. According to the mitochondrial gene cytochrome *b*, *Mystromys* is similarly cladistically removed from *Calomyscus* and instead is sister genus to *Petromyscus* (Jansa et al., 1999). The ancestral stock of *Mystromys* and its close relative *Proodontomys* is yet to be identified, although Tong and Jaeger (1993) speculated that *Mystromys* was derived from *Democricetodon*, identified from the middle Miocene of East Africa. Mein et al. (2000*b*), however, disputed that origin and hypothesized that *Mioharimys*, late Miocene of Namibia, represents the ancestral group from which living *Mystromys* evolved.

Mystromys Wagner, 1841. Gelehrte Anz. I. K. Bayer. Akad. Wiss., München, 12(54), col. 434.
    TYPE SPECIES: *Mystromys albipes* Wagner, 1841 (= *Otomys albicaudatus* A. Smith, 1834).
    COMMENTS: Gross morphology of male accessory glands described by Voss and Linzey (1981). Gastric anatomy, development, and physiology thoroughly investigated and considered with respect to systematics, function, and diet (Maddock and Perrin, 1981, 1983; Perrin and Curtis, 1980; Perrin and Maddock, 1983). Spermatozoal morphology described by Breed (1995*d*), who noted that sperm head ultrastructure generally resembles that of the cricetine *Mesocricetus*.

Mystromys albicaudatus (A. Smith, 1834). S Afr. Quart. J., 2:148.
    COMMON NAME: African White-tailed Rat.
    TYPE LOCALITY: S Africa, Eastern Cape Prov., Albany Dist.
    DISTRIBUTION: South Africa (relict population in Western Cape Prov., lowlands of Eastern Cape Prov., and highveld of Free State, Gauteng and Mpumalanga Provinces, and NW KwaZulu-Natal), Lesotho, and S Swaziland.
    STATUS: IUCN – Endangered.
    SYNONYMS: *albicaudatus* (Desmarest, 1822) [Meester et al. (1986) described origin and possible status of this name]; *albicaudatus* (Smith, 1834); *albipes* (Wagner, 1841); *antiquus* Broom, 1948 [probably *nomen nudum*; see Avery, 1998]; *fumosus* Thomas and Schwann, 1905); *hauslichtneri* Broom 1937; *lanuginosa* (Lichtenstein, 1842).
    COMMENTS: Reviewed by de Graaff (1997*z*), Meester et al. (1986), and Skinner and Smithers (1990). The taxon *longicaudatus* was described as a species of *Mystromys* (Noack, 1887) and listed that way by G. M. Allen (1939), but Misonne (1966) determined the holotype to be an immature example of *Mastomys natalensis*, an identification accepted by de Graaff (1981) and Meester et al. (1986), and verified by multivariate analysis (Van der Straeten and Robbins, 1997, who noted the specimen is a young adult). See Meester et al. (1986) for type localities and taxonomic references of all described forms; Grubb (2001; see also Opinion 2005 of the International Commission

on Zoological Nomenclature, 2002*b*) suggested that *albicaudatus* Geoffroy, 1803, is a senior homonym of *Otomys albicaudatus* A. Smith, 1834, a possibility that should be formally resolved. Found in fynbos, succulent Karoo, nama-karoo, grassland, arid savanna, and savanna woodland (Mugo et al., 1995).

Fossil relatives include *Mystromys pocockei*, and *Proodontomys cookei*, represented from Pliocene-Pleistocene australopithecine sites in South Africa (Avery, 1998; Denys, 1991; Pocock, 1987). Late Pleistocene fossils had been identified as the extinct M. *hausleitneri*, but Lavocat (1956) doubted that they could be separated from extant M. *albicaudatus*, a synonymy that Avery (1995, 1998) substantiated with new material from Gladysvale and Swartkrans caves. *Proodontomys cookei* has been collected with M. *albicaudatus* in those late Pleistocene sediments, and while the latter still occurs in South Africa, the former apparently became extinct between 1 and 0.7 million years ago (Avery, 1998). "*Mystromys*" also recorded from the Pleistocene of Namibia (Senut et al., 1992).

Subfamily Nesomyinae Major, 1897. Proc. Zool. Soc. Lond., 1897:718.
SYNONYMS: Brachytarsomyes Ellerman, 1941; Brachyuromyes Ellerman, 1941; Eliuri Ellerman, 1941; Gymnuromyinae Ellerman, 1941; Nesomyinae Major, 1897 (Nesomyidae Tullberg, 1899; Nesomyini Vorontsov, 1959).
COMMENTS: Emended definition—Muroid rodents in which the jugal is large, spanning most of the middle zygomatic arch (Major, 1897); the tongue retains three circumvallate papillae (Tullberg, 1899; Vorontsov, 1967); and the enamel face of the lower incisors bears two low and inconspicuous ridges, close-set in parallel and positioned mediolaterally (confirmed by MDC in all nine genera).

Group exceedingly diverse morphologically (see Carleton and Musser, 1984:341-2), but the above characters, in combination, uniquely define nesomyines among assemblages of living muroids. Jugal size and number of circumvallate papillae are probably primitive conditions, but the acquisition of various lower incisor ornamentations is thought derived (Flynn et al., 1985; L. D. Martin, 1980). Similar longitudinal incisor ridges are also found in certain cricetomyines among extant groups, and in the extinct Miocene African genera *Afrocricetodon*, *Notocricetodon*, and *Protarsomys* (Flynn et al., 1985; Lavocat, 1973). The morphological diversity of Madagascar's native rodents has questioned their monophyletic origin and prompted numerous classificatory arrangements (see discussions in Carleton and Musser, 1984; Jansa and Carleton, 2003*a*). Proponents of a single ancestral origin have usually arranged nesomyines as a subfamily or tribe within Cricetidae (e.g., Miller and Gidley, 1918; Simpson, 1945; Vorontsov, 1959) or as a subfamily within a broadly defined Nesomyidae, which includes African groups such as cricetomyines, tachyoryctines, and *Mystromys* (Chaline et al., 1977; Lavocat, 1978). Ellerman (1941, 1949) was convinced that nesomyines are polyphyletic and dispersed the genera among five subfamilies of Muridae *sensu lato*.

Recent studies of DNA-DNA hybridization and genetic sequence data, although not finally resolving all aspects of the debate, persuasively demonstrate that Ellerman's radical classificatory treatment is untenable. Surveys of few nesomyine taxa, using DNA hybridization or nuclear genes, have supported monophyly of Nesomyinae (DuBois et al., 1996; Michaux and Catzeflis, 2000; Michaux et al., 2001*b*); whereas, results based on dense taxonomic sampling, using a mitochondrial gene, have disclosed paraphyly, although most nesomyine genera were associated closely within two clades (Jansa et al., 1999). According to phylogenetic interpretations of nuclear genes, Michaux et al. (2001*b*) depicted Nesomyinae (two genera sampled) as a basal clade with other African groups (Cricetomyinae, Dendromurinae, Mystromyinae), a result that supplies some empirical support for Lavocat's (1973, 1978) concept of a family Nesomyidae that embraces the archaic remnants of an African radiation dating to the early Miocene or late Oligocene (also see Carleton and Musser, 1984:344).

Living members endemic to Madagascar, known only from late Quaternary and subfossil Holocene deposits (McKenna and Bell, 1997). One Miocene occurrence, described from Africa, however, appreciably extends the fossil history of the group, if the identification proves true. Lavocat (1978) viewed *Protarsomys*, early Miocene of Kenya, as close to the ancestry of Malagasy Nesomyinae, and Chaline et al. (1977) emphasized that relationship by placing the Miocene fossil in synonymy under extant *Macrotarsomys*, as did F. Petter (1990). Others have questioned so close an affinity and specifically disputed their generic equivalence (Carleton and Goodman, 1996:246-249, 252). *Macrotarsomys* represents a morphology that evolved in situ, coincident with formation of the island's dry western landscapes. Whether *Protarsomys* is a distant phylogenetic relative to extant nesomyines must await additional critical review of Miocene taxa and-or future discoveries. *Brachyuromys*, otherwise known only from the Recent of Madagascar, was mistakenly recorded from the late Miocene of Namibia (Conroy et al., 1992), together with *Myocricetodon*, *Notocricetodon*, and *Protarsomys*, but this fossil was subsequently reidentified as an extinct spalacid genus, *Harasibomys* (Mein et al., 2000a). Thomas (1915) in erecting the replacement name *Majoria* (for *Myoryctes* Major, 1908) uncritically stated that it represented "a fossil Madagascan rodent," an association just as uncritically perpetuated by others (Simpson, 1945; Vorontsov, 1959; Carleton and Musser, 1984); the holotype is an innominate bone of *Plesiorycteropus*, a member of the extinct order Bibymalagasia (see MacPhee, 1994).

Ellerman (1949) provided the first critical synopsis of nesomyine taxa, with subsequent descriptions and taxonomic arrangements updated by F. Petter (1972c, 1975a). Resurgence in field and museum studies over the past decade has vastly improved taxonomic, distributional, and ecological knowledge of nesomyines (Carleton, 1994; Carleton and Goodman, 1996, 1998, 2000; Carleton and Schmidt, 1990; Goodman and Carleton, 1996, 1998; Goodman and Soarimalala, 2002; Goodman et al., 1996, 1999b; Ryan et al., 1993; Soarimalala et al., 2001), as summarized for all species in Goodman and Benstead (2003). Conservation concerns associated with introduced *Rattus* populations and their impact upon native rodents discussed by Goodman (1995).

Brachytarsomys Günther, 1875. Proc. Zool. Soc. Lond., 1875:79.
> TYPE SPECIES: *Brachytarsomys albicauda* Günther, 1875.
> COMMENTS: The incipient prismatic condition of the dentition persuaded Ellerman (1941) to classify *Brachytarsomys* as a primitive arvicoline. Specific discrimination and distributions reviewed by Carleton and Goodman (2003a).

Brachytarsomys albicauda Günther, 1875. Proc. Zool. Soc. Lond., 1875:80.
> COMMON NAME: White-tailed Antsangy.
> TYPE LOCALITY: Madagascar, between "Tamatave and Morondava;" type locality requires formal restriction, which may be possible with archival research because the species occurs only in the east.
> DISTRIBUTION: Rain forest at middle elevations, E Madagascar.
> STATUS: IUCN – Lower Risk (lc).

Brachytarsomys villosa F. Petter, 1962. Mammalia, 26:570.
> COMMON NAME: Hairy-tailed Antsangy.
> TYPE LOCALITY: Description based on a specimen held in captivity in the "Vivarium de Tsimbazaza;" type locality requires formal restriction.
> DISTRIBUTION: So far recorded only from the western slopes of the Anjanaharibe-Sud Massif, N Madagascar.
> COMMENTS: Named as a subspecies by F. Petter (1962a) based on a zoo specimen, but Carleton and Schmidt (1990) suggested that *villosa* is a distinct species. Rediscovered in the wild by S. M. Goodman et al. (2001b) and contrasted to *B. albicauda*; distributional limits unknown.

Brachyuromys Major, 1896. Ann. Mag. Nat. Hist., ser. 6, 18:322.
> TYPE SPECIES: *Brachyuromys ramirohitra* Major, 1896.
> COMMENTS: Ellerman (1941) allied *Brachyuromys* as a tribe within Tachyoryctinae, an affinity earlier considered plausible by Major (1897). Cladistic interpretation of cytochrome *b* sequences instead indicates close kinship with *Nesomys* (Jansa et al., 1999). Specific discrimination and distributions reviewed by Jansa and Carleton (2003b).

Brachyuromys betsileoensis (Bartlett, 1880). Proc. Zool. Soc. Lond., 1879:770 [1880].
> COMMON NAME: Lesser Short-tailed Rat.

TYPE LOCALITY: Madagascar, "S.E. Betsileo."

DISTRIBUTION: Central Highlands and its eastern fringes, ca. 900-2450 m, Madagascar.

STATUS: IUCN – Lower Risk (nt).

COMMENTS: Described as a species of *Nesomys* but referred to *Brachyuromys* by Major (1896). Occurs sympatrically with *B. ramirohitra* (Goodman and Rasolonandrasana, 2001).

Brachyuromys ramirohitra Major, 1896. Ann. Mag. Nat. Hist., ser. 6, 18:323.

COMMON NAME: Greater Short-tailed Rat.

TYPE LOCALITY: Madagascar, Fianarantsoa Prov., 6 hours SE Fandriana, Ampitambe forest.

DISTRIBUTION: Isolated records, ca. 900-1990 m, in highlands of N and C Madagascar.

STATUS: IUCN – Lower Risk (nt).

COMMENTS: Extent poorly documented, but recent records paint a broader distribution in the Central Highlands (Soarimalala et al., 2001) as well as its presumably isolated occurrence in the Northern Highlands (Jansa and Carleton, 2003*b*).

Eliurus Milne-Edwards, 1885. Ann. Sci. Nat., Zool. Paleontol. (Paris), 20: Art. 1 bis.

TYPE SPECIES: *Eliurus myoxinus* Milne-Edwards, 1885.

COMMENTS: Genus arrayed as a monotypic tribe within Murinae by Ellerman (1941), an affinity confuted by molecular data (Dubois et al., 1996; Jansa et al., 1999). Ellerman (1949) set the precedent for recognizing only two species, the large polytypic *E. myoxinus* and the small monotypic *E. minor*, an arrangement later followed by F. Petter (1972*c*, 1975*a*) and others (Corbet and Hill, 1991; Honacki et al., 1982). Revised by Carleton (1994), who recognized 8 species and summarized their morphological identification and distributions; interspecific relationships based on cladistic interpretation of cytochrome *b* sequences studied by Jansa et al. (1999). Review of specific taxonomy and distributions, along with key to identification, provided by Carleton (2003).

Eliurus antsingy Carleton, Goodman, and Rakotondravony, 2001. Proc. Biol. Soc. Wash., 114:974.

COMMON NAME: Tsingy Tufted-tailed Rat.

TYPE LOCALITY: Madagascar, Toliara Prov., Antsingy Forest, near Bekopaka; about 19°07.5'S, 44°49.0'E.

DISTRIBUTION: So far known from only two localities in dry forest peculiar to karst topography (tsingy), W Madagascar.

COMMENTS: Specific relationships unclear; although found in W Madagascar, *E. antsingy* is apparently not closely related to the more widely distributed western species, *E. myoxinus* (Carleton et al., 2001).

Eliurus ellermani Carleton, 1994. Am. Mus. Novit., 3087:39.

COMMON NAME: Ellerman's Tufted-tailed Rat.

TYPE LOCALITY: Madagascar, Toamasina Prov., near Hiaraka, about 18 km ESE Maroantsetra, 850 m; about 15°30'S, 49°56'E (as amended by Carleton and Goodman, 1998:181).

DISTRIBUTION: Known only from two widely separated localities in NE Madagascar.

COMMENTS: Status uncertain; provisionally retained as distinct from the morphologically similar *E. tanala* (see Carleton and Goodman, 1998).

Eliurus grandidieri Carleton and Goodman, 1998. Fieldiana Zool., N.S., 90:165.

COMMON NAME: Grandidier's Tufted-tailed Rat.

TYPE LOCALITY: Madagascar, Antsiranana Prov., Reserve Speciale d'Anjanaharibe-Sud, 11 km WSW Befingitra, 1550 m; 14°44.5'S, 49°27.5'E.

DISTRIBUTION: Middle to upper montane forest, 1250-1875 m, in the Northern Highlands and northern portion of the Central Highlands, E Madagascar.

COMMENTS: Certain morphological features suggest that the species is most closely related to *E. petteri*.

Eliurus majori Thomas, 1895. Ann. Mag. Nat. Hist., ser. 6, 16:164.

COMMON NAME: Major's Tufted-tailed Rat.

TYPE LOCALITY: Madagascar, Fianarantsoa Prov., Ambohimitambo forest, 4500 ft (1372 m).

DISTRIBUTION: Middle to upper montane forest, 1000-2000 m, E Madagascar, from Montagne d'Ambre in the north to the Anosyenne Mountains in the south.

STATUS: IUCN – Endangered.

COMMENTS: Morphologically similar to and presumably closely related with *E. penicillatus* (Carleton, 1994).

Eliurus minor Major, 1896. Ann. Mag. Nat. Hist., ser. 6, 18:462.
COMMON NAME: Lesser Tufted-tailed Rat.
TYPE LOCALITY: Madagascar, Fianarantsoa Prov., Ampitambe.
DISTRIBUTION: Humid forests of E Madagascar, near sea level to 1875 m.
STATUS: IUCN – Lower Risk (lc).
COMMENTS: Co-occurs regionally with most other species of *Eliurus*. Extensive size variation and broad altitudinal occurrence suggest a species composite.

Eliurus myoxinus Milne-Edwards, 1885. Ann. Sci. Nat., Zool. Paleontol. (Paris), 20: Art. 1 bis.
COMMON NAME: Western Tufted-tailed Rat.
TYPE LOCALITY: Madagascar, Toliara Prov., Tsilambana.
DISTRIBUTION: Dry deciduous forest and xerophilous habitats, W Madagascar, from the vicinity of Ankarafantsika to the island's southernmost tip (Carleton et al., 2001: Fig. 3).
STATUS: IUCN – Lower Risk (lc).
COMMENTS: Although endemic to western biotopes, the range of *E. myoxinus* closely approaches that of eastern species in the extreme south (Carleton et al., 2001; Goodman et al., 1999). Sister species of *E. minor* according to phylogenetic interpretation of mitochondrial DNA sequences (Jansa et al., 1999).

Eliurus penicillatus Thomas, 1908. Ann. Mag. Nat. Hist., ser. 8, 2:453.
COMMON NAME: White-tipped Tufted-tailed Rat.
TYPE LOCALITY: Madagascar, Fianarantsoa Prov., Ampitambe.
DISTRIBUTION: Known only from the type locality.
STATUS: IUCN – Critically Endangered.
COMMENTS: Morphologically like *E. majori* except for possession of a white-tipped caudal tuft. Tentatively retained as a species by Carleton (1994); specific status maintained based on morphometric comparisons with *E. majori* and *E. tanala* by Carleton and Goodman (2000).

Eliurus petteri Carleton, 1994. Am. Mus. Novit., 3087:37.
COMMON NAME: Petter's Tufted-tailed Rat.
TYPE LOCALITY: Madagascar, Toamasina Prov., 8 km from Fanovana.
DISTRIBUTION: Restricted region in foothills, 450-530 m, EC Madagascar.
COMMENTS: To date, the recent biological surveys have not uncovered a wider distribution of *E. petteri* in eastern forests.

Eliurus tanala Major, 1896. Ann. Mag. Nat. Hist., ser. 6, 18:462.
COMMON NAME: Tanala Tufted-tailed Rat.
TYPE LOCALITY: Madagascar, Fianarantsoa Prov., 30 mi (48 km) S Fianarantsoa, near Vinanitelo.
DISTRIBUTION: Middle elevations in humid forest, 775-1625 m, from Northern Highlands to the S Anosyenne Mountains, E Madagascar.
STATUS: IUCN – Lower Risk (lc).
COMMENTS: Distribution overlaps that of its morphologically similar congener *E. webbi* at lower elevations. Geographic variation evaluated by Carleton and Goodman (1998).

Eliurus webbi Ellerman, 1949. Families and Genera of Living Rodents, 3(App. II):163.
COMMON NAME: Webb's Tufted-tailed Rat.
TYPE LOCALITY: Madagascar, Fianarantsoa Prov., 20 mi (32 km) S Farafangana.
DISTRIBUTION: Lowland rainforest, sea level-1150 m, E Madagascar, from Montagne d'Ambre in the north to southernmost reaches of the Anosyenne and Vohimena Mountains, including littoral forests in the extreme southeast.
STATUS: IUCN – Lower Risk (nt).
COMMENTS: Geographic variation evaluated by Carleton and Goodman (2000). Sister species of *E. tanala* according to phylogenetic interpretation of mitochondrial DNA sequences (Jansa et al., 1999).

Gymnuromys Major, 1896. Ann. Mag. Nat. Hist., ser. 6, 18:324.
TYPE SPECIES: *Gymnuromys roberti* Major, 1896.
COMMENTS: Ellerman (1941) created the subfamily Gymnuromyinae in recognition of the distinctive features of this species, which he believed to be derived from a *Nesomys*-like ancestor. Molecular data fail to support this independent status and instead generally relate the genus to other nesomyines (Jansa et al., 1999). Contrasts with *Eliurus* and distribution reviewed by Carleton and Goodman (2003*b*).

Gymnuromys roberti Major, 1896. Ann. Mag. Nat. Hist., ser. 6, 18:324.
COMMON NAME: Voalavoanala.
TYPE LOCALITY: Madagascar, Fianarantsoa Prov., Ampitambe forest.
DISTRIBUTION: E Madagascar humid forest, 500-1625 m, from the Northern Highlands, along the eastern flank of the central highlands, to the southern limits of the Anosyenne Mountains.
STATUS: IUCN – Vulnerable.
COMMENTS: Some have voiced concern that populations of *G. roberti* are being supplanted by introduced *Rattus* (see Carleton and Schmidt, 1990; Goodman, 1995).

Hypogeomys A. Grandidier, 1869. Rev. Mag. Zool. Paris, ser. 2, 21:338.
TYPE SPECIES: *Hypogeomys antimena* A. Grandidier, 1869.
COMMENTS: Retained by Ellerman (1941) within Cricetinae as a genus of obscure relationships. Sister genus to the *Macrotasomys-Monticolomys* clade according to phylogenetic assessment of mitochondrial DNA sequences (Jansa et al., 1999). The fossil *Hypogeomys boulei* G. Grandidier (1912) proved to be an example of *Plesiorycteropus*, member of the extinct order Bibymalagasia (see MacPhee, 1994). Ecology, behavior, and conservation status reviewed by Sommer (2003).

Hypogeomys antimena A. Grandidier, 1869. Rev. Mag. Zool. Paris, ser. 2, 21:339.
COMMON NAME: Votsovotsa.
TYPE LOCALITY: Madagascar, Toliara Prov., between the banks of the Tsiribihina and Andranomena Rivers.
DISTRIBUTION: Narrow coastal zone in deciduous dry forest with sandy soils, WC Madagascar.
STATUS: IUCN – Endangered.
COMMENTS: Although its present range is greatly restricted, *H. antimena* was more widespread in SW Madagascar until the late Holocene, 1400 years before present (Goodman and Rakotodravony, 1996). Ecology, breeding biology, and conservation status of extant populations addressed by Cook et al. (1991) and Sommer (1997, 2001). Most closely related to *H. australis*, known as subfossils from cave deposits in C and extreme S Madagascar, and documented as a valid species by Goodman and Rakotodravony (1996), who discussed its possible survival into the period of the island's early human colonization.

Macrotarsomys Milne-Edwards and G. Grandidier, 1898. Bull. Mus. Hist. Nat. Paris, ser. 1, 4:179.
TYPE SPECIES: *Macrotarsomys bastardi* Milne-Edwards and G. Grandidier, 1898.
COMMENTS: Retained within Cricetinae by Ellerman (1941) but more closely related to other nesomyines, in particular *Monticolomys* (Carleton and Goodman, 1996; Jansa et al., 1999). Lavocat (1978), Chaline et al. (1977), and F. Petter (1990) aligned *Macrotarsomys* with the Kenyan Miocene fossil *Protarsomys*, a relationship questioned by Carleton and Schmidt (1990) and Carleton and Goodman (1996). Specific discrimination, distributions, and biology reviewed by Carleton and Goodman (2003*c*).

Macrotarsomys bastardi Milne-Edwards and G. Grandidier, 1898. Bull. Mus. Hist. Nat. Paris, ser. 1, 4:179.
COMMON NAME: Lesser Big-footed Mouse.
TYPE LOCALITY: Madagascar, Fianarantsoa Prov., east of Ihosy River, near Ravori; see Carleton and Schmidt (1990:14).
DISTRIBUTION: Dry deciduous forest and open savannah in NW, W, and S Madagascar, from the vicinity of Ankarafantsika to the island's southernmost tip.

STATUS: IUCN – Lower Risk (lc).
SYNONYMS: *occidentalis* Ellerman, 1949.
COMMENTS: Ellerman (1949) described *occidentalis* as a subspecies; its status has yet to be critically assessed.

Macrotarsomys ingens F. Petter, 1959. Mammalia, 23:140.
COMMON NAME: Greater Big-footed Mouse.
TYPE LOCALITY: Madagascar, Mahajanga Prov., Ankarafantsika Reserve, near Ampijoroa.
DISTRIBUTION: Known only from the type locality and its vicinity, NW Madagascar.
STATUS: IUCN – Critically Endangered.
COMMENTS: Occurs sympatrically with *M. bastardi*.

Monticolomys Carleton and Goodman, 1996. Fieldiana Zool., N.S., 85:233.
TYPE SPECIES: *Monticolomys koopmani* Carleton and Goodman, 1996.
COMMENTS: Sister genus to *Macrotarsomys* according to morphological (Carleton and Goodman, 1996) and molecular (Jansa et al., 1999) information.

Monticolomys koopmani Carleton and Goodman, 1996. Fieldiana Zool., N.S., 85:235.
COMMON NAME: Koopman's Montane Voalavo.
TYPE LOCALITY: Madagascar, Antananarivo Prov., Manjakatompo, 1800 m; ca. 19°20′S, 47°26′E.
DISTRIBUTION: Spottily distributed in sclerophyllous montane forest, 1625-2000 m, of the Central and Southern Highlands, E Madagascar.

Nesomys Peters, 1870. Sitzb. Ges. Naturf. Fr., Berlin, p. 54.
TYPE SPECIES: *Nesomys rufus* Peters, 1870.
SYNONYMS: *Hallomys* Jentink, 1879.
COMMENTS: Major (1897) synonymized Jentink's (1879a) *Hallomys* (type species = *H. audeberti*) under *Nesomys* and allocated the genus to the Cricetinae, as did Ellerman (1941). Sister genus of *Brachyuromys* according to phylogenetic interpretation of mitochondrial DNA sequences (Jansa et al., 1999). Ellerman (1941, 1949) accepted *audeberti* and *lambertoni* as species as described, but F. Petter (1972c, 1975a) arranged them as subspecies of *N. rufus*, synonymies observed in systematic checklists (Corbet and Hill, 1991; Honacki et al., 1982; Musser and Carleton, 1993). Carleton and Schmidt (1990) indicated that each is morphologically highly distinctive, and Ryan et al. (1993) encountered *N. audeberti* and *N. rufus* in sympatry. Distributions, ecology, and behavior reviewed by Ryan (2003).

Nesomys audeberti (Jentink, 1879). Notes Leyden Mus., 1879:107.
COMMON NAME: White-bellied Nesomys.
TYPE LOCALITY: "Maisine and Savary.—N. E. Madagascar;" type locality requires formal restriction.
DISTRIBUTION: Low to middle elevation rain forest, sea level-1000 m, from vicinity of Antongil Bay to Vohimena Mtns, E Madagascar.
COMMENTS: Carleton and Schmidt (1990:18) discussed the indeterminate nature of Audebert's collecting localities.

Nesomys lambertoni G. Grandidier, 1928. Bull. Acad. Malgache, Nouv. Ser., 11:95.
COMMON NAME: Western Nesomys.
TYPE LOCALITY: Madagascar, Toliara Prov., vicinity of Maintirano (as amended by F. Petter, 1962a:571).
DISTRIBUTION: Documented only from the type locality, WC Madagascar, but certainly more widely distributed in the west.
COMMENTS: So far known by only three specimens.

Nesomys rufus Peters, 1870. Sitb. Ges. Naturf. Fr. Berlin, p. 55.
COMMON NAME: Rufous Nesomys.
TYPE LOCALITY: Madagascar, Antsiranana Prov., Vohima.
DISTRIBUTION: Broadly distributed in montane forest, 900-2300 m, from the Northern Highlands to the southern Anosyenne Mtns, E Madagascar.
STATUS: IUCN – Lower Risk (lc).

Voalavo Carleton and Goodman, 1998. Fieldiana Zool., N.S., 90:182.
> TYPE SPECIES: *Voalavo gymnocaudus* Carleton and Goodman, 1998.
> COMMENTS: Sister genus to *Eliurus* based on morphological features (Carleton and Goodman, 1998). Generic status questioned by phylogenetic study of cytochrome *b* data, in which *Voalavo gymnocaudus* is grouped as sister species to *Eliurus grandidieri*, the two forming a basal clade to all other *Eliurus* (Jansa et al., 1999).

Voalavo gymnocaudus Carleton and Goodman, 1998. Fieldiana Zool., N.S., 85:182.
> COMMON NAME: Naked-tailed Voalavo.
> TYPE LOCALITY: Madagascar, Antsiranana Prov., Reserve Speciale d'Anjanaharibe-Sud, 12.2 km WSW Befingitra, 1950 m; 14°44.8'S, 49°26.0'E.
> DISTRIBUTION: Upper montane and sclerophyllous forest of the Anjanaharibe-Sud and Marojejy massifs, 1250-1950 m, N Madagascar; may occur on other mountains in the Northern Highlands.

Subfamily Petromyscinae Roberts, 1951. Mammals of South Africa, p. 434.
> COMMENTS: Emended definition—Small-bodied (76-112 mm), terrestrial muroids with large ears, short limbs, short hind feet and digits, hairy tail (caudal hairs as long as 2-3 scale rows) equal to or moderately longer than head and body, and two pairs inguinal mammae; front feet with stubby thumb and four clawed digits (two outer nearly as long as the two inner); hind foot with five clawed digits (first very short, next three subequal in length, fifth slightly shorter) and full complement of plantar pads (hypothenar much smaller than thenar); pelage long, soft and silky; rostrum long and slender, interorbit broad and smooth, braincase wide and flat without temporal ridging; zygomatic plate wide with prominent projecting anterior spine (deep zygomatic notch) and inconspicuous masseteric tubercle; incisive foramina narrow and reaching anterior margins of M1s or extending beyond them, posterior palatine foramina at middle of M2s; bony palate wide and projecting past posterior margin of M3s to form a prominent shelf confluent with long, very wide, and slightly excavated parapterygoid fossae; mesopterygoid fossa very narrow, breached by spacious sphenopalatine vacuities; alisphenoid strut wide, robust, and appressed against braincase; subsquamosal foramen small in young animals, closed in adults, its former position marked by prominent hamular process; carotid circulatory partially derived (large stapedial foramen, no sphenofrontal foramen or squamosal-alisphenoid groove; pattern 2 as per Carleton, 1980, and Voss, 1988); bulla moderately inflated; mandible with delicate coronoid processes below which incisor capsule forms low mound; upper incisors strongly ophisthodont, enamel faces smooth; lower incisors ungrooved; molars brachydont with uppers three-rooted and lowers two-rooted; anterocone(id) of M1 and m1 large and entire; cusps longitudinally connected by short crests; large cingular lingual cusp adjacent to protocone on M1 and M2 and coalesces with wear, anteroloph and mesoloph absent, anterolophid absent, short mesolophid connecting large mesostylid with entoconid on m1 and m2; wide posterior cingulum (posteroloph) connects posterior margins of metacone and hypocone on M1 and M2, prominent posterior cingulum on m1 and m2; M2 two-thirds size of M1, M3 very small, half the size of M2 with simple C-shaped occlusal pattern unlike that of M2; (drawn from study of specimens in AMNH and USNM, along with information from F. Petter, 1967*b*, and Carleton and Musser, 1984; see dental terminology in Carleton and Musser, 1989). Contents—*Petromyscus* Thomas, 1926.
>
> Based on the lingual cusps on their upper molars, Hayman (1962*a*) regarded *Delanymys* and *Petromyscus* as dendromurines. Lavocat (1964) considered their cusp patterns as structural links between *Mystromys* and the dendromurines and allied them with cricetids, a view shared by F. Petter (1966*c*), W. Verheyen (1965*b*), and earlier by Ellerman (1941). Whereas these authors regarded the accessory lingual cusp to be an elaboration of the cingulum, Petter (1967*b*) considered it to represent the protocone, the new cusp having developed from the longitudinal crest (mure). Petter therefore united *Delanymys* and *Petromyscus* in Roberts' (1951) Petromyscinae and stressed their distant relationship to true dendromurines. Aside from the shared lingual cusps, *Delanymys* and *Petromyscus* have little in common and their union as sister taxa in the same subfamily has been challenged (Carleton and Musser, 1984; Denys, 1994*c*). Here we disassociate the

two and treat *Petromyscus* as the sole known member of Petromyscinae (see account of Delanymyinae). According to phyletic interpretation of cytochrome *b* data (Jansa et al., 1999), the genus is closely related to *Mystromys* and both form a clade that is distantly removed from the single dendromurine sampled (*Steatomys*).

Petromyscus has not yet been convincingly linked by fossils to a possible ancestral group, although both Tong and Jaeger (1993) and Mein et al. (2000*b*) believed that it represents a lineage derived from Miocene myocricetodontines. Isolated molars identified as "Petromyscinae" have been uncovered from middle Miocene sediments of Namibia and as *Petromyscus* from late Miocene beds in the same region (Senut et al., 1992). The latter teeth have been recently reidentified as *Harimyscus hoali*, a form regarded as the earliest known petromyscine (Mein et al., 2000*b*, 2004). Namibia today comprises the greater part of the distribution of three extant species of *Petromyscus*.

Petromyscus Thomas, 1926. Ann. Mag. Nat. Hist., ser. 9, 17:179.

TYPE SPECIES: *Praomys collinus* Thomas and Hinton, 1925.

COMMENTS: Most faunal lists, either with or without reservations, have recognized only two species (Ellerman et al., 1953; Meester et al., 1986; Misonne, 1974). We retain our earlier arrangement (Musser and Carleton, 1993), based on study of museum specimens and original descriptions, and continue to recognize four morphologically distinctive species, a conclusion also recorded by Skinner and Smithers (1990). Comprehensive systematic revision should be undertaken.

Petromyscus barbouri Shortridge and Carter, 1938. Ann. S. Afr. Mus., 32:288.

COMMON NAME: Barbour's Pygmy Rock Mouse.

TYPE LOCALITY: South Africa, W Northern Cape Prov., Little Namaqualand, Witwater, Kamiesberg, 3500-3800 ft (1067-1158 m).

DISTRIBUTION: Known only from rocky areas in the Springbok and Kamiesberg regions and Loeriesfontein area in Little Namaqualand, W Northern Cape Prov., South Africa; limits unknown.

STATUS: IUCN – Least Concern.

COMMENTS: Reviewed by Skinner and Smithers (1990) and de Graaff (1997*ii*). Although listed as a subspecies of *P. collinus* (Meester et al., 1986), *P. barbouri* can be separated by the diagnostic short and bicolored tail, as noted by Shortridge and Carter (1938), its smaller skull and relatively shorter rostrum, much shorter molar rows, and lack of postaxillary teats.

Petromyscus collinus (Thomas and Hinton, 1925). Proc. Zool. Soc. Lond., 1925:237.

COMMON NAME: Pygmy Rock Mouse.

TYPE LOCALITY: Namibia, Damaraland, Karibib, northwest of Windhoek, 3142 ft (958 m).

DISTRIBUTION: SW Angola (along the Inselberge belt between Namib and mopani areas; Crawford-Cabral, 1966*b*, 1998) and Kaokoveld region in N Namibia, south through Namibia, to W Limpopo of South Africa south of the Orange River in Goodhouse and Pella areas; also recorded from Western Cape Prov.

STATUS: IUCN – Lower Risk (lc).

SYNONYMS: *bruchus* (Thomas and Hinton, 1925); *capensis* Shortridge and Carter, 1938; *kaokoensis* Roberts, 1938; *kurzi* Lehmann, 1955; *namibensis* Roberts, 1948; *rufus* Lundholm, 1955; *variabilis* Lundholm, 1955.

COMMENTS: Reviewed by Skinner and Smithers (1990) and de Graaff (1997*jj*). Roberts (1951) listed *capensis*, known only from Goodhouse, as a species, and Meester et al. (1986) treated it as a full synonym of *P. collinus barbouri*. In arranging *capensis* as a subspecies of *P. collinus*, however, Shortridge and Carter (1938) properly reflected its affinity and appreciated that its morphology is unlike *P. barbouri*, which they described in the same paper and knew well. Roberts (1951) also treated *bruchus* (S Namibia, Great Brukkaros Mtn) as a separate species, merging the northern *shortridgei* as a subspecies. The latter is clearly a species separate from *P. collinus* (see comments under *shortridgei*), but the status of *bruchus* will have to be illuminated in a revision; the few specimens we have seen from Great Brukkaros Mtn are examples of *P. collinus*.

Petromyscus monticularis (Thomas and Hinton, 1925). Proc. Zool. Soc. Lond., 1925:238.

COMMON NAME: Brukkaros Pygmy Rock Mouse.

TYPE LOCALITY: Namibia, Great Namaqualand, Great Brukkaros Mtn, near Berseba.

DISTRIBUTION: S Namibia (vicinity of type locality and south of there between Aus region in the west and South African border near Rietfontein area in the east) and N South Africa (extreme Northern Cape Prov. on south bank of the Orange River at Augrabies Falls; USNM 452333).

STATUS: IUCN – Lower Risk (nt).

COMMENTS: Geographic limits of this distinctive species have yet to be determined. Reviewed by de Graaff (1997kk), who claimed that it was known only from vicinity of the Great Brukkaros Mtn but speculated that the range may extend southward into the Northern Cape Prov.

Petromyscus shortridgei Thomas, 1926. Proc. Zool. Soc. Lond., 1926:302.

COMMON NAME: Shortridge's Pygmy Rock Mouse.

TYPE LOCALITY: Extreme S Angola, Ruacana Falls, 3350 ft (1021 m).

DISTRIBUTION: W and S Angola and N Namibia (S to Erongo Mtns and Okahandja region); limits unknown.

STATUS: IUCN – Lower Risk (lc).

COMMENTS: Reviewed by Skinner and Smithers (1990) and de Graaff (1997ll). Roberts (1951) treated *shortridgei* as a subspecies of *P. bruchus*, and Meester et al. (1986) listed it as a subspecies of *P. collinus*. However, Thomas' species description correctly expressed the distinctness of the animal; also see Schlitter (*in* Meester et al., 1986) and Skinner and Smithers (1990). The larger size of *P. shortridgei*, darker fur with a less silky texture, and lack of postaxillary teats clearly separate it from *P. collinus*, a judgement based upon AMNH and FMNH specimens, which also inform our estimate of its geographic range. Crawford-Cabral (1998) recorded *P. shortridgei* only from the vicinity of Ruacana Falls, S Angola near the Namibia border, and speculated that the specimen (AMNH 81915) from farther north at Coporolo represented *P. collinus*. External and cranial morphology of the Coporolo specimen falls within the range of variation in *P. shortridgei* (as identified by Hill and Carter, 1941:104), not *P. collinus*. The two species overlap in distribution in SW Angola and N Namibia but have not been collected at the same place.

Family Cricetidae Fischer, 1817. Mém. Soc. Imp. Nat. Moscow, 5:372.

COMMENTS: The cricetid-murid question has persisted as the dominant theme, or uncertainty, involving the higher level classification of muroid rodents over the past century (see summaries in Carleton, 1980, and Carleton and Musser, 1984): in essence, should the various subfamilies be approximately equally apportioned between Cricetidae and Muridae (Miller and Gidley, 1918; Simpson, 1945) or should most be placed under an all encompassing Muridae (Alston, 1876; Ellerman, 1940; Thomas, 1896—even these studies, however, accorded the fossorially specialized forms separate familial rank; i.e., the spalacines, rhizomyines, and myospalacines in various combinations). While the inclusive view of Muridae has gained acceptance over the latter half of the 20th century (Hershkovitz, 1962; Hooper and Musser, 1964a; McKenna and Bell, 1997), there are notable departures (Chaline et al., 1977; Lavocat, 1978), and Reig (1980, 1981, 1984), in particular, has steadfastly espoused recognition of a separate Cricetidae, the contents of which correspond in large part to the taxa covered here. Principal subfamilial exceptions to Reig's coverage are exclusion of certain African clades (Cricetomyinae, Dendromurinae, Nesomyinae), here arranged in a broadly defined Nesomyidae (as per Lavocat, 1973, 1978), and inclusion of Arvicolinae and Lophiomyinae, each of which Reig regarded as separate families. Accepting those adjustments, monophyly of the family finds general support in phylogenetic evaluation of genetic sequence data (Dubois et al., 1999; Michaux and Catzeflis, 2000; Michaux et al., 2001b; Robinson et al., 1997). Nevertheless, this cladistic hypothesis of Cricetidae should be viewed as provisional, and taxonomic sampling must be considerably expanded to reinforce its validity, notably with addition of groups such as Lophiomyinae and many more cladistically older members drawn from Cricetinae, Sigmodontinae, and Tylomyinae.

Although monophyletic on molecular trees, unambiguous diagnosis of so large and heterogeneous an assemblage would prove challenging with the morphological character base currently referenced in phylogenetic studies. We note that all members possess a biserial arrangement of the molar cusps, with retention of a longitudinal connection (mure/murid) among them and formation of a discrete anterocone(id) on the first molars. Such character states are likely ancestral for the superfamily, as are other traits exhibited by many cricetid species, in some cases a majority of them (three-rooted upper and two-rooted lower molars; possession of mesolophs(ids); malleus configuration parallel; tegmen tympani conformation; complete carotid circulatory pattern; entepicondylar foramen present; stomach unilocular-hemiglandular; full complement of male reproductive glands). However, parallel evolution of derived conditions for each of these characters is apparent at lower taxonomic levels within Cricetidae (subfamilies and tribes) and among other families of Muroidea as currently understood.

The fossil record of "cricetids" is comparatively rich in taxa, deep in time, and broad in geographic representation, and their phylogenetic systematics is correspondingly complicated. Several groups—e.g., cricetodontines, eucricetodontines, paracricetodontines, and cricetopines—have been variously acknowledged as families separate from Cricetidae; McKenna and Bell (1997) recapitulated family-group rankings used for such extinct cricetids and retained most as tribes or subfamilies within Muridae *sensu lato*. Also see overview by Carleton and Musser (1984) and regional treatments by Freudenthal et al. (1992), Hugueney (1999), Kälin (1999), Korth (1994), Lavocat (1978), Martin (1980), Mein and Freudenthal (1971), and Rümmel (1999). The evolutionary fate of many of these extinct groups as progenitor to living cricetid assemblages is generally unclear and-or much disputed; where some phyletic link is plausibly documented, usually not until Miocene in age, we mention those groups in the subfamily comments. Geological range from the early Oligocene to Recent of Europe, late Eocene to Recent of Asia, late Miocene to Recent of Africa, late Eocene to Recent of North America, and early Pliocene to Recent of South America (McKenna and Bell, 1997).

Subfamily Arvicolinae Gray, 1821. Lond. Med. Repos., 15:303.

SYNONYMS: Alticoli Gromov, 1977; Arvicolidae Gray, 1821 (Arvicolina Bonaparte, 1837; Arvicolini Giebe, 1855; Arvicolinae Baird, 1857; Arvicolae Winge, 1887; Arvicolini Kretzoi, 1955, unnecessary naming); Braminae Miller and Gidley, 1918; Clethrionomyini Hooper and Hart, 1962 (Clethrionomyi Gromov, 1977; Clethriomyina Pavlinov and Rossolimo, 1987); Dicrostonychini Kretzoi, 1955 (Dicrostonyxini Gromov, 1972; Dicrostonychinae Chaline, 1973; Dicrostonychina Pavlinov and Rossolimo, 1987); Dolomyinae Chaline, 1975; Ellobiusini Gill, 1872, justified emendation by Pavlinov et al., 1995a (Ellobiinae Gill, 1872, not Ellobiinae Adams, 1858; Ellobii Weber, 1928; Ellobiini Simpson, 1945); Fibrini Mehely, 1914; Lagurini Kretzoi, 1955 (Lagurina Pavlinov and Rossolimo, 1987); Lemnina Gray, 1825 (Lemmi Miller, 1896; Lemmini Simpson, 1945; Lemminae Kretzoi, 1955); Microtidae Cope, 1891 (Microtinae Miller, 1896; Microti Miller, 1896; Microtini Simpson, 1945); Myodini Kretzoi, 1969; Neofibrini Hooper and Hart, 1962; Ondatrini Gray, 1825 (Ondatrini Kretzoi, 1955, unnecessary naming; Ondatrinae Repenning, 1982; Ondatrina Pavlinov et al., 1995); Phenacomyini Zagorodnyuk, 1990 (*nomen nudum*); Pitymyini Repenning, 1983; Pliomyini Kretzoi, 1969 (Pliomyini Chaline, 1975, unnecessary naming; Pliomyi Gromov, 1977); Pliophenacomyini Repenning, Fejfar, and Heinrich, 1990; Prometheomyinae Kretzoi, 1955 (Prometheomyini Hooper and Hart, 1962; Prometheomyina Pavlinov et al., 1995); Synaptomyini Koenigswald and L. D. Martin, 1984.

COMMENTS: See Kretzoi (1955, 1962, 1969) for family-group priority of Arvicolidae Gray, 1821, instead of Microti Miller, 1896. Although some have intentionally maintained the latter name (e.g., Repenning, 1992, 1998), a group concept of arvicoline rodents, recognized as Arvicolinae, actually had emerged long prior to Miller's (1896) seminal monograph (e.g., Baird, 1857; Coues, 1874; Murray, 1866; Alston, 1876; Lataste, 1887). Carleton and Musser (1984) generally defined and reviewed the limits and contents of the subfamily. Hinton's (1926a) classic monograph, although never completed, still remains the most authoritative systematic, morphological, and biogeographic review for many genera. Synthetic taxonomic treatments are available for broad regions, including the Palearctic (Agadzhanyan and Yatsenko, 1984; Corbet, 1978c, 1984; Ellerman and Morrison-Scott, 1951), Eurasia (Gromov and Erbajeva, 1995; Gromov and Polyakov, 1977; Ognev, 1963,

1964; Pavlinov and Rossolimo, 1987; Pavlinov et al., 1995a), Europe (Mitchell-Jones et al., 1999; Niethammer and Krapp, 1982a), and North America (Hall, 1981; Hall and Cockrum, 1953; Wilson and Ruff, 1999). Biochronology, paleogeography, and paleoecology of the arvicoline radiation comprehensively reviewed by Repenning (1990), Repenning et al. (1990, 1998), Fejfar and Repenning (1992), Montuire et al. (1997), and Chaline et al. (1999, and references therein). Dating of the early arvicoline phylogenesis, about 5-6 million years ago, is in reasonable agreement whether based on paleontological or molecular-clock perspectives (Chaline et al., 1999; Conroy and Cook, 1999).

The earliest indisputable arvicolines are represented by early Pliocene fossils in Holarctic North America, Europe and NW Asia (the primitive *Prosomys*; Chaline et al., 1999; Fejfar, 1990a; Repenning, 1998, 2003, as *Promimomys*). Southern Asia was not populated by voles until the late Pliocene, presumably derived from immigration of a European *Mimomys* stock (Kotlia, 1994; Kotlia and Koenigswald, 1992); early Pliocene dispersion of *Mimomys* to North America, via Beringia, important to the diversification of New World arvicolines (Repenning, 2003). Several independent lineages of cricetids with rooted, slightly hypsodont molars appeared in the late Miocene of Europe, Asia, and North America and survived into the Pleistocene in Europe; these "microtoid cricetids," such as *Baranomys* and *Microtoscoptes*, are believed to precede the appearance of true voles and were replaced by them (see review by Fejfar, 1999a). Although some have treated certain of these genera as Arvicolinae (e. g., Gromov and Polyakov, 1977; Kretzoi, 1969; Repenning, 1998; Repenning et al., 1990; Zheng and Li, 1990), most have recognized them as dentally progressive cricetids that exploited a graminivorous niche (see review by Chaline et al., 1999). Arvicolines are convincingly derived from a cricetid ancestral stock (Gromov and Polyakov, 1977; Kretzoi, 1955; Michaux et al., 2001b), yet whether their phylogenetic roots are embedded in the microtoid cricetids or another cricetid lineage is unknown. Chaline et al. (1999:242) summarized the uncertainty of the paleontological evidence: "there is a wide array of cricetids with arvicoline features but it is currently impossible to specify their involvement in the origin of arvicolines."

Broad, multispecies surveys have been undertaken on morphological and biochemical systems of arvicolines that bear on issues of their phenetic divergence and phylogenetic relationships. E.g., comparative and functional studies of the dentition (Abramson, 1993; Brunet-Lecomte and Chaline, 1992; Bustos, 2002; Contoli, 1993; Hinton, 1926a; Koenigswald, 1980, 1982; Miller, 1896); of the cranium (Courant et al., 1997; Gromov, 1990; Kratochvíl, 1982; Pietsch, 1980); of middle ear anatomy (Hooper, 1968; Pavlinov, 1984a); of cutaneous and subcutaneous glands (Quay, 1954, 1968; Sokolov and Dzhemukhadze, 1991); of the arterial system (Durán et al., 1998); of myology (Kesner, 1980, 1986; Repenning, 1968; Stein, 1986, 1987); of the digestive tract (Carleton, 1981; Quay, 1954; Vorontsov, 1979); of reproductive structures (Anderson, 1960; Hooper and Hart, 1962; Niethammer, 1972). Chromosomal comparisons, both standard and banded karyotypes, are prevalent in recent systematic studies (Ashley and Fredga, 1994; Burgos et al., 1989; Modi, 1987; Radosavlievic et al., 1990; Zagorodnyuk, 1990, 1991c; Zima and Kral, 1984a); especially see Zagorodnyuk (1992c), who tabulated chromosomal traits of 63 species of Arvicolini and discussed their taxonomic and geographic patterns. Molecular studies have broadly addressed phylogenetic questions, including allozyme variation (Chaline and Graf, 1988; Gill et al., 1987; Graf, 1982; Mezhzherin et al., 1993, 1995; Moore and Janecek, 1990), albumin evolution (Nikoletopoulos et al., 1992), and DNA, whether hybridization comparisons (Catzeflis, 1990; Catzeflis et al., 1987; Din et al., 1993), interspersed repetitive elements (Modi, 1996; Vanlerberghe et al., 1993), or mitochondrial and nuclear gene sequences (Conroy and Cook, 1999; Martin et al., 2000; Michaux and Catzeflis, 2000; Michaux et al., 2001b). Although morphological data has historically formed the foundation for defining Arvicolinae, this body of molecular data robustly supports the subfamily's monophyly.

Notwithstanding the proliferation of family-group names, practically a one-to-one correspondence with recognized genera, suprageneric relationships remain somewhat ambiguous—e.g., compare the tribal contents of Ognev (1963), Hooper and Hart (1962), Kretzoi (1969), Gromov and Polyakov (1977), and Repenning (1992). The instability of tribal limits is mirrored in the irresolvable polytomies disclosed in phylogenetic evaluation of mitochondrial DNA sequences, suggesting rapid taxonomic diversification over a short time period (Conroy and Cook, 1999). In general, we follow McKenna and Bell (1997) for tribal affiliations (Table 1) and provide explanations for departures from their taxonomic scheme. We observe Ketzoi's (1969) Myodini, type genus *Myodes*, for most arvicolines with rooted molars and simple molar patterns. In the following generic and specific remarks, therefore, readers should understand that wherever we use *Myodes* or Myodini the cited publications generally refer to *Clethrionomys* (or *Evotomys*) and Clethrionomyini.

Alticola Blanford, 1881. J. Asiat. Soc. Bengal, 50:96.
    TYPE SPECIES: *Arvicola stoliczkanus* Blanford, 1875.
    SYNONYMS: *Aschizomys* Miller, 1899; *Platycranius* Kastschenko, 1901.
    COMMENTS: Myodini. Or placed in subtribe Myodina, Prometheomyini (Pavlinov and Rossolimo, 1998; Pavlinov et al., 1995a). *Alticola* is broadly related to *Myodes*, an affinity early acknowledged by Hooper and Hart (1962), who associated *Alticola* with *Clethrionomys* (= *Myodes*), *Eothenomys*, *Hyperacrius*, *Dinaromys*, and *Phenacomys* in Clethrionomyini, a grouping later supported by Gromov and Polyakov (1977) and Mezhzherin et al. (1995). Appendicular myological and osteological traits reinforce the monophyly of *Alticola* and its close association with *Myodes* and *Eothenomys* (Stein, 1987). We follow Gromov and Polyakov (1977) and others (Hille and Stubbe, 1996; Pavlinov and Rossolimo, 1987; Pavlinov et al., 1995a) who recognize the subgenera *Alticola*, *Aschizomys*, and *Platycranius*.

    Alticola is implicated as polyphyletic in DNA-DNA hybridization studies that disclose *Alticola macrotis* as sister species to *Myodes rufocanus* and *A. argentatus* as sister species to *M. rutilus-M. glareolus* (Gileva et al., 1989). Phylogenetic analyses of mitochondrial genes also support closer relationship of *A. macrotis* to species of *Myodes* (Conroy and Cook, 1999). Since only two of the 11 species of *Alticola* have been thus far surveyed in molecular studies, its paraphyly relative to *Myodes* will require broader taxonomic sampling. Schwarz (1939) reviewed the Himalayan species, and cytogenetic results are documented by Hielscher et al. (1992). Phylogenetic analysis of electrophoretic data by Hille and Stubbe (1996) supported relationships among *A. argentatus*, *A. barakshin*, and *A. semicanus*, as proposed by Rossolimo and Pavlinov (1992).

Alticola albicaudus (True, 1894). Proc. U. S. Natl. Mus., 17:12.
    COMMON NAME: White-tailed Mountain Vole.
    TYPE LOCALITY: India, Baltistan, Braldu Valley.
    DISTRIBUTION: Himalayan portions of Baltistan (Braldu Valley, Nahr Nulla) and Ladakh (Phyang Nulla), NW India (Rossolimo and Pavlinov, 1992; Rossolimo et al., 1994).
    STATUS: IUCN – Lower Risk (nt).
    SYNONYMS: *acmaeus* Schwarz, 1939.
    COMMENTS: Subgenus *Alticola*, *A. roylei-A. argentatus* species group (Rossolimo and Pavlinov, 1992). Although usually included in *A. roylei* (Corbet, 1978c; Ellerman and Morrison-Scott, 1951; Gromov and Polyakov, 1977), Hinton (1926a) earlier pointed out the diagnostic specific traits of *albicaudus*. Rossolimo and Pavlinov (1992) verified this ranking, fully redescribed the species, and contrasted it with morphologically similar forms. Schwarz's (1939) *acmaeus* represents another population of *A. albicaudus* (Rossolimo and Pavlinov, 1992; our study of holotype).

Alticola argentatus (Severtzov, 1879). Izv. Soc. Nat. Anthrop. Etnogr., 8, 2:82.
    COMMON NAME: Silver Mountain Vole.
    TYPE LOCALITY: Tajikistan, Pamir Mountains, Murgab District, Alichur.
    DISTRIBUTION: Tien Shan mountains from Xinjiang in NW China (Ma et al., 1987); southwest through mountains of E Kazakhstan (Dzhungarskiy Alatau and other highlands) and Kyrgyzstan (Talasskiy Alatau and other local ranges); to Pamir Mtns of S Kyrgyzstan, Tajikistan, NW India (Agrawal, 2000, discussed under *A. blanfordi*), and the Hindu Kush of NW Pakistan and N Afghanistan; not present in Tibet or Himalayas.
    STATUS: IUCN – Lower Risk (lc).
    SYNONYMS: *alaica* Rosanov, 1935; *argurus* (Thomas, 1909); *blanfordi* (Scully, 1880); *gracilis*

Kashkarov, 1923; *lahulius* Hinton, 1926; *leucurus* (Severtsov, 1873) [not Gerbe, 1852];
longicauda Kashkarov, 1923; *longicaudata* Ognev, 1950; *parvidens* Schlitter and Setzer,
1973; *phasma* Miller, 1912; *rosanovi* Ognev, 1940; *saurica* Afanasiev and Bazhanov,
1948; *shnitnikovi* Ognev, 1940; *severtzovi* (Tichomirov and Korchagin, 1889); *subluteus*
Thomas, 1914; *tarasovi* Rossolimo and Pavlinov, 1992; *villosa* Kashkarov, 1923;
worthingtoni Miller, 1906.

COMMENTS: Subgenus *Alticola*, *A. roylei-A. argentatus* species group. Included under *A. roylei*
by Ellerman and Morrison-Scott (1951) and Corbet (1978c), but separated as a species
by Rossolimo (1989) and Rossolimo and Pavlinov (1992). Rossolimo and Pavlinov
(1992) speculated that *A. argentatus* descended from a Himalayan form morphologically similar to *A. roylei*, and in turn was the ancestor to northern *A. tuvinicus*. Trait-
frequency data of m1s and M3s among 10 geographic samples documented by
Tokmergenov (1992); M3 variation analyzed with geometric techniques by Pavlinov
et al. (1994). See Rossolimo et al. (1994) for recognized subspecies.

Alticola barakshin Bannikov, 1947. Bull. Moscow Soc. Nat., Biol., 52, 4:217.
COMMON NAME: Gobi Altai Mountain Vole.
TYPE LOCALITY: S Mongolia, Gobi Altai, Gurvan Saihan Ridge, Dzun Saihan.
DISTRIBUTION: Low to middle altitudes in Tuva region (Kyzyl Valley), Russia; southward
through Gobi and Mongol Altais, rocky outcrops over transAltai Gobi Desert and
Barun Khurai Valley, to S Mongolia and adjacent China (Hou et al., 1995).
STATUS: IUCN – Lower Risk (lc).
COMMENTS: Subgenus *Alticola*. Included in *A. stoliczkanus* by Corbet (1978c) and Pavlinov
and Rossolimo (1987). Subsequently revised as species and morphologically
contrasted with the geographically adjacent *A. semicanus* in N Mongolia and
A. stoliczkanus in China (Rossolimo et al., 1988, 1994; Rossolimo and Pavlinov, 1992;
those authors described the range overlap of *A. barakshin* and *A. tuvinicus* and the
possibile contact between the former and *A. argentatus*. Chromosomal data provided
by Yatsenko (1980).

Alticola lemminus (Miller, 1898). Proc. Acad. Nat. Sci. Philadelphia, 1898:369.
COMMON NAME: Lemming Mountain Vole.
TYPE LOCALITY: Russia, Siberia, Bering Strait, Plover Bay, Kelsey Station.
DISTRIBUTION: NE Siberia from Chukotskiy (Chukotka) Peninsula to Kamchatka, westward
through the Kolyma Plateau (Khrebei Kolymskiv) to N, C, and S Yakutskaya (the River
Lena basin from the Laptev Sea Coast to the Olekma River).
STATUS: IUCN – Lower Risk (lc).
SYNONYMS: *lemniscus* (Satunin, 1908); *vicina* Portenko, 1963; *yakutensis* Vasil'eva, 1993
[*nomen nudum*].
COMMENTS: Subgenus *Aschizomys*. Stature as species and genus-group allocation highly
varied. Retained in *Aschizomys* as genus by Ellerman (1941). Corbet (1978c) transferred
lemminus to *Eothenomys*, but Russian workers continue to refer it to *Alticola* (Gromov
and Polyakov, 1977; Ognev, 1964; Pavlinov and Rossolimo, 1987). Earlier, Hinton
(1926a:279) recognized *Aschizomys* but speculated that the species is a member of the
Myodes rufocanus group; Miller (1940a:94) identified the holotype as "nothing more
than an alcohol-discolored specimen of the extreme East Asian representative of
Myodes rufocanus," an opinion followed by Ellerman and Morrison-Scott (1951). Some
have treated *lemminus* as a subspecies of *A. macrotis* (Bolshakov et al., 1985; Gromov
and Baranova, 1981; Gromov and Erbajeva, 1995; Gromov and Polyakov, 1977),
whereas others have maintained the two as separate (Musser and Carleton, 1993;
Pavlinov and Rossolimo, 1987; Pavlinov et al., 1995a). Based on renewed inspection of
the holotype, we endorse Ognev's (1964) allocation of *lemminus* to *Alticola*, subgenus
Aschizomys, and reemphasize its impressive diagnostic features as a species.

The status of included populations requires additional study. Significant
chromosomal and morphological differences between samples from Chukotka and
Yakutskaya led Bykova et al. (1978) to speculate that "*lemminus*" is a composite of two
species. Vasil'eva's (1999:113) analysis of non-metric cranial traits indicated that
"*A. lemminus* probably includes a complex of distinct forms," and interpreted samples

from the Chukotka Peninsula, N Yakutskaya, and S Yakutskaya as each representing well defined subspecies. Her evaluation uncovered two major lineages, corresponding to the allopatric ranges of *lemminus* and *macrotis* as given here, a division also reflected in a molecular study (Rybnikov et al., 1986); finer divergence patterns within the two major clusters prompted Vasil'eva to suggest "that the taxa and populations under consideration are currently subjected to the process of speciation *in statu nascendi*." Nikanorov (2000) included *A. lemminus* among mammals that occur on the Kamchatka Peninsula.

Alticola macrotis (Radde, 1862). Reise in den Suden von Ost-Sibierien, 1:196.
COMMON NAME: Large-eared Mountain Vole.
TYPE LOCALITY: Russia, S Siberia, S Krasnoyarsk Krai, Vostochnyy Sayan Mtns.
DISTRIBUTION: Altai of extreme NW Xinjiang and S Siberia, eastward through Tuvinskaya (Tuva region) and Sayan Mtns, to highlands in Lake Baikal region.
STATUS: IUCN – Lower Risk (lc).
SYNONYMS: *altaica* Vinogadov, 1933; *fetisovi* Galkina and Epifantseva, 1988; *vinogradovi* Rasorenova, 1933.
COMMENTS: Subgenus *Aschizomys*. Closely related to the northern *A. lemminus* (see above), and either not allocated to subgenus (Ellerman, 1941), placed it in the subgenus *Alticola* (Ognev, 1964), or considered under the subgenus *Aschizomys* (Gromov and Erbajeva, 1995; Gromov and Polyakov, 1977; Pavlinov and Rossolimo, 1987). Galkina and Epifantseva (1988) described *fetisovi* as species, but others have included it in *A. macrotis* or considered its status unclear (Vasil'eva, 1999); Pavlinov et al. (1995*a*) and Pavlinov and Rossolimo (1998) persuasively treated *fetisovi* as a synonym. Chromosomal variation reported by Bolshakov et al. (1985).

Alticola montosa (True, 1894). Proc. U. S. Natl. Mus., 17:11.
COMMON NAME: Kashmir Mountain Vole.
TYPE LOCALITY: N India, C Kashmir, 11,000 ft (3353 m).
DISTRIBUTION: Jammu and Kashmir, ca. 2450-4000 m (Agrawal, 2000; Hinton, 1926*a*; Rossolimo and Pavlinov, 1992).
STATUS: IUCN – Vulnerable.
SYNONYMS: *imitator* (Bonhote, 1905).
COMMENTS: Subgenus *Alticola*, *A. roylei-A. argentatus* species group (Rossolimo and Pavlinov, 1992). Usually incorporated in *A. roylei* (Corbet, 1978*c*; Gromov and Polyakov, 1977), but the diagnostic traits of *montosa* clearly distinguish it from geographically adjacent *roylei*, as Hinton (1926*a*) long ago noted and Rossolimo and Pavlinov (1992) verified. Reviewed by Agrawal (2000).

Alticola olchonensis Litvinov, 1960. Zool. Zhurn., 39(12):1889.
COMMON NAME: Lake Baikal Mountain Vole.
TYPE LOCALITY: S Russia, Irkutskaya Oblast, Olkhon Isl, EC Lake Baikal.
DISTRIBUTION: Endemic to Olkhon and Ogoi Isls, Lake Baikal.
SYNONYMS: *baicalensis* Litvinov, 1961.
COMMENTS: Subgenus *Aschizomys*. Originally described as a species, later included in *A. tuvinicus* (Pavlinov and Rossolimo, 1987; Rossolimo, 1988; Rossolimo and Pavlinov, 1992), and eventually reinstated as species (Pavlinov and Rossolimo, 1998; Pavlinov et al., 1995*a*). Gromov and Polyakov (1977) suggested that *olchonensis* belonged in *Aschizomys*, as did Litinov (1960), but all subsequent revisions and checklists have associated *olchonensis* with the subgenus *Alticola*. Pavlinov (2002, in litt.) indicated that *olchonensis* does not belong in *A. tuvinicus* and should be returned to *Aschizomys* following the original describer. Both its specific status and subgeneric relationships deserve reassessment.

Alticola roylei (Gray, 1842). Ann. Mag. Nat. Hist. [ser. 1], 10:265.
COMMON NAME: Royle's Mountain Vole.
TYPE LOCALITY: India, Kumaon.
DISTRIBUTION: W Himalayas, 2600-3900 m, from Kulu Valley in Himachal Pradesh to N Kumaon in Uttar Pradesh, N India (Agrawal, 2000).

STATUS: IUCN – Lower Risk (nt).

SYNONYMS: *cautus* Hinton, 1926.

COMMENTS: Subgenus *Alticola*. Once considered the broadest-ranging species of *Alticola* in central Asia, encompassing *A. argentatus* and many of its synonyms (Corbet, 1978*c*; Ellerman, 1961; Ellerman and Morrison-Scott, 1951). With removal of the latter (Rossolimo, 1989; Rossolimo and Pavlinov, 1992), the geographic and morphological definition of *A. roylei* conforms to that presented by Hinton (1926*a*) and Ellerman (1941).

Alticola semicanus (G. M. Allen, 1924). Am. Mus. Novit., 133:6.

COMMON NAME: Mongolian Mountain Vole.

TYPE LOCALITY: Mongolia, SE Khangai Mtns, upper reaches of Ongyin Gol River "Sain Noin Khan."

DISTRIBUTION: S Tuva region, Russia, throughout most of N and C Mongolia to adjacent Nei Mongol, N China.

STATUS: IUCN – Lower Risk (lc).

SYNONYMS: *alleni* Argyropulo, 1933.

COMMENTS: Subgenus *Alticola*. Originally described as a subspecies of *Microtus worthingtoni* and listed as a subspecies of *Alticola worthingtoni* by Ellerman (1941); later synonymized with *A. roylei* (Ellerman and Morrison-Scott, 1951; Corbet, 1978*c*) or *A. argentatus* (see taxonomic history in Rossolimo and Pavlinov, 1992). Reinstatement as a species (Pavlinov and Rossolimo, 1987) vindicated by the revisions of Rossolimo et al. (1988, 1994), and Rossolimo and Pavlinov (1992). The geographic range of *A. semicanus* approaches but does not overlap *A. tuvinicus* and *A. bararshin*; Rossolimo and Pavlinov (1992) morphologically contrasted the latter two and *A. semicanus*.

Alticola stoliczkanus (Blanford, 1875). J. Asiat. Soc. Bengal, 44:107.

COMMON NAME: Stoliczka's Mountain Vole.

TYPE LOCALITY: NW India, Ladakh (Kashmir), Kuenlun Mtns, Nubra Valley (as restricted by neotype designated by Rossolimo and Pavlinov, 1992:172).

DISTRIBUTION: N India (Jammu, Kashmir; Himachal Pradesh; Agrawal, 2000), Nepal and N Sikkim, northward through Xizang (Tibet) to Kunlun Shan in S Xinjiang and Qinghai, N China (Zhang et al., 1997); probably does not extend farther north than the Nan Shan and Qilian Shan (Rossilimo and Pavlinov, 1992; Rossilimo et al., 1994).

STATUS: IUCN – Lower Risk (lc) as *A. stoliczkanus* and *A. stracheyi*.

SYNONYMS: *acrophilus* (Miller, 1899); *bhatnagari* Biswas and Khajuria, 1955; *cricetulus* (Miller, 1899); *kaznakovi* (Satunin, 1903); *lama* (Barret-Hamilton, 1900); *nanschanicus* (Satunin, 1903); *stracheyi* (Thomas, 1880).

COMMENTS: Subgenus *Alticola*. The taxon *stracheyi* was conventionally included in *A. stoliczkanus* (Schwarz, 1939; Ellerman and Morrison-Scott, 1951; Gromov and Polyakov, 1977; Corbet, 1978*c*; Rossolimo and Pavlinov, 1992) until reinstated as a species by Feng et al. (1986), as earlier arranged by Hinton (1926*a*) and Ellerman (1941). Rossolimo and Pavlinov (1992) tentatively recognized three geographic clusters: one centered in Kashmir (typical *stoliczkanus*), another along S Himalayan slopes from Nepal to Sikkim (*bhatnagari*), and the last ranging over Tibet and Kuen Lun Shan (*lama*). We follow their synonymy but emphasize that inquiries into the taxonomic significance of this geographic variation are warranted.

Alticola strelzowi (Kastchenko, 1899). Izv. Imp. Tomsk. Univ., 16:50.

COMMON NAME: Strelzow's Mountain Vole.

TYPE LOCALITY: Russia, Altai Krai, Altai Mtns, near Lake Teniga.

DISTRIBUTION: Altai of NW Mongolia, Siberia, and Xinjiang in NW China (Ma et al., 1987), west through Kazakhstan to Karaganda region.

STATUS: IUCN – Lower Risk (lc).

SYNONYMS: *depressus* (Ognev, 1944); *desertorum* (Kastschenko, 1901); *desertorum* Ognev, 1950.

COMMENTS: Subgenus *Platycranius*. Citations and synonyms are discussed by Pavlinov and Rossolimo (1987) and Gromov and Erbajeva (1995).

Alticola tuvinicus Ognev, 1950. [Mammals of USSR and Adjacent Countries], 7:520.
  COMMON NAME: Tuva Mountain Vole.
  TYPE LOCALITY: Russia, Tuvinskaya (Tuva), Barun-Kemtchik District, Kyzyl Mozhalyk
      (= Kyzyr).
  DISTRIBUTION: Discontinuous in the Altai of Russia and NW Mongolia, east through Tuva
      region and N part of Khubsugul Lake Valley, to SW shore of Lake Bailkal (Rossolimo
      et al., 1988; Rossolimo and Pavlinov, 1992).
  STATUS: IUCN – Lower Risk (lc).
  SYNONYMS: *khubsugulensis* Litvinov, 1973; *kosogol* Litvinov, 1973.
  COMMENTS: Subgenus *Alticola*, *A. roylei-A. argentatus* species group (Rossolimo and Pavlinov,
      1992). Originally described as a species, as regarded by Pavlinov and Rossolimo (1987),
      then synonymized under the ubiquitous *A. roylei* (Corbet, 1978c), or listed as a
      subspecies of *A. argentatus* (Gromov and Erbajeva, 1995). Thoroughly revised by
      Rossolimo et al. (1988) and Rossolimo and Pavlinov (1992), who recognized *kosogol*
      and *olchonensis* as subspecies (latter now a species; see account). Rossolimo and
      Pavlinov (1992) noted that *A. tuvinicus* has the northernmost distribution of any in
      the subgenus *Alticola* and overlaps the range of *A. barakshin* in NW Mongolia and
      SW Tuva; southern range is separated from *A. argentatus* by the broad Zayson
      depression between Kazakhstan and NW China. They considered *A. tuvinicus* to be
      descended from a form similar to *A. argentatus*, its closest relative.

Arborimus Taylor, 1915. Proc. California Acad. Sci., ser. 4, 5:119.
  TYPE SPECIES: *Phenacomys longicaudus* True, 1890.
  SYNONYMS: *Paraphenacomys* Repenning and Grady, 1988.
  COMMENTS: Phenacomyine (see remarks under *Phenacomys*). Described as a subgenus of
      Phenacomys and conventionally recognized as such or as a complete synonym (Carleton
      and Musser, 1984; Hall, 1981; Howell, 1926). Evidence for generic stature marshalled by
      Johnson (1968, 1973), but opinion on the validity of this rank has vacillated in recent
      systematic works (yes, as per George, 1999, Jones et al., 1997, and Musser and Carleton,
      1993; no, as per McKenna and Bell, 1997, Repenning and Grady, 1988, and Verts and
      Carraway, 1998). Whether *Arborimus* is most closely related to *Phenacomys* or to some
      other arvicoline has not been cladistically substantiated with sampling that includes
      problematic species like *albipes* and other archaic arvicolines; the methodologies and
      critical information bases are here, but an informative sampling design must be
      assembled. Specific and subspecific classification basically set forth by Howell (1926) and
      Hall and Cockrum (1953), as part of *Phenacomys*, and by Johnson and George (1991).

Arborimus albipes (Merriam, 1901). Proc. Biol. Soc. Wash., 14:125.
  COMMON NAME: White-footed Vole.
  TYPE LOCALITY: USA, California, Humboldt Co., Humboldt Bay, redwood forest near Arcata.
  DISTRIBUTION: Pacific coastal zone south of Columbia River, from W Oregon to extreme
      NW California, USA.
  STATUS: IUCN – Data Deficient.
  COMMENTS: More generalized terrestrial form and habits have suggested a closer relationship
      to *Phenacomys intermedius* (e.g., Hall, 1981, placed *albipes* and *intermedius* together in
      subgenus *Phenacomys*). Johnson and Maser (1982) enumerated character states that
      instead support closer congruence of *albipes* with species of *Arborimus*. However, see
      Repenning and Grady (1988), who diagnosed the subgenus *Paraphenacomys* of
      Phenacomys to contain *albipes*, which they viewed as more distantly related to the sister
      species *intermedius* and *longicaudus*; *Paraphenacomys* is known by fossil forms from the
      late Pliocene of North America and Beringian Asia (Repenning and Grady, 1988;
      Repenning et al., 1987). See Verts and Carraway (1995, Mammalian Species, 494).

Arborimus longicaudus (True, 1890). Proc. U. S. Natl. Mus., 13:303.
  COMMON NAME: Red Tree Vole.
  TYPE LOCALITY: USA, Oregon, Coos Co., Marshfield.
  DISTRIBUTION: Coastal area and Western Cascade Mtns, W Oregon (Verts and Carraway,
      1998:Fig. 11-106), USA.

STATUS: IUCN – Lower Risk (lc).

SYNONYMS: *silvicola* (A. B. Howell, 1921).

COMMENTS: Although Howell (1926) recognized *silvicola* as a nominal species, subsequent research has favored its synonymy under *A. longicaudus* (Johnson, 1968), where it has been maintained as a subspecies (Hall, 1981; Johnson and George, 1991). Formerly included populations in California assigned to the new species *A. pomo* (see next account). See Hayes (1996, Mammalian Species, 532).

Arborimus pomo Johnson and George, 1991. Los Angeles Co. Nat. Hist. Mus., Contr. Sci., 429:12.

COMMON NAME: Sonoma Tree Vole.

TYPE LOCALITY: USA, California, Sonoma Co., 0.8 km N Jenner, Jenner Ridge; 38°27′N, 123°06′W.

DISTRIBUTION: Coastal coniferous forests of NW California, south of Klamath Mtns as far as Sonoma Co., USA.

STATUS: IUCN – Data Deficient.

COMMENTS: Closely related to *A. longicaudus* (Johnson and George, 1991). See Adam and Hayes (1998, Mammalian Species, 593).

Arvicola Lacepede, 1799. Tab. Div. Subd. Orders Genres Mammifères, p. 10.

TYPE SPECIES: *Mus amphibius* Linnaeus, 1758.

SYNONYMS: *Alviceola* de Blainville, 1817; *Hemiotomys* de Sélys Longchamps, 1836; *Ochetomys* Fitzinger, 1867; *Paludicola* Blasius, 1857 [not Wagner, 1830, or Hodgson, 1837]; *Praticola* Fatio, 1867 [not Swainson, 1837].

COMMENTS: Arvicolini, subtribe Arvicolina (Pavlinov et al., 1995a). *Arvicola* is phylogenetically close to *Microtus* (Burgos et al., 1989; Chaline and Graf, 1988; Graf, 1982; Mezhzherin et al., 1993). Excludes the North American *Microtus richardsoni* (see that account), placed in *Arvicola* by Hooper and Hart (1962).

Heinrich (1990) hypothesized that *Arvicola* evolved from the extinct *Mimomys*, a view already presented by Hinton (1926a). Many studies summarize the rich European fossil history, with differing emphases on the transition leading from Pliocene *Mimomys* or *Cromeromys* to the early Pleistocene *Mimomys savini*, and eventually to modern species of *Arvicola* (Chaline, 1990; Chaline et al., 1999; Desclaux et al., 2000; Maul et al., 2000; Neraudeau et al., 1995; Rekovets, 1990). The European middle Pleistocene *A. mosbachensis* is the oldest species of *Arvicola*, as known to date, from which the living forms can be derived (Agadzhanian, 2000; Maul et al., 2000; Rekovets, 1990). In broad phylogeographic studies of *A. sapidus* and *terrestris* (here = *A. amphibius*), Taberlet et al. (1998) and Hewitt (1999) identified S European peninsulas (Iberian, Italian, and Balkan) as major ice-age refugia from which genetically distinct taxa emerged. Kolfschoten (1992) presented another view of postglacial immigrations, extinctions, and recolonizations, as based on the *Arvicola* fossil record in NW Europe.

Panteleyev's (2000) monograph exhaustively treats *Arvicola*'s systematics and phylogeny, morphology and physiology, ecology and behavior, and agricultural and epidemiological impacts. The number of species recognized has varied from one (*terrestris*—Ellerman and Morrison-Scott, 1951), to four (*amphibius, sapidus, scherman*, and *terrestris*—Hinton, 1926a) or seven (*amphibius, illyricus, italicus, musignani, sapidus, scherman*, and *terrestris*—Miller, 1912a), but usually just two (*sapidus* and *terrestris*—e.g., Corbet et al., 1970; Corbet, 1978c, 1984; Corbet and Hill, 1991; Gromov and Polyakov, 1977; Musser and Carleton, 1993). Here, we acknowledge the strongly differentiated fossorial form, *A. scherman*, as a third species, along with *A. amphibius* and *A. sapidus*, although we anticipate that future revisionary research will undoubtedly converge towards Miller's (1912a) recognition of biodiversity.

Arvicola amphibius (Linnaeus, 1758). Syst. Nat., 10th ed., 1:61.

COMMON NAME: Eurasian Water Vole.

TYPE LOCALITY: England.

DISTRIBUTION: Europe (excluding C and S Spain but including N Spain and N Portugal) east through Siberia to Lena River Basin (Yakutskaya); from Arctic Sea south to Lake Baikal and N Tien Shan Mtns of NW China (Xinjiang) through NW Iran, Iraq, N Israel,

Caucasus, and Turkey (Harrison and Bates, 1991; Lay, 1967); also in Great Britain except Ireland (Corbet, 1978c).

STATUS: IUCN – Lower Risk (lc) as *A. terrestris*.

SYNONYMS: *abbotti* Hinton, 1910; *abrukensis* Reinwaldt, 1927; *americana* Gray, 1842; *antiquus* Pomel, 1853; *aquaticus* (Cuvier, 1817); *aquaticus* (Billberg, 1827); *argyropus* Cabrera, 1901; *armenius* (Thomas, 1907); *ater* (Billberg, 1827); *ater* Macgillivray, 1832; *bactonensis* Hinton, 1926; *barabensis* (Heptner, 1948); *brigantium* Thomas, 1928; *cantiana* Hinton, 1910; *caucasicus* Ognev, 1933; *cernjavskii* Petrov, 1949; *chosaricus* Alexandrova, 1976; *cubanensis* Ognev, 1933; *destructor* Savi, 1839; *djukovi* Ognev and Formosov, 1927; *ferrugineus* Ognev, 1933; *fuliginosus* de Sélys Longchamps, 1845 [*nomen nudum*]; *gracilis* Heller, 1955; *greenii* Hinton, 1926; *hintoni* Aharoni, 1932; *hunasensis* Carls, 1986; *hyperryphaeus* (Heptner, 1948); *illyricus* (Barrett-Hamilton, 1899); *italicus* Savi, 1839; *jacutensis* Ognev, 1933; *jenissijensis* Ognev, 1933; *kalmankensis* Za—igin, 1980; *karatshaicus* (Heptner, 1948); *korabensis* Martino, 1937; *kuruschi* Heptner and Formosov, 1928; *kuznetzovi* Ognev, 1933; *littoralis* (Billberg, 1827); *martinoi* Petrov, 1949; *moenana* Heller, 1969; *meridionalis* Ognev, 1922; *minor* de Sélys Longchamps, 1845 [*nomen nudum*]; *musignani* de Sélys Longchamps, 1839; *nigricans* de Sélys Longchamps, 1845 [*nomen nudum*]; *obensis* Egorin, 1939; *ognevi* Turov, 1926; *pallasi* Ognev, 1913; *pallasi* Ognev, 1913; *paludosus* (Linnaeus, 1771); *persicus* de Filippi, 1865; *pertinax* Savi, 1839; *praeceptor* Hinton, 1926; *reta* Miller, 1910; *rufescens* (Satunin, 1908) [not de Sélys Longchamps, 1836]; *scythicus* Thomas, 1914; *stankovici* Petrov, 1949; *tanaitica* Kalabuchov and Raevsky, 1930; *tataricus* Ognev, 1933; *taurica* Ognev, 1923; *terrestris* (Linnaeus, 1758); *turovi* Ognev, 1933; *uralensis* Egorin, 1940; *variabilis* Ognev, 1933 [not Rörig and Börner, 1905]; *volgensis* Ognev, 1933; *weinheimensis* Heller, 1952.

COMMENTS: Linnaeus' *amphibius* and *terrestris*, both proposed in 1758 on the same page, are now considered conspecific by most researchers, but which name should be properly used is unsettled. Corbet (1978c:105) noted that "*amphibius* should have priority (presumably following Blasius (1857) as first reviser). Although strictly correct this is contrary to long-established usage and would cause considerable confusion and ambiguity" (also see discussion in Corbet et al., 1970:315). The usage is not so long established since the two forms were considered separate species through the middle 1900s (Ellerman, 1941; Hinton, 1926a; Miller, 1910, 1912a), and in the first work that considered them as conspecific, Blasius (1857) placed *terrestris* as a subjective synonym of *A. amphibius*. The reminder of Blasius' role as first revisor dates from Van den Brink (1967), who employed *A. amphibius* as the valid name as have other systematists (e.g., Panteleyev, 2000; Zagorodnyuk, 1992c, 2000). In our view, confusion and ambiguity will be lessened only by using the name combination that is acknowledged as "strictly correct," *A. amphibius*.

The larger issue involves the homogeneity of populations arranged here under one nominal species. The exceptionally robust water voles in the English Isles (*amphibius*) have been maintained as specifically distinct from populations on the European continent (*terrestris*) (Hinton, 1926a; Miller, 1910, 1912a). The status of *italicus* also deserves critical review: e.g., Miller (1912a) had considered it a species, and Taberlet et al. (1998), using phylogeographic analyses of cytochrome *b*, found that samples from N Italy (*italicus*) are more genetically divergent than *terrestris* (here = *A. amphibius*) is from *scherman*. Denser geographic sampling and more critical analyses of morphologies and molecules are warranted.

European populations are generally reviewed by Reichstein (1982b) and Mitchell-Jones et al. (1999), and those of Russia and adjacent territories by Gromov and Erbajeva (1995). Faunal studies and checklists have proliferated to augment regional knowledge of the water vole's morphology, distribution, and biology: Scotland (Stewart et al., 1999); Portugal (Ramalhinho and Mathias, 1988); N Spain (Ventura, 1992, 1993a, b, 1993-1994; Ventura and Gosalbez, 1989, 1992b; Castien and Gosalbez, 1992); Italy (Amori et al., 1999; Cantini, 1991); Switzerland (Hausser, 1995); N Germany (Dolch et al., 1994); Netherlands (Pelzers, 1992); E Baltic region (Miljutin, 1997, 1998; Timm et al., 1998); Sumava Mtns region, SW Bohemia (Andera and

Èervený, 1994); Slovenia (Kryštufek, 1991); Slovakia (Danko, 1994; Mošanský, 1994; Stanko, 1995; Stanko and Mošanský, 1994, 2000; Stanko et al., 2000); Czech Republic (Šmaha, 1996); greater Serbia and Montenegro (Petrov, 1992); N Israel (Qumsiyeh, 1996); Iran, Iraq, and nearby regions (Harrison and Bates, 1991); and China (Zhang et al., 1997).

Taxonomic and distributional studies have broadly covered morphological variability (Kratochvíl, 1980, 1983; Nikolaeva, 1982; Ventura, 1991). Morphometric analyses document variation along altitudinal transects and within geographic regions (Kratochvíl, 1981b, 1983), and taxonomically illuminate variation among samples from the Iberian Peninsula and C Europe (Ventura and Gosálbez, 1989) and that in Traika Depression of Bulgaria (Mitev and Miteva, 1991). Ventura and Sans-Fuentes (1997) documented geographic variation in non-metric trait frequencies in SW Europe. Molar patterns have been studied in several contexts: dental ontogeny and its significance for interpreting relationships of fossil and living species (Kratochvíl, 1980); concordance with subspecies recognized in the Pyrenees and Iberian region (Ventura, 1991); and correlation with geography, age, and diet (Nikolaeva, 1982). Ventura et al. (1993) described abdominal arterial configuration of Spanish water voles and compared it with other muroids. Variation in cranial size and shape in relation to fluctuations in population density and environment explored by Galaktionov (1995) and Kovaleva et al. (1996). Corbet et al. (1970) morphometrically demonstrated that only one kind of *Arvicola* occurs in the British Isles and considered it to be conspecific with continental populations (*terrestris*).

Several fossil species have been described from the middle to late Pleistocene (*abbotti, antiquus, bactonensis, cantiana, chosaricus, gracilis, greenii, hunasensis, kalmankensis, praeceptor*) and have been treated as extinct subspecies (Hutterer and Koenigswald, 1993; Kolfschoten, 1990; see review in Zagorodnyuk, 2000; see Maul et al., 2000, for a different view). Kratochvíl (1980, 1981b) regarded the Pleistocene *cantiana* to be merely part of the chronocline of *A. amphibius*. European Pleistocene records are provided by Kowalski (2001).

Although Neolithic samples document the presence of *Arvicola* on Sicily 3000-4000 years ago, none was uncovered in an exhaustive survey of barn owl pellets, leading Catalisano and Sarà (1995) to conclude its recent extinction.

Arvicola sapidus Miller, 1908. Ann. Mag. Nat. Hist., ser. 8, 1:195.
COMMON NAME: Southwestern Water Vole.
TYPE LOCALITY: Spain, Burgos Prov., Santo Domingo de Silos.
DISTRIBUTION: Portugal, Spain, and France.
STATUS: IUCN – Lower Risk (nt).
SYNONYMS: *musiniani* (Lataste, 1884); *tenebricus* Miller, 1908.
COMMENTS: Although arranged as a subspecies of *terrestris* by Ellerman and Morrison-Scott (1951), *A. sapidus* is strongly differentiated in morphological, morphometric, and chromosomal traits (Panteleyev, 1996; Reichstein, 1982a; Ventura and Sans-Fuentes, 1997). Phylogenetic analyses of cytochrome-*b* segregated *A. sapidus* from European samples of *A. amphibius* at a high level of percent sequence divergence (Taberlet et al., 1998, as *terrestris*).

Reviewed by Hinton (1926a), Corbet et al. (1970), Corbet (1978c), Reichstein (1982a), and Mitchell-Jones et al. (1999). New information covering morphology, discrimination from *A. amphibius*, and ecology supplied by Garde et al. (1993), Garde and Escala (1994, 1996, 1999), Ventura (2000), Ventura and Gosalbez (1990, 1992a), and Ventura et al. (1989, 1994). Population in NE Spain documented by owl pellet remains (Torre et al., 1996); distribution in N Spain and considerable range overlap with *A. amphibius* covered by Castien and Gosalbez (1992).

Arvicola scherman (Shaw, 1801). Gen. Zool. II, Pt. 1, p. 75.
COMMON NAME: Montane Water Vole.
TYPE LOCALITY: Germany, Strassburg.
DISTRIBUTION: Mountains of N Spain, through C Europe (S Netherlands to SC France and eastwards to Slovakia), to C Romania (Panteleyev, 2000:Fig. 2).

SYNONYMS: *albus* (Bechstein, 1801); *argentoratensis* Desmarest, 1822; *buffonii* (Fischer, 1829); *cantabriae* Ventura and Gosálbez, 1989; *canus* (Bechstein, 1801); *castaneus* de Sélys Longchamps, 1845 [*nomen nudum*]; *exilis* Lydekker, 1910; *exitus* Miller, 1910; *gutsulius* Zagorodnyuk, 2000; *monticola* de Sélys Longchamps, 1838; *minor* (Leske, 1779) [listed as *nomen dubium* by Miller, 1912*a*, and Hinton, 1926*a*]; *niger* de Sélys Longchamps, 1845 [*nomen nudum*]; *schermaus* (Hermann, 1804).

COMMENTS: Earlier recognized as a species (Trouessart, 1910; Miller, 1912*a*; Hinton, 1926*a*; Ognev, 1950) until Ellerman and Morrison-Scott (1951) reassigned *scherman* as another subspecies of an all-embracing *terrestris*, a classification observed through the late 1900s (Gromov and Polyakov, 1977; Corbet, 1978*c*, 1984; Honacki et al., 1982; Corbet and Hill, 1991; Musser and Carleton, 1993). However, the two long known ecological morphotypes clearly correspond to two biological species: an amphibious form (*A. amphibius*) that is widely distributed in Eurasia, and a smaller fossorial species (*A. scherman*) isolated in certain European mountains (Alps, Carpathians, Cantabrian, Massif Central, Pyrenees). They contrast in body mass, pelage coloration, social behavior, mating system, use of space, cranial size and shape, and incisor protrusion (LaVille, 1989; Mitchell-Jones et al., 1999; Panteleyev, 1996; Sausy, 2000; Warmerdam, 1982). In the Netherlands and Belgium, Warmerdam (1982) meticuously characterized the geographic complementarity and morphometric discrimination of the two forms, as did Panteleyev (1996) in a geographically broader synopsis of the two. Although both authors ultimately retained the fossorial *scherman* as a distinctive race of *terrestris* (here = *A. amphibius*), theirs is an exceptionally robust application of subspecies usage. Based on this persuasive body of literature, coupled with examinations of AMNH and USNM series, we can only endorse the recent views of Panteleyev (2000) and Zagorodnyuk (1992*b*, *c*, 2000) in acknowledging *scherman* as a species. Genetic distances based on allozymic analysis are slightly less between samples of *A. scherman* and *A. amphibius* than are any pair-wise comparisons among 10 *Microtus* species sampled by Mezhzherin et al. (1993).

Blanfordimys Argyropulo, 1933. Z. Säugetierk., 8:182.

TYPE SPECIES: *Microtus bucharensis* Vinogradov, 1930.

COMMENTS: Arvicolini, subtribe Arvicolina (Pavlinov et al., 1995*a*). Originally proposed as a subgenus of *Microtus*, a ranking traditionally acknowledged by Russian authors (Golenishchev and Sablina, 1991; Gromov and Erbajeva, 1995; Gromov and Polyakov, 1977; Ognev, 1964; Pavlinov and Rossolimo, 1987). In other taxonomic variations, Corbet (1978*c*) assigned *afghanus* and *bucharensis*, the type-species of *Blanfordimys*, to the genus *Pitymys*, and Chaline (1974) placed it in *Neodon*, subgenus *Microtus*. Ellerman (1941, 1948) considered the diagnostic traits of *afghanus* so impressive that he recognized the genus, an appreciation shared by many others (Ellerman and Morrison- Scott, 1951; Musser and Carleton, 1993; Pavlinov and Rossolimo, 1998; Pavlinov et al., 1995*a*; Zagorodnyuk, 1990).

Blanfordimys is emphasized as *Allophaiomys*-like in retaining certain primitive traits, notably the simple M3 and m1 patterns and a high diploid number (Nadachowski and Zagorodnyuk, 1996). *Allophaiomys* is a late Pliocene-Pleistocene complex thought to be ancestral to *Microtus* and closely related genera (Chaline et al., 1999; R. A. Martin, 1995; Repenning, 1992). These primitive features, coupled with their localized ranges along the southern margin of the Palearctic arvicoline distribution, suggest that the species are Pleistocene relicts. Zagorodnyuk (1992*c*) even treated *afghanus* and *bucharensis* as species of *Allophaiomys* because the karyotype of *afghanus* (2n = 58) resembles that (2n = 62) hypothesized as "one of the first steps in the karyotype differentiation of Arvicolini" (Nadachowski and Zagorodnyuk, 1996:390). Although the molars of *bucharensis* and *Allophaiomys* are similar, those of *afghanus* are more elaborate, approaching the morphology of some *Neodon* (see that account). In other Holarctic regions, *Allophaiomys* is thought to have evolved into various subgenera of *Microtus*, either directly into Nearctic *Pitymys* and Palearctic *Terricola* or through a morphologically intermediate stage resembling *Lasiopodomys* (Chaline et al., 1999; Repenning, 1992); however, Nadachowski and Zagorodnyuk (1996) postulated a slower evolutionary rate in S Asia, those populations surviving as *Blanfordimys* and other *Allophaiomys*-like forms (also see *Neodon* and *Phaiomys*).

   Blanfordimys has inflated auditory bullae and mastoid regions, the latter so enlarged that they nearly project beyond the occipital condyles and believed to be a derived configuration that occurs in no other arvicoline (Gromov and Polyakov, 1977; however, *Microtus paradoxus* exhibits comparable bullar and mastoidal inflation). The combination of derived (bullar and mastoid inflation) and primitive (M3 and m1 patterns) traits separates *Blanfordimys* from *Neodon*, to which it is otherwise closely similar in molar patterns (especially *N. irene* and *N. juldaschi*); from *Phaiomys leucurus* (also differing by its less robust cranium, longer incisive foramina, and greater variation and more elaborate configuration in M3 and m1 patterns); and from *Proedromys*, *Lasiopodomys*, and subgenera of *Microtus* (including *Pitymys* and *Terricola*). Generic isolation of *Blanfordimys* defines an explicit kinship hypothesis relative to *Microtus* and those other genera that should be critically tested with molecular data and an array of morphological traits, not solely molar occlusal patterns. Except for the allozymic survey of Mezhzherin et al. (1993), which nested *afghanus* among the 10 species of *Microtus* sampled, other multi-character and multi-specific inquiries are unavailable.

Blanfordimys afghanus (Thomas, 1912). Ann. Mag. Nat. Hist., ser. 8, 9:349.
   COMMON NAME: Afghan Vole.
   TYPE LOCALITY: Afghanistan, Badkhiz, Gulran.
   DISTRIBUTION: High steppes and semi-desert in S Turkmenistan, Uzbekistan, Tajikistan, and C Afghanistan; isolated population in Great Balkhan Mtns on E coast of Caspian Sea.
   STATUS: IUCN – Lower Risk (lc).
   SYNONYMS: *afganensis* Agadzhanyan and Yatsenko, 1984; *balchanensis* (Heptner and Shukurov, 1950); *dangarinensis* Golenishchev and Sablina, 1991.
   COMMENTS: Taxonomy and distribution of Afghanistan populations reported by Niethammer (1970) and Hassinger (1973), karyotype (2n = 58, FN = 60 or 61) documented by Lyapunova and Zagorodnyuk (1990), and morphometric and karyological analyses provided by Golenishchev and Sablina (1991), who recognized three subspecies.

Blanfordimys bucharensis (Vinogradov, 1930). Rukovodstvok opredelenlyu gryzunov Srednei Azii [Key to Determine Rodents of Central Asia], p. 45.
   COMMON NAME: Bucharian Vole.
   TYPE LOCALITY: Tajikistan, Zeravshan Range, 8 km S Pendzhikent, near village of Zivan, 2200 m.
   DISTRIBUTION: Mountains of SW Tajikistan, possibly N Afghanistan; limits unresolved.
   STATUS: IUCN – Lower Risk (nt).
   SYNONYMS: *bucharicus* (Vinogradov, 1931); *davydovi* Golenishchev and Sablina, 1991.
   COMMENTS: Reviewed by Gromov and Erbajeva (1995). Ellerman (1941) listed *bucharensis* as species, but it is usually included in *afghanus* (Corbet, 1978*c*; Ellerman and Morrison-Scott, 1951; Ognev, 1964; Pavlinov and Rossolimo, 1987). Morphometric analyses and chromosomal complement (2n = 48, FN = 52) indicate that *bucharensis* is distinct, with two morphological and geographical components, one newly described as the subspecies *davydovi* (Golenishchev and Sablina, 1991). See Pavlinov and Rossolimo (1987:195) for use of *bucharensis* instead of *bucharicus*.

Caryomys Thomas, 1911. Abstr. Proc. Zool. Soc. Lond., 1911:4.
   TYPE SPECIES: *Microtus* (*Eothenomys*) *inez* Thomas, 1908.
   COMMENTS: Myodini. Thomas (1911*c*, *d*) proposed *Caryomys* as a subgenus of *Microtus* to contain the Chinese species *eva*, *inez*, and *nux* (now included in *inez*; see below). Hinton (1923) at first elevated *Caromys* to genus but later (1926*a*) included it in *Evotomys* (= *Myodes*) because he considered the holotypes of *eva*, *inez*, and *nux* to be young examples of *E. rufocanus shanseius*, a synonymy followed by others (Ellerman, 1941; Ellerman and Morrison-Scott, 1951; Gromov and Polyakov, 1977). A. B. Howell (1929), however, realigned *Caryomys* as a subgenus of *Microtus* and adamantly declared *inez* to be valid and different from any *Clethrionomys* (= *Myodes*); G. M. Allen (1940) concurred in recognizing

eva and *inez* as species but in the subgenus *Caryomys* of *Eothenomys*. While the specific validity of both is currently accepted (Corbet and Hill, 1992; Kaneko, 1992*c*), *Caryomys* has
either remained sequestered in *Eothenomys* (Corbet, 1978*c*; Musser and Carleton, 1993;
McKenna and Bell, 1997; Pavlinov et al., 1995*a*) or ignored (Corbet and Hill, 1992).

Recent generic resurrection of *Caryomys* issues primarily from the karyotypic study of
Ma and Jiang (1996). Both *eva* and *inez* have 2n = 54 with mostly telocentric pairs; all
Myodes and *Eothenomys* sampled have 2n = 56, with one metacentric pair (Ando et al.,
1988; Gamperl, 1982; Harada et al., 1991; Iwasa et al., 1999*a*, *b*; Kaneko et al., 1998; Ma
and Jiang, 1996; Modi and Gamperl, 1989; Obara et al., 1995; Sokolov et al., 1990;
Yoshida et al., 1989). In pelage texture and coloration, teat number (four versus eight in
Myodes), and rootless molars, *eva* and *inez* resemble *Eothenomys* (see G. M. Allen, 1940).
Their molars, however, match species of *Myodes*, or as Thomas (1911*d*:175) characterized
their occlusal patterns, "the teeth with the triangles nearly all closed [like *Myodes*], instead
of being mostly open and connected with each other [as in *Eothenomys*]" (also see G. M.
Allen, 1940; Hinton, 1923, 1926*a*).

Karyotypes, an *Eothenomys*-like external morphology and rootless molars, and a
Myodes-like occlusal pattern identify *eva* and *inez* as a monophyletic group (genus), at
which rank *Caryomys* was reviewed by Ye et al. (2002). According to these data, inclusion
of *eva* and *inez* in either *Eothenomys* or *Myodes* would make the latter polyphyletic and un-
diagnosable; the monophyly of *Caryomys* and its phylogenetic separation from the latter
require critical evaluation.

Caryomys eva (Thomas, 1911). Abstr. Proc. Zool. Soc. Lond., 1911(90):4.

> COMMON NAME: Eva's Red-backed Vole.
>
> TYPE LOCALITY: China, Gansu (Kansu), SE of Tauchow, 10,000 ft (3028 m).
>
> DISTRIBUTION: China, mountains of S Gansu and adjoining Shaanxi, Sichuan, and Hubei,
> 2400-3600 m; Zhang et al. (1997) included Ningxia and Qinghai provinces in the
> distribution.
>
> STATUS: IUCN – Lower Risk (lc) as *Eothenomys eva*.
>
> SYNONYMS: *alcinous* (Thomas, 1911); *aquilus* (G. M. Allen, 1912).
>
> COMMENTS: Listed as a member of *Evotomys* or *Clethrionomys rufocanus* by Hinton (1926*a*),
> Ellerman and Morrison-Scott (1951), and Gromov and Polyakov (1977). Correctly
> acknowledged and revised as a species (of *Eothenomys*) by G. M. Allen (1940), Corbet
> (1978*c*), Corbet and Hill (1992), and Kaneko (1992*c*), as followed by Musser and
> Carleton (1993) and Pavlinov et al. (1995*a*). Phallic morphology described by Yang
> et al. (1992) and contrasted with species of *Eothenomys*. Corbet and Hill (1992) noted
> that the southern populations (*alcinous*) have much darker upperparts and underparts
> than the slight frosting of typical *eva*. Kaneko's (1991) morphometric analysis
> associated the holotypes of *alcinous* and *aquilus* among samples of *eva*.

Caryomys inez (Thomas, 1908). Abstr. Proc. Zool. Soc. Lond., 1908(63):45.

> COMMON NAME: Inez's Red-backed Vole.
>
> TYPE LOCALITY: China, Shanxi (Shansi), mtns 12 mi (19 km) NW Kolanchow, 7000 ft
> (2134 m).
>
> DISTRIBUTION: N Sichuan and SE Shaanxi through Shanxi provinces, China, possibly farther
> east.
>
> STATUS: IUCN – Lower Risk (lc) as *Eothenomys inez*.
>
> SYNONYMS: *nux* (Thomas, 1910).
>
> COMMENTS: Following Hinton (1926*a*), *inez* was synonymized with *Clethrionomys rufocanus*
> (Ellerman and Morrison-Scott, 1951; Gromov and Polyakov, 1977), but its distinctive
> status is well documented, typically within *Eothenomys* (G. M. Allen, 1940; Corbet,
> 1978*c*; Corbet and Hill, 1992; Kaneko, 1992*c*). Hinton (1926*a*) and Ellerman and
> Morrison-Scott (1951) allocated *nux* to *Clethrionomys rufocanus shanseius*, but
> G. M. Allen (1940) treated it as a subspecies of *Eothenomys inez*. Kaneko's (1991)
> morphometric analysis identified the holotype of *nux* as an example of *inez*, as
> concluded earlier by A. B. Howell (1929). Kaneko restricted the range to Shaanxi and
> Shanxi provinces, but Zhang et al. (1997) mapped a far broader distribution.

Chionomys Miller, 1908. Ann. Mag. Nat. Hist., ser. 8, 1:97.

TYPE SPECIES: *Arvicola nivalis* Martins, 1842.

COMMENTS: Arvicolini, subtribe Arvicolina (Pavlinov et al., 1995*a*). We provisionally retain *Chionomys* in Arvicolini, following Gromov and Polyakov (1977), but stress that its tribal affinities are unresolved. Others have viewed the genus as a member of Myodini, as based on allozymic data (Mezhzherin et al., 1995) or on the stratigraphic sequence of fossils that suggest common ancestry with *Myodes* (Kretzoi, 1969; Chaline, 1987).

Miller (1912*a*), although describing *Chionomys* as a genus, later employed it as subgenus, a status that became entrenched in the literature (Chaline et al., 1999; Corbet, 1978*c*; Krapp, 1982*a*; Neuhäuser, 1936) with rare dissent (e.g., Gromov and Polyakov, 1977). A diverse information base, however, depicts *Chionomys* as a lineage apart from *Microtus* (Chaline and Graf, 1988; Graf, 1982; Kryštufek, 1999*c*; Mezhzherin et al., 1993, 1995; Nadachowski, 1990*a*; Pavlinov and Rossolimo, 1987, 1998; Pavlinov et al., 1995*a*; Zagorodnyuk, 1990). Van der Meulen (1978) considered *Suranomys* to be a junior synonym, but the type species of *Suranomys* (= *Microtus malei*) is related to the *Microtus oeconomus* group, not to *Chionomys* (Nadachowski, 1990*a*). New World *Microtus longicaudus* was referred to *Chionomys* by Anderson (1960), but a variety of data sources allies the former with *Microtus* proper (Chaline and Graf, 1988; Conroy and Cook, 1999, 2000; Graf, 1982).

Gromov and Polyakov (1977) provided detailed morphological and geographical descriptions of *Chionomys* and its three species, which they viewed as distinctive montane forms with ecologies convergent to those of *Alticola* and *Dinaromys*. Karyotypic variation is reported by Sablina et al. (1988) and Zima and Kral (1984*a*). Nadachowski (1991) discussed the systematics, geographic variation and evolution of *Chionomys* species based on dental traits. Kryštufek (1999*c*:335) remarked that snow voles "live in a narrow band of the south-western Palaearctic which is approximately 5500 km long and up to 2500 km broad." All three species are sympatric in the Pontic Mtns, NE Turkey and the Caucasus (Kryštufek, 1999*c*; Nadachowski, 1990*a*), and they also occurred there during the Pleistocene (Nadachowski and Baryshnikov, 1991). Nadachowski (1990*a*, 1991) suggested that *C. nivalis* evolved in Europe by the late Pleistocene and the others evolved in the Caucasus-Near East region, with *C. roberti* splitting from a *gud-roberti* ancestral stock by middle Pleistocene. Kryštufek (1999*c*) considered this hypothesized speciation in light of Pleistocene sea level oscillations around the Bosphorous landbridge and their vicariant effect (see Hosey, 1982).

Chionomys gud (Satunin, 1909). Izv. Kavkas. Mus., 4:272.

COMMON NAME: Gudaur Snow Vole.

TYPE LOCALITY: Georgia, Caucasus Mtns, Gudaur, near Krestovskii Pass.

DISTRIBUTION: Caucasus in Russia, Georgia, and Azerbaijan, and the E Black Sea Mtns of NE Turkey.

STATUS: IUCN – Lower Risk (nt).

SYNONYMS: *gotschobi* (Shidlovsky, 1919) [*nomen nudum*]; *ighesicus* (Ellerman and Morrison-Scott, 1951); *lasistanius* (Neuhäuser, 1936); *lghesicus* (Shidlovsky, 1919); *lucidus* (Shidlovsky, 1919) [*nomen nudum*]; *nenjukovi* Formosov, 1931; *oseticus* (Shidlovsky, 1919).

COMMENTS: Ellerman and Morrison-Scott (1951) and Corbet (1978*c*) listed *gotschobi* and *lghesicus* as synonyms of *nivalis*, but Pavlinov and Rossolimo (1987, 1998) and Pavlinov et al. (1995*a*) included them in *C. gud*. Dental traits indicate a closer relationship to *C. roberti* than to *C. nivalis* (Nadachowski, 1991). In Turkey, the species has been collected at the same localities as *C. nivalis* but prefers more mesic habitats (Kryštufek, 1999*c*).

Chionomys nivalis (Martins, 1842). Rev. Zool. Paris, p. 331.

COMMON NAME: European Snow Vole.

TYPE LOCALITY: Switzerland, Berner Oberland, Faulhorn.

DISTRIBUTION: Mountains of S Europe (Pyrenees, the Alps, the Carpathians, Balkan mountains, Pindhos Range and Mt Olimbos), east to Turkey, W Caucasus (Armenia, Georgia), Lebanon, W Syria, Zagros and Elburz Mtns of W and N Iran, and Kopet Dag of S Turkmenistan.

STATUS: IUCN – Lower Risk (nt).

SYNONYMS: *abulensis* (Agacino, 1936); *aleco* Paspalev, Martino and Peshev, 1952; *alpinus* (Wagner, 1843); *appenninicus* (Dal Piaz, 1929); *aquitanius* (Miller, 1908); *cedrorum* Spitzenberger, 1973; *dementievi* (Heptner, 1939); *hermonis* (Miller, 1908); *lebruni* (Crespon, 1844); *leucurus* (Gerbe, 1852) [not Blyth, 1863]; *loginovi* (Ognev, 1950); *malyi* (Bolkay, 1925); *mirhanreini* (Schäfer, 1935); *nivicola* (Schinz, 1845); *olympius* (Neu-häuser, 1936); *petrophilus* (Wagner, 1853); *pontius* (Miller, 1908); *radnensis* (Ehik, 1942); *satunini* (Shidlovsky, 1919); *spitzenbergerae* Nadachowski, 1990; *trialeticus* (Shidlovsky, 1919); *ulpius* (Miller, 1908); *wagneri* Martino, 1940.

COMMENTS: European populations reviewed by Krapp (1982*a*) and Mitchell-Jones et al. (1999), those in Russia and nearby regions by Gromov and Erbajeva (1995). Other regional and geopolitical accounts of distribution and taxonomy include the Navarra region, N Spain (Castien and Gosalbez, 1992, as *Microtus nivalis*); Switzerland (Hausser, 1995; Maurizio, 1994); Italy (Amori et al., 1999), especially N Italy (Cantini, 1991; Cresti et al., 1994; Locatelli and Paolucci, 1996*a*, *b*; Paolucci et al., 1993); Alpic, Dinaric, and Shara-Pindic mountain systems in Serbia and Montenegro (Petrov, 1992); Slovakia (Mošanský, 1994; Stanko and Mošanský, 1994, 2000); Slovenia (Kryštufek, 1991); Turkey (Kryštufek, 1999*c*); Greece (Niethammer, 1987*b*); the Middle East (Harrison and Bates, 1991); and Iran (Lay, 1967).

Kratochvíl (1981*a*) analyzed intraspecific variation among Carpathian samples and reviewed European and Turkish subspecies. The subspecies *spitzenbergerae* denotes a population from S Turkey that was previously misidentified by Spitzenberger (1971*b*) as *C. gud*, which is known only from NE Turkey where it is sympatric with *C. nivalis* (Nadachowski, 1990*b*). Association of the overlooked *appenninicus* is discussed by Kryštufek (1997). Vertical distribution in Serbia and Montenegro documented by Kryštufek and Kovacic (1989); geographic variation within Austria and Serbia and Montenegro by Kryštufek (1990); distribution in Portugal, Spain and France and morphometric distinctions from regional *Microtus* by Madureira (1983). Chromosomal variation in Bulgarian populations reported by Peshev and Belcheva (1979); Zima et al. (1997*a*) documented karyotypes from a Macedonian population and reviewed chromosomal data from other geographic areas. Allozymic variation within N Italy and Israel suggests that the Israeli population from Mt Hermon (*hermonis*) is a separate species (Filippucci et al., 1991). Nadachowski (1992) analyzed dental patterns of Quaternary and extant populations to reconstruct timing and routes of past migrations; he also mapped the insular montane distribution from Spain to Iran, the scientific names applied to each segment, and Quaternary fossil sites. Although now absent from the British Isles, *C. nivalis* occurred in S England during the last Pleistocene glaciation (Kowalski, 1967; Nadachowski, 1991).

Chionomys roberti (Thomas, 1906). Ann. Mag. Nat. Hist., ser. 7, 17:418.

COMMON NAME: Robert's Snow Vole.

TYPE LOCALITY: Turkey, Sumela (=Meryemana), south of Trebizond (Kryštufek, 1999*c*).

DISTRIBUTION: Forests in W Caucasus (Georgia and Azerbaijan), and in E Black Sea Mtns, NE Turkey.

STATUS: IUCN – Lower Risk (nt).

SYNONYMS: *circassicus* (Heptner, 1948); *occidentalis* (Turov, 1928); *personatus* Ognev, 1924; *pshavus* (Shidlovski, 1919); *turovi* (Hoffmeister, 1949).

COMMENTS: Distributional and biological aspects of Turkish populations reported by Kryštufek (1999*c*), who noted slight chromosomal differences between those and Caucasus populations.

Dicrostonyx Gloger, 1841. Gemein Hand.- Hilfsbuch. Nat., 1:97.

TYPE SPECIES: *Mus hudsonius* Pallas, 1778.

SYNONYMS: *Borioikon* Poliakov, 1881; *Cuniculus* Wagler, 1830 [not of Brisson, 1762, Gronovius, 1763, or Mayer, 1790]; *Misothermus* Hensel, 1855; *Tylonyx* Schulze, 1897.

COMMENTS: Dicrostonychini. *Dicrostonyx* was initially grouped with other lemmings following Miller's (1896) classic Lemmi-Microti division (e.g., Ellerman, 1941; Hinton, 1926; Ognev, 1963; Simpson, 1945). A large and diverse information base, however, requires its tribal separation from the true lemmings (Lemmini) and suggests that the cladistic origin of *Dicrostonyx* dates to the earliest radiation of arvicolines (Carleton, 1981; Chaline and Graf, 1988; Conroy and Cook, 1999; Gromov and Polyakov, 1977; Hinton, 1926*a*; Hooper and Hart, 1962; Kretzoi, 1969; Mezhzherin et al., 1995; Modi, 1987, 1996). Various aspects of taxonomy, karyology, distribution, and ecology are summarized by Stenseth and Ims (1993). Pliocene-Pleistocene changes in molar complexity traced by Agadzhanyan (1986); Quaternary fossils place the generic occurrence in North America far to the south of its present distribution, e.g., in Wyoming, Iowa, and Maryland (Graham and Lundelius, 1994; Mead and Mead, 1989).

The simple viewpoint of a single circumpolar species, *D. torquatus*, as advocated by Ognev (1963) and Rausch (1953, 1963), was unsettled by karyological and breeding studies in recent decades that implied a geographically partitioned superspecies complex (Chernyavskii and Kozlovskii, 1980; Gileva, 1983; Krohne, 1982; Rausch, 1977; Rausch and Rausch, 1972). The occurrence of collared lemmings in quite different tundra biotopes (e.g., see Youngman, 1975) alone disputed the existence of a single species within North America. Later authorities either have listed the North American karyotypic variants as species (Baker et al., 2003*b*; Corbet and Hill, 1991; Honacki et al., 1982; Jones et al., 1986, 1997; Musser and Carleton, 1993); or continued to recognize most as subspecies of *D. groenlandicus*, retaining only *D. exsul* on St. Lawrence Isl and *D. hudsonius* on the Ungava Peninsula as separate (Hall, 1981); or modified Hall's treatment to yield two New World species, *D. hudsonius* and *D. groenlandicus*, the latter embracing *exsul* and *vinogradovi* on Wrangel Isl (Jarrell and Fredga, 1993).

The kinds of studies needed to settle this indecision emerged in the 1990s (Borowik and Engstrom, 1993; Eger, 1995; Ehrich et al., 2000; Engstrom et al., 1993; Fedorov et al., 1999*a*; Van Wynsberghe and Engstrom, 1992). While the framework for taxonomic understanding is thus much improved, straightforward declarations of synonymy of studied taxa would help much to signal systematic conclusions. As indicated in 1993, the specific recognition of *D. groenlandicus*, *D. hudsonius*, *D. richardsoni*, and *D. unalascensis* is defensible on morphological grounds, and such an interpretation is generally consistent with these recent research contributions. The distinctiveness of other North American taxa, namely *kilangmiutak* and *rubricatus*, has not been sustained by this body of research, which points to their synonymy under *D. groenlandicus* (Eger, 1995; Ehrich et al., 2000). We continue to provisionally retain as species *D. nelsoni* and *D. nunatakensis* (specimens of latter not seen), examples of which have yet to be critically addressed in the on-going systematic review of *Dicrostonyx*; nor is the evidence for merging *D. vinogradovi* as another form of *D. groenlandicus*, as advanced by Jarrell and Fredga (1993), sufficiently convincing to date.

Dicrostonyx groenlandicus (Traill, 1823). *In* Scoresby, J. Voy. to Northern Whale-Fishery . . . , p. 416.
COMMON NAME: Nearctic Collared Lemming.
TYPE LOCALITY: Greenland, Jamesons Land.
DISTRIBUTION: N Greenland and Queen Elizabeth Isls, islands in the District of Franklin, and Southampton Isl; N North America above treeline, from NE District of Keewatin, Canada, to N Alaska, USA.
STATUS: IUCN – Lower Risk (lc) as *D. groenlandicus*, *D. kilangmiutak*, and *D. rubricatus*.
SYNONYMS: *alascensis* Stone, 1900; *clarus* Handley, 1953; *kilangmiutak* Anderson and Rand, 1945; *lentus* Handley, 1953; *rubricatus* (Richardson, 1889).
COMMENTS: In their broad specific concept, Jarrell and Fredga (1993) viewed all North American taxa, except *D. hudsonius*, as junior synonyms, including the insular forms *exsul* and *vinogradovi*. Morphological, distributional, breeding, chromosomal, and-or molecular evidence, albeit uneven and incomplete, persuades us to maintain *D. nelsoni*, *D. nunatakensis*, *D. richardsoni*, *D. unalascensis*, and *D. vinogradovi* as distinct (see those accounts).

Standard and banded chromosomal comparisons reported by Borowik and Engstrom (1993), who supported the synonymy of *clarus* and *lentus*. Engstrom et al. (1993) reported *kilangmiutak* (2n = 47-50) as karyotypically separable from *D. groenlandicus* (2n = 38-44), calling them "cytospecies," but found the two to be only marginally differentiated in mitochrondrial DNA sequences; samples drawn from the range of *kilangmiutak* are morphometrically (Eger, 1995) and genetically (Ehrich et al., 2000) unremarkable in comparisons with *D. groenlandicus* proper. The form *rubricatus* (Beringian distribution) shares a distinctive XY-autosomal fusion pattern with *D. groenlandicus* (Pearyland distribution), different from that found in *D. richardsoni* (Borowik and Engstrom, 1993); levels of differentiation in cranial form (Eger, 1995) and nucleotide sequences (Ehrich et al., 2000) provide no persuasive evidence for specific separation of *rubricatus*, although the latter study did disclose weakly defined western and eastern clades divided by the Mackenzie River.

Dicrostonyx hudsonius (Pallas, 1778). Nova Spec. Quad. Glir. Ord., p. 208.
>   COMMON NAME: Ungava Collared Lemming.
>   TYPE LOCALITY: Canada, Labrador.
>   DISTRIBUTION: NE Labrador, N Quebec, and Belcher Isls in Hudson Bay, Canada.
>   STATUS: IUCN – Lower Risk (lc).
>   COMMENTS: Unbanded karyotype resembles that of *D. richardsoni* (see Krohne, 1982); lack of XY-autosomal fusion believed to be primitive within the genus (Borowik and Engstrom, 1993). Relatively early origination, possibly well before the Wisconsin glaciation, proposed from collared lemming stock isolated in periglacial tundra to the southeast of the continental ice sheet (Borowik and Engstrom, 1993; Eger, 1995). Geographic and nongeographic variation evaluated by Eger (1995).

Dicrostonyx nelsoni Merriam, 1900. Proc. Wash. Acad. Sci., 2:25.
>   COMMON NAME: Nelson's Collared Lemming.
>   TYPE LOCALITY: USA, Alaska, Norton Sound, St. Michael.
>   DISTRIBUTION: W Alaska and Alaska Peninsula, USA.
>   STATUS: IUCN – Data Deficient as *D. exsul*, Lower Risk (lc) as *D. nelsoni*.
>   SYNONYMS: *exsul* G. M. Allen, 1919; *peninsulae* Handley, 1953.
>   COMMENTS: Rausch and Rausch (1972) noted successful mating among F1 progeny of *D. nelsoni* and collared lemmings on St. Lawrence Isl (*exsul*), while Jarrell and Fredga (1993) reported meiotic incompatibility and reduced fertility between stocks of *D. nelsoni* and *rubricatus* (= *D. groenlandicus*). In view of the complex appearance of Beringia refugia in space and time and the possibility of isolation from northern or southern stocks (e.g., see MacDonald and Cook, 1996), the homogeneity of populations in western Alaska (*D. nelsoni*) and those of *D. groenlandicus* should be empirically demonstrated, including samples from Beringean landbridge islands (*exsul* and *D. vinogradovi*), the Alaska Peninsula (*peninsulae*), and Aleutian chain (*D. unalascensis*).

Dicrostonyx nunatakensis Youngman, 1967. Proc. Biol. Soc. Wash., 80:31.
>   COMMON NAME: Ogilvie Mountains Collared Lemming.
>   TYPE LOCALITY: Canada, Yukon Territory, Ogilvie Mtns, 20 mi (32 km) S Chapman Lake, 5500 ft (1646 m); 64°35'N, 138°13'W.
>   DISTRIBUTION: Known only from the Ogilvie Mtns, NC Yukon Territory, Canada.
>   STATUS: IUCN – Data Deficient.
>   COMMENTS: As remarked by Youngman (1967, 1975), this form contrasts markedly with nearby *rubricatus* and *kilangmiutak* (both = *D. groenlandicus*); tentatively retained as a species by Honacki et al. (1982), Musser and Carleton (1993), and Jones et al. (1997). The geographic isolation of this form, in rocky alpine tundra south of the High Arctic tunda zone, invites testing of refugial hypotheses using multiple data sets and applying a phylogeographic approach.

Dicrostonyx richardsoni Merriam, 1900. Proc. Wash. Acad. Sci., 2:26.
>   COMMON NAME: Richardson's Collared Lemming.
>   TYPE LOCALITY: Canada, Manitoba, Fort Churchill.
>   DISTRIBUTION: W coast of Hudson Bay west to vicinity of Great Slave Lake, District of MacKenzie, Canada; extent of westward distribution unknown.
>   STATUS: IUCN – Lower Risk (lc).
>   COMMENTS: Autosomal polymorphisms and distinctive sex chromosomes based on standard karyotypes reported by Van Wynsberghe and Engstrom (1992), who retained

richardsoni as distinct from *D. groenlandicus* and postulated its differentiation in a SC periglacial refugium. Strong divergence of mitochrondrial DNA genotypes, based on restriction fragment analysis, indicates that the origination of *D. richardsoni* predated the Wisconsin glaciation (Engstrom et al., 1993).

Dicrostonyx torquatus (Pallas, 1778). Nova Spec. Quad. Glir. Ord., p. 206.
COMMON NAME: Palearctic Collared Lemming.
TYPE LOCALITY: Russia, Siberia, mouth of Ob River.
DISTRIBUTION: Palearctic tundra from White Sea, W Russia, to Chukotski Peninsula, NE Siberia, and Kamchatka (Nikanorov, 2000); including Novaya Zemlya and New Siberian isls, Arctic Ocean (Corbet, 1978*c*; Jarrell and Fredga, 1993:Fig. 5).
STATUS: IUCN – Lower Risk (lc).
SYNONYMS: *chionopaes* G. M. Allen, 1914; *lenae* (Kerr, 1792); *lenensis* (Pallas, 1779); *pallida* (Middendorff, 1853); *ungulatus* (Von Baer, 1841).
COMMENTS: Once believed to encompass most or all New World populations (e.g., Rausch, 1953, 1963*b*), but karyotypic and breeding evidence (summarized by Jarrell and Fredga, 1993) supports the strict application of *D. torquatus* for only Eurasian populations (also see Fedorov et al., 1999*a*). Chromosomal traits of populations from the Polar Urals (*torquatus*), and those from the Laptev Sea coast, and Rautan Isl off the coast of the Chukotka Peninsula (*chionopaes*) are similar, and crosses between these two subspecies yield fertile progeny (Gileva, 1980). Unusual sex-chromosome constitution and other chromosomal information summarized by Gileva et al. (1980), Gileva (1983), and Zima and Král (1984). Phylogeographic clades identified from restriction-site analysis of mitochrondrial DNA are mostly congruent with chromosomal races (Fedorov et al., 1999*a*). Although absent from Great Britain's modern fauna, the species occurred there during Pleistocene and Last Glacial (Late Palaeolithic) times (Sutcliffe and Kowalski, 1976; Yalden, 1999).

Dicrostonyx unalascensis Merriam, 1900. Proc. Wash. Acad. Sci., 2:25.
COMMON NAME: Unalaska Collared Lemming.
TYPE LOCALITY: USA, Alaska, Umnak Isl.
DISTRIBUTION: Umnak and Unalaska isls of Aleutian Archipelago, Alaska, USA.
STATUS: IUCN – Data Deficient.
SYNONYMS: *stevensoni* Nelson, 1929.
COMMENTS: Conventionally viewed as a form of *D. groenlandicus* (Hall, 1981; Hall and Cockrum, 1953; Jarrell and Fredga, 1993); retained as a species based on its distinctive pelage, craniodental form, and lack of specialized nival pelage and foreclaws, traits thought to be unique among *Dicrostonyx* (Gilmore, 1933; Musser and Carleton, 1993; Nelson, 1929; Rausch and Rausch, 1972). The most cranially divergent among the North American forms that Eger (1995) included in her morphometric analyses.

Dicrostonyx vinogradovi Ognev, 1948. Zveri S.S.S.R. i prilezhashchikh stran: Gryzuny (prodolzhenie) [Mammals of the U.S.S.R. and adjacent countries], 6:509.
COMMON NAME: Wrangel Island Collared Lemming.
TYPE LOCALITY: Russia, SE Siberia, Wrangel Isl (Os. Vrangelya), off coast of Anadyr region.
DISTRIBUTION: Known only from the type locality.
STATUS: IUCN – Critically Endangered.
COMMENTS: Included in *D. torquatus* by Corbet (1978*c*), but Chernyavskii and Kozlovskii (1980) interpreted chromosomal data (2n = 28), morphological traits, and infertility with Siberian populations to validate *vinogradovi* as a separate species. Jarrell and Fredga (1993) speculated that such hybrid depression is unlikely with Alaskan populations having similar derived karyotypes (fewest telocentric chromosomes) and so synonymized *vinogradovi* under *D. groenlandicus sensu lato*; mitochondrial DNA haplotypes also link the Wrangel Isl form with specimens (provenience and taxon not specified) from Alaska (Fedorov, 1999; Fedorov et al., 1999*a*). Given the antiquity and endemism of *Lemmus* populations on Wrangel Isl (*L. portenkoi*, see Chernyavskii et al., 1993), the island's remoteness from the Alaskan mainland, and the complexity of Beringian landscapes, this proposed synonymy deserves confirmation with broader taxonomic sampling and other data.

Dinaromys Kretzoi, 1955. Acta Geol. Acad. Sci. Hung., 3:351.

TYPE SPECIES: *Microtus* (*Chionomys*) *marakovici* Bolkay, 1924 (= *Microtus bogdanovi* Martino, 1922).

COMMENTS: Pliomyini. Formerly referenced as *Dolomys* until Kretzoi (1955; also see Corbet, 1978c) explained the correct usage of *Dinaromys* for *bogdanovi*. Past tribal associations emphasize the archaic characteristics of this enigmatic genus, allocated to a monotypic subfamily Dolomyinae (Chaline, 1975), or to Ondatrini (see Corbet, 1978c), Clethrionomyini (Gromov and Polyakov, 1977; Hooper and Hart, 1962; McKenna and Bell, 1997), or Prometheomyini (Pavlinov et al., 1995a). Koenigswald (1980) uncovered no close resemblance between its molar enamel microstructure and that of any living arvicoline and suggested relationship with an extinct species of *Propliomys* (= *Pliomys*), a late Pliocene genus also placed in Pliomyini by Kretzoi (1969). Chaline et al. (1999) actually derived *Dinaromys* from the Pliocene *Pliomys hungaricus*. Kretzoi's (1969) referral of *Dinaromys* to the Pliomyini (or subtribe Pliomyi as per Gromov and Polyakov, 1977) reflects a stronger phyletic hypothesis (also see Chaline et al., 1999, and Zagorodnyuk, 1990). Phylogenetic analyses drawing upon molecular information would be enlightening.

Dinaromys bogdanovi (Martino, 1922). Ann. Mag. Nat. Hist., ser. 9, 9:413.

COMMON NAME: Balkan Snow Vole.

TYPE LOCALITY: W Serbia and Montenegro, Rijeka Prov., Montenegro, Cetinje.

DISTRIBUTION: Isolated pockets in karst mountains, sea level to 2200 m, of W Balkans— Dinaric Alps in Croatia, Bosnia and Herzegovina, Montenegro and Kosovo; and the Šara-Pindus Mtns of Macedonia; probably occurs in Albania and Greece.

STATUS: IUCN – Lower Risk (nt).

SYNONYMS: *coeruleus* Mirić, 1960; *grebenscikovi* (Martino, 1935); *korabensis* (Martino, 1937); *longipedis* Dulic and Vidimic, 1967; *marakovici* (Bolkay, 1924); *preniensis* (Martino, 1940); *trebevicensis* Gligić, 1959.

COMMENTS: Petrov (1992) and Mitchell-Jones et al. (1999) provided informative reviews. Zoogeographic aspects discussed by Petrov (1979); distribution in Montenegro and its zoological significance reported by Kryštufek and Vohralík (1992). Chromosomal data presented by Zima and Kral (1984a) and Zima et al. (1997a). Eight subspecies have been recognized, forming two groups that are distinguished by M1 patterns and genetic divergence (Mitchell-Jones et al., 1999); Kryštufek et al. (2000b) provided information on age determination and molar structure. The extant species is closely related to two Pleistocene species—*D. dalmatinus* from N Italy, Serbia and Montenegro, and S Greece (Petrov and Todorovic, 1982) and *D. topachevskii* from Uzbekistan (Nesin and Skorik, 1989)—and to the late Pliocene *D. allegranzii* from NE Italy (Sala, 1996). Middle Pleistocene fossils of *D. bogdanovi* in N Italy (Zanalda, 1994) underscore the relictual character of the modern range.

Ellobius Fischer, 1814. Zoognosia, 3:72.

TYPE SPECIES: *Mus talpinus* Pallas, 1770.

SYNONYMS: *Afganomys* Topachevski, 1965; *Afghanomys* Baryshnikov and Baranova, 1983; *Chthonergus* Nordmann, 1839; *Lemmomys* Lesson, 1842; *Myospalax* Blyth, 1846 [not Laxmann, 1769, or Hermann, 1783].

COMMENTS: Ellobiusini (emended by Pavlinov et al., 1995a, because Ellobiini Gill 1872 is a junior homonym of Ellobiinae Adams, 1858, whose type genus *Ellobium* is a mollusk). Gromov and Polyakov (1977) excluded *Ellobius* from arvicolines; Pavlinov and Rossolimo (1987) viewed the genus as Cricetidae *incertae sedis*, questioning whether it belonged in Arvicolinae or Cricetinae; and Gromov and Erbajeva (1995) included it in Cricetinae. Most workers, however, have recognized *Ellobius*, albeit highly specialized morphologically, as the only extant member of a tribe within Arvicolinae (Corbet, 1978c; Hooper and Hart, 1962; Kretzoi, 1969; McKenna and Bell, 1997; Pavlinov and Rossolimo, 1998; Pavlinov et al., 1995a; Topachevskii and Rekovets, 1982), an affiliation supported by DNA sequence (Just et al., 1995) and allozymic analyses (Mezhzherin et al., 1995). Ultrastructure and evolution of sex chromosomes discussed by Kolomiets et al. (1991). Two subgenera are recognized, *Ellobius* and *Afganomys*; Zagorodnyuk (1990) listed the latter as a genus. Topachevskii and Rekovets (1982) documented morphological trends and interspecific interpretations from the late Pliocene to Recent. The present range of *Ellobius* is only part of a wider distribution that once embraced Israel and North Africa in the middle to late Pleistocene (Jaeger, 1988*a*).

Ellobius alaicus Vorontsov et al., 1969. *In* Vorontsov (ed.), [The Mammals: Evolution, karyology, taxonomy, fauna], Novosibirsk, p. 127.
COMMON NAME: Alai Mole Vole.
TYPE LOCALITY: Kyrgyzstan, Alai Valley, between Sary-Tashem and Bardabo, 3300 m.
DISTRIBUTION: Recorded only from the Alai Mtns, S Kyrgyzstan.
STATUS: IUCN – Endangered.
COMMENTS: Subgenus *Ellobius*. Included, with reservation, in *E. talpinus* by Corbet (1978*c*), an assignment contradicted by chromosomal data and breeding results that reveal its closer relationship to *E. tancrei*, with which it is parapatric (Corbet, 1984; Lyapunova et al., 1990, and references therein); provisionally placed in that species by Pavlinov and Rossolimo (1987).

Ellobius fuscocapillus Blyth, 1843. J. Asiat. Soc. Bengal, 11:887.
COMMON NAME: Southern Mole Vole.
TYPE LOCALITY: Pakistan, Baluchistan Region, Quetta Div., Quetta.
DISTRIBUTION: E Iran, Afghanistan, W Pakistan, and S Turkmenistan in the Kopet Dag Mtns; outlying population in E Turkmenistan, 500 km northeast of the central range (Marochkina, 1996).
STATUS: IUCN – Lower Risk (lc).
SYNONYMS: *farsistani* Ugarov, 1928; *intermedius* Scully, 1887.
COMMENTS: Subgenus *Afganomys*. Chromosomal data presented by Vorontsov et al. (1980) and Lyapunova et al. (1980); in contrast to *E. lutescens* and *E. tancrei*, *E. fuscocapillus* has the standard mammalian XX/XY sex chromosomes (Just et al., 1995). *Ellobius fuscocapillus* does not occur today in the S Levant (Israel), but fossils document its former presence during Pleistocene intervals, about 220,000-110,000 and 80,000-60,000 years ago (see Tchernov, 1992, 1994, and references therein).

Ellobius lutescens Thomas, 1897. Ann. Mag. Nat. Hist., ser. 6, 20:308.
COMMON NAME: Transcaucasian Mole Vole.
TYPE LOCALITY: Turkey, Kurdistan, Van.
DISTRIBUTION: S Caucasus south through E Turkey and NW Iran.
STATUS: IUCN – Lower Risk (lc).
SYNONYMS: *legendrei* Goodwin, 1940; *woosnami* Thomas, 1905.
COMMENTS: Subgenus *Afganomys*. Treated by Corbet (1978*c*) as a synonym of *E. fuscocapillus* but shown to be a distinct species (Corbet, 1984; Vorontsov et al., 1980, and references therein), as earlier listed by Ellerman and Morrison-Scott (1951). The low diploid number (17) and "weird" mode of sex determination (both sexes XO, X unpaired in male meiosis, apparent lack of genomic differentiation between males and females) continue to stimulate inquiry (Baumstart et al., 2001; Just et al., 1995, 2002; Vogel et al., 1988, 1998; Zima and Kral, 1984*a*; and references cited in each). Vogel et al. (1998) concluded that the entire Y chromosome is lost in *E. lutescens*, and Baumstart et al. (2001) found evidence for a yet unknown gene responsible for testes determination. After extensive synthesis of the literature and equivocal search for other genes effecting gonadal differentiation, Just et al. (2002) remarked that the mechanism of sex determination in *E. lutescens* "remains bizarre." Vogel et al. (1998) speculated on the origin of *E. lutescens* from an *Ellobius* species with XX males, which they proposed as a very recent event in evolutionary time. Geographic distribution, morphology, and karyotype of population in eastern Turkey reported by Cokun (2001).

Ellobius talpinus (Pallas, 1770). Nova Comm. Acad. Sci. Petropoli, 14, I:568.
COMMON NAME: Northern Mole Vole.
TYPE LOCALITY: Russia, W Bank of Volga River, between Kuibyshev (= Samara) and Kostychi.
DISTRIBUTION: Steppes of S Ukraine and Crimea, east through Kazakhstan to N of Balkhash Lake, and in Turkmenistan.

STATUS: IUCN – Lower Risk (lc).

SYNONYMS: *ciscaucasicus* Sviridenko, 1936 [*nomen nudum*]; *murinus* (Pallas, 1770); *rufescens* (Eversmann, 1850); *tanaiticus* Zubko, 1940; *transcaspiae* Thomas, 1912.

COMMENTS: Subgenus *Ellobius*. Chromosomal polymorphism among samples from the Pamir-Alai Mtns analyzed by Lyapunova et al. (1980); other chromosomal data summarized by Zima and Kral (1984*a*). Cytological determination of Turkmenian populations as *E. talpinus* and comparisons with *E. tancrei* provided by Yakimenko and Lyapunova (1986); distribution and habitat preference in the Crimea detailed by Tovpinets (1993); in E Turkmenistan, *E. talpinus* and *E. tancrei* are separated by the Amudarya River (Marochkina, 1996).

Ellobius tancrei Blasius, 1884. Zool. Anz., 7:197.

COMMON NAME: Eastern Mole Vole.

TYPE LOCALITY: Kazakhstan, Zaysan Lake Valley, Kendyrlik (= Przevalskoie).

DISTRIBUTION: E Turkmenistan (Yakimenko and Lyapunova, 1986; Marochkina, 1996) and Uzbekistan, east through E Kazakhstan, to Mongolia and adjacent China in NW Xinjiang, Nei Mongol, N Shaanxi, N Gansu, and Ningxia (Zhang et al., 1997, as *E. talpinus*).

STATUS: IUCN – Lower Risk (lc).

SYNONYMS: *albicatus* Thomas, 1912; *coenosus* Thomas, 1912; *fusciceps* Thomas, 1909; *fuscipes* Vinogradov, 1936; *kastschenkoi* Thomas, 1912; *larvatus* G. M. Allen, 1924; *ognevi* Dukelsky, 1927; *orientalis* G. M. Allen, 1924; *ursulus* Thomas, 1912.

COMMENTS: Subgenus *Ellobius*. Corbet (1978*c*) included *tancrei* in *E. talpinus*, but the former is morphologically and chromosomally distinct and its geographic range is allopatric to that of *E. talpinus* (see Corbet, 1984; Pavlinov and Rossolimo, 1987; Yakimenko and Lyapunova, 1986). Chromosomal contrasts with *E. talpinus* reported by Yakimenko and Lyapunova (1986) and with *E. alaicus* by Lyapunova et al. (1990). This species has a 2n = 32-54, with an XX sex chromosome combination in both males and females, and like *E. lutescens*, has lost its Y chromosome (Just et al., 1995; Vogel, 1998).

Eolagurus Argyropulo, 1946. Vestn. Akad. Nauk Kazakh. SSR, 7-8:44.

TYPE SPECIES: *Georychus luteus* Eversmann, 1840.

COMMENTS: Lagurini (see account of *Lagurus*). Also placed variously in subtribe Lagurina, Prometheomyini (Pavlinov and Rossolimo, 1998; Pavlinov et al., 1995*a*), or in Arvicolini (McKenna and Bell, 1997). Corbet (1978*c*) viewed *Eolagurus* as part of *Lagurus*, but subsequent authorities have considered them separate genera (Corbet and Hill, 1991; Gromov and Erbajeva, 1995; Gromov and Polyakov, 1977; Pavlinov and Rossolimo, 1987, 1998; Pavlinov et al., 1995*a*; Zagorodnyuk, 1990). Descent of *Eolagurus* from the early Pleistocene, its inferred derivation from a rooted Pliocene ancestor, and elaboration of the enamel schmelzmuster summarized by Koenigswald and Tesakov (1997).

Taxonomic confusion surrounding origin and application of the generic name discussed by Pavlinov and Rossolimo (1987). Pavlinov (2002, in litt.) explained that Argyropulo (1946) based the genus on *luteus* and *przewalskii* but did not explicitly designate a type species; justification for regarding *luteus* as the type species is Argyropulo's passing comment that *przewalskii* might be a geographical race of *luteus*, suggesting its fixation by original monotypy. Pavlinov and Rossolimo (1987) pointed out that the first unambiguous validation of *Eolagurus* may originate from Corbet's (1978*c*:116) statement that it was intended as a subgenus for *luteus*.

Eolagurus luteus (Eversmann, 1840). Bull. Soc. Nat. Moscow, p. 25.

COMMON NAME: Yellow Steppe Lemming.

TYPE LOCALITY: Kazakhstan, NW of Aral Sea.

DISTRIBUTION: Dry steppes, semideserts, and stable sand dunes in Zaysan Lake basin of E Kazakhstan, NW Xinjiang, and W Mongolia north of the Gobi Desert (Gromov and Erbajeva, 1995; Gromov and Polyakov, 1997; Ma et al., 1987).

STATUS: IUCN – Lower Risk (cd).

SYNONYMS: *gromovi* Topatchevski, 1963; *praeluteus* Schevtschenko, 1965; *volgensis* Alexandrova, 1976.

COMMENTS: Gromov and Polyakov (1977) and Gromov and Erbajeva (1995) excellently covered morphology, geographic distribution, and phylogeny; cranial and dental morphology also described by Hinton (1926a). The three synonyms were originally appied as species to samples of *Eolagurus* from early and middle Pleistocene sediments, but are now regarded as extinct subspecies of *E. luteus* (Gromov and Polyakov, 1977).

Eolagurus przewalskii (Büchner, 1889). Wiss. Res. Przewalski Cent.-Asien, Reisen, Zool., I: (Säugeth.), p. 127.
COMMON NAME: Przewalski's Steppe Lemming.
TYPE LOCALITY: China, Qinghai, Tsaidam region, shore of Iche-zaidemin Nor.
DISTRIBUTION: Montane meadows and river banks, from S Xinjiang and N Xizang, W China, east through Quinghai and N Gansu to W and S Mongolia and Nei Mongol; limits unknown.
STATUS: IUCN – Lower Risk (lc).
COMMENTS: A distinctive species united with *E. luteus* by Corbet (1978c) but treated as separate by most systematists (G. M. Allen, 1940; Corbet and Hill, 1991; Musser and Carleton, 1993; Pavlinov et al., 1995a). Gromov and Polyakov's (1977) redescription of *E. przewalskii* and illustrations of crania and dentitions unequivocally discriminate the two, whose ranges overlap in the lakes region of W Mongolia.

Eothenomys Miller, 1896. N. Am. Fauna, 12:45.
TYPE SPECIES: *Arvicola melanogaster* Milne-Edwards, 1871.
SYNONYMS: *Anteliomys* Miller, 1896.
COMMENTS: Myodini. Closely related to *Alticola*, *Caryomys*, *Hyperacrius*, and *Myodes*, all of which are usually placed in Clethrionomyini (Gromov and Polyakov, 1977; Hooper and Hart, 1962; Koenigswald, 1980) or the subtribe Myodina, Prometheomyini (Pavlinov and Rossolimo, 1998; Pavlinov et al., 1995a). Under *Eothenomys*, Corbet (1978c) included *Phaulomys* (placed by Ye et al., 2002, in *Clethrionomys* and by Japanese systematists in *Eothenomys*), *Aschizomys* (listed here in *Alticola*), and *Caryomys* (treated as a separate genus). *Anteliomys* was proposed as a subgenus of *Microtus* (Hinton, 1923), then tranferred to *Evotomys* (Hinton, 1926a), and finally treated as a subgenus of *Eothenomys* (Osgood, 1932; G. M. Allen, 1940); recently, Ye et al. (2002) also observed subgeneric usage for *Anteliomys*, although they suggested its reinstatement as genus. Phallic morphology of *E. melanogaster* and *E. miletus* is described by Yang et al. (1992) and contrasted with *Caryomys eva*.

Molar patterns and the lack of pectoral and postaxillary mammae are among the traits distinguishing *Eothenomys* from *Myodes* (see account of *M. shanseius*), but most systematists emphasize the degree of molar hypsodonty in adults, rootless (evergrowing) in *Eothenomys* versus rooted in *Myodes* (Corbet and Hill, 1992; Hinton, 1926a; Kaneko, 1990, 1992c, 1996b). Rootless molars, however, have apparently been acquired independently in some lineages of *Myodes* (see that account). The endemic species on the Korean Peninsula (revised by Kaneko, 1990, as *E. regulus*), the N Chinese *shanseius* (transferred to *Eothenomys* by Kaneko, 1992c), and all the Japanese endemics formerly considered in *Eothenomys* or *Phaulomys* (*andersoni*, *imaizumii*, and *smithii*) have rootless molars, and their allocation to *Eothenomys* has not been questioned until recently (all have also been regarded as species of *Clethrionomys*, here = *Myodes*; see Ellerman and Morrison-Scott, 1951). Phylogenetic analyses of mitochrondrial and nuclear ribosomal DNA convincingly demonstrated that these Korean and Japanese endemics are closely related to species of *Myodes* and form a monophyletic group that excludes *Eothenomys melanogaster* (Suzuki et al., 1999b; Wakana et al., 1996).

Does *Eothenomys* simply represent a southern continental extension of *Myodes*? The Palearctic distribution of *Myodes* mostly lies above 45°N latitude, with southeastern projections onto the Korean Peninsula, adjacent mainland China, and the main Japanese isls (see Gromov and Erbajeva, 1995). In contrast, species belonging to *Eothenomys*, as we view it, are distributed in the mountains of S China, south of about 40°N latitude, with limited extensions into NE India, N and WC Burma, N Thailand, NW Vietnam (west of the Red River), and probably EC Burma (east of the Salween River), and N Laos. In phylogenetic analyses of mitochondrial and ribosomal DNA sequences, *melanogaster* (type species of *Eothenomys*) was nested between *rutilus* (type species of *Myodes*) and the monophyletic *Myodes rufocanus* cluster (*andersoni, smithii, imaizumii, regulus, rex*) (Suzuki et al., 1999*b*). Systematists should examine other species of *Eothenomys* in a phylogenetic frame-work and revisit the significance of rootless versus rooted molars in our generic con-structs.

Corbet and Hill (1992:401) wrote of *Eothenomys* that "The southern forms are badly in need of revision, especially to establish the limits of *E. melanogaster* in which Ellerman & Morrison-Scott (1951) included a great diversity of forms that seem unlikely to be con-specific." Thanks to Kaneko's careful studies (1992*c*, 1996*b*, 2002), most of those species have been delineated by diagnostic morphological traits and their geographic distribu-tions empirically documented. Kaneko's revisions and the species we recognize here (based on specimen examination) in some ways reflect the treatment of Chinese *Eothen-omys* by G. M. Allen (1940), which for the time was an insightful portrayal of species di-versity. Consult also the review by Ye et al. (2002), which derives from an earlier revision that we have not seen.

Anteliomys is sometimes used as a subgenus but authors have differently allocated the species (compare G. M. Allen, 1940, and Pavlinov et al., 1995*a*). We arrange the species into two groups, based largely on molar patterns as described by Kaneko (1996*b*): *E. mela-nogaster* group (*E. cachinus, E. melanogaster, E. miletus*) and *E. chinensis* group (*E. chinensis, E. custos, E. olitor, E. proditor, E. wardi*). Kaneko (1996*b*:106) also recognized two groups of species occurring in Sichuan and Yunnan: one consists of the five species in his *E. chi-nensis* group; the other "is the *E. melanogaster* group, which includes *confinii, eleusis, fidelis, miletus* and *mucronatus*." Kaneko (2002) has recently revised this latter complex, but our interpretation of species limits and distributions, based upon specimen study and pertinent literature, differs somewhat from his. These two assemblages generally mirror former subgeneric groups; their apparent monophyly should be tested with analyses of a variety of morphological, chromosomal, and molecular data. *Eothenomys* dates from the late Pliocene in the Sichuan-Guizhou region, S China, where it is also represented by Pleistocene fossils identified as extinct *E. praechinensis* and living *E. chinensis* (both in *E. chinensis* species group), and by samples from late Pliocene and Pleistocene cave sedi-ments of extant *E. melanogaster* (Zheng, 1993; Zheng and Li, 1990).

Eothenomys cachinus (Thomas, 1921). J. Bombay Nat. Hist. Soc., 27:504.

COMMON NAME: Kachin Red-backed Vole.

TYPE LOCALITY: NE Burma, Kachin State, Imaw Bum, 9000 ft (2743 m).

DISTRIBUTION: Montane forest, 2300-3200 m, west of the Salween River Valley in NE Burma (Thomas, 1921*g*, the holotype; Anthony, 1941), and adjacent NW Yunnan (holotype of *confinii* Hinton, 1923); range may extend farther south to extreme W Yunnan and EC Burma.

SYNONYMS: *confinii* Hinton, 1923.

COMMENTS: *E. melanogaster* species group. Described by Thomas (1921*g*) as a species but sub-sequently included either in *E. melanogaster* (Hinton, 1923; Ellerman, 1941, 1961; Ellerman and Morrison-Scott, 1951; Corbet, 1978*c*; Musser and Carleton, 1993; Pavlinov et al., 1995*a*) or *E. miletus* (Corbet and Hill, 1992). Comparisons of Anthony's (1941) large AMNH sample, obtained at the type locality of *cachinus* and vicinity, with *E. miletus* and *E. melanogaster* proper confirm Thomas' view that *cachinus* is a distinc-tive species. Adult *E. cachinus* are large-bodied and long-tailed, with soft, long and thick fur, the upperparts bright tawny brown and the underparts gray washed with hues ranging from pale buff to ochraceous (the only species in the *E. melanogaster* group with such bright venters). They are about the same body size as the large *E. miletus*, but have longer tails (mean = 50.9 mm, range = 43-61 mm, *n* = 105) than *E. miletus* (43.6 mm, 35-48 mm, *n* = 39) and *E. melanogaster* (36.1 mm, 31-42 mm, *n* = 34); no other species in the *E. melanogaster* group has such a long tail. Compared with *E. miletus*, cranium of *E. cachinus* is slightly smaller in length and width dimen-sions and strikingly much lower in profile (flattened compared with the high, arched and stocky cranium of *E. miletus*); M3 has four lingual salient angles (like *E. miletus*), but also four labial angles in 95 of 109 specimens (three in *E. miletus*). Occlusal pattern consists of a single anterior lamina, two inverted chevron-shaped laminae, and an irregularly-shaped heel, while all other forms have only two anterior laminae and a heel with an angular labial projection. Thomas (1921g) clearly described the M3 pattern in *cachinus*, and Hinton (1923) used it to separate *cachinus* from other forms of *melanogaster* in his key to *Eothenomys*.

We synonymize the NW Yunnan *confinii* with *E. cachinus* in contrast to its peripatetic association under *E. eleusis* (G. M. Allen, 1940; Zhang et al., 1997), *E. melanogaster* (Corbet and Hill, 1992), *E. mucronatus* (Kaneko, 2002), or *E. miletus* (Ye et al., 2002). Hinton's (1923) description of the pelage and tail length (59 mm) matches that of *E. cachinus*, his cranial differences between *confinii* and *miletus* precisely distinguish *E. cachinus* from *E. miletus*, and *confinii* too was obtained west of the Salween River Valley adjacent to the region where Anthony (1941) worked. Kaneko (2002) relegated both *cachinus* and *confinii*, along with *miletus*, to *E. mucronatus* (here considered a junior synonym of *E. melanogaster*; see below). Examination of the same specimens studied by Kaneko (2002) supplies no morphological evidence for considering the three taxa as members of a single species. Specimens of *E. cachinus* (including *confinii*) and *E. miletus* can be consistently separated by cranial conformation and relative tail length (comparing specimens of similar age), as previously indicated.

In NE Burma, Anthony (1941) encountered *E. cachinus* (identified as *E. melanogaster cachinus*) in heavy forest, 2300-3200 m, and found the smaller *E. melanogaster* (reported as *E. m. libonotus*) at lower elevations, 1200-2750, in open habitats (meadows, near cultivation, and shrubby growth along streams). This region of Burma also harbors two endemic murines, *Niviventer brahma* (Musser, 1970b, 1973a) and an undescribed species of *Niviventer* (Musser and Lunde, ms). Curiously, Zhang et al. (1997) listed *cachinus* both as a separate species and as a synonym of *E. melanogaster*. The specimens that Ellerman (1961) recorded from N Burma are *E. melanogaster*.

Eothenomys chinensis (Thomas, 1891). Ann. Mag. Nat. Hist., ser. 6, 8:117.
  COMMON NAME: Sichuan Red-Backed Vole.
  TYPE LOCALITY: China, W Sichuan Province, Kia-ting-fu (= Leshan; see Kaneko, 1996b for additional information).
  DISTRIBUTION: Recorded only between 1500 and 3000 m "on both sides of the River Datu He near Omei Shan, Sichuan Province at 29-30° N" (Kaneko, 1996b:104, 105).
  STATUS: IUCN – Lower Risk (lc).
  SYNONYMS: *tarquinus* (Thomas, 1912).
  COMMENTS: *E. chinensis* species group. The type-species of *Anteliomys* (Hinton, 1926a) and recognized as a species since its description by Thomas in the usual 20th-century systematic compendia. This large-bodied species, with the longest tail of any *Eothenomys* (up to 76 mm), is allopatric to the ranges of *E. custos*, *E. proditor*, and *E. wardi*, and reaches lower altitudes (see Kaneko, 1996b). Kaneko (1996b) arranged *tarquinus* as a synonym.

Eothenomys custos (Thomas, 1912). Ann. Mag. Nat. Hist., ser. 8, 9:517.
  COMMON NAME: Southwest China Red-backed Vole.
  TYPE LOCALITY: China, Yunnan, A-tun-tsi (= Dequen Xian; see Kaneko, 1996b for additional information), 11,500-12,500 ft (3505-3810 m).
  DISTRIBUTION: Recorded only "from the extreme north-west of Yunnan, the Likiang Range, the loop of the Jinsha Jiang River, and from central Sichuan," 2500-4800 m (Kaneko, 1996b:108).
  STATUS: IUCN – Lower Risk (lc).
  SYNONYMS: *cangshanensis* Wang and Li, 2000; *hintoni* Osgood, 1932; *ninglangensis* Wang and Li, 2000; *rubelius* Hinton, 1932; *rubellus* (G. M. Allen, 1924).
  COMMENTS: *E. chinensis* species group. Listed as a species of *Anteliomys* by Hinton (1926a). Whether in *Eothenomys* or *Anteliomys*, this distinct Chinese endemic has always been recognized as a species in faunal reports and checklists. Kaneko (1996b) explained why the holotypes of *hintoni* and *rubelius* represent examples of *E. custos*; inclusion of *cangshanensis* and *ninglangensis* follows Ye et al. (2002).

Eothenomys melanogaster (Milne-Edwards, 1871). Nouv. Arch. Mus. Hist. Nat. Paris, Bull., 7:93.
  COMMON NAME: Père David's Red-backed Vole.
  TYPE LOCALITY: China, W Sichuan, Moupin.

DISTRIBUTION: Mountains of SE China in S Anhui, Zhejiang, Fujian (AMNH, MCZ, USNM), NE Jiangxi, N Guangodong, W Hubei (MCZ), Ghizhou (FMNH, USNM), Sichuan (AMNH, FMNH, MCZ, USNM), S Gansu, and W Yunnan (AMNH, FMNH, USNM) (as per specimens examined, G. M. Allen, 1940, and Zhang et al., 1997). Also in Taiwan (M.-J. Yu, 1996); NE India (Mishmi Hills in Arunachal Pradesh; Agrawal, 2000); N Burma (Ellerman, 1961, identified as *cachinus*; BMNH, FMNH); NE Burma and Chin Hills, EC Burma (Anthony, 1941; AMNH); N Thailand (summit Doi Inthanon, Chiengmai Prov.; J. T. Marshall, Jr., 1977a; MCZ, USNM); and extreme NW Vietnam west of the Red River (Osgood, 1932; Dang et al., 1994; FMNH, MCZ). Known altitudinal range 700-3000 m (Kaneko, 2002).

STATUS: IUCN – Lower Risk (lc).

SYNONYMS: *aurora* (G. M. Allen, 1912); *bonzo* (Cabrera, 1922); *chenduensis* Wang and Li, 2000; *colurnus* (Thomas, 1911); *eleusis* (Thomas, 1911); *kanoi* Tokuda, 1937; *libonotus* Hinton, 1923; *mucronatus* (G. M. Allen, 1912); *yingjiangensis* Wang and Li, 2000.

COMMENTS: *E. melanogaster* species group. Corbet (1978c) suspected that more than one species is represented in what he identified as *E. melanogaster*, and Corbet and Hill (1992:401) posed their treatment of the species as "very tentative." After checking large museum series, extracting *E. cachinus* and *E. miletus* along with their synonyms, we believe most synonyms listed here actually belong with *E. melanogaster*, allocations in most part concordant with Kaneko's (2002) revision. Specimens of *E. melanogaster* are generally small-bodied with small skulls, have dark brown to blackish upperparts, slate gray underparts (washed with buff or brown in some specimens), and possess a short to medium tail (21-42 mm) relative to body length. Cranial size varies geographically, with specimens from Sichuan and Yunnan averaging smaller than those from E China, Burma, N Thailand, and NW Vietnam. Three lingual salient angles on each M3 were considered to be typical for *E. melanogaster* (G. M. Allen, 1940; Corbet and Hill, 1992), but number of lingual angles varies within the species, as Anthony (1941) noted for his large sample from NE Burma and confirmed by survey of M3s in AMNH and FMNH material: NE Burma, 65 have four angles, 52 have three; WC Burma, one with four angles, seven with three; Sichuan, nine with four angles, 37 with three; Ghizhou, two with four angles; Yunnan, 25 with four angles; Fujian, four with three salient angles; all specimens from N Thailand and NW Vietnam uniformly have four lingual angles. Furthermore, range in expression of the fourth lingual angle varies from a prominent projection, through progressive size reductions, to undetectable. AMNH and FMNH specimens from Yunnan, identified as *E. eleusis* by G. M. Allen because they have long tails and four lingual angles, are examples of *E. melanogaster*. Taxonomic significance of the variation in M3 lingual angles, tail length, and cranial size, as outlined here and discussed by Kaneko (2002), should be assessed by careful morphometric and molecular analyses.

The taxa *bonzo, colurnus, kanoi, libonatus,* and *mucronatus* have been traditionally included in *E. melanogaster* (G. M. Allen, 1940; Corbet, 1978c; Hinton, 1923). Ye et al. (2002) listed *chenduensis* as a form of *E. melanogaster* and *yingjiangensis* as a subspecies of *E. eleusis*; Corbet and Hill (1992) also excluded *eleusis* from *E. melanogaster*. Kaneko (2002) used *mucronatus* as the oldest name for populations we identify as *E. miletus*, also including *chenduensis*; neither he nor we have have examined the holotype of *chenduensis* and its referral here is provisional. Allocation of the problematic taxa *aurora, eleusis,* and *mucronatus* requires comment.

Described as a species from W Hubei (G. M. Allen, 1912; see Kaneko, 2002:62, for type locality details), G. M. Allen (1940) later arranged *aurora* as a subspecies of *E. miletus* (curiously, Allen was unaware that *aurora* had priority, an oversight noted by Ellerman and Morrison-Scott, 1951:668). Included in *E. melanogaster* in most subsequent checklists and faunal studies, excepting Ye et al. (2002), who listed it as a subspecies of *E. eleusis*. The type series of *aurora* (MCZ 7788, holotype) consists of individuals that are larger than most *E. melanogaster*, but they resemble in size and pelage coloration *colurnus* from E China and samples from farther south in Burma, N Thailand, and NW Vietnam, which all have greater average cranial dimensions than do most samples from Sichuan and Yunnan (study of MCZ specimens). The W Hubei series consists of adults and younger animals, all with four lingual angles on M3; the adults had been originally identified on skin tags as *aurora*, the younger individuals as *melanogaster*. The oldest adults of *aurora* are only slightly smaller than *mucronatus* from W Sichuan.

Like *aurora*, G. M. Allen (1912) had originally described *mucronatus* as a species but subsequently (1940:809) considered it a full synonym of *E. melanogaster melanogaster*. Although larger than average for *E. melanogaster*, all specimens in the type series of *mucronatus* have three lingual angles, as is usual for *E. melanogaster*. In addition to the holotype, another adult from the type locality had been identified as *mucronatus* but a younger specimen was referred to *E. melanogaster*. While recognizing that the holotype of *mucronatus* (MCZ 7789) is slightly larger than typical *E. melanogaster* from Sichuan, it does not possess the even larger and stockier skull with high braincase that characterizes *E. miletus*; contrary to Kaneko's (2002) revision, we refer *mucronatus* to *E. melanogaster*, following G. M. Allen (1940). Kaneko also identified samples from C Burma, N Thailand, and N Vietnam as *E. mucronatus*, treating *libonotus* as its synonym. To us, material from mountains outside of China represents isolated populations of *E. melanogaster* that average slightly larger in size.

The taxon *eleusis* is based upon a sample from N Yunnan, 21 miles E Chao-tung-fu (= Zhaotong on modern maps, 27°20'N, 103°50' E). Described as a subspecies of *E. melanogaster* by Thomas (1911*e*), as also ranked by Hinton (1923, 1926*a*), and conventionally included therein (Corbet, 1978*c*; Ellerman, 1941; Ellerman and Morrison-Scott, 1951; Musser and Carleton, 1993; Pavlinov et al., 1995*a*). G. M. Allen (1940) regarded *eleusis* as a valid species because of its longer tail and M3 with four lingual salient angles, a status recently followed by others (Corbet and Hill, 1992; Wang, 2003; Ye et al., 2002; Zhang et al., 1997). Fur coloration, body length (less than 100 mm), and skull dimensions (occipitonasal length less than 25 mm) of the type series of *eleusis* match *E. melanogaster* (see G. M. Allen, 1940; Thomas, 1911*e*), but the tails are significantly longer (*eleusis*—mean = 48.9 mm, range = 46-55, *n* = 10; *melanogaster*—36.1 mm, 31-42 mm, *n* = 34) and every M3 has four lingual salient angles. Although tail length approaches *E. cachinus*, no other features are common to that species, which appears to be restricted to mountains west of the Salween River Valley, more than four degrees longitude west of Chao-tung-fu. Most *E. melanogaster* have only three lingual salient angles on the M3, but the number of angles varies within and among geographic samples (Kaneko, 2002; our observations). As Kaneko (2002) demonstrated, the type series of *eleusis* cannot be separated from other *E. melanogaster* except for their long tails; either these were inaccurately measured or the original sample represents the extreme in tail length for *E. melanogaster*.

Liu (1994) used bacular dimensions as useful for dividing Sichuan specimens into age groups. Conventional karyotype of Taiwan sample (2n = 56, FN = 58) like that characterizing species of *Myodes* (Harada et al., 1991), which prompted Ye et al. (2002) to exclude it from *Eothenomys* but leave its generic assignment as indeterminate. Ecology, altitudinal distribution and genetic substructure of Taiwanese populations extensively documented by H.-T. Yu (1993, 1994, 1995).

Eothenomys miletus (Thomas, 1914). Ann. Mag. Nat. Hist., ser. 8, 14:474.
    COMMON NAME: Yunnan Red-backed Vole.
    TYPE LOCALITY: SW China, W Yunnan, 10 m W Yang-pi, 7000 ft (2134 m).
    DISTRIBUTION: Mountains of SW Sichuan and W to C Yunnan east of the Salween River Valley. S China (as per specimens examined in AMNH, BMNH, FMNH, MCZ, and USNM); limits unresolved (in Ghizhou according to Zhang et al., 1997, but verification needed).
    SYNONYMS: *fidelis* Hinton, 1923.
    COMMENTS: *E. melanogaster* species group, its largest-bodied member (Kaneko, 1996*b*). Described as a subspecies of *melanogaster*, a placement commonly observed throughout the 1900s, less commonly regarded as distinct (G. M. Allen, 1940; Corbet and Hill, 1992; Ye et al., 2002; Zhang et al., 1997). The account of *miletus* by Thomas (1914*c*) and of *fidelis* by Hinton (1923) described examples of the same species, as G. M. Allen (1940) later perceived; pelage coloration and measurements of each holotype fall within the variation observed in the large AMNH and FMNH series from

Yunnan and W Sichuan. Kaneko (2002) used *mucronatus* as the oldest name for this species and included *libonotus* from NW India as well as samples from C Burma, N Thailand, and N Vietnam; we identify this material as *E. melanogaster* (see account).

Eothenomys miletus occurs in the same regions as *E. melanogaster* but is usually larger in body size and cranial dimensions with a strikingly higher cranium; a tawny brown to reddish brown dorsal coat that is soft, thick and long (shorter coat, velvety in texture, dark brown to melanistic in most *melanogaster*); and gray underparts either frosted or washed with brown to ochraceous hues (stark slate gray in *melanogaster*, infrequently washed with brown or buff). Particular samples of *E. melanogaster* (*aurora*, *colurnus*, and *mucronatus*, and those from N Burma, Thailand, and NW Vietnam) may overlap *E. miletus* in skull length and width, but they lack its deep and robust cranial conformation and large bulla. While an M3 with four lingual salient angles (*E. miletus*) versus three (*E. melanogaster*) has been used to separate them (G. M. Allen, 1940; Corbet and Hill, 1992; Hinton, 1923, 1926a), this character works only for specimens of the latter from Sichuan (see *E. melanogaster* account); furthermore, not all *E. miletus* possess four lingual angles—of 71 specimens from Yunnan and Sichuan (AMNH, FMNH), 64 have four angles but seven exhibit only three. Both species are recorded from the same localities in Yunnan (G. M. Allen, 1912, 1940; our study) but precise microhabitat information is unavailable. Wu et al. (1989) and Ye et al. (2002) recorded the 2n as 56, with all acrocentric elements.

Eothenomys olitor (Thomas, 1911). Abstr. Proc. Zool. Soc. Lond., 1911(100):50.
>   COMMON NAME: Black-eared Red-backed Vole.
>   TYPE LOCALITY: China, Yunnan, Chao-tung-fu (= Zhaotong Xian), 6700 ft [2042 m] (Kaneko, 1999b, provided additional information).
>   DISTRIBUTION: NE Yunnan at the type locality and SW Yunnan between the Mekong and Salween River valleys, 1800-3350 m (Kaneko, 1996b).
>   STATUS: IUCN – Lower Risk (lc).
>   SYNONYMS: *hypolitor* Wang and Li, 2000.
>   COMMENTS: *E. chinensis* species group. The smallest *Eothenomys* in body size. Hinton (1926a) allocated *olitor* to the subgenus *Eothenomys*, as did Corbet and Hill (1992), but others have placed it in *Anteliomys* (G. M. Allen, 1940; Gromov and Polyakov, 1977) or in the subgenus *Caryomys* (Pavlinov et al., 1995a). G. M. Allen (1940) extended the range to W Yunnan, but Kaneko (1996b) explained why this record is not *olitor*; inclusion of *hypolitor* observes the summary of Ye et al. (2002).

Eothenomys proditor Hinton, 1923. Ann. Mag. Nat. Hist., ser. 9, 11:152.
>   COMMON NAME: Yulongxuen Red-backed Vole.
>   TYPE LOCALITY: China, Yunnan, Lichiang Range (=Yulongxuen), 13,000 ft (Kaneko, 1996b, provided additional information).
>   DISTRIBUTION: "Restricted to the border between Sichuan and Yunnan, at around 27-28°N and 100-102°E, and that it lives in meadows and in rocky areas" (Kaneko, 1996b:109); known altitudinal range 2500-4200 m.
>   STATUS: IUCN – Lower Risk (lc).
>   COMMENTS: *E. chinensis* species group. Morphologically a giant version of *E. olitor* (Kaneko, 1996b). Treated as a species of *Eothenomys* (G. M. Allen, 1940; Corbet and Hill, 1992; Hinton, 1926a), but placed in *Anteliomys* (Gromov and Polyakov, 1977), and in subgenus *Caryomys* (Pavlinov et al., 1995a). In his account of *E. proditor*, Osgood (1932) explained why *Anteliomys* deserves at most subgeneric status within *Eothenomys*. Karyotype of *E. proditor* (2n = 32, FN = 56) differs from other *Eothenomys* so far reported (Yang et al., 1998), and Ye et al. (2002) would exclude it from the genus.

Eothenomys wardi (Thomas, 1912). Ann. Mag. Nat. Hist., ser. 8, 9:516.
>   COMMON NAME: Ward's Red-backed Vole.
>   TYPE LOCALITY: China, NW Yunnan, Chamutong (= Tra-mu-tang), west of Atunsi, 13,000 ft (Kaneko, 1996b, provided additional information).
>   DISTRIBUTION: Extreme NW Yunnan in the Mekong and Salween valleys, 2400-4250 m (Corbet and Hill, 1992; Kaneko, 1996b).
>   COMMENTS: *E. chinensis* species group. Placed in the genus *Anteliomys* by Hinton (1923).

Described as a species by Thomas, recognized as such by Hinton (1923, 1926a) and Ellerman (1941), but usually included in *E. chinensis* (G. M. Allen, 1940; Ellerman and Morrison-Scott, 1951; Musser and Carleton, 1993; Pavlinov et al., 1995a). Corbet and Hill (1992) and Kaneko (1996b), however, followed Thomas because *wardi* has a much shorter tail and hind feet, and appreciably smaller auditory bulla than specimens of *chinensis*, and is latitudinally isolated from that species.

Hyperacrius Miller, 1896. N. Am. Fauna, 12:54.
  TYPE SPECIES: *Arvicola fertilis* True, 1894.
  COMMENTS: Myodini. Assigned to Clethrionomyini by Hooper and Hart (1962) and Gromov and Polyakov (1977); placed in subtribe Myodina, Prometheomyini, by Pavlinov et al. (1995a) and Pavlinov and Rossolimo (1998). Phylogenetically near *Alticola* but more fossorial (Corbet, 1978c; Corbet and Hill, 1992). Taxonomy, geographic distribution, and subspecific classification monographed by Phillips (1969).

Hyperacrius fertilis (True, 1894). Proc. U. S. Natl. Mus., 17:10.
  COMMON NAME: Subalpine Kashmir Vole.
  TYPE LOCALITY: India, Kashmir, Pir Panjal Mtns, 8500 ft (2591 m).
  DISTRIBUTION: Subalpine scrub and meadows, 2450-3600 m, N India (Jammu and Kashmir) and N Pakistan.
  STATUS: IUCN – Lower Risk (lc).
  SYNONYMS: *aitchisoni* (Miller, 1897); *brachelix* (Miller, 1899); *zygomaticus* Phillips, 1969.
  COMMENTS: Indian populations reviewed by Agrawal (2000).

Hyperacrius wynnei (Blanford, 1881). J. Asiat. Soc. Bengal, 49:244.
  COMMON NAME: Conifer Kashmir Vole.
  TYPE LOCALITY: Pakistan, Murree, 7000 ft (2134 m) (as fixed by lectotype selection by Phillips, 1969:462).
  DISTRIBUTION: Coniferous forests and associated grasslands, 1850-3050 m, in N India (Jammu and Kashmir) and Pakistan (Murree Hills in the lower Kahgan Valley E of Indus River, and west of the Indus in Swat).
  STATUS: IUCN – Lower Risk (lc).
  SYNONYMS: *traubi* Phillips, 1969.
  COMMENTS: Indian populations reviewed by Agrawal (2000) and Pakistan by Phillips (1969).

Lagurus Gloger, 1841. Gemein Hand.-Hilfsbuch. Nat., 1:97.
  TYPE SPECIES: *Mus lagurus* Pallas, 1773.
  SYNONYMS: *Eremiomys* Poliakov, 1881; *Eremomys* Heude, 1898; *Lagurodon* Kretzoi, 1956; *Laguropsis* Kretzoi, 1956; *Prolagurus* Kormos, 1938.
  COMMENTS: Lagurini. Closely related to *Eolagurus*, which together constitute the extant members of Lagurini (Gromov and Polyakov, 1977; Hooper and Hart, 1962; Zagorodnyuk, 1990). Also arranged as subtribe Lagurina within Prometheomyini (Pavlinov and Rossolimo, 1998; Pavlinov et al., 1995a), Clethrionomyini (Mezhzherin et al., 1995), or Arvicolini (McKenna and Bell, 1997). Pavlinov and Rossolimo (1987) clarified the status of *Mus lagurus* as the type species, instead of *L. migratorius* as misindicated by Corbet (1978c). Koenigswald and Tesakov (1997) documented the enamel microstructure of *Lagurus* and related its significance for reconstructing relationships among extant and fossil lagurines, from the Pliocene to Recent. Excludes North American *Lemmiscus curtatus* (see that account).

Lagurus lagurus (Pallas, 1773). Reise Prov. Russ. Reichs., 2:704.
  COMMON NAME: Steppe Vole.
  TYPE LOCALITY: Kazakhstan, mouth of Ural River.
  DISTRIBUTION: Steppes, mountains, and northern deserts from Ukraine through Kazakhstan to W Altai steppes of S Russia, adjacent W Mongolia, and NW China (NW Xinjiang; Zhang et al., 1997).
  STATUS: IUCN – Lower Risk (lc).
  SYNONYMS: *abacanicus* Serebrennikov, 1929; *agressus* Serebrennikov, 1929; *altorum* Thomas,

1912; *major* Zazhigin, *in* Gromov and Polyakov, 1977; *migratorius* Gloger, 1841; *occidentalis* Migulin, 1938; *saturatus* Ognev, 1950 [*nomen nudum*].

COMMENTS: Morphology, geographic range, and phylogenetic history described by Gromov and Polyakov (1977) and Pavlinov et al. (1995a). Chromosomal data reviewed by Zima and Kral (1984a). Using geometric morphometry, Courant et al. (1997) noted cranial shape convergence between *L. lagurus* and lemmings (*Lemmus* and *Dicrostonyx*) and discussed ecological parallels between these two groups. Although the present range of *L. lagurus* does not extend west of Ukraine, the species occurred in Europe and S England during the last interglacial and glacial cycle (Kowalski, 1967; Sutcliffe and Kowalski, 1976).

Lasiopodomys Lataste, 1887. Ann. Mus. Civ. Stor. Nat. Genova, 2a, 4:268.

TYPE SPECIES: *Arvicola brandtii* Radde, 1861.

SYNONYMS: *Lemmimicrotus* Tokuda, 1941.

COMMENTS: Arvicolini. Although systematists agree that *Lasiopodomys* belongs in this tribe, they have disputed its generic status. G. M. Allen (1940) treated *Lasiopodomys* as a full synonym of *Phaiomys*, included in *Microtus* as a subgenus. Others have relegated it to a separate subgenus of *Microtus* (Corbet, 1978c; Corbet and Hill, 1991; Ellerman and Morrison-Scott, 1951). Hinton (1926a) enumerated the many features that isolate *Lasiopodomys*: very short pinnae; elongate front claws; claw rather than nail on thumb; plantar surfaces densely furred; three labial salient angles on a simple M3, not four on a relatively elaborate molar; anterolabial margin of M3 concave, not angular; cusps elongate, not triangular; and m1 cap with only lingual secondary wing, not labial and lingual wings (Gromov and Polyakov, 1977, offered additional traits). Both neontologists and paleontologists have broadly acknowledged Hinton's treatment (Ellerman, 1941; Gromov and Erbajeva, 1995; Gromov and Polyakov, 1977; Pavlinov and Rossolimo, 1987, 1998; Pavlinov et al., 1995a; Repenning, 1992; Repenning et al., 1990; Smorkacheva et al., 1990; Zagorodnyuk, 1990; Zheng and Li, 1990). In an allozymic analysis of Palaearctic voles, Mezhzherin et al. (1993) documented membership of *L. brandtii* in the same clade as *Microtus fortis* and *M. gregalis*. Except for their study, the allocation of *Lasiopodomys* to *Microtus* has not issued from a data-rich and taxonomically broad phylogenetic study; until such evidence prescribes otherwise, *Lasiopodomys* should be retained as a genus following Hinton (1926a).

Gromov and Polyakov (1977) presciently described the two extant species as remnants of an ancient group that was more abundant in the past, a view consistent with that of Repenning (1992). Based on M3 and m1 patterns, he allocated several extinct species to *Lasiopodomys*, dating from the early Pleistocene in Eurasia (1.5-1.2 million years ago), Beringia (1.6-1.3 million years ago), and North America west of the Rockies (1.2 million years ago), and from the late Pleistocene (850,000 years ago) in North America east of the Rockies (also Zheng and Li, 1990). Repenning (1992) placed the evolution of *Lasiopodomys* from *Allophaiomys* in Eurasia (about 1.5 million years ago), and both genera reached North America at different times (see also Chaline, 1999, for discussion of *Allophaiomys*); in neither region did *Lasiopodomys* give rise to any *Microtus*. Other authorities do not share Repenning's notions about the pivotal evolutionary significance of the *Lasiopodomys* morphotype, and instead recognize only a great range of variation in *Allophaiomys*, from which most members of Arvicolini were derived (see Agusti, 1991; Chaline et al., 1999; R. A. Martin, 1987, 1989b, 1995).

Lasiopodomys brandtii (Radde, 1861). Melanges Biol. Acad. St. Petersbourg, 3:683.

COMMON NAME: Brandt's Vole.

TYPE LOCALITY: NE Mongolia, near Tarei-Nor (G. M. Allen, 1940).

DISTRIBUTION: Mongolia and adjacent Transbaikalia, Russia; Nei Mongol, Jilin, and Hebei provinces, NE China (Zhang et al., 1997).

STATUS: IUCN – Lower Risk (lc).

SYNONYMS: *aga* (Kastschenko, 1912); *hangaicus* Bannikov, 1948; *warringtoni* (Miller, 1913).

COMMENTS: G. M. Allen (1940) provided a detailed description of the species; phallic morphology documented by Yang et al. (1992).

Lasiopodomys fuscus (Büchner, 1889). Wiss. Res. Przewalski Cent.-Asien Reisen, Zool., I:(Säuget.), p. 125.
    COMMON NAME: Smokey Vole.
    TYPE LOCALITY: China, S Qinghai Prov., Yushu Autonomous Prefecture, Zhidoi Co., upper reaches of Zi Qu (also as Zhi Qu or Tongtian He) River, a tributary of the Yangtze; 33°40'N, 96°15'E (as restricted by lectotype designation by Hoffmann, 1996a).
    DISTRIBUTION: E Tibetan Plateau, western Qinghai Prov., C China; limits unresolved.
    STATUS: IUCN – Lower Risk (nt).
    COMMENTS: Included in *Phaiomys leucurus*, whether as *Pitymys* or *Microtus* (Corbet, 1978c; Ellerman and Morrison-Scott, 1951; Pavlinov et al., 1995a; Zhang and Wang, 1963), but separated as a species and transferred to *Lasiopodomys* by Zheng and Wang (1980; also Musser and Carleton, 1993). USNM specimens of *fuscus* verify its membership in *Lasiopodomys*, not *Phaiomys*. Hoffmann (1996a) carefully described old expedition routes and collecting sites, selected a lectotype of *fuscus*, and documented sympatry between it and the wider ranging *Phaiomys leucurus*.

Lasiopodomys mandarinus (Milne-Edwards, 1871). Rech. Hist. Nat. Mammifères, p. 129.
    COMMON NAME: Mandarin Vole.
    TYPE LOCALITY: China, Shanxi (Shansi), probably near Saratsi.
    DISTRIBUTION: NE and C China (Nei Mongol, Liaoning, Beijing, Shaanxi, Shanxi, Henan, Jiangsu, and Anhui; Zhang et al., 1997); N Mongolia; Transbaikal region and E and SE Siberia of Russia; Korea (Won and Smith, 1999); range limits uncertain.
    STATUS: IUCN – Lower Risk (lc).
    SYNONYMS: *faeceus* (G. M. Allen, 1924); *jeholensis* (Mori, 1939); *johannes* (Thomas, 1910); *kishidai* (Mori, 1930); *mandrianus* (Miller, 1896); *pullus* (Miller, 1911); *vinogradovi* (Fetisov, 1936).
    COMMENTS: External, cranial, dental, and distributional characteristics presented by G. M. Allen (1940). G-banding of sex chromosomes and its significance described by Zhu et al. (1994). Structural and functional comparisons of the masticatory complex in *L. mandarinus* and *Myospalax* undertaken by Li and Wang (1999). Aspects of reproductive biology documented by Zorenko et al. (1994).

Lemmiscus Thomas, 1912. Ann. Mag. Nat. Hist., ser. 8, 9:401.
    TYPE SPECIES: *Arvicola curtata* Cope, 1868.
    COMMENTS: Arvicolini. Named as a subgenus of *Lagurus* to segregate New World sagebrush voles from Old World steppe voles. Davis (1939) underscored the morphological separation between New and Old World forms and raised *Lemmiscus* to a genus, a view endorsed by Carleton's (1981) study of gastric anatomy. Subsequent faunal studies and checklists have variously listed *Lemmiscus* as a genus (Carleton and Musser, 1984; Corbet and Hill, 1991; Gromov and Polyakov, 1977; Musser and Carleton, 1993) or as a subgenus of *Lagurus* (Hall, 1981; Honacki et al., 1982). Certain morphological traits associate *Lemmiscus* with *Microtus* (Carleton, 1981; Davis, 1939), a relationship supported by phylogenetic evaluation of long repetitive DNA segments (Modi, 1996); chromosomal banding patterns provided little resolution of its affinity (Modi, 1987). Some paleontologists continue to view New World sagebrush voles as lagurines that migrated to North America in the Pleistocene (Chaline, 1985; Chaline et al., 1999); Repenning (1992), however, considered the dental similarities between *Lemmiscus* and *Microtus* sufficiently impressive to warrant their joint placement in a tribe Microtini. Taxonomic representation necessary to discriminate the migration scenario from autochthonous differentiation in the New World has yet to be realized in recent molecular studies.

Lemmiscus curtatus (Cope, 1868). Proc. Acad. Nat. Sci. Philadelphia, 20:2.
    COMMON NAME: Sagebrush Vole.
    TYPE LOCALITY: USA, Nevada, Esmeralda Co., Mt Magruder, Pigeon Spring.
    DISTRIBUTION: Sagebrush steppe and desert from S Alberta and SE Saskatchewan, Canada, south to EC California and NW Colorado (Fitzgerald et al., 1994), including the Columbia Basin of interior Oregon (Verts and Carraway, 1998) and Washington, USA.
    STATUS: IUCN – Lower Risk (lc).

SYNONYMS: *artemisiae* (Anthony, 1913); *decurtata* (Coues, 1877); *intermedius* (Taylor, 1911); *levidensis* Goldman, 1941; *orbitus* (Dearden and Lee, 1955); *pallidus* (Merriam, 1888); *pauperrimus* (Cooper, 1868).

COMMENTS: See Carroll and Genoways, 1980 (Mammalian Species, 124, as *Lagurus curtatus*).

Lemmus Link, 1795. Beitr. Naturgesch., 1(2):75.

TYPE SPECIES: *Mus lemmus* Linnaeus, 1758.

SYNONYMS: *Brachyurus* Fischer, 1813; *Hypudaeus* Illiger, 1811; *Lemnus* Gray, 1825; *Lemmus* Rochebrune, 1843; *Mirus* Brunner, 1938 [not Albers, 1850]; *Miromus* Brunner, 1951.

COMMENTS: Lemmini. Nominative genus of Miller's (1896) classic tribe Lemmi, then including *Dicrostonyx* (see that account). Distinctiveness still recognized within a tribe, including *Myopus* and *Synaptomys*, a clade believed to represent an early line of arvicoline evolution (Abramson, 1993; Carleton, 1981; Chaline and Graf, 1988; Graf, 1982; Gromov and Polyakov, 1977; Hinton, 1926*a*; Hooper and Hart, 1962; Jarrell and Fredga, 1993; Koenigswald, 1980). The monophyly of true lemmings, excluding *Dicrostonyx*, is also supported by recent cladistic evaluations of allozymes (Mezhzherin et al., 1995), nuclear repetitive DNA elements (Modi, 1996), and mitochrondrial DNA sequences (Conroy and Cook, 1999). *Myodes* Pallas, 1811, usually placed in the synonymy of *Lemmus* (e.g., Hall, 1981; Hinton, 1926*a*; Miller, 1896), is the oldest name for *Clethrionomys* (see that account); also see account of *Myopus*, once treated as a subgenus of *Lemmus*.

Old World taxa covered by Corbet (1978, 1984), Gromov and Polyakov (1977), Pavlinov and Rossolimo (1987), and Gromov and Erbajeva (1995); biology, distributions, and taxonomic arrangements of New and Old World species summarized by Stenseth and Ims (1993). Fossil history reviewed by Koenigswald and L. D. Martin (1984) and Fejfar and Repenning (1998); zoogeography discussed by Rausch and Rausch (1975*b*). All sampled *Lemmus* have 2n = 50, with autosomes similar to those in *Synaptomys* (Jarrell and Fredga, 1993; Zima and Kral, 1984*a*). In an important series of reports, Fedorov (1999) and Fedorov et al. (1999*a, b*) assessed mitochondrial DNA variation among co-occurring samples of *Lemmus* and *Dicrostonyx* to explore circumpolar phylogeographic patterns and discussed the geographic incongruence of genetic differentiation within the two lemming genera.

The alpha taxonomy of true lemmings is complicated. The genus is clearly more diverse than the single Holarctic species proposed by Sidorowicz (1960, 1964). Two (Pokrovski et al., 1984; Rausch and Rausch, 1975*b*), three (Musser and Carleton, 1993; Pavlinov and Rossolimo, 1987), four (Gromov and Erbajeva, 1995; Honacki et al., 1982; Jarrell and Fredga, 1993; Pavlinov and Rossolimo, 1998; Pavlinov et al., 1995*a*), or five (Corbet and Hill, 1991) species have been recognized, and the senior names employed and synonymies allocated among them are inconsistent or contradictory. We distill this information and evidence supplied in recent studies as the basis for recognizing five allopatric species.

Lemmus amurensis Vinogradov, 1924. Ann. Mag. Nat. Hist., ser. 9, 14:186.

COMMON NAME: Amur Brown Lemming.

TYPE LOCALITY: Russia, Siberia, Pikan, on Zeya River, a tributary of the Amur River.

DISTRIBUTION: Larch taiga of E Siberia; from the Arctic coast between the Lena and Kolyma Rivers southeastward onto the Kamchatka Peninsula, and southward through the Verkhoyansk and Cherskogo Mtns and the Omolon River to the upper Amur River basin and region east of Lake Baikal; also on islands in the New Siberian Arch. (Novosibirskiye Ostrova) (Chernyavskii et al., 1980, 1993; Federov, 1999*a*; Jarrell and Fredga, 1993:Fig. 2).

STATUS: IUCN – Lower Risk (lc).

SYNONYMS: *chrysogaster* J. A. Allen, 1903; *flavescens* Vinogradov, 1925; *ognevi* Vinogradov, 1933; *xanthotrichus* Vinogradov, 1925 [*nomen nudum*, see Ellerman and Morrison-Scott, 1951:656].

COMMENTS: Revised by Chernyavskii et al. (1980); additional chromosomal data analyzed by Gileva et al. (1984). The distinctiveness of *L. amurensis* has been reaffirmed by Chernyavskii et al. (1993), who studied karyotypes, craniodental traits, and pelage coloration and reported its range extension in the Kamchatka Peninsula. Fedorov et al. (1999*b*) similarly demonstrated its genetic isolation relative to *L. trimucronatus* and

L. sibiricus based on mitochondrial DNA analyses. Their study highlights the bio-geographic importance of the Lena-Kolyma catchment, with *L. trimucronatus* ranging east of the Kolyma River, *L. sibiricus* ranging to the west of the Lena River, and *L. amurensis* found in-between; discontinuities in mitochondrial DNA sequences among samples of *Dicrostonyx* are also localized in the region of the Kolyma and Lena Rivers (Federov, 1999; Federov et al., 1999*a*).

Includes *chrysogaster*, a form sometimes viewed as a species (e.g., Gileva et al., 1984; Gromov and Erbajeva, 1995). In identifying their Chukotskiy sample as the species *chrysogaster*, Gileva et al. (1984) inadvertently confused understanding of the taxonomy and ranges of these lemmings (see remarks under *L. sibiricus* and *L. trimucronatus*). The type locality of *chrysogaster* is Gichiga, a place near the west coast of the Okhotsk Sea (Ellerman and Morrison-Scott, 1951:656) and at the margin of the known range of *L. amurensis*, and we refer *chrysogaster* to that species. Pavlinov and Rossolimo (1987) retained, with reservation, *chrysogaster* in the synonymy of *L. sibiricus*, which then was thought to occur in the region as far as the Kamchatka Peninsula.

Lemmus lemmus (Linnaeus, 1758). Syst. Nat., 10th ed., 1:59.
  COMMON NAME: Norway Lemming.
  TYPE LOCALITY: Sweden, Lappmark.
  DISTRIBUTION: Mountains of Scandinavia and tundra from W Norway and Sweden to Kola (Kolskiy) Peninsula at W margin of the White Sea (NW Russia).
  STATUS: IUCN – Lower Risk (lc).
  SYNONYMS: *borealis* Nilsson, 1829; *iterator* (Gistel, 1850); *norvegicus* Desmarest, 1822.
  COMMENTS: European populations reviewed by Tast (1982*a*) and Mitchell-Jones et al. (1999); Russia segment reviewed by Gromov and Erbajeva (1995). Phylogenetic analysis of cytochrome *b* haplotypes has revealed appreciable nucleotide divergence between Scandinavian and Siberian samples (Fedorov et al., 1999*b*); that genetic characterization, in combination with unique pelage coloration, supports the specific integrity of *L. lemmus* relative to the geographically allopatric *L. sibiricus*. Although now absent from the British Isles, the species occurred there during the Pleistocene and Late Glacial (Late Palaeolithic) times (Sutcliffe and Kowalski, 1976; Yalden, 1999).

Lemmus portenkoi Tchernyavsky, 1967. Zool. Zhur., 46:12.
  COMMON NAME: Wrangel Island Lemming.
  TYPE LOCALITY: Russia, SE Siberia Magadanskaya Oblast, Ostrov Vrangelya (Wrangel Isl).
  DISTRIBUTION: Endemic to Wrangel Isl.
  COMMENTS: Jarrell and Fredga (1993) included *portenkoi* in *L. sibiricus*, observing their convention of synonymizing insular taxa with mainland species. However, Chernyavskii et al. (1993) interpreted the cytogenetic and morphological peculiarities of the Wrangel Isl form to reflect an old Palearctic origin and regarded it as a separate species. Although Wrangel Isl lies close to the Chukotskiy mainland, where *L. trimucronatus* occurs, cytochrome *b* sequence data instead link the island populations with those farther west that we recognize as *L. amurensis* (Fedorov et al., 1999*b*); the percent sequence divergence between the Wrangel Isl lemming and *L. amurensis* approximates that distinguishing *L. lemmus* and *L. sibiricus*, which together with the information provided by Chernyavskii et al. (1993), lends support to its recognition as a species. Wrangel Isl provides habitat for another endemic lemming, *Dicrostonyx vinogradovi* (see that account), yet that species is more closely related to Nearctic, not Eurasian, forms according to mitochondrial DNA haplotypes (Federov, 1999; Fedorov et al., 1999*a*). During the Pleistocene, the island merged with the Beringian land mass, including the mainland east of the Kolyma River stretching into Alaska; Fedorov et al. (1999*b*) developed historical arguments to explain the isolation of brown lemming populations on Wrangel Isl in the course of their range expansions and contractions within Beringia.

Lemmus sibiricus (Kerr, 1792). *In* Linnaeus, Anim. Kingdom, p. 241.
  COMMON NAME: Siberian Brown Lemming.
  TYPE LOCALITY: Russia, Yamalo-Nenetskaya Nats. Okr., between Polar Ural Mtns and lower course of Ob River.

DISTRIBUTION: Palearctic tundra landscapes—from Arkhangel region on eastern border of White Sea, W Russia, eastward to W border of the Lena River.

STATUS: IUCN – Lower Risk (lc).

SYNONYMS: *bungei* Vinogradov, 1924; *kittlitzi* (Brandt, 1845); *kittlitzi* (Middendorff, 1853); *minor* (Pallas, 1811) [*nomen nudum*, as discussed by Ellerman and Morrison-Scott, 1951:656]; *migratorius* (Illiger, 1815); *novosibiricus* Vinogradov, 1924; *obensis* Brants, 1827; *paulus* G. M. Allen, 1914.

COMMENTS: This species was formerly defined broadly to encompass not only most Palearctic forms but also North American taxa here retained as *L. trimucronatus* (see next). Within Eurasia, the geographic range was thought to extend eastward beyond the Lena River to the Kolymskaya region, E Siberia (e.g., Jarell and Fredga, 1993); however, phylogeographic patterns based on cytochrome *b* sequences substantiate a pronounced division between samples west (*L. sibiricus*) and east (*L. amurensis*, see above) of the Lena River (Fedorov et al., 1999*b*), a frontier coincident with significant haplotype discontinuities among samples of *Dicrostonyx* (Fedorov et al., 1999*a*). Fedorov et al. (1999*b*) interpreted the level of sequence divergence to indicate cladogenesis prior to the last glaciation and implicated the formation of montane and continental ice sheets near the Lena River as the historical barrier driving this divergence. *Lemmus sibiricus* proper is genetically closely related to *L. lemmus* and possibly referrable to that species (Fedorov et al., 1999*b*; Jarrell and Fredga, 1993).

Lemmus trimucronatus (Richardson, 1825). *In* Parry, Journal of a second voyage , App.:309.

COMMON NAME: Nearctic Brown Lemming.

TYPE LOCALITY: Canada, District of Mackenzie, Point Lake.

DISTRIBUTION: N Chukotskiy region in far NE Siberia (coastal region east of Kolyma River, not inland); in North America, from W Alaska east to Baffin Isl and Hudson Bay, and south in the Rocky Mtns to C British Columbia, Canada; also Nunivak and St. George isls in the Bering Sea, Pribilof Isls, and Canadian Archipelago (Jarrell and Fredga, 1993:Fig. 2).

SYNONYMS: *alascensis* Merriam, 1900; *harroldi* Swarth, 1931; *helvolus* (Richardson, 1828); *minusculus* Osgood, 1904; *nigripes* (True, 1894); *phaiocephalus* Manning and Macpherson, 1958; *subarticus* Bee and Hall, 1956; *yukonensis* Merriam, 1900.

COMMENTS: Rausch (1953) proposed the synonymy of *trimucronatus* and *nigripes* under Old World *L. sibiricus*, a taxonomic arrangement elaborated by Rausch and Rausch (1975*b*) and maintained in subsequent faunal and systematic works (Banfield, 1974; Hall, 1981; Honacki et al., 1982; Jones et al., 1986, 1997; Musser and Carleton, 1993). As now understood, *L. trimucronatus* is the only true lemming to exhibit a recent transberingian geographic distribution; see Chernyavskii et al. (1993) and Federov et al. (1999*a*), who speculated about Beringian history and possible dispersion pathways for lemmings. North American subspecies revised by Davis (1944) and retained as such by Hall and Cockrum (1953) and Hall and Kelson (1959); North American populations reviewed by Batzli (1999). Corbet and Hill (1991) continued to recognize the St. George Isl form *nigripes* as a species.

Among Eurasian *Lemmus*, hybridization results (Pokrovski et al., 1984), meiotic inquiries (Gileva et al., 1984), and gene sequences (Federov et al., 1999*a*) have underscored the sharp genetic discontinuity of populations that inhabit the Chukotskiy region, NE Siberia. Gileva (1983) and Gileva et al. (1984) represented the cytogenetic peculiarities of the Chukotskiy population (which they identified as *chrysogaster*) as an independent species (along with *L. amurensis*, *L. lemmus*, and *L. sibiricus*) and suggested that it may prove conspecific with North American *L. trimucronatus*. Chernyavskii et al (1993) agreed based on identical karyotypes obtained from their Chukotskiy and North American samples. These data are collectively compelling, and we view the North American and Chukotskiy populations as a separate entity for which the oldest name is *L. trimucronatus*, a species distinct from the strictly Palearctic *L. sibiricus*. By interesting contrast, mitochondrial DNA haplotypes from samples of *Dicrostonyx* in Chukotskiy relate them to W Siberian populations, not Nearctic taxa (Fedorov, 1999; Fedorov et al., 1999*a*).

Microtus Schrank, 1798. Fauna Boica, 1(1):72.

TYPE SPECIES: *Microtus terrestris* Schrank, 1798 (= *Mus arvalis* Pallas, 1778).

SYNONYMS: *Agricola* Blasius, 1857; *Alexandromys* Ognev, 1914; *Ammomys* Bonaparte, 1831; *Arbusticola* Shidlovsky, 1919; *Arvalomys* Chaline, 1974; *Aulacomys* Rhoads, 1894; *Bicunedens* Hodgson, 1863; *Campicola* Schulze, 1890 [not Swainson, 1827]; *Campicoloma* Strand, 1928; *Chilotus* Baird, 1857; *Euarvicola* Acloque, 1899; *Herpetomys* Merriam, 1898; *Iberomys* Chaline, 1972; *Isodelta* Cope, 1871; *Meridiopitymys* Chaline, 1974; *Micrurus* Major, 1877 [not Ehrenberg, 1831]; *Mynomes* Rafinesque, 1817; *Oecomicrotus* Rabeder, 1981; *Orthriomys* Merriam, 1898; *Pallasiinus* Kretzoi, 1964; *Parapitymys* Chaline, 1978; *Pedomys* Baird, 1857; *Pinemys* Lesson, 1836; *Pitymys* McMurtrie, 1831; *Psammomys* Le Conte, 1830 [not Cretzschmar, 1828], *Steneocranius* Trouessart, 1904; *Stenocranius* Kastschenko, 1901; *Sumeriomys* Argyropulo, 1933; *Suranomys* Chaline, 1972; *Sylvicola* Fatio, 1867 [not Harris, 1782, or Humphrey, 1797]; *Terricola* Fatio, 1867 [not Fleming, 1828]; *Tetramerodon* Rhoads, 1894; *Tibercola* Koenigswald, Fejfar, and Tchernov, 1992; *Tyrrhenicola* Major, 1905.

COMMENTS: Arvicolini. Nowhere are the explosiveness and recency of arvicoline evolution more dramatically highlighted than by the inconsistency of systematic treatment of genus-group taxa to be subsumed by *Microtus*. Little consensus exists concerning the morphological limits or monophyly of many of these taxa, a situation that partly reflects the overly narrow reliance of our classifications on dental characters undergoing rapid change (see Guthrie, 1971; Koenigswald, 1980). The recency of speciation is another contributory factor—Conroy and Cook (2000a) dated the major pulse of diversification within *Microtus* as only 1.3 million years ago. See accounts of *Blanfordimys*, *Chionomys*, *Lasiopodomys*, *Neodon*, *Phaiomys*, and *Proedromys*, often included in *Microtus* but which are here treated as genera.

North American forms revised by Bailey (1900) and taxonomy updated by Hall and Cockrum (1953) and Hall (1981); many aspects of anatomy, paleontology, taxonomy, and zoogeography covered in Tamarin (1985). See R. A. Martin (1995) for comments on the usage of *Pedomys* and *Pitymys* as subgenera of *Microtus* and classification of fossil forms. The more diverse Palearctic fauna is covered in several comprehensive synopses (Corbet, 1978c; Gromov and Erbajeva, 1995; Gromov and Polyakov, 1977; Meyer et al., 1996; Mitchell-Jones et al., 1999; Niethammer and Krapp, 1982a; Ognev, 1963b, 1964; Pavlinov et al., 1995a). See Pozdnyakov (1996) for summary of morphological and chromosomal variation within the subgenus *Alexandromys* and arrangement of species groups. Fossil *Microtus* are known from the late Pliocene of Eurasia and North America (Chaline et al., 1999; McKenna and Bell, 1997; Repenning, 1992; Van der Muelen, 1978). The genus is generally thought to have been derived from the Pliocene *Allophaiomys*, either directly for some North American and European species (Chaline et al., 1999; Repenning, 1992) or indirectly through a morphological intermediate similar to *Lasiopodomys* (Repenning, 1992); Kotlia (1994), on the other hand, suggested direct descent from the Pliocene *Mimomys* in S Asia.

The Holarctic distribution of *Microtus* (see map in Gromov and Erbajeva, 1995) has spawned two principal scenarios of zoogeographic interpretation. The traditional notion has emphasized multiple intercontinental dispersals and broad transcontinental distributions of subgenera, a viewpoint more often propounded by paleontologists (R. A. Martin, 1974, 1987; Repenning, 1980, 1983, 1992, 1998; Repenning et al., 1990; Van der Muelen, 1978). An alternative hypothesis stresses one or two intercontinental dispersions and extensive regional cladogenesis; such a viewpoint has gained more credence, emerging from studies of allozymes (Chaline and Graf, 1988; Graf, 1982; Moore and Janecek, 1990), chromosomes (Zagorodnyuk, 1990), mitochondrial DNA (Conroy and Cook, 1999, 2000a; Conroy et al., 2001), and fossils (Brunet-Lecomte and Chaline, 1991, 1992; Chaline, 1974; Chaline et al., 1999; Kryštufek et al., 1996; R. A. Martin, 1995). According to these collective results, pitymyine forms in the Old World (*Terricola*) differentiated independently of those in the New World (*Pedomys*, *Pitymys*); New World common voles (*Mynomes*) are cladistically separated from Old World subgenera such as *Microtus* proper, *Alexandromys*, and *Agricola*; and the invasion of the New World semiaquatic niche is recognized (subgenus *Aulacomys*) as independent of the Old World water-vole radiation (genus *Arvicola*). See especially Conroy and Cook (2000a) for review of these hypotheses and comments on taxonomy and biogeography.

Microtus abbreviatus Miller, 1899. Proc. Biol. Soc. Wash., 13:13.

    COMMON NAME: Insular Vole.

    TYPE LOCALITY: USA, Alaska, Bering Sea, Hall Isl.

    DISTRIBUTION: Hall and St. Matthew Isls, Bering Sea (Alaska, USA).

    STATUS: IUCN – Data Deficient.

    SYNONYMS: *fisheri* Merriam, 1900.

    COMMENTS: An insular relative of *M. miurus* of the Alaskan mainland, generally retained as a
        species (Fedyk, 1970; Jones et al., 1997; Rausch and Rausch, 1968) but separate status
        questioned (Conroy and Cook, 2000*a*). Also see accounts of *M. gregalis* and *M. miurus*.

Microtus agrestis (Linnaeus, 1761). Fauna Suecica, 2nd ed., p. 11.

    COMMON NAME: Field Vole.

    TYPE LOCALITY: Sweden, Uppsala.

    DISTRIBUTION: Britain and nearby small islands (except Ireland); continental distribution
        extends from Scandinavia and Baltic region (Miljutin, 1997, 1998; Timm et al., 1998)
        east through Siberia to the Lena River; in the south from N and C Portugal, Pyrennes of
        N Spain (Brunet-Lecomte, 1991; Castiens and Gosalbez, 1992; Torre et al., 1996),
        France, Belgium, Netherlands (Lange, 1992), Germany (Dolch et al., 1994), Switzer-
        land (Hausser, 1995), Austria, Czech Republic (Andera and Èervený, 1994; Šmaha,
        1996), Slovakia (Danko, 1994; Kminiak, 1996; Mošanský, 1994; Stanko, 1995; Stanko
        and Mošanský, 1994, 2000; Stanko et al., 2000), N Italy (Amori et al., 1999; Locatelli
        and Paolucci, 1996*a*), Poland, Hungary, Slovenia (Kryštufek, 1991), east through
        N Croatia and N Bosnia and Herzegovina (Petrov, 1992), N Serbia, Romania, Ukraine,
        and Kazhakstan to S Urals, Altai Mtns, NW China (NW Xinjiang; Zhang et al., 1997),
        and Lake Baikal region (Corbet, 1978*c*; Krapp and Niethammer, 1982; Mitchell-Jones
        et al., 1999).

    STATUS: IUCN – Lower Risk (lc).

    SYNONYMS: *agrestoides* Hinton, 1910; *angustifrons* (Fatio, 1905); *arcturus* Thomas, 1912; *ar-
        gyropoli* Ognev, 1944; *argyropuli* Ognev, 1950; *argyropuloi* Ognev, 1952; *armoricanus*
        Heim de Balsac and Beaufort, 1966; *bailloni* (de Sélys Longchamps, 1841); *britannicus*
        (de Sélys Longchamps, 1847); *bucklandii* (Giebel, 1847); *carinthiacus* Kretzoi, 1958;
        enez-groezi Heim de Balsac and Beaufort, 1966; *estiae* Reinwaldt, 1927; *exsul* Miller,
        1908; *fiona* Montagu, 1922; *gregarius* (Linneaus, 1766); *hirta* (Bellamy, 1839); *insul*
        Lydekker, 1909; *insularis* (Nilsson, 1844); *intermedia* (Bonaparte, 1845); *latifrons* (Fatio,
        1905); *levernedii* (Crespon, 1844); *luch* Barrett-Hamilton and Hinton, 1913; *mac-
        gillivrayi* Barrett-Hamilton and Hinton, 1913; *mial* Barrett-Hamilton and Hinton,
        1913; *mongol* Thomas, 1911; *neglectus* (Jenyns, 1841); *nigra* (Fatio, 1869); *nigricans*
        (Kerr, 1792); *ognevi* Scalon, 1935; *orioecus* Cabrera, 1924; *pannonicus* Ehik, 1924; *pallida*
        Melander, 1938; *punctus* Montagu, 1923; *rozianus* (Bocage, 1865); *rufa* (Fatio, 1900);
        scaloni Heptner, 1948; *tridentinus* Dal Piaz, 1924; *wettsteini* Ehik, 1928.

    COMMENTS: Subgenus *Microtus*, *agrestis* species group *sensu* Zagorodnyuk (1990). Considered
        a member of subgenus *Microtus* by Pavlinov et al. (1995*a*), Gromov and Erbajeva
        (1995), and Meyer et al. (1996), but placed in subgenus *Agricola* by Zagorodnyuk
        (1990). Although once regarded as conspecific with North American *M. pennsylvanicus*
        on morphological grounds (Klimkiewicz, 1970), chomosomal differences indicate
        those similarities to represent convergence, not phylogenetic alliance (Vorontsov and
        Lyapunova, 1976; Zagorodnyuk, 1990). Phylogenetic inference based on cytochrome
        b sequences also widely separated *M. agrestis* from any North American *Microtus* (Con-
        roy and Cook, 2000*a*), and allozymic analysis isolated it by large genetic distances
        from nine other Eurasian *Microtus* sampled (Mezhzherin et al., 1993).

        Distribution in Portugal, Spain and France and morphometric discrimination from
        other *Microtus* and *Chionomys* in the region documented by Madureira (1983). Kooij
        et al. (1997) employed multivariate analyses to identify dentaries extracted from owl
        pellets as either *M. agrestis* or *M. arvalis*; Brunet-Lecomte et al. (1996) confidently dis-
        criminated the two species based on analysis of m1s; Kapischke (1992) recognized six
        distinct molar surface patterns in samples from Germany. Chromosomal data re-
        viewed by Zima and Kral (1984*a*). Almaça (1993) assessed variation in pelage color,
        morphometrics, and dental traits among samples from N Portugal, Massif Central of

France, and French lowlands and found little support for traditional subspecies but detected clinal variation between highland and lowland populations. Sequence analysis of mitochondrial DNA subdivided Eurasian *M. agrestis* into three allopatric clades, suggesting extended isolation in different glacial refugia (Jaarola and Searle, 2002): western (W and C Europe, spreading from the Carpathians); eastern (Lithuania to C Asia, originating from the S Urals or Caucasus); and southern (Portugal to Hungary, expanding from the Iberian Peninsula and perhaps representing a separate species).

Microtus anatolicus Kryštufek and Kefelioğlu, 2002. Bonn. Zool. Beitr., 50:8.
  COMMON NAME: Anatolian Vole.
  TYPE LOCALITY: Turkey, Konya, Cihanbeyli, Yapali köyü.
  DISTRIBUTION: Known only from the type locality and vicinity.
  COMMENTS: Subgenus *Microtus*, *socialis* species group. When Kefelioğlu and Kryštufek (1999) first identified this 2n = 60 population, they refrained from designating it a new species until the characterization of *M. irani* was improved; such was later accomplished by Kryštufek and Kefelioğlu (2002), who described significant morphological traits that distinguish it from their new species, *M. anatolicus* (no chromosomal data are available for *M. irani* from its type locality). *Microtus anatolicus* is chromosomally close to Turkish samples of *M. socialis* (2n = 62, FN = 60), but Kryštufek and Kefelioğlu (2002) claimed that "categorical cranial differences provide strong evidence against restricting the new species to a merely Robertsonian population of *M. socialis*."

Microtus arvalis (Pallas, 1778). Nova Spec. Quadr. Glir. Ord., p. 78.
  COMMON NAME: Common Vole.
  TYPE LOCALITY: Russia, Leningrad Oblast, Pushkin-town (as restricted by neotype selection by Malygin and Yatsenko, 1986; formerly as "Germany," e.g., Ellerman and Morrison-Scott, 1951).
  DISTRIBUTION: NE Portugal and C and N Spain (Brunet-Lecomte, 1991; Castien and Gosalbez, 1992; Gonzalez-Esteban et al., 1995) through France, Belgium, Netherlands (Jonkers, 1992, mapped distributional changes between 1850 and 1988), Germany (Dolch et al., 1994), Switzerland (Hausser, 1995; Maurizio, 1994), N Italy (Amori et al., 1999; Bigini and Turini, 1995; Cantini, 1991, Paolucci et al., 1993), Austria, Hungary, Czech Republic (Andera and Èervený, 1994; Šmaha, 1996), Slovakia (Danko, 1994; Kminiak, 1996; Mošanský, 1994; Stanko, 1995; Stanko and Mošanský, 1994, 2000; Stanko et al., 1994, 2000), Poland, the Baltic region (Miljutin, 1997, 1998; Timm et al., 1998), and Denmark (but not most of Fennoscandia), and eastward to C and S Urals in Sverdlovsk, Chelyabinsk, and Orenburg districts in Russia (Gileva et al., 1996); south through Slovenia (Kryštufek, 1991), Serbia and Montenegro (Petrov, 1991), Romania, N and W Bulgaria; east through Russia from Krym (Crimea) and E Ukraine through Siberia to the upper Yenesei River; south through NW Mongolia, NW China (NW Xinjiang; Zhang et al., 1997), the Altai Mtns and Kazakhstan (Kovalskaya, 1994), to the Caucasus, N and E Turkey (Kryštufek and Vohralík, 2001; Pamukoglu and Albayrak, 1996), NW Iran and east through the Elburz Mtns to N Khorassan Prov. of NE Iran (Lay, 1967; type series of *khorkoutensis* and holotype of *hyrcania*; see below). Also insular populations on the Orkney Isls (Channel Isls, but not the British Isles), and Yeu (France).
  STATUS: IUCN – Lower Risk (lc) as *M. arvalis* and *M. obscurus*.
  SYNONYMS: *albus* (Bechstein, 1801); *arvensis* (Schinz, 1840); *angularis* Miller, 1908; *assimilis* (Rörig and Börner, 1905) [*nomen nudum*]; *assimilis* Miller, 1912; *asturianus* Miller, 1908; *ater* (de Sélys Longchamps, 1845) [*nomen nudum*]; *brevirostris* Ognev, 1924; *calypsus* Montagu, 1923; *campestris* (Blasius, 1853); *caucasicus* Satunin, 1896; *cimbricus* Stein, 1931; *contigua* (Rörig and Börner, 1905); *corneri* Hinton, 1910; *cunicularius* (Ray, 1847); *depressa* (Rörig and Börig, 1905) [*nomen nudum*]; *depressa* Miller, 1912; *duplicatus* Rörig and Börner, 1905) [*nomen nudum*]; *duplicatus* Miller, 1912; *flava* (Fatio, 1905); *fulva* (Fatio, 1869) [not Millet, 1828]; *fulvus* Geoffroy, 1803 [not Desmarest, 1816]; *fulvus* (Miller, 1912) [not Millet, 1828 or Fatio, 1869; see Ellerman and Morrison-Scott, 1951:697 for status of this name]; *galliardi* (Fatio, 1905); *ghalgai* (Krassovsky, 1929) [*no-*

men nudum; see Pavlinov et al., 1995*a*]; *grandis* Martino and Martino, 1948; *gudauricus* Ognev, 1929; *hawelkae* Bolkay, 1925; *heptneri* Hamar, 1963; *hyrcania* Goodwin, 1940; *igmanensis* Bolkay, 1919; *incertus* (de Sélys Longchamps, 1841); *incognitus* Stein, 1931; *iphigeniae* Heptner, 1946; *khorkoutensis* Goodwin, 1940; *macrocranius* Ognev, 1924; *meldensis* Delost, 1955; *meridianus* Miller, 1908; *mystacinus* (de Filippi, 1865); *obscurus* (Eversmann, 1841); *orcadensis* Millais, 1904; *oyaensis* Heim de Balsac, 1940; *principalis* (Rörig and Börig, 1905) [*nomen nudum*]; *principalis* Miller, 1912; *rhodopensis* Heinrich, 1936; *ronaldshaiensis* Hinton, 1913; *rousiensis* Hinton, 1913; *rufescentefuscus* (Schinz, 1845); *ruthenus* Ognev, 1950; *sandayensis* Millais, 1905; *sarnius* Miller, 1909; *simplex* (Rörig and Börig, 1905) [*nomen nudum*]; *simplex* Miller, 1912; *terrestris* (Schrank, 1798) [not Linnaeus, 1758]; *transcaucasicus* Ognev, 1924; *transuralensis* Serebrennikov, 1929; *variabilis* (Rörig and Börner, 1905) [*nomen nudum*]; *variabilis* Miller, 1912; *vulgaris* (Desmarest, 1822); *westrae* Miller, 1908. Some synonyms listed and much of the southern and eastern distribution outlined for *M. arvalis* by Ellerman and Morriscon-Scott (1951) and Corbet (1978*c*) actually refer to *M. levis* (formerly *rossiaemeridionalis*); see that account.

COMMENTS: Subgenus *Microtus*, *arvalis* species group (Pavlinov and Rossolimo, 1987, 1998; Pavlinov et al., 1995*a*; Zagorodnyuk, 1990). Most closely related to *M. levis* (formerly *rossiaemeredionalis*), *M. transcaspicus*, and *M. ilaeus* (Meyer et al., 1996). Phylogenetic analysis of cytochrome *b* sequences reinforces the strong alliance between *M. arvalis* and *M. levis* (Conroy and Cook, 2000*a*).

Monograph edited by Sokolov and Bashenina (1994) encapsulates karyology, geographic distribution, taxonomy, morphology, ecology, and contrasts with *M. levis* (as *rossiaemeridionalis*). So broadly a distributed species has spawned numerous karyotypic studies on chromosomal variation, genomic mapping, and interspecific comparisons (Baskevich, 1996*b*; Burgos et al., 1989; Gileva et al., 1996; Mazurok et al., 1996*a*; Mitev and Mitev, 1991*c*; Zima and Kral, 1984*a*; Zima et al., 1997*a*). Morphological variation among samples from NE Spain documented by Gosalbez and Sans-Coma (1977). Distribution in Portugal, Spain and France and morphometric discrimination from species of *Microtus* and *Chionomys* within the region documented by Madureira (1983). Kooij et al. (1997) identified dentaries extracted from owl pellets collected in France, Germany, and the Netherlands as either *M. arvalis* or *M. agrestis* by multivariate analyses, and Brunet-Lecomte et al. (1996) documented differences in m1s between the two species. Molar variability of German samples and its significance documented by Kapischke (1997). Berry (1996) provided an excellent taxonomic review of populations on the Orkney Isls; Ligtvoet and Wijngaarden (1994) traced the rapid colonization by *M. arvalis* of a Netherlands landbridge island recently rejoined to the mainland and displacement of its indigenous occupant, *M. oeconomus*.

Goodwin (1940) described *khorkoutensis* (Khorkout Range near Dasht, NE Iran) as a subspecies of *M. arvalis*, and Corbet (1978*c*) questionably included it in *M. transcaspicus* (also in *arvalis* species group). Goodwin's holotype (AMNH 88764) and referred specimens (AMNH 88762, 88763) are much too small for *transcaspicus* and differ in size and qualitative features from the only other vole in the region, *M. paradoxus* (*socialis* species group). Our comparisons indicate that cranial and external characteristics of *khorkoutensis* fit within the range of variation in *M. arvalis*.

Goodwin (1940) also described *hyrcania* (Gouladah, NE Iran) as a distinct species, but Musser and Carleton (1993) and Kryštufek and Kefelioğlu (2002) erroneously placed it in *M. socialis*. The holotype and only known specimen, however, lacks the woolly and buffy brownish gray upperparts of *M. socialis* (soft and brownish in *hyrcania*), its white front and hind feet (brown in *hyrcania*), short tail (longer in *hyrcania*), larger cranium and relatively larger bullae (small bullae in *hyrcania*). Although a young adult, its external and skull features are nearly identical to those of a young adult *khorkoutensis*. Both it and *hyrcania* morphologically resemble specimens of *M. arvalis* from W Kazakhstan and Kyrgyzstan, not the larger, white-footed and short-tailed *M. socialis* from Kazakhstan and NW Iran.

The taxon *obscurus* is either included in *M. arvalis* (e.g., Baskevich, 1996*b*; Corbet, 1978*c*; Gromov and Erbajeva, 1995; Meyer et al., 1996, 1997; Mezhzherin et al., 1993;

Pavlinov and Rossolimo, 1987, 1998; Pavlinov et al., 1995*a*) or treated as a separate species based primarily on chromosomal morphology and hybridization results (Malygin and Luis, 1996; Zagorodnyuk, 1991*a*, *b*; and cited references). True *arvalis* has 2n = 46, FN = 84, that currently identified as *M. obscurus* or the *"obscurus"* form of *M. arvalis* has 2n = 46, FN = 72 or 74, and their hybrids demonstrate low fertility (Malygin and Luis, 1996). In Europe and N Russia, *obscurus* ranges to the east and south of the distribution of *M. arvalis* (see map in Zagorodnyuk, 1991*a*), and Meyer et al. (1997) documented their close approach to one another, separated by only 24 kms in Russia. Golenishchev et al. (2001) have uncovered a natural but narrow intergradation zone in which gene flow between the two parapatric karyomorphs approximates that among populations of the same karyotype.

Microtus bavaricus Konig, 1962. Senckenberg. Biol., 43:2.
> COMMON NAME: Bavarian Pine Vole.
> TYPE LOCALITY: Germany, Bavarian Alps, Garmisch-Partenkirchen, 730 m.
> DISTRIBUTION: Germany, Bavarian Alps (N margin of E Alps at type locality and across the border in Tyrol region of Austria; Haring et al., 2000; Konig, 1982).
> STATUS: IUCN – Data Deficient.
> COMMENTS: Subgenus *Terricola*, *subterraneus* species group *sensu* Chaline et al. (1988) and Pavlinov et al. (1995*a*). Taxonomic history reviewed by Haring et al. (2000) and Spitzenberger et al. (2000). The morphological separation of *bavaricus* from its close relatives, *M. multiplex* and *M. liechtensteini*, was confirmed by Spitzenberger et al. (2000) in a multivariate analysis of m1s. In another morphometric analysis of m1s, *M. bavaricus* (only three teeth) was phenetically closest to *M. tatricus* (Brunet-Lecomte and Nadachowski, 1998), but phylogenetic analysis of DNA sequences (mitochondrial control region, *CR*) discounts close affinity with *M. tatricus* (Haring et al., 2000) and instead allies *M. bavaricus* with *M. liechtensteini* (see account). *Microtus bavaricus* is extinct at the type locality (Haring et al., 2000; Spitzenberger et al., 2000), but still lives in the N Tyrol region of the Alps in E Austria (Haring et al., 2000).

Microtus brachycercus Lehmann, 1961. Zool. Anz., 167:223.
> COMMON NAME: Calabria Pine Vole.
> TYPE LOCALITY: S Italy, Calabrian Peninsula, Camigliatello Silano.
> DISTRIBUTION: Calabrian Peninsula of S Italy (Krapp, 1982*c*; Galleni et al., 1998).
> COMMENTS: Subgenus *Terricola*, *savii* species group (Chaline et al., 1988; Zagorodnyuk, 1990). Initially described as a subspecies of *M. savii* (Lehmann, 1961) and usually retained in that species (Corbet, 1978*c*; Krapp, 1982*c*; Krapp and Winking, 1976; Mitchell-Jones et al., 1999; Pavlinov et al., 1995*a*). Karyotype of samples from Calabrian region is similar to those from elsewhere on the Italian Peninsula and Sicily (2n = 54), but the X chromosome is larger and submetacentric (small metacentric in *M. savii*) and the Y is twice the size (Galleni et al., 1992, 1998; Niethammer, 1981). F1 hybrids obtained between *savii* and *brachycercus*, but F2 offspring were not produced because F1 males are sterile; those data, the gonosomal differences, and morphological traits (Lehmann, 1961) prompted Galleni et al. (1994, 1998) to recognize *brachycercus* as a species endemic to the Calabrian Peninsula of S Italy.

Microtus breweri (Baird, 1857). Mammalia, *in* Repts. U.S. Expl. Surv., 8(1):525.
> COMMON NAME: Beach Vole.
> TYPE LOCALITY: USA, Massachusetts, Muskeget Isl, off Nantucket.
> DISTRIBUTION: Known only from the type locality.
> STATUS: IUCN – Lower Risk (nt).
> COMMENTS: Subgenus *Mynomes*, *pennsylvanicus* species group *sensu* Zagorodnyuk (1990). An insular vicariant of *M. pennsylvanicus*, the two are inseparable karyotypically (Fivush et al., 1975; Modi, 1986), marginally distinct electrophoretically (Kohn and Tamarin, 1978), but morphologically sharply discrete (Bailey, 1900; Miller, 1896; Moyer et al., 1988). Although posited as conspecific with *M. pennsylvanicus* (e.g., Corbet and Hill, 1991; Jones et al., 1986; Modi, 1986; Whitaker and Hamilton, 1998), Moyer et al. (1988) mustered convincing evidence for the retention of *breweri* as a species. See Tamarin and Kunz (1974, Mammalian Species, 45).

Microtus cabrerae Thomas, 1906. Ann. Mag. Nat. Hist., ser. 7, 17:576.

COMMON NAME: Cabrera's Vole.

TYPE LOCALITY: Spain, Madrid Prov., Sierra de Guadarrama, near Rascafria.

DISTRIBUTION: Fragmented range in Spain and Portugal; and; subfossils document past occurrence of *M. cabrerae* in French side of the Pyrenees, which is outside its modern range (Corbet, 1984; Fernández-Salvador, 1998, 2000; Mitchell-Jones et al., 1999; Niethammer, 1982*f*).

STATUS: IUCN – Lower Risk (nt).

SYNONYMS: *dentatus* Miller, 1910.

COMMENTS: Subgenus *Microtus*, *agrestis* species group (Pavlinov and Rossolimo, 1987, 1998; Pavlinov et al., 1995*a*). Placed in subgenus *Agricola* by Zagorodnyuk (1990); Chaline (1974) used the subgenus *Iberomys*, a name superceded by *Agricola*, for *dentatus* and *brecciensis*. Biochemical and chromosomal data reported by Millet et al. (1982), Burgos et al. (1989), and Jimenez et al. (1991); other chromosomal data contributed by Zima and Kral (1984*a*); molecular and cytogenetic characteristics of satellite DNA and heterochromatin reported by Modi (1993). Biological and morphometric traits of Spanish samples and their significance evaluated by Ventura et al. (1998); occurrence in Spain, Portugal and France (subfossils) and morphometric discrimination from other *Microtus* and *Chionomys* in the same region documented by Madureira (1983). Closely related to the Pleistocene *M. brecciensis*, recorded from Spain, SE France, and NW Italy (Niethammer, 1982*f*).

Microtus californicus (Peale, 1848). Mammalia *in* Repts. U.S. Expl. Surv., 8:46.

COMMON NAME: California Vole.

TYPE LOCALITY: USA, California, Santa Clara Co., vicinity of San Francisco Bay, San Francisquito Creek near Palo Alto (as fixed by Kellogg, 1918:5).

DISTRIBUTION: Oak woodlands and grasslands of Pacific coast, from SW Oregon through California, USA, to N Baja California Norte, México.

STATUS: U.S. ESA – Endangered; IUCN – Vulnerable as *M. c. mohavensis* and *M. c. scirpensis*, Data Deficient as *M. c. stephensi*, Lower Risk (nt) as *M. c. vallicola*, otherwise Lower Risk (lc).

SYNONYMS: *aequivocatus* Osgood, 1928; *aestuarinus* Kellogg, 1918; *constrictus* Bailey, 1900; *edax* (Le Conte, 1853); *eximius* Kellogg, 1918; *grinnelli* Huey, 1931; *halophilus* Von Bloeker, 1937; *hyperuthrus* Elliot, 1903; *kernensis* Kellogg, 1918; *mariposae* Kellogg, 1918; *mohavensis* Kellogg, 1918; *neglectus* Kellogg, 1918; *paludicola* Hatfield, 1935; *perplexabilis* Grinnell, 1926; *sanctidiegi* Kellogg, 1918 [replacement name for *neglectus* Kellogg, 1918]; *sanpabloensis* Thaeler, 1961; *scirpensis* Bailey, 1900; *stephensi* Von Bloeker, 1932; *trowbridgii* (Baird, 1857); *vallicola* Bailey, 1898.

COMMENTS: Broadly affiliated with other North American *Microtus*, but evidence for nearest specific relative contradictory (compare assessments of Anderson, 1959; Conroy and Cook, 2000*a*; Hooper and Hart, 1962; Moore and Janecek, 1990). Zagorodnyuk (1990) acknowledged enigmatic phyletic stature as sole member of *californicus* species group, subgenus *Mynomes*. Geographic races delineated by Kellogg (1918); Gill (1980) recorded instances of sterility in hybrids between *M. c. californicus* and *M. c. stephensi*.

Microtus canicaudus Miller, 1897. Proc. Biol. Soc. Wash., 11:67.

COMMON NAME: Gray-tailed Vole.

TYPE LOCALITY: USA, Oregon, Polk Co., Willamette Valley, McCoy.

DISTRIBUTION: Willamette Valley, NW Oregon (Verts and Carraway, 1998:Fig. 11-111) and adjacent Washington, USA.

STATUS: IUCN – Lower Risk (lc).

COMMENTS: Subgenus *Mynomes*. Reduced to a subspecies of *M. montanus* by Hall and Kelson (1951); resurrected to specific status based on karyotypic and electrophoretic evidence (Hsu and Johnson, 1970; Johnson, 1968; Modi, 1986). Sibling species to *M. montanus* (Hoffmann and Koeppl, 1985; Modi, 1986, 1987; Zagorodnyuk, 1990) or to *M. townsendii* (Conroy and Cook, 2000*a*; Conroy et al., 2001). See Verts and Carraway (1987, Mammalian Species, 267).

Microtus chrotorrhinus (Miller, 1894). Proc. Boston Soc. Nat. Hist., 26:190.

COMMON NAME: Rock Vole.

TYPE LOCALITY: USA, New Hampshire, Coos Co., Mt Washington, head of Tuckerman's Ravine, 5300 ft (1615 m).

DISTRIBUTION: S Labrador south through S Quebec and Ontario, Canada, to NE Minnesota, N New York, and N New England states, USA; isolated segments in the C (S New York, NE Pennsylvania) and S Appalachian Mtns (W Maryland to E Tennessee and W North Carolina).

STATUS: IUCN – Data Deficient as *M. c. ravus*, Lower Risk (nt) as *M. c. carolinensis*, otherwise Lower Risk (lc).

SYNONYMS: *carolinensis* Komarek, 1932; *ravus* Bangs, 1898.

COMMENTS: Although conventionally viewed as closely related to (Anderson, 1960), if not conspecific with (Hall and Kelson, 1959), *M. xanthognathus*, morphological, chromosomal, and molecular information reveals their more distant kinship (Bailey, 1900; Conroy and Cook, 2000*a*; Guilday, 1982; R. L. Martin, 1973, 1979; Rausch and Rausch, 1974). The expanded definition of the subgenus *Aulacomys* (= *M. chrotorrhinus*, *M. longicaudus*, *M. richardsoni*, *M. xanthognathus*) as conceived by Zagorodnyuk (1990) is decidedly polyphyletic on parsimony and likelihood trees generated from mito- chondrial DNA sequences (Conroy and Cook, 2000*a*). Genic variation evaluated by Kilpatrick and Crowell (1985). See Kirkland and Jannett (1982, Mammalian Species, 180).

Microtus clarkei (Hinton, 1923). Ann. Mag. Nat. Hist., ser. 9, 11:158.

COMMON NAME: Clarke's Vole.

TYPE LOCALITY: China, Yunnan, divide between the Kuikiang and Salween Rivers, 11,000 ft (3353 m); 28°N.

DISTRIBUTION: High mountains, ca. 3300-4300 m, of W and S Yunnan (G. M. Allen, 1940; Zhang et al., 1997), SE Xizang (Feng et al., 1986, as *M. millicens*), SE Tibet (China) and N Burma (Ellerman, 1961; also FMNH 40960-40963).

STATUS: IUCN – Lower Risk (nt) as *Volemys clarkei*.

COMMENTS: Subgenus *Alexandromys*. Placed in *Volemys* by Zagorodnyuk (1990) as the only member of the *clarkei* species group, an arrangement followed by Musser and Carleton (1993) and Pavlinov et al. (1995*a*). Hinton (1923), however, compared *M. clarkei* with *M. calamorum*, now a synonym of *M. fortis*, and thought the two to be closely related, an alliance also implied by G. M. Allen (1940). Specimen examination leads us to concur: molar patterns are very similar in *M. clarkei* and *M. fortis* (see illustrations in Hinton, 1923), as are general skull conformation and body and tail proportions. In contrast to species of *Volemys*, *M. clarkei* lacks a low, smooth cranium, has smaller auditory bullae, and possesses M1-2 and m1-2 occlusal patterns like those of most other *Microtus*, not the configurations unique to *Volemys* (see that account). Although morphologically close to *M. fortis*, both Hinton's (1923) and G. M. Allen's (1940) comparisons indicate that *M. clarkei* is a separate species. Their typical ecological associations are distinctive and geographic ranges allopatric: *M. clarkei* in coniferous forests and alpine meadows in the high mountains of W Yunnan, N Burma, and SE Tibet; *M. fortis* along lowland lake shores and riverbanks, its closest occurrence in C Guizhou (Zhang et al., 1997:237, mapped and vouchered localities of the two).

Microtus daghestanicus (Shidlovsky, 1919). Raboty Zemskoi Opytnoi Stantsi, 2:12.

COMMON NAME: Caucasus Pine Vole.

TYPE LOCALITY: Russia, Daghestan, Caucasus Mtns, Karda.

DISTRIBUTION: N Caucasus Mtns (S Russia) and S Caucasus from Georgia south to S Armenia and Azerbaijan (Achverdjan et al., 1992) and in adjacent E Black Sea Mtns of NE Turkey (Kryštufek and Vohralík, 2001); possibly occurs in NW Iran.

STATUS: IUCN – Lower Risk (nt) as *M. nasarovi*, Lower Risk (lc) as *M. doghestanicus*.

SYNONYMS: *intermedius* Shidlovsky, 1919 [not Bonaparte, 1845, or Taylor, 1911]; *nasarovi* (Schidlovsky, 1938); *suramensis* Heptner, 1948 [see Corbet, 1978*c*, for origin of this replacement name].

COMMENTS: Subgenus *Terricola*, *subterraneus* species group (Pavlinov and Rossolimo, 1998;

Pavlinov et al., 1995*a*; Zagorodnyuk, 1990). Included in *Pitymys subterraneus* by
Ellerman and Morrison-Scott (1951) or in *P. majori* by Corbet (1978*c*). However,
Kratochvíl and Kral (1974), Baskevich et al. (1984), Achverdjan et al. (1992), and
Baskevich (1997) provided evidence that supports *daghestanicus* as a separate species, a
stature broadly endorsed (Gromov and Erbajeva, 1995; Pavlinov and Rossolimo, 1987,
1998; Pavlinov et al., 1995*a*). Closely related to *M. subterraneus* and *M. majori* (Baskevich, 1997; Macholán et al., 2001; Mezhzherin et al., 1995). Zima and Kral (1984*a*)
reviewed chromosomal information. Bashkevich (1997) contributed other karyotypic
and spermatozoal analyses and identified morphological traits potentially useful for
discriminating *M. daghestanicus* and *M. majori* in sympatry. Comparisons in molar size
variation among Turkish *M. daghestanicus, M. majori*, and *M. subterraneus* described by
Kryštufek and Vohralík (2004).

Within *M. daghestanicus*, although all geographic samples possess FN = 58, those
from Armenia and Azerbaijan exhibit 11 karyomorphs ranging from 2n = 38 to 54 and
those from the N Caucasus uniformly have 2n = 54 (Achverdjan et al., 1992).
Achverdjan et al. (1992) contrasted this pattern with *M. majori* (2n = 54, FN = 60) from
the Caucasus and *M. schelkovnikovi* (2n = 54, FN = 62) from SE Azerbaijan, speculating
that pine vole (subgenus *Terricola*) evolution in the Caucasus either resulted in species
with the same 2n but different fundamental numbers (*M. majori* and *M. schelkovnikovi*)
or produced the same FN but variable diploid counts (*M. daghestanicus*). The taxon
nasarovi has been included in *M. majori* (Corbet, 1978*c*) or listed as a separate species
(Pavlinov and Rossolimo, 1987; Pavlinov et al., 1995*a*; Zagorodnyuk, 1990). Its
chromosome complement (2n = 38 or 42, FN = 58) falls within the range of karyomorphs described for *M. daghestanicus*, prompting Achverdjan et al. (1992) and
Gromov and Erbajeva (1995) to include it as a subjective synonym.

Microtus dogramacii Kefelioğlu and Kryštufek, 1999. J. Nat. Hist., 33:301.
COMMON NAME: Doğramaci's Vole.
TYPE LOCALITY: NC Turkey, Amasya, Suluova, Boyali köyü, 900 m; 35°36′E, 41°40′N.
DISTRIBUTION: Recorded only from two places on the Anatolian Plateau, the type locality
    and near Konya in the south (Kefelioğlu and Kryštufek, 1999).
COMMENTS: Subgenus *Microtus, socialis* species group. A distinctive species similar in body
    size and morphology to Turkish *M. socialis* but different in a combination of external
    and cranial proportions and in karyotype (2n = 48, FN = 46, 48, 50 in *M. dogramacii*;
    2n and FN = 62 in *M. socialis*). Kefelioğlu and Kryštufek (1999) and Kryštufek and
    Kefelioğlu (2002) provided chromosomal, phenetic, and morphometric analyses that
    critically contrast *M. dogramacii* with *M. anatolicus* and Turkish *M. socialis* and
    M. guentherii. Çolak et al. (1997) identified *M. irani* from SE Turkey with 2n = 46;
    Kryštufek and Vohralík (2001) suggested that their sample may be close to
    M. dogramacii, which is chromosomally polymorphic.

Microtus duodecimcostatus (de Selys-Longchamps, 1839). Rev. Zool. Paris, p. 8.
COMMON NAME: Mediterranean Pine Vole.
TYPE LOCALITY: France, Gard, Montpellier.
DISTRIBUTION: S Portugal, Spain (except NW region; Brunet-Lecomte, 1991; Castiens and
    Gosalbez, 1992; Torre et al., 1996), and SE France.
STATUS: IUCN – Lower Risk (lc).
SYNONYMS: *centralis* (Miller, 1908); *flavescens* (Cabrera, 1924); *fuscus* (Miller, 1908); *ibericus*
    (Gerbe, 1854); *pascuus* (Miller, 1911); *provincialis* (Miller, 1909); *regulus* (Miller, 1908).
COMMENTS: Subgenus *Terricola, duodecimcostatus* species group (Chaline et al., 1988;
    Pavlinov et al., 1995*a*). Reviewed by Niethammer (1982*i*) and Mitchell-Jones et al.
    (1999). Mathias (1996) morphometrically analyzed cranial variation within *M. duo-
    decimcostatus* and compared it with *M. lusitanicus*; craniometric analyses contrasting
    M. duodecimcostatus with *M. gerbii* and *M. lusitanicus* documented by Spitz (1978).
    Chromosomal data summarized by Zima and Kral (1984*a*); meiotic behavior of sex
    chromosomes reported by Carnero et al. (1991). Garcia (1992) reported complete
    albinism in a population from SW Spain, the only instance of abnormal pelage
    coloration so far known. Presence of both *M. duodecimcostatus* and *M. lusitanicus*

recorded from Upper Palaeolithic to Neolithic cave sediments in central Portugal; only the latter is present in the modern fauna (Brunet-Lecomte and Povoas, 1993). The proficient swimming ability of *M. duodecimcostatus* may partly explain its wide geographic distribution that spans rivers which form barriers to the dispersal of other small-bodied fossorial mammals (Giannoni et al., 1994).

Microtus evoronensis Kovalskaya and Sokolov, 1980. Zool. Zh., 59:1410.
COMMON NAME: Evoron Vole.
TYPE LOCALITY: Russia, Khabarovsk Krai, Lake Evoron (= Zvoron) basin, Devyatka River.
DISTRIBUTION: Known only from the type locality (Meyer et al., 1996).
STATUS: IUCN – Critically Endangered.
COMMENTS: Subgenus *Alexandromys*, *maximowiczii* species group (Pavlinov and Rossolimo, 1998; Pavlinov et al., 1995a; Zagorodnyuk, 1990). Placed in subgenus *Microtus* by Pavlinov and Rossolimo (1987) and Meyer et al. (1996). Considered closely related to *M. mujanensis* and *M. maximowiczii* (Meyer, 1983); close kinship to *M. mujanensis* sustained by molar variation within the *M. maximowiczii* species group (Pozdnyakov, 1993) and by data accumulated from morphological, karyological, and hybridization studies (Meyer et al., 1996).

Microtus felteni Malec and Storch, 1963. Senckenberg. Biol., 44:171.
COMMON NAME: Balkan Pine Vole.
TYPE LOCALITY: Macedonia, Pelister Mtns, near Trnovo-Magarevo.
DISTRIBUTION: Endemic to the Balkan region: S Serbia, Macedonia (Zima et al., 1997a), Albania, and N Greece; Niethammer (1982h, 1987b); Mitchell-Jones et al. (1999).
STATUS: IUCN – Lower Risk (nt).
COMMENTS: Subgenus *Terricola*, *savii* species group (Chaline et al., 1988; Pavlinov et al., 1995a). Reviewed by Niethammer (1982h) and Mitchell-Jones et al. (1999). Chromosomal data consolidated by Zima and Kral (1984a); biochemical comparisons among *M. felteni*, *M. subterraneus*, and other species conducted by Gill et al. (1987).

Microtus fortis Büchner, 1889. Wiss. Res. Przewalski's Cent.- Asien. Reisen, Zool., I:(Säugeth.), p. 99.
COMMON NAME: Reed Vole.
TYPE LOCALITY: China, Nei Mongol, Ordos Desert, Huang Ho Valley, Sujan.
DISTRIBUTION: Lowlands of Transbaikalia and Amur region (Kovalskaya et al., 1988; Meyer et al., 1996), south through E and C China (Heilongjiang and Nei Mongol; south through Jilin, Liaoning, and Shandong to Ningxia, Shaanxi, Gansu, NE Sichuan, and C Guizhou; east through N Guangxi and Hunan to N Jiangxi, Fujian, Zhejiang, Jiangsu, and S Anhui; Zhang et al., 1997); also on Sakhalin Isl (Dobson, 1994; Voronov, 1992) and the Korean Peninsula (Won and Smith, 1999).
STATUS: IUCN – Lower Risk (lc).
SYNONYMS: *calamorum* Thomas, 1902; *dolichocephalus* Mori, 1930; *fujianensis* Hong, 1981; *michnoi* Kastschenko, 1910; *pelliceus* Thomas, 1911; *superus* Thomas, 1911; *uliginosus* James and Johnson, 1955.
COMMENTS: Subgenus *Alexandromys* (Zagorodnyuk, 1990) or subgenus *Microtus* (Gromov and Polyakov, 1977; Meyer et al., 1996; Pavlinov and Rossolimo, 1987). Most researchers view this species as phylogenetically close to *M. mongolicus* (Meyer, 1983; Meyer et al., 1996; Radjabli et al., 1984). Zagorodnyuk (1990) listed *M. fortis* as the only member of its species group and related to the *M. middendorffii* group, which contains *M. middendorffii*, *M. miurus*, *M. mongolicus*, and *M. sachaliensis*. This association is supported by allozymic analysis (Mezhzherin et al., 1993); data integrated from chromosomal, morphological and hybridization studies (Meyer et al., 1996); and phylogenetic analysis of cytochrome *b*, also including *M. montebelli*, *M. kikuchii*, and *M. oeconomus* (Conroy and Cook, 2000a). Chromosomal variation and its significance assessed by Kovalskaya et al. (1988, 1991).
  Kaneko and Hasegawa (1995) reported *M. fortis* and *M. oeconomus* from late Pleistocene cave deposits on Miyako Isl in the Ryukyu Isls, and claimed that the late Pleistocene samples from continental China, other Ryukyu Isls, and Honshu, Japan identified in the literature as *brandtioides*, *epiratticeps*, and *complicidens* (not necessarily the holotypes, which were not studied) actually represent *M. fortis* (modern range does

not include Japan or the Ryukyu Isls). According to Zheng and Li (1990), specimens identified as *brandtioides*, *epiratticeps*, and *complicidens* from China are a mixed assemblage. At Choukoutien, e.g., the oldest fossil (early Pleistocene) is true *brandtioides*, fragments from other sites are nearly indistinguishable from living *Lasiopodomys brandti*, specimens called *epiratticeps* may represent more than a single species, *complicidens* is in a different genus (*Hexianomys*), and *M. fortis* is present but had been misidentified as *brandtioides*. Zheng and Li (1990) regarded true *brandtioides* to be a species of *Lasiopodomys*, as did Repenning (1992).

Microtus gerbei (Gerbe, 1879). Le Naturaliste, 1:51.
> COMMON NAME: Pyreneean Pine Vole.
> TYPE LOCALITY: France, Loire-Inferieure, Dreneuf.
> DISTRIBUTION: SW France north to the Loire, and south through the Pyrenees Mtns of France and N Spain (Castien and Gosalbez, 1992; Krapp, 1982*d*; Mitchell-Jones et al., 1999).
> STATUS: IUCN – Lower Risk (lc).
> SYNONYMS: *brunneus* (Miller, 1908); *planiceps* (Miller, 1908); *pyrenaicus* (de Sélys Longchamps, 1847).
> COMMENTS: Subgenus *Terricola*, *savii* species group (Chaline et al., 1988; Zagorodnyuk, 1990; Pavlinov et al., 1995*a*). Reviewed by Krapp (1982*d*) under the name *pyrenaicus*; according to Spitz (1978), the latter is a *nomen dubium* and *gerbei* assumes priority, an opinion currently maintained (Mitchell-Jones et al., 1999). Corbet (1978*c*) included *gerbei* in *M. savii*, but chromosomal and craniometric studies underscore its specific integrity (Kratochvíl and Kral, 1974; Spitz, 1978). Chromosomal data (as *pyrenaicus*) summarized by Zima and Kral (1984*a*).
> > Brunet-Lecomte and Chaline (1993) and Brunet-Lecomte et al. (1995) identified *gerbei* and *pyrenaicus* as subspecies based on morphometric analysis of molar morphology. Another multivariate analysis of dental measurements phenetically placed *M. gerbei* (as *pyrenaicus*) between *M. duodecimcostatus* and *M. subterraneus* and disclosed minor morphological differences among the subspecies *pyrenaicus*, *planiceps*, and *brunneus* (Brunet-Lecomte et al., 1994). Those authors also hypothesized the gradual evolution of *M. gerbei* from middle Pleistocene *M. mariaclaudiae*.

Microtus gregalis (Pallas, 1779). Nova Spec. Quad. Glir. Ord., p. 238.
> COMMON NAME: Narrow-headed Vole.
> TYPE LOCALITY: Russia, Siberia, E of Chulym River.
> DISTRIBUTION: Discontinuous distribution in four regions. Largest range is in forests or steppes from Volga River eastward through Kazakhstan, across the Pamirs, Tien Shan and Altai Mtns, NW China (NW Xinjiang), N Mongolia, and Transbaikalia to Amur area and NE China (Heilongjiang, Nei Mongol, Hebei, and Henan; Zhang et al., 1997); another area is to the north in the Lena River Basin; last two are farther north in the Siberian tundra, where one area stretches from the Kolyma River area west to Taymyr Peninsula, the other from mouth of the Ob River to the White Sea. Absent from the British Isles.
> STATUS: IUCN – Lower Risk (lc).
> SYNONYMS: *angelicus* Hinton, 1910; *angustus* Thomas, 1908; *brevicauda* Kastschenko, 1901; *buturlini* (Ognev, 1922); *castaneus* Kashkarov, 1923; *dolguschini* Afanasiev, 1939; *dukelskiae* Ognev, 1950; *egorovi* Baranov and Feigin, 1980; *eversmanni* (Poljakov, 1881); *kossogolicus* (Ognev, 1923); *kriogenicus* Rekovets, 1978; *major* (Ognev, 1923); *montosus* Argyropulo, 1932; *nordenskioldi* (Poljakov, 1881); *pallasii* Kastschenko, 1901 [*nomen nudum*]; *raddei* (Poljakov, 1881); *ravidulus* Miller, 1899; *sirtalaensis* Yung, 1966; *slowzowi* (Poljakov, 1881); *talassicus* Heptner, 1948; *tarbagataicus* Ognev, 1944; *tianschanicus* Buchner, 1889; *tundrae* Ognev, 1944; *unguiculatus* (Vinogradov, 1935); *zachvatkini* Heptner, 1945.
> COMMENTS: Subgenus *Stenocranius*, the only included species (Gromov and Polyakov, 1977; Pavlinov and Rossolimo, 1987, 1998; Pavlinov et al., 1995*a*; Zagorodnyuk, 1990). Corbet (1978*c*) claimed that *M. abbreviatus* on Hall and St Matthews Isls in the Bering Sea and *M. miurus* in Alaska are closely related vicariant species; based on morphologi-

cal and zoogeographic criteria, Rausch (1964) considered North American *miurus* to be conspecific with Asian *gregalis*. Ample studies and an array of data convincingly refute this connection and reveal the morphological similarities between *M. gregalis* and *M. miurus* as convergent (Conroy and Cook, 2000*a*; Fedyk, 1970; Vorontsov and Lyapunova, 1976; Zagorodnyuk, 1990; also see *M. miurus*). Allozymic analysis by Mezhzherin et al. (1993) placed *M. gregalis* in a clade with *M. oeconomus*, *M. midden-dorffii*, *M. fortis*, and *Lasiopodomys brandtii*, but their taxon sampling was limited.

Intraspecific chromosomal variation among Mongolian samples documented by Kovalskaya (1989) and earlier chromosomal data presented by Zima and Kral (1984*a*). Occurrence on the Svjatoj Nos peninsula and isthmus in Lake Baikal documented by Reiter et al. (1995). Dupal (1998) reported chronoclinal changes in m1s from ancestral *M. hintoni* of the early Pleistocene, through the middle Pleistocene *M. gregaloides*, to living *M. gregalis* in Russia; transformations in crown length can be correlated with three trends associated with geographical and physical environmental gradients. An early origin of *M. gregalis* is consistent with its phylogenetic position basal to both Eurasian and North American species, as suggested by analysis of cytochrome *b* sequences (Conroy and Cook, 2000*a*). The taxa *egorovi* and *kriogenicus*, based on late Pleistocene fossils, were described as subspecies of *M. gregalis* (Dupal, 1998). Although not part of the modern fauna on the British Isles, *M. gregalis* occurred there during Late Glacial times in the Pleistocene (Kowalski, 1967; Yalden, 1999).

Microtus guatemalensis Merriam, 1898. Proc. Biol. Soc. Wash., 12:108.
COMMON NAME: Guatemalan Vole.
TYPE LOCALITY: Guatemala, Huehuetenango Dept., Todos Santos, 10,000 ft (3048 m).
DISTRIBUTION: Highland meadows of C Chiapas, México, and C Guatemala.
STATUS: IUCN – Lower Risk (lc).
COMMENTS: Type species of *Herpetomys*, usually placed in *Microtus*, with (Bailey, 1900) or without (Hall, 1981) subgeneric division; or placed in *Pitymys* when used as a genus (Honacki et al., 1982; R. A. Martin, 1974). Sister species to *M. oaxacensis* according to cladistic interpretation of molar characters (R. A. Martin, 1995) and cytochrome b sequence data (Conroy et al., 2001). The latter evidence indicates that *Herpetomys* should be considered a junior synonym of *Pitymys*, whether used as subgenus or genus.

Microtus guentheri (Danford and Alston, 1880). Proc. Zool. Soc. Lond., 1880:62.
COMMON NAME: Guenther's Vole.
TYPE LOCALITY: Turkey, Maras Prov., Taurus Mtns, near Maras (= Marash).
DISTRIBUTION: Lowlands and foothills of SE Europe (Niethammer, 1982*e*) in S Serbia and Macedonia (Petrov, 1992), SE Bulagaria, Greece (Vohralík, 1992), and Turkish Thrace (also on isls of St. Thomas and Lesbos); extends across Anatolia, except E Black Sea Mtns (Kefelioğlu and Kryštufek, 1999; Kryštufek and Vohralík, 2001), into coastal region of Syria, Lebanon and C Israel near Jaffa (Qumsiyeh, 1996); NE Iraq and NW Iran (south to Alïgûrdarz in Lurestan Prov and eastward to Tehran Prov; specimens identified by Kryštufek and Kefelioğlu, 2002, as "non-*socialis*" voles); also on Cyrenaican Plateau and adjacent coastal plain of N Libya (Ranck, 1968); Kryštufek (in litt., 2002) outlined the range in the Near East.
STATUS: IUCN – Lower Risk (nt).
SYNONYMS: *ankaraensis* Yiğit and Çolak, 2002; *hartingi* Barrett-Hamilton, 1903; *lydius* Blackler, 1916; *macedonicus* Kretzoi, 1964; *machintoni* Bate, 1937; *martinoi* Petrov, 1939; *mustersi* Hinton, 1926; *philistinus* Thomas, 1917; *shevketi* Neuhäuser, 1936; *strandzensis* Markov, 1960. See Golenishchev et al. (2000*b*) for map of most type localities.
COMMENTS: Subgenus *Microtus*, *socialis* species group (Zagorodnyuk, 1990) or subgenus *Sumeriomys* (Gromov and Polyakov, 1977; Pavlinov and Rossolimo, 1987; Pavlinov et al., 1995*a*). Corbet (1978*c*) initially included *guentheri* in *M. socialis* but later (1984) recognized its specific distinctness, a status earlier demonstrated by Neuhäuser (1936), Felten et al. (1971), and Morlok (1978). Harrison and Bates (1991) continued to unite *guentheri* with *M. socialis*. C- and G-banding chromosomal patterns in SE Bulgarian reported by Belcheva et al. (1980), and standard karyotype documented for SE Turkey by

Çolak et al. (1997); molecular and cytogenetic characteristics of satellite DNA and heterochromatin reported by Modi (1993). Greater breadth of chromosomal data integrated by Zima and Kral (1984a) and Zima (in litt., 2002); 2n = 54, FN = 52-56 in *M. guentheri* and 2n and FN = 62 for *M. socialis*. Cranial traits clearly distinguish *M. guentheri* from *M. socialis* (Kock and Nader, 1983; Kock et al., 1972); Zykov and Zagorodnyuk (1988) morphometrically separated it from *M. socialis* and *M. paradoxus*; and Kefelioğlu and Kryštufek (1999) contrasted it with Turkish *M. socialis* and other species using chromosomal and multivariate analyses. Allozymic analysis (40 loci) employing samples of *M. guentheri*, *M. arvalis*, and *Chionomys nivalis* revealed strong, species-level divergence among the three (Markov et al., 1995). Closest relatives of *M. guentheri* are *M. socialis*, *M. paradoxus*, and apparently *M. qazvinensis* (see those accounts). *Microtus paradoxus*, recorded from the mountains of NE Iran and adjacent Kopet Dag Mtns in S Turkmenistan, is similar to *M. guentheri* in body size and pelage coloration and texture. Whether it represents the easternmost distribution of *M. guentheri* or a separate species requires resolution, as does the status of other taxa listed within the synonymy.

Morphological and geographic definition of the species has been unclear due partly to substantial geographic variation and partly to its confusion with *M. irani* (Kock and Nader, 1983). Yiğit and Çolak (2002) morphologically and morphometrically examined Turkish samples and concluded that two species are present: *M. guentheri* in SE Turkey and *M. lydius* (type locality Izmur; Blackler, 1916), with *M. l. lydius* in coastal W Turkey and the larger-bodied *M. l. ankaraensis* throughout C Anatolia. Chromosomal traits do not separate *M. lydius* from *M. guentheri*, as was previously noted by Kefelioğlu (1995) and Çolak et al. (1997), but they have been considered separable based on bacular morphology and pelage coloration (brown versus yellowish brown). Kryštufek and colleagues (in litt., 2002) revisited the problem by analyzing 13 samples from Macedonia, Greece, throughout Turkey, Lebanon, Israel, NW Iran, and N Libya, including holotypes of *hartingi*, *lydius*, *philistinus*, *mustersi* and topotypes of *martinoi* (= *macedonicus*), *ankaraensis*, and *guentheri*; their preliminary results revealed two indistinct clusters, joined by intermediate samples and suggesting continuous morphological variation among the taxa represented. They regarded these "intergrading samples as belonging to *guentheri*, which is supported by available evidence on the karyotype."

The taxon *mustersi* (Hinton, 1926b), described as distinct from but morphologically closely related to *philistinus* from Israel (= *M. guentheri*), is the only living arvicoline in Africa, known by only a few specimens collected at or near the type locality. Described as a species and often recognized as such (G. M. Allen. 1939; Ellerman, 1941; Ranck, 1968); or included in *M. guentheri* (Ellerman and Morrison-Scott, 1951; Toschi, 1954), *M. socialis* (Corbet, 1978c; Harrison and Bates, 1991), or *M. irani* (Musser and Carleton, 1993). Libyan voles are a zoogeographic enigma, "particularly since they are not known from coastal Egypt, the only conceivable dispersal route," wrote Ranck (1968:73), who speculated that the population is a relict "from the period when the more boreal climates of the Pleistocene prevailed over North Africa. The Cyrenaican Plateau by virtue of its higher elevation has retained at least a vestige of these boreal elements, whereas the remainder of coastal Africa has become much drier and more sparsely vegetated and no longer contains habitats suitable for voles." The morphological identity of *mustersi* with *M. guentheri* should be tested by chromosomal and molecular analyses. Recorded in Middle Palaeolithic deposits in Lebanon, where *M. guentheri* is now common (Kersten, 1992), perhaps reaching the E Mediterranean region from the north during middle Pleistocene (Tchernov, 1988). Bate (1937) described *machintoni* based on fossils from Palestine, but Tchernov (1968) considered it part of a chronocline within *M. guentheri*.

Microtus ilaeus Thomas, 1912. Ann. Mag. Nat. Hist., ser. 8, 9:348.
COMMON NAME: Kazakhstan Vole.
TYPE LOCALITY: Kazakhstan, Semirechyia, Djarkent, banks of Ussek River.
DISTRIBUTION: W Uzbekistan just south of the Aral Sea, eastward through S Kazakhstan nearly to border south of Lake Balkhash (Meyer et al., 1996:128) and into NW Xinjiang Prov., NW China (Zhang et al., 1997).

STATUS: IUCN – Lower Risk (lc) as *M. kirgisorum*.

SYNONYMS: *igromovi* Meir, Golenischev, Radjably, and Sablina, 1996; *ileos* Vinogradov, 1930; *innae* Ognev, 1950; *kirgisorum* (Ognev, 1950).

COMMENTS: Subgenus *Microtus*, *arvalis* species group (Meyer et al., 1996; Zagorodnyuk, 1990). Described as a subspecies of *M. arvalis* and later elevated to species (Meyer, 1983; Meyer et al., 1981), as now observed (Corbet, 1984; Gromov and Erbajeva, 1995; Musser and Carleton, 1993; Pavlinov and Rossolimo, 1987, 1995; Pavlinov et al., 1995a). Kovalskaya (1994) documented its distribution in E Kazakhstan, reported probable sympatry between *M. kirgisorum* and *M. arvalis "obscurus"* in the Ili River valley, and noted that the region circumscribed the type locality of Thomas' *ilaeus*. Meyer et al. (1996) revised *M. ilaeus*, synonymizing *kirgisorum* and describing *igromovi* as a subspecies; Malygin and Luis (1996) associated *innae* with *M. ilaeus*. Chromosomal data (2n = 54, FN = 78) presented by Orlov et al. (1983), Kovalskaya (1994), and Meyer et al. (1996) in phylogenetic studies of the *M. arvalis* group. Histoenzymatic indices of specific skin glands also support the specific status of *M. ilaeus* (Sokolov and Dzhemukhadze, 1995, as *kirgisorum*). The species closely resembles the southern *M. transcaspicus* in pelage coloration, large body size, and karyotype (2n = 52 for *M. transcaspicus*) and is also closely allied to *M. levis* (Meyer et al., 1996).

Microtus irani Thomas, 1921. J. Bombay Nat. Hist. Soc., 27:580.

COMMON NAME: Iranian Vole.

TYPE LOCALITY: WC Iran, NW Fars Prov, Shiraz, 5200 ft (see Kryštufek and Kefelioğlu, 2002).

DISTRIBUTION: Recorded only from the type locality.

STATUS: IUCN – Lower Risk (lc).

COMMENTS: Subgenus *Microtus*, *socialis* species group (Zagorodnyuk, 1990). Most closely related to *M. paradoxus*, *M. socialis*, and *M. guentheri*. Described by Thomas (1921h) as a species known only from Shiraz, as recognized by Ellerman and Morrison-Scott (1951), but included in *M. socialis* by Ellerman (1941), Corbet (1978c), and Harrison and Bates (1991). Although Kock et al. (1972) and Kock and Nader (1983) reaffirmed the species integrity of *M. irani*, they uncritically expanded its geographic range to extend from Israel to W Iran, a definition later perpetuated by others (Pavlinov and Rossolimo, 1987, 1998; Zagorodnyuk, 1990; Musser and Carleton, 1993). Kryštufek and Kefelioğlu (2002), however, documented *M. socialis* proper as extending south into W Iran, along the Zagros Mtns to a locality 38 km west of Shiraz, where samples maintain an identity markedly unlike the type series of *irani*. Kock and Nader (1983) also considered most records of *M. guentheri* from Syria, Lebanon, and Israel to be *M. irani* and that *philistinus* (Thomas, 1917d; type locality, Ekron, SE Jaffa, Israel) may be an older name; Musser and Carleton (1993) retained Kock and Nader's (1983) distributional range but used *philistinus*.

In their revision of the *socialis* species group in the Near East, Kryštufek and Kefelioğlu (2002) defined the morphological and distributional limits of true *M. socialis* (see that account), identified a second group of "non-*socialis*" voles in Iran (later identified as *M. guentheri* fide Kryštufek, in litt., 2002; see above account), and excluded *irani* from both. Kryštufek and Kefelioğlu (2002) redescribed *M. irani* based on the holotype and three original topotypes (Thomas, 1921h, had recorded six in his type series, but they located only four in the BMNH), contrasted the species with the "non-*socialis*" group from Iran and Turkey, reaffirmed the identity of *philistinus* with *M. guentheri* and its separation from *M. irani*, and regarded the latter as "an independent species, known solely from its type locality." The phylogenetic alliance of *M. irani* to other members of the *M. socialis* group needs reevaluation, using a larger sample from the type locality and accessing chromosomal and molecular data.

Microtus kikuchii Kuroda, 1920. Dobuts. Zasshi, 32:40-41.

COMMON NAME: Taiwan Vole.

TYPE LOCALITY: Taiwan, Mt Morrison, 10,000 ft (3048 m).

DISTRIBUTION: Highlands of Taiwan, usually above 3000 m (Lin et al., 1987; H.-T. Yu, 1993; M.-J. Yu, 1996).

STATUS: IUCN – Vulnerable as *Volemys kikuckii*.

COMMENTS: Subgenus *Alexandromys*. Zagorodnyuk (1990) listed this species as the only member of the *kikuchii* species group, genus *Volemys*, an arrangement followed by Musser and Carleton (1993); treated as a species of *Microtus* by Corbet and Hill (1992), or listed in *Microtus*, subgenus *Volemys*, by Pavlinov et al. (1995a). Contrast in cranial conformation between *M. kikuchii* and *Volemys musseri*, as well as their striking coronal dissimilarities, earlier emphasized by Lawrence (1982); Corbet and Hill (1992) speculated that *M. kikuchii* is closely allied to *M. fortis*, a kinship certainly indicated by molar patterns. Phylogenetic analysis of cytochrome *b* sequences arranged *M. kikuchii* as sister species of *M. oeconomus*, in a clade that includes *M. fortis*, *M. middendorffii*, and *M. montebelli* (Conroy and Cook, 2000a); furthermore, the standard karyotype of *M. kikuchii* (2n = 30, FN = 52) is similar to those of *M. montebelli* and *M. oeconomus* (Harada et al., 1991). Alliance with *M. oeconomus* and its kin is zoogeographically sensible since it and *M. fortis* apparently had a wider geographic distribution during the Pleistocene (including the Ryukyu Isls). Information about the holotype of *M. kikuchii* and type locality amplified by Jones (1975). Cranial and dental morphometry reported by Kaneko (1987). Ecology, altitudinal distribution, and genetic substructure among populations exhaustively covered by H.-T. Yu (1993, 1994, 1995).

Microtus levis Miller, 1908. Ann. Mag. Nat. Hist., ser. 8, 1:197.

COMMON NAME: East European Vole.

TYPE LOCALITY: Romania, Prahova, Gageni, in foothills of S Carpathian Mtns north of Bucharesti (after Ellerman and Morrison-Scott, 1951:698).

DISTRIBUTION: S Finland and Baltic region (Miljutin, 1997, 1998; Timm et al., 1998), eastwards through S Urals (Gileva et al., 1996) and Novosibirsk suburbs in W Siberia (Yakimenko and Kryukov, 1997) to SW margin of Lake Baikal, south through the Caucasus (Baskevich, 1996b) and probably NW Iran, and across the Ukraine in the north and Turkey in the south (Kryštufek and Vohralík, 2001) to E Slovakia (Mošanský, 1994), E and S Romania (Zima et al., 1981), Bulgaria, Greece, Macedonia and S Serbia and Montenegro (Petrov, 1992), and E Albania; accidentally introduced to the Arctic Svalbard Archipelago (Fredga et al., 1990; Yoccoz et al., 1993; as *epiroticus*). Meyer et al. (1996) provide the best map.

STATUS: IUCN – Lower Risk (lc) as *Microtus rossiaemeridionalis*.

SYNONYMS: *caspicus* Ognev, 1950; *epiroticus* Ondrias, 1966; *ghalgai* Krassovsky, 1929; *muhlisi* Neuhäuser, 1936; *relictus* Neuhäuser, 1936; *rhodopensis* Heinrich, 1936; *rossiaemeridionalis* Ognev, 1924; *subarvalis* Meyer, Orlov, and Skholl, 1972 [not Heller, 1930].

COMMENTS: Subgenus *Microtus*, *arvalis* species group (Pavlinov and Rossolimo, 1987, 1998; Pavlinov et al., 1995a; Zagorodnyuk, 1990). Populations referable to this E European species (2n = 54, FN = 56), first separated from *M. arvalis* by Meyer et al. (1972), have been listed or discussed as *subarvalis* (Corbet, 1978c), *epiroticus* (Petrov and Ruzic, 1982), or *rossiaemeridionalis* (Sokolov and Bashenina, 1994). Nomenclatural usage reviewed by Fredga (1995); Masing (1999) consolidated reports over the last 30 years spread under those epithets. The senior status of *levis* remains provisional.

 Microtus rossiaemeridionalis is accepted by most researchers (Gromov and Erbajeva, 1995; Mitchell-Jones et al., 1999; Zagorodnyuk, 1991a), following the assertions of Malygin and Yatsenko (1986), but others still prefer *subarvalis* (Mazurok et al., 1995) or *epiroticus* (Vohralík and Sofianidou, 1992a). Vohralík and Sofianidou (1992a:364), while accepting Malygin and Yatsenko's conclusions, noted that "a finite solution concerning the correct name of this vole cannot be made until the vole forms described earlier from northern Iran (above all *M. mystacinus*) will be revised. That is why we prefer to use the commonly used name *M. epiroticus* in the meantime." Zagorodnyuk (1991b) illustrated and described cranial traits useful for distinguishing *rossiaemeridionalis* from *M. arvalis*, and Masing (1999) applied Zagorodnyuk's characters and new ones to distinguish large series of "*rossiaemeridionalis*" (2n = 54) and *M. arvalis* (2n = 46) with known karyotypes. He also confirmed that the type series of Miller's (1908) *M. levis* (included in *M. arvalis* by Ellerman and Morrison-Scott, 1951) exhibited features typical of "*rossiaemeridionalis*" but not *M. arvalis*. Masing's (1999) identity is convincing and liberates vole researchers from "the mouthful name *M. rossiaemeridionalis*" (Fredga, 1995:95). Whether *levis* is the oldest valid name awaits identity of *mystacinus* (de Filippi, 1865; see Ellerman and Morrison-Scott, 1951:698) from the Elburz Mtns in N Iran, as Vohralík and Sofianidou (1992*a*) had cautioned. Lay (1967) firmly associated the holotype and cotypes of *mystacinus* with *M. arvalis*, not *M. socialis* which occurs in the same region, but *rossiaemeridionalis* had not yet been appreciated as distinct. Reexamination of the holotype of *mystacinus* is needed.

Abundant cytological and morphological comparisons with the *obscurus* form of *M. arvalis* underscore the specific distinctness of *M. levis* (Gavrila et al., 1986; Gilvea et al., 1996; Kral et al., 1981; Kratochvíl, 1982*b*; Meyer et al., 1972; Naumova et al., 1990; Zagorodnyuk, 1991*a, b*; Zima et al., 1991), as do analyses of allozymes (Mezhzherin et al., 1993) and cytochrome *b* sequences (Conroy and Cook, 2000*a*). High-resolution G-banding relates *M. levis* most closely to *M. transcaspicus* within the *M. arvalis* species group (Mazurok et al., 1996*b*), an affinity supported by phylogenetic analyses of chromosomal, morphological, and hybridization data (Meyer et al., 1996). Regional chromosomal variation reported for W Siberia (Yakimenko and Kryukov, 1997), the Caucasus (Baskevich, 1996*b*), and Urals (Gileva et al., 1996); chromosomal and subchromosomal mapping documented by Mazurok et al. (1995). Albumin evolution using microcomplement fixation compared with *M. thomasi* (Nikoleto-poulos et al., 1992). Arvicoline examplar used to quantify gene expression patterns, development, and topography in the evolution of mammalian teeth (Jernvall et al., 2000). *Microtus arvalis* and *M. levis* broadly overlap in distribution (Meyer et al., 1996:129; Zagorodnyuk, 1991*b*:30) and occur sympatrically in a few regions (Meyer et al., 1996; Tikhonov et al., 1998). Gromov and Erbajeva (1995) included *caspicus* as a subspecies of *M. arvalis*, but Malygin and Luis (1996) demonstrated its identity with *M. levis*.

Microtus liechtensteini (Wettstein, 1927). Anz. Akad. Wiss., Wien, 20:2.

COMMON NAME: Liechtenstein's Pine Vole.

TYPE LOCALITY: Croatia, northwestern segment of Dinaric Alps, Velebit Mtns, summit of Mali Rajinac, 1699 m (Brunet-Lecomte and Kryštufek, 1993).

DISTRIBUTION: SE Alps in N Italy (east of Adige River valley in Trentino), S and C Austria, and Slovenia; south to the NW Dinaric Alps in Croatia; southern isolates in Pannonian Plain in Croatia, C Bosnia and Herzegovina, and Mt Tara in W Serbia (Brunet-Lecomte and Kryštufek, 1993; Haring et al., 2000).

SYNONYMS: *petrovi* Kryštufek, 1983.

COMMENTS: Subgenus *Terricola*, *M. subterraneus* species group (placed in the *duodecimcostatus* species group by Pavlinov et al., 1995*a*). Listed as a subspecies of *Pitymys subterraneus* (Ellerman and Morrison-Scott, 1951), included in *M. multiplex* (Krapp, 1982*b*; Petrov, 1992), or viewed as a species (Corbet, 1978*c*; Kryštufek, 1983; Petrov and Zivkovic, 1971). Taxonomic treatment of *liechtensteini* reviewed by Brunet-Lecomte and Kryštufek (1993) and Spitzenberger et al. (2000). Two distinctive karyotypes exist, one (2n = 46-48, FN = 48-50) characteristic of *M. multiplex* (Brunet-Lecomte and Volobouev, 1994; Graf and Meylan, 1980), the other (2n = 46, FN = 46) referrable to *M. liechtensteini* (Graf and Meylan, 1980; Storch and Winking, 1977). Chromosomal data and hybridization reports have questioned the specific status of *liechtensteini* (Graf and Meylan, 1980; Storch and Winking, 1977; Zima and Kral, 1984*a*). Brunet-Lecomte and Volobouev (1994), however, considered the chromosomal rearrange-ments to cytogenetically isolate *liechtensteini* from *M. multiplex*. Separate specific status is also supported by comparative morphometric analyses of m1s (Brunet-Lecomte and Kryštufek, 1993; Spitzenberger et al., 2000) and phylogenetic analysis of DNA sequences that include *M. multiplex*, *M. subterraneus*, *M. bavaricus*, and *M. liechtensteini* (Haring et al., 2000). These sequence data also suggest an early divergence of *M. subter-raneus* relative to the *M. multiplex-M. liechtensteini* clade, but cannot discount the past occurrence or continuing sporadic hybridization between the latter and *M. subter-raneus*. Haring et al. (2000) further speculated on glacial refugia: *M. bavaricus* in the N Alps, *M. liechtensteini* in the Dinaric Mtns, and *M. multiplex* in the SW Alps. In Slovenia, *M. liechtensteini* and *M. subterraneus* are mostly parapatric and can be cleanly distinguished by morphometric analysis (Kryštufek, 1991). Brunet-Lecomte and Kryštufek's (1993) morphometric study identified *petrovi* as a subspecies of

M. liechtensteini, not *M. multiplex*. Tarsal glands compared with those of *M. subterraneus* by Hrabe (1977).

Microtus limnophilus Büchner, 1889. Wiss. Res. Przewalski Cent.-Asien. Reis. Zool., I: (Säugeth.), p. 110.
COMMON NAME: Lacustrine Vole.
TYPE LOCALITY: China, Qinghai.
DISTRIBUTION: NC China (from N Sichuan, E Qinghai, Gansu, and Shaanxi northeast through Ningxia to C Nei Mongol and NW Xinjiang; Zhang et al., 1997, as *M. oeconomus*) to W Mongolia.
STATUS: IUCN – Lower Risk (lc).
SYNONYMS: *flaviventris* Satunin, 1903; *malcolmi* Thomas, 1911; *malygini* Courant et al., 1999.
COMMENTS: Subgenus *Alexandromys* (including subgenus *Pallasiinus sensu* Zagorodnyuk, 1990, and Pavlinov et al., 1995*a*). Formerly included in *M. oeconomus* (Corbet, 1978*c*; Ellerman and Morrison-Scott, 1951), but Malygin et al. (1990, and references therein) separated *limnophilus* as a parapatric species in W Mongolia; further evidence of interspecific distinction and of intraspecific karyotypic variation presented by Courant et al. (1999). Correlation and interdependence among non-metric traits with sex, age, and body size discussed by Markowski (1995).

Microtus longicaudus (Merriam, 1888). Am. Nat., 22:934.
COMMON NAME: Long-tailed Vole.
TYPE LOCALITY: USA, South Dakota, Custer Co., Black Hills, Custer, 5500 ft (1676 m).
DISTRIBUTION: Rocky Mountains and adjacent foothills, from E Alaska and N Yukon, south through British Columbia and SW Alberta, Canada, to E California and W Colorado; including Pacific coastal taiga to N California; disjunct southern pockets in S California, Arizona, and New Mexico, USA.
STATUS: IUCN – Data Deficient as *M. l. bernardinus*, *M. l. coronarius*, and *M. l. leucophaeus*, otherwise Lower Risk (lc).
SYNONYMS: *abditus* A. B. Howell, 1923; *alticola* (Merriam, 1890); *angusticeps* Bailey, 1898; *angustus* Hall, 1931; *baileyi* Goldman, 1938; *bernardinus* Merriam, 1908; *cautus* J. A. Allen, 1899; *coronarius* Swarth, 1911; *halli* Hayman and Holt, 1941 [replacement name for *angustus* Hall, 1931]; *incanus* Lee and Durrant, 1960; *latus* Hall, 1931; *leucophaeus* (J. A. Allen, 1894); *littoralis* Swarth, 1933; *macrurus* Merriam, 1898; *mordax* (Merriam, 1891); *sierrae* Kellogg, 1922; *vellerosus* J. A. Allen, 1899.
COMMENTS: Sometimes viewed as a Nearctic member of *Chionomys* (Anderson, 1959), or allocated to subgenus *Microtus* (Chaline, 1974; Hall, 1981), or to subgenus *Aulacomys* (Zagorodnyuk, 1990). Although strongly differentiated relative to other North American *Microtus* (e.g., Hooper and Hart, 1962; Modi, 1987; Moore and Janecek, 1990), the phyletic affinity of *M. longicaudus* lies with this complex and not Old World *Chionomys* (Chaline and Graf, 1988; Conroy and Cook, 2000*a*; Gromov and Polyakov, 1977; Zagorodnyuk, 1990).

Extensive karyotypic (Judd and Cross, 1980) and molecular (Conroy and Cook, 2000*b*) variation reported, albeit not necessarily concordant, which invites continued taxonomic investigation. Well-defined geographic groupings of cytochrome *b* haplotypes interpreted in light of Pleistocene climatic changes, possible refugia, and likely isolation during northward reexpansion along different colonizing routes (Conroy and Cook, 2000*b*). Finley and Bogan (1995) discussed the inconsistent usage and problematic identification of subspecies (*alticola*, *mordax*, and *longicaudus*) in NW Colorado, an area which tellingly circumscribes some of the most distinctive populations (Southern Rockies Clade) identified by Conroy and Cook (2000*b*) and whose genetic divergence approaches that of other *Microtus* species. Includes *coronarius*, which Jones et al. (1986) viewed as an insular derivative and subspecies of *M. longicaudus*; MacDonald and Cook (1996) continued to list *coronarius* as species and advised broader comparisons with *M. longicaudus* populations to vindicate its status. See Smolen and Keller (1987, Mammalian Species, 271).

Microtus lusitanicus (Gerbe, 1879). Rev. Mag. Zool. Paris, Ser. 3, 7:44.
COMMON NAME: Lusitanian Pine Vole.

TYPE LOCALITY: Portugal.

DISTRIBUTION: Portugal, N and C Spain (Brunet-Lecomte, 1991; Castien and Gosalbez, 1992), and SW France.

STATUS: IUCN – Lower Risk (lc).

SYNONYMS: *depressus* (Miller, 1908); *gerritmilleri* Kretzoi, 1958; *hurdanensis* (Agacino, 1938); *mariae* Major, 1905; *pelandonius* (Miller, 1908).

COMMENTS: Subgenus *Terricola*, *duodecimcostatus* species group (Chaline et al., 1988). Graf (1982) and Chaline and Graf (1988) apparently used *mariae* for this species, which reflected Winking's (1976) opinion in separating the latter from *M. duodecimcostatus*; Niethammer (1982*j*) explained the priority of *lusitanicus*. Chromosomal data reviewed by Zima and Kral (1984*a*). Morphometric analyses of variation within *M. lusitanicus* and multivariate contrasts with *M. duodecimcostatus* and-or *M. gerbii* conducted by Spitz (1978) and Mathias (1996). Brunet-Lecomte and Povoas (1993) successfully separated isolated molars of *M. lusitanicus* and *M. duodecimcostatus* using morphometric analyses of extant and Upper Palaeolithic-Neolithic samples in C Portugal. The fossorial *M. lusitanicus* is an adept swimmer (Giannoni et al., 1993), like its fossorial relative *M. duodecimcostatus* (Giannoni et al., 1994), an ability facilitating dispersal in mountainous habitats dissected by numerous streams.

Microtus majori Thomas, 1906. Ann. Mag. Nat. Hist., ser. 7, 17:419.

COMMON NAME: Major's Pine Vole.

TYPE LOCALITY: Turkey, Trabzon Prov., Sumelas (= Meryemana), 30 mi (48 km) S Trabzon.

DISTRIBUTION: NE Turkey (humid forest along southern shore of Black Sea east of type locality; Kryštufek and Vohralík, 2001; Macholán et al., 2001*a*), N Caucasus (Russia), S Caucasus (Georgia, Armenia, and W Azerbaijan), and NW Iran.

SYNONYMS: *ciscaucasicus* (Ognev, 1924); *colchicus* Shidlovsky, 1919; *dinniki* Satunin, 1903 [*nomen nudum*]; *labensis* Heptner, 1948; *rubelianus* Shidlovsky, 1919; *transcaucasicus* (Khatukhov and Tembotov, 1982); *vinogradovi* Sviridenko, 1936 [*nomen nudum*; not of Fetisov, 1936].

COMMENTS: Subgenus *Terricola*, *subterraneus* species group (Chaline et al., 1988) or *majori* species group (Pavlinov and Rossolimo, 1998; Pavlinov et al., 1995*a*; Zagorodnyuk, 1990). Morphologically and genetically closely related to *M. subterraneus* and *M. daghestanicus* (Baskevich, 1997; Macholán et al., 2001*a*; Mezhzherin et al., 1995). Geographic range once thought to extend into the Balkans, S Greece, and W Turkey (Kivanc, 1986; Niethammer, 1987*b*), but a diverse character base aligns those populations with *M. subterraneus* and excludes *M. majori* from the European fauna (Kryštufek et al., 1994).

Reviewed by Storch (1982) and comparative chromosomal data summarized by Zima and Kral (1984*a*). Macholán et al. (2001*a*) documented karyotypic and allozymic contrasts between Turkish and European *M. subterraneus* and Turkish *M. majori*; their ranges approach closely in NW Turkey but sympatry not yet demonstrated (see account of *M. subterraneus*). Karyotypic and spermatozoal comparisons among *M. majori* (2n = 54, FN = 60), *M. daghestanicus* (2n = 54, FN = 58), and *M. subterraneus* (2n = 52, FN = 60) validated the species status of each, indicated closer relationship between *M. daghestanicus* and *M. subterraneus*, and identified traits useful for separating sympatric samples of the morphologically similar *M. majori* and *M. daghestanicus* (Baskevich, 1997). Invariant karyotype of *M. majori* from the Caucasus contrasted with highly variable 2n (38-54) and conservative FN (58) of *M. daghestanicus* by Achverdjan et al. (1992), who discussed their significance to pine vole evolution. Comparisons in molar size variation among Turkish *M. majori*, *M. subterraneus*, and *M. daghestanicus* elucidated by Kryštufek and Vohralík (2004).

Microtus maximowiczii (Schrenk, 1859). Reisen und Forsch., *in* Säugeth. Amurlande St. Petersburg., 1:140.

COMMON NAME: Maximowicz's Vole.

TYPE LOCALITY: Russia, Chita Oblast, upper Amur region, mouth of Omutnaya River.

DISTRIBUTION: E shore of Lake Baikal to upper Amur region (Meyer et al., 1996), E Mongolia, and NE China (Heilongjiang, Nei Mongol, Jilin, Hebei, and N Shaanxi; Zhang et al., 1997).

STATUS: IUCN – Lower Risk (lc).

SYNONYMS: *gromovi* Vorontsov, Boeskorov, Ljapunova, and Revin, 1988; *ungurensis* Kastschenko, 1912.

COMMENTS: Subgenus *Alexandromys*, *maximowiczii* species group (Pavlinov and Rossolimo, 1998; Pavlinov et al., 1995*a*; Zagorodnyuk, 1990). Included in subgenus *Microtus* by Gromov and Polyakov (1977), Pavlinov and Rossolimo (1987), and Meyer et al. (1996). Karyotypic, morphological and hybridization evidence closely relates *M. maximowiczii* to *M. mujanensis* and *M. evorensis* (Meyer et al., 1996). Reviewed by Orlov and Kovalskaya (1978) and Meyer (1983). Populations in Russia revised by Meyer et al. (1996), who attributed the two synonyms. Additional intraspecific chromosomal polymorphisms analysed by Kovalskaya (1977). Molar variation in the *M. maximowiczii* species group, as analyzed by Pozdnyakov (1993), reveals significant differences between *maximowiczii* from the Amur region and other geographic samples.

Microtus mexicanus (Saussure, 1861). Rev. Mag. Zool. Paris, Ser. 2, 13:3.

COMMON NAME: Mexican Vole.

TYPE LOCALITY: México, Puebla, Volcán de Orizaba.

DISTRIBUTION: Patchy occurrence in mountains from extreme S Utah and extreme S Colorado (Fitzgerald et al., 1994:9-102) to C Arizona and New Mexico (Frey and LaRue, 1993, for new records), USA; south in Sierra Madres through interior México to C Oaxaca.

STATUS: U.S. ESA and IUCN—Endangered as *M. m. hualpaiensis*; IUCN – Vulnerable as *M. mexicanus* and *M. mongollonensis hualpaiensis*.

SYNONYMS: *fulviventer* Merriam, 1890; *fundatus* Hall, 1948; *guadalupensis* Bailey, 1902; *hualpaiensis* Goldman, 1938; *madrensis* Goldman, 1938; *mogollonensis* (Mearns, 1890); *navaho* Benson, 1934; *neveriae* Hooper, 1955; *ocotensis* T. Alvarez and Hernández-Chávez, 1993; *phaeus* (Merriam, 1892); *salvus* Hall, 1948; *subsimus* Goldman, 1938.

COMMENTS: Interspecific affinities unclear: affiliated with *Microtus sensu stricto*, in particular *M. californicus* (Anderson, 1959, 1960; Conroy and Cook, 2000*a*; Conroy et al., 2001) or *M. montanus* and *M. pennsylvanicus* (Moore and Janecek, 1990); associated with the subgenus *Pitymys* by Hooper and Hart (1962); chromosomal banding data uninformative (Modi, 1987); also see Hoffmann and Koeppl (1985). Morphometric, electrophoretic, and karyotypic variation among isolated populations in SW USA studied by Wilhelm (1982); nongeographic variation of US and Mexican populations evaluated by Frey and Moore (1990); geographic variation of populations in C México assessed by Alvarez and Hernández-Chávez (1993), who considered the regional subspecies as valid (*fundatus*, *mexicanus*, and *salvus*) and described a new one (*ocotensis*).

Critical overhaul of the *mexicanus* complex is needed. Musser (1964) reduced the Oaxacan form *fulviventer* to a subspecies of *M. mexicanus*; whereas, Frey and LaRue (1993) and Frey (1999) elevated populations in the SW USA (*mogollonensis*) as distinct, arguing the truly superficial basis for its original synonymy (Bailey, 1932) and noting earlier reports of slight chromosomal differences between US and Mexican populations (Judd, 1980). Fitzgerald et al. (1994) and Saldaña-DeLeon and Jones (1998), on the other hand, retained *mogollonensis* as a subspecies for Colorado populations; furthermore, the interpretation of extensive and relatively recent post-Pleistocene dispersion of *M. mexicanus* into the SW USA (Davis and Callahan, 1992) questions the length of isolation of these populations and the potential for their differentiation as a separate species. A phylogeographic approach, integrating morphological and genetic variation across the collective range of these taxa, would offer a firmer basis for delineating species limits and diagnosing them.

Microtus middendorffii (Poljakov, 1881). Mem. Acad. Imp. Sci. St. Petersbourg, 39:70.

COMMON NAME: Middendorff's Vole.

TYPE LOCALITY: Russia, Krasnoyarsk Krai, Taimyr Peninsula.

DISTRIBUTION: N Siberia, from the Polar Ural Mtns east through the Yenisey River basin past the N Lena River region to the N Kolyma River area, south in Urals to 62N, and in the Lena Valley near Yakutsk (Meyer et al., 1996).

STATUS: IUCN – Lower Risk (lc) as *M. middendorffii* and *M. hyperboreus*.

SYNONYMS: *hyperboreus* Vinogradov, 1934; *obscurus* (Middendorff, 1853) [not Eversmann, 1841]; *ryphaeus* Heptner, 1948; *swerevi* Scalon, 1935; *tasensis* Skalon, 1935; *uralensis* Skalon, 1935 [not Poliakoff, 1881].

COMMENTS: Subgenus *Alexandromys* (Pavlinov et al., 1995a; Pavlinov and Rossolimo, 1998). Also variously included in subgenus *Pallasiinus* (Zagorodnyuk, 1990) or subgenus *Microtus* (Gromov and Erbajeva, 1995; Gromov and Polyakov, 1977; Meyer et al., 1996). Meyer (1983) considered *M. middendorffii* to be phylogenetically distant from other northern and central Asian species of *Microtus*, but close relationship to *M. oeconomus* broadly supported by analyses of allozymes (Mezhzherin et al., 1993), karyology and morphology (Meyer et al., 1996), and DNA sequences (Conroy and Cook, 2000a). Vorontsov and Lyapunova (1976) considered *M. middendorffii* (and *hyperboreus*) to be chromosomally closely related to North American *M. miurus*, but cytochrome *b* sequences group *M. middendorffii* with *M. fortis* and Palaearctic species, not *M. miurus* and North American endemics (Conroy and Cook, 2000a). Karyotype documented by Boyeskorov et al. (1993, as *hyperboreus*).

The taxon *hyperboreus* has been widely treated as a separate species (Gromov and Erbajeva, 1995; Gromov and Polyakov, 1977; Meyer, 1983; Musser and Carleton, 1993; Pavlinov and Rossolimo, 1987; Pavlinov et al., 1995a). Corbet (1978c) included *hyperboreus* in *M. middendorffii* based on Gileva's (1972) demonstration of their complete interfertility. Meyer et al. (1996) comprehensively marshalled chromosomal, morphological, and hybridization data that sustain inclusion of *hyperboreus* as a subspecies in *M. middendorffii*, along with typical *middendorffii* and *ryphaeus*.

Microtus miurus Osgood, 1901. N. Am. Fauna, 21:64.

COMMON NAME: Singing Vole.

TYPE LOCALITY: USA, Alaska, Cook Inlet, Turnagain Arm, head of Bear Creek in mountains near Hope City.

DISTRIBUTION: Alaska, USA, except the central portion and Alaska Peninsula; eastwards through much of Yukon Territory to westernmost Northwest Territories and extreme NW British Columbia, Canada.

STATUS: IUCN – Lower Risk (lc).

SYNONYMS: *andersoni* Rand, 1945; *cantator* Anderson, 1947; *muriei* Nelson, 1931; *oreas* Osgood, 1907; *paneaki* Rausch, 1950.

COMMENTS: Subgenus indeterminate. Classically treated as a member of the subgenus *Stenocranius* (e.g., Hall, 1981; Miller and Kellogg, 1955), an association that purported close affinity to Old World *M. gregalis* (see that account); synonymized with same to form a Holarctic species by Rausch (1964) and Rausch and Rausch (1968). Other data convincingly argue their specific distinctiveness and distant kinship (Anderson, 1960; Fedyk, 1970; Gromov and Polyakov, 1977; Vorontsov and Lyapunova, 1986), and phylogenetic studies of chromosomes (Zagorodnyuk, 1990) and mitochondrial DNA sequences (Conroy and Cook, 2000a) require the removal of *M. miurus* (along with *M. abbreviatus*) from *Stenocranius* proper. Zagorodnyuk (1990) had emphasized this divergence by placing *M. miurus* in the subgenus *Alexandromys*, *M. middendorffii* species group, along with *M. mongolicus* and *M. sachalinensis*. According to the results of Conroy and Cook (2000a), *M. miurus*, and its sister species *M. abbreviatus*, originated early within the radiation of North American *Microtus*, apart from Palearctic species.

Microtus mongolicus (Radde, 1861). Melange Biol. Acad. St. Petersbourg, 3:681.

COMMON NAME: Mongolian Vole.

TYPE LOCALITY: Russia, Transbaikalia (Chitinskaya Oblast), Omutnaya River, a tributary to the Amur River.

DISTRIBUTION: Transbaikalia, Mongolia, and NE China (N Nei Mongolia, Heilongjiang, Jilin).

STATUS: IUCN – Lower Risk (lc).

SYNONYMS: *baicalensis* Fetisov, 1941; *poljakovi* Kastschenko, 1901; *xerophilus* Skalon, 1936.

COMMENTS: Subgenus *Alexandromys*, *middendorffii* species group (Zagorodnyuk, 1990) or *mongolicus* species group (Pavlinov and Rossolimo, 1998; Pavlinov et al., 1995a).

Meyer et al. (1996) placed *M. mongolicus* in the subgenus *Microtus*, and their integrated study of karyotypes, morphology, and hybridization reveals *M. mongolicus* to be closely related to *M. fortis*, by implication a member of the *maximowiczii* species group. Karyotypic analyses by Orlov et al. (1983), Yatsenko et al. (1980), and Radjabli et al. (1984) include phylogenetic inferences among Asian species of *Microtus*.

Microtus montanus (Peale, 1848). Mammalia *in* Repts. U.S. Expl. Surv., 8:44.

COMMON NAME: Montane Vole.

TYPE LOCALITY: USA, California, Siskiyou Co., headwaters of Sacramento River, near Mt Shasta.

DISTRIBUTION: Cascade, Sierra Nevada, and Rocky Mountain ranges: SC British Columbia, Canada, south to EC California, S Utah, and NC New Mexico, USA; disjunct populations in S Nevada, EC Arizona, and WC New Mexico (Frey et al., 1995).

STATUS: IUCN – Data Deficient as *M. m. codiensis* and *M. m. zygomaticus*, Vulnerable as *M. m. fucosus* and *M. m. nevadensis*, Lower Risk (nt) as *M. m. arizonensis* and *M. m. rivularis*, otherwise Lower Risk (lc).

SYNONYMS: *amosus* Hall and Hayward, 1941; *arizonensis* Bailey, 1898; *canescens* Bailey, 1898; *caryi* Bailey, 1917; *codiensis* S. Anderson, 1954; *dutcheri* Bailey, 1898; *fucosus* Hall, 1935; *fusus* Hall, 1938; *longirostris* (Baird, 1857); *micropus* Hall, 1935; *nanus* (Merriam, 1891); *nevadensis* Bailey, 1898; *nexus* Hall and Hayward, 1941; *pratincolus* Hall and Kelson, 1951; *rivularis* Bailey, 1898; *undosus* Hall, 1935; *yosemite* Grinnell, 1914; *zygomaticus* S. Anderson, 1954.

COMMENTS: Subgenus *Mynomes*, *montanus* species group (Zagorodnyuk, 1990). Geographic variation and subspecific classification assessed by Anderson (1959), who viewed *M. oeconomus* as its sister species. Hooper and Hart (1962) arranged *M. montanus* with *M. pennsylvanicus* and *M. townsendii*, a general relationship corroborated by karyotypic and genic analyses (Modi, 1987; Moore and Janecek, 1990). Two karyotypic morphs reported by Judd et al. (1980), who raised the question of their specific distinction. See Sera and Early (2003, Mammalian Species, 716).

Microtus montebelli (Milne-Edwards, 1872). Rech. Hist. Nat. Mammifères, p. 285.

COMMON NAME: Japanese Grass Vole.

TYPE LOCALITY: Japan, Honshu Isl, Fusiyama.

DISTRIBUTION: Endemic to the Japanese islands of Honshu, Sado, and Kyushu; not currently found on Hokkaido, Shikoku or the Ryukyu Isl (Dobson, 1994; Kaneko, 1994) but may have occurred on Shikoku in the Pleistocene (Kawamura, 1988).

STATUS: IUCN – Lower Risk (lc).

SYNONYMS: *brevicorpus* Tokuda, 1933; *hatanedzumi* (Sasaki, 1904).

COMMENTS: Subgenus *Alexandromys* (including *Pallasiinus*, as per Zagorodnyuk, 1990, and Pavlinov et al., 1995*a*) or subgenus *Microtus* (Gromov and Polyakov, 1977). Phylogenetic interpretation of cytochrome *b* sequences allies *M. montebelli* in a tight clade with *M. fortis*, *M. middendorffii*, *M. kikuchii*, and *M. oeconomus* (Conroy and Cook, 2000*a*); karyotypic information also indicates special relationship to *M. oeconomus* (Borodin et al., 1997; Zagorodnyuk, 1990). Other chromosomal data summarized by Tsuchiya (1981) and Satoh and Obara (1995). Gastric musculature documented by Ohuchi et al. (1992). Corbet (1978*c*) included Sikotan Isl in the Kuriles within the species range, but this record is erroneous (Kawamura, 1988, and references therein).

Kawamura (1988, 1991, 1994) exhaustively summarized molar variation of Holocene and middle-late Pleistocene samples and defined Quaternary species endemic to Japan. *Microtus montebelli*, now the only surviving Japanese endemic, then occurred with populations of *M. epiratticepoides* and *M.* cf. *brandtioides* (perhaps = *M. fortis*; see Kaneko and Hasegawa, 1995), both of which disappeared between 15,000 and 10,000 years before present.

Microtus mujanensis Orlov and Kovalskaya, 1978. Zool. Zh., 57:1224.

COMMON NAME: Muya Valley Vole.

TYPE LOCALITY: Russia, Buryat, Bauntovski Dist., Vitim River Basin, Muya Valley.

DISTRIBUTION: Known only from vicinity of the type locality (Meyer et al., 1996).

STATUS: IUCN – Critically Endangered.

COMMENTS: Subgenus *Alexandromys*, *maximowiczii* species group (Pavlinov and Rossolimo, 1998; Pavlinov et al., 1995*a*; Zagorodnyuk, 1990). A member of subgenus *Microtus* according to Pavlinov and Rossolimo (1987) and Meyer et al. (1996). Analysis of molar variation in the *M. maximowiczii* species group indicates close relationship between *M. mujanensis* (2n = 38, FN = 50-53) and *M. evoronensis* (2n = 40, FN = 58) (Pozdnyakov, 1993), as does integration of morphological, karyological, and hybridization data (Meyer, 1983; Meyer et al., 1996).

Microtus multiplex (Fatio, 1905). Arch. Sci. Phys. Nat. Geneve, Ser. 4, 19:193.
   COMMON NAME: Alpine Pine Vole.
   TYPE LOCALITY: Switzerland, Ticino Canton, near Lugano.
   DISTRIBUTION: Western and Central Alps in E France, Switzerland (Maurizio, 1994; Hausser, 1995), far W Austria, and N Italy, including N Italian Apennines (Amori et al., 1999; Cantini, 1991; Paolucci et al., 1994). The Adige River Valley (Trentino region) in the Alps of NE Italy is apparently the boundary between ranges of *M. multiplex* (to the west) and *M. liechtensteini* (eastward); see Brunet-Lecomte and Volobouev (1994), Haring et al. (2000), and Spitzenberger et al. (2000).
   STATUS: IUCN – Lower Risk (lc).
   SYNONYMS: *druentius* (Miller, 1911); *fatioi* (Mottaz, 1909); *leponticus* Thomas, 1906; *niethammeri* Brunet-Lecomte and Volobouev, 1994; *orientalis* Dal Piaz, 1924; *vuillemeyi* Brunet-Lecomte, Chaline, and Campy, 1993.
   COMMENTS: Subgenus *Terricola*, *subterraneus* species group (Chaline et al., 1988; Pavlinov et al., 1995*a*) or *multiplex* species group (Zagorodnyuk, 1990). Most closely related to *M. bavaricus* and *M. liechtensteini* (see latter account for discussion of relationships). Brunet-Lecomte and Volobouev (1994) and Brunet-Lecomte (1995) identified populations from the W French Alps and Rhone Valley, at the western margin of the geographic range, as the subspecies *niethammeri* based on m1 morphology and chromosomes. The taxon *vuillemeyi* was described as a fossil subspecies from the Pleistocene of E France (Brunet-Lecomte et al., 1993).

Microtus oaxacensis Goodwin, 1966. Am. Mus. Novit., 2243:1.
   COMMON NAME: Tarabundí Vole.
   TYPE LOCALITY: México, Oaxaca, Ixtlán Dist., near Vista Hermosa, Tarabundí Ranch, 5000 ft (1524 m).
   DISTRIBUTION: Eastern slopes, Sierra de Juárez, 1600-2500 m, NC Oaxaca, México.
   STATUS: IUCN – Lower Risk (nt).
   COMMENTS: Subgenus *Pitymys*. Differentially diagnosed with respect to *M. mexicanus* and *M. umbrosus* by Goodwin (1966); description amplified and resemblance to *M. guatemalensis* noted by Jones and Genoways (1967), an affinity supported by their dental and genetic similarity (Conroy et al., 2001; R. A. Martin, 1995). Grouped with other relictual "pitymyine" species found at southern edge of the Nearctic *Microtus* distribution by Hoffmann and Koeppl (1985), a biogeographic interpretation supported by Conroy et al. (2001). Karyotype reported by Cervantes et al. (1997), who noted its derived condition compared with other New World *Microtus*. Altitudinally parapatric distribution with respect to *M. mexicanus* illuminated by Sánchez H. et al. (1996). See Frey and Cervantes (1997, Mammalian Species, 556).

Microtus ochrogaster (Wagner, 1842). *In* Schreber, Die Säugethiere, Suppl. 3:592.
   COMMON NAME: Prairie Vole.
   TYPE LOCALITY: USA, Indiana, Posey Co., New Harmony (as fixed by Bole and Moulthrop, 1942:157).
   DISTRIBUTION: Northern and Central Great Plains—EC Alberta to S Manitoba, Canada; south to N Texas Panhandle (Choate and Killebrew, 1991), SW Oklahoma (Smith, 1992), and Arkansas; eastwards to C Tennessee, westernmost West Virginia, and W Ohio, USA; relictual populations in C Colorado, N New Mexico, and coastal prairies (*ludovicianus*) of SW Louisiana and adjacent Texas, USA.
   STATUS: IUCN – Lower Risk (lc).
   SYNONYMS: *austerus* (Le Conte, 1853); *cinnamonea* (Baird, 1858); *haydenii* (Baird, 1857);

ludovicianus Bailey, 1900; *minor* (Merriam, 1888); *ohioensis* Bole and Moulthrop, 1942; *similis* Severinghaus, 1977; *taylori* Hibbard and Rinker, 1943.

COMMENTS: Subgenus *Pedomys*, which has been employed as a subgenus of *Microtus* (Bailey, 1900; R. A. Martin, 1995; Van der Meulen, 1978), as a full synonym of *Microtus* (Hall, 1981), or as a synonym of *Pitymys* whether ranked as a genus (Repenning, 1983) or subgenus (Hooper and Hart, 1962; Zagorodnyuk, 1990). The purported close affinity of *M.* (*Pedomys*) *ochrogaster* with North American pitymyine species (subgenus *Pitymys*) has received little support from allozymic and gene-sequence studies (Chaline and Graf, 1988; Conroy and Cook, 2000*a*; Conroy et al., 2001; Moore and Janacek, 1990). R. A. Martin (1995) retained the two as indigenous North American subgenera based on cladistic analysis of dental traits.

Geographic variation over the Central Great Plains studied by Choate and Williams (1978). The strong morphometric segregation of *minor* from other *M. ochrogaster* advises renewed scrutiny of its status (Severinghaus, 1977); also see remarks by R. A. Martin (1991) on the occurrence of "two size classes of *M. ochrogaster*" in certain late Pleistocene deposits and report of extensive heterochromatin variability by Modi (1993). Extralimital late Pleistocene records summarized by R. A. Martin (1991) and late Holocene occurrence reported for NW Wyoming (Barnosky, 1994); chronological phyletic sequence of fossil taxa leading to *M. ochrogaster* postulated by R. A. Martin (1995). Includes *ludovicianus*, an isolated form of the coastal prairies formerly considered a separate species and now apparently extinct (Lowery, 1974). See Stalling (1990, Mammalian Species, 355).

Microtus oeconomus (Pallas, 1776). Reise Prov. Russ. Reichs., 3:693.

COMMON NAME: Root Vole.

TYPE LOCALITY: Russia, Siberia, Ishim Valley.

DISTRIBUTION: Holarctic in tundra, northern taiga, and grassy meadows. In Palearctic, from Fennoscandia across N European Russia and Siberia to Kamchatka Peninsula (Nikanorov, 2000) and borderlands of Bering Sea; south through through Baltic region (Miljutin, 1997, 1998; Timm et al., 1998) to NE Germany (Dolch, 1994), Poland (Salata-Pilacinska, 1990), Belarus, Ukraine, N Kazakhstan, transbaikal region (Reiter et al., 1995, documented occurrence on Svjatoj Nos peninsula and isthmus in Lake Baikal), N Mongolia, NE China (N Nei Mongol), and discontinuously in the Ussuri region bordering the Japanese Sea, including, Sakhalin and Kurile Isls; relict populations in Netherlands (Ligtvoet, 1992; Ligtvoet and Wijngaarden, 1994), S Norway and C Sweden, Finnish Baltic Coast, E Austria, SW Slovakia (Mošanský, 1994; Stanko and Mošanský, 2000), and W Hungary; absent from the British Isls but present on St. Lawrence Isl in Bering Sea. In Nearctic, from Alaska through Yukon Territory, eastwards to W Northwest Territories, and southwards to extreme NW British Columbia and Alexander Archipelago, Alaska (see records by MacDonald and Cook, 1996:579).

STATUS: IUCN – Critically Endangered as *M. o. arenicola*, Vulnerable as *M. o. mehelyi*, Data Deficient as *M. o. amakensis*, *M. o. elymocetes*, *M. o. innuitus*, *M. o. popofensis*, *M. o. punukensis*, and *M. o. sitkensis*, otherwise Least Concern.

SYNONYMS: *altaicus* Ognev, 1944; *amakensis* Murie, 1930; *anikini* Egorin, 1939; *arenicola* (de Sélys Longchamps, 1841); *dauricus* Kastschenko, 1910; *elymocetes* Osgood, 1906; *endoecus* Osgood, 1909; *epiratticeps* Young, 1934; *finmarchicus* Siivonen, 1967; *flaviventris* Satunin, 1903; *gilmorei* Setzer, 1952; *hahlovi* Skalon, 1935; *innuitus* Merriam, 1900; *kadiacensis* Merriam, 1897; *kamtschatica* (Pallas, 1779); *karaginensis* Kostenko, 1984 [*nomen nudum*]; *karaginensis* Kostenko, 1989; *kharanurensis* Courant et al., 1999; *kjusjurensis* Koljuschev, 1935; *koreni* G. M. Allen, 1914; *macfarlani* Merriam, 1900; *medius* (Nilsson, 1844); *mehelyi* Ehik, 1928; *montiumcaelestinum* Ognev, 1944; *naumovi* Stroganov, 1936; *operarius* (Nelson, 1893); *ouralensis* (Poliakov, 1881); *petshorae* Ognev, 1944; *popofensis* Merriam, 1900; *punukensis* Hall and Gilmore, 1932; *ratticeps* (Keyserling and Blasius, 1841); *shantaricus* Ognev, 1929; *sitkensis* Merriam, 1897; *stimmingi* (Nehring, 1899); *suntaricus* Dukelski, 1928; *tschuktschorum* Miller, 1899; *uchidae* Kuroda, 1924; *unalascensis* Merriam, 1897; *uralensis* (Poljakov, 1881); *yakutatensis* Merriam, 1900.

COMMENTS: Subgenus *Alexandromys*. Earlier included in the subgenus *Pallasiinus* (Gromov and Erbajeva, 1995; Pavlinov and Rossolimo, 1998; Pavlinov et al., 1995a; Zagorodnyuk, 1990), also containing *M. montebelli* and *M. limnophilus*, or arranged in subgenus *Microtus* (Gromov and Polyakov, 1977; Pavlinov and Rossolimo, 1987). Based on morphometric, chromosomal and hybridization analyses, Meyer et al. (1996) placed *M. oeconomus* in a clade containing *M. middendorffii*, *M. mongolicus*, *M. fortis*, *M. sachalinensis*, *M. maximowiczii*, *M. jujanensis*, and *M. evoronensis*; according to phylogenetic analysis of cytochrome *b* sequences, member of a clade that includes the Taiwanese *M. kikuchii* as sister-species, along with *M. montebelli*, *M. middendorffii*, and *M. fortis* (Conroy and Cook, 2000a). Close alliance of *M. oeconomus* and *M. montebelli* is also supported by pairing patterns of X-Y chromosome (Borodin et al., 1997).

Homogeneity of Old and New World populations averred by Zimmerman (1942) and sustained by subsequent studies (Nadler et al., 1976, 1978; Ognev, 1964; Rausch, 1953; Conroy and Cook, 2000a). Interpretations of mitochondrial DNA led Conroy and Cook (2000a) to propose that *M. oeconomus* represented one of only two colonizations of North America by Old World stocks, the earlier one radiating into the North American endemics. Allozymic and chromosomal differentiation among subspecies surrounding the Bering Strait minimal, also suggesting a relatively late entry of the species into North America and rapid expansion of populations once arrived (Lance and Cook, 1998).

Geographic variation and subspecies of Nearctic populations reviewed by Paradiso and Manville (1961), populations in Russia and adjacent regions by Gromov and Erbajeva (1995), and European populations by Tast (1982b) and Mitchell-Jones et al. (1999). Chromosomal data evaluated by Zima and Kral (1984a); Modi (1987) and Lance and Cook (1998) documented recent chromosomal results (uniformly 2n = 30, FN = 54). Synaptic association of sex chromosomes and its evolutionary significance discussed by Ashley and Fredga (1994). Gastrointestinal morphology of Chinese sample and its adaptive significance reported by Wang et al. (1995). Interdependence and correlation among non-metric characters with sex, age, and body size evaluated by Markowski (1995). Molar patterns intensively investigated, addressing frequency and interpretation of morphotypes in large isolated sample from W Poland (Rachowiak, 1993), assessment of variation in extant samples against which patterns in extinct species can be judged (Angermann, 1984), correlation of variation with latitude and altitude (Pozdnyakov and Litvinov, 1994), and departures from bilateral symmetry (Kovaleva et al., 2002).

Kaneko and Hasegawa (1995) identified *M. oeconomus* from a late Pleistocene cave on Miyako Isl, Ryukyu Isls, an occurrence far outside the modern and Pleistocene distribution as known by other fossils. They also allocated Pleistocene *epiratticeps* as another synonym of *M. oeconomus*. The taxon *corneri*, based on Pleistocene (Late Glacial) cave deposits in the British Isles, testifies the former presence of *M. oeconomus* (Berry, 1996; Yalden, 1999).

Ognev (1964) identified *ratticeps* Keyserling and Blasius, 1841, as the proper name applicable to this species, and suggested that *oeconomus* Pallas, 1776, was based upon an example of *M. gregalis*. Ellerman and Morrison-Scott (1951:705), however, noted that *kamtschaticus* Pallas, 1779, would be the oldest name, not *ratticeps*, if Ognev is correct (also see Hall, 1981:805). While most systematists continue to utilize *oeconomus* Pallas, 1776, for reasons of familiarity, *ratticeps* has been used in the past (G. M. Allen, 1940) and even recently (Berry, 1996). This nomenclatural uncertainty requires formal resolution.

Microtus oregoni (Bachman, 1839). J. Acad. Nat. Sci. Philadelphia, 8:60.

  COMMON NAME: Creeping Vole.
  TYPE LOCALITY: USA, Oregon, Clatsop Co., Astoria.
  DISTRIBUTION: Moist coniferous forest seres of Pacific Northwest, from SW British Columbia, Canada, south to NW California, USA.
  STATUS: IUCN – Lower Risk (lc).
  SYNONYMS: *adocetus* Merriam, 1908; *bairdii* Merriam, 1897; *cantwelli* Taylor, 1920; *morosus* Elliot, 1899; *serpens* Merriam, 1897.

COMMENTS: Subgenus *Mynomes*, sole member of *oregoni* species group (Zagorodnyuk, 1990). Type species of *Chilotus*, conventionally recognized as a subgenus of *Microtus* (see Anderson, 1959, 1960; Bailey, 1900), and occasionally including Old World *M. socialis* as a comember (Chaline, 1974; Ognev, 1964) or not (Anderson, 1959). Hooper and Hart (1962), however, found the morphological evidence insufficient to warrant subgeneric segregation of *oregoni* from North American species of *Microtus*, a viewpoint sustained by genetic distance comparisons (Moore and Janecek, 1990) and reflected in the classification of Zagorodnyuk (1990); sister species to *M. longicaudus* according to mitochondrial DNA sequence data (Conroy and Cook, 2000a) or to the *M. chrotorrhinus-M. pinetorum* clade according to repetitive DNA fragments (Modi, 1996). Diploid number and sex-determining mechanism unique among North American *Microtus* studied (Ohno et al., 1963). See Carraway and Verts, 1985 (Mammalian Species, 233).

Microtus paradoxus (Ognev and Heptner, 1928). Zool. Anz., 75:263.
  COMMON NAME: Paradox Vole.
  TYPE LOCALITY: S Turkmenistan, Kopet Dag Mtns, Chuli, near Ashkhabad.
  DISTRIBUTION: Kopet Dag Mtns, S Turkmenistan, and mountains of NE Iran in N Khorassan Prov. (near Dasht, Gowadoh, Dergematie, Kaur, and Gorgan; series in AMNH and USNM; Kryštufek and Kefelioğlu, 2002); may also occur in N Afghanistan, identified as *M. guentheri* (Hassinger, 1973), but the two cited specimens need to be reexamined.
  COMMENTS: Subgenus *Microtus*, *socialis* species group (Zagorodnyuk, 1990). Placed in subgenus *Sumeriomys* by Gromov and Erbajeva (1995). Described as a species, later relegated to *M. socialis* (Corbet, 1978c; Ellerman, 1941; Ellerman and Morrison-Scott, 1951; Pavlinov and Rossolimo, 1987). Specific status reasserted by Zykov and Zagorodnyuk (1988), whose morphometric analysis distinguished it from both *M. socialis* (karyotype also 2n = 62, FN = 60, but much larger in body and cranial dimensions) and *M. guentheri* (2n = 54, FN = 56); Golenishchev et al. (2002) amplified and endorsed their action. Goodwin (1940) had identified Iranian series from Dasht, Gowadoh and Dergematie as *M. socialis paradoxus*, the name combination at the time. Kryštufek and Kefelioğlu (2002) and Kryštufek (in litt., 2002) demonstrated that those samples are not the smaller-bodied *M. socialis*, and although morphologically similar to the "non-*socialis*" group in Iran, they differ strikingly in certain cranial traits and likely represent *M. paradoxus*. Although examining Goodwin's AMNH specimens and the large USNM series leads us to concur, we suspect that *M. guentheri* and *M. paradoxus* will prove to be more closely related than now appreciated; the nature of the genetic relationship between the two warrants further study.

Microtus pennsylvanicus (Ord, 1815). *In* Guthrie, New Geogr., Hist., Comml., Grammar, Philadelphia, 2nd ed., 2:292.
  COMMON NAME: Meadow Vole.
  TYPE LOCALITY: USA, Pennsylvania, "meadows below Philadelphia."
  DISTRIBUTION: Meadowlands interspersed across boreal and mixed coniferous-deciduous biomes of North America: C Alaska to Labrador, including Newfoundland and Prince Edward Isl, Canada; south in Rocky Mountains to N New Mexico, in Great Plains to N Kansas (see Frey and Moore, 1990), and in Appalachians and along eastern seaboard to N Georgia and South Carolina, USA; outlier populations in W New Mexico and peninsular Florida, USA, and in N Chihuahua, México.
  STATUS: U.S. ESA—Endangered as *M. p. dukecampbelli*; IUCN – Vulnerable as *M. p. dukecampbelli*, Lower Risk (nt) as *M. p. admiraltiae*, *M. p. kincaidi*, *M. p. provectus*, and *M. p. shattucki*, not evaluated as *M. p. chihuahuensis*, otherwise Lower Risk (lc).
  SYNONYMS: *acadicus* Bangs, 1897; *admiraltiae* Heller, 1909; *alborufescens* (Emmons, 1840); *alcorni* Baker, 1951; *aphorodemus* Preble, 1902; *arcticus* Cowan, 1951; *aztecus* (J. A. Allen, 1893); *chihuahuensis* Bradley and Cockrum, 1968; *copelandi* Youngman, 1967; *dekayi* (Audubon and Bachman, 1854); *drummondii* (Audubon and Bachman, 1853); *stonei* J. A. Allen, 1899; *dukecampbelli* Woods, Post, and Kilpatrick, 1982; *enixus* Bangs, 1896; *finitus* S. Anderson, 1956; *fontigenus* Bangs, 1896; *fulva* (Audubon and Bachman, 1841); *funebris* Dale, 1940; *hirsutus* (Emmons, 1840); *inspectus* (J. A. Allen, 1899); *insperatus* (J. A. Allen, 1894); *insularis* Bailey, 1898; *kincaidi* Dalquest, 1941; *labradorius*

Bailey, 1898; *longipilis* (Baird, 1857); *magdalenensis* Youngman, 1967; *microcephalus* (Rhoads, 1894); *modestus* (Baird, 1857); *nasuta* (Audubon and Bachman, 1841); *nesophilus* Bailey, 1898 [replacement name for *insularis* Bailey, 1898]; *nigrans* Rhoads, 1897; *noveboracensis* (Rafinesque, 1820); *oneida* (DeKay, 1842); *palustris* (Harlan, 1825); *pratensis* (Rafinesque, 1817); *provectus* Bangs, 1908; *pullatus* S. Anderson, 1956; *riparius* (Ord, 1825); *rubidus* Dale, 1940; *rufescens* (DeKay, 1842); *rufidorsum* (Baird, 1857); *shattucki* Howe, 1901; *tananaensis* Baker, 1951; *terraenovae* (Bangs, 1894); *uligocola* S. Anderson, 1956; *wahema* Bailey, 1920.

COMMENTS: Subgenus *Mynomes*, *pennsylvanicus* species group (Zagorodnyuk, 1990). Proposed as conspecific with Old World *M. agrestis* by Klimkiewicz (1970), but G-banded chromosomal differences support their recognition as distinct species (Modi, 1987; Vorontsov and Lyapunova, 1986). Aside from the probable insular derivative *M. breweri* (see that account), *pennsylvanicus* is closely related to *M. montanus* and *M. townsendii* among New World species (see Conroy and Cook, 2000*a*; Hooper and Hart, 1962; Modi, 1987; Moore and Janecek, 1990). Insular form *provectus* relegated to subspecific status by Chamberlain (1954) and Moyer et al. (1988), and *nesophilus* by Jones et al. (1986). Regional studies of variation undertaken (e.g., Anderson, 1956; Anderson and Hubbard, 1971; Weddle and Choate, 1983), but comprehensive review of entire species warranted. Late Pleistocene and Holocene vegetational and climatic changes discussed by Woods et al. (1982) apropos isolation of the Florida population (*dukecambelli*) and other arvicoline austral relicts. Intricate study of M3 variation in fossil and extant samples conducted by Barnosky (1993), who attempted to infer microevolutionary processes from broad stratigraphic and geographic patterns. See Reich (1981, Mammalian Species, 159).

Microtus pinetorum (Le Conte, 1830). Ann. Lyc. Nat. Hist., 3:133.

COMMON NAME: Woodland Vole.

TYPE LOCALITY: USA, Georgia, Liberty Co., old LeConte plantation near Riceboro (as interpreted by Bailey, 1900:63).

DISTRIBUTION: Temperate deciduous forest zone of E USA—eastern shoreline from S Maine to NC Florida, west to C Wisconsin and E Texas; isolated population (*auricularis*) on the Edwards Plateau, C Texas, may be extinct (Goetze, 1998).

STATUS: IUCN – Lower Risk (lc).

SYNONYMS: *apella* (Le Conte, 1853); *auricularis* Bailey, 1898; *carbonarius* (Handley, 1952); *kennicottii* (Baird, 1857); *nemoralis* Bailey, 1898; *parvulus* (A. H. Howell, 1916); *scalopsoides* (Audubon and Bachman, 1841); *schmidti* (Jackson, 1941).

COMMENTS: Type species of *Pitymys*, a taxon sometimes accorded generic status, particularly by European mammalogists and paleontologists (e.g., Corbet, 1978*c*; Koenigswald, 1980; Repenning, 1983, 1992; Van der Meulen, 1978). *Pitymys* is more often viewed as a subgenus of *Microtus* (e.g., Chaline et al., 1999; Gromov and Polyakov, 1977; Hall, 1981; Miller and Kellogg, 1955), a relationship and ranking convincingly endorsed by a variety of morphological, chromosomal, and molecular studies (Carleton, 1981; Conroy and Cook, 2000*a*; Conroy et al., 2001; Hooper and Hart, 1962; Modi, 1996; Moore and Janacek, 1990; Zagorodnyuk, 1990).

Van der Meulen (1978) regarded *nemoralis* and *parvulus* as species distinct from *pinetorum*, an evaluation shared by Repenning (1983) and Brunet-LeComte and Chaline (1992) for *nemoralis*; R. A. Martin (1991) did not view *parvulus* as separate. Whitaker and Hamilton (1998), on the other hand, regarded both taxa, and other eastern forms, as inseparable from *M. p. pinetorum*. Such disparate conclusions cry out for critical examination with a broader array of information bases, such as chromosomal and genetic surveys (e.g., Moore and Janecek, 1990; J. W. Wilson, 1984). *Pitymys hibbardi*, a late Pleistocene Floridian form, synonymized with *M. pinetorum* by R. A. Martin (1995). See Smolen (1981, Mammalian Species, 147).

Microtus qazvinensis Golenishchev, 2003. Russian J. Theriol., 1:118.

COMMON NAME: Qazvin Vole.

TYPE LOCALITY: Iran, Qazvin Prov., 65 km S Qazvin City, Bu'in-Zahra; 49°58′N, 35°39′E.

DISTRIBUTION: Known only from the type locality, N Iran.

COMMENTS: Subgenus *Microtus, socialis* species group (our allocation), or subgenus *Sumeriomys* (Golenishchev et al., 2002). A close relative of *M. guentheri* distinguished by pelage coloration, bacular morphology, and more complex occlusal pattern; F1 hybrids of *M. guentheri* and *M. qazvinensis* proved sterile (Golenishchev et al., 2002*b*). Relationship of *M. qazvinensis* to populations of *M. guentheri* will remain obscure until much-needed revision of the latter is available (see above account).

Microtus quasiater (Coues, 1874). Proc. Acad. Nat. Sci. Philadelphia, 26:191.

COMMON NAME: Jalapan Vole.

TYPE LOCALITY: México, Veracruz, Jalapa.

DISTRIBUTION: Eastern slopes of the Sierra Madre Oriental, 500-2150 m, SE San Luis Potosi to N Oaxaca, México (Ramírez-Pulido et al., 1991:Fig. 1).

STATUS: IUCN – Lower Risk (lc).

COMMENTS: Subgenus *Pitymys*. Named as a variety of *M. pinetorum* but specific status affirmed thereafter (Bailey, 1900; Hall and Cockrum, 1953; Repenning, 1983). Typically viewed as most closely related to *pinetorum* in subgenus or genus *Pitymys* (Anderson, 1960; Hall and Cockrum, 1953; van der Meulen, 1978), a convention challenged by Repenning (1983) and Moore and Janecek (1990), who supported sister-group kinship between *ochrogaster* and *quasiater*. Closest kinship with *M. pinetorum*, however, reaffirmed in phylogenetic analysis of cytochrome *b* sequences that included *M. ochrogaster* and 16 other species of North American *Microtus* (Conroy et al., 2001). Nongeographic variation, distribution, and natural history summarized by Ramírez-Pulido et al. (1991); nonbanded karyotype reported by Cervantes et al. (1994) and compared with other Mexican species.

Microtus richardsoni (DeKay, 1842). Zoology of New York, Part I, Mammals, p. 91.

COMMON NAME: North American Water Vole.

TYPE LOCALITY: Canada, Alberta Prov., vicinity of Jasper House (as interpreted by Bailey, 1900:60).

DISTRIBUTION: Wet subalpine and alpine meadows of Rocky Mountains, from S British Columbia and Alberta, Canada, to W Wyoming and C Utah, USA; and of Cascade Mountains, from SW British Columbia south through WC Oregon.

STATUS: IUCN – Lower Risk (lc).

SYNONYMS: *arvicoloides* (Rhoads, 1894); *macropus* (Merriam, 1891); *myllodontus* Rasmussen and Chamberlain, 1959; *principalis* Rhoads, 1895.

COMMENTS: Senior synonym of the type (= *arvicoloides*) of *Aulacomys*. In early classifications (Bailey, 1900; Miller, 1896), *richardsoni* was placed with Old World water voles in the subgenus *Arvicola* (including *Aulacomys*) of *Microtus*, a relationship reaffirmed by Hooper and Hart (1962) and followed by others, with *Arvicola* employed either at the subgeneric (Hall, 1981) or generic level (Jones et al., 1975). Substantial evidence argues the retention of *richardsoni* within *Microtus* and the restriction of *Arvicola* to Old World forms (Carleton, 1981; Conroy and Cook, 2000*a*; Gromov and Polyakov, 1977; Hinton, 1926*a*; Jannett, 1997; Koenigswald, 1980; Repenning, 1980; Zagorodnyuk, 1990). Hoffmann and Koeppel (1985) further suggested the autochthonous origin of *M. richardsoni* in North America from an early stock that also gave rise to *M. xanthogna-thus*, an idea consistent with the biochemical similarity of *richardsoni* to certain New World *Microtus* (Nadler et al., 1978) and with Zagorodnyuk's (1990) usage of the subgenus *Aulacomys*; cytochrome *b* sequence data, although also supporting a North American origin, disclose nearest kinship to *M. (Pitymys) pinetorum* (Conroy and Cook, 2000*a*). See Ludwig (1984, Mammalian Species, 223).

Microtus sachalinensis Vasin, 1955. Zool. Zh., 34:427.

COMMON NAME: Sakhalin Island Vole.

TYPE LOCALITY: Russia, Sakhalin Isl, Poronaysk region, Olen River.

DISTRIBUTION: Endemic to Sakhalin Isl (Dobson, 1994).

STATUS: IUCN – Lower Risk (nt).

COMMENTS: Subgenus *Alexandromys*, *middendorffii* species group (Zagorodnyuk, 1990) or *maximowiczii* species group (Pavlinov and Rossolimo, 1998; Pavlinov et al., 1995*a*). The latter species group association more accurately reflects data synthesized from

karyological, morphological, and hybirdization studies, which phylogenetically link
M. sachalinensis most closely to *M. maximowiczii* within the subgenus *Alexandromys*
(Meyer et al., 1996).

Microtus savii (de Selys-Longchamps, 1838). Rev. Zool. Paris, p. 248.
COMMON NAME: Savi's Pine Vole.
TYPE LOCALITY: Italy, near Pisa.
DISTRIBUTION: Italian peninsula (except NE and Calabrian Peninsulan in the south) and
Sicily (Amori et al., 1999; Cerone and Aloise, 1994; Cresti et al., 1994; Paolucci et al.,
1993), South Tessin in S Switzerland (Hausser, 1995), and marginally in SE France;
thought to occur on Elba but that record is unconfirmed; sea level to 2000 m (Cheylan,
1991).
STATUS: IUCN – Lower Risk (lc).
SYNONYMS: *nebrodensis* Mina-Palumbo, 1868; *niethammericus* Contoli, 2003; *selysii* (Gerbe,
1852); *tolfetanus* Contoli, 2003.
COMMENTS: Subgenus *Terricola*, *savii* species group (Chaline et al., 1988; Zagorodnyuk,
1990). Revised by Contoli (2003) and Krapp and Winking (1976) and reviewed by
Corbet (1978c, 1984), Krapp (1982c), and Mitchell Jones et al. (1999). Chromosomal
data consolidated by Zima and Kral (1984a); karyology of Italian samples documented
by Galleni et al. (1992, 1994, 1998). Closely related to the living *M. brachycercus* on the
Calabrian Peninsula (see species account) and two extinct species: *M. henseli* from
Sardinia and Corsica (Brunet-Lecomte and Chaline, 1991; Krapp, 1982c) and *M. melitensis* from Malta (Brunet-Lecomte and Chaline, 1991; Burgio and Kotsakis, 1986).

Microtus schelkovnikovi (Satunin, 1907). Izv. Kavkas. Mus., 3, 1:242.
COMMON NAME: Schelkovnikov's Pine Vole.
TYPE LOCALITY: SE Azerbaijan, Talysk Mtns, near village of Dzhi.
DISTRIBUTION: S Azerbaijan (Talysk Mtns) and NW Iran (Elburz Mtns), in mountains
bounding southern margin of Caspian Sea.
STATUS: IUCN – Lower Risk (lc).
SYNONYMS: *dorothea* (Ellerman, 1948).
COMMENTS: Subgenus *Terricola*, sole member of *schelkovnikovi* species group (Zagorodnyuk,
1990), or the *majori* species group with connate *M. majori* (Pavlinov et al., 1995a;
Pavlinov and Rossolimo, 1998). Although lumped under *Pitymys subterraneus* by
Ellerman Morrison-Scott (1951), its specific validity is well documented (Kratochvíl,
1970; Kratochvíl and Kral, 1974). Chromosomal data summarized by Zima and Kral
(1984a). Karyotypic contrasts between Caucasus *M. schelkovnikovi* (2n = 54, FN = 62),
M. majori (2n = 54, FN = 60), and *M. daghestanicus* (2n = 38-54, FN = 58) and inferences
on pine vole chromosomal evolution (subgenus *Terricola*) in the Caucasus offered by
Achverdjan et al. (1992); see account of *M. daghestanicus*. The name *kaznakovi* was
thought to be potentially valid for this species (see reference in Corbet, 1984), but
Pavlinov and Rossolimo (1987) avoided its use, and Rossolimo and Pavlinov (1992)
and Rossolimo et al. (1994) placed it in the synonymy of *Alticola stoliczkanus*.

Microtus schidlovskii Argyropulo, 1933. Z. Säugetierk., 8:182.
COMMON NAME: Schidlovsky's Vole.
TYPE LOCALITY: Transcaucasia, NW Armenia, Leninakan Dist., "vicinity of Nalband railway
station, 35 km E from Leninakan" (Golenishchev et al., 2002:53), 1200 m.
DISTRIBUTION: Transcaucasia, recorded from W Armenia and Georgia (Ahverdyan et al.,
1991b); limits unresolved.
SYNONYMS: *colchicus* Argyropulo, 1932 [not Shidlovsky, 1919]; *goriensis* Argyropulo, 1935.
COMMENTS: Subgenus *Microtus*, *socialis* species group (Zagorodnyuk, 1990). Placed in
subgenus *Sumeriomys* by Gromov and Erbajeva (1995). Long included in *M. socialis*
(Ellerman and Morrison-Scott, 1951; Pavlinov and Rossolimo, 1987; Pavlinov et al.,
1995a). Pavlinov and Rossolimo (1998) and Kryštufek and Kefelioğlu (2002) also
questioned the separate status for *schidlovskii*, and Pavlinov and Rossolimo (1998)
placed it in the synonymy of *M. socialis* with reservation. Reinstated as species by
Ahverdyan et al. (1991a, 1991b), who demonstrated that hybridization between
schidlovskii (2n and FN = 60-62) and *M. socialis* (2n and FN = 62) produced only sterile

F1s and analyzed supportive chromosomal and morphological data. Species status reaffirmed by Golenishchev et al. (2002) based on bacular morphology and hybridization trials. Gromov and Erbajeva (1995) listed *goriensis* (a replacement name for *colchicus*) as a subspecies.

Microtus socialis (Pallas, 1773). Reise Prov. Russ. Reichs., 2:705.

COMMON NAME: Social Vole.

TYPE LOCALITY: Kazakhstan, probably Gur'evsk Oblast (Gur'ev Dist.) between Volga and Ural Rivers (Kryštufek and Vohralík, 2001).

DISTRIBUTION: Palearctic steppe from Dneper River and Crimea east to Lake Balkhash and NW Xinjiang in China (Zhang et al., 1997), south through Caucasus and E half of Turkey (Kizilirmak River may be W boundary; Kefelioğlu and Kryštufek, 1999; Kryštufek and Kefelioğlu, 2002) to NW Syria, Lebanon (restricted to Mt Lebanon), N Iraq, and NW Iran (Kryštufek and Kefelioğlu, 2002, mapped range in the Near East).

STATUS: IUCN – Lower Risk (lc).

SYNONYMS: *aristovi* Golenishchev, 2002; *astrachanensis* (Erxleben, 1777); *binominatus* Ellerman, 1941; *bogdoensis* Wang and Ma, 1982; *gravesi* Goodwin, 1934; *nikolajevi* Ognev, 1950; *parvus* Satunin, 1901; *satunini* Ognev, 1924 [not Shidlovksy, 1919]; *syriacus* (Brants, 1827); *zaitsevi* Golenishchev, 2002.

COMMENTS: Subgenus *Microtus*, *socialis* species group (Zagorodnyuk, 1990). Subgenus *Sumeriomys* according to Pavlinov and Rossolimo (1987, 1998), Gromov and Polyakov (1977), Gromov and Erbajeva (1995), and Pavlinov et al. (1995a). Chromosomal data reviewed by Zima and Kral (1984a) and Golenishchev et al. (2002); Kefelioğlu and Kryštufek (1999) reported karyotypes for Turkish samples (2n and FN = 62). Harrison and Bates (1991) included *guentheri* and *irani* under *M. socialis* because they could not discriminate the three but provided no substantiating data (see *M. guentheri* and *M. irani*). Range of *M. socialis* proper has been obscured because researchers indiscriminantly lumped all voles more or less matching the description of *socialis* (Corbet, 1978c; Ellerman and Morrison-Scott, 1951; Harrison and Bates, 1991; Lay, 1967; Ognev, 1964); however, at least seven (*M. anatolicus*, *M. dogramacii*, *M. guentheri*, *M. irani*, *M. paradoxus*, *M. qazvinensis*, *M. schidlovskii*) have been separated (Golenishchev et al., 2002; Kefelioğlu and Kryštufek, 1999; Kryštufek and Kefelioğlu, 2002). Golenishchev et al. (2002) described *aristovi* and *zaitsevi* as subspecies.

Most taxonomic problems lie with populations in the Near East, where Kryštufek and Kefelioğlu (2002) have resolved species limits by using discriminant function analysis to identify *M. socialis* in Iran, the Caucasus, Lebanon, Syria, and Turkey and to distinguish it from the "non-socialis" complex (see *M. guentheri*). They also considered *hyrcania* as an outlying population of *M. socialis*, but its holotype is an example of *M. arvalis* (see account). The holotype of *gravesi* and four other AMNH specimens in the type series are examples of *M. socialis*. Hassinger (1973) recorded *M. socialis* from N Afghanistan, an unlikely occurrence, but those two specimens, originally identified as *M. guentheri*, should be reexamined to verify the identity as that species or *M. paradoxus* (see account).

Microtus subterraneus (de Selys-Longchamps, 1836). Essai Monogr. sur les Campagnols des Env. de Liege., 10.

COMMON NAME: Common Pine Vole.

TYPE LOCALITY: Belgium, Liege, Waremme.

DISTRIBUTION: Atlantic coasts of N and C France through S Netherlands (Ligtvoet and Straetmans, 1992), across C Europe (SW Bohemia [Andera and Èervený, 1994]; Switzerland [Maurizio, 1994; Hausser, 1995]; Slovakia [Danko, 1994; Kminiak, 1996; Mošanský, 1994; Stanko, 1995; Stanko and Mošanský, 1994, 2000; Stanko et al., 2000]; Czech Republic [Šmaha, 1996]; Italy [Amori et al., 1999; Locatelli and Paolucci, 1996a]) to Ukraine and the Don River (Gromov and Erbajeva, 1995), south through Romania and Balkan region into N and SE Greece, and Turkey (mountains in N, W and S but not on Anatolian Plateau; Kryštufek and Vohralík, 2001) eastward to south of Trabzon (Macholán et al., 2001a); isolated populations in Russia (near St Petersburg;

Zagorodnyuk, 1989) and Estonia (Miljutin, 1997, 1998; Timm et al., 1998). Absent from Mediterranean coast and islands.

STATUS: IUCN – Lower Risk (lc).

SYNONYMS: *atratus* (Stein, 1931); *brauneri* Martino, 1926; *capucinus* (Miller, 1908); *dacius* (Miller, 1908); *dinaricus* Kretzoi, 1959; *ehiki* (Wettstein, 1927); *fingeri* (Neuhäuser, 1936); *fusca* (Fatio, 1900) [not Buchner, 1889]; *grafi* Brunet-Lecomte, Nadarowski, and Chaline, 1992; *hercegovinensis* (Martino, 1940); *hungaricus* (Ehik, 1926); *incertoides* (Wettstein, 1927); *incertus* (Sélys-Longchamps, 1841); *klozeli* (Ehik, 1942); *kupelwieseri* (Wettstein, 1925); *martinoi* (Ehik, 1935); *matrensis* (Ehik, 1930); *mustersi* (Martino, 1937) [not Hinton, 1926]; *neglectus* Petrov, 1992 [*nomen nudum*]; *neuhauseri* (Martino and Paspalev, 1955); *nyirensis* (Ehik, 1930); *rufescente-fuscus* (Schinz, 1845); *rufofuscus* (Schinz, 1845) [*nomen nudum*]; *subterraneoides* Petrov, 1992 [*nomen nudum*]; *transsylvanicus* (Ehik, 1924); *transvolgensis* (Schaposchnikov and Schanev, 1958); *ukrainicus* (Vinogradov, 1922); *wettsteini* (Ehik, 1926); *zimmermanni* (Matschie, 1924).

COMMENTS: Subgenus *Terricola*, *subterraneus* species group (Chaline et al., 1988; Zagorodnyuk, 1990). Reviewed by Niethammer (1982g). Karyotypic information available in Sablina et al. (1989), Zima (1986), Zima and Kral (1984a), and Zima et al. (1997a); biochemical comparisons among M. *subterraneus*, M. *felteni*, and other voles implemented by Gill et al. (1987). Synonyms follow Zagorodnyuk (1989), who speculated that the Romanian *dacius* is a separate species, but Brunet-Lecomte et al. (2001) regarded the minor cytogenetic differences as inconclusive. The taxon *grafi* was described as a species from late Pleistocene fossils from Bulgaria (Brunet-Lecomte et al., 1992) but is now regarded as a chronological subspecies of M. *subterraneus* (Brunet-Lecomte et al., 2001). Present in the British Isles during early Pleistocene interglacials (Yalden, 1999).

Regional studies have addressed the discrimination of M. *subterraneus*. Pelage and vibrissal contrasts with M. *tatricus* documented by Kratochvíl (1969). Distribution and morphological variation reported by Zagorodnyuk (1989, 1992a) in taxonomic overview of species in subgenus *Terricola*; he clarified distribution of the 2n = 52 and 2n = 54 chromosomal morphs in the Baltic region and morphometrically contrasted M. *subterraneus* with M. *agrestis*, M. *arvalis*, M. *levis* (formerly *rossiaemeridionalis*), and M. *oeconomus* (Zagorodnyuk and Mezhzherin, 1992). Morphometric analysis of m1s vindicated specific separation of M. *subterraneus* from M. *multiplex* and M. *liechtensteini* (Brunet-Lecomte and Kryštufek, 1993), a result also indicated by sequence analysis of the mitochondrial control region (Haring et al., 2000). Morphometric distinction between M. *subterraneus* and M. *liechtensteini* and their distributions in Slovenia discussed by Kryštufek (1991). Comparisons in molar size variation among Turkish M. *subterraneus*, M. *majori*, and M. *daghestanicus* presented by Kryštufek and Vohralík (2004). Vohralík and Sofianidou (1992a) identified M. *subterraneus* from SE Greece (Thrace region) but expressed skepticism about the reliability of size distinctions used to segregate it from M. *majori*. Macholán et al. (2001a) utilized karyotypic and allozymic analyses to document the distributions of the 2n = 52 and 2n = 54 karyomorphs of M. *subterraneus* and their chromosomal and genetic distinctions from M. *majori* (2n = 54); low genetic distance between the two suggests a recent divergence, only 170-350 thousand years ago. Their ranges closely approach one another in NE Turkey but sympatry not yet demonstrated (Macholán et al., 2001a).

The substantial genetic and morphological variation of M. *subterraneus* throughout its range, especially the Balkans (Kryštufek et al., 1994), prompted Macholán et al. (2001a:40) to propose this view: "taking into account close karyotypic and allozymic relationships (and the fact that some intraspecific genetic distances within M. *subterraneus* are higher than those between M. *subterraneus* and M. *majori*), we cannot rule out a hypothesis that M. *subterraneus* is paraphyletic, with M. *majori* (and, by the same token, M. *daghestanicus*) being one of its constituents." Further research should explore their reservations.

Microtus tatricus Kratochvíl, 1952. Acta Acad. Sci. Nat. Moravo-Siles., 24:155-194.

COMMON NAME: Tatra Vole.

TYPE LOCALITY: Czech Republic, Poprad Dist., Velka Studena Dolina valley, High Tatra Mtns.

DISTRIBUTION: W and E Carpathian Mtns, 800-2350 m; isolated populations in montane
    spruce forests and meadows of Tatra Mtns between Czech Republic, Slovakia
    (Mošanský, 1994) and S Poland, Pilsko Mtn, and Beslksid Ziwiecki Mtns; also
    W Ukraine and N Romania (Zagorodnyuk, 1988; Zagorodnyuk and Zima, 1992);
    possibly also in the S Carpathians, Romania (Zagorodnyuk and Zima, 1992).
STATUS: IUCN – Lower Risk (nt).
SYNONYMS: *zykovi* Zagorodnyuk, 1989.
COMMENTS: Subgenus *Microtus*. Formerly placed in subgenus *Terricola*, *subterraneus* species
    group (Chaline et al., 1988; Pavlinov and Rossolimo, 1998; Pavlinov et al., 1995*a*). A
    distinctive species reviewed by Niethammer (1982*l*) and Zagorodnyuk and Zima
    (1992); karyotype with 2n = 32, lower than any species of *Terricola* (Zima and Kral,
    1984*a*). Zagorodnyuk (1989) systematically reviewed species in the subgenus *Terricola*
    and described *zykovi*, now recognized as the subspecies occurring in the E Carpathians
    (Mitchell-Jones et al., 1999). Presence in those mountains originally based on
    morphology until Zagorodnyuk and Zima (1992) added chromosomal evidence to
    conclusively identify *M. tatricus*. Zagorodnyuk et al. (1992) also documented
    M. tatricus and *M. subterraneus* in the E Carpathian Mtns in the Ukraine, corrected
    misidentifications as *M. agrestis* or *M. arvalis*, and described diagnostic morphological
    traits.
    Zagorodnyuk and Zima (1992) proposed that the pitymyan rhombus in *M. tatricus*,
    a cardinal dental trait for assigning it to the subgenus *Terricola*, was acquired
    independently and considered the species to be related to those in subgenus *Microtus*.
    Haring et al. (2000:237) corroborated such a relationship, using DNA sequence
    analyses that disclosed *M. tatricus* "as a distinct lineage that split off prior to the
    radiation of the *multiplex* complex, perhaps shortly after the separation of the
    M. subterraneus lineage." Zagorodnyuk and Zima (1992) propounded a biogeographic
    scenario to explain the Pleistocene differentiation of *M. tatricus* and its isolation in the
    Carpathians. Martínková and Dudich (2003) consolidated all distribution records and
    discussed the strong insularity of the species' distribution.

Microtus thomasi (Barrett-Hamilton, 1903). Ann. Mag. Nat. Hist., ser. 7, 11:306.
    COMMON NAME: Thomas's Pine Vole.
    TYPE LOCALITY: Montenegro, Vranici.
    DISTRIBUTION: Endemic to SW Balkans, from the Neretva River in Herzegovina through
        Montenegro, Albania and mainland Greece to the Peloponessos (including Euboea
        Isl).
    STATUS: IUCN – Lower Risk (nt).
    SYNONYMS: *atticus* (Miller, 1910); *byroni* (Bolkay, 1926).
    COMMENTS: Subgenus *Terricola*, *duodecimcostatus* species group (Chaline et al., 1988;
        Pavlinov et al., 1995*a*) or sole member of *thomasi* species group (Zagorodnyuk, 1990).
        Generally reviewed by Niethammer (1982*k*) and Petrov (1992); taxonomic history
        reviewed by Brunet-Lecomte and Nadachowski (1994). Karyotype is polymorphic,
        2n = 40 to 44 and FN = 42 to 46 (Giagia-Athanasopoulou et al., 1995; Mitchell-Jones
        et al., 1999; Zima and Kral, 1984*a*). Based on morphometric analysis of m1s,
        M. thomasi is strongly divergent from six other species examined in subgenus *Terricola*
        (Brunet-Lecomte and Nadachowski, 1994), a finding that supports species-group
        isolation (Zagorodnyuk, 1990). Synonymy of *atticus* confirmed by microcomplement
        fixation comparisons of albumin (Nikoletopoulos et al., 1992), m1 morphometry
        (Brunet-Lecomte and Nadachowski, 1994), and morphometric and allozymic
        variation (Tsekoura et al., 2002).

Microtus townsendii (Bachman, 1839). J. Acad. Nat. Sci. Philadelphia, 8:60.
    COMMON NAME: Townsend's Vole.
    TYPE LOCALITY: USA, Oregon, Multnomah Co., Wappatoo (Sauvie) Isl, lower Columbia River
        near mouth of Willamette River (as located by Bailey, 1900:46).
    DISTRIBUTION: Wet meadows and marshes of Pacific Northwest, from extreme SW British
        Columbia, Canada, to NW California, USA, including Vancouver and neighboring
        islands.

STATUS: IUCN – Lower Risk (cd) as *M. t. cowani*, otherwise Lower Risk (nt).

SYNONYMS: *cowani* Guiguet, 1955; *cummingi* Hall, 1936; *laingi* Anderson and Rand, 1943; *occidentalis* (Peale, 1848); *pugeti* Dalquest, 1940; *tetramerus* (Rhoads, 1894).

COMMENTS: Subgenus *Mynomes*, *pennsylvanicus* species group (Zagorodnyuk, 1990). Broadly allied with *M. pennsylvanicus* and *M. montanus* (Anderson, 1959; Hooper and Hart, 1962; Modi, 1987); sister species to *M. canicaudus* according to phyletic reconstructions of cytochrome *b* sequences (Conroy and Cook, 2000*a*). See Cornely and Verts (1988, Mammalian Species, 325).

Microtus transcaspicus Satunin, 1905. Izv. Kavkas. Mus., 2:57.

COMMON NAME: Middle East Vole.

TYPE LOCALITY: Turkmenistan, Kopet Dag Mtns, Chuli Valley, near Ashkhabad.

DISTRIBUTION: Dry montane steppe habitats on isolated mountains from N slopes of Kopet-Dag Mtns in S Turkmenistan (Meyer et al., 1996), mountains in E Iran in the NE (Khorassan Prov, 5 km N Kashmar, USNM) and S (Kuh-e Laleh-Zar and Kuh-e Hazar Mtns south of Kerman; Roguin, 1988), and the Hindu Kush of N Afghanistan (Ellerman, 1948; Parvan Prov, Shibar Pass, FMNH).

STATUS: IUCN – Endangered as *M. kermanensis*, Lower Risk (lc) as *M. transcaspicus*.

SYNONYMS: *kermanensis* Roguin, 1988.

COMMENTS: Subgenus *Microtus*, *arvalis* species group (Pavlinov and Rossolimo, 1987, 1998; Pavlinov et al., 1995*a*; Zagorodnyuk, 1990). Based on banding chromosomal data, Mazurok et al. (1996*b*) closely allied *M. transcaspicus* with *M. levis* among species in the *M. arvalis* group, an association also supported by allozymic data (Mezhzherin et al., 1993) and syntheses of karyology, morphology, and hybridization (Meyer et al., 1996). Taxonomy reviewed by Malygin (1978), Meyer et al. (1981, 1996), and Meyer (1983).

Roguin (1988) described *kermanensis* as a species, recognized as part of the *M. arvalis* group (Musser and Carleton, 1993; Pavlinov et al., 1995*a*; Zagorodnyuk, 1990). His description and specimen measurements fall within the range of variation typical of *M. transcaspicus* (see review of *M. transcaspicus* in Turkmenistan by Meyer et al., 1996), as do those large-bodied specimens of "*M. arvalis*" from N Afghanistan reported by Niethammer (1970) and Hassinger (1973). *Microtus transcaspicus* is a valid species diagnosed in part by its very large size and diploid number (adult occipitonasal length = 28-32 mm, 2n = 52) compared with the smaller-bodied *M. arvalis* (occipitonasal length = 22-28 mm, 2n = 46). *Microtus arvalis* does occur in the mountains of N Iran but has not been recorded so far east at those latitudes.

Microtus umbrosus Merriam, 1898. Proc. Biol. Soc. Wash., 12:107.

COMMON NAME: Zempoaltépec Vole.

TYPE LOCALITY: México, Oaxaca, Mt Zempoaltépec, 8200 ft (2500 m).

DISTRIBUTION: Cerro Zempoaltépec, 1800-3000 m, NC Oaxaca, México.

STATUS: IUCN – Lower Risk (lc).

COMMENTS: Type species and sole member of *Orthriomys*, a taxon typically identified as a subgenus of *Microtus* (Anderson, 1959; Bailey, 1900; Hall and Cockrum, 1953; R. A. Martin, 1995; Zagorodnyuk, 1990). Nonbanded karyotype reported by Cervantes et al. (1994), who noted its agreement with the complement hypothesized as ancestral for Arvicolinae. Viewed as cladistically basal to other North American pitymyines (subgenera *Herpetomys*, *Pedomys*, and *Pitymys*) based on phylogenetic study of molar traits (R. A. Martin, 1995); more removed from pitymyine species according to mitochrondrial gene sequences and weakly associated with *M. chrotorrhinus* (Conroy et al., 2001). Conroy et al. (2001) suggested that *M. umbrosus* represents an independent invasion of southern latitudes, the other consisting of species related to *M. pinetorum*, subgenus *Pitymys* (*M. guatemalensis*, *M. oaxacensis*, *M. quasiater*). See Frey and Cervantes (1997, Mammalian Species, 555).

Microtus xanthognathus (Leach, 1815). Zool. Misc., 1:60.

COMMON NAME: Taiga Vole.

TYPE LOCALITY: Canada, Manitoba, Hudson Bay.

DISTRIBUTION: Western boreal taiga zone, from EC Alaska to W Northwest Territories, southeastwards to C Alberta and W coast of Hudson Bay, Canada.

STATUS: IUCN – Lower Risk (lc).

COMMENTS: Relationships obscure—apparently not a sister species to *M. chrotorrhinus* as once believed (see that account and Lidicker and Yang, 1986) nor closely related to *M. richardsoni*, subgenus *Aulacomys*, as arranged by Zagorodnyuk (1990; see Conroy and Cook, 2000a). Karyotype reported and affinities discussed by Rausch and Rausch (1974). Late Pleistocene distribution reached C and E USA, as far south as S Nebraska, N Arkansas, and Virginia (Graham and Lundelius, 1994).

Myodes Pallas, 1811. Zoographia Rosso Asiatica, I:173.

TYPE SPECIES: *Mus rutilus* Pallas, 1779 (as subsequently designated by Lataste, 1883b:349).

SYNONYMS: *Clethrionomys* Tilesius, 1850; *Craseomys* Miller, 1900; *Evotomys* Coues, 1874; *Glareomys* Rasorenova, 1952; *Neoaschizomys* Tokuda, 1935; *Phaulomys* Thomas, 1905.

COMMENTS: In the early systematic literature, red-backed voles were commonly listed under *Evotomys* (e.g., Hinton, 1926a; Miller, 1896, 1924; Miller and Rehn, 1901; Trouessart, 1880-1881), until Palmer (1928) argued the priority of *Clethrionomys* Tilesius, 1850. Most subsequent systematic authorities and faunal works throughout the 1900s have adopted Palmer's correction in recognizing *Clethionomys* as the valid name (Corbet, 1978c; Corbet and Hill, 1991; Ellerman, 1941; Gromov and Erbajeva, 1995; Hall, 1981; McKenna and Bell, 1997; Miller and Kellogg, 1955; Mitchell-Jones et al., 1999; Musser and Carleton, 1993; Wilson and Ruff, 1999). In his deliberate search of the early European literature, Palmer (1928) had anticipated that he would discover an older synonym for *Evotomys*.

Regrettably, Palmer overlooked Pallas's (1811) *Myodes* and Lataste's (1883b) subsequent designation of *Mus rutilus* as its type species, the same one that he selected for *Clethrionomys*. Pallas (1811) included 10 species in his *Myodes* but did not specify a type species, a common omission for the era. In his studies of the voles of France, Lataste (1883a:324) had discussed the origin and contents of Pallas's *Myodes*, noted (1883b:348-349) the junior status of North American *Evotomys* as then newly named by Coues (1874), and finally (1883b:349) clearly indicated *rutilus* as type species of *Myodes* to preserve the genus-group name. Lataste reiterated his restricted usage of *Myodes*, as subgenus, and the type status of *rutilus* in his 1887 classification of New and Old World *Microtus*. Some modern workers have accordingly employed *Myodes* as the oldest name for red-backed voles (Heller, 1968; Kretzoi, 1964, 1969; Kretzoi and Vertes, 1965a, 1965b; Naglov, 1998; Zagorodnyuk, 1990), and others have acknowledged it as probably valid while using *Clethrionomys* for reasons of familiarity (Gromov and Polyakov, 1977; Pavlinov and Rossolimo, 1987, 1998; Pavlinov et al., 1995a).

How *Myodes* became so long misassociated as a junior synonym of *Lemmus* is pertinent to consideration of its status. Whereas some early European workers narrowed the definition of Pallas' *Myodes* to identify red-backed voles (Sélys Longchamps, 1839), North American taxonomists had used the genus to embrace brown lemmings (Baird, 1857; Coues, 1874). Lataste (1883a, b; 1887) frequently mentioned Sélys Longchamps as the prior authority for applying *Myodes* to red-backed voles (*glareolus* and *rutilus*), but took the critical additional step in designating *Mus rutilus* as its type species. Baird (1857) expressly followed Keyserling and Blasius (1840) for precedence in applying *Myodes* to North American brown lemmings (all as *M. obensis*). The latter's classificatory listing of European birds and mammals arranged four species within *Myodes* (*lemmus*, *obensis*, *lagurus*, and *torquatus*), but they did not note a type species. The latter meaning was the one emphasized by Miller (1896:15), who stated "Since *Myodes* contained species of exactly the same modern genera as *Lemmus* Link and no groups not included in the latter, the name is a synonym of *Lemmus*." In other words, he based their synonymy on the equivalence of contents, as he understood generic limits in 1896, not on the principle of typification (Article 61; ICZN, 1999). Miller (1896:19-24) elsewhere summarized Lataste's 1883 papers and even presented his composition of *Myodes* (as a subgenus of *Microtus*) that included only *rutilus* and *glareolus*. Palmer (1904:439) also had knowledge of Lataste's nomenclatural action with regard to *Myodes* ("type said to be *M. rutilus*!"), yet confoundingly omitted any mention of it in his (1928) subsequent fixation of *Mus rutilus* in order to employ *Clethrionomys* instead of *Evotomys*. Miller's interpretation was formalized by Hinton (1926a:210), who considered *Mus lemmus* to be the type species of *Myodes* and thereby reinforced its junior synonymy under *Lemmus*. In light of Lataste's (1883*b*) legitimate indication of *Mus rutilus* as type species for *Myodes* (Article 69.1), Hinton's passing listing of *Mus lemmus* as its type is invalid to establish the equivalence of *Myodes* with *Lemmus* (Article 70.2). Unfortunately, Hinton's incorrect synonymy was perpetuated in the influential classifications of Ellerman (1941) and Simpson (1945) and became broadly adopted thereafter, especially by North American authors but not universally by European workers.

To our knowledge, three genus-group taxa are founded on *Mus rutilus* Pallas as type species: *Myodes* Pallas (1811), as subsequently designated by Lataste (1883*b*); *Clethrionomys* Tilesius (1850), as subsequently designated by Palmer (1928); and *Evotomys* Coues (1874), as originally designated by Coues. Where two or more genus-group taxa have the same type species, their names are considered to be objective synonyms (Article 61.3.3; ICZN, 4th Ed., 1999) and the senior among them assumes priority. Pavlinov et al. (1995*a*) questionably indicated *Myodes* as a *nomen oblitum* in the synonymy of *Clethrionomys*. The ICZN (1999), however, explains that both conditions of Article 23.9.1 must be jointly satisfied to reverse the principle of priority: the senior synonym cannot have been used as a valid name after 1899 (*nomen oblitum*), and the junior synonym must have been used extensively as the prevailing name (also see, 1999:111). Although the second condition supports retention of *Clethrionomys*, the first inescapably demands recognition of *Myodes*. In the following species comments, and elsewhere, readers should understand that wherever we use *Myodes* the supporting literature generally refers to *Clethrionomys* or *Evotomys*.

Usage of the family-group name is not so clearcut. Hooper and Hart (1962) named Clethrionomyini for those arvicolines with rooted molars and simple occlusal patterns. Later, Kretzoi (1969) coined Myodini for this tribe, given his recognition of Lataste's (1883*b*) type species fixation and apparently unaware of Hooper and Hart's action. Both Clethrionomyini (Gromov and Polyakov 1977; McKenna and Bell, 1997) and Myodini (Pavlinov and Rossolimo, 1998; Pavlinov et al., 1995*a*; Zagorodnyuk, 1990) have been recognized in recent classifications that employ arvicoline tribal ranks. The ICZN does not require replacement of an older family-group name when its type genus is reallocated as junior synonym of some other genus, upon which another family-group name is based (Article 40.1); however, maintenance of the junior family-group name may be defended in instances where a valid subsequent fixation had been overlooked and instability or confusion could likely ensue (Articles 41, 65.2.2). In view of the non-universal employment of Clethrionomyini versus Myodini and to lessen such potential confusion, we recommend usage of the junior name Myodini. The case must be ultimately decided by petition to the Commission.

We much regret that Miller, Hinton, and Palmer failed to appreciate Lataste's (1883*b*) nomenclatural action affecting the status of *Myodes*. Over 50 years passed between description of *Evotomys* (Coues, 1874) and its replacement by *Clethrionomys* (Palmer, 1928). Just 36 years later, Kretzoi (1964) reminded the systematic community of Lataste's type-species selection, reestablished the priority of *Myodes*, and listed *Clethrionomys* and *Evotomys* among its junior synonyms. With the advent of the global internet and greatly enhanced communication among scientists, we doubt that such oversights, issuing from the inertia of provincially familiar taxonomies, can be so long perpetuated.

Myodini. *Myodes* is closely related to *Eothenomys* (Koenigswald, 1980; Kretzoi, 1969), if not congeneric with it as suggested by some authors (Corbet, 1978*c*, 1984; Hooper and Hart, 1962; and see discussion under *Eothenomys*). Their close phylogenetic alliance is also supported by results of chromosomal analyses (Iwasa et al., 1999*b*) and DNA-DNA hybridization (Din et al., 1993).

Broad taxonomic reviews are provided by Aimi (1980), Corbet (1978*c*, 1984), and Gromov and Polyakov (1977). Aimi also retraced the complex taxonomic history of species, especially Japanese forms, variously allocated to *Evotomys*, *Clethrionomys*, *Anteliomys*, *Phaulomys*, or *Eothenomys*. Gromov and Erbajeva (1995) mapped the generic range and reviewed morphological and geographic contrasts among species within Russia and adjacent regions. All species of *Myodes* so far examined have 2n = 56, with 26 pairs of telocentric and one pair of metacentric chromosomes (Gamperl, 1982; Iwasa, 1998; Iwasa et al., 1999; Kaneko et al., 1998; Kartavtseva et al., 1998; Ma and Jiang, 1996; Modi and Gamperl, 1989; Nadler et al., 1976; Obara et al., 1995; Sokolov et al., 1990; Vorontsov et al., 1978; Yoshida et al., 1989). Electrophoretic data presented by Nadler et al. (1978), Chaline and Graf (1988), and Mezhzherin and Serbenyuk (1992); variation in mitochondrial DNA control region examined by Matson and Baker (2001). Fossil *Myodes* date to the late Pliocene of China (Zheng, 1993), early Pleistocene of Europe, late Pliocene of Ukraine (Topachevskii et al., 1998), and middle Pleistocene of North America (McKenna and Bell, 1997). Kawamura (1988) interpreted relationships among European and Japanese species based on living and Pleistocene samples.

Myodes andersoni (Thomas, 1905). Abstr. Proc. Zool. Soc. Lond., 1905(23):18.

COMMON NAME: Anderson's Red-backed Vole.

TYPE LOCALITY: Japan, Honshu, Iwate Prefecture, Tsunagi, near Morioka.

DISTRIBUTION: Known only from C and N Honshu Isl (Kaneko, 1994).

STATUS: IUCN – Vulnerable as *Phaulomys andersoni*.

SYNONYMS: *niigatae* (Anderson, 1909).

COMMENTS: Originally named as a species of *Evotomys*, later included in the synonymy of *C. rufocanus smithii* (Ellerman and Morrison-Scott, 1951), and eventually revised by Aimi (1980) as a species of *Eothenomys*. Subsequently arranged as a species of *Phaulomys* (Musser and Carleton, 1993), *Eothenomys* (Kaneko, 1994), or *Clethrionomys* (Pavlinov et al., 1995a). Evidence for treatment under *Myodes* is explained in account of *M. smithii*. Species reviewed by Kaneko (1994, as *Eothenomys*).

Kaneko et al. (1992a) documented vertical distributions, zone of overlap (650-1325 m), and morphological separation of *M. andersoni* (alpine habitats generally above 1000 m) and *M. smithii* in C Honshu (also see Kimura et al., 1994, 1999). Honshu specimens of problematic identification can be confidently associated using Y-chromosome morphology, certain external proportions and cranial traits, and mammae formula (Iwasa, 2000; Iwasa and Tsuchiya, 2000). G-band chromosomal homologies between *M. andersoni* and *M. rufocanus* documented by Obara (1986). Junior synonymy of *niigatae*, occasionally treated as specifically distinct, clarified by Aimi (1980) based on morphological traits. His conclusion is strongly corroborated by cross-breeding experiments (Kitahara and Kimura, 1995), analyses of mitochondrial and nuclear ribosomal DNA (Suzuki et al., 1999b), and C-banding comparisons of the Y chromosome (Iwasa and Tsuchiya, 2000); however, Kaneko (1994) continued to recognize *niigatae* as a valid species occurring in C Honshu.

Myodes californicus (Merriam, 1890). N. Am. Fauna, 4:26.

COMMON NAME: Western Red-backed Vole.

TYPE LOCALITY: USA, California, Humboldt Co., Eureka.

DISTRIBUTION: Coastal coniferous forest from the Columbia River south through W Oregon (Verts and Carraway, 1998:Fig. 11-97) to NW California, USA.

STATUS: IUCN – Lower Risk (lc) as *Clethrionomys californicus*.

SYNONYMS: *mazama* (Merriam, 1897); *obscurus* (Merriam, 1897).

COMMENTS: The name *occidentalis* was formerly applied to this species (e.g., Hall and Cockrum, 1953), but populations north of the Columbia River, which include *occidentalis* and *caurinus*, have been reassigned to *M. gapperi* (Cowan and Guiguet, 1965; Johnson and Ostenson, 1959). See Alexander and Verts, 1992 (Mammalian Species, 406).

Myodes centralis (Miller, 1906). Ann. Mag. Nat. Hist., ser. 7, 17:373.

COMMON NAME: Tien Shan Red-backed Vole.

TYPE LOCALITY: Kazakhstan, W Tien Shan Mtns, Koksu valley, 9000 ft (2743 m).

DISTRIBUTION: Known only from Tien Shan in Kazakhstan, Kyrgyzstan, and adjacent NW Xinjiang, China (Zhang et al., 1997, as *C. frater*).

STATUS: IUCN – Lower Risk (lc) as *Clethrionomys centralis*.

SYNONYMS: *frater* (Thomas, 1908).

COMMENTS: A distinctive species related to *M. glareolus*. Recognized as valid by Hinton (1926a) and most subsequent systematists (Corbet, 1984; Gromov and Erbajeva, 1995; Pavlinov and Rossolimo, 1987; Pavlinov et al., 1995a), but Corbet (1978c) included it in *M. glareolus*. Some Russian (Gromov and Polyakov, 1977) and Chinese (Ma and Jiang, 1996; Ma et al., 1987; Zhang et al., 1997) researchers still use *frater* for the species or retain both as distinct (Wang, 2003). Chromosomal data recorded by Vorontsov et al. (1978), Sokolov et al. (1990), and Ma and Jiang (1996). Not surprisingly, allozymic analyses cluster *M. centralis* and *M. glareolus* apart from other continental species (Mezhzherin and Serbenyuk, 1992).

Myodes gapperi (Vigors, 1830). Zool. J., 5:204.
>   COMMON NAME: Southern Red-backed Vole.
>   TYPE LOCALITY: Canada, Ontario, between York (= Toronto) and Lake Simcoe.
>   DISTRIBUTION: Most of Canada from N British Columbia to Labrador, excluding Newfoundland; south in the Appalachians to N Georgia and NW South Carolina (Laerm et al., 1995), in the Great Plains to N Iowa, and in the Rockies to C New Mexico and EC Arizona, USA; extralimital isolates in NW and E Pennsylvania and S New Jersey.
>   STATUS: IUCN – Data Deficient as *M. g. solus* (sic), Lower Risk (nt) as *C. g. maurus*, otherwise Lower Risk (lc).
>   SYNONYMS: *arizonensis* (Cockrum and Fitch, 1952); *athabascae* (Preble, 1908); *brevicaudus* (Merriam, 1891); *carolinensis* (Merriam, 1888); *cascadensis* (Booth, 1945); *caurinus* (Bailey, 1898); *fuscodorsalis* (J. A. Allen, 1894); *galei* (Merriam, 1890); *gaspeanus* (Anderson, 1943); *gauti* (Cockrum and Fitch, 1952); *hudsonius* (Anderson, 1940); *idahoensis* (Merriam, 1891); *limitis* (Bailey, 1913); *loringi* (Bailey, 1897); *maurus* (Kellogg, 1939); *nivarius* (Bailey, 1897); *occidentalis* (Merriam, 1890); *ochraceus* (Miller, 1894); *pallescens* (Hall and Cockrum, 1940) [a replacement name for *rufescens* Smith, 1940]; *paludicola* (Doutt, 1941); *phaeus* (Swarth, 1911); *proteus* (Bangs, 1897); *pygmaeus* (Rhoads, 1894); *rhoadsii* (Stone, 1893); *rufescens* (Smith, 1940); *rupicola* (Poole, 1949); *saturatus* (Rhoads, 1894); *soleus* (Hall and Cockrum, 1952); *stikinensis* (Hall and Cockrum, 1952); *uintaensis* (Doutt, 1941); *ungava* (Bailey, 1897); *wrangeli* (Bailey, 1897).
>   COMMENTS: Close relationship to, but genetic segregation from, *M. rutilus* supported by allozymic data (Mezhzherin and Serbenyuk, 1992; Nadler et al., 1978), although some had earlier suggested, without presentation of data, that *gapperi* and *rutilus* are conspecific (Bee and Hall, 1956; Youngman, 1975). In laboratory crosses with Eurasian *M. glareolus*, the partially infertile hybrids led Grant (1974) to view the two as semispecies of recent divergence. In a narrower taxonomic survey, Nadler et al. (1978) viewed Old World *rufocanus* as closely related to the New World *gapperi-rutilus* complex, but broader analysis of allozyme polymorphisms indicated the *gapperi-rutilus* clade to share common ancestry with *glareolus-centralis*, not *rufocanus* (Mezhzherin and Serbenyuk, 1992).
>   Most highly variable in gastric morphology among species of *Myodes* studied by Carleton (1981) and differing from *M. rutilus*. See MacDonald and Cook (1996:579) for the need to clarify specific and intraspecific identities of populations occurring in the Alexander Archipelago and Alaska Panhandle. Genetic structuring and heterozygosity patterns studied in populations isolated in the S Appalachians by Reese et al. (2001). See account of *M. californicus* for allocation of *occidentalis* and *caurinus* to *M. gapperi*. See Merritt (1981, Mammalian Species, 146).

Myodes glareolus (Schreber, 1780). Die Säugethiere, 4:680.
>   COMMON NAME: Bank Vole.
>   TYPE LOCALITY: Denmark, Lolland Isl.
>   DISTRIBUTION: W Palaearctic forests from France and Scandinavia to Lake Baikal, south to N Spain, N Italy (isolated montane populations farther south), the Balkans (but not most of Greece), W and N Turkey, N Kazakhstan and the Altai and Sayan Mtns; also occurs on Britain and SW Ireland.
>   STATUS: IUCN – Lower Risk (lc).
>   SYNONYMS: *alstoni* (Barrett-Hamilton and Hinton, 1913); *arvalis* (Geoffroy, 1803) [not Desmarest, 1816; see Ellerman and Morrison-Scott, 1951:663, for status]; *bernisi* (Rey, 1972); *bicolor* (Fatio, 1862); *bosniensis* (Martino, 1945); *britannicus* (Miller, 1900); *caesarius* (Miller, 1908); *cantueli* (Saint Girons, 1969); *curcio* (Lehman, 1961); *devius* (Stroganov, 1948); *erica* (Barret-Hamilton, 1913); *fulvus* (Millet, 1828); *garganicus*

(Hagen, 1958); *gorka* (Montagu, 1923); *hallucalis* (Thomas, 1906); *harrisoni* (Hinton, 1926); *helveticus* (Miller, 1900); *hercynicus* (Mehlis, 1831); *insulaebellae* (Heim de Balsac, 1940); *intermedius* (Burg, 1923); *istericus* (Miller, 1909); *italicus* (Dal Piaz, 1924); *jurassicus* (Burg, 1923); *kennardi* (Hinton, 1926); *makedonicus* (Felten and Storch, 1965); *minor* (Kerr, 1792); *nageri* (Schinz, 1845); *norvegicus* (Miller, 1900); *ognevi* (Serebrennikov, 1927); *petrovi* (Martino, 1945); *pirenaica* (Cabrera, 1924); *pirinus* (Wolf, 1940); *ponticus* (Thomas, 1906); *pratensis* (Baillon, 1834); *pratensis* (Bell, 1837) [not Baillon, 1834]; *reinwaldti* (Hinton, 1921); *riparia* (Yarrell, 1832) [not Ord, 1825]; *rubidus* (Baillon, 1834); *rufescens* (de Sélys Longchamps, 1836); *ruttneri* (Wettstein, 1926); *saianicus* (Thomas, 1911); *sibiricus* (Egorin, 1936) [not Poliakoff, 1881]; *skomerensis* (Barrett-Hamilton, 1903); *sobrus* (Montagu, 1923); *suecicus* (Miller, 1900); *tomensis* (Heptner, 1948); *variscicus* (Wettstein, 1954); *vasconiae* (Miller, 1900); *vesanus* (Hinton, 1926); *wasjuganensis* (Egorin, 1939).

COMMENTS: European populations reviewed by Viro and Niethammer (1982) and Mitchell-Jones et al. (1999), those in Russia and adjacent regions by Gromov and Erbajeva (1995). Regional reports covering range, taxonomy, and ecology are available for: N Turkey (Pamukoglu and Albayrak, 1996), Greece (Thrace; Vohralík and Sofianidou, 1992a), Serbia and Montenegro (Kryštufek and Vohralík, 1992; Petrov, 1992), Slovenia (Kryštufek, 1991), NE Spain (Torre et al., 1996) and Navarra region of N Spain (Castien and Gosalbez, 1992), Netherlands (Bergers and Bussink, 1995; Mostert, 1992a), E Baltic region (Miljutin, 1997, 1998; Timm et al., 1998), Italia (Amori et al., 1999; Cantini, 1991; Cerone and Aloise, 1994; Cresti et al., 1994; Locatelli and Paolucci, 1996a), Slovakia (Danko, 1994; Kminiak, 1996; Mošanský, 1994; Stanko, 1995; Stanko and Mošanský, 1994, 2000; Stanko et al., 1994, 2000), Czech Republic (Šmaha, 1996), N Germany (Dolch et al., 1994), Sumava Mtns of SW Bohemia (Andera and Èervený, 1994), and Switzerland (Hausser, 1995; Maurizio, 1994).

Morphologically and biochemically, *M. glareolus* is closely allied with *M. centralis* (Mezhzherin and Serbenyuk, 1992). Evolutionary relationships among British samples reported by Steven (1953), biochemical differentiation among populations over short geographic distances presented by Leitner and Hartl (1988). Univariate and multivariate distillations of metric and non-metric traits of Austrian samples, as integrated with allozymic and mitochondrial DNA analyses, point to distinctive eastern and western moieties: "the Eastern Alps were colonized in postglacial times by two different phylogenetic lineages (*nageri* from the west, *glareolus* or *istericus* from the east) that probably had split during the Würm glaciation" (Spitzenberger et al., 1999:69). Y-chromosome polymorphisms encountered at the same localities in Serbia and Montenegro thought to indicate postglacial secondary contact (Vujošević and Blagojević, 1997). Normal synaptic association between gonosomes in *M. glareolus* documented by Ashley and Fredga (1994) in a phylogenetic context. M3 variation among extant and fossil samples analyzed and interpreted by Bauchau and Chaline (1987). Based on huge samples from the Czech Republic, Zejda et al. (1994) concluded that modal differences in m1 and M3 patterns mirrored genetic divergence among populations where samples are large and age groups are considered. In a homogeneous population, Plakchotnikova et al. (1992) uncovered no significant differences in molar dimensions between right and left sides or between males and females. Istomin (1994a, b) tabulated the infrequent occurrence of polydonty and interpreted variation in other cranial traits across fragmented habitats within the S Taiga, NW Russia. Literák and Zejda (1995) described color mutations in wild populations in the Czech Republic. Bank voles represent a recent introduction to Ireland, not recorded until 1964; Smiddy and Sleeman (1994) surveyed their range and rate of expansion in County Cork, SW Ireland, presumably from a small founder population as indicated by mitochrondrial DNA analyses (Ryan et al., 1996).

Myodes imaizumii (Jameson, 1961). Pacific Sci., 15:599.

COMMON NAME: Imaizumi's Red-backed Vole.

TYPE LOCALITY: Japan, Honshu, Kii Peninsula, Wakayama Prefecture, Nachi.

DISTRIBUTION: Recorded only from the S end of the Kii Peninsula.

COMMENTS: Described as a species of *Clethrionomys*, *imaizumii* was demoted to a subspecies of *andersoni* by Aimi (1967) and afterwards placed in full synonymy of *Eothenomys andersoni* (Aimi, 1980; Musser and Carleton, 1993; Pavlinov et al., 1995*a*). See *M. smithii* account for association of *imaizumii* with *Myodes* rather than *Eothenomys*. Kitahara (1995) exhaustively studied molar patterns and craniometric data of C Honshu voles and also concluded that the Kii populations represent *M. andersoni*; lab-breeding trials between *imaizumii* and *andersoni* have produced fertile hybrids (Kitahara and Kimura, 1995). Suzuki et al. (1999*b*) postulated a biogeographic scenario to explain the southward extension of *andersoni* into the Kii Peninsula, followed by their isolation, as *imaizumii*, when climates warmed and *M. smithii* expanded from west to the east.

Detailed chromosomal (Iwasa et al., 1999*b*) and molecular studies (Suzuki et al., 1999*b*), however, just as convincingly attest the absence of genetic interchange between *andersoni* and *imaizumii*. As noted by Iwasa et al. (1999), *imaizumii* is now geographically widely isolated from *andersoni* and their ability to produce fertile hybrids in the wild is untestable. MtDNA sequences portray *imaizumii* as unique, but its rDNA profile reflects a mixture of those characterizing *M. andersoni* and *M. smithii*, leading Suzuki et al. (1999*b*:519) to wonder whether "ancestral populations of *smithii-andersoni* were polymorphic at each of the variable restriction sites and fixation of variants progressed during speciation, except for *E. imaizumii* in which the polymorphic status has been preserved until now." Pending completion of such on-going studies, we follow Kaneko (1994) in retaining *imaizumii* as species based on its distinctive molecular traits and isolated geographic range.

Myodes regulus (Thomas, 1907). Proc. Zool. Soc. Lond., 1906:863 [1907].
    COMMON NAME: Korean Red-backed Vole.
    TYPE LOCALITY: Korea, Mingyong, 110 mi (177 km) SE Seoul.
    DISTRIBUTION: Korean Peninsula (Won and Smith, 1999, as *Eothenomys*).
    STATUS: IUCN – Lower Risk (lc) as *Eothenomys regulus*.
    COMMENTS: Usually included in *M. rufocanus* (G. M. Allen, 1940; Ellerman and Morrison-Scott, 1951; Gromov and Polyakov, 1977; Hinton, 1926*a*; also see references in Kaneko, 1990) but treated as a separate species by Corbet (1978*c*). In a study of red-backed voles from Russia, NE China, and Korea, Kaneko (1990) thoroughly documented morphological distinctions between *M. rufocanus* and *M. regulus*, considering the latter to be a Korean endemic and suggesting (p. 129) that "the true geographical demarcation line between the two species lies on the western and southern boundary of the Kaima Plateau, North Korea." Kaneko also discovered that adult and old *regulus* lack molar roots, a key character for separating *Myodes* from *Eothenomys*, and so transferred the Korean endemic to *Eothenomys*.

Subsequent mitochondrial and nuclear ribosomal DNA analyses conclusively demonstrate that *M. regulus* is distinct but phylogenetically close to *M. rufocanus*, not typical *Eothenomys* (Suzuki et al., 1999*b*, whose study included *melanogaster*, the type species of *Eothenomys*; Wakana et al., 1996). Wakana et al. (1996:22) noted that the "absence of rooting of the molars in the Korean vole is a characteristic that may have developed within a short period of evolutionary time in the Korean population." The G-banding pattern of *regulus* is "essentially identical" to that of *M. rufocanus*, both differing from *M. glareolus*, but Y-chromosome morphology implicates closer relationship between *M. regulus* and *M. smithii* (Iwasa et al., 1999*a*), a Japanese vole also having unrooted molars. As understated by Iwasa et al. (1999*a*:40), phylogenetic affinities among *rufocanus*, *regulus*, and *smithii* "are extremely complicated." To this complicated mix must be added *M. shanseius* (also rootless), which is so morphologically similar to *M. regulus* that the same specimens have been variously identified as one or the other. They differ only in length of tail, interorbital breadth, and M3 occlusal patterns; as currently known, their distributions are allopatric, perhaps separated by the Liao He River in Liaoning Prov. (Kaneko, 1992*c*). The genetic integrity of *M. regulus* relative to *M. shanseius* similarly warrants additional study.

Myodes rex (Imaizumi, 1971). J. Mammal. Soc. Japan, 6:99.
    COMMON NAME: Hokkaido Red-backed Vole.

TYPE LOCALITY: Japan, Rishiri Isl (off NW coast of Hokkaido), Mt Rishiri, Kanrosen.
DISTRIBUTION: Hokkaido and small islets of Rishiri and Rebun off NW coast of Hokkaido
    (Kaneko et al., 1998:25, 28).
STATUS: IUCN – Lower Risk (nt).
SYNONYMS: *montanus* (Imaizumi, 1972).
COMMENTS: Treated as a distinct species (Abe, 1973*a*, *b*, 1984) until Aimi (1980) included *rex*
    in *M. rufocanus*, an allocation followed by Corbet (1978*c*) and Musser and Carleton
    (1993). Kaneko and Sato (1993), however, presented morphological traits that
    distinguish the two as species on Rishiri Isl (also Kaneko et al., 1998) and demon-
    strated their sympatry and habitat affinities. Subsequent evaluation of ribosomal and
    mitochondrial DNA sequences clearly sustains the specific integrity of *M. rex* (Suzuki
    et al., 1999*b*; Wakana et al., 1996). Imaizumi's (1971) description of *M. rex*, differential
    comparisons with *M. rufocanus*, and records of habitat segregation (*M. rufocanus* in
    open grassy fields, *M. rex* in coniferous forest) are lucid and thorough.
        Imaizumi (1972) described *montanus* (Hidaka Mtns, Hokkaido) as the other species
    in his *M. rex* group. A wealth of morphological (Abe, 1973*a*, *b*, 1984; Aimi, 1980),
    chromosomal (Kashiwabara and Onoyama, 1988; Tsuchiya, 1981), allozymic (Yoshida
    et al., 1989), and DNA sequence data (Suzuki et al., 1999*b*; Wakana et al., 1996)
    represents *montanus* as another population of *M. rex* on Hokkaido. Reports of *M. rex* on
    Shikotan and Shibotsu Isls in the S Kurils and on Sakhalin Isl (Kostenko and Allenova,
    1978) require verification (Kaneko et al., 1998:25).
        Imaizumi (1971, 1972) regarded *M. rex* to be closely related to *M. rufocanus* but
    more primitive in external, cranial, and dental traits. His view is broadly consistent
    with phylogenetic interpretations of mitochondrial and ribosomal DNA sequences
    and biogeographic reconstructions of species diversification (Wakana et al., 1996). At
    present, *M. rex*, *M. rufocanus* and *M. rutilus* occur on Hokkaido (Kaneko, 1994; Kaneko
    et al., 1998; Wakana et al., 1996).

Myodes rufocanus (Sundevall, 1846). Ofv. K. Svenska Vet.-Akad. Forhandl. Stockholm, 3:122.
    COMMON NAME: Gray Red-backed Vole.
    TYPE LOCALITY: Sweden, Lappmark.
    DISTRIBUTION: N Palearctic from Scandinavia through Siberia to Kamchatka (Nikanorov,
        2000), Sakhalin, and Taraku (south of Shikotan Isl in the S Kuril Isls) in Russia, south to
        S Ural Mtns, the Altai Mtns, Transbaikal, N China (NW Xinjiang in the west, and Nei
        Mongol, Liaoning, Jilin, and Heilongjiang in the NE), N portion of Korean peninsula,
        N Japan (Hokkaido and the offshore islets of Rishiri, Daikoku, Teuri, and Yagishiri),
        and the S Kurile Isls of Kunashiri, Shikotan, Shibotsu and others.
    STATUS: IUCN – Lower Risk (lc).
    SYNONYMS: *akkeshii* (Imaizumi, 1949); *arsenjevi* (Dukelsky, 1928); *bargusinensis* (Turov, 1924);
        bedfordiae (Thomas, 1905); *bromleyi* Kostenko [date unknown, *nomen nudum*; see
        Gromov and Erbajeva, 1995]; *changbaishanensis* (Jang, Ma, and Luo, 1993); *irkutensis*
        (Ognev, 1924); *kamtnschaticus* (Poliakov, 1881); *kolymensis* (Ognev, 1922); *kurilensis*
        (Tokuda, 1932); *latastei* (J. A. Allen, 1903); *microtinus* (Kuzyakin, 1963) [*nomen nudum*];
        siberica (Poliakov, 1881); *sikotanensis* (Tokuda, 1935); *wosnessenskii* (Poliakov, 1881);
        yesomontanus (Kishida, 1931).
    COMMENTS: Chromosomal and molecular evidence relates *M. rufocanus* most closely to
        Korean *M. regulus* and Japanese *M. rex*, *M. andersoni*, *M. smithii*, and *M. imaizumii*
        (Iwasa et al., 1999*a*; Suzuki et al., 1999*b*; Wakana et al., 1996); it is also mor-
        phologically similar to the Chinese *M. shanseius* (see that account). Allozymic
        (Mezhzherin and Serbenyuk, 1992) and chromosomal data (Sokolov et al., 1990) re-
        veal *M. rufocanus* as highly differentiated from other *Myodes* sampled (*M. centralis*,
        M. glareolus, *M. gapperi*, *M. rutilus*). DNA/DNA hybridization and homologous land-
        mark analyses also distantly isolate *M. rufocanus* from *M. glareolus* and New World
        M. rutilus and *M. gapperi* (Din et al., 1993). *Myodes rufocanus* and *M. rex* have rooted
        molars (in adults), but their purported close relatives—*M. andersoni*, *M. smithii*, and
        M. imaizumi from the main Japanese islands, and *M. shanseius* and *M. regulus* from
        mainland China and the Korean Peninsula, respectively—have rootless molars. These
        data collectively presume that aquisition of ever-growing molars occurred indepen-

dently from a rooted *rufocanus*-like ancestor in insular regions and on the continent, a distributional pattern in a significant dental innovation unlike that observed among other species of *Myodes*.

Morphological variation broadly assessed by Kaneko (1990), who documented morphological distinctions between *M. rufocanus* and the Korean endemic *M. regulus* (as *Eothenomys*). Geographic distribution of chromosomal variation across continental E Russia, Sakhalin Isl, and other small islands assessed by Kartavtseva et al. (1998), who recorded a conservative 2n (56) but a variable FN (56-59); Hokkaido sample also chromosomally similar to continental species (Obara et al., 1995) ; phylogenetic analyses of chromosomes and allozymes undertaken by Yoshida et al. (1989) for Japanese populations. Contrasting phylogeographic patterns, based on cytochrome *b* versus *Sry* gene, documented among populations from E Russia, Sakhalin Isl, and Hokkaido and offshore islands (Iwasa et al., 2000). M3 variation reported by Abe (1982). Cranial skeleton and myology associated with incisal biting and mastication examined by Satoh (1997, 1998, 1999) and compared with the murine *Apodemus speciosus*; cardiac musculature of the pulmonary vein and vena cava described by Endo and Yanagawa (1994). Ishibashi et al. (1992) used DNA fingerprinting to track individual relatedness during cyclic fluctuations in population density so common to *M. rufocanus*.

European distributions reviewed by Henttonen and Viitala (1982), Korean by Won and Smith (1999). Occurrence on the Svjatoj Nos peninsula and isthmus in Lake Baikal documented by Reiter et al. (1995). Kaneko et al. (1998) extensively covered the systematics and distribution of *M. rufocanus* on Hokkaido; the species occurs on Haikkado and those offshore islands separated only by narrow and shallow straits, but no extant population occupies Oshima Isl, separated from the coast of Hokkaido by a wide and deep strait (Iwasa et al., 2000). In the E Palearctic, the range of *M. rufocanus* is separated from Chinese species of *Eothenomys* by the Gobi Desert (Kaneko, 1992*c*).

Tokuda (1935) described *sikotanensis* (Shikotan Isl in the Kurile islands) and proposed the genus *Neoaschizomys* to contain it (in 1941, he transferred these to *Clethrionomys*). Imaizumi (1949) described *akkessi* from Daikoku Isl (near SE Hokkaido) as a subspecies of *sikotanensis*; the *nomen nudum* status of *microtinus*, associated with *sikotanensis*, is discussed by Gromov and Polyakov (1977) and Pavlinov and Rossolimo (1987). Although *sikotanensis* is frequently recognized as a distinct species (Gromov and Erbajeva, 1995; Gromov and Polyakov, 1977; Imaizumi, 1971, 1972; Musser and Carleton, 1993; Pavlinov and Rossolimo, 1987, 1998; Pavlinov et al., 1995*a*), chromosomal and molecular samples from these islands, and other reported insular occurrences, are unremarkably aligned with *M. rufocanus* (Mezhzherin and Serbenyuk, 1992; Sokolov et al., 1990; Wakana et al. 1996). Kaneko et al. (1998) reported that the holotype of *sikotanensis* is lost but original molar illustrations, especially the M3, match those of *M. rufocanus*, not *M. rex*; they too concluded that *sikotanensis* is another insular example of *M. rufocanus*, a view consistent with the vole's overwater dispersal ability (Iwasa et al. 2000).

Myodes rutilus (Pallas, 1779). Nova Spec. Quadr. Glir. Ord., p. 246.

COMMON NAME: Northern Red-backed Vole.

TYPE LOCALITY: Russia, Siberia, center of Ob River delta.

DISTRIBUTION: Holarctic—in Old World, from N Scandinavia east to Chukotski Peninsula, and south to N Kazakhstan, Transbaikalia, Mongolia, N China (NW Xinjiang in the west, Nei Mongol and Ningxia in northcentral, and Jilin and Heilongjiang in the northeast; Zhang et al., 1997), extreme N Korean Peninsula (Won and Smith, 1999), and Sakhalin and Hokkaido Isls (Dobson, 1994; Henttonen and Peiponen, 1982; Nikanorov, 2000); St. Lawrence Isl, Bering Sea; in New World, from Alaska east to Hudson Bay, and south to N Alaska Panhandle, N British Columbia, and NE Manitoba, Canada.

STATUS: IUCN – Lower Risk (lc).

SYNONYMS: *alascensis* (Miller, 1898); *albiventer* (Hall and Gilmore, 1932); *amurensis* (Schrenk, 1859); *baikalensis* (Ognev, 1924); *dawsoni* (Merriam, 1888); *dorogostaiskii* (Vinogradov, 1933) [*nomen nudum*]; *finmarchius* (Siivonen, 1967); *glacialis* (Orr, 1945); *hintoni* (Vinogradov, 1933) [*nomen nudum*]; *hintoni* (Zolotarev, 1936); *insularis* (Heller, 1910);

jacutensis (Vinogradov, 1927); *jochelsoni* (J. A. Allen, 1903); *laticeps* (Ognev, 1924); *latigriseus* (Argyropulo and Afanasiev, 1939); *lenaensis* (Koljuschev, 1936); *mikado* (Thomas, 1905); *mollessonae* (Kastschenko, 1910); *narymensis* (Argyropulo and Afanasiev, 1939); *orca* (Merriam, 1900); *otus* (Turov, 1924); *parvidens* (Ognev, 1924); *platycephalus* (Manning, 1957); *rjabovi* (Beljaeva, 1953); *rossicus* (Dukelsky, 1928); *russatus* (Radde, 1862); *salairicus* (Egorin, 1936); *tugarinovi* (Vinogradov, 1933) [*nomen nudum*]; *tundrensis* (Bolshakov and Schwarz, 1965); *uralensis* (Vinogradov, 1933) [*nomen nudum*]; *uralensis* (Koljusch, 1936); *vinogradovi* (Naumov, 1933); *volgensis* (Kaplanov and Raevsky, 1928); *washburni* (Hanson, 1952); *watsoni* (Orr, 1945).

COMMENTS: Conspecificity of Old (*rutilus*) and New World (*dawsoni*) populations advanced by Rausch (1953) and corroborated by subsequent studies (e.g., Nadler et al., 1976, 1978; Rausch and Rausch, 1975a). Phylogenetic interpretation of allozymes indicates that *M. rutilus* is most closely related to *M. gapperi*, the two forming a monophyletic group along with *M. glareolus-M. centralis* but excluding *M. rufocanus* (Mezhzherin and Serbenyuk, 1992). Analyses of autosomes (Iwasa et al., 1999b) and mitochondrial and nuclear ribosomal DNA (Suzuki et al., 1999b; Wakana et al., 1996) also emphasize the phyletic distance between *M. rutilus* and *M. rufocanus* and its allies (*M. andersoni, M. smithii, M. rex, M. regulus*).

North American populations revised, as *Clethrionomys dawsoni*, by Orr (1945), and, as *C. rutilus* by Manning (1956). European populations reviewed by Henttonen and Peiponen (1982) and Mitchell-Jones et al. (1999); those in Russia and adjacent regions by Gromov and Erbajeva (1995). M3 variation and its systematic implications evaluated by Nakatsu (1982) for Japanese populations. Iwasa et al. (1999b) contrasted the sex chromosomes and pachytene synapses of Hokkaido *M. rutilus* with Hokkaido *M. rufocanus* and Japanese species of *Eothenomys* (here = *Myodes*); karyotypes of *M. rutilus* and *M. rufocanus* from Hokkaido are essentially identical and closely resemble mainland species of *Myodes* (Obara et al., 1995). Phylogeography of *M. rutilus* in NE Asia, and comparisons with *M. rufocanus* which has a similar geographic range, assessed by mitochondrial cytochrome *b* sequences (Iwasa et al., 2002). Occurrence on the Svjatoj Nos peninsula and isthmus in Lake Baikal documented by Reiter et al. (1995). In Hokkaido, Japan, *M. rutilus* occurs with *M. rufocanus* and *M. rex* (see those accounts and that of *M. gapperi*).

Myodes shanseius (Thomas, 1908). Proc. Zool. Soc. Lond., 1908:643.

COMMON NAME: Shanxi Red-backed Vole.

TYPE LOCALITY: China, Shanxi, Chao Cheng Shan, "Mo-er-Shan" (= Mt Nanyanshan or Mt Guandi Shan), 9296 ft (2789 m); 37°54′N, 111°30′E (as fixed by Kaneko, 1992c:95).

DISTRIBUTION: NC China: C Nei Mongol, S Gansu, N Shanxi, N Shaanxi, Beijing, and Hebei provinces (Kaneko, 1992c; Zhang et al., 1997).

STATUS: IUCN – Lower Risk (lc) as *Eothenomys shanseius*.

SYNONYMS: *jeholicus* (Kuroda, 1939).

COMMENTS: Usually regarded as a subspecies of *rufocanus* (G. M. Allen, 1940; Ellerman, 1941; Ellerman and Morrison-Scott, 1951; Gromov and Polyakov, 1977; Hinton, 1926a; Howell, 1929; Wang, 2003; Ye et al., 2002). Because adults possess rootless molars, Corbet (1978c) reassociated *shanseius* with *Eothenomys*, an allocation followed by Corbet and Hill (1992), Musser and Carleton (1993), and Pavlinov et al. (1995a). Karyotype matches that of other *Myodes* (Ma and Jiang, 1996). Species revised by Kaneko (1992c), as *Eothenomys shanseius*, who judged the holotype of *jeholicus* to be a young example.

Pelage texture, hair length, and color pattern in *shanseius*, however, are typical of species of *Myodes*, strikingly unlike the dark brown, short-furred *Eothenomys* (comparative series in AMNH and USNM). Like other *Myodes*, female *shanseius* have four pairs of mammae (one pectoral, one axillary, and two inguinal; USNM 155055 and 175527, both lactating adults), compared with only two inguinal pairs in *Eothenomys*. Occlusal patterns in *shanseius* also resemble those of *Myodes*: particularly the five closed, alternate triangles on m1 (six opposite and medially confluent in *Eothenomys*); four closed, alternate triangles on m2 (four opposite and confluent in *Eothenomys*); and M3 with two closed, alternate triangles (two opposite and confluent in *Eothenomys*). We hypothesize that the rootless condition in *shanseius* was acquired independently, as is thought to have evolved in the Korean *M. regulus* and species endemic to Japan (*M. andersoni*, *M. smithii*, *M. imaizumii*), all of which are closely related to *M. rufocanus* (see those accounts). Evaluation of this generic reallocation using other taxonomic information is needed, including critical exemplar species of both *Myodes* and *Eothenomys*.

Geographic distribution of *M. shanseius* is allopatric to that of the Korean endemic *M. regulus* (see that account), which it morphologically resembles (Kaneko, 1992c). The *M. shanseius*-*M. regulus* complex represents a segment of *Myodes* that occurs south and east of the Gobi desert in China and eastward onto the Korean Peninsula; although those populations have acquired rootless molars, they appear to be phylogenetically closer to rhizodont *M. rufocanus*.

Myodes smithii (Thomas, 1905). Ann. Mag. Nat. Hist., ser. 7, 15:493.

COMMON NAME: Smith's Red-backed Vole.
TYPE LOCALITY: Japan, Honshu, Hyogo Prefecture, Kobe.
DISTRIBUTION: Japanese islands of Dogo, Honshu, Shikoku, and Kyushu (Kaneko, 1992b, 1994).
STATUS: IUCN – Lower Risk (lc) as *Phaulomys smithii*.
SYNONYMS: *kageus* (Imaizumi, 1957); *okiensis* (Tokuda, 1932).
COMMENTS: Type species of *Phaulomys* Thomas (1905b), diagnosed as a subgenus of *Evotomys*. Imaizumi (1949) transferred *Phaulomys* to *Eothenomys*, an action largely ignored (e.g., Ellerman and Morrison-Scott, 1951) until adopted by Corbet (1978c) and Aimi (1980). *Phaulomys* was reinstated as a genus, containing *andersoni* and *smithii*, by Kawamura (1988), who thought the two extant species with rootless molars (a key trait of *Eothenomys*) were derived from the middle Pleistocene *japonicus*, which has rooted molars. Molar and external traits also led Tanaka (1971) to recognize *Phaulomys* as a genus distinct from *Eothenomys*, primarily because *smithii* combined characteristics of both *Myodes* and *Eothenomys*. Chromosomal analyses of *smithii* suggest that *Myodes* and *Phaulomys* are derived from a common ancestor (Ando et al., 1988).

We (1993:532) previously recognized *Phaulomys* because "The removal of Japanese *Phaulomys* from *Eothenomys*, whose species-diversity centers in the mountains of China, and its proposed association with *Clethrionomys* [= *Myodes*] is zoogeographically plausible and outlines a precise hypothesis that can be tested by analyses of other data sets." In just such a taxonomically and geographically broad study, using mitochondrial and nuclear ribosomal DNA, Suzuki et al. (1999b) demonstrated: 1) that *andersoni*, *imaizumii*, and *smithii* are closely related to Japanese and Asian *M. rufocanus* and to Korean *M. regulus*; 2) that this clade is phyletically remote from *M. rutilus* and *Eothenomys melanogaster*; and 3) that no monophyletic group corresponding to *Phaulomys* as a separate genus is supported. They concluded (1999b:520) that the allocations of Japanese red-backed voles to *Myodes*, *Eothenomys*, or *Phaulomys* should be reevaluated. The incorporation of *smithii* into *Myodes* had earlier been urged by Yoshida et al. (1989), based on chromosomal and allozymic data, and was so arranged by Pavlinov et al. (1995a). The molecular relationships reported by Suzuki et al. imply that the evergrowing molars in *andersoni* and *smithii* are independently derived from a rooted *Myodes* ancestor endemic to Japan, as earlier hypothesized by Kawamura (1988).

Species revised, as *Eothenomys*, and many name combinations traced by Aimi (1980). Morphology, altitudinal and geographic distribution, taxonomic comparisons, and nomenclatural history meticulously investigated by Kaneko (1992b, 1994, 1996a). Synonymy of *kageus* demonstrated by Aimi (1980), Kaneko (1985) and Ando et al. (1988); later Kaneko (1994, as *Eothenomys*) recognized *kageus* as occurring in E Honshu and *smithii* in W Honshu. The taxon *okiensis*, described as a subspecies of *rufocanus*, was placed in synonymy by Aimi (1980). Chromosomal data widely available, in comparisons among arvicoline species and within samples of *smithii* (Ando et al., 1988, 1991; Iwasa and Tsuchiya, 2000; Iwasa et al., 1999a, b; Tsuchiya, 1981; Vorontsov et al., 1980; Yoshida et al., 1989). See Kawamura (1991, 1994) for discussion of Holocene occurrence in archeological sites on Honshu, Shikoku, and Kyushu. Both *M. smithii* and *M. andersoni* are discussed by Dobson (1994, as *Phaulomys*) in context of elucidating patterns of distribution in Japanese land mammals.

Myopus Miller, 1910. Smithson. Misc. Coll., 52:497.

> TYPE SPECIES: *Myodes schisticolor* Lilljeborg, 1844.
>
> COMMENTS: Lemmini. Conventionally treated as a genus until Chaline (1972) regarded the molar differences between *schisticolor* and *Lemmus* to reflect species-level distinctions (also Chaline and Mein, 1979; Chaline et al., 1989). Koenigswald and L. D. Martin (1984) also cited fundamental molar similarity for their arrangement of *Myopus* as a subgenus of *Lemmus*. Others, however, have maintained their generic segregation (Gromov and Erbajeva, 1995; Jarrel and Fredga, 1993; Niethammer and Henttonen, 1982; Pavlinov and Rossolimo, 1987, 1998; Pavlinov et al., 1995*a*), as we do here. Although *Lemmus* and *Myopus* share certain dental resemblances, they are readily distinguished by karyotype, a profoundly different sex chromosome mechanism, body size, fur coloration, morphology (feet, skull and eyes), habitat, and behavior (Jarrell and Fredga, 1993); such contrasts have not been addressed in a phylogenetic context along with other forms comprising these genus-group taxa. In the absence of such studies, the phenotypically restricted character set mustered from a paleontological perspective is insufficient to falsify the hypothesis that *schisticolor* represents a monophyletic group separate from species of *Lemmus*. Jarrell and Fredga (1993) noted that the mostly biarmed karyotype of *Myopus* could be derived from a *Lemmus*-like karyotype by numerous centric fusions. Known from late Pleistocene of Asia (McKenna and Bell, 1997).

Myopus schisticolor (Lilljeborg, 1844). Ofv. K. Svenska Vet.-Akad. Forhandl. Stockholm, I, p. 33.

> COMMON NAME: Wood Lemming.
>
> TYPE LOCALITY: Norway, Gulbrandsdal, N end of Mjosen, near Lillehammer.
>
> DISTRIBUTION: Coniferous taiga zone from Norway and Sweden through Siberia to Kolyma River and Kamchatka (Nikanorov, 2000), south to the Altai Mtns, N Mongolia and NE China (N Heilongjiang and N Nei Mongol; Zhang et al., 1997), and the Sikhote Alin Range (Corbet, 1978*c*); also a southern isolate in the Ural Mtns, near source of Ural River, about 450 km south of previously recorded limit at 58°N.
>
> STATUS: IUCN – Near Threatened.
>
> SYNONYMS: *middendorfii* Vinogradov, 1922; *morulus* Hollister, 1912; *saianicus* Hinton, 1914; *thayeri* G. M. Allen, 1914; *vinogradovi* Skalon and Raevski, 1940.
>
> COMMENTS: Described in detail by Miller (1912*a*) and Hinton (1926*a*); reviewed by Gromov and Polyakov (1977) and Corbet (1978*c*, 1984). Intraspecific chromosomal variation and extraordinary genetic system of sex determination leading to female-biased sex ratio documented by Gropp et al. (1976), Kozlovskij (1986), Fredga et al. (1976, 1977, 1993), and Gileva and Fedorov (1991). Departures from the usual 2n = 32 documented for Siberian samples (Gileva et al., 1983; Kozlovsky, 1985), but the chromosomal difference does not signify more than a single species (Jarrell and Fredga, 1993). Genetic variation is low in populations from Western Siberia and Scandinavia (Fedorov, 1990, 1993; Fredga et al., 1993). Using isozyme analysis, Fedorov et al. (1994) recorded significant genetic divergence between the Fennoscandian, Western Siberian, and Eastern Siberian regions but little differentiation among populations within each. Age and geographic variation among Norwegian populations assessed by Kratochvíl (1979).
>
> Courant et al. (1997) discovered cranial shape convergence between *M. schisticolor* and certain voles and discussed possible ecological influences for such parallel shape changes in distantly related arvicolines. Regarded as rare, but Emelyanova (1994) found wood lemmings to be a dominant species in small mammal communities of the taiga zone, NE Siberia, forming the primary food of hawk owls.

Neodon Horsfield, 1841. A Catalogue of the Mammalia in the Museum of the Hon. East-India Company:145-146 (as corrected by Kaneko and Smeenk, 1996; not Hodgson, 1849, as entrenched in the literature).

> TYPE SPECIES: *Neodon sikimensis* Horsfield, 1841.
>
> SYNONYMS: *Bicunedens* Hodgson, 1863.
>
> COMMENTS: Arvicolini. Included in Pitymyini by Repenning et al. (1990) and Repenning (1992). Maintained as a genus by some specialists (Ellerman, 1941; Hinton, 1923, 1926*a*; Zagorodnyuk, 1990, 1992*c*), as a subgenus of *Pitymys* by others (Corbet, 1978*c*; Ellerman,

1941; Ellerman and Morrison-Scott, 1951), or a subgenus of *Microtus* (G. M. Allen, 1940; Gromov and Erjabeva, 1995; Gromov and Polyakov, 1977; Musser and Carleton, 1993; Pavlinov et al., 1995a). Hinton (1923, 1926a) enumerated cranial traits that characterize *Neodon* and considered it to be closely related to *Pitymys*, primarily based on the m1 occlusal pattern; species of Nearctic *Pitymys* and Palearctic *Terricola* (here treated as subgenera of *Microtus*) and *Neodon sikimensis* all possess this pitymyine m1 configuration (as defined by Repenning, 1992:65). Both *N. sikimensis* and pitymyine forms also have an elongate and complex M3, generally with four lingual salient angles in contrast to the simpler pattern seen in *Phaiomys leucurus* (see Hinton, 1923, 1926a; Nadachowski and Zagorodnyuk, 1996; Repenning, 1992). Hinton (1923) included *forresti, irene, oniscus* (= *irene*), and *carruthersi* (= *juldaschi*) in *Neodon*. Occlusal patterns of their m1s vary, some similar to the pattern in *N. sikimensis*, others less complex and generally, but not in detail, resembling the *Allophaiomys*-like configuration as seen in *Blanfordimys* and *Phaiomys* (specimens in AMNH, FMNH, and USNM; also see Nadachowski and Zagorodnyuk, 1996). All have a more elaborate M3 than that in *Phaiomys leucurus*, and none possesses the deep and stout cranium like that species or its short, constricted incisive foramina (see account of *P. leucurus*).

Species of *Neodon* (along with *Blanfordimys* and *Phaiomys*) are characterized as Pleistocene relicts because their molar occlusal patterns resemble the extinct *Allophaiomys*. Further, one species (*N. juldaschi*) has a primitive karyotype similar to that hypothesized as ancestral to Arvicolini (2n = 56, Chaline and Matthey, 1971; or 2n = 54, Zagorodnyuk, 1992c). Thus, Nadachowski and Zagorodnyuk (1996:387) viewed these voles as "*Allophaiomys*-like" species that represent "Pleistocene relicts or a return to an initial type," with ranges in C and S Asia at the austral periphery of the Palearctic arvicoline distribution. We provisionally retain *Neodon* as a lineage independent of *Phaiomys* and *Microtus* until phylogenetic relationships can be analyzed by a suite of information bases, including morphological traits other than molar patterns.

Neodon forresti Hinton, 1923. Ann. Mag. Nat. Hist., ser. 9, 11(9):156.
COMMON NAME: Forrest's Mountain Vole.
TYPE LOCALITY: S China, NW Yunnan, Divide between Mekong and Yangtze Rivers (27°30′ N), 11,000-12,000 ft (3350-3660 m).
DISTRIBUTION: Extreme NW Yunnan, China (Hinton, 1923), and northernmost Burma (Ellerman, 1961), 3350-3660 m.
COMMENTS: Morphologically close to *N. irene* but body size larger, pelage longer and darker (Hinton, 1923). Monographed as species by G. M. Allen (1940), but he noted that *forresti* may be only a southern subspecies of *irene*, as later recognized by Ellerman (1947a, 1961); *forresti* was subsequently arranged as a subspecies of *N. sikimensis* (Weigel, 1969). Intergradation between *forresti* and *irene* has never been demonstrated. None of the 112 specimens of *irene* we studied from NW Sichuan (see measurements in Lawrence, 1982), or those documented by G. M. Allen from S Gansu and N Yunnan, overlap in size with specimens of *forresti* of comparable age.

Neodon irene (Thomas, 1911). Abstr. Proc. Zool. Soc. Lond., 1911(90):5.
COMMON NAME: Irene's Mountain Vole.
TYPE LOCALITY: China, Sichuan, Tatsienlu.
DISTRIBUTION: High mountains in the Chinese provinces of E Qinghai, S Gansu, W Sichuan, NE Xizang, and NW Yunnan (G. M. Allen, 1940; Feng et al., 1986; Zhang et al., 1997).
STATUS: IUCN – Lower Risk (lc) as *Microtus irene*.
SYNONYMS: *oniscus* Thomas, 1911.
COMMENTS: Member of the *sikimensis* species group (Zagorodnyuk, 1990). Ellerman (1941, 1947b, 1961) and Ellerman and Morrison-Scott (1951) listed *irene* as a valid species, but Weigel (1969) allocated it to *Pitymys sikimensis* as did Corbet (1978c), citing Gruber's (1969) "detailed study" as substantiation. However, the union of *irene* and *sikimensis* is insufficiently demonstrated by Gruber's study, whose data actually underscore the trenchant differences in size and dentition between the two. Ellerman (1947b, 1961) highlighted these marked differences in body size and m1 patterns in his keys and descriptions, contrasts reinforced by Feng et al. (1986), who recorded *irene* from

NE Tibet and *sikimensis* just to the west, without evidence of character overlap. Lawrence (1982) recorded localities and habitat information for a large AMNH series of *N. irene* collected in Qionglai Shan (2900-3660 m), Sichuan.

Neodon juldaschi (Severtzov, 1879). Sap. Turk. Otd. Obsh.Lubit. Estestv., 1:63.
COMMON NAME: Juniper Mountain Vole.
TYPE LOCALITY: NE Tajikistan, Pamir Mtns, Kara-Kul Lake basin, near Aksu.
DISTRIBUTION: W Tien Shan (Talasskiy Altai), south to Gissar and Turkestan ranges in E Tajikistan and SW Kyrgyzstan and east though the Pamir Mtns in E Tajikistan to NW Xizang (Tibet; Feng et al, 1986) and SW Xinjiang (Zhang et al., 1997); farther south in N Pakistan (Roberts, 1977) and NE Afghanistan (Hassinger, 1973; Niethammer, 1970).
STATUS: IUCN – Lower Risk (lc) as *Microtus juldaschi*.
SYNONYMS: *carruthersi* Thomas, 1909; *pamirensis* Miller, 1899; *thalassensis* Sludsky, 1988 [*nomen nudum*; see Pavlinov and Rossolimo, 1998]; *yuldaschi* (Severtzov, 1879) [original spelling but alteration to *juldaschi* unclear; see Pavlinov et al., 1995a].
COMMENTS: M3 occlusal patterns range from slightly more elaborate than *Allophaiomys* to a configuration resembling *N. sikimensis* (see Hinton, 1923; Nadachowski and Zagorodnyuk, 1996). The taxon *carruthersi* was treated as a separate species (Ellerman, 1941; Ellerman and Morrison-Scott, 1951; Ognev, 1964) until synonymized with *juldaschi* (Corbet, 1978c), a synonymy thereafter observed (Gromov and Erbajeva, 1995; Gromov and Polyakov, 1977; Pavlinov and Rossolimo, 1987, 1998; Pavlinov et al., 1995a; Musser and Carleton, 1993). Bolshakov and Pokrovskii (1969) questioned the morphological distinction of *carruthersi* and *juldaschi* and demonstrated unhindered hybridization between the two with completely fertile offspring. Gromov and Polyakov (1977), however, remained skeptical, noting that Bolshakov and Pokrovskii's sample of *carruthersi* was not topotypic (Tajikistan, Gissar Mtns, 100 miles east of Samarkand, 9000-10,000 ft) but from the Macha River in the Zeravshan Range, just north of the Gissar Range. Because the specimens used by Bolshakov and Pokrovskii originated from the Maikhur Ravine in the Gissar Valley, and because the type locality is simply given as the Gissar Mtns, we find it reasonable to accept their sample as representative of *carruthersi* (Ironically, their breeding stock of *juldaschi* originated from Chechekta in the Pamirs, about 60 km south of the type locality, and drew no criticism).

Gileva and Pokrovskii (1970; and cited references) incorporated past chromosomal results and documented only slight differences between *carruthersi* from the Gissar Range (2n = 54, FN = 56-58) and *M. juldaschi* from Chechekta (2n = 54, FN = 60-61). Subsequently, Gileva et al. (1982) substantially broadened their geographic representation, including material from the type locality of *juldaschi*, and concluded that populations in the Pamirs (*juldaschi*) and those from the Gissar and Turkestan ranges (*carruthersi*) formed one group distinct from another complex in the W Tien Shan; whether the two groups represented different species or just geographic variants was not stated (Gromov and Erbajeva, 1995, regarded all as one species). While the status of the western Tien Shan population invites resolution, the published evidence portrays *carruthersi* and *juldashi* as examples of the same biological entity.

Neodon sikimensis (Horsfield, 1841). A Catalogue of the Mammalia in the Museum of the Hon. East-India Company:145-146 (as corrected by Kaneko and Smeenk, 1996; not Hodgson, 1849, as entrenched in the literature).
COMMON NAME: Sikkim Mountain Vole.
TYPE LOCALITY: India, Sikkim.
DISTRIBUTION: Himalayas from W to E Nepal (based on FMNH series), through NE India (Sikkim and Darjeeling Dist.) to E Bhutan (Ellerman, 1947a, 1961); also SC and E Xizang (Tibet; Feng et al., 1986; Zhang et al., 1997).
STATUS: IUCN – Lower Risk (lc) as *Microtus sikimensis*.
SYNONYMS: *thricolis* (Gray, 1863).
COMMENTS: Geographic distribution of *N. sikimensis* on the E Tibetan Plateau is parapatric with the smaller-bodied and morphologically different *N. irene* (see account), once

included in *N. sikimensis*. Banded karyotype (2n = 48, FN = 56) described by Mekada et al. (2002), who postulated its derivation from that characterizing *N. juldaschi* (as *Microtus carruthersi*). Indian population reviewed by Agrawal (2000); aspects of habitat, individual and age variation reported by Abe (1971) and Gruber (1969).

Neofiber True, 1884. Science, 4:34.
TYPE SPECIES: *Neofiber alleni* True, 1884.
SYNONYMS: *Schistodelta* Cope, 1899.
COMMENTS: Neofibrini. A relictual form once more widespread in the E USA (Frazier, 1977; Hibbard and Dalquest, 1973). Although sometimes placed with *Ondatra* in Ondatrini (Chaline and Mein, 1979; Repenning et al., 1990), various morphological features suggest a more distant relationship (Carleton, 1981; Hooper and Hart, 1962; Koenigswald, 1980), as does some fossil evidence (R. A. Martin, 1975; Zakrzewski, 1974). Alternatively, *Neofiber* either has been cited as Arvicolinae *incertae sedis* (Gromov and Polyakov, 1977; Koenigswald, 1980; Kretzoi, 1969) or affiliated with *Arvicola* and *Microtus* in a broadly-defined Arvicolini (Zagorodnyuk, 1990). Also see remarks under *Ondatra*.

Neofiber alleni True, 1884. Science, 4:34.
COMMON NAME: Round-tailed Muskrat.
TYPE LOCALITY: USA, Florida, Brevard Co., Georgiana.
DISTRIBUTION: Most of peninsular Florida, extralimitally in extreme SE Georgia and Florida panhandle, USA.
STATUS: IUCN – Lower Risk (nt).
SYNONYMS: *apalachicolae* Schwartz, 1953; *exoristus* Schwartz, 1953; *nigrescens* A. H. Howell, 1920; *struix* Schwartz, 1952.
COMMENTS: Geographic races initially delimited by Schwartz (1953), but Whitaker and Hamilton (1998) recommended *nigrescens* and *struix* as synonyms of *N. a. alleni*. See Birkenholz (1972, Mammalian Species, 15).

Ondatra Link, 1795. Beytrage zur Naturgeschichte, 1(2):76.
TYPE SPECIES: *Castor zibethicus* Linnaeus, 1766.
SYNONYMS: *Fiber* G. Cuvier, 1800; *Moschomys* Billberg, 1827; *Mussacus* Oken, 1816; *Neondatra* Hibbard, 1937; *Pliopotamys* Wilson, 1933; *Simotes* Fischer, 1817.
COMMENTS: Ondatrini. Assigned either as sole tribal member (Hooper and Hart, 1962; Koenigswald, 1980) or with *Neofiber* (Chaline and Mein, 1979; Repenning et al., 1990); or placed in subtribe Ondatrina, Arvicolini (Pavlinov et al., 1995*a*). Based on allozymic analysis, Mezhzherin et al. (1995) recognized Ondatrini as the first group to have diverged from the basal arvicoline stock during the late Miocene. Evidence from fossil history and morphology indicates that the two genera of "muskrats" are more distantly related than their inclusion in the same tribe would connote (Carleton, 1981; Hooper and Hart, 1962; Koenigswald, 1980; R. A. Martin, 1975, 1979); analysis of long interspersed nucleotide elements, on the other hand, unites *Ondatra* and *Neofiber* as sister genera (Modi, 1996). Originated from the Pliocene form *Pliopotamys minor* (Eshelman, 1975; Zakrzewski, 1974); Pliocene-Pleistocene changes in dentition and size ("chronomorphs") leading to extant species are traced by R. A. Martin (1996), who reallocated *Pliopotamys* as junior synonym.

Ondatra zibethicus (Linnaeus, 1766). Syst. Nat., 12th ed., 1:79.
COMMON NAME: Common Muskrat.
TYPE LOCALITY: E Canada.
DISTRIBUTION: North America, north to the treeline, including Newfoundland; south to the Gulf of México, Rio Grande and lower Colorado River valleys. Introduced to Czech Republic in 1905 and now widespread in the Palearctic, including C and N Europe, most of Ukraine, Russia, and Siberia, adjacent parts of Mongolia and scattered throughout China, NE Korea, and Honshu Isl, Japan (Gromov and Erbajeva, 1995, for Eurasian range; Mitchell-Jones et al., 1999, for European range); also into southernmost Argentina (Galliari et al., 1996; Olrog and Lucero, 1981).
STATUS: IUCN – Data Deficient as *O. z. ripensis*, otherwise Lower Risk (lc).
SYNONYMS: *albus* (Sabine, 1823); *americana* Tiedemann, 1808; *aquilonius* (Bangs, 1899);

bernardi Goldman, 1932; *cinnamominus* (Hollister, 1910); *hudsonius* (Preble, 1902); *goldmani* Huey, 1938; *maculosa* (Richardson, 1829); *macrodon* (Merriam, 1897); *mergens* (Hollister, 1910); *niger* (Fitzinger, 1867); *niger* (Brass, 1911); *nigra* (Richardson, 1829); *obscurus* (Bangs, 1894); *occipitalis* (Elliot, 1903); *osoyoosensis* (Lord, 1863); *pallidus* (Mearns, 1890); *ripensis* (Bailey, 1902); *rivalicius* (Bangs, 1895); *spatulatus* (Osgood, 1900); *varius* (Fitzinger, 1867); *zalophus* (Hollister, 1910).

COMMENTS: Subspecific classification revised, under the name *Fiber*, by Hollister (1911); Whitaker and Hamilton (1998) regarded *macrodon* as inseparable from *O. z. zibethicus*. Comprehensive summaries of systematics, ecology, population biology, and economic status provided by Pietsch (1982), Perry (1982), and Sokolov and Lavrov (1993). Correlation between masticatory muscles and cranial architecture monographed by Vendeloo (1953). Viriot et al. (1993) used digital imaging to analyze molar ontogeny and wear effects in extant samples; the technique may quantitatively refine the evolutionary chronocline as traced in the stratigraphic record (e.g., Chaline et al., 1999; R. A. Martin, 1996).

History of introductions, mostly for fur farming, and early population spread in Eurasia and USA reviewed by Storer (1937) and Willner (1984). Recent faunal studies—including records of introduction, distribution, and population expansion—collectively underscore the muskrat's pervasive exploitation of Eurasian aquatic habitats: Netherlands (Hoeve and Wijlaars, 1992); Sumava Mtn region, SW Bohemia (Andera and Èervený, 1994); Switzerland (Hausser, 1995); Italy (Amori et al., 1999; Andreotti et al., 2001); Slovakia (Mošanský, 1994; Stanko and Mošanský, 2000); Czech Republic (Šmaha, 1996); Serbia and Montenegro (Petrov, 1992); Slovenia (Kryštufek, 1991); E Baltic region (Miljutin, 1997, 1998; Timm et al., 1998); Korea (Won and Smith, 1999); Svjatoj Nos peninsula and isthmus in Lake Baikal (Reiter et al., 1995); Kamchatka region (Nikanorov, 2000); Russia and adjacent regions (Gromov and Erbajeva, 1995); China (Zhang et al., 1997); and Japan (Kaneko, 1994). See Willner et al. (1980, Mammalian Species, 141).

Phaiomys Blyth, 1863. J. Asiatic Soc. Bengal, 32(1):89.

TYPE SPECIES: *Phaiomys leucurus* Blyth, 1863.

COMMENTS: Arvicolini. Variably recognized as a genus (Ellerman, 1941; Hinton, 1923, 1926*a*; R. A. Martin, 1987, 1989*b*; Repenning, 1992; Repenning et al., 1990); a subgenus of *Microtus* (G. M. Allen, 1940; Ellerman and Morrison-Scott, 1951; Musser and Carleton, 1993; Ognev, 1950); a subgenus of *Pitymys* (Corbet, 1978*c*; Ellerman, 1947*a*, 1961; Zheng and Wang, 1980); or a subgenus of *Neodon* (Gromov and Polyakov, 1977; Zagorodnyuk, 1990, 1992*c*). Nadachowski and Zagorodnyuk (1996) identified *Phaiomys leucurus* as a Pleistocene relict because its M3 and m1 occlusal patterns are simple and closely similar to the extinct *Allophaiomys*. Chaline (1987:253) proclaimed that "an isolate of the ancestral *Allophaiomys pliocaenicus* stayed in the Himalayas and survives as *Phaiomys leucurus*." R. A. Martin (1987, 1989*b*) thought the m1 patterns in *Allophaiomys pliocaenicus* and *Phaiomys* to be so similar that he included the former as subgenus of the latter, and later (1995) acknowledged the differences between the two at the species level. Kormos (1932), however, who named and first described *Allophaiomys*, noted cranial, mandibular, and minor dental traits that contrast the two genera, and Repenning (1992) described other minute but significant dental differences between them.

Hinton (1926*a*:48) believed that *P. leucurus* is "closely related to *Arvicola*, and seems like the latter to be descended from some close ally of *Mimomys*." This relationship was formalized by Kretzoi (1955; also Repenning et al., 1990, and Repenning, 1992), who included *Phaiomys*, *Arvicola*, and *Allophaiomys*, all with rootless molars, in the Arvicolini (not as used here), along with *Mimomys* and other fossil genera that possess rooted molars and a range in hypsodonty. *Phaiomys* can be derived from *Allophaiomys*, which in turn was presumably descended "from a still inadequately identified *Mimomys* lineage" (Chaline et al., 1999). Both M3 and m1 occlusal patterns are also closely similar in *Phaiomys* and *Arvicola* (see Hinton, 1926*a*; Rekovets, 1990); moreover, the cranium of *Phaiomys* is stout and rugged, with very short and posteriorly constricted incisive foramina, a conformation similar to species of *Arvicola* but unlike any *Microtus*, and in particular unlike *Neo-*

don with which *Phaiomys* has been associated. Repenning explicitly denied any close tie between *Phaiomys* and *Neodon*, allocating the latter to Pitymyini and thereby viewing *Phaiomys* and *Neodon* as independent lineages derived from ancestral *Allophaiomys*. Along with the molar differences, other external and cranial contrasts—compared with *Neodon*, *Phaiomys* has longer and stouter foreclaws, a robust and deeper cranium, much shorter and posteriorly constricted incisive foramina, inflated auditory bulla, moderately inflated mastoid region, and robust mandible with prominent alveolar processes—support a view of two genera.

The hypothesis that *Phaiomys* is a monophyletic lineage (genus) independent of the radiation of species-groups within *Microtus* draws attention to its morphological connection to ancestral *Allophaiomys*, as argued by paleontologists, and provides a taxonomic framework to explore its postulated alliance with *Arvicola*, other relictual groups (*Blanfordimys* and *Neodon*), and species of *Microtus* using molecular analyses and a broader variety of morphological data than molar patterns. *Phaiomys leucurus* occurs only at high altitudes in the rugged Himalayas and Tibetan Plateau at the periphery of the Palearctic range of Arvicolini, a region Hinton (1923:155) emphasized as "that great refuge for archaic Microtines formed by the highlands of Central and South-eastern Asia."

Phaiomys leucurus (Blyth, 1863). J. Asiat. Soc. Bengal, 32:89.

COMMON NAME: Blyth's Mountain Vole.

TYPE LOCALITY: NW India, Ladakh, near Lake Chomoriri (= Tsomoriri).

DISTRIBUTION: Chinese provinces of S Xinjiang (Zhang et al., 1997), S Qinghai (Zheng and Wang, 1980) and Xizang (Feng et al., 1986) on the Tibetan Plateau (Hoffmann, 1996*a*), and high altitudes in the Himalayas west to NW India (Jammu, Kashmir, and Himachal Pradesh, above 4500 m; Agrawal, 2000).

STATUS: IUCN – Lower Risk (lc) as *Microtus leucurus*.

SYNONYMS: *blythi* (Blanford, 1875); *everesti* Thomas and Hinton, 1922; *petulans* (Wroughton, 1911); *strauchi* (Büchner, 1889); *tsaidamensis* (Satunin, 1903); *waltoni* (Bonhote, 1905); *zadoensis* (Zheng and Wang, 1980).

COMMENTS: Revised by Zheng and Wang (1980); allocation of *tsaidamensis* to *P. leucurus* documented by Hoffmann (1996*a*). Hoffmann provided a detailed historical review of expeditionary routes and collection sites of *P. leucurus* on the Tibetian Plateau, and recorded places of sympatry with *Lasiopodomys fuscus* (formerly included in *leucurus*). Feng et al. (1986) also discussed and mapped distribution on the Tibetan Plateau (molar rows they illustrate are either incorrectly drawn or from another vole). Zheng and Wang (1980) recognized *leucurus*, *waltoni*, and *zadoensis* as subspecies. The older *leucurus* Gerbe, 1852, is a synonym of *nivalis*, a species transferred from *Microtus* to *Chionomys* (see that account). Should *nivalis* ever be returned to *Microtus*, the name *blythi* would have to replace *leucurus*, a doubtful eventuality considering the great phylogenetic distance between *Chionomys* and species of *Microtus*.

Phenacomys Merriam, 1889. N. Am. Fauna, 2:32.

TYPE SPECIES: *Phenacomys intermedius* Merriam, 1889.

SYNONYMS: *Propliophenacomys* L. D. Martin, 1975.

COMMENTS: Phenacomyini. Apart from *Arborimus*, nearest generic kin uncertain—placed as Arvicolinae *incertae sedis* (Chaline et al., 1999; Gromov and Polyakov, 1977); or with Phenacomyini, including *Arborimus* (Zagorodnyuk, 1990); Arvicolini, including *Phaiomys* and certain extinct genera (Repenning et al., 1990); or Myodini (McKenna and Bell, 1997). The rooted molars and lack of cement in reentrant angles are plesiomorphic traits that suggest an early differentiation of *Phenacomys* within the arvicoline radiation, and paleontologists have proposed its origin from a lineage of Beringian *Mimomys* in the early Pliocene (Repenning and Grady, 1988; Repenning et al. 1987). Early cladistic separation is also suggested by phylogenetic analysis of highly repetitive DNA (LINE-1) elements, in which *Phenacomys* forms an unresolved basal trichotomy with *Dicrostonyx* and a third branch subtending seven other genera surveyed (Modi, 1996). Revised by Howell (1926) and Hall and Cockrum (1953), then including species assigned to *Arborimus* (see that account).

We acknowledge phenacomyine in an informal sense because Phenacomyini, as coined by Zagorodnyuk (1990:27), is a *nomen nudum* and unavailable. Zagorodnyuk used the formal name in a paragraph listing of arvicoline tribes and member genera, without indicating its status as new and lacking any statement of differentiation. In view of the old and pronounced phyletic separation of *Phenacomys* suggested by the above studies, taxonomic sampling should be broadened and the family-group clade properly named and diagnosed should such a conclusion prove sustainable.

Phenacomys intermedius Merriam, 1889. N. Am. Fauna, 2:32.
COMMON NAME: Western Heather Vole.
TYPE LOCALITY: Canada, British Columbia Prov., 20 mi (32 km) NNW Kamloops.
DISTRIBUTION: NW British Columbia and SW Alberta, Canada, south to N New Mexico, C Utah, and N California, USA; disjunct populations in EC California and W Nevada, USA.
STATUS: IUCN – Lower Risk (lc).
SYNONYMS: *celsus* A. B. Howell, 1923; *constablei* J. A. Allen, 1899; *laingi* Anderson, 1942; *levis* A. B. Howell, 1923; *olympicus* Elliot, 1899; *oramontis* Rhoads, 1895; *orophilus* Merriam, 1891; *preblei* Merriam, 1897; *pumilus* (Elliot, 1899); *truei* J. A. Allen, 1894.
COMMENTS: Howell (1926) originally recognized three species (*intermedius*, *mackenzii*, and *ungava*) of heather voles, later reduced to two (*intermedius* and *ungava* including *mackenzii*) by Anderson (1942, 1947). Crowe (1943) further lumped all under *intermedius* based on suspected intergrades from SW Alberta, and the recognition of a single species has been generally followed (e.g., Banfield, 1974; Corbet and Hill, 1991; Hall and Cockrum, 1953) but not exclusively so (Cowan and Guiguet, 1965; Miller and Kellogg, 1955; Peterson, 1966). Foster and Peterson (1961) questioned Crowe's appreciation of age-effects in his identification of the alleged intergrades between *intermedius* and *ungava*. As remarked by Cowan and Guiguet (1965), the matter of their synonymy "requires more detailed examination before a decision can be reached," an appraisal which stands equally valid today. See McAllister and Hoffmann (1988, Mammalian Species, 305, including *ungava*).

Phenacomys ungava Merriam, 1889. N. Am. Fauna, 2:35.
COMMON NAME: Eastern Heather Vole.
TYPE LOCALITY: Canada, Quebec Prov., Fort Chimo, near Ungava Bay.
DISTRIBUTION: S Yukon across much of Canada to E Labrador; southwards skirting the E Rocky Mountains to S Alberta, along the N Great Lakes and lower St. Lawrence River, SE Quebec.
STATUS: IUCN – Lower Risk (lc).
SYNONYMS: *celatus* Merriam, 1889; *crassus* Bangs, 1900; *latimanus* Merriam, 1889; *mackenzii* Preble, 1902; *soperi* Anderson, 1942.
COMMENTS: See remarks under *P. intermedius* and McAllister and Hoffmann (1988) on proper usage of *ungava* for this form. Distributional limits in late Pleistocene much farther south, as far as N Arkansas, C Tennessee, and Virginia (Graham and Lundelius, 1994).

Proedromys Thomas, 1911. Abstr. Proc. Zool. Soc. Lond., 1911(90):4.
TYPE SPECIES: *Proedromys bedfordi* Thomas, 1911.
COMMENTS: Arvicolini. Although included in *Microtus* by Ellerman and Morrison-Scott (1951) and Corbet (1978c), most systematists have maintained *Proedromys* as genus (G. M. Allen, 1940; Gromov and Polyakov, 1977; McKenna and Bell, 1997; Musser and Carleton, 1993; Pavlinov et al., 1995a; Wang et al., 1966; Zhang et al., 1997). The taxon's diagnostic traits, as enumerated by Thomas (1911d— massive cranium with wide, heavy, and grooved upper incisors and remarkably short lower incisors, and molar peculiarities), identify it as an independent lineage derived from some as yet unidentified ancestral arvicoline stock. Gromov and Polyakov (1977) considered *Proedromys* a relict of unidentifiable affinites, and Repenning (1992:65), based on molar occlusal patterns, speculated that "an origin out of *Allophaiomys* or early *Lasiopodomys* seems apparent but is not as yet documented."

Proedromys bedfordi Thomas, 1911. Abstr. Proc. Zool. Soc. Lond., 1911(90):4.
COMMON NAME: Duke of Bedford's Vole.

TYPE LOCALITY: China, S Gansu, 60 mi (97 km) SE Minchow.

DISTRIBUTION: Recorded from S Gansu and N Sichuan (Wang et al., 1966; Zhang et al., 1997), China.

STATUS: IUCN – Lower Risk (nt).

COMMENTS: The only extant species of the genus. Pleistocene fragments from Shanxi, Hebei, and Shandong provinces have been identified as *P.* cf. *bedfordi* (Zheng and Li, 1990).

Prometheomys Satunin, 1901. Zool. Anz., 24:572.

TYPE SPECIES: *Prometheomys schaposchnikowi* Satunin, 1901.

COMMENTS: Prometheomyini. The only extant member of an archaic line that most specialists isolate as a tribe within Arvicolinae (Gromov and Polyakov, 1977; Hooper and Hart, 1962; Kretzoi, 1955, 1969; McKenna and Bell, 1997; Musser and Carleton, 1993; Pavlinov and Rossolimo, 1987; Zagorodnyuk, 1990). Repenning et al. (1990), however, aligned *Prometheomys* with another presumed relic, *Ellobius*, in Prometheomyinae. A third view was offered by Pavlinov et al. (1995a) and Pavlinov and Rossolimo (1998), who sequestered *Prometheomys* within its own subtribe of a broadly defined Prometheomyini (also *Alticola*, *Clethrionomys* [= *Myodes*], *Dicrostonyx*, *Dinaromys*, *Eolagurus*, *Eothenomys*, *Hyperacrius*, and *Lagurus* in other subtribes).

Fossil occurrences date from the late Pleistocene of W Asia (Agadzhanyan, 1993; McKenna and Bell, 1997). *Stachomys*, which ranged widely from Germany to the Lake Baikal region during the middle and late Pliocene, may be the progenitor of *Prometheomys*, which survived as a relict in the Caucasus Mtns (Agadzhanyan, 1993).

Prometheomys schaposchnikowi Satunin, 1901. Zool. Anz., 24:572.

COMMON NAME: Long-clawed Mole Vole.

TYPE LOCALITY: Georgia (Gruzinskaya), Caucasus Mtns, Gudaur, S of Krestovyi Pass, 6500 ft (as given by Ognev, 1963b).

DISTRIBUTION: Alpine zone of Caucasus Mtns, Georgia, and extreme NE Turkey (E Black Sea Mtns; Kryštufek and Vohralík, 2001).

STATUS: IUCN – Lower Risk (lc).

COMMENTS: Cranial and dental morphology detailed by Hinton (1926a); chromosomal data summarized by Zima and Kral (1984a); karyotype in sample from NE Turkey reported by Colak et al. (1999).

Synaptomys Baird, 1857. Mammalia *in* Repts. U.S. Expl. Surv., 8(1):558.

TYPE SPECIES: *Synaptomys cooperi* Baird, 1857.

SYNONYMS: *Kentuckomys* Koenigswald and L. D. Martin; 1984; *Metaxyomys* Zakrzewski, 1972; *Mictomys* True, 1894; *Praesynaptomys* Kowalski, 1977.

COMMENTS: Lemmini. Many taxonomic characters associate *Synaptomys* with the true lemmings (*Lemmus* and *Myopus*) in a clade, usually regarded as Lemmini and believed to represent an early line of arvicoline evolution (Carleton, 1981; Chaline and Graf, 1988; Conroy and Cook, 1999; Graf, 1982; Gromov and Polyakov, 1977; Hinton, 1926a; Hooper and Hart, 1962; Koenigswald, 1980). Fossil history reviewed by Koenigswald and L. D. Martin (1984), Abramson (1993), Fejfar and Repenning (1998), Kowalski (2001), and Martin et al. (2003). The last authors segregated *Synaptomys* (and *Mictomys* as genus), together with a European fossil (*Tobienia*), as Synaptomyini, apart from Lemmini (*Lemmus*, *Myopus*, and *Plioctomys*), both groups thought to be descendants from an ancestral *Mimomys* stock in the early Pliocene.

Although described as a genus, Miller (1896) arranged *Mictomys* as a subgenus of *Synaptomys*, as conventionally recognized by neontologists in the 1900s (Hall, 1981; Honacki et al., 1982; Howell, 1927; Musser and Carleton, 1993), albeit not uniformly (Jarrell and Fredga, 1993). Paleontologists, on the other hand, have emphasized the dental contrasts of *Synaptomys* and *Mictomys* at the generic level (Fejfar and Repenning, 1998; Koenigswald and L. D. Martin, 1984; Kretzoi, 1969; Repenning and Grady, 1988), albeit not uniformly (Abramson, 1993). While certain early Pliocene European taxa have been linked with the origin of *Synaptomys sensu lato* (see Fejfar and Repenning, 1998, Chaline et al., 1999, and Martin et al., 2003 for reviews), the divergence of *borealis* and *cooperi* is thought to have occurred in North America, whether dated to a common ancestor in the late Pliocene (Repenning and Grady, 1988) or one in the middle Pleistocene (Fejfar and Repenning, 1998). As understood, the cladistic dichotomy and its recency seem equivocal for treating *Mictomys* and *Synaptomys* as separate genera; that possibility should be explored using other information and a broad sampling of New and Old World lemmings, one that includes other archaic forms outside of Lemmini, such as *Arvicola*, *Dinaromys*, *Phenacomys*, and *Prometheomys*. Current specific and subspecific classification (e.g., Hall, 1981) essentially framed by Howell (1927).

Synaptomys borealis (Richardson, 1828). Zool. J., 3:517.
    COMMON NAME: Northern Bog Lemming.
    TYPE LOCALITY: Canada, District of Mackenzie, Great Bear Lake, Fort Franklin.
    DISTRIBUTION: Patchy occurrence in boreal habitats, from Alaska to N Washington, USA, eastwards across much of interior Canada to Labrador; disjunct range segment from Gaspe Peninsula, Quebec, to C New Hampshire, USA.
    STATUS: IUCN – Lower Risk (nt).
    SYNONYMS: *andersoni* J. A. Allen, 1903; *artemisiae* Anderson, 1932; *bullatus* Preble 1902; *chapmani* J. A. Allen, 1903; *dalli* Merriam, 1896; *innuitus* (True, 1894); *medioximus* Bangs, 1900; *smithi* Anderson and Rand, 1943; *sphagnicola* Preble, 1899; *truei* Merriam, 1896; *wrangeli* Merriam, 1896.
    COMMENTS: Subgenus *Mictomys*. Pleistocene records establish the species in the Great Basin, far to the south of its current range (see Mead et al., 1992).

Synaptomys cooperi Baird, 1857. Mammalia *in* Repts. U.S. Expl. Surv., 8(1):558.
    COMMON NAME: Southern Bog Lemming.
    TYPE LOCALITY: USA, New Hampshire, Carroll Co., Jackson (as fixed by Bole and Moulthrop, 1942:146).
    DISTRIBUTION: Midwestern and E USA through SE Canada, including Nova Scotia and Cape Breton Isl; as far south as W North Carolina and NE Arkansas; outlying populations in SW Kansas, W Nebraska, and the Dismal Swamp region of SE Virginia-NE North Carolina.
    STATUS: IUCN – Extinct as *S. c. paludis* and *S. c. relictus*, otherwise Lower Risk (lc).
    SYNONYMS: *fatuus* Bangs, 1896; *gossii* (Coues, 1877); *helaletes* Merriam, 1896; *jesseni* Long, 1987; *kentucki* Barbour, 1956; *paludis* Hibbard and Rinker, 1942; *relictus* Jones, 1958; *saturatus* Bole and Moulthrop, 1942; *stonei* Rhoads, 1893.
    COMMENTS: Subgenus *Synaptomys*. Geographic diversification within the species evaluated by Wetzel (1955), who refined the subspecific arrangement. Geographic variation among populations in the C Great Plains studied by Wilson and Choate (1997), who retained subspecies defined within the region and commented upon the conservation status of the relictual races (*paludis* and *relictus*) isolated at the western margin of the species distribution; for new range reports and biogeographic discussion of populations (*helaletes*) at the eastern periphery of the species distribution, see Lee and Clark (1993) and Clark et al. (1993). Using landmark data, Courant et al. (1997) demonstrated stronger convergence in cranial shape between *S. cooperi* and surface dwelling voles like *Myodes* rather than its lemming relatives. See Linzey (1983, Mammalian Species, 210).

Volemys Zagorodnyuk, 1990. Vestn. Zool., 2:28.
    TYPE SPECIES: *Microtus musseri* Lawrence, 1982.
    COMMENTS: Arvicolini. *Volemys* circumscribes two species restricted to the alpine-subalpine zone in S China that may be part of an older fauna (Lawrence, 1982). Both have traditionally been placed in *Microtus*, although their phylogenetic relationships were regarded as obscure or equivocal (G. M. Allen, 1940; Zagorodnyuk, 1990). Corbet and Hill (1992) did not recognize *Volemys*, and Pavlinov et al. (1995a) arranged it as a subgenus of *Microtus*. In his generic description, Zagorodnyuk (1990) added two other species: Taiwanese *kikuchii* and *clarkei* from the mountains of Yunnan, N Burma, and SE Tibet. Phylogenetic analysis of cytochrome *b* sequences demonstrated that *kikuchii* is sister-species of *M. oeconomus* (Conroy and Cook, 2000a), within a clade embracing *M. monte-belli*, *M. middendorffii*, and *M. fortis*, all members of *Microtus*, subgenus *Alexandromys* (see

account of *M. kikuchii*). In reassessing the characteristics of *clarkei* (see that account), we find that it too is morphologically unlike *V. millescens* and *V. musseri* and instead resembles *M. fortis*, another species in the subgenus *Alexandromys*.

Although Zagorodnyuk (1990) formally proposed *Volemys*, its diagnostic traits had been previewed by Lawrence (1982), who concluded that *musseri* and *millescens* do not morphologically fit with other species-groups of *Microtus*. Those traits are long tail relative to head and body; smooth and flattened cranium; slim dentary with a low ramus; inflated auditory bullae; M1-2 with a large posterolingual triangle confluent with the opposite labial triangle and forming an inverted chevron-shaped lamina (in *V. musseri*; the configuration characterizes only the M2 in *V. millescens*); m1 with only four closed triangles anterior to the posterior lamina, the anterior triangles confluent and with the large anteroconid cap; anterior lamina of each m2 is chevron-shaped and formed by confluent lingual and labial triangles. *Volemys millescens* is smaller than *V. musseri* and differs in pelage coloration, but is similar in molar patterns except for the lack of an M1 posterolingual triangle. Their dental patterns are unique among species of *Microtus*, and in combination with the distinctive cranial and mandibular conformations and external proportions, isolate *millicens* and *musseri* from any other species-group of *Microtus* or any other genus of Arvicolini.

Species of *Volemys*, along with *Neodon irene* and certain *Eothenomys*, are the only voles recorded from the high mountains of W Sichuan, China, where no species of true *Microtus* has been found. *Microtus fortis* does occur in SE Sichuan but inhabits river valleys at low altitudes, and *M. clarkei* is known only from above 3400 m in W Yunnan and N Burma (see those accounts).

Volemys millicens (Thomas, 1911). Abstr. Proc. Zool. Soc. Lond., 1911(100):49.
    COMMON NAME: Sichuan Vole.
    TYPE LOCALITY: China, NW Sichuan, Weichoe, Si-ho River Valley, 12,000 ft (3658 m).
    DISTRIBUTION: Known only from the type locality and SE Xizang, Tibet (Feng et al., 1986).
    STATUS: IUCN – Lower Risk (lc).
    COMMENTS: Detailed descriptions provided by Thomas (1911e), G. M. Allen (1940), and
        Ellerman (1961); Lawrence (1982) described traits that distinguish it from *V. musseri*.
        Gromov and Polyakov (1977) provisionally allocated *millicens* to subgenus *Neodon* of
        Microtus. Feng et al. (1986) identified ten specimens from SE Xizang (Tibet) as *millicens*,
        although their measurements average larger than those of the type series (compare
        Feng et al., 1986:396, with Lawrence, 1982:16). All of the Tibetan specimens are
        morphologically similar to *V. millescens*, not *V. musseri*, according to Darrin Lunde
        (pers. comm., 2004) who examined the sample at IZAS in Beijing. The locality of
        millicens mapped by Zhang et al. (1997) in S Yunnan needs verification; the site is more
        plausible for *Microtus clarkei*.

Volemys musseri (Lawrence, 1982). Am. Mus. Novit., 2745:6.
    COMMON NAME: Marie's Vole.
    TYPE LOCALITY: China, W Sichuan, Qionglai Shan, 30 mi (48 km) W Wenquan, 9000 ft
        (2743 m).
    DISTRIBUTION: Qionglai Shan, W Sichuan, 2318-3660 m.
    STATUS: IUCN – Lower Risk (lc).
    COMMENTS: Lawrence (1982) located the type locality of *V. musseri* about 35 miles
        southwest from that of *V. millicens*, the two separated by the Chehshieh Shan. The 45
        examples of *V. musseri* were trapped in the same lines as *Neodon irene* and *Eothenomys
        melanogaster* (Lawrence, 1982).

Subfamily Cricetinae Fischer, 1817. Mém. Soc. Imp. Nat. Moscow, 5:372.
    SYNONYMS: Cricetini Fischer, 1817 (Cricetinorum Fischer, 1817; Cricetina Gray, 1825;
        Cricetinae Murray, 1866; Cricetidae Rochebrune, 1883; Criceti Winge, 1887; Cricetoidea
        Thaler, 1966); Ischymomyini Topachevskii, 1992.
    COMMENTS: Carleton and Musser (1984) diagnosed the subfamily using morphological traits
        and reviewed general characters, major fossil groups, and past association with New
        World sigmodontines. Living hamsters form a monophyletic group bounded by

unambiguously derived morphological traits. Their cladistic integrity is reinforced by phylogenetic analysis of nuclear gene sequences (Michaux and Catzeflis, 2000; Michaux et al., 2001*b*), which associate hamsters as a monophyletic clade within a major lineage that includes Neotominae and Arvicolinae (and Myospalacinae, but see that subfamily), apart from other muroid groups.

Corbet (1978*c*:88) noted that "generic divisions within the group are rather unstable and a fresh, comprehensive classification is required," a conclusion we echo after studying specimens and literature. One of these problematic divisions is *Cricetulus*, and Corbet (1978*c*:90) acknowledged that some species are frequently placed in *Allocricetulus* and *Tscherskia*, but "pending a review of generic classification in the subfamily as a whole," he preferred to include those taxa in *Cricetulus*. The highly distinctive and diagnosable morphologies of these forms as much argue the opposite: to retain them as separate genera until a systematic revision is performed, credible relationships are established, and genus-group synonymies are demonstrated. In cranial traits, *Allocricetulus* closely resembles *Cricetus*, not *Cricetulus*, and the latter is more like *Phodopus* than either *Tscherskia* or *Allocricetulus* (specimens studied in AMNH, FMNH, MCZ, and USNM). Gromov and Erbajeva (1995), Pavlinov and Rossolimo (1987, 1998), Pavlinov et al. (1995*a*), and Wang (2003) also treated *Allocricetulus*, and *Tscherskia* as genera separate from *Cricetulus*. Broad regional treatises condense much systematic and biological information on hamsters, including identification keys, in the Palearctic (Corbet, 1978*c*, 1984; Ellerman and Morrison-Scott, 1951), Europe (Mitchell-Jones et al., 1999; Niethammer, 1982*a, d*), Russia (Gromov and Erbajeva, 1995), China (Wang, 2003), and Indomalayan region (Corbet and Hill, 1992). Chromosomal information for European species recorded by Zima and Kral (1984*a*). Zagorodnyuk (1992*b*) reviewed taxonomic status of scientific names associated with samples from the Ukraine.

Antiquity of Cricetinae extends to the middle Miocene of N Africa, early Miocene of Europe, late Miocene of Asia, and Pliocene in the Mediterranean region. The seven extant genera are a remnant of a Neogene radiation resulting in a plethora of species in 16 extinct genera (extracted from Fahlbusch, 1996; Kälin, 1999; Kowalski, 2001; McKenna and Bell, 1997).

Allocricetulus Argyropulo, 1932. Trudy Zoologich. Inst. AN SSSR, 1:242 (*fide* Pavlinov, 2002, in litt.).
    TYPE SPECIES: *Cricetus eversmanni* Brandt, 1859.
    COMMENTS: Considered valid as a subgenus of *Cricetulus* by Ellerman (1941, 1961) and
        Ellerman and Morrison-Scott (1951).

Allocricetulus curtatus (G. M. Allen, 1925). Am. Mus. Novit., 179:3.
    COMMON NAME: Mongolian Hamster.
    TYPE LOCALITY: China, W Nei Mongol, Iren Dabasu (= Ehrlien).
    DISTRIBUTION: Mongolian steppes north of the Altai and in adjacent China in NW Xinjiang,
        N Gansu, N Ningxia (Qin, 1991), Anhui (Liu et al., 1985), and Nei Mongol; see Zhang
        et al. (1997, mapped as *C. eversmanni*).
    STATUS: IUCN – Lower Risk (lc).
    COMMENTS: Although some systematists have treated *curtatus* as a subspecies of *A. evers-*
        manni (G. M. Allen, 1940; Ellerman and Morrison-Scott, 1951; Ma et al., 1987), most
        have continued to recognize it as a separate species (Corbet, 1978*c*; Corbet and Hill,
        1991; Gromov and Erbajeva, 1995; Musser and Carleton, 1993; Pavlinov and
        Rossolimo, 1987, 1998; Pavlinov et al., 1995*a*). A fresh perspective is required to
        validate its specific status.

Allocricetulus eversmanni (Brandt, 1859). Melanges. Biol. Acad. St. Petersbourg, p. 210.
    COMMON NAME: Eversmann's Hamster.
    TYPE LOCALITY: Russia, Orenburg Oblast, near Orenburg.
    DISTRIBUTION: Steppes of N Kazakhstan from Volga River to the upper Irtysh River at Zaysan
        Lake in E Kazakhstan; also N Xinjiang Prov., NW China (Wang, 2003).
    STATUS: IUCN – Lower Risk (lc).
    SYNONYMS: *beljawi* (Argyropulo, 1933); *belajevi* (Selevin, 1934); *beljaevi* (Kuznetzov, 1944);
        microdon (Ognev, 1925); *pseudocurtatus* Vorontsov and Kryukova, 1969 [*nomen nudum*].

COMMENTS: See Pavlinov and Rossolimo (1987) for allocation of *pseudocurtatus*.

Cansumys G. M. Allen, 1928. J. Mammal., 9:244.
> TYPE SPECIES: *Cansumys canus* G. M. Allen, 1928.
> COMMENTS: Formerly considered a synonym of *Cricetulus* (Corbet, 1978c; Ellerman, 1941;
> > Ellerman and Morrison-Scott, 1951) or *Tscherskia* (Carleton and Musser, 1984; Pavlinov
> > and Rossolimo, 1987). Ross (1988) underscored the combination of unusual pelage
> > pattern, large body size, long hairy tail, and high-crowned (but rooted) selenodont-like
> > molars that makes *Cansumys* unique among cricetines, and subsequent workers have
> > supported its obvious generic separation from *Cricetulus* and *Tscherskia* (Corbet and Hill,
> > 1992; Musser and Carleton, 1993; Pavlinov et al., 1995a). Zhang et al. (1997) recognized
> > *canus* but retained *Cansumys* in *Cricetulus*, without explanation.

Cansumys canus G. M. Allen, 1928. J. Mammal., 9:245.
> COMMON NAME: Gansu Hamster.
> TYPE LOCALITY: China, S Gansu Prov., Jonê (=Choni or Cho-Ni), "around 2500 m" (Corbet
> > and Hill, 1992:393).
> DISTRIBUTION: Known only from the type locality and possibly Henan Prov., NC and
> > EC China; limits unresolved.
> STATUS: IUCN – Lower Risk (lc).
> COMMENTS: Viewed either as a form of *Cricetulus triton* (Corbet, 1978c; Ellerman, 1941;
> > Ellerman and Morrison-Scott, 1951) or as a valid species of *Cansumys* (Corbet and Hill,
> > 1991, 1992; Musser and Carleton, 1993; Pavlinov et al., 1995a). Zhang et al. (1997) and
> > Wang (2003) included Shaanxi Prov. within the distribution of *canus* (as a species of
> > *Cricetulus* or *Cansumys*, respectively) because they regarded *ningshaanensis* as its
> > subspecies; in fact, Song (1985) had described *ningshaanensis* from Shaanxi as a
> > subspecies of *Tscherskia triton*. The tail in *ningshaanensis* has a longer white segment
> > and is longer relative to head and body as compared with *triton*, but no other features
> > in the original description implicate *ningshaanensis* as a form of *canus*. *Cansumys canus*
> > possesses long semi-hypsodont molar rows (6.4-6.6 mm; based on the holotype and
> > two topotypes in AMNH and FMNH), but *ningshaanensis* has shorter toothrows
> > (4.7-5.7 mm), well within the range characteristic of *T. triton*. The occurrence of
> > *C. canus* in Henan Prov., as reported by Lu and Wang (1996), also requires
> > reverification.

Cricetulus Milne-Edwards, 1867. Ann. Sci. Nat. (Paris), 7:376.
> TYPE SPECIES: *Cricetulus griseus* Milne-Edwards, 1867 (= *Mus barabensis* Pallas, 1773).
> SYNONYMS: *Allocricetus* Schaub, 1930; *Cricetinus* Zdansky, 1928; *Moldavimus* Samson and
> > Radulesco, 1973; *Urocricetus* Satunin, 1903.
> COMMENTS: See comments under subfamily. Genus documented from the late Miocene of
> > Europe and early Pliocene of Asia (McKenna and Bell, 1997).

Cricetulus alticola Thomas, 1917. Ann. Mag. Nat. Hist., ser. 8, 19:455.
> COMMON NAME: Ladak Dwarf Hamster.
> TYPE LOCALITY: India, Ladak (Kashmir), Shushul, 13,500 ft (4115 m).
> DISTRIBUTION: NE India in Jammu and Ladak (Agrawal, 2000), Nepal (Lim and Ross, 1992),
> > and western part of Tibetan Plateau (Xizang) in China (Feng et al., 1986).
> STATUS: IUCN – Lower Risk (lc).
> COMMENTS: Conventionally recognized as a species (Corbet, 1978c; Corbet and Hill, 1991;
> > Ellerman, 1941, 1961; Ellerman and Morrison-Scott, 1951; Musser and Carleton, 1993;
> > Pavlinov et al., 1995a), but Feng et al. (1986) considered *alticola* to be a subspecies of
> > *C. kamensis* that occurs in W Xizang Prov. (an allocation followed by Zhang et al.,
> > 1997, and Wang, 2003).

Cricetulus barabensis (Pallas, 1773). Reise Prov. Russ. Reichs., 2:704.
> COMMON NAME: Striped Dwarf Hamster.
> TYPE LOCALITY: Russia, W Siberia, banks of Ob River, Kasmalinskii Bor (village in Altai Mtns;
> > *fide* Pavlinov, 2002, in litt).
> DISTRIBUTION: Steppes of S Siberia from Irtysh River to Ussuri region, south through

Transbaikalia to Mongolia (Gromov and Erbajeva, 1995), NE China (Zhang et al., 1997), and Korean Peninsula (Won and Smith, 1999).

STATUS: IUCN – Lower Risk (lc).

SYNONYMS: *ferrugineus* Argyropulo, 1941; *fumatus* Thomas, 1909; *furunculus* (Pallas, 1779); *griseus* (Milne-Edwards, 1867) [not Kashkarov, 1923]; *manchuricus* Mori, 1930 [see Kaneko and Maeda, 2002]; *mongolicus* (Thomas, 1888); *obscurus* (Milne-Edwards, 1867); *pseudogriseus* Iskhakova, 1974 [*nomen nudum*]; *pseudogriseus* Orlov and Iskhakova, 1975; *tuvinicus* Iskhakova, 1974 [see Corbet, 1984, and Pavlinov et al., 1995*a*, for status of this name]; *xinganensis* Wang, 1980.

COMMENTS: Status of *griseus* remains unsettled: originally described as a species (Milne-Edwards, 1867), later submerged as a subspecies of *C. barabensis* (G. M. Allen, 1940; Corbet, 1978*c*; Ellerman, 1941; Ellerman and Morrison-Scott, 1951), then reelevated as distinct (Corbet and Hill, 1991; Malygin et al., 1992; Orlov and Iskhakova, 1975; Pavlinov and Rossolimo, 1987; Wang, 2003). Orlov and Iskhakova (1975) had specifically separated *griseus* (2n = 22, FN = 38) from *C. barabensis* (2n = 20, FN = 38) and described the closely related *C. pseudogriseus* (2n = 24, FN = 38) based only on chromosomal traits. Kral et al. (1984) had difficulty in karyotypically characterizing the three because of the extensive homology among chromosomal arms and questioned their distinctiveness. Corbet (1978*c*) also discussed the problem and included *pseudogriseus* and *griseus* in *C. barabensis*, as did Pavlinov et al. (1995*a*) and Pavlinov and Rossolimo (1998). Malygin et al. (1992), however, reaffirmed the chromosomal and specific distinction of *griseus* and *pseudogriseus* from *C. barabensis*, and Gromov and Erbajeva (1995) arranged *pseudogriseus* as a species.

Cricetulus kamensis (Satunin, 1903). Ann. Zool. Mus. St. Petersbourg, 7:574.

COMMON NAME: Tibetan Dwarf Hamster.

TYPE LOCALITY: China, NE Tibet (Xizang Prov.), Mekong Dist., Moktschjun River.

DISTRIBUTION: China, Tibetan Plateau (SE Xinjiang, Xizang, and adjacent parts of Qinghai and Gansu; Zhang et al., 1997).

STATUS: IUCN – Lower Risk (lc).

SYNONYMS: *kozlovi* Satunin, 1903; *lama* Bonhote, 1905; *tibetanus* Thomas, 1922.

COMMENTS: Some workers would include *alticola* as a subspecies (Feng et al., 1986; Wang, 2003; Zhang et al., 1997; see account of *C. alticola*). G. M. Allen (1940) treated *kozlovi* as a synonym of *C. barabensis*, and *lama* was listed as a species by Ellerman (1941) and Ellerman and Morrison-Scott (1951). See Wang and Zheng (1973) for the association of *kozlovi*, *lama*, and *tibetanus* as subjective synonyms of *C. kamensis*.

Cricetulus longicaudatus (Milne-Edwards, 1867). Rech. Hist. Nat. Mammifères, p. 13.

COMMON NAME: Long-tailed Dwarf Hamster.

TYPE LOCALITY: China, N Shanxi (Shansi), near Saratsi.

DISTRIBUTION: Altai and Tuva regions of Russia and Kazakhstan (Gromov and Erbajeva, 1995), NW China (Xinjiang), Mongolia, and adjacent Chinese regions in Nei Mongol, Hebei, Beijing, Tianjin, Henan, N Shanxi, Ningxia, N Sichuan, and N Xizang (Tibet); see Feng et al. (1986), Qin (1991), Corbet and Hill (1992), Wang (2003), and Zhang et al. (1997).

STATUS: IUCN – Lower Risk (lc).

SYNONYMS: *andersoni* Thomas, 1908; *chiumalaiensis* Wang and Cheng, 1973; *dichrootis* Satunin, 1902; *griseiventris* (Satunin, 1903) [not Thomas, 1917]; *kozhantschikovi* Vinogradov, 1927; *nigrescens* G. M. Allen, 1925.

COMMENTS: G. M. Allen (1940) listed *dichrootis* as a synonym of *C. barabensis*, but Corbet (1978*c*) included it in *C. longicaudatus*. Revised by Wang and Zheng (1973).

Cricetulus migratorius (Pallas, 1773). Reise Prov. Russ. Reichs., 2:703.

COMMON NAME: Gray Dwarf Hamster.

TYPE LOCALITY: W Kazakhstan, lower Ural River.

DISTRIBUTION: SE Greece, NW Romania, SE Bulgaria, and S European Russia (Mitchell-Jones et al., 1999); eastwards through Kazakhstan to S Mongolia and N China (Xinjiang, Nei Mongol, Ningxia, Gansu, and Qinghai; Qin, 1991; Wang, 2003; Zhang et al., 1997); southwards through Turkey (Kryštufek and Vohralík, 2001; Osborn, 1965; Pamukoglu and Albayrak, 1996) and Transcaucasia to Israel (Mendelssohn and Yom-Tov, 1999; Qumsiyeh, 1996), Jordan, Lebanon, Iraq, Iran (Lay, 1967), Afghanistan (Hassinger, 1973), Pakistan (Roberts, 1977), and N India (Jammu and Kashmir; Agrawal, 2000).

STATUS: IUCN – Lower Risk (nt).

SYNONYMS: *accedula* (Pallas, 1779); *arenarius* (Pallas, 1773); *atticus* Nehring, 1902; *bellicosus* Scharleman, 1915; *caesius* Kashkarov, 1923; *cinerascens* (Wagner, 1848); *cinereus* Kashkarov, 1926; *coerulescens* (Severtzov, 1879); *elisarjewi* Afanasiev, 1953; *falzfeini* Matschie, 1918; *fulvus* (Blanford, 1875); *griseiventris* (Thomas, 1917) [not Satunin, 1902]; *griseus* (Kashkarov, 1923) [not Milne-Edwards, 1867]; *isabellinus* (de Filippi, 1865); *murinus* (Severtzov, 1876); *myosurus* Argyropulo, 1932 [*nomen nudum*]; *neglectus* Ognev, 1916; *ognevi* Argyropulo, 1932 [*nomen nudum*]; *ognevi* Argyropulo, 1941; *pamirensis* Ognev, 1923 [*nomen nudum*]; *phaeus* (Pallas, 1779); *pulcher* Ognev, 1924; *sviridenkoi* Pidoplitschka, 1928 [*nomen nudum*]; *tauricus* Satunin, 1908 [*nomen nudum*]; *vernula* Thomas, 1917; *zvierezombi* Pidoplitschka, 1928. (See Zagordnyuk, 1992*b*, for comments regarding scientific names applied to Ukranian samples).

COMMENTS: Major compendia summarize the systematic biology of the species in Europe and Russia (Gromov and Erbajeva, 1995; Mitchell-Jones et al., 1999; Niethammer, 1982*d*). Allozymic variation between *isabellinus* from Kopet Dag Mtns and *bellicosus* from Ukranian steppes discussed by Mezhzherin (2001*b*). *Cricetulus migratorius* has occurred in the southern Levant since 80,000-70,000 years before present and apparently replaced two fossil species (recorded as *Allocricetus*), *C. magnus* (70,000-60,000 years before present) and *C. jesreelicus* (not recorded later than 120,000 years before present) (Tchernov, 1992, 1994, and cited references).

Cricetulus sokolovi Orlov and Malygin, 1988. Zool. Zh., 67:305.

COMMON NAME: Sokolov's Dwarf Hamster.

TYPE LOCALITY: W Mongolia, Bayan Hongor, SW shore Orog Nuur Lake (*fide* Pavlinov, 2002, in litt).

DISTRIBUTION: W and S Mongolia, C Nei Mongol of N China.

STATUS: IUCN – Lower Risk (lc).

COMMENTS: A species well defined by chromosomal and pelage traits (Malygin et al., 1992; Orlov and Malygin, 1988). Samples from Mongolia had been identified as *C. obscurus* (Kral et al., 1984, and references therein; Orlov and Malygin, 1988); true *obscurus* is a form of *C. barabensis*.

Cricetus Leske, 1779. Anfansgr. Naturg., 1:168.

TYPE SPECIES: *Mus cricetus* Linnaeus, 1758.

SYNONYMS: *Hamster* Lacepède, 1799; *Heliomys* Gray, 1873.

COMMENTS: Only one living species, but others are represented by fossils in an evolutionary radiation that dates from the middle Miocene of N Africa and late Miocene of Europe (McKenna and Bell, 1997).

Cricetus cricetus (Linnaeus, 1758). Syst. Nat., 10th ed., 1:60.

COMMON NAME: Common Hamster.

TYPE LOCALITY: Germany.

DISTRIBUTION: Belgium across C Europe, W Siberia, and N Kazakhstan to the upper Yenesei and Altai region and NW China (NW Xinjiang; Wang, 2003; Zhang et al., 1997); see Mitchell-Jones et al. (1999) for former European range.

STATUS: IUCN – Lower Risk (lc).

SYNONYMS: *albus* Fitzinger, 1867; *babylonicus* Nehring, 1903; *canescens* Nehring, 1899; *frumentarius* Pallas, 1811; *fulvus* (Bechstein, 1801); *fuscidorsis* Argyropulo, 1932 [*nomen nudum*]; *fuscidorsis* Argyropulo, 1936; *germanicus* (Kerr, 1792); *jeudii* (Gray, 1873); *latycranius* Ognev, 1923; *nehringi* Matschie, 1901; *niger* Fitzinger, 1867; *niger* Bogdanov, 1871 [*nomen nudum*]; *niger* (Simroth, 1906); *nigricans* (Lacépède, 1799) [not Brandt, 1832]; *polychroma* Krulikovski, 1916; *rufescens* Nehring, 1899; *stavropolicus* Satunin, 1907; *tauricus* Ognev, 1924; *tomensis* Ognev, 1924; *varius* Fitzinger, 1867; *vulgaris* Geoffroy, 1803 [see Ellerman and Morrison-Scott, 1951:629].

COMMENTS: Mitchell-Jones et al. (1999) recounted the drastic reduction in W Europe due to persecution and hunting. Other regional taxonomic synopses and chronicles of the species' decline are available for the Netherlands (Lenders and Pelzers, 1982; Pelzers and Lenders, 1992), Sumava Mtns of SW Bohemia (Andera and Èervený, 1994), Northrhine-Westphalia in Germany (Hutterer and Geiger-Roswora, 1997), Slovakia (Mošanský, 1994; Stanko and Mošanský, 2000), Slovenia (Kryštufek, 1991), Serbia and Montenegro (Petrov, 1992), and Czech Republic (Šmaha, 1996). Morphological variability among European samples evaluated by Grulich (1987a, b, 1991, and references therein). Recent reports emphasize conservation status and reintroduction attempts for various European countries (Godmann and Kasabi, 2001; Hellwig, 2001; Jordan, 2001; Losinger, 2001; Mercelis, 2001; Schreiber, 2001; Ulbrich and Kayser, 2001). Once present in the British Isles during the Pleistocene (Kowalski, 1967; Yalden, 1999).

Mesocricetus Nehring, 1898. Zool. Anz., 21:49.

TYPE SPECIES: *Cricetus nigricans* Brandt, 1832 (= *Cricetus raddei* Nehring, 1894).

SYNONYMS: *Mediocricetus* Nehring, 1898 [*nomen nudum*]; *Semicricetus* Nehring, 1898 [*nomen nudum*; see Pavlinov et al., 1995a].

COMMENTS: Ellerman and Morrison-Scott (1951) recognized one extant species with six subspecies. Morphological and karyological differences, however, led Hamer and Schutowa (1965) to recognize four genetically isolated species (*auratus, brandti, newtoni,* and *raddei*). Popular checklists have recognized either three living species (Corbet, 1978c; Corbet and Hill, 1991) or four living and one recently extinct (Musser and Carleton, 1993; Pavlinov et al., 1995a). Pieper (1984) described *M. rathgeberi* based on Holocene fossils from the Greek island of Armathia (off coast of Kasos Isl between Kriti and Rodhos). Known from the late Pliocene of Asia and late Pleistocene of Europe (Kowalski, 2001; McKenna and Bell, 1997).

Mesocricetus auratus (Waterhouse, 1839). Proc. Zool. Soc. Lond., 1839:57.

COMMON NAME: Golden Hamster.

TYPE LOCALITY: Syria, Aleppo.

DISTRIBUTION: Vicinity of type locality and SE Turkey.

STATUS: IUCN – Endangered.

COMMENTS: Lyman and O'Brien (1977) discussed the geographic range of *auratus* and evidence for segregating it from *M. brandti*; also see Kryštufek and Vohralík (2001). Based upon chromosomal data, Doğramaci et al. (1994b) recorded *M. auratus* from SE Turkey, and Yiğit et al. (2000b) later identified only one sample from 10 km E Kilis Prov., SE Turkey, after intensive survey efforts. They recorded *M. brandti* from throughout Turkey and contrasted its morphology and chromosomes (2n = 42, FN = 82 and 84) with those of *M. auratus* (2n = 44, FN = 82). Adler (1948) and Murphy (1985) recorded the origin of the laboratory stocks of *M. auratus* and history of their distributions to various laboratories in France, India, and the United States. Coronary arteries are described by Sans-Coma et al. (1993) and infraorbital glands by Kühnel (1983).

Mesocricetus brandti (Nehring, 1898). Zool. Anz., 21:331.

COMMON NAME: Brandt's Hamster.

TYPE LOCALITY: Georgia, near Tbilisi.

DISTRIBUTION: Anatolian Turkey and Black Sea Mtns in N Turkey, east into the Caucasus (Armenia, Georgia, Azerbaijan; Gromov and Erbajeva, 1995), and south to NW Iran (Lay, 1967), N Iraq, Syria, Lebanon, and N Israel (Qumsiyeh, 1996, as *auratus*).

STATUS: IUCN – Lower Risk (lc).

SYNONYMS: *koenigi* Nehring, 1898.

COMMENTS: Corbet (1978c) reservedly included *brandti* in *M. auratus*, following the traditional view (Ellerman, 1941; Ellerman and Morrison-Scott, 1951). Hamer and Schutowa (1965) earlier established the specific integrity of *M. brandti* and Lyman and O'Brien (1977) exhaustively revised the species; additional chromosomal data provided by Fang and Jagiello (1992). Population and range in C Anatolian Turkey documented by Spitzenberger (1972) and summarized by Kryštufek and Vohralík

(2001). Yiğit et al. (2000*b*) broadly inventoried *M. brandti* in Turkey and contrasted its morphology and karyotype with those of *M. auratus* in SE Turkey (see above account).

Mesocricetus newtoni (Nehring, 1898). Zool. Anz., 21:329.
COMMON NAME: Romanian Hamster.
TYPE LOCALITY: Bulgaria, Kolarovgrad (= Schumla or Shumen).
DISTRIBUTION: Restricted to right side of Danube River in SE Romania and N Bulgaria.
STATUS: IUCN – Vulnerable.
COMMENTS: Systematic summaries and comparisons with other species are provided by Niethammer (1982*c*) and Mitchell-Jones et al. (1999); 2n = 38 (Raicu and Bratosin, 1965). Hosey (1982) postulated origin of this species around 20,000-10,000 years before present, resulting from dispersion of Near East *Mesocricetus* across a Bosphorus land bridge and subsequent isolation due to post-Pleistocene sea level increase.

Mesocricetus raddei (Nehring, 1894). Zool. Anz., 18:148.
COMMON NAME: Ciscaucasian Hamster.
TYPE LOCALITY: Russia, N Caucasus, Dagestan, Samur River.
DISTRIBUTION: Russia, steppes along N slopes of Caucasus from Dagestan to Don River and Sea of Azov.
STATUS: IUCN – Lower Risk (lc).
SYNONYMS: *avaricus* Ognev and Heptner, 1927; *nigricans* (Brandt, 1832) [not Lacépède, 1799]; *nigriculus* Nehring, 1898.
COMMENTS: Convincingly separated as a species on morphological and chromosomal grounds by Hamer and Schutowa (1965).

Phodopus Miller, 1910. Smithson. Misc. Coll., 52:498.
TYPE SPECIES: *Cricetulus bedfordiae* Thomas, 1908 (= *Cricetulus roborovskii* Satunin, 1903).
SYNONYMS: *Cricetiscus* Thomas, 1917.
COMMENTS: Chromosomal data reported by Spyropoulos et al. (1982) and Schmid et al. (1986). Fossil documentation only for the Pleistocene of Eurasia (McKenna and Bell, 1997).

Phodopus campbelli (Thomas, 1905). Ann. Mag. Nat. Hist., ser. 7, 15:322.
COMMON NAME: Campbell's Desert Hamster.
TYPE LOCALITY: NE Mongolia, Shaborte (see Pavlinov and Rossolimo, 1987, for additional comments).
DISTRIBUTION: Transbaikalia in Russia, Mongolia, and adjacent China (W Xinjiang, N Hebei, and N Nei Mongol).
STATUS: IUCN – Lower Risk (lc).
SYNONYMS: *crepidatus* Hollister, 1912; *tuvinicus* Orlov and Iskharova, 1974 [see Pavlinov and Rossolimo, 1987:171].
COMMENTS: Included in *P. sungorus* by some researchers (G. M. Allen, 1940; Corbet, 1978*c*; Ellerman, 1941; Ellerman and Morrison-Scott, 1951; Gromov and Erbajeva, 1995; Wang, 2003; Zhang et al., 1997), but regarded as distinct by others (Musser and Carleton, 1993; Pavlinov and Rossolimo, 1987, 1998; Pavlinov et al., 1995*a*). Separation of *campbelli* is supported by chromosomal differences (Safronova et al., 1992).

Phodopus roborovskii (Satunin, 1903). Ann. Zool. Mus. St. Petersbourg, 7:571.
COMMON NAME: Roborovski's Desert Hamster.
TYPE LOCALITY: China, Nan Shan Mtns, upper part of Shargol Dzhin River.
DISTRIBUTION: Tuva (Russia) and E Kazakhstan; W and S Mongolia; adjacent regions of China from NW Xinjiang east through N Gansu, N Qinghai, Ningxia, N Shaanxi, N Shanxi, and Nei Mongolia to Liaoning and Jilin (Ma et al., 1987; Qin, 1991; Wang, 2003; Zhang et al., 1997).
STATUS: IUCN – Lower Risk (lc).
SYNONYMS: *bedfordiae* (Thomas, 1908); *praedilectus* Mori, 1930 [see Kaneko and Maeda, 2002]; *przewalskii* Vorontsov and Kriukova, 1969.
COMMENTS: The status of *przewalskii* is discussed by Corbet (1978*c*) and Pavlinov and Rossolimo (1987). G. M. Allen (1940) treated *bedfordiae* as a separate species, but most

have included it in *P. roborovskii* (Corbet, 1978*c*; Ellerman, 1941; Ellerman and Morrison-Scott, 1951; Musser and Carleton, 1993; Pavlinov and Rossolimo, 1987; Pavlinov et al., 1995*a*).

Phodopus sungorus (Pallas, 1773). Reise Prov. Russ. Reichs., 2:703.
    COMMON NAME: Striped Desert Hamster.
    TYPE LOCALITY: E Kazakhstan, 100 km west of Semipalatinsk, near Grachevsk.
    DISTRIBUTION: E Kazakhstan and SW Siberia.
    STATUS: IUCN – Lower Risk (lc).
    COMMENTS: Safronova et al. (1992) referenced chromosomal data to justify the separation of *campbelli* from *P. sungorus*.

Tscherskia Ognev, 1914. Moskva Dnev. Zool. otd. obsc. liub. jest., 2:102.
    TYPE SPECIES: *Tscherskia albipes* Ognev, 1914 (= *Cricetulus triton* de Winton, 1899).
    SYNONYMS: *Asiocricetus* Kishida, 1929.
    COMMENTS: Although many have included *Tscherskia* in *Cricetulus* (G. M. Allen, 1940; Corbet, 1978*c*; Corbet and Hill, 1992; Ellerman, 1941; Ellerman and Morrison-Scott, 1951; Zhang et al., 1997), its distinctive morphology suggests distant relationship to the dwarf hamsters and generic segregation (Carleton and Musser, 1984; Gromov and Erbajeva, 1995; Musser and Carleton, 1993; Pavlinov and Rossolimo, 1987; Pavlinov et al., 1995*a*). Oldest records date to Late Pliocene of Europe (McKenna and Bell, 1997).

Tscherskia triton (de Winton, 1899). Proc. Zool. Soc. Lond., 1899:575.
    COMMON NAME: Greater Long-tailed Hamster.
    TYPE LOCALITY: China, N Shantung.
    DISTRIBUTION: Upper Ussuri, Russia; NE China from Heilongjiang and Nei Mongol southeast through Jilin, Liaoning, Hebei, Shandong, Henan, and Anhui (Liu et al., 1985) and west through Shanxi to Shaanxi (north and south of Qinling Mtns) (Wang, 2003; Zhang et al., 1997); also Korean Peninsula (Won and Smith, 1999).
    STATUS: IUCN – Lower Risk (lc).
    SYNONYMS: *albipes* Ognev, 1914; *arenosus* (Mori, 1939) [see Kaneko and Maeda, 2002]; *bampensis* (Kishida, 1929) [see Kaneko and Maeda, 2002]; *collinus* (G. M. Allen, 1925); *fuscipes* (G. M. Allen, 1925); *incanus* (Thomas, 1908); *meihsienensis* (Ho, 1935); *nestor* (Thomas, 1907); *ningshaanensis* Song, 1985; *yamashinai* (Kishida, 1929) [see Kaneko and Maeda, 2002].
    COMMENTS: G. M. Allen (1940) thoroughly redescribed the species, but whether one or more species are represented among the named forms remains to be resolved. Two sibling chromosomal species have been recognized (*albipes* and *triton*) but soon after refuted based on additional chromosomal data (see Corbet, 1984, and references therein). Song (1985) proposed *ningshaanensis* for a sample of *T. triton* from Shaanxi, but Wang (2003) and Zhang et al. (1997) listed it as a subspecies of *Cansumys canus* (see that account). Karyotypes and B chromosomes from several Chinese samples described by J.-W. Wang et al. (1999). A related fossil species, *T. rusa*, has been described from Holocene material in NW Iran (Storch, 1974), far outside the range of extant *Tscherskia*.

Subfamily Lophiomyinae Milne-Edwards, 1867. Arch. Mus. Hist. Nat. Paris. Memoires C, III, p. 81-116.
    SYNONYMS: Lophiomides Milne-Edwards, 1867 (Lophiomyidae Gill, 1872; Lophiomyoidea Gill, 1872; Lophiomyinae Thomas, 1896).
    COMMENTS: See Carleton and Musser (1984) for diagnosis, general characterisitics, habits, and habitat. The unique morphological adaptations of *Lophiomys* have always been recognized at either the family or subfamily level. Past estimates of phylogenetic relationships, or rather uncertainty, were expressed by arranging *Lophiomys* in its own family (Alston, 1876; Ellerman, 1941; Gill, 1872; Reig, 1981; Tullberg, 1899) or in a subfamily of either Muridae (Carleton and Musser, 1984; Musser and Carleton, 1993; Thomas, 1896; Weber, 1904; Winge, 1924), Cricetidae (G. M. Allen, 1939; Chaline et al., 1977; Corbet, 1978*c*; Miller and Gidley, 1918; Simpson, 1945), or Nesomyidae (Lavocat, 1973).
        Lavocat (1973) regarded *Lophiomys* as a possible derivative from the Miocene Af-

rocricetodontinae. Wahlert (1984) postulated a close phylogenetic alliance, based upon dental morphology, between *Lophiomys* and *Cricetops dormitor* from the early Oligocene of Mongolia and Kazakhstan, allocated both to Lophiomyinae, and considered the subfamily closely related to Cricetinae. Close relationship between *Lophiomys* and *Cricetops* has been disproven (Carrasco and Wahlert, 1999), and *Cricetops* and the early Miocene *Enginia* from Turkey are now placed in the Cricetopsinae (McKenna and Bell, 1997), excluding *Lophiomys*. The independent acquisition of dental similarities between extant *Lophiomys* and *Cricetops* is indicated by lophiomyine fossils from Spain and Morocco. *Protolophiomys ibericus*, late Miocene of S Spain (Aguilar and Thaler, 1987), combines some primitive cranial and dental states with the peculiar pebble-textured cranium like extant *Lophiomys*, and this combination also characterizes *Lophiomys maroccanus* from the Pliocene of Morocco (Aguilar and Michaux, 1989-1990). Moroccan studies have also yielded isolated molars identified as *Lophiomys* sp. from Miocene-Pliocene (Geraads, 1998) and Pliocene-Pleistocene sediments (Aguilar and Michaux, 1989-1990).

Although extant *Lophiomys* have complex molars (Wahlert, 1984), they are less so in the extinct forms, which resemble Miocene cricetids such as *Megacricetodon* and *Democricetodon*. The latter first appears in early Miocene sediments of Europe, is among the most common cricetids uncovered in the middle Miocene, and is regarded as the ancestor for many later generic lineages (Kälin, 1999). *Democricetodon* has also been documented in the middle Miocene of East Africa, and Tong and Jaeger (1993) speculated that it represents the African ancestor of *Lophiomys*. Whether derived from an African or European ancestral stock, available dental evidence links *Lophiomys* with cricetids, not nesomyids, possibly originating from an ancestor resembling *Democricetodon* (Carleton and Musser, 1984; Tong and Jaeger, 1993). A test of this hypothesis with broader cladistic analyses of other information bases is welcomed.

Topachevskii and Skorik (1984) described *Microlophiomys*, based upon isolated first upper and lower molars recovered from late Miocene sediments in Ukraine, as a member of Lophiomyinae. Its elaborate occlusal surfaces recall those of extant *Lophiomys*, but no other evidence so far confirms its allocation to the subfamily. The M1 (but not m1) also resembles that of extant *Mesocricetus*, and considering the spectacular parallelism in dental patterns among cricetids, one could interpret *Microlophiomys* as a derived form more closely related to Eurasian cricetines.

Lophiomys Milne-Edwards, 1867. L'Institut, Paris, 35:46.
    TYPE SPECIES: *Lophiomys imhausii* Milne-Edwards, 1867.
    SYNONYMS: *Phractomys* Peters, 1867; *Phragmomys* Peters, 1867.
    COMMENTS: Closest relatives are the fossil taxa noted in subfamily account.

Lophiomys imhausi Milne-Edwards, 1867. L'Institut, Paris, 35:46.
    COMMON NAME: Maned Rat.
    TYPE LOCALITY: Somalia ("Probably from African coast opposite Aden, where it was purchased," G. M. Allen, 1939:315; also see Thomas, 1910c:222).
    DISTRIBUTION: E Sudan, Eritrea, Ethiopia, Djibouti, Somalia, Kenya, NE Uganda, and W Tanzania (Kock and Künzel, 1999); sea level to 3300 m in Ethiopia (Yalden et al., 1976), in lowland forests in Djibouti (Pearch et al., 2001), but apparently restricted to mountain forest in Kenya and Uganda (Delany, 1975; Clausnitzer and Kityo, 2001; Hollister, 1919).
    STATUS: IUCN – Lower Risk (lc).
    SYNONYMS: *aethiopicus* (Peters, 1867); *bozasi* Oustalet, 1902; *hindei* Thomas, 1910; *ibeanus* Thomas, 1910; *smithi* Rhoads, 1896; *testudo* Thomas, 1905; *thomasi* Heller, 1912.
    COMMENTS: Thomas (1910c) recognized four species of *Lophiomys*, but Ellerman (1940:636) noted that "Thomas evidently came to the conclusion that all the East African 'species' were one, as there is a note in his tracts to this effect. I am inclined to go further and think that until more material comes to hand all forms must be treated as races of the earliest name *imhausi*." G. M. Allen (1939:315) independently noted that "All the recognizable forms are doubtless races of *L. imhausi*." Ellerman's view prevails today and has yet to be tested by careful taxonomic revision. Distribution in the isolated Harenna Forest, S Ethiopia, documented by Lavrenchenko (2000); range and habitat

in Djibouti reported by Pearch et al. (2001); Kock and Künzel (1999) included a detailed map and list of all collection localities covering known range of the species, which is disjunct. No records exist in the Danakil Desert between Djibouti-Sudan and Ethiopia, or in the expansive arid region between Ethiopia and Kenya; whether these gaps are real or an artifact of collection inactivity is unknown. Dor's (1966) record of cranial fragments, dated to the 11th century from the Judean desert near the Dead Sea, may represent an import, and references to a range in Arabia (Kingdon, 1990) are unsubstantiated by recent collections (Kock and Künzel, 1999).

Subfamily Neotominae Merriam, 1894. Proc. Philadelphia Acad. Sci, 1894:228.

SYNONYMS: Baiomyini new tribe (see *Baiomys* account); Neotominae Merriam 1894 (Neotomini Vorontsov, 1959); Ochrotomyini new tribe (see *Ochrotomys* account); Onychomyini Vorontsov, 1959; Peromyscini Hershkovitz, 1966*b*; Reithrodontomyini Vorontsov, 1959.

COMMENTS: Merriam's (1894) definition of the subfamily included North American woodrats and certain South American fossils with high-crowned molars (*Ptyssophorus* and *Tretomys*, now considered synonyms of *Reithrodon* by Pardiñas, 2000*a*). Usage as a formal subfamily was observed (e.g., Miller and Rehn, 1901; Miller, 1912*b*) until Miller and Gidley (1918) considered the genera to be members of a diverse Cricetinae, as did Ellerman (1940) and Simpson (1945). A broadened family-group concept reemerged in an informal way as the "neotomine-peromyscines" (Hooper, 1960; Hooper and Musser, 1964*a*; Carleton, 1980) and was eventually nomenclaturally recognized as distinct from sigmodontines, whether as a tribe (Hershkovitz, 1966*b*, as Peromyscini) or subfamily (Reig, 1980, 1981, as Neotominae). Phylogenetic diagnosis and cladistic demonstration of neotomine monophyly remain ambiguous based on taxonomically broad surveys of morphological traits (Carleton, 1973, 1980; Steppan, 1995; Voss and Linzey, 1981) or cytochrome *b* data (D'Elía et al., 2003), but not other mitochondrial genes (Engel et al., 1998) or mitochondrial and nuclear genes in combination (D'Elía, 2003). Coupled with this uncertainty is that of the sister-group relationship between neotomines and sigmodontines (see D'Elía, 2000, for review), as assumed in evolutionary narratives (e.g., Hershkovitz, 1966*b*; Marshall, 1979; Patterson and Pascual, 1972) and early interpretations of phylogeny (Hooper and Musser, 1964*a*). Other cognate possibilities, such as arvicolines and Old World cricetines, are variously implicated in phylogenetic studies of morphology (Carleton, 1980), of DNA-DNA hybridization (Catzeflis et al., 1993), and of mitochondrial and nuclear DNA sequences (D'Elía, 2003; D'Elía et al., 2003; Engel et al., 1998; Michaux et al., 2001*b*).

The ancestry of extant neotomines has been loosely connected to *Copemys* (Jacobs and Lindsay, 1984; Slaughter and Ubelaker, 1984), a middle Miocene-early Pliocene North American cricetid that has been variously invoked as progenitor of *Peromyscus* (Lindsay, 1972), *Onychomys* (Jacobs, 1977*b*), and *Bensonomys* (Baskin, 1978). The morphological limits and specific contents of *Copemys* are poorly understood, however, and its evolutionary relationships and biogeographic origin have been subject to several, sometimes contradictory, interpretations (see commentary and references in Carleton and Musser [1984:306-308], Baskin [1986:296], and Korth [1994:231-232]). Examples of extant genera are known from the late Miocene (*Neotoma* and *Peromyscus*), and others appear in early Pliocene strata (*Baiomys, Onychomys, Reithrodontomys*) (Carleton and Eshelman, 1979; Korth, 1994; Packard, 1960; Zakrzewski, 1993).

Periodic synopses of systematic understanding and distributions provided by Miller and Rehn (1901), Miller (1924), Hall and Kelson (1959), and Hall (1981). Recent faunal treatises update and summarize, in varying detail, aspects of natural history, distribution, biogeography, and specific and-or infraspecific taxonomy: North America (Baker et al., 2003*b*; Jones et al., 1997; Wilson and Ruff, 1999) and the E USA (Whitaker and Hamilton, 1998); Honduras (Marineros and Gallegos, 1998) and Central America (Reid, 1997); México (Ceballos et al., 2002*a*; Ramírez-Pulido et al., 1996), NW México (Alvarez-Castañeda and Cortés-Calva, 1999), and the states of Baja California (Hafner and Riddle, 1997), Chiapas (Espinoza M. et al., 1999*a, b*), Jalisco (Guerrero Vázquez et al., 1995), México (Ramírez-Pulido et al., 1995), Morelos (Alvarez-Castañeda, 1996; Alvarez et al. 1998), Sonora (Caire, 1997), and Quintana Roo (Pozo de la Tijera and Escobedo Cabrera, 1999).

Baiomys True, 1894. Proc. U. S. Natl. Mus. (1893), 16:758 [1894].

TYPE SPECIES: *Hesperomys taylori* Thomas, 1887.

COMMENTS: Baiomyini new tribe. Type genus—*Baiomys* True, 1894. Definition—small terrestrial muroid rodents with tail shorter than head-and-body (Hooper, 1972; Osgood, 1909; Packard, 1960); hindfoot narrow with digits I and V short relative to II-IV, plantar pads 1 and 4 proximally positioned compared with 2-3 (Carleton, 1980; Osgood, 1909); cranium small but stoutly built, rostrum short, interorbit and braincase lacking ridges; interparietal compressed, wedge-shaped, not contacting squamosal; carotid circulation derived (character state 3 per Carleton, 1980); tegmen tympani adnate to squamosal (Voss, 1993), alisphenoid strut present; M3 cylindrical, less than -½ size of M2; vertebral column with 13 thoracic and 6 lumbar vertebrae, 1st rib articulating 7th cervical and 1st thoracic, humerus lacking entepicondylar foramen (Carleton, 1980); trochlear process of calcaneum distal (Carleton, 1980); two complete and five incomplete transverse palatal ridges (Carleton, 1980); stomach unilocular-hemiglandular, caecum a short simple sac, gall bladder present (Carleton, 1973, 1980); ampullary glands simple and preputial glands large (Arata, 1964; Carleton et al., 1975); baculum and phallus short and broad, with a shallow crater and terminal urinary meatus, dorsal and ventral lappets absent (Blair, 1942; Hooper, 1960); voice a staccato trill, males with prominent midventral glands (Blair, 1941; Hooper and Carleton, 1976). Contents—*Baiomys* True, 1894; *Scotinomys* Thomas, 1913.

Relegated to a subgenus of *Peromyscus* by Osgood (1909) but reinstated as a separate genus by Miller (1912b), a status reinforced by Blair (1942) and Hooper (1958). Relationship to South American forms (i.e., *Calomys*) advanced by Packard (1960), but other studies have disclosed cognate affinity to *Scotinomys* (Carleton, 1980; Carleton et al., 1975; D'Elía, 2003; Engel et al., 1998; Hooper, 1960; Hooper and Musser, 1964a). Revised by Packard (1960), who delineated subspecific ranges and also summarized fossil taxa (early Pliocene through Pleistocene) within a phylogenetic context. Karyology evaluated by Yates et al. (1979) and biochemical variation by Calhoun et al. (1989). Hershkovitz (1962) transferred the South American form *hummelincki*, erroneously described under *Baiomys*, to the phyllotine genus *Calomys*.

Baiomys musculus (Merriam, 1892). Proc. Biol. Soc. Wash., 7:170.

COMMON NAME: Southern Pygmy Mouse.

TYPE LOCALITY: México, Colima, Colima.

DISTRIBUTION: SW Nayarit and C Veracruz, México, to NW Nicaragua, excluding Yucatán Peninsula and Caribbean tropical lowlands.

STATUS: IUCN – Lower Risk (lc).

SYNONYMS: *brunneus* (J. A. Allen and Chapman, 1897); *grisescens* Goldman, 1932; *handleyi* Packard, 1958; *infernatis* Hooper, 1952; *nebulosus* Goodwin, 1959; *nigrescens* (Osgood, 1904); *pallidus* Russell, 1952; *pullus* Packard, 1958.

COMMENTS: See Packard and Montgomery (1978, Mammalian Species, 102).

Baiomys taylori (Thomas, 1887). Ann. Mag. Nat. Hist., ser. 5, 19:66.

COMMON NAME: Northern Pygmy Mouse.

TYPE LOCALITY: USA, Texas, Duval Co., San Diego.

DISTRIBUTION: SE Arizona and SW New Mexico (see Stuart and Scott, 1992), SW Oklahoma (see Tumlison et al., 1993), and E Texas (see Roberts et al., 1997), USA, south to Michoacán, C Hidalgo, and C Veracruz, México.

STATUS: IUCN – Lower Risk (lc).

SYNONYMS: *allex* (Osgood, 1904); *analogus* (Osgood, 1909); *ater* Blossom and Burt, 1942; *canutus* Packard, 1960; *fuliginatus* Packard, 1960; *paulus* (J. A. Allen, 1903); *subater* (Bailey, 1905).

COMMENTS: Range expanding northwardly and westwardly in Texas and Oklahoma as documented by recent collections (L. Choate et al., 1990; L. Choate and Jones, 1998; Roberts et al., 1997; Tumlison et al., 1993). See Eshelman and Cameron (1987, Mammalian Species, 285).

Habromys Hooper and Musser, 1964. Occas. Pap. Mus. Zool., Univ. Michigan, 635:12.
  TYPE SPECIES: *Peromyscus lepturus* Merriam, 1898.
  COMMENTS: Reithrodontomyini. Genus circumscribes species originally placed with the *Peromyscus mexicanus* complex, subgenus *Peromyscus* (Osgood, 1909). Hooper and Musser (1964*b*) acknowledged their distinctive morphology as the subgenus *Habromys* of *Peromyscus* (and Hooper, 1968), a clade that Carleton (1980, 1989) viewed at the generic level. Systematic evidence weakly supports common ancestry of *Habromys* with *Neotomodon* and/or *Podomys* (Carleton, 1980; Hooper and Musser, 1964*b*; Stangl and Baker, 1984*b*). Male reproductive tract examined by Hooper (1958), Hooper and Musser (1964*b*), and Linzey and Layne (1969, 1974). Species definitions revised and vouchered distributions summarized by Carleton et al. (2002).

Habromys chinanteco (Robertson and Musser, 1976). Occas. Pap. Mus. Nat. Hist., Univ. Kansas, 47:1.
  COMMON NAME: Chinanteco Deermouse.
  TYPE LOCALITY: México, Oaxaca, north slope Cerro Pelón, 31.6 km S Vista Hermosa, 2650 m.
  DISTRIBUTION: The type locality and its vicinity on gulf-facing slopes, 2080-2650 m, of the Sierra de Juárez, NC Oaxaca, México.
  STATUS: IUCN – Lower Risk (lc).
  COMMENTS: A small-bodied species morphologically similar to *H. simulatus* and sympatric with *H. lepturus* (see Robertson and Musser, 1976). Possible status as a junior synonym of *H. simulatus* needs resolution (see Carleton et al., 2002).

Habromys delicatulus Carleton, Sánchez, and Urbano Vidales, 2002. Proc. Bio. Soc. Wash., 115:491.
  COMMON NAME: Delicate Deermouse.
  TYPE LOCALITY: México, México, Municipio Jilotepec, Dexcaní Alto, 2 km E and 3.5 km S Jilotepec, Cañada de la Ermita, 2570 m; 19°56′N, 99°30′W.
  DISTRIBUTION: Known only from the type locality.
  COMMENTS: Smallest species of *Habromys* currently known.

Habromys ixtlani (Goodwin, 1964). Am. Mus. Novit., 2183:3.
  COMMON NAME: Ixtlán Deermouse.
  TYPE LOCALITY: México, Oaxaca, District of Ixtlán, Cerro Machín, 5 mi (8 km) NE Macuiltianguis, 9000 ft (2743 m).
  DISTRIBUTION: Upper elevations of Sierra de Juárez, 2350-3150 m, NC Oaxaca, México.
  COMMENTS: Originally described as a species but regarded as a well marked subspecies of *H. lepturus* by Musser (1969). Morphological differentiation from *H. lepturus* and other *Habromys* reevaluated by Carleton et al. (2002), who returned *ixtlani* to specific rank.

Habromys lepturus (Merriam, 1898). Proc. Biol. Soc. Wash., 12:118.
  COMMON NAME: Zempoaltepec Deermouse.
  TYPE LOCALITY: México, Oaxaca, Cerro Zempoaltépec, 8200 ft (2500 m).
  DISTRIBUTION: Humid montane forest on Sierra de Zempoaltépec, 2500-3000 m, NC Oaxaca, México.
  STATUS: IUCN – Lower Risk (nt).
  COMMENTS: Formerly included *ixtlani* as a subspecies (Musser, 1969).

Habromys lophurus (Osgood, 1904). Proc. Biol. Soc. Wash., 17:72.
  COMMON NAME: Crested-tailed Deermouse.
  TYPE LOCALITY: Guatemala, Huehuetenango Dept., Todos Santos, 10,000 ft (3048 m).
  DISTRIBUTION: Highlands, 1950-3110 m, of Chiapas, México, C Guatemala, and NW El Salvador.
  STATUS: IUCN – Lower Risk (lc).
  COMMENTS: Although Robertson and Musser (1976) noted much diversity among their samples of *H. lophurus*, Carleton et al. (2002) interpreted their range of variation as consonant with recognition of a single species.

Habromys simulatus (Osgood, 1904). Proc. Biol. Soc. Wash., 17:72.
  COMMON NAME: Jico Deermouse.
  TYPE LOCALITY: México, Veracruz, near Jico, 6000 ft (1830 m).

DISTRIBUTION: Eastern middle slopes of Sierra Madre Oriental, 1830-2200 m, from S Hidalgo and C Veracruz to extreme NW Oaxaca, México.

STATUS: IUCN – Endangered.

COMMENTS: Distribution summarized by Carleton et al. (2002) but enhancement of range limits needed. For example, the status of the form reported as *Peromyscus* affinity *simulatus* from the Sierra de Taxco, N Guerrero (León Paniagua and Romo Vázquez, 1993), needs clarification; the presence of *H. simulatus* proper in the Cordillera Transvolcanica seems zoogeographically implausible.

Hodomys Merriam, 1894. Proc. Acad. Nat. Sci. Philadelphia, 46:232.

TYPE SPECIES: *Neotoma alleni* Merriam, 1892.

COMMENTS: Neotomini. Maintained as a genus (e.g., Goldman, 1910; Ellerman, 1941) until arranged as a subgenus of *Neotoma* by Burt and Barkalow (1942). Carleton (1980) expressed relationships and morphological differentiation of *Hodomys* at the generic rank, perhaps closely related to *Xenomys* (also see Carleton, 1973; Hooper, 1960; and Schaldach, 1960); parsimony and likelihood evaluations of cytochrome *b* data also support this rank and interpretation of kinship (Edwards and Bradley, 2002*b*).

Hodomys alleni (Merriam, 1892). Proc. Biol. Soc. Wash., 7:168.

COMMON NAME: Allen's Woodrat.

TYPE LOCALITY: México, Colima, Manzanillo.

DISTRIBUTION: S Sinaloa to Oaxaca; interior México along basins of Río Balsas to C Puebla and Río Tehuacán to N Oaxaca.

STATUS: IUCN – Lower Risk (nt).

SYNONYMS: *elattura* Osgood, 1904; *guerrerensis* Goldman, 1938; *vetulus* Merriam, 1894.

COMMENTS: Kelson (1952) placed *vetulus* as a subspecies of *N. alleni*. See Genoways and Birney, 1974 (Mammalian Species, 41, as *Neotoma alleni*), who first reported the karyotype (2n=48).

Isthmomys Hooper and Musser, 1964. Occas. Pap. Mus. Zool., Univ. Michigan, 635:12.

TYPE SPECIES: *Megadontomys flavidus* Bangs, 1902.

COMMENTS: Reithrodontomyini. Species associated here were originally classified in *Megadontomys*, used either as a genus or as a subgenus of *Peromyscus* (Osgood, 1909). *Isthmomys* was later diagnosed as a subgenus of *Peromyscus* (Hooper and Musser, 1964*b*) and maintained at this rank by Hooper (1968); diagnosis emended and accorded generic status by Carleton (1980, 1989). Sister-group relationship with *Megadontomys* proposed by Carleton (1980) but unsupported by chromosomal banding data (Stangl and Baker, 1984*b*). Aspects of morphology considered by Carleton (1973, 1980), Hooper and Musser (1964*b*), and Linzey and Layne (1969, 1974); karyology by Stangl and Baker (1984*b*).

Isthmomys flavidus (Bangs, 1902). Bull. Mus. Comp. Zool., 9:27.

COMMON NAME: Yellow Deermouse.

TYPE LOCALITY: Panamá, Chiriquí Prov., Volcán de Chiriquí, Boquete, 4000 ft (1219 m).

DISTRIBUTION: Intermediate elevations in W Panamá (Chiriquí region) and on Azuero Peninsula (see Handley, 1966*a*).

STATUS: IUCN – Lower Risk (lc).

Isthmomys pirrensis (Goldman, 1912). Smithson. Misc. Coll., 60(2):5.

COMMON NAME: Mt Pirri Deermouse.

TYPE LOCALITY: Panamá, Darién Prov., Mt Pirri, headwaters of Río Limón, 4500 ft (1372 m).

DISTRIBUTION: Easternmost Panamá and Serranía del Darién, extreme NW Colombia (see Alberico et al., 2000).

STATUS: IUCN – Lower Risk (nt).

COMMENTS: Whether *I. pirrensis* is a junior synonym of *I. flavidus*, as opined by Hooper (1968), has yet to be assessed.

Megadontomys Merriam, 1898. Proc. Biol. Soc. Wash., 12:115.

TYPE SPECIES: *Peromyscus thomasi* Merriam, 1898.

COMMENTS: Reithrodontomyini. Used variably as a genus until Osgood (1909) stabilized its taxonomic ranking as a subgenus of *Peromyscus*, and so followed by Hooper and Musser (1964*b*) and Hooper (1968). Carleton (1980, 1989) viewed the relationships and differentiation of *Megadontomys* at the generic level (but see Rogers, 1983). Aspects of morphology studied by Carleton (1973, 1980), Hooper and Musser (1964*b*), and Linzey and Layne (1969, 1974). Karyological affinities evaluated by Rogers (1983), Rogers et al. (1984), and Stangl and Baker (1984). Werbitsky and Kilpatrick (1987) reported relatively low levels of genetic similarity among the nominal forms.

Megadontomys cryophilus (Musser, 1964). Occas. Pap. Mus. Zool., Univ. Michigan, 636:13.
    COMMON NAME: Oaxacan Big-toothed Deermouse.
    TYPE LOCALITY: México, Oaxaca, Distrito de Ixtlán, 13 mi (21 km) NE Llano de las Flores, south slope Cerro Pelón, 9200 ft (2804 m).
    DISTRIBUTION: Cloud and pine-oak forests in highlands of N Oaxaca, México.
    STATUS: IUCN – Lower Risk (nt).
    COMMENTS: Named as a subspecies of *Peromyscus thomasi*. Citing morphological traits and genetic differentiation reported by Werbitsky and Kilpatrick (1987), Carleton (1989) arranged *cryophilus* as a species. Recent elevational surveys by Briones-Salas et al. (2001) in the Sierra Mazteza and Sánchez-Cordero (2001) in the Sierra Mixteca amplify the geographic occurrence of this species in Oaxaca.

Megadontomys nelsoni (Merriam, 1898). Proc. Biol. Soc. Wash., 12:116.
    COMMON NAME: Nelson's Big-toothed Deermouse.
    TYPE LOCALITY: México, Veracruz, Jico, 6000 ft (1830 m).
    DISTRIBUTION: E slopes of Sierra Madre Oriental, from SE Hidalgo to C Veracruz, México.
    STATUS: IUCN – Lower Risk (lc).
    COMMENTS: Relegated to a subspecies of *Peromyscus thomasi* by Musser (1964); reinstated to species rank by Carleton (1989).

Megadontomys thomasi (Merriam, 1898). Proc. Biol. Soc. Wash., 12:116.
    COMMON NAME: Thomas's Big-toothed Deermouse.
    TYPE LOCALITY: México, Guerrero, mountains near Chilpancingo, 9700 ft (2957 m).
    DISTRIBUTION: High Sierra Madre del Sur of Guerrero, México.
    STATUS: IUCN – Lower Risk (lc).
    COMMENTS: Formerly encompassed *cryophilus* and *nelsoni* as subspecies (Musser, 1964); see above accounts.

Nelsonia Merriam, 1897. Proc. Biol. Soc. Wash., 11:277.
    TYPE SPECIES: *Nelsonia neotomodon* Merriam, 1897.
    COMMENTS: Neotomini. Viewed as a basal member of clade including *Neotoma* and related genera (Carleton, 1980; Engel et al., 1998; Hooper, 1954, 1960). Phylogenetic significance of banded karyotype discussed by Engstrom and Bickham (1983). Revised initially by Hooper (1954), who recognized a single species, and later by Engstrom et al. (1992), who resurrected *N. goldmani* as distinct from *N. neotomodon*.

Nelsonia goldmani Merriam, 1903. Proc. Biol. Soc. Wash., 16:80.
    COMMON NAME: Goldman's Diminutive Woodrat.
    TYPE LOCALITY: México, Michoacán, Mt Tancítaro.
    DISTRIBUTION: Cordillera Transvolcanica, México, from Colima and S Jalisco eastwards through N Michoacán to N Estado de México.
    STATUS: IUCN – Lower Risk (lc).
    SYNONYMS: *cliftoni* Genoways and Jones, 1968.
    COMMENTS: Engstrom et al. (1992) retained *cliftoni* as a valid subspecies.

Nelsonia neotomodon Merriam, 1897. Proc. Biol. Soc. Wash., 11:278.
    COMMON NAME: Western Diminutive Woodrat.
    TYPE LOCALITY: México, Zacatecas, mountains near Plateado, 8200 ft (2500 m).
    DISTRIBUTION: Sierra Madre Occidental from S Durango to N Jalisco and Aguascalientes, México.
    STATUS: IUCN – Lower Risk (lc).

Neotoma Say and Ord, 1825. J. Acad. Nat. Sci. Philadelphia, 4:345.

TYPE SPECIES: *Mus floridana* Ord, 1818.

SYNONYMS: *Homodontomys* Goldman, 1910; *Parahodomys* Gidley and Gazin, 1933; *Parneotoma* Hibbard, 1967; *Teanopus* Merriam, 1903; *Teonoma* Gray, 1843.

COMMENTS: Neotomini. Phylogenetic relationships of the genus considered by Hooper and Musser (1964*a*), Carleton (1980), and Edwards and Bradley (2002*b*). Anatomical systems described by Arata (1964), Burt and Barkalow (1942), Carleton (1973, 1980), Hooper (1960), and Howell (1926); fossil taxa (Miocene-Recent) and trends in dental evolution reviewed by Zakrzewski (1993). Karyotypic variation and evolution assessed by Mascarello and Hsu (1976) and Koop et al. (1985); multispecific surveys of molecular variation and its systematic implications covered by Planz et al. (1996), Edwards and Bradley (2001, 2002*a*, *b*), and Edwards et al. (2001), especially for temperate forms.

Revised by Goldman (1910), then including only *Homodontomys*, *Teonoma*, and the nominate subgenus. Burt and Barkalow (1942) established the prevailing subgeneric framework (e.g., Hall, 1981), also relegating *Hodomys* and *Teanopus* to subgenera. Carleton (1973, 1980) reinstated *Hodomys* as a genus (see above account), an action supported by phylogenetic analysis of cytochrome *b* sequences (Edwards and Bradley, 2002*b*). Traditional species groups within the subgenus *Neotoma* (e.g., Burt and Barkalow, 1942; Goldman, 1910) are undergoing critical reassessment; see Birney (1976), Mascarello (1978), Planz et al. (1996), Edwards et al. (2001), and Edwards and Bradley (2002*a*, *b*) for evolving views on interspecific affinities.

Neotoma albigula Hartley, 1894. Proc. California Acad. Sci., Ser. 2, 4:157.

COMMON NAME: White-throated Woodrat.

TYPE LOCALITY: USA, Arizona, Pima Co., vicinity of Fort Lowell, near Tucson.

DISTRIBUTION: SW Colorado and W New Mexico west of the Río Grande to SE California, USA, south to N Sinaloa and S Chihuahua west of the Río Conchos, México, including islands in the Sea of Cortez (see Edwards, et al. 2001:Fig. 4).

STATUS: IUCN – Endangered as *N. varia*, Lower Risk (lc) as *N. albigula*.

SYNONYMS: *angusticeps* Merriam, 1894; *brevicauda* Durrant 1934; *cumulator* Mearns, 1897; *grandis* Elliot, 1904; *laplataensis* F. W. Miller, 1933; *mearnsi* Goldman, 1915; *melanura* Merriam, 1894; *seri* Townsend, 1912; *sheldoni* Goldman, 1915; *varia* Burt, 1932; *venusta* True, 1894.

COMMENTS: Subgenus *Neotoma*, *floridana* species group (*sensu* Edwards and Bradley, 2002*b*). Closely related to *N. floridana* and *N. micropus* (Birney, 1976; Hooper, 1960; Planz et al., 1996), the three considered semispecies by Zimmerman and Nejtek (1977). Formerly included populations recognized as *N. leucodon* (see below) based on DNA restriction-site (Planz et al., 1996) and gene-sequence (Edwards et al., 2001) investigations. Patterns of conventional morphometric variation (Rogers and Schmidly, 1981), however, do not intelligibly correspond to the specific limits as drawn by Edwards et al. (2001); denser geographic sampling across the river barriers identified by those authors is warranted to bolster evidence for specific separation and to refine distributional limits. Sister species to the *N. floridana*-*N. magister* clade based on parsimony and likelihood distillations of cytochrome *b* sequences (Edwards et al., 2001; Edwards and Bradley, 2002*b*).

Subspecific taxonomy updated by Hall and Genoways (1970), but continued designation of racial divisions is uninformative in light of recent specific changes. Attention should be devoted to the validity of synonyms here attributed following the study of Edwards et al. (2001). Includes *varia* from Isla Datil (Turner), Sea of Cortez, formerly treated as a species but considered inseparable from *N. albigula* by Bogan (1997), who retained the form as a subspecies. See Macêdo and Mares (1988, Mammalian Species, 310).

Neotoma angustapalata Baker, 1951. Univ. Kansas Mus. Nat. Hist. Misc. Publ., 5:217.

COMMON NAME: Tamaulipan Woodrat.

TYPE LOCALITY: México, Tamaulipas, 70 km (by highway) S Ciudad Victoria and 6 km W Panamerican Highway, El Carrizo.

DISTRIBUTION: SW Tamaulipas and adjacent San Luis Potosí, México.

STATUS: IUCN – Lower Risk (lc).

COMMENTS: Subgenus *Neotoma*, *mexicana* species group (*sensu* Edwards and Bradley, 2002*b*). Specific status maintained by Birney (1973) but level of relationship to *N. mexicana* or to *N. micropus* unclear.

Neotoma anthonyi J. A. Allen, 1898. Bull. Am. Mus. Nat. Hist., 10:151.
COMMON NAME: Anthony's Woodrat.
TYPE LOCALITY: México, Baja California Norte, Todos Santos Isl.
DISTRIBUTION: Known only from the type locality.
STATUS: IUCN – Endangered.
COMMENTS: Subgenus *Neotoma*, *lepida* species group (*sensu* Edwards and Bradley, 2002*b*). Listed as nominal species of *N. lepida* group by Goldman (1932); also see Mascarello (1978). Not collected since 1910; judged "extinct or very close to extinction" by Mellink (1992:139) probably due to the introduction of domestic cats. See Cortés-Calva et al. (2001, Mammalian Species, 663).

Neotoma bryanti Merriam, 1887. Am. Nat., 21:191.
COMMON NAME: Bryant's Woodrat.
TYPE LOCALITY: México, Baja California Norte, Cedros Isl.
DISTRIBUTION: Known only from the type locality.
STATUS: IUCN – Endangered.
COMMENTS: Subgenus *Neotoma*, *lepida* species group (*sensu* Edwards and Bradley, 2002*b*). Listed as nominal species of *N. lepida* group by Goldman (1932); also see Mascarello (1978). Genetically unremarkable compared with samples of *N. lepida* from Baja California and may represent the oldest name available for those populations from deserts of coastal California and the Baja peninsula (see Patton and Alvarez-Castañeda, 2005, and remarks under *N. lepida*). See Alvarez-Castañeda and Yensen (1999, Mammalian Species, 619).

Neotoma bunkeri Burt, 1932. Trans. San Diego Soc. Nat. Hist., 7:181.
COMMON NAME: Coronados Island Woodrat.
TYPE LOCALITY: México, Baja California Sur, Coronados Isl; 26°06′N, 111°18′W.
DISTRIBUTION: Known only from the type locality.
STATUS: IUCN – Endangered.
COMMENTS: Subgenus *Neotoma*, *lepida* species group (*sensu* Edwards and Bradley, 2002*b*). Likely conspecific with *N. lepida* according to Mascarello (1978). Not collected since 1932; judged "extinct for years or even decades" by Smith et al. (1993:152), probably due to a combination of habitat destruction and domestic cat predation (also see Alvarez-Castañeda and Ortega-Rubio, 2003).

Neotoma chrysomelas J. A. Allen, 1908. Bull. Am. Mus. Nat. Hist., 24:653.
COMMON NAME: Nicaraguan Woodrat.
TYPE LOCALITY: Nicaragua, Matagalpa Dept., Matagalpa.
DISTRIBUTION: NW Nicaragua, Honduras.
STATUS: IUCN – Lower Risk (lc).
COMMENTS: Subgenus *Neotoma*, *mexicana* species group (*sensu* Edwards and Bradley, 2002*b*). Hall (1981) suggested that *chrysomelas* is conspecific with *N. mexicana*, a proposal that should be considered apropos of a much needed revision of the latter.

Neotoma cinerea (Ord, 1815). *In* Guthrie, New Geogr., Hist., Comml., Grammar, Philadelphia, 2nd ed., 2:292.
COMMON NAME: Bushy-tailed Woodrat.
TYPE LOCALITY: USA, Montana, Cascade Co., Great Falls.
DISTRIBUTION: SE Yukon and westernmost Northwest Territories, south through Alaskan Panhandle, British Columbia and W Alberta, Canada; in W USA, from Washington to W Dakotas as far south as EC California, S Nevada, N Arizona, and NW New Mexico.
STATUS: IUCN – Lower Risk (lc).
SYNONYMS: *acraea* (Elliot, 1904); *alticola* Hooper, 1940; *apicalis* Elliot, 1903; *arizonae* Merriam, 1893; *cinnamomea* J. A. Allen, 1895; *columbiana* Elliot, 1899; *drummondii* (Richardson, 1828); *fusca* True, 1894; *grangeri* J. A. Allen, 1894; *lucida* Goldman, 1917;

macrodon Kelson, 1949; *occidentalis* Baird, 1855; *orolestes* Merriam, 1894; *pulla* Hooper, 1940; *rupicola* J. A. Allen, 1894; *saxamans* Osgood, 1900.

COMMENTS: Subgenus *Teonoma*. Type species of *Teonoma*, conventionally acknowledged as a distinctive subgenus of *Neotoma* (e.g., Burt and Barkalow, 1942; Goldman, 1910; Hall, 1981; Hooper, 1960). Cladistically basal to all other species of *Neotoma* surveyed by Planz et al. (1996) and Edwards and Bradley (2002b); the latter authors discussed the possible generic elevation of *Teonoma* but demurred pending cladistic examination of *N.* (*Teanopus*) *phenax*. Late Quaternary changes in body size as correlated with temperature fluctuations documented by Smith et al. (1995). See Smith (1997, Mammalian Species, 564).

Neotoma devia Goldman, 1927. Proc. Biol. Soc. Wash., 40:205.
COMMON NAME: Arizona Woodrat.
TYPE LOCALITY: USA, Arizona, Painted Desert, Tanner Tank, 5200 ft (1585 m).
DISTRIBUTION: W Arizona, USA, east and south of the Colorado River; NW Sonora, México.
STATUS: IUCN – Lower Risk (lc).
SYNONYMS: *aureotunicata* Huey, 1937; *auripila* Blossom, 1933; *bensoni* Blossom, 1933; *flava* Benson, 1935; *harteri* Huey, 1937.
COMMENTS: Subgenus *Neotoma*, *lepida* species group (*sensu* Edwards and Bradley, 2002b). Once ranked as a subspecies of *N. lepida*, from which several data sources support its specific-level divergence (Koop et al., 1985; Mascarello, 1978; Patton and Alvarez-Castañeda, 2005). But see Hoffmeister (1986), who disputed the specific status of *devia* based on his interpretation of morphological integration among Arizona populations. Unequivocal discrimination from *N. lepida*, verification of species-group synonyms, delimitation of geographic range, and delineation of subspecies, if defensible, all require further attention. Planz (1999) noted only *auripila* as synonym; others provisionally listed here conform to the geographic range as depicted by Riddle et al. (2000b:Fig. 1).

Neotoma floridana (Ord, 1818). Bull. Sci. Soc. Philom. Paris, 1818:181.
COMMON NAME: Eastern Woodrat.
TYPE LOCALITY: USA, Florida, Duval Co., St. Johns River, near Jacksonville.
DISTRIBUTION: SC and SE USA (see Monty et al., 1995, for range enhancement in Illinois), from EC Colorado to C Texas, eastwards to the Atlantic seaboard, from S North Carolina to peninsular Florida; isolated population on Florida Keys (*smalli*).
STATUS: U.S. ESA – Endangered as *N. f. smalli*; IUCN – Endangered *N. f. smalli*, Lower Risk (nt) as *N. f. baileyi* and *N. f. haematoreia*, otherwise Lower Risk (lc).
SYNONYMS: *attwateri* Mearns, 1897; *baileyi* Merriam, 1894; *campestris* J. A. Allen, 1894; *haematoreia* A. H. Howell, 1934; *illinoensis* A. H. Howell, 1910; *osagensis* Blair, 1939; *rubida* Bangs, 1898; *smalli* Sherman, 1955.
COMMENTS: Subgenus *Neotoma*, *floridana* species group (*sensu* Edwards and Bradley, 2002b). Hybridization with *N. micropus* possible but introgression along narrow contact zone judged insubstantial (Birney, 1973). See Planz et al. (1996) and Edwards et al. (2001) for studies of kinship and phylogeography based on DNA information. Whitaker and Hamilton (1998) implied that eastern populations (except *smalli*) are all synonymous with the nominate subspecies, a possibility consistent with the study of Edwards et al. (2001), who noted well-marked western and eastern clades among the subspecies sampled, coincident with the Mississippi River; denser geographic sampling is required. Few records from the Edwards Plateau, C Texas, reviewed by Goetze (1998); genetic differentiation and status of populations in S Illinois assessed by Monty et al. (2003). Formerly included *magister* as a subspecies (see below). See Wiley (1980, Mammalian Species, 139).

Neotoma fuscipes Baird, 1857. Mammalia *in* Repts. U.S. Expl. Surv., 8(1):495.
COMMON NAME: Dusky-footed Woodrat.
TYPE LOCALITY: USA, California, Sonoma Co., Petaluma.
DISTRIBUTION: Coastal and Cascade mountains, W Oregon, southwards to Inner Coastal Range, WC California, and N Sierra Nevadas, EC California, USA.

STATUS: U.S. ESA – Endangered as *N. f. riparia*; IUCN – Critically Endangered as *N. f. riparia*, Data Deficient as *N. f. annectens*, otherwise Lower Risk (lc).

SYNONYMS: *affinis* Elliot, 1898; *annectens* Elliot, 1898; *bullatior* Hooper, 1939; *monochroura* Rhoads, 1894; *perplexa* Hooper, 1938; *riparia* Hooper, 1938; *splendens* True, 1894.

COMMENTS: Subgenus *Neotoma*, *lepida* species group (*sensu* Edwards and Bradley, 2002*b*). Type species of *Homodontomys*. Sister-group relationship subject to conflicting interpretations, either *N. cinerea* (Carleton, 1980; Cudmore, 1986; Koop et al., 1985) or the *N. lepida* complex (Edwards and Bradley, 2002*b*); the former kinship supports synonymy of *Homodontomys* under subgenus *Teonoma*, the latter under subgenus *Neotoma* as enacted by Burt and Barkalow (1942). Subspecific classification revised by Hooper (1938). Patterns of morphological and genetic variation broadly evaluated by Matocq (2003), who demonstrated specific level divergence of southern populations and reinstated those as *N. macrotis* (see below). See Carraway and Verts (1991, Mammalian Species, 386).

Neotoma goldmani Merriam, 1903. Proc. Biol. Soc. Wash., 16:48.
COMMON NAME: Goldman's Woodrat.
TYPE LOCALITY: México, Coahuila, Saltillo, 5000 ft (1524 m).
DISTRIBUTION: Mexican Plateau, 1160-2320 m, from SE Chihuahua to S San Luis Potosí and N Querétaro, México.
STATUS: IUCN – Lower Risk (lc).
COMMENTS: Subgenus *Neotoma*, *floridana* species group (*sensu* Edwards and Bradley, 2002*b*). Chomosomal formula viewed as primitive for the genus (Harris and McCullough, 1988; Lee and Elder, 1977). Species group affinity variously interpreted, associated with *N. lepida* (Goldman, 1910) or *N. albigula* (Rainey and Baker, 1955); cytochrome *b* evaluations decidedly relate *N. goldmani* to the *N. albigula-N. floridana* clade (Edwards et al., 2001; Edwards and Bradley, 2002*b*). See Hrachovy et al. (1996, Mammalian Species, 545).

Neotoma lepida Thomas, 1893. Ann. Mag. Nat. Hist., ser. 6, 12:235.
COMMON NAME: Desert Woodrat.
TYPE LOCALITY: USA, "Simpson's Route" between Camp Floyd (= Fairfield), Utah and Carson City, Nevada (as restricted by Goldman, 1932:61).
DISTRIBUTION: SE Oregon to C Utah and WC Colorado, south through Nevada, NW Arizona, and S California, USA, to S Baja California Sur, México.
STATUS: IUCN – Data Deficient as *N. l. intermedia*, otherwise Lower Risk (lc).
SYNONYMS: *abbreviata* Goldman, 1909; *arenacea* J. A. Allen, 1898; *aridicola* Huey, 1957; *bella* Bangs, 1899; *californica* Price, 1894; *desertorum* Merriam, 1894; *egressa* Orr 1934; *felipensis* Elliot, 1903; *gilva* Rhoads, 1894; *grinnelli* Hall, 1942; *insularis* Townsend, 1912; *intermedia* Rhoads, 1894; *latirostra* Burt, 1932; *marcosensis* Burt, 1932; *marshalli* Goldman, 1939; *molagrandis* Huey, 1945; *monstrabilis* Goldman, 1932; *nevadensis* Taylor, 1910; *notia* Nelson and Goldman, 1931; *nudicauda* Goldman, 1905; *perpallida* Goldman, 1909; *petricola* von Bloeker, 1938; *pretiosa* Goldman, 1909; *ravida* Nelson and Goldman, 1931; *sanrafaeli* Kelson, 1950; *sola* Merriam, 1894; *vicina* Goldman, 1909.
COMMENTS: Subgenus *Neotoma*, *lepida* species group (*sensu* Edwards and Bradley, 2002*b*). Even with removal of *N. devia*, "*lepida*" may still represent a composite of two species (Mascarello, 1978; Riddle et al., 2000*b*). As underscored in the phylogeographic study of Patton and Alvarez-Castañeda (2005), based on cytochrome *b* data, populations from the coastal region of W California and the Baja California peninsula (= *intermedia*) are more genetically differentiated from *lepida* proper than are those of *N. devia*. Taxonomic stature of these coastal-peninsular populations, and the possible nomenclatural priority of the insular form *bryanti* for this complex, await integrated morphological and molecular confirmation (see Patton and Alvarez-Castañeda, 2005). As with *N. devia*, explicit allocation of species-group synonyms is needed. See Verts and Carraway (2002, Mammalian Species, 699).

Neotoma leucodon Merriam, 1894. Proc. Bio. Soc. Wash., 9:120.
COMMON NAME: White-toothed Woodrat.

TYPE LOCALITY: México, San Luis Potosí, San Luis Potosí.

DISTRIBUTION: SE Colorado, E New Mexico, and W Texas, USA, south in the Mexican Plateau to N Jalisco and N México.

SYNONYMS: *durangae* J. A. Allen, 1903; *latifrons* Merriam, 1894; *melas* Dice, 1929; *montezumae* Goldman, 1905; *robusta* Blair, 1939; *subsolana* Alvarez, 1962; *warreni* Merriam, 1908; *zacatecae* Goldman, 1905.

COMMENTS: Subgenus *Neotoma*, *micropus* species group (*sensu* Edwards and Bradley, 2002b). Rearranged as a subspecies of *N. albigula* by Goldman (1910), but specific status defended by Edwards et al. (2001) based on phylogenetic analysis of cytochrome-*b* sequences. Hybridization (reported as *N. albigula*) suspected with *N. micropus* in Colorado (Finley, 1958) and apparent intergradation of the two in Coahuila (Anderson, 1969); sister species to *N. micropus* supported by mitochondrial DNA data (Edwards et al., 2001). With changes in specific definitions and geographic ranges, meaningful delimitation of subspecies is currently unwarranted, as noted by Edwards et al. (2001); provisional allocation of other species-group epithets not sampled by those authors should be confirmed.

Neotoma macrotis Thomas, 1893. Ann. Mag. Nat. Hist, ser.6, 12:234.

COMMON NAME: Big-eared Woodrat.

TYPE LOCALITY: USA, California, San Diego Co., San Diego.

DISTRIBUTION: Santa Lucia Range, WC California, and Sierra Nevada, EC California, southwards to N Baja California Norte, México; isolated population in SE California.

STATUS: IUCN – Data Deficient as *N. fuscipes luciana*.

SYNONYMS: *cnemophila* Elliot, 1904; *dispar* Merriam, 1894; *luciana* Hooper, 1938; *martirensis* Orr, 1934; *mohavensis* Elliot, 1904; *simplex* True, 1894; *streatori* Merriam, 1894.

COMMENTS: Subgenus *Neotoma*, *lepida* species group (*sensu* Edwards and Bradley, 2002b). Intergradation with *N. fuscipes* maintained by Hooper (1938), who retained *macrotis* and related populations as subspecies of the former (e.g., see Hall, 1981). Purported hybridization zones reexamined by Matocq (2003), who uncovered pronounced and concordant differentiation in morphology, microsatellite genotypes, and mitochondrial DNA sequences between northern (*fuscipes*) and southern (*macrotis*) populations in the mountains flanking the Sacramento-San Joaquin Valley and elevated the latter set to a species. Matocq (2002) elaborated upon phylogeographic patterns of inter- and intraspecific diversification in the context of glacial and hydrographic changes that occurred in the Sierra Nevadas during the late Pliocene-Pleistocene.

Neotoma magister Baird, 1857. Mammals, *in* Repts. Expl. Surv. 8:498.

COMMON NAME: Allegheny Woodrat.

TYPE LOCALITY: USA, Pennsylvania, cave near Carlisle (Cumberland Co.) and Harrisburg (Dauphin Co.).

DISTRIBUTION: Allegheny Mountains, E USA, along a southwesterly tract from extreme SE New York and NW New Jersey to N Alabama and NW Georgia; isolated pockets in S Indiana and S Ohio.

STATUS: IUCN – Lower Risk (nt).

SYNONYMS: *pennsylvanica* Stone, 1893.

COMMENTS: Subgenus *Neotoma*, *floridana* species group (*sensu* Edwards and Bradley, 2002b). Originally described as a species, retained as such by Goldman (1910), later placed as a subspecies of *N. floridana* by Burt and Barkalow (1942). Birney (1976), however, noted that *magister* may prove distinct from *N. floridana*, an assessment borne out by subsequent morphological and genetic studies (Edwards and Bradley, 2001; Hayes and Harrison, 1992; Hayes and Richmond, 1993; Planz et al., 1996). Historical biogeography and timing of divergence from *N. floridana* postulated by Hayes and Harrison (1992) and Edwards and Bradley (2001).

Neotoma martinensis Goldman, 1905. Proc. Biol. Soc. Wash., 18:28.

COMMON NAME: San Martín Woodrat.

TYPE LOCALITY: México, Baja California Norte, San Martín Isl.

DISTRIBUTION: Known only from the type locality.

STATUS: IUCN – Endangered.

COMMENTS: Subgenus *Neotoma*, *lepida* species group (*sensu* Edwards and Bradley, 2002*b*). Listed as nominal species of *N. lepida* group by Goldman (1932); also see Mascarello (1978). Not collected since 1963, possibly extinct due to domestic cat predation (see Alvarez-Castañeda and Cortés-Calva, 1999). See Cortés-Calva et al. (2001, Mammalian Species, 657).

Neotoma mexicana Baird, 1855. Proc. Acad. Nat. Sci. Philadelphia, 7:333.

COMMON NAME: Mexican Woodrat.

TYPE LOCALITY: México, Chihuahua, mountains near Chihuahua.

DISTRIBUTION: SE Utah and C Colorado, USA, southwards through W and interior México, to highlands of Guatemala, El Salvador, and W Honduras.

STATUS: IUCN – Lower Risk (lc).

SYNONYMS: *atrata* Burt, 1939; *bullata* Merriam, 1894; *chamula* Goldman, 1909; *distincta* Bangs, 1903; *eremita* Hall, 1955; *fallax* Merriam, 1894; *ferruginea* Tomes, 1862; *fulviventer* Merriam, 1894; *griseoventer* Dalquest, 1951; *inopinata* Goldman, 1933; *inornata* Goldman, 1938; *isthmica* Goldman, 1904; *madrensis* Goldman, 1905; *navus* Merriam, 1903; *ochracea* Goldman, 1905; *orizabae* Merriam, 1894; *parvidens* Goldman, 1904; *picta* Goldman, 1904; *pinetorum* Merriam, 1893; *scopulorum* Finley, 1953; *sinaloae* J. A. Allen, 1898; *solitaria* Goldman, 1905; *tenuicauda* Merriam, 1892; *torquata* Ward, 1891; *tropicalis* Goldman, 1904; *vulcani* Sanborn, 1935.

COMMENTS: Subgenus *Neotoma*, *mexicana* species group (*sensu* Edwards and Bradley, 2002*b*). Goldman (1910) recognized the synonyms listed here under seven species, an early view on diversity uncritically overturned during the broad-brushed application of the biological species concept (Hall, 1955; Hooper, 1955). However, a composite of three or more species, e.g., as indicated by the comments of Sánchez-Hernández et al. (1999) on the morphological distinctiveness of *tenuicauda* and *torquata* in Michoacán, and by the findings of Edwards and Bradley (2002*a*) on the large genetic distances recorded among US and Méxican populations. Using mitochondrial DNA sequences, Edwards and Bradley (2002*b*) subsequently elevated two epithets, *isthmica* and *picta*, to species, but the few specimens examined and very limited geographic representation of each raise questions about their eventual application to the genetic groups identified. The status of these forms, and others such as *ferruginea*, *sinaloae*, and *torquata*, merits reconsideration in the context of a full systematic revision that treats *N. angustapalata* and *N. chrysomelas*, that broadens sampling to include pivotal older names, and that integrates morphological and molecular information.

Nearly all species-group epithets are recognized as subspecies (e.g., Hall, 1955, 1981), a formality that is unwarranted given the stronger need for basic specific revision. Using allozymic data, Sullivan (1994) studied genetic divergence and biogeographic patterns among montane populations in Arizona and New Mexico; low-elevation records in Utah and their implications for inter-mountain colonization and Great Basin biogeography discussed by Grayson et al. (1996). See Cornely and Baker (1986, Mammalian Species, 262).

Neotoma micropus Baird, 1855. Proc. Acad. Nat. Sci. Philadelphia, 7:333.

COMMON NAME: Southern Plains Woodrat.

TYPE LOCALITY: México, Tamaulipas, Charco Escondido.

DISTRIBUTION: SE Colorado and SW Kansas through W Texas and most of New Mexico, USA; south in México to N Chihuahua, E San Luis Potosí, and S Tamaulipas.

STATUS: IUCN – Lower Risk (lc).

SYNONYMS: *canescens* J. A. Allen, 1891; *leucophaea* Goldman, 1933; *littoralis* Goldman, 1905; *planiceps* Goldman, 1905; *surberi* Elliot, 1899.

COMMENTS: Subgenus *Neotoma*, *micropus* species group (*sensu* Edwards and Bradley, 2002*b*). Limited hybridization documented with *N. floridana* in Oklahoma (Birney, 1973) and believed probable with *N. leucodon* (reported as *N. albigula*) in Colorado (Finley, 1958) and Coahuila (Anderson, 1969); also see commentary under *N. albigula*, *N. floridana*, and *N. leucodon*. Genetically distinct from but generally related to the *N. albigula* complex (*sensu* Planz et al., 1996), in particular *N. leucodon* (*micropus* species group *sensu* Edwards et al., 2001). See Braun and Mares (1989, Mammalian Species, 330).

Neotoma nelsoni Goldman, 1905. Proc. Biol. Soc. Wash., 18:29.
COMMON NAME: Nelson's Woodrat.
TYPE LOCALITY: México, Veracruz, Perote, 7800 ft (2377 m).
DISTRIBUTION: Known only from the type locality.
STATUS: IUCN – Endangered.
COMMENTS: Subgenus *Neotoma*. Affiliated with *N. albigula* but stature as a separate species uncertain (see Hall and Genoways, 1970). The removal, as *N. leucodon*, of southern populations formerly associated with *albigula* necessitates reevalution of the status of *nelsoni* and its affinity with repect to the *floridana* and *micropus* species groups (see Edwards and Bradley, 2002*b*).

Neotoma palatina Goldman, 1905. Proc. Biol. Soc. Wash., 18:27.
COMMON NAME: Bolaños Woodrat.
TYPE LOCALITY: México, Jalisco, Bolaños, 2800 ft (853 m).
DISTRIBUTION: EC Jalisco, México.
STATUS: IUCN – Lower Risk (nt).
COMMENTS: Subgenus *Neotoma*. Specific distinctiveness from *N. albigula* reasserted by Hall and Genoways (1970), but see above comments under *N. nelsoni*.

Neotoma phenax (Merriam, 1903). Proc. Biol. Soc. Wash., 16:81.
COMMON NAME: Sonoran Woodrat.
TYPE LOCALITY: México, Sonora, Río Mayo, Camoa.
DISTRIBUTION: SW Sonora and NW Sinaloa, México.
STATUS: IUCN – Lower Risk (lc).
COMMENTS: Subgenus *Teanopus*. Type species of *Teanopus*, a taxon recognized as genus until reduced to a subgenus by Burt and Barkalow (1942). Karyotype interpreted as highly derived (Koop et al., 1985); highly differentiated from and cladistically basal to most species surveyed by Planz et al. (1996) using mitochondrial DNA restriction site analyses. See Jones and Genoways (1978, Mammalian Species, 108).

Neotoma stephensi Goldman, 1905. Proc. Biol. Soc. Wash., 18:32.
COMMON NAME: Stephen's Woodrat.
TYPE LOCALITY: USA, Arizona, Mohave Co., Hualapai Mtns, 6300 ft (1920 m).
DISTRIBUTION: Extreme SC Utah, N Arizona, and NW New Mexico, USA.
STATUS: IUCN – Lower Risk (lc).
SYNONYMS: *relicta* Goldman, 1932.
COMMENTS: Subgenus *Neotoma*, *lepida* species group (*sensu* Edwards and Bradley, 2002*b*). Morphological separation from *N. lepida* and geographic variation reviewed by Hoffmeister and de la Torre (1960). Sister taxon to *N. mexicana sensu lato* based on phallic morphology (Hooper, 1960) and restriction-site analysis of mitochondrial DNA (Planz et al., 1996), but associated with the *lepida* complex based on cytochrome *b* sequence data (Edwards and Bradley, 2002*b*). See Jones and Hildreth (1989, Mammalian Species, 328).

Neotomodon Merriam, 1898. Proc. Biol. Soc. Wash., 12:127.
TYPE SPECIES: *Neotomodon alstoni* Merriam, 1898.
COMMENTS: Reithrodontomyini. Conventionally ranked as a genus, later judged closely related to *Peromyscus sensu lato* (Hooper and Musser, 1964*b*). Synonymy as a subgenus of *Peromyscus* advocated by Yates et al. (1979) and J. C. Patton et al. (1981), whereas Carleton (1980, 1989) retained *Neotomodon* as a genus. Various kinds of evidence suggest the phyletic association of *Neotomodon*, *Podomys*, and perhaps *Habromys* (Carleton, 1980; Hooper and Musser, 1964*b*; Linzey and Layne, 1969; J. C. Patton et al., 1981; Stangl and Baker, 1984), although the cladistic results among these studies conflict in detail and on conclusions about rank.

Neotomodon alstoni Merriam, 1898. Proc. Biol. Soc. Wash., 12:128.
COMMON NAME: Volcano Deermouse.
TYPE LOCALITY: México, Michoacán, Nahuatzen, 8500 ft (2591 m).

DISTRIBUTION: Endemic to the Cordillera Transvolcanica, México, from WC Michoacán eastwards to C Veracruz.

STATUS: IUCN – Lower Risk (lc).

SYNONYMS: *orizabae* Merriam, 1898; *perotensis* Merriam, 1898.

COMMENTS: Williams and Ramírez-Pulido's (1984) evaluation of geographic variation disclosed no basis for subspecific divisions. See Williams et al. (1985, Mammalian Species, 242, as *Peromyscus alstoni*).

Ochrotomys Osgood, 1909. N. Am. Fauna, 28:222.

TYPE SPECIES: *Arvicola nuttalli* Harlan, 1832.

COMMENTS: Ochrotomyini new tribe. Type genus—*Ochrotomys* Osgood, 1909. Definition— medium-sized semiarboreal muroid rodents with tail about equal to head-and-body (Packard, 1969); hindfoot short and broad, digit V nearly as long as II-IV, plantar pads large and close set; carotid circulation complete (character state 0 per Carleton, 1980); skull with amphoral interorbit and smooth braincase; zygomatic plate narrow, no formation of dorsal notch; tegmen tympani adnate to squamosal (Voss, 1993), alisphenoid strut present; molars more brachyodont than *Peromyscus*, enamel thick, accessory lophs(ids) well developed (Hooper, 1957); m3 slightly smaller than m2, enteroconid and posterolophid usually discrete elements; vertebral column with 13 thoracic and 6 lumbar vertebrae, 1st rib articulating 7th cervical and 1st thoracic (Carleton, 1980); humerus lacking entepicondylar foramen (Manville, 1961); three complete and four incomplete transverse palatal ridges (Carleton, 1980); gastric glandular epithelium somewhat reduced and incisura angularis moderately deep (Carleton, 1973), gall bladder absent (Carleton, 1980); preputial glands large, ampullary glands absent (Arata, 1964); baculum and phallus short and wide, spination pronounced, urethral process present, dorsal and ventral lappets lacking (Blair, 1942; Hooper, 1958); 2n=52, G-banding patterns highly derived (Engstrom and Bickham, 1982; Patton and Hsu, 1967). Contents— *Ochrotomys* Osgood, 1909.

Although named as a subgenus of *Peromyscus*, the distant kinship and generic segregation of *Ochrotomys* have been repeatedly sustained (Blair, 1942; Carleton, 1980; Hooper, 1958; Hooper and Musser, 1964b; Patton and Hsu, 1967). Informally grouped with "peromyscines" by Carleton (1980), but his own results (see Carleton, 1989:115), together with karyological and molecular studies (Engel et al., 1998; Engstrom and Bickham, 1982), suggest earlier origination within the neotomine radiation.

Ochrotomys nuttalli (Harlan, 1832). Mon. Am. J. Geol. Nat. Sci. Philadelphia, p. 446.

COMMON NAME: Golden Mouse.

TYPE LOCALITY: USA, Virginia, Norfolk Co., Norfolk.

DISTRIBUTION: SE USA, from SE Missouri across to E West Virginia and S Virginia, south to E Texas, the Gulf coast, and C Florida.

STATUS: IUCN – Lower Risk (lc).

SYNONYMS: *aureolus* (Audubon and Bachman, 1841); *flammeus* (Goldman, 1941); *floridanus* Packard, 1969; *lewisi* (A. H. Howell, 1939); *lisae* Packard, 1969.

COMMENTS: Subspecific classification revised by Packard (1969), but Whitaker and Hamilton (1998) saw no basis for recognizing any. See Linzey and Packard (1977, Mammalian Species, 75).

Onychomys Baird, 1857. Mammalia *in* Repts. U.S. Expl. Surv., 8(1):458.

TYPE SPECIES: *Hypudaeus leucogaster* Wied-Neuwied, 1841.

COMMENTS: Reithrodontomyini. While Vorontsov (1959) arranged *Onychomys* as sole member of its own tribe, apart from Reithrodontomyini, a robust body of data now supports its close phyletic affinity with *Peromyscus* and related genera (Allard and Honeycutt, 1991; Carleton, 1980; Hooper and Musser, 1964b; Stangl and Baker, 1984b; Sullivan et al., 1995), in particular *Osgoodomys* (Engel et al., 1998), and recommends the synonymy of these two family-group taxa (Reithrodontomyini having line priority).

Revised by Hollister (1914), with later regional reviews by Engstrom and J. Choate (1979), Riddle and J. Choate (1986), and Van Cura and Hoffmeister (1966). Status of fossil forms (early Pliocene-Holocene) and affinity with living species evaluated by Carleton and Eshelman (1979). Chromosomal evolution among three species investigated by Baker et al. (1979), allozymic differentiation by Sullivan et al. (1986), and molecular relationships by Allard and Honeycutt (1991), Riddle and Honeycutt (1990), and Sullivan et al. (1995). Historical biogeography considered by Riddle and Honeycutt (1990) and Riddle (1995).

Onychomys arenicola Mearns, 1896. Preliminary diagnosis of new mammals from the Mexican border of the U. S., p. 3 (preprint of Proc. U. S. Natl. Mus., 19:137-140).
COMMON NAME: Chihuahuan Grasshopper Mouse.
TYPE LOCALITY: USA, Texas, El Paso Co., 6 mi (10 km) above El Paso.
DISTRIBUTION: Chihuahuan Desert: SE Arizona, SC New Mexico, and W Texas, USA, south to Aguascalientes, San Luis Potosí, and W Tamaulipas, México.
STATUS: IUCN – Lower Risk (lc).
SYNONYMS: *canus* Merriam, 1904; *surrufus* Hollister, 1914.
COMMENTS: Placed in full synonymy of *O. t. torridus* by Hollister (1914). Sympatry with *O. torridus* and karyotypic discrimination reported by Hinesley (1979), who raised *O. arenicola* to species (also see Baker et al., 1979). Sullivan et al. (1986) and Riddle and Honneycutt (1990) viewed *O. arenicola* and *O. leucogaster* as sister taxa, but Allard and Honeycutt (1991) depicted the cladistic topography, based on ribosomal DNA analyses, as (*O. arenicola* (*O. leucogaster-O. torridus*)). Synonymy follows Riddle (1999), who did not indicate subspecies; specimen-based amplification of geographic range still highly welcomed.

Onychomys leucogaster (Wied-Neuwied, 1841). Reise in Nord-America, 2:99.
COMMON NAME: Northern Grasshopper Mouse.
TYPE LOCALITY: USA, North Dakota, Oliver Co., Mandan village near Fort Clark.
DISTRIBUTION: S Alberta, S Saskatchewan, and SW Manitoba, Canada, south through much of Great Plains and Great Basin region of USA, to NC Sonora and N Tamaulipas, México.
STATUS: IUCN – Data Deficient as *O. l. durranti*, otherwise Lower Risk (lc).
SYNONYMS: *albescens* Merriam, 1904; *arcticeps* Rhoads, 1898; *breviauritus* Hollister, 1913; *brevicaudus* Merriam, 1891; *capitulatus* Hollister, 1913; *durranti* Riddle and J. Choate, 1986; *fuliginosus* Merriam, 1890; *fuscogriseus* Anthony, 1913; *longipes* Merriam, 1889; *melanophrys* Merriam, 1889; *missouriensis* (Audubon and Bachman, 1851); *pallescens* Merriam, 1890; *pallidus* Herrick, 1885; *ruidosae* Stone and Rehn, 1903; *utahensis* Goldman, 1939.
COMMENTS: Geographic variation and subspecific taxonomy reviewed for C Great Plains (Engstrom and J. Choate, 1979) and for Great Basin region (Riddle and J. Choate, 1986). Biogeographic implications of intraspecific differentiation explored by Riddle and J. Choate (1986) and Riddle et al. (1993). See McCarty (1978, Mammalian Species, 87).

Onychomys torridus (Coues, 1874). Proc. Acad. Nat. Sci. Philadelphia, 26:183.
COMMON NAME: Southern Grasshopper Mouse.
TYPE LOCALITY: USA, Arizona, Graham Co., Camp Grant.
DISTRIBUTION: C California, S Nevada, and extreme SW Utah, USA, south to N Baja California Norte, Sonora, and N Sinaloa, México.
STATUS: IUCN – Data Deficient as *O. t. ramona* and *O. t. tularensis*, otherwise Lower Risk (lc).
SYNONYMS: *clarus* Hollister, 1913; *knoxjonesi* Hollander and Willig, 1992; *longicaudus* Merriam, 1889; *macrotis* Elliot, 1903; *perpallidus* Mearns, 1896; *pulcher* Elliot, 1904; *ramona* Rhoads, 1893; *tularensis* Merriam, 1904; *yakiensis* Merriam, 1904.
COMMENTS: See comments under *O. arenicola*. See McCarty (1975, Mammalian Species, 59, including *arenicola*).

Osgoodomys Hooper and Musser, 1964. Occas. Pap. Mus. Zool., Univ. Michigan, 635:12.
TYPE SPECIES: *Peromyscus banderanus* J. A. Allen, 1897.
COMMENTS: Reithrodontomyini. Diagnosed as a subgenus of *Peromyscus* by Hooper and Musser (1964b) and retained there by Hooper (1968); proposed as a distinct genus by Carleton (1980). Anatomy of male reproductive system detailed by Linzey and Layne (1969, 1974).

Relationships assessed based on chromosomal banding (Rogers et al., 1984), protein electrophoresis (Schmidly et al., 1985), and mitochondrial DNA sequencing (Engel et al., 1998).

Osgoodomys banderanus (J. A. Allen, 1897). Bull. Am. Mus. Nat. Hist., 9:51.

COMMON NAME: Osgood's Deermouse.

TYPE LOCALITY: México, Nayarit, Valle de Banderas.

DISTRIBUTION: Coastal plain of S Nayarit to S Guerrero, interior of Michoacán and Guerrero along the basin of the Río Balsas, México.

STATUS: IUCN – Lower Risk (lc).

SYNONYMS: *vicinior* Osgood, 1904.

COMMENTS: The forms *angelensis*, *coatlanensis*, and *sloeops* had been mistaken as subspecies of *O. banderanus* and were reassigned to *Peromyscus mexicanus* by Musser (1969).

Peromyscus Gloger, 1841. Gemein Hand.-Hilfsbuch. Nat., 1:95.

TYPE SPECIES: *Peromyscus arboreus* Gloger, 1841 (= *Mus leucopus* Rafinesque, 1818).

SYNONYMS: *Haplomylomys* Osgood, 1904; *Sitomys* Fitzinger, 1867; *Vesperimus* Coues, 1874; *Trinodontomys* Rhoads, 1894.

COMMENTS: Reithrodontomyini. The *Drosophila* of North American mammalogy—the alpha-level classification of the genus has been revised three times (Osgood, 1909; Hooper, 1968; Carleton, 1989) and its biology and evolution have been twice monographed (King, 1968; Kirkland and Layne, 1989). Multispecies surveys have broadly sampled morphology of the genus (Carleton, 1973, 1980; Hooper, 1957, 1958; Hooper and Musser, 1964*b*; Linzey and Layne, 1969, 1974), its karyology (Robbins and Baker, 1981; L. W. Robbins et al., 1983; Rogers et al., 1984; Stangl and Baker, 1984*b*), biochemical variation (Avise et al., 1974, 1979; Brownell, 1983; Fuller et al., 1984; J. C. Patton et al., 1981; Rogers and Engstrom, 1992; Schmidly et al., 1985; Zimmerman et al., 1978), and geographical ecology (Glazier, 1980). See especially Greenbaum et al. (1994) for compilation of chromosomal banding data on the genus (2n = 48 in all species) and review of its cytosystematic applications.

Greater emphasis on phylogenetic systematics has altered Osgood's (1909) original generic scope. *Baiomys* and *Ochrotomys* have been removed and ranked as separate genera (Carleton, 1980; Hooper, 1958; Hooper and Musser, 1964*b*). Taxa arranged as subgenera by Hooper (1968—*Habromys*, *Isthmomys*, *Megadontomys*, *Osgoodomys*, and *Podomys*) have been considered genera by Carleton (1980) but not others (Rogers, 1983; Stangl and Baker, 1984*b*). Expansion of the generic limits *sensu* Hooper (1968) has been advocated to encompass *Neotomodon* (Stangl and Baker, 1984*b*; Yates et al., 1979) and perhaps *Onychomys* (Stangl and Baker, 1984). Nomenclatural resolution of these alternative proposals awaits definitive study. *Haplomylomys* has been used as a subgenus to contain the *californicus* and *eremicus* species groups, all others being assigned to the subgenus *Peromyscus*; Carleton (1989) emphasized species group assemblages rather than subgenera.

Major subdivisions of *Peromyscus* have received added revisionary attention, especially the *eremicus* (Avise et al., 1974; Lawlor, 1971*a*, *b*; Riddle et al., 2000*a*, *c*), *maniculatus* (Allard et al., 1987; Gunn and Greenbaum, 1986; Hogan et al., 1993, 1997), *boylii* (Avise et al., 1974; Bradley and Schmidly, 1987; Bradley et al., 1989; Carleton, 1977, 1979; De-Walt et al., 1993*b*; Rennert and Kilpatrick, 1986, 1987; Schmidly, 1973; Schmidly et al., 1988; Smith, 1990; Sullivan et al., 1991, 1997; Tiemann-Boege et al., 2000), *truei* (DeWalt et al., 1993*b*; Janecek, 1990; Modi and Lee, 1984; Schmidly, 1973; Zimmerman et al., 1975), and *mexicanus* (Huckaby, 1980; Musser, 1971; Rogers and Engstrom, 1992; Smith et al., 1986) species groups. See Tiemann-Boege et al. (2000) for commentary on membership in the *aztecus*, *boylii*, and *truei* species groups; and Hafner et al. (2001) for proposed affinities of insular taxa in the Sea of Cortez with species of the *boylii*, *eremicus*, or *maniculatus* species groups.

Peromyscus is thought to have evolved from the Miocene *Copemys* (Lindsay, 1972), a poorly characterized genus variously proposed as also ancestral to *Onychomys* (Jacobs, 1977) and *Bensonomys* (Baskin, 1978). Fossil species that are indisputably assigned to *Peromyscus* date from the early Pliocene (lower Blancan) of North America (e.g., Korth, 1994), but certain forms described from the late Miocene (Clarendonian-Hemphillian) have been assigned to *Copemys* or to *Peromyscus* (e.g., compare Shotwell, 1967*a*, and Hibbard, 1968, versus Lindsay, 1972, and Korth, 1994). By Pleistocene and Holocene times, the genus is well represented in the fossil record, including examples of living species (Graham and Lundelius, 1994). The possible fossil occurrence of *Peromyscus* from the late Pleistocene of Ecuador, as first reported by Fejfar et al. (1993), was later described as a new genus (*Copemyodon*) of the "Copemyne-Peromyscine group" (Fejfar et al., 1995). Two fossil species described from Pleistocene-Late Pleistocene deposits in the Channel Isls, off S California, may have survived initial human contact; these large tetralophodont forms, *P. anyapahensis* White (1966) and *P. nesodytes* Wilson (1936), were principally compared with *P. californicus* and deserve further study to ascertain their degree of differentiation and relationships to living species.

Peromyscus attwateri J. A. Allen, 1895. Bull. Am. Mus. Nat. Hist., 7:330.
> COMMON NAME: Texas Deermouse.
> TYPE LOCALITY: USA, Texas, Kerr Co., Turtle Creek.
> DISTRIBUTION: Edwards Plateau (see Goetze, 1998) and E Llano Estacado (see L. Choate, 1997) of C and N Texas, eastwards through SW and E Oklahoma, to SE Kansas, SW Missouri, and NW Arkansas, USA.
> STATUS: IUCN – Lower Risk (lc).
> SYNONYMS: *bellus* Bangs, 1896; *cansensis* Long, 1961; *laceyi* Bailey, 1905.
> COMMENTS: *P. truei* species group. Cognate relationship to *P. difficilis* (and *nasutus*) suggested by allozymic data (Janecek, 1990; Sullivan et al. 1991), a relationship bolstered by restriction-site (DeWalt et al., 1993b) and gene-sequence analyses (Tiemann-Boege et al., 2000) of mitochondrial DNA. Classified as a geographic race of *P. boylii* (Osgood, 1909) until rediagnosed as a species by Schmidly (1973). Biochemical evolution studied by Kilpatrick (1984); morphometric variation among Arkansas populations evaluated by Sugg et al. (1990). Populations found along the eastern escarpment of the Llano Estacado, N Texas, reidentified as *P. attwateri*, not *P. boylii* (L. Choate, 1997). See Schmidly (1974, Mammalian Species, 48).

Peromyscus aztecus (Saussure, 1860). Rev. Mag. Zool. Paris, Ser. 2, 12:105.
> COMMON NAME: Aztec Deermouse.
> TYPE LOCALITY: México, Veracruz, vicinity of Mirador, about 3800 ft ([~1158 m] as designated by Osgood, 1909:156).
> DISTRIBUTION: Humid montane and cloud forests, 800-3140 m, from C Veracruz and C Guerrero, through Oaxaca and E Chiapas, México, to Guatemala, Honduras, and El Salvador.
> STATUS: IUCN – Lower Risk (lc).
> SYNONYMS: *cordillerae* Dickey, 1928; *evides* Osgood, 1904; *hondurensis* Goodwin, 1941; *oaxacensis* Merriam, 1898; *yautepecus* Goodwin, 1955.
> COMMENTS: *P. aztecus* species group. Traditionally viewed as a subspecies of *P. boylii* (Osgood, 1909; Hall and Kelson, 1959). Species status recognized by Alvarez (1961); morphological recognition, distribution, and synonymies clarified by Musser (1969) and Carleton (1979). Systematic relationships examined by Bradley and Schmidly (1987), Smith (1990), Sullivan and Kilpatrick (1991), and Sullivan et al. (1997). Formerly included *P. hylocetes* (see account); other junior synonyms (*evides*, *hondurensis*, *oaxacensis*) have been treated as species (Hooper, 1968; Osgood, 1909) and their inclusion under *P. aztecus* deserves further scrutiny, especially those populations that occupy highlands south of the Isthmus of Tehuantepec (e.g., see Sullivan et al., 1997, 2000). Segregated together with *P. hylocetes*, *P. spicilegus*, and *P. winkelmanni* as the *aztecus* species group (Carleton, 1989; Sullivan et al., 1997; Tiemann-Boege et al., 2000). See Vázquez et al. (2001, Mammalian Species, 649).

Peromyscus beatae Thomas, 1903. Ann. Mag. Nat. Hist., ser. 7, 11:485.
> COMMON NAME: Orizaba Deermouse.
> TYPE LOCALITY: México, Veracruz, Mt Orizaba.
> DISTRIBUTION: C Veracruz and C Guerrero to Oaxaca and Chiapas, México, southwards through highlands of Guatemala, El Salvador, and Honduras.
> SYNONYMS: *sacarensis* Dickey, 1928.
> COMMENTS: *P. boylii* species group. Formerly arranged as a synonym of *P. boylii levipes*

(Hooper, 1968; Osgood, 1909) or *P. levipes* (Carleton, 1989). Specific status, including *sacarensis* as junior synonym and recognized subspecies, reasserted based on karyotypic, morphologic, and genetic evidence (Bradley and Schmidly, 1987; Bradley et al., 1989, 2000; Houseal et al. 1987; Schmidly et al., 1988). Sister taxon of *P. levipes*; taxonomic level of differentiation of populations west (*beatae*) and east (*sacarensis*) of the Isthmus of Tehuantepec requires additional study (see Bradley et al., 2000).

Peromyscus boylii (Baird, 1855). Proc. Acad. Nat. Sci. Philadelphia, 7:335.
COMMON NAME: Brush Deermouse.
TYPE LOCALITY: USA, California, Eldorado Co., Middle Fork of American River, near Auburn.
DISTRIBUTION: Much of SW USA, from N California to westernmost Oklahoma, south to N Baja California Norte and Trans-Pecos Texas (see Bradley et al., 1999a), along the Sierra Madre Occidental and W Central Plateau to Queretaro and W Hidalgo, México.
STATUS: IUCN – Lower Risk (lc).
SYNONYMS: *gaurus* Elliot, 1903; *glasselli* Burt, 1932; *major* (Rhoads, 1893); *metallicola* Elliot, 1904; *parasiticus* Elliot, 1904; *pinalis* (Miller, 1893); *robustus* (J. A. Allen, 1893); *rowleyi* (J. A. Allen, 1893); *utahensis* Durrant, 1946.
COMMENTS: *P. boylii* species group. Many taxa previously consigned as subspecies or full synonyms by Osgood (1909) or Hall (1981) have been elevated to species: namely *P. attwateri* (see Schmidly, 1973), *P. aztecus* (see Alvarez, 1961; Carleton, 1979; Hooper, 1968), *P. baetae* (see Bradley and Schmidly, 1987; Schmidly et al., 1988), *P. levipes* (see Schmidly et al., 1988), *P. madrensis* (see Carleton, 1977; Carleton et al., 1982), *P. simulus* (see Carleton, 1977), *P. sagax* (see Bradley et al., 1996a), and *P. spicilegus* (see Carleton, 1977). Others have been realigned with other species—*cordillerae* and *evides* under *P. aztecus* (Carleton, 1979); *ambiguus* under *P. levipes* (Castro-Campillo et al, 1999; Schmidly et al., 1988); *sacarensis* under *P. baetae* (Bradley et al., 2000); and *penicillatus* under *P. nasutus* (Diersing, 1976).

These changes in status and synonymy of taxa once associated under one species *sensu* Osgood (1909) involve the systematics of at least three species groups by current understanding. Chromosomal and molecular information suggests that yet other forms in W México await differential characterization as species (see overview by Tiemann-Boege et al., 2000). Karyotypic variation of *P. boylii* and its kin investigated by Houseal et al. (1987), Lee et al. (1972), and Schmidly and Schroeter (1974); biochemical and molecular variation by Avise et al. (1974), Kilpatrick and Zimmerman (1975), Rennert and Kilpatrick (1986, 1987), Sullivan et al. (1991), Tiemann-Boege et al. (2000); morphological variation by Bradley and Schmidly (1987), Carleton (1977, 1979), and Schmidly et al. (1988).

Peromyscus bullatus Osgood, 1904. Proc. Biol. Soc. Wash., 17:63.
COMMON NAME: Perote Deermouse.
TYPE LOCALITY: México, Veracruz, Perote.
DISTRIBUTION: Known from only two localities in Veracruz, EC México.
STATUS: IUCN – Endangered.
COMMENTS: *P. truei* species group. Retained as a species since its discovery (Carleton, 1989; Hoffmeister, 1951; Osgood, 1909), although Hooper (1968) suspected that it would prove to be a subspecies of *P. truei*, a possibility that has yet to be addressed.

Peromyscus californicus (Gambel, 1848). Proc. Acad. Nat. Sci. Philadelphia, 4:78.
COMMON NAME: California Deermouse.
TYPE LOCALITY: USA, California, Monterey Co., Monterey.
DISTRIBUTION: C and S California, USA, excluding San Joaquin Valley, to NW Baja California Norte, México.
STATUS: IUCN – Lower Risk (lc).
SYNONYMS: *benitoensis* Grinnell and Orr, 1934; *insignis* Rhoads, 1895; *mariposae* Grinnell and Orr, 1934; *parasiticus* (Baird, 1857).
COMMENTS: *P. californicus* species group. Electrophoretic and morphometric variation investigated by Smith (1979), who retained only a northern (*californicus*) and southern (*insignis*) subspecies. See Merritt (1978, Mammalian Species, 85).

Peromyscus caniceps Burt, 1932. Trans. San Diego Soc. Nat. Hist., 7:174.

COMMON NAME: Monserrat Island Deermouse.

TYPE LOCALITY: México, Baja California Sur, Monserrat Isl; 25°38′N, 111°02′W.

DISTRIBUTION: Known only from Monserrat Isl.

STATUS: IUCN – Vulnerable.

COMMENTS: *P. eremicus* species group. Although membership in *Haplomylomys* is generally recognized (Avise et al., 1971; Burt, 1932; Hooper, 1968), nearest kin and specific stature unresolved. Affinity to *P. eva* highlighted by Lawlor (1971*b*), who suggested its possible ranking as a subspecies thereof, but later (1983) maintained it as an insular species. Strongly and about equally differentiated from examples of *P. eva*, *P. eremicus*, and *P. fraterculus* based on allozymic comparisons by Avise et al. (1974). Listed, without explanation, in the *crinitus* species group by Hall (1981). Reassigned as a subspecies of *P. fraterculus* based on mitochondrial DNA similarity by Hafner et al. (2001). Nevertheless, the bacular contrasts to *P. eremicus-P. fraterculus*, as recorded by Lawlor (1971*b*), are impressive and recommend additional morphological and molecular comparisons that include Burt's (1932) original type series, reference samples of *P. eva* and *P. fraterculus*, and the recent vouchers utilized by Hafner et al. (2001). See Alvarez-Castañeda et al. (1998, Mammalian Species, 602).

Peromyscus crinitus (Merriam, 1891). N. Am. Fauna, 5:53.

COMMON NAME: Canyon Deermouse.

TYPE LOCALITY: USA, Idaho, Jerome Co., Shoshone Falls, north side of Snake River.

DISTRIBUTION: E Oregon and SW Idaho, south through Nevada and parts of Utah and W Colorado, USA, to EC Baja California Norte and NW Sonora, México.

STATUS: IUCN – Lower Risk (lc).

SYNONYMS: *auripectus* (J. A. Allen, 1893); *delgadilli* Benson, 1940; *disparilis* Goldman, 1932; *doutii* Goin, 1944; *pallidissimus* Huey, 1931; *pergracilis* Goldman, 1939; *peridoneus* Goldman, 1937; *petraius* Elliot, 1904; *rupicolus* Benson, 1940; *scitulus* Bangs, 1899; *scopulorum* Benson, 1940; *stephensi* Mearns, 1897.

COMMENTS: *P. crinitus* species group. Revised by Osgood (1909); subspecific taxonomy updated by Hall and Hoffmeister (1942). Initially placed in subgenus *Haplomylomys* (Osgood, 1909); later transferred to subgenus *Peromyscus* (Hooper and Musser, 1964*b*). Karyotype viewed as primitive for the genus (Greenbaum and Baker, 1978; Stangl and Baker, 1984*b*). Population on Isla San Luís Gonzaga (*pallidissimus*), N Sea of Cortez, genetically similar to mainland *P. crinitus* and retained as subspecies (Hafner et al., 2001). A composite of at least two species, the status of long-tailed forms (e.g., *disparilis* and *delgadilli*) in particular meriting reassessment. See Johnson and Armstrong (1987, Mammalian Species, 287).

Peromyscus dickeyi Burt, 1932. Trans. San Diego Soc. Nat. Hist., 7:176.

COMMON NAME: Dickey's Deermouse.

TYPE LOCALITY: México, Baja California Sur, Tortuga Isl; 27°21′N, 111°54′W.

DISTRIBUTION: Known only from Tortuga Isl.

STATUS: IUCN – Endangered.

COMMENTS: *P. eremicus* species group. Genically similar to *P. eremicus* and *P. merriami* (Avise et al., 1974) and presumably originated from an *eremicus*-like progenitor (Lawlor, 1983), but gene sequence data decisively portray mice on Tortuga Isl as derivatives of *P. merriami* (Hafner et al., 2001). The last authors have recommended the recognition of *dickeyi* as a subspecies of *P. merriami*. Cranium and fur traits of *dickeyi* do closely resemble those of *P. merriami* but the bullae appear smaller and the tail relatively shorter, diagnostic features noted by Burt (1932). A sounder footing for its synonymy and possible subspecific retention should involve broader morphological and molecular surveys within the range of *P. merriami*, together with the original type series of the island form. See Cortés-Calva and Alvarez-Castañeda (2001, Mammalian Species, 659).

Peromyscus difficilis (J. A. Allen, 1891). Bull. Am. Mus. Nat. Hist., 3:298.

COMMON NAME: Southern Rock Deermouse.

TYPE LOCALITY: México, Zacatecas, Sierra de Valparaíso.

DISTRIBUTION: W Chihuahua and SE Coahuila, south to C Oaxaca, México.

STATUS: IUCN – Lower Risk (lc).

SYNONYMS: *amplus* Osgood, 1904; *felipensis* Merriam, 1898; *petricola* Hoffmeister and de la Torre, 1959; *saxicola* Hoffmeister and de la Torre, 1959.

COMMENTS: *P. truei* species group. Revised by Hoffmeister and de la Torre (1961) to include *nasutus* (see species account below) and *comanche* (transferred to *P. truei*—Schmidly, 1973). Karyologic and biochemical evidence has suggested that *P. difficilis* and *P. nasutus* are sibling species (Avise et al., 1979; Zimmerman et al., 1975, 1978); classified as such by Carleton (1989), but see Janecek (1990), who continued to view *nasutus* populations as part of *P. difficilis*. DeWalt et al. (1993*b*) also noted the somewhat low levels of genetic divergence among *attwateri*, *difficilis*, and *nasutus* (e.g., less than *P. gratus* versus *P. truei*) but retained them as species, as did Tiemann-Boege et al. (2000). The matter of their synonymy merits reconsideration and decisive resolution. Gaona (1997) evaluated sex and age variation of populations in Veracruz and Puebla.

Peromyscus eremicus (Baird, 1857). Mammalia *in* Repts. U.S. Expl. Surv., 8(1):479.

COMMON NAME: Cactus Deermouse.

TYPE LOCALITY: USA, California, Imperial Co., Old Fort Yuma, Colorado River opposite Yuma, Arizona.

DISTRIBUTION: SE California, S Nevada, and SW Utah east to Trans-Pecos Texas, USA; south along mainland coast to C Sinaloa, and on the Mexican Plateau to N Zacatecas and S San Luis Potosí, México.

STATUS: IUCN – Lower Risk (lc).

SYNONYMS: *alcorni* Anderson, 1972; *anthonyi* (Merriam, 1887); *arenarius* Mearns, 1896; *avius* Osgood, 1909; *cedrosensis* J. A. Allen, 1898; *cinereus* Hall, 1931; *collatus* Burt, 1932; *insulicola* Osgood, 1909; *papagensis* Goldman, 1917; *phaeurus* Osgood, 1904; *polypolius* Osgood, 1909; *pullus* Blossom, 1933; *sinaloensis* Anderson, 1972; *tiburonensis* Mearns, 1897.

COMMENTS: *P. eremicus* species group. Once included *P. merriami* (see Hoffmeister and Lee, 1963*a*), *P. eva* (see Lawlor, 1971*b*), and *P. fraterculus* (see Riddle et al., 2000*a*, *c*). Even following removal of these several species confused under *eremicus*, mitochondrial DNA variation delineates well-marked W (Sonoran) and E (Chihuahuan) lineages whose taxonomic status should be explored (see Walpole et al., 1997; Riddle et al., 2000*c*). Lawlor (1971*a*) relegated the population on Isla Turner, *collatus*, to subspecies, a relationship and ranking bolstered by molecular analyses (Hafner et al., 2001), but Hall (1981) and Alvarez-Castañeda and Cortés-Calva (1999) continued to list it as a species. Relationships of *P. eremicus* to various insular forms in Gulf of California elucidated first by Lawlor (1971*a*, *b*, 1983) and later Hafner et al. (2001). Genetic variation broadly surveyed by Avise et al. (1974), Walpole et al. (1997), and Riddle et al. (2000*c*). See Veal and Caire (1979, Mammalian Species, 118).

Peromyscus eva Thomas, 1898. Ann. Mag. Nat. Hist., ser. 7, 1:44.

COMMON NAME: Southern Baja Deermouse.

TYPE LOCALITY: México, Baja California Sur, San José del Cabo.

DISTRIBUTION: Most of Baja California Sur and Carmen Isl, México.

STATUS: IUCN – Lower Risk (lc).

SYNONYMS: *carmeni* Townsend, 1912.

COMMENTS: *P. eremicus* species group. Ranked by Osgood (1909) as a race of *P. eremicus*, but Lawlor (1971*b*), noting morphological differences and sympatry of the two, resurrected *P. eva* as a species, including the population on nearby Isla Carmen as a subspecies (also see Hafner et al., 2001). Specific differentiation sustained by electrophoretic and gene-sequence analyses (Avise et al., 1974; Riddle et al., 2000*a*, *c*); sister species to *P. fraterculus* (Riddle et al., 2000*c*), another member of the Baja California Peninsular Desert fauna (e.g., Hafner and Riddle, 1997; Riddle et al., 2000*a*, *b*). See Alvarez-Castañeda and Cortés-Calva (2003, Mammalian Species, 738).

Peromyscus fraterculus (Miller, 1892). Amer. Nat., 26:261.

COMMON NAME: Northern Baja Deermouse.

TYPE LOCALITY: USA, California, San Diego Co., Dulzura.

DISTRIBUTION: SW California, USA, through Baja California Norte, to EC Baja California Sur, México.

SYNONYMS: *herroni* (Rhoads, 1893); *homochroia* Elliot, 1903; *nigellus* (Rhoads, 1894); *propinquus* J. A. Allen, 1898.

COMMENTS: *P. eremicus* species group. Long considered a geographic race of *P. eremicus* following the revision of Osgood (1909; e.g., also and Hall, 1981, Lawlor, 1971*a*). Studies of allozymic (Avise et al., 1974) and mitochondrial DNA variation (Walpole et al., 1997) within *eremicus* as conceived by Osgood have intimated that western and eastern moieties are specifically distinct. Riddle et al. (2000*a, c*) have confirmed this pronounced genetic divergence, reinstated *fraterculus* to species, demonstrated its common ancestry with *P. eva*, and interpreted its Pleistocene origination in a phylogeographic context. Synonyms follow Riddle et al. (2000*c*).

Peromyscus furvus J. A. Allen and Chapman, 1897. Bull. Am. Mus. Nat. Hist., 9:201.
COMMON NAME: Blackish Deermouse.
TYPE LOCALITY: México, Veracruz, 1.5 mi (2.4 km) E Jalapa, 4400 ft (1341 m).
DISTRIBUTION: Wet and cool forests along E flanks of Sierra Madre Oriental from extreme SE San Luis Potosí to NC Oaxaca, México.
STATUS: IUCN – Lower Risk (lc).
SYNONYMS: *angustirostris* Hall and Alvarez, 1961; *latirostris* Dalquest, 1950.
COMMENTS: *P. furvus* species group. Morphological limits, geographic variation, and distribution reviewed by Huckaby (1980), nongeographic variation by Martínez-Coronel et al. (1997). Full synonyms include *latirostris* (see Hall, 1971) and *angustirostris* (see Musser, 1964), a conclusion supported by electrophoretic study of allozymes (Harris and Rogers, 1999). Specific homogeneity of taxon questioned by phylogeographic analysis of cytochrome *b* sequences (Harris et al., 2000), which suggests that the status and identity of populations in northernmost Oaxaca be reconsidered. Considered a member of the *mexicanus* group by Hooper (1968); segregated as *P. furvus* species group, tentatively including *P. mayensis* and *P. ochraventer*, by Carleton (1989). Although well differentiated from other *mexicanus*-group forms, based on very limited taxonomic sampling, kinship with *P. ochraventer* is more distant according to mitochondrial DNA analyses (Harris et al., 2000).

Peromyscus gossypinus (Le Conte, 1853). Proc. Acad. Nat. Sci. Philadelphia, 6:411.
COMMON NAME: Cotton Deermouse.
TYPE LOCALITY: USA, Georgia, Liberty Co., near Riceboro, probably Le Conte Plantation.
DISTRIBUTION: SE USA, from SE Oklahoma, extreme S Illinois (see Feldhammer et al., 1998) and SE Virginia, southwards, skirting the S Appalachians, to E Texas, the Gulf States, and peninsular Florida.
STATUS: U.S. ESA – Endangered as *P. g. allapaticola*; IUCN – Extinct as *P. g. restrictus*, Vulnerable as *P. g. allapaticola*, otherwise Lower Risk (lc).
SYNONYMS: *allapaticola* Schwartz, 1952; *anastasae* Bangs, 1898; *cognatus* (Le Conte, 1855); *insulanus* Bangs, 1898; *megacephalus* (Rhoads, 1894); *mississippiensis* Rhoads, 1896; *nigriculus* Bangs, 1896; *palmarius* Bangs, 1896; *restrictus* A. H. Howell, 1939; *telmaphilus* Schwartz, 1952.
COMMENTS: *P. leucopus* species group. Lectotype based on MCZ specimen collected by Le Conte designated by Helgen (in Helgen and McFadden, 2001). Hybridization in lab documented with *P. leucopus* (Bradshaw, 1968), but genetic integrity of *P. gossypinus* is strongly corroborated in most field studies (Engstrom et al., 1982; Price and Kennedy, 1980; L. W. Robbins et al., 1985); limited hybridization of the two may occur in S Illinois (Barko and Feldhammer, 2002). Subspecific realignments based on morphometric and allozymic variation of insular and continental populations presented by Boone et al. (1993); in another survey, Boone et al. (1999) documented broad geographic patterns of genetic variation and discussed their general lack of correspondence to subspecific boundaries. See Wolfe and Linzey (1977, Mammalian Species, 70).

Peromyscus grandis Goodwin, 1932. Am. Mus. Novit., 560:4.
COMMON NAME: Large Deermouse.

TYPE LOCALITY: Guatemala, Alta Verapaz Dept., Finca Concepción, 3 mi (5 km) S San Miguel Tucurú, 3750 ft (1143 m).

DISTRIBUTION: S Alta Verapaz and NE Baja Verapaz, Guatemala; limits of distribution unknown.

STATUS: IUCN – Lower Risk (nt).

COMMENTS: *P. mexicanus* species group. Maintained as a species by Huckaby (1980). The level of relationship of this species to the allopatric forms *P. guatemalensis* and *P. zarhynchus* requires investigation.

Peromyscus gratus Merriam, 1898. Proc. Biol. Soc. Wash., 12:123.

COMMON NAME: Saxicoline Deermouse.

TYPE LOCALITY: México, Distrito Federal, Tlalpan.

DISTRIBUTION: SW New Mexico, USA, south from W Chihuahua and SE Coahuila, through interior México to C Oaxaca.

STATUS: IUCN – Lower Risk (lc).

SYNONYMS: *erasmus* Finley, 1952; *gentilis* Osgood, 1904; *pavidus* Elliot, 1903; *zapotecae* Hooper, 1957; *zelotes* Osgood, 1904.

COMMENTS: *P. truei* species group. Mexican populations revised, as part of *P. truei*, by Hoffmeister (1951). Divergent karyotypes and genetic distances had questioned unity of *P. truei* (Lee et al., 1972; Zimmerman et al., 1978); sympatry documented in New Mexico by Modi and Lee (1984), who raised *P. gratus* to a species (also see DeWalt et al., 1993b; Janecek, 1990; Tiemann-Boege et al., 2000).

Peromyscus guardia Townsend, 1912. Bull. Am. Mus. Nat. Hist., 31:126.

COMMON NAME: La Guarda Deermouse.

TYPE LOCALITY: México, Baja California Norte, Ángel de la Guarda Isl, 29°33'N, 113°35'W.

DISTRIBUTION: Isls of Ángel de la Guarda, Granito, and Mejía, N Gulf of California, México.

STATUS: IUCN – Lower Risk (nt).

SYNONYMS: *harbisoni* Banks, 1967; *mejiae* Burt, 1932.

COMMENTS: *P. eremicus* species group. Status and relationships with respect to *P. eremicus* and *P. interparietalis* illuminated by Brand and Ryckman (1969), Lawlor (1971a), and Avise et al. (1974); also see remarks on those studies by Hafner et al. (2001:785). Formerly included *P. interparietalis* (see Banks, 1967, and Lawlor, 1971a). Extinction or critical endangerment of various island populations, principally due to domestic cat introduction, evaluated by Mellink et al. (2002) and Alvarez-Castañeda and Ortega-Rubio (2003).

Peromyscus guatemalensis Merriam, 1898. Proc. Biol. Soc. Wash., 12:118.

COMMON NAME: Guatemalan Deermouse.

TYPE LOCALITY: Guatemala, Huehuetenango Dept., Todos Santos, 10,000 ft (3048 m).

DISTRIBUTION: Intermediate to high elevations in mountains of S Chiapas, México, and SW Guatemala.

STATUS: IUCN – Lower Risk (lc).

COMMENTS: *P. mexicanus* species group. Morphological variation and distribution clarified by Huckaby (1980). Level of relationship to morphologically similar allopatric forms *P. grandis* and *P. zarhynchus* unresolved. Formerly included *altilaneus* Osgood, 1904, here removed to *P. mexicanus* (see that account).

Peromyscus gymnotis Thomas, 1894. Ann. Mag. Nat. Hist., ser. 6, 14:365.

COMMON NAME: Naked-eared Deermouse.

TYPE LOCALITY: "Guatemala."

DISTRIBUTION: Pacific coastal plain and adjacent foothills, near sea level-1675 m, from S Chiapas, México, to S Nicaragua, west of Lake Nicaragua (after Huckaby, 1980; Jones and Yates, 1983; J. G. Owen et al., 1990).

STATUS: IUCN – Lower Risk (lc).

SYNONYMS: *allophylus* Osgood, 1904.

COMMENTS: *P. mexicanus* species group. Formerly classified as a subspecies of *P. mexicanus* (Osgood, 1909); Musser (1971) elevated *gymnotis* to species and pointed out the junior status of *allophylus*. Differentiation from *P. mexicanus* sustained by Huckaby (1980) and Jones and Yates (1983), who also amplified the southern range and morphological recognition of *P. gymnotis*. As noted by Carleton (1989), the northward distribution in the region of the Isthmus of Tehuantepec now needs clarification, particularly in light of the report of a *mexicanus* sample from near Berriozábal, Chiapas, that genetically links with *P. gymnotis* (Rogers and Engstrom, 1992).

Peromyscus hooperi Lee and Schmidly, 1977. J. Mammal., 58:263.

COMMON NAME: Hooper's Deermouse.

TYPE LOCALITY: México, Coahuila, 2.5 mi (4 km) W, 21 mi (34 km) S Ocampo, 3500 ft (1067 m).

DISTRIBUTION: Grassland transition zone, 1000-2000 m, from C Coahuila to northernmost Zacatecas and San Luis Potosí, México.

STATUS: IUCN – Lower Risk (lc).

COMMENTS: *P. hooperi* species group. Allocated, with reservation, to the subgenus *Peromyscus* as sole member of *P. hooperi* species group (Schmidly et al., 1985). See Alvarez-Castañeda (2002, Mammalian Species, 709).

Peromyscus hylocetes Merriam, 1898. Proc. Biol. Soc. Wash., 12:124.

COMMON NAME: Transvolcanic Deermouse.

TYPE LOCALITY: México, Michoacán, Patzcuaro, 7000 ft (2134 m).

DISTRIBUTION: Endemic to the Cordillera Transvolcanica, 1980-3085 m, from WC Jalisco eastwards to Distrito Federal and N Morelos, México.

COMMENTS: *P. aztecus* species group. Described and early maintained as a species (Merriam, 1898; Osgood, 1909), suggested as conspecific with *oaxacensis* (Hooper, 1968), and subsequently synonymized as one of five montane subspecies of *P. aztecus*, including *oaxacensis* (Carleton, 1979). Cladistic interpretation of cytochrome *b* sequence data indicates the earlier separation of a *hylocetes* lineage relative to other populations of *P. aztecus* (Sullivan and Kilpatrick, 1991; Sullivan et al., 1997, 2000), and genetic divergence levels favor its reinstatement as a species.

Peromyscus interparietalis Burt, 1932. Trans. San Diego Soc. Nat. Hist., 7:175.

COMMON NAME: San Lorenzo Deermouse.

TYPE LOCALITY: México, Baja California Norte, San Lorenzo Sur Isl; 28°36′N, 112°51′W.

DISTRIBUTION: North and South San Lorenzo Isls, and Salsipuedes Isl, N Gulf of California, México.

STATUS: IUCN – Endangered.

SYNONYMS: *lorenzi* Banks, 1967; *ryckmani* Banks, 1967.

COMMENTS: *P. eremicus* species group. Considered distinct from *P. guardia* and *P. eremicus* (*sensu lato*) by Banks (1967) and Lawlor (1971a; also see Avise et al., 1974, and Brand and Ryckman, 1969). Marginally differentiated from *P. eremicus* based on all characters surveyed by Lawlor (1971a) but retained as species; phyletic relations, as divulged by mitochondrial DNA (COIII) sequences, indicate derivation from a *P. eremicus* (*sensu stricto*) stock and suggest only subspecific differentiation from those mainland populations (Hafner et al., 2001).

Peromyscus keeni (Rhoads, 1894). Proc. Acad. Nat. Sci. Philadelphia, 46:258.

COMMON NAME: Northwestern Deermouse.

TYPE LOCALITY: Canada, British Columbia, Queen Charlotte Isls, Graham Isl, Massett.

DISTRIBUTION: S Alaska, W British Columbia, and W Washington, west of the Coastal and Cascade mountain ranges and including continental shelf islands and archipelagos.

STATUS: IUCN – Lower Risk (lc) as *P. oreas* and *P. sitkensis*.

SYNONYMS: *algidus* Osgood, 1909; *beresfordi* Guiguet, 1955; *cancrivorus* McCabe and Cowan, 1945; *carli* Guiguet, 1955; *doylei* McCabe and Cowan, 1945; *hylaeus* Osgood, 1908; *interdictus* Anderson, 1932; *isolatus* Cowan, 1935; *macrorhinus* (Rhoads, 1894); *maritimus* McCabe and Cowan, 1945; *oceanicus* Cowan, 1935; *oreas* Bangs, 1898; *pluvialis* McCabe and Cowan, 1945; *prevostensis* Osgood, 1901; *rubriventer* McCabe and Cowan, 1945; *sartinensis* Guiguet, 1955; *sitkensis* Merriam, 1897; *triangularis* Guiguet, 1955.

COMMENTS: *P. maniculatus* species group. Relegated to a subspecies of *P. maniculatus* by

Osgood (1909) and thereafter maintained as a junior synonym (Hall, 1981; Hooper, 1968; Musser and Carleton, 1993). Specific status, synonymies, and distribution with respect to *P. maniculatus* clarified by Hogan et al. (1993), who reassigned *oreas* and *sitkensis*, taxa formerly regarded as species, to subspecific rank; much morphological, genetic, and karyological information has been reported under those names (e.g., Allard and Greenbaum, 1988; Allard et al. 1987; Calhoun and Greenbaum, 1991; Cowan and Guiguet, 1965; Gunn and Greenbaum, 1986; Osgood, 1909; Pengilly et al., 1983; Sheppe, 1961; Sullivan et al., 1990; Thomas, 1973). About half of the synonyms listed are associated on the basis of distributional proximity and literature reports (see Hogan et al., 1993:829), a zoogeographically plausible supposition that deserves firsthand specimen examination and empirical ratification. Reconstruction of historical distributions to account for divergence of *P. keeni* and *P. maniculatus* considered by Zheng et al. (2003), who also discussed the status of *sitkensis*.

Peromyscus leucopus (Rafinesque, 1818). Am. Mon. Mag., 3:446.

COMMON NAME: White-footed Deermouse.

TYPE LOCALITY: USA, Kentucky, Ballard Co., near mouth of Ohio River (as restricted by Osgood, 1909:115-116).

DISTRIBUTION: S Alberta and to S Ontario, Quebec and Nova Scotia, Canada; throughout much of C and E USA, excluding Florida; southwards to N Durango and along Caribbean coast to Isthmus of Tehuantepec and NW Yucatán Peninsula, México.

STATUS: IUCN – Data Deficient as *P. l. ammodytes*, otherwise Lower Risk (lc).

SYNONYMS: *affinis* (J. A. Allen, 1891); *ammodytes* Bangs, 1905; *arboreus* Gloger, 1841; *aridulus* Osgood, 1909; *arizonae* (J. A. Allen, 1894); *brevicaudus* Davis, 1939; *campestris* (Le Conte, 1853); *canus* Mearns, 1896; *castaneus* Osgood, 1904; *caudatus* Smith, 1939; *cozulmelae* Merriam, 1901; *easti* Paradiso, 1960; *emmonsi* (DeKay, 1840); *flaccidus* J. A. Allen, 1903; *fuscus* Bangs, 1905; *incensus* Goldman, 1942; *lachiguiriensis* Goodwin, 1956; *mearnsii* (J. A. Allen, 1891); *mesomelas* Osgood, 1904; *michiganensis* (Audubon and Bachman, 1842); *minnesotae* Mearns, 1901; *musculoides* Merriam, 1898; *myoides* (Gapper, 1830); *noveboracensis* (Fischer, 1829); *ochraceus* Osgood, 1909; *texanus* (Woodhouse, 1853); *tornillo* Mearns, 1896.

COMMENTS: *P. leucopus* species group. Revised by Osgood (1909). Morphometric discrimination of *P. leucopus* from *P. maniculatus* studied in central US and New England (J. Choate, 1973; J. Choate et al., 1979); Rich et al. (1996) combined salivary amylase genotypes with discriminant function analyses to confidently separate these morphologically similar species (also see account of *P. gossypinus*). Distinctive eastern (*leucopus*) and southwestern (*texanus*) cytotypes reported (Baker et al., 1983), with introgression across hybrid zone in central Oklahoma intensively studied (Nelson et al., 1987; Schmidt, 1999; Simmons et al., 1992; Stangl, 1986; Stangl and Baker, 1984*a*; Van Den Bussche et al., 1993*b*); possible mechanisms of heterochromatic chromosomal change that would produce the different cytotypes examined by Bowers et al. (1998).

Morphometric variation examined over microgeographic scale in various regions that provide insight to understanding macrogeographic patterns of size, including NC Kansas (Tolliver et al., 1987), SW Tennessee (Elrod and Kennedy, 1995), and Texas (Owen, 1989). Size and chromatic variation of series in the Llano Estacado, N Texas, and its relationship to recognized subspecies, *texanus* and *tornillo*, discussed by L. Choate (1997). See Long and Long (1993) and Kilpatrick et al. (1994) for local range expansions within Wisconsin and Maine, respectively. Genetic variation investigated at different spatial scales and habitats (Browne, 1977; Schnake-Greene et al., 1990); variation in microsatellite DNA repeats as potential markers to determine mating systems, gene flow, and demic relationships elucidated by Schmidt (1999). Whitaker and Hamilton (1998) regarded *noveboracensis* and the nominate subspecies as broadly intergrading and synonymized the two. See Lackey et al. (1985, Mammalian Species, 247).

Peromyscus levipes Merriam, 1898. Proc. Biol. Soc. Wash., 12:123.

COMMON NAME: Nimble-footed Deermouse.

TYPE LOCALITY: México, Tlaxcala, Mt Malinche, 8400 ft (2560 m).

DISTRIBUTION: C Nuevo León and W Tamaulipas, in and along the Sierra Madre Oriental, to C Veracruz, Tlaxcala, Distrito Federal, and Morelos, México.

STATUS: IUCN – Lower Risk (lc).

SYNONYMS: *ambiguus* Alvarez, 1961.

COMMENTS: *P. boylii* species group. Systematic studies have confirmed sympatry of *P. boylii rowleyi* and *P. levipes* in Queretaro and Hidalgo and supported its elevation to species (Houseal et al., 1987; Rennert and Kilpatrick, 1987; Schmidly et al., 1988; Sullivan et al., 1991). The taxon *baetae* was provisionally retained by Carleton (1989) under *P. levipes*, whereas Schmidly et al. (1988) viewed it as a cryptic species distinct from *P. levipes*, a treatment borne out by subsequent studies (see above account). Includes *ambiguus* as a definable subspecies based on morphometric and gene-sequence evidence (Castro-Campillo et al., 1999); geographic and nongeographic variation, subspecific distributions, and nomenclatural changes summarized by Castro-Campillo et al. (1999).

Peromyscus madrensis Merriam, 1898. Proc. Biol. Soc. Wash., 12:16.

COMMON NAME: Tres Marías Deermouse.

TYPE LOCALITY: México, Nayarit, Tres Marías Isls, María Madre Isl.

DISTRIBUTION: Tres Marías Isls, México.

STATUS: IUCN – Vulnerable.

COMMENTS: *P. boylii* species group. Treated as a subspecies of *P. boylii* by Osgood (1909) but considered distinct by Carleton (1977) in a revision of *boylii* group forms in W México. Viewed as closely related to *P. simulus* by Carleton et al. (1982), but cladistically basal to most other *boylii*-group species in trees generated from cytochrome *b* data (Tiemann-Boege et al., 2000). Perhaps extirpated on María Magdalena Isl, where *Rattus rattus* is now abundant (Carleton et al., 1982; Wilson, 1991).

Peromyscus maniculatus (Wagner, 1845). Arch. Naturgesch., 11, 1:148.

COMMON NAME: North American Deermouse.

TYPE LOCALITY: Canada, Labrador, Moravian settlements.

DISTRIBUTION: Panhandle of Alaska and across N Canada, south through most of continental USA, excluding the SE and E seaboard, to southernmost Baja California Sur and to NC Oaxaca, México; including many landbridge islands.

STATUS: IUCN – Lower Risk (nt) as *P. m. anacapae* and *P. m. clementis*, otherwise Lower Risk (lc).

SYNONYMS: *abietorum* Bangs, 1896; *akeleyi* Elliot, 1899; *alpinus* Cowan, 1937; *anacapae* von Bloeker, 1942; *angustus* Hall, 1932; *anticostiensis* Moulthrop, 1937; *arcticus* (Coues, 1877); *arcticus* (Mearns, 1890); *argentatus* Copeland and Church, 1906; *artemisiae* (Rhoads, 1894); *assimilis* Nelson and Goldman, 1931; *austerus* (Baird, 1855); *bairdii* (Hoy and Kennicott, 1857); *bairdii* (Coues, 1877); *balaclavae* McCabe and Cowan, 1945; *blandus* Osgood, 1904; *borealis* Mearns, 1911 [renaming of *arcticus* Mearns, 1890]; *canadensis* (Miller, 1893); *catalinae* Elliot, 1903; *cineritius* J. A. Allen, 1898; *clementis* Mearns, 1896; *coolidgei* Thomas, 1898; *deserticolus* (Mearns, 1890); *dorsalis* Nelson and Goldman, 1931; *dubius* J. A. Allen, 1898; *elusus* Nelson and Goldman, 1931; *eremus* Osgood, 1909; *exiguus* J. A. Allen, 1898; *exterus* Nelson and Goldman, 1931; *fulvus* Osgood, 1904; *gambelii* (Baird, 1857); *georgiensis* Hall, 1938; *geronimensis* J. A. Allen, 1898; *gracilis* (Le Conte, 1855); *gunnisoni* Goldman, 1937; *hollisteri* Osgood, 1909; *hueyi* Nelson and Goldman, 1932; *imperfectus* Dice, 1925; *inclarus* Goldman, 1939; *insolatus* (Rhoads, 1894); *labecula* Elliot, 1903; *luteus* Osgood, 1905; *magdalenae* Osgood, 1909; *margaritae* Osgood, 1909; *martinensis* Nelson and Goldman, 1931; *medius* Mearns, 1896; *nebrascensis* (Coues, 1877); *nebrascensis* (Mearns, 1890); *nubiterrae* Rhoads, 1896; *oresterus* Elliot, 1903; *osgoodi* Mearns 1911 [renaming of *nebrascensis* Mearns, 1890]; *ozarkiarum* Black, 1935; *pallescens* J. A. Allen, 1896; *perimekurus* Elliot, 1903; *plumbeus* C. F. Jackson, 1939; *rubidus* Osgood, 1901; *rufinus* (Merriam, 1890); *sanctaerosae* von Bloeker, 1940; *santacruzae* Nelson and Goldman, 1931; *saturatus* Bangs, 1897; *saxamans* McCabe and Cowan, 1945; *serratus* Davis, 1939; *sonoriensis* (Le Conte, 1853); *streatori* Nelson and Goldman, 1931; *subarcticus* J. A. Allen, 1899; *thurberi* (J. A. Allen, 1893); *umbrinus* Miller, 1897.

COMMENTS: *P. maniculatus* species group. A broadly distributed and morphologically highly variable form once confused with many taxa now reallocated to *P. leucopus* (see Osgood, 1909). Formerly included long-tailed populations in NW North America recently separated as *P. keeni*. Status and relationships of *P. maniculatus* have been extensively addressed with regard to *P. keeni* (or as its junior synonyms *oreas* and *sitkensis*—Allard et al., 1987; Allard and Greenbaum, 1988; Gunn and Greenbaum, 1986; Hogan et al., 1993, 1997; Zheng et al., 2003), *P. melanotis* (Bowers, 1974; Bowers et al., 1973; Greenbaum and Baker, 1978), *P. polionotus* (Avise et al., 1979; Hogan et al., 1997; Robbins and Baker, 1981), and *P. sejugis* (Avise et al., 1979; Hafner et al., 2001; Hogan et al., 1997; Lawlor, 1983); see those accounts.

Regional studies of geographic variation have clarified distributions and realigned certain subspecific ranges: e.g., populations in N Wisconsin and on islands in the N Great Lakes (Long and Long, 1993); those in S Maine (Kilpatrick et al., 1994); those in Arizona (Hoffmeister, 1986); those inhabiting the Llano Estacado, N Texas and vicinity (Cooper et al., 1993); and those at the southern end of the Mexican Plateau, C México (Nanez-Jiminez and Martínez-Coronel, 1995). Differences in craniodental shape according to dietary consistency investigated using laboratory animals by Myers et al. (1996), who discussed their bearing on systematic interpretations among wild populations. Probable extinction of San Roque Isl population (*cineritius*) reported by Alvarez-Castañeda and Cortes-Calva (1999).

Even after removal of long-tailed populations in NW North America (i.e., *P. keeni*), appreciable variation in morphological, chromosomal, and biochemical data still cautions that more than one species is represented among the short-tailed (e.g., *bairdii*) and long-tailed (e.g., *gracilis*, *nubiterrae*) morphotypes in C and NE North America (Bradshaw and Hsu, 1972; Calhoun et al., 1988; Koh and Peterson, 1983; Lansman et al., 1983; Myers Uncie et al., 1998). Trees derived from mitochondrial DNA sequences represent *maniculatus* as paraphyletic with respect to *P. keeni* and *P. sejugis* (Hogan et al. 1997). Meaningful delineation of subspecies can only be achieved within the context of this much needed revision; see Hall (1981) for conventional arrangement of races (excepting those now removed to *P. keeni*).

Peromyscus mayensis Carleton and Huckaby, 1975. J. Mammal., 56:444.
COMMON NAME: Mayan Deermouse.
TYPE LOCALITY: Guatemala, Huehuetenango Dept., about 7 km NW Santa Eulalia, Yaiquich, 2950 m.
DISTRIBUTION: Known only from the type locality and close vicinity, WC Guatemala.
STATUS: IUCN – Endangered.
COMMENTS: *P. furvus* species group. Nearest specific relatives obscure; provisionally aligned with the *P. furvus* species group (Carleton, 1989).

Peromyscus megalops Merriam, 1898. Proc. Biol. Soc. Wash., 12:119.
COMMON NAME: Broad-faced Deermouse.
TYPE LOCALITY: México, Oaxaca, La Cieneguilla Ranch, near Santa María Ozolotepec, 10,000 ft (3048 m).
DISTRIBUTION: Humid forests in mountains of C Guerrero, S and NC Oaxaca, México.
STATUS: IUCN – Lower Risk (lc).
SYNONYMS: *auritus* Merriam, 1898; *comptus* Merriam, 1898.
COMMENTS: *P. megalops* species group. Following Osgood (1909), Hall (1981) continued to maintain *melanurus* as a subspecies, a form that Huckaby (1980) documented as a species distinct from *P. megalops*. Distribution and ecology reviewed by Musser (1964); Briones-Salas et al. (2001) recently documented the species in the Sierra Mazteca, NC Oaxaca. Carleton (1989) observed Osgood's (1909) earlier arrangement of a *megalops* species group (including *P. melanocarpus* and *P. melanurus*) apart from the *mexicanus* species group; segregation from *mexicanus* complex receives support from genic data, but not including *melanocarpus* (Rogers and Engstrom, 1992).

Peromyscus mekisturus Merriam, 1898. Proc. Biol. Soc. Wash., 12:124.
COMMON NAME: Puebla Deermouse.
TYPE LOCALITY: México, Puebla, Chalchicomula, 8400 ft (2560 m).

DISTRIBUTION: SE Puebla, México.

STATUS: IUCN – Vulnerable.

COMMENTS: *P. melanophrys* species group. A distinctive species known only by two specimens, the type and one from Tehuacán, Puebla (Hooper, 1947). Provisionally retained with the *melanophrys* species group (Carleton, 1989).

Peromyscus melanocarpus Osgood, 1904. Proc. Biol. Soc. Wash., 17:73.

COMMON NAME: Black-wristed Deermouse.

TYPE LOCALITY: México, Oaxaca, Cerro Zempoaltépec, above Yacochi.

DISTRIBUTION: Humid montane forests of Sierras de Zempoaltépec, Juárez, and Mazteca, ca. 1500-2500 m, NC Oaxaca, México.

STATUS: IUCN – Lower Risk (lc).

COMMENTS: Species group indeterminate. Restricted range and distinctive morphology substantiated by Huckaby (1980); recently documented in the Sierra Mazteca, NC Oaxaca, by Briones-Salas et al. (2001). Age, sex, and individual variation assessed by Cervantes et al. (1993*b*). Nearest specific relatives uncertain: initially segregated in the *megalops* species group apart from the *mexicanus* complex by Osgood (1909) and later Carleton (1989); or arranged within an expanded *mexicanus* species group by Hooper (1968) and Huckaby (1980). Neither of these associations finds support in phenetic and cladistic analyses of allozymes (Rogers and Engstrom, 1992), a result that leaves the species group assignment of *melanocarpus* as *incertae sedis*. See Rickart and Robertson (1985, Mammalian Species, 241).

Peromyscus melanophrys (Coues, 1874). Proc. Acad. Nat. Sci. Philadelphia, 26:181.

COMMON NAME: Plateau Deermouse.

TYPE LOCALITY: México, Oaxaca, Santa Efigenia.

DISTRIBUTION: S Durango and Coahuila, south through interior México to Chiapas.

STATUS: IUCN – Lower Risk (lc).

SYNONYMS: *coahuilensis* Baker, 1952; *consobrinus* Osgood, 1904; *micropus* Baker, 1952; *xenurus* Osgood, 1904; *zamorae* Osgood, 1904.

COMMENTS: *P. melanophrys* species group. Revised by Baker (1952) to include *xenurus*, which Osgood (1909) had retained as a species. Relationships evaluated by Schmidly et al. (1985) and Stangl and Baker (1984*b*). The level of differentiation of *micropus* in Jalisco and Nayarit deserves reconsideration.

Peromyscus melanotis J. A. Allen and Chapman, 1897. Bull. Am. Mus. Nat. Hist., 9:203.

COMMON NAME: Black-eared Deermouse.

TYPE LOCALITY: México, Veracruz, Las Vigas, 8000 ft (2438 m).

DISTRIBUTION: Cordillera Transvolcanica in C México (E Jalisco to C Veracruz), northwards along Sierra Madre Oriental to S Nuevo León and along Sierra Madre Occidental to W Chihuahua; isolated populations in SE Arizona, USA.

STATUS: IUCN – Lower Risk (lc).

SYNONYMS: *cecilii* Thomas, 1903; *zamelas* Osgood, 1904.

COMMENTS: *P. maniculatus* species group. Genetic, karyotypic, and geographic variation studied by Bowers et al. (1973), who included populations in S Arizona previously identified as *P. maniculatus rufinus*. However, see Hoffmeister (1986) who disputed their conclusions and maintained all Arizona populations as *P. maniculatus*. Bowers (1974) discerned no basis for subspecific divisions; geographic and nongeographic variation of populations inhabiting the Cordillera Transvolcanica evaluated by Martínez-Coronel et al. (1991).

Peromyscus melanurus Osgood, 1909. N. Am. Fauna, 28:215.

COMMON NAME: Black-tailed Deermouse.

TYPE LOCALITY: México, Oaxaca, below Pluma Hidalgo, 3000 ft (914 m).

DISTRIBUTION: Pacific slopes of the Sierra Madre del Sur of Oaxaca, México.

STATUS: IUCN – Vulnerable.

COMMENTS: *P. megalops* species group. Although named as a subspecies of *P. megalops*, Huckaby (1980) substantiated the specific status and range of *melanurus*. Banded karyotype described by Smith et al. (1986).

Peromyscus merriami Mearns, 1896. Preliminary diagnoses of new mammals from the Mexican border of the United States, p. 2; preprint of Proc. U. S. Nat. Mus., 19:138.

COMMON NAME: Merriam's Deermouse.

TYPE LOCALITY: México, Sonora, Sonoyta, on Sonoyta River.

DISTRIBUTION: SC Arizona, USA, through W Sonora to C Sinaloa, México.

STATUS: IUCN – Lower Risk (lc).

SYNONYMS: *goldmani* Osgood, 1904.

COMMENTS: *P. eremicus* species group. Osgood (1909) placed *merriami* under *P. eremicus*, a synonymy overturned by Hoffmeister and Lee (1963a), who documented their sympatric occurrence and morphological differentiation. Geographic variation evaluated by Hoffmeister and Diersing (1973), who retained *goldmani* as a subspecies; Lawlor (1971b) did not. Generally related to *P. eremicus sensu lato* (Avise et al., 1974; Lawlor, 1971a) and to *P. pembertoni* (Lawlor, 1971a, 1983); based on mitochondrial DNA sequences, viewed as cognate species to *P. eremicus sensu stricto*, which together form a sister group to the clade *P. eva-P. fraterculus* (Riddle et al., 2000a, c). Includes the insular form *dickeyi* according to mitochondrial DNA (COIII) comparisons (Hafner et al., 2001).

Peromyscus mexicanus (Saussure, 1860). Rev. Mag. Zool. Paris, Ser. 2, 12:103.

COMMON NAME: Mexican Deermouse.

TYPE LOCALITY: México, Veracruz, 10 km E Mirador (as restricted by Dalquest, 1950:8).

DISTRIBUTION: In México, along the Atlantic coast from S San Luis Potosí to the Isthmus of Tehuantepec, and along the Pacific coast, from the Guerrero-Oaxaca border to C Chiapas; upper foothills and middle-elevation mountains in Guatemala, through El Salvador, Honduras, and Nicaragua, to highlands in Costa Rica and W Panamá (Chiriquí region).

STATUS: IUCN – Lower Risk (lc).

SYNONYMS: *altilaneus* Osgood, 1904; *angelensis* Osgood, 1904; *azulensis* Goodwin, 1956; *cacabatus* Bangs, 1902; *coatlanensis* Goodwin, 1956; *hesperus* Harris, 1940; *nicaraguae* J. A. Allen, 1908; *nudipes* (J. A. Allen, 1891); *orientalis* Goodwin, 1938; *orizabae* Merriam, 1898; *philombrius* Dickey, 1928; *putlaensis* Goodwin, 1964; *salvadorensis* Dickey, 1928; *sloeops* Goodwin, 1955; *teapensis* Osgood, 1904; *tehuantepecus* Merriam, 1898; *totontepecus* Merriam, 1898; *tropicalis* Goodwin, 1932.

COMMENTS: *P. mexicanus* species group. Geographic range, variation, and taxonomic synonymy summarized by Huckaby (1980), the species construct followed here. Includes forms formerly viewed as members of *Osgoodomys banderanus* (*angelensis, coatlanensis,* and *sloeops*—see Musser, 1969), *P. guatemalensis* (*tropicalis*—see Musser, 1969), and *P. megalops* (*azulensis*—see Huckaby, 1980); and see account of *P. gymnotis,* which had been arranged as a race of *P. mexicanus.* Euchromatic banding patterns identical in *P. mexicanus* and related species so far examined (Smith et al., 1986), and levels of genic divergence are similarly unremarkable (Rogers and Engstrom, 1992).

The conspecific status of populations now arranged under *P. mexicanus* direly needs corroboration. For one, the status of *altilaneus,* described by Osgood (1904) from Todos Santos, Guatemala, also the type locality of *P. guatemalensis,* remains problematic. The taxon was synonymized under *guatemalensis* by Huckaby (1980; also see Carleton and Huckaby, 1975) but listed as a species by Hall (1981), who followed Osgood (1909) and Hooper (1968). Carleton (1989) noted the cranial resemblance of *altilaneus* to *P. mexicanus,* and while a high-elevation occurrence for the latter, the type locality of *altilaneus* does fall within the elevational range as known for *P. mexicanus sensu lato* (Huckaby, 1980). Critical testing of this provisional synonymy is required. For another, southern populations identified as *nudipes* have been arranged as a synonym by some (Carleton, 1989; Huckaby, 1980) or considered a distinct species by others (Hooper, 1968; Osgood, 1909), including recent studies (Rogers and Engstrom, 1992; Smith et al., 1986). Based on specimens from a single locality, the latter studies leave unanswered the status of other regional epithets (*cacabatus* Bangs, *hesperus* Harris, and *orientalis* Goodwin), the distributional extent of *nudipes,* assuming that this is the applicable name, and whether *P. mexicanus* proper occurs within the same region. While we expect that a separate *mexicanus*-like species will be shown to inhabit these

Mesoamerican southern highlands, its documentation will require denser geographic and altitudinal sampling and more critical analyses than so far mustered. For a third, the status of *angelensis*, found along the lower and dry Pacific-facing slopes in Oaxaca, deserves investigation; reassigned from *banderanus* to *mexicanus* (Musser, 1969), these populations contrast in pelage color and cranial form with those inhabiting higher elevations in Oaxaca, notably *totontepecus* (see discussion in Carleton, 1989:89). Lastly, see remarks above under *P. gymnotis* on the need to clarify distributions and species-group synonyms of populations occurring in the lowlands of the Isthmus of Tehuantepec.

Peromyscus nasutus (J. A. Allen, 1891). Bull. Amer. Mus. Nat. Hist., 3:229.
COMMON NAME: Northern Rock Deermouse.
TYPE LOCALITY: USA, Colorado, Larimer Co., Estes Park.
DISTRIBUTION: C Colorado and SE Utah, south through New Mexico and Trans-Pecos Texas, USA, to NW Coahuila, México.
STATUS: IUCN – Lower Risk (lc).
SYNONYMS: *griseus* Benson, 1932; *penicillatus* Mearns, 1896.
COMMENTS: *P. truei* species group. Considered a distinct species (Osgood, 1909) until synonymized under *P. difficilis* by Hoffmeister and de la Torre (1961). Separated as a sibling species of *P. difficilis* by Zimmerman et al. (1975, 1978), as also recognized by Carleton (1989); however, the weak genic divergence persuaded Janecek (1990) to retain *nasutus* within *P. difficilis*. DeWalt et al. (1993b) also uncovered low levels of genetic divergence among *attwateri*, *difficilis*, and *nasutus* (e.g., less than *P. gratus* versus *P. truei*) but retained them as species, as did Tiemann-Boege et al. (2000) based on a broader survey of cytochrome *b* sequences among members of the *boylii* and *truei* species groups. Also see remarks under *P. difficilis*. Includes *penicillatus*, which Osgood (1909) had misclassified under *P. boylii* (see Diersing, 1976); distributional records in the Davis Mtns, W Texas, reviewed by Bradley et al. (1999b).

Peromyscus ochraventer Baker, 1951. Univ. Kansas Mus. Nat. Hist. Misc. Publ., 5:213.
COMMON NAME: El Carrizo Deermouse.
TYPE LOCALITY: México, Tamaulipas, El Carrizo, 70 km (by highway) S Ciudad Victoria and 6 km W Panamerican Highway, 2800 ft (853 m).
DISTRIBUTION: Moist forests of S Tamaulipas and adjacent San Luis Potosí, México.
STATUS: IUCN – Lower Risk (lc).
COMMENTS: Species group indeterminate. Revised by Huckaby (1980) as part of the *mexicanus* species group (*sensu* Hooper, 1968), an assignment not supported by allozymic analyses (Rogers and Engstrom, 1992). Departs from conservative karyotypic pattern exhibited by species of the *mexicanus* group (Robbins and Baker, 1981; Smith et al., 1986). Provisionally assigned to the *furvus* species group by Carleton (1989), but this association too is questioned by cytochrome *b* data, which indicate *P. ochraventer* to be strongly divergent genetically from both *P. furvus* and *mexicanus*-group species (*P. mexicanus*, *P. melanocarpus*) (Harris et al., 2000).

Peromyscus pectoralis Osgood, 1904. Proc. Biol. Soc. Wash., 17:59.
COMMON NAME: White-ankled Deermouse.
TYPE LOCALITY: México, Querétaro, Jalpan.
DISTRIBUTION: SE New Mexico and C Texas, USA, south along the Mexican Plateau and Cordillera Oriental to N Jalisco and Hidalgo, México.
STATUS: IUCN – Lower Risk (lc).
SYNONYMS: *collinus* Hooper, 1952; *eremicoides* Osgood, 1904; *laceianus* Bailey, 1906.
COMMENTS: Species group indeterminate. Although traditionally arranged as a member of the *boylii* species group (Carleton, 1989; Hooper, 1968; Osgood, 1909), a robust variety of other systematic data leaves the matter of its affinity as unresolved (e.g., Rogers et al., 1984; Sullivan and Kilpatrick, 1991; Tiemann-Boege et al., 2000). Revised by Schmidly (1972), who recognized three subspecies that correspond to patterns of genetic differentiation (Kilpatrick and Zimmerman, 1976); consanguinity of included populations nevertheless suspect (see Avise et al., 1974). See Schmidly (1974, Mammalian Species, 49).

Peromyscus pembertoni Burt, 1932. Trans. San Diego Soc. Nat. Hist., 7:176.

COMMON NAME: Pemberton's Deermouse.

TYPE LOCALITY: México, Sonora, San Pedro Nolasco Isl; 27°58′N, 111°24′W.

DISTRIBUTION: Known only from the type locality.

STATUS: IUCN – Extinct.

COMMENTS: *P. eremicus* species group. Derivation from a *P. merriami*-like ancestor postulated by Lawlor (1971*a*). Probably extinct—see Lawlor (1983) and Alvarez-Castañeda and Ortega-Rubio (2003). See Alvarez-Castañeda and Cortés-Calva (2003, Mammalian Species, 734).

Peromyscus perfulvus Osgood, 1945. J. Mammal., 26:299.

COMMON NAME: Tawny Deermouse.

TYPE LOCALITY: México, Michoacán, 10 km W Apatzingán, 1040 ft (317 m).

DISTRIBUTION: Coastal lowlands of Jalisco and Colima, along Río Balsas to interior Michocán and northernmost Guerrero, México.

STATUS: IUCN – Lower Risk (lc).

SYNONYMS: *chrysopus* Hooper, 1955.

COMMENTS: *P. melanophrys* species group. Placed with *P. melanophrys* group by Hooper (1968), an association supported by electrophoretic data (Schmidly et al., 1985). The geographic range and ecological affinity of *P. perfulvus* and *Osgoodomys banderanus* are tightly congruent and suggest similar biogeographic histories.

Peromyscus polionotus (Wagner, 1843). Arch. Naturgesch., 9, 2:52.

COMMON NAME: Oldfield Deermouse.

TYPE LOCALITY: USA, Georgia.

DISTRIBUTION: SE USA, from SC Tennessee to W South Carolina, south through Alabama and Georgia, to panhandle and N peninsular Florida, USA.

STATUS: U.S. ESA – Endangered as *P. p. allophrys*, *P. p. ammobates*, *P. p. peninsularis*, *P. p. phasma*, and *P. p. trissyllepsis*, Threatened as *P. p. niveiventris*; IUCN – Extinct as *P. p. decoloratus*, Critically Endangered as *P. p. trissyllepsis*, Endangered as *P. p. allophrys*, *P. p. ammobates*, *P. p. peninsularis*, and *P. p. phasma*, Lower Risk (nt) as *P. p. leucocephalus* and *P. p. niveiventris*, otherwise Lower Risk (lc).

SYNONYMS: *albifrons* Osgood, 1909; *allophrys* Bowen, 1968; *ammobates* Bowen, 1968; *arenarius* Bangs, 1898; *baliolus* Bangs, 1898; *colemani* Schwartz, 1954; *decoloratus* A. H. Howell, 1939; *griseobracatus* Bowen, 1968; *leucocephalus* A. H. Howell, 1920; *lucubrans* Schwartz, 1954; *niveiventris* (Chapman, 1889); *peninsularis* A. H. Howell, 1939; *phasma* Bangs, 1898; *rhoadsi* Bangs, 1898; *subgriseus* (Chapman, 1893); *sumneri* Bowen, 1968; *trissyllepsis* Bowen, 1968.

COMMENTS: *P. maniculatus* species group. Revised by Osgood (1909). Pleistocene origination of subspecies hypothesized by Bowen (1968); intraspecific genetic differentiation surveyed by Selander et al. (1971). See commentary on subspecific variation and ranges by Whitaker and Hamilton (1998), who considered *colemani* and *subgriseus* as inseparable from the nominate subspecies. Membership within *maniculatus* group and relationships to other *Peromyscus* assessed by Avise et al. (1979), Greenbaum et al. (1978), Rogers et al. (1984), and Stangl and Baker (1984*b*).

Peromyscus polius Osgood, 1904. Proc. Biol. Soc. Wash., 17:61.

COMMON NAME: Chihuahuan Deermouse.

TYPE LOCALITY: México, Chihuahua, Colonia García.

DISTRIBUTION: WC Chihuahua, México.

STATUS: IUCN – Vulnerable.

COMMENTS: Species group indeterminate. A distinctive species of enigmatic relationships: initially placed with the *truei* group (Osgood, 1909) but provisionally reassigned to the *boylii* group by Hoffmeister (1951; also see Bradley and Schmidly, 1987, and commentary by Anderson, 1972:348). Allozymic distance data weakly support the latter association (Bradley et al., 1996; Kilpatrick and Zimmerman, 1975; Schmidly et al., 1985), but mitochondrial DNA (cytochrome *b*) interpretations of its kinship are inconclusive (Tiemann-Boege et al., 2000).

Peromyscus pseudocrinitus Burt, 1932. Trans. San Diego Soc. Nat. Hist., 7:173.
COMMON NAME: Coronados Deermouse.
TYPE LOCALITY: México, Baja California Sur, Coronados Isl; 26°06′N, 111°18′W.
DISTRIBUTION: Known only from the type locality.
STATUS: IUCN – Critically Endangered.
COMMENTS: *P. eremicus* species group. Tentatively allocated to *crinitus* species group by Hooper (1968); affinities to *P. eremicus* asserted by Lawlor (1971*a*) and believed derived therefrom (Lawlor, 1983). See Alvarez-Castañeda (1998, Mammalian Species, 601).

Peromyscus sagax Elliot, 1903. Field Columbian Mus. Publ. 71, Zool. Ser., 3:142.
COMMON NAME: Michoacán Deermouse.
TYPE LOCALITY: México, Michoacán, La Palma.
DISTRIBUTION: NC Michoacán, México; limits uncertain.
COMMENTS: Species group indeterminate. Recognition long obscured as a full synonym first of *P. truei gratus* (Osgood, 1909) and later of *P. boylii levipes* (Hoffmeister, 1946), but morphometric, karyotypic, and electrophoretic information demonstrates the form's specific separation from *P. boylii*, *P. levipes*, and *P. gratus* (Bradley et al., 1996). Gene-sequence data indicate that *P. sagax* is as much differentiated from the *boylii* complex as from the *aztecus* and *truei* species groups (Tiemann-Boege et al., 2000).

Peromyscus sejugis Burt, 1932. Trans. San Diego Soc. Nat. Hist., 7:171.
COMMON NAME: Santa Cruz Deermouse.
TYPE LOCALITY: México, Baja California Sur, Santa Cruz Isl; 25°17′N, 110°43′W.
DISTRIBUTION: Santa Cruz and San Diego Isls, S Gulf of California, México.
STATUS: IUCN – Lower Risk (lc).
COMMENTS: *P. maniculatus* species group. Highly differentiated morphologically and allozymically from mainland *P. maniculatus* (Avise et al., 1974, 1979; Burt, 1932), though considered a derivative from a *maniculatus*-like ancestor (Lawlor, 1983), a relationship supported by sequence data from several mitochondrial DNA genes (Hafner et al., 2001; Hogan et al., 1997). Banded karyotype described by L. R. Smith et al. (2000), who reported a unique heterochromatin addition in *sejugis* as compared with *P. maniculatis* and recommended retention of the former as species. Hafner et al. (2001:787), on the other hand, viewed the sequence divergence between the two as minimal and predicted that additional study "will demonstrate that *P. sejugis* should be included as a subspecies of *P. maniculatus.*" Ultimate decision on its status will hinge upon a broader context of taxonomic inquiry that includes peninsular populations of *P. maniculatus* (*coolidgei*) and those along the northern Pacific coast (*P. keeni*), as noted by Hogan et al. (1997). See Alvarez-Castañeda (2001, Mammalian Species, 658).

Peromyscus simulus Osgood, 1904. Proc. Biol. Soc. Wash., 17:64.
COMMON NAME: Sinaloan Deermouse.
TYPE LOCALITY: México, Nayarit, San Blas.
DISTRIBUTION: Coastal plain and lower river valleys, sea level-200 m, from WC Sinaloa through W Nayarit, México.
STATUS: IUCN – Lower Risk (nt).
COMMENTS: *P. boylii* species group. Previously classified as a geographic race of *P. boylii* (Hall, 1981; Osgood, 1909); elevated to species by Carleton (1977). Nongeographic and geographic variation evaluated by Schmidly and Bradley (1995), who provided new distributional records. Status and interspecific relationships within the *boylii* group addressed by Carleton et al. (1982), Bradley and Schmidly (1987), Sullivan et al. (1991), and Tiemann-Boege et al. (2000). See Roberts et al. (2001, Mammalian Species, 669).

Peromyscus slevini Mailliard, 1924. Proc. California Acad. Sci., ser. 4, 12:1221.
COMMON NAME: Catalina Deermouse.
TYPE LOCALITY: México, Baja California Sur, Santa Catalina Isl, 17 mi (27 km) NE Punta San Marcial, 25°43′50″N.
DISTRIBUTION: Known only from the type locality.

STATUS: IUCN – Critically Endangered.

COMMENTS: Species group indeterminate. Proposed derivation from a *maniculatus*-like ancestor (Lawlor, 1983) has been questioned on morphological (Carleton, 1989), chromosomal (L. R. Smith et al., 2000), and genetic grounds (Hogan et al., 1997; Hafner et al., 2001). The latter molecular studies conflict in representing *slevini* either as a species cladistically basal to the *P. leucopus-P. maniculatus* species groups (Hogan et al., 1997) or as another possible synonym of *P. fraterculus* (Hafner et al., 2001). Examination of vouchers used by Hafner et al. (2001) reveals those specimens (CIB 703-708, 711, 714-717) to be examples of *P. fraterculus*, not the species described by Mailliard (1924) and later reported by Burt (1934) based on his 1931 fieldwork (M. Carleton and T. Lawlor, in litt.). Those used by Hogan et al. (1997) and L. R. Smith et al. (2000) do prove to be *P. slevini* proper (specimens in TCWC). As noted by Carleton (1989), the cranium of *P. slevini* exhibits resemblances to certain *Peromyscus* found on the Mexican mainland, explicitly with members of the *P. melanophrys* species group. The status and relationships of *P. slevini* and the identities of deermice currently inhabiting Santa Catalina warrant further investigation. See Alvarez-Castañeda and Cortés-Calva (2002, Mammalian Species, 705).

Peromyscus spicilegus J. A. Allen, 1897. Bull. Am. Mus. Nat. Hist., 9:50.

COMMON NAME: Gleaning Deermouse.

TYPE LOCALITY: México, Jalisco, Mascota, Mineral San Sebastián.

DISTRIBUTION: Low to middle elevations, 15-1980 m, along the W flanks of the Sierra Madre Occidental, México, from SE Sonora and extreme SW Chihuahua to NE Colima and WC Michoacán.

STATUS: IUCN – Lower Risk (lc).

COMMENTS: *P. aztecus* species group. Arranged as a subspecies of *P. boylii* by Osgood (1909; and Hall, 1981). Morphological, chromosomal, molecular, and distributional data support its recognition as a species having closer affinity to *P. aztecus* and its allies (see Bradley and Schmidly, 1987; Carleton, 1977, 1979; Carleton et al., 1982; Hooper, 1968; Sullivan and Kilpatrick, 1991; Sullivan et al., 1997; Tiemann-Boege et al., 2000). Includes populations in Michoácan once identified as *evides* (see Carleton, 1977; Smith et al., 1989); Bradley et al. (1996) substantially extended the northern occurrence of the species as far as SE Sonora. Variation with altitude investigated by Sánchez-Cordero and Villa-Ramírez (1988); geographic and nongeographic variation evaluated by Bradley et al. (1996*b*), who regarded patterns and levels of interpopulational divergence as insufficient to delimit subspecies. See Roberts et al. (1998, Mammalian Species, 596).

Peromyscus stephani Townsend, 1912. Bull. Am. Mus. Nat. Hist., 31:126.

COMMON NAME: San Esteban Deermouse.

TYPE LOCALITY: México, Sonora, San Esteban Isl, 28°34′N, 113°21′W.

DISTRIBUTION: Known only from Isla San Esteban.

STATUS: IUCN – Endangered.

COMMENTS: *P. boylii* species group. Lawlor (1971*a*) demonstrated the *boylii*-like traits of *P. stephani*, which was previously classified in the subgenus *Haplomylomys* (Hooper, 1968), and favored its recognition as species. Electrophoretic investigations have sustained this affinity and status (Avise et al., 1974), but mitochondrial COIII gene sequences imply subspecific recognition within *P. boylii* (Hafner et al., 2001). In a denser taxonomic survey of *boylii* group forms, using the cytochrome *b* gene and including several subspecies of *P. boylii sensu stricto*, Tiemann-Boege et al. (2000) viewed *P. stephani* as sister taxon to the *P. boylii-P. simulus* clade and argued its status as species, an interpretation that best reflects the information sources so far consulted.

Peromyscus stirtoni Dickey, 1928. Proc. Biol. Soc. Wash., 41:5.

COMMON NAME: Stirton's Deermouse.

TYPE LOCALITY: El Salvador, La Unión Dept., Río Goascorán, 13°30′ N, 100 ft (30.5 m).

DISTRIBUTION: Intermittently found in dry to semiarid lowlands of SE Guatemala, SE El Salvador, Honduras, and W Nicaragua.

STATUS: IUCN – Lower Risk (lc).

COMMENTS: *P. mexicanus* species group. Although Hooper (1968) questioned the specific recognition of *P. stirtoni*, others have substantiated its distinctive morphology and habitat (Huckaby, 1980; Jones and Yates, 1983). Association with the *mexicanus* complex considered tentative by Carleton (1989), but G-banded karyotype identical to that reported for the core members of the *mexicanus* group (Peppers et al., 1999). Late Holocene remains from Guanacaste Prov., Costa Rica, have been interpreted as evidence of recent range contraction (Woodman, 1988), but new records of the species from nearby SW Nicaragua as much suggest inadequacy of basic field survey in parts of Central America (Woodman et al., 2002). See Jones (1990, Mammalian Species, 361).

Peromyscus truei (Shufeldt, 1885). Proc. U. S. Natl. Mus., 8:407.
COMMON NAME: Piñon Deermouse.
TYPE LOCALITY: USA, New Mexico, McKinley Co., Fort Wingate.
DISTRIBUTION: USA, SW and C Oregon to W and SE Colorado, south to N Baja California Norte (México), Arizona, and New Mexico; isolated populations in N Texas (*comanche*) and S Baja California Sur (*lagunae*).
STATUS: IUCN – Lower Risk (nt) as *P. t. comanche*, otherwise Lower Risk (lc).
SYNONYMS: *chlorus* Hoffmeister, 1941; *comanche* Blair, 1943; *dyselius* Elliot, 1898; *gilberti* (J. A. Allen, 1893); *hemionotis* Elliot, 1903; *lagunae* Osgood, 1909; *lasius* Elliot, 1904; *martirensis* (J. A. Allen, 1893); *megalotis* Merriam, 1890; *montipinoris* Elliot, 1904; *nevadensis* Hall and Hoffmeister, 1940; *preblei* Bailey, 1936; *sequoiensis* Hoffmeister, 1941.
COMMENTS: *P. truei* species group. Revised by Hoffmeister (1951), who included Mexican populations subsequently recognized as *P. gratus* (see above species account). Schmidly (1973) reallocated *comanche* as a geographic variant of *P. truei*, a synonymy supported by an array of information (e.g., DeWalt et al., 1993b; Janecek, 1990; Modi and Lee, 1984); localized distribution and conservation status of this subspecies addressed by Yancey et al. (1996). Geographic variation in pelage color among Great Basin populations evaluated by Carraway and Verts (2002), who discussed results apropos recognized subspecies. Range extension and locality records in Oregon verified by Carraway et al. (1993) and Verts and Carraway (1998:Fig. 11-79). See Hoffmeister (1981, Mammalian Species, 161).

Peromyscus winkelmanni Carleton, 1977. Occas. Pap. Mus. Zool., Univ. Michigan, 675:2.
COMMON NAME: Coalcomán Deermouse.
TYPE LOCALITY: México, Michoacán, 6.3 mi (= 10.1 km, by road) WSW Dos Aguas, 8000 ft (2438 m).
DISTRIBUTION: Isolated localities in the Sierra de Coalcomán, Michoacán, and in the Sierra Madre del Sur, Guerrero, México.
STATUS: IUCN – Lower Risk (nt).
COMMENTS: *P. aztecus* species group. Karyotype reported by Smith et al. (1989); relationships evaluated by Bradley and Schmidly (1987), Sullivan and Kilpatrick (1991), and Sullivan et al. (1997). The occurrence of the species in Guerrero, as reported in several studies (e.g., Smith et al., 1989; Sullivan and Kilpatrick, 1991; Sullivan et al., 1997), seems biogeographically implausible; identification of these vouchers should be reconfirmed (We were unsuccessful in trying to locate them). Removed to *aztecus* species group by Carleton (1989), an allocation supported by other evidence (e.g., Sullivan et al., 1997; Tiemann-Boege et al., 2000).

Peromyscus yucatanicus J. A. Allen and Chapman, 1897. Bull. Am. Mus. Nat. Hist., 9:8.
COMMON NAME: Yucatán Deermouse.
TYPE LOCALITY: México, Yucatán, Chichén Itzá.
DISTRIBUTION: N Yucatán Peninsula, México.
STATUS: IUCN – Lower Risk (lc).
SYNONYMS: *badius* Osgood, 1904.
COMMENTS: *P. mexicanus* species group. The notion that *P. yucatanicus* is most closely related to *P. mexicanus* (e.g., Hooper, 1968; Lawlor, 1965) has not found support from morphological information (Carleton, 1973; Huckaby, 1980); the weak differentiation in chromosomes and allozymes offers little resolution of their level of kinship, as is

true among *mexicanus*-group species in general (Rogers and Engstrom, 1992; Rogers et al., 1984; Smith et al., 1986). Huckaby (1980) followed Lawlor (1965) in treating *P. yucatanicus* as monotypic; Hall (1981) accepted Osgood's (1909) subspecific arrangement. See Young and Jones (1983, Mammalian Species, 196).

Peromyscus zarhynchus Merriam, 1898. Proc. Biol. Soc. Wash., 12:117.
COMMON NAME: Chiapan Deermouse.
TYPE LOCALITY: México, Chiapas, mountains above Tumbalá, 5500 ft (1076 m).
DISTRIBUTION: Middle to high elevation cloud forest, 1000-2900 m, in C and SE Chiapas, México.
STATUS: IUCN – Vulnerable.
SYNONYMS: *cristobalensis* Merriam, 1898.
COMMENTS: *P. mexicanus* species group. Distribution and morphological variation reviewed by Huckaby (1980). Allozymic (Rogers and Engstrom, 1992) and chromosomal (Smith et al., 1986) comparisons convey little differentiation of *P. zarhynchus* from other members of the *mexicanus* group. The recent report (Horath and Navarrete-Gutierrez, 1997) of *P. zarhynchus* from SE Chiapas, near the range of *P. guatemalensis*, invites additional study of their relationship and taxonomic status. See McClellan and Rogers (1997, Mammalian Species, 562).

Podomys Osgood, 1909. N. Am. Fauna, 28:226.
TYPE SPECIES: *Hesperomys floridanus* Chapman, 1889.
COMMENTS: Reithrodontomyini. Named as a subgenus of *Peromyscus* by Osgood (1909) and maintained as such by Hooper (1968). Carleton (1980, 1989) argued for generic recognition, a ranking disputed by others (Rogers et al., 1984; Stangl and Baker, 1984*b*). Various morphological features suggest relationship to *Neotomodon* and/or *Habromys* (Carleton, 1980; Hooper and Musser, 1964*b*; Linzey and Layne, 1969). Accessory reproductive glands and spermatozoan morphology described by Linzey and Layne (1969, 1974).

Podomys floridanus (Chapman, 1889). Bull. Am. Mus. Nat. Hist., 2:117.
COMMON NAME: Florida Deermouse.
TYPE LOCALITY: USA, Florida, Alachua Co., Gainesville.
DISTRIBUTION: Allopatric segments in panhandle and peninsular Florida, USA.
STATUS: IUCN – Vulnerable. Considered threatened by Florida agencies due to disappearance of scrub habitat (see Layne, 1990).
COMMENTS: Smith et al. (1973) noted low heterozygosity levels within populations and high genetic similarity among them.

Reithrodontomys Giglioli, 1874. Bull. Soc. Geogr. Ital., Roma, 11:326.
TYPE SPECIES: *Reithrodon megalotis* Baird, 1857.
SYNONYMS: *Aporodon* Howell, 1914; *Cudahyomys* Hibbard, 1944; *Ochetodon* Coues, 1874.
COMMENTS: Reithrodontomyini. Genus viewed as closely related to *Peromyscus*, whether defined broadly (Hooper and Musser, 1964*b*) or narrowly (Carleton, 1980). Cladistic interpretations of banded chromosomes have not corroborated so close an affinity (Rogers et al., 1984; Stangl and Baker, 1984*b*), but those of mitochondrial and nuclear gene sequences continue to support near kinship (Allard and Honeycutt, 1991; D'Elía, 2003; Engel et al., 1998; Smith and Patton, 1999), although the sampling of critical exemplars in the latter studies can be much improved.
    Alpha taxonomy revised by Allen (1895), Howell (1914), and Hooper (1952), the last of whom framed the currently-used subgeneric division (*Aporodon* and *Reithrodontomys*) and species groups (*megalotis*, *fulvescens*, *mexicanus*, and *tenuirostris*). For comparative studies of morphology, see Arata (1964), Carleton (1973, 1980), and Hooper (1952, 1959); of karyology, see Carleton and Myers (1979), Engstrom et al. (1981), Hood et al. (1984), and Robbins and Baker (1980); of allozymic variation, see Arellano et al. (2003), Arnold et al. (1983), and Nelson et al. (1984); of gene-sequence data, see Bell et al. (2001). In general, the aforementioned studies and information sources lack the taxonomic breadth or data structure appropriate to critically test the phyletic validity of the subgeneric dichot-

omy and all four species groups proposed by Hooper (1952). We repeat his intrageneric classification as the last synoptic treatment, bearing that caveat in mind: e.g., Carleton and Myers (1979) recommended research emphasis upon species-group associations and their interrelationships instead of the two subgenera. Distributions and species limits of all species groups in Middle America require renewed systematic attention. A key to the species is found in Spencer and Cameron (1982); ecogeographic distributions of Mexican species reviewed by Sánchez (1993).

Reithrodontomys brevirostris Goodwin, 1943. Am. Mus. Novit., 1231:1.
COMMON NAME: Short-nosed Harvest Mouse.
TYPE LOCALITY: Costa Rica, Alajuela Prov., canyons above Villa Quesada, 5000 ft (1524 m).
DISTRIBUTION: Allopatric populations in highlands of NC Nicaragua (see Jones and Genoways, 1970) and C Costa Rica.
STATUS: IUCN – Lower Risk (lc).
SYNONYMS: *nicaraguae* Jones and Genoways, 1970.
COMMENTS: Subgenus *Aporodon*, *mexicanus* species group. Status of the Nicaraguan form, which Jones and Genoways considered describing as a species, should be reevaluated. See Jones and Baldassarre (1982, Mammalian Species, 192).

Reithrodontomys burti Benson, 1939. Proc. Biol. Soc. Wash., 52:147.
COMMON NAME: Sonoran Harvest Mouse.
TYPE LOCALITY: México, Sonora, Río Sonora, Rancho de Costa Rica.
DISTRIBUTION: WC Sonora to WC Sinaloa, México.
STATUS: IUCN – Lower Risk (lc).
COMMENTS: Subgenus *Reithrodontomys*, *megalotis* species group. A relict species having close affinity to *R. montanus* (Hooper, 1952).

Reithrodontomys chrysopsis Merriam, 1900. Proc. Biol. Soc. Wash., 13:152.
COMMON NAME: Volcano Harvest Mouse.
TYPE LOCALITY: México, México, Volcán Popocatépetl, 11,500 ft (3505 m).
DISTRIBUTION: Cordillera Transvolcanica, SE Jalisco to WC Veracruz, México.
STATUS: IUCN – Lower Risk (lc).
SYNONYMS: *colimae* Merriam, 1901; *orizabae* Merriam, 1901; *perotensis* Merriam, 1901; *tolucae* Merriam, 1901.
COMMENTS: Subgenus *Reithrodontomys*, *megalotis* species group.

Reithrodontomys creper Bangs, 1902. Bull. Mus. Comp. Zool., 39:39.
COMMON NAME: Talamancan Harvest Mouse.
TYPE LOCALITY: Panamá, Chiriquí Prov., Volcán de Chiriquí, 11,000 ft (3353 m).
DISTRIBUTION: Upper elevations, 1300-3350 m, in the Cordilleras Tilarán, Central, and Talamanca, Costa Rica, to Chiriquí region, W Panamá.
STATUS: IUCN – Lower Risk (lc).
COMMENTS: Subgenus *Aporodon*, *tenuirostris* species group. See Hopp and Rogers (1994, Mammalian Species, 482).

Reithrodontomys darienensis Pearson, 1939. Not. Naturae, Acad. Nat. Sci. Philadelphia, 6:1.
COMMON NAME: Darien Harvest Mouse.
TYPE LOCALITY: Panamá, Darién Prov., Santa Cruz de Cana, upper Río Tuyra, 2000 ft (610 m).
DISTRIBUTION: E Panamá, including Azuero Peninsula, and perhaps adjacent Colombia.
STATUS: IUCN – Lower Risk (lc).
COMMENTS: Subgenus *Aporodon*, *mexicanus* species group. Hooper (1952) viewed *R. darienensis* as closely related to *R. gracilis*.

Reithrodontomys fulvescens J. A. Allen, 1894. Bull. Am. Mus. Nat. Hist., 6:319.
COMMON NAME: Fulvous Harvest Mouse.
TYPE LOCALITY: México, Sonora, Oposura, 2000 ft (610 m).
DISTRIBUTION: SC Arizona, NC, S and E Texas, to SW Missouri and W Mississippi, USA; south through much of México, to W Nicaragua; excluding Yucatán Peninsula and Caribbean coastal lowlands.
STATUS: IUCN – Lower Risk (lc).

SYNONYMS: *amoenus* (Eliot, 1905); *aurantius* J. A. Allen, 1895; *canus* Benson, 1939; *chiapensis* A. H. Howell, 1914; *chrysotis* Elliot, 1899; *difficilis* Merriam, 1901; *griseoflavus* Merriam, 1901; *helvolus* Merriam, 1901; *inexspectatus* Elliot, 1903; *infernatis* Hooper, 1950; *intermedius* J. A. Allen, 1895; *laceyi* J. A. Allen, 1896; *meridionalis* Anderson and Jones, 1960; *mustelinus* A. H. Howell, 1914; *nelsoni* A. H. Howell, 1914; *tenuis* J. A. Allen, 1899; *toltecus* Merriam, 1901; *tropicalis* Davis, 1944.

COMMENTS: Subgenus *Reithrodontomys*, *fulvescens* species group. Chromosomal complement (2n = 50) interpreted as primitive for the genus (Robbins and Baker, 1980). Genetically strongly divergent from six other species of the subgenus *Reithrodontomys* based on cytochrome *b* sequences (Bell et al., 2001). Westerly dispersion along grassy railway borders onto Llano Estacado, NC Texas, documented by Yancey and Jones (1997); also see Peppers et al. (1998) for records of northward extension in Texas. Appreciable fixed allelic differences detected among certain samples from the US and Mexico advise the need for taxonomic review of included synonyms (Arellano et al., 2003; Arnold et al., 1983). See Spencer and Cameron (1982, Mammalian Species, 174).

Reithrodontomys gracilis J. A. Allen and Chapman, 1897. Bull. Am. Mus. Nat. Hist., 9:9.
COMMON NAME: Slender Harvest Mouse.
TYPE LOCALITY: México, Yucatán, Chichén Itzá.
DISTRIBUTION: Yucatán Peninsula and coastal Chiapas, México, south along Pacific watershed to WC Costa Rica (see Reid and Langtimm, 1993).
STATUS: IUCN – Lower Risk (lc).
SYNONYMS: *anthonyi* Goodwin, 1932; *harrisi* Goodwin, 1945; *insularis* Jones, 1964; *pacificus* Goodwin, 1932.
COMMENTS: Subgenus *Aporodon*, *mexicanus* species group. In the allozymic study of Arellano et al. (2003), samples of *R. gracilis* emerged as paraphyletic with regard to *R. spectabilis* on Cozumel Isl; their results identify the need for critical review of the *gracilis-spectabilis* complex. See Young and Jones (1984, Mammalian Species, 218).

Reithrodontomys hirsutus Merriam, 1901. Proc. Wash. Acad. Sci., 3:553.
COMMON NAME: Hairy Harvest Mouse.
TYPE LOCALITY: México, Jalisco, Ameca, 4000 ft (1219 m).
DISTRIBUTION: SC Nayarit and NW Jalisco, México.
STATUS: IUCN – Lower Risk (nt).
SYNONYMS: *levipes* Merriam, 1901.
COMMENTS: Subgenus *Reithrodontomys*, *fulvescens* species group.

Reithrodontomys humulis (Audubon and Bachman, 1841). Proc. Acad. Nat. Sci. Philadelphia, 1:97.
COMMON NAME: Eastern Harvest Mouse.
TYPE LOCALITY: USA, South Carolina, Charleston Co., vicinity of Charleston.
DISTRIBUTION: SE USA, from E Oklahoma and E Texas eastwards to the Atlantic seaboard, from S Maryland to peninsular Florida.
STATUS: IUCN – Lower Risk (lc).
SYNONYMS: *carolinensis* (Audubon and Bachman, 1842); *dickinsoni* Rhoads, 1895; *impiger* Bangs, 1898; *lecontii* (Audubon and Bachman, 1842); *merriami* J. A. Allen, 1895; *virginianus* A. H. Howell, 1940.
COMMENTS: Subgenus *Reithrodontomys*, *megalotis* species group. The standard and banded karyotypes depart markedly from other species of the *megalotis* group, and substantial variation in fundamental numbers has been reported among populations (Bradley et al, 1988; Carleton and Myers, 1979; Engstrom et al., 1981; Robbins and Baker, 1980). Genetically strongly divergent from six other species of the subgenus *Reithrodontomys* based on cytochrome *b* sequences (Bell et al., 2001). Whitaker and Hamilton (1998) regarded the subspecies *virginianus* and *humulis* as broadly intergrading and synonymized the former; L. Choate and Jones (1998) mentioned possible westward expansion of *R. h. merriami* in Oklahoma. See Stalling (1997, Mammalian Species, 565).

Reithrodontomys megalotis (Baird, 1857). Mammalia *in* Repts. U.S. Expl. Surv., 8(1):451.
COMMON NAME: Western Harvest Mouse.

TYPE LOCALITY: México-USA boundary region, between Janos, Chihuahua, and San Luis Springs, Grant Co., New Mexico.

DISTRIBUTION: SC British Columbia and SE Alberta, Canada; through much of W and NC USA; south to N Baja California Norte and through interior México to C Oaxaca.

STATUS: IUCN – Lower Risk (lc).

SYNONYMS: *alticolus* Merriam, 1901; *amoles* A. H. Howell, 1914; *arizonensis* J. A. Allen, 1895; *aztecus* J. A. Allen, 1893; *caryi* A. H. Howell, 1935; *catalinae* (Elliot, 1904); *cinereus* Merriam, 1901; *deserti* J. A. Allen, 1895; *distichlis* von Bloeker, 1937; *dychei* J. A. Allen, 1895; *hooperi* Goodwin, 1954; *klamathensis* Merriam, 1899; *limicola* von Bloeker, 1932; *longicaudus* (Baird, 1857); *nebrascensis* J. A. Allen, 1895; *nigrescens* A. H. Howell, 1914; *pallidus* Rhoads, 1893; *pectoralis* Hanson, 1944; *peninsulae* (Elliot, 1903); *ravus* Goldman, 1939; *santacruzae* Pearson, 1951; *saturatus* J. A. Allen and Chapman, 1897; *sestinensis* J. A. Allen, 1903.

COMMENTS: Subgenus *Reithrodontomys*, *megalotis* species group. Formerly included *R. zacatecae*, which Hood et al. (1984) identified as a separate species. The substantial chromosomal variation reported (see Engstrom et al., 1981) cautions that yet other species are lumped under *R. megalotis*. Multivariate discrimination from *R. montanus* in Kansas demonstrated by Hoofer et al. (1999). Recent range expansions into NE Illinois (see Pigage and Pigage, 1994) and within N Texas (Yancey and Jones, 1997) reported; L. Choate (1997) discussed the clinal intergradation between the subspecies *aztecus* and *megalotis* on the Llano Estacado, Texas-New Mexico; southern range extension on Baja Peninsula reported by Rios and Alvarez-Castañeda (2002). See Webster and Jones (1982, Mammalian Species, 167).

Reithrodontomys mexicanus (Saussure, 1860). Rev. Mag. Zool. Paris, Ser. 2, 12:109.

COMMON NAME: Mexican Harvest Mouse.

TYPE LOCALITY: México, Veracruz, Mirador (as restricted by Hooper, 1952:140).

DISTRIBUTION: S Tamaulipas and WC Michoacán, México, south through Middle American highlands to W Panamá; Andes of W Colombia and N Ecuador.

STATUS: IUCN – Lower Risk (lc).

SYNONYMS: *cherrii* (J. A. Allen, 1891); *costaricensis* J. A. Allen, 1895; *garichensis* Enders and Pearson, 1940; *goldmani* Merriam, 1901; *howelli* Goodwin, 1932; *jalapae* Merriam, 1901; *lucifrons* A. H. Howell, 1932; *milleri* J. A. Allen, 1912; *minusculus* A. H. Howell, 1932; *ocotepequensis* Goodwin, 1937; *orinus* Hooper, 1949; *potrerograndei* Goodwin, 1945; *riparius* Hooper, 1955; *scansor* Hooper, 1950.

COMMENTS: Subgenus *Aporodon*, *mexicanus* species group. A species composite requiring detailed revision, as underscored by the large genetic distances disclosed among Middle American populations and the polyphyly of nominal "*mexicanus*" in the resultant cladogram (Arellano et al., 2003).

Reithrodontomys microdon Merriam, 1901. Proc. Wash. Acad. Sci., 3:548.

COMMON NAME: Small-toothed Harvest Mouse.

TYPE LOCALITY: Guatemala, Huehuetenango Dept., Todos Santos, 10,000 ft (3048 m).

DISTRIBUTION: Isolated pockets in highlands of N Michoacán and Distrito Federal, N Oaxaca, and C and S Chiapas (see Espinosa M. et al., 1999a), México, and WC Guatemala.

STATUS: IUCN – Lower Risk (nt).

SYNONYMS: *albilabris* Merriam, 1901; *wagneri* Hooper, 1950.

COMMENTS: Subgenus *Aporodon*, *tenuirostris* species group. Included forms doubtfully conspecific; e.g., in cladistic reconstructions based on allozymic data, Guatemalan and Mexican samples of *R. microdon* are paraphyletic with respect to *R. tenuirostris* (Arellano et al., 2003).

Reithrodontomys montanus (Baird, 1855). Proc. Acad. Nat. Sci. Philadelphia, 7:335.

COMMON NAME: Plains Harvest Mouse.

TYPE LOCALITY: USA, Colorado, Saguache Co., upper part of the San Luis Valley (as restricted by Allen, 1895:125; also see Armstrong's, 1972:190, lucid summary of the contradictory references to the type locality).

DISTRIBUTION: High Plains of C USA, from W South Dakota and E Wyoming to EC Texas and extreme SE Arizona; NE Sonora and Chihuahua to N Durango, México.
STATUS: IUCN – Lower Risk (lc).
SYNONYMS: *albescens* Cary, 1903; *griseus* Bailey, 1905.
COMMENTS: Subgenus *Reithrodontomys*, *megalotis* species group. Multivariate discrimination from *R. megalotis* in Kansas demonstrated by Hoofer et al. (1999). New records on Edwards Plateau, EC Texas, documented by Goetze et al. (1993). See Wilkins (1986, Mammalian Species, 257).

Reithrodontomys paradoxus Jones and Genoways, 1970. Occas. Pap. W. Found. Vert. Zool., 2:12.
COMMON NAME: Nicaraguan Harvest Mouse.
TYPE LOCALITY: Nicaragua, Carazo Dept., 3 mi (5 km) NNW Diriamba, about 660 m.
DISTRIBUTION: Isolated records from SW Nicaragua and WC Costa Rica.
STATUS: IUCN – Lower Risk (nt).
COMMENTS: Subgenus *Aporodon*, *mexicanus* species group. Morphologically closest to *R. brevirostris*; extent of distribution unknown. See Jones and Baldassarre (1982, Mammalian Species, 192).

Reithrodontomys raviventris Dixon, 1908. Proc. Biol. Soc. Wash., 21:197.
COMMON NAME: Salt-marsh Harvest Mouse.
TYPE LOCALITY: USA, California, San Mateo Co., Redwood City.
DISTRIBUTION: Salt marshes around San Francisco Bay, California, USA.
STATUS: U.S. ESA – Endangered; IUCN – Endangered as *R. r. raviventris*, Lower Risk (cd) as *R. r. halicoetes*.
SYNONYMS: *halicoetes* Dixon, 1909.
COMMENTS: Subgenus *Reithrodontomys*, *megalotis* species group. Morphological and chromosomal differentiation between the subspecies *raviventris* and *halicoetes* thought to represent terminal stages of speciation (Fisler, 1965; Shellhammer, 1967). Chromosomal banding, allozymic, and DNA data reveal sister-group relationship to *R. montanus* (Bell et al., 2001; Hood et al., 1984; Nelson et al., 1984). See Shellhammer (1982, Mammalian Species, 169).

Reithrodontomys rodriguezi Goodwin, 1943. Am. Mus. Novit., 1231:1.
COMMON NAME: Rodriguez's Harvest Mouse.
TYPE LOCALITY: Costa Rica, Cartago Prov., Volcán de Irazú, 9400 ft (2865 m).
DISTRIBUTION: Patchily recorded in the Cordilleras Central and Talamancae, C Costa Rica (see McPherson, 1985; Timm et al., 1989).
STATUS: IUCN – Vulnerable.
COMMENTS: Subgenus *Aporodon*, *tenuirostris* species group.

Reithrodontomys spectabilis Jones and Lawlor, 1965. Univ. Kansas Publ., Mus. Nat. Hist., 16:413.
COMMON NAME: Cozumel Harvest Mouse.
TYPE LOCALITY: México, Quintana Roo, Isla Cozumel, 2.5 km N San Miguel.
DISTRIBUTION: Restricted to Cozumel Isl, México.
STATUS: IUCN – Endangered.
COMMENTS: Subgenus *Aporodon*, *mexicanus* species group. A large insular species, posited to share common ancestry with *R. gracilis* (Jones and Lawlor, 1965), a relationship generally supported by allozymic data (Arellano et al., 2003). See Jones (1982, Mammalian Species, 193).

Reithrodontomys sumichrasti (Saussure, 1861). Rev. Mag. Zool. Paris, Ser. 2, 13:3.
COMMON NAME: Sumichrast's Harvest Mouse.
TYPE LOCALITY: México, Veracruz, Mirador (as restricted by Hooper, 1952:72).
DISTRIBUTION: Allopatric segments in Middle American highlands: SW Jalisco and S San Luis Potosí to C Guerrero and EC Oaxaca, México; C Chiapas, México, to NC Nicaragua; C Costa Rica to W Panamá.
STATUS: IUCN – Lower Risk (lc).
SYNONYMS: *alleni* A. H. Howell, 1914; *australis* J. A. Allen, 1895; *dorsalis* Merriam, 1901; *luteolus* A. H. Howell, 1914; *modestus* Thomas, 1907; *nerterus* Merriam, 1901; *otus*

Merriam, 1901; *rufescens* J. A. Allen and Chapman, 1897; *seclusus* Hall and Villa, 1949; *underwoodi* Goodwin, 1937; *vulcanius* Bangs, 1902.

COMMENTS: Subgenus *Reithrodontomys*, *megalotis* species group. Reconstruction of historical biogeography based on mitochondrial DNA sequence data intimates that "*sumichrasti*" as defined by Hooper (1952) is a composite of two or more species (Sullivan et al., 2000), a finding that should be expanded within the context of a full revision. Sister species (*sensu lato*) to the *R. megalotis-R. zacatecae* clade based on cytochrome *b* data (Bell et al., 2001).

Reithrodontomys tenuirostris Merriam, 1901. Proc. Wash. Acad. Sci., 3:547.

COMMON NAME: Narrow-nosed Harvest Mouse.

TYPE LOCALITY: Guatemala, Huehuetenango Dept., Todos Santos, 10,000 ft (3048 m).

DISTRIBUTION: Mountains of Chiapas, México, and C Guatemala (see Arellano and Rogers, 1994:Fig. 3; Rogers et al., 1983).

STATUS: IUCN – Lower Risk (lc).

SYNONYMS: *aureus* Merriam, 1901.

COMMENTS: Subgenus *Aporodon*, *tenuirostris* species group. Karyotype similar to those of *R. creper* and *R. mexicanus* (Rogers et al., 1983). See Arellano and Rogers (1994, Mammalian Species, 477).

Reithrodontomys zacatecae Merriam, 1901. Proc. Wash. Acad. Sci., 3:557.

COMMON NAME: Zacatecan Harvest Mouse.

TYPE LOCALITY: México, Zacatecas, Sierra de Valparaíso.

DISTRIBUTION: W Chihuahua to WC Michoacán, México.

STATUS: IUCN – Lower Risk (nt).

SYNONYMS: *obscurus* Merriam, 1901.

COMMENTS: Subgenus *Reithrodontomys*, *megalotis* species group. Hooper (1952) noted the phenetic distinction of *zacatecae* from *R. megalotis saturatus* in Jalisco and Michoacán but elected to retain it as a subspecies. Karyotypic divergence of *zacatecae* reported by Hood et al. (1984), who proposed its elevation to species; status sustained by genetic distance levels, as sister species to *R. megalotis* (Bell et al., 2001). Distributional limits, morphological discrimination, and assignment of species-group synonyms deserve amplification.

Scotinomys Thomas, 1913. Ann. Mag. Nat. Hist., ser. 8, 11:408.

TYPE SPECIES: *Hesperomys teguina* Alston, 1877.

COMMENTS: Baiomyini. Until diagnosed by Thomas, species associated here had been affiliated with *Akodon* (e.g., Bangs, 1902) and later the genus with akodontines (Vorontsov, 1959). Common ancestry with *Baiomys* inferred on the basis of shared morphological traits (Carleton, 1980; Carleton et al., 1975; Hooper, 1960; Hooper and Musser, 1964*b*) and phylogenetic evaluation of mitochondrial and nuclear genes (D'Elía, 2003; Engel et al., 1998), a relationship weakened by karyological banding data (Rogers and Heske, 1984). Also see comments under *Baiomys*. Revised by Hooper (1972); for other aspects of biology and systematics, see Carleton et al. (1975), Hooper and Carleton (1976), and Rogers and Heske (1984).

Scotinomys teguina (Alston, 1877). Proc. Zool. Soc. Lond., 1876:755 [1877].

COMMON NAME: Short-tailed Singing Mouse.

TYPE LOCALITY: Guatemala, Alta Verapaz Dept., Cobán.

DISTRIBUTION: Intermediate elevations of Middle America from E Oaxaca, México, to W Panamá.

STATUS: IUCN – Lower Risk (lc).

SYNONYMS: *apricus* (Bangs, 1902); *cacabatus* Goodwin, 1945; *endersi* Goodwin, 1946; *episcopi* Enders and Pearson, 1939; *escazuensis* Goodwin, 1945; *garichensis* Enders and Pearson, 1939; *irazu* (J. A Allen, 1904); *leridensis* Enders and Pearson, 1939; *rufoniger* Sanborn, 1935; *stenopygius* Buchanan and Howell, 1967; *subnubilis* Goldman, 1935.

COMMENTS: Level of differentiation and status of populations north (*teguina*) and south (*irazu*) of the Nicaraguan Depression merit reconsideration. Hooper (1972) recognized four subspecies.

Scotinomys xerampelinus (Bangs, 1902). Bull. Mus. Comp. Zool., 39:41.

> COMMON NAME: Long-tailed Singing Mouse.
> TYPE LOCALITY: Panamá, Chiriquí Prov., Volcán de Chiriquí, 10,300 ft (3139 m).
> DISTRIBUTION: High elevations in Cordilleras Central and Talamancae of Costa Rica to Volcán Chiriquí region in W Panamá.
> STATUS: IUCN – Lower Risk (lc).
> SYNONYMS: *harrisi* Goodwin, 1945; *longipilosus* Goodwin, 1945.
> COMMENTS: Hooper (1972) synonymized *harrisi* and *longipilosus*, forms described as species, under *S. xerampelinus* and deemed the intraspecific variation as insufficient to retain subspecies.

Xenomys Merriam, 1892. Proc. Biol. Soc. Wash., 7:160.

> TYPE SPECIES: *Xenomys nelsoni* Merriam, 1892.
> COMMENTS: Neotomini. Retained as a genus since its description. Among woodrats believed to share common ancestry with *Hodomys* (Carleton, 1980); banded karyotype reported by Haiduk et al. (1988), who noted its resemblance to those of *Onychomys* and *Peromyscus*; restriction-site analyses of mitochrondrial DNA portrayed *Xenomys* and *Ototylomys* as a clade basal to nine *Neotoma* species studied, including *N. cinerea* (subgenus *Teonoma*) and *N. phenax* (subgenus *Teanopus*) (Planz et al., 1996). Sister-group kinship with *Hodomys* and generic rank reaffirmed based on broad molecular survey (cytochrome *b*) of *Neotoma* species, including *N.* (*Teonoma*) *cinerea* (Edwards and Bradley, 2002*b*).

Xenomys nelsoni Merriam, 1892. Proc. Biol. Soc. Wash., 7:161.

> COMMON NAME: Magdalena Woodrat.
> TYPE LOCALITY: México, Colima, Hacienda Magdalena (= Pueblo Juárez, as per Schaldach, 1960), between Ciudad Colima and Manzanillo.
> DISTRIBUTION: Tropical coastal lowlands, sea level-450 m, of Colima and SW Jalisco, México.
> STATUS: IUCN – Lower Risk (nt).
> COMMENTS: Most natural history information contained in the reports of Schaldach (1960) and Ceballos (1990). See Ceballos et al. (2002*b*, Mammalian Species No. 704).

Subfamily Sigmodontinae Wagner, 1843. *In* Schreber, Die Säugethiere, Suppl., 3:398.

> SYNONYMS: Akodontini Vorontsov, 1959; Hesperomyinae Murray, 1866 (Hesperomyes Winge, 1887; Hesperomyidae Ameghino, 1889; Hesperomyini, Simpson, 1945); Ichthyomyini Vorontsov, 1959; Oryzomyini Vorontsov, 1959; Phyllotiini Vorontsov, 1959 (Phyllotini Reig, 1980); Reithrodonini Vorontsov, 1959 (Reithrodontini Reig, 1980; Reithrodontina, Steppan, 1995); Scapteromyini Massoia, 1979; Sigmodontes Wagner, 1843 (Sigmodontinae Thomas, 1896; Sigmodonini Vorontsov, 1959; Sigmodontini Hershkovitz, 1966); Thomasomyini Steadman and Ray, 1982; Wiedomyini Reig, 1980; Zygodontomyini Eisenberg, 1989 [*nomen nudum*].
> COMMENTS: Priority of family-group name Sigmodontinae, dating from Sigmodontes Wagner, 1843, established by Hershkovitz (1966*c*) and Reig (1980). Taxonomic and nomenclatural histories of many forms compiled by Tate (1932*a-h*). Comprehensive and influential alpha-level classifications presented by Gyldenstolpe (1932), Ellerman (1941), and Cabrera (1961). The studies of Carleton (1980), Gardner and Patton (1976), Hershkovitz (1962, 1966*b, c*), Hooper and Musser (1964), and Reig (1980, 1981, 1984, 1987) contain information on higher-level relationships and classificatory arrangements. For critical overviews of phylogenetic and biogeographic issues, see Hershkovitz (1966*b*, 1972), Reig (1981, 1984, 1986), D'Elía (2000, 2003), and Pardiñas et al. (2002). Also see commentaries for Neotominae and Tylomyinae, whose genera were formerly arranged within Sigmodontinae, with or without indication of tribal distinction (Carleton and Musser, 1984; McKenna and Bell, 1997; Musser and Carleton, 1993).
>
> For a paleontological background, see Baskin (1978, 1986), Jacobs and Lindsay (1984), Marshall (1979), Reig (1978, 1984), Pardiñas (1999), Pardiñas et al. (2002), Slaughter and Ubelaker (1984), and Simpson (1980). Especially see Pardiñas (1995, 2000*a*) for redeterminations of Ameghino's (1889) many descriptions from Pliocene and Pleistocene beds in Argentina, and Pardiñas et al. (2002) for an encyclopedic review of South American fossil taxa and their bearing on the current systematics and biogeographic history of Sigmodon-

tinae. Within South America, earliest known examples of living genera (*Auliscomys*, *Necromys*, and possibly *Reithrodon*) date from the early Pliocene (Montehermosan Formation), and by the late Pliocene, a greater diversity of genera, living and extinct (*Abrothrix*, *Akodon*, *Cholomys*, *Dankomys*, *Graomys*, *Panchomys*, *Scapteromys*, *Wiedomys*), has been recorded (Reig, 1978, 1981, 1994; Pardiñas, 1997, 2000a; Pardiñas and Tonni, 1998; Quintana, 2002). By Pleistocene times, many fossils of living genera and species are known, mainly from deposits in Argentina and Brazil (Pardiñas, 1999, 2000a; Pardiñas et al., 2002). A late Pleistocene site in Ecuador contains not only a representative composition of the extant fauna (*Akodon*, *Anotomys*, *Microryzomys*, *Phyllotis*, *Sigmodon*, *Thomasomys*) but also a new genus (*Copemyodon*) believed to represent a member of the "Copemyne-Peromyscine group" (Fejfar et al., 1993, 1996). In North America, *Sigmodon* and its possible ancestor *Prosigmodon* are known from early Pliocene (late Hemphillian) sites in the S USA and N México (Jacobs and Lindsay, 1981; Lindsay and Jacobs, 1985; R. A. Martin, 1979). Certain low-crowned tetralophodont forms from the Pliocene of North America (*Bensonomys* and *Symmetrodontomys*) have been interpreted as primitive members of Phyllotini and Akodontini (Baskin, 1978, 1986; Czaplewski, 1987; Korth, 1994; Lindsay and Jacobs, 1985), affiliations dismissed by Reig (1980, 1984) as dental convergence. These contrasting views, and their critical bearing on sigmodontine (and neotomine) phylogeny and biogeography, have yet to be satisfactorily reconciled with fresh empirical analyses and from disinterested viewpoints. Using molecular divergence estimates (assuming an *Akodon*-*Necromys* split at 3.5 million years ago), Smith and Patton (1999) placed the major radiation of sigmodontines within South America at 10-14 million years ago, an interval harmonious with a middle Miocene appearance and well before late Pliocene landbridge formation, as earlier advocated by Hershkovitz (1966b, 1972d) and Reig (1980, 1984).

Following Thomas (1906d, 1916c, 1917c), sigmodontine genera have been informally or formally arranged into tribes (see discussions in Carleton and Musser, 1989; D'Elía, 2000; Hershkovitz, 1966c; Reig, 1981, 1984; Smith and Patton, 1999; Voss, 1993). Confirmatory evidence for the monophyly and formal employment of these suprageneric groupings is vastly improved—especially for the Akodontini (D'Elía, 2003; D'Elía et al., 2003; Reig, 1987; Smith and Patton, 1991, 1993; Steppan, 1995), Ichthyomyini (Voss, 1988), Oryzomyini (Myers et al., 1995; Patton and da Silva, 1995; Smith and Patton, 1999; Steppan, 1995; Voss and Carleton, 1993; Weksler, 2003), and Phyllotini (Braun, 1993; Olds and Anderson, 1989; Ortiz et al., 2000b; Spotorno et al., 2001; Steppan, 1993, 1995)—but much probing phylogenetic investigation of this kind is yet required.

The tribal affiliations cited here basically conform to Reig (1980, 1981, 1984), as modified by Smith and Patton (1999). Notably, we continue to maintain the Thomasomyini (= *Aepeomys*, *Chilomys*, *Rhipidomys*, and *Thomasomys*) as distinct from the Oryzomyini (see Thomas, 1906d, 1917c; Hershkovitz, 1966c; Carleton and Musser, 1989; Voss and Carleton, 1993); furthermore, we resurrect Vorontsov's (1959) Reithrodontini for the (*Euneomys*-(*Neotomys*-*Reithrodon*)) clade (see comments under *Reithrodon*). An oxymycterine group, as informally used by Hershkovitz (1966c), was never named, and evidence for such a tribe as phyletically distinct from Akodontini has not been forthcoming (Barrantes et al., 1993; D'Elía, 2003; D'Elía et al, 2003; Reig, 1987; Smith and Patton, 1999). Similarly, the scapteromyines of Hershkovitz (1966c), formalized as Scapteromyini by Massoia (1979b; also Reig, 1980, 1981, 1984, and McKenna and Bell, 1997), appear to be polyphyletic and cladistically interspersed among akodontines in molecular studies to date (D'Elía, 2003; D'Elía et al., in press; Smith and Patton, 1999). On the other hand, a congruent body of evidence reveals that certain genera currently aligned within Akodontini, here parenthetically identified as the Southern Andean clade, may eventually warrant tribal recognition (Barrantes et al., 1993; D'Elía, 2003; D'Elía et al., 2003; Reig, 1986, 1987; Smith and Patton, 1999; Spotorno et al., 1990); this possibility should be pursued and framed within a full and proper diagnosis. An intractable residuum of genera, many of them the "plesiomorphic Neotropical muroids" enumerated by Voss (1993), so far refuses to disclose their genealogical secrets, whether based on morphological or molecular information, and are noted as Sigmodontinae *incertae sedis* (see Table 1).

Faunal treatises and annotated checklists are becoming more available, providing taxonomic, distributional, and bibliographic compilations for Sigmodontinae, on both continental (Eisenberg, 1989; Eisenberg and Redford, 1999; Emmons and Feer, 1997; Redford and Eisenberg, 1992) and regional scales (Argentina—Braun and Mares [1995*b*], Chebez and Massoia [1996], Díaz [2000], Díaz and Barquez [1999], Galliari et al. [1996], Heinonen Fortabat and Chebez [1997], Mares et al. [1981*b*, 1989*b*, 1997]; Pardiñas et al. [2003*b*]; Bolivia—Anderson [1993, 1997], Anderson et al. [1993], Salazar-Bravo et al. [2003]; Brazil—Avila-Pires [1994], Fonseca et al. [1996]; Chile— Muñoz-Pedreros [2000]; Colombia—Alberico et al. [2000]; Ecuador—Tirira [1999, 2000]; Panamá—Mendez [1993]; Paraguay—Gamarra de Fox and Martin [1996]; Perú—Pacheco et al. [1995]; Surinam— Husson [1978]; Uruguay—González [2000*b*, 2001], Mones [2001], Mones and Philippi [1992]; Venezuela—Linares [1998]). The amount of taxonomic detail, extent of specimen documentation, and primary literature attribution presented in the aforementioned works vary greatly.

Abrawayaomys Souza Cunha and Cruz, 1979. Bol. Mus. Biol. Prof. Mello-Leitao Zool., 96:2.
    TYPE SPECIES: *Abrawayaomys ruschii* Cunha and Cruz, 1979.
    COMMENTS: Sigmodontinae *incertae sedis*. Diagnostic traits seem to combine aspects of *Neacomys*, *Oryzomys*, and *Akodon*, and both Reig (1987) and Smith and Patton (1999) acknowledged the enigmatic affinities of *Abrawayaomys* as uncertain. Certain cranial features of *Abrawayaomys* suggest an archaic "thomasomyine," perhaps distantly related to the other endemic genera of SE Brazil (e.g., see comments under *Delomys*).

Abrawayaomys ruschii Cunha and Cruz, 1979. Bol. Mus. Biol. Prof. Mello-Leitao Zool., 96:2.
    COMMON NAME: Ruschi's Rat.
    TYPE LOCALITY: Brazil, Espírito Santo, Forno Grande, Castelo.
    DISTRIBUTION: Known from scattered localities in Espírito Santo and Minas Gerais, Brazil, and Misiones Prov., Argentina (see Massoia et al., 1991).
    STATUS: IUCN – Endangered.

Abrothrix Waterhouse, 1837. Proc. Zool. Soc. Lond., 1837:21.
    TYPE SPECIES: *Mus longipilis* Waterhouse, 1837.
    SYNONYMS: *Chroeomys* Thomas, 1916; *Habrothrix* Wagner, 1843 [misspelling].
    COMMENTS: Akodontini (S Andean clade). An assemblage of species whose level of taxonomic recognition has come full circle: early employed as a genus (Gyldenstolpe, 1932; Thomas, 1916); later viewed as subgenus of *Akodon* (Cabrera, 1961; Ellerman, 1941; Musser and Carleton, 1993; Reig, 1978, 1981, 1987) or even a full synonym of *Microxus* (Hershkovitz, 1966*c*); and recently reinstated to genus based on congruence of morphological, karyological, and genetic traits that set it apart from typical *Akodon* (Apfelbaum and Reig, 1989; Barrantes et al., 1993; Gallardo et al., 1988; Smith and Patton, 1993, 1999; Spotorno et al., 1990). See Reig (1987), Spotorno et al. (1990), and Barrantes et al. (1993) for a review of the treatment of *Abrothrix* and its specific contents. Sister genus to the long-clawed akodonts (*Chelemys*, etc.) according to cladistic interpretations of allozymic and cytochrome *b* data (Barrantes et al., 1993; D'Elía, 2003; D'Elía et al., 2003; Smith and Patton, 1993, 1999).
        Working with N species of the S Andean clade (i.e., *jelskii* and *andinus*), Patton et al. (1989; and later Smith and Patton, 1991, 1993, and Patton and Smith, 1992) conveyed their genetic divergence and cladistic separation by employing the local genus-group taxon *Chroeomys* (type species in Perú). Described as a genus (Thomas, 1916*c*), *Chroeomys*, like *Abrothrix*, was afterwards conventionally viewed as a subgenus of *Akodon* (Cabrera, 1961; Ellerman, 1941; Gardner and Patton, 1976; Reig, 1987) until the studies of Patton and Smith. Although Smith and Patton (1999) have continued to recognize *Chroeomys* as a genus (including *jelskii* but not *andinus* which they returned to *Abrothrix*), their own gene-sequence analyses consistently nest *jelskii* among species of *Abrothrix*, not basal to them. These results, together with the persuasive body of evidence cited above, recommend the junior synonymy of *Chroeomys*.

Abrothrix andinus (Philippi, 1858). Arch. Naturgesh., 23(1):77.
    COMMON NAME: Andean Akodont.
    TYPE LOCALITY: Chile, Santiago Prov., Altos Andes.

DISTRIBUTION: Altiplano, generally above 3500 m, from SC Perú, through extreme
W Bolivia, to NW Argentina and C Chile (as far as 34°S; Muñoz Pedreros, 2000).

STATUS: IUCN – Lower Risk (lc) as *Chroeomys andinus*.

SYNONYMS: *cinnamomea* (Philippi, 1896); *dolichonyx* (Philippi, 1896); *gossei* (Thomas, 1920);
jucundus (Thomas, 1913); *polius* Osgood, 1944.

COMMENTS: Anderson (1997) employed *dolichonyx* as subspecies for Bolivian populations
but emphasized the need for fresh evaluation of intraspecific variation.

Abrothrix hershkovitzi (Patterson, Gallardo, and Freas, 1984). Fieldiana Zool., N. S., 23:8.

COMMON NAME: Hershkovitz's Akodont.

TYPE LOCALITY: Chile, Magallanes Prov., Isla Capitán Aracena, head of Bahía Morris, 60 m;
54°14'S, 71°30'W.

DISTRIBUTION: Outer islands of the Chilean Archipelago.

STATUS: IUCN – Lower Risk (lc) as *Akodon hershkovitzi*.

COMMENTS: A member of *A.* (*Akodon*) after Patterson et al. (1984) and also Muñoz Pedreros
(2000); subgenus or genus *Abrothrix* according to Reig (1987) and others (e.g., Pearson
and Smith, 1999; Smith and Patton, 1999). Pearson and Smith (1999) suspected that
hershkovitzi will prove to be another synonym of *A. olivaceus*, including *xanthorhinus*;
this possibility should be demonstrated rather than assumed in view of the
morphological separation presented in its diagnosis (Patterson et al., 1984).

Abrothrix illuteus Thomas, 1925. Ann. Mag. Nat. Hist., ser. 9, 15:582.

COMMON NAME: Gray Akodont.

TYPE LOCALITY: Argentina, Tucumán Prov., Sierra de Aconquija, 3000-4000 m.

DISTRIBUTION: NW Argentina (Jujuy, Tucumán, and Catamarca; see Díaz, 2000; Mares et al.,
1997).

STATUS: IUCN – Lower Risk (lc) as *Akodon illuteus.*.

COMMENTS: Variably placed with *Abrothrix*, usually as a subgenus (Cabrera, 1961; Reig, 1987),
or with *Akodon sensu stricto* (Gardner and Patton, 1976). Morphological and karyological
resemblances to other species of *Abrothrix* affirmed by Liascovich et al. (1989).

Abrothrix jelskii (Thomas, 1894). Ann. Mag. Nat. Hist., ser. 6, 14:360.

COMMON NAME: Ornate Akodont.

TYPE LOCALITY: Perú, Junín Dept., Junín.

DISTRIBUTION: Altiplano, about 2200-5000 m, from C Perú (see Arana-Cardó and Ascorra,
1994) to W Bolivia and NW Argentina.

STATUS: IUCN – Lower Risk (lc) as *Chroeomys jelskii*.

SYNONYMS: *bacchante* (Thomas, 1902); *cayllomae* (Thomas, 1901); *cruceri* (Thomas, 1901);
inambarii (Thomas, 1901); *inornatus* (Thomas, 1917); *ochrotis* (Sanborn, 1947);
pulcherrimus (Thomas, 1897); *pyrrhotis* (Thomas, 1894); *scalops* (Thomas, 1884); *sodalis*
(Thomas, 1913).

COMMENTS: Revised by Sanborn (1947b); Anderson (1997) observed Sanborn's subspecific
boundaries for Bolivian samples. Patton and Smith (1992) questioned whether
chromatically highly differentiated northern and southern populations really belong
to a single species.

Abrothrix lanosus (Thomas, 1897). Ann. Mag. Nat. Hist., ser. 6, 20:218.

COMMON NAME: Woolly Akodont.

TYPE LOCALITY: Chile, Magallanes Prov., Isla Madre de Dios, Seno Monteith (as corrected by
Pardiñas, in litt.).

DISTRIBUTION: Southernmost Chile and Argentina, as far north as NW Santa Cruz Prov. (see
Galliari and Pardiñas, 1999).

STATUS: IUCN – Lower Risk (lc) as *Akodon lanosis*.

COMMENTS: Relegated to a subspecies of *A. longipilis* by Mann (1978) but others have
affirmed its specific status (Osgood, 1943; Yañez et al., 1978). Earlier associated with
Microxus (e.g., Gyldenstolpe, 1932) but later allocated to subgenus *Abrothrix* of *Akodon*
(Osgood, 1943; Reig, 1987); maintained with *Akodon* by Muñoz Pedreros (2000).

Abrothrix longipilis (Waterhouse, 1837). Proc. Zool. Soc. Lond., 1837:16.

COMMON NAME: Long-haired Akodont.

TYPE LOCALITY: Chile, Coquimbo Prov., Coquimbo.

DISTRIBUTION: C to S Chile and Argentina.

STATUS: IUCN – Lower Risk (lc) as *Akodon longipilis*.

SYNONYMS: *angustus* (Thomas, 1927); *apta* (Osgood, 1943); *brachytarsus* (Philippi, 1900); *castaneus* Osgood, 1943; *francei* (Thomas, 1908); *fusco-ater* (Philippi, 1900); *hirtus* (Thomas, 1895); *melampus* (Philippi, 1900); *modestior* Thomas, 1919; *moerens* Thomas, 1919; *nubila* Thomas, 1929; *porcinus* (Philippi, 1858); *suffusus* (Thomas, 1903).

COMMENTS: Species closely related to, perhaps conspecific with *sanborni* (see Osgood, 1943; Pine et al., 1979); also see *A. lanosus*. Pearson (1984) assigned *Chelemys angustus* to synonymy of *A. longipilis*; other synonyms follow Osgood (1943) and Reig (1987). Subspecific distributions in Chile sketched by Muñoz Pedreras (2000); easternmost records in Patagonian Argentina (Río Negro and Chubut Provs.) reported by Teta et al. (2002).

Abrothrix markhami (Pine, 1973). Ann. Inst. Patagonia, 4(1-3):423-426.

COMMON NAME: Wellington Akodont.

TYPE LOCALITY: Chile, Magallanes Prov., Isla Wellington, about 1.2 km NW Puerto Eden.

DISTRIBUTION: Isla Wellington, S Chile.

STATUS: IUCN – Lower Risk (lc) as *Akodon markhami*.

COMMENTS: Considered a species of the subgenus *Akodon* by its describer (also Muñoz Pedreros, 2000); referred to *Abrothrix* by Smith and Patton (1999).

Abrothrix olivaceus (Waterhouse, 1837). Proc. Zool. Soc. Lond., 1837:16.

COMMON NAME: Olive-colored Akodont.

TYPE LOCALITY: Chile, Aconcagua Prov., Valparaíso.

DISTRIBUTION: N Chile, south through C Chile and bordering area of westernmost Argentina, to Patagonian region of S Chile and Argentina, including Tierra del Fuego.

STATUS: IUCN – Lower Risk (lc) as *Akodon olivaceus*, *Akodon mansoensis*, and *Akodon xanthorhinus*.

SYNONYMS: *atratus* (Philippi, 1900); *beatus* Thomas, 1919; *brachiotis* (Waterhouse, 1837); *brevicaudatus* (Philippi, 1872); *canescens* (Waterhouse, 1837); *chonoticus* (Philippi, 1900); *foncki* (Philippi, 1900); *germaini* (Philippi, 1900); *infans* (Philippi, 1900); *landbecki* (Philippi, 1900); *lepturus* (Philippi, 1900); *llanoi* Pine, 1976; *longibarbus* (Philippi, 1900); *macronychos* (Philippi, 1900); *mansoensis* (De Santis and Justo, 1980); *mochae* (Philippi, 1900); *nasica* (Philippi, 1900); *nemoralis* (Philippi, 1900); *pencanus* (Philippi, 1900); *psilurus* (Philippi, 1900); *renggeri* (Waterhouse, 1839); *ruficaudus* (Philippi, 1900); *senilis* (Philippi, 1900); *trichotis* (Philippi, 1900); *vinealis* (Philippi, 1900); *xanthopus* (Philippi, 1900); *xanthorhinus* (Waterhouse, 1837).

COMMENTS: Systematists have either viewed *olivaceus* and *xanthorhinus* as distinct species, based on their apparent morphological and ecological separation in local settings (Osgood, 1943; Patterson et al., 1984; Pearson, 1995); or treated *olivaceus* as a polytypic species, also embracing southern *xanthorhinus*, based on broader regional trends of morphological and morphometric variation (Mann, 1978; Yañez et al., 1979). In an allozymic study drawing upon limited geographic sampling, Barrantes et al. (1993) depicted *xanthorhinus* as more differentiated from *A. longipilis* than is *olivaceus* and represented the three as species. Recently, detailed transects in two regions where Valdivian rainforest (*olivaceus* morphotype) grades into Patagonian steppe (*xanthorhinus* morphotype) have disclosed evidence of demic intergradation, results that strongly argue their synonymy (Smith et al., 2001; Pearson and Smith, 1999). Smith et al. (2001) elaborated a biogeographic hypothesis to explain the morphological and ecological diversification within *A. olivaceus* during the late Pleistocene.

Type locality and karyotype of *xanthorhinus* discussed by Patterson et al. (1984), who judged the form *infans* Philippi (1900), synonymized under *Akodon xanthorhinus* by Osgood (1943), as a *nomen dubium* and considered *llanoi* Pine (1976b) as a subjective synonym of *Akodon x. xanthorhinus*. Also includes *mansoensis*, described as a species but whose distinctiveness was questioned by Reig (1987), Monjeau et al. (1994), and Pearson (1995); the form is morphologically and genetically inseparable from *A. olivaceus* proper (Pearson and Smith, 1999). Pearson and Smith (1999) implicated *hershko-*

vitzi as yet another junior synonym of *A. olivaceus*. Although Smith et al. (2001) offered a trinomial arrangement for the samples they studied, the many synonymies recently enacted or suggested recommend a fresh evaluation of intraspecific variation to reinforce formal recognition of racial divisions. Attribution of Philippi's numerous epithets follows Osgood (1943). See Lozada et al. (1996, Mammalian Species, 540—in part, as *Abrothrix xanthorhinus*).

Abrothrix sanborni Osgood, 1943. Field Mus. Nat. Hist. Publ., Zool. Ser., 30:194.
COMMON NAME: Sanborn's Akodont.
TYPE LOCALITY: Chile, Chiloé Prov., south end of Isla de Chiloé, mouth of Río Inio.
DISTRIBUTION: S Chile and possibly adjacent Argentina.
STATUS: IUCN – Lower Risk (lc) as *Akodon sanborni*.
COMMENTS: Perhaps conspecific with *A. longipilis*, a possibility that requires formal documentation (see Pearson, 1995; Pine et al., 1979).

Aepeomys Thomas, 1898. Ann. Mag. Nat. Hist., ser. 7, 1:452.
TYPE SPECIES: *Oryzomys lugens* Thomas, 1896.
COMMENTS: Thomasomyini. Synonymized under *Thomasomys* by Osgood (1933c), an arrangement followed by Ellerman (1941), Cabrera (1961), and Handley (1976); generic status maintained by Gyldenstolpe (1932), Gardner and Patton (1976), and Musser and Carleton (1993). Morphological recognition contrasted to *Thomasomys sensu stricto* by Ochoa G. et al. (2001); karyology reported by Gardner and Patton (1976) and Aguilera et al. (1994, 2000). Diagnosis and specific contents amended by Voss et al. (2002), who removed *Aepeomys fuscatus* J. A. Allen (1912) to *Handleyomys* and *Aepeomys vulcani* Thomas (1898) to *Thomasomys* (see those accounts).

Aepeomys lugens (Thomas, 1896). Ann. Mag. Nat. Hist., ser. 6, 18:306.
COMMON NAME: Mérida Aepeomys.
TYPE LOCALITY: Venezuela, Mérida State, La Loma del Morro, 3000 m.
DISTRIBUTION: Mérida Andes of Venezuela to Andean Ecuador.
STATUS: IUCN – Lower Risk (lc).
SYNONYMS: *ottleyi* (Anthony, 1932).
COMMENTS: Karyotype (2n = 28) reported by Aguilera et al. (2000), who noted its strong divergence from other Venezuelan series (2n = 44) later described as *A. reigi* (see next). Voss et al. (2002) confirmed the allocation of *ottleyi* as an example of *A. lugens*.

Aepeomys reigi Ochoa G., Aguilera, Pacheco, and Soriano, 2001. Z. Saugetierk., 66:230.
COMMON NAME: Reig's Aepeomys.
TYPE LOCALITY: Venezuela, Lara State, Parque Nacional Yacambu, 17 km SE Sanare, El Blanquito, 1600 m; 09°40′N, 69°37′W.
DISTRIBUTION: Venezuelan Andes (Lara and Trujillo), 1600-3230 m, W Venezuela.
COMMENTS: Morphologically similar to *A. lugens* but size larger and chromosomal formula very different, 2n =44 versus 2n = 28 in *A. lugens* (Aguilera et al., 1994, 2000).

Akodon Meyen, 1833. Verhandl. Kais. Leop.-Carol. Akad. Wiss., 16(2):599.
TYPE SPECIES: *Akodon boliviensis* Meyen, 1833.
SYNONYMS: *Acodon* Agassiz, 1846; *Axodon* Giebel, 1855; *Chalcomys* Thomas, 1916; *Hypsimys* Thomas, 1918; *Microxus* Thomas, 1909; *Plectomys* Borchert and Hansen, 1983 [*nomen nudum* as per Hershkovitz, 1990c].
COMMENTS: Akodontini. The morphotypical akodontine genus stands at the nexus of a host of specific-and generic-level taxonomic problems, but a review of all classificatory variations would be more confusing than enlightening. The nomenclatural history of the following genus-group taxa has been significantly, at times confusingly, intertwined with that of *Akodon*: *Abrothrix*, *Bolomys*, *Chalcomys*, *Chroeomys*, *Deltamys*, *Hypsimys*, *Microxus*, *Necromys*, *Thalpomys*, and *Thaptomys* (especially see Cabrera, 1961; Ellerman, 1941; Gyldenstolpe, 1932; Massoia and Pardiñas, 1993; Reig, 1984, 1987; Tate, 1932g; Thomas, 1916c). See accounts of *Abrothrix* (including *Chroeomys*), *Deltamys*, *Necromys* (including *Bolomys*), *Thalpomys*, and *Thaptomys* for history and arguments on their generic status. Divorced of these taxa, those *Akodon* species thus far surveyed in allozymic and gene-

sequence studies represent a monophyletic clade on resultant trees (Barrantes et al., 1993; D'Elía, 2003; D'Elía et al., 2003; Smith and Patton, 1993, 1999). The formation of species groups is a taxonomic edifice still under construction, with contradictory views over membership apparent in initial attempts (e.g., Hershkovitz, 1990c; Myers, 1989; Myers and Patton, 1989; Myers et al., 1990; Rieger et al., 1995) and many species unassociated for want of convincing study.

Renewed interest in basic revisionary study is rendering species definitions and distributional limits of akodonts less imposing (see especially Apfelbaum and Reig, 1989; Hershkovitz, 1990c; Myers, 1989; Myers and Patton, 1989; Myers et al., 1990; Patton and Smith, 1992; Patton et al., 1989; Reig, 1987). Distributions, morphological traits, karyotypes, and nomenclatural problems of Peruvian species reviewed by Patton and Smith (1992). Although much improved, inconsistent taxonomic treatment of many species-group epithets and unnamed populations with distinctive karyotypes (e.g., Geise et al., 2001; Pardiñas et al., 2003a; Silva and Yonenaga-Yassuda, 1998) advise continued intense alpha-level attention, including regular citation of voucher specimens. Galliari et al. (1996) highlighted names that have appeared in the literature (*diminutus*, *massoiai*, *minoprioi*, *sarmientoi*), but were never properly described and are considered *nomina nuda* (also see Braun et al., 2000).

Akodon aerosus Thomas, 1913. Ann. Mag. Nat. Hist., ser. 8, 11:406.
COMMON NAME: Yungas Akodont.
TYPE LOCALITY: Ecuador, Tungurahua Prov., upper Río Pastaza, Mirador, 1500 m.
DISTRIBUTION: Upper montane forests along E Andean slopes, 1200-2000 m, in Ecuador, Perú, and C Bolivia (Anderson, 1997; Patton and Smith, 1992).
STATUS: IUCN – Lower Risk (lc).
SYNONYMS: *baliolus* Osgood, 1915.
COMMENTS: Type species of *Chalcomys*, usually placed as a synonym of the subgenus *Akodon* (Cabrera, 1961; Reig, 1987). Formerly included as a subspecies of *A. urichi* by Cabrera (1961), but Gardner and Patton (1976), noting the pronounced difference in diploid number, listed the two as separate species; we include *urichi* under *Necromys* (see that account). Probably more than one species yet masquerades under the name of *A. aerosus* (see Patton and Smith, 1992; Patton et al., 1990); status of *surdus* as a possible junior synonym needs resolution (see Patton and Smith, 1992). Anderson (1997) recognized *baliolus* as a valid subspecies for Bolivian populations.

Akodon affinis (J. A. Allen, 1912). Bull. Am. Mus. Nat. Hist., 31:89.
COMMON NAME: Cordillera Occidental Akodont.
TYPE LOCALITY: Colombia, Valle del Cauca Dept., San Antonio, near Cali, 8000 ft (2438 m).
DISTRIBUTION: Cordillera Occidental of W Colombia.
STATUS: IUCN – Lower Risk (lc).
SYNONYMS: *tolimae* J. A. Allen, 1913.
COMMENTS: Subgenus *Akodon*. Tentatively affiliated with *Microxus* by Gyldenstolpe (1932), but Anthony (1924) had intimated its membership in *Akodon* and Cabrera (1961) arranged it as such.

Akodon albiventer Thomas, 1897. Ann. Mag. Nat. Hist., ser. 6, 20:217.
COMMON NAME: White-bellied Akodont.
TYPE LOCALITY: Argentina, Salta Prov., Bajo Río Cachí.
DISTRIBUTION: SE Perú, through SW Bolivia, to N Argentina and extreme NE Chile.
STATUS: IUCN – Lower Risk (lc).
SYNONYMS: *berlepschii* Thomas, 1898.
COMMENTS: Subgenus *Akodon*. Sometimes referenced as a member of *Bolomys* (Bianchi et al., 1971; Gardner and Patton, 1976), but its inclusion, together with *berlepschii*, within *Akodon* proper is more strongly supported based on morphlogical comparisons and molecular analyses (D'Elía et al., 2003; Pine et al., 1979; Reig, 1987). Anderson (1997) retained *berlepschii* as a valid subspecies.

Akodon aliquantulus Díaz, Barquez, Braun, and Mares, 1999. J. Mammal., 80:788.
COMMON NAME: Diminutive Akodont.

TYPE LOCALITY: Argentina, Tucumán Prov., Tafí Viejo Dept., Las Aguitas, Cumbres del Tafícillo, 1700 m; 26°42'S, 65°22'W.

DISTRIBUTION: Known only from the type locality.

COMMENTS: Subgenus *Akodon*, *A. boliviensis* species group (Myers et al., 1990). A small species related to *A. puer* (here = *A. lutescens*) but morphological differentiation from that form unimpressive based on multivariate ordinations and univariate overlap (Díaz et al., 1999:Fig. 2, Table 1). The differing treatments of species limits and affinities among these small akodonts (especially *aliquantulus*, *alterus*, *boliviensis*, *caenosus*, *lutescens*, *spegazzinii*, *subfuscus*, and *tucumanensis*—see Hershkovitz, 1990c, vis a vis Myers et al., 1990, and overview by Díaz et al., 1999) signal the need for concerted regional examination using multiple data sources, larger sample sizes, specimens from intermediate regions, and detailed transects across habitat types, e.g., following the revisionary model of Myers et al. (1990).

Akodon azarae (Fischer, 1829). Synopsis Mamm., p. 325.

COMMON NAME: Azara's Akodont.

TYPE LOCALITY: Argentina, Entre Ríos Prov., about 30°30'S latitude between the Uruguay and Paraná Rivers.

DISTRIBUTION: NE and EC Argentina, Paraguay, Uruguay, and extreme S Brazil.

STATUS: IUCN – Lower Risk (lc).

SYNONYMS: *arenicola* (Waterhouse, 1837); *bibianae* Massoia, 1971; *hunteri* Thomas, 1917.

COMMENTS: Subgenus *Akodon*, *A. boliviensis* species group (Hershkovitz, 1990c). Olrog and Lucero (1981) maintained *arenicola* as a species distinct from *A. azarae*, but karyotypic and other data support their union (Vitullo et al., 1986; Ximenez et al., 1972). Species documented in the early-middle Pleistocene of Argentina by Pardiñas (1993), who discussed its biogeographic implications.

Akodon bogotensis (Thomas, 1895). Ann. Mag. Nat. Hist., ser. 6, 16:369.

COMMON NAME: Bogotá Akodont.

TYPE LOCALITY: Colombia, Cundinamarca Dept., Bogotá region, 8750 ft (2667 m).

DISTRIBUTION: Andes of NW Venezuela, E and C Colombia.

STATUS: IUCN – Lower Risk (lc).

COMMENTS: This form has been conventionally treated as a member of *Microxus* (e.g., Cabrera, 1961). As noted by Patton et al. (1989), the representation of *Microxus* as generically distinct from *Akodon* is more often based on traits of the Andean species *bogotensis* (e.g., Linares, 1998), not the type species of *Microxus* (= *Akodon mimus*; see below). If not a *Microxus*, the distinctiveness of *bogotensis* remains an issue (e.g., see Reig, 1987; Ventura et al., 2000), and like those authors, we are impressed by the non-*Akodon* characteristics of the species. Attention to other possible phylogenetic connections may prove fruitful: e.g., Pérez-Zapata et al. (1992) have identified certain similarities between *bogotensis* and *Podoxomys roraimae*. Male reproductive glands described by Voss and Linzey (1981), who noted that the species possessed a single pair of ventral prostates compared with two in most akodonts dissected.

Akodon boliviensis Meyen, 1833. Verhandl. Kais. Leop.-Carol. Akad. Wiss., 16(2):600, pl. 43, fig. 1.

COMMON NAME: Bolivian Akodont.

TYPE LOCALITY: Perú, Puno Dept., Pichu-Pichun, 14,000 ft (location clarified by Myers et al., 1990:49).

DISTRIBUTION: Altiplano of SE Perú to SC Bolivia.

STATUS: IUCN – Lower Risk (lc).

SYNONYMS: *pacificus* Thomas, 1902.

COMMENTS: Subgenus *Akodon*, *A. boliviensis* species group (Myers et al., 1990). Karyology and geographic variation evaluated by Myers et al. (1990), who more narrowly defined the morphological and distributional boundaries of *A. boliviensis*, excluding *spegazzinii* and *subfuscus*, forms which had been arranged as subspecies (e.g., Cabrera, 1961). Southern range limits poorly understood (see Myers et al., 1990).

Akodon budini (Thomas, 1918). Ann. Mag. Nat. Hist., ser. 9, 1:191.

COMMON NAME: Budin's Akodont.

TYPE LOCALITY: Argentina, Jujuy Prov., Leon, 1500 m.
DISTRIBUTION: Mountains of NW Argentina and SC Bolivia (see Emmons, 1997).
STATUS: IUCN – Lower Risk (lc).
SYNONYMS: *deceptor* (Thomas, 1921).
COMMENTS: Type species of *Hypsimys*, which Cabrera (1961), Reig (1987), and Smith and
          Patton (1999) recognized as a subgenus of *Akodon*; listed as a genus by Galliari et al.
          (1996) without comment. Karyotype reported by Vitullo et al. (1986) and biochemical
          divergence by Apfelbaum and Reig (1989).

Akodon cursor (Winge, 1887). E Museo Lundii, 1(3):25.
COMMON NAME: Cursorial Akodont.
TYPE LOCALITY: Brazil, Minas Gerais, Lagoa Santa, Rio das Velhas.
DISTRIBUTION: Atlantic Forest formations in SE Brazil (Bahia to Paraná) and perhaps
          NE Argentina (Misiones Prov., see Pardiñas et al., 2003*a*); limits uncertain.
STATUS: IUCN – Lower Risk (lc).
COMMENTS: Subgenus *Akodon*, *A. mollis* species group (Hershkovitz, 1990*c*) or *A. cursor*
          species group (Geise et al., 2001; Rieger et al., 1995). Earlier recognized as a subspecies
          of *arviculoides* (e.g., Cabrera, 1961; Gyldenstolpe, 1932), which Reig (1978, 1987)
          reallocated to *Bolomys lasiurus* (see *Necromys*). Species characterized by an unusually
          low diploid number (2n = 14-15) and exhibiting substantial chromosomal variation
          (Fagundes et al., 1998; Sbalqueiro and Nascimento, 1996); however, as noted by Geise
          et al. (2001), application of *cursor* to this karyomorph (2n = 14-15, FN = 18-20) requires
          critical study of type specimens since the *A. montensis* karyomorph (2n = 24, FN = 42)
          occurs in sympatry at the type locality. Phylogeographic structure poorly defined
          based on cytochrome *b* haplotypes (Geise et al., 2001). Chromosomal, allozymic, and
          genetic studies have referenced an unnamed form from Pernambuco and Paraíba that
          is closely related to *A. cursor* (Geise et al., 2001; Rieger et al., 1995). Formerly included
          montensis (see below).

Akodon dayi Osgood, 1916. Field Mus. Nat. Hist. Publ., Zool. Ser., 10:208.
COMMON NAME: Dusky Akodont.
TYPE LOCALITY: Bolivia, Cochabamba Dept., Todos Santos, Río Chaparé.
DISTRIBUTION: Lowland to intermediate elevations (up to 2450 m), S Pando to WC Santa
          Cruz, Bolivia (Anderson, 1997).
STATUS: IUCN – Lower Risk (lc).
COMMENTS: Subgenus *Akodon*, *A. varius* species group (Myers, 1989). Interpreted as a
          subspecies of *A. tapirapoanus* (= *Bolomys lasiurus*) by Cabrera (1961); considered a
          species closely related to *A. toba* by Myers (1989).

Akodon dolores Thomas, 1916. Ann. Mag. Nat. Hist., ser. 8, 18:324.
COMMON NAME: Córdoba Akodont.
TYPE LOCALITY: Argentina, Córdoba Prov., near Villa Dolores, Yacanto, 900 m.
DISTRIBUTION: C Argentina (Catamarca, Córdoba, and Santiago del Estero).
STATUS: IUCN – Lower Risk (lc).
COMMENTS: Subgenus *Akodon*, *A. varius* species group (Myers, 1989). Considered closely
          related to *A. molinae* (Bianchi et al., 1979; Myers, 1989; Wittouck et al., 1995), if not
          conspecific with it (Hershkovitz, 1990*c*), but karyological differences (2n = 34-40 in
          A. dolores versus 42-44 in *A. molinae*) and generally non-overlapping distributions
          suggest distinctive species, as clarified by Tiranti (1998*b*).

Akodon fumeus Thomas, 1902. Ann. Mag. Nat. Hist., ser. 7, 9:137.
COMMON NAME: Smoky Akodont.
TYPE LOCALITY: Bolivia, Cochabamba Dept., Río Secure, El Choro, 3500 m.
DISTRIBUTION: E Andean slopes, about 1000-3500 m, of SE Perú, WC Bolivia (Anderson,
          1997), and NW Argentina (Jujuy, see Díaz and Barquez, 1999).
STATUS: IUCN – Lower Risk (lc).
COMMENTS: Subgenus *Akodon*, *A. fumeus* species group (Myers and Patton, 1989). Treated as a
          subspecies of *A. mollis* (Cabrera, 1961; Gyldenstolpe, 1932; Hershkovitz, 1990*c*); viewed
          as a distinct species more closely related to *A. kofordi* by Myers and Patton (1989).

Akodon iniscatus Thomas, 1919. Ann. Mag. Nat. Hist., ser. 9, 3:205.
COMMON NAME: Patagonian Akodont.
TYPE LOCALITY: Argentina, Chubut Prov., Valle del Lago Blanco, Koslowsky region, 100 m.
DISTRIBUTION: WC to EC Argentina (Buenos Aires Prov., see Pardiñas and Galliari, 1999), as far south as Santa Cruz Prov.
STATUS: IUCN – Lower Risk (lc).
SYNONYMS: *collinus* Thomas, 1919; *nucus* Thomas, 1926.
COMMENTS: Subgenus *Akodon*. The form *nucus* has been variously listed as a species (Galliari et al., 1996; Hershkovitz, 1990*c*; Reig, 1987), but Barros et al. (1990) and Pearson (1995) retained it as a subspecies pending critical revision of this complex. Karyotype (2n = 33-34) reported by Barros et al. (1990), who noted its extensive banding homologies with samples of *A. puer* (here = *A. lutescens*).

Akodon juninensis Myers, Patton, and Smith, 1990. Misc. Publ. Mus. Zool., Univ. Michigan, 177:41.
COMMON NAME: Junín Akodont.
TYPE LOCALITY: Perú, Junín Dept., 22 km (by road) N La Oroya (at junction of Hwy 3 to Junín and Hwy 20 to Tarma), 4040 m.
DISTRIBUTION: E and W Andean slopes, above 2700 m, of C Perú, south along W slopes to Dept. Ayacucho.
STATUS: IUCN – Lower Risk (lc).
COMMENTS: Subgenus *Akodon*, *A. boliviensis* species group (Myers et al., 1990).

Akodon kofordi Myers and Patton, 1989. Occas. Pap. Mus. Zool., Univ. Michigan, 721:14.
COMMON NAME: Koford's Akodont.
TYPE LOCALITY: Perú, Puno Dept., 9 km (by road) N Limbani, Agualani, 2840 m.
DISTRIBUTION: Depts. Cusco and Puno, SE Perú, 2750 to 2900 m, and Cochabamba Dept., C Bolivia, 1833-3700 m.
STATUS: IUCN – Lower Risk (lc).
COMMENTS: Subgenus *Akodon*, *A. fumeus* species group (Myers and Patton, 1989). Distribution and ecological associations amplified by Salazar-Bravo et al. (2002*b*).

Akodon latebricola (Anthony, 1924). Am. Mus. Novit., 139:3.
COMMON NAME: Ecuadorean Akodont.
TYPE LOCALITY: Ecuador, Tungurahua Prov., east of Ambato on Río Cusutagua, Hacienda San Francisco, 8000 ft (2438 m).
DISTRIBUTION: Cordillera Oriental, ca. 2400-3840, NC Ecuador (Pichincha and Tungurahua Provs.).
STATUS: IUCN – Lower Risk (lc).
COMMENTS: Subgenus *Microxus* (see remarks under *Akodon bogotensis* and *A. mimus*). Anthony (1924) described the species based on a single specimen under *Microxus*, as genus, and critically compared it with *bogotensis*. Additional specimens reported, range extended northwardly, and morphological description expanded by Voss (2003).

Akodon leucolimnaeus Cabrera, 1926. Rev. Chilena Hist. Nat., 30:320.
COMMON NAME: Catamarca Akodont.
TYPE LOCALITY: Argentina, Catamarca Prov., Laguna Blanca, 3100 m.
DISTRIBUTION: NW Argentina; limits unknown.
COMMENTS: Subgenus *Akodon*, *A. boliviensis* species group (Galliari et al., 1996). Reassigned as a subspecies of *lactens* (Gyldenstolpe, 1932; Cabrera, 1961) and thereafter associated as a member of *Bolomys* (Reig, 1987; Hershkovitz, 1990*c*). Reassigned as a species of *Akodon* by Galliari and Pardiñas (1995), its status considered provisional pending an emended diagnosis and clarification of its relationship.

Akodon lindberghi Hershkovitz, 1990. Fieldiana Zool., N.S., 57:16.
COMMON NAME: Lindbergh's Akodont.
TYPE LOCALITY: Brazil, Distrito Federal, 20 km NW Brasília, Parque Nacional de Brasília, Matosa, 1100 m.
DISTRIBUTION: Cerrado habitat, C and SE Brazil.
STATUS: IUCN – Vulnerable.

COMMENTS: Subgenus *Akodon*, *A. boliviensis* species group (Hershkovitz, 1990*c*). Morphology, karyotype, and habitat described by Hershkovitz (1990*c*); additional karyotypic and distributional information reported by Svartman and Almeida (1994) and Geise et al. (1996, 2001). Hershkovitz (1990*c*) discussed the *nomen nudum* *Plectomys paludicola* and its possible equivalence to the species *A. lindberghi*.

Akodon lutescens J. A. Allen, 1901. Bull. Am. Mus. Nat. Hist., 14:46.
COMMON NAME: Altiplano Akodont.
TYPE LOCALITY: Perú, Puno Dept., Tirapata, 15,000 ft (4572 m).
DISTRIBUTION: High altiplano of C Perú (Puno), through W Bolivia, to NW Argentina.
STATUS: IUCN – Lower Risk (lc) as *A. puer.*
SYNONYMS: *caenosus* Thomas, 1918; *puer* Thomas, 1902.
COMMENTS: Subgenus *Akodon*, *A. boliviensis* species group (Myers et al., 1990). Anderson (1997) noted the earlier availability of *lutescens* for this species. Karyologic and morphometric variation presented, as *A. caenosus*, by Bárquez et al. (1980). In their revision of the *boliviensis* group, Myers et al. (1990) referred *caenosus*, maintained as a species by Cabrera (1961), and *lutescens*, classified as a subspecies of *A. andinus* by Cabrera (1961), to subspecies of *A. puer* (also see Vitullo et al., 1986). Hershkovitz (1990*c*) listed *caenosus* as a species, as did Mares et al. (1997), but Anderson (1997) retained it, along with *lutescens* and *puer*, as formal subspecies. Also see remarks under *A. aliquantulus*.

Akodon mimus (Thomas, 1901). Ann. Mag. Nat. Hist., ser. 7, 7:183.
COMMON NAME: Hocicudo-like Akodont.
TYPE LOCALITY: Perú, Puno Dept., Limbane, 2600 m.
DISTRIBUTION: Upper E Andean slopes, about 2000-3700 m, from SE Perú (Puno Dept.; Patton and Smith, 1992) to WC Bolivia (Cochabamba Dept.; Anderson, 1997).
STATUS: IUCN – Lower Risk (lc).
COMMENTS: Type species of *Microxus*, a taxon arranged either as a subgenus of *Akodon* (Cabrera, 1961), as a distinct akodontine genus (Ellerman, 1941; Gyldenstolpe, 1932; Reig, 1987; Thomas, 1916*c*), or as a synonym of the "oxymycterine" genus *Abrothrix* (Hershkovitz, 1966*c*). Allozymic and cytochrome *b* evidence reveals that *mimus* is phyletically closer to species of *Akodon sensu stricto* than to those of *Abrothrix* (including *Chroeomys*) and *Bolomys* (= *Necromys*) and doubtfully warrants even subgeneric recognition based on these data (see Patton and Smith, 1992; Patton et al., 1989; Smith and Patton, 1991, 1993).

Akodon molinae Contreras, 1968. Zool. Platense, 1(2):9-12.
COMMON NAME: Molina's Akodont.
TYPE LOCALITY: Argentina, Buenos Aires Prov., Partido de Villarino, Laguna Chasicó.
DISTRIBUTION: S Argentina (Mendoza to Buenos Aires Provs., south to Chubut).
STATUS: IUCN – Lower Risk (lc).
COMMENTS: Subgenus *Akodon*, *A. varius* species group (Myers, 1989). Chromosomal polymorphism (2n = 42-44) and distributional extent with regard to *A. dolores* (also see remarks therein) reported by Tiranti (1998*b*).

Akodon mollis Thomas, 1894. Ann. Mag. Nat. Hist., ser. 6, 14:363.
COMMON NAME: Soft-furred Akodont.
TYPE LOCALITY: Perú, Piura Dept., Tumbes.
DISTRIBUTION: Pacific lowlands and adjoining montane regions of N Ecuador to WC Perú.
STATUS: IUCN – Lower Risk (lc).
SYNONYMS: *altorum* Thomas, 1913; *fulvescens* Hershkovitz, 1940.
COMMENTS: Subgenus *Akodon*, *A. mollis* species group. Extensive altitudinal range and substantial interpopulational differentiation suggest a composite of two or more species (Patton and Smith, 1992; Smith and Patton, 1993; study of specimens in AMNH); status of highland form *altorum* as synonym questioned by Voss (2003). Also see remarks under *A. fumeus*, once included as a subspecies.

Akodon montensis Thomas, 1913. Ann. Mag. Nat. Hist., ser. 8, 11:405.
COMMON NAME: Montane Akodont.

TYPE LOCALITY: Paraguay, Paraguarí Dept., Sapucái.

DISTRIBUTION: E Paraguay (Gamarra de Fox and Martin, 1996), NE Argentina (Pardiñas et al., 2003*a*), and SE Brazil (Rio de Janeiro to Rio Grande do Sul, including gallery forest settings in Minas Gerais and Goiás; Geise et al., 2001; Rieger et al., 1995).

COMMENTS: Subgenus *Akodon*, *A. cursor* species group (Rieger et al., 1995). Named as a subspecies of *A. arviculoides*, a ranking observed by Gyldenstolpe (1932) and Cabrera (1961). Morphologically similar to *A. cursor*, under which it has been included (e.g., Musser and Carleton, 1993; Reig, 1987; Ximénez and Langguth, 1970), but karyotypically, allozymically, and genetically distinct from that species (D'Elía et al., 2003; Geise et al., 1998, 2001; Rieger et al., 1995). Where *A. cursor* and *A. montensis* occur in proximity, the latter species is found at higher elevations, 800 m and above (Geise et al., 2001); past reports for this species in Uruguay (e.g., Ximénez and Langguth, 1970) have proven to be *A. reigi* (González et al., 1998). Pardiñas et al. (2003*a*) reviewed the taxon's involved historical treatment and summarized evidence for its discrimination from *A. cursor* (also see that account).

Akodon mystax Hershkovitz, 1998. Bonn Zool. Beitr., 47:220.

COMMON NAME: Caparaó Akodont.

TYPE LOCALITY: Brazil, Minas Gerais, Pico da Bandeira, W slope Mt Caparaó, Arrozal, 2300 m.

DISTRIBUTION: Known only from highland localities, 2000-2700 m, in Minas Gerais and Rio de Janeiro, SE Brazil.

COMMENTS: Subgenus *Akodon*, *A. boliviensis* species group (Hershkovitz, 1990*c*) or *A. cursor* species group (Geise et al., 2001). Described as related to *A. sanctipaulensis*, *A. lindberghi*, and *A. azarae* (Hershkovitz, 1998), but gene-sequence data suggest closer affinity to *A. cursor-A. montensis* (Geise et al., 2001). Status with respect to *A. paranaensis* and to *A. reigi*, which are morphologically very similar and have the same diploid number (Christoff et al., 2000; González et al., 1998), should be reviewed and their morphological definition and distributional limits as three separate species verified.

Akodon neocenus Thomas, 1919. Ann. Mag. Nat. Hist., ser. 9, 3:213.

COMMON NAME: Neuquén Akodont.

TYPE LOCALITY: Argentina, Neuquén Prov., upper Río Negro, Río Limay.

DISTRIBUTION: WC Argentina; limits unknown.

STATUS: IUCN – Lower Risk (lc).

COMMENTS: Subgenus *Akodon*, *A. varius* species group (Myers, 1989). A subspecies of *A. varius* according to Cabrera (1961); a species more closely related to *A. dolores* or *A. toba* according to Myers (1989).

Akodon oenos Braun, Mares, and Ojeda, 2000. Z. Saugetierk., 65:218.

COMMON NAME: Monte Akodont.

TYPE LOCALITY: Argentina, Mendoza Prov., Dept. La Valle, La Pega; 32°48'S, 68°40'W.

DISTRIBUTION: Known only from the type locality and vicinity, 600-1200 m, WC Argentina.

COMMENTS: Subgenus *Akodon*, *A. varius* species group (Myers, 1989). A medium-size species compared with *A. molinae*. Braun et al. (2000) noted that their new species corresponds to a *nomen nudum* (*minoprioi*) that had appeared in a meeting abstract (also see Galliari et al., 1996).

Akodon orophilus Osgood, 1913. Field Mus. Nat. Hist. Publ., Zool. Ser., 10:98.

COMMON NAME: Utcubamba Akodont.

TYPE LOCALITY: Perú, Amazonas Dept., Leimabamba, Alto Utcubamba, 2400 m.

DISTRIBUTION: Upper E Andean slopes, C Perú.

STATUS: IUCN – Lower Risk (lc).

SYNONYMS: *orientalis* Osgood, 1913.

COMMENTS: Subgenus *Akodon*. Although initially described as a subspecies of *A. mollis*, Osgood (1943:197) later reinstated *orophilus* as a distinct species, at which rank it has since remained (Cabrera, 1961; Reig, 1987; Patton and Smith, 1992). Formerly included *torques* as a subspecies (Cabrera, 1961).

Akodon paranaensis Christoff, Fagundes, Sbalqueiro, Mattevi, and Yonenaga-Yassuda, 2000. J. Mammal., 81:844.

COMMON NAME: Paraná Akodont.

TYPE LOCALITY: Brazil, Paraná State, Piraquara, Estação Ecologica Canguiri.

DISTRIBUTION: NE Argentina (Misiones Prov.), SE Brazil (Paraná and Rio Grande do Sul).

COMMENTS: Subgenus *Akodon*. Populations assigned to this species had been formerly identified as *A. serrensis* (e.g., Liascovich and Reig, 1989), but karyotypic and morphometric separation from that form clarified by Christoff et al. (2000). Similar in size to members of the *A. mollis* species group (*sensu* Hershkovitz, 1990c), but Christoff et al. (2000) left the assignment of *A. paranaensis* as species group indeterminate pending additional study of its relationship. Also see remarks under *A. mystax* and involved discussion by Pardiñas et al. (2003a) on the tenuous basis for segregating *paranaensis* and *reigi*; resolution is needed.

Akodon pervalens Thomas, 1925. Ann. Mag. Nat. Hist., ser. 9, 15:579.

COMMON NAME: Tarija Akodont.

TYPE LOCALITY: Bolivia, Tarija Dept., Caraparí, 1000 m.

DISTRIBUTION: SC Bolivia, 900-2100 m (presumably into N Argentina).

COMMENTS: Subgenus *Akodon*. Status and relationship variously recognized: named as a subspecies of *A. sylvanus* by Thomas (1925; as followed by Musser and Carleton, 1993); reallocated to a subspecies of *A. varius* by Cabrera (1961); retained as a nominal species similar to *A. cursor* by Myers (1989); placed as a synonym of *A. serrensis* by Hershkovitz (1990c). As dryly noted by Anderson (1997), who also listed *pervalens* as species, the want of unanimity in its treatment underlines the need for taxonomic resolution.

Akodon reigi González, Langguth, and Oliveira, 1998. Comun. Zool. Mus. Hist. Nat. Montevideo, 12(191):2.

COMMON NAME: Reig's Akodont.

TYPE LOCALITY: Uruguay, Lavalleja Dept., Paso Averias; 33°60'S, 54°40'W.

DISTRIBUTION: Gallery forests of E Uruguay and extreme S Brazil (Rio Grande do Sul).

COMMENTS: Subgenus *Akodon*, *A. cursor* species group *sensu* Rieger et al. (1995). Principally contrasted to *A. cursor* and *A. serrensis*; see remarks by González et al. (1998) on karyotype (2n = 44) and misattributed reports of this new species as *A. serrensis*. Status and distributional limits with regard to *A. mystax* and especially *A. paranaensis* deserve careful review (see those accounts and discussion by Pardiñas et al., 2003a).

Akodon sanctipaulensis Hershkovitz, 1990. Fieldiana Zool., N. S., 57:23.

COMMON NAME: São Paulo Akodont.

TYPE LOCALITY: Brazil, São Paulo, Primeiro Morro.

DISTRIBUTION: Serra do Mar, SE Brazil.

STATUS: IUCN – Lower Risk (nt).

COMMENTS: Subgenus *Akodon*, *A. boliviensis* species group (Hershkovitz, 1990c). Morphological separation from *A. cursor* and *A. serrensis* amplified by Christoff et al. (2000).

Akodon serrensis Thomas, 1902. Ann. Mag. Nat. Hist., ser. 7, 9:61.

COMMON NAME: Serra do Mar Akodont.

TYPE LOCALITY: Brazil, Paraná State, Serra do Mar, Roca Nova, 1000 m.

DISTRIBUTION: SE Brazil; distributional limits uncertain.

STATUS: IUCN – Lower Risk (lc).

SYNONYMS: *leucogula* Miranda Ribeiro, 1905.

COMMENTS: Subgenus *Akodon*, *A. mollis* species group (Hershkovitz, 1990c). Recognized as a species since its description (Cabrera, 1961; Gyldenstolpe, 1932; Moojen, 1952). Most strongly differentiated, based on cytochrome *b* sequences, of six Brazilian *Akodon* species examined by Geise et al. (2001). The single example of *A. serrensis* studied by D'Elía et al. (2003) is genetically (cytochrome *b*) most similar to *Thaptomys nigrita*, a finding that recommends expansion of sampling and review of the taxon's assignment within *Akodon* proper.

New records of sympatry with *A. cursor* reported by Hershkovitz (1998), who amplified the discrimination of these two morphologically similar species (also see Liascovich and Reig, 1989). Karyotypic and morphometric separation from *A. paranaensis* illuminated by Christoff et al. (2000) and from *A. reigi* by González et al. (1998); strong genetic (cytochrome *b*) divergence from members of *A. cursor* species group confirmed by Geise et al. (2001). Hershkovitz (1990*c*) listed *leucogula* as a species, but Christoff et al. (2000) demonstrated its synonymy under *A. serrensis* and uncovered no basis for subspecific recognition. Morphometric variation in a local sample (Paraná, Brazil) investigated by Kosloski (1997), whose finding of two distinct phenetic groupings may relate to one of the species recently described from the region (e.g., *A. paranaensis* or *A. reigi*).

Akodon siberiae Myers and Patton, 1989. Occas. Pap. Mus. Zool., Univ. Michigan, 720:4.
COMMON NAME: Cochabamba Akodont.
TYPE LOCALITY: Bolivia, Cochabamba Dept., 28 km (by road) W Comarapa, 2800 m; 17°51′S, 64°40′W.
DISTRIBUTION: Known only from the vicinity of the type locality, about 1800-3000 m, Cochabamba and W Santa Cruz Depts., C Bolivia (Anderson, 1997).
STATUS: IUCN – Vulnerable.
COMMENTS: Subgenus *Hypsimys*. Morphology and chromosomes suggest that *A. siberiae* is closely related to *A. budini*, the type species of *Hypsimys* (see Myers and Patton, 1989).

Akodon simulator Thomas, 1916. Ann. Mag. Nat. Hist., ser. 8, 18:335.
COMMON NAME: White-throated Akodont.
TYPE LOCALITY: Argentina, Tucumán Prov., San Pablo, Villa Nouges, 1200 m.
DISTRIBUTION: E lower Andean slopes, 650-2400 m, from SC Bolivia (Tarija; Anderson, 1997) to NW Argentina (Jujuy to Catamarca; Mares et al., 1997; Ortiz et al., 2000*a*).
STATUS: IUCN – Lower Risk (lc).
SYNONYMS: *glaucinus* Thomas, 1919; *tartareus* Thomas, 1919.
COMMENTS: Subgenus *Akodon*, *A. varius* species group (Myers, 1989). Named as a species but later placed as a subspecies of *A. varius* (Cabrera, 1961; Thomas, 1926). Returned to species rank by Myers (1989), who emphasized the need for continued study of their level of differentiation; he recognized *glaucinus* and *tartareus* as distinctive subspecies. Karyology and morphometrics reviewed, as *A. varius*, by Bárquez et al. (1980); age variation in a large Argentinian sample analysed by Eschevarria (1995).

Akodon spegazzinii Thomas, 1897. Ann. Mag. Nat. Hist., ser. 6, 20:216.
COMMON NAME: Spegazzini's Akodont.
TYPE LOCALITY: Argentina, Salta Prov., lower Río Cachí.
DISTRIBUTION: E Andean slopes, 400 to 1000 m, NW Argentina.
STATUS: IUCN – Lower Risk (lc).
SYNONYMS: *alterus* Thomas, 1919; *tucumanensis* J. A. Allen, 1901.
COMMENTS: Subgenus *Akodon*, *A. boliviensis* species group (Myers et al., 1990). Formerly considered a subspecies of *A. boliviensis* (Cabrera, 1961); Myers et al. (1990) segregated *spegazzinii* as a species, retaining *tucumanensis* as a subspecies. Karyology and morphometrics reviewed, as *A. boliviensis tucumanensis*, by Barquez et al. (1980). Cabrera (1961) placed *alterus* in full synonymy with *A. boliviensis tucumanensis*; karyotypes of *alterus* and *tucumanensis* reported as identical (2n = 40, FN = 40) by Blaustein et al. (1992), who also discussed their weak allozymic and morphological divergence. Mares et al. (1997) listed *alterus* and *tucumanensis* as species principally based on their different ecological associations. Another complex of small akodonts in need of revision; also see remarks under *A. aliquantulus*.

Akodon subfuscus Osgood, 1944. Field Mus. Nat. Hist. Publ., Zool. Ser., 29:195.
COMMON NAME: Puno Akodont.
TYPE LOCALITY: Perú, Puno Dept., drainage of Río Inambari, Limbani, 9000 ft (2743 m).
DISTRIBUTION: E Andean upper slopes, about 2000-4500 m, from SC Perú (Puno) to Bolivia (Cochabamba) (see Anderson, 1997:Fig. 717; Patton and Smith, 1992:Fig. 3).
STATUS: IUCN – Lower Risk (lc).
SYNONYMS: *arequipae* Myers, Patton, and Smith, 1990.

COMMENTS: Subgenus *Akodon*, *A. boliviensis* species group (Myers et al., 1990). Described as a subspecies of *A. boliviensis*; raised to specific level and another subspecies named by Myers et al. (1990).

Akodon surdus Thomas, 1917. Smithson. Misc. Coll., 68(4):2.
COMMON NAME: Slate-bellied Akodont.
TYPE LOCALITY: Perú, Cuzco Dept., Huadquiña, 5000 ft (1524 m).
DISTRIBUTION: E Andes of SE Perú.
STATUS: IUCN – Lower Risk (lc).
COMMENTS: Subgenus *Akodon*. Retained as a species by Reig (1987), Myers (1989), and Hershkovitz (1990c), but considered a possible synonym of *A. aerosus* by Patton and Smith (1992); reconciliation of these differing treatments needed.

Akodon sylvanus Thomas, 1921. Ann. Mag. Nat. Hist., ser. 9, 7:184.
COMMON NAME: Woodland Akodont.
TYPE LOCALITY: Argentina, Jujuy Prov., Sierra de Santa Barbara, Sunchal, 1200 m.
DISTRIBUTION: NW Argentina.
STATUS: IUCN – Lower Risk (lc).
COMMENTS: Subgenus *Akodon*. Retained as a species until Cabrera (1961) synonymized it as a race of *A. azarae*; Myers (1989) tentatively listed *sylvanus* and *azarae* as separate species. Formerly included *pervalens* (see that account).

Akodon toba Thomas, 1921. Ann. Mag. Nat. Hist., ser. 9, 7:178.
COMMON NAME: Toba Akodont.
TYPE LOCALITY: Paraguay, Presidente Hayes Dept., northern Chaco, Jesematathla, 100 m.
DISTRIBUTION: Chaco of W Paraguay, SE Bolivia, and contiguous N Argentina.
STATUS: IUCN – Lower Risk (lc).
COMMENTS: Subgenus *Akodon*, *A. varius* species group (Myers, 1989). Relegated to a subspecies of *A. varius* by Cabrera (1961); karyotypic and morphological discrimination from *varius* proper summarized by Myers (1989).

Akodon torques (Thomas, 1917). Smithson. Misc. Coll., 68:3.
COMMON NAME: Cloud Forest Akodont.
TYPE LOCALITY: Perú, Cuzco Dept., Machu Picchu, 10,000 ft (3048 m).
DISTRIBUTION: E Andean cloud forest, 2000-3500 m, SE Perú.
STATUS: IUCN – Lower Risk (lc).
COMMENTS: Subgenus *Akodon*. Relegated to a subspecies of *A. orophilus* (e.g., Cabrera, 1961), but recent studies have revealed that *torques* is a genetically distinctive form more closely related to *A. mollis* (Patton and Smith, 1992; Patton et al., 1989, 1990). Described as a species of *Microxus* but considered a member of *Akodon* proper by later authors (Ellerman, 1941; Thomas, 1927), placement vindicated by allozymic and cytochrome *b* data (Patton et al., 1989; Smith and Patton, 1993).

Akodon varius Thomas, 1902. Ann. Mag. Nat. Hist., ser. 7, 9:134.
COMMON NAME: Variable Akodont.
TYPE LOCALITY: Bolivia, Cochabamba Dept., Cochabamba, 2400 m.
DISTRIBUTION: E Andean slopes, 2000-3000 m, of W Bolivia.
STATUS: IUCN – Lower Risk (lc).
COMMENTS: Subgenus *Akodon*, *A. varius* species group. Cabrera (1961) included as subspecies *neocenus*, *simulator*, and *toba*, forms here arranged as species according to the preliminary revision of Myers (1989).

Amphinectomys Malygin, 1994 (in Malygin et al., 1994). Zool. Zhur., 73:198.
TYPE SPECIES: *Amphinectomys savamis* Malygin, 1994 (in Malygin et al., 1994).
COMMENTS: Oryzomyini. A water rat morphologically similar to *Nectomys* but with more expansive interdigital webbing and broader interorbital region (Malygin et al., 1994). The new genus was differentially diagnosed with respect to nearby populations of *Nectomys squamipes sensu* Hershkovitz (1944), now well known to be a heterogeneous composite (see below); the relationship and generic status of *savamis* should be confirmed within a broader sampling of *Nectomys* species and other oryzomyines. The karyotype (2n = 52;

Malygin et al., 1994) fits within the range of variation documented for species of *Nectomys* (2n = 38-59; e.g., Andrades-Miranda et al., 2001*b*), but critical banding comparisons are needed; genetically closely related to *Nectomys squamipes* based on nuclear gene (IRBP) sequences (Weksler, 2003). Anatomy of male reproductive tract reported by Malygin and Rosmiarek (1997) and compared with other oryzomyines.

Amphinectomys savamis Malygin, 1994 (in Malygin et al., 1994). Zool. Zhur., 73:203.
    COMMON NAME: Ucayali Water Rat.
    TYPE LOCALITY: Perú, Loreto Dept., Requena Prov., right bank Río Ucayali, 7 km E Henaro Errera; 04°55'S, 73°45'W.
    DISTRIBUTION: Known only by the holotype.

Andalgalomys Williams and Mares, 1978. Ann. Carnegie Mus., 47:197.
    TYPE SPECIES: *Andalgalomys olrogi* Williams and Mares, 1978.
    COMMENTS: Phyllotini. Includes forms formerly associated with *Graomys*; morphological and karyotypic traits summarized by Olds et al. (1987), who emended the generic diagnosis. Closely related to *Graomys* (e.g., Olds et al., 1987), but recent phylogentic colloquy has centered on the monophyly and ergo the generic validity of the taxon. Braun (1993) and Steppan (1993, 1995), using morphological characters, found that exemplars of *Andalga-lomys* associated among those of *Graomys*; the former author retained *Andalgalomys* as genus whereas the latter formally merged it under *Graomys* (in a reanalysis, Steppan and Sullivan, 2000, provisionally retained both as genera). Taxonomically narrow (Anderson and Yates, 2000) and broad (D'Elía et al., 2003) surveys of cytochrome *b* sequences do convey the monophyly of *Andalgalomys*, well delineated from species of *Graomys*, on resultant trees.

    The issue has become more complex commensurate with the emerging appreciation of taxonomic diversity within this phyllotine assemblage, expressed at both the generic (*Salinomys* Braun and Mares, 1995; *Tapecomys* Anderson and Yates, 2000) and specific (*Andalgalomys roigi* Mares and Braun, 1996) levels, and the shifting snapshots of cladistic relationship as these new forms are serially considered (e.g., compare Braun and Mares, 1995, with Anderson and Yates, 2000, and Steppan and Sullivan, 2000). As is sometimes the case, the conflicting nomenclatural recommendations rest variably on the scope of sampled species, the character bases referenced, and the requisite weight to be accorded recent statistical methodologies vis a vis differential diagnosis (see Mares and Braun [2000*b*] and Steppan and Sullivan [2000] for an engaging representation of issues and viewpoints). Given the nascent stage of our systematic understanding, Anderson and Yates (2000) offered a reasoned overview of the problems to be resolved and a reasonable provisional taxonomy that we follow here.

Andalgalomys olrogi Williams and Mares, 1978. Ann. Carnegie Mus., 47:203.
    COMMON NAME: Olrog's Pericote.
    TYPE LOCALITY: Argentina, Catamarca Prov., 15 km (on route 62) W Andalgala, west bank Río Amanao.
    DISTRIBUTION: Known only from the vicinity of the type locality.
    STATUS: IUCN – Vulnerable.

Andalgalomys pearsoni (Myers, 1977). Occas. Pap. Mus. Zool., Univ. Michigan, 676:1.
    COMMON NAME: Pearson's Pericote.
    TYPE LOCALITY: Paraguay, Boquerón Dept., 410 km (by road) NW Villa Hayes.
    DISTRIBUTION: Chaco of W Paraguay and SE Bolivia (Santa Cruz Dept.).
    STATUS: IUCN – Lower Risk (lc).
    SYNONYMS: *dorbignyi* Olds, Anderson, and Yates, 1987.
    COMMENTS: Diploid count differs by one pair (76 versus 78) in the two subspecies (Myers, 1977; Olds et al., 1987).

Andalgalomys roigi Mares and Braun, 1996. J. Mammal., 77:929.
    COMMON NAME: Roig's Pericote.
    TYPE LOCALITY: Argentina, San Luis Prov., 8 km W La Botija, Pampa de las Salinas, 510 m; 36°12'27"S, 66°39'35"W.

DISTRIBUTION: Semiarid thorn scrub from Catamarca Prov. south to San Luis Prov. (Mares et al., 1997), NC Argentina.

COMMENTS: Mares and Braun (1996) contrasted their new species with *A. olrogi* and *A. pearsoni*.

Andinomys Thomas, 1902. Proc. Zool. Soc. Lond., 1902(1):116.

TYPE SPECIES: *Andinomys edax* Thomas, 1902.

COMMENTS: Phyllotini. Revised by Hershkovitz (1962). Standard karyotype reported and compared to other phyllotines by Pearson and Patton (1976) and Simonetti and Spotorno (1980). Based on homologous chromosomal bands, Spotorno et al. (1994) suggested the primitive nature of *Andinomys* and its early divergence within the phyllotine radiation; cladistic studies of morphological traits indicate a later genesis, in a clade variously containing *Chinchillula*, *Irenomys*, and *Punomys* (Braun, 1993; Steppan, 1995).

Andinomys edax Thomas, 1902. Proc. Zool. Soc. Lond., 1902(1):116.

COMMON NAME: Andean Mouse.

TYPE LOCALITY: Bolivia, Potosí Dept., between Potosí and Sucre, El Cabrado, 3700 m.

DISTRIBUTION: Altiplano of extreme S Perú (Puno) and N Chile (Pine et al., 1979; Spotorno, 1976), through WC Bolivia (Anderson, 1997), to NW Argentina (Jujuy to La Rioja; Díaz and Barquez, 1999; Mares et al., 1997; Ortiz et al., 2000a).

STATUS: IUCN – Lower Risk (lc).

SYNONYMS: *lineicaudatus* Yepes, 1935.

COMMENTS: Díaz and Barquez (1999) remarked on subspecific distribution in N Argentina and urged revision to resolve whether two species are represented.

Anotomys Thomas, 1906. Ann. Mag. Nat. Hist., ser. 7, 17:86.

TYPE SPECIES: *Anotomys leander* Thomas, 1906.

COMMENTS: Ichthyomyini. Formerly included *trichotis* (Handley, 1976), which Voss (1988) removed to the new genus *Chibchanomys*. Phylogenetic relationships studied by Voss (1988).

Anotomys leander Thomas, 1906. Ann. Mag. Nat. Hist., ser. 7, 17:87.

COMMON NAME: Ecuadoran Ichthyomyine.

TYPE LOCALITY: Ecuador, Pichincha Prov., Volcán Pichincha, 11,500 ft (3505 m).

DISTRIBUTION: N Ecuador at high elevations.

STATUS: IUCN – Endangered.

COMMENTS: Diploid number (2n = 92) reported by Gardner (1971); taxonomy and distribution reviewed by Voss (1988).

Auliscomys Osgood, 1915. Field Mus. Nat. Hist. Publ., Zool. Ser., 10:190.

TYPE SPECIES: *Reithrodon pictus* Thomas, 1884.

SYNONYMS: *Maresomys* Braun, 1993.

COMMENTS: Phyllotini. Diagnosed as a subgenus of *Phyllotis* and variably recognized afterwards at that rank (Osgood, 1947; Pearson, 1958) or as a genus (Gyldenstolpe, 1932; Thomas, 1926a). Recent evidence supports the monophyly and probable earlier divergence of *Auliscomys* (Pearson and Patton, 1976; Simonetti and Spotorno, 1980), usually as closely related to *Galenomys* (Braun, 1993; Steppan, 1993, 1995; Steppan and Sullivan, 2000). Includes *Maresomys* whose type species (= *boliviensis*) Braun (1993) interpreted as sister group to *Galenomys*, but which Steppan (1995) viewed as cognate to other species of *Auliscomys* and placed as synonym of the latter. The critical nodes on which these nomenclatural decisions hinge reflect slightly different coding schemes of morphological traits and rely upon consensus depictions of numerous equal-length trees; as such, the issue invites additional testing with other information sources. Anderson (1997) used *Maresomys* as a subgenus of *Auliscomys*. Alpha systematics revised by Pearson (1958) and Hershkovitz (1962) as part of *Phyllotis*. Banded karyotypes and chromosomal evolution evaluated and compared with *Loxodontomys micropus* by Walker and Spotorno (1992).

Auliscomys boliviensis (Waterhouse, 1846). Proc. Zool. Soc. Lond., 1846:9.

COMMON NAME: Bolivian Pericote.

TYPE LOCALITY: Bolivia, Potosí Dept., "a few leagues" south of Potosí, 12,000 ft (3658 m).
DISTRIBUTION: Altiplano from S Perú (Arequipa Dept.) to extreme N Chile and WC Bolivia (3450-4770 m).
STATUS: IUCN – Lower Risk (lc).
SYNONYMS: *flavidior* (Thomas, 1902).
COMMENTS: Type species of *Maresomys* Braun (1993).

Auliscomys pictus (Thomas, 1884). Proc. Zool. Soc. Lond., 1884:457.
COMMON NAME: Colorful Pericote.
TYPE LOCALITY: Perú, Junín Dept., Junín, 13,700 ft (4176 m).
DISTRIBUTION: High Andes, ca. 3500-4800 m, from C Perú (Ancash Dept.) to NW Bolivia (La Paz Dept.).
STATUS: IUCN – Lower Risk (lc).
SYNONYMS: *decoloratus* (Osgood, 1915).

Auliscomys sublimis (Thomas, 1900). Ann. Mag. Nat. Hist., ser. 7, 6:467.
COMMON NAME: Lofty Pericote.
TYPE LOCALITY: Perú, Arequipa Dept., Rinconado Malo Pass, between Cailloma and Calalla, 18,000 ft (5486 m).
DISTRIBUTION: Altiplano from S Perú (Ayacucho Dept.), through W Bolivia (3800-4740 m) and adjacent Chile, to NW Argentina (Jujuy and Salta; Díaz, 1999).
STATUS: IUCN – Lower Risk (lc).
SYNONYMS: *leucurus* (Thomas, 1919).
COMMENTS: Anderson (1997) retained subspecific divisions for Bolivian populations.

Bibimys Massoia, 1979. Physis, sec. C., 38(95):2.
TYPE SPECIES: *Bibimys torresi* Massoia, 1979.
COMMENTS: Akodontini. Diagnosed as a third genus of scapteromyines, along with *Kunsia* and *Scapteromys* (see Hershkovitz, 1966c), and so maintained by Reig (1980, 1981, 1984, 1986). More closely related to *Akodon* and kin than to *Kunsia-Scapteromys* according to parsimony distillations of cytochrome *b* sequence data (D'Elía et al., in press). In analyses of mitochondrial and nuclear genes, D'Elía (2003) recognized *Bibimys* as one of five major clades within Akodontini.
Generic diagnosis emended, alpha taxonomy reviewed, and known fossil and recent occurrences consolidated by D'Elía et al. (in press). They provisionally retained three species, given the inadequacies of sample size and geographic representation, but acknowledged that morphological, karyotypic, and genetic evidence for their separation is unpersuasive. Karyology reported by Dyzenchauz and Massarini (1999) and Gonçalves et al. (in press). Fossil records (early Pleistocene to Holocene) of Argentina and Brazil summarized by Pardiñas (1996), who recharacterized the dental characteristics of the genus and compared them with *Kunsia* and *Scapteromys*.

Bibimys chacoensis (Shamel, 1931). J. Wash. Acad. Sci., 21:247.
COMMON NAME: Chacoan Akodont.
TYPE LOCALITY: Argentina, Chaco Prov., Las Palmas.
DISTRIBUTION: NE Argentina.
STATUS: IUCN – Lower Risk (lc).
COMMENTS: Described as a species of *Akodon*; transferred to *Bibimys* by Massoia (1980a), who noted the need to clarify the differentiation of this form from *B. torresi*.

Bibimys labiosus (Winge, 1887). E Museo Lundii, 1(3):25.
COMMON NAME: Lagoa Santa Akodont.
TYPE LOCALITY: Brazil, Minas Gerais State, Pleistocene cave deposits near Lagoa Santa.
DISTRIBUTION: Minas Gerais, Brazil.
STATUS: IUCN – Lower Risk (lc).
COMMENTS: Described as a species of *Scapteromys*, referred to *Akodon* by Hershkovitz (1966c), and confirmed as a species of *Bibimys* by Massoia (1980a). Although described as a fossil (Winge, 1887) from the Lagoa Santa caves, the species survives in the region of Minas Gerais (D'Elía et al., in press; Paglia et al., 1995). Diploid number (2n = 70) same

as *B. torresi* but fundamental numbers differ (FN = 76 versus 80; Gonçalves et al., in press).

Bibimys torresi Massoia, 1979. Physis, sec. C., 38(95):3.
COMMON NAME: Torres' Akodont.
TYPE LOCALITY: Argentina, Buenos Aires Prov., Paraná River delta at the confluence of Arroyo Las Piedras with Arroyo Cucarachas, Campana, Estación Experimental del INTA, Canal 6 (as refined by Pardiñas, 1996).
DISTRIBUTION: EC Argentina.
STATUS: IUCN – Lower Risk (nt).
COMMENTS: Karyotype (2n = 70) reported by Dyzenshauz and Massarini (1999) and contrasted to species of *Scapteromys*.

Blarinomys Thomas, 1896. Ann. Mag. Nat. Hist., ser. 6, 18:310.
TYPE SPECIES: *Oxymycterus breviceps* Winge, 1887.
COMMENTS: Akodontini. Sister genus to *Brucepattersonius* according to phylogenetic evaluations of cytochrome *b* sequences (D'Elía et al., 2003; Smith and Patton, 1999) or to the clade *Brucepattersonius-Lenoxus* using cytochrome *b* and IRBP (D'Elía, 2003).

Blarinomys breviceps (Winge, 1887). E Museo Lundii, 1(3):34.
COMMON NAME: Blarinine Akodont.
TYPE LOCALITY: Brazil, Minas Gerais, Rio das Velhas, Pleistocene cave deposits near Lagoa Santa.
DISTRIBUTION: Atlantic Forest region of SE Brazil (Bahia to São Paulo; Silva et al., 2003) and NE Argentina (Misiones Prov.; Massoia, 1993).
STATUS: IUCN – Lower Risk (nt).
COMMENTS: Based on compilation of old and new records, Silva et al. (2003) suggested that *B. breviceps* is extinct at the type locality, now set within Cerrado vegetation, and confined to the more mesic Atlantic Forest zone. See Matson and Abravaya (1977, Mammalian Species, 74).

Brucepattersonius Hershkovitz, 1998. Bonn. Zool. Beitr., 47:227.
TYPE SPECIES: *Brucepattersonius soricinus* Hershkovitz, 1998.
COMMENTS: Akodontini. Morphological traits resemble those of *Oxymycterus*, considered its probable nearest generic relative by Hershkovitz (1998). Parsimony and likelihood trees derived from cytochrome *b* data instead most closely relate the genus to *Blarinomys*, within a clade that includes *Lenoxus*, *Kunsia*, and *Scapteromys* (D'Elía et al., 2003; Smith and Patton, 1999). With recognition of this distinctive genus, attention should be directed to documenting the status and distributions of the newly described forms, most known only from their type locality and based on differentiating traits that intimate age and-or individual variation.

Brucepattersonius albinasus Hershkovitz, 1998. Bonn. Zool. Beitr., 47:235.
COMMON NAME: White-nosed Akodont.
TYPE LOCALITY: Brazil, Minas Gerais State, Parque Nacional de Caparaó, Pico de Bandeira, 2700 m.
DISTRIBUTION: Known only from the type locality.

Brucepattersonius griserufescens Hershkovitz, 1998. Bonn. Zool. Beitr., 47:233.
COMMON NAME: Gray-bellied Akodont.
TYPE LOCALITY: Brazil, Minas Gerais State, Parque Nacional de Caparaó, Terreirao, 2400 m
DISTRIBUTION: Humid montane forest and scrub, 1300-2700 m, E Minas Gerais and Espírito Santo to Rio de Janeiro, SE Brazil.
COMMENTS: Distribution amplified and karyotype reported by Bonvicino et al. (1998b), who noted that it (2n = 52, FN = 52-53) differs from all *Oxymycterus* species so far examined (2n = 54, FN = 64; see Bonvicino et al., 1998b).

Brucepattersonius guarani Mares and Braun, 2000. Occas. Pap. Sam Noble Oklahoma Mus. Nat. Hist., 9:9.
COMMON NAME: Guaraní Akodont.

TYPE LOCALITY: Argentina, Misiones Prov., Guaraní Dept., 6 km NE (by Hwy 2) junction of Hwy 2 and Arroyo Paraíso, 360 m.

DISTRIBUTION: Known only from the type locality.

Brucepattersonius igniventris Hershkovitz, 1998. Bonn. Zool. Beitr., 47:232.

COMMON NAME: Red-bellied Akodont.

TYPE LOCALITY: Brazil, São Paulo State, Iporanga (Petar) State Park, 200 m; 24°35'S, 48°35'W.

DISTRIBUTION: Known only from the type locality.

Brucepattersonius iheringi (Thomas, 1896). Ann. Mag. Nat. Hist., ser. 6, 18:308.

COMMON NAME: Ihering's Akodont.

TYPE LOCALITY: Brazil, Rio Grande do Sul State, Rio dos Linos, Taquara do Mundo Novo.

DISTRIBUTION: NE Argentina (Misiones) and SE Brazil.

STATUS: IUCN – Lower Risk (lc) as *Oxymycterus iheringi*.

COMMENTS: A distinctive species of varied generic associations—included in *Microxus* or *Akodon* (e.g., Avila-Pires, 1994; Cabrera, 1961; Ellerman, 1941; Gyldenstolpe, 1932; Moojen, 1952); reallocated to *Oxymycterus* by Massoia (1963*b*) and thereafter classified (Corbet and Hill, 1991; Honacki et al., 1982; Musser and Carleton, 1993); recently assigned to the new genus *Brucepattersonius* by Hershkovitz (1998), an allocation seconded by Bonvicino et al. (1998*b*). Morphological recognition amplified by Massoia and Fornes (1969).

Brucepattersonius misionensis Mares and Braun, 2000. Occas. Pap. Sam Noble Oklahoma Mus. Nat. Hist., 9:7.

COMMON NAME: Misiones Akodont.

TYPE LOCALITY: Argentina, Misiones Prov., Guaraní Dept., ca. 2 km W Parque Provincial Moconá, junction of Hwy 21 and Arroyo Oveja Negra.

DISTRIBUTION: Known only from the type locality.

Brucepattersonius paradisus Mares and Braun, 2000. Occas. Pap. Sam Noble Oklahoma Mus. Nat. Hist., 9:3.

COMMON NAME: Arroyo of Paradise Akodont.

TYPE LOCALITY: Argentina, Misiones Prov., Guaraní Dept., junction of Hwy 2 and Arroyo Paraíso, 197 m.

DISTRIBUTION: Known only from the type locality.

Brucepattersonius soricinus Hershkovitz, 1998. Bonn. Zool. Beitr., 47:232.

COMMON NAME: Soricine Akodont.

TYPE LOCALITY: Brazil, São Paulo State, Ribeirão Fundo, 30 m; 24°15'S, 47°45'W.

DISTRIBUTION: Known only from several localities in São Paulo, SE Brazil.

Calomys Waterhouse, 1837. Proc. Zool. Soc. Lond., 1837:21.

TYPE SPECIES: *Mus bimaculatus* Waterhouse, 1837 (= *Mus laucha* Fischer, 1814).

SYNONYMS: *Hesperomys* Waterhouse, 1839.

COMMENTS: Phyllotini. Generally viewed as a primitive clade relative to other phyllotine genera (Hershkovitz, 1962; Pearson and Patton, 1976; Reig, 1986), a viewpoint corroborated by phylogenetic analyses of trait data, both morphological (Braun, 1993; Steppan, 1993; 1995) and molecular (Smith and Patton, 1999). Those morphological studies differ in disclosing the basal position of *Calomys* spp. as monophyletic (Braun, 1993) or paraphyletic (Steppan, 1993); gene-sequence data convincingly ratify the taxon's monophyly (Salazar-Bravo et al., 2001, 2002*a*). Salazar-Bravo et al. (2001) dated the initial cladogenesis among *Calomys* species to ca. 9 million years ago, possibly spurred by the early development of South American grasslands, and reviewed hypotheses and data that concern the Great American Interchange and areas of origination. Known from Pleistocene of Argentina (e.g., Pardiñas, 1999). Status of North American Miocene-Pliocene form *Bensonomys* as a subgenus of *Calomys* (Baskin, 1978, 1986; Czaplewski, 1987; Korth, 1994; Lindsay and Jacobs, 1985) or as a merely dentally similar neotomine (Reig, 1980, 1984) needs decisive resolution, as it bears importantly on larger issues of New World rodent evolution and biogeography. McKenna and Bell (1997) assigned *Bensonomys* to Neotomini, within Sigmodontinae *sensu lato*.

Generic revision by Hershkovitz (1962), who reduced the previously recognized 10-15 species (e.g., Cabrera, 1961; Ellerman, 1941) to only four, two of which, *laucha* and *callosus*, have been revealed as species complexes (Bonvicino and Almeida, 2000; Corti et al., 1987; Massoia et al., 1968; Olds, 1988; Pearson and Patton, 1976; Reig, 1986; Williams and Mares, 1978). See Bonvicino and Almeida (2000) and Bonvicino et al. (2003c) for compilation of the prolific karyological literature on the genus and correction of misidentified voucher specimens. Interspecific genealogies and species groups inferred from morphology (Olds, 1988), banded chromosomes (Espinosa et al., 1997; Vitullo et al., 1990), and cytochrome *b* sequences (Salazar-Bravo et al., 2001) are substantially incongruent. The following recognition of species attempts to integrate recent morphological, karyotypic, and molecular studies (principally, Bonvicino and Almeida, 2000; Olds, 1988; Salazar-Bravo et al., 2001). Nonetheless, the continued listing of *Calomys* specimens as species indeterminate (e.g., Anderson, 1997; Salazar-Bravo et al., 2001) and reports of new karyotypic variants (Espinosa et al., 1997; Lima and Kasahara, 2001) caution that considerable alpha taxonomic investigation is yet required.

Calomys boliviae (Thomas, 1901). Ann. Mag. Nat. Hist., ser. 7, 8:253.

COMMON NAME: Bolivian Laucha.

TYPE LOCALITY: Bolivia, La Paz Dept., Río Solocame, 1200 m.

DISTRIBUTION: E Andean slopes, ca. 600-2700 m, in WC Bolivia and NW Argentina.

STATUS: IUCN – Lower Risk (lc).

SYNONYMS: *fecundus* (Thomas, 1926).

COMMENTS: Conventionally arranged as a subspecies or synonym of *C. callosus* (Cabrera, 1961; Hershkovitz, 1962). Use here follows informal listing of Reig (1986), who included *fecundus* as a subjective synonym, an attribution supported by their morphological similarity (Olds, 1988). Anderson (1997), following Olds (1988), treated *boliviae* and *fecundus* as synonyms of *C. venustus*, type locality in Córdoba, Argentina. The specific segregation of *fecundus* (2n = 54), as based strictly on S Bolivian and NW Argentinian samples, from *C. venustus* (2n = 56) is sustained by differentiation in karyotypes and mitochondrial DNA sequence analyses (Salazar-Bravo et al., 2001, 2002a); however, the proper application of *boliviae* as senior synonym or its recognition as another valid species distinct from *fecundus* deserves concrete demonstration, as does its distributional extent with regard to *C. venustus* in NW Argentina.

Calomys callidus (Thomas, 1916). Ann. Mag. Nat. Hist., ser. 8, 17:182.

COMMON NAME: Reclusive Laucha.

TYPE LOCALITY: Argentina, Corrientes Prov., Goya, 600 ft (183 m).

DISTRIBUTION: EC Argentina and E Paraguay.

STATUS: IUCN – Lower Risk (lc).

COMMENTS: Diagnosed as a subspecies of *C. venustus* and recognized as such (e.g., Cabrera, 1961) until Hershkovitz (1962) placed both in synonymy with *C. callosus callosus*. Specific status of *callidus* affirmed by Corti et al. (1987), who summarized its distribution, karyotypic traits, and morphometric discrimination; banded chromosomal comparisons with *C. venustus* and systematic commentary provided by Vitullo et al. (1990).

Calomys callosus (Rengger, 1830). Naturgesch. Säugeth. Paraguay, p. 231.

COMMON NAME: Big Laucha.

TYPE LOCALITY: Paraguay, Neembucú Dept., banks of Río Paraguay about 27°S latitude (refined as opposite mouth of Río Bermejo by Hershkovitz, 1962:172, and restricted as vicinity of Ciudad Pilar by Contreras, 1992:2).

DISTRIBUTION: Dry and subhumid areas in E Bolivia, N Argentina, Paraguay, and contiguous WC Brazil; isolated records in SE and E Brazil (see Salazar-Bravo et al., 2002a).

STATUS: IUCN – Lower Risk (lc).

SYNONYMS: *muriculus* (Thomas, 1921).

COMMENTS: According to Hershkovitz (1962), *callosus* also encompassed *boliviae* (see species account), *callidus* (see species account), *fecundus* (see *C. boliviae*), and *venustus* (see species account). Others have referenced, without explanation, *muriculus* (Williams and Mares, 1978) as a separate species, although Reig (1986) did not; specimens collected near the type locality of *muriculus* are interspersed among samples of *C. callosus* proper in cytochrome *b* sequence analyses (Salazar-Bravo et al., 2002*a*). Especially see Salazar-Bravo et al. (2002*a*:194-195) for karyotypic amendment of *C. callosus* (2n = 50, FN = 66; not 2n = 36, FN = 48 as reported by Pearson and Patton, 1976). Also see Contreras (1992) for review of conflicting interpretations of the type locality and the need to obtain topotypic material to confirm the identity of the species as currently understood. His recommendation is especially prescient today since the restricted type locality of *C. callosus* is extralimital to the distribution of the species as currently circumscribed (Salazar-Bravo, 2002*a*:Fig. 1).

Calomys expulsus (Lund, 1841). K. Danske Vidensk. Selsk. Afhandl., 8:280.
    COMMON NAME: Caatinga Laucha.
    TYPE LOCALITY: Brazil, Minas Gerais State, Rio das Velhas, Lagoa Santa.
    DISTRIBUTION: Caatinga and Cerrado formations from Pernambuco southwestwardly through Goiás, C Brazil.
    COMMENTS: Morphologically similar to but karyotypically strongly divergent (2n = 66, FN = 68) from *C. callosus* (Bonvicino and Almeida, 2000), the species with which it had been previously synonymized (Hershkovitz, 1962; Musser and Carleton, 1993). Morphometric, karyotypic, distributional, and ecological distinctions from *C. tener* summarized by Bonvicino and Almeida (2000); those authors also noted that past faunal and ecological reports on *Calomys* from C Brazil, identified as *C. callosus* (e.g., Mares et al., 1981*a*, 1989*a*), probably refer to this species and-or to *C. tener*, which may occur sympatrically in ecotonal settings between the Atlantic Forest and Cerrado biomes. Ontogeny of cranial shape variation analyzed by Hingst-Zaher et al. (2000).

Calomys hummelincki (Husson, 1960). Stud. Faun. Curacao Carib. Isl., 43:34.
    COMMON NAME: Hummelinck's Laucha.
    TYPE LOCALITY: Netherlands West Indies, Curaçao, Klein Santa Martha.
    DISTRIBUTION: Llanos of NE Colombia (La Guajira), N and C Venezuela, and the continental-shelf islands Curaçao and Aruba.
    STATUS: IUCN – Lower Risk (lc).
    COMMENTS: Originally named as a species of *Baiomys*; reidentified as *Calomys* and included in *C. laucha* by Hershkovitz (1962), who considered the Venezuelan populations to be introductions. Indigenous status and specific distinctiveness reasserted by Handley (1976) and underscored by subsequent research (Garcia et al., 1999; Martino and Capanna, 2002; Pérez-Zapata et al., 1987); biogeographic scenario for derivation of *C. hummelincki* from southern populations of *Calomys* developed by Martino et al. (2002). Cytogenetics, allozymic and morphometric variation, and natural history monographed by Martino (2000); distribution discussed by Voss (1991*a*) in the context of biogeographic comparisons with other small mammals inhabiting nonforest vegetation in N South America.

Calomys laucha (Fischer, 1814). *In* Eschwege, J. Brasilien, Neue Bibliothek. Reisenb., 15(2):209.
    COMMON NAME: Little Laucha.
    TYPE LOCALITY: Paraguay, Central Dept., vicinity of Asunción (as restricted by Hershkovitz, 1962:153).
    DISTRIBUTION: Dry biotopes in SC Bolivia (Chaco), N and EC Argentina, W Paraguay, WC and extreme S Brazil, and Uruguay (González et al., 1995).
    STATUS: IUCN – Lower Risk (lc).
    SYNONYMS: *bimaculatus* (Waterhouse, 1837); *bonariensis* (Osgood, 1933); *dubius* (Fischer, 1829); *gracilipes* (Waterhouse, 1837); *pusillus* (Philippi, 1858).
    COMMENTS: Formerly included *hummelincki* (See Handley, 1976), *musculinus* (see Massoia et al., 1968), and *tener* (see species account). Attribution of *pusillus* follows Hershkovitz (1962; also see Osgood, 1943:239); that of *bimaculatus* as per Brum-Zorilla et al. (1990). Allozymic variability and differentiation among samples in EC Argentina assayed by Chiappero et al. (2002) and Gardenal et al. (2002).

Calomys lepidus (Thomas, 1884). Proc. Zool. Soc. Lond., 1884:454.

　　COMMON NAME: Graceful Laucha.

　　TYPE LOCALITY: Perú, Junín Dept., Junín.

　　DISTRIBUTION: Altiplano of C Perú, through W Bolivia (2950-4820 m), to NE Chile and NW Argentina (Jujuy, Salta, and Tucúman; Díaz, 1999; Ortiz et al., 2000a).

　　STATUS: IUCN – Lower Risk (lc).

　　SYNONYMS: *argurus* (Thomas, 1919); *carilla* (Thomas, 1902); *ducilla* (Thomas, 1901); *marcarum* (Thomas, 1917); *montanus* (Sanborn, 1950).

　　COMMENTS: Revised by Hershkovitz (1962). Populations in Jujuy, NW Argentina, exhibit a different chromosomal complement (2n = 44, FN = 68), as reported by Espinosa et al. (1997), from those described from Peru (2n = 36, FN = 68) by Pearson and Patton (1976); Espinosa et al. (1997) offered a Robertsonian explanation to account for such the karyotypic differences between chromosomal races of a single species, but also acknowledged that two species may be involved. Anderson (1997) regarded *carillus* and *ducillus* as valid subspecies in the Bolivian Andes.

Calomys musculinus (Thomas, 1913). Ann. Mag. Nat. Hist., ser. 8, 11:138.

　　COMMON NAME: Drylands Laucha.

　　TYPE LOCALITY: Argentina, Jujuy Prov., Maimará, 2230 m.

　　DISTRIBUTION: Arid habitats over wide elevations in WC Bolivia (see Anderson, 1997: Fig. 738), W Paraguay, and Argentina (Jujuy and Salta as far south as Chubut; Díaz, 2000; Mares et al., 1997; Ortiz et al., 2000a; Pearson, 1995); limits uncertain.

　　STATUS: IUCN – Lower Risk (lc).

　　SYNONYMS: *cordovensis* (Thomas, 1916); *cortensis* (Thomas, 1920); *murillus* (Thomas, 1916).

　　COMMENTS: Included in *C. laucha* by Hershkovitz (1962) but morphological and karyotypic evidence supports its specific status (Corti et al., 1987; Massoia et al., 1968). Sister species to *C. lepidus* according to phylogenetic interpretation of cytochrome *b* sequences (Salazar-Bravo et al., 2001). Morphometric variation among select phyto-geographic samples examined by Provensal and Polop (1993); allozymic variability among samples in EC Argentina investigated by Gardenal et al. (2002) and contrasted with sympatric *C. laucha*. Synonymy follows Massoia et al. (1968) and Contreras and Rosi (1980), who considered *murillus* a subspecies of *C. musculinus*; others have treated the former as a species (Olds, 1988; Reig, 1986). In the molecular study of Salazar-Bravo et al. (2001), it is noteworthy that their sample of *C. musculinus* from Buenos Aires Province, presumably referable to *murillus* proper (type locality La Plata, Buenos Aires Prov.), was genetically like population samples drawn from NW Argentina, presumably referable to *musculinus* proper. Until the issue can be substantively resolved, we continue to view *murillus* as a subjective synonym.

Calomys sorellus (Thomas, 1900). Ann. Mag. Nat. Hist., ser. 7, 6:297.

　　COMMON NAME: Peruvian Laucha.

　　TYPE LOCALITY: Perú, Libertad Dept., 8 mi (13 km) S Huamachuco, 3500 m.

　　DISTRIBUTION: Peruvian Andes, above 2000 m, from Libertad to Puno Depts.

　　STATUS: IUCN – Lower Risk (lc).

　　SYNONYMS: *frida* (Thomas, 1917); *miurus* (Thomas, 1926).

　　COMMENTS: Classified as a subspecies of *C. lepidus* (Cabrera, 1961); discrimination from and sympatry with *C. lepidus* documented by Hershkovitz (1962), who relegated *frida* and *miurus* to synonymy under *C. sorellus* (also see Pearson and Patton, 1976).

Calomys tener (Winge, 1887). E Museo Lundii, 1(3):15.

　　COMMON NAME: Delicate Laucha.

　　TYPE LOCALITY: Brazil, Minas Gerais State, Rio das Velhas, Lagoa Santa.

　　DISTRIBUTION: Atlantic Forest region and habitats bordering the Cerrado, SE Brazil (S Goiás, Minas Gerais, São Paulo), and in NE Argentina (Misiones Prov., Massoia, 1988) and E Bolivia (Anderson, 1997); range limits need refinement.

　　STATUS: IUCN – Lower Risk (lc).

　　COMMENTS: Arranged as a subspecies of *C. laucha* by Hershkovitz (1962), but others have retained *tener* and *C. laucha* as distinct species (Cabrera, 1961; Mares et al., 1989a; Moojen, 1952; Olds, 1988). Karyotypic identity (2n = 66, FN = 66) clarified by

Bonvicino and Almeida (2000), who expanded upon its morphological and ecological separation from *C. expulsus*. Anderson (1997) considered identification of the Bolivian specimens as tentative and advised further review of series presently allocated to *C. callosus*.

Calomys tocantinsi Bonvicino, Lima, and Almeida, 2003. Rev. Bras. Zool., 20:301.
COMMON NAME: Tocantins Laucha.
TYPE LOCALITY: Brazil, Tocantins State, 33 km SW Formoso do Araguaia, Rancho Beira Rio; 11°47'S, 49°45'W.
DISTRIBUTION: Cerrado habitat, states of Mato Grosso and Tocantins, C Brazil.
COMMENTS: A large species, similar in size to *C. callosus* and *C. expulsus* but morphometrically and karyotypically (2n = 46, FN = 66) well differentiated from those species and from *C. tener* (Bonvicino et al., 2003c). The new form was first detected based on its distinctive karyotype (Lima and Kasahara, 2001).

Calomys venustus (Thomas, 1894). Ann. Mag. Nat. Hist., ser. 6, 14:359.
COMMON NAME: Córdoba Laucha.
TYPE LOCALITY: Argentina, Córdoba Prov., Cosquín.
DISTRIBUTION: WC Argentina; limits uncertain.
COMMENTS: Earlier retained as a species as described (Cabrera, 1961; Gyldenstolpe, 1932), but swept under Hershkovitz's (1962) broad concept of *callosus* and provisionally retained as such by Musser and Carleton (1993). Olds (1988; also Anderson, 1997) recognized *C. venustus* as separate from *C. callosus* proper, and defined it broadly to include *boliviae* and *fecundus* in Bolivia and NW Argentina (also see account of *C. boliviae*); others have more narrowly applied the name to the C Argentinian populations (e.g., Salazar-Bravo et al., 2001, 2002a; Vitullo et al., 1990) as distinct from *fecundus* (here = *C. boliviae*). Banded chromosomal comparisons with *C. callidus* provided by Vitullo et al. (1990), who considered the two, along with *fecundus* (here = *C. boliviae*) to form a closely related group, as did Espinosa et al. (1997). Salazar-Bravo et al. (2002a) similarly recognized a *callosus-venustus* clade, also embracing *callidus* and *fecundus* as species, although the taxa represented in these studies are not entirely equivalent.

Chelemys Thomas, 1903. Ann. Mag. Nat. Hist., ser. 7, 12:242.
TYPE SPECIES: *Akodon megalonyx* Waterhouse, 1845.
COMMENTS: Akodontini (S Andean clade). Named as a subgenus of *Akodon*, later ranked as a genus (Gyldenstolpe, 1932; Reig, 1987; Thomas, 1927) or consolidated under *Notiomys*, together with *Geoxus* (Cabrera, 1961; Ellerman, 1941; Osgood, 1925, 1943). Trees derived from cytochrome *b* data consistently disclose close kinship among these three long-clawed, semifossorial taxa (and *Pearsonomys*—Smith and Patton, 1999), a finding that is cladistically harmonious with their monophyly as perceived by Osgood (1925), who arranged the species under the one genus *Notiomys*. However, morphological differentiation among them is pronounced, as emphasized by Thomas (1927), Gyldenstolpe (1932), Pearson (1984), and Reig (1987). The latter two authors enumerated diagnostic traits and amplified their morphological definition as genera; also see Patterson (1992b) for other morphological and morphometric comparisons and a key to the genera and species of long-clawed akodonts. *Chelemys angustus* Thomas, 1927, is based on a specimen of *Akodon longipilis* according to Pearson (1984).

Chelemys delfini (Cabrera, 1905). Rev. Chil. Hist. Nat., 9:15.
COMMON NAME: Magellanic Long-clawed Akodont.
TYPE LOCALITY: Chile, Magallanes Prov., Straits of Magellan, Punta Arenas.
DISTRIBUTION: Southernmost Chile.
COMMENTS: Taxonomic status and distributional extent require refinement—treated as a subspecies of *C. megalonyx* by Mann (1978) and Tamayo and Frassinetti (1980); retained as a species by Osgood (1943) and Reig (1987). Johnson et al. (1990) recognized *C. delfini* and *C. macronyx* as definable forms, separable by body size, in the Patagonian region of extreme S Chile.

Chelemys macronyx (Thomas, 1894). Ann. Mag. Nat. Hist., ser. 6, 14:362.
> COMMON NAME: Andean Long-clawed Akodont.
> TYPE LOCALITY: Argentina, Mendoza Prov., near Volcán Peteroa (not Fort San Rafael, as corrected by Pearson and Lagiglia, 1992).
> DISTRIBUTION: S Andes along Chile-Argentina boundary, about 34°S latitude south to Straits of Magellan. Distribution in C Chile augmented by Pine et al. (1979) and in Argentina by Pearson (1995) and Teta et al. (2002).
> STATUS: IUCN – Lower Risk (lc).
> SYNONYMS: *alleni* (Osgood, 1925); *connectens* (Osgood, 1925); *fumosus* Thomas, 1927); *vestitus* (Thomas, 1903).
> COMMENTS: Treated as a subspecies of *C. megalonyx* by Mann (1978) and Tamayo and Frassinetti (1980); recognized as species by Osgood (1943) and Pearson (1984, 1995). Osgood (1943) designated the skin of *Notiomys connectens* as the holotype of this composite specimen (skull an *Abrothrix*) and synonymized it with *macronyx vestitus*.

Chelemys megalonyx (Waterhouse, 1845). Proc. Zool. Soc. Lond., 1844:154 [1845].
> COMMON NAME: Large Long-clawed Akodont.
> TYPE LOCALITY: Chile, Valparaíso Prov., Lake Quintero.
> DISTRIBUTION: C Chile, coastal region from Coquimbo Prov. south to Cautín Prov.
> STATUS: IUCN – Lower Risk (lc).
> SYNONYMS: *microtis* (Philippi, 1900); *niger* (Philippi, 1872); *scalops* (Gay, 1847).
> COMMENTS: Tamayo and Frassinetti (1980) listed *delfini* (see above account) as a subspecies of *C. megalonyx*.

Chibchanomys Voss, 1988. Bull. Am. Mus. Nat. Hist., 188:321.
> TYPE SPECIES: *Ichthyomys trichotis* Thomas, 1897.
> COMMENTS: Ichthyomyini. Described as a species of *Ichthyomys*, *trichotis* was thereafter associated first with *Rheomys* (Cabrera, 1961; Tate, 1932h) and then *Anotomys* (Handley, 1976). Generic distinctiveness and phylogenetic relationships substantiated by Voss (1988); also see Jenkins and Barnett (1997). Anderson (1997) reported a single specimen of *Chibchanomys* species indeterminate from NW Bolivia (La Paz).

Chibchanomys orcesi Jenkins and Barnett, 1997. Bull. Nat. Hist. Mus. Lond., 63:124.
> COMMON NAME: Las Cajas Ichthyomyine.
> TYPE LOCALITY: Ecuador, Azuay Prov., Las Cajas, Lake Luspa, 3700 m; 02°50′S, 79°30′W.
> DISTRIBUTION: Páramo, 3100-4000 m, of the Las Cajas Plateau, S Ecuador.
> COMMENTS: Although exhibiting certain similarities to *Neusticomys*, the brunt of apomorphic traits substantiates the placement of *orcesi* in *Chibchanomys*. Phylogenetic relationships to other ichthyomyine genera examined by Jenkins and Barnett (1997); behavioral and ecological observations supplied by Barnett (1997).

Chibchanomys trichotis (Thomas, 1897). Ann. Mag. Nat. Hist., ser. 6, 20:220.
> COMMON NAME: Chibchan Ichthyomyine.
> TYPE LOCALITY: Colombia, "W. Cundinamarca."
> DISTRIBUTION: Andean highlands in far W Venezuela, Colombia, and C Perú.
> STATUS: IUCN – Lower Risk (nt).
> COMMENTS: Provenience of type, taxonomy, and distribution reviewed by Voss (1988).

Chilomys Thomas, 1897. Ann. Mag. Nat. Hist., ser. 6, 19:500.
> TYPE SPECIES: *Oryzomys instans* Thomas, 1895.
> COMMENTS: Thomasomyini. Arranged with oryzomyines (Reig, 1984) or suggested as more closely related to thomasomyines (Musser and Carleton, 1993); molecular analyses (cytochrome *b*) support the latter tribal affinity, in a clade composed of (*Rhipidomys* (*Chilomys-Thomasomys*)) (D'Elía et al., 2003; Smith and Patton, 1999). Aspects of morphology reported by Carleton (1973) and Voss and Linzey (1981).

Chilomys instans (Thomas, 1895). Ann. Mag. Nat. Hist., ser. 6, 16:368.
> COMMON NAME: Andean Chilomys.

TYPE LOCALITY: Colombia, Cundinamarca Dept., Bogotá region, Hacienda de La Selva, 4600 ft (1402 m).

DISTRIBUTION: N Andes, from C Ecuador (Musser et al., 1998:106), through C and N Colombia, to W Venezuela (Handley, 1976; Linares, 1998).

STATUS: IUCN – Lower Risk (lc).

SYNONYMS: *fumeus* Osgood, 1912.

COMMENTS: Osgood (1912) described *fumeus* (type locality: Colombia, Norte de Santander, Páramo de Tamá) as a species distinct from *C. instans*; however, Cabrera (1961) relegated it to subspecific status, as also observed by Linares (1998), and the genus has thereafter been viewed as monotypic (e.g., Corbet and Hill, 1991; Honacki et al., 1982; Musser and Carleton, 1993). Osgood's employment of subspecies was especially robust so that he chose to diagnose *fumeus* as a species is by itself instructive. In view of the other valid species described from the Páramo de Tamá region (e.g., *Oligoryzomys griseolus, Thomasomys hylophilus*), the status of *fumeus* invites another appraisal.

Chinchillula Thomas, 1898. Ann. Mag. Nat. Hist., ser. 7, 1:280.

TYPE SPECIES: *Chinchillula sahamae* Thomas, 1898.

COMMENTS: Phyllotini. Revised by Hershkovitz (1962); karyology reviewed by Pearson and Patton (1976).

Chinchillula sahamae Thomas, 1898. Ann. Mag. Nat. Hist., ser. 7, 1:280.

COMMON NAME: Achallo.

TYPE LOCALITY: Bolivia, La Paz Dept., 50 km N Mt Sajama, Esperanza, Pacajes, 4200 m.

DISTRIBUTION: Altiplano region of S Perú, W Bolivia (above 4000 m), and N Chile.

STATUS: IUCN – Lower Risk (lc).

COMMENTS: The presence of this species in NW Argentina, as reported in many faunal works and checklists (Cabrera, 1961; Díaz, 2000; Musser and Carleton, 1993), is not based on confirmed identification of any vouchered specimen (see Galliari et al., 1996; Ortiz et al., 2000a).

Delomys Thomas, 1917. Ann. Mag. Nat. Hist., ser. 8, 20:196.

TYPE SPECIES: *Hesperomys dorsalis* Hensel, 1873.

COMMENTS: Sigmodontinae *incertae sedis*. *Delomys* has variously stood as a genus (Avila-Pires, 1960c; Gyldenstolpe, 1932; Tate, 1932f) or merged under *Thomasomys*, with or without subgeneric division (Ellerman, 1941; Moojen, 1952; Osgood, 1933d). As remarked by Osgood (1933d), Thomas in part created *Delomys* to taxonomically underscore the geographic disjunction between thomasomyines in SE Brazil and the diverse radiation of typical *Thomasomys* in the N Andes (e.g., see Reig, 1986). Defining morphological traits consolidated and contrasted with types species of *Thomasomys* and *Oryzomys* by Voss (1993); chromosomal variation reported by Zanchin et al. (1992b) and Bonvicino and Geise (1995). While the morphological differentiation of *Delomys* is comparable to that of other sigmodontine genera, its cladistic stature with regard to Andean thomasomyines— and that of other SE Brazilian endemics like *Juliomys, Rhagomys, Phaenomys*, and *Wilfredomys* (oryzomyine-thomasomyine plesions *sensu* Voss, 1993)—must await broader-based phylogenetic inquiry. In one such preliminary view using cytochrome *b* sequence data, *Delomys* erratically formed a basal clade with various other tribes depending upon the analysis (Smith and Patton, 1999); those authors arranged the genus as one of several "additional unique lines."

Number of recognized species given as four (Moojen, 1952), three (Bonvicino and Geise, 1995; Reig, 1986), two (Ellerman, 1941; Gyldenstolpe, 1932; Hershkovitz, 1998; Voss, 1993), or one (Cabrera, 1961; Corbet and Hill, 1991; Honacki et al., 1982). The three recognized here integrates the revision of Voss (1993) and karyological results of Zanchin et al. (1992b) and Bonvicino et al. (1995). Avila-Pires (1960c) associated *Calomys plebejus* Winge, 1887, as another form of *Delomys*; Voss (1993) accepted the generic assignment but regarded *plebejus* as a *nomen dubium*, considering its equivalence to any of the living species as indeterminate.

Delomys collinus Thomas, 1917. Ann. Mag. Nat. Hist., ser. 8, 20:197.
 COMMON NAME: Montane Delomys.
 TYPE LOCALITY: Brazil, Rio de Janeiro State, Itatiaya, 4800 ft (1463 m).
 DISTRIBUTION: Middle to upper elevations, 1000-2700 m, in disjunct mountain ranges
  (E Minas Gerais, Espírito Santo and Rio de Janeiro), SE Brazil.
 COMMENTS: Voss (1993) noted the usual absence of pectoral mammae in this form, but
  absent clear mensural differentiation, elected to retain it under *D. dorsalis*, perhaps as a
  subspecies as originally placed by Thomas (1917*c*); whereas, Bonvicino and Geise
  (1995) obtained different fundamental numbers from *collinus* samples compared with
  those of *D. dorsalis* (2n = 82 in both) and recommended specific recognition. The
  concordance of these independent traits, and others, should be verified with larger
  samples as a basis for reinforcing their status as distinct species.

Delomys dorsalis (Hensel, 1873). Abh. König. Akad. Wiss. Berlin, 1872:42.
 COMMON NAME: Striped Delomys.
 TYPE LOCALITY: Brazil, Rio Grande do Sul State, Taquara (as restricted by Avila-Pires,
  1994:372).
 DISTRIBUTION: Atlantic Forest region of SE Brazil (Minas Gerais and Espírito Santo to
  Rio Grande do Sul) and NE Argentina (Misiones).
 STATUS: IUCN – Lower Risk (lc).
 SYNONYMS: *lechei* (Trouessart, 1904); *obscura* (Leche, 1886) [not *obscurus* Wagner, 1843].

Delomys sublineatus (Thomas, 1903). Ann. Mag. Nat. Hist., ser. 7, 12:240.
 COMMON NAME: Pallid Delomys.
 TYPE LOCALITY: Brazil, Espírto Santo State, Engenheiro Reeve (= Rive), 500 m.
 DISTRIBUTION: Atlantic Forest region of SE Brazil (Minas Gerais and Espírito Santo to Santa
  Catarina).
 STATUS: IUCN – Lower Risk (lc).

Deltamys Thomas, 1917. Ann. Mag. Nat. Hist., ser. 8, 20:98.
 TYPE SPECIES: *Deltamys kempi* Thomas, 1917.
 COMMENTS: Akodontini. Described as a genus, but usually treated as a subgenus of *Akodon* (e.g.,
  Cabrera, 1961; Ellerman, 1941; Musser and Carleton, 1993; Reig, 1987; Smith and Patton,
  1999), though not uniformly (Gyldenstolpe, 1932; Massoia, 1980*b*). Status of *Deltamys*
  recently argued as a genus (Bianchini and Delupi, 1994) and diagnosis so emended
  (González and Massoia, 1995). In cladistic analyses of allozymes (Barrantes et al., 1993)
  and gene sequences (D'Elía, 2003; D'Elía et al., 2003), *D. kempi* emerges as a basal sister-
  group to other *Akodon sensu stricto*. By itself, this cladistic dichotomy is ambiguous with
  regard to its ranking as merely distinctive subgenus or separate genus; the issue deserves
  further consideration based on a broader sampling of akodonts and other information.

Deltamys kempi Thomas, 1917. Ann. Mag. Nat. Hist., ser. 8, 20:98.
 COMMON NAME: Kemp's Akodont.
 TYPE LOCALITY: Argentina, Buenos Aires Prov., Río Paraná Delta, La Plata Estuary, Isla Ella,
  1 m.
 DISTRIBUTION: EC Argentina, Uruguay, and SE Brazil (Rio Grande do Sul—González and
  Massoia, 1995; also see D'Elía et al., 2003:Fig. 1).
 STATUS: IUCN – Lower Risk (lc) *as Akodon kempi*.
 SYNONYMS: *langguthi* Massoia, 1980 [*nomen nudum*]; *langguthi* González and Massoia, 1995.
 COMMENTS: Distribution and ecology reviewed by Massoia (1964); karyotype (2n = 35-38,
  FN = 38) and unique sex-determining mechanism described by Sbalqueiro et al. (1984)
  and Castro et al. (1991). Status of populations exhibiting different karyotypes (Castro
  et al., 1991) warrants further investigation. See González and Pardiñas (2002,
  Mammalian Species, 711).

Eligmodontia F. Cuvier, 1837. Ann. Sci. Nat. (Paris), ser. 2, 7:169.
 TYPE SPECIES: *Eligmodontia typus* F. Cuvier, 1837.
 SYNONYMS: *Elimodon* Fitzinger, 1867; *Eligmodon* Wagner, 1841; *Heligmodontia* Aggasiz, 1846.
 COMMENTS: Phyllotini. Morphologically similar to *Calomys* (Hershkovitz, 1962; Williams and

Mares, 1978), but cladistic interpretations of morphological and molecular data represent the genus as more highly derived, usually as a basal clade to *Phyllotis* and various kin depending upon the taxa sampled (Braun, 1993; Engel et al., 1998; Spotorno et al., 2001; Steppan, 1993; Smith and Patton, 1999).

Revised by Hershkovitz (1962), who synonymized all forms as *E. typus*, but morphological, chromosomal, and distributional evidence supports greater species diversity (e.g., Kelt et al., 1991; Mann, 1978; Mares et al., 1989*b*; Osgood, 1943; Zambelli et al., 1992). The thrust of recent revisionary research has overwhelmingly reinforced the latter view (Hillyard et al., 1997; Sikes et al., 1997; Sportorno et al., 1994; Tiranti, 1997) and collectively indicates the number of species as at least four. Others are occasionally listed (Braun, 1993; Díaz and Barquez, 1999; Galliari et al., 1996) or noted as species indeterminate (Mares et al., 1997), a situation that underscores the need for continued revision. Tiranti (1997) viewed the northern (*E. moreni-E. puerulus*) and southern (*E. morgani-E. typus*) pairs as sister clades and offered a Plio-Pleistocene biogeographic interpretation to explain their phyletic diversification.

Eligmodontia moreni (Thomas, 1896). Ann. Mag. Nat. Hist., ser. 6, 18:307.
COMMON NAME: Monte Laucha.
TYPE LOCALITY: Argentina, La Rioja Prov., Chilecito, 1200 m.
DISTRIBUTION: Atlantic-facing Andean slopes at intermediate elevations, NW Argentina (Salta to Neuquén Prov.); limits poorly documented.
STATUS: IUCN – Lower Risk (lc).
COMMENTS: Mares et al. (1981*b*, 1989*b*) noted the morphological, distributional, and ecological distinction of *E. moreni* and *E. puerulus* in Salta Prov., Argentina, and Spotorno et al. (1994) substantiated their pronounced karyotypic divergence (2n = 34 versus 2n = 50, respectively).

Eligmodontia morgani J. A. Allen, 1901. Bull. Am. Mus. Nat. Hist., 14:409.
COMMON NAME: Western Patagonian Laucha.
TYPE LOCALITY: Argentina, Santa Cruz Prov., basaltic canyons 50 mi (80 km) SE Lago Buenos Aires.
DISTRIBUTION: W Patagonian region of S Argentina (C Neuquén and SC Río Negro Provs. south to NW Santa Cruz; see Hillard et al., 1997:Fig. 2) and adjacent Chile.
STATUS: IUCN – Lower Risk (lc).
COMMENTS: This species epithet was used by Kelt et al. (1991) for the 2n = 32 chromosomal form as compared with 2n = 44 in *E. typus* proper; the same contrast holds for other populations in sympatry or parapatry in SC Argentina (Tiranti, 1997; Zambelli et al., 1992). Multivariate analysis of craniodental variables underscores the strong morphological divergence of specimens with known karyotype and cytochrome *b* haplotype (Hillard et al., 1997; Sikes et al., 1997). Historical biogeography with regard to *E. typus* discussed by Hillard et al. (1997), who elaborated upon the complementarity of their present-day distributions to biotic provinces within Patagonia.

Eligmodontia puerulus (Philippi, 1896). Anal. Mus. Nac. Chile, Zool. Ent., 13a:20.
COMMON NAME: Altiplano Laucha.
TYPE LOCALITY: Chile, Antofagasta Prov., San Pedro de Atacama, 3223 m.
DISTRIBUTION: Altiplano, usually above 3000 m, of extreme S Perú, through NE Chile and WC Bolivia, to NW Argentina.
STATUS: IUCN – Lower Risk (lc).
SYNONYMS: *hirtipes* (Thomas, 1902); *jacunda* Thomas, 1919; *marica* Thomas, 1918; *tarapacensis* Mann, 1945.
COMMENTS: Reduced to a subspecies of *E. typus* by Hershkovitz (1962), but morphological (Mann, 1978; Osgood, 1943) and karyotypic (Kelt et al., 1991; Ortells et al., 1989; Spotorno et al., 1994) data sustain the specific recognition of *E. puerulus*. Further systematic review is required: e.g., Braun (1993) and Díaz and Barquez (1999) viewed *hirtipes* as a separate species but Anderson (1997) did not; and Galliari et al. (1996) and Mares et al. (1997) listed *marica* as distinct.

Eligmodontia typus F. Cuvier, 1837. Ann. Sci. Nat. (Paris), ser. 2, 7:169.
>   COMMON NAME: Eastern Patagonian Laucha.
>   TYPE LOCALITY: Argentina, "Buenos Aires."
>   DISTRIBUTION: E Patagonian region of S Argentina, Buenos Aires and La Pampa Provs.
>   southwards to Santa Cruz.
>   STATUS: IUCN – Lower Risk (lc).
>   SYNONYMS: *elegans* (Waterhouse, 1837); *hypogaeus* (Cabrera, 1934); *pamparum* Thomas,
>   1913.
>   COMMENTS: Hershkovitz (1962:185) discussed the uncertainty of the type locality (also see
>   Ortells et al., 1989:138), which should be formally restricted in order to settle the
>   morphological identification of *E. typus* and promote a critical definition of the species
>   within the genus. See account of *E. morgani* for references on the morphometric,
>   karyotypic, and molecular differentiation of the two species where their ranges
>   approach one another or overlap. Cabrera (1961) referred *Graomys hypogaeus* Cabrera
>   (1934) as a full synonym of *Phyllotis griseoflavus medius*, Massoia (1976) placed it in
>   synonymy under *E. typus sensu lato*, and Williams and Mares (1978) questioned
>   whether the type is a composite specimen; critical restudy of the holotype and new
>   field collections are required to finally ascertain its status.

Euneomys Coues, 1874. Proc. Acad. Nat. Sci. Philadelphia, 26:185.
>   TYPE SPECIES: *Reithrodon chinchilloides* Waterhouse, 1839.
>   SYNONYMS: *Chelemyscus* Thomas, 1925.
>   COMMENTS: Reithrodontini. Sister genus to *Neotomys-Reithrodon* based on cladistic evaluation
>   of morphological characters (Ortiz et al., 2000b; Pardiñas, 1997; Steppan, 1995; Steppan
>   and Sullivan, 2000). See remarks under *Reithrodon* on tribal affiliation. Relatively few
>   specimens exist to substantiate the specific taxonomy and distribution of the genus.
>   Revised by Hershkovitz (1962), who acknowledged several nominal species but indicated
>   their probable synonymy under *E. chinchilloides*, as Mann (1978) later formalized for
>   Chilean populations. Yañez et al. (1987) summarized extant specimen data for *Euneomys*
>   and similarly concluded that only one species is represented. Others have recognized two
>   or more (e.g., Gyldenstolpe, 1932; Osgood, 1943; Pearson and Christie, 1991; Pine et al.,
>   1979; Reise and Gallardo, 1990). The latter arrangement is more nearly correct but must
>   be documented by rigorous revisions.

Euneomys chinchilloides (Waterhouse, 1839). Zool. Voy. H. M. S. "Beagle," Mammalia, p. 72.
>   COMMON NAME: Tierra del Fuego Euneomys.
>   TYPE LOCALITY: Chile, Magallanes Prov., Tierra del Fuego, Straits of Magellan, south shore
>   near E entrance.
>   DISTRIBUTION: Isla Grande de Tierra del Fuego, neighboring islands, and southernmost
>   Chile (Magallanes); limits uncertain.
>   STATUS: IUCN – Lower Risk (lc).
>   SYNONYMS: *ultimus* Thomas, 1916.
>   COMMENTS: Hershkovitz (1962) had suggested that other nominal species would all prove to
>   be synonyms of *E. chinchilloides*. His supposition has been taxonomically observed
>   (Mann, 1978; Yañez et al., 1987) but misrepresents species diversity in the genus (see
>   Pearson and Christie, 1991; Reise and Gallardo, 1990). Includes *petersoni* (see account
>   below) according to Pearson and Christie (1991), Pearson (1995), and Reise and
>   Gallardo (1990), but this synonymy requires more persuasive documentation. Also
>   includes *noei* (see *E. mordax* account) according to Reise and Gallardo (1990), but this
>   referral is incorrect.

Euneomys fossor (Thomas, 1899). Ann. Mag. Nat. Hist., ser. 7, 4:280.
>   COMMON NAME: Burrowing Euneomys.
>   TYPE LOCALITY: Argentina, "Prov. Salta."
>   DISTRIBUTION: Known only from the type locality, which is questionable as *Euneomys* is
>   otherwise unrecorded this far north in Argentina (see Mares et al., 1989b).
>   COMMENTS: Holotype of *Reithrodon fossor* Thomas, 1899, type species of *Chelemyscus*
>   Thomas, 1925, is a composite, the designated type skull referable to *Euneomys*

(Osgood, 1943:164; Pearson, 1984:231). Status is doubtful —"the name *Chelemyscus fossor* should be attached to an appropriate species of *Euneomys* by some future revisor" (Pearson, 1984:231). Galliari et al. (1996) considered it a *nomen dubium*.

Euneomys mordax Thomas, 1912. Ann. Mag. Nat. Hist., ser. 8, 10:410.
> COMMON NAME: Large Euneomys.
> TYPE LOCALITY: Argentina, Fort San Rafael; location uncertain, probably not the San Rafael in Mendoza Prov. (see remarks by Pearson and Christie, 1991:126, and Reise and Gallardo, 1990:79).
> DISTRIBUTION: WC Argentina (Mendoza and Neuquen Provs.) and adjacent region of Chile (Santiago, Bío-Bío, and Araucanía Provs.), apparently at higher elevations (1740-3000 m); range limits unknown.
> STATUS: IUCN – Lower Risk (lc).
> SYNONYMS: *noei* Mann, 1944.
> COMMENTS: Clearly a species distinct from the *chinchilloides-petersoni* complex as conveyed by its large size, singular pattern of incisor grooving, and chromatic differences (Pearson and Christie, 1991; Reise and Gallardo, 1990). Pearson and Christie (1991) treated *noei* as conspecific with *E. mordax*, but Reise and Gallardo (1990) considered it a subjective synonym of *E. chinchilloides*. The former allocation is supported by the sulcation of the upper incisors (a wide, shallow trough about centrally positioned) that characterizes the holotype (USNM 391817); although a subadult, the dimensions of the molars, interorbit, and braincase describe a large animal, one fully as robust as *E. mordax* (e.g., examples from La Parva; USNM 399400, 399401). At the type locality (Valle de la Junta, Lo Valdes, 2500 m), the holotype was collected with another specimen (USNM 391818) that is referable to the *chinchilloides-petersoni* complex, as suggested by its incisor groove (a narrow crease situated mediolaterally), smaller size of those same cranial dimensions, and narrower breadth of the incisors and distal nasals. Pearson and Christie (1991) recorded other sites of sympatry or parapatry between *E. mordax* and the *chinchilloides-petersoni* complex (all identified as *E. chinchilloides* by them) in Chile (La Parva, vicinity of Paso Pino Hachado) and Argentina (vicinity of Copahue, Cueva Traful). Karyotype (2n = 42, FN = 66) reported by Reise and Gallardo (1990) based on a voucher from Paso Pino Hachado; its identity should be reconfirmed.

Euneomys petersoni J. A. Allen, 1903. Bull. Am. Mus. Nat. Hist., 19:192.
> COMMON NAME: Peterson's Euneomys.
> TYPE LOCALITY: Argentina, Santa Cruz Prov., upper Río Chico, near the Cordilleras.
> DISTRIBUTION: WC Argentina (Neuquen Prov.) and C Chile (Santiago Prov.) southwards to extreme S Argentina and adjacent Chile, excluding Tierra del Fuego; limits uncertain.
> STATUS: IUCN – Lower Risk (lc).
> SYNONYMS: *dabbeni* Thomas, 1919.
> COMMENTS: Relegated to a subspecies or full synonym of *E. chinchilloides* by most systematists (Gyldenstolpe, 1932; Hershkovitz, 1962; Mann, 1978; Muñoz Pedreros, 2000; Pearson, 1995; Pearson and Christie, 1991; Reise and Gallardo, 1990; Yañez et al., 1987). While examples of both *E. chinchilloides* (e.g., FMNH 50600, 50601, 50736; USNM 482138-482140) and *E. petersoni* (e.g., FMNH 50583, 50584-50593, 50595-50599; USNM 84197, 84200, 84202) possess upper incisors with distinct mediolateral grooves, the series otherwise differ in size and color, abrupt contrasts over relatively short geographic distances that persuaded Osgood (1943) to maintain each as species. A strong size separation is actually conveyed in the morphometric analysis of Reise and Gallardo (1990:Fig. 3), which employed all variables and in which samples of *chinchilloides* proper are non-overlapping in multivariate space (Note that certain operational taxonomic units defined by those authors are undoubtedly species composites, a situation that affects measures of intra-sample covariation and compromises statistics of inter-sample dispersion). As implied by the range limits, we tentatively assign those northern samples that co-occur with *E. mordax* to *E. petersoni*, but, as noted by Pine et al. (1979), they do not convincingly fit with either *E. chinchilloides* or *E. petersoni* as known by populations in S Chile and Argentina. Until such

problems and differences can be resolved in the context of a substantive generic revision, using larger samples and other kinds of data, we continue to follow Osgood (1943). Karyotype (2n = 36, FN = 66) reported by Reise and Gallardo (1990, as *E. chinchilloides*).

Galenomys Thomas, 1916. Ann. Mag. Nat. Hist., ser. 8, 17:143.

TYPE SPECIES: *Phyllotis garleppi* Thomas, 1898.

COMMENTS: Phyllotini. Named as a subgenus of *Euneomys*, later elevated to genus (Thomas, 1926*a*), and reassociated as a subgenus of *Phyllotis* (Ellerman, 1941; Osgood, 1947; Corbet and Hill, 1991). Revised by Hershkovitz (1962), who considered *Galenomys* a distinct genus, as did Pearson (1958); phylogenetic evaluation of morphological characters depicts *Galenomys* as closely related to species of *Ausliscomys* (Braun, 1993; Steppan, 1995).

Galenomys garleppi (Thomas, 1898). Ann. Mag. Nat. Hist., ser. 7, 1:279.

COMMON NAME: Garlepp's Pericote.

TYPE LOCALITY: Bolivia, La Paz Dept., northeast of Mt Sajama, Esperanza, 4140 m.

DISTRIBUTION: Altiplano, generally above 3000 m, of S Perú, N Chile, and adjacent Bolivia.

STATUS: IUCN – Lower Risk (lc).

Geoxus Thomas, 1919. Ann. Mag. Nat. Hist., ser. 9, 3:207.

TYPE SPECIES: *Oxymycterus valdivianus* Philippi, 1858.

COMMENTS: Akodontini (S Andean clade). Revised by Osgood (1925, 1943) as part of *Notiomys*— see remarks under *Chelemys*. Sister genus to *Pearsonomys* according to maximum parsimony and likelihood analyses of cytochrome *b* sequences (Smith and Patton, 1999).

Geoxus valdivianus (Philippi, 1858). Arch. Naturgesch., 24(1):303.

COMMON NAME: Valdivian Long-clawed Akodont.

TYPE LOCALITY: Chile, Valdivia Prov.

DISTRIBUTION: C and S Chile (see Saavedra and Simonetti, 2001, for northern range extension), including Mocha and Chiloe Isls, to Straits of Magellan, and S Argentina (Neuquén to Santa Cruz Provs.).

STATUS: IUCN – Lower Risk (lc).

SYNONYMS: *araucanus* (Osgood, 1925); *bicolor* (Osgood, 1943); *bullocki* (Osgood, 1943); *chiloensis* (Osgood, 1925); *fossor* Thomas, 1919; *michaelseni* (Matschie, 1898); *microtis* (J. A. Allen, 1903).

COMMENTS: Pearson (1984) noted the uncertainty of the generic association and specific status of *michaelseni*, which has been arranged as a subspecies (Osgood, 1943; Patterson, 1992*b*) or later listed as species (Reig, 1987). The taxon's status and relationship clearly merit further study. Subspecific distributions in Chile sketched by Muñoz Pedreros (2000).

Graomys Thomas, 1916. Ann. Mag. Nat. Hist., ser. 8, 17:141.

TYPE SPECIES: *Mus griseoflavus* Waterhouse, 1837.

SYNONYMS: *Bothriomys* Ameghino, 1889.

COMMENTS: Phyllotini. Doubtfully retained as a genus distinct from *Phyllotis* by Ellerman (1941), later included in *Phyllotis* as a formal subgeneric division (Osgood, 1947; Pearson, 1958) or not (Hershkovitz, 1962). Thomas (1919) remained firm regarding the generic distinctiveness of *Phyllotis* and *Graomys*, and the latter was so reinstated based on karyological data (Pearson and Patton, 1976) and other traits (see Olds and Anderson, 1989; Reig, 1978). Generic stature uniformly reinforced in broad phylogenetic studies of phyllotine taxa, using either morphological characters (Braun, 1993; Steppan, 1993, 1995) or molecular data (Anderson and Yates, 2000). Also see remarks under *Andalgalomys*, which Steppan (1995) included as a subjective synonym of *Graomys*.

Williams and Mares (1978) transferred *Graomys pearsoni* Myers, 1977, to *Andalgalomys*, and Massoia (1976) synonymized *G. hypogaeus* Cabrera, 1934, with *Eligmodontia typus*. Pardiñas (1995) identified the synonymy of *Bothriomys* with *Graomys*, contra Herskovitz (1962) who placed it under *Euneomys* (McKenna and Bell [1997] placed it under both

Graomys and *Euneomys*), and recommended retention of *Graomys* as *nomen protectum*, it being the familiar and long used genus-group name for these phyllotine rodents.

Graomys centralis (Thomas, 1902). Ann. Mag. Nat. Hist., ser. 7, 9:240.

COMMON NAME: Central Pericote.

TYPE LOCALITY: Argentina, Córdoba Prov., Cruz del Eje, 600 m.

DISTRIBUTION: C Argentina (La Rioja, Catamarca, and Córdoba Provs.); distributional limits uncertain.

COMMENTS: Described as a subspecies of *griseoflavus* and early recognized at that rank (Cabrera, 1961; Ellerman, 1941; Gyldenstolpe, 1932); later merged as a full synonym of *Phyllotis g. griseoflavus* by Hershkovitz (1962). Substantial divergence in allozymes, genomic content, and karyotypes, together with laboratory breeding results, collectively supports the specific status of populations referrable to *centralis* (Ramírez et al., 2001; Theiler and Blanco, 1996; Theiler and Gardenal, 1994; Theiler et al., 1999; Zambelli et al., 1994), as known from C Argentina. However, whether *centralis* will prove to be the oldest name to apply to the 2n = 42 karyomorph, versus 2n = 36-38 in *G. griseoflavus*, or whether other names should be added to its synonymy, will require broader analyses of populations of *Graomys*, especially those yet included within *G. griseoflavus*; see Tiranti (1998*a*) for a thoughtful overview of the geographic ranges of different karyomorphs and the nomenclatural issues to be illuminated in order to apply the proper specific epithets.

Graomys domorum (Thomas, 1902). Ann. Mag. Nat. Hist., ser. 7, 9:132.

COMMON NAME: Pale Pericote.

TYPE LOCALITY: Bolivia, Cochabamba Dept., Tapacari, 3000 m.

DISTRIBUTION: E slopes of Andes in SC Bolivia (600-3700 m) and NW Argentina.

STATUS: IUCN – Lower Risk (lc).

SYNONYMS: *taterona* Thomas, 1926.

COMMENTS: Maintained as species by Cabrera (1961), but ranked as a subspecies of *griseoflavus* by Hershkovitz (1962). Chromosomal divergence reported by Pearson and Patton (1976), who reinstated *domorum* as a species (also see Olds et al., 1987). Reig (1978) followed Cabrera (1961) in allocating *taterona* to *G. domorum*, and Anderson (1997) maintained the latter as a definable subspecies.

Graomys edithae Thomas, 1919. Ann. Mag. Nat. Hist., ser. 9, 3:495.

COMMON NAME: Otro Cerro Pericote.

TYPE LOCALITY: Argentina, Catamarca Prov., Otro Cerro, 3000 m.

DISTRIBUTION: Known only from the type locality.

STATUS: IUCN – Lower Risk (lc).

COMMENTS: Status uncertain—placed as a full synonym of *griseoflavus medius* (Cabrera, 1961); retained as a nominal species of *Phyllotis* (Hershkovitz, 1962) or of *Graomys* (Gyldenstolpe, 1932; Mares et al., 1997; Myers, 1977; Williams and Mares, 1978); or considered a *nomen dubium* (Galliari et al., 1996). Thomas' (1919) measurements certainly characterize *G. edithae* as a smaller form than *cachinus* and *medius*, taxa collected at lower elevations in the region of Otro Cerro and now regarded as examples of the larger *G. griseoflavus*. Reinvestigation of original type series, along with altitudinal transects across vegetation zones in the region, would shed much light. Using an old map (1893) of Catamarca Province, as cited by Cabrera (1961:487), Pardiñas (pers. com.) located the type locality Otro Cerro in the Sierra de Ambato, Capayán Dept., ca. 31.5 km W Huillapima and 16.5 km NNW Chumbicha; formal amendment should be effected in a revisionary context.

Graomys griseoflavus (Waterhouse, 1837). Proc. Zool. Soc. Lond., 1837:28.

COMMON NAME: Common Pericote.

TYPE LOCALITY: Argentina, Río Negro Prov., mouth of Río Negro.

DISTRIBUTION: SC Bolivia, W Paraguay, and nearby Brazil, south through W Argentina to S Chubut Prov.

STATUS: IUCN – Lower Risk (lc).

SYNONYMS: *cachinus* (J. A. Allen, 1901); *chacoensis* (J. A. Allen, 1901); *lockwoodi* Thomas, 1918; *medius* Thomas, 1919.

COMMENTS: See remarks under *G. centralis* and *G. domorum*, taxa formerly included under *G. griseoflavus* as revised by Hershkovitz (1962). Pardiñas (1995) reallocated the late Pleistocene forms *Bothriomys catenatus* and *Oxymicterus impexus*, described by Ameghino (1889), as full synonyms of *G. griseoflavus*; these reallocations and other fossil sites in Buenos Aires establish the broader distribution of the species in the late Pleistocene.

Handleyomys Voss, Gómez-Laverde, and Pacheco, 2002. Am. Mus. Novit., 3373:5.

TYPE SPECIES: *Aepeomys fuscatus* J. A. Allen, 1912.

COMMENTS: Oryzomyini. The two species, *fuscatus* and *intectus*, composing this new genus had been recognized as problematic members within their genera of original description, *Aepeomys* and *Oryzomys* respectively (e.g., Gardner and Patton, 1976; Musser and Carleton, 1993; Ochoa et al., 2001). Discrimination from those genera, morphological description, distribution, and ecological information supplied by Voss et al. (2002). Females lack pectoral mammae as common to most oryzomyine species (see Voss and Carleton, 1993), but *Handleyomys* possesses other synapomorphies that cladistically affiliate it with that tribe, a relationship supported by molecular evaluation (Weksler, 2003). Nearest relatives remain uncertain (Voss et al., 2002); affiliated with certain Middle American *Oryzomys* according to nuclear DNA sequence data (Weksler, 2003).

Handleyomys fuscatus (J. A. Allen, 1912). Bull. Am. Mus. Nat. Hist., 31:89.

COMMON NAME: Dusky-footed Handley's Mouse.

TYPE LOCALITY: Colombia, Valle del Cauca Dept., San Antonio, near Cali, 7000 ft (2134 m).

DISTRIBUTION: Cordillera Occidental, 1700-2580 m, W Colombia.

STATUS: IUCN – Lower Risk (lc) as *Aepomys fuscatus*.

COMMENTS: Formerly ranked as a species of *Thomasomys* (Ellerman, 1941), subspecies of *Aepeomys lugens* (Cabrera, 1961), or usually a species of *Aepeomys* (Gardner and Patton, 1976; Musser and Carleton, 1993). Karyotype highly divergent among oryzomyines and thomasomyines reported by Gardner and Patton (1976). Specific status reinforced by Voss et al. (2002), who morphologically and morphometrically contrasted it to *H. intectus*.

Handleyomys intectus (Thomas, 1921). Ann. Mag. Nat. Hist., ser. 9, 8:356.

COMMON NAME: White-footed Handley's Mouse.

TYPE LOCALITY: Colombia, Antioquia Dept., 8 km E Medellín, Santa Elena, 9000 ft (2743 m).

DISTRIBUTION: Cordillera Central, 1500-2800 m, C Colombia.

STATUS: IUCN – Lower Risk (lc) as *Oryzomys intectus*.

COMMENTS: A species whose distinctive traits have confusingly invoked its past association with *Oryzomys balneator* or *Melanomys* (Gyldenstolpe, 1932), *Nectomys* (Ellerman, 1941), or the subgenus *Oryzomys* (Cabrera, 1961). Reallocated to *Handleyomys* and compared with *H. fuscatus* by Voss et al. (2002).

Holochilus Brandt, 1835. Mem. Acad. Imp. Sci. St. Petersbourg, ser. 6, 3(2):428.

TYPE SPECIES: *Holochilus sciureus* Wagner, 1842 (by subsequent designation - see comments below).

SYNONYMS: *Holochilomys* Brandt, 1855.

COMMENTS: Oryzomyini. Arranged by Hershkovitz (1955a) as one of four genera of sigmodont rodents. Based on reproductive anatomy, Hooper and Musser (1964) remarked that *Holochilus* may represent a "well differentiated oryzomyine rather than a sigmodont." Retained, with *Sigmodon* proper, in the tribe Sigmodontini by Reig (1984, 1986); transferred to Oryzomyini by Voss and Carleton (1993), who associated *Holochilus* within a tetralophodont clade that includes *Pseudoryzomys* and *Lundomys* (also see Carleton and Olson, 1999). Membership in Oryzomyini sustained by taxonomically broad, cladistic studies of morphological, karyological, and molecular characters (Baker et al., 1983; Malygin and Rosmiarek, 1997; Smith and Patton, 1999; Steppan, 1995; Weksler, 2003).

We follow the proposal of Voss and Abramson (1999) for designating *H. sciureus*

Wagner, 1842, as the type species of *Holochilus* in order to continue traditional usage of the genus for a myomorphous sigmodontine. As they explain, the type species as heretofore given, *Mus* (*Holochilus*) *leucogaster* Brandt, 1835, was regrettably listed by Miller and Rehn (1901:89) without specimen examination and is based on an example of the echimyid *Trinomys*. *Mus simpsonii* Philippi, 1900, occasionally listed as a species of *Holochilus* (e.g., Gyldenstolpe, 1932), was reidentified as an example of *Rattus norvegicus* by Osgood (1943). See Musser et al. (1998:280-284) for the contorted taxonomic history of *Mus physodes* Brants, 1827 (or Lichtenstein, 1827), formerly associated with *Holochilus*, whether as species (Ellerman, 1941; Gyldenstolpe, 1932) or as junior synonym of *H. brasiliensis* (Cabrera, 1961; Hershkovitz, 1955*a*).

Genus revised by Hershkovitz (1955*a*), who consolidated 13 nominal species (e.g., Ellerman, 1941) under *H. brasiliensis* and diagnosed a new one, *H. magnus*; the latter has been removed to the genus *Lundomys* (Voss and Carleton, 1993). Subsequent studies have revealed that *brasiliensis* of Hershkovitz (1955*a*) is a composite of three or more species (Aguilera and Pérez-Zapata, 1989; Gardner and Patton, 1976; Massoia, 1980*a*, 1981; Reig, 1986). In general, evidence for the species acknowledged below issues from more localized studies that lack the persuasion of a full generic review; these names, distributions, and synonyms must therefore be accepted as tentative pending such a synoptic revision. Our association of synonyms basically observes the species classification of Massoia (1980*a*). Known from the middle Pleistocene, Bolivia, based on the fossil species, *H. primigeneus*, which Steppan (1996) described as possibly ancestral to living *Holochilus*; also early through late Pleistocene of Argentina, as *H. brasiliensis* (Pardiñas, 1999).

Holochilus brasiliensis (Desmarest, 1819). Nouv. Dict. Hist. Nat., Nouv. ed., 29:62.
>   COMMON NAME: Brazilian Marsh Rat.
>   TYPE LOCALITY: Brazil, Minas Gerais State, Lagoa Santa (as restricted by Hershkovitz, 1955*a*:662).
>   DISTRIBUTION: SE Brazil, Uruguay, and EC Argentina.
>   STATUS: IUCN – Lower Risk (lc).
>   SYNONYMS: *anguyu* (Brandt, 1835); *cancellinus* Wagner, 1843; *darwini* Thomas, 1897; *leucogaster* (Brandt, 1835); *vulpinus* (Brants, 1827).
>   COMMENTS: Morphological definition and geographic range reduced in scope by Massoia (1980*a*, 1981), who segregated *H. sciureus* as a species distinct from *H. brasiliensis*. González (2000*b*) retained *vulpinus* as a subspecies for Uruguayan populations.

Holochilus chacarius Thomas, 1906. Ann. Mag. Nat. Hist., ser. 7, 18:446.
>   COMMON NAME: Chacoan Marsh Rat.
>   TYPE LOCALITY: Paraguay, Concepción Dept., E Chaco, one "league" NW Concepción, 500 ft (152 m).
>   DISTRIBUTION: Paraguay and NE Argentina.
>   STATUS: IUCN – Lower Risk (lc).
>   SYNONYMS: *balnearum* Thomas, 1906.
>   COMMENTS: Considered distinct by Massoia (1980*a*), who treated *balnearum* as a synonym; Reig (1986) mentioned *balnearum* as another good species. Highly variable in diploid (2n = 48-56) and fundamental numbers (Nachman, 1992*b*; Nachman and Myers, 1989; Vidal et al., 1976). Nachman (1992*a*) recognized Paraguayan samples (as *H. brasiliensis chacarius*) as karyotypically distinguishable and specifically distinct from Argentinian populations (as *H. vulpinus*).

Holochilus sciureus Wagner, 1842. Arch. Naturgesch., ser. 8, 1:16.
>   COMMON NAME: Amazonian Marsh Rat.
>   TYPE LOCALITY: Brazil, Minas Gerais State, Rio São Francisco.
>   DISTRIBUTION: Broad reaches of Orinoco and Amazon River basins: Venezuela (including an isolated locality in the Maracaibo Basin, NW of the Andes; see Linares, 1998:Fig. 169), Guianas, N and C Brazil, and Amazonian regions of Colombia, Ecuador, Perú, and Bolivia.
>   STATUS: IUCN – Lower Risk (lc).
>   SYNONYMS: *amazonicus* Osgood, 1915; *berbericensis* Morrison-Scott, 1937; *guianae* Thomas,

1901; *incarum* Thomas, 1920; *multannus* Ameghino, 1889; *nanus* Thomas, 1897; *venezuelae* J. A. Allen, 1904.

COMMENTS: Discrimination from *H. brasiliensis* documented by Massoia (1980a, 1981), who raised *sciureus* to specific rank. Many forms swept under Hershkovitz's (1955a) concept of *brasiliensis* actually belong to this "species," which itself is a composite—e.g., Amazonian populations characterized by a 2n = 55-56 (Patton et al., 2000), but Venezuelan populations characterized by 2n = 44 (Aguilera and Perez-Zapata, 1989). The latter authors, and Aguilera et al. (1993), have recognized *H. venezuelae* as a species, but Linares (1998) continued to employ *H. sciureus* for those same series, acknowledging *venezuelae* as a subspecies. In addition to *venezuelae*, Reig (1986) listed *amazonicus* and *guianae* as probable valid species. The persisting disagreement over number of valid species, uncertain correspondence of karyotypic variants to definable morphologies, and vagueness of distributional limits will only be illuminated by wholesale generic revision. Synonymy of *multannus* asserted by Massoia and Pardiñas (1993).

Ichthyomys Thomas, 1893. Proc. Zool. Soc. Lond., 1893:337.
　　TYPE SPECIES: *Ichthyomys stolzmanni* Thomas, 1893.
　　COMMENTS: Ichthyomyini. Revised by Voss (1988), who defined four species among the eight described taxa and evaluated their phylogenetic relationships.

Ichthyomys hydrobates (Winge, 1891). Vidensk. Medd. Nat. Foren. Kjobenhavn, ser. 5, 3:20.
　　COMMON NAME: Silver-bellied Ichthyomyine.
　　TYPE LOCALITY: Venezuela, Mérida State, Sierra de Mérida. The holotype may have originated from the vicinity of Mérida, 1600-1700 m (Voss, 1988).
　　DISTRIBUTION: Andes of W Venezuela, Colombia, and N Ecuador.
　　STATUS: IUCN – Lower Risk (nt).
　　SYNONYMS: *nicefori* Thomas, 1924; *soderstromi* Tate, 1931.
　　COMMENTS: Taxonomy and distribution reviewed by Voss (1988), who retained *nicefori* and *soderstromi* as subspecies.

Ichthyomys pittieri Handley and Mondolfi, 1963. Acta Biol. Venezuela, 3:417.
　　COMMON NAME: Pittier's Ichthyomyine.
　　TYPE LOCALITY: Venezuela, Aragua State, Rancho Grande, near headwaters of Río Limón.
　　DISTRIBUTION: Cordillera Central, N Venezuela (see Linares, 1998:Fig. 164).
　　STATUS: IUCN – Vulnerable.
　　COMMENTS: Taxonomy reviewed by Voss (1988).

Ichthyomys stolzmanni Thomas, 1893. Proc. Zool. Soc. Lond., 1893:339.
　　COMMON NAME: Stolzmann's Ichthyomyine.
　　TYPE LOCALITY: Perú, Junín Dept., Chanchamayo, near Tarma, 923 m.
　　DISTRIBUTION: E Ecuador and Perú.
　　STATUS: IUCN – Lower Risk (lc).
　　SYNONYMS: *orientalis* Anthony, 1923.
　　COMMENTS: Taxonomy and distribution reviewed by Voss (1988), who retained *orientalis* as a subspecies.

Ichthyomys tweedii Anthony, 1921. Am. Mus. Novit., 20:1.
　　COMMON NAME: Tweedy's Ichthyomyine.
　　TYPE LOCALITY: Ecuador, El Oro Prov., Río Amarillo, Portovelo, 2000 ft (610 m).
　　DISTRIBUTION: W Ecuador and C Panamá.
　　STATUS: IUCN – Lower Risk (lc).
　　SYNONYMS: *caurinus* Thomas, 1924.
　　COMMENTS: Taxonomy and distribution reviewed by Voss (1988), who placed *caurinus* in full synonymy.

Irenomys Thomas, 1919. Ann. Mag. Nat. Hist., ser. 9, 3:201.
　　TYPE SPECIES: *Mus tarsalis* Philippi, 1900.
　　COMMENTS: Sigmodontinae *incertae sedis*. Usually grouped with phyllotines (Olds and

Anderson, 1989; Vorontsov, 1959), a tribal affiliation supported by cladistic analyses of morphological traits (Braun, 1993; Steppan, 1995) but not of cytochrome *b* sequences (D'Elía et al., 2003; Smith and Patton, 1999).

Irenomys tarsalis (Philippi, 1900). Ann. Mus. Nac. Chile Zool., 14:10.
> COMMON NAME: Large-footed Irenomys.
> TYPE LOCALITY: Chile, Valdivia Prov., Fundo San Juan, near La Unión.
> DISTRIBUTION: C and S Chile (see Saavedra and Simonetti, 2000, for northern range extension), including Chiloe and Guaitecas Isls, and adjacent Argentina (Neuquén to Chubut Provs.—see Pardiñas et al., 2003*b*; Pearson, 1995).
> STATUS: IUCN – Lower Risk (lc).
> SYNONYMS: *longicaudatus* (Philippi, 1900).
> COMMENTS: Muñoz Pedreros (2000) maintained *longicaudatus*, along with nominate *tarsalis*, as subspecies for Chilean populations.

Juliomys González, 2000. Com. Zool. Mus. Hist. Nat. Montevideo, 12(196):3.
> TYPE SPECIES: *Thomasomys pictipes* Osgood, 1933.
> COMMENTS: Sigmodontinae *incertae sedis*. Osgood (1933*b*) described *pictipes* as another species of *"Thomasomys"* endemic to the Atlantic Forest region of SE South America. Provisionally reassigned to *Wilfredomys* by Musser and Carleton (1993) on the assertions of Osgood (1933*b*) and Pine (1980*b*) that *pictipes* is most closely related to *Thomasomys oenax*, type species of Avila-Pires' (1960*b*) *Wilfredomys*. With description of *Juliomys*, all of the principal thomasomyine morphological themes in SE South America have now been generically isolated. However, González's (2000*a*) diagnosis serves mainly to repeat or reinforce contrasts between the species *pictipes* and *oenax*, whose differentiation is already well established (Osgood, 1933*b*; Pine, 1980). The broader issue of cladistic relations among these SE endemics still requires elucidation to adjudge generically meaningful clades and tribal memberships among them. In phylogenetic reconstructions based on cytochrome *b* sequences, the genus emerges as monophyletic (using *J. pictipes*, *J. rimofrons*, and an undescribed species from São Paulo, but not *Wilfredomys oenax*), strongly divergent from Thomasomyini proper, and variably affiliated basally with different sigmodontines (D'Elía et al., 2003; Oliveira and Bonvicino, 2002; Smith and Patton, 1999); suprageneric affinities unresolved. Also see remarks under *Delomys*.

Juliomys pictipes (Osgood, 1933). Field Mus. Nat. Hist. Publ., Zool. Ser., 20(2):11.
> COMMON NAME: Contreras' Juliomys.
> TYPE LOCALITY: Argentina, Misiones Prov., Río Paraná, 100 mi (161 km) S Río Iguazú, Caraguatay.
> DISTRIBUTION: NE Argentina and SE Brazil (São Paulo and Santa Catarina).
> STATUS: IUCN – Lower Risk (lc) as *Wilfredomys pictipes*.
> COMMENTS: See González (2000*a*:4) for remarks on location and habitat of the type locality. Karyotype (2n = 36, FN = 34) reported (as *Wilfredomys pictipes*) by Bonvicino and Otazu (1999), who noted many banding patterns congruent with *Rhipidomys* species and generally weak homologies with those of *Delomys*.

Juliomys rimofrons Oliveira and Bonvicino, 2002. Acta Theriol., 47(3):310.
> COMMON NAME: Cleft-headed Juliomys.
> TYPE LOCALITY: Brazil, Minas Gerais State, Itamonte Municipality, Serra da Mantiqueira, east of Pedra Furada, Brejo da Lapa, 2000 m; 22°21'S, 44°44'W.
> DISTRIBUTION: Known only from the type locality, SE Brazil.
> COMMENTS: Similar to *J. pictipes*, but *J. rimofrons* possesses longer pelage, fewer chromosomes (2n = 20, FN = 34), and a carotid arterial circulation with a supraorbital branch (Oliveira and Bonvicino, 2002).

Juscelinomys Moojen, 1965. Rev. Brasil. Biol., 25:281.
> TYPE SPECIES: *Juscelinomys candango* Moojen, 1965.
> COMMENTS: Akodontini. Reig (1987:36) viewed *Juscelinomys* as "an akodontine closely related to *Oxymycterus* and *Lenoxus*," an assessment consistent with initial parsimony analyses using cytochrome *b* data (Emmons, 1999*b*). In evaluations of cytochrome *b* and IRBP

sequences, *Juscelinomys* is portrayed as sister genus to *Oxymycterus* and more distantly related to *Lenoxus* (D'Elía, 2003). Morphological discrimination from *Oxymycterus* and *Brucepattersonius* amplified by Emmons (1999b). Collection of additional series is necessary to corroborate the distinction of the described tribal species and to amplify their distributions.

Moojen (1965) provisionally referred Winge's (1887) *Oxymycterus talpinus* to *Juscelinomys*. Langguth (in Fonseca et al., 1996) commented that the form, known to date only as a subfossil, should not be acknowledged as a recent species (e.g., Musser and Carleton, 1993), but in view of the rediscoveries of other species described first as fossils from the Lagoa Santa deposits (e.g.: *Bibimys labiosus*, *Blarinomys breviceps*, *Lundomys molitor*, *Pseudoryzomys simplex*) or those named subfossils listed as synonyms of living species (e.g.: *principalis* of *Kunsia tomentosus*, *coronatus* of *Oryzomys russatus*), such a conclusion may be premature (also see Voss and Myers, 1991). In any case, the possible status of *talpinus* as senior synonym for other *Juscelinomys* taxa should await critical comparisons and analyses of the type and all known specimens.

Juscelinomys candango Moojen, 1965. Rev. Brasil. Biol., 25:281.
COMMON NAME: Candango Akodont.
TYPE LOCALITY: Brazil, Federal District, Brasília, Parque Zoobotanico, 1030 m.
DISTRIBUTION: C Brazil.
STATUS: IUCN – Lower Risk (nt).
COMMENTS: Morphology of type series amplified by Emmons (1999b) and compared with two new species (see next accounts).

Juscelinomys guaporensis Emmons, 1999. Am. Mus. Novit., 3280:4.
COMMON NAME: Guaporé Akodont.
TYPE LOCALITY: Bolivia, Santa Cruz Dept., Velasco Prov., Parque Nacional Noel Kempff Mercado, Flor de Oro, 210 m; 13°33.10'S, 61°00.51'W.
DISTRIBUTION: Known only from the type locality.
COMMENTS: The type and single known specimen is cranially similar to *J. candango* but smaller and has different colored pelage.

Juscelinomys huanchacae Emmons, 1999. Am. Mus. Novit., 3280:2.
COMMON NAME: Huanchaca Akodont.
TYPE LOCALITY: Bolivia, Santa Cruz Dept., Velasco Prov., Serrania de Huanchaca, Parque Nacional Noel Kempff Mercado, Campamento "Huanchaca II," 700 m; 14°31.42'S, 60°44.37'W.
DISTRIBUTION: Known only from the type locality.
COMMENTS: The type and single known specimen is the largest individual recorded for the genus.

Kunsia Hershkovitz, 1966. Z. Säugetierk., 31(2):112.
TYPE SPECIES: *Mus tomentosus* Lichtenstein, 1830.
COMMENTS: Akodontini. Species formerly included in *Scapteromys* until set apart in *Kunsia* by Hershkovitz (1966c), who arranged both genera in the "scapteromyine group," which he viewed as closely related to oxymycterines. Formal tribal segregation of the two genera, later including *Bibimys*, advanced by Massoia (1979b) and Reig (1980, 1981). Although the probable sister genus of *Scapteromys*, parsimony and likelihood analyses of cytochrome *b* data represent the two as nested within the akodontine radiation and distantly related to *Bibimys* (D'Elía et al., 2003; Smith and Patton, 1999). D'Elía (2003) identified the pair *Kunsia-Scapteromys* as one of five major clades within Akodontini. Karyology summarized by Gardner and Patton (1976). Confirmation of species limits, junior synonyms, and geographic ranges overdue.

Kunsia fronto (Winge, 1887). E Museo Lundii, 1(3):44.
COMMON NAME: Fossorial Kunsia.
TYPE LOCALITY: Brazil, Minas Gerais State, Rio das Velhas, Pleistocene cave deposits ("Lapa da Escrivania no. 5") near Lagoa Santa.
DISTRIBUTION: Isolated localities in NE Argentina, and EC Brazil; range poorly documented.

STATUS: IUCN – Vulnerable.

SYNONYMS: *chacoensis* (Gyldenstolpe, 1932); *planaltensis* Avila-Pires, 1972.

COMMENTS: Taxonomy and distribution reviewed by Avila-Pires (1972).

Kunsia tomentosus (Lichtenstein, 1830). Darst. Säugeth., 7(15):33.

COMMON NAME: Woolly Kunsia.

TYPE LOCALITY: Brazil, SE area along the Río Uruguay (as restricted by Hershkovitz, 1966c:120).

DISTRIBUTION: NE Bolivia (Beni and Santa Cruz Depts.—Anderson, 1993, 1997) and WC Brazil (Mato Grosso); Pleistocene cave samples in Minas Gerais, Brazil.

STATUS: IUCN – Vulnerable.

SYNONYMS: *gnambiquarae* (Mirando Ribeiro, 1914); *principalis* (Lund, 1840).

COMMENTS: Rare in collections; range inadequately known. See Massoia and Fornes (1965) and Hershkovitz (1966c) for justification of synonymy.

Lenoxus Thomas, 1909. Ann. Mag. Nat. Hist., ser. 8, 4:236.

TYPE SPECIES: *Oxymycterus apicalis* J. A. Allen, 1900.

COMMENTS: Akodontini. Viewed as closely related to *Oxymycterus* by Reig (1987), but electrophoretic data reveal *Lenoxus* as cladistically removed from *Oxymycterus* proper (Patton et al., 1989) and gene sequences indicate closer relationship to a clade including *Blarinomys*, *Brucepattersonius*, *Kunsia*, and *Scapteromys* (D'Elía, 2003; D'Elía et al., 2003; Smith and Patton, 1999).

Lenoxus apicalis (J. A. Allen, 1900). Bull. Am. Mus. Nat. Hist., 13:224.

COMMON NAME: White-tailed Akodont.

TYPE LOCALITY: Perú, Puno Dept., Valley of Río Inambari, Santo Domingo Mine, 6000 ft (1830 m).

DISTRIBUTION: Cloud forest of E Andean slopes, about 1500-2500 m, in SE Perú and WC Bolivia (La Paz Dept.).

STATUS: IUCN – Lower Risk (nt).

SYNONYMS: *boliviae* Sanborn, 1950.

COMMENTS: Anderson (1997) used *boliviae* as subspecies for the Bolivian populations.

Loxodontomys Osgood, 1947. J. Mamm., 28:172.

TYPE SPECIES: *Mus micropus* Waterhouse, 1837.

COMMENTS: Phyllotini. Diagnosed as a subgenus of *Phyllotis* and maintained at that rank (Pearson, 1958) or as a full synonym (Hershkovitz, 1962). Transferred to *Auliscomys*, employed as a genus, by Simonetti and Spotorno (1980). Based on morphological studies, however, the species do not cladistically nest within *Auliscomys* or *Phyllotis* proper but instead are associated with a clade containing *Reithrodon* and variably other genera (Braun, 1993; Spotorno et al., 1998; Steppan, 1993, 1995).

Loxodontomys micropus (Waterhouse, 1837). Proc. Zool. Soc. Lond., 1837:17.

COMMON NAME: Southern Pericote.

TYPE LOCALITY: Argentina, Santa Cruz Prov., "interior plains of Patagonia" at 50°S latitude, near the Río Santa Cruz (probably near La Argentina as per Hershkovitz, 1962:392).

DISTRIBUTION: S Andes of Chile and Argentina, from about 38°S latitude to Straits of Magellan.

STATUS: IUCN – Lower Risk (lc) as *Auliscomys micropus*.

SYNONYMS: *alsus* (Thomas, 1919); *fumipes* (Osgood, 1943).

COMMENTS: Revised by Pearson (1958); banded karyotype published by Spotorno and Walker (1979).

Loxodontomys pikumche Spotorno, Cofre, Manriquez, Vilina, Marquet, and Walker, 1998. Rev. Chilena Hist. Nat., 71:362.

COMMON NAME: Pikumche Pericote.

TYPE LOCALITY: Chile, Region Metropolitana, Cajón del Río Maipo, Cruz de Piedra, 55 km S de la Central Hidroelectrica de Las Melosas, 24°50 m; 34°10'S, 69°58'W.

DISTRIBUTION: C Chile; limits unknown.

COMMENTS: Northern sister species to *L. micropus* and allopatric with it as so far known, the two differing in chromosomal and morphological traits (Spotorno et al., 1998).

Lundomys Voss and Carleton, 1993. Am. Mus. Novit., 3085:5.
    TYPE SPECIES: *Hesperomys molitor* Winge, 1887.
    COMMENTS: Oryzomyini. A tetralophodont form related to *Holochilus*, *Noronhomys*, and *Pseudoryzomys* (Carleton and Olson, 1999; Voss and Carleton, 1993; Weksler, 2003).

Lundomys molitor (Winge, 1887). E Museo Lundii, 1(3):14.
    COMMON NAME: Lund's Amphibious Rat.
    TYPE LOCALITY: Brazil, Minas Gerais State, cave chamber ("Lapa da Escrivania Nr. 5") near Lagoa Santa.
    DISTRIBUTION: SE Brazil and Uruguay.
    STATUS: IUCN – Lower Risk (lc) as *Holochilus magnus*.
    SYNONYMS: *magnus* Hershkovitz, 1955.
    COMMENTS: Species originally described based on fossils from Recent cave deposits in Minas Gerais, Brazil. Extant populations (described as *Holochilus magnus* Hershkovitz, 1955) apparently restricted to Rio Grande do Sul, Brazil, and Uruguay (González, 2000*b*; Olrog and Lucero, 1981), but Pleistocene fossils known from EC Argentina (Pardiñas, 2000*b*). Banded karyotypic comparisons with *H. brasiliensis* presented by Freitas et al. (1983*a*, as *H. magnus*); synonymy of *magnus* demonstrated by Voss and Carleton (1993); González (2000*b*) recognized *magnus* as a subspecies for Uruguayan populations.

Megalomys Trouessart, 1881. Le Naturaliste, 1:357.
    TYPE SPECIES: *Mus pilorides* (Desmarest, 1826 = *Mus desmarestii* Fischer, 1829).
    SYNONYMS: *Moschomys* Trouessart, 1903 [unnecessary substitute for *Megalomys*; not *Moschomys* Billberg, 1827 = *Ondatra*]; *Moschophoromys* Elliot, 1904 [unnecessary substitute for *Moschomys*].
    COMMENTS: Oryzomyini. Derived postcranial traits of *Megalomys* are consistent with its association within Sigmodontinae *sensu stricto* (lack of entepicondylar foramen, tuberculum of 1st rib contacting both the 1st thoracic and 7th cervical vertebrae; as per *M. desmarestii*, BMNH 1850.11.30.6); other cranial characters more narrowly point to its membership within Oryzomyini (long palate with prominent posterolateral pits, absence of alisphenoid strut; as per *M. luciae*, BMNH 53.12.16.2, and *M. desmarestii*, BMNH 50.11.30.6 and 55.12.24.201), as diagnosed by Voss and Carleton (1993). A combination of features suggests that the close relatives of *Megalomys* may be sought among certain derived oryzomyines, such as *Oryzomys sensu stricto*, or *Nectomys* (e.g., thenar and hypothenar pads present but barely indicated; molars brachyodont and pentalophodont, with M1/m1 four-rooted; supraorbital and temporal ridging well developed and reflected dorsally; carotid circulation derived).
        Named as a subgenus of the all-inclusive *Hesperomys*, later associated with *Holochilus* by Thomas (1884) or *Oryzomys* by Major (1901), and subsequently recognized as an oryzomyine genus endemic to the Lesser Antilles (Ellerman, 1941; Hall, 1981; Steadman and Ray, 1982; Tate, 1932*c*). Does not include the extinct Galápagos endemic *curoi*, described as a species of *Megalomys* by Niethammer (1964); see Steadman and Ray (1982) and account of *Megaoryzomys*. Morphology redescribed and taxonomy revised by Ray (1962) as a subgenus of *Oryzomys*, following the undeveloped opinions of Osgood (1947) and Hershkovitz (1948*b*). The four nominal species (*audreyae* Hopwood, 1926; *curazensis* Hooijer, 1959; *desmarestii* Fischer, 1829; *luciae* Forsyth Major, 1901) are extinct, possibly all due to anthropogenic causes, but *M. desmarestii* and *M. luciae*, listed below, persisted into the middle, perhaps late, 1800s (Ray, 1962; Woods, 1989). Known from Pleistocene-Holocene fossils, Pre-Columbian midden deposits, and a few modern specimens (Ray, 1962). See G. M. Allen (1942) for compilation of meagre historical accounts.

Megalomys desmarestii (Fischer, 1829). Synopsis Mamm., p. 316.
    COMMON NAME: Desmarest's Pilorie.
    TYPE LOCALITY: West Indes, Lesser Antilles, Martinique.

DISTRIBUTION: Known only from Martinique.

STATUS: IUCN – Extinct.

SYNONYMS: *pilorides* (Desmarest, 1826) [preoccupied by *Mus pilorides* Pallas, 1778, a composite]; *piloris* (Major, 1901).

COMMENTS: As compared with *M. luciae* (BMNH 53.12.16.2, the holotype), the cranium of *M. desmarestii* (BMNH 50.11.30.6, 55.12.24.201) is generally larger in all dimensions, its m3 larger relative to m2, and the upper incisors yellowish-orange instead of plain orange.

Megalomys luciae (Major, 1901). Ann. Mag. Nat. Hist., ser. 7, 7:206.

COMMON NAME: Santa Lucian Pilorie.

TYPE LOCALITY: West Indes, Lesser Antilles, Santa Lucia.

DISTRIBUTION: Known only from Santa Lucia.

STATUS: IUCN – Extinct.

COMMENTS: Provisionally retained as species by Ray (1962), who suspected that improved samples would demonstrate conspecificity with *M. desmarestii*.

Megaoryzomys Lenglet and Coppois, 1979. Bull. Acad. R. Belgique, Classe des Sciences, Sér 5., 65:635.

TYPE SPECIES: *Megalomys curioi* Niethammer, 1964.

COMMENTS: Sigmodontinae *incertae sedis*. Originally described under *Megalomys*, an oryzomyine taxon endemic to the Lesser Antilles. Morphological contrasts with *Megalomys* and biogeographic implausibilty of that identity addressed by Steadman and Ray (1982), who instead emphasized its similarities to large species of extant *Thomasomys* and *Rhipidomys* and formally allied it under Thomasomyini. While not an oryzomyine per se, as demonstrated by Steadman and Ray, the relationships and tribal affiliation of the unfortunately christened *Megaoryzomys* deserve reconsideration within a broader sampling of New World cricetids and from a cladistic perspective (also see remarks and comparisons by Hutterer and Oromi, 1993, and Leo L. and Gardner, 1993). Hutterer and Oromi (1993) considered that the presence and distinctiveness of *Megaoryzomys* suggested three independent colonizations of the Galapágos Isls by cricetid rodents, *Nesoryzomys* and *Oryzomys* comprising the other two.

Megaoryzomys curioi (Niethammer, 1964). Mammalia, 28:596.

COMMON NAME: Galapágos Giant Rat.

TYPE LOCALITY: Ecuador, Galápagos Isls, Indefatigable Isl (= Santa Cruz), volcanic tube 3 km N Academy Bay (as corrected by Hutterer and Oromi, 1993:67).

DISTRIBUTION: Known only from late Quaternary and subfossil cave deposits (precise age indeterminate) on Santa Cruz Isl.

STATUS: Dead but not forgotten.

COMMENTS: Lectotype designated by Steadman and Ray (1982:6). See Hutterer and Oromi (1993) for commentary on varied renditions of the type locality, morphological redescription based on new samples, and discussion of the taxon's very recent extinction given its recovery with bones of *Nesoryzomys* and *Mus*.

Melanomys Thomas, 1902. Ann. Mag. Nat. Hist., ser. 7, 10:248.

TYPE SPECIES: *Oryzomys phaeopus* Thomas, 1894 (= *Hesperomys caliginosus* Tomes, 1860).

COMMENTS: Oryzomyini. Usually arranged as a subgenus of *Oryzomys* since Goldman's (1918) revision (e.g., Cabrera, 1961; Ellerman, 1941; Hall, 1981; Reig, 1986; Tate, 1932*e*). J. A. Allen (1913), however, provided morphological criteria, as contrasted to the type species (= *Mus palustris*) of *Oryzomys*, defending his retention of *Melanomys* as a genus, as did Gyldenstolpe (1932); its synonymy in *Oryzomys* proper deserves more rigorous, character-based, phylogenetic substantiation. The taxon has not been revised; the three nominal species listed stem from Cabrera (1961), who acknowledged his solely literature-based interpretation as provisional; Allen (1913) had recognized nine and Gyldenstolpe (1932) listed eight. A morphologically distinctive and distributionally circumscribed taxon ripe for revisionary and biogeographic study.

Melanomys caliginosus (Tomes, 1860). Proc. Zool. Soc. Lond., 1860:263.

COMMON NAME: Dusky Melanomys.

TYPE LOCALITY: Ecuador, Esmeraldas Prov., Esmeraldas.

DISTRIBUTION: Central American lowlands from easternmost Honduras through Panamá; in South America, N and W Colombia, including the Chocó (see Cadena et al., 1998), to SW Ecuador and NW Venezuela.

STATUS: IUCN – Lower Risk (lc).

SYNONYMS: *affinis* (J. A. Allen, 1912); *buenavistae* J. A. Allen, 1913; *chrysomelas* (J. A. Allen, 1897); *columbianus* (J. A. Allen, 1899); *idoneus* (Goldman, 1912); *lomitensis* J. A. Allen, 1913; *monticola* (J. A. Allen, 1912); *obscurior* (Thomas, 1894); *olivinus* (Thomas, 1902); *oroensis* J. A. Allen, 1913; *phaeopus* (Thomas, 1894); *tolimensis* J. A. Allen, 1913; *vallicola* J. A. Allen, 1913.

COMMENTS: Chromosomal complement described by Gardner and Patton (1976).

Melanomys robustulus Thomas, 1914. Ann. Mag. Nat. Hist., ser. 8, 14:243.

COMMON NAME: Robust Melanomys.

TYPE LOCALITY: Ecuador, Morona-Santiago Prov., Gualaquiza, 2500 ft (762 m).

DISTRIBUTION: SE Ecuador.

STATUS: IUCN – Lower Risk (lc).

COMMENTS: A distinct species judged from our study of specimens in the AMNH and BMNH.

Melanomys zunigae (Sanborn, 1949). Publ. Mus. Hist. Nat., Javier Prado, Zool., 1(3):2.

COMMON NAME: Zuniga's Melanomys.

TYPE LOCALITY: Perú, Lima Dept., Lomas de Atocongo.

DISTRIBUTION: WC Perú.

STATUS: IUCN – Lower Risk (lc).

Microakodontomys Hershkovitz, 1993. Fieldiana Zool., N.S., 75:2.

TYPE SPECIES: *Microakodontomys transitorius* Hershkovitz, 1993.

COMMENTS: Oryzomyini. Provisionally left as Sigmodontinae *incertae sedis* by Hershkovitz (1993:2), who remarked that "Overall resemblance to *Microryzomys* and *Oligoryzomys* suggests differentiation from an oryzomyine stock in a scrub brush habitat transitional between forest and savanna." To us, the holotype displays features that clearly ally it with Oryzomyini *sensu* Voss and Carleton (1993). Founded on a single young specimen with damaged skull, the diagnostic traits of this new genus and species, which "could be mistaken for a well-differentiated species of *Oligoryzomys*" (Hershkovitz, 1993:4), should be verified with improved series from the type locality.

Microakodontomys transitorius Hershkovitz, 1993. Fieldiana Zool., N.S., 75:2.

COMMON NAME: Transitional Colilargo.

TYPE LOCALITY: Brazil, Federal District, Parque Nacional de Brasília, ca. 20 km NW Brasília, 1100 m; 15°47'S, 47°55'W.

DISTRIBUTION: Known only from the type locality.

Microryzomys Thomas, 1917. Smithson. Misc. Coll., 68:1.

TYPE SPECIES: *Hesperomys minutus* Tomes, 1860.

SYNONYMS: *Thallomyscus* Thomas, 1926.

COMMENTS: Oryzomyini. Named as a subgenus of *Oryzomys* and either retained as such (Cabrera, 1961; Ellerman, 1941; Osgood, 1933a) or placed in synonymy with *Oligoryzomys* (Gyldenstolpe, 1932; Tate, 1932e; Thomas, 1926). Junior synonymy of *Thallomyscus* established by Osgood (1933a). Raised to genus by Carleton and Musser (1984) and later revised by them (1989). Cladistically primitive member of Oryzomyini, sharing certain morphological traits with *Oligoryzomys* (Carleton and Musser, 1989; Carleton and Olson, 1999); sister genus to *Neacomys* according to gene sequence studies (Myers et al., 1995; Patton and da Silva, 1995; Smith and Patton, 1999) or to the *Neacomys-Oligoryzomys* clade according to allozymic analyses (Dickerman and Yates, 1995).

Microryzomys altissimus (Osgood, 1933). Field Mus. Nat. Hist. Publ., Zool. Ser., 20:5.

COMMON NAME: Páramo Colilargo.

TYPE LOCALITY: Perú, Pasco Dept., La Quinua, mountains north of Cerro de Pasco, 11,600 ft (3536 m).

DISTRIBUTION: Subalpine and páramo formations, mostly 2500-4000 m, in the Andes of Colombia, Ecuador, and Perú.

STATUS: IUCN – Lower Risk (lc).

SYNONYMS: *chotanus* Hershkovitz, 1940; *hylaeus* Hershkovitz, 1940.

COMMENTS: Diagnosed as a subspecies of *Oryzomys minutus* but elevated to species by Hershkovitz (1940), who named two additional subspecies. Subspecific arrangement followed by Cabrera (1961) but none retained by Carleton and Musser (1989).

Microryzomys minutus (Tomes, 1860). Proc. Zool. Soc. Lond., 1860:215.

COMMON NAME: Montane Colilargo.

TYPE LOCALITY: Ecuador, Chimborazo Prov., probably near Pallatanga.

DISTRIBUTION: Lower montane to subalpine forest, mostly 2000-3500 m, from N Venezuela, through Colombia, Ecuador and Perú, to WC Bolivia.

STATUS: IUCN – Lower Risk (lc).

SYNONYMS: *aurillus* (Thomas, 1917); *dryas* (Thomas, 1898); *fulvirostris* (J. A. Allen, 1912); *humilior* (Thomas, 1898).

COMMENTS: Formerly included *altissimus* as a subspecies by Osgood (1933*a*), who also treated *aurillus*, *humilior*, and *fulvirostris* as subspecies. No races were deemed diagnosable by Carleton and Musser (1989).

Neacomys Thomas, 1900. Ann. Mag. Nat. Hist., ser. 7, 5:153.

TYPE SPECIES: *Hesperomys spinosus* Thomas, 1882.

COMMENTS: Oryzomyini. Sister genus to *Microryzomys* according to gene sequence studies (Myers et al., 1995; Patton and da Silva, 1995; Smith and Patton, 1999; Weksler, 2003) or to *Oligoryzomys* according to allozymic analyses (Dickerman and Yates, 1995). Recent taxonomic descriptions and detailed regional comparisons have markedly improved understanding of species diversity and their morphological discrimination (Patton et al., 2000; Voss et al., 2001). Delineation of yet other species and enhancement of distributional limits still required (e.g., see Malygin and Rosmiarek, 1996; Patton et al., 2000). Preliminary molecular geographic patterns discerned by Patton et al. (2000) offer a template for future taxonomic studies.

Neacomys dubosti Voss, Lunde, and Simmons, 2001. Bull. Am. Nat. Hist., 263:78.

COMMON NAME: Dubost's Neacomys.

TYPE LOCALITY: French Guiana, Paracou.

DISTRIBUTION: SE Surinam, French Guiana, and Amapá, Brazil.

COMMENTS: Morphological comparisons with *N. guianae*, *N. paracou*, and *N. tenuipes* provided by Voss et al. (2001). Known sympatrically with *N. guianae* and *N. paracou*.

Neacomys guianae Thomas, 1905. Ann. Mag. Nat. Hist., ser. 7, 16:310.

COMMON NAME: Guianan Neacomys.

TYPE LOCALITY: Guyana, Demerara River, 120 ft (37 m).

DISTRIBUTION: Guianas, S Venezuela, and N Brazil.

STATUS: IUCN – Lower Risk (lc).

COMMENTS: Morphological comparisons with *N. dubosti*, *N. paracou*, and *N. tenuipes* provided by Voss et al. (2001). Known sympatrically with *N. dubosti* and *N. paracou*.

Neacomys minutus Patton, da Silva, and Malcolm, 2000. Bull. Am. Mus. Hist., 244:105.

COMMON NAME: Minute Neacomys.

TYPE LOCALITY: Brazil, Amazonas State, left bank Rio Juruá, Altimira; 6°35'S, 68°54'W.

DISTRIBUTION: So far known from the central and lower drainage of the Rio Juruá, W Brazil; range limits require amplification.

COMMENTS: Morphological, karyological (2n = 35-36), and molecular differentiation from *N. musseri* and *N. spinosus* presented by Patton et al. (2000). Known sympatrically with *N. spinosus*.

Neacomys musseri Patton, da Silva, and Malcolm, 2000. Bull. Am. Mus. Hist., 244:98.

COMMON NAME: Musser's Neacomys.

TYPE LOCALITY: Perú, Cusco Dept., Manu Biosphere Reserve, 72 km (by road) NE Paucartambo at km 152, 1460 m.

DISTRIBUTION: So far known from the headwaters of the Rio Juruá, SE Perú and extreme W Brazil; range limits require amplification.

COMMENTS: Morphology, karyology (2n = 34), and genetic differentiation from *N. minutus* and *N. spinosus* presented by Patton et al. (2000). Known sympatrically with *N. spinosus*.

Neacomys paracou Voss, Lunde, and Simmons, 2001. Bull. Am. Nat. Hist., 263:81.
COMMON NAME: Paracou Neacomys.
TYPE LOCALITY: French Guiana, Paracou.
DISTRIBUTION: Guianan subregion of Amazonia—SE Venezuela, through Guyana, Surinam, and French Guiana, to Amapá, Brazil, and southwards to Amazonas and Pará, Brazil.
COMMENTS: Morphological comparisons with *N. dubosti*, *N. guianae*, and *N. tenuipes* provided by Voss et al. (2001). Known sympatrically with *N. dubosti* and *N. guianae*.

Neacomys pictus Goldman, 1912. Smithson. Misc. Coll., 60:2.
COMMON NAME: Painted Neacomys.
TYPE LOCALITY: Panamá, Darién Prov., Cana, 1800 ft (549 m).
DISTRIBUTION: Known only from easternmost Panamá.
STATUS: IUCN – Lower Risk (nt).
COMMENTS: Cabrera (1961) allocated *pictus* to subspecific standing under *N. tenuipes*, and Handley (1966) concurred, but examination of series in USNM suggests that the relationship and status of this gracile form bear reexamination.

Neacomys spinosus (Thomas, 1882). Proc. Zool. Soc. Lond., 1882:105.
COMMON NAME: Common Neacomys.
TYPE LOCALITY: Perú, Amazonas Dept., Huambo, 3700 ft (1128 m).
DISTRIBUTION: WC Brazil to Andean foothills and lowlands of SE Colombia, E Ecuador, E Perú, and N and C Bolivia.
STATUS: IUCN – Lower Risk (lc).
SYNONYMS: *amoenus* Thomas, 1903; *carceleni* Hershkovitz, 1940; *typicus* Thomas, 1900.
COMMENTS: Karyotype (2n = 64) compared with other oryzomyines by Gardner and Patton (1976). Morphological, karyological, and molecular differentiation from *N. minutus* and *N. musseri* presented by Patton et al. (2000). Lawrence (1941) recognized the nominate form and *carceleni* as subspecies, a division supported by the limited samples genetically analyzed by Patton et al. (2000). Known sympatrically with *N. minutus* and *N. musseri*.

Neacomys tenuipes Thomas, 1900. Ann. Mag. Nat. Hist., ser. 7, 5:153.
COMMON NAME: Narrow-footed Neacomys.
TYPE LOCALITY: Colombia, Cundinamarca Dept., Bogotá region, Guaquimay.
DISTRIBUTION: W and NC Colombia, N Venezuela.
STATUS: IUCN – Lower Risk (lc).
SYNONYMS: *pusillus* J. A. Allen, 1912.
COMMENTS: Named and recognized (e.g., Ellerman, 1941) as a subspecies of *N. spinosus* until elevated to specific status by Lawrence (1941).

Necromys Ameghino, 1889. Actas Acad. Nac. Cienc. Córdoba, 6:120.
TYPE SPECIES: *Necromys conifer* Ameghino, 1889.
SYNONYMS: *Bolomys* Thomas, 1916; *Cabreramys* Massoia and Fornes, 1967.
COMMENTS: Akodontini. Type species of this neglected genus, based on a fossil form, incorrectly synonymized under the phyllotine *Calomys callosus* by Hershkovitz (1962), an allocation followed by McKenna and Bell (1997), but identification as akodontine and generic priority over *Bolomys* earlier reestablished by Massoia (1985). Diagnostic dental traits, lectotype selection, nomenclatural history, and allocation of fossil and living species provided by Massoia and Pardiñas (1993). Those authors referred to *conifer* as a *nomen dubium*, a misleading choice of words because the meaning clearly intended in their discussion is that the fossil is not certainly identifiable with a living species, namely *N. benefactus* which today occurs in the same region as *conifer*.
Allocation of *amoenus* to *Necromys*, and by extension the junior synonymy of *Bolomys*,

is weakened by certain phylogenetic iterations of cytochrome *b* sequence data (Smith and Patton, 1993) but not others (D'Elía, 2003; D'Elía et al., 2003; Smith and Patton, 1999). Whether species in high Andean grasslands (*Bolomys*) and those in the S lowland savannas (*Necromys*) form monophyletic, diagnosable clades with respect to one another and to *Akodon* proper will require broader sampling of critical taxa. The issue parallels the evolutionary and systematic questions involving Andean *Thomasomys* and those 'thomasomyine' forms isolated in SE South America (see remarks under *Delomys*). Sister genus to the clade *Akodon-Thaptomys* (Smith and Patton, 1993, 1999), to *Thaptomys* (D'Elía et al., 2003), or to *Thalpomys* (D'Elía, 2003) according to gene sequence studies.

Contains species conventionally known in the recent literature under *Bolomys*, a taxon recognized either as a subgenus of *Akodon* (Cabrera, 1961; Ellerman, 1941) or as an akodontine genus (Gyldenstolpe, 1932; Musser and Carleton, 1993; Reig, 1987; Thomas, 1916). Emendation of diagnostic traits, synonymy of *Cabreramys*, and provisional list of species presented by Reig (1987). Complex taxonomic history and diagnosis further refined based on Bolivian species by Anderson and Olds (1989) and on Argentinian and Uruguayan forms by Galliari and Pardiñas (2000). Chromosomal traits and invalid karyotypic references reviewed by Maia and Langguth (1981) and Reig (1987). Species recognized here largely observe the generic review of Galliari and Pardiñas (2000).

Necromys amoenus (Thomas, 1900). Ann. Mag. Nat. Hist., ser. 7, 6:468.
COMMON NAME: Pleasant Akodont.
TYPE LOCALITY: Perú, Arequipa Dept., Calalla, Río Colca, near Sumbay, 3500 m.
DISTRIBUTION: Altiplano grasslands, above 3200 m, of SE Perú and W Bolivia, as far south as Tarija Dept. (Anderson, 1993).
STATUS: IUCN – Lower Risk (lc) as *Bolomys amoenus*.
COMMENTS: Morphological definition as type species of *Bolomys* Thomas, 1916, clarified by Reig (1987) and Anderson and Olds (1989). Additional Bolivian localities reported by Salazar-Bravo et al. (2002*b*).

Necromys benefactus (Thomas, 1919). Ann. Mag. Nat. Hist., ser. 9, 3:214.
COMMON NAME: Argentine Akodont.
TYPE LOCALITY: Argentina, Buenos Aires Prov., Bonifacio (= Laguna Alsina as per Galliari and Pardiñas, 2000), 50 m.
DISTRIBUTION: Isolated populations in Córdoba, La Pampa, and Buenos Aires provinces, EC Argentina.
COMMENTS: Relegated to a subspecies of *obscurus* by Cabrera (1961) and Reig (1978), but Massoia and Fornes (1967) considered the two as distinct species. Distribution documented based on vouchered specimens and clear morphological segregation from *N. obscurus* underscored by Galliari and Pardiñas (2000). The latter authors regarded the fossil *conifer* Ameghino as the probable senior synonym for this species but elected to retain *benefactus* Thomas as the valid name for reasons of familiarity; the possibility of their synonymy requires firm demonstration, with replacement by the senior name if so.

Necromys lactens (Thomas, 1918). Ann. Mag. Nat. Hist., ser. 9, 1:188.
COMMON NAME: White-chinned Akodont.
TYPE LOCALITY: Argentina, Jujuy Prov., León, about 1500 m.
DISTRIBUTION: E Andean highlands, about 2100-4000 m, of SC Bolivia (see Anderson and Olds, 1989) and NW Argentina (Jujuy, Salta, Catamarca, and Tucumán Provs.; see Ortiz et al., 2000*a*).
STATUS: IUCN – Lower Risk (lc) as *Bolomys lactens*.
SYNONYMS: *negrito* (Thomas, 1926); *orbus* (Thomas, 1919).
COMMENTS: Galliari and Pardiñas (1995) reidentified *leucolimnaeus*, previously considered a subspecies of *lactens* (Cabrera, 1961; Gyldenstolpe, 1932), as a species of *Akodon* and questioned (2000:226) the synonymy of *negrito* and *orbus*, as attributed by Cabrera (1961), as "dubious." Capllonch et al. (1997) had resurrected *N. orbus* based on a single specimen from Tucumán Prov., but Pardiñas and Galliari (1998*b*) considered the action to be inadequately documented.

Necromys lasiurus (Lund, 1840). K. Dansk. Vid. Selsk. Naturv. Math. Afhandl., 8:50.

COMMON NAME: Hairy-tailed Akodont.

TYPE LOCALITY: Brazil, Minas Gerais State, Rio das Velhas, Lagoa Santa.

DISTRIBUTION: C Brazil south of the Amazon River, extreme SE Perú (as per Pacheco et al., 1995), and perhaps NE Argentina (see Galliari et al., 1996); limits uncertain.

STATUS: IUCN – Lower Risk (lc) as *Bolomys lasiurus*.

SYNONYMS: *arviculoides* (Wagner, 1842); *brachyurus* (Wagner, 1845); *fuscinus* (Thomas, 1897); *lasiotus* (Lund, 1838); *orobinus* (Wagner, 1842); *pixuna* (Moojen, 1943); *renggeri* (Pictet, 1844).

COMMENTS: Revised as part of *Zygodontomys* (Hershkovitz, 1962); inclusion of *lasiurus* within *Bolomys* (or *Akodon*), not *Zygodontomys*, supported by karyological and morphological information (see Gardner and Patton, 1976; Maia and Langguth, 1981; Voss and Linzey, 1981). Geographic variation assessed, as *Bolomys lasiurus*, by Macêdo and Mares (1987), who retained only *fuscinus* and *lasiurus* as subspecies. The validity of listed synonyms very much needs confirmation: e.g., the taxon *arviculoides* is considered a species by some (Cabrera, 1961; Gyldenstolpe, 1932; Ximénez & Langguth, 1970), or reallocated to *lasiurus* by others (Macêdo and Mares, 1987; Reig, 1978, 1987), while Galliari and Pardiñas (2000) passingly mention it as another species related to *N. obscurus*. Also see following account of *N. lenguarum*, included by some (e.g., Macêdo and Mares, 1987) as another junior synonym; morphological discrimination and distributional limits of the two forms require refinement.

Necromys lenguarum (Thomas, 1898). Ann. Mag. Nat. Hist., ser. 7, 2:271.

COMMON NAME: Paraguayan Akodont.

TYPE LOCALITY: Paraguay, Presidente Hayes Dept., Chaco Boreal, Waikthlatingwaialwa.

DISTRIBUTION: Chacoan formations of E Bolivia, W Paraguay, and possibly N Argentina (see Galliari et al., 1996).

SYNONYMS: *tapiropoanus* (J. A. Allen, 1916).

COMMENTS: Taxonomic status unresolved—demoted to subspecies under *Akodon obscurus* (Cabrera, 1961) or placed in full synonymy with *Bolomys l. lasiurus* (Macêdo and Mares, 1987), or generally if equivocally recognized as a species (Anderson, 1997; Anderson and Olds, 1989; Galliari and Pardiñas, 2000; Reig, 1987). Anderson (1997) maintained *tapiropoanus*, type locality in Mato Grosso, W Brazil, as a subspecies for Bolivian populations. Also see remarks under *N. lasiurus*.

Necromys obscurus (Waterhouse, 1837). Proc. Zool. Soc. Lond., 1837:16.

COMMON NAME: Dark-furred Akodont.

TYPE LOCALITY: Uruguay, Maldonado Dept., Maldonado.

DISTRIBUTION: Isolated populations along coastal areas in S Uruguay (*obscurus*) and Buenos Aires Prov., EC Argentina (*scagliarum*).

STATUS: IUCN – Lower Risk (lc) as *Bolomys obscurus*.

SYNONYMS: *scagliarum* Galliari and Pardiñas, 2000.

COMMENTS: Type species of *Cabreramys* (Massoia and Fornes, 1967), a taxon considered a junior synonym of *Bolomys* by Reig (1978, 1987) and of *Necromys* by Massoia and Pardiñas (1993). Formerly contained *benefactus* as a subspecies (see above account). Geographic variation and allopatric distribution in light of late Pleistocene climatic changes discussed by Galliari and Pardiñas (2000). The strong morphometric differentiation between and probable long-term isolation of the named populations, as demonstrated by the latter authors, recommend reconsideration of their status using other information sources.

Necromys punctulatus (Thomas, 1894). Ann. Mag. Nat. Hist., ser 6, 14:361.

COMMON NAME: Ecuadoran Akodont.

TYPE LOCALITY: Ecuador.

DISTRIBUTION: Indeterminate area of E Ecuador and perhaps Colombia.

STATUS: IUCN – Lower Risk (lc) as *Bolomys punctulatus*.

COMMENTS: Described as a species of *Akodon* but referred to a subspecies of *Zygodontomys brevicauda* by Hershkovitz (1962). Thomas (1916c) had earlier appreciated the morphological resemblance between *amoenus*, type species of *Bolomys*, and *punctulatus*; Voss (1991*b*) later amplified the *Bolomys*-like traits of *punctulatus*, provisionally retaining it as a species and noting its similarity to *lasiurus*. The enigmatic distribution of the few fragmentary specimens assignable to *punctulatus*, which originate from a region outside of the core geographic range of *Bolomys* (= *Necromys*), is discussed by Voss (1991*b*).

Necromys temchuki (Massoia, 1982). Hist. Nat., Corrientes, 2(11):91.

COMMON NAME: Temchuk's Akodont.

TYPE LOCALITY: Argentina, Misiones Prov., Depto. Capital, Costa del Arroyo Zaimán, Villa Miguel Lanús.

DISTRIBUTION: NE Argentina (Misiones, Corrientes, Formosa, and Chaco Provs.; see Contreras, 1982).

STATUS: IUCN – Lower Risk (lc) as *Bolomys temchuki*.

SYNONYMS: *elioi* (Contreras, 1982); *liciae* (Contreras, 1982); *temchuki* (Massoia, 1980) [*nomen nudum*].

COMMENTS: Although the authorship is sometimes dated to Massoia, 1980*b* (Musser and Carleton, 1993; Honacki et al, 1982), Massoia had then mentioned the name (then spelled *temchucki*) in the context of a preliminary note; no holotype or museum repository was designated, and descriptive and comparative information is scant. As with another preliminary name mentioned in the same paper (*Deltamys kempi langguthi*, later described formally by González and Massoia, 1995), the availability of *temchuki* should properly date from Massoia's (1982) explicit diagnosis of the species. As remarked by Galliari et al. (1996), the relationships and discrimination of *temchuki* with regard to *N. lasiurus*, also known from the same general region, merit reexamination and clear documentation.

Necromys urichi (J. A. Allen and Chapman, 1897). Bull. Am. Mus. Nat. Hist., 9:19.

COMMON NAME: Northern Akodont.

TYPE LOCALITY: Trinidad, Caparo.

DISTRIBUTION: Trinidad and Tobago, highlands of N and S Venezuela, E Colombia, N Brazil.

STATUS: IUCN – Lower Risk (lc) as *Akodon urichi*.

SYNONYMS: *chapmani* (J. A. Allen, 1913); *meridensis* (J. A. Allen, 1904); *saturatus* (Tate, 1939); *tobagensis* (Goodwin, 1962); *venezuelensis* (J. A. Allen, 1899).

COMMENTS: Although traditionally arranged as a species of *Akodon* (Cabrera, 1961; Musser and Carleton, 1993), subgenus *Chalcomys* according to Thomas (1916*c*), *urichi* is here assigned to *Necromys* following those molecular studies that disclose its close relationship to *amoenus* and *lasiurus* apart from representative species of *Akodon* proper (D'Elía, 2003; D'Elía et al., 2003; Smith and Patton, 1993, 1999). Formerly included *aerosus* as a subspecies (Cabrera, 1961), but karyotypic (Gardner and Patton, 1976) and gene sequence (D'Elía et al., 2003; Smith and Patton, 1993, 1999) data demonstrate their distant relationship (see account of *Akodon aerosus*).

Morphometric variation of Venezuelan populations studied by Ventura et al. (2000), who recognized *meridensis*, *saturatus*, and *venezuelensis* as subspecies along with the nominate form. Linares (1998) arranged the populations in S Venezuela as a species (*saturatus*) distinct from those in the Andes (*urichi*). To date, craniodental morphometrics (Ventura et al., 2000) and cytochrome *b* comparisons (Smith and Patton, 1999) provide no compelling evidence for doing so, but the biogeographic plausibility of such a divergence merits further attention.

Nectomys Peters, 1861. Abh. König. Akad. Wiss. Berlin, 1860:151 [1861].

TYPE SPECIES: *Mus squamipes* Brants, 1827.

SYNONYMS: *Potamys* Liais, 1872.

COMMENTS: Oryzomyini. A broad and diverse character base indelibly and uniformly pictures the phyletic heritage of *Nectomys* as oryzomyine (Baker et al., 1983; Dickerman and Yates, 1995; Hershkovitz, 1944, 1960; Hooper and Musser, 1964*a*; Myers et al., 1995; Patton and da Silva, 1995; Smith and Patton, 1999; Steppan, 1995; Voss and Carleton, 1993). Hershkovitz (1944) included *Sigmodontomys* as a subgenus; Gardner and Patton (1976) urged removal of *Sigmodontomys* to *Oryzomys*, and Musser and Carleton (1993) provisionally elevated it to genus.

Regional studies are beginning to uncover the species diversity masked by Hershkovitz's (1944) concept of a single, pan-Amazonian species, *N. squamipes* (e.g., Andrades-Miranda et al., 2001*b*; Gómez-LaVerde et al., 1999; Patton et al., 2000; Voss et al., 2001). Integration of these regional perspectives nevertheless warrants continued alpha-level revisionary investigation to voucher geographic ranges and to properly associate junior synonyms.

Nectomys apicalis Peters, 1861. Abhandl. Preuss. Akad. Wiss., 1860:152 [1861].

COMMON NAME: Western Amazonian Nectomys.

TYPE LOCALITY: Ecuador, Napo Prov., Tena, 512 m (as restricted by Hershkovitz, 1944:53).

DISTRIBUTION: Westernmost Brazil (Acre and Amazonas), and contiguous lowlands and Andean foothills of C and E Ecuador, E Perú, and WC Bolivia (see Anderson, 1997: Fig. 685; Patton et al., 2000:Fig. 78); distributional limits uncertain.

SYNONYMS: *fulvinus* Thomas, 1897; *garleppii* Thomas, 1899; *montanus* Hershkovitz, 1944; *napensis* Hershkovitz, 1944; *saturatus* Thomas, 1897; *vallensis* Hershkovitz, 1944.

COMMENTS: Argued by Patton et al. (2000) as the oldest name for those water rat populations in western Amazonia with a 2n = 38-42 karyotype, including *garleppii*, which Reig (in Honacki et al., 1982) and Barros et al. (1992) recognized as a species. More than a single species is likely represented among the listed synonyms, whether they belong to one of the other species recognized here or to some other species yet to be determined.

Nectomys magdalenae Thomas, 1897. Ann. Mag. Nat. Hist., ser. 6, 19:499.

COMMON NAME: Magdalena Nectomys.

TYPE LOCALITY: Colombia, Cundinamarca Dept., lowlands near Magdalena River.

DISTRIBUTION: Basins of the Magdalena and Cauca Rivers, N Colombia.

SYNONYMS: *grandis* Thomas, 1897.

COMMENTS: Although considered a subspecies of *N. squamipes* by Hershkovitz (1944), Gómez-Laverde et al. (1999) elevated *magdalenae* to species based on its karyotype (2n = 34), which is uniquely divergent from those populations so far reported from Amazonia and SE Brazil (2n = 38-59; see other accounts). Reig (in Honacki et al., 1982) had considered *grandis* and *magdalenae* as possible synonyms of *apicalis*, but the distinctive karyotype and emerging biogeographic appreciation of the Transandean lowlands as a region of specific differentiation warrant provisional species recognition. In spite of the page priority of *grandis* over *magdalenae*, both names dating from Thomas 1897, Gómez et al. (1999) elected *magdalenae* as the valid name because the animal karyotyped originated near the type locality and since the International Code of Zoological Nomenclature (International Commission on Zoological Nomenclature, 1999) does not require strict adherence to page precedence, acceding to the judgement of first revisors. The soundness of the species definition, junior synonymy, and distributional extent should be confirmed with wider geographic sampling and additional analyses of karyotypic and other data.

Nectomys palmipes J. A. Allen and Chapman, 1893. Bull. Am. Mus. Nat. Hist., 5:209.

COMMON NAME: Trinidad Nectomys.

TYPE LOCALITY: Trinidad, Victoria County, Princes Town.

DISTRIBUTION: Isl of Trinidad and nearby region of NE Venezuela; limits of distribution unknown.

STATUS: IUCN – Lower Risk (lc).

SYNONYMS: *tatei* Hershkovitz, 1948.

COMMENTS: Arranged by Hershkovitz (1944) as one of many subspecies of *N. squamipes*. Barros et al. (1992) reinstated *palmipes* to species based on its inordinately low diploid number (2n = 16-17) as compared to other populations of *Nectomys*, which range from 2n = 38 to 59 (Barros et al., 1992; Gardner and Patton, 1976). Linares (1998) retained *palmipes* as a subspecies of *N. squamipes*; junior status of *tatei* identified by Voss et al. (2001).

Nectomys rattus (Pelzeln, 1883). Verhl. kaiserl.-konigl. zool.-bot. Gesellsch., Wien, 33 (Suppl.):73.

COMMON NAME: Amazonian Nectomys.

TYPE LOCALITY: Brazil, Amazonas State, right bank of upper Rio Negro, Marabitanas, 100 m.

DISTRIBUTION: Amazonia—E Colombia, NW and S Venezuela, Guianas, N and C Brazil, and perhaps lowlands of E Perú; distributional limits need specimen-based verification.

STATUS: IUCN – Critically Endangered as *N. parvipes*.

SYNONYMS: *amazonicus* Hershkovitz, 1944; *mattensis* Thomas, 1903; *melanius* Thomas, 1910; *parvipes* Petter, 1979; *tarrensis* Hershkovitz, 1948.

COMMENTS: Most forms included here were once treated as subspecies of *N. squamipes* following Hershkovitz (1944). Diagnosis emended as the species *N. melanius* and morphology contrasted to *N. palmipes* and *N. squamipes sensu stricto* by Voss et al. (2001); they mentioned that *rattus* Pelzeln (1883) may be the proper name to use for this species and provided information on the type's existence and its identity as a *Nectomys* (2001:98-99, footnote). Bonvicino (in Andrades-Miranda et al., 2001*b*) has employed the name *rattus* as senior synonym for populations with 2n = 52-54 (Baker et al., 1983; Barros et al., 1992; Voss et al., 2001), a usage acknowledged by Andrades-Miranda et al. (2001*b*); whereas, Patton et al. (2000) suggested *mattensis* as the oldest name applicable to this widespread karyotypic variant. The provisional basis for the senior name, included synonyms, and geographic range cobbled together here is patently clear and must be confirmed by more robust studies that include topotypic material. Bonvicino (1999) reassigned *Nectomys parvipes* Petter (1979) as another species of *Sigmodontomys*, but Voss et al. (2001) considered the type to represent a small individual of *melanius* (here = *N. rattus*); a third and final determination is required.

Nectomys squamipes (Brants, 1827). Het Geslacht der Muizen, p. 138.

COMMON NAME: Atlantic Forest Nectomys.

TYPE LOCALITY: Brazil, São Paulo State, São Sebastião (as restricted by Hershkovitz, 1944:32).

DISTRIBUTION: SE Brazil (Pernambuco to Rio Grande do Sul), NE Argentina (Misiones Prov.), and E Paraguay.

STATUS: IUCN – Lower Risk (lc).

SYNONYMS: *aquaticus* (Lund, 1841); *brasiliensis* (Pictet and Pictet, 1844); *olivaceus* Hershkovitz, 1944; *pollens* Hollister, 1914; *robustus* (Burmeister, 1854).

COMMENTS: Hershkovitz (1944) arrayed most nominal taxa of water rats as subspecies of *N. squamipes*, a polytypic view maintained by Cabrera (1961). Gardner and Patton (1976) intimated the mixed specific composition of "*squamipes*," and Reig (in Honacki et al., 1982; 1986) offered enumerations of valid species and probable synonymies. Formerly included *apicalis*, *magdalenae*, *palmipes*, and *rattus* (see those accounts). The narrower definition of *N. squamipes* now coalescing (that is, containing those populations with 2n = 56-59—Andrades et al., 2001*b*; Barros et al., 1992; Bonvicino et al., 1996) corresponds to a species restricted to the Atlantic Forest region, overlapping with *N. rattus* in riverine settings of the Cerrado-Caatinga biomes. See Ernest (1986, Mammalian Species, 265).

Neotomys Thomas, 1894. Ann. Mag. Nat. Hist., ser. 6, 14:346.

TYPE SPECIES: *Neotomys ebriosus* Thomas, 1894.

COMMENTS: Reithrodontini. Although grouped with sigmodont rodents by Hershkovitz (1955*a*), other studies have convincingly linked the genus with phyllotines (Olds and Anderson, 1989; Pearson and Patton, 1976; Spotorno et al., 2001), usually as sister genus of *Reithrodon* (Braun, 1993; Steppan, 1995) and in a clade including *Euneomys* and certain fossil genera (Ortiz et al., 2000*b*; Pardiñas, 1997; Steppan and Pardiñas, 1998; Steppan and Sullivan, 2000). See remarks under *Reithrodon* on tribal affiliation.

Neotomys ebriosus Thomas, 1894. Ann. Mag. Nat. Hist., ser. 6, 14:348.

COMMON NAME: Red-nosed Neotomys.

TYPE LOCALITY: Perú, Junín Dept., Vítoc Valley.

DISTRIBUTION: Altiplano grasslands and shrubby steppe, mostly 2500-4500 m, from C Perú (Junín), south through northernmost Chile and W Bolivia, to NW Argentina (see Anderson, 1997; Bárquez, 1983; Pardiñas and Ortiz, 2001; Sanborn, 1947*a*).

STATUS: IUCN – Lower Risk (lc).

SYNONYMS: *vulturnus* Thomas, 1921.

COMMENTS: Revised by Sanborn (1947*a*), who reduced *vulturnus* to a subspecies of
N. ebriosus, as recognized by Anderson (1997) for Bolivian populations. Distribution
augmented by Pearson (1951), Pine et al. (1979), and Pardiñas and Ortiz (2001). First
fossil occurrences documented from the late Pleistocene-Holocene of NW Argentina
by Pardiñas and Ortiz (2001), who discussed the paleoecological significance of the
lower altitudes recorded.

Nesoryzomys Heller, 1904. Proc. California Acad. Sci., 3:241.

TYPE SPECIES: *Nesoryzomys narboroughi* Heller, 1904 (= *Oryzomys indefessus* Thomas, 1899).

COMMENTS: Oryzomyini. Realigned as a subgenus of *Oryzomys* by Ellerman (1941), following
the comments of Goldman (1918). Morphological, genic, and karyological information,
however, sustains the generic separation of *Nesoryzomys* (Beaufort, 1963; Gardner and
Patton, 1976; Patton and Hafner, 1983; Smith and Patton, 1999) and intimates that the
group is an old Galapagos immigrant, originating *ca*. 3-3.5 million years ago (see Patton
and Hafner, 1983). See Key and Heredia (1994) and Dowler et al. (2000) for recent assessments of conservation status, commensal rodent introductions, and natural history of
populations on the various islands.

Nesoryzomys darwini Osgood, 1929. Field Mus. Nat. Hist. Publ., Zool. Ser., 17:23.

COMMON NAME: Darwin's Nesoryzomys.

TYPE LOCALITY: Ecuador, Galápagos Archipelago, Santa Cruz Isl, Academia Bay.

DISTRIBUTION: Santa Cruz (= Indefatigable) Isl.

STATUS: IUCN – Extinct. Probably extinct, last recorded in 1930 (see Dowler et al., 2000;
Patton and Hafner, 1983).

Nesoryzomys fernandinae Hutterer and Hirsch, 1979. Bonn. Zool. Beitr., 30:278.

COMMON NAME: Fernandina Nesoryzomys.

TYPE LOCALITY: Ecuador, Galápagos Archipelago, Fernandina Isl.

DISTRIBUTION: Known only from the type locality.

STATUS: IUCN – Vulnerable.

COMMENTS: Name based on material recovered from fresh owl pellets; morphological
recognition expanded by Dowler and Carroll (1996) and Dowler et al. (2000) based on
recently collected specimens from rediscovered population and compared with
sympatric examples of *N. indefessus narboroughi*. Believed closely related to *N. darwini*
(see Hutterer and Hirsch, 1979).

Nesoryzomys indefessus (Thomas, 1899). Ann. Mag. Nat. Hist., ser. 7, 4:280.

COMMON NAME: Santa Cruz Nesoryzomys.

TYPE LOCALITY: Ecuador, Galápagos Archipelago, Santa Cruz (= Indefatigable) Isl, Academia
Bay.

DISTRIBUTION: Santa Cruz, Baltra (= South Seymour), and Fernandina (= Narborough) Isls.

STATUS: IUCN – Extinct as *N. indefessus*, Near Threatened as *N. narboroughi*. *Nesorysomys i.
indefessus* is probably extinct, none documented since 1934 (see Patton and Hafner,
1983). Populations of *N. i. narboroughi* on Fernandina Isl, which lacks commensal
Rattus and *Mus*, appear stable (see Dowler et al., 2000; Patton and Hafner, 1983).

SYNONYMS: *narboroughi* Heller, 1904.

COMMENTS: Patton and Hafner (1983:539) recommended that *indefessus*, *narboroughi*, and
swarthi are "best considered races of a single species, which differ primarily in pelage
color." Their analyses sustain this conclusion with regard to *indefessus* and *narboroughi*
but not the strong craniodental differentiation demonstrated for *N. swarthi*.

Nesoryzomys swarthi Orr, 1938. Proc. California Acad. Sci., 23(21):304.

COMMON NAME: Santiago Nesoryzomys.

TYPE LOCALITY: Ecuador, Galápagos Archipelago, San Salvador Isl, Sullivan Bay.

DISTRIBUTION: Known only from San Salvador (= Santiago, James) Isl.

STATUS: IUCN – Vulnerable.

COMMENTS: Status viewed as an insular race of *N. indefessus* by Patton and Hafner (1983), but Orr (1938) underscored the trenchant diagnostic traits that separate *N. swarthi* from both *N. indefessus* and *narboroughi*. Extant populations thought to exist as of 1965 (see Peterson, 1966*b*) and recently confirmed in sympatry with *Rattus* and *Mus* (Dowler et al., 2000).

Neusticomys Anthony, 1921. Am. Mus. Novit., 20:2.
> TYPE SPECIES: *Neusticomys monticolus* Anthony, 1921.
> SYNONYMS: *Daptomys* Anthony, 1929.
> COMMENTS: Ichthyomyini. Phylogenetic relationships studied by Voss (1988), who allocated *Daptomys* as a synonym. Distribution and morphological traits of lowland species summarized by Voss et al. (2001). Each species is generally known from few specimens, the animals rarely encountered and difficult to collect; as such, species diversity may be underestimated and ranges of known forms are minimal approximations.

Neusticomys monticolus Anthony, 1921. Am. Mus. Novit., 20:2.
> COMMON NAME: Montane Ichthyomyine.
> TYPE LOCALITY: Ecuador, Pichincha Prov., Cordillera Occidental, Nono Farm, "San Francisco," 10,500 ft (3200 m).
> DISTRIBUTION: Andes, ca. 1800-3750 m, of W Colombia and N Ecuador.
> STATUS: IUCN – Lower Risk (lc).
> COMMENTS: Taxonomy and distribution reviewed by Voss (1988), who noted the probably lower elevation of the type locality at 8580 ft (2615 m).

Neusticomys mussoi Ochoa G. and Soriano, 1991. J. Mammal., 72:97.
> COMMON NAME: Musso's Ichthyomyine.
> TYPE LOCALITY: Venezuela, Táchira State, 14 km SE Pregonero, Río Potosí, Paso Hondo, 1050 m.
> DISTRIBUTION: Known only from the type locality, far W Venezuela.
> STATUS: IUCN – Endangered.
> COMMENTS: The two known specimens represent the least aquatically specialized ichthyomyine species described thus far.

Neusticomys oyapocki (Dubost and Petter, 1978). Mammalia, 42:436.
> COMMON NAME: Guianan Ichthyomyine.
> TYPE LOCALITY: French Guiana, Trois-Sauts, near the banks of the Oyapock River; 02°10'N, 53°11'W.
> DISTRIBUTION: Known only from three localities in French Guiana (Voss et al., 2001) and one in Amapá, NE Brazil (Nunes, 2002).
> STATUS: IUCN – Endangered.
> COMMENTS: Based on a single specimen originally referred to *Daptomys*. Maintained as a species by Voss (1988); additional material reported by Voss et al. (2001) and Nunes (2002).

Neusticomys peruviensis (Musser and Gardner, 1974). Am. Mus. Novit., 2537:7.
> COMMON NAME: Peruvian Ichthyomyine.
> TYPE LOCALITY: Perú, Loreto Dept., Balta, 300 m; 10°08'S, 17°13'W.
> DISTRIBUTION: Known only from two localities in lowland E Perú (see Pacheco and Vivar, 1996; Pacheco et al., 1993).
> STATUS: IUCN – Endangered.
> COMMENTS: Described as a form of *Daptomys* and maintained as a species by Voss (1988).

Neusticomys venezuelae (Anthony, 1929). Am. Mus. Novit., 383:2.
> COMMON NAME: Venezuelan Ichthyomyine.
> TYPE LOCALITY: Venezuela, Sucre State, 15 mi (24 km) W Cumanacoa, headwaters of Río Neverí, 2400 ft (732 m).
> DISTRIBUTION: E and S Venezuela, Guyana.
> STATUS: IUCN – Endangered.
> COMMENTS: Type species of *Daptomys*, considered a junior synonym of *Neusticomys* by Voss (1988).

Noronhomys Carleton and Olson, 1999. Am. Mus. Novit., 3256:9.
  TYPE SPECIES: *Noronhomys vespuccii* Carleton and Olson, 1999.
  COMMENTS: Oryzomyini. An extinct insular endemic and tetralophodont form most closely
    related to *Holochilus* according to cladistic interpretation of craniodental characters
    (Carleton and Olson, 1999).

Noronhomys vespuccii Carleton and Olson, 1999. Am. Mus. Novit., 3256:10.
  COMMON NAME: Vespucci's Rodent.
  TYPE LOCALITY: Brazil, Territorio de Fernando de Noronha, Ilha Fernando de Noronha,
    beach dunes (late Quaternary) near Ponta de Santo Antonio; 03°50′S, 32°24′W.
  DISTRIBUTION: Known only from the type locality.
  COMMENTS: Variation and morphological discrimination from *Holochilus* and *Lundomys*
    provided by Carleton and Olson (1999). Fragmentary historical archives tenuously
    indicate that this species was encountered by Amerigo Vespucci during his 1503
    voyage to the New World and was extirpated soon thereafter (see Carleton and Olson,
    1999:44-49).

Notiomys Thomas, 1890. *In* Milne-Edwards, Mission Sci. Cap. Horn, 1882-3, 6, Mamm., p. 23.
  TYPE SPECIES: *Hesperomys* (*Notiomys*) *edwardsii* Thomas, 1890.
  COMMENTS: Akodontini (S Andean clade). Alpha systematics revised by Osgood (1925), who
    viewed *Chelemys* and *Geoxus* as synonyms (see comments under those genera and in
    Pearson, 1984, and Reig, 1987). Sister genus to *Chelemys* according to maximum
    parsimony and likelihood analyses of cytochrome *b* sequences (Smith and Patton, 1999).

Notiomys edwardsii (Thomas, 1890). *In* Milne-Edwards, Mission Sci. Cap. Horn, 1882-3, 6, Mamm.,
    p. 24.
  COMMON NAME: Edward's Long-clawed Akodont.
  TYPE LOCALITY: Argentina, Santa Cruz Prov., south of Santa Cruz, Corpen Aike (as refined by
    Pardiñas and Galliari, 1998*a*:124).
  DISTRIBUTION: S Argentina, from S Río Negro Prov. to C Santa Cruz Prov. (see Pardiñas and
    Galliari, 1998*a*:Fig. 5).
  STATUS: IUCN – Lower Risk (lc).
  COMMENTS: Known by few specimens; morphology and natural history amplified by
    Pearson (1984, 1995) and distribution by Pardiñas and Galliari (1998*a*) and Teta et al.
    (2002).

Oecomys Thomas, 1906. Ann. Mag. Nat. Hist., ser. 7, 18:444.
  TYPE SPECIES: *Rhipidomys benevolens* Thomas, 1901 (= *Hesperomys bicolor* Tomes, 1860).
  COMMENTS: Oryzomyini. Diagnosed as a subgenus of *Oryzomys* to segregate arboreal, pencil-
    tailed sigmodontines with a long palate from *Rhipidomys*, under which many of the
    species included here were first described. Thereafter treated alternatively as a subgenus of
    Oryzomys (Ellerman, 1941; Goldman, 1918) or as full genus (Gyldenstolpe, 1932; Thomas,
    1917*c*) until Hershkovitz's (1960) revision stabilized its ranking as a subgenus (e.g.,
    Cabrera, 1961; Hall, 1981). Systematists have recently acknowledged the morphological
    and karyotypic distinctiveness of *Oecomys* at the generic level (Andrades-Miranda et al.,
    2001*b*; Carleton and Musser, 1984; Gardner and Patton, 1976; Reig, 1984, 1986); species
    so far surveyed genetically are reciprocally monophyletic with other oryzomyine genera
    (Smith and Patton, 1999; Weksler, 2003) but broader phylogenetic substantiation is
    desirable.
      Revised by Hershkovitz (1960), who consolidated some 25 species (e.g., Ellerman,
    1941; Gyldenstolpe, 1932) into just two, *bicolor* and *concolor*. Although this gross
    underestimation of specific diversity within *Oecomys* has been intimated by other authors
    (e.g., Gardner and Patton, 1976; Reig, 1986), it has yet to be documented within a full
    taxonomic revision. The species identified here mainly repeat those compiled by Musser
    and Carleton (1993), based on a (still) unfinished revision; while all of these will stand as
    valid, some, such as *O. bicolor* and *O. trinitatis*, are undoubtedly composites even now.
    Morphological and karyotypic comparisons of species in W Brazil (Patton et al., 2000)
    and French Guiana (Voss et al., 2001) underscore the far greater local biodiversity that

was masked by Hershkovitz's (1960) interpretation and supply regional glimpses of the species richness within the Neotropical realm yet to be fully understood.

Oecomys auyantepui Tate, 1939. Bull. Am. Mus. Nat. Hist, 76:193.
COMMON NAME: Guianan Oecomys.
TYPE LOCALITY: Venezuela, Bolívar State, Mt Auyán-Tepuí, 1100m.
DISTRIBUTION: SC Venezuela eastwards through the Guianas to Amapá, Brazil and southwards to Amazonas, Brazil, north of Amazon River.
COMMENTS: Synonymized under *concolor* by Hershkovitz (1960) and under *paricola* by Musser and Carleton (1993). Morphological distinctiveness elaborated by Voss et al. (2001) and contrasted with *O. paricola*, its probable sister taxon.

Oecomys bicolor (Tomes, 1860). Proc. Zool. Soc. Lond., 1860:217.
COMMON NAME: White-bellied Oecomys.
TYPE LOCALITY: Ecuador, Morona-Santiago Prov., Gualaquiza, Río Gualaquiza, 885 m.
DISTRIBUTION: E Panamá to W Colombia and Ecuador; Venezuela, Guianas, N and C Brazil; Amazonian drainage of Bolivia, Perú, Ecuador, and Colombia.
STATUS: IUCN – Lower Risk (lc).
SYNONYMS: *benevolens* (Thomas, 1901); *dryas* (Thomas, 1900); *endersi* Goldman, 1933; *florenciae* J. A. Allen, 1916; *milleri* J. A. Allen, 1916; *nitedulus* Thomas, 1910; *occidentalis* (Hershkovitz, 1960); *phelpsi* Tate, 1939; *rosilla* (Thomas, 1904); *trabeatus* G. M. Allen and Barbour, 1923.
COMMENTS: Synonymy of *O. phelpsi* Tate attributed by Musser and Patton (1989). A morphologically and genetically distinct member of the *bicolor* complex, species identity uncertain, was collected at localities nearby those of *O. bicolor* proper in the Rio Juruá basin, W Brazil (Patton et al., 2000).

Oecomys catherinae Thomas, 1909. Ann. Mag. Nat. Hist., ser. 8, 4:234.
COMMON NAME: Atlantic Forest Oecomys.
TYPE LOCALITY: Brazil, Santa Catarina State, Joinville.
DISTRIBUTION: Atlantic Forest Region of SE Brazil, Bahia to Santa Catarina, and along riverine forest into Cerrado and Caatinga regions; limits poorly documented.
SYNONYMS: *bahiensis* (Hershkovitz, 1960) [replacement name for *Mus cinnamomeus* Pictet and Pictet, 1844, preoccupied by *Mus cinnamomeus* Lichtenstein, 1830, a *Proechimys*]; *cinnamomeus* (Pictet and Pictet, 1844).
COMMENTS: Member of the *O. trinitatis* group but size larger than *O. trinitatis* proper, pelage more luxuriant, and supraorbital shelves more pronounced.

Oecomys cleberi Locks, 1981. Bol. Mus. Nac., Zool., 300:1.
COMMON NAME: Cleber's Oecomys.
TYPE LOCALITY: Brazil, Federal District, Universidade de Brasília, Fazenda Agua Limpa; 45°54'W, 15°57'S.
DISTRIBUTION: Known only from the type locality.
STATUS: IUCN – Endangered.
COMMENTS: Allied to *O. bicolor* or *O. paricola*; status and distributional extent require study.

Oecomys concolor (Wagner, 1845). Arch. Naturgesch., 11:147.
COMMON NAME: Unicolored Oecomys.
TYPE LOCALITY: Brazil, Amazonas State, Rio Curicuriari, a tributary of the upper Rio Negro, below São Gabriel.
DISTRIBUTION: S Venezuela south of the llanos (see Linares, 1998:Fig 142), NW Brazil, E Colombia, and NE Bolivia (as per Anderson, 1997); range limits poorly documented.
STATUS: IUCN – Lower Risk (lc).
SYNONYMS: *marmorsurus* (Thomas, 1899).
COMMENTS: *Oecomys concolor* proper has a restricted distribution compared with Hershkovitz's (1960) broad view of its contents; most of those forms are actually synonyms of the very different species *O. flavicans*, *O. roberti*, *O. superans*, or *O. trinitatis*. Characters of the skin and skull indicate that *O. concolor* is closely related to *O. mamorae*.

Oecomys flavicans (Thomas, 1894). Ann. Mag. Nat. Hist., ser. 6, 14:351.
    COMMON NAME: Tawny Oecomys.
    TYPE LOCALITY: Venezuela, Mérida State, Mérida, 1600 m.
    DISTRIBUTION: Coastal Range and Cordillera de Mérida of N and W Venezuela (see Linares,
        1998:Map 140), west to Sierra de Santa Marta of NE Columbia, perhaps including the
        Cordillera Oriental.
    STATUS: IUCN – Lower Risk (lc).
    SYNONYMS: *illectus* (Bangs, 1896); *mincae* J. A. Allen, 1913.

Oecomys mamorae (Thomas, 1906). Ann. Mag. Nat. Hist., ser. 7, 18:445.
    COMMON NAME: Marmore Oecomys.
    TYPE LOCALITY: Bolivia, Cochabamba Dept., upper Río Mamoré, Mosetenes.
    DISTRIBUTION: N and C Bolivia, N Paraguay, and WC Brazil.
    STATUS: IUCN – Lower Risk (lc).

Oecomys paricola (Thomas, 1904). Ann. Mag. Nat. Hist., ser. 7, 14:194.
    COMMON NAME: Brazilian Oecomys.
    TYPE LOCALITY: Brazil, Pará State, Igarapé-Assu, 50 m.
    DISTRIBUTION: C Brazil south of the Amazon River; range limits uncertain.
    STATUS: IUCN – Lower Risk (lc).
    COMMENTS: Formerly included *auyantepui* according to Musser and Carleton (1993), that
        form elevated to species by Voss et al. (2001).

Oecomys phaeotis (Thomas, 1901). Ann. Mag. Nat. Hist., ser. 7, 7:181.
    COMMON NAME: Dusky Oecomys.
    TYPE LOCALITY: Perú, Puno Dept., upper Río Inambari, Sagrario, 1000 m.
    DISTRIBUTION: E slopes of Peruvian Andes; limits unresolved.
    STATUS: IUCN – Lower Risk (lc).
    COMMENTS: Part of the *bicolor* or *paricola* complex.

Oecomys rex Thomas, 1910. Ann. Mag. Nat. Hist., ser. 8, 6:504.
    COMMON NAME: Regal Oecomys.
    TYPE LOCALITY: Guyana, Demerara Dist., Supenaam River.
    DISTRIBUTION: Far E Venezuela (Bolívar; see Linares, 1998), Guianas, and NE Brazil north of
        the Amazon (Amapá and Amazonas).
    STATUS: IUCN – Lower Risk (lc).
    SYNONYMS: *regalis* (Hershkovitz, 1960) [an unjustified replacement name].

Oecomys roberti (Thomas, 1904). Proc. Zool. Soc. Lond., 1903(2):237 [1904].
    COMMON NAME: Robert's Oecomys.
    TYPE LOCALITY: Brazil, Mato Grosso State, Santa Anna da Chapada (= Chapada Dos
        Guimarães), 800 m.
    DISTRIBUTION: S Venezuela, Guianas, and Amazonian region of W Brazil, E Perú, and
        extreme N Bolivia.
    STATUS: IUCN – Lower Risk (lc).
    SYNONYMS: *guianae* Thomas, 1910; *tapajinus* Thomas, 1909.

Oecomys rutilus Anthony, 1921. Am. Mus. Novit., 19:4.
    COMMON NAME: Reddish Oecomys.
    TYPE LOCALITY: Guyana, Mazaruni-Potaro Dist., Kartabo.
    DISTRIBUTION: Extreme E Venezuela, Guyana, Surinam, and French Guiana to Amazonas,
        Brazil (see Voss et al., 2001:Fig. 60).
    STATUS: IUCN – Lower Risk (lc).
    COMMENTS: Morphological separation from *O. bicolor* and distribution amplified by Voss
        et al. (2001).

Oecomys speciosus (J. A. Allen and Chapman, 1893). Bull. Am. Mus. Nat. Hist., 5:212.
    COMMON NAME: Savannah Oecomys.
    TYPE LOCALITY: Trinidad, Princes Town.
    DISTRIBUTION: Savannahs of NE Colombia, C and N Venezuela, and Trinidad.
    STATUS: IUCN – Lower Risk (lc).

SYNONYMS: *caicarae* J. A. Allen, 1913; *trichurus* (J. A. Allen, 1899).

Oecomys superans Thomas, 1911. Ann. Mag. Nat. Hist., ser. 8, 8:250.
COMMON NAME: Large Oecomys.
TYPE LOCALITY: Ecuador, Pastaza Prov., Río Bobonaza, Canelos, 2100 ft (640 m).
DISTRIBUTION: Lower Andean slopes of E Colombia, Ecuador, and Perú, including contiguous lowlands of W Amazonia (see Patton et al., 2000).
STATUS: IUCN – Lower Risk (lc).
SYNONYMS: *melleus* Anthony, 1924; *palmeri* Thomas, 1911.

Oecomys trinitatis (J. A. Allen and Chapman, 1893). Bull. Am. Mus. Nat. Hist., 5:213.
COMMON NAME: Long-furred Oecomys.
TYPE LOCALITY: Trinidad, Princes Town.
DISTRIBUTION: Tropical rainforests from SW Costa Rica to C Brazil, including Guianas, Trinidad and Tobago; E Andean slopes of WC Colombia to SC Perú.
STATUS: IUCN – Lower Risk (lc).
SYNONYMS: *frontalis* (Goldman, 1912); *fulviventer* (J. A. Allen, 1899); *helvolus* (J. A. Allen, 1913); *klagesi* (J. A. Allen, 1904); *osgoodi* Thomas, 1924; *palmarius* (J. A. Allen, 1899); *splendens* Hayman, 1938; *subluteus* (Thomas, 1898); *tectus* (Thomas, 1901); *vicencianus* (J. A. Allen, 1913).
COMMENTS: The synonyms listed here form the core of Hershkovitz's (1960) *concolor*, but their consanguinity is highly doubtful.

Oligoryzomys Bangs, 1900. Proc. New England Zool. Club, 1:94.
TYPE SPECIES: *Oryzomys navus* Bangs, 1899 (= *Hesperomys fulvescens* Saussure, 1860).
COMMENTS: Oryzomyini. Described as a subgenus of *Oryzomys* and usually recognized as such (Ellerman, 1941; Hall, 1981; Reig, 1984; Tate, 1932*e*) or as a genus (Contreras and Berry, 1983; Gyldenstolpe, 1932), with *Microryzomys* as a full synonym (Gyldenstolpe, 1932; Tate, 1932) or not (Cabrera, 1961). Diagnosis emended at the generic level and morphology contrasted with *Microryzomys* and *Oryzomys sensu stricto* by Carleton and Musser (1989). Monophyly of the genus supported by electrophoretic (Dickerman and Yates, 1995) and mitochondrial and nuclear gene-sequence data (Myers et al., 1995; Smith and Patton, 1999; Weksler, 2003), in a clade also containing *Microryzomys* and *Neacomys*. Attempts to define species groups have so far produced conflicting results and should be viewed as preliminary forays to stimulate critical phylogenetic study (e.g., see Andrades-Miranda et al., 2001*a*; Bonvicino and Weksler, 1998; Carleton and Musser, 1989; Dickerman and Yates, 1995; Myers and Carleton, 1981; Myers et al., 1995).
    Species-level revisions and range documentation much needed: estimates range from one (Hershkovitz, 1966*c*) to 30 (Tate, 1932*e*), usually around 12 (Cabrera, 1961). Regional studies have typically recorded two to four species in sympatry or parapatry (Bonvicino and Weksler, 1998; Carleton and Musser, 1995; Contreras and Berry, 1983; Massoia, 1973; Myers and Carleton, 1981; Olds and Anderson, 1987). Chromosomes of many species described and compared in Andrades-Miranda et al. (2001*a*), Bonvicino and Weksler (1998), Espinosa and Reig (1991), Gallardo and Patterson (1985), Gardner and Patton (1976), and Myers and Carleton (1981); published karyotypes summarized for all South American species by Andrades-Miranda et al. (2001*a*). The 18 species recognized here generally observe the preliminary review of Carleton and Musser (1989), as amplified by later descriptions. A tally of 25-30 species remains probable as extrapolated from the following observations: the Andean complex has yet to be critically addressed; definition of *O. fulvescens sensu stricto* and its valid synonyms is still lacking; investigations continue to list many *Oligoryzomys* as species indeterminate (e.g., Anderson, 1997; Andrades-Miranda et al., 2001*a*; Carleton and Musser, 1989). Inclusion of vouchered material from type localities will prove crucial to advancing comprehension of specific diversity and stabilizing nomenclatural usage within the genus.

Oligoryzomys andinus (Osgood, 1914). Field Mus. Nat. Hist. Publ., Zool. Ser., 10:156.
COMMON NAME: Andean Colilargo.
TYPE LOCALITY: Perú, La Libertad Dept., upper Río Chicama, Hacienda Llagueda, 6000 ft (1830 m).

DISTRIBUTION: W Perú and WC Bolivia; geographic and altitudinal limits uncertain.

STATUS: IUCN – Lower Risk (lc).

COMMENTS: Karyotype reported by Gardner and Patton (1976). Recorded as sympatric with *O. arenalis* in Lima Dept., Perú (Arana-Cardó and Ascorra, 1994).

Oligoryzomys arenalis (Thomas, 1913). Ann. Mag. Nat. Hist., ser. 8, 12:571.

COMMON NAME: Sandy Colilargo.

TYPE LOCALITY: Perú, Lambayeque Dept., Eten, 10 m.

DISTRIBUTION: Arid and semiarid coastal plain of Perú.

STATUS: IUCN – Lower Risk (lc).

COMMENTS: Level of relationship to *O. fulvescens* warrants clarification.

Oligoryzomys brendae Massoia, 1998. 2nd Congreso Argentino de Zoonosis y 1st Congreso Argentino y Latinamericano de Enfermedades Emergentes, Buenos Aires, p. 243.

COMMON NAME: Brenda's Colilargo.

TYPE LOCALITY: Argentina, Tucumán Prov., Depto. Tafí Viejo, Cerro San Javier, 1000 m.

DISTRIBUTION: Tucumán, Salta, and Catamarca Provs., NW Argentina.

COMMENTS: Compared with *O. flavescens* and *O. longicaudatus* by Massoia (1998) but details for recognition skimpy; additional study required to illuminate status, discrimination from other regional forms, and distribution.

Oligoryzomys chacoensis (Myers and Carleton, 1981). Misc. Publ. Mus. Zool., Univ. Michigan, 161:19.

COMMON NAME: Chacoan Colilargo.

TYPE LOCALITY: Paraguay, Boquerón Dept., km 419 along Trans Chaco Hwy, northwest of Villa Hayes.

DISTRIBUTION: Dryer habitats of SE Bolivia, W Paraguay, SW Brazil, and N Argentina.

STATUS: IUCN – Lower Risk (lc).

COMMENTS: Karyotype reported by Myers and Carleton (1981); morphometric variation by Myers and Carleton (1981) and Olds and Anderson (1987). Status with regard to *O. andinus* questioned by Carleton and Musser (1989), but genetic data underscore the specific distinctiveness of the two (Myers et al., 1995).

Oligoryzomys delticola (Thomas, 1917). Ann. Mag. Nat. Hist., ser. 8, 20:96.

COMMON NAME: Large Colilargo.

TYPE LOCALITY: Argentina, Buenos Aires Prov., Delta del Paraná, Isla Ella, 1 m.

DISTRIBUTION: EC Argentina, Uruguay, and S Brazil (Rio Grande do Sul).

STATUS: IUCN – Lower Risk (lc).

COMMENTS: A large-bodied species that karyotypically and morphometrically resembles *O. eliurus* and *O. nigripes*; the level of differentiation among the three and their appropriate taxonomic status require additional investigation (see comparisons and commentary in Andrades-Miranda et al. [2001a], Bonvicino and Weksler [1998], Espinoza and Reig [1991], and Myers and Carleton [1981]).

Oligoryzomys destructor (Tschudi, 1844). Fauna Peruana, 1:182.

COMMON NAME: Tschudi's Colilargo.

TYPE LOCALITY: Perú, Huanuco Dept., haciendas along the Río Chinchao, 900-1000 m (as discussed by Hershkovitz, 1940:81; formal fixation necessary).

DISTRIBUTION: E Andean slopes of S Colombia, through Ecuador, Perú, and WC Bolivia (see Anderson, 1997:Fig. 691), to NW Argentina (Tucumán Prov. as per OMNH 1062, 1106, 1140, 2205, 2239, 2734).

STATUS: IUCN – Data Deficient.

SYNONYMS: *maranonicus* (Osgood, 1914); *melanostoma* (Tschudi, 1844); *spodiurus* (Hershkovitz, 1940); *stolzmanni* (Thomas, 1894).

COMMENTS: Considered a subspecies of *O. longicaudatus* by Gyldenstolpe (1932) and Cabrera (1961); provisionally resurrected as species by Carleton and Musser (1989). Karyotype reported by Gardner and Patton (1976) as *Oryzomys longicaudatus* variant (4). Distributional limits, validity of included synonyms, and relationships to other Andean species all require detailed investigation.

Oligoryzomys eliurus (Wagner, 1845). Arch. Naturgesch., 11:147.
COMMON NAME: Brazilian Colilargo.
TYPE LOCALITY: Brazil, São Paulo State, Ytararé.
DISTRIBUTION: C and SE Brazil; range limits uncertain.
STATUS: IUCN – Lower Risk (lc).
SYNONYMS: *pygmaeus* (Wagner, 1845); *utiaritensis* (J. A. Allen, 1916).
COMMENTS: Suggested as conspecific with *O. nigripes* (see Myers and Carleton, 1981); also see
        comments under *O. delticola*. Diploid number (2n = 62) reported as the same as
        O. nigripes but with lower fundamental numbers (FN = 64-66 versus 78-82), based on
        specimens from Goías, C Brazil, where the two species were recorded in sympatry
        (Andrades-Miranda et al., 2001a). This karyotype resembles that earlier reported by
        Bonvicino and Weksler (1998) as *O. fornesi* from Goías, also there recorded in sympatry
        with *O. nigripes*. The latter authors instead attributed a high-FN karyotype from São
        Paulo (as reported by Yonenaga et al., 1976) to *O. eliurus*. Reconciliation of these
        apparent discrepancies is needed as are vouchered documentation of the species'
        distributional limits and its clear morphological discrimination from *O. nigripes*.

Oligoryzomys flavescens (Waterhouse, 1837). Proc. Zool. Soc. Lond., 1837:19.
COMMON NAME: Flavescent Colilargo.
TYPE LOCALITY: Uruguay, Maldonado Dept., Maldonado.
DISTRIBUTION: E Paraguay (Bonvicino and Weksler, 1998), SE Brazil (Rio Grande do Sul to
        Bahia), Uruguay, and N to SC Argentina; range limits uncertain.
STATUS: IUCN – Lower Risk (lc).
SYNONYMS: *antoniae* (Massoia, 1979); *occidentalis* Contreras and Rosi, 1980.
COMMENTS: Chromosomal variation reported and taxonomic implications discussed by
        Sbalqueiro et al. (1991); allozymic polymorphisms and gene flow among populations
        distributed along the lower Río Paraná, EC Argentina, discussed by Chiappero et al.
        (1997). Morphometrically (Myers and Carleton, 1981) and karyotypically (Sbalqueiro
        et al., 1991) similar to *O. fornesi*; their level of differentiation deserves further
        illumination (see next). In their informal review, Carleton and Musser (1989)
        acknowledged three *flavescens*-like forms from Peru, Bolivia, and Brazil; these morphs
        should be critically identified, and named as necessary, as part of a much-needed
        revision of *flavescens*.

Oligoryzomys fornesi (Massoia, 1973). Rev. Invest. Agropec. Ser. 1, Biol. Prod. Anim., 10:22.
COMMON NAME: Fornes' Colilargo.
TYPE LOCALITY: Argentina, Formosa Prov., Naineck.
DISTRIBUTION: NE Argentina, E Paraguay, and SC Brazil (Goiás to Paraíba; Bonvicino and
        Weksler, 1998).
COMMENTS: Viewed as a synonym of *O. microtis* (Carleton and Musser, 1989; Olds and
        Anderson, 1987), an assignment contradicted by morphometric (Bonvicino and
        Weksler, 1998) and cytochrome *b* (Myers et al., 1995) data. Bonvicino and Weksler
        (1998) recorded the species as present in Brazil, but Andrades-Miranda et al. (2001a)
        did not. Bonvicino and Weksler (1998) reidentified certain Paraguayan specimens
        reported as *O. fornesi* by Myers and Carleton (1981) as *O. flavescens* based on possession
        of lower diploid (62) and fundamental (64) numbers and certain cranial shape
        differences. Myers and Carleton interpreted these minor chromosome differences as
        populational variation (2n = 62-66, FN = 64-68), nor can we readily grasp the
        purported shape differences from Bonvicino and Weksler's principle component
        results. Also see comments on *O. flavescens*.

Oligoryzomys fulvescens (Saussure, 1860). Revue Mag. Zool. Paris, ser. 2, 12:102.
COMMON NAME: Fulvous Colilargo.
TYPE LOCALITY: México, Veracruz State, Orizaba (as restricted by Merriam, 1901a:295).
DISTRIBUTION: W and E versants of S México, through Central America, to Ecuador, N and
        C Venezuela, Guianas, and northernmost Brazil in South America.
STATUS: IUCN – Lower Risk (lc).
SYNONYMS: *costaricensis* (J. A. Allen, 1893); *delicatus* (J. A. Allen and Chapman, 1897);
        fulvescens (Saussure, 1860); *engraciae* (Osgood, 1945); *lenis* (Goldman, 1915); *mayensis*

(Goldman, 1918); *messorius* (Thomas, 1901); *munchiquensis* (J. A. Allen, 1912); *navus* (Bangs, 1899); *nicaraguae* (J. A. Allen, 1910); *pacificus* (Hooper, 1952); *tenuipes* (J. A. Allen, 1904).

COMMENTS: All Middle American forms retained as subspecies by some (see Hall, 1981; Jones and Engstrom, 1986). Morphometric variation among Middle American forms evaluated by Carleton and Musser (1995) and their morphology contrasted with *O. vegetus*. Reported karyotypic variation intimates that more than one species occurs among the listed synonyms (Andrades-Miranda et al., 2001*a*; Gardner and Patton, 1976; Haiduk et al., 1979), whose status, particularly the South American taxa provisionally associated by Carleton and Musser (1995), warrants additional study. Linares (1998), e.g., recognized all Venezuelan populations (except *O. griseolus*) as *O. fulvescens*; whereas, Andrades-Miranda et al. (2001*a*) listed specimens from Roraima, Brazil, as *O.* cf. *messorius*, a form described from Guyana; and see Voss et al. (2001) for another regional perspective regarding discrimination and specific assignment of Guianan samples vis a vis lowland populations in Middle America (*O. fulvescens*) and N Brazil (*O. microtis*).

Oligoryzomys griseolus (Osgood, 1912). Field Mus. Nat. Hist. Publ., Zool. Ser., 10:49.
   COMMON NAME: Grizzled Colilargo.
   TYPE LOCALITY: Venezuela, Táchira State, upper Río Táchira, west of Páramo de Tamá, 6000-7000 ft (1830-2130 m).
   DISTRIBUTION: Táchira Andes of extreme W Venezuela and Cordillera Oriental of E Colombia.
   STATUS: IUCN – Lower Risk (lc).
   COMMENTS: As noted by Osgood (1912) and verified by specimens in USNM, this distinctive form contrasts sharply with neighboring populations of *O. fulvescens* (e.g., *navus* and *tenuipes*) and instead resembles *O. vegetus* of lower Middle America.

Oligoryzomys longicaudatus (Bennett, 1832). Proc. Zool. Soc. Lond., 1832:2.
   COMMON NAME: Long-tailed Colilargo.
   TYPE LOCALITY: Chile, Valparaíso Prov. (as suggested by Osgood, 1943:143).
   DISTRIBUTION: N Chile and NW Argentina, southwards along Andes to approximately to 50°S latitude; extralimital in EC Argentina (Buenos Aires Prov.); range limits uncertain.
   STATUS: IUCN – Lower Risk (lc).
   SYNONYMS: *agilis* (Philippi, 1900); *amblyrrhynchus* (Philippi, 1900); *araucanus* (Philippi, 1900); *commutatus* (Philippi, 1900); *coppingeri* (Thomas, 1881); *diminutivus* (Philippi, 1900); *dumetorum* (Philippi, 1900); *exiguus* (Philippi, 1900); *glaphyrus* (Philippi, 1900); *macrocercus* (Philippi, 1900); *malaenus* (Philippi, 1900); *melanizon* (Philippi, 1900); *mizurus* (Thomas, 1916); *nigribarbis* (Philippi, 1900); *pampanus* (Massoia, 1973); *pernix* (Philippi, 1900); *peteroanus* (Philippi, 1900); *philippii* (Landbeck, 1858); *saltator* (Philippi, 1900).
   COMMENTS: Formerly encompassed most Andean populations of *Oligoryzomys* (see Cabrera, 1961), including *destructor* and *magellanicus* (see those specific accounts). Morphometric variation among Chilean populations investigated by Gallardo and Palma (1990), who placed *philippii* in full synonymy with *O. longicaudatus* and removed *magellanicus* to specific status. Populations of *O. longicaudatus* (OMNH 2412, 2423, 2445, 2471) and *O. destructor* (see Distribution in that account) apparently approach one another in Tucumán Prov., NW Argentina. That in Buenos Aires Prov., EC Argentina, described as *longicaudatus pampanus* by Massoia (1973), is extralimital to the principally Andean distribution of the species as currently understood, but other isolated occurrences have been reported from intermediate provinces (La Pampa and Río Negro—De Santis et al., 2001). Relationships, validity of included synonyms, and level of divergence from allopatric forms like *O. destructor* and *O. nigripes* warrant intensive study (see commentary in Carleton and Musser [1989], Mares et al. [1997], and Pardiñas et al. [2003*b*]). Attribution of numerous Philippi (1900) epithets follows Osgood (1943).

Oligoryzomys magellanicus (Bennett, 1836). Proc. Zool. Soc. Lond., 1835:191 [1836].
   COMMON NAME: Patagonian Colilargo.

TYPE LOCALITY: Chile, Magallanes Prov., Straits of Magellan, Port Famine.

DISTRIBUTION: S Patagonian region of Chile and Argentina, including Tierra del Fuego.

STATUS: IUCN – Lower Risk (lc).

COMMENTS: Karyotypic, morphometric, and phallic differentiation from *O. longicaudatus* supports the specific recognition of *O. magellanicus* (Gallardo and Palma, 1990; Gallardo and Patterson, 1985).

Oligoryzomys microtis (J. A. Allen, 1916). Bull. Am. Mus. Nat. Hist., 35:525.

COMMON NAME: Small-eared Colilargo.

TYPE LOCALITY: Brazil, Amazonas State, north bank of lower Rio Solimões, 70 km WSW Manaus, Manacaparú (as restricted by Voss et al., 2001:118-119).

DISTRIBUTION: Amazon Basin of Brazil and contiguous lowlands of Perú, Bolivia, and Paraguay.

STATUS: IUCN – Lower Risk (lc).

SYNONYMS: *chaparensis* (Osgood, 1916); *mattogrossae* (J. A. Allen, 1916).

COMMENTS: Karyotype reported by Gardner and Patton (1976) as *Oryzomys longicaudatus* variant (2) for Peruvian sample; slightly different diploid and fundamental numbers documented for specimens from Amapá, Brazil (Andrades-Miranda et al., 2001*a*). Gene sequence analyses indicate that *O. microtis* and *O. fornesi* represent separate species groups (Myers et al., 1995), not populations of a single species (Carleton and Musser, 1989; Olds and Anderson, 1987). With removal of *fornesi* from synonymy and fixation of the type locality, specific definition and distributional limits require amendment; also see Voss et al. (2001) for comparisons with *O. fulvescens* and comments on taxonomy.

Oligoryzomys nigripes (Olfers, 1818). *In* Eschwege, J. Brasilien, Neue Bibliothek. Reisenb., 15:209.

COMMON NAME: Black-footed Colilargo.

TYPE LOCALITY: Paraguay, Paraguarí Dept., Ybycui National Park, 85 km SSE Atyra (as restricted by neotype designation by Myers and Carleton, 1981:14).

DISTRIBUTION: E Paraguay, N Argentina, and Atlantic Forest region, C and SE Brazil (Rio Grande do Sul to Paraíba, interior to Goiás and Distrito Federal; also Ilha Grande; see Andrades-Miranda et al., 2001*a*, and Bonvicino and Weksler, 1998).

STATUS: IUCN – Lower Risk (lc).

SYNONYMS: *tarsonigro* (Fischer, 1814).

COMMENTS: Neotype designated, diagnosis emended, and karyotype described (2n = 62, FN = 80) by Myers and Carleton (1981), who regarded *Mus tarsonigro* Fischer, 1814, as a *nomen oblitum*; further, they recommended that *Mus longitarsus* Rengger, 1830, which could apply to either *O. microtis* or *O. nigripes*, be considered a *nomen dubium*. Distributional extent complex; in SE Brazil, predominantly a sylvan animal whose range may penetrate drier Cerrado and Caatinga biomes along riparian forests (Bonvicino and Weksler, 1998); also found on offshore island, Ilha Grande, Rio de Janeiro (L. G. Pereira et al., 2001). Pattern of microgeographic variation in craniodental variables evaluated in E Paraná, Brazil (as *Oryzomys nigripes*), by Santos (1997). Relationship to forms such as *delticola*, *eliurus*, and *longicaudatus* warrants additional investigation (also see comments under *O. delticola*).

Oligoryzomys stramineus Bonvicino and Weksler, 1998. Z. Saugetierk., 63:98.

COMMON NAME: Straw-colored Colilargo.

TYPE LOCALITY: Brazil, Goiás State, Terezina de Goiás, 24 km N Terezina, 15 km SW Rio Paraná (tributary of the upper Rio Tocantins), Fazenda Vao dos Bois, road GO-118, km 275, 424 m; 13°34'29"S, 47°10'57"W.

DISTRIBUTION: Cerrado (Goiás and Minas Gerais) and Caatinga (Paraíba and Pernambuco) formations of NE and C Brazil (see Bonvicino and Weksler, 1998:Fig. 7).

COMMENTS: A large-bodied form sympatric with *O. fornesi* or *O. nigripes*. Morphometric, karyological (2n = 52, FN = 68), and distributional comparisons with *O. chacoensis*, *O. delticola*, and *O. nigripes* provided by Bonvicino and Weksler (1998); other karyological comparisons by Andrades-Miranda et al. (2001*a*).

Oligoryzomys vegetus (Bangs, 1902). Bull. Mus. Comp. Zool., 39:35.
> COMMON NAME: Sprightly Colilargo.
> TYPE LOCALITY: Panamá, Chiriqui Prov., Volcán de Chiriqui, Boquete, 4000 ft (1219 m).
> DISTRIBUTION: Lower montane and montane wet forests, 840-3000 m, in C Costa Rica and
> W Panamá.
> STATUS: IUCN – Lower Risk (nt).
> SYNONYMS: *creper* (Goodwin, 1945); *reventazoni* (Goodwin, 1945).
> COMMENTS: Relegated to a subspecies of *O. fulvescens* by Goldman (1918) and so arranged
> thereafter (e.g., Hall, 1981). Bangs (1902), however, correctly recognized the sympatry
> of his new species with *O. fulvescens*, and Carleton and Musser (1995) clarified its
> morphological definition, regional synonymies, and geographic range.

Oligoryzomys victus (Thomas, 1898). Ann. Mag. Nat. Hist., ser. 7, 1:178.
> COMMON NAME: St. Vincent Colilargo.
> TYPE LOCALITY: Lesser Antilles, Saint Vincent.
> DISTRIBUTION: Known only from the type locality.
> STATUS: IUCN – Endangered.
> COMMENTS: Earlier classifications listed *O. victus* as an *Oryzomys* of uncertain affinity
> (Ellerman, 1941; Goldman, 1918), and Hall and Kelson (1959) erroneously placed it
> with their "*tectus* group" (= *Oecomys*). Thomas (1898a), and later Ray (1962),
> emphasized its alliance with species of *Oligoryzomys*. Known only by the holotype;
> presumably extinct (see Ray, 1962).

Oryzomys Baird, 1857. Mammalia *in* Repts. U.S. Expl. Surv., 8(1):458.
> TYPE SPECIES: *Mus palustris* Harlan, 1837.
> SYNONYMS: *Macruroryzomys* Hershkovitz, 1948 [*nomen nudum*]; *Micronectomys* Hershkovitz,
> 1948 [*nomen nudum*].
> COMMENTS: Oryzomyini. A taxonomically complex and nomenclaturally confused taxon
> whose definition was successively broadened by Goldman (1918), Tate (1932d, e), and
> Ellerman (1941). By the time of Cabrera (1961), the genus embraced as subgenera
> *Melanomys*, *Microryzomys*, *Nesoryzomys*, *Oecomys*, and *Oligoryzomys*—a polyphyletic
> agglomeration of taxa as evolutionarily divergent from one another, and from *Oryzomys*
> *sensu stricto*, as *Neacomys* is from *Nectomys*, forms traditionally accorded generic status (see
> Carleton and Musser, 1989:52). Others have recognized some or all of these as genera
> (e.g., Carleton and Musser, 1984, 1989; Gardner and Patton, 1976; Gyldenstolpe, 1932;
> Reig, 1981, 1984; Thomas, 1917c), as we continue to do here; see appropriate generic
> accounts for taxonomic histories.
> Notwithstanding the elevation of the aforementioned genus-group taxa, the generic
> boundary as denotatively conveyed by the species listed here remains a polyphyletic
> shell. The species may be morphologically and-or geographically sorted into the
> following groups whose relationship to one another and to other oryzomyine genera is
> incompatible with their continued association under a single genus: Albigularis complex
> (*albigularis*, *auriventer*, *caracolus*, *devius*, *keaysi*, *levipes*, *meridensis*); Balneator (*balneator*);
> Chapmani complex (*chapmani*, *saturatior*); Hammondi (*hammondi*); Megacephalus
> complex (*megacephalus* [*laticeps*, *megacephalus*, *perenensis*, *suazensis*] and *yunganus* [*tatei*,
> *yunganus*] species groups); Nitidus complex (Transandean [*alfaroi*, *bolivaris*, *melanotis*,
> *rhabdops*, *rostratus*, and *talamancae*] and Amazonian-Atlantic Forest [*emmonsae*, *kelloggi*?,
> *lamia*, *legatus*, *macconnelli*, *nitidus*, and *russatus*] species groups); North and Central
> American *Oryzomys sensu stricto* (*couesi*, *dimidiatus*, *gorgasi*, *nelsoni*, and *palustris*); and the
> very heterogeneous residuum of South American "*Oryzomys*" (*angouya*, *galapagoensis*,
> *maracajuensis*, *marinhus*, *polius*, *scotti*, *subflavus*, and *xanthaeolus*). See discussions in
> Carleton and Musser (1989), Carleton and Olson (1999), and Voss et al. (2002) on the
> need to objectively represent the monophyly and diagnosis of *Oryzomys sensu stricto* as
> elementary to improving comprehension of orzyomyine phylogeny. See Weksler (2003)
> for molecular evidence that pointedly underscores the polyphyletic contents of the genus
> as currently arranged.
> Karyotypic information for many species supplied by Baker et al. (1983), Gardner and
> Patton (1976), and Haiduk et al. (1979); multispecific molecular surveys and extended

taxonomic commentary provided by Patton et al. (2000), Bonvicino et al. (2001), and Weksler (2003); for morphological surveys, see Carleton (1973, 1980), Hooper and Musser (1964a), Musser et al. (1998), Voss and Linzey (1981), and Voss et al. (2001); for overview of habitat and distribution, see Sánchez-Cordero and Valadez-Azua (1989). Much basic alpha-level revision yet required.

Oryzomys albigularis (Tomes, 1860). Proc. Zool. Soc. Lond., 1860:264.

COMMON NAME: White-throated Oryzomys.

TYPE LOCALITY: Ecuador, Chimborazo Prov., Pallatanga, 4950 ft (1509 m).

DISTRIBUTION: Montane forests of N and W Venezuela, easternmost Panamá, Andes of Colombia and Ecuador, to N Perú.

STATUS: IUCN – Lower Risk (lc).

SYNONYMS: *childi* Thomas, 1895; *maculiventer* J. A. Allen, 1891; *moerex* Thomas, 1914; *oconnelli* J. A. Allen, 1913; *pectoralis* J. A. Allen, 1912; *pirrensis* Goldman, 1913; *villosus* J. A. Allen, 1899.

COMMENTS: Hershkovitz's (1944) footnoted listing of specific synonyms of *O. albigularis* set the precedent for Cabrera's (1961) arrangement of the South American forms as subspecies, a viewpoint reiterated in regional studies (e.g., Handley, 1966a, 1976). Gardner and Patton (1976) demonstrated the composite nature of Hershkovitz's (1944) and Cabrera's (1961) concept of *albigularis*; however, the determination of priority and refinement of distributions require much museum-based research. Here we follow the taxonomy of Gardner and Patton (1976), Patton et al. (1990), and Márquez et al. (2000) in recognizing *O. auriventer, O. caracolus, O. devius, O. keaysi, O. levipes,* and *O. meridensis* as separate species (see those accounts). Gardner and Patton (1976) reassociated Cabrera's (1961) name-combination *O. a. boliviae* as a junior synonym of *O. nitidus*. Potential and actual distribution in Venezuela evaluated using GIS techniques by Anderson (2003b).

Oryzomys alfaroi (J. A. Allen, 1891). Bull. Am. Mus. Nat. Hist., 3:214.

COMMON NAME: Alfaro's Oryzomys.

TYPE LOCALITY: Costa Rica, Alajuela Prov., San Carlos.

DISTRIBUTION: Lowland to lower montane forests from S Tamaulipas and Oaxaca, México, through Middle America, to W Colombia and Ecuador.

STATUS: IUCN – Lower Risk (lc).

SYNONYMS: *agrestis* Goodwin, 1959; *dariensis* Goldman, 1915; *gloriaensis* Goodwin, 1956; *gracilis* Thomas, 1894; *incertus* J. A. Allen, 1908; *intagensis* Hershkovitz, 1940; *palatinus* Merriam, 1901; *palmirae* J. A. Allen, 1912.

COMMENTS: Goldman (1918) forged a broad definition of the species, expanded more so by Hall and Kelson (1959), that encompassed many forms previously treated as distinct (e.g., Merriam, 1901a). We recognize *chapmani, rhabdops,* and *saturatior* as species (revision in progress). *Oryzomys alfaroi* proper may be more closely related to *melanotis-rostratus* than to the *chapmani-saturatior* group. Karyotype reported by Haiduk et al. (1979) and Engstrom (1984).

Oryzomys angouya (Fischer, 1814). Zoognosia. Tabulis synopticis, illustrata, p. 71.

COMMON NAME: Angouya Oryzomys.

TYPE LOCALITY: Paraguay, Misiones Dept., east of the Río Paraguay, 2.7 km (by road) N San Antonio (as fixed by neotype designation by Musser et al., 1998).

DISTRIBUTION: SE Brazil, NE Argentina, and E Paraguay.

STATUS: IUCN – Lower Risk (lc) as *O. buccinatus* and *O. ratticeps*.

SYNONYMS: *angouya* (Desmarest, 1819); *anguya* (Rengger, 1830); *buccinatus* (Olfers, 1818); *leucogaster* (Wagner, 1845); *paraganus* Thomas, 1924; *ratticeps* (Hensel, 1872); *rex* (Winge, 1887); *tropicius* Thomas, 1924.

COMMENTS: *Oryzomys angouya* is the oldest available name based on Azara's (1801) "Rat troisieme, ou Rat Angouya," a species conventionally known as *O. buccinatus* and-or *O. ratticeps* (Avila-Pires, 1960a; Cabrera, 1961; Ellerman, 1941; Gyldenstolpe, 1932; Hershkovitz, 1959a; Moojen, 1952; Musser and Carleton, 1993; Thomas, 1924b). Morphology characterized, neotype designated, and synonyms attributed by Musser et al. (1998). Slight karyotypic variation, reflecting supernumerary chromosomes, reported for specimens from Rio de Janeiro, Brazil (2n = 80-82, FN = 88-90; L. G. Pereira et al., 2001).

Oryzomys auriventer Thomas, 1890. Ann. Mag. Nat. Hist., ser. 4, 7:379.
COMMON NAME: Golden-bellied Oryzomys.
TYPE LOCALITY: Ecuador, Tungurahua Prov., upper Río Pastaza, Mirador, 1500 m.
DISTRIBUTION: E Ecuador and N Perú.
STATUS: IUCN – Lower Risk (lc).
SYNONYMS: *nimbosus* Anthony, 1926.
COMMENTS: A subspecies of *O. albigularis* as recognized by Cabrera (1961); considered distinct by Gardner and Patton (1976).

Oryzomys balneator Thomas, 1900. Ann. Mag. Nat. Hist., ser. 7, 5:273.
COMMON NAME: Ecuadoran Oryzomys.
TYPE LOCALITY: Ecuador, Tungurahua Prov., upper Río Pastaza, 20 mi (32 km) E Baños, Mirador, 1500 m.
DISTRIBUTION: E and S Ecuador, N Perú; range extent uncertain.
STATUS: IUCN – Lower Risk (lc).
SYNONYMS: *hesperus* Anthony, 1924.
COMMENTS: Not a member of *Oryzomys* sensu stricto, but relationships obscure. In many features, *balneator* resembles species of *Microryzomys*: all are small, soft-furred oryzomyines with a long, scantily-haired tail; delicate cranium with rounded, featureless braincase and small ectotympanic bullae; narrow zygomatic plate with a shallow dorsal notch; complete carotid circulatory pattern (stapedial foramen and squamosal-alisphenoid groove present) and alisphenoid strut absent; brachyodont, pentalophodont molars with bifurcated anterocone, uppers with three roots and lowers two; and conspicuous capsular process on the mandible. Although cognizant of such similarities when composing the *Microryzomys* revision (Carleton and Musser, 1989), equally numerous differences persuaded us not to include *balneator* within the genus. The cranium of *balneator* is larger and differently proportioned than both *M. altissimus* and *M. minutus*; its interorbit is much broader and bears a slight post-orbital shelf; the zygoma are convergent anteriorly, not squared; incisive foramina are exceptionally short, terminating well anterior to the M1s; anteroconid is a single cone, not bifurcated as in *Microryzomys*; the hind feet are comparatively big, more elongate over the metatarsal section, digit V short, and the plantar pad conformation correspondingly shifted. Phylogenetic studies that include representatives of such genera as *Microryzomys*, *Oligoryzomys*, and *Neacomys* should profitably illuminate the relationships of *balneator* and its generic assignment.

Oryzomys bolivaris J. A. Allen, 1901. Bull. Am. Mus. Nat. Hist., 14:405.
COMMON NAME: Long-whiskered Oryzomys.
TYPE LOCALITY: Ecuador, Bolívar Prov., Porvenir, 1800 m.
DISTRIBUTION: Lowland evergreen to lower montane cloud forest from E Honduras, through E Nicaragua, Costa Rica and Panamá, to W Colombia and WC Ecuador (see Musser et al., 1998:Fig. 50); about sea level to 1800 m.
STATUS: IUCN – Lower Risk (lc).
SYNONYMS: *alleni* Goldman, 1915; *bombycinus* Goldman, 1912; *castaneus* J. A. Allen, 1901; *orinus* Pearson, 1939; *rivularis* J. A. Allen, 1901.
COMMENTS: Names conventionally applied to this species include *bombycinus*, as reviewed by Pine (1971), and *rivularis*, as discussed by Gardner and Patton (1976). Priority of *bolivaris* established by Musser et al. (1998), who summarized the morphological variation, geographic distribution, and synonymy of the species. Pine (1971) retained subspecies (under the name *bombycinus*), but Musser et al. (1998) judged intraspecific variation too insubstantial to circumscribe geographic races.

Oryzomys caracolus Thomas, 1914. Ann. Mag. Nat. Hist., ser. 8, 14:242.
COMMON NAME: Costa Central Oryzomys.
TYPE LOCALITY: Venezuela, Distrito Federal, Cerro del Avila, Galiparé, 6000 ft (1830 m).
DISTRIBUTION: Costa Central cordilleras, NC Venezuela (Aragua, Miranda, Distrito Federal).

COMMENTS: Although long ranked as a species (Ellerman, 1941; Gyldenstolpe, 1932), *caracolus* was considered a synonym of *O. capito* (= *O. megacephalus*) by Hershkovitz (1960) and Cabrera (1961) and of *O. albigularis* by Musser and Carleton (1993). Karyotypic and morphometric assessments among Venzuelan populations led Aguilera et al. (1995) and Márquez et al. (2000) to view *O. caracolus* and *O. meridensis* as distinct species. Rivas and Pefaur (1999*a*) obtained similar but not wholly concordant geographic patterns of morphometric differentiation and elected to maintain the taxa as subspecies of *O. albigularis*. These differing interpretations advise continued study of populations in the Venezuelan Andes and verification of their taxonomic status and distributions.

Oryzomys chapmani Thomas, 1898. Ann. Mag. Nat. Hist., ser. 7, 1:179.
   COMMON NAME: Chapman's Oryzomys.
   TYPE LOCALITY: México, Veracruz State, Jalapa, 4400 ft (1341 m).
   DISTRIBUTION: Cloud forest elevations of Sierra Madre Oriental (Tamaulipas to Veracruz), Sistema Montañosa (N Oaxaca), and Sierra Madre del Sur (S Oaxaca and Guerrero), México.
   STATUS: IUCN – Lower Risk (lc).
   SYNONYMS: *caudatus* Merriam, 1901; *dilutior* Merriam, 1901; *guerrerensis* Goldman, 1915; *huastecae* Dalquest, 1951.
   COMMENTS: Relegated to a subspecies of *O. alfaroi* by Goldman (1918). Goodwin (1969) recognized *caudatus* as distinct from *O. alfaroi* in N Oaxaca, but, as earlier arranged by Merriam (1901), *O. chapmani* has priority for this Mexican species. We view *O. saturatior* as the vicariant relative of *O. chapmani* (revision in progress). Karyotype reported, as *O. caudatus*, by Haiduk et al. (1979).

Oryzomys couesi (Alston, 1877). Proc. Zool. Soc. Lond., 1876:756 [1877].
   COMMON NAME: Coues' Oryzomys.
   TYPE LOCALITY: Guatemala, Alta Verapaz Dept., Cobán.
   DISTRIBUTION: Extreme S Texas, USA; México, excluding NC plateau region, south through most of Central America (see Platt et al., 2000, for Belize records), to NW Colombia (see Hershkovitz, 1987); including Jamaica, Isla Cozumel, and allopatric populations in S Baja California Sur (*peninsulae*) and WC Sonora (*lambi*).
   STATUS: IUCN – Lower Risk (lc).
   SYNONYMS: *albiventer* Merriam, 1901; *antillarum* Thomas, 1898; *apatelius* Elliot, 1904; *aquaticus* J. A. Allen, 1891; *aztecus* Merriam, 1901; *azuerensis* Bole, 1937; *bulleri* J. A. Allen, 1897; *cozumelae* Merriam, 1901; *crinitus* Merriam, 1901; *fulgens* Thomas, 1893; *gatunensis* Goldman, 1912; *goldmani* Merriam, 1901; *jalapae* J. A. Allen and Chapman, 1897; *lambi* Burt, 1934; *mexicanus* J. A. Allen, 1897; *molestus* Elliot, 1903; *peninsulae* Thomas, 1897; *peragrus* Merriam, 1901; *pinicola* A. Murie, 1932; *regillus* Goldman, 1915; *richardsoni* J. A. Allen, 1910; *richmondi* Merriam, 1901; *rufinus* Merriam, 1901; *rufus* Merriam, 1901; *teapensis* Merriam, 1901; *zygomaticus* Merriam, 1901.
   COMMENTS: Retained as a species by Goldman (1918) until Hall (1960) considered it only subspecifically distinct from *O. palustris*. Benson and Gehlbach (1979) returned *O. couesi* to specific status based on morphological contrasts with *O. p. texensis* in supposed area of intergradation; protein electrophoresis corroborates their lack of hybridization and genetic distinctiveness in sympatry (Schmidt and Engstrom, 1994). Karyotype reported by Benson and Gehlbach (1979) and Haiduk et al. (1979); morphometric comparisons with *O. palustris* by Humphrey and Setzer (1989). Alvarez-Castañeda (1994) discussed the indigenous status and possible extinction of the population (*peninsulae*) isolated near the tip of Baja California.
       Following Hall's (1960) example, other insular or localized subspecies—namely, *antillarum, azuerensis, cozumelae, fulgens, gatunensis,* and *peninsulae* (see Handley, 1966*a*; Hershkovitz, 1971; Jones and Lawlor, 1965)—were swept under *O. palustris*; they are here included in *O. couesi* because of geographic proximity. A composite of at least three species, the Mesoamerican populations referred to "*couesi*," as well as *O. dimidiatus* and *O. gorgasi*, should be critically reviewed (also see Sánchez et al., 2001).

Oryzomys curasoae McFarlane and Debrot, 2001. Caribbean J. Sci., 37:182.

COMMON NAME: Curaçao Oryzomys.

TYPE LOCALITY: Venezuela, Curaçao Isl, fissure 30 m below north face of Tafelberg Santa Barbara.

DISTRIBUTION: Known only from Curaçao Isl.

STATUS: Extinct; found in old owl-pellet deposits together with *Rattus rattus*, indicating that extinction transpired some time after initial European contact in 1499 (McFarlane and Debrot, 2001).

COMMENTS: Described as belonging to *Oryzomys*, subgenus *Oecomys*, but the long incisive foramina, zygomatic notch, and deep posterolateral palatal pits of the figured specimen indicate an example of *Oryzomys sensu stricto*; generic confirmation is required.

Oryzomys devius Bangs, 1902. Bull. Mus. Comp. Zool., 39:34.

COMMON NAME: Talamancan Oryzomys.

TYPE LOCALITY: Panamá, Chiriqui Prov., Volcán de Chiriqui, Boquete, 5000 ft (1524 m).

DISTRIBUTION: Highlands of Costa Rica and westernmost Panamá.

STATUS: IUCN – Lower Risk (lc).

COMMENTS: Maintained as a species until relegated to synonymy under *O. albigularis* by Handley (1966), emulating the treatment of South American *albigularis*-like forms by Cabrera (1961). Gardner (1983*a*), however, continued to rank *devius* as a species; its status must be evaluated within a revisionary context of the entire *albigularis* complex.

Oryzomys dimidiatus (Thomas, 1905). Ann. Mag. Nat. Hist., ser. 7, 15:586.

COMMON NAME: Nicaraguan Oryzomys.

TYPE LOCALITY: Nicaragua, Zelaya Dept., Río Escondido, 7 mi (11 km) below Rama.

DISTRIBUTION: SE Nicaragua.

STATUS: IUCN – Lower Risk (nt).

COMMENTS: Described as a species of *Nectomys* and included therein as *incertae sedis* by Hershkovitz (1944). Later designated as the type species of *Micronectomys*, subgenus *Oryzomys*, by Hershkovitz (1948*b*), who afterwards (1970) acknowledged the taxon as a *nomen nudum* (also see Pine and Wetzel, 1975:653). Morphological features resemble those of the *O. couesi-palustris* complex (Hershkovitz, 1970), a relationship supported by Sánchez et al. (2001). Only two specimens known (see Jones and Engstrom, 1986).

Oryzomys emmonsae Musser, Carleton, Brothers, and Gardner, 1998. Bull. Am. Mus. Nat. Hist., 236:233.

COMMON NAME: Emmons' Oryzomys.

TYPE LOCALITY: Brazil, Pará State, E bank Rio Xingu, 52 km SSW Altamira, below 100 m; 03°39'S, 52°22'W.

DISTRIBUTION: EC Brazil, south of the Rio Amazonas between the lower reaches of the Xingu and Tocantins Rivers.

COMMENTS: A member of the *O. nitidus* complex (Musser et al., 1998; Patton et al., 2000). Sympatry with *O. macconnelli* and *O. megacephalus* reported by Musser et al. (1998); range limits require further documentation.

Oryzomys galapagoensis (Waterhouse, 1839). Zool. Voy. H. M. S. "Beagle," Mammalia, p. 66.

COMMON NAME: Galapagos Oryzomys.

TYPE LOCALITY: Ecuador, Galápagos Isls, Chatham Isl.

DISTRIBUTION: San Cristobal (= Chatham) and Sante Fe (= Barrington) Isls.

STATUS: IUCN – Extinct as *O. g. galapagoensis*, Vulnerable as *G. g. bauri*.

SYNONYMS: *bauri* J. A. Allen, 1892.

COMMENTS: Cabrera (1961) assigned *bauri* as a synonym of *O. galapagoensis*, an action more fully documented by Patton and Hafner (1983). Related to *Oryzomys xanthaeolus* complex on mainland South America (Gardner and Patton, 1976; Patton and Hafner, 1983). Extirpated from San Cristobal Isl (*galapagoensis* proper) but populations (*O. g. bauri*) still inhabit Sante Fe Isl (see Dowler et al., 2000; Patton and Hafner, 1983).

Oryzomys gorgasi Hershkovitz, 1971. J. Mammal., 52:700.

COMMON NAME: Gorgas' Oryzomys.

TYPE LOCALITY: Colombia, Antioquia Dept., basin of Río Atrato, Loma Teguerre, just below and opposite Sautata (Chocó), 1 m; 7°54'N, 77°W.

DISTRIBUTION: Lowlands of NW Colombia and NW Venezuela as so far known.

STATUS: IUCN – Critically Endangered.

COMMENTS: Known only by the holotype until new records from Venezuela reported by Sánchez et al. (2001); those authors supplemented the taxon's description, argued its specific status in differential comparisons with *O. couesi, O. dimidiatus,* and *O. palustris,* and reaffirmed its membership in *Oryzomys sensu stricto.*

Oryzomys hammondi (Thomas, 1913). Ann. Mag. Nat. Hist., ser. 8, 12:570.

COMMON NAME: Hammond's Oryzomys.

TYPE LOCALITY: Ecuador, Pichincha Prov., Mindo, 4213 ft (1284 m).

DISTRIBUTION: NW Ecuador.

STATUS: IUCN – Lower Risk (lc).

COMMENTS: Described as a species of *Nectomys* and included therein as *incertae sedis* by Hershkovitz (1944). Later designated as the type species of *Macruroryzomys,* subgenus *Oryzomys,* by Hershkovitz (1948b), who afterwards (1970) acknowledged the taxon as a *nomen nudum* (also see Pine and Wetzel, 1975:653). Distributional limits poorly defined and affinities obscure, perhaps related to the extirpated Antillean form *Megalomys* (Ray, 1962).

Oryzomys keaysi J. A. Allen, 1900. Bull. Am. Mus. Nat. Hist., 13:225.

COMMON NAME: Keays's Oryzomys.

TYPE LOCALITY: Perú, Puno Dept., valley of upper Río Inambari, Inca Mines (= Santo Domingo), 6000 ft (1830 m).

DISTRIBUTION: Montane rainforest of E Peruvian Andes.

STATUS: IUCN – Lower Risk (lc).

SYNONYMS: *obtusirostris* J. A. Allen, 1900.

COMMENTS: A subspecies of *O. albigularis sensu* Cabrera (1961). Genetically divergent from *O. albigularis* and *O. levipes;* contiguously allopatric to the latter (see Patton et al., 1990).

Oryzomys lamia Thomas, 1901. Ann. Mag. Nat. Hist., ser. 7, 8:528.

COMMON NAME: Buffy-sided Oryzomys.

TYPE LOCALITY: Brazil, Minas Gerais State, along Rio Jordão, a small tributary of the Rio Paranaíba, 800 m.

DISTRIBUTION: Limited to Cerrado in Minas Gerais and Goiás states, C Brazil (see Bonvicino et al., 1998a).

STATUS: IUCN – Lower Risk (lc). Considered vulnerable by Bonvicino et al. (1998a).

COMMENTS: Retained as a species (e.g., Gyldenstolpe, 1932; Musser and Carleton, 1993) until reallocated to the synonymy of *O. russatus* by Musser et al. (1998). However, clear morphometric differentiation (Musser et al., 1998) and divergent karyotype (2n = 58, FN = 82; Bonvicino et al., 1998a) from *O. russatus* (2n = 80, FN = 86; Silva et al., 2000) recommend specific recognition. Allied to *O. intermedius* (= *O. russatus*) according to Thomas (1901b) and with the *O. nitidus* group *sensu* Musser et al. (1998).

Oryzomys laticeps (Lund, 1840). Preprint of 1841 K. Dansk. Vidensk. Selskav Afhandl., Kjobenhavn, 8:279.

COMMON NAME: Atlantic Forest Oryzomys.

TYPE LOCALITY: Brazil, Minas Gerais State, valley of the Rio das Velhas, Lagoa Santa.

DISTRIBUTION: Atlantic Forest region of SE Brazil.

STATUS: IUCN – Lower Risk (nt) as *O. oniscus.*

SYNONYMS: *oniscus* Thomas, 1904; *saltator* (Winge, 1887).

COMMENTS: Lectotype selected, synonyms identified, and morphological separation from *O. megacephalus* and related oryzomyines elaborated by Musser et al. (1998); includes *oniscus,* previously recognized as a separate species (e.g., Gyldenstolpe, 1932; Moojen, 1952; Musser and Carleton, 1993). Cladistic interpretation of cytochrome *b* data reveals a sister-species relationship between *O. laticeps* and *O. perenensis* (Patton et al., 2000).

Oryzomys legatus Thomas, 1925. Ann. Mag. Nat. Hist., ser. 9, 15:577.
    COMMON NAME: Tarija Oryzomys.
    TYPE LOCALITY: Bolivia, Tarija Dept., Caraparí, 1000 m.
    DISTRIBUTION: E Andean slopes of SC Bolivia and NW Argentina.
    STATUS: IUCN – Lower Risk (lc).
    COMMENTS: Considered conspecific with *O. capito* (= *O. megacephalus*) (Cabrera, 1961; Hershkovitz, 1960), viewed as a probable synonym of *O. nitidus* (Gardner and Patton, 1976), or retained as a species (Gyldenstolpe, 1932; Mares et al., 1989b; Massoia, 1975; Musser and Carleton, 1993). Musser et al. (1998) acknowledged the morphological distinctiveness of *legatus* but ultimately arranged it within a broadly defined *O. russatus*, a related form of the *nitidus* species group. This proposed relationship finds no support from trees derived from cytochrome *b* sequences (Patton et al., 2000), in which specimens of *legatus* fall among those of *O. nitidus* proper, not with *O. russatus* from SE Brazil. In view of the pronounced morphometric differentiation of *legatus* and *nitidus* in SC Bolivia (Musser et al., 1998:208-218), we follow Patton et al. (2000) in regarding both as species and stress that the lack of reciprocal monophyly between the respective data sets invites further study.

Oryzomys levipes Thomas, 1902. Ann. Mag. Nat. Hist., ser. 7, 9:129.
    COMMON NAME: Nimble-footed Oryzomys.
    TYPE LOCALITY: Perú, Puno Dept., Limbane, 2200 m.
    DISTRIBUTION: Cloud forest of SE Perú to WC Bolivia (see Anderson, 1997:Fig. 696).
    STATUS: IUCN – Lower Risk (nt).
    COMMENTS: Synonymized under *O. albigularis keaysi* by Cabrera (1961). Genetically divergent from and altitudinally parapatric to *O. keaysi* in Perú (see Patton et al., 1990).

Oryzomys macconnelli Thomas, 1910. Ann. Mag. Nat. Hist., ser. 8, 6:186.
    COMMON NAME: MacConnell's Oryzomys.
    TYPE LOCALITY: Guyana, Demerara Dist., Supenaam River, a tributary of the Lower Essequibo.
    DISTRIBUTION: Tropical evergreen rain forest, sea level to 1524 m, of SC Colombia, E Ecuador and Perú, eastwards to S Venezuela, Guianas, and N Brazil (see Musser et al., 1998: Fig. 78).
    STATUS: IUCN – Lower Risk (lc).
    SYNONYMS: *incertus* J. A. Allen, 1913; *mureliae* J. A. Allen, 1915 [replacement name for *incertus*].
    COMMENTS: Included in *O. capito* (= *O. megacephalus*) by Hershkovitz (1960) but retained as species by Cabrera (1961). Morphological and karyotypic basis for species recognition reinforced by Pine (1973), Gardner and Patton (1976), and Husson (1978) (and see comments under *O. megacephalus*). Musser et al. (1998) contributed a synopsis of its distribution, chromosomal variation, and morphological discrimination from related oryzomyines; differences in karyotype and cytochrome b haplotypes (Patton et al., 2000) intimate that *macconnelli* as presently defined is a composite of two species. Phylogeographic relationships among populations of *O. macconnelli* and between *O. macconnelli* and other *nitidus*-group species examined by Costa (2003).

Oryzomys maracajuensis Langguth and Bonvicino, 2002. Arq. Mus. Nac., Rio de Janeiro, 60:292.
    COMMON NAME: Maracaju Oryzomys.
    TYPE LOCALITY: Brazil, Matto Grosso do Sul State, Municipality Maracaju, Fazenda da Mata; 21°38'S, 55°09'W.
    DISTRIBUTION: Definitely known only from the type locality, S Brazil, but probably occurs in E Paraguay; limits unknown.
    COMMENTS: *O. subflavus* species group. Similar to *O. subflavus* but larger, with longer fur and different karyotype (2n = 56, FN = 58); collected sympatrically with *O. scotti*.

Oryzomys marinhus Bonvicino, 2003. Mamm. Biol., 68:84.
    COMMON NAME: Marinho's Oryzomys.
    TYPE LOCALITY: Brazil, Goiás State, Jaborandi Municipality, Fazenda Sertão do Formoso (formerly Fazenda Jucurutu), 775 m; 14°40'20"S, 45°49'71"W.

DISTRIBUTION: Known only from the type locality in the Cerrado, SC Brazil.

COMMENTS: *O. subflavus* species group. Size larger than *O. subflavus*, with better developed supraorbital ridges and distinctive karyotype (2n = 56, FN = 54).

Oryzomys megacephalus (Fischer, 1814). Zoognosia. Tabulis synopticis, illustrata, p. 71.

COMMON NAME: Azara's Broad-headed Oryzomys.

TYPE LOCALITY: Paraguay, Canendiyu Dept., east of Río Paraguay, 13.3 km (by road) N Curuguaty, 255 m; 24°31'S, 55°42'W (as fixed by neotype designation by Musser et al., 1998:252).

DISTRIBUTION: Lowland tropical rainforests of E Amazonia—E and S Venezuela, Guianas, N and C Brazil, E Paraguay; including Trinidad; western limits indeterminate.

STATUS: IUCN – Lower Risk (lc).

SYNONYMS: *capito* (Olfers, 1818); *cephalotes* (Desmarest 1819); *goeldi* Thomas, 1897; *modestus* J. A. Allen, 1899; *velutinus* J. A. Allen and Chapman, 1893.

COMMENTS: *Oryzomys megacephalus* (Fischer, 1814) is the oldest available name based on Azara's (1801) "Rat second, ou Rat a grosse tete," a form conventionally known as *O. capito* (Olfers, 1818) in the middle 1900s (e.g., Cabrera, 1961; Hall, 1981; Hershkovitz, 1960). Taxonomic understanding of this species was obscured by the infelicitous footnote of Hershkovitz (1960:544), who suggested the synonymy of some 20 taxa under *O. capito*, a passing opinion formally expanded by Cabrera (1961) and followed by other authors (e.g., Hall, 1981; Handley, 1966a, 1976). The extremeness of this viewpoint was exposed by the karyotypic study of Gardner and Patton (1976), a watershed paper that has sparked critical review, still on-going, of this diverse species complex.

As a result, the following forms considered synonyms by Hershkovitz (1960) and/or Cabrera (1961) are now acknowledged as distinct species or as synonyms of other species (see separate accounts): *O. bolivaris, boliviae* (= *O. nitidus*), *castaneus* (= *O. bolivaris*), *O. caracolus, intermedius* (= *O. russatus*), *O. laticeps, O. legatus, magdalenae* (= *O. talamancae*), *O. macconnelli, medius* (= *O. talamancae*), *mollipilosus* (= *O. talamancae*), *O. nitidus, oniscus* (= *O. laticeps*), *O. perenensis, rivularis* (= *O. bolivaris*), *sylvaticus* (= *O. talamancae*), *O. talamancae*, and *O. yunganus*. Nomenclatural priority, neotype designation, synonyms, and morphological definition of *O. megacephalus* proper addressed by Musser et al. (1998); phylogeographic relationships and other morphometric comparisons supplied by Patton et al. (2000); banded chromosomal comparisons with related species presented by Volobouev and Aniskin (2000). Sister species to *O. laticeps* (Bonvicino and Martins Moreira, 2001; Costa, 2003) or the *O. laticeps-O. perenensis* clade (Patton et al., 2000) based on cytochrome *b* sequence data.

Oryzomys melanotis Thomas, 1893. Ann. Mag. Nat. Hist., ser. 6, 11:404.

COMMON NAME: Black-eared Oryzomys.

TYPE LOCALITY: México, Jalisco State, Mineral San Sebastián.

DISTRIBUTION: Low to intermediate elevations of W México, from S Sinaloa to SW Oaxaca.

STATUS: IUCN – Data Deficient.

SYNONYMS: *colimensis* Goldman, 1918.

COMMENTS: Revised by Goldman (1918), who recognized *melanotis* and *rostratus* as separate species within a *melanotis* species group. Hooper (1953) viewed the geographic complementarity of the two as a subspecific pattern, and so they were recognized by Hall and Kelson (1959) and Hall (1981). Engstrom (1984) returned *rostratus* to a separate species based on a robust variety of data interpreted within a zoogeographic context.

Oryzomys meridensis Thomas, 1894. Ann. Mag. Nat. Hist., ser. 6, 14:351.

COMMON NAME: Mérida Oryzomys.

TYPE LOCALITY: Venezuela, Mérida State, Mérida.

DISTRIBUTION: Sierra de Mérida, W Venezuela.

COMMENTS: Reallocated as a subspecies of *O. albigularis* by Cabrera (1961), but extensive chromosomal rearrangements suggest specific status (Aguilera et al., 1995). Karyotypic and morphometric differentiation from Venzuelan populations identified as *O. caracolus* assessed by Aguilera et al. (1995) and Márquez et al. (2000); also see remarks under the latter. Nongeographic variation in crania and dentitions evaluated (as *O. albigularis*) by Rivas and Pefaur (1999*b*).

Oryzomys nelsoni Merriam, 1898. Proc. Biol. Soc. Wash., 12:15.
COMMON NAME: Nelson's Oryzomys.
TYPE LOCALITY: México, Nayarit State, María Madre Isl.
DISTRIBUTION: Known only from the type locality.
STATUS: IUCN – Extinct; known only by the 4 specimens of the type series and presumed to be extinct (see Wilson, 1991).
COMMENTS: Allied to *O. couesi*, under which Hershkovitz (1971) listed it as a subspecies. Unquestionably a species distinct from mainland *O. couesi* and *O. palustris* as others have recognized (Goldman, 1918; Hall, 1981). See Alvarez-Castañeda and Méndez (2003, Mammalian Species, 735).

Oryzomys nitidus (Thomas, 1884). Proc. Zool. Soc. Lond., 1884:452.
COMMON NAME: Elegant Oryzomys.
TYPE LOCALITY: Perú, Junín Dept., valley of Río Tulumayo, 10 km S San Ramón, Amable Maria, 2000 ft (as located by Gardner and Patton, 1976:42).
DISTRIBUTION: Lowland rain forest and Andean foothills, 50-1985 m, of E Perú, E Bolivia, and WC Brazil (Acre and Mato Grosso) (See Musser et al., 1998:Fig. 79; Patton et al., 2000).
STATUS: IUCN – Lower Risk (lc).
SYNONYMS: *boliviae* Thomas, 1901.
COMMENTS: A species formerly included in *O. capito* (= *O. megacephalus*) by Cabrera (1961) or affiliated with *O. alfaroi* by Hershkovitz (1966*c*). Gardner and Patton (1976) provided karyotypic and morphological evidence justifying the specific distinction of *O. nitidus* from both and listed *boliviae*, *intermedius*, and *legatus* as likely synonyms. Distribution, synonymy, and morphological recognition clarified by Musser et al. (1998), who arrayed *O. nitidus* with *O. emmonsae*, *O. macconnelli*, and *O. russatus* as a species group; their monophyletic association is generally supported by cytochrome *b* data but cladistic details differ depending on the study (Bonvicino and Martins Moreira, 2001; Patton et al., 2000). Detailed karyotypic comparisons of *O. nitidus* with other *nitidus*-group species presented by Silva et al. (2000); limited resemblance of homologous chromosomal bands indicates that *O. nitidus* is distantly related to the *O. megacephalus*-*O. yunganus* assemblage (Volobouev and Aniskin, 2000).

Oryzomys palustris (Harlan, 1837). Am. J. Sci., 31:385.
COMMON NAME: Marsh Oryzomys.
TYPE LOCALITY: USA, New Jersey, Salem Co., "Fastland," near Salem.
DISTRIBUTION: SE USA—SE Kansas, S Illinois, and S New Jersey, south, avoiding the Appalachian Mtns, to E and coastal Texas, along the Gulf coast, and into peninsular Florida and lower Florida Keys.
STATUS: U.S. ESA—Endangered in the lower Florida Keys (west of the Seven Mile Bridge) as *O. p. natator*; IUCN – Data Deficient as *O. p. natator*, otherwise Lower Risk (lc).
SYNONYMS: *argentatus* Spitzer and Lazell 1978; *coloratus* Bangs, 1898; *natator* Chapman, 1893; *oryzivora* (Bachman, 1854); *planirostris* Hamilton, 1955; *sanibeli* Hamilton, 1955; *texensis* J. A. Allen, 1894.
COMMENTS: Formerly included *couesi* and related forms as subspecies (see account of *O. couesi*). Geographic variation evaluated by Humphrey and Setzer (1989), who acknowledged the nominate and only one other subspecies, *O. p. natator*; Whitaker and Hamilton (1998), on the other hand, regarded the island taxa *planirostris* and *sanibeli* as valid subspecies and viewed *natator* as indistinguishable from nominate *palustris*, observing their philosophical emphasis of overwater gaps as "primary isolating mechanisms" rather than renewed data analyses. Overwater dispersal capability studied by Forys and Dueser (1993) in the context of understanding differentiation of populations on near-shore islands. Recoveries from sub-Recent archeological and cave excavations document the species substantially to the north (E Nebraska to SW Pennsylvania) of its present range limits (see Graham and Lundelius, 1994; Richards, 1980).

The form *argentatus* was described as a species from the lower Florida Keys (Goodyear and Lazell, 1986; Spitzer and Lazell, 1978), later synonymized under *O. palustris natator* (Humphrey and Setzer, 1989), and pointedly reinstated as a species (Goodyear, 1991). Its status continues to oscillate in faunal works (or reviews thereof), unilluminated by fresh data or reanalysis (e.g., a species, as per Lazell, 1993, and Whitaker and Hamilton, 1998; not a species, as per Brown, 1997, and Rose and Dueser, 1999). We continue to follow Humphrey and Setzer (1989) because of their better measurement protocol, larger sample sizes, broader geographic and taxonomic scope, stronger analytical rigor that accounts sources of nongeographic variation, and firmer grasp of the biological and paleogeographic contexts. As an anecdotal aside, based on few specimens reviewed and not to be confused with scientific process, we cannot separate the crania of *argentatus* from those of *O. palustris* with the same confidence that one can sort *O. palustris* from *O. couesi* or *O. nelsoni*. In light of the subspecific endemism displayed among other Florida-Key mammals, the status of *argentatus* merits further study drawing upon genetic evidence; the investigatory tools and data sources appropriately sensitive to this taxonomic question are at hand and merely await application. See Wolfe (1982, Mammalian Species, 176).

Oryzomys perenensis J. A. Allen, 1901. Bull. Am. Mus. Nat. Hist., 14:406.
   COMMON NAME: Western Amazonian Oryzomys.
   TYPE LOCALITY: Perú, Junín Dept., Perené, 800 m.
   DISTRIBUTION: E Andean foothills and western margins of the Amazon Basin, including C and SE Colombia, E Ecuador, E Perú, E Bolivia, and WC Brazil (see Patton et al., 2000:Fig. 96); eastern limits indeterminate.
   COMMENTS: Morphometric and karyotypic differentiation of western Amazonian populations (*perenensis*) from eastern ones (*O. megacephalus sensu stricto*) recognized by Musser et al. (1998), who doggedly elected to retain the former as a synonym of the latter pending vouchered demonstration of sympatry. Genetic divergence persuasively documented by Patton et al. (2000), who reinstated *O. perenensis* as a species, as have others (Bonvicino and Martins Moreira, 2001; Voss et al., 2001). Eastern distributional limits, however, remain obscure and undocumented by specimens; given the morphological ambiguity of certain series from middle Amazonas (see Musser et al., 1998), the stature and geographic limits of *perenensis* should continue to be refined with regard to western populations of *O. megacephalus*.

Oryzomys polius Osgood, 1913. Field Mus. Nat. Hist. Publ., Zool. Ser., 10:97.
   COMMON NAME: Marañon Oryzomys.
   TYPE LOCALITY: Perú, Amazonas Dept., mountains east of Balsas, Tambo Carrizal, 5000 ft (1524 m).
   DISTRIBUTION: NC Perú (Amazonas, Cajamarca, Piura).
   STATUS: IUCN – Lower Risk (lc).
   COMMENTS: Affinities obscure; compared to *O. xanthaeolus* by Osgood (1913). A valid and very distinctive species, all locality records so far confined to the dry lowlands of the upper Río Marañon basin, east of the N Peruvian Andes.

Oryzomys rhabdops Merriam, 1901. Proc. Wash. Acad. Sci., 3:292.
   COMMON NAME: Highland Oryzomys.
   TYPE LOCALITY: Guatemala, Quezaltenango Dept., Calel, 10,000 ft (3048 m).
   DISTRIBUTION: Highlands of S Chiapas, México, and C Guatemala.
   STATUS: IUCN – Lower Risk (lc).
   SYNONYMS: *angusticeps* Merriam, 1901.
   COMMENTS: Allocated to subspecies of *O. alfaroi* by Goldman (1918), as conventionally observed in later systematic works (e.g., Hall, 1981). Merriam's form is a valid species, having a much longer tail, longer pelage, and larger cranium than true *O. alfaroi*.

Oryzomys rostratus Merriam, 1901. Proc. Wash. Acad. Sci., 3:293.
   COMMON NAME: Long-nosed Oryzomys.
   TYPE LOCALITY: México, Puebla State, Metlaltoyuca.
   DISTRIBUTION: Deciduous and evergreen tropical forests from C Tamaulipas to Oaxaca and

Yucatán Peninsula, México, through Guatemala, El Salvador, and Honduras (see Lee and Bradley, 1992), to S Nicaragua (see Jones and Engstrom, 1986).

STATUS: IUCN – Lower Risk (lc).

SYNONYMS: *carrorum* Lawrence, 1947; *megadon* Merriam, 1901; *salvadorensis* Felton, 1958; *yucatanensis* Merriam, 1901.

COMMENTS: Formerly considered a subspecies of *O. melanotis* (Hall, 1981; Hooper, 1953), Engstrom (1984) effectively argued the specific distinctiveness of *O. rostratus* (also see account of *O. melanotis*). Homogeneity of populations assigned to *O. rostratus* warrants additional study.

Oryzomys russatus (Wagner, 1848). Abh. Math.-Phys. Classe K. B. Akad. Wiss. (Munchen), 5:312.

COMMON NAME: Russet Oryzomys.

TYPE LOCALITY: Brazil, São Paulo State, Ipanema.

DISTRIBUTION: E Paraguay, SE Brazil (Bahia to Rio Grande do Sul), and NE Argentina (as per Massoia, 1975; Musser et al., 1998).

STATUS: IUCN – Lower Risk (lc) as *O. intermedius* and *O. kelloggi*.

SYNONYMS: *coronatus* (Winge, 1887); *intermedia* (Leche, 1886); *kelloggi* Avila-Pires, 1959; *moojeni* Avila-Pires, 1959; *physodes* (Brants, 1827) [preoccupied by *Mus physodes* Olfers, 1818].

COMMENTS: The oldest available name for the form conventionally identified in the literature as *intermedius*, either used as a subspecies of *O. capito* (= *O. megacephalus*) (Cabrera, 1961) or as a distinct species (Gardner and Patton, 1976; Moojen, 1952; Musser and Carleton, 1993). Nomenclatural priority of *russatus* Wagner, lectotype designation, morphological definition, and distribution presented by Musser et al. (1998). A member of the *O. nitidus* group *sensu* Musser et al. (1998); most closely related to *O. emmonsae* and *O. macconnelli* based on evaluations of cytochrome *b* sequence data (Bonvicino and Martin Moreira, 2001; Patton et al., 2000). Extensive karyotypic variation advises that other species remain to be identified among the populations referred to *russatus* (see Silva et al., 2000).

Oryzomys saturatior Merriam, 1901. Proc. Wash. Acad. Sci., 3:290.

COMMON NAME: Cloud Forest Oryzomys.

TYPE LOCALITY: México, Chiapas State, Tumbalá, 5000 ft (1524 m).

DISTRIBUTION: Cloud forest elevations from S Oaxaca and Chiapas, México, through Guatemala, Honduras, and El Salvador, to NC Nicaragua.

STATUS: IUCN – Lower Risk (lc).

SYNONYMS: *hylocetes* Merriam, 1901.

COMMENTS: Described as a subspecies of *O. chapmani* (Merriam, 1901a) and later retained as a subspecies of *O. alfaroi* (Goldman, 1918; Hall and Kelson, 1959). The divergence of *O. chapmani* and *O. saturatior*, whose populations are confined to wet montane forest separated by the Isthmus of Tehuantepec, suggests that they are sister species.

Oryzomys scotti Langguth and Bonvicino, 2002. Arq. Mus. Nac., Rio de Janeiro, 60:290.

COMMON NAME: Lindbergh's Oryzomys.

TYPE LOCALITY: Brazil, Goiás State, Municipality Corumbá de Goiás, Morro dos Cabelundos; 15°54'S, 48°48'W.

DISTRIBUTION: Cerrado of Matto Grosso do Sul to Goiás and W Minas Gerais, C Brazil (Langguth and Bonvicino, 2002:Fig. 1).

COMMENTS: *O. subflavus* species group. Similar to *O. subflavus* but size smaller, dorsal-ventral contrast less defined, and karyotype different (2n = 58, FN = 70-72); reported as "*O. subflavus* variant 4" by Bonvicino et al. (1999).

Oryzomys seuanezi Weksler, Geise, and Cerqueira, 1999. Zool. Jour. Linnean Soc., 125:454.

COMMON NAME: Seuánez's Oryzomys.

TYPE LOCALITY: Brazil, Rio de Janeiro State, Casimiro de Abreu Municipality, Fazenda União, 50 m; 22°25'S, 42°02'W.

DISTRIBUTION: Atlantic Forest, SE Brazil.

COMMENTS: Morphometric and karyological comparisons with other Brazilian members of the *O. megacephalus* complex provided by Weksler et al. (1999); a sister species to

O. *oniscus* (or *O. laticeps sensu* Musser et al., 1998). The level of differentiation of this species from Lund's (1840) *laticeps*, identified by Musser et al. (1998) as the senior synonym for the large *megacephalus*-like form of the Atlantic Forest region, deserves further examination.

Oryzomys subflavus (Wagner, 1842). Arch. Naturgesh., 8(1):362.
COMMON NAME: Flavescent Oryzomys.
TYPE LOCALITY: Brazil, Minas Gerais State, Lagoa Santa (as restricted by Cabrera, 1961:396).
DISTRIBUTION: Lowland forests of E Brazil (Ceará and Río Grande do Norte southwestwardly to São Paulo), E Bolivia (less than 500 m, per Anderson, 1997), and extreme SE Peru (as *buccinatus*, per Pacheco et al., 1995); limits uncertain.
STATUS: IUCN – Lower Risk (lc).
SYNONYMS: *vulpinoides* (Schinz, 1845) [a renaming of *vulpinus* Lund, preoccupied by *Mus vulpinus* Brants, 1827]; *vulpinus* (Lund, 1840).
COMMENTS: Morphology characterized, holotype identified, and synonyms attributed by Musser et al. (1998); definition further restricted and karyotype (2n = 54, FN = 62), based on topotypical specimens, reported by Langguth and Bonvicino (2002). Hershkovitz (1960) placed *catherinae* Thomas, and *rex* Thomas, as synonyms of *O. subflavus*; however, both are forms of *Oecomys* (see that account). The supposed distribution of *O. subflavus* in the Guianas (e.g., Honacki et al., 1982) issued from this erroneous allocation of names (e.g., not recorded by Husson, 1978, or Voss et al., 2001). Age and secondary sexual variation in a large sample (Perambuco, Brazil) investigated by Brandt and Pessóa (1994). Extensive karyotypic (2n = 46-58, FN = 56-72) and molecular variation indicates the presence of several species within the nominal taxon (Andrades-Miranda et al., 2002*b*; Bonvicino et al., 1999; Bonvicino and Martins Moreira, 2001:Fig. 4, which see for geographic occurrences of these genetic entities), and documentation of this diversity initiated by the descriptive studies of Langguth and Bonvicino (2002) and Bonvicino (2003); see accounts of *O. maracajuensis, O. marinhus,* and *O. scotti*. The distribution of *O. subflavus* proper may conform to the Atlantic Forest region and along gallery forests into the Cerrado; with description of the new species, status of populations allocated to *subflavus* outside of this region (e.g., Anderson, 1997; Gamarra de Fox and Martin, 1996; Pacheco et al., 1995) should be reassessed.

Oryzomys talamancae J. A. Allen, 1891. Proc. U. S. Natl. Mus., 14:193.
COMMON NAME: Transandean Oryzomys.
TYPE LOCALITY: Costa Rica, Limón Prov., Talamanca.
DISTRIBUTION: Forested lowlands, sea level to 1525 m, from NW Costa Rica, through Panamá, to W and NC Colombia, W Ecuador, and N Venezuela (see Musser et al., 1998:Fig. 66).
STATUS: IUCN – Lower Risk (lc).
SYNONYMS: *carrikeri* J. A. Allen, 1908; *magdalenae* J. A. Allen, 1899; *medius* Robinson and Lyon, 1901; *mollipilosus* J. A. Allen, 1899; *panamensis* Thomas, 1901; *sylvaticus* Thomas, 1900; *villosus* J. A. Allen, 1899.
COMMENTS: Considered a junior synonym of *O. capito* (= *O. megacephalus*) (Hershkovitz, 1960), an opinion that led to its arrangement as a subspecies thereof (Hall, 1981; Handley, 1966*a*). Species status maintained by Gardner (1983*a*) and reasserted based on G-banding comparisons with *capito* (= *O. megacephalus*) by Pérez-Zapata and Aguilera (1996). Morphological recognition, distributional limits, karyotypic variation, and synonymies amplified by Musser and Williams (1985) and Musser et al. (1998).

Oryzomys tatei Musser, Carleton, Brothers, and Gardner, 1998. Bull. Am. Mus. Nat. Hist., 236:100.
COMMON NAME: Tate's Oryzomys.
TYPE LOCALITY: Ecuador, Tungurahua Prov., Palmera, 4000 ft (1219 m).
DISTRIBUTION: Known from only three localities along E Andean foothills, 1128-1524 m, of C Ecuador.
COMMENTS: A close relative of *O. yunganus* (see Musser et al., 1998).

Oryzomys xanthaeolus Thomas, 1894. Ann. Mag. Nat. Hist., ser. 6, 14:354.
COMMON NAME: Yellowish Oryzomys.
TYPE LOCALITY: Perú, Piura Dept., Tumbez.
DISTRIBUTION: WC Ecuador to W Perú.
STATUS: IUCN – Lower Risk (lc).
SYNONYMS: *baroni* J. A. Allen, 1897; *ica* Osgood, 1944.
COMMENTS: Probable mainland relative of *O. galapagoensis* (Gardner and Patton, 1976;
    Patton and Hafner, 1983). Nongeographic variation in sample from NW Perú
    evaluated by Lavrenchenko (1994), who documented pronounced sexual dimorphism
    and commented on its evolutionary significance. Stomach morphology and diet
    reported by Guabloche et al. (2002) for Peruvian population. Possible species status of
    ica deserves evaluation.

Oryzomys yunganus Thomas, 1902. Ann. Mag. Nat. Hist., ser. 7, 9:130.
COMMON NAME: Amazonian Oryzomys.
TYPE LOCALITY: Bolivia, Cochabamba Dept., Charuplaya, 1350 m.
DISTRIBUTION: Evergreen rain forest of Amazonia, sea level-2000 m, from Guianas and
    S Venezuela (Ochoa et al., 1988) to C Brazil, including lowlands and Andean foothills
    of C Colombia, E Ecuador, E Perú, and N Bolivia (see Musser et al., 1998:Fig. 14).
STATUS: IUCN – Lower Risk (lc).
COMMENTS: Considered a subspecies of *O. capito* (= *O. megacephalus*) by Cabrera (1961);
    specific differences illuminated by Gardner and Patton (1976) and sustained by strong
    genetic divergence therefrom (Patton et al., 2000). Morphological definition,
    localization of type locality, and distributional limits provided by Musser et al. (1998),
    who noted substantial morphometric differentiation between western and eastern
    populations that may reflect separate species; also see Voss et al. (2001) for additional
    comments on this variation and for other comparisons with *O. megacephalus*.

Oxymycterus Waterhouse, 1837. Proc. Zool. Soc. Lond., 1837:21.
TYPE SPECIES: *Mus nasutus* Waterhouse, 1837.
COMMENTS: Akodontini. Studies of electromorphic and gene-sequence variation decisively
    support the monophyly of *Oxymycterus* (minus *iheringi*, see *Brucepattersonius*) and its tribal
    placement as an akodontine (D'Elía, 2003; D'Elía et al., 2003; Hinojosa et al., 1987;
    Hoffmann et al., 2002; Patton et al., 1989; Smith and Patton, 1993, 1999); according to
    cytochrome *b* data, a sister group to the *Necromys-Thaptomys-Akodon* clade (Smith and
    Patton, 1999) or to *Juscelinomys* (D'Elía, 2003; D'Elía et al., 2003). Studies addressing the
    morphological definition of *Oxymycterus* include Carleton (1973), Hooper and Musser
    (1964a), Hinojosa et al. (1987), Vorontsov (1967), and Voss and Linzey (1981). Diploid
    number of species thus far karyotyped (ca. 8) have proven to be identical (2n = 54,
    FN = 64) (Bonvicino et al., 1998b; L. G. Pereira et al., 2001; Vitullo et al., 1986).
    Specific taxonomy needs full published revision, including emended diagnoses of
    valid taxa, objective criteria for attributed synonyms, and vouchered documentation of
    ranges. Nominal species here recognized discordantly assimilate Cabrera's (1961)
    classification, Reig's (1987) provisional tally of valid forms (based on his examination of
    BMNH types), and recent descriptions by Hinojosa et al. (1987), Hershkovitz (1994,
    1998), and Hoffmann et al. (2002). Generic characters, nomenclatural foundation of
    described forms, and natural history reports were compiled by Hershkovitz (1994).

Oxymycterus akodontius Thomas, 1921. Ann. Mag. Nat. Hist., ser. 9, 8:615.
COMMON NAME: Akodont-like Hocicudo.
TYPE LOCALITY: Argentina, Jujuy Prov., 20 km E Tilcara, Higuerilla, 2000 m.
DISTRIBUTION: NW Argentina.
STATUS: IUCN – Lower Risk (lc).
COMMENTS: Perhaps conspecific with *O. paramensis* according to Cabrera (1961), Vitullo
    et al. (1986), and Reig (1987); retained as species by Galliari et al. (1996).

Oxymycterus amazonicus Hershkovitz, 1994. Fieldiana Zool., N.S., 79:23.
COMMON NAME: Amazonian Hocicudo.
TYPE LOCALITY: Brazil, Pará State, right bank lower Rio Tapajós, Fordlândia; 03°40'S, 55°30'W.

DISTRIBUTION: Lower Amazon Basin, south of the Rio Amazonas between the Rios Tocantins and Madeira, C Brazil, as least as far south as NW Mato Grosso (per additional localities reported by Musser et al., 1998:239).

COMMENTS: Considered a small size-class species, principally compared with *O. nasutus*; sister species to *O. delator* according to phylogenetic analysis of cytochrome *b* data (Hoffmann et al., 2002).

Oxymycterus angularis Thomas, 1909. Ann. Mag. Nat. Hist., ser. 8, 4:237.

COMMON NAME: Angular Hocicudo.
TYPE LOCALITY: Brazil, Pernambuco State, São Lourenço, 30 m.
DISTRIBUTION: Extreme E Brazil (Alagoas, Ceará, Pernambuco).
STATUS: IUCN – Lower Risk (lc).
COMMENTS: A large form maintained as a species since its description (e.g., Cabrera, 1961; Moojen, 1952).

Oxymycterus caparoae Hershkovitz, 1998. Bonn. Zool. Beitr., 47:244.

COMMON NAME: Mt Caparaó Hocicudo.
TYPE LOCALITY: Brazil, Minas Gerais State, Parque Nacional de Caparaó, Arrozal, 2400 m.
DISTRIBUTION: Humid montane forest and secondary vegetation, 1800-2700 m, E Minas Gerais and Espírito Santo to Rio de Janeiro, SE Brazil (see Bonvicino et al., 1998*b*).
COMMENTS: A medium species similar to *O. paramensis* and *O. nasutus*; level of differentiation from latter should be reevaluated in the context of a full generic revision.

Oxymycterus dasytrichus (Schinz, 1821). Cuvier's Das Thierreich, 1:288.

COMMON NAME: Atlantic Forest Hocicudo.
TYPE LOCALITY: Brazil, Bahia State, Rio Mucurí (as restricted by Cabrera, 1961:468).
DISTRIBUTION: Atlantic Forest region of SE Brazil (Bahia to São Paulo; see Hoffmann et al., 2002, L. G. Pereira et al., 2001).
SYNONYMS: *dasytrichos* (Wied-Neuwied, 1826); *dosytrichos* (Schinz, 1821); *rostellatus* (Wagner, 1842).
COMMENTS: Original spelling as *Mus dosytrichos*, emended to *dasytrichus* by Tate (1932*g*:17). Arranged as a full synonym of *O. rufus* by Ellerman (1941) or recognized subspecies by Cabrera (1961), and so associated by Musser and Carleton (1993). Fonseca et al. (1996) and L. G. Pereira et al. (2001) listed *dasytrichus* as species, and confirmatory data (mitochrondrial DNA) demonstrating its clear separation from *O. rufus* presented by Hoffmann et al. (2002). Karyotype (2n = 54, FN = 64) reported by L. G. Pereira et al. (2001). The taxon *rostellatus* is listed as a species by Fonseca et al. (1996).

Oxymycterus delator Thomas, 1903. Ann. Mag. Nat. Hist., ser. 7, 11:489.

COMMON NAME: Paraguayan Hocicudo.
TYPE LOCALITY: Paraguay, Paraguarí Dept., Sapucaí.
DISTRIBUTION: E Paraguay, SC Brazil.
STATUS: IUCN – Lower Risk (lc).
COMMENTS: Provisionally retained as a species by Reig (1987), who questioned its distinction from *O. rufus*. Genetically well delineated from *O. rufus* and sister species to *O. amazonicus* according to phylogenetic analysis of cytochrome *b* data (Hoffmann et al., 2002). Distributional limits uncertain: not recorded for the Brazilian fauna by Fonseca et al. (1996) but mapped as broadly distributed in SC Brazil by Hoffmann et al. (2002).

Oxymycterus hiska Hinojosa, Anderson, and Patton, 1987. Am. Mus. Novit., 2898:14.

COMMON NAME: Small Hocicudo.
TYPE LOCALITY: Perú, Puno Dept., 14 km W Yanahuaya, 2210 m; 14°19'S, 69°21'W.
DISTRIBUTION: E Andean slopes in SE Peru and NW Bolivia (La Paz and Cochabamba Depts.), 610-3500.
STATUS: IUCN – Vulnerable
COMMENTS: Smallest species of *Oxymycterus*, morphologically resembling *O. hucucha* but genetically closer to *O. paramensis* (Hoffmann et al., 2002). Bolivian records originally reported as *O. paramensis* (Anderson, 1997) and reidentified by Oliveira (in Salazar-Bravo et al., 2002*b*).

Oxymycterus hispidus Pictet, 1843. Mem. Soc. Phys. Hist. Nat. Geneve, 10:212.
COMMON NAME: Hispid Hocicudo.
TYPE LOCALITY: Brazil, Bahia State, Bahia.
DISTRIBUTION: SE Brazil.
STATUS: IUCN – Lower Risk (lc).
SYNONYMS: *hispidulus* (Schinz, 1845) [unjustified renaming of *hispidus* Pictet].
COMMENTS: A large-sized species traditionally thought to include *judex, misionalis* and
   quaestor as synonyms (Cabrera, 1961; Musser and Carleton, 1993; Reig, 1987); see
   account of *O. quaestor*. Emended diagnosis and documentation of geographic range
   required.

Oxymycterus hucucha Hinojosa, Anderson, and Patton, 1987. Am. Mus. Novit., 2898:15.
COMMON NAME: Quechuan Hocicudo.
TYPE LOCALITY: Bolivia, Cochabamba Dept., 28 km (by road) W Comarapa, 2800 m; 17°51'S,
   64°40'W.
DISTRIBUTION: Known only from the type locality and vicinity, about 2600-3000 m,
   C Bolivia.
STATUS: IUCN – Vulnerable.
COMMENTS: Morphologically similar to *O. hiska*.

Oxymycterus inca Thomas, 1900. Ann. Mag. Nat. Hist., ser. 7, 6:298.
COMMON NAME: Inca Hocicudo.
TYPE LOCALITY: Perú, Junín Dept., Ucayali watershed, Río Perené, 800 m.
DISTRIBUTION: SC Perú to N Bolivia, as far south as Cochabamba and W Santa Cruz Depts.
   (see Anderson, 1997:Fig. 727).
STATUS: IUCN – Lower Risk (lc).
SYNONYMS: *doris* Thomas, 1916; *iris* Thomas, 1901; *juliacae* J. A. Allen, 1900.
COMMENTS: Both Cabrera (1961) and Reig (1987) concurred in citing the listed synonyms as
   subspecies. Anderson (1997) outlined subspecific boundaries of *doris* and *iris* in
   Bolivia; distributional extent and delineation of subspecies in northern sector of range
   unclear.

Oxymycterus josei Hoffmann, Lessa, and Smith, 2002. J. Mammal., 83:411.
COMMON NAME: Cook's Hocicudo.
TYPE LOCALITY: Uruguay, Maldonado Dept., Balneario Las Flores, west margin of Arroyo
   Tarariras.
DISTRIBUTION: SW Uruguay, south of the Río Negro.
COMMENTS: A medium-sized species that is morphologically similar to *O. nasutus*, with
   which it occurs sympatrically; sister species to *O. rufus* according to phylogenetic
   analysis of cytochrome *b* data (Hoffmann et al., 2002).

Oxymycterus nasutus (Waterhouse, 1837). Proc. Zool. Soc. Lond., 1837:16.
COMMON NAME: Long-nosed Hocicudo.
TYPE LOCALITY: Uruguay, Maldonado Dept., Maldonado.
DISTRIBUTION: Uruguay and adjacent SE Brazil (Rio Grande do Sul to São Paulo).
STATUS: IUCN – Lower Risk (lc).
COMMENTS: Comparisons with *O. iheringi* provided by Massoia and Fornes (1969). They
   (and Vitullo et al., 1986) further noted the morphological distinction of *O. nasutus*
   from *O. rufus*, under which it had been ranked as a subspecies (Cabrera, 1961);
   pronounced divergence from the latter sustained by genetic (cytochrome *b*) and
   morphometric analyses (Hoffmann et al., 2002).

Oxymycterus paramensis Thomas, 1902. Ann. Mag. Nat. Hist., ser. 7, 9:139.
COMMON NAME: Yungas Hocicudo.
TYPE LOCALITY: Bolivia, Cochabamba Dept., upper Río Securé, Choquecamate, 4000 m.
DISTRIBUTION: Middle to upper E Andean slopes, ca. 1000-4000 m, in SE Perú, WC Bolivia,
   and NW Argentina.
STATUS: IUCN – Lower Risk (lc).
SYNONYMS: *jacentior* Thomas, 1925; *nigrifrons* Osgood, 1944.

COMMENTS: Parenthetically mentioned as a subspecies of *rutilans* (= *O. rufus*) by Hershkovitz (1966*c*) but most systematists have recognized *paramensis* as a species (Anderson, 1997; Cabrera, 1961; Gyldenstolpe, 1932; Honacki et al., 1982; Musser and Carleton, 1993). Allozymic (Hinososa et al., 1987) and gene-sequence (Hoffmann et al., 2002) analyses support sibling kinship with *O. hiska*, and the latter study underscores its genetic differentiation from *O. rufus*. Musser and Carleton (1993) suggested that placement of *jacentior* with this species merits reconsideration; Cabrera (1961) retained it, *nigrifrons*, and the nominate form as discernable subspecies, as did Anderson (1997:Fig. 728) for Bolivian populations.

Oxymycterus quaestor Thomas, 1903. Ann. Mag. Nat. Hist., ser. 7, 11:226.

COMMON NAME: Quaestor Hocicudo.

TYPE LOCALITY: Brazil, Paraná State, Serra do Mar, Roça Nova.

DISTRIBUTION: NE Argentina (Misiones), SE Brazil (N Rio Grande do Sul to Rio de Janeiro), and perhaps extreme E Paraguay.

SYNONYMS: *judex* Thomas, 1909; *misionalis* Sanborn, 1931.

COMMENTS: *Oxymycterus quaestor*, *judex*, and *misionalis* were all described as species, but Cabrera (1961) and Reig (1987) considered them to be subspecific synonyms of *O. hispidus*, as subsequently observed (Corbet and Hill, 1991; Musser and Carleton, 1993). Recent checklists have referenced some or all as species, without explanation (e.g., Fonseca et al., 1996; Galliari et al., 1996; Heinonen Fortabat and Chebez, 1997). Hoffmann et al. (2002) too acknowledged the three as species; however, their analytical samples are substantially congruent in morphometric space (CVA based on 29 craniodental measurements), a posteriori misclassification among the three is relatively high, and molecular samples of two (*judex* and *quaestor*) form a clade that appears comparable to intraspecific sequence divergence levels presented for other species (e.g., *O. dasytrichus*, *O. delator*, and *O. rufus*). We provisionally retain the three as one species, pending unambiguous argument for their specific status.

Oxymycterus roberti Thomas, 1901. Ann. Mag. Nat. Hist., ser. 7, 8:530.

COMMON NAME: Robert's Hocicudo.

TYPE LOCALITY: Brazil, Minas Gerais State, Rio Jordão, Paranaíba, 700-900 m.

DISTRIBUTION: Vicinity of type locality; range limits uncertain.

STATUS: IUCN – Lower Risk (lc).

Oxymycterus rufus (Fischer, 1814). Zoognosia, 3:71.

COMMON NAME: Rufous Hocicudo.

TYPE LOCALITY: Argentina, Entre Ríos Prov., 32.5°S latitude along lower Río Paraná (as restricted by Hershkovitz, 1994:36, but see comments).

DISTRIBUTION: EC Argentina.

STATUS: IUCN – Lower Risk (lc).

SYNONYMS: *platensis* Thomas, 1914; *rutilans* (Olfers, 1818).

COMMENTS: Priority of *Mus rufus* Fischer, 1814, over *Mus rutilans* Olfers, 1818, both based on Azara's (1801) descriptions, deserves formal stabilization as does clarification of the morphological identity of the species with Azara's "Rat cinquieme ou rat roux." Similarly, decisive fixation of the type locality is required. Cabrera (1961) cited Rengger (1830) for the placement of the type locality near Asunción, Paraguay, but Rengger's basis for this interpretation is unclear since Azara (1802) mentioned collecting the species in an arroyo at 32.5 degrees, presumably S latitude. Hershkovitz (1994), perhaps following Musser and Carleton's observations (1993:727), restricted the type locality as given above; Galliari et al. (1996) instead noted that the type locality should be restricted to the vicinity of San Ignacio Guazú, Paraguay. Vitullo et al. (1986) supported the union of *O. rufus* and *platensis* as one species based on their chromosomal similarity, a synonymy that warrants reassessment given the karyotypic monomorphism that apparently characterizes the genus (Bonvicino et al., 1998*b*). Vicariant sister species of *O. josei* in Uruguay according to phylogenetic analysis of cytochrome *b* data (Hoffmann et al., 2002).

Paralomys Thomas, 1926. Ann. Mag. Nat. Hist., ser. 9, 17:315.

    TYPE SPECIES: *Phyllotis gerbillus* Thomas, 1900.

    COMMENTS: Phyllotini. Although named as a genus and early recognized as such (Cabrera, 1961; Gyldenstolpe, 1932; Pearson, 1958), *Paralomys* was later submerged as a subgenus of *Calomys* (Ellerman, 1941) or as a subgenus or full synonym of *Phyllotis* (Hershkovitz, 1962; Osgood, 1947; Pearson and Patton, 1976). Uncertainty over its monophyletic union with those species and taxonomic rank has been revived by broad phylogenetic studies of morphological characters that suggest the distant relationship of *gerbillus* and *amicus* to other *Phyllotis* (Braun, 1993; Steppan, 1995). Braun (1993) elevated *Paralomys* to its former generic status, but Steppan (1995) reservedly maintained it as a synonym for want of strongly supported cladistic structure and demonstable monophyly of *gerbillus-amicus* in the various trees generated (also see Steppan and Sullivan, 2000). Compared with other *Phyllotis*, we are impressed by the very brachyodont molars of *gerbillus*; their more cuspidate construction, somewhat resembling *Eligmodontia* more than *Phyllotis*, as remarked by Thomas (1926*a*); the small size of the m1 anteroconid, only weakly demarcated from the metaconid; along with other traits as enumerated by Braun (1993). According to such characters, and following Thomas (1926*a*), we view *amicus* as more closely related to other *Phyllotis*, an interpretation supported by phylogenetic analysis of abbreviated cytochrome *b* sequences (Spotorno et al., 2001). Further investigation using other information sources is urged to clarify relationships among *gerbillus*, those small northern species of *Phyllotis*, and other brachyodont phyllotines such as *Calomys* and *Eligmodontia*.

Paralomys gerbillus (Thomas, 1900). Ann. Mag. Nat. Hist., ser. 7, 5:151.

    COMMON NAME: Gerbilline Pericote.

    TYPE LOCALITY: Perú, Piura Dept., Piura, 30 m.

    DISTRIBUTION: Sechura Desert, NW Perú.

    STATUS: IUCN – Lower Risk (lc) as *Phyllotis gerbillus*.

Pearsonomys Patterson, 1992. Zool. J. Linn. Soc., 106:132.

    TYPE SPECIES: *Pearsonomys annectens* Patterson, 1992.

    COMMENTS: Akodontini (S Andean clade). A long-clawed akodont with long pinnae, morphologically most similar to *Geoxus* and probably its closest generic relative (Patterson, 1992*b*), a relationship affirmed by molecular data (Smith and Patton, 1999).

Pearsonomys annectens Patterson, 1992. Zool. J. Linn. Soc., 106:136.

    COMMON NAME: Pearson's Long-clawed Akodont.

    TYPE LOCALITY: Chile, Valdivia Prov., Region de Los Lagos, 42 km N and slightly E Valdivia, near Mehuín, 100 m; 39°26'S, 73°10'W.

    DISTRIBUTION: Known only from the type locality and vicinity but may inhabit other areas of the Valdivian temperate rainforest zone, WC Chile.

    STATUS: IUCN – Data Deficient.

    COMMENTS: Another species endemic to the austral *Nothofagus* temperate rainforests, highlighting the importance of the S Andean region in understanding biotic genesis and sigmodontine biogeography (see Patterson, 1992*b*).

Phaenomys Thomas, 1917. Ann. Mag. Nat. Hist., ser. 8, 20:196.

    TYPE SPECIES: *Oryzomys ferrugineus* Thomas, 1894.

    COMMENTS: Sigmodontinae *incertae sedis* per Smith and Patton (1999) or Oryzomyini *sensu lato* per Reig (1981). Another of the SE Brazilian endemics whose systematic relationships are minimally understood. Bonvicino et al. (2001) reported on a recently collected specimen, documenting its karyotype (2n = 78, FN = 114), redescribing the taxon's morphology, and comparing it with Andean Thomasomyini and other Atlantic Forest thomasomyine-like genera. The authors noted that morphological characters reinforce the plesiomorphic nature of the genus, thereby offering little insight to relationships, while the chromosomal complement demonstrates affinity with species of *Delomys*.

Phaenomys ferrugineus (Thomas, 1894). Ann. Mag. Nat. Hist., ser. 6, 14:352.

    COMMON NAME: Rusty Phaenomys.

TYPE LOCALITY: Brazil, Rio de Janeiro State, Rio de Janeiro.

DISTRIBUTION: Known only from a restricted area in the Serra do Mar, Rio de Janeiro and São Paulo states, S Brazil.

STATUS: IUCN – Endangered.

COMMENTS: Known museum records (total = 11) consolidated by Maia Vaz (2000), who noted that the species has not been collected within the last sixty years and commented on its conservation status. An individual collected in 1998, from Rio de Janeiro, brings the total to 12 and indicates that the species can persist even in slightly degraded habitat (Bonvicino et al., 2001).

Phyllotis Waterhouse, 1837. Proc. Zool. Soc. Lond., 1837:27.

TYPE SPECIES: *Mus darwini* Waterhouse, 1837.

COMMENTS: Phyllotini. Pearson (1958) utilized *Auliscomys*, *Graomys*, and *Loxodontomys* as distinctive subgenera, whereas Hershkovitz (1962) placed all three, plus *Paralomys*, in complete synonymy—see respective accounts for their reinstatement as genera. Full generic revisions initiated by Pearson (1958) and Hershkovitz (1962), followed by extensive refinement of species definitions and distributions as a result of multifaceted research over the past 25 years—see especially Pearson (1972), Pearson and Patton (1976), Spotorno (1976), Spotorno and Walker (1983), Walker et al. (1984), and Steppan (1998). Steppan (1998) discussed alternative interpretations of interspecific relationships and patterns of historical biogeography within the genus.

Phyllotis amicus (Thomas, 1900). Ann. Mag. Nat. Hist., ser. 7, 5:355.

COMMON NAME: Peruvian Pericote.

TYPE LOCALITY: Perú, Cajamarca Dept., Tolón, 100 m.

DISTRIBUTION: Coast and lower Pacific slopes of W Perú, Depts. Piura (see Lavrentchenko and Dmitriev, 1994) to Ayacucho.

STATUS: IUCN – Lower Risk (lc).

SYNONYMS: *maritimus* Thomas, 1900; *montanus* Thomas, 1900.

COMMENTS: Sister species of *P. limatus* according to phylogenetic analysis of abbreviated cytochrome *b* sequences (Spotorno et al., 2001). The basis for the reported presence of *P. amicus* in Argentina (Heinonen Fortabat and Chebez, 1997; Massoia and Donadío, 1990) has been reidentified as an example of *Oligoryzomys chacoensis* (Galliari et al., 1996).

Phyllotis andium Thomas, 1912. Ann. Mag. Nat. Hist., ser. 8, 10:409.

COMMON NAME: Andean Pericote.

TYPE LOCALITY: Ecuador, Cañar Prov., Cañar, 8500 ft (2591 m).

DISTRIBUTION: E and W Andean slopes from C Ecuador (Tungurahua) to C Perú (Lima).

STATUS: IUCN – Lower Risk (lc).

SYNONYMS: *fruticicolus* Anthony, 1922; *melanius* Thomas, 1913; *stenops* Osgood, 1914; *tamborum* Osgood, 1914.

Phyllotis bonariensis Crespo, 1964. Neotropica, 10:99.

COMMON NAME: Bonaerense Pericote.

TYPE LOCALITY: Argentina, Buenos Aires Prov., Depto. Tornquist, Abra de la Ventana, 500 m.

DISTRIBUTION: EC Argentina; limits uncertain.

STATUS: IUCN – Lower Risk (nt).

COMMENTS: Described as a subspecies of *P. darwini* but considered distinct by Reig (1978) and Galliari et al. (1996); determination of status and distribution requires fresh investigation.

Phyllotis caprinus Pearson, 1958. Univ. California Publ. Zool., 56:435.

COMMON NAME: Capricorn Pericote.

TYPE LOCALITY: Argentina, Jujuy Prov., Tilcara, 8000 ft (2438 m).

DISTRIBUTION: Upper slopes of E Andes, about 2100-3750 m, from SC Bolivia to northernmost Argentina.

STATUS: IUCN – Lower Risk (lc).

COMMENTS: Demoted to a subspecies of *P. darwini* by Hershkovitz (1962), but distributional and morphological evidence substantiates the specific recognition of *P. caprinus* (Pearson, 1958).

Phyllotis darwini (Waterhouse, 1837). Proc. Zool. Soc. Lond., 1837:28.
COMMON NAME: Darwin's Pericote.
TYPE LOCALITY: Chile, Coquimbo Prov., Coquimbo.
DISTRIBUTION: Coastal WC Chile.
STATUS: IUCN – Lower Risk (lc).
SYNONYMS: *boedeckeri* (Philippi, 1900); *campestris* (Philippi, 1900); *dichrous* (Philippi, 1900); *fulvescens* Osgood, 1943; *griseoflavus* (Philippi, 1900); *illapelinus* (Philippi, 1900); *megalotis* (Philippi, 1900); *melanotis* (Philippi, 1900); *melanonotus* (Philippi and Landbeck, 1858); *mollis* (Philippi, 1900); *platytarsus* (Philippi, 1900); *segethi* (Philippi, 1900).
COMMENTS: *Phyllotis darwini sensu stricto* has been revealed as a geographically restricted species following a quarter-century of researches upon its polytypic predecessor, which was believed to range over much of the S Andes. See accounts of *P. bonaeriensis*, *P. caprinus*, *P. definitus*, *P. limatus*, *P. magister*, *P. osgoodi*, *P. wolffsohni*, and *P. xanthopygus*, all previously arrayed as subspecies of *P. darwini* by Pearson (1958) and/or Hershkovitz (1962), for references addressing their specific status.

Phyllotis definitus Osgood, 1915. Field Mus. Nat. Hist. Publ., Zool. Ser., 10:189.
COMMON NAME: Ancash Pericote.
TYPE LOCALITY: Perú, Ancash Dept., Mácate, 9000 ft (2743 m).
DISTRIBUTION: Andes of Ancash, Perú.
STATUS: IUCN – Lower Risk (lc).
COMMENTS: Included in *P. magister* by Pearson (1958) and in *P. darwini* by Hershkovitz (1962) but distinctive karyotype is unlike either of those species (see Pearson, 1972).

Phyllotis haggardi Thomas, 1908. Ann. Mag. Nat. Hist., ser. 7, 2:270.
COMMON NAME: Ecuadoran Pericote.
TYPE LOCALITY: Ecuador, Pichincha Prov., Mt Pichincha, above Quito, 3400-4000 m.
DISTRIBUTION: Andes of C Ecuador.
STATUS: IUCN – Lower Risk (lc).
SYNONYMS: *elegantulus* Thomas, 1913; *fuscus* Anthony, 1924.

Phyllotis limatus Thomas, 1912. Ann. Mag. Nat. Hist., ser. 8, 10:407.
COMMON NAME: Lima Pericote.
TYPE LOCALITY: Perú, Lima Dept., Chosica, 3000 ft (914 m).
DISTRIBUTION: Western coast and contiguous Pacific slopes of the Andes, sea level- 4000 m, from WC Perú to N Chile (see Steppan, 1998:Fig. 5).
COMMENTS: Described as a subspecies of *P. darwini* by Thomas (1912) and therewith ranked by others (Cabrera, 1961; Hershkovitz, 1962; Pearson, 1958). Steppan (1998) defended the specific status of *limatus*, including populations confused under *P. xanthopygus rupestris*, based on congruent morphological and molecular data; he further noted appreciable differentiation between northern and southern populations of *P. limatus* but declined to recognize these formally as subspecies. Sister species of *P. xanthopygus* according to maximum parsimony analyses of cytochrome *b* sequences (Steppan, 1998), perhaps arising by peripheral isolation from a western lineage of *P. xanthopygus rupestris* (Kuch et al., 2002). Late Quaternary occurrence of the species in the S Atacama Desert, slightly to the south of its current range, documented by mitochondrial DNA recovered from ancient rodent midden (Kuch et al., 2002).

Phyllotis magister Thomas, 1912. Ann. Mag. Nat. Hist., ser. 8, 10:406.
COMMON NAME: Majestic Pericote.
TYPE LOCALITY: Perú, Arequipa Dept., Arequipa, 2300 m.
DISTRIBUTION: Upper Pacific slopes of Andes from C Perú (see Arana-Cardó and Ascorra, 1994) to N Chile.
STATUS: IUCN – Lower Risk (lc).
COMMENTS: Hershkovitz (1962) treated *magister* as a subspecies of *P. darwini*, a rank inconsistent with records of sympatry (Pearson, 1958) and karyotypic differences

(Pearson, 1972; Pearson and Patton, 1976). Pearson (1958) arranged *definitus* as a subspecies of *P. magister* but later (1972) supported its specific status. Intestinal area and volume compared with *P. xanthopygus* and related to dietary differences (Lagos and Bozinovic, 1999).

Phyllotis osgoodi Mann, 1945. Biologica, 2:81.
COMMON NAME: Osgood's Pericote.
TYPE LOCALITY: Chile, Tarapacá Prov., Parinacota.
DISTRIBUTION: Altiplano of NE Chile.
STATUS: IUCN – Lower Risk (lc).
COMMENTS: Relegated to synonymy under *P. darwini* (Hershkovitz, 1962; Pearson, 1958) but considerable evidence supports its validity as a species (Spotorno, 1976; Spotorno and Walker, 1979, 1983).

Phyllotis osilae J. A. Allen, 1901. Bull. Am. Mus. Nat. Hist., 14:44.
COMMON NAME: Bunchgrass Pericote.
TYPE LOCALITY: Perú, Puno Dept., Osila (= Asillo), 17 mi (27 km) ENE Ayaviri, 13,000 ft (3962 m) (as interpreted by Pearson, 1958:426).
DISTRIBUTION: Upper Andean slopes on Atlantic drainage, from SC Perú (Cuzco), through W Bolivia (1700-4900 m), to N Argentina (Catamarca).
STATUS: IUCN – Lower Risk (lc).
SYNONYMS: *lutescens* Thomas, 1902; *nogalaris* Thomas, 1921; *phaeus* Osgood, 1944; *tucumanus* Thomas, 1912.
COMMENTS: All-acrocentric karyotype (2n = 68) interpreted as ancestral state of the genus (Pearson and Patton, 1976). In Bolivia, Anderson (1997) recognized *osilae* (*lutescens* as a full synonym) and *phaeus* as subspecies.

Phyllotis wolffsohni Thomas, 1902. Ann. Mag. Nat. Hist., ser. 7, 9:131.
COMMON NAME: Wolffsohn's Pericote.
TYPE LOCALITY: Bolivia, Cochabamba Dept., Tapacarí, 9900 ft (3018 m).
DISTRIBUTION: Upper E Andean slopes in WC Bolivia, 1300-3875 m (see Anderson, 1997: Fig. 750).
STATUS: IUCN – Lower Risk (lc).
COMMENTS: Hershkovitz (1962) synonymized *wolffsohni* as a subspecies of *P. darwini*, a relationship at variance with morphological and karyotypic information (Pearson, 1958; Pearson and Patton, 1976).

Phyllotis xanthopygus (Waterhouse, 1837). Proc. Zool. Soc. Lond., 1837:28.
COMMON NAME: Yellow-rumped Pericote.
TYPE LOCALITY: Argentina, Santa Cruz Prov., coast of Santa Cruz.
DISTRIBUTION: WC Perú, in and along the Andes to Santa Cruz Prov., S Argentina, and adjacent Magallanes Prov., S Chile; latitudinal (15-51S) and altitudinal ranges (sea level-5600 m) exceptional as currently defined (see Kramer et al., 1999:Fig. 3, for extent of geographic races).
STATUS: IUCN – Lower Risk (lc).
SYNONYMS: *abrocodon* Thomas, 1926; *arenarius* Thomas, 1902; *capito* (Philippi, 1860); *chilensis* Mann, 1945; *glirinus* (Philippi, 1896); *lanatus* (Philippi, 1896); *oreigenus* Cabrera, 1926; *posticalis* Thomas, 1912; *ricardulus* Thomas, 1919; *rupestris* (Gervais, 1841); *vaccarum* Thomas, 1912; *wolffhuegeli* Mann, 1944.
COMMENTS: Viewed as a geographic race of *P. darwini* by Pearson (1958), Cabrera (1961), and Hershkovitz (1962). Morphometric, electrophoretic, karyotypic, and molecular differentiation supports the specific recognition of *P. xanthopygus* (Spotorno and Walker, 1983; Steppan, 1998; Walker et al., 1984); genetic diversity among and within isolated Patagonian demes examined by Kim et al. (1998), who postulated the early Pleistocene origin of haplotype lineages and highlighted the vicariant influence of the Río Chubut. Steppan (1998) noted the need for continued systematic examination of the populations now assigned to the highly variable *xanthopygus*: e.g., *chilensis* is sometimes treated as a species (Anderson, 1997); *rupestris* and *vaccarum* (Spotorno et al., 2001) appear distinct from *xanthopygus* in phylogenetic analysis of abbreviated

cytochrome *b* sequences (Spotorno et al., 2001); *xanthopygus* paraphyletic with respect to *P. limatus* in broader molecular survey of nominal subspecies (Kuch et al., 2002). See Kramer et al. (1999, Mammalian Species, 617).

Podoxymys Anthony, 1929. Am. Mus. Novit., 383:4.
> TYPE SPECIES: *Podoxymys roraimae* Anthony, 1929.
> COMMENTS: Akodontini. Systematics little known aside from original comparisons, report on gastric morphology by Carleton (1973), brief comments of Reig (1987), and new data on systematics and karyology (2n = 16) by Pérez-Zapata et al. (1992).

Podoxymys roraimae Anthony, 1929. Am. Mus. Novit., 383:4.
> COMMON NAME: Roraima Akodont.
> TYPE LOCALITY: Guyana, Mazaruni-Potaro Dist., summit of Mt Roraima, 8600 ft (2621 m).
> DISTRIBUTION: Guyana and adjacent portions of Venezuela and Brazil (as per Fonseca et al., 1996). Probably occurs on other neighboring tepuis within Venezuela (Linares, 1998).
> STATUS: IUCN – Lower Risk (nt).
> COMMENTS: Known only by the original type series of 5 specimens and a single specimen reported by Pérez-Zapata et al. (1992).

Pseudoryzomys Hershkovitz, 1962. Fieldiana Zool., 31:208.
> TYPE SPECIES: *Oryzomys wavrini* Thomas, 1921 (= *Hesperomys simplex* Winge, 1887).
> COMMENTS: Oryzomyini. Associated with phyllotines by Hershkovitz (1962) and viewed as the tribe's most ancient clade by Braun (1993). However, excluded from Phyllotini by Olds and Anderson (1989), and oryzomyine characteristics illuminated by Voss and Myers (1991). Formally allocated to Oryzomyini as the tribal diagnosis was phylogenetically recast by Voss and Carleton (1993); the morphology of the glans penis also supports this tribal assignment (Langguth and Silva Neto, 1993) as do other morphological traits (Steppan, 1995). See Voss and Myers (1991:418) on the formal availability of the genus-group name.

Pseudoryzomys simplex (Winge, 1887). E Museo Lundii, 1(3):11.
> COMMON NAME: False Oryzomys.
> TYPE LOCALITY: Brazil, Minas Gerais State, near Lagoa Santa.
> DISTRIBUTION: EC Bolivia, NE Argentina (Chaco, Formosa, Sante Fe; Pardiñas et al., in litt.), and W Paraguay, to E Brazil (Pernambuco).
> STATUS: IUCN – Lower Risk (lc).
> SYNONYMS: *wavrini* (Thomas, 1921); *reigi* Pine and Wetzel, 1975.
> COMMENTS: Formerly classified as *Oryzomys incertas sedis* (Cabrera, 1961) until genus diagnosed by Hershkovitz (1962). Lectotype designated, synonymy presented, karyotype and distribution discussed by Voss and Myers (1991).

Punomys Osgood, 1943. J. Mammal., 24:369.
> TYPE SPECIES: *Punomys lemminus* Osgood, 1943.
> COMMENTS: Generic-level affinities uncertain (Osgood, 1943): left as Sigmodontinae *incertae sedis* by some (Hershkovitz, 1962; Reig, 1980, 1981, 1984; Smith and Patton, 1999; Steppan, 1995) or formally assigned to Tribe Phyllotini by others (Braun, 1993; McKenna and Bell, 1997; Olds and Anderson, 1989; Vorontsov, 1959). Pacheco and Patton (1995) provided a synopsis of distributional records, specific discrimination, and natural history.

Punomys kofordi Pacheco and Patton, 1995. Z. Saugetierk., 60:86.
> COMMON NAME: Eastern Puna Mouse.
> TYPE LOCALITY: Perú, Puno Dept., 13 mi (20.8 km) ENE Crucero, Lago Aricoma, 15,000 ft (4550 m); 14°17'S, 69°47'W.
> DISTRIBUTION: Cordillera Oriental, above 4500 m, S Perú.

Punomys lemminus Osgood, 1943. J. Mammal., 24:369.
> COMMON NAME: Western Puna Mouse.
> TYPE LOCALITY: Perú, Puno Dept., San Antonio de Esquilache, 4500 m.
> DISTRIBUTION: Cordillera Occidental, above 4400 m, S Perú.

STATUS: IUCN – Lower Risk (lc).

Reithrodon Waterhouse, 1837. Proc. Zool. Soc. Lond., 1837:29.

TYPE SPECIES: *Reithrodon typicus* Waterhouse, 1837.

SYNONYMS: *Proreithrodon* Ameghino, 1889; *Ptyssophorus* Ameghino, 1889; *Tretomys* Ameghino, 1889.

COMMENTS: Reithrodontini. Classically arranged by Hershkovitz (1955*a*) with sigmodont rodents, but a compelling body of morphological evidence points to its general kinship with phyllotines (Braun, 1993; Olds and Anderson, 1989; Ortiz et al., 2000*b*; Pardiñas, 1997; Pardiñas and Galliari, 2001; Pearson and Patton, 1976; Steppan, 1993, 1995). In their rediagnosis of Phyllotini, Olds and Anderson (1989) had identified a morphologically well-defined complex that they designated the "*Reithrodon*-Group," comprised of *Euneomys*, *Neotomys*, and *Reithrodon*. Such an assemblage, along with certain extinct genera, has consistently appeared in phylogenetic studies based on morphological characters (Ortiz et al., 2000*b*; Pardiñas, 1997; Steppan, 1993, 1995; Steppan and Sullivan, 2000). Cladistic interpretations of mitochondrial genes, however, dispute tribal alignment of *Reithrodon* with phyllotines, or with *Sigmodon*, and inconsistently represent its phyletic affinity (D'Elía et al., 2003; Engel et al., 1998; Smith and Patton, 1999); samples of *Euneomys* and *Neotomys* were not represented in these studies. Whether inclusion of the latter genera will bring the "*Reithrodon*-Group" back within Phyllotini *sensu lato* (i.e., a subtribe Reithrodontina as per Steppan, 1995), or whether the phyletic remoteness of *Reithrodon* and kin will be sustained invites further investigation. Vorontsov (1959) had previously recognized a formal tribal separation of *Reithrodon* from Phyllotini; availability of the family-group name issues from his action. Attribution of fossil genera described by Ameghino (1889) follows Pardiñas (2000*a*).

Reithrodon auritus (Fischer, 1814). Zoognosia, 3:71.

COMMON NAME: Hairy-soled Conyrat.

TYPE LOCALITY: Argentina, Buenos Aires Prov., pampas south of Buenos Aires, south bank of the Río de la Plata (as restricted by Hershkovitz, 1959*a*:349).

DISTRIBUTION: Isolated records in the Pampean region, N and C Argentina; more or less continuously distributed over the Patagonian region, sea level-3000 m, from C Argentina and adjacent Chile through Tierra del Fuego (see Ortiz et al., 2000; Pardiñas and Galliari, 2001:Fig. 3).

STATUS: IUCN – Lower Risk (lc).

SYNONYMS: *auritus* (Desmarest, 1819); *caurinus* Thomas, 1920; *cuniculoides* Waterhouse, 1837; *evae* Thomas, 1927; *flammarum* Thomas, 1912; *hatcheri* J. A. Allen, 1903; *marinus* Thomas, 1920; *obscurus* J. A. Allen, 1903; *pachycephalus* (Philippi, 1900); *pampanus* Thomas, 1916; *physodes* (Illiger, 1815) [*nomen nudum*]; *physodes* (Olfers, 1818); *spegazzinii* (Ameghino, 1889).

COMMENTS: See the weighty Mammalian Species account (No. 664) of Pardiñas and Galliari (2001) for discrimination from *R. typicus*, remarks on the type locality, review of fossil synonyms and numerous paleontological occurrences (middle Pliocene-Holocene), and distribution of the four conventionally recognized subspecies (Cabrera, 1961), "whose boundaries are highly hypothetical." Pardiñas (1995) allocated the late Pleistocene *Bothriomys spegazzinii*, described by Ameghino (1889) from Argentinian sediments, as a synonym of *R. auritus*. Reports of the species from the Malvinas (= Falkland) Isls remain unconfirmed (see Pardiñas and Galliari, 2001). Also see account of *R. typicus*, formerly included as a subspecies.

Reithrodon typicus Waterhouse, 1837. Proc. Zool. Soc. Lond., 1837:30.

COMMON NAME: Naked-soled Conyrat.

TYPE LOCALITY: Uruguay, Maldonado Dept., Maldonado.

DISTRIBUTION: EC Argentina (Entre Ríos and Corrientes), Uruguay, and extreme S Brazil (Rio Grande do Sul).

SYNONYMS: *currentium* Thomas, 1920.

COMMENTS: Although early maintained as distinct from *R. auritus* (Ellerman, 1941; Gyldenstolpe, 1932), Osgood (1943) placed *typicus* and all conyrats in a single species, and the genus thereafter was typically viewed as monotypic, whether as *R. physodes* Olfers, 1818 (Cabrera, 1961; Corbet and Hill, 1991; Hershkovitz, 1955*a*, 1959; Honacki et al., 1982) or *R. auritus* Fischer, 1814 (Musser and Carleton, 1993; Reig, 1978). Specific status reaffirmed by Ortells et al. (1988), who noted substantial differences in diploid and fundamental numbers between *R. auritus* proper (2n = 34) and *typicus* (2n = 28), as earlier reported by Freitas et al. (1983*b*, as *auritus*).

Rhagomys Thomas, 1917. Ann. Mag. Nat. Hist., ser. 8, 20:192.
TYPE SPECIES: *Hesperomys rufescens* Thomas, 1886.
COMMENTS: Generic affinities uncertain: originally allied with *Oryzomys-Oecomys* (Thomas, 1917*c*), or included among oryzomyine genera (Tate, 1932*f*), or listed as Sigmodontinae *incertae sedis* (McKenna and Bell, 1997; Reig, 1980, 1984; Smith and Patton, 1999). Taxonomic history reviewed and genus critically recharacterized by Luna and Patterson (2003), who considered the biogeographic implications of a new *Rhagomys* in SE Peru, far away from the Atlantic Forest region where its congenor and other oryzomyine-thomasomyine endemic genera are found.

Rhagomys longilingua Luna and Patterson, 2003. Fieldiana Zool., N.S., 101:3.
COMMON NAME: Long-tongued Rhagomys.
TYPE LOCALITY: Perú, Cusco Dept., Paucartambo Prov., Manu Biosphere Reserve, along Río Cosñipata, below "Suecia," 1900 m; 13°06.032′S, 71°34.125′W.
DISTRIBUTION: Dense forest of E Andean slopes, 450-2100 m, SE Perú (Cusco and Madre de Dios Depts.).
COMMENTS: Similar to *R. rufescens* in unique possession (among Sigmodontinae) of a hallux with nail and in most verifiable cranial traits, but fur of *R. longilingua* spiny and inter-orbit strongly beaded; so far known by only three specimens (Luna and Patterson, 2003).

Rhagomys rufescens (Thomas, 1886). Ann. Mag. Nat. Hist., ser. 5, 17:250.
COMMON NAME: Rufescent Rhagomys.
TYPE LOCALITY: Brazil, Rio de Janeiro State, Rio de Janeiro.
DISTRIBUTION: Known only from the type locality.
STATUS: IUCN – Critically Endangered.
COMMENTS: Only two specimens exist, one with the skull now missing. Morphology redescribed and skull of holotype illustrated by Luna and Patterson (2003).

Rheomys Thomas 1906. Ann. Mag. Nat. Hist., ser. 7, 17:421.
TYPE SPECIES: *Rheomys underwoodi* Thomas, 1906.
SYNONYMS: *Neorheomys* Goodwin, 1959.
COMMENTS: Ichthyomyini. All named forms suggested as conspecific under *trichotis* by Hershkovitz (1966*c*) but Voss (1988) recognized four species. Formerly included *trichotis* (Tate, 1932), a species later designated as the type of *Chibchanomys* (Voss, 1988). *Neorheomys*, diagnosed as a subgenus (Goodwin, 1959*a*), was placed in full synonymy by Voss (1988). Separation from *Ichthyomys* questioned by Ellerman (1941) and Hall (1981); relationships, generic stature, and Mesoamerican endemism illuminated by Voss (1988).

Rheomys mexicanus Goodwin, 1959. Am. Mus. Novit., 1967:4.
COMMON NAME: Mexican Ichthyomyine.
TYPE LOCALITY: México, Oaxaca State, Miahuatlán Dist., San José Lachiguirí, 4000 ft (1219 m).
DISTRIBUTION: Oaxaca, México.
STATUS: IUCN – Lower Risk (nt).
COMMENTS: Type species of Goodwin's (1959*a*) *Neorheomys*. Viewed as closely related to *R. underwoodi* (Voss, 1988).

Rheomys raptor Goldman, 1912. Smithson. Misc. Coll., 60:7.
COMMON NAME: Goldman's Ichthyomyine.
TYPE LOCALITY: Panamá, Darién Prov., Cerro Pirre, near headwaters of Río Limón, 4500 ft (1372 m).

DISTRIBUTION: Isolated segments in highlands of C Costa Rica and W Panamá (*hartmanni*), and in E Panamá (*raptor*).

STATUS: IUCN – Lower Risk (lc).

SYNONYMS: *hartmanni* Enders, 1939.

COMMENTS: Treated as a subspecies of *Rheomys trichotis* (Cabrera, 1961), but species status affirmed by Voss (1988). Enders (1939) described *hartmanni* as a species; recognized as such (e.g., Hall, 1981) until Voss (1988) allocated the form to a subspecies of *raptor*. Status of the morphologically distinctive *hartmanni*, isolated in the Talamancan highlands, deserves reconsideration using other information sources.

Rheomys thomasi Dickey, 1928. Proc. Biol. Soc. Wash., 41:11.

COMMON NAME: Thomas' Ichthyomyine.

TYPE LOCALITY: El Salvador, San Miguel Dept., Finca San Felipe, Cerro Cacaguatique, 3500 ft (1067 m).

DISTRIBUTION: Highlands of S México (Chiapas), Guatemala, and El Salvador.

STATUS: IUCN – Lower Risk (lc).

SYNONYMS: *chiapensis* Hooper, 1947; *stirtoni* Dickey, 1928.

COMMENTS: Of two named subspecies, Voss (1988) retained *stirtoni* as valid in addition to the nominate form.

Rheomys underwoodi Thomas, 1906. Ann. Mag. Nat. Hist., ser. 7, 17:422.

COMMON NAME: Underwood's Ichthyomyine.

TYPE LOCALITY: Costa Rica, Cartago Prov., near Tres Ríos.

DISTRIBUTION: Highlands of C Costa Rica and W Panamá.

STATUS: IUCN – Lower Risk (lc).

COMMENTS: Closely related to *R. mexicanus* (Voss, 1988).

Rhipidomys Tschudi, 1845. Untersuchungen uber die Fauna peruana (Therologie), p. 183.

TYPE SPECIES: *Hesperomys leucodactylus* Tschudi, 1845.

COMMENTS: Thomasomyini. Tribe (1996) explained why the genus-group name *Rhipidomys* is available from Tschudi (1845), not 1844 as conventionally cited (e.g., Cabrera, 1961; Ellerman, 1941). External and cranial morphology characterized and contrasted with superficially similar species of *Oecomys* by Tribe (1996), Patton et al. (2000), and Voss et al. (2001). Sister genus to the *Chilomys-Thomasomys* clade based on phylogenetic evaluations of mitochondrial DNA sequences from 2-6 species surveyed (D'Elía et al., 2003; Smith and Patton, 1999). As noted by Tribe (1996), monophyly of the associated species within a single genus has yet to be convincingly demonstrated; relationships among the distinctive morphologies identifiable within the taxon may involve other thomasomyines and eventual rearrangement of generic boundaries. Tribe designated those morphologies as three formal "sections": Fulviventer (*R. caucensis, R. fulviventer, R. venustus, R. wetzeli*); Leucodactylus (*R. austrinus, R. couesi, R. emiliae, R. gardneri, R. latimanus, R. leucodactylus; R. macrurus, R. mastacalis, R. modicus, R. nitela, R. ochrogaster, R. venezuelae*); and Macconnelli (*R. macconnelli*).

No published revisionary standard is available. Species and synonyms acknowledged herein in general observe our earlier listing (1993), as based on examination of holotypes and series in AMNH, BMNH, and USNM, and the unpublished thesis of Tribe (1996), which is the best synthesis of valid species and their distributional limits based on exhaustive examination of museum specimens. Nonetheless, the distributions of many species remain inadequately documented, and the continued listing of *Rhipidomys* as species indeterminate or undescribed species (e.g., Anderson, 1997; Costa, 2003; L. G. Pereira et al., 2001; Tribe, 1996) concedes our incomplete knowledge of their alpha-level systematics. Volobouev and Catzeflis (2000) summarized karyotypic data for the genus and discussed mechanisms of cytogenetic change; Andrades-Miranda (2002) provided additional karyotypic records and also summarized reports for the genus; Costa (2003) analyzed phylogeographic patterns among select species distributed in the Amazon Basin and Atlantic Forest.

Rhipidomys austrinus Thomas, 1921. Ann. Mag. Nat. Hist., ser. 9, 7:183.

COMMON NAME: Southern Andean Rhipidomys.

TYPE LOCALITY: Argentina, Jujuy Prov., Sierra de Santa Bárbara, El Sunchal, 1200 m.
DISTRIBUTION: Lower E Andean slopes and foothills, about 350-1750 m, from La Paz Dept., WC Bolivia, south to Jujuy Prov., NW Argentina.
STATUS: IUCN – Lower Risk (lc).
SYNONYMS: *collinus* Thomas, 1925.
COMMENTS: *R. leucodactylus* section *sensu* Tribe (1996). Variously thought to intergrade with *R. leucodactylus* (e.g., Cabrera, 1961; Mares et al., 1989*b*) or with *R. couesi* (e.g., Anderson, 1997). Retained as species by Musser and Carleton (1993) and by Tribe (1996), who characterized its morphology and vouchered its geographic occurrence.

Rhipidomys caucensis J. A. Allen, 1913. Bull. Am. Mus. Nat. Hist., 32:601.
COMMON NAME: Colombian Rhipidomys.
TYPE LOCALITY: Colombia, Cauca Dept., Cerro Munchique, 8225 ft (2507 m).
DISTRIBUTION: Middle to high elevations, 2200-3500 m, W Andes of Colombia.
STATUS: IUCN – Lower Risk (nt).
COMMENTS: *R. fulviventer* section *sensu* Tribe (1996). A small species morphologically similar to *R. wetzeli* (Tribe, 1996).

Rhipidomys couesi (J. A. Allen and Chapman, 1893). Bull. Am. Mus. Nat. Hist., 5:214.
COMMON NAME: Coues' Rhipidomys.
TYPE LOCALITY: Trinidad, 7 mi (11 km) SE Princes Town and 12 mi (19 km) N of the southern coast.
DISTRIBUTION: Moist lowland forest that follows an arc from Trinidad, through N and WC Venezuela, to C Colombia (Meta).
STATUS: IUCN – Lower Risk (lc).
SYNONYMS: *cumananus* Thomas, 1900.
COMMENTS: *R. leucodactylus* section *sensu* Tribe (1996). Cabrera (1961) identified *couesi* as a subspecies of *R. sclateri*. As emphasized by its original describer Thomas (1887*b*), *sclateri* closely resembles Peruvian *R. leucodactylus*, and, in both Venezuela (see Handley, 1976) and Perú, *R. couesi* and *R. leucodactylus* are morphologically distinct.

Rhipidomys emiliae (J. A. Allen, 1916). Bull. Am. Mus. Nat. Hist., 35:525.
COMMON NAME: Eastern Amazon Rhipidomys.
TYPE LOCALITY: Brazil, Pará State, Rio Moju.
DISTRIBUTION: E Amazonia, principally in Pará and Mato Grosso east of the Rio Xingu and as far south as Serra do Roncador, C Brazil.
COMMENTS: *R. leucodactylus* section *sensu* Tribe (1996). Described as a species of *Oecomys*, transferred to *Rhipidomys* by Goodwin (1953), and thereafter associated as a form of *R. mastacalis* (Cabrera, 1961; Musser and Carleton, 1993) until reinstated as a species by Tribe (1996). Distributional occurrence and morphological discrimination with respect to *R. nitela* to the west and *R. macrurus* to the east require amplification. Sympatry reported with *R. nitela* at Serra do Roncador (Tribe, 1996).

Rhipidomys fulviventer Thomas, 1896. Ann. Mag. Nat. Hist., ser. 6, 18:304.
COMMON NAME: Tawny-bellied Rhipidomys.
TYPE LOCALITY: Colombia, Cundinamarca Dept., Aguadulce, 2400 ft (732 m).
DISTRIBUTION: Isolated populations in the Andes of SW and C Colombia, and W and NE Venezuela.
STATUS: IUCN – Lower Risk (lc).
SYNONYMS: *elatturus* Osgood, 1914; *similis* J. A. Allen, 1912; *tenuicauda* (J. A. Allen, 1899).
COMMENTS: *R. fulviventer* section *sensu* Tribe (1996). Listed as a subspecies of *R. latimanus* by Cabrera (1961) but clearly a species separate from the *latimanus-venezuelae* complex in Venezuela (Handley, 1976; Tribe, 1996). Includes *venustus* as a subspecies according to Tribe (1996) but not Linares (1998). To us, the considerable variation among Venezuelan series (USNM) indicates the presence of at least two species: *venustus* proper in the Mérida Andes and *fulviventer sensu lato* in the coastal mountains (*tenuicauda*) and Páramo de Tamá (*elatturus*). Whether the latter two small forms are conspecific with *fulviventer* proper, as provisionally arranged here, invites further study using other information bases. The highly disjunct distribution of *fulviventer*, as conveyed by the

listed synonyms, must in part reflect the need for additional field inventory. Equally plausible, revisionary investigation may uncover complementary, geographically replacing, specific ranges among these synonyms; the status and relationships of *similis* and *tenuicauda* are foremost candidates for such renewed attention. Compared with *R. wetzeli* by Gardner (1989).

Rhipidomys gardneri Patton, da Silva, and Malcolm, 2000. Bull. Am. Mus. Nat. Hist., 244:165.

COMMON NAME: Gardner's Rhipidomys.

TYPE LOCALITY: Perú, Madre de Dios Dept., Reserva Cusco Amazónico, 14 km E Puerto Maldonado, left (= north) bank of Río Madre de Dios, 200 m; 12°33'S, 69°03'W.

DISTRIBUTION: Westernmost Brazil (Acre) and lowlands of SE Perú, perhaps including the valley of the Río Ucayali.

COMMENTS: *R. leucodactylus* section *sensu* Tribe (1996). Morphology and karyotype contrasted to *R. leucodactylus* by Patton et al. (2000).

Rhipidomys latimanus (Tomes, 1860). Proc. Zool. Soc. Lond., 1860:213.

COMMON NAME: Broad-footed Rhipidomys.

TYPE LOCALITY: Ecuador, Chimborazo Prov., Pallatanga, 1485 m.

DISTRIBUTION: Mid-elevation Andean forests, about 450-2200 m, of C and W Colombia, C Ecuador, and extreme N Perú; allopatric populations in easternmost Panamá.

STATUS: IUCN – Vulnerable as *R. scandens*, Lower Risk (lc) as *R. latimanus*.

SYNONYMS: *cocalensis* J. A. Allen, 1912; *microtis* Thomas, 1896; *mollissimus* J. A. Allen, 1912; *pictor* Thomas, 1904; *quindianus* J. A. Allen, 1913; *scandens* Goldman, 1913.

COMMENTS: *R. leucodactylus* section *sensu* Tribe (1996). Includes *scandens* Goldman, previously recognized as a species (e.g., Handley, 1966a; Musser and Carleton, 1993) but which Tribe (1996) placed in full synonymy under *R. l. latimanus*. Cabrera (1961) also considered *fulviventer* and *venustus* as subspecies of *R. latimanus*, but these two forms are clearly distinct from the *latimanus-venezuelae* complex in northern South America (Handley, 1976; Tribe, 1996). Karyotype reported by Gardner and Patton (1976).

   Handley (1976) suggested that Andean populations in E Colombia and N Venezuela, which he identified as *R. venezuelae*, may form the eastern component of *R. latimanus*, and Tribe (1996) formally arranged the former as a subspecies of the latter. Tribe's morphometric analyses, however, disclose appreciable differentiation between the taxa; his results, and the biogeographic complexity documented for other groups distributed over these same Andean ranges, persuade us to provisionally retain the two as separate species. While undoubtedly closely related, the issue of their synonymy merits unambiguous verification with other data.

Rhipidomys leucodactylus (Tschudi, 1845). Untersuchungen uber die Fauna peruana (Therologie), p. 183.

COMMON NAME: White-footed Rhipidomys.

TYPE LOCALITY: Perú, Region Andrés Avelino Cáceres (former Junín Dept.), Montaña de Vítoc area (as restricted by Tribe, 1996:201).

DISTRIBUTION: Guianas, S Venezuela, N and C Brazil, Ecuador, Perú, and WC Bolivia (see Anderson, 1993); extralimital in W Ecuador and NW Perú.

STATUS: IUCN – Lower Risk (lc).

SYNONYMS: *aratayae* Guillotin and Petter, 1984; *bovallii* Thomas, 1911; *equatoris* Thomas, 1915; *goodfellowi* Thomas, 1900; *lucullus* Thomas, 1911; *rex* Thomas, 1927; *sclateri* (Thomas, 1887).

COMMENTS: *R. leucodactylus* section *sensu* Tribe (1996). The homogeneity of junior epithets and populations gathered under *R. leucodactylus*, especially those west of the Andes, should be tested with other data and new collections. Thomas (1887b), e.g., named *sclateri* as a species from British Guiana and viewed it as the eastern counterpart of Peruvian *R. leucodactylus*; Guillotin and Petter (1984), however, described *aratayae* as a subspecies of *R. leucodactylus* from French Guiana; Voss et al. (2001) provisionally accepted both as examples of *R. leucodactylus*. Morphology and karyotype of W Brazil samples characterized by Patton et al. (2000), who compared the species with *R. gardneri*.

Rhipidomys macconnelli de Winton, 1900. Trans. Linn. Soc. Lond, 8:52.
> COMMON NAME: Tepui Rhipidomys.
> TYPE LOCALITY: Venezuela, Bolívar State, Mt Roraima, 8700 ft (2652 m).
> DISTRIBUTION: Highlands of S Venezuela (Bolívar and Amazonas) and neighboring parts of N Brazil and W Guyana.
> STATUS: IUCN – Lower Risk (lc).
> SYNONYMS: *subnubis* Tate, 1939.
> COMMENTS: *R. macconnelli* section *sensu* Tribe (1996). Sometimes misallocated as a species of *Thomasomys* (e.g., Gyldenstolpe, 1932) but see Hershkovitz (1959*b*). Tribe (1996:194) viewed *macconnelli* "as a possible basal offshoot of the genus."

Rhipidomys macrurus (Gervais, 1855). Expédition dans les parties centrales de l'Amerique dus Sud, Zoologie, Tome 1:111.
> COMMON NAME: Long-tailed Rhipidomys.
> TYPE LOCALITY: Brazil, Goiás State, Crixás.
> DISTRIBUTION: Gallery and semideciduous forests of the cerrado-caatinga biomes (Ceará southwestwards to E Mato Grosso and Minas Gerais), C Brazil.
> SYNONYMS: *cearanus* Thomas, 1910.
> COMMENTS: *R. leucodactylus* section *sensu* Tribe (1996). Initially reassociated as a subspecies of *R. venezuelae* (Gyldenstolpe, 1932) and later of *R. mastacalis* (Cabrera, 1961; Musser and Carleton, 1993). Resurrected as a species by Tribe (1996), containing those populations with 2n = 44 and low fundamental numbers, 48-51 (Svartman and Almeida, 1993; Zanchin et al., 1992*a*). The high-FN specimens reported by Zanchin et al. (1992*a*) as the species *cearanus* were reidentified by Tribe (1996) as *R. mastacalis*.

Rhipidomys mastacalis (Lund, 1840). K. Dansk. Vid. Selsk. Naturv. Math. Afhandl., p. 24.
> COMMON NAME: Atlantic Forest Rhipidomys.
> TYPE LOCALITY: Brazil, Minas Gerais State, Rio das Velhas, Lagoa Santa.
> DISTRIBUTION: Atlantic Forest region, SE Brazil (Pernambuco to São Paulo).
> STATUS: IUCN – Lower Risk (lc).
> SYNONYMS: *leucodactylus* (Wagner, 1845) [not Tschudi, 1845]; *maculipes* (Pictet and Pictet, 1844).
> COMMENTS: *R. leucodactylus* section *sensu* Tribe (1996). Comprehending the definition and distribution of Lund's *mastacalis* stands at the center of systematic problems involving medium-sized rhipidomys with whitish venters. As broadly conceived by Cabrera (1961), the form contained, as subspecies or complete synonyms, the following taxa now associated with other species: *R. emiliae*, *R. macrurus*, *R. nitela* (including *fervidus* and *yuruanus*), *tenuicauda* (= *R. fulviventer*), and *R. venezuelae* (see individual accounts). The restricted concept of the species, as currently understood, circumscribes yet another form endemic to the Atlantic Forest zone, SE Brazil. Although the morphological distinction between *R. macrurus* and *R. mastacalis* is not well marked (Tribe, 1996), chromosomal studies consistently reveal populations with low (48-52) versus high (74-80) fundamental numbers, respectively (2n = 44 in all; Andrades-Miranda et al., 2002*a*; Svartman and Almeida, 1993; Zanchin et al., 1992*a*). The distributions of the two in SE Brazil appear to intertwine in complex fashion; other species are likely involved as noted by Tribe (1996), who cited many specimens as *cf. macrurus* or as species indeterminate.

Rhipidomys modicus Thomas, 1926. Ann. Mag. Nat. Hist., ser. 9, 18:161.
> COMMON NAME: Peruvian Rhipidomys.
> TYPE LOCALITY: Perú, San Martín Dept., Puca Tambo, about 50 mi (80 km) E Chachapoyas, 5100 ft (1555 m).
> DISTRIBUTION: Middle elevations, 700-1800 m, of E Andean slopes, C Perú.
> COMMENTS: *R. leucodactylus* section *sensu* Tribe (1996). Arranged as a subspecies of *R. leucodactylus* by Cabrera (1961) and as a subjective synonym of *R. couesi* by Musser and Carleton (1993). Specific status reasserted and diagnosis emended by Tribe (1996), who recorded its sympatry with examples of *R. leucodactylus*.

Rhipidomys nitela Thomas, 1901. Ann. Mag. Nat. Hist., ser. 7, 8:148.
   COMMON NAME: Guianan Rhipidomys.
   TYPE LOCALITY: Guyana, Rupununi Dist., Kanuku Mountains, Kwaimatta, 240 ft (73 m).
   DISTRIBUTION: Amazonian lowlands in S Venezuela, Guianas, and NC Brazil. The reports of
         the species in Bolivia (Anderson, 1997) and Colombia (Tribe, 1996) cannot be
         confirmed, and the species appears to be confined to Amazonia east of the Rios Negro-
         Madeira as so far documented (see Voss et al., 2001).
   STATUS: IUCN – Lower Risk (lc).
   SYNONYMS: *fervidus* Thomas, 1904; *milleri* J. A. Allen, 1913; *yuruanus* J. A. Allen, 1913.
   COMMENTS: *R. leucodactylus* section *sensu* Tribe (1996). Previously listed either as a race of
         R. venezuelae (Gyldenstolpe, 1932) or *R. mastacalis* (Cabrera, 1961), but distinct from
         true *venezuelae* in Venezuela (see Handley, 1976, who used the species name *mastaca-
         lis*) and from the *mastacalis* complex in EC Brazil (Tribe, 1996). Morphometric and
         chromosomal variation cautions that populations arranged as *nitela* represent a
         composite of two species (Tribe, 1996; Volobouev and Catzeflis, 2000); Andrades-
         Miranda et al. (2002*a*) considered the 2n = 48 and FN = 68 karyotype to represent
         R. nitela proper.

Rhipidomys ochrogaster J. A. Allen, 1901. Bull. Am. Mus. Nat. Hist., 14:43.
   COMMON NAME: Buff-bellied Rhipidomys.
   TYPE LOCALITY: Perú, Puno Dept., valley of Río Inambari, Inca Mines (= Santo Domingo),
         6000 ft (1830 m).
   DISTRIBUTION: Known only from the type locality and vicinity, SE Perú.
   STATUS: IUCN – Lower Risk (nt).
   COMMENTS: *R. leucodactylus* section *sensu* Tribe (1996). Retained as a species until judged the
         same as *R. l. leucodactylus* by Cabrera (1961). J. A. Allen (1901*b*), however, pointedly
         contrasted his new species with Peruvian *R. leucodactylus*; Musser and Carleton (1993)
         and Tribe (1996) retained it as species. Geographic range and closest specific relative
         uncertain.

Rhipidomys venezuelae Thomas, 1896. Ann. Mag. Nat. Hist., ser. 6, 18:303.
   COMMON NAME: Venezuelan Rhipidomys.
   TYPE LOCALITY: Venezuela, Mérida State, Mérida, 1630 m.
   DISTRIBUTION: Mountains of N and W Venezuela, including Trinidad and Tobago, and
         E Colombia.
   STATUS: IUCN – Lower Risk (lc).
   SYNONYMS: *tobagi* Goodwin, 1961.
   COMMENTS: *R. leucodactylus* section *sensu* Tribe (1996). This form may represent
         the E Andean complement of *R. latimanus* (see that account). The provisional
         association of *R. nitela tobagi* Goodwin, heretofore placed as a synonym of *R. nitela*
         (e.g., Musser and Carleton, 1993; Tribe, 1996), follows the observations of Voss et al.
         (2001).

Rhipidomys venustus Thomas, 1900. Ann. Mag. Nat. Hist., ser. 7, 5:152.
   COMMON NAME: Mérida Rhipidomys.
   TYPE LOCALITY: Venezuela, Mérida, Las Vegas del Chama, 1400 m.
   DISTRIBUTION: Isolated populations in mountains of W and N Venezuela.
   STATUS: IUCN – Lower Risk (lc).
   COMMENTS: *R. fulviventer* section *sensu* Tribe (1996). Placed as a subspecies of *R. latimanus* by
         Cabrera (1961) but distinct from the *latimanus-venezuelae* complex in Venezuela
         (Handley, 1976; Tribe, 1996); or placed as a subspecies if *R. fulviventer* by Tribe (1996).
         See comments in that account.

Rhipidomys wetzeli Gardner, 1989. *In* Eisenberg, Advances in Neotropical Mammalogy, p. 417.
   COMMON NAME: Wetzel's Rhipidomys.
   TYPE LOCALITY: Venezuela, Territorio Federal Amazonas, Cerro de la Neblina, Camp VII,
         1800 m; 00°50′40″N, 65°58′10″W.
   DISTRIBUTION: Highlands in S Venezuela (Bolívar and Amazonas) and N Brazil (as per
         Fonseca et al., 1996).

STATUS: IUCN – Lower Risk (lc).

COMMENTS: *R. fulviventer* section *sensu* Tribe (1996). A small species, contrasted to *R. fulviventer* by Gardner (1989) and morphometrically well differentiated from the *fulviventer-venustus* complex (Tribe, 1996).

Salinomys Braun and Mares, 1995. J. Mammal., 76:505.

TYPE SPECIES: *Salinomys delicatus* Braun and Mares, 1995.

COMMENTS: Phyllotini. Sister taxon to the *Andalgalomys-Graomys* clade according to phylogenetic interpretation of morphological traits (Braun and Mares, 1995) or only to *Andalgalomys* per cytochrome *b* data (Anderson and Yates, 2000). Also see remarks under *Andalgalomys*.

Salinomys delicatus Braun and Mares, 1995. J. Mammal., 76:514.

COMMON NAME: Delicate Salinomys.

TYPE LOCALITY: Argentina, San Luis Prov., Ayacucho Dept., 23 km N Route 20, Pampa de Las Salinas, La Botija, 1300 ft (396 m).

DISTRIBUTION: Scrublands associated with salt flats, 380-412 m, WC Argentina (San Juan and San Luis).

Scapteromys Waterhouse, 1837. Proc. Zool. Soc. Lond., 1837:20.

TYPE SPECIES: *Mus tumidus* Waterhouse, 1837.

COMMENTS: Akodontini. Hershkovitz (1966c) viewed *Scapteromys* as the cognate of *Kunsia*, a close kinship supported by phylogenetic investigations of gene-sequence data, but not including *Bibimys* as third member of a scapteromyine tribe *sensu* Massoia (1979b, 1980a) (see D'Elía, 2003; D'Elía et al. 2003; Smith and Patton, 1999). Hershkovitz (1966c) removed *tomentosus* and *fronto* to *Kunsia*, and Massoia (1980a) transferred *chacoensis* and *labiosus* to *Bibimys*. Karyology summarized by Gardner and Patton (1976) and Brum-Zorilla et al. (1986). Fossil occurrences (late Pliocene through Holocene) documented by Reig (1994) and Pardiñas (1996, 1999). Generic revision required to consolidate the morphological basis for specific discrimination and to vouch geographic occurrences. Also see accounts of *Bibimys* and *Kunsia*.

Scapteromys aquaticus Thomas, 1920. Ann. Mag. Nat. Hist., ser. 9, 5:477.

COMMON NAME: Argentine Swamp Rat.

TYPE LOCALITY: Argentina, Buenos Aires Prov., Isla Ella.

DISTRIBUTION: EC Argentina and E Paraguay.

COMMENTS: Retained as a species (e.g., Cabrera, 1961; Gyldenstolpe, 1932) until Massoia and Fornes (1964) realigned *aquaticus* as a subspecies of *S. tumidus*, a reallocation commonly observed in later systematic works (Corbet and Hill, 1991; Hershkovitz, 1966c; Honacki et al., 1982; Musser and Carleton, 1993). Karyotypic differences consistently disclosed between *aquaticus* (2n = 32) and *tumidus* (2n = 24) have persuaded others to retain the two as distinct species (Brum et al., 1973; Brum-Zorilla et al., 1986; Gardner and Patton, 1976; Gentile de Fronza, 1970), and they are represented as genetically separable in phylogenetic studies of cytochrome *b* data (D'Elía et al., 2003, in press).

Scapteromys tumidus (Waterhouse, 1837). Proc. Zool. Soc. Lond., 1837:15.

COMMON NAME: Uruguay Swamp Rat.

TYPE LOCALITY: Uruguay, Maldonado Dept., Maldonado.

DISTRIBUTION: Southernmost Brazil (Rio Grande do Sul) and Uruguay.

STATUS: IUCN – Lower Risk (lc).

COMMENTS: Distribution and morphological identity reviewed by Massoia and Fornes (1964) and Hershkovitz (1966c), including *aquaticus*. Review of museum specimens indicates that only a single species, *S. tumidus*, is present in Uruguay (González, 1994). Also see remarks under *S. aquaticus*.

Scolomys Anthony, 1924. Am. Mus. Novit., 139:1.

TYPE SPECIES: *Scolomys melanops* Anthony, 1924.

COMMENTS: Sigmodontinae *incertae sedis*. Listed in tribe Oryzomyini by Reig (1984), a tribal affinity supported by Voss and Carleton (1993) and Patton and da Silva (1995). Unlike other oryzomyines, however, specimens of *Scolomys* lack pectoral mammae (e.g., Gyldenstolpe, 1932; Patton and da Silva, 1995), and subsequent phyletic studies of mitochondrial DNA (cytochrome *b*) do not support close affinity with representatives of that tribe (D'Elía et al., 2003; Smith and Patton, 1999); discussion of these results and their bearing on tribal membership discussed by Gómez-Laverde et al. (2004). Defining generic traits emended and contrasted with those of *Neacomys* by Patton and da Silva (1995) and Patton et al. (2000). Distributional records, karyotypes, morphological variation, and natural history contained in the generic reviews of Patton and da Silva (1995) and Gómez-Laverde et al. (2004).

Scolomys melanops Anthony, 1924. Am. Mus. Novit., 139:2.

COMMON NAME: Short-nosed Scolomys.

TYPE LOCALITY: Ecuador, Pastaza Prov., Mera, 3800 ft (1158 m).

DISTRIBUTION: E Ecuador and NE Perú (Gómez-Laverde et al., 2004:Fig.1; Hice, 2001).

STATUS: IUCN – Endangered.

COMMENTS: Discrimination from *S. ucayalensis* amplified by Gómez-Laverde et al. (2004).

Scolomys ucayalensis Pacheco, 1991. Publ. Mus. Hist. Nat., Ser. A Zool., Univ. Nac. Mayor de San Marcos, 37:1.

COMMON NAME: Long-nosed Scolomys.

TYPE LOCALITY: Perú, Loreto Dept., 2.8 km E Jenaro Herrera, right bank of Río Ucayali, 135 m; 79°39'W, 04°52'S.

DISTRIBUTION: S Colombia, NE Peru, and westernmost Brazil (Acre and Amazonas) (Gómez-Laverde et al., 2004:Fig. 1).

STATUS: IUCN – Endangered.

SYNONYMS: *juruaense* Patton and da Silva 1995.

COMMENTS: Based on qualitative traits, morphometric analyses, and cytochrome *b* sequences, Gómez-Laverde et al. (2004) considered the recently described *juruaense* as insufficiently differentiated to warrant specific separation.

Sigmodon Say and Ord, 1825. J. Acad. Nat. Sci. Philadelphia, 4(2):352.

TYPE SPECIES: *Sigmodon hispidus* Say and Ord, 1825.

SYNONYMS: *Deilemys* Saussure, 1860; *Lasiomys* Burmeister, 1854; *Sigmomys* Thomas, 1901.

COMMENTS: Sigmodontini. Hershkovitz (1955*a*) arranged *Sigmodon* with *Holochilus*, *Neotomys*, and *Reithrodon* as the sigmodont group, but other evidence has eroded the tribal-level affinity of each to *Sigmodon* (see those generic accounts) and leaves the genus as the sole living tribal member. Cladistic isolation from other sigmodontine genera, typically as a basal lineage, affirmed in broad taxonomic surveys of phallic morphology (Hooper and Musser, 1964*a*) and mitochondrial genes (D'Elía et al., 2003; Engel et al., 1998; Smith and Patton, 1999). Ambiguous evidence for geographic origin, North versus South America, discussed by Voss (1992) and Peppers et al. (2002).

Sigmomys, type species *Sigmodon alstoni*, has been variously treated as a distinct genus (Ellerman, 1941; Gyldenstolpe, 1932; Handley, 1976), a subgenus of *Sigmodon* (Husson, 1978), or a full synonym (Cabrera, 1961; Hershkovitz, 1955*a*; Voss, 1992). Peppers et al. (2002) discussed the possible recognition of *alstoni* within a monotypic genus, subgenus, or simply species group; in view of its morphological distinctiveness (Voss, 1992) and cladistic dichotomy relative to all other *Sigmodon* species (Peppers et al., 2002), the formal rank of subgenus seems appropriate.

North American forms revised by Bailey (1902), with subsequent partial revisions by Baker (1969), Zimmerman (1970), Voss (1992), Carleton et al. (1999), and Peppers and Bradley (2000); further alpha-taxonomic study dearly needed. Standard karyology summarized by Zimmerman (1970) and Carleton et al. (1999); banded karyotypes by Elder (1980) and Elder and Lee (1985); albumin differentiation by Fuller et al. (1984); and mitochondrial DNA variation by Peppers and Bradley (2000) and Peppers et al. (2002). Skin-and-skull morphology (Baker, 1969; Bailey, 1902), karyology (Elder and Lee, 1985; Zimmerman, 1970), and molecular data (Peppers et al., 2002) have yielded conflicting pictures of species-group associations. Species groups observed here follow the last study, which broadly treats most species under the revised understanding of *S. hispidus*.

Sigmodon alleni Bailey, 1902. Proc. Biol. Soc. Wash., 15:112.
>   COMMON NAME: Allen's Cotton Rat.
>   TYPE LOCALITY: México, Jalisco State, Mascota, Mineral San Sebastián.
>   DISTRIBUTION: W México, from S Sinaloa to S Oaxaca.
>   STATUS: IUCN – Lower Risk (lc).
>   SYNONYMS: *guerrerensis* Nelson and Goldman, 1933; *macdougalli* Goodwin, 1955; *macrodon*
>   Goodwin, 1955; *vulcani* J. A. Allen, 1906.
>   COMMENTS: Subgenus *Sigmodon*, *S. hispidus* species group. Sister species to *S. hirsutus*
>   according to phylogenetic interpretation of cytochrome *b* data (Peppers and Bradley,
>   2000). The forms *guerrerensis* and *vulcani* appear to be proper synonyms of *S. alleni*, but
>   the status of the Oaxacan populations described as *macdougalli* and *macrodon* deserves
>   renewed study and confirmation. Formerly included *planifrons* according to Baker's
>   (1969) revision, here regarded as a species. See Shump and Baker (1978, Mammalian
>   Species, 95).

Sigmodon alstoni (Thomas, 1881). Proc. Zool. Soc. Lond., 1880:691 [1881].
>   COMMON NAME: Alston's Cotton Rat.
>   TYPE LOCALITY: Venezuela, Sucre State, Cumaná.
>   DISTRIBUTION: Intermittently distributed in savannas over NE Colombia, N and E Vene-
>   zuela, Guyana, Surinam, and N Brazil (regional aspects of range amplified by Husson,
>   1978; Linares, 1998; and Voss, 1992).
>   STATUS: IUCN – Lower Risk (lc).
>   SYNONYMS: *savannarum* (Thomas, 1901); *venester* (Thomas, 1914).
>   COMMENTS: Subgenus *Sigmomys*. Suspected to be "individuals of South American
>   representatives of *S. hispidus* with grooved incisors" (Hershkovitz, 1955a:647), but
>   distantly related to other species of *Sigmodon* according to phylogenetic interpretation
>   of cytochrome *b* sequences (Peppers and Bradley, 2000; Peppers et al., 2002). Voss
>   (1992) discerned no morphological or morphometric foundation for delineating
>   subspecies, but Linares (1998) recognized *savannarum* and *venester*, along with the
>   nominate form, as races for Venezuelan populations.

Sigmodon arizonae Mearns, 1890. Bull. Am. Mus. Nat. Hist., 2:287.
>   COMMON NAME: Arizona Cotton Rat.
>   TYPE LOCALITY: USA, Arizona, Yavapai Co., 3 mi (5 km) SE Camp Verde, Bell's Ranch
>   (as restricted by Hoffmeister, 1986).
>   DISTRIBUTION: Extreme SE California, SC Arizona, and extreme SW New Mexico, USA;
>   south along coastal plain and adjoining foothills, sea level-1900 m, in W México to
>   S Nayarit.
>   STATUS: IUCN – Extinct as *S. a. arizonae*, Lower Risk (nt) as *S. a. plenus*, otherwise Lower Risk
>   (lc).
>   SYNONYMS: *cienegae* A. B. Howell, 1919; *jacksoni* Goldman, 1918; *major* Bailey, 1902; *plenus*
>   Goldman, 1928.
>   COMMENTS: Subgenus *Sigmodon*, *S. hispidus* species group. Zimmerman (1970) recognized
>   the chromosomal and morphological differences that support the specific separation
>   of *S. arizonae* from *S. hispidus*, under which it had been arranged as a subspecies (Bailey,
>   1902; Mearns, 1890). Distribution and differentiating traits amplified by Severinghaus
>   and Hoffmeister (1978), Elder and Lee (1985), Hoffmeister (1986), Carleton et al.
>   (1999), and Frey et al. (2002). Association of *jacksoni*, known only by the type speci-
>   men, as a junior synonym has been questioned (Carleton et al., 1999; Hoffmeister,
>   1986). Hall (1981) retained all named forms as subspecies, a classificatory formality
>   unwarranted without fresh studies of populational variation across the species' range.
>   Probable sister species of *S. mascotensis* according to a variety of data (Carleton et al.,
>   1999; Elder, 1980; Peppers and Bradley, 2000; Peppers et al., 2002; Zimmerman, 1970).

Sigmodon fulviventer J. A. Allen, 1889. Bull. Am. Mus. Nat. Hist., 2:180.
>   COMMON NAME: Tawny-bellied Cotton Rat.
>   TYPE LOCALITY: México, Zacatecas State, Zacatecas.
>   DISTRIBUTION: SE Arizona, WC New Mexico, and SW Texas (see Stangl, 1992), USA, south
>   through interior México to Guanajuato and NW Michoacán.

STATUS: IUCN – Extinct as *S. f. goldmani*, otherwise Lower Risk (lc).

SYNONYMS: *dalquesti* Stangl, Jr., 1992; *goldmani* Bailey, 1913; *melanotis* Bailey, 1902; *minimus* Mearns, 1894; *woodi* M. C. Gardner, 1948.

COMMENTS: Subgenus *Sigmodon*, *S. fulviventer* species group. Sister species to *S. leucotis* according to phylogenetic interpretation of cytochrome *b* data (Peppers and Bradley, 2000; Peppers et al., 2002). See Baker and Shump (1978, Mammalian Species, 94).

Sigmodon hirsutus (Burmeister, 1854). Abhandlungen Naturforschender Gesellschaft Halle, Sitzungberichte, 2:16.

COMMON NAME: Southern Cotton Rat.

TYPE LOCALITY: Venezuela, Zulia State, Maracaibo.

DISTRIBUTION: Nicaragua to C Panamá, Central America; in South America to N Colombia and N Venezuela (see Linares, 1998:Fig. 171); northern limits uncertain.

SYNONYMS: *austerulus* Bangs, 1902; *bogotensis* J. A. Allen, 1897; *borucae* J. A. Allen 1897; *chiriquensis* J. A. Allen, 1904; *griseus* J. A. Allen, 1908; *sanctaemartae* Bangs, 1898.

COMMENTS: Subgenus *Sigmodon*, *S. hispidus* species group. Central and South American taxa included as subspecies of *S. hispidus* (Cabrera, 1961; Hall and Kelson, 1959), observing the polytypic species template handed down from Bailey (1902). Genetic separation, inferred from cytochrome *b* sequence data, of southern populations revealed by Peppers and Bradley (2000), who reinstated *hirsutus* as species; sister species to *S. alleni* based on those data (and Peppers et al., 2002). Among the species-group synonyms, the South American taxa *bogotenis* and *sanctaemartae*, which have yet to be surveyed genetically, warrant additional attention; Gyldenstolpe (1932) and Ellerman (1941) listed both as species.

Sigmodon hispidus Say and Ord, 1825. J. Acad. Nat. Sci. Philadelphia, 42:354.

COMMON NAME: Hispid Cotton Rat.

TYPE LOCALITY: USA, Florida, St. Johns River.

DISTRIBUTION: SE USA, from S Nebraska to C Virginia and south to SE Arizona and peninsular Florida; NW Chihuahua to N Tamaulipas, south through interior México at least to C Zacatecas and W San Luis Potosí (southern limits with respect to *S. mascotensis* and *S. toltecus* need reverification).

STATUS: IUCN – Lower Risk (nt) as *S. h. eremicus* and *S. h. insulicola*, otherwise Lower Risk (lc).

SYNONYMS: *alfredi* Goldman and M. C. Gardner, 1947; *berlandieri* Baird, 1855; *confinis* Goldman, 1918; *eremicus* Mearns, 1897; *exsputus* G. M. Allen, 1920; *floridanus* A. H. Howell, 1943; *insulicola* A. H. Howell, 1943; *komareki* M. C. Gardner, 1948; *littoralis* Chapman, 1889; *pallidus* Mearns, 1897; *solus* Hall, 1951; *spadicipygus* Bangs, 1898; *texianus* (Audubon and Bachman, 1853); *virginianus* M. C. Gardner, 1946.

COMMENTS: Subgenus *Sigmodon*, *S. hispidus* species group. The morphological limits of *hispidus*, as set forth by Bailey (1902) and expanded by others (Cabrera, 1961; Hall, 1981; Hall and Kelson, 1959; Miller, 1924), were first uncovered as composite by Zimmerman (1970), an insight supported and extended by others (Carleton et al., 1999; Peppers and Bradley, 2000; Severinghaus and Hoffmeister, 1978; Voss, 1992). The much reified definition of *S. hispidus* corresponds to a species restricted to open landscapes in the SC and SE USA and NC México (see Carleton et al., 1999; Peppers and Bradley, 2000). See comments under *S. arizonae*, *S. hirsutus*, *S. inopinatus*, *S. mascotensis*, *S. peruanus*, *S. toltecus*, and *S. zanjonensis*, forms formerly ranked as subspecies.

Given the substantial reconstitution of *S. hispidus*, conventional recognition of subspecies (e.g., Hall, 1981) seems pointless without fresh assessment of geographic variation. Also see remarks by L. Choate (1997) on the subspecies occurring in the Llano Estacado, New Mexico-Texas, by Goetze (1998) for those on the Edwards Plateau, C Texas, and by Whitaker and Hamilton (1998) on SE populations and recognizable subspecies. Aberrant somatic chromosomal variation within local populations and its significance for characterizing interpopulational karyotypic differences discussed by Berend (1993). See Cameron and Spencer (1981, Mammalian Species, 158).

Sigmodon inopinatus Anthony, 1924. Am. Mus. Novit., 114:3.
COMMON NAME: Ecuadorian Cotton Rat.
TYPE LOCALITY: Ecuador, Chimborazo Prov., Mt Chimborazo, Urbina, 11,400 ft (3474 m).
DISTRIBUTION: Known only from high Andes in Azuay and Chimborazo provinces, Ecuador.
STATUS: IUCN – Lower Risk (lc).
COMMENTS: Subgenus *Sigmodon* (*S. fulviventer* species group?). Lumped under *S. hispidus* by
Cabrera (1961), following the footnoted opinion of Hershkovitz (1955a), but
distinctive morphology reasserted and species reinstated by Voss (1992).

Sigmodon leucotis Bailey, 1902. Proc. Biol. Soc. Wash., 15:115.
COMMON NAME: White-eared Cotton Rat.
TYPE LOCALITY: México, Zacatecas, Sierra de Valparaíso, 2653 m.
DISTRIBUTION: Interior México, from SW Chihuahua and S Nuevo Leon to C Oaxaca.
STATUS: IUCN – Lower Risk (lc).
SYNONYMS: *alticola* Bailey, 1902; *amoles* Bailey, 1902.
COMMENTS: Subgenus *Sigmodon*, *S. fulviventer* species group. See Shump and Baker (1978,
Mammalian Species, 96).

Sigmodon mascotensis J. A. Allen, 1897. Bull. Am. Mus. Nat. Hist., 9:54.
COMMON NAME: West Mexican Cotton Rat.
TYPE LOCALITY: México, Jalisco State, Mascota, Mineral San Sebastián, 3300 ft (1006 m).
DISTRIBUTION: W México, from S Nayarit and SW Zacatecas southwards to extreme
SW Chiapas; along interior arid basins as far east as W Hidalgo, W Puebla, and
NW Oaxaca (see Carleton et al., 1999:Fig. 16).
STATUS: IUCN – Lower Risk (lc).
SYNONYMS: *atratus* Hall, 1949; *colimae* J. A. Allen, 1897; *inexoratus* Elliot, 1903; *ischyrus*
Goodwin, 1956; *obvelatus* Russell, 1952; *tonalensis* Bailey, 1902.
COMMENTS: Subgenus *Sigmodon*, *S. hispidus* species group. Zimmerman (1970) removed
mascotensis from *S. hispidus* and reinstated it as species, a change supported by
Severinghaus and Hoffmeister (1978) and Elder and Lee (1985). Revised by Carleton
et al. (1999), who elaborated upon its morphological separation from *S. arizonae* (see
above account) and *S. hispidus* and identified other junior synonyms mistakingly
allocated to *S. hispidus*; see those authors for remarks on broad patterns of intraspecific
variation, which do not concord with conventional subspecies boundaries (e.g., Hall,
1981).

Sigmodon ochrognathus Bailey, 1902. Proc. Biol. Soc. Wash., 15:115.
COMMON NAME: Yellow-nosed Cotton Rat.
TYPE LOCALITY: USA, Texas, Brewster Co., Chisos Mountains, 8000 ft (2438 m).
DISTRIBUTION: SE Arizona, extreme SW New Mexico, and Transpecos, Texas, USA, south to
C Durango, México; northern outlier population may persist in Guadalupe Mtns,
Transpecos, Texas (see Stangl and Dalquest, 1991).
STATUS: IUCN – Lower Risk (lc).
SYNONYMS: *baileyi* J. A. Allen, 1903; *madrensis* Goldman and M. C. Gardner, 1947; *montanus*
Benson, 1940.
COMMENTS: Subgenus *Sigmodon*, *S. hispidus* species group. Basal clade within the *S. hispidus*
assemblage according to phylogenetic interpretations of cytochrome *b* data (Peppers
et al., 2002), not with *S. fulviventer* and kin (Baker, 1969). Findley and Jones (1960)
detected no consistent pattern of size and pelage color variation to justify recognition
of subspecies, a conclusion sustained by the low genetic distances recorded among
populations (Carroll et al., 2002). See Baker and Shump (1978, Mammalian Species,
97).

Sigmodon peruanus J. A. Allen, 1897. Bull. Am. Mus. Nat. Hist., 9:118.
COMMON NAME: Peruvian Cotton Rat.
TYPE LOCALITY: Perú, La Libertad Dept., Trujillo, 200 ft (61 m).
DISTRIBUTION: Pacific coastal plain and contiguous Andean foothills of W Ecuador and
NW Perú.
STATUS: IUCN – Lower Risk (lc).

SYNONYMS: *chonensis* J. A. Allen, 1913; *lonnbergi* Thomas, 1921; *puna* J. A. Allen, 1903; *simonsi* J. A. Allen, 1901.

COMMENTS: Subgenus *Sigmodon*, *S. fulviventer* species group. Lumped under *S. hispidus* by Cabrera (1961) following the footnoted opinion of Hershkovitz (1955a); reinstated as a species by Voss (1992) and morphologically contrasted to that species and other South American taxa. The provocative molecular results presented by Peppers et al. (2002) phyletically link this South American endemic with Middle American *S. fulviventer* and *S. leucotis*.

Sigmodon planifrons Nelson and Goldman, 1933. Proc. Biol. Soc. Wash., 46:197.
COMMON NAME: Miahuatlán Cotton Rat.
TYPE LOCALITY: México, Oaxaca State, Juquila, 5000 ft (1524 m).
DISTRIBUTION: As so far known, limited to Sierra de Miahuatlán, S Oaxaca, México.
SYNONYMS: *minor* Goodwin, 1955; *setzeri* Goodwin, 1959 [replacement name for *minor*].
COMMENTS: Subgenus *Sigmodon* (*S. hispidus* species group?). Described as a species but relegated to a subspecies of *S. alleni* by Baker (1969). As noted by Nelson and Goldman (1933b), the diminutive size exhibited by the type series, as well as the material and sympatry later reported by Goodwin (1955b, 1969), suggests that the form is distinct from *S. alleni*; renewed study of the Oaxacan taxa named by Goodwin (1955b) will prove critical to much-needed revision of the *S. alleni* complex (see discussion in Carleton et al., 1999).

Sigmodon toltecus (Saussure, 1860). Rev. Mag. Zool. Paris, Ser. 2, 12:98.
COMMON NAME: Toltec Cotton Rat.
TYPE LOCALITY: México, Veracruz, near Mirador (as amended by Dalquest, 1953:163).
DISTRIBUTION: Foothills and lowlands of E México, across the Isthmus of Tehuantepec, into the Yucatán Peninsula and N Guatemala; range limits require vouchered documentation.
SYNONYMS: *fervidus* Lydekker, 1904 [lapsus for *furvus*]; *furvus* Bangs, 1903; *microdon* Bailey, 1902; *saturatus* Bailey, 1902.
COMMENTS: Subgenus *Sigmodon*, *S. hispidus* species group. Ranked as a subspecies of a broadly defined *S. hispidus* by Bailey (1902) and so classified thereafter (Hall and Kelson, 1959; Hall, 1981). Specific rank defended based on mitochrondrial DNA evaluations that disclose strong genetic divergence and cladistic separation from *S. hispidus* proper (Peppers and Bradley, 2000). Molecular studies have so far sampled *toltecus* and *furvus*; other species-group synonyms provisionally observe Carleton's et al. (1999) narrative outline of the *toltecus* complex; their synonymy needs confirmation. Sister species to the *S. alleni-S. hirsutus* clade according to phylogenetic interpretation of cytochrome *b* data (Peppers and Bradley, 2000; Peppers et al., 2002).

Sigmodon zanjonensis Goodwin, 1932. Am. Mus. Novit., 528:1.
COMMON NAME: Montane Cotton Rat.
TYPE LOCALITY: Guatemala, Quezaltenango Dept., Zanjón, 9000 ft (2743 m).
DISTRIBUTION: Highlands of Chiapas, México, and Guatemala.
SYNONYMS: *villae* Goodwin, 1958.
COMMENTS: Subgenus *Sigmodon* (*S. hispidus* species group?). Described as species, but soon afterwards relegated to subspecies by Goodwin (1934) and thereafter maintained (Hall and Kelson, 1959; Hall, 1981). With dismantling of the polyphyletic mishmash of *hispidus*, the status and relationship of these highland Central American taxa remain uncertain, as noted by Carleton et al. (1999). The geographic and altitudinal ranges as illustrated in conventional maps (e.g., Hall, 1981) are improbably broad and likely include lowland populations referrable to *S. toltecus* and possibly *S. hirsutus*; limits require critical analysis and vouchered documentation. In certain cranial and pelage traits, examples of *S. zanjonensis* resemble *S. alleni*; such impressions require empirical testing.

Sigmodontomys J. A. Allen, 1897. Bull. Am. Mus. Nat. Hist., 9:38.
TYPE SPECIES: *Sigmodontomys alfari* J. A. Allen, 1897.
COMMENTS: Oryzomyini. Named as a genus but later viewed as a synonym of *Nectomys*

(Ellerman, 1941; Gyldenstolpe, 1932), usually ranked as subgenus (Hershkovitz, 1944). Hershkovitz (1944:71) stressed the weakness of their association: "The apparent relationship of *Sigmodontomys* to *Nectomys* . . . is probably attributable to an independant development from the common oryzomyine stock rather than to a divergence from a more recent *Nectomys*-like stock," an assessment later corroborated by other data (Gardner and Patton, 1976; Hooper and Musser, 1964*a*). Gardner and Patton (1976) formally transferred *Sigmodontomys* to a subgenus of *Oryzomys*, and we continue to list it as genus, while urging the need to refine its specific contents and their relationships to *Oryzomys* proper. Cadena et al. (1998) reported a specimen as *Sigmodontomys* species indeterminate from the Chocó, SW Colombia.

Sigmodontomys alfari J. A. Allen, 1897. Bull. Am. Mus. Nat. Hist., 9:39.
COMMON NAME: Short-tailed Sigmodontomys.
TYPE LOCALITY: Costa Rica, Limón Prov., 3 mi (5 km) E Guápiles, Jiménez, 700 ft (213 m).
DISTRIBUTION: Lowland forest from E Honduras to Panamá; C and W Colombia to NW Venezuela and NW Ecuador (see Musser et al., 1998:Fig. 51).
STATUS: IUCN – Lower Risk (lc).
SYNONYMS: *barbacoas* (J. A. Allen, 1916); *efficax* Goldman, 1913; *esmeraldarum* Thomas, 1901; *ochraceus* (J. A. Allen, 1908); *ochrinus* (Thomas, 1921); *russulus* (Thomas, 1897).
COMMENTS: Karyotype reported by Gardner and Patton (1976). Musser et al. (1998) summarized locality records and compared them with other oryzomyines having trans-Andean distributions.

Sigmodontomys aphrastus (Harris, 1932). Occas. Pap. Mus. Zool., Univ. Michigan, 248:5.
COMMON NAME: Long-tailed Sigmodontomys.
TYPE LOCALITY: Costa Rica, San José Prov., San Joaquin de Dota, 4000 ft (1219 m).
DISTRIBUTION: Known only from the type locality, Chiriqui Prov. in W Panamá, and Pichincha Prov. in NC Ecuador (Voss, 1988:423).
STATUS: IUCN – Critically Endangered.
COMMENTS: Usually listed as a species of *Oryzomys* of uncertain relationship (Hall, 1981). Assignment to *Sigmodontomys* tentative following the observations of Ray (1962). Distribution meagerly documented, known by only four specimens.

Tapecomys Anderson and Yates, 2000. J. Mammal., 81:21.
TYPE SPECIES: *Tapecomys primus* Anderson and Yates, 2000.
COMMENTS: Phyllotini. A large phyllotine, as big as *Andinomys edax*, but morphologically, chromosomally, and genetically well differentiated from that species. Conflicting views of relationships and validity of generic status obtained in phylogenetic analyses—sister genus to the *Andalgalomys-Salinomys* clade according to parsimony analyses based on cytochrome *b* sequences (Anderson and Yates, 2000; D'Elía et al., 2003), or a species of *Graomys* (including *Andalgalomys*) as suggested by consensus trees based on morphological characters (Steppan, 1993; Steppan and Sullivan, 2000). Also see remarks under *Andalgalomys*.

Tapecomys primus Anderson and Yates, 2000. J. Mammal., 81:21.
COMMON NAME: Primordial Tapecua.
TYPE LOCALITY: Bolivia, Tarija Dept., Tapecua, 1500 m; 21°26'S, 63°55'W.
DISTRIBUTION: So far known only from the type locality, SE Bolivia.

Thalpomys Thomas, 1916. Ann. Mag. Nat. Hist., ser. 8, 18:339.
TYPE SPECIES: *Thalpomys lasiotis* Thomas, 1916 (not *Mus lasiotis* of Lund, 1841).
COMMENTS: Akodontini. Recognized as a genus, as initially diagnosed, by Gyldenstolpe (1932); reclassified as a subgenus of *Akodon* by Ellerman (1941) and so observed by Cabrera (1961), or considered a subjective synonym of *Bolomys* (Reig, 1987). See Hershkovitz (1990*a*) for availability of genus-group name, its differentiating characters from typical *Akodon*, and definition of included species.

Thalpomys cerradensis Hershkovitz, 1990. J. Nat. Hist., 24:777.
COMMON NAME: Cerrado Akodont.

TYPE LOCALITY: Brazil, Distrito Federal, Parque Nacional de Brasília, 1100 m.
DISTRIBUTION: Cerrado of C Brazil.
STATUS: IUCN – Lower Risk (lc).

Thalpomys lasiotis Thomas, 1916. Ann. Mag. Nat. Hist., ser. 8, 18:339.
COMMON NAME: Hairy-eared Akodont.
TYPE LOCALITY: Brazil, Minas Gerais State, Lagoa Santa.
DISTRIBUTION: Cerrado of C Brazil.
STATUS: IUCN – Lower Risk (lc).
SYNONYMS: *reinhardti* (Langguth, 1975).
COMMENTS: For synonymy of the unnecessary replacement name *Akodon reinhardti*, see Hershkovitz (1990*a*).

Thaptomys Thomas, 1916. Ann. Mag. Nat. Hist., ser. 8, 18:339.
TYPE SPECIES: *Hesperomys subterraneus* Hensel, 1873 (= *Mus nigrita* Lichtenstein, 1829).
COMMENTS: Akodontini. Following its description as an akodontine genus, the taxon has been usually retained as a subgenus of *Akodon* (Cabrera, 1961; Ellerman, 1941; Massoia, 1963*a*) or even a full synonym of the nominate subgenus (Reig, 1987). Recently, Hershkovitz (1990*a*, 1998) has reasserted the singular morphological traits of the taxon as "definitely non-*Akodon*" and concurred with Thomas' (1916*c*) generic segregation. Cladistic interpretations of cytochrome *b* sequences portray *nigrita* as genetically highly divergent from *Akodon* proper, either as sister group to that genus (Smith and Patton, 1999), to *Necromys* (D'Elía et al., 2003), or to the clade *Necromys-Thalpomys* (D'Elía, 2003).

Thaptomys nigrita (Lichtenstein, 1829). Darst. Säugeth., 7:pl. 35, fig. 1.
COMMON NAME: Ebony Akodont.
TYPE LOCALITY: Brazil, Rio de Janeiro State, vicinity of Rio de Janeiro.
DISTRIBUTION: SE Brazil (Bahia to Rio Grande do Sul), E Paraguay, and NE Argentina (Misiones).
STATUS: IUCN – Lower Risk (lc) as *Akodon nigrita*.
SYNONYMS: *fuliginosus* (Wagner, 1845); *henseli* (Leche, 1896); *orycter* (Lund, 1841); *subterraneus* (Hensel, 1873).
COMMENTS: Massoia (1963*a*) clarified the identity of *subterraneus*, the type of *Thaptomys*, as a synonym of the earlier named *nigrita* and amplified the distribution of the species.

Thomasomys Coues, 1884. Am. Nat., 18:1275.
TYPE SPECIES: *Hesperomys cinereus* Thomas, 1882.
SYNONYMS: *Erioryzomys* Bangs, 1900; *Inomys* Thomas, 1917.
COMMENTS: Thomasomyini. A speciose and taxonomically complex genus whose nomenclatural history is intertwined with *Aepeomys*, *Delomys*, and *Wilfredomys*, taxa that have been used as subgenera (e.g., Ellerman, 1941; Cabrera, 1961) or as genera (see their accounts). Comparative anatomical studies involving some *Thomasomys* include Carleton (1973), Hooper and Musser (1964*a*), and Voss and Linzey (1981); chromosomal numbers of several species published by Gardner and Patton (1976), Gómez-Laverde et al. (1997), and Aguilera et al. (2000). Based on 6-7 species surveyed, sister genus to *Chilomys* according to phylogenetic interpretations of mitochondrial DNA sequences (D'Elía et al., 2003; Smith and Patton, 1999). Revisionary studies required to critically overhaul species taxonomy and to place ranking of genus-group taxa in a phylogenetic context. Lacking a current revisionary standard, species recognized here basically follow Cabrera (1961), supplemented by our own museum examinations and subsequent descriptions.

Thomasomys apeco Leo L. and Gardner, 1993. Proc. Biol. Soc. Wash., 106:417.
COMMON NAME: Apeco Thomasomys.
TYPE LOCALITY: Perú, San Martín Dept., Parque Nacional Río Abiseo, ca. 25 km NE Pataz, Valle de Los Chochos, 3280 m.
DISTRIBUTION: Known only from the type locality and vicinity, in upper montane forest, 3250-3380 m, NC Perú.
COMMENTS: The largest living thomasomyine as so far known; differentially compared with *T. aureus* and *Megaoryzomys curioi*, an extinct Pleistocene form from the Galapagos Isls.

Thomasomys aureus (Tomes, 1860). Proc. Zool. Soc. Lond., 1860:219.
>COMMON NAME: Golden Thomasomys.
>TYPE LOCALITY: Ecuador, Chimborazo Prov., "Pallatanga," 4950 ft (1509 m). See discussions by Gardner (1983*b*) and Voss (2003) on indeterminate location of Pallatanga, a place below the altitudinal range documented for *T. aureus*; amendment of type locality needed.
>DISTRIBUTION: Andean forests, about 2400-4000 m, from far W Venezuela (see Linares, 1998:Fig. 152) and E Colombia, through Ecuador and Peru, to WC Bolivia (see Anderson, 1997:Fig. 702); vouchered documentation of range needed.
>STATUS: IUCN – Lower Risk (lc).
>SYNONYMS: *altorum* J. A. Allen, 1914; *princeps* (Thomas, 1895).
>COMMENTS: Highly differentiated morphologically from other species of *Thomasomys* (see Carleton, 1973; Hooper and Musser, 1964*a*; Voss and Linzey, 1981). Karyotype reported by Gardner and Patton (1976). Based on the *Thomasomys* material in the BMNH, Ellerman (1941) had recognized an *aureus* group, consisting of large species and also including *nicefori*, *popayanus*, and *praetor*; and Cabrera (1961) later consolidated most of these as subspecies within a highly variable *T. aureus*. The indiscriminate lumping prompted Gardner and Romo R. (1993) to refer to the populations as the "*T. aureus* complex," which they suspected to consist of three or more valid species. Voss (2003) provided essential traits for separating *T. aureus* proper from *T. popaynus* and *T. praetor* (see those accounts).

Thomasomys baeops (Thomas, 1899). Ann. Mag Nat. Hist., ser. 7, 3:152.
>COMMON NAME: Short-faced Thomasomys.
>TYPE LOCALITY: Ecuador, El Oro Prov., Chilla Valley, Río Pita, 3500 m.
>DISTRIBUTION: W Andes of Ecuador.
>STATUS: IUCN – Lower Risk (lc).

Thomasomys bombycinus Anthony, 1925. Am. Mus. Novit., 178:1.
>COMMON NAME: Silky Thomasomys.
>TYPE LOCALITY: Colombia, Antioquia Dept., Paramillo, 12,500 ft (3810 m).
>DISTRIBUTION: Cordillera Occidental of Colombia.
>STATUS: IUCN – Lower Risk (lc).

Thomasomys caudivarius Anthony, 1923. Am. Mus. Novit., 55:4
>COMMON NAME: White-tipped Thomasomys.
>TYPE LOCALITY: Ecuador, El Oro Prov., Cordillera de Chilla, Taraguacocha, 10,750 ft (3277 m).
>DISTRIBUTION: Andes of C and S Ecuador (Bolivar and El Oro Provs. as per Luna and Pacheco, 2002).
>COMMENTS: Described as a species and considered a large member of the "*cinereus* group" by Anthony (1923); relegated to a subspecies of *T. cinereus* by Cabrera (1961). Tirira (1999) check-listed *caudivarius* as a valid species, a status reaffirmed by Luna and Pacheco (2002).

Thomasomys cinereiventer J. A. Allen, 1912. Bull. Am. Mus. Nat. Hist., 31:80.
>COMMON NAME: Ashy-bellied Thomasomys.
>TYPE LOCALITY: Colombia, Cauca Dept., 40 mi (64 km) W Popayán, 10,340 ft (3152 m).
>DISTRIBUTION: Upper Andean elevations in Colombia.
>STATUS: IUCN – Lower Risk (lc).
>SYNONYMS: *contradictus* Anthony, 1925; *dispar* Anthony, 1925.
>COMMENTS: Formerly included *erro* as a subspecies (Cabrera, 1961); see below.

Thomasomys cinereus (Thomas, 1882). Proc. Zool. Soc. Lond., 1882:108.
>COMMON NAME: Olive-gray Thomasomys.
>TYPE LOCALITY: Perú, Cajamarca Dept., Cutervo, 9200 ft (2804 m).
>DISTRIBUTION: N Perú.
>STATUS: IUCN – Lower Risk (lc).
>COMMENTS: Formerly included *caudivarius* as subspecies according to Cabrera (1961).

Thomasomys cinnameus Anthony, 1924. Am. Mus. Novit., 139:5.
COMMON NAME: Cinnamon-colored Thomasomys.
TYPE LOCALITY: Ecuador, Tungurahua Prov., east of Ambato, Hacienda San Francisco, 8000 ft (2438 m).
DISTRIBUTION: Cordillera Oriental, ca. 2400-3800 m, NC Ecuador (Tungurahua and Napo Provs.).
COMMENTS: A small form described as a species by Anthony (1924), later treated as a synonym of *T. gracilis* (Cabrera, 1961; Musser and Carleton, 1993), but resurrected by Voss (2003), who amplified its unique traits as compared with *T. gracilis* and *T. hudsoni*.

Thomasomys daphne Thomas, 1917. Smithson. Misc. Coll., 68(4):2.
COMMON NAME: Daphne's Thomasomys.
TYPE LOCALITY: Perú, Cuzco Dept., Ocabamba Valley, 9100 ft (2774 m).
DISTRIBUTION: E Andean slopes from S Perú to WC Bolivia (see Anderson, 1997:Fig. 703).
STATUS: IUCN – Lower Risk (lc).
SYNONYMS: *australis* Anthony, 1925.
COMMENTS: Anderson (1997) maintained *australis* as the subspecies for Bolivian populations.

Thomasomys eleusis Thomas, 1926. Ann. Mag. Nat. Hist., ser. 9, 17:614.
COMMON NAME: Peruvian Thomasomys.
TYPE LOCALITY: Perú, Amazonas Dept., mountains east of Balsas, Tambo Jenes, 12,000 ft (3658 m).
DISTRIBUTION: NC Perú.
STATUS: IUCN – Lower Risk (lc).

Thomasomys erro Anthony, 1926. Am. Mus. Novit., 240:3
COMMON NAME: Wandering Thomasomys.
TYPE LOCALITY: Ecuador, Napo Prov., headwaters of Río Suno, a small tributary of the upper Río Napo, upper slopes of Volcán Sumaco, about 8000-9000 ft (2438-2743 m).
DISTRIBUTION: Cordillera Oriental, 2400-3600 m, NC Ecuador.
COMMENTS: Anthony (1926) considered *T. erro* to be a highly distinctive species that was possibly derived from *T. cinereiventer*; Cabrera (1961) formalized the latter opinion as the trinomial *T. cinereiventer erro*, a synonymy followed by Musser and Carleton (1993) but not Tirira (1999). New records of *T. erro* reported by Voss (2003), who enumerated morphological traits that clearly discriminate the two forms as species.

Thomasomys gracilis Thomas, 1917. Smithson. Misc. Coll., 68(4):2.
COMMON NAME: Gracile Thomasomys.
TYPE LOCALITY: Perú, Cuzco Dept., Machu Picchu, 12,000 ft (3658 m).
DISTRIBUTION: Andes of SE Perú, about 2750-4300 m.
STATUS: IUCN – Lower Risk (lc).
COMMENTS: Formerly included *cinnameus* and *hudsoni* as subspecies or referred synonyms (Cabrera, 1961; Musser and Carleton, 1993, respectively), forms reinstated to species by Voss (2003).

Thomasomys hudsoni Anthony, 1923. Am. Mus. Novit., 55:3.
COMMON NAME: Hudson's Thomaomys.
TYPE LOCALITY: Ecuador, Azuay Prov., Bestión, 10,100 ft (3078 m).
DISTRIBUTION: Known only from the type locality, S Ecuador.
COMMENTS: Considered a subspecies of *T. gracilis* by Cabrera (1961); differentiating features contrasted with *T. cinnameus* and *T. gracilis* and returned to species rank by Voss (2003); known only by the holotype.

Thomasomys hylophilus Osgood, 1912. Field Mus. Nat. Hist. Publ., Zool. Ser., 10(5):50.
COMMON NAME: Woodland Thomasomys.
TYPE LOCALITY: Colombia, Norte de Santander Dept., Upper Río Tachira, Páramo de Tamá, 7500 ft (2286 m).
DISTRIBUTION: Cordillera Oriental, E Colombia, and Cordillera de Mérida, W Venezuela (see Linares, 1998:Fig. 150).
STATUS: IUCN – Lower Risk (lc).

Thomasomys incanus (Thomas, 1894). Ann. Mag. Nat. Hist., ser. 6, 14:350.
COMMON NAME: Black-eared Thomasomys.
TYPE LOCALITY: Perú, Junín Dept., Vítoc Valley.
DISTRIBUTION: Andes of C Perú.
STATUS: IUCN – Lower Risk (lc).
SYNONYMS: *fraternus* Thomas, 1927.
COMMENTS: Type species of *Inomys*.

Thomasomys ischyrus Osgood, 1914. Field Mus. Nat. Hist. Publ., Zool. Ser., 10(12):162.
COMMON NAME: Long-tailed Thomasomys.
TYPE LOCALITY: Perú, Amazonas Dept., 65 km E Chachapoyas, near Uchco, Tambo Almirante, 5000 ft (1524 m).
DISTRIBUTION: N to C Perú.
STATUS: IUCN – Lower Risk (lc).

Thomasomys kalinowskii (Thomas, 1894). Ann. Mag. Nat. Hist., ser. 6, 14:349.
COMMON NAME: Kalinowski's Thomasomys.
TYPE LOCALITY: Perú, Junín Dept., Vítoc Valley.
DISTRIBUTION: Andes of C Perú.
STATUS: IUCN – Lower Risk (lc).
COMMENTS: Chromosomal formula reported by Gardner and Patton (1976).

Thomasomys ladewi Anthony, 1926. Am. Mus. Novit., 239:1.
COMMON NAME: Ladew's Thomasomys.
TYPE LOCALITY: Bolivia, La Paz Dept., Río Aceromarca, 10,800 ft (3292 m).
DISTRIBUTION: Known only from a restricted region in La Paz Dept., Andes of NW Bolivia.
STATUS: IUCN – Lower Risk (lc).

Thomasomys laniger (Thomas, 1895). Ann. Mag. Nat. Hist., ser. 6, 16:59.
COMMON NAME: Soft-furred Thomasomys.
TYPE LOCALITY: Colombia, Cundinamarca Dept., Bogotá Region, 8750 ft (2667 m).
DISTRIBUTION: Andes of C Colombia and adjacent W Venezuela (see Linares, 1998:Fig. 149).
STATUS: IUCN – Lower Risk (lc).
SYNONYMS: *emeritus* Thomas, 1916.
COMMENTS: Chromosomal, mensural, and ecological discrimination from *T. niveipes* in Colombia presented by Gómez-Laverde et al. (1997); karyotypic formula for Venezuelan populations reported by Aguilera et al. (2000).

Thomasomys macrotis Gardner and Romo R. 1993. Proc. Biol. Soc. Wash., 106:762.
COMMON NAME: Large-eared Thomasomys.
TYPE LOCALITY: Perú, San Martín Dept., Parque Nacional Río Abiseo, ca. 30 km NE Los Alisos, Puerta del Monte, 3250 m.
DISTRIBUTION: Known only from the Pampa del Cuy Valley, in upper montane elfin forest, 3250-3380 m, NC Perú.
COMMENTS: A big-eared and relatively short-tailed thomasomyine, somewhat larger than *T. aureus* but craniodentally more similar to *T. ischyrus* and kin.

Thomasomys monochromos Bangs, 1900. Proc. New England Zool. Club, 1:97.
COMMON NAME: Unicolored Thomasomys.
TYPE LOCALITY: Colombia, Magdalena Dept., Sierra Nevada de Santa Marta, Páramo de Macotama, 11,000 ft (3353 m).
DISTRIBUTION: Extreme NE Colombia.
STATUS: IUCN – Lower Risk (nt).
COMMENTS: Included as a subspecies of *T. laniger* by Cabrera (1961); karyotype reported by Gardner and Patton (1976) as a species without comment.

Thomasomys niveipes (Thomas, 1896). Ann. Mag. Nat. Hist., ser. 6, 18:305.
COMMON NAME: White-footed Thomasomys.
TYPE LOCALITY: Colombia, Cundinamarca Dept., La Hoya del Barro (spelling corrected per Gómez-Laverde et al., 1997).
DISTRIBUTION: Depts. Boyacá and Cundinamarca, above 2900 m, C Colombia.

STATUS: IUCN – Lower Risk (lc).

COMMENTS: Included as a subspecies of *T. laniger* by Cabrera (1961) but returned to species status by Musser and Carleton (1993) based on their own specimen comparisons. Sympatry with and chromosomal (2n = 24, FN = 42), mensural, and ecological discrimination from *T. laniger* (2n = 40, FN = 40) documented by Gómez-Laverde et al. (1997), who provided additional information on the location and habitat of the type locality.

Thomasomys notatus Thomas, 1917. Smithson. Misc. Coll., 68(4):2.

COMMON NAME: Dusky-footed Thomasomys.

TYPE LOCALITY: Perú, Cuzco Dept., Torontoy, 9500 ft (2896 m).

DISTRIBUTION: SE Perú.

STATUS: IUCN – Lower Risk (nt).

COMMENTS: Standard karyotype (2n = 44, FN = 44) reported by Gardner and Patton (1976).

Thomasomys onkiro Luna and Pacheco, 2002. J. Mammal., 83:835.

COMMON NAME: Ashaninka Thomasomys.

TYPE LOCALITY: Perú, Cuzco Dept., La Convención Prov., Vilcabamba Cordillera, between the Ene and Urubamba Rivers, 325 km NW Cuzco, 3350 m; 11°39'36"S, 73°40'02"W.

DISTRIBUTION: Known only from elfin forest at the type locality, S Peru.

COMMENTS: A medium-sized species differentially compared with *T. caudivarius* and *T. silvestris* and morphologically most similar to the latter; presumed to be a relict restricted to the Cordillera Vilcabamba (Luna and Pacheco, 2002).

Thomasomys oreas Anthony, 1926. Am. Mus. Novit., 239:2.

COMMON NAME: Montane Thomasomys.

TYPE LOCALITY: Bolivia, La Paz Dept., Cocopunco, 10,000 ft (3048 m).

DISTRIBUTION: SC Perú (Pacheco et al., 1993) and WC Bolivia (La Paz; Anderson, 1997).

STATUS: IUCN – Lower Risk (lc).

Thomasomys paramorum Thomas, 1898. Ann. Mag. Nat. Hist., ser. 7, 1:453.

COMMON NAME: Páramo Thomasomys.

TYPE LOCALITY: Ecuador, Chimborazo Dept., páramo south of Mt Chimborazo.

DISTRIBUTION: Páramo and páramo-forest ecotone of high Andes, Ecuador.

STATUS: IUCN – Lower Risk (lc).

COMMENTS: Ecological information presented by Voss (2003).

Thomasomys popayanus J. A. Allen, 1912. Bull. Am. Mus. Nat. Hist., 31:81.

COMMON NAME: Popayán Thomasomys.

TYPE LOCALITY: Colombia, Cauca Dept., Cordillera Occidental, 40 mi (64 km)W Popayán, 10,340 ft (3152 m).

DISTRIBUTION: Andes of W and C Colombia.

SYNONYMS: *nicefori* Thomas, 1921.

COMMENTS: Reduced to a subspecies of *T. aureus* by Ellerman (1941), a synonymy repeated by others (Cabrera, 1961; Musser and Carleton, 1993). Species stature, provisionally including *nicefori*, argued by Voss (2003).

Thomasomys praetor (Thomas, 1900). Ann. Mag. Nat. Hist., ser. 7, 5:354.

COMMON NAME: Cajamarca Thomasomys.

TYPE LOCALITY: Perú, Cajamarca Dept., páramo between Cajamarca and San Pablo, 4000 m.

DISTRIBUTION: NW Perú; limits uncertain.

COMMENTS: Another of the large *Thomasomys* relegated to a subspecies of *T. aureus* by Cabrera (1961). Morphological contrasts sketched by Voss (2003), who acknowledged *T. praetor* as species.

Thomasomys pyrrhonotus Thomas, 1886. Ann. Mag. Nat. Hist., ser. 5, 18:421.

COMMON NAME: Reddish-backed Thomasomys.

TYPE LOCALITY: Perú, Cajamarca Dept., Río Malleta, Tambillo, 5800 ft (1768 m).

DISTRIBUTION: Andes of S Ecuador and NW Perú.

STATUS: IUCN – Lower Risk (lc).

SYNONYMS: *auricularis* Anthony, 1923.

COMMENTS: Morphology redescribed and additional specimens reported by Pine (1980*b*). Tirira (1999) listed *auricularis* as a valid species.

Thomasomys rhoadsi Stone, 1914. Proc. Acad. Nat. Sci. Philadelphia, 66:12.
COMMON NAME: Rhoads's Thomasomys.
TYPE LOCALITY: Ecuador, Pichincha Prov., Mt Pichincha, Hacienda Garzón, 10,500 ft (3200 m).
DISTRIBUTION: Andes of NC Ecuador.
STATUS: IUCN – Lower Risk (lc).
SYNONYMS: *fumeus* Anthony, 1924.
COMMENTS: Need to confirm junior status of *fumeus* urged by Voss (2003).

Thomasomys rosalinda Thomas and St. Leger, 1926. Ann. Mag. Nat. Hist., ser. 9, 18:345.
COMMON NAME: Rosalinda's Thomasomys.
TYPE LOCALITY: Perú, Amazonas Dept., Goncha, 8500 ft (2591 m).
DISTRIBUTION: NC Perú.
STATUS: IUCN – Lower Risk (lc).

Thomasomys silvestris Anthony, 1924. Am. Mus. Novit., 114:2.
COMMON NAME: Sylvan Thomasomys.
TYPE LOCALITY: Ecuador, Pichincha Prov., Santo Domingo trail on western slope of Mt Corazón, Las Máquinas, 7000 ft (2134 m).
DISTRIBUTION: Andes of NC Ecuador (Pichincha Prov.).
STATUS: IUCN – Lower Risk (lc).
COMMENTS: Relegated to a subspecies of *T. ischyrus* by Cabrera (1961), but explicitly distinguished from that form by Anthony (1924); accepted as a valid species by Tirira (1999), Luna and Pacheco (2002), and Voss (2003).

Thomasomys taczanowskii (Thomas, 1882). Proc. Zool. Soc. Lond., 1882:109.
COMMON NAME: Taczanowski's Thomasomys.
TYPE LOCALITY: Perú, Cajamarca Dept., Río Malleta, Tambillo, 5800 ft (1768 m).
DISTRIBUTION: Isolated localities in NW Perú and WC Bolivia (see Anderson, 1993).
STATUS: IUCN – Lower Risk (lc).
COMMENTS: Karyology reviewed by Gardner and Patton (1976). Anderson (1997) regarded the assignment of Bolivian specimens as tentative.

Thomasomys ucucha Voss, 2003. Am. Mus. Novit., 3421:10.
COMMON NAME: Ucucha Thomasomys.
TYPE LOCALITY: Ecuador, Napo Prov., valley of Río Papallacta, about 3-5 km (by trail) NNW Papallacta, 11,100 ft (3384 m).
DISTRIBUTION: Crest of Cordillera Oriental, 3350-3700 m, NC Ecuador (Napo and Pichincha Provs.).
COMMENTS: A medium-sized species with procumbent incisors contrasted principally with *T. hylophilus* (Voss, 2003).

Thomasomys vestitus (Thomas, 1898). Ann. Mag. Nat. Hist., ser. 7, 1:454.
COMMON NAME: Mérida Thomasomys.
TYPE LOCALITY: Venezuela, Mérida, Río Milla, 1630 m.
DISTRIBUTION: Mérida Andes, ca. 1600-2400 m, W Venezuela (see Linares, 1998:Fig. 151).
STATUS: IUCN – Lower Risk (lc).
COMMENTS: Karyotype described by Aguilera et al. (2000), who noted its general conformity to the condition supposed as primitive for the genus.

Thomasomys vulcani (Thomas, 1898). Ann. Mag. Nat. Hist., ser. 7, 1:452.
COMMON NAME: Pichincha Thomasomys.
TYPE LOCALITY: Ecuador, Pichincha Prov., Mt Pichincha, 12,000 ft (3658 m).
DISTRIBUTION: Ecuadoran Andes, range uncertain.
COMMENTS: Described as a species of *Aepeomys* and thereafter classified as a subspecies or synonym of *lugens*, whether listed under *Aepeomys* (Musser and Carleton, 1993) or *Thomasomys* (Cabrera, 1961). Pacheco (in Voss et al., 2002) considered *vulcani* to be a form of *Thomasomys* proper, its status and relationships yet to be resolved.

Wiedomys Hershkovitz, 1959. Proc. Biol. Soc. Wash., 72:5.

TYPE SPECIES: *Mus pyrrhorhinos* Wied-Neuwied, 1821.

COMMENTS: Wiedomyini. Tribe formally diagnosed by Reig (1980) to contain a fossil genus, *Cholomys*, recovered from E Argentina, and the problematic form *pyrrhorhinos*, which has been variously classified as a species of *Oryzomys* or *Thomasomys* (see Tate, 1932*f*; Osgood, 1933*d*; and Hershkovitz, 1959*b*). Genus documented, as *W. marplatensis*, from the late Pliocene (Sanandresian) of Argentina (Quintana, 2002). Noting phallic similarities between *W. pyrrhorhinos* and certain phyllotines, Langguth and Silva Neto (1993) considered it an early offshoot of the South American sigmodontine radiation; based on a broader survey of morphological traits, Steppan (1995:60) considered its possible derivation from "a basal 'thomasomyine' grade." Although these somewhat conflicting assessments perhaps argue for continued tribal segregation, more definitive interpretation of the genus' phylogenetic position is desirable.

Wiedomys pyrrhorhinos (Wied-Neuwied, 1821). Reise nach Brasilien, 2:177.

COMMON NAME: Red-nosed Wiedomys.

TYPE LOCALITY: Brazil, Bahia State, caatingas along the Riacho da Ressaca, between the farms Tamboril and Ilha (as clarified by Avila-Pires, 1965).

DISTRIBUTION: SE Brazil, from Ceará to Rio Grande do Sul.

STATUS: IUCN – Lower Risk (lc).

COMMENTS: Locality records clarified by Pine (1980*b*).

Wilfredomys Avila-Pires, 1960. Bol. Mus. Nac., Nov. Ser., Rio de Janeiro, 220:3.

TYPE SPECIES: *Thomasomys oenax* Thomas, 1928.

COMMENTS: Sigmodontinae *incertae sedis*. Genus diagnosed to encompass another problematic thomasomyine-like species from the Atlantic Forest region; see Osgood (1933*b*) and Pine (1980*b*) for historical reviews of the generic affiliations of the type species. The categorical recognition of this taxon, conventionally treated as a subgenus of *Thomasomys* (e.g., Carleton and Musser, 1984; Corbet and Hill, 1991; Pine, 1980*b*), will depend on phylogenetic studies involving other SE Brazilian endemics (also see comments under *Delomys* and *Juliomys*).

Wilfredomys oenax (Thomas, 1928). Ann. Mag. Nat. Hist., ser. 10, 1:154.

COMMON NAME: Rufous-nosed Wilfredomys.

TYPE LOCALITY: Brazil, Rio Grande do Sul State, Saõ Lourenço.

DISTRIBUTION: SE Brazil to N and C Uruguay.

STATUS: IUCN – Lower Risk (lc).

COMMENTS: Morphology redescribed by Pine (1980*b*) and critically compared with *Juliomys pictipes* by González (2000*a*).

Zygodontomys J. A. Allen, 1897. Bull. Am. Mus. Nat. Hist., 9:38.

TYPE SPECIES: *Oryzomys cherriei* J. A. Allen, 1895 (= *Oryzomys brevicauda* J. A. Allen and Chapman, 1893).

COMMENTS: Oryzomyini. *Zygodontomys* was considered a member of the phyllotine group by Hershkovitz (1962), an affinity disputed by others (Hooper and Musser, 1964*a*; Olds and Anderson, 1989; Pearson and Patton, 1976), or as Sigmodontinae *incertae sedis* (Reig, 1984; Voss, 1991*a*). Classification in Oryzomyini formally proposed by Voss and Carleton (1993), a tribal relationship supported by Steppan (1995) and Bonvicino et al. (2003*d*). First revised by Hershkovitz (1962), who included South American populations now allocated to *Bolomys* (= *Necromys*) *lasiurus* (see Maia and Langguth, 1981; Reig, 1987; Voss and Linzey, 1981). Generic definition emended and species taxonomy revised by Voss (1991*a*), as followed here; in addition to the two species that Voss recognized, Reig (*in* Reig et al., 1990*a*) considered *microtinus* as valid, a possibility that deserves further exploration. Karyology reported by Gardner and Patton (1976, Costa Rica and Colombia), Reig et al. (1990*a*, Venezuela), and Mattevi et al. (2002, Brazil); all karyotypes summarized as well as new variants reported by Bonvicino et al. (2003*d*).

Zygodontomys brevicauda (J. A. Allen and Chapman, 1893). Bull. Am. Mus. Nat. Hist., 5:215.

COMMON NAME: Short-tailed Zygodont.

TYPE LOCALITY: Trinidad, Princes Town.

DISTRIBUTION: Savannas from SE Costa Rica through Panamá, Colombia, Venezuela, and the Guianas, to Brazil north of the Amazon River; including Trinidad and Tobago and smaller continental-shelf islands adjacent Panamá and Venezuela.

STATUS: IUCN – Lower Risk (lc).

SYNONYMS: *cherriei* (J. A. Allen, 1895); *fraterculus* J. A. Allen, 1913; *frustrator* (J. A. Allen and Chapman, 1897); *griseus* J. A. Allen, 1913; *microtinus* (Thomas, 1894); *reigi* Tranier, 1976; *sanctaemartae* (J. A. Allen, 1899); *seorsus* Bangs, 1901; *soldadoensis* Goodwin, 1965; *stellae* Thomas, 1899; *thomasi* J. A. Allen, 1901; *tobagi* Thomas, 1900; *ventriosus* Goldman, 1912.

COMMENTS: Geographic variation evaluated by Voss (1991*a*), who recognized *cherriei* and *microtinus*, along with the nominate form, as subspecies. Formerly included *punctulatus*, which Voss (1991*b*) reidentified as a member of *Bolomys* (= *Necromys*). Karyotypic variation of Brazilian populations reported by Mattevi et al. (2002), who questioned whether *brevicauda* as currently arranged is a composite; Bonvicino et al. (2003*d*) reached the same conclusion based on evaluation of cranial traits and karyotypes.

Zygodontomys brunneus Thomas, 1898. Ann. Mag. Nat. Hist., ser. 7, 2:269.

COMMON NAME: Brown Zygodont.

TYPE LOCALITY: Colombia, "El Saibal, W. Cundinamarca" (indeterminate location of type locality discussed by Voss, 1991*a*).

DISTRIBUTION: Intermontane valleys of N Colombia, 350-1300 m.

STATUS: IUCN – Lower Risk (lc).

SYNONYMS: *borreroi* (Hernandez-Camacho, 1957).

COMMENTS: Relegated to a subspecies of *Z. brevicauda* by Gyldenstolpe (1932) and Hershkovitz (1962), but specific status documented by Voss (1991*a*).

Subfamily Tylomyinae Reig, 1984. Revista Brasil. Genet., 7(2):338.

SYNONYMS: Tylomyinae Reig, 1984 (Tylomyini McKenna and Bell, 1997).

COMMENTS: Emended definition—Medium to large-sized arboreal cricetid rodents with tail slightly longer than head and body; four mammae arranged as two inguinal pairs; hindfoot short and broad, digit V nearly equal to II-IV, plantar pads large and closely approximate, ungual tufts present; interorbit cuneate, supraorbital shelves pronounced, dorsally reflected, continuing (in adults) as well-defined temporal ridges; interparietal conspicuous, long and wide, laterally contacting squamosal; zygomatic plate relatively narrow, dorsal notch absent or weakly suggested; tegmen tympani adnate to squamosal (Voss, 1993); alisphenoid strut present, postglenoid foramen tiny, subsquamosal fenestra absent, hamular process undefined (Carleton, 1980); palatal conformation short-wide (Hershkovitz, 1962), parapterygoid fossa shallow and relatively narrow; mesopterygoid fossa typically fully ossified, sphenopalatine vacuities absent or inconspicuous slits (Carleton, 1980); molars brachyodont, strongly cuspidate with principal cones opposite, mesoloph(id) and other accessory enamel ridges well developed; M2 four-rooted, m3 large and in size and coronal topography resembling m2; 1st rib articulating only 1st thoracic vertebra, humerus with entepicondylar foramen, trochlear process of calcaneum broad and positioned proximally (Carleton, 1980); stomach unilocular-hemiglandular, gall bladder absent, caecum long and elaborately infolded (Carleton, 1973, 1980); glans penis short and wide with rudimentary to deep crater, spines large and widely spaced, baculum shorter than glans, with moderate cartilaginous spine but without lateral bacular digits (Hooper, 1960; Hooper and Musser, 1964*a*). Contents—*Nyctomys* Saussure, 1860; *Otonyctomys* Anthony, 1932; *Ototylomys* Merriam, 1901; *Tylomys* Peters, 1866.

Vorontsov (1959) arranged these genera as members of Oryzomyini (including "thomasomyines"), within a broadly defined Cricetinae. Hooper (1960) first advanced a family-group concept for these Middle American endemics, then comprising only *Tylomys* and *Ototylomys* which he associated as one group of four within North American genera having a "simple" phallus, the neotomine-peromyscines (= Neotominae *sensu* Reig, 1980, and others). Subsequently viewed either as early offshoots of a lineage leading to *Neotoma* (Hooper and Musser, 1964*a*) or as an older clade, possibly including *Nyctomys*, basal to the divergence of neotomine-peromyscines and sigmodontines (Carleton, 1980). Reig (1984) nomenclaturally formalized the rank as subfamily and listed its generic contents; McKenna and Bell (1997) used it as tribe within Sigmodontinae *sensu lato*.

Although tylomyines can be morphologically circumscribed, as here defined, abundant traits indicate that *Tylomys-Ototylomys* (Tylomyini; see emended definition under *Tylomys* account) and *Nyctomys-Otonyctomys* (Nyctomyini new tribe; see *Nyctomys* account) are themselves distantly related (see below). While various kinds of evidence do point to the antiquity of certain of these genera (Arata, 1964; Carleton, 1980; D'Elía et al., 2003; Engel et al., 1998; Haiduk et al., 1988; Hooper and Musser, 1964a; Sarich, 1985; Steppan, 1995; Voss and Linzey, 1981), no recent study has mustered the taxonomic sampling needed to critically test the several questions of monophyly at issue, whether early diverging members of Neotomini (Hooper and Musser, 1964a), a primitive clade apart from Neotominae and Sigmodontinae (Carleton, 1980; Reig, 1984), or another phyletic topography as yet unrecognized in our classification.

Nyctomys Saussure, 1860. Rev. Mag. Zool. Paris, ser. 2, 12:106.

TYPE SPECIES: *Hesperomys sumichrasti* Saussure, 1860.

COMMENTS: Nyctomyini new tribe. Type genus—*Nyctomys* Saussure, 1860. Definition— Medium-sized tylomyine rodents with rich tawny to reddish-brown pelage, tail densely furred and noticeably penicillate; cranium with short, stocky rostrum and arched dorsal profile; jugal thin but a constant middle element of zygomatic arch; carotid circulation complete (*Otonyctomys*) or derived (most *Nyctomys*); basihyal with small, knobby entoglossal process; vertebral column with 13 thoracics and 6 lumbars (Carleton, 1980); M1 oval, anterocone narrow and undivided, M3 ovate, smaller than M2; two complete and five incomplete transverse palatal ridges (Carleton, 1980); accessory reproductive glands with two pairs of preputials, ampullaries loose and filiform, dorsal prostates strongly modified, anterior prostates and vesiculars absent (Voss and Linzey, 1981); glans penis with urethral process and dorsal papilla (Hooper and Musser, 1964a). Contents— *Nyctomys* Saussure, 1860; *Otonyctomys* Anthony, 1932.

Placed with thomasomyine group by Hershkovitz (1944, 1962, 1966c) based on its pentalophodont molars and short-wide palate. Numerous studies have supported removal of *Nyctomys* from thomasomyines proper and even questioned its placement within Sigmodontinae *sensu stricto* (Arata, 1964; Carleton, 1980; Haiduk et al., 1988; Hooper and Musser, 1964a; Steppan, 1995; Voss and Linzey, 1981). Phylogenetic interpretation of mitochondrial DNA sequences also suggest an ancient New World lineage, basal to Sigmodontinae or to Cricetinae-Sigmodontinae depending upon the trees generated (Engel et al., 1998).

Nyctomys sumichrasti (Saussure, 1860). Rev. Mag. Zool. Paris, ser. 2, 12:107.

COMMON NAME: Sumichrast's Vesper Rat.

TYPE LOCALITY: Mexico, Veracruz, Uvéro, 20 km NW Santiago Tuxtla (as restricted by Alvarez, 1963:583).

DISTRIBUTION: Lowland and lower montane forests from S Jalisco and S Veracruz, Mexico, south to C Panama, excluding the Yucatán Peninsula.

STATUS: IUCN – Lower Risk (lc).

SYNONYMS: *colimensis* Laurie, 1953; *costaricensis* Goldman, 1937; *decolorus* (True, 1894); *florencei* Goldman, 1937; *nitellinus* Bangs, 1902; *pallidulus* Goldman, 1937; *salvini* (Tomes, 1862); *venustulus* Goldman, 1916.

COMMENTS: Nongeographic variation of Nicaraguan sample evaluated by Genoways and Jones (1972b); specimens from Michoacán documented by Sánchez-Hernández et al. (1999). Standard and banded karyotypes described by Lee and Elder (1977) and Haiduk et al. (1988), respectively. No current revisionary standard available. Based on examination of USNM series, populations can be sorted according to concordant variation in carotid circulatory pattern and molar-root number: those to the north and west of the Isthmus of Tehuantepec (*sumichrasti, pallidulus*) possess a complete pattern and three-rooted M2; whereas, those farther south and east of the Isthmus (*costaricensis, decolorus, florencei, nitellinus, venustulus*) exhibit a derived carotid condition and four-rooted M2. The existence of a geographic pattern should be confirmed with larger series, especially samples from Chiapas and Guatemala (*salvini*), and its taxonomic significance explored with other data sources.

Otonyctomys Anthony, 1932. Am. Mus. Novit., 586:1.
    TYPE SPECIES: *Otonyctomys hatti* Anthony, 1932.
    COMMENTS: Nyctomyini. Although singularly distinctive, many morphological traits associate *Otonyctomys* with *Nyctomys* as its closest relative (see above and Hooper and Musser, 1964*a*), but its phylogenetic position and other aspects of systematic biology have received little attention.

Otonyctomys hatti Anthony, 1932. Am. Mus. Novit., 586:1.
    COMMON NAME: Yucatán Vesper Rat.
    TYPE LOCALITY: Mexico, Yucatán, Chichén Itzá.
    DISTRIBUTION: Yucatán Peninsula, Mexico, south to N Belize and NE Guatemala (Peten Dept.).
    STATUS: IUCN – Lower Risk (lc).
    COMMENTS: Range wholly allopatric to that of its probable sister taxon *Nyctomys*; new records in Quintana Roo documented by Aranda et al. (1997).

Ototylomys Merriam, 1901. Proc. Wash. Acad. Sci., 3:561.
    TYPE SPECIES: *Ototylomys phyllotis* Merriam, 1901.
    COMMENTS: Tylomyini. Emended Definition—Large-sized tylomyine rodents, pelage somber gray to grayish-brown, tail nearly naked; cranium with moderately long, tapering rostrum and flat dorsal profile; jugal absent or tiny and irregularly formed, zygomatic processes of maxillary and squamosal in contact; carotid circulation derived (character state 3 per Carleton, 1980); basihyal with large, attenuate entoglossal process, vertebral column with 14-15 thoracics and 6 lumbars (Carleton, 1980); M1 rectangular, anterocone broad and deeply bifurcate, lingual and labial conules subequal to primary cusps, M3 rectangular and resembling M2 in size and enamel configuration; three complete and five-nine incomplete transverse palatal ridges (Carleton, 1980); accessory reproductive glands with one pair of preputials, ampullaries compact and coiled, vesiculars present and highly modified (Arata, 1964; Lawlor, 1969), ductus deferens with distal ampulla (Arata, 1964; Carleton, 1980); urethral process absent (*Ototylomys*) or rudimentary (*Tylomys*), dorsal papilla absent (Hooper, 1960). Contents—*Ototylomys* Merriam, 1901; *Tylomys* Peters, 1866. Closedly related to *Tylomys* according to morphological features (Carleton, 1980; Hooper, 1960; Lawlor, 1969; Merriam, 1901*b*) and to albumin immunological distances (Sarich, 1985).

Ototylomys phyllotis Merriam, 1901. Proc. Wash. Acad. Sci., 3:562.
    COMMON NAME: Big-eared Climbing Rat.
    TYPE LOCALITY: Mexico, Yucatán, Tunkás.
    DISTRIBUTION: C Costa Rica north to Yucatán Peninsula, S Tabasco, and N Chiapas, Mexico; isolated record from NC Guerrero, Mexico.
    STATUS: IUCN – Lower Risk (lc).
    SYNONYMS: *affinis* Laurie, 1953; *australis* Osgood, 1931; *brevirostris* Laurie, 1953; *connectens* Sanborn, 1935; *fumeus* J. A. Allen, 1908; *guatemalae* Thomas, 1909; *phaeus* Merriam, 1901.
    COMMENTS: Distribution and geographic variation examined by Lawlor (1969), who retained *australis* and *connectens* as subspecies. See Lawlor (1982, Mammalian Species, 181).

Tylomys Peters, 1866. Monatsb. K. Preuss. Akad. Wiss. Berlin, 1866:404.
    TYPE SPECIES: *Hesperomys nudicaudus* Peters, 1866.
    COMMENTS: Tylomyini. Closely related to *Ototylomys* (see previous account). Phylogenetic evaluations of cytochrome *b* sequences arrange *Tylomys* either as sister-group to Sigmodontinae (D'Elía et al., 2003) or as a member of a polytomy that includes Arvicolinae-Neotominae-Sigmodontinae (D'Elía et al., in press). Nominal species listed follow Hall (1981:626), who remarked that "Study may show that some of the species are only subspecies." Such taxonomic study is overdue.

Tylomys bullaris Merriam, 1901. Proc. Wash. Acad. Sci., 3:561.
COMMON NAME: Chiapan Climbing Rat.
TYPE LOCALITY: Mexico, Chiapas, Tuxtla Gutierrez.
DISTRIBUTION: Known only from the type locality.
STATUS: IUCN – Critically Endangered.

Tylomys fulviventer Anthony, 1916. Bull. Am. Mus. Nat. Hist., 35:366.
COMMON NAME: Fulvous-bellied Climbing Rat.
TYPE LOCALITY: Panama, Darien Prov., Tacarcuna, 4200 ft (1280 m).
DISTRIBUTION: Easternmost Panama.
STATUS: IUCN – Lower Risk (nt).
COMMENTS: Possibly a subspecies of *T. mirae* according to Cabrera (1961) or a synonym of
T. panamensis according to Handley (1966a).

Tylomys mirae Thomas, 1899. Ann. Mag. Nat. Hist., ser. 7, 4:278.
COMMON NAME: Mira Climbing Rat.
TYPE LOCALITY: Ecuador, Imbabura Prov., Rio Mira, Paramba.
DISTRIBUTION: W Colombia and NW Ecuador.
STATUS: IUCN – Lower Risk (lc).
SYNONYMS: *bogotensis* Goodwin, 1955.

Tylomys nudicaudus (Peters, 1866). Monatsb. K. Preuss. Akad. Wiss. Berlin, 1866:404.
COMMON NAME: Peter's Climbing Rat.
TYPE LOCALITY: Guatemala.
DISTRIBUTION: C Guerrero and C Veracruz, Mexico, south to S Nicaragua, excluding Yucatan
Peninsula.
STATUS: IUCN – Lower Risk (lc).
SYNONYMS: *gymnurus* Villa, 1941; *microdon* Goodwin, 1955; *villai* Schaldach, 1966.
COMMENTS: Goodwin (1934) passingly noted that La Primavera (Dept. Alto Verapaz) may be
the type locality, a restriction which should be formally considered by a future revisor.
Goodwin (1969) relegated *gymnurus* to a subspecies of *T. nudicaudus*.

Tylomys panamensis (Gray, 1873). Ann. Mag. Nat. Hist., ser. 4, 12:417.
COMMON NAME: Panamá Climbing Rat.
TYPE LOCALITY: Panamá.
DISTRIBUTION: Easternmost Panamá (see Goldman, 1920).
STATUS: IUCN – Vulnerable
COMMENTS: Handley (1966a) suggested that *fulviventer* and *watsoni* may prove to be junior
synonyms of *T. panamensis*.

Tylomys tumbalensis Merriam, 1901. Proc. Wash. Acad. Sci., 3:560.
COMMON NAME: Tumbalá Climbing Rat.
TYPE LOCALITY: Mexico, Chiapas, Tumbalá.
DISTRIBUTION: Known from isolated localites in Chiapas, S México (see Espinoza M. et al.,
1999a, b).
STATUS: IUCN – Critically Endangered.
COMMENTS: Recorded with *T. nudicaudus* in reserves of Chiapas (Espinoza M. et al., 1999a,
b).

Tylomys watsoni Thomas, 1899. Ann. Mag. Nat. Hist., ser. 7, 4:278.
COMMON NAME: Watson's Climbing Rat.
TYPE LOCALITY: Panamá, Chiriqui Prov., Volcan de Chiriqui, Bogava (= Bugaba), 800 ft
(244 m).
DISTRIBUTION: Costa Rica and W Panamá.
STATUS: IUCN – Lower Risk (lc).

Family Muridae Illiger, 1811. Abhandl. K. Akad. Wiss., Berlin for 1804-11, p. 46, 129.
COMMENTS: Phylogenetic analyses of mitochondrial and nuclear gene sequences support the
monophyly of Gerbillinae and Murinae (Adkins et al., 2001; Montgelard et al., 2002;
Robinson et al., 1997); Deomyinae, Gerbillinae, and Murinae (Catzeflis et al., 1995; Con-

roy and Cook, 1999; Dubois et al., 1999; Hänni et al., 1995; Martin et al., 2000); and those three subfamilies plus Otomyinae (Jansa and Weksler, 2004; Michaux and Catzeflis, 2000; Michaux et al., 2001*b*). Until analyses of molecular data recovered a monophyletic clade consisting of *Acomys*, *Lophuromys*, *Uranomys*, and *Deomys* (Deomyinae), the first three genera had been listed in classifications and checklists under Murinae. They share many of the diagnostic cranial traits associated with murines and the defining "upper first molars with three cusps in the anterior two transverse chevrons" and positions of the two lingual cusps relative to the adjacent central cusps (Carleton and Musser, 1984; Flynn et al., 1985:605). *Deomys* was either viewed as the sole member of Deomyinae or an unusual dendromurine (see *Deomys* account). Gerbils have been treated as a separate family within Muroidea or subfamily within Cricetidae (separated from Muridae) or more encompassing Muridae (see Gerbillinae introduction). Although strongly differentiated in external and cranial characters among muroids, most gerbil species lack longitudinal enamel crests on their molars, as found in cricetids, and resemble murines in this derived trait, which has been the primary basis for postulating a sister-group relationship between Gerbillinae and Murinae (Flynn et al., 1985; Tong and Jaeger, 1993).

Despite past treatments of otomyines as an independent family (Roberts, 1951), subfamily of Nesomyidae (Chaline et al., 1977; Lavocat, 1978) or Cricetidae (Misonne, 1974), or in an inclusive Muridae generally comparable to Muroidea (Carleton and Musser, 1984; Ellerman, 1941; Thomas, 1896), morphological information from fossil and living samples, along with sequences from mitochondrial and nuclear genes, unequivocally link Vlei and Whistling Rats with African murines (see Otomyinae introduction). Including Otomyinae within a restricted Muridae (separate from Cricetidae) was already formulated by Tullberg (1899), Miller and Gidley (1918), and Simpson (1945). Cranial, dental, and mandibular conformation of otomyines lacks derived traits that link them with any other African clade, living or fossil, and resembles the cranial configuration of arvicanthine genera (especially *Arvicanthis*, *Pelomys*, and *Mylomys*). Furthermore, there is an impression that enlarged M3s (with four to 10 laminae) and m1s (four to seven laminae), diagnostic of otomyines, are not found elsewhere among murids and are only seen in nonmurid groups (Misonne, 1969). The species of late Miocene *Microtia* from the Gargano fissure fillings in southern Italy are typical European Neogene murines, possibly derived from *Stephanomys* or *Apodemus*, with enlarged M3s and m1s accommodating four to seven transverse, tubercular and chevron-like laminations (Freudenthal, 1976; Freudenthal and Suárez, 1999; Jaeger and Hartenberger, 1989; Millien and Jaeger, 2001; Millien-Parra, 2000*b*; Zafonte and Masini, 1992). The laminae of otomyines are flat and lack indication of the murine triserial pattern and cuspidation (except in early fossils) but the proliferation of laminae in *Microtia* and otomyines suggests a deep shared genetic heritage.

We isolate the African *Leimacomys* in its own subfamily to highlight the enigmatic nature of its mosaic character pattern and uncertain phylogenetic affinities. Usually associated with dendromurines (see Leimacomyinae introduction), its cranial configuration is unlike any genus in that group and instead resembles murines, gerbils, and deomyines such as *Lophuromys*. Molar occlusal patterns were thought to match those in the dendromurine *Steatomys*, but in many ways are more similar to gerbillines (Denys et al., 1995) or extinct myocricetodontines (C. Denys, in litt., 2001). While nesting Leimacomyinae within Muridae seems to reflect morphology, its status as an independent clade, or alternatively as a member of either Deomyinae or Gerbillinae, will require testing with a broader survey of African genera in Muridae and Nesomyidae in which the focus is on morphological traits other than just dentition, and the incorporation of DNA gene sequences.

Evolutionary origin of the Muridae as constituted here may be traced back through extinct myocricetodontines (early Miocene and late Miocene-early Pliocene in Arabian Peninsula, early to late Miocene in Pakistan, middle Miocene of China, middle to late Miocene in North and East Africa, early to middle Miocene of Turkey, and late Miocene-early Pliocene in Spain; Qiu, 2001; Wessels, 1996) to early Miocene cricetids such as *Spanocricetodon* in Pakistan and and the morphologically similar *Notocricetodon* of Africa (de Bruijn et al., 1996). Tong and Jaeger (1993) speculated that myocricetodontines are paraphyletic, producing both gerbils and murines, which may be a sister-group to African dendromurines and cricetomyines; they estimated divergence of gerbillines-murines from dendromurines-cricetomyines at 18 million years before present. Such a relationship is reflected in the phylogenetic analyses of nuclear IRBP gene sequences by Jansa and Weksler (2004) in which those deomyines, gerbillines, and otomyines sampled (here = Muridae) form a major sister-group to exemplars of nesomyines, dendromurines, cricetomyines, mystromyines, and petromyscines (here = Nesomyidae).

Subfamily Deomyinae Thomas, 1888. Proc. Zool. Soc. Lond., 1888:132.

SYNONYMS: Acomyinae Dubois, Catzeflis, and Beintema, 1999 [*nomen nudum*]; Acomyinae Michaux and Catzeflis, 2000 [*nomen nudum*]; *Deomyes* Thomas, 1888 (Deomyinae Lydekker, 1889; Deomyinae Ellerman, 1941).

COMMENTS: Up to 1999 the clade containing *Acomys* was informally referred to as "acomyines" (Hänni et al., 1995; E. Verheyen et al., 1995, 1996). Dubois et al. (1999:181-182) wrote that "*Acomys* and *Uranomys* constitute a monophyletic clade at the subfamily level, denoted 'Acomyinae' and noted "throughout this paper, we will use the term Acomyinae for the clade containing *Acomys* and *Uranomys*." Later Michaux and Catzeflis (2000:286) indicated that "Following Hänni *et al.* (1995, p. 132), we name 'acomyines' or [provisionally] 'Acomyinae' as the clade containing the genera *Acomys*, *Deomys*, *Lophuromys*, and *Uranomys*." Acomyinae as used by Dubois et al. is unavailable because it was unaccompanied "by a description or definition that states in words characters that are purported to differentiate the taxon" or "a bibliographic reference to such a published statement," which is stipulated by the fourth edition of the International Code of Zoological Nomenclature (International Commission on Zoological Nomenclature, 1999: Articles 13.1, 13.2), and Acomyinae as proposed by Michaux and Catzeflis cannot formally stand as a family-group name because its formation does not satisfy requirements described in articles 16.1 ("Every new name published after 1999, including new replacement names (nomina nova), must be explicitly indicated as intentionally new" and 16.2 ("a new family-group name published after 1999 must be accompanied by citation of the name of the type genus") of the fourth edition of the Code (ICZN, 1999).

Lydekker (1889:1418) wrote that "The subfamily *Deomyinae* is only known by *Deomys*, of the Congo Valley, which has upper molars intermediate in structure between those of the preceding [*Hesperomys*] and following subfamilies [Murinae]." The earliest reference, however, to a formal group equivalent to subfamily rank is Thomas's (1888c) *Deomyes*. After describing *Deomys ferrugineus*, Thomas wrote that "In the complete systematic arrangement of the Muridae, therefore, we shall have to look upon *Deomys* as forming by itself a special section, the *Deomyes*, intermediate between the *Mures* [modern Murinae] and *Criceti* [equivalent to modern New and Old World Cricetidae and Malagasy Nesomyinae]." Ellerman (1941:317) proposed Deomyinae, mostly because "the zygomatic plate and infraorbital foramen. . . appear unique," listed Thomas's *Deomyes* as a synonym, but was apparently unaware of Lydekker's use. Because *Deomys*, the type genus for Deomyes, which was clearly used by Thomas to designate a suprageneric rank, is a member of the clade also containing *Acomys*, *Lophuromys*, and *Uranomys*, Deomyinae is available as the formal name of the subfamily.

From the time that Dollmann described *Uranomys* in 1909, researchers have appreciated the morphological resemblance between it, *Acomys*, and *Lophuromys*, especially in pelage texture, palatal configuration, and molar occlusal patterns (Denys et al., 1992; Denys and Michaux, 1992; Heim de Balsac, 1963; Heim de Balsac and Lamotte, 1958; Heller, 1911; Hinton, 1921; Ingoldby, 1929; Misonne, 1969; Rosevear, 1969; Thomas, 1910*d*). But it has been the accumulation of nonmorphological evidence that has really revealed the monophyletic unity of these three genera and their phylogenetic isolation from the rest of Murinae. Immunological experiments (Fraguedakis-Tsolis et al, 1993; Hammer et al., 1987; Montgelard, 1992, including *Uranomys*; Sarich, 1985) and analysis of amino acid sequences of insulin (Graur, 1994) identified *Acomys* as being as distant from Murinae as are some of the other subfamilies of Muridae (Wilson et al., 1987). These results, and the bond between *Acomys* and *Uranomys*, were also indicated by chromosomal data (Viegas-Péquignot et al., 1986), analyses of allozymic variation (Bonhomme et al., 1985), and early DNA/DNA hybridization experiments (Catzeflis, 1990). A subsequent DNA-DNA hybridization study by Chevret et al. (1993*a*) sampled *Acomys*, *Uranomys*, and *Lophuromys*, and identified them as a monophyletic clade, a conclusion supported by analysis of albumin immunology (Watts and Baverstock, 1995*b*). Phylogenetic analyses of mitochondrial 12S and rRNA sequences (Hänni et al., 1995), nuclear pancreatic ribonuclease A gene sequences (Dubois et al., 1999), and mtDNA cytochrome *b* sequences (E. Verheyen et al., 1995) of the combination *Acomys-Uranomys* or *Acomys-Lophuromys* added additional molecular evidence for the reality of the clade. Analysis of highly repeated sequences of nuclear genome (LINES) identified a DNA element, *Lx*, which is unique to Murinae but absent from *Acomys*, *Uranomys*, and *Lophuromys* (Furano et al., 1994; Pascale et al., 1990; Usdin et al., 1995). Samples of *Deomys* were ultimately added to those of the other three genera and incorporated in phylogenetic analyses of sequences from mtDNA cytochrome *b* (E. Verheyen et al., 1996, *Lophuromys* and *Deomys*); the nuclear protein-coding gene LCAT (Michaux and Catzeflis, 2000); LCAT in combination with von Willebrand Factor (vWF) (Michaux et al., 2001); and the combination of DNA/DNA hybridization experiments, complete mitochondrial 12S and rRNA, and nuclear LCAT. Results strongly corroborated the distant relationship of *Acomys*, *Uranomys*, and *Lophuromys* to the rest of murid rodents and demonstrated unequivocally their close phylogenetic association with *Deomys*. A contrary view based on phylogenetic analyses of complete cytochrome *b* sequences by Martin et al. (2000) left the position of *Acomys* unresolved relative to the other murine genera examined. Chevret and Hänni (1994) provided an excellent review of the molecular and biochemical results and their limitations involving *Acomys* and its relatives.

Systematists agreed that *Acomys* and the other deomyines should be excluded from Murinae but had not identified a sister group (Chevret and Hänni, 1994; Dubois et al., 1999). DNA/DNA hybridization experiments (Catzeflis, 1990; Chevret et al., 1993*a*) and analysis of DNA sequences from the fifth exon of the KBP gene (Agulnik and Silver, 1996) indicated *Acomys*, *Uranomys*, and *Lophuromys* to be phylogenetically closer to gerbillines than murines. Watts and Baverstock (1995*b*), however, suggested the *Acomys* clade to possibly be the sister group to the rest of Murinae. Michaux and Catzeflis (2000:289) thought their molecular results provisionally confirmed a sister group relationship between Gerbillinae, Deomyinae, and Murinae, "implying a later separation between them with regard to the other subfamilies of Muridae." The reconstruction of phylogenetic relationships based on analyses of combined LCAT and vWF sequences, "provides a clear and strong picture: acomyines [=deomyines] cluster with gerbillines. . ., and these two lineages are sister to murines" and "According to the molecular-clock analysis, the separation between the three subfamilies appeared 17.9-20.8 MYA [Early? Miocene]", which "was characterized by changes in climate which favored the spread in Europe, Africa, and America of allochthonous rodent groups probably coming from Asia" (Michaux et al., 2001:2026). The sister-group relationship of gerbillines and deomyines and their inclusion with murines in a monophyletic clade is substantiated by analyses of nuclear IRBP sequences (Jansa and Weksler, 2004). However, results combining DNA/DNA hybridization experiments with analyses of mitochondrial and nuclear gene sequences defined a clade formed of gerbils, murines, and deomyines but could not decipher the relationships among these three subfamiles, and Chevret et al. (2001) suggested that using other molecules (mainly protein-coding genes that evolve slowly), repeated elements, and a variety of morphological data may resolve sister-group affinities among the three subfamilies.

Reproductive biology of *Acomys* is special among murines (Dieterlen, 1961, 1962, 1963). Spermatozoal morphology described for samples of *Acomys*, *Lophuromys*, and *Uranomys*, seems to support their isolation from the rest of the genera sampled in Murinae (Baskevich and Lavrenchenko, 1995; Breed, 1995*d*), and Breed (1995*d*:418) suggested that the *Acomys-Uranomys* lineage "diverged from the base of the Murinae-Otomyine clade." Breed (1995*a*) also noted that spermatozoal morphology of *Lophuromys* is "totally different" from that in *Acomys* and *Uranomys* as well as the other African murines sampled. A resistant-fit theta-rho analysis surveying magnitude and direction of positional changes of homologous landmarks on M1s grouped *Acomys* with Murinae and was equivocal in evaluating relationships among *Acomys*, *Uranomys*, and *Lophuromys* (Xu et al., 1996). To date, no spermatozoal traits or other morphological characters have been identified as

synapomorphies to supplement the biochemical and molecular data to define Deomyinae (Chevret et al., 2001).

Acomys I. Geoffroy, 1838. Ann. Sci. Nat. Zool. (Paris), ser. 2, 10:126.
   TYPE SPECIES: *Mus cahirinus* É. Geoffroy, 1803.
   SYNONYMS: *Acanthomys* Lesson, 1842 [not Gray, 1867, or Tokuda, 1941]; *Acosminthus* Gloger,
       1841; *Peracomys* F. Petter and Roche, 1981; *Subacomys* Denys, Gautun, Tranier, and
       Volobouev, 1994.
   COMMENTS: A brief list of species and subspecies made by Setzer (1975); the arrangement of spe-
       cies provided by Musser and Carleton (1993) based on review of literature and study of
       specimens; chromosomal reviews by Matthey (1965*a*, *b*, 1968), Volobouev et al. (1991),
       Sokolov et al. (1992, 1993), and Denys et al. (1994) and regional chromosomal studies of
       species (cited in the species accounts); study of molar occlusal patterns (Denys et al.,
       1994; F. Petter, 1983), and spermatozoal morphology (Baskevich and Lavrenchenko,
       1995; Breed, 1995*a*); biochemical and molecular analyses of large species-clusters (Barome
       et al., 1998, 2000; Janecek et al., 1991) and smaller assemblages cited in the species ac-
       counts; regional systematic revisions (referenced in the species accounts); and review of
       geographic distributions (Bates, 1994) have all contributed to the present understanding
       of species-limits within *Acomys*, their geographic ranges, and phylogenetic relationships.
       Regrettably, there is still no comprehensive taxonomic revision of the genus incorporat-
       ing morphological as well as chromosomal and molecular data. Extant species are sorted
       into three subgenera (*Acomys*, *Peracomys*, and *Subacomys*), but these groupings require re-
       assessment by systematic revision. Van der Straeten (1994) provided a bibliography of *Ac-
       omys* from Africa and the Middle East that is coded by species and subjects. Our allocation
       of species to subgenera follows Denys et al. (1994).
       According to Denys (1990*b*), the earliest fossil representatives of *Acomys* were found in
       Southern African sediments at least 4.5 million years old (early Pliocene; see also Avery,
       1998, 2000, and Senut et al., 1992). Its evolutionary history undoubtedly extends back in
       time even farther because Geraads (2001) described *Preacomys*, considered a "forerunner"
       of *Acomys*, based upon isolated molars from late Miocene sediments in Ethiopia
       (Chorora). Late Miocene sediments in Namibia have also yielded samples representing
       Preacomys, one of the three species documented similar if not identical with the Ethiopian
       form (Mein et al., 2004). Barome et al. (2001) presented two biogeographic hypotheses for
       the center of species' origin and their dispersion. One recognizes origin of *Acomys* in
       southern Africa and dispersal northwards along with subsequent evolution into species.
       The other suggests that East Africa (Ethiopia and Kenya) is the center, with dispersion and
       attendant speciation in one direction to the Mediterranean region, and in another to
       southern Africa. In either scenario, "The speciation process could also have been caused
       by a progressive fragmentation of the distribution area, resulting from the formation of
       the Rift Valley in its south-western part, or from the extension of tropical forest zones di-
       viding the savannah domain into mosaic blocks during climatic fluctuations and isolat-
       ing the different species by vicariance" (Barome et al., 2001).

Acomys airensis Thomas and Hinton, 1921. Nov. Zool. Tring, 28:8
   COMMON NAME: Western Saharan Spiny mouse.
   TYPE LOCALITY: Niger, Mt Baguezan, Asben (Aïr), 5200 ft (1585 m).
   DISTRIBUTION: Recorded from S Mauritania, Mali, Niger, and Chad (Bâ et al., 2001; Barome
       et al., 1998, 2000; Dobigny et al., 2001*a*, *b*, 2002*b*, 2003; Le Berre, 1990; Tranier et al.,
       1999).
   COMMENTS: Subgenus *Acomys*. Initially described as a species but later included in *A. cinera-
       ceus* (F. Petter, 1954; Rosevear, 1969), *A. dimidiatus* (Setzer, 1975), or *A. cahirinus*
       (Musser and Carleton, 1993; Rosevear, 1969). Tranier (1975*b*) first reported its
       karyotype (2n = 40), and subsequent comparisons among samples of *airensis*,
       dimidiatus, and *cahirinus* using chromosomal traits highlighted *airensis* as a species
       relative to the other two (Tranier et al., 1999; Volobouev et al., 1991, 1996). Phylo-
       genetic analyses of mtDNA cytochrome *b* sequences provided concordant results
       (Barome et al., 2000, 2001*a*, *b*; Dobigny et al., 2001*a*). Denys et al. (1994) described the
       dental traits distinguishing *A. airensis* from *A. cahirinus*, *A. chudeaui*, and *A. seurati*. In

his initial chromosomal report, Tranier (1975b) placed *A. airensis* in the *A. cahirinus* species group, a judgement reinforced by subsequent analyses of chromosomes (Volobouev et al., 2002b) and mtDNA cytochrome *b* sequences (Barome et al., 2000, who included it in an *A. cahirinus-A. dimidiatus* group); this assemblage also contains *A. johannis*, *A. minous*, *A. nesiotes*, *A. cilicicus*, and an unnamed species from Burkina Faso.

Meinig (2000) obtained five specimens from S Mali that are probably *A. airensis*, although they could be examples of *A. johannis*. Dobigny et al. (2001a) documented a sample of *A. airensis* from the Adrar des Iforas massif in NE Mali and reported the karyotype as 2n = 44 and 46, FNa = 66, which expands the previously recorded diploid number of the species (2n = 40 and 43, see references in Dobigny et al., 2001a, b; or 2n = 42, FN = 68, Volobouev et al., 2002b). Analysis of cytochrome *b* sequences from the NE Mali specimens positively identified them as *A. airensis*. The records from S Mauritania exhibited 2n = 40, NFa = 66, which is typical for *A. airensis* (Bâ et al., 2001). Although *A. airensis* has been caught in rocky habitats, it is often collected in gardens, grain storage units, and huts, in Mauritania, Mali, and Niger (Bâ et al., 2001; Dobigny et al., 2001a).

Acomys cahirinus (É. Geoffroy, 1803). Cat. Mam. Mus. Natl. Hist. Nat., Paris, p. 195.

COMMON NAME: Northeast African Spiny Mouse.

TYPE LOCALITY: Egypt, Cairo.

DISTRIBUTION: NE Africa: Libya, Egypt (Osborn and Helmy, 1980), Sinai Peninsula (Saleh and Basuony, 1998), N Sudan, Ethiopia (identified by chromosomal traits; Sokolov et al., 1992, 1993), and Djibouti (Pearch et al., 2001); see Bates (1994); W and S distributional limits unresolved.

STATUS: IUCN – Lower Risk (lc).

SYNONYMS: *albigena* (Heuglin, 1877); *helmyi* Osborn, 1980; *hunteri* de Winton, 1901; *megalodus* Setzer, 1959; *nubicus* (Heuglin, 1877); *sabryi* Kershaw, 1922; *viator* Thomas, 1902.

COMMENTS: Subgenus *Acomys*. Musser and Carleton (1993) presented a broad definition of *A. cahirinus* that included some of the species listed here (*airensis*, *chudeaui*, *dimidiatus* and its synonyms, *johannis* and *seurati*) and had a geographic range extending from NW Africa to S Pakistan (as mapped by Bates, 1994, who followed Musser and Carleton, 1993). A growing body of published research, mostly by French scientists, has rendered Musser and Carleton's view simplistic and an underestimate of the actual number of species. As currently understood, the geographic distribution of *A. cahirinus* centers in NE Africa, and the species is distinguished from *A. airensis*, *A. chudeaui*, *A. dimidiatus*, *A. johannis*, and *A. seurati* by dental characters (Denys et al., 1994), chromosomal traits (Volobouev et al., 1991, 1996, 2002b; 2n = 36 for *A. cahirinus*, 2n from 38 to 68 for the other species), and mtDNA cytochrome *b* sequences (Barome et al., 2000, 2001a, b); see accounts of those species.

Morphological and geographic definition of *A. cahirinus* remains incomplete; some of the taxa listed here as synonyms may belong to other species. El Ashmawy (1990), for example, documented different G- and C-banding patterns in *cahirinus* and *hunteri* from Egypt. Osborn and Helmy (1980), however, noted the ample evidence of intergradation between *hunteri* (described from NE Sudan) and *cahirinus* but no indication that *hunteri* intergrades with the more southern *cineraceus*. We provisionally include *viator* (type locality NC Libya) in the synonymy of *A. cahirinus*, but Osborn and Helmy (1980) recorded smaller measurements for that form in Egypt (occurs only in extreme SW corner of the country) compared with all the other samples of *A. cahirinus*, and Denys et al. (1994) described the molar morphology of *viator* as distinct from that of *A. cahirinus*. Ranck (1968) provided detailed descriptions of pelage and habitat for samples from Libya that he assigned to *viator*. Whether the latter represents the easternmost segment of *A. cahirinus*, is a separate species, or a population of some other *Acomys* (the Algerian *seurati*, for example) will have to be determined by cytogenetic, molecular, and additional morphological inquiries.

Setzer (1959:95) claimed melanistic *A. cahirinus* to be "almost exclusively commensal" or, if not actually caught in houses, collected in cultivated fields adjacent to villages. Osborn and Helmy (1980:298) noted that melanism is common in the Egyptian population "because it is the commonest mouse in buildings and houses. Some have been taken in gardens, date groves, and rocky hills and cliffs bordering the Nile Delta and Valley." The rat is also numerous in tombs and temples where the melanism becomes diluted.

Acomys chudeaui Kollman, 1911. Abh. Bayer Akad. Wiss. Math. Natürwiss., Kl. 3:195.

COMMON NAME: Chudeau's Spiny Mouse.

TYPE LOCALITY: Mauritania, Atar, SW of Biskra.

DISTRIBUTION: Western Sahara: Recorded from C Morocco (mapped by Aulagnier and Thevenot, 1986, as *cahirinus*) and W Mauritania (Benazzou, 1983; Le Berre, 1990); geographic limits unresolved.

COMMENTS: Subgenus *Acomys*. Usually listed in the synonymy of *A. cahirinus* (e. g., Musser and Carleton, 1993), but chromosomal data (Benazzou, 1983; 2n = 40 for *chudeaui*, 2n = 36 for *cahirinus*) has been used to support species status (Le Berre, 1990; Le Berre and Le Guelte, 1990) as a member of the *A. cahirinus-A. dimidiatus* group (Benazzou, 1983). Denys et al. (1994) described distinguishing dental differences between *A. chudeaui*, *A. cahirinus*, and the Algerian *A. seurati*, suggesting they reflected differences among species. The precise phylogenetic relationships among *A. chudeaui*, *A. seurati*, and *A. airensis* (in which 2n varies from 40 to 46) require resolution.

Acomys cilicicus Spitzenberger, 1978. Ann. Nat. Hist. Mus. Wien, 81:444.

COMMON NAME: Turkish Spiny Mouse.

TYPE LOCALITY: Asiatic Turkey, Vil Mersin, 17 km E Silifke.

DISTRIBUTION: Known only from the type locality (Bates, 1994).

STATUS: IUCN – Critically Endangered.

COMMENTS: Subgenus *Acomys*. Corbet (1984) commented on this species. Karyotype documented by Macholán et al. (1995). Zima et al. (1999b) considered the karyotype to be closely similar to that of *A. nesiotes* from Cyprus, stated that karyotypes of *A. cilicicus* and *A. cahirinus* from Egypt were "almost identical" (p. 151), and suggested "that *A. cahirinus*, *A. cilicicus*, and *A. nesiotes* represent a separate group of karyotypically closely related species of Eastern Mediterranean spiny mice." This notion is corroborated by phylogenetic analyses of mtDNA cytochrome *b* sequences (Barome et al., 1998, 2000) and breeding experiments (Frynta and Sádlová, 1998). Kivanç et al. (1997b) reported the same karyotype, but also looked at phallic morphology and concluded *A. cilicicus* was distinct because of its unique combination of phallic and other morphological traits in combination with a karyotype nearly identical to that of *A. cahirinus*. See accounts of *A. minous* and *A. nesiotes*.

Acomys cineraceus Heuglin, 1877. Reise in Nordost-Afrika, 2:70.

COMMON NAME: Gray Spiny Mouse.

TYPE LOCALITY: EC Sudan, "Eastern Sennaar and about Kalabat" (G. M. Allen, 1939).

DISTRIBUTION: C and S Sudan, N Uganda, C and S Ethiopia (specimens in USNM), and Djibouti (Pearch et al., 2001); distributional limits unresolved.

STATUS: IUCN – Lower Risk (lc).

SYNONYMS: *cinerascens* Heuglin, 1877; *hawashensis* Frick, 1914; *hystrella* Heller, 1911; *intermedius* Wettstein, 1916; *johannis* Thomas, 1912; *lowei* Setzer, 1956; *witherbyi* de Winton, 1901 [Dieterlen, in litt.].

COMMENTS: Subgenus *Acomys*. Formerly included in *A. cahirinus* (F. Petter, 1983; Setzer, 1975), but Dieterlen (in litt.) noted that *A. cineraceus* is a distinct species and one of four (*A. wilsoni*, *A. percivali*, and *A. cahirinus*) occurring in Sudan. Separation of *cineraceus* from *A. cahirinus* is supported by chromosomal data (2n = 48 or 50 for *cineraceus*, 2n = 36 for *cahirinus*; Kunze et al., 1999b) and analysis of pericentric satellite DNA (Kunze et al., 1999a). F. Petter (1983) recognized *witherbyi* as a species, and reported that it coexists with a member of the *cahirinus-dimidiatus* complex in Sudan. Both F. Petter (1983) and Denys et al. (1994) did not associate *lowei* (W Sudan) with *A. cineraceus*, but with *A. mullah* (see that account). In morphology, *A. cineraceus* closely resembles *A. kempi* (our study of specimens), which ranges from S Ethiopia and Somalia to NW Tanzania; systematic revision would reveal whether each is a species, or simply represents a population of one species. Limits of *A. cineraceus*'s geographic range is unresolved, especially its W segment. Musser and Carleton (1993) extended the distribution west to Burkina Faso and Ghana, but at least two species occupy that wide range west of Sudan, *Acomys airensis* (2n = 40 to 46) and *A. johannis* (2n = 66 or 68); see those accounts. From their study of external and dental morphology of holotypes included in *A. cineraceus*, Denys et al. (1994:225) noted their "very superficial comparison shows that probably different species have been grouped together, and that there is a need for a revision of this group." We agree.

Acomys dimidiatus (Cretzschmar, 1826). Rüppel Atlas, 37, taf. 13, fig. a.
   COMMON NAME: Eastern Spiny Mouse.
   TYPE LOCALITY: Egypt, Sinai.
   DISTRIBUTION: Sinai Peninsula of Egypt (Saleh and Basuony, 1998, as *cahirinus*), Jordan, Israel (Mendelssohn and Yom-Tov, 1999, as *cahirinus*), Lebanon, Syria, Saudi Arabia, Yemen (Al-Jumaily, 1998, as *cahirinus*), Oman, United Arab Emirates (Stuart and Stuart, 1995, as *cahirinus*), S Iraq, S Iran, and S Pakistan. The range (see Bates, 1994) is basically east of the distribution of the morphologically similar North African *A. cahirinus*.
   SYNONYMS: *carmeliensis* Haas, 1952; *flavidus* Thomas, 1917; *hispidus* (Brandts, 1827); *homericus* Thomas, 1923; *megalotis* (Lichtenstein, 1829); *whitei* Harrison, 1980.
   COMMENTS: Subgenus *Acomys*. *Acomys dimidiatus* is morphologically very similar to *A. cahirinus* and with few exceptions (e. g., Ellerman, 1941; Morrison-Scott, 1939; Setzer, 1959, 1975) has usually been listed in the synonymy of that species (Ellerman and Morrison-Scott, 1951; Ellerman et al., 1953; Musser and Carleton, 1993). Volobouev et al. (1996:217) correctly noted "The difficulty to find diagnostic morphological characters is especially notable in the *cahirinus-dimidiatus* group in which a number of forms and their taxonomic rank are subject to incessant revision." The chromosomal differences "provide a strong cytogenetic isolation" demonstrating the species status of *dimidiatus* compared with *cahirinus*, a conclusion they had proposed in an earlier study contrasting chromosomal features among *cahirinus*, *dimidiatus*, and *airensis* (Volobouev et al., 1991). That separation is also indicated by comparative karyological studies (Kunze et al., 1999*b*; Volobouev et al., 2002*b*), phylogenetic analyses based on mtDNA cytochrome *b* gene sequences (Barome et al., 2000, 2001*a*, *b*), pericentric satellite DNA (Kunze et al., 1999*b*), and comparative study of dental traits among species of *Acomys* (Denys et al., 1994). See Al-Saleh (1988), Barome et al. (2001*a*), Macholán et al. (1995), Qumsiyeh et al. (1986), Sokolov et al. (1992, 1993), Volobouev et al. (1991), and references cited therein for additional documentation of chromosomal data and its significance. Most samples of *A. dimidiatus* have 38 chromosomes, but the diploid number is 36 in some populations on the Sinai Peninsula and in Israel, and there is a narrow hybridization zone on the E Sinai (Nevo, 1989; Wahrman and Goitein, 1972; reported as *cahirinus*). Kronfield et al. (1994) reported ecological characteristics of *A. dimidiatus* (as *cahirinus*) and *A. russatus* where they occur together in S Israel.
      Haas's (1952) *carmeliensis* was described as a species of *Acomys* and based upon fragments from prehistoric (Natufian-Neolithic) sediments of Abu Usba Cave in the Mount Carmel region of Israel. Tchernov (1968) regarded the specimens as inseparable from contemporary *A. cahirinus* in Israel, which was the name applied to the Israeli populations now called *A. dimidiatus*.

Acomys ignitus Dollman, 1910. Ann. Mag. Nat. Hist., ser. 8, 6:229.
   COMMON NAME: Fiery Spiny Mouse.
   TYPE LOCALITY: Kenya, Voi.
   DISTRIBUTION: Usambara Mtns in NE Tanzania (specimens in FMNH), and Kenya (F. Petter, 1983, reports the species from Somalia but gives no exact locality). See Bates (1994); range limits unknown.
   STATUS: IUCN – Lower Risk (lc).
   COMMENTS: Subgenus *Acomys*. F. Petter (1983) recognized *ignitus* as a valid species, clashing with Setzer (1975), who regarded it as part of *A. dimidiatus*. F. Petter's action reflects reality, reinforcing the views of Hollister (1919) and Ellerman (1941), who recognized

ignitus as a separate species, but associated the names *pulchellus*, *kempi*, and *montanus* either as subspecies or direct synonyms (we arrange those names with *A. kempi*; see that account). Janecek et al. (1991) considered *ignitus* distinct and phylogenetically closely related to *A. cahirinus*, based on genetic data. Their conclusions, along with morphological data (Denys et al., 1994; our examination of skins and skulls) and analysis of mtDNA cytochrome *b* sequences (Barome et al. 1998, 2000, 2001*a*) define the specific integrity of *ignitus*.

Acomys johannis Thomas, 1912. Ann. Mag. Nat. Hist., ser. 8, 9:272.
COMMON NAME: Johan's Spiny Mouse.
TYPE LOCALITY: N Nigeria, Bauchi Plateau Kabwir (Rosevear, 1969, provided more information about holotype and origin of common name).
DISTRIBUTION: Burkina Faso, S Niger, Nigeria, N Benin, and N Cameroon; limits of geographic range unresolved.
COMMENTS: Subgenus *Acomys*. Formerly included in *A. cahirinus* (Setzer, 1975) or *A. cineraceus* (Musser and Carleton, 1993). Recently, Sicard and Tranier (1996) provided a detailed report on the geographic distribution of three pelage color phenotypes of *Acomys* occurring in Burkina Faso (all with 2n = 66 or 68; also see Gautun et al., 1986), assigned them to *A. johannis*, and contrasted their external, cranial, and dental morphology with *A. chudeaui* (2n = 42), *A. airensis* (2n = 40 to 46) and the Algerian *A. seurati* (2n = 38). Using mtDNA cytochrome *b* sequences, Barome et al. (2000) identified the specimens from Burkina Faso as *Acomys* sp. and specimens from Niger, Benin, Cameroon, and Niger as *A. johannis*. This species is a member of the *A. cahirinus-A. dimidiatus* group (Barome et al., 2000) and different from either *A. cahirinus* or *A. cineraceus*, but it needs to be more clearly diagnosed and its geographic range delineated.

Sicard and Tranier (1996) were impressed with the close similarity in cranial and dental traits between *A. johannis* from Burkina Faso and *A. mullah* from Ethiopia and Somalia. This observation may be significant because Sokolov et al. (1992, 1993) karyotyped specimens of *Acomys* from the Ethiopian Rift Valley. Several samples characterized as 2n = 36 were determined to be *A. cahirinus*, and another sample with 2n = 68 was identified only as *Acomys* sp. If the latter is *A. mullah*, which occurs in Ethiopia, it would corroborate the morphological alliance suggested by Sicard and Tranier. In addition to *A. cahirinus* and *A. mullah*, only *A. cineraceus* (2n = 48 or 50; Kunze et al., 1999*a*, *b*), *A. wilsoni* (2n = 50; Matthey, 1968), and *A. percivali* (karyotype unknown) have been recorded from Ethiopia.

Acomys kempi Dollman, 1911. Ann. Mag. Nat. Hist., ser. 8, 8:125.
COMMON NAME: Kemp's Spiny Mouse.
TYPE LOCALITY: Kenya, Chanler Falls, N Guaso Nyiro.
DISTRIBUTION: S Ethiopia, S Somalia, Kenya and NE Tanzania (samples in FMNH and USNM); also see Bates (1994); limits unknown.
STATUS: IUCN – Lower Risk (lc).
SYNONYMS: *montanus* Heller, 1914; *pulchellus* Dollman, 1911.
COMMENTS: Subgenus *Acomys*. Originally described by Dollman as a subspecies of *A. ignitus* and listed that way by Ellerman (1941) and Hollister (1919), but considered a subspecies of *A. cahirinus* by Setzer (1975). Treated as a species by Janecek et al. (1991) with closest evolutionary ties to *A. cahirinus*. The morphological characteristics and geographic range of *kempi* may represent the eastern segment of *A. cineraceus*. Hollister (1919) correctly explained why *pulchellus* is a synonym of *A. kempi*; we include *montanus* based upon our studies.

Acomys louisae Thomas, 1896. Ann. Mag. Nat. Hist., ser. 6, 18:269.
COMMON NAME: Louise's Spiny Mouse.
TYPE LOCALITY: Somalia, 40 mi (64.4 km) south of Berbera.
DISTRIBUTION: Somalia and Djibouti; see Bates (1994); limits unknown.
STATUS: IUCN – Lower Risk (lc).
SYNONYMS: *umbratus* Thomas, 1923.
COMMENTS: Subgenus *Peracomys*. The type-species of subgenus *Peracomys* (F. Petter and

Roche, 1981); distinction from other *Acomys* confirmed by Denys et al. (1994). F. Petter (1983) included *umbratus* in the species. Pearch et al. (2001) listed *A. louisae* for the Djibouti fauna following Scaramella et al. (1974).

Acomys minous Bate, 1906. Proc. Zool. Soc. Lond., 1905(2):321 [1906].

COMMON NAME: Crete Spiny Mouse.

TYPE LOCALITY: Greece, Crete Isl, Kanea.

DISTRIBUTION: Endemic to Crete (Greece); see Bates (1994).

STATUS: IUCN – Vulnerable.

COMMENTS: Subgenus *Acomys*. Treated as a species by Dieterlen (1978*a*), Corbet and Hill (1991), Cheylan (1991), and Denys et al. (1994). Chromosomal banding pattern documented by Kunze et al. (1999*b*:228), who noted the close similarity between karyotypes of Egyptian *A. cahirinus* and *A. minous*, and speculated that "*Acomys* may have reached Crete by ship with humans," an assumption concordant "with the lack of fossil *Acomys* records in Crete" (see Dieterlen, 1978, and discussion and references in Barome et al. [2001*a*]). This hypothesis was challenged by Barome et al. (2001*a*) based on mtDNA cytochrome *b* sequences. Their sample of *A. minous* was composed of two lineages, one (group A) clustered with Egyptian *A. cahirinus* and *A. nesiotes* from Cyprus, the other (group B) formed a clade with the Turkish *A. cilicicus*. Cytochrome *b* sequences were closely similar among the four species, the maximum nucleotide divergence equivalent to intraspecific divergence found in other species of *Acomys*, and Barome et al. (2001*a*:44) concluded that *minous*, *nesiotes*, and *cilicicus* are very closely related to *A. cahirinus*, "and the low sequence divergence among them suggests that they could belong to the same species, an hypothesis that must be tested with cytogenetics, morphology, and hybridization." Acknowledging the lack of fossil *Acomys* from Crete and the current belief that it was inadvertently introduced to Crete by humans, Barome et al. noted that the cytochrome *b* diversity found in *A. minous* was more ancient than the time of human colonization with common ancestry of the sequences extending back to the middle Pleistocene (0.7 to 0.1 million years ago). The polymorphism does not represent diversification following a human introduction but was present in the population from which the Crete animals were derived. Whether the Crete population represents immigration from two distinct populations or a single variable one will have to be determined by surveying polymorphisms of *Acomys* from Egypt, Cyprus, and Turkey. Volobouev et al. (2002*b*:266) do not recognize the distinctness of *A. minous* relative to *A. cahirinus*, stating that the minor chromosomal difference between the two (Kunze et al., 1999*b*) "is not sufficient to cause reproductive isolation."

Acomys mullah Thomas, 1904. Ann. Mag. Nat. Hist., ser. 7, 14:103.

COMMON NAME: Mullah Spiny Mouse.

TYPE LOCALITY: Ethiopia.

DISTRIBUTION: Ethiopia and Somalia; see Bates (1994); limits unknown.

STATUS: IUCN – Lower Risk (lc).

SYNONYMS: *brockmani* Dollman, 1911.

COMMENTS: Subgenus *Acomys*. F. Petter (1983) recognized *brockmani* as a valid species, suggesting it might be referable to *mullah*, which is the older name. These two forms are characterized by large molar rows, also diagnostic of *lowei*, which F. Petter implicitly associated with both *mullah* and *brockmani* as separate species, as did Denys et al. (1994). Ellerman (1941) also recognized *mullah* and *brockmani* as species, but Setzer (1975) arranged *lowei* as a subspecies of *A. cahirinus*, Dieterlen (in litt.) regarded it as a synonym of *A. cinerasceus*, and Setzer (1975) treated *mullah* and *brockmani* as subspecies of *A. dimidiatus*. Yalden et al. (1976) listed *mullah* as a synonym of *A. cahirinus*. Musser and Carleton (1993) followed Dieterlen in associating *lowei* with *A. cinerasceus*, an identity we retain pending a systematic revision of that species (see account).

Acomys nesiotes Bate, 1903. Ann. Mag. Nat. Hist., ser. 7, 11:565.

COMMON NAME: Cyprus Spiny Mouse.

TYPE LOCALITY: Cyrus, Kernyia Hills, near Dikomo.

DISTRIBUTION: Cyprus; see in Bates (1994).

STATUS: IUCN – Data Deficient.

COMMENTS: Subgenus *Acomys*. Originally described as a species, subsequently listed as a subspecies of *A. dimidiatus* (Ellerman, 1941) or included in *A. cahirinus* (Corbet, 1978c), but again treated as a species (Spitzenberger, 1978b). Standard karyotype (2n=38) described by Zahavi and Wahrman (1956) and Zima et al. (1999b); chromosomal banding pattern similar to the Turkish *A. cilicicus* and Egyptian *A. cahirinus* (Zima et al., 1999b). This association corroborated by mtDNA cytochrome *b* sequences (Barome et al., 2001a) and breeding experiments (Frynta and Sádlová, 1998); the Crete, Cyprus, and Turkish species join *A. cahirinus* in a monophyletic clade to the exclusion of that formed by samples of *A. dimidiatus* or other species of *Acomys* sampled. See accounts of *A. cilicicus* and *A. minous* for additional discussion.

Acomys percivali Dollman, 1911. Ann. Mag. Nat. Hist., ser. 8, 8:126.

COMMON NAME: Percival's Spiny Mouse.

TYPE LOCALITY: Kenya, Chanler Falls, Nyiro.

DISTRIBUTION: S Sudan (east of White Nile), E Uganda, SW Ethiopia, and Kenya; see Bates (1994).

STATUS: IUCN – Lower Risk (lc).

COMMENTS: Subgenus *Acomys*. Treated as a synonym of *kempi*, which in turn was included within *A. cahirinus* by Setzer (1975). However, Hollister (1919) recognized *percivali* as a species based upon many specimens he examined, as did Ellerman (1941) and F. Petter (1983). Hubert (1978b) identified specimens from Ethiopia as *A. percivali*, and Neal (1983) used *percivali* as a species in comparing breeding patterns between it and *A. wilsoni* in C Kenya. Janecek et al. (1991) regarded *percivali* as the species genetically most closely related to *A. wilsoni*, and Denys et al. (1994) noted the nearly identical molar morphology in the two species. Matthey (1968) recorded a karyotype (2n = 36, FN = 68) for what he identified as *A. percivali* from Ethiopia, but Sokolov et al. (1992, 1993) regarded Matthey's results as applying to *A. cahirinus*. See account of *A. wilsoni*.

Acomys russatus (Wagner, 1840). Abh. Akad. Wiss. Münchin, 3:195.

COMMON NAME: Golden Spiny Mouse.

TYPE LOCALITY: Egypt, Sinai.

DISTRIBUTION: Egypt east of the Nile (Osborn and Helmy, 1980), S and E Sinai (Haim and Tchernov, 1974; Nevo, 1989; Saleh and Basuony, 1998), Jordan, Israel (Mendelssohn and Yom-Tov, 1999), Saudi Arabia, N Yemen (Al-Jumaily, 1998), and Oman (Harrison and Bates, 1991); see Bates (1994).

STATUS: IUCN – Lower Risk (lc).

SYNONYMS: *aegyptiacus* Bonhote, 1912; *affinis* Gray, 1843; *harrisoni* Atallah, 1970; *lewisi* Atallah, 1967.

COMMENTS: Subgenus *Acomys*. Qumsiyeh et al. (1986) retained *lewisi* as a species because fur color and bacular morphology of *lewisi* are distinctive compared with *A. russatus* (Atallah, 1967), even though the karyotype of *lewisi* from Jordan is indistinguishable from *A. russatus*. However, based on morphological evidence, *lewisi* was included in *A. russatus* by other systematists (Corbet, 1978c; Harrison and Bates, 1991; Osborn and Helmy, 1980), an allocation also supported by genetic data (Janecek et al., 1991). See Nevo (1985, 1989, and references therein) for additional chromosomal data and its significance.

    Among species of *Acomys*, *A. russatus* is very distinctive in its molar morphology (Denys et al., 1994) and chromosomal traits (2n = 66), an isolation bolstered by phylogenetic analysis of mtDNA cytochrome *b* sequences, which does not associate it closely with any other single species or any species in the *A. cahirinus-A. dimidiatus* group (Barome et al., 2000, 2001a, b). *Acomys russatus* and *A. dimidiatus* are sympatric in Israel and Arabia, and their similarities and differences in ecology, physiology, and activity patterns where they coexist in Israel have been extensively documented (see references in Nevo, 1989).

Acomys seurati Heim de Balsac, 1936. Bull. Biol. Fr. Belg., 21:356, Fig. 6(4); 389, Fig. 15(15).

COMMON NAME: Seurat's Spiny Mouse.

TYPE LOCALITY: S Algeria, Iniker, A'hagger (Ahaggar Mtns; Ellerman and Morrison-Scott, 1951; Ranck, 1968).

DISTRIBUTION: Recorded only from mountains of S Algeria, in Hoggar, Mouydir and Tassili n'Ajjers (Bates, 1994; Kowalski and Rzebik-Kowalska, 1991).

COMMENTS: Subgenus *Acomys*. Described as a species by Heim de Balsac, but listed as a synonym of *A. cahirinus* by Ellerman and Morrison-Scott (1951) and Musser and Carleton (1993); however, the karyotype is different (2n = 38, FN = 68 compared with 2n = 36, FN = 68 for *A. cahirinus*) and the two can be distinguished by molar morphology (Denys et al., 1994:237). Kowalski and Rzebik-Kowalska (1991) provided references to distributional records, ecology, and morphology of the species.

Acomys spinosissimus Peters, 1852. Reise nach Mossambique, Säugeth., p. 160.
    COMMON NAME: Southern African Spiny Mouse.
    TYPE LOCALITY: Mozambique, Tette and Buio.
    DISTRIBUTION: NE Tanzania (Amani; series in USNM) and EC Tanzania (Kilosa and Morogoro regions; specimens in FMNH and MNHN), SE Dem. Rep. Congo, Zambia, Malawi (Chitaukali et al., 2001, on Nyika Plateau; Denys et al., 1999), Zimbabwe, E Botswana, C Mozambique, and N and NW South Africa; see Bates (1994).
    STATUS: IUCN – Lower Risk (lc).
    SYNONYMS: *selousi* de Winton, 1896; *transvaalensis* Roberts, 1926.
    COMMENTS: Subgenus *Acomys*. Revised by Dippenaar and Rautenbach (1986). Genetic data indicated that *A. spinosissimus* should be placed in a species-group separate from other *Acomys* (Janecek et al., 1991). Barome et al. (2001*b*) reported morphological and molecular analyses of *A. spinosissimus* from Tanzania and Malawi compared with samples from farther south in the range. Their results not only demonstrated the monophyly of all geographic samples (from Tanzania, Malawi, Mozambique, Zimbabwe, and NE South Africa) of *A. spinosissimus*, but also the distinctiveness of this species compared with South African *A. subspinosus* and other northern species (*A. wilsoni*, *A. russatus*, *A. ignitus*, *A. airensis*, *A. cahirinus*, and *A. dimidiatus*; see also Barome et al., 2000). Reviewed by de Graaf (1997*b*).

Acomys subspinosus (Waterhouse, 1838). Proc. Zool. Soc. Lond., 1837:104 [1838].
    COMMON NAME: Cape Spiny Mouse.
    TYPE LOCALITY: South Africa, Western Cape Province, Cape of Good Hope.
    DISTRIBUTION: South Africa; restricted to Western Cape Province (Bates, 1994; Dippenaar and Rautenbach, 1986; Skinner and Smithers, 1990).
    STATUS: IUCN – Least Concern.
    COMMENTS: Subgenus *Subacomys*. Revised by Dippenaar and Rautenbach (1986). Musser and Carleton (1993) noted that *A. subspinosus* should be in a species-group by itself, according to the results of genetic analysis by Janecek et al. (1991). Its separation from other *Acomys* is also indicated by phylogenetic analyses of mtDNA cytochrome *b* sequences (Barome et al., 2000, 2001*b*). Their study of cranial and dental morphology prompted Denys et al. (1994) to propose the subgenus *Subacomys* for *A. subspinosus*. Reviewed by de Graaff (1997*c*).

Acomys wilsoni Thomas, 1892. Ann. Mag. Nat. Hist., ser. 6, 10:22.
    COMMON NAME: Wilson's Spiny mouse.
    TYPE LOCALITY: Kenya, Mombasa.
    DISTRIBUTION: S Sudan, S Ethiopia, S Somalia, Uganda, Kenya, and south to EC Tanzania (Kondoa; specimens in AMNH); see Bates (1994); limits unknown.
    STATUS: IUCN – Lower Risk (lc).
    SYNONYMS: *ablutus* Dollman, 1911; *argillaceus* Hinton and Kershaw, 1920; *boronei* De Beaux, 1934; *enid* St. Leger, 1932; *nubilus* Dollman, 1914.
    COMMENTS: Subgenus *Acomys*. Formerly included in *A. subspinosus* by Setzer (1975) but considered a species by most systematists (Corbet and Hill, 1991; Ellerman, 1941; Hollister, 1919; Matthey, 1968; Musser and Carleton, 1993; F. Petter, 1983; F. Petter and Roche, 1981; Rupp, 1980; Yalden et al., 1976). Genetically most similar to *A. percivali* (Janecek et al., 1991) with which it occurs sympatrically (Neal, 1983, for example, studied the breeding patterns in both species from the same locality in

C Kenya); Denys et al. (1994) noted a close similarity in molar patterns between the two species. Specimens of *nubilus* are larger and longer-tailed than *wilsoni* and may represent a separate species; de Beaux's (1934) description of *boronei* recalls *nubilus*. Those two names possibly represent the closely related *A. percivali*, which has a longer tail relative to head and body length than does *A. wilsoni*. The need to revise the *A. wilsoni-A. percivali* species complex was stressed by Denys et al. (1994). Matthey (1968) documented chromosomal traits (2n = 50). Our study of specimens indicates *A. wilsoni* (and *A. percivali*) to be very different from species in the *A. cahirinus-A. dimidiatus* group, and all other species of *Acomys*, a conclusion reinforced by phylogenetic analyses of mtDNA cytochrome *b* gene sequences among species of *Acomys* (Barome et al., 2000).

Deomys Thomas, 1888. Proc. Zool. Soc. Lond., 1888:130.

TYPE SPECIES: *Deomys ferrugineus* Thomas, 1888.

COMMENTS: Described by Thomas (1888c:132) as a unique member of a "special section" within Muridae (*Deomyes*) with molar traits intermediate between cricetines and murines. A year later, Lydekker (1889) referred *Deomys* to the Deomyinae; subsequently, Thomas (1896) placed *Deomys* in the Dendromurinae, and with the exception of Ellerman, (1941), there it remained until the 1990s (see historical discussions in Rosevear, 1969, and Carleton and Musser, 1984). The morphological traits of *Deomys* do not fit well in the Dendromurinae (Carleton and Musser, 1984; Rosevear, 1969), which influenced Ellerman (1941) to propose Deomyinae (primarily because of its peculiar and morphologically primitive conformation of the zygomatic plate) to contain the genus (Ellerman listed Thomas's *Deomyes* as a synonym and was apparently unaware of Lydekker's Deomyinae). Recent DNA-DNA hybridization experiments (Denys et al., 1995) and phylogenetic analyses of mtDNA cytochrome *b* sequences (E. Verheyen et al., 1995), nuclear protein-coding DNA sequences (LCAT, Michaux and Catzeflis, 2000; LCAT and vWF, Michaux et al., 2001), and the combination of DNA/DNA hybridization and mitochondrial and nuclear DNA sequences (Chevret et al., 2001) consistently placed *Deomys* in the same clade as *Acomys*, *Lophuromys*, and *Uranomys*. Phylogenetic analyses of dental traits, however, did not group *Deomys* with the acomyines (Denys et al., 1995). Sperm morphology of *Deomys* is highly derived compared with species of Dendromurinae and indicates it does not belong with that subfamily (Breed, 1995d), but Breed did not include it within the *Acomys-Uranomys* group he defined using spermatozoal characters. *Deomys ferrugineus* feeds mostly on insects, millipedes, centipedes, slugs and crustacea (Genest-Villard, 1980) and its unique zygomasseteric morphology and molar topography may reflect adaptations to that diet. In reporting their results based on mtDNA sequences, E. Verheyen et al. (1996:784) suggested that "the clustering of *Deomys* with the *Lophuromys* clade may be less surprising than has been suggested elsewhere (Denys et al., 1995). The *Deomys* skull is not only very distinct from that of other dendromurines, but its zygomatic plate and internal organs (stomach) have been reported to resemble the same features in *Lophuromys*." Even Ellerman (1941:317) admitted that the zygomatic plate of *Deomys* resembled that of *Lophuromys*. Osgood (1936:238) described how the pelage of *Deomys* has a "more superficial resemblance to *Acomys* than to any other well-known genus," pointed to configuration of auditory bulla and petrosal as being similar in *Deomys* and *Lophuromys*, and noted that Tullberg (1899) commented on the similarity in anterior portion of skull and internal organs with *Lophuromys*.

Deomys ferrugineus Thomas, 1888. Proc. Zool. Soc. Lond., 1888:130.

COMMON NAME: Congo Forest Rat.

TYPE LOCALITY: Lower Congo.

DISTRIBUTION: Uganda, Rwanda, Dem. Rep. Congo, SW Central African Republic, S Cameroon, Gabon, Republic of Congo, Equatorial Guinea (including Bioko).

STATUS: IUCN – Lower Risk (lc).

SYNONYMS: *christyi* Thomas, 1915; *poensis* Eisentraut, 1965; *vandenberghei* Rahm and W. Verheyen, 1960.

COMMENTS: Hatt (1940), Rosevear (1969), and Eisentraut (1973) provided informative descriptions and reviews. Lemire (1966) reported myological and skeletal interaction in

mastication. Eisentraut (1973) provided color plate contrasting the reddish brown upperparts of mainland samples with the dark brown dorsum of *poensis* from Bioko.
Rosevear (1969) described the range in variation in coloration of dorsal pelage (intense
reddish brown to reddish brown with a blackish rump to dark brown or blackish over
most of back and rump; Eisentraut's plate illustrated this variation) and the meaningless attempts to attach the names listed in synonymy to parts of that range. Kerbis Peterhans and Patterson (1995:329) included *D. ferrugineus* in their African guild of
"waders" along with species of *Malacomys* and *Colomys goslingi*: "Without swimming
and while perched on elongate hind feet, all three consume insects and other small animals in shallow forest streams and pools."

Lophuromys Peters, 1874. Monatsb. K. Preuss. Akad. Wiss. Berlin, p. 234.

TYPE SPECIES: *Lasiomys afer* Peters, 1866 (= *Mus sikapusi* Temminck, 1853).

SYNONYMS: *Kivumys* Dieterlen, 1987; *Lasiomys* Peters, 1866 [not Burmeister, 1854]; *Neanthomys*
Toschi, 1946.

COMMENTS: Because *Lasiomys* is preoccupied by Burmeister's *Lasiomys*, Peters proposed
Lophuromys with the same type species. Dieterlen (1976b, 1987) arranged the species into
subgenera *Lophuromys* and *Kivumys*; the latter apparently represents a separate
monophyletic group (containing *L. luteogaster, L. medicaudatus,* and *L. woosnami*) that
may eventually be elevated to generic rank. *Neanthomys* was based on a specimen missing
the tail; see Lavrenchenko et al. (1998b) for additional discussion and references. Most
changes from the 10 or 11 species recognized by G. M. Allen (1930) and Ellerman (1941)
to the five listed by Misonne (1974), the 10 in Musser and Carleton (1993), and the 21
acknowledged here result from the contributions of Dieterlen (1976b, 1987), F. Petter
(1972d), W. Verheyen et al. (1996, 1997, 2000, 2002) and Lavrechenko et al. (1998b); the
overall systematic revision of *Lophuromys* is being continued by W. Verheyen and his
colleagues and will provide future information about diversity of species in the genus,
their geographic distributions, and phylogenetic relationships.

Lophuromys aquilus (True, 1892). Proc. U. S. Nat. Mus., 15:460.

COMMON NAME: Dark-colored Brush-furred Rat.

TYPE LOCALITY: East Africa, Tanzania, Mt Kilimanjaro, 8000 ft (2650 m; coordinates given by
W. Verheyen et al., 2002).

DISTRIBUTION: Lowland and montane habitats from NE Angola (Crawford-Cabral, 1998)
throughout the Dem. Rep. Congo, Rwanda (Corti et al., 2000; Monfort, 1992) and
Burundi (Maddalena et al., 1989) to Uganda (Delaney, 1975), Kenya (Hollister, 1919;
except montane habitats on Aberdare Range and Mt Kenya; see account of *L. zena*),
and south through Tanzania (Swinnerton and Hayman, 1951; except Mt Meru; see
account of *L. verhageni*) to Malawi (Ansell and Dowsett, 1988), N Zambia (Ansell,
1978), and NE Mozambique (Smithers and Lobão Tello, 1976); recorded as either
L. aquilus or *L. flavopunctatus* in faunal accounts.

STATUS: IUCN – Data Deficient as *L. cinereus*.

SYNONYMS: *cinereus* Dieterlen and Gelmroth, 1974; *laticeps* Thomas and Wroughton, 1907;
major Thomas and Wroughton, 1907; *margarettae* Heller, 1912; *rita* Dollman, 1910;
rubecula Dollman, 1909.

COMMENTS: Subgenus *Lophuromys, L. aquilus* species group (W. Verheyen et al., 2002).
Described as a species, which was recognized in early checklists (G. M. Allen, 1939;
Ellerman, 1940d), but subsequently included in *L. flavopunctatus* (Misonne, 1974;
Musser and Carleton, 1993). Analyses of morphometric traits and mtDNA cytochrome
b sequences by W. Verheyen et al. (2002) demonstrated the conspecificity of the synonyms with *L. aquilus* and definition of the species relative to others in the *L. flavopunctatus* complex of species. Geographic samples of *L. aquilus* form a monophyletic
clade most closely related to a clade containing the Ethiopian *L. brunneus* and *L. chrysopus*, and Kenyan *L. zena*, both clusters considered to be members of the *L. aquilus* species group (W. Verheyen et al., 2002). Maddalena et al. (1989) reported chromosomal
data for Burundi sample. Three-dimensional geometric morphometrics was used by
Corti et al. (2000) to look at patterns of growth and cranial size and shape in *L. aquilus*
(reported as *flavopunctatus*) compared with *L. sikapusi* and *L. woosnami*. Stanley et al.

(2000) documented specimens from the Eastern Arc Mtns of Tanzania. Altitudinal distribution on Ugandan slopes of Ruwenzori Mtns reviewed by Kerbis Peterhans et al. (1998) in context of surveying small mammal distributions in those highlands (but samples from Ruwenzori Mtns are apparently a separate species in the *L. aquilus* group; W. Verheyen et al., 2002). Habitat and altitudinal distribution on Ugandan slopes of Mt Elgon recorded by Clausnitzer and Kityo (2001) and Clausnitzer et al. (2003; both as *L. flavopunctatus*).

The holotype of *rubecula* is from the Kenyan slopes of Mt Elgon. In the morphometric analysis by W. Verheyen et al. (2002), it clusters with holotypes of the Ethiopian *L. chrysopus*, and Cameroon *L. dieterleni* and *L. eisentrauti*, suggesting close affinity; however, the rest of their sample from Mt Elgon bunches with those identified as *L. aquilus*. Dieterlen and Gelmroth (1974) described *L. cinereus* from Marais Mukaba in Parc National du Kahuzi-Biega (Dem. Rep. Congo), which was recorded only from the type locality. It is morphologically closely similar to what they called *L. flavopunctatus*, and possibly only a gray morph of that species: "Es ist nicht auszuschließen, daß die vier als *Lophuromys cinereus* beschriebenen Stücke graue Farbmutanten von *L. flavopunctatus* sind" (Dieterlen, 1987:188). Identity of the holotype of *cinereus* with *L. aquilus* was confirmed by W. Verheyen et al. (2002).

Lophuromys angolensis W. Verheyen, Dierckx, and Hulselmans, 2000. Bull. Inst. Roy. Sci. Nat. Belgique, Biol., 70:255
COMMON NAME: Angolan Brush-furred Rat.
TYPE LOCALITY: SW Dem. Rep. Congo, Mbwambala, 500 m (W. Verheyen et al., 2000, provided additional data).
DISTRIBUTION: Lowland tropical evergreen rain forest and mountain forest (disturbed and highly disturbed secondary and primary formations) along rim of C rainforest block in SW Dem. Rep. Congo (500 m), and W highlands of Angola (1000-2600 m); see W. Verheyen et al. (2000).
COMMENTS: Subgenus *Lophuromys*, *L. sikapusi* species group (W. Verheyen et al., 2000). W. Verheyen et al. (2000) documented significant morphometric differences between *L. angolensis* and samples of *L. sikapusi* and *L. ansorgei* as revealed by univariate and multivariate analyses, but noted the lack of qualitative morphological distinctions between *L. angolensis* and the other two. Reviewed by Crawford-Cabral (1998) as *L. sikapusi*.

Lophuromys ansorgei de Winton, 1896. Proc. Zool. Soc. Lond., 1896:607.
COMMON NAME: Ansorge's Brush-furred Rat.
TYPE LOCALITY: Kenya, Nyanza Province, Mumia's.
DISTRIBUTION: West Africa from lower reaches of the Congo River in E Dem. Rep. Congo (Boma, Kinshasa) and possibly S Cameroon, and in E Africa from Uganda through W Kenya to N Tanzania.
SYNONYMS: *manteufeli* Matschie, 1911; *pyrrhus* Heller, 1911.
COMMENTS: Subgenus *Lophuromys*, *L. sikapusi* species complex (W. Verheyen et al., 2000). Usually included in *L. sikapusi*, but regarded as a separate species by W. Verheyen et al. (1997, 2000). The species has yet to be fully revised and its geographic range delineated; the distribution recorded here is extracted from W. Verheyen et al. (1997, 2000). Geographic range of *L. ansorgei* either extends from East Africa along the northern and southern rim of the Congolese Central Forest Block to the lower reaches of the Congo River, or the range along the lower Congo is a remnant of a once more expansive distribution (W. Verheyen et al., 2000). Dieterlen (1987) discussed *ansorgei* and *pyrrhus* as subspecies of *L. sikapusi*.

Lophuromys brevicaudus Osgood, 1936. Zool. Ser. Field Mus. Nat. Hist., 20:241.
COMMON NAME: Short-tailed Brush-furred Rat.
TYPE LOCALITY: Ethiopia, Arusi (=Arsi) Province, Chilalo Mtns, Mt Albasso.
DISTRIBUTION: Endemic to SC Ethiopian highlands on east side of Ethiopian Rift Valley between 2400 and 3750 m; recorded from the Chilalo and Gedeb Mtns by Osgood (1936) and from the Bale Mtns by Lavrenchenko et al. (1998*b*).
COMMENTS: Subgenus *Lophuromys*, *L. flavopunctatus* species group (Lavrenchenko et al.,

1998*b*, 2000). Described as a distinctive species (Osgood, 1936), but later included in *L. flavopunctatus* (Misonne, 1974; Yalden et al., 1976, 1996). Allozymic (Lavrenchenko et al., 2000), chromosomal (Aniskin et al., 1997), and morphological and morphometric data (Lavrenchenko et al., 1998*b*; W. Verheyen et al., 2002) separate *brevicaudus* as a species. Lavrenchenko et al. (1998*b*) noted that *L. brevicauda* is diurnal and one of the most common small mammals occurring in the *Erica-Hypericum* heath zone (3170-3750 m) in the Harenna Forest on southern slopes of the Bale Mtns. At about 3500 m, *L. brevicauda*'s range narrowly overlaps lower margin of the altitudinal distribution of *L. melanonyx*. At the lower end of *L. brevicauda*'s altitudinal range, between 2400 and 2760 m in the *Schefflera-Hagenia* belt on southern slope of Bale Massif, it coexists with *L. chrysopus* (Lavrenchenko, 2000; Lavrenchenko et al., 1998*b*).

Allozymic data (Lavrenchenko et al., 2000) indicate that *L. melanonyx* diverged basally to the sister species *L. brevicaudus* and *L. chrysopus*. *Lophuromys chrysopus*, *L. brevicaudus*, and *L. melanonyx* "occur in the Bale Mountains, replacing each other in the different altitudinal belts (tropical forest–heathland–afro-alpine zone) each time with a small overlap suggesting an adaptive pattern of speciation" (Lavrenchenko et al., 1998*b*). Multivariate analyses of morphometric data (Bekele and Corti, 1994) from seven geographic samples of *L. flavopunctatus* and *L. melanonyx* uncovered three groups: *L. megalonyx*; a second cluster of samples from N, C, and W Ethiopia; and a third assemblage from the Bale Mtns. Lavrenchenko et al. (1998*b*) thought the Bale Mtns sample to be a mixture of *L. brevicauda* and *L. chrysopus*.

Lophuromys brevicaudus, along with *L. flavopunctatus* and *L. melanonyx*, are Ethiopian endemics and members of the strictly Ethiopian *L. flavopunctatus* species group. The other two Ethiopian endemic *Lophuromys* (*brunneus* and *chrysopus*) apparently form a clade more closely related to the *L. aquilus* species group than to *L. brevicaudus* and its allies (W. Verheyen et al., 2002).

Lophuromys brunneus Thomas, 1906. Ann. Mag. Nat. Hist., ser. 7, 18:305.

COMMON NAME: Thomas's Ethiopian Brush-furred Rat.

TYPE LOCALITY: S Ethiopia, Manno-Jimma, 4200 ft (1280 m; W. Verheyen et al., 2002, provide coordinates).

DISTRIBUTION: Endemic to S Ethiopia on highlands west of the Rift Valley, from the Simien Mtns in the north to Manno-Jimma in the south (see specimens examined in W. Verheyen et al., 2002).

SYNONYMS: *simensis* Osgood, 1936.

COMMENTS: Subgenus *Lophuromys*, *L. aquilus* species group (W. Verheyen et al., 2002). Originally described as a subspecies of *L. aquilus*, later allocated to *L. flavopunctatus* as a probable synonym of *L. f. zaphiri* (Osgood, 1936), and included in *L. flavopunctatus* until now (Misonne, 1974; Musser and Carleton, 1993; Yalden et al., 1976, 1996). Morphometric analyses by W. Verheyen et al. (2002) indicated *brunneus* to be a species distinct from all the other Ethiopian *Lophuromys* except the Ethiopian *L. chrysopus* to which it is closely related, and that *simensis* is likely a synonym. Lavrenchenko et al. (1998*b*) speculated that *simensis* may represent a separate species in the Simien Mtns. See account of *L. chrysopus*.

Lophuromys chrysopus Osgood, 1936. Zool. Ser. Field Mus. Nat. Hist., 20:242.

COMMON NAME: Ethiopian Forest Brush-furred Rat.

TYPE LOCALITY: Ethiopia, Sidamo Province, Allata.

DISTRIBUTION: Endemic to Ethiopia in montane forests of the E and W plateaus (separated by the Ethiopian Rift Valley); probably occurs in most forests of the SW mountains (Lavrenchenko et al., 1998*b*).

COMMENTS: Subgenus *Lophuromys*, *L. aquilus* species group (W. Verheyen et al., 2002). Described as a subspecies of *L. aquilus* by Osgood (1936) but subsequently included in *L. flavopunctatus* (Yalden et al., 1976, 1996). Allozymic, chromosomal, as well as morphological and morphometric data (Aniskin et al., 1997; Lavrenchenko et al., 1998*b*, 2000; W. Verheyen et al., 2002) distinguish *chrysopus* as a separate species from both *L. flavopunctatus* and *L. brevicaudus*. The chromosomal composition alone signaled the distinctness of *chrysopus* (2n = 54, FN = 60, as contrasted with 2n = 68, FN = 78 for *bre*-

vicaudus; and 2n = 68 for *flavopunctatus*; Aniskin et al., 1997). Differences between samples from E and W plateaus in morphometrics, pelage coloration, and distribution of B-chromosomes suggest some geographic differentiation in *L. chrysopus* (Lavrenchenko et al., 1998*b*). Morphometric analyses place *L. chrysopus* in the same cluster as the Ethiopian *L. brunneus* and montane Kenyan *L. zena*, and separate from the group formed by geographic samples of *L. aquilus* (W. Verheyen et al., 2002). Those authors also noted that holotypes of Cameroon *eisentrauti* and *dieterleni*, Mt Elgon *rubecula*, and Kenyan *zena* cluster with samples of *L. chrysopus*, suggesting all represent the same species, a possibility requiring testing by analysis of morphometric, chromosomal, and molecular variation in highland samples from the region between Ethiopia and the Cameroon and Kenyan mountains. In the Harenna forest block on S slopes of Bale Mtns, *L. chrysopus* is one of the most abundant rodents in tropical evergreen rainforest between 1550 and 2760 m and is sympatric with the heathland *L. brevicauda* from 2400 to 2760 m (Lavrenchenko, 2000; Lavrenchenko et al., 1998*b*).

Lophuromys dieterleni W. Verheyen, Hulselmans, Colyn, and Hutterer, 1997. Bull. Inst. Roy. Sci. Nat. Belgique, Biol., 67:173.
    COMMON NAME: Mount Oku Brush-furred Rat.
    TYPE LOCALITY: West Cameroon, Bamenda-Banso highlands, border of the crater lake on Mt Oku, 2100 m (see W. Verheyen et al., for additional information).
    DISTRIBUTION: Recorded by five specimens collected in 1967 from the type locality, an isolated patch of forest on Mt. Oku (W. Verheyen et al., 1997).
    COMMENTS: Subgenus *Lophuromys*, *L. aquilus* species group (W. Verheyen et al., 2002). Apparently phylogenetically most closely allied with *L. eisentrauti* from nearby Mt Lefo (W. Verheyen et al., 1997). Holotypes of *L. eisentrauti*, *L. dieterleni*, Mt Elgon *L. rubecula*, and montane Kenyan *L. zena* cluster with samples of *L. chrysopus*, suggesting all represent the same species, a possibility requiring testing by analyses of morphometric, chromosomal, and molecular variation in geographic samples from the region between Ethiopia and the Cameroon and Kenyan highlands. Morphometric and morphological comparisons between *L. dieterleni* and samples of *L. eisentrauti*, *L. sikapusi*, *L. flavopunctatus*, and *L. ansorgei* documented by W. Verheyen et al. (1997). An example of *L. sikapusi* has been collected at the base of Mt Oku and that species may prove to be sympatric with *L. dieterleni* (W. Verheyen et al., 1997). *Lophuromys dieterleni*, three murine rodents (*Lamottemys okuensis*, *Lemniscomys mittendorfi*, and *Hylomyscus grandis*), and a golden mole (*Chrysochloris balsaci*) are endemic to Mt Oku (W. Verheyen et al., 1997).

Lophuromys dudui W. Verheyen, Hulselmans, Dierckx, and E. Verheyen, 2002. Bull. Inst. Roy. Sci. Nat. Belgique, Biol., 72:147.
    COMMON NAME: Dudu's Brush-furred Rat.
    TYPE LOCALITY: N Dem. Rep. Congo, Orientale (Haut-Zaïre), right bank Congo River near Kisangani, Masako Forest Reserve, "rainforest of the Tshoporiver," 440 m, 00°36'N, 25°13'E (W. Verheyen et al., 2002:147).
    DISTRIBUTION: Dem. Rep. Congo, lowland tropical evergreen rainforest on right side of Congo River from Kisangani area east to W foothills of the Western Rift Mtns, from the Garamba-Blukwa-Djugu region in NE Dem. Rep. Congo to Irangi in the south (details and localities provided by W. Verheyen et al., 2002).
    COMMENTS: Subgenus *Lophuromys*, *L. aquilus* species group (W. Verheyen et al., 2002). A distinct species defined by clear morphometric and molecular traits. Populations on left bank of the Congo River near Kisangani represent *L. aquilus*.

Lophuromys eisentrauti Dieterlen, 1978. Bonn. Zool. Beitr., 29:296.
    COMMON NAME: Mount Lefo Brush-furred Rat.
    TYPE LOCALITY: West Cameroon, Bamenda-Banso highlands, Mt Lefo, 2550 m (Hutterer et al., 1992*a*; W. Verheyen et al., 1997).
    DISTRIBUTION: An isolated patch of montane forest on Mt Lefo and so far recorded only from the type locality.
    COMMENTS: Subgenus *Lophuromys*, *L. aquilus* species group (W. Verheyen et al., 2000). Dieterlen (1979*b*) described *eisentrauti* as a subspecies of *L. sikapusi* but Hutterer et al.

(1992a) were so impressed by the small body size of *eisentrauti* ("a dwarf form") relative to that of *L. sikapusi* from the surrounding region that they raised it to the rank of species. A morphometric revision of *Lophuromys* from the mountain forest archipelago of W Cameroon and SE Nigeria corroborated the specific status of *eisentrauti* and identified *L. dieterleni*, endemic to nearby Mt Oku in the Gotel Mtns, as its closest phylogenetic relative (W. Verheyen et al., 1997). See account of *L. dieterleni*.

Lophuromys flavopunctatus Thomas, 1888. Proc. Zool. Soc. Lond., 1888:14.

COMMON NAME: Ethiopian Buff-spotted Brush-furred Rat.

TYPE LOCALITY: Ethiopia, Shoa (=Shewa) Province, probably obtained at Ankober, 100 mi. (161 km) NE Addis Ababa (Thomas, 1888d), 3000 m (coordinates given by W. Verheyen et al., 2002).

DISTRIBUTION: Endemic to Ethiopian plateau west of the Rift Valley (Osgood, 1936; W. Verheyen et al., 2002).

STATUS: IUCN – Lower Risk (lc).

SYNONYMS: *giaquintoi* (Toschi, 1946); *zaphiri* Thomas, 1906.

COMMENTS: Subgenus *Lophuromys*, *L. flavopunctatus* species group (W. Verheyen et al., 2002). Originally described as a species, which later came to embrace Ethiopian *brunneus*, *simensis*, and *zaphiri* (Osgood, 1936; G. M. Allen, 1939; Ellerman, 1941), and eventually a dozen taxa representing a geographic range with an isolated segment in highlands of Ethiopia, and a vast distribution outside of that region, extending from NE Angola through the Congo basin to Uganda, Kenya, Tanzania, and south to Malawi, N Zambia and N Mozambique (Dieterlen, 1976b; Misonne, 1974; Musser and Carleton, 1993). The peculiar range and appreciable morphological variation among samples prompted Musser and Carleton (1993) to suggest that more than one species was present, which has been demonstrated by recent analyses of allozymic (Lavrenchenko et al., 2000), chromosomal (Aniskin et al., 1997), and morphological, morphometric, and molecular data (Lavrenchenko et al., 1998b; W. Verheyen et al., 2002). What was once considered a single species is now a complex containing at least 11 species, with *L. flavopunctatus* being restricted to the Ethiopian plateau west of the Rift Valley (W. Verheyen et al., 2002), as Osgood (1936) had earlier defined it. Association of the two synonyms with *L. flavopunctatus* demonstrated by Lavrenchenko et al. (1998b) and W. Verheyen et al. (2002). Closest relatives of *L. flavopunctatus* are the Ethiopian endemics *L. brevicaudus* and *L. melanonyx*; all other species in the complex are members of the *L. aquilus* species group (W. Verheyen et al., 2002). See account of *L. brevicaudus*.

Lophuromys huttereri W. Verheyen, Colyn, and Hulselmans, 1996. Bull. Inst. Roy. Sci. Nat. Belgique, Biol., 66:255.

COMMON NAME: Hutterer's Brush-furred Rat.

TYPE LOCALITY: N Dem. Rep. Congo, near village of Yaenero on left side of Congo River, 450 m (see W. Verheyen et al., 1996, for additional information).

DISTRIBUTION: N Dem. Rep. Congo; equatorial lowland tropical rain forest between 300 and 450 m on left side of Congo River between Lualaba and Lomami Rivers, and west of the Lomami River to Ndele.

COMMENTS: Subgenus *Lophuromys*; closely related to *L. nudicaudus*, which is found on the right side of the Congo River (see that account). Morphological and morphometric contrasts among *L. huttereri*, *L. nudicaudus*, and *L. sikapusi* documented by W. Verheyen et al. (1996), who also mapped the distribution and discussed how the range of *L. huttereri* is concordant with the South-Central zoogeographic faunal region.

Lophuromys luteogaster Hatt, 1934. Am. Mus. Novit., 708:4.

COMMON NAME: Buff-bellied Brush-furred Rat.

TYPE LOCALITY: Dem. Rep. Congo, Orientale (Haute-Zaire), Ituri Dist., Medje.

DISTRIBUTION: NE and E Dem. Rep. Congo (Medje, Irangi, Bafwasende, and Tungula; Dieterlen, 1987); limits unknown.

STATUS: IUCN – Lower Risk (lc).

COMMENTS: Subgenus *Kivumys*. W. Verheyen et al. (1996:250) referred to a "*L.woosnami-luteogaster* species-complex." Redescribed by W. Verheyen (1964c), who noted that skull of holotype is an *L. aquilus* (discussed as *flavopunctatus*) but the skin does represent the species named by Hatt, and both skin and skull of the paratype are *luteogaster*; earlier, Osgood (1936:244) had identified the holotype as being a mismatch: "Although the skin is unique among *Lophuromys*, the skull is perfectly normal for a member of the *aquilus* group." Additional descriptions and references to this poorly known species were summarized by Dieterlen (1975, 1976b, 1987).

Lophuromys medicaudatus Dieterlen, 1975. Bonn. Zool. Beitr., 26:295.

COMMON NAME: Western Rift Brush-furred Rat.
TYPE LOCALITY: Dem. Rep. Congo, Kivu region, Lemera-Nyabutera.
DISTRIBUTION: Montane forests of Kivu highlands in E Dem. Rep. Congo and Rwanda above 2000 m (Dieterlen, 1987); a montane Western Rift endemic.
STATUS: IUCN – Lower Risk (nt).
COMMENTS: Subgenus *Kivumys*. Information summarized by Dieterlen (1975, 1976b, 1987).

Lophuromys melanonyx F. Petter, 1972. Mammalia, 36:177.

COMMON NAME: Black-clawed Brush-furred Rat.
TYPE LOCALITY: Ethiopia, Bale Dist., Dinshu.
DISTRIBUTION: S and C Ethiopia, west of the Ethiopian Rift Valley (Dieterlen, 1987), and in the Bale Mtns, east of the Rift Valley, 3100-4050 m.
STATUS: IUCN – Lower Risk (nt).
COMMENTS: Subgenus *Lophuromys*, *L. flavopunctatus* species group (Lavrenchenko et al., 1998b, 2000). Endemic to Afro-alpine moorland of Ethiopia, where it shares habitat with the other moorland specialists *Stenocephalemys albocaudata*, *Arvicanthis blicki*, *Tachyoryctes macrocephalus*, and *Otomys typus* (Demeter and Topal, 1982; Rupp, 1980; Yalden, 1988; Yalden and Largen, 1992; Yalden et al., 1976). Lower portion of *L. melanonyx*'s range overlaps that of *L. brevicauda* (Lavrenchenko et al., 1998b). Statistical summary of external measurements for large series of *L. melanonyx* reported by Sillero-Zubiri (1995a). Karyotype documented by Aniskin et al. (1997) and Corti et al. (1995). Habitat preferences, abundance, and biomass of this species along with others in the Bale Mtns and their relevance to the endangered Ethiopian wolf (*Canis simensis*) documented by Sillero-Zubiri et al. (1995c). See account of *L. brevicaudus*.

Lophuromys nudicaudus Heller, 1911. Smithson. Misc. Coll., 56(17):11.

COMMON NAME: Fire-bellied Brush-furred Rat.
TYPE LOCALITY: Cameroon, Bula country, Efulen (see Rosevear, 1969, for information about the holotype).
DISTRIBUTION: Lowland tropical evergreen rainforest, usually below 700 m, from S Cameroon, Equatorial Guinea (including Bioko), Gabon, Republic of Congo, SW Central African Republic, and N Dem. Rep. Congo (Orientale), between the Ubangui River in the west and the Aruwimi River in the east; Basoko, 01°13′ N, 23°35′ E, near the confluence of the Aruwimi River with the Congo, is the easternmost record. Despite intensive survey, *L. nudicaudus* is not found on the right side of the Congo River between mouth of the Aruwimi River and Kisangani to the southeast); elsewhere it occurs only on right side of Congo River. Distribution concordant with the West-Central and Cameroon zoogeographic faunal regions (W. Verheyen et al., 1996).
STATUS: IUCN – Lower Risk (lc).
SYNONYMS: *afer* (Tullberg, 1893) [not Peters, 1866: see Rosevear, 1969:304]; *naso* Thomas, 1911; *parvulus* Eisentraut, 1965; *tullbergi* Matschie, 1911 [see Rosevear, 1969].
COMMENTS: Subgenus *Lophuromys*. Most closely related to *L. huttereri* from the left side of the Congo River in N Dem. Rep. Congo (W. Verheyen et al., 1996). Included in *L. sikapusi* by Misonne (1974), but is a distinct species; *L. sikapusi* and *L. nudicaudus* occur together in the same region in W Africa. Rosevear (1969) discussed traits distinguishing the two species and the taxonomic history of samples in West Africa. W. Verheyen et al. (1996) summarized taxonomic history of the two species (and allocations of the taxa associated with them), and morphological and morphometric analyses that distinguish them. Dieterlen (1978b) explained why *naso* is a synonym. Eisentraut (1965) described *parvulus* as a subspecies of *L. sikapusi* but Rosevear (1969) included it in *L. nudicaudus*. W. Verheyen et al. (1996) endorsed the synonymy of *naso* with

L. nudicaudus and synonymized *parvulus* with *tullbergi*; they were reluctant to include the latter in *L. nudicaudus* until biochemical and other data could be analyzed, despite the apparent union of *tullbergi* with *L. nudicaudus* in their multivariate analysis. Karyotype of Cameroon specimens was described by W. Verheyen and Van der Straeten (1980).

Lophuromys rahmi Verheyen, 1964. Rev. Zool. Bot. Afr., 69:206.
COMMON NAME: Rahm's Brush-furred Rat.
TYPE LOCALITY: Dem. Rep. Congo, Kivu, Bogamanda near Lemera.
DISTRIBUTION: Montane forest in E Dem. Rep. Congo (Kivu) and Rwanda above 1800 m; limits unknown, probably a montane Western Rift endemic.
STATUS: IUCN – Lower Risk (nt).
COMMENTS: Subgenus *Lophuromys*. Distributional and biological information summarized by W. Verheyen (1964a), Van der Straeten and W. Verheyen (1983), and Dieterlen (1976b, 1987).

Lophuromys roseveari W. Verheyen, Hulselmans, Colyn, and Hutterer, 1997. Bull. Inst. Roy. Sci. Nat. Belgique, Biol., 67:167.
COMMON NAME: Mount Cameroon Brush-furred Rat.
TYPE LOCALITY: West Cameroon, Mt Cameroon, Musake, 1850-2200 m (see W. Verheyen et al., 1997, for additional information).
DISTRIBUTION: Endemic to Mt Cameroon, 1000-3100 m.
COMMENTS: Subgenus *Lophuromys*, *L. sikapusi-ansorgei* species group (W. Verheyen et al., 1997). Rosevear (1969:305) had one example of this species and described it under "*Lophuromys* sp." W. Verheyen et al. (1997:172) concluded that *L. roseveari* is distinct from *L. sikapusi* inhabiting lowland forests and the other montane forest patches in the montane forest archipelago of the adjacent Nigeria-Cameroon region. They cautioned that "in view of the considerable morphometrical differences also encountered between the studied OTU's [Gabon, Republic of Congo, Central African Republic, N Cameroon, representing *sikapusi*], we expect that the exact systematical position of the Mount Cameroun form will only become clear within a continent-wide revision of the *Lophuromys sikapusi-ansorgei* complex." Other small mammals also recorded only from montane habitats on Mt Cameroon are the shrews *Sylvisorex morio* and *Crocidura eisentrauti* and the otomyine *Otomys burtoni* (W. Verheyen et al., 1997).

Lophuromys sikapusi (Temminck, 1853). Esquisses Zool. sur la Côte de Guine, p. 160.
COMMON NAME: Rusty-bellied Brush-furred Rat.
TYPE LOCALITY: Ghana, Dabacrom.
DISTRIBUTION: From Sierra Leone through Ghana and S Nigeria to Republic of Congo, Gabon, Cameroon, and Central African Republic; eastern limits unresolved (see W. Verheyen et al., 1997, 2000, for range east of Ghana).
STATUS: IUCN – Lower Risk (lc).
SYNONYMS: *afer* (Peters, 1866) [not Tullberg, 1893].
COMMENTS: Subgenus *Lophuromys*, *L. sikapusi* species complex. Phylogenetically most closely allied with *L. angolensis* and *L. ansorgei* (see those accounts). Inclusion of *afer* in *L. sikapusi* documented by W. Verheyen et al. (1997). West African samples reviewed by Rosevear (1969). Chromosomal information summarized by Gautun et al. (1986). Grubb et al. (1998) reviewed Ghana and Sierra Leone populations. Three-dimensional geometric morphometrics was employed to investigate growth patterns and differences in size and shape of the skull in samples of *L. sikapusi* from Côte d'Ivoire, *L. aquilus* (discussed under *flavopunctatus*) from Rwanda, and *L. woosnami* (Corti et al., 2000).

Lophuromys verhageni W. Verheyen, Hulselmans, Dierckx, and E. Verheyen, 2002. Bull. Inst. Roy. Sci. Nat. Belgique, Biol., 72:153.
COMMON NAME: Verhagen's Brush-furred Rat.
TYPE LOCALITY: N Tanzania, Mt. Meru, 2600 m (coordinates given by W. Verheyen et al., 2002).
DISTRIBUTION: Endemic to Mt Meru, 2600-3050 m (see W. Verheyen et al., 2002).

COMMENTS: Subgenus *Lophuromys*, *L. aquilus* species group (W. Verheyen et al., 2002). Defined by morphometric data and mtDNA cytochrome *b* sequences, and phylogenetically most closely related to *L. aquilus*.

Lophuromys woosnami Thomas, 1906. Ann. Mag. Nat. Hist., ser. 7, 18:146.
COMMON NAME: Woosnam's Brush-furred Rat.
TYPE LOCALITY: Uganda, Ruwenzori East, Mubuku Valley, 6000 ft (1830 m).
DISTRIBUTION: E Dem. Rep. Congo, W Uganda, Rwanda, and Burundi; a montane Western Rift endemic.
STATUS: IUCN – Lower Risk (lc).
SYNONYMS: *prittiei* Thomas, 1911.
COMMENTS: Subgenus *Kivumys*. The impressive genetic differences between this species and *L. flavopunctatus* prompted W. Verheyen et al. (1986) to suggest that the two belong in at least separate subgenera, reflecting Dieterlen's (1976*b*) earlier conclusion based on morphology. Later, Dieterlen (1987) proposed the subgenus *Kivumys* for the species. Chromosomal data for sample from Burundi reported by Maddalena et al. (1989). Corti et al. (2000) used three-dimensional geometric morphometrics to study cranial growth patterns, size, and shape in samples representing *L. aquilus* (reported as *flavopunctatus*), *L. sikapusi*, and *L. woosnami*. Growth patterns were parallel among the species, no significant sexual dimorphism in shape was discovered, and the primary shape difference set *L. aquilus* and *L. sikapusi* apart from *L. woosnami*, reflecting the chromosomal data and Dieterlen's subgeneric partitioning. Altitudinal distribution on Ugandan slopes of Ruwenzori Mtns reviewed by Kerbis Peterhans et al. (1998) in context of surveying small mammal distributions in those highlands.

Lophuromys zena Dollman, 1909. Ann. Mag. Nat. Hist., ser. 8, 4:550.
COMMON NAME: Zena's Brush-furred Rat.
TYPE LOCALITY: East Africa, S Kenya, near Nyeri, east side of Aberdare Range, 9581 ft (2920 m; coordinates given by W. Verheyen et al., 2002).
DISTRIBUTION: Recorded only from montane habitats in the Aberdare Range and on Mt Kenya in S Kenya (W. Verheyen et al., 2002).
COMMENTS: Subgenus *Lophuromys*, *L. aquilus* species group (W. Verheyen et al., 2002). Originally described as a species, later treated as a subspecies of *L. aquilus* (G. M. Allen, 1939; Ellerman, 1941), and finally included in synonymy of *L. flavopunctatus* (Misonne, 1974; Musser and Carleton, 1993). Recent analyses of morphometric data and mtDNA cytochrome *b* sequences distinguish the montane Aberdare Range and Mt Kenya samples as a separate and distinct species from all the other samples formerly included in either *L. aquilus* or *L. flavopunctatus*, and identifies samples from lower slopes of those mountains as *L. aquilus* (W. Verheyen et al., 2002). Morphometric analyses place *L. zena* in a clade with Ethiopian *L. chrysopus* and *L. brunneus* separate from the clade containing geographic samples of *L. aquilus* (W. Verheyen et al., 2002). Holotypes of *L eisentrauti*, *L. dieterleni*, Mt Elgon *L. rubecula*, and *L. zena* cluster with samples of *L. chrysopus*, suggesting all represent the same species, a possibility requiring testing by analyses of morphometric, chromosomal, and molecular variation in geographic samples from the region between Ethiopia and the Cameroon and Kenyan highlands (W. Verheyen et al., 2002).

Uranomys Dollman, 1909. Ann. Mag. Nat. Hist., ser. 8, 4:551.
TYPE SPECIES: *Uranomys ruddi* Dollman, 1909.
COMMENTS: Dollman (1909) noted that pelage traits of *Uranomys* indicated close relationship to *Lophuromys* but cranial structure resembled the Ethiopian *Muriculus* (related to *Mus*). In its combination of external, cranial, and dental traits, *Uranomys* is phylogenetically distant from *Muriculus* but is closely related to *Lophuromys* and *Acomys* (Denys and Michaux, 1992; Denys et al., 1992; Heim de Balsac, 1963; Heim de Balsac and Lamotte, 1958; Heller, 1911; Hinton, 1921; Ingoldby, 1929; Misonne, 1969; Rosevear, 1969; Thomas, 1910*d*). Cladistic affinity of the three is also substantiated by the extensive biochemical, molecular, and chromosomal studies cited in the subfamily account. A cladistic analysis of morphological traits by Hutterer et al. (1988) aligned *Uranomys* and *Acomys* with the

extinct *Malpaisomys* from the Canary Isls, but Montgelard (1992) presented immunological data demonstrating that *Malpaisomys* is phylogenetically more closely related to *Mus* than to either *Uranomys* or *Acomys*. G. M. Allen (1939) listed seven species of *Uranomys* but noted that six of them would probably prove to be subspecies of *U. ruddi*, which is how the genus is currently treated by others (Lavrenchenko, 1993; Musser and Carleton, 1993; W. Verheyen, 1964b), but not here because we are not recognizing subspecies.

Uranomys ruddi Dollman, 1909. Ann. Mag. Nat. Hist., ser. 8, 4:552.

COMMON NAME: Rudd's Bristle-furred Rat.

TYPE LOCALITY: Kenya, Mt Elgon, Kirui, 6000 ft (1830 m).

DISTRIBUTION: Savannas in Senegal (Duplantier and Granjon, 1992), Guinea (Ziegler et al., 2002), Côte d'Ivoire, Sierra Leone, Ghana (Decher and Bahian, 1999; Ryan and Attuquayefio, 2000), Togo, N Nigeria, N Cameroon, NE Dem. Rep. Congo, SW Ethiopia (Lavrenchenko, 1993; Yalden et al., 1996), Uganda (Delany, 1975; Hollister, 1919), Kenya, Tanzania (W. Verheyen, pers. comm.), C Mozambique Smithers and Lobão Tello, 1976), Malawi (Ansell and Dowsett, 1988), and SE Zimbabwe (Smithers and Wilson, 1979); see G. M. Allen (1939), Lavrenchenko (1993), Misonne (1974), and Rosevear (1969); limits unknown.

STATUS: IUCN – Lower Risk (lc).

SYNONYMS: *acomyoides* Ingoldby, 1929; *foxi* Thomas, 1912; *oweni* Thomas, 1910; *shortridgei* Hayman, 1953; *tenebrosus* Hinton, 1921; *ugandae* Heller, 1911; *woodi* Hinton, 1921.

COMMENTS: Chromosomal data presented by Viegas-Péquignot et al. (1983) in context of chromosomal phylogeny of selected murines; protein electrophoretic data analysed by Iskandar and Bonhomme (1984). Karyotype of Senegal sample reported by Granjon et al. (1992). Karyotypes are geographically variable: 2n = 50 in Senegal, 2n = 58 in Côte d'Ivoire, and 2n = 52 in Central African Republic, suggesting that more than one species is represented. Furthermore, specimens of *shortridgei* are darker than the others and have much larger molars. Ingoldby (1929) was impressed with the traits of *acomyoides*, which he named as a species, and he pointed out the orthodont configuration of its incisors, which contrasted with the proodont facies of all the other named forms. The significance of geographic variation in fur color, craniodental traits, and chromosomal characteristics has to be assessed in a systematic revision of the genus, which should also employ data from DNA sequences.

　　The Ethiopian record is documented by Lavrenchenko (1993), who also provided range map of the species and description of the phallus. Spermatozoal morphology described by Baskevich and Lavrenchenko (1995) and Breed (1995a). Grubb et al. (1998) reviewed populations in Ghana and Sierra Leone. Range in Southern African Sub-region mapped by de Graaff (1997p), who also provided short review.

Subfamily Gerbillinae Gray, 1825. Ann. Philos., n.s., 10:342.

SYNONYMS: Ammodillini Pavlinov, 1981; Desmodilliscina Pavlinov, 1982; Gerbillina Gray, 1825 (Gerbillidae De Kay, 1842; Gerbillinae Alston, 1876; Gerbillina Pavlinov, 1982; Gerbillini Pavlinov, 1982); Gerbillurina Pavlinov, 1982; Merionina Brandt, 1844 (Merionides Giebel, 1855; Merionidinae Schmidtlein, 1893; Merioninae Heptner, 1933); Pachyuromyina Pavlinov, 1982; Rhombomyinae Heptner, 1933 (Rhombomyini Pavlinov and Rossolimo, 1987; Rhombomyina Pavlinov, Dubrovskii, Rossolimo, and Potapova, 1990); Taterillinae Chaline, Mein and F. Petter, 1977 (Taterillina Pavlinov, 1982).

COMMENTS: Whether viewed as a subfamily of Cricetidae or Muridae (Alston, 1876; Carleton and Musser, 1984; Ellerman, 1941; McKenna and Bell, 1997; Miller and Gidley, 1918; Musser and Carleton, 1993; Simpson, 1945; Thomas, 1896) or a separate family (Chaline et al., 1977; Pavlinov et al., 1990, 1995a; Reig, 1980; Wessels, 1998, 1999), species of gerbils form a distinct group defined by a suite of derived morphological traits. In a series of reports, Pavlinov (1980a, 1981a, b, 1982a, 1984b, 1985, 1986, 1987, 2001) presented analyses of skeletal, dental, and male genital characters of gerbils and developed a hypothesis of their phylogeny and a classification. The monograph by Pavlinov et al. (1990) represents a culmination of these efforts, in which the phylogeny, species classification, morphology, ecology, and geographical distribution of the genera and species in Gerbillinae are comprehensively reviewed. They recognized the family Gerbillidae, with two subfamilies. Taterillinae contains tribes Taterillini (*Tatera*, *Taterillus*, and *Gerbilliscus*) and Gerbillurini (*Gerbillurus* and *Desmodillus*). Gerbillinae is composed of Gerbillini (*Dipodillus*, *Gerbillus*, *Monodia*, and *Microdillus*), Desmodilliscini (*Desmodilliscus*), Pachyuromyini (*Pachyuromys*), and Rhombomyini (*Sekeetamys*, *Meriones*, *Brachiones*, *Psammomys*, and *Rhombomys*). Pavlinov et al. (1990) were uncertain about the subfamily allocation of Ammodillini (*Ammodillus*). An excellent short review of character distribution within gerbils that is the basis for the classification of Pavlinov et al. (1990) was provided by Pavlinov (2001). We observe their subfamilies as tribes and their tribes as subtribes.

Cytogenetic data for the subfamily summarized by Qumsiyeh and Schlitter (1991) and Viegas-Péquignot et al. (1986). Rates of protein, chromosomal, and morphological evolution in four genera reported by Qumsiyeh and Chesser (1988). Chromosomal and biochemical results of several species and genera documented and discussed in a phylogenetic context by Benazzou et al. (1982*a*, *b*, 1984) and Qumsiyeh (1986). Problems in using Robertsonian rearrangements to determine monophyly were elaborated by Qumsiyeh et al. (1987) using species of *Gerbilliscus* and *Gerbillurus*. Other chromosomal reports significant to gerbilline systematics are those by Gamperl and Vistorin (1980), Ratomponirina et al. (1986, 1989), and the references cited in the various species accounts below. Anatomy, physiology, adaptive significance, and evolution of the middle and inner ear of gerbillines documented by Lay (1972) and Pavlinov (1988, 2001); significance of acoustic emissions and morphology of cochlea in context of adaptation and systematics reported by Bridelance (1987) and Plassmann et al. (1987); the relationship of acoustical adaptations in relation to steppe and desert environments summarized by Petter et al. (1984); and significance of the relation between size of pinna and auditory bulla elaborated by Pavlinov and Rogovin (2000). Breed (1995*d*) described spermatozoal morphology of *Gerbilliscus leucogaster* and *Gerbillurus paeba*, regarded them as highly derived and unlike the spermatozoa of other muroid genera, but drew no phylogenetic inferences. Comparative study of male genital morphology of species in 12 genera and its taxonomic significance documented by Pavlinov (1986).

The origin and evolution of North African gerbils discussed by Tong (1989) whose results were presented in a phylogenetic classification of genera and compared with other classifications based on morphological, chromosomal, and biochemical data. Pavlinov (2001) contrasted Tong's view with the classification by Pavlinov et al. (1990) and pointed out disagreements. Cockrum and Setzer (1976) reviewed holotypes and type localities associated with North African species. Palearctic species reviewed by Corbet (1978*c*, 1984); Russian species checklisted by Pavlinov and Rossolimo (1987) and reviewed by Gromov and Erbajeva (1995). Eurasian species listed by Pavlinov et al. (1995*a*). Zoogeography (patterns of geographic distributions, correlation with bioclimates, definition of faunal categories, and historical origin of the fauna) of Moroccan gerbils presented by Aulagnier (1991). Patterns of geographic variation in relative importance of gerbillines across the "Great Palaearctic Desert Belt" using biogeographic and ecological approaches reviewed and quantified by Shenbrot and Krasnov (2001).

Morphology clearly defines Gerbillinae (Carleton and Musser, 1984; Pavlinov et al., 1990). Molecular data does also, and indicates that phylogenetic affinities of gerbils are with murines and deomyines. Analyses of the nuclear protein-coding gene LCAT sequences (Michaux and Catzeflis, 2000), and LCAT in combination with sequences of the von Willebrand Factor (Michaux et al., 2001) placed gerbils as a sister-group with deomyines, in a clade also containing murines. This topography is also consistent with results of analyses of complete mtDNA cytochrome *b* (Martin et al., 2000) and nuclear IRBP sequences (Jansa and Weksler, 2004); the Murinae-Gerbillinae link was also revealed by analyses of DNA sequences from the nuclear genes GHR and BRCA1 (Adkins et al., 2003). From their molecular-clock estimate, Michaux et al. (2001) suggested the three families diverged 20.8-17.9 million years ago, with deomyines and gerbils separating 18.5-16.5 million years ago, events transpiring in the early Miocene. Such divergence times are not unreasonable in context of known evolutionary history of these groups. The earliest gerbil fossils come from the late Miocene of Eurasia and North Africa (Wessels, 1998), and those of its sister-group, the myocricetodontines, extend back to the early Miocene of

Turkey, Pakistan, and Saudi Arabia, middle Miocene of China and Africa, and Late Miocene of Spain and Namibia (Jaeger, 1977*a,b*; Lindsay, 1994; Mein et al., 2004; Qiu, 2001; Wessels, 1996, 1998, 1999). Current consensus suggests the Gerbillinae to have evolved from myocricetodontines (de Bruijn and Whybrow, 1994; Jaeger, 1977*b*; Lindsay, 1994; Tong, 1989; Wessels, 1998), whether viewed as a family containing Gerbillinae (Wessels, 1996), a subfamily by itself (Tong and Jaeger, 1993) or within Gerbillidae (Ameur-Chehbeur, 1991; Mein et al., 2000*b*, 2004; Tong, 1989; Wessels, 1998, 1999), or a tribe within Gerbillinae (McKenna and Bell, 1997). Tong and Jaeger (1993) would also derive murines from an early ancestral myocricetodontine with the split between the two groups occurring 16 million years ago (middle Miocene). The earliest deomyines are represented by fossils of *Acomys* from the early Pliocene (Denys, 1990*b*) and *Preacomys* from the late Miocene (Geraads, 2001; Mein et al., 2004); earlier deomyines or fossils actually linking deomyines with gerbils have yet to be discovered.

Ammodillus Thomas, 1904. Ann. Mag. Nat. Hist., ser. 7, 14:102.

TYPE SPECIES: *Gerbillus imbellis* de Winton, 1898.

COMMENTS: Tribe Ammodillini. Reviewed by Pavlinov (1981*a*), who was so impressed with its unique morphological traits that he proposed the monotypic tribe Ammodillini. The genus and single species were reviewed by Pavlinov et al. (1990), who expressed uncertainty whether *Ammodillus* should be placed in subfamily Taterillinae or Gerbillinae.

Ammodillus imbellis (de Winton, 1898). Ann. Mag. Nat. Hist., ser. 7, 1:249.

COMMON NAME: Ammodile.

TYPE LOCALITY: Somalia, Goodar.

DISTRIBUTION: Somalia (Roche and Petter, 1968) and E Ethiopia (Yalden et al., 1976).

STATUS: IUCN – Vulnerable.

COMMENTS: Reviewed by Roche and Petter (1968) and Pavlinov et al. (1990). Chromosomal traits documented by Capanna and Merani (1981).

Brachiones Thomas, 1925. Ann. Mag. Nat. Hist., ser. 9, 16:548.

TYPE SPECIES: *Gerbillus przewalskii* Büchner, 1889.

COMMENTS: Tribe Gerbillini, Rhombomyina. More closely related to *Meriones* than to any other genera; "just a specialized relative of *Meriones*" (Pavlinov, 2001).

Brachiones przewalskii (Büchner, 1889). Wiss. Res. Przewalski Cent.-Asian Reisen., Zool., I:(Säugeth.), p. 51.

COMMON NAME: Przewalski's Jird.

TYPE LOCALITY: China, Xinjiang, Lob Nor.

DISTRIBUTION: China, deserts from N Xinjiang through N Gansu to W Nei Mongol north of the Tien Shan (see Wang, 2003, and Zhang et al., 1997).

STATUS: IUCN – Lower Risk (lc).

SYNONYMS: *arenicolor* Miller, 1900; *callichrous* Heptner, 1934.

COMMENTS: Reviewed by Corbet (1978*c*) and Pavlinov et al. (1990).

Desmodilliscus Wettstein, 1916. Anz. Akad. Wiss. Wien, 53:153.

TYPE SPECIES: *Desmodilliscus braueri* Wettstein, 1916.

COMMENTS: Tribe Gerbillini, Subtribe Desmodilliscina. Evolutionary history as represented by fossils extends back to the Holocene of East Africa (Denys, 1999).

Desmodilliscus braueri Wettstein, 1916. Anz. Akad. Wiss. Wien, 53:153.

COMMON NAME: Pouched Gerbil.

TYPE LOCALITY: Sudan, south of El Obeid.

DISTRIBUTION: Recorded from Sahelian savanna in N and C Sudan, N Cameroon, S Niger, N Nigeria, C Mali, N Burkina Faso (Gautun et al., 1985), Senegal (Ba et al., 2000; Duplantier and Granjon, 1992), and W Mauritania (overall distributions reviewed by Hutterer and Dieterlen [1986] and Dobigny et al. [2002*b*]).

STATUS: IUCN – Lower Risk (lc).

SYNONYMS: *buchanani* Thomas and Hinton, 1920; *fuscus* Setzer, 1969.

COMMENTS: Taxonomically reviewed by Setzer (1969), who recognized three subspecies. A subsequent analysis of geographic distribution and variation in morphometric traits was provided by Hutterer and Dieterlen (1986), who noted the lack of significant sexual and geographic variation in the species. Notes on behavior, distribution, and population dynamics offered by Poulet (1984). Karyotype (2n = 76, FNa = 104) from Niger (Dobigny et al., 2002a) differs in diploid number (2n = 78) from a Senegal sample (Granjon et al., 1992).

Desmodillus Thomas and Schwann, 1904. Abstr. Proc. Zool. Soc. Lond., 1904(2):6.
   TYPE SPECIES: *Gerbillus auricularis* Smith, 1834.
   COMMENTS: Tribe Taterillini, Subtribe Gerbillurina. A Southern African endemic represented by fossils through the Holocene to early Pliocene (Langebaanweg) of South Africa (Avery, 1998; Denys, 1999) and the Pleistocene of Namibia (Senut et al., 1992).

Desmodillus auricularis (Smith, 1834). S. Afr. Quart. J., Ser. 2, 2:160.
   COMMON NAME: Cape Short-tailed Gerbil.
   TYPE LOCALITY: South Africa, Little Namaqualand, Kamiesberg (Meester et al., 1986).
   DISTRIBUTION: South Africa (Northern Cape and Western Cape Provs., SW Free State, and Gauteng), S Botswana, Namibia, and SW Angola (see Perrin, 1997a; Skinner and Smithers, 1990; Crawford-Cabral, 1998, reviewed and mapped the Angolan records).
   STATUS: IUCN – Lower Risk (lc).
   SYNONYMS: *brevicaudatus* (F. Cuvier, 1838); *caffer* (Wagner, 1842); *hoeschi* Lehmann, 1955; *pudicus* Dollman, 1910; *robertsi* Lundholm, 1955; *shortridgei* Lundholm, 1955; *wolfi* Lehmann, 1955.
   COMMENTS: Taxonomy and geographic distribution summarized by Meester et al. (1986). Ecology and distribution reviewed by Griffin (1990). Chromosomal traits documented by Qumsiyeh (1986) and compared with cytogenetic traits of southern African species of *Gerbillurus* and *Gerbilliscus*. Significance of differences in behavior patterns (Dempster et al., 1993) and acoustic repertoire (Dempster and Perrin, 1994) documented between *D. auricularis* and species of *Gerbilliscus*.

Dipodillus Lataste, 1881. Le Naturalliste (Paris), 1:506.
   TYPE SPECIES: *Gerbillus* (*Dipodillus*) *simoni* Lataste, 1881.
   SYNONYMS: *Petteromys* Pavlinov, 1982.
   COMMENTS: Tribe Gerbillini, Subtribe Gerbillina. The species included here, all with hairless plantar surfaces, were all discussed by Lay (1983) under *Gerbillus*. Originally described as a subgenus of *Gerbillus*, *Dipodillus* has subsequently been treated that way (Ellerman, 1941; Ellerman and Morrison-Scott, 1951; Lay, 1983; Musser and Carleton, 1993) or as a genus (G. M. Allen, 1939; Corbet, 1978c; Osborn and Helmy, 1980; Pavlinov, 2001; Pavlinov et al., 1990; F. Petter, 1975b; Qumsiyeh and Schlitter, 1991). We follow F. Petter, 1975b, Pavlinov et al. (1990) and Pavlinov (2001) in separating species of *Dipodillus* from those in *Gerbillus*. Osborn and Helmy (1980) provided tables and illustrations of morphological traits distinguishing the Egyptian species.

Dipodillus bottai (Lataste, 1882). Le Naturaliste (Paris), 4:36.
   COMMON NAME: Botta's Dipodil.
   TYPE LOCALITY: Sudan, Sennar.
   DISTRIBUTION: Recorded only from Sudan and Kenya.
   STATUS: IUCN – Lower Risk (lc) as *Gerbillus bottai*.
   COMMENTS: Subgenus *Petteromys*. F. Petter (1975b) listed *harwoodi* and *bottai* as species and considered *luteolus* as a synonym of *D. campestris*. Kock (1978) included *luteolus* and *harwoodi* in *D. bottai*. Lay (1983:339) claimed the holotype of *harwoodi* lacks an accessory tympanum but holotype of *luteolus* has an accessory tympanum, and cannot represent the same species. Lay had not examined the holotype of *bottai* and rendered the conclusion that "because of the middle ear characters one of the two forms listed by Kock (1978) cannot be synonymous, perhaps neither are." He preferred to regard *D. bottai* "as valid and monotypic pending revision," and listed *harwoodi* as a species

and *luteolus* as a synonym of *D. stigmonyx*. Pavlinov et al. (1990) reviewed the species and discussed *harwoodi*, *luteolus*, and *stigmonyx* under *D. bottai*.

Dipodillus campestris (Loche, 1867). Expl. Sci. Alg. Zool. Mamm., p. 106.

COMMON NAME: North African Dipodil.

TYPE LOCALITY: Algeria, Constantine Province, Philipeville.

DISTRIBUTION: N Africa, from Morocco to Egypt and Sudan.

STATUS: IUCN – Critically Endangered as *Gerbillus quadrimaculatus*, Lower Risk (lc) as
  G. campestris.

SYNONYMS: *brunnescens* (Ranck, 1968); *cinnamomeus* Cabrera, 1922; *dodsoni* Thomas, 1913;
  gerbii (Loche, 1867) [*nomen nudum*, see Cockrum and Setzer, 1976]; *haymani* (Setzer,
  1958); *hilda* (Thomas, 1918); *minutus* (Loche, 1867) [see Cockrum and Setzer, 1976];
  patrizii de Beaux, 1932; *quadrimaculatus* (Lataste, 1882); *riparius* Cabrera, 1922; *rozsikae*
  Thomas, 1908; *somalicus* Thomas, 1910; *venustus* (Sundevall, 1843); *wassifi* Setzer,
  1958.

COMMENTS: Subgenus *Petteromys*. Nineteen species-group names have been associated with
  D. campestris by different authors in various combinations, as summarized by Lay
  (1983), who also noted that most opinions lacked supportive evidence and that some
  of the synonyms are unidentifiable or *nomina nuda*. Lay (1983) remarked that
  although most authors followed Ellerman and Morrison-Scott (1951) in listing
  quadrimaculatus equivocally under *G. nanus*, it should be kept separate until a revision
  is available. Pavlinov et al. (1990) synonymized it with *Dipodillus campestris* and we
  follow their allocation. Musser and Carleton (1993) had allocated *hilda* to *G. nanus*,
  but it belongs here (Aulagnier, in litt.). Those names listed here are probably correctly
  associated with *D. campestris*, but the species requires refined definition through
  careful systematic revision. Different geographical populations were reviewed by
  Aulagnier and Thevenot (1986, Morocco) Kowalski and Rzebik-Kowalska (1991,
  Algeria), Ranck (1968, Libya), and Osborn and Helmy (1980, Egypt). Benazzou and
  Zyadi (1990) conducted a biometric study analyzing variation among Moroccan
  populations. Dobigny et al. (2001*a, b*) documented chromosomal traits (2n = 56,
  FNa = 68) from Adrar des Iforas in NE Mali, which are similar to karyotypes reported
  from the Maghreb and S Niger (Dobigny et al., 2002*b*), and from Tunisia (Chetoui
  et al., 2002). Significance of APRT gene sequences in illuminating substitution rate
  variation among *D. campestris* and other muroids presented by Fieldhouse et al.
  (1997). Cockrum and Setzer (1976:643) clarified the author and date of publication of
  campestris (also see Kowalski and Rzebik-Kowalska, 1991). Reviewed by Pavlinov et al.
  (1990). Fragments identified as *D. campestris* are reported from probable middle
  Pleistocene sites in Morocco (Amani and Geraads, 1993; Geraads, 1994), and *D. biber-
  soni* was described (as *Gerbillus*) from late Pliocene fossils in Moroccan strata and is
  considered closely related to *D. campestris* (Geraads, 1995).

Dipodillus dasyurus (Wagner, 1842). Arch. Naturgesch., 8:20.

COMMON NAME: Wagner's Dipodil.

TYPE LOCALITY: Sinai.

DISTRIBUTION: Arabian Peninsula, Iraq, Syria, Jordan, Lebanon, Israel, Sinai, E desert of
  Egypt (see Harrison and Bates, 1991); also recorded from Turkey (Yiğit et al., 1997).

STATUS: IUCN – Lower Risk (lc) as *Gerbillus dasyurus*.

SYNONYMS: *dasyroides* Nehring, 1901; *gallagheri* (Harrison, 1971); *leosollicitus* (Lehmann,
  1966); *palmyrae* (Lehman, 1966).

COMMENTS: Subgenus *Petteromys*. Review of taxonomy, morphology, and distribution
  provided by Harrison and Bates (1991). Those authors also listed *lixa* as a synonym of
  D. dasyurus, but the holotype, a young animal, has an accessory tympanum and bare-
  soled hind feet, which is uncharacteristic of *D. dasyurus* but does suggest alliance with
  Gerbillus nanus (Lay, 1983). Egyptian population reviewed by Osborn and Helmy
  (1980), that on the Sinai Peninsula by Saleh and Basuony (1998), Yemen segment
  covered by Al-Jumaily (1998). Morphological, chromosomal, and ecological
  characteristics of the Turkish population documented by Yiğit et al., 1997*a*), of the
  Israeli and Jordanian segments by Qumsiyeh (1996) and Mendelssohn and Yom-Tov

(1999). Reviewed by Pavlinov et al. (1990). Fossils identified as *D. dasyurus* have been found at intermittent levels in the early to late Pleistocene of Israel (Tchernov, 1986, 1992, 1994, and references cited in those reports).

Dipodillus harwoodi (Thomas, 1901). Ann. Mag. Nat. Hist., ser. 7, 8:275.
COMMON NAME: Harwood's Dipodil.
TYPE LOCALITY: Kenya, Lake Naivasha.
DISTRIBUTION: Kenya and Tanzania (see F. Petter, 1975*b*, and Swynnerton and Hayman, 1951).
STATUS: IUCN – Lower Risk (lc) as *Gerbillus harwoodi*.
SYNONYMS: *luteus* Dollman, 1914.
COMMENTS: Subgenus *Petteromys*. Defined as a distinct species by Roche (1975). The relationship between this species and *D. bottai* requires resolution; see comments under account of *D. bottai* and the discussion by Pavlinov et al. (1990).

Dipodillus jamesi (Harrison, 1967). Mammalia, 31:383.
COMMON NAME: James's Dipodil.
TYPE LOCALITY: Tunisia, between Bou Ficha and Enfidaville.
DISTRIBUTION: Tunisia.
STATUS: IUCN – Lower Risk (lc) as *Gerbillus jamesi*.
COMMENTS: Subgenus *Petteromys*. Recognized as valid by F. Petter (1975*b*), Lay (1983), and Pavlinov et al. (1990).

Dipodillus lowei (Thomas and Hinton, 1923). Proc. Zool. Soc. Lond., 1923:261.
COMMON NAME: Lowe's Dipodil.
TYPE LOCALITY: Sudan, Jebel Marra.
DISTRIBUTION: Known only from the type locality.
STATUS: IUCN – Critically Endangered as *Gerbillus lowei*.
COMMENTS: Subgenus *Petteromys*. Synonymized with *D. campestris* by F. Petter (1975*b*) without supporting evidence, but should be kept separate pending revision of the genus (Lay, 1983). Associated with *D. campestris* by Pavlinov et al. (1990).

Dipodillus mackilligini Thomas, 1904. Ann. Mag. Nat. Hist., ser. 7, 14:158.
COMMON NAME: Mackilligin's Dipodil.
TYPE LOCALITY: Egypt, Wadi Alagi, E desert of Nubia.
DISTRIBUTION: E desert of S Egypt (see Osborn and Helmy, 1980) and probably adjacent Sudan.
STATUS: IUCN – Lower Risk (lc) as *Gerbillus mackillingini*.
COMMENTS: Subgenus *Petteromys*. Although some authors have placed this species with *D. nanus* (Corbet, 1978*c*; Ellerman and Morrison-Scott, 1951; F. Petter, 1975*b*), Osborn and Helmy (1980) demonstrated its specific distinction. Reviewed by Pavlinov et al. (1990).

Dipodillus maghrebi (Schlitter and Setzer, 1972). Proc. Biol. Soc. Wash., 84:387.
COMMON NAME: Maghreb Dipodil.
TYPE LOCALITY: Morocco, Fes Province, 15 km WSW Taounate (see Lay, 1983, and map in Aulagnier and Thevenot, 1986).
DISTRIBUTION: Recorded only from the Fes region (between the coast and tip of Middle Atlas Mtns) in N Morocco where it is abundant (Aulagnier et al., 1993).
STATUS: IUCN – Vulnerable as *Gerbillus maghrebi*.
COMMENTS: Subgenus *Dipodillus*. A distinctive Moroccan endemic unrelated to *D. simoni* as Schlitter and Setzer (1972) proposed, but instead phylogenetically linked to *D. campestris* according to Lay (1983). Aulagnier et al. (1993) documented detailed cranial and dental comparisons between samples of *D. maghrebi* and those of sympatric *D. campestris*. Reviewed by Pavlinov et al. (1990).

Dipodillus rupicola (Granjon, Aniskin, Volobouev, and Sicard, 2002). J. Zool. Lond., 256: 183.
COMMON NAME: Rupicolous Dipodil.
TYPE LOCALITY: Sahelian region of N Mali, 12 km east of Mopti, Emnal'here, right bank of

main course of Niger River (14°28′26″ N, 04°05′16″ W); Granjon et al. (2002*a*) provided details of trap location and habitat.

DISTRIBUTION: Recorded only from the type locality.

COMMENTS: Subgenus *Petteromys*. A distinctive species (2n = 52, FNa = 72) that is morphologically adapted to climb about on rocky outcrops and may be restricted to such habitats, which is unusual among species of *Dipodillus* and *Gerbillus*. Granjon et al. (2002*a*) described the species, compared it with others, and provided observations on habitat and behavior. The Rupicolous Dipodil is one of nine species of *Gerbillus* and *Dipodillus* occurring in a small area of the West African sahelian region located between SE Mauritania, N Mali, and W Niger. *Dipodillus rupicola* was found in close proximity to *G. tarabuli*, *G. nigeriae*, and *G. henleyi*, but only *D. rupicola* was caught on the rocks along with a species of *Acomys*, another rupicolous muroid. Granjon et al. (2002*a*) described *rupicola* as a *Gerbillus* but in the classification of Pavlinov et al. (1990), which we follow here, it would be placed in subgenus *Petteromys* of *Dipodillus*; in morphology and body size, *D. rupicola* is most like *D. campestris*, but differs strikingly in karyotype.

Dipodillus simoni Lataste, 1881. Le Naturaliste (Paris), 3:499.

COMMON NAME: Simon's Dipodil.

TYPE LOCALITY: Algeria, Oued Magra.

DISTRIBUTION: Along the coast of Egypt (west of Nile Delta) and NE Libya, coastal and inland in NW Libya and Tunisia, and high plateau region of the Atlas in Algeria and E Morocco (see Cockrum et al., 1976).

STATUS: IUCN – Lower Risk (lc) as *Gerbillus simoni*.

SYNONYMS: *kaiseri* (Setzer, 1958).

COMMENTS: Subgenus *Dipodillus*. A distinctive species (Lay, 1983; Pavlinov et al., 1990) revised by Cockrum et al. (1976*a*) and reviewed by Pavlinov et al. (1990). Regional reviews are available for Egypt (Osborn and Helmy, 1980), Libya (Ranck, 1968), and Algeria (Kowalski and Rzebik-Kowalska, 1991). Although samples from coastal Egypt (*kaiseri*) have, on average, slightly longer tails, darker pelage, and smaller auditory bulla than do those from the western portion of the range (*simoni*), the variation is clinal east to west, prompting Cockrum et al. (1976*a*) to treat all populations as a single species. This interpretation was endorsed by Osborn and Helmy (1980) from their study of Egyptian samples and comparisons with data in the literature, and supports Wassif's (1956, 1960) earlier conclusion that *kaiseri* is synonymous with *D. simoni*. Chromosomal data summarized by Qumsiyeh and Schlitter (1991); karyotype (2n = 60, FN = 72) of Tunisian sample described by Chetoui et al. (2002).

Dipodillus somalicus (Thomas, 1910). Ann. Mag. Nat. Hist., ser. 8, 5:197.

COMMON NAME: Somalian Dipodil.

TYPE LOCALITY: Somalia, Upper Sheikh.

DISTRIBUTION: Somalia (Lay, 1983) and Djibouti (Pearch et al., 2001).

STATUS: IUCN – Lower Risk (lc) as *Gerbillus somalicus*.

COMMENTS: Subgenus *Petteromys*. Included in *D. campestris* by F. Petter (1975*b*) and Pavlinov et al. (1990), but treated by Lay (1983) as a distinct species pending taxonomic revision.

Dipodillus stigmonyx (Heuglin, 1877). Reise in Nordost-Afrika, 2:78.

COMMON NAME: Khartoum Dipodil.

TYPE LOCALITY: Sudan, Khartoum.

DISTRIBUTION: Sudan.

STATUS: IUCN – Lower Risk (lc) as *Gerbillus stigmonyx*.

SYNONYMS: *luteolus* (Thomas, 1901).

COMMENTS: Subgenus *Petteromys*. Listed as a synonym of *G. campestris* by F. Petter (1975*b*); retained as separate by Lay (1983) pending revisionary study.

Dipodillus zakariai Cockrum, Vaughn, and Vaughn, 1976. Mammalia, 40:320.

COMMON NAME: Kerkennah Islands Dipodil.

TYPE LOCALITY: N Africa, Tunisia, Kerkennah Isls, 0.5 km E Kellabine, about 42 km east of Sfax, Tunisia.

DISTRIBUTION: Recorded only from the Kerkennah Isls, off the Tunisian coast opposite the coastal city of Sfax.

COMMENTS: Subgenus *Dipodillus*. Lay (1983) included *zakariai* in *D. simoni* but provided no reasons for doing so. Pavlinov et al. (1990) recognized it as a species. Cockrum et al. (1976*a*) provided adequate morphological data to support the hypothesis that *zakariai* is an insular species related to mainland *D. simoni*.

Gerbilliscus Thomas, 1897. Proc. Zool. Soc. Lond., 1897:433.

TYPE SPECIES: *Gerbillus boehmi* (Noack, 1887).

SYNONYMS: *Taterona* Wroughton, 1917.

COMMENTS: Tribe Taterillini, Subtribe Taterillina. Usually included as a subgenus of *Tatera* (Musser and Carleton, 1993, for example). Pavlinov (1981*b*, 2001) and Pavlinov et al. (1990) regarded true *Tatera* to consist only of the Asian species *T. indica*, initially placed all the African species in subgenera *Gerbilliscus* and *Taterona* (Pavlinov, 1981*b*), then separated *Gerbilliscus* as a genus (which included *Taterona* as a subgenus), and identified *Taterillus* as its closest relative. Pavlinov's phylogenetic analysis reflects the distinctive traits associated with auditory structures and occlusal pattern of m1 separating Asian from all African species and we follow his arrangement. African *Gerbilliscus* is also distinguished from Asian *Tatera* by humerus morphology (entepicondylar foramen present in *Gerbilliscus*, absent in *Tatera*; Bates, 1988) and karyotypes (2n = 36-52 in *Gerbilliscus*, 2n = 68 for *Tatera*; Rao et al., 1968; Qumsiyeh and Schlitter, 1991; Yosida, 1981; Yiğit et al., 2001).

Taxonomic revisions of various inclusiveness (when *Gerbilliscus* was included in *Tatera*) were provided by Pirlot (1955) and Bates (1985, 1988). Davis (1975*a*) reviewed the genus and, aside from *C. boehmi*, clustered all the species either in an *afra* group or *robusta* group. Chromosomal information for some species was reported by Matthey and Petter (1970) and summarized by Qumsiyeh and Schlitter (1991). Chromosomal contrasts among Southern African species of *Gerbilliscus*, *Gerbillurus*, and *Desmodillus* documented by Qumsiyeh (1986). Results of craniometric studies of Angolan *Gerbilliscus* (as *Tatera*) were presented by Crawford-Cabral (1988) and Crawford-Cabral and Pacheco (1991). Neal (1982, as *Tatera*) provided contrasts in reproductive characteristics of *G. nigricaudus*, *G. robustus*, and *G. validus*. Significance of divergence in acoustic repertoire, ultrasonic vocalizations and associated behavior, and overall behavioral patterns among selected species of *Gerbilliscus*, *Gerbillurus*, and *Desmodillus* in different combinations reported by Dempster and Perrin (1991, 1994) and Dempster et al. (1991, 1992, 1993).

Evolutionary history as indicated by fossils (reported as *Tatera*) extends to early Pliocene and Pleistocene in East and South Africa (Avery, 1998, 2000; Denys, 1987*a*, 1989*b*, 1990*a*; Sabatier, 1982; Wessels, 1998) and Pleistocene in Namibia (Senut et al., 1992). *Protatera*, from the late Miocene of North Africa and late Miocene-early Pliocene of Spain, is the oldest member of Taterillini in Africa and may be ancestral to *Gerbilliscus* (Jaeger, 1977*b*; see reviews by Denys, 1999, and Wessels, 1998; reported as *Tatera*).

Gerbilliscus afra (Gray, 1830). Spicil. Zool., p. 10.

COMMON NAME: Cape Gerbil.

TYPE LOCALITY: South Africa, Cape of Good Hope, vicinity of Cape Town (as restricted by Meester et al., 1986).

DISTRIBUTION: Endemic to the Cape Macchia zone, Western Cape Province, South Africa (see Meester et al., 1986; Perrin, 1997*g*; Skinner and Smithers, 1990); recorded from fynbos and succulent Karoo biomes (Mugo et al., 1995).

STATUS: IUCN – Least Concern as *Tatera afra*.

SYNONYMS: *africanus* (Cuvier, 1838); *caffer* (Wagner, 1842); *gilli* (Roberts, 1929); *schlegelii* (Smuts, 1832).

COMMENTS: Subgenus *Taterona*. Taxonomy and distribution summarized by Meester et al. (1986), who listed the species in the *G. afra* group. Pre- and postmating isolation in karyotypically similar *G. afra* and *G. brantsii* is documented by Dempster (1996), who suggested "despite the lack of chromosomal divergence often associated with specia-tion events in rodents, postmating isolation through hybrid disadvantage and possibly male sterility is operating." Reviewed by Perrin (1997*g*).

Gerbilliscus boehmi (Noack, 1887). Zool. Jahrb. Syst., 2:241.

COMMON NAME: Boehm's Gerbil.

TYPE LOCALITY: Dem. Rep. Congo, Katanga Province, Marungu, Qua Mpala (see Ansell, 1978, and Hill and Carter, 1941, for placement of type locality in S Dem. Rep. Congo; it has also been identified as N Zambia [G. M. Allen, 1939; Bates, 1988]).

DISTRIBUTION: E Angola (Crawford-Cabral, 1998, discussed the possible Angolan record), S Dem. Rep. Congo (Hatt, 1940*a*), N Zambia (Ansell, 1978), Malawi (Ansell and Dowsett, 1988), Tanzania (Swynerton and Hayman, 1951), Kenya (Hollister, 1919), and Uganda (Delany, 1975).

STATUS: IUCN – Lower Risk (lc) as *Tatera boehmi*.

SYNONYMS: *fallax* (Thomas and Schwann, 1904); *fraterculus* (Thomas, 1898); *varia* (Heller, 1910).

COMMENTS: Subgenus *Gerbilliscus*. Reviewed by Davis (1975*a*) and Pavlinov et al. (1990); northeast African population revised by Bates (1988). The double-grooved incisors and fringed, white-tipped tail of *G. boehmi* are unique among species of *Gerbilliscus*, and these traits prompted Thomas (1896) to propose *Gerbilliscus* as a subgenus of *Tatera* to contain *boehmi*.

Gerbilliscus brantsii (Smith, 1836). Rept. Exped. Exploring Central Africa, p. 43.

COMMON NAME: Highveld Gerbil.

TYPE LOCALITY: South Africa, Ladybrand, E Free State Province, near Lesotho border (see Meester et al., 1986, for details).

DISTRIBUTION: Subtropical and wooded grasslands of South Africa (most of country except southern region; see Perrin, 1997*h*; Skinner and Smithers, 1990; and P. J. Taylor, 1998), W Zimbabwe (Smithers and Wilson, 1979), Botswana (Smithers, 1971), C and E Namibia (Skinner and Smithers, 1990), S Angola (Crawford-Cabral, 1998), and SW Zambia (Ansell, 1978).

STATUS: IUCN – Lower Risk (lc) as *Tatera brantsii*.

SYNONYMS: *breyeri* (Roberts, 1926); *draco* (Wroughton, 1906); *griquae* (Wroughton, 1906); *humpatensis* (Hill and Carter, 1937); *joanae* (Thomas, 1926); *maccalinus* (Sundevall, 1847); *maputa* (Roberts, 1936); *miliaria* (Wroughton, 1906); *montanus* (A. Smith, 1842); *namaquensis* (Shortridge and Carter, 1938); *natalensis* (Roberts, 1929); *perpallida* (Dollman, 1910); *ruddi* (Wroughton, 1906); *tongensis* (Roberts, 1931).

COMMENTS: Subgenus *Taterona*. Taxonomy and distribution summarized by Meester et al. (1986), who assigned the species to the *G. afra* group. Geographic variation in protein and enzyme markers among samples from Lesotho was reported by Maurer et al. (1976). Pre- and postmating isolation in karyotypically identical *G. afra* and *G. brantsii* documented by Dempster (1996; see account of *G. afra*). The Angolan *humpatensis* was described as a species (Hill and Carter, 1937) but subsequently listed as a subspecies of *G. brantsii* (F. Petter, 1975*b*) or a possible subspecies of *G. leucogaster* (Crawford-Cabral, 1986*a*), or simply included in *G. brantsii* (Pavlinov et al., 1990). Crawford-Cabral (1988, 1998) and Crawford-Cabral and Pacheco (1991) treated *humpatensis* as a relict species surviving only in the Humpata highlands, a view requiring better documentation. There is significant geographic variation in chromatic and morphological traits within what is now called *G. brantsii* and some populations are isolated (F. Petter, 1975*b*); perhaps *humpatensis* refers to one of those isolated fragments but its status can only be documented by a careful revisionary study incorporating samples of *G. brantsii* from throughout its geographic range. Southern African population reviewed by Perrin (1997*h*) and Skinner and Smithers (1990).

Gerbilliscus guineae (Thomas, 1910). Ann. Mag. Nat. Hist., ser. 8, 5:351.

COMMON NAME: Guinean Gerbil.

TYPE LOCALITY: Guinea-Bissau, Gunnal.

DISTRIBUTION: From Gambia and Senegal (Duplantier and Granjon, 1992), through Guinea (Ziegler et al., 2002), Sierra Leone, and S Mali (Meinig, 2000) to Burkina Faso and Ghana.

STATUS: IUCN – Lower Risk (lc) as *Tatera guineae*.

SYNONYMS: *picta* (Hayman, 1936).

COMMENTS: Subgenus *Taterona*. Reviewed by Rosevear (1969). Shown to be morphologically distinct from *G. robustus* by Bates (1985) and Pavlinov (1997). Gautun et al. (1985) provided chromosomal information. Ghana and Sierra Leone populations reviewed by Grubb et al. (1998), who explained why *picta* should be included in *G. guineae*.

Gerbilliscus inclusus (Thomas and Wroughton, 1908). Proc. Zool. Soc. Lond., 1908:169.
COMMON NAME: Gorongoza Gerbil.
TYPE LOCALITY: Mozambique, Gorongoza Dist., Tambarara.
DISTRIBUTION: E Zimbabwe (Smithers and Wilson, 1979), C Mozambique (north of Zambezi River; Smithers and Lobao Tello, 1976) to NE Tanzania (Swynnerton and Hayman, 1951, as *T. leucogaster cosensi*).
STATUS: IUCN – Lower Risk (lc) as *Tatera inclusa*.
SYNONYMS: *cosensi* (Kershaw, 1921); *pringlei* (Hubbard, 1970).
COMMENTS: Subgenus *Taterona*. Taxonomy and geographic range summarized by Meester et al. (1986), who listed the species in *G. afra* group. Southern African population reviewed by Perrin (1997*i*) and Skinner and Smithers (1990).

Gerbilliscus kempi (Wroughton, 1906). Ann. Mag. Nat. Hist., ser. 7, 17:375.
COMMON NAME: Northern Savanna Gerbil.
TYPE LOCALITY: Nigeria, Aguleri.
DISTRIBUTION: Throughout Subsaharan northern savannas from Senegal (Duplantier and Granyon, 1992, as *gambiana*) and Guinea (Mt Nimba) through Burkina Faso, S Mali (Meinig, 2000), Nigeria, S Niger, N Cameroon, S Chad, Central African Republic, S Sudan, S Ethiopia, Uganda, and NE Dem. Rep. Congo to SW Kenya (west of Eastern Rift Valley); see Bates (1988).
STATUS: IUCN – Lower Risk (lc) as *Tatera kempi*.
SYNONYMS: *beniensis* (Hatt, 1935); *benvenuta* (Hinton and Kershaw, 1920); *dichrura* (Thomas, 1915); *dundasi* (Wroughton, 1909); *flavipes* (Allen, 1914); *gambiana* (Thomas, 1910); *giffardi* (Wroughton, 1906); *hopkinsoni* (Thomas, 1911); *lucia* (Hinton and Kershaw, 1920); *nigrita* (Wroughton, 1906); *ruwenzorii* (Thomas and Wroughton, 1910); *smithi* (Wroughton, 1909); *soror* (Allen, 1914); *welmani* (St. Leger, 1929).
COMMENTS: Subgenus *Taterona*. Bates (1988) treated *kempi* as a subspecies of *G. validus* (as did D. H. S. Davis, 1975*a*), distinguishing it from *validus* by the different conformation of the anterior lamina (anteroconid) on first lower molar. The pattern in *kempi* is characteristic of all samples from the northern savanna (Senegal to SW Kenya), the configuration in *validus* typifies specimens from the southern savanna (Angola to S Kenya east of Rift Valley), a dichotomy also documented by F. Petter (1959) and D. H. S. Davis (1966, 1975*a*). For Bates, the two populations are geographically discrete and best treated as distinct subspecies, "although the exact rank in the taxonomic hierarchy of such allopatric populations is difficult to determine" (Bates, 1988:277). Current evidence supports the hypothesis that each is a species, an arrangement that can be tested with other kinds of data and additional samples from relevant locations. Rosevear (1969) recognized *kempi*, *hopkinsoni*, and *welmanni* as separate species within West Africa, but commented that *hopkinsoni* and *welmanni* were probably only subspecifically distinct; Pavlinov et al. (1990) listed these names in the synonymy of *G. validus*. Chromosomal data reported from samples collected in Burkina Faso (Gautun et al., 1985, as *hopkinsoni*), Mt Nimba in Guinea (Gautun et al., 1986, as either *kempi* or *hopkinsoni*), Benin (Codjia et al., 1994), Niger (Dobigny et al., 2002*b*, as *Tatera gambiana*), and Ethiopia (Bulatova et al., 2002, as *Tatera valida*). Populations from Ghana, Sierra Leone, and Gambia reviewed by Grubb et al. (1998). Morphometric and ecological data recorded by Ryan and Attuquayefio (2000) for sample from S Ghana, and ecological information documented by Decher and Bahian (1999) for population on Accra Plains of S Ghana. Swimming ability tested by Duplantier and Bâ (2001, as *gambiana*) in context of assessing capability for colonizing islands off coast of Senegal and crossing inland waterways.

Gerbilliscus leucogaster (Peters, 1852). Bericht Verhandl. K. Preuss., Akad. Wiss., Berlin, 17:274.
COMMON NAME: Bushveld Gerbil.
TYPE LOCALITY: Mozambique, north of Zambezi River, Mesuril (as restricted by Davis, 1949:1004).

DISTRIBUTION: N and W South Africa (see Perrin, 1997*f*; Skinner and Smithers, 1990; P. J. Taylor, 1998), Mozambique (including islands in the Bazaruto Arch. off the coast of S Mozambique; Downs and Wirminghaus, 1997, Smithers and Lobao Tello, 1976), Zimbabwe (Smithers and Wilson, 1979), Botswana, Namibia (Griffin, 1990), Malawi (Denys et al., 1999), Zambia (Ansell, 1978), Angola (Crawford-Cabral, 1998), SW Tanzania (Davis, 1975*a*), and S Dem. Rep. Congo (Davis, 1975*a*).

STATUS: IUCN – Lower Risk (lc) as *Tatera leucogaster*.

SYNONYMS: *angolae* (Wroughton, 1906); *bechuanae* (Wroughton, 1906); *beirae* (Roberts, 1951); *beirensis* (Roberts, 1929); *kaokoensis* Lehmann, 1955; *limpopoensis* (Roberts, 1929); *littoralis* (Roberts, 1929); *lobengulae* (de Winton, 1898); *mashonae* (Wroughton, 1906); *mitchelli* (Roberts, 1929); *ndolae* (Kershaw, 1922); *nigrotibialis* (Monard, 1933); *nyasae* (Wroughton, 1906); *panja* (Wroughton, 1906); *pestis* (Roberts, 1929); *pretoriae* (Roberts, 1929); *salsa* (Wroughton, 1906); *schinzi* (Noack, 1889); *shirensis* (Wroughton, 1906); *stellae* (Wroughton, 1906); *tenuis* (Peters, 1852); *tzaneenensis* (Roberts, 1929); *waterbergensis* (Roberts, 1938); *zuluensis* (Roberts, 1931).

COMMENTS: Subgenus *Taterona*. Taxonomy and distribution summarized by Meester et al. (1986), who included this species in the *G. robustus* group. Nongeographic variation in Botswana sample analyzed by Swanepoel et al. (1979). Discriminant function analyses contrasting Angolan *G. validus* and *G. leucogaster* documented by Crawford-Cabral and Pacheco (1991). Southern African Subregion populations reviewed by Perrin (1997*f*) and Skinner and Smithers (1990).

Gerbilliscus nigricaudus (Peters, 1878). Monatsb. K. Preuss. Akad. Wiss., Berlin, 1879:200 [1878].

COMMON NAME: Black-tailed Gerbil.

TYPE LOCALITY: Kenya, Taita, Ndi.

DISTRIBUTION: Ethiopia (Duckworth et al., 1993), Kenya, Somalia, and Tanzania. Apparently does not occur west of the Eastern Rift Valley (see Bates, 1988).

STATUS: IUCN – Lower Risk (lc) as *Tatera nigricaudata*.

SYNONYMS: *nyama* (Dollman, 1911); *percivali* (Heller, 1914).

COMMENTS: Subgenus *Taterona*. Revised by Bates (1988), who summarized distributional and ecological information, and reviewed by Pavlinov et al. (1990). The taxa *bayeri* and *bodessae*, included in this species by Davis (1975*a*), were transferred to *G. robustus* by Bates (1988; see that account).

Gerbilliscus phillipsi (de Winton, 1898). Ann. Mag. Nat. Hist., ser. 7, 1:253.

COMMON NAME: Phillips's Gerbil.

TYPE LOCALITY: Somalia, Hanka Dadi.

DISTRIBUTION: Somalia and the Rift Valley in Ethiopia and Kenya (see Bates, 1988), westward to Senegal and Mali (see Pavlinov, 1997).

STATUS: IUCN – Lower Risk (lc) as *Tatera phillipsi*.

SYNONYMS: *bodessana* (Frick, 1914); *minusculus* Osgood, 1936; *umbrosa* (Dollman, 1912).

COMMENTS: Subgenus *Taterona*. Revised by Bates (1988), who summarized distributional and ecological information, and reviewed by Pavlinov et al. (1990). Sympatric with *G. robustus*. The form *minusculus* was included here by Pavlinov et al. (1990) but designated *incertae sedis* by Bates (1988) because all specimens he examined from the type locality, including topotypes, were juveniles and impossible to identify as either *G. phillipsi* or *G. robusta*. A discriminant function analysis of cranial traits was used by Pavlinov (1997) to define the specific distinctness of *G. robustus*, *G. philippsi*, and *G. guineae* and also to demonstrate that holotypes of *bodessana* and *minusculus* clustered with examples of *G. phillipsi*.

Gerbilliscus robustus (Cretzschmar, 1826). In Rüppell, Atlas Reise Nordl. Afrika, Zool. Säugeth., 1:75.

COMMON NAME: Fringe-tailed Gerbil.

TYPE LOCALITY: Sudan, Ambukol.

DISTRIBUTION: Burkina Faso, Chad, Sudan, Ethiopia, Somalia, Uganda, Kenya, and Tanzania (see Bates, 1985, 1988).

STATUS: IUCN – Lower Risk (lc) as *Tatera robusta*.

SYNONYMS: *bayeri* (Lönnberg, 1918); *bodessae* (Frick, 1914); *iconica* (Dollman, 1911); *loveridgei* (Hatt, 1935); *macropus* (Heuglin, 1877); *mombasae* (Wroughton, 1906);

muansae (Matschie, 1911); *murinus* (Sundevall, 1842); *pothae* (Heller, 1910); *shoana* (Wroughton, 1906); *swaythlingi* (Kershaw, 1921); *taylori* (Hatt, 1935); *vicinus* (Peters, 1878).

COMMENTS: Subgenus *Taterona*. Revised by Bates (1988), who summarized chromosomal, distributional and ecological data, and reviewed by Pavlinov et al. (1990). Pavlinov et al. (1990) allocated *bayeri* and *bodessana* to *G. nigricaudus*, following Davis (1975a), but Bates (1988) assigned them to the synonymy of *G. robustus*. Pavlinov (1997) used discriminant analysis of cranial variables to verify the distinctness of *G. robustus* compared with *G. guineae* and *G. phillipsi* and showed that the holotype of *bodessana* clustered with *G. phillipsi* and those of *bodessae* and *taylori* grouped with samples of *G. robustus*. Bulatova et al. (2002) provided chromosomal information for Ethiopian sample.

Gerbilliscus validus (Bocage, 1890). J. Sci. Math. Phys. Nat. Lisboa, 2(5):6.
COMMON NAME: Southern Savanna Gerbil.
TYPE LOCALITY: Angola, Rio Cuando, Ambaca, Quissange, Caconda.
DISTRIBUTION: Southern savanna from Angola (Crawford-Cabral, 1998) through S Dem. Rep. Congo, Zambia, and SW Tanzania to S Kenya (between the Rift Valley and coast).
STATUS: IUCN – Lower Risk (lc) as *Tatera valida*.
SYNONYMS: *liodon* (Thomas, 1902); *neavei* (Wroughton, 1907); *taborae* (Kershaw, 1921).
COMMENTS: Subgenus *Taterona*. Revised by Bates (1988) as *Tatera validus validus*. Allopatric with and closely related to the West African *G. kempi* (see that account). Discriminant function analyses comparing Angolan *G. validus* and *G. leucogaster* provided by Crawford-Cabral and Pacheco (1991).

Gerbillurus Shortridge, 1942. Ann. S. Afr. Mus., 36(1):52.
TYPE SPECIES: *Gerbillus vallinus* Thomas, 1918.
SYNONYMS: *Paratatera* Petter, 1983; *Progerbillurus* Pavlinov, 1982.
COMMENTS: Tribe Taterillini, Subtribe Gerbillurina. Sometimes included in *Gerbillus* (see references in Schlitter et al., 1984), but concordance of morphological, allozymic, and chromosomal data supports the monophyly of *Gerbillurus* and indicates its close relationship to the *robusta* group of *Taterillus* (Qumsiyeh et al., 1987, and references therein). Pavlinov (1987), Pavlinov et al. (1990), and Pavlinov (2001), however, considered *Gerbillurus* to be a sister-species of *Desmodillus* and to form a monophyletic tribe. Comparisons in thermal parameters, macro-and micro-environments, and interspecific aggression among four sympatric species of *Gerbillurus* were documented by Downs and Perrin (1989, 1990) and Dempster and Perrin (1990) in context of adaptive significance and phylogenetic relationships. Divergence in acoustic repertoire, ultrasonic vocalizations and associated behavior, and general behavior patterns among four species of *Gerbillurus*, one *Gerbilliscus*, and *Desmodillus* in different combinations provided by Dempster and Perrin (1991, 1994) and Dempster et al. (1991, 1992, 1993). Standard karyotypic data reported by Schlitter et al. (1984), a hypothesis of phylogenetic relationships among the four species, derived from chromosomal data, was advanced by Qumsiyeh et al. (1991), and chromosomal comparisons among southern African species of *Gerbillurus*, *Gerbilliscus*, and *Desmodillus* was documented by Qumsiyeh (1986). Reviewed by Davis (1975a). *Gerbillurus* is represented by samples from the late Pliocene of Namibia (Senut et al., 1992) and South Africa, and Holocene of East Africa (Denys, 1999).

Gerbillurus paeba (A. Smith, 1836). Rept. Exped. Exploring Central Africa, app., p. 43.
COMMON NAME: Paeba Hairy-footed Gerbil.
TYPE LOCALITY: South Africa, W Eastern Cape Province, Vryberg (as restricted by Roberts, 1951:401).
DISTRIBUTION: South Africa (western half and Limpopo; see Perrin, 1997c, Skinner and Smithers, 1990), W Mozambique, W Zimbabwe, Botswana, Namibia (Griffin, 1990), and SW Angola (Crawford-Cabral, 1998).
STATUS: IUCN – Lower Risk (lc).
SYNONYMS: *broomi* (Thomas, 1918); *calidus* (Thomas, 1918); *coombsi* (Roberts, 1929); *exilis* (Shortridge and Carter, 1938); *infernus* (Lundholm, 1955); *kalaharicus* (Roberts, 1932);

leucanthus (Thomas, 1927); *mulleri* (Roberts, 1946); *oralis* (Thomas and Hinton, 1925); *swakopensis* (Roberts, 1951); *swalius* (Thomas and Hinton, 1925); *tenuis* (A. Smith, 1842).

COMMENTS: Subgenus *Progerbillurus*. Meester et al. (1986) and Skinner and Smithers (1990) summarized South African taxonomy and distribution. Based on analyses of chromosomal data, Qumsiyeh (1986) hypothesized that *G. paeba* and *G. vallinus* evolved from a common ancestor. Significance of variation in cranial size and shape between *G. paeba* and *Desmodillus auricularis* in sympatry is documented by Matson and Christian (1996). Southern African Subregion population reviewed and mapped by Perrin (1997*c*).

Gerbillurus setzeri (Schlitter, 1973). Bull. S. California Acad. Sci., 72:13.
COMMON NAME: Setzer's Hairy-footed Gerbil.
TYPE LOCALITY: Namibia, Gobabeb, 1 mi (1.6 km) E Namib Desert Research Station.
DISTRIBUTION: Namib Desert, from the Namib-Naukluft Natl. Park north through Namibia to extreme SW Angola (Crawford-Cabral, 1998; Meester et al., 1986; Skinner and Smithers, 1990).
STATUS: IUCN – Lower Risk (lc).
COMMENTS: Subgenus *Gerbillurus*. Interpretation of chromosomal data indicated *G. setzeri* and *G. tytonis* are closely related (Qumsiyeh et al., 1991). Reviewed by Perrin (1997*e*) and Griffin (1990), who also mentioned an undescribed form (*G.* cf. *setzeri*) coexisting with *G. setzeri*.

Gerbillurus tytonis (Bauer and Niethammer, 1960). Bonn. Zool. Beitr., (1959), 10:255 [1960].
COMMON NAME: Dune Hairy-footed Gerbil.
TYPE LOCALITY: Namibia, Namib Desert, Sossusvlei.
DISTRIBUTION: Namibia (Sossusvlei, Sandwich Harbour and Bobabeb, Namib Desert, and Farm Canaan near diamond region of Namibia).
STATUS: IUCN – Lower Risk (lc).
COMMENTS: Subgenus *Paratatera*. Taxonomic history summarized by Meester et al. (1986); distribution and habitat reviewed by Griffin (1990), Perrin (1997*b*), and Skinner and Smithers (1990).

Gerbillurus vallinus (Thomas, 1918). Ann. Mag. Nat. Hist., ser. 9, 2:148.
COMMON NAME: Brush-tailed Hairy-footed Gerbil.
TYPE LOCALITY: South Africa, Northern Cape Province, Bushmanland, Kenhart, Tuin (see Meester et al., 1986).
DISTRIBUTION: From South Africa (Northern Cape Province) northwest through Namibia towards Brukaros-Karas Mtns and C Namib Desert.
STATUS: IUCN – Lower Risk (lc).
SYNONYMS: *seeheimi* Lundholm, 1955.
COMMENTS: Subgenus *Gerbillurus*. Taxonomy summarized by Meester et al. (1986); ecology and range reviewed by Griffin (1990) and Skinner and Smithers (1990). Reviewed and mapped by Perrin (1997*d*).

Gerbillus Desmarest, 1804. Tabl. Méth. Hist Nat., *in* Nouv. Dict. Hist. Nat., 24:22.
TYPE SPECIES: *Gerbillus aegyptius* Desmarest, 1804 (= *Dipus gerbillus* Olivier, 1801).
SYNONYMS: *Endecapleura* Lataste, 1882; *Hendecapleura* Lataste, 1894; *Monodia* Heim de Balsac, 1943.
COMMENTS: Tribe Gerbillini, Subtribe Gerbillina. *Gerbillus* has never been adequately revised. Lay (1983) summarized taxonomic difficulties attendant with determining if the genus was monophyletic, discussed significant character complexes that would bear on any scheme to separate species into subgenera (nature of plantar surfaces, size of auditory bulla, relative tail length, dental traits, accessory tympanum, and karyotype), provided an annotated checklist of the species he considered valid, and a list of named forms with literature citations and type localities (including geographic coordinates). Lack of concordance among the suites of characters discouraged Lay from allocating species to higher categories conventionally recognized as either subgenera or genera (see Pavlinov et al., 1990, for example), and he felt compelled to recognize a single genus without subgenera

until the group was systematically re-evaluated. Several taxonomic revisions, pertinent faunal stuides (referred to in the accounts of species), reviews, and the important monograph by Pavlinov et al. (1990) have emerged since 1983; these in combination with our study of specimens and Lay's taxonomic overview form the basis for our summary, which can be viewed as a working hypothesis of specific diversity in *Gerbillus*. In extracting the species usually associated with *Dipodillus*, we follow Pavlinov et al. (1990) and Pavlinov (2001) in recognizing those species as a separate cluster (*Dipodillus*) related to species of *Gerbillus* and depart from Lay's arrangement. Pavlinov (in litt., 2002), however, stressed that "*Dipodillus* and *Gerbillus* could be both paraphyletic and the subgenera in each of them are certainly grades and not clades." Within *Gerbillus*, our listing of species under subgenera *Gerbillus* or *Hendecapleura* generally echos the treatments by Pavlinov et al. (1990) and F. Petter (1975b).

An excellent account of species occurring east of the Euphrates River was provided by Lay and Nadler (1975). The distribution of six species endemic to North Africa was mapped and discussed by Cheylan (1990) in the context of assessing endemism and speciation of Mediterranean mammals. Tranier and Julien-Laferriere (1990) commented on suggested identities of several species from North Africa. Qumsiyeh and Schlitter (1991) summarized chromosomal data; subsequent documentation of chromosomal complements and their significance to reconstructing phylogenetic relationships among species of *Gerbillus* will be found in some of the references cited in various accounts of species. A contribution to allozymic variation within and between species was offered by Nevo (1982). Evolutionary tendencies within *Gerbillus*, as reflected in dental traits, were discussed by F. Petter (1973c), and significance of variation in incisor microstructure within the genus was reported by Flynn (1982c). Species of *Gerbillus* were included by Bonhomme et al. (1985) and Pascale et al. (1990) in their assessment of phylogenetic relationships among muroid rodents with electrophoretic and DNA-sequence data. Volobouev et al. (1995b:258) characterized and isolated two repeated DNA sequences in *G. nigeriae*, *G. aureus*, and *G. nanus* and pointed out that ". . . in *Gerbillus* species, chromosomal evolution involves rearrangements of both repeated sequences and other chromosomal structures." Osborn and Helmy (1980) provided clear morphological comparisons among species of *Gerbillus* occurring in Egypt. Granjon et al. (2002a) pointed out that seven species of *Gerbillus* (*G.* cf *nancillus*, *G. henleyi*, *G. nanus*, *G. gerbillus*, *G. nigeriae*, *G. tarabuli*, and *G.* cf *pyramidum*) and two of *Dipodillus* (*D. campestris* and *D. rupicola*) can be unambiguously identified in the West African Subsaharan sahelian region between southeastern Mauritania, northern Mali, and western Niger.

Evolutionary history as documented by fossils dates to early Pleistocene of China (Lucas, 2001), late Pleistocene of C India (Patnaik, 1995) and middle Pliocene to early Pleistocene of Africa (Denys, 1989b, 1999; Tong, 1989; Wesselman, 1984; Wessels, 1998). The earliest member of Gerbillini and relative of *Gerbillus* is *Debruijnimys* from early Pliocene of Spain and late Pliocene of North Africa (Castillo and Agustí, 1996; Wessels, 1998).

Gerbillus acticola Thomas, 1918. Ann. Mag. Nat. Hist., ser. 9, 2:147.
    COMMON NAME: Berbera Gerbil.
    TYPE LOCALITY: Somalia, Berbera.
    DISTRIBUTION: Somalia.
    STATUS: IUCN – Lower Risk (lc).
    COMMENTS: Subgenus *Gerbillus*. Lay (1983:337) explained why F. Petter's (1975b) synonomy of *G. alticola* with *G. pyramidum* should not be followed: ". . . the nearest documented occurrence of *G. pyramidum* is over 1300 kilometers distant in Sudan." Lay claimed that *acticola* ". . . should be regarded as distinct pending revision and karyotypic analysis." Pavlinov et al. (1990) included *acticola* in *G. pyramidum* but with a question.

Gerbillus agag Thomas, 1903. Proc. Zool. Soc. Lond., 1903:296.
    COMMON NAME: Agag Gerbil.
    TYPE LOCALITY: Sudan, W Kordofan, Agageh Wells.
    DISTRIBUTION: Mali, N Nigeria, and Niger to Chad, Sudan and Kenya.
    STATUS: IUCN – Critically Endangered as *G. cosensis* and *G. dalloni*, Lower Risk (lc) as *G. agag*.

SYNONYMS: *cosensi* Dollman, 1914; *dalloni* Heim de Balsac, 1936; *maradius* Kock, 1978.

COMMENTS: Subgenus *Gerbillus*. Taxa listed above as synonyms have been associated with *G. agag* (Kock, 1978; Pavlinov et al., 1990; F. Peter, 1975*b*), but Lay (1983) found the limits of this species impossible to define based on published analyses. He regarded it as monotypic pending systematic revision. The taxon *dalloni* was united with *G. agag* by F. Petter (1975*b*) without supporting documentation, ". . . but the type localities are separated by more than 2300 km" (Lay, 1983:340), and *dalloni* should ". . . be regarded as a valid species pending revision." Although F. Petter (1975*b*;9) included *cosensi* as a synonym of *G. agag*, he also noted it "could be a valid species." Kock (1978) described *maradius* as a subspecies of *G. agag*. The morphological, chromosomal, molecular, and geographic limits of *G. agag* clearly require resolution but until then we group together the taxa treated as species by Lay (1983), an arrangement found also in Pavlinov et al. (1990).

Gerbillus amoenus (de Winton, 1902). Ann. Mag. Nat. Hist., ser. 7, 9:46.

COMMON NAME: Pleasant Gerbil.

TYPE LOCALITY: Egypt, Giza Province.

DISTRIBUTION: Recorded only from Egypt and Libya.

STATUS: IUCN – Lower Risk (lc).

COMMENTS: Subgenus *Hendecapleura*. Included in *G. nanus* by Corbet (1978*c*), but convincingly separated and shown to be a distinctive species by Osborn and Helmy (1980), who reviewed the Egyptian population. Lay (1983) speculated that *G. amoenus* may range across Tunisia and Algeria to Mauritania, noted its past associations with *Dipodillus dasyurus* and *D. campestris*, and advised future comparison with *Gerbillus nanus*. Among Egyptian species of *Gerbillus*, *G. amoenus* is morphologically most closely related to *G. nanus* and *G. henleyi* (Osborn and Helmy, 1980). Ranck (1968) reviewed the Libyan populations and recorded significant geographic variation. Reviewed by Pavlinov et al. (1990).

Gerbillus andersoni de Winton, 1902. Ann. Mag. Nat. Hist., ser. 7, 9:45.

COMMON NAME: Anderson's Gerbil.

TYPE LOCALITY: Egypt, E Alexandria, Mandara.

DISTRIBUTION: SW Jordan and Israel (Qumsiyeh, 1996; Mendelssohn and Yom-Tov, 1999), Egypt (Sinai Peninsula and Nile Delta south to El Faiyum as mapped for *G. a. andersoni* by Osborn and Helmy, 1980:120), Libya and Tunisia (Cockrum et al., 1976*b*; Lay, 1983).

STATUS: IUCN – Vulnerable as *G. allenbyi* and *G. bonhotei*, Lower Risk (lc) as *G. andersoni*.

SYNONYMS: *allenbyi* Thomas, 1918; *blanci* Cockrum, Vaughn, and Vaughn, 1976; *bonhotei* Thomas, 1919; *eatoni* Thomas, 1902; *inflatus* Ranck, 1968; *versicolor* Ranck, 1968.

COMMENTS: Subgenus *Gerbillus*. The forms *allenbyi*, *inflatus*, and *bonhotei* have all been listed as synonyms of *G. andersoni* (e.g., Cockrum et al., 1976*b*; Harrison and Bates, 1991; Osborn and Helmy, 1980; Pavlinov et al., 1990, 1995*a*), but Lay (1983) argued that current evidence does not support the union of these forms and that *allenbyi* and *bonhotei* should be treated as species pending a revision. He also suggested that *blanci* and *eatoni* be tentatively associated with *G. andersoni*. We acknowledge the need for a critical new look at geographic variation in *G. andersoni* but include *allenbyi*, *bonhotei* and the other taxa listed above as synonyms following Pavlinov et al. (1990). Furthermore, Cockrum et al. (1976*b*) provided credible evidence for uniting *allenbyi*, *eatoni* and other taxa with *G. andersoni*, as did Qumsiyeh (1996) for including *allenbyi*. Harrison and Bates (1991) recorded *bonhotei* from Jordan, an occurrence requiring corroboration according to Lay (1983). Qumsiyeh (1996) reviewed range, habitat, and biology of populations in Jordan and Israel.

Gerbillus aquilus Schlitter and Setzer, 1972. Proc. Biol. Soc. Wash., 86:167.

COMMON NAME: Swarthy Gerbil.

TYPE LOCALITY: Iran, 60 km W Kerman.

DISTRIBUTION: SE Iran, W Pakistan, S Afghanistan (see Lay and Nadler, 1975).

STATUS: IUCN – Lower Risk (lc).

SYNONYMS: *subsolanus* Schlitter and Setzer, 1973.

COMMENTS: Subgenus *Gerbillus*. Originally described as a subspecies of *G. cheesmani* but Lay and Nadler (1975) presented data supporting its status as a separate and distinctive species. Schlitter and Setzer (1973) described *subsolanus* as a subspecies of *G. cheesmani*.

Gerbillus brockmani (Thomas, 1910). Ann. Mag. Nat. Hist., ser. 8, 5:420.
    COMMON NAME: Brockman's Gerbil.
    TYPE LOCALITY: Somalia, Burao, 85 mi (137 km) S Berbera.
    DISTRIBUTION: Somalia.
    STATUS: IUCN – Lower Risk (lc).
    COMMENTS: Subgenus *Hendecapleura*. Petter (1975b) placed *brockmani* in synonymy with *G. nanus*, but no evidence indicates that *G. nanus* occurs anywhere remotely near Somalia (see Lay, 1983). The phylogenetic affinities of *G. brockmani* require clarification.

Gerbillus burtoni F. Cuvier, 1838. Trans. Zool. Soc. Lond., 2:145, pl. 25.
    COMMON NAME: Burton's Gerbil.
    TYPE LOCALITY: Sudan, Dharfur.
    DISTRIBUTION: Known only from the type locality.
    STATUS: IUCN – Critically Endangered.
    COMMENTS: Usually treated as a synonym of *G. pyramidum* (Pavlinov et al., 1990; also see references in Lay, 1983), but Lay (1983) mustered diagnostic cranial traits seen in the original cranial illustrations that appear to distinguish them. Until a critical revision of the *G. pyramidum* complex defines the status of *burtoni*, we retain it as a species following Lay.

Gerbillus cheesmani Thomas, 1919. J. Bombay Nat. Hist. Soc., 26:748.
    COMMON NAME: Cheesman's Gerbil.
    TYPE LOCALITY: Iraq, Lower Euphrates, near Basra.
    DISTRIBUTION: SW Iran, C and S Iraq, Saudi Arabia, Oman, North Yemen, South Yemen, and Kuwait (see Al-Jumaily, 1998; Harrison and Bates, 1991; Lay and Nadler, 1975), and S Jordan (Qumsiyeh, 1996).
    STATUS: IUCN – Lower Risk (lc).
    SYNONYMS: *arduus* Cheesman and Hinton, 1924; *maritimus* Sanborn and Hoogstraal, 1953.
    COMMENTS: Subgenus *Gerbillus*. Reviewed by Lay and Nadler (1975), Lay (1983), Harrison and Bates (1991), and Pavlinov et al. (1990). Chromosomal polymorphism among samples from Kuwait was reported by Badr and Asker (1980). Detailed comparisons between Qatarian *G. cheesmani* and *G. nanus* in cranial morphology discussed by Madkour (1984).

Gerbillus dongolanus (Heuglin, 1877). Reise in Nordost-Afrika, 2:79.
    COMMON NAME: Dongola Gerbil.
    TYPE LOCALITY: Sudan, Dongola.
    DISTRIBUTION: Known only from the type locality.
    STATUS: IUCN – Lower Risk (lc).
    COMMENTS: Subgenus *Gerbillus*. Lay (1983) noted that several authors have synonymized this taxon with *G. pyramidum* but always without confirmatory documentation. Until such data is available, the species should be considered valid, excluded from the taxonomic morass that has characterized *G. pyramidum*. Pavlinov et al. (1990) also included *dongolanus* with *G. pyramidum*.

Gerbillus dunni Thomas, 1904. Ann. Mag. Nat. Hist., ser. 7, 14:101.
    COMMON NAME: Dunn's Gerbil.
    TYPE LOCALITY: Somalia, Gerlogobi.
    DISTRIBUTION: Ethiopia, Somalia, Djibouti (Pearch et al., 2001).
    STATUS: IUCN – Lower Risk (lc).
    COMMENTS: Subgenus *Gerbillus*. The conspecificity of *dunni* with *G. latastei* was suggested, without supporting evidence, by Cockrum (1977), but Lay (1983) noted that the latter's geographic range lies more than 4000 km from that of *G. dunni*. Yalden et al. (1996) included *dunni* in *G. pulvinatus*. Chromosomal data documented by Capanna and Merani (1981).

Gerbillus famulus Yerbury and Thomas, 1895. Proc. Zool. Soc. Lond., 1895:551.
> COMMON NAME: Black-tufted Gerbil.
> TYPE LOCALITY: South Yemen, Aden, Lehej.
> DISTRIBUTION: Endemic to South Yemen and North Yemen (see Harrison and Bates, 1991:272).
> STATUS: IUCN – Lower Risk (lc).
> COMMENTS: Subgenus *Hendecapleura*. A large and elegant gerbil whose morphology, geographic range, and ecology were elucidated by Harrison and Bates (1991); also reviewed by Pavlinov et al. (1990).

Gerbillus floweri Thomas, 1919. Ann. Mag. Nat. Hist., ser. 9, 3:559.
> COMMON NAME: Flower's Gerbil.
> TYPE LOCALITY: Egypt, Sinai, S of El Arish.
> DISTRIBUTION: Known only from the type locality.
> STATUS: IUCN – Critically Endangered.
> COMMENTS: Subgenus *Gerbillus*. Usually considered a synonym of *G. pyramidum*, but Lay (1983) explained why it should not be united with that species (in his comments under *G. pyramidum*). Reviewed by Pavlinov et al. (1990).

Gerbillus garamantis Lataste, 1881. Le Naturaliste, 3:507.
> COMMON NAME: Algerian Gerbil.
> TYPE LOCALITY: Algeria, Ouargla, Sidi Roueld.
> DISTRIBUTION: Algeria.
> STATUS: IUCN – Lower Risk (lc).
> COMMENTS: Subgenus *Hendecapleura*. Most workers have included *garamantis* in *G. nanus* (e.g., Kowalski and Rzebik-Kowalska, 1991; Pavlinov et al., 1990; F. Petter, 1975*b*; Qumsiyeh, 1996), but never with supporting evidence. Matthey (1954) recorded a 2n = 54 for a sample without designated locality that was identified as *garamantis*. Lay (1983:342) noted that 2n = 52 is usually associated with *G. nanus* and that *G. garamantis* be recognized "provisionally pending confirmation and clarification of Matthey's results." The species was reviewed as a population of *G. nanus* by Kowalski and Rzebik-Kowalska (1991), who were uneasy with including *garamantis* in *G. nanus* and did not believe that the same species extended from Morocco to Pakistan. They also noted that *garamantis* would be the oldest name for North African populations if determined to be different from those in Asia. See account of *G. nanus*.

Gerbillus gerbillus (Olivier, 1801). Bull. Sci. Soc. Philom. Paris, 2:121.
> COMMON NAME: Lesser Egyptian Gerbil.
> TYPE LOCALITY: Egypt, Giza Province.
> DISTRIBUTION: From SW Jordan and S Israel through Egypt, N Sudan (Yalden et al., 1996) and Djibouti to Morocco and SW Mauritania; also N and S Mali, N Niger, and N Chad (see Aulagnier and Thevenot, 1986; Corbet, 1978*c*; Dobigny et al., 2002*b*; Granjon et al., 2002*b*; Harrison and Bates, 1991:283; Mendelssohn and Yom-Tov, 1999; Osborn and Helmy, 1980:131; Pearch et al., 2001; Qumsiyeh, 1996).
> STATUS: IUCN – Lower Risk (lc).
> SYNONYMS: *aegyptius* Desmarest, 1804; *aeruginosus* Ranck, 1968; *asyutensis* Setzer, 1960; *discolor* Ranck, 1968; *foleyi* Heim de Balsac, 1936; *hirtipes* Lataste, 1881; *longicaudus* (Wagner, 1843); *psammophilous* Ranck, 1968; *sudanensis* Setzer, 1956.
> COMMENTS: Subgenus *Gerbillus*. Geographic portions reviewed by Ranck (1968, Libya), Corbet (1978*c*), Lay (1983), Kowalski and Rzebik-Kowalska (1991, Algeria), Osborn and Helmy, 1980, Egypt), Djibouti (Pearch et al., 2001), Saleh and Basuony (1998, Sinai Peninsula), Qumsiyeh (1996, Israel and Jordan) and Harrison and Bates (1991, Arabian Peninsula). Meinig (2000) reported *G. gerbillus* from an isolated sand-belt in S Mali. In 1983, Lay drew attention to the lack of inquiry into variation in this species, which has such an extensive range; that complaint stands today and the species needs careful taxonomic review. The form *hirtipes* was synonymized with *G. gerbillus* by Cockrum (1976*a*), but because of his inadequate documentation, Lay (1983) was reluctant to accept this union. Cockrum's evidence is scanty, but we are swayed by Kowalski and Rzebik-Kowalska's (1991) argument for merging *hirtipes* with *G. gerbillus*. Lay (1983)

listed *aegyptius* as *incertae sedis*, but Ellerman and Morrison-Scott (1951) and Corbet (1978c) placed it in the synonymy of *G. gerbillus*. Lay (1983) also treated *longicaudus* as *incertae sedis*, but Cockrum and Setzer (1976) quoted Thomas as having seen the holotype in Munich and identifying it as *G. gerbillus*. The origin of multiple sex chromosomes in this species was discussed by Wahrman et al. (1983); karyotype (2n = 42/43, FNa = 72) similar in samples from Niger, Mauritania, Algeria, Tunisia, Morocco, Israel, and Egypt (see references in Dobigny et al., 2002b). Reviewed by Pavlinov et al. (1990), who also treated the above-listed taxa as synonyms. Ecology and aspects of membership in small mammal community of coastal SW Mauritania reported by Granjon et al. (2002b).

Gerbillus gleadowi Murray, 1886. Ann. Mag. Nat. Hist., ser. 5, 17:246.
  COMMON NAME: Indian Hairy-footed Gerbil.
  TYPE LOCALITY: Pakistan, Upper Sind, Rohri Dist, Mirpur-Drahrki Taluka, 15 mi (24 km) SW Rehti, Beruto.
  DISTRIBUTION: NW India (Rajasthan and Gujarat; see Agrawal, 2000), sand dunes along Indus Valley of Pakistan (see Lay and Nadler, 1975).
  STATUS: IUCN – Lower Risk (lc).
  COMMENTS: Subgenus *Gerbillus*. A distinctive species defined by diagnostic morphological and chromosomal data (Lay and Nadler, 1975). Reviewed by Pavlinov et al. (1990). Agrawal (2000) and Chakraborty and Agrawal (2000) reviewed habitat and distribution for Indian populations.

Gerbillus grobbeni Klaptocz, 1909. Zool. Jahrb., Syst., 27:252.
  COMMON NAME: Grobben's Gerbil.
  TYPE LOCALITY: Libya, Cyrenaica, Dernah.
  DISTRIBUTION: Known only from the type locality.
  STATUS: IUCN – Critically Endangered.
  COMMENTS: Subgenus *Hendecapleura*. Petter (1975b) and Corbet (1978c) included *grobbeni* in *G. nanus*, but Lay (1983) agreed with Ranck (1968) in retaining *grobbeni* as a species until its status is clarified .

Gerbillus henleyi (de Winton, 1903). Novit. Zool., 10:284.
  COMMON NAME: Pygmy Gerbil.
  TYPE LOCALITY: Egypt, Wadi Natron, Zaghig.
  DISTRIBUTION: From C Morocco (south of Atlas Mtns) and Algeria through N Africa to Sinai Peninsula, NE Sudan, and Djibouti (Perch et al., 2001); then through Israel and Jordan (Qumsiyeh, 1996; Mendelssohn and Yom-Tov, 1999), with scattered records in W Saudi Arabia, N Yemen, and Oman (Harrison and Bates); also recorded from Burkina Faso (Maddalena et al., 1988; Volobouev et al., 1995a), Niger (Dobigny et al., 2002b), and N Senegal (Bâ et al., 2000; Duplantier and Granjon et al., 1992; Duplantier et al. 1991a).
  STATUS: IUCN – Lower Risk (lc).
  SYNONYMS: *jordani* (Thomas, 1918); *makrami* (Setzer, 1958); *mariae* (Bonhote, 1910).
  COMMENTS: Subgenus *Hendecapleura*. Broad segments of the species reviewed by Ranck (1968, Libya), Osborn and Helmy (1980, Egypt), Qumsiyeh (1996, Israel and Jordan), Aulagnier and Thevenot (1986, Morocco), Harrison and Bates (1991, Arabian Peninsula), Pearch et al. (2001, Djibouti), Al-Jumaily (1998, Yemen), and Kowalski and Rzebik-Kowalska (1991, Algeria). The occurrences in Burkina Faso and Senegal were postulated to reflect the southward expansion of Saharan environments (Duplantier et al., 1991a). R- and C-banding of chromosomes of *G. henleyi*, *G. nanus*, and *G. poecilops* was documented by Volobouev et al (1995a:60), who concluded that the three species form "a natural group deriving from the same ancestor, which may have been poorly adapted to dry conditions." Additional chromosomal data from Niger sample documented by Dobigny et al. (2002b). The record from Djibouti is the southernmost on the African continent (Pearch et al., 2001). Reviewed by Pavlinov et al. (1990).

Gerbillus hesperinus Cabrera, 1936. Bol. Real. Soc. Esp. Hist. Nat., p. 365.
>COMMON NAME: Moroccan Gerbil.
>TYPE LOCALITY: Morocco, Mogador (= Essouira).
>DISTRIBUTION: Coastal Morocco north of High Atlas Mtns (see Aulagnier and Thevenot, 1986).
>STATUS: IUCN – Lower Risk (lc).
>COMMENTS: Subgenus *Gerbillus*. See the references cited by Lay (1983) for characters defining this distinctive Moroccan endemic. Lay (1975) also documented the striking morphological and chromosomal contrasts between *G. hesperinus* and Moroccan *G. hoogstraali* and *G. occiduus*. Also reviewed by Aulagnier and Thevenot (1986) and Pavlinov et al. (1990).

Gerbillus hoogstraali Lay, 1975. Fieldiana Zool., 65:90.
>COMMON NAME: Hoogstraal's Gerbil.
>TYPE LOCALITY: Morocco, 7 km S Taroudannt.
>DISTRIBUTION: Recorded only from Morocco in the Souss Plain (between High Atlas and Anti Atlas Mtns) and the type locality (Aulagnier, in litt.; see map in Aulagnier and Thevenot, 1986).
>STATUS: IUCN – Critically Endangered.
>COMMENTS: Subgenus *Gerbillus*. Diagnostic features along with morphological and chromosomal comparisons between *G. hoogstraali* and Moroccan *G. hesperinus* and *G. occiduus* presented by Lay (1975). Associated with *G. latastei* by Pavlinov et al. (1990). Reviewed by Aulagnier and Thevenot (1986).

Gerbillus latastei Thomas and Trouessart, 1903. Bull. Soc. Zool. France, 28:172.
>COMMON NAME: Lataste's Gerbil.
>TYPE LOCALITY: Tunisia, Kebili.
>DISTRIBUTION: Tunisia and Libya (Lay, 1983, suspected the species occurs in Algeria but Kowalski and Rzebik-Kowalska, 1991, did not record it from there).
>STATUS: IUCN – Lower Risk (lc).
>SYNONYMS: *aureus* Setzer, 1956; *favillus* Setzer, 1956; *nalutensis* Ranck, 1968.
>COMMENTS: Subgenus *Gerbillus*. The three synonyms were definitely merged with *G. latastei* by Cockrum (1977), who suggested that *bonhotei*, *dunni*, *perpallidus*, *riggenbachi*, and *rosalinda* are possible synonyms but admitted that more material, particularly chromosomal data, is needed to resolve their relationships. Lay (1983) discussed the basis for considering each of these five as separate species. Chromosomal traits (2n = 74, FN = 94-97) of Tunisian sample described by Chetoui et al. (2002). The Libyan populations (under *G. aureus*) were reviewed by Ranck (1968). Reviewed by Pavlinov et al. (1990).

Gerbillus mauritaniae (Heim de Balsac, 1943). Bull. Mus. Hist. Nat. Paris, 15:287.
>COMMON NAME: Mauritanian Gerbil.
>TYPE LOCALITY: Mauritania, Aouker Region, south of Archane Titarek.
>DISTRIBUTION: Known only by the holotype, which is apparently lost, from the type locality.
>STATUS: IUCN – Critically Endangered.
>COMMENTS: Gerbillini. Type species of *Monodia* (Heim de Balsac, 1943). According to Lay (1983), F. Petter (1975*b*), and Roche (1975), the diagnostic traits of *Monodia* (three large independent tubercles covered with short and stiff hair in metatarsal region with remainder of sole naked, M3 with small posterior cusp) do not warrant generic distinction. Pavlinov et al. (1990) recognized *Monodia* as a genus but with a question. Pavlinov (in litt., 2001) explained that he examined a locality sample of pigmy gerbils from Mauritania in USNM and recognized two different mandible morphologies within it. One of them is typical *Gerbillus* and another he thought to be atypical of *Gerbillus* and superficially resembling that of *Desmodilliscus*; the two kinds were figured in Pavlinov et al. (1990:42, fig. 16). He supposed the atypical mandible to represent Heim de Balsac's *Monodia*, and so resurrected the genus as distinct in Pavlinov et al. (1990). We examined the same Mauritanian series, which does consist of two entities, three examples of a very small-bodied species and six of an even smaller one. Given the vagueness of Heim de Balsac's description, apparent lack of type material, and

need for careful revision of pygmy gerbils in West Africa, it is uncertain which series represents *mauritaniae*. General mandibular shape of both is similar and resembles mandibles of other small *Gerbillus*, such as *G. henleyi* or the larger *G. nanus*. Each of the sympatric Mauritanian species has a naked plantar surface, accessory tympanum, and multichambered mastoid bullae with the same number of compartments as in many *Gerbillus* proper. Until specimens can accurately be identified as *mauritaniae* and their diagnostic traits revealed, we depart from the treatment by Pavlinov et al. (1990) and submerge *Monodia* into *Gerbillus*, without even subgeneric status, as indicated by other gerbil taxonomists.

Gerbillus mesopotamiae Harrison, 1956. J. Mammal., 37:417.
   COMMON NAME: Mesopotamian Gerbil.
   TYPE LOCALITY: Iraq, southwest of Faluja, W bank of Euphrates River, near Amiriya.
   DISTRIBUTION: Iraq and SW Iran in valleys of the Tigris, Euphrates, and Karun Rivers (see Lay and Nadler, 1975).
   STATUS: IUCN – Lower Risk (lc).
   COMMENTS: Subgenus *Hendecapleura*. Originally described as a subspecies of *Dipodillus dasyurus*, but later regarded as a distinctive species defined and reviewed by Lay and Nadler (1975), Harrison and Bates (1991), and Pavlinov et al. (1990).

Gerbillus muriculus (Thomas and Hinton, 1923). Proc. Zool. Soc. Lond., 1923:263.
   COMMON NAME: Darfur Gerbil.
   TYPE LOCALITY: Sudan, Darfur, Madu, 80 mi (129 km) NE El Fasher.
   DISTRIBUTION: Sudan.
   STATUS: IUCN – Lower Risk (lc).
   COMMENTS: Subgenus *Hendecapleura*. Lay (1983) regarded this species as valid pending revision. Reviewed by Pavlinov et al. (1990).

Gerbillus nancillus Thomas and Hinton, 1923. Proc. Zool. Soc. Lond., 1923:260.
   COMMON NAME: Sudan Gerbil.
   TYPE LOCALITY: Sudan, Plains of Darfur, 45 mi (72 km) N El Fasher.
   DISTRIBUTION: Sandy Sudanian and Sahelian savannas in Sudan (near El Fasher), Egypt (Wadi Umm-Ashera, Nasser Lake Shore), Chad, Niger, and Mali.
   STATUS: IUCN – Lower Risk (nt).
   COMMENTS: Subgenus *Gerbillus*. Possibly a distinct species (Lay, 1983), but poorly known. Discussed by Pavlinov et al. (1990). Karyotype from Niger sample (2n = 56, FNa = 54) described by Dobigny et al. (2002*b*). The record from Egypt is based on two specimens collected by Dr. E. Emel'anova and identified by Pavlinov (in litt., 2002), those from Chad, Niger, and Mali are documented by Dobigny et al., 2002*b*).

Gerbillus nanus Blanford, 1875. Ann. Mag. Nat. Hist., ser. 4, 16:312.
   COMMON NAME: Baluchistan Gerbil.
   TYPE LOCALITY: Pakistan, Gedrosia (see Lay, 1983).
   DISTRIBUTION: An extensive range from the Baluchistan region of NW India (Rajasthan and Gujarat; see Agrawal, 2000), Pakistan, S Afghanistan, and Iran through the Arabian Peninsula, Iraq, Jordan, Israel, and North Africa to Morocco (SE side of Atlas Mtns; see Aulagnier and Thevenot, 1986) and Mauritania (Granjon et al., 1997*a*, 2002*b*), and south in the Sahara to at least Niger (Dobigny et al., 2002*b*), NE Mali (Dobigny et al., 2001*a, b*); see excellent map in Lay and Nadler (1975) portraying range east of the Euphrates River.
   STATUS: IUCN – Lower Risk (lc).
   SYNONYMS: *arabium* (Thomas, 1918); *indus* Thomas, 1920; *lixa* Yerbury and Thomas, 1895; *mimulus* (Thomas, 1902); *setonbrownei* Harrison, 1968.
   COMMENTS: Subgenus *Hendecapleura*. Regional reviews of the species provided by Lay and Nadler (1975, Iraq to India), Agrawal (2000) and Chakraborty and Agrawal (2000; NW India), Qumsiyeh (1996, Israel and Jordan), Mendelssohn and Yom-Tov (1999, Israel), Harrison and Bates (1991, Arabia), Al-Jumaily (1998, Yemen), Osborn and Helmy (1980, Egypt), Ranck (1968, Libya), Kowalski and Rzebik-Kowalska (1991, Algeria), Aulagnier and Thevenot (1986, Morocco), Granjon et al. (1997*a*, 2002*b*,

SW Mauritania), Dobigny et al. (2001*a, b* N Mali), and Shenbrot and Krasnov (1997, Israel). Lay (1983) remarked that *G. nanus* and *G. amoenus* share several morphological and chromosomal traits and that the nature of their relationship should be explored by careful revision. Cranial morphology of Qatarian *G. nanus* and *G. cheesmani* contrasted by Madkour (1984). Analysis of R- and C-banding of chromosomes by Volobouev et al. (1995*a*) of *G. nanus*, *G. henleyi*, and *G. poecilops* revealed that the three species form a natural phylogenetic group derived from the same ancestor. Furthermore, Volobouev et al. (1995*a*:60) concluded that "*Gerbillus nanus* probably represents a series of forms presently undergoing speciation and showing incipient karyotype and morphological differentiation." They also pointed out the vast geographic distribution of *G. nanus* and described three morphological and two chromosomal types discovered in their samples, two from Arabia and one from Pakistan: "a small form with large bullae and some submetacentric chromosomes (Al Hofuf), a large form with large bullae and acrocentric chromosomes (Taif), and finally, a large form from Pakistan with small bullae and acrocentric chromosomes." Dobigny et al. (2001*a*) documented chromosomal traits of *G. nanus* from NE Mali, and Dobigny et al. (2002*b*) described karyotype from Niger samples. Reviewed by Pavlinov et al. (1990). The Algerian *garamantis* is usually included in *G. nanus* (Pavlinov et al., 1990; F. Petter, 1975*b*), but we list it as a separate species following Lay (1983); see account of *G. garamantis*.

Gerbillus nigeriae Thomas and Hinton, 1920. Novit. Zool., 27:317.
    COMMON NAME: Nigerian Gerbil.
    TYPE LOCALITY: Nigeria, Farniso, near Kano.
    DISTRIBUTION: Recorded from Niger, N Nigeria, Burkina Faso, and SW Mauritania.
    STATUS: IUCN – Lower Risk (lc).
    COMMENTS: Subgenus *Gerbillus*. Lay (1983) maintained this species as valid even though it has been synonymized with *G. agag* by some workers (Pavlinov et al., 1990; F. Petter, 1975*b*). Complex chromosomal polymorphism characterizes *G. nigeriae* (Volobouev et al., 1988*b*), and left-handed Z-DNA was detected in metaphasic chromosomes in one sample (Viegas-Péquignot et al., 1982). Karyotypic sampling from Niger, Burkina Faso, Mali, and Mauritania exhibit impressive variation in 2n (60-74) and FNa (116-114) (Dobigny et al., 2002*b*; Gautun et al., 1985; Granjon et al. 1997*a*; Tranier, 1975*a*; Volobouev et al., 1988*b*). Ecology and membership in small mammal community of coastal SW Mauritania documented by Granjon et al. (2002*b*). *Gerbillus nigeriae* is regarded as one of the nine unambiguously identified species of gerbils occurring in the West African Sahelian zone (Granjon et al., 2002*a*).

Gerbillus occiduus Lay, 1975. Fieldiana Zool., 65:94.
    COMMON NAME: Occidental Gerbil.
    TYPE LOCALITY: Morocco, Aoreora, 80 km WSW Goulimine.
    DISTRIBUTION: Coastal Morocco from south of Anti Atlas to Tarfaya (Aulagnier, in litt.; also see Aulagnier and Thevenot, 1986).
    STATUS: IUCN – Critically Endangered.
    COMMENTS: Subgenus *Gerbillus*. Lay (1983) noted that the karyotype of this Moroccan endemic was distinct from that of *G. andersoni*, and provided detailed morphological and chromosomal contrasts between *G. occiduus* and Moroccan *G. hoogstraali* and *G. hesperinus*. Pavlinov et al. (1990) included *occiduus* in *G. pyramidum*. Reviewed by Aulagnier and Thevenot (1986).

Gerbillus perpallidus Setzer, 1958. J. Egypt Publ. Health Assoc., 33:221.
    COMMON NAME: Pale Gerbil.
    TYPE LOCALITY: Egypt, Bir Victoria.
    DISTRIBUTION: N Egypt, west of the Nile River.
    STATUS: IUCN – Lower Risk (lc).
    COMMENTS: Subgenus *Gerbillus*. Treated as a species by most authors except Cockrum (1977) who placed it in *G. latastei*. See Lay (1983), Osborn and Helmy (1980), and Pavlinov et al. (1990) for reviews of this distinctive species.

Gerbillus poecilops Yerbury and Thomas, 1895. Proc. Zool. Soc. Lond., 1895:549.
 COMMON NAME: Large Aden Gerbil.
 TYPE LOCALITY:Yemen, Aden, Lahej.
 DISTRIBUTION:Yemen and SW Saudi Arabia.
 STATUS: IUCN – Lower Risk (nt).
 COMMENTS: Subgenus *Hendecapleura*. A valid species reviewed by Harrison and Bates (1991)
  and Pavlinov et al. (1990). Volobouev et al. (1995*a*) analyzed R- and C-banding of
  chromosomes from *G. poecilops*, *G. nanus*, and *G. henleyi* and concluded that all three
  were derived from the same ancestor. Furthermore, *G. poecilops* is the most plesiomorphic and appears to be a "living fossil" isolated in the coastal mountain ranges of
  the Red Sea in Yemen and SW Saudi Arabia.

Gerbillus principulus (Thomas and Hinton, 1923). Proc. Zool. Soc. Lond., 1923:262.
 COMMON NAME: Principal Gerbil.
 TYPE LOCALITY: Sudan, Jebel Meidob, El Malha.
 DISTRIBUTION: Known only from the type locality.
 STATUS: IUCN – Critically Endangered.
 COMMENTS: Subgenus *Hendecapleura*. This species has been associated with *G. nanus* or
  G. watersi, but Lay (1983) regarded it as valid pending systematic revision. See the
  discussion in Pavlinov et al. (1990).

Gerbillus pulvinatus Rhoads, 1896. Proc. Acad. Nat. Sci. Philadelphia, p. 537.
 COMMON NAME: Rhoads's Gerbil.
 TYPE LOCALITY: Ethiopia, Lake Rudolf, Rusia.
 DISTRIBUTION: Endemic to NE Africa in Ethiopia, Eritrea, and Djibouti.
 STATUS: IUCN – Critically Endangered as *G. bilensis*, Lower Risk (lc) as *G. pulvinatus*.
 SYNONYMS: *bilensis* Frick, 1914.
 COMMENTS: Subgenus *Gerbillus*. Lay (1983:339) questioned the synonymy of *bilensis* with
  pulvinatus as allocated without comment by F. Petter (1975*b*): "Inasmuch as these two
  forms were described from opposite ends of the rift valley in Ethiopia that are separated by at least 700 km and because they have not been critically compared, *G. bilensis*
  should be regarded as distinct pending revision." Until that comparison is available we
  include *bilensis* in *G. pulvinatus*. Yalden et al. (1996) would also include *dunni* as a
  synonym of *G. pulvinatus*. Records from Djibouti are documented by Pearch et al.
  (2001), who also provided a good description of the species.

Gerbillus pusillus Peters, 1878. Monatsb. K. Preuss. Akad. Wiss. Berlin, 1879:201 [1878].
 COMMON NAME: Least Gerbil.
 TYPE LOCALITY: Kenya, Ndi and Kitui.
 DISTRIBUTION: Tanzania (Swynnerton and Hayman, 1951), Kenya, Ethiopia, and S Sudan
  (Dieterlen and Nikolaus, 1985).
 STATUS: IUCN – Lower Risk (lc) as *G. diminutus*, *G. percivali*, *G. ruberrimus*, and *G. pusillus*.
 SYNONYMS: *diminutus* Dollman, 1911; *percivali* Dollman, 1914; *ruberrimus* Rhoads, 1896.
 COMMENTS: Subgenus *Hendecapleura*. Roche (1975) united these three synonyms with
  G. pusillus based upon his study of original descriptions and specimens, and F. Petter
  (1975*b*) suggested that all three probably represented *G. pusillus*. Lay (1983), however,
  listed *G. pusillus* and all three synonyms as separate monotypic species pending a
  revision. We prefer Roche's conclusion as the hypothesis to be tested because it was
  derived from specimen examination. Chromosomal data documented by Capanna
  and Merani (1981). Reviewed by Pavlinov et al. (1990), who also accepted Roche's
  synonyms.

Gerbillus pyramidum Geoffroy, 1803. Cat. Mam. Mus. Natl. Hist. Nat., Paris, p. 202.
 COMMON NAME: Greater Egyptian Gerbil.
 TYPE LOCALITY: Egypt, Giza Province.
 DISTRIBUTION: Documented from Egypt, Nile delta and valley south to Sudan (N Sudan,
  oases of Western Desert and SE Eastern Desert, and Khartoum region in EC Sudan (see
  Osborn and Helmy, 1980:97); specimens with same karyotype have been collected

from the Mauritanian coast, N Mali, and W Niger, so actual range may extend from coastal Mauritania to Egypt and Sudan (Dobigny et al., 2001*a, b*, 2002*b*).

STATUS: IUCN – Lower Risk (lc).

SYNONYMS: *elbaensis* Setzer, 1958; *gedeedus* Osborn and Helmy, 1980.

COMMENTS: Subgenus *Gerbillus*. Many authors have viewed this species as ranging extensively from the Sinai throughout North Africa to Morocco and Senegal, and containing many synonyms. Lay (1983), however, discussed the taxonomic quagmire, noting that most synonyms formerly associated with *G. pyramidum* probably did not represent that species or are presently unidentifiable, and described the wide chromosomal variation in specimens from different geographic regions. "Inasmuch as a distinctive karyotype of 2n = 38 with all biarmed elements has been reported for topotypical material and such a karyotype has not been reported from any locality outside Egypt, it should be assumed that *G. pyramidum* inhabits only the region mapped by Osborn and Helmy" (Lay, 1983:346). Tawill and Niethammer (1989) discussed the morphological and chromosomal identification of a sample from Khartoum as most probably *G. pyramidum*. The chromosomal and morphometric analyses of North African samples usually identified as *G. pyramidum* reinforced the view that it has been recorded only from Egypt and Sudan (Granjon et al., 1999). Dobigny et al. (2001*a*), however, reported samples with 2n = 38 from coastal Mauritania, NE Mali, and NW Niger that are "probably referable to *G. pyramidum*" (see also Dobigny et al., 2002*b*), and Granjon et al. (2002*a*) identified *G.* cf. *pyramidum* as one of the many species of *Gerbillus* and *Dipodillus* from the same general region encompassing the subsahelian region of Mauritania, Mali, and W Niger in West Africa. Apparently the species has a wider distribution than realized, a range that overlaps that of its close relative, *G. tarabuli*. Aspects of *G. pyramidum* are reviewed by Pavlinov et al. (1990). For publication from Geoffroy, 1803 see Opinion 2005 of the International Commission on Zoological Nomenclature (2002*b*).

Gerbillus rosalinda St. Leger, 1929. Ann. Mag. Nat. Hist., ser. 10, 4:295.

COMMON NAME: Rosalinda Gerbil.

TYPE LOCALITY: Sudan, Kordofan, Abu Zabad, 145 km SW El Obeid.

DISTRIBUTION: Sudan.

STATUS: IUCN – Lower Risk (lc).

COMMENTS: Subgenus *Gerbillus*. Both F. Petter (1975*b*) and Lay (1983) listed this species as distinct pending revision of the genus. Reviewed by Pavlinov et al. (1990).

Gerbillus syrticus Misonne, 1974. Bull. Inst. R. Sci. Nat. Belg., 50:1.

COMMON NAME: Sand Gerbil.

TYPE LOCALITY: Libya, 12 km N Nofilia.

DISTRIBUTION: Known only from the type locality.

STATUS: IUCN – Critically Endangered.

COMMENTS: Subgenus *Hendecapleura*. Relationship of *G. syrticus* has to be evaluated in a taxonomic revision in Lay's opinion (1983), but Pavlinov et al. (1990) included it in *G. henleyi* with a question. We list it as a separate species pending revisionary study of the genus.

Gerbillus tarabuli Thomas, 1902. Proc. Zool. Soc. Lond., 1902:5.

COMMON NAME: Tarabul's Gerbil.

TYPE LOCALITY: Libya, Sebha.

DISTRIBUTION: Ranges from N Senegal and S Mauritania (Bâ et al., 2001; Dobigny et al., 2002*b*; Duplantier et al., 1991*a*; Duplantier and Granjon, 1992; Granjon et al., 1992, as cf *pyramidum*; Granjon et al., 2002*b*) westward to the Cyrenaican Plateau of Libya (Ranck, 1968) and the Tibesti Mtns of Chad (Setzer and Ranck, 1971; Granjon et al., 1999:305).

STATUS: IUCN – Lower Risk (nt) as *G. riggenbachi*, Lower Risk (lc) as *G. tarabuli*.

SYNONYMS: *hamadensis* Ranck, 1968; *riggenbachi* Thomas, 1903; *tibesti* Setzer and Ranck, 1971.

COMMENTS: Subgenus *Gerbillus*. Usually listed as a synonym of *G. pyramidum* (see Pavlinov et al., 1990), Lay et al. (1975) noted that *tarabuli* could be distinguished by morphological traits. Future inquiry, according to Lay (1983:347) ". . . should examine the possibility that the 2n = 40 . . .forms reported from Tunisia, Algeria, Morocco and Senegal . . .are conspecific and may be referable to *G. tarabuli*." Granjon et al. (1999) have examined the distribution of samples with the 2n = 40 complement and using analyses of G- and C-banding and morphometric data redefined the species limits of *G. tarabuli* and its geographic distribution, as outlined above. Diploid number and other chromosomal traits confirmed by Chetoui et al. (2002) in Tunisian sample. Lay (1983) felt *hamadensis*, named as a subspecies of *G. pyramidum* by Ranck (1968), should provisionally be included in *G. tarabuli*. Granjon et al. (1999) explained why *riggenbachi* (formerly recorded from the type locality at Rio de Oro in Western Sahara and south on the coast at Lagwera, near border with Mauritania (see map in Aulagnier and Thevenot, 1986) and N Senegal (Duplantier et al., 1991a) is a synonym of *G. tarabuli*; *riggenbachi* has been included in either *G. pyramidum* (F. Petter, 1975b) or *G. latastei* (Cockrum, 1977), but Lay (1983:347) contended that it ". . . should be regarded as distinct pending comprehensive revision. . .," an arrangement followed by Aulagnier and Thevenot (1986). Granjon et al. (1999) also implicitly included *tibesti* (from the Tibesti Mtns of Chad) in *G. tarabuli*; *tibesti* was originally described as a subspecies of *G. pyramidum* but Lay (1983:346) noted "It is unclear whether *G. p. tibesti* represents this form, another species or is distinct." Musser and Carleton (1993) listed it as a separate species pending revisionary studies of North African *Gerbillus*. The record from Mauritania is documented by Bâ et al. (2001) and was identified by its karyotype (2n = 40, FN = 74), which is typical of the species. See Dobigny et al. (2002b) for additional chromosomal information.

Gerbillus tarabuli is most closely related to *G. pyramidum* (2n = 38) from Egypt and Sudan, and Granjon et al. (1999) suggested the distribution of *G. tarabuli* to be separated from the range of *G. pyramidum* in Egypt by the Cyrenaican Plateau and northern region of the Libyan desert. Those geomorphic features may separate Egyptian *G. pyramidum* from *G. tarabuli*, but apparently the former also occurs in the Sahelian region of SE Mauritania, N Mali, and W Niger (Granjon et al., 2002a). Dobigny et al. (2001a) documented *G. tarabuli* from N Mali, recorded the diploid number as 40, reported a species from the same area with a 2n of 38 that is probably *G. pyramidum*, and recorded that species (identifications based on karyotypes) from the region outlined by Granjon et al. (2002a). Ecology and membership in community of small mammals in coastal SW Mauritania reported by Granjon et al. (2002b). See account of *G. pyramidum*.

Gerbillus vivax (Thomas, 1902). Proc. Zool. Soc. Lond., 1902:8.

COMMON NAME: Vivacious Gerbil.
TYPE LOCALITY: Libya, Sebha.
DISTRIBUTION: Libya.
STATUS: IUCN – Lower Risk (lc).
COMMENTS: Subgenus *Hendecapleura*. Variously placed in either *G. dasyurus* (Ellerman and Morrison-Scott, 1951), *G. amoenus* (Ranck, 1968), or *G. nanus* (F. Petter, 1975b; Corbet, 1978c), but Lay (1983) disputed its association with *G. dasyurus* (holotype of *vivax* possesses an accessory tympanum and bare feet, traits uncharacteristic of *G. dasyurus*) and urged that *vivax* be retained as a species until its relationship to *G. amoenus* and *G. nanus* is assessed by systematic revision. Pavlinov et al. (1990) included *vivax* in *G. amoenus* following Ranck (1968) and Osborn and Helmy (1980).

Gerbillus watersi de Winton, 1901. Novit. Zool., 8:399.

COMMON NAME: Waters's Gerbil.
TYPE LOCALITY: Sudan, Upper Nile, Shendi.
DISTRIBUTION: Endemic to NE Africa in Sudan, Somalia, and Djibouti.
STATUS: IUCN – Lower Risk (lc) as *G. juliani* and *G. watersi*.
SYNONYMS: *juliani* (St. Leger, 1935).
COMMENTS: Subgenus *Hendecapleura*. Listed both as a subspecies of *G. nanus* or as a valid species by F. Petter (1975b) in the same report. The species should be considered distinct until revisionary studies advise otherwise (Lay, 1983). Roche and Petter (1968)

reviewed *juliani* under *Monodia*, but F. Petter (1975*b*) later synonymized it with *G. watersi* without supporting evidence. Roche (1975) provided evidence for uniting *juliani* with *G. watersi*, an action endorsed by Pavlinov et al. (1990). Lay (1983), however, recognized *juliani* as valid pending revision of the genus. Pearch et al. (2001) documented records from Djibouti, provided a good description of the species, comparing it with *G. henleyi* from the same country. Reviewed by Pavlinov et al. (1990).

Meriones Illiger, 1811. Prodr. Syst. Mamm. Avium., p. 82.

TYPE SPECIES: *Mus tamariscinus* Pallas, 1773.

SYNONYMS: *Cheliones* Thomas, 1919; *Idomeneus* Schultze, 1900; *Meraeus* Billberg, 1828 [*nomen nudum*]; *Pallasiomys* Heptner, 1933; *Parameriones* Heptner, 1937 [not Tchernov and Chetboun, 1984].

COMMENTS: Tribe Gerbillini, Subtribe Rhombomyina. No modern systematic revision is available for *Meriones*. The early revision by Chaworth-Musters and Ellerman (1947), as updated and modified by Ellerman and Morrison-Scott (1951), Corbet (1978*c*), and Pavlinov et al. (1990), represents the most current review of species. Additional taxonomic, distributional, and evolutionary views are found in the taxonomic reports and regional faunal studies cited throughout the accounts below. Most workers agree on definitions of the species we list here, but careful systematic revision will probably uncover a greater number of species. Chromosomal data concerning *Meriones* was summarized by Nadler and Lay (1967) in the context of assessing relationships among the species and subsequent chromosomal information was summarized by Qumsiyeh and Schlitter (1991). Additional chromosomal data and its significance to understanding phylogenetic relationships among six species of *Meriones* reported by Benazzou et al. (1982*a*, *b*). Lay and Nadler (1969) summarized laboratory hybridization attempts among several species of *Meriones*. Records of species from Iran and Pakistan reported by Lay et al. (1970). Yiğit et al. (1997*b*) reviewed ranges, diagnostic traits, and other characteristics of Turkish species. Comparisons of male genital morphology by Pavlinov (1986) pointed to the phylogenetic isolation of *M. tamariscinus* among the species of *Meriones*. Gromov and Erbajeva (1995) documented morphological and distributional contrasts among species of *Meriones* in Russia and Sokolov and Neronov (1993) provided a three-volume bibliography covering research dealing with *Meriones* that includes distribution maps for species in that country.

Pavlinov et al. (1990) recognized four subgenera in *Meriones* (*Meriones*, *Parameriones*, *Pallasiomys*, and *Cheliones*), and our subgeneric allocation of species generally follows their treatment. Evolutionary history as indicated by fossils first encountered in late Pliocene-early Pleistocene sediments of North Africa (Tong, 1998; Wessels, 1999), Holocene of East Africa (Denys, 1999), early Pleistocene of Israel (Tchernov, 1986, discussed under *Parameriones*), and middle Pleistocene strata in China (Lucas, 2001). *Meriones* is thought to have evolved from the extinct *Pseudomeriones*, the earliest known gerbilline, recorded from the early part of late Miocene in Turkey and Afghanistan, and later in late Miocene and Pliocene in those two countries and Spain, Greece, N Iran/ Turkmenistan, and China (Wessels, 1998).

Meriones arimalius Cheesman and Hinton, 1924. Ann. Mag. Nat. Hist., ser. 9, 14:554.

COMMON NAME: Arabian Jird.

TYPE LOCALITY: Saudi Arabia, Yabrin (Jabrin), Djebel Agoula.

DISTRIBUTION: "Northern sands of the Rub al Khali in Saudi Arabia and Oman" (Harrison and Bates, 1991:297).

STATUS: IUCN – Endangered.

COMMENTS: Subgenus *Pallasiomys*. Ellerman and Morrison-Scott (1951) listed *arimalius* as a valid species, but it was later included in *M. libycus* (Corbet, 1978*c*; Harrison and Bates, 1991). Pavlinov et al. (1990:294) reinstated *arimalius* as a separate species and reviewed its salient characters. Even from the terse description of its diagnostic traits provided by Harrison and Bates (1991:297), who recognized the form as a subspecies of *M. libycus*, it is evident that *arimalius* is morphologically different from populations of *lybicus* north of it in Saudi Arabia. The species was also considered distinct by Nadler and Lay (1967).

Meriones chengi Wang, 1964. Acta Zootaxon. Sinica, 1:9.

COMMON NAME: Cheng's Jird.

TYPE LOCALITY: China, N Xinjiang (Sinkiang), Da-Ho-Yien, Turfan.

DISTRIBUTION: Recorded from several localities in a small area of N Xinjiang (see Wang, 2003, and Zhang et al., 1997).

STATUS: IUCN – Critically Endangered.

COMMENTS: Subgenus *Pallasiomys*. Wang (1964) considered this species, based on a series of adult and immature specimens, to be most closely related to *M. meridianus*, which also occurs in Xinjiang Province (Ma et al., 1987). Pavlinov et al. (1990, 1995a) questionably included *chengi* in the synonymy of *M. meridianus*. The relationship of *chengi* to the latter species needs to be assessed by revision of *Meriones*, especially the *M. meridianus* complex.

Meriones crassus Sundevall, 1842. K. Svenska Vet. Akad., Ser. 3, p. 233.

COMMON NAME: Sundevall's Jird.

TYPE LOCALITY: Egypt, Sinai, Fount of Moses (Ain Musa).

DISTRIBUTION: Across North Africa from Morocco through Niger, Sudan, and Egypt to Israel, Jordan, Syria, SE Anatolia (Turkey), Saudi Arabia, Iraq, Iran, and Afghanistan.

STATUS: IUCN – Lower Risk (lc).

SYNONYMS: *asyutensis* Setzer, 1961; *charon* Thomas, 1919; *ismahelis* Cheesman and Hinton, 1924; *longifrons* Lataste, 1884; *pallidus* Bonhote, 1912; *pelerinus* Thomas, 1919; *perpallidus* Setzer, 1961; *swinhoei* Scully, 1881; *tripolius* Thomas, 1919.

COMMENTS: Subgenus *Pallasiomys*. Reviewed by Corbet (1978c) and Pavlinov et al. (1990). Regional reviews are available for Morocco (Aulagnier and Thevenot, 1986), Algeria (Kowalski and Rzebik-Kowalska, 1991), Libya (Ranck, 1968), Egypt (Osborn and Helmy, 1980), Sinai Peninsula (Saleh and Basuony, 1998), Israel and Jordan (Mendelssohn and Yom-Tov, 1999; Qumsiyeh, 1996), SE Turkey (Yiğit et al., 1997b, 1998b), the Arabian Peninsula (Harrison and Bates, 1991), Iran (Lay, 1967; Morshed and Patton, 2002), and Afghanistan (Hassinger, 1973). See Koffler (1972, Mammalian Species, 9). Yiğit et al. (1998b) contrasted the distribution, morphology and chromosomes between *M. crassus* (2n = 60, FN = 76) and *M. meridianus* (2n = 50, FNa = 78) in E Turkey. See Dobigny et al. (2002b) for additional karyotypic information and references.

Meriones dahli Shidlovsky, 1962. Definition Rodents Zakabkaziya, p. 115.

COMMON NAME: Dahl's Jird.

TYPE LOCALITY: Armenia, Sadarak steppe, foothills of Vardanis (Saraibulak) Ridge.

DISTRIBUTION: Local sandy habitats in Armenia.

STATUS: IUCN – Endangered.

COMMENTS: Subgenus *Pallasiomys*. Included in *M. meridianus* by Corbet (1978c) but shown to be a separate species by Dyatlov and Avanyan (1987), whose results were based on morphological, biochemical, and chromosomal traits, as well as interbreeding experiments. Reviewed by Pavlinov et al. (1990).

Meriones grandis Cabrera, 1907. Bol. Real Soc. Española Hist. Nat., Madrid, 7:175.

COMMON NAME: Moroccan Jird.

TYPE LOCALITY: Morocco, Marrakesh.

DISTRIBUTION: Mediterranean littoral from Morocco through N Algeria to Tunisia (see Pavlinov, 2000).

COMMENTS: Subgenus *Pallasiomys*. Moroccan population mapped and reviewed by Aulagnier and Thevenot (1986, as *M. shawi*). Pavlinov et al. (1990) recognized *grandis* as a species, but Aulagnier and Thevenot (1986) included it in *M. shawi*. Geographic ranges of *M. shawi* and *M. grandis* broadly overlap through E Morocco, N Algeria, and Tunisia, and examples of both species have been collected at a few locations; see Pavlinov (2000) who revised the species and documented morphometric and other traits distinguishing it from *M. shawi*. Fragments identified as *M. shawi* are reported from possible middle Pleistocene sediments at Jebel Irhoud in Morocco (Amani and Geraads, 1993); whether these fossils are *shawi* or *grandis* has to be determined.

Meriones hurrianae Jordon, 1867. Mamm. India, p. 186.

COMMON NAME: Indian Desert Jird.

TYPE LOCALITY: India, Hurriana Dist.

DISTRIBUTION: Primarily in Thar Desert in SE Iran, Pakistan, and NW India (Haryana, Rajasthan, and Gujarat; see Agrawal, 2000).

STATUS: IUCN – Lower Risk (lc).

SYNONYMS: *collinus* (Thomas, 1919).

COMMENTS: Subgenus *Cheliones*. Pakistan populations reviewed by Roberts (1977), those in India by Agrawal (2000) and Chakraborty and Agrawal (2000). Hassinger (1973) discussed old records of the species from Afghanistan as probably erroneous. Reviewed by Pavlinov et al. (1990).

Meriones libycus Lichtenstein, 1823. Verz. Doublet. Zool. Mus. Univ. Berlin, p. 5.

COMMON NAME: Libyan Jird.

TYPE LOCALITY: "Libische Wuste" (Libyan Desert), as restricted by lectotype designation by Pavlinov (1982c:1767); usually listed as Egypt, near Alexandria, due to the interpretation of Lichtenstein's type locality by Chaworth-Musters and Ellerman (1947:485).

DISTRIBUTION: North Africa from Western Sahara (Rio de Oro) and Morocco to Egypt, through Saudi Arabia, Jordan, Iraq, Syria, Iran, Afghanistan, and east through Turkmenistan, Uzbekistan, and S Kazakhstan to W China (Xinjiang). A record from SE Anatolia has not been confirmed (Kryštufek and Vohralík, 2001).

STATUS: IUCN – Lower Risk (lc).

SYNONYMS: *afghanus* Pavlinov and Rossolimo, 1987; *amplus* Ranck, 1968; *aquilo* Thomas, 1912; *caucasicus* (Satunin, 1896) [accidental renaming of *caucasius*]; *caucasius* Brandt, 1855; *caudatus* Thomas, 1919; *collium* Severtzov, 1873; *confalonieri* de Beaux, 1931; *edithae* Cheesman and Hinton, 1924; *erythrourus* (Gray, 1842); *evelynae* Cheesman and Hinton, 1924; *eversmanni* (Bogdanov, 1889); *farsi* Schlitter and Setzer, 1973; *gaetulus* Lataste, 1882; *guyonii* (Loche, 1867); *heptneri* Argyropulo, 1936 [*nomen nudum*, Pavlinov et al., 1990]; *intermedius* Gromov, 1952; *iranensis* Goodwin, 1939; *luridus* Ranck, 1968; *marginae* (Heptner, 1933); *mariae* Cabrera, 1907; *maxeratis* (Heptner, 1933); *melanurus* Rüppell, 1842; *oxianus* (Heptner, 1933); *renaultii* (Loche, 1867); *schousboeii* (Loche, 1867); *schwarzovi* Toktosunov, 1977 [*nomen nudum*]; *sogdianus* (Heptner, 1933); *syrius* Thomas, 1919; *tuareg* Thomas, 1925; *turfanensis* (Satunin, 1903).

COMMENTS: Subgenus *Pallasiomys*. Reviewed by Corbet (1978c) and Pavlinov et al. (1990). Regional studies cover populations in Morocco (Aulagnier and Thevenot, 1986), Algeria (Kowalski and Rzebik-Kowalska, 1991), Libya (Ranck, 1968, as *caudatus*), Egypt (Osborn and Helmy, 1980), Jordan (Qumsiyeh, 1996), Arabian Penninsula (Harrison and Bates, 1991), Iran (Lay, 1967; Morshed and Patton, 2002), Afghanistan (Hassinger, 1973), the Pribalkhashye region of S Kazakhstan (Burdelov et al., 1993, as *erythrourus*), Russia (Gromov and Erbajeva, 1995), and the Xinjiang Province of W China (Ma et al., 1987; Wang, 2003). Comparative craniometric analyses between Moroccan samples of *M. libycus* and *M. shawi* obtained in sympatry were reported by Zaime and Pascal (1988). Morphological and karyotypic contrasts between these same two species as well as laboratory hybridization experiments, were recorded by Lay and Nadler (1969). In North Africa, *M. libycus* inhabits the Sahara desert, but does extend to the Mediter-ranean in Morocco, Algeria, and Libya where it overlaps the distribution of *M. shawi*, which is primarily Mediterranean littoral (Lay and Nadler, 1969; Zaime and Pascal, 1988). Intrapopulation polymorphism of the 13th heterochromatin chromosome among samples of *M. libycus* from Kazakhstan and its significance for reconstructing possible dispersal routes into C Asia is reported by Korobitsina and Kartavtseva (1992). Citations for synonyms among Russian samples were supplied by Pavlinov and Rossolimo (1987); *afghanus* is proposed as a new subspecies in that checklist. Schlitter and Setzer (1973) described *farsi* as a subspecies of *M. erythrourus*. Lay and Nadler (1969) also clarified why *caudatus*, used by Ranck (1968) as a species name for Libyan samples, simply refers to *M. libycus*. Corbet (1978c:127), apparently unaware of the report by Lay and Nadler (1969), followed Ranck and listed *caudatus* as a species, but cautioned that *caudatus* may be ". . . conspecific with *M. libycus* and that the Libyan forms assigned by Ranck to *M. libycus* should really be allocated to *M. shawi.*" This is correct and restates the past confusion that has ". . . clouded the taxonomy of *M. shawi* and *M. libycus* because of uncertainty concerning the number of species in this complex and their nomenclature" (Lay and Nadler, 1969:44). Pavlinov (1982c) documented the identity of *erythrourus* and *caudatus* with *M. libycus*, but Zhang et al. (1997) still recognized *erythrourus* as a separate species.

Meriones meridianus (Pallas, 1773). Reise Prov. Russ. Reichs., p. 702.
 COMMON NAME: Midday Jird.
 TYPE LOCALITY: Se Russia, Astrakhanskaya Oblast, Dosang (as restricted by Heptner, *in* Vinogradov et al., 1936, not Chaworth-Musters and Ellerman, 1947; see Pavlinov and Rossolimo, 1987).
 DISTRIBUTION: From Lower Don River and north of the Caucasus to Mongolia and N China (provinces of Nei Mongol, Hebei, Henen, Shanxi, N Shaanxi, Xinjiang, Gansu, NE Qinghai, and Ningxia; see Ma et al. [1987], Wang [2003], and Zhang et al. [1997]), south to E Turkey, E Iran and N Afghanistan. The isolated segment in Armenia mentioned by Corbet (1978c) refers to *M. dahli* (see that account).
 STATUS: IUCN – Lower Risk (lc).
 SYNONYMS: *auceps* Thomas, 1908; *brevicaudatus* (Milne-Edwards, 1867); *buechneri* Thomas, 1909; *cryptorhinus* (Blanford, 1875); *fulvus* Eversmann, 1848; *heptneri* (Kuznetzov, 1944) [not of Argyropulo, 1944]; *jei* Wang, 1964; *karelini* (Kolossow, 1935); *lepturus* (Büchner, 1889); *littoralis* (Heptner, 1927) [*nomen nudum*]; *massagetes* (Heptner, 1933); *muleiensis* Wang, 1981; *nogaiorum* (Heptner, 1927); *penicilliger* (Heptner, 1933); *psammophilus* (Milne-Edwards, 1871); *roborowskii* (Büchner, 1889); *shitkovi* (Heptner, 1933); *tropini* Kartavtseva and Korobitsyna, 1986 [*nomen nudum*]; *urianchaicus* (Vinogradov, 1927); *uschtaganicus* (Rall, 1940); *zhitkovi* (Heptner, 1936).
 COMMENTS: Subgenus *Pallasiomys*. Intraspecific chromosomal variation among Russian samples reported by Korobitsyna and Kartavtseva (1988). Utilizing several sets of data, including results from hybridization studies, Dyatlov and Avanyan (1987) tested conspecificity of the subspecies *meridianus*, *nogaiorum*, and *dahli*, and concluded that *nogaiorum* should be considered a semispecies and *dahli* a species (see that account). We retain *nogaiorum* in *M. meridianus* pending unequivocal results demonstrating its evolutionary status. Phallic morphology described by Yang et al. (1992) and contrasted with *M. unguiculatus*. Yiğit et al. (1998b) contrasted geographic range, morphology and karyotypes between samples of *M. meridianus* and *M. crassus* in E Turkey. Reviewed by Pavlinov et al. (1990). The segment in E Anatolia (Turkey) was reviewed by Yiğit et al. (1997b, 1998b). Range in Russia, taxonomy, and other characteristics reviewed by Gromov and Erbajeva (1995), who listed both *nogaiorum* and *dahli* in the synonymy of *M. meridianus*.

Meriones persicus (Blanford, 1875). Ann. Mag. Nat. Hist., ser. 4, 16:312.
 COMMON NAME: Persian Jird.
 TYPE LOCALITY: Iran, Kohrud, north of Isfahan.
 DISTRIBUTION: Iran, adjacent regions of Transcaucasia, Turkey (E Anatolia), Iraq, Turkmenistan, Afghanistan and Pakistan (west of Indus River).
 STATUS: IUCN – Lower Risk (lc).
 SYNONYMS: *ambrosius* Thomas, 1919; *baptistae* Thomas, 1920; *gurganensis* Goodwin, 1939; *rossicus* Heptner, 1931; *suschkini* (Kashkarov, 1925).
 COMMENTS: Subgenus *Parameriones*. Regional studies available for the Middle East (Harrison and Bates, 1991), Pakistan (Roberts, 1977), Afghanistan (Hassinger, 1973), Iran (Lay, 1967; Morshed and Patton, 2002), E Anatolian Turkey (Kryštufek and Vohralík, 2001), and Russia (Gromov and Erbajeva, 1995). Yiğit and Çolak (1999) recorded 2n = 42, FN = 78 for samples from E Turkey. Reviewed by Pavlinov et al. (1990).

Meriones rex Yerbury and Thomas, 1895. Proc. Zool. Soc. Lond., 1895:552.
 COMMON NAME: King Jird.
 TYPE LOCALITY: Yemen, Lahej, near Aden.

DISTRIBUTION: Yemen highlands in SW Arabian Peninsula, from Mecca in Saudi Arabia to near Aden in Yemen (see Al-Jumaily [1998] and Harrison and Bates [1991:289]).

STATUS: IUCN – Lower Risk (lc).

SYNONYMS: *buryi* Thomas, 1902; *philbyi* (Morrison-Scott, 1939).

COMMENTS: Subgenus *Parameriones*, according to Nadler and Lay (1967) and Harrison and Bates (1991), but not allocated to a subgenus by Pavlinov et al. (1990), who reviewed the species. A distinctive jird that ". . . has evolved in the special and peculiar environment of the Yemen highlands. . ." (Harrison and Bates, 1991:291). Karyotype (2n = 38, FN = 74) documented by Al-Saleh and Khan (1987).

Meriones sacramenti Thomas, 1922. Ann. Mag. Nat. Hist., ser. 9, 10:552.

COMMON NAME: Buxton's Jird.

TYPE LOCALITY: Israel, 10 mi (16 km) S Beersheba.

DISTRIBUTION: A small range in Israel (on coastal plain south of the Yarqon River and in the N Negev) and NE Sinai Peninsula of Egypt (Harrison and Bates, 1991; Osborn and Helmy, 1980).

STATUS: IUCN – Endangered.

SYNONYMS: *legeri* Aharoni, 1932.

COMMENTS: Subgenus *Pallasiomys*. A distinctive Israeli and Sinai endemic, reviewed by Osborn and Helmy (1980), Pavlinov et al. (1990), Harrison and Bates (1991), Qumsiyeh (1996), and Mendelssohn and Yom-Tov (1999). Apparently most closely related to *M. shawi* and *M. libycus* in its morphology (see Harrison and Bates, 1991, and references therein).

Meriones shawi (Duvernoy, 1842). Mem. Soc. Sci. Nancy, 3:22.

COMMON NAME: Shaw's Jird.

TYPE LOCALITY: Algeria, Oran.

DISTRIBUTION: Mediterranean littoral from E Morocco through N Algeria, Tunisia, Libya, and Egypt to N Sinai, never found more than about 150 mi (240 km) inland (Lay and Nadler, 1969, and references therein; Pavlinov, 2000).

STATUS: IUCN – Lower Risk (lc).

SYNONYMS: *albipes* Lataste, 1882; *auratus* Ranck, 1968; *auziensis* Lataste, 1882; *azizi* Setzer, 1956; *crassibulla* Lataste, 1885; *isis* Thomas, 1919; *laticeps* Lataste, 1885; *longiceps* Lataste, 1885; *richardii* (Loche, 1867); *savii* (Loche, 1867); *sellysii* (Pomel, 1856); *trouessarti* Lataste, 1882.

COMMENTS: Subgenus *Pallasiomys*. A distinctive species that ranges mostly north of *M. libycus* but is sympatric with it in several regions (Lay and Nadler, 1969; Zaime and Pascal, 1988). The distribution maps of Algerian (Kowalski and Rzebik-Kowalska, 1991) and Egyptian (Osborn and Helmy, 1980) ranges of *M. shawi* and *M. libycus* illustrate their geographic relationships–mostly parapatric, but sympatric near the coast. *Meriones shawi* and *M. libycus* are often confused in museum collections and published reports (see the reviews by Lay and Nadler, 1969, and Pavlinov et al., 1990). The taxon *grandis* has traditionally been included in *M. shawi* but Pavlinov et al. (1990) reviewed *grandis* as a species and Pavlinov (2000) recorded the morphological traits distinguishing *M. grandis* from *M. shawi*; he also documented broad overlap between the two species in E Morocco, N Algeria, and Tunisia (see account of *M. grandis*).

Meriones tamariscinus (Pallas, 1773). Reise Prov. Russ. Reichs., 2:702.

COMMON NAME: Tamarisk Jird.

TYPE LOCALITY: Kazakhstan, Saraitschikowski (= Saraichik).

DISTRIBUTION: N Caucasus and Kazakhstan to the Altai Mtns, and through N Xinjiang and Nei Mongol of China (Wang, 2003; Zhang et al., 1997).

STATUS: IUCN – Lower Risk (lc).

SYNONYMS: *ciscaucasicus* (Satunin, 1903); *collium* Severtsov, 1873; *jaxartensis* (Ognev and Heptner, 1928); *kokandicus* Heptner, 1933; *montanus* Severtsov, 1873 [*nomen nudum*, Pavlinov et al., 1990]; *satschouensis* (Satunin, 1903); *tamaricinus* (Pallas, 1779).

COMMENTS: Subgenus *Meriones*. Reviewed by Corbet (1978c) and Pavlinov et al. (1990); regional reviews offered by G. M. Allen (1940, China), and Ma et al. (1987, Xinjiang). Range in Russia, taxonomy, contrasts with other species and additional characteristics reviewed by Gromov and Erbajeva (1995). The restriction to subgenus *Meriones* reflects the very distinctive male genital morphology of *M. tamariscinus* compared with other species of *Meriones* (Pavlinov, 1986).

Meriones tristrami Thomas, 1892. Ann. Mag. Nat. Hist., ser. 6, 9:148.

COMMON NAME: Tristram's Jird.

TYPE LOCALITY: Israel, Dead Sea region.

DISTRIBUTION: From Israel, Lebanon, and W Jordan to Turkey (Kryštufek and Vohralík, 2001; Yiğit et al., 1998*a*), Syria, N Iraq, NW Iran, and Transcaucasia (see Harrison and Bates, 1991:294).

STATUS: IUCN – Lower Risk (lc).

SYNONYMS: *blackleri* Thomas, 1903; *bodenheimeri* Aharoni, 1932; *bogdanovi* Heptner, 1931; *intraponticus* Neuhäuser, 1936; *kariateni* Aharoni, 1932; *kilisensis* Yiğit and Çolak, 1998; *lycaon* Thomas, 1919; *qatafensis* Haas, 1951.

COMMENTS: Subgenus *Pallasiomys*. Treated as a subspecies of *M. shawi* by Ellerman and Morrison-Scott (1951). The species was generally reviewed by Corbet (1978*c*) and Pavlinov et al. (1990) and regionally reviewed by Harrison and Bates (1991, Arabia), Qumsiyeh (1996) and Mendelssohn and Yom-Tov (1999, Israel and Jordan), Misonne (1957, Syria), Lay (1967, C Asia), Yiğit et al. (1998*a*, Turkey), and Gromov and Erbajeva (1995, Russia). Chromosomal polymorphism and its significance among Transcaucasian samples were reported by Korobitsyna and Korablev (1980). Populations in Turkey have a stable diploid number (72) but the fundamental number varies between 76 in W Turkey and 82 in samples from the rest of Turkey (a diagnostic traits of *M. t. kilisensis* is its FN of 78; Yiğit and Çolak, 1998). *Meriones tristrami* has resided in the S Levant, as documented by fossils, since at least 160,000 years before present (see reviews and references cited by Tchernov, 1992, 1994). Saleh and Basuony (1998) claimed the Sinai Peninsula was the westernmost part of *M. tristrami*'s range but provided no documentation; we cannot locate records west of Israel.

Meriones unguiculatus (Milne-Edwards, 1867). Ann. Sci. Nat. Zool., ser. 7, 5:377.

COMMON NAME: Mongolian Jird.

TYPE LOCALITY: China, N Shanxi, 10 mi (16 km) NE of Tschang-Kur, Eul-che san hao (= Ershi san hao).

DISTRIBUTION: Mongolia, and adjacent regions of Siberia (Transbaikalia) and of China from E Gansu, N Ningxia, N Shaanxi, N Shanxi, and Hebei, through C and N Nei Mongol and Liaoning (see Gromov and Erbajeva [1995], Wang [2003], and Zhang et al. [1997]).

STATUS: IUCN – Lower Risk (lc).

SYNONYMS: *chihfengensis* Mori, 1939 [see Kaneko and Maeda, 2002]; *koslovi* (Satunin, 1903); *kurauchii* Mori, 1930; *selenginus* Heptner, 1949.

COMMENTS: Subgenus *Pallasiomys*. Reviewed by G. M. Allen (1940), Corbet (1978*c*), and Pavlinov et al. (1990). Corbet also included Xinjiang in the distribution of the species, but Ma et al. (1987) did not record it there. Spicer and Schulte (1994) documented cochlear structure at the cellular level. This study is an example of the many undertaken on this species, as this is the gerbil usually sold as pets and used in medical laboratories (Turton, 1984). Laboratory colonies were derived from twenty pairs captured in the Amur River basin in 1935, initially taken to Japan from which a colony was shipped to the United States, and from there others were distributed to Europe (Turton, 1984). Yang et al. (1992) described phallic morphology and contrasted it with that of *M. meridianus*. Cao et al. (1995) reported histology of the glans penis in *M. unguiculatus* and contrasted the pattern and morphology of its epidermal spines with *M. meridianus* and *Rhombomys opimus*, and commented upon the significance of the spines for discriminating among species. See Gulotta (1971, Mammalian Species, 3).

Meriones vinogradovi Heptner, 1931. Zool. Anz., 94:122.

COMMON NAME: Vinogradov's Jird.

TYPE LOCALITY: Iran, Persian Azarbaijan.

DISTRIBUTION: E Anatolian Turkey, N Syria, N Iran, and Armenia and Azerbaijan (see Harrison and Bates, 1991).

STATUS: IUCN – Lower Risk (lc).

COMMENTS: Subgenus *Pallasiomys*. A distinctive species (Pavlinov and Rossolimo, 1987; Pavlinov et al., 1995*a*) that was reviewed by Pavlinov et al. (1990) and Harrison and Bates (1991). The Iranian population was reviewed by Lay (1967), the Syrian by Misonne (1957), Turkish by Yiğit et al. (1997*b*), and the segment in Russia by Gromov and Erbajeva (1995).

Meriones zarudnyi Heptner, 1937. Byull. Moscow Ova. Ispyt. Prir. Otd. Biol., 46:19.
COMMON NAME: Zarudny's Jird.
TYPE LOCALITY: Turkmenistan, Kushka (Afghanistan-Turkmenistan border).
DISTRIBUTION: NE Iran, N Afghanistan, and S Turkmenistan.
STATUS: IUCN – Endangered.
COMMENTS: Subgenus *Pallasiomys*. Reviewed by Corbet (1978*c*) and Pavlinov et al. (1990) and listed by Pavlinov and Rossolimo (1987) as a distinctive species. The Afghanistan population was reviewed by Hassinger (1973), the E Anatolian by Yiğit et al. (1997), and segment in Turkmenistan by Gromov and Erbajeva (1995).

Microdillus Thomas, 1910. Ann. Mag. Nat. Hist., ser. 8, 5:197.
TYPE SPECIES: *Gerbillus peeli* de Winton, 1898.
COMMENTS: Tribe Gerbillini, Subtribe Gerbillina. A close relative of *Gerbillus* and *Dipodillus* (Pavlinov et al., 1990).

Microdillus peeli (de Winton, 1898). Ann. Mag. Nat. Hist., ser. 7, 1:250.
COMMON NAME: Somali Pygmy Gerbil.
TYPE LOCALITY: Somalia, Eyk.
DISTRIBUTION: Recorded only from Somalia (see Roche and F. Petter, 1968).
STATUS: IUCN – Lower Risk (nt).
COMMENTS: Reviewed by Roche and F. Petter (1968) and Pavlinov et al. (1990).

Pachyuromys Lataste, 1880. Le Naturaliste, 2(40):313.
TYPE SPECIES: *Pachyuromys duprasi* Lataste, 1880.
COMMENTS: Tribe Gerbillini, Subtribe Pachyuromyina (sole member).

Pachyuromys duprasi Lataste, 1880. Le Naturaliste, 2(40):313.
COMMON NAME: Fat-tailed Jird.
TYPE LOCALITY: Algeria, Laghouat.
DISTRIBUTION: N Sahara desert from Western Sahara and Morocco to N Egypt.
STATUS: IUCN – Lower Risk (lc).
SYNONYMS: *faroulti* Thomas, 1920; *natronensis* de Winton, 1903.
COMMENTS: The Moroccan population was mapped by Aulagnier and Thevenot (1986), Algerian population reviewed by Kowalski and Rzebik-Kowalska (1991), the Libyan segment by Ranck (1968), and the Egyptian by Osborn and Helmy (1980). Chromosomal data summarized by Qumsiyeh and Schlitter (1991).

Psammomys Cretzschmar, 1828. *In* Rüppell, Atlas Reise Nordl. Afr., Zool. Säugeth., p. 56.
TYPE SPECIES: *Psammomys obesus* Cretzschmar, 1828.
SYNONYMS: *Parameriones* Tchernov and Chetboun, 1984 [not of Heptner, 1937].
COMMENTS: Tribe Gerbillini, Subtribe Rhombomyina. Reviewed by Corbet (1978*c*, 1984). Evolutionary history as indicated by fossils extends to Late Pleistocene of Asia (McKenna and Bell, 1997).

Psammomys obesus Cretzschmar, 1828. *In* Rüppell, Atlas Reise Nordl. Afr., Zool. Säugeth., p. 58, pl. 22.
COMMON NAME: Fat Sand Rat.
TYPE LOCALITY: Egypt, Alexandria.
DISTRIBUTION: In North Africa from Morocco (Aulagnier and Thevenot, 1986) and Algeria (Kowalski and Rzebik-Kowalska, 1991), through Tunisia and coastal region of Egypt (Osborn and Helmy, 1980), into Syria, Jordan, Israel, and parts of Arabian Peninsula (Harrison and Bates, 1991); also on coast of Sudan (Corbet, 1978*c*).
STATUS: IUCN – Lower Risk (lc).

SYNONYMS: *algiricus* Thomas, 1902; *dianae* Morrison-Scott, 1939; *elegans* Heuglin, 1877; *nicolli* Thomas, 1908; *roudairei* Lataste, 1881; *terraesanctae* Thomas, 1902; *tripolitanus* Thomas, 1902.

COMMENTS: Electrophoretic, chromosomal, and morphological traits were analyzed by Qumsiyeh and Chesser (1988) in the context of assessing evolutionary change among four genera of gerbils. Reviewed by Corbet (1978c) and Pavlinov et al. (1990). Arabian population reviewed by Harrison and Bates (1991), Israeli and Jordanian by Qumsiyeh (1996) and Mendelssohn and Yom-Tov (1999), Egyptian by Osborn and Helmy (1980), Algerian by Kowalski and Rzebik-Kowalska (1991), and Moroccan by Aulagnier and Thevenot (1986).

Psammomys vexillaris Thomas, 1925. Ann. Mag. Nat. Hist., ser. 9, 16:198.
    COMMON NAME: Lesser Sand Rat.
    TYPE LOCALITY: Libya, Tripolitania Province, Bu Ngem (Bondjem).
    DISTRIBUTION: Algeria, Tunisia, and Libya.
    STATUS: IUCN – Lower Risk (lc).
    SYNONYMS: *edusa* Thomas, 1925.
    COMMENTS: Sometimes included in *P. obesus* (Corbet, 1978c; Pavlinov et al., 1990, 1995a) but Ranck (1968) regarded it as a separate species. Cockrum et al. (1977) enumerated the morphological and chromosomal traits distinguishing the two. Kowalski and Rzebik-Kowalska (1991) did not recognize this species in Algeria and discussed only *P. obesus*. We recognize *vexillaris* as distinct pending a new look at geographic variation in morphological, chromosomal, and molecular traits among populations of *Psammomys*.

Rhombomys Wagner, 1841. Gelehrte Anz. I. K. Bayer Akad. Wiss., München, 12, 52:421.
    TYPE SPECIES: *Rhombomys pallidus* Wagner, 1841 (= *Meriones opimus* Lichtenstein, 1823).
    SYNONYMS: *Amphiaulacomys* Lataste, 1882; *Pliorhombomys* Fokanov, 1964 [*nomen nudum*]; *Pliorhombomys* Fokanov, 1976.
    COMMENTS: Tribe Gerbillini, Subtribe Rhombomyina. Reviewed by Corbet (1978c). The inclusion of *Pliorhombomys* in *Rhombomys* was documented by Pavlinov (1992b). Fossils indicate an evolutionary history extending to late Pliocene in Asia (McKenna and Bell, 1997).

Rhombomys opimus (Lichtenstein, 1823). Naturh. Abh. Eversmann's Reise, p. 122.
    COMMON NAME: Great Gerbil.
    TYPE LOCALITY: Kazakhstan, Kzyl-Ordinskaya, KaraKumy Desert (see Pavlinov and Rossolimo, 1987).
    DISTRIBUTION: From S Mongolia and N China (Xinjiang, Gansu, Nei Mongolia) to Kazakhstan, Iran, Afghanistan, and SW Pakistan (Corbet, 1987c, Ma et al., 1987; Wang, 2003; Zhou et al., 2000).
    STATUS: IUCN – Lower Risk (lc).
    SYNONYMS: *alaschanicus* Matschie, 1911; *dalversinicus* Kashkarov, 1926; *fumicolor* Heptner, 1933; *giganteus* Buchner, 1889; *major* Burdelov, 1989; *minor* Burdelov, 1989; *nigrescens* Satunin, 1903; *pallidus* Wagner, 1841; *pevzovi* Heptner, 1939; *sargadensis* Heptner, 1939; *sodalis* Goodwin, 1939.
    COMMENTS: Reviewed by Pavlinov et al. (1990). Regional reviews cover Pakistan (Roberts, 1977), Iran (Lay, 1967), Afghanistan (Hassinger, 1973), Russia and adjacent regions (Gromov and Erbajeva, 1995), and N China (Xinjiang, Gansu, and Nei Mongolia; Ma et al., 1987; Zhang et al., 1997; Zhou et al., 2000). *Rhombomys opimus* is unique among gerbils in having hypsodont, evergrowing (rootless) molars with cement in the reentrant angles, but Pavlinov (1982b, 1996) found a few old individuals with rooted molars from several localities in Iran; however, this discovery does not help clarify taxonomic status of the species. Chromosomal data summarized by Qumsiyeh and Schlitter (1991). Morphology and schmelzmuster of rootless molars in relation to jaw motion documented by Koenigswald et al. (1994) and compared to 24 other extant rodent genera. Phallic morphology described by Yang et al. (1992). Age identification and population structure of Chinese population reported by Zhou et al. (2002), who

also mapped the nine different dry habitats in N China occupied by *R. opimus*, noting that its distribution has been altered there through human activities.

Sekeetamys Ellerman, 1947. Proc. Zool. Soc. Lond., 1947-1948(117):271 [1947].

TYPE SPECIES: *Gerbillus calurus* Thomas, 1892.

COMMENTS: Tribe Gerbillini, Subtribe Rombomyina. Reviewed by Corbet (1978*c*), who also discussed the past allocation of *calurus* to either *Meriones* or *Gerbillus*. Based primarily upon morphology, Pavlinov et al. (1990), thought *Sekeetamys* was most closely related to *Meriones* and *Brachiones*, but Tong (1989) aligned it with *Microdillus* and *Gerbillus*. One of four gerbil genera in which electrophoretic, chromosomal, and morphological traits were examined to assess rates of evolutionary change (Qumsiyeh and Chesser, 1988).

Sekeetamys calurus (Thomas, 1892). Ann. Mag. Nat. Hist., ser. 6, 9:76.

COMMON NAME: Bushy-tailed Jird.

TYPE LOCALITY: Egypt, Sinai, near Tor.

DISTRIBUTION: Restricted to rocky and cliff habitat from E Egypt (east side of the Nile) through Sinai, SE Israel and SW Jordan into C Saudi Arabia (Harrison and Bates, 1991; Mendelssohn and Yom-Tov, 1999; Osborn and Helmy, 1980; Qumsiyeh, 1996).

STATUS: IUCN – Lower Risk (lc).

SYNONYMS: *makrami* Setzer, 1961.

COMMENTS: Although researchers now agree that *calurus* merits generic separation, in the past it has been regarded as a species of *Gerbillus*, *Dipodillus*, or *Meriones* (see references in Corbet, 1978*c*, and Osborn and Helmy, 1880). Morphology, taxonomy, and ecology summarized by Osborn and Helmy (1980) for Egyptian population, Qumsiyeh (1996) for populations in Israel and Jordan, and Harrison and Bates (1991) for Arabian Peninsula. Distribution on Sinai Peninsula recorded by Saleh and Basuony (1998). Chromosomal data summarized by Qumsiyeh and Schlitter (1991). Reviewed in detail by Pavlinov et al. (1990).

Tatera Lataste, 1882. Le Naturaliste, Paris 2:126.

TYPE SPECIES: *Dipus indicus* Hardwicke, 1807.

COMMENTS: Tribe Taterillini, Subtribe Taterillina. According to Pavlinov et al. (1990) and Pavlinov (2001), Taterillini also includes the African *Taterillus* and *Gerbilliscus* (once included in *Tatera;* see generic account of the latter). Tong (1989) also included *Gerbillurus* in this group and did not separate *Gerbilliscus* from *Tatera*.

Evolutionary history of *Tatera* dates from *T. pinjoricus*, which is represented by isolated molars from late Pliocene Siwalik beds of NW India (Patnaik, 1997), which appears to be a primitive phylogenetic ally of living *T. indica* (Flynn et al., 2003). Patnaik (1997) considered *Tatera pinjoricus* to be dentally similar to the late Miocene *Abudhabia* sp. from Pakistan, *A. baynunensis* from the United Arab Emirates (de Bruijn and Whybrow, 1994) and the early to late Pliocene *A. kabulense* from NW India and N Afghanistan (see review by Wessels, 1998). In addition to those species, *A. pakistanensis* has been described from the Siwalik late Miocene of Pakistan (Flynn and Jacobs, 1999), *A. radinsky* from the Pliocene of NE Afghanistan (Flynn et al., 2003), and the first non-Asian record (*Abudhabia* sp.) was documented by Winkler (2003) from the late Miocene of N Kenya. Rather than considering *Abudhabia* to be directly related to *Tatera*, Flynn et al. (2003:279) suggested ". . . that *Abudhabia* is a taterilline related to *Taterillus*, *Gerbilliscus*, and *Taterona* in Africa, plus *Tatera* in Asia. . . ," a hypothesis testable ". . . through a survey of osteological features across the group, a basis for phylogenetic analysis enriched by the fossil record." *Abudhabia* is significant not only to understanding evolutionary origin of taterillines but Gerbillinae in general "because its dental characteristics are intermediate between the Myocricetodontinae and the Gerbillinae" (de Bruijn and Whybrow, 1994:414). Detailed comparisons between the Afghanistan Pliocene *Abudhabia radinskyi* and modern *Tatera indica* provided by Flynn et al. (2003).

Tatera indica (Hardwicke, 1807). Trans. Linn. Soc. Lond., 8:279.

COMMON NAME: Indian Gerbil.

TYPE LOCALITY: India, United Province, between Benares and Hardwar.

DISTRIBUTION: An extensive range from SE Anatolia in Turkey (Yiğit et al., 2001), Syria, Iraq, and Kuwait through Iran, Afghanistan, and Pakistan into most of Indian Peninsula north to the Terai region of S Nepal; also Sri Lanka (see Bates, 1988).

STATUS: IUCN – Lower Risk (lc).

SYNONYMS: *bailwardi* Wroughton, 1906; *ceylonica* Wroughton, 1906; *cuvieri* (Waterhouse, 1838); *dunni* Wroughton, 1917; *hardwickei* (Gray, 1843); *monticola* Wroughton, 1906; *otarius* (Cuvier, 1838); *persica* Wroughton, 1906; *pitmani* Cheesman, 1921; *scansa* Wroughton, 1906; *sherrini* Wroughton, 1917; *taeniurus* (Wagner, 1843).

COMMENTS: Revised by Bates (1988), who recognized three distinctive subspecies. Regional reviews include the segments from Arabian Peninsula (Harrison and Bates, 1991), SE Turkey (Yiğit et al., 2001), Syria (Misonne, 1957), Iran (Lay, 1967; Morshed and Patton, 2002), Afghanistan (Hassinger, 1973), Pakistan (Roberts, 1977), and India (Agrawal, 2000; Chakraborty and Agrawal, 2000). This is the only species that Pavlinov (1981*b*) and Pavlinov et al. (1990) allocated to *Tatera*. Karyotype and chromosomal polymorphism of Indian samples documented by Rao et al. (1968) and Yosida (1981); chromosomal data summarized by Qumsiyeh and Schlitter (1991), and karyotype of Turkish population (2n = 68, FN = 84) documented by Yiğit et al. (2001). Ecology and distribution in the Aravalli Ranges in Rajasthan, India documented by Prakash et al. (1995*a*, *b*). Late Pleistocene fossils from C India described as *T.* cf *indica* by Patnaik (1995).

Taterillus Thomas, 1910. Ann. Mag. Nat. Hist., ser. 8, 6:222.

TYPE SPECIES: *Gerbillus emini* Thomas, 1892.

SYNONYMS: *Taterina* Wettstein, 1916.

COMMENTS: Tribe Taterillini, Subtribe Taterillina. Robbins (1971) proposed a new dental terminology for the genus based on a large sample of *T. gracilis*, analysed (1973) nongeographic variation drawn from the same sample, and summarized (1977) morphometric and chomosomal differentiation among the seven species that he considered valid (only *T. petteri* and *T. tranieri* have since been added). Chromosomal data were reported for some species by Matthey and F. Petter (1970) and karyotypic information was summarized by Qumsiyeh and Schlitter (1991). Like many genera of African muroid rodents, *Taterillus* requires critical systematic revision to determine species definitions and their distributional limits. Several of the species now recognized are morphologically closely similar to one another, prompting many workers to consider them "sibling" or cryptic species (Dobigny et al., 2002*a*, 2003; Sicard et al., 1988). Recently, Volobouev and Granjon (1996:47) documented a sex-chromosome system in West African *T. arenarius*, summarized all chromosome data for the species of *Taterillus* (and provided references), and noted that "These data clearly show that karyotype evolution in *Taterillus* is still continuing. Taking into account the apparently frequent occurrence of tandem translocations and pericentric inversions in the evolution of *Taterillus* and the important role of these rearrangements in the establishment of reproductive isolation . . ., the existence of other chromosomal races and species is extremely likely." This prediction was recently realized by discovery of two new cytotypes ("new cryptic species"), each reproductively isolated from one another, in samples from the Lake Chad area in W Chad and SE Niger (Dobigny et al., 2002*a*), and a new species (*T. tranieri*) from Mali and Mauritania diagnosed primarily by karyotype (Dobigny et al., 2003). Apparently the species of *Taterillus* in West Africa (*T. arenarius*, *T. gracilis*, *T. petteri*, *T. pygargus*, and *T. tranieri*) comprise a monophyletic complex of sibling species defined by the same autosome-gonosome translocation, which has not been seen in species from Central and East Africa (Dobigny et al., 2003; Volobouev and Granjon, 1996). Dobigny et al. (2003:300) use "karyospecies" to describe these "cryptic" species "that can only be unambiguously characterized as good biological species by means of their karyotype." Evolutionary history as documented by fossils extends back to late Pliocene of subsaharan Africa (Denys, 1987*a*, 1989*b*, 1999; see review by Wessels, 1998).

Taterillus arenarius Robbins, 1974. Proc. Biol. Soc. Wash., 87:399.

COMMON NAME: Robbins's Tateril.

TYPE LOCALITY: Mauritania, Trarza Region, Tiguent.

DISTRIBUTION: N Sahelian savanna and subdesert from Mauritania through Mali to Niger (see Robbins, 1974); eastern limits unknown although Sicard et al. (1988) believed the species to be confined to the left bank of the Niger River.

STATUS: IUCN – Lower Risk (lc).

COMMENTS: In a detailed multivariate analysis, Robbins (1974) compared his new species with samples of *T. gracilis* and *T. pygargus*. The latter and *T. arenarius* are sympatric in S Mauritania. Results from a chromosome banding study (R- and C-bands) of *T. arenarius* as reported by Volobouev and Granjon (1996:45), "revealed the presence of an XX/XY$_1$Y$_2$ sex-chromosome system in the karyotype [2n = 30/31, FNa = 36], as found previously in three other congeneric species." Ecology and membership in small mammal community of SW Mauritania discussed by Granjon et al. (2002*b*).

Taterillus congicus Thomas, 1915. Ann. Mag. Nat. Hist., ser. 8, 16:147.

COMMON NAME: Congo Tateril.

TYPE LOCALITY: Dem. Rep. Congo, Orientale, Upper Uele River (Welle), Poko.

DISTRIBUTION: Cameroon, Chad, Central African Republic, Dem. Rep. Congo, Sudan, Uganda.

STATUS: IUCN – Lower Risk (lc).

SYNONYMS: *clivosus* Thomas and Hinton, 1923.

COMMENTS: Assignment of *clivosus* follows Robbins (1977).

Taterillus emini (Thomas, 1892). Ann. Mag. Nat. Hist., ser. 6, 9:78.

COMMON NAME: Emin's Tateril.

TYPE LOCALITY: Uganda, Wadelai.

DISTRIBUTION: Sudan, SW Ethiopia (Yalden et al., 1996), Uganda, NW Kenya, NE Dem. Rep. Congo.

STATUS: IUCN – Lower Risk (lc).

SYNONYMS: *anthonyi* Hatt, 1934; *butleri* Wroughton, 1910; *gyas* Thomas, 1918.

COMMENTS: Allocation of the synonyms to this species follows Robbins (1977).

Taterillus gracilis (Thomas, 1892). Ann. Mag. Nat. Hist., ser. 6, 9:77.

COMMON NAME: Gracile Tateril.

TYPE LOCALITY: Gambia (see Robbins, 1974).

DISTRIBUTION: N Nigeria, Niger, S Mali, and Burkina Faso to Gambia and Senegal (see Dobigny et al., 2002*b*; Duplantier and Granjon, 1992; Grubb et al., 1998; Meinig, 2000).

STATUS: IUCN – Lower Risk (lc).

SYNONYMS: *angelus* Thomas and Hinton, 1920; *meridionalis* Robbins, 1974; *nigeriae* Thomas, 1911.

COMMENTS: Chromosomal data (2n = 36/37, FNa = 42-44) for samples from Burkino Faso and Senegal were reported by Gautun et al. (1985) and Volobouev and Granjon (1996); Dobigny et al. (2002*b*) recorded a slightly greater range in diploid and fundamental numbers in Niger samples. Rosevear (1969) included *angelus* in this species, but Sicard et al. (1988) discussed its possible specific status. The form *nigeriae* was provisionally considered valid by Rosevear (1969) but was treated as a subspecies of *T. gracilis* by Robbins (1974).

Taterillus harringtoni (Thomas, 1906). Ann. Mag. Nat. Hist., ser. 7, 18:303.

COMMON NAME: Harrington's Tateril.

TYPE LOCALITY: Ethiopia, east of Lake Turkana (Rudolf), near Mutti Galeb.

DISTRIBUTION: Central African Republic, Sudan, Ethiopia, Somalia, E Uganda, Kenya, and Tanzania.

STATUS: IUCN – Lower Risk (lc).

SYNONYMS: *illustris* Dollman, 1911; *kadugliensis* (Wettstein, 1916); *lorenzi* (Wettstein, 1916); *lowei* Dollman, 1914; *melanops* Allen, 1912; *meneghettii* Toschi, 1946; *nubilus* Dollman, 1911; *osgoodi* Wroughton, 1910; *perluteus* Thomas and Hinton, 1923; *rufus* (Wettstein, 1916); *tenebricus* Dollman, 1911; *zammarani* de Beaux, 1922.

COMMENTS: Allocation of synonyms follows Robbins (1977). Karyotypes of samples from Ethiopia and Kenya (2n = 44, FNa = 62, 64) summarized by Volobouev and Granjon (1996).

Taterillus lacustris (Thomas and Wroughton, 1907). Ann. Mag. Nat. Hist., ser. 7, 19:37.

COMMON NAME: Lake Chad Tateril.

TYPE LOCALITY: Nigeria, Lake Chad (= Kaddai).

DISTRIBUTION: NE Nigeria and Cameroon.

STATUS: IUCN – Lower Risk (lc).

COMMENTS: Rosevear (1969) synonymized *lacustris* with *T. gracilis*, but both Robbins (1974, 1977) and F. Petter (1975*b*) treated it as a distinct species. Analysis of serum proteins pointed to a closer relationship of *T. lacustris* to *T. pygargus* rather than *T. gracilis* (Tranier et al., 1974).

Taterillus petteri Sicard, Tranier, and Gautun, 1988. Mammalia, 52:188.

COMMON NAME: Petter's Tateril.

TYPE LOCALITY: Burkina Faso (Upper Volta), Oudalan Province, near Oursi Pond; 14°38'N, 00°26'W (Sicard et al., 1988:188).

DISTRIBUTION: Sahelian savanna of E Burkina Faso and W Niger west of Niger River.

STATUS: IUCN – Lower Risk (lc).

COMMENTS: Ansell (1989) maintained the availability of *petteri* should properly date from its first checklist citation (Gautun et al., 1985), not from its later formal description (Sicard et al., 1988), but Sicard et al. identified the holotype and museum in which it is stored and provided a differential diagnosis. According to Sicard et al. (1988), *T. petteri* is confined to the loop of the Niger River and is parapatric with *T. gracilis*, from which it differs in morphology, ecology, biochemistry, and physiology. In the same paper, they stated that *T. petteri* may actually be conspecific with *angelus* (from Gambia), although they acknowledged not examining the holotype of the latter, and they speculated that *angelus* and *lacustris* may prove to be the same but, because the type of the former is young and that of the latter is old, the relationship is difficult to demonstrate. Karyotype (2n = 18/19, FNa = 28) is distinctive compared with some other West African *Taterillus*; closest karyological relative is apparently *T. tranieri* (Dobigny et al., 2003).

Taterillus pygargus (F. Cuvier, 1838). Trans. Zool. Soc. Lond., 2:142.

COMMON NAME: Senegal Tateril.

TYPE LOCALITY: Senegal, probably St. Louis (as suggested by Robbins, 1977:191).

DISTRIBUTION: Gambia, Senegal, S Mauritania, W and S Mali (Meinig, 2000), and S Niger (Dobigny et al., 2002*b*).

STATUS: IUCN – Lower Risk (lc).

COMMENTS: Cuvier's *pygargus* was included in *Gerbillus pyramidum*, but the holotype is an example of *Taterillus* (see F. Petter et al., 1972, and F. Petter, 1975*b*, and references therein). Apparently unaware of F. Petter's observation, Lay (1983) listed *pygargus* as a species of *Gerbillus* known only from the type locality, which he thought was Egypt. Karyotype (2n = 22/23, FNa = 38-40) described in detail and contrasted with two newly described cytotypes representing two species (2n = 22/23, FN = 40; 2n = 24/25, FN = 44; identified as *Taterillus* sp. 1 and *Taterillus* sp. 2) by Dobigny et al. (2002*a*). Additional chromosomal information for Niger samples of *T. pygargus* provided by Dobigny et al. (2002*b*).

Taterillus tranieri Dobigny, Granjon, Aniskin, Ba, and Volobouev. 2003 Mammal. Biol., 68:301.

COMMON NAME: Tranieri's Tateril.

TYPE LOCALITY: W Mali, Dilly (15°01'129"N, 07°40'084"W); see Dobigny et al. (2003) for details.

DISTRIBUTION: Recorded from Sahelian savanna of W Mali and SE Mauritania (see Dobigny et al., 2003, for locality coordinates and habitat).

COMMENTS: Karyotype (2n = 14/15, FNa = 22-24) exhibits the lowest diploid number among gerbillines and differs from its chromosomally closest ally, *T. petteri* (2n = 18/19, FNa = 28), by "two telomere-telomere translocations accompanied by two centromere activations/inactivations, one non-reciprocal translocation, and three pericentric inversion" (Dobigny et al., 2003:309). No consistent qualitative differences in external and cranial morphological or pelage coloration could be dectected between *T. tranieri* and four other West African species: *T. arenarius*, *T. gracilis*, *T. petteri*, and *T. pygargus*.

Subfamily Leimacomyinae new subfamily.

COMMENTS: Type genus–*Leimacomys* Matschie, 1893. Definition–medium-sized terrestrial, insectivorous muroid with a scantily haired (caudal hairs length of one tail scale), tapering tail much shorter than length of head and body (TL/HBL = 32%), four digits and rudimentary fifth on front foot, five on hind foot, claws nearly straight, and dense pelage without guard hairs; cranium robust, rostrum moderately long and wide, and tapering distally in lateral perspective; interorbit wide; prominent supraorbital ridges border wide wedge-shaped postorbital region merging with wide braincase outlined by weak temporal ridging; occiput squarish, interparietal wide and narrow; broad zygomatic plate with prominent anteriorly projecting spine (deep zygomatic notch), masseteric tubercle weakly developed; zygomatic arches stout and not bowed outward; braincase moderately deep, with convex dorsal outline, and overhangs occipital condyles; expansive parietals form dorsal sides of braincase along with squamosal; incisive foramina narrow and very long, extending to lingual roots of first molar and close to posterior palatine foramina; bony palate wide, and posterior to third molar merges with adjacent pterygoids to form shelf roofing more than half of mesopterygoid fossa (extending almost to basisphenoid-presphenoid suture); auditory bullae small; dentary with high ramus, elongate coronoid process, deep emargination between condyloid and angular processes, incisor alveolus ending in prominent knob anterior to root of coronoid process, not extending into condyloid process; upper incisors slender, proodont, shallowly grooved on anterior enamel faces, those surfaces oriented as the arc of a circle (closely similar to form in *Tateomys*; see illustration in Musser and Durden, 2002); lower incisors very slender; molars with primitive number of roots; upper molar rows diverging anteriorly; upper molars with simple laminae, most formed by broad coalesence of two cusps (outlines of individual cusps on anterocone and posterior lamina of M1, and posterior lamina of M2 not evident), second lamina on M1 and anterior lamina on M2 chevron-shaped and formed by three broadly coalesced cusps (a cingular lingual cusp, protocone, and paracone); posterior cingulum, cusp t7, cingular cusplets, anterolabial cusp on M2 and M3, longitudinal grooves separating cusp rows, and mures (longitudinal crests) absent from upper molars; anterolabial and posterolabial cusplets, and posterior cingula absent from lower molars; anteroconid broadly fused with chevron-shaped anterior lamina; M3 and m3 moderately reduced compared to first and second molars (Misonne, 1966; Rosevear, 1969; Dieterlen, 1976a; Denys et al., 1995; Denys, 1993). Contents–only the type genus.

Compared with *Steatomys* and *Lophuromys* by Matschie (1893) who placed *Leimacomys* in Dendromurinae, kept there by G. M. Allen (1939), but transferred to Murinae by Ellerman (1940) and Simpson (1945), and subsequently regarded as an enigmatic dendromurine (Carleton and Musser, 1984; Rosevear, 1969). Conformation of body and skull is murine or deomyine, not dendromurine; cranial configuration is strikingly similar to the deomyine *Lophuromys*, but the resemblance has been interpreted as convergent, not phylogenetic, reflecting similar terrestrial and insectivorous habits (Dieterlen, 1976a). Nothing about *Leimacomys*'s external and cranial morphology suggests affinity with dendromurines (Misonne, 1966; Rosevear, 1969); only molar occlusal patterns, which have been interpreted as resembling those of *Steatomys* (Dieterlen, 1976a; Misonne, 1966; presence of a lingual cusp attached to second row of cusps on M1 and anterior cusp row on M2), have kept *Leimacomys* in Dendromurinae (Denys, 1993). Phylogenetic analysis of primarily dental traits incorporating dendromurines, deomyines, cricetomyines, petromyscines, gerbillines, *Mystromys* and *Macrotarsomys* by Denys et al. (1995) placed *Leimacomys* in a clade with gerbils and *Mystromys* that was isolated from the dendromurine clade containing *Dendromus*, *Steatomys*, *Malacothrix*, and *Megadendromus*, and suggests that the lingual cusp is independently derived in *Leimacomys*. The affinity with gerbils and *Mystromys* reflects shared simple molar occlusal patterns, but details of the pattern (lingual cusp present, longitudinal crests absent) differ and external and cranial traits of *Leimacomys* are unlike either gerbils or *Mystromys*. Phylogenetically isolated from living dendromurines, murines, gerbillines, and other muroid groups, "*Leimacomys* pourrait représenter le dernier élément vivant d'une lignée de rongeurs Muroidea isolée depuis longtemps et endémique du bloc forestier de l'Afrique occidentale" (Denys, 1993:617).

Leimacomys Matschie, 1893. Sitzb. Ges. Naturf. Fr. Berlin, p. 107.

TYPE SPECIES: *Leimacomys büttneri* Matschie, 1893.

SYNONYMS: *Limacomys* Lydekker, 1894 (see Rosevear, 1969:487).

COMMENTS: Rosevear (1969) provided a particularly good summary of past taxonomic allocations of *Leimacomys* and the nature of the types, which consist of an imperfect skin and a specimen preserved in fluid; only one cranium and mandible have been located, but the different wear on upper and lower molars indicates the two elements to be mismatched.

Leimacomys büttneri Matschie, 1893. Sitzb. Ges. Naturf. Fr. Berlin, p. 109.

COMMON NAME: Büttner's African Forest Mouse.

TYPE LOCALITY: Togo, Bismarckburg, 710 m (see Denys, 1993; Misonne, 1966; and Rosevear, 1969 for additional data).

DISTRIBUTION: Recorded only from the type locality; may also occur in adjacent high forest of Ghana (Grubb et al., 1998).

STATUS: IUCN – Data Deficient.

COMMENTS: Still known only by two specimens obtained in 1890 (Dieterlen, 1976*a*; Misonne, 1966; Rosevear, 1969). The species has not been collected for nearly a century and has been either declared extinct, regarded as critically endangered, or believed to still exist (see discussion and references in Grubb et al., 1998:189). Recent mammal surveys of Togo have not included forest habitats (Grubb et al., 1998) and so are uninformative about the present existence of *L. buettneri* in that country. The type locality is in high forest and the species is insectivorous (Dieterlen, 1976*a*); otherwise, its ecology and actual geographic distribution are a mystery. Insectivorous muroids are notoriously difficult to capture with usual kinds of traps and require pitfall and other techniques. Focused survey of long duration in forests of Togo and nearby in Ghana should be conducted to assess the status of this unique rodent.

Subfamily Murinae Illiger, 1811. Abh. Phys. Klasse K.-Preuss Akad. Wiss. Berlin, for 1804-11, p. 46, 129 [1811].

SYNONYMS: Anisomyes Ellerman, 1941 (Anisomyini Lidicker and Brylski, 1987); Arvicanthini Ducroz, Volobouev, and Granjon, 2001 [*nomen nudum*]; Coniluridae Dahl, 1897 (Conilurini Lee, Baverstock, and Watts, 1981); Hydromyina Gray, 1825 (Hydromyinae Alston, 1876; Hydromyes Winge, 1887; Hydromyini Lee, Baverstock and Watts, 1981); Murina Illiger, 1811 (Murina Gray, 1825; Murinae Murray, 1866; Murini Winge, 1887; Mures Winge, 1887); Phloeomyinae Alston, 1876 (Phloeomyini Tullberg, 1899); Pseudomyinae Simpson, 1961; Rattidae Burnett, 1830; Uromyini Lee, Baverstock, and Watts, 1981; Rhynchomyinae Thomas, 1897.

COMMENTS: Diagnosis employing molar and cranial traits, and enumeration of external, cranial, postcranial, dental, reproductive, and arterial characteristics presented by Carleton and Musser (1984). Contents of subfamily generally as presented by them except that *Acomys*, *Lophuromys*, and *Uranomys*, formerly considered murines (e.g., Carleton and Musser, 1984; Ellerman, 1941; Misonne, 1969; Musser and Carleton, 1993), are excluded and treated under subfamily Deomyinae. Murinae is characterized by a cohesive cluster of external, cranial, postcranial, dental, reproductive, and arterial characteristics (Carleton and Musser, 1984), but derived molar conditions form the primary basis for defining the subfamily. Two neomorphic cusps, the anterostyle (t1) and enterostyle (t4), are present on the lingual border of M1 and form two chevron-shaped, transverse lamina; both upper and lower molars lack longitudinal enamel crests (mures/ids) between lamina; and cusps on the lower molars are positioned opposite one another (Flynn et al., 1985; Jacobs et al., 1989; Freudenthal and Martin Suarez, 1999). Other derived cranial features include their modified carotid circulatory pattern (sphenofrontal foramen and squamosal-alisphenoid groove absent; stapedial foramen present) and a reduced tegmen tympani that does not contact the posterior squamosal (Bugge 1970; Carleton and Musser, 1984; our unpublished survey). Cladistic integrity of Murinae is also supported by results of mitochondrial and nuclear gene sequences, although generic sampling is still limited and some inquiries have used only *Mus* and *Rattus* to represent this huge subfamily (Adkins et al., 2001, 2003; Catzeflis et al., 1995; Conroy and Cook, 1999; DeBry and Sagel, 2001; Dubois et al., 1996,

1999; Fieldhouse et al., 1997; Furano et al., 1994; Graur, 1994; Hänni et al., 1995; Huchon et al., 1999; Jansa et al., 1999; Jansa and Weksler, 2004; Kass et al., 1997; Larizza et al., 2002; Martin et al., 2000; Michaux and Catzeflis, 2000; Michaux et al., 2001; Montgelard et al., 2002; Pascale et al., 1990; Robinson et al., 1997; Usdin et al., 1995; E. Verheyen et al., 1995; Verneau et al., 1997, 1998).

No general tribal arrangement of genera is available except for Australian and New Guinea groupings (Conilurini and Hydromyini for Australian species, Watts and Aslin, 1981; Hydromyini, Watts and Baverstock, 1994a, which includes Conilurini and Hydromyini of Watts and Aslin, 1981; and Anisomyini for some New Guinea genera, Lidicker and Brylski, 1987, Watts and Baverstock, 1994a), and African Arvicanthini (Ducroz et al., 2001). Attempts to reflect intergeneric relationships have been made by some muroid specialists and checklist compilers who separated genera into formal and informal categories. Alston used Hydromyinae, Phloeomyinae, and Murinae (although he included what are now non-murine forms) for genera we place in Murinae. Tullberg (1899) classified genera into Murini and Phloeomyini, and also included Otomyini within Muridae. Thomas (1896) listed Hydromyinae, Rhynchomyinae, and Phloeomyinae as coequal with Murinae in a larger Muridae. Miller and Gidley (1918) also recognized Hydromyinae, Phloeomyinae, and Murinae. Tate (1936:505) allocated murines to three groups: 1) those with "simple *Rattus*-like molar teeth," generally the Murinae of Thomas (1896); 2) genera with "complexly folded molars," the Phloeomyinae; and 3) murines "with specialized multi-rooted molars having basin-like depressions with raised edges, a definite tendency for non-development or loss of the third molars, and a trend in the direction of an aquatic habitus," the Hydromyinae. Ellerman (1941) recognized four clusters: 1) *Anisomys* Group (Anisomyes), containing only the New Guinea *Anisomys*, a cluster Ellerman thought might deserve subfamily rank; 2) *Mures* Group, containing most murine genera; 3) Rhynchomyinae for the Philippine *Rhynchomys*; and 4) the Hydromyinae, which resembled Tate's group. Simpson (1945) listed genera in his Muridae under Murinae, Phloeomyinae, Rhynchomyinae, and Hydromyinae (and also included Dendromurinae and Otomyinae). Misonne (1969) sorted all genera into four Divisions with subgroups: *Lenothrix-Parapodemus* Division containing the *Lenothrix* and *Parapodemus* groups; 2) *Arvicanthis* Division; 3) *Rattus* Division, which contains the *Praomys*, *Maxomys*, *Rattus*, *Uromys*, and *Mus* groups; and 4) Basin shaped molars Division, composed of Hydromyinae, Rhynchomyinae, and a third group composed of *Echiothrix*, *Macruromys*, and *Crunomys*. Chaline et al. (1977) allocated genera to either Murinae or Hydromyinae. The checklist by Pavlinov et al. (1995a) segregated genera into primary groups containing sections: 1) *Micromys* Group, with *Pithecheir*, *Batomys*, and *Micromys* Sections; 2) *Apodemus* Group, containing the *Apodemus*, *Tokudaia*, and *Mus* Sections; 3) *Arvicanthis* Group; 4) *Rattus* Group, composed of *Rattus*, *Dacnomys*, *Praomys*, and *Uromys* Sections; 5) *Phloeomys* Group; 6) *Rhynchomys* Group, which includes the *Chrotomys* and *Rhynchomys* Sections; and 7) *Incertae sedis*, composed of the *Crunomys*, *Nesokia*, *Melasmothrix*, and *Hydromys* Sections. We have borrowed Misonne's (1960) primary category and arranged genera in 29 Divisions, each presumably monophyletic (Table 1). A few are defined by published diagnostic morphological and/or molecular traits and could be treated as formal tribes; contents of others derive from our unpublished research. Placement of some genera is provisional and integrity of those clusters require testing with a greater range of morphological and molecular data than is now available. Each of seven Divisions contains only a single genus because no reliable information is presently available that would support allocation to a larger group; some of these genera may have no close living phylogenetic relatives. Reasons supporting our dispositions are explained in the generic accounts. Future research results may indicate some genera be reallocated, and some of our smaller Divisions be combined with others to form larger monophyletic clades. Various phylogenetic analyses of mtDNA cytochrome *b*, 12S and 16S rRNA sequences, as well as nuclear IRBP sequences, for example, associate the African *Aethomys*, *Micaelamys*, *Grammomys*, *Hybomys*, *Dasymys*, *Lemniscomys*, *Rhabdomys*, *Desmomys*, *Pelomys*, *Mylomys*, and *Arvicanthis* either in a single clade or in smaller clusters within a large African clade (Castiglia et al., 2003; Ducroz et al., 2001; Lecompte, 2003), which would consolidate our *Aethomys*, *Arvicanthis*, *Dasymys*, and *Hybomys* Divisions. Results from albumin immunology suggested that those murine genera sampled form large monophyletic clades reflecting geographic origins–New Guinean, Australasian, Southeast Asian, and African (Watts and Baverstock, 1995b, 1996). DNA/DNA hybridization experiments (Chevret, 1994), analyses of mtDNA complete cytochrome *b* sequences (Lecompte, 2003; Lecompte et al., 2002b) and combination of cytochrome *b* and nuclear IRBP gene sequences (Lecompte, 2003), by contrast, indicate the African murines to be paraphyletic and reflect at least four separate colonization events.

Comparative chromosomal data provided in phylogenetic framework and other contexts for European species (Zima and Kral, 1984a), some Asian groups (Cao and Tran, 1984; Gadi and Sharma, 1983; Raman and Sharma, 1977; Rickart and Musser, 1993), some African species (Robbins and Baker, 1978), and murines in general (Viegas-Péquignot et al., 1983, 1985, 1986). Phylogenetic relationships among murines inferred from biochemical sources include the studies of allozymic variation (Bonhomme et al., 1985; Iskandar and Bonhomme, 1984), microcomplement fixation of albumin (Watts and Baverstock, 1995b, 1996), DNA-DNA hybridization experiments (Catzeflis, 1990; Catzeflis et al., 1987), amplification of ancient murine Lx family of long interspersed repeated DNA (L1, Line-1) and L1 retrotransposons (Furano et al., 1994; Pascale et al., 1990; Usdin et al., 1995; Verneau et al., 1997, 1998), and several mitochondrial and nuclear gene sequence studies (see above citations and those in subfamily accounts). Morphological bases for phylogenetic interpretation provided by relationship between body mass, testes mass, and sperm size in murines (Breed and Taylor, 2000); variation in sperm head morphology (Breed, 1991, 2004); comparative hair morphology (Keogh, 1985), soft palate topography (Eisentraut, 1969a), and digestive system anatomy (Perrin and Curtis, 1980). The discrepancy between the relatively rapid divergence of murine taxa as revealed by fossils and the much slower rates indicated by molecular data is reconciled by Jaeger et al. (1986) by postulating accelerated rates of evolution for certain proteins and a higher rate of nucleotide substitution in murines than ordinarily seen in other eutherians.

Taxonomy and geographic distributions of African murines are embedded in a vast literature. Phylogenetic affinities among African taxa have been based on DNA/DNA hybridization experiments (Chevret, 1994), microcomplement fixation of albumin (Watts and Baverstock, 1995a), spermatozoal morphology (Baskevich and Lavrenchenko, 1995; Breed, 1995a, d), and all the chromosomal, molecular, morphological, and morphometric data referenced in the accounts of African species. Watts and Baverstock (1995a:431) viewed the evolutionary picture of African murines in the framework of "a period of rapid radiation from a single ancestor, beginning 8-10 million years ago and still continuing. As a result the systematics of these groups is confusing and will take the concerted efforts of workers in many different disciplines to clarify." Based on mtDNA cytochrome *b* and 12S and 16S rRNA sequences, Ducroz et al. (2001) identified an African murine clade (containing *Aethomys*, *Grammomys*, *Hybomys*, *Dasymys*, *Lemniscomys*, *Desmomys*, *Rhabdomys*, *Pelomys*, *Mylomys*, and *Arvicanthis*), which they isolated as Arvicanthini, and which partly accords with the evolutionary reconstruction of Watts and Baverstock. Chevret's (1994) results from DNA/DNA hybridization experiments, however, indicate African murines to be paraphyletic, as does Lecompte's (2003) analyses of combined mtDNA cytochrome *b* and nuclear IRBP sequences. Lecompte also recovered an arvicanthine group, and identified three other clades: *Praomys* group (*Myomyscus*, *Stenocephalemys*, *Praomys*, *Mastomys*, *Heimyscus*, *Hylomyscus*, *Zelotomys*, and *Colomys*), a *Mus* (*Nannomys*) cluster, and a clade containing only *Malacomys*.

Late Miocene is the earliest that murines are recorded from African strata (Geraads, 2001; Jaeger, 1977b; Mein et al., 1993, 2004; Winkler, 2003). Whether African murines were derived from a single or several ancestral groups or originated by one or more immigrations cannot be resolved by the meager number of Miocene taxa recovered from African sediments. Southern Asia is vaguely indicated as the source area, with arrivals in Africa occurring in middle to late Miocene-early Pliocene (Jacobs, 1985; Winkler, 1994, 2002). Late Miocene is the earliest documentation for *Progonomys* and *Paraethomys* in North Africa (Jaeger, 1977b; Mein et al., 1993) and *Progonomys* and *Saidomys* in Kenya (Winkler, 2003; *Karnimata* was used, but that is a synonym of *Progonomys*, according to Mein et al., 1993). Murines from earlier Miocene beds have been found only in N Pakistan, and among them is the middle Miocene *Antemus*, currently regarded as the earliest murine (Freudenthal and Martin Suarez, 1999; Jacobs and Downs, 1994). Fossils of living endemic African murines first appear in the late Miocene of Namibia and Pliocene elsewhere on the continent (Denys, 1999; Mein et al., 2004; Winkler, 2002; see generic accounts). *Mastomys* and *Arvicanthis*, now mostly occurring in Africa, are each represented by an extinct species that lived in Israel during the Pleistocene (Tchernov, 1968, 1996), and *Pelomys* by a Pliocene species on the Mediterranean island of Rhodes (de Bruijn et al., 1996). See Denys (1999) for a review of fossil African murine faunas and past habitats in relationship to the modern fauna and distributions.

All native New Guinea and Australian rodents are members of Murinae. Classical compendia based upon morphological and distributional studies for murines of New Guinea and adjacent archipelagos remain those of Tate (1951) and Laurie and Hill (1954), which have been largely supplanted by current faunal treatises (Flannery, 1995*a*, *b*) and taxonomic reports cited in the various generic and species accounts. Reports relevant to understanding phylogenetic relationships among New Guinea endemics documented comparative phallic morphology (Lidicker, 1968), chromosomal information (Donnellan, 1987), variation in immunological distances assessed by microcomplement fixation of albumin (Watts and Baverstock, 1994*a*), and sperm morphology (Breed and Aplin, 1994). Evolution of the spermatozoon in New Guinea and Australian murines combined is treated by Breed (1997), who compared the range in sperm head morphology with the phylogenetic frameworks constructed by Watts et al. (1992) for Australian murines and Watts and Baverstock (1994*a*) for New Guinea species based on analyses of albumin immunology. Swann et al. (2002) explored the cDNA nucleotide sequence encoding the ZPC protein of selected species of *Pseudomys*, *Notomys*, and *Hydromys*, and its significance. Albumin immunological evidence derived from New Guinea marsupials was used to determine time and mode of origin of the terrestrial fauna and is pertinent to inquiries into historical origin of the endemic murines (Aplin et al., 1993). Aplin (2005) has reviewed Australian-New Guinea murine diversity, phylogeny and biogeography.

Earliest Australian record of murines consists of more than a dozen species recovered from Pliocene strata, among them representatives of *Leggadina*, *Pseudomys*, and *Zyzomys* (Aplin, 2005; Godthelp, 1990; Rich et al., 1991). Such a high diversity of fossil taxa suggests arrival of murines in Australia either before the Pliocene, possibly 7 mya (Godthelp, 1990) or later, only about 5 mya (early Pliocene; Aplin, 2005; K. Aplin, in litt., 2005). That murines are immigrants to Australia is indicated by the lack of fossils before early Pliocene times (Aplin, 2005; Pledge, 1992:140; Rich, 1991; Rich et al., 1991). In the rich assemblage of local faunas from the Riversleigh region in NW Queensland, for example, in which fossils have been retrieved from late Oligocene to Pleistocene strata, earliest murines were recovered only from late Pliocene sediments, about 3 million years old (Aplin, 2005; Archer et al., 1991; Rich et al., 1991). Possible earliest record of murines on New Guinea still consists only of a sliver of incisor enamel from the Otibanda Formation, 3.0-3.3 myo (late Pliocene; Aplin, 2005; Flannery, 1995*a*; Pledge, 1992; Rich et al, 1991).

Phylogenetic relationships among Australian murines and between them and non-Australian species have been inferred from chromosomal data (Baverstock et al., 1977*c-e*, 1983*a*, *b*), microcomplement fixation of albumin to measure immunological distances (Baverstock et al., 1977*a*, *b*, 1980; Watts et al., 1992, Watts and Baverstock, 1994*a*, *b*; 1995*b*), and spermatozoal and male reproductive tract morphology (Breed, 1980, 1983, 1984, 1985, 1986, 1990, 1997, 2000; Breed and Sarafis, 1978, 1983). In addition to the large taxonomic literature (see species accounts), important regional publications summarize their history of discovery, distribution, conservation, and status in South Australia (Robinson et al., 2000, which includes summary of deposits from which subfossils were collected), Victoria (Menkhorst, 1995*a*; Seebeck, 1995*a*; Seebeck and Menkhorst, 2000), New South Wales (Dickman, 1994; Dickman et al., 1993, 2000*a*), Queensland (Dickman et al, 2000*b*), arid Northern Territory (Cole and Woinarski, 2000), monsoonal tropics of Northern Territory (Woinarski, 2000), Western Australia (How et al., 2001; Morris, 2000), Tasmania (Hocking and Driessen, 2000), Cape Range peninsula of NW Australia (Baynes and Jones, 1993), and Australia in general (Strahan, 1995). Other topics covered in general summaries focusing on native murines or including them in broader mammalian surveys are patterns and cases of extinction and decline in murines (Smith and Quin, 1996), extinctions on Australian islands with discussion of causes and conservation implications (Burbidge and Manly, 2002), factors influencing species richness on Australian islands (Burbidge et al., 1997), and how range changes in tropical NE Australia may be attributed to late Quaternary climatic changes (Winter, 1997).

Molecular data indicate that phylogenetic affinities of murines are with gerbils and deomyines. Analyses of the nuclear protein-coding LCAT sequences (Michaux and Catzeflis, 2000), and LCAT in combination with sequences of the von Willebrand Factor (Michaux et al., 2001) clusters gerbils with deomyines, which form a sister-group to murines in a monophyletic clade. This topography is also consistent with results of analyses of complete mtDNA cytochrome *b* (Martin et al., 2000) and nuclear IRBP sequences (Jansa and Weksler, 2004); the Murinae-Gerbillinae link was also revealed by analyses of DNA sequences from the nuclear genes GHR and BRCA1 (Adkins et al., 2003). From their molecular-clock estimate, Michaux et al. (2001) suggested the three subfamilies diverged from the muroid stock 20.8-17.9 million years ago (early Miocene), which is earlier than documented by the first definite murine fossils.

"What is the ancestor of the true Murinae?" (de Bruijn et al., 1996:255) is still unanswered and the subject of active inquiry. The earliest molars currently generally accepted as murine belong to *Antemus chinjiensis* from 13.75 million-year-old Siwalik strata in N Pakistan (Freudenthal and Martin Suárez, 1999; Jacobs and Downs, 1994). Originally described as the earliest and most primitive (dentally) murine or murid (Flynn et al., 1985; Jacobs, 1977, 1978; Jacobs et al., 1989, 1990), *A. chinjiensis* was regarded as a dendromurine by Brandy (1981). Earlier records of isolated molars identified as *A. primitivus* from Pakistan and *A. thailandicus* from Thailand have since been removed to *Potwarmus*, a dendromurine (Lindsay, 1988) or myocricetodontine (Tong and Jaeger, 1993). Jacobs and Downs (1994) recorded a morphological transition in Siwalik stratigraphy from "morphologically certain *Potwarmus*" (14.3 million years ago) to "certain *Antemus*" (*A. chinjiensis*,13.75 million years ago) and from that genus to *Progonomys* or *Progonomys*-like forms between 12.5 and 11.8 million years ago; Siwalik *Mus auctor*, considered the earliest *Mus* by Jacobs (1978), was derived from *Progonomys* at 5.7 million years ago. "The series *Potwarmus-Antemus-Progonomys-Mus* appears to be one long sequence of anagenesis" (Jacobs and Downs, 1994:155; which is an oversimplification considering that several different late Miocene lineages, including one very similar to *Mus*, have been identified in Europe; Mein et al., 1993). However, Freudenthal and Martin Suárez (1999:403) cautioned that "This does not necessarily mean that *Antemus* is derived from *Potwarmus*, and that Muridae are derived from the Dendromurinae, but it may mean that we have come as close as possible to the origin of the Muridae, and that taxons [sic] more primitive than *Antemus* will be arranged in other families than the Muridae." For de Bruijn et al. (1996:253), morphology of M1 in *Antemus*, *Potwarmus* and similar forms (usually considered myocricetodontines) represent an evolutionary grade, and ". . . assignment to subfamily . . . remains more or less a matter of taste at this stage." The cladistic position of *Potwarmus*, or forms with similar dental morphology, may be significant to understanding origin of the Murinae. If that genus is a myocricetodontine and represents a morphological transitory stage between that large Asian and African group and murines, then the strong phylogenetic link between gerbillines and murines indicated by molecular data is reflected in the fossil record because the myocricetodontines are the postulated ancestral group from which gerbils evolved (see Gerbillinae introduction).

While the phylogenetic affinity of *Antemus* is equivocal to some, extinct *Progonomys* or *Progonomys*-like taxa are universally accepted as the earliest forms exhibiting "an essentially modern grade of murine dental morphology" (Jacobs and Downs, 1994:151), and the first murines to have migrated out of Asia (presumably in the Late Miocene) "to arrive in Europe and Anatolia were at the *Progonomys* stage-of-evolution" (de Bruijn et al., 1996:255); *Progonomys*-like morphology has also been identified in late Miocene deposits of North Africa (Jaeger, 1977b; Mein et al., 1993) and East Africa (Winkler, 2003, as *Karnimata*). Mein et al. (1993:62), however, pointed out that most of the early murine samples identified as *Progonomys* in the literature are not that genus but represent a cluster of separate lineages; by the early Vallesian of Europe (late Miocene, 11-10 million years ago), for example, several murine lineages were already present representing *Apodemus*, *Progonomys*, a *Mus*-like form, a *Parapodemus*-like morph, and an unidentified murine, and none "of these lineages has an ancestor-descendant relationship with any of the other ones, but they have a common origin." Fossil links between the diverse living Indo-Australian endemic murines and the late Miocene Siwalik forms have yet to be discovered; the oldest in that region are Pliocene and either too fragmentary to be identified, represent extant genera, or are closely related to them (Chaimanee, 1998; Godthelp, 1990, 1997; Pledge, 1992).

Murine rodents are indigenous to Africa (but not Madagascar), Europe (excluding Ireland), Middle East, Arabian Peninsula, N Eurasia, Indomalayan region, Ceylon, Taiwan, Hainan, Japan, Ryukyu Isls, Philippine Isls, and archipelagos stretching from the Sunda Shelf through Moluccas to the New Guinea area, Australia, and Tasmania. Occurrence of some species outside their natural ranges likely result from intentional or inadvertent anthropogenic processes: an *Apodemus* on Iceland; two *Bandicota* in the Indomalayan region; several *Rattus* and *Mus* in the Indoaustralian region and Pacific Isls; and *Mus musculus*, *Rattus rattus*, and *R. norvegicus* nearly worldwide outside of Asia.

Abditomys Musser, 1982. Am. Mus. Novit., 2730:3.

    TYPE SPECIES: *Rattus latidens* Sanborn, 1952.

    COMMENTS: Rattus Division. Monotypic member of the Philippine New Endemic cluster (Musser and Heaney, 1992).

Abditomys latidens (Sanborn, 1952). Fieldiana Zool., 33:125.

    COMMON NAME: Luzon Broad-toothed Rat

    TYPE LOCALITY: Philippines, Luzon Isl, Mountain Province, Mt Data, 7500 ft (2286 m).

    DISTRIBUTION: Greater Luzon Faunal Region; endemic to C and N Luzon where it is know only by a specimen from Laguna Province and another from Mountain Province (Heaney et al., 1998; Musser, 1982*a*).

    STATUS: IUCN – Lower Risk (lc).

    COMMENTS: A close phylogenetic relative of *Tryphomys adustus*, another Luzon endemic (Musser and Heaney, 1992). Taxonomy and morphological data provided by Musser (1982*a*), who also reviewed the scanty available ecological information. One of the few Indoaustralian murines to bear a nail-like claw on the hallux instead of a claw.

Abeomelomys Menzies, 1990. Science in New Guinea, 16:133.

    TYPE SPECIES: *Melomys sevia* Tate and Archbold, 1935.

    COMMENTS: *Pogonomys* Division. Monotypic member of the New Guinea Old Endemics and a close morphological relative of *Pogonomelomys* (our research). Our allocation of *Abeomelomys* to the *Pogonomys* cluster is provisional pending a wider phylogenetic inquiry into New Guinea murines incorporating molecular and qualitative morphological data. Sperm morphology of *Abeomelomys* is distinctive and unlike that typical of *Uromys* and its allies in our *Uromys* Division (Breed and Aplin, 1994) or *Coccymys* and *Pogonomelomys* with which *Abeomelomys* has been associated in the past (Breed and Aplin, 1994; Musser and Breed, ms).

Abeomelomys sevia (Tate and Archbold, 1935). Am. Mus. Novit., 803:3.

    COMMON NAME: Papuan Abeomelomys.

    TYPE LOCALITY: Papua New Guinea, Morobe Province, Huon Peninsula, Cromwell Range, Sevia, 1400 m.

    DISTRIBUTION: Papua New Guinea; disjunct montane populations in Central Cordillera of Papua New Guinea from Star Mtns (Telefomin area) in the west to Wau region of Morobe Province in the east, and isolated population in the Cromwell Range on Huon Peninsula; 1400-3100 m (Flannery, 1995*a*; Menzies, 1990).

    STATUS: IUCN – Lower Risk (lc) as *Pogonomelomys sevia*.

    SYNONYMS: *tatei* (Hinton, 1943).

    COMMENTS: Originally described as a species of *Melomys* (Tate and Archbold, 1935) and subsequently transferred to *Pogonomelomys* (Rümmler, 1938) where it remained (Laurie and Hill, 1954; Tate, 1951) until *sevia* was made the type species of

Abeomelomys (Menzies, 1990). Morphology of *sevia* has always been considered distinctive compared to *mayeri* and *bruijnii*, the two species of *Pogonomelomys* (Flannery, 1990b; Tate, 1951), but whether it should be separated from that genus remains a question because the traits used by Menzies to diagnose *Abeomelomys* simply duplicated the diagnostic morphological characters of *sevia* and did not identify *Abeomelomys* as a separate monophyletic group in any phylogenetic sense. Furthermore, Menzies (1990:134) distinguished *A. sevia* from the two species of *Pogonomelomys* by only its "grey-based ventral fur and the relatively long incisive foramina." Subsequently, Menzies (1996) also cited Breed and Aplin's (1994) description of spermatozoal morphology of *sevia* as corroborating evidence for its generic status, but Breed and Aplin did not have comparative samples of *Pogonomelomys*, and primitive features characterize spermatozoal morphology of *sevia*. Menzies (1990) did not recognize *tatei* because in his opinion it was indistinguishable from *sevia*, but no data was presented to support this synonymy.

Aethomys Thomas, 1915. Ann. Mag. Nat. Hist., ser. 8, 16:477.

TYPE SPECIES: *Epimys hindei* Thomas, 1902.

COMMENTS: *Aethomys* Division. An isolated member of an apparently monophyletic Subsaharan murine radiation as estimated by albumin immunology (Watts and Baverstock, 1995a), but included in an *Arvicanthis* Division by Misonne (1969), who focused on molar occlusal patterns. Analyses of mtDNA sequences (cytochrome *b* and 12S and 16S gene fragments) placed *Aethomys* as the basal member of an African murine clade containing *Grammomys, Hybomys, Dasymys, Leminiscomys, Rhabdomys, Desmomys, Pelomys, Mylomys*, and *Arvicanthis* (Ducroz et al., 2001). The affinity with *Arvicanthis, Hybomys*, and *Grammomys* is supported by DNA/DNA hybridization (Chevret, 1994). *Aethomys* was originally proposed as subgenus of *Epimys* (=*Rattus*), then elevated to generic rank (see G. M. Allen, 1939:267). The genus was first reviewed by Ellerman (1941), then D. H. S. Davis (1975b), and most recently Chimimba (1998) and Chimimba et al. (1999; also see below), but some species-complexes still require careful systematic revision, particularly taxa extralimital to the Southern African Subregion, to understand species-diversity and their geographic distributions, intraspecific variation, and interspecific phylogenetic relationships. The large-bodied, short-tailed members of the *A. kaiseri* complex (*A. hindei, A. kaiseri, A. thomasi*, and *A. stannarius*), for example, may comprise a monophyletic unit, is considered such by Crawford-Cabral (1998, 1999; as a superspecies) based upon morphology, but integrity of the grouping requires testing by analyses of molecular and other kinds of data.

Micaelamys, diagnosed as a subgenus of *Rattus* by Ellerman (1941) primarily by dental traits, has been traditionally and is currently used as a subgenus for *A. granti* and *A. namaquensis*, with few exceptions (Senut et al., 1992). The two species are set apart from all other *Aethomys* by length of tail relative to body length and pelage coloration (D. H. S. Davis, 1975b), cranial morphology (Chimimba, 1997; Chimimba et al., 1999), dental traits (Ellerman, 1941; Misonne, 1969), chromosomal characteristics (Matthey, 1954, 1958, 1964; Visser and Robinson, 1986), spermatozoal structure and bacular features (Visser and Robinson, 1987). Results from analysis of microcomplement fixation of albumin indicated *A. namaquensis* (subgenus *Micaelamys*) to be farther from *A. chrysophilus* (subgenus *Aethomys*) in immunological distance units than would be expected for congeneric species (Watts and Baverstock, 1995a). These results were reinforced by phylogenetic analyses of mtDNA cytochrome *b*, 12S and 16S ribosomal RNA sequences, which separated *A. namaquensis* and *A. chrysophilus* into separate clades, each associated with species in other genera of the African murines sampled (Ducroz et al., 2001). A more recent phylogenetic analyses of cytochrome *b* sequences by Castiglia et al. (2003b) placed *A. chrysophilus* and *A. kaiseri* in a monophyletic clade near *Grammomys dolichurus* in all trees but excluded *A. namaquensis*, associating it with either *Lemniscomys* or basal to members of an arvicanthine clade. We treat *Micaelamys* (containing *M. granti* and *M. namaquensis*) as a distinct genus: chromosomal, morphological, and molecular data clearly define it as a separate monophyletic cluster that may or may not be closely related to species of *Aethomys* (at least those used in the studies cited above). We also include *Micaelamys* with *Aethomys* in the same Division, but that allocation is provisional. In phylogenetic analyses of mtDNA cytochrome *b* sequences, for example, *Aethomys* (represented by *A. chrysophilus* and *A. kaiseri*) is closest to *Grammomys*, while *Micaelamys* (*M. namaquensis*) joins either *Lemniscomys* or is basal to an arvicanthine clade depending on the analyses (Castiglia et al., 2003*b*; Ducroz et al., 2001). Analyses of combined mtDNA cytochrome *b*, 12S and 16S ribosomal RNA sequences identified *A. chrysophilus* and *G. dolichurus* as a single clade, and placed *M. namaquensis* basal to that clade, a *Dasymys/Hybomys* clade, and an arvicanthine clade (Ducroz et al., 2001). Unequivocal phylogenetic affinity of *Aethomys* and *Micaelamys* within the evolutionary radiation of African murines has yet to be uncovered, and until such resolution we isolate them in their own Division. Phylogenetic analyses employing a broader sampling of not only the speciose *Aethomys* but of endemic African murine genera using molecular, chromosomal, and morphological data sets may resolve the phylogenetic relationships of *Aethomys* and *Micaelamys*. Based on their results from cytochrome *b* sequences, Castiglia et al. (2003*b*) estimated time of divergence between *Aethomys* (sampling *A. chrysophilus* and *A. kaiseri*) and *Grammomys dolichurus* as about 8 million years ago.

Chromosomal information available for various species (which includes those in *Aethomys* and *Micaelamys*: Baker et al., 1988*c*; Castiglia et al., 2003*b*; Matthey, 1954, 1958, 1964; Viegas-Péquignot et al., 1986; Visser and Robinson, 1986). Chimimba and Dippenaar (1994) morphometrically assessed sexual dimorphism and age variation in *A. chrysophilus* and *M. namaquensis* (the latter as *Aethomys*) and subsequently used statistical procedures to select the most discriminating set of morphometric characters to use in systematically revising several species of *Aethomys* (Chimimba and Dippenaar, 1995). Study of the correlation between diet and molar cusp patterns in *A. chrysophilus* and *M. namaquensis* by Denys (1994*a*, the latter as *Aethomys*) allowed her to determine possible diets of extinct Pliocene species. Spermatozoal morphology of several species described by Breed (1995*a*), and used in survey of relationship among body mass, testes mass, and sperm size within murine rodents (Breed and Taylor, 2000). Morphometric and morphological definitions of southern African species, along with a taxonomic synthesis for this group of *Aethomys* (which included those we place in *Micaelamys*), provided by Chimimba (1998) and Chimimba et al. (1999). Assessments of intraspecific variation and its significance in several southern African species produced by Chimimba (2000, 2001*a*, *b*) and Chimimba et al. (1999).

Evolutionary history of *Aethomys* (not *Micaelamy*) as documented by fossils extends back to the Pliocene and Pleistocene of East Africa (Denys, 1987*b*, 1994*a*; Jaeger, 1976; Jaeger and Wesselman, 1976; Wesselman, 1984), early Pliocene-late Pleistocene of South Africa (Avery, 1995, 1998, 2000; Denys, 1990*c*), and late Miocene of Namibia (Senut et al., 1992); the Namibian specimens represent the earliest occurrence of a large-bodied murine in southern Africa; see review by Denys (1999) and Mein et al. (2004).

Aethomys bocagei (Thomas, 1904). Ann. Mag. Nat. Hist., ser. 7, 13:416.
  COMMON NAME: Bocage's Aethomys.
  TYPE LOCALITY: Angola, Pungo Andongo.
  DISTRIBUTION: Recorded only from NW Angola (Crawford-Cabral, 1998) where it occurs in savannas interspersed with forest; geographic boundaries unresolved.
  STATUS: IUCN – Lower Risk (lc).
  COMMENTS: Morphology very similar to *A. silindensis* (see that account). Apparently uncommon (Crawford-Cabral, 1998).

Aethomys chrysophilus (de Winton, 1897). Proc. Zool. Soc. Lond., 1896:801 [1897].
  COMMON NAME: Red Veld Aethomys.
  TYPE LOCALITY: E Zimbabwe, Mashonaland, Mazoe (see Chimimba, 1998:430).
  DISTRIBUTION: From SE Kenya south through Tanzania, Malawi (Denys et al., 1999), Zambia, S Angola (Crawford-Cabral, 1998), N Namibia, N and E Botswana, Zambia, Mozambique, and NE South Africa in a narrow band bordering Zimbabwe, Botswana, and Mozambique, generally the course of the Limpopo River; range abstracted from Linzey et al. (2003).
  STATUS: IUCN – Lower Risk (lc).
  SYNONYMS: *acticola* (Thomas and Wroughton, 1908); *alticola* Lydekker, 1910; *imago* Thomas, 1927; *singidae* Kershaw, 1923; *voi* Osgood, 1910.

COMMENTS: Previously, data from chromosomes, hemoglobin electromorph mobility and banding patterns, and spermatozoal and bacular morphologies indicated that populations traditionally identified as *A. chrysophilus* in southern Africa consisted of two groups: *chrysophilus* and another species unidentified by scientific name (Breed, 1995*a*; Breed et al., 1988; Gordon and Rautenbach, 1980; Gordon and Watson, 1986; Visser and Robinson, 1986, 1987); the two are claimed to be indistinguishable in external morphology. This other species has recently been identified as *A. ineptus*, which, in addition to the traits listed above, can be distinguished by mtDNA cytochrome *b* sequences (Linzey, et al., 2003).

The two species were thought to be extensively sympatric in the Southern African Subregion (Taylor, 2000*b*), a conclusion based upon morphometric analyses of cytogenetically identified voucher specimens and material of unknown cytotypes (Chimimba, 1998; Chimimba et al., 1999). Using positively identified specimens, however, Linzey et al. (2003) provided an altered hypothesis of distributions that indicates their geographic ranges are mostly parapatric with a narrow zone of overlap and sympatry (a site in Limpopo Province of South Africa is the only confirmed record of syntopy) along lowlands of the Limpopo River drainage in NE South Africa. *Aethomys ineptus* occurs in the remainder of C, N, and E South Africa (see that account), and of the two, *A. chrysophilus* is the only one to range extensively outside that country. Samples from NW Tanzania have 2n = 50 and were identified as *Aethomys* cf. *chrysophilus* (Fadda et al., 2001*b*), which is the same 2n characterizing southern African samples of *A. chrysophilus*; 2n = 44 is diagnostic for *A. ineptus* (Chimimba, 1998; Chimimba et al., 1999). With the revised definition of *A. chrysophilus* offered by Linzey et al. (2003), the intraspecific analysis of samples from southern Africa suggesting the presence of two subspecies (Chimimba, 2000) requires reevaluation.

Recently, Castiglia et al. (2003*b*) characterized Tanzanian *A. chrysophilus* by chromosomal (C- and G-banding) and molecular traits (complete mtDNA cytochrome *b* sequence) and compared results with similar data for Zambian *A. kaiseri*. The 2n = 50 shared by the two species also characterizes *A. bocagei* (Matthey, 1954; Visser and Robinson, 1986), and Castiglia et al. (2003*b*:85) suggested this is "... the ancestral chromosomal number..." for *Aethomys*. Divergence between the two species from a common ancestor occurred 5-4 million years ago, as derived from the cytochrome *b* sequence analyses.

Non-geographic variation associated with sex and age in *A. chrysophilus* reported by Chimimba and Dippenaar (1994). Several regional studies reported distributional, ecological, and taxonomic information for samples from Mt Kilimanjaro (Grimshaw et al., 1995), Bazaruto Arch. off the coast of S Mozambique (Downs and Wirminghaus, 1997), and the Southern African Subregion (de Graaff, 1997*y*).

Aethomys hindei (Thomas, 1902). Ann. Mag. Nat. Hist., ser. 7, 9:218.
COMMON NAME: Hinde's Aethomys.
TYPE LOCALITY: Kenya, Machakos.
DISTRIBUTION: N Cameroon, S Chad, Central African Republic, N and NE Dem. Rep. Congo, S Sudan, SW Ethiopia (Bekele and Schlitter, 1989; Yalden et al., 1996), Uganda, Kenya, and Tanzania (no farther south than Muanza); southern limits unresolved. Range abstracted mostly from D. H. S. Davis (1975*b*) and Denys and Tranier (1992).
STATUS: IUCN – Lower Risk (lc).
SYNONYMS: *alghazal* Wroughton, 1907; *centralis* Heller, 1914; *helleri* Hollister, 1918; *medicatus* Wroughton, 1909; *norae* Wroughton, 1909.
COMMENTS: Originally described as a species but later incorrectly arranged as a subspecies of *A. kaiseri* (e.g., Hollister, 1919; Swynnerton and Hayman, 1951), with which it is sympatric. Actual geographic range of *A. hindei* is unresolved because many series in museum collections and reported in the literature are misidentified as *A. kaiseri*. D. H. S. Davis (1975*b*) recognized two morphologically distinctive populations as subspecies, one to the east of the Eastern Rift Valley (*hindei*), and the other to the west (*medicatus*). Recent preliminary comparisons of cranial and dental traits among geographic samples of *A. hindei* along with multivariate analyses prompted Denys and Tranier (1992) to suggest that *A. hindei* likely consists of several species or distinctive

geographic populations of one species: *hindei*, which would be restricted to Kenya; *alghazal* from N Cameroon, Central African Republic, NE Dem. Rep. Congo, and N Uganda; and *medicatus* occurring in S Chad, Central African Republic, C Uganda, and SW Kenya. The complex requires critical revisionary study to determine significance of appreciable geographic variation in morphological traits.

Aethomys ineptus (Thomas and Wroughton, 1908). Proc. Zool. Soc. Lond., 1908:546.
COMMON NAME: Tete Veld Aethomys.
TYPE LOCALITY: Mozambique, Zambezi River, Tette (= Tete); see Chimimba (1998:432) for details.
DISTRIBUTION: Documented from NE South Africa (Mpumalanga, Gauteng, North West, Limpopo, Northern Cape, and KwaZulu-Natal Provs.; from about 25°30′ S in the north southward to the Durban region) and S Mozambique; distribution in this region unresolved (Linzey et al., 2003).
SYNONYMS: *capricornis* Roberts, 1926; *fouriei* Roberts, 1946; *harei* Roberts, 1946; *magalakuini* Roberts, 1926; *pretoriae* (Roberts, 1913); *tongensis* Roberts, 1931; *tzaneenensis* Jameson, 1909.
COMMENTS: Separation of *A. ineptus* from *A. chrysophilus* documented by Chimimba (1998), Chimimba et al. (1999), and Linzey et al. (2003). Geographic range of *A. ineptus* was thought to incorporate parts of Namibia, Botswana, and Zimbabwe (Chimimba, 1998; Chimimba et al., 1999; Taylor, 2000b), but recent documentation of positively identified specimens indicates its distribution to be largely parapatric with that of *A. chrysophilus*, which occurs in South Africa only along the NE margin approximating the Limpopo River drainage, but has an extensive range northward to SE Kenya (Linzey et al., 2003). Redefinition of *A. ineptus* by Linzey et al. (2003) prompts a reevaluation of the clinal pattern of cranial size within the species proposed by Chimimba (2001b). See account of *A. chrysophilus*.

Aethomys kaiseri (Noack, 1887). Zool. Jahrb. Syst., 2:228.
COMMON NAME: Kaiser's Aethomys.
TYPE LOCALITY: Dem. Rep. Congo, Katanga Marungu.
DISTRIBUTION: SW Uganda, S Kenya, Rwanda, S and E Dem. Rep. Congo, W and SW Tanzania, Malawi, Zambia, and NC Angola; range mostly derived from D. H. S. Davis (1975b).
STATUS: IUCN – Lower Risk (lc).
SYNONYMS: *amalae* Dollman, 1914; *hintoni* Hatt, 1934; *manteufeli* Matschie, 1911; *pedester* Thomas, 1911; *turneri* Heller, 1914; *vernayi* Hill and Carter, 1937; *walambae* Wroughton, 1907.
COMMENTS: Significance of regional morphological variation needs to be assessed by a systematic revision. Westernmost population is that in Angola ("... the Cuanza-Luando Mesopotamia, the Cuango River and the Southern Lunda Province"; Crawford-Cabral, 1999:180), referred to as *A. k. vernayi*, which seems to be isolated from the primary range of *A. kaiseri* (Hill and Carter [1941] treated *vernayi* as a species); genetic intergradation between populations of *vernayi* and *kaiseri* has yet to be documented (Crawford-Cabral, 1998, 1999). Crawford-Cabral (1998) reviewed Angolan population, and Grimshaw et al. (1995) discussed Mt Kilimanjaro records. Cytogenetics (C- and G-banding) and complete mtDNA cytochrome *b* sequence of Zambian *A. kaiseri* characterized and constrasted with samples of *A. chrysophilus* by Castiglia et al. (2003b); see account of the latter.

Aethomys nyikae (Thomas, 1897). Proc. Zool. Soc. Lond., 1897:431.
COMMON NAME: Nyika Aethomys.
TYPE LOCALITY: N Malawi, Nyika Plateau (probably on lower slopes; Ansell and Ansell, 1973; Ansell and Dowsett, 1988; Ansell, 1989b).
DISTRIBUTION: Known only from NE Zambia (1978), Malawi (Ansell and Dowsett, 1988; Ansell, 1989b), NE Angola (Crawford-Cabral, 1998), and S Dem. Rep. Congo (Hatt, 1934b); see Davis (1975b).
STATUS: IUCN – Lower Risk (lc).
SYNONYMS: *dollmani* Hatt, 1934.
COMMENTS: Delany (1975) considered *nyikae* to be a synonym of *A. kaiseri*, but others

(D. H. S. Davis, 1975*b*; Ansell, 1978, 1989*b*) correctly treated it as a separate species. Ansell (1978) documented geographic range and discussed identity of *dollmani*; inclusion of that taxon, which is based upon samples from S Dem. Rep. Congo (Hatt, 1934*b*), is provisional (Davis, 1975*b*). The published record from Eastern Ngorima Reserve in E Zimbabwe (Davis, 1975b), represented by the only specimen (now lost) from south of the Zambezi River, is probably based on a juvenile *A. silindensis* (Chimimba, 1998:434) and is likely the basis for de Graaff's (1997*x*) indication that *A. nyikae* occurs in the Southern African Subregion. All current evidence indicates the species does not occur in southern Africa south of the Zambezi-Cunene Rivers (C. T. Chimimba, 2002, in litt.). A specimen collected at Duque de Bragança, just north of the Central Angolan Plateau, may represent *A. nyikae* or an unnamed species related to it (Crawford-Cabral, 1999). *Aethomys nyikae* and *A. kaiseri* are sympatric in Zambia and probably E Angola (Crawford-Cabral, 1998).

Aethomys silindensis Roberts, 1938. Ann. Transvaal Mus., 19:245.
  COMMON NAME: Seilinda Aethomys.
  TYPE LOCALITY: E Zimbabwe, Mt Seilinda, Chirinda Forest (see Chimimba, 1998:433, for additional data).
  DISTRIBUTION: E Zimbabwe (Chimimba et al., 1999:508; Skinner and Smithers, 1990:277).
  STATUS: IUCN – Lower Risk (lc).
  COMMENTS: For more than 30 years known only by two specimens from the type locality, but material has now been obtained from the nearby Ngorima Reserve in Melsetter Dist. and Stapleford in Umtali Dist., and the species may also extend north along the escarpment in E Zimbabwe and the adjacent W Vila Pery Dist. in Mozambique (Chimimba, 1998; Skinner and Smithers, 1990). Particular derived external, cranial, and dental traits phylogenetically tie *A. silindensis* to the Angolan *A. bocagei*. *Aethomys silindensis* is morphologically very distinctive compared with other species in the genus occurring in the Southern African Subregion (Chimimba et al., 1999). Reviewed by de Graaff (1997*u*).

Aethomys stannarius (Thomas, 1913). Ann. Mag. Nat. Hist., ser. 8, 11:482.
  COMMON NAME: West African Aethomys.
  TYPE LOCALITY: N Nigeria, Bauchi Province, Kabwir.
  DISTRIBUTION: N Nigeria to W Cameroon.
  STATUS: IUCN – Lower Risk (nt).
  COMMENTS: A West African endemic that was originally described as a species and recognized as such (G. M. Allen, 1939; Ellerman, 1941; Rosevear, 1969), subsequently included in *A. hindei* as a subspecies (D. H. S. Davis, 1975*b*), then treated again as a distinctive species (Denys and Tranier, 1992; Hutterer and Joger, 1982; Musser and Carleton, 1993). Rosevear (1969) provided an excellent and comprehensive description. Member of the murine fauna endemic to West Africa (see account of *Grammomys buntingi*).

Aethomys thomasi (de Winton, 1897). Ann. Mag. Nat. Hist., ser. 6, 20:327.
  COMMON NAME: Thomas's Aethomys.
  TYPE LOCALITY: Angola, Galanga.
  DISTRIBUTION: Endemic to Central Plateau of Angola; see Crawford-Cabral, 1999:190).
  STATUS: IUCN – Lower Risk (lc).
  COMMENTS: A morphologically distinctive Angolan endemic related to *A. kaiseri* and the "eastern African *A. hindei-A. medicatus* complex of species" (Crawford-Cabral, 1999:179). Although Crawford-Cabral (1986*a*) suspected *A. kaiseri* to overlap with *A. thomasi* in the central highlands, he has recently shown the two species to have parapatric distributions in C Angola (Crawford-Cabral, 1999), and outlined an hypothesis describing origin of *A. thomasi* within the context of zoogeographic origins and current distributions for other members of the *A. kaiseri* complex (Crawford-Cabral, 1999:191).

Anisomys Thomas, 1904. Proc. Zool. Soc. Lond., 1903(2):199 [1904].
  TYPE SPECIES: *Anisomys imitator* Thomas, 1904.
  COMMENTS: *Pogonomys* Division. Monotypic member of the New Guinea Old Endemics

(Musser, 1981c). Lidicker (1968) documented phallic morphology of *Anisomys imitator* and other New Guinea endemic murines, and concluded that the phallic morphology of *Anisomys* retained a high proportion of ancestral states. Breed and Aplin (1994:26) regarded morphology of the sperm head of *Anisomys* to be plesiomorphic for Muridae and the genus to be one of the "earliest offshoots of the Australo-Papuan murid radiation." *Anisomys* was used as the type genus of Anisomyini by Lidicker and Brylski (1987), but Ellerman (1941) had already isolated the genus as Anisomyes, which he thought probably merited subfamily rank. Watts and Baverstock (1994b), based on microcomplement fixation of albumin, retained Anisomyini and also included *Chiruromys, Hyomys, Macruromys, Mallomys, Coccymys,* and *Pogonomys* in it.

Anisomys imitator Thomas, 1904. Proc. Zool. Soc. Lond., 1903:200 [1904].

COMMON NAME: Uneven-toothed Rat.

TYPE LOCALITY: Papua New Guinea, Central Province, Aroa River, Avera.

DISTRIBUTION: Forested mountain backbone of mainland New Guinea from the Weyland Range in the west to the eastern margin of the Owen Stanley Range in the east; also in the Huon Peninsula; sea level to 3500 m.

STATUS: IUCN – Lower Risk (lc).

COMMENTS: Flannery (1990b, 1995a) provided description of the species and revealing distribution map. Musser and Lunde (ms) documented its altitudinal distributions in the Snow Mtns of Prov. of Papua (= Irian Jaya) and Mt Dayman region of eastern Papua New Guinea. Chromosomal data reported by Donnellan (1987).

Anonymomys Musser, 1981. Bull. Am. Mus. Nat. Hist., 168:300.

TYPE SPECIES: *Anonymomys mindorensis* Musser, 1981.

COMMENTS: *Dacnomys* Division. Contrasted with Indomalayan and Sulawesian murines by Musser (1981c) and Musser and Newcomb (1983). Morphology and distribution reviewed and compared with other Philippine murines by Musser and Heaney (1992). Except that *Anonymomys* has no close phylogenetic links to any of the Philippine Old Endemics or New Endemics in the *Rattus* Division (Musser and Heaney, 1992), its affinities are unresolved. Of the two published speculations, a tie to Bornean and Sulawesi *Haeromys* (Musser and Newcomb, 1983) or primitive relative of Indomalayan *Niviventer* and its phylogenetic allies (Musser, 1981b), the latter is a better estimate based upon interpretation of available morphological data: "*Anonymomys* may be distantly related to *Niviventer* and its relatives, perhaps a species with many primitive characters that is part, possibly a remnant, of an early stock from which *Niviventer, Chiromyscus, Leopoldamys,* . . . and *Dacnomys* evolved" (Musser, 1981b). We provionally include *Anonymomys* in the *Dacnomys* Division until additional information, including chromosomes and analyses of gene sequences, is marshalled to test the suggested alternative alliances. Within the context of phylogenetic relationships among endemic Philippine murines, *Anonymomys* is an Indosundaic element on Mindoro Isl and shares this distinction with an undescribed species of *Maxomys* (specimens examined by Musser), and *Rattus mindorensis*. The other indigenous Mindoran murines are species of *Apomys* and *Chrotomys,* which are strictly Philippine in geographic and phylogenetic origin, and also members of the Philippine Old Endemic assemblage.

Anonymomys mindorensis Musser, 1981. Bull. Am. Mus. Nat. Hist., 168:300.

COMMON NAME: Mindoro Forest Anonymomys.

TYPE LOCALITY: Philippines, Mindoro Isl, Halcon Range, Ilong Peak, 4500 ft (1370 m).

DISTRIBUTION: Greater Mindoro Faunal Region: endemic to Mindoro and recorded only from primary mountain forest on Mount Halcon although probably occurs throughout the island where montane forest formations persist.

STATUS: IUCN – Vulnerable.

COMMENTS: The original series of this arboreal montane rat consists of three specimens collected in 1954 (Musser, 1981b); three additional examples were collected on northern flanks of Mount Halcon during 1992 (in CMNH). *Anonymomys mindorensis* joins *Apomys gracilirostris, Chrotomys mindorensis, Rattus mindorensis,* and an undescribed species of *Maxomys* as Mindoran endemics.

Apodemus Kaup, 1829. Skizz. Entwickel.-Gesch. Nat. Syst. Europ. Thierwelt, 1:154.

TYPE SPECIES: *Mus agrarius* Pallas, 1771.

SYNONYMS: *Alsomys* Dukelski, 1928; *Karstomys* Martino, 1939; *Nemomys* Thomas, 1924; *Petromys* Martino, 1934; *Sylvaemus* Ognev, 1924.

COMMENTS: *Apodemus* Division. Diagnosed by Niethammer (1978*d*) using skeletal and soft tissues, body size, and dental traits, and by Martín Suárez and Mein (1998) using dental characters only. Recognized species have been allocated among the subgenera *Apodemus*, *Sylvaemus*, *Alsomys*, and *Karstomys* (Corbet, 1978*c*; Zimmermann, 1962), but whether these names designate monophyletic clusters and should be retained as subgenera or instead raised to generic rank remains to be answered by critical systematic revision of the entire group, which is currently unavailable. Most taxonomists appreciate at least the subgeneric validity of *Apodemus* and *Sylvaemus*; some suggest *Sylvaemus* should be raised to generic rank because of its great morphological and genetic divergence from *Apodemus* (Britton-Davidian et al., 1991; Mezhzherin and Zykov, 1991); others either treat *Sylvaemus* as a separate genus or suggest it should be recognized at that rank (Bonhomme et al., 1985; Chelomina, 1998; Chelomina et al., 1998*b*; Filippucci et al., 1996; Mezhzherin, 1997*a*; Pavlinov and Rossolimo, 1998; Pavlinov et al., 1995*a*; Zagorodnyuk, 1992*b*, 1993; Zagorodnyuk et al., 1997); and even recognition of *Apodemus*, *Alsomys*, and *Sylvaemus* as genera (with *Karstomys* as a subgenus of the latter) has been advocated (Mezhzherin, 1997*a*). Pavlinov et al. (1995*a*) and Pavlinov and Rossolimo (1998) listed *Alsomys* as a subgenus of *Apodemus*, and *Karstomys* as a subgenus of genus *Sylvaemus*. Phylogenetic analysis of complete mtDNA cytochrome *b* sequences identified *A. agrarius*, *A. mystacinus*, *A. uralensis*, *A. flavicollis*, *A. sylvaticus*, and *A. alpicola* as a monophyletic clade (Martin et al., 2000); the latter four form a monophyletic subgroup (subgenus *Sylvaemus*), and *A. agrarius* (subgenus *Apodemus*) and *A. mystacinus* (subgenus *Karstomys*) fall outside that cluster, a pattern taxonomically interpreted by employing three subgenera (Martin et al., 2000). But Martin et al. also noted that the genetic distances of species in subgenus *Sylvaemus* compared to *A. agrarius* and *A. mystacinus* was of the same magnitude as the *Sylvaemus* cluster was to other murine genera sampled (*Micromys*, *Mus*, *Rattus*, and *Acomys*). Results from analyses of combined nuclear IRBP and mtDNA cytochrome *b* and 12S rRNA sequences indicated *Apodemus* to be monophyletic and consisting of *Apodemus* (*A. semotus*, *A. peninsulae*, and *A. agrarius*) and *Sylvaemus* (*A. sylvaticus*, *A. alpicola*, and *A. flavicollis*) groups, with *A. mystacinus* being related to *Sylvaemus*, although with weak support and possibly forming a third group, *Karstomys* (Michaux et al., 2002*a*). Using morphological and published genetic information, Musser et al. (1996) defined three groups into which the species discussed in the accounts that follow are allocated: *Apodemus* Group (*A. agrarius*, *A. chevrieri*, *A. speciosus*, *A. peninsulae*, *A. latronum*, *A. draco*, and *A. semotus*); *Sylvaemus* Group (*A. sylvaticus*, *A. flavicollis*, *A. uralensis*, *A. mystacinus*, *A. epimelas*, *A. alpicola*, *A. witherbyi*, *A. hyrcanicus*, *A. ponticus*, *A. rusiges*, and *A. pallipes*); and *A. argenteus* as the sole member of the third group. The partitioning has generally been corroborated by analyses of protein electrophoretic results (Filippucci et al., 2002) and mitochondrial and nuclear gene sequences (Chelomina, 1998; Chelomina et al., 1998*b*; Liu et al., 2004; Michaux et al., 2002*a*; Serizawa et al., 2000). Liu et al. (2004) and Serizawa et al. (2000) diverged from this phyletic trichotomy by isolating *A. gurkha* as the sole member of a fourth group, which we recognize here, along with the other three groups, and at the same time abandon subgeneric designations. These studies underline the cladistic diversity within what is now called *Apodemus*, identify at least four monophyletic groups, and undermine past efforts to cram all species into either *Apodemus* or *Sylvaemus*, which besides being a futile endeavor does not reflect phylogenetic reality.

Examples of morphological, chromosomal, and molecular inquires bearing on systematics of *Apodemus* are (also consult those references cited in species accounts): phallic comparisons among European (Williams et al., 1980) and Chinese (Liu et al., 2000; Yang and Fang, 1988) species; geometric morphometry of upper molars in European species (Janzekovic and Kryštufek, 2004); taxonomic differences in testes size among European species (Kratochvíl, 1971); comparative chromosomal studies among European (Bekasova et al., 1980; Soldatovic et al., 1975; Vujosevic et al., 1984; Zima et al., 1997*a*) and Japanese (Tsuchiya, 1981) species; presence or absence of B chromosomes and their possible bio-

logical and taxonomic significance (Zima and Macholán, 1995); electrophoretic variations of allozymes among a variety of species in systematic context (Darviche et al., 1979,
and references therein; Filippucci, 1992; Filippucci et al., 2002; Fraguedakis-Tsolis et al.,
1983; Gemmeke, 1980; Gill et al., 1987; Macholán et al., 2001b; Mezhzherin et al., 1992;
Verimli et al., 2001); electrophoretic, karyological, and morphological distinctions
among several species (Britton-Davidian et al., 1991; Mezhzherin, 1997a; Vorontsov
et al., 1989); differences in restriction endonuclease of nuclear DNA used to assess relationships among three Austrian species (Csaikl et al., 1990) and determine genetic distances and reconstruct phylogenetic tree for six species (Chelomina, 1993a, b); evolution
of microsatellite alleles in four species (Makova et al., 2000); DNA-DNA hybridization results from analyses of three species (Catzeflis, 1990; Catzeflis et al., 1987); evolutionary
classification of European members of subgenus *Sylvaemus* based on allozymic and chromosomal data (Orlov et al., 1996a, b); relationships among EuroCaucasian and Asian species as assessed by mtDNA cytochrome *b* sequences (Chelomina et al., 1998b), restriction
analysis of total nuclear DNA (Chelomina, 1998), and a combination of cranial morphometrics, genotypic and allele frequencies of allozymes, and typing of mtDNA cytochrome
b restriction fragment patterns and sequences (Hille et al., 2002); analyses of mtDNA
cytochrome *b* divergence among western European species (Reutter et al., 2003); isozymic, chromosomal, and molecular divergence among Caucasian species (Chelomina,
1998a); differentiation of ribosomal DNA restriction sites among seven species (Suzuki
et al., 1990); reconstructing evolutionary history and phylogenetic relationships among
eastern Asian species derived from mtDNA cytochrome *b* and nuclear IRBP sequences (Serizawa et al., 2000; Suzuki et al., 2003; also see accounts of species); phylogeny and taxonomy of 15 species emphasizing Chinese species (Liu et al., 2004); and phylogeny of *Apodemus* focusing on subgenus *Sylvaemus* using nuclear IRBP gene sequences and cytochrome *b* and 12S rRNA mitochondrial markers (Michaux et al., 2002a). The inconsistency of molecular evolutionary rates affecting allozyme divergence within *Apodemus* is
illuminated in an intelligent discussion by Hartl et al. (1992).

Numerous taxonomic provincial studies described morphological and other distinctions among sympatric or allopatric species of *Apodemus*; examples are a study of five species from Bulgaria (Popov, 1981); four species from Germany (Feiler and Tegcgn, 1998),
lower Danube River region (Fedorchenko and Zagorodniuk, 1994), Serbia and Montenegro (Todorovi et al., 1974), W Turkey (Filippucci et al. 1996), E Turkey, W Armenia, and
N Iran (Frynta et al., 2001); three species from Poland (Ruprecht, 1979), the Baltic region
(Zagorodnyuk and Mezhzherin, 1992), and Daghestan (Lavrenchenko and Likhnova,
1995); and two species from Korea (Koh, 1988; Park et al., 1990). Vorontsov et al. (1989)
described the presence of at least five species in the Caucasus, some of which at the time
were distinguished only by biochemical traits; Vorontsov et al. (1992) recently identified
and defined four species from there, and Mezhzherin et al. (1992) compared the Tran-
caucasian assemblage with European species using morphology and allozymic data.
Tchernov (1979, 1994) reported on polymorphism, size trends and Pleistocene paleoclimatic responses of three Israeli species in an evolutionary context. Geographic ranges
of various Palaearctic species mapped by Panteleyev (1998). Palaearctic species reviewed
by Corbet (1978c, 1984) and Kobayashi (1985), N Eurasian by Mezhzherin (1997a), and
Chinese by Xia (l984, 1985). Liu et al. (2002) provided review of taxonomy and phylogenetic relationships among the Chinese species they recognize.

Evolutionary history of *Apodemus* is documented back to late Pliocene in NW India
(Kotlia, 1992), early Pliocene in S and N China (Cai and Qiu, 1993; Qiu and Storch, 2000;
Wu and Flynn, 1992; Zheng, 1993) and late Miocene (early Vallesian, 11.1-9.7 million
years ago) of S and C Europe (Freudenthal and Martín Suárez, 1999; Martín Suàrez and
Mein, 1998). The genus was formerly regarded as no older than late Turolian (8.7-7.5 million years ago) and to have been derived from *Progonomys* (then thought to be the oldest
European murine) through *Parapodemus*. Recent studies indicate that European fossil species of *Apodemus* exhibit their diagnostic derived dentition when first encountered in the
Miocene, the genus is older than European *Progonomys* (Freudenthal and Martín Suárez,
1999) or any other European murine, and represents an independent lineage "of unknown origin" (Martín Suárez and Mein, 1998; but see de Bruijn et al., 1999, for another

interpretation). Only *Progonomys* from the Siwaliks of Pakistan is older, the earliest record of fossils identified as that genus coming from strata dated at about 11.8 million years ago (Jacobs and Downs, 1994; Jacobs et al., 1990), or "*Progonomys*-like species, or indeed true *Progonomys*" at 12.3 million years ago (Pilbeam et al., 1996:103). Whether *Apodemus* diverged from this earlier Asian *Progonomys* stock (in which case its presence in Europe represents westward dispersal during the Miocene) or evolved independently from an ancient Asian murine progenitor is unknown. Certainly *Apodemus* (and perhaps *Mus*; see that generic account) has the longest range in time of any known murine (Martín Suárez and Mein, 1998).

Apodemus agrarius (Pallas, 1771). Reise Prov. Russ. Reichs., 1:454.

COMMON NAME: Striped Field Mouse.

TYPE LOCALITY: Russia, Ulianovsk Obl., middle Volga River, Ulianovsk (formerly Simbirsk).

DISTRIBUTION: Palearctic and Oriental regions in two disjunct segments. One from C Europe (from S Finland, Baltic region, through Poland and C Germany to NE Italy and through the Balkan countries to Greece, Romania, Bulgaria, and Turkish Thrace; Mitchell-Jones et al., 1999; Petrov, 1992; Vohralík, 1992; Vohralík and Sofianidou, 1992*b*) east to the Caucasus, parts of Kazakhstan and Kyrgyzstan, and west of Lake Baikal region and adjacent Mongolia and NW China (Xinjiang). The other portion from the Amur River region in far E Russia extending through Korea (including some offshore islands) and westward through China from northern reaches of Nei Mongol and Heilongjiang in the northeast to W Yunnan (Wang, 2003; Zhang et al., 1997); also in the small Senkaku Isl group of Japan (Kaneko, 1994) and on Taiwan (M.-J. Yu, 1996). Expansion of geographic range due to anthropogenic pressures reported by Karaseva et al. (1992).

STATUS: IUCN – Lower Risk (lc).

SYNONYMS: *albostriatus* (Bechstein, 1801); *caucasicus* Kuznetzov, 1944; *chejuensis* Johnson and Jones, 1955; *coreae* Thomas, 1908; *gloveri* Kuroda, 1939 [see Kaneko and Maeda, 2002]; *harti* (Thomas, 1898); *henrici* Lehmann, 1970; *insulaemus* Tokuda, 1939 [*nomen nudum*; see Kaneko and Maeda, 2002]; *insulaemus* Tokuda, 1941 [see Kaneko and Maeda, 2002]; *istrianus* Kryštufek, 1985; *kahmanni* Malec and Storch, 1963; *karelicus* Ehrström, 1914; *maculatus* (Bechstein, 1801); *mantchuricus* (Thomas, 1898); *nicolskii* Charlemagne, 1933 [see Zagorodnyuk, 1992*b*]; *nikolskii* Migouline, 1927; *ningpoensis* (Swinhoe, 1870); *ognevi* Johansen, 1923; *pallescens* Johnson and Jones, 1955; *pallidior* Thomas, 1908; *pratensis* (Ockskay, 1831); *rubens* (Oken, 1816); *septentrionalis* Ognev, 1924; *tianschanicus* Ognev, 1940; *volgensis* Kuznetzov, 1944.

COMMENTS: *Apodemus* group. Most closely related to *A. chevrieri* as reflected by morphology (see that account). Phylogenetic analyses of mtDNA cytochrome *b* and nuclear IRBP sequences clearly identified *A. chevrieri* and *A. agrarius* as sister-species that are part of a larger clade composed of *A. peninsulae*, *A. draco*, *A. semotus*, *A. latronum*, and *A. speciosus*, which represents an evolutionary radiation separate from that producing the Japanese *A. argenteus* and Nepalese *A. gurkha* (Liu et al., 2004; Suzuki et al., 2003). Noteworthy morphological studies bearing on systematics of European populations are: seasonal changes in gut morphology, and the relationships between allozyme heterozygosity and gut morphology (Borkowska, 1995; Borkowska and Ratkiewicz, 1996); research on the thymus (Bazan-Kubik and Skrzypiec, 1992); and comparisons between postnatal development of German *A. agrarius* and *A. sylvaticus* (Pelz et al., 1996). Age and geographic variation in Eurasian, Chinese, Korean, Polish, Bulgarian, and Yugoslavian populations documented by results from morphometric analyses (Koh, 1983, 1991; Koh and Tikhonova, 1998; Koh et al., 1997, 1998*a*, *b*; Kryštufek, 1985*b*; Liu et al., 1991; Sikorski, 1982). Karyological data reported by Kang and Koh (1976), Koh (1982), Lungeanu et al. (1986), J.-X. Wang et al. (1993), Boyeskorov et al. (1992), and Yiğit et al. (2000*a*). Chromatic, morphological, and biochemical information for Chinese and European populations reported in context of assessing relationships among geographic samples (Hille and Meinig, 1996; Wang You-zhi, 1985*b*; Yang and Lu, 1998; Zhao and Lu, 1986). Analyses of mitochondrial and nuclear ribosomal DNA in samples from N China and Korea reported by Han et al. (1996) and Wang and Hung (1997). Other inquiries focusing on Korean populations have produced informative morphometric and chromosomal analyses (Koh, 1988), microgeographic genetic variation detected by mtDNA sequence divergence (Kim et al., 2000), and spermatozoal morphology and function (Lee and Son, 1997). Based on its large body size and mtDNA genotype, Koh (1991) and Koh and Yoo (1992) suggested that the population from Cheju Isl (Korea) represented the distinct species *A. chejuensis* (recognized as an endemic by Won and Smith, 1999, in their checklist of mammals of the Korean Peninsula), but in later analyses treated it as a morphologically distinctive insular population of *A. agrarius* (Koh et al., 1998a). An evaluation of subspecies in E China provided by Liu et al. (1991).

Included in Mezhzherin's (1997a) revision of N Eurasian *Apodemus*; European population reviewed by Böhme (1978b) and Mitchell-Jones et al. (1999), that ranging over Russia and adjacent regions by Gromov and Erbajeva (1995). Innumerable past regional faunal accounts covering distribution, ecology, genetics, and sometimes taxonomy have included or focused on *A. agrarius*, including these more current reports for: N Germany (Dolch et al., 1994), Austria (Bauer and Spitzenberger, 1996; Spitzenberger, 1997; documentation of first record for country), Italy (Amori et al., 1999; Paolucci et al., 1993), E Baltic region (Miljutin, 1997, 1998; Timm et al., 1998), NW Lithuania (Juškaitis and Baranauskas, 2001). Slovakia (Danko, 1994; Kminiak, 1996; Mošanský, 1994; Stanko, 1995; Stanko and Mošanský, 2000; Stanko et al., 1994, 2000; Uhrin and Benda, 2000), Czech Republic (Bárta and Benda, 1998; Polechová and Graciasová, 2000; Zima and Anděra, 1996), Translvanian Romania (Istrate, 1998), Serbia and Montenegro (Petrov, 1992), Slovenia (Kryštufek, 1991), Grecian Thrace (Vohralík, 1992), Greece and S Balkans (Vohralík and Sofianidou, 1992b), Bulgaria (Peshev, 1996), Turkish Thrace (Kryštufek and Vohralík, 2001), Korea (Won and Smith, 1999), China and Korea (Musser et al., 1996), Taiwan (Adler, 1995; H.-T. Yu, 1995), and Japan (Kaneko, 1994).

The low interpopulation genetic divergence among European populations (Hille and Meinig, 1996) suggests a recent colonization of that region. Sediments dated 5500 years BC have yielded the earliest European fossils (Toškan and Kryštufek, in press), thus bolstering the supposition that *A. agrarius* is an Asiatic immigrant and a relatively new member of the European fauna (Kowalski, 2001; Martín Suárez and Mein, 1998).

Apodemus alpicola Heinrich, 1952. J. Mammal., 33:260.
    COMMON NAME: Alpine Field Mouse.
    TYPE LOCALITY: Allgäu, Osterachtal, S Germany.
    DISTRIBUTION: Endemic to NW parts of the Alps: S Germany, Austria (Bauer and
        Spitzenberger, 1996; Spitzenberger and Englisch, 1996), Liechtenstein, Switzerland
        (Hausser, 1995; Margry, 1996; Maurizio, 1994), N Italy (Amori et al., 1999), and
        SE France; altitudinal range 550-2000 m (Mitchell-Jones et al., 1999).
    STATUS: IUCN – Data Deficient.
    SYNONYMS: *alpinus* Heinrich, 1951.
    COMMENTS: *Sylvaemus* group. Originally described as a high-altitude form of *A. flavicollis*,
        but recognized as distinct by Storch and Lütt (1989), who reviewed morphology and
        distribution of *S. alpicola* and noted that it occurred syntopically with *A. sylvaticus* and
        A. flavicollis. The specific integrity of *S. alpicola* has been confirmed by allozymic and
        gene sequence studies (Filippucci, 1992; Filippucci et al., 2002; Liu et al., 2004; Michaux et al., 2002a; Reutter et al., 2003; Serizawa et al., 2000; Suzuki et al., 2003; Vogel et al., 1991). Filippucci's (1992) report of allozymic variation at 28-33 loci suggested that *A. alpicola* may be most closely related to *A. uralensis*, which occurs from E Europe to NW China (similar results are reported for Turkish samples by Filippucci et al., 1996), and that *A. uralensis*, *A. flavicollis*, *A. alpicola*, and *A. hermonensis* (= *A. witherbyi*) separated recently from a common ancestor. The sister-species relationship between *A. alpicola* and *A. uralensis* was corroborated by Michaux et al.'s (2002a) phylogenetic analyses of the mtDNA cytochrome *b* and 12S rRNA sequences, and Reutter et al.'s (2003) analyses of cytochrome *b* haplotype divergences, but not by complete mtDNA cytochrome *b* sequences (Liu et al., 2004). A discriminant function analysis of cranial and dental measurements for distinguishing samples of *A. sylvaticus*, *A. flavicollis*, and *A. alpicola* provided by Reutter et al. (1999), who also documented variation in banded chromosomes between *A. alpicola* and *A. uralensis* (Reutter et al., 2001). Michaux et al.

(2001a) and Reutter et al. (2002) described a relatively simple process where species-specific primers derived from mtDNA cytochrome *b* sequences can be used to rapidly identify specimens of *A. alpicola*, *A. flavicollis*, and *A. sylvaticus* in regions of sympatry or syntopy. Included in Mezhzherin's (1997a) review of N Eurasian *Apodemus*, and also reviewed by Mitchell-Jones et al. (1999). Musser et al. (1996) listed specimens and discussed subgeneric allocation.

Apodemus argenteus (Temminck, 1844). *In* Siebold, Temminck, and Schlegel, Fauna Japonica, Arnz et Socii, Lugduni Batavorum, p. 51.

COMMON NAME: Small Japanese Field Mouse.

TYPE LOCALITY: Japan.

DISTRIBUTION: Endemic to Japan (Dobson, 1994); the four main islands (Hokkaido, Honshu, Shikoku, and Kyushu) along with some smaller ones (Abe and Ishii, 1987; Corbet, 1978c:136; Kaneko, 1994).

STATUS: IUCN – Lower Risk (lc).

SYNONYMS: *celatus* (Thomas, 1906); *geisha* (Thomas, 1905); *hokkaidi* (Thomas, 1906); *sagax* Thomas, 1908; *tanei* Kuroda, 1924 [see Kaneko and Maeda, 2002]; *yakui* (Thomas, 1906).

COMMENTS: *argenteus* group. Zimmermann (1962) included *A. argenteus* in the subgenus *Alsomys*, but this allocation was questioned by Corbet (1978c:136). Based on external, cranial, dental, and chromosomal features, Musser et al. (1996:184) could not place *A. argenteus* in any existing subgenus of *Apodemus* and wrote that the species "needs to be compared with other species in the genus within a revisionary study that focuses on phylogenetic analyses of morphological and biochemical characters before we can identify its nearest phyletic affinity." Comparison of genomes among several species of *Apodemus*, as assessed by differentiation of ribosomal DNA restriction sites (Suzuki et al., 1990), restriction-fragment-length polymorphism (RFLP) of nuclear DNA (Chelomina, 1998; Chelomina et al., 1995), mtDNA cytochrome *b* sequences (Chelomina et al., 1998b; Liu et al., 2004) and phylogenetic analyses of mitochondrial cytochrome *b* and nuclear IRBP gene sequences (Serizawa et al., 2000; Suzuki et al., 2003), indicates that *A. argenteus* differs from all other European and Asian species analyzed, and represents an ancient independent lineage, one of the three initial evolutionary radiations leading to the modern Asian *Apodemus* fauna (Liu et al., 2004; Serizawa et al., 2000; Suzuki et al., 2003). Its small body size, arboreal adaptations, and distinctive morphology compared with all the other species of *Apodemus* were also recognized by Thomas (1905b), who described *A. argenteus* under the name *Micromys geisha*, and proposed *hokkaidi*, *yakui*, and *celatus* as subspecies of *M. geisha* (see synonymy in Ellerman and Morrison-Scott, 1951:570). Ellerman (1941:96) also recognized a "*geisha* group." Kawamura (1989:85; also see Kawamura, 1991) studied molars and cranial fragments of *A. argenteus* from middle and late Pleistocene sediments as well as Holocene and Recent material, and suggested that the "species is relatively primitive in dental morphology and possibly near to ancestral forms of the genus *Apodemus*."

　　Other studies pertinent to systematics of *A. argenteus* are: electrophoretic analyses of 17 enzymes reported in context of biochemical systematics of Japanese *Apodemus* (Saitoh et al., 1989), chromosomal and morphometric comparisons between *A. argenteus* and other Japanese *Apodemus* (Vorontsov et al., 1977a), polymorphic microsatellite DNA markers identified in *A. argenteus* that can be used to investigate genetic variation within the species (Ohnishi et al., 1998), study of nonrandom distribution of sister chromatid exchanges (Satoh and Obara, 1995), topographic distribution (Kaneko, 1992a), variation in molar size among samples from the Japanese Oki Isls (Sakai et al., 1997), comparisons with *A. sylvaticus* and *A. speciosus* in adaptive latitudinal trends in mandible shape (Renaud and Michaux, 2003), inter- and intraspecific patterns of morphological variation in sympatric *A. argenteus* and *A. speciosus* and its importance in insular isolation and biogeographic gradients (Renaud and Millien, 2001), and significance of lower incisor size and shape in ecological and taxonomic inquiries (Millien-Parra, 2000a). Reviewed by Musser et al. (1996) and Kaneko (1994). Because the type series of *Mus argenteus* Temminck, 1844 is composite, a lectotype was chosen by Smeenk et al. (1982), which stabilizes the name; year of publication of *argenteus* is usually listed as 1845 but the description appeared in 1844; see Holthuis and Sakai (1970).

Apodemus chevrieri (Milne-Edwards, 1868). Rech. Hist. Nat. Mammifères, p. 288.
COMMON NAME: Chevrier's Field Mouse.
TYPE LOCALITY: China, Sichuan, Moupin.
DISTRIBUTION: WC China; from Shaanxi and S Gansu through W Hubei, Sichuan, and Guizhou to W Yunnan (Corbet and Hill, 1992; Zhang et al., 1997).
STATUS: IUCN – Lower Risk (lc).
SYNONYMS: *fergussoni* Thomas, 1911.
COMMENTS: *Apodemus* group. Sister-species to *A. agrarius* (see that account). Included in *A. agrarius* by G. M. Allen (1940), Ellerman and Morrison-Scott (1951), and Corbet (1978c), but was earlier listed as a species by Ellerman (1941). Currently recognized as a species distinct from *A. agrarius* (Corbet and Hill, 1992; Koh, 1991; Liu et al., 2002; Musser and Carleton, 1993; Musser et al., 1996; Wang You-zhi, 1985b; Y. Wang, 2003; Xia, 1984; Yang and Fang, 1988). Sympatric with *A. agrarius* in Sichuan and Ghizhou provinces (Wang You-zhi, 1985b; Xia, 1985). On Wuliang Mtn in C Yunnan, *A. chevrieri* occurs between 1800 and 2300 m where it overlaps with *A. draco*, which is more common at higher altitudes up to 2860 m (Jiang and Wang, 2000). Reviewed and contrasted with *A. agrarius* by Musser et al. (1996).
　　　Evolutionary history in the Sichuan-Guizhou region of S China extends back to early Pleistocene (Zheng, 1993).

Apodemus draco (Barrett-Hamilton, 1900). Proc. Zool. Soc. Lond., 1900:418.
COMMON NAME: South China Field Mouse.
TYPE LOCALITY: S China, NW Fujian, Kuatun.
DISTRIBUTION: China (Fujian, Jiangxi, Zhejiang, Anhui, Henan, Hubei, and Hunan in the east through Shanxi, Shaanxi, Ninxia, Gansu and Guangxi to Guizhou, Sichuan, Yunnan, and SE Xizang in the west), N and EC Burma (Kachin and Chin states, respectively), and NE India (Arunachal Pradesh; USNM 564492, 564493); see Musser et al. (1996), Agrawal (2000), Corbet and Hill (1992), Feng et al. (1986), and Zhang et al. (1997).
STATUS: IUCN – Lower Risk (lc).
SYNONYMS: *argenteus* (Swinhoe, 1870) [not Temminck, 1844]; *badius* (Swinhoe, 1870) [not Blyth, 1859]; *ilex* Thomas, 1922; *orestes* Thomas, 1911.
COMMENTS: *Apodemus* Group. Included in subgenus *Alsomys* by Zimmerman (1962) and Pavlinov et al. (1995a) but placed in the "*Apodemus* Group" by Musser et al. (1996), which contains the Eastern Asian species Zimmerman placed in *Alsomys* as well as *A. agrarius* and *A. chevrieri*, which are in subgenus *Apodemus*. Originally described as a subspecies of *A. sylvaticus*, and listed that way by some workers (G. M. Allen, 1940; Ellerman and Morrison-Scott, 1951), but correctly listed as a separate species by others (Corbet, 1978c; Corbet and Hill, 1991; Ellerman, 1941; Xia, 1985), who also thought there was a close relationship between *A. draco* and *A. argenteus*, which is unsupported by cranial and dental morphology (Musser et al., 1996). Corbet and Hill (1992) recognized *orestes* as a separate species sympatric or possibly parapatric with *A. draco*, and their treatment is followed in recent regional faunal reports (Agrawal, 2000, for India, and Wang, 2003, for China) and Jiang and Wang (2000), who analyzed morphometric variation. Musser et al. (1996) found no evidence for this arrangement in their study of large samples from W China, and *orestes* is also included in *A. draco* by Liu et al. (2002, 2004). The Taiwanese *A. semotus* is the closest phylogenetic relative of *A. draco*. Karyotype reported by Chen et al. (1996). Analysis of protein electrophoresis aligned *A. draco* with *A. latronum* rather than *A. chevrieri*, in Bing et al.'s (1996) study of these three species, an alliance also reflected by morphology. Phylogenetic analyses of mtDNA cytochrome *b* and nuclear IRBP sequences placed *A. draco*, *A. latronum*, and *A. semotus* in the same clade that is part of a larger evolutionary radiation (containing the Asian *A. agrarius*, *A. chevrieri*, *A. peninsulae*, and *A. speciosus*) separate from the early and independent radiations that produced the Japanese *A. argenteus* and Nepalese

A. gurkha (Liu et al., 2004; Suzuki et al., 2003). Reviewed by Musser et al. (1996). Evidence from complete mtDNA cytochrome *b* sequences prompted Liu et al. (2004:12) to recognize *ilex* (W Yunnan) as a separate species, but noted that ". . . further detailed morphological and molecular analyses are needed to confirm its species level status."

Apodemus epimelas (Nehring, 1902). Sitz. Ber. Ges. Nat. Fr. Berlin, 1902:2

COMMON NAME: Western Broad-toothed Field Mouse.

TYPE LOCALITY: SE Greece, Parnassós, Agoriana.

DISTRIBUTION: W and S Balkans in W Croatia, W Bosnia and Herzegovina, S Serbia and Montenegro, Albania, Macedonia, W Bulgaria and Greece (the European portion of *mystacinus* mapped by Mitchell-Jones et al. [1999]; also see Peshev [1996], Petrov [1992], Prigioni [1996], and Vohralík [1992], all reported as *mystacinus*), and Adriatic islands of Korula and Mljet (Mitchell-Jones et al., 1999).

COMMENTS: *Sylvaemus* group. A close relative of *A. mystacinus*, which has been considered the sole member of *Karstomys*, although Musser et al. (1996) placed it in their *Sylvaemus* Group, a relationship weakly to strongly corroborated by allozymic studies (Filippucci et al., 2002; Mezhzherin, 1997a) and phylogenetic analyses of nuclear IRBP and mtDNA cytochrome *b* and 12S rRNA sequences (Michaux et al., 2002a, 2005). Described as a species and listed that way by Ellerman (1941) but subsequently treated as a subspecies of *A. mystacinus* (Ellerman and Morrison-Scott, 1951; Corbet, 1978c; Musser and Carleton, 1993; Pavlinov et al., 1995a). Based on morphology and biochemical genetic variation data, Mezhzherin (1997a) separated *epimelas* as a species separate from *A. mystacinus*, an action supported by phylogenetic analyses of protein electrophoresis of 28-38 gene loci (Filippucci et al., 2002) and mtDNA cytochrome *b* and 12S rRNA sequences (Michaux et al., 2002a). That the Balkan *epimelas* was a different species than the Anatolian *mystacinus* was indicated in Mitchell-Jones et al. (1999), and the basic discriminatory data was provided earlier by Storch (1977; who also wrote the account in Mitchell-Jones et al., 1999) who documented the significant differences between the two in occlusal patterns of M1 and M2 (posterior cingulum is free in most *epimelas*, but coalesced with cusp t8 in *mystacinus*); we verified these contrasts in series from Greece, Montenegro, and Turkey in USNM. Karyotype of Macedonian sample recorded and compared with published karyotypes from other regions (Zima et al., 1997a, as *mystacinus*). Greek samples of *A. epimelas* (recorded as *mystacinus*), *A. sylvaticus*, and *A. flavicollis* contrasted using data from electrophoretic and immunological sources (Fraguedakis-Tsolis et al., 1983).

Apodemus epimelas may represent a population of *A. mystacinus* that dispersed across a Pleistocene Bosporous land bridge (between Balkan Peninsula and Turkish Thrace plus NW Anatolian Turkey) 20,000-10,000 years before present and achieved genetic isolation after post-Pleistocene sea level rise reconnected the Black and Marmara Seas through the Bosporous channel (Hosey, 1982). Michaux et al. (2002a) suggested the isolation of Balkan and Near East groups of *A. mystacinus* and the subsequent differentiation of *A. epimelas* resulted from late Pliocene or early Pleistocene cooling intervals associated with low population densities.

Apodemus flavicollis (Melchior, 1834). Dansk. Staat. Norg. Pattedyr, p. 99.

COMMON NAME: Yellow-necked Field Mouse.

TYPE LOCALITY: Denmark, Sieland Isl.

DISTRIBUTION: S England and Wales; on the continent from N Spain through most of Europe to 64E N in S Finland and S Sweden, and south to S Italy, Balkan region, and Greece (Mitchell-Jones et al., 1999; Van der Straeten, 1977a); east through Belarus and Ukraine to Urals; throughout Turkey (Filippucci et al., 1996; Frynta et al., 2001; Kryštufek and Vohralík, 2001; Macholán et al., 2001b; Pamukoglu and Albayrak, 1996; specimens in USNM) and east to W Armenia (Frynta et al., 2001; Macholán et al., 2001), Zagros Mtns of W Iran (Kryštufek, 2002a; Macholán et al, 2001; USNM material), and south to Syria, Lebanon, and C and N Israel (Filippucci, 1992; Filippucci et al., 2002; Qumsiyeh, 1996:289). The species also occurs on some NE Aegean Isls off W coast of Turkey (Özkan and Kryštufek, 1999) and S Greece (Vohralík et al., 1996), Adriatic islands (Mitchell-Jones et al., 1999), and Corfu in the Mediterranean

(Cheylan, 1991); Corbet (1978c:134) and Mezhzherin (1997a:35) mapped overall range. Absent from most of Iberian Peninsula and W France, Iceland, Ireland, Balearic Isls, and islands of Sardinia, Corsica, Sicily, and Cyprus; also does not occur in most of the Caucasus (except WArmenia) where it is apparently replaced by the closely related *A. ponticus*.

STATUS: IUCN – Lower Risk (lc) as *A. flavicollis* and *A. arianus*.

SYNONYMS: *arianus* (Blanford, 1881); *brauneri* Martino, 1927; *cellarius* (Fischer, 1866); *dietzi* Kahmann, 1964; *erythronotus* (Blanford, 1875) [not Temminck, 1844]; *fennicus* (Hilzheimer, 1911); *geminae* Lehmann, 1961; *levantinus* Bate, 1942; *princeps* (Barrett-Hamilton, 1900); *samariensis* Ognev, 1923; *saturatus* Neuhäuser, 1936; *stankovici* Martino and Martino, 1937; *tauricus* (Pallas, 1811); *typicus* (Barrett-Hamilton, 1900); *wintoni* (Barrett-Hamilton, 1899).

COMMENTS: *Sylvaemus* group. Geographic variation of E European samples in the context of testing correlation between phenotypic and genotypic variations of certain morphological characters documented by Orlov and Okulova (2001). Additional research using a variety of data sources that provide comparative information among populations of *A. flavicollis* are: B chromosome polymorphism among populations (Vujosevic et al., 1991), and presence or absence of B chromosomes and the attendant biological significance (Blagojevi and Vujoševi, 1995; Macholán and Zima, 1997; Ramalhinho and Libois, 2002; Vujoševi and Blagojevi, 1995, 2000; Zima et al., 1999a, 2003); chromosomal morphologies of Macedonian (Zima et al., 1997a) and Baltic (Boyeskorov et al., 1992) samples; genetic variation in a fluctuating Polish population (Wójcik, 1993); and mtDNA cytochrome *b* haplotype divergence among W European samples (Reutter et al., 2003). Phylogeographic history (using mtDNA cytochrome *b* sequences) of *A. flavicollis* in Europe and the Near and Middle East in the context of determining Quaternary glacial refugia and postglacial recolonization events reported by Michaux et al. (2004b).

Contrasts and similarities between *A. flavicollis* and *A. sylvaticus* are available based upon: chromosomal morphology, allozymic variation, and mitochondrial and nuclear DNA sequences (Bellinvia et al., 1999; Britton-Davidian et al., 1991; Chelomina, 1998; Chelomina et al., 1998a, b; Debrot and Mermod, 1977; Filippucci, 1992; Filippucci et al., 1996, 2002; Gemmeke, 1980, 1981b; Hirning et al., 1989; Macholán et al., 2001b; Michaux et al., 2002a, 2004a]; Mezhzherin, 1997a; Mezhzherin and Zykov, 1991; Mezhzherin et al., 1992; Nadjafova et al., 1993; Orlov et al., 1996a, b; Reutter et al., 2003; Serizawa et al., 2000; Suzuki, et al., 1990; Tegelstrom and Jaarola, 1989); genetic structure of populations living in four different biotypes (Amori et al., 2002b); morphometric analyses, in context of evolutionary divergence, morphometric identification, and habitat selection (Amori and Contoli, 1994; Dolan and Yates, 1981; Feiler and Tegegn, 1998; Fernendes et al., 1991; Fielding, 1966; Hedges, 1969; Kryštufek and Stojanovski, 1996; Mezhzherin and Lashkova, 1992; Nores, 1988; Panzironi et al., 1993; Popov, 1993; Van der Straeten, 1976b; Van der Straeten and Van der Straeten-Harrie, 1977); dental variation (Filippucci et al., 1996; Tvrtkovi, 1976); temporal variation of non-metric traits (Sádlová and Frynta, 2001); and habitat in Norway (Lura et al., 1995). Differences in blood serum proteins between *A. flavicollis* and *A. witherbyi* (reported as *hermonensis*) documented by Verimli et al. (2000). Based on allozyme and chromosome data, Orlov et al. (1996a, b) placed *A. flavicollis* and *A. ponticus* in the superspecies *A. flavicollis*. The species was a part of Mezhzherin's (1997a) revision of N Eurasian *Apodemus* based upon morphology and molecular data.

The inclusion of *arianus*, along with *erythronotus* (the name *arianus* replaced; see Kryštufek, 2000), in the above synonymy follows Kryštufek's (2002a) identification of the holotype. Musser and Carleton (1993) had used *arianus* as the oldest name for the species we now list as *A. witherbyi* (see that account). The taxon *stankovici*, described by Martino and Martino (1937) as a subspecies of *A. sylvaticus* from the Korab Mtns of W Macedonia, was treated as an endemic Balkan species by Petrov (1993/94), who considered *A. flavicollis* and *A. stankovici* to be narrowly sympatric but found in different habitats. In Petrov's view, the range of *A. stankovici* would be concordant with other mammalian Balkan relicts, such as the vole, *Dinaromys bogdanovi*, and the mole,

Talpa stankovici. However, morphometric analysis of SW Balkan samples of *A. flavicollis* and *A. sylvaticus* along with the holotype and paratypes of *stankovici* by Kryštufek and Stojanovski (1996) demonstrated the conspecificity of the latter with *A. flavicollis*. There may be another species (in addition to *A. flavicollis* and *A. sylvaticus*) in the SW Balkan Mtns, but it is not *stankovici* (B. Kryštufek, 2002, in litt.).

Bate's (1942*b*) *levantinus*, based upon Palaeolithic specimens from Israel and described as a species morphologically similar to *A. flavicollis* (see expanded description in Tchernov, 1968), is now regarded as synonymous with *A. flavicollis* (Tchernov, 1986) and documents the evolutionary history of the species, which now lives in Israel, back to early Pleistocene in that country (see also Tchernov, 1996).

European populations reviewed by Niethammer (1978*b*) and Mitchell-Jones et al. (1999). Among the numerous regional reports detailing geographic range, taxonomy, habitat, and other biological aspects of *A. flavicollis* are the following more recent contributions for: Norway (Lura et al., 1995); Netherlands, where the species is known only from the southern tip (Bergers and Foppen, 1992; Thissen and Hollander, 1996); Belgium (Libois, 1996; Van der Straeten, 1976*b*); Germany (Alf et al., 1997; Berger and Feldmann, 1997; Dolch et al. 1994; Niedenführ and Rathke, 1996); Austria (Bauer and Spitzenberger, 1969); Switzerland (Hausser, 1995; Maurizio, 1994); Slovakia (Danko, 1994; Kminiak, 1996; Mošanský, 1994; Stanko, 1995; Stanko and Mosansky, 1994, 2000; Stanko et al., 1994, 2000); Czech Republic (Anděra and ervený, 1994; Šmaha, 1996; Zima and Anděra, 1996); Translvanian Romania (Istrate, 1998); Bulgaria (Chassovnikarova and Markov, 1999; Peshev, 1996); Italy (Amori et al., 1999, *2002a, b*; Cagnin et al., 1996; Cantini, 1991; Cerone and Aloise, 1994; Cresti et al., 1992; Locatelli and Paolucci, 1996*a*); E Baltic region (Miljutin, 1997, 1998; Timm et al., 1998); Lithuania (Juškaitis and Baranauskas, 2001); Serbia and Montenegro (Petrov, 1992); Slovenia (Kryštufek, 1991); Albania (Prigioni, 1969); SE Greece (Thrace; Vohralík, 1992); Turkey (Kryštufek and Vohralík, 2001); Armenia (Gromov and Erbajeva, 1995); and Israel (Qumsiyeh, 1966).

USNM 369845, a young adult from Kordestan Province (near Marivan), Zagros Mtns, is one of the few records of *A. flavicollis* from Iran, where the species has been recorded only in the Zagros Mtns. It was taken at the same place on the day before USNM 369846, an *A. witherbyi*, was collected. The two species in Iran are regularly misidentified in museum collections. Iranian *A. flavicollis* has paler dorsal pelage than Turkish samples, whitish gray underparts (white in *A. witherbyi*), a round pectoral patch (streak in *A. witherbyi*), molar occlusal patterns that are typical of Turkish *A. flavicollis* (Filippucci et al., 1996), M1-3 length = 4.2 mm (usually less than 4.0 in *A. witherbyi*), and large bullae (5.2 mm; 4.2-4.8 mm in USNM Iranian *A. witherbyi*, usually 4.5 mm).

From 1948 (following Heptner, 1948*b*) until the 1990s, some authors (e. g., Ondrias, 1966, Fedorchenko and Zagorodnyuk, 1994, Mezhzherin, 1997*a*) have substituted the earlier *tauricus* Pallas, 1811 (type locality is mountains of Crimea), for *A. flavicollis*. Corbet (1978c:134), however, explained that *flavicollis* "preponderates in the very extensive ecological literature on the species. It therefore seems desirable to retain the name *flavicollis*, and the inadequacy of the original description of *tauricus*, which amounts only to 'multo major et elegantissimi velleris' (relative to *A. sylvaticus*), would seem to justify its rejection as not certainly determinable." Pavlinov and Rossolimo (1987) listed *tauricus* as *nomen oblitum* and Pavlinov et al. (1995*a*) treated it the same way but with a question. Kryštufek (2002*a*) countered by demonstrating that *tauricus* has been used as a senior synonym of *flavicollis* after 1899 (thus invalidating its status as a *nomen oblitum*), that *flavicollis* is the only *Apodemus* found in the Crimean Mtns (which negates Corbet's concern), and that *tauricus* is potentially valid for replacing *flavicollis*. Zagorodnyuk (1992*b*:45) offered "*Sylvaemus taurica* (Pall.)" as a new combination. The nomenclatural issue requires resolution.

Apodemus gurkha Thomas, 1924. J. Bombay Nat. Hist. Soc., 29:888.
COMMON NAME: Nepalese Field Mouse.
TYPE LOCALITY: Nepal, Gorkha, Laprak.

DISTRIBUTION: Endemic to coniferous forest in C Nepal between 2200 and 3600 m (Martens and Niethammer, 1972).

STATUS: IUCN – Lower Risk (lc).

COMMENTS: *gurkha* group. Usually considered a member of subgenus *Alsomys* (Pavlinov et al., 1995a), but placed in an *Apodemus* Group by Musser et al. (1996), which corresponds to subgenus *Apodemus*. Phylogenetic analyses of mtDNA cytochrome *b* and nuclear IRBP sequences identified *A. gurkha* as distinct from any other species sampled (*A. argenteus, A. speciosus, A. peninsulae, A. agrarius,A. chevrieri, A. semotus, A. draco, A. latronum, A. sylvaticus, A. flavicollis,* and *A. alpicola*), and to represent an ancient lineage, one of the first to have diverged from an ancestral stock (*A. argenteus* is the other ancient line), and one that is older than the evolutionary origins of *A. agrarius* and its East Asian allies and the European species of *Apodemus* (Serizawa et al., 2000; Suzuki et al., 2003). Reviewed and contrasted with *A. sylvaticus* by Martens and Niethammer (1972). Chromosomal data reported and compared with other *Apodemus* by Gemmeke and Niethammer (1983). Reviewed by Musser et al. (1996) and others who also recognized *gurkha* as a distinct species (Agrawal, 2000; Corbet, 1978c; Corbet and Hill, 1991, 1992). Narrowly sympatric and syntopic with *A. pallipes*, which occurs primarily at higher altitudes (Corbet and Hill, 1992; Martens and Niethammer, 1972; reported as *A. sylvaticus* in both reports). *Apodemus gurkha*, with its restricted endemic range and apparently ancient origin, and *A. pallipes*, which follows the Himalayas westward to the Pamirs, N Pakistan, and the Hindu Kush of N Afghanistan and is part of the *Sylvaemus* radiation, are the only *Apodemus* currently recorded from Nepal.

Apodemus hyrcanicus Vorontsov, Boyeskorov, and Mezhzherin, 1992. Zool. Zh., 71:127.

COMMON NAME: Hyrcanian Field Mouse.

TYPE LOCALITY: Azerbaijan, Caucasus, Astarinski (R-NO), Hirkauski Preserve, "Piavolil" Dist., 450 m.

DISTRIBUTION: E Caucasus, where it is found in the low mountain broadleaf forests of the Talysh region of SE Azerbaijan (Gromov and Erbajeva, 1995; Vorontsov et al., 1992:124). Then ranges eastward in the Hyrcanian forests (deciduous formations in which *Quercus, Fagus,* and *Carpinus* predominate) along southern border of the Caspian Sea in N Iran (on coast and northern slopes of Elburez Mtns) from Astara on the west coast (USNM 354778); through Sama (FMNH 97462, 97464), Sari (USNM 369857) and Amol (USNM 369858) on southern border; to Dar Kaleh (AMNH series) on SE coast; then extends east of Caspian Sea on northern slopes of Elburez Mtns in Khorassan Province of NE Iran through the Gorgan (large FMNH and USNM samples), Gouladah (AMNH material), Shah Pasand (USNM), and Bojnurd (USNM) regions to vicinity of Dasht (series in AMNH), which is easternmost record of species (at 37°19' N/56°01' E); possibly also occurs in SW Turkmenistan on eastern border of Caspian Sea. Altitudinal range, sea level to 1830 m.

STATUS: IUCN – Data Deficient.

COMMENTS: *Sylvaemus* group. *Apodemus hyrcanicus* was originally characterized by Vorontsov et al. (1989) using four allozyme loci and subsequently defined by Mezhzherin et al. (1992) and Vorontsov et al. (1992) with 40 loci in addition to chromosomal and morphological characters. Macholán et al. (2001) have recently identified (using allozyme and morphological data) *A. hyrcanicus* from two localities in N Iran, and our identification of specimens in AMNH, FMNH, and USNM uncovered the range throughout N Iran between the Elburez Mtns and Caspian Sea and east into NE Iran. Frynta et al. (2001) provided the first multivariate analysis of morphological traits for *A. hyrcanicus*, contrasting specimens from N Iran with samples of *A. flavicollis, A.witherbyi* (reported as *arianus*), and *A. uralensis* from W Iran, Armenia, and E Turkey. Included in Mezhzherin's (1997a) study of N Eurasian *Apodemus*. Analysis of 18-38 loci by Filippucci et al. (2002) revealed *A. hyrcanicus* to be phylogenetically closest to *A. uralensis* among the ten species sampled.

Throughout its range in Iran, *A. hyrcanicus* is not sympatric with either *A. flavicollis* or *A. witherbyi*, the only other species recorded for the country. The former is known only from the Zagros Mtns in W Iran, but the latter ranges widely in rocky and arid habitats. Goodwin (1940) collected both species in the vicinity of Dasht, but *A. hyr-*

canicus (reported as *arianus*) was caught at 3000 ft (914 m) in deciduous forest and
A. witherbyi (reported as *A. sylvaticus chorassanicus*) at 5000 ft (1524 m) on rocky, arid
nonforested slopes.

Apodemus latronum Thomas, 1911. Abstr. Proc. Zool. Soc. Lond., 100:49.
  COMMON NAME: Large-eared Field Mouse.
  TYPE LOCALITY: China, W Szechwan, Tatsienlu.
  DISTRIBUTION: S China (E Xizang, Yunnan, Sichuan, and Qinghai) and N Burma (Ellerman,
      1961; Feng et al., 1986; Musser et al., 1996; Wang, 2003; Zhang et al., 1997).
  STATUS: IUCN – Lower Risk (lc).
  COMMENTS: *Apodemus* group. Usually placed in subgenus *Alsomys* (Pavlinov et al., 1995a);
      included in an *Apodemus* group by Musser et al. (1996). Treated as a subspecies of
      A. draco by Feng et al. (1986), but *A. latronum* is a separate species that is sympatric
      with *A. draco* in Sichuan, Yunnan, and NE Burma (Musser et al., 1996). Karyotype re-
      ported by Chen et al. (1996). Analysis of protein electrophoresis aligns *A. latronum*
      with *A. draco* and not *A. chevrieri*, the only three species in the study (Bing et al., 1996),
      an alliance also reflected by morphology. Phylogenetic analyses of mtDNA cyto-
      chrome *b* and nuclear IRBP sequences united *A. latronum* in a clade with *A. draco* and
      A. semotus that is part of a larger lineage containing *A. agrarius*, *A. chevrieri*, *A. penin-
      sulae*, and *A. speciosus* (Suzuki et al., 2003; also see account of *A. draco*). Musser et al.
      (1996) reviewed taxonomic history, morphology, relationships, and listed localities
      and specimens from China (Sichuan and Yunnan) and NE Burma. Evolutionary his-
      tory extends back to the early Pleistocene as documented by fossils from cave sedi-
      ments in the Sichuan-Guizhou region of S China (Zheng, 1993) and fissure deposits in
      Shandong (Zheng et al., 1997).

Apodemus mystacinus (Danford and Alston, 1877). Proc. Zool. Soc. Lond., 1877:279.
  COMMON NAME: Eastern Broad-toothed Field Mouse.
  TYPE LOCALITY: Turkey, Adana Province, Bulgar Dagh Mt, Zebil.
  DISTRIBUTION: On some Aegean islands (Rhodes, Crete, Corfu, and other inshore islands;
      Cheylan, 1991; Özkan and Kryštufek et al., 1999; Storch, 1977); eastward through
      Turkey (Felten et al., 1973; Kryštufek and Vohralík, 2001; Pamukoglu and Albayrak,
      1996) to S Georgia in Caucasus (Gromov and Erbajeva, 1995); south and east to N and
      C Israel, Lebanon, NW Jordan (Benda and Sádlová, 1999; Mendelssohn and Yom-Tov,
      1999; Qumsiyeh, 1996; Tchernov, 1979, who discussed extant and Pleistocene
      samples), Syria (Shehab et al., 1999), and N Iraq; see non-European distribution in
      Corbet (1978c), Niethammer (1978a), Mitchell-Jones (1999), and review of
      distribution in the Near East by Harrison and Bates (1991).
  STATUS: IUCN – Lower Risk (lc).
  SYNONYMS: *euxinus* G. Allen, 1914; *pohlei* Aharoni, 1932; *rhodius* Festa, 1914; *smyrnensis*
      (Thomas, 1903).
  COMMENTS: *Sylvaemus* group. Reviewed by Storch (1977) and Niethammer (1978a), who in-
      cluded *krkensis* as a subspecies; that form, however, is a color phase of *A. sylvaticus* (see
      that account). Usually regarded as sole member of subgenus *Karstomys*, but included in
      a *Sylvaemus* Group by Musser et al. (1996); placed in subgenus *Karstomys* of genus *Syl-
      vaemus* by Mezhzherin (1997a) and Pavlinov et al. (1995a), but in an "*Apodemus*
      group" by Liu et al. (2004). Electrophoretic analyses of allozyme variation indicated
      A. mystacinus (and *A. epimelas*) to be a distinct species justifiably placed within sub-
      genus *Sylvaemus* (Britton-Davidian et al., 1991; Filippucci, 1992; Filippucci et al., 2002;
      Gemmeke, 1980; Mezhzherin et al., 1992). Data from nuclear IRBP and mtDNA
      cytochrome *b* and 12S rRNA sequences also indicated weak affinity with *Sylvaemus*
      (Michaux et al., 2002a), as did analyses of cytochrome *b* haplotype divergences (Reut-
      ter et al., 2003), but not complete mtDNA cytochrome *b* sequences (Liu et al., 2004).
      Vorontsov et al. (1989) provided chromosomal data in context of defining species of
      Apodemus in the Caucasus. Formerly included the Balkan *epimelas*, which is now
      treated as a separate species (see that account). Sequences from mtDNA cytochrome *b*
      and D-loop and the IRBP nuclear gene identified two distinct clades within *A. my-*

stacinus, one occupying NW and E Turkey, and the Near East, the other inhabiting SW Turkey and the island of Crete (Michaux et al., 2005).

Remains of *A. mystacinus* were uncovered from subfossil cave deposits in S and E Anatolia (Corbet and Morris, 1967; Kock et al., 1972) as well as middle Pleistocene (Storch, 1975) and Holocene (Besenecker et al., 1972) sediments on the Aegean island of Khios. Fossils are also common in middle Palaeolithic assemblages from Lebanon (Kersten, 1992) and date back to middle Pleistocene in other parts of the Levant (Tchernov, 1968, 1986, 1992, 1994).

Apodemus pallipes (Barrett-Hamilton, 1900). Proc. Zool. Soc. Lond., 1900:417.

COMMON NAME: Himalayan Field Mouse.

TYPE LOCALITY: E Tajikistan, Pamir Altai, Surhad Wahkan; see Barrett-Hamilton (1900:417) and Mezhzherin (1997*a*:38) for condition of holotype.

DISTRIBUTION: Pamirs and adjacent mountainous extensions in S Kyrgyzstan and Tajikistan (Mezhzherin, 1997*a*); south through the Hindu Kush and auxiliary high ranges of N Afghanistan from Herat Province in the west (FMNH specimens) to Kabul region (FMNH) and provinces of Badakhshan (FMNH) and Konar (FMNH 103601) in the east; Pamirs and Himalayas in N Pakistan in districts of Chitral (material in AMNH), Swat (USNM series), Dir (USNM), Hazara (large USNM samples, including USNM 411143 and 411144, which are topotypes of *pentax*), Gilgit (USNM specimens), and Baltistan (USNM); Jammu-Kashmir region of NW India (USNM and AMNH material); eastward through Himalayas to C Nepal (specimens in FMNH) and possibly SW Xizang (Tibet). Altitudinal range, 1465-3965 m, with most records from above 2440 m.

STATUS: IUCN – Lower Risk (lc) as *Apodemus wardi*.

SYNONYMS: *bushengensis* Zheng, 1979; *pentax* (Wroughton, 1908); *wardi* (Wroughton, 1908).

COMMENTS: *Sylvaemus* group. Originally described as a subspecies of *Mus sylvaticus*, retained as a subspecies of *Apodemus sylvaticus* (Ellerman, 1941; Ellerman and Morrison-Scott, 1951), treated as a synonym of that species (Corbet, 1978*c*; Pavlinov and Rossolimo, 1987), and finally included in *A. uralensis* (Musser and Carleton, 1993; Pavlinov and Rossolimo, 1998; Pavlinov et al., 1995*a*). Mezhzherin (1997*a*), who examined the holotype, used *pallipes* for a species phylogenetically related to *A. uralensis*, but larger in body size with paler dorsum, more white on underparts, and occurring farther south in the Pamirs of Kyrgyzstan and Tajikistan, and mountainous N Afghanistan and Pakistan. Many of the specimens in the institutions cited above, especially those from high altitudes (Kashmir, for example), conform to original descriptions of *pallipes* (Barrett-Hamilton, 1990) and *wardi* from Kashmir (Wroughton, 1908*b*); specimens from lower altitudes tend to have darker pelage, matching *pentax* (Wroughton, 1908*b*) from Thandiani in N Pakistan. Our study of large series in AMNH, FMNH, and USNM (including types and topotypes) supports Mezhzherin's view, and allowed us to expand the range eastward to C Nepal. Musser and Carleton's (1993) account of *A. wardi* generally corresponds to present definition of *A. pallipes*. It is this species that was documented as marginally sympatric and syntopic with the C Nepalese *A. gurkha* but generally occurring at higher altitudes (Corbet and Hill, 1992; Martens and Niethammer, 1972; as *A. sylvaticus* in both reports). In Suzuki et al.'s (2003) analyses of DNA sequences, their specimen of *"A. wardi"* from Mt Nilgiri, Nepal refers to *A. pallipes*. *Apodemus pallipes* is sympatric with *A. rusiges* in the Jammu-Kashmir region of NW India and N Pakistan (see that account), and is apparently the only *Apodemus* occurring elsewhere (except C Nepal) within the range outlined above. That distribution is apparently centered in the Pamirs and follows the Hindu Kush, subsidiary ranges radiating to the southwest from the Pamirs, and the Himalayas extending to the east.

Apodemus pallipes is smaller in body size than either *A. gurkha* or *A. rusiges*; its dorsal coat ranges from pale buffy brown to drab brownish gray, ventral fur whitish gray or solid white without a pectoral wash; tail either coequal or longer than head and body, rarely reaching 117 mm in adults (most fall within 100-110 mm); skull length rarely greater than 27 mm, length of molar row 3.6-4.0 (3.8 mm most frequent). *Apodemus pallipes* is not the same as *A. witherbyi*, which occurs in S Pakistan, Iran, and farther west (see that account), but the two species closely resemble each other in body size and relative tail length, and some samples are similar in fur coloration. The stephanodont pattern on M1 characterizing most *A. witherbyi* (Filippucci et al., 1996) occurs at a significantly lower frequency in *A. pallipes*. If this morphological resemblance reflects phylogeny, it would explain the similarity in protein polymorphism between Nepalese samples identified as *wardi* and Iranian samples (allozymic data originally reported by Darviche et al, 1979), and the striking differences between them and samples of true *A. sylvaticus*, which indicated that the Nepalese and Iranian specimens were more closely related to each other than to *A. sylvaticus* and were unlikely to be geographic representatives of the latter (Gemmeke and Niethammer, 1982). Apparently *A. pallipes* is also closely related to *A. uralensis* (see that account).

Two other names may belong in synonymy. The Tibetan *bushengensis* from SW Xizang was described and subsequently listed as a subspecies of *A. sylvaticus* (Feng et al., 1986); we provisionally allocate it to *A. pallipes* until specimens can be examined. Blanford's (1879) *Mus sublimus* from Kashmir (Ladakh, west of Pankong Lake, 13,000 ft [3962 m]) had provisionally been associated with *M. musculus* (e.g., Ellerman, 1961), but the holotype is a young *Apodemus* (J. T. Marshall, Jr., 1977*b*:213; Wroughton, 1920). Marshall equated it with *wardi* (= *A. pallipes*), but it could also be a young *A. rusiges*; the description of *sublimus* predates both names. Because the holotype is young and the skull is lost (J. T. Marshall, Jr., 1977*b*), the specimen may not be identifiable. We suggest the name be treated as *nomen dubium*, at least until another attempt is made to identify Blanford's specimen and resolve the issue.

Apodemus peninsulae (Thomas, 1907). Proc. Zool. Soc. Lond., 1906:862 [1907].

COMMON NAME: Korean Field Mouse.

TYPE LOCALITY: Korea, 110 mi SE of Seoul, Mingyoung.

DISTRIBUTION: A northern segment in S Siberia from Altai Mtns in the west to Ussuri region in the east (Gromov and Erbajeva, 1995; Reiter et al., 1995, who documented occurrence on Svjatoj Nos peninsula and isthmus in Lake Baikal), and the Russian island of Sakhalin (Abe et al., 1996) and Japanese island of Hokkaido (Dobson, 1994; Kaneko, 1994); an eastern arm ranging south through E Mongolia and NE China (Heilongjiang, Jilin, Liaoning, Hebei, and E Nei Mongol), and the Korean Peninsula (Won and Smith, 1999); then extending westward through N China (Shanxi, Shaanxi, SE Gansu to SE Qinghai), and south through SW Sichuan to NW Yunnan and E Xizang (Tibet). We are unaware of any records south of NW Yunnan and west of about 92° E longitude. Musser and Carleton (1993) incorrectly included the Chinese province of Xinjiang within the range, but only *A. agrarius* and *A. uralensis* (recorded as *A. sylvaticus tsherga*) are known from that region (Ma et al., 1987). The distribution outlined here is based upon our study of museum specimens (Musser et al., 1996) and the references cited above, and is similar to the distribution described by Wang (2003) and Zhang et al. (1997).

STATUS: IUCN – Lower Risk (lc).

SYNONYMS: *giliacus* (Thomas, 1907); *major* (Radde, 1862) [not Pallas, 1779, or Brants, 1827]; *majusculus* (Turov, 1924); *nigritalus* Hollister, 1913; *praetor* Miller, 1914; *qinghaiensis* Feng, Zheng and Wu, 1983; *rufulus* (Dukelski, 1928); *sowerbyi* Jones, 1956; *tscherga* (Kastchenko, 1899).

COMMENTS: *Apodemus* group. Usually considered a member of subgenus *Alsomys* (Pavlinov et al., 1995*a*) but placed in subgenus *Apodemus* by Pavlinov and Rossolimo (1987) and Mezhzherin and Zykov (1991), and included in an *Apodemus* Group by Musser et al. (1996) based upon morphology, which has been consistently corroborated by analyses of allozymes and sequences from mtDNA cytochrome *b* and nuclear DNA (Chelomina, 1998; Chelomina et al., 1998*b*; Filippucci et al., 2002; Liu et al, 2004; Mezhzherin, 1997*a*; Michaux et al., 2002*a*; Seriwaza et al., 2000; Suzuki et al., 2003). The Chinese *qinghaiensis* was described as a subspecies (Feng et al., 1983). Corbet (1978*c*) listed *nigritalus* as a synonym of *A. sylvaticus*, but the holotype is *A. peninsulae* (Musser and Carleton, 1993; Musser et al., 1996; Pavlinov and Rossolimo, 1987). *Apodemus peninsulae* is sympatric with the smaller-bodied *A. uralensis* in the Altai region, which Hollister (1913*b*) had noted (but under different names; see Ellerman and Morrison-Scott, 1951:567). In a revision of N Eurasian species of *Apodemus*, Mezhzherin (1997*a*) used the older name *major* instead of *peninsulae*, but *major* is twice preoccupied (see Ellerman and Morrison-Scott, 1951:566).

Available chromosomal information consists of karyotype description (Chen et al., 1996), and B-chromosomal analyses and references to chromosomal studies of *A. peninsulae* (Bekasova et al., 1980; Borisov and Malygin, 1991; Karamysheva et al., 2002; Kartavtseva et al., 2000; Kolomiets et al., 1988). Biochemical systematics and biogeographical implications in reference to *A. speciosus* and the Hokkaido *giliacus* reported by Saitoh et al. (1989), who also reviewed the contrasting treatment of *giliacus* as either a subspecies of *A. peninsulae* or a separate species. Molecular and chromosomal variation along with its geographic significance investigated (under the name *Alsomys major*) by Mezhzherin (2001*a*). Korean samples have been the subject of studies involving variation of mtDNA restriction fragments (Koh et al., 1995*b*) and morphometric and chromosomal analyses (Koh, 1988); morphometric variation among Chinese and Korean samples elucidated by Koh and Lee (1994) and Koh et al. (1996). The latter authors included a sample from NE Xinjiang in their study, but its much smaller body size compared with their other samples of true *peninsulae* from E China and Korea suggests their Xinjiang sample represents the small-bodied *A. uralensis*, which occurs in the region. Phylogenetic analyses of mtDNA pcytochrome *b* and nuclear IRBP sequences isolated *A. peninsulae* as a separate lineage in a larger clade containing *A. speciosus*, an *A. agrarius-A. chevrieri* cluster, and a *A. draco-A. semotus-A. latronum* group, which represents an evolutionary lineage separate from the more ancient radiations producing the Nepalese *A. gurkha* and Japanese *A. argenteus* (Suzuki, et al., 2003). Liu et al. (2004) could not resolve phylogenetic affinities among *A. peninsulae, A. speciosus, A. chevreri,* and *A. agrarius* using complete mtDNA cytochrome *b* sequences. Musser et al. (1996) outlined the pattern of geographic variation derived from qualitative inspection of their material, which is similar to that revealed by multivariate analysis of morphometric variation (Koh and Lee, 1994; Koh et al., 1996).

The taxon *tscherga* has been treated as a subspecies of *A. sylvaticus* (Ellerman and Morrison-Scott, 1951; Liu et al., 2002; Ma et al., 1987; Pavlinov and Rossolimo, 1987) or a synonym of either *A. uralensis* (Musser and Carleton, 1993) or *A. peninsulae* (Corbet, 1978*c*). Mezhzherin (1997*a*) explained that its original description suggested a composite of *A. sylvaticus* and *A. peninsulae*, but Kuznetzov studied the type series and identified all the specimens as *A. peninsulae* (or *A. major*, the name preferred by the Russians). Fragments identified as *A. peninsulae* have been recovered from late Pleistocene cave sediments in the Sichuan-Quizhou region of S China (Zheng, 1993) and fissure fillings in the Shandong area (Zheng et al., 1997).

Apodemus ponticus (Sviridenko, 1936). Abs. Works Zool. Inst. Moscow State Univ., 3:103.

COMMON NAME: Caucasus Field Mouse.

TYPE LOCALITY: N Caucasus, Chernomorski Dist., (Black Sea) Olgino Village.

DISTRIBUTION: Endemic to the Caucasus, from shore of Azov Sea through the Caucasus (S Russia, Georgia, Azerbaijan); see Mezhzherin (1991, 1997*a*), Orlov et al. (1996*a*, *b*), and Vorontsov et al. (1992).

STATUS: IUCN – Lower Risk (lc).

SYNONYMS: *argyropuli* Ellerman and Morrison-Scott, 1951; *argyropuloi* Heptner, 1948; *brevicauda* (Sviridenko, 1936); *parvus* Argyropulo, 1941 [not Bechstein, 1796]; *persicus* Gromov, 1963 [see Mezhzherin, 1991]; *planicola* (Sviridenko, 1936); *samaricus* (Sviridenko, 1936).

COMMENTS: *Sylvaemus* group. The synonyms have been associated with *A. flavicollis* (Ellerman and Morrison-Scott, 1951; Harrison and Bates, 1991), but they represent samples of *A. ponticus*, which is apparently a distinct species (Bobrinskii et al., 1944) defined by chromosomal, allozymic, molecular, and morphological traits that is closely related to *A. flavicollis* (Chelomina, 1998; Chelomina et al., 1998*a*, *b*; Mezhzherin, 1991, 1997*a*; Orlov et al., 1996*a*, *b*; Reutter et al., 2003; Vereshchagin, 1967:510; Vorontsov et al., 1992; Mezhzherin's 1991 report is the most comprehensive). *Apodemus flavicollis* is absent from most of the Caucasus, and has been reported only from Armenia (Frynta et al., 2001; Macholán et al., 2001*b*). The names *argyropuli* Ellerman and Morrison-

Scott, 1951, and *argyropuloi* Heptner, 1948 (see Harrison and Bates, 1991), were proposed to replace *parvus*. Included in Mezhzherin's (1997*a*) revision of N Eurasian *Apodemus*. Contrasted with *A. arianus* and *A. argenteus* by restriction analysis of nuclear DNA (Chelomina et al., 1995), and with *A. witherbyi* (as *A. fulvipectus*) and *A. uralensis* by comparison of isozymic, chromosomal, and molecular divergence (Chelomina et al., 1998*a*), and with European and Georgian species by a combination of methodological approaches (Hille et al., 2002). Placed in the superspecies *A. flavicollis* by Orlov et al. (1996*a*, *b*). Reviewed by Gromov and Erbajeva (1995, as *fulvipectus*).

Apodemus rusiges Miller, 1913. Proc. Biol. Soc. Wash., 26:81.

COMMON NAME: Kashmir Field Mouse.

TYPE LOCALITY: N India, C Kashmir ("Central Kashmir and the Pir Panjal" True, 1894).

DISTRIBUTION: Himalayas in Northern Area of NE Pakistan east of Indus River Valley (Baltistan and Hazara districts; large series in USNM) and Kashmir and Jammu region of NW India (large samples in USNM, including cotypes of *griseus*); altitudinal range, 1980-3350 m.

STATUS: IUCN – Lower Risk (lc).

SYNONYMS: *griseus* (True, 1894) [not Mina Palumbo, 1868].

COMMENTS: *Sylvaemus* group. Endemic to the Himalayas of N Pakistan (east of the Indus Valley) and NW India. Originally described as *Mus arianus griseus* (True, 1894), subsequently listed as a subspecies of either *A. flavicollis* (Ellerman, 1941, 1961; Ellerman and Morrison-Scott, 1951) or *A. sylvaticus* (Corbet, 1978*c*; Wroughton, 1920), placed in the synonymy of *A. sylvaticus* (Corbet and Hill, 1992), and finally regarded as a synonym of *A. sylvaticus wardi* (Agrawal, 2000). *Apodemus sylvaticus* does not occur in the Himalayas (see that account), and *wardi* is a synonym of *A. pallipes* (see that account). The latter and *A. rusiges* are the only *Apodemus* recorded from the Himalayas of N Pakistan and NW India and their ranges broadly overlap (we have not seen samples collected from the same place). *Apodemus rusiges* is larger than *A. pallipes* and further differs in having a much longer tail relative to length of head and body (often reaching 120 mm; coequal or not as relatively long in *A. pallipes*), dark brownish gray upperparts (pale buffy brown to drab grayish brown), larger skull (up to 30 mm long; rarely more than 27 mm in *A. pallipes*) with longer rostrum, and longer molar row (3.8-4.2 mm; rarely greater than 3.8 mm in *A. pallipes*); also see description in Wroughton (1908*b*).

Apodemus flavicollis, with which *rusiges* had also been associated, has not been found east of the Zagros Mtns of W Iran, and has smaller average cranial dimensions but larger bullae and a conspicuously higher braincase (small bullae and low cranium in *A. rusiges*), buffy pectoral suffusion that extends across chest in some specimens (no pectoral markings in *A. rusiges*), brighter upperparts, and shorter tail relative to head and body.

Distribution described here derives from the large series in USNM. Other smaller collections examined do not have specimens from beyond that range. Ellerman (1961) identified *rusiges* from the Kumaun region of N India, which is farther east than the Jammu-Kashmir area and adjacent to Nepal, but without studying the specimens we cannot determine if they are *A. rusiges* or *A. pallipes*. *Apodemus rusiges* may not occur west of the Indus Valley in N Pakistan. The USNM surveys collected large series of *A. rusiges* and *A. pallipes* east of the deep Indus River Valley in N Pakistan, but only *A. pallipes* was obtained west of the Indus.

Apodemus semotus Thomas, 1908. Ann. Mag. Nat. Hist., ser. 8, 1:44.

COMMON NAME: Taiwan Field Mouse.

TYPE LOCALITY: Taiwan (M.-j. Yu, 1996).

DISTRIBUTION: Endemic to montane habitats on Taiwan (H.-T. Yu, 1994).

STATUS: IUCN – Lower Risk (nt).

COMMENTS: *Apodemus* group. Usually included in subgenus *Alsomys* (Pavlinov et al, 1995*a*); placed in an *Apodemus* Group by Musser et al. (1996) based on morphology, which is consistent with analyses of allozymic data and mitochondrial and nuclear DNA sequences (Filippucci et al., 2002; Liu et al., 2004; Michaux et al., 2002*a*; Serizawa et al.,

2000; Suzuki et al., 2003); however, Chelomina et al. (1998b) grouped *A. semotus* with *A. argenteus* using mtDNA cytochrome *b* sequences. Originally described as a species, but subsequently treated as a subspecies of *A. sylvaticus* (Ellerman, 1949; Ellerman and Morrison-Scott, 1951), then either regarded as a species closely related in morphology to mainland *A. draco* (Corbet, 1978c) or questionably included in that species (Corbet and Hill, 1992), and finally recognized as a separate specific entity (Musser and Carleton, 1993; Liu et al., 2002; Musser et al., 1996; Wang, 2003; molecular reports cited above). Included in studies on ecology, altitudinal distributions, and patterns of diversification and genetic population structure of small Taiwanese mammals (Adler, 1996; H.-T. Yu, 1993, 1994, 1995). Relationship of montane *A. semotus* to lowland *A. agrarius*, based upon allozymic data, documented by H.-T. Yu (1995).

Apodemus speciosus (Temminck, 1844). *In* Siebold, Temminck and Schlegel, Fauna Japonica, Arnz et Socii, Lugduni Batavorum, p. 52.

COMMON NAME: Large Japanese Field Mouse.

TYPE LOCALITY: Japan.

DISTRIBUTION: Endemic to Japan (Dobson, 1994); found on the four larger islands (Hokkaido, Honshu, Shikoku, and Kyushu) and some smaller islands, but south only to Yakushima (Corbet, 1978c; Abe and Ishii, 1987; Kaneko, 1994; Tsuchiya, 1974).

STATUS: IUCN – Lower Risk (lc).

SYNONYMS: *ainu* (Thomas, 1906); *dorsalis* Kuroda, 1924 [see Kaneko and Maeda, 2002]; *insperatus* Kuroda, 1938 [see Kaneko and Maeda, 2002]; *miyakensis* Imaizumi, 1969 [see Kaneko and Murakami, 1996, and Kaneko and Maeda, 2002]; *navigator* (Thomas, 1906); *sadoensis* Tokuda, 1939 [*nomen nudum*; see Kaneko and Maeda, 2002]; *sadoensis* Tokuda, 1941 [see Kaneko and Maeda, 2002]; *tusimaensis* Tokuda, 1939 [*nomen nudum*; see Kaneko and Maeda, 2002]; *tusimaensis* Tokuda, 1941 [see Kaneko and Maeda, 2002].

COMMENTS: *Apodemus* group. Usually included in subgenus *Alsomys* (e.g., Pavlinov et al., 1995a) but placed in subgenus *Apodemus* by Pavlinov and Rossolimo (1987) and Mezhzherin and Zykov (1991); considered a member of an *Apodemus* Group by Musser et al. (1996), which is consistent with analyses of allozymic data and mtDNA cytochrome *b* and nuclear DNA sequences (Chelomina, 1998; Chelomina et al., 1998b; Liu et al., 2004; Mezhzherin, 1997a; Serizawa et al., 2000; Suzuki et al., 2003). Phylogenetic analyses of mtDNA cytochrome *b* and nuclear IRBP sequences identified *A. speciosus* as an independent lineage within a radiation that includes the separate lineage of *A. peninsulae*, the *A. agrarius-A. chevrieri* clade, and *A. draco-A. semotus-A. latronum* cluster; this radiation is separate from those two producing the Nepalese *A. gurkha* and Japanese *A. argenteus* (Suzuki et al., 2003; also see Liu et al., 2004). Sympatric with *A. peninsulae* on Hokkaido. Geographic variation extensively analyzed by Imaizumi (1962). Some of the synonyms (*ainu*, *navigator*, and *miyakensis*) have been listed as separate species (Vorontsov et al., 1977a; also see discussion in Corbet, 1978c), but we follow Corbet (1978c) who included them in a single species. This action was also reflected by Tsuchiya (1974), who divided the names into two groups of *A. speciosus* based on biochemical and chromosomal evidence. Saitoh et al. (1989) provided additional biochemical information in context of systematic comparisons among Japanese *Apodemus*. Evolution of restriction sites of ribosomal DNA in natural populations reported by Suzuki et al. (1994b). Inter- and intraspecific patterns of morphological variation in sympatric *A. speciosus* and *A. argenteus* and its importance in insular isolation and biogeographic gradients reported by Renaud and Millien (2001). Analysis of the cranial and mandibular skeletal and myological morphology associated with incisal biting and mastication reported by Satoh (1997, 1998, 1999) and contrasted with the arvicoline *Myodes rufocanus*. Significance of lower incisor shape and size related to ecological and taxonomic inquiries investigated by Millien-Parra (2000a). Adaptive latitudinal trends in mandible shape contrasted with *A. sylvaticus* and *A. argenteus*. Reviewed by Kawamura (1989, 1991, 1994), Kaneko (1994), Musser et al., (1996), and partially by Mezhzherin (1997a). Catalog of specimens in National Science Museum, Tokyo, provided by Endo et al. (2002).

Fossils extend the history of *A. speciosus* in Japan back to the middle Pleistocene and Kawamura (1989:57) noted that "the temporal morphological changes since the middle Pleistocene are generally slight." Citation for the original description is usually cited as 1845, but was published in 1844; see Holthuis and Sakai (1970).

Apodemus sylvaticus (Linnaeus, 1758). Syst. Nat., 10th ed., 1:62.

COMMON NAME: Long-tailed Field Mouse.

TYPE LOCALITY: Sweden, Uppsala (neotype designated by Zagorodnyuk [1993]).

DISTRIBUTION: European and N African: Europe north to Scandinavia; south to NW Turkey (Thrace and NW Anatolia; Filippucci et al., 1996; Pamukoglu and Albayrak, 1996; specimens in USNM); and east to C Belarus, E Ukraine, and closely adjacent W Russia, which is the easternmost limit of the species (Zagorodnyuk, 1993); see maps in Zagorodnyuk et al. (1997:39), Mezhzherin (1997a:33), Niethammer (1978c:341), and Mitchell-Jones et al. (1999:275). Range in N Africa extends from Atlas Mtns in Morocco east across Algiers to Tunisia (Aulagnier, 1991; Aulagnier and Thevenot, 1986; Kock and Felton, 1980). Also found on Iceland; Britain, Ireland, and numerous nearby islands; Aegean islands (Kryštufek, 2002a; Özkan and Kryštufek, 1999); some islands in the Tuscan Arch. (De Marinis et al., 1996); Sardinia, Corsica (Masseti, 1993), and other Mediterranean islands (Alcover and Gosalbez, 1988; Amori, 1993; Amori and Masseti, 1996; Cheylan, 1991).

STATUS: IUCN – Lower Risk (lc).

SYNONYMS: *albus* (Bechstein, 1801); *algirus* (Pomel, 1856); *alpinus* (Burg, 1921); *bergensis* (Krausse, 1921); *butei* Hinton, 1914; *callipides* (Cabrera, 1907); *candidus* (Bechstein, 1796); *celticus* (Barrett-Hamilton, 1900); *chamaropsis* (Levaillant, 1857); *charkovensis* (Migulin, 1936); *clanceyi* Harrison, 1947; *creticus* Miller, 1910 (see Kryštufek, 2002a); *cumbrae* Hinton, 1914; *dichruroides* Miric, 1960; *dichrurus* (Rafinesque, 1814) [not Cabrera, 1921]; *dichrurus* Cabrera, 1921 [not Rafinesque, 1814]; *eivissensis* Alcover, 1977; *fiolagan* Hinton, 1914; *flaviventris* (Petrov, 1943); *flavobrunneus* (Hilzheimer, 1911); *fridariensis* (Kinnear, 1906); *frumentariae* Sans-Coma and Kahmann, 1977; *ghia* Montagu, 1923; *grandiculus* Degerbol, 1939; *granti* Hinton, 1914; *griseus* (Mina Palumbo, 1868); *hamiltoni* Hinton, 1914; *hayi* (Waterhouse, 1838); *hebridensis* (de Winton, 1895); *hermani* Felten and Storch, 1970; *hessei* Miric, 1960; *hirtensis* (Barrett-Hamilton, 1899); *ifranensis* Saint Girons and Bree, 1962; *ilvanus* Kahmann and Niethammer, 1971; *intermedius* (Bellamy, 1839); *isabellinus* (Mina Palumbo, 1868); *islandicus* (Thienemann, 1824); *krkensis* Miric, 1968; *larus* Montagu, 1923; *leptodus* Kretzoi, 1956; *leucocephalus* (Bechstein, 1796); *maclean* Hinton, 1914; *maximus* (Burg, 1925); *milleri* de Beaux, 1926; *nesiticus* Warwick, 1940; *niger* (Bechstein, 1796); *parvus* (Bechstein, 1793) [not Argyropulo, 1941], *pecchioli* (Pecchioli, 1844); *reboudia* (Loche, 1867); *rufescens* Saint Girons and Bree, 1962; *spadix* Fritsche, 1934; *thuleo* Hinton, 1919; *tirae* Montagu, 1923; *tural* Montagu, 1923; *varius* (Bechstein, 1796); *vohlynensis* (Migulin, 1938) [Charlemagn, 1936, proposed the name but it is a *nomen nudum*; see Pavlinov and Rossolimo, 1987; Zagorodnyuk, 1992b].

COMMENTS: *Sylvaemus* group. The geographic range as mapped by Corbet (1978c) east of Belarus and E Ukraine reflects distributions of other species (*A. uralensis, A. witherbyi, A. pallipes,* and *A. rusiges*) once included within *A. sylvaticus;* "*A. sylvaticus* is virtually absent from the entire area of the Middle East" (Macholán et al., 2001b:806). Contrary to published records, *A. sylvaticus* is not part of the modern Israeli fauna (Filippucci et al., 1989). Tchernov (1979, 1986, 1994, 1996), however, maintains it is and is also represented there by fossils from middle to late Pleistocene cave sediments (late Acheullan to Upper Mousterian), and from 40,000 to 10,000 years B.C. During that Pleistocene interval, *A. mystacinus* was found along with *A. sylvaticus* in the same caves, but *A. flavicollis* was usually absent (Tchernov, 1979). Tchernov's "*sylvaticus*" probably represents *A. wltherbyi* (B. Kryštufek, 2002, in litt.; see account of *A. witherbyi*), which along with *A. mystacinus* and *A. flavicollis* are part of the modern Israeli fauna (see accounts of latter two). An Arabian Peninsula record at Qatar is based on *Mus* (Kock and Nader, 1990). See Kowalski (2001) for the inclusion of Hungarian *leptodus*.

The population of *A. sylvaticus* on Corsica arrived there (presumably by passive human transport from peninsular Italy through the Tyrrhenian islands, Michaux et al.,

1996*a*) at the beginning of the third millennium B.C. and occurred together with the endemic murine *Rhagamys orthodon* and arvicoline *Microtus henseli*, both of which vanished by the end of the first millenium B.C. (Vigne, 1992; Libois et al., 1993; and references cited in those reports). The population on Sicily represents a Holocene anthropogenic introduction (about 750,000 years ago); its origin is not peninsular Italy or W Europe but possibly North Africa or the E Mediterranean basin (Michaux et al., 1998). Allozymic variation among European and North African samples of *A. sylvaticus* (Filippucci, 1992; Filippucci et al., 2002; Libois et al., 2001) indicated high genetic homogeneity among samples of North African populations and recent derivation from SW Europe most likely through anthropogenic introduction, results corroborated by Michaux et al. (2003) based on phylogenetic analyses of mtDNA cytochrome *b* sequences. North African populations are no older than Holocene (Kowalski and Rzebik-Kowalska, 1991) or late Pleistocene (Ouahbi et al., 2001). Dobson (1998, 2000) claimed *A. sylvaticus* is unknown from the Maghreb fossil record prior to the Iron Age (about 3000-2300 years before present) and its presence in North Africa is likely the result of human intervention, but Ouahbi et al. (2001) documented it from late Pleistocene sediments in N Morocco.

Allozymic (Byrne et al., 1990; Fernandes et al., 1991; Gemmeke, 1981*a*) and morphometric (Alcantara, 1991; Murbach, 1979) intrapopulational analyses provided results of differentiation within *A. sylvaticus* and change in its evolutionary history (see Berry, 1973, and references therein); other allozymic and morphometric analyses of European populations presented by Gemmeke (1981*a*) and Nauroz (1984); high genetic diversity within samples from W Europe as measured by mtDNA cytochrome *b* sequences documented by Reutter et al. (2003); and phylogeography based on study of mtDNA cytochrome *b* sequences, which suggested postglacial colonizations of all of Europe from Iberian and S France refuge regions (Michaux et al., 2002*a*, 2003). Several studies have focused on aspects relevant to systematics of *A. sylvaticus*: morphometric and molecular analyses of samples from Mediterranean region in the context of taxonomy and biogeography (Libois et al., 1993; Michaux et al., 1996*a*, *b*, 1998; Sarà and Casamento, 1995*b*), adaptive latitudinal trends in mandible shape (Renaud and Michaux, 2003), insular gigantism in the Mediterranean (Alcover and Gosalbez, 1988; Libois and Fons, 1990, and references cited therein; Michaux et al., 2002*b*), B chromosomes in samples from Czech Republic (Zima et al., 1997*d*), sex-chromosome heterochromatin variation (Nová et al., 2002), epigenetic characteristics and divergence among Bulgarian populations (Markov and Chassovnikarova, 1999), morphometric analyses of samples from Iberian Peninsula (González-Esteban et al., 1996), variability in allometric relationships between body mass and skull size in Spanish samples (Alcantara et al., 1991), and karyotypes of Macedonian sample (Zima et al., 1997*a*). Conspecificity of *krkensis* with *A. sylvaticus* documented by Williams et al. (1980) and Dolan and Yates (1981). The name *charkovensis*, usually associated with *A. uralensis* (e. g., Mezhzherin, 1997*b*; Zagorodnyuk, 1992*b*), was identified by Zagorodnyuk (1993) as referring to the easternmost population of *A. sylvaticus*; he also selected a lectotype. Based only on chromosomal data, Orlov et al. (1996*a*, *b*) regarded *vohlynensis* as a separate species within the *A. sylvaticus* superspecies, and one that occurs in Central Europe, the Balkans, and east to the Dnepr basin. Distribution and population density of *A. sylvaticus* at the easternmost limits of its range (E Ukraine) and comparisons with *A. uralensis* documented by Naglov (1995). The range expansion of *A. sylvaticus* into forested tracts that were originally clay semidesert in the Trans-Volga region outlined by Bykov (1990). See account of *A. flavicollis* for reports documenting morphological, chromosomal, and molecular contrasts between it and the closely related and morphologically similar *A. sylvaticus*.

European populations reviewed by Niethammer (1978*c*) and Mitchell-Jones et al. (1999); N Eurasian segment by Mezhzherin (1997*a*, *b*). Literature covering ecological aspects of *A. sylvaticus* is voluminous, extending back for decades, and only a few current regional reports focusing on distribution, taxonomy, habitat, and other aspects are listed here for populations in: Pyrenees and NE Spain (Castién and J. Gosálbez, 1992; Torre et al., 1996), Portugal (Santos-Reis and Mathias, 1996), Netherlands

(Thissen and Hollander, 1996; Wammes, 1992a), Belgium (Libois, 1996), N Germany (Dolch et al., 1994), Slovakia (Danko, 1994; Kminiak, 1996; Mošanský, 1994; Stanko and Mosansky, 2000; Stanko et al., 1994), Czech Republic (And—ra and ervený, 1994; Smaha, 1996; Zima and And—ra, 1996), Austria (Bauer and Spitzenberger, 1996), Switzerland (Hausser, 1995; Maurizio, 1994), Italy (Amori et al., 1999, 2002b; Cagnin et al., 1996; Cantini, 1991; Cresti et al., 1994; Cerone and Aloise, 1994; Locatelli and Paolucci, 1996), E Baltic region (Miljutin, 1997, 1998; Timm et al., 1998), Serbia and Montenegro (Petrov, 1992), Albania (Prigioni, 1969); Bulgaria (Peshev, 1996), Translvanian Romania (Istrate, 1998), Slovenia (Kryštufek, 1991), Thrace (Greece; Vohralík, 1992) and NE Turkey (Kryštufek and Vohralík, 2001). North African populations reviewed by Kock and Felten (1980) and Kowalski and Rzebik-Kowalska (1991); the latter authors also point out that *algirus* Pomel, 1856 and *chamaeropsis* Loche, 1867 may refer to species of *Mus* or *Gerbillus* rather than *Apodemus*. Abundance of Algerian populations in different habitats compared with *Mus spretus* reported by Khidas et al. (2002).

Apodemus uralensis (Pallas, 1811). Zoogr. Rosso-Asiat., 1:168.
COMMON NAME: Herb Field Mouse.
TYPE LOCALITY: Russia, S Ural Mtns.
DISTRIBUTION: C Europe (E Germany S Poland, Czech Republic, Slovakia, NE Austria, N Hungary, south to E Croatia, N Serbia and Montenegro, Bulgaria and N Romania); Baltic region (NW Lithuania, Latvia and Estonia); east through W Russia and Ukraine to E Kazakhstan (Djarkent, USNM specimens including 155471, the holotype of *microtis*), Siberian Altai in S Russia (USNM), NW China (Xinjiang; Ma et al., 1987, as *tscherga*; USNM 259545 from Aksu), and the Altai of Mongolia; south in the Caucusus and throughout N Turkey (see Filippucci et al., 1996; Kryštufek and Vohralík, 2001; Mezhzherin, 1997a:36; Mitchell-Jones et al., 1999; Frynta et al., 2001; Steiner, 1978, under *A. microps*; Zhang et al., 1997, as *A. sylvaticus*); E and S boundary of range in C Asia unresolved.
STATUS: IUCN – Lower Risk (lc).
SYNONYMS: *baessleri* (Dahl, 1919) [see Zagorodnyuk, 1992b]; *balchanensis* (Kashkarov, 1981) [*lapsus calami* according to Pavlinov et al., 1995a]; *cimrmani* Vohralík, 2002; *ciscaucasicus* (Ognev, 1924); *kastschenkoi* Kuznetzov, 1932; *major* (Severtsov, 1873) [not Pallas, 1779, or Radde, 1862]; *microps* Kratochvíl and Rosicky, 1952; *microtis* Miller, 1912; *mosquensis* (Ognev, 1913); *nankiangensis* Wang, 1964; *pallidus* Kashkarov, 1926; *parvulus* Mosanský, 1994 [*nomen nudum* according to Vohralík, 2002]; *tokmak* (Severtzov, 1873) [*nomen nudum*].
COMMENTS: *Sylvaemus* group. Previously listed as a subspecies of *A. sylvaticus* (Corbet, 1978c; Ellerman and Morrison-Scott, 1951; Pavlinov and Rossolimo, 1987), but now recognized as the oldest name for a distinctive small-bodied species formerly called *A. microps* (Musser and Carleton, 1993; Pavlinov and Rossolimo, 1998; Pavlinov et al., 1995a). Synonymy of European *microps* with Asian *uralensis* has been documented by results of a variety of molecular and morphological studies (Bellinvia et al., 1999; Filippucci et al., 1996; Macholán et al., 2001b; Mezhzherin, 1997a; Mezhzherin and Mikhailenko, 1991; Mezhzherin and Zykov, 1991; Mezhzherin et al., 1992; Reutter et al., 2003). Mezhzherin (1997b:309) summarized historical literature indicating the presence of a species different from *A. sylvaticus* that corresponded to *A. uralensis*. Pertinent systematic treatments cover chromosomal and allozymic variation (as *microps*; Gemmeke, 1983; Mezhzherin, 1997b; Vorontsov et al., 1989, 1992); morphological data presented in context of distinguishing species of *Apodemus* (Tvrtkovic and Dzukic, 1977); variation in banded chromosomes between *A. uralensis* (as *microps*) and *A. alpicola* (Reutter et al., 2001); morphology (skin, skull, and dentition) of northern Turkish samples (Filippucci et al., 1996), and results of discriminant function analyses separating samples of E Turkish *A. uralensis* from Turkish and Iranian *A. flavicollis*, *A. witherbyi* (reported as *hermonensis*), and *A. hyrcanicus* (Frynta et al., 2001); comparison of isozymic, chromosomal, and molecular divergence among *A. uralensis*, *A. witherbyi* (reported as *A. fulvipectus*), and *A. ponticus* (Chelomina et al., 1998a); and morphological traits distinguishing *A. uralensis* and *A. sylvaticus* in E Ukraine (Naglov, 1995). Electrophoretic analysis of allozymic data indicated a recent separation of

A. flavicollis, A. alpicola, A. uralensis, and *A. witherbyi* (reported as *hermonensis*) from a common ancestor (Filippucci, 1992; Filippucci et al., 1966), and a closer link between *A. uralensis* and *A. hyrcanicus* than between *A. uralensis* and *A. flavicollis, A. sylvaticus,* or *A. witherbyi* (reported as *hermonensis*), the other species sampled (Macholán et al, 2001*b*). However, analyses of sequences from mtDNA cytochrome *b* and 12S rRNA identify *A. uralensis* as a sister-species to *A. alpicola* (Michaux et al., 2002*a*), as did cytochrome *b* haplotype divergences (Reutter et al., 2003), but not complete mtDNA cytochrome *b* sequences (Liu et al., 2004). Allozymic variation and morphology suggest a close alliance with *A. pallipes* from the Pamirs and Himalayas (Mezhzherin, 1997*a*). In Reutter et al's (2003) analyses of cytochrome *b* haplotype divergences, a a sample from GenBank labeled "*sylvaticus* 2" from Pakistan (specimen in FMNH) used by Jansa et al. (1999) and Liu et al. (2004) clustered with samples of *A. uralensis*; "*sylvaticus* 2" represents *A. pallipes* as that is the most common species in Pakistan and the one usually found in museum collections (Musser's identifications). Results derived from "*sylvaticus* 2" reinforces the close phylogenetic allicance between *A. pallipes* and *A. uralensis* indicated by Mezhzherin's (1997*a*) study; this alliance is also substantiated by qualitative observation of specimens (Musser's research), and phylogenetic analyses of mtDNA cytochrome *b* and nuclear IRBP sequences (Suzuki et al., 2003; sample identified as *A. wardi,* but see account of *P. pallipes*). Hille et al. (2002) summarized the various cladistic trees involving *A. uralensis* and other *Apodemus,* highlighting similarities and incongruences among them.

Species-boundaries are far from resolved. Using only chromosomal banding data, Orlov et al. (1996*a, b*) separated *mosquensis* and *ciscaucasicus* as separate species within what they referred to as the superspecies *Sylvaemus uralensis*; two forms of *A. uralensis* from the Caucasus were also detected by Hille et al. (2002). Based on morphological and allozymic data, Mezhzherin (1997*a*) recognized a clade containing *uralensis, kastschenkoi,* and *pallipes* (which had been listed as a synonym of *A. uralensis*; e.g., Musser and Carleton, 1993) and applied each name to a separate species. Mezhzherin explained that *kastschenkoi* was proposed by Kuznetzov for the small-bodied species in the Siberian Altai that had been identified as *tscherga.* The original description of *tscherga* suggested a composite of *A. sylvaticus* and *A. peninsulae,* but Kuznetzov studied the type series and identified all members as *A. peninsulae* (Russians use *A. major* instead of *peninsulae*). Mezhzherin noted that *kastschenkoi* has dark brown upperparts, gray belly, and is readily distinguished from *A. uralensis,* and that the names *microtis, tokmak,* and *balchaschensis* are probable synonyms of *A. uralensis.* We are not impressed with the degree of genetic distance separating Mezhzherin's samples of *A. uralensis* and *kastschenkoi* or the slight morphological differences between specimens we studied from the Siberian Altai and those representing *A. uralensis* from the west. On the other hand, species-status for *pallipes,* which occurs to the south in the Pamirs and Himalayas, and is larger-bodied with generally paler upperparts and white or whitish gray underparts (see that account), is supported by genetic-distance measurements and morphology, as Mezhzherin indicated. The NW Chinese *nankiangensis* (from Aksu in Xinjiang) was described as a subspecies of *A. sylvaticus* and compared with *tscherga* from farther north in Xinjiang and the Siberian Altai from which it differs primarily in color (sandy brown upperparts instead of grayish brown; Corbet, 1984; Wang, 1964); we include it as a synonym of *A. uralensis.* An isolated population in NW Bohemia in the Czech Republic was extensively documented by Vohralík (2002) under *cimrmani* (as a new subspecies of *A. microps*).

European populations reviewed by Steiner (1978, under *microps*) and Mitchell-Jones et al. (1999), Russian range and taxonomy reviewed by Gromov and Erbajeva (1995). Recent faunal studies incorporating distribution, taxonomy, ecology, and other information are available for: E Germany (Feiler and Tegegn, 1998), Austria (Bauer and Spitzenberger, 1996, as *microps*), Baltic region (Miljutin, 1997, 1998; Timm et al., 1998; Zagorodnyuk and Mezhzherin, 1992, as *microps*), Lithuania (Juškaitis et al., 2001), Serbia and Montenegro (Petrov, 1992, as *microps*), E Carpathians (Kyselyuk, 1993, as *microps*), Czech Republic (Krška, 1996, under *microps*; Vohralík, 2002, as *microps*; Zima and And—ra, 1996), Slovakia (Mosansky, 1994; Stanko and

Mosansky, 1994, 2000; Stanko et al., 1994), Bulgaria (Peshev, 1996, as *microps*), and NW China (Ma et al., 1987).

Apodemus witherbyi (Thomas, 1902). Ann. Mag. Nat. Hist., ser. 7, 10:490.

COMMON NAME: Steppe Field Mouse.

TYPE LOCALITY: S Iran, Fars Province, Shul (see Lay, 1967, for coordinates).

DISTRIBUTION: Plains, mountain and plateau steppes, and highland semideserts (not found in desert depressions) from E of the Dnepr River in the S Ukraine, Crimea, N Caucasus, S Caucasus (Georgia, Armenia, and Azerbaijan), Anatolian Turkish steppe and Bozcaada Isl (Kryštufek and Vohralík, 2001; specimens in USNM); south to N Israel and NW Jordan (Benda and Sádlová, 1999); through most of C and N Iran in provinces of Azarbayjan (FMNH and USNM material), Kordestan (FMNH and USNM), Ilam (series in FMNH), Lorestan (large samples in AMNH, FMNH, and USNM), Isfahan (FMNH material), Fars (FMNH), Semnan (FMNH 97469), Tehran (FMNH 341459), C and E Mazandaran (FMNH and USNM), N and E Khorasan (AMNH, FMNH, and USNM material); Kopet-Dag Mtns of SW Turkmenistan; and eastward in WC Pakistan, (about 90 km NE Quetta, USNM specimens); probably also occurs in Afghanistan, NE Iraq, and Lebanon and adjacent SW Syria. Distribution, which is southeast of range of *A. sylvaticus*, derived from specimens in AMNH, FMNH, and USNM and published reports (Filippucci et al., 1989; Mezhzherin, 1997a:33; Mezhzherin and Zagorodnyuk, 1989; Zagorodnyuk et al., 1997:39).

STATUS: IUCN – Endangered as *A. hermonensis*, Lower Risk (lc) as *A. fulvipectus*.

SYNONYMS: *caessareanus* Bate, 1942; *chorassanicus* (Ognev and Heptner, 1928); *falzfeini* Mezhzherin and Zagorodnyuk, 1989 [see Zagorodnyuk, 1992*b*]; *fulvipectus* (Ognev, 1924); *hermonensis* Filippucci, Simson, and Nevo, 1989; *iconicus* Heptner, 1948; *kilikiae* Kretzoi, 1964; *planicola* (Sviridenko, 1936); *saxatilis* Krassovsky, 1929; *saxatilis* (Sviridenko, 1936); *tauricus* (Barret-Hamilton, 1900) [not Pallas, 1811].

COMMENTS: *Sylvaemus* group. The taxon *witherbyi* was originally described as a subspecies of *Mus sylvaticus*, subsequently arranged as a subspecies of *A. sylvaticus* (Ellerman, 1941), treated as a synonym of *A. sylvaticus arianus* (Ellerman and Morrison-Scott, 1951) or *A. sylvaticus* (Corbet, 1978*c*), and listed as a synonym of *A. arianus* (Musser and Carleton, 1993; Pavlinov et al., 1995*a*). Zagorodnyuk (1996*a*) finally identified the holotype as an example of *A. uralensis*. This is the species identified as *A. arianus* by Musser and Carleton (1993), who separated it from *A. sylvaticus* on the basis of its distinctive pelage and smaller body size. Recently, Zagorodnyuk (1996*a*), Zagorodnyuk et al. (1997), and Mezhzherin (1997*a*) elucidated the morphological and distributional boundaries of *A. witherbyi* (as *arianus*) and their definition incorporates the names and ranges of *fulvipectus*, *falzfeini*, and *hermonensis*, each of which has been treated as a species by various authors (Chelomina et al., 1998*a*; Filippucci et al. 1989, 1996; Mezhzherin and Zagorodnyuk, 1989; Musser and Carleton, 1993). Mezhzherin and Zagorodnyuk (1989) described *falzfeini* as a species, which O. Rossolimo (in litt.) considered to be the same as *A. fulvipectus*. Subsequent results from genetic studies led Mezhzherin and Zykov (1991) to treat *falzfeini* as identical with *chorassanicus*, which was originally described as a subspecies of *A. sylvaticus*; the identity of *chorassanicus* with *fulvipectus* and *hermonensis* is supported by allozymic data (see Filippucci et al., 2002:407). Vorontsov et al. (1992) recognized *fulvipectus* as the oldest name for this species; it was one of the distinct electrophoretic forms in the Caucasus revealed by Vorontsov et al. (1989). *Apodemus hermonensis* was described from the alpine "tragacanthic" belt at about 2000 m on Mt Hermon and was considered to be morphologically and genetically closely related to *A. flavicollis*, which it displaces on Mt Hermon (and possibly also Lebanon and Antilebanon mountain ranges) at elevations above 1900 m (Filippucci et al., 1989). Based on allozymic and morphological data, *hermonensis* was subsequently identified from Anatolian Turkey and N Iran and suspected to be synonymous with *fulvipectus* (Filippucci et al., 1996; Macholán et al., 2001*b*); Michaux et al. (2002*a*), however, retained the two as separate species based upon analysis of mtDNA cytochrome *b* sequences.

Kryštufek (2002*a*) has convincingly demonstrated that the holotype of *arianus* is an example of *A. flavicollis*, that the holotype of *tauricus* Barrett-Hamilton, 1900, is the same as *hermonensis*, and is an older name but preoccupied by *tauricus* Pallas, 1811, which is associated with *A. flavicollis*. He suggested that *iconicus* Heptner, 1948 was the oldest available name for the species, but also noted that the holotype of *witherbyi* Thomas, 1902 might be identical with *hermonensis*. The holotype is similar to *hermonensis* in all external, cranial, and dental traits (including the stephanodont pattern on M1), except for its longer bullae (4.8 mm) and shorter molar row (3.4 mm), and Kryštufek was reluctant to declare *witherbyi* and *hermonensis* as being the same. Two specimens we studied from the Zagros Mtns of W Iran, USNM 369849 (Baneh, Kordistan Province) and USNM 350595 (Borujerd region, Luristan Province), have all the chromatic and morphological attributes of *hermonensis*; molar row is 3.5 mm in the former, bullar length is 4.8 mm in the latter. Like these two, the holotype of *witherbyi* is at one end of the range of variation in lengths of molar row and bullae. We use *witherbyi* as the oldest name for what has been called *arianus* (Musser and Carleton, 1993; Zagorodnyuk et al., 1997; Mezhzherin, 1997a), *hermonensis*, *fulvipectus* or *iconicus* (Filippucci et al., 1996; Kryštufek and Vohralík, 2001); no other species of *Apodemus* occurs in the Fars region, and all of its other traits match those of *hermonensis* and the other material we assemble here.

Apodemus witherbyi may be the species represented by the sample from Qazvin, N Iraq, that Darviche et al. (1979) separated electrophoretically from *A. sylvaticus* and *A. flavicollis*, but still considered closer to the latter. A close alliance between *A. flavicollis* and *A. witherbyi* is supported by other studies employing protein electrophoresis (Filippucci et al., 1989, 2002; Macholán et al., 2001b; Mezhzherin, 1997a) and discriminant function analysis of morphometric traits (Frynta et al., 2001), but not mtDNA cytochrome *b* gene sequences in which *A. witherbyi* is isolated within subgenus *Sylvaemus* (Michaux et al., 2002a; reported as *A. hermonensis*). Included in Mezhzherin's (1997a) revision of N Eurasian *Apodemus*. Compared (as *A. fulvipectus*) with *A. ponticus* and *A. argenteus* by restriction analysis of nuclear DNA (Chelomina et al., 1995), and with *A. ponticus* and *A. uralensis* by comparison of isozymic, chromosomal and molecular divergence (Chelomina et al., 1998a). Phylogenetic affinity of *A. witherbyi* requires sharper resolution, particularly concerning the nature of the relationship with *A. pallipes* from the Pamirs and Himalayas and the Himalayan *A. rusiges*.

Easternmost record of *A. witherbyi* is from the Quetta region of WC Pakistan, and the four specimens from there match those in the large samples from Iran in chromatic traits, cranial morphology, and molar occlusal patterns. In Iran, *A. witherbyi* is the most widespread *Apodemus*, occurs syntopically with *A. flavicollis* in the Zagros Mtns, and is altitudinally parapatric with *A. hyrcanicus* (see those accounts); these are the only three *Apodemus* currently recorded from Iran. Lay (1967:186) noted that a bright, small *Apodemus*, and a larger, darker species occur in N Iran, and considered them ecological subspecies; the brighter one is *A. witherbyi*, the darker *A. hycanicus* (see that account).

Apodemus witherbyi, *A. mystacinus*, and *A. flavicollis* are the only *Apodemus* that now occur in Israel (see accounts of latter two). Tchernov (1968, 1979, 1986, 1996) has identified fossilized *Apodemus* recovered from Pleistocene Israeli cave sediments. Three size classes of molars are represented. Tchernov identified the largest as *A. mystacinus*, the next in size as *A. flavicollis*, and the smallest as *A. sylvaticus* (Tchernov, 1986:263), which he regarded as part of the modern Israeli fauna. He (Tchernov, 1986) also synonymized Bate's (1942b) Pleistocene *A. caesareanus* with *A. sylvaticus*. The latter, however, does not occur in Israel or anywhere in the Middle East (Filippucci et al., 1989; Macholán et al., 2001b), and Kryštufek (2002, in litt.) thinks the smallest morphotype in the Pleistocene samples represents *A. witherbyi*, not *A. sylvaticus*. We agree. Certainly the illustrations of *caesareanus* molar rows (Tchernov, 1968:52, 54) exhibit some of the traits characteristic of *A. witherbyi* and the molars are similar in dimensions (Filippucci et al., 1996; Kryštufek, 2002a).

Apomys Mearns, 1905. Proc. U.S. Natl. Mus., 28:455.

TYPE SPECIES: *Apomys hylocetes* Mearns, 1905.

COMMENTS: *Chrotomys* Division. At one time *Apomys* was included in *Rattus*, but is a distinct genus and forms a monophyletic group within the assemblage of Philippine Old Endemics (Musser and Heaney, 1992). Taxonomic history of the genus and preliminary systematic revision provided by Musser (1982*b*), who separated species into an *Apomys datae* Group and *Apomys abrae-hylocetes* Group, which has been confirmed by phylogenetic analyses of mitochindrial DNA cytochrome *b* sequences (Steppan et al., 2003); additional taxonomic notes and phylogenetic relationships outlined by Musser and Heaney (1992). Phylogenetic analyses of complete mtDNA cytochrome *b* sequences for 13 of the 16 genera of endemic Philippine murines united *Apomys* within a clade containing *Archboldomys*, *Chrotomys* (along with *Celaenomys*), and *Rhynchomys*, which are also Old Endemics (Jansa and Heaney, 2001). This alliance is supported by data from chromosomes (Rickart and Heaney, 2002) and morphology (Musser and Heaney, 1992). Five undescribed species are recorded from Negros, Panay, Sibuyan, Camiguin, and Mindoro islands (Heaney et al., 1998); actual distribution in archipelago and number of species still unknown. A sample from Negros has 2n = 30, FN = 50, one of the lowest diploid numbers recorded for Indoaustralian murines (Rickart and Musser, 1993).

Phylogenetic analyses of mtDNA cytochrome *b* sequences for ten species of *Apomys* (six described, four undescribed) presented by Steppan et al. (2003). Their results confirm the monophyly of *Apomys*, and its endemicity to the Philippines, suggest that the genus originated on Luzon Isl, and imply its diversification within the archipelago during the Pliocene. Some speciation events occurred on Luzon or Mindanao, ". . . the two largest, oldest, and most topographically complex islands" (Steppan et al., 2003:699), one speciation event is associated with vicariance related to Pleistocene sea-level fluctuations, but most speciation events are associated with dispersal to newly developed oceanic islands.

Apomys abrae (Sanborn, 1952). Fieldiana Zool., 33(2):133.
    COMMON NAME: Luzon Cordillera Apomys.
    TYPE LOCALITY: Philippines, Luzon Isl, Abra Province, Abra, 3500 ft (1067 m).
    DISTRIBUTION: Greater Luzon Faunal Region. Endemic to Central Cordillera of N Luzon (Benguet, Ilocos Norte, and Mountain provinces), 1060-2500 m (Musser, 1982b; Heaney et al., 1998).
    STATUS: IUCN – Lower Risk (lc).
    COMMENTS: Included within the *Apomys abrae-hylocetes* Group by Musser (1982*b*).

Apomys datae (Meyer, 1899). Abh. Mus. Dresden, ser. 7, 7:25.
    COMMON NAME: Northern Luzon Apomys.
    TYPE LOCALITY: Philippines, N Luzon Isl, Lepanto.
    DISTRIBUTION: Greater Luzon Faunal Region. Most records come from highlands in N Luzon where it inhabits primary montane forest in the Sierra Madre and Central Cordillera, 760-2500 m (Heaney et al., 1998; Musser, 1982*b*), but the species has also been taken near sea level on W coast of Ilocos Norte Province (AMNH 252474, 252475).
    STATUS: IUCN – Lower Risk (lc).
    SYNONYMS: *major* Miller, 1910.
    COMMENTS: Member of the *Apomys datae* Group according to Musser (1982*b*), who also documented the inclusion of *major*; the Mindoran *A. gracilirostris* is the only other member of the *A. datae* group (see account below). Standard karyotype (2n = 44, FN = 54) closely similar to *A. musculus* and the shrew rats *Chrotomys* (including *Celaenomys*) and *Rhynchomys* and strikingly different from other species of *Apomys* sampled (Rickart and Heaney, 2002). The karyotype of *A. datae*, along with a primitive pattern of cephalic arterial circulation, suggest it is the most primitive species within *Apomys*. It, along with *A. gracilirostris*, also form the basal clade within the genus as revealed by phylogenetic analysis of mtDNA cytochrome *b* sequences (Steppan et al., 2003)

Apomys gracilirostris Ruedas, 1995. Proc. Biol. Soc. Wash., 108:305.
    COMMON NAME: Large Mindoro Apomys.
    TYPE LOCALITY: Philippines, Mindoro Isl, Mindoro Occidental Province, Municipality of San Teodoro, North Ridge approach to Mt Halcon, 1580 m, 13°16′48″N, 121°59′19″E.
    DISTRIBUTION: Greater Mindoro Faunal Region. A Mindoro endemic recorded only from montane forest on northern flank of Mt Halcon between 1250 and 1950 m, but may

occur throughout the Mindoran highlands still covered by mountain forest (Ruedas, 1995).

STATUS: IUCN – Vulnerable.

COMMENTS: *Apomys datae* Group. In morphology of rostrum and incisors, *A. gracilirostris* is unique among species of *Apomys*, but phylogenetic analysis of mtDNA cytochrome *b* sequences placed it with the Luzon *A. datae* (Steppan et al., 2003). Large body size, long rostrum relative to rest of cranium, and primitive pattern of cephalic arterial supply also tie *A. gracilirostris* to *A. datae* (Ruedas, 1995; Musser's study of the Mindoran specimens). Presence of *A. gracilirostris* on Mindanao likely reflects dispersal to that island, probably from an ancestral stock on Luzon (Steppan et al., 2003). Discovery of *A. gracilirostris* accentuates the dual biogeographic nature of Mindoro's murine fauna: *Chrotomys mindorensis*, *Apomys musculus*, *A. gracilirostris*, and *Rattus everetti* have Philippine affinities; *Rattus mindorensis*, an undescribed species of *Maxomys* (specimens examined by Musser), and *Anonymomys mindorensis* are related to Indochinese and Sundaic faunas.

Apomys hylocetes Mearns, 1905. Proc. U.S. Natl. Mus., 28:456.

COMMON NAME: Mindanao Mossy Forest Apomys.

TYPE LOCALITY: Philippines, Mindanao Isl, Mt Apo, 6000 ft (1830 m).

DISTRIBUTION: Greater Mindanao Faunal Region. Endemic to Mindanao in primary montane forest between 1900 and 2800 m; probably widespread on high peaks of the island (Heaney et al., 1998.

STATUS: IUCN – Lower Risk (lc).

SYNONYMS: *petraeus* Mearns, 1905.

COMMENTS: Included within *Apomys abrae-hylocetes* Group by Musser (1982*b*), who also explained why *petraeus* is a synonym. Standard karyotype (2n = 48, FN = 56) of *A. hylocetes* is strikingly dissimilar to that of sympatric *A. insignis* (2n = 36, FN = 36), but this phyletic estimate is contradicted by phylogenetic analyses of mtDNA cytochrome *b* sequences, which indicates they are sister species and most closely related to an undescribed species from the islands of Leyte and Biliran and an undescribed species on Camiguin Isl (Steppan et al., 2003). An hypothesis explaining the evolutionary origins of these four species is presented by Steppan et al. (2003).

Apomys insignis Mearns, 1905. Proc. U.S. Natl. Mus., 28:459.

COMMON NAME: Mindanao Montane Forest Apomys.

TYPE LOCALITY: Philippines, S Mindanao Isl, Mt Apo, 6000 ft (1830 m).

DISTRIBUTION: Endemic to the Greater Mindanao Faunal Region from islands of Dinagat and Mindanao in primary and secondary forests, 900-2800 m (Heaney et al., 1998).

STATUS: IUCN – Lower Risk (lc).

SYNONYMS: *bardus* Miller, 1910.

COMMENTS: Included within *Apomys abrae-hylocetes* Group by Musser (1982*b*), who explained why *bardus* is a synonym. See account of *A. hylocetes*.

Apomys littoralis (Sanborn, 1952). Fieldiana Zool., 33(2):134.

COMMON NAME: Mindanao Lowland Apomys.

TYPE LOCALITY: Philippines, Mindanao Isl, Bugasan, Cotabato, 50 ft (15 m).

DISTRIBUTION: Endemic to the Greater Mindanao Faunal Region on islands of Biliran, Bohol, Leyte, and Mindanao (Musser and Heaney, 1992; Rickart et al., 1993; Heaney et al., 1998). Formerly recorded from Negros (Musser, 1982*b*), but that sample represents an undescribed species (Heaney et al., 1998).

STATUS: IUCN – Lower Risk (lc).

COMMENTS: Included within *Apomys abrae-hylocetes* Group by Musser (1982*b*). This species is known by a juvenile holotype with damaged skull from Mindanao and several referred series from Leyte and nearby smaller islands (Heaney et al., 1998). Distributional and ecological data provided by Rickart et al. (1993). The standard karyotype reported for a sample from Leyte (2n = 44, FN = 88; Rickart and Musser, 1993), which expresses the highest FN recorded for Indoaustralian murines, actually comes from an undescribed species (Rickart and Heaney, 2002; Steppan et al., 2003).

Apomys microdon Hollister, 1913 (not Peters, 1852). Proc. U.S. Natl. Mus., 46:327.

COMMON NAME: Small Luzon Apomys.

TYPE LOCALITY: Philippines, Cataduanes Isl, Biga.

DISTRIBUTION: Endemic to the Greater Luzon Faunal Region and widespread on Luzon in secondary lowland forest and primary montane forest (Heaney et al., 1991; Heaney et al., 1998; Heaney et al., 1999; Musser and Heaney, 1992). Formerly thought to also occur on Leyte and Dinagat (Musser, 1982*b*), but those samples have been referred to *A. littoralis.*

STATUS: IUCN – Lower Risk (lc).

SYNONYMS: *hollisteri* (Ellerman, 1949).

COMMENTS: Included within *Apomys abrae-hylocetes* Group by Musser (1982*b*). Phylogenetic analyses of mtDNA cytochrome *b* sequences identifies *M. musculus*, an undescribed species from Sibuyan Isl and Negros Isl, and another undescribed species from Sibuyan Isl as the closest phylogenetic allies to *A. microdon* (Steppan et al., 2003).

    Ellerman (1949) provided *hollisteri* as a replacement name for *Apomys microdon* because it was preoccupied by *Mus microdon* Peters (1852), which is a form of *Mastomys*, and Ellerman had included both *Apomys* and *Mastomys* in *Rattus*. Corbet and Hill (1992:379) used *hollisteri* as the correct name stating that *microdon* was permanently invalid following Article 59(b) of the 3rd edition of the International Code of Zoological Nomenclature (International Commission on Zoological Nomenclature, 1985*d*), which states that a "junior secondary homonym replaced before 1961 is permanently invalid." But according to article 59.3 of the 4th edition of the Code (ICZN, 1999), a "junior secondary homonym replaced before 1961 is permanently invalid unless the substitute name is not in use and the relevant taxa are no longer considered congeneric, in which case the junior homonym is not to be rejected on grounds of that replacement." Except for Corbet and Hill's account, *microdon* is the name that has been in use, especially for the last 20 years, in all the ecological, biological, biogeographical, and taxonomic literature covering Philippine mammals. Additionally, all modern practicing rodent systematists agree that Hollister's *microdon* and Peters's *microdon* refer to species in different genera (*Apomys* and *Mastomys*, respectively) that are in unrelated (except as members of Murinae) phylogenetic clades. We continue to follow Musser (1982*b*) in using *microdon*. Corbet and Hill (1992:379) correctly noted that *A. microdon* is closely related to *A. insignis* (an alliance documented by Musser, 1982*b*) and also stated that *A. microdon* "could conceivably be considered conspecific with *A. insignis*" but morphological evidence does not support this supposition (Musser, 1982*b*).

Apomys musculus Miller, 1911. Proc. U.S. Natl. Mus., 38:403.

COMMON NAME: Least Philippine Apomys.

TYPE LOCALITY: Philippines, Luzon Isl, Benguet, Baguio, Camp John Hay, 5000 ft (1524 m).

DISTRIBUTION: Philippines: Dinagat, S and N Luzon, and Mindoro (Heaney et al., 1998).

STATUS: IUCN – Lower Risk (lc).

COMMENTS: Included within *Apomys abrae-hylocetes* Group by Musser (1982*b*). Standard karyotype (2n = 42, FN = 52; Rickart and Musser, 1993) very similar to that of *A. datae* (Rickart and Heaney, 2002). Three species may be represented in what is now called *musculus*: typical *musculus* from N Luzon, a separate species in S Luzon, and another on Mindoro (Heaney, pers. comm.). Altitudinal distribution and ecological data summarized by Balete and Heaney (1997) and Heaney et al. (1999). Judged by phylogenetic analyses of mtDNA cytochrome *b* sequences, *A. musculus* is most closely related to two undescribed species and *A. microdon* (see that account).

Apomys sacobianus Johnson, 1962. Proc. Biol. Soc. Wash., 75:318.

COMMON NAME: Long-nosed Luzon Apomys.

TYPE LOCALITY: Philippines, Luzon Isl, Pampanga Province, Sacobia River, Clark Air Base.

DISTRIBUTION: Greater Luzon Faunal Region. Endemic to Luzon and recorded only from primary forest at the type locality; limits unresolved.

STATUS: IUCN – Vulnerable.

COMMENTS: Included within the *Apomys abrae-hylocetes* Group by Musser (1982*b*), but allied to the *Apomys datae* Group by Ruedas (1995) because he thought the holotype (USNM 304352) and another specimen he identified as *A. sacobianus* (USNM 557717) exhibited the primitive murine cephalic arterial pattern typical of *A. datae*. Musser (1982*b*), however, described a derived pattern for *A. sacobianus*, which is typical for species in the *A. abrae-hylocetes* Group, and our restudy of the holotype reaffirms those observations. *Apomys sacobianus* is morphologically closely related to the N Luzon *A. abrae*, being a slightly larger-bodied, grayer version occurring in the S lowlands (Musser, 1982*b*); whether it represents a larger-bodied geographic variant of *A. abrae* or a separate species will have to be determined by study of specimens from intermediate localities and altitudes.

Heaney et al. (1998) recorded *A. sacobianus* from Isabela, Mountain, and Zambales provinces on Luzon, each represented by a single specimen. He (L. Heaney, in litt., 2002) tentatively determined the Isabela and Zambales examples to be *A. sacobianus*, but acknowledged that the specimens require restudy to verify the identifications. USNM 557717 is from Mountain Province (Kiangan, Duwit, 4000 ft [1219 m]), has the derived carotid circulatory pattern typical of *A. abrae-A. hylocetes* (scheme B in Musser, 1982*b*:6), but is much larger than *A. sacobianus* in skull length (40.0 mm versus 38 mm for *A. sacobianus*) and length of M1-3 (8.0 mm versus 7.1 mm), and likely represents a separate undescribed species; no example of *A. abrae*, which also occurs in Mountain Province, or any other species of *Apomys* comes close to it in size (Musser, 1982*b*:table 4). Until identity of the other two specimens from Isabela and Zambales provinces are re-evaluated, we consider *A. sacobianus* to be represented only by the holotype.

Archboldomys Musser, 1982. Bull. Am. Mus. Nat. Hist., 174:30.
TYPE SPECIES: *Archboldomys luzonensis* Musser, 1982.
COMMENTS: *Chrotomys* Division. Member of the Philippine Old Endemics (Musser and Heaney, 1992). Based strictly upon morphological data, Musser and Heaney (1992) supposed *Archboldomys* to be most closely related to *Crunomys*, but phylogenetic analyses of complete mtDNA cytochrome *b* sequences for 13 of the 16 genera of Philippine murines indicate *Archboldomys* is a member of a clade that includes *Apomys*, *Chrotomys* (including *Celaenomys*), and *Rhynchomys* but not *Crunomys* (Jansa and Heaney, 2001), and chromosomal data does not support a close alliance between *Archboldomys* and *Crunomys* (Rickart and Musser, 1993; Rickart and Heaney, 2002). Reviewed by Rickart et al. (1998).

Archboldomys luzonensis Musser, 1982. Bull. Am. Mus. Nat. Hist., 174:30.
COMMON NAME: Isarog Shrew Mouse.
TYPE LOCALITY: Philippines, SE Luzon Isl, Camarines Sur Province, Mt Isarog, 6560 ft (2000 m).
DISTRIBUTION: Greater Luzon Faunal Region where it is endemic to montane habitat on Mt Isarog in SE peninsula of Luzon (Heaney et al., 1999; Rickart et al., 1991).
STATUS: IUCN – Endangered.
COMMENTS: Morphological descriptions and comparisons with *Crunomys* and Sulawesi shrew rats provided by Musser (1982*c*). Altitudinal distribution and ecologial notes in Rickart et al. (1991), Balete and Heaney (1997), and Heaney et al. (1999). Standard karyotype and G-banding patterns are distinctive (2n = 26, FN = 43): its low 2n, high number of bi-armed chromosomes relative to telocentric elements, and peculiar pattern of the sex chromosomes define a highly derived karyotype compared to those characteristic of other Old Endemics sampled, but is uninformative in assessing the phylogenetic relationships of *A. luzonensis* among Philippine murines (Rickart and Musser, 1993; Rickart and Heaney, 2002). Morphology suggests a tie to Philippine *Crunomys* (Musser and Heaney, 1992), but molecular data indicated close phylogenetic alliance with species of *Apomys*, *Chrotomys* (including *Celaenomys*), and *Rhynchomys* and no close association with species of *Crunomys* (Jansa and Heaney, 2001).

Archboldomys musseri Rickart, Heaney, Tabaranza, Jr., and Balete, 1998. Fieldiana Zool. n.s., 89:17.
COMMON NAME: Sierra Madre Shrew Mouse.
TYPE LOCALITY: Philippines, N Luzon Isl, Cagayan Province, Callao Municipality, Sierra Madre Range, Mt Cetaceo, 1650 m, 17°42'N, 122°02'E.

DISTRIBUTION: Greater Luzon Faunal Region where it is known from the type locality and the Balbalasang region of Kalinga Province at 1900 m (Rickart and Heaney, 2002), but may occur elsewhere in montane forest throughout N Luzon (Rickart et al., 1998).

COMMENTS: Documented by two specimens from the type locality and a sample from Kalinga Province referred to *A. musseri*. Standard karyotype (2n estimated as 44 with mostly telocentric elements) strikingly different from that of *A. luzonensis* (Rickart and Heaney, 2002) and may turn out to be indistinguishable or closely similar to those of *Chrotomys silaceus*, *C. gonzalesi*, and *Rhynchomys isarogensis* (karyotypes described by Rickart and Musser, 1993, and Rickart and Heaney, 2002).

Arvicanthis Lesson, 1842. Nouv. Tabl. Regn. Anim. Mammalifères, p. 147.

TYPE SPECIES: *Lemmus niloticus* É. Geoffroy Saint-Hilaire, 1803.

SYNONYMS: *Isomys* Sundevall, 1843.

COMMENTS: *Arvicanthis* Division. Morphological features tie *Arvicanthis* to species of *Pelomys*, *Mylomys*, *Rhabdomys*, and *Lemniscomys* (Musser, 1987*b*), which has been confirmed by results from mtDNA sequences (cytochrome *b*, 12S and 16S rRNA genes), with the addition of *Desmomys*, by Ducroz et al. (2001). Their results indicate three distinct lineages: one containing *Arvicanthis*, *Mylomys*, and *Pelomys*; another with *Desmomys* and *Rhabdomys*; and a third containing only *Lemniscomys*. The Asian *Golunda*, provisionally (Musser, 1987*b*) or certainly (Jacobs, 1978; Misonne, 1969; Sabatier, 1982) allied with this clade, is not part of it according to molecular data (Ducroz et al., 2001; Lecompte, 2003; see *Golunda* account).

　　　Opinions on the number of species in *Arvicanthis* have varied from one (Misonne, 1974) to several (G. M. Allen, 1939; Corbet and Hill, 1991; Dollman, 1911; Ellerman, 1941; Musser and Carleton, 1993; Rousseau, 1983). The first and still useful review of *Arvicanthis* was made by Dollman (1911). Several recent attempts to assess variability in the genus using breeding studies (F. Petter et al., 1969; Philippi, 1994), morphometric analyses (Bekele et al., 1993; Corti and Fadda, 1996; Fadda and Corti, 1998, 2001; Rousseau, 1983), spermatozoal morphology (Baskevich and Lavrenchenkio, 1995), chromosomal evidence (Baskevich and Lavrenchenko, 2000; Capanna and Civitelli, 1988; Castiglia et al., 2003*a*; Civitelli et al., 1995*a*; Corti et al., 1996*b*; Garagna et al., 1999; Orlov et al., 1992*a*; Volobouev et al., 1987, 1988*a*, 2002*a*, *b*; and references cited there), electrophoretic patterns of blood proteins (Kaminski et al., 1984), allozymic variation (Capula et al., 1997; Milishnikov et al., 1992), analysis of mitochondrial (cytochrome *b*, 12S and 16S rRNA) gene sequences (Ducroz et al., 1998, 2001), and a combination of data sources (Capanna et al., 1996*b*; Ducroz et al., 1997) have provided insights into species-diversity within *Arvicanthis*, as have results obtained based upon our examination of museum specimens (focusing on qualitative traits as well as external and craniodental measurements), and study of the discussions by Dollman (1911), Hollister (1919), Osgood (1936), and Corbet and Yalden (1972). Results of the careful chromosomal, molecular, and interbreeding studies by Volobouev et al. (1988*a*, 2002*a*, *b*), Ducroz (1998), and Ducroz et al. (1997, 1998) have been especially significant in uncovering the species diversity of *Arvicanthis* in West Africa.

　　　Of the seven species we list here, *A. abyssinicus* and *A. blicki* are endemic to the Ethiopian Plateau and easily diagnosed and recognizable. Two, *A. ansorgei* and *A. rufinus*, have been recorded only from West Africa and the full extent of their geographic distributions have yet to be resolved. The East African *A. nairobae* is distinctive but its actual range and phylogenetic relationship to other species requires investigation. The small-bodied *A. neumanni* is also endemic to East Africa and exhibits significant geographic variation possibly reflecting the presence of another species in the southern part of its range. *Arvicanthis niloticus* has the most expansive geographic distribution of any *Arvicanthis* and as defined here may still consist of more than one entity. Several samples of *Arvicanthis* have karyotypes different from any reported for other species and remain taxonomically unidentified. Capanna and Civitelli (1988) recorded a karyotype with 2n = 44 from a sample they identified as *A. niloticus* from Somalia, which contrasts with the 2n = 62 of *A. niloticus* or documented diploid number for any other sample of *Arvicanthis*. Two unidentified Ethiopian samples were reported by Orlov et al. (1992*a*): 2n = 60, FN = 78 (S Ethiopia) and 2n = 56, FN = 60

(SW Ethiopia); the former might apply to *A. nairobae*, which occurs in the same region of S Ethiopia (Fadda and Corti, 2001); Tanzanian specimens of *A. nairobae*, for example, exhibit 2n = 62, FN = 78 (Castiglia et al., 2003*a*; Fadda et al., 2001*b*).

Living species of *Arvicanthis* occur only in subsaharan Africa and the SW portion of the Arabian peninsula, but the genus is represented in middle to late Pleistocene cave deposits in Israel by *A. ectos* (Bate, 1942*b*; Tchernov, 1968), and is not recorded as fossils in cave sites younger than 80,000-70,000 years before present (Tchernov, 1986, 1992, 1994, and references cited in therein). Isolated molars identified as *A. niloticus* were obtained from late Pleistocene sediments in N Morocco, which is outside the modern range of the genus (Ouahbi et al., 2001), and from the late Pleistocene of East Africa (Jaeger, 1976; Wesselman, 1984; see review by Denys, 1999). *Arvicanthis* sp. was recovered from late Pleistocene beds in Tunisia (Mein and Pickford, 1992). The earliest fossil representing *Arvicanthis* comes from 3-million year old sediments (late Pliocene) in the Omo Valley of Ethiopia (Wesselman, 1984). Based upon complete mtDNA cytochrome *b* sequences, Ducroz et al. (1998) estimated *Arvicanthis* and *Lemniscomys* diverged from a common ancestor 5.5-7.7 million years ago. The split between *Arvicanthis* and *Mylomys* + *Pelomys* would have occurred between 5 and 6.3 millions years ago (Ducroz et al., 2001). Subsequently the North African species may have split from the other *Arvicanthis* 4.3-6.2 million years ago, and the separation of *A. niloticus*, *A. abyssinicus*, *A. blicki*, and *A. neumanni* was a Plio-Pleistocene event (Capanna et al., 1996*b*; see extended discussion in Ducroz et al., 1998, 2001).

Arvicanthis abyssinicus (Rüppell, 1842). Mus. Senckenberg., 3:104.
  COMMON NAME: Ethiopian Arvicanthis.
  TYPE LOCALITY: Ethiopia, Simien Province, Simien Mtns, Entschetqab (see Osgood, 1936:252).
  DISTRIBUTION: Ethiopian Plateau between 1300-3400 m (Yalden et al., 1976).
  STATUS: IUCN – Lower Risk (lc).
  SYNONYMS: *fluvicinctus* Osgood, 1936; *rufodorsalis* (Heuglin, 1877); *saturatus* Dollman, 1911.
  COMMENTS: An Ethiopian endemic with 2n = 62, FN = 68 (Corti et al., 1996; Orlov and Bulatova, 1997; Orlov et al., 1992*a*). Historically the species was perceived to embrace forms occurring from Ethiopia south to Zambia (G. M. Allen, 1939; Dollman, 1911; Ellerman, 1941), but Osgood (1936:252) discussed several of *A. abyssinicus*'s diagnostic features, noting that it "is not unlikely that it is confined to Ethiopia and at least some of the forms of Kenya and Uganda which have been associated with it will need other allocation." This view has been reinforced by analyses of morphometric (Bekele et al., 1993; Corti and Fadda, 1996; Fadda and Corti, 2001; Rousseau, 1983), chromosomal (Baskevich and Lavrenchenko, 2000; Corti et al., 1996*b*; Orlov et al., 1991*a*), and mtDNA cytochrome *b* sequences (Ducroz et al., 1998), as well as our examination of specimens. Yalden and Largen (1992) speculated that *A. abyssinicus* may also occur in Kenya, Uganda, and elsewhere but no specimens or any other kind of data suggests its distribution extends beyond the Ethiopian Plateau.

  Arvicanthis abyssinicus and *A. blicki* are the only species of *Arvicanthis* endemic to the Ethiopian Plateau; both are part of middle and high altitude grassland and moorland mammal communities also unique to the plateau (Demeter and Topal, 1982; Rupp, 1980; Yalden, 1988); are more closely related to each other than to any other species of the genus, as judged by morphological, allozymic, and chromosomal traits (Capanna et al., 1996; Capula et al., 1997; Musser and Carleton, 1993); and may have diverged from a common ancestor during the Pleistocene (Capanna et al., 1996*b*). Both *A. abyssinicus* and *A. blicki* are close phylogenetic allies of *A. niloticus* (which occurs in Ethiopia as well as elsewhere) and the three constitute a monophyletic group (Ducroz et al., 1998; Corti et al., 1996*b*). *Arvicanthis neumanni* is another member of the Ethiopian fauna but does not occur on the plateau and also ranges far beyond that country's boundaries.

  Rousseau (1983) regarded *mearnsi*, described as a subspecies of *A. abyssinicus* by Frick (1914), to be a subspecies of *A. abyssinicus*, but its qualitative dental and chromatic characteristics fall within the range of variation seen in *A. niloticus*, as Osgood (1936) and Yalden et al. (1976) noted (discussed by them under either *lacernatus* or *dembeensis*).

Arvicanthis ansorgei Thomas, 1910. Ann. Mag. Nat. Hist., ser. 8, 5:353.

COMMON NAME: Sudanian Arvicanthis.

TYPE LOCALITY: Bissau Guinea, Gunnal.

DISTRIBUTION: Grasslands and bush in Sudanian savannas from Gambia (Fadda and Corti, 2001) and S Senegal (Casamance region south of Gambia River; Granjon et al., 1992) through S Mali, Burkina-Faso, and Niger (Dobigny et al., 2002*b*; Ducroz et al., 1998; Volobouev et al., 2002*a*) to S Chad (Fadda and Corti, 2001); limits of range unresolved.

COMMENTS: 2n = 62, FNa = 74-76 (Volobouev et al., 2002*a*). This is the species originally separated from samples of *A. niloticus* by its chromosomal features and identified as ANI-3 (Volobouev et al., 1987, 1988*a*; Ducroz et al., 1997). Subsequent analyses incorporating chromosomal refinements, DNA/DNA hybridization, cross-breeding experiments in the laboratory, three-dimensional geometric morphometrics, and mtDNA cytochrome *b* sequences (Ducroz, 1998; Ducroz et al., 1997, 1998; Fadda and Corti, 2001; Volobouev et al., 2002*a*), demonstrated that *ansorgei* is not a geographic form of *A. niloticus* but a separate species. The two are sympatric along the southern course and in most of the Inner delta of Niger River in Mali (Volobouev et al., 2002*a*), but elsewhere the "... distribution of the two species is largely parapatric and follows the latitudinal patterns of the West-African biogeographical domains, which are related to the latitudinal patterns of annual rainfuall in this region" (Sicard et al., 2004:5). The range of *A. ansorgei* is generally south of *A. niloticus*, which inhabits the more arid Sahelian and N Sudanian savannas, and north of *A. rufinus*, found in the more humid S Sudanian-Guinean savannas and adjacent evergreen forest belt. In Senegal, *A. ansorgei* occurs only south of the Gambia River while *A. niloticus* inhabits drier habitats north of the Gambia (Granjon et al., 1992). Phylogenetic analysis of mtDNA cytochrome *b* sequences by Ducroz et al. (1998) indicated *A. ansorgei* to be more closely related to samples from East Africa (Tanzania and SW Uganda) than to those from Central Africa (Benin and Central African Republic), a discordance with similarities based upon Procrustes distances, which readily distinguished West African from East African assemblages (Fadda and Corti, 2001).

The holotype of *ansorgei* was not used in any of the analyses cited above, but it is the only taxon described from a place close to where samples of the ANI-3 chromosomal type were collected, and Volobouev et al. (2002*a*) thought it reasonable to use *ansorgei* as the correct name; no synonyms have been identified.

Arvicanthis blicki Frick, 1914. Ann. Carnegie Mus., 9:20.

COMMON NAME: Blick's Arvicanthis.

TYPE LOCALITY: Ethiopia, South Chilalo Mtns, Hora Mt base camp, 9000 ft (2743 m).

DISTRIBUTION: Ethiopia; between 2750 and 4050 m from plateau on E side of Ethiopian Rift Valley (Yalden et al., 1976; Yalden et al., 1996).

STATUS: IUCN – Lower Risk (lc).

COMMENTS: An Ethiopian endemic with 2n = 48, FNa = 68 (Corti et al., 1996*b*; Lavrenchenko et al., 1997) that is a characteristic diurnal member of the Afro-Alpine moorland zone above 3500 m (Demeter and Topal, 1982; Rupp, 1980; Yalden, 1988; Yalden et al., 1976). In morphology, ecological and geographic distribution, and habits, *A. blicki* is a very distinctive species, as recognized by Dorst (1972). Judged by similarities in molar occlusal topography and dorsal fur patterning, chromosomal traits, and allozymic data, *A. blicki* is more closely related to *A. abyssinicus* than to any other species of *Arvicanthis* and together with that species and *A. niloticus* forms a monophyletic clade (see account of *A. abyssinicus*). Statistical summary of external measurements for a large sample reported by Sillero-Zubiri et al. (1995*a*). Habitat preferences, abundance, and biomass of rodents in the Bale Mtns, including *A. blicki*, and relevance to the endangered *Canis simensis* recorded by Sillero-Zubiri (1995*c*).

Arvicanthis nairobae J. A. Allen, 1909. Bull. Am. Mus. Nat. Hist., 26:168.

COMMON NAME: East African Arvicanthis.

TYPE LOCALITY: Kenya, Nairobi.

DISTRIBUTION: Recorded from S Ethiopia (Sidamo) through Kenya in the Rift Valley south to the Dodoma region in EC Tanzania; western and southern limits unresolved (based largely upon our study of specimens; the Ethiopian record and some specimens from W Kenya were identified using discriminant function analysis by Fadda and Corti, 2001). An unidentified sample of *Arvicanthis* from SW Uganda clusters with Tanzanian samples of *A. nairobae* in the analysis of mtDNA cytochrome *b* sequences by Ducroz et al. (1998), which indicates the species may range farther west than described here.

STATUS: IUCN – Lower Risk (lc).

SYNONYMS: *chanleri* Dollman, 1911; *nubilans* Wroughton, 1909; *pallescens* Dollman, 1914; *praeceps* Wroughton, 1909; *rumruti* Dollman, 1911; *virescens* Heller, 1914.

COMMENTS: 2n = 62, FN = 78 (Castiglia et al., 2003a; Fadda et al., 2001). The name *nairobae* is the oldest applicable to samples from Kenya and E Tanzania containing animals smaller in body size and generally brighter and buffier in pelage tones and hues than those larger and darker specimens we have identified as *A. niloticus* from W Tanzania and Uganda. Some specimens of *A. nairobae* closely resemble those of *A. neumanni* in pelage coloration, and could be mistaken for it, but are larger in body size. Corbet and Yalden (1972) even suggested that *chanleri* might represent *A. neumanni* (they used *somalicus*), but the holotype is larger and fits within the range of variation seen among samples of *A. nairobae* (our study of holotypes and larger samples in AMNH, BMNH, and USNM). Records of sympatry between *A. nairobae* and *A. neumanni* are documented by series from Mount Lololokwi ("an isolated mountain east of the Mathews Range, about midway between Mount Kenia and Mount Marsabit," Hollister, 1919; specimens in USNM), and the Dodoma region of Tanzania (samples in AMNH), and the two are probably broadly sympatric throughout their ranges in Kenya and Tanzania.

The morphological and geographic relationships between *A. nairobae* and *A. niloticus* require resolution, which could be accomplished by critical systematic review of the *A. niloticus-A. nairobae* complex. For example, *A. nairobae* is found with *A. neumanni* in the Dodoma region of Tanzania and they are the only species of *Arvicanthis* (represented by specimens) from E Tanzania. In our samples from the Tabora area to the west of Dodoma we could detect only one species, which is larger in body size and has much darker pelage than samples from Dodoma. It is this species (which we identify as *A. niloticus*) that ranges south into Zambia and north through W Tanzania and Uganda. According to the geometric morphometrical analysis by Fadda and Corti (2001), however, two morphological types occur in that region, which they designated as *A. sp.* and *A sp.1*; a third form, *A. sp2*, had a geographic distribution similar to *A. nairobae* and included the holotype. Studying the distribution of characters from samples collected along a transect between the Dodoma and Tabora regions would likely elucidate character and geographic relationships among the different populations.

Ducroz et al. (1998) provided another view of possible relationships between *A. nairobae* and *A. niloticus* in their phylogenetic analysis of mtDNA cytochrome *b* sequences. Samples of *Arvicanthis* from West Africa identified by Volobouev et al. (2002a) as *A. ansorgei*, those from Central African Republic and Benin (the latter identified as *A. rufinus*; Volobouev et al., 2002a), and samples from East Africa representing *A. nairobae* formed a cluster ("clade 1") separate from that composed of *A. niloticus*, *A. abyssinicus*, and *A. neumanni* ("clade 2"), which is congruent with chromosomal data (e.g., Castiglia et al., 2003a) because "clade 1" is composed of species with a high FN. Within "clade 1" the West African lineage was phylogenetically more closely allied to the East African linage than to the samples from Benin and Central African Republic. Analyses of cytogenetic data has affirmed the specific integrity of *A. nairobae* and its closer phylogenetic relationship to West African *A. ansorgei* and *A. rufinus* than to the East African lineage consisting of *A. niloticus*, *A. nairobae*, *A. abyssinicus*, and *A. blicki* (Castiglia et al., 2003a). The inclusion of *rumruti* and *nubilans* in the synonymy above is based upon their identification as *A. nairobae* by Fadda and Corti's (2001) morphometric analysis.

Arvicanthis neumanni (Matschie, 1894). Sitz.-Ber. Ges. Naturf. Fr. Berlin, 1894: 204.

COMMON NAME: Neumann's Arvicanthis.

TYPE LOCALITY: C. Tanzania, Kondoa District, Barungi.

DISTRIBUTION: N and E Rift Valleys of Ethiopia (Yalden et al., 1976, 1996), Somalia, extreme SE Sudan (Dieterlen and Nikolaus, 1985), and south through Kenya in and east of the Rift Valley (Dollman, 1911; Hollister, 1919) to C and EC Tanzania on both sides of the Rift, including base of Mt Kilimanjaro (Grimshaw et al., 1995).

STATUS: IUCN – Lower Risk (lc) as *A. somalicus*.

SYNONYMS: *reptans* Dollman, 1911; *somalicus* Thomas, 1903.

COMMENTS: 2n = 62, FN = 66 or 67 for Ethiopia (Baskevich and Lavrenchenko, 2000); 2n = 53-54, FN = 62 for Tanzania (Castiglia et al., 2003*a*; Fadda and Corti, 2001), but more precisely 2n = 53-54 due to a Robertsonian fusion and derived from a larger sample (Fadda et al., 2001*b*). Under the name *somalicus* Thomas (1903*a*), *A. neumanni* has been treated as a species (G. M. Allen, 1939; Dollman, 1911; Ellerman, 1941; Hollister, 1919), and with an exception or two (e.g., Misonne, 1974), retains that status (Corti and Fadda, 1996; Ducroz et al., 1998; Fadda and Corti, 2001; Musser and Carleton, 1993; Rousseau, 1983; Yalden et al., 1976). Both reliable published records (see above) and specimens we examined document the northern and central range of *A. neumanni*. Known southern limits of the species are defined by samples from the Mawele region south of Tabora (Mwanasomano's, 31 mi [50 km] S Tabora) in C Tanzania and southeast of there at Kilosa in EC Tanzania (specimens in MCZ). The species is sympatric with both *A. niloticus* and *A. nairobae* (see those accounts). G. M. Allen and Loveridge (1933:117) recorded three specimens from Mwanza on the southern margin of Lake Victoria in Tanzania under the name *muansae* thinking they represented topotypes of that form, but the holotype of *muansae* from Mwanza is very large with dark pelage (Matschie, 1911) and is an example of *A. niloticus*, while their sample is the much smaller and paler *A. neumanni*; Mwanza is probably another site of sympatry.

Specimens from the southern margin of the range in Tanzania (Dodoma) were used in the geometric morphometric analysis by Fadda and Corti (2001) who noted they had small skulls as in *A. neumanni* from Kenya, Somalia, Ethiopia, but were shaped differently, indicating they represented either a separate species or a distinctive geographic variant within *A. neumanni*. The large series we examined from the region, including material from Dodoma, exhibit the small size and pelage characteristics common to samples from farther north in Kenya, Ethiopia, and Somalia; it would not be surprising to discover geographic variation within *A. neumanni* considering its extensive East African range. Alternately, if the Tanzanian population is a different species (the karyotype of *A. neumanni* from specimens collected elsewhere in Tanzania is different from Ethiopian samples; Fadda and Corti, 2001), *A. neumanni* would be the correct name for it, and *A. somalicus* would be the species with a range through Kenya to Somalia, S Ethiopia, and SE Sudan. After reporting karyotypes of their samples from NE Tanzania, Fadda et al. (2001*b*) compared the material to large series from the central plateau of Tanzania and relevant holotypes, concluding "that further analysis is needed before deciding on the exact systematic position of the different *Arvicanthis* populations of this part of Tanzania."

Spermatozoal morphology of *A. neumanni* was described by Baskevich and Lavrenchenko (1995, as *somalicus*) who noted that it is the most distinctive compared to other species of *Arvicanthis* sampled (*A. abyssinicus*, "*dembeensis*" and "*A.* sp."). Phylogenetic analysis of mtDNA cytochrome *b* sequences indicated *A. neumanni* (Tanzanian samples) is the basal member of a clade also containing *A. abyssinicus* and *A. niloticus* (Ducroz et al., 1998). Cytogenetic data also places those three species, along with *A. blicki*, in the same phylogenetic lineage (Castiglia et al., 2003*a*).

Arvicanthis niloticus (É. Geoffroy, 1803). Cat. Mam. Mus. Natl. Hist. Nat., Paris, p. 186.

COMMON NAME: African Arvicanthis.

TYPE LOCALITY: Egypt.

DISTRIBUTION: Grassland and bush of Sahelian and northern Sudanian savannas, steppe, and semidesert in subsaharan N Africa, primarily in anthropogenic habitats, from Senegal (north of Gambia River), S Mauritania eastward through Mali, Burkina Faso, C and S Niger (Dobigny et al., 2002*a*), and C and S Chad to Sudan and western half and EC portion of Ethiopia (Bâ et al., 2001; Corti et al., 1996*b*, as *dembeensis*; Dobigny et al., 2001*a*, *b*; Granjon et al., 1992, 2002*b*; Osborn and Hemy, 1980;); north along the

Nile Valley through Egypt; south through NE Dem. Rep. Congo, Uganda, S Burundi, and W Tanzania into E Zambia, where the population is isolated from the nearest one in SW Tanzania (Ansell, 1978). There is a record from N Malawi (1 skull only, living animals never found; Ansell and Dowsett, 1988), and the species also occurs in SW Arabian Peninsula (SW Yemen and W Oman; Al-Jumaily, 1998; Snowden et al., 2000; and references cited therein). The only truly Saharan record is from SE Algeria (Hoggar), and Dobigny et al. (2001a:216) speculated that *A. niloticus* "may persist across the whole Sahara, but only in some massifs or wadies where ecological conditions are more of the subdesert and/or steppe types than purely arid. Gardens also seem to be a good refuge for this species, and its commensalism may also explain its maintaining across the Sahara, via relictual populations." In Niger, the species was frequently encountered "in most wild bushy areas, gardens, villages and towns" (Dobigny et al., 2002b:498).

STATUS: IUCN – Lower Risk (lc).

SYNONYMS: *centralis* Dollman, 1911; *centrosus* Hollister, 1916; *dembeensis* (Rüppell, 1842); *discolor* (Wagner, 1842); *jebelae* Heller, 1911; *kordofanensis* Wettstein, 1916; *luctuosus* Dollman, 1911; *major* (Sundevall, 1843); *mearnsi* Frick, 1914; *minor* (Sundevall, 1843); *muansae* Matschie, 1911; *naso* Pocock, 1934; *nubilans* Wroughton, 1909; *ochropus* (Heuglin, 1877); *pelliceus* Thomas, 1928; *raffertyi* Frick, 1914; *reichardi* (Noack, 1887); *rhodesiae* St. Leger, 1932; *rossii* de Beaux, 1925; *rubescens* Wroughton, 1909; *solatus* Thomas, 1925; *tenebrosus* Kershaw, 1923; *testicularis* (Sundevall, 1843); *variegatus* (Lichtenstein, 1823); *zaphiri* Dollman, 1911.

COMMENTS: 2n = 62, FNa = 62 or 64 (Corti et al., 1996b, as *dembeensis*; Volobouev et al., 2002a). This is the species identified by the chromosomal cytotype ANI-1 (Volobouev et al., 1988a; see review in Ducroz et al., 1997) with two different forms identifying a different FN: ANI-1a (FNa = 62) in samples east of Mali and Burkina Faso to Ethiopia and Egypt, including the type locality; and ANI-1b (FNa = 64), found west of there (Volobouev et al., 2002a). Specimens with FNa = 66 were identified as *dembeensis* by Orlov et al. (1992a) and others (see references in Baskevich and Lavrenchenko, 2000). Closest phylogenetic relationships of *A. niloticus* appears to be with the Ethiopian *A. abyssinicus*, judged by analysis of chromosomal, allozymic, and morphometric data along with mtDNA cytochrome *b* sequences (Capanna et al., 1996b; Capula et al., 1997; Corti et al., 1996b; Ducroz et al., 1998; Fadda and Corti, 2001; *niloticus* was reported as *dembeensis* in most of these reports).

In 1993, Musser and Carleton reported dissatisfaction with their definition of *A. niloticus* (which included *A. ansorgei* and *A. rufinus*), noting appreciable intra- and interpopulational variation in body size, pelage coloration, and cytological and biochemical traits, and speculated that more than one species may be present in the complex, as also suggested by the morphometric (Rousseau, 1983), chromosomal (Viegas-Péquignot et al., 1983; Volobouev et al., 1987) and electrophoretic (Kaminski et al., 1984) analyses available at the time. Subsequent analyses of a combination of data sets has been used to separate *A. ansorgei* and *A. rufinus* from *A. niloticus* (see those accounts) and identify a third yet unnamed species from the Central African Republic (Ducroz, 1998; Ducroz et al., 1997, 1998; Fadda and Corti, 2001; Volobouev et al., 2002a). Furthermore, the geometric morphometric analysis by Fadda and Corti (2001) has questioned the identity of samples from Uganda and E Tanzania that we consider to be *A. niloticus* so our definition of *A. niloticus* may still be paraphyletic; additional study of samples from Uganda and E Tanzania (which should include the NE Zambian population) employing data from morphology, chromosomes, and gene sequences is certainly warranted.

Our inclusion of two synonyms require amplification. One is *dembeensis* (type locality, Deraske, Lake Tana, N Ethiopia), which has been recognized as a distinct species occurring in Ethiopia and nearby regions (Demeter, 1983; Yalden et al., 1976, 1996); and treated as a distinct species in recent results of chromosomal, molecular, and morphometric analyses (Baskevich and Lavrenchenko, 1994, 2000; Bekele et al., 1993; Bulatova et al., 2002; Capanna et al., 1996b; Capula et al., 1997; Corti and Fadda, 1996; Corti et al., 1996b; Orlov et al., 1992a). Thomas (1910b) had once associated *dem-*

beensis with *Desmomys*, noting that it had alredy been referred to *Mus*, *Arvicanthis*, *Golunda*, *Pelomys*, and *Oenomys*. Later, however, he (Thomas, 1928*a*) referred *dembeensis* to *Arvicanthis*, a view confirmed by Osgood (1936) and reinforced by Dieterlen (1974). The name *lacernatus* was once used in place of *dembeensis* for the Ethiopian samples of *Arvicanthis* (G. M. Allen, 1939; Corbet and Yalden, 1972; Osgood, 1936) but cranium of the holotype is an example of *Meriones* (Yalden et al., 1976). *Arvicanthis dembeensis* was thought to be a separate species because Ethiopian samples were originally contrasted by Corbet and Yalden (1972, under the name *lacernatus*) with *abyssinicus*, which they treated as a subspecies of *A. niloticus*; however, *abyssinicus* is a distinctive species (see that account). From our study of specimens, we observed that the diagnostic features attributed to *dembeensis* also described the morphology in samples of *A. niloticus* from Egypt and westward throughout the steppes, semidesert, and N Sudanian belt as outlined above in the distribution. For that reason, Musser and Carleton (1993) included *dembeensis* within the synonymy of *A. niloticus* for which we were accused of confusing its specific rank (Baskevich and Lavrenchenko, 2000). We are not confused nor are Ducroz et al. (1999:114), who noted the small genetic divergence between their samples of *dembeensis* and Egyptian *A. niloticus* and the lack of chromosomal differentiation between the two (Capanna et al., 1996*b*; Corti et al., 1996*b*; Volobouev et al., 1988*a*), and thus declared "there is no convincing evidence for the recognition of *A. dembeensis* as a distinct species and suggest referring it to as *A. niloticus* until conflicting evidence is produced." J.-F. Ducroz (in litt., 2001) now has additional information from Sudanese samples "clearly showing that there is a continuity from the genetic or karyologic point of view from Lower Egypt to Ethiopia, thus strengthening the point that there is no reason to distinguish *dembeensis*."

The other name is *testicularis* (type locality, Bahr-el-Abiad, N Sudan). Samples identified as *A. testicularis* were thought to occur together with samples of *A. niloticus* at several places in Uganda (Delany, 1975; Dollman, 1911; Hollister, 1919) and at Khartoum in Sudan (Fadda and Corti, 1998), but after study of large series from Uganda and Sudan we could not distinguish two kinds at any of the localities (Musser and Carleton, 1993) and found no morphological or distributional reasons why *testicularis* should not be included in *A. niloticus*. In a study of geographic variation among samples of *Arvicanthis* along the Nile Valley from Cairo to S Sudan using geometric morphometric techniques, Fadda and Corti (1998) recognized *A. niloticus* in Egypt and in Sudan at the Blue Nile and Khartoum, and *A. testicularis* at Khartoum and elsewhere in Sudan, but "museum labels were used for identification" (p. 105). They identified a "general trend in size decrease from North to South" (p. 108), and the suggestion of "an opposite clinal trend in shape" (p. 109), and claimed their scatterplot of first two principal compoments to demonstrate two clear clusters, which we could not appreciate and are only evident if the specimens were preidentified from their skin labels as to either *niloticus* or *testicularis*. If there are compelling data indicating specific integrity for *testicularis* they have not been published. J.-F. Ducroz (in litt., 2001) informed us of his "additional data from Sudan clearly showing . . . a continuity from the genetic or karylogic point of view from Lower Egypt to Ethiopia, thus strengthening the point that there is no reason to distinguish . . . *testicularis* as a distinct species." Philippi (1994) has demonstrated normal fertility of F1 and "backcross" individuals between samples of *A. niloticus* from Egypt and Sudan.

Corbet (1984) questioned including the S Arabian *naso* with *A. niloticus*, but current morphometric analysis supports that identity (Fadda and Corti, 2001). Verifiable sympatry with other species of *Arvicanthis* has been recorded: with *A. ansorgei* along the southern course and in most of the Inner delta of Niger River in Mali; and with *A. neumanni* in the Rift Valley of Ethiopia (*niloticus* was recorded as either *dembeensis* or *lacernatus*; Corbet and Yalden, 1972; Demeter, 1983) and the Mawele region (Mwanasomano's, 30 mi [48 km] S Tabora) of C Tanzania (specimens in MCZ). Allopatry or parapatry, however, is usual. In Ethiopia, *A. niloticus* occurs up to about 2000 ft (610 m) and is replaced at higher elevations by *A. abyssinicus* and *A. blicki* (Rupp, 1980; Yalden et al., 1976); in Kenya and Tanzania, *A. niloticus* ranges west of the E Rift Valley and is allopatric with *A. nairobae*, found in and east of the Rift as we have defined the species.

Arvicanthis rufinus (Temminck, 1853). Esquisses sur la côte de Guiné:163.

COMMON NAME: Guinean Arvicanthis.

TYPE LOCALITY: S Ghana, Elmina.

DISTRIBUTION: Guinean and southern Sudanian deciduous forest and woodland savannas, and clearings in adjacent evergreen forest belt from Sierra Leone through Benin and Ghana to S Nigeria (Ducroz et al., 1998; Fadda and Corti, 2001; Volobouev et al., 2002*a*).

SYNONYMS: *mordax* Thomas, 1911; *occidentalis* Wroughton, 1906; *setosus* Thomas, 1905.

COMMENTS: 2n = 62, FNa = 76 (Volobouev et al., 2002*a*). This is the species separated from *A. niloticus* and identified as chromosomal type ANI-4, and the subject of studies incorporating data from chromosomes, allozymes, mtDNA cytochrome *b* sequences, and morphometrics (Capanna et al., 1996*b*; Capula et al., 1997; Civitelli et al., 1995; Corti et al., 1996*b*; Ducroz et al., 1998; Fadda and Corti, 2001; Garagna et al., 1999; Volobouev et al., 2002*a*). This species and *A. blicki* are the largest in body size among the species of *Arvicanthis* (Fadda and Corti, 2001). Phylogenetic analysis of mtDNA cytochrome *b* sequences by Ducroz et al. (1998) indicated that *A. rufinus* is most closely related to samples from the Central African Republic (2n = 58, FNa = 70), an affinity also supported by chromosomal data even though the karyotypes seem very different (Volobouev et al., 1987, 2002*a*), but contradicted by three-dimensional geometric morphometric analysis (Fadda and Corti, 2001). All available data, including morphometrics (Fadda and Corti, 2001) point to the Central African Republic sample as representing a distinct species, but additional and larger samples from Central Africa must be analyzed before a name can be applied to it (Volobouev et al., 2002*a*).

Only the holotype of *mordax* was included in the morphometric study by Fadda and Corti (2001) and shown to match specimens with the cytotype ANI-4; the other scientific names associated with this species are based on holotypes collected from places within the known range of ANI-4 and were not employed in any of the analyses cited above. No other species of *Arvicanthis* is known to be sympatric with *A. rufinus* so the assignment of these names seems reasonable.

Bandicota Gray, 1873. Ann. Mag. Nat. Hist., ser. 4, 12:418.

TYPE SPECIES: *Mus giganteus* Hardwicke, 1804 (= *Mus indicus* Bechstein, 1800).

SYNONYMS: *Gunomys* Thomas, 1907.

COMMENTS: *Rattus* Division. *Nesokia* has been considered the closest relative of *Bandicota* (Corbet and Hill, 1992; Misonne, 1969: Musser and Brothers, 1994; Niethammer, 1977; Radtke and Niethammer, 1984/85; Wroughton, 1908; Watts and Baverstock, 1994*b*), and outside of that alliance, its phylogenetic position among murines is nearest *Rattus*. Misonne (1969) was unsure where to place *Bandicota* except he thought it unrelated to *Rattus*; but Niethammer (1977), Pradhan and Bhagwat (1990), Gadi and Sharma (1983), and Gemmeke and Niethammer (1984) concluded from morphological, chromosomal, and allozymic data that *Rattus* is close to *Bandicota*, a conclusion also endorsed by Musser and Brothers (1994), who reviewed the problem and evidence aligning the two genera. Analyses of microcomplement fixation of albumin (Watts and Baverstock, 1994*b*) and DNA sequences of LINE-1 elements (Verneau et al., 1997, 1998) also placed *Bandicota* within a *Rattus* clade, as did DNA/DNA hybridization results (Chevret, 1994 [cited in Verneau et al., 1997). Although allozymic data and albumin immunology indicate *Bandicota* could be united with *Nesokia*, Corbet and Hill (1992:352) noted that the "cranial and dental differences are as great as between many genera of Murinae," and that the union would be nomenclaturally confusing since *Bandicota indica* "would become *Nesokia indica* (Bechstein) and the present *Nesokia indica* (Gray) would become *N. hardwickei* (Gray)." However, the nomenclatural issue should not prevent the assimilation of *Bandicota* into *Nesokia* if future analyses of a larger suite of morphological traits, along with mitochondrial and nuclear gene sequences, identifies species of each group as members of the same monophyletic clade. Revision of genus in the context of identifying Thailand samples provided by Musser and Brothers (1994). Spermatozoal morphology of *B. bengalensis* resembles that in species of *Rattus* and many other murines and may represent the ancestral configuration; it is strikingly different from spermatozoal structure of

B. indica and *B. savilei*, which is apparently unlike spermatozoa of any other eutherian described to date (Breed, 1993, 1998). Enamel microstructure of incisors and molars and its significance documented by Patnaik (2002).

Insight into evolutionary history of *Bandicota* comes from fossils uncovered in late Pliocene Swalik strata in NW India that represent at least two species (*Bandicota* sp. and *B. sivalensis*; Patnaik, 1997, 2001). *Bandicota* cf *bengalensis* was identified from late Pleistocene fragments collected in C India (Patnaik, 1995). Middle Pleistocene cave sediments in C Thailand yielded isolated molars of *B. indica* and *B. savilei*; the latter was also discovered in cave deposits of the same age in peninsular Thailand (Chaimanee, 1998).

Bandicota bengalensis (Gray, 1835, *in* 1830-1835). Illustr. Indian Zool., pl. 21 (see Ellerman and Morrison-Scott, 1955).

COMMON NAME: Lesser Bandicoot Rat.

TYPE LOCALITY: India, Bengal.

DISTRIBUTION: Probable natural range extends from N and SE Pakistan (the Punjab and Sind, respectively; Roberts, 1997) through most of India (Agrawal, 2000), Sri Lanka, S lowlands of Nepal, and Bangladesh east to Burma. Introduced to Penang Isl off the W coast of Malay Peninsula (Chasen, 1936), the Aceh region of N Sumatra and E Java (Kloss, 1921; Musser and Newcomb, 1983), and Saudi Arabia (Kock et al., 1990). Corbet and Hill (1992) noted a report of *B. bengalensis* from Patta Isl, Kenya, but did not know if the population was established. Indomalayan distribution mapped by Musser and Brothers (1994:7).

STATUS: IUCN – Lower Risk (lc).

SYNONYMS: *barclayanus* (Anderson, 1878); *blythianus* (Anderson, 1878); *daccaensis* (Tytler, 1854); *dubius* (Kelaart, 1850); *gracilis* (Nehring, 1902); *insularis* Phillips, 1936; *kok* (Gray, 1837); *lordi* (Wroughton, 1908); *morungensis* (Hodgson, *in* Horsfield, 1855); *plurimammis* (Hodgson, in Horsfield, 1855); *providens* (Elliot, 1839); *sindicus* (Wroughton, 1908); *sundavensis* (Kloss, 1921); *tarayensis* (Hodgson, *in* Horsfield, 1855); *varillus* (Thomas, 1907); *varius* (Thomas, 1907); *wardi* (Wroughton, 1908).

COMMENTS: The most morphologically divergent of the species now placed in *Bandicota*; so impressive are the differences with the other species that *B. bengalensis* has been placed in its own genus, *Gunomys* (see Wroughton, 1908a). Morphological contrasts with *B. indica* and *B. savilei* documented by Musser and Brothers (1994). Geographic variation and one view of subspecies presented by Agrawal and Chakraborty (1976). Chromosomal data reported by Sharma and Raman (1971, 1973), Gadi and Sharma (1983), and Dubey and Raman (1992). Lekagul and Felten (1989) recognized *varius* as a distinct species in Thailand, but that record was based on specimens of *B. savilei* (Musser and Brothers, 1994). Whether some of the many names associated with *B. bengalensis* identify diagnosable geographic entities is unresolved because no "careful systematic study of morphological and geographic variation among all the samples now identified as *B. bengalensis* is available" (Musser and Brothers, 1994:6). Prakash et al. (1995a, b) recorded specimens and habitat in Abu Hill, Aravalli Ranges in Rajasthan, India, and Chakraborty and Agrawal (2000) documented habitat and distribution in Gujarat State. Indian population reviewed by Agrawal (2000).

Bandicota indica (Bechstein, 1800). *In* Pennant, Allgemeine Ueber Vierfuss. Thiere, 2:497.

COMMON NAME: Greater Bandicoot Rat.

TYPE LOCALITY: India, Pondicherry.

DISTRIBUTION: Extends from throughout most of India (Agrawal, 2000), Sri Lanka, Bangladesh, lowlands of Nepal through Burma, S China (Yunnan, S Sichuan, Guizhou, Guangxi, Guangdong, Fujian, Sichuan, Jiangxi, and Hong Kong Isls; Wang, 2003, and Zhang et al., 1997), Taiwan (Wang, 2003, and M.-J. Yu, 1996), Thailand (J. T. Marshall, Jr., 1977a; Robinson et al., 1995), Laos (Aplin et al., 2003b; Smith et al., In Press), Cambodia (Aplin et al., 2003b, c) and Vietnam (Dang et al., 1994; also Cat Ba Isl, off coast of N Vietnam, Kuznetsov, 2000). Introduced into Kedah and Perlis regions of Malay Peninsula (Harrison, 1956; J. T. Marshall, Jr., 1977a) as well as Java (Musser and Newcomb, 1983). Its spotty distribution may reflect other geographic introductions (Taiwan for example); "since it is commensal, large, and delicious to eat, this bandicoot may have been spread by man in comparatively recent times" (J. T. Marshall, Jr., 1977a:428). Corbet and Hill (1992:352) included Pakistan within the range but Roberts (1977, 1997) did not record it from there and we cannot locate any specimens from that country. Indomalayan range mapped by Musser and Brothers (1994).

STATUS: IUCN – Lower Risk (lc).

SYNONYMS: *bandicota* (Bechstein, 1800) [see Ellerman, 1941]; *elliotanus* (Anderson, 1878); *eloquens* (Kishida, 1926) [see Kaneko and Maeda, 2002]; *gigantea* (Hardwicke, 1804); *jabouillei* Thomas, 1927; *macropus* (Hodgson, 1845); *malabarica* (Shaw, 1801); *maxima* Pradhan, Mondal, Bhagwat, and Agrawal, 1993; *mordax* Thomas, 1916; *nemorivaga* (Hodgson, 1836); *perchal* (Shaw, 1801); *setifera* (Horsfield, 1824); *siamensis* Kloss, 1919; *sonlaensis* Dao, 1975; *taiwanus* (Tokuda, 1939) [*nomen nudum*, see Kaneko and Maeda, 2002]; *taiwanus* (Tokuda, 1941) [not Horikawa, 1929; see Kaneko and Maeda, 2002].

COMMENTS: Usually sympatric with one or other of *B. bengalensis* and *B. savilei* in the Indomalayan region (Musser and Brothers, 1994). In C Burma all three species occur sympatrically in the rainfed rice production environment (K. Aplin, in litt., 2004). A careful systematic revision is necessary to assess the significance of morphological and biochemical variation within *B. indica*. Pradhan et al. (1989), for example, reported that *B. gigantea* is specifically distinct from *B. indica*, citing differences in body size, and haemoglobin and eye lens proteins. However, because the skull of the holotype of *gigantea* is broken and diagnostic traits of *gigantea* could not be confirmed, Pradhan et al. (1993) described this taxon as *B. maxima*, diagnosed primarily by large size, cuticular pattern of dorsal hairs, and certain cranial proportions. Agrawal (2000:151) noted that the difference in cranial proportions between *B. indica* and *B. maxima* "appears to be nothing but individual variation due to growth changes" and that the difference in hair texture needs to be tested by study of more specimens of different sex and age groups collected from different geographic localities; Agrawal included *maxima* in *B. indica*. This action was taken earlier by Chakraborty and Chakraborty (1991), who have provided the only decent study of intraspecific geographic variation within *B. indica*. They concluded that the morphological differences between *indica* and *gigantea* (= *maxima*) simply reflected normal size variation within any large geographic sample of the species and the documented biochemical differences needed to be tested by sampling throughout the range of the species, not just from the Bombay-Pune region. Chakraborty and Chakraborty (1991) did uncover one significant pattern of geographic variation. Specimens from peninsular India (typical *indica*) have significantly longer nasals (absolutely and relative to skull length) than do samples from NE India (*nemorivaga*). This pattern was detected independently by Corbet and Hill (1992) who noted that the animals with longer skulls and nasals occurred on Sri Lanka as well as peninsular India and were allopatric to the range of samples from NE India and eastward in the Indomalayan region. Chakraborty and Chakraborty (1991) treated these two populations as subspecies because of the considerable overlap they found in other characters. If there are two species within what is now called *B. indica*, it will be reflected in the differences between these two groups, not between *indica* and *gigantea*.

Morphologically, *B. indica* is more closely related to *B. savilei* than to *B. bengalensis* (Musser and Brothers, 1994), but based on gel electrophoretic comparisons, closer to *Nesokia* than to other species of *Bandicota* (Radtke and Niethammer, 1984/85). Chromosomal data for Thai samples provided by Markvong et al. (1973) and for Indian samples by Gadi and Sharma (1983). Li et al. (1998) documented variation in mtDNA sequences from three sites in Guangdong, China, with the intent to compare the results with other species of murines. Ecological relationships with species of *Maxomys*, *Mus*, *Leopoldamys*, and *Niviventer* in S Yunnan reported by Wu et al. (1996). Habitat use on Taiwan documented by Adler (1995), and habitat and distribution in Gujarat State of W India recorded by Chakraborty and Agrawal (2000).

Bandicota savilei Thomas, 1916. J. Bombay Nat. Hist. Soc., 24:641.

COMMON NAME: Savile's Bandicoot Rat.

TYPE LOCALITY: Burma, Mt Popa, about 2500 ft (760 m).

DISTRIBUTION: C Burma, Thailand (throughout the country north of the Isthmus of Kra and south of the Isthmus to the southern end of peninsular Thailand), Vietnam (Dang et al., 1994), Cambodia (Aplin et al., 2003*b*; also specimens in FMNH); and an unconfirmed photographic record from S Laos (Aplin et al., 2003*b*).

STATUS: IUCN – Lower Risk (lc).

SYNONYMS: *bangchakensis* Boonsong and Felten, 1989; *curtata* Thomas, 1929; *giaraiensis* Dao and Cao, 1990; *hichensis* Dao, 1961.

COMMENTS: Occurs sympatrically with *B. indica* in C Burma (K. Aplin, in litt., 2003), C and S Thailand, and S and N Vietnam and C Cambodia (Aplin et al., 2003*b*); found with *B. bengalensis* and *B. indica* in C Burma where in some places *B. savilei* lives in fields and *B. bengalensis* in village houses, and at other sites both are found together in rainfed paddy fields (Aplin, in litt., 2004; Musser and Brothers, 1994). *Bandicota savilei* is a distinctive species as Thomas (1916*d*) pointed out; unfortunately it was later treated as a subspecies of *B. indica* (Ellerman, 1961; Ellerman and Morrison-Scott, 1951). Specimens are usually misidentified as *B. bengalensis*, but in external, cranial, and dental morphology, *B. savilei* can easily be distinguished from *B. bengalensis* and more closely resembles *B. indica* (Musser and Brothers, 1994). Aplin et al. (2003*b*) described external morphological criteria for distinguishing *B. indica* from *B. savilei*, based upon large samples collected in sympatry in Cambodia and Vietnam.

Electrophoretic data, however, indicated a closer relationship with *B. bengalensis* than with *B. indica* (Radtke and Niethammer, 1984[85]). *Bandicota indica* and *B. savilei* share a unique spermatozoal structure (Breed, 1993, 1998). Chromosomal information reported by Markvong et al. (1973). Lekagul and Felten (1989) described *bangchakensis* as a species; *hichensis* and *giaraiensis* were proposed as subspecies of *B. bengalensis* (Dao, 1961; Dao and Cao, 1990) and listed in the synonymy of that species by Corbet and Hill (1992). Musser and Brothers (1994) explained why these three names represent samples of *B. savilei* and provided a revision of the species.

Batomys Thomas, 1895. Ann. Mag. Nat. Hist., ser. 6, 16:162.

TYPE SPECIES: *Batomys granti* Thomas, 1895.

SYNONYMS: *Mindanaomys* Sanborn, 1953.

COMMENTS: *Phloeomys* Division. A Philippine Old Endemic (Musser and Heaney, 1992). Using morphological data, Musser and Heaney (1992) placed *Batomys* in the same monophyletic group as *Crateromys* and *Carpomys* and among some alternative hypotheses about relationships suggested that *Phloeomys* may also be a member of this group. Phylogenetic analyses of complete mtDNA cytochrome *b* sequences for 13 of the 16 genera of Philippine murines bring together *Batomys*, *Crateromys*, and *Phloeomys* in a clade separate from those formed by other Old Endemic genera; species of *Carpomys* were not sampled (Jansa and Heaney, 2001). Revised by Musser et al. (1998*a*).

Batomys dentatus Miller, 1911. Proc. U.S. Natl. Mus., 38:400.

COMMON NAME: Large-toothed Batomys.

TYPE LOCALITY: Philippines, N Luzon Isl, Benguet, Haights-in-the-Oaks, 7000 ft (2134 m).

DISTRIBUTION: Greater Luzon Faunal Region. Endemic to mountain forest on N Luzon and known only from the type locality.

STATUS: IUCN – Data Deficient.

COMMENTS: Still represented only by the holotype (Musser et al., 1998*a*).

Batomys granti Thomas, 1895. Ann. Mag. Nat. Hist., ser. 6, 16:162.

COMMON NAME: Luzon Batomys.

TYPE LOCALITY: Philippines, N Luzon Isl, Mountain Province, Mt Data, 7000 ft (2134 m).

DISTRIBUTION: Greater Luzon Faunal Region. Recorded only from mountain forests of Luzon Isl, Mt Data in north and Mt Isarog in SE peninsula; probably more widespread in mountains of N Luzon (Heaney et al., 1998).

STATUS: IUCN – Lower Risk (lc).

COMMENTS: Morphological features indicate *B. granti* to be more closely related to *B. salomonseni* than to other species of *Batomys* (Musser et al., 1998*a*). Karyotype (2n = FN = 52) indistinguishable between samples from S and N regions of Luzon, and from *B. salomonseni* (Rickart and Heaney, 2002; Rickart and Musser, 1993). Altitudinal range and ecological information provided by Heaney et al. (1999).

Batomys russatus Musser, Heaney, and Tabaranza, Jr. 1998. Am. Mus. Novit., 3237:34.
COMMON NAME: Russet Batomys.
TYPE LOCALITY: Philippines, Dinagat Isl (Surigao del Norte Province), Plaridel Municipality, Libjo (=Albor).
DISTRIBUTION: Greater Mindanao Faunal Region. Endemic to Dinagat Isl in lowland tropical evergreen rainforest.
COMMENTS: Sympatric with *B. salomonseni*. Unique among all species of living Murinae in retaining a complete cephalic arterial pattern (the primitive configuration for muroid rodents) as indicated by the presence of a sphenofrontal foramen, squamosal-alisphenoid groove, large stapedial foramen, and groove scoring ventral surface of posterolateral pterygoid plate (Musser et al., 1998*a*). *Batomys russatus* is part of a small cluster of mammals endemic to Dinagat (which includes *Crateromys australis* and *Podogymnura aureospinula*).

Batomys salomonseni (Sanborn, 1953). Vidensk. Medd. Nat. Foren. Kjobenhavn, 115:287.
COMMON NAME: Mindanao Batomys.
TYPE LOCALITY: Philippines, Mindanao Isl, Bukidnon Province, Mt Katanglad, 1600 m.
DISTRIBUTION: Greater Mindanao Faunal Region on Isls of Mindanao, Dinagat, Biliran, and Leyte; tropical lowland evergreen rainforest to montane forest formations.
STATUS: IUCN – Lower Risk (lc).
COMMENTS: Originally described as the only species in *Mindanaomys* (Sanborn, 1953), but that genus considered inseparable from *Batomys* (Misonne, 1969; Musser and Heaney, 1992; Musser et al., 1998*a*). Morphologically more closely related to *B. granti* than to *B. dentatus* or *B. russatus* (Musser et al., 1998*a*), and the standard karyotype (2n = FN = 52) is indistinguishable from that of *B. granti* (Rickart and Heaney, 2002; Rickart and Musser, 1993). Distributional and ecological information summarized by Rickart et al. (1993) and Heaney et al. (1999).

Berylmys Ellerman, 1947. Proc. Zool. Soc. Lond., 1947-1948(117):261 [1947].
TYPE SPECIES: *Epimys manipulus* Thomas, 1916.
COMMENTS: *Rattus* Division. Originally proposed as a subgenus of *Rattus* by Ellerman (1947*b*) for *B. manipulus* and *B. berdmorei*, but elevated to generic rank in a revision by Musser and Newcomb (1983). They also recorded taxonomic history of names and groups associated with *Berylmys*, reported past evolutionary histories of species (centered in Indochina), and found that *Berylmys* was dentally similar to *Niviventer*, *Maxomys*, and *Leopoldamys*, but shared some derived cranial characters with *Rattus* and that phylogenetic relationships were still unknown. Sperm morphology united *Berylmys* with *Sundamys*, *Rattus*, and *Leopoldamys* (Breed and Yong, 1986), but that union was based on a shared spermatozoal form which is probably primitive. Analyses of microcomplement fixation of albumin (Watts and Baverstock, 1994*a*) and DNA sequences from LINE-1 retrotransposons (Usdin et al., 1995; Verneau et al., 1997, 1998) placed *Berylmus* (represented by *B. bowersi*) in a *Rattus* group, as did cladistic analysis of molar traits (Chaimanee, 1998). Musser and Newcomb (1983) summarized chromosomal information; additional chromosomal data for Vietnam samples reported by Bulatova et al. (1992).

Berylmys berdmorei (Blyth, 1851). J. Asiat. Soc. Bengal, 20:173.
COMMON NAME: Berdmore's Berylmys.
TYPE LOCALITY: S Burma (S Tenasserim), Mergui.
DISTRIBUTION: S China (S Yunnan; Wang, 2003; Yang and Wu, 1979; Zhang et al., 1997), S Burma, Thailand (J. T. Marshall, Jr., 1977*a*; Robinson et al., 1995), Cambodia, N and C Laos (Aplin et al., 2003*c*; Smith et al., In Press), and S Vietnam (Dang et al., 1994; Kuznetsov, 2000); also on Con Dao Isl (= Con Son Isl). Details of overall range are documented by Musser and Newcomb (1983) and Corbet and Hill (1992).
STATUS: IUCN – Lower Risk (lc).
SYNONYMS: *magnus* (Kloss, 1916); *mullulus* (Thomas, 1916).

COMMENTS: *Berylmys berdmorei* is the only species of *Berylmys* recorded from a small island (Con Dao, S Vietnam) off continental Indochina. Evolutionary history extends back to the middle Pleistocene, based on molars recovered from cave sediments in C Thailand (Chaimanee, 1998).

Berylmys bowersi (Anderson, 1879). Anat. Zool. Res., Yunnan, p. 304.
COMMON NAME: Bower's Berylmys.
TYPE LOCALITY: China, Yunnan Province, Kakhyen Hills, Hotha, 4500 ft (1370 m).
DISTRIBUTION: NE India (Agrawal, 2000), N and C Burma, S China (Yunnan, Guangxi, Fujiau, and S Anhui; Liu et al., 1985), N and peninsular Thailand, N Laos, Vietnam (Dang et al., 1994), Malay Peninsula, and NW Sumatra (Medan); see Musser and Newcomb (1983) and Corbet and Hill (1992).
STATUS: IUCN – Lower Risk (lc).
SYNONYMS: *ferreocanus* (Miller, 1900); *kennethi* (Kloss, 1919); *lactiiventer* (Kloss, 1919); *latouchei* (Thomas, 1897); *totipes* (Dao, 1966); *wellsi* (Thomas, 1921).
COMMENTS: Once relegated to *Rattus* in either subgenus *Stenomys* (Ellerman, 1947a) or *Bullimus* (Misonne, 1969); see Musser and Newcomb (1983) for taxonomic history. Not known to occur on small islands off continental margin; *B. bowersi* is the only species of *Berylmys* found on Malay Peninsula and a large island of the Sunda Shelf (Sumatra). Spermatozoal morphology and its significance documented by Breed and Yong (1986). Feng et al. (1986) identified a sample from SE Xizang (Tibet, China) as *B. bowersi* (which was accepted by Wang, 2003) but whether it is that species or *B. mackenziei* is unclear. Wang (2003) and Zhang et al. (1997) described an expansive distribution for *B. bowersi* in S China but because they do not recognize *B. mackenziei*, which also occurs in the same region, it is unclear whether their records represent just *B. bowersi* or both species. Fossils recovered from early to late Pleistocene cave strata in the Sichuan-Guizhou region of S China were identified as *B. bowersi* (Zheng, 1993).

Berylmys mackenziei (Thomas, 1916). J. Bombay Nat. Hist. Soc., 24:40.
COMMON NAME: Mackenzie's Berylmys.
TYPE LOCALITY: Burma, Chin Hills, 50 mi (80 km) west of Kindat.
DISTRIBUTION: NE India in Meghalaya (Shillong and Cherrapunji), Nagaland, Mizoram (Lushai Hills), and Manipur (Bishenpur and Tamenglong); C and S Burma, China (Sichuan), and S Vietnam; see Musser and Newcomb (1983) and Agrawal (2000).
STATUS: IUCN – Lower Risk (lc).
SYNONYMS: *fea* (Thomas, 1916).
COMMENTS: Specimens of this distinctive species are usually misidentified as *B. bowersi*; see detailed comparisons in Musser and Newcomb (1983).

Berylmys manipulus (Thomas, 1916). J. Bombay Nat. Hist. Soc., 24, 3:413.
COMMON NAME: Manipur Berylmys.
TYPE LOCALITY: C Burma, Kabaw Valley, Kampat, 20 mi (32 km) west of Kindat.
DISTRIBUTION: India in Assam (Golaghat), Nagaland (Kekrima and Naga Hills), and Manipur (Bishenpur, Senapati, and Imphal); N and C Burma, and S China (W Yunnan); distribution extracted from Agrawal (2000), Musser and Newcomb (1983), Wang (2003), Yang and Wu (1979), and Zhang et al. (1997).
STATUS: IUCN – Lower Risk (lc).
SYNONYMS: *kekrimus* Roonwal, 1948.
COMMENTS: The taxon *kekrimus* was described as a subspecies and based on specimens from Nagaland, but Agrawal (2000), who reviewed the Indian segment of *B. manipulus*, could find no morphological basis for recognizing any significant differences between the Nagaland sample and typical *B. manipulus*.

Bullimus Mearns, 1905. Proc. U.S. Natl. Mus., 28:450.
TYPE SPECIES: *Bullimus bagobus* Mearns, 1905.
COMMENTS: *Rattus* Division. Although described as a distinctive genus by Mearns, *Bullimus* has been treated as a subgenus of *Rattus* in most taxonomic works (e.g., Ellerman, 1941; Misonne, 1969; Simpson, 1945). Generically reinstated by Musser (1982c) and Musser and Newcomb (1983) and rediagnosed by Musser and Heaney (1992), who also

documented taxonomic history and past associations with other murine genera. *Bullimus* belongs to the group of Philippine New Endemics (Musser and Heaney, 1992), and phylogenetic analyses of complete mtDNA cytochrome *b* sequences for 13 of the 16 genera of endemic Philippine murines placed it in a clade with *Rattus everetti*, *Limnomys*, and *Tarsomys* (Jansa and Heaney, 2001), which are also New Endemics. The association of *Bullimus* with genera in a *Rattus* Division is also supported by albumin immunology (Watts and Baverstock, 1994*b*) and chromosomal data (Rickart and Heaney, 2002; Rickart and Musser, 1993). Revised by Rickart et al. (2002).

Bullimus bagobus Mearns, 1905. Proc. U.S. Natl. Mus., 28:450.
COMMON NAME: Mindanao Bullimus.
TYPE LOCALITY: Philippines, Mindanao Isl, Mt Apo, Todaya, 4000 ft (1220 m).
DISTRIBUTION: Greater Mindanao Faunal Region in the Philippines. Isls of Samar, Calicoan, Leyte, Dinagat, Siargao, Mindanao, Bohol, and Maripipi, but probably occurs on other islands in the Greater Mindanao Faunal Region; altitudinal range 200-1800 m.
STATUS: IUCN – Lower Risk (lc).
SYNONYMS: *barkeri* (Johnson, 1946); *rabori* (Sanborn, 1952).
COMMENTS: Morphology, ecology and distribution documented and contrasted with other species of *Bullimus* by Rickart et al., 1993, 2002). Standard karyotype (2n = 42, FN = 58) described by Rickart and Musser (1993).

Bullimus gamay Rickart, Heaney, and Tabaranza, Jr., 2002. J. Mammal., 83:427.
COMMON NAME: Camiguin Bullimus.
TYPE LOCALITY: Philippines, Camiguin Isl, Camiguin Province, Mt Timpoong, 2 km N, 6.5 km W Mahinog, 1275 m, 09°11′ N/124°43′ E.
DISTRIBUTION: Endemic to Camiguin Isl in the Bohol Sea, 8 km north of NC Mindanao; found in remaining forests on the island, 900-1475 m.
COMMENTS: Each of the other species of *Bullimus* is endemic to one of the great Pleistocene land masses. Although adjacent to Mindanao Isl, Camiguin Isl is not part of the Greater Mindanao land mass of the Pleistocene (Rickart et al., 2002). Isolated by a deep-water channel, it remained separate from Mindanao during Pleistocene sea level fluctuations. *Bullimus gamay* is a member of a rich mammalian fauna native to Camiguin of which 19 also occur on Mindanao: "proximity to a large island with a rich fauna (Mindanao) has allowed some colonization to occur, but isolation by deep water has been sufficient to promote speciation among some non-volant mammals" (Rickart et al., 2002). Presumably *B. gamay* is more closely related to *B. bagobus*, the species on Mindanao and other islands once part of Pleistocene Greater Mindanao, than to *B. luzonicus*, which is endemic to Pleistocene Greater Luzon. An undescribed species of *Apomys* is the only other murine endemic to Camiguin Isl (Heaney and Tabaranza, Jr., ms).

Bullimus luzonicus (Thomas, 1895). Ann. Mag. Nat. Hist., ser. 6, 16:163.
COMMON NAME: Luzon Bullimus.
TYPE LOCALITY: Philippines, Luzon Isl, Lepanto, Mt Data, 8,000 ft (2440 m).
DISTRIBUTION: Greater Luzon Faunal Region in the Philippines. Endemic to Luzon and represented by samples from scattered localities over the island, 200-2440 m.
STATUS: IUCN – Lower Risk (lc).
COMMENTS: A poorly known species documented by few samples. Specimens from S Luzon are discussed by Heaney et al. (1999). Morphological and distributional comparisons with *B. bagobus* and *B. gamay* provided by Rickart et al. (2002).

Bunomys Thomas, 1910. Ann. Mag. Nat. Hist., ser. 8, 6:508.
TYPE SPECIES: *Mus coelestis* Thomas, 1896.
SYNONYMS: *Bunomys* Grandidier, 1905 [not Thomas, 1910]; *Frateromys* Sody, 1941.
COMMENTS: *Rattus* Division. Usually considered a Sulawesi endemic, although the Nusa Tenggaran *naso* has been referred to *Bunomys* (Corbet and Hill, 1992). Recognized as a genus until the 1930s (Tate, 1936), subsequently included in *Rattus* (Ellerman, 1941; Simpson, 1945; Laurie and Hill, 1954; Misonne, 1969), finally treated as separate genus (Musser and Newcomb, 1983; Musser, 1984, 1987*a*, 1991; Musser and Carleton, 1993;

Corbet and Hill, 1991, 1992; Pavlinov et al., 1995a). Phylogenetic relationships unresolved, but member of Sulawesi New Endemics. Spermatozoal morphology distinctive although some species resemble the spermatozoal configuration in *Rattus*, which reflects shared primitive features (Breed and Musser, 1991). Kitchener et al. (1991a) considered *Bunomys* a very close relative of *Paulamys*, a Flores endemic; both are ecological if not morphological equivalents (Kitchener et al., 1998; Musser et al., 1986). Analysis of microcomplement fixation of albumin (Baverstock and Watts, 1994b) indicated *Bunomys* (based on *B. chrysocomus*) belongs in a "*Rattus*-like" clade; may be most closely related to *Rattus timorensis*, and *Komodomys rintjanus* in Nusa Tenggara (Lesser Sunda Isls); and could be polyphyletic, although their division of species is discordant from the two groups identified below based on morphometric traits. Geographic and altitudinal distributions of species and association with different forest formations summarized by Musser and Dagosto (1987), Musser (1991), and Musser and Holden (1991). Sulawesi species can be morphologically divided into two clusters, the *B. chrysocomus* Group (*B. chrysocomus*, *B. coelestis*, and *B. prolatus*), and the *B. fratrorum* Group (*B. andrewsi*, *B. fratrorum*, *B. penitus*, and an undescribed species from C Sulawesi). Information in the following accounts derive from published sources and a revision of the genus by Musser (ms), in which the new species from C Sulawesi will be described. Grandidier (1905:50) proposed *Bunomys* for a Madagascar rat, and Sody (1941:260) used *Frateromys*, but the former is a *nomina nudum*.

Bunomys andrewsi (J. A. Allen, 1911). Bull. Am. Mus. Nat. Hist., 30:366.
　　COMMON NAME: Andrew's Bunomys.
　　TYPE LOCALITY: Indonesia, NE Sulawesi, Buton Isl.
　　DISTRIBUTION: Sulawesi; central core, SE and SW peninsulas; primarily found in lowland tropical evergreen rainforest but reaches lower montane rainforest in some places.
　　STATUS: IUCN – Lower Risk (nt) as *B. heinrichi*, otherwise Lower Risk (lc).
　　SYNONYMS: *adspersus* (Miller and Hollister, 1921); *heinrichi* (Tate and Archbold, 1935); *inferior* (Tate and Archbold, 1935).
　　COMMENTS: *Bunomys fratrorum* Group. Recognized as a distinct species in a "*Rattus chrysocomus* Group" (Tate, 1936; Ellerman, 1941), then in a "*coelestis* group" of *Rattus* (Ellerman, 1949a), subsequently treated as member of subgenus *Rattus* (Laurie and Hill, 1954), and currently included in *Bunomys* (Musser, 1981c; Musser and Newcomb, 1983). In morphology and pelage features, *B. andrewsi* is most closely related to *B. fratrorum* of the NW peninsula and *B. penitus* in the central core and SE peninsula. The taxon *heinrichi* has been treated as a species (Musser, 1991; Musser and Holden, 1991; Musser and Carleton, 1993), although originally described as a subspecies of *Rattus penitus* and listed that way by Ellerman in 1941; later arranged as a subspecies of *Rattus adspersus* (Ellerman, 1949a; Laurie and Hill, 1954), or included within *Bunomys penitus* (Musser, 1981c; Musser and Newcomb, 1983) or *Frateromys penitus* (Sody, 1941). Karyotype summarized by Rickart and Musser (1993). Considerable geographic variation in craniodental measurements typifies *B. andrewsi*, but its significance will have to be determined by study of larger samples from more localities than are currently available.

Bunomys chrysocomus (Hoffmann, 1887). Abh. Zool. Anthrop.-Ethnology Mus. Dresden, 3:17.
　　COMMON NAME: Common Bunomys.
　　TYPE LOCALITY: Indonesia, N Sulawesi, Minahassa.
　　DISTRIBUTION: Sulawesi; throughout island generally between 200 and 1500 m, reaching 2200 m in some regions.
　　STATUS: IUCN – Lower Risk (lc).
　　SYNONYMS: *brevimolaris* (Tate and Archbold, 1935); *koka* Tate and Archbold, 1935; *nigellus* (Miller and Hollister, 1921); *rallus* (Miller and Hollister, 1921).
　　COMMENTS: Core of *B. chrysocomus* Group. Placed in a "*Rattus chrysocomus* Group" (Tate, 1936), in subgenus *Rattus* (Ellerman, 1941), then subgenus *Maxomys* of *Rattus* (Ellerman 1949a: Laurie and Hill, 1954, but later treated as a *Bunomys* (Musser, 1981c, 1991; Musser and Newcomb, 1983; Musser and Carleton, 1993); Corbet and Hill,

1992). Chromosomes described by Duncan (1976) and karyotypic data summarized and compared with other murines by Rickart and Musser (1993).

On the NE peninsula, *B. chrysocomus* is sympatric with *B. fratrorum*; in the central core of the island it is broadly sympatric with an undescribed species at middle altitudes and marginally sympatric with *B. andrewsi* at low elevations and *B. penitus* in montane forest. No extant specimens of *B. chrysocomus* have been collected from lowlands of the SW peninsula, partly because most of the original forest has been removed and partly because the remaining tracts have not been adequately surveyed, but the species once occurred in the deforested areas as represented by subfossils (Musser, ms).

Bunomys coelestis (Thomas, 1896). Ann. Mag. Nat. Hist., ser. 6, 18:248.
COMMON NAME: Lampobatang Bunomys.
TYPE LOCALITY: Indonesia, SW Sulawesi, Bonthain Peak (Gunung Lampobatang), 6000 ft (1830 m).
DISTRIBUTION: Sulawesi; endemic to the slopes of Gunung Lampobatang at tip of SW peninsula; found in montane forest formations between 1800 and 2500 m.
STATUS: IUCN – Endangered.
COMMENTS: *Bunomys chrysocomus* Group. More closely related to *B. chrysocomus* than to any other species of *Bunomys*. Retained as a species of *Bunomys* by Tate (1936), but subsequently included in a "*coelestis* group" of *Rattus* (Ellerman, 1941, 1949a), listed as a species in subgenus *Rattus* (Laurie and Hill, 1954), or synonymized under *Bunomys chrysocomus* (Musser, 1981c; Musser and Newcomb, 1983). Finally reinstated as a species restricted to montane forest on Gunung Lampobatang (Musser, 1991; Musser and Holden, 1991). Tate and Archbold (1935) described *koka* from mountains of the SE peninsula as a subspecies of *B. coelestis*, but that is a sample of *B. chrysocomus*.

Bunomys fratrorum (Thomas, 1896). Ann. Mag. Nat. Hist., ser. 6, 18:246.
COMMON NAME: Northeastern Peninsula Bunomys.
TYPE LOCALITY: Indonesia, NE Sulawesi, Rurukan, 3500 ft (1067 m).
DISTRIBUTION: Sulawesi; Endemic to NE region of N peninsula.
STATUS: IUCN – Lower Risk (nt).
COMMENTS: *Bunomys fratrorum* Group. Despite Thomas' clear declaration that *fratrorum* was distinct from *chrysocomus*, an evaluation also supported by Tate (1936) and Sody (1941), *fratrorum* was listed as a synonym of *B. chrysocomus* by Ellerman (1949a) and Laurie and Hill (1954). The two are sympatric on the NE peninsula (Musser, 1970d, 1981c, 1991; Musser and Newcomb, 1983; Musser, ms). In contrast to C Sulawesi where three species of the *B. fratrorum* species-complex are altitudinally distributed in a pattern of parapatric ranges, *B. fratrorum* is the only representative on the NE peninsula where it inhabits tropical evergreen rainforest formations from lowlands to high in mountains. Sody (1941) designated *fratrorum* as the type species of *Frateromys*.

Bunomys penitus (Miller and Hollister, 1921). Proc. Biol. Soc. Wash., 34:72.
COMMON NAME: Montane Bunomys.
TYPE LOCALITY: Indonesia, C Sulawesi, Gunung Lehio, above 6000 ft (1830 m).
DISTRIBUTION: Sulawesi; endemic to montane forest formations in central core and SE peninsula.
STATUS: IUCN – Lower Risk (lc).
SYNONYMS: *sericatus* (Miller and Hollister, 1921).
COMMENTS: *Bunomys fratrorum* Group. Originally described as a species of *Rattus*, then listed as a species in the *chrysocomus* Group of *Rattus* (Tate, 1936; Ellerman, 1941), treated as a subspecies of *Rattus adspersus* (Laurie and Hill, 1954), and included in *Frateromys* by Sody (1941). *Bunomys penitus* is distinctive in morphology and confined to lower and upper montane evergreen rainforest (Musser, 1987a, 1991; Musser and Dagosto, 1987; Musser and Holden, 1991; Musser and Newcomb, 1983; Musser, ms). Karyotypic data summarized by Rickart and Musser (1993).

In the C part of Sulawesi, *B. penitus* inhabits montane forest formations, is replaced by an undescribed species in tropical evergreen rainforest at middle elevations, and by

B. *andrewsi* below that species in lowland forests. No records indicate that any species in the B. *fratrorum* species-group are sympatric, but B. *chrysocomus* of the B. *chrysocomus* Group is sympatric and syntopic with all three of the B. *fratrorum*-group species (Musser, ms).

Bunomys prolatus Musser, 1991. Am. Mus. Novit., 3001:4.
>   COMMON NAME: Tambusisi Bunomys.
>   TYPE LOCALITY: Indonesia, C Sulawesi, Gunung Tambusisi, 6000 ft (1830 m).
>   DISTRIBUTION: Known only from 6000 ft (1830 m) on Gunung Tambusisi, Sulawesi.
>   STATUS: IUCN – Endangered.
>   COMMENTS: *Bunomys chrysocomus* Group. Morphologically most closely related to
>   B. *chrysocomus*, with which it is nearly sympatric (both were caught at the same
>   elevation but in different habitats; Musser, 1991).

Carpomys Thomas, 1895. Ann. Mag. Nat. Hist., ser. 6, 16:161.
>   TYPE SPECIES: *Carpomys melanurus* Thomas, 1895.
>   COMMENTS: *Phloeomys* Division. A Philippine Old Endemic in the same monophyletic cluster as
>   *Crateromys* and *Batomys* (Musser and Heaney, 1992). Species of *Carpomys* are endemic to
>   N Luzon and still represented only by the samples described by Thomas (1895) and
>   recorded by Sanborn (1952*a*) and Largen (1985). Both species are apparently restricted to
>   tropical upper montane forest (mossy forest), have long tails relative to head and body
>   densely covered with long hairs, and short and broad feet. Conformation of tail and hind
>   feet suggests the two species are arboreal; their molar occlusal patterns (see illustrations in
>   Musser and Heaney, 1992) indicate a diet of leaves, flowers, and possibly fruit.

Carpomys melanurus Thomas, 1895. Ann. Mag. Nat. Hist., ser. 6, 16:162.
>   COMMON NAME: Large Luzon Carpomys.
>   TYPE LOCALITY: Philippines, NW Luzon Isl, Mt Data, 7000-8000 ft (2130-2440 m).
>   DISTRIBUTION: Greater Luzon Faunal Region in the Philippines. Known only from Mt Data,
>   but probably occurs in other forested highlands of N Luzon.
>   STATUS: IUCN – Data Deficient.

Carpomys phaeurus Thomas, 1895. Ann. Mag. Nat. Hist., ser. 6, 16:162.
>   COMMON NAME: Small Luzon Carpomys.
>   TYPE LOCALITY: Philippines, NW Luzon Isl, Mt Data, 7000-8000 ft (2130-2440 m).
>   DISTRIBUTION: Greater Luzon Faunal Region in the Philippines. Recorded only from Mt Data
>   and Mt Kapilingan (Sanborn, 1952*a*), but likely occurs elsewhere in mountain forests
>   of N Luzon.
>   STATUS: IUCN – Data Deficient.

Chiromyscus Thomas, 1925. Proc. Zool. Soc. Lond., 1925:503.
>   TYPE SPECIES: *Mus chiropus* Thomas, 1891.
>   COMMENTS: *Dacnomys* Division. Reviewed by Musser (1981*b*), who regarded *Chiromyscus* as
>   morphologically closely related to *Niviventer*.

Chiromyscus chiropus (Thomas, 1891). Ann. Mus. Civ. Stor. Nat. Genova, ser. 2, 10:884.
>   COMMON NAME: Indochinese Chiromyscus.
>   TYPE LOCALITY: E Burma, Karin Hills.
>   DISTRIBUTION: SW China (S Yunnan; Wang, 2003), E Burma, N Thailand (J. T. Marshall, Jr.,
>   1977*a*), Vietnam (Dang et al., 1994), Laos (Aplin et al., 2003*c*; Smith et al., In Press;
>   AMNH 272902); see Musser (1981*b*) and Corbet and Hill (1992).
>   STATUS: IUCN – Lower Risk (lc).
>   COMMENTS: Reviewed by Corbet and Hill (1992) and Musser (1981*b*). This arboreal species is
>   morphologically similar to *Niviventer langbianis* (Musser and Lunde, ms) and easily
>   confused with it, but generally has a longer molar row, higher supraorbital and
>   temporal cranial ridging, a bicolored or mottled tail, more expansive orange pattern
>   on upperparts, and a nail-like claw on each hallux instead of a small claw.

Chiropodomys Peters, 1869. Monatsb. K. Preuss. Akad. Wiss. Berlin, p. 448.
>   TYPE SPECIES: *Mus gliroides* Blyth, 1856.
>   SYNONYMS: *Insulaemus* Taylor, 1934.
>   COMMENTS: *Micromys* Division. Revised by Musser (1979). Among Asian murines, phylogenetic
>   position of *Chiropodomys* is ambiguous. Musser and Newcomb (1983) postulated a phylo-
>   genetic link between *Chiropodomys* and *Hapalomys*, which was also indicated by Misonne
>   (1969) and Chaimanee (1998) based upon molar occlusal patterns, and a weaker tie to
>   *Haeromys*. Ellerman (1949a;132) saw a "close relationship" between *Chiropodomys* and
>   *Vandeleuria*. Spermatozoal morphology of *Chiropodomys* resembles that described for spe-
>   cies of *Hapalomys*, *Maxomys*, and *Haeromys*, and is also similar to *Mus* (Breed and Musser,
>   1991; Breed and Yong, 1986). Musser (1979) and Musser and Newcomb (1983) recorded
>   other published opinions about relationships of *Chiropodomys*. No data supports the in-
>   clusion of *Chiropodomys* within the Phloeomyinae along with *Coryphomys*, *Lenomys*,
>   *Pogonomys*, *Chiruromys*, *Mallomys*, *Phloeomys*, and *Crateromys*, which is where Tate (1936)
>   and Simpson (1945) listed it (Ellerman, 1949a:132; our studies).
>
>   Evolutionary history of *Chiropodomys* has been confined to the Indomalayan region
>   (as defined by Corbet and Hill, 1992). The genus extends back to late Pliocene in S China
>   where *C. primitivus* has been described from isolated molars recovered from cave sedi-
>   ments in the Sichuan-Guizhou region (Zheng, 1993), and *C. gliroides* is represented by
>   molars from middle or late Pleistocene cave layers in Guangxi Province of S China (Chen
>   et al., 2002). In Thailand, the extant *C. gliroides* is recorded back to late Pliocene, and the
>   extinct *C. maximus* is represented by isolated molars from middle Pleistocene deposits
>   (Chaimanee, 1998). On the Sunda Shelf, *C. gliroides* has been identified from isolated mo-
>   lars recovered from early Pleistocene beds on Java (Van der Meulen and Musser, 1999).

Chiropodomys calamianensis (Taylor, 1934). Monogr. Bur. Sci. Manila, 30:470.
>   COMMON NAME: Palawan Pencil-tailed Tree Mouse.
>   TYPE LOCALITY: Philippines, Calamian Isls, Busuanga Isl, Minuit, sea level.
>   DISTRIBUTION: Endemic to Greater Palawan Faunal Region on islands Busuanga, Balabac,
>   and Palawan (politically part of the Philippines but faunistically an extension of the
>   Sunda Shelf).
>   STATUS: IUCN – Lower Risk (lc).
>   COMMENTS: Originally described as the type species of *Insulaemus* (Taylor, 1934). In
>   morphology and geographic continuity, *C. calamianensis* is most closely related to
>   *C. major* of mainland Borneo (Musser, 1979). *Chiropodomys calamianensis* joins
>   *Maxomys panglima* and *Palawanomys furvus* as the only recorded murines endemic to
>   the Greater Palawan Faunal Region.

Chiropodomys gliroides (Blyth, 1856). J. Asiat. Soc. Bengal, 24:721.
>   COMMON NAME: Indomalayan Pencil-tailed Tree Mouse.
>   TYPE LOCALITY: NW India (Assam), Khasi Hills, Cherrapunji.
>   DISTRIBUTION: Documented from W China (S Guangxi and Yunnan; Wang, 2003, and
>   Zhang et al., 1997), NE India (Assam; Agrawal, 2000), Burma, Thailand (J. T. Marshall,
>   Jr. 1977a; Robinson et al., 1995), Laos (Smith et al., In Press), Vietnam (Dang et al.,
>   1994; Lunde et al., 2003b), Malay Peninsula (Medway, 1969), S Sumatra, Pulau Nias,
>   Kepulauan Tujuh, Kepulauan Natuna, Java, and Bali; range primarily extracted from
>   Corbet and Hill (1992), Musser (1979), and Wu and Deng (1984); probably also occurs
>   on other small islands of the Sunda Shelf and in Cambodia.
>   STATUS: IUCN – Lower Risk (lc).
>   SYNONYMS: *ana* Thomas and Wroughton, 1909; *jingdongensis* Wu and Deng, 1984; *niadis*
>   Miller, 1903; *peguensis* (Blyth, 1859); *penicillatus* Peters, 1868.
>   COMMENTS: Variation among some samples in certain morphological features significantly
>   correlated with geography both on mainland Indochina and islands of the Sunda Shelf
>   (Musser, 1979). One example of that variation is in the cranial traits distinguishing
>   populations south of the Isthmus of Kra (10E N latitude) from those occurring to the
>   north. Another is reflected in Wu and Deng's (1984) description of *C. jingdongensis*,
>   based on a small sample from W Yunnan, in which a few cranial dimensions average
>   slightly larger than most samples of *C. gliroides* from elsewhere in its range, but are not

otherwise significantly different (Corbet and Hill, 1992, and our study of specimens; however, Zheng et al. [1997] and Wang [2003] continued to recognize *jingdongensis* as a separate species). Chromosomal data reported by Yong (1973, 1983), J. T. Marshall, Jr. (1977*a*), and Tsuchiya et al. (1979). Dang et al. (1994) mapped localities in C Vietnam but the species has also been collected farther north in Ha Tinh Province and near the Chinese Border in Ha Giang Province (Lunde et al., 2003*b*; specimens in AMNH and IEBR).

Chiropodomys karlkoopmani Musser, 1979. Bull. Am. Mus. Nat. Hist., 162(6):389.
COMMON NAME: Koopman's Pencil-tailed Tree Mouse.
TYPE LOCALITY: Indonesia, Kepulauan Mentawai, Pulau Pagai Utara.
DISTRIBUTION: Known only from Pagai (Musser, 1979), and Siberut (Jenkins and Hill, 1982) in the Mentawai Isls.
STATUS: IUCN – Endangered.
COMMENTS: The largest in body size of any described *Chiropodomys*, and possibly most closely related to *C. major*. Part of the rodent fauna endemic to the Mentawai Arch. (see account of *Leopoldomys siporanus*).

Chiropodomys major Thomas, 1893. Ann. Mag. Nat. Hist., ser. 6, 11:344.
COMMON NAME: Greater Pencil-tailed Tree Mouse.
TYPE LOCALITY: Malaysia (Borneo), Sarawak, Sadong.
DISTRIBUTION: Borneo: recorded only from Sarawak and Sabah (Musser, 1979), but probably also occurs in Kalimantan; found between 900 and 1500 m on Mt Kinabalu in Sabah (Corbet and Hill, 1992; Md Nor, 2001).
STATUS: IUCN – Lower Risk (lc).
SYNONYMS: *legatus* Thomas, 1911; *pictor* Thomas, 1911.
COMMENTS: Most closely related to *C. calamianensis* in morphology and geography. Member of a group of murines endemic to Borneo that includes *Chiropodomys muroides*, *C. pusillus*, *Haeromys margaretae*, *Maxomys alticola*, *M. baeodon*, *M. ochraceiventer*, *Niviventer rapit*, *Pithecheirops otion*, and *Rattus baluensis*.

Chiropodomys muroides Medway, 1965. J. Malay. Branch R. Asiat. Soc., 36(3):133.
COMMON NAME: Gray-bellied Pencil-tailed Tree Mouse.
TYPE LOCALITY: Malaysia (Borneo), Sabah, Gunung Kinabalu, Bundu Tuhan, 4000 ft (1220 m).
DISTRIBUTION: Known only by a few specimens from Gunung Kinabalu (1100-1220 m; Md Nor, 2001) and N Kalimantan (Long Petak), but probably occurs elsewhere on Borneo.
STATUS: IUCN – Lower Risk (nt).
COMMENTS: The smallest in body size of any known species of *Chiropodomys*; its closest phylogenetic allies may be *C. gliroides* and *C. pusillus*.

Chiropodomys pusillus Thomas, 1893. Ann. Mag. Nat. Hist., ser. 6, 11:345.
COMMON NAME: Lesser Pencil-tailed Tree Mouse.
TYPE LOCALITY: Malaysia (Borneo), Sabah, Gunung Kinabalu, 1000 ft (305 m).
DISTRIBUTION: Recorded only from Sabah, Sarawak, and S Kalimantan (Musser, 1979), but probably occurs throughout Borneo.
STATUS: IUCN – Lower Risk (lc).
COMMENTS: Thomas described *pusillus* as a species, but Musser (1979) treated it as a distinctive subspecies of *C. gliroides*. However, the few known specimens of *pusillus* represent a population of small-bodied mice in which the range of variation of most dimensions are outside of the range recorded for all other samples of *C. gliroides*. On Mt Kinabalu in Sabah, *C. pusillus* is found between 300 and 1220 m (Md Nor, 2001; recorded as *gliroides*). Member of the endemic Bornean murine fauna (see account of *C. major*).

Chiruromys Thomas, 1888. Proc. Zool. Soc. Lond., 1888:237.
TYPE SPECIES: *Chiruromys forbesi* Thomas, 1888.
COMMENTS: *Pogonomys* Division. Member of the New Guinea Old Endemics (Musser, 1981*c*).

This group of arboreal species was united with *Pogonomys* as a subgenus (Laurie and Hill, 1954; Tate, 1951; Thomas, 1897*a*) until the chromosomal and morphometric study by Dennis and Menzies (1979) demonstrated how different *Chiruromys* is compared to *Pogonomys*. The closest phylogenetic relative was thought to be *Pogonomys*, as assessed by morphology (for example, Tate, 1951); analysis of microcomplement fixation of albumin (Watts and Baverstock, 1994*a*) tends to support this alliance and unites *Chiruromys* with *Pogonomys, Hyomys, Macruromys, Mallomys, Coccymys*, and *Anisomys* in the same clade. Tate included *Chiruromys* (as a subgenus of *Pogonomys*) in the Phloeomyinae, which included *Chiropodomys, Crateromys, Lenomys, Mallomys*, and *Phloeomys* (an arrangement followed by Simpson, 1945), but, except for its link to *Mallomys*, no evidence supports this allocation (Ellerman, 1949*a*; our research). Flannery (1995*a*) provided photographs of most of the species and summarized distributional and biological information. *Chiruromys forbesi* is found on the mainland and adjacent D'Entrecasteaux Isls, and two species have been recorded only from mainland Papua. A fourth undescribed species is endemic to the Louisiade Arch. (Musser and Lunde, in ms.).

Chiruromys forbesi Thomas, 1888. Proc. Zool. Soc. Lond., 1888:239.

> COMMON NAME: Forbes's Chiruromys.
>
> TYPE LOCALITY: Papua New Guinea, Central Province, Astrolabe Range, Sogeri, 1500 ft (458 m).
>
> DISTRIBUTION: Papua New Guinea; endemic to mainland of SE Papua New Guinea, sea level to 700 m. Not recorded west of Oomsis Creek in valley of Markham River (see map of this region in Brass, 1959), and extends eastward to Bara Bara near Milne Bay (Thomas, 1897*a*); also on D'Entrecasteaux Isls (Goodenough, Fergusson, and Normanby). Occurs in lowlands (sea level to 700 m) on the mainland, but up to 1300 m on Goodenough Isl and nearly 900 m on Normanby Isl (Flannery, 1995a, and specimens in AMNH). Dennis and Menzies (1979) included the Louisiade Isls in the range, but samples in AMNH of *Chiruromys* from there are not *forbesi* but examples of a separate undescribed species (Musser and Lunde, ms).
>
> STATUS: IUCN – Lower Risk (lc).
>
> SYNONYMS: *major* (Tate and Archbold, 1935); *mambatus* (Thomas, 1920); *pulcher* Thomas, 1895; *satisfactus* (Tate and Archbold, 1935); *shawmayeri* (Laurie, 1952); *vulturnus* (Thomas, 1920).
>
> COMMENTS: Significance of the appreciable geographic variation in body size and other traits recently assessed by Musser and Lunde (ms) who provided results of multivariate analysis of morphometric variation in samples from the mainland of E Papua and the three largest D'Entrecasteaux Isls (Goodenough, Fergusson and Normanby). Within the island samples, that from Goodenough averages larger in cranial dimensions than the other two, and as a group the three insular samples average larger than those from mainland Papua in both cranial and dental dimensions. The names *satisfactus* and *major* were applied to samples from Goodenough Isl (Tate and Archbold, 1935), and *shawmayeri* to specimens from Fergusson Isl (Laurie, 1952).
>
> The contrast between mainland and insular populations reflects some evolutionary divergence in the island populations after separation from the mainland at the end of the Pleistocene when sea levels rose. The D'Entrecasteaux Isls are on the continental shelf and have likely been either connected to or separated from mainland Papua at several times during glacial and interglacial Pleistocene intervals, which provided intermittent periods of gene flow between island and mainland populations. Islands in the Louisiade Arch., far eastward of the mainland, are surrounded by deeper seas and past connection with the mainland was either tenuous or nonexistent, which is reflected in the significant differentiation of the populations of *Chiruromys* (an undescribed species; Musser and Lunde, in ms.) and *Melomys* (undescribed species related to *M. lutillus*; see that account) that are endemic to Sudest Isl in the Louisiades.

Chiruromys lamia (Thomas, 1897). Ann. Mus. Civ. Stor. Nat. Genova, 18:615.

> COMMON NAME: Broad-headed Chiruromys.
>
> TYPE LOCALITY: Papua New Guinea, Central Province, lower Kemp Welch River, Ighibirei (see Thomas, 1897*a*:4, for details).

DISTRIBUTION: Mainland Papua New Guinea; known only from the Owen Stanley Range in SE Papua, from 1200 to 2300 m (Flannery, 1995a; specimens in AMNH, BMNH).

STATUS: IUCN – Lower Risk (lc).

SYNONYMS: *kagi* (Tate, 1951).

COMMENTS: Occurs in the same general mainland region as *C. forbesi*, but replaces it at higher elevations. Specimens and habitat on eastern flanks of Mount Dayman discussed by Cole et al. (1997).

Chiruromys vates (Thomas, 1908). Ann. Mag. Nat. Hist., ser. 8, 2:495.

COMMON NAME: Lesser Chiruromys.

TYPE LOCALITY: Papua New Guinea, Central Province, Upper St. Joseph's River (= Angabunga River), Madeu, 50 m NE Hall Sound, 2000-3000 ft (610-915 m).

DISTRIBUTION: Mainland Papua New Guinea; S side of Central Cordillera, from Lake Murray area in the Trans-Fly region (Western Province) east to the Kokoda Gap region near type locality; sea level to 1500 m (Flannery, 1990b, 1995a).

STATUS: IUCN – Lower Risk (lc).

Chrotomys Thomas, 1895. Ann. Mag. Nat. Hist., ser. 6, 16:161.

TYPE SPECIES: *Chrotomys whiteheadi* Thomas, 1895.

SYNONYMS: *Celaenomys* Thomas, 1898.

COMMENTS: *Chrotomys* Division. Philippine Old Endemic shrew rats unrelated to shrew rats and shrew mice endemic to Australia and New Guinea (Musser and Heaney, 1992). Phylogenetic analyses of complete mtDNA cytochrome *b* sequences for 13 of the 16 genera of endemic Philippine murines places *Chrotomys* (along with *Celaenomys*) in the same clade as *Apomys*, *Archboldomys*, and *Rhynchomys*, which are also Old Endemics (Jansa and Heaney, 2001); karyotypic variation supports this alliance (Rickart and Heaney, 2002). Actual distribution in archipelago and number of species in genus still unknown; an undescribed species has been found in lowland forest on Sibuyan Isl (Heaney et al., 1999).

Chrotomys gonzalesi Rickart and Heaney, 1991. Proc. Biol. Soc. Wash., 104:389.

COMMON NAME: Isarog Chrotomys.

TYPE LOCALITY: Philippines, SE Luzon Isl, Camarines Sur Province, W slope Mt Isarog, 4 km N, 21 km E Naga, 1350 m.

DISTRIBUTION: Greater Luzon Faunal Region in the Philippines. Known only from Mt Isarog, Luzon, in lower and upper montane evergreen forest formations, 1350-1750 m (Heaney et al., 1998, 1999).

STATUS: IUCN – Critically Endangered.

COMMENTS: Morphology, ecology, and comparisons with other species of *Chrotomys* provided by Rickart and Heaney (1991) and Rickart et al. (1991). Altitudinal distribution and ecology summarized by Heaney et al. (1999). Standard karyotype (2n = 44, FN = 52; Rickart and Musser, 1993) indistinguishable from those of *Rhynchomys isarogensis* and *Chrotomys silaceus* (Rickart and Heaney, 2002; the latter recorded as *Celaenomys*).

Chrotomys mindorensis Kellogg, 1945. Proc. Biol. Soc. Wash., 58:123.

COMMON NAME: Lowland Chrotomys.

TYPE LOCALITY: Philippines, Mindoro Isl, 3 mi (5 km) SSE of San Jose (Central), 200 ft (61 m).

DISTRIBUTION: Lowlands (sea level to 1000 m) of N Luzon and Mindoro (Heaney et al., 1998).

STATUS: IUCN – Lower Risk (lc).

COMMENTS: Considered specifically distinct from *C. whiteheadi* by Musser et al. (1982b). Larger series and analyses of both morphological and molecular data are needed to test the hypothesis that the populations on Mindoro (Greater Mindoro Faunal Region) and Luzon (Greater Luzon Faunal Region) represent the same species.

Chrotomys silaceus (Thomas, 1895). Ann. Mag. Nat. Hist., ser. 6, 16:161.

COMMON NAME: Blazed Luzon Chrotomys.

TYPE LOCALITY: Philippines, N Luzon Isl, Mt Data, Lepanto, 8000 ft (2440 m).

DISTRIBUTION: Greater Luzon Faunal Region in the Philippines. Endemic to montane forest in N Luzon; known from Benguet and Kalinga provinces (Heaney et al., 1998; Largen, 1985; Rickart and Heaney, 2002; Sanborn, 1952a), 1800-2500 m, but probably lives in montane forest formations elsewhere on N Luzon.

STATUS: IUCN – Lower Risk (lc) as *Celaenomys silaceus*.

COMMENTS: Standard karyotype (2n = 44, FN = 52) is indistinguishable from those of *Rhynchomys isarogensis* and *Chrotomys gonzalesi* (Rickart and Heaney, 2002). Made the type species of *Celaenomys*, a genus separated from *Chrotomys* by its lack of M3 (Thomas, 1898b). *Celaenomys* has been recognized in past compendia (Ellerman, 1941; Heaney et al., 1998; Musser and Heaney, 1992), but recently synonymized with *Chrotomys* because M3 occurs in one or both molar rows in some specimens (Corbet and Hill, 1992; Musser's observations). A suite of highly distinctive and derived external and cranial traits unites *silaceus* with other species of *Chrotomys* to the exclusion of the other Philippine shrew rats, *Archboldomys*, *Crunomys*, and *Rhynchomys* (Musser and Heaney, 1992).

Chrotomys whiteheadi Thomas, 1895. Ann. Mag. Nat. Hist., ser. 6, 16:161.

COMMON NAME: Montane Chrotomys.

TYPE LOCALITY: Philippines, N Luzon Isl, Mountain Province, Mt Data, 8000 ft (2440 m).

DISTRIBUTION: Greater Luzon Faunal Region in the Philippines. Endemic to N Luzon in the Central Cordillera of Benguet, Mountain, and Kalinga provinces, 1000-1500 m (Heaney et al., 1999; Rickart and Heaney, 2002).

STATUS: IUCN – Vulnerable.

COMMENTS: One specimen is recorded from Irisan, Benguet Province (Hollister, 1913b); a small series comes from Kalinga Province (L. Heaney, in litt., 2002), and all other examples are from the type locality (Largen, 1985; Sanborn, 1952a; Thomas, 1895). Standard karyotype (2n = 38, FN = 52) differs from *Rhynchomys isarogensis*, *Chrotomys gonzalesi*, and *C. silaceus* (2n = 44, recorded as *Celaenomys*) by three Robertsonian translocations (Rickart and Heaney, 2002).

Coccymys Menzies, 1990. Science in New Guinea, 16:132.

TYPE SPECIES: *Pogonomelomys ruemmleri* Tate and Archbold, 1941.

COMMENTS: *Pogonomys* Division. A New Guinea Old Endemic. Analysis of immunological distances by Watts and Baverstock (1994a) for New Guinea murines indicated *Coccymys* (represented by *C. ruemmleri*) showed no clear affinities with any other genus but appeared to be a distinctive member of their "*Anisomys*" clade (our *Pogonomys* Division). Sperm morphology of *C. ruemmleri* described by Breed and Aplin (1994) and its significance as a guide to relationships of *Coccymys* within endemic New Guinea murines outlined by Breed (1997). Our allocation to the *Pogonomys* Division is provisional pending phylogenetic analyses incorporating a wide array of morphological data sets combined with a range of different gene sequences.

Coccymys albidens (Tate, 1951). Bull. Am. Mus. Nat. Hist., 97:286.

COMMON NAME: White-toothed Coccymys.

TYPE LOCALITY: New Guinea, Prov. of Papua (= Irian Jaya), Snow Mtns, 15 mi (24 km) N Mt Wilhelmina, Lake Habbema, 3225 m.

DISTRIBUTION: New Guinea, Prov. of Papua (= Irian Jaya); known only from the N slopes of the Snow Mtns (Pegunungan Maoke) at the type locality and downslope at 2800 m (9 km NE Lake Habbema). Possibly endemic to the Snow Mtns; all the past collecting activity in the mountains of Papua New Guinea have not discovered the species living there.

STATUS: IUCN – Endangered.

COMMENTS: Still represented only by six specimens collected in 1938. Three of these form the type series, one of the other three had been misidentified as "*Pogonomys sylvestris*" and two were misidentified as "*Pogonomelomys ruemmleri*". Originally described as a species of *Melomys* (Tate, 1951), but Musser and Carleton (1993) placed *albidens* in *Coccymys* because so many of its traits are unlike those defining species within *Melomys* and some characteristics are more similar to those defining *reummleri*. This conclusion was reached independently by Flannery (1990*b*) and Menzies (1990), although neither of them formally allocated *albidens* and *ruemmleri* to the same genus. Further study has revealed that *albidens* and *ruemmleri* seem to form a monophyletic group, as defined by morphology, but are separated by a suite of external, cranial, and dental traits (Musser and Lunde, in ms.).

Coccymys ruemmleri (Tate and Archbold, 1941). Am. Mus. Novit., 1101:6.

    COMMON NAME: Rümmler's Coccymys.

    TYPE LOCALITY: New Guinea, Prov. of Papua (= Irian Jaya), Snow Mtns (Pengunungan Maoke), N slope Mt Wilhelmina, Lake Habbema, 3225 m.

    DISTRIBUTION: New Guinea; Central Cordillera from Mt Wilhelmina in the Snow Mtns (Pegunungan Maoke) of Prov. of Papua (= Irian Jaya) to Mt Saint Mary in Central Province of Papua New Guinea; 2000-4050 m; Flannery (1995*a*) and Musser and Lunde (ms). Apparently absent from the Owen Stanley Range in E Papua New Guinea.

    STATUS: IUCN – Lower Risk (lc).

    SYNONYMS: *shawmayeri* (Hinton, 1943).

    COMMENTS: The species *ruemmleri* was originally described as a *Pogonomelomys* (Tate and Archbold, 1941), but because its morphology is so different from other species in that genus Tate (1951) placed *ruemmleri* in a group within *Pogonomelomys* separate from *mayeri* and *bruijnii*, the species considered typical of the genus. The distinctiveness of *ruemmleri* was reinforced by Lidicker's (1968) study of phallic morphology; others have noted that it was not part of the same monophyletic group containing the other species of *Pogonomelomys* (for example, Flannery, 1990*b*). Finally, Menzies (1990) made *ruemmleri* the type species of *Coccymys*. Before the reports of Lidicker, Flannery, and Menzies, the unique character of *ruemmleri* had been ascertained by Jack Mahoney, who died before he could finish his revision of the group. The form *shawmayeri* was described by Hinton (1943), who considered it a remarkable species of *Rattus* unlike any of the New Guinea species and possibly closely related to Nepal and Sikkim endemics. An undescribed species of *Coccymys* occurs in the Owen Stanley Range (Musser and Lunde, in ms.).

Colomys Thomas and Wroughton, 1907. Ann. Mag. Nat. Hist., ser. 7, 19:379.

    TYPE SPECIES: *Colomys goslingi* Thomas and Wroughton, 1907.

    COMMENTS: *Colomys* Division. *Nilopegamys* has been included in *Colomys* (Hayman, 1966) but was recently separated from it (Kerbis Peterhans and Patterson, 1995). Using dental characteristics, Misonne (1969) placed *Colomys* in a *Parapodemus* group of his *Lenothrix-Parapodemus* Division with an inferred relationship to *Malacomys*, but that affinity is currently rejected (Dieterlen, 1983; E. Lecompte, 2002*b*; results of our research). Molar occlusal patterns of *Colomys* resemble those of the African *Zelotomys* (Misonne, 1969, and verified by our observations), a shared-derived morphological feature concordant with results from phylogenetic analyses of complete mtDNA cytochrome *b* sequences (Lecompte et al., 2002*b*) and nuclear IRBP gene sequences (Lecompte, 2003), which indicates *Colomys* to be sister to *Zelotomys*. Those molecular data align *Colomys* and *Zelotomys* with *Myomyscus verreauxii* in a monophyletic cluster embedded in a larger group composed of *Praomys*, *Mastomys*, *Hylomyscus*, *Heimyscus*, other *Myomyscus*, and *Stenocephalemys*; partial 16S rRNA mtDNA sequences also associated *Colomys* with *Praomys*, *Mastomys*, and *Hylomyscus* (Fadda et al., 2001*a*).

Colomys goslingi Thomas and Wroughton, 1907. Ann. Mag. Nat. Hist., ser. 7, 19:380.

    COMMON NAME: African Wading Rat.

    TYPE LOCALITY: Dem. Rep. Congo, Orientale, Uele River, Gambi.

    DISTRIBUTION: Recorded from Liberia (Lofa), Cameroon, NE Angola, NW Zambia, Dem. Rep. Congo, Ruanda, Uganda, E Kenya, S Sudan, and W Ethiopia (range mostly abstracted from Dieterlen, 1983); limits unresolved.

    STATUS: IUCN – Lower Risk (lc).

    SYNONYMS: *bicolor* Thomas, 1912; *denti* St. Leger, 1930; *eisentrauti* Dieterlen, 1983; *ruandensis* Dieterlen, 1983.

    COMMENTS: Study of external, cranial, and dental variation in context of taxonomic revision provided by Dieterlen (1983). Recent research by Rainer Hutterer and his colleagues (R. Hutterer, in litt., 1999) is uncovering a much greater range in morphological variation within and among geographic samples of *Colomys* than has been documented, which indicates the current definition of *C. goslingi* to be a composite of several separate species.

Colomys goslingi is most commonly found along banks of small and shallow flowing streams and pools in tropical evergreen rainforests (Dieterlen, 1983; Hatt, 1940a), but has also been taken along streams in grassland far from forest (Hayman, 1966). Hayman supposed *Colomys* to "be a relict forest form" and its survival in regions where forests have retreated or been cleared "may be due to its aquatic specialisation which enabled it to continue to exist wherever any permanent water with sufficient bordering vegetational cover may provide its basic requirements" (p. 36). The species is a carnivore preying on limnetic macroinvertebrates and small vertebrates, and possesses special morphological and neurological adaptations as well as behavioral repertoire that are correlated with extracting such a diet from streams and pools (Dieterlen and Statzner, 1981; Stephan and Dieterlen, 1982; Kerbis Peterhans and Patterson, 1995). *Colomys* wades through the stream, using its front feet to sift mud and debris and its vibrissae–which are spread out over the water surface–to detect prey (*colo* is Latin for separating or sifting so *Colomys* means sifting mouse). *Colomys goslingi* is a member of an African guild of waders that includes species of *Malacomys* and *Deomys ferrugineus*: "Without swimming and while perched on elongate hind feet, all three consume insects and other small animals in shallow forest streams and pools" (Kerbis Peterhans and Patterson, 1995:329). Comparisons with Neotropical ichthyomyines provided by Voss (1988) and Kerbis Peterhans and Patterson (1995). Crawford-Cabral (1998) reviewed and mapped the few Angolan records. Habitat and altitudinal distribution on Ugandan slopes of Mt Elgon reported by Clausnitzer and Kityo (2001).

Conilurus Ogilby, 1838. Trans. Linn. Soc. Lond., 18:124.
    TYPE SPECIES: *Conylurus constructor* Ogilby, 1837 (= *Hapalotis albipes* Lichtenstein, 1829).
    SYNONYMS: *Conylurus* Ogilby, 1837; *Hapalotis* Lichtenstein, 1829.
    COMMENTS: *Pseudomys* Division. Member of the Australian Old Endemics (Musser, 1981c) that is phylogenetically closely allied to *Mesembriomys* (Watts et al., 1992). *Conilurus* was also considered a phylogenetic ally of *Leporillus* (Watts et al., 1992), but that conclusion was based on one-way microcomplement fixation of albumin reactions and is unsupported by more recent analyses using DNA sequences that group *Leporillus* with *Pseudomys* and its relatives (K. Alpin, in litt., 2004). Watts and Baverstock (1994a) included *Conilurus* within a larger clade, the Hydromyini (incorporating Conilurini where *Conilurus* has usually been placed; Baverstock, 1984), which encompassed members of our *Hydromys, Xeromys, Pseudomys*, and *Uromys* Divisions (the "Australasian clade" of Watts and Baverstock, 1995b, 1996). Mahoney and Richardson (1988:154) cataloged taxonomic, distributional, and biological references, and explained that the name *Conilurus* should be maintained because it is in general use and the valid older name *Conylurus* is an unused senior synonym.

Conilurus albipes (Lichtenstein, 1829). Darst. Säugeth., 6, 2 unno. text pages and pl. 29.
    COMMON NAME: White-footed Conilurus.
    TYPE LOCALITY: SE Australia (Neuholland); see Mahoney and Richardson (1988:155).
    DISTRIBUTION: Australia; once occurred in a strip from SE Queensland through coastal New South Wales and Victoria into SE South Australia (Robinson et al., 2000; Watts and Aslin, 1981:130); see Williams and Menkhorst (1995a) for past distribution in Victoria.
    STATUS: IUCN – Extinct.
    SYNONYMS: *constructor* (Ogilby, 1837); *destructor* Palmer, 1897.
    COMMENTS: Living *C. albipes* have not been encountered by naturalists for over a century and the species is apparently extinct (Mahoney and Richardson, 1988; Watts and Aslin, 1981; Dixon, 1995a).

Conilurus penicillatus (Gould, 1842). Proc. Zool. Soc. Lond., 1842:12.

COMMON NAME: Brush-tailed Conilurus.

TYPE LOCALITY: Australia, Northern Territory, Port Essington, sea shore; see Mahoney and Richardson (1988:155).

DISTRIBUTION: Australia; northern coastal area of Northern Territory and adjacent islands (Melville Isl, Bathurst Isl, Groote Eylandt, Sir Edward Pellew Group, and Wellesley Isl, and the extreme NE Western Australia (Kemper, 1995a:553; Watts and Aslin, 1981: 133); SC Papua New Guinea in Morehead region (Flannery, 1995a; Waithman, 1979).

STATUS: IUCN – Lower Risk (lc).

SYNONYMS: *hemileucurus* (Gould, 1858); *melanura* (Gray, 1844); *melibius* Thomas, 1921; *randi* Tate and Archbold, 1938.

COMMENTS: Analyses of external morphology of glans penis and spermatozoal structure provided by Breed (1984), Breed and Sarafis (1978), and Morrissey and Breed (1982). Kemper and Schmitt (1992) reported multivariate study of external and cranial morphology among samples from N Australian and SC New Guinea in the context of assessing geographic variation and zoogeography; populations on Mellville and Bathurst Isls and SC New Guinea are morphometrically the most distinct compared with mainland Australian populations. Studies of chromosomal features (Baverstock et al., 1977c, 1983b), electrophoretic data (Baverstock et al., 1981), phallic morphology (Lidicker and Brylski, 1987), and dental traits (Misonne, 1969) supported the hypothesis that *C. penicillatus* is phylogenetically closely related to *Mesembriomys gouldi*, and microcomplement fixation data placed it in the same clade close to species of *Leporillus* and *Mesembriomys* (Watts et al., 1992; but see generic comments). *Conilurus penicillatus* is one of the few Old Endemic Australian murines that also occurs in the Trans-Fly region of SC New Guinea from where it is still represented by only two specimens (Flannery, 1995ba:258), and is one of 17 Australian mammalian species that are found only in that part of the island (Norris and Musser, 2001).

A morphologically distinctive *penicillatus*-sized form of *Conilurus* is present in Recent cave deposits in the Christmas Creek area, inland of Townsville, N Queensland (K. Aplin, in litt., 2004). This form differs from both typical *penicillatus* and *randi*; further comparisons are needed with *hemileucurus*, the holotype of which was collected "at an unknown locality by the A. C. Gregory's Expedition to North Australia in 1855-1856" (Mahoney and Richardson, 1988:156). This expedition visited localities not too far from Christmas Creek.

Coryphomys Schaub, 1937. Verh. Naturf. Ges. Basel, 48:2.

TYPE SPECIES: *Coryphomys buehleri* Schaub, 1937.

COMMENTS: *Pogonomys* Division. Simpson (1945) listed *Coryphomys*, along with *Lenomys*, *Pogonomys*, *Chiropodomys*, *Mallomys*, *Phloeomys*, and *Crateromys* in the Phloeomyinae, but no data supports association with any of those genera except possibly *Pogonomys* and *Mallomys*. Based upon our study, we suggest that *Coryphomys* is part of an early radiation of New Guinea endemics that include the extant species in our *Pogonomys* Division, which is comparable to the Anisomyini as outlined by Watts and Baverstock (1994b).

Coryphomys buehleri Schaub, 1937. Verh. Naturf. Ges. Basel, 48:2.

COMMON NAME: Buhler's Coryphomys.

TYPE LOCALITY: Indonesia, Nusa Tenggara, W Timor, cave deposit near Nikiniki.

DISTRIBUTION: Recorded only from Timor.

STATUS: Extinct.

COMMENTS: Known only by subfossil fragments collected at the type locality and from sites in E Timor (Glover, 1986). Related to endemic New Guinea murines and not to species of *Papagomys*, *Hooijeromys*, *Komodomys*, or *Paulamys* on Flores (Musser's research). *Coryphomys buehleri* is one of four species, each in its own genus, of giant rats (three have yet to be named and described) endemic to Timor; all are known only by subfossils (Glover, 1986) and may be related to the New Guinea Old Endemics we place in the *Pogonomys* Group.

Crateromys Thomas, 1895. Ann. Mag. Nat. Hist., ser. 6, 16:163.

> TYPE SPECIES: *Phloeomys schadenbergi* Meyer, 1895.
>
> COMMENTS: *Phloeomys* Division. A Philippine Old Endemic in the same monophyletic group as *Batomys* and *Carpomys* (Musser and Heaney, 1992). Reviewed by Musser and Gordon (1981) and Musser et al. (1985). Phylogenetic analyses of complete mtDNA cytochrome *b* sequences for 13 of the 16 genera of endemic Philippine murines indicates *Crateromys* belongs in a clade with *Batomys* and *Phloeomys* (Jansa and Heaney, 2001). The tie to *Phloeomys* was suggested an an alternative hypothesis by Musser and Heaney (1992). Except for the phylogenetic link to *Phloeomys*, no data supports a close relationship between *Crateromys* and *Coryphomys*, *Lenomys*, *Pogonomys*, *Chiruromys*, *Chiropodomys*, and *Mallomys*, which, along with *Crateromys* and *Phloeomys*, were brought together in the Phloeomyinae by Tate (1936), an arrangement followed by Simpson (1945). Placed by Misonne (1969) in a *Lenothrix* Group of a more inclusive *Lenothrix-Parapodemus* Division, and in a *Batomys* Section of a larger *Micromys* Group by Pavlinov et al. (1995a). Actual distribution in the archipelago and number of species in the genus have yet to be determined.

Crateromys australis Musser, Heaney, and Rabor, 1985. Am. Mus. Novit., 2821:3.

> COMMON NAME: Dinagat Crateromys.
>
> TYPE LOCALITY: Philippines, Surigao del Norte Province, Dinagat Isl, Loreto Municipality, Balitbiton.
>
> DISTRIBUTION: Greater Mindanao Faunal Region in the Philippines. Endemic to Dinagat Isl and known only from the type locality in tropical lowland evergreen rainforest (Heaney et al., 1999).
>
> STATUS: IUCN – Endangered.
>
> COMMENTS: Known only by the holotype. People living on the nearby island of Siargao reported seeing an arboreal rat resembling *C. australis* (Oliver et al, 1993). Pelage texture and coloration of *C. australis* is strikingly different from the other species of *Crateromys*, in some ways recalling species of *Batomys*; its molar occlusal patterns also resemble those characterizing *Batomys*. The latter much smaller in body size, and terrestrial. A larger sample of *C. australis* is required from which data can be derived for analyses of morphological and molecular traits to test the following phylogenetic alternatives: 1) the species is a *Crateromys* unlike any other in the genus, 2) a giant arboreal *Batomys*, 3) a phylogenetic link between *Crateromys* and *Batomys*, or 4) member of a separate clade.

Crateromys heaneyi Gonzales and Kennedy, 1996. J. Mammal., 76:26.

> COMMON NAME: Panay Crateromys.
>
> TYPE LOCALITY: Philippines, W Panay Isl, Antique Province, San Remigio Municipality, Barangay Maytawis, SW side of greater Mt Baloy, 11°01′N, 122°14′E.
>
> DISTRIBUTION: Greater Negros-Panay Faunal Region in the Philippines. Endemic to Panay Isl in lowland tropical primary and secondary evergreen rainforest formations (Gonzales and Kennedy, 1996; Schweigert, 1998).
>
> STATUS: IUCN – Endangered.
>
> COMMENTS: Schweigert (1998) provided distributional and ecological notes. Morphological features indicate a closer relationship to *C. schadenbergi* than to other species of *Crateromys* (Gonzales and Kennedy, 1996; Musser's study of specimens).

Crateromys paulus Musser and Gordon, 1981. J. Mammal., 62:515.

> COMMON NAME: Ilin Crateromys.
>
> TYPE LOCALITY: Philippines, Mindoro Occidental Province, Ilin Isl.
>
> DISTRIBUTION: Greater Mindoro Faunal Region in the Philippines. Known only from the type locality.
>
> STATUS: IUCN – Critically Endangered.
>
> COMMENTS: Still represented only by the holotype. The species is probably extinct on Ilin (Pritchard, 1989), but people living on nearby Mindoro Isl report an animal resembling *C. paulus* living in lowland forest (Oliver et al., 1993).

Crateromys schadenbergi (Meyer, 1895). Abh. Mus. Dresden, 6:1.

COMMON NAME: Luzon Crateromys.

TYPE LOCALITY: Philippines, N Luzon Isl, Mountain Province, Mt Data.

DISTRIBUTION: Greater Luzon Faunal Region in the Philippines. Endemic to the mountains of N Luzon (Heaney et al., 1998; Oliver et al., 1993; Sanborn, 1952a).

STATUS: IUCN – Vulnerable.

COMMENTS: The largest-bodied of all the species of *Crateromys* and an arboreal inhabitant of oak-pine forest where it builds stick-nests in tree crowns for shelter (Oliver et al., 1993; Heaney et al., 1998). In many ways, *C. schadenbergi* is an ecological and morphological equivalent of large-bodied tropical tree squirrels (Sciuridae), which do not occur in the Greater Luzon Faunal Region.

Cremnomys Wroughton, 1912. J. Bombay Nat. Hist. Soc., 21:340.

TYPE SPECIES: *Cremnomys cutchicus* Wroughton, 1912.

COMMENTS: *Millardia* Division. An Indian subcontinental endemic that was incorporated into subgenus *Rattus* (Ellerman, 1941), then arranged as a valid subgenus within *Rattus* (Ellerman, 1961; Ellerman and Morrison-Scott, 1951), and finally reinstated as a distinctive genus related to *Millardia* (Gadi and Sharma, 1983; Misonne, 1969; Raman and Sharma, 1977) and *Madromys* (see that account).

Evolutionary history as revealed by fossils from Swalik strata in NW India can be traced back to the early Pliocene (identified as *Cremnomys* sp. and *C.* cf. *cutchicus*) and may have been derived from a *Progonomys*-like form, which is one of the earliest undisputed Miocene murines (Patnaik, 1997, 2001; discussed as *Karnimata*, which Mein et al., 1993, synonymized with *Progonomys*). Lower molars from late Pleistocene cave deposits in the Sichuan-Guizhou region of S China were identified as *?Cremnomys* sp. by Zheng (1993), which if correct, reflects a wider range for the genus during the Pleistocene.

Cremnomys cutchicus Wroughton, 1912. J. Bombay Nat. Hist. Soc., 21:340.

COMMON NAME: Cutch Cremnomys.

TYPE LOCALITY: India, Gujarat, Kutch, Dhonsa.

DISTRIBUTION: An Indian peninsular endemic: Kutch, Kathiawar, S Rajputana, Gujarat, and Bihar in NW India; Mysore, Bellary, and Eastern Ghats in S Peninsula (Agrawal, 2000; Chakraborty and Agrawal, 2000; Corbet and Hill, 1992).

STATUS: IUCN – Lower Risk (lc).

SYNONYMS: *australis* Thomas, 1916; *caenosus* Thomas, 1916; *leechi* Harrison, 1974; *medius* Thomas, 1916; *siva* Thomas, 1916; *rajput* Thomas, 1916.

COMMENTS: Ellerman (1961) suggested *australis* to be a valid species, but Agrawal (2000) did not find any significant patterns in geographic variation of morphological traits and treated all the named subspecies, including *australis*, as synonyms of *C. cutchicus*; Prakash et al. (1995a) synonymized *rajput* with *medius*. Chromosomal information and its significance reported by Raman and Sharna (1977), Sharma and Gadi (1977), Gadi and Sharma (1983), Rishi and Puri (1984), and Sobti and Gill (1984). Reviewed by Corbet and Hill (1992) and Agrawal (2000). Prakash et al. (1995a, b) documented distribution and ecology in Aravalli Ranges in Rajasthan, India, and Chakraborty and Agrawal (2000) recorded ecology, taxonomy, and distribution of samples from Gujarat State, NW India.

Cremnomys elvira (Ellerman, 1946). Ann. Mag. Nat. Hist., ser. 11, 13:207.

COMMON NAME: Elvira Cremnomys.

TYPE LOCALITY: India, E Ghats, Tamil Nadu, Salem Dist., Kurumbapatti.

DISTRIBUTION: Another Indian peninsular endemic; known only from SE India at the type locality in Tamil Nadu (Agrawal, 2000; Corbet and Hill, 1992).

STATUS: IUCN – Vulnerable.

COMMENTS: Still only represented by few specimens from region of the type locality. Reviewed by Agrawal (2000).

Crossomys Thomas, 1907. Ann. Mag. Nat. Hist., ser. 7, 20:70.

TYPE SPECIES: *Crossomys moncktoni* Thomas, 1907.

COMMENTS: *Hydromys* Division. Member of New Guinea Old Endemics (Musser, 1981c). Based on phallic morphology, Lidicker (1968) speculated that *C. moncktoni* is not closely related to *Hydromys*, an affinity supported by microcompliment fixation of albumin that instead clusters *Crossomys* with *Leptomys*, *Xeromys*, and *Pseudohydromys* (which includes *Mayermys* and *Neohydromys*), an assemblage that albumin immunology indicated is part of a larger clade containing members of our *Hydromys*, *Pseudomys*, and *Uromys* Divisions, an "Australasian clade" (Watts and Baverstock, 1994a, 1995b, 1996). Despite Lidicker's evaluation and the alliance suggested by albumin immunology, we place *Crossomys* in the *Hydromys* Division along with *Hydromys*, *Microhydromys*, and *Parahydromys*. Overall morphology of *Crossomys*, including a shared primitive cephalic arterial pattern, resembles *Hydromys* and its allies more closely than it does any genus in the *Xeromys* Division. Unpublished phylogenetic analyses of mtDNA cytochrome *b* sequences placed *Crossomys* closer to *Hydromys* than to *Leptomys* (K. Aplin, in litt., 2004). Results of the data from microcomplement fixation of albumin (Watts and Baverstock, 1994a) may not be a reliable indicator of affinities in this case. *Crossomys* is clearly distant from *Hydromys* and *Parahydromys*, but the albumin immunological distances within each group were nearly as great as those between the the group with *Hydromys* and *Parahydromys* and the cluster containing *Xeromys*, *Crossomys*, *Leptomys*, and *Pseudohydromys* (K. Aplin, in litt., 2004). *Hydromys*, as represented by *H. chrysogaster*, is an amphibious generalist, utilizing a range of aquatic and semiaquatic prey, from arthropods and mollusks to small vertebrates (Olsen, 1995). *Crossomys moncktoni* is highly specialized, both in external and cranial morphology for feeding apparently exclusively on aquatic insects, mostly nymphs and larvae (Voss, 1988). Among amphibious muroids, each prey regime exemplified by *Hydromys leucogaster* and *Crossomys* is associated with particular external and cranial morphologies (Voss, 1988). In many respects, the cranial specializations of *Crossomys* form an enlarged version of the cranial conformation characterizing *Hydromys habbema* (see Flannery, 1995a:plate 24), which also preys predominantly but not exclusively on aquatic insect larvae and nymphs (Voss, 1988).

Crossomys moncktoni Thomas, 1907. Ann. Mag. Nat. Hist., ser. 7, 20:71.

COMMON NAME: Earless New Guinea Water Rat.

TYPE LOCALITY: New Guinea, SE Papua New Guinea, Central Province, Brown River, Serigina, 4500 ft (1370 m).

DISTRIBUTION: Papua New Guinea; known only from Papua New Guinea Central Cordillera and mountains on Huon Peninsula, 1000-2700 m (Flannery, 1990b:191, 1995a).

STATUS: IUCN – Lower Risk (lc).

COMMENTS: Flannery (1990b, 1995b) provided photographs and summaries of distributional and biological information for this amphibious rat. Chromosomal data reported by Donnellan (1987). Comparisons with *Hydromys habbema*, Neotropical ichthyomyines, and other small, semiaquatic carnivorous mammals made by Voss (1988). Leary and Seri (1997) reported specimens from Mt. Sisa in the Central Cordillera. *Crossomys moncktoni* is part of a cluster of Old Endemics apparently confined to E New Guinea (the others are *Leptomys elegans*, *L. ernstmayeri*, *L. signatus*, *Pseudohydromys fuscus*, *P. murinus*, *Abeomelomys sevia*, all species of *Chiruromys*, *Hyomys goliath*, *Protochromys fellowsi*, *Melomys dollmani*, *Paramelomys gressitti*, *P. levipes*, *P. moncktoni*, *Pogonomelomys bruijni*, and *Pogonomys championi*).

Crunomys Thomas, 1897. Trans. Zool. Soc. Lond., 14(6):393.

TYPE SPECIES: *Crunomys fallax* Thomas, 1897.

COMMENTS: *Crunomys* Division. Revised by Musser (1982c) and reviewed by Rickart et al. (1998). An Old Endemic of the Philippines (Musser and Heaney, 1992) and Sulawesi (Musser, 1981c; Musser and Durden, 2002). Thomas (1898b) placed the type species in Hydromyinae, but after describing *C. melanius* was unsure whether *Crunomys* should be placed in Hydromyinae or Murinae (Thomas, 1907c); Ellerman (1941) and Misonne (1969) thought it to be murine. Using morphological data, Musser and Heaney (1992) postulated a close phylogenetic link between *Crunomys* and the Philippine *Archboldomys*, but phylogenetic analyses of complete mtDNA cytochrome *b* sequences for 13 of 16 genera of endemic Philippine murines indicated that Philippine *Crunomys* is not part of the clade containing *Archboldomys* or any other Philippine endemic but instead is more closely related to Sundaic genera (Jansa and Heaney, 2001), a phyletic pattern also supported by chromosomal morphology (Rickart and Heaney, 2002). *Crunomys* may belong to an ancient group that diverged early from a Miocene *Progonomys*-like murine progenitor.

Crunomys celebensis Musser, 1982. Bull. Am. Mus. Nat. Hist., 174:16.
COMMON NAME: Sulawesi Shrew Mouse.
TYPE LOCALITY: Indonesia, C Sulawesi, near Tomado, 3500 ft (1067 m).
DISTRIBUTION: C Sulawesi; known only from tropical lowland evergreen rainforest formations in mountain valley of Danau Lindu and upper drainage of Sungai Miu in the Kulawi region.
STATUS: IUCN – Endangered.
COMMENTS: A small-bodied, dark-furred terrestrial insectivore recorded by only three specimens. Morphologically very distinct from Philippine species of *Crunomys*, and without close relatives on Sulawesi except for possibly *Sommeromys*, another Sulawesian endemic (Musser, 1982c; Musser and Durden, 2002). Spermatozoal morphology distinctive (Breed and Musser, 1991), but unrevealing in assessing phylogenetic relationships. Compared with Philippine species of *Crunomys* by Rickart et al. (1998); morphology reviewed and contrasted with *Sommeromys* by Musser and Durden (2002; see that account).

Crunomys fallax Thomas, 1897. Trans. Zool. Soc. Lond., 14:394.
COMMON NAME: Luzon Shrew Mouse.
TYPE LOCALITY: Philippines, NC Luzon Isl, Isabella Province, Sierra Madre Range, 1000 ft (305 m).
DISTRIBUTION: Greater Luzon Faunal Region. Known only from the type locality.
STATUS: IUCN – Critically Endangered.
COMMENTS: Recorded only by the holotype. Morphologically more similar to the Philippine *C. melanius* and *C. suncoides* than to the Sulawesian *C. celebensis* (Musser, 1982c; Rickart et al., 1998). Two specimens reported as *C. fallax* by Danielsen et al. (1994) and Mallari and Jensen (1993) that were trapped in mossy forest in the Sierra Madre Range are actually *Archboldomys musseri* (Rickart et al., 1998:22).

Crunomys melanius Thomas, 1907. Proc. Zool. Soc. Lond., 1907:141.
COMMON NAME: Mindanao Shrew Mouse.
TYPE LOCALITY: Philippines, Mindanao Isl, Davao Province, Mt Apo, 3000 ft (915 m).
DISTRIBUTION: Greater Mindanao Faunal Region on islands of Mindanao and Leyte; also on Camiguin Isl; altitudinal range, near sea level to 1550 m (Rickart et al., 1998; Heaney et al., 1998).
STATUS: IUCN – Lower Risk (lc).
SYNONYMS: *rabori* Musser, 1982.
COMMENTS: The taxon *rabori* was shown to be an old adult example of *C. melanius* (Rickart et al., 1998), as Musser (1982c) suspected.

Crunomys suncoides Rickart, Heaney, Tabaranza, Jr., and Balete, 1998. Fieldiana Zool., ns, 89:8.
COMMON NAME: Katanglad Shrew Mouse.
TYPE LOCALITY: Philippines, Mindanao Isl, Bukidnon Province, 18.5 km S, 4 km E Camp Phillips, 2250 m, 08°09'N, 124°51'E.
DISTRIBUTION: Greater Mindanao Faunal Region. Known only from the type locality but may occur elsewhere on Mindanao in primary tropical upper montane rainforest (Rickart et al., 1998).
COMMENTS: Recorded only by the holotype. On Mt Katanglad, the montane *C. suncoides* is separated by about 700 m from the upper altitudinal range limits of *C. melanius*, which also occurs elsewhere in lowlands near sea level (Rickart et al., 1998). Standard karyotype (2n = FN = 36) substantially different from any other Philippine Old Endemic (Rickart et al., 1998; Rickart and Heaney, 2002).

Dacnomys Thomas, 1916. J. Bombay Nat. Hist. Soc., 24(3):404.
>   TYPE SPECIES: *Dacnomys millardi* Thomas, 1916.
>   COMMENTS: *Dacnomys* Division. Reviewed by Musser (1981*b*). Closest phylogenetic alliance may be with *Niviventer*, particularly with members of the *N. andersoni* group (Musser, 1981*b*).

Dacnomys millardi Thomas, 1916. J. Bombay Nat. Hist. Soc., 24(3):404.
>   COMMON NAME: Millard's Dacnomys.
>   TYPE LOCALITY: India, West Bengal, near Darjeeling, Gopaldhara, 3440 ft (1050 m).
>   DISTRIBUTION: E Nepal (specimens in FMNH), NE India (West Bengal, Nagaland, and Arunachal Pradesh; Agrawal, 2000), N Laos, NW Vietnam (only west of the Red River; specimens in FMNH and MVZ), and S China (S and NW Yunnan; C. Li et al., 1987; Wang, 2003; Zhang et al., 1997); probably occurs over a wider geographic range (N Burma, for example; Musser, 1981*b*).
>   STATUS: IUCN – Lower Risk (lc).
>   SYNONYMS: *ingens* Osgood, 1932; *wroughtoni* Thomas, 1922.
>   COMMENTS: The few known specimens show appreciable variation in body size, but as Corbet and Hill (1992:360) noted, "too few specimens are available to assess the overall variation." Reviewed by Osgood (1932), Agrawal (2000), Corbet and Hill (1992), and Musser (1981*b*).

Dasymys Peters, 1875. Monatsb. K. Preuss. Akad. Wiss., Berlin, p. 12.
>   TYPE SPECIES: *Dasymys gueinzii* Peters, 1875 (= *Mus incomtus* Sundevall, 1847; see G. M. Allen, 1939).
>   COMMENTS: *Dasymys* Division. Closest phylogenetic allies unidentified. Many pelage, cranial, and dental traits of *Dasymys* suggest alliance with *Aethomys*, but Misonne (1969) considered *Dasymys* to be the most phylogenetically isolated of endemic African murines. In their analysis based on microcomplement fixation of albumin, Watts and Baverstock (1995*a*) placed *Dasymys* as the sister group to their African, Australasian, and Southeast Asian clades. Despite the seemingly isolated position of *Dasymys*, and its apparent slow rate of albumin evolution, Watts and Baverstock were reluctant to highlight the genus as a separate clade and provisionally included it within their African group. Analyses of combined mtDNA (cytochrome *b*) and 12S and 16S rRNA gene fragments by Ducroz et al. (2001) placed *Dasymys* near *Hybomys* in a clade basal to an arvicanthine clade (containing *Lemniscomys*, *Desmomys*, *Rhabdomys*, *Pelomys*, *Mylomys*, and *Arvicanthis*); an *Aethomys chrysophilus*/*Grammomys dolichurus* clade was situated basal to the *Dasymys*/*Hybomys* cluster. Ducroz et al.'s (2001) analysis using only cytochrome *b* sequences placed *Dasymys* basal to a cluster comprised of *Hybomys*, otomyines, *Grammomys*, and *Aethomys*, a *Micaelamys* clade, and the arvicanthine cluster. The association with *Grammomys*, *Aethomys*, *Hybomys*, and *Arvicanthis* is supported by DNA/DNA hybridization (Chevret, 1994). Lecompte's (2003) phylogenetic analyses of mtDNA cytochrome *b* and nuclear IRBP sequences identified a clade containing *Dasymys*, *Hybomys*, *Grammomys*, and some arvicanthines (*Arvicanthis*, *Lemniscomys*, and *Pelomys*). Castiglia et al.'s (2003*b*) phylogenetic analyses of mtDNA cytochrome *b* sequences revealed two patterns. In one tree *Dasymys* clustered with *Hybomys* in a clade that was basal to a clade containing arvicanthines (*Desmomys*, *Rhabdomys*, *Pelomys*, *Mylomys*, *Arvicanthis*, *Lemniscomys*) *and Micaelamys*; in another tree, *Dasymys* is basal to a clade containing otomyines with the arvicanthines basal to the *Dasymys* and otomyine clusters. Such molecular studies promise insight but generic and species sampling is still limited; until better resolution is achieved we restrict *Dasymys* to its own division. Chromosomal data summarized by Carleton and Martinez (1991). Sperm morphology similar to that of Australian 'hydromyines' (Breed, 1995*a*). Measurements, localities with coordinates, and other information listed for holotypes of *Dasymys* taxa by W. Verheyen et al. (2003:53-54).
>   Species diversity within *Dasymys* has shifted from Ellerman's (1941:121) view, "It appears very unlikely that there is more than one species of this genus. . .," to the recognition of the nine species covered here, which is still likely an underestimate. Our accounts derive primarily from study of museum specimens and the taxonomic reviews by Carleton and Martinez (1991) and W. Verheyen et al. (2003). Wetland habitats combined with

dense ground cover are associated with the presence of *Dasymys* and, as Carleton and Martinez (1991:429) noted, the ". . . inherently patchy nature of such wetland environments may account for the interrupted distribution among *Dasymys* populations . . ."; past fragmentation of such moist habitats with subsequent genetic isolation of populations may partly explain the species diversity that today characterizes the genus. At least two extinct species have been identified from early Pleistocene beds of South Africa (Avery, 1998, 2000), and the genus is represented by Pleistocene fossils from Namibia (Senut et al., 1992); see review by Denys (1999).

Dasymys alleni Lawrence and Loveridge, 1953. Bull. Mus. Comp. Zool., Harvard, 110:53.
    COMMON NAME: Glover Allen's Dasymys.
    TYPE LOCALITY: SW Tanzania, Rungwe Mtn, Ilolo (09°10'S, 33°37'E, as per W. Verheyen et al., 2003).
    DISTRIBUTION: Apparently endemic to southern part of the Eastern Arc Mtns (Mt. Rungwe) and the mountainous western rim of the Rift Valley in Dem. Rep. Congo from the Kivu Volcanos in the north to the Kalemi (Albertville) region in the south (W. Verheyen et al., 2003), and probably farther south in the Marungu Mtns.
    COMMENTS: Member of the *D. incomtus* species complex (W. Verheyen et al., 2003). Originally described as a very distinctive subspecies of *D. incomtus* (Lawrence and Loveridge, 1953) and usually treated that way (e.g., Musser and Carleton, 1993), but recently reinstated as a separate species based on multivariate analyses of craniometric data in combination with mtDNA cytochrome *b* sequences (W. Verheyen et al., 2003).

Dasymys cabrali W. Verheyen, Hulselmans, Dierckx, Colyn, Leirs, and E. Verheyen, 2003. Bull. Inst. Roy. Sci. Nat. Belgique, Biol., 73:39.
    COMMON NAME: Crawford-Cabral's Dasymys.
    TYPE LOCALITY: NE Namibia, Groot Fontein Dist., near the Okavango River (Omatoka Junction), 1080 m (17°56'S, 20°25'E, as per W. Verheyen et al., 2003).
    DISTRIBUTION: Apparently endemic to the Okavango Basin (W. Verheyen et al., 2003). The only specimens explicitly assigned to *D. cabrali* come from the type locality and Caprivi Strip of NW Namibia, but records from SE Angola (Crawford-Cabral, 1998), SW Zambia (Ansell, 1978), and N Botswana (Smithers, 1971) possibly represent this species.
    COMMENTS: Samples from the range of *D. cabrali* were originally regarded as examples of *D. nudipes* (e.g., Musser and Carleton, 1993), but those from the Omatoka Junction and Caprivi Strip in N Namibia were recently shown to be craniometrically very distinct from central Angolan *D. nudipes* (W. Verheyen et al., 2003). Crawford-Cabral (1998:64) had noted that possibly ". . . the specimens from Okavango and adjacent countries of Zambia and Angola represent a different subspecies," and W. Verheyen et al. (2003) interpreted the multivariate distance in their discriminant function analyses to reflect two separate species, *D. nudipes* from the central highlands of Angola, and *D. cabrali* from the Okavango Basin.

Dasymys foxi Thomas, 1912. Ann. Mag. Nat. Hist., ser. 8, 9:685.
    COMMON NAME: Fox's Dasymys.
    TYPE LOCALITY: Nigeria, Panyam, 4000 ft (1220 m); 09°27'N, 09°52'E (W. Verheyen et al., 2003).
    DISTRIBUTION: Endemic to the Jos Plateau in Nigeria.
    STATUS: IUCN – Lower Risk (nt).
    COMMENTS: Member of the *D. incomtus* species complex (W. Verheyen et al., 2003). Originally described as a species, but subsequently treated as a subspecies of *D. incomtus* (Ellerman, 1941; Happold, 1987; Misonne, 1974; Rosevear, 1969) until the review by Carleton and Martinez (1991), who contrasted *foxi*'s distinctive morphological and distributional traits (larger body size is one of these) with the smaller-bodied *D. rufulus* (see account), the other and more common species in West Africa. Multivariate analyses of craniometric data by W. Verheyen et al. (2003:48) clustered a sample from Pulima, NW Ghana, with that from Panyam on the Jos Plateau, and W. Verheyen et al. speculated that the species of *Dasymys* living ". . . in the Guinean and Sudanese savannahs [of West Africa] should probably all be referred to

the *foxi* taxon. . . ." This conclusion conflicts with results of the multivariate analyses of cranial and dentail measurements reported by Carleton and Marinez (1991) that showed a clear separation of samples collected in Liberia (including type series of *rufulus*), Ivory Coast, Ghana (including a large sample from Pulima), Togo, and Nigeria (west of the Jos Plateau) from a large sample near Panyam on the Jos Plateau. Those analyses, along with their qualitative inspection of specimens, identified the smaller-bodied *D. rufulus* as occurring from Sierra Leone east to W Nigeria, and and larger-bodied *D. foxi* as restricted to the Jos Plateau.

Dasymys incomtus (Sundevall, 1847). Ofv. K. Svenska Vet.-Akad. Forhandl, Stockholm, 1846, 3:120 [1847].

COMMON NAME: Common Dasymys.

TYPE LOCALITY: South Africa, "Caffraria prope Portum Natal," (= Durban, KwaZulu-Natal); 29°15′S, 31°01′E (W. Verheyen et al., 2003).

DISTRIBUTION: South Africa (de Graaff, 1997*g*; Skinner and Smithers, 1990; P. J. Taylor, 1998), Malawi (Ansell and Dowsett, 1988), Zambia (Ansell, 1978), Zimbabwe (Smithers and Wilson, 1979), C and S Mozambique (Smithers and Lobao Tello, 1976), Angola (Crawford-Cabral, 1998), Dem. Rep. Congo, Uganda (Delaney, 1975), Kenya (Hollister, 1919), Tanzania (Swynnerton and Hayman, 1951), Ethiopia, S Sudan (Setzer, 1956); limits unknown. Also recorded from N Botswana (Smithers, 1971), but unclear whether these records represent *D. incomtus*, *D. cabrali*, or both species.

STATUS: IUCN – Data Deficient.

SYNONYMS: *bentleyae* (Thomas, 1892); *capensis* Roberts, 1936; *edsoni* Hatt, 1934; *fuscus* de Winton, 1897; *griseifrons* Osgood, 1936; *gueinzii* Peters, 1875; *helukus* Heller, 1910; *longipilosus* Eisentraut, 1963; *medius* Thomas, 1906; *nigridius* Hollister, 1916; *orthos* Heller, 1911; *palustris* Setzer, 1956; *savannus* Heller, 1911; *shawi* Kershaw, 1924.

COMMENTS: Despite recent attempts at systematic revision (e. g., W. Verheyen et al., 2003), morphological and geographic definition of *D. incomtus* remains intractable and probably a complex of several species. W. Verheyen et al. (2003), for example, considered *bentleyae* to be one of the synonyms of *D. incomtus* in one part of their report (p. 36) but elsewhere treated it as a separate species with an expansive range ". . . consisting mainly of the fringes of the lowland rain forest between the Atlantic coast and the western rift . . .," and ". . . in the fringes of the western forest block and in the region adjacent to the highlands of the western flank of Lake Malawi" (p. 48). *Dasymys incomtus*, according to W. Verheyen et al. (2003:48), has the widest range, covering ". . . the moist woodlands of western, northcentral and northeastern Africa, . . . the moist woodlands of the north western and eastern part of southern Africa and . . . approximately the western half of the inland of Kenya and Tanzania. . . ." Chromosomal variation from South African specimens reported by Gordon (1991:413) who noted it "is equivocal whether the different chromosomal forms represent distinct species characterized by fixed rearrangements or a chromosomally polymorphic species." Allozymic and cranial differences are also concordant with the chromosomal distinctions (2n = 38, FN = 44 from KwaZulu-Natal; 2n = 46, FN = 44 from Northern Province) between the populations (Mullin et al., 2002). Non-geographic morphometric variation in samples from Southern Africa subregion reported by Mullin et al. (2001). Spermatozoon morphology documented by Breed (1995*a*) and Breed and Pillay (1999). A morphometric analysis by Crawford-Cabral and Pacheco (1989) suggested that Angolan *bentleyae* is a separate species, but later Crawford-Cabral (1998) listed it as a subspecies of *D. incomtus* and noted the wide gap in geographic range between the Angolan populations and those of typical *D. incomtus*.

Recorded from fynbos, grassland, and savanna woodlands biomes in Southern African Subregion (Mugo et al., 1995); those populations reviewed by Skinner and Smithers (1990) and de Graaff (1997*g*). Documented from the isolated Harenna Forest of S Ethiopia by Lavrenchenko (2000) who claimed it was a new record for the entire region of the Ethiopian Rift Valley, but Yalden et al. (1996) had already recorded collections from the Didessa River valley and the Jimma region in S Ethiopia. Habitat and altitudinal distribution on Ugandan slopes of Mt Elgon reported by Clausnitzer and Kityo (2001), and in SW Uganda by Lunde and Sarmiento (2002).

Dasymys montanus Thomas, 1906. Ann. Mag. Nat. Hist., ser. 7, 18:143.

> COMMON NAME: Ruwenzori Dasymys.
>
> TYPE LOCALITY: Uganda, Ruwenzori East, Mubuku Valley, 12,500 ft (3810 m); 00°22′N, 30°00′E (W. Verheyen et al., 2003).
>
> DISTRIBUTION: Known only from Ruwenzori Mtns, Uganda, between 2600 and 3810 m (Kerbis Peterhans et al., 1998); a montane Western Rift endemic.
>
> STATUS: IUCN – Vulnerable.
>
> COMMENTS: Usually included in *D. incomtus* (Delany, 1975), but distinguished from that species by its long, fine fur that is very dark over upperparts and dark gray washed with buff on underparts; very short tail; short rostrum; low, squat cranium; wide zygomatic breadth (Thomas, 1906*a*; specimens in BMNH examined by Musser). *Dasymys montanus* is replaced by *D. incomtus* at lower altitudes on the E slopes of the Ruwenzoris, and both are recorded from 2600 m, the lowest point for *D. montanus* and highest for *D. incomtus* (Kerbis Peterhans et al., 1998). Based on multivariate analyses of craniometric traits derived from holotypes, W. Verheyen et al. (2003) suggested that *montanus* and *medius* (holotype from Ruwenzori Mtns at 1800 m) represented the same population and that both taxa were synonyms of *D. incomtus*. Our study of holotypes and other series, including the material reported by Kerbis Peterhans et al. (1998), indicates *D. montanus* to be a separate species from the lower-altitude *D. incomtus* (*medius*).

Dasymys nudipes (Peters, 1870). Jorn. Sci. Math., Phys. Nat., Lisboa, ser. 1, 3:126.

> COMMON NAME: Angolan Dasymys.
>
> TYPE LOCALITY: Angola, Humpata, 15°01′S, 13°21′E (based on neotype designated by W. Verheyen et al., 2003:38).
>
> DISTRIBUTION: Apparently endemic to the Angolan Plateau (Huambo highlands) in WC Angola (the range in WC Angola, not that indicated in SE Angola, mapped by Crawford-Cabral, 1998:154); see W. Verheyen et al. (2003).
>
> STATUS: IUCN – Data Deficient.
>
> COMMENTS: Treated by G. M. Allen (1939), Hill and Carter (1941), and Roberts (1951) as a species, but included in *D. incomtus* by Ellerman (1941) and most later writers of lists (for example, Ansell, 1978; Meester et al., 1986; Misonne, 1974). Crawford-Cabral (1983) recorded sympatry between *D. nudipes* and *D. incomtus* in Angola, and our survey of series from Chitau identified as *D. nudipes* by Hill and Carter (1941:98) revealed that it consists of both *D. nudipes* and *D. incomtus*, qualitative observations supported by multivariate analyses of morphometric traits (Crawford-Cabral and Pacheco, 1989; W. Verheyen et al., 2003). Specimens from Lukolela (Équateur) and Luluabourg (Kasaï-Occidental), in Dem. Rep. Congo, have been included in *D. nudipes* (Crawford-Cabral, 1983), but these are examples of *D. incomtus* (in AMNH). We examined the holotype and only specimen of *edsoni*, described as a subspecies of *D. nudipes* from Lukolela, middle Dem. Rep. Congo (Hatt, 1934*b*), and it too is an example of *D. incomtus*. Samples from the Okavango Basin, formerly considered examples of *D. nudipes*, represent *D. cabrali* (see that account).

Dasymys rufulus Miller, 1900. Proc. Wash. Acad. Sci., 2:639.

> COMMON NAME: West African Dasymys.
>
> TYPE LOCALITY: Liberia, Mt Coffee (06°30′N, 10°35′W, as per W. Verheyen et al., 2003).
>
> DISTRIBUTION: Senegal (Duplantier and Granjon, 1992; Granjon et al., 1992, as *incomtus*), Guinea (Ziegler et al., 2002), Sierra Leone, Liberia, Ivory Coast, Ghana, Togo, Benin, and W Nigeria (see Carleton and Martinez, 1991:429).
>
> STATUS: IUCN – Data Deficient.
>
> COMMENTS: Originally described as a species, but subsequently treated as a subspecies of *D. incomtus* (Ellerman, 1941; Happold, 1987; Misonne, 1974; Rosevear, 1969) until Carleton and Martinez (1991) used multivariate analyses to discriminate the smaller-bodied *D. rufulus* from the larger-bodied Nigerian *D. foxi* (see that account). Multivariate analyses of craniometric data by W. Verheyen et al. (2003) along with mtDNA cytochrome *b* sequences also identified *rufulus* as a separate species and prompted W. Verheyen et al. (2003:48) to speculate that its geographic range is

"... more or less limited to the enclosed savannahs near the coast," and that the populations found in the "the fringes of the rainforest and the adjacent guinean savannahs may be a new taxon related to the *bentleyae* group. These conclusions conflict with those of Carleton and Martinez (1991) who identified only *D. rufulus* in West Africa (except on the Jos Plateau where *D. foxi* occurs). Grubb et al. (1998) reviewed populations in Gambia, Sierra Leone, and Ghana. Chromosomal data from Senegal samples (2n = 36, FN = 48) documented by Granjon et al. (1992, as *incomtus*).

Dasymys rwandae W. Verheyen, Hulselmans, Dierckx, Colyn, Leirs, and E. Verheyen, 2003. Bull. Inst. Roy. Sci. Nat. Belgique, Biol., 73:45.
COMMON NAME: Rwandan Dasymys.
TYPE LOCALITY: NW Rwanda, Virunga Volcanoes (Nyungwe Forest), Kinigi, 2250 m (01°26'S, 29°36'E, as per W. Verheyen et al., 2003).
DISTRIBUTION: Probably endemic to the Virunga Volcanoes forming eastern rim of the Rift Valley in Rwanda (W. Verheyen et al., 2003).
COMMENTS: Member of the *D. incomtus* species complex, and diagnosed by multivariate analyses of craniodental measurements and complete mtDNA cytochrome *b* sequences relative to examples of *D. alleni* and *D. bentleyae* (W. Verheyen et al., 2003). Specimens from mountains to the south in W Burundi are not *D. rwandae* but *D. bentleyae* judged from craniometric data.

Dasymys sua W. Verheyen, Hulselmans, Dierckx, Colyn, Leirs, and E. Verheyen, 2003. Bull. Inst. Roy. Sci. Nat. Belgique, Biol., 73:41.
COMMON NAME: Tanzanian Dasymys.
TYPE LOCALITY: E Tanzania, Mbete (06°52'S, 37°41'E, as per W. Verheyen et al., 2003:41), near Morogoro, on flanks of the Uluguru Range (Kitundu Forest), 1540 m.
DISTRIBUTION: Known only from upland plains in the Morogoro region of EC Tanzania, 400-1600 m (W. Verheyen et al., 2003).
COMMENTS: Member of the *D. incomtus* species complex. Defined by discriminant function analyses using craniometric traits as well as complete mtDNA cytochrome *b* sequences in comparison with samples of *D. rufulus* and the *bentleyae* portion of the *D. incomtus* species complex (W. Verheyen et al., 2003).

Dephomys Thomas, 1926. Ann. Mag. Nat. Hist., ser. 9, 17:177.
TYPE SPECIES: *Mus defua* Miller, 1900.
COMMENTS: *Hybomys* Division. Another genus founded by Thomas and either relegated to *Rattus* as a subgenus or combined with *Stochomys*, which was then treated as a subgenus of either *Rattus* or *Aethomys* (D. H. S. Davis, 1965; Rosevear, 1969; Van der Straeten, 1984). The generic integrity of *Dephomys* was recognized by Misonne (1969) and Rosevear (1969), who also suggested it was not especially closely related to *Stochomys*, the conventional view at the time. Some molar traits and multivariate analysis of morphological variation suggest a close relationship to *Hybomys*, and particularly to subgenus *Hybomys* rather than subgenus *Typomys* (Misonne, 1969; Van der Straeten, 1984); other qualitative characters, however, support a distant affinity. Careful analyses of a suite of morphological features coupled with DNA sequences employing samples of *Dephomys* and other African forest murine genera is required to uncover the phylogenetic alliance of *Dephomys* relative to *Hybomys* and *Stochomys*.

Dephomys defua (Miller, 1900). Proc. Wash. Acad. Sci., 2:635.
COMMON NAME: Defua Dephomys.
TYPE LOCALITY: Liberia, Mt Coffee.
DISTRIBUTION: Specimens are from Sierra Leone, Guinea (Mt Nimbo), Liberia, Côte d'Ivoire, and Ghana (derived from references cited below and specimens we examined).
STATUS: IUCN – Lower Risk (lc).
COMMENTS: Part of the murine fauna endemic to West Africa (see account of *Grammomys buntingi*). Taxonomy, morphology, range, and habits reviewed by Rosevear (1969) and Van der Straeten (1984). Grubb et al. (1998) reviewed populations in Ghana and Sierra Leone.

Dephomys eburneae (Heim de Balsac and Bellier, 1967). Mammalia, 31:157.

COMMON NAME: Ivory Coast Dephomys.

TYPE LOCALITY: Côte d'Ivoire, Lamto.

DISTRIBUTION: Records are from Côte d'Ivoire and Liberia (specimens in USNM); may occur in SW Ghana (Grubb et al., 1998).

STATUS: IUCN – Lower Risk (lc).

COMMENTS: Originally described as a subspecies of *defua* (see discussion in Rosevear, 1969), but shown to be a separate species by Van der Straeten (1984). Part of the murine fauna endemic to West Africa (see account of *Grammomys buntingi*). Chromosomal data from Côte d'Ivoire sample reported by Tranier and Dosso (1979).

Desmomys Thomas, 1910. Ann. Mag. Nat. Hist., ser. 8, 5:284.

TYPE SPECIES: *Pelomys harringtoni* Thomas, 1902.

COMMENTS: *Arvicanthis* Division. Listed as a genus by G. M. Allen (1939), but usually treated as a subgenus of *Pelomys* (Corbet and Hill, 1991; Ellerman, 1941; Rupp, 1980; Yalden et al., 1976). Most of the diagnostic traits described by Thomas (1910*b*) are outside the range of morphological variation seen among species of *Pelomys*. Our study of specimens revealed that general external traits and cranial conformation of *D. harringtoni* resemble species of *Mylomys* and *Pelomys*, but that *Desmomys* has its own derived dental patterns (ridge-like cusp t9 connecting central cusp t8 with labial cusp t6 on M1 and M2, ridge-like cusp t7 on M2). Chromosomal data also support the extraction of *D. harringtoni* from *Pelomys* (Capanna et al., 1996*a*). Analyses of mtDNA sequences of cytochrome *b*, along with 12S and 16S ribosomal rRNA gene fragments support membership of *Desmomys* in an *Arvicanthis* Group and indicate it to be in the same lineage as *Rhabdomys*, which is distinct from a lineage containing *Lemniscomys*, and another formed by *Arvicanthis*, *Mylomys*, and *Pelomys* (Ducroz et al., 2001). See account of *Mylomys* for status of *rex*, which is usually placed in *Desmomys* (Yalden et al., 1976).

Desmomys harringtoni (Thomas, 1902). Proc. Zool. Soc. Lond., 1902:313.

COMMON NAME: Harrington's Desmomys.

TYPE LOCALITY: Ethiopia, W Shoa, Kutai, Katchisa.

DISTRIBUTION: Ethiopian plateau between 1500 and 3300 m, east and west of Rift Valley (Lavrenchenko, 2003; Rupp, 1980; Yalden and Largen, 1992; Yalden et al., 1976, 1996).

STATUS: IUCN – Lower Risk (lc).

COMMENTS: Another Ethiopian endemic that is apparently semi-arboreal (Yalden et al., 1976). Chromosomal data (2n = 52, FN = 78) reported by Capanna et al. (1996*a*), Lavrenchenko et al. (1989), and Orlov and Bulatova (1997). Protein variation in geographic samples documented by Milishnikov et al. (1992), and spermatozoal morphology described by Baskevich and Lavrenchenko (1995). Reviewed as an Ethiopian endemic by Yalden and Largen (1992, as a *Pelomys*). Distribution is allopatric to *D. yaldeni* (Lavrenchenko, 2003).

Desmomys yaldeni Lavrenchenko, 2003. Bonn. zool. Beitr., 50:320.

COMMON NAME: Yalden's Desmomys.

TYPE LOCALITY: Ethiopia, Sheko Forest, 07°04'N, 35°30'E, 1930 m.

DISTRIBUTION: Recorded from humid afromontane forest in SW Ethiopia (Sheko Forest and Gore), 1800-1930 m (Lavrenchenko, 2003).

COMMENTS: Differs from allopatric *D. harringtoni* by smaller size, pelage coloration, and details of the karyotype, 2n = 52, FN = 62 (Lavrenchenko, 2003).

Diomys Thomas, 1917. J. Bombay Nat. Hist. Soc., 25:203.

TYPE SPECIES: *Diomys crumpi* Thomas, 1917.

COMMENTS: *Millardia* Division. Misonne (1969) suggested that *Diomys* is related to *Chiromyscus*, *Dacnomys*, and *Niviventer* (Misonne used the name *Maxomys* for this group), but Musser and Newcomb (1983) hypothesized that *Millardia*, *Cremnomys*, and other Indian genera may be more closely allied to *Diomys*, a view held by Ellerman (1947*a*) who wrote that it is "a pro-odont offshoot of *Millardia* which resembles it in having very long palatal foramina and a long palate, also in the shortened fifth hind-toe."

Diomys crumpi Thomas, 1917. J. Bombay Nat. Hist. Soc., 25:203.
>COMMON NAME: Crump's Diomys.
>TYPE LOCALITY: India, Bihar, Hazaribagh Dist., Paresnath Hills.
>DISTRIBUTION: Recorded from NE India (Bihar and Manipur; Agrawal, 2000), SW Nepal, and N Burma (see Musser and Newcomb, 1983); limits unknown.
>STATUS: IUCN – Lower Risk (nt).
>COMMENTS: Reviewed by Ingles et al. (1980), Corbet and Hill (1992), and Musser and Newcomb (1983). Indian population reviewed by Agrawal (2000). The species has never been collected again from the type locality, a place where *D. crumpi* may not occur because the original description was based upon a broken skull mismatched with a skin of *Millardia meltada* from Paresnath Hills, Behar (Agrawal, 2000; Corbet and Hill, 1992).

Diplothrix Thomas, 1916. J. Bombay Nat. Hist. Soc., 24:404.
>TYPE SPECIES: *Lenothrix legata* Thomas, 1906.
>COMMENTS: *Rattus* Division. Member of a clade containing genera more closely related to *Rattus* than to other murines. Endemic, along with *Tokudaia*, to the Ryukyu Isls between the Tokara and Kerama Straits. Suzuki et al. (2000:23) noted that "the genetic constitution of the rodents in the Okinawa area is distinct from that of the rodents in the surrounding areas, including the Japanese mainland and Taiwan. . . . in the region of the Central Ryukyus, mammals with a unique and ancient origin have inhabited a small geographic region."

Diplothrix legata (Thomas, 1906). Ann. Mag. Nat. Hist., ser. 7, 17:88.
>COMMON NAME: Ryukyu Islands Tree Rat.
>TYPE LOCALITY: Japan, Ryukyu Isls, Amami-oshima Isl.
>DISTRIBUTION: Japan, Ryukyu Isls of Amami-oshima, Tokun-oshima, and Okinawa (known by modern specimens only in north, but by Quaternary fossils from farther south on island, and from Miyako Isl, about 250 km SW of Okinawa; see Kawamura, 1989, 1991, 1994).
>STATUS: IUCN – Endangered.
>SYNONYMS: *okinavensis* (Namie, 1909) [see Kaneko and Maeda, 2002].
>COMMENTS: Historical allocation of *legata* with either *Lenothrix* or *Rattus* is reviewed by Kawamura (1989). Phylogenetic relationship–discerned from molar occlusal patterns, cranial morphology, and body form–is close to *Rattus* and far from *Lenothrix*; "phylogeny of this unique genus will be sufficiently understood, when the fossil murids from China, India and Southeast Asia will be investigated in detail" (Kawamura, 1989:110). Analyses of mtDNA cytochrome *b* and nuclear IRBP gene sequences support morphological data in clustering *D. legata* with the species of *Rattus* used (*R. norvegicus*, *R. argentiventer*, and *R. rattus*) and not with the other murines sampled (species of *Apodemus*, *Micromys*, *Mus*, and *Tokudaia*), and support the hypothesis that *D. legata* is a survivor from Pliocene and early Pleistocene immigrant ancestral population to the Ryukyus (Suzuki et al., 2000). Chromosomal data reviewed by Tsuchiya (1981). A review by Kaneko (1994) includes beautiful color photographs of live rats showing their long well-haired tail and long body guard hairs extending far beyond the dorsal coat, which are features common to many species of arboreal murines, particularly those in the *Rattus* Division.

Echiothrix Gray, 1867. Proc. Zool. Soc. Lond., 1867:599.
>TYPE SPECIES: *Echiothrix leucura* Gray, 1867.
>SYNONYMS: *Craurothrix* Thomas, 1896.
>COMMENTS: *Echiothrix* Division. Thomas (1898*b*:397) explained why *Echiothrix* is the proper name and not *Craurothrix*, which he earlier proposed thinking *Echiothrix* was preoccupied. Based upon shared cranial features, Thomas (1898*b*) thought *Echiothrix* to be related to the Philippine *Rhynchomys* and placed them in the Rhynchomyinae. Later workers disagreed (see summaries in Musser, 1969*b*, 1990), and, except to recognize that *Echiothrix* is an Old Endemic of Sulawesi (Musser, 1981*c*), no one has discovered the closest phylogenetic ally of this highly specialized, terrestrial vermivore. Spermatozoal mor-

phology resembles that of *Margaretamys*, *Maxomys*, and in some aspects even *Rattus*, but is ambiguous in illuminating phylogenetic alliances (Breed and Musser, 1991), as is the meager chromosomal data for the genus (Musser, 1990). External, cranial, and dental morphology reviewed by Musser (1969*b*, 1990).

Miller and Hollister (1921*a*) described two species in addition to *E. leucura*, but these were included in the latter by Laurie and Hill (1954) and Corbet and Hill (1992), although Musser and Carleton (1993) proclaimed that a revision of the genus is required to determine if N and C Sulawesian samples represent one or two species. Recent multivariate analyses and study of qualitative dental traits indicate that available samples represent two species: one in the NE tip of the N peninsula, the other on the NC part of the N Peninsula and in the central core of Sulawesi (Musser, ms). This biogeographic pattern, one species at the tip of the NE peninsula and a related species restricted to the north and western portions of the N peninsula and the C part of the island, is common to species-pairs in several genera of endemic Sulawesian murines (for example, *Taeromys*, *Rattus xanthurus* Group, and *Bunomys*).

Echiothrix centrosa Miller and Hollister, 1921. Proc. Biol. Soc. Wash., 34:67.

COMMON NAME: Central Sulawesi Echiothrix.

TYPE LOCALITY: Indonesia, C Sulawesi, Winatu (01°34' S, 119°59' E), 2500 ft (760 m).

DISTRIBUTION: Sulawesi: NC portion of N peninsula and central core of islands, sea level to 975 m in lowland tropical evergreen rainforest; southern limits unknown.

SYNONYMS: *brevicula* Miller and Hollister, 1921.

COMMENTS: A distinct species distinguished from *E. leucura* by its shorter cranium and mandible, and shorter molar rows with simpler occlusal patterns. The pelage coloration and size features used by Miller and Hollister (1921*a*) to distinguish *brevicula*, which is also based upon a sample from C Sulawesi, are not diagnostic. The derived karyotype of male documented by Musser (1990, as *leucurus*): 2n = 40, FN = 75, consisting of two telocentric pairs, three metacentric pairs, and the rest submetacentric. Stomach morphology described and contrasted with the Sulawesian shrew rats *Melasmothrix* and *Tateomys*, and the insectivorous *Sommeromys* (Musser and Durden, 2002).

Echiothrix leucura Gray, 1867. Proc. Zool. Soc. Lond., 1867:600.

COMMON NAME: Northern Sulawesi Echiothrix.

TYPE LOCALITY: Indonesia, Sulawesi, NE tip of N peninsula, Manado (= Menado), 01°30'N/124°50'E, coastal plain near sea level (as restricted by Tate, 1936:586).

DISTRIBUTION: Sulawesi: NE tip of the N peninsula, sea level to 1100 m in lowland tropical evergreen rainforest (Musser, 1990; Musser, ms).

STATUS: IUCN – Lower Risk (lc).

COMMENTS: A significantly larger relative of *E. centrosa*, with somewhat more elaborate molar occlusal patterns. Gray (1867*a*) thought the holotype came from Australia, but Jentink (1883) speculated the species would prove to be found only on Sulawesi. Laurie and Hill (1954) listed the type locality as probably N Sulawesi, but Tate (1936) had already restricted it to the NE tip of the N peninsula by designating Manado.

Eropeplus Miller and Hollister, 1921. Proc. Biol. Soc. Wash., 34:94.

TYPE SPECIES: *Eropeplus canus* Miller and Hollister, 1921.

COMMENTS: *Pithecheir* Division. Ellerman (1941) allied *Eropeplus* closely with *Rattus*, but its nearest phylogenetic relative is Sulawesian *Lenomys*, an affinity supported by external, cranial, and spermatozoal characters (Breed and Musser, 1991; Musser, 1981*c*; Tate, 1936). Spermatozoal morphology of *Eropeplus* and *Lenomys* is unique among Sulawesian taxa sampled (Breed and Musser, 1991). See *Lenomys* account.

Eropeplus canus Miller and Hollister, 1921. Proc. Biol. Soc. Wash., 34:94.

COMMON NAME: Sulawesi Soft-furred Rat.

TYPE LOCALITY: Indonesia, C Sulawesi, Gunung Lehio, above 6000 ft (1830 m).

DISTRIBUTION: C Sulawesi; known only by small samples from a few places in montane tropical rainforest formations (Musser, 1970*d*; Musser and Holden, 1991), 1800-2300 m; probably occurs throughout central core of the island in montane forest habitats.

STATUS: IUCN – Endangered.

COMMENTS: Monotypic. Nearly indistinguishable from *Lenomys meyeri* in fur coloration and texture, tail color pattern, length of fur covering over base of tail, cranial conformation, and spermatozoal configuration, but differs sharply in molar occlusal patterns (Musser, 1981c).

Golunda Gray, 1837. Mag. Nat. Hist. [Charlesworth's], 1:586.

TYPE SPECIES: *Golunda ellioti* Gray, 1837.

COMMENTS: *Golunda* Division. Reviewed and compared with *Hadromys* and *Mylomys* by Musser (1987b), who noted that its dental similarity with the latter, and *Pelomys*, was probably convergent. Analyses of mtDNA gene sequences (cytochrome *b*, 12S and 16S rRNA fragments) are ambiguous in ascertaining the phylogenetic affinity of *Golunda* except to clearly refute the hypothesis presented by some (e. g., Jacobs, 1978; Misonne, 1969) that it and *Mylomys* are closely related and derived from a *Pelomys*-like ancestor (Ducroz et al., 2001). As Ducroz et al. (2001:198) noted, "Further studies including a larger sample of African murine taxa will be necessary to evaluate the precise place of this genus." Recent analysis of nuclear IRBP gene sequences divorces *Golunda* from alliances with any of the sampled African genera, especially arvicanthines (Lecompte, 2003). We isolate *Golunda* in its own division until its relationship to other murines is better understood; molecular comparisons with living Asian *Hadromys* would be especially illuminating. Furthermore, three hypotheses derived from fossil samples have been proposed and require testing. First, *Golunda* originated in Africa and migrated to Asia in late Pliocene (Jacobs, 1978; Patnaik, 2001). Second, *Golunda* is one of the lineages evolving out of an African arvicanthine group that migrated to Asia during early Pliocene (Cheema, et al., 2003). Finally, *Golunda* evolved from late Miocene Asian *Parapelomys* (Patnaik, 1997), which in turn was derived from the earlier Asian *Karnimata* (Jacobs, 1978; Jacobs and Downs, 1994; *Karnimata* = *Progonomys*, according to Mein et al., 1993), pointing to *Golunda* as not only an Asian endemic but derived from an endemic Miocene Asian fauna. Pliocene fragments identified as a species of *Golunda* have been recorded from Ethiopia, but Musser (1987b) explained why they do not represent this genus. All current information about evolutionary history of *Golunda* substantiates its endemism to the Indian subcontinent. Enamel microstructure of incisors and molars and its significance documented by Patnaik (2002). Isolated molars from Siwalik strata document *Golunda*'s presence back to the early Pliocene of NW India (*G. tatroticus*) and middle Pliocene-early Pleistocene of NW India and N Pakistan (*G. kelleri* and *G.* sp.); see Cheema et al. (1997, 2003); Kotlia (1992); Patnaik (1997, 2001); Gupta and Prasad (2001); Jacobs (1978).

Golunda ellioti Gray, 1837. Mag. Nat. Hist. [Charlesworth's], 1:586.

COMMON NAME: Indian Bush Rat.

TYPE LOCALITY: India, Dharwar.

DISTRIBUTION: SE Iran (Misonne, 1990), Pakistan (Roberts, 1977, 1997), Nepal (Ellerman, 1961), N and NE India south through Indian peninsula to Sri Lanka (Agrawal, 2000).

STATUS: IUCN – Lower Risk (lc).

SYNONYMS: *bombax* Thomas, 1923; *coenosa* Thomas, 1923; *coffaeus* Kelaart, 1850; *coraginis* Thomas, 1923; *gujerati* Thomas, 1923; *hirsutus* (Elliot, 1839); *limitaris* Thomas, 1923; *myothrix* (Hodgson, 1845); *newara* (Kelaart, 1850); *paupera* Thomas, 1923; *watsoni* (Blanford, 1876).

COMMENTS: Agrawal (2000) reviewed the Indian population and could find no significant geographic variation in fur coloration or other morphological features. Ecology and distribution in the Aravalli Ranges in Rajasthan, India, documented by Prakash et al. (1995a, b), in Western Ghats of S India by Chandrasekar-Rao and Sunquist (1996), and in Gujarat State of NW India by Chakraborty and Agrawal (2000).

Grammomys Thomas, 1915. Ann. Mag. Nat. Hist., ser. 8, 16:150.

TYPE SPECIES: *Mus dolichurus* Smuts, 1832.

COMMENTS: *Oenomys* Division. A distinctive genus as asserted by Ellerman (1941) and other workers (e.g., Hutterer and Dieterlen, 1984; Misonne, 1969; Rosevear, 1969), and not part of *Thamnomys* with which it has often been united as a subgenus (G. M. Allen, 1939; Hatt,

1940*b*; Hollister, 1919; Misonne, 1974; F. Petter and Tranier, 1975). Morphological and chromosomal similarities exist with *Thallomys* (Olert et al., 1978). Analysis of micro-complement fixation of albumin associated *Grammomys* with *Lemniscomys*, *Pelomys*, *Rhabdomys*, and *Thallomys* (Watts and Baverstock, 1995*a*). Research using mitochondrial gene sequences (DNA cytochrome *b*, 12S and 16S rRNA fragments) placed *Grammomys* next to *Aethomys* within an African murine clade consisting of *Hybomys*, *Dasymys*, *Lemniscomys*, *Rhabdomys*, *Desmomys*, *Pelomys*, *Mylomys*, and *Arvicanthis* (Ducroz et al., 2001). Partly reviewed by F. Petter and Tranier (1975) and Hutterer and Dieterlen (1984), who provided morphological, distributional, and chromosomal comparisons.

Two of the species listed below, *G. kuru* and *G. poensis* (both formerly included in *G. rutilans*; see Musser and Carleton, 1993, for example), along with an undescribed species from Mt Oku in W Cameroon, form a monophyletic group separate from the other species of *Grammomys* and will eventually be placed in a new genus (R. Hutterer, pers. comm., 2002). The taxa *kuru* and *poensis* (and *rutilans*, which is unavailable) have been considered members of *Thamnomys* (G. M. Allen, 1939; Ellerman, 1941; Hutterer and Dieterlen, 1984), but external, cranial, and dental morphology is more similar to that characterizing species of *Grammomys* (where it was listed by D. H. S. Davis, 1965, and Misonne, 1974), and we include them within that genus pending publication of Hutterer's revision (he also provided most of the information for the two accounts). Fossils identified as *Grammomys* come from Pliocene-Pleistocene strata of East Africa and South Africa (Avery, 2000; Jaeger, 1976; Jaeger and Wesselman, 1976; see review by Denys, 1999) and Pleistocene sediments of Namibia (Senut et al., 1992).

Grammomys aridulus Thomas and Hinton, 1923. Proc. Zool. Soc. Lond., 1923:268.
COMMON NAME: Arid Woodland Grammomys.
TYPE LOCALITY: Sudan, Darfur, Wadi Aribo, Kulme.
DISTRIBUTION: WC Sudan (Dieterlen and Nikolaus, 1985).
STATUS: IUCN – Lower Risk (lc).
COMMENTS: Usually either listed as a subspecies of *G. macmillani* (G. M. Allen, 1939; Ellerman, 1941; Setzer, 1956) or included in *G. dolichurus* (Misonne, 1974), but considered a distinct species by Hutterer and Dieterlen (1984).

Grammomys buntingi (Thomas, 1911). Ann. Mag. Nat. Hist., ser. 8, 7:381.
COMMON NAME: Bunting's Grammomys.
TYPE LOCALITY: Liberia, Bassa, Gonyon.
DISTRIBUTION: Zone of high forest, coastal scrub or Guinea woodland in West Africa from Senegal (Duplantier and Granjon, 1992), Sierra Leone (Grubb et al., 1998), and Guinea to Côte d'Ivoire and Liberia; see Rosevear (1969).
STATUS: IUCN – Lower Risk (lc).
COMMENTS: Except for Misonne (1974), who included it in *G. dolichurus*, *buntingi* has always been listed or discussed as a species of *Grammomys* (G. M. Allen, 1939; Ellerman, 1941; Hutterer and Dieterlen, 1984; F. Petter and Trainer, 1975; Rosevear, 1969). Apparently *G. buntingi* is a member of a suite of species endemic to West Africa that includes *Dephomys defua*, *D. eburneae*, *Hybomys planifrons*, *H. trivirgatus*, *Hylomyscus baeri*, *Lemniscomys bellieri*, *Malacomys cansdalei*, *M. edwardsi*, *Oenomys ornatus*, *Praomys daltoni*, *P. derooi*, and *P. rostratus* (Carleton and Robbins, 1985; Hutterer and Dieterlen, 1984).

Grammomys caniceps Hutterer and Dieterlen, 1984. Stuttg. Beitr. Naturk., A, 374:12.
COMMON NAME: Gray-headed Grammomys.
TYPE LOCALITY: Kenya, Malindi.
DISTRIBUTION: N Kenya and S Somalia.
STATUS: IUCN – Lower Risk (lc).
COMMENTS: Karyotype of Kenyan sample and its significance documented by Hutterer and Dieterlen (1984), and chromosomal variation among individuals from S Somalia described and discussed by Roche et al. (1984). The latter authors assigned their sample to the *G. dolichurus* group and, independent of Hutterer and Dieterlen (1984), noted the distinctive quality of the Somalian species as indicated by chromosomal evidence.

Grammomys cometes (Thomas and Wroughton, 1908). Proc. Zool. Soc. Lond., 1908:549.
    COMMON NAME: Mozambique Grammomys.
    TYPE LOCALITY: Mozambique, Inhambane.
    DISTRIBUTION: From Pirie Forest (NW of King William's Town) in SE Eastern Cape Province
        of South Africa north through KwaZulu-Natal and Limpopo provinces of that country
        into E Zimbabwe (Melsetter and Umtali districts) and Mozambique south of the
        Zambezi River (de Graaff, 1981, 1997*h*; Meester et al., 1986; Skinner and Smithers,
        1990; Smither and Tello, 1976); an inhabitant of the savanna woodland biome in
        southern Africa (Mugo et al., 1995).
    STATUS: IUCN – Lower Risk (lc).
    SYNONYMS: *silindensis* Roberts, 1938.
    COMMENTS: The geographic range of *G. cometes* has been outlined as extending north from
        South Africa through East Africa to S Sudan (Hutterer and Dieterlen, 1984), but
        pending revisionary study of the genus we restrict it to the E segment of the Southern
        African Subregion south of the Zambezi River (similar to the range mapped by Skinner
        and Smithers, 1990:225), and consider samples north of that river to be *G. ibeanus* (see
        that account). We studied the holotype of *cometes* and the other specimens in the type
        series noted by Thomas and Wroughton (1908); these animals are on average larger
        and have more highly inflated bullae than do those from north of the Zambesi River.
        Ansell (1978) and Ansell and Dowsett (1988) assigned samples from Zambia and
        Malawi to *cometes*, but were also impressed with the chromatic and morphological
        contrast between them and the holotype from Inhambane. The specimen from the
        Pirie Forest (in AMNH) represents a range extension south of KwaZulu-Natal. That
        population was discussed by P. J. Taylor (1998) and Taylor et al. (1994*a*), who noted
        sympatry between *G. cometes* and *G. dolichurus* at some localities and at others a
        gradation of the diagnostic traits usually used to distinguish the two species. Whether
        those characters have limited discriminatory use in these areas or hybridization is
        occurring is unclear; the two species require taxonomic study in that part of southern
        Africa.

Grammomys dolichurus (Smuts, 1832). Enumer. Mamm. Capensium, p. 38.
    COMMON NAME: Common Grammomys.
    TYPE LOCALITY: South Africa, near Cape Town.
    DISTRIBUTION: From Nigeria east to S Ethiopia; then south through N Dem. Rep. Congo,
        Uganda (Delany, 1975; Clausnitzer and Kityo, 2001, discussed distribution and habitat
        on Ugandan slopes of Mt Elgon), Kenya (Hollister, 1919), Tanzania (Swynnerton and
        Hayman, 1951; Grimshaw et al., 1995, distribution on Mt Kilimanjaro; Stanley et al.,
        2002, presence in Gonja Forest Reserve), and C and S Malawi (Ansell and Dowsett,
        1988) to N and E South Africa (from Limpopo Province along coast through KwaZulu-
        Natal and Eastern Cape provinces to Port Elizabeth; de Graaff, 1981, 1997*i*; P. J. Taylor,
        1998), E Zimbabwe, and Mozambique (Smithers and Lobao Tello, 1976); and west
        through Zambia (except in northeast on Nyika Plateau; Ansell, 1978) to Angola
        (Crawford-Cabral, 1998); limits of geographic range unresolved.
    STATUS: IUCN – Lower Risk (lc).
    SYNONYMS: *angolensis* Hill and Carter, 1937; *arborarius* (True, 1892) [not of Peters, 1852];
        baliolus (Osgood, 1910); *discolor* (Thomas, 1910); *elgonis* (Thomas, 1910); *insignis*
        (Dollman, 1911); *littoralis* (Heller, 1912); *polionops* (Osgood, 1910); *surdaster* (Thomas
        and Wroughton, 1908); *tongensis* Roberts, 1931.
    COMMENTS: Number of scientific names reflects morphological and chromosomal variation
        correlated with geography and suggests more than one species is represented (Hutterer
        and Dieterlen, 1984; Meester et al., 1986); the complex requires careful revision. For
        example, specimens of true *dolichurus* from South Africa have duller pelage and more
        inflated bullae than animals from East and West Africa; should these prove to be
        diagnostic specific differences, the northern populations should be identified as
        G. surdaster. The Ethiopian records are based upon a specimen from Kefa (in USNM)
        and one documented by Duckworth et al. (1993), but not those recorded by Yalden et
        al. (1976), which represent other species (Hutterer and Dieterlen, 1984).

Grammomys dryas (Thomas, 1907). Ann. Mag. Nat. Hist., ser. 7, 19:123.
    COMMON NAME: Albertine Rift Grammomys.
    TYPE LOCALITY: "Kenya Colony" (Uganda), Ruwenzori East, 6000-7000 ft (1830-2130 m).
    DISTRIBUTION: A montane Western Rift endemic: Ruwenzoris and Kivu region in Uganda
        and Dem. Rep. Congo (AMNH and BMNH specimens), NW Burundi (specimens in
        FMNH).
    STATUS: IUCN – Lower Risk (lc).
    COMMENTS: Originally described as a species of *Thamnomys* by Thomas (1907a) and listed
        that way by G. M. Allen (1939). Ellerman (1941), however, treated it as valid species of
        Grammomys, which, in the absence of a critical systematic revision of the genus, best
        expresses current knowledge. Thomas (1907a) noted the diagnostic mammary count
        in *G. dryas*, which, in combination with cranial traits, set it apart from other described
        forms (our study of specimens). See Kerbis Peterhans et al. (1998) for review of
        altitudinal distribution of *G. dryas* in context of the entire Ruwenzori small mammal
        fauna.

Grammomys gigas (Dollman, 1911). Ann. Mag. Nat. Hist., ser. 8, 7:527.
    COMMON NAME: Mount Kenya Grammomys.
    TYPE LOCALITY: Kenya, Mt Kenya, Solai, 9000 ft (2743 m).
    DISTRIBUTION: Known only from the vicinity of Mt Kenya.
    STATUS: IUCN – Endangered.
    COMMENTS: Recorded only by the holotype. Recognized as a species in most lists
        (G. M. Allen, 1939; Ellerman, 1941). Hutterer and Dieterlen (1984) insisted the species
        has to be recognized because of its large teeth, an opinion we share derived from our
        study of the holotype; but the possibility that it is simply a large individual of
        G. ibeanus is a hypothesis worth testing.

Grammomys ibeanus (Osgood, 1910). Field Mus. Nat. Hist. Publ., Zool. Ser., 10:8.
    COMMON NAME: East African Grammomys.
    TYPE LOCALITY: Kenya, Molo.
    DISTRIBUTION: From extreme NE Zambia (Nyika Plateau; Ansell, 1978) and Malawi (Ansell
        and Dowsett, 1988) north through highlands of E Tanzania (Stanley et al., 1998; and
        specimens in MCZ) and Kenya to S Sudan (Hollister, 1919; Hutterer and Dieterlen,
        1984).
    STATUS: IUCN – Lower Risk (lc).
    COMMENTS: Morphological and geographic definition of *G. ibeanus* is unsatisfactory,
        particularly the extent of its distribution in Tanzania. Hutterer and Dieterlen (1984)
        treated *ibeanus* as a form of *G. cometes*, but the striking morphological distinctions
        between samples of *ibeanus* and the type series of *cometes* prompted our specific
        ranking of *ibeanus* (see also account of *G. cometes*).

Grammomys kuru (Thomas and Wroughton, 1907). Ann. Mag. Nat. Hist., ser. 7, 19:381.
    COMMON NAME: Eastern Rainforest Grammomys.
    TYPE LOCALITY: NE Dem. Rep. Congo, Orientale, Angu, Uele River.
    DISTRIBUTION: Probably most of the E Congo Basin in S Central African Republic, Dem. Rep.
        Congo, forest patches in W Uganda (Delany, 1975, as *rutilans*), and possibly Republic
        of Congo; limits unresolved.
    SYNONYMS: *centralis* (Dollman, 1914).
    COMMENTS: Phylogenetically closely related to *G. poensis* but differs in its smaller skull and
        shorter molar rows. The karyotype 2n = 50 described by Matthey (1963) for a specimen
        captured in Republic of Congo probably refers to *G. kuru* and contrasts with the
        2n = 36 reported from the Côte d'Ivoire (see account of *G. poensis*). The taxon *kuru* has
        been treated as a synonym of *G. rutilans* (D. H. S. Davis, 1965; Musser and Carleton,
        1993) or sometimes listed as a species (e.g., Ellerman, 1941), but with provisions (Hatt,
        1940b; Hutterer and Dieterlen, 1984; Thomas, 1915); our study of the holotype
        revealed it is a very young adult. Measurements of the molar rows are like those in
        samples from E Dem. Rep. Congo and unlike samples of the larger *G. poensis* (see
        discussion in Hatt, 1940b).

Grammomys macmillani (Wroughton, 1907). Ann. Mag. Nat. Hist., ser. 7, 20:504.
> COMMON NAME: Macmillan's Grammomys.
> TYPE LOCALITY: Ethiopia, north of Lake Rudolf, Wouida.
> DISTRIBUTION: Sierra Leone, Liberia, Central African Republic, S Sudan, S Ethiopia, N Dem.
>> Rep. Congo, Kenya, Uganda (including Bugala Isl in Lake Victoria), Tanzania, Malawi,
>> Mozambique, and E Zimbabwe; limits unresolved.
> STATUS: IUCN – Lower Risk (lc).
> SYNONYMS: *callithrix* (Hatt, 1934); *erythropygus* Setzer, 1956; *gazellae* (Thomas, 1910); *oblitus*
>> (Osgood, 1910); *ochraceus* (G. M. Allen, 1912); *usambarae* (Matschie, 1915); *vumbaensis*
>> Roberts, 1938; *vumbensis* G. M. Allen, 1939.
> COMMENTS: Hutterer and Dieterlen (1984) provided historical association of the name with
>> other taxa, and summary of range as they knew it. Chromosomal variation (under
>> name of *gazellae*) described by Civitelli et al. (1989). Spermatozoal morphology
>> documented by Breed (1995*a*, as *gazellae*). The records from Sierra Leone, Tanzania,
>> Malawi, and Mozambique are based on samples in AMNH, MCZ, and USNM and
>> represent significant range extensions beyond that outlined by Hutterer and Dieterlen
>> (1984). Judged by Roberts's (1938) description and our study of one of the specimens
>> in his original series (MCZ 24046), *vumbaensis* from E Zimbabwe (Vumba and Mount
>> Selinda) belongs in the synonymy of *G. macmillani* rather than *G. dolichurus* where
>> Smithers and Wilson (1979) and Meester et al. (1986) listed it. Matschie's (1915)
>> *usambarae* from N Tanzania is also likely an example of *G. macmillani* because of its
>> small size and short molar rows. Distribution in the Eastern Arc Mtns of Tanzania
>> documented by Stanley et al. (1998), and in SW Uganda by Lunde and Sarmiento
>> (2002).

Grammomys minnae Hutterer and Dieterlen, 1984. Stuttg. Beitr. Naturk., A, 374:10.
> COMMON NAME: Ethiopian Grammomys.
> TYPE LOCALITY: S Ethiopia, Sidamo Province, edge of Bulcha Forest, 1800 m.
> DISTRIBUTION: S Ethiopia; limits unknown.
> STATUS: IUCN – Lower Risk (nt).
> COMMENTS: Recorded only from two localities. About the same body size as *G. dolichurus*
>> and *G. macmillani* but has a very different karyotype (2n = 32, FN = 64; 2n = 52, FN = 66
>> in *G. dolichurus* and 2n = 68-76 in *G. macmillani*). Reviewed as an Ethiopian endemic
>> by Yalden and Largen (1992; also see Yalden et al., 1996).

Grammomys poensis (Eisentraut, 1965). Zool. Jb. Syst., Jena, 92:26.
> COMMON NAME: Western Rainforest Grammomys.
> TYPE LOCALITY: Equatorial Guinea, Bioko, Moca, 1200 m.
> DISTRIBUTION: Western tropical rain forest blocks and outlying patches: from Guinea
>> (Mt Nimba) eastward through Côte d'Ivoire, Ghana (Grubb et al., 1998, as *rutilans*),
>> and Togo to S Nigeria, then south through Cameroon, Equatorial Guinea (including
>> Bioko; Eisentraut, 1965), and Gabon to N Angola (Crawford-Cabral, 1998, as *rutilans*).
>> This outline is provisional; samples from Gabon and N Angola should be reexamined
>> to determine if they are this species or the smaller *G. kuru*.
> STATUS: IUCN – Lower Risk (lc) as *G. rutilans*.
> SYNONYMS: *rutilans* (Peters, 1876) [not Olfers, 1818].
> COMMENTS: The species is better known as *Grammomys rutilans* (Peters, 1876; see Musser and
>> Carleton, 1993, for example). *Mus rutilans* Peters, 1876, however, is a junior synonym
>> of *Mus rutilans* Olfers, 1818, a South American sigmodontine cricetid, which has been
>> used in the combination *Oxymycterus rutilans* (Olfers, 1818) by Hershkovitz (1959*a*),
>> Vaz-Ferreira (1960), Reig (1964, 1965), Carleton (1973), and Voss and Linzey (1981).
>> Hershkovitz (1994) recently placed *rutilans* in the synonymy of *Oxymycterus rufus*
>> (Fischer, 1814). *Mus rutilans* Peters, 1876, therefore, is not available; *poensis*, described
>> as a subspecies of *G. rutilans* by Eisentraut (1965) refers to the same genus and species,
>> and is the valid species name for the taxon. The karyotype (2n = 36) reported by
>> Tranier and Dosso (1979) for an individual captured in Côte d'Ivoire refers to
>> *G. poensis*. Phylogenetically most closely related to *G. kuru* (see that account).

Hadromys Thomas, 1911. J. Bombay Nat. Hist. Soc., 20:999.

TYPE SPECIES: *Mus humei* Thomas, 1886.

COMMENTS: *Hadromys* Division. Usually considered closely allied to *Arvicanthis* and its relatives, especially *Golunda* (Misonne, 1969), but a combination of primitive and derived cranial and dental traits divorces *Hadromys* from that group: best hypothesis now available on phylogenetic affinities would be to derive the species of *Hadromys* from some late Miocene Asian ancestor, possibly a species of *Karnimata* (= *Progonomys*, according to Mein et al., 1993), and to consider any resemblance to the *Arvicanthis* cluster a reflection of convergent evolution (Musser, 1987b). No fossils have been described that suggest phylogenetic links between *Hadromys* and either Asian *Golunda* or extinct Asian genera such as *Saidomys*, *Parapelomys*, and *Dilatomys* (see descriptions of those genera in Brandy, 1981; Gupta and Prasad, 2001; Jacobs, 1978; Ôen, 1983) that have been labeled "Asian arvicanthines" (e.g., Cheema et al., 2003)

Hadromys loujacobsi, known by fossil fragments from Late Pliocene and early Pleistocene sediments in Punjab region of N Pakistan, is a more derived (dentally) relative of living *H. humei* (Cheema et al., 2003; Musser, 1987b). *Prohadromys varavudhi*, described from isolated molars found in Thailand (late Pliocene to early Pleistocene) is dentally closely related to the species of *Hadromys*, but has less specialized molars reflected by their lower coronal heights and more arcuate cusp rows (Chaimanee, 1998). Something with that kind of molar configuration could be a direct potential ancestor of *Hadromys*. An earlier dental fragment from middle Pliocene of Pakistan was identified as *Hadromys* sp. and is also less derived than *H. loujacobsi* (Cheema et al., 1997). Available data indicate that *H. humei* and its extant and extinct relatives once occurred over a wide area of the Indomalayan region and may have originated and evolved in what is now Indochina (Chaimanee and Jaeger, 2000b).

Hadromys humei (Thomas, 1886). Proc. Zool. Soc. Lond., 1886:63.

COMMON NAME: Hume's Hadromys.

TYPE LOCALITY: India, Manipur, Moirang.

DISTRIBUTION: NE India (Manipur and NW Assam; Musser, 1987b; Agrawal, 2000).

STATUS: IUCN – Lower Risk (lc).

COMMENTS: Reviewed by Musser (1987a). The species occurred throughout Thailand to Songkla Province in peninsular Thailand at the Thai-Malay border (south of the present Isthmus of Kra, 10°30' N) during middle Pleistocene, about 137,000 years ago, judged by fossilized molars identified as *H. humei* (Chaimanee and Jaeger, 2000b). During that earlier time Thailand, and most likely other parts of the Indomalayan region may have been dominated by drier and cooler climates and extensive savanna environments similar to the habitats in which *H. humei* is found today. In the context of this past distribution, the present range of *H. humei* appears to be relictual (Chaimanee and Jaeger, 2000b).

Hadromys yunnanensis Yang and Wang, 1987. Acta Theriologica Sinica, 7:46.

COMMON NAME: Yunnan Hadromys.

TYPE LOCALITY: China, W Yunnan Province, Ruili County, Hulan (24°09' N, 97°51' E), 1300 m.

DISTRIBUTION: Recorded only from Ruili County in W Yunnan, 970-1300 m (Yang and Wang, 1987).

COMMENTS: Originally described as a distinctive subspecies of *H. humei*, but the Yunnan animal is easily distinguished by its much larger body size (123-140 mm as opposed to 98-120 mm in *H. humei*), relatively shorter tail, pure white underparts (grayish white tinged with pale or rich buff in *H. humei*), significantly longer diastema (8.1-8.5 mm versus 6.7-7.8 mm in *H. humei*), and shorter palate relative to skull length. These contrasts support the hypothesis that *yunnanensis* represents a separate species, a conclusion also suggested by Corbet and Hill (1992) who were impressed by the differences. In his checklist of Chinese mammals, Wang (2003) treated *yunnanensis* as a species.

Haeromys Thomas, 1911. Ann. Mag. Nat. Hist., ser. 8, 7:207.

TYPE SPECIES: *Mus margarettae* Thomas, 1893.

COMMENTS: *Micromys* Division. A Sundaic and Sulawesian endemic. Cranial, dental, and spermatozoal morphology suggested a distant phylogenetic link to *Chiropodomys* (Breed and Musser, 1991; Musser and Newcomb, 1983). This alliance was indicated by Thomas (1893:346) in describing *margarettae*: "Skull with a very peculiar and noteworthy resemblance to that of *Chiropodomys* agreeing with that of *Ch. gliroides* so closely that it is not until a close examination is made that the differences become apparent." Chromosomal data is ambiguous in assessing *Haeromys*'s closest relatives (Musser, 1990). A new species from C Sulawesi will be described within the context of a revision of *Haeromys* being prepared by Musser (ms).

Haeromys margarettae (Thomas, 1893). Ann. Mag. Nat. Hist., ser. 6, 11:346.

COMMON NAME: Margaret's Haeromys.

TYPE LOCALITY: Malaysia, Sarawak, Penrisen Hills.

DISTRIBUTION: Borneo; recorded only from the type locality and Sabah (Chasen and Kloss, 1932).

STATUS: IUCN – Vulnerable.

COMMENTS: Known only by the holotype (BMNH 93.4.1.55) and a specimen from Bettotan in Sabah (RMBR 4.7900); Medway's (1977) record from East Kalimantan is a nestling of *Sundamys muelleri* (ZMO 8226, identified by Musser).

Haeromys minahassae (Thomas, 1896). Ann. Mag. Nat. Hist., ser. 6, 18:247.

COMMON NAME: Lowland Sulawesi Haeromys.

TYPE LOCALITY: Indonesia, Sulawesi, NE tip of N peninsula, Minahassa region, Rurukan (01°21′N/124°52′E), 1000 ft (305 m).

DISTRIBUTION: Sulawesi; recorded only from tropical lowland evergreen rain forest at the NE tip of the N peninsula and central core between 75 and 1000 m (Musser, 1990; Musser, ms).

STATUS: IUCN – Lower Risk (lc).

COMMENTS: External, cranial, and dental features distinguish *H. minahassae* and an undescribed Sulawesian species from those on Borneo and Palawan (Musser, ms). Karyotype is primitive (2n = 48, FN = 54) consisting of two pairs of metacentrics and the rest telocentrics (Musser, 1990); spermatozoal morphology is distinctive (Breed and Musser, 1991). Of the two species of *Haeromys* endemic to Sulawesi, an undescribed form occupies montane forest formations and its close relative *H. minahassae* occurs at lower altitudes in lowland evergreen rain forest; the two are also distinguished by pelage coloration and cranial traits (Musser, ms).

Haeromys pusillus (Thomas, 1893). Ann. Mag. Nat. Hist., ser. 6, 11:232.

COMMON NAME: Sundaic Haeromys.

TYPE LOCALITY: Malaysia, Sabah, Mt Kinabalu.

DISTRIBUTION: Documented (Musser, ms) by six examples from Borneo (Sarawak, Sabah, E and SW Kalimantan) and two from the Greater Palawan Faunal Region of the Philippines (one from Palawan, the other from Calauit Isl).

STATUS: IUCN – Vulnerable.

COMMENTS: The two Philippine specimens have paler underparts on head and body and dorsal metatarsal surfaces than does the Bornean sample and also differ slightly in dental traits, but because there are so few specimens from either Borneo or the Greater Palawan Faunal Region, the significance of this variation has yet to be determined. Most closely related to *H. margarettae*, and included, incorrectly in our view, in that species by Payne et al. (1985).

Hapalomys Blyth, 1859. J. Asiat. Soc. Bengal, 28:296.

TYPE SPECIES: *Hapalomys longicaudatus* Blyth, 1859.

COMMENTS: *Micromys* Division. Reviewed by Musser (1972). One of the few murine genera with representatives in both Indochina and on the Sunda Shelf (Musser and Newcomb, 1983). Although molar occlusal patterns are highly derived, they are more similar to those of *Chiropodomys* than to any other extant Asian murine (Chaimanee, 1998; Misonne, 1969), a suggested alliance supported by shared derived traits of the feet, digits, and skull (Musser and Newcomb, 1983).

Musser and Newcomb (1983) suggested the evolutionary history of *Hapalomys* to have been confined to Indochina, a view now confirmed by Pleistocene fossils that document a small radiation in the region probably originating in the Pliocene. Isolated molars from early Pleistocene cave sediments in the Sichuan-Guizhou region of S China have been described as *H. eurycidens*, *H. angustidens*, and *H. gracilis* (Zheng, 1993); *H. khaorupchangi* is represented by molars from early and middle Pleistocene cave deposits on peninsular Thailand (Chaimanee, 1998). A hypothesis of cladistic relationships among three of these extinct and the two living species, based upon molar traits, is provided by Chaimanee (1998).

Hapalomys delacouri Thomas, 1927. Proc. Zool. Soc. Lond., 1927:55.
COMMON NAME: Lesser Marmoset Rat.
TYPE LOCALITY: S Vietnam, Dakto.
DISTRIBUTION: S China (Hainan Isl and S Guangxi on the mainland; Musser [1972], Wang [2003], Zhang et al. [1997]), N Laos (Musser, 1972), and C Vietnam (Dang et al., 1994); limits unknown.
STATUS: IUCN – Lower Risk (nt).
SYNONYMS: *marmosa* G. M. Allen, 1927; *pasquieri* Thomas, 1927.
COMMENTS: Corbet and Hill (1992) recognized two groups on the mainland separated by degree of tail pilosity and some cranial and dental dimensions, but the significance of these differences needs to be assessed by study of larger series; *H. delacouri* is still represented only by a few specimens. The species has been infrequently encountered by collectors, is probably restricted to a special habitat (bamboo, for example), has a patchy distribution, and will likely be found to occur over a broader range than is now indicated by extant specimens.

   Isolated molars have been uncovered from Holocene cave sediments in the Sichuan-Guizhou region of S China (Zheng, 1993, as *H.* cf. *delacouri*) and middle or late Pleistocene cave deposits in Guangxi Province of S China (Chen et al., 2002) and NE Thailand (Chaimanee, 1998, as *H. delacouri*); it would not be surprising to find the species still living in N Thailand and other parts of S China.

Hapalomys longicaudatus Blyth, 1859. J. Asiat. Soc. Bengal, 28:296.
COMMON NAME: Greater Marmoset Rat.
TYPE LOCALITY: Burma, Tenasserim, Sitang River Valley.
DISTRIBUTION: SW China (W Yunnan; Wang, 2003), SE Burma (Ellerman, 1961), SW and peninsular Thailand (J. T. Marshall, Jr., 1977a; Robinson et al., 1995), and Malay Peninsula (Medway, 1978; specimens in USNM); limits unknown.
STATUS: IUCN – Lower Risk (nt).
COMMENTS: Karyotype uninformative about phylogenetic relationships (Yong et al., 1982). Spermatozoal morphology similar to that of *Chiropodomys* and *Maxomys* and also resembling the basic structure found in species of *Mus* and *Apodemus* (Breed and Yong (1986). Reviewed by Musser (1972). Evolutionary history of *H. longicaudatus* extends back to middle Pleistocene as documented by isolated molars recovered from cave sediments in peninsular and NW Thailand (Chaimanee, 1998).

Heimyscus Misonne, 1969. Mus. Roy. l'Afrique Cent., Tervuren, Zool., no. 172:125.
TYPE SPECIES: *Hylomyscus fumosus* Brosset, Dubost, and Heim de Balsac, 1965.
COMMENTS: *Stenocephalemys* Division. A distinctive genus formerly phylogenetically allied with *Hylomyscus*. DNA/DNA hybridization experiments, for example, place *Heimyscus fumosus* closer to *Hylomyscus stella* than to the species of *Praomys*, *Myomyscus*, and *Mastomys* examined (Chevret et al., 1994), which is reflected by nuclear IRBP sequences in which *Heimyscus fumosus* joins *H. stella* and *H. parvus* in a monophyletic group (Lecompte, 2003). Analyses of complete mtDNA cytochrome *b* sequences, however, indicate *Heimyscus* to be phylogenetically divergent from *Hylomyscus* and apparently more closely related to *Praomys* (Lecompte et al., 2002b).

Heimyscus fumosus (Brosset, Dubost, and Heim de Balsac, 1965). Biologia Gabonica, 1:154.
COMMON NAME: Smokey Heimyscus.
TYPE LOCALITY: Gabon, Makokou.
DISTRIBUTION: Recorded only from the type locality, S Cameroon (Robbins et al., 1980),

SW Gabon (specimens in FMNH, USNM), Republic of Congo (Granjon, 1991), and
S Central African Republic (F. Petter and Genest, 1970; Malcom and Ray, 2000); known
distribution is in lowland forests between the Sanaga River in the west and the
Oubangui-Congo River in the east (Nicolas et al., 2003).

STATUS: IUCN – Lower Risk (lc).

COMMENTS: Originally described as a species of *Hylomyscus*, but differs significantly from
any species in that genus in morphological and chromosomal traits (Misonne, 1969;
Robbins et al., 1980), ecology (Duplantier, 1989), and complete mtDNA cytochrome *b*
sequences (Lecompte et al., 2002*b*). Chevret et al. (1994), however, suggested that the
degree of nucleotide substitutions indicated *fumosus* could remain in *Hylomyscus*
under subgenus *Heimyscus*. Chromosomal traits (2n = 40, FNa = 48) described by
Lecompte (2003). Geographical distribution and morphological variation based upon
239 specimens summarized by Nicolas et al. (2003).

Hybomys Thomas, 1910. Ann. Mag. Nat. Hist., ser. 8, 5:85.

TYPE SPECIES: *Mus univittatus* Peters, 1876.

SYNONYMS: *Typomys* Thomas, 1911.

COMMENTS: *Hybomys* Division. The species have traditionally been arranged into two groups
and reflected taxonomically as either subgenera (*Hybomys* and *Typomys*; G. M. Allen,
1939; Carleton and Robbins, 1985; Ellerman, 1941; Misonne, 1969) or genera (Thomas,
1911*b*; see Rosevear's [1969] excellent exposition). Morphometric analyses of cranial and
dental dimensions by Van der Straeten (1984) showed a closer alliance between the
subgenus *Hybomys* and *Dephomys*, a West African forest endemic, than with *Typomys*, and
prompted Van der Straeten to argue strongly for recognizing *Typomys* as a separate genus.
But his conclusion was founded on multivariate analyses of continuous variables and
ignored discrete character-trait differences; his tree is a phenogram of shape-size
similarity, not a cladogram of shared-derived characters hypothesizing phylogenetic
affinity. Carleton and Robbins (1985:983) corroborated the dichotomy between *Hybomys*
and *Typomys* using a suite of qualitative features and chromosomal data, but commented
that whether "*Typomys* and *Hybomys* merit generic segregation must await an evaluation
of character variation among *Hybomys* and its near relatives." We also note that *Hybomys*
and *Typomys* share several traits setting them apart from *Dephomys* (dorsal striping, no
pronounced guard hairs, narrower hind foot with short fifth digit, no hypothenar pad,
tail shorter than head and body, weaker temporal ridging, shorter incisive foramina,
deeper zygomatic notch, larger auditory bullae, and cusp t3 of M1 larger and separate
from cusp t2). Additional research is needed to resolve the phylogenetic relationships
among *Hybomys*, *Typomys*, and other genera of African rainforest murines. Analysis of
microcomplement fixation of albumin grouped *Hybomys univittatus* with *Stochomys*
(Watts and Baverstock, 1995*a*), but other genera and species critical to resolving the
Typomys-Hybomys problem were not included in their inquiry. Analyses of mtDNA
cytochrome *b* and 12S and 16S rRNA gene fragments placed *Hybomys* with *Dasymys* in an
African murine clade containing *Aethomys*, *Grammomys*, *Lemniscomys*, *Desmomys*,
Rhabdomys, *Pelomys*, *Mylomys*, and *Arvicanthis* (Ducroz et al., 2001). The association of
Hybomys with *Dasymys*, *Grammomys*, and arvicanthines is also supported by DNA/DNA
hybridization studies (Chevret, 1994), analyses of nuclear IRBP gene sequences
(Lecompte, 2003), and another study of mtDNA cytochrome *b* sequences (Castiglia et al.,
2003*b*). Carleton and Robbins (1985) reviewed patterns of morphometric differentiation,
qualitative features, chromosomal data, geographic distributions, and ecological and
zoogeographical associations of several species, focusing on West African forms but
discussing taxonomic problems within the *H. univittatus* complex.

Hybomys badius Osgood, 1936. Zool. Ser. Field Mus. Nat. Hist., 20:254.

COMMON NAME: Cameroon Highland Hybomys.

TYPE LOCALITY: Cameroon, southwest slope Mt Cameroon, 5800 ft (1768 m).

DISTRIBUTION: Mt Cameroon and Mtns Lefo and Oku in the Bamenda-Banso highlands,
W Cameroon.

STATUS: IUCN – Endangered as *H. eisentrauti*.

SYNONYMS: *eisentrauti* Van der Straeten and Hutterer, 1986.

COMMENTS: Subgenus *Hybomys*. We use *badius* as the oldest name for *eisentrauti*, described by Van der Straeten and Hutterer (1986), whose definition is based mostly on its geographic separation from *H. lunaris*, as Van der Straeten et al. (1986) perceived that species, not on differentiating character data. Their multivariate analysis revealed three clusters of specimen scores, one representing *H. basilii* from Bioko, another "*H. lunaris*" from E Dem. Rep. Congo and Rwanda (not true *lunaris*; see below), and a third *H. univittatus* collected in S Cameroon. Specimens of *eisentrauti* lie close to examples of *badius* and both sets of scores fall within their "*H. lunaris*." Curiously and confusingly, they treated Osgood's *badius* as a subspecies of *H. univittatus*, even though those specimen scores are well separated from the *H. univittatus* cluster. The analysis of Van der Straeten and Hutterer provides no clearcut basis for recognition of the populations on Mt Cameroon (*badius*) as specifically distinct from those on Mt Lefo and Mt Oku (*eisentrauti*). Assessing the level of relationship between populations from the E margin of the Central African Forest Block, called "*lunaris*," and those from the W Cameroon highlands, *H. badius* (including *eisentrauti*), will require a fresh systematic review of the *H. univittatus* complex (see below).

Hybomys basilii Eisentraut, 1965. Zool. Jahrb. Syst., 92:20.
   COMMON NAME: Bioko Hybomys.
   TYPE LOCALITY: Equatorial Guinea, Bioko, Mocatal, 1200 m.
   DISTRIBUTION: Endemic to Bioko.
   STATUS: IUCN – Lower Risk (nt).
   COMMENTS: Subgenus *Hybomys*. Originally described as a subspecies of *H. univittatus*, but raised to specific rank by Van der Straeten (1985) based on morphometric analyses contrasting it with samples identified as *H. univittatus* and *H. lunaris* (Van der Straeten, 1985; Van der Straeten and Hutterer, 1986).

Hybomys lunaris (Thomas, 1906). Ann. Mag. Nat. Hist., ser. 7, 18:145.
   COMMON NAME: Ruwenzori Hybomys.
   TYPE LOCALITY: Uganda, Ruwenzori East, Mubuku Valley, 6000 ft (1830 m).
   DISTRIBUTION: Ruwenzori Mtns, W Uganda.
   STATUS: IUCN – Lower Risk (lc).
   COMMENTS: Subgenus *Hybomys*. Originally described as a subspecies of *H. univittatus* by Thomas, *lunaris* was raised to a species based upon chromosomal traits (Verheyen and Van der Straeten, 1985) and morphometric analyses (Van der Straeten, 1985; Van der Straeten et al., 1986). Van der Straeten et al. (1986) identified the species as occurring in NE and E Dem. Rep. Congo, W Uganda, and Rwanda, but their samples from these areas are not representative of true *lunaris*, a morphology so far known only from the Ruwenzori Mtns. Verheyen and Van der Straeten (1985) documented 2n = 48 for a Rwandan sample they labeled *H. lunaris*, which contrasted with 2n = 44 from their Cameroon material they identified as *H. univittatus*. However, Carleton and Robbins (1985) recorded a 2n = 48 from northwest of Dongila, the type locality of *H. univittatus* and considered that karyotype typical of the latter. Carleton and Robbins (1985:985) explained that "Verheyen and Van der Straeten (1985) assumed that *lunaris* applied to their Rwandan locality because it was the geographically closest epithet available for easternmost populations of *Hybomys*. Thomas (1906) described *lunaris* from an intermediate elevation on the northeastern slope of Mt Ruwenzori, the Mubuku Valley, Uganda, and his diagnosis clearly depicts a small, delicate form of *Hybomys*. In fact, our two Ugandan examples [USNM 526765 and 526766] from Kanyawara, a place near the type-locality, precisely fit Thomas' description in their coloration, delicate skull, and small size. . . . These two specimens contrast sharply with the larger robust skulls seen in our examples from Uwinka, Rwanda [USNM 340830-340834], the same locality where Verheyen and Van der Straeten obtained karyotypic preparations. Morphologically and chromosomally, the karyotyped *Hybomys* from Gabon, Rwanda, and Dem. Rep. Congo, reported herein and by Verheyen and Van der Straeten (1985), conform closely and appear to represent but one species, *univittatus*." Carleton and Robbins (1985) suggested that *lunaris* proper is a valid species, and Kerbis Peterhans et al. (1998) restricted *H. lunaris* to the Ruwenzori Mtns. The "*lunaris*" populations

identified by Verheyen and Van der Straeten (1985; also Van der Straeten et al., 1986) may prove to be a new species, but their separation from *H. badius* and *H. univittatus* will require concerted revision and critical association of distinctive morphologies to type specimens.

Hybomys planifrons (Miller, 1900). Proc. Wash. Acad. Sci., 2:641.
COMMON NAME: Liberian Forest Hybomys.
TYPE LOCALITY: Liberia, Mt Coffee.
DISTRIBUTION: Liberian Forest Zone: N and E Sierra Leone, Liberia, S half of Guinea (Barnett et al., 1996; Ziegler et al., 2002), and W Côte d'Ivoire west of the Sassandra River; Carleton and Robbins (1985:990).
STATUS: IUCN – Lower Risk (lc).
COMMENTS: Subgenus *Typomys*. Part of the murine fauna endemic to West Africa (see account of *Grammomys buntingi*). Taxonomic status, phylogenetic relationship, and significance of distribution in Liberian forest refuge reviewed by Carleton and Robbins (1985). Additional data reported by Gautun et al. (1986). Grubb et al. (1998) reviewed Sierra Leone populations, and Barnett et al. (1996) extended its range in Guinea to the Kounounkan Massif.

Hybomys trivirgatus (Temminck, 1853). Esquisses Zool. sur la Côte de Guine, p. 159.
COMMON NAME: West African Hybomys.
TYPE LOCALITY: Ghana (= Gold Coast), Dabocrom.
DISTRIBUTION: From Sierra Leone west to SW Nigeria west of the Niger River (Carleton and Robbins, 1985:990).
STATUS: IUCN – Lower Risk (lc).
SYNONYMS: *pearsei* Ingoldby, 1929 [justified emendation of *pearcei* by Rosevear, 1969:378].
COMMENTS: Subgenus *Typomys*. Part of the murine fauna endemic to West Africa (see account of *Grammomys buntingi*). Morphology, phylogenetic affinities, and geographic range reviewed by Carleton and Robbins (1985). Grubb et al. (1998) reviewed populations in Ghana and Sierra Leone. Neither Rosevear (1969) nor Carleton and Robbins (1985) uncovered evidence for subspecific differentiation of *pearsei* from SW Nigeria.

Hybomys univittatus (Peters, 1876). Monatsb. K. Preuss. Akad. Wiss., Berlin, p. 479.
COMMON NAME: Peters's Hybomys.
TYPE LOCALITY: Gabon, Dongila.
DISTRIBUTION: From SE Nigeria (on E side of Cross River), through Cameroon, Equatorial Guinea, Gabon, Republic of Congo, S Central African Republic, Dem. Rep. Congo, and extreme NW Zambia to S Uganda and W Rwanda (Carleton and Robbins, 1985:990).
STATUS: IUCN – Lower Risk (lc).
SYNONYMS: *rufocanus* (Tullberg, 1893).
COMMENTS: Subgenus *Hybomys*. Review of species and comparisons with *H. trivirgatus* and *H. planifrons* reported by Carleton and Robbins (1985). The degree of morphological variation seen among populations of *H. univittatus* suggests that it is a composite of morphologically similar species. One problem has been inadvertently identified by Van der Straeten et al. (1986), who separated *lunaris* from *H. univittatus*. Their morphometric analysis, and similar results reproduced in other reports (Van der Straeten, 1985; Van der Straeten and Hutterer, 1986), clearly distinguish specimens from S Cameroon identified as *H. univitattus* from samples obtained in E Dem. Rep. Congo and Rwanda, which they identified as *H. lunaris*. Real *lunaris*, however, is a species of small body size and more delicate build, to date known only from the Ruwenzori highlands (see that account). The "*lunaris*" of Van der Straeten and colleagues is a different species from *H. univitattus*, which appears to inhabit the Central African Forest Block (Carleton and Robbins, 1985), but there is no available scientific name that can be attached to it. To adequately define this eastern species will require a fresh systematic inquiry into qualitative and quantitative relationships among samples of the *H. univittatus* complex (see comments under *H. badius* and *H. lunaris*). Habitat and locality in Kalinzu Forest of SW Uganda recorded by Lunde and Sarmiento (2002).

Hydromys E. Geoffroy, 1804. Bull. Sci. Soc. Philom. Paris, 3(93):353.

TYPE SPECIES: *Hydromys chrysogaster* E. Geoffroy, 1804.

SYNONYMS: *Baiyankamys* Hinton, 1943.

COMMENTS: *Hydromys* Division. Member of the Australian and New Guinea Old Endemics (Musser, 1981*c*). Considered an Australian member of a restricted Hydromyini by Baverstock (1984), not the more inclusive Hydromyini of Watts and Baverstock (1994*a*). Analysis of allozymic variation supported a weak link between *H. chrysogaster* and *Xeromys* (Baverstock et al., 1981), but albumin immunology (Watts and Baverstock, 1994*a*, 1996) and spermatozoal morphology (Breed and Aplin, 1994; Breed, 1997), using *H. chrysogaster*, strongly allied *Hydromys* with New Guinea *Parahydromys*, and the Australian *Leggadina*, and placed the three in a larger clade containing members of our *Xeromys*, *Pseudomys*, and *Uromys* Divisions, which Watts and Baverstock (1994*a*) defined as the Hydromyini (or "Australasian clade") to the exclusion of strictly New Guinea species in our *Pogonomys* Division (Anisomyini of Watts and Baverstock, 1994*a*) and *Lorentzimys* Division. However, analyses of multiple mitochondrial and nuclear DNA sequences indicate *Leggadina* to be closer to a *Pseudomys/Notomys* clade than to *Hydromys* (F. Ford, in litt., 2004; see *Leggadina* account). Comparisons with Neotropical ichthyomyines made by Voss (1988). Flannery (1990*b*, 1995*a*) provided photographs and distributional and biological summaries of species. Phallic morphology of *H. chrysogaster* and *H. habbema* described by Lidicker (1968). Mahoney (1968) explained why *Baiyankamys* is a synonym of *Hydromys*.

Hydromys chrysogaster E. Geoffroy, 1804. Bull. Sci. Soc. Philom. Paris, 93:354.

COMMON NAME: Common Water Rat.

TYPE LOCALITY: Australia, Tasmania, Bruny Isl (see Mahoney and Richardson, 1988:157).

DISTRIBUTION: Australia: freshwater lakes and rivers as well as swamp, salt marsh, and supralittoral habitats (absent from C Australian region); also found on Tasmania and numerous smaller islands off the coast of Australia (Friend and Thomas, 1990; Robinson et al., 2000; Rounsevell et al., 1991; Seebeck, 1995*b*; Watts and Aslin, 1981:67); Kai Isls and Aru Isls. New Guinea: throughout most of the island from sea level to 1900 m (Flannery, 1990*b*:188; 1995*a*:237;). Also on the Melanesian and Wallacean islands of Goodenough, Yapen, Biak, Kiriwina, Fergusson, Normanby, and Obi (Flannery, 1995*b*).

STATUS: IUCN – Lower Risk (lc).

SYNONYMS: *apicalis* Kuhl, 1820; *beccarii* Peters, 1874; *caurinus* Thomas, 1909; *esox* Thomas, 1906; *flaviventer* Owen, 1840-1845; *fuliginosus* Gould, 1853; *fulvogaster* Jourdan, 1837; *fulvolavatus* Gould, 1853 *fulvoventer* Cuvier, 1837; *grootensis* Troughton, 1935; *illuteus* Thomas, 1922; *lawnensis* Troughton, 1935; *leucogaster* Geoffroy, 1804; *longmani* Thomas, 1923; *lutrilla* Gould, 1853; *melicertes* Thomas, 1921; *moae* Troughton, 1935; *nauticus* Thomas, 1921; *oriens* Troughton, 1937; *reginae* Thomas and Dollman, 1909.

COMMENTS: Chromosomal data presented by Baverstock et al. (1977*c*, 1983*b*). Morphology of spermatozoa and male reproductive tract discussed in context of comparative study of Australian murines (Breed, 1984, 1986; Breed and Sarafis, 1978; Morrissey and Breed, 1982). References to distributional, taxonomic, and biological literature for Australia cataloged by Mahoney and Richardson (1988:156). Significance of variation in body size and pelage color needs to be assessed in context of careful systematic revision of the species; more than one species likely exists among available samples (the single specimen from Obi Isl, for example, likely represents a separate species, as does a very small-bodied specimen from highland oak forest on Goodenough Isl, which may be a montane endemic distinct from *H. chrysogaster* in the lowlands; K. Helgen, in litt., 2003). Australian population reviewed by Olsen (1995), New Guinea by Flannery (1995*a*). Leary and Serl (1997) reported specimens from Mt Sisa in Papua New Guinea. Aplin et al. (1999) reported a specimen of this species from a Holocene archaeological site on the Ayamaru Plateau, central Bird's Head Peninsula of West Papua.

Hydromys habbema Tate and Archbold, 1941. Am. Mus. Novit., 1101:3.

COMMON NAME: New Guinea Mountain Water Rat.

TYPE LOCALITY: New Guinea, Prov. of Papua (= Irian Jaya), 15 km N Mt Wilhelmina, Lake Habbema, 3225 m.

DISTRIBUTION: New Guinea, Prov. of Papua (= Irian Jaya), known only from the type locality and NE slope of Mt Wilhelmina between 3560 and 3600 m (Tate, 1951:227); limits unknown.

STATUS: IUCN – Lower Risk (nt).

COMMENTS: Reviewed and mapped by Flannery (1995a).

Hydromys hussoni Musser and Piik, 1982. Zool. Meded. Leiden, 56:157.

COMMON NAME: Husson's Water Rat.

TYPE LOCALITY: New Guinea, Prov. of Papua (= Irian Jaya), Wissel Lakes, Lake Paniai, Enarotali, 1765 m (see Musser and Piik, 1982, for other details).

DISTRIBUTION: New Guinea; known only from the Wissel Lakes region in the W foothill margin of the Snow Mtns (Pegunungan Maoke) in Prov. of Papua (= Irian Jaya) (Flannery, 1990b:186, 1995a; Musser and Piik, 1982:156); limits unresolved.

STATUS: IUCN – Lower Risk (nt).

COMMENTS: Overall morphology and anatomical proportions of *H. hussoni* are like those of *H. chrysogaster*, although it is much smaller than that species (Musser and Piik, 1982). Ziegler (1984) identified a specimen as *H. hussoni* obtained from Bianyik, 5 km south of Maprik in lowlands (213 m) of the Torricelli Mtns on N coast of East Sepik Province in Papua New Guinea. This specimen is about the same body size as the series from Enarotali, has smaller skull and bullae, shorter molar rows (Ziegler, 1984) and to us is not an example of *H. hussoni*, but represents a separate undescribed species (K. Helgen, in litt., 2004, who examined the specimen, concurs).

Hydromys neobritannicus Tate and Archbold, 1935. Am. Mus. Novit., 803:8.

COMMON NAME: New Britain Water Rat.

TYPE LOCALITY: Bismarck Arch., New Britain Isl, Wide Bay, Balayang, Bainings.

DISTRIBUTION: Endemic to New Britain Isl.

STATUS: IUCN – Vulnerable.

COMMENTS: Included in *H. chrysogaster* by Ellerman (1941) and Ziegler (1982b), but recognized as a New Britain endemic by Flannery and White (1991) and should retain this status until significance of its diagnostic traits can be assessed in a systematic revision of the large-bodied forms of *Hydromys*, as Tate (1951:236) already noted. Reviewed by Flannery (1995b).

Hydromys shawmayeri (Hinton, 1943). Ann. Mag. Nat. Hist., ser. 11, 10:552.

COMMON NAME: Shaw Mayer's Water Rat.

TYPE LOCALITY: Papua New Guinea, SE Bismarck Range, Purari-Ramu Divide, Baiyanka, 6500 ft (1980 m).

DISTRIBUTION: New Guinea; recorded only along the Papua New Guinea Central Cordillera from Hagen Range in the west to Mt Kaindi area in the east (localities 1-4 that were mapped as "*H. habbema*" in Musser and Piik, 1982:156).

STATUS: IUCN – Lower Risk (nt).

COMMENTS: Although identified as *H. habbema*, specimens from Papua New Guinea were noted to exhibit significant morphological differences from true *H. habbema* in Prov. of Papua (= Irian Jaya), not only in body size but also in certain cranial proportions (Mahoney, 1968; Musser and Piik, 1982; Tate, 1951). These and other contrasts support the hypothesis of eastern (*H. shawmayeri*) and western (*H. habbema*) montane species, a pattern that is not uncommon to other New Guinea murines (e. g., *Pseudohydromys murinus* and *P. occidentalis*). Mahoney (1968) discussed the problems associated with the original holotype of *shawmayeri*, which was described as a species of *Baiyankamys*. Reviewed by Flannery (1995a).

Hylomyscus Thomas, 1926. Ann. Mag. Nat. Hist., ser. 9, 17:174.

TYPE SPECIES: *Epimys aeta* Thomas, 1911.

COMMENTS: *Stenocephalemys* Division. Taxonomic, distributional, and biological summaries of West African forms provided by Rosevear (1969). Morphometric, distributional, and chromosomal data for some species in Cameroon reported by Eisentraut (1969, 1973) and

Robbins et al. (1980). Species from Côte d'Ivoire reviewed by Heim de Balsac and Aellen (1965). Distributions of species listed below based primarily on study of museum specimens. See Rosevear (1969) and Robbins et al. (1980) for taxonomic history of the alternating use of *Hylomyscus* as a genus or subgenus. DNA/DNA hybridization results set *Hylomyscus* well apart from a cluster formed by species of *Praomys*, *Myomyscus*, and *Mastomys* (Chevret et al., 1994); analysis of microcomplement fixation of albumin brings *Praomys*, *Mastomys*, *Myomyscus*, and *Hylomyscus* together in a clade with the last separated from the other three (Watts and Baverstock, 1997*a*); and monophyly of *Hylomyscus* within a monophyletic *Praomys* group (= *Stenocephalemys* Division) is supported by cladistic analyses of variation in skeletal and dental traits (Lecompte et al., 2002*a*), complete mitochondrial cytochrome *b* (Lecompte et al., 2000*b*), partial 16S rRNA mitochondrial gene sequences (Fadda et al., 2001*a*), and nuclear IRBP gene sequences (Lecompte, 2003). Van der Straeten and Robbins (1997) examined most holotypes of the *Mastomys-Praomys-Myomys-Hylomyscus* complex and sorted them into groups by principal component analyses of continuous character variation; these generally brought together all holotypes traditionally associated with *Hylomyscus* in a cluster separate from those representing other genera.

Hylomyscus aeta (Thomas, 1911). Ann. Mag. Nat. Hist., ser. 8, 7:591.
    COMMON NAME: Beaded Hylomyscus.
    TYPE LOCALITY: Cameroon, Bitye, Ja River.
    DISTRIBUTION: Central forest block from Equatorial Guinea (including Bioko) to Cameroon, Gabon, Central African Republic, Republic of Congo, Dem. Rep. Congo, W Uganda, and NW Burundi; limits unknown (distribution based primarily on specimens examined in various institutions).
    STATUS: IUCN – Lower Risk (lc).
    SYNONYMS: *laticeps* Osgood, 1936; *shoutedeni* (Dollman, 1914); *weileri* (Lönnberg and Gyldenstolpe, 1925).
    COMMENTS: *Hylomyscus aeta* is easily distinguished from all other species of *Hylomyscus* by its distinct supraorbital shelves. Hatt (1940*a*) tentatively included *aeta*, *shoutedeni*, and *weileri* in *H. carillus*, but our study of samples and holotypes does not support his arrangement. Variation in body size exists among geographic samples but its significance has not been assessed by careful study. Swynnerton and Hayman's (1951:316) report of *H. aeta* in the Uluguru Mtns, EC Tanzania, is apparently based on Allen and Loveridge's (1933) listing of *H. weileri*, now considered a synonym of *H. aeta*; these specimens (series in MCZ) are all examples of the *H. denniae* complex. Chromosomal data reported by Robbins et al. (1980). Eisentraut (1969*b*) described *grandis* from Mt Oku as a subspecies of *H. aeta* but Hutterer et al. (1992*a*) treated it as a separate species (see that account). Discussed by Eisentraut (1973) in the context of faunal evolution in West Africa. Taken on the ground and on vines and branches above ground in the Kalinzu Forest of SW Uganda (Lunde and Sarmiento, 2002).

Hylomyscus alleni (Waterhouse, 1838). Proc. Zool. Soc. Lond., 1837:77 [1838].
    COMMON NAME: Allen's Hylomyscus.
    TYPE LOCALITY: Equatorial Guinea, Bioko, Gulf of Guinea.
    DISTRIBUTION: Equatorial Guinea (including Bioko) and West Africa from Guinea (Mt Nimba) to Gabon and Cameroon; limits unknown.
    STATUS: IUCN – Lower Risk (lc).
    SYNONYMS: *canus* Sanderson, 1940; *montis* Eisentraut, 1969; *simus* G. M. Allen and Coolidge, 1930.
    COMMENTS: We do not follow Rosevear (1969) in restricting *H. alleni* to Bioko (= Fernando Poo), and agree with Eisentraut (1969*b*) and Robbins et al. (1980) in recognizing the species in West Africa (based on our study of Eisentraut's material and other specimens). Robbins et al. (1980) reported sympatry between *alleni* and *stella* in S Cameroon, as well as cranial and chromosomal distinctions. However, the morphological differences they noted are slight and variable even within a single sample, the karyotypes all have 2n = 46 and only differ slightly in FN (68 versus 70), and they did not demonstrate whether the contrasts represented intra- or interpopulational

variation. The chromosomal differences are not impressive considering that a sample referred to *H. stella* from Burundi had 2n = 48 and FN= 86 (Maddalena et al., 1989). Some authors (Brosset et al., 1965; Heim de Balsac and Aellen, 1965) have regarded *simus* as the species distributed throughout West Africa and into Angola, and *stella* in C Africa, but we follow Rosevear (1969) in not being able to distinguish any sample that can be called *simus*. The definitions of both *alleni* and *stella* on continental Africa need clarification, as Rosevear (1969) so cogently elaborated. Populations in Ghana and Sierra Leone reviewed by Grubb et al. (1998). Morphometric and ecological data recorded by Ryan and Attuquayefio (2000) and ecological data discussed by Decher and Bahian (1999) for samples from the Accra Plains in S Ghana. Information covering vertical distribution in S Côte d'Ivoire (under *H. simus*) recorded by Adam (1977). Discussed by Eisentraut (1973) in the context of a retrospective on faunal evolution in West Africa.

Hylomyscus baeri Heim de Balsac and Aellen, 1965. Biologia Gabonica, 1:175.
  COMMON NAME: Baer's Hylomyscus.
  TYPE LOCALITY: Côte d'Ivoire, Adiopodoume.
  DISTRIBUTION: Recorded only from Côte d'Ivoire, Ghana, and Sierra Leone.
  STATUS: IUCN – Lower Risk (lc).
  COMMENTS: Part of the murine fauna endemic to West Africa (see account of *Grammomys buntingi*). Distributional and morphometric data summarized by Robbins and Setzer (1979); Ghana and Sierra Leone records reviewed by Grubb et al. (1998).

Hylomyscus carillus (Thomas, 1904). Ann. Mag. Nat. Hist., ser. 7, 13:418.
  COMMON NAME: Angolan Hylomyscus.
  TYPE LOCALITY: Angola, Andongo, Pungo, 1200 m.
  DISTRIBUTION: WC Angola; limits unknown (Crawford-Cabral, 1998, claimed the species occurs in SE Dem. Rep. Congo).
  STATUS: IUCN – Lower Risk (lc).
  COMMENTS: Associated with either *H. aeta* (Hatt, 1940a) or *H. alleni* (G. M. Allen, 1939), the affinities of *carillus* may be closer to *H. stella* (based on our study of the holotype and series from Pungo in BMNH). We have identified three distinct species of *Hylomyscus* occurring in Angola: *H. carillus*, an unnamed population related to East African *H. denniae*, and *H. stella* (specimens in AMNH, BMNH, and FMNH). Crawford-Cabral (1998) reviewed and mapped records he identified as *H. carillus* and regarded this as the only species in Angola; some of his records likely represent the other two species.

Hylomyscus denniae (Thomas, 1906). Ann. Mag. Nat. Hist., ser. 7, 18:144.
  COMMON NAME: Montane Hylomyscus.
  TYPE LOCALITY: Uganda, Mubuku Valley, Ruwenzori East, 7000 ft (2134 m).
  DISTRIBUTION: Montane forest islands in WC Angola, extreme E Dem. Rep. Congo, Uganda, W Rwanda, and Kenya through Tanzania to NE Zambia.
  STATUS: IUCN – Lower Risk (lc).
  SYNONYMS: *anselli* Bishop, 1979; *endorobae* (Heller, 1910); *vulcanorum* (Lönnberg and Gyldenstolpe, 1925).
  COMMENTS: The species was superficially reviewed by Hatt (1940a) and somewhat more thoroughly by Bishop (1979). The records of *H. carillus* from Chitau and Hanha in WC Angola (Hill and Carter, 1941:98) are based upon examples of what we identify here as *H. denniae*; additional samples (in FMNH) were collected in the same region on Mt Moco and Mt Soque. Although far west from the nearest records of *H. denniae*, most morphological characteristics of the Angolan series fall within the range of variation among samples now defined as that species. That variation, however, is appreciable, especially in body size, and its significance in determining whether one or more species is present in what is now regarded as *H. denniae* has to be assessed by systematic revision; certainly more than one species is represented. The distribution of *H. denniae*-like populations in afromontane habitats of East Africa and the Angolan plateau parallels the pattern seen in *Rhabdomys dilectus* (see account). Stanley et al. (1998) discussed distribution in the Eastern Arc Mtns of Tanzania. Altitudinal distribution of *H. denniae* on Ugandan slopes of Ruwenzori Mtns reviewed by Kerbis Peterhans et al. (1998), who

regarded true *H. denniae* as a Ruwenzori endemic. Habitat and altitudinal distribution on Ugandan slopes of Mt Elgon documented by Clausnitzer and Kityo (2001).

In their principle components analysis of holotypes, Van der Straeten and Robbins (1997) interpreted the score for the holotype of *endorobae* as falling within the *Praomys* cluster. But that specimen is part of a large series used by Heller to describe *endorobae* and in a preliminary multivariate analysis generated by Carleton, the entire series forms a swarm overlapping the cluster of *denniae* from Ruwenzori. The holotype is a large individual and it is not surprising that in a principle components analysis of holotypes alone it would fall near or among small holotypes of *Praomys*. The holotype of *endorobae* also has all the qualitative traits we associate with *Hylomyscus* and not *Praomys*: scarcely expressed cusps t3 and t9 (the latter just a slight spur off cusp t8), zygomatic plate relatively narrow and the notch shallow, and digit 5 of hindfoot about as long as digits 2-4.

Hylomyscus grandis Eisentraut, 1969. Z. Säugetierk., 34:300.
> COMMON NAME: Mount Oku Hylomyscus.
> TYPE LOCALITY: West Cameroon, Banso Highlands, Lake Oku on Mt Oku, 2100 m.
> DISTRIBUTION: Recorded only from upper slopes of Mt Oku (Eisentraut, 1969b, 1973).
> COMMENTS: Still known only by the four specimens collected by Eisentraut. He (Eisentraut, 1969b) described *grandis* as a subspecies of *H. aeta*, but it has a larger skull and significantly longer molar rows and is a separate species (Hutterer et al., 1992a; our study of specimens at ZFMK). Eisentraut collected examples of *H. alleni* at the same altitude at Lake Oku (Eisentraut, 1969b; our identifications of his specimens). In addition to *H. grandis*, other species of small mammals endemic to Mount Oku are the muroid rodents *Lophuromys dieterleni*, *Lamottemys okuensis*, and *Lemniscomys mittendorfi* and the golden mole *Chrysochloris balsaci*.

Hylomyscus parvus Brosset, Dubost, and Heim de Balsac, 1965. Biologia Gabonica, 1:149.
> COMMON NAME: Lesser Hylomyscus.
> TYPE LOCALITY: Gabon, Belinga, 800 m.
> DISTRIBUTION: N and E Dem. Rep. Congo, S Central African Republic, N Gabon, and S Cameroon (Dudu et al., 1989).
> STATUS: IUCN – Lower Risk (lc).
> COMMENTS: A distinctive species discussed by Dudu et al. (1989). Regularly trapped, although in small numbers, in Makokou, Gabon (Duplantier, 1989).

Hylomyscus stella (Thomas, 1911). Ann. Mag. Nat. Hist., ser. 8, 7:590.
> COMMON NAME: Stella Hylomyscus.
> TYPE LOCALITY: E Dem. Rep. Congo, Ituri Forest, between Mawambi and Avakubi.
> DISTRIBUTION: From S and SE Nigeria, Gabon, Cameroon, Republic of Congo, Central African Republic, S Sudan to Dem. Rep. Congo, N Angola, Uganda, and W Kenya to EC Tanzania, Burundi, and Rwanda; limits unknown.
> STATUS: IUCN – Lower Risk (lc).
> SYNONYMS: *kaimosae* (Heller, 1912).
> COMMENTS: The records from Sudan and N Angola are documented by specimens in FMNH; those from Tanzania are in BMNH. A specimen from the Gotel Mtns in SE Nigeria may represent the most northern population in Cameroon and Nigeria (Hutterer et al., 1992a). Some E African samples were reviewed by Bishop (1979). Using allozymic data, Iskandar et al. (1988) documented two species occurring together in NW Gabon. One is *H. stella* but they could not place a name on the other. Recognizable morphological variation exists among the samples of *H. stella* and this significance has to be assessed by critical systematic revision. The variation in 2n and FN among samples identified as *H. stella* was documented by Robbins et al. (1980) and Maddalena et al. (1989). Fadda et al. (2001a) used *kaimosae* as a species without explanation. Included by Eisentraut (1973) in his discussion of faunal evolution in West Africa. Documented from the Kalinzu Forest of SW Uganda by Delany (1975) and Lunde and Sarmiento (2002). Ecological data (demographic and spatial parameters during 16 months) in Gabon summarized by Duplantier (1989).

Hyomys Thomas, 1904. Proc. Zool. Soc. Lond., 1903(2):198 [1904].

TYPE SPECIES: *Hyomys meeki* Thomas, 1904 (= *Mus goliath* Milne-Edwards, 1900).

COMMENTS: *Pogonomys* Division. Member of the New Guinea Old Endemics (Musser, 1981*c*). Phallic morphology documented by Lidicker (1968). Distributional and biological data summarized by Flannery (1990*b*, 1995*a*). Analysis of immunological distances by Watts and Baverstock (1994*a*) placed *Hyomys* in a clade with *Chiruromys*, *Pogonomys*, *Anisomys*, *Coccymys*, *Mallomys*, and *Macruromys*, but could not identify clear affinity with any particular genus within that group. Based upon their primitive sperm morphologies, Breed and Aplin (1994:26) speculated that *Hyomys*, along with *Anisomys*, "may represent some of the earliest offshoots of the Australo-Papuan radiation." Whether only one or more species are present in *Hyomys* has never been satisfactorily resolved (Flannery, 1990*b*; Rümmler, 1938; Tate, 1951), but our examination of museum specimens revealed the two species listed below.

Hyomys dammermani Stein, 1933. Z. Säugetierk., 8:95.

COMMON NAME: Western Hyomys.

TYPE LOCALITY: New Guinea, Prov. of Papua (= Irian Jaya), Weyland Range, Kunupi Mtn.

DISTRIBUTION: New Guinea; from the Vogelkop region (Arfak Mtns), Weyland Range, and Snow Mtns in Prov. of Papua (= Irian Jaya) east along Central Cordillera to Schrader Range and the Mt Hagen and Nondugl region in Papua New Guinea; to the east it is apparently replaced by *H. goliath*.

STATUS: IUCN – Lower Risk (lc).

COMMENTS: Originally described as a subspecies of *H. meeki* (= *H. goliath*; Rummler, 1938; Stein, 1933), *H. dammermani* is a small-bodied species with only traces of white wisps about the ears.

Hyomys goliath (Milne-Edwards, 1900). Bull. Mus. Hist. Nat., Paris, 6:165.

COMMON NAME: Eastern Hyomys.

TYPE LOCALITY: Papua New Guinea, Central Province, highlands of Aroa River basin.

DISTRIBUTION: Papua New Guinea; Central Cordillera from the Kratke Mtns in the north and Mt Sisa in the south (Leary and Seri, 1997) eastward to Mt Dayman at the end of the Owen Stanley Range, south to mountains of Milne Bay Province, and mountains of Huon Peninsula.

STATUS: IUCN – Lower Risk (lc).

SYNONYMS: *meeki* Thomas, 1904; *strobilurus* Rümmler, 1933.

COMMENTS: This is the larger-bodied species with prominent white auricular tufts that is identified as *meeki* in the older literature.

Kadarsanomys Musser, 1981. Zool. Verhandel., 189:5.

TYPE SPECIES: *Rattus canus sodyi* Bartels, 1937.

COMMENTS: *Rattus* Division. The only murine genus endemic to Java (Musser and Newcomb, 1983; Musser, 1986).

Kadarsanomys sodyi (Bartels, 1937). Treubia, 16:45.

COMMON NAME: Javan Bamboo Rat.

TYPE LOCALITY: Indonesia, W Java, Gunung Pangrango-Gede, 1000 m.

DISTRIBUTION: Java.

STATUS: IUCN – Lower Risk (nt).

COMMENTS: Morphology, natural history, and comparisons with *Rattus* and *Lenothrix* reported by Musser (1981*a*); the species is arboreal and was collected from nests in bamboo. Represented only by modern series collected in W Java during 1933-1935, and subfossil fragments from C and E Java (Musser and Newcomb, 1983). Among Sundaic murines, *Kadarsanomys* has no close phylogenetic allies, but some cranial and dental traits indicate a distant relationship with *Rattus* (Musser and Newcomb, 1983), and especially the Sulawesi *R. xanthurus* Group. *Kadarsanomys sodyi* joins *Sundamys maxi*, *Mus vulcani*, *Niviventer lepturus*, *Maxomys bartelsii*, and *Pithecheir melanurus* as a member of the murine fauna endemic to Java (Musser, 1986).

Komodomys Musser and Boeadi, 1980. J. Mammal., 61:397.
> TYPE SPECIES: *Rattus rintjanus* Sody, 1941.
> COMMENTS: *Rattus* Division. An endemic of Nusa Tenggara, Indonesia. Although cranial conformation is distinctive, dental morphology allies *Komodomys* to *Papagomys*, another endemic of Nusa Tenggara found only on Flores Isl (Musser, 1981*c*; Musser and Boeadi, 1980). Membership in a *Rattus* Division is supported by cranial and dental morphology (Musser and Newcomb, 1983) along with albumin immunology (Watts and Baverstock, 1994*b*); the albumin data ties *Komodomys* more closely to Sulawesian *Bunomys* than to *Papagomys*. Recent unpublished allozyme electrophoresis analysis demonstrates a close relationship between *K. rintjanus* and *Rattus timorensis* (see that account). Curiously, Pavlinov et al. (1995*a*) listed *Komodomys*, along with *Papagomys* in a *Pithecheir* Section of a more inclusive *Micromys* Group.

Komodomys rintjanus (Sody, 1941). Treubia, 18:310.
> COMMON NAME: Nusa Tenggara Komodomys.
> TYPE LOCALITY: Indonesia, Nusa Tenggara, Pulau Rintja, Lohoboeaja.
> DISTRIBUTION: Nusa Tenggara: islands of Rintja, Padar, and Flores; probably occurs on other islands in the Lesser Sunda chain (e.g., Komodo).
> STATUS: IUCN – Vulnerable.
> COMMENTS: Populations on Rintja and Padar are represented by modern samples, that from Flores by subfossil fragments recovered from sediments 4000-3000 years old (Musser, 1981*c*; undescribed material studied by Musser).

Lamottemys F. Petter, 1986. Cimbebasia, Ser. A, 8:98.
> TYPE SPECIES: *Lamottemys okuensis* F. Petter, 1986.
> COMMENTS: *Oenomys* Division. A distinctive genus whose closest phylogenetic relative is probably *Oenomys* (Dieterlen and Van der Straeten, 1988; F. Petter, 1986).

Lamottemys okuensis F. Petter, 1986. Cimbebasia, Ser. A, 8:98.
> COMMON NAME: Mount Oku Lamottemys.
> TYPE LOCALITY: W Cameroon, Mt Oku.
> DISTRIBUTION: Known only by a few specimens from Mt Oku (Dieterlen and Van der Straeten, 1988; F. Petter, 1986).
> STATUS: IUCN – Endangered.
> COMMENTS: Palatal ridges described and contrasted with *Oenomys hypoxanthus* by Fülling (1992).

Leggadina Thomas, 1910. Ann. Mag. Nat. Hist., ser. 8, 6:606.
> TYPE SPECIES: *Mus forresti* Thomas, 1906.
> COMMENTS: *Pseudomys* Division. Sometimes included in *Pseudomys*, but a distinctive genus considered to be an Australian Old Endemic (Musser, 1981*c*). Usually included in the Conilurini (Baverstock, 1984), but data from microcomplement fixation of albumin indicated that *L. forresti* is not closely related to *Pseudomys* or any of the other Australian Old Endemics (Watts et al., 1992), but clusters with *Hydromys* (Australia and New Guinea) and the New Guinea endemic *Parahydromys* (Watts and Baverstock, 1994*a*, 1996), an association seemingly supported by spermatozoal morphology (Breed and Aplin, 1994; Breed, 1997). However, a recently completed phylogenetic analysis of multiple mitochondrial and nuclear genes in a wide selection of "conilurines" (in the sense of Baverstock, 1984) shows that *Leggadina* is much closer to a *Pseudomys/Notomys* clade than to *Hydromys* (F. Ford, in litt., 2004). Compared with DNA sequences, especially from a combination of nuclear and mitochondrial genes, the albumin immulogical data, as well as sperm morphology, are apparently unreliable indicators for detecting deep phylogenetic relationships. *Leggadina* and other members in the *Pseudomys* Division are part of a larger clade that includes genera in our *Hydromys*, *Xeromys*, and *Uromys* Divisions, an assemblage that Watts and Baverstock (1994*a*) defined as Hydromyini (or the "Australasian clade," Watts and Baverstock, 1995*b*, 1996) as opposed to a strictly New Guinean Anisomyini (our *Pogonomys* Division). Some species of *Pseudomys*, namely *P. delicatulus* and *P. hermannsburgensis*, are often included in *Leggadina* (Lidicker and Brylski, 1987) but only *L. forresti*

and *L. lakedownensis* belong here (Mahoney and Posamentier, 1975; Mahoney and Richardson, 1988; Watts and Aslin, 1981).

Morphology of male reproductive tract, external anatomy of glans penis, and spermatozoal structure documented in context of comparative study of Australian murines by Breed (1980, 1984, 1986), Morrissey and Breed (1982), and Lidicker and Brylski (1987). Biochemical and chromosomal data discussed by Baverstock et al. (1976*a*, 1981, 1983*b*). Taxonomic, distributional, and biological references for species catalogued by Mahoney and Richardson (1988). Moro et al. (1998) analysed variation in mtDNA cytochrome *b* sequences with particular focus on Western Australian populations. Cooper et al. (2003*a*) clarified the wider distributions of the two species based on much newly collected material and a combination of allozyme electrophoresis and morphological analysis. Earliest record of *Leggadina* comes from late Pliocene sediments (Aplin, 2005; Godthelp, 1997).

Leggadina forresti (Thomas, 1906). Abstr. Proc. Zool. Soc. Lond., 1906(32):6.
  COMMON NAME: Forrest's Leggadina.
  TYPE LOCALITY: Australia, Northern Territory, Alexandria (for additional information, see Mahoney and Richardson, 1988:158).
  DISTRIBUTION: Inland Australia; W half of Queensland, NW New South Wales, N South Australia, S Northern Territory, and localities in far eastern part of Western Australia (Banks and Rushton, 1998; Reid and Morton, 1995; Robinson et al., 2000; Watts and Aslin, 1981). Populations in Pilbara region, including Thevenard Isl, once considered to be *L. forresti*, are now referred to *L. lakedownensis* (Cooper et al., 2003*a*; Moro et al., 1998).
  STATUS: IUCN – Lower Risk (nt).
  SYNONYMS: *berneyi* (Troughton, 1936); *messorius* (Thomas, 1925); *waitei* (Troughton, 1932).
  COMMENTS: Reviewed by Watts and Aslin (1981), Reid and Morton (1995), and Cooper et al. (2003*a*).

Leggadina lakedownensis Watts, 1976. Trans. R. Soc. S. Aust., 100:105.
  COMMON NAME: Lakeland Downs Leggadina.
  TYPE LOCALITY: Australia, Queensland, Lakeland Downs, 110 km north of Cooktown.
  DISTRIBUTION: NE and N Australia; recorded from inland localities in C Queensland and coastal regions in far N Queensland, in subtropical region of Northern Territory, and Kimberley and Pilbara regions (including Thevanard Isl) of Western Australia (Cooper et al., 2003*a*).
  STATUS: IUCN – Lower Risk (nt).
  COMMENTS: A distinctive species distinguished from its close relative *L. forresti* by a suite of morphological, biochemical, and chromosomal traits (Baverstock et al., 1976*a*; Cooper et al., 2003*a*; Moro et al., 1998; Watts, 1976). Kidney structure described by Moro (2000). Moro et al. (1998) reported variation in mtDNA cytochrome *b* sequenes between the Pilbara (including Thevanard Isl) and the Kimberley and argued for taxonomic distinction of these populations. Cooper et al. (2003*a*) undertook a broader analysis based on much newly collected material and a combination of allozyme electrophoresis and morphological analysis. Although regional morphological differences were revealed, the various populations are genetically similar from the Pilbara region through to the Northern Territory. Populations in Queensland may be genetically more distinct; however, insufficient samples were available to quantify the extent of divergence. The Thevanard Isl population is larger in all dimensions than any in the nearby Pilbara region but with minimal genetic differentiation.

Lemniscomys Trouessart, 1881. Bull. Soc. Etudes Sci. Angers, 10:124.
  TYPE SPECIES: *Mus barbarus* Linnaeus, 1766.
  COMMENTS: *Arvicanthis* Division. Morphological data clusters *Lemniscomys* with species of *Arvicanthis*, *Pelomys*, *Mylomys*, and *Rhabdomys* (Musser, 1987*a*), which is corroborated by analysis of mitochondrial gene sequences (DNA cytochrome *b*, 12S and 16S rRNA gene fragments), and also includes *Desmomys* (Ducroz et al., 2001). The sequence data of Ducroz et al. also indicates that within this arvicanthine cluster, *Lemniscomys* is the only member of a lineage separate from that comprising *Desmomys* and *Rhabdomys*, and from

another containing *Arvicanthis*, *Mylomys*, and *Pelomys*. Analysis of microcomplement fixation of albumin associates *Lemniscomys* with *Pelomys*, *Rhabdomys*, *Grammomys*, and *Thallomys* (Watts and Baverstock, 1995*a*). Chromosomal data has been summarized for several species (Fadda et al., 2001*b*; Filippucci et al., 1986; Gautun et al., 1985). Spermatozoal morphology documented for *L. striatus* and *L. zebra* (Breed, 1995*a*; reported as *barbarus*) and *L. macculus* (Baskevich and Lavrenchenko, 1995). Evolutionary history of *Lemniscomys* can be traced back to late Pliocene of East Africa and the genus is represented in several Pleistocene and Holocene sites in East and South Africa (Avery, 2000; Denys, 1999).

Lemniscomys barbarus (Linnaeus, 1766). Syst. Nat., 12th ed., 1:addenda.

COMMON NAME: Barbary Lemniscomys.

TYPE LOCALITY: Morocco (= "Barbaria," see discussion in Carleton and Van der Straeten, 1997).

DISTRIBUTION: "Coastal region of Morocco, Algeria, and Tunisia, northwest and north of the Atlas Mountains" (Carleton and Van der Straeten, 1997:665); endemic to the Barbarian region (Maghreb) of NW Africa.

STATUS: IUCN – Least Concern.

SYNONYMS: *ifniensis* Morales Agacino, 1935.

COMMENTS: Member of the *L. barbarus* group, which also contains *L. zebra* and *L. hoogstraali*. Carleton and Van der Straeten (1997) provided a comprehensive systematic revision of this assemblage. Results of their morphometric and color pattern analyses demonstrated that the populations formerly constituting *L. barbarus* (e. g., as listed by Musser and Carleton, 1993) actually consists of *L. barbarus*, which is endemic to NW Africa, and *L. zebra*, stretching across Subsaharan Africa from Senegal in the west to Kenya and Tanzania in the east. Chromosomal data for an Algerian sample documented by Filippucci et al. (1986). Fragments identified as *L. barbarus* are reported from middle Pleistocene sediments at Jebel Irhoud in Morocco (Amani and Geraads, 1993).

Lemniscomys bellieri Van der Straeten, 1975. Rev. Zool. Afr., 89:906.

COMMON NAME: Bellier's Lemniscomys.

TYPE LOCALITY: Côte d'Ivoire, Ayeremou (= Lamto).

DISTRIBUTION: Guinea and Doka woodland of C Guinea (Ziegler et al., 2002), Côte d'Ivoire and Ghana (Van der Straeten, 1975); possibly occurs in Sierra Leone (Grubb et al., 1998).

STATUS: IUCN – Data Deficient.

COMMENTS: Part of the murine fauna endemic to West Africa (see account of *Grammomys buntingi*). Karyological and morphometric comparisons between *L. bellieri* and *L. striatus* reported by Van der Straeten and Verheyen (1978*a*). A relative of *L. macculus*, according to Van der Straeten (1975), who also recorded chromosomal information (Van der Straeten, 1977*b*); Matthey (1954) and Tranier and Gautun (1979) provided additional chromosomal data. This close morphological alliance between *L. bellieri* and *L. macculus* is supported by analyses of mitochondrial gene sequences (cytochrome *b*, 12S and 16S rRNA gene fragments) by Ducroz et al. (2001), who also noted the weak molecular differentiation between them to be less than between their geographic samples of *L. striatus*, suggesting *bellieri* and *macculus* to represent a single species.

Lemniscomys griselda (Thomas, 1904). Ann. Mag. Nat. Hist., ser. 7, 13:414.

COMMON NAME: Griselda's Lemniscomys.

TYPE LOCALITY: Angola, Jinga country, Muene Coshi.

DISTRIBUTION: Known only from Angola as the species is currently defined; Crawford-Cabral (1998) reviewed and mapped Angolan records and thought the range may extend into Dem. Rep. Congo and Zambia. Because there is currently no unambiguous morphological, chromosomal, or molecular definition of *L. griselda*, its geographic distribution is impossible to define.

STATUS: IUCN – Lower Risk (lc).

COMMENTS: Morphometrically related to *L. rosalia*, *L. roseveari*, and *L. linulus* (Van der Straeten (1980*a*, *b*). Before Van der Straeten (1980*b*) described *roseveari* and removed *rosalia* from *L. griselda*, that species was considered to have a wide distribution extending from Angola through southern Africa and up E Africa to S Kenya. Results of Van der Straeten's (1980*b*) multivariate analyses of craniodental measurements produced three slightly overlapping clusters of samples. Specimens from Balovale, Zambia, were described as *L. roseveari*; most population samples representing several described subspecies formed another cluster, which was identified as *L. rosalia*; and the third group represented *L. griselda*. No qualitative traits were used to distinguish the three. Furthermore, except for *L. roseveari*, geographic ranges of the other two species were not outlined. We have examined large series of specimens in the *L. griselda* complex from Angola, Zambia, and southern Africa (in AMNH and USNM), are impressed with the range of intra-and intersample variation in craniodental dimensions, and had difficulty sorting the specimens into Van der Straeten's three groups. There is a very large series from Balovale, Zambia, in AMNH, which were not examined by Van der Straeten and we assume represent his *L. roseveari*. These are somewhat smaller and brighter than the slightly larger and darker Angolan *L. griselda*, but the chromatic and craniodental differences between the Balovale series and all other samples Van der Straeten would identify as *rosalia* are unimpressive (our examination was restricted to inspection and not morphometric analyses). Also uneasy about Van der Straeten's definition of *L. griselda*, Crawford-Cabral (1998:71) advised "that the taxonomic status (species *versus* subspecies) of the Southern African forms of this complex be re-evaluated." We agree. While provisionally accepting Van der Straeten's results here, we await a rigorous new revision of the *L. griselda* complex incorporating samples from throughout its geographic range and clear definition of geographic distributions for whatever species are recognized.

Lemniscomys hoogstraali Dieterlen, 1991. Bonn. Zool. Beitr., 42:11.

COMMON NAME: Hoogstraal's Lemniscomys.

TYPE LOCALITY: Sudan, Upper Nile Prov, Paloich, 12 mi (19 km) N Niayok (Carleton and Van der Straeten, 1997 provided additional information).

DISTRIBUTION: Recorded only from the type locality.

STATUS: IUCN – Data Deficient.

COMMENTS: Still known only by the holotype. A member of the *Lemniscomys barbarus* group. Dieterlen (1991) summarized knowledge of this species, and Carleton and Van der Straeten (1997) provided a current review. Its diagnostic dorsal pattern of stripes and large body size relative to *L. zebra*, which occurs all around it, prompted Carleton and Van der Straeten (1997:667) to retain *L. hoogstraali* as a distinct species, but they cautioned that "More and larger series must be assembled to rigorously evaluate the characters and status of Dieterlen's new form, so far known only by the holotype, its condition imperfect."

Lemniscomys linulus (Thomas, 1910). Ann. Mag. Nat. Hist., ser. 8, 6:429.

COMMON NAME: Senegal Lemniscomys.

TYPE LOCALITY: Senegal (= French Gambia), Gamon.

DISTRIBUTION: Sudan savanna and forest clearings in Senegal and Côte d'Ivoire (Van der Straeten, 1980*a*); may also occur in Gambia and Ghana (Grubb et al., 1998).

STATUS: IUCN – Data Deficient.

COMMENTS: Once treated as a subspecies of *L. griselda* (G. M. Allen, 1939), but now considered a separate West African species morphologically related to *L. griselda* (Van der Straeten, 1980*a*). Member of murine assemblage endemic to West Africa (see account of *Grammomys buntingi*).

Lemniscomys macculus (Thomas and Wroughton, 1910). Trans. Zool. Soc. Lond., 19:515.

COMMON NAME: Buffoon Lemniscomys.

TYPE LOCALITY: Uganda, SE Ruwenzori, Mokia.

DISTRIBUTION: Recorded from Savannahs of NE Dem. Rep. Congo, S Sudan, Ethiopia, Uganda, and Kenya; limits unknown.

STATUS: IUCN – Lower Risk (lc).

SYNONYMS: *akka* (Thomas, 1915).

COMMENTS: Reviewed by Van der Straeten and Verheyen (1979*a*). The species is often confused with *L. striatus*, but occurs sympatric with it in East Africa (Hollister, 1919) and Ethiopia (Yalden et al., 1996). Bulatova et al. (2002) recorded 2n = 56, FN = 60 for an Ethiopian sample. Considered very close to *L. bellieri* in morphological and molecular characters and possibly conspecific with it (Ducroz et al., 2001; see account of *L. bellieri*).

Lemniscomys mittendorfi Eisentraut, 1968. Bonn. Zool. Beitr., 19:7.
COMMON NAME: Mittendorf's Lemniscomys.
TYPE LOCALITY: West Cameroon, Banso Highlands, Lake Oku on Mt Oku, 2100 m.
DISTRIBUTION: Known only from the type locality.
STATUS: IUCN – Endangered.
COMMENTS: Described as a subspecies of *L. striatus*, included in that species by Misonne (1974), but treated as separate by Van der Straeten and Verheyen (1980), who also suggested it has morphometric affinities with *L. macculus* and *L. bellieri*. Fülling (1992) contrasted samples of *L. mittendorfi* and *L. striatus* and documented differences in morphology of palatal ridges and karyotype (2n = 56, FN = 66-72 for *L. mittendorfi*; 2n = 44, FN = 72 for *L. striatus*). Eisentraut (1973) discussed the species and *L. striatus* in the context of a monograph of the West African fauna, in which he provided a color plate contrasting upperparts of *L. striatus* and *L. mittendorfi*. In addition to *L. mittendorfi*, other Mt Oku endemics are the muroid rodents *Hylomyscus grandis*, *Lamottemys okuensis*, and *Lophuromys dieterleni* and the golden mole *Chrysochloris basaci*.

Lemniscomys rosalia (Thomas, 1904). Ann. Mag. Nat. Hist., ser. 7, 13:414.
COMMON NAME: Single-Striped Lemniscomys.
TYPE LOCALITY: Tanzania, Nguru Mtns, Monda.
DISTRIBUTION: N Namibia, South Africa (KwaZulu-Natal, Mpumalanga, Limpopo, and North West provinces; de Graaff, 1997*d*; P. J. Taylor, 1998), E Swaziland, Zimbabwe, C and N Botswana, Mozambique, Zambia, Malawi, Tanzania, and S Kenya; range inadequately resolved.
STATUS: IUCN – Lower Risk (lc).
SYNONYMS: *calidior* (Thomas and Wroughton, 1908); *dorsalis* (Smith, 1845) [not Fischer, 1814]; *fitzsimonsi* Roberts, 1915; *maculosus* (Osgood, 1910); *mearnsi* Heller, 1914; *phaeotis* (Thomas, 1910); *sabiensis* Roberts, 1946; *sabulatus* Thomas, 1927; *spinalis* Thomas, 1916 [replacement for *dorsalis* A. Smith, 1845, preoccupied by *dorsalis* G. Fischer, 1814]; *zuluensis* Roberts, 1931.
COMMENTS: Once included in *L. griselda* (G. M. Allen, 1939), but now considered a distinct species most closely related to *L. griselda* and *L. roseveari* (Van der Straeten, 1980*b*); see account of *L. griselda*. Karyotype (2n = 54, FN = 64) for samples from NE Tanzania reported by Fadda et al. (2001*b*). Populations in Southern African Subregion reviewed by de Graaff (1997*d*).

Lemniscomys roseveari Van der Straeten, 1980. Ann. Cape Prov. Mus. Nat. Hist., 13(5):55.
COMMON NAME: Rosevear's Lemniscomys.
TYPE LOCALITY: Zambia, Zambezi (= Balovale), 1015 m (see Van der Straeten, 1980*b*, for additional data).
DISTRIBUTION: Known only from the type locality and Solwezi in Zambia.
STATUS: IUCN – Lower Risk (nt).
COMMENTS: Morphometrically most closely related to *L. rosalia* and *L. griselda* (Van der Straeten, 1980*b*); see account of *L. griselda*.

Lemniscomys striatus (Linnaeus, 1758). Syst. Nat., 10th ed., 1:62.
COMMON NAME: Typical Lemniscomys.
TYPE LOCALITY: "India" (= Sierra Leone; see G. M. Allen, 1939:394).
DISTRIBUTION: From Guinea (Ziegler et al., 2002), Sierra Leone and Ghana (Grubb et al., 1998), and Burkina Faso west to Ethiopia (Yalden et al, 1996), and south into N Angola (Crawford-Cabral, 1998) and through Kenya (Hollister, 1919), Uganda (Delany, 1975), Rwanda, E Dem. Rep. Congo, and Tanzania (Swynnerton and Hayman, 1951; Grimshaw et al., 1995, reviewed Mt Kilimanjaro records) into NE Zambia (Ansell, 1978) and N Malawi (Ansell and Dowsett, 1988).

STATUS: IUCN – Lower Risk (lc).

SYNONYMS: *ardens* (Thomas, 1910); *dieterleni* Van der Straeten, 1976; *fasciatus* (Wroughton, 1906); *luluae* Matschie, 1926; *lynesi* Thomas and Hinton, 1923; *massaicus* (Pagenstecher, 1885); *micropus* (Heller, 1911); *orientalis* (Desmarest, 1819) [not Cretzschmar, 1826]; *orientalis* Hatt, 1935; *pulchella* (Gray, 1864); *pulcher* (Wroughton, 1906); *spermophilus* Heller, 1912; *venustus* (Thomas, 1911); *versustus* (Thomas, 1911); *wroughtoni* (Thomas, 1910).

COMMENTS: Reviewed by Van der Straeten and Verheyen (1980). Carleton and Van der Straeten (1997) discussed and identified the holotype of Linnaeus's *Mus striatus*. Chromosomal information reported by Matthey (1959), Van der Straeten (1977b), and Van der Straeten and Verheyen (1979a). Van der Straeten (1976a) described *dieterleni* as a distinctive subspecies of *L. striatus* occurring in the Lake Kivu region of E Dem. Rep. Congo. He (1981) also documented the identity of *venustus* as representing a population of *L. striatus*. Morphology of palatal ridges and karyotypes contrasted with *L. mittendorfi* by Fülling (1992); see account of *L. mittendorfi*. Ecological and other data for populations from S Ghana reported by Ryan and Attuquayefio (2000) and Decher and Bahian (1999). Hutterer et al. (1992a) noted that *L. striatus* is common in Nigeria and Cameroon. Kerbis Peterhans et al. (1998) reviewed altitudinal distribution on Ugandan slopes of Ruwenzori Mtns. Isolated teeth from Pleistocene deposits in East Africa have been identified as *L.* aff. *striatus* (Wesselman, 1984).

Lemniscomys zebra (Heuglin, 1864). Beit. Zool. Cent.–Afrika's Leopoldina, 31:10.

COMMON NAME: Heuglin's Lemniscomys.

TYPE LOCALITY: Sudan, Bahr el Ghazal (Carleton and Van der Straeten, 1997, provided additional information).

DISTRIBUTION: "Grassy woodlands and savannas south of the Sahara Desert, from Senegal in the west to southern Sudan in the east, southwards through northeastern-most Zaire [Dem. Rep. Congo], northern Uganda and western Kenya, to northcentral Tanzania;" sea level to 1220 m (Carleton and Van der Straeten, 1997:669). Not yet documented from Ethiopia, although the species probably occurs in the W region of that country (Carleton and Van der Straeten, 1997).

STATUS: IUCN – Least Concern.

SYNONYMS: *albolineatus* (Osgood, 1910); *convictus* (Osgood, 1910); *dunni* (Thomas, 1903); *manteufeli* Matschie, 1911; *nigeriae* (Thomas, 1912); *nubalis* Thomas and Hinton, 1923; *olga* Thomas and Hinton, 1921; *orientalis* Hatt, 1935 [not Desmarest, 1819]; *oweni* Thomas, 1911; *spekei* (de Winton, 1897).

COMMENTS: A member, along with *L. barbarus* and *L. hoogstraali*, of the *L. barbarus* group. Results of morphometric analyses by Carleton and Van der Straeten (1997) distinguished this species from *L. barbarus*, which is endemic to the Barbarian province (Maghreb) of Northwest Africa. See their comprehensive revision for details of taxonomy, geographic distribution, geographic character variation, and selection of a lectotype (p. 669). Karyotypes documented (usually as *L. barbarus*) from several geographic samples (Fadda et al., 2001b, NE Tanzania; Gautun et al., 1986, Burkina Faso; Dobigny et al., 2002b, Niger; Matthey, 1954, Côte d'Ivoire; Van der Straeten, in Carleton and Van der Straeten, 1997, Cameroon), all of which have the same diploid number as *L. barbarus* (2n = 54). Grubb et al. (1998) reviewed populations from Ghana and Gambia. Morphometric and ecological data for sample from S Ghana recorded by Ryan and Attuquayefio (2000, as *barbarus*).

Lenomys Thomas, 1898. Trans. Zool. Soc. Lond., 14:409.

TYPE SPECIES: *Mus meyeri* Jentink, 1879.

COMMENTS: *Pithecheir* Division. A Sulawesi endemic reviewed by Musser (1970d, 1981c, 1984). Tate (1936), followed by Simpson (1945), listed *Lenomys* as a member of the Phloeomyinae along with *Coryphomys*, *Chiruromys*, *Pogonomys*, *Mallomys*, *Phloeomys*, *Chiropodomys*, and *Crateromys*, but no evidence supports this arrangement (Ellerman, 1949a; our studies). External, cranial, and spermatozoal traits tied *Lenomys* phylogenetically close to *Eropeplus*, another Sulawesian endemic (Breed and Musser, 1991; Musser, 1981c). Both genera may be most closely allied to the Sundaic *Lenothrix* and *Pithecheir* (Musser's unpublished research), an alliance already suggested by Misonne (1969) based on molar occlusal patterns. *Lenomys* and *Eropeplus* were placed in a *Pithecheir* Section of a larger *Micromys* Group by Pavlinov et al. (1995*a*). Reviewed by Musser (1981*c*).

Lenomys meyeri (Jentink, 1879). Notes Leyden Mus., 1:12.
COMMON NAME: Meyer's Lenomys.
TYPE LOCALITY: Indonesia, Sulawesi, NE peninsula, Menado.
DISTRIBUTION: Recorded only from N, C, and SW Sulawesi.
STATUS: IUCN – Lower Risk (lc).
SYNONYMS: *lampo* Tate and Archbold, 1935; *longicaudus* Miller and Hollister, 1921.
COMMENTS: Represented by a small extant series from the N peninsula and C core, and by both modern specimens and subfossils from the SW peninsula of Sulawesi (Musser, 1970*d*; 1984). A second undescribed species is known from a subfossil fragment collected in the SW arm of Sulawesi (Musser and Holden, 1991; Musser, ms). Flannery (1995*b*) identified a large-bodied rat, with shaggy pelage and long bicolored tail, from Pulau Sangir (off coast of NE peninsular tip of Sulawesi) as *Lenomys meyeri*, but the specimen represents an undescribed species related to mainland *Rattus xanthurus* (Musser, ms); *Lenomys* has yet to be collected on any island off the coast of Sulawesi.

Lenothrix Miller, 1903. Proc. U.S. Natl. Mus., 26(1317):466.
TYPE SPECIES: *Lenothrix canus* Miller, 1903.
COMMENTS: *Pithecheir* Division. Chromosomal and biochemical data suggested a close phylogenetic link between *Lenothrix* and *Niviventer* (Chan et al., 1979), and a high divergence from species of *Maxomys*, which were once placed in *Lenothrix* (Gadi and Sharma, 1983). In its cranial and dental morphology, *Lenothrix* is unlike any species in those genera; derived molar occlusal patterns are much like those in *Pithecheir* and *Lenomys* (Chaimanee, 1998; Misonne, 1969; Musser and Newcomb, 1983), but spermatozoal conformation is highly divergent from any Sundaic murine. In sum, *Lenothrix* is a Sundaic endemic characterized by many primitive external, cranial, dental, and chromosomal features and a few derived dental and spermatozoal traits; despite several claims, its phylogenetic relationships still require illumination (see discussions in Chaimanee, 1998; Breed and Yong, 1986; Musser, 1981*a*, *b*, *c*; Musser and Newcomb, 1983). Its inclusion here in a *Pithecheir* Division reflects an hypothesis that has to be tested by analyses of molecular and a range of morphological data. *Lenothrix* was included in a *Pithecheir* Section of a larger *Micromys* Group by Pavlinov et al. (1995*a*). *Lenothrix* may be a member of one of the earliest groups to have diverged from the core murine lineage as represented by a *Progonomys*-like ancestor (Chaimanee, 1998). Sundaic *Pithecheir* and *Pithecheirops*, along with Sulawesian *Eropeplus*, *Lenomys*, and *Margaretamys*, are likely members of the same cluster.

Lenothrix canus Miller, 1903. Proc. U.S. Natl. Mus., 26(1317):466.
COMMON NAME: Sundaic Lenothrix.
TYPE LOCALITY: Indonesia, Pulau Tuangku (west of Sumatra).
DISTRIBUTION: Malay Peninsula, Penang Isl, Tuangku Isl, and Borneo (Sarawak, Sabah, and SW Kalimantan); Musser (1981*a*) and references cited below.
STATUS: IUCN – Lower Risk (lc).
SYNONYMS: *malaisia* (Kloss, 1931).
COMMENTS: Historical allocations of *canus* to species-clusters now defined as *Rattus*, *Diplothrix*, or *Maxomys*, as well as comparisons between *L. canus* and other Sundaic endemics documented by Musser (1981*a*, *b*) and Musser and Newcomb (1983). It is an arboreal species occupying lowland and foothill secondary and primary forests on the Malay Peninsula (Muul and Lim, 1971), and has been taken at 550 m on the slopes of Mt Kinabalu in Sabah (Md Nor, 2001). The Kalimantan record, from Gunung Palung National Park, was noted by Blundell (1996:258) and verified by A. J. Gorog (pers. comm.), who examined the specimen.

Leopoldamys Ellerman, 1947. Proc. Zool. Soc. Lond., 1947-1948(117):267 [1947].
TYPE SPECIES: *Mus sabanus* Thomas, 1887.

COMMENTS: *Dacnomys* Division. Definition and contrasts with *Rattus* and *Niviventer* provided by Musser (1981*b*), who also reviewed morphological, chromosomal, and distributional characteristics. Additional chromosomal data for Vietnam samples provided by Bulatova et al. (1992). *Leopoldamys* is dentally similar to *Berylmys*, *Maxomys*, and *Niviventer*. Sperm morphology unites *Leopoldamys* with *Berylmys*, *Sundamys*, and *Rattus* (Breed and Yong, 1986), but alliance is based on shared spermatozoal form that is likely primitive. Chromosomal traits suggested *Leopoldamys* is more closely related to *Bandicota*, *Berylmys*, *Nesokia*, *Rattus*, and *Sundamys*, than to *Lenothrix*, *Maxomys*, or *Niviventer* (Gadi and Sharma, 1983). Best estimates of relationships are derived from molecular and morphological sources. Allozymic and morphological data clearly separates *Leopoldamys* from *Rattus* (Chan et al., 1979; Musser, 1981*b*; Musser and Newcomb, 1983). Cladistic analysis of DNA sequences of LINE-1 elements placed *Leopoldamys* and *Niviventer* as sister-genera in a clade separate from that containing *Rattus*, *Berylmys*, *Bandicota*, and *Sundamys* (members of our *Rattus* Division), and another clade containing only *Maxomys* (Verneau et al., 1997, 1998), results also reflected in studies of albumin immunology (Watts and Baverstock, 1994*b*), DNA/DNA hybridization assays (Chevret, 1994 [cited in Verneau et al., 1997]; Ruedas and Kirsch, 1997), and generally in cranial and dental traits (Musser and Newcomb, 1983).

   Leopoldamys is represented by fossils identified as *L. edwardsioides* from early Pleistocene cave sediments in the Sichuan-Guizhou where it was replaced at later horizons in the early Pleistocene by *L. edwardsi* (Zheng, 1993); *L.* sp. is represented by molars recovered from cave strata in Guangxi Province of S China (Chen et al., 2002). Isolated molars recovered from middle Pleistocene cave sediments in Thailand have been identified as *L. sabanus*, and teeth from late Pliocene and early Pleistocene horizons were described as *L. minutus*, a possible ancestor of extant *L. edwardsi* (Chaimanee, 1998).

Leopoldamys ciliatus (Bonhote, 1900). Proc. Zool. Soc. Lond., 1900:879.
   COMMON NAME: Sundaic Mountain Leopoldamys.
   TYPE LOCALITY: Malaysia, Perak (Malay Peninsula), Gunung Inas.
   DISTRIBUTION: Montane forest formations on Malay Peninsula (usually above 1000 m; Medway, 1969; Yong, 1970) and mountain backbone of Sumatra (usually above 1000 m; Chasen, 1940; Robinson and Kloss, 1916; Miller, 1942); range also derived from study of specimens in ANSP, AMNH, BMNH, RMBR, RMNH, and USNM.
   SYNONYMS: *setiger* (Robinson and Kloss, 1916).
   COMMENTS: Originally described as a species of *Mus* and associated with *M. edwardsi* by Bonhote (1900; see history in Yong, 1970), *ciliatus* was subsequently arranged as a subspecies of *edwardsi*, whether associated with *Rattus* or *Leopoldamys* (Chasen, 1940; Corbet and Hill, 1992; Ellerman, 1941; Musser, 1981*b*; Musser and Carleton, 1993; Osgood, 1932). Morphological, ecological, behavioral, reproductive, chromosomal, and genic contrasts between Malayan *L. ciliatus* (as *edwardsi*) and *L. sabanus* documented by Yong (1970). The two species occur altitudinally parapatric on the Peninsula, but have been taken at the same place in the Sumatran highlands (Miller, 1942; he recorded three specimens, all identified as *setiger* [= *L. ciliatus*], but only ANSP 20354 is that species, the other two being *L. sabanus* [ANSP 20352, 20353]; Musser's identifications). Uniting Sumatran populations (*setiger*) with those on Malay Peninsula within the same species requires testing by analyses of multi-trait morphological data and molecular sequences.

Leopoldamys edwardsi (Thomas, 1882). Proc. Zool. Soc. Lond., 1882:587.
   COMMON NAME: Edward's Leopoldamys.
   TYPE LOCALITY: China, mountains of W Fujian (probably Kuatun).
   DISTRIBUTION: NW India (W Bengal, Arunachal Pradesh, Meghalaya, and Nagaland; Agrawal, 2000; Ellerman, 1961); N Burma (Anthony, 1941; Ellerman, 1961), S and C China (to S Anhui, including Hainan Isl; G. M. Allen [1940] Liu et al. [1985], Wang [2003], Wu et al. [1996]), N Laos (Ellerman, 1961; Osgood, 1932), N and C Vietnam (Dang et al., 1994; Lunde et al., 2003*b*; Osgood, 1932), and isolated montane population in N Thailand (Phu Kadeung Plateau, Loei Province; J. T. Marshall, Jr., 1977*a*). Range (excluding *ciliatus* and *milleti*) mostly extracted from Musser (1981*b*),

Corbet and Hill (1992), and specimens examined in AMNH, BMNH, FMNH, IEBR, MVZ, RMBR, and USNM.

STATUS: IUCN – Lower Risk (lc).

SYNONYMS: *garonum* (Thomas, 1921); *gigas* (Satunin, 1903); *hainanensis* (Xu and Yu, 1985); *listeri* (Thomas, 1916); *melli* (Matschie, 1922).

COMMENTS: Requires taxonomic revision. Association of the synonyms with *L. edwardsi* essentially derives from Osgood's (1932:311) suggestion, which has yet to be tested by careful systematic study. Osgood included the Malayan and Sumatran *ciliatus* (including *setiger*), along with the S Vietnam *milleti*, as forms of *L. edwardsi*, but we list them as separate species. In addition to the synonyms assembled by Osgood (1932) and Musser (1981b), Xu and Yu (1985) described *hainanensis* from Hainan Isl. Those specimens have significantly smaller external, cranial, and dental measurements than characterize other samples of *L. edwardsi* and if adults may represent a separate species. Phallic morphology described by Yang and Fang (1988) in context of assessing phylogenetic relationships among Chinese murines. Ecological relationships with species of *Bandicota*, *Mus*, *Maxomys*, *Niviventer*, and *Rattus* in S Yunnan reported by Wu et al. (1996).

A large series (in AMNH and IEBR) collected between 200 and 1200 m in the N Truong Son Range along the N Vietnam-Laos border in Hà Tinh Province of Vietnam resembles *L. edwardsi* in pelage coloration but approaches *L. sabanus* in relative tail length and other features. Comparison of their mtDNA cytochrome *b* sequences with sequences from *L. sabanus* and *L. edwardsi* taken farther north in Vietnam isolates the Hà Tinh sample as a separate species for which no name is currently available (Gorog et al. 2004; J. L. Patton, in litt., 1999). How this population phylogenetically is related to samples now identified as *L. edwardsi* from N Laos and Thailand as well as NE India and N Burma requires systematic resolution. Fossils identified as *L. edwardsi* come from early Pleistocene to Recent cave sediments in the Sichuan-Guozhou region of S China (Zheng, 1993).

Leopoldamys milleti (Robinson and Kloss, 1922). Ann. Mag. Nat. Hist., ser. 9, 9:94.

COMMON NAME: Millet's Leopoldamys.

TYPE LOCALITY: S Vietnam, Lâm Dğng Province, Langbian Mtns (= Lam Vien Plateau), Dà Lat, 5000 ft (1524 m).

DISTRIBUTION: Documented only from the Langbian highlands in the Dà Lat region; limits unresolved.

COMMENTS: Originally described as "a remarkably distinct race" of *Rattus edwardsi* (Robinson and Kloss, 1922:94) and retained as such whether *edwardsi* was associated with *Rattus* or *Leopoldamys* (Corbet and Hill, 1992; Ellerman, 1941; Musser, 1981b; Musser and Carleton, 1993; Osgood, 1932). The few representatives of *milleti*, however, contrast with *L. edwardsi* by their very dark dorsal pelage and larger bullae. Robinson and Kloss (1922:94) noted that color and pelage texture of the holotype "suggested relationship with the *bowersi* [= *Berylmys*] group of rats," and a recently trapped specimen was initially identified by the collector as a possible new species of *Berylmys* (examined by Musser). Recognizing *milleti* as a species highlights its distinctive morphology and zoogeographic distribution. The nature of its relationship to the northern *Leopoldamys edwards* and Sundaic *L. ciliatus* requires resolution by revision of the *L. edwardsi* complex employing morphological and molecular data sets. The murine *Rattus osgoodi* is also recorded only from the Langbian highlands (see that account), and *Maxomys moi* occurs there, but its range also extends to nearby S Laos and the Vietnamese highlands north of the Langbian Plateau (to Quàng Tri Province).

Leopoldamys neilli (J. T. Marshall, Jr., 1976). Family Muridae: rats and mice, p. 485. Privately printed by Government Printing Office, Bangkok.

COMMON NAME: Neill's Leopoldamys.

TYPE LOCALITY: Thailand, Saraburi Prov, Kaengkhoi Dist., "outside the entrance to the bat cave, half-way up the face of a wooded limestone cliff, 200 meters altitude."

DISTRIBUTION: N and SW Thailand north of the peninsular region (J. T. Marshall, Jr.,1977a; study of specimens in AMNH, BMNH, and USNM); limits unresolved (may also occur in adjacent Burma and Vietnam).

STATUS: IUCN – Endangered.

COMMENTS: Documented by few specimens: three samples are from wooded limestone cliffs in Saraburi Province and adjacent Kanchanaburi Province in the southwest (J. T. Marshall, Jr., 1977a; specimens in AMNH and USNM), three come from lowland bamboo forest in a different part of Kanchanaburi Province (Wiles, 1981), and a fourth consists of owl pellet fragments collected in Loei Province in the north (BMNH 94.374). Phylogenetic relationship of *L. neilli* to other extant members of *Leopoldamys* requires resolution by systematic revision employing morphological and molecular data. See Musser (1981b:237) for discussion of original publication.

Leopoldamys sabanus (Thomas, 1887). Ann. Mag. Nat. Hist., ser. 5, 20:269.

COMMON NAME: Indomalayan Leopoldamys.

TYPE LOCALITY: Malaysia, Sabah (N Borneo), Gunung Kinabalu.

DISTRIBUTION: SE Bangladesh (Chittagong Hill Tracts, AMNH 251694), Thailand (J. T. Marshall, Jr., 1977a; Robinson et al., 1995), Vietnam from the north in Tuyên Quang Province to the south in Ninh Thuân Province (Dang et al., 1994; Osgood, 1932; Van Peenen et al., 1969; and four islands off the coast; Kuznetsov, 2000), Laos (Aplin et al., 2003c; Osgood, 1932; Smith et al., In Press; Van Peenen et al., 1969), S and SW Cambodia (Elephant Mtns, specimens in FMNH; Cardamom Mtns, A. Smith, in litt.), S Burma and most islands in Mergui Arch., Malay Peninsula, Sumatra, Java, Borneo, and smaller islands on the Sunda Shelf except Bali; northern limits unresolved. Range mostly extracted from Musser (1981c), Corbet and Hill (1992), and study of museum specimens in AMNH, BMNH, FMNH, IEBR, MVZ, MZB, RMBR, RMNH, and USNM.

STATUS: IUCN – Lower Risk (lc).

SYNONYMS: *balae* (Miller, 1903); *bunguranensis* (Chasen, 1935); *clarae* (Miller, 1913) [not Rümmler, 1935]; *dictatorius* (Chasen, 1940); *fremens* (Miller, 1902); *heptneri* (Dao, 1961), *herberti* (Kloss, 1916); *insularum* (Miller, 1913); *lancavensis* (Miller, 1900); *lucas* (Miller, 1903); *luta* (Miller, 1913); *macrourus* (Jentink, 1879); *mansalaris* (Lyon, 1916); *masae* (Miller, 1903); *matthaeus* (Miller, 1903); *mayapahit* (Robinson and Kloss, 1919); *nasutus* (Lyon, 1911); *revertens* (Robinson and Kloss, 1922); *salanga* (Chasen, 1940); *stentor* (Miller, 1913); *strepitans* (Miller, 1900); *stridens* (Miller, 1903); *stridulus* (Miller, 1903); *tapanulius* (Lyon, 1916); *tersus* (Thomas and Wroughton, 1909); *tuancus* (Lyon, 1916); *ululans* (Robinson and Kloss, 1916); *vociferans* (Miller, 1900).

COMMENTS: Appreciable morphological variation exists between samples from north and south of Isthmus of Kra, and among insular samples from the Sunda Shelf; systematic revision using a suite of morphological traits and molecular data is required to assess whether variation is characteristic of one or several species (Musser, 1981b). Recent phylogenetic analyses of mtDNA cytochrome *b* sequences by Gorog et al. (2004) identified separate lineages, each from Borneo, Sumatra, and the Malay Peninsula forming an unresolved trichotomy, and a Vietnam cluster basal to that trichotomy consisting of a paraphyletic pattern of Vietnamese *L. sabanus* and *L. edwardsi*. Gorog et al. (2004) suggested that *L. sabanus* evolved on the Indochinese mainland and the vicariance patterns revealed by molecular analyses ". . . likely have their roots in the Pliocene fragmentation of the Sunda block. . ." rather than widespread dispersal across the late Pleistocene Sunda Shelf and subsequent isolation with increasing sea levels. An early preglacial presence of *Leopoldamys* in the Indochinese region is supported by late Pliocene to middle Pleistocene fossils (Chaimanee, 1998). Usually found in lowland rainforests, infrequently in montane habitats, but does reach 3100 m on the slopes of Mt. Kinabalu in Sabah (Md Nor, 2001) and 1300 m in some highlands on Malay Peninsula (Yong, 1970). Aspects of ultrastructure of pineal gland morphology described by Pévet and Yadav (1980) in comparative context. The name *macrourus* has priority over *sabanus*, but is based on a specimen of uncertain origin (Musser, 1981b).

Leopoldamys siporanus (Thomas, 1895). Ann. Mus. Civ. Stor. Nat. Genova, 34:11.

COMMON NAME: Mentawai Archipelago Leopoldamys.

TYPE LOCALITY: Indonesia, Kepulauan Mentawai, Pulau Sipora.

DISTRIBUTION: Endemic to Mentawai Arch.; islands of Siberut, Sipora, North Pagai, and South Pagai.

STATUS: IUCN – Vulnerable.

SYNONYMS: *soccatus* (Miller, 1903).

COMMENTS: Originally described as a species, subsequently recognized as such by Ellerman (1941), but later treated as a subspecies of either *Rattus sabanus* (Chasen, 1940) or *R. edwardsi* (Ellerman, 1961). In its morphology, *L. siporanus* is distinct from either of those species. Where it fits within the web of phylogenetic relationships among species of *Leopoldamys* has yet to be determined. *Leopoldamys siporanus* might be closely allied to one of the extant species, but because evolutionary history of the genus is documented back to late Pliocene in Indochina (Chaimanee, 1998), *L. siporanus* might represent an early lineage isolated in the Mentawai Arch. and within a phylogenetic pattern of relationships. Systematic revision of *Leopoldamys* using a suite of morphological character sets along with molecular data would be revealing. *Leopoldamys siporanus* joins *Maxomys pagensis*, *Chiropodomys karlkoopmani*, *Rattus lugens*, *Iomys sipora*, *Hylopetes sipora*, *Petinomys lugens*, *Callosciurus melanogaster*, *Sundasciurus fraterculus*, and *Lariscus obscurus* as part of the rodent fauna endemic to the Mentawai Arch.

Leporillus Thomas, 1906. Ann. Mag. Nat. Hist., ser. 7, 17:83.

TYPE SPECIES: *Hapalotis apicalis* Gould, 1853.

COMMENTS: *Pseudomys* Division. Member of the Australian Old Endemics (Musser, 1981c:167), which includes the Conilurini where Baverstock (1984) listed *Leporillus*. Conilurines, however, are currently treated as part of a larger clade, the Hydromyini, which also includes members we place in the *Hydromys*, *Xeromys*, *Pseudomys*, and *Uromys* Divisions (the "Australasian clade" of Watts and Baverstock, 1994a, 1995b, 1996). Mahoney and Richardson (1988) cataloged taxonomic, distributional, and biological references. Species of *Leporillus* are the ecological and superficially morphological counterparts of North American woodrats (*Neotoma*). Australian researchers are now surveying stick-rat middens and gathering megafossils and pollen to use in assessing environmental history and human impact in Australian deserts, just as North American woodrat middens have been studied for the same reason (Pearson, 1999; Pearson and Dodson, 1993). Pearson et al. (2001) extracted hair, teeth, and bone from middens of *L. apicalis* and *L. conditor* to determine former distribution and community composition of mammal and nonmammal species in arid Australia, some of which are no longer present in the areas where middens are currently found. Earliest record of *Leporillus* comes from late Pliocene sediments (see review in Aplin, 2005).

Leporillus apicalis (Gould, 1853). Proc. Zool. Soc. Lond., 1853:126.

COMMON NAME: Lesser Stick-nest Rat.

TYPE LOCALITY: Australia, South Australia; see Mahoney and Richardson (1988:159).

DISTRIBUTION: Australia: once found in S Northern Territory, C and S Western Australia, South Australia, and western parts of New South Wales and Victoria, but now presumed to be extinct; extent of former range indicated by specimens caught in late 1800s and early 1900s and distribution of empty stick nests (Mahoney and Richardson, 1988:159; Robinson, 1995a:558; Robinson et al., 2000; Watts and Aslin, 1981:152; L. M. Williams, 1995a).

STATUS: IUCN – Extinct (but see comments below).

COMMENTS: The last specimen of *L. apicalis* was collected near Mt Crombie in 1933, but there are recent reports from Western Australia of stick nests in caves with fresh green vegetation woven into their structure, a tantalizing sign that the species may not be extinct (Robinson, 1995a). Former distribution, along with detailed ecological information, summarized by Copley (1999).

Leporillus conditor (Sturt, 1848). Narr. Exped. C. Aust., 1:120.

COMMON NAME: Greater Stick-nest Rat.

TYPE LOCALITY: Australia, New South Wales, Polia area, about 45 miles (72 km) from Laidley Ponds; see Mahoney and Richardson (1988:160).

DISTRIBUTION: Australia; once ranged on mainland from lower Darling River to Nullarbor Plain in New South Wales, probably NW Victoria, South Australia, and SE corner of Western Australia, and presumed to be extinct; living population on Franklin Isl in Nuyt's Arch. of W South Australia (Mahoney and Richardson, 1988:160; Robinson et al., 2000; Watts and Aslin, 1981:147; L. M. Williams, 1995*b*); rats from a captive breeding program have been released onto Reevesby, Salutation, and St. Peter Isls in S Australia (Robinson, 1995*b*). Former distribution and detailed ecological information summarized by Copley (1999).

STATUS: CITES – Appendix I; U.S. ESA – Endangered; IUCN – Endangered.

SYNONYMS: *jonesi* Thomas, 1921.

COMMENTS: Phylogenetic significance of spermatozoal morphology reported by Breed and Sarafis (1978) and Breed (1997); chromosomal morphology described by Baverstock et al. (1977*c*). Electrophoretic data indicated *L. conditor* is phylogenetically closely allied to *Pseudomys* (Baverstock et al., 1981), but information from analyses of phallic and dental morphology placed *L. conditor* in same monophyletic groups as *Conilurus* and *Mesembriomys*, to the exclusion of *Pseudomys* (Lidicker and Brylski, 1987; Misonne, 1969). Ellis (1995*a*) reviewed distribution of *L. conditor* in the Mootwingee National Park of W New South Wales based on historical records and subfossils. Reviewed by Watts and Aslin (1981) and Robinson (1995*b*).

Leptomys Thomas, 1897. Ann. Mus. Civ. Stor. Nat. Genova, 18:610.

TYPE SPECIES: *Leptomys elegans* Thomas, 1897.

COMMENTS: *Xeromys* Division. Member of the New Guinea Old Endemics (Musser, 1981*c*). Microcomplement fixation of albumin studies (Watts and Baverstock, 1994*a*, 1995*b*, 1996) and spermatozoal morphology (Breed, 1997; Breed and Aplin, 1994) support membership of *Leptomys* in a larger clade containing members of our *Hydromys*, *Xeromys*, *Pseudomys*, and *Uromys* Divisions (the "Australasian clade" of Watts and Baverstock, 1995*b*, 1996), excluding the New Guinea endemics in the *Pogonomys* Division (Anisomyini of Watts and Baverstock, 1994*a*) and *Lorenztimys* Division. A derived cephalic arterial pattern, along with other morphological features, is shared by *Leptomys*, *Pseudohydromys* (which includes *Mayermys* and *Neohydromys*), and *Lorentzimys* (Musser and Heaney, 1992). Certain phallic traits (Lidicker and Brylski, 1987), also united *Leptomys* and *Lorentzimys*, but that association conflicts with analyses of albumin immunology (Watts and Baverstock, 1994) and spermatozoal structure (Breed and Aplin, 1994). Chromosomal data provided by Donnellan (1987), but whether the sampled species was *L. elegans* or *L. ernstmayri* is unclear.

Most lists and faunal studies published since 1951 recognized only one species in *Leptomys* (Flannery, 1990*b*; Laurie and Hill, 1954; Menzies and Dennis, 1979; Tate, 1951), but Rümmler's (1938) revision in which he identified two species (*L. elegans* and *L. ernstmayri*), Tate and Archbold's (1938) description of a third (*L. signatus*), and two more being described elsewhere (Helgen, Musser and Lunde, in ms.) accurately reflect the known diversity (Musser's study of specimens in AMNH and BMNH). Except for a population in the Arfak Mtns in western Prov. of Papua (= Irian Jaya), all known records of *Leptomys* come from eastern or Papua New Guinea.

Leptomys elegans Thomas, 1897. Ann. Mus. Civ. Stor. Nat. Genova, 18:610.

COMMON NAME: Elegant Leptomys.

TYPE LOCALITY: Papua New Guinea (= "British New Guinea"), Central Province, Astrolabe Range behind Port Moresby. The specimen was collected by Dr. Lamberto Loria between 1890 and 1893 and Tate (1951:223) claimed that "Loria collected in the Astrolabe Range behind Port Moresby," but Laurie and Hill, 1954:132 noted that no exact locality was published. Thomas (1897*a*:4), however wrote that Loria's localities "are mostly between the Owen Stanley Range and the sea, in or near the watershed of the Kemp Welch river," an area that would include the Astrolabe Range and adjacent Sogeri Plateau.

DISTRIBUTION: Papua New Guinea; known only by specimens from area of Mount Sisa below 1500 m (Dwyer, 1984; K. Helgen, in litt., 2004) and the Kikori River Basin at 450 m (Leary and Seri, 1997) in Southern Highlands Province, the Wharton Range at 1253 m and Astrolabe Range at 410-520 m (near and northwest of Port Moresby), and Mt Dayman at 700 m in the Manau Range; limits unknown (Flannery, 1995a). Flannery (1995a) also recorded three specimens collected at 800 m on Mount Victory in Oro Province in the E peninsula.

STATUS: IUCN – Critically Endangered.

COMMENTS: Poorly represented in museum collections. Leary and Seri (1997) collected three from the Kikori River Basin in S Papua, one from 450 m, the other from 1280-1300 m. They grumbled about Musser and Carleton's (1993) reference to original published descriptions as a revision of *Leptomys* and that these primary literature sources were unclear in distinguishing *L. elegans* from *L. ernstmayri*, but were still able to determine their three specimens to be the stockier species as Flannery (1995a) and Rümmler (1938) had described *L. elegans*. Cole et al. (1997) obtained *Leptomys* from the E flank of Mt Dayman in E Papua and interestingly had no problem separating their specimens as different from *L. elegans* using the primary literature referenced by Musser and Carleton (1993).

Leptomys ernstmayri Rummler, 1932. Das Aquarium, 6:134.

COMMON NAME: Ernst Mayr's Leptomys.

TYPE LOCALITY: Papua New Guinea, Morobe Province, Huon Peninsula, Saruwaged Mtns, Ogeramnang, 1785 m.

DISTRIBUTION: Papua New Guinea (above 1500 m in Astrolabe Range west to Kratke Mtns and Purosa region in Eastern Highlands Province, then northeast to mountains on Huon Peninsula; probably occurs farther west in Papuan Central Cordillera, but limits unknown) Reviewed by Flannery (1995a) and Helgen, Musser, and Lunde (in ms.).

STATUS: IUCN – Lower Risk (lc).

COMMENTS: A distinctive species distinguished from *L. elegans* by the diagnostic morphological traits and altitudinal distributions that were carefully enumerated by Rümmler (1932). It is much smaller in body size and replaces *L. elegans* at high altitudes in tropical montane rainforest formations; in the Astrolabe Range, for example, *L. ernstmayri* displaces *L. elegans* at elevations above 1000 m (Musser's study of samples at AMNH). Usually misidentified as *L. elegans* in museum collections. Rümmler (1938) included the Arfak Mtns in extreme western Prov. of Papua (= Irian Jaya) within the range of *L. ernstmayri*, but the sampe from there represents a separate undescribed species, as does material from the eastern end of New Guinea in the Mount Dayman region (Helgen, Musser, and Lunde, in ms.).

Leptomys signatus Tate and Archbold, 1938. Am. Mus. Novit., 982:2.

COMMON NAME: Fly River Leptomys.

TYPE LOCALITY: Papua New Guinea, Western Province, Sturt Isl Camp, Fly River (north bank), near sea level, 08°15'S, 142°15'E.

DISTRIBUTION: Known only from the type locality along lower Fly River; limits unknown.

STATUS: IUCN – Critically Endangered.

COMMENTS: The diagnostic traits Tate and Archbold (1938) used to separate this species from the other two (short, dense fur and gray upperparts with a white blaze on forehead) identify a lowland population of *Leptomys* that shows no morphological intergradation with samples of the other three species of *Leptomys* (Musser and Lunde, ms). *Leptomys signatus* is still known only by the four examples in the type series.

Limnomys Mearns, 1905. Proc. U.S. Natl. Mus., 28:451.

TYPE SPECIES: *Limnomys sibuanus* Mearns, 1905.

COMMENTS: *Rattus* Division. Taxonomic history and past erroneous association with *Rattus* reviewed by Musser (1977b), and Musser and Heaney (1992). Revised by Musser and Heaney (1992), who considered *Limnomys* a montane forest member of the Philippine New Endemics. Phylogenetic analyses of complete mtDNA cytochrome *b* sequences for 13 of the 16 genera of endemic Philippine rodents indicates that species of *Limnomys* are sister-taxa to *Tarsomys*, and in a clade containing species of *Rattus everetti* and *Bullimus* (Jansa and Heaney, 2001), which are also New Endemics. This alliance along with membership in the *Rattus* Division corroborated by chromosomal data (Rickart and Heaney, 2002).

Limnomys bryophilus Rickart, Heaney, and Tabaranza, 2003. J. Mammal., 84:1445.

COMMON NAME: Gray-bellied Limnomys.

TYPE LOCALITY: Philippines, NC Mindanao, Bukidnon Province, Mt Kitanglad Range, 18.5 km S, 4 km E Camp Phillips, 2250 m (08°09′30″N, 124°51′E).

DISTRIBUTION: Greater Mindanao Faunal Region in the Philippines. Recorded only from Mt Kitanglad Range between 2250 and 2800 m primary tropical upper montane rain-fores and transition between upper and lower montane forest; may occur elsewhere in montane forests on Mindanao (Rickart et al.,2003).

COMMENTS: A larger-bodied, longer-tailed close phylogenetic ally of *L. sibuanus* with grayish rather than white underparts. The latter occurs in lower montane forest on the Mt Kitanglad Range, is replaced at higher elevations by *L. bryophilus*, but the two are sympatric at the type locality (Rickart et al.,2003). Karyotype indistinguishable from that of *L. sibuanus* (Rickart and Heaney, 2002).

Limnomys sibuanus Mearns, 1905. Proc. U.S. Natl. Mus., 28:452.

COMMON NAME: White-bellied Limnomys.

TYPE LOCALITY: Philippines, SE Mindanao, Davao City Province, Mt Apo, 6600 ft (2012 m).

DISTRIBUTION: Greater Mindanao Faunal Region in the Philippines. Endemic to Mindanao Isl between 1900 and 2800 m in primary tropical lower and upper montane rainforest (recorded from Mt Apo, Mt Malindang, and Mt Katanglad; Musser and Heaney, 1992; Musser, 1994; Heaney, 2001).

STATUS: IUCN – Lower Risk (lc).

SYNONYMS: *mearnsi* Hollister, 1913.

COMMENTS: Standard karyotype (2n = 42, FN = 61/62) described by Rickart and Heaney (2002). See account of *L. bryophilus*.

Lorentzimys Jentink, 1911. Nova Guinea, 9:166.

TYPE SPECIES: *Lorentzimys nouhuysi* Jentink, 1911.

COMMENTS: *Lorentzimys* Division. Member of the New Guinea Old Endemics (Musser, 1981*c*). Past estimates of cladistic affinities ranged from enigmatic (Misonne, 1969) to alliance with either *Leggadina* (Simpson, 1961) or *Haeromys* (Ellerman, 1941). Analyses of phallic traits (Lidicker and Brylski, 1987) indicated that *Lorentzimys* and *Leptomys* are related, a hypothesis also supported by molar occlusal patterns and a shared derived cephalic arterial configuration (Musser's study of material in AMNH). Analysis of immunological distances by Watts and Baverstock (1994*a*, 1996), however, isolated *Lorentzimys* in its own clade to the exclusion of the other New Guinea and Australian genera sampled, which included *Leggadina* and *Leptomys*. *Lorentizimys* remains a New Guinea endemic whose closest phylogenetic alliance has yet to be uncovered.

Lorentzimys nouhuysi Jentink, 1911. Nova Guinea, 9:166.

COMMON NAME: Long-footed Tree Mouse.

TYPE LOCALITY: New Guinea, Prov. of Papua (= Irian Jaya), Noord (= Lorentz River), Bivak 2, 400 m.

DISTRIBUTION: New Guinea; Records are from S lowlands and middle altitudes, throughout Central Cordillera from N slopes of the Snow Mtns of Prov. of Papua (= Irian Jaya) in the west to Mt Dayman (Papua New Guinea) in the east, and the Torricelli Mtns (Flannery, 1995*a*:280); 30-2700 m; limits unknown.

STATUS: IUCN – Lower Risk (lc).

SYNONYMS: *alticola* Tate and Archbold, 1941.

COMMENTS: Photograph, distributional and biological information provided by Flannery (1990*b*:197). Phallic morphology described by Lidicker (1968). Some authors recognized *alticola* as a species (Lidicker, 1968; Lidicker and Brylski, 1987; Tate, 1951), and Flannery (1995*a*:280) noted that "*alticola* differs in a number of ways from the typical form, and further research is needed to determine its true status." Chromosomal data (2n = 46) reported by Donnellan (1987).

Macruromys Stein, 1933. Z. Säugetierk., 8:94.

TYPE SPECIES: *Macruromys elegans* Stein, 1933.

COMMENTS: *Pogonomys* Division. Member of the New Guinea Old Endemics (Musser, 1981*c*), and most closely related to *Chiruromys*, *Pogonomys*, *Coccymys*, *Hyomys*, *Anisomys*, and *Mallomys* as indicated by analysis of immunological distances (Watts and Baverstock, 1994*a*). Distributional and biological information summarized by Flannery (1990*b*; 1995*a*).

Macruromys elegans Stein, 1933. Z. Säugetierk., 8:95.
COMMON NAME: Lesser Macruromys.
TYPE LOCALITY: New Guinea, Prov. of Papua (= Irian Jaya), Weyland Range, Mt Kunupi, 1400-1800 m.
DISTRIBUTION: Known only from the type locality.
STATUS: IUCN – Critically Endangered.
COMMENTS: Possibly endemic to Weyland Range. Still represented only by the type series, three females and a male collected in October, 1931 (Flannery, 1995a).

Macruromys major Rümmler, 1935. Z. Säugetierk., 10:105.
COMMON NAME: Greater Macruromys.
TYPE LOCALITY: Papua New Guinea, Eastern Highlands Province, Kratke Mtns, Buntibasa Dist, 4000-5000 ft (1220-1524 m).
DISTRIBUTION: New Guinea; known from a few mid-altitude localities (1200-1500 m) on N slopes of Snow Mtns in Prov. of Papua (= Irian Jaya) and Central Cordillera of Papua New Guinea to the Eastern Highlands Province, southeast to Mt. Simpson (Flannery, 1990*b*; 1995*a*; Musser and Lunde, ms).
STATUS: IUCN – Endangered.
COMMENTS: Phallic morphology and its systematic significance described by Lidicker (1968). The scanty distributional and ecological information associated with *M. major* summarized by Flannery (1995*a*).

Madromys Sody, 1941. Treubia, 18 (part 2):258.
TYPE SPECIES: *Mus blanfordi* Thomas, 1881.
COMMENTS: *Millardia* Division. After being named a species of *Mus*, *blanfordi* was transferred to *Rattus* and finally to *Cremnomys* (see comments in following account). To Ellerman (1961:605), *blanfordi* was a *Rattus* in subgenus *Rattus*, but "an aberrant species. . . it stands rather well apart from typical *Rattus*, and is probably allied to *Cremnomys*." Misonne (1969:126) studied dental traits and included *blanfordi* in *Cremnomys*, but noted that "it could be made as well a genus on its own." Corbet and Hill (1992:349) acknowledged that "The inclusion of *blanfordi* in *Cremnomys* is supported by data on chromosomes. . . but is nevertheless very dubious." The extraction of *blanfordi* from *Cremnomys* was also supported by Agrawal (2000), who suggested that "*blanfordi* may be given an independent status as a genus equivalent to *Millardia*, *Cremnomys*, and *Rattus*." We agree. Unfortunately, Sody diagnosed *Madromys* only by its three pairs of mammae, a trait also characteristic of *Cremnomys* and some other tropical Asian murines. However, *blanfordi* differs from all species of *Cremnomys* by a suite of characters. Among them are its much larger body size; tail that is brown for proximal two-thirds and all white distally, well-haired and tufted at the tip (monocolor or dark brown above and mottled below in *Cremnomys*); longer braincase relative to rostrum, its dorsal outline more like an elongate triangle rather than vase-shaped; much wider zygomatic plate relative to cranial size, with deeper zygomatic notch and more pronounced, projecting anterodorsal shoulder; very large, globular and inflated bullae (small and uninflated in *Cremnomys*); and the molar occlusal traits described by Misonne (1969:126); these contrasts are based upon our study of specimens at AMNH, BMNH, and FMNH, and most holotypes representing taxa in *Cremnomys*. Among Indomalayan murines, *Madromys* is phylogenetically closest to *Cremnomys*, *Millardia*, and *Diomys* and unrelated to *Rattus* (based upon morphology and chromosomal data). Fossils identified as morphologically similar to *M. blanfordi* come from late Pliocene Siwalik sediments in NW India, and the species may have ancestral roots in the late Miocene *Progonomys* complex of N India and N Pakistan (Patnaik, 1997).

Madromys blanfordi (Thomas, 1881). Ann. Mag. Nat. Hist., ser. 5, 7:24.
COMMON NAME: Blanford's Madromys.

TYPE LOCALITY: India, Madras, Kadapa.

DISTRIBUTION: Endemic to Sri Lanka and Peninsular India, from southern provinces north to Bihar in the east and near Bombay in the west (Agrawal, 2000).

STATUS: IUCN – Lower Risk (lc) as *Cremnomys blanfordi*.

COMMENTS: Described as a species of *Mus* (Thomas, 1881), then allocated to *Rattus* (Ellerman, 1941, 1961), *blanfordi* is phylogenetically distant from any species in that genus and is related to species in *Cremnomys*, a conclusion based on dental morphology (Misonne, 1969) and chromosomal evidence (Gadi and Sharma, 1983; Raman and Sharma, 1977; Rao and Lakhotia, 1972). See discussion in *Madromys* above. Chromosomal number and configuration of *M. blanfordi* very similar to that recorded for *C. cutchicus* and *C. elvira* (Raman and Sharma, 1977), but differing in amount of C-band-positive constitutive heterochromatin (Sharma and Gadi, 1977). Ecology and occurrence in Western Ghats of S India documented by Chandrasekar-Rao and Sunquist (1996).

Malacomys Milne-Edwards, 1877. Bull. Sci. Soc. Philom. Paris, ser. 6, 12:10.

TYPE SPECIES: *Malacomys longipes* Milne-Edwards, 1877.

COMMENTS: *Malacomys* Division. Study of dental features led Misonne (1969:106) to write that "*Malacomys* stands apart from all the other African genera by its unusual dental characters. This is an advanced genus which cannot be clearly related to any other one." *Malacomys* was included in allozymic analyses, which were inconclusive as to detecting its closest relative (Bonhomme et al., 1985). DNA/DNA hybridization experiments (Chevret, 1994) clustered *Malacomys longipes* as one branch of a trifurcation with the second leading to a "*Praomys* group" (*Praomys* + *Myomys* + *Mastomys* + *Hylomyscus*) and the third to a "*Mus* group" (*Mus*, including subgenera *Pyromys*, *Nannomys*, and *Coelomys*). External, cranial, and dental traits suggest *Malacomys* is phylogenetically allied with *Praomys*, particularly through what appears to be the morphologically annectant *P. lukolelae* and *P. verschureni* (our study of specimens; see account of *P. lukolelae*). This view is countered by the cladistic analyses of morphological traits in species of *Praomys*, *Hylomyscus*, *Mastomys*, and *Myomyscus* (reported as *Myomys*) where *Malacomys* was used as an outgroup and was consistently excluded from a monophyletic "*Praomys* group" (Lecompte et al., 2002*a*), and "was always very distant from the '*Praomys* group', even when not designed as outgroup in parsimony or any other type of analyses" (L. Granjon, in litt., 2002). *Malacomys* is also distant to monophyletic *Praomys* group in preliminary results from phylogenetic analyses of nuclear IRBP gene sequences (E. Lecompte, in litt., 2002). Analyses of a broader range of genes along with denser taxon sampling that includes *Malacomys* would be welcome and may provide a sharper picture of its phyletic affinity within a reconstruction of phylogenetic relationships among African murines.

Two revisions (Rautenbach and Schlitter, 1978; Van der Straeten and Verheyen, 1979*b*) provided somewhat different evaluations of morphological variation and its interpretation in context of species-diversity. The morphological distinctions between *M. edwardsi* and *M. longipes* were documented more than six decades ago by Hayman (1935). Geographic distributions outlined below come mostly from the two revisions.

Species of *Malacomys* are referred to as African swamp rats (Wilson and Cole, 2000), but they do not inhabit swamps. Their range of habitats is exemplified by the sympatric range of *M. edwardsi* and *M. cansdalei* in Ghana. Both eat slugs, earthworms, and plant parts, and require primary tropical evergreen rainforest, but *M. edwardsi* is found in higher, well drained ground, and *M. cansdalei* prefers damper places—valley bottoms, moist lower slopes, stream banks, and wet mud (Grubb et al., 1998; Happold, 1987). *Malacomys edwardsi* has been caught on moist soil near a river in SW Guinea (Barnett et al., 1996). In the Kalinzu Forest of SW Uganda, *M. longipes* was collected on dry ground around stumps in logged forest (Lunde and Sarmiento, 2002), and Granjon (1991) trapped it in a variety of habitats in Republic of Congo, from degraded primary forest to swamp forest. In their experience, Kerbis Peterhans and Patterson (1995) noted that *M. longipes* was usually captured in shallow forest streams or along their margins (also indicated by Hoffmann, 1997). They included species of *Malacomys*, *Colomys goslingi*, and *Deomys ferrugineus* in an African guild of waders (see account of *Colomys*).

Malacomys cansdalei Ansell, 1958. Ann. Mag. Nat. Hist., ser. 13, 1:342.
COMMON NAME: Cansdale's Malacomys.
TYPE LOCALITY: Ghana, Oda.
DISTRIBUTION: Forest zone of S Ghana, S Côte d'Ivoire, and E Liberia.
STATUS: IUCN – Lower Risk (lc).
SYNONYMS: *giganteus* Bellier and Gautun, 1968.
COMMENTS: Part of the murine fauna endemic to West Africa (see account of *Grammomys buntingi*). Rautenbach and Schlitter (1978) considered *cansdalei* a subspecies of *M. longipes*, but Van der Straeten and Verheyen (1979*b*) demonstrated its specific distinction. Grubb et al. (1998) reviewed the Ghana population.

Malacomys edwardsi Rochebrune, 1885. Bull. Sci. Soc. Philom. Paris, ser. 7, 9:87.
COMMON NAME: Edwards' Malacomys.
TYPE LOCALITY: Liberia Mellacoree River (= Melikhoure River, Guinea according to Rautenbach and Schlitter, 1978:410).
DISTRIBUTION: Sierra Leone, Guinea, Liberia, and southern regions of Côte d'Ivoire, Ghana, and Nigeria.
STATUS: IUCN – Lower Risk (lc).
COMMENTS: Member of the murine fauna endemic to West Africa (see account of *Grammomys buntingi*). Chromosomal data reported by Matthey (1958) and Van der Straeten and Verheyen (1979*b*). Grubb et al. (1998) reviewed populations from Ghana and Sierra Leone, and Barnett et al. (1996) recorded the species from the Kounounkan Massif in SW Guinea. See Happold (1987) for ecology and distribution in Nigeria.

Malacomys longipes Milne-Edwards, 1877. Bull. Sci. Soc. Philom. Paris, ser. 6, 12:10.
COMMON NAME: Common Malacomys.
TYPE LOCALITY: Gabon, Gaboon River (vicinity of Ogooue, Gabon; see Rautenbach and Schlitter, 1978:414).
DISTRIBUTION: From SE Nigeria, C and S Cameroon, Equatorial Guinea (including Bioko; Eisentraut, 1965), and Gabon eastward through Republic of Congo, S Central African Republic, Dem. Rep. Congo to S Sudan, Uganda, and Rwanda; also to the south in NW Zambia and NE Angola (Crawford-Cabral, 1998).
STATUS: IUCN – Lower Risk (lc).
SYNONYMS: *australis* Ansell, 1958; *centralis* de Winton, 1897; *wilsoni* Thomas, 1916.
COMMENTS: Karyotype of sample from Côte d'Ivoire is 2n = 48, FN = 52 (Fna = 48), with the autosomal complement consisting of all telocentric pairs except for a single pair of small metacentrics, a composition regarded by Viegas-Péquignot et al. (1983) to be close to the presumed ancestral complement for murines. A sample from Gabon exhibits an identical karyotype (Lecompte, 2003).

Mallomys Thomas, 1898. Novit. Zool., 5:1.
TYPE SPECIES: *Mallomys rothschildi* Thomas, 1898.
SYNONYMS: *Dendrosminthus* de Vis, 1907.
COMMENTS: *Pogonomys* Division. Member of the New Guinea Old Endemics (Musser, 1981*c*). Tate (1936) included *Mallomys*, along with *Chiropodomys*, *Crateromys*, *Lenomys*, *Phloeomys*, *Chiruromys*, and *Pogonomys* in the Phloeomyinae, an arrangement followed by Simpson (1945), but no data support such an allocation (Ellerman, 1949*a*; our research). Using microcomplement fixation of albumin, Watts and Baverstock (1994*a*) placed *Mallomys* in a clade with *Anisomys*, *Pogonomys*, *Chiruromys*, *Coccymys*, *Hyomys*, and *Macruromys* (their Anisomyini, our *Pogonomys* Division), but noted that its karyotype (2n = 48) and derived sperm morphology (Breed, 1991) suggest affinities with *Hydromys* and members of genera we place in the *Pseudomys*, *Xeromys*, and *Uromys* Divisions, which form a clade (their Hydromyini) separate from the New Guinea endemics in the *Pogonomys* Division. Careful phylogenetic analyses of other data sets (morphological and molecular) are needed to more clearly resolve the cladistic relationship of *Mallomys*. Revised by Flannery et al. (1989), but actual number of species in genus still unresolved. A separate yet undescribed species occurs on the Arfak Mtns in Prov. of Papua (= Irian Jaya) (K. Helgen, in litt., 2003). Aplin et al. (1999) reported material probably referable to this same taxon from a late

Pleistocene archaeological site on the Ayamaru Plateau, central Bird's Head Peninsula of Prov. of Papua (= Irian Jaya).

Mallomys aroaensis (De Vis, 1907). Ann. Queensl. Mus., 7:11.
COMMON NAME: De Vis's Mallomys.
TYPE LOCALITY: Papua New Guinea, Central Province, head of Aroa River (Flannery et al., 1989, provided additional data).
DISTRIBUTION: New Guinea; in Prov. of Papua (= Irian Jaya) in the Arfak Mtns of Vogelkop region and on the S side of the Snow Mtns (Pegunungan Maoke), then throughout the Central Cordillera of Papua New Guinea to Mt Simpson in the east; also highlands of Huon Peninsula; 1100-2700 m (Flannery et al., 1989:95; Flannery, 1995*a*).
STATUS: IUCN – Lower Risk (lc).
SYNONYMS: *hercules* Thomas, 1912.
COMMENTS: Broadly sympatric with *M. rothschildi* along the Central Cordillera in Papua New Guinea (Flannery, 1995*a*). Cole et al. (1997) discussed a specimen obtained on E flanks of Mt Dayman.

Mallomys gunung Flannery, Aplin, and Groves, 1989. Rec. Aust. Mus., 41:101.
COMMON NAME: Alpine Mallomys.
TYPE LOCALITY: New Guinea, Prov. of Papua (= Irian Jaya), 2 km E Mt Wilhelmina, 3800 m (Flannery et al., 1989, gave details).
DISTRIBUTION: New Guinea, Prov. of Papua (= Irian Jaya), Snow Mtns (Pegunungan Maoke); known only from vicinity of the type locality and Mt Carstenz (Flannery et al., 1989:97; Flannery, 1995*a*:286).
STATUS: IUCN – Critically Endangered.
COMMENTS: A very distinctive species known only by a few specimens collected between 3500 and 4050 m. On the N slopes of the Snow Mtns it is replaced at lower elevations by *M. istapantap* and *M. rothschildi*.

Mallomys istapantap Flannery, Aplin, and Groves, 1989. Rec. Aust. Mus., 41:96.
COMMON NAME: Subalpine Mallomys.
TYPE LOCALITY: Papua New Guinea, Mt Hagen Dist., Korelum (Flannery et al., 1989, provided additional information).
DISTRIBUTION: New Guinea; high altitudes along Central Cordillera from Mt Victoria in Owen Stanley Range of E Papua New Guinea to Bele River region in Snow Mtns of Prov. of Papua (= Irian Jaya).
STATUS: IUCN – Lower Risk (nt).
COMMENTS: Recorded from the Snow Mtns (Pegunungan Maoke), where it has been taken on the N slopes at the Bele River and on the S slopes near Tembagapura, and the Central Cordillera in Papua New Guinea from Mt Giluwe in the west to Mt Victoria in the east; 2200-3850 m (Flannery, 1995*a*:289).

Mallomys rothschildi Thomas, 1898. Novit. Zool., 5:2.
COMMON NAME: Rothschild's Mallomys.
TYPE LOCALITY: Papua New Guinea, Central Province, Owen Stanley Range, between Mt Musgrave and Mt Scratchley, 5000-6000 ft (1524-1830 m); see Flannery et al. (1989).
DISTRIBUTION: New Guinea; Central Cordillera, from Weyland Range in W Prov. of Papua (= Irian Jaya) to Owen Stanley Range in Papua New Guinea; 1550-3700 m (Flannery, 1995*a*:290; Flannery et al., 1989:93).
STATUS: IUCN – Lower Risk (lc).
SYNONYMS: *argentata* Rothschild and Dollman, 1932; *weylandi* Rothschild and Dollman, 1932.
COMMENTS: *Mallomys rothschildi* and *M. aroaensis* are thought to be sympatric at a few localities (Flannery, 1995*a*; Flannery et al., 1989), but the morphological variation present among samples requires re-examination to test whether it reflects presence of one or two species.

Malpaisomys Hutterer, Lopez-Martinez, and Michaux, 1988. Palaeovertebrata, 18:246.
TYPE SPECIES: *Malpaisomys insularis* Hutterer, Lopez-Martinez, and Michaux, 1988.

COMMENTS: *Oenomys* Division. Represented only by late Pleistocene and Holocene samples. A phylogenetic analysis by Hutterer et al. (1988) using 27 morphological traits placed *Malpaisomys* in a monophyletic group with the extinct *Stephanomys* (late Miocene to late Pliocene of Europe and N Africa) and the deomyines *Acomys* and *Uranomys*, to the exclusion of *Rattus* and *Mus*. A subsequent cladistic study with 17 cranial and 23 dental characters supported a very close phylogenetic association between *Malpaisomys* and *Stephanomys*, and an alliance between those two genera and *Acomys* on one hand, and *Thamnomys* and *Oenomys* on the other (López-Martínez et al., 1998). A study of antibody reaction against albumin of *Malpaisomys*, *Acomys*, *Uranomys*, and *Mus* (bone extracts were used for *Malpaisomys*) indicated *Malpaisomys* to be allied with *Mus* to the exclusion of *Acomys* and *Uranomys* (Montgelard, 1992). We would like to see results from a phylogenetic study incorporating a greater range of murine genera and genes; until then we view the hypothesis by López-Martínez et al. (1998) to best estimate phylogenetic affinities of *Malpaisomys* and allocate it to our *Oenomys* Division. Certainly molar occlusal patterns are similar to those characterizing genera in that cluster.

Two species of *Canariomys*, also probably in the *Oenomys* Division, along with *Malpaisomys*, represent the known murine fauna once endemic to the Canary Isls and now apparently extinct (Boye et al., 1992; Hutterer et al., 1988; Lopez-Martinez and Lopez-Jurado, 1987; Michaux et al., 1996c). *Canariomys bravoi* is represented only by Pleistocene and possibly Holocene specimens from Tenerife Isl. There are abundant remains from several archaeological sites (Michaux et al., 1996c; Alcover et al., 1998); some specimens have been[14]C-dated as 12,230 " 140 years before present (Michaux et al., 1996c). This is much earlier than human arrival to the Canary Islands, about 500 BC. The ^{14}C date, however, does not indicate date of extinction for *C. bravoi* and humans may have been contemporaneous because recently a mummified specimen has been found (Alcover et al., 1998). *Canariomys tamarani*, from Gran Canaria Isl (Lopez-Martinez and Lopez-Jurado, 1987), was terrestrial, herbivorous and is known only by specimens from the Holocene to pre-Hispanic epoch (about 2000 years before present). Fossils have been found associated with remains of goats, dogs, and *Mus musculus* in sediments dated at 130 years BC so *C. tamarani* and humans overlapped for a time (Lopez-Martinez and Lopez-Jurado, 1987). While *Malpaisomys* was small (estimated weight about 40 grams; Boye et al., 1992), the two species of *Canariomys* were large rats, up to one kilogram or more (Michaux et al., 1996c). The two genera are not sympatric on any island. *Malpaisomys* has been found only on the three eastern isls (Fuerteventura, Lanzarote, and Graciosa); the two large central isls (Tenerife and Gran Canaria) are provenances of *Canariomys*. No endemic murines have been discovered on the western isls. A cladistic analysis by Lopez-Martinez and Lopez-Jurado (1987) suggested that *C. bravoi* and *C. tamarani* were in the same monophyletic group, a cluster related to the African *Pelomys-Arvicanthis* assemblage. However, a phylogenetic analysis by Hutterer et al. (1988) indicated that each species of *Canariomys* was in a different monophyletic cluster, *C. bravoi* with *Grammomys dolichurus* and *Oenomys hypoxanthus*, and *C. tamarani* with *Arvicanthis niloticus* and *Rattus rattus*. Careful reexamination of specimens will be necessary to resolve these contrasting views.

Lopez-Martinez and Lopez-Jurado (1987) regarded the two species of *Canariomys* as an emigrant relict from an African Miocene fauna. *Malpaisomys*, also a member of the *Oenomys* Division, likely represents a separate emigrant invasion of the isls (R. Hutterer, in litt., 2003). Possibly the closest phylogenetic relatives of these three species will prove to be extinct murines represented by Miocene and Pliocene fossils.

Malpaisomys insularis Hutterer, Lopez-Martinez, and Michaux, 1988. Palaeovertebrata, 18:246.
    COMMON NAME: Lava Mouse.
    TYPE LOCALITY: Canary Isls, Fuerteventura Isl, Cueva Villaverde near La Oliva, at stratigraphic level dated about 1070 years before present (see Hutterer et al., 1988, for additional information).
    DISTRIBUTION: Isls of Fuerteventura, Lanzorote, and Graciosa in the Canary Isls (Hutterer et al., 1988).
    STATUS: IUCN – Extinct.
    COMMENTS: *Malpaisomys insularis* has been recorded from sediments dated from 25,000 and 32,000 years before present up to historical times when the species became extinct,

sometime between 800 years before present and now (Hutterer et al., 1988; Michaux et al., 1991). Reconstruction and study of the postcranial skeleton suggest the species was adapted to living in lava fields (Boye et al., 1992). Boye et al. (1992) reported that about 2000 years before present, *Mus musculus* was apparently casually imported to the islands by humans who also arrived then, and from that time to the historical period, populations of *M. insularis* declined and were progressively replaced by house mice. This interaction between lava mice and house mice is the hypothesized causal reason for extinction of *M. insularis*. *Malpaisomys insularis* is part of a mammalian fauna endemic to the E Canary Isls that includes the shrew *Crocidura canariensis* (Hutterer et al., 1987*a*; Michaux et al., 1991; Boye et al., 1992).

Mammelomys Menzies, 1996. Aust. J. Zool., 44:383.
  TYPE SPECIES: *Melomys rattoides* Thomas, 1922.
  COMMENTS: *Pogonomys* Division. Member of New Guinea Old Endemics. Originally included in *Melomys*. Generic diagnosis based primarily on morphometric analyses (Menzies, 1996), some qualitative cranial and dental features, extremely long hind feet with very short first digit, single pair of teats (not found in any other Papuan murine), and spermatozoal morphology (Breed and Aplin, 1994). The monophyly of *rattoides* and *lanosus* relative to *Melomys*, *Uromys*, and their close relatives in our *Uromys* Division is strongly supported by albumin immunology, which indicated they are not part of the core species defining either *Uromys* or *Melomys*, and judged from immunological distances (Watts and Baverstock, 1994*a*) and sperm morphology (Breed and Aplin, 1994) could not properly be placed in either of those genera. Our study of specimens in AMNH and BMNH revealed that the two species share a derived cephalic arterial circulation and its osseous reflection in the cranium (see Musser and Heaney, 1992; Musser et al., 1998) that is not found in any other genus in our *Uromys* Division (traits not mentioned by Menzies, 1996). Our inclusion of *Mammelomys* in the *Pogonomys* Division is provisional, but concordant with sperm morphology (Breed and Aplin, 1994), and a better estimate of its cladistic relationships than is placement in a group close to *Melomys*. *Mammelomys* contains two species, both revised by Menzies (1996).

Mammelomys lanosus (Thomas, 1922). Ann. Mag. Nat. Hist., ser. 9, 9:263.
  COMMON NAME: Highland Mammelomys.
  TYPE LOCALITY: New Guinea, Prov. of Papua (= Irian Jaya), Doormanpad-bivak (3°30′S, 138°30′E), 2400 m.
  DISTRIBUTION: Mountain forests (above 1000-1500 m) of N New Guinea; from the type locality in Prov. of Papua (= Irian Jaya) eastward along Central Cordillera through the Telefomin region, Schrader Range, Kratke Mtns, and Mt Hagen region to the Wau area of Papua New Guinea; also from Cyclops Range in the northern coastal mountains; not recorded from Huon Peninsula.
  STATUS: IUCN – Lower Risk (lc).
  SYNONYMS: *shawmayeri* Rümmler, 1935.
  COMMENTS: Once treated as a subspecies of *Paramelomys levipes*, *lanosus* is a separate species (Flannery, 1990*b*; Menzies, 1996). It is unrelated to *P. levipes*, but instead is phylogenetically allied with *Mammelomys rattoides*, a larger-bodied species that displaces it in lower altitudes. This close association was also reflected in Menzies' (1990) similarity coefficient analysis and his later systematic revision (Menzies, 1996). Flannery (1990*b*) listed *shawmayeri* as a synonym of *M. rattoides*, but holotype is an example of *M. lanosus* (Menzies, 1996; Musser's study of type series in BMNH).

Mammelomys rattoides (Thomas, 1922). Ann. Mag. Nat. Hist., ser. 9, 9:263.
  COMMON NAME: Lowland Mammelomys.
  TYPE LOCALITY: New Guinea, NW Prov. of Papua (= Irian Jaya), Mamberano River, Pionier-bivak (2°20′S, 138°0′ E), 200 ft (61 m).
  DISTRIBUTION: New Guinea; N slopes and lowlands from the Mamberano River in Prov. of Papua (= Irian Jaya) through lower slopes north of Idenberg River to the Telefomin region in West Sepik Province of Papua New Guinea (Flannery and Seri, 1990) and eastward to the Gogol River in Madang Province; between sea level and 1500 m

(Menzies, 1996); also on Yapen Isl in NE Prov. of Papua (= Irian Jaya) (Flannery, 1995b; AMNH 160268 and 222193, BMNH 46.643); limits of range unknown on mainland.

STATUS: IUCN – Lower Risk (lc).

COMMENTS: Distribution based on specimens we studied in AMNH and BMNH, and Menzie's (1996) account. Tate's (1951:290) record of M. rattoides from Cyclops Mtns is based on M. lanosus. Closest relative is M. lanosus, which occurs in montane forests at higher altitudes than M. rattoides (see account of M. lanosus).

Margaretamys Musser, 1981. Bull. Am. Mus. Nat. Hist., 168(3):275.

TYPE SPECIES: *Mus beccarii* Jentink, 1880.

COMMENTS: *Pithecheir* Division. Morphological, chromosomal, and distributional features monographed by Musser (1981b). Spermatozoal morphology distinctive and similar among the three species (Breed and Musser, 1991). Member of the Sulawesi Old Endemics (Musser, 1981c); nearest phylogenetic ally unknown, but probably more closely related to Sulawesi *Lenomys* and *Eropeplus* and Sundaic *Lenothrix* than any other murine (Musser's unpublished research): "Possibly the three species of *Margaretamys* represent fragments of an early murid stock of which *Lenothrix* is a more primitive remnant" (Musser, 1981b: 325). Pavlinov et al. (1995a) listed *Margaretamys* in a *Pithecheir* Section of a more inclusive *Micromys* Group. All three species of *Margaretamys* are arboreal and found only in primary forest formations.

Margaretamys beccarii (Jentink, 1880). Notes Leyden Mus., 2:11.

COMMON NAME: Spiny Lowland Margaretamys.

TYPE LOCALITY: Indonesia, Sulawesi, NE peninsula, Menado-Langowan.

DISTRIBUTION: NE peninsula and C Sulawesi in lowland tropical evergreen rain forest, near sea level to1000 m.

STATUS: IUCN – Lower Risk (lc).

SYNONYMS: *leucopus* (Jentink, 1879) [not Gray, 1867]; *thysanurus* (Sody, 1932).

COMMENTS: History of incorrect taxonomic allocations of *beccarii* recorded by Musser (1971e, 1981b). Insufficient samples exist to determine any morphometric or qualitative differences between populations in central core of the island and the NE peninsula. The only one of the three *Margaretamys* with spinous pelage.

Margaretamys elegans Musser, 1981. Bull. Am. Mus. Nat. Hist., 168:286.

COMMON NAME: Elegant Margaretamys.

TYPE LOCALITY: Indonesia, C Sulawesi, Gunung Nokilalaki, 6500 ft (1980 m).

DISTRIBUTION: Originally recorded only from 1600-2285 m in montane rain forest on Gunung Nokilalaki, but a specimen (AMNH 267763) was obtained from farther south in the Mamasa Area (02°56′S/119°22′E) of C Sulawesi. Probably occurs on mountains elsewhere in C core of the island.

STATUS: IUCN – Vulnerable.

COMMENTS: The largest in body size of the three species and the most squirrel-like in habitus.

Margaretamys parvus Musser, 1981. Bull. Am. Mus. Nat. Hist., 168:294.

COMMON NAME: Lesser Margaretamys.

TYPE LOCALITY: Indonesia, C Sulawesi, Gunung Nokilalaki, 7400 ft (2256 m).

DISTRIBUTION: Known only from 1830-2286 m in montane rain forest on Gunung Nokilalaki, but probably occurs elsewhere in mountainous C Sulawesi.

STATUS: IUCN – Vulnerable.

COMMENTS: The smallest of the three species and the most divergent in cranial and dental traits.

Mastacomys Thomas, 1882. Ann. Mag. Nat. Hist., ser. 5, 9:413.

TYPE SPECIES: *Mastacomys fuscus* Thomas, 1882.

COMMENTS: *Pseudomys* Division. Included in *Pseudomys* by Musser and Carleton (1993), who followed the action of Watts et al. (1992). Mahoney and Richardson (1988) and Happold (1995) maintained *Mastacomys* separate from *Pseudomys*. A recently completed DNA sequencing study by F. Ford (James Cook University, Townsville), utilizing several different mitochondrial and nuclear genes, strongly supports the inclusion of *fuscus* within *Pseudomys* as currently constituted. The diversity within *Pseudomys*, however, is such as to warrant recognition of several genera including *Mastacomys* (F. Ford, in litt., 2004). Mahoney and Richardson (1988) cataloged distributional, taxonomic, and biological references. Member of Australian Old Endemics (Musser, 1981*c*; Watts and Aslin, 1981).

Mastacomys fuscus Thomas, 1882. Ann. Mag. Nat. Hist., ser. 5, 9:413.

COMMON NAME: Broad-toothed Mastacomys.

TYPE LOCALITY: Australia, Tasmania; see Mahoney and Richardson (1988:160) for details.

DISTRIBUTION: Australia; modern records from E New South Wales, E Victoria, and Tasmania, but late Pleistocene-Holocene fragments indicated range once included Kangaroo Isl, Carrieton, and Naracoorte in South Australia (Archer et al., 1984; Happold, 1995; Pledge, 1990; Robinson et al., 2000); see Menkhorst (1995*b*) for the range in Victoria, and Rounsevell et al. (1991) for a summary of Tasmanian records.

SYNONYMS: *brazenori* Ride, 1956; *castaneus* (Higgins and Petterd, 1884) [not Waterhouse, 1843]; *mordicus* Thomas, 1922; *wombeyensis* Ride, 1956.

COMMENTS: The form *fuscus* and the other taxa listed above were all described and revised under *Mastacomys* (Ride, 1956; Wakefield, 1972*b*), and this genus has always been recognized (Watts and Aslin, 1981; Mahoney and Richardson, 1988). However, chromosomal morphology (Baverstock et al., 1977*c*), G-banding homologies (Baverstock et al., 1983*b*), allozymic data (Baverstock et al., 1981), and phallic morphology (Lidicker and Brylski, 1987) linked *fuscus* with some species of *Pseudomys*, and Watts et al. (1992) united *fuscus* with that genus. Sperm head structure reported by Breed (1984), and variation in external morphology of glans penis documented by Morrissey and Breed (1982). Taxonomic, distributional, and biological references cataloged by Mahoney and Richardson (1988:160); ecological and distributional information summarized by Watts and Aslin (1981) and Happold (1995).

Mastomys Thomas, 1915. Ann. Mag. Nat. Hist., ser. 8, 16:477.

TYPE SPECIES: *Mus coucha* Smith, 1834.

SYNONYMS: *Myomys* Thomas, 1915.

COMMENTS: *Stenocephalemys* Division. Historical taxonomic use of *Mastomys* as a genus or subgenus of either *Rattus* or *Praomys* summarized by Meester et al. (1986), Chevret et al. (1994), and Granjon et al. (1997*b*). Morphological (Lecompte et al., 2002*a*; Van der Straeten, 1979; Van der Straeten and Robbins, 1997), protein electrophoretic analyses (Iskandar and Bonhomme, 1984), DNA/DNA hybridization experiments (Chevret et al., 1994), analyses of complete mtDNA cytochrome *b* sequences (Lecompte et al., 2002*b*), and chromosomal data (Britton-Davidian et al., 1995; Granjon et al., 1997*b*) support union of most of the species listed below in a monophyletic group different from those clades containing species of *Praomys*, *Myomyscus*, and *Hylomyscus*. *Mastomys pernanus* is the exception (see that account). Important to studies of specific diversity, biogeography, population and community ecology, and medicine, species of *Mastomys* have been examined in several contexts (see reviews by Granjon et al., 1997*b*; Keogh and Price, 1981; Leirs, 1992; Skinner and Smithers, 1990; and Taylor, 2000*b*). Until the late 1990s, definition of species by chromosomal and biochemical traits from some regions (Duplantier et al., 1990*a*; Green et al., 1980; Hubert et al., 1983) proceeded faster than definitions based on morphology, resulting in a new view of species-diversity in the genus, but also an ignorance of morphological limits of those species and their real geographic distributions. Recently, however, integration of chromosomal, molecular, and morphometric data has allowed sharper resolution of species limits, research summarized by Granjon et al. (1997*b*), who also presented the most current and best review of *Mastomys* systematics. Except for adding new data and various comments, we follow their conception of species contents.

Monophyly of *M. erythroleucus*, *M. natalensis*, *M. huberti*, and *M. coucha* is supported by cladistic analyses of chromosomal traits (Britton-Davidian et al., 1995; Volobouev et al., 2002*b*) and variation in skeletal, dental, and soft tissue characters (Lecompte et al., 2002*a*). Analyses of complete mtDNA cytochrome *b* sequences support monophyly of the same four species and adds *M. kollmannspergeri* (reported as *verheyeni*) to that cluster (Lecompte et al., 2002*b*). Lavrenchenko et al. (1992) provided a review of chromosomal and biochemical polymorphism (including hemoglobin electrophoresis) in samples of *Mastomys* and the significance of these data in distinguishing species and other topics. Pattern of karyotype evolution among several *Mastomys*, its correlation with phylogenetic patterns indicated by analyses of gene sequences, and relevance to cladogenesis in the genus discussed by Volobouev et al. (2002*b*). Spermatozoal morphology of selected species documented by Breed (1995*a*) and Baskevich and Lavrenchenko (1995). Analysis of microcomplement fixation of albumin grouped *Mastomys* with *Otomys*, *Myomys* (=*Myomyscus*), *Hylomyscus*, and *Praomys* (Watts and Baverstock, 1995*a*).

Robbins and Van der Straeten (1989) and Van der Straeten and Robbins (1997) studied holotypes that are the basis of most names associated with the *Mastomys-Myomys* (= *Myomyscus*)-*Hylomyscus-Praomys* group, and identified the taxa they definitely regard as representing *Mastomys* but unfortunately did not allocate any of them to species. One of these names is *Mus colonus* Brants (1827), which was made the type species of *Myomys* by Thomas (1915). Because some workers claimed *colonus* was unidentifiable and therefore *Myomys* is invalid, *Myomyscus* was proposed by Shortridge (1942) to replace *Myomys*. The problem and historical opinions have been thoroughly reviewed (Roberts, 1951; Ellerman et al., 1953; Rosevear, 1969; Van der Straeten and Verheyen, 1978*b*; Meester et al., 1986). Recently, Van der Straeten and Robbins (1997) located the specimen (an adult male) described by Brants, which represents the holotype of *colonus*, and to them the skull and teeth "shows all *Mastomys* characters" (p. 153); they did not assign the holotype to a definite species of *Mastomys*. Brants (1827) did not indicate the type locality, but Van der Straeten and Robbins reported that "Zondagsrivier" is written on the specimen label and that place is listed as "Uitenhage, Port Elizabeth (33°46′S, 25°24′E)" in the gazetteer they consulted. This is far eastward of the geographic range of *Myomyscus verreauxii*, very close to if not within the western margin of the distribution of *M. coucha*, but not within the range of *M. natalensis* (as mapped by Skinner and Smithers, 1990, and de Graaff, 1997*q*; but see Taylor, 2000*b*). Musser studied the specimen set aside by Van der Straeten in Berlin and agrees with the identification by Van der Straeten and Robbins. It is not related to *Myomyscus verreauxii* as Roberts (1951) indicated (he treated it as a species of *Myomyscus*, but did not have the benefit of examining the holotype), but is a *Mastomys*.

Identifying the holotype of *colonus* as a *Mastomys* affects the taxonomy of *Mastomys* and *Myomys* in three ways. First, *Myomys* has to be arranged as a synonym of *Mastomys* (both were named by Thomas and preliminarily diagnosed on the same page with *Myomys* on the line above *Mastomys*). Second, *Myomys* must be replaced by *Myomyscus* to cover the species formerly assigned to it. Finally, *colonus* may be the earliest name for what is now called *M. coucha*. Here we allocate *Myomys* to the synonymy of *Mastomys* but await future study to determine the specific identity of the holotype of *colonus*. That could be accomplished by including the holotype in a multivariate discriminant function analysis among *Mastomys* samples from southern Africa, and by extracting a piece of dry tissue, bone or tooth from it for analysis by mitochondrial or nuclear DNA sequencing.

Despite accumulation of systematic information through the last decade, the species of *Mastomys* still require careful taxonomic revision; geographic ranges have to be resolved for most species, and associating the many scientific names proposed with currently defined species remains a vexing problem. Living species of *Mastomys* occur primarily in subsaharan Africa (isolated population in WC Morocco is the North African exception) and the genus is represented in late Pliocene, early Pleistocene, and Holocene deposits of East and South Africa (Avery, 1998, 2000; Denys, 1987*a*, 1999; Jaeger, 1976; Wesselman. 1984), but at least three once occurred in the S Levant (Israel). *Mastomys batei*, described from middle and Late Pleistocene (Bate, 1942*b*; Tchernov, 1968), is absent from sediments dated later than 80,000-70,000 years before present (Tchernov, 1986, 1992, 1994, and references cited therein). Fossils identified as *Mastomys galilensis* and *M. levantinus* were recovered from the early Pleistocene 'Ubeidiya Formation in the Central Jordan Valley of Israel (Tchernov, 1986).

Mastomys awashensis Lavrenchenko, Likhnova, and Baskevich (in Lavrenchenko et al., 1998*a*).
    Z. Säugetierk., 63:44.

COMMON NAME: Awash Mastomys.

TYPE LOCALITY: Ethiopia, "collected at the bank of the Awash River near Koka Lake" (Lavrenchenko et al., 1998a:46), 08°23′N 39°09′E.

DISTRIBUTION: Apparently endemic to the Ethiopian Rift Valley where it "is confined to a small part of the Upper Awash Valley" (Lavrenchenko et al., 1998a:48). All known examples were captured at the E bank of Koka Lake and Awash National Park where they inhabit Awash riverbank vegetated by *Acacia-Commiphora* thornbush with high grass, and adjacent agricultural habitats.

COMMENTS: This species occurs with *M. natalensis* and *M. erythroleucus* in the same region. Karyotype of *M. awashensis* (2n = 32, FNa = 54, Lavrenchenko et al., 1998a; or 2n = 32, FNa = 52, Volobouev et al., 2002b) is dissimilar to that of *M. erythroleucus* (2n = 38, FN = 50) and resembles karyotype of *M. natalensis* from the Ethiopian Rift Valley (2n = 32, FNa = 54-56), but the two are distinguished by frequency of metacentric and submetacentric elements, form of the Y-chromosome, and C-banding pattern (Baskevich and Orlov, 1993; Lavrenchenko et al., 1998a). Analyses of allozymic and morphometric data indicate *M. awashensis* is more closely related to *M. erythroleucus* than to *M. natalensis*, but *M. awashensis* and *M. natalensis* are more similar in chromosomal traits and configuration of tail scales; in glans penis morphology, *M. awashensis* differs from either of the other species, which are similar in that character complex (Lavrenchenko and Baskevich, 1996; Lavrenchenko et al., 1998a). *Mastomys awashensis* shares a common fixed hemoglobin electromorph with South African *M. coucha*, different from that found in Ethiopian *M. natalensis* and *M. erythroleucus* (Lavrenchenko et al., 1992, 1998a). Those latter authors noted that the discordance among morphometric, genital, molecular, and chromosomal character suites in distinguishing samples of the three Ethiopian species may reflect a mosaic pattern of evolution. Character discordance, however, has not been an unappreciated reality to generations of working systematists, as many morphologically similar sets of species can only be distinguished by a combination of traits.

Yalden and Largen (1992) claimed that mammalian endemics of Ethiopia are associated with open habitats at high altitude or with forest blocks and are not found in the Awash Valley and other dry lowlands. Lavrenchenko et al. (1998a:49) used the occurrence of *M. awashensis* and a species of *Acomys* (uniquely defined by chromosomal traits according to Sokolov et al., 1993) in the upper Awash Valley as exceptions to Yalden and Largen's proclamation, noting that the valley "with its unique rodent fauna is an integral part of the Ethiopian region with high faunistic diversity and endemism."

Mastomys coucha (Smith, 1834). Rept. Exped. Exploring Central Africa, p. 43.

COMMON NAME: Southern African Mastomys.

TYPE LOCALITY: South Africa, Northern Cape Province, between Orange River and Tropic of Capricorn (see Meester et al., 1986:286).

DISTRIBUTION: Endemic to Southern African Subregion: South Africa (provinces of Eastern and Northern Cape, KwaZulu-Natal, Free State, Gauteng, Mpumalanga, and S and W Limpopo; also in Lesotho), S and W Zimbabwe, C Namibia (Granjon et al., 1997b; Skinner and Smithers, 1990; P. J. Taylor, 1998).

STATUS: IUCN – Lower Risk (lc).

SYNONYMS: *bradfieldi* Roberts, 1926; *breyeri* (Roberts, 1915); *limpopoensis* (Roberts, 1914); *marikquensis* (Smith, 1836); *sicialis* Shortridge, 1934; *silaceus* (Wagner, 1842); *socialis* (Roberts, 1913).

COMMENTS: Characterized by 2n = 36, FNa = 52-56, and a distinctive hemoglobin electromorph, *M. coucha* occurs sympatrically with *M. natalensis* in some areas but allopatrically in other regions of the Southern African Subregion (de Graaff, 1997r); the latter is distinguished by a different hemoglobin pattern (Green et al., 1980), 2n = 32, FN = 54 (Volobouev et al., 2002b), and three isozyme markers (Smit et al., 2001). The two species also differ in cranial, phallic, and spermatozoal morphology as well as reproductive behavior, growth patterns, ultrasonic vocalizations, and pheromones (Breed, 1995a; Dippenaar et al., 1993; Jackson and van Aarde, 2003; Lavrenchenko and Baskevich, 1996; Skinner and Smithers, 1990; Taylor, 2000b; and references cited therein), and can also be separated by principal component analysis of cranial and dental measurements (Dippenaar et al., 1993). Although the two are widely sympatric in southern Africa, their geographic ranges appear concordant with rainfall patterns, with *M. coucha* predominating in drier habitats characteristic of southwestern Africa and *M. natalensis* inhabiting more mesic regions in the east (Taylor, 2000*b*, and references cited therein). Phylogenetic analyses of chromosomal data indicate *M. coucha* is more closely related to *M. huberti* and *M. natalensis* than to *M. erythroleucus* (Britton-Davidian et al., 1995). Karyotype of *M. coucha* is similar to *M. shortridgei* from NE Namibia and NW Botswana, but sex chromosomes differ and the two species can be distinguished by spermatozoal morphology (see references in Granjon et al., 1997*b*). Synonyms listed are only those pertaining to samples from South Africa for reasons explained by Meester et al. (1986); of these, Robbins and Van der Straeten (1989) regarded *marikquensis* to be a *Myomys* (although Van der Straeten and Robbins, 1997, did not include it in their more recent allocation of holotypes to *Myomys*), and Robbins (in Meester et al., 1986) claimed it may be a species distinct from *Myomyscus verreauxii*. Reviewed by Granjon et al. (1997*b*) and de Graaff (1997*r*).

Mastomys erythroleucus (Temminck, 1853). Esq. Zool. sur la Côte de Guine, p. 160.
    COMMON NAME: Reddish-white Mastomys.
    TYPE LOCALITY: Guinea.
    DISTRIBUTION: Mostly Subsaharan N Africa; an isolated population in WC Morocco (Aulagnier, 1991; Aulagnier and Thevenot, 1986); main range in Gambia, Senegal, Guinea, Ghana, Sierra Leone, Côte d'Ivoire, Burkina Faso, Mali, Niger, Benin, Nigeria, Cameroon, Central African Republic, Sudan, Ethiopia, E. Dem. Rep. Congo, Burundi, and W Uganda. See Granjon et al. (1997*b*), Volobouev et al. (2001), and Ziegler et al. (2002).
    STATUS: IUCN – Lower Risk (lc).
    SYNONYMS: *calopus* (Cabrera, 1906); *gambianus* (Thomas, 1911); *peregrinus* (de Winton, 1898).
    COMMENTS: Chromosomal (2n = 38, FNa = 50-56) and electrophoretic data as well as fur color (reddish brown dorsum, whitish venter) set this species apart from other *Mastomys* with which it occurs (Baskevich and Orlov, 1993; Duplantier et al. 1990*a*, *b*; Granjon et al., 1997*b*; Lavrenchenko et al., 1992; Volobouev et al., 2001; and references cited therein). The geographic range described by Granjon et al. (1997*b*) is primarily derived from the distribution of samples with 2n = 38 (see references to chromosomal reports cited by Granjon et al., 1997*b*). Codjia et al. (1996) documented chromosomal distinctions between *M. erythroleucus* and *M. natalensis* from S Benin. *Mastomys erythroleucus* is sympatric with *M. huberti* and *M. natalensis* in Senegal (Duplantier et al., 1990*a*, 1996). Reproductive distinctions among *M. erythroleucus*, *M. natalensis*, and *M. huberti* in Senegal documented by Duplantier et al. (1996). In Senegal, *M. erythroleucus* is a generalist and inhabits a variety of habitats, but *M. huberti* is restricted to humid natural or cultivated areas, and *M. natalensis* is found only in villages (Duplantier et al., 1996). Ethiopian populations of *M. natalaensis* and *M. erythroleucus* exhibited similar ecological ranges (Bulatova et al., 2002). Phylogenetic analysis of chromosomal data indicates that *M. erythroleucus*, although still a member of *Mastomys* and not *Myomys* or *Praomys*, is set apart from *M. natalensis*, *M. huberti*, and *M. coucha*, the other species analyzed (Britton-Davidian et al., 1995). Mitochondrial DNA cytochrome *b* sequences identifies *M. erythroleucus* as sister group to *M. natalensis* (Lecompte et al., 2002*b*). Granjon et al. (1997*b*) provided an excellent review of *M. erythroleucus* and noted that while no single body or cranial measurement will unambiguously discriminate among *M. erythroleucus*, *M. natalensis*, and *M. huberti* in Senegal, discriminant function analyses of cranial, mandibular, and dental measurements do separate the three species. Lavrenchenko and Baskevich (1996) documented differences in spermatozoal morphology distinguishing Ethiopian *M. erythroleucus* and *M. natalensis*, and Bulatova et al. (2002) recorded chromosomal contrasts; the two species are sympatric in the Gambela region. Reviews covering taxonomy, distribution, ecology, and other subjects are available for Morocco (Aulagnier, 1991; Aulagnier and Thevenot, 1986); Gambia, Ghana, and Sierra Leone

(Grubb et al., 1998); S Ghana (Ryan and Attuquayefio, 2000) and Accra Plains of S Ghana (Decher and Bahian, 1999).

More than one species is likely represented in what is now called *M. erythroleucus*. In an analysis of R- and C-bands of samples with 2n = 38, Volobouev et al. (2001) reported that the FN varied from 40 to 60, a result of pericentric inversion polymorphism involving 3-12 chromosome pairs. Presuming that introgressive hybridization is interrupted among populations differing from one another by 3-5 to 11-12 pericentric inversions, Volobouev et al. hypothesized the presence of three "cryptic" species. One (designated MER-1 with FN = 50-56) occurs throughout Subsaharan Africa and includes karyotyped samples from Senegal, Côte d'Ivoire, Mali, Benin, Cameroon, Dem. Rep. Congo, Sudan, and most likely Ethiopia and the isolated Moroccan population; this is *M. erythroleucus*. The second species (MER-2 with FN = 40-41) is represented by samples from Chad and Sudan (see account of *M. kollmannspergeri*), and the third "tentative species" (MER-3 with FN = 59-60) consists of samples from E Dem. Rep. Congo, W Uganda, and possibly Mali and Chad. Volobouev and his colleagues are currently testing this hypothesis by analyses of molecular traits, and have also identified a new 2n = 38 chromosomal species from Niger (MER-4, FN = 52).

The taxa *calopus* and *peregrinus* are based upon Moroccan specimens and Aulagnier and Thevenot (1986) allocated them to the synonymy of *M. erythroleucus*, the only species of *Mastomys* recorded from that country. Wroughton (1908b) claimed *gambianus* was a synonym of *erythroleucus* and Grubb et al. (1998) treated it as such. At the present time we cannot allocate the following names identifying holotypes collected from within what is now accepted as the range of *M. erythroleucus* and *M. natalensis*. **Sudan**: *agurensis* Setzer, 1956; *azrek* (Wroughton, 1911); *blainei* (Wroughton, 1907); *kulmei* Setzer, 1956; *limbatus* (Wagner, 1845); *macrolepis* (Sundevall, 1843); *marrensis* Setzer, 1956. Holotypes of *azrek*, *blainei*, and *macrolepis* from S Sudan may be examples of *M. kollmannspergeri* (Denys et al., 2002; see that account). **Ethiopia:** *gardulensis* (Frick, 1914; a species of *Praomys* according to Yalden et al., 1975, 1996, but holotype clusters with those of other *Mastomys* in the morphometric analysis of Van der Straeten and Robbins, 1997); *lateralis* (Heuglin, 1877); *tacaziena* (Heuglin, 1877). The taxon *fuscirostris* Wagner, 1845 (type locality, Sennaar, Sudan) was associated with *Myomys albipes* by G. M. Allen (1939) and recognized as *Praomys albipes fuscirostris* by Setzer (1956). The holotype has never been examined or identified. It cannot be an example of *albipes*, which is an Ethiopian endemic that has never been collected outside of that country, and we suspect the name should be associated as a species or synonym with *Mastomys* and provisionally list it here until the holotype can be located and identified.

Mastomys huberti (Wroughton, 1909). Ann. Mag. Nat. Hist., ser. 8, 1:255.

COMMON NAME: Hubert's Mastomys.

TYPE LOCALITY: N Nigeria, Zungeru.

DISTRIBUTION: Recorded from S coastal Mauritania (Granjon et al., 1997a, 2002b), Senegal, Mali, Burkina-Faso, and N Nigeria (Granjon et al., 1997b); limits unresolved.

COMMENTS: Characterized by 2n = 32, FNa = 44-46 (Hubert et al., 1983; Duplantier et al., 1990a; Britton-Davidian et al., 1995) or 2n = 32, FNa = 42 (Volobouev et al., 2002b). This is not the same species as *M. natalensis*, which is also characterized by 2n = 32, but a different FN (52-54; Britton-Davidian et al., 1995), hemoglobin pattern (C. B. Robbins et al., 1983), and apparently serum proteins (Robbins and Van der Straeten, 1989). Duplantier et al. (1990a, b) contrasted results of protein electrophoresis and chromosomal traits derived from Senegalese samples of *M. huberti*, *M. natalensis*, and *M. erythroleucus*. Reproductive distinctions among *M. huberti*, *M. erythroleucus*, and *M. natalensis* in Senegal are documented by Duplantier et al. (1996). Phylogenetic analysis of chromosomal data by Britton-Davidian et al. (1995) indicated that *M. huberti* is more closely related to *M. natalensis* than to *M. coucha* and *M. erythroleucus*, the other two species analyzed.

Qumsiyeh et al. (1990) reported a 2n = 32, FNa = 50-54 from Kenyan samples they considered to be different from that of *M. natalensis* and regarded them as the same species having a similar karyotype from Somalia reported by Capanna et al. (1982), who identified their material as *M. huberti*. Qumsiyeh et al. also claimed (without examining holotypes) that *hildebrandtii* (Ndi, Taita, Kenya, is type locality) is an older name for *huberti*; their allocation was followed by Musser and Carleton (1993). Granjon et al. (1997*b*) explained that the karyotype characteristic of *M. huberti* (2n = 32, FNa = 44-46) has not been found in samples from East Africa and is restricted to West Africa, and the samples reported by Qumsiyeh et al. (1990) do not represent *M. huberti*. Granjon et al. also noted (p. 8) "The existence of an East African species distinct from *M. natalensis*, and that may be *M. hildebrandtii* is not sufficiently supported to date." Van der Straeten and Robbins (1997) considered the holotype of *hildebrandtii* to be a *Mastomys*, not an example of *Praomys*, but did not allocate it to species. We place *hildebrandtii* in the synonymy of *M. natalensis* (see also Lecompte, 2003).

Mastomys kollmannspergeri (Petter, 1957). Mammalia, 21:129.

COMMON NAME: Kollmannsperger's Mastomys.

TYPE LOCALITY: N Niger, Aïr region, Titintarat (= Tchet-en-Taghat; Dobigny et al., 2002*b*), 150 km N Agadez (Petter, 1957), "near the wells of Tedjidda-n-Tesemt" (Rosevear, 1969:423) (= Teguidda-n-Tessoumt; Dobigny et al., 2002*b*).

DISTRIBUTION: N Niger (Dobigny et al., 2002*b*), NE Nigeria and N Cameroon (Robbins and Van der Straeten, 1989), Chad (Denys et al., 2002), and S Sudan (Volobouev et al., 2001); limits unresolved.

STATUS: IUCN – Lower Risk (nt) as *M. verheyeni*.

SYNONYMS: *verheyeni* Robbins and Van der Straeten, 1989.

COMMENTS: Originally described as a subspecies of *M. natalensis* (Petter, 1957) from N Niger, chromosomal data documented by Dobigny et al. (2002*b*) from samples near the type locality of *kollmannspergeri* exhibit the same configuration as the chromosomal species that Volobouev et al. (2001) identified as MER-2 (2n = 38, FNa = 40-41) from Chad (Denys et al., 2002) and S Sudan, and Dobigny et al. (2002*b*) provisionally referred their specimens to *M. kollmannspergeri*. Robbins and Van der Straeten (1989) named and described the same species as *M. verheyeni* with a range restricted to the savannas surrounding Lake Chad in NE Nigeria and N Cameroon. It was reviewed under that name by Granjon et al. (1997*b*). Conspecificity of *verheyeni* with *M. kollmannspergeri* is supported by phylogenetic analyses of mtDNA cytochrome *b* and nuclear IRBP gene sequences (L. Granjon, in litt., who noted the molecular work was done by E. Lecompte, 2003), as well as morphometric and cytogenetic analyses (Denys et al., 2002). Study of mitochondrial cytochrome *b* sequences places *M. kollmannspergeri* (reported under *verheyeni*) as sister to a closely related cluster containing *M. coucha*, *M. huberti*, *M. natalensis*, and *M. erythroleucus* (Lecompte et al., 2002*b*). Petter's *kollmannspergeri* may not be the oldest name for the species. Holotypes of *azrek*, *blainei*, and *macrolepis* (see account of *M. erythroleucus* for names and dates) from S Sudan may be examples of the same species (C. Denys, in litt., 2002); if so, one of these older names will replace *kollmannspergeri* (Denys et al., 2002).

Mastomys natalensis (Smith, 1834). S. Afr. Quart. J., ser. 2, 2:156.

COMMON NAME: Natal Mastomys.

TYPE LOCALITY: South Africa, KwaZulu-Natal, Port Natal (= Durban).

DISTRIBUTION: Widespread in subsaharan Africa except for SW portion of continent (see Skinner and Smithers, 1990, and de Graaff, 1997*q*, for range in the Southern African Subregion; see Granjon et al., 1997*b*, for generalized map of overall distribution).

STATUS: IUCN – Lower Risk (lc) as *Myomys fumatus*, *Mastomys hildebrandtii* and *M. natalensis*.

SYNONYMS: *caffer* (Smith, 1834); *cuninghamei* (Wroughton, 1908); *durumae* (Heller, 1912); *effectus* (Dollman, 1911); *evelyni* (Dollman, 1911); *fumatus* (Peters, 1878); *gardulensis* (Frick, 1914); *fusca* (Bocage, 1890); *hildebrandtii* (Peters, 1878); *illovoensis* (Jentink, 1909); *ismaillae* (Heller, 1914); *itigiensis* Hatt, 1935; *kerensis* (Heuglin, 1877); *komatiensis* Roberts, 1926; *longicaudatus* (Noack, 1887); *microdon* (Peters, 1852); *muscardinus* (Wagner, 1843) [see Meester et al., 1986, for reason behind this allocation]; *neumanni* (Heller, 1912); *ovamboensis* Roberts, 1926; *pallida* (Dollman, 1914); *panya* (Heller, 1910); *rufa* (Bocage, 1890); *somereni* (Kershaw, 1923); *tana* (True, 1893); *tinctus* (Heller, 1918); *ugandae* (de Winton, 1897); *victoriae* (Matschie, 1911); *zuluensis* (Thomas and Schwann, 1905).

COMMENTS: Characterized by 2n = 32, FNa = 52-54 (Britton-Davidian et al., 1995; Green et al., 1980; Volobouev et al., 2002b) and a distinctive hemoglobin electromorph (Green et al., 1980; C. B. Robbins et al. 1983). Samples with these chromosomal features have also been found in Senegal (Duplantier et al., 1990a, b; Granjon et al., 1996); Côte d'Ivoire, Central African Republic, Republic of Congo, and Chad (Matthey, 1955, 1966a, b); Sierra Leone and Burundi (C. B. Robbins et al., 1983); S Benin (Codjia et al., 1996); Burkina Faso, Mali, Niger, and Nigeria (Granjon et al., 1997b; Hubert et al., 1983); Somalia (Capanna et al., 1982); Ethiopia (Baskevich and Orlov, 1993; Lavrenchenko et al., 1998a); Tanzania (Fadda et al., 2001; Leirs, 1992; Leirs et al., 1989, 1993); Zimbabwe (Lyons et al., 1980); Namibia (Hallett, 1979), and South Africa (Granjon et al., 1996; Green et al., 1980;). Using chromosomal data, laboratory crosses, and principal component analysis of cranial and dental measurements, Granjon et al. (1996) demonstrated that samples of *M. natalensis* from South Africa were conspecific with those obtained from Senegal. Musser and Carleton (1993) had speculated that *M. natalensis* "probably occurs in Angola, S Zaire [Dem. Rep. Congo], Zambia, Malawi, Mozambique, and farther north in Tanzania and perhaps may even range more extensively in West Africa, but at this time samples from those regions have not been identified by linking chromosomal and biochemical data to morphology. Realizing this problem, some regional faunal accounts provisionally list specimens under *M. natalensis* (e.g., Ansell and Dowsett, 1988, for Malawi mammals)." Data now available indicate that the species occurs throughout subsaharan Africa (except SE southern Africa; Granjon et al., 1997b; Skinner and Smithers, 1990), and may be "the most widely distributed mammal of Africa." (Granjon et al., 1997b).

The following studies focus on systematics and distribution of *M. natalensis*: reproductive distinctions among *M. natalensis*, *M. erythroleucus*, and *M. huberti* in Senegal (Duplantier et al., 1996); chromosomal contrasts between *M. natalensis* and *M. erythroleucus* from S Benin (Codjia et al., 1996); phylogenetic analysis of chromosomal data that indicated closer relationship between *M. natalensis* and *M. huberti* than with *M. coucha* and *M. erythroleucus* (Britton-Davidian et al., 1995); data on trapping and recover from owl pellets in N Malawi (Denys et al., 1999), and regarded as the most common species trapped in S Mali (Meinig, 2000); distribution in the Eastern Arc Mtns and Gonja Forest Reserve of Tanzania (Stanley et al., 1998, 2002); and occurrence in the Bazaruto Arch. off the coast of S Mozambique (Downs and Wirminghaus, 1997). Population in KwaZulu-Natal Province of South Africa extensively discussed by P. J. Taylor (1998), and that in Southern African subregion reviewed by de Graaff (1997q). See account of *M. coucha* for contrasts with that southern African species.

Synonyms based on South Africa samples are those listed by Meester et al. (1986). All other synonyms refer to samples described from Angola, Uganda, Kenya, and Tanzania (see G. M. Allen, 1939, and Crawford-Cabral, 1998), regions where other species closely related to *M. natalensis* (*M. coucha*, *M. huberti*, and *M. erythroleucus*) are not presently known to occur. Holotypes that are the basis for most of these names have been identified as examples of *Mastomys* by Van der Straeten and Robbins (1997) but were not allocated to species. Eventually, our allocations will have to be verified by a systematic revision of *M. natalensis* that includes accurate identification of holotypes. The identity of *longicaudatus*, originally described as a species of *Mystromys*, was first documented by Misonne (1966). Other possible synonyms are listed in account of *M. erythroleucus*. The taxon *fumatus* has traditionally been regarded as a species of *Myomys* (see Musser and Carleton, 1993), but the holotype is a young example of *Mastomys* (Van der Straeten and Robbins, 1997; Musser's examination of holotype); see account of *Myomyscus brockmani*.

Tanzanian populations of *M. natalensis* have been the focus of a three-year study by Leirs (1992) and Leirs et al. (1990, 1993, 1996) resulting in exhaustive information on various aspects of population ecology. Based upon their morphometric and chromosomal study of 6083 specimens from localities throughout Tanzania, they concluded *M. natalensis*–not *M. coucha*, *M. huberti*, or *M. erythroleucus*—is the most common species and that seasonal variation in cranial dimensions within populations is greater than geographic variation among them.

Mastomys pernanus (Kershaw, 1921). Ann. Mag. Nat. Hist., ser. 9, 8:568.

COMMON NAME: Dwarf Mastomys.

TYPE LOCALITY: SW Kenya, Amala (Mara) River; see G. M. Allen (1939:406) and Misonne and Verschuren (1964:655) for details.

DISTRIBUTION: Published records are from SW Kenya, N Tanzania, and Rwanda (Misonne and Verschuren, 1964; see map inVan der Straeten, 1999:228).

STATUS: IUCN – Data Deficient.

COMMENTS: Known by very few specimens. Ellerman (1941) listed *pernanus* as a distinctive species of *Rattus* in the subgenus *Mastomys*, but earlier G. M. Allen (1939) had treated it as a species of *Myomys*. Robbins and Van der Straeten (1989) claimed this a possibility. Misonne and Verschuren (1964, 1966b), however, discussed the problem and concluded that *pernanus* should be associated with *Mastomys*, an arrangement endorsed by Van der Straeten (1999). Van der Straeten (1999) studied and measured all known specimens (except those from Rwanda extracted from raptor pellets and a skull from Tanzania reported by Misonne and Verschuren, 1964, which could not be found in the Koninklijk Belgisch Instituut voor Natuurwetenschappen at Brussels). His principal component analysis contrasting two specimens of *M. pernanus* with holotypes of taxa in *Praomys*, *Mastomys*, *Myomys*, *Myomyscus*, and *Hylomyscus* indicated *M. pernanus* could be a dwarf *Mastomys*. Van der Straeten (1999) and Granjon et al. (1997b) suggested that molecular data might provide sharper resolution of *M. pernanus*'s affinities. Recent analyses of complete mtDNA cytochrome *b* (Lecompte et al., 2002b) and nuclear IRBP gene (Lecompte, 2003) sequences did not cluster *M. pernanus* with the other species of *Mastomys* tested (*M. coucha*, *M. huberti*, *M. erythroleucus*, *M. natalensis*, and *M. kollmannspergeri* [reported as *verheyeni*]), but aligned it with *Heimyscus* and species of *Hylomyscus*. Keeping *pernanus* within *Mastomys* renders that genus polyphyletic, but extracting it will require additional analyses using a broader array of genes to determine if *pernanus* is the sole member of a separate genus related to *Hylomyscus* or to be included within that genus (see discussion in Lecompte, 2003).

Mastomys shortridgei (St. Leger, 1933). Proc. Zool. Soc. Lond., 1933:411.

COMMON NAME: Shortridge's Mastomys.

TYPE LOCALITY: N Namibia, Grootfontein Dist., Okavango-Omatako junction.

DISTRIBUTION: Extreme NW Botswana and NE Namibia (Caprivi area) in the region of the confluence of Okavango and Kwito Rivers where it apparently inhabits reedbeds and swamp grasses (de Graaff, 1981:216; Skinner and Smithers, 1990:270); also E Angola, where the species is associated with marshes (Crawford-Cabral, 1998).

STATUS: IUCN – Lower Risk (lc).

SYNONYMS: *legerae* (Ellerman, Morrison-Scott, and Hayman, 1953).

COMMENTS: Karyotype (2n = 36, FNa = 50) similar to that of *M. coucha* (Granjon et al., 1997b). Taxonomy reviewed by Meester et al. (1986:285). Some workers considered *shortridgei* closely related to or the same as *M. angolensis* (see references cited in Meester et al., 1986). That species has five pairs of teats and most authors attributed the same number to *M. shortridgei*, but the latter has eight pairs (see references in Granjon et al., 1997b). This is also a significant distinction for the generic allocation of *shortridgei* because Granjon et al. (1997b) employed more than ten teats as one of the diagnostic traits of *Mastomys* and included *shortridgei* in the genus but excluded *angolensis*. A principal component analysis based on holotypes of the *Mastomys-Praomys-Myomys-Hylomyscus* complex also nested that of *shortridgei* deep within the cluster identified as *Mastomys* (Van der Straeten and Robbins, 1997). *Mastomys shortridgei* occurs sympatrically with *M. natalensis* (de Graaff, 1997q, 1997s; Skinner and Smithers, 1990). Reviewed by de Graaff (1997s) and Granjon et al. (1997b), who also cited references describing chromosomal and spermatozoal distinctions between *M. shortridgei* and *M. coucha*.

Maxomys Sody, 1936. Natuurk. Tidjschr. Ned.-Ind., 96:55.

TYPE SPECIES: *Mus bartelsii* Jentink, 1910.

COMMENTS: *Maxomys* Division. Definition and contents of *Maxomys* provided by Musser et al. (1979), who discussed historical allocations of species to groups of *Rattus* and summarized chromosomal information. The exclusion of *Maxomys* from *Rattus* also supported by al-lozymic data (Chan et al., 1979), albumin immunology (Watts and Baverstock, 1994*b*), and DNA/DNA hybridization studies (Ruedas and Kirsch, 1997). Morphological analyses indicated that *Maxomys* shares all its dental derived molar traits with *Berylmys*, *Mus*, and *Niviventer*, and some of its external features with *Mus* (Musser and Newcomb, 1983). Esti-mate of relationships based on karyotypic data placed *Maxomys* close to *Niviventer* and *Lenothrix* and far from *Rattus* (Gadi and Sharma, 1983). Spermatozoal morphology of Ma-layan *Maxomys* and one Sulawesian species is similar to *Chiropodomys* and *Hapalomys*, a configuration also like that seen in *Mus* and *Apodemus* (Breed and Musser, 1991; Breed and Yong, 1986). Sperm conformation in the other species of Sulawesian *Maxomys* were unlike the Malayan species and resembled that of the Sulawesian endemic *Margaretamys* (Breed and Musser, 1991). In their study of DNA/DNA hybridization, Ruedas and Kirsch (1997) sampled five of the 19 species of *Maxomys*, all Sundaic (*M. ochraceiventer*, *M. white-headi*, *M. rajah*, *M. bartelsii*, and *M. surifer*); they formed a monophyletic clade relative to other Indomalayan genera sampled (*Rattus*, *Sundamys*, *Niviventer*, and *Leopoldamys*); whether all other species of *Maxomys*, especially those endemics on Sulawesi, are also members of this clade remains to be determined. Albumin immunology indicates *Max-omys* to be "a separate lineage within South-east Asian murines not particularly close to any other" (Watts and Baverstock, 1994*b*), and analysis of DNA sequence of LINE-1 ele-ments places *Maxomys* in a clade separate from that containing *Leopoldamys* and *Niviven-ter* (members of our *Dacnomys* Division), and another formed by *Rattus*, *Berlymys*, *Sun-damys*, and *Bandicota*, genera in our *Rattus* Division (Verneau et al., 1997, 1998). We isolate *Maxomys* in its own division, which highlights current estimates of its phylogene-tic position within the evolutionary diversity of extant Indomalayan murines.

Closest relative of *Maxomys* may be the extinct *Ratchaburimys ruchae*, described from jaw fragments and isolated molars recovered from late Pliocene and early Pleistocene cave sediments in S Thailand north of the Isthmus of Kra (10°30′ N) (Chaimanee, 1998). Aside from late Pliocene and Pleistocene examples of Thai *M. surifer* (see that account), the only other fossils determined as *Maxomys* come from early Pleistocene beds in E Java (Van der Meulen and Musser, 1999). There is an undescribed species of *Maxomys* from Mindoro Isl in the Philippines (see accounts of *Anonymomys mindorensis* and *Apomys gracilirostris*) and two from C Sulawesi (Musser, ms).

Maxomys alticola (Thomas, 1888). Ann. Mag. Nat. Hist., ser. 6, 2:408.
COMMON NAME: Bornean Mountain Maxomys.
TYPE LOCALITY: Malaysia, Sabah (N Borneo), Mt Kinabalu ("from a high level").
DISTRIBUTION: Known only from two mountains in Sabah, N Borneo: from 1070 to 3360 on Mt Kinabalu (Md Nor, 2001), and also on Gunung Trus Madi.
STATUS: IUCN – Endangered.
SYNONYMS: *kinabaluensis* (Chasen, 1937).
COMMENTS: The name *alticola* was once applied to populations from lower elevations on Gunung Kinabalu, the Malay Peninsula, and Sumatra (Chasen, 1940), but was shown to be the name for a distinct species endemic to mountains of Sabah (Medway, 1964, 1977). Reviewed by Corbet and Hill (1992). Part of a suite of murine species endemic to Borneo (Musser, 1986; also see account of *Chiropodomys major*).

Maxomys baeodon (Thomas, 1894). Ann. Mag. Nat. Hist., ser. 6, 14:452.
COMMON NAME: Small Bornean Maxomys.
TYPE LOCALITY: Malaysia, Sabah (N Borneo), Mt Kinabalu.
DISTRIBUTION: Known only from a few scattered localities in Sabah and Sarawak (N Borneo); occurs between 900 and 1400 m on Mt Kinabalu in Sabah (Md Nor, 2001).
STATUS: IUCN – Endangered.
SYNONYMS: *trachynotus* (Cabrera, 1920).
COMMENTS: A very distinctive species still represented in museum collections by few specimens (Musser et al., 1979). The record from Gunung Palung National Park in SW Kalimantan (Blundell, 1996) is erroneous and represents misidentified *M. white-headi* (A. J. Gorog, pers. comm., 2001). Part of the suite of murines endemic to Borneo (Musser, 1986; also see account of *Chiropodomys major*).

Maxomys bartelsii (Jentink, 1910). Notes Leyden Mus., 33:69.
   COMMON NAME: Bartels's Javan Maxomys.
   TYPE LOCALITY: Indonesia, W Java, Gunung Pangerango-Gede, 6000 ft (1830 m).
   DISTRIBUTION: Endemic to montane forests on volcanos of W and C Java (Van Peenen et al.,
      1974).
   STATUS: IUCN – Lower Risk (lc).
   SYNONYMS: *obscuratus* (Bartels, 1938); *tjibunensis* (Sody, 1933).
   COMMENTS: Morphology, chromosomes, and geographic distribution reviewed by Van
      Peenen et al. (1974); skull and dentition illustrated in Musser and Newcomb (1983).
      Reviewed also by Corbet and Hill (1992). Member of a suite of murine rodents endemic
      to Java (Musser, 1986; see account of *Kadarsanomys sodyi*).

Maxomys dollmani (Ellerman, 1941). Families and Genera of Living Rodents, 2:218.
   COMMON NAME: Dollman's Sulawesi Maxomys.
   TYPE LOCALITY: Indonesia, C Sulawesi, Quarles Mtns, Rantekaroa, 6000 ft (1830 m).
   DISTRIBUTION: Type locality and Gunung Tanke Salokko in the SE peninsula of Sulawesi;
      known only from high elevations in montane forests of the SE peninsula and S region
      of the C core of the island.
   STATUS: IUCN – Vulnerable.
   COMMENTS: Originally described by Ellerman as a subspecies of *Rattus hellwaldii*, but shown
      to be a distinct species by Musser (1969c), and morphologically allied to *M. hellwaldii*
      (Musser, 1991). *Maxomys dollmani* is scansorial and a smaller-bodied version of an
      undescribed species collected from cool and wet primary forest in the mountains of
      C Sulawesi between 854 and 1460 m. That new species is sympatric with *M. hellwaldii*
      in the lower part of its altitudinal range and occurs with *M. musschenbroekii*
      throughout the altitudinal distribution. The geographic pattern exhibited by the
      C highlands species and *M. dollmani* (the former in the C core of the island, the latter
      on the SE peninsula and S portion of the C core) is common to some sets of species in
      Taeromys and the *Rattus xanthurus* Group (see those accounts).

Maxomys hellwaldii (Jentink, 1878). Notes Leyden Mus., 1:11.
   COMMON NAME: Hellwald's Sulawesi Maxomys.
   TYPE LOCALITY: Indonesia, Sulawesi, NE peninsula, Menado.
   DISTRIBUTION: Throughout Sulawesi in tropical lowland evergreen rain forest, sea level to
      3300 ft (1006 m; Musser and Holden, 1991; Musser, ms).
   STATUS: IUCN – Lower Risk (lc).
   SYNONYMS: *cereus* (Miller and Hollister, 1921); *griseogenys* (Sody, 1941); *localis* (Miller and
      Hollister, 1921).
   COMMENTS: Among Sulawesian representatives of *Maxomys*, *M. hellwaldii* is most closely
      related to *M. dollmani* and two undescribed species in highlands of C Sulawesi (Musser,
      1991; Musser, ms). Spermatozoal morphology unlike species of Malayan *Maxomys*
      studied, and more similar to members of *Margaretamys*, another Sulawesian endemic
      (Breed and Musser, 1991). There is appreciable morphological variation among
      samples from different regions of the island, which may reflect presence of more than
      one species (Musser, ms). Stomach morphology described and contrasted with other
      Sulawesian endemics by Musser and Durden (2002).

Maxomys hylomyoides (Robinson and Kloss, 1916). J. Str. Br. Roy. Asiat. Soc., 73:273.
   COMMON NAME: Sumatran Mountain Maxomys.
   TYPE LOCALITY: Indonesia, W Sumatra, Korinchi Peak, 7300 ft (2225 m).
   DISTRIBUTION: Endemic to montane forests in mountainous backbone of W Sumatra.
   STATUS: IUCN – Lower Risk (nt).
   COMMENTS: At one time listed as a subspecies of *M. alticola* (Chasen, 1940), but later
      reinstated as a distinctive species (Medway, 1964; Musser et al., 1979). Member of a
      suite of murines endemic to Sumatra that includes *Maxomys inflatus*, *Mus crociduroides*,
      Rattus korinchi, and *R. hoogerwerfi* (Musser, 1986).

Maxomys inas (Bonhote, 1906). Proc. Zool. Soc. Lond., 1906:9.
   COMMON NAME: Malayan Mountain Maxomys.

TYPE LOCALITY: Malaysia, Perak (Malay Peninsula), Gunung Inas.

DISTRIBUTION: Endemic of Malay Peninsula (and possibly peninsular Thailand south of Isthmus of Kra) in montane forests, rarely occurring below 3000 ft (914 m; Medway, 1969).

STATUS: IUCN – Lower Risk (lc).

COMMENTS: Once treated as a subspecies of Bornean *M. alticola* (Chasen, 1940), but identified as a separate species by Medway (1964). A close phylogenetic ally to *M. whiteheadi*, as indicated by morphological, biochemical, chromosomal, and spermatozoal traits (Breed and Yong, 1986; Chan et al., 1978, 1979; Yong, 1969). *Maxomys inas* joins *Niviventer cameroni* and *Pithecheir parvus* as the three living endemic murines on the Malay Peninsula south of the Isthmus of Kra (Musser, 1986; see account of *Niviventer cameroni*).

Maxomys inflatus (Robinson and Kloss, 1916). J. Str. Br. Roy. Asiat. Soc., 73:273.

COMMON NAME: Broad-nosed Sumatran Maxomys.

TYPE LOCALITY: Indonesia, W Sumatra, Korinchi Peak, Sungei Kumbang, 4700 ft (1433 m).

DISTRIBUTION: Endemic to mountain forests of W Sumatra.

STATUS: IUCN – Lower Risk (lc).

COMMENTS: One of the most distinctive species of *Maxomys*, as reflected by its wide rostrum and inflated nasolacrimal capsules (Corbet and Hill, 1992; Musser et al., 1979). Part of the suite of murine species endemic to Sumatra (Musser, 1986; see account of *Maxomys hylomyoides*).

Maxomys moi (Robinson and Kloss, 1922). Ann. Mag. Nat. Hist., ser. 9, 9:95.

COMMON NAME: Indochinese Mountain Maxomys.

TYPE LOCALITY: S Vietnam, Lâm Dǧng Province, Langbian Mtns (= Lam Vien Plateau), Arbre Broyé (= Tram Hanh), 5400 ft (1646 m).

DISTRIBUTION: Recorded only from highlands of S Vietnam in Quàng Tri, Quàng Nam, and Lâm Dǧng provinces (Dang et al., 1994; Van Peenen et al., 1969) and adjacent S Laos on Plateau Bolovens; 190-1500 m; does not occur in the Annamite Mtns along the Laos-Vietnam border north of Quàng Tri Province. Distribution based primarily upon study of specimens in AMNH, BMNH, FMNH, and USNM.

STATUS: IUCN – Lower Risk (lc).

COMMENTS: Originally described as a species by Robinson and Kloss (1922), later listed as a subspecies of either *Maxomys surifer* or *Rattus coxingi* (=*Niviventer coninga*), but finally reinstated as a distinct species (Musser et al., 1979; Van Peenen et al., 1969). Closest phylogenetic relative is *M. surifer*, an estimate based on cranial, dental, and chromosomal traits shared by both species (Bulatova et al., 1992; Duncan and Van Peenen, 1971; Musser et al., 1979).

Maxomys musschenbroekii (Jentink, 1878). Notes Leyden Mus., 1:10.

COMMON NAME: Musschenbroek's Sulawesi Maxomys.

TYPE LOCALITY: Indonesia, Sulawesi, NE peninsula, Menado.

DISTRIBUTION: Throughout Sulawesi in most forest formations (Musser, 1991; Musser and Holden, 1991).

STATUS: IUCN – Lower Risk (lc).

SYNONYMS: *aspinatus* (Tate and Archbold, 1935); *lalawora* (Sody, 1941); *tetricus* (Miller and Hollister, 1921).

COMMENTS: Considered conspecific with Sundaic *M. whiteheadi* by Ellerman and Morrison-Scott (1951), but *musschenbroekii* is a separate species endemic to Sulawesi (Medway, 1977; Musser, 1991). Spermatozoal morphology more similar to that described for Malayan species of *Maxomys* than to sperm of other sampled Sulawesian species (Breed and Musser, 1991). Occurs sympatrically with all other species of Sulawesian *Maxomys*, and does not show significant morphological variation among samples from different island regions (Musser, ms).

Maxomys ochraceiventer (Thomas, 1894). Ann. Mag. Nat. Hist., ser. 6, 14:451.

COMMON NAME: Ochraceous-bellied Bornean Maxomys.

TYPE LOCALITY: Malaysia, Sabah (N Borneo), Gunung Kinabalu, below 1700 m (Md Nor, 2001).

DISTRIBUTION: Sabah and Sarawak (N Borneo), and East Kalimantan (E Borneo); apparently restricted to hills (Medway, 1964).

STATUS: IUCN – Lower Risk (lc).

SYNONYMS: *perasper* (Shamel, 1940).

COMMENTS: Once treated as a lower altitudinal subspecies of the higher montane *M. alticola* on Borneo (Chasen, 1940), but the two species are sympatric at 1067 m on slopes of Gunung Kinabalu (Medway, 1977). Reviewed by Corbet and Hill (1992). Member of the cluster of murine species endemic to Borneo (Musser, 1986; also see account of *Chiropodomys major*).

Maxomys pagensis (Miller, 1903). Smithson. Misc. Coll., 45:39.

COMMON NAME: Mentawai Archipelago Maxomys.

TYPE LOCALITY: Indonesia, Kepulauan Mentawai, Pulau Pagai Selattan (S Pagai Isl), off coast W Sumatra.

DISTRIBUTION: Endemic to islands of South Pagai, North Pagai, Sipora, and Siberut in Mentawai Arch.

STATUS: IUCN – Lower Risk (nt).

COMMENTS: Usually listed as a subspecies of *M. surifer* (Chasen, 1940), but treated as a species by Musser et al. (1979) and retained as such by Corbet and Hill (1992). Closest phylogenetic relative is probably *M. surifer*. Part of the rodent fauna endemic to Mentawai Arch. (see account of *Leopoldamys siporanus*).

Maxomys panglima (Robinson, 1921). Ann. Mag. Nat. Hist., ser. 9, 7:235.

COMMON NAME: Palawan Maxomys.

TYPE LOCALITY: Philippines, Palawan Isl.

DISTRIBUTION: Greater Palawan Faunal Region; endemic to Balabac, Palawan, Busuanga, Calauit, and Culion Isls; politically part of Philippines, but faunistically an extension of the Sunda Shelf. Recorded from primary and secondary tropical lowland evergreen rainforest from sea level to 1000 m (Heaney et al., 1998).

STATUS: IUCN – Lower Risk (nt).

SYNONYMS: *palawanensis* (Taylor, 1934).

COMMENTS: Treated in the past as a subspecies of *M. surifer*, but morphological features support its independence as a species; uncertain whether *M. panglima* is more closely related to *M. surifer* or *M. rajah* (Musser et al., 1979). Joins *Palawanomys furvus* and *Chiropodomys calamianus* as the only recorded murines endemic to the Greater Palawan Faunal Region.

Maxomys rajah (Thomas, 1894). Ann. Mag. Nat. Hist., ser. 6, 14:451.

COMMON NAME: Rajah Sundaic Maxomys.

TYPE LOCALITY: Malaysia, Sarawak (N Borneo), Gunung Batu Song.

DISTRIBUTION: Endemic to the Sunda Shelf; Peninsular Thailand south of Isthmus of Kra, Malay Peninsula, Riau Arch., Sumatra, amd Borneo; absent from Java and Bali.

STATUS: IUCN – Lower Risk (lc).

SYNONYMS: *hidongis* (Kloss, 1921); *lingensis* (Miller, 1900); *pellax* (Miller, 1900); *similis* (Robinson and Kloss, 1916).

COMMENTS: Insular distribution similar in broad outline to that of *M. whiteheadi*. Once regarded as conspecific with *M. surifer* (Ellerman and Morrison-Scott, 1951; Misonne, 1969), but occurs sympatrically with that species, and although samples of each are regularly misidentified, the two differ in a suite of morphological, ecological, behavioral, and biochemical traits, as well as albumin immunology (Chan et al., 1979; Corbet and Hill, 1992; Musser et al., 1979; Yong, 1972; Watts and Baverstock, 1994b).

Maxomys surifer (Miller, 1900). Proc. Biol. Soc. Wash., 13:148.

COMMON NAME: Indomalayan Maxomys.

TYPE LOCALITY: Peninsular Thailand, Trang.

DISTRIBUTION: Indochina; S Burma, Thailand (J. T. Marshall, Jr., 1977a; Robinson et al., 1995), Laos, S and SW Cambodia (Elephant Mtns, specimens in FMNH; Cardamom Mtns, A. Smith, in litt., 2002), throughout Vietnam (Dang et al., 1994; Van Peenen et al., 1969; specimens in AMNH and IEBR; including Thom and PhuQuoc Isls off the

S coast of Vietnam, see Kuznetsov, 2000), and extreme S Yunnan near Laotian border (specimens in IZAS, D. Lunde, in litt., 2004; Wang, 2003; Wu et al., 1996); and the Sunda Shelf (Peninsular Thailand, Malay Peninsula, Borneo, Sumatra, Java, and many smaller islands); in addition to references cited above, range is generally extracted from Corbet and Hill (1992) and Musser et al. (1979).

STATUS: IUCN – Lower Risk (lc).

SYNONYMS: *anambae* (Miller, 1900); *antucus* (Lyon, 1916); *aoris* (Robinson, 1912); *banacus* (Lyon, 1916); *bandahara* (Robinson, 1921); *bentincanus* (Miller, 1903); *binominatus* (Kloss, 1915); *butangensis* (Miller, 1900); *carimatae* (Miller, 1906); *casensis* (Miller, 1903); *catellifer* (Miller, 1903); *changensis* (Kloss, 1916); *connectens* (Kloss, 1916); *domelicus* (Miller, 1903); *eclipsis* (Kloss, 1916); *finis* (Kloss, 1916); *flavidulus* (Miller, 1900); *flavigrandis* (Kloss, 1911); *grandis* (Kloss, 1911); *koratis* (Kloss, 1919); *kramis* (Kloss, 1919); *kutensis* (Kloss, 1916); *leonis* (Robinson and Kloss, 1911); *luteolus* (Miller, 1903); *mabalus* (Lyon, 1916); *manicalis* (Robinson and Kloss, 1914); *microdon* (Kloss, 1908) [not Peters, 1852]; *muntia* (Chasen, 1940); *natunae* (Chasen, 1940); *pelagius* (Kloss, 1916); *pemangilis* (Robinson, 1912); *perflavus* (Lyon, 1911); *pidonis* (Chasen, 1940); *pinacus* (Lyon, 1916); *puket* (Chasen, 1940); *ravus* (Robinson and Kloss, 1916); *saturatus* (Lyon, 1911); *serutus* (Miller, 1906); *siarma* (Kloss, 1919); *solaris* (Sody, 1934); *spurcus* (Robinson and Kloss, 1914); *telibon* (Chasen, 1940); *tenebrosus* (Kloss, 1916); *ubecus* (Lyon, 1911); *umbridorsum* (Miller, 1903); *verbeeki* (Sody, 1930).

COMMENTS: The only species of *Maxomys* with a range encompassing Indochinese and Sundaic faunal regions. The two groups of samples differ in morphological features, and the significance of this variation should be determined in a careful systematic revision of the genus. *Berylmys bowersii*, *Chiropodomys gliroides*, and *Leopoldamys sabanus* have roughly concordant geographic ranges and demonstrate similar patterns of geographic morphological variation (Musser and Newcomb, 1983; Musser et al., 1979). Virtually every insular sample from the Sunda Shelf has been given a scientific name, but any significance of insular variation has yet to be assessed. Corbet and Hill (1992) detected a geographic pattern in pelage coloration, but whether it is concordant with distribution of morphological and molecular traits has yet to be determined. Phylogenetic analyses of mtDNA cytochrome *b* sequences by A. J. Gorog et al. (2004) identified six distinct lineages in what is now defined as *M. surifer* associated with 1) Java, 2) Sumatra, 3) Borneo, 4) the Malay Peninsula, 5) S Vietnam, and 6) NE Vietnam. The Javan and Sumatra lineages formed one clade, which is sister to the Bornean clade, the Malay Peninsula lineage is sister to the Java/Sumatra and Bornean clades, and the two Vietnamese lineages are basal to all those other lineages with the southern Vietnamese cluster siser to the clades from the Malay Peninsula and Sundanese islands. Gorog et al. (2004) suggested that *M. surifer* evolved on the Indochinese mainland and the vicariance patterns revealed by molecular analyses ". . . likely have their roots in the Pliocene fragmentation of the Sunda block. . ." rather than widespread dispersal across the late Pleistocene Sunda Shelf and subsequent isolation with increasing sea levels. The early presence of *Maxomys* in the Indomalaysian region is supported by late Pliocene to middle Pleistocene fossils of *Maxomys* and related forms (Chaimanee, 1998; van der Meulen and Musser, 1999).

Comparative spermatozoal morphology documented by Breed and Yong (1986), chromosomal and biochemical data summarized by Chan et al. (1979) in phylogenetic context, and chromosomal data for Vietnam samples reported by Bulatova et al. (1992). Md Nor (1996) recorded its distribution and ecology on the small islands off the tip of N Sabah. Generally a lowland species, *M. surifer* reaches 1680 m on the slopes of Mt Kinabalu in Sabah (Md Nor, 2001, and references cited therein). Ecological relationships with species of *Bandicota*, *Mus*, *Rattus*, *Leopoldamys*, and *Niviventer* in S Yunnan near Laotian border discussed by Wu et al. (1996). *Maxomys surifer* is the only *Maxomys* occurring in China (S Yunnan). Wang (2003), however, recognized *M. rajah* from S Yunnan and *M. musschenbroekii* from Guizhou Province, but the former is a Sundaic endemic and the latter lives only on Sulawesi (see those accounts). Isolated molars identified as *M. surifer* have been recovered from late Pliocene to Holocene cave sediments in Thailand (Chaimanee, 1998).

Maxomys wattsi Musser, 1991. Am. Mus. Novit., 3001:20.

COMMON NAME: Watts's Sulawesi Maxomys.

TYPE LOCALITY: Indonesia, C Sulawesi, Gunung Tambusisi, Tambusisi Damar, 4700 ft (1432 m).

DISTRIBUTION: Known only from 4700-6000 ft (1432-1830 m) on Gunung Tambusisi, C Sulawesi.

STATUS: IUCN – Endangered.

COMMENTS: A distinctive species that is morphologically isolated from all other Sulawesian *Maxomys* except possibly *M. musschenbroekii*; assessing its phylogenetic relations to other species in the genus will require systematic revision of *Maxomys* (Musser, 1991).

Maxomys whiteheadi (Thomas, 1894). Ann. Mag. Nat. Hist., ser. 6, 14:452.

COMMON NAME: Whitehead's Sundaic Maxomys.

TYPE LOCALITY: Malaysia, Sabah (N Borneo), Gunung Kinabalu.

DISTRIBUTION: Peninsular Thailand south of Isthmus of Kra, Malay Peninsula, Sumatra, Borneo, and various adjacent islands; absent from Java and Bali; see Corbet and Hill (1992) and Musser et al. (1979).

STATUS: IUCN – Lower Risk (lc).

SYNONYMS: *asper* (Miller, 1900); *batamanus* (Lyon, 1907); *batus* (Miller, 1911); *coritzae* (Sody, 1941); *klossi* (Bonhote, 1906); *mandus* (Lyon, 1908); *melanurus* (Shamel, 1940); *melinogaster* (Cabrera, 1920); *perlutus* (Thomas, 1911); *piratae* (Chasen, 1940); *subitus* (Chasen, 1940).

COMMENTS: A Sundaic endemic. Ellerman and Morrison-Scott (1951) included *whiteheadi* in the Sulawesi *M. musschenbroekii*, but the two are separate species (Chasen, 1940; Medway, 1977; Musser, 1991; Musser et al., 1979; Tate, 1936). Spermatozoal morphology (Breed and Yong, 1986) and data from biochemical, morphological, and cytological studies (Chan et al., 1978, 1979; Yong, 1969) pointed to a close relationship between *M. whiteheadi* and the Malayan *M. inas*. Md Nor (1996) documented occurrence of *M. whiteheadi* on the small islands off the tip of N Sabah. Generally a lowland species on Borneo, it does reach 2100 m on the slopes of Mt Kinabalu (Md Nor, 2001, and references cited therein). Noticeable geographic variation in pelage coloration and aspects of cranial morphology exists among insular samples on the Sunda Shelf, but its significance has yet to be determined. However, phylogenetic analyses of mtDNA cytochrome *b* sequences by Gorog et al. (2004) identified a clade containing two separate lineages on Borneo and another clade containing samples from Sumatra and the Malay Peninsula. Significance of this pattern is discussed by Gorog et al. (2004) in the context of determining whether these lineages represent deep prePleistocene vicariant events or broad Pleistocene migrations over the Sunda Shelf associated with formation of land bridges and insular isolation with subsequent rising sea level. Reviewed by Corbet and Hill (1992).

Melasmothrix Miller and Hollister, 1921. Proc. Biol. Soc. Wash., 34:93.

TYPE SPECIES: *Melasmothrix naso* Miller and Hollister, 1921.

COMMENTS: *Melasmothrix* Division. Member of Sulawesian Old Endemics (Musser, 1981c). Closely related to *Tateomys* as judged by morphological, ecological, and spermatozoal characters (Breed and Musser, 1991; Musser, 1982c). Both genera are highly specialized diurnal (*Melasmothrix*) or nocturnal (*Tateomys*) vermivores whose phylogenetic affinities have yet to be determined. However, their molar occlusal patterns (Musser, 1982c) are somewhat specialized versions of those characterizing Miocene *Progonomys*-like morphologies, and the shrew rats may have evolved from one of the earliest divergences from a *Progonomys*-like stock.

Melasmothrix naso Miller and Hollister, 1921. Proc. Biol. Soc. Wash., 34:93.

COMMON NAME: Diurnal Sulawesian Shrew Rat.

TYPE LOCALITY: Indonesia, C Sulawesi, Rano Rano, 6000 ft (1830 m); see Musser (1982c) for additional information.

DISTRIBUTION: Sulawesi; known only from tropical upper montane rain forest at Rano Rano and on Gunung Nokilalaki, 1950-2286 m; probably occurs in upper montane forest on other mountains in C core of the island.

STATUS: IUCN – Endangered.

COMMENTS: Stomach morphology described and contrasted with other Sulawesi shrew rats and the insectivorous *Sommeromys macrorhinos* by Musser and Durden (2002). Other morphological aspects, along with altitudinal distribution, diet, and ecology of this terrestrial, diurnal, and primarily vermivorous shrew rat reviewed by Musser (1982c).

Melomys Thomas, 1922. Ann. Mag. Nat. Hist., ser. 9, 9:261.

TYPE SPECIES: *Uromys rufescens* Alston, 1877.

COMMENTS: *Uromys* Division. Member of the Australian and New Guinea region Old Endemics (Musser, 1981c, 1982b). Data from microcomplement fixation of albumin indicated Australian *Melomys* is closely related to *Uromys* and in the same monophyletic group with *Mesembriomys, Leporillus, Conilurus,* and *Zyzomys* (Watts et al., 1992). These genera join other Australian and some New Guinea genera to form an "Australasian clade" as estimated by albumin immunology (Watts and Baverstock, 1994a, 1995b, 1996). Menzies (1996) has revised the New Guinea species that were formerly included in *Melomys*. His results considerably contracted the scope of the genus by raising subgenus *Paramelomys* to generic rank, separating *lanosus* and *rattoides* as a separate genus (see account of *Mammelomys*) and placing *fellowsi* in its own genus (see account of *Protochromys*). Mahoney and Richardson (1988) cataloged taxonomic, distributional, and biological references to Australian species; Flannery (1995a, b) summarized the same for many New Guinea *Melomys*.

Melomys aerosus (Thomas, 1920). Ann. Mag. Nat. Hist., ser. 9, 6:428.

COMMON NAME: Dusky Seram Melomys.

TYPE LOCALITY: Indonesia, Pulau Seram, Mt Manusela, 6000 ft (1830 m).

DISTRIBUTION: Endemic to Seram Isl, 1200-1830 m.

STATUS: IUCN – Lower Risk (nt).

COMMENTS: Tate (1951:292) suggested a relationship between *M. aerosus* and the New Guinea *M. levipes* group, but cranial morphology indicates *M. aerosus* to be more closely related to Australian species of *Melomys*, especially *M. cervinipes*, than to the endemic *Melomys* of New Guinea (Musser's study of specimens in BMNH). Menzies (1996:418) noted that *M. aerosus* "has the long incisive foramina characteristic of the *cervinipes* division of *Melomys*" but "does not fit comfortably in any group" Reviewed by Flannery (1995a) and Helgen (2003b).

Melomys arcium (Thomas, 1913). Ann. Mag. Nat. Hist., ser. 8, 12:214.

COMMON NAME: Rossel Island Melomys.

TYPE LOCALITY: Papua New Guinea, Louisiade Arch., Rossel Isl.

DISTRIBUTION: Endemic to Rossel Isl, 50-700 m.

COMMENTS: Originally described as a species of *Uromys*, treated as a subspecies of *M. leucogaster* by Rümmler (1938), Musser and Carleton (1993), and Flannery (1995b), but retained as a species by Laurie and Hill (1954) and Menzies (1996). *Melomys arcium* is closely related to *M. leucogaster* (Menzies, 1996, and our study of the holotype and AMNH 159597-159600), and Flannery (1995b:135) wrote that although closely related to *M. leucogaster*, it may represent "a distinct species." Flannery (1995b:135) noted that most islands off the E coast of Papua New Guinea in Milne Bay have been heavily surveyed and neither *M. leucogaster* nor *M. arcium* has been found, "Therefore it seems likely that it is found only on Rossel among the Milne Bay Islands. This is very unusual, because Rossel is the most distant of the Louisiade Group from New Guinea. Presumably, it did exist previously on other islands of the group, but has become extinct on all except Rossel."

Melomys bannisteri Kitchener and Maryanto, 1993. Rec. West. Aust. Mus., 16:428.

COMMON NAME: Great Key Island Melomys.

TYPE LOCALITY: Indonesia, Maluku Tengah, Pulau Kai Besar (Great Key Isl), 2 km W Fakoi, 200 m (see Kitchener and Maryanto, 1993b, for details).

DISTRIBUTION: Recorded only from Pulau Kai Besar in Kepulauan Kai (Ewab), between Seram Isl and the Aru Isls.

COMMENTS: Morphologically similar to *M. lutillus*. Flannery (1995*b*:138) included *bannisteri* in *M. lutillus* because "it is clearly part of the *M. lutillus* group," and also noted that "the confused taxonomic relationships of the *M. lutillus* group make it difficult to assess the validity of this taxon." *Melomys bannisteri* is one average larger than either *M. lutillus* or *M. frigicola* on New Guinea, and much larger than *M. burtoni* from the Trans-Fly region (*muscalis*); compare the measurements listed by Kitchener and Maryanto (1993*b*:431) with those of *M. lutillus* and *M. frigicola* presented by Menzies (1996:397). Kitchener and Maryanto's taxon should be highlighted as a species until its diagnostic characteristics can be more critically assessed in the context of a revision of these small-bodied *Melomys*. The Kai Isls are in deep water and not on the continental shelf connecting Australia and New Guinea beneath the Arafura Sea, and any endemic species such as *M. bannisteri* may be the product of a longer evolutionary history in isolation than is true of faunas on continental shelf islands associated with Australia and New Guinea.

Melomys bougainville Troughton, 1936. Rec. Aust. Mus., 19:344.

COMMON NAME: Bougainville Island Melomys.

TYPE LOCALITY: Solomon Isls, Bougainville Isl, Buin.

DISTRIBUTION: Recorded only from islands of Buka, Bougainville, and Choiseul in northern Solomon Arch. (Flannery and Wickler, 1990; Flannery, 1995*b*).

STATUS: IUCN – Lower Risk (nt).

COMMENTS: Although historically treated as a subspecies of *M. rufescens*, *M. bougainville* is a separate species known by small samples of extant and archaeological specimens (Flannery and Wickler, 1990). Most closely related to *M. rufescens*, *M. matambuai*, and *M. paveli* (Helgen, 2003*b*; Menzies, 1996).

Melomys burtoni (Ramsay, 1887). Proc. Linn. Soc. N.S.W., ser. 2, 2:531.

COMMON NAME: Grassland Melomys.

TYPE LOCALITY: Australia, Western Australia, near Derby; Mahoney and Richardson (1988:161) provided details.

DISTRIBUTION: Australia; along the coast "from just south of the New South Wales-Queensland border, north to the tip of Cape York, and in coastal areas of the Northern Territory and north-eastern Western Australia" (Watts and Aslin, 1981:84; Kerle, 1995*c*:633); also found on many offshore islands. SC New Guinea (recorded only from the Trans-Fly region).

STATUS: IUCN – Lower Risk (lc).

SYNONYMS: *albiventer* Kellogg, 1945; *australius* Thomas, 1924; *callopes* Finlayson, 1943; *froggatti* Troughton, 1937; *insulae* Troughton and Le Souef, 1929; *littoralis* (Lönnberg, 1916); *melicus* (Thomas, 1913); *mixtus* Troughton, 1935; *murinus* (Thomas, 1913); *muscalis* (Thomas, 1913).

COMMENTS: Results of chromosomal and electrophoretic studies reported by Baverstock et al. (1977*c*, 1980, 1981, 1983*b*). Anatomy of male reproductive tract and spermatozoal morphology presented by Breed and Sarafis (1978), Morrissey and Breed (1982), and Breed (1984, 1986). Tate (1951), Mahoney and Richardson (1988), and Musser and Carleton treated populations from Australian and New Guinea region as a single species, and Musser and Carleton (1993) noted that the complex needs revision using data from morphological, chromosomal, and molecular sets to assess significance of the variation apparent both among samples from New Guinea and between those from New Guinea and Australia. In the absence of such a revisionary study, Flannery (1995*a, b*) treated the New Guinea samples as a separate species, *M. lutillus*. Recently, Menzies (1996) has provided a revision of the New Guinea segment, as *M. lutillus*, recognizing *M. l. lutillus* as occurring in E Papua, *M. l. muscalis* as confined to the Trans-Fly savannas, and samples from the Snow Mtns of Prov. of Papua (= Irian Jaya) as a separate species, *M. frigicola*. Menzies (1996:380) realized the close relationship between Australian and New Guinea populations, writing "There is a possibility that New Guinean *lutillus*, especially the Trans-Fly population, *muscalis*, could be conspecific with the tropical Australian *burtoni*." Here we incorporate the Trans-Fly segment (*muscalis*) into Australian *M. burtoni*, and recognize *M. lutillus* and *M. frigicola* as occurring elsewhere on New Guinea (see those accounts). This seems a reasonable hypothesis to test in the context of a needed revision incorporating samples from both Australia and New Guinea. If *muscalis* truly represents the Austral *M. burtoni*, this would be another example of an Australian element also occurring only in the Trans-Fly region of S New Guinea (Norris and Musser, 2001). Australian segment reviewed by Watts and Aslin (1981) and Kerle (1995*c*).

Melomys capensis Tate, 1951. Bull. Am. Mus. Nat. Hist., 97:295.
   COMMON NAME: Cape York Melomys.
   TYPE LOCALITY: Australia, Queensland, Nesbit River, Rocky Scrub (east of Coen), 1500 ft (457 m).
   DISTRIBUTION: Australia, Queensland, Iron and McIlwraith Ranges of Cape York (north of Cooktown).
   STATUS: IUCN – Lower Risk (lc).
   COMMENTS: Originally described as a subspecies of *M. cervinipes* by Tate (1951:295), but separated as a distinct species by Baverstock et al. (1980) based on differences in blood proteins; otherwise, *M. capensis* and *M. cervinipes* are similar in morphological traits and body size (Watts and Aslin, 1981:82). Other allozymic results presented by Baverstock et al. (1981). Reviewed by Leung (1995), who also presented the first detailed ecological study of the species (Leung, 1999*a*).

Melomys caurinus (Thomas, 1921). Treubia, 2:112.
   COMMON NAME: Short-tailed Talaud Melomys
   TYPE LOCALITY: Indonesia, Kepulauan Talaud, Pulau Karakelang.
   DISTRIBUTION: Recorded only by four specimens from Pulau Karakelang and Pulau Salebabu (MZB) in the Talaud Isls (Helgen, 2003*b*).
   COMMENTS: Morphologically very similar to *M. leucogaster* (Menzies, 1996; our own study of the specimens discussed by Flannery, 1995*b*). Originally described as a species, but Rümmler (1938) and Ellerman (1941) treated *caurinus* as a subspecies of *M. leucogaster*, Musser and Carleton (1993) listed it in the synonymy of that species, and Laurie and Hill (1954) arranged it as a subspecies of *M. fulgens*. Tate (1951) recognized *caurinus* as a distinct species as have Flannery (1995*b*) and Menzies (1996). It is sympatric with *M. talaudium*, which has a longer tail relative to head and body length, suggesting that *M. caurinus* may be terrestrial and *M. talaudium* arboreal (Flannery, 1995*b*; Thomas, 1921*i*). Except for tail length these two species are hardly distinguishable in cranial and dental features and to Flannery (1995*b*:132) "there seems little doubt that these two species have evolved in the Talaud Isls from a common ancestor."

Melomys cervinipes (Gould, 1852). Mamm. Aust., pt. 4, 3:pl. 14.
   COMMON NAME: Fawn-footed Melomys.
   TYPE LOCALITY: Australia, Queensland, Stradbrook Isl (as indicated by Thomas' lectotype designation; see Mahoney and Richardson, 1988:162).
   DISTRIBUTION: Australia; extant range is closed forest and more open habitat along the E Australian coast from Cooktown region of Cape York in Queensland south to Gosford area of New South Wales (Watts and Aslin, 1981:79; Redhead, 1995*a*:636). Late Pleistocene specimens indicate distribution once extended farther south to the Pyramids Cave region in Victoria (Wakefield, 1972*a*).
   STATUS: IUCN – Lower Risk (lc).
   SYNONYMS: *banfieldi* (De Vis, 1907); *bunya* Tate, 1951; *eboreus* Thomas, 1924; *limicauda* Troughton, 1935; *pallidus* Troughton and Le Souef, 1929.
   COMMENTS: Anatomy of male reproductive tract and spermatozoa reported by Breed and Sarafis (1978), Morrissey and Breed (1982), and Breed (1984, 1986). Chromosomal morphology, G-banding homologies, and results of electrophoretic analyses presented by Baverstock et al. (1977*c*, 1980, 1981, 1983*b*), who (1980) reported that *M. cervinipes* is phylogenetically close to *M. capensis* but electrophoretically distant, having experienced a rapid rate of protein evolution relative to *M. capensis* and *M. burtoni*. Reviewed by Watts and Aslin (1981) and Redhead (1995*a*).

Melomys cooperae Kitchener, in Kitchener and Maryanto, 1995. Rec. West. Aust. Mus., 17:43.
COMMON NAME: Yamdena Island Melomys.
TYPE LOCALITY: Indonesia, Maluku Tenggara, Tanimbar Isl Group, Yamdena Isl,
1 km S Kebun Lorulun, *c*. 20 km N Saumlaki, 200 m.
DISTRIBUTION: Recorded only from Yamdena Isl in the Tanimbar Group of islands.
COMMENTS: A member of what Menzies (1996) called the *Melomys rufescens* Group.
Kitchener and Maryanto (1995*a*) compared this species with *M. leucogaster*, *M. rubicola*,
M. arcium, *M. fulgens*, *M. talaudium*, *M. caurinus*, and *M. rufescens*.

Melomys dollmani Rümmler, 1935. Z. Säugetierk., 10:106.
COMMON NAME: Dollman's Melomys.
TYPE LOCALITY: Papua New Guinea, Eastern Highlands Province, Kratke Mtns, 1200-1500 m.
DISTRIBUTION: Papua New Guinea; above 1200 m along Central Cordillera in the Eastern
Highlands (Graina area, Wau region, Kratke Mtns, Upper Ramu River Plateau, Okapa
area) to slopes of Mt Hagen (Tomba); also apparently Mt Sisa (see Flannery et al., 1994);
does not occur east of the Okapa area on mountains of the E peninsula of Papua.
COMMENTS: Although usually listed as a subspecies or synonym of *M. rufescens* (Flannery,
1990*b*; Laurie and Hill, 1954; Tate, 1951), *M. dollmani* is a separate species. It has been
collected at the same place as *M. rufescens* in the Kratke Mtns, on the Upper Ramu
River Plateau, in the Okapa area, and at Tomba (specimens in AMNH and BMNH). This
is the same species listed as *M. gracilis* by Musser and Carleton (1993). Musser recently
reexamined the specimens described as *gracilis* from the Owen Stanley Range, both by
qualitative observation and discriminant function analysis, and agrees with Menzies'
(1996) allocation of the taxon to *M. rufescens*. However, *M. dollmanni* is not the same as
gracilis. In all multivariate analyses it is set well apart from all the other clusters
conforming to different geographic groups of *M. rufescens* (Musser, ms). Compared
with *M. rufescens* from the same area, *M. dollmani* has a significantly longer head and
body, hind foot, and much longer tail; longer skull and incisive foramina, wider
interorbit and mastoid breadth, shorter molar row, conspicuously narrower zygomatic
plate, lower mandibular ramus; woollier pelage, and 1-3 caudal hairs per scale (a single
hair per scale in all *M. rufescens* examined, including *gracilis*). The presence of two
species of *rufescens*-like *Melomys* on Mt Hagen and Mt Sisa was was also demonstrated
by analysis of allozymic variation (Flannery et al., 1994).

Melomys fraterculus (Thomas, 1920). Ann. Mag. Nat. Hist., ser. 9, 6:428.
COMMON NAME: Manusela Melomys.
TYPE LOCALITY: Indonesia, Pulau Seram, Mt Manusela, 6000 ft (1830 m).
DISTRIBUTION: Endemic to Seram Isl.
STATUS: IUCN – Lower Risk (nt).
COMMENTS: Still known only by the type series (2 specimens in BMNH; Helgen, 2003*b*).
Originally described as a species of *Uromys*, then placed in *Pogonomelomys* by Rümmler
(1938), kept there by Tate (1951), but returned to *Melomys* by Laurie and Hill (1954).
Based on our study of specimens, *M. fraterculus* shares many derived cranial features
with the Australian *M. cervinipes* complex and may be more closely related to the
indigenous Australian *Melomys* than to those on New Guinea. Menzies (1990:134),
however, wrote that "Without more material and additional study it is difficult to say
more about it. For the present it is best left in *Melomys*," and for Flannery (1995*b*:133)
"It may ultimately be necessary to create a new genus for it." Helgen (2003*b*) promised
a review in a forthcoming report.

Melomys frigicola Tate, 1951. Bull. Am. Mus. Nat. Hist., 97:303.
COMMON NAME: Snow Mountains Grassland Melomys.
TYPE LOCALITY: Indonesia, Prov. of Papua (= Irian Jaya), Snow Mtns (Pegunungan Maoke),
Bele River, 18 km N Lake Habbema, 2200 m.
DISTRIBUTION: Known only from the N slopes of the Snow Mtns and nearby Baliem River,
1600-2800 m; probably occurs elsewhere in Pegunungan Maoke.
COMMENTS: Originally described as a subspecies of *M. lutillus* by Tate (1951), retained that
way by Laurie and Hill (1954) and Flannery (1995*a*), but included among the
synonyms of *M. burtoni* by Mahoney and Richardson (1988) and Musser and Carleton

(1993). Menzies (1996) elevated *frigicola* to species rank based on results of morphometric analysis. Most closely related to *M. lutillus* of New Guinea and *M. burtoni* of Australia and the Trans-Fly region of New Guinea (see those accounts).

Melomys fulgens (Thomas, 1920). Ann. Mag. Nat. Hist., ser. 9, 6:426.

COMMON NAME: Seram Long-tailed Melomys.

TYPE LOCALITY: Indonesia, S Pulau Seram, Teluk Taluti (=Teloeti Bay).

DISTRIBUTION: Recorded only from the southern coast of Seram Isl.

COMMENTS: Still known only by the holotype and one other specimen (in BMNH; Helgen, 2003*b*). Originally described as a *Uromys*, arranged as a subspecies of *M. leucogaster* by Rümmler (1938) and Ellerman (1941), placed in synonymy of that species by Musser and Carleton (1993), but treated as a distinct species by Tate (1951), Laurie and Hill (1954), Flannery (1995*b*), and Menzies (1996). Although superficially very similar in its morphology to *M. leucogaster*, it has a much longer tail relative to length of head and body that is scaleless and calloused near the tip, indicating dorsal prehensility, an adaptation absent in the shorter-tailed *M. leucogaster*. Reviewed by Flannery (1995*b*) and Helgen (2003*b*).

Melomys howi Kitchener, in Kitchener and Suyanto, 1996. Rec. West. Aust. Mus., 18:113.

COMMON NAME: Riama Island Melomys.

TYPE LOCALITY: Indonesia, Maluka Tenggara, Tanimbar Isls, Pulau Riama, sea level (see Kitchener and Suyanto, 1996, for additional information).

DISTRIBUTION: Known only from Riama, a small island off the west coast of the larger Pulau Selaru in the Tanimbar Isls.

COMMENTS: In its morphology and body size, *M. howi* is closely related to the New Guinea *M. lutillus*; its description recalls those in samples from Woodlark, Misima, and Sudest islands east of the Papua New Guinea mainland, which resemble mainland *M. lutillus* but are larger in body size. Those samples, the specimens from Riama Isl, and series of *M. lutillus* from mainland New Guinea need to be reanalyzed to determine how many species are actually present in the complex and their relationship to one another. *Melomys howi* is not closely related to *M. cooperae* from the large Pulau Yamdena in the Tanimbar Isls; that distinct species is a member of the New Guinea *M. rufescens-M. leucogaster* complex. Kitchener and Suyanto (1996) provided habitat and other information about *M. howi*.

Melomys leucogaster (Jentink, 1908). Nova Guinea, 9:3.

COMMON NAME: White-bellied Melomys.

TYPE LOCALITY: New Guinea, S Prov. of Papua (= Irian Jaya), Lorentz River, Alkmaar, 300 m.

DISTRIBUTION: New Guinea; primarily S side of Central Cordillera from the type locality in Prov. of Papua (= Irian Jaya) east through Papua New Guinea to the Mori River on the mainland (see Flannery, 1995*a*:296), sea level to 1400 m (Menzies, 1996); also recorded from Jayapura (Hollandia) on north coast of Prov. of Papua (= Irian Jaya) (Tate, 1951), Yule Isl off the coast of E Papua New Guinea in Central Province, and small treeless offshore islands (Flannery, 1995*a*).

STATUS: IUCN – Lower Risk (lc).

SYNONYMS: *latipes* Tate and Archbold, 1935.

COMMENTS: A close morphological relative of *M. rufescens* and placed in "The *rufescens* Division" of *Melomys* by Menzies (1996). Tate (1951:307) recorded four specimens (AMNH152842-152845) from Hollandia (Jayapura) on the north coast of Prov. of Papua (= Irian Jaya, Indonesia), which is way outside of the known range of *M. leucogaster*, and thought by Menzies (1996:405) to be an error. Musser examined the rats, which are examples of *M. leucogaster*, and there is no indication in the catalog or other archival documents that they could have been collected elsewhere and mislabeled.

Melomys lutillus (Thomas, 1913). Ann. Mag. Nat. Hist., ser. 8, 12:216.

COMMON NAME: Papua Grassland Melomys.

TYPE LOCALITY: SE Papua New Guinea, Angabunga River, Owgarra.

DISTRIBUTION: Discontinuous range from the northern lowlands of Prov. of Papua (= Irian Jaya) throughout Papua New Guinea in foothills and higher in Central Cordillera, from sea level to 1500 m on mainland; not in the Trans-Fly region (Menzies, 1996). Also recorded from Woodlark Isl (AMNH 159593 and 159594), Misima Isl (AMNH 159591, 159592, and 190514), and the Conflict Isls (sample in BMNH) in the Louisiade Arch. east of mainland E Papua (Flannery, 1995*b*).

SYNONYMS: *hintoni* Rümmler, 1935.

COMMENTS: Menzies (1996) outlined the range as E New Guinea; we extend it along the N coast to the Lake Sentani region in Prov. of Papua (= Irian Jaya), the type locality of *hintoni*. Menzies did not allocate *hintoni* to any of the subspecies of *M. lutillus* he recognized. Menzies included *muscalis* from the Trans-Fly region as a distinct subspecies of *M. lutillus*, but we place it in the Australian *M. burtoni* (see that account). The specimens from Woodlark and Misima Isls are slightly larger in body size than mainland samples of *M. lutillus*. There are three specimens from Sudest Isl in the Louisiade Arch. (AMNH 159595, 159596, 190566) that represent an undescribed species. Although morphologically related to mainland *M. lutillus*, they are much larger in body size with conspicuously longer molar rows, larger even than the animals on Woodlark and Misima.

Melomys matambuai Flannery, Colgan, and Trimble, 1994. Proc. Linn. Soc. N.S.W., 114:39.

COMMON NAME: Manus Island Melomys.

TYPE LOCALITY: Papua New Guinea, Bismarck Arch., Admirality Isls, Manus Isl.

DISTRIBUTION: Known only by two specimens from Manus Isl.

COMMENTS: A member of the "*Melomys rufescens* Division" (Menzies, 1996) and "very distinct, both in its biochemistry and morphology, and . . . presumably a long-isolated species" (Flannery, 1995*b*:140). *Melomys matambuai* is most closely related to *M. paveli* on Seram Isl., *M. bougainville* on Manus Isl., and the wide-ranging *M. rufescens* (Helgen, 2003*b*; see account of *M. paveli*).

Melomys obiensis (Thomas, 1911). Ann. Mag. Nat. Hist., ser. 8, 7:208.

COMMON NAME: Obi Island Melomys.

TYPE LOCALITY: Indonesia, Malukus, Pulau Obi.

DISTRIBUTION: Endemic to Obi and Bisa, Isls, south of Halmahera (Flannery, 1995*b*; Helgen, 2003*b*).

STATUS: IUCN – Lower Risk (nt).

COMMENTS: Judged by morphological traits, possibly a close relative of *M. fraterculus* from Seram, and more closely related to Australian *Melomys cervinipes* than to any New Guinea species, an observation gleaned from Musser's study of specimens and earlier recorded by Tate (1951:297).

Melomys paveli Helgen, 2003. J. Zool. London, 261:168.

COMMON NAME: Pavel's Seram Melomys.

TYPE LOCALITY: Indonesia, coast of S Pulau Seram, Piliana, 400 m (see Helgen, 2003*b*, for coordinates and additional information).

DISTRIBUTION: Known only from the type locality.

COMMENTS: Represented only by the holotype. Member of the *M. rufescens* group as defined by Menzies (1996) and originally described as a subspecies of *M. rufescens* that in body size and pelage coloration is most similar to *M. r. niviventer* occurring in forest and savannas of the lower Fly and Digul Rivers in S New Guinea (see Menzies, 1996, and Tate, 1951 for characteristics of *niviventer*). A better hypothesis recognizes *paveli* as a species related to *M. rufescens*, *M. bougainville*, and *M. matambuai* (see those accounts), for Helgen (2003*b*:169) noted that *paveli* "differs in several ways from other mosaic-tailed rats placed in *M. rufescens* and can probably be considered as distinctive as *M. matambuai* or *M. bougainville*, both of which are currently separated from *M. rufescens* as distinct species." Analyses of additional specimens of *paveli* using morphological traits and gene sequences may resolve the question of whether the resemblance between it and *M. r. niviventer* reflects close genetic alliance as an island population of *M. rufescens* or traits independently acquired during evolution in insular isolation.

Melomys rubicola Thomas, 1924. Ann. Mag. Nat. Hist., ser. 9, 13:298.

COMMON NAME: Bramble Cay Melomys.

TYPE LOCALITY: Australia, Queensland, Torres Strait, Bramble Cay, about 9ES, 144EE; see Mahoney and Richardson (1988:163).

DISTRIBUTION: Australia; endemic to Bramble Cay at the extreme N end of the Great Barrier Reef of Queensland (Limpus et al., 1983).

STATUS: IUCN – Critically Endangered.

COMMENTS: Studies of blood proteins and morphology suggested *M. rubicola* was closely related to *M. capensis*, which is endemic to Cape York in N Queensland (Limpus et al., 1983). Structure of sperm head described by Breed (1984). Reviewed by Watts (1995g).

Melomys rufescens (Alston, 1877). Proc. Zool. Soc. Lond., 1877:124.

COMMON NAME: Black-tailed Melomys.

TYPE LOCALITY: "Duke of York Isl or adjacent parts of New Britain or New Ireland" (Tate, 1951:304).

DISTRIBUTION: New Guinea; throughout the island continent, from the Vogelkop in Prov. of Papua (= Irian Jaya) to the E end of Papua New Guinea, coastal lowlands to high altitudes in mountains (at least 2000 m). Also on islands of New Britain, New Ireland, Mioko, and Lamassa in the Bismarck Arch.; three large islands in E Prov. of Papua (= Irian Jaya), Yapen, Waigeo, and Salawati; on islands of Blup Blup and Karkar, off northern coast of Papua New Guinea; and on Sideia Isl, off coast of SE Papua New Guinea (Emmons and Kinbag, 2002; Flannery, 1995b; Flannery and White, 1991; Menzies, 1996); Kitchener and Maryanto (1995a:49) recorded it from Wokam Isl in the Aru Isl Group.

STATUS: IUCN – Lower Risk (lc) as *M. gracilis* and *M. rufescens*.

SYNONYMS: *calidior* (Thomas, 1911); *gracilis* (Thomas, 1906); *hageni* Troughton, 1937; *musavora* (Ramsay, 1877); *niviventer* Tate, 1951; *sexplicatus* (Jentink, 1907); *stalkeri* (Thomas, 1904); *wisselensis* Menzies, 1996.

COMMENTS: The significant geographic variation in morphological traits present among samples allows recognition of four distinct groups; *rufescens* from N and W New Guinea and the Bismarck Arch., *niviventer* from Fly River drainage, *stalkeri* from E Papua New Guinea, and *hageni* from the Eastern Highlands (based on Musser's study of specimens in AMNH and BMNH), which is generally concordant with the subspecies recognized by Menzies (1996) in his revision of New Guinea *Melomys*. Menzies (1996) also described *wisselensis* from the Wissel Lakes region in Prov. of Papua (= Irian Jaya) as a subspecies of *M. rufescens*. Morphologically and likely phylogenetically, *M. rufescens* is related to *M. paveli* on Seram Isl, *M. matambuai* on Manus Isl, and *M. bougainville* on the northern Solomon islands, and according to Helgen (2003b:170), all four "... undoubtedly share a recent common ancestry. . . ."

Melomys spechti Flannery and Wickler, 1990. Aust. Mammal., 13:130.

COMMON NAME: Buka Island Melomys.

TYPE LOCALITY: Solomon Isls, Buka Isl, Kilu Rockshelter.

DISTRIBUTION: Recorded only from Buka Isl.

COMMENTS: A distinctive species known only by archaeological fragments (Flannery and Wickler, 1990).

Melomys talaudium (Thomas, 1921). Ann. Mag. Nat. Hist., ser. 9, 7:248.

COMMON NAME: Long-tailed Talaud Melomys.

TYPE LOCALITY: Indonesia, Kepulauan Talaud, Pulau Salebabu, Lirung.

DISTRIBUTION: Recorded only from the islands of Karakelang and Salebabu in the Talaud Isls (specimens in BMNH and MZB; Helgen, 2003b).

COMMENTS: Morphologically similar to *M. leucogaster* (Menzies, 1996; our own study of the specimens discussed by Flannery, 1995b). Originally described as a species, but Rümmler (1938) and Ellerman (1941) treated *talaudium* as a subspecies of *M. leucogaster*, Musser and Carleton (1993) listed it in the synonymy of that species, and Laurie and Hill (1954) arranged it as a subspecies of *M. fulgens*. Tate (1951) recognized *talaudium* as a distinct species as have Flannery (1995b) and Menzies (1996). It is sympatric with *M. caurinus*, which has a shorter tail relative to head and body length, suggesting that *M. caurinus* may be terrestrial and *M. talaudium* arboreal (Flannery, 1995b; Thomas, 1921i). Except for tail length these two species are hardly distinguishable in cranial and dental features and to Flannery (1995b:132) "there seems little doubt that these two species have evolved in the Talaud islands from a common ancestor."

Mesembriomys Palmer, 1906. Proc. Biol. Soc. Wash., 19:97.

> TYPE SPECIES: *Mus hirsutus* Gould, 1842 (= *Hapalotis gouldii* Gray, 1843).
>
> SYNONYMS: *Ammomys* Thomas, 1906 [not Bonaparte, 1831].
>
> COMMENTS: *Pseudomys* Division. Member of the Australian Old Endemics (Musser, 1981c:167), which includes the Conilurini where Baverstock (1984) placed *Mesembriomys*, and closely related to *Leporillus* and *Conilurus* (Watts et al., 1992). Analyses of immunological distances (Watts and Baverstock, 1994b, 1996) and sperm morphology (Breed and Aplin, 1994; Breed, 1997) support the inclusion of *Mesembriomys* in large clade that includes members of our *Hydromys*, *Xeromys*, *Pseudomys*, and *Uromys* Divisions from Australia and New Guinea (the "Australasian clade" of Watts and Baverstock, 1995b, 1996), but not our strictly New Guinea *Pogonomys* Division or *Lorentzimys* Division. Mahoney and Richardson (1988:164) cataloged taxonomic, distributional, and biological references.

Mesembriomys gouldii (Gray, 1843). List Specimens Mamm. Coll. Br. Mus., p. 116.

> COMMON NAME: Black-footed Mesembriomys.
>
> TYPE LOCALITY: Australia, Northern Territory, Port Essington; see details in Mahoney and Richardson (1988:163).
>
> DISTRIBUTION: Australia; N Western Australia, N Northern Territory, N Queensland, Melville Isl, and Bathurst Isl (Friend, 1991; Watts and Aslin, 1981).
>
> STATUS: IUCN – Lower Risk (nt).
>
> SYNONYMS: *hirsutus* (Gould, 1842) [not Elliot, 1839]; *melvillensis* Hayman, 1936; *rattoides* Thomas, 1924.
>
> COMMENTS: Analyses of chromosomal and electrophoretic data (Baverstock et al., 1977a, c, 1981, 1983b) as well as phallic and dental morphology (Lidicker and Brylski, 1987; Misonne, 1969) indicated *M. gouldii* is phylogenetically most closely related to species of *Conilurus*. Reviewed by Watts and Aslin (1981) and Friend and Calaby (1995).

Mesembriomys macrurus (Peters, 1876). Monatsb. K. Preuss. Akad. Wiss. Berlin, p. 355.

> COMMON NAME: Golden-backed Mesembriomys.
>
> TYPE LOCALITY: Australia, Western Australia, "at a small mainland creek, Mermaid Strait" (Mahoney and Richardson, 1988:164).
>
> DISTRIBUTION: N Western Australia and N Northern Territory (Watts and Aslin, 1981:128); probably extinct in NW central region of Western Australia (Mahoney and Richardson, 1988:164). Possibly S New Guinea (Flannery, 1995a:65).
>
> STATUS: IUCN – Vulnerable.
>
> SYNONYMS: *boweri* (Ramsay, 1887).
>
> COMMENTS: Reviewed by Watts and Aslin (1981) and McKenzie and Kerle (1995), who discussed decline of the species in some regions and other areas where secure populations persist.

Micaelamys Ellerman, 1941. Families and Genera of Living Rodents, 2:170.

> TYPE SPECIES: *Mus granti* Wroughton, 1908.
>
> COMMENTS: *Aethomys* Division. Formerly included in *Aethomys* as a subgenus, but a variety of data sets ranging from morphological to molecular indicate the two species discussed below belong in a monophyletic group, *Micaelamys*, separate from that containing species of *Aethomys* (see generic account of *Aethomys*). References cited in the following accounts treated *granti* and *namaquensis* as species of *Aethomys*. Including *Micaelamys* with *Aethomys* in the same Division is provisional. In phylogenetic analyses of mtDNA cytochrome *b* sequences, *Aethomys* (represented by *A. chrysophilus* and *A. kaiseri*) is closest to *Grammomys*, while *Micaelamys* (*M. namaquensis*) joins other African genera depending on the analyses (Castiglia et al., 2003b; Ducroz et al., 2001). Phylogenetic analyses employing a broader sampling of not only species in *Aethomys* but of endemic African murine genera using molecular, chromosomal, and morphological data sets may resolve the phylogenetic relationships of *Aethomys* and *Micaelamys*. Evolutionary history as

documented by fossils extends back to the late Pliocene of South Africa (*"A. cf. namaquensis"*; Denys, 1990*c*), and Pleistocene of Namibia (*"Micaelamys"*; Senut et al., 1992).

Micaelamys granti (Wroughton, 1908). Ann. Mag. Nat. Hist., ser. 8, 1:257.

COMMON NAME: Grant's Micaelamys.

TYPE LOCALITY: South Africa, NE Western Cape Province, Deelfontein, north of Richmond, (see Chimimba, 1998:435, for details).

DISTRIBUTION: South Africa, known only from the fynbos, succulent karoo, and nama-Karoo biomes (Mugo et al., 1995) in SC South Africa (the Great Karoo region; Chimimba et al., 1999:508; de Graaff, 1997*w*; Skinner and Smithers, 1990:279).

STATUS: IUCN – Least Concern as *Aethomys granti*.

COMMENTS: Meester et al. (1986:292) and Chimimba (1998:435) provided historical taxonomic allocations of *granti*, which ranged from *Myomys*, through *Rattus* and *Mastomys*, and finally to *Micaelamys* as either genus or subgenus of *Aethomys*. Reviewed and compared with *M. namaquensis*, its closest relative, by Skinner and Smithers (1990) and also reviewed by de Graaff (1997*w*). Analysis of geographic variation (Chimimba et al., 1998) suggested a clinal pattern (southwesterly-northeasterly) with size of cranium negatively correlated with longitude and positively with latitude. No significant steps in the clinal variation were detected and subspecies were not recognized.

Micaelamys namaquensis (A. Smith, 1834). S. Afr. Quart. J., 2:160.

COMMON NAME: Namaqua Micaelamys.

TYPE LOCALITY: South Africa, S Western Cape Province, Little Namaqualand, Cape of Good Hope (restricted to Witwater by Shortridge, 1942).

DISTRIBUTION: E Angola (Crawford-Cabral, 1998), South Africa (except parts of Western, Northern, and Eastern Cape provinces, coastal KwaZulu-Natal Province, and Namib Desert; de Graaff, 1997*v*; P. J. Taylor, 1998), Botswana, Zimbabwe, S and C Mozambique (absent from central and coastal regions), S Malawi, and SE Zambia. Range abstracted from Skinner and Smithers (1990:278) and Chimimba et al. (1999:507).

STATUS: IUCN – Lower Risk (lc) as *Aethomys namaquensis*.

SYNONYMS: *arborarius* Peters, 1852; *auricomis* de Winton, 1897; *avarillus* Thomas and Wroughton, 1908; *avunculus* (Thomas, 1904); *calarius* Thomas, 1926; *capensis* Roberts, 1926; *centralis* Schwann, 1906; *drakensbergi* Roberts, 1926; *epupae* Von Lehmann, 1975; *grahami* Roberts, 1915; *klaverensis* Roberts, 1926; *lechochloides* Roberts, 1926; *lehocla* A. Smith, 1836; *longicaudatus* Von Lehmann, 1955; *monticularis* Jameson, 1909; *namibensis* Roberts, 1946; *phippsi* Hill and Carter, 1937; *siccatus* Thomas, 1926; *waterbergensis* Roberts, 1938.

COMMENTS: Originally described as a species of *Gerbillus* (Gerbillinae), subsequently treated as a species of *Aethomys* (G. M. Allen, 1939), *Thallomys* (Ellerman, 1941) or *Rattus* in subgenus *Praomys* (Ellerman et al., 1953). There is appreciable variation in body size and pelage coloration among geographic samples, but past systematic studies do not recognize subspecies or significant clinal patterns of variation (Chimimba, 1998; Chimimba et al., 1999). A recent intraspecific morphometric analysis across a more comprehensive geographic region in southern Africa suggested recognition of four subspecies (Chimimba, 2001*a*) in which distributional limits coincide with major phytogeographical zones. Integrity of these subspecific units, however, requires independent testing with molecular data. Non-geographic variation due to sex and age reported by Chimimba and Dippenaar (1994). Of all the species of either *Micaelamys* or even *Aethomys*, *M. namaquensis* has the most extensive geographic distribution and is sympatric with all the other species of *Aethomys* and *Micaelamys* occurring in the Southern African Subregion (Chimimba et al., 1999). Reviewed by Meester et al. (1986), Skinner and Smithers (1990), Chimimba (1998), Chimimba et al. (1999), and de Graaff (1997*v*).

Microhydromys Tate and Archbold, 1941. Am. Mus. Novit., 1101:2.

TYPE SPECIES: *Microhydromys richardsoni* Tate and Archbold, 1941.

COMMENTS: *Hydromys* Division. Member of the New Guinea Old Endemics (Musser, 1981*c*). Reviewed by Flannery (1989).

Microhydromys musseri Flannery, 1989. Proc. Linn. Soc. N.S.W., 111:216.

COMMON NAME: Torricelli Mountains Shrew Mouse.

TYPE LOCALITY: Papua New Guinea, West Sepik Province, Torricelli Mtns, Mt Somoro, 1350 m (see Flannery, 1989, for details).

DISTRIBUTION: Recorded only from the type locality.

STATUS: IUCN – Lower Risk (lc).

COMMENTS: Represented only by the holotype. A very distinct species that is part of a highland fauna endemic to the N coastal ranges of Papua New Guinea (see account of *Paraleptomys rufilatus*). Its cranial and dental morphology is most like *M. richardsoni*, but "both possess a number of independent specializations not seen in the other" suggesting that they "have been evolving separately for a considerable period of time" (Flannery, 1989:220). Flannery preferred to include *musseri* in *Microhydromys* rather than its own monotypic genus because doing so would highlight its apparent close morphological affinity with *richardsoni*, at least "until relationships within the group are clarified" (p. 220), but a new genus will likely have to be erected for the species.

Microhydromys richardsoni Tate and Archbold, 1941. Am. Mus. Novit., 1101:2.

COMMON NAME: Richardson's Shrew Mouse.

TYPE LOCALITY: New Guinea, Prov. of Papua (= Irian Jaya), Snow Mtns (Pegunungan Maoke), Idenburg River, 4 km SW Bernhard Camp, 850 m.

DISTRIBUTION: New Guinea; scattered localities in hill forest from type locality in Prov. of Papua east to Sogeri in Port Moresby region (Flannery, 1995a:248).

STATUS: IUCN – Lower Risk (nt).

COMMENTS: The smallest-bodied of New Guinea's endemic murids, the only one among members of the *Hydromys* and *Xeromys* Divisions with longitudinally grooved upper incisors, and known only by five specimens (Flannery, 1995a). Distributional and biological data summarized by Flannery (1989, 1990b, 1995a). In its morphological structure, *M. richardsoni* is a miniature version of *Hydromys* but with grooved upper incisors and strictly terrestrial rather than amphibious habitus. It also possesses the primitive cephalic arterial pattern, a conformation shared with species of *Xeromys*, *Crossomys*, *Hydromys*, *Paraleptomys*, and *Parahydromys*, but not with *Leptomys* or the terrestrial New Guinea shrew mice in *Pseudohydromys* (which includes *Mayermys* and *Neohydromys*) that exhibit a derived configuration. Whether the same pattern is present in *M. musseri* has yet to be determined.

Micromys Dehne, 1841. *Micromys agilis*, Ein Neues Säugetier der Fauna von Dresden, p. 1.

TYPE SPECIES: *Micromys agilis* Dehne, 1841 (= *Mus minitus* Pallas, 1771).

COMMENTS: *Micromys* Division. Generic diagnosis based upon molar traits provided by Storch (1987). Molar morphology indicates a close relationship with members of Misonne's (1969) *Progonomys* group (within a more inclusive *Lenothrix-Parapodemus* Division), and phylogenetic relationships assessed by microcomplement fixation of albumin (Watts and Baverstock, 1995b) pointed to the Asian *Vandeleuria* (also a member of Misonne's *Progonomys* cluster) as the closest living relative of *Micromys*, an alliance hinted at by Jüdes's (1981) interpretation of chromosomal data, and explicitly indicated by Ellerman (1949:132) based upon morphology. Phylogenetic analyses of complete mtDNA cytochrome *b* sequences by Martin et al. (2000) could not resolve the phylogenetic position of *Micromys* relative to other murines and deomyines sampled (*Apodemus*, *Mus*, *Rattus*, *Acomys*), nor could analyses of sequences from the nuclear LCAT (Robinson et al., 1997); neither of these studies included *Vandeleuria*. *Micromys* possesses a quality of the *Lx* family of long interspersed repeated DNA compared with other extant murines sampled (*Mus*, *Praomys*, *Rattus*, *Bandicota*, *Arvicanthis*, *Mastomys*, *Hylomyscus*, *Aethomys*, *Dasymys*, and *Apodemus*) that indicates it to be the most divergent of those genera and an outgroup to them (Furano et al., 1994). Those authors speculated that *Micromys* diverged early from the core murine lineage, an observation consistent with scnDNA hybridization data (Catzeflis et al., 1992), analyses of sequences from nuclear IRBP gene, and mtDNA cytochrome *b* and 12S rRNA (Michaux et al., 2002a), and the early presence of the genus and phylogenetic history in the Neogene. *Micromys* may have been the first lineage to have diverged from a common ancestral *Progonomys*-like form approximately 12 million years ago (Furano et al., 1994). Based upon her cladistic analysis of molar traits, Chaima-nee (1998) suggested the Indomalayan *Vandeleuria*, *Lenothrix*, and *Pithecheir* are also members of this primary divergent lineage.

The early evolutionary origin of *Micromys* is reflected in its phylogenetic history documented by fossils representing ten extinct species: one from the late Miocene of Nei Mongol (Storch, 1987), seven from the late Miocene, Pliocene, and Pleistocene of Europe, one of which occurs in Europe and Nei Mongol (Qiu and Storch, 2000; Weerd, 1979; see Storch and Dahlmann, 1995, who reviewed all the fossil species of *Micromys*), one from the early to late Pliocene of S China (Cai and Qiu, 1993; Wu and Flynn, 1992), and another from the early Pleistocene of S China (Zheng, 1993). "The fossil *Micromys* species do not represent a single lineage which finally leads to the living *M. minutus*. . . . the genus underwent radiations and shows a rather complex phylogenetic history" (Storch and Dahlmann, 1995:128). Only one living species is recognized by most workers, but whether extant samples represent one or more species has yet to be resolved by critical systematic revision.

Micromys minutus (Pallas, 1771). Reise Prov. Russ. Reichs., 1:454.
    COMMON NAME: Harvest Mouse.
    TYPE LOCALITY: Russia, Ulyanovsk. Obl. Middle Volga River, Simbirsk (now Ulyanovsk).
    DISTRIBUTION: From NW Spain through most of Europe (including Thrace region of Turkey; Kryštufek and Vohralík, 2001, but missing from the Alps, Portugal, and most of Sweden, Norway, Italy, and Spain; Mitchell-Jones et al., 1999), across Siberia to Ussuri region and Korea (Won and Smith, 1999), north to about 65°E in European Russia and Yakutia, south to N edge of Caucasus and N Mongolia (Gromov and Erbajeva, 1995); isolated ranges in NW China (Xinjiang) and throughout S and NE China (from SE Xizang in west to Heilongjiang and Nei Mongol in far northeast; Wang, 2003, and Zhang et al., 1997); south to NW Vietnam (Dang et al., 1994), N Burma (Anthony, 1941; Ellerman, 1961), and NE India (Meghalaya and Nagaland; Agrawal, 2000). Island distributions include Britain; Texel, Terschelling, and Ameland off coast of Netherlands in Wadden Sea (Mostert, 1992*b*; Naber, 1982); Japan (Honshu, Shikoku, Kyushu, and Tsushima), Quelpart Isl (Korea), and Taiwan (M.-J. Yu, 1996); see Corbet (1978*c*) for details, and map by Gromov and Erbajeva (1995). Possibly introduced to the Japanese Isls because no fossils have ever been found while all the Japanese endemic muroids are represented by Pleistocene and Holocene fossils (Kowalski and Hasegawa, 1976). The species also occurs in Great Britain where it was also probably introduced (Sutcliffe and Kowalski, 1976; Yalden, 1999).
    STATUS: IUCN – Lower Risk (nt).
    SYNONYMS: *agilis* Dehne, 1841; *aokii* Kuroda, 1922 [see Kaneko and Maeda, 2002]; *arundinaceus* (Petenyi, 1882); *arvensis* (Leach, 1816) [*nomen nudum*]; *avenarius* (Wolf, 1794); *batarovi* (Kastschenko, 1910); *berezowskii* Argyropulo, 1929; *brauneri* Martino, 1930; *campestris* (Desmarest, 1822); *danubialis* Simonescu, 1971; *erythrotis* (Blyth, 1856); *fenniae* (Hilzheimer, 1911); *flavus* (Kerr, 1792); *hertigi* Johnson and Jones, 1955; *hondonis* Kuroda, 1933 [see Kaneko and Maeda, 2002]; *japonicus* Thomas, 1906; *kastschenkoi* Charlamagne, 1915 [see Zagorodnyuk, 1992*b*]; *kurodai* Mori, 1942 [see Kaneko and Maeda, 2002]; *kytmanovi* (Kastschenko, 1910); *mehelyi* Bolkay, 1925; *meridionalis* (Costa, 1844); *messorius* Kerr, 1792; *minatus* (Schinz, 1840); *minimus* (White, 1789); *oryzivorus* (de Sélys-Longchamps, 1841); *parvulus* (Hermann, 1804) [not Mosanský, 1994]; *pendulinus* (Hermann, 1804); *pianmaensis* Peng, 1981; *pratensis* (Ockskay, 1831); *pumilus* (F. Cuvier, 1842); *pygmaeus* (Milne-Edwards, 1872); *sareptae* (Hilzheimer, 1911); *shenshiensis* Li, Wu, and Shao, 1965; *soricinus* (Hermann, 1780); *subobscurus* Fritsche, 1934; *takasagoensis* Tokuda, 1939 [*nomen nudum*; see Kaneko and Maeda, 2002]; *takasagoensis* Tokuda, 1941 [see Kaneko and Maeda, 2002]; *triticeus* (Boddaert, 1785); *typicus* (Barrett-Hamilton, 1899); *ussuricus* (Barrett-Hamilton, 1899); *zhenjiangensis* Huang, 1989.
    COMMENTS: Reviewed by Corbet (1978*c*, 1984). Chromosomal information reported by Jüdes (1981), Zima (1983), Lungeanu et al. (1984), Solleder et al. (1984), and Schmid et al. (1987). Results of morphometric analyses of selected European samples and its significance to applying subspecific names to the geographic variation reported by

Kratochvíl and Simionescu (1983). Alveolar pattern of molar roots and comparisons with those of *Rattus* and *Apodemus* reported by Gallego (1974). Phallic morphology of Chinese samples described by Yang and Fang (1988) in context of assessing relationships among Chinese murines. Haffner (1996) described a tendon-locking mechanism in *M. minutus* that is engaged when the middle phalanx is bent so that less muscular energy is expended when twigs or stalks are grasped. This structure should be compared with digits in species of the related *Vandeleuria* and *Vernaya*.

European populations reviewed by Böhme (1978*a*) and Mitchell-Jones et al. (1999). Indomalayan populations reviewed by Corbet and Hill (1992), who discussed the morphological differences between European and Indomalayan samples. Regional reports covering range, ecology, and other biological aspects available for S Norway (Kooij et al., 2001) and SW Sweden, where it was introduced by humans and is now possibly extinct (Hansson and Fredga, 1996); N Germany (Dolch et al, 1994); Austria (Bauer and Spitzenberger, 1996); Slovakia (Danko, 1994; Kminiak, 1996; Stanko, 1995; Stanko and Mošanský, 1994, 2000; Stanko et al., 2000); Czech Republic (And—ra and ervený, 1994; Miloš, 1994; Smaha, 1996; Zima and And—ra, 1996); Translvanian Romania (Istrate, 1998); Slovenia (Kryštufek, 1991); SE Greece (Thrace; Vohralík and Sofianidou, 1992*a*); Albania (Prigioni, 1969); Serbia and Montenegro (Kryštufek and Kovacic, 1984; Petrov, 1992); Bulgaria (Peshev, 1996; Vohralík, 1985); N Spain in Navarra region (Castién and Gosálbez, 1992); N Italy (Agnelli and Lazzeretti, 1995, who also documented identity of *Mus meridionalis* with *M. minutus*; Amori et al., 1999; Paolucci et al., 1993); Baltic region (Miljutin, 1997, 1998; Timm et al., 1998;); Lithuania (Juškaitis and Baranauskas, 2001); Netherlands (Mostert, 1992*b*; Thissen and Hollander, 1996); Belgium (Libois, 1996); Great Britain (Perrow and Jowitt, 1995, who also described decline and future of populations); and Japan (Dobson, 1994; Kaneko, 1994). European late Pleistocene and Pleistocene records reviewed by Kowalski (2001).

Millardia Thomas, 1911. J. Bombay Nat. Hist. Soc., 20:998.

TYPE SPECIES: *Golunda meltada* Gray, 1837.

SYNONYMS: *Grypomys* Thomas, 1911; *Guyia* Thomas, 1917; *Millardomys* Sody, 1941.

COMMENTS: *Millardia* Division. Listed as a genus by Ellerman in 1941, but later as a subgenus of *Rattus* (Ellerman, 1961). By 1969, *Millardia* was again treated as a genus and thought to be closely related to the Indian *Cremnomys* (Misonne, 1969). Subsequent analyses of morphological features and particularly chromosomal traits have demonstrated the great phylogenetic distance between *Rattus* and *Millardia* (Gadi and Sharma, 1983; Mishra and Dhandra, 1975; Raman and Sharma, 1977). Cytogenetic analyses resulted in a phylogenetic hypothesis isolating *Millardia* and *Cremnomys* from other Asian genera (Gadi and Sharma, 1983), which is corroborated by analyses of mitochondrial sequences (DNA cytochrome *b*, 12S and 16S rRNA gene fragments), in which Ducroz et al. (2001:200) found no evidence of close relationship between *Millardia* and African arvicanthines (contradicting results from DNA/DNA hybridization; Chevret et al., 1994) or between *Millardia* and any murine genera tested; *Millardia* was basal in nearly all their phylogenetic reconstructions. This position is supported by albumin immunology suggesting that "*Millardia* appears to be a monogeneric clade arising early in the history of the murines" (Watts and Baverstock, 1995*b*). That ancient lineage also contains *Cremnomys*, *Madromys*, and probably *Diomys*, all Indian subcontinent endemics. An affinity with the African *Praomys* has been suggested (Corbet and Hill, 1992; Misonne, 1969) but not substantiated by phylogenetic analyses of morphological or molecular traits (e. g., Lecompte, 2003). Agrawal (2000) characterized *Millardia* and reviewed its taxonomic history; also reviewed by Corbet and Hill (1992). Enamel microstructure of incisors and molars and its significance presented by Patnaik (2002).

Two species of *Millardia*, until 1982 thought to be strictly an Asian genus, were recorded from the late Pliocene of Ethiopia (Sabatier, 1982). One of them was reidentified as an *Acomys* (Denys, 1990*a*); identity of the other needs to be reevaluated, but to us its molar occlusal patterns also resemble those of *Acomys*. Outside of the supposed African record, evolutionary history of *Millardia* (identified as cf *Millardia* or *M.* cf. *meltada*) is documented from early Pliocene to late Pleistocene strata only on the Indian subcontinent

(NW India and N Pakistan; Gupta and Prasad, 2001; Musser, 1987*b*; Patnaik, 1995, 1997, 2001).

Millardia gleadowi (Murray, 1886). Proc. Zool. Soc. Lond., 1885:809 [1886].
   COMMON NAME: Sand-colored Metad.
   TYPE LOCALITY: Pakistan, Clifton Plain, Karachi.
   DISTRIBUTION: S and C Pakistan on west side of Indus River (Roberts, 1977, 1997) and adjacent NW India (Agrawal, 2000).
   STATUS: IUCN – Lower Risk (lc).
   COMMENTS: A distinctive species morphologically unlike any other *Millardia* and adapted to semideserts of shifting sand dunes, clay flats, and rocky hillsides. Raman and Sharma (1977) reported essentially no similarity between karyotypes of *M. meltada* and *M. gleadowi*. Its inclusion in *Millardia* should be reexamined. Indian population reviewed by Agrawal (2000); ecology and distribution in Gujarat State of NW India recorded by Chakraborty and Agrawal (2000).

Millardia kathleenae Thomas, 1914. J. Bombay Nat. Hist. Soc., 23:29.
   COMMON NAME: Burmese Metad.
   TYPE LOCALITY: Burma, Pagan.
   DISTRIBUTION: Apparently endemic to C Burma (Ellerman, 1961).
   STATUS: IUCN – Lower Risk (lc).
   COMMENTS: Sody (1941) proposed the genus *Millardomys* for this species. Aplin (in litt., 2004) trapped it at Mt Popa and near Pagan, found it absent from surrounding areas, and suspected it to be locally common in the central dry zone of Burma.

Millardia kondana Mishra and Dhanda, 1975. J. Mammal., 56:76.
   COMMON NAME: Large Metad.
   TYPE LOCALITY: India, Maharashtra State, Poona Dist, Sinhgarh (18°23′N, 73°42′E).
   DISTRIBUTION: Endemic to India; known only from the Sinhgarh Plateau in the Maharashtra region.
   STATUS: IUCN – Endangered.
   COMMENTS: Morphological comparisons between this distinctive species and *M. gleadowi*, *M. kathleenae*, and *M. meltada* reported by Mishra and Dhanda (1975). Reviewed by Agrawal (2000).

Millardia meltada (Gray, 1837). Ann. Mag. Nat. Hist., [ser. 1], 1:586.
   COMMON NAME: Common Metad.
   TYPE LOCALITY: India, S Mahratta, Dharwar.
   DISTRIBUTION: Sri Lanka; Indian Peninsula west to Gujarat and Rajasthan, north to Himachal Pradesh, and east to West Bengal; E Pakistan; and Terai region of Nepal (Agrawal, 2000, Chakraborty and Agrawal, 2000; Corbet and Hill, 1992; Ellerman, 1961; Rana, 1985).
   STATUS: IUCN – Lower Risk (lc).
   SYNONYMS: *comberi* (Wroughton, 1907); *dunni* Thomas, 1917; *lanuginosus* (Elliot, 1839); *listoni* (Wroughton, 1907); *mettada* (Wroughton, 1907); *pallidior* Ryley, 1914; *singuri* Mandal and Ghosh, 1981.
   COMMENTS: Cytogenetics of this species is the subject of a substantial body of literature (Nanda and Raman, 1981; Raman and Sharma, 1977; Sobti and Gill, 1984; Yosida, 1978*a*). Mandal and Ghosh (1981) described *singuri* as a subspecies of *M. meltada*. Subspecies formerly recognized in *M. meltada* (Ellerman, 1961) were based upon differences in fur coloration, but Agrawal (2000) noted the lack of significant geographic pattern in color variation as well as external and cranial dimensions among samples and treated the taxa as synonyms. Habitat and distribution on Abu Hill in the Aravalli Range in Rajasthan, India documented by Prakash et al. (1995*a*, 1995*b*), and in Gujarat State of NW India by Chakraborty and Agrawal (2000). Reviewed by Agrawal (2000).

Muriculus Thomas, 1903. Proc. Zool. Soc. Lond., 1902(2):314 [1903].
   TYPE SPECIES: *Mus imberbis* Rüppell, 1842.

COMMENTS: *Mus* Division. When Thomas proposed *Muriculus* he suggested it might be related to *Lophuromys*, but Osgood (1936) found little affinity between the two genera and instead noted a closer morphological relationship with *Mus* and *Zelotomys*. The close tie to *Mus* is real (Ellerman, 1941), and Misonne (1974) noted that *Muriculus* might be merged with *Mus*; however, the middorsal stripe and morphological specializations of rostrum, mandible, incisors, and increased expanse of incisor enamel associated with pronounced proodonty, are not part of the character suite defining *Mus* and set *Muriculus* apart as a distinctive genus.

Muriculus imberbis (Rüppell, 1842). Mus. Senckenberg., 3:110.
    COMMON NAME: Ethiopian Striped Mouse.
    TYPE LOCALITY: N Ethiopia, Simien, 3300 m.
    DISTRIBUTION: Endemic to mountains of Ethiopia on both sides of the Rift Valley, 1900-3400 m (Rupp, 1980).
    STATUS: IUCN – Vulnerable.
    SYNONYMS: *chilaloensis* Osgood, 1936.
    COMMENTS: Member of a unique rodent fauna endemic to high mountains of Ethiopia. Osgood (1936) described *chilaloensis* as a subspecies of *M. imberbis*. Yalden and Largen (1992) reviewed the species as an Ethiopian endemic, which is infrequently encountered by collectors (Yalden et al., 1996).

Mus Linnaeus, 1758. Syst. Nat., 10th ed., 1:59.
    TYPE SPECIES: *Mus musculus* Linnaeus, 1758.
    SYNONYMS: *Budamys* Kretzoi and Vertes, 1967; *Coelomys* Thomas, 1915; *Drymomys* Tschudi, 1844; *Gatamiya* Deraniyagala, 1966; *Hylenomys* Thomas, 1925; *Leggada* Gray, 1837; *Leggadilla* Thomas, 1914; *Musculus* Rafinesque, 1814; *Mycteromys* Robinson and Kloss, 1918; *Nannomys* Peters, 1876; *Oromys* Robinson and Kloss, 1916 [not Leidy, 1853]; *Pseudoconomys* Rhoads, 1896; *Pyromys* Thomas, 1911; *Tautatus* Kloss, 1917.
    COMMENTS: *Mus* Division. Extant species of *Mus* are contained in subgenera *Coelomys*, *Mus*, *Nannomys*, and *Pyromys*, each diagnosed by a suite of discrete morphological traits (see J. T. Marshall, Jr., 1977*b*, 1986, for diagnoses of *Coelomys*, *Mus*, and *Pyromys*; morphological characters distinguishing each subgenus are listed by Chevret et al., 2003), morphometric features (Macholán, 2001), and biochemical characteristics (Bonhomme, 1986; She et al., 1990). *Nannomys*, *Pyromys*, and *Coelomys* alternatively have been treated as genera (Bonhomme, 1986; She et al., 1990). The DNA-DNA hybridization and morphological study by Catzeflis and Denys (1992) using selected species of subgenera *Nannomys*, *Pyromys*, *Coelomys*, and *Mus* indicated the two African species sampled are more closely allied to *Mus* than to other African and European non-*Mus* genera sampled, and at least as close to subgenus *Mus* as the subgenera *Coelomys* and *Pyromys*: *Nannomys* "appears to be an offshoot of the Asian mice radiation, and this split could have occurred without important morphological changes at the beginning of the Pliocene" (p. 228). Their study reinforced the inclusion of the suite of endemic African species in *Mus*. Results of phylogenetic analysis using mitochondrial 12S rDNA gene sequences, which focused on the relationships of Sumatran *Mus crociduroides*, also demonstrated the monophyly of *Mus* compared with other murine genera (Sourrouille et al., 1995), as did albumin immunology (Watts and Baverstock, 1995*b*), and analyses of sequences from six genes "representing paternally, maternally, and biparentally inherited regions of the genome" (Lundrigan et al., 2002:410). Recent analyses of combined morphological traits, DNA/DNA hybridization, and mitochondrial 12S rRNA sequences indicate subgenera *Mus*, *Nannomys*, *Pyromys*, and *Coelomys* represent four distinctive monophyletic groups (Chevret et al., 2003; also see review by Guénet and Bonhomme, 2003). Results of Lundrigan et al.'s (2002) analyses also recognized the same four clades, but at the same time indicated all four to be in a larger monophyletic cluster relative to *Hylomyscus*, *Mastomys*, and *Rattus*, which were used to root the phylogenies. Lundrigan et al. (2002:424) noted that a denser sampling of *Mus* species and murine genera are needed to rigorously test the monophyly of *Mus*, ". . . but for now there is no phylogenetic justification for excluding *Pyromys*, *Coelomys*, or *Nannomys* from the genus *Mus*." Relevance of metrical, chromosomal, and allozymic variation to systematics of *Mus*, particulary the impulse to elevate subgenera to genera, was dis-

cussed by Corbet (1990). Enamel microstructure of incisors and molars in *Mus* and its significance documented by Patnaik (2002).

Reviews of variable depth and content are available for Asian (J. T. Marshall, Jr., 1977*b*) and European species (J. T. Marshall, Jr., 1981, 1986, 1998; J. T. Marshall, Jr. and Sage, 1981; Bonhomme et al., 1984; Gerasimov et al., 1990; She et al., 1990), and "wild mice" as a popular mammalian model (Guénet and Bonhomme, 2003). Appraisal of fossil evidence and morphological traits for European species summarized by Thaler (1986). A key to European house mice (*M. musculus*, *M. spretus*, *M. macedonicus*, and *M. spicilegus*) based mostly on cranial and dental traits provided by Macholán (1996*a*), along with summaries of diagnostic characters, habitats, and geographic ranges. Morphometric analysis of 12 population samples representing those four species were also documented by Macholán (1996*b*), and he provided results of multivariate analysis of morphometric variation in Asian species of subgenus *Mus* and two species of *Nannomys* and contrasted it with the phylogenies reconstructed from allozymic analyses (Macholán, 2001). Most African species require careful systematic review (see accounts of species in subgenus *Nannomys*); Ansell (in Meester et al., 1986:280) believed the African segment of the genus to be "over-split," but our study of specimens suggests a current underestimate of actual species-diversity.

Reconstructing evolutionary relationships among species is a prime research goal of *Mus* systematists and they are using phylogenetic analyses of an increasing variety of molecular data (sequences from a range of mitochondrial and nuclear genes) along with sophisticated chromosomal information to test the current phylogenetic picture based upon morphological traits. Much of the molecular data is concordant with morphology in supporting current estimate of species affinities, but not all. Nishioka et al. (1993) surveyed a repetitive Y chomosomal sequence (1.1 kb long and designated 142.4) among species of *Mus* in subgenera *Mus*, *Pyromys*, *Coelomys*, and *Nannomys* and found a restriction fragment length polymorphism that was different in *M. m. musculus* and *M. m. domesticus*. Among the other species sampled, however, the accumulation patterns of this repetitive sequence did not correlate with the phylogenetic relationships of species as determined by morphological and other molecular traits, and Nishioka et al. concluded that the 142.4 related sequences are evolutionarily unstable.

Phylogenetic relationships of *Mus* are unclear. Study of molar traits led Misonne (1969) to place the genus in his *Rattus* Division, and to suggest an affinity with the deomyines *Acomys* and *Uranomys*. This grouping is uncorroborated when tested with molecular data, which isolates the latter two genera as a separate group not closely related to any murines (see account of Deomyinae). Cladistic analyses of complete mtDNA cytochrome *b* sequences (Martin et al., 2000) and sequences from nuclear IRBP gene and the cytochrome *b* and 12S rRNA mitochondrial markers (Michaux et al., 2002*a*) indicate a tighter link between *Mus* and *Apodemus* and its relatives than between *Mus* and *Rattus*, results corroborated by other molecular studies (see references in Michaux et al., 2002*a*). Generic sampling is not dense in these analyses, and the suggested *Mus-Apodemus* alliance must be tested by wider taxon and geographic sampling. Albumin immunology, for example, pointed to *Mus* as "a distinct monogeneric clade arising from the central murine stem . . . at about the same time as *Apodemus* and the African group [*Rhabdomys*, *Grammomys*, *Arvicanthis*, *Thallomys*, *Praomys*, *Hybomys*, *Stochomys*, *Hylomyscus*, *Pelomys*, *Aethomys*, *Lemniscomys*, and *Otomys*]" (Watts and Baverstock, 1995*b*:112), which may be a more realistic estimate of evolutionary origin. A species resembling the M1 occlusal pattern of *Mus* comes from the same C European beds containing the oldest European murines (early Vallesian, 11.1-9.7 million years ago) and is regarded as related to *Mus*; it, *Apodemus*, *Progonomys*, and two other groups comprise the earliest European murine lineages (Freudenthal and Martín Suárez, 1999; Mein et al., 1993). The fate of the *Mus*-like form in Europe is unknown because true *Mus* is absent from the later European Miocene and Pliocene sediments and appears only in the Pleistocene (Auffray et al., 1990*c*; Kowalski, 2001), but Mein et al. (1993) noted the very close similarity between the Vallesian "*Mus*" and Siwalik *Progonomys*, from which Jacobs (1978) and others (Jacobs and Downs, 1994) would derive the earliest *Mus*. This phyletic ambiguity surrounding *Mus* prompts us to isolate it in its own group. Most systematists have interpreted, as all the molecular work has reinforced, the phylogenetic relationship between *Mus* and *Rattus* to be distant, and perhaps instead of focusing interminably and tediously on the imaginary *Mus-Rattus* split (dredged up in most paleontological and molecular works focusing on murine systematics; see also Adkins et al., 2003), researchers can instead look for data signals identifying the closest allies of *Mus* and time of its split from the ancestral murine stock.

Mus auctor, from late Miocene sediments in the Siwaliks of Pakistan, is claimed to be the earliest record of the genus (5.7 million years ago; Jacobs, 1978; Jacobs and Downs, 1994; Patnaik, 1997; Patnaik et al., 1996). But at that time the *Mus* dental specializations are already apparent and indicate an earlier evolutionary origin (see discussion in Chaimanee, 1998), which may be found in molar samples from early Vallesian (11-10 million years ago) of Europe considered closely related to *Mus* in morphology, if not a species of *Mus* (Mein et al., 1993; Suárez and Mein, 1998; those latter authors even noted that *Mus*, along with *Apodemus*, have the longest time range of any murine, referring to the Vallesian form). Jacobs and Downs (1994) derive *Mus* from Asian *Progonomys debruijni* or a species similar to it (but Mein et al., 1993, excluded *debruijni* from *Progonomys*). During Pliocene times *Mus* samples come from C Asia and Africa (see review in Auffray et al., 1990), and by the Plio-Pleistocene, subgenus *Mus* is represented by a cluster of species from N India that are dentally similar to the living *M. booduga* and *M. terricolor* from the Indian subcontinent (Gupta and Prasad, 2001; Kotlia, 1992, 1995; Patnaik, 1997, 2001; Patnaik et al., 1993, 1996). Recently, a 2-million year old skull (late Pliocene) has been found representing a species (*M. linnaeus*) that might be ancestral to *M. musculus*, which in the eyes of Patnaik et al. (1996) support the out-of-India origin of house mice. Boursot et al. (1996:406), e.g., speculated that "This species appears to have originated in the N Indian subcontinent, from where it colonised the Middle East," which is the hypothesis also favored by Din et al. (1996), by not by Prager et al. (1998). Elsewhere in Asia, Zheng (1993) recorded *Mus* sp. from late Pliocene cave sediments of the Sichuan-Guizhou region in S China, and Zheng et al. (1997) documented *M. musculus* from Pleistocene fissure deposits in the Shandong region. Chaimanee (1998:214) documented four extant Thai species based upon molars recovered from late Pliocene and Pleistocene cave sediments. In her opinion, "Southeast Asia is the center of evolution of [*Mus*] with four sympatric species occurring [in Thailand] since the Pleistocene." Examples of species in subgenus *Mus* have also been recovered from early Pleistocene sediments on Java (Van der Meulen and Musser, 1999).

The extinct *Mus minotaurus* is represented by only teeth and mandibular fragments "from a cave deposit in a small limestone ridge between Canea and Suda" on Crete Isl in the Mediterranean (Bate, 1942*a*:46); age range probably late Pleistocene to Holocene (found in Neolithic levels; Mayhew, 1977). This is one of the large-bodied species in the genus, approximating the size of the extant Indomalayan *Mus shortridgei* (Bate, 1942*a*). Kuss and Misonne (1968) suggested a close relationship between *M. minotaurus* and the Indomalayan *M. pahari*, *M. mayori* and *M. shortridgei* as well as the African *M. bufo*. Mayhew (1977, 1996) had described *M. bateae*, having molars smaller than *M. minotaurus* but larger than *M. musculus*, from probable middle Pleistocene sediments on Crete and postulated that it and *M. minotaurus* are part of a single lineage derived from an ancestral population resembling *M. musculus*. Both *M. minotaurus* and *M. bateae* are found in the higher biostratigraphic depositional levels on Crete, always in the zone above those containing the extinct endemic murine *Kritimys* (Dermitzakis and de Vos, 1987). *Mus minotaurus* was still present in late cave levels along with traces of human activities and probably became extinct after the arrival of humans (about 8000 years before present) and the subsequent introduction of *Rattus* and *Mus musculus*, which ultimately provided devastating competition (Dermitzakis and de Vos, 1987).

Earliest records of *Mus* from Subsaharan Africa come from late Pliocene sediments, and samples of the genus have also been obtained from Pleistocene and Holocene strata (Avery, 1998, 2000; Denys, 1999; Jaeger, 1976; Jaeger and Wesselman, 1976; Senut et al., 1992). Geraads (1998) described a Miocene-Pliocene species from Morocco that may be the earliest known *Mus* recorded from Africa. Mein and Pickford (1992) discussed species recovered from Pleistocene beds in Tunisia.

Mus baoulei (Vermeiren and Verheyen, 1980). Rev. Zool. Afr., 94:573.
    COMMON NAME: Baoule's Mouse.

TYPE LOCALITY: Côte d'Ivoire, Lamto.

DISTRIBUTION: Known only from Côte d'Ivoire and E Guinea (Grubb et al., 1998, guessed the species may reach Ghana or Sierra Leone).

STATUS: IUCN – Lower Risk (lc).

COMMENTS: Subgenus *Nannomys*. Occurs sympatrically with *M. minutoides* and *M. setulosus*, morphology closely similar to species in the *M. sorella* group (Vermeiren and Verheyen, 1980). The diagnostic traits reported for *M. baoulei* by Vermeiren and Verheyen (1980) are those that set *M. sorella* apart from other species of *Mus* (Verheyen, 1965a). Judged by their description, *M. baoulei* is distinguished from *M. sorella* by smaller size, a contrast that also exists between *M. sorella* and *M. neavei* (see that account). The relationship of *baoulei* to other members of the *M. sorella* group requires fresh assessment.

Mus booduga (Gray, 1837). Mag. Nat. Hist. [Charlesworth's], 1:586.

COMMON NAME: Little Indian Field Mouse.

TYPE LOCALITY: India, S Mahratta.

DISTRIBUTION: Sri Lanka, Peninsular India (north to Jammu and Kashmir; Agrawal, 2000; Chakraborty and Agrawal, 2000), Bangladesh (Posamentier, 1989), S Nepal and C Burma (Corbet and Hill, 1992), and Pakistan (Roberts, 1977, 1997; J. T. Marshall, Jr., 1998).

STATUS: IUCN – Lower Risk (lc).

SYNONYMS: *albidiventris* Blyth, 1852 [not Burg, 1923]; *fulvidiventris* Blyth, 1852; *lepidoides* (Fry, 1931); *weragami* (Deraniyagala, 1965).

COMMENTS: Subgenus *Mus*. Revised by J. T. Marshall, Jr. (1977b). Results from chromosomal analyses reported by Sen and Sharma (1983) and Sharma et al. (1986) in context of evolutionary divergence relative to other species of *Mus*. Chromosomal information along with molecular data (allozymes, serum proteins, mtDNA sequences) used by Sharma (1996) in a comparative study with other species of *Mus*. Corbet and Hill (1992) suspected the species to be widespread but its distribution poorly known because of confusion with the morphologically similar *M. terricolor*. They listed *terricolor* Blyth, 1851 and *beavanii* Peters, 1866 as synonyms of *M. booduga*, but the former is the oldest name for *M. dunni* and the latter is a synonym of *terricolor*. Agrawal (2000) reviewed the Indian populations and treated *dunni* and *terricolor* as synonyms of *M. booduga*. In a morphometric study of samples from the C Punjab region of N Pakistan, Rana et al. (1998) uncovered two forms, one possibly representing *M. terricolor* (reported as *dunni*), the other *M. booduga*. Overlap in some features, however, prohibited a clear taxonomic decision and the authors stressed the need for breeding and cytogenetic studies of additional samples. Dental patterns of *M. booduga* and *M. terricolor* are closely similar to species of Plio-Pleistocene *Mus* from the Indian subcontinent (Patnaik et al., 1993). See account of *M. terricolor* for additional comparisons with *M. booduga*.

Mus bufo (Thomas, 1906). Ann. Mag. Nat. Hist., ser. 7, 18:145.

COMMON NAME: Toad Mouse.

TYPE LOCALITY: Uganda, Ruwenzori East, 6000 ft (1830 m).

DISTRIBUTION: E Dem. Rep. Congo (Kivu region), adjacent Uganda, Rwanda, and Burundi; a montane Western Rift endemic.

STATUS: IUCN – Lower Risk (lc).

SYNONYMS: *ablutus* G. M. Allen and Loveridge, 1939; *wambutti* (Lönnberg and Gyldenstolpe, 1925).

COMMENTS: Subgenus *Nannomys*. In body size and morphology, *M. bufo* superficially resembles the large-bodied *M. triton*, but they are distinguished by dental traits and tail length (F. Petter and Matthey, 1975) as well as karyotypes (Robbins and Baker, 1978), and occur together in the Kivu region of E Dem. Rep. Congo (specimens in AMNH). Electrophoretic analysis of 19 protein enzymes at 24 loci indicated *M. bufo* to be more closely related to *M. gratus* (= *M. minutoides*) than to *M. triton* (Van Rompaey et al., 1984). Chromosomal data for Burundi samples reported by Maddalena et al. (1989). Altitudinal distribution on Ugandan slopes of Ruwenzori Mtns reviewed by Kerbis

Peterhans et al. (1998); documented from Kalinzu Forest in SW Uganda by Lunde and Sarmiento (2002), and from Kibale Forest by Hoffmann (1997).

Mus callewaerti (Thomas, 1925). Ann. Mag. Nat. Hist., ser. 9, 15:668.
COMMON NAME: Callewaert's Mouse.
TYPE LOCALITY: S Dem. Rep. Congo, Kasaï-Occidental, Kananga (= Luluabourg), 610 m.
DISTRIBUTION: Recorded only from NE and C Angola (Crawford-Cabral, 1998), S and W Dem. Rep. Congo; limits unknown.
STATUS: IUCN – Lower Risk (lc).
COMMENTS: Subgenus *Nannomys*. Thomas described this species as a member of the genus *Hylenomys*, which is now united with *Mus* (Hill and Carter, 1941; Misonne, 1965a) as a synonym of *Nannomys*. It is the largest-bodied of any of the African *Mus* (F. Petter and Matthey, 1975), has ivory incisors and very large auditory bullae, and in external features resembles *M. triton*, which misled Hatt (1940a) into treating *callewaerti* as a subspecies of *triton*. Misonne (1965a) summarized distributional and other information.

Mus caroli Bonhote, 1902. Novit. Zool., 9:627.
COMMON NAME: Ryukyu Mouse.
TYPE LOCALITY: Japan, Ryukyu (= Liukiu) Isls, Okinawa Isl.
DISTRIBUTION: Natural range probably from Ryukyu Isls (Kaneko, 1994) to Taiwan (M.-J. Yu, 1996, and Wang, 2003), S China (Fujian, Guangdong, Guizhou, Guangxi, Yunnan, Hainan Isl; Wang, 2003, and Zhang et al., 1997) and Hong Kong (Chandrasekar-Rao and Musser, 1993), Vietnam (Dang et al., 1994; also Cat Ba Isl, off north coast; Kuznetsov, 2000), Laos (Aplin et al., 2003c; Smith et al., In Press), Cambodia, and Thailand (N of Isthmus of Kra; J. T. Marshall, Jr., 1977a). Also recorded from Malay Peninsula (S Kedah State), Sumatra, Java, Madura, and Flores Isls in Nusa Tenggara, all places where it was likely inadvertently introduced (Musser and Newcomb, 1983).
STATUS: IUCN – Lower Risk (lc).
SYNONYMS: *boninensis* Kishida, 1926 [*nomen nudum*]; *boninensis* Kuroda, 1930; *formosanus* Kuroda, 1925; *kurilensis* Kuroda, 1924; *ouwensi* Kloss, 1921.
COMMENTS: Subgenus *Mus*. For synonyms see Kaneko and Maeda (2002). Morphologically similar to but easily distinguished (especially by cranial traits) from *M. cervicolor* (Macholán, 2001; J. T. Marshall, Jr., 1977b); member of a clade containing *M. cookii* and *M. cervicolor* as assessed by sequences of several different genes (Graur, 1994; Lundrigan et al., 2002). This alignment also supported by combined analyses of morphological traits, DNA/DNA hybridization, and mitochondrial 12S rRNA sequences (Chevret et al, 2003). Multivariate morphometric analysis by Macholán (2001) of Thai and Vietnamese samples suggested significant morphometric differences between populations from the two geographic regions and Macholán speculated that the Mekong River may impede gene flow between them. Habitat use on Taiwan reported by Adler (1995). Reviewed by Corbet and Hill (1992), who doubted that *ouwensi* is synonymous with *M. caroli*. Sterile hybrid females were obtained from crosses between *M. caroli* and *M. musculus* (West et al., 1978). Genetic variation across mainland Southeast Asian range surveyed by Terashima et al. (2003). Aplin et al. (2003c) illustrated a specimen of *M. caroli* collected near Hmawbe in C Burma (the same taxon was subsequently collected near Mawlamyine, Mon State; K. Aplin, in litt., 2004). Molecular analysis in the laboratory of Dr. H. Suzuki has confirmed the general relationship of this taxon to *M. caroli* (K. Aplin, in litt., 2004). The Burmese population, however, is much larger-bodied than typical *M. caroli* from farther east.

Musser and Carleton (1993:623) listed *kakhyenensis*, described by Anderson (1879) from a specimen collected in the Kakhyen Hills of W Yunnan, in the synonymy of *M. caroli*. It is an older name than *caroli*, J. T. Marshall, Jr. (1977b:211, 1998:53) examined the holotype (body in fluid, skull missing) but without the skull could not identify it to species, declared the name indeterminate, and urged its suppression. Aplin (in litt., 2004) noted that "this seems insufficient grounds given the dual possibilities of recollecting at a known type locality and/or DNA recovery from an extant holotype." Molars recovered from cave sediments in Thailand document occurrence of *M. caroli* back to late Pliocene in that country (Chaimanee, 1998).

Mus cervicolor Hodgson, 1845. Ann. Mag. Nat. Hist., [ser. 1], 15:268.
  COMMON NAME: Fawn-colored Mouse.
  TYPE LOCALITY: Nepal.
  DISTRIBUTION: Indigenous range from N India (Jammu, Kashmir, Uttar Pradesh, Sikkim,
      West Bengal, Meghalaya, and Manipur; Agrawal, 2000) and Nepal east through Burma,
      Thailand (J. T. Marshall, Jr., 1977a; Robinson et al., 1995), Laos (Aplin et al., 2003c;
      Smith et al., In Press), Cambodia, Vietnam (Dang et al., 1994), and S China (Yunnan;
      Wang, 2003, and Zhang et al., 1997); see overall distribution in J. T. Marshall, Jr.
      (1977a). Also recorded from Sumatra and Java where it has likely been inadvertently
      introduced (Musser and Newcomb, 1983). Mandal and Ghosh (1984) reported a
      specimen from South Andaman Isl in the Andaman Isls; whether representative of
      natural range or anthropogenic introduction is unknown ("most likely introduced,"
      J. T. Marshall, Jr., in litt., 2004). Roberts (1977, 1997) reported the species from C and
      S Pakistan but that identification requires verification.
  STATUS: IUCN – Lower Risk (lc).
  SYNONYMS: *annamensis* (Robinson and Kloss, 1922); *cunicularis* Blyth, 1855; *imphalensis*
      (Roonwal, 1948); *nitidulus* Blyth, 1859; *popaeus* (Thomas, 1919); *strophiatus* Hodgson,
      1845.
  COMMENTS: Subgenus *Mus*. Closely related to *Mus caroli* in morphology (Macholán, 2001); a
      member of a clade containing *M. cookii* and *M. caroli* as assessed by sequences from
      several different genes (Graur, 1994; Lundrigan et al., 2002), and combined analyses of
      morphological traits, DNA/DNA hybridization, and mitochondrial 12S rRNA
      sequences (Chevret et al., 2003). J. T. Marshall, Jr. (1977b) recognized two subspecies in
      Thailand, the large-bodied *T. c. popaeus* found in forests, and the smaller-bodied
      T. c. cervicolor inhabiting ricefields, with some intermediates in forest near Tak,
      NW Thailand. Macholán's (2001) multivariate analysis of morphometric traits could
      not distinguish samples of the two kinds, and he noted their respective ranges formed
      a mosaic defined by habitat discontinuities; the two subspecies are also indistinguish-
      able by molecular markers (Auffray et al., 2003). Reviewed by Corbet and Hill (1992).
      Mus cervicolor has resided in Thailand since middle Pleistocene, judged from isolated
      molars recovered from cave deposits (Chaimanee, 1998).

Mus cookii Ryley, 1914. J. Bombay Nat. Hist. Soc., 22:663.
  COMMON NAME: Cook's Mouse.
  TYPE LOCALITY: N Burma, Shan States, Gokteik, 2133 ft (650 m).
  DISTRIBUTION: India (disjunct, one part in S Peninsular India, the other in the northeast;
      Agrawal, 2000), Nepal through Burma and S China (SW Yunnan; Wang, 2003; Zhang
      et al., 1997, as *M. famulus cookii*) to N and C Thailand (J. T. Marshall, Jr., 1977a;
      Robinson et al., 1995), Laos (Smith et al., In Press), and Vietnam (Dang et al., 1994).
  STATUS: IUCN – Lower Risk (lc).
  SYNONYMS: *darjilingensis* Hodgson, 1849 [holotype is example of *M. cookii*, but name is a
      nomen nudum; J. T. Marshall, Jr., 1977b, 1998]; *nagarum* (Thomas, 1921); *palnica*
      (Thomas, 1923); *rahengis* (Kloss, 1920); *thai* (Kloss, 1917).
  COMMENTS: Subgenus *Mus*. Revised by J. T. Marshall, Jr. (1977b) and reviewed by Corbet and
      Hill (1992). Analysis of sequences from the *Sry* gene indicates close alliance with *M. cervi-*
      color (Graur, 1994), and combined analyses of morphological traits, DNA/DNA
      hybridization, and mitochondrial 12S rRNA sequences bring *M. cooki* together with
      M. cervicolor and *M. caroli* in a clade separate from a clade of European species (Chevret
      et al., 2003). Except for *M. musculus, M. cooki* is the only species of *Mus* common to
      peninsular India and Southeast Asia, but phylogenetic relationships between populations
      in these two regions have not been critically examined and perhaps more than one species
      is present in what is now defined as *M. cooki*. Agrawal (2000), for example, recognized
      typical *cooki* with a skull longer than 23 mm, and *M. c. nagrum* with skull length less than
      23 mm. J. T. Marshall, Jr. (1977b:202) described two size classes in sympatry in NE India
      and Thailand, but noted "intermediates" from Burma. Significance of this variation
      requires review by restudy of museum specimens and analyses of DNA sequences from
      new material. Isolated molars from Thailand caves identified as *M. cookii* indicate the
      species has existed in the region since late Pliocene (Chaimanee, 1998).

Mus crociduroides (Robinson and Kloss, 1916). J. Str. Br. Roy. Asiat. Soc., 73:271.
 COMMON NAME: Sumatran Shrewlike Mouse.
 TYPE LOCALITY: Indonesia, W Sumatra, Korinchi Peak, 10,000 ft (3050 m).
 DISTRIBUTION: Upper montane rain forest in mountain chain along W Sumatra.
 STATUS: IUCN – Lower Risk (lc).
 COMMENTS: Subgenus *Coelomys*. A distinct montane species endemic to the mountains of
  W Sumatra (Musser, 1986; Musser and Newcomb, 1983). Listed by Chasen (1940) as a
  species of *Mycteromys*, which also contained the Javan *M. vulcani*, the closest relative of
  M. crociduroides. Phylogenetic analysis of mitochondrial 12S rRNA sequences clustered
  M. crociduroides with *M. pahari* (Catzeflis and Denys, 1992; Chevret et al., 2003),
  reinforcing the alliance indicated by morphological traits (Chevret et al., 2003;
  J. T. Marshall, Jr., 1977*b*) – both are in subgenus *Coelomys*. Member of the suite of
  murines endemic to Sumatra (Musser, 1986; also see account of *Maxomys hylomyoides*).

Mus famulus Bonhote, 1898. J. Bombay Nat. Hist. Soc., 12:99.
 COMMON NAME: Servant Mouse.
 TYPE LOCALITY: S India, Tamil Nadu, Nilgiri Hills, Coonoor, 5000 ft (1524 m).
 DISTRIBUTION: An Indian endemic recorded only from the Western Ghats (= Sahyadris) in
  tropical evergreen rain forest covering the Nilgiri Hills in SW peninsular India, about
  1500 m (Agrawal, 2000; Corbet and Hill, 1992).
 STATUS: IUCN – Endangered.
 COMMENTS: Subgenus *Mus*. Revised by J. T. Marshall, Jr. (1977*b*); reviewed by Agrawal (2000)
  and Corbet and Hill (1992). Originally considered a member of subgenus *Coelomys*
  along with *M. vulcani*, *M. crociduroides*, *M. mayori*, and *M. pahari* (J. T. Marshall, Jr.,
  1977*b* [Marshall acknowledges his mistake and now supports the subgeneric
  allocation identified here; in litt., 2004]. However, recent analyses of morphological
  traits, DNA/DNA hybridization, and mitochondrial 12S rRNA sequences indicates
  close relationship with European (*M. spicilegus*, *M. spretus*, and *M. musculus*) and Asian
  (*M. cervicolor*, *M. cookii*, and *M. caroli*) clades within subgenus *Mus* (Chevret et al., 2003;
  Guénet and Bonhomme, 2003); molecular data place *M. famulus* as sister to first
  M. fragilicauda and then the European clade (see review by Guénet and Bonhomme,
  2003), but morphology nests it within the Asian clade. The murine *Vandeleuria
  nilagirica* is also recorded only from the Nilgiri Hills; *Rattus satarae* and the lion-tailed
  macaque (*Macaca silenus*) occur there also but have a more extensive range northward
  in the Western Ghats to which they are endemic.

Mus fernandoni (Phillips, 1932). Spolia Zeylan., 16:325.
 COMMON NAME: Ceylon Spiny Mouse.
 TYPE LOCALITY: Sri Lanka, Mulhalkelle Dist., Kubalgamuwa, 3000 ft (915 m).
 DISTRIBUTION: Endemic to Sri Lanka.
 STATUS: IUCN – Lower Risk (lc).
 COMMENTS: Subgenus *Pyromys*. Revised by J. T. Marshall, Jr. (1977*b*); reviewed by Corbet and
  Hill (1992) and Phillips (1980).

Mus fragilicauda Auffray, Orth, Catalan, Gonzalez, Desmarais, and Bonhomme, 2003. Zoologica
 Scripta, 32:121.
 COMMON NAME: Sheath-Tailed Mouse.
 TYPE LOCALITY: SC Thailand, Nahkon Ratchasima (Khorat) Province, Wang Nam Yen Dist.,
  Ban Nong Sanga (14°32'33"N, 101°57'44"E).
 DISTRIBUTION: Recorded from the type locality and Tumbon, both places 50 km SSE of
  Khorat. Either this taxon or a close relative (DNA sequences are weakly divergent) is
  present in Sekong Province of Laos (K. Aplin, in litt., 2004). Additional field surveys
  and reexamination of museum specimens may reveal an even wider range.
 COMMENTS: Subgenus *Mus*. Known by 21 specimens collected in dry grass and patches of
  pygmy bamboo along roadsides or dikes bordering dry ricefields; *M. cervicolor* and
  M. caroli were taken at the same localities. *Mus fragilicauda* closely resembles *M. cervicolor* in morphology, differing in some cranial measurements and qualitative traits, fur
  texture, and its thinner skin that is easily pulled from the tail when mice are handled
  (Auffray et al., 2003); the two also differ in chromosomal features. Analyses of DNA

sequences indicates *M. fragilicauda* to be cladistically aligned most closely first with Indian *M. famulus* and then with European *M. musculus, M. macedonicus, M. spicilegus,* and *M. spretus,* but not closely with Asian *M. cervicolor, M. caroli,* or *M. cookii* (Guénet and Bonhomme, 2003). This recently described species was not identified in the field but in the laboratory after exhaustive morphological, chromosomal, and molecular analyses.

Mus goundae F. Petter and Genest, 1970. Mammalia, 34:455.
COMMON NAME: Gounda Mouse.
TYPE LOCALITY: Central African Republic, vicinity of Gounda River (F. Petter, 1981*b*, provided coordinates).
DISTRIBUTION: Recorded only from the vicinity of the type locality (Jotterand, 1972); limits unknown.
STATUS: IUCN – Vulnerable.
COMMENTS: Subgenus *Nannomys*. F. Petter (1981*b*) treated *M. goundae* as a species related to others in the *M. sorella* group, but the nature of that alliance remains unresolved (see account of *M. sorella*). Chromosomal data reported by Jotterand (1972).

Mus haussa (Thomas and Hinton, 1920). Novit. Zool., 27:319.
COMMON NAME: Hausa Mouse.
TYPE LOCALITY: Nigeria, Farniso.
DISTRIBUTION: Senegal (Duplantier and Granjon, 1992) and S Mauritania through Mali, Côte d'Ivoire, Burkina Faso, Ghana, S Niger (Dobigny et al., 2002*b*), and Benin to N Nigeria; limits undocumented.
STATUS: IUCN – Lower Risk (lc).
COMMENTS: Subgenus *Nannomys*. Body size, pelage color and pattern, and other morphological traits of *M. haussa* are very similar to those of *M. tenellus* (F. Petter, 1963*c*, 1972*a*; Rosevear, 1969), and F. Petter (1969) considered combining them; relationship between the two needs to be assessed by systematic revision of the *M. tenellus* complex. Chromosomal data (2n = 31-34, FN = 38) reported by Dobigny et al. (2002*b*), Jotterand (1972), and Matthey (1967*a*). Ghana population reviewed by Grubb et al. (1998). Abundant in owl pellets reported from S Mali (Meinig, 2000).

Mus indutus (Thomas, 1910). Ann. Mag. Nat. Hist., ser. 8, 5:89.
COMMON NAME: Desert Pygmy Mouse.
TYPE LOCALITY: N South Africa, Northern Cape Province, Molopo River, west of Morokwen.
DISTRIBUTION: N South Africa, W Zimbabwe, Botswana, C and N Namibia (de Graaff, 1997*n*:143; Skinner and Smithers, 1990:264); also S Angola (Crawford-Cabral, 1998).
STATUS: IUCN – Lower Risk (lc).
SYNONYMS: *deserti* (Thomas, 1912); *pretoriae* (Roberts, 1926); *valschensis* (Roberts, 1926).
COMMENTS: Subgenus *Nannomys*. Skinner and Smithers (1990) and Meester et al. (1986) discussed the morphological and chromosomal distinctions between *M. indutus* and *M. minutoides*. Skinner and Smithers (1990) also summarized biological data, and Meester et al. (1986) provided citations for synonyms. Definition of this species is ambiguous (Meester et al., 1986; Skinner and Smithers, 1990). Meester et al. (1986) included the Angolan *sybilla* in *M. indutus,* but our study of the holotype revealed that *sybilla* belongs with *M. minutoides*.

Mus macedonicus Petrov and Ruzic, 1983. Proc. Fauna SR Serbia, Serbian Acad. Sci. and Arts, Belgrade, 2:177.
COMMON NAME: Macedonian Mouse.
TYPE LOCALITY: Macedonia, near Valandovo.
DISTRIBUTION: Mediterranean environments in the Balkan Peninsula (Macedonia, Bulgaria south of the Stara Planina Mtns to Greece and some Aegean islands; Mitchell-Jones et al., 1999:285; Peshev, 1996; Petrov, 1992), Turkey, Transcaucasia (Gromov and Erbajeva, 1995, as *abbotti*), N and W Iran (see map in J. T. Marshall, Jr. 1998), Syria, Jordan, Lebanon, and Israel (Auffray et al., 1988, 1990*b*); see Macholán (1996*a*) for distributional details. Recorded also from Cyprus (Cheylan, 1991, as *abbotti*; Cucchi et al., 2002).

STATUS: IUCN – Lower Risk (lc).

SYNONYMS: *camini* Bate, 1942; *makovensis* Orlov, Nadjafova, and Bulatova, 1992; *spretoides* Bonhomme, Catalan, Britton-Davidian, Chapman, Moriwaki, Nevo, and Thaler, 1984 [*nomen nudum*; J. T. Marshall, Jr., 1998].

COMMENTS: Subgenus *Mus*. Originally described as a subspecies of *M. hortulanus* (Petrov and Ruzic, 1983), but now recognized as a separate species (Boursot et al., 1993; Bonhomme, 1986; Macholán, 1996*a*, 1996*c*; J. T. Marshall, Jr., 1998; Sage, et al., 1993; and references cited therein). *Mus macedonicus* was first recognized as a distinct species by Kratochvil (1986), who monographed it as *M. abbotti*, the name still used by some researchers (Cheylan, 1991; Mezhzherin and Kotenkova, 1992; Orlov et al., 1992*b*, Gromov and Erbajeva, 1995, e. g.), but the holotype of *abbotti* is an example of *M. musculus domesticus* (Boursot et al., 1993; J. T. Marshall, Jr., 1998). The taxon *tataricus*, another synonym of *M. musculus domesticus*, has also been used by Russian researchers for *M. macedonicus* (Mezhzherin and Kotenkova, 1992; Kotenkova and Bulatova, 1994; also see discussion in Mitchell-Jones et al., 1999:284). Orlov et al. (1992*b*) described *makovensis* as a subspecies of *M. abbotti*.

    Mus macedonicus occurs sympatrically, but never syntopically, with *M. musculus domesticus* (Auffray et al., 1990*b*; Macholán, 1996*a*; Petrov and Ruzic, 1985:234), and is sympatric with *M. spicilegus* in a narrow zone along the Black Sea (Boursot et al., 1993 and references therein). Results of allozymic and morphometric analyses of *M. macedonicus* (some reported as *spretoides*) in context of phylogenetic studies provided by Bonhomme et al. (1984), Auffray et al. (1990*b*), Gerasimov et al. (1990), and She et al. (1990). Forms a clade with *M. musculus*, *M. spicilegus*, and *M. spretus* based on analyses of DNA sequences from several different genes (Auffrey et al., 2003; Graur, 1994; Larizza et al., 2002; Lundrigan et al., 2002; Prager et al., 1996, 1998). Autosomal chromosome complement of *M. macedonicus* is the same as *M. musculus* and related species (2n = NF = 40), but has distinctive cytogenetic markers (Ivanitskaya et al., 1996*b*). Morphometric and immunological comparisons between Macedonian *M. macedonicus* and samples of *M. musculus* from Greek Isls in the Aegean and Ionian seas reported by Chondropoulos et al. (1995). A deeply divergent mitochondrial clade has been recongized recently by Orth et al. (2002) in Israel, suggesting several glacial refuges south of the Caucasus.

    The species was recorded from Cyprus by Cheylan (1991, as *abbotti*) who cited Spitzenberger (1978*b*), but she documented only *M. musculus*, and in a more exhaustive study of *Mus* populations from the Aegean and Ionian Isls, Chondropoulos et al. (1995) identified *M. musculus domesticus* as the only *Mus* now living on Cyprus. However, Cucchi et al. (2002) demonstrate that *M. m. domesticus* and *M. macedonicus* now live on Cyprus and have been there from at least the 9th millenium BC. Whether the latter is a survivor from an indigenous Pleistocene fauna on Cyprus or introduced inadvertently by the first agro-pastoral societies migrating from the mainland is unknown (Cucchi et al., 2002).

    Fossil history of indigenous Israeli *M. macedonicus* (discussed as *spretoides*) relative to colonization and origin of commensalism in *M. musculus* reported by Auffray et al. (1988, 1990*b*). *Mus macedonicus* now occurs in the Levantine region, has been there since at least the late Acheulian (late middle Pleistocene), 160,000 years ago (Tchernov, 1992, 1996), and may have been present by the early Pleistocene, 1.4 million years ago (Tchernov, 1996, as *M.* cf. *macedonicus*). Bate (1942*b*) described *M. camini* from late Pleistocene (Acheulian segment of the Palaeolithic) Tabun Cave deposits in Palestine, but Tchernov (1968, 1986) identified it as an ancient part of the molar variation within the range shown by the chronocline of hundreds of specimens of identified as *M. musculus* found in sediments from the Acheulian to present in Israel. Reexamination of Tchernov's material revealed two lineages (Auffray et al., 1988). The oldest is represented by specimens from late Pleistocene Achulian sediments to the youngest level surveyed, which is about 10,000 years old; these samples are *M. macedonicus*, which lived in the Levant even earlier (Tchernov, 1996) and occurs there today (Auffray et al., 1990*b*). The other lineage represents *M. musculus*, which was found only in the youngest level, indicating recent colonization of the Levant by the house mouse no older than about 10,000 years ago. Molars of *camini* come from middle and late Pleistocene levels in Tabun Cave and likely represent *M. macedonicus* as implied by Tchernov (1996).

Mus mahomet Rhoads, 1896. Proc. Acad. Nat. Sci. Philadelphia, p. 532.
COMMON NAME: Mahomet Mouse.
TYPE LOCALITY: SC Ethiopia, Sheikh Mahomet.
DISTRIBUTION: Ethiopian highlands, 1500-3400 m (Lavrenchenko, 2000; Rupp, 1980; Yalden et al., 1976, 1996; specimens in FMNH), SW Uganda and SW Kenya (specimens in USNM); limits unknown.
STATUS: IUCN – Lower Risk (lc).
SYNONYMS: *emesi* Heller, 1911.
COMMENTS: Subgenus *Nannomys*. Heller (1911) described *emesi* as a subspecies of *M. musculoides*, but Hollister (1919:96) and Hatt (1940a) treated it as a distinct species. Musser's study of Hatt's specimens from NE Dem. Rep. Congo (in AMNH) revealed they consisted of *M. minutoides* and *M. sorella*. The holotype of *emesi* and most of Hollister's other examples from Uganda do represent a species distinct from *M. minutoides*; in morphology and chromatic traits, we cannot distinguish the series of *emesi* from the large samples of *mahomet* collected by Osgood in Ethiopia. *Mus mahomet* is sympatric with *M. minutoides* in Uganda and Kenya and narrowly sympatric or closely parapatric with *M. setulosus* in Ethiopia (Yalden et al., 1976; specimens in FMNH). Yalden and Largen (1992) reviewed the species as an Ethiopian endemic. Karyotype (2n = FN = 36) documented by Orlov and Bulatova (1997), spermatozoal morphology described by Baskevich and Lavrenchenko (1995). Yalden et al. (1976:30) suspected that *kerensis*, currently included in synonymy of *Mastomys natalensis*, might be the correct name for *M. mahomet*.

Mus mattheyi F. Petter, 1969. Mammalia, 33:118.
COMMON NAME: Matthey's Mouse.
TYPE LOCALITY: Ghana, Accra.
DISTRIBUTION: Recognized only from the type locality.
STATUS: IUCN – Data Deficient.
COMMENTS: Subgenus *Nannomys*. This form has been recorded from Senegal, Côte d'Ivoire, Burkina Faso, and Ghana (Duplantier and Granjon, 1992; Grubb et al., 1998; F. Petter, 1969; F. Petter et al., 1971; F. Petter and Matthey, 1975), but after studying large series (in MNHN and USNM) of the *M. haussa* and *M. minutoides* complexes from West Africa, we are unable to assign anything to *M. mattheyi*. The morphological traits used to distinguish *M. mattheyi* from *M. haussa* by F. Petter (1969) and F. Petter and Matthey (1975) vary in a continuous fashion from typical *M. haussa* morphology to that considered diagnostic for *M. mattheyi*. The status of *M. mattheyi* needs to be reassessed in a taxonomic revision of the group. A chromosomal complement of 2n = 36, FN = 36 characterizes samples identified as *M. matheyi* and is considered primitive for African *Mus* (Jotterand-Bellomo, 1986; F. Petter, 1969).

Mus mayori (Thomas, 1915). J. Bombay Nat. Hist. Soc., 23:415.
COMMON NAME: Mayor's Mouse.
TYPE LOCALITY: Sri Lanka, Central Mtns, Pattipola, 6200 ft (1890 m).
DISTRIBUTION: Endemic to rainforests of Sri Lanka.
STATUS: IUCN – Lower Risk (nt).
SYNONYMS: *pococki* Ellerman, 1947 [not Tichomirow and Kortchagin, 1889].
COMMENTS: Subgenus *Coelomys*. Recognized by Ellerman (1941:234) as the only species in genus *Coelomys*, but morphological and molecular traits support inclusion of *M. mayori* in a clade also containing *M. crociduroides*, *M. pahari*, *and M. vulcani*, and point to *Coelomys* as a highly distinct subgenus of *Mus* (Chevret et al., 2003; J. T. Marshall, Jr., 1977b). Phillips (1980) recognized highland *mayori* and lowland *pococki* as subspecies, and summarized distributional and biological information for both. J. T. Marshall, Jr. (1977b) and Corbet and Hill (1992) discussed variation in pelage coloration and texture within each form, and neither regarded those traits to reflect significant intraspecific entities.

Mus minutoides Smith, 1834. S. Afr. Quart. J., ser. 2, 2:157.

COMMON NAME: Southern African Pygmy Mouse.

TYPE LOCALITY: South Africa, S Western Cape Province, Cape Town.

DISTRIBUTION: Southern African Subregion: Zimbabwe, Mozambique (south of the Zambezi River), southern and eastern regions of South Africa (S Northern Cape, Western Cape, and Eastern Cape provinces; KwaZulu-Natal, Lesotho, Free State, C and E Limpopo provinces), and Swaziland; northern limits unresolved (de Graaff, 1997o; Meester et al., 1986:283; Skinner and Smithers, 1990:264).

STATUS: IUCN – Lower Risk (lc).

SYNONYMS: *marica* (Thomas, 1910); *minimus* (Peters, 1852) [not White, 1789]; *umbratus* (Thomas, 1910).

COMMENTS: Subgenus *Nannomys*. Relationship of this species to *M. musculoides* has to be assessed by careful systematic revision of the *minutoides-musculoides* complex (see following account). Until such a study reveals otherwise, we view *M. minutoides* as occurring only in the Southern African Subregion, as the range of the species is outlined by de Graaf (1997o) and Skinner and Smithers (1990). Pertinent reviews covering taxonomy, morphology, distribution, and biology are available for populations in Zimbabwe (Smithers and Wilson, 1979), Mozambique (south of the Zambezi River; Smithers and Lobao Tello, 1976), KwaZulu-Natal in South Africa (P. J. Taylor, 1998); and the Southern African Subregion (de Graaff, 1997o; Skinner and Smithers, 1990).

Mus musculoides Temminck, 1853. Esquisses Zool. Sur la Côte de Guine, p. 161.

COMMON NAME: Subsaharan Pygmy Mouse.

TYPE LOCALITY: West Africa, "Côte de Guine."

DISTRIBUTION: Subsaharan Africa (including Ethiopia and Somalia) southward to contact with *M. minutoides* (see preceeding account).

STATUS: IUCN – Critically Endangered as *M. kasaicus*, Lower Risk (lc) as *M. musculoides*.

SYNONYMS: *bella* (Thomas, 1910); *enclavae* Heller, 1911; *gallarum* (Thomas, 1910); *gondokorae* Heller, 1911; *grata* (Thomas and Wroughton, 1910); *kasaica* (Cabrera, 1924); *paulina* (Thomas, 1918); *petila* Hollister, 1916; *soricoides* Heller, 1914; *sungarae* Heller, 1911; *sybilla* (Thomas, 1918); *vicina* (Thomas, 1910).

COMMENTS: Subgenus *Nannomys*. Whether samples reflect only one or a complex of species is unresolved. Meester et al. (1986:283) noted that Van der Straeten "regards *minutoides* as a complex of different species and considers East and West African taxa different from those occurring in Southern Africa." F. Petter and Matthey (1975:3) recognized only *M. minutoides*, noting that among all taxa referable to that species ". . . it is still impossible to recognize those which morphologically merit specific rank." Karyotypes, however, are more revealing, and by using them F. Petter and Matthey (1975) could distinguish the typical South African *minutoides* (2n = 18-19), populations from all of West Africa and a part of Central Africa (2n =18-34), and populations from southern East Africa (2n = 30). The taxon *grata* (or *gratus*) is often listed as a separate species (Hatt, 1940a; Hollister, 1919; F. Petter and Matthey, 1975). Unresolved also is the geographic distributionof *M. musculoides*, and the nature of the biological relationship between it and *M. minutoides*. Samples from Angola and Zambia, for example, have not been critically studied to determine whether only one or both species are present (e.g., F. Crawford-Cabral, 1998; Petter and Matthey, 1975). The considerable chromosomal variation among samples from West Africa was documented by Jotterand (1972), Jotterand-Bellomo (1984, 1986), and Matthey (1967b) under the identification of *minutoides/musculoides*, a label that reflects current understanding of specific limits in this complex. Analyses of sequences from six genes indicate subgenus *Nannomys*, as represented by *M. musculoides*, is sister-group to subgenus *Pyromys*, which in turn phylogenetically connects with subgenus *Mus* (Lundrigan et al., 2002; as *minutoides* from Kenya).

Pertinent reviews covering taxonomy, morphology, distribution, and biology are available for populations in Senegal (Duplantier and Granjon, 1992); Gambia, Sierra Leone, and Ghana (Grubb et al., 1998); Accra Plains of S Ghana (Decher and Bahian, 1999); Nigeria (Happold, 1987); West Africa in general (Rosevear, 1969); S Sudan (Setzer, 1956); Uganda and Kenya (Delany, 1975; Hollister, 1919); Tanzania (Swynnerton and Hayman, 1951; Stanley et al., 1998; 2000, 2002); Zambia (Ansell, 1978); Malawi (Ansell and Dowsett, 1988); Dem. Rep. Congo (Misonne, 1974; specimens in AMNH); Republic of Congo (Dowsett and Granjon, 1991); Angola (Crawford-Cabral, 1998). Neither Yalden et al. (1976, 1996) nor Rupp (1980) recorded *M. minutoides* from Ethiopia, but we have examined many specimens from that country (in BMNH and FMNH).

The taxon *kasaica* (holotype from Dem. Rep. Congo, Kasaï Occidental Province, Kananga [= Luluabourg]) was originally described as a subspecies of *Leggada bella* (Cabrera, 1924), but subsequently treated as a species of *Mus* belonging to the *M. sorella* group (F. Petter, 1981*b*), an arrangement provisionally followed by Musser and Carleton (1993). We place it in the synonymy of *M. minutoides* after Musser's study of the holotype graciously sent to him by Dr. Josefina Barreiro of the Museo Nacional de Sciencias Naturales, Madrid. The skin and skull exhibit all the morphological traits characteristic of *M. musculoides* and none of the diagnostic features peculiar to species in the *M. sorella* group.

Mus musculus Linnaeus, 1758. Syst. Nat., 10th ed., 1:62.

COMMON NAME: House Mouse.

TYPE LOCALITY: Sweden, Uppsala County, Uppsala.

DISTRIBUTION: Spread over the world's continents and islands (except Antarctica) through its close association with humans (Ellerman and Morrison-Scott, 1951; J. T. Marshall, Jr., 1998); in some areas restricted to human dwellings and habitats maintained by human activity; sometimes feral where introduced; and maintaining natural, wild populations in other regions. Distributional summaries available for Europe (Mitchell-Jones et al., 1999), Italy (Amori et al., 1999; Andreotti et al., 2001), Eolian Arch. (Cristaldi and Amori, 1988), Netherlands (Wammes, 1992*b*), Greek island of Astpálaia (Angelici et al., 1992) and other Greek Isls (Chondropoulos et al., 1995), Balearic Isls (Alcover and Gosalbez, 1988), Slovakia (Mosansky, 1994; Stanko, 1995; Stanko and Mosansky, 1994, 2000), Czech Republic (Smaha, 1996), Transylvanian Romania (Istrate, 1998), Baltic region (Miljutin, 1998; Timm et al., 1998), Russia and adjacent regions (Gromov and Erbajeva, 1995), Svjatoj Nos peninsula and isthmus in Lake Baikal (Reiter et al., 1995) and Kamchatka region (Nikanorov, 2000) in Russia, Korea (Won and Smith, 1999), Philippines (Heaney et, al., 1998), India (Agrawal, 2000), Vietnam (Dang et al., 1994), China (Zhang et al., 1997), Australia (Redhead et al., 1991; Singleton, 1995; Watts and Aslin, 1981), New Guinea (Flannery, 1995a), New Zealand (Murphy and Pickard, 1990), Mariana Isls (Stinson, 1994), Hawaii (Tomich, 1986), and Africa (Ansell, 1978; Ansell and Dowsett, 1988; Aulagnier and Thevenot, 1986; Crawford-Cabral, 1998; Dobigny et al., 2002*b*; de Graaff, 1981; Duplantier et al., 1997; Grubb et al., 1998; Osborn and Helmy, 1980; Ranck, 1968; Skinner and Smithers, 1990).

STATUS: IUCN – Lower Risk (lc).

SYNONYMS: *albicans* Billberg, 1827; *amurensis* Argyropulo, 1933; *arenarius* Migulin, 1938; *bicolor* Tichomirow and Kortchagin, 1889; *borealis* Ognev, 1924; *decolor* Argyropulo, 1932; *funereus* Ognev, 1924; *gansuensis* Satunin, 1902; *germanicus* Noack, 1918; *gilvus* Petényi, 1882; *hanuma* Ognev, 1948; *hapsaliensis* Reinwaldt, 1927; *helvolus* Fitzinger, 1867 [*nomen nudum*]; *heroldii* Krausse, 1922; *hortulanus* Nordmann, 1840; *kambei* Kishida and Mori, 1931 [*nomen nudum*]; *kuro* Kuroda, 1940 [see Kaneko and Maeda, 2002]; *longicauda* Mori, 1939 [see Kaneko and Maeda, 2002]; *manchu* Thomas, 1909; *mongolium* Thomas, 1908; *niveus* Billberg, 1827; *nogaiorum* Heptner, 1934; *orii* Kuroda, 1924 [see Kaneko and Maeda, 2002]; *oxyrrhinus* Kashkarov, 1922; *pachycercus* Blanford, 1875; *polonicus* Niezabitowsky, 1934; *raddei* Kastschenko, 1910; *rotans* Fortuyn, 1912; *rufiventris* Argyropulo, 1932; *sareptanicus* Hilzheimer, 1911; *severtzovi* Kashkarov, 1922; *solymarensis* Kretzoi, *in* Jánossy, 1986 [*nomen nudum* according to Kowalski, 2001]; *striatus* Billberg, 1827; *synanthropus* Kretzoi, 1965; *takagii* Kishida and Mori, 1931 [*nomen nudum*]; *takayamai* Kuroda, 1938 [see Kaneko and Maeda, 2002]; *tomensis* Kastschenko, 1899; *utsuryonis* Mori, 1938 [see Kaneko and Maeda, 2002]; *variabilis* Argyropulo, 1933; *vinogradovi* Argyropulo, 1933; *wagneri* Eversmann, 1848; *yamashinai*

Kuroda, 1934 [see Kaneko and Maeda, 2002]; *yesonis* Kuroda, 1928 [see Kaneko and Maeda, 2002]; **bactrianus** Blyth, 1846; **castaneus** Waterhouse, 1843; *albertisii* Peters and Doria, 1881; *bieni* Young, 1934; *canacorum* Revilloid, 1914; *commissarius* Mearns, 1905; *dubius* Hodgson, 1845 [not Fischer, 1829]; *dunckeri* Mohr, 1923; *fredericae* Sody, 1933; *manei* Gray, 1843 [*nomen nudum*]; *manei* Kelaart, 1852; *mohri* Ellerman, 1941; *momiyamai* Kuroda, 1920 [see Kaneko and Maeda, 2002]; *mystacinus* Mohr, 1923 [not Danford and Alston, 1877]; *nipalensis* Hodgson, 1841 [*nomen nudum*]; *rama* Blyth, 1865; *sinicus* Cabrera, 1922; *taitensis* Zelebor, 1869 [probably *nomen nudum*]; *taiwanus* Horikawa, 1929 [not Tokuda, 1941; see Kaneko and Maeda, 2002]; *tytleri* Blyth, 1859; *urbanus* Hodgson, 1845; *viculorum* Anderson, 1879; **domesticus** Schwarz and Schwarz, 1943 [not Rutty, 1772, a *nomen nudum*, but conserved as *domesticus* Schwarz and Schwarz, 1943; see explanation and references in J. T. Marshall, Jr., 1998, and ICZN, 1990]; *abbotti* Waterhouse, 1837; *adelaidensis* Gray, 1841; *airolensis* Burg, 1921; *albidiventris* (Burg, 1923) [not Blyth, 1852]; *albinus* Minà Palumbo, 1868; *albus* Bechstein, 1801; *ater* Fraipont, 1907 [*nomen nudum*; not Millais, 1905]; *azoricus* Schinz, 1845; *brevirostris* Waterhouse, 1837; *deserti* (Loche, 1867) [see Cockrum and Setzer, 1976]; *candidus* Laurent, 1937 [not Bechstein, 1796]; *caudatus* Martino, 1934; *corsicus* Kratochvil, 1986; *faeroensis* Clarke, 1904; *far* Cabrera, 1921; *flavescens* Fischer, 1872 [not Elliot, 1839, or Waterhouse, 1837]; *flavus* Bechstein, 1801 [not Kerr, 1792]; *formosovi* Heptner, 1930; *gentilis* Brants, 1827; *gerbillinus* Blyth, 1853; *helgolandicus* Zimmerman, 1953; *helviticus* Burg, 1923; *homourus* Hodgson, 1845; *indianus* Wied, 1862; *jalapae* J. A. Allen and Chapman, 1897; *jamesoni* Krausse, 1921; *kalehpeninsularis* Goodwin, 1940; *lundii* Fitzinger, 1867 [*nomen nudum*]; *makovensis* Orlov, Nadjafova, and Bulatova, 1992; *maculatus* Bechstein, 1801; *major* Severtzov, 1873 [not Brants, 1827, or Pallas, 1779]; *melanogaster* Minà Palumbo, 1868; *microdontoides* Noack, 1889; *modestus* Wagner, 1842; *muralis* Barrett-Hamilton, 1899; *mykinessiensis* Degerbol, 1940; *nattereri* Fitzinger, 1867 [*nomen nudum*]; *niger* Bechstein, 1801 [not Bechstein, 1796]; *nudoplicatus* Gaskoin, 1856; *pallescens* Heuglin, 1877; *parvulus* Tschudi, 1844 [not Hermann, 1804, or Mosanský, 1994]; *percnonotus* Moulthrop, 1942; *peruvianus* Peale, 1848; *poschiavinus* Fatio, 1869; *praetextus* Brants, 1827; *rubicundus* Minà Palumbo, 1868; *simsoni* Higgins and Petterd, 1883; *subcaeruleus* Fritsche, 1928 [not Lesson, 1842]; *subterraneus* Montessus, 1899; *tataricus* Satunin, 1908; *theobaldi* Blyth, 1853; *orientalis* Cretzschmar, 1826 [not Desmarest, 1819]; *vignaudii* Des Murs and Prévost, 1850; **gentilulus** Thomas, 1919. **Unassigned**: *albula* Kishida, 1924 [Japan; see Kaneko and Maeda, 2002]; *cinereomaculatus* Fitzinger, 1867 [Europe, *nomen nudum*]; *molossinus* Temminck, 1844 [Japan; holotype is hybrid between *castaneus* and *musculus*; J. T. Marshall, Jr., 1998]; *nordmanni* Keyserling and Blasius, 1840 [*nomen nudum*]; *reboudi* Loche, 1867 [Lataste, 1883*a*, and Cabrera, 1923, identified this as a house mouse; J. T. Marshall, Jr. treated it as a synonym of *domesticus* but noted from the original description that the tail is too long for *M. spretus* and the eye was gerbil-like; Kowalski and Rzebik-Kowalska, 1991, claimed the holotype to be lost and the name should be treated as *nomen dubium*]; *tantillus* G. M. Allen, 1927 [holotype is a hybrid between *musculus* and *castaneus*; J. T. Marshall, Jr., 1998]; *varius* Fitzinger, 1867 [not Bechstein, 1796; Europe, *nomen nudum*]; *yonakuni* Kuroda, 1924 [S Ryukyu Isls; description seems to indicate hybrid between *castaneus* and *musculus* and such a phenetic mixture is reflected in specimens from Okinawa identified by J. T. Marshall, Jr.; see also Kaneko and Maeda, 2002]. J. T. Marshall, Jr. (1998) is the only systematist who has studied all available holotypes and read all original descriptions of taxa associated with house mice (except *arenarius*, which is included in *musculus* by Zagorodnyuk, 1996*b*, and *modestus*, listed as a synonym by Meester et al., 1986). Allocation of synonyms follows J. T. Marshall, Jr. (1998), whose report should be consulted for type locality, assignment, and remarks for each name, and a map of 77 continental type localities. Some of the synonyms may eventually be differently allocated once the range of *domesticus* and status of *bacterianus* are resolved (see below).

Two taxa have been erroneously associated with *Mus* and the identity of another is uncertain. Musser and Carleton (1993) listed *molissimus* (Dehne, 1855) in the synonymy of *M. musculus*, but J. T. Marshall, Jr. (1998) claimed the original description is that of a dormouse, *Eliomys* or *Muscardinus*. Hodgson's (1845:267) *"Mus? hydrophilus"* has been listed as *incertae sedis* near *Mus musculus* accounts (Ellerman and Morrison-Scott, 1951; Ellerman, 1961), but Hodgson described an animal much too large to be any species of *Mus* known to occur in Nepal, and he grouped it with the species of rats, not the mice he described. Identity of the Taiwanese *Mus kagii* Kuraoka, 1912, has yet to be determined (Kaneko and Maeda, 2002).

Three names listed in the synonymy were proposed for fossils. Young (1934:80) described *"Mus musculus* var. *bieni"* from "Pleistocene" of Choukoutien, China. Pei (1936) studied more than 200 samples he identified as *M. musculus* from another Choukoutien locality and remarked that Young's specimens fell well within the size range of typical *M. musculus*. Late Pleistocene Hungarian fossils were described by Kretzoi as the species *synanthropus* (in Kretzoi and Vértes, 1965*a*) and *solymarensis* (probably proposed by Kretzoi and listed in Jánossy, 1986) in subgenus *Budamys* of *Mus*. Both names are listed as synonyms of *M. musculus* by Kowalski (2001).

COMMENTS: Subgenus *Mus*. Schwarz and Schwarz (1943) provided a revision that was followed with minor changes by Ellerman and Morrison-Scott (1951). This arrangement was criticized by Jones and Johnson (1965:394) who found that specimens from Asia that they studied "bear little or no relation to this idealized classification." Subsequent treatments of this group were presented by J. T. Marshall, Jr. (1977*b*, 1981, 1986, 1998) and J. T. Marshall, Jr. and Sage (1981). The most recent classification combined biochemical analyses of European, Asian, and African mice (Bonhomme et al., 1984; Boursot et al., 1993, 1996; Prager et al., 1998). The translation of these results, as well as incorporation of morphological data, into a new classification of *Mus musculus* and its allies (J. T. Marshall, Jr, 1998), and the allocation of the many names to *M. musculus*, are surprisingly concordant with the treatment of Schwarz and Schwarz (1943). Some of the scientific names listed under *M. musculus* by those authors, but now placed elsewhere, are here associated with *M. spicilegus*, *M. spretus*, *M. macedonicus*, and *M. caroli* (see those accounts).

There are three distinctive and well known groups of house mice (Bonhomme et al., 1994; Boursot et al., 1993, 1996; Din et al., 1996; J. T. Marshall, Jr., 1998; Sage et al., 1993; Yonekawa et al., 1994), a fourth not as well documented (Din et al., 1996), and a fifth recently identified (Prager et al., 1996). Among *Mus* researchers, the three usual complexes (*castaneus*, *domesticus*, and *musculus*) are either recognized as species (J. T. Marshall, Jr., 1998; J. T. Marshall, Jr. and Sage, 1981; Prager et al., 1998; Sage et al., 1993), or subspecies as we list them here (see discussions in Boursot et al., 1993, 1996; Yonekawa et al., 1994; Din et al., 1996; Prager et al., 1998); the fourth group (*bactrianus*) is generally treated as a subspecies (Din et al., 1996); the fifth (*gentilulus*) is considered to be a possible species (Prager et al., 1998). Analyzing mtDNA by restriction enzymes and sequencing in worldwide samples of *M. musculus* led Yonekawa et al. (1994) to recognize the same three subspecies described above and a fourth, *M. m. bactrianus*, in N India, Pakistan, and Afghanistan. They assigned subspecific rather than specific status to these four geographic entities "because they have not been sexually isolated from each other under natural conditions" (p. 34). Presence of four identifiable geographic groups connected by significant gene flow among them is also indicated by analysis of variation of ribosomal RNA (Suzuki and Kurihara, 1994). Whether viewed as species or subspecies, each group is characterized by distinctive genetic and morphological traits. In their phylogenetic analyses of sequences from the mtDNA control region, Prager et al. (1998) identified four major lineages with *M. m. domesticus* as sister group to *M. m. castaneus*, *M. m. musculus*, and *M. m. gentilulus*, and the latter being sister to *M. m. castaneus* and *M. m. musculus*.

Mus musculus musculus ranges from C Europe and Scandinavia through E Europe, Ukraine, Turkmenistan, SW Georgia, NC Iran, through N Afghanistan (north of the Hindu Kush) and N Asia to Manchuria, Korea, and Japan (Prager et al., 1998). Introduced pockets of this form occur within the ranges of the other two groups (in the Nile delta, for example; J. T. Marshall, Jr., 1998).

Mus musculus domesticus occurs in N Africa and ranges in Eurasia from W Europe (and most Mediterranean islands) through S Eurasia to the Caucasus and eastward through Iran, Afghanistan and Pakistan to N India and Nepal. It is this form that was spread inadvertently by European colonization throughout the New World, numerous islands, Australia, and probably southern Africa (see Meester et al., 1986, for discussion about the possible subspecies in southern Africa). Auffray et al. (1988) hypothesized the origin of commensalism in *M. m. domesticus* to have taken place in the Fertile Crescent. Prager et al. (1998) identified samples from Afghanistan as *musculus* and *castaneus*, and specimens from Pakistan, N India, and Nepal as *castaneus*; the actual range of *domesticus* in these regions requires clarification.

Mus musculus castaneus extends from Central Asia (C Afghanistan south of Hindu Kush, Pakistan, peninsular and NE India, E and C Nepal) to throughout SE Asia; it has spread to Taiwan and eastward through the Moluccas to New Guinea; also on the Mariana Isls (J. T. Marshall, Jr., 1998; Prager et al., 1998). Populations on Japan (traditionally referred to as *molossinus*) consist of either *musculus*, *castaneus*, or hybrids between the two (Yonekawa et al., 1988).

The fourth and lesser known group, *M. m. bactrianus* from Afghanistan, was first identified by Yonekawa et al. (1981) and Bonhomme et al. (1984) using mitochondrial and nuclear gene sequences. In their analysis of nuclear gene sequences of several samples from the Indian subregion, Din et al. (1996:535) would restrict the range from that proposed by Yonekawa et al. (1994), writing that "*M. m. bactrianus* defined on the basis of the different mitochondrial. . . and nuclear genes. . . from the region of Kabul in Afghanistan. . . has a very distinct nuclear gene composition. . . and probably belongs to a geographically confined group of populations living in the valleys of Afghanistan where contact with the neighbouring regions is limited by the high mountain ranges which surround the country" and "can no longer be considered to be representative of the other *Mus musculus* populations from the central area of the species range [Pakistan and peninsular India]." Prager et al. (1998:835) noted that *M. m. bactrianus* is "the least well defined and characterized" of the five groups "and it is not known whether it is a cohesive genetic entity." Kandahar in SE Afghanistan is type locality of *bactrianus*.

The fifth group, *M. m. gentilulus*, was recently identified by analyses of sequences from the mtDNA control region (mtDNA) in a sample from Yemen; the mice are also distinctive in their small body size (Prager et al., 1998). This form may have a more expansive distribution and Prager et al. (1998:856) noted the desirability of genetically sampling populations from elsewhere on the Arabian Peninsula, "all along the northern shores of the Persian Gulf and the Gulf of Oman, and also the Horn of Africa and adjacent regions." Similar mitochondrial sequences and Y chromosomal traits that characterize *M. m. gentilulus* have recently been discovered in samples of *M. musculus* from Madagascar (Duplantier et al., 2002). Those authors suggest this kinship supports a hypothesis of colinization by human importation from the Arabian Peninsula to islands along East Africa, including Madagascar, a colonization route also claimed for the shrew *Suncus murinus* (Hutterer and Tranier, 1990).

Mus m. musculus and *M. m. domesticus* meet and hybridize through a narrow zone (less than 50 km) in C Europe (from the Jutland Peninsula to Bulgaria), which has been the focus of extensive genetic, parasitological, and some morphological and behavioral studies (Agulnik et al., 1993; Alibert et al., 1994; Auffray et al., 1996; Britton-Davidian et al., 2002; Garagna et al., 1992; Lazarová, 1999; Macholán and Zima, 1994; Macholán et al., 2003; Moulia et al., 1991; Munclinger et al., 2002; Niwa-Kawakita, 1994; Prager et al, 1993; Sage et al., 1986*b*; Smadja and Ganem, 2002; Turker et al., 1993; also see the reviews and citations referenced by Boursot et al. [1993], J. T. Marshall, Jr. [1998], Mezhzherin et al. [1998], and Sage et al. [1993]; Mitchell-Jones et al. [1999] provided vivid depictions of the European ranges of *domesticus* and *musculus*), and in a much wider zone (more than 300 km) in the Transcaucasus region (Mezhzherin et al., 1998; Orth et al., 1996). The hybrid zone between *M. m. domesticus* and *M. m. castaneus* may be in NW India (J. T. Marshall, Jr., 1998), but that contact requires critical analysis (Agrawal [2000:121] listed the Indian populations of *M. m. domesticus*, which he referred to as *praetextus* and *homourus*, as "outdoor subspecies" and *M. m. castaneus* as an "indoor subspecies"). In their analysis of nuclear gene sequences, Din et al. (1996) documented samples from Tehran in N Iran as having a nuclear gene pool containing elements of *domesticus* and mice from N India and Pakistan, which strengthens the hypothesis of a series of intergrading populations between N India and the periphery of the *domesticus* range in Europe. *Mus m. musculus* and *M. m. castaneus* contact each other in a wide zone of intergradation in N China (more than 1000 km in latitude), in the main Japanese Isls, and on Okinawa (Bonhomme et al., 1989; Boursot et al., 1993, and references cited therein; J. T. Marshall, Jr., 1998; Mezhzherin et al., 1998; Yonekawa et al., 1994).

Samples of *M. musculus* have been the focus of numerous morphometric, chromosomal, and molecular studies undertaken in context of phylogenetic inquiries bearing upon discrimination and relationships among populations of *M. musculus* and between it and other species of *Mus* as well as general treatises on speciation and phylogeny. Some of the contributions published during the last two decades that also summarized and cited earlier reports are Evans (1981), Martin et al. (2000), Foster et al. (1981), Capanna (1982), Bonhomme et al. (1984), Potter et al. (1986), Giagia et al. (1987), Hubner (1988), Nishioka (1987), Winking et al., (1988), Britton-Davidian (1990), Corti and Ciabatti (1990), Garagna et al. (2002), Gerasimov et al. (1990), She et al. (1990), Capanna and Corti (1991), Searle (1991), Viroux and Bauchau (1992), Karn and Dlouhy (1991), Bush and Paigen (1992), Fraguedakis-Tsolis (1992), Mathias and Ramalhino (1992), Mezhzherin and Kotenkova (1992), Ritte et al. (1992), Scriven and Bauchau (1992), Ganem (1993), Garagna et al. (1993), Hauffe and Searle (1993), Korobitsyna et al. (1993), Said et al. (1993), Searle et al. (1993), Capanna and Redi (1994), Capanna et al. (1994), Giménez and Bidau (1994), Miyashita et al. (1994), Nagamine et al. (1994), Ohtsuka et al. (1996), Tokumitsu and Ogawa (1994), Chondropoulos et al. (1996), Ganem et al. (1996), Macholán (1996*c*), Naruse et al. (1996), Smets et al. (1996), Garagna et al. (1997), Said et al. (1993, 1999), Britton-Davidian et al. (2000), Gündüz et al., 2000; Huang et al. (2001), Leamy et al., (2002), Karn et al. (2002), and Wallace et al. (2002).

A highlight among the many chromosomal studies is the exceptionally variable karyotype characteristic of *M. m. domesticus* throughout its range compared with other subspecies, especially *M. m. musculus* (Britton-Davidian et al., 2000; Capanna, 1982; Capanna et al., 1994; Chondropoulos et al., 1996; Hauffe and Searle, 1993; Muñoz-Muñoz et al., 2003; Said et al., 1993, 1999; Nachman and Searle, 1995, provided an excellent synopsis). In addition to the standard karyotype of all acrocentric chromosomes (2n = 40), numerous populations of *domesticus* exhibit metacentric chromosomes derived from Robertsonian fusion mutations. Such mutations are absent from most populations of *M. m. musculus*, and when present occur at very low frequencies (Zima and Macholán, 1989; Zima et al., 1990).

Recently, initial sequencing and comparative analysis of the mouse genome has been documented using laboratory strain C57BL/6J, and revealing about 30,000 genes, with over 90% of mouse and human genomes "reflecting segments in which the gene order in the most recent common ancestor has been conserved in both species" (Mouse Genome Sequencing Consortium, 2002).

Analysis of DNA sequences from a set of genes places *M. musculus* in the same clade as *M. spretus*, *M. macedonicus*, and *M. spicilegus* (Graur, 1994; Larizza et al., 2002; Lundrigan et al., 2002; Martin et al., 2000; Prager et al., 1996, 1998). Combined analyses of morphological traits, DNA/DNA hybridization, and mitochondrial 12S rRNA sequences align *M. musculus* with *M. spicilegus*, *M. macedonicus*, and *M. spretus* in a clade separate from an Asian clade composed of *M. caroli*, *M. cervicolor*, and *M. cooki* (Auffrey et al., 2003; Chevret et al., 2003; also supported by cranial morphology, J. T. Marshall, Jr., 1977*b*, and in litt., 2004). Osteometric comparisons between subfossil remains from Lavezzi Isl off the coast of S Corsica and recent Corsican populations reported by Vigne et al. (1993). Analyses of mtDNA variation among samples from Japanese islands reveal a polyphyletic origin of Japanese *M. musculus* derived from *musculus*, *castaneus*, and *domesticus* strains (Bonhomme et al., 1989; Yonekawa et al., 1994; also confirmed by specimens in USNM and original descriptions, J. T. Marshall, Jr., in litt.). These results are also supported by a morphometric study of insular variation of *M. musculus* from various Japanese islands (Takada et al., 1994). Morphometric variation in samples from the Japanese Izu Isls south of Tokyo Bay documented by Takada et al. (1999), and their relationship to populations on the Ogasawara Isls elucidated by Takada et al. (2002). Morphometric and immunological relationships among samples from Greek Isls in the Ionian and Aegean seas indicate that all Greek mainland and insular samples are *M. musculus domesticus*, the same subspecies that occurs in W Europe, North Africa, Turkey, and eastward (Chondropoulos et al., 1995). A taxonomic assessment of house mice from the Ukraine, Moldavia, and European sector of Russia by Zagorodnyuk (1996*b*) reinforced the specific status of *M. spicilegus* (including *sergii* and *makovensis*) and recognized the following subspecies of *Mus musculus*: *musculus* (which includes *hapsaliensis*, *polonicus*, *borealis*, and *funerus*), *formosovi* (includes *tataricus*), *wagneri* (including *nogaiorum*, *bicolor*, *wagneri*, and *sareptanicus*), and *hortulanus* (includes *nordmanni* and *arenarius*). Geographical distribution of hemoglobin *Hbb* haplotypes in samples from E Asia and correlation with subspecies defined by morphology discussed by Kawashima et al. (1995). A comprehensive summary of genetic map for *Mus musculus* and its current applications and future prospects presented by Copeland et al. (1993). Taxonomic results of morphometric and genetic characteristics of samples from China, Korea, Taiwan, and Japan identify several geographic groups of *M. musculus* in the region (Tsuchiya et al., 1994).

Two important reviews are critical to understanding evolution of house mice. Boursot et al. (1993) discussed phylogenetic position of *M. musculus* within the genus, origin and radiation of the species, range expansions and secondary contacts of subspecies, origin and consequences of commensalism, and chromosomal evolution. Sage et al. (1993) summarized biosystematics and evolution of the house mouse complex, and use of house mice as models in studies of hybrid zone biology, chromosomal evolution and speciation, and testing phylogenetic reconstruction. In other pertinent reviews, Thaler (1986) gave an appraisal of fossil evidence and morphological traits of *M. musculus* and other European species of *Mus*. Kotenkova and Bulatova (1994) edited contributions covering evolution and systematics of house mice, geographic distributions, isolating mechanisms, intraspecific genetic polymorphisms, karyotype variability, and behaviour. Bonhomme et al. (1994) presented a synthetic view of genetic differentiation within *M. musculus* and the hypothesis that the complex formed a ring species (Din et al., 1996, did not think the evidence sufficient to test this idea). Moriwaki et al. (1994) documented genetic variation and its geographical distribution, immunology, chromosomes and genetic mapping, and establishment of wild-derived laboratory strains. Silver (1995) provided a useful compendium for systematists and laboratory researchers. Finally, Berry (1984) reviewed general biology, domestication, history in biological research, and other topics.

Mitochondrial DNA and nuclear gene sequences have been used to determine the evolutionary origin and radiation of *M. musculus*, and results are the basis of two models. One identifies the N Indian subcontinent as the place where *M. musculus* likely evolved (Boursot et al., 1996; Din et al., 1996), a region also known for the greatest diversity of species in subgenus *Mus* recorded by Plio-Pleistocene fossils (Patnaik et al., 1993, 1996). Genetic data indicated that ranges of *musculus*, *castaneus*, and *domesticus* likely "correspond to three distinct possible paths of expansion from the Indian cradle" (Boursot et al., 1993:128, and references cited therein), a conclusion based on the greater genetic variability in samples from the Indian subcontinent compared with the genetic variation in the peripheral regions of its range (Din et al., 1996). Boursot et al. (1996) also championed this view (their scenario, summarized in fig. 4, p. 405, exemplifies the hypothesis), and they look upon *M. m. bactrianus* as an isolate in mountainous Afghanistan derived from the ancestral population in the N Indian subcontinent. The other model was proposed by Prager et al. (1998:835): "an origin in west-central Asia . . . and the sequential spreading of mice first to the southern Arabian Peninsula, thence eastward and northward into south-central Asia, and later from south-central Asia to north-central Asia (and thence into most of northern Eurasia) and to southeastern Asia."

"One of the most characteristic features of the house mouse life history is probably its commensalism in relation to humans. The worldwide colonization by this species is mainly due to passive transport by humans and is a consequence of its ecological dependence on them" (Boursot et al., 1993:135). *Mus m. castaneus* is found only in buildings but the other two subspecies occur in a wide variety of anthropogenic and wild environments. Analyses, using palaeontological and archaeozoological approaches, of the colonization process of W Eurasia and the origin of commensalism were presented by Auffray et al. (1988, 1990*a, c*) and Auffray and Britton-Davidian (1992). Cucchi et al. (2000) documented unintentional anthropogenic introduction to Cyprus during the 9th and 8th millenia BC. Jaarola et al. (1999) outlined colonization history of Fennoscandia by house mice. The living population of *M. m. domesticus* in the Maderian Arch., 600 km off the Atlantic coast of NW Africa, is also represented by subfossil specimens (Mathias and Mira, 1992; Pieper, 1981) indicating presence before Portuguese discovery of the archipelago in 1419. Analyses of mitochondrial D-loop sequences by Gündüz et al. (2001) indicated the source area for the mice was probably the population of *domesticus* from N Europe; the Vikings may have inadvertently brought the mice in the ninth century, although no historical evidence exists of their visit. Their study is an example of molecular data providing clues to human historical colonization events rather than the opposite — using human history to infer colonization history of commensals.

Populations in nearly all of the United States descended from the European *M. musculus domesticus*, probably in the 16th century, but a population around Lake Casitas, about five miles (8 km) from the Pacific coast in S California, possesses certain genic traits and retroviruses not found in *M. m. domesticus* but identical to those in populations of the Asian *M. m. castaneus* (Gardner et al., 1991). Immigrant Chinese laborers who worked on railroads and ranches in the Lake Casitas area in the middle 1800s were likely accompanied by *castaneus* on the voyages from China to S California (Gardner et al., 1991).

The commensal proclivity of *M. musculus* ultimately led to its domestication by humans, first as pets and later as laboratory animals. The mitochondrial genome and most of the nuclear genome of classical laboratory inbred strains of *M. musculus* originated from the European *M. m. domesticus* (see review and references in Berry, 1984, Bonhomme et al., 1987, Flegr et al., 1994, and Nadeau, 2002; and J. T. Marshall, Jr., in litt., 2004, notes that laboratory mice match W European *domesticus* in body size, tail length and conformation of zygomatic plate) but questions still exist about genealogy of the strains and their relationships to one another and to wild forms. This is a significant problem because laboratory strains constitute the "universal mammalian model" (Flegr et al., 1994: 33) or "the mammalian archetype in genetic studies" (Bonhomme et al., 1987:52), and knowledge of their origins and genetic relationships among them is critical to interpreting experimental results in laboratories or phylogenetic comparisons between inbred strains and wild populations of *M. musculus* and other species. Bonhomme et al. (1987:53) discussed the polyphyletic origin of laboratory mice and noted that "any one of the alleles present in inbred strains could have come from *domesticus*, *musculus* or *castaneus*," and they are "a mosaic of genomes coming from different taxonomic origins," a conclusion echoed by Sage et al. (1993; see also Blank et al., 1986; Nishioka, 1987). In a phylogenetic analysis of mitochondrial D-loop sequences in laboratory and wild strains, Flegr et al. (1994) reported that some of the strains they used were more similar to *M. m. musculus* than to *M. m. domesticus*. Moriwaki (1994) provided an informative diagram illustrating the genetic differentiation and geographic ranges of subspecies groups of *M. musculus* and their relevance to the origin of laboratory strains. According to Moriwaki (p. xxiii) "the genetic background of today's laboratory mice is mainly derived from that of the European *domesticus* subspecies group and a little from that of some Asian mice, probably Japanese fancy mice belong to the *musculus* subspecies group." Aligning the genome sequence from the C57BL/6J strain with sequences from other laboratory strains suggested that the genomes of inbred strains used "are mosaics with the vast majority of segments derived from *domesticus* and *musculus* sources" (Wade et al., 2002; Nadeau, 2002).

Mus neavei (Thomas, 1910). Ann. Mag. Nat. Hist., ser. 8, 5:90.
  COMMON NAME: Neave's Mouse.

TYPE LOCALITY: SE Zambia, E Loangwe Dist., Petauke, 2400 ft (732 m; Ansell, 1978, provided coordinates).

DISTRIBUTION: E Dem. Rep. Congo, SE Zambia (Ansell, 1978), S Zimbabwe, Limpopo Province of South Africa, W Mozambique, and S Tanzania; (range derived from Meester et al., 1986:282, and our study of material in AMNH, BMNH, and USNM). Distributional limits undocumented; supposed records of *M. neavei* from Malawi represent other species (Ansell and Dowsett, 1988).

STATUS: IUCN – Lower Risk (lc).

COMMENTS: Subgenus *Nannomys*. Originally described as a species, *neavei* was later treated as a subspecies of *M. sorella* (Verheyen, 1965a), an arrangement accepted by Ansell (1978), Meester et al. (1986), and Skinner and Smithers (1990). F. Petter (1981b), however, pointed out that while a member of the *M. sorella* group, *neavei* should be treated as a separate species; in morphology and body size it appears to be close to *M. oubanguii* (F. Petter, 1981b). Our study (series in AMNH, BMNH, and USNM) corroborates F. Petter's view. *Mus neavei* is a distinct species and easily distinguished from *M. sorella* by its richer tawny fur, much smaller size, more delicate cranium, and shorter molar rows (3.0-3.2 mm in seven examples of *M. neavei*, 3.2-3.7 mm in nine *M. sorella*). How *M. neavei* is related to *M. oubanguii* and the small-bodied *M. baoulei* (both in the *M. sorella* group) is unresolved.

Mus orangiae (Roberts, 1926). Ann. Transvaal Mus., 11:251.

COMMON NAME: Orange Mouse.

TYPE LOCALITY: South Africa, N Free State, Kruisementifontein, Viljoensdrift, near Vereeniging.

DISTRIBUTION: South Africa; N Free State (Vermeiren and Verheyen, 1983).

STATUS: IUCN – Least Concern.

COMMENTS: Subgenus *Nannomys*. Either treated as a species possibly allied to *M. setzeri* (Vermeiren and Verheyen, 1983) or listed as a subspecies of *M. minutoides* (Meester et al., 1986; Skinner and Smithers, 1990). The definition of this form and its phylogenetic relationship to other southern African species require resolution.

Mus oubanguii F. Petter and Genest, 1970. Mammalia, 34:454.

COMMON NAME: Oubangui Mouse.

TYPE LOCALITY: Central African Republic, La Maboke, Ippy, Bangassou (F. Petter, 1981b, provided coordinates).

DISTRIBUTION: Recorded only from Central African Republic (savanna north of Oubangui River); see Jotterand (1972:332).

STATUS: IUCN – Vulnerable.

COMMENTS: Subgenus *Nannomys*. Sympatric with *M. setulosus* and *M. minutoides* (F. Petter and Genest, 1970), but a phylogenetic member of the *M. sorella* group, according to F. Petter (1981b), who also noted that its morphology, except for a dental trait, is similar to that of *M. neavei* (see that account). Chromosomal information, in context of understanding chromosomal evolution among species of African *Mus*, documented by Jotterand (1972) and Jotterand-Bellomo (1984, 1986).

Mus pahari Thomas, 1916. J. Bombay Nat. Hist. Soc., 24:414.

COMMON NAME: Indochinese Shrewlike Mouse.

TYPE LOCALITY: India, Sikkim, Batasia, 6000 ft (1830 m).

DISTRIBUTION: From NE India (Sikkim, West Bengal, Arunachal Pradesh, Assam, Nagaland, Meghalaya, and Mizoram; Agrawal, 2000) and Bhutan through N Burma, S China (SE Xizang, Yunnan, S Sichuan, Guizhou, and Guangxi; Wang [2003], Zhang et al. [1997], Wu et al. [1996], FMNH 40710), Thailand (J. T. Marshall, Jr., 1977a; Robinson et al., 1996), Cardamom Mtns of SW Cambodia (A. Smith, in litt.), Laos, and C and N Vietnam (Dang et al., 1994; specimens in AMNH and IEBR); see J. T. Marshall, Jr. (1977a) and Corbet and Hill (1992).

STATUS: IUCN – Lower Risk (lc).

SYNONYMS: *gairdneri* (Kloss, 1920); *jacksoniae* (Thomas, 1921); *meator* (G.M. Allen, 1927); *mocchauensis* Dao, 1978.

COMMENTS: Subgenus *Coelomys*. Dao (1978) described *mocchauensis* as a subspecies of

M. pahari. Chromosomal features of Thai samples reported by Gropp et al. (1973); 2n = FN = 48. Relationships within *Mus* assessed by sequences of the *Sry* gene (Graur, 1994); forms clade basal to species in subgenera *Nannomys*, *Pyromys*, and *Mus* according to study of *Sry* and five other genes (Lundrigan et al., 2002) and DNA/DNA hybridization experiments (Chevret et al., 2003). The close association between *M. pahari* and *M. crociduroides* and placement in subgenus *Coelomys* is supported by analysis of mitochondrial 12S rRNA sequences (Chevret et al., 2003). Those two species and *M. mayori* belong in a clade as indicated by analyses of morphological traits (Chevret et al., 2003; J. T. Marshall, Jr., 1977*b*). Sequences of APRT gene used to illuminate substitution rate variation among *M. pahari*, *M. spicilegus*, and other muroids (Fieldhouse et al., 1997). Ecology of population in S Yunnan and comparisons with species of *Rattus*, *Bandicota*, *Maxomys*, *Leopoldamys*, and *Niviventer* from the same forests reported by Wu et al. (1996).

Documented in middle and late Pleistocene cave sediments from the Sichuan-Guizhou region (Zheng, 1993) and Guangxi (Chen et al., 2002) of S China. Isolated molars from cave deposits in Thailand indicates the species was present in that region since the middle Pleistocene (Chaimanee, 1998).

Mus phillipsi Wroughton, 1912. J. Bombay Nat. Hist. Soc., 21:772.
COMMON NAME: Phillips's Mouse.
TYPE LOCALITY: India, Central Province, Nimur Dist., Asirgarh, 1500 ft (458 m).
DISTRIBUTION: S and WC Peninsular India (Rajasthan, Gujarat, Madhya Pradesh, Andra Pradesh, Karnataka, and Tamil Nadu; Agrawal, 2000).
STATUS: IUCN – Lower Risk (lc).
SYNONYMS: *siva* (Thomas and Ryley, 1913); *surkha* (Wroughton and Ryley, 1913).
COMMENTS: Subgenus *Pyromys*. An Indian endemic revised by J. T. Marshall, Jr. (1977*b*) and reviewed by Agrawal (2000). Its occurrence and ecology in the Aravalli ranges of Rajasthan State documented by Prakash et al. (1995*a*, *b*, *c*), and in Gujarat State by Chakraborty and Agrawal (2000).

Mus platythrix Bennett, 1832. Proc. Zool. Soc. Lond., 1832:121.
COMMON NAME: Flat-haired Mouse.
TYPE LOCALITY: Peninsular India, Dukhun (=Deccan).
DISTRIBUTION: Peninsular India north to West Bengal (Agrawal, 2000; Corbet and Hill, 1992).
STATUS: IUCN – Lower Risk (lc).
SYNONYMS: *bahadur* (Wroughton and Ryley, 1913); *grahami* (Ryley, 1913); *hannyngtoni* (Ryley, 1913).
COMMENTS: Subgenus *Pyromys*. Close relationship with *M. saxicola*, also in subgenus *Coelomys*, substantiated by analyses of DNA/DNA hybridization and mitochondrial 12S rRNA sequences, but not morphological traits (Chevret et al., 2003). May be more than one species within present definition of *M. platythrix*. Samples from the north-west have 2n = 30, those from S India have 2n = 26, but "the limits of these forms and any correlation with morphological characters have not been defined" (Corbet and Hill, 1992:330). Another Indian endemic revised by J. T. Marshall, Jr. (1977*b*) and reviewed by Agrawal (2000). Occurrence and ecology in the Aravalli ranges of Rajasthan State documented by Prakash et al. (1995*a*, *b*, *c*), and in the Western Ghats of S India by Chandrasekar-Rao and Sunquist (1996). Roberts (1977, 1997) reported *M. platythrix* from S Pakistan, but that record probably represents *M. saxicola*, which occurs in India and S Pakistan (Corbet and Hill, 1992:328).

Mus saxicola Elliot, 1839. Madras J. Litt. Sci., 10:215.
COMMON NAME: Saxicolous Mouse.
TYPE LOCALITY: India, Madras.
DISTRIBUTION: India (disjunct distribution mapped by Agrawal, 2000), S Nepal, and S Pakistan.
STATUS: IUCN – Lower Risk (lc).
SYNONYMS: *cinderella* (Wroughton, 1912); *gurkha* (Thomas, 1914); *khumbuensis* Biswas and Khajuria, 1968; *priestlyi* (Thomas, 1911); *pygmaeus* Biswas and Khajuria, 1955 [not Milne-Edwards, 1874]; *ramnadensis* Bentham, 1908; *sadhu* Wroughton, 1911.

COMMENTS: Subgenus *Pyromys*. Indian populations reviewed by Agrawal (2000), who recognized three subspecies based primarily upon the chromosomal traits originally documented by Rishi and Puri (1978; 2n = 22-26). Corbet and Hill (1992) expressed the need to confirm the inclusion of *gurkha*, which has soft fur, in the spinous-furred *M. saxicola*. J. T. Marshall, Jr. (1998:63) examined the holotype of *pygmaeus* and identified it as a nestling *M. saxicola*; Musser and Carleton (1993) had listed it as a synonym of *M. musculus*. Sister-group to members of subgenus *Mus* as assessed by analysis of sequences from six genes (Lundrigan et al., 2002). Analyses of DNA/DNA hybridizations and mitochondrial 12S rRNA sequences but not morphological traits support close relationship between *M. saxicola* and *M. platythrix*, also in subgenus *Pyromys* (Chevret et al., 2003). Occurrence and ecology in the Aravalli ranges of Rajasthan State reported by Prakash et al. (1995*a*, *b*, *c*) and in Gujarat State of NW India by Chakraborty and Agrawal (2000).

Mus setulosus Peters, 1876. Monatsb. K. Preuss. Akad. Wiss. Berlin, p. 480.
   COMMON NAME: Peters's Mouse.
   TYPE LOCALITY: Cameroon, Victoria.
   DISTRIBUTION: From Senegal (Duplantier and Granjon, 1992), Guinea (Mt Nimba) and Sierra Leone; eastward through Liberia, Côte d'Ivoire, Ghana, Togo, Benin, Nigeria, Cameroon, Gabon, Central African Republic, N Dem. Rep. Congo (Orientale), S Sudan, WC and S Ethiopia; to N Uganda and W Kenya (range documented by Grubb et al., 1998; Rosevear, 1969; F. Petter and Genest, 1970; our study of samples in AMNH, BMNH, FMNH, and USNM).
   STATUS: IUCN – Lower Risk (lc).
   SYNONYMS: *pasha* (Thomas, 1910); *proconodon* (Thomas, 1910).
   COMMENTS: Subgenus *Nannomys*. A distinct species sometimes confused with *M. minutoides*, which occurs over approximately the same region (Rosevear, 1969). Multivariate analysis of morphometric traits by Macholán (2001) indicated that *M. setulosus* and *M. minutoides* were distantly related. Both *pasha* (Thomas, 1910*a*) and *proconodon* (Rhoads, 1896) were originally described as species; Osgood (1936) associated *pasha* with *M. proconodon*, and we agree with his identification. F. Petter and Matthey (1975) regarded *pasha* as a species, noting that it might be referrable to *M. setulosus*. Both Osgood (1936) and Yalden et al. (1976) recognized *proconodon* as a species endemic to Ethiopia. Our study of Osgood's specimens, some of which are near-topotypes, revealed that their morphological traits fell within the range of variation typical of *M. setulosus*. Our identification was foreshadowed by F. Petter and Matthey (1975), who cited the range of *M. setulosus* to include Ethiopia, based on a letter from J. Prevost. Chromosomal data for samples from West Africa documented by Jotterand (1972), Jotterand-Bellomo (1981, 1986), and Matthey (1964).

Mus setzeri F. Petter, 1978. Mammalia, 42:377.
   COMMON NAME: Setzer's Mouse.
   TYPE LOCALITY: Botswana, 82 km west of Mohembo (near Namibian border).
   DISTRIBUTION: NE Namibia, NW and S Botswana, and W Zambia (Vermeiren and Verheyen, 1983).
   STATUS: IUCN – Lower Risk (lc).
   COMMENTS: Subgenus *Nannomys*. A unique desert species reviewed by Vermeiren and Verheyen (1983), Meester et al. (1986), Skinner and Smithers (1990), and de Graaff (1997*k*).

Mus shortridgei (Thomas, 1914). J. Bombay Nat. Hist. Soc., 23:30.
   COMMON NAME: Shortridge's Mouse.
   TYPE LOCALITY: Burma, Mt Popa, 4961 ft (1512 m).
   DISTRIBUTION: Burma (Ellerman, 1961), Thailand (J. T. Marshall, Jr., 1977*a*), SW Cambodia (A. Smith, in litt.), C Laos (Smith et al., In Press), and NW Vietnam (Dao, 1978); see J. T. Marshall, Jr. (1977*a*:431).
   STATUS: IUCN – Lower Risk (lc).
   SYNONYMS: *nghialoensis* Dao, 1966.
   COMMENTS: Subgenus *Pyromys*. Dao (1966) described *nghialoensis* as a subspecies of

M. platythrix, which at the time embraced the Indochinese *shortridgei*. Chromosomal composition of Thai samples, which demonstrate a complex polymorphism, reported by Gropp et al. (1973). Evolutionary history extends back to the late Pliocene of Thailand, based on isolated molars recovered from cave sediments (Chaimanee, 1998).

Mus sorella (Thomas, 1909). Ann. Mag. Nat. Hist., ser. 8, 4:548.

COMMON NAME: Thomas's Mouse.

TYPE LOCALITY: W Kenya, Mt Elgon, Kirui, 6000 ft (1830 m).

DISTRIBUTION: Documented by specimens from E Cameroon, EC Angola, NE and SE Dem. Rep. Congo, Uganda, Kenya, and N Tanzania (F. Petter, 1981*b*; Verheyen, 1965*a*; specimens examined in AMNH, BMNH, CM, and USNM); range limits unresolved.

STATUS: IUCN – Lower Risk (lc).

SYNONYMS: *acholi* Heller, 1911; *wamae* Heller, 1911.

COMMENTS: Subgenus *Nannomys*. Closest relatives are *M. baoulei*, *M. goundae*, *M. neavei*, and *M. oubanguii*; F. Petter (1981*b*) placed these (except *M. baoulei*) together in the *M. sorella* group. He also recognized *wamae* and *acholi* as species in the *sorella* complex, but after examining holotypes and other specimens we agree with Verheyen (1965*a*), who united them with *M. sorella*. There are at least two morphologically distinct species in the group, *M. sorella* and *M. neavei* (see that account), but the nature of their phylogenetic relationship to other forms in this complex needs to be assessed by critical systematic review. Reidentification of museum specimens might also help resolve boundaries of geographic ranges. The specimens from Angola, for example, were originally identified by Hill and Carter (1941) as *M. bella*. The southern African range mapped by de Graaff (1997*m*) reflects *M. neavi* and not *M. sorella*.

Mus spicilegus Petényi, 1882. Termeszetrajzi Fuzetek, Budapest, 5:114.

COMMON NAME: Mound-building Mouse.

TYPE LOCALITY: Hungary, Budapest, Rakos Plains.

DISTRIBUTION: Lowlands of Austria, S Slovakia (Bauer et al., 1998; Stollmann and Macholán, 1999), Hungary, Romania, Serbia and Montenegro, Albania, Greece, N Bulgaria (Peshev, 1996), and steppes of Moldavia and S Ukraine (see references cited below).

STATUS: IUCN – Lower Risk (nt).

SYNONYMS: *adriaticus* Kryštufek and Macholán, 1998; *mehelyi* Bolkay, 1925; *petenyi* Kryzhov, 1936 [see Zagorodnyuk, 1992*b*]; *sergii* Valch, 1927 [see Zagorodnyuk, 1992*b*]. Ellerman and Morrison-Scott (1951:608) and J. T. Marshall, Jr. (1998) listed *acervator*, *acervifex*, *canicularius*, and *caniculator* as alternative names proposed by Petényi in the same report where he described *M. spicilegus*; all are *nomina nuda*.

COMMENTS: Subgenus *Mus*. This is the mouse that constructs soil-covered storage mounds of grain, and was formally known as *M. hortulanus* (Corbet, 1984; Gromov and Erbajeva, 1995); however, the holotype of *hortulanus* is an example of *M. musculus* so the earliest name for the species is *spicilegus* (Gerasimov et al., 1990; J. T. Marshall, Jr., 1998; Zagorodnyuk, 1996*b*). Sympatric and frequently syntopic with *M. m. musculus* (Mitchell-Jones et al., 1999). Results of morphometric and allozymic analyses reported by Bonhomme et al. (1984), Petrov and Ruzic (1985), Gerasimov et al. (1990), She et al. (1990), and Lyalyukhina et al. (1991). Forms a clade with *M. musculus*, *M. macedonicus*, and *M. spretus* as indicated by analyses of DNA sequences from several different genes (Graur, 1994; Larizza et al., 2002; Lundrigan et al., 2002; Martin et al., 2000). Combined analyses of morphological traits, DNA/DNA hybridization, and mitochondrial 12S rRNA brings *M. spicilegus* together with *M. spretus*, *M. macedonicus* and *M. musculus* in a clade distinct from an Asian clade consisting of *M. caroli*, *M. cervicolor*, and *M. cooki* (Auffrey et al., 2003; Chevret et al., 2003). Other cytogenetic and biochemical contrasts between the species (reported as *hortulanus*) and *M. musculus* recorded by Bulatova and Kotenkova (1990) and Yakimenko et al. (1990). Distribution of the p53 pseudogene as another molecular trait separating *M. spicilegus* from *M. musculus* documented by Ohtsuka et al. (1996). Phylogenetic analysis of mitochondrial D-loop sequences also distinguishes *M. spicilegus* from *M. spretus* and *M. musculus* and indicates a separation time between *M. spicilegus* and *M. musculus* of about 3 million years ago (Flegr et al., 1994). Phylogenetic analyses of complete mtDNA cytochrome *b* sequences by Martin et al. (2000) clustered *M. spicilegus* closer to *M. musculus* than to *M. spretus*. Morphometric revision of samples from throughout the range of *M. spicilegus* and contrasts with *M. macedonicus* in the context of identifying a sample from the Adriatic coast of Serbia and Montenegro were reported by Kryštufek and Macholán (1998) who described it as a subspecies, *M. s. adriaticus*. In an extension of this study, the occurrence of *M. spicilegus* in Albania and Greece and comparison with *M. macedonicus* were documented in some detail by Macholán and Vohralík (1997). Sequences of APRT gene were used to illuminate substitution rate variation among *M. spicelegus*, *M. pahari*, and other muroids (Fieldhouse et al., 1997). Zagorodnyuk and Berezovsky (1994) reported aspects of taxonomy, geographic distribution, and ecology of *M. spicilegus* in the steppes and forest-steppes in the districts of Odessa, Nikolayev, Kirovograd, Vinnitsa, and Cherkassy districts across S Ukraine. Gromov and Erbajeva (1995) reviewed Russian segment (as *hortulanus*). Bauer et al. (1998), Stollmann (1998), and Stollmann and Macholán documented the Slovakian record. A four-year study of the species in Austria covering morphological traits, ecology, and population biology is reported by Unterholzner and Willenig (2000). European population reviewed by Mitchell-Jones et al. (1999).

Bauer (2000) noted that *M. spicilegus* is an endemic of SE Europe and, based on his review of the evolutionary time-table generated by allozymic and DNA data, speculated that the species evolved between middle and late Pleistocene near the Black Sea, possibly near the Danube delta. "*M. spicilegus* in SE Central Europe is so totally dependent on man-made habitats and continuous human intervention, that it well might be a rather late (neolithic or even later) immigrant" (Bauer, 2000:101). Bauer also provided an extensive review of the documented evolutionary history of this species and *Mus musculus*.

Mus spretus Lataste, 1883. Actes Soc. Linn. De Bordeaux, ser. 7, 4:27.

COMMON NAME: Western Mediterranean Mouse.

TYPE LOCALITY: Algeria, Oued Magra, between M'sila and Barika, north of Hodna.

DISTRIBUTION: Natural grasslands and agricultural fields in Western Mediterranean climatic zone of W Europe (S France, Spain, Portugal, Balearic Isls; Mitchell-Jones et al., 1999: 291) and North Africa (Morocco, Algeria, Tunisia, and Libya); see J. T. Marshall, Jr. (1981; 1998) and Macholán (1996a).

STATUS: IUCN – Lower Risk (lc).

SYNONYMS: *caoccii* Krausse, 1919; *hispanicus* Miller, 1909; *lusitanicus* Miller, 1909; *lynesi* Cabrera, 1923; *mogrebinus* Cabrera, 1911; *parvus* Alcover, Gosalbez, and Orsini, 1985 [not Beckstein, 1796]; *rifensis* Cabrera, 1923.

COMMENTS: Subgenus *Mus*. See J. T. Marshall, Jr. (1998) for identification of synonyms. Throughout its range, *M. spretus* is sympatric but not syntopic with *M. musculus domesticus* (Boursot et al., 1993; Mitchell-Jones et al., 1999). The species has been the subject of various inquiries covering biometrical and morphological analyses (Darviche and Orsini, 1982; Engels, 1980, 1983b; Gerasimov et al., 1990; Palomo, 1988; Palomo et al., 1983; Vargas, et al., 1984); chromosomal and electrophoretic studies (Cano et al., 1984; Engels, 1983a; Matsuda and Chapman, 1992; Traut et al., 1992); distribution of the p53 pseudogene as another molecular trait distinguishing *M. spretus* from *M. spicilegus* and *M. musculus* (Ohtsuka et al., 1996); phylogenetic analysis of mitochondrial D-loop sequences to also distinguish *M. spretus* from *M. spicilegus* and *M. musculus* (Flegr et al., 1994); variability in mtDNA that revealed two genetically distinct phylogenetic groups within *M. spretus* (Boursot, et al., 1985); contrasts between *M. spretus* and *M. musculus* (*domesticus*) using DNA sequence of LINE-1 elements (Casavant and Hardies, 1994); phylogenetic relationships assessed by sequences from several different genes that placed *M. spretus* in a clade with *M. musculus*, *M. macedonicus*, and *M. spicilegus* (Graur, 1994; Larizza et al., 2002; Lundrigan et al., 2002; Martin et al., 2000; Prager et al., 1996, 1998); combined analyses of morphological traits, DNA/DNA hybridization, and mitochondrial 12S rRNA sequences join *M. spretus* with *M. spicilegus*, *M. macedonicus*, and *M. musculus* in a clade separate from an Asian clade composed of *M. caroli*, *M. cervicolor*, and *M. cooki* (Auffray et al., 2003; Chevret et al., 2003); and differences between *M. spretus* and *M. musculus* in thermoregulatory capabilities

(Gorecki et al., 1990). Evidence indicates that *M. spretus* forms a sister group to *M. spicilegus*, *M. macedonicus*, and all taxa embraced by *Mus musculus* (Prager et al., 1998). Alcover et al. (1985) described *parvus* as a subspecies of *M. spretus*. European population reviewed by Mitchell-Jones et al. (1999), Balearic Isls by Alcover et al. (1985) and Alcover and Gosalbez (1988), and Moroccan by Aulagnier and Thevenot (1986). Relationship between habitats and abundance in Algeria reported by Khidas et al. (2002). Torre et al. (1996) analyzed population in NE Spain by studying owl pellet remains.

   Mus spretus has been identified from middle Pleistocene beds at Jebel Irhoud in Morocco (Amani and Geraads, 1993), from late Pleistocene in the N Rif Mtns of N Morocco (Ouahbi et al., 2001), and Pleistocene sediments of Tunisia (Mein and Pickford, 1992). Dobson (1998, 2000) postulated that the species is native to the Maghreb of North Africa and may have been transported by humans, presumably inadvertently, to SW Europe, which J. T. Marshal, Jr. (in litt., 2004) finds unlikely because *M. spretus* is found only in "wild" outdoor habitats.

Mus tenellus (Thomas, 1903). Proc. Zool. Soc. Lond., 1903(1):298.
   COMMON NAME: Delicate Mouse.
   TYPE LOCALITY: Sudan, Blue Nile, Roseires.
   DISTRIBUTION: Sudan (Setzer, 1956), S Ethiopia (below 2000 m; Rupp, 1980), S Somalia, and south through Kenya to C Tanzania (Dodoma), including lower flanks of Mt Kilimanjaro (Grimshaw et al., 1995); range based upon our study of material at AMNH, BMNH, CNHM, MCZ, AND USNM; distributional limits unresolved.
   STATUS: IUCN – Lower Risk (lc).
   SYNONYMS: *aequatorius* Setzer, 1953; *delamensis* Setzer, 1956; *gerbillus* (G. M. Allen and Loveridge, 1933); *suahelica* (Thomas, 1910).
   COMMENTS: Subgenus *Nannomys*. Reviewed by F. Petter (1972*a*) and Yalden et al. (1976). Morphologically and ecologically closely similar to *M. haussa* (see that account). The southernmost record is based on holotype of *gerbillus* (G. M. Allen and Loveridge, 1933), which is an example of *M. tenellus*; *aequatorius*, *delamensis*, and *suahelica* also represent that species (our study of holotypes). Yalden et al. (1976) listed *gallarum* as a synonym of *M. tenellus*, but the holotype is a *M. minutoides*. Judged by our studies of museum specimens, most published Ethiopian records of *M. tenellus* are actually *M. minutoides*, but we have not seen the samples from SW Ethiopia referenced by Yalden et al. (1996).

Mus terricolor Blyth, 1851. J. Asiat. Soc. Bengal, 20:172.
   COMMON NAME: Earth-Colored mouse.
   TYPE LOCALITY: S India, Bengal, neighborhood of Calcutta.
   DISTRIBUTION: Indigenous to peninsular India and Nepal (J. T. Marshall, Jr., 1998), Bangladesh (Comilla District, Chittagong Province; Alpin et al., 2003*c*; K. Alpin, in litt., 2004), and Pakistan (J. T. Marshall, Jr., 1998); occurs also in Medan region of N Sumatra (Indonesia) where it was probably inadvertently introduced (Musser and Newcomb, 1983, recorded as *dunni*).
   STATUS: IUCN – Lower Risk (lc).
   SYNONYMS: *beavanii* Peters, 1866; *dunni* Wroughton, 1912.
   COMMENTS: Subgenus *Mus*. Formerly referred to as *M. dunni* (J. T. Marshall, Jr., 1977*b*, 1986), but *terricolor* is the older name (J. T. Marshall, Jr., 1977*b*, 1998:80, and in litt., 1989). Chromosomal results presented by Sharma et al. (1986, under *dunni*) and Boyeskorov et al. (1997, as *dunni*) in context of evolutionary divergence from other species of *Mus*. Closely related to *Mus booduga*. Both species have 2n = 40, with all telocentric chromosomes, but *M. booduga* has a slightly smaller Y chromosome. All populations sampled show the same karyotype; *M. terricolor*, however, has a large submetacentric X and telocentric Y. Furthermore, different populations of *M. terricolor* have three different karyotypes (formally labelled I, II, and III) indicating various stages of evolutionary differentiation in which heterochromatin may be important in the speciation process (Bahadur and Sharma, 1995; Sharma et al., 2002). In a combined chromosomal and molecular study (allozymes, serum proteins, and mitochondrial

cytochrome *b* DNA sequences), Sharma (1996) concluded the *M. booduga-M. terricolor* lineage probably evolved simultaneously with the *M. caroli-M. cookii-M. cervicolor* lineage and *M. terricolor* itself is in an active phase of speciation. Balajee and Sharma (1994) found *M. musculus*-like AT-rich heterochromatin in *M. booduga* and the three chromosomal forms of *M. terricolor*. Patnaik et al. (1993) contrasted dental traits and cranial features between *M. booduga* and *M. terricolor* and compared dental patterns of those two with molars of six species of *Mus* represented by late Pliocene-Early Pleistocene fossils from the N portion the Indian subcontinent, concluding that the fossil *Mus* are dentally closely related to the extinct species and reflect an early diversification in the evolutionary history of the genus. The multivariate morphometric analysis by Macholán (2001) suggested *M. terricolor* to be distantly related to the two other species in subgenus *Mus* that were examined, *M. caroli* and *M. cervicolor*. Occurrence and ecology of *M. terricolor* in the Aravalli ranges of Rajasthan State documented by Prakash et al. (1995*a*, *b*, *c*), and in Gujarat State, NW India by Chakraborty and Agrawal (2000). The Bangladesh record was confirmed from comparisons with Indian and Nepalese samples using DNA sequences, all of which were very similar (K. Aplin, in litt., 2004).

Mus triton (Thomas, 1909). Ann. Mag. Nat. Hist., ser. 8, 4:548.
COMMON NAME: Gray-bellied Mouse.
TYPE LOCALITY: Kenya, Mt Elgon, Kirui, 6000 ft (1830 m).
DISTRIBUTION: N and E Dem. Rep. Congo, Uganda (Delany, 1975), Kenya (Hollister, 1919), S Ethiopia (Lavrenchenko, 2000; Yalden et al., 1996), Tanzania (Grimshaw et al., 1995; Stanley et al., 1998; Swynnerton and Hayman, 1951), Malawi (Ansell, 1989*b*; Ansell and Dowsett, 1988), Tete Dist. of Mozambique (de Graaff, 1997*h*; Smithers and Lobao Tello, 1976), Zambia (Ansell, 1978), and C and NE Angola (Crawford-Cabral, 1998).
STATUS: IUCN – Lower Risk (lc).
SYNONYMS: *birungensis* (Lönnberg and Gyldenstolpe, 1925); *fors* (Thomas, 1909); *imatongensis* Setzer, 1953; *murilla* (Thomas, 1910); *naivashae* (Heller, 1910).
COMMENTS: Subgenus *Nannomys*. Listed as a questionable synonym of *M. mahomet* by Yalden et al. (1976:30), who were unsure about the equivalence of *mahomet* and *triton* and merely noted that Ethiopian samples previously identified as *triton* were really *mahomet*. Lavrenchenko (2000) claimed that *M.* cf *triton* from the Ethiopian Harenna Forest, along with Ethiopian *M. mahomet*, belong to the same cytotaxonomic group, which excludes true *M. triton*; Yalden et al. (1996) noted that *M. triton* and *M. mahomet* are sympatric in S Ethiopia but the former lives in forest, the latter in grassy forest clearings. The description of *Mus birungensis* (Lonnberg and Gyldenstolpe, 1925) mirrors the range of variation of *M. triton* in samples (in AMNH) we have examined from the Kivu region of E Dem. Rep. Congo. Considerable chromosomal polymorphism has been reported in samples identified as *M. triton* (Robbins and Baker, 1978). Altitudinal distribution on Ugandan slopes of Ruwenzori Mtns reviewed by Kerbis Peterhans et al. (1998). Extant southern limit of species is Zambia and Tete Dist. of Mozambique (about 17ES), but it was present in KwaZulu-Natal, South Africa up to about 60,000 years ago (Avery, 1991).

Mus vulcani (Robinson and Kloss, 1919). Ann. Mag. Nat. Hist., ser. 9, 4:378.
COMMON NAME: Javan Shrew-like Mouse.
TYPE LOCALITY: Indonesia, W Java, Gunung Gede, Gandang Badak, 7900 ft (2408 m).
DISTRIBUTION: Endemic to upper montane tropical evergreen rain forests in mountains of W Java.
STATUS: IUCN – Lower Risk (nt).
COMMENTS: Subgenus *Coelomys*. Originally described as a subspecies of *Mycteromys crociduroides* (Chasen, 1940), but *vulcani* is a separate species (based upon Musser's study of specimens of *vulcani* and *crociduroides*). Both *M. vulcani* and *M. crociduroides* are the only living native *Mus* recorded from islands on the Sunda Shelf (Musser, 1986; Musser and Newcomb, 1983). Part of the suite of murines endemic to Java (see account of *Kadarsanomys sodyi*). Curiously, Wang (2003) listed *M. vulcani* as part of the Chinese fauna from S Yunnan.

Mylomys Thomas, 1906. Ann. Mag. Nat. Hist., ser. 7, 18:224.

> TYPE SPECIES: *Mylomys cuninghamei* Thomas, 1906 (= *Golunda dybowskii* Pousargues, 1893).
>
> COMMENTS: *Arvicanthis* Group. The Indian *Golunda* and *Mylomys* are usually considered close relatives of each other, but Musser (1987b) discussed traits that indicated their distant relationship and a closer phylogenetic alliance between *Mylomys* and *Pelomys*, which is substantiated by mtDNA sequences of cytochrome *b* and 12S and 16S rRNA gene fragments (Ducroz et al., 2001). Those latter two genera are members of a division that includes *Arvicanthis*, *Lemniscomys*, and *Rhabdomys* as assessed by morphological data (Musser, 1987b). The molecular results of Ducroz et al. also indicate that *Desmomys* belongs with this assemblage, and that *Mylomys*, *Pelomys*, and *Arvicanthis* form one lineage, *Desmomys* and *Rhabdomys* a second, and a third contains only *Lemniscomys*.

Mylomys dybowskii (Pousargues, 1893). Bull. Soc. Zool. Fr., 18:163.

> COMMON NAME: Common Mylomys.
>
> TYPE LOCALITY: Central African Republic, Kemo River.
>
> DISTRIBUTION: Guinea (Mt Nimba), Côte d'Ivoire, Ghana (Grubb et al., 1998), S Cameroon, Republic of Congo, Central African Republic, W, N and E Dem. Rep. Congo, Rwanda, Tanzania (Swynnerton and Hayman, 1951), Kenya (Hollister, 1919), Uganda (Delany, 1975), and S Sudan (Setzer, 1958); abstracted from Rosevear (1969), Misonne (1974), many faunal reports, and specimens examined in AMNH, BMNH, FMNH, and USNM.
>
> STATUS: IUCN – Lower Risk (lc).
>
> SYNONYMS: *alberti* Thomas, 1915; *christyi* Thomas, 1917; *cuninghamei* Thomas, 1906; *lowei* Hayman, 1935; *lutescens* Thomas, 1915; *massaicus* Lönnberg, 1916; *rex* (Thomas, 1906); *richardi* Hayman, 1939; *roosevelti* (Heller, 1910).
>
> COMMENTS: Hatt (1940a) noted that the cotypes of *dybowskii* are examples of *Mylomys*, not *Pelomys* under which the name had been listed (Ellerman, 1941), and selected a lectotype. The identity was verified by F. Petter (1962b), who also thought there were two species in the genus, *M. dybowskii* and the West African *M. lowei*, named by Hayman (1935) as a subspecies of *M. cuninghamei*. Rosevear (1969) followed F. Petter in recognizing *M. lowei* and in turn was followed by Grubb et al. (1998). Hayman (1935:934), however, regarded all forms to be part of a single species: "In considering what value to place on this Ashanti form [*lowei*] it has seemed best to regard all the known *Mylomys* as a subspecies of the original *cuninghamei* Thos. From that small and rather dark form, through the middle-sized *lutescens* and *christyi* to the large brightly-coloured races *roosevelti*, *alberti*, and the present form, the differences lie chiefly in comparative size and intensity of colouring. In my opinion the characters of *lowei* are not sufficient to justify full specific differentiation." Significance of geographic variation in chromatic and morphological traits has yet to be assessed by critical systematic revision; whether the genus is monotypic, contains the two species listed here, or consists of more than two is unresolved. Chromosomal data available for samples from Central African Republic (Matthey, 1970) and Mt Nimba (Guinea; Gautun et al., 1986).

Mylomys rex (Thomas, 1906). Ann. Mag. Nat. Hist., ser. 7, 18:304.

> COMMON NAME: Ethiopian Mylomys.
>
> TYPE LOCALITY: C Ethiopia, Kaffa, Charada Forest.
>
> DISTRIBUTION: Recorded only from the type locality.
>
> COMMENTS: The taxon *rex*, represented only by the holotype (a skin without skull), was described by Thomas (1906c) as a species of *Arvicanthis*, but later "provisionally considered as a giant member of *Desmomys*" (Thomas, 1916a:68). Dieterlen (1974) challenged the validity of *rex*, but Yalden et al. (1976) pointed out the features distinguishing the holotype from samples of *D. harringtoni*, and treated *rex* as another distinctive species endemic to Ethiopia. Musser and Carleton (1993:630) wrote that "Our study of the holotype skin reveals it to be a large and probably old adult of *Mylomys* that is not as brightly pigmented as most samples of that genus. Whether the holotype actually came from Ethiopia, or represents a separate species of *Mylomys* are unknown; we provisionally list *rex* in the synonymy of *M. dybowskii*." Yalden et al. (1996) were content to adopt this provisional arrangement, but explained there was

no doubt the specimen was collected in S Ethiopia, and that it "may indeed be specifically distinct from *M. dybowskii*, since the latter is generally considered to be a savanna form while the type of *rex* was apparently obtained in tropical deciduous forest at an altitude of 1800 m" and the speculation "that *rex* could be one of Ethiopia's very few lowland forest endemics . . . may yet be proved correct, once further specimens become available." Lavrenchenko (2002) endorsed the possibility that *rex* may be a distinct species. We treat *rex* as a species, which is better than hiding a possible endemic in the synonymy of *M. dybowskii*; study of additional specimens will reveal its actual relationship to that widespread species.

Myomyscus Shortridge, 1942. Ann. S. Afr. Mus., 36:93.

  TYPE SPECIES: *Mus verroxii* (= *verreauxi*) A. Smith, 1834.

  COMMENTS: *Stenocephalemys* Division. *Myomys* has traditionally been used for the species we list under *Myomyscus*, but the holotype of *colonus* (Brants, 1827), which is the type species of *Myomys*, is an example of *Mastomys* (Van der Straeten and Robbins, 1997), making *Myomys* a synonym of that genus. Shortridge's (1942) *Myomyscus* is the next available name (see generic account of *Mastomys*). The species *ruppi* and *albipes* were listed in *Myomys* by Musser and Carleton (1993), but are here added to the Ethiopian *Stenocephalemys* (see accounts of those species). They also placed *daltoni* and *derooi* in *Myomys*, but here we transfer them to *Praomys* (see those accounts).

  Even as reconstituted here, *Myomyscus* remains polyphyletic. Recent analyses of complete mtDNA cytochrome *b* sequences, for example, places *M. verreauxii* (type-species of the genus) as sister to *Colomys* and *Zelotomys* (although with weak support), *brockmani* as sister group to *Mastomys* and *Stenocephalemys*, and *yemeni* loosely and anciently related to *Praomys verschureni* (Lecompte et al., 2002*b*). By contrast, analyses of nuclear IRBP gene sequences cluster *M. brockmani* and *M. yemeni* and indicate it to be a sister group to *Stenocephalemys* (E. Lecompte, in litt., 2002); the IRBP sequences support the monophyletic cluster of *M. verreauxii*, *Colomys*, and *Zelotomys*. Cladistic analyses of morphological traits demonstrate *Myomyscus* (reported as *Myomys*) to be a sister-genus of *Mastomys*, with the two species sampled (*fumatus* [=*brockmani*] and *daltoni* [here placed in *Praomys*]) typical of tree savannas while species of *Mastomys* occupy grass savannas (Lecompte et al., 2002*a*). Chromosomal and immunological data related *Myomyscus brockmani* (reported as *fumatus*) to species of *Mastomys* (Qumsiyeh et al., 1990), morphometric analyses placed *M. verreauxii* closer to *Mastomys* than to other species of *Myomyscus* (Van der Straeten and Dieterlen, 1983; Van der Straeten and Robbins, 1997), and some researchers (e.g., Misonne, 1969, 1974; Qumsiyeh et al., 1990) included *Myomys* (now a synonym of *Mastomys*) and *Myomyscus* within *Praomys*. Analysis of microcomplement fixation of albumin grouped *Myomyscus verreauxii* (recorded as *Myomys*) with *Mastomys*, *Praomys*, and *Hylomyscus* (Watts and Baverstock, 1995*a*). Van der Straeten (in litt., 1994) considered *Myomyscus* to contain only *verreauxii*, a hypothesis consistent with molecular analyses (Lecompte et al., 2002*b*; Lecompte, in litt., 2002). We retain *Myomyscus* separate from *Mastomys* and *Praomys*, which accords with current research results (Lecompte et al., 2002*a, b*), and would welcome future morphological and molecular phylogenetic analyses incorporating a wide range of morphological traits and different genes along with denser taxon sampling that might finally define the contents of *Myomyscus* and test the cladistic affinity of *M. verreauxii* with species of *Colomys* and *Zelotomys* as indicated by mitochondrial and nuclear DNA sequences.

Myomyscus angolensis (Bocage, 1890). J. Sci. Math., Phys. Nat., Lisboa, ser. 2, 2:12.

  COMMON NAME: Angolan Myomyscus.

  TYPE LOCALITY: Angola, Capangombe, interior of Moçâmedes District, 527 m (additional information provided by Crawford-Cabral, 1989*b*, 1998).

  DISTRIBUTION: W Angola, primarily on the Angolan Plateau (Crawford-Cabral, 1998).

  STATUS: IUCN – Lower Risk (lc) as *Mastomys angolensis*.

  SYNONYMS: *angolae* (Crawford-Cabral, 1989).

  COMMENTS: Our inclusion of this species in *Myomyscus* is provisional. It has been placed in *Rattus* (Ellerman, 1941), *Myomyscus* (or *Myomys*) (G. M. Allen, 1939; D. H. S. Davis, 1965; Hill and Carter, 1941), *Praomys* (see references in Crawford-Cabral, 1989*b*), or re-

garded as a species of *Mastomys* (Crawford-Cabral, 1989*b*, 1998; Misonne, 1974). Ansell (1978) and Ellerman et al. (1953) thought it morphologically similar to *M. shortridgei*, which they treated as a subspecies of *M. angolensis*. In their review of *Mastomys*
systematics, Granjon et al. (1997*b*) transferred *angolensis* to *Myomys* and retained *shortridgei* in *Mastomys*. Incorporation of *M. angolensis* in phylogenetic analyses derived
from morphological and molecular data is needed to clarify its relationship. *Myomyscus angolensis* is either sympatric (Hill and Carter, 1941; Crawford-Cabral, 1983,
1998) or altitudinally parapatric (specimens in FMNH from Mt Soque) with *Mastomys*
natalensis in Angola.

    In a report devoted to the identity of Bocage's *Mus angolensis*, Crawford-Cabral
(1989*b*) noted that the type series was destroyed by fire in the Lisbon Museum in 1978.
According to the original description the specimens represented a species with a tail
much longer than head and body, white feet, soft and thick fur, and five pairs of teats,
all characters of other species in *Myomyscus*. Crawford-Cabral considered Angolan
specimens collected after 1890 outside of the Capangombe region and identified as *angolensis* to be another species. He proposed *Praomys angolae* for this rat and considered
it a *Mastomys*, not an example of *Myomyscus*. In our view he simply renamed Bocage's
angolensis.

Myomyscus brockmani (Thomas, 1908). Ann. Mag. Nat. Hist., ser. 7, 18:298.

    COMMON NAME: Brockman's Myomyscus.

    TYPE LOCALITY: Somalia (British Somaliland), Upper Sheikh.

    DISTRIBUTION: Primarily tree savannas from EC Tanzania north through Kenya (Hollister,
        1919) and N Uganda (Delany, 1975) into Somalia, SE Ethiopia (Yalden et al., 1976,
        1996; specimens in CM), and S Sudan (Setzer, 1956); W and S limits unknown.

    SYNONYMS: *allisoni* (Hayman, 1960); *niveiventris* (Osgood, 1910); *oweni* Setzer, 1956;
        subfuscus (Osgood, 1910); *ulae* (Heller, 1910).

    COMMENTS: Chromosomal and immunological data indicate close relationship with species
        of *Mastomys* (Qumsiyeh et al., 1990), an alliance also suggested by analyses of mtDNA
        cytochrome *b* sequences (Lecompte et al., 2002*b*). Analyses of nuclear IRBP gene
        sequences, however, indicates *brockmani*, along with *M. yemeni*, to be closely allied to
        species of *Stenocephalemys* and should be included in that genus (Lecompte, 2003).
        Spermatozoal morphology described by Baskovich and Lavrenchenko (1995, as
        fumatus). This species has traditionally been known as *M. fumatus* (Musser and
        Carleton, 1993), but Van der Straeten and Robbins (1997) demonstrated the holotype
        of *fumatus* to be an example of *Mastomys* and not *Myomys* (Musser independently
        examined the holotype and identified it as a *Mastomys*). We include *fumatus* in the
        synonymy of *Mastomys natalensis* (see that account). The next oldest name available
        for this species of *Myomyscus* is Thomas's (1908) *brockmani*. Yalden et al. (1976)
        included *tana* in the synonymy but the holotype is a *Mastomys* (Van der Straeten and
        Robbins, 1997).

Myomyscus verreauxii (Smith, 1834). S. Afr. Quart. J., 2:156.

    COMMON NAME: Verreaux's White-footed Rat.

    TYPE LOCALITY: South Africa, Western Cape Province, Cape of Good Hope, near Cape Town.

    DISTRIBUTION: South Africa, Western Cape Province, from Olifants River in the west to
        Nature's Valley, Plettenberg Bay in the east (de Graaff, 1981:218; Skinner and Smithers,
        1990:271).

    STATUS: IUCN – Least Concern.

    SYNONYMS: *verreauxi* (Sclater, 1901); *veroxii* (Smith, 1834).

    COMMENTS: A distinctive South African endemic confined to the fynbos biome (Mugo et al.,
        1995). Taxonomy reviewed by Meester et al. (1986); distributional and biological
        information provided by de Graaff (1981) and Skinner and Smithers (1990). Roberts
        listed measurements and treated *verreauxii* as a subspecies of "*Myomys colonus*."
        Analyses of complete mtDNA cytochrome *b* sequences (Lecompte et al., 2002*b*) and
        nuclear IRBP gene sequences (Lecompte, 2003) separates *M. verreauxii* from *M.
        brockmani* and *M. yemeni* (*M. angolensis* has yet to be included in molecular inquiries)
        and clusters it with *Colomys* and *Zelotomys*, providing an hypothesis that should be

tested with morphological data and sequences from a wider range of genes. Ultimately, *verreauxii* may represent the only species of *Myomyscus*. Reviewed by de Graaff (1997*t*). Earliest fossils of *Myomyscus*, presumably either *M. verreauxii* or something related to it, come from late Pliocene deposits of South Africa (see review by Denys, 1999).

Myomyscus yemeni Sanborn and Hoogstraal, 1953. Fieldiana Zool., 34:241.
    COMMON NAME: Yemen White-footed Rat.
    TYPE LOCALITY: Yemen, Kariet Wadi Dhahr, six miles (9.7 km) northwest of San'a, 6400 ft (1950 m).
    DISTRIBUTION: Recorded only from N Yemen and SW Saudi Arabia (see Harrision and Bates, 1991:249; also references in Al-Jumaily, 1998, for Yemen population).
    STATUS: IUCN – Lower Risk (lc) as *Myomys yemeni*.
    COMMENTS: Originally described by Sanborn and Hoogstraal (1953) as a subspecies of *Myomys fumatus* (= *brockmani*), the diagnostic traits of *yemeni* are outside the range of variation recorded for any sample of *M. brockmani*. Our study of holotype and specimens of *M. yemeni* and *M. brockmani* at FMNH revealed that *M. yemeni* is much larger in body, cranial, and dental dimensions than *M. brockmani* (no overlap in length of molar rows, for example), with paler pelage and significantly larger ears and auditory bullae (both absolutely and relative to body size). The morphological attributes of *M. yemeni* define a distinctive species; however, its phylogenetic relationships to other species in the genus, or within the *Stenocephalemys* Division, have yet to be fully resolved. Analyses of mtDNA cytochrome *b* sequences, for example, suggests *M. yemeni* to be more closely related to *Praomys verschureni* than to any other species we retain in *Myomyscus* (Lecompte et al., 2002*b*), but nuclear IRBP gene sequences indicates *M. yemeni*, along with *M. brockmani*, to form a sister group to *Stenocephalemys* and should be included in that genus (Lecompte, 2003). Chromosomal traits (2n = FNa = 36) described by Lecompte (2003). Reviewed (as *fumatus*) by Harrison and Bates (1991) and the Yemen population by Al-Jumaily (1998).

Nesokia Gray, 1842. Ann. Mag. Nat. Hist., [ser. 1], 10:264.
    TYPE SPECIES: *Arvicola indica* Gray, 1830.
    SYNONYMS: *Erythronesokia* Khajuria, 1981; *Nesocia* Blanford, 1891; *Spalacomys* Peters, 1860.
    COMMENTS: *Rattus* Division. Early placed with the Philippine *Phloeomys* in the Phloeomyinae (Alston, 1876), but phylogenetically most closely allied to *Bandicota* among living murines (Corbet and Hill, 1992; Jansa et al., 1999; Musser and Brothers, 1994; Watts and Baverstock, 1994*b*); see generic account of *Bandicota*. Of the two living species in the genus, *N. indica* is fossorial, *N. bunnii* is amphibious. Earliest record of *Nesokia* is *N. panchkulaensis* from early Pleistocene Pinjor beds in the Siwaliks of N India (Raghavan, 1989).

Nesokia bunnii (Khajuria, 1981). Bull. Nat. Hist. Res. Centre, 7:162.
    COMMON NAME: Long-tailed Nesokia.
    TYPE LOCALITY: SE Iraq, Basra Province, Al-Qurna.
    DISTRIBUTION: Recorded only from marshes at the confluence of Tigris and Euphrates Rivers in SE Iraq. Possibly occurs in Al-Hawizeh marsh to the east straddling the Iraq-Iran border.
    STATUS: IUCN – Lower Risk (nt).
    COMMENTS: Originally described under genus *Erythronesokia* by Khajuria (1981), but shown to be a distinctive species of *Nesokia* by Al-Robaae and Felten (1990). The meager available ecological information about this unique *Nesokia* is reported by Khajuria (1981) and Al-Robaae and Felten (1990). Because the vast wetlands of southern Iraq were largely destroyed by the Saddam Hussein regime during the last decade (Associated Press, 2003; UNEP, 2003), current population status of *N. bunnii* in Iraq is unknown.

Nesokia indica (Gray, 1830, *in* 1830-1835). Illustr. Indian Zool., pl. xi (see Ellerman and Morrison-Scott, 1955, and Corbet and Hill, 1992).
    COMMON NAME: Short-tailed Nesokia.
    TYPE LOCALITY: India (uncertain).

DISTRIBUTION: Modern range covers Bangladesh, N India (Bihar, West Bengal, Punjap, Haryana, Delhi, Uttar Pradesh, Gujarat, and Rajasthan; Agrawal, 2000; Chakraborty and Agrawal, 2000), Pakistan, Afghanistan, Iran, Iraq, Syria, Saudi Arabia, Israel-Jordan, NE Egypt, NW China (Xinjiang, south of Tian Shan; Wang, 2003, and Zhang et al., 1997), Turkmenistan, Uzbekistan, and Tajikistan.

STATUS: IUCN – Lower Risk (lc).

SYNONYMS: *bacheri* Nehring, 1897; *bailwardi* Thomas, 1907; *beaba* Wroughton, 1908; *boettgeri* Radde and Walter, 1889; *brachyura* Büchner, 1889; *buxtoni* Thomas, 1919; *chitralensis* Schlitter and Setzer, 1973; *dukelskiana* Heptner, 1928; *griffithi* Horsfield, 1851; *hardwickei* (Gray, 1837); *huttoni* (Blyth, 1846); *indicus* (Peters, 1860); *insularis* Goodwin, 1940; *legendrei* Goodwin, 1939; *myosura* (Wagner, 1845); *satunini* Nehring, 1899; *scullyi* Wood-Mason, 1876; *suilla* Thomas, 1907.

COMMENTS: Reviewed by Corbet (1978c) and Corbet and Hill (1992). Chromosomal data in different contexts reported by Thelma and Rao (1982), Gadi and Sharma (1983), Rao et al. (1983), Juyal et al. (1989), and Dubey and Raman (1992). External, cranial, and dental morphology, along with albumin immunology supports a close phylogenetic relationship with *Bandicota* (Misonne, 1969; Musser and Brothers, 1994; Niethammer, 1977; Watts and Baverstock, 1994b; Wroughton, 1908a), and electrophoretic comparisons of eight loci indicated a sister-species alliance with *B. indica* (Radtke and Niethammer, 1984[1985]). Chromosomal traits are closely similar in *N. indica* and *Bandicota bengalensis* (Gadi and Sharma, 1983). Substantial morphological variation is present among geographic samples of *N. indica*, and careful systematic revision is required to determine whether this variation represents one or more species. Geographic reviews covering taxonomy, ecology, and distribution available for India (Agrawal, 2000), Pakistan (Roberts, 1977, 1997), Israel and Jordan (Mendelssohn and Yom-Tov, 1999; Qumsiyeh, 1996), Syria (Kock and Nader, 1983; Misonne, 1957), Iran (Lay, 1967), Egypt (Osborn and Helmy, 1980), and regions adjacent to Russia (Gromov and Erbajeva, 1995). Inclusion of SE Anatolia (Turkey) in a distribution map of *N. indica* was probably based on nearby Syrian records and the species has yet to be found in Turkey (Kryštufek and Vohralík, 2001). Fossils from late Pleistocene sites in Egypt and N Sudan are beyond modern range in NE Africa (Osborn and Helmy, 1980).

Nesoromys Thomas, 1922. Ann. Mag. Nat. Hist., ser. 9, 9:263.

TYPE SPECIES: *Stenomys ceramicus* Thomas, 1920.

COMMENTS: *Rattus* Division. Rümmler (1938) united *Nesoromys* with *Stenomys* explaining only that the differences between it and other species in *Stenomys* was insufficient to recognize a separate genus for *ceramicus*; he considered *ceramicus* to represent *Stenomys*, a group primarily New Guinean in distribution, on Seram. The species, however, exhibits a combination of primitive and highly derived traits not seen in other species of *Stenomys* or *Rattus* (Aplin et al., 2003b; Misonne, 1969). While *ceramicus* does bear a superficial resemblance to New Guinea *R. niobe* (which Rümmler placed in *Stenomys*) in body size, pelage coloration and texture, and general cranial conformation, it differs considerably from any other endemics in the Indomaylan, Moluccan, and Australia-New Guinea regions by its combination of very specialized bony palate (extremely broad and projecting way beyond posterior margins of molar rows), short incisive foramina that end well anterior to molar rows, and shallow pterygoid fossae, prompting Ellerman (1941: 258) to admonish that ". . . until intermediate forms are discovered between this genus and *Rattus* in characters of the palate, it must stand as a very distinct genus." We agree and retain the genus (as did Ellerman [1949a], Laurie and Hill [1954], Misonne [1969], and Simpson [1945]) with its morphologically distinctive species until a relationship is demonstrated otherwise by phylogenetic analyses employing morphology and DNA sequences from a broad range of Indo-Australian murines.

Nesoromys ceramicus (Thomas, 1920). Ann. Mag. Nat. Hist., ser. 9, 6:425.

COMMON NAME: Seram Island Mountain Rat.

TYPE LOCALITY: Indonesia, Moluccas, Pulau Seram, Gunung Manusela, 6000 ft (1830 m).

DISTRIBUTION: Endemic to montane forests of Pulau Seram.

STATUS: IUCN – Lower Risk (lc) as *Stenomys ceramicus*.

COMMENTS: Morphologically superficially similar to the *Rattus niobe* complex of New Guinea, but may not be closely related to that group (Ellerman, 1941; Flannery, 1995*b*). Originally described as a species of *Nesoromys* (Thomas, 1920*c*), but subsequently merged with *Stenomys* (Rümmler, 1938; Musser and Carleton, 1993). Known only by the type series consisting of three specimens (in BMNH) and an additional example in the Western Australia Museum (Helgen, 2003*b*).

Nilopegamys Osgood, 1928. Field Mus. Nat. Hist., Zool. Ser., 12:185.

TYPE SPECIES: *Nilopegamys plumbeus* Osgood, 1928.

COMMENTS: *Colomys* Division. Described as a genus by Osgood (1928) and recognized as such by G. M. Allen (1939) and Ellerman (1941), but subsequently included in *Colomys* by Hayman (1966) and most checklists published after his report covering African rodents, including the revision of *Colomys* by Dieterlen (1983) in which *plumbeus* was maintained as a subspecies of *C. goslingi*, the allocation recommended by Hayman (1966). Kerbis Peterhans and Patterson (1995) reviewed the taxonomic history and resurrected *Nilopegamys* within a context of comparison with *Colomys* and other African muroids.

Nilopegamys plumbeus Osgood, 1928. Field Mus. Nat. Hist., Zool. Ser., 12:185, pl. 15.

COMMON NAME: Ethiopian Amphibious Rat.

TYPE LOCALITY: NW Ethiopia, Gojjam, Little Abbai, a small stream that is a tributary of the Blue Nile (11°7'20"N, 37°12'40"W); see Kerbis Peterhans and Patterson (1995) for additional information.

DISTRIBUTION: Recorded only from the type locality.

STATUS: IUCN – Critically Endangered.

COMMENTS: Represented only by the type specimen, which is morphologically similar to *Colomys*. Kerbis Peterhans and Patterson (1995) enumerated the traits they considered to distinguish the two genera and noted that *N. plumbeus* has the largest cranial capacity and foramen magnum area of any African muroid sampled, close only to *Colomys*, *Malacomys*, and *Deomys*. Amphibious rodents exhibit large brains relative to body size with an enlarged medulla oblongata, which is correlated with an expansive foramen magnum (Voss, 1988). Kerbis Peterhans and Patterson (1995) contended that *N. plumbeus* has more morphological adaptations reflecting an amphibious life style than any other African muroid, including *C. goslingi*. That *plumbeus* represents a monophyletic group separate from the species of *Colomys*, however, is not universally accepted and the significance of the traits enumerated by Kerbis Pererhans and Patterson (1995) distinguishing *Nilopegamys* and *Colomys* is being tested by Hutterer and his colleagues within the context of a systematic revision of *Colomys* (R. Hutterer, in litt., 1999).

Niviventer J. T. Marshall, Jr., 1976. Family Muridae: rats and mice. Government Printing Office, Bangkok, p. 402 (See Musser, 1981*b*, for discussion of original citation).

TYPE SPECIES: *Mus niviventer* Hodgeson, 1836.

COMMENTS: *Dacnomys* Division. Diagnosed and contrasted with other Indo-Sundaic genera by Musser (1981*b*), who also reviewed morphological, chromosomal, and distributional information. Additional chromosomal data are available for Vietnamese (Bulatova et al., 1992; Baskevich and Kuznetsov, 2000), Taiwanese (H.-T. Yu et al., 1996) and Chinese species (Wang et al., 1997).

Closest phylogenetic relatives are Indochinese *Chiromyscus* and *Dacnomys*; among Sundaic genera, *Niviventer* shares dental derivations with *Berylmys*, *Leopoldamys*, and *Maxomys* (Musser, 1981*b*; Musser and Newcomb, 1983). Analyses of chromosomal data postulated similarities among *Niviventer*, *Lenothrix*, and possibly *Maxomys*, and an origin from a common ancestor (Gadi and Sharma, 1983). Analyses of allozymic and morphological data for Malay Peninsula species demonstrated substantial separation from *Rattus*, with which *Niviventer* had been merged (Ellerman, 1941, 1961; see history in Musser, 1981*b*), and alliance with *Lenothrix* in protein variation, but equivocal affinities in morphological context (Chan et al., 1979). Spermatozoal morphology equivocal in assessing phylogenetic relationships (Breed and Yong, 1986). Cladistic analysis of DNA sequence from LINE-1 elements placed *Niviventer* and *Leopoldamys* together in a clade (our *Dacnomys* Division) separate from another containing *Rattus*, *Berylmys*, *Sundamys*, and *Bandicota* (our *Rattus* Division), and a third with only *Maxomys* (Verneau et al., 1997, 1998). These kinship patterns are also supported by albumin immunology (Watts and Baverstock, 1994*b*), DNA/DNA hybridization (Chevret, 1994 [cited in Verneau, 1997]), and generally by cranial and dental traits (Musser and Newcomb, 1983). Pavlinov et al. (1995*a*) listed *Niviventer* in a *Dacnomys* Section of a more inclusive *Rattus* Group. Phallic morphology of three Chinese taxa described by Yang and Fang (1988) in context of assessing phylogenetic relationships among Chinese murines.

Recent multivariate analyses of morphometric traits for samples of *Niviventer* (Musser and Lunde, ms) substantiate the limits of 11 species (*N. andersoni*, *N. brahma*, *N. coninga*, *N. cremoriventer*, *N. culturatus*, *N. eha*, *N. excelsior*, *N. hinpoon*, *N. langbianis*, *N. lepturus*, and *N. niviventer*) that were considered clearly defined using morphological criteria (Corbet and Hill, 1992; Musser, 1981*b*; Musser and Newcomb, 1983), and helped formulate definitions of *N. confucianus*, *N. fulvescens*, and *N. tenaster* in the Indochinese region and *N. cameroni*, *N. rapit*, and *N. fraternus* on the Sunda Shelf. The analyses also identify an undescribed species from N Burma and a new genus related to *Niviventer* from C Laos; descriptions of these taxa are being prepared by Musser and colleagues.

Niviventer is represented by fossils found in China and Thailand. Evolutionary history of *N. andersoni* and *N. confucianus* in S China extends back to early Pleistocene as documented by fossils recovered from cave sediments in the Sichuan-Guizhou region (Zheng, 1993); fragments found in the same region, but at late Pliocene to early Pleistocene horizons were described as *N. preconfucianus* (Zheng, 1993). Both *N. confucianus* and *N. preconfucianus* were discovered in Pleistocene fissure deposits in the Shandong region, with the latter found in earlier horizons (Zheng et al., 1997). Late or middle Pleistocene cave layers in Guangxi Province of S China yielded three molar classes, identified as *Niviventer* sp. 1-3 (Chen et al., 2002). From middle Pleistocene cave sediments in Thailand come examples of extant *N. fulvescens* and extinct *N. gracilis* (Chaimanee, 1998). The earliest record of *Niviventer* is from 4.5 million-year-old strata (early Pliocene) in the Yushe Basin of Shanxi Province, which indicates an early origin for the *Dacnomys* Division (L. J. Flynn, in litt., 2003).

Niviventer andersoni (Thomas, 1911). Abstr. Proc. Zool. Soc. Lond., 1911(90):4.
  COMMON NAME: Anderson's Niviventer.
  TYPE LOCALITY: SW China, Sichuan, Omi San, 6000 ft (1830 m).
  DISTRIBUTION: SW China; E Xizang (Tibet), Yunnan, W Sichuan, N Ghizhou and S Shaanxi (Feng et al., 1986; Musser and Chiu, 1979; Wang, 2003).
  STATUS: IUCN – Lower Risk (lc).
  SYNONYMS: *lushuiensis* Wu and Wang, 2002 [*nomen nudum*].
  COMMENTS: Systematically revised and contrasted with *N. excelsior* and *N. confucianus* by Musser and Chiu (1979). Closest relative is apparently *N. excelsior*; the two species are set apart from other *Niviventer* by primitive traits they share, the two form a tight cluster reflected in multivariate analysis of morphometric traits (Musser and Lunde, ms), and both are isolated in the high mountains of S China (Musser, 1981*b*). This hypothesized phyletic alliance and morphological isolation from other species of *Niviventer* should be tested by analyses of molecular data. Anthony (1941) recorded *N. andersoni* from NE Burma but that series represents an undescribed species of *Niviventer* not closely related to *N. andersoni* (Musser and Lunde, ms). In his checklist of Chinese mammals, Wang (2003:204) listed *lushuiensis* as simply Wu *et* Wang, subsp. Nov. 2002," a subspecies of *N. andersoni*, but the name has yet to be published with a diagnosis and identification of holotype (D. Lunde, in litt., 2004).

Niviventer brahma (Thomas, 1914). J. Bombay Nat. Hist. Soc., 23:232.
  COMMON NAME: Brahman Niviventer.
  TYPE LOCALITY: NE India, NE Arunachal Pradesh, Anzong Valley in Mishmi Hills, 6000 ft (1830 m).
  DISTRIBUTION: NE India (NE Arunachal Pradesh), N Burma (Adung Valley and Nyetmaw River region), and SW China (NW Yunnan; Wang, 2003).
  STATUS: IUCN – Lower Risk (lc).

COMMENTS: Morphological and geographic limits outlined by Musser (1970*b*, 1973*a*, 1981*b*), who also reported that the species is represented by few specimens, and is most closely related to *N. eha* in morphological features, a conclusion echoed by Corbet and Hill (1992). We have not examined the material upon which the Chinese record is based.

Niviventer cameroni (Chasen, 1940). Bull. Raffles Mus., 15:176.
COMMON NAME: Cameron Highlands Niviventer.
TYPE LOCALITY: Malay Peninsula, Pahang, Cameron Highlands, 5000 ft (1524 m).
DISTRIBUTION: Recorded only from mountain forest in the Cameron Highlands, 5000-6600 ft (1524-2012 m; specimens in BMNH, RMBR, and USNM).
COMMENTS: Originally described as a subspecies of *Rattus rapit* (= *Niviventer rapit*) but differs from *N. rapit* in being much larger in all cranial and dental dimensions, especially the longer molar row, and lacking a conspicuously tufted tail. Whether *N. cameroni* is phylogenetically more closely related to *N. rapit* than to other species on the Sunda Shelf and in the Indochinese region needs to be determined using other morphological suites and gene sequences. *Niviventer cameroni* joins *Maxomys inas* and *Pithecheir parvus* as the three endemic murines occurring on the Malay Peninsula south of the Isthmus of Kra.

Niviventer confucianus (Milne-Edwards, 1871). Nouv. Arch. Mus. Hist. Nat. Paris, 7, Bull.:93.
COMMON NAME: Confucian Niviventer.
TYPE LOCALITY: China, Szechwan, Moupin.
DISTRIBUTION: N Burma and mainland China (from Yunnan and W Sichuan west to Fujian and north to Jilin Province; Wang, 2003, and Zhang et al., 1997); also mountains of NW Thailand (summit Doi Inthanon, Chiengmai Province) and extreme NW Vietnam (summit Mt Fan Si Pan west of the Red River); may also be found on summits of mountains in N Laos; not recorded from islands off coast of China. Because of past confusion of *N. confucianus* with *N. tenaster*, *N. niviventer*, and *N. fulvescens*, the range outlined here is derived primarily from Musser's identification of specimens in AMNH, BMNH, FMNH, MCZ, MVZ, and USNM.
STATUS: IUCN – Lower Risk (lc).
SYNONYMS: *canorus* (Thomas, 1911); *chihliensis* (Thomas, 1917); *deqinensis* Deng and Wang, 2000; *elegans* (Shih, 1931); *littoreus* (Cabrera, 1922); *luticolor* (Thomas, 1908); *mentosus* (Thomas, 1916); *naoniuensis* (Zhang and Zhao, 1984); *sacer* (Thomas, 1908); *sinianus* (Shih, 1931); *yajiangensis* Deng and Wang, 2000; *yushuensis* (Wang and Zheng, 1981); *zappeyi* (G. M. Allen, 1912).
COMMENTS: Usually included in *N. niviventer*, but that allocation is not supported by present evidence (Abe, 1983; Corbet and Hill, 1992; Musser, 1981*b*; Musser and Lunde, ms). *Niviventer confucianus* is basically endemic to SE portion of the Palearctic region east of the Himalayas and generally north of the Tropic of Cancer; south of that line, this northern species has been found only on mountaintops in NW Thailand and NW Vietnam. Sympatric with *N. fulvescens* and an undescribed larger-bodied species in N Burma, with *N. fulvescens* in S China, and with *N. andersoni* and *N. confucianus* in highlands of Sichuan. *Niviventer confucianus* is replaced westward in the Himalayas by the smaller-bodied *N. niviventer* and in mountains south of the Tropic of Cancer by the larger-bodied and montane *N. tenaster* (Musser and Lunde, ms). In NW Vietnam, *N. confucianus* is found near summit of Mt Fan Si Pan with *N. tenaster* and *N. fulvescens* occuring at lower altitudes (Musser and Lunde, ms). Westward range boundary is unresolved. Dimensions of Chakraborty's (1975) specimen of *"Rattus niviventer niviventer"* from W Bhutan are too large for *N. niviventer*, match those typical of *N. confucianus*, and may represent that species, as could the samples from NE India treated as *N. niviventer* by Agrawal (2000). If so, the westward range of *N. confucianus* would extend to Bhutan but no farther; another look at the Bhutan and NE Indian series would resolve the issue (see account of *N. niviventer*).

Morphometric analyses (Musser and Lunde, ms) indicate *N. confucianus* to be most closely related to *N. tenaster*, an alliance supported by analysis of mtDNA cytochrome *b* sequences (J. L. Patton, in litt., 2000). Geographic variation in body size and pelage coloration is conspicuous within *N. confucianus*, but its significance has yet to be re-vealed by careful systematic revision. Wang and Zheng (1981) reported results from a systematic study under the name of *Rattus niviventer*, and described *yushuensis* as a sub-species. Zhang and Zhao (1984) proposed *naoniuensis* as a subspecies. Deng et al. (2000) diagnosed *yajiangensis* from W Sichuan and *deqinensis* from NW Yunnan as sub-species of *Niviventer confucianus*. Deng et al. (2000) also suggested that ten subspecies be recognized in *N. confucianus*; one of these, *lotipes*, represents *N. tenaster* (see that ac-count), and another, *culturatus*, is treated here as a separate species (see account).

Chromosomal characteristics documented for Thai (Markvong et al., 1973) and Chinese (Wang et al., 1997) samples. Ecological relationship with *N. fulvescens* and other murines in S Yunnan reported by Wu et al. (1996). Morphology of digestive tract related to energy metabolism in *N. confucianus* (reported as *Rattus niviventer confu-cianus*) versus *Rattus norvegicus* presented by Bao et al. (1998)

Niviventer coninga (Swinhoe, 1864). Proc. Zool. Soc. Lond., 1864:185.
    COMMON NAME: Spiny Taiwan Niviventer.
    TYPE LOCALITY: Taiwan.
    DISTRIBUTION: Endemic to Taiwan, below about 2000 m and common at 1300 m (H.-T. Yu, 1993, 1994; M.-J. Yu, 1996).
    STATUS: IUCN – Lower Risk (nt) as *N. coxinga*.
    SYNONYMS: *coxinga* (Swinhoe, 1871); *coxingi* (Thomas, 1892) [unjustified emendation].
    COMMENTS: Corbet and Hill (1992:364) explained why *coninga* instead of *coxingi* is the correct name for this species. Musser (1981*b*) included a population from NE Burma, but that is a separate undescribed species; current morphometric analyses indicates *N. coninga* is not closely related to any species of *Niviventer* on the mainland (Musser and Lunde, ms), but analysis of mtDNA cytochrome *b* sequences ties *N. coninga* to *N. tenaster* within a clade also containing *N. confucianus* and *N. culturatus* (J. L. Patton, in litt., 2000). Chromosomal data documented by H.-T. Yu et al. (1996) and contrasted with karyotype of Taiwanese *N. culturatus*, which indicates that *N. coninga* and and Taiwanese *N. culturatus* are not sister-species but represent independent invasions of Taiwan from the Asian mainland, a pattern also supported by allozymic data (H.-T. Yu, 1995). Ecology, altitudinal distribution, and genetic population structure documented by H.-T. Yu (1993, 1994, 1995) and Adler (1995, 1996).

Niviventer cremoriventer (Miller, 1900). Proc. Biol. Soc. Wash., 13:144.
    COMMON NAME: Sundaic Arboreal Niviventer.
    TYPE LOCALITY: S Peninsular Thailand, Trang Province, Trang (07°30′N, 99°18′E).
    DISTRIBUTION: Peninsular Thailand and Malaya and some offshore islands; Mergui Arch., Anambas Isls; Sumatra, and smaller islands of Nias, Billiton and Banka; Borneo and some offshore islands, Java, and Bali; see Musser (1973*c*) and Corbet and Hill (1992). Does not occur north of the Isthmus of Kra (10°30′N).
    STATUS: IUCN – Lower Risk (lc).
    SYNONYMS: *barussanus* (Miller, 1911); *cretaceiventer* (Robinson and Kloss, 1919); *flaviventer* (Miller, 1900); *gilbiventer* (Miller, 1903); *kina* (Bonhote, 1903); *malawali* (Chasen and Kloss, 1932); *mengurus* (Miller, 1911); *solus* (Miller, 1913); *spatulatus* (Lyon, 1911); *sumatrae* (Bartels, 1937).
    COMMENTS: Revised by Musser (1973*c*, 1981*b*); a Sundaic endemic. Considerable variation in body size exists among samples, with those in small island populations being larger than individuals in populations on larger islands and the Malay Peninsula (Musser, 1973*c*). Tuen et al. (2000) documented *N. cremoriventer* from peninsular Mt Santubong in W Sarawak, and Md Nor (1996) recorded its distribution and ecology on the small islands off N tip of Sabah. This generally lowland species reaches 1530 m on the slopes of Mt Kinabalu in Sabah (Md Nor, 2001, and references cited therein). *Niviventer cremoriventer* joins *N. rapit* as the only two species in the genus occurring on Borneo.

Niviventer culturatus (Thomas, 1917). Ann. Mag. Nat. Hist., ser. 8, 20:198.
    COMMON NAME: Soft-furred Taiwan Niviventer.

TYPE LOCALITY: Taiwan, Mt Arizan, 8000 ft (2440 m).

DISTRIBUTION: Endemic to highlands of Taiwan, 2000-3000 m (H.-T. Yu, 1994; M.-J. Yu, 1996).

STATUS: IUCN – Lower Risk (nt).

COMMENTS: Historically listed as a species (Ellerman, 1941), a subspecies of *N. niviventer* (Ellerman, 1961; Ellerman and Morrison-Scott, 1951; Wang and Zheng, 1981), or included in *N. confucianus* (Musser, 1981*b*; Corbet and Hill, 1992). *Niviventer culturatus* is a distinctive insular form that in morphology superficially resembles mainland *N. confucianus*, but differs sufficiently that it should be treated as a species, as revealed by multivariate analysis of morphometric traits by Musser and Lunde (ms), which also clusters *N. culturatus* with the Javan *N. lepturus*, Chinese *N. andersoni* and *N. excelsior*, and Burmese *N. brahma*, but not with *N. confucianus*. Analyses of mitochondrial cytochrome *b* sequences also highlights the specific integrity of *N. culturatus* but places it with *N. confucianus*, *N. coninga*, and *N. tenaster* in a monophyletic clade (J. L. Patton, in litt., 2000; *N. langbianis* and *N. fulvescens* were the only other *Niviventer* sampled). Chromosomal data documented by H.-T. Yu et al. (1996) and compared with karyotype of *N. coninga* (see account of that species). Ecology, distribution and abundance along an elevational gradient, and genetic population structure documented by H.-T. Yu (1993, 1994, 1995); microhabitat use reported by Adler (1996).

Niviventer eha (Wroughton, 1916). J. Bombay Nat. Hist. Soc., 24:428.

COMMON NAME: Smoke-bellied Niviventer.

TYPE LOCALITY: India, Sikkim, Lachen, 8800 ft (2680 m).

DISTRIBUTION: Recorded from C and E Nepal, India (West Bengal, Sikkim, and N Assam; Agrawal, 2000, and Ellerman, 1961), N Burma (Musser, 1970*b*), and S China (S Xizang and W Yunnan; Wang, 2003) in montane forests.

STATUS: IUCN – Lower Risk (lc).

SYNONYMS: *ninus* (Thomas, 1922).

COMMENTS: Reviewed by Musser (1970*b*) and Corbet and Hill (1992). Shared morphological traits and proportions support the hypothesis of close phylogenetic relationship between *N. eha* and *N. brahma* (Musser, 1981*b*; Corbet and Hill, 1992), but this alliance should be tested by analyses of other morphological traits and gene sequences. Anthony reviewed the population from N Burma, and G. M. Allen (1940) discussed distribution and morphological traits of population in Yunnan. Zhang et al. (1997) mapped the species as occurring east of Yunnan in the provinces of Guizhou and Guangxi, but those records are so far out of the documented biogeographical range that they require verification. Biology, habitats, and aspects of morphological variation for samples from C and E Nepal reported by Abe (1971) and Gruber (1969).

Niviventer excelsior (Thomas, 1911). Abstr. Proc. Zool. Soc. Lond., 1911(90):4.

COMMON NAME: Sichuan Niviventer.

TYPE LOCALITY: SW China, W Sichuan, Tatsienlu, 9000 ft (2744 m).

DISTRIBUTION: SW China (W Sichuan, NW and C Yunnan); see Musser and Chiu (1979) and Wang (2003).

STATUS: IUCN – Lower Risk (lc).

SYNONYMS: *tengchongensis* Deng and Wang, 2002 [*nomen nudum*].

COMMENTS: A montane southwestern Chinese endemic apparently morphologically most closely related to *N. andersoni* and occurring sympatrically with it and *N. confucianus* (Musser, 1981*b*; Musser and Chiu, 1979). Early to middle Pleistocene cave sediments in the Sichuan Guizhou region of S China have yielded fossils identified as *N. cf excelsior* (Zheng, 1993). See account of *N. andersoni*. In his checklist of Chinese mammals, Wang (2003:203) listed *tengchongensis* as simply "Deng *et* Wang, subsp. Nov. 2002," a subspecies of *N. excelsior*, but the taxon has yet to be published with a diagnosis and identification of holotype (D. Lunde, in litt., 2004).

Niviventer fraternus (Robinson and Kloss, 1916). J. Strs. Br. Roy. Asiat. Soc., 73:373.

COMMON NAME: Montane Sumatran Niviventer.

TYPE LOCALITY: Indonesia, W Sumatra, Korinchi Peak, 4700 ft (1432 m).

DISTRIBUTION: Montane forest formations along mountainous backbone of W Sumatra.

SYNONYMS: *atchinensis* (Sody, 1941).

COMMENTS: Originally described as *Epimys fraternus*, but subsequently treated as a subspecies of *Rattus rapit* (Chasen, 1940), *R. orbus* (Ellerman, 1941), or *Niviventer rapit* (= Chasen's *R. rapit*; Musser, 1970b; Corbet and Hill, 1992); the latter authorities also included the Bornean *rapit* and Malay Peninsula *cameroni* as subspecies. *Niviventer fraternus* is about same size or larger than *N. rapit*, but smaller in body size than *N. cameroni*, without a tufted tail, and has a wider zygomatic plate than either *N. cameroni* or *N. rapit*. Multivariate analyses of morphometric traits cluster *N. fraternus* tightly with *N. fulvescens* and its allies and not with either *N. cameroni* or *N. rapit* (Musser and Lunde, ms). Member of a murine cluster endemic to Sumatra (Musser, 1986; also see account of *Maxomys hylomyoides*).

Niviventer fulvescens (Gray, 1847). Cat. Hodgson Coll. Br. Mus., p. 18.

COMMON NAME: Indomalayan Niviventer.

TYPE LOCALITY: Nepal.

DISTRIBUTION: From S Himalayas (Nepal and N India; see Agrawal, 2000, for Indian distribution; Corbet and Hill, 1992 extend range to N Pakistan, but we have not seen any specimens from that region) through Bangladesh, S China (Wang, 2003; including Hainan Isl and Hong Kong), and Indochina (including Con Son Isl and several other islands off coast of Vietnam; Kuznetsov, 2000; Musser and Lunde, ms) to the Sunda Shelf on Peninsular Thailand (and offshore Koh Chang), Malay Peninsula (and offshore Koh Samui), Sumatra, Java, and Bali; absent from Borneo and other islands on the Sunda Shelf. Because samples of *N. fulvescens* have been confused with those of *N. confucianus*, *N. niviventer*, and *N. tenaster* in museum collections and the literature, the range outlined here derives primarily from Musser's identification of specimens in AMNH, ANSP, BMNH, FMNH, HUNHM, MCZ, MNHN, MZB, RMBR, RMNH, USNM, and ZFMK.

STATUS: IUCN – Lower Risk (lc).

SYNONYMS: *baturus* (Sody, 1932); *besuki* (Sody, 1931); *blythi* (Kloss, 1917); *bukit* (Bonhote, 1903); *caudatior* (Hodgson, 1849); *cinnamomeus* (Blyth, 1859) [not Pictet, 1844]; *condorensis* (Kloss, 1926); *flavipilis* (Shih, 1930); *gracilis* (Miller, 1913); *huang* (Bonhote, 1905); *jacobsoni* (Bartels, 1937); *jerdoni* (Blyth, 1863); *lepidus* (Miller, 1913); *lepturoides* (Sody, 1934); *lieftincki* (Chasen, 1939); *ling* (Bonhote, 1905); *marinus* (Kloss, 1916); *mekongis* (Robinson and Kloss, 1922); *minor* (Shih, 1930); *octomammis* (Gray, 1863); *orbus* (Robinson and Kloss, 1914); *pan* (Robinson and Kloss, 1914); *temmincki* (Kloss, 1921); *treubii* (Robinson and Kiloss, 1919); *vulpicolor* (G. M. Allen, 1926); *wongi* (Shih, 1931).

COMMENTS: Some authors have referred to populations on Bali, Java, Sumatra, Malay Peninsula, and peninsular Thailand as either *N. rapit* (Chasen, 1940) or *N. bukit* (Corbet and Hill, 1992; J. T. Marshall, Jr., 1977a; Musser, 1981b), and those occurring north of the Isthmus of Kra as *N. fulvescens*, pending taxonomic revision of the group. Abe's (1983) study, which focused on Thai samples, and recent morphometric analyses combining Indochinese and Sundaic samples (Musser and Lunde, ms) support the hypothesis that specimens of *bukit* represent *N. fulvescens* (Abe, 1983), an arrangement reflecting the earlier view of Ellerman (1941) and particularly Osgood (1932:305): "The relationship of *fulvescens* to southern forms is obvious in several instances, especially in that of *R. f. bukit* which can at most be no more than a subspecies." This hypothesis will require additional testing with other data sets, including DNA sequences. The morphometric study by Musser and Lunde (ms) also demonstrates that *N. fulvescens* is part of a tight cluster containing the Thai endemic *N. hinpoon*, and the Sumatran *N. fraternus*, another phylogenetic pattern that will have to be tested by analyses of other kinds of morphological traits and gene sequences. Analysis of mtDNA cytochrome *b* sequences indicateed *N. fulvescens* to be phyletically closer to a monophyletic clade containing *N. tenaster*, *N. coninga*, *N. confucianus*, and *N. culturatus* than is *N. langbianis* (J. L. Patton, in litt., 2000; no other *Niviventer* were sampled). Neithammer and Martens (1975) regarded *fulvescens* to be only a color morph of *N. niviventer* in Nepal, but the two species are readily distinguished by pelage coloration and external, cranial, and dental dimensions (Abe, 1977; Musser and Lunde, ms).

If Indochinese and Sundaic populations all represent *N. fulvescens*, then that species is the only member of the genus with a geographic distribution encompassing the Indochinese mainland and some islands on the Sunda Shelf; other murines with roughly equivalent ranges are *Berylmys bowersii, Chiropodomys gliroides, Leopoldamys sabanus*, and *Maxomys surifer* (Musser and Newcomb, 1983; but see accounts of *L. sabanus* and *M. surifer*). Spermatozoal morphology of Malay Peninsula sample described by Breed and Yong (1986, as *bukit*) in comparative context. Chromosomal data for samples from Vietnam, Thailand, Malaya, Java, and Hong Kong summarized by Baskevich and Kuznetsov (2000). Biology, habitat, morphological variation, and taxonomy described by Abe (1971, 1977) for samples from C Nepal. Ecological relationships with *N. confucianus* and other murines in S Yunnan reported by Wu et al. (1996). Vietnamese records and other information provided by Dang et al. (1994), Kuznetsov (2000) and Lunde et al. (2003*b*). Fossils identified as *N. fulvescens* were recovered from late Pleistocene cave sediments in the Sichuan-Guizhou region of S China (Zheng, 1993).

Niviventer hinpoon (J. T. Marshall, Jr., 1976). Family Muridae: rats and mice, p. 459. Privately printed by Government Printing Office, Bangkok.
   COMMON NAME: Limestone Niviventer.
   TYPE LOCALITY: Thailand, Saraburi Province, Kaengkhoi Dist., "outside the entrance to the bat cave, half-way up the face of a wooded limestone cliff, 200 meters altitude."
   DISTRIBUTION: Endemic to Korat Plateau in Thailand (J. T. Marshall, Jr., 1977*a*; specimens in AMNH and USNM).
   STATUS: IUCN – Lower Risk (nt).
   COMMENTS: See Musser (1981*b*) for discussion of original citation. A very distinctive species (the only one with buffy underparts) that is morphometrically most closely related to *N. fulvescens* (Musser and Lunde, ms).

Niviventer langbianis (Robinson and Kloss, 1922). Ann. Mag. Nat. Hist., ser. 9, 9:96.
   COMMON NAME: Indochinese Arboreal Niviventer.
   TYPE LOCALITY: S Vietnam, Lâm Dǎng Province, Langbian Mtns (= Lam Vien Plateau) Langbian Peak, 1800-2300 m.
   DISTRIBUTION: Tropical evergreen rainforest formations in Indochina north of the Isthmus of Kra: recorded from NE India (Aranachal Pradesch; USNM 564482; Musser, 1973*c*), Burma (Musser, 1973*c*), Thailand north of Isthmus of Kra (J. T. Marshall, Jr., 1977*a*), Cardamom Mtns of SW Cambodia (A. Smith, in litt., 2002), Laos (Musser, 1973*c*), and Vietnam (Dang et al., 1994; Lunde et al., 2003*b*; specimens in AMNH and IEBR). See maps in Musser (1973*c*) and Corbet and Hill (1992). Sketch of range derived from Musser (1973*c*) and Musser's study of recently collected material in AMNH and IEBR; see also Lunde et al., 2003*b*).
   STATUS: IUCN – Lower Risk (lc).
   SYNONYMS: *indosinicus* (Osgood, 1932); *quangninhensis* (Dao and Cao, 1990); *vientianensis* (Bourret, 1942).
   COMMENTS: Morphological limits and comparisons with *N. cremoriventer*, which is similar in some external and cranial traits, were reported by Musser (1973*c*, 1981*b*); also reviewed by Corbet and Hill (1992). Dao and Cao (1990) described *quangninhensis* as a subspecies of *Rattus cremoriventer* (= *N. cremoriventer*). *Niviventer cremoriventer* occurs in Indochina only south of the Isthmus of Kra (see that account), but *N. langbianis* is still referred to as either *Rattus cremoriventer* or *Niviventer cremoriventer* in most of the current Vietnamese (Dang et al., 1994) and Chinese (Zhang et al., 1997) literature. One out of the several distinct traits characterizing *N. langbianis* is its monocolored brown tail, which is the feature used by most researchers to separate the species from specimens of *N. fulvescens*, but the tail ranges from bicolored to monocolored brown in *N. fulvescens* and those with brown tails are often misidentified as *N. langbianis*. Even examples of Indochinese *Rattus*, which generally have unpatterned monocolored brown tails, have been identified as *N. langbianis* (Musser's observations from study of museum collections). Throughout its range, *N. langbianis* is found with *N. fulvescens* in the same habitats. We have depended solely upon our study of specimens (except for the Cambodian sample) for the distribution outlined above. Kuznetsov (2000)

recorded the species (as *cremoriventer*) from three islands off the coast of N Vietnam; Wang (2003) listed it from Yunnan Province in W China (as *N. cremoriventer indosinicus*); and Zhang et al. (1997) mapped records in the provinces of Sichuan, Guizhou, and Yunnan in S China; identity of these specimens, however, has to be verified, especially those from Sichuan and Guizhou, which seem too far north and may represent *N. confucianus* with monocolored brown tails. *Niviventer langbianis* has been collected in far N Vietnam west and east of the Red River (Lunde et al., 2003*b*; Musser, 1973*c* and specimens in AMNH, IEBR, and MVZ). NE India and N Vietnam are the northern extremes of *N. langbianis* documented by specimens; whether this arboreal species also occurs in S China where tropical evergreen rainforests have not been destroyed has yet to be determined. Karyotype of sample from Dalat Plateau in Vietnam recorded by Baskevich and Kuznetsov (2000).

Niviventer lepturus (Jentink, 1879). Notes Leyden Mus., 2:17.
> COMMON NAME: Montane Javan Niviventer.
> TYPE LOCALITY: Indonesia, W Java, Gunung Gede.
> DISTRIBUTION: Endemic to montane forest in W and C Java (based upon specimens in AMNH, BMNH, MVZ, RMBR, and RMNH).
> STATUS: IUCN – Lower Risk (lc).
> SYNONYMS: *fredericae* (Sody, 1931); *maculipectus* (Sody, 1934).
> COMMENTS: Reviewed and compared with other species of *Niviventer* by Musser (1981*b*). Originally described as a species, subsequently kept that way in a "*lepturus* Group" by Ellerman (1941:197), then arranged as a subspecies of *rapit* (Chasen, 1940; Sody, 1941), and finally recognized again as a separate species and distinctive member of the endemic Javan murine fauna (Corbet and Hill, 1992; Musser, 1986; Musser and Newcomb, 1983; see account of *Kadarsanomys sodyi*). Multivariate analyses of morphometric traits cluster *N. lepturus* with the Taiwanese *N. culturatus*, Chinese *N. andersoni* and *N. excelsior*, and Burmese *N. brahma*, and not with any of the other species of *Niviventer* occurring on the Sunda Shelf (Musser and Lunde, ms).

Niviventer niviventer (Hodgson, 1836). J. Asiat. Soc. Bengal, 5:234.
> COMMON NAME: Himalayan Niviventer.
> TYPE LOCALITY: Nepal, Katmandu.
> DISTRIBUTION: Himalayas in montane habitats from N Pakistan (USNM 359793), through NW India (Uttar Pradesh), and Nepal to NE India in Sikkim and higher altitudes of West Bengal (Darjiling Dist.); 1800-3600 m (Abe, 1977; Agrawal, 2000). Range derived from Musser's study of specimens (in BMNH, HUNHM, ROM, USNM, and ZFMK), which coincides with that outlined by Corbet and Hill (1992) but not Agrawal (2000; see below).
> STATUS: IUCN – Lower Risk (lc).
> SYNONYMS: *lepcha* (Wroughton, 1916); *monticola* (Ghose, 1964); *niveiventer* (Blanford, 1891).
> COMMENTS: Recognized as a species since Ellerman (1941); *N. niviventer* later came to embrace populations extending from Nepal through Indochina, Malay Peninsula, and some islands on the Sunda Shelf (Ellerman, 1961; Ellerman and Morrison-Scott, 1951; Niethammer and Martens, 1975); some of the names associated with *N. niviventer* in those reports now identify separate species or are synonyms of others (*N. confucianus*, *N. culturatus*, *N. fulvescens*, and *N tenaster*; see those accounts). "There is yet no convincing evidence that the Nepalese populations are the same as those from areas farther east in northern Burma and China" (Musser, 1981*b*:253), or from Thailand (Abe, 1983:160), or anywhere else in Indochina or on the Sunda Shelf. Multivariate analysis of morphometric traits incorporating samples of most named forms in *Niviventer* isolates *N. niviventer* from clusters containing the four species listed above (Musser and Lunde, ms).
>
> Our study of specimens in museum collections indicates the geographic distribution of *N. niviventer* to be strictly Himalayan as outlined by Corbet and Hill (1992). Agrawal (2000) included Bhutan, NE India (Assam, Nagaland, Arunachal Pradesh, Meghalaya, and Manipur), and Burma within the range of *N. niviventer*, but material we have seen from Burma is either *N. confucianus* (in the north) or *N. tenaster* (WC and

S Burma; see those accounts), and samples from NE India are *N. fulvescens*; we have not seen Agrawal's material in ZSI. Agrawal noted that specimens from NE India are larger than those from Nepal and Sikkim, and we suspect they may represent *N. confucianus*, which is a larger animal (Musser and Lunde, ms), or even misidentified *N. fulvescens*. To finally resolve easternmost limits of *N. niviventer* and westernmost of *N. confucianus* will require reexamination of Agrawal's series from NE India. Biology, habitats, morphological variation, and taxonomy reported by Abe (1971, 1977) for samples of *N. niviventer* from C Nepal. See account of *N. confucianus*.

Niviventer rapit (Bonhote, 1903). Ann. Mag. Nat. Hist., ser. 7, 11:123.

COMMON NAME: Montane Bornean Niviventer.

TYPE LOCALITY: Malaysia, Sabah (N Borneo), Gunung Kinabalu.

DISTRIBUTION: Highlands of Borneo (specimens in BMNH, MZB, RMBR UMMZ, and USNM); found between 940 and 3360 m on the slopes of Mt Kinabalu in Sabah (Md Nor, 2001), and probably occurs throughout Borneo in mountain forests. In 1998, a large sample was collected by A. J. Gorog on Bukit Baka in Bukit Raya National Park in SW Kalimantan (specimens in UMMZ and MZB).

STATUS: IUCN – Lower Risk (lc).

COMMENTS: Reviewed and contrasted with other species of *Niviventer* by Musser (1981b), who arranged *cameroni* from the Malay Peninsula and *fraternus* from Sumatra as subspecies. Musser and Carleton (1993) noted that the "hypothesis that only one species is involved in such a disjunct insular distribution requires evaluation in context of systematic revision of *Niviventer*." Recent unpublished multivariate analyses of morphometric variation among samples of all species of *Niviventer* indicates that the three taxa are morphologically highly distinctive and that each should be treated as a separate species (Musser and Lunde, ms). In those analyses, *N. rapit* is isolated among species of *Niviventer*, does not cluster with either *N. cameroni* or *N. fraternus*, and differs from the latter two by its long and tufted tail, short molar rows, small bullae, and short and wide incisive foramina, among other traits. Member of the suite of murine species endemic to Borneo (Musser, 1986; see also account of *Chiropodomys major*).

Niviventer tenaster (Thomas, 1916). Ann. Mag. Nat. Hist., ser. 8, 17:425.

COMMON NAME: Indochinese Mountain Niviventer.

TYPE LOCALITY: S Burma (Kayin State), Mulayit (also spelled Mooleyit, Mulaiyit, Muleyit) Taung (= peak) (16°11′N/98°32′E), 5000-6000 ft (1525-1830 m).

DISTRIBUTION: Mountains of WC (Mt Victoria) and S (Mulayit Taung) Burma, NW Thailand (Doi Pui and Doi Suthep, Chiengmai Province), S Cambodia (Elephant Mtns), S Laos, Vietnam, and in China in the Tengchong region of W Yunnan (east of the Salween River bordering N Burma; specimens in IZAS, D. Lunde, in litt., 2004); also on Hainan Isl off southern coast of China; probably occurs in mountains of N Laos. Range primarily derived from specimens identified by Musser in AMNH, BMNH, FMNH, IEBR, USNM, and ZMA; also see Lunde et al. (2003b).

STATUS: IUCN – Lower Risk (lc).

SYNONYMS: *champa* (Robinson and Kloss, 1922); *lotipes* (G. M. Allen, 1926).

COMMENTS: Originally described as a species, *tenaster* was later arranged as a subspecies of *N. cremoriventer* (Ellerman, 1941, 1961; Ellerman and Morrison-Scott, 1951) until the type specimens were shown to represent a separate species (Musser, 1973c, 1981b). The large-bodied and montane *N. tenaster* is morphometrically most closely related to *N. confucianus* among species of *Niviventer* (Musser and Lunde, ms). Analysis of mtDNA cytochrome *b* sequences points to *N. coninga* as the close ally of *N. tenaster*, and clusters both within a monophyletic clade also containing *N. confucianus* and *N. culturatus* (J. L. Patton, in litt., 2000; *N. fulvescens* and *N. langbianis* were the only other *Niviventer* sampled). *Niviventer tenaster* is restricted to montane habitats in Indochina mostly beyond the range of *N. confucianus*, and is sympatric in some places with *N. fulvescens* and *N. langbianis* (Lunde et al., 2003b; Musser, 1981b; specimens in AMNH and IEBR). In NW Vietnam, *N. tenaster* occurs on the same mountain (Fan Si Pan) as *N. confucianus*, but at lower montane altitudes (Musser and Lunde, ms). Mountains in N Burma, where *N. tenaster* might also be expected, are instead occupied by *N. confucianus*,

N. fulvescens and an undescribed species larger in body size than *N. tenaster*, darker in fur coloration, and morphometrically unrelated (Musser and Lunde, ms). Robinson and Kloss (1922) described *champa* as a subspecies of *Rattus bukit*, and *lotipes* was proposed as a subspecies of *Rattus confucianus* by G. M. Allen (1926); morphometric analyses closely cluster large samples from type localities of both taxa (which includes holotypes) with samples of *N. tenaster* from elsewhere in the range (Musser and Lunde, ms). Karyotype of sample from Dalat Plateau in Vietnam recorded by Baskevich and Kuznetsov (2000).

Notomys Lesson, 1842. Nouv. Tabl. Regne Anim. Mammifères, p. 129.

    TYPE SPECIES: *Dipus mitchellii* Ogilby, 1838.

    SYNONYMS: *Ascopharynx* Waite, 1900; *Podanomalus* Waite, 1898; *Thylacomys* Waite, 1898.

    COMMENTS: *Pseudomys* Division. Member of the Australian Old Endemics (Musser, 1981*c*), which includes the Conilurini in which Baverstock (1984) placed *Notomys*. Gross and microscopical anatomy of neck glands described by Watts (1975); morphological variation in female reproductive tract documented by Breed (1985); morphology of male reproductive tract, glans penis, and spermatozoa described by Breed (1980, 1984, 1986), Breed and Sarafis (1978), and Morrissey and Breed (1982); results of electrophoretic studies presented by Baverstock et al. (1977*b*, 1981); chromosomal evolution and G-banding homologies addressed by Baverstock et al. (1977*c*, *e*, 1983*b*). Mahoney and Richardson (1988) cataloged references to taxonomy, distribution, and biology of the species.

    Members of *Notomys* form a monophyletic group diagnosed by a suite of distinctive morphological and genic traits; closest phylogenetic relatives are species of *Pseudomys* (see Lidicker and Brylski, 1987, and references therein; Watts et al., 1992); that genus, *Notomys*, and other Australian Old Endemics have been assembled in an expanded Hydromyini, which contains both Australian and some New Guinea genera and is supported by albumin immunology and spermatozoal morphology (Watts and Baverstock, 1995*b*, 1996).

    There is an undescribed species of large-bodied *Notomys* (Great Hopping Mouse) known from Recent and complete remains collected in the Flinders Ranges and Davenport Range of South Australia; it may have become extinct in the Flinders Ranges between 1850 and 1900 (Flannery, 1995*d*; Robinson et al., 2000; Watts and Aslin, 1981). K. Aplin (in litt., 2004) noted that many of the historically recorded species of *Notomys* declined so quickly that they are currently known from only a few museum specimens, often in poor condition and from widely scattered localities. Aplin views the current taxonomy of *Notomys* to seriously underestimate the real historic diversity, and states that the true diversity will only be revealed by a combination of 1) systematic survey of subfossil assemblages, 2) comprehensive DNA sampling of historical specimens, and 3) selective DNA sampling of subfossils to link the two.

Notomys alexis Thomas, 1922. Ann. Mag. Nat. Hist., ser. 9, 9:316.

    COMMON NAME: Spinifex Hopping Mouse.

    TYPE LOCALITY: Australia, Northern Territory, 35 miles (56.3 km) SW of Alroy, 800 ft (244 m); see Mahoney and Richardson (1988:166).

    DISTRIBUTION: Australia; Western Australia, Northern Territory, South Australia, and W Queensland (Breed, 1995*b*:568; Robinson et al., 2000; Watts and Aslin, 1981:109).

    STATUS: IUCN – Lower Risk (lc).

    SYNONYMS: *everardensis* Finlayson, 1940; *reginae* Troughton, 1936.

    COMMENTS: Of all the species of *Notomys*, *N. alexis* has the most extensive geographic range (Watts and Aslin, 1981). Variation in sperm head morphology documented by Breed and Sarafis (1983) and its phylogenetic significance discussed by Breed (1997). Studies on sperm storage in vas deferens and comparison with other Notomys provided by Peirce et al. (2003). Digestive tract traits and relation to diet documented by Murray et al. (1994). Reviewed by Watts and Aslin (1981) and Breed (1995*b*).

Notomys amplus Brazenor, 1936. Mem. Nat. Mus. Melb., 9:7.

    COMMON NAME: Short-tailed Hopping Mouse.

    TYPE LOCALITY: Australia, Northern Territory, Charlotte Waters.

DISTRIBUTION: Australia; S Northern Territory and N South Australia (see Watts and Aslin, 1981:104); recorded as subfossils from South Australia (Robinson et al., 2000).

STATUS: IUCN – Extinct.

COMMENTS: Known by only two extant specimens from the type locality (Watts and Aslin, 1981) and a skin collected during the last century from Burt Plain near Alice Springs (in the Australian Museum; T. Flannery, in litt., 2002), but also represented by owl pellet deposits from Flinders Ranges of South Australia (Dixon, 1995b); apparently extinct (Mahoney and Richardson, 1988; Watts and Aslin, 1981). Reviewed by Watts and Aslin (1981b) and Dixon (1995d).

Notomys aquilo Thomas, 1921. Ann. Mag. Nat. Hist., ser. 9, 8:540.

COMMON NAME: Northern Hopping Mouse.

TYPE LOCALITY: Australia, Queensland, Cape York.

DISTRIBUTION: Australia; N Queensland and N Northern Territory (Groote Eylandt and N Arnhem Land); Watts and Aslin (1981:113).

STATUS: U.S. ESA – Endangered; IUCN – Lower Risk (nt).

SYNONYMS: *carpentarius* Johnson, 1959.

COMMENTS: Found only in three coastal or near-coastal areas of ridges around the Gulf of Carpentaria (Flannery, 1995c). Distribution, habitat, and conservation status of *N. aquilo* documented by Woinarski et al. (1999).

Notomys cervinus (Gould, 1853). Proc. Zool. Soc. Lond., 1851:127 [1853].

COMMON NAME: Fawn Hopping Mouse.

TYPE LOCALITY: Australia, "Interior of South Australia" (as restricted by Thomas's lectotype designation; see Mahoney and Richardson, 1988:167).

DISTRIBUTION: Australia; SW Queensland, NW New South Wales, South Australia, and S Northern Territory (Ellis, 1993; Robinson et al., 2000; Watts, 1995a:574; Watts and Aslin, 1981:99).

STATUS: IUCN – Lower Risk (nt).

SYNONYMS: *aistoni* Brazenor, 1934.

COMMENTS: For date of publication see Mahoney and Richardson (1988:167). Reviewed by Watts and Aslin (1981) and Watts (1995a).

Notomys fuscus (Jones, 1925). Rec. S. Aust. Mus., 3:3.

COMMON NAME: Dusky Hopping Mouse.

TYPE LOCALITY: Australia, South Australia, Ooldea Dist. (as restricted by lectotype designation; see Mahoney and Richardson, 1988:168).

DISTRIBUTION: Inland Australia; SE Western Australia, S Northern Territory, South Australia, and SW Queensland (Robinson et al., 2000; Watts and Aslin, 1981:115; Watts, 1995b:576). Also in W New South Wales (Watts, in litt.).

STATUS: IUCN – Vulnerable.

SYNONYMS: *eyreius* Finlayson, 1960; *filmeri* Mack, 1961.

COMMENTS: Reviewed by Watts and Aslin (1981) and Watts (1995b). Distribution, habitat, and conservation status documented by Moseby et al. (1999).

Notomys longicaudatus (Gould, 1844). Proc. Zool. Soc. Lond., 1844:104.

COMMON NAME: Long-tailed Hopping Mouse.

TYPE LOCALITY: Australia, Western Australia, Moore River (as restricted by Thomas's lectotype designation; see Mahoney and Richardson, 1988:168).

DISTRIBUTION: Australia: Western Australia and Northern Territory (Watts and Aslin, 1981:107; Dixon, 1995c:577); extinct but recorded as subfossils from South Australia (Robinson et al., 2000). "Given the wide range of locations that live animals were collected from, and the increasing number of locations where its remains have been found, this species must have once occupied much of the arid and semi-arid zones of western and central Australia." (Ellis, 1995:40).

STATUS: IUCN – Extinct.

SYNONYMS: *sturti* Thomas, 1921.

COMMENTS: No living animals have either been seen or trapped since 1901, and the species is apparently extinct (Watts and Aslin, 1981). Dixon (1995c:578) noted that "Once widespread throughout arid and semiarid country where the vegetation included acacia and eucalypt woodlands, hummock grassland and low shrubland, the Long-tailed Hopping-mouse is now unknown and possibly extinct." Male reproductive anatomy and spermatozoal morphology described by Breed (1990).

Notomys macrotis Thomas, 1921. Ann. Mag. Nat. Hist., ser. 9, 8:538.

COMMON NAME: Big-eared Hopping Mouse.
TYPE LOCALITY: Australia, Western Australia, Moore River.
DISTRIBUTION: Known only from the type locality (Dixon, 1995*d*:578).
STATUS: IUCN – Extinct.
SYNONYMS: *megalotis* Iredale and Troughton, 1934.
COMMENTS: Represented only by the holotype and paratype from "Australia" (Mahoney, 1975). Apparently extinct. Closest phylogenetic relative is probably *Notomys cervinus* (Mahoney, 1975). Reviewed by Watts and Aslin (1981) and Dixon (1995*d*).

Notomys mitchellii (Ogilby, 1838). Lond. Edinb. Philos. Mag. J. Sci., 12:96.

COMMON NAME: Mitchell's Hopping Mouse.
TYPE LOCALITY: Australia, Victoria, about 12 km SE Lake Boga (see Mahoney and Richardson, 1988:169).
DISTRIBUTION: Australia; S Western Australia, S South Australia, and W Victoria (Bennett and Lumsden, 1995:213; Robinson et al., 2000; Watts, 1995*c*:579; Watts and Aslin, 1981:118); once occurred in SW New South Wales but is now apparently extinct there (Mahoney and Richardson, 1988:170).
STATUS: IUCN – Lower Risk (lc).
SYNONYMS: *alutacea* Brazenor, 1934; *gouldi* (Gould, 1863); *macropus* Thomas, 1921; *richardsonii* (Gray, 1844).
COMMENTS: The largest living species of *Notomys*; reviewed by Watts and Aslin (1981) and Watts (1995*c*).

Notomys mordax Thomas, 1922. Ann. Mag. Nat. Hist., ser. 9, 9:317.

COMMON NAME: Darling Downs Hopping Mouse.
TYPE LOCALITY: Australia, SE Queensland, Darling Downs.
DISTRIBUTION: Known only from the type locality.
STATUS: IUCN – Extinct.
COMMENTS: Still represented only by the the skull of the holotype (Mahoney, 1977). Apparently extinct (Mahoney and Richardson, 1988; Watts, 1995*d*; Watts and Aslin, 1981). Provenance of the skull was questioned by Mack (1961), but Mahoney (1977) claimed there were no adequate reasons for doubting that it came from Darling Downs and considered *N. mordax* a valid species closely related to *N. mitchelli*. Watts and Aslin (1981:121) remarked that "it is not possible to be sure that this one skull really represents a distinct species, or whether it is simply that of a large specimen of Mitchell's hopping-mouse." K. Aplin (in litt., 2004) wrote that a subfossil sample obtained recently from near Coonabarabran (NE New South Wales) contains good material of a *Notomys* similar to *N. mitchelli* from W Victoria, but with slightly smaller teeth (*N. mitchelli* has smaller molars than exhibited by the larger-toothed *N. mordax*; Mahoney, 1977). Habitat around the site is not dissimilar to that of Darling Downs. Various possibilities are being considered: 1) that the Coonabarabran material is referable to *mordax*, with the holotype of *mordax* being an individual with unusually large teeth; 2) that two species of *Notomys* were found in this wider region, namely *mordax* and a smaller, *mitchelli*-like taxon; and 3) that all of the N New South Wales and SE Queensland *Notomys*, including the holotype of *mordax*, are referable to *N. mitchelli*. Aplin also noted a possible recent sight record of a live hopping mouse from the Pilliga region (NE New South Wales); this record is now being seriously investigated by personnel of the New South Wales National Parks Service.

Oenomys Thomas, 1904. Ann. Mag. Nat. Hist., ser. 7, 13:416.

TYPE SPECIES: *Mus hypoxanthus* Pucheran, 1855.
SYNONYMS: *Aenomys* Kershaw, 1923.
COMMENTS: *Oenomys* Division. A phylogenetic analysis of 17 cranial and 23 dental traits related

Oenomys most closely to African *Thamnomys* (Lopez-Martinez et al., 1998), a link previously suggested by Hatt (1940*a*), and *Lamottemys*. A morphometric study using a Fourier analysis applied to outlines of first upper and lower molars underscored the existence of the two species discussed below and the presence of possibly two other undescribed species (Renaud, 1999). DNA/DNA hybridization experiments placed *Oenomys* in a clade with *Arvicanthis*, *Dasymys*, *Aethomys*, and *Hybomys* (Chevret, 1994). Isolated molars of *Oenomys* have been uncovered from Pliocene and Pleistocene sediments in East Africa (Jaeger, 1976; Sabatier, 1982; Wesselman, 1984; see review by Denys, 1999).

Oenomys hypoxanthus (Pucheran, 1855). Rev. Mag. Zool. Paris, ser. 2, 7:206.
COMMON NAME: Common Oenomys.
TYPE LOCALITY: Gabon.
DISTRIBUTION: Tropical forest block from S Nigeria south to W and NE Angola (Crawford-Cabral, 1998), and east across Dem. Rep. Congo (incl. islands of Zaire River between Kisangani and Kinshasa; Colyn and Dudu, 1986) to Rwanda, Burundi, and Uganda; also isolated forest patches in S Sudan, SW Ethiopia (Yalden et al., 1996), Kenya and W Tanzania (see section of map east of Ghana in Dieterlen and Rupp, 1976).
STATUS: IUCN – Data Deficient.
SYNONYMS: *albiventris* Eisentraut, 1968; *anchietae* (Bocage, 1890); *bacchante* (Thomas, 1903); *editus* Thomas and Wroughton, 1910; *marungensis* (Noack, 1887); *moerens* Thomas, 1911; *oris* Thomas, 1911; *rufinus* (Matschie, 1895) [not Temminck, 1855]; *talangae* Setzer, 1956; *unyori* (Thomas, 1903); *vallicola* Heller, 1914.
COMMENTS: Another African species showing appreciable geographic variation in fur color and body size (Dieterlen and Rupp, 1976; Rosevear, 1969; Thomas, 1915). Chromosomal data reported by Matthey (1963, 1967*a*) and Maddalena et al. (1989). Palatal ridges of sample from Mount Oku described and contrasted with *Lamottemys okuensis* (Fülling, 1992). Altitudinal distribution on Ugandan slopes of Ruwenzori Mtns reviewed by Kerbis Peterhans et al. (1998).

Oenomys ornatus Thomas, 1911. Ann. Mag. Nat. Hist., ser. 8, 7:378.
COMMON NAME: West African Oenomys.
TYPE LOCALITY: Ghana, Bibianaha, near Dunkwa.
DISTRIBUTION: Records are from SE Guinea (Mt Nimba; Tranier and Gautun, 1979), E Sierra Leone, and S Ghana (Grubb et al., 1998); probably also occurs in Côte d'Ivoire and Liberia in suitable forest habitats.
STATUS: IUCN – Lower Risk (lc).
COMMENTS: Originally described by Thomas (1911*b*) as a species, and subsequently either listed that way (G. M. Allen, 1939; Ellerman, 1941) or included in *O. hypoxanthus* (Misonne, 1974). Rosevear (1969), however, recognized *ornatus* as a distinctive subspecies of *O. hypoxanthus*, and Tranier and Gautun (1979) reinstated its specific uniqueness as indicated by chromosomal, morphological, and distributional attributes. Populations in Sierra Leone and Ghana reviewed by Grubb et al. (1998). Member of murine fauna endemic to West Africa (see account of *Grammomys buntingi*).

Palawanomys Musser and Newcomb, 1983. Bull. Am. Mus. Nat. Hist., 174:335.
TYPE SPECIES: *Palawanomys furvus* Musser and Newcomb, 1983.
COMMENTS: *Rattus* Division. Phylogenetic affinities unclear, but among murines native to the Sunda Shelf, *Palawanomys* appears morphologically most closely allied to the cluster of genera that includes *Rattus*; broader regional comparisons required before more precise affinities can be determined (Musser and Newcomb, 1983).

Palawanomys furvus Musser and Newcomb, 1983. Bull. Am. Mus. Nat. Hist., 174:335.
COMMON NAME: Palawan Mountain Rat.
TYPE LOCALITY: Philippines, Palawan Isl, Brooke's Point Municipality, Mt Mantalingajan, 4500 ft (1370 m).
DISTRIBUTION: Greater Palawan Faunal Region (politically part of the Philippines but faunistically an insular extension of the Sunda Shelf). Known only from the type locality.

STATUS: IUCN – Endangered.

COMMENTS: Recorded only by four melanistic examples, possibly highly restricted in distribution (Heaney et al., 1998; Musser and Newcomb, 1983). One of three murines endemic to the Greater Palawan Faunal Region (*Maxomys panglima* and *Chiropodomys calamianus* are the other two).

Papagomys Sody, 1941. Treubia, 18:322.

TYPE SPECIES: *Mus armandvillei* Jentink, 1892.

COMMENTS: *Rattus* Division. Formerly thought closely related to *Mallomys* but has little affinity to that New Guinea Old Endemic and instead is phylogenetically related to the living *Komodomys rintjanus* and Pleistocene *Hooijeromys nusatenggara*, both endemics of Nusa Tenggara (Musser, 1981c) and members of the *Rattus* Division. This association is indicated by cranial and dental morphology (Musser and Newcomb, 1993) along with albumin immunology (Watts and Baverstock, 1994b). Pavlinov et al. (1995a) listed *Papagomys* and *Komodomys* in a *Pithecheir* Section of a more inclusive *Micromys* Group. *Hooijeromys* is the only rat found in late Pleistocene sediments on Flores (0.8-0.7 million years ago) and is part of an extinct depauperate insular fauna that also consisted of the living *Varanus komodoensis* (Komodo dragon), the elephantid *Stegodon florensis*, and *Homo erectus* (Sondaar et al., 1994; Van den Bergh et al., 2001). The two species of *Papagomys* were reviewed by Musser (1981c).

Papagomys armandvillei (Jentink, 1892). Weber's Zool. Ergebn., 3:79, pl. 5.

COMMON NAME: Armandville's Papagomys.

TYPE LOCALITY: Indonesia, Nusa Tenggara (Lesser Sunda Isls), Pulau Flores.

DISTRIBUTION: Recorded only from Flores Isl.

STATUS: IUCN – Vulnerable.

SYNONYMS: *besar* Hooijer, 1957; *verhoeveni* Hooijer, 1957.

COMMENTS: Known by extant specimens as well as subfossil fragments (3000-4000 years old).

Papagomys theodorverhoeveni Musser, 1981. Bull. Am. Mus. Nat. Hist., 169:95.

COMMON NAME: Theodore Verhoeven's Papagomys.

TYPE LOCALITY: Indonesia, Nusa Tenggara (Lesser Sunda Isls), Pulau Flores, Menggarai Province, Liang Toge, a cave near Warukia, 1 km south of Lepa.

DISTRIBUTION: Known only from Flores Isl.

STATUS: IUCN – Extinct (but see comments).

COMMENTS: Recorded only as subfossil fragments (3000-4000 years old), but possibly still living on Flores.

Parahydromys Poche, 1906. Zool. Anz., 30:326.

TYPE SPECIES: *Limnomys asper* Thomas, 1906.

SYNONYMS: *Drosomys* Thomas, 1906; *Limnomys* (Thomas, 1906) [not Mearns, 1905].

COMMENTS: *Hydromys* Division. Member of the New Guinea Old Endemics (Musser, 1981c). Thought to be closely related to species of *Hydromys* judged by cranial and dental morphology (Tate, 1951; Flannery, 1989; Musser's study of specimens) phallic traits (Lidicker, 1968), immunological distances (Watts and Baverstock, 1994), and spermatozoal structure (Breed, 1997; Breed and Aplin, 1994). Based upon albumin immunology and spermatozoal traits, *Parahydromys* joins *Hydromys* and *Leggadina* in a larger clade containing members of our *Xeromys*, *Pseudomys*, and *Uromys* Divisions (the "Australasian clade," Watts and Baverstock, 1996), and not the clades formed by our *Pogonomys* or *Lorentzimys* Divisions (Watts and Baverstock, 1994a, 1995b, 1996; Breed and Aplin, 1994; Breed, 1997). However, analyses of multiple mitochondrial and nuclear DNA sequences disassociate *Leggadina* from *Parahydromys* and *Hydromys* and cluster it with a *Pseudomys-Notomys* clade (F. Ford, in litt., 2004; see *Leggadina* generic account).

Parahydromys asper (Thomas, 1906). Ann. Mag. Nat. Hist., ser. 7, 17:326.

COMMON NAME: New Guinea Waterside Rat.

TYPE LOCALITY: Papua, New Guinea, Central Province, Owen Stanley Range, Richardson Range, Mt Gayata, 2000-4000 m.

DISTRIBUTION: New Guinea; Central Cordillera from the Arfak Mountains and Weyland Range in Prov. of Papua (= Irian Jaya) to E flanks of the Owen Stanley Range in Papua New Guinea, as well as Huon Peninsula (Flannery, 1995a:251).

STATUS: IUCN – Lower Risk (lc).

COMMENTS: Flannery (1990b, 1995a) reviewed the little distributional and biological information recorded for the species, which unlike species of *Hydromys*, is not amphibious. Donnellan (1987) reported chromosomal data. Cole et al. (1997) discussed a specimen from the Mt Dayman region at the E end of the species' range. Aplin et al. (1999) reported a specimen of this species from a Holocene archaeological site on the Ayamaru Plateau, central Bird's Head Peninsula of Prov. of Papua (= Irian Jaya).

Paraleptomys Tate and Archbold, 1941. Am. Mus. Novit., 1101:1.

TYPE SPECIES: *Paraleptomys wilhelmina* Tate and Archbold, 1941.

COMMENTS: *Hydromys* Division. Member of the New Guinea Old Endemics (Musser, 1981c). Based on phallic and some cranial traits (Lidicker, 1968; Tate, 1951), *Paraleptomys* has traditionally been phylogenetically linked to *Leptomys*, which is more closely allied to *Pseudohydromys* (which includes *Mayremys* and *Neohydromys*), and *Xeromys* than to any member of the *Hydromys* Division (see account of *Leptomys*). Although cranial conformation is similar in *Paraleptomys* and *Leptomys*, the former is more like *Hydromys* in possessing a primitive murine cephalic arterial pattern (derived configuration in *Leptomys*), lacking third upper and lower molars (retained in *Leptomys*), and having basined occlusal patterns on the remaining molars (much more elaborate occlusal surfaces in *Leptomys*). We suggest that *Paraleptomys* is a terrestrial representative of our *Hydromys* Division (*Hydromys, Parahydromys, Microhydromys,* and *Crossomys*), and not our *Xeromys* Division (*Xeromys, Leptomys,* and *Pseudohydromys*); this hypothesis requires testing with data from additional morphological and molecular sources.

Paraleptomys rufilatus Osgood, 1945. Fieldiana Zool., 31:1.

COMMON NAME: Northern Paraleptomys.

TYPE LOCALITY: New Guinea, Prov. of Papua (= Irian Jaya), Cyclops Mtns, Mt Dafonsero, 4700 ft (1432 m).

DISTRIBUTION: NC New Guinea; known only from montane forest formations (1200-1800 m) on N coastal ranges in Prov. of Papua (= Irian Jaya) on Mt. Dafonsero, Cyclops Mtns (Flannery, 1995a; Osgood, 1945) and adjacent Papua New Guinea on Mt. Somoro, Torricelli Mtns (Flannery, 1995a:253), and Mt. Menawa, Bewani Mtns (specimens in BBM, K. Helgen, pers. comm.).

STATUS: IUCN – Lower Risk (lc).

COMMENTS: A distinct species (Flannery, 1990b; Osgood, 1945) that along with the murine *Microhydromys musseri*, monotreme *Zaglossus attenboroughi*, and marsupials *Dendrolagus scottae* and *Petaurus abidi* is endemic to the northern coastal ranges (Flannery, 1990b, 1995a).

Paraleptomys wilhelmina Tate and Archbold, 1941. Am. Mus. Novit., 1101:1.

COMMON NAME: Central Cordilleran Paraleptomys.

TYPE LOCALITY: New Guinea, Prov. of Papua (= Irian Jaya), near Mt Wilhelmina, Snow Mtns (Pegunungan Maoke), 9 km NE Lake Habbema, 2800 m.

DISTRIBUTION: C New Guinea; known only from N slopes of Snow Mtns between Idenburg River and Mt Wilhelmina in Prov. of Papua (= Irian Jaya) (Tate and Archbold, 1941) and the Tifalmin Valley in W Papua New Guinea (Flannery and Seri, 1990:189; Flannery, 1995a); altitudinal range, 1800-2800 m (specimens in AMNH).

STATUS: IUCN – Vulnerable.

COMMENTS: A common part of the montane small mammal fauna on north slopes of the Snow Mtns (Musser and Lunde, ms). More than two species may be represented in available samples: series from above 2200 m on north slopes of the Snow Mtns and Tifalmin Valley represent *P. wilhelmina*, but specimens from below 2200 m on north slopes of the Snow Mtns likely sample a morphologically closely related but separate species (K. Helgen, in litt., 2003).

Paramelomys Rümmler, 1936. Z. Säugetierk., 10:248.

TYPE SPECIES: *Uromys levipes* Thomas, 1897.

COMMENTS: *Uromys* Division. Originally described as a subgenus of *Melomys* by Rümmler (1936) and either retained as a synonym of that genus (Musser and Carleton, 1993; Tate, 1951) or ignored (Laurie and Hill, 1954) until Menzies (1996) rediagnosed *Paramelomys* and raised it to generic rank (using principally multivariate analyses of continuous variation in morphological traits). Included are nine species whose shared traits, according to Menzies, isolate them as a monophyletic group endemic to New Guinea and separate from the species of *Melomys*. Analysis of albumin immunology of some species also supports their exclusion from *Melomys* (Watts and Baverstock, 1994a). All species revised by Menzies (1996).

Paramelomys gressitti Menzies, 1996. Aust. J. Zool., 44:407.

COMMON NAME: Gressitt's Paramelomys.

TYPE LOCALITY: Papua New Guinea, Morobe Province, Wau region, near summit of Mt Kaindi, 2300 m.

DISTRIBUTION: Known only from 2300 m on Mt Kaindi and 2400 m on Mt Garaina (80 km from Mt Kaindi), both in Morobe Province. In addition to the five specimens in the type series listed by Menzies (1996:407), five others have also been collected on Mt Kaindi (USNM 357443-357445, 357488, and 357489).

COMMENTS: A very distinctive species restricted to montane forest formations and apparently morphologically related to *P. moncktoni* and *P. lorentzii* (Menzies, 1996).

Paramelomys levipes (Thomas, 1897). Ann. Mus. Civ. Stor. Nat. Genova, 18:617.

COMMON NAME: Papuan Lowland Paramelomys.

TYPE LOCALITY: New Guinea, Papua New Guinea, Central Province, Sogeri Plateau, Haveri, 700 m (additional information provided by Laurie and Hill, 1954:121, and Menzies, 1989).

DISTRIBUTION: Papua New Guinea, S Central Province; the only specimens we have seen are from the Sogeri Plateau and Astrolabe Range near Port Moresby, below about 700 m. Leary and Seri (1997) reported two examples at Omo (06°58'41" S 144°18'15" E), 170 m in the Kikori River Basin in S Papua New Guinea. Menzies (1996:415) gave the range as "Lowlands of eastern New Guinea, probably not over 1200 m. Lack of specimens prevents defining precise distribution."

STATUS: IUCN – Lower Risk (lc) as *Melomys levipes*.

COMMENTS: A lectotype was designated by Menzies (1989), but Rümmler (1938) had already indicated which of Thomas' two cotypes should be considered the holotype. Our examination of series indicated that true *P. levipes* is documented by the holotype and Tate's (1951:291) series from Baruari and Itiki in the Astrolabe Range (other material Tate listed as *P. levipes* are examples of either *P. platyops* or *P. mollis*). All other records usually associated with *P. levipes* (Laurie and Hill, 1954; Tate, 1951) represent either *Mammelomys lanosus* and *M. rattoides* or *Paramelomys lorentzii* and *P. mollis*. In morphology and altitudinal distribution, *P. levipes* is morphologically very similar to *P. naso* from SW lowlands of Prov. of Papua (= Irian Jaya) and may represent the SE Papuan New Guinea relative of that species. A skull only from New Britain (AMNH 194397) is cranially and dentally similar to *P. levipes*, but has slightly wider molars. Flannery (1995b) also examined that specimen and referred it to *P. levipes*. Neither Flannery nor we are sure about this identification; better material and a larger series are needed to determine actual identity of the skull.

Paramelomys lorentzii (Jentink, 1908). Nova Guinea, 9:3.

COMMON NAME: Lorentz's Paramelomys.

TYPE LOCALITY: New Guinea, Prov. of Papua (= Irian Jaya), Lorentz River, Resi Camp, 900 m.

DISTRIBUTION: New Guinea; specimens are from lowlands along the south side of Central Cordillera, from Mimika River in SW Prov. of Papua (= Irian Jaya) to middle altitudes and lowlands of the upper and middle Fly River; sea level to 780 m (based on specimens examined in AMNH and BMNH); limits unknown.

STATUS: IUCN – Lower Risk (lc) as *Melomys lorentzii*.

COMMENTS: Usually listed as a subspecies of *Melomys levipes* [= *Peramelomys lorentzii*] (Laurie and Hill, 1954; Tate, 1951), *lorentzii* is a distinct species morphologically related to *P. moncktoni* (Menzies, 1996, and our observations). *P. lorentzii* and *P. moncktoni* have not yet been recorded as sympatric, but they both occur along the Fly River in S Papua New Guinea. *Paramelomys lorentzii* was collected by members of the Archbold 1936-1937 Expedition along the upper Fly River and downriver to Lake Daviumbu in the middle Fly River; *P. moncktoni* was collected opposite Sturt Isl on the lower Fly River, about 115 direct air miles (185 km) southeast of Lake Daviumbu.

Paramelomys mollis (Thomas, 1913). Ann. Mag. Nat. Hist., ser. 8, 12:210.

COMMON NAME: Montane Soft-furred Paramelomys.

TYPE LOCALITY: New Guinea, Prov. of Papua (= Irian Jaya), Nassau Range, upper Utakwa River, S slope Mount Carstenz, Camp Padang, 6c, 5500 ft (1676 m).

DISTRIBUTION: New Guinea; scattered in montane forest throughout Central Cordillera from Arfak Mtns in Vogelkop and Weyland Range of Prov. of Papua (= Irian Jaya) to Mt Dayman at end of the Owen Stanley Range in extreme E Papua New Guinea; above 1200 m to 2500 m or more; not recorded from N coastal ranges or the Huon Peninsula (Flannery, 1995a; Menzies, 1996; and specimens we studied).

STATUS: IUCN – Lower Risk (lc) as *Melomys mollis*.

SYNONYMS: *arfakianus* (Rümmler, 1935); *clarae* (Rümmler, 1935); *meeki* (Rümmler, 1935); *stevensi* (Rümmler, 1935); *weylandi* (Rümmler, 1935).

COMMENTS: Specimens representing *P. mollis* have been identified as *P. levipes* in the literature (usually as *Melomys levipes*), and the synonyms listed here have also been associated with that species. However, judged by Musser's study of specimens in AMNH and BMNH, *P. mollis* is a distinct species tied to montane forest formations and differs from *P. levipes* in habitat and morphology. Specimens from W Prov. of Papua (= Irian Jaya) are more darkly pigmented than those from E Papua New Guinea, but series from Papua New Guinea Central and Eastern highlands bridge this chromatic gap. Menzies's (1996:416) concept of the species is much the same as ours (Musser and Carleton, 1993) and he noted, "As this rat has a wide altitudinal range. . . and occurs on discontinuous mountain ranges, some geographical variation is to be expected."

Paramelomys moncktoni (Thomas, 1904). Ann. Mag. Nat. Hist., ser. 7, 14:399.

COMMON NAME: Monckton's Paramelomys.

TYPE LOCALITY: Papua New Guinea, Northern Province, NE coast, 8°30'S, 148°20'E (Kumusi River).

DISTRIBUTION: Papua New Guinea; reliable records are from coastal plains and foothills (sea level to 1400 m) of SE Papua New Guinea, from Wewak on the NE coast eastward to S lowlands where westernmost record is from Sturt Isl camp on the lower Fly River (from our study of specimens, and Menzies, 1996); also on Sideia Isl, just off the E coast near Milne Bay. Flannery (1995b) recorded a specimen from Yapen Isl in NW Prov. of Papua (= Irian Jaya) and mapped two localities on the mainland of SW Prov. of Papua (= Irian Jaya) (Flannery, 1995a); these should be reexamined to determine if the former is a *P. platyops* and the latter *P. lorentzii*, or if indeed they are *P. moncktoni*.

STATUS: IUCN – Lower Risk (lc) as *Melomys moncktoni*.

SYNONYMS: *sturti* (Tate, 1951).

COMMENTS: Horizontal and altitudinal distributions of this species have been misunderstood due to incorrect identifications of specimens. *Paramelomys moncktoni* was thought to have a primarily southern distribution from Prov. of Papua (= Irian Jaya) to Papua New Guinea, and to extend up into moss forest (see map and discussion in Flannery, 1990b:225), but all samples from moss forest represent other species. Examples of *P. moncktoni* come only from the restricted range described above (based upon series in AMNH and BMNH; also see Menzies, 1996).

The forms *intermedius* and *shawi* are usually associated with *P. moncktoni* (Flannery, 1990b; Laurie and Hill, 1954; Tate, 1951, under *Melomys*), but the holotype of *shawi* is a *P. rubex* and the type series of *intermedius* belongs to *P. platyops* (Musser's study of specimens and Menzies, 1996). Musser and Carleton (1993) incorrectly allocated *sturti* to *P. lorentzii*, and Menzies (1996) correctly reassociated it with *P. moncktoni*. Possibly the E New Guinea ecological and phylogenetic equivalent of the SW New Guinea *P. lorentzii*.

Paramelomys naso (Thomas, 1911). Ann. Mag. Nat. Hist., ser. 8, 7:386.

COMMON NAME: Long-nosed Paramelomys.

TYPE LOCALITY: New Guinea; SW Prov. of Papua (= Irian Jaya), foothills of Nassau Range, Whitewater Camp on the Kafari (=Kaparé) River (see Laurie and Hill, 1954:121).

DISTRIBUTION: New Guinea; lowlands of SW Prov. of Papua (= Irian Jaya) and Weyland Range, limits unknown on mainland; also on Wokam Isl in the Aru Isl Group (Menzies, 1995; Flannery, 1995b, identified the specimen from Wokam as *P. lorentzii*).

COMMENTS: Treated as a synonym of *Melomys levipes lorentzii* by Rümmler (1938) and as a subspecies of *M. levipes* by Tate (1951) and Laurie and Hill (1954). Musser and Carleton (1993) incorrectly included *naso* in *P. lorentzii* but the two are sympatric (Menzies, 1996). Morphologically and probably phylogenetically most closely related to *P. levipes* in the lowlands of SE Papua New Guinea. An adult from the Weyland Range (AMNH 101957) collected at 5000 ft (1524 m), much higher than any other record, has all the characteristics of *P. naso* (including one hair per tail scale); it was collected along with examples of montane *P. mollis* and *P. rubex*.

Paramelomys platyops (Thomas, 1906). Ann. Mag. Nat. Hist., ser. 7, 17:327.

COMMON NAME: Common Lowland Paramelomys.

TYPE LOCALITY: SE Papua New Guinea, Central Province, head of Aroa River.

DISTRIBUTION: New Guinea, throughout lowlands and mid-mountain altitudes on the mainland except seasonally dry savanna forests of the southern lowlands; altitudinal range from sea level to 1500 m. Also occurs on islands of Yapen and Biak in Prov. of Papua (= Irian Jaya); New Britain in the Bismarck Arch.; and the islands of Normanby, Fergusson, and Goodenough in the D'Entrecasteaux Arch. (Flannery, 1995a; Menzies, 1996; specimens in AMNH).

STATUS: IUCN – Lower Risk (lc) as *Melomys platyops*.

SYNONYMS: *fuscus* (Rümmler, 1935); *intermedius* (Rümmler, 1935); *jobiensis* (Rümmler, 1935); *mamberanus* (Sody, 1937).

COMMENTS: This species was thought to range primarily throughout N New Guinea (see map in Flannery, 1990b:224) but our reidentification of museum specimens and holotypes, along with Menzies's revision (1996) reveals otherwise. The form *intermedius* was originally described as a subspecies of *Melomys moncktoni* (see Rümmler, 1938), but members of type series from Utakwa River (type locality) in SW Prov. of Papua (= Irian Jaya) have only one hair per scale (all *P. moncktoni* have three hairs per scale), as does *P. platyops*, and their cranial, dental, and other external traits are also characteristic of *P. platyops*, not *P. moncktoni*. Geographic variation in body size exists among samples of *P. platyops*, especially lowland versus highland samples (Menzies, 1996, and our observations) and mainland versus island populations (Flannery, 1995b), and a careful systematic revision of the species is required to assess its significance. Leary and Seri (1997) reported specimens and habitat information in the Kikori River Basin of S Papua New Guinea.

Paramelomys rubex (Thomas, 1922). Ann. Mag. Nat. Hist., ser. 9, 9:263.

COMMON NAME: Mountain Paramelomys.

TYPE LOCALITY: New Guinea, Prov. of Papua (= Irian Jaya), Mamberano River, Doormanpad-bivak, 1410 m.

DISTRIBUTION: New Guinea; montane forests in the Central Cordillera from the Arfak Mtns on the Vogelkop of W Prov. of Papua (= Irian Jaya) to Mt Dayman at the end of the Owen Stanley Range in E Papua New Guinea; also in the N coastal Torricelli Mtns and Huon Peninsula; altitudinal range, 900-3000 m (specimens we examined; Menzies, 1996; see map in Flannery, 1995a).

STATUS: IUCN – Lower Risk (lc) as *Melomys rubex*.

SYNONYMS: *alleni* (Rümmler, 1935); *arfakiensis* (Rümmler, 1935); *clarus* (Rümmler, 1935); *pohlei* (Rümmler, 1935); *rutilus* (Rümmler, 1935); *shawi* (Tate and Archbold, 1935); *stresemanni* (Rümmler, 1935); *tafa* (Tate and Archbold, 1935).

COMMENTS: The many names applied to *P. rubex* reflect morphological variation among samples that is concordant with interrupted highland distributions (Menzies, 1996, and our study of specimens in various museums). Menzies (1996) revised the species and earlier (Menzies, 1974) commented on the status of some names associated with *P. rubex*.

Paramelomys steini (Rümmler, 1935). Z. Säugetierk., 10:111.
COMMON NAME: Stein's Paramelomys.
TYPE LOCALITY: E Prov. of Papua (= Irian Jaya), Sumuri Mtn, Weyland Mtns, 2000-2600 m.
DISTRIBUTION: Recorded only from the type locality.
COMMENTS: Similar to *P. rubex* in fur coloration, but averages larger in most dimensions except for its shorter and broader hind feet (Menzies, 1996). *Paramelomys rubex* occurs at lower elevations (originally reported as *shawi* from 1200 m) in the Weyland Mtns.

Paruromys Ellerman, 1954. *In* Laurie and Hill, List of land mammals of New Guinea, Celebes and adjacent islands, p. 117.
TYPE SPECIES: *Rattus dominator* Thomas, 1921.
COMMENTS: *Rattus* Division. Described by Ellerman as a subgenus of *Rattus*, but now recognized as distinct genus (Musser and Newcomb, 1983; Musser, 1984). Membership in the *Rattus* Division is supported by morphological data (Musser and Newcomb, 1983) and albumin immunology (Watts and Baverstock, 1994*b*). Morphological comparisons with Sundaic *Sundamys*, also in the *Rattus* Division, recorded by Musser and Newcomb (1983).

Paruromys dominator (Thomas, 1921). Ann. Mag. Nat. Hist., ser. 9, 7:244.
COMMON NAME: Giant Sulawesi Rat.
TYPE LOCALITY: Indonesia, N Sulawesi, Minahassa, Mt Masarang, 4000 ft (1220 m).
DISTRIBUTION: Sulawesi; Tropical evergreen lowland and montane rainforests throughout the island, from sea level to tree-line.
STATUS: IUCN – Endangered as *P. ursinus*, otherwise Lower Risk (lc).
SYNONYMS: *frosti* (Ellerman, 1949); *ursinus* (Sody, 1941).
COMMENTS: Musser and Newcomb (1983) reviewed the taxonomic allocations of *dominator* from the time it was originally described as a species of *Rattus* (Thomas, 1921*a*), through its allocation to genus *Taeromys* (Sody, 1941) and use as type-species of subgenus *Paruromys* in *Rattus* (Ellerman, *in* Laurie and Hill, 1954), up to its inclusion in subgenus *Bullimus* in *Rattus* (Misonne, 1969). Spermatozoal morphology of *P. dominator* is unlike species of *Rattus* or any other species for which data from spermatozoal morphology are available (Breed and Musser, 1991), and stomach morphology is also unique among sampled murines (Musser and Durden, 2002). Musser (1971*b*) documented the association of Ellerman's *frosti* with *P. dominator*. Sody's (1941) *ursinus*, based upon specimens from the upper slopes of Gunung Lampobatang on the SE peninsula of Sulawesi, was described as a subspecies of *Taeromys dominator*, later included in the synonymy of that species (Musser, 1984), and subsequently recognized as a separate species (Musser and Holden, 1991; Downing et al., 1998). Recent multivariate analysis of cranial and dental traits by Musser (ms.), however, support the earlier inclusion of *ursinus* in *P. dominator*. *Paruromys dominator*, *Maxomys muschenbroekii*, and *Rattus hoffmanni* are the only Sulawesian murines occurring over the entire island in most forest formations (Musser and Holden, 1991).

    Paruromys dominator is represented by subfossil fragments (Holocene) from lowlands of the SW peninsula (Musser, 1984) and by two lower molars recovered from late Pliocene-early Pleistocene sediments in the Walanae Formation (Downing et al., 1998). This is the earliest record of a murine from Sulawesi and joins a distinctive extinct Pleistocene fauna consisiting of giant tortoises, crocodiles, the pig *Celebochoerus*, and the elephants *Stegodon sompoensis* and *Elephas celebensis*; except for *Paruromys*, this unique assemblage is unlike the modern fauna (see review by Van den Bergh et al., 2001).

Paulamys Musser, 1986. *In* Musser et al., Am. Mus. Novit., 2850:2.
TYPE SPECIES: *Floresomys naso* Musser, 1981.
SYNONYMS: *Floresomys* Musser, 1981 [not Fries et al., 1955].
COMMENTS: *Rattus* Division. Phylogenetically most closely related to Sulawesian *Bunomys* as assessed by morphology (Kitchener et al., 1991*a*), and the Timorese *Rattus timorensis* as

inferred from albumin immunology (Watts and Baverstock, 1994b); included in *Bunomys* by Corbet and Hill (1992). This postulated affinity between *Paulamys* and *Bunomys* needs testing by study of more extant specimens of *P. naso* in a phylogenetic context that would also compare the sample with species of native New Guinea, Nusa Tenggara, and Australian *Rattus*, along with Sulawesi *Bunomys*. Musser's *Floresomys* is invalid because of Fries et al.'s (1955) *Floresomys*, an Eocene geomyoid (McKenna and Bell, 1997:181).

Paulamys naso (Musser, 1981). Bull. Am. Mus. Nat. Hist., 169:112.
COMMON NAME: Paula's Long-nosed Rat.
TYPE LOCALITY: Indonesia, Nusa Tenggara (Lesser Sunda Isls), Pulau Flores, Menggarai Province, Liang Toge, cave near Warukia, 1 km S Lepa.
DISTRIBUTION: Recorded only on Flores Isl.
STATUS: IUCN – Extinct (but see comments).
COMMENTS: Originally described from subfossil fragments (Musser, 1981c; Musser et al., 1986), but an extant specimen collected in tropical rainforest at 1600 m from SC Flores was referred to this species by Kitchener et al. (1991a), who suggested it is closely related to Sulawesian *Bunomys*. Apparently *P. naso* is not uncommon in forested habitats between 1000 and 2000 m on Gunung Ranaka in West Flores (Kitchener and Yani, 1998; Kitchener et al., 1998).

Pelomys Peters, 1852. Bericht Verhandl. K. Preuss. Akad. Wiss. Berlin, 17:275.
TYPE SPECIES: *Mus (Pelomys) fallax* Peters, 1852.
SYNONYMS: *Komemys* De Beaux, 1924.
COMMENTS: *Arvicanthis* Division. Definition and phylogenetic position of *Pelomys* need to be reassessed in context of a systematic revision of arvicanthine murines. In overall morphology, the genus is most closely related to *Mylomys* and *Desmomys*, which in turn are members of a group also containing species of *Arvicanthis*, *Lemniscomys*, and *Rhabdomys* (Musser, 1987b), an alliance corroborated by mtDNA sequences of cytochrome *b* and 12S and 16S rRNA gene fragments (Ducroz et al., 2001). The latter study indicated *Pelomys* to be most closely related to *Mylomys* (Ducroz et al., 2001) and some authors regard them as congeneric (e. g., Heim de Balsac and Bellier, 1967). Analysis of microcomplement fixation of albumin groups *Pelomys* with *Lemniscomys*, *Rhabdomys*, *Grammomys*, and *Thallomys* (Watts and Baverstock, 1995a). *Komemys* is usually treated as a subgenus for *P. hopkinsi* and *P. isseli* (e.g., Delany, 1975), but we recognize it here as a synonym of *Pelomys*. *Pelomys* is represented by Pliocene and Pleistocene fossils from East Africa (Jaeger, 1976; Jaeger and Wesselman, 1976; Wesselman, 1984), South Africa (de Graaff, 1961), Algeria (de Bruijn et al., 1970; Jaeger, 1975), and outside of Africa on Mediterranean Rhodes Isl (de Bruijn et al., 1970, 1996); see review by Denys (1999). The early Pliocene record of *Pelomys* from Afghanistan (Sen et al., 1979) is an example of the extinct *Parapelomys* (Brandy et al., 1980).

Pelomys campanae (Huet, 1888). Le Naturaliste, ser. 2, 10(31):143.
COMMON NAME: Angolan Pelomys.
TYPE LOCALITY: W Angola, Landana.
DISTRIBUTION: W Angola (Crawford-Cabral, 1983) and W Dem. Rep. Congo.
STATUS: IUCN – Lower Risk (lc).
COMMENTS: A distinctive species occurring either sympatrically or parapatrically with *P. fallax* in parts of its range (Crawford-Cabral, 1983).

Pelomys fallax (Peters, 1852). Bericht Verhandl. K. Preuss. Akad. Wiss. Berlin, 17:275.
COMMON NAME: East African Pelomys.
TYPE LOCALITY: Mozambique, Caya Dist., Zambezi River.
DISTRIBUTION: Savanna habitats from S Kenya (Hollister, 1919) and SW Uganda (Delany, 1975) through Tanzania (Swynnerton and Hayman, 1951), E and S Dem. Rep. Congo, Angola (Crawford-Cabral, 1998), Zambia (Ansell, 1978), Malawi (Ansell and Dowsett, 1988), and Mozambique (Smithers and Lobao Tello, 1976), to E and NW Zimbabwe (Smithers and Wilson, 1979) and N Botswana (Smithers, 1971).
STATUS: IUCN – Lower Risk (lc).
SYNONYMS: *australis* Roberts, 1913; *concolor* Heller, 1912; *frater* Thomas, 1904; *insignatus*

Osgood, 1910; *iridescens* Heller, 1912; *luluae* Matschie, 1926; *rhodesiae* Roberts, 1929; *vumbae* Roberts, 1946.

COMMENTS: Appreciable variation in size and fur color exists between samples from Angola and Zambia and those from the rest of the geographic range of *P. fallax*, suggesting more than one species may be present in this complex. No extant records are from South Africa but the species occurred in KwaZulu-Natal more than 17,000 years before present when the region was covered with deciduous woodland instead of thornveld (Avery, 1991). Southern African Subregion population reviewed by de Graaff (1981, 1997*a*) and Skinner and Smithers (1990).

Pelomys hopkinsi Hayman, 1955. Rev. Zool. Bot. Afr., 52:323.
COMMON NAME: Hopkins' Pelomys.
TYPE LOCALITY: SW Uganda, Kigezi, Rwamachuchu.
DISTRIBUTION: Rwanda, Uganda, and SW Kenya (Bekele and Schlitter, 1989).
STATUS: IUCN – Vulnerable.
COMMENTS: Morphologically similar to *P. isseli*, but significant distinguishing traits suggest that *hopkinsi* and *isseli* should be viewed as separate species (Bekele and Schlitter, 1989).

Pelomys isseli (de Beaux, 1924). Ann. Mus. Civ. Stor. Nat. Genova, 51:207.
COMMON NAME: Lake Victoria Pelomys.
TYPE LOCALITY: Uganda, Lake Victoria, Kome Isl.
DISTRIBUTION: Uganda; endemic to islands of Kome, Bugala, and Bunyama in Lake Victoria (Bekele and Schlitter, 1989; Delany, 1975).
STATUS: IUCN – Vulnerable.
COMMENTS: Morphologically closely related to *P. hopkinsi*.

Pelomys minor Cabrera and Ruxton, 1926. Ann. Mag. Nat. Hist., ser. 9, 17:601.
COMMON NAME: Least Pelomys.
TYPE LOCALITY: Dem. Rep. Congo, Kasaï-Occidental, Kananga (= Luluabourg).
DISTRIBUTION: NE Angola (Crawford-Cabral, 1998), NW Zambia (Ansell, 1978), S and E Dem. Rep. Congo, and W Tanzania (Swynnerton and Hayman, 1951).
STATUS: IUCN – Lower Risk (lc).
COMMENTS: A distinct species; aspects of its morphology resemble some species of *Lemniscomys* (e.g., *L. griselda*).

Phloeomys Waterhouse, 1839. Proc. Zool. Soc. Lond., 1839:108.
TYPE SPECIES: *Mus* (*Phloeomys*) *cumingi* Waterhouse, 1839.
COMMENTS: *Phloeomys* Division. Part of the Philippine Old Endemics; phylogenetic relationships relative to genera in other areas of Indo-Australian region unresolved (Musser and Heaney, 1992). Originally included with the Asian *Nesokia* (in reality, a phylogenetic close relative of *Bandicota* and *Rattus*) in the Phloeomyinae by Alston (1976) because of their similar laminar molar occulusal patterns, but listed as the only genus in that subfamily by Thomas (1896). Considered a member of Phloeomyinae by Tate (1936) and Simpson (1945), along with *Chiropodomys, Coryphomys, Crateromys, Lenomys, Mallomys,* and *Pogonomys,* but except for *Crateromys,* no data supports such an allocation (Ellerman, 1949*a*; our research). Phylogenetic analyses of complete mitochondrial cytochrome b sequences for 13 of the 16 genera of endemic Philippine murines join *Phloeomys, Crateromys,* and *Batomys* in a clade separate from other Old Endemic genera (Jansa and Heaney, 2001).

Phloeomys cumingi (Waterhouse, 1839). Proc. Zool. Soc. Lond., 1839:108.
COMMON NAME: Southern Luzon Phloeomys.
TYPE LOCALITY: Philippines, Luzon Isl.
DISTRIBUTION: Greater Luzon Faunal Region on S Luzon, Marinduque, and Catanduanes Isls (Heaney et al., 1991, 1998; Musser and Heaney, 1992; Oliver et al., 1993).
STATUS: IUCN – Vulnerable.
SYNONYMS: *albayensis* (Elera, 1895) [*nomen nudum* according to Ellerman, 1941:293]; *elegans* (Cabrera, 1901).

COMMENTS: Sometimes considered conspecific with *P. pallidus*, but *P. cumingi* is a distinct species (Corbet and Hill, 1992; Musser and Heaney, 1992; Thomas, 1898*b*). Standard karyotype (2n = 44, FN = 66) reported by Rickart and Musser (1993), G-banding pattern described by Rickart and Heaney (2002). Elevational and ecological summary for S Luzon population provided by Heaney et al. (1999).

Phloeomys pallidus Nehring, 1890. Sitzb. Ges. Naturf. Fr. Berlin, p. 106.
    COMMON NAME: Northern Luzon Phloeomys.
    TYPE LOCALITY: Philippines, Luzon Isl.
    DISTRIBUTION: Greater Luzon Faunal Region. Endemic to N and C Luzon; widespread in primary and secondary tropical forests from sea level to at least 2000 m (see Oliver et al. [1993] and Heaney et al. [1998], and references cited therein).
    STATUS: IUCN – Lower Risk (nt).
    COMMENTS: Standard karyotype (2n = 40, FN = 60) described by Jotterand-Bellomo and Schauenberg, 1988, as *P. cumingi*); a minimum of two Robertsonian and three non-Robertsonian events separate karyotypes of the two *Phloeomys* (Rickart and Musser, 1993). This species is becoming common in zoos.

Pithecheir Lesson, 1840. Species des Mammifères, p. 264.
    TYPE SPECIES: *Pithecheir melanurus* Lesson, 1840.
    SYNONYMS: *Pithechir* Müller, 1840 [*nomen nudum*; see Ellerman and Morrison-Scott, 1951, and Corbet and Hill, 1992].
    COMMENTS: *Pithecheir* Division. The name is often attributed to Cuvier 1833 (e.g., Ellerman, 1941; Musser and Carleton, 1993), "which was a vernacular description of 'Pithecheir melanure' and has no validity" (Ellerman and Morrison-Scott, 1955:39). Reviewed by Emmons (1993*b*) in context of describing Bornean *Pithecheirops*. *Pithecheir* is an endemic of the Sunda Shelf. Musser and Newcomb (1983) suggested on the basis of cranial and dental traits that *Pithecheir* is distantly related to *Lenothrix*, which is supported by molar occlusal patterns (Misonne, 1969) but unsupported by chromosomal data (Yong et al., 1982) and spermatozoal morphology (Breed and Yong, 1986). Chaimanee's (1998) cladistic analysis of molar traits, however, placed *Pithecheir*, *Lenothrix*, and *Vandeleuria* in the same clade and speculated that it, along with the Palearctic *Micromys*, may have been the earliest evolutionary cluster to have diverged from the main murine lineage as represented by a *Progonomys*-like ancestor. Ellerman (1949:132) had also indicated a close relationship between *Vandeleuria* and *Pithecheir*. Except that it is unquestionably morphologically and phylogenetically most closely related to the Bornean *Pithecheirops*, alliances to Indomalayan genera other than *Lenothrix* have yet to be uncovered, and the association of *Pithecheir* with *Lenothrix* and *Vandeleuria* has to be tested by analyses of molecular and additional morphological data. Members of our *Pithecheir* Division are Sundaic (*Lenothrix*, *Pithecheir* and *Pithecheirops*) or Sulawesian (*Eropeplus*, *Lenomys* and *Margaretamys*) and may prove to be most closely allied to members of the *Micromys* Division (*Micromys*, *Chiropodomys*, *Vandeleuria*, *Haplomys*, *Vernaya*, and *Haeromys*), which are Palearctic, Indomalayan, and Sulawesian in geographic distributions. Pavlinov et al. (1995*a*) listed *Pithecheir* in a *Pithecheir* Section of a more inclusive *Micromys* Group.
    Present and past distribution and radiation of species in *Pithecheir* have centered on the Sunda Shelf south of the Isthmus of Kra (10°30' N). In addition to the living species, isolated molars identified as the extinct *P. peninsularis* were recovered from early and middle Pleistocene cave sediments in peninsular Thailand south of the Isthmus (Chaimanee, 1998).

Pithecheir melanurus Lesson, 1840. Species des Mammifères, p. 265.
    COMMON NAME: Javan Pithecheir.
    TYPE LOCALITY: Indonesia, Java (Bengal or Sumatra have also been proposed as type localities; see Corbet and Hill, 1992, for history and restriction to Java).
    DISTRIBUTION: Java only (see Musser, 1982*d*:76).
    STATUS: IUCN – Lower Risk (nt).
    SYNONYMS: *melanurus* Müller, 1840 [*nomen nudum*; see Corbet and Hill, 1992].
    COMMENTS: Represented by few specimens collected before World War II. Bartels' (1937)

observations on habitat and habits remain the most complete report of this Javan endemic, which is a member of the suite of murine rodents unique to Java (Musser, 1986; see account of *Kadarsanomys sodyi*).

Pithecheir parvus Kloss, 1916. J. Fed. Malay St. Mus., 6:250.
    COMMON NAME: Malay Peninsula Pithecheir.
    TYPE LOCALITY: Malay Peninsula, Selangor, Bukit Kutu, near Kuala Kubu, 3,400 ft (1035 m).
    DISTRIBUTION: Malay Peninsula (Pahang and Selangor).
    STATUS: IUCN – Lower Risk (lc).
    COMMENTS: Originally described as a subspecies of *P. melanurus*, but *parvus* is a distinctive species endemic to Malay Peninsula (Musser and Newcomb, 1983; Muul and Lim, 1971). Karyotype has same diploid number as *Hapalomys longicaudus*, but is uninformative about inferring phylogenetic relationships (Yong, et al., 1982). Spermatozoal morphology very distinctive (no apical hook), unlike that of any other Sundaic endemic (Breed and Yong, 1986), and resembling sperm of Sulawesian *Lenomys* and *Eropeplus* (Breed and Musser, 1991). Joins *Maxomys inas* and *Niviventer cameroni* in being the only recorded endemic murines on the Malay Peninsula south of the Isthmus of Kra (Musser, 1986; see account of *Niviventer cameroni*). Distribution of *P. parvus* has been restricted to south of the Isthmus of Kra since early Pleistocene. Isolated molars identified as this species have been recovered from early and middle Pleistocene cave sediments in peninsular Thailand south of the Isthmus (Chaimanee, 1998).

Pithecheirops Emmons, 1993. Proc. Biol. Soc. Wash., 106:752.
    TYPE SPECIES: *Pithecheirops otion* Emmons, 1993.
    COMMENTS: *Pithecheir* Division. Closely related to *Pithecheir* and morphologically similar to species in that genus except for small details of molar occlusal patterns, and its much smaller and uninflated bullae, all of which were noted in the original description and comparisons. The contrasting middle ear features not clearly emphasized are: blade of manubium nearly parallel in *Pithecheirops* (perpendicular in *Pithecheir*), orbicularis apophysis present (absent in *Pithecheir*), pars flaccida present (absent in *Pithecheir*), and as a consequence of the latter, the posterodorsal rim of ectotympanic is incomplete (complete in *Pithecheir*). These states in *Pithecheirops* are likely plesiomorphic relative to those in *Pithecheir*, which agrees with the difference in extent of bullar inflation between the two. Whether these comparisons contrast a more ancestral species with derived ones within a monophyletic group (single genus), or whether *Pithecheirops* actually represents a different monophyletic clade (two genera) will have to be determined by surveying adults within a phylogenetic study employing as many data sets as possible and a broad range of Indomalayan genera.

Pithecheirops otion Emmons, 1993. Proc. Biol. Soc. Wash., 106:753.
    COMMON NAME: Bornean Pithecheirops.
    TYPE LOCALITY: Malaysia, Sabah (N Borneo), Danum Valley Field Centre, 150 m (see Emmons, 1993*b*, for more information).
    DISTRIBUTION: Known only from the type locality.
    COMMENTS: Described from a single juvenile collected in "dense viny roadside secondary brush on an abandoned logging road" in what was primary lowland dipterocarp forest, which still persists "within 600 m and is the dominant vegetation type of the entire surrounding region" (Emmons, 1993*b*:753). Member of a group of murines endemic to Borneo (see account of *Chiropodomys major*).

Pogonomelomys Rümmler, 1936. Z. Säugetierk., 11:248.
    TYPE SPECIES: *Melomys mayeri* Rothschild and Dollman, 1932.
    COMMENTS: *Pogonomys* Division. Member of the New Guinea Old Endemics (Musser, 1981*c*). Revised by Menzies (1990). Originally described as a subgenus of *Melomys*, but now regarded as a distinct genus (Laurie and Hill, 1954; Menzies, 1990; Tate, 1951). Its phylogenetic relationships to *Melomys* and other related genera in the *Uromys* cluster still require resolution by multi-trait morphological and molecular inquiries. Our examina-

tion of skins and skulls, however, suggests *Pogonomelomys* is not closely related to any member in the *Uromys* Division. Spermatozoal morphology is also unlike that of *Uromys* and most of its allies (Musser and Breed, ms), but resembles that of some of the genera we list in our *Pogonomys* Division (Breed and Aplin, 1994; Musser and Breed, ms), where we provisionally place *Pogonomelomys*. Aplin et al. (1999) reported on material of a likely undescribed species of *Pogonomelomys* from Holocene archaeological sites on the Ayamaru Plateau, central Bird's Head Peninsula of Prov. of Papua (= Irian Jaya).

Pogonomelomys bruijni (Peters and Doria, 1876). Ann. Mus. Civ. Stor. Nat. Genova, 8:336.
    COMMON NAME: Bruijn's Pogonomelomys.
    TYPE LOCALITY: W New Guinea, Prov. of Papua (= Irian Jaya), Pulau Salawati, off W coast of Vogelkop.
    DISTRIBUTION: New Guinea; known only by 11 specimens from type locality and Vogelkop mainland in Prov. of Papua (= Irian Jaya); lower Fly River, and low altitudes on Mt Bosavi and Mt Sisa in Papua New Guinea; not recorded higher than about 60 m (Menzies, 1990; Flannery, 1995a:312).
    STATUS: IUCN – Critically Endangered.
    SYNONYMS: *brassi* Tate and Archbold, 1941.
    COMMENTS: The lowland, larger-bodied morphological and phylogenetic counterpart of the highland *P. mayeri*. Aplin et al. (1999) reported material of this species from late Pleistocene to mid-Holocene archaeological sites on the Ayamaru Plateau, central Bird's Head Peninsula of Prov. of Papua (= Irian Jaya).

Pogonomelomys mayeri (Rothschild and Dollman, 1932). Abstr. Proc. Zool. Soc. Lond., 1932(353):14.
    COMMON NAME: Shaw Mayer's Pogonomelomys.
    TYPE LOCALITY: New Guinea, Prov. of Papua (= Irian Jaya), Weyland Range, Gebroeders Mtns, 5000 ft (1524 m).
    DISTRIBUTION: New Guinea; mountains from Weyland Range in Prov. of Papua (= Irian Jaya) to Wau region and Huon Peninsula in Morobe Province of Papua New Guinea; not known from the Vogelkop in Prov. of Papua (= Irian Jaya) or the Owen Stanley Range in E Papua New Guinea; 400-1500 m.
    STATUS: IUCN – Lower Risk (lc).
    COMMENTS: Phallic morphology described by Lidicker (1968). Specimens of *P. mayeri* and *P. bruijni* were collected on Mt Sisa in the Kikori River Basin of S Papua New Guinea (Leary and Seri, 1997), but altitudinal relationships between the two there have not been published.

Pogonomys Milne-Edwards, 1877. C. R. Acad. Sci. Paris, 85:1081.
    TYPE SPECIES: *Mus* (*Pogonomys*) *macrourus* Milne-Edwards, 1877.
    COMMENTS: *Pogonomys* Division. Member of the New Guinea and Australian Old Endemics (Musser, 1981c). Analysis of immunological distances by Watts and Baverstock (1994a) indicated *Pogonomys* to be related to *Chiruromys*, *Anisomys*, *Coccymys*, *Hyomys*, and possibly *Mallomys*. Tate's (1936) inclusion of *Pogonomys* (with *Chiruromys* as a subgenus) in the Phloeomyinae along with *Chiropodomys*, *Crateromys*, *Lenomys*, *Mallomys*, and *Phloeomys* (followed by Simpson, 1945) has no merit except for the link with *Mallomys* (Ellerman, 1949a; our research). A chromosomal and morphometric study, which separated species of *Pogonomys* from those of *Chiruromys*, was offered by Dennis and Menzies (1979). Additional chromosomal data reported by Donnellan (1987). *Pogonomys fergussoniensis* is endemic to the D'Entrecasteaux Isls; all other species are recorded only from mainland localities on New Guinea and Australia.
        We list five species of *Pogonomys*, another undescribed species is endemic to the Snow Mtns in Prov. of Papua (= Irian Jaya) (Musser and Lunde, in ms.), and there is possibly a seventh that has yet to be named occurring in NE coastal Queensland (in rainforests of Cape York Peninsula and farther south in the wet tropics between Cooktown and Townsville; Watts and Aslin, 1981, and Winter and Whitford, 1995). Mahoney and Richardson (1988:170) catalogued taxonomic, distributional, and biological references for the Australian sample, which was identifed as *P. mollipilosus* by Watts and Aslin (1981) and reviewed

under that name by Winter and Whitford (1995). The holotype of *mollipilosus*, however, was obtained near Daru on the south coast of the Trans-Fly region of S New Guinea and is an example of *P. macrourus*, which is known only from mainland New Guinea (see account of that species). The Australian *Pogonomys* has a much larger body and longer tail than does *P. macrourus* (compare measurements for the Australian sample listed in Winter and Whitford, 1995:643, with those of New Guinea *P. macrourus* given by Flannery, 1995*a*), dark brownish gray upperparts and pure white underparts (bright reddish brown dorsal fur in *P. macrourus*), and does not appear to represent *P. macrourus*. Its body size, tail length, and fur coloration recall the New Guinea *P. loriae*, a generally montane inhabitant (see account of that species), but it averages smaller in those external dimensions (contrast measurements listed by Winter and Whitford with those for *P. loriae* presented by Flannery, 1995*a*:316). Furthermore, it seems biogeographically implausible that *P. loriae* also occurs in NE Queensland. No other species of nonvolant mammal that is endemic to the Australian-New Guinea region exhibits such a distribution. The typically Australian species that also occur in New Guinea are known only from the Trans-Fly region, not the Central Cordillera or outlying mountain ranges (Norris and Musser, 2001). The Australian *Pogonomys* is most likely a separate species from the New Guinea representatives, and its alliance needs to be determined by careful comparison with samples of *P. macrourus* and *P. loriae*; at that time perhaps the holotype of *mollipilosus* should be reexamined. We are satisfied that it represents *P. macrourus* (as is explained in the account of that species), but Dennis and Menzies (1979:322) were not as sure and cautioned that "The possibility that *mollipilosus* is an Australian species occurring in New Guinea only in the lower Fly River region cannot be ignored."

Pogonomys championi Flannery, 1988. Rec. Aust. Mus., 40:333.
    COMMON NAME: Champion's Pogonomys.
    TYPE LOCALITY: Papua New Guinea, Sandaun Provomce, Telefomin Valley, Ofektaman, 1400 m (see Flannery, 1988, for additional information).
    DISTRIBUTION: Papua New Guinea; known only from Telefomin and Tifalmin valleys between 1400 and 2300 m (Flannery, 1988).
    STATUS: IUCN – Vulnerable.
    COMMENTS: Apparently ecologically replaces the smaller-bodied *P. sylvestris*, but is morphologically similar to *P. macrourus* (Musser and Lunde, ms.).

Pogonomys fergussoniensis Laurie, 1952. Bull. British Mus. (Nat. Hist.) Zool., 1:299.
    COMMON NAME: D'Entrecasteaux Archipelago Pogonomys.
    TYPE LOCALITY: Papua New Guinea, D'Entrecasteaux Isls, Fergusson Isl, Awabula District, Saibutu (Laurie gave "Faralulu district" and "Taibutu", which are incorrect).
    DISTRIBUTION: Recorded (only by five specimens) from Fergusson Isl (holotype and paratype, Laurie, 1952:299, and AMNH 157601), Goodenough Isl (AMNH 157613), and Normanby Isl (AM M20309, Flannery, 1995*b*) in the D'Entrecasteaux group off the NE end of Papua New Guinea.
    COMMENTS: Recognized as a species in Laurie and Hill's (1954) checklist, but treated as a distinct subspecies of *P. loriae* by Dennis and Menzies (1979) in their morphometric revision of *Pogonomys* and *Chiruromys*, and finally included in *P. loriae* (Flannery, 1990*b*, 1995*a*; Musser and Carleton, 1993). Flannery (1995*b*:149) listed it under *P. loriae* in his account of the *Pogonomys* on the D'Entrecasteaux Isls, but noted that "*fergussoniensis* is very distinctive, being very large and having a reddish rather than grey dorsum. I suspect that further studies will reveal it to represent a distinct species." *Pogonomys fergussoniensis* can be distinguished from all samples of *P. loriae* by its significantly larger body and skull size (see measurements in Dennis and Menzies, 1979:320), brownish red upperparts, buffy gray underparts, and sleek fur (dark brownish gray dorsal coat that is thick and woolly and white underparts in *P. loriae*). Furthermore, Dennis and Menzies (1979) found appreciable separation between samples of *fergussoniensis* and *loriae* in their discriminant analysis results.

Pogonomys loriae Thomas, 1897. Ann. Mus. Civ. Stor. Nat. Genova, 18:613.
    COMMON NAME: Loria's Pogonomys.

TYPE LOCALITY: Papua New Guinea, Central Province, mountains behind Astrolabe Range, near Mt Wori, Haveri, 700 m (Laurie and Hill, 1954:96, provided details).

DISTRIBUTION: New Guinea; Recorded only from the Vogelkop and Weyland Range in Prov. of Papua (= Irian Jaya), but widespread in Papua New Guinea from the Telefomin area in the west through the Central Cordillera to the E flanks of the Owen Stanley Range in the east; also found in the N coastal Torricelli Mtns, the Huon Peninsula and the upper Fly River drainage; 100-3000 m (Flannery, 1995a and Dennis and Menzies, 1979:328).

STATUS: IUCN – Lower Risk (lc).

SYNONYMS: *dryas* Thomas, 1904.

COMMENTS: Until the revision by Dennis and Menzies (1979), this species was known as *P. mollipilosus* (Tate, 1951), but the holotype of that taxon is a *P. macrourus* (see account of that species). Dennis and Menzies (1979) also included *fergussoniensis* from the D'Entrecasteaux Isls, which we separate as a distinct species (see that account). *Pogonomys loriae* has traditionally been considered to be montane, occurring "generally over 1500 m," (Dennis and Menzies, 1979:330), which is generally accurate but there are interesting exceptions. Five specimens listed by Tate (1951) from the upper Fly River near the foothills of the Central Cordillera were collected at about 100 m, which is the lowest record of the species. Flannery (1995a:316) noted that "In northern New Guinea it is restricted to a narrow elevational band of between approximately 200 and 800 metres. Above this, it is replaced by *Pogonomys macrourus*." This is unusual because elsewhere *P. macrourus* is found in lowlands and mid-elevations and is usually replaced at higher elevations by *P. loriae*. Most of the distributional records of *P. loriae* are from Papua New Guinea. The species has not been recorded from the Snow Mtns (Pegunungan Maoke), that vast Central Cordillera of Prov. of Papua (= Irian Jaya), despite intensive collecting activity in places. For example, members of the 1938-1939 Archbold Expedition to the N slope of the Snow Mtns obtained about 2400 mammals representing 68 indigenous species of echidna, marsupials, bats, and rodents, but did not encounter *P. loriae* (Musser and Lunde, ms). If the absence of *P. loriae* from the Snow Mtns is real and not an artifact of collecting, its overall distribution would be roughly concordant with that described for *P. sylvestris* (Musser and Lunde, ms; see account of that species). Cole et al. (1997) reported *P. loriae* to be common on the E flanks of Mt Dayman, noting that *P. loriae* was usually found up to 1600 m and replaced above that altitude by *P. sylvestris*, although they overlap between 1000 and 2000 m. There is much to learn about altitudinal and geographic distribution of *P. loriae* throughout New Guinea.

Pogonomys macrourus (Milne-Edwards, 1877). C. R. Acad. Sci. Paris, 85:1081.

COMMON NAME: Chestnut Pogonomys.

TYPE LOCALITY: New Guinea, Prov. of Papua (= Irian Jaya), Vogelkop, Arfak Mtns, Amberbaki.

DISTRIBUTION: New Guinea; throughout lowland and midmontane forests from sea level to 1800 m (Flannery, 1995a:319); also recorded from Yapen and New Britain (Rümmler, 1938; Flannery, 1995b); represented in the Trans-fly region (near Daru) only by the holotype of *mollipilosus*.

STATUS: IUCN – Lower Risk (lc).

SYNONYMS: *derimapa* Tate and Archbold, 1935; *huon* Tate and Archbold, 1935; *lepidus* Thomas, 1897; *mollipilosus* (Peters and Doria, 1881).

COMMENTS: Phallic morphology described by Lidicker (1968). There is appreciable geographic variation in cranial and dental dimensions and proportions among samples of *P. macrourus* (Musser and Lunde, ms) and its significance should be assessed in a fresh taxonomic revision of the species. The holotype of *mollipilosus* is a young adult that Tate (1951:280), who examined it, associated with the holotypes of *loriae* and *dryas*, although he had "some doubt of the absolute identity of these with *mollipilosus*." However, Dennis and Menzies (1979) and Flannery (in litt.), who also studied the holotype, thought it an example of *P. macrourus*, the New Guinea species with reddish brown upperparts and smaller body size and shorter tail length than *P. loriae* (which has dark brownish gray upperparts tinged with buff and white underparts). Tate wrote that the dorsal coat of the holotype is reddish brown and the ventral coat creamy white; the color is typical of *P. macrourus*, and Tate's cranial and dental measurements (p. 376) are closer to samples of that species than to *P. loriae*. In their morphometric study, Dennis and Menzies (1979) thought the holotype fell between the clusters of *P. macrourus* and *P. loriae* in their discriminant function analysis, but because the specimen came from lowlands of the Trans-Fly region (near Daru) they provisionally identified it as *P. macrourus*. They also noted (p. 322) that Oldfield Thomas had written on the tag attached to the holotype, "Is a *Uromys* and I think is the same as *Pogonomys macrourus* M. Edw." Thomas had described *loriae* and *dryas* and knew the difference between the large-bodied, brownish gray rat those names represent and the smaller-bodied animal with reddish brown fur we know as *P. macrourus*. The scatter plot of canonical variate scores presented by Dennis and Menzies (1979:321) can also be interpreted differently. We see the score for the holotype of *mollipilosus* as forming one end of the *macrourus* cluster and not intermediate in position. Determining the identity of *mollipilosus* is not a trivial exercise because that name has been used for the population of *Pogonomys* endemic to the Cape York Peninsula of NE Australia (see generic discussion).

Pogonomys sylvestris Thomas, 1920. Ann. Mag. Nat. Hist., ser. 9, 6:534.
    COMMON NAME: Gray-bellied Pogonomys.
    TYPE LOCALITY: Papua New Guinea, Morobe Province, Rawlinson Mtns, 1500 m.
    DISTRIBUTION: New Guinea; in Prov. of Papua (= Irian Jaya) known only from the Arfak Mountains in the Vogelkop region, but widespread in mountainous Papua New Guinea, from about the Mendi region east through the mountains (including those on the Huon Peninsula) to the western margin of the Owen Stanley Range at elevations between 1300 and 2800 m. It has not been taken on the N coastal ranges and is conspicuously absent from the mountain valleys in the Telefomin area of S Sandaun Province in W Papua (where *P. championi* occurs), and from the Snow Mountains (Pegunungan Maoke) in Prov. of Papua (= Irian Jaya), which is occupied by an undescribed species related to *P. sylvestris* (Musser and Lunde, ms.).
    STATUS: IUCN – Lower Risk (lc).
    COMMENTS: This is the smallest in body size of any species of *Pogonomys*. The few localities from the Central Cordillera in Prov. of Papua (= Irian Jaya) attributed to *P. sylvestris* (Flannery, 1995a, for example) represent an undescribed species (Musser and Lunde, ms.), and *P. sylvestris* is apparently replaced in the Telefomin area by *P. championi* (Flannery, 1988). Cole et al. (1997) reported that *P. sylvestris* is common in the Mt Dayman region at the eastern margin of the Owen Stanley Range.

Praomys Thomas, 1915. Ann. Mag. Nat. Hist., ser. 8, 15:4
    TYPE SPECIES: *Epimys tullbergi* Thomas, 1894.
    SYNONYMS: *Berberomys* Jaeger, 1975.
    COMMENTS: *Stenocephalemys* Division. *Myomyscus* (or *Myomys*), *Mastomys*, and *Hylomyscus* have been united with *Praomys* as subgenera (D. H. S. Davis, 1965; Misonne, 1974), but are treated as separate genera here and by other workers such as Rosevear (1969), who also reviewed the taxonomic history of some species in *Praomys* as well as its generic status relative to the other genera allied with it. Cladistic analyses of morphological variation has also supported the monophyly of *Mastomys*, *Myomyscus*, and *Hylomyscus* relative to *Praomys*, but also revealed that the latter is paraphyletic, consisting of a monophyletic *P. tullbergi* group that is separate from the other *Praomys* sampled (members of the *P. jacksoni* and *P. delectorum* clusters), which may have to be contained in a separate genus (Lecompte et al., 2002a). This pattern has been confirmed by analyses of complete mtDNA cytochrome *b* sequences (Lecompte et al., 2002b). Van der Straeten and Dieterlen (1987) and Van der Straeten and Dudu (1990) provided brief reviews of the historical and current allocations of forms to the *P. tullbergi*, *P. jacksoni*, and *P. delectorum* complexes. Chromosomal and biochemical traits reviewed or referenced by Qumsiyeh et al. (1990), who combined *Myomys* (= *Myomyscus*) and *Mastomys* with *Praomys*. Additional chromosomal information recorded by Maddalena et al. (1989). Although variously claimed to be closely related to *Mus* and *Apodemus* (Sarich, 1985), *Millardia* (Misonne, 1969), or *Rattus*

(Ellerman, 1941), results of DNA/DNA hybridization (Chevret et al., 1994) and analysis of microcomplement fixation of albumin (Watts and Baverstock, 1995a) demonstrate distant affinity to those genera and instead collectively group *Praomys* with *Mastomys*, *Myomys* (=*Myomyscus*), and *Hylomyscus*. *Praomys* is the core of a "*Praomys* group" as reported in the literature also containing *Heimyscus*, *Hylomyscus*, *Mastomys*, *Myomyscus*, and *Stenocephalemys* (for example, Lecompte et al., 2001, 2002a, b), which is equivalent to our *Stenocephalemys* Division.

Defining limits of species within *Praomys* has been the research focus of several mammalian systematists, among them Eric van der Straeten and his colleagues, with whom he has published a series of informative reports covering *Praomys* taxonomy and increasing our knowledge about the number of species and their characteristics. Van der Straeten and Dudu (1990) and Van der Straeten and Kerbis Peterhans (1999) recognized four species groups. One is the *P. jacksoni* complex, which includes the species *degraaffi, jacksoni, minor, montis, mutoni, peromyscus, sudanensis, viator,* and at least one undescribed species. Of these, only *degraaffi, minor,* and *mutoni* have been adequately defined and their geographic ranges documented; the four others are treated here as synonyms of *P. jacksoni*. A second is the *P. tullbergi* complex containing *misonnei, morio, rostratus, tullbergi, petteri, hartwigi,* and *obscurus* (Hutterer et al., 1992a; Lecompte et al., 1999; Van der Straeten et al., 2003). A third is the *P. delectorum* complex and consisting of *delectorum, melanotus, octomastis,* and *taitae*. The latter three taxa have yet to be differentially diagnosed and their geographic ranges unambiguously defined; we include them here under *P. delectorum*. The fourth is the *P. lukolelae* complex encompassing *lukolelae* (treated as a species of *Malacomys* by Musser and Carleton, 1993) "and probably *verschureni* (described as *Malacomys*)" (Van der Straeten and Kerbis Peterhans, 1999:89). Van der Straeten and Kerbis Peterhans (1999:89) commented that "each of these species-complexes consists of several well-defined and closely related species. Differences between these species-complexes are also well marked. We believe that elevation of these species-complexes to the status of genera should be considered. For example, the morphological differences between the *P. tullbergi* species-complex and the *P. jacksoni* species-complex are equal to those between *Lemniscomys* and *Rhabdomys*." This is a hypothesis that has to be tested with morphological and molecular data within a framework surveying all African murine genera, not just *Praomys*. Cladistic analyses of morphological variation does support the generic separation of the *P. tullbergi* group (Lecompte et al., 2002a) from the *P. jacksoni* cluster, as does phylogenetic analyses of complete mtDNA cytochrome *b* sequences (Lecompte et al., 2002b). We add a fifth group to *Praomys* composed of *P. daltoni* and *P. derooi*, which are usually allocated to *Myomys* (= *Myomyscus*; Musser and Carleton, 1993; Rosevear, 1969; Van der Straeten, 1979; Van der Straeten and Verheyen, 1978b; see those accounts)

Principle components analysis of most holotypes associated with *Praomys* was presented by Van der Straeten and Robbins (1997); the specimens tended to form a cluster separate from those containing holotypes of *Mastomys* and *Hylomyscus*. Lecompte et al. (2001) offered a taxonomic key to the species of *Praomys* (except *P. lukolelae* and *P. verschureni*), and summaries of their geographical distributions and chromosomal traits.

Extant species of *Praomys* now occur primarily in Subsaharan forest and tree savanna habitats, but the genus is represented by *P. skouri* from late Pliocene sediments in Morocco, when the environment may have been similar to present conditions in East and South Africa (Geraads, 1995; Geraads et al., 1998), and by *P. darelbeidae* from the middle Pleistocene of that country, a time of open, dry environments (Geraads, 1994). *Praomys skouri* has also been found in early Pleistocene of Tunisia, along with several other extinct *Praomys* at different Pleistocene intervals (Mein and Pickford, 1992). *Berberomys* was proposed as a subgenus for the Moroccan *P. skouri*, but its diagnostic traits also apply to an extinct species described as a *Mastomys* recovered from sediments at Olduvai Gorge in Tanzania (Geraads, 1995). Late Pliocene of Ethiopia (hadar) is the oldest documented Subsaharan record of *Praomys* (Denys, 1999; Sabatier, 1982), but apparently strata in Chad have yielded an even older occurrence (early Pliocene) about 5 million years before present (Denys, pers. comm., in Lecompte, 2003).

Praomys daltoni (Thomas, 1892). Ann. Mag. Nat. Hist., ser. 6, 10:181.
    COMMON NAME: Dalton's Praomys.

TYPE LOCALITY: West Africa, Senegal, Niokolo Koba National Park (see discussion in Grubb et al., 1998:197).

DISTRIBUTION: Sudanian and Sahelian zones (see Dobigny et al. 2002b:500) from Gambia and Senegal through Guinea (Ziegler et al., 2002), Sierra Leone, N Côte d'Ivoire, Mali, Burkina Faso, Niger, Ghana, Togo, Benin, Nigeria, S Chad and Central African Republic to SW Sudan; E limits unresolved.

STATUS: IUCN – Lower Risk (lc) as *Myomys daltoni*.

SYNONYMS: *butleri* (Wroughton, 1907); *ingoldbyi* (Ellerman, 1941); *saturatus* (Ingoldby, 1929) [not Lyon, 1911]; *tuareg* (Braestrup, 1935).

COMMENTS: Reviewed by Rosevear (1969) and Van der Straeten and Verheyen (1978b). Although its geographic range allopatrically complements the distribution of *Myomyscus brockmani*, *P. daltoni* is not conspecific with that E African species, a conclusion based on our study of specimens and analyses of morphological (Lecompte et al., 2001a) and molecular (Lecompte et al., 2002b) data. DNA/DNA hybridization results clustered *daltoni* with species of *Praomys* (Chevret et al., 1994) as did phylogenetic analyses of morphometric traits (Van der Straeten, 1979), complete mtDNA cytochrome *b* sequences (Lecompte et al., 2002b), and nuclear IRBP gene data (E. Lecompte, in litt., 2002); however, partial 16S rRNA mitochondrial sequences pointed to *daltoni* as sister group to species of *Stenocephalemys*. Van der Straeten (in litt., 1994) regarded *derooi* and *daltoni* to be closely related to *Praomys tullbergi*. Chromosomal morphology documented for samples from Côte d'Ivoire (Matthey, 1964), Senegal (Granjon et al., 1992), and NE Mali (Dobigny et al., 2001a, b; Volobouev et al., 2002b); 2n = 36, FNa = 34 for all samples. The taxon *tuareg* was described as a subspecies of *Grammomys macmillani*, but listed by Rosevear (1969) as a species of *Grammomys* of doubtful validity, and finally identified by Braestrup and Hutterer (1985) as a possibly distinct subspecies of *P. daltoni*. Setzer (1956) retained *butleri*, known only by the holotype collected in SW Sudan, as a species, but its morphological traits, judged by his description, are those of *P. daltoni*. Van der Straeten and Verheyen (1978) documented the identity of *ingoldbyi* with *P. daltoni*. Grubb et al. (1998) reviewed populations in Gambia, Sierra Leone, and Ghana. Senegal samples documented by Duplantier and Granjon (1992) and Granjon et al. (1992). Niger records documented by Dobigny et al. (2002b). Trapped and recovered from owl pellets in S Mali (Meinig, 2000). Two specimens taken in NE Mali inside the city of Kidal and on the Adrar des Iforas massif are the northernmost records of the species, and Dobigny et al. (2001a:216) speculated that *P. daltoni* is usually found in Sudano-Sahelian habitats and its occurrence in the Adrar des Iforas "supports the hypothesis that this massif shelters a relictual fauna, isolated since the recent southward progression of the Sahara desert" and noted that "the survival of this species in this area is favoured by its commensalism." L Granjon (in litt., 2002) confirmed that *P. daltoni* is widely distributed in Mali, "outdoors in the South, indoors in the North."

Praomys degraaffi Van der Straeten and Kerbis Peterhans, 1999. S. Afr. J. Zool., 34:81.

COMMON NAME: De Graaff's Praomys.

TYPE LOCALITY: Burundi, Nyamuugari, 2200 m.

DISTRIBUTION: Recorded only from moist montane forest covering the Albertine Rift Mtns in Burundi, Rwanda and Uganda (Van der Straeten and Kerbis Peterhans, 1999, who also indicated the species to be "restricted" to high forested elevations in the Albertine Rift).

STATUS: IUCN – Near Threatened.

COMMENTS: A distinct and one of the more easily identified species in the *P. jacksoni* complex; distinguished from *P. jacksoni* by a suite of chromosomal traits (2n = 26, FN = 24), fur coloration, morphometric data, and number of teats. Where the altitudinal ranges of *P. degraaffi* and *P. jacksoni* narrowly overlap, the former is usually found at higher elevations.

Praomys delectorum (Thomas, 1910). Ann. Mag. Nat. Hist., ser. 8, 6:430.

COMMON NAME: East African Praomys.

TYPE LOCALITY: S Malawi, Mlanji Plateau, 5500 ft (1675 m).

DISTRIBUTION: High plateaus and isolated mountains from NE Zambia (Nyika Plateau, Maku-
tus, and Mafingas; Ansell, 1978) and Malawi (Nyika Plateau; Ansell and Dowsett, 1988),
through Tanzania (Swynnerton and Hayman, 1951) to SE Kenya (Hollister, 1919).

STATUS: IUCN – Lower Risk (lc).

SYNONYMS: *melanotus* G. M. Allen and Loveridge, 1933; *octomastis* Hatt, 1940; *taitae* (Heller,
1912).

COMMENTS: Of the synonyms belonging here, *taitae* was described as a species (Heller, 1912)
and recognized as such by Hollister (1919) and Swynnerton and Hayman (1951),
melanotus was described as a form of *P. tullbergi* (G. M. Allen and Loveridge, 1933), and
octomastis was presented as a subspecies of *P. jacksoni* (Hatt, 1940b). Demeter and
Hutterer (1986) suggested that *taitae* is synonymous with *Hylomyscus denniae*, but it is
not, judging from our study of specimens and relevant holotypes. Van der Straeten
and Kerbis Peterhans (1999) recognized *melanotus*, *octomastis*, and *taitae* as valid
species in the *P. delectorum* complex, but presented no substantiating evidence
defining their morphological and distributional limits. A sample of *taitae* had 2n = 48
(Matthey, 1965a). Distribution in the Eastern Arc Mtns of Tanzania discussed by
Stanley et al. (1998) and on slopes of Mt Kilimanjaro by Grimshaw et al. (1995).

Praomys derooi Van der Straeten and Verheyen, 1978. Z. Säugetierk., 43:33.

COMMON NAME: Deroo's Praomys.

TYPE LOCALITY: Togo, Borgou, 160 m.

DISTRIBUTION: Known only from E and S Ghana, Togo, Benin, and W Nigeria (see Van der
Straeten and Verheyen, 1978b).

STATUS: IUCN – Lower Risk (lc) as *Myomys derooi*.

COMMENTS: Part of the murine fauna endemic to West Africa (see account of *Grammomys
buntingi*). Described as a savanna species found living in and around human dwellings.
Grubb et al. (1998) reviewed the Ghana population, and Decher et al. (1997) recorded
two animals from the Accra Plains region in SE Ghana, one caught in a house, the
other near a house. Phylogenetic relationships of this species are with *P. daltoni*; the
two are closely similar in pelage coloration and texture (except that underparts are
gray in most *derooi* and white in most *daltoni*, but we have seen series of USNM
P. daltoni ranging from gray to white), external, cranial, and dental morphology (our
observations), habitat (see account of *P. daltoni*), and variation in mtDNA cytochrome
b sequences (Lecompte et al., 2002b) and nuclear IRBP gene sequences (E. Lecompte, in
litt., 2002). The morphological and molecular resemblances are so close we suspect the
samples of *derooi* simply represent local populations of *P. daltoni*. Even Van der
Straeten's (1979) canonical analyses of cranial and dental measurements exposed
broad overlap between his samples of *P. daltoni* and *P. derooi*. Ziegler et al. (2002)
discussed an example of *P. derooi* from C Guinea collected in a village, but the record is
so far from the known, well documented range that we suspect it represents a
misidentification of *P. daltoni*.

Praomys hartwigi Eisentraut, 1968. Bonn. Zool. Beitr., 19:8-11.

COMMON NAME: Hartwig's Praomys.

TYPE LOCALITY: Cameroon, Bamenda Plateau, Mt Oku, Lake Oku, 2100 m (coordinates
provided by Ansell, 1989:56).

DISTRIBUTION: Recorded only in isolated mountain forests along the mountain chain on the
Bamenda Plateau in W Cameroon (Mt Bambuto, Mt Lefo and Mt Oku; Hutterer et al.,
1992a; Van der Straeten, in litt., 1994).

STATUS: IUCN – Endangered.

COMMENTS: Lecompte et al. (1999) regarded *P. hartwigi* as a member of the *P. tullbergi*
complex, which has been confirmed by phylogenetic analyses of morphological traits
(Lecompte et al., 2002a) and complete mtDNA cytochrome *b* sequences (Lecompte
et al., 2002b). Fülling (1992) contrasted morphology of palatal ridges between samples
of *P. hartwigi* (2 + 7 = 9) and *P. jacksoni* (2 + 5 = 7) collected from slopes of Mt Oku. Morphologically and phylogenetically allied to *P. obscurus* (see that account), which is
separated from *P. hartwigi* by about 100 km in the Gotel Mtns on the Mambilla Plateau
to the northeast of the Bamenda Plateau (Hutterer et al., 1992a).

Praomys jacksoni (de Winton, 1897). Ann. Mag. Nat. Hist., ser. 6, 20:318.

COMMON NAME: Jackson's Praomys.

TYPE LOCALITY: Uganda, Entebbe.

DISTRIBUTION: C Nigeria through Cameroon and Central African Republic to S Sudan, Dem. Rep. Congo, NE Angola (Crawford-Cabral, 1998), Uganda, Rwanda, Kenya, and southward through E Tanzania to N and E Zambia; the most expansive geographic range of any species in *Praomys* (Lecompte et al., 2001).

STATUS: IUCN – Data Deficient as *P. jacksoni*, *P. montis*, and *P. peromyscus*.

SYNONYMS: *montis* (Thomas and Wroughton, 1910); *peromyscus* (Hollister, 1919); *sudanensis* Setzer, 1956; *viator* (Thomas, 1911).

COMMENTS: 2n = 28, FN = 30 (Matthey, 1959). At one time listed as a subspecies of *P. tullbergi* (e.g., Hollister, 1919), *jacksoni* is a distinct species occurring sympatrically with *P. tullbergi* (G. M. Allen, 1939; Ansell, 1978; Van der Straeten and Dieterlen, 1987; Van der Straeten and Dudu, 1990; Lecompte et al., 2001). The problem of identifying the holotype of *P. jacksoni*, as well as current names associated with the species, was reviewed and discussed by Van der Straeten and Dieterlen (1987), Van der Straeten and Dudu (1990), and Van der Straeten and Kerbis Peterhans (1999). Van der Straeten and Dudu (1990) regarded *montis* and *peromyscus* as valid species and Van der Straeten and Kerbis Peterhans (1999) claimed that *montis*, *peromyscus*, *sudanensis*, and *viator*, all names that we place in synonymy, are separate species. The canonical analyses presented by those authors, however, employing series representing these taxa and other distinct species of *Praomys* (*P. degraaffi* and *P. mutoni*), do not indicate any significant separation between clusters identified as *montis*, *peromyscus*, *sudanensis*, and *viator* and topotypes of *jacksoni*. *Praomys jacksoni* as we list it here may be composite, but until the forms representing potential species (*viator*, *montis*, *peromyscus*, *sudanensis*) have been clearly diagnosed and their geographic ranges described we include them here under *P. jacksoni*.

Several pertinent reports cover: craniometrical comparisons among four samples of *P. jacksoni* collected at different altitudes (850-3300 m) in the Kivu region of E Dem. Rep. Congo (Van der Straeten and Dieterlen, 1992); altitudinal distribution of *P. j. montis* on Ugandan slopes of Ruwenzori Mtns (Kerbis Peterhans, 1998, who regarded *montis* as a Ruwenzori endemic); specimens from Kalinzu forest of SW Uganda that were collected on the forest floor and on woody vines 1.5 m above the ground (Lunde and Sarmiento, 2002); samples from mountain chain in W Cameroon and SE Nigeria reported as "*Praomys* cf. *jacksoni*" (Hutterer et al., 1992a); habitat and distribution on Ugandan slopes of Mt Elgon (Clausnitzer and Kityo, 2001); and samples from Kibale Forest in Uganda (Hoffmann, 1997). Ziegler et al. (2002) listed *P. jacksoni* from C Guinea but the voucher material comes from so far west of the nearest documented occurrence (Nigeria) that we are dubious of its identification.

Praomys lukolelae (Hatt, 1934). Am. Mus. Novit., 708:13.

COMMON NAME: Lukolela Praomys.

TYPE LOCALITY: Dem. Rep. Congo, Equateur, Lukolela.

DISTRIBUTION: N Dem. Rep. Congo; recorded from the type locality and Kisangani region in Orientale (specimens in UCA).

STATUS: IUCN – Near Threatened as *Malacomys lukolelae*.

COMMENTS: F. Petter (1975c) and Chevret et al. (1994) recorded *P. lukolelae* from the Central African Republic, but the specimens we have seen from that series represent an undescribed species of *Praomys* in the *P. tullbergi* group (see also Lecompte et al., 2001, 2002a, b) that has now been described as *P. petteri* (see that account). Up to 1990, *P. lukolelae* was represented only by the three specimens in the type series (Van der Straeten and Dudu, 1990), but we have examined 90 specimens from the Kisangani region that are morphologically closely similar to the three examples from Lukolela. Hatt (1934b) described *lukolelae* as a subspecies of *Praomys tullbergi*, but its long and very slim hind feet, tip of short fifth digit extending only to base of digital pad of fourth digit, very large ears, cranial conformation, and molar dental patterns identify the specimens as a separate species. Citing those same traits, Musser and Carleton (1993) placed *lukolelae* in *Malacomys*, noting its close morphological similarity with verschureni, which had been described as a species of *Malacomys* and was still associated with that genus in the early 1990s. Compared with *Praomys*, the species of *Malacomys* tend to have a small or no subsquamosal fenestra (usually large in *Praomys*), no cusp t3 on second upper molar (usually present), no anterolabial or posterolabial cusplets on first and second lower molars (present in most specimens), and the three central digits of the hind foot very long relative to the two outer digits (shorter in *Praomys*). In addition to the external traits mentioned above, the type series of *lukolelae* and specimens from Kisangani exhibit subsquamosal variation similar to that in *Malacomys*, most specimens lack a cusp t3 on the second upper molar, most do not have anterolabial and posterolabial cusplets on the lower molars, and configuration of the digits is like that in *Malacomys*.

Van der Straeten and Dudu (1990) had recognized a *Praomys lukolelae* group containing only that species and stated that *Malacomys verschureni* was related to the *P. lukolelae* complex. By 1999, Van der Straeten and Kerbis Peterhans (1999:89) recognized a "*lukolelae* species-complex enclosing *lukolelae* and probably *verschureni* (described as *Malacomys*)." Van der Straeten's view received support from Denys, who was studying the *lukolelae* complex and wrote in 1995 (in litt.) that her study of cranial and dental traits would align *lukolelae* with *Praomys* rather than *Malacomys*. We defer to our colleagues' allocations and list both species in *Praomys* but stress the need for critical phylogenetic study of *lukolelae* and *verschureni* in relation to the species of *Malacomys* and *Praomys* and reevaluation of generic boundaries. In our study of this group we appreciated the distinctions between species of *Praomys* and *Malacomys* but also detected close morphological alliance between them, in particular the *P. tullbergi* complex and *Malacomys*, with *P. lukolelae* and *P. verschureni* nearly bridging the gap in several diagnostic characters. This assessment, however, clashes with cladistic analyses of morphological traits (Lecompte et al., 2002*a*), sequences from complete mtDNA cytochrome *b* (Lecompte et al., 2002*b*) and nuclear IRBP gene sequences (E. Lecompte, in litt., 2002), which do not cluster *Malacomys* even close to any genus in *Stenocephalemys* Division or any subgroup of *Praomys*.

In the morphometric analysis of type specimens of *Praomys*, *Mastomys*, *Myomys* (*Myomyscus*), and *Hylomyscus* by Van der Straeten and Robbins (1997), the holotype of *lukolelae* clustered with holotypes of taxa associated with *Praomys*, which to them endorsed the association of *lukolelae* with *Praomys* and not *Malacomys*. But this connection is unwarranted without comparison with holotypes representing species of *Malacomys*, taxa omitted from their analysis.

Praomys minor Hatt, 1934. Am. Mus. Novit., 708:11.
    COMMON NAME: Least Praomys.
    TYPE LOCALITY: Dem. Rep. Congo, Equateur, Lukolela.
    DISTRIBUTION: Recorded only from the type locality.
    STATUS: IUCN – Vulnerable
    COMMENTS: Still only known by the three specimens in Hatt's sample. Originally described as a subspecies of *P. jacksoni* (Hatt, 1934*b*), and then treated as a subspecies of *P. tullbergi* (F. Petter, 1975*c*), *minor* is a distinct species in the *P. jacksoni* complex (Van der Straeten and Dieterlen, 1987; Van der Straeten and Dudu, 1990; Van der Straeten and Kerbis Peterhans, 1999).

Praomys misonnei Van der Straeten and Dieterlen, 1987. Stuttg. Beitr. Naturk. ser. A, 402:3.
    COMMON NAME: Misonne's Praomys.
    TYPE LOCALITY: E Dem. Rep. Congo, Kivu region, Irangi.
    DISTRIBUTION: N and E Dem. Rep. Congo.
    STATUS: IUCN – Lower Risk (lc).
    COMMENTS: Sympatric with *P. jacksoni* at the type locality and with *P. jacksoni* and *P. mutoni* in Orientale (Van der Straeten and Dudu, 1990). Regarded as related to *P. tullbergi* by Van der Straeten and Dieterlen (1987; Van der Straeten and Dudu, 1990; Dudu et al., 1997). Examples of *misonnei* we examined from the Ituri Forest in E Dem. Rep. Congo and Gamangui in the Orientale Province, where *P. jacksoni* was also trapped, are morphologically very similar to *P. tullbergi*; the possibility that *misonnei* simply

represents populations of *P. tullbergi* at the eastern margins of its geographic range needs to be considered in any systematic revision of the complex. They are very closely associated in phylogenetic analyses of morphological traits (Lecompte et al., 2002*a*) and complete mtDNA cytochrome *b* sequences (Lecompte et a., 2002*b*). Qumsiyeh et al. (1990) identified Kenyan samples as *P. misonnei*, and their chromosomal and electrophoretic characteristics relate the samples to what they identified as *Praomys hildebrandti*; identity of the Kenyan material should be reassessed (Van der Straeten and Kerbis Peterhans, 1999, examined five of the specimens used by Qumsiyeh et al. in their chromosomal and electrophoretic analyses and identified them as members of the *P. jacksoni* complex, yet Lecompte et al., 2001, claimed that Van der Straeten examined the Kenyan specimens and considered them to be *P. misonnei*). Dudu et al. (1997) provided the first information on reproductive biology in samples from the Masako forest near Kisangani, Dem. Rep. Congo.

Praomys morio (Trouessart, 1881). Bull. Soc. Etudes Sci. Angers, 10:121.

COMMON NAME: Cameroon Praomys.

TYPE LOCALITY: Cameroon, Mt Cameroon, 7000 ft (2135 m); see Rosevear (1969:399).

DISTRIBUTION: Mt Cameroon and the mountainous island of Bioko (Equatorial Guinea), essentially the distribution outlined by Eisentraut (1970).

STATUS: IUCN – Vulnerable.

SYNONYMS: *maura* (Gray, 1862) [not Waterhouse, 1839].

COMMENTS: 2n = 34 (Matthey, 1965*a*). A member of the *P. tullbergi* complex. Musser and Carleton (1993) restricted *P. morio* to Mt Cameroon, although Eisentraut (1970) recorded it from the island of Bioko, and F. Petter (1965) discussed samples from the Central African Republic. Musser and Carleton also noted that "The species requires definition; alleged distinctions between it and *P. tullbergi* may not reflect specific differences (Hutterer, in litt.). Our study revealed that series from outside of Mt Cameroon identified as *morio* are either *tullbergi* or an undescribed species of *Praomys* (the series from Central African Republic, for example [see *P. petteri* below])." Van der Straeten (in litt., 1994) wrote us that "The morphological differences between *morio* and *tullbergi* are clear and were described in detail by Eisentraut. I obtained the same results using all *Praomys* specimens collected by Eisentraut in Bioko and on Mount Cameroon in a principal component analysis. All this is in agreement with the breeding experiments carried out by Eisentraut." Lecompte et al. (1999) redescribed the holotype of *morio*, providing illustrations of the skull and molar rows along with cranial and dental measurements. Morphological traits of this holotype place it among the species in the *P. tullbergi* complex as that group is defined by Van der Straeten and Dieterlen (1987) and Van der Straeten and Dudu (1990), an arrangement supported by cladistic analyses of morphological traits (Lecompte et al., 2002*a*).

Praomys mutoni Van der Straeten and Dudu, 1990. *In* Peters and Hutterer (eds.), Vertebrates in the tropics, Museum Alexander Koenig, Bonn, p. 75.

COMMON NAME: Riverine Praomys.

TYPE LOCALITY: N Dem. Rep. Congo, Orientale, Batiabongena (Masako Forest Reserve), 00°36′N, 25°13′E.

DISTRIBUTION: Recorded only from the type locality.

STATUS: IUCN – Lower Risk (nt).

COMMENTS: A tropical evergeen riverine forest species morphologically related to *P. jacksoni* and occurring sympatrically with it (Van der Straeten and Dudu, 1990). Dudu et al. (1997) documented data covering reproductive biology for this distinctive species. Mutoni is a Swahili word for riverbank or wet area (mto or muto = river), and refers to the typical biotype where the specimens were trapped (Van der Straeten and Dudu, 1990).

Praomys obscurus Hutterer and Dieterlen, 1992. Bonn. Zool. Beitr., 43:402.

COMMON NAME: Gotel Mountain Praomys.

TYPE LOCALITY: SE Nigeria, Mambilla Plateau, Gotel Mtns, Gangirwal, 2300 m (coordinates and habitat information provided by Hutterer et al., 1992*a*).

DISTRIBUTION: Recorded only from fern-grassland, swamp and gallery forest, and along forest streams at the type locality and nearby Chappal Waddi.

STATUS: IUCN – Endangered.

COMMENTS: Hutterer et al. (1992a) described *obscurus* as a subspecies of *P. hartwigi*. Those authors were also impressed with the morphological distinctions between typical *P. hartwigi* from Mt. Oku and *obscurus* from the Gotel Mtns and were unsure whether to treat the latter as a subspecies of *P. hartwigi* (Dieterlen's opinion) or a separate species (Hutterer's view). We treat *obscurus* as a separate species because contrasted with *P. hartwigi* it has much darker pelage, larger body measurements, but significantly smaller cranial and dental measurements (Hutterer et al., 1992a); this proposal was recently advocated by Hutterer (in litt., 2003). *Praomys obscurus* is morphologically and phylogenetically most closely related to *P. hartwigi* (see that account), which occurs in mountain forest on the Bamenda Plateau to the southwest of the Gotel Mtns; both species are members of the *P. tullbergi* complex (Hutterer et al., 1992a).

Praomys petteri Van der Straeten, Lecompte, and Denys, 2003. Bonn. zool. Beitr., 50:333.

COMMON NAME: Petter's Praomys.

TYPE LOCALITY: Central African Republic, Boukoko, 03°54′ N, 17°56′ E.

DISTRIBUTION: Tropical lowland rainforest in S Cameroon, S Central African Republic, and S Republic of Congo.

COMMENTS: One of the largest-bodied species of *Praomys* in the *P. tullbergi* group. Populations formerly included in *P. morio* (F. Petter, 1965; Matthey, 1965), *P. lukolelae* (F. Petter, 1975; Genest-Villard, 1980; Granjon, 1991; Chevret et al., 1994), or considered distinct but unnamed (Musser and Carleton, 1993). Phylogenetic affinities discussed by Van der Straeten et al. (2003), who compared *P. petteri* with samples of *P. rostratus*, *P. misonnei*, *P. verschureni*, and *P. lukolelae*. Locally sympatric with either *P. tullbergi* or *P. jacksoni* in Republic of Congo (Granjon, 1991).

Praomys rostratus (Miller, 1900). Proc. Wash. Acad. Sci., 2:637.

COMMON NAME: West African Praomys.

TYPE LOCALITY: Liberia, Mt Coffee.

DISTRIBUTION: Recorded only from forest in Liberia, C Guinea (Ziegler et al., 2002), Mt Nimba region of Guinea, and Côte d'Ivoire; limits unresolved.

STATUS: IUCN – Lower Risk (lc).

COMMENTS: 2n = 34 (Gautun et al., 1986). Member of the *P. tullbergi* complex. Originally described as a subspecies of *tullbergi*, but Van der Straeten and Verheyen (1981) distinguished *rostratus* from *tullbergi* by its greater body size, noted that both kinds were sympatric, and raised *rostratus* to specific rank. A similar distribution of body sizes, as well as different ecologies, were claimed by Gautun et al. (1986) from Mt Nimba, and they also separated their samples into either *P. tullbergi* or *P. rostratus*. However, the existence of a larger-bodied species in W Africa related to *P. tullbergi* requires verification because published data separating the two is not convincing according to L. Granjon (in litt., 2002): "The text by Gautun et al. (1986) is not clear, and in Mali we now have a good sample of 2n = 34 (all acrocentric autosomes) *Praomys* from various localities with measurements covering the entire range of the two species as analyzed by Van der Straeten and Verheyen (1981), in which no individual measurement would distinguish between the two species." If proved a distinct species, *P. rostratus* would be part of the murine fauna endemic to W Africa (see account of *Grammomys buntingi*).

Praomys tullbergi (Thomas, 1894). Ann. Mag. Nat. Hist., ser. 6, 13:205.

COMMON NAME: Tullberg's Praomys.

TYPE LOCALITY: Ghana, Ashanti, Wasa, Ankober River.

DISTRIBUTION: Forest and Guinea woodland from Gambia River in the west through Cameroon to N and E Dem. Rep. Congo; Bioko (Equatorial Guinea); also NW Angola (Crawford-Cabral, 1998; specimens in FMNH) and Kakamega Forest Reserve in NW Kenya (specimen in USNM); limits unknown.

STATUS: IUCN – Lower Risk (lc).

SYNONYMS: *burtoni* (Thomas, 1892) [not Ramsay, 1887].

COMMENTS: Reviewed by Rosevear (1969), Van der Straeten and Verheyen (1981), and more recent reports cited in the above species accounts. Morphologically closely related to

P. misonnei, P. morio, and *P. rostratus.* Grubb et al. (1998) reviewed populations in the Gambia, Sierra Leone, and Ghana. Ecological information recorded for population on Accra Plains of S Ghana by Decher and Bahian (1999). A karyotype of 2n = 34 (all acrocentric autosomes) was recorded by Granjon et al. (1992) from Senegal, which is similar to karyotypes obtained from samples collected in Côte d'Ivoire and Central African Republic (Matthey, 1958). Additional comments on Senegal distribution given by Duplantier and Granjon (1992). Barnett et al. (1996) documented its occurrence on the Kounounkan Massif in SW Guinea. See account of *P. rostratus.*

Praomys verschureni (Verheyen and Van der Straeten, 1977). Rev. Zool. Afr., 91:739.
COMMON NAME: Verschuren's Praomys.
TYPE LOCALITY: Dem. Rep. Congo, Orientale, Mamiki.
DISTRIBUTION: Known only from NE Dem. Rep. Congo (eastern edge of Central African high forest block).
STATUS: IUCN – Near Threatened as *Malacomys verschureni.*
COMMENTS: Described as a species of *Malacomys* and all published information referenced under that taxonomic combination. Original description was based on one specimen (Verheyen and Van der Straeten, 1977), and the species is still represented by very few examples (Dieterlen and Van der Straeten, 1984; Robbins and Van der Straeten, 1982). Chromosomal information reported by Robbins and Van der Straeten (1982). Analyses of complete mtDNA cytochrome *b* sequences placed *P. verschureni* near *Myomyscus yemeni* and not species of *Praomys* (Lecompte et al., 2002*b*), but nuclear IRBP gene sequences alligned it with species of *Praomys* (E. Lecompte, in litt., 2002). See account of *P. lukolelae.*

Protochromys Menzies, 1996. Aust. J. Zool., 44:416.
TYPE SPECIES: *Melomys fellowsi* Hinton, 1943.
COMMENTS: *Uromys* Division. Member of the New Guinea Old Endemics that is morphologically and morphometrically closely related to *Melomys* and *Paramelomys* (Menzies, 1996).

Protochromys fellowsi (Hinton, 1943). Ann. Mag. Nat. Hist., ser. 11, 10:554.
COMMON NAME: Papuan Protochromys.
TYPE LOCALITY: Papua New Guinea, Purari-Ramu Divide, Baiyanka, 8000 ft (2440 m).
DISTRIBUTION: Papua New Guinea; known only from a few places at high elevations (usually above 2000 m) in the Central Cordillera, from the Porgera area in the west (143EE) eastward to Mt Wilhelm in the Bismarck Range in the east (see Flannery, 1995*a*:292); may range farther west and east in Papua New Guinea (Menzies, 1996).
STATUS: IUCN – Vulnerable as *Melomys fellowsi.*
COMMENTS: Originally described as a very distinctive species of *Melomys* that resembles *M. rufescens* and *M. leucogaster* in some morphometric features (Menzies, 1990). Its inclusion in *Melomys* has always been questioned by systematists familiar with the group. Recently, Menzies (1996) demonstrated that while *fellowsi* shared traits with both *Melomys* and *Paramelomys*, it should be separated from both groups of species because of its pale (unpigmented to pale yellow) incisors, narrow zygomatic plate, and alisphenoid strut. Its separation as a different monophyletic entity will have to be tested by phylogenetic analyses of the *Melomys-Paramelomys* complex that also incorporate molecular data and a broader array of morphological traits.

Pseudohydromys Rümmler, 1934. Z. Säugetierk., 9:47.
TYPE SPECIES: *Pseudohydromys murinus* Rümmler, 1934.
SYNONYMS: *Mayermys* Laurie and Hill, 1954; *Neohydromys* Laurie, 1952.
COMMENTS: *Xeromys* Division. Member of the New Guinea Old Endemics (Musser, 1981*c*). Species of *Pseudohydromys, Neohydromys,* and *Mayermys* are all terrestrial, primarily insectivorous, small-bodied shrew mice living in mountain forests of New Guinea. They share dense, velvety fur, generally similar cranial conformation (see photographs in Laurie and Hill [1954] and Flannery [1995*a*:plates 22-23]), spacious postglenoid and middle lacerate foramina, derived configuration of cephalic arterial pattern (Musser and Heaney, 1992), loss of third molars (*Pseudohydromys, Neohydromys*) or second and third

molars (*Mayermys*), dentary shape and degree of penetration of incisor alveolus, extent of enamel relative to dentine on upper and lower incisors, similar phallic morphology (Lidicker, 1968) and spermatozoal structure (Breed and Aplin, 1994), and no divergence in immunological distance (albumins indistinguishable among the three as assessed by microcomplement fixation of albumin; Watts and Baverstock, 1994*a*). Species in the three genera clearly form a tight monophyletic group that should be taxonomically expressed by uniting *Neohydromys* and *Mayermys* with *Pseudohydromys*. In proposing *Neohydromys*, Laurie (1952:311) compared it to *Microhydromys*, another small-bodied shrew mouse, which has no close morphological relationship to *Pseudohydromys*, *Neohydromys*, or even *Mayermys* (see generic account of *Microhydromys*). Morphological contrasts between *Neohydromys fuscus* and *Pseudohydromys murinus* presented by Laurie are of the magnitude typical of interspecific variation seen in other murine genera. With its flatter cranium and loss of second and third molars, *Mayermys ellermani* stands apart from the other species, but these are autapomorphies within the monophyletic cluster. Judged by albumin immunology (Watts and Baverstock, 1994*a*, 1996) and spermatozoal morphology (Breed, 1997; Breed and Aplin, 1994), *Pseudohydromys* is part of a larger clade containing species in our *Hydromys*, *Xeromys*, *Pseudomys*, and *Uromys* Divisions (the "Australasian clade" of Watts and Baverstock, 1995*b*, 1996).

Pseudohydromys ellermani (Laurie and Hill, 1954). List of Land Mammals of New Guinea, Celebes and Adjacent Islands, 1952, p. 134.

COMMON NAME: Shaw Mayer's Shrew Mouse.

TYPE LOCALITY: Papua New Guinea, Chimbu Province, Bismarck Range, N slopes Mt Wilhelm, 8000 ft (2440 m).

DISTRIBUTION: New Guinea: known from montane forest localities scattered along the Central Cordillera from Porokma in the Lake Habbema area (Snow Mtns) of Prov. of Papua (= Irian Jaya) through the Telefomin region in W Papua New Guinea eastward to the Wau area and all the way to near Agaun near the E portion of the Owen Stanley Range (Flannery 1995*a*:245, as *Mayermys*).

STATUS: IUCN – Vulnerable as *Mayermys ellermani*.

COMMENTS: The type species of *Mayermys* (Laurie and Hill, 1954). Flannery (1990*b*:183, 1995*a*) provided a photograph of the animal and a summary of distributional and biological data. This small-bodied mouse is unique among all living muroid rodents in having only four minute molars (12 is usual, three in each quadrant of the jaw). An undescribed species related to *P. ellermani* occurs in the SE peninsula of Papua New Guinea (K. Helgen, in litt., 2003). Of the species in *Pseudohydromys*, *P. ellermani* is the only one with with a distribution in both western and eastern portions of the Central Cordillera.

Pseudohydromys fuscus (Laurie, 1952). Bull. Br. Mus. (Nat. Hist.) Zool., 1:311.

COMMON NAME: Mottled-tailed Shrew Mouse.

TYPE LOCALITY: Papua New Guinea, Chimbu Province, N slopes Mt. Wilhelm, 9000-10,000 ft (2744-3048 m).

DISTRIBUTION: Papua New Guinea: Known from Lake Louise (Sandaun Province), Mt Erimbari and Mt Wilhelm (Chimbu Province), and Mt Kaindi (Morobe Province) above 2400 m (Flannery, 1995*a*, as *Neohydromys*).

STATUS: IUCN – Vulnerable as *Neohydromys fuscus*.

COMMENTS: Type species of *Neohydromys*. The little information about habitat and diet is reviewed by Flannery (1995*a*). Except for its smaller molars (relative to cranial size) and slightly proodont upper incisors, *P. fuscus* closely resembles *P. murinus* in pelage coloration and texture, body size, and craniodental traits (Musser's observations). *Pseudohydromys fuscus*, along with *P. murinus*, are known only from the eastern Central Cordillera of New Guinea.

Pseudohydromys murinus Rümmler, 1934. Z. Säugetierk., 9:48.

COMMON NAME: Eastern New Guinea Shrew Mouse.

TYPE LOCALITY: Papua New Guinea, Morobe Province, Wau area, Mt. Missim, 7000 ft (2135 m).

DISTRIBUTION: Papua New Guinea; known only from several localities around the Wau area in the east and the vicinity of Mt. Wilhelm in the west (Laurie, 1952; Flannery, 1995*a*).

STATUS: IUCN – Critically Endangered.

COMMENTS: The scanty information available for distribution, ecology, and diet is summarized by Flannery (1995a).

Pseudohydromys occidentalis Tate, 1951. Bull. Am. Mus. Nat. Hist., 97:224.

COMMON NAME: Western New Guinea Shrew Mouse.

TYPE LOCALITY: New Guinea, Prov. of Papua (= Irian Jaya), north of Mt Wilhelmina, Snow Mtns (Pegunungan Maoke), Lake Habbema, 3225 m.

DISTRIBUTION: New Guinea; known only from the area around Lake Habbema and slopes of Mt Wilhelmina in Prov. of Papua (= Irian Jaya) (Tate, 1951:225); apparently a Snow Mtn endemic.

STATUS: IUCN – Vulnerable.

COMMENTS: *Pseudohydromys occidentalis* and its eastern relative *P. murinus* are found only in tropical upper montane rainforest. Other than being montane and primarily insectivorous, little is recorded about its natural history (Flannery, 1995a). Records from the Star Mtns and Victor Emmanuel Range in W Papua New Guinea attributed to *P. occidentalis* (Flannery, 1995a:256) actually represent a separate morphologically distinctive species (K. Helgen, in litt., 2003).

Pseudomys Gray, 1832. Proc. Zool. Soc. Lond., 1832:39.

TYPE SPECIES: *Pseudomys australis* Gray, 1832.

SYNONYMS: *Gyomys* Thomas, 1910; *Paraleporillus* Martinez and Lidicker, 1971; *Thetomys* Thomas, 1910.

COMMENTS: *Pseudomys* Division. Member of the Australian Old Endemics (Musser, 1981c), part of which is the Conilurini where Lee et al. (1981) and Baverstock (1984) placed *Pseudomys*. The genus is now regarded by some as part of an expanded Hydromyini (which includes Australian and New Guinea genera in our *Hydromys*, *Uromys*, and *Xeromys* Divisions), an "Australasian clade" defined by similarities in albumin immunology and spermatozoal structure (Watts and Baverstock, 1994a, 1995b, 1996). Taxonomic, distributional, and biological references to all species cataloged by Mahoney and Richardson (1988).

Data from several character suites have been used to estimate relationships among species of *Pseudomys*: anatomy of male and female reproductive tracts (Breed, 1980, 1985, 1986); phallic morphology (Lidicker and Brylski, 1987; Morrissey and Breed, 1982); spermatozoal morphology (Breed, 1983, 1984, 1997; Breed and Sarafis, 1978); electrophoretic variation (Baverstock et al., 1977a, 1981); albumin immunology (Watts et al., 1992); and chromosomes (Baverstock et al., 1977c, 1983b). Despite such different approaches, no study estimating phylogenetic relationships is available that integrates all these data with information from skins, skulls, and dentitions. Consequently, opinion still ranges from including all the distinctive species groups in one genus (*Pseudomys*), to allocating them to three genera (*Pseudomys*, *Gyomys*, and *Leggadina*), to merging *Mastacomys* and *Leporillus* with *Pseudomys* (see discussions in Baverstock et al., 1981; Breed, 1983; Lidicker and Brylski, 1987; Watts and Aslin, 1981; and Watts et al., 1992). Watts et al. (1992) discussed the futility of attempting to split *Pseudomys* and they regarded it as a single but complex genus in which results from albumin immunology comfirmed its monophyly relative to the other Australian murines except *Mastacomys*, which they merged with *Pseudomys*. In contrast, Mahoney and Richardson (1988:160) and Happold (1995) retained *Mastacomys* as a separate genus, a course we follow. A recently completed molecular phylogenetic study by F. Ford (James Cook University), utilizing a combination of mitochondrial and nuclear genes, provides strong evidence that *Pseudomys* is paraphyletic with respect to both *Mastacomys* and *Notomys* (F. Ford, in litt., 2004). Ford's phylogeny is currently being integrated with all other available morphological, cytological and immunological datasets, and with new observations on cranial and dental morphology (F. Ford and K. Aplin, in litt., 2004), to produce a new generic arrangement for this key group of Australian Old Endemics. The use of plantar pad patterns to help identify three species of *Pseudomys* in contrast to *Mus* and *Rattus* is described by Cooper (1993, 1994). Anatomy of mammary glands in two species is described by Griffiths and Simms (1993).

There are at least two undescribed species of *Pseudomys*. One (Basalt Plains Mouse) is represented by hundreds of specimens "collected from owl-roost deposits and Aboriginal middens on the basalt plains of western Victoria, particularly in lava tubes associated with dormant volcanos" (Flannery, 1995e:619). It may have survived into the nineteenth century and vanished when its natural habitat of tussock grassland was destroyed by sheep. The other is morphologically similar to but distinct from *P. australis* and represented by subfossils from S South Australia (Watts and Aslin, 1981); the name *auritus*, now listed as one of the synonyms under *P. australis*, may apply to this taxon (Robinson et al., 2000). Van Dyck (1997) reported another potentially undescribed living species from Queensland west of the Great Dividing Range from near Cloncurry to Camooweal. On DNA evidence, this population was found to be very close to *P. johnsoni* (F. Ford, in litt., 2004). *Pseudomys vandyki*, the earliest Australian record of *Pseudomys*, is represented by molars from probable late Pliocene sediments (Aplin, 2005; Godthelp, 1990).

K. Aplin (in litt., 2004) noted that many of the historically recorded species of *Pseudomys* declined so quickly that they are currently known from only a few museum specimens, often in poor condition and from widely scattered localities. Aplin views the current taxonomy of *Pseudomys* to seriously underestimate the real historic diversity, and states that the true diversity will only be revealed by a combination of 1) systematic survey of subfossil assemblages, 2) comprehensive DNA sampling of historical specimens, and 3) selective DNA sampling of subfossils to link the two.

Pseudomys albocinereus (Gould, 1845). Proc. Zool. Soc. Lond., 1845:78.
COMMON NAME: Ash-gray Pseudomys.
TYPE LOCALITY: Australia, Western Australia, "scrubby plains near Perth" (as restricted by Thomas' lectotype designation; see Mahoney and Richardson, 1988:171).
DISTRIBUTION: Australia, SW Western Australia (from Shark Bay area southeast to Israelite Bay); also found on islands of Bernier, Dorre, Shark Bay, and Woody (Watts and Aslin, 1981:196, and Morris, 1995:584).
STATUS: IUCN – Lower Risk (lc).
SYNONYMS: *squalorum* (Thomas, 1907).
COMMENTS: Analysis of phallic morphology suggested *P. albocinereus* belongs in group with *P. fumeus* and *P. shortridgei* (Lidicker and Brylski, 1987), but electrophoretic data placed it in a cluster containing *P. apodemoides* and seven other species, excluding *P. fumeus* and *P. shortridgei* (Baverstock et al., 1981). Dental traits suggested *P. albocinereus* is closely related to the Pliocene *P. vandycki*, and if this resemblance reflects monophyly, the two species form a distinct group within *Pseudomys* (Godthelp, 1990). But Watts (in litt.) wrote us that "virtually all data supports close relationships between *P. apodemoides* and *P. albocinereus*. Relationships beyond this are any one's guess." See also Watts et al. (1992).

Pseudomys apodemoides Finlayson, 1932. Trans. R. Proc. Soc. S. Aust., 56:170.
COMMON NAME: Silky Pseudomys.
TYPE LOCALITY: Australia, Southern Australia, Coombe.
DISTRIBUTION: Australia; Mallee-heath of SE South Australia and W Victoria (Murray-Darling Basin); see maps in Cockburn (1995a:586), Menkhorst (1995c:214), and Watts and Aslin (1981:199, only the portion in SE South Australia and W Victoria); also see Robinson et al. (2000).
STATUS: IUCN – Lower Risk (lc).
COMMENTS: Phylogenetic relationships are equivocal to some (see discussion in Lidicker and Brylski, 1987:635, and references therein), but not to other workers (see preceding account).

Pseudomys australis Gray, 1832. Proc. Zool. Soc. Lond., 1832:39.
COMMON NAME: Plains Pseudomys.
TYPE LOCALITY: Australia, New South Wales, SW side of Liverpool Plains; see Mahoney and Richardson (1988:172).
DISTRIBUTION: Australia; New South Wales, S Queensland, South Australia, and S Northern Territory; late Pleistocene to Recent remains from W Victoria (Breed and Head, 1991; Robinson et al., 2000; Watts and Aslin, 1981; Williams and Menkhorst, 1995b); probably extinct in New South Wales (Mahoney and Richardson, 1988:172), where the

species has not been found alive for more than 100 years (Ellis, 1995); present range summarized by Watts (1995*e*).

STATUS: IUCN – Vulnerable.

SYNONYMS: *auritus* Thomas, 1910; *flavescens* Troughton, 1936; *lineolatus* (Gould, 1845); *minnie* Troughton, 1932; *murinus* (Gould, 1845); *stirtoni* (Martinez and Lidicker, 1971).

COMMENTS: Microscopic structure of hooks on sperm head reported by Flaherty and Breed (1982), and protein composition of ventral processes on the sperm head and its significance documented by Breed et al. (2000). Phallic information suggested *P. australis* is related to *P. gouldii, P. higginsi,* and *P. nanus,* to the exclusion of other species of *Pseudomys* (Lidicker and Brylski, 1987:635); electrophoretic data (Baverstock et al., 1981) and spermatozoal morphology (Breed, 1983) concordant with this association. Reviewed by Watts and Aslin (1981) and Watts (1995*e*). Comparative ecology of two populations in N South Australia is reported by Brandle and Moseby (1999), and distribution, ecology, and conservation status are documented by Brandle et al. (1999).

Pseudomys bolami Troughton, 1932. Rec. Aust. Mus., 18:292.

COMMON NAME: Bolam's Pseudomys.

TYPE LOCALITY: Australia, South Australia, Ooldea.

DISTRIBUTION: Australia, S South Australia and S Western Australia (Kitchener, 1985:216, and Watts, 1995*f*:588; Robinson et al., 2000); possibly once occurred in W Victoria (Menkhorst and Williams, 1995).

STATUS: IUCN – Lower Risk (lc).

COMMENTS: Originally described by Troughton (1932*a*) as a subspecies of *P. hermannsburgensis,* but distinguished from it and redescribed by Kitchener et al. (1984*a*), who also reported sympatry of both species at Goongarrie, Western Australia. Reviewed by Watts (1995*f*).

Pseudomys calabyi Kitchener and Humphreys, 1987. Rec. West. Aust. Mus., 13:296.

COMMON NAME: Kakadu Pebble-mound Pseudomys.

TYPE LOCALITY: Australia, Northern Territory, Uranium Development Project Falls, 100 m.

DISTRIBUTION: Australia, recorded only from a small area in N part of Northern Territory (Woinarski et al., 1995*a*).

COMMENTS: Kitchener and Humphreys (1987) proposed *calabyi* as a distinctive subspecies of *P. laborifex* (see also Mahoney and Richardson, 1988:177), but Woinarski et al. (1995*a*) claimed that study of more material collected between 1988 and 1990 confirms its specific status. The small range of *P. calabyi* is allopatric to that of *P. laborifex.*

Pseudomys chapmani Kitchener, 1980. Rec. West. Aust. Mus., 8:405.

COMMON NAME: Western Pebble-mound Mouse.

TYPE LOCALITY: Australia, Western Australia, Pilbara Dist., East Hammersley Range, West Angelas Mine Site (Kitchener, 1980, provided additional information).

DISTRIBUTION: Australia, NW Western Australia; extant specimens known only from Pilbara Dist. (Kitchener, 1985:216), but distribution of pebble mounds indicates range once extended through Gascoyne to Murchison Dist. with S limit near Mileura, N limit the Great Sandy Desert, and E limit the Gibson Desert (Dunlop and Pound, 1981; Start and Kitchener, 1995).

STATUS: IUCN – Lower Risk (lc).

COMMENTS: This species builds pebble mounds and is sympatric with *P. hermannsburgensis* (Kitchener, 1980), which does not construct such mounds (Dunlop and Pound, 1981), but phylogenetically most closely allied to *P. johnsoni,* another species that constructs pebble mounds (Kitchener, 1985). The use of pebble mounds as indicators of the presence of *P. chapmani* is documented by Anstee (1996). Reviewed by Start and Kitchener (1995).

Pseudomys delicatulus (Gould, 1842). Proc. Zool. Soc. Lond., 1842:13.

COMMON NAME: Delicate Pseudomys.

TYPE LOCALITY: Australia, Northern Territory, Port Essington (as restricted by Thomas's lectotype designation; see Mahoney and Richardson, 1988:173).

DISTRIBUTION: Torresian distribution in N Australia; N coastal region from near Port

Hedland in Western Australia to Bundaberg area in Queensland, including some nearshore islands (Watts and Aslin, 1981:188; Braithwaite and Covacevich, 1995:593). SC Papua New Guinea, Morehead region on the trans-Fly plains (Flannery, 1995a; Waithman, 1979).

STATUS: IUCN – Lower Risk (nt).

SYNONYMS: *mimulus* Thomas, 1926; *pumilus* Troughton, 1936.

COMMENTS: Electrophoretic data and spermatozoal morphology supported a close relationship between *P. delicatulus*, *P. novaehollandiae*, and *P. pilligaensis* (Breed, 1983; Briscoe et al., 1981). Sperm head morphology documented by Breed (2000) in context of distinguishing *P. delicatulus* from *P. patrius*, and revealing significant geographic variation within the former. Reviewed by Watts and Aslin (1981) and Braithwaite and Covacevich (1995). There is no evidence of range reduction in Australia from the time of European settlement (Braithwaite and Covacevich, 1995). Palmer (2001) described some new collection sites of *P. delicatulus* in Queensland representing range extensions. *Pseudomys delicatulus* is one of 17 Australian species of mammals that are also found in the Trans-Fly region of SC New Guinea and nowhere else on the island (Norris and Musser, 2001).

Pseudomys desertor Troughton, 1932. Rec. Aust. Mus., 18:293.

COMMON NAME: Desert Pseudomys.

TYPE LOCALITY: Australia, "Central Australia" (see Mahoney and Richardson, 1988:174).

DISTRIBUTION: Arid and semiarid tropical regions of Australia; Western Australia, South Australia, Northern Territory, Queensland, New South Wales, and possibly Victoria; also recorded from Bernier Isl, Western Australia (Kerle, 1995a:595; Mahoney and Richardson, 1988:174); Robinson et al. (2000) summarized the distribution in South Australia; L. M. Williams (1995c) discussed the possible distribution in Victoria.

STATUS: IUCN – Lower Risk (nt).

SYNONYMS: *murrayensis* (Krefft, 1862); *subrufus* (Krefft, 1862).

COMMENTS: Spermatozoal morphology similar to *P. australis* and many other species of *Pseudomys* (Breed, 1983). Both names listed above are unused senior synonyms of *desertor* and should not be resurrected (Mahoney and Richardson, 1988:174). Reviewed by Kerle (1995a). Read et al. (1999) documented the distribution, ecology, and current status of *P. desertor* in South Australia.

Pseudomys fieldi (Waite, 1896). Rept. Horn Sci. Exped. Cent. Aust., Zool., 2:403.

COMMON NAME: Shark Bay Pseudomys.

TYPE LOCALITY: Australia, S Northern Territory, Alice Springs.

DISTRIBUTION: Australia. The only natural living population occurs on Bernier Isl in Shark Bay, Western Australia; some animals from there were translocated to Doole Isl in Exmouth Gulf in 1993 (Morris and Robinson, 1995).

STATUS: CITES – Appendix I as *P. praeconis*; U.S. ESA – Endangered as *P. fieldi* and *P. praeconis*; IUCN – Critically Endangered as *P. fieldi*, Vulnerable as *P. praeconis*.

SYNONYMS: *praeconis* Thomas, 1910.

COMMENTS: Clustered with most other species of *Pseudomys*, judged by electrophoretic data (Baverstock et al., 1981, reported under *praeconis*). In the 1980s, *fieldi* was thought to be represented only by the holotype collected at Alice Springs in Northern Territory in 1895 and *praeconis* was known only from Bernier Isl. While listing *fieldi* as a species, Watts and Aslin (1981:171) wrote that "It is difficult to determine whether or not this represents a distinct species or a rather aberrant specimen of some other species." Subfossil samples have now been discovered along the west coast south of Shark Bay, and through Western Australia (the upper Gascoyne, northern Goldfields, and Gibson Desert) to the S region of Northern Territory. Study of this material and the holotypes of *fieldi* and *praeconis* indicates that all the samples represent the same species and that it once had an extensive mainland distribution before European settlement (Morris and Robinson, 1995).

Pseudomys fumeus Brazenor, 1934. Mem. Nat. Mus. Melb., 8:158.

COMMON NAME: Smoky Pseudomys.

TYPE LOCALITY: Australia, Victoria, Turton's Pass, Otway Forest.

DISTRIBUTION: Australia, Victoria (Cockburn, 1995*b*:599; Menkhorst, 1995*d*:219; Watts and Aslin, 1981:202). Range during late Pleistocene extended into E New South Wales (Wakefield, 1972*a*).

STATUS: U.S. ESA – Endangered; IUCN – Vulnerable.

COMMENTS: Electrophoretic data separated *P. fumeus* from other species of *Pseudomys* (Baverstock et al., 1981); phallic morphology suggested a close tie between *P. fumeus*, *P. albocinereus*, and *P. shortridgei*, a group for which the generic name *Gyomys* is available (Lidicker and Brylski, 1987:635); but spermatozoal structure tied *P. fumeus* with most other species of *Pseudomys* (Breed, 1983). Reviewed by Watts and Aslin (1981) and Cockburn (1995*b*).

Pseudomys glaucus Thomas, 1910. Ann. Mag. Nat. Hist., ser. 8, 6:609.

COMMON NAME: Blue-gray Pseudomys.

TYPE LOCALITY: Australia, S Queensland.

DISTRIBUTION: Australia, Murray-Darling basin in New South Wales and S Queensland (Mahoney and Richardson, 1988:175).

STATUS: IUCN – Critically Endangered. Possibly extinct (Mahoney and Richardson, 1988:175).

COMMENTS: Questionably listed as a synonym of *P. apodemoides* by Watts and Aslin (1981), and and sometimes considered synonymous with that species (Cockburn, 1995*a*), but no data supports such an association (Aplin, in litt., 2004; our survey of relevant literature).

Pseudomys gouldii (Waterhouse, 1839). Zool. Voy. H.M.S. "Beagle," Mammalia, 2:67.

COMMON NAME: Gould's Pseudomys.

TYPE LOCALITY: Australia, New South Wales, north of Hunter River (as restricted by Thomas' lectotype designation; see Mahoney and Richardson, 1988:175).

DISTRIBUTION: Australia; range based on Recent and subfossil specimens includes E South Australia and New South Wales (Dixon, 1995*e*); subfossils of either *P. gouldii* or *P. fieldi* (the two are difficult to distinguish) are from localities widespread across South Australia (Robinson et al., 2000.

STATUS: U.S. ESA – Endangered; IUCN – Extinct.

SYNONYMS: *rawlinnae* Troughton, 1932.

COMMENTS: For full citation and other information see Mahoney and Richardson (1988:175). Apparently extinct (Mahoney and Richardson, 1988:175); no live animals seen or collected since the middle 1850s (Watts and Aslin, 1981:169). Phallic morphology indicated *gouldii* is clustered with *P. australis*, *P. higginsi*, and *P. nanus* (Lidicker and Brylski, 1987). Reviewed by Watts and Aslin (1981) and Dixon (1995*e*).

Pseudomys gracilicaudatus (Gould, 1845). Proc. Zool. Soc. Lond., 1845:77.

COMMON NAME: Eastern Chestnut Pseudomys.

TYPE LOCALITY: Australia, Queensland, Darling Downs, Oakey Creek; see Mahoney and Richardson (1988:175).

DISTRIBUTION: Australia; modern range along the eastern coast from Townsville in N Queensland to Sydney area in New South Wales; subfossil specimens from farther south in New South Wales (Mahoney and Posamentier, 1975) and from S Victoria (Fox, 1995*a*:601; Watts and Aslin, 1981:180; Watts and Tweedie, 1992).

STATUS: IUCN – Lower Risk (lc).

SYNONYMS: *ultra* Troughton, 1939.

COMMENTS: Phylogenetically closely related to *P. nanus*, an estimate based on spermatozoal morphology (Breed, 1983), electrophoretic data (Baverstock et al., 1977*a*, 1981), and morphology of skin, skull, and teeth (Watts and Aslin, 1981). Fox (1995*a*) provided a short recent review.

Pseudomys hermannsburgensis (Waite, 1896). Rept. Horn Sci. Exped. Cent. Aust., Zool., 2:405.

COMMON NAME: Sandy Inland Pseudomys.

TYPE LOCALITY: Australia, Northern Territory, George Gill Range (as restricted by lectotype designation; see Mahoney and Richardson, 1988:176).

DISTRIBUTION: Australia; arid parts of Western Australia, South Australia, Northern Territory,

W Queensland, W New South Wales, and NW Victoria (Breed, 1995*c*:604; Robinson et al., 2000; Watts and Aslin, 1981:190; see also the range retraction by Ellis, 1995*b*).
STATUS: IUCN – Lower Risk (lc).
SYNONYMS: *brazenori* Troughton, 1937.
COMMENTS: Electrophoretic data (Baverstock et al., 1981) and spermatozoal morphology (Breed, 1983) clustered *P. hermannsburgensis* with most other species in *Pseudomys*, but phallic morphology interpreted by Lidicker and Brylski (1987) indicated closer affinity to species they placed in *Leggadina*. Digestive tract traits in relation to diet documented by Murray et al. (1994). Reviewed by Breed (1995*c*).

Pseudomys higginsi (Trouessart, 1897). Cat. Mamm. Viv. Foss., 1:473.
COMMON NAME: Long-tailed Pseudomys.
TYPE LOCALITY: Australia, Tasmania, Kentishbury; see Mahoney and Richardson (1988:177).
DISTRIBUTION: Australia; extant population known only from Tasmania (Rounsevell et al., 1991); represented on mainland in Victoria and E New South Wales by late Pleistocene samples (Wakefield, 1972*b*).
STATUS: IUCN – Lower Risk (lc).
SYNONYMS: *australiensis* Wakefield, 1972; *leucopus* (Higgins and Petterd, 1881) [not Rafinesque, 1818].
COMMENTS: Spermatozoal morphology (Breed, 1983) and electrophoretic data (Baverstock et al., 1981) placed *P. higginsi* with most other species of *Pseudomys*, but phallic anatomy (Lidicker and Brylski, 1987) clustered *P. higginsi* with *P. australis, P. gouldii,* and *P. nanus.* Wakefield (1972*b*) described *australiensis* as a subspecies based on late Pleistocene fossils. Reviewed by Watts and Aslin (1981) and Green (1995).

Pseudomys johnsoni Kitchener, 1985. Rec. West. Aust. Mus., 12:208.
COMMON NAME: Central Pebble-mound Pseudomys.
TYPE LOCALITY: Australia, C Northern Territory, Kurundi Station, Kurinelli Mine, 150 m (Kitchener, 1985, provided additional information); also Queensland west of the Great Dividing Range, from near Cloncurrys to Camooweal (Van Dyck, 1997, as potentially undescribed species).
DISTRIBUTION: Australia; originally reported from small area in arid C Northern Territory (Kerle, 1995*b*:608; Kitchener, 1985:216). Recent discovery of *P. johnsoni* in Queensland highlights lack of detailed survey data for many parts of northern Australia.
STATUS: IUCN – Lower Risk (nt).
COMMENTS: Closest phylogenetic relative is apparently *P. chapmani*, which occurs in NW Western Australia (Kitchener, 1985).

Pseudomys laborifex Kitchener and Humphreys, 1986. Rec. West. Aust. Mus., 12:420.
COMMON NAME: Kimberley Pseudomys.
TYPE LOCALITY: Australia, N Western Australia, Kimberley Region, Mitchell Plateau, adjacent to Camp Creek, 270 m (Kitchener and Humphreys, 1986:421, provided more information, including description of habitat at type locality).
DISTRIBUTION: Australia, N Western Australia and N Northern Territory (Kitchener and Humphreys, 1986:430, 1987:292).
STATUS: IUCN – Lower Risk (lc).
COMMENTS: Reviewed by Kitchener (1995).

Pseudomys nanus (Gould, 1858). Proc. Zool. Soc. Lond., 1858:242.
COMMON NAME: Western Chestnut Pseudomys.
TYPE LOCALITY: Australia, Western Australia, Victoria Plains (as restricted by Thomas' lectotype designation; see Mahoney and Richardson, 1988:177).
DISTRIBUTION: Australia; W coast of Western Australia (also Barrow Isl), between Port Hedland and the Barkly Tableland in NE Western Australia, N Northern Territory, and NW Queensland (also South-West Isl in Gulf of Carpentaria); Watts and Aslin (1981: 177); once ranged through W part of Western Australia (Robinson, 1995*c*:610).
STATUS: IUCN – Lower Risk (nt).
SYNONYMS: *ferculinus* (Thomas, 1902).
COMMENTS: Type species of *Thetomys*. A close relative of *P. gracilicaudatus* to the exclusion of

other species of *Pseudomys*, judged by analyses of electrophoretic data (Baverstock et al., 1977*a*, 1981); but a member of the group that includes only *P. australis*, *P. gouldi*, and *P. higginsi* based on phallic morphology (Lidicker and Brylski, 1987); however, not distinctive relative to most other species in the genus as judged by spermatozoal form (Breed, 1983). Anatomy of mammary glands described by Griffiths and Simms (1993). Reviewed by Watts and Aslin (1981) and Robinson (1995*c*).

Pseudomys novaehollandiae (Waterhouse, 1843). Proc. Zool. Soc. Lond., 1842:146 [1843].
COMMON NAME: New Holland Pseudomys.
TYPE LOCALITY: Australia, New South Wales, upper Hunter River, Yarrundi (as restricted by Thomas' lectotype designation; see Mahoney and Richardson, 1988:178).
DISTRIBUTION: Australia; Coastal region of E New South Wales, S Victoria, and N Tasmania (Kemper, 1995*b*:611; Rounsevell et al., 1991:711; Seebeck and Menkhorst, 1995:222; Watts and Aslin, 1981:193; Wilson, 1994:47, 1996:32).
STATUS: U.S. ESA – Endangered; IUCN – Lower Risk (lc).
COMMENTS: Type species of *Gyomys*. External, cranial, and dental morphology, along with electrophoretic data and spermatozoal anatomy, pointed to a close relationship between *P. novaehollandiae* and *P. pilligaensis* (Breed, 1983; Briscoe et al., 1981). Wilson and Roede (1997) described how populations can be sampled using hair sampling tubes. Reviewed by Watts and Aslin (1981) and Kemper (1995*b*). Distribution and its correlation with vegetation type in Victoria documented by Lock and Wilson (1999).

Pseudomys occidentalis Tate, 1951. Bull. Am. Mus. Nat. Hist., 97:246.
COMMON NAME: Western Pseudomys.
TYPE LOCALITY: Australia, Western Australia, Tambellup.
DISTRIBUTION: Australia; extant range in SW Western Australia, subfossil specimens indicate species extended along S coastline to Kangaroo Isl off coast of South Australia (Kitchener, 1992; Robinson et al., 2000; Watts and Aslin, 1981:205).
STATUS: U.S. ESA and IUCN – Endangered.
COMMENTS: Considered "rare and likely to become extinct," but range was contracting before arrival of Europeans (Watts and Aslin, 1981:205). Electrophoretic data clustered *P. occidentalis* with all other *Pseudomys* analyzed except *P. fumeus*, *P. gracilicaudatus*, *P. nanus*, and *P. shortridgei* (Baverstock et al., 1981). Reviewed by Whisson and Kitchener (1995).

Pseudomys oralis Thomas, 1921. Ann. Mag. Nat. Hist., ser. 9, 8:621.
COMMON NAME: Hastings River Pseudomys.
TYPE LOCALITY: Australia; exact place unknown, but likely located in NE New South Wales or SE Queensland (Mahoney and Richardson, 1988:179).
DISTRIBUTION: Australia; extant specimens from NE New South Wales and SE Queensland, but late Pleistocene fossils are from farther south in New South Wales and E Victoria (Gynther and O'Reilly, 1995; Meek and Triggs, 1999; Poole, 1994; Read, 1993; Watts and Aslin, 1981:170).
STATUS: IUCN – Endangered.
COMMENTS: Originally described as a subspecies of *P. australis*. Once considered to be the least known and rarest of *Pseudomys* (Watts and Aslin, 1981), but is less rare than thought although extremely local in distribution (Kirkpatrick, 1995). Body size and geographic distribution documented by Read (1993)

Pseudomys patrius (Thomas and Dollman, 1909). Proc. Zool. Soc. Lond., 1908:791.
COMMON NAME: Eastern Pebble-mound Pseudomys.
TYPE LOCALITY: Australia; Queensland, Mount Inkerman; see Mahoney and Richardson (1988:179).
DISTRIBUTION: Australia; coastal region of NE Queensland, along the Gread Dividing Range from near Townsville to Kilkivan (Van Dyck, 1997).
STATUS: IUCN – Vulnerable.
COMMENTS: Treated as a distinct species by Fox and Briscoe (1980) and Mahoney and Richardson (1988:179), but earlier judged by Mahoney to be a synonym of *P. delicatulus* (Kitchener, 1985:218), where it was placed by Watts and Aslin (1981) and

Braithwaite and Covacevich (1995). Recently *patrius* was returned to the status of a species (Van Dyck, 1997), an action reinforced by Breed's (2000) study of sperm head morphology: "*P. patrius*. . . has a sperm head morphology that differs markedly in its structural organisation from the spermatozoon of all *P. delicatulus* so far examined."

Pseudomys pilligaensis Fox and Briscoe, 1980. Aust. Mamm., 3:112.
    COMMON NAME: Pilliga Pseudomys.
    TYPE LOCALITY: Australia, New South Wales, Merriwindi State Forest, 3 km west of Pilliga-Baradine Road., Cumberdeen Road (Fox and Briscoe, 1980, provided additional information).
    DISTRIBUTION: Australia, N New South Wales, collected from a few localities within the Pilliga Scrub (Fox and Briscoe, 1980:119).
    STATUS: IUCN – Vulnerable.
    COMMENTS: Chromosomal morphology described by Fox and Briscoe (1980). Morphological and electrophoretic data supported a close phylogenetic relationship of *P. pilligaensis* to *P. delicatulus* and *P. novaehollandiae* (Briscoe et al., 1981), which was reinforced by spermatozoal morphology (Breed, 1983). Reviewed by Fox (1995*b*). Recent DNA sequencing studies by F. Ford reveal a complex mosaic of *delicatulus* and *novaehollandiae* genotypes and phenotypes within the geographic area occupied by '*pillagaensis*,' raising the distinct possibility that the latter 'taxon' actually identifies a zone of hybrid interaction between *delicatulus* and *novaehollandiae* (F. Ford, in litt., 2004).

Pseudomys shortridgei (Thomas, 1907). Proc. Zool. Soc. Lond., 1906:765 [1907].
    COMMON NAME: Heath Pseudomys.
    TYPE LOCALITY: Australia, SW Western Australia, neighborhood of Woyerling Reserve, 973 ft (297 m); additional information in Mahoney and Richardson (1988:180).
    DISTRIBUTION: Australia; S Western Australia and SW Victoria (Grampian Mtns and Portland areas); see maps in Watts and Aslin (1981:185) and Menkhorst (1995*e*:223); living specimen collected in 1967 from South Australia where the species is otherwise represented only by subfossils (Robinson et al., 2000).
    STATUS: U.S. ESA – Endangered; IUCN – Lower Risk (cd).
    COMMENTS: Thought to be extinct in Western Australia (Watts and Aslin, 1981), but recently rediscovered there (Baynes et al., 1987). Level of DNA sequence divergence between Western and South Australian populations examined by Cooper et al. (2003*b*). Electrophoretic data suggested *P. shortridgei* is phylogenetically isolated from all other species of *Pseudomys* (Baverstock et al., 1981); spermatozoal morphology unlike most other *Pseudomys* but similar to that of *P. delicatulus*, *P. novaehollandiae*, and *P. pilligaensis* (Breed, 1983); phallic anatomy linked *P. shortridgei* to *P. albocinereus* and *P. fumeus*, a group that could be generically recognized by calling it *Gyomys* (Lidicker and Brylski, 1987:635). Gross and microscopic anatomy of the gastrointestinal tract documented by Meulman et al. (1999). Reviewed by Watts and Aslin (1981) and Cockburn (1995*c*).

Rattus Fischer, 1803. Natl. Mus. Nat. Paris, 2:128.
    TYPE SPECIES: *Mus decumanus* Pallas, 1779 (see Hollister, 1916*b*; = *Mus norvegicus* Berkenhout, 1769).
    SYNONYMS: *Acanthomys* (Gray, 1867) [not Lesson, 1842, or Tokuda, 1941]; *Christomys* Sody, 1941; *Cironomys* Sody, 1941; *Epimys* Trouessart, 1881; *Geromys* Sody, 1941; *Mollicomys* Sody, 1941; *Octomys* (Sody, 1941) [not Thomas, 1922]; *Pullomys* Sody, 1941; *Rattus* Frisch, 1775; *Stenomys* Thomas, 1910; *Togomys* Dieterlen, 1989.
    COMMENTS: *Rattus* Division. *Rattus* Frisch, 1775, is unavailable. Corbet and Hill (1992:334) noted that Fischer's original spelling is *Ruttus* but ". . . there is no evidence within this publication that the spelling *Ruttus* was an error. However it has, very sensibly, been universally accepted as an error for *Rattus* and it would serve no useful purpose to revert to *Ruttus*." Sody (1941) proposed the genera *Christomys*, *Cironomys*, *Geromys*, *Mollicomys*, *Octomys*, and *Pullomys* for various species we list in *Rattus*; Corbet and Hill (1992) doubted the validity of *Mollicomys* and *Octomys* since they are based on mammary formulae shared with other murines (but see generic account of *Taeromys*). *Togomys* is based on *R. exulans*

(Dieterlen, in Ansell, 1989). Taxonomic changes altering the definition of *Rattus* as understood by systematists working in the middle 1900s (Chasen, 1940; Tate, 1936; Ellerman, 1941, 1949a, 1961; Ellerman and Morrison-Scott, 1951; Laurie and Hill, 1954; Simpson, 1945) have been documented and summarized by several specialists (Misonne, 1969; Musser, 1981b; Musser and Newcomb, 1983; Musser and Holden, 1991; Musser and Heaney, 1992). Species in many Asian and African genera recognized today were once embraced by the broad definition of *Rattus* (e. g., Chasen, 1940; Ellerman, 1941, 1949a; Ellerman et al., 1953): *Cremnomys*, *Millardia*, *Madromys*, *Berylmys*, *Sundamys*, *Kadarsanomys*, *Komodomys*, *Maxomys*, *Leopoldamys*, *Niviventer*, *Srilankamys*, *Margaretamys*, *Bunomys*, *Paruromys*, *Taeromys*, *Apomys*, *Bullimus*, *Limnomys*, *Tarsomys*, *Aethomys*, *Micaelamys*, *Praomys*, *Mastomys*, *Hylomyscus*, and *Myomyscus*. A few of these genera are phylogenetically allied to species currently arrayed in *Rattus* (members of *Rattus* Division; see Table 1), but others represent very distant lineages having separate origins from the early ancestral murine stock.

Even divorced of species in these genera, *Rattus* remains a heterogeneous accumulation of species. The wide range of morphological variation present disallows an objective generic diagnosis and indicates the present contents to consist of several monophyletic clusters that may or may not prove to be embraced by a single genus. This polyphyletic pattern is reflected in our sorting of species into the following six species groups, and a seventh assemblage containing unaffiliated species for want of adequate information or study.

1. *Rattus norvegicus* species group. *Rattus norvegicus*, the type species of the genus, is traditionally placed in subgenus *Rattus* with *R. rattus* and its close allies (e. g., Ellerman, 1941, Ellerman, 1949a), but the two diverge significantly in morphology (Musser's studies), allozymic variation, albumin immunology, and L1 (LINE-1) retrotransposons and amplification events, and mtDNA cytochrome *b* sequences (Aplin, in litt., 2004; Chan, 1977; Chan et al., 1979; Baverstock et al., 1983a, c, 1986; Verneau et al., 1997, 1998; Watts, in litt.). *Rattus nitidus* and *R. pyctoris* also belong here as indicated by molecular data and some morphological traits (see those accounts).

2. *Rattus exulans* species group. Another species usually included within subgenus *Rattus* along with *R. argentiventer*, *R. nitidus*, *R. norvegicus*, *R. rattus*, and others that have, in the past, formed the core of that group (e.g., Ellerman, 1941; Misonne, 1969). Some electrophoretic and chromosomal data have supported this allocation (Chan, 1977; Chan et al., 1979; Raman and Sharma, 1977), but a larger body of morphological, other biochemical data, mtDNA cytochrome *b* sequences, and L1 (LINE-2) retrotransposons and amplification events indicate that *R. exulans* is phylogenetically distant from species in our *R. rattus* group as well as from those in our *R. norvegicus* group (K. Aplin, in litt., 2004; Gemmeke and Niethammer, 1984; Medway and Yong, 1976; Pasteur et al., 1982; Verneau et al., 1997, 1998). Based on analyses of mtDNA cytochrome *b* sequences, *R. exulans* is as phylogenetically distant from the *R. rattus* group as are the *R. leucopus* (New Guinea-Moluccan lineage) and *R. fuscipes* (Australian lineage) assemblages from the *R. rattus* cluster (K. Aplin, in litt., 2004).

3. *Rattus rattus* species group. Species usually regarded to represent the core of species related to *R. rattus* (*R. adustus*, *R. andamanensis*, *R. argentiventer*, *R. baluensis*, *R. blangorum*, *R. burrus*, *R. hoffmanni*, *R. koopmani*, *R. losea*, *R. lugens*, *R. mindorensis*, *R. mollicomulus*, *R. osgoodi*, *R. palmarum*, *R. rattus*, *R. satarae*, *R. stoicus*, *R. simalurensis*, *R. tanezumi*, *R. tawitawiensis*, and *R. tiomanicus*; Musser and Holden, 1991). Whether or not this cluster, *R. norvegicus*, and *R. exulans*, form the monophyletic contents of *Rattus* proper is a possibility that has to be assessed by systematic revision and phylogenetic study (Musser and Heaney, 1985, 1992; Musser and Holden, 1991). Comparative chromosomal data for many of these species summarized by Musser and Holden (1991) and Rickart and Musser (1993), additional chromosomal descriptions for Vietnam samples reported by Bulatova et al. (1992), and silver-stained karyotypes of Chinese species compared by Chen et al. (1992). Schwarz and Schwarz (1967) offered an idiosyncratic revision of the group.

4. *Rattus fuscipes* species group. Native Australian species (*R. colletti*, *R. fuscipes*, *R. lutreolus*, *R. sordidus*, *R. tunneyi*, and *R. villosissimus*), revised by Taylor and Horner (1973) and reviewed by Watts and Aslin (1981); one (*R. sordidus*) also occurs on New Guinea. Al-

lozymic variation and chromosomal studies indicate that the species form a monophyletic cluster to the exclusion of either *R. rattus* or *R. norvegicus* (Baverstock et al., 1977*d*, 1983*a*, 1986), but "the extent of biochemical divergence between the Australian *Rattus* on the one hand, and either *R. rattus* or *R. norvegicus* on the other is no greater than that between *R. rattus* and *R. norvegicus*, thus supporting the notion that the Australian forms do indeed belong to the genus *Rattus*" (Baverstock et al., 1986:29). Baverstock et al. (1983*a*, 1986) proposed a phylogenetic hypothesis based on cladistic analyses of electrophoretic, immunologic, and chromosomal data, and another proposed set of relationships assessed by isozyme electrophoresis (Baverstock et al., 1986). Seddon and Baverstock (2000) compared evolutionary history of the second exon of RT1.Ba and intron b with the hypothesis suggested by Baverstock et al. (1986). Mahoney and Richardson (1988) cataloged taxonomic, distributional, and biological references.

Preliminary phylogenetic analyses of mtDNA cytochrome *b* sequences from exemplars in the *R. fuscipes*, *R. leucopus*, and *R. xanthurus* groups indicate that the *R. fuscipes* cluster (Australian endemics) is sister-group to the *R. leucopus* assemblage (New Guinea and Moluccan endemics), and that *R. xanthurus* and its allies (Sulawesi endemics) is sister to those two assemblages (K. Aplin et al., 2004). In Aplin's view, members of the *R. fuscipes* and *R. leucopus* groups each may represent ". . . discrete radiations albeit closely related and with some exchange across the Torres Strait." Should some supraspecific nomenclatural designation be warranted, Sody's 1941) *Geromys* is available for the Australian lineage, and Thomas's (1910*e*) *Stenomys* for the New Guinea-Mollucan radiation.

5. *Rattus leucopus* species group. Species indigenous to New Guinea and adjacent archipelagos (*R. arfakiensis*, *R. arrogans*, *R. giluwensis*, *R. jobiensis*, *R. leucopus*, *R. mordax*, *R. niobe*, *R. novaeguineae*, *R. omichlodes*, *R. praetor*, *R. richardsoni*, *R. steini*, *R. vandeuseni*, and *R. verecundus*); one (*R. leucopus*) also occurs in NE Australia. Most have been subjects of systematic revisions (Taylor et al., 1982, 1983). Known endemics of the Moluccas (*R. elaphinus*, *R. feliceus*, and two undescribed Moluccan species [Musser and Holden, ms]) belong in this group. Included here also are species Musser and Carleton (1993) and Flannery (1995*a*) placed in *Stenomys*. All former *Stenomys* are members of a *Rattus* clade as determined by albumin immunology (Watts and Baverstock, 1994*b*, 1995*b*, 1996; *R. niobe* was the only species sampled), skin and skull morphology (specimens in AMNH and BMNH), and spermatozoal traits (Breed and Aplin, 1994). Multivariate analyses of morphometric traits brought together *R. niobe*, *R. verecundus*, and *R. richardsoni* in a cluster separate from other native species of New Guinea *Rattus* (Taylor et al., 1982), but phylogenetic analyses of mtDNA cytochrome *b* sequences indicate *R. verecundus* to be very closely related to *R. leucopus* (K. Aplin, in litt., 2004) and Aplin (2005) has suggested that *Stenomys* should be subsumed again within *Rattus*. Rümmler (1938), on morphological grounds, appreciated the alliance between *leucopus* and *verecundus* and treated them as species of *Stenomys*, along with *R. niobe* and the Seramese *Nesoromys ceramicus*. Flannery (1990*b*, 1995*a*, *b*) provided photographs of animals along with distributional and biological summaries.

6. *Rattus xanthurus* species group. Species native to Sulawesi, adjacent Pulau Peleng, and Pulau Sangir in Kepulauan Sangir off tip of NE peninsula of Sulawesi (*R. bontanus*, *R. marmosurus*, *R. pelurus*, *R. salocco*, *R. xanthurus*, and an undescribed species from Kepulauan Sangir, off tip of NE Sulawesi). Musser and Holden (1991) speculated this monophyletic cluster may not belong in *Rattus*, but preliminary phylogenetic analyses of mtDNA cytochrome *b* sequences indicates the Sulawesi endemics are sister to the *R. leucopus* (New Guinea and Moluccas) and *R. fuscipes* (Australia) groups, and all three fall within the bounds of *Rattus* if such distinctive clusters as the *R. rattus* and *R. exulans* groups are included (K. Aplin, in litt., 2004).

7. *Rattus* species group unresolved. Phylogenetic affinities of *R. annandalei*, *R. enganus*, *R. everetti*, *R. hainaldi*, *R. hoogerwerfi*, *R. korinchi*, *R. macleari*, *R. montanus*, *R. morotaiensis*, *R. nativitatis*, *R. ranjiniae*, *R. sanila*, and *R. timorensis* are unresolved; some will eventually be excluded from the genus (see individual species accounts).

Reports presenting nonmorphological information pertinent to understanding evolutionary relationships among species of *Rattus* include chromosomal data (Baverstock et al., 1883*a*; Capanna and Corti, 1991; Gadi and Sharma, 1983; Milyutin, 1990; Raman and Sharma, 1977; Rickart and Musser, 1993; Yosida, 1980), allozymic variation (Baverstock et al., 1986; Watts and Baverstock, 1994*b*), and L1 (LINE-1) retrotransposons and amplification events (Verneau et al., 1997, 1998); additional studies are referenced in the species accounts. Mahoney and Richardson (1988) summarized taxonomic, distibutional, and biological references for the Australian species.

Descriptions of nine undescribed species of *Rattus* are being prepared by various researchers. One is from Pulau Bisa (Bisa Rat) in Kepulauan Obi of the Moluccas that is represented only by a skull (Flannery, 1995*b*). Another was collected on Mt Kemenagi in the Kikori River Basin of S Papua and the Mt Karimui area (Leary and Seri, 1997). Two additional new species were found in archaeological deposits on Morotai Isl in the N Moluccas (Flannery et al., 1998; Helgen, 2003*b*), a fifth (possibly not referable to *Rattus*) has been found on Timor by K. Helgen (see account of *R. timorensis*), two occur on Pulau Taliabu in the Sula Arch. (Musser, 1981*c*; Musser and Holden, ms), one is known from upland habitat near Loei in N Thailand (Aplin, in litt., 2004), and the last, a chromosomally and morphologically distinct member (a close relative of *R. villosissimus*) of the *R. fuscipes* group, is found in C Queensland, Australia (Aplin, in litt., 2004).

Evolutionary history of *Rattus* extends to late Pliocene in Asia. Two species, identified only as *Rattus* sp., are represented by fossils recovered from middle to late Pleistocene cave sediments in the Sichuan-Guizhou region (Zheng, 1993) and Guangxi Province (Chen et al., 2002) of S China. In Thailand, the extant *R. andamanensis* (reported as *sikkimensis*) has been documented back to early Pleistocene, the extinct *R. jaegeri* is represented by fragments from late Pliocene and early Pleistocene cave sediments, and specimens identified as *Rattus* sp. come from middle Pleistocene strata (Chaimanee, 1998). See also Chaimanee and Jaeger (2000*a*) for a look at diversity of Thai *Rattus* during the Plio-Pleistocene. The extinct *R. trinilensis* and at least five other species of *Rattus* have been recovered from Pleistocene sediments in C and E Java and represent the earliest records of the genus on the Sunda Shelf (already present by 1.6 million years ago; Musser, 1982*d*; Van der Meulen and Musser, 1999). In the Hondo region of Japan, *R. norvegicus* is represented by Holocene and late Pleistocene fossils, and a form identified as *R.* aff. *norvegicus* from the middle Pleistocene that may have been ancestral to *R. norvegicus* in the Japanese Isls (Kowalski and Hasegawa, 1976; Kawamura, 1989). The extinct *R. haasi* comes from late Acheulian (middle Pleistocene) cave sediments in Israel (Tchernov, 1968, 1996). *Rattus casimcensis* and *R. dobrogicus* are represented by fragments from the late Pleistocene of Romania (Kowalski, 2001).

Some fossils originally identified as *Rattus* are examples of other genera. The m1 reported as "cf. *Rattus*" from the Chinese Pliocene of Shaanxi (Jacobs and Li, 1982) is an example of extinct Chinese *Saidomys* (Cai and Qiu, 1993). *Rattus* sp., also based upon an m1, was recorded from Plio-Pleistocene Pinjor beds in the Siwaliks of NW India (Gaur, 1986), but the cusp pattern resembles those in *Cremnomys* and *Millardia* rather than *Rattus* (Musser's observations). Finally, late Pleistocene isolated molars from N Pakistan Siwaliks identified as "cf. *Rattus*" by Jacobs (1978) are examples of *Hadromys loujacobsi* (Musser, 1987*b*). No fossils of true *Rattus* have yet been recovered from Plio-Pleistocene Siwalik strata in either N Pakistan or NW India (Musser, 1987*b*; Patnaik, 2000, 2001).

Rattus adustus Sody, 1940. Treubia, 17:397.

> COMMON NAME: Burnished Enggano Rat.
> TYPE LOCALITY: Indonesia, Pulau Enggano, off the coast of W Sumatra (and off the continental shelf), Kiojoh, sea level.
> DISTRIBUTION: Pulau Enggano.
> STATUS: IUCN – Vulnerable.
> COMMENTS: *Rattus rattus* species group. Still known only by the holotype. Although Sody (1940) described *adustus* as a species, he later listed it as a subspecies of *R. rattus* in a section also containing *lugens* and *mentawai*, populations endemic to the Mentawai Isls (Sody, 1941). In morphology and geographic proximity, *R. adustus* is related to *R. lugens*, the two are allies of *R. simalurensis* from the Simalur Arch., and all three share close kinship with Sundaic *R. tiomanicus* (Musser and Heaney, 1985; Musser, ms.). *Rattus adustus*, *R. enganus* (see that account), and *R. tiomanicus* are the only native *Rattus* (and the sole native murines) on Pulau Enggano.

Rattus andamanensis (Blyth, 1860). J. Asiat. Soc. Bengal, 29:103.

COMMON NAME: Indochinese Forest Rat.

TYPE LOCALITY: India, Andaman Isls, South Andaman Isl.

DISTRIBUTION: S China (Yunnan, Guangxi, Fujian, islands of Hong Kong and Hainan), Vietnam (including four coastal islands; Kuznetsov, 2000, recorded as *koratensis*), Laos, Cambodia, Thailand (including Koh Klum off SE Thailand in the Gulf of Siam), C and N Burma, NE India (Sikkim, N West Bengal, Arunachal Pradesh, Nagaland, Meghalaya), Bhutan, and E Nepal. Not recorded from the mainland of peninsular Thailand south of Isthmus of Kra (10°E, 30′ N), but occurs on four islands (Koh Tau, Koh Phangan, Koh Samui, and Koh Kra) off the coast well south of the Isthmus (see Musser and Heaney, 1985). Also on the Andaman Isls (islands of North Andaman, Interview, Middle Andaman, Long, Henry Lawrence, Havelock, South Andaman, and Little Andaman) and Car Nicobar, northernmost of the Nicobar Isls. Limits in NE India and Nepal unresolved. Distribution based on specimens examined by Musser).

STATUS: IUCN – Vulnerable as *R. sikkimensis*

SYNONYMS: *burrulus* (Miller, 1902); *flebilis* (Miller, 1902); *hainanicus* G. M. Allen, 1925; *holchu* Chaturvedi, 1965; *klumensis* (Kloss, 1916); *koratensis* Kloss, 1919; *kraensis* (Kloss, 1916); *remotus* (Robinson and Kloss, 1914); *sikkimensis* Hinton, 1919; *yaoshanensis* Shih, 1930.

COMMENTS: *Rattus rattus* species group. Usually listed as *sikkimensis*, which was described as a subspecies of *R. rattus* (Hinton, 1919a), then arranged as a synonym of *R. r. brunneusculus* (Ellerman, 1961), later identified as *R. sikkimensis* (Musser and Newcomb, 1983; Musser and Heaney, 1985; Musser and Carleton, 1993), and discussed under *R. remotus* by Corbet and Hill (1992), who noted that *remotus* is an older name than *sikkimensis*. South Vietnamese samples have also been described under *R. sladeni* (Van Peenen et al., 1969), N Vietnamese series as *R. koratensis* (Dao, 1985), and Thai samples under *R. koratensis* and *R. remotus* (J. T. Marshall, Jr., 1977a).

The oldest name for the species is *andamanensis*, which Musser and Lunde (ms) determined by study of the holotype and large series (in USNM, KNMB, and ZSI) collected on the Andaman Isls. A multivariate analysis of morphometric traits has verified that the Andaman Isls populations and those on mainland Indochina are the same species (Musser and Lunde, ms). Corbet and Hill (1992:344) included *macmillani* (Hinton, 1919) as a synonym with question, but the holotype is an example of *Rattus tanezumi* (Musser's study of the holotype at BMNH). Those authors also listed *klumensis* and *kraensis* in the synonymy of *R. rattus*, but the respective holotypes represent *R. andamanensis* (Musser's study of holotypes and other specimens). Shih's (1930) *yaoshanensis* from Guangzi Province of S China was described as a subspecies of *Niviventer confucianus* and is listed that way by Corbet and Hill (1992), but the holotype (which Musser has not examined) is too large for *confucianus* as are its cranial and dental dimensions; measurement values and pelage coloration fall within the range of morphological variation exhibited by Chinese samples of *R. andamanensis* (a young adult from the type series sent to AMNH is an example of *R. andamanensis*). A sample from Car Nicobar was described as *R. rattus holchu* by Chaturvedi (1965), but Miller (1902) had already described *burrulus* from the same island (thought to be conspecific with *R. burrus* by Corbet and Hill, 1992; see that account), and all are examples of *R. andamanensis* (Musser's study of the respective holotypes and other specimens). While *R. andamanensis* is sympatric with *R. tanezumi* on mainland Indochina and many offshore islands in the South China Sea, it and *R. palmarum* are the only *Rattus* found on Car Nicobar, and *R. andamanensis* occurs with *R. stoicus* on the Andaman Isls from which *R. tanezumi* is absent (a sample from Barren Isl in the Andamans described as *atratus* by Miller [1902] and two [BMNH 10.7.26.2 and 10.7.26.3] from South Brother Isl, northeast of Little Andaman Isl, represent introduced *R. rattus*; Musser's study of holotype of *atratus* in USNM, and BMNH specimens). Holotypes tied to the synonyms for *R. andamanensis* document a large-bodied species of *Rattus* with rich brown upperparts, white underparts, long tail relative to head and body length, 12 teats, robust skull, small bullae, and large molars. Chromosomal comparisons between *R. andamanensis* (reported as *koratensis*) and other species of *Rattus* from S Vietnam reported by Baskevich and Kuznetsov (1998).

Rattus annandalei (Bonhote, 1903). *In* Annandale, Fasciculi Malayenses, Zool., pt. I:30.
COMMON NAME: Annandale's Sundaic Rat.
TYPE LOCALITY: Malaysia (Malay Peninsula), S Perak, Sungkei.
DISTRIBUTION: Malay Peninsula, Singapore, E Sumatra, and islands of Padang and Rupat off
the coast of E Sumatra (see Musser and Newcomb, 1983:515, and references therein).
STATUS: IUCN – Lower Risk (lc).
SYNONYMS: *bullatus* (Lyon, 1908); *villosus* (Lyon, 1908).
COMMENTS: *Rattus* species group unresolved. A Sundaic endemic; superficially resembles
Sundamys muelleri and some species of *Rattus* in primitive external, cranial, and dental
features, but in other specialized traits *R. annandalei* is unlike any species of *Sundamys*
and may eventually be removed from *Rattus* (Musser and Newcomb, 1983).

Rattus arfakiensis Rümmler, 1935. Z. Säugetierk., 10:118.
COMMON NAME: Vogelkop Mountain Rat.
TYPE LOCALITY: New Guinea, NW Prov. of Papua (= Irian Jaya), Vogelkop Peninsula, Arfak
Mtns, 2000 m.
DISTRIBUTION: Recorded only by the holotype from the Arfak Mtns.
COMMENTS: *Rattus leucopus* species group; member of the *R. niobe* complex. Usually
associated with *R. niobe* as a subspecies (Laurie and Hill, 1954; Tate, 1951) or one of the
synonyms of *R. niobe arrogans* (Taylor et al., 1982). Its brown upperparts and long
incisive foramina do not fit neatly into either *R. niobe*, *R. pococki*, or *R. arrogans*. We
provisionally recognize *arfakiensis* as a species until a fresh assessment is made of
individual and geographic variation among samples of the *R. niobe* complex. A larger
series of *arfakiensis* is required for study to determine its relationship to populations of
this complex in the Central Cordillera of Prov. of Papua (= Irian Jaya) and Papua New
Guinea.

Rattus argentiventer (Robinson and Kloss, 1916). J. Strs. Br. Roy. Asiat. Soc., 73:274.
COMMON NAME: Ricefield Rat.
TYPE LOCALITY: Indonesia, west coast of Sumatra, Pasir Ganting.
DISTRIBUTION: Indochina: Thailand and Koh Samui off the east coast of peninsular Thailand
(J. T. Marshall, Jr., 1977*a*), Cambodia, C Laos (Smith et al., In Press), and Vietnam
(Dang et al., 1994) in Indochina (including islands of Cham and Thô Chu off the coast
of S Vietnam; Kuznetsov, 2000). Sunda Shelf: Malay Peninsula, Sumatra, Java, Borneo,
Kangean Isl, and Bali. Nusa Tenggara (Lesser Sunda Isls): islands of Lombok, Sumbawa,
Sangeang, Komodo, Rintja, Flores, Adonara, Lembata, Alor, Sumba, Timor, and
Tanimbar. East of Indochina and the Sunda Shelf: Philippines (Cebu, Luzon, Mindoro,
Negros, and Mindanao Isls; Heaney et al., 1998); Sulawesi (Musser and Holden, 1991);
and one place and date of collection in New Guinea (Musser, 1973*b*; Taylor et al.,
1982). See Musser (1973*b*) and Maryanto (2003) for details of range. Corbet and Hill
(1992), followed by Helgen (2003*b*), recorded the species from Seram in the Moluccas,
but we cannot locate any specimens that would substantiate that occurrence, and
Maryanto's (2003) taxonomic revision of Indonesian populations does not include
Seram within the distribution.
STATUS: IUCN – Lower Risk (lc).
SYNONYMS: *bali* Kloss, 1921; *brevicaudatus* Horst and Raadt, 1918; *chaseni* Sody, 1941;
hoxaensis Dao, 1960; *kalimantanensis* Maryanto, 2003; *pesticulus* Thomas, 1921;
saturnus Sody, 1941; *umbriventer* Kellogg, 1945.
COMMENTS: *Rattus rattus* species group. The incorrect historical association of *argentiventer* as
a subspecies of *Rattus rattus* summarized by Musser (1973*b*). Judged by its close morphological alliance with species that Ellerman (1941) placed in subgenus *Rattus*, which
are mostly mainland Asian in origin, and its peculiar Indo-australian geographic distribution that is discordant with ranges of endemic species, *R. argentiventer* was likely
inadvertently introduced into the highly distinctive murine faunas of the Sunda Shelf,
Philippines, Sulawesi, Nusa Tenggara, and New Guinea from its Indochinese home-
land, possibly with the spread of rice culture (Musser, 1973*b*; Musser and Holden,
1991; Musser and Newcomb, 1983; Taylor et al., 1982). *Rattus hoxaensis* from C Vietnam, described by Dao (1960), represents *R. argentiventer* (Musser's identification of

specimens in ZMVNU that Dao himself identified as *hoxaensis*); reasons for allocating the other synonyms explained in Musser (1973*b*). Geographic variation in external and cranial traits for populations from the Malay Peninsula, islands on the Sunda Shelf, Sulawesi, and Nusa Tenggara presented by Maryanto (2003). Pertinent ecological aspects of populations presented in a comparative context documented for populations on W Java (Tristiani et al., 1998, 2000; Leung and Sudarmaji, 1999). Usually found in lowlands, *R. argentiventer* reaches 1646 m on the slopes of Mt. Kinabalu in Sabah (Md Nor, 2001), and was caught (in 1991) in irrigated rice paddies at Nenas, highest village (1100 m) on Gunung Mutis, Timor (K. Aplin, in litt., 2004).

Rattus arrogans Thomas, 1922. Ann. Mag. Nat. Hist., ser. 9, 9:263.

COMMON NAME: Western New Guinea Mountain Rat.

TYPE LOCALITY: New Guinea, Prov. of Papua (= Irian Jaya), N slopes Snow Mtns (Pegunungan Maoke), Doormanpad-bivak, 2400 m, 03°30'S, 138°30'E (Laurie and Hill, 1954).

DISTRIBUTION: Western New Guinea; recorded from the central mountainous backbone of Prov. of Papua (= Irian Jaya; Pegunungan Maoke) eastward to the Star Mtns in W Papua New Guinea (longitude 141°E), 2200-4050 m (some of the material cited by Taylor et al., 1982 as *Rattus niobe arrogans*; Flannery and Seri, 1990; specimens in AMNH).

SYNONYMS: *haymani* (Ellerman, 1941); *klossi* Thomas, 1913 [not Bonhote, 1906].

COMMENTS: *Rattus leucopus* species group; member of the *R. niobe* complex. Taylor et al., (1982) treated *arrogans* as a subspecies of *R. niobe* and western montane counterpart of the eastern *R. n. niobe*. Our examination of hundreds of specimens in AMNH led us to a different interpretation. There are at least three species in what is currently regarded as *R. niobe*. One is *R. niobe*, which we have not been able to identify outside of Papua New Guinea. In Prov. of Papua (= Irian Jaya), two kinds occur on N slopes of the Snow Mtns. Between 1500 and 2150 m is a dark blackish brown (upperparts and underparts) species closely similar in fur coloration to most Papuan *R. niobe*, but averages larger in external and most cranial dimensions. We provisionally apply *pococki* to this species, as did Tate (1951:340, as *Rattus niobe pococki*) who examined the same material. The zone between 2200 and 3900 m is occupied by a different species that has brown instead of blackish pelage, even larger dimensions, a larger skull with more inflated braincase, and wider molars. Thomas's (1922*a*, *b*) and Tate's (1951) description of *arrogans* fit these specimens (type locality is in same region but farther north), and they were also assigned to *arrogans* by Tate (1951, as a subspecies of *niobe*). Tate's (1951) description of the holotype of *haymani* (replacement name for Thomas's, 1913*b*, *klossi*) suggests it is a synonym. Flannery (1990*b*:241) noted that unpublished results from biochemical studies indicated that two species are present in what Taylor et al. (1982) defined as a single entity, one at high altitudes and the other at lower, but did not indicate which region of New Guinea. According to K. Aplin (in litt., 2004), those results came from samples collected by him on Mt. Karimui (northern Papua New Guinea in Chimbu Province) in 1984. He encountered two morphologically (in external and cranial traits) distinct forms overlapping at about 1400 m. His allozyme study confirmed they were also genetically distinct (more so than *R. leucopus* and *R. verecundus* that were included on the same gels). Flannery and Seri (1990) also recognized a high-altitude form they labelled *arrogans* and a middle-altitude entity they identified as *pococki* in the Central Cordillera of West Sepik Province in northwestern Papua New Guinea. Whether these names identify the same species that live on northern slopes of the Snow Mtns remains to be determined.

Rattus baluensis (Thomas, 1894). Ann. Mag. Nat. Hist., ser. 6, 14:454.

COMMON NAME: Kinabalu Rat.

TYPE LOCALITY: Malaysia, Sabah (N Borneo), Mt Kinabalu, 8000 ft (2438 m).

DISTRIBUTION: Known only between 1524 and 3360 m on slopes of Mt Kinabalu, N Borneo (Md Nor, 2000; Musser, 1986).

STATUS: IUCN – Endangered.

COMMENTS: *Rattus rattus* species group. Taxonomic history of past association of *baluensis* with *Rattus rattus* documented by Musser (1986), who also noted that its closest relative is probably *R. tiomanicus*, which occurs in lowlands of Borneo and on many islands on the Sunda Shelf. Member of the suite of murine species endemic to Borneo (Musser, 1986; see also account of *Chiropodomys major*).

Rattus blangorum Miller, 1942. Proc. Acad. Nat. Sci. Philadelphia, 94:145.

COMMON NAME: Aceh Rat.

TYPE LOCALITY: Indonesia, N Sumatra, Aceh foothills of Gunung Leuser, near Blangnanga Base Camp, 3600 ft (1097 m; see Miller, 1942, for details).

DISTRIBUTION: Known only from the type locality.

COMMENTS: *Rattus rattus* species group. Still represented only by the holotype and another specimen (ANSP 20348, 20349). Originally described as a species (Miller, 1942), but later included in *R. tiomanicus* (Musser and Califia, 1982; Corbet and Hill, 1992; Musser and Carleton, 1993). Recent multivariate analyses employing morphometric traits of Sundaic samples of the *R. tiomanicus* complex indicate *blangorum* to be distinct from all other insular samples in that assemblage, which is partly a reflection of its small body size compared to the usual size range in *R. tiomanicus* (Musser, ms.). Member of the suite of murine species endemic to Sumatra (Musser, 1986, and account of *Maxomys hylomyoides*).

Rattus bontanus Thomas, 1921. Ann. Mag. Nat. Hist., ser. 9, 7:246.

COMMON NAME: Southwestern Xanthurus Rat.

TYPE LOCALITY: Indonesia, SW Sulawesi, Mt Bonthain (Gunung Lampobatang), 2000 ft (610 m).

DISTRIBUTION: SW Sulawesi: 600-2500 m on the slopes of Gunung Lampobatang and adjacent coastal lowlands.

STATUS: IUCN – Vulnerable as *R. bontanus*, Lower Risk (nt) as *R. foramineus*.

SYNONYMS: *foramineus* Sody, 1941.

COMMENTS: *Rattus xanthurus* species group. Sody (1941) questionably included this species in *Taeromys*, Laurie and Hill (1954) and Musser (1984) treated it as a subspecies of *R. xanthurus*, but Musser and Holden (1991) contended that it is a distinct species most closely related to *R. foramineus*, which occurs in coastal lowlands of the southern end of the SW peninsula of Sulawesi and is represented only by four modern specimens (Sody, 1941) and subfossil fragments from a few localities (Musser, 1984; Musser and Holden, 1991). Recent morphometric analyses reveal that *foramineus* is a lowland population of *R. bontanus*, which has external and morphological traits that distinguish it from any other species in the *R. xanthurus* group of Sulawesi (Musser, ms). Sody described *pelurus* from Pulau Peleng as a subspecies of *R. foramineus*, but the Peleng Isl rat is a separate species (Musser and Holden, 1991; see that account).

Rattus burrus (Miller, 1902). Proc. U. S. Natl. Mus., 24:768.

COMMON NAME: Nicobar Archipelago Rat.

TYPE LOCALITY: India, Nicobar Isls, Trinkat Isl.

DISTRIBUTION: Islands of Trinkat, Little Nicobar, and Great Nicobar in the Nicobar Arch.; apparently absent from Car Nicobar, northernmost of the Nicobar Isls, which is inhabited by populations of *R. andamanensis* and *R. palmarum* (see those accounts).

STATUS: IUCN – Vulnerable.

SYNONYMS: *burrescens* (Miller, 1902).

COMMENTS: *Rattus rattus* species group. Recent morphometric analyses indicate *R. burrus* to be most closely related to *R. simalurensis* from the Simalur Arch., *R. lugens* from the Mentawai Arch., and *R. adustus* from Pulau Enggano. Except for larger body size, morphology of the four species resembles that characterizing the *R. tiomanicus* complex from the Sunda Shelf (Musser, 1986; Musser, ms.). *Rattus burrus* is also morphologically similar to *R. palmarum* from Car Nicobar, the northernmost of the Nicobar Isls (see that account), but differs in being smaller in body size (e. g., greatest skull length = 41.3-46.7 mm for 12 adult *R. burrus*, 49.0-54.0 mm for three adult *R. palmarum*), with softer pelage. Whether *burrus* is a separate species or simply a smaller-bodied insular variant of the larger-bodied and coarse-furred *R. palmarum* warrants careful study. Both taxa may be Nicobar endemics, but their status will have to be tested by a systematic revision of the *R. tiomanicus* complex. Corbet and Hill (1992) questionably included *burrus* and *burrescens* as synonyms of *R. tiomanicus* and regarded burrulus, which we treat as a synonym of *R. andamanensis* (see that account), as possibly conspecific with *burrus*. No other species of native *Rattus* is sympatric with *R. burrus*, but examples of *R. tanezumi* have been collected from Great Nicobar (holotype of *pulliventer* Miller, 1902) and Little Nicobar (BMNH 20.3.1907) Isls, and *R. rattus* is documented from Nancowry Isl (USNM 111800); both species were introduced.

Rattus colletti (Thomas, 1904). Novit. Zool., 11:599.

COMMON NAME: Australian Dusky Rat.

TYPE LOCALITY: Australia, Northern Territory, South Alligator River (see Mahoney and Richardson, 1988).

DISTRIBUTION: Australia; known only from the coastal floodplains of the Northern Territory, the most restricted range of all the native Australian *Rattus* (Watts and Aslin, 1981; C. K. Williams, 1995).

STATUS: IUCN – Lower Risk (lc).

COMMENTS: *Rattus fuscipes* species group. Arranged as a subspecies of *R. sordidus* by Taylor and Horner (1973), but subsequently treated as a separate species (Mahoney and Richardson, 1988; Watts and Aslin, 1981; C. K. Williams, 1995). Although diploid counts are very different, *Rattus colletti* (2n = 42) is genetically very close to *R. villosissimus* (2n = 50); the two hybridize in the laboratory, but the offspring exhibit severely reduced fertility (Baverstock et al., 1983a, 1986). An undescribed species related to *R. colletti* and *R. villosissimus* is known from a small area in C Queensland (Aplin, in litt., 2004). Fertile hybrids have also been obtained between laboratory crosses of *R. colletti* and *R. tunneyi* (Baverstock et al., 1983a).

Rattus elaphinus Sody, 1941. Treubia, 18:307.

COMMON NAME: Sula Archipelago Rat.

TYPE LOCALITY: Indonesia, Kepulauan Sula, Pulau Taliabu (east of Pulau Peleng and Sulawesi).

DISTRIBUTION: Known only from Pulau Taliabu, Indonesia (Musser and Holden, 1991) and adjacent Pulau Manggole in the Sula Arch., but absent from nearby Pulau Sanana in the same archipelago, where only the introduced *R. tanezumi* and *R. exulans* are found (Flannery, 1995b).

STATUS: IUCN – Vulnerable.

COMMENTS: *Rattus leucopus* species group. A morphologically distinctive species that is likely most closely related to native *Rattus* on the Moluccas and New Guinea. Amplified description and comparisons with Sulawesian *R. hoffmanni* and *R. koopmani* from Peleng Isl provided by Musser and Holden (1991).

Rattus enganus (Miller, 1906). Proc. U.S. Natl. Mus., 30:821.

COMMON NAME: Enggano Island Rat.

TYPE LOCALITY: Indonesia, Pulau Enggano (southwest of Sumatra and off the continental margin).

DISTRIBUTION: Known only from Pulau Enggano.

STATUS: IUCN – Critically Endangered.

COMMENTS: *Rattus* species group unresolved. Represented only by the holotype. Some cranial, dental, and external features resemble those in *R. macleari* from Christmas Isl and *R. xanthurus* from Sulawesi, but the holotype of *R. enganus* is morphologically distinctive; determining its phylogenetic affinities within *Rattus* will require more specimens from Pulau Enggano and revisionary study of the genus (Musser and Newcomb, 1983). In addition to *R. enganus* and *R. adustus*, there is a single specimen of *R. tiomanicus* (USNM 140975) collected on Pulau Enggano.

Rattus everetti (Günther, 1879). Proc. Zool. Soc. Lond., 1879:75.

COMMON NAME: Philippine Forest Rat.

TYPE LOCALITY: Philippines, N Mindanao Isl (see Heaney and Rabor, 1982, for details).

DISTRIBUTION: Endemic and widespread in the Philippines except for the Greater Palawan and Sulu faunal regions and the Batanes-Babuyan groups. Islands of Luzon, Catanduanes, Mindoro, Sibuyan, Ticao, Camiguin, Samar, Calicoan, Leyte, Dinagat, Siargao, Mindanao,

Basilan, Bohol, Biliran, Marinduque, Panay, and Maripipi (Heaney et al., 1998); probably occurs on other islands in the Philippine Arch. (Musser and Heaney, 1992).

STATUS: IUCN – Lower Risk (lc).

SYNONYMS: *albigularis* (Mearns, 1905); *gala* (Miller, 1910); *tagulayensis* (Mearns, 1905); *tyrannus* (Miller, 1910).

COMMENTS: *Rattus* species group unresolved. Member of the Philippine New Endemics (Musser and Heaney, 1992). May be more than one species represented in insular samples; the *everetti* complex needs critical systematic revision. Corbet and Hill (1992), for example, recognized *tyrannus* from Ticao and Negros Isls as a separate species. Cranial, dental, and spermatozoal characters elaborated by Breed and Musser (1991) and Musser and Heaney (1992). Not closely related to other species of *Rattus* endemic to the Philippines and likely should be generically separated. Phylogenetic analyses of complete DNA mitochondrial cytochrome *b* sequences for 13 of the 16 genera of endemic Philippine murines place *R. everetti* as either basal to other species of *Rattus* or near the clade containing species of *Bullimus*, *Tarsomys*, and *Limnomys* (Jansa and Heaney, 2001; Heaney, pers. comm.), which are also New Endemics. Standard karyotype (2n = 42, FN = 64) described by Rickart and Musser (1993).

Rattus exulans (Peale, 1848). Mammalia *in* Repts. U.S. Expl. Surv., 8:47.

COMMON NAME: Pacific Rat.

TYPE LOCALITY: Society Isls, Tahiti Isl (France).

DISTRIBUTION: E Bangladesh, C and S Burma, Thailand, Laos, Cambodia, C and S Vietnam, Yongxing Isl in the Xisha Arch (in South China Sea southeast of Hainan Isl between 16° and 18°N; Wang, 2003), E Taiwan and Miyakojima Isl in S Ryukyus (Motokawa et al., 2001*a*), Sundaic region (incl. Mentawai Isls, and islands of Enggano, Nias, and Simeulule), Christmas Isl (Gibson-Hill, 1947), Sulawesi, Philippines (Heaney et al., 1998), Moluccas (Flannery, 1995*b*), and Nusa Tenggara (Lesser Sunda Isls); New Guinea Region (Taylor et al., 1982; Flannery, 1995*a*), SW Pacific Isls (Flannery, 1995*b*) Adele and Murray Isls off the coast of NW and NE Australia (not recorded from mainland; Mahoney and Richardson, 1988; Taylor and Horner, 1973; Watts, 1995*h*; Watts and Aslin, 1981), Micronesia, New Zealand (Atkinson and Moller, 1990), Polynesia (including Caroline Isls; Buden, 1996*a*, 1996*b*), Hawaiian Isls (Tomich, 1986), and Easter Isl. Not documented from Andaman or Nicobar Isls (Chaturvedi, 1980; Musser's research), despite assertion of Wodzicki and Taylor (1984), who otherwise adequately summarized general distribution; details recorded by Musser and Newcomb (1983) and Corbet and Hill (1992). Range also based upon our study of specimens in several museums (also see Matisoo-Smith et al., 1998).

STATUS: IUCN – Lower Risk (lc).

SYNONYMS: *aemuli* (Thomas, 1896); *aitape* Troughton, 1937; *apicus* (Mearns, 1905); *basilanus* (Hollister, 1913); *bocourti* (Milne-Edwards, 1872); *browni* (Alston, 1877); *buruensis* (J. A. Allen, 1911); *calcis* (Hollister, 1911); *clabatus* (Lyon, 1906); *concolor* (Blyth, 1859); *echimyoides* (Ramsay, 1877); *ephippium* (Jentink, 1880); *equile* Robinson and Kloss, 1927; *eurous* Miller and Hollister, 1921; *gawae* Troughton, 1845; *hawaiiensis* Stone, 1917; *huegeli* (Thomas, 1880); *jessook* (Jentink, 1879); *lassacquerei* Sody, 1933; *leucophaetus* (Hollister, 1913); *luteiventris* (J. A. Allen, 1910); *malengiensis* Sody, 1941; *manoquarius* Sody, 1934; *maorium* (Hutton, 1870); *mayonicus* (Hollister, 1913); *melanoderma* (Dieterlen, 1986); *meringgit* Sody, 1941; *micronesiensis* Tokuda, 1933 [see Kaneko and Maeda, 2002]; *negrinus* (Thomas, 1898); *obscurus* (Miller, 1900) [not Waterhouse, 1837]; *ornatulus* (Hollister, 1913); *otteni* Kopstein, 1931; *pantarensis* (Mearns, 1905); *praecelsus* Troughton, 1937; *pullus* (Miller, 1901); *querceti* (Hollister, 1911); *raveni* Miller and Hollister, 1921; *rennelli* Troughton, 1945; *schuitemakeri* Sody, 1933; *solatus* Kellogg, 1945; *stragulum* (Robinson and Kloss, 1916); *suffectus* Troughton, 1937; *surdus* (Miller, 1903); *tibicen* Troughton, 1937; *todayensis* (Mearns, 1905); *vigoratus* (Hollister, 1913); *vitiensis* (Peale, 1848); *vulcani* (Mearns, 1905); *wichmanni* (Jentink, 1890).

COMMENTS: *Rattus exulans* species group. Inadvertent or intentional human introduction or possibly natural rafting is responsible for most of the Pacific insular occurrences (Langdon, 1995; Matisoo-Smith and Robins, 2004; Matisoo-Smith et al., 1998; Roberts,

1991), and for distributions on islands and archipelagos outside of mainland SE Asia, the region where the species may have originated (Musser and Newcomb, 1983). Matisoo-Smith et al. (1998) claimed that *R. exulans* was among the plant and animal species carried by ancestral Polynesians in their colonizing canoes. They analyzed mtDNA sequences in an array of Polynesian samples, concluding that the sequences "prove to be valuable genetic markers for tracing the migration routes and movement of the first humans entering the remote Pacific" (p. 15149). A subsequent study by Matisoo-Smith and Robins (2004) documents evidence for origins and dispersals of Polynesians derived from mtDNA phylogenies of *R. exulans*. Standard karyotype (2n = 42, FN = 60) for Philippines, Indomalayan region, Oceania, and Papua New Guinea summarized by Rickart and Musser (1993), and for Taiwan sample by Motokawa et al. (2001*a*).

In a significant report on variation in skull size among *R. exulans*, *R. rattus*, and *R. norvegicus* in New Zealand and other Pacific islands, Yoram et al. (1999) documented increase in skull size with latitude for *R. exulans* but not for the other two. Other relationships between skull size and island area as well as species sympatry are also presented; all aspects bear on understanding insular morphological variation in these species, especially *R. exulans*, and their taxonomy. Reliability of radiocarbon dating skeletal remains of New Zealand *R. exulans* discussed by Hedges (2000) and Higham and Petchey (2000), a test between radiocarbon dates and those obtained by Carbon 14 accelerator mass spectrometry using *R. exulans* reported by Holdaway and Beavan (1999), and possible role of burial contamination or diet in radiocarbon anomalies documented by Beavan-Athfield and Sparks (2001*a*, 2001*b*; also see references cited there). Reliability of dating techniques bears significantly on first colonization dates of *R. exulans* to New Zealand. Data presented by Holdaway (1996:226) indicates the Pacific rat was established on both main islands nearly 2000 years ago, much earlier than the time established for human settlement (about 850 years ago) based on unequivocal evidence. Holdaway reasoned that *R. exulans* was introduced by "transient human visitors, who either left immediately or quickly died out," and noted that for "more than 1,000 years after its arrival, the Pacific rat was the sole exotic predator in New Zealand." Whether *R. exulans* was spread by human transport or natural rafting throughout the Pacific is a controversial topic, and Langdon's (1995) discussion of rodent colonization of Easter Isl is relevant in this context (see also Matisoo-Smith and Robins, 2004; Matisoo-Smith et al., 1998).

Rattus feliceus Thomas, 1920. Ann. Mag. Nat. Hist., ser. 9, 6:423.
COMMON NAME: Spiny Seram Island Rat.
TYPE LOCALITY: Indonesia, Kepulauan Maluku, Pulau Seram, Gunung Manusela, 6000 ft (1830 m).
DISTRIBUTION: Known only from Seram Isl, sea level to 1830 m (Helgen, 2003*b*).
STATUS: IUCN – Vulnerable.
COMMENTS: *Rattus leucopus* species group. Amplified description and comparison with *R. koopmani* from Peleng Isl provided by Musser and Holden (1991), who also discussed past subspecific allocations of *feliceus* to other species of *Rattus*. Phylogenetic affinities of *R. feliceus* are probably with species of *Rattus* endemic to New Guinea region (Musser and Holden, 1991), an alliance based upon morphology that should be tested with molecular data. Reviewed by Flannery (1995*b*) and Helgen (2003*b*).

Rattus fuscipes (Waterhouse, 1839). Zool. Voy. H.M.S. "Beagle," Mammalia, p. 66.
COMMON NAME: Australian Bush Rat.
TYPE LOCALITY: Australia, Western Australia, Albany, "Little Grove" on Princess Royal Harbor, 4 mi (6 km) S Mt Melville (as restricted by neotype designation by Taylor and Horner, 1973; also see Mahoney and Richardson, 1988).
DISTRIBUTION: Coastal, subcoastal, and offshore islands of SW Western Australia; S coast from Eyre Peninsula in South Australia (Robinson et al., 2000) to W Victoria; coastal and subcoastal Victoria from Otway Peninsula north to near Rockhampton in Queensland; coastal Queensland from Townsville to Cooktown (Lunney, 1995*a*:652; Seebeck, 1995*c*:225; Taylor and Horner, 1973:15).

STATUS: IUCN – Lower Risk (lc).

SYNONYMS: *assimilis* (Gould, 1858); *brazenori* Tate, 1940; *coracius* Thomas, 1923; *glauerti* Thomas, 1926; *greyii* (Gray, 1841); *manicatus* (Gould, 1858); *mondraineus* Thomas, 1921; *murrayi* Thomas, 1923; *peccatus* Troughton, 1937; *pelori* Finlayson, 1960; *ravus* Brazenor, 1936.

COMMENTS: *Rattus fuscipes* species group. Taylor and Horner (1973) suggested, on morphological grounds, that the Queensland population of *R. fuscipes* (*coracius*) has a common ancestry with Queensland *R. leucopus*, a hypothesis reasserted by Taylor et al. (1982, 1983). This relationship, however, is unsupported by either chromosomal (Dennis and Menzies, 1978) or allozymic data (Baverstock et al., 1983a, 1986). Reviewed by Taylor and Horner (1973), Watts and Aslin (1981), Mahoney and Richardson (1988), and Lunney (1995a). See Taylor and Calaby (1988a, Mammalian Species, 298). The name *manicatus* was included in the synonymy of *Rattus fuscipes* by Taylor and Horner (1973), but listed as *incertae sedis* by Mahoney and Richardson (1988:192).

Rattus giluwensis Hill, 1960. J. Mammal., 41:277.

COMMON NAME: Mount Giluwe Rat.

TYPE LOCALITY: NE New Guinea, Papua, Mt Giluwe, 11,000-12,000 ft (3350-3660 m).

DISTRIBUTION: Recorded only from Mt Giluwe in Papua New Guinea and adjoining highlands, 2195-3660 m (Taylor et al., 1982).

STATUS: IUCN – Lower Risk (nt).

SYNONYMS: *melanurus* Laurie and Hill, 1954 [not Shamel, 1940].

COMMENTS: *Rattus leucopus* species group. Originally described as *R. ruber melanurus* (Laurie and Hill, 1954:112), but the name is preoccupied by *Rattus melanurus* Shamel, 1940, which refers to a sample of *Maxomys whiteheadi*. A morphologically distinctive species whose phylogenetic affinities are equivocal (Taylor et al., 1982, 1983).

Rattus hainaldi Kitchener, How, and Maharadatunkamsi, 1991. Rec. West. Aust. Mus., 15:557.

COMMON NAME: Hainald's Flores Island Rat.

TYPE LOCALITY: Indonesia, Nusa Tenggara, Pulau Flores, Gunung Ranakah, 1300 m, above Kampong Robo, Desa Longko, 8 km SSE Ruteng.

DISTRIBUTION: Known only from Flores Isl.

STATUS: IUCN – Lower Risk (nt).

COMMENTS: *Rattus* species group unresolved. Described from two specimens (Kitchener et al., 1991c), and later found to be common in forest habitats (Kitchener and Yani, 1998; Kitchener et al., 1998). Phylogenetic affinities uncertain; placement in *Rattus* originally provisional (Kitchener et al., 1991), results from albumin immunology suggest transfer to either *Bunomys* or *Komodomys* (Watts and Baverstock, 1994b), but electrophoretic data indicate *R. hainaldi* to be equidistant from the *Komodomys-Rattus timorensis* clade of Nusa Tenggara clade and true *Rattus* (K. Aplin, in litt., 2004).

Rattus hoffmanni (Matschie, 1901). Abh. Senckenb. Naturforsch. Ges., 25:284.

COMMON NAME: Hoffmann's Sulawesi Rat.

TYPE LOCALITY: Indonesia, NE Sulawesi, Minahassa region.

DISTRIBUTION: Sulawesi; throughout island except upper slopes of Gunung Lampobatang at the end of the SW peninsula; also on Pulau Malenge in Kepulauan Togian (Musser and Holden, 1991).

STATUS: IUCN – Lower Risk (lc).

SYNONYMS: *biformatus* Sody, 1941; *celebensis* (Hoffmann, 1887) [not Gray, 1867]; *linduensis* Miller and Hollister, 1921; *mengkoka* Tate and Archbold, 1935; *mollicomus* Miller and Hollister, 1921; *tatei* Ellerman, 1941.

COMMENTS: *Rattus rattus* species group. Morphological, chromosomal, distributional, and ecological boundaries of species elaborated by Musser and Holden (1991), who also documented taxonomic history. Stomach morphology described and contrasted with other Sulawesian endemics by Musser and Durden (2002). Closest relative of *R. hoffmanni* is *R. mollicomulus* from Gunung Lampobatang, Sulawesi. *Rattus koopmani*, endemic to Pulau Peleng, is also likely closely related to *R. hoffmanni*. Phylogenetic affinities of this cluster to other species of *Rattus* are unresolved.

Rattus hoogerwerfi Chasen, 1939. Treubia, 17(3):496.

COMMON NAME: Hoogerwerf's Sumatran Rat.

TYPE LOCALITY: Indonesia, W Sumatra, Aceh, Gunung Leuser, Blang Kedjeren, 2900 ft (885 m).

DISTRIBUTION: Recorded from only between 885 and 2835 m in foothills and upper slopes of Gunung Leuser, Sumatra (Musser, 1986).

STATUS: IUCN – Vulnerable.

COMMENTS: *Rattus* species group unresolved. Very distinctive and possibly not a member of *Rattus*. Morphology and comparisons with other species documented by Musser and Newcomb (1983) and Musser (1986). Although phylogenetic affinities with species of *Rattus* from Sulawesi, the Philippines, and Christmas Isl have been claimed (Miller, 1942), and with *R. korinchi* (Chasen, 1939), no compelling evidence links *R. hoogerwerfi* with any other species (Musser, 1986). Member of the suite of murine species endemic to Sumatra (see Musser, 1986, and account of *Maxomys hylomyoides*).

Rattus jobiensis Rümmler, 1935. Z. Säugetierk., 10:116.

COMMON NAME: Yapen Island Rat.

TYPE LOCALITY: New Guinea, Prov. of Papua (= Irian Jaya), Pulau Yapen in Teluk Cendrawasih (Geelvinck Bay).

DISTRIBUTION: New Guinea, Prov. of Papua (= Irian Jaya), recorded only from islands of Yapen, Owi, Supiori, and Biak in Geelvinck Bay (Flannery, 1995*b*:154; Taylor et al., 1982:262).

STATUS: IUCN – Lower Risk (nt).

SYNONYMS: *biakensis* Troughton, 1945; *owiensis* Troughton, 1946.

COMMENTS: *Rattus leucopus* species group. Originally described as a subspecies of *R. leucopus*, *jobiensis* is a distinctive, separate species that may be more closely related to Moluccan endemics than to native mainland New Guinea species (Taylor et al., 1982). Reviewed by Flannery (1995*b*).

Rattus koopmani Musser and Holden, 1991. Bull. Am. Mus. Nat. Hist., 206:389.

COMMON NAME: Koopman's Peleng Island Rat.

TYPE LOCALITY: Indonesia, Kepulauan Banggai, Pulau Peleng (1°23′S, 123°14′E), which is separated from mainland Sulawesi by deepwater Selat Peleng.

DISTRIBUTION: Known only from Peleng Isl.

STATUS: IUCN – Lower Risk (nt).

COMMENTS: *Rattus rattus* species group. Represented only by the holotype. In many characters, *R. koopmani* resembles the Sulawesian *R. hoffmanni* and bears the same degree of relationship to that species as the Peleng *R. pelurus* does to mainland *R. xanthurus*. More specimens are needed to assess morphological variation and better estimate phylogenetic affinity (Musser and Holden, 1991). Occurs with *R. pelurus*, a member of the *R. xanthurus* group; no other species of *Rattus* have been found on Pulau Peleng, but past biological surveys have been limited.

Rattus korinchi (Robinson and Kloss, 1916). J. Str. Br. Roy. Asiatic Soc., 73:275.

COMMON NAME: Sumatran Mountain Rat.

TYPE LOCALITY: Indonesia, W Sumatra, Propinsi Jambi, Gunung Kerinci, Sungai Kring, 7300 ft (2225 m).

DISTRIBUTION: Recorded only from Gunung Kerinci and Gunung Talakmau in W Sumatra (Musser, 1986:4).

STATUS: IUCN – Lower Risk (lc).

COMMENTS: *Rattus* species group unresolved. Revised by Musser (1986). Represented by very few specimens, and morphologically unlike any other described species of *Rattus*. Member of the cluster of murine species endemic to Sumatra (Musser, 1986; see also account of *Maxomys hylomyoides*).

Rattus leucopus (Gray, 1867). Proc. Zool. Soc. Lond., 1867:598.

COMMON NAME: Cape York Rat.

TYPE LOCALITY: Australia, Queensland, Cape York (as restricted by Thomas' lectotype selection; see Mahoney and Richardson, 1988:183).

DISTRIBUTION: Australia, Queensland: one population ranges from the tip of Cape York south down E side of the peninsula to vicinity of Coen, another from region of Cooktown south along the coast to Tully; all records are east of the Great Dividing Range (Moore and Leung, 1995; Taylor and Horner, 1973; Watts and Aslin, 1981). New Guinea: widespread in lowlands south of Central Cordillera, in N and S lowland regions fringing the Owen Stanley Range in E Papua New Guinea (Flannery, 1995a; Taylor et al., 1982:232); also Wokam Isl in the Aru Isls (Flannery, 1995b). Altitudinal range, sea level to 1200 m (Flannery, 1995a).

STATUS: IUCN – Lower Risk (lc).

SYNONYMS: *cooktownensis* Tate, 1951; *dobodurae* Troughton, 1946; *mcilwraithi* Tate, 1951; *personata* (Krefft, 1867); *ratticolor* (Jentink, 1908); *ringens* (Peters and Doria, 1881); *terra-reginae* (Alston, 1879).

COMMENTS: *Rattus leucopus* species group. This species and *R. sordidus* are the only two native *Rattus* occurring on both New Guinea and the NE coastal region of Australia (Taylor et al., 1982). Morphologically related to other species of *Rattus* native to New Guinea (Taylor et al., 1982). Morphological data interpreted by Taylor and Horner (1973) to indicate close affiliation between *R. leucopus* and *R. fuscipes* from coastal Queensland; allozymic data discordant with this view (Baverstock et al., 1983a, 1986). Australian segment reviewed by Watts and Aslin (1981), Mahoney and Richardson (1988), and Moore and Leung (1995); New Guinea segment reviewed by Flannery (1995a). Leung (1999b) provided the first detailed ecological study of *R. leucopus* in Australia. Leary and Seri (1997) discussed specimens taken in the Kikori River Basin of S Papua.

Rattus losea (Swinhoe, 1871). Proc. Zool. Soc. Lond., 1870:637 [1871].

COMMON NAME: Losea Rat.

TYPE LOCALITY: Taiwan.

DISTRIBUTION: Taiwan (M.-J. Yu, 1996), Pescadores Isls, S China (Fujian, Guangdong, Guangxi, Jiangxi, Guizhou, Chongqing, E Sichuan, S Shaanxi, Hainan Isl, Hong Kong; Wang, 2003), Vietnam (Dang et al., 1994), C and S Laos (Smith et al., In Press), Thailand (excluding peninsular Thailand; J. T. Marshall, Jr., 1977a; Robinson et al., 1995), and S and SW Cambodia (specimens in FMNH and MNHN; A. Smith in litt., 2000); details of range reported by Musser and Newcomb (1985) and also based upon our study of other specimens not recorded in that report.

STATUS: IUCN – Lower Risk (lc).

SYNONYMS: *exiguus* Howell, 1927; *sakeratensis* Gyldenstolpe, 1917.

COMMENTS: *Rattus rattus* species group. Known morphological, geographical, and altitudinal boundaries; correctly and incorrectly associated scientific names; and geographic variation reviewed by Musser and Newcomb (1985). They also suggested that *R. losea* is morphologically and probably phylogenetically closest to *R. osgoodi* from the highlands of S Vietnam and phylogenetically linked to *R. argentiventer*; see also Corbet and Hill (1992:341). Chromosomal contrasts between *R. losea* and three other species of *Rattus* from S Vietnam documented by Baskevich and Kuznetsov (1998). Karyotype from Taiwanese sample described by H.-T. Yu et al. (1996), who noted that the chromosomal complement is the same as that recorded for Thai samples except for differences in the X chromosome. Additional chromosomal data summarized by Rickart and Musser (1993). Adler (1995) documented habitat use on Taiwan.

Rattus lugens (Miller, 1903). Smithson. Misc. Coll., 45:33.

COMMON NAME: Mentawai Archipelago Rat.

TYPE LOCALITY: Indonesia, Kepulauan Mentawai, Pulau Utara Pagai (North Pagai Isl).

DISTRIBUTION: Islands of Siberut, Sipora, Pagai Utara, and Pagai Selattan in the Mentawai Arch. off coast of SW Sumatra; these islands lie on a slender extension of the continental shelf.

STATUS: IUCN – Lower Risk (lc).

SYNONYMS: *mentawai* Chasen and Kloss, 1928.

COMMENTS: *Rattus rattus* species group. Chasen (1940) listed *lugens* and *mentawai* each as a subspecies of *R. rattus*, but the Mentawai endemic is a morphologically distinctive species most closely related to *R. adustus* from Enggano Isl and *R. simalurensis* from the adjacent Simalur Arch., and both those two species along with *R. lugens* appear to be morphologically related to the *R. tiomanicus* complex (Musser, 1986; Musser and Califia, 1982; Musser and Heaney, 1985; Musser, ms.). Member of the murine fauna endemic to Mentawai Arch. (see account of *Leopoldamys siporanus*).

Rattus lutreolus (J. E. Gray, 1841). *In* G. Gray, App. C *in* Jour. Two Exped. Aust., II:409.
  COMMON NAME: Australian Swamp Rat.
  TYPE LOCALITY: Australia, New South Wales, Hunter River, Moscheto Isl (as restricted by Thomas' lectotype selection; see Mahoney and Richardson, 1988:184).
  DISTRIBUTION: Tasmania (Rounsevell et al., 1991); coastal and subcoastal habitats from vicinity of Adelaide in SE South Australia east and north through S Victoria, New South Wales to SE Queensland, with isolated populations in N Queensland; also on Kangaroo Isl off coast of South Australia (Lunney, 1995b:656; Robinson et al., 2000; Seebeck, 1995d:227; Taylor and Horner, 1973:53; Watts and Aslin, 1981:231).
  STATUS: IUCN – Lower Risk (lc).
  SYNONYMS: *cambricus* Troughton, 1937; *imbil* Troughton, 1937; *lacus* Tate, 1951; *pachyurus* (Higgins and Petterd, 1884); *petterdi* (Trouessart, 1904); *tetragonurus* (Higgins and Petterd, 1884); *vellerosus* (Gray, 1847); *velutinus* (Thomas, 1882).
  COMMENTS: *Rattus fuscipes* species group. Watts and Aslin (1981) provided comprehensive discussion of the species. Gross and microscopic anatomy of gastrointestinal tract described by Meulman et al. (1999). Reviewed by Mahoney and Richardson (1988) and Lunney (1995b). See Taylor and Calaby (1988b, Mammalian Species, 299).

Rattus macleari (Thomas, 1887). Proc. Zool. Soc. Lond., 1887:513.
  COMMON NAME: Maclear's Christmas Island Rat.
  TYPE LOCALITY: Christmas Isl (Australia).
  DISTRIBUTION: Endemic to Christmas Isl, 320 km south of Java in the Indian Ocean; thought to be extinct by 1908 (Andrews, 1909) and now considered extirpated (Flannery, 1990c), with time of extinction between 1901 and 1904 (Pickering and Norris, 1996).
  STATUS: IUCN – Extinct.
  COMMENTS: *Rattus* species group unresolved. Ellerman (1941) first listed the species as the only member of "*macleari*" group in subgenus *Rattus*, then placed it and *R. nativitatis* in same group within subgenus *Stenomys* of *Rattus* (Ellerman, 1949a). Chasen (1940) thought *R. macleari* to be nearest *Sundamys muelleri*, but in their comparisons, Musser and Newcomb (1983) found no support for this alliance. Misonne (1969) included *R. macleari* in subgenus *Rattus*; at the other extreme, Sody (1941) proposed genus *Christomys* to contain it. In the original description, Thomas (1887c) indicated *R. macleari* belonged to a group that included *Taeromys celebensis*, *Lenomys meyeri*, *Rattus everetti*, and *R. xanthurus* (he treated all as *Rattus*); of these, only *R. xanthurus* resembles *R. macleari* (Musser and Newcomb, 1983). Phylogenetic relationships remain unresolved, but Musser (1986) suggested *R. macleari* should be compared with a group of species that includes *R. annandalei*, *R. enganus*, *R. korinchi*, *R. montanus*, *R. nativitatis*, and *R. xanthurus*, none of them members of subgenus *Rattus* but distantly related to it.
    Until the introduction of *Rattus rattus* in 1899 (apparently from a cargo of hay carried by the S. S. Hindustan; Pickering and Norris, 1996), *R. macleari* and *R. nativitatis* were the only rats living on Christmas Isl. The possible role of disease transferred from *R. rattus* to *R. macleari* in abetting its extinction is discussed by Pickering and Norris (1996), who also identified specimens with color patterns they interpreted to reflect hybridization between the introduced species and *R. macleari*. All specimens, however, can be identified as either *R. rattus* or *R. macleari* and provide no evidence of hybridization (Musser and Norris, ms.), which is consistent with the distant phyletic relationship between the two as assessed by morphological traits. Whatever the causes of extinction (and introduced infectious diseases may have played a role) hybridization was not one of them.

Rattus marmosurus Thomas, 1921. Ann. Mag. Nat. Hist., ser. 9, 7:246.
  COMMON NAME: Marmoset Xanthurus Rat.

TYPE LOCALITY: Indonesia, NE Sulawesi, Minahassa, Gunung Masarang, 2000 ft (610 m).

DISTRIBUTION: Northern arm and central core of Sulawesi. Absent from SE peninsula where it is replaced by *R. salocco* and SW peninsula where *R. bontanus* occurs. Altitudinal range, lowlands to mountain summits in a variety of forest formations.

STATUS: IUCN – Lower Risk (lc).

SYNONYMS: *facetus* Miller and Hollister, 1921; *tondanus* Sody, 1932.

COMMENTS: *Rattus xanthurus* species group. Although *marmosurus* has been listed as a subspecies of *R. xanthurus* (Ellerman, 1941; Laurie and Hill, 1954), most researchers besides Thomas have recognized its specific uniqueness (Misonne, 1969; Musser, 1971*e*, 1984; Musser and Holden, 1991; Sody, 1941; Tate, 1936). Furthermore, the two are sympatric in NE Sulawesi (Musser, 1971*e*). Sody (1941) questionably included *marmosurus* in *Taeromys*, but it is a member of the *Rattus xanthurus* group. Musser (1971*e*) explained why *tondanus*, from the NE arm, is a synonym; *facetus* is based on a juvenile from the central core of the island. Musser and Carleton (1993) confined the species to NE region of the northern peninsula of Sulawesi, but recent morphometric analyses unequivocally indicate that it is *R. xanthurus* that is restricted to the NE arm and *R. marmosurus* that has the wider distribution on the island (Musser, ms).

Rattus mindorensis (Thomas, 1898). Trans. Zool. Soc. Lond., 14:402.

COMMON NAME: Mindoro Mountain Rat.

TYPE LOCALITY: Philippines, Mindoro Isl, Mt Dulangan, 5000 ft (1524 m).

DISTRIBUTION: Greater Mindoro Faunal Region. Endemic to highlands of Mindoro (Philippines).

STATUS: IUCN – Vulnerable.

SYNONYMS: *picinus* (Hollister, 1913).

COMMENTS: *Rattus rattus* species group. Phylogenetically distant from other species of *Rattus* endemic to the Philippines (Musser and Heaney, 1992). Morphologically closely related to *Rattus tiomanicus*, which is native to Malay Peninsula and islands on the Sunda Shelf, and possibly only an insular variant of that species (Musser and Califia, 1982; Musser and Heaney, 1985; Musser, ms). Musser (1977*b*) explained why *picinus* is a synonym. Murine fauna on Mindoro consists of Sundaic and Philippine elements, and *R. mindorensis* is a member of the former (see account of *Anonymomys*).

Rattus mollicomulus Tate and Archbold, 1935. Am. Mus. Novit., 802:4.

COMMON NAME: Lampobatang Sulawesi Rat.

TYPE LOCALITY: Indonesia, Sulawesi, SW Sulawesi, Gunung Lampobatang, Wawokaraeng, 1500 m.

DISTRIBUTION: Known only from higher slopes of Gunung Lampobatang, Sulawesi (Musser and Holden, 1991).

STATUS: IUCN – Vulnerable.

COMMENTS: *Rattus rattus* species group. Morphological and distributional limits outlined by Musser and Holden (1991), who also provided past historical allocations of the name. Closest relative is *R. hoffmanni*, which occurs in lowlands of the SW peninsula and throughout the rest of Sulawesi (see that account).

Rattus montanus Phillips, 1932. Ceylon J. Sci., Sec. B, 16:323.

COMMON NAME: Sri Lankan Mountain Rat.

TYPE LOCALITY: Sri Lanka (Ceylon), West Haputale, Ohiya, 6000 ft (1830 m).

DISTRIBUTION: Known only from the type locality, Horton Plains at 2135 m, and Nuwara Eliya at 1830 m in primary montane forests of Central and Uva Provinces of Sri Lanka (Phillips, 1980).

STATUS: IUCN – Critically Endangered.

COMMENTS: *Rattus* species group unresolved. A montane endemic on Sri Lanka and morphologically so unlike other species of *Rattus* that it should probably be removed from the genus, despite McKay's (1984) assertion that it is nothing more than a large form of *R. rattus* (see Musser, 1986:22, and Corbet and Hill, 1991:346). Like *R. annandalei*, *R. hoogerwerfi*, *R. korinchi*, *R. macleari*, *R. nativitatis*, and members of the *R. xanthurus* group, the Ceylon endemic seems isolated within the morphological boundaries of *Rattus* as presently understood (Musser, 1986).

Rattus mordax (Thomas, 1904). Ann. Mag. Nat. Hist., ser. 7, 14:398.
   COMMON NAME: Eastern New Guinea Rat.
   TYPE LOCALITY: Papua New Guinea, Kumusi River (see Tate, 1951:333, for comments).
   DISTRIBUTION: Papua New Guinea; Huon Peninsula and both sides of Owen Stanley Range
      on mainland, sea level to 2750 m (Taylor et al., 1982; Flannery, 1995a). Also recorded
      from Goodenough, Fergusson, and Normanby Isls in the D'Entrecasteaux Arch.;
      Woodlark Isl in the Trobriand Arch.; Misima and Sudest in the Louisiade Arch.; the
      Conflict Group of islands; and Sideia Isl, off the tip of E Papua (Flannery, 1995b).
   STATUS: IUCN – Lower Risk (lc).
   SYNONYMS: *fergussoniensis* Laurie, 1952.
   COMMENTS: *Rattus leucopus* species group. Originally described by Thomas as a species,
      mordax had been treated as a subspecies of either *leucopus*, *ringens*, or *ruber* until it was
      reinstated by Taylor et al. (1982). Reviewed by Flannery (1995a).

Rattus morotaiensis Kellogg, 1945. Proc. Biol. Soc. Wash., 58:66.
   COMMON NAME: Halmahara Rat.
   TYPE LOCALITY: Indonesia, Moluccas, Pulau Morotai, off N coast of Pulau Halmahera.
   DISTRIBUTION: Endemic to the Halmahara Isls; recorded from Morotai, Halmahara, and
      Batjan Isls (Flannery, 1995b).
   STATUS: IUCN – Lower Risk (lc).
   COMMENTS: *Rattus* species-group unresolved. Arboreal, and the only recorded murine
      endemic to the Halmahara Isls. Kellogg (1945) regarded *R. morotaiensis* to be more
      closely related to the New Guinea species ("*R. ringens* group") than to Sundaic species
      ("*Rattus rajah* group"), but morphology of the Halmahara rat indicates otherwise
      (K. Aplin, in litt., 2004). It is characterized by a suite of distinctive external traits (very
      spiny fur, long slightly tufted tail), cranial conformation, and molar structure
      (posterior cingula on upper molars, large peg-like anterolabial and anterolingual cusps
      on m1, wide cingular margins on m2, and crenulated enamel on all molars) not found
      in any other species of *Rattus* from New Guinea, the Moluccas, or anywhere else
      (Musser and Holden, ms). Reviewed by Flannery (1995b).

Rattus nativitatis (Thomas, 1889). Proc. Zool. Soc. Lond., 1888:533 [1889].
   COMMON NAME: Christmas Island Rat.
   TYPE LOCALITY: Christmas Isl (Australia).
   DISTRIBUTION: Endemic to Christmas Isl, 320 km south of Java in the Indian Ocean;
      suspected to be extinct by 1908 (Andrews, 1909) and is now considered extinct
      (Flannery, 1990c).
   STATUS: IUCN – Extinct.
   COMMENTS: *Rattus* species group unresolved. For Thomas (1888b), the morphology of
      R. nativitatis distanced it from any other described species of *Rattus*. Ellerman (1941)
      first listed the species as the only member of the "*nativitatis*" group in subgenus *Rattus*,
      then placed it and *R. macleari* in same group within subgenus *Stenomys* of *Rattus*
      (Ellerman, 1949a). Chasen (1940) thought *R. nativitatis* to be without close relatives in
      Malaysia, but Misonne (1969) placed it close to *rajah* in the subgenus *Leopoldamys* of
      Rattus, an allocation rejected by Musser (1981b) and Musser and Newcomb (1983).
      Four hypotheses about phylogenetic position of *R. nativitatis* require testing: 1) it is
      most closely related to *R. macleari*, the other endemic on Christmas Isl; 2) it is not
      related to *R. macleari* but to other species of *Rattus*; 3) it is a member of *Rattus* but
      phylogenetically isolated from all other species; 4) it is not even a member of *Rattus*.

Rattus niobe (Thomas, 1906). Ann. Mag. Nat. Hist., ser. 7, 17:327.
   COMMON NAME: Eastern New Guinea Mountain Rat.
   TYPE LOCALITY: Papua New Guinea, Angabunga River, Owgarra, 2750 m.
   DISTRIBUTION: Montane habitat in Papua New Guinea: "from the southeastern extremity of
      New Guinea northward and westward to 141EE longitude; also in the Saruwaged
      Mountains of the Huon Peninsula " (Taylor et al., 1982:193), 762-4000 m.
   STATUS: IUCN – Lower Risk (lc) as *Stenomys niobe*.
   SYNONYMS: *rufulus* Thomas, 1922; *stevensi* Rümmler, 1935.
   COMMENTS: *Rattus leucopus* species group. Our view of *R. niobe* corresponds to Papuan *R. n.*

niobe as defined by Taylor et al. (1982). They recognized Central Cordillera populations in Prov. of Papua (= Irian Jaya) and a population in the Arfak Mtns of the Vogelkop Peninsula as *R. n. arrogans*, which we divided into *R. arrogans*, *R. pococki*, and *R. arfakiensis* (see those accounts). Chromosomal morphology presented by Dennis and Menzies (1978). Phallic anatomy described by Lidicker (1968). See account of *R. arrogans*.

Rattus nitidus (Hodgson, 1845). Ann. Mag. Nat. Hist., [ser. 1], 15:267.

COMMON NAME: White-footed Indochinese Rat.

TYPE LOCALITY: Nepal.

DISTRIBUTION: Mainland Southeast Asia from S China (SE Xixang, Yunnan, Sichuan, Guizhou, Hunna, Guangxi, Guangdong, Fujian, Jiangxi, Zhejiang, Shanghai, Jiangsu, Anhui, S Shaanxi, SE Gansu, and Hainan Isl; Wang, 2003), Vietnam (including the coastal islands of Cat Ba and Thô Chu; Kuznetsov, 2000), Laos, N Thailand, Burma, Bangladesh, Nepal, Bhutan, and N India (Uttar Pradesh, Sikkim, West Bengal, Arunachal Pradesh, Meghalaya, Tripura, Mizoram, and Manipur; Agrawal, 2000); this distribution probably represents the indigenous range (Musser and Holden, 1991). Records east of the continental shelf are from C Sulawesi, Luzon Isl in the Philippines (recorded only from Benguet Province, Heaney et al., 1998; Musser, 1977*a*; Musser and Holden, 1991), Pulau Seram in the Moluccas, the Vogelkop Peninsula of the Prov. of Papua (= Irian Jaya), and the Palau Isls (east of the Philippines; Barbehenn, 1974); this distribution likely represents introductions mediated by human agency (Musser and Holden, 1991).

STATUS: IUCN – Lower Risk (lc).

SYNONYMS: *aequicaudalus* (Hodgson, 1849); *guhai* (Nath, 1952); *horeites* (Hodgson, 1845); *manuselae* Thomas, 1920; *obsoletus* Hinton, 1919; *rahengis* Kloss, 1919; *ruber* (Jentink, 1880); *rubricosa* (Anderson, 1879); *subditivus* Miller and Hollister, 1921; *vanheurni* Sody, 1933.

COMMENTS: *Rattus norvegicus* species group. Association of synonyms documented by Corbet and Hill (1992), Ellerman (1941), Khajuria et al. (1977), J. T. Marshall, Jr. (1977*a*, *b*), Musser (1981*c*), Musser and Holden (1991), and Taylor et al. (1982). Phallic morphology described by Yang and Fang (1988) in context of assessing phylogenetic relationships among Chinese murines. Yang et al. (1994) and Zeng et al. (1996*a*, *b*, 1999) reported various aspects of population ecology for Chinese populations, which are among the few studies of this kind covering *R. nitidus*. In pristine environments, *R. nitidus* lives in forested habitats along streams and readily enters water (field observations by G. Musser and D. Lunde in N Vietnam), which may account for its dense pelage resembling that of *R. norvegicus*, also a forager in streams and lakes. The names *manuselae* (Seram), *ruber* and *vanheurni* (Vogelkop Peninsula), and *subditivus* (Sulawesi) were applied to specimens thought to represent native species (Musser, 1877*a*; Musser and Holden, 1991).

Analyses of DNA sequences from L1 (LINE-1) elements identified a specimen of *Rattus* from Pulau Seram (Gunung Binaiya) in the Malukus as *R. rattus moluccarius* (Usdin et al., 1995) or *R. cf. moluccarius* (Verneau et al., 1997, 1998) that has close molecular similarities to *R. norvegicus*. However, the holotype and type series of *moluccarius* is from nearby Pulau Buru and represent a large-bodied population of *R. tanezumi* (Musser, 1970*a*, discussed under *R. rattus*). The voucher specimen (not examined by us) is probably an example of *R. nitidus*, which occurs on Seram "... in many habitat types over a broad altitudinal range" (Helgen, 2003*b*:170). The type series of *manuselae*, for example, was collected at 1830 m, and Helgen (2003*b*) listed samples taken from 350 m to 1830 m; *R. norvegicus* has not been recorded from Seram, and if it has been introduced there would likely be restricted to port towns (see account of *R. norvegicus*).

The phylogenetic association of *R. nitidus* with *R. norvegicus* indicated by DNA sequences from L1 elements is substantiated by recent analyses of mtDNA cytochrome *b* sequences, which aligns samples of *R. nitidus* (and *R. pyctoris*; see account) with *R. norvegicus* (K. Aplin, in litt., 2004). Morphology also supports this alliance. Both *R. nitidus* and *R. norvegicus* have dense and soft fur, six pairs of teats, and an upper M1 in which the anterolabial cusp (t3) on the anterior lamina is missing or undetectable due to its coalesence with the adjacent central cusp.

Rattus norvegicus (Berkenhout, 1769). Outlines of the Natural History of Great Britain and Ireland, 1:5.

COMMON NAME: Brown Rat.

TYPE LOCALITY: Great Britain.

DISTRIBUTION: Original distribution assumed to be SE Siberia, N China (Heilongjiang), and Hondo region (islands of Honshu, Shikoku, and Kyushu; see Dobson, 1994) of Japan (Jones and Johnson, 1965; Kowalski and Hasegawa, 1976; Kawamura, 1989), but introduced worldwide where it is more common in colder climates of higher N and S latitudes (Kucheruk, 1990); in warmer regions and tropics restricted to habitats highly modified by humans (e. g., sewers, buildings, wharves, breakwaters, ports, and large cities; Johnson, 1962a, Corbet and Hill, 1992). Considered extinct in Norway (Syvertsen et al., 1996).

STATUS: IUCN – Lower Risk (lc).

SYNONYMS: *aquaticus* (Rutty, 1772); *albinus* (Donaldson, 1912); *albus* Hatai, 1907 [not of Fitzinger, 1867; see Kaneko and Maeda, 2002]; *americanus* (de Kay, 1842); *caraco* (Pallas, 1778); *caspius* (Oken, 1816); *cauquenensis* (Philippi, 1900); *decaryi* (Grandidier, 1934); *decumanoides* (Hodgson, 1841) [*nomen nudum*]; *decumanus* (Pallas, 1779); *discolor* (Noack, 1918); *fossilis* (Ameghino, 1889) [see Massoia and Pardiñas, 1993]; *fossor* (Walker, 1808); *griseipectus* (Milne-Edwards, 1872); *hibernicus* (Thompson, 1837) [see McKee, 1981]; *hoffmanni* (Trouessart, 1904) [not of Matschie, 1901]; *humiliatus* (Milne-Edwards, 1868); *hybridus* (Bechstein, 1800); *insolatus* Howell, 1927; *javanus* (Hermann, 1804); *kurodobu* Kuroda, 1953 [unavailable, see Kaneko and Maeda, 2002]; *leucosternum* (Rüppell, 1842); *lutescens* (Gay, 1848); *magnirostris* (Mearns, 1905); *major* (Hoffmann, 1887); *maniculatus* (Wagner, 1848); *maurus* (Waterhouse, 1839); *migrans* (Zimmermann, 1777); *orii* Kuroda, 1952 [see Kaneko and Maeda, 2002]; *otomoi* Yamada, 1930; *ouangthomae* (Milne-Edwards, 1871); *plumbeus* (Milne-Edwards, 1874); *praestans* (Trouessart, 1904); *primarius* Kastschenko, 1912; *shirokuma* Kuroda, 1953 [unavailable; see Kaneko and Maeda, 2002]; *simpsoni* (Philippi, 1900); *socer* Miller, 1914; *sowerbyi* Howell, 1928; *suffureoventris* Kuroda, 1952 [see Kaneko and Maeda, 2002]; *surmulottus* (Severinus, 1779); *tamarensis* (Higgins and Petterd, 1883).

COMMENTS: *Rattus norvegicus* species group. Latitudinal geographic variation among Chinese populations reported by Wu (1982). Samples from native Asian, free-living introduced, and laboratory populations have been the subject of numerous morphological (e.g., Bugge, 1970; Cheng and Yen, 1993; Greene, 1935; Kiliaridis et al., 1996; Kimura et al., 1994, 1996; Millien-Parra, 2000a; Naftel et al., 1999; Piette and Lametschwandtner, 1995; Puzachenko and Lapshov, 1994; Sakamoto, 1996), physiological, chromosomal and molecular (many are summarized in Yosida, 1980, and Levan et al., 1990; see also Behboudi et al., 2002; Belcheva et al., 1992; Koh et al., 1995; Robinson, 1984; Rothenburg et al., 2002) studies, which have produced among the mass of data a gene map of *R. norvegicus* (Levan et al., 1990). Results of attempted hybridizations between *R. norvegicus* and different forms of *R. rattus* summarized by Yosida (1980). Assessment and review of swimming capability in context of potential to colonize tropical oceanic islands provided by Spennemann and Rapp (1980). Monographic treatment edited by Sokolov and Karasjova (1990) covers systematics, geographic distribution, ecology, behavior, and practical importance generally worldwide but with a focus on populations in Russia. Overall review of systematics reported by Schwarz and Schwarz (1967) and Miljutin (1990), who also included *R. norvegicus* in treatises covering ecomorphology and ecological strategies of Baltic rodents (Miljutin, 1997, 1998). Phylogenetic relationships to other members of subgenus *Rattus* equivocal (e.g., contrast Chan et al., 1979, with Pasteur et al., 1982), but morphological, isozymic, and molecular data indicate significant phylogenetic divergence between *R. norvegicus* and *R. rattus* (Baverstock et al., 1983a, b, 1986; Chan, 1977; Verneau et al., 1997, 1998). Phallic morphology of Chinese samples described by Yang and Fang (1988) in context of assessing phylogenetic relationships among Chinese murines. Significance of the ecological distributions between *R. norvegicus* and *R. rattus* in the Caucasus investigated by Kalinin (1993), and in the Galápagos Isls by Key and Woods (1996). Contrasts in surface topography of hind foot plantar pads between *R. norvegicus*, *R. argentiventer*,

R. rattus, and *R. exulans* documented by Yabe et al. (1998); the first two are primarily terrestrial and have low and inconspicuous lamellae on the pads, but the last two are terrestrial and arboreal and their pads are adorned with well developed lamellae.

European populations reviewed Becker (1978*b*) and Mitchell-Jones et al. (1999); range, taxonomy, and other characteristics of species in Russia and adjacent regions summarized by Gromov and Erbajeva (1995); taxonomy and distribution in Indo-malayan region presented by Corbet and Hill (1992); North American taxonomy and range outlined by Hall (1981); Chilean by Osgood (1943); and Australian populations, which are restricted to major coastal cities and ports, reviewed by Mahoney and Richardson (1988) and Watts (1995*i*). Summaries of geographic distribution, habitat, ecology, interaction with other mammalian species as well as humans, and sometimes history of colonization, are available in varying depth of coverage for Hawaiian Isls (Tomich, 1986); New Zealand (Moors, 1990); New Guinea (Flannery, 1995*a*; Taylor et al., 1982); Moluccas and SW Pacific Isls (Flannery, 1995*b*); Micronesia (Barbehenn, 1974; J. T. Marshall, Jr., 1961); Philippines (Heaney et al., 1998); Japan (Kaneko, 1994); Korea (Jones and Johnson, 1965; Won and Smith, 1999); Russia on the Svjatoj Nos peninsula and isthmus in Lake Baikal where the species is common in most human settlements (Reiter et al., 1995), the Kamchatka region (Nikanorov, 2000), SE European part of Russia (Varshavsky et al., 1989), and from Russia to Europe and Asia in general with particular regions detailed (Kucheruk, 1900); China (Wang, 2003; Zhang et al., 1997); Taiwan (Adler, 1995; M.-J. Yu, 1996); large cities, ports, and river mouths in India (Hossack, 1907) where it used to be common in Calcutta but is being gradually replaced by *Bandicota bengalensis* (Agrawal, 2000); Afghanistan (Hassinger, 1973); Iran (Lay, 1967); Turkey, Gökçeada Isl, and Cyprus (Kryštufek and Vohralík, 2001; Yi—it et al., 1998); Serbia and Montenegro (Petrov, 1992); Slovenia (Kryštufek, 1991); Albania (Prigioni, 1996); Transylvanian Romania (Istrate, 1998); Bulgaria (Mitev and Miteva, 1991*a*; Peshev, 1996); Slovakia (Danko, 1994; Kminiak, 1996; Mosansky, 1994; Stanko and Mosansky, 2000); Austria (Bauer and Spitzenberger, 1996); Czech Republic (Smaha, 1996; Zima and And—ra, 1996); Baltic region (Timm et al., 1998); Italy (Amori et al., 1999; Andreotti et al., 2001); Netherlands (Blaaderen and Bosman, 1992; Thissen and Hollander, 1969); Belgium (Libois, 1996); France (Beaufort et al., 1996; Vincent, 2001); Portugal (Santos-Reis and Mathias, 1996); Spanish Balearic Isls (Alcover and Gosalbez, 1988); Madagascar (Duplantier and Duchemin, 2003); Morocco (Aulagnier and Thevenot, 1986); Libya (Ranck, 1968); Egypt (Osborn and Helmy, 1980); Ghana and Sierra Leone (Grubb et al., 1998); Angola, where the species is restricted to coastal settlements (Crawford-Cabral, 1998); South Africa, where it is unknown outside of ports and larger coastal towns (de Graaff, 1981). Usually found only in seaports in West Africa and thought incapable of penetrating inland, *R. norvegicus* was found in the inland city of Bamako in S Mali by Meinig (2000). Meinig noted that fresh water, a necessity for *R. norvegicus*, is available all year in irrigation channels and canals and assumed that the brown rat was inadvertently transported to Bamako along the only railway line, which originates in the port city of Dakar, Senegal. Colonization of the New World by *R. norvegicus* reviewed by Armitage (1993).

Rattus norvegicus is thought to be native to the Hondo region of Japan because it is represented there by Holocene and late Pleistocene fossils along with a form identified as *R.* aff. *norvegicus* from the middle Pleistocene that may have been ancestral to *R. norvegicus* in the Japanese Isls; the rest of the endemic Japanese rodent fauna is also represented by Pleistocene and Holocene fossils (Kowalski and Hasegawa, 1976; Kawamura, 1989). Late Pleistocene-Holocene records of *R. norvegicus* come from cave deposits in the Sichuan-Guizhou region of China (Zheng, 1993), which adds credance to the conventional view that the original range probably also included N China (particularly in Heilongjiang Province) and SE Siberia (see references under Distribution). From their indigenous range, *R. norvegicus* is claimed to have reached Europe by the 18th century, presumably through Russia (Robinson, 1984), and the British Isles during the same period (Yalden, 1999). Miljutin (1983:62) noted that while some authorities defended a much earlier invasion of Europe ("long before the beginning of our era"), geographical and historical data proves *R. norvegicus* could not have reached Europe "before the discovery of the Indian sea-route by Vasco da Gama, i.e. before the year of 1499." By at least 1745, the Norway rat had reached American shores, but the principal thrust of the invasion occurred between the 1760s and 1770s, coinciding with the massive wave of British immigration to the region of what would become the original 13 English colonies (Armitage, 1993).

Metamorphosis of *R. norvegicus* "from evil harbinger of pestilence to hero of modern medicine" is described by Clause (1993:330) who related how rat breeding and husbandry were a part of the research programs at several institutions during the early 1900s, but that it was at the Wistar Institute in Philadelphia where "the concept of the rat as an instrument for scientific research was most clearly articulated and that the engineering of a superior animal was most vigorously promoted as an objective of the institution's scientific program." Creation, maintenance, and distribution of the famous Wistar Rats as "standardized animals" are cogently reviewed by Clause (1993).

Rattus novaeguineae Taylor and Calaby, 1982 (*in* Taylor et al., 1982, Bull. Am. Mus. Nat. Hist., 173:259).

COMMON NAME: Papua New Guinea Rat.

TYPE LOCALITY: Papua New Guinea, Kalolo Creek, 1070 m.

DISTRIBUTION: Recorded only from C Papua New Guinea "from Kassam westward to Karimui in the north, and southward to Koranga, at altitudes ranging from 740 to 1525 m" (Taylor et al., 1982:259), and from Wasi Falls and Mt Sisa in the Kikori River Basin of S Papua New Guinea (Leary and Seri, 1997); see Taylor et al. (1982) and Flannery (1995a).

STATUS: IUCN – Lower Risk (lc).

COMMENTS: *Rattus leucopus* species group. Virtually all taxonomic, distributional, and ecological information associated with *R. novaeguineae* is contained in the reports by Flannery (1995a), Leary and Seri (1997), and Taylor et al. (1982).

Rattus omichlodes (Misonne, 1979). Bull. Inst. Roy. Sci. Nat. Belgium (Biologie), 51:6.

COMMON NAME: Arianus' New Guinea Mountain Rat.

TYPE LOCALITY: New Guinea, Prov. of Papua (= Irian Jaya), W Snow Mtns (Pegunungan Maoke), Ertsberg (04°04'S, 137°07'E), 3400 m.

DISTRIBUTION: Recorded only from western region of Snow Mtns at Ertsberg (3400 m) and Carstensz Peak area, 2950-3950 m (Flannery, 1995a:340).

COMMENTS: *Rattus leucopus* species group. Regarded by Taylor et al. (1982) as identical to *R. richardsoni* after examining two paratypes of *omichlodes* (KNMB 4030, 4031), a synonymy followed by Musser and Carleton (1993) who had not studied the type series. Misonne (1979), however, had collected *R. richardsoni* at the same place but in a different habitat and his comparison with *omichlodes* clearly indicated the two are different species. Flannery (1995a;340) revived the latter after finding it in alpine scrub with samples of *R. richardsoni* and *R. arrogans* (recorded as *R. niobe*) in Meren Valley in the Carstensz Peak region; at lower elevations, "[*Rattus*] *omichlodes* occupies boggy alpine heath while [*Rattus*] *niobe* [= *R. arrogans*] occurs in mossy forest." Flannery had also examined the holotype. Misonne (1979:9) noted that *R. arrogans* (reported as *R. niobe*) was common in the forest and lower margins of alpine grassland, that *R. richardsoni* seemed to be restricted to bushes and grasses above treeline, and *R. omichlodes* to be still more restricted and found only "in shallow marshes with high grass and small bushes, above treeline." Our recent examination of paratypes of *omichlodes* (KNMB 4030, 4031) and comparisons with samples in AMNH of *R. richardsonii* and *R. arrogans* corroborate Misonne's original discovery and Flannery's action. *Rattus omichlodes* is a different entity than *R. richardsoni*, but its relationship to *R. arrogans* has to be resolved in comprehensive morphometric and molecular analyses. From our inspection of specimens, we could not separate the paratypes of *omichlodes* from the four specimens Flannery sent us from 3950 m in Barren Valley below Meran Glacier, Tembagapura Area, in which he thought two of the specimens represented *niobe* and two *omichlodes*. Flannery's material and the paratypes are thick-furred small-bodied rats the size of *R. niobe* and much smaller than the large *R. arrogans* from N slope of the Snow Mtns. This entire complex should be reexamined to redefine *omichlodes* in relation to the large *R. arrogans* and series from the S flanks of the Snow Mtns. Our recognition of *R. omichlodes* is provisional pending a fresh review of the *R. niobe* complex.

Rattus osgoodi Musser and Newcomb, 1985. Am. Mus. Novit., 2814:18.

COMMON NAME: Osgood's Vietnamese Rat.

TYPE LOCALITY: S Vietnam, Lâm Dğng Province, Langbian Peak, 5000 ft (1524 m; see Musser and Newcomb, 1985:18, for details).

DISTRIBUTION: Recorded only from two localities on Langbian (= Lam Vien) Plateau in the Dà Lat region of S Vietnam, 900-2000 m (Musser and Newcomb, 1985).

STATUS: IUCN – Lower Risk (lc).

COMMENTS: *Rattus rattus* species group. A Lang Bian Plateau endemic morphologically and probably phylogenetically related to *R. losea* (Musser and Newcomb, 1985). Among the smallest in body size of the *R. rattus* species group.

Rattus palmarum (Zelebor, 1869). Reise Oesterr. Fregatte Novara. Zool., I(Wirbelthiere), I(Säugeth.), p. 26.

COMMON NAME: Car Nicobar Rat.

TYPE LOCALITY: India, Nicobar Isls, Car Nicobar (see below).

DISTRIBUTION: Nicobar Isls, Car Nicobar.

STATUS: IUCN – Vulnerable.

SYNONYMS: *novarae* (Fitzinger, 1861); *palmarum* (Fitzinger, 1861); both names are *nomina nuda* (Miller, 1902:759).

COMMENTS: *Rattus rattus* species group. A distinctive species known only by four specimens in the original series (NMW B26, B27, 21497, and 27027; Musser and Heaney, 1985; Musser and Newcomb, 1983). Because Zelebor (1869) provided no exact collection site, Musser and Heaney (1985) stated that the island was unknown where the rats were obtained. In response, Dr. K. Bauer wrote Musser the following: "It is true, that Zelebor (1869) gave no more exact information. But K. Scherzer, the 'Historiographer' of the expedition did. In his detailed three volume report 'Reise der Österreichischen Fregatte Novara um die Erde in den Jahren 1857, 1858, 1859 unter den Befehlen des Commodore R. von Wüllerstort-Urbair,' published in Vienna 1861/62 (and later in Italian and English translations), he summarized the scientific results. And in connection with a sketch of the fauna found on the Nicobar Islands (vol. 2:71/72) he states (in translation): 'In mammals, all islands of the group are poor. We only found 8 species . . . and two different murids (*Mus*). One of these, nearly as big as the Norway Rat, we saw only on Car-Nicobar and Sambelong (= Great Nicobar); . . . always in the crowns of cocos palms, very fast, difficult to see and shoot, doing heavy damage to cocos plantations. . . . A second, in size like our Black Rat, lives on Car-Nicobar in holes in the ground that it shares in total peace with a (terrestrial) crab (*Gecarcinus*).' . . . This history gives detailed information on spots visited and time spent there. . . . Careful study of this text might well give further hints, but already from a hurried overview it seems quite certain, that the *palmarum* sample came from Car-Nicobar."

Rattus palmarum is large (head and body length for three adults = 225-240 mm, greatest skull length = 49.0-54.0 mm) and has a moderately long tail (220-231 mm); brownish, dense, and rough dorsal coat with long guard hairs; white underparts; ten pairs of mammae, and a large, robust cranium with thick and high dorsolateral ridges. The skull is a larger, more robust version of that seen in *R. burrus* from the other Nicobar Isls, *R. simalurensis* from the Simalur Group and *R. lugens* from the Mentawai Arch. It may be most closely related to *R. burrus*, which occurs on Trinkat, Little Nicobar, and also Great Nicobar Isls but not Car Nicobar, and it may be *R. palmarum* and *R. burrus* that Scherzer saw on Great Nicobar. *Rattus andamanensis* occurs on Car Nicobar (see that account). It is about the same body size as Scherzer's "Black Rat," and the larger-bodied *R. palmarum*, which also occurs on that island, matches the "Norway Rat" in body size.

Rattus pelurus Sody, 1941. Treubia, 18:308.

COMMON NAME: Peleng Island Xanthurus Rat.

TYPE LOCALITY: Indonesia, Kepulauan Banggai, Pulau Peleng (1°23'S, 123°14'E), east of Sulawesi.

DISTRIBUTION: Known only from Peleng Isl.

STATUS: IUCN – Vulnerable.

COMMENTS: *Rattus xanthurus* species group. Originally described by Sody as a subspecies of *R. foramineus*, but provisionally treated as an insular population of *R. xanthurus* by Musser (1984), and finally reviewed as a distinctive species by Musser and Holden (1991), who also placed it in the *R. xanthurus* group. Recent morphometric analyses corroborated the separation of *pelurus* as a different species from mainland *R. xanthurus* (Musser, ms).

Rattus pococki (Ellerman, 1941). Family and Genera of Living Rodents, 2:206.

COMMON NAME: Pocock's New Guinea Highland Rat.

TYPE LOCALITY: New Guinea, Prov. of Papua (= Irian Jaya), Weyland Range, Mt Sumuri, 2500 m.

DISTRIBUTION: New Guinea, middle altitudes along Central Cordillera of Prov. of Papua (= Irian Jaya), 1500-2500 m.

SYNONYMS: *clarae* Rümmler, 1935 [not Miller, 1913].

COMMENTS: *Rattus leucopus* species group; member of the *R. niobe* complex. *Rattus pococki* is replaced at higher elevations in the Central Cordillera (Pegunungan Maoke) of Prov. of Papua (= Irian Jaya) by *R. arrogans* (see that account). Relationship to *R. niobe* of Papua New Guinea remains to be determined.

Rattus praetor (Thomas, 1888). Ann. Mag. Nat. Hist., [ser. 1], 2:158.

COMMON NAME: Large New Guinea Spiny Rat.

TYPE LOCALITY: Solomon Isls, Guadalcanal Isl, Aola.

DISTRIBUTION: New Guinea; on the mainland north of the Central Cordillera from Vogelkop Peninsula throughout N New Guinea to about the Sepik-Ramu drainage in Papua New Guinea, and south of the Cordillera to W Prov. of Papua (= Irian Jaya); sea level to 1900 m; absent from the Trans-Fly region and E Papua New Guinea. Living populations also recorded from Manus Isl in Admiralty Isls; Bougainville and Guadalcanal Isls in the Solomon Isls; Karkar Isl, Blup Blup Isl, Bat Isl, New Britain Isl, and New Ireland Isl in the Bismarck Arch.; and Salawati and Gebe islands off the western tip of Prov. of Papua (= Irian Jaya); found as fossils only on Nissan and Tikopia islands in the Solomons (Emmons and Kinbag, 2001; Flannery, 1995*a*, *b*; Flannery and White, 1991; Taylor et al., 1982). Also recorded from Fiji Isl as a subfossil (K. Aplin, in litt., 2004).

STATUS: IUCN – Lower Risk (lc).

SYNONYMS: *bandiculus* Thomas, 1922; *coenorum* Thomas, 1922; *mediocris* Troughton, 1936; *purdiensis* Troughton, 1946; *sansapor* Troughton, 1946; *tramitius* Thomas, 1922; *utakwa* Ellerman, 1941.

COMMENTS: *Rattus leucopus* species group. In Tate's (1951) monograph, this species was listed as a subspecies of *R. ruber*; the holotype of *ruber* is an example of the introduced *R. nitidus* (Taylor et al., 1983, summarized the taxonomic history). According to Flannery (1995*b*:159), *R. praetor* "has been introduced prehistorically into most of its insular distribution, reaching New Ireland by 3500 year ago. . . and the Solomon Islands soon after." Reviewed by Flannery (1995*a*, *b*).

Rattus pyctoris (Hodgson, 1845). Ann. Mag. Nat. Hist., [ser. 1], 15:267.

COMMON NAME: Himalayan Rat.

TYPE LOCALITY: Nepal.

DISTRIBUTION: Mountains of SE Kazakhstan, Kyrgyzstan, E Uzbekistan, and Tajikistan (Kucheruk, 2000); EC Iran (NE Kerman Province; FMNH series); N Afghanistan (Hassinger, 1973; Niethammer and Martens, 1975; BMNH and FMNH specimens); N Pakistan (Akhtar, 1959; USNM material); N India (Agrawal, 2000; large series in BMNH, FMNH, and USNM); Nepal (Ellerman, 1961; Niethammer and Martens, 1975; BMNH, FMNH, and USNM specimens); and S China (Yunnan, Sichuan, and Guangdong; AMNH and FMNH material).

STATUS: IUCN – Lower Risk (lc) as *R. turkestanicus*.

SYNONYMS: *celsus* G. M. Allen, 1926; *gilgitianus* Akhtar, 1959; *khumbuensis* Biswas and Khajuria, 1955; *rattoides* (Hodgson, 1845) [not Pictet and Pictet, 1844]; *shigarus* (Miller, 1913); *turkestanicus* (Satunin, 1903); *vicerex* (Bonhote, 1903).

COMMENTS: *Rattus norvegicus* species group. Despite past use of *rattoides* Hodgson, 1845 (Caldarini et al., 1989; Corbet, 1978c; Ellerman, 1961; Ellerman and Morrison-Scott, 1951), it is preoccupied by *rattoides* Pictet and Pictet, 1844, a synonym of *R. rattus* (Schlitter and Thonglongya, 1971). The name *turkestanicus* (type-locality = Kyrgyzstan, Oshskaya Obl., Lenniskii p-h, Arslanbob) has generally replaced *rattoides* (Corbet and Hill, 1992; Musser and Carleton, 1993; Musser and Newcomb, 1985; Niethammer and Martens, 1975) for the species but should be abandoned for *pyctoris*, which is the oldest name (Musser and Carleton, 1993). Hodgson's (1845) *pyctoris* has historically been treated as a synonym of either *R. rattus* (Ellerman, 1941) or *R. nitidus* (Corbet, 1978c; Ellerman, 1961; Ellerman and Morrison-Scott, 1951). Musser studied the holotype of *pyctoris* (BMNH 45.1.8.381), the skin and skull of a young adult. The skin is in poor condition, and posterior part of the cranium and posterior portions of the dentaries are missing. However, the skin is sufficiently intact to determine that the fur is shaggy and brown over the upperparts, in contrast to *R. nitidus*, which has dark grayish brown upperparts and dense, soft pelage. The rostrum is wide and short (narrow and slender in *R. nitidus*), molars are chunky and wide (thinner and gracile in *R. nitidus*), and cusp t3, although small, is present on M1 (missing or so coalesced with cusp t2 that it is usually undetectable). Musser found the holotype of *pyctoris* to closely resemble the holotype of Hodgson's *rattoides* and to be inseparable from BMNH Nepalese specimens usually identified as either *rattoides* or *turkestanicus*.

Three distinctive morphological, chromosomal, and geographic forms are included under *R. pyctoris* (Caldarini et al., 1989; Capanna and Corti, 1991; Niethammer and Martens, 1975) and were first recognized by Hinton (1922), who treated all as species (*R. turkestanicus*, *R. vicerex*, and *R. rattoides*) but also suggested that each may instead be a well-differentiated subspecies. Subsequently, *R. pyctoris* and *R. vicerex* were reported to occur sympatrically in Jammu and Kashmir regions of NW India (Chakraborty, 1983; Agrawal, 2000). The latter is said to be distinguished from *R. pyctoris* by its shorter tail (shorter than head and body, conspicuously longer in *R. pyctoris*) and fewer teats (four pairs versus six pairs in *R. turkestanicus*); most of the specimens are young and their identification as a species separate from *R. pyctoris* has to be verified within the framework of a careful systematic treatment (Musser could not distinguish two species in the large series from N India and N Pakistan examined at BMNH and USNM). Such an inquiry should also examine significance of variation in pelage coloration (Agrawal, 2000; Corbet and Hill, 1992; our study) and chromosomal morphology (see above) among geographic samples, and determine whether the three groups represent separate species or geographic variants.

Distribution of *R. pyctoris* according to habitat types in SE Kazakhstan documented by Kucheruk (2000, as *R. turkestanicus*), who also described habitat interaction between that species and *R. norvegicus* and *R. rattus*. In contrast to those two, populations of *R. pyctoris* are less dependent upon anthropogenic habitats and its natural range has not been enlarged by following human settlement. Comparisons among *R. pyctoris*, *R. norvegicus*, and *R. rattus* in mandible shape and its potential taxonomic significance documented by Lapshov (1992, as *turkestanicus*). Russian range, taxonomy, and other characteristics reviewed by Gromov and Erbajeva (1995, as *turkestanicus*).

Including *R. pyctoris* in the same clade as *R. norvegicus* and *R. nitidus* is based primarily upon recent analyses of mtDNA cytochrome *b* sequences (K. Aplin, in litt., 2004). *Rattus pyctoris* has shaggy, dense fur and six pairs of teats, traits also characterizing *R. norvegicus*. The M1 of *R. pyctoris* also has a reduced anterolabial cusp (t3) relative to the adjacent two cusps forming the anterior lamina; cusp t3 is usually absent or undetectable in most examples of *R. norvegicus* and *R. nitidus*, but prominent in all other species of *Rattus*.

Rattus ranjiniae Agrawal and Ghosal, 1969. Proc. Zool. Soc. Calcutta, 22:41.

COMMON NAME: Ranjini's Rat.

TYPE LOCALITY: SW Indian Peninsula, India, Kerala State, Trivandrum.

DISTRIBUTION: Recorded only by specimens collected in rice fields from the districts of Trichur, Alleppy, and Trivandrum in the state of Kerala.

STATUS: IUCN – Vulnerable.

COMMENTS: *Rattus* species group unresolved. Still represented by few specimens. Information about the species is meager and contained in the original description and the review by Agrawal (2000). Study of two paratypes kindly loaned to Musser by Dr. S. Chakraborty revealed that *R. ranjiniae* is characterized by large claws relative to body size, very long and slender hind feet, large body size, long molar rows, small bullae, narrow incisive foramina, and a short bony plate that does not extend past the third molars. These traits combine in a morphology that is unique compared with all other species now placed in *Rattus* (a conclusion also endorsed by Corbet and Hill, 1992). Phylogenetic relationships of *R. ranjiniae* are unknown; this species probably should be removed from *Rattus*.

Rattus rattus (Linnaeus, 1758). Syst. Nat., 10th ed., 1:61.

COMMON NAME: Roof Rat.

TYPE LOCALITY: Sweden, Uppsala County, Uppsala.

DISTRIBUTION: Native to Indian Peninsula (Niethammer, 1975) and introduced worldwide in the temperate zone and parts of the tropical and subantarctic zones (Alcover and Gosalbez, 1988; Al-Jumaily, 1998; Amoti et al., 1999; Andreotti et al., 2001; Armitage, 1994; Aulagnier and Thevenot, 1986; Becker, 1978a; Blaaderen, 1992; Corbet and Hill, 1992; Crawford-Cabral, 1998; de Graaff, 1981; de Roguin, 1991; Dieterlen, 1979; Dobigny et al., 2002b; Downs and Wirminghaus, 1997; Duplantier et al., 1991b, 1997; Endepols et al., 2001; Gromov and Erbajeva, 1995; Grubb et al., 1998; Hall, 1981; Harrison and Bates, 1991; Helgen, 2001; Innes, 1990; Johnson, 1962a, b; Key et al., 1998; Kryštufek, 1991; Kryštufek and Vohralík, 2001; Kucheruk, 1994; Lay, 1967; Le Berre, 1990; Mahoney and Richardson, 1988; Meinig, 2000; Mitchell-Jones, et al., 1999; Niethammer, 1975; Osborn and Helmy, 1980; Osgood, 1943; Petrov, 1992; Prakash et al., 1995a, 1995b; Qumsiyeh, 1996; Ranck, 1968; Roberts, 1977, 1997; Smaha, 1996; Stanley et al., 1998, 2000, 2002; Taylor and Horner, 1973; Taylor et al., 1982; Tomich, 1986; Twigg, 1992; Vincent, 2001; Vohralík and Andera, 2000; Watts, 1995j; Yosida, 1980; Yosida et al., 1985). Southernmost limit is the subantarctic Macquarie Isl where *R. rattus* was introduced by sealers during the 19th century (Pye et al., 1999).

STATUS: IUCN – Lower Risk (lc).

SYNONYMS: *aequicaudalis* (Hodgson, 1845); *aethiops* (Philippi, 1900); *albiventer* (Jentink, 1909); *albiventris* (Jentink, 1909); *albus* (Fitzinger, 1867) [not of Hatai, 1907]; *alexandrino-rattus* (Fatio, 1902); *alexandrinus* (É. Geoffroy, 1803); *alexandrinus* (Desmarest, 1819); *arboreus* (Horsfield, 1851); *arboricola* (Gould, 1863); *asiaticus* (Gray, 1837); *ater* (Fitzinger, 1867); *ater* (Millais, 1905); *atratus* (Miller, 1902) [not of Philippi, 1900]; *atridorsum* (Miller, 1903); *auratus* (Grandidier, 1899); *beccarii* (Peters and Doria, 1881) [not Jentink, 1881]; *brahminicus* (Lloyd, 1909); *brookei* (Crew, 1923); *caeruleus* (Lesson, 1842); *caledonicus* (Wagner, 1842); *ceylonus* (Kelaart, 1850); *chionogaster* (Cabrera, 1921), *chionogaster* (Lönnberg and Mjoberg, 1916); *coquimbensis* (Philippi, 1900); *crassipes* (Blyth, 1859); *cyaneus* (Philippi, 1895) [not of Molina, 1782]; *doboensis* (de Beaufort, 1911); *domesticus* (Fitzinger, 1857); *doriae* (Trouessart, 1897); *erythronotus* (Temminck, 1844); *flavescens* (Elliot, 1839); *flavigaster* (Heuglin, 1861); *flaviventris* (Brants, 1827); *frugivorus* (Rafinesque, 1814); *fuliginosus* (Bonaparte, 1833); *fulvaster* (Fitzinger, 1867); *fuscus* (Fitzinger, 1857); *galapagoensis* (Waterhouse, 1830); *girensis* Hinton, 1918; *griseocaeruleus* (Higgins and Petterd, 1883); *indicus* (É. Geoffroy, 1803) [not Bechstein, 1800]; *infralineatus* (Blylth, 1863) [*nomen nudum*]; *insularis* (Waterhouse, 1838) [not Phillips, 1936]; *intermedius* (Ninni, 1882); *jacobiae* (Waterhouse, 1838); *jujensis* (Lönnberg, 1916); *jurassicus* (Burg, 1921); *kandianus* (Kelaart, 1850); *kandiyanus* (Kelaart, 1887); *keelingensis* Tate, 1950; *kelaarti* (Wroughton, 1915); *kijabius* (J. A. Allen, 1909); *latipes* (Bennett, 1835); *leucogaster* (Pictet, 1841); *longicaudus* Mori, 1937 [see Kaneko and Maeda, 2002]; *muansae* (Matschie, 1911); *narbadae* Hinton, 1918; *nemoralis* (Selys Longchamps, 1841) [not Blyth, 1851]; *nericola* Cabrera, 1921; *novaezelandiae* (Buller, 1871); *osorninus* (Philippi, 1900); *personatus* (Krefft, 1867); *picteti* (Schinz, 1845); *rattiformis* (Matschie, 1915); *rattoides* (Pictet and Pictet, 1844) [not of Hodgson, 1845]; *rufescens* (Gray, 1837); *ruthenus* Ognev and Stroganov, 1936; *saltuum* (Philippi, 1900); *samharensis* (Heuglin, 1877); *setosus* (Lund, 1841); *siculae* (Lesson,

1827); *subcaeruleus* (Lesson, 1842); *subrufus* (Philippi, 1900); *sueirensis* Cabrera, 1921; *sylvestris* (Pictet, 1841); *tectorum* (Savi, 1825); *tetragonurus* (Kelaart, 1850); *tettensis* (Peters, 1852); *tompsoni* (Ramsay, 1881); *variabilis* (Higgins and Petterd, 1883); *varius* (Fitzinger, 1867) *wroughtoni* Hinton, 1919.

COMMENTS: *Rattus rattus* species group. Numerous cytogenetic studies focusing on the *R. rattus* complex, as summarized by Baverstock et al. (1983*c*), Bekasova and Mezhova (1983), Niethammer (1975), and Yosida (1980), have revealed two basic groups of populations. The Oceanian or European type has 2n = 38 (40 in some), the Asian type is characterized by 2n = 42; the two are also distinguished by biochemical features (Baverstock et al., 1983*c*) as well as morphological traits (Schwabe, 1979). An ongoing taxonomic study of the *Rattus rattus* complex by K. Aplin and H. Suzuki and their collaborators, has confirmed that separateness of the 2n = 38 populations (based on samples from four continents, but not yet including Indian samples) from the various East and Southeast Asian populations (Aplin et al., 2003*c*). Moreover, their studies indicate that the widespread 'introduced' populations are genetically all very similar, consistent with progressive lineage restriction as a result of multiple founder effects (K. Aplin, in litt., 2004). Where the Asian type is indigenous, the Oceanian form is restricted to ports or on ships in harbor. Both chromosomal kinds apparently occur together without evidence of interbreeding on the Polynesian island of Fiji (Yosida et al., 1985), but do hybridize in the laboratory (usually producing sterile offspring) and on the South Pacific islands of Chichijima and Eniwetok (with apparent introgression). The biological status of the two kinds is best summarized by Baverstock et al. (1983*c*:978), who noted that if "the chromosomal, electrophoretic and laboratory hybridization data are considered together, it seems that the 2n = 38 and 2n = 42 forms are best considered as incipient species. Where they meet, they may introgress, become sympatric without interbreeding or one may replace the other depending upon the prevailing biological conditions," a view earlier espoused by Capanna (1974). *Rattus rattus* is the name for the 2n = 38/40 group, and we list it as a species separate from the 2n = 42 form, for which the oldest name is apparently *R. tanezumi* (see that account).

Populations on Ceylon are unique among *R. rattus* in that most highland samples exhibit 2n = 40 ("Ceylonese type"; found nowhere else within the range of the species), lowland samples have 2n = 38, and hybrids (2n = 39) occur in both regions (Yosida et al., 1979). Detailed analyses of chromosomal and biochemical traits have been used to characterize these Ceylonese populations, extent of hybridization in the laboratory and field, and evolutionary origin of the 2n = 40 configuration in context of phylogenetic relationship to typical peninsular Indian *R. rattus* (Yosida, 1977, 1978*b*, 1979, 1980; Yosida et al., 1974).

Historically, three color phases have been considered as subspecies with their own geographic distributions: *alexandrinus* (grayish brown dorsum, pure white venter), *frugivorus* (grayish brown dorsum and gray through grayish buff to slate-gray venter), and *rattus* (ranging from grayish black upperparts merging into slate-gray underparts to completely melanistic coats). Nearly 140 years ago, de L'isle (1865) determined through breeding experiments that all three morphs interbreed freely, can even be found in the same litter, and signal color polymorphism within a single species, not intraspecific geographic variation, results substantiated by subsequent research and observations (Feldman, 1926; Grubb et al., 1998, and references cited therein; Hinton, 1918; Johnson, 1946*b*, 1962*a*; Jones and Johnson, 1965; Rosevear, 1969; Tomich, 1968, 1986; Tomich and Kami, 1966; see color plate in Innes, 1990). Prakash et al. (1995*a*) recorded ecology and distribution of *R. rattus alexandrinus* from the Aravalli Ranges in Rajasthan, India, could not distinguish differences between what had been called *rufescens* and the "*alexandrinus*" color phase of *R. rattus*, and synonymized the two.

Allocation of five synonyms requires explanation. Musser and Califia (1982) listed *keelingensis* as a synonym of *diardii*, which is included in *R. tanezumi* (see that account); Musser reexamined the holotype and part of the type series collected from Pulau Tikus in the Cocos-Keeling Isls and now realizes they are introduced *R. rattus*. Both 2n = 42 and 2n = 38 have been attributed to specimens identified as Indian *wroughtoni* from the Mysore region (Gadi and Sharma, 1983; Lakhotia et al., 1973; Raman and Sharma, 1977); the holotype and paratypes are examples of *R. rattus* (Corbet, in litt., 2003; Musser's study of types and material at BMNH), which accords with 2n = 38 recorded for most of the specimens; the individuals with 2n = 42 are likely *R. satarae* (see that account). Holotypes of Hinton's (1918) Indian *girensis* and *narbadae* and most of the specimens in BMNH identified as these taxa represent *R. rattus* (Musser's study of the specimens); material in the BMNH identified by Hinton as *arboreus* also sample populations of *R. rattus*.

Several recent studies elucidate regional colonization history, geographic distribution, and systematics of *R. rattus*: genetic diversity within and between W Mediterranean island populations (Cheylan et al., 1998; Cheylan, 1999); morphometric analysis of population from Congreso Isl, and comparison to nearby Iberian mainland populations (Ventura and López-Fuster, 2000); osteometric analysis of archaeological samples from the XIV and XVIIth centuries on Lavezzi Isl (off the coast of S Corsica), and comparison with recent samples from the island and Corsica (Vigne et al., 1993); analyses of chromosomes and mtDNA restriction patterns comparing samples from islands in the Mediterranean and Atlantic with mainland Europe and Africa (Libois et al., 1996); morphometric study comparing samples from the Azores with a continental sample (Ramalhinho et al., 1996); analysis and interpretation of archaeological samples from medieval (XIIIth Century) Portugal (Morales and Rodriguez, 1997); historical interpretation of presence in Eolian Arch. (Cristaldi and Amori, 1988); genetic and morphological divergence among introduced populations in the Galápagos Arch. (Patton et al., 1975); systematics of Turkish samples based upon chromosomal and morphological traits (Yi—it et al., 1998); assessment and review of swimming ability in context of potential for colonizing tropical oceanic islands (Spennemann and Rapp, 1989); review of Korean population (Jones and Johnson, 1965; Won and Smith, 1999); status and role in Baltic region (Miljutin, 1997, 1998; Timm et al., 1998).

Rattus rattus is unknown in Europe before the Holocene (Kowalski, 2001), and may have reached Mediterranean Europe by pre-Roman times (4000-2300 years BC) along ancient trade routes from India through the Indus Valley and Mesopotamia (Armitage, 1994; Toškan and Kryštufek, in press). The species already occupied (apparently as a commensal) coastal regions of N Israel by the Epipalaeolithic, about 15,000 years ago (Tchernov, 1968, 1996), and so was part of the Levant fauna since just before end of the Pleistocene. There is also a record of *R. rattus* from a cold interval in middle Pleistocene cave sediments of Turkish Thrace (Santel and Koenigswald, 1998). By the fourth century AD, the roof rat was in Britain and associated with Roman occupation (Armitage, 1994; Yalden, 1999). The earliest known record of *R. rattus* in the Americas comes from Haiti, at the site of *La Navidad*, the "first Spanish settlement in the New World, established by Columbus on December 26th, 1492"; by the late 1500s, the species was established along coasts of North, Central, and South America (Armitage, 1993: 174-175).

Rattus rattus, not the Asian *R. tanezumi*, also occurs on Madagascar (Duplantier et al., 2003; Musser's research). The species may have accompanied the first human immigrants to Madagascar about 2000 years ago with speculated transport through Zanzibar and Comoros islands to Madagascar, the same sea route used by Arab traders that facilitated the introduction of *Suncus murinus* (Duplantier and Duchemin, 2003; Hutterer and Tranier, 1990). *Rattus rattus* presently occurs everywhere on Madagascar, ". . . homes, fields, and forests, from sea level up to elevations of more than 2400 m" (Duplantier and Duchemin, 2003:1191). Its distribution over the island and along altitudinal transects, as well as associations with different forest formations and anthropogenic habitats has been extensively documented by recent faunal surveys (Carleton and Goodman, 2000; Goodman et al., 1996a, b; Goodman and Carleton, 1996, 1998; Stephenson, 1995). Relationship between the introduced *R. rattus* and nesomyine rodents, tenrecs and small-bodied lemurs in the context of protecting that endemic and unique assemblage (particularly competition between nesomyines and *R. rattus*, and possible elimination of the former by the introduced murine) is explicitly addressed by Goodman (1995) and Ramanamanjato and Ganzhorn (2001). *Rattus*

rattus, not any of the native nesomyines, is also a reservoir for plague, which has been present in Madagascar for at least a century, and is restricted to two foci above 800 m (Duplantier et al., 2003).

Rattus richardsoni (Tate, 1949). Am. Mus. Novit., 1421:1.

COMMON NAME: Richardson's New Guinea Mountain Rat.

TYPE LOCALITY: New Guinea, Prov. of Papua (= Irian Jaya), northern slopes of Snow Mtns (Pegunungan Maoke), near Lake Habbema, 3225 m.

DISTRIBUTION: New Guinea, Prov. of Papua (= Irian Jaya); recorded from three areas high in the Snow Mtns: Lake Habbema region, Mt Wilhelmina, and Mt Jaya (Carstensz Range), 3225-4500 m (Flannery, 1995a; Taylor et al., 1982).

STATUS: IUCN – Lower Risk (lc) as *Stenomys richardsoni*.

COMMENTS: *Rattus leucopus* species group. Described as an unusual and distinctive member of *Rattus*, but later provisionally associated with a "*niobe* group" (Tate, 1951), a relationship supported by multivariate analyses of morphometric variables (Taylor et al., 1982). Most specimens are from tussock grassland and tundra-like habitat (mainly rock or gravel interspersed with mats of herbs and grass tufts), regions that are cold and wet throughout the year (Flannery, 1995a; Taylor et al., 1982). Taylor et al. (1982) included *omichlodes*, but Flannery (1995a) correctly removed it from synonymy (see that account).

Rattus salocco Tate and Archbold, 1935. Am. Mus. Novit., 802:7.

COMMON NAME: Southeastern Xanthurus Rat.

TYPE LOCALITY: Indonesia, SE peninsula of Sulawesi, Pegunungan Mekongga (= Mengkoka), Tanke Salokko (highest point on Pengunungan), 03°35′S/121°15′E, 1500 m.

DISTRIBUTION: SE Peninsula; recorded from Tanke Salokko and in lowlands at Mowewe (03°57′S/121°43′E), 300 m.

SYNONYMS: *orientalis* (Revilliod, 1911) [not Desmarest, 1819, or Cretzschmar, 1826].

COMMENTS: *Rattus xanthurus* species group. Described as a species by Tate and Archbold, but subsequently usually included in *R. xanthurus* (e. g., Musser and Carleton, 1993). Musser (ms) demonstrated that the species is most closely related to *R. marmosurus* of C and N Sulawesi, and documented the synonymy of *orientalis*. See account of *Taeromys arcuatus* for information on the biogeographic pattern exhibited by this and other species pairs on Sulawesi.

Rattus sanila Flannery and White, 1991. Nat. Geog. Res. Explor., 7:102.

COMMON NAME: New Ireland Forest Rat.

TYPE LOCALITY: Bismarck Arch., New Ireland, Balof, site 2.

DISTRIBUTION: Apparently endemic to New Ireland.

COMMENTS: *Rattus* species group unresolved. Represented only by subfossil fragments dated at 3000 years before present and older (late Pleistocene to mid-Holocene), but may still occur in primary forest, which has not been adequately sampled. Flannery and White (1991) described *sanila* as a subspecies of *R. mordax*, but most dental measurements of *sanila* exceed and do not overlap those of even the largest known *R. mordax*, suggesting the former to be a separate species, which even Flannery and White acknowledged (also see Flannery, 1995b). Recent taxonomic and stratigraphic analysis of rodent remains from New Ireland and Manus Isl has confirmed the specific status of *sanila* (K. P. Aplin and M. Leavesley, in litt., 2004). Although originally phylogenetically allied with *R. mordax* (Flannery and White, 1991), K. Aplin (in litt., 2004) noted that upper molars of *R. sanila* ". . . are very complex in comparison with typical New Guinea *Rattus*," and he suspects the species may be a relict of an archaic, and original, dispersal of ancestral *Rattus* stock to New Guinea and Australia.

Rattus satarae Hinton, 1918. J. Bombay Nat. Hist. Soc., 26:87.

COMMON NAME: Sahyadris Forest Rat.

TYPE LOCALITY: SW Peninsular India, E margin of Western Ghats (= Sahyadris), Maharashtra State, Satara Dist., Ghatmatha (Crestline), 2000 ft (610 m).

DISTRIBUTION: Endemic to the Western Ghats and occurring in tropical evergreen rain forest. Records are from N part of the Western Ghats at and near the type locality

(Hinton, 1918; Tiwari et al., 1971) and Nilgiri Hills (Tamil Nadu State) in the south (Verneau et al., 1997), which approximates the extent of tropical evergreen rain forest along the SW mountainous margin of SW peninsular India (see Morley, 2000:162); altitudinal range, 1500-2150 m (from data with specimens). Although wet, tropical forests once occurred along the length of the Western Ghats (Subramanyam and Nayer, 1974), they are presently reduced to small, isolated patches through conversion to agriculture and plantations (Chandrasekar-Rao and Sunquist, 1996); these forest islands between Satara region and Nilgiri Hills probably also contain *R. satarae*.

COMMENTS: A distinctive species with golden brown upperparts, white underparts, soft and dense fur, a very long tail relative to head and body length, and 2n = 42. Originally described as a subspecies of *R. rattus* by Hinton (1918:88), who noted that it was "distinguished from all other Indian subspecies by its peculiar skull and relatively long tail." Sympatric with *R. rattus* (*wroughtoni*, 2n = 38) from which it is distinguished by its fur texture, dorsal coloration, very long tail, and cranial, molecular, and biochemical traits (Verneau et al., 1997:425; Usdin et al., 1995). Chromosomal studies had identified *R. rattus wroughtoni* with 2n = 42 and *R. r. rufescens* having 2n = 38 from S Western Ghats (Lakhotia et al., 1973; Gadi and Sharman, 1983). Samples used by Verneau et al. (1997) from the Nilgiri Hills consisted of both taxa. Gordon Corbet (in litt., 2003) examined their material and wrote that the "holotype and two paratypes of *wroughtoni* show all the cranial characters of the 2n = 38 group; hence this name cannot be applied to the form with 42 chromosomes," and noted that the holotype of *satarae* "agrees well with the karyotyped 2n = 42 group from S India in 10 out of 14 cranial characters, including the most important ones mentioned in the diagnosis." Because two distinct species of *Rattus* occur in the S Western Ghats, ecological information gathered from rats living there identified as only *R. r. wroughtoni* (e.g., Boshell-M and Rajagopalan, 1968; Chandrasekar-Rao and Sunquist, 1996; Rajagopalan, 1970) is not as useful as it could have been. *Rattus satarae* joins the other murines *Mus famulus* and *Vandeleuria nilagirica*, and the lion-tailed macaque (*Macaca silenus*) in being endemic to tropical rainforests along the Western Ghats.

Rattus simalurensis (Miller, 1903). Proc. U.S. Natl. Mus., 26:458.

COMMON NAME: Simalur Archipelago Rat.

TYPE LOCALITY: Indonesia, Sumatra, Pulau Simeulue (Simalur).

DISTRIBUTION: Simalur Isl and nearby islands of Siumat, Lasia, and Babi.

STATUS: IUCN – Lower Risk (lc).

SYNONYMS: *babi* Lyon, 1916; *lasiae* Lyon, 1916.

COMMENTS: *Rattus rattus* species group. The form *simalurensis* and its two synonyms were each listed as a separate subspecies of *R. rattus* by Chasen (1940), but *simalurensis* is distinct from *R. rattus* and in morphology represents larger-bodied island variants of *R. tiomanicus* (Musser, 1986; Musser and Califia, 1982; Musser and Heaney, 1985); *simalurensis*, *babi*, and *lasiae* were included in *R. tiomanicus* by Corbet and Hill (1992). Whether populations on the Simalur islands are distinct species or insular representatives of *R. tiomanicus* that occur off the margin of the continental shelf will have to be determined by systematic revision of the *R. tiomanicus* complex. Recent morphometric analyses indicate *R. simalurensis* to be more closely tied to *R. burrus* (Nicobar Isls), *R. lugens* (Mentawai Arch.), and *R. adustus* (Pulau Enggano) than to *R. tiomanicus* (Musser, ms).

Rattus sordidus (Gould, 1858). Proc. Zool. Soc. Lond., 1857:242 [1858].

COMMON NAME: Canefield Rat.

TYPE LOCALITY: Australia, Queensland, open plains Darling Downs (as restricted by Thomas' lectotype designation; see Mahoney and Richardson, 1988:187).

DISTRIBUTION: Australia: E coast from tip of Cape York to NE New South Wales, and some offshore islands (see Watts and Aslin, 1981:239, and Redhead, 1995b:662). New Guinea: lowlands south of Central Cordillera from Dobodura in E Papua New Guinea west and north to Koembe in Prov. of Papua (= Irian Jaya); sea level to 670 m (see Taylor et al., 1983:265, and Flannery, 1995a:335); also on Yule Isl, off the coast of SE Papua (Flannery, 1995b).

STATUS: IUCN – Lower Risk (nt).

SYNONYMS: *aramia* Troughton, 1937; *brachyrhinus* Tate and Archbold, 1935; *bunae* Troughton, 1946; *conatus* Thomas, 1923; *gestri* (Thomas, 1897); *gestroi* [misspelling of *gestri* first appearing in Laurie and Hill, 1954]; *youngi* Thomas, 1926.

COMMENTS: *Rattus fuscipes* species group. One of the two species of native *Rattus* in the New Guinea-Australian region that occurs on both land masses. On morphological evidence, Taylor and Horner (1973) arranged *villosissimus* and *colletti* as subspecies of *R. sordidus*. Later evaluations, however, using chromosomal, biochemical, and hybridization data, suggested the three should be viewed as separate species in the same monophyletic cluster (Baverstock et al., 1977*d*, 1983*a*, 1986), which is the way they are treated in current faunal accounts and catalogs (Mahoney and Richardson, 1988; Redhead, 1995*b*; Watts and Aslin, 1981).

Rattus steini Rümmler, 1935. Z. Säugetierk., 10:115.

COMMON NAME: Stein's New Guinea Rat.

TYPE LOCALITY: New Guinea, Prov. of Papua (= Irian Jaya), Weyland Range, Mt Kunupi.

DISTRIBUTION: Mostly mid-montane elevations along Central Cordillera of the Prov. of Papua (= Irian Jaya) to C Papua New Guinea, as well as highlands along N coast and on Huon Peninsula; 20-2800 m; absent from the Vogelkop Peninsula, Trans-Fly region, and E Papua (see Taylor et al., 1982:243, and Flannery, 1995*a*:337).

STATUS: IUCN – Lower Risk (lc).

SYNONYMS: *baliemensis* Taylor and Calaby, 1982; *försteri* (Rümmler, 1935); *hageni* Troughton, 1937; *rosalinda* Hinton, 1943.

COMMENTS: *Rattus leucopus* species group. Until Taylor et al.'s (1982) revision, the distinctness of *R. steini* was obscured by its historical allocation at different times as a subspecies of *R. leucopus*, *R. mordax*, *R. ringens* (= *R. leucopus*), *R. ruber* (= *R. nitidus*), and *Stenomys verecundus*.

Rattus stoicus (Miller, 1902). Proc. U.S. Natl. Mus., 24:759.

COMMON NAME: Andaman Archipelago Rat.

TYPE LOCALITY: India, Andaman Isls, Henry Lawrence Isl.

DISTRIBUTION: Islands of Henry Lawrence, Little Andaman, and South Andaman in the Andaman Arch. (Musser and Newcomb, 1983).

STATUS: IUCN – Vulnerable.

SYNONYMS: *rogersi* (Thomas, 1907); *taciturnus* (Miller, 1902).

COMMENTS: *Rattus rattus* species group. Despite past confusion with *Sundamys muelleri*, *R. stoicus* is defined by a unique set of derived and primitive features and is endemic to the Andaman Isls (Musser and Heaney, 1985; Musser and Newcomb, 1983). Its closest phylogenetic ally within the *R. rattus* species complex has yet to be determined (Musser, 1986).

Rattus tanezumi Temminck, 1844, *In* Siebold, Temminck, and Schlegel. Fauna Japonica, Arnz et Socii, Lugduni Batavorum, p. 51.

COMMON NAME: Oriental House Rat.

TYPE LOCALITY: Japan, possibly from near Nagasaki on Kyushu Isl (see Jones and Johnson, 1965).

DISTRIBUTION: Apparently indigenous to SE Asia, from E Afghanistan (Niethammer and Martens, 1975) through C and S Nepal (below about 2000 m), Bhutan, N India, N Bangladesh and NE India into S and C China (including Hainan Isl), Korea, and mainland Indochina (including offshore islands) south to Isthmus of Kra; also probably native to Mergui Arch.; not found in the Andaman Isls and most of the Nicobar Isls. Whether native or introduced to Taiwan and Japan is unknown (but see Yosida and Harada, 1985; Japanese distribution reviewed by Kaneko, 1994). Most likely introduced to the Malay Peninsula and islands on the Sunda Shelf (Medway and Yong, 1976) and nearby archipelagos just off of the Shelf, including the Mentawais (Musser and Califia, 1982; Musser and Newcomb, 1983) and Nicobar Isls (the only museum records are the holotype of *pulliventer* from Great Nicobar Isl described by Miller, 1902, and BMNH 20.3.1907 from Little Nicobar Isl; Musser's identifications). Certainly introduced to the Philippines (Musser, 1977*a*; Heaney et al., 1998, summarized records), Sulawesi (Musser and Holden, 1991), and numerous islands east through the Moluccas and Nusa Tenggara (Flannery, 1995*b;* Musser, 1970*a*, 1972, 1981*c*) to W New Guinea (Flannery, 1995*a*; Sody, 1941), and farther east through Micronesia to islands of Eniwetok and Fiji (Johnson, 1962*a*, *b*), but not to the Samoas where *R. rattus* occurs (Yosida et al., 1985).

STATUS: IUCN – Lower Risk (lc).

SYNONYMS: *alangensis* Chasen, 1937; *amboinensis* Laurie and Hill, 1954; *argyraceus* Sody, 1941; *auroreus* Sody, 1941; *barussanoides* Sody, 1941; *benguetensis* [Hollister, 1913; see Musser, 1977*a*]; *bhotia* Hinton, 1918; *brevicaudus* Kuroda, 1952 [see Kaneko and Maeda, 2002]; *brevicaudus* Chakraborty, 1975; *brunneus* (Hodgson, 1845); *brunneusculus* (Hodgson, 1845); *bullocki* Roonwal, 1948; *canna* (Swinhoe, 1871); *coloratus* (Hollister, 1913); *dammermani* Thomas, 1921; *dentatus* (Miller, 1913); *diardii* (Jentink, 1880); *exsul* (Miller, 1913); *flavipectus* (Milne-Edwards, 1872); *fortunatus* (Miller, 1913); *gangutrianus* Hinton, 1919; *germaini* (Milne-Edwards, 1872); *griseiventer* (Bonhote, 1903); *insulanus* (Miller, 1913); *kadanus* Chasen, 1937; *kelleri* (Mearns, 1905); *khyensis* Hinton, 1919; *kurokuma* Kuroda, 1953 [unavailable; see Kaneko and Maeda, 2002]; *kramensis* Kloss, 1919; *lalolis* Tate and Archbold, 1935; *lanensis* Kloss, 1919; *lontaris* Chasen 1937; *macmillani* Hinton, 1919; *makassarius* Sody, 1941; *makensis* Kloss, 1916; *mansorius* Johnson, 1962; *masaretes* Sody, 1937; *mesanis* Kloss, 1919; *mindanensis* (Mearns, 1905); *moheius* Chasen, 1937; *molliculus* Robinson and Kloss, 1922; *moluccarius* Sody, 1933; *neglectus* (Jentink, 1880); *nemoralis* (Blyth, 1851) [not Selys Longchamps, 1841]; *obiensis* Sody, 1941; *ouangthomae* (Milne-Edwards, 1872); *palelae* Miller and Hollister, 1921; *palembang* Tate and Archbold, 1935; *panjius* Chasen, 1937; *pannellus* (Miller, 1913); *pannosus* (Miller, 1900); *pelengensis* Sody, 1941; *pipidonis* Chasen, 1937; *poenitentiarii* (Kloss, 1915); *portus* (Kloss, 1915); *povolny* Niethammer and Martens, 1975; *pulliventer* (Miller, 1902); *rangensis* (Kloss, 1916); *robiginosus* (Hollister, 1913); *robinsoni* Chasen, 1940; *robustulus* (Blyth, 1859); *sakisimana* Tokuda, 1939 [*nomen nudum*; see Kaneko and Maeda, 2002]; *samati* Sody, 1932; *santalum* Sody, 1932; *sapoensis* Sody, 1941; *satarae* Hinton, 1918; *septicus* Sody, 1933; *shirokuma* Kuroda, 1953 [unavailable; see Kaneko and Maeda, 2002]; *sladeni* (Anderson, 1879); *sumbae* Sody, 1930; *tablasi* Taylor, 1934; *talaudensis* Sody, 1941; *tatkonensis* Hinton, 1910; *thai* Kloss, 1917; *tikos* Hinton, 1919; *tistae* Hinton, 1918; *toxi* Sody, 1941; *turbidus* (Miller, 1913); *yeni* Dao, 1960; *yunnanensis* (Anderson, 1879); *zamboangae* (Mearns, 1905).

COMMENTS: *Rattus rattus* species group. The authority is usually cited as 1845, but was published in 1844 (Holthuis and Sakai, 1970). The name *tanezumi* is the oldest for the 2n = 42 group of Asian houserats that is distinguished from the 2n = 38/40 *R. rattus* not only by chromosomal characters but also morphological and biochemical traits (see account of *R. rattus*). The probable indigenous range is generally north and east of peninsular India (judged by Musser's study of material in AMNH, BMNH, FMNH, MZB, RMNH, USNM, and geographic distribution of karyotypes with 2n = 42; Duncan and Van Peenen, 1971; Gadi and Sharma, 1983; Markvong et al., 1973; Niethammer, 1975; Niethammer and Martens, 1975; Ray-Chaudhuri and Pathak, 1970; Raman and Sharma, 1974, 1977; Gill and Gupta, 1989; Tripathy et al., 1985; Yong, 1969; Yosida et al., 1971). In the southern portion of the Western Ghats (along SW peninsula in Karnataka and Maharashtra States) at Sagar, three individuals trapped in forest exhibited 2n = 42, and three from a nearby "domestic site" had 2n = 38. The former diploid number was attributed to *R. r. wroughtoni* and the latter to *R. r. rufescens* (Lakhotia et al., 1973; Pradhan et al., 1991), but holotypes and paratypes of *wroughtoni* represent *R. rattus* (which includes *rufescens* as a synonym), and the 2n = 42 sample is *R. satarae* (see that account). Dark-bellied populations identified as *R. rattus rufescens* are found occurring together with white-bellied rats in India, which prompted Tiwari et al. (1972) to recognize *rufescens* as a separate species, but it is simply a phenotypic morph of *R. rattus* found in the same populations containing the white-bellied morph (see account of *R. rattus*). Distributional and ecological relationships between *R. rattus* and *R. tanezumi* along the N portion of the Indian subcontinent where indigenous ranges are either parapatric or overlap still requires resolution (see discussion in Corbet and Hill, 1992).

More than one species may be present in our concept of *R. tanezumi*, but careful taxonomic revision using morphological, chromosomal and molecular characters will have to test such an hypothesis. Verneau et al. (1997), for example, split Vietnamese *R. flavipectus* from *R. tanezumi* because of a difference in Fundamental Numbers of karyotypes. Wang (2003) recognized *yunnanensis* as a separate species, but no data substantiates that treatment (see G. M. Allen, 1940, and Osgood, 1932). He also listed *brunneusculus* as a species with *sladeni*, *molliculus*, *canna*, *tistae*, and *hainanicus* as subspecies. We associate the first four taxa with *R. tanezumi* and *hainanicus* with *R. andamanensis* (see that account). An extensive taxonomic survey of these populations, based on a combination of DNA sequencing and morphological studies, is currently underway by K. Aplin, H. Suzuki and their various collaborators (Aplin et al., 2003*b*; K. Aplin, in litt., 2004). Their preliminary analysis, using the mitochondrial cytochrome *b* gene, suggests that the 2n = 42 Asian group includes two distinct taxa, with a level of divergence equal to or greater that characterizes the 2n = 38 cluster. Aplin et al. (2003*b*:489) summarized the fundamental division as between ". . . an endemic Southeast Asian taxon (recorded from Vietnam, Cambodia and southern Laos) and a northern and South Asian taxon (recorded thus far from Japan, Hong Kong, northern Vietnam, northern Laos and Bangladesh). These taxa are probably regionally sympatric in parts of Vietnam and Laos." Their work also suggests that island populations off the Sunda Shelf contain a complex admixture of both major haplotype groups, suggestive of prehistoric introductions from multiple areas with subsequent hybridization (K. Aplin, in litt., 2004). Aplin et all. (2003*c*:489) suggested that ". . . it would be premature to speculate on the appropriate names for each of the two Asian taxa," and caution that ". . .the apparent extension of the north Asian taxon (which probably includes Japanese *tanezumi*) through to the Indian subcontinent raises the possibility that *tanezumi* may not be the earliest available name for this group."

Females in most populations of *R. tanezumi* have five pairs of mammae (one pectoral, one postaxillary, one abdominal, and two inguinal), but in some populations the postaxillary pair is twinned with the teats close together (less than 1 cm apart, usually 5 mm). This configuration is different from the six pairs of teats characterizing certain species of *Rattus* in the *R. rattus* species group (e. g., *R. andamanensis*) in which there are two distinct pairs of postaxillary teats, usually 1 cm or more apart. We have not seen such variation in *R. rattus*.

Phallic morphology of Chinese *flavipectus* described by Yang and Fang (1988) in context of assessing phylogenetic relationships among Chinese murines. Standard karyotype of Philippine samples described by Rickart and Musser (1993). Taxonomic, chromosomal, and ecological aspects of Chinese populations (under *flavipectus*) described by Zhang et al. (2000). Md Nor (1996) documented its distribution and ecology on the small islands off the northern tip of Sabah. The species does not occur on Christmas Isl or Madagascar (*R. rattus* was introduced on those islands), or in the Andaman and Nicobar Isls except for Greater Nicobar Isl (see above); *R. rattus* from Barren Isl is the only recorded introduction of a commensal *Rattus* in the Andaman Isls (see account of *R. andamanensis*).

Rattus tawitawiensis Musser and Heaney, 1985. Am. Mus. Novit., 2818:5.
    COMMON NAME: Tawitawi Forest Rat.
    TYPE LOCALITY: Philippines, Sulu Arch., Tawitawi Isl, Batu Batu.
    DISTRIBUTION: Greater Sulu Faunal Region. Known only from Tawitawi Isl in S part of Sulu Arch.
    STATUS: IUCN – Vulnerable.
    COMMENTS: *Rattus rattus* species group. Still recorded by only three specimens. Phylogenetic affinities unclear; in some traits resembles Sulawesian *R. hoffmanni* and members of the *Rattus palmarum-burrus-simalurensis-lugens* clusters on the Nicobar, Simalur, and Mentawai islands more closely than any described *Rattus* from the Indo-Australian region (Musser and Heaney, 1985).

Rattus timorensis Kitchener, Aplin, and Boeadi, 1991. Rec. West. Aust. Mus., 15:446.
    COMMON NAME: Timor Forest Rat.

TYPE LOCALITY: Indonesia, Nusa Tenggara, Timor, Gunung Mutis, 1900 m, 7 km E Desa Nenas.

DISTRIBUTION: Known only from the type locality.

STATUS: IUCN – Data Deficient.

COMMENTS: *Rattus* species-group unresolved. Represented only by the holotype. Some of the large series of subfossil fragments collected in E Timor by Glover (1986) may be this species. Phylogenetic affinities unknown and possibly not even a member of *Rattus* (Kitchener et al., 1991*b*). Albumin immunological analysis suggests a close phyletic affinity with *Bunomys chrysocomus* from Sulawesi and *Komodomys rintjanus* from Nusa Tenggara (Watts and Baverstock, 1994*b*); the close alliance between *rintjanus* and *timorensis*, and their great phylogenetic distance from *Rattus*, is supported by a recent unpublished allozyme electrophoresis study (K. Aplin, in litt., 2004). Recently, K. Helgen (in litt., 2003) collected an extant specimen from a forest fragment on Timor that is not *R. timorensis*, but an undescribed species that is also represented by large series in Glover's subfossil collections.

Rattus tiomanicus (Miller, 1900). Proc. Wash. Acad. Sci., 2:212.

COMMON NAME: Malaysian Field Rat.

TYPE LOCALITY: Malaysia, Pahang, Tioman Isl, off the east coast Malay Peninsula.

DISTRIBUTION: Endemic to the Sunda Shelf and some offshore islands. Records on the Shelf are from peninsular Thailand south of Isthmus of Kra (10°30′N), the Malay Peninsula, Sumatra, Java, Bali, Borneo, Palawan (see Heaney et al., 1998, for records from the Palawan Faunal Region), and many smaller islands. Off the Sunda Shelf, *R. tiomanicus* is documented from Enggano Isl, southwest of Sumatra, and Maratua Arch., east of Borneo (Musser and Califia, 1982; Musser and Heaney, 1985).

STATUS: IUCN – Lower Risk (lc).

SYNONYMS: *ambersoni* Schwarz and Schwarz, 1967; *banguei* Chasen and Kloss, 1932; *batin* Robinson, 1916; *delirius* Sody, 1941; *ducis* (Lyon, 1911); *fulmineus* (Miller, 1913); *generatius* Sody, 1941; *jalorensis* (Bonhote, 1903); *jarak* (Bonhote, 1905); *jemuris* Chasen and Kloss, 1931; *julianus* (Miller, 1903); *kabanicus* Hill, 1960; *kunduris* Chasen and Kloss, 1931; *lamucotanus* (Lyon, 1911); *lasurius* Sody, 1941; *luxuriosus* Chasen, 1935; *maerens* (Miller, 1911); *mangalumis* Kloss, 1931; *mara* (Miller, 1913); *pauper* (Miller, 1913); *payanus* Chasen and Kloss, 1931; *pemanggis* Chasen, 1940; *perhentianus* Chasen, 1940; *pharus* Hill, 1960; *piperis* Schwarz and Schwarz, 1967; *rhionis* (Thomas and Wroughton, 1909); *roa* (Miller, 1913); *roquei* Sody, 1929; *rumpia* (Robinson and Kloss, 1911); *sabae* Medway, 1965; *sebasianus* Sody, 1941; *siantanicus* (Miller, 1900); *sribuatensis* Hill, 1960; *tambelanicus* (Miller, 1900); *tenggolensis* Yong, 1971; *terutavensis* Hill, 1960; *tingius* (Miller, 1913); *tua* (Miller, 1913); *vernalus* Sody, 1940; *viclana* (Miller, 1913).

COMMENTS: *Rattus rattus* species group. Reviewed by Musser and Califia (1982), who also summarized and provided references documenting the incorrect historical association of *tiomanicus* and the other synonyms listed here as subspecies of *R. rattus*. They also pointed out that careful study of inter-island variation among named forms of the *R. tiomanicus* complex is necessary before relationships among the insular populations can be discerned; more than one species may be represented in what is now viewed as *R. tiomanicus*. Miller's (1942) *blangorum* from the Aceh region of N Sumatra, for example, was originally described as a species, then included in *R. tiomanicus* (Musser and Califia, 1982), and is here reinstated as a separate species (see account of *R. blangorum*).

　　Rattus mindorensis from Mindoro Isl in the Philippines; *R. simalurensis* from the islands of Babi, Lasia, Siumat, and Simalule, off the northwest coast of Sumatra; *R. lugens* from the Mentawai Arch., and *R. adustus* from Pulau Enggano (all islands off W coast of Sumatra), and *R. burrus* from some of the Nicobar Isls, are morphologically similar to the *R. tiomanicus* complex and should be considered part of it (Musser, 1986; Musser and Heaney, 1985). Whether they are species or island forms of *R. tiomanicus* can be determined within the framework of a systematic revision of the *R. tiomanicus* complex employing morphological and molecular data.

　　Distribution and ecology of *R. tiomanicus* on the small islands off the tip of N Sabah documented by Md Nor (1996). It is found primarily in the lowlands on Borneo but reaches 1700 m on slopes of Mt Kinabalu in Sabah (Md Nor, 2001, and references cited therein).

Rattus tunneyi (Thomas, 1904). Novit. Zool., 11:223.

COMMON NAME: Australian Pale Field Rat.

TYPE LOCALITY: Australia, Northern Territory, Mary River.

DISTRIBUTION: Australia; NE and SW Western Australia, Northern Territory, E Queensland, and NE New South Wales. Also occurs on offshore islands. Recorded historically (holotype of *austrinus* probably from Kangaroo Isl; see Mahoney and Richardson, 1988:188), but otherwise only by subfossils (Robinson et al., 2000) from South Australia. Extant range vastly reduced from former distribution (see Taylor and Horner, 1973:89; Watts and Aslin, 1981; Braithwaite and Baverstock, 1995:663).

STATUS: IUCN – Lower Risk (nt).

SYNONYMS: *apex* Troughton, 1939; *austrinus* Thomas, 1921; *culmorum* (Thomas and Dollman, 1909); *dispar* Brazenor, 1936; *melvilleus* Thomas, 1921; *vallesius* Thomas, 1921; *woodwardi* (Thomas, 1908).

COMMENTS: *Rattus fuscipes* species group. This species hybridized in the laboratory with *R. colletti* (Baverstock et al., 1983a, 1986). Revised by Taylor and Horner (1973) and reviewed by Watts and Aslin (1981), Mahoney and Richardson (1988), and Braithwaite and Baverstock (1995).

Rattus vandeuseni (Taylor and Calaby, 1982). *In* Taylor et al., Bull. Am. Mus. Nat. Hist., 173:211.

COMMON NAME: Van Deusen's New Guinea Mountain Rat.

TYPE LOCALITY: Papua New Guinea, Maneau Range, N slope Mt Dayman, Middle Camp, 1540 m.

DISTRIBUTION: Recorded only from the Mt Dayman region at the E end of the Owen Stanley Range in Papua New Guinea, 1500-1540 m (Musser and Carleton, 1993; Musser and Lunde, ms).

STATUS: IUCN – Endangered as *Stenomys vandeuseni*.

COMMENTS: *Rattus leucopus* species group. Cole et al. (1997) reported one example from an E ridge of Mt Dayman (identified by comparing it with two specimens from the type series) and noted that Flannery had indicated to them that most *R. vandeuseni* (in AM) from that area were from 1300 m. Flannery (1995a) also reported specimens from the Agaun Valley at 300 m (according to Cole et al., 1997), but it is unclear if these are from 300 m (in which case they should be reexamined to determine if they are really the lowland *R. verecundus*) or the specimens from 1300 m. We can only vouch that the type series, which we studied, represents a montane species morphologically distinguishable from the lower altitudinal *R. verecundus*.

Taylor and Calaby described *vandeuseni* as a subspecies of *R. verecundus* (Taylor et al., 1982), but its distinctive morphology and habitat indicate otherwise, as the describers even suggested, and as Flannery (1990b:245) speculated. *Rattus verecundus* occurs on Mt Dayman in rain forest at 700 m and is replaced at 1540 m in montane oak forest by *R. vandeuseni*, a relationship that is similar to the distributions of *Leptomys elegans* (rain forest) and *Leptomys* sp. (oak forest) on Mt Dayman (Musser and Lunde, ms).

Rattus verecundus (Thomas, 1904). Novit. Zool., 11:598.

COMMON NAME: New Guinea slender Rat.

TYPE LOCALITY: Papua New Guinea, Central Province, Aroa River, Avera, 200 m.

DISTRIBUTION: New Guinea; from the Vogelkop region in W Prov. of Papua (= Irian Jaya) to easternmost Papua New Guinea; nearly all records are from Papua New Guinea; altitudinal range, 150 to 2750 m (Taylor et al., 1982:204, and Flannery, 1995a:345).

STATUS: IUCN – Lower Risk (lc) as *Stenomys verecundus*.

SYNONYMS: *mollis* Rümmler, 1935; *tomba* (Laurie, 1952); *unicolor* Rümmler, 1935.

COMMENTS: *Rattus leucopus* species group. Revised by Taylor et al. (1982), who recognized three subspecies: *mollis*, *unicolor*, and *verecundus*. Chromosomal morphology discussed by Dennis and Menzies (1978). Leary and Seri (1997) described the distribution in the Kikori River Basin in S Papua where *R. verecundus* occurs sympatrically with *R. niobe* at the upper altitudinal part of its range and *Rattus leucopus* at the lower end.

Rattus villosissimus (Waite, 1898). Proc. R. Soc. Victoria, 10:125.

COMMON NAME: Australian Long-haired Rat.

TYPE LOCALITY: Australia, Queensland, "from probably the vicinity of Goonhaghooheeny Billabong, Cooper Creek" (Mahoney and Richardson, 1988:188).

DISTRIBUTION: Australia; broad inland range from NW Western Australia through Northern Territory into most of Queensland and N South Australia and N New South Wales (see Watts and Aslin, 1981:245); fossils indicate a past broader range once extended across the Nullarbor Plain to the Great Australian Bight (Watts and Aslin, 1981; Watts, 1995*k*). Range in South Australia summarized by Robinson et al. (2000).

STATUS: IUCN – Lower Risk (lc).

SYNONYMS: *longipilis* (Gould, 1854) [not Waterhouse, 1837]; *profusus* Thomas, 1921.

COMMENTS: *Rattus fuscipes* species group. Geographic range is allopatric to the coastal *R. sordidus* in Queensland and *R. colletti* in Northern Territory (see Taylor and Horner, 1973:72). The three species are closely related; *villosissimus* was treated as a subspecies of *R. sordidus* by Taylor and Horner (1973), but is considered genically closer to *colletti* by Baverstock et al. (1983*a*, 1986); see accounts of *R. sordidus* and *R. colletti*. An undescribed species related to *R. villosissimus* and *R. colletti* is known from a small area in C Queensland (Aplin, in litt., 2004). Analyses of electrophoretic data by Gemmeke and Niethammer (1984) indicated *R. villosissimus* to be greatly separated from *R. argentiventer*, *R. exulans*, *R. norvegicus*, and *R. tiomanicus*, and closer to species of *Bandicota* and *Maxomys*. Reviewed by Mahoney and Richardson (1988), Watts and Aslin (1981), and Watts (1995*k*). Occurrence of *R. villosissimus* in Mootwingee National Park of W New South Wales documented by Ellis (1995*a*) based on skulls and mummified remains. Effects of inbreeding on skeletal development reported by Lacy and Horner (1996).

Rattus xanthurus (Gray, 1867). Proc. Zool. Soc. Lond., 1867:598.

COMMON NAME: Northeastern Xanthurus Rat.

TYPE LOCALITY: Indonesia, NE Sulawesi, Tondano, 3600 ft (1100 m).

DISTRIBUTION: NE Sulawesi only, from vicinity of Teluk Kuandang (0°50′N, 122°52′E) eastward in NE region of N peninsula of Sulawesi.

STATUS: IUCN – Lower Risk (lc).

SYNONYMS: *faberi* (Jentink, 1883); *paraxanthus* (Sody, 1941).

COMMENTS: *Rattus xanthurus* species group. See Musser (1971*c-e*, 1984) for justification behind allocation of synonyms. One of the larger-bodied members of this complex, which includes *R. bontanus*, *R. marmosurus*, *R. salocco*, and *R. pelurus*. Occurs sympatrically with *R. marmosurus* on the NE Peninsula. Sody (1941) included *xanthurus* in *Taeromys*. Although the *R. xanthurus* group may eventually be removed from *Rattus* (Musser and Holden, 1991), it does not belong in *Taeromys* (Musser, ms).

Rhabdomys Thomas, 1916. Ann. Mag. Nat. Hist., ser. 8, 18:69.

TYPE SPECIES: *Mus pumilio* Sparrman, 1784.

COMMENTS: *Arvicanthis* Division. Morphological traits place *Rhabdomys* in a group containing species of *Arvicanthis*, *Lemniscomys*, *Mylomys*, and *Pelomys* (Musser, 1987*b*), which is corroborated by analysis of mitochondrial gene sequences (cytochrome *b*, 12S and 16S rRNA gene fragments), and also includes *Desmomys* (Ducroz et al., 2001). The sequence data of Ducroz et al. also indicate that within this arvicanthine cluster, *Rhabdomys* and *Desmomys* are members of a lineage separate from that comprised of only *Lemniscomys*, and from another containing *Arvicanthis*, *Mylomys*, and *Pelomys*. Analysis of microcomplement fixation of albumin groups *Rhabdomys* with *Lemniscomys*, *Pelomys*, *Grammomys*, and *Thallomys* (Watts and Baverstock, 1995*a*).

When Wroughton (1905*b*) reviewed *R. pumilio*, he distinguished four morphological groups, each with different forms, and although unsure about the taxonomic status to give the forms, thought each group represented a separate species. Distribution of character variation, however, forced him to conclude that (p. 630) "in view of the absolute identity of pattern, the variability of coloration, and the difficulty of deciding the inter-relationship of the different forms, the simpler and safer way is to call them all subspecies of the original species *pumilio*." Wroughton's view prevails today. Checklists (G. M. Allen, 1939; Ellerman, 1941; Ellerman et al., 1953), faunal studies (e.g., Ansell, 1978; Ansell and

Dowsett, 1988; Roberts, 1951; Skinner and Smithers, 1990; Smithers, 1971), a study of possible influences of climate on length of tail (Coetzee, 1970), and preliminary studies on geographic variation (see Meester et al., 1986:275) have not critically analyzed patterns of variation in pelage coloration and pattern along with morphology to assess whether only one or several species exist. Interpopulation breeding studies (Pillay, 2000) and allozymic analyses (Mahida et al., 1999) of South African populations have provided new comparative data, but these results are ambiguous in ascertaining whether more than one species is represented among the samples. Analysis of the allozymic data resulted in ranges of genetic similarities suggesting different subspecies were involved, if supported by other evidence, which Mahida et al. (1999) promised to provide derived from DNA sequencing and cytogenetic studies.

Analyses by Rambau et al. (2003;564) have provided that evidence in the form of mtDNA cytochrome *b* sequences and cytogenetic data. Two major and highly divergent lineages were identified, one inhabiting mesic regions of southern Africa, characterized by dark gray or dark brown fur and 2n = 46 or 48. The other occupies xeric habitats, and has much paler pelage and 2n = 48. High genetic distance, ecological difference, and differences in mating behavior between the two lineages indicate two species: *R. pumilio* in xeric grasslands, *R. dilectus* living in mesic environments. Origin of the two may have been facilitated by survival of an ancestral population in two refugia during climatic oscillations in the Pliocene, one in the mesic regions of NE Africa, the other in drier environments of southern Africa. Two species had already been recognized in Angola by Hill and Carter (1941:102); *R. pumilio* from the C and S highlands, and *R. bechuanae* form the arid southwestern desert just north of Namibia. They noted the lack of intergradation between the two kinds and we have not found any evidence in their material (in AMNH) that such intergradation exists. The samples from the highlands are likely *P. dilecuts*, the desert sample is *P. pumilio* (see accounts below).

Several studies are pertinent to systematics of *Rhabdomys*, but were published before the resolution of two species. Morphology of digestive system in relation to diet and evolution described by Perrin and Curtis (1980). In a study of size variation (using greatest length of skull) among samples from across southern Africa, Yom-Tov (1993) discovered a positive correlation between size and mean minimum temperature of the coldest month. He also determined that samples of *Rhabdomys* from the zone of sympatry with the *Lemniscomys griselda* complex, which are also diurnal herbivores but much larger in body size, are significantly smaller than samples from regions in which *Lemniscomys* does not occur. Yom-Tov suggested character release to be a primary factor in determining body size of *Rhabdomys* in southern Africa. Spermatozoal morphology documented by Breed (1995a). Populations in Southern African Subregion reviewed by de Graaff (1981, 1997e) and Skinner and Smithers (1990).

Rhabdomys is represented by isolated molars from the Pleistocene of Namibia (Senut et al., 1992); early Pleistocene Swartkrans and Sterkfontein cave sediments in South Africa (Avery, 1998, 2000). The earliest record is from Langebaanweg in South Africa and may be five million years old (early Pliocene; Denys, 1999).

Rhabdomys dilectus (de Winton, 1897). Proc. Zool. Soc. Lond., 1896:803.

COMMON NAME: Mesic Four-striped Grass Rat.

TYPE LOCALITY: NE Zimbabwe, Mashonaland, Mazowe.

DISTRIBUTION: E South Africa (Rambau et al., 2003), E Zimbabwe (Smithers and Wilson, 1979), WC Mozambique (Smithers and Lobao Tello, 1976), Malawi (Nyika Plateau and Mulanje Massif; Ansell and Dowsett, 1988), NE Zambia (Nyika Plateau; Ansell, 1978), SE Dem. Rep. Congo (Kasaki, Marungu Mtns, 2300 m; Hatt, 1940b), highlands in Tanzania (Grimshaw et al., 1995; Shore and Garbett, 1991; Swynnerton and Hayman, 1951; Vesey-Fitzgerald, 1966), Kenya (Hollister, 1919), E Uganda (Mt. Elgon; Clausnitzer, 2001; Clausnitzer and Kityo, 2001; Delany, 1975), and S and C Angola (Crawford-Cabral, 1998; Hill and Carter, 1941); all referred to as *P. pumilio* except Rambau et al. (2003).

SYNONYMS: *algoae* Roberts, 1946; *angolae* (Wroughton, 1905); *bethuliensis* Roberts, 1946; *chakae* (Wroughton, 1905); *cradockensis* Roberts, 1946; *diminutus* (Thomas, 1893); *griquoides* Roberts, 1946; *moshesh* (Wroughton, 1905); *nyasae* (Wroughton, 1905); *vaalensis* Roberts, 1946.

COMMENTS: Definition of *R. dilectus* generally follows Rambau et al. (2003), who sampled populations in South Africa, W Zimbabwe, Nyika Plateau in Malawi, and Mt. Elgon in Uganda. The species occurs in mesic grasslands and savannas in southern Africa (probably the more mesic of the six biotic zones of southern Africa in which *Rhabdomys* occurs; Yom-Tov, 1993) but north of that region (in SE Demo Rep. Congo, NE Zambia, Malawi, Tanzania, Kenya, and Uganda) is restricted to disjunct montane savannas (e. g., Delany, 1975; Hollister, 1919; and other references cited above for regions north of South Africa, Zimbabwe, and Mozambique).

    Two species occur in Angola. *Rhabdomys pumilio* occupies only the southwestern Namib desert of Angola (see account below); it is their sample from here that Carter and Hill (1941:102) identified as *R. bechuanae*. The other is represented by populations occurring over most of the Angolan Plateau. Rambau et al. (2003) did not sample Angola, but we assume these highland populations to represent *R. dilectus*. This afromontane distribution, an isolated population in Angola and isolated populations along highlands of East Africa, is concordant with the distributional pattern seen in *Hylomyscus denniae* (see account).

Rhabdomys pumilio (Sparrman, 1784). K. Svenska Vet.-Akad. Handl. Stockholm, p. 236.
COMMON NAME: Xeric Four-striped Grass Rat.
TYPE LOCALITY: South Africa, S Western Cape Province, east of Knysna, Tsitsikamma Forest, Slangrivier.
DISTRIBUTION: Xeric grasslands and savannas from W South Africa north through Namibia and C and S Botswana to SW Angola.
STATUS: IUCN – Data Deficient.
SYNONYMS: *bechuanae* (Thomas, 1893); *cinereus* (Thomas and Schwann, 1904); *deserti* (Dollman, 1910); *donavani* (Lesson, 1827); *fouriei* Roberts, 1946; *griquae* (Wroughton, 1905); *intermedius* (Wroughton, 1905); *lineatus* (F. Cuvier, 1829); *major* (Brants, 1827); *meridionalis* (Wroughton, 1905); *namaquensis* Roberts, 1946; *namibensis* Roberts, 1926; *orangiae* Roberts, 1946; *prieskae* Roberts, 1946; *septemvittatus* (Schinz, 1845); *typicus* (Sclater, 1899); *vittatus* (Wagner, 1842).
COMMENTS: As defined by Rambau et al. (2003), *P. pumilio* is endemic to the Southern African Subregion. Rambau et al. did not sample populations from Angola or Botswana, but we regard Hill and Carter's (1941:102) record of "*R. bechuanae*" from the Namib desert in southwestern Angola as an extension of *P. pumilio* through Namibia (see accounts above). Most of Botswana is covered by arid savanna, and records of *Rhabdomys* from there (see de Graaff, 1997e; Smithers, 1971) are likely *P. pumilio* and not *P. dilectus*, which occupies mesic grasslands and savannas (Rambau et al., 2003).

Rhagamys Major, 1905. Geol. Mag., Dec. 5, 2:503.
TYPE SPECIES: *Mus orthodon* Hensel, 1856.
COMMENTS: *Apodemus* Division. Redescribed by Schaub (1938), revised by Brandy (1978) and Martín Suárez and Mein (1998). Consists of two extinct species known only from the Mediterranean islands of Corsica and Sardinia. Martín Suárez and Mein (1998:92) explained that "When Forsyth Major (1905) created the genus *Rhagamys* he considered it to be closer to the group *Mäuse* (that is what we now call *Apodemus* and *Micromys*) than to the group *Ratten* (what we call *Mus* and *Rattus*)."

Rhagamys orthodon (Hensel, 1856). Z. Dtsch. Geol. Ges.:281.
COMMON NAME: Tyrrhenian Field Rat.
TYPE LOCALITY: Unknown (see Alcover et al., 1998), presumably from somewhere on either Corsica or Sardinia.
DISTRIBUTION: Recorded only from Corsica and Sardinia (Kotsakis, 1980; Caloi et al., 1986; Martín Suárez and Mein, 1998; Alcover et al., 1998; E. Pereira et al., 2001).
STATUS: Extinct.
COMMENTS: An extinct insular species that was large in body size (up to 50 grams; Libois et al., 1993) compared with its continental ancestors, and with bulky, high-crowned molars. Represented only by fossils recovered from late Pleistocene to Holocene sediments (Sondaar et al., 1984; Sondaar, 2000; Amori, 1993; Vigne, 1988; E. Pereira et al.,

2001). A smaller species, *minor*, known by fossils from early and middle Pleistocene deposits on both islands, was described as a species of *Rhagamys* (Brandy, 1978; E. Pereira et al., 2001; Sondaar, 2000) but later transferred to *Rhagapodemus*. This extinct genus dates back to late Miocene, and all the other species are represented by fossils from the Mediterranean continental mainland (Martín Suárez and Mein, 1998); *Rhagapodemus minor* has never been found in late Pleistocene horizons on Corsica or Sardinia, the time zone during which *Rhagamys orthodon* is first recorded. *Rhagamys orthodon* coexisted with the introduced *Apodemus sylvaticus* at the beginning of the third millennium but vanished at the end of the first millennium BC (Libois et al., 1993). Alcover et al. (1998) assumed that all mammalian species living on islands during late Pleistocene and Holocene survived until the arrival of humans and, except for a small cervid on Corsica, the other "Late Pleistocene mammals co-existed for several centuries with the exotic species introduced by man, from the beginning of his colonization of the Tyrrhenian islands" (Masseti, 1993:211; see also Vigne, 1992). The more than 25 species of extant mammals now found on Corsica and Sardinia represent gradual introductions by humans. However, before humans colonized the islands, about 9000 years ago (Mesolithic), the endemic fauna was unbalanced as is typical of insular assemblages and consisted of large soricids (*Episoriculus similis* on Sardinia iand *E. corsicanus* on Corsica), a talpid (*Talpa tyrrhenica*), a fox-sized canid (*Cynotherium sardous*), an otter (*Cyrnaonyx majori*), a small cervid (*Megaloceros cazioti*), a pika (*Prolagus sardus*), a vole (*Microtus henseli*), and *Rhagamys orthodon*. *Megaloceros* was exterminated soon after human arrival, but the shrew, vole, pika and Tyrrhenian field rat coexisted with humans until between 2000 and 1000 years ago (between Roman time and Middle Ages) and disappeared sometime between the arrival of *Rattus rattus* (during the Roman period) and the present (Vigne, 1990; Vigne and Marinval-Vigne, 1991; Vigne, 1992). The extinction of the four small mammal species "may be correlated with drastic deforestation" and "a major increase in agriculture, the effects of which were probably aggravated by the appearance of *Rattus rattus*, an aggressive competitor" (see Vigne, 1992:93, for comprehensive review and pertinent literature references).

Martín Suárez and Mein (1998) described molars of *Rhagamys orthodon* as large and highly derived compared with its relatives in *Rhagapodemus*, a consequence of insular evolution. They speculated *R. orthodon* possibly to have been derived from *Rhagapodemus* or even *Apodemus*, but lack of pertinent fossils prevents resolution of its evolutionary origin. Except for the hypsodonty exhibited by *R. orthodon*, Barrett-Hamilton (1900:422) thought its molar occlusal patterns were similar to *A. sylvaticus* and *A. agrarius*, "and may have been a direct offshoot from a common stock."

Rhynchomys Thomas, 1895. Ann. Mag. Nat. Hist., ser. 6, 16:160.

TYPE SPECIES: *Rhynchomys soricoides* Thomas, 1895.

COMMENTS: *Chrotomys* Division. Member of the Philippine Old Endemics (Musser and Heaney, 1992). Considered the only genus in Rhynchomyinae by Thomas (1896), who later included *Echiothrix* (Thomas, 1898b). Phylogenetic analyses of complete mtDNA cytochrome b sequences for 13 of the 16 genera of endemic Philippine murines place *Rhynchomys* in a clade containing *Apomys*, *Chrotomys* (including *Celaenomys*), and *Archboldomys* (Jansa and Heaney, 2001), which is supported by chromosomal data (Rickart and Heaney, 2002). The species of Philippine shrew rats in *Rhynchomys*, *Chrotomys* (including *Celaenomys*), and *Archboldomys* are unrelated to the other groups of shrew rats endemic to either Sulawesi or the Australia-New Guinea region (Musser and Heaney, 1992).

Rhynchomys isarogensis Musser and Freeman, 1981. J. Mammal., 62:154.

COMMON NAME: Isarog Rhynchomys.

TYPE LOCALITY: Philippines, Camarines Sur Province, SE Peninsula of Luzon Isl, Mt Isarog, 5000 ft (1524 m).

DISTRIBUTION: Greater Luzon Faunal Region. Known only from Mt Isarog, Luzon, in montane forest formations from 1125 to 1750 m (Heaney et al., 1998; Rickart et al., 1991).

STATUS: IUCN – Vulnerable.

COMMENTS: Distributional and ecological information summarized by Heaney et al. (1999). Standard karyotype (2n = 44, FN = 52; Rickart and Musser, 1993; Rickart and Heaney, 2002) indistinguishable from those of *Chrotomys silaceus* (recorded as *Celaenomys*) and *C. gonzalesi* (Rickart and Heaney, 2002).

Rhynchomys soricoides Thomas, 1895. Ann. Mag. Nat. Hist., ser. 6, 16:160.
    COMMON NAME: Northern Luzon Rhynchomys.
    TYPE LOCALITY: Philippines, N Luzon Isl, Mountain Province, Mt Data, 8000 ft (2440 m).
    DISTRIBUTION: Greater Luzon Faunal Region; recorded only from Mt Data in the Central Cordillera (Largen, 1985; Sanborn, 1952a; Thomas, 1895) and Kalinga Province (Rickart, in litt.), but probably occurs elsewhere in montane forest formations in N Luzon.
    STATUS: IUCN – Lower Risk (lc).

Solomys Thomas, 1922. Ann. Mag. Nat. Hist., ser. 9, 9:261.
    TYPE SPECIES: *Uromys sapientis* Thomas, 1902.
    SYNONYMS: *Unicomys* Troughton, 1935.
    COMMENTS: *Uromys* Division. Member of the New Guinea region Old Endemics (Musser, 1981c). Included in *Melomys* by Ellerman (1941:226), then transferred to *Uromys* by Tate (1951:312), but finally recognized again as a distinct genus (Flannery and Wickler, 1990; Laurie and Hill, 1954:128). Traditionally allied with *Uromys* and *Melomys* (Menzies, 1990; Misonne, 1969; Simpson, 1961; Tate, 1951), and analysis of albumin immunology linked *Solomys* "closely with a group of *Melomys* species represented by *M. cervinipes*" (Watts and Baverstock, 1994a:301). Sperm head morphology, however, is unusual among murines and unlike that characteristic of species in *Uromys* and *Melomys* (Breed, 1997; Breed and Aplin, 1994). Membership of *Solomys* in the *Uromys* cluster should be tested by phylogenetic analyses of morphological and molecular data.
    Species of *Solomys* are arboreal and endemic to the Solomon Isls. There are no specimens, either living or fossil, from Malaita and San Cristobal Isls in the S Solomons, but stories told by local people to Flannery suggest that *Solomys* once also occurred on both (Flannery, 1995b). This is also probably the large rat discussed by Kwa'ioloa and Burt (2001).

Solomys ponceleti (Troughton, 1935). Rec. Aust. Mus., 19:260.
    COMMON NAME: Poncelet's Solomys.
    TYPE LOCALITY: Solomon Isls, Bougainville Isl, about 10 mi (16 km) inland from Buin.
    DISTRIBUTION: Solomon Isls; endemic to islands of Buka, Bougainville, and Choiseul (Flannery, 1995b; Flannery and Wickler, 1990); sample from Buka Isl represented only by fossils.
    STATUS: IUCN – Endangered.
    COMMENTS: An arboreal species that builds stick nests high in large tall trees and known by a few extant specimens and archaeological fragments (Flannery et al., 1988; Flannery and Wickler, 1990). Reviewed by Flannery (1995b).

Solomys salamonis (Ramsay, 1883). Proc. Linn. Soc. N.S.W., 7:43.
    COMMON NAME: Florida Island Solomys.
    TYPE LOCALITY: Solomon Isls, Florida Isl in the Nggela Group (Flannery and Wickler, 1990:13).
    DISTRIBUTION: Recorded only from Florida Isl in the Solomon Arch. (Flannery and Wickler, 1990).
    STATUS: IUCN – Vulnerable.
    COMMENTS: Originally described as a *Mus*, the species was transferred to *Uromys* (Rümmler, 1938; Tate, 1951), but then correctly placed in *Solomys* by Troughton (1936). Holotype of *S. salamonis* is an adult male and was originally preserved in alcohol with the skull extracted, but the body has been lost and only the skull now represents the species (Flannery, 1995b).

Solomys salebrosus Troughton, 1936. Rec. Aust. Mus., 19:436.
    COMMON NAME: Bougainville Island Solomys.

TYPE LOCALITY: Solomon Isls, Bougainville Isl.

DISTRIBUTION: Recorded from the islands of Buka, Bougainville, and Choiseul in Solomon Arch.; sample from Buka Isl represented only by fossils (Flannery, 1995*b*).

STATUS: IUCN – Lower Risk (nt).

COMMENTS: Known by extant specimens and archaeological fragments (Flannery and Wickler, 1990). Reviewed by Flannery (1995*b*).

Solomys sapientis (Thomas, 1902). Ann. Mag. Nat. Hist., ser. 7, 9:44.

COMMON NAME: Isabel Island Solomys.

TYPE LOCALITY: Solomon Isls, Santa Isabel Isl.

DISTRIBUTION: Recorded only from Santa Isabel Isl (Flannery and Wickler, 1990).

STATUS: IUCN – Vulnerable.

COMMENTS: Represented by extant specimens; no archaeological fragments have been uncovered. Reviewed by Flannery (1995*b*).

Solomys spriggsarum Flannery and Wickler, 1990. Aust. Mamm., 13:133.

COMMON NAME: Buka Island Solomys.

TYPE LOCALITY: Solomon Isls, Buka Isl, Kilu rockshelter.

DISTRIBUTION: Endemic to Buka Isl.

COMMENTS: Known only by subfossil archaeological material (Flannery and Wickler, 1990).

Sommeromys Musser and Durden, 2002. Am. Mus. Novit., 3368:4.

TYPE SPECIES: *Sommeromys macrorhinos* Musser and Durden, 2002.

COMMENTS: *Crunomys* Division. Monotypic and represented only by the holotype. Morphology of skin, skull, dentition, and stomach described and contrasted with other murines, particularly the Sulawesian shrew rats (*Melasmothrix*, *Tateomys*, and *Echiothrix*) and Sulawesi and Philippine shrew mice (*Crunomys*) by Musser and Durden (2002). Certain derived cranial osteological and cephalic arterial complexes suggest affinity with *Crunomys*; otherwise, *Sommeromys* has no close morphological or phylogenetic affinity with any other described Sulawesian endemic. It may be a highly specialized relict of a group that diverged early from the Miocene *Progonomys*-like murine ancestral stock.

Sommeromys macrorhinos Musser and Durden, 2002. Am. Mus. Novit., 3368:7.

COMMON NAME: Sommer's Sulawesi Rat.

TYPE LOCALITY: Indonesia, C Sulawesi, Gunung Tokala, 02°13'S, 120°04'E, 2400 m (see Musser and Durden, 2002:7, for additional information).

DISTRIBUTION: Recorded only from tropical upper montane rain forest at the type locality.

COMMENTS: Gracile, insectivorous, nocturnal, scansorial, and arboreal. The species is unique among murines in morphology of its highly specialized rostral region, tail, feet and digits (Musser and Durden, 2002). Member of montane forest species known only from mountainous C Sulawesi (Musser and Durden, 2002; Musser and Holden, 1991).

Spelaeomys Hooijer, 1957. Zool. Meded. Leiden, 35:306.

TYPE SPECIES: *Spelaeomys florensis* Hooijer, 1957.

COMMENTS: *Pogonomys* Division. Reviewed by Musser (1981*c*), who also recorded past opinions about phylogenetic affinities of *Spelaeomys*, noted that it is not closely related to the other Nusa Tenggara endemics *Hooijeromys*, *Komodomys*, *Papagomys*, or *Paulamys*, and hypothesized that "*Spelaeomys* belongs with the old native genera of New Guinea, possibly Australia, and likely Timor." Molar occlusal patterns and configuration of unreported cranial fragments (studied by Musser) resemble a giant version of species in New Guinea *Pogonomys* and *Chiruromys*. But additional data is required to test this notion of cladistic membership with the New Guinea endemics in the *Pogonomys* Division. The alternative hypothesis relates *Spelaeomys* to members of the *Pithecheir* Division, which is centered on the Sunda Shelf and Sulawesi. Two genera in that Division, Sundaic *Lenothrix* and Sulawesian *Lenomys*, also have molar occlusal patterns resembling those in *Spelaeomys* (see illustrations in Musser, 1981*c*).

Spelaeomys florensis Hooijer, 1957. Zool. Meded. Leiden, 35:306.

COMMON NAME: Flores Island Spelaeomys.

TYPE LOCALITY: Indonesia, Nusa Tenggara, Palau Flores, Manggarai Province, Liang Toge, a cave near Warukia, 1 km S Lepa.

DISTRIBUTION: Known only from Flores Isl.

STATUS: IUCN – Extinct.

COMMENTS: A large-bodied, probably arboreal frugivore and folivore represented only by subfossil fragments (3000-4000 years old). The species may still live on Flores, possibly other nearby islands, and should be sought in any remnants of tropical evergreen rainforest.

Srilankamys Musser, 1981. Bull. Am. Mus. Nat. Hist., 168:268.

TYPE SPECIES: *Rattus ohiensis* Phillips, 1929.

COMMENTS: *Dacnomys* Division. A montane insular relict with possible phylogenetic ties to *Chiromyscus* and *Niviventer*, but not to *Rattus* (Musser, 1981*b*). The genus also exhibits some morphological traits characteristic of *Maxomys*; its inclusion in the *Dacnomys* Division has to be tested by broader character analyses that include chromosomal and molecular data.

Srilankamys ohiensis (Phillips, 1929). Ceylon J. Sci., Sec. B, 15:167.

COMMON NAME: Phillips's Srilankamys.

TYPE LOCALITY: Sri Lanka (Ceylon), Ohiya, W Haputale, 6000 ft (1830 m).

DISTRIBUTION: Primary lowland tropical and montane evergreen rainforest formations on mountains in Uva and Central Provinces of Sri Lanka between 915 and 2135 m (Phillips, 1980).

STATUS: IUCN – Lower Risk (nt).

COMMENTS: Musser (1981*b*) described the history of past and incorrect allocations of *ohiensis* to groups that are now recognized as *Apomys, Lenothrix, Leopoldamys, Maxomys, Niviventer*, and *Rattus*.

Stenocephalemys Frick, 1914. Ann. Carnegie Mus., 9:7.

TYPE SPECIES: *Stenocephalemys albocaudata* Frick, 1914.

COMMENTS: *Stenocephalemys* Division. An Ethiopian endemic that is phylogenetically related to species of *Praomys, Mastomys, Myomyscus, Heimyscus*, and *Hylomyscus* (Lecompte et al., 2002*b*). Significance of analyses of craniodental and phallic structure, allozymes, chromosomal data, partial mitochondrial 16S rRNA and complete mtDNA cytochrome *b* gene sequences, and geometric morphometrics from samples of *S. albipes, S. albocaudata*, and *S. griseicauda* (Corti et al., 1999; Fadda and Corti, 2000; Fadda et al., 2001*a*; Lavrenchenko et al., 1999, 2000; Lecompte et al., 2002*b*) are discussed in accounts that follow. Statistical summaries of external measurements from large series of *S. albocaudata* and *S. griseicauda* reported by Sillero-Zubiri (1995*a*), and the habitat preferences of these two species, abundance, and biomass in relation to predation by the endangered Ethiopian wolf (*Canis simensis*) are documented by Sillero-Zubiri (1995*c*). Spermatozoal morphology described by Baskevich and Lavrenchenko (1995) and Breed (1995*a*). Isolated molars from late Miocene Ethiopian sediments (Chorora) were identified by Geraads (2001) as possibly being *Stenocephalemys*, and the genus is also represented by Pleistocene remains recovered in Ethiopia (Jaeger, 1979); see review by Denys (1999).

Stenocephalemys albipes (Rüppell, 1842). Mus. Senckenberg., 3:107.

COMMON NAME: White-footed Stenocephalemys.

TYPE LOCALITY: Ethiopia, Massawa.

DISTRIBUTION: Ethiopia; endemic to Ethiopian Plateau between 800-3300 m (Van der Straeten and Dieterlen, 1983; Yalden and Largen, 1992; Yalden et al., 1976, 1996).

STATUS: IUCN – Lower Risk (nt) as *Myomys albipes*.

SYNONYMS: *alettensis* (Frick, 1914); *ankoberensis* (Frick, 1914); *leucopus* (Fitzinger, 1867); *minor* (Heuglin, 1877).

COMMENTS: Usually listed as either a species of *Myomys* (G. M. Allen, 1939), *Myomyscus* (D. H. S. Davis, 1965), or *Praomys* (Yalden et al., 1976, 1996). Morphometric traits related *albipes* closely to species of *Stenocephalemys* and what was described as *Praomys ruppi* (Van der Straeten and Dieterlen, 1983). Furthermore, qualitative external, cranial (see Rupp, 1980: Fig. 5), and molar (see illustrations in Misonne, 1969) features of *albipes* are more similar to species of *Stenocephalemys* than to other species in *Myomys* (= *Myomyscus*). Musser and Carleton (1993) noted that "The phylogenetic significance of this morphologically annectant relationship of *M. albipes* between other *Myomys* and *Stenocephalemys* needs to be assessed by taxonomic revision of both groups." Analyses of chromosomal data (Corti et al., 1995, 1999; Lavrenchenko et al., 1999), allozymic information (Lavrenchenko et al., 1999, 2000), partial 16S rRNA mitochondrial gene sequences (Fadda et al., 2001*a*), and complete DNA mitochondrial cytochrome *b* sequences (Lecompte et al., 2002*b*) clustered *S. albipes*, *S. albocaudata*, and *S. griseicauda* in a monophyletic group that excluded species of *Praomys* and *Myomyscus* (reported as *Myomys*) sampled, either *P. daltoni* or *M. brockmani* (reported as *fumatus*). The molecular data also suggested that *S. griseicauda* is more closely related to *S. albipes* than to *S. albocaudata*. To explain differences in size and shape of *S. albipes*, *S. ruppi*, *S. albocaudata*, *and S. griseicauda*, Fadda and Corti (2000) studied three-dimensional geometric morphometrics of these species, contrasted them with *Myomyscus brockmani* (reported as *fumatus*), and demonstrated a positive correlation between altitude and size and shape of skull. Shape differences clustered *S. albipes* and *M. brockmani* in a separate group from *S. albocaudata* and *S. griseicauda*, which is discordant with the monophyly indicated by chromosomal and molecular data. Fadda and Corti suggested that the strong correlation with altitude and associated environmental changes may be the causal factor for the shape changes. Both *S. albocaudata* and *S. griseicauda* inhabit treeless Afro-alpine moorland. Rodents with extremely narrow interorbits, expansive zygomatic arches, and other shape changes in the orbital region allowing a more dorsal position of eyes (stenocephaly) are adapted to survive in open country where they can scan the sky more efficiently for raptors, their primary predators (F. Petter, 1972*d*). The cranial similarity between *S. albocaudata* and *S. griseicauda* may reflect adaptations to similar habitats rather than phylogenetic relationship. The *Myomyscus*-like cranium of *S. albipes* is probably a convergent adaptation to forested habitats. The distribution of *S. albipes* in relation to forest blocks documented by Bekele and Corti (1997:287): "The species prefers the dense forests that are progressively shrinking, so perhaps it can be used as a reliable indicator of forest block reduction on an historical basis."

The holotype of *ankoberensis* (which we examined) did not fit comfortably in any principal component cluster of holotypes associated with *Praomys*, *Mastomys*, *Hylomyscus*, *Myomys*, and *Myomyscus* reported by Van der Straeten and Robbins (1997) because it is an example of *S. albipes*, which was not included in their analysis. The two Sudanese specimens reported by Setzer (1956) as *P. albipes fuscirostris* are not *albipes* but may be associated with *Mastomys* as either a species or synonym (see account of *Mastomys erythroleucus*). Pertinent reports cover allozyme variation within geographic samples of *S. albipes* (Milishnikov et al., 1992), postnatal development and reproduction in captive-bred *S. albipes* (Bekele, 1995), and distribution in the isolated Harenna Forest of S Ethiopia (Lavrenchenko, 2000).

Stenocephalemys albocaudata Frick, 1914. Ann. Carnegie Mus., 9:8.

COMMON NAME: White-tailed Stenocephalemys.

TYPE LOCALITY: S Ethiopia, Chilalo Mtns, Inyala Camp.

DISTRIBUTION: Ethiopia; endemic to the eastern plateau (eastern side of Rift Valley), 3000-4050 m.

STATUS: IUCN – Lower Risk (nt).

COMMENTS: Inhabits the Afro-alpine moorland where it occurs with *Arvicanthis blicki*, *Lophuromys melanonyx*, and *Tachyoryctes macrocephalus*, which are also moorland specialists (Rupp, 1980; Yalden, 1988; Yalden and Largen, 1992). Chromosomal data reported by Corti et al. (1995, 1999) and Lavrenchenko et al. (1999). Yalden et al. (1996) cited Schlitter who claimed he caught the species on Mt Entoto (north of Addis Ababa), which would extend the range to the highlands west of the Rift Valley but that record requires verification.

Stenocephalemys griseicauda F. Petter, 1972. Mammalia, 36:171.

COMMON NAME: Gray-tailed Stenocephalemys.

TYPE LOCALITY: Ethiopia, Bale Mtns, Dinsho.

DISTRIBUTION: Ethiopia; S and N highlands, 2400-3900 m (Demeter and Topal, 1982; Yalden, 1988; Yalden and Largen, 1992).

STATUS: IUCN – Lower Risk (nt).

COMMENTS: Another Ethiopian mountain endemic that overlaps altitudinally with *S. albocaudata* but occupies bushy areas rather than moorland (Yalden, 1988). Chromosomal data documented by Corti et al. (1999) and Lavrenchenko et al. (1999). Distribution in isolated Harenna Forest of S Ethiopia documented by Lavrenchenko (2000). Yalden et al. (19965) cited Schlitter as collecting this species and *S. albocaudata* on Mt Entoto north of Addis Ababa, which would be west of the Rift Valley, but those records have yet to be documented by publication.

Stenocephalemys ruppi (Van der Straeten and Dieterlen, 1983). Ann. Mus. R. Afr. C., 237:121.

COMMON NAME: Rupp's Stenocephalemys.

TYPE LOCALITY: Ethiopia, Bonke, north of Bulta, 2800-3200 m.

DISTRIBUTION: Ethiopia; known only from Bonke and Bulta in the Gamo Gofa region of SW Ethiopia, 2700-3200 m (Rupp, 1980; Van der Straeten and Dieterlen, 1983; Yalden and Largen, 1992)

STATUS: IUCN – Vulnerable as *Myomys ruppi*.

COMMENTS: Originally described as a species of *Praomys* based on material collected by Rupp (1980), who illustrated the skull as "*Praomys albipes*, stenocephaler Typ" (p. 92). Musser and Carleton (1993) wrote that "*M. ruppi* combines morphological features of both *M. albipes* and *Stenocephalemys*, an observation reinforced by morphometric analyses (Van der Straeten and Dieterlen, 1983). Were it not for the long tail of *M. ruppi* (a trait shared with *M. albipes*), the species could as easily be included within *Stenocephalemys*." In their canonical analysis of *ruppi*, *albipes*, and *S. griseicauda*, Van der Straeten and Dieterlen (1983) derived scores for *ruppi* that fell between those representing the other two species, but portrayed *ruppi* as being overall more closely related to *S. griseicauda*. Fadda and Corti (2000) disagreed with that assessment, noting that in their study "*ruppi* shares centroid size values and shape changes that are typical of *M. albipes*," indicated they have unpublished evidence suggesting divergence within "Ethiopian *Myomys*" (presumably *albipes*), and considered the relationship of the Bonke population as unresolved. We await their results but meanwhile all available data about *ruppi* support its transfer from *Myomys* (=*Myomyscus*) to *Stenocephalemys*.

Stochomys Thomas, 1926. Ann. Mag. Nat. Hist., ser. 9, 17:176.

TYPE SPECIES: *Dasymys longicaudatus* Tullberg, 1893.

COMMENTS: *Hybomys* Division. Listed as a genus by G. M. Allen (1939), but allocated to *Rattus* as a subgenus by Ellerman (1941). D. H. S. Davis (1965) placed *Stochomys* in *Aethomys* as a subgenus, which reflected Thomas' (1915) early allocation of *longicaudatus* to subgenus *Aethomys*. Misonne (1969) and Rosevear (1969), however, reinstated the generic status of *Stochomys*, correctly noting that it was distinctive and not closely related to *Rattus*. It has been phylogenetically associated with *Dephomys* (D. H. S. Davis, 1965; Misonne, 1969; an alliance supported by our study of specimens), but Van der Straeten (1984) considered it unrelated to that genus, as reflected by multivariate analysis of morphometric variation. His conclusion, however, derives from analysis of continuous variables and neglected discrete differences in character traits, and the resulting tree is a phenogram of shape-size similarity, not a cladogram reflecting shared-derived characters hypothesizing phylogenetic affinity. Features of pelage, skull, and dentition characterizing *Stochomys* are similar to those found in some species of *Aethomys*, which prompted Visser and Robinson (1986) to suggest that both belong in the same monophyletic clade in which *Stochomys* is the high forest partner of savanna *Aethomys*. Analysis of microcomplement fixation of albumin, however, grouped *Stochomys* and *Hybomys* together to the exclusion of *Aethomys* (Watts and Baverstock, 1995a, b), a pairing suggested by Misonne (1969) who examined molar occlusal patterns, but not by chromosomal data (Viegas-Péquignot et al., 1986). Spermatozoal morphology is primitive for murines (Breed, 1995a). Analyses of APRT gene sequences from *Stochomys* were ambiguous in assessing its cladistic position but did illuminate significance of substitution rate variation among *Stochomys* and other muroids

(Fieldhouse et al., 1997). Additional inquiry into the phylogenetic affinity of *Stochomys* is warranted, employing broader generic sampling and analyses of mitochondrial and nuclear gene sequences.

Stochomys longicaudatus (Tullberg, 1893). Nova Acta Reg. Soc. Sci. Uppsala, ser. 3, 16:36.
    COMMON NAME: Target Rat.
    TYPE LOCALITY: Cameroon.
    DISTRIBUTION: Tropical evergreen forest; recorded from Togo, S Nigeria, Central African Republic, Cameroon, Gabon, Dem. Rep. Congo, and Uganda.
    STATUS: IUCN – Lower Risk (lc).
    SYNONYMS: *hypoleucus* (Pucheran, 1855) [not Sundevall]; *ituricus* (Thomas, 1915); *sebastianus* (de Winton, 1897).
    COMMENTS: Rosevear (1969) described the historical allocations of *longicaudatus* to *Aethomys*, *Dasymys*, *Epimys*, *Mus*, *Rattus*, and *Stochomys*. Morphometric analyses, which resulted in distinguishing two distinct subspecies, reported by Van der Straeten (1984).

Sundamys Musser and Newcomb, 1983. Bull. Am. Mus. Nat. Hist., 174:401.
    TYPE SPECIES: *Mus muelleri* Jentink, 1879.
    COMMENTS: *Rattus* Division. Endemic to the Malay Peninsula and islands on the Sunda Shelf. Revised by Musser and Newcomb (1983), whose analysis of morphological traits placed it phylogenetically closer to *Rattus* and *Kadarsanomys* than to other Sundaic genera. The close relationship to *Rattus* is corroborated by albumin immunology (Watts and Baverstock, 1994*b*, 1995*b*, 1996), spermatozoal morphology (Breed and Yong, 1986), and analyses of LINE-1 retrotransposons (Verneau et al., 1997, 1998).

Sundamys infraluteus (Thomas, 1888). Ann. Mag. Nat. Hist., ser. 6, 2:409.
    COMMON NAME: Mountain Sundamys.
    TYPE LOCALITY: Malaysia, Sabah (N Borneo), Kinabalu.
    DISTRIBUTION: Gunung Kinabalu, 920-2930 m (Md Nor, 2001), and Gunung Trus Madi in Sabah; W mountain chain of Sumatra, 700-2400 m.
    STATUS: IUCN – Lower Risk (lc).
    SYNONYMS: *atchinus* (Miller, 1942).
    COMMENTS: A montane species related to *S. muelleri*, which occurs at middle altitudes and lowlands; narrowly sympatric with it on both N Borneo and Sumatra. Specimens from each island differ slightly in pelage tone and craniodental dimensions and proportions, but the sample from Sumatra (*atchinus*) is small and significance of the differences between it and the much larger series from Sabah requires study of more material from Sumatra (Musser and Newcomb, 1983).

Sundamys maxi (Sody, 1932). Natuurh. Maandbl. Maastricht., 21:157.
    COMMON NAME: Javan Sundamys.
    TYPE LOCALITY: Indonesia, Java, Bandung, Cibuni.
    DISTRIBUTION: Known only from W Java, 900-1350 m.
    STATUS: IUCN – Endangered.
    COMMENTS: Described as a species of *Rattus* (Sody, 1932), later treated as a subspecies of *R. infraluteus* (Chasen, 1940), but shown to be a very distinctive species and member of the suite of endemic Javanese murines (Musser and Newcomb, 1983; see account of *Kadarsanomys sodyi*). Still only represented by specimens collected between 1932 and 1935 by Max Bartels Jr. Phylogenetic relationship may be closer to *S. muelleri* than to *S. infraluteus* as judged by morphology. This is the only species of *Sundamys* known to occur on Java and the only one with a derived cephalic arterial pattern.

Sundamys muelleri (Jentink, 1879). Notes Leyden Mus., 2:16.
    COMMON NAME: Müller's Sundamys.
    TYPE LOCALITY: Indonesia, W Sumatra, Batang Singgalang, Padang Highlands.
    DISTRIBUTION: Endemic to large and small islands and peninsula on the Sunda Shelf south of the Isthmus of Kra (10°30'N): SW peninsular Burma, peninsular Thailand, Malay Peninsula, Sumatra, Borneo, and Palawan; and on the smaller islands of Siantan

(Anamba Isl), many of the Riau Isls, Tuangku and Bangkuru (Banjak Isls), Mansalar (W Sumatra), Pinie, Tanahmasa and Tanahbala (Batu Isls), Banka, Bunguan and Serasan (Natuna Isls), Karimata Isl (SW Borneo), Sebuku (SE Borneo), Balembangan and Banggi (N Borneo), and Balabac, Culion, and Busuanga (Palawan Faunal Region). Older published records from islands off Sunda Shelf (Nicobars, for example) proved to represent other species (all but the Karimata records are documented in the gazetteer and range maps in Musser and Newcomb, 1983; specimens in UMMZ and MZB were obtained on Karimata Isl by A. J. Gorog in 2000). Does not occur on Java.

STATUS: IUCN – Lower Risk (lc).

SYNONYMS: *balabagensis* (Sanborn, 1952); *balmasus* (Lyon, 1916); *borneanus* (Miller, 1913); *campus* (Robinson and Kloss, 1916); *chombolis* (Lyon, 1909); *crassus* (Lyon, 1911); *credulus* (Chasen, 1940); *culionensis* (Sanborn, 1952); *domitor* (Miller, 1903); *firmus* (Miller, 1902); *foederis* (Robinson and Kloss, 1911); *integer* (Miller, 1901); *otiosus* (Chasen, 1935); *pinatus* (Lyon, 1916); *pollens* (Miller, 1913); *potens* (Miller, 1913); *sebucus* (Lyon, 1911); *terempa* (Chasen and Kloss, 1928); *valens* (Miller, 1913); *validus* (Miller, 1900); *victor* (Miller, 1913); *virtus* (Lyon, 1916); *waringensis* (Sody, 1941).

COMMENTS: Morphological, chromosomal, and spermatozoal data suggested *S. muelleri* is a distant relative of *Rattus* but in the same monophyletic clade (Breed and Yong, 1986; Musser and Newcomb, 1983; Watts and Baverstock, 1994*b*). Like *Maxomys surifer*, which also has an extensive insular distribution on the Sunda Shelf (see that account), nearly every island sample of *S. muelleri* was described as a separate subspecies, some based on real differences in pelage coloration and body size, others diagnosed primarily by their insular distribution (see history in Musser and Newcomb, 1983). There are detectable differences among some samples. Populations from SW Burma, Peninsular Thailand, and the Malay Peninsula, for example, are significantly larger in body size than those from elsewhere on the Sunda Shelf and may be a separate species (Musser and Newcomb, 1983). Phylogenetic analyses of gene sequences from insular and peninsular samples of *S. muelleri* may illuminate significance and historical origin of the documented morphological variation.

Occurrence of *S. muelleri* on peninsular Mt Santubong in W Sarawak recorded by Tuen et al. (2000), and its distribution and ecology on the small islands off the tip of N Sabah documented by Md Nor (1996). Generally a lowland species, *S. muelleri* reaches 1650 m on slopes of Mt Kinabalu in Sabah (Md Nor, 2001, and citations therein).

Taeromys Sody, 1941. Treubia, 18:260.

TYPE SPECIES: *Mus* (*Gymnomys*) *celebensis* Gray, 1867.

SYNONYMS: *Arcuomys* Sody, 1941.

COMMENTS: *Rattus* Division. A Sulawesi endemic. Formerly included in *Rattus* by Ellerman (1949*a*), and in subgenus *Bullimus* of *Rattus* by Misonne (1969), but considered a distinct genus by Musser (1981*c*, 1984, 1987) and Musser and Newcomb (1983); albumin immunology nestles *Taeromys* in a *Rattus* clade (also containing *Sundamys*, *Bunomys*, *Komodomys*, *Stenomys*, *Bullimus*, *Bandicota*, *Nesokia*, *Paruromys*, *Papagomys*, and *Berylmys*; Watts and Baverstock, 1994*b*, 1995*b*, 1996). Partially reviewed and contrasted with Sundaic *Sundamys* by Musser and Newcomb (1983). Marked interspecific contrasts in spermatozoal morphology evident among species and reflected by differences in cranial and dental traits (Breed and Musser, 1991). Sody (1941:260) proposed nine new genera, each diagnosed by mammary formula. Among these were *Taeromys* and *Arcuomys*, which Corbet and Hill (1992:287) regarded as doubtfully valid ". . . since they were based solely upon mammary formulae shared with other genera proposed on the same page." But Sody diagnosed his genera in a comparative context, listing the mammary formulae for all the new genera, and the names seem to satisfy the conditions of Article 13 in the fourth edition of the International Code of Zoological Nomenclature (International Commission on Zoological Nomenclature, 1999). Corbet and Hill (1992:287) also noted that *Arcuomys* had line priority over *Taeromys*, but the latter can be viewed as having priority following Musser (1981*c*:137) as "first reviser." Because Sody's diagnosis only highlighted mammary formula, we provide the following emended diagnosis for *Taeromys*: A genus of Murinae (as that subfamily was defined by Carleton and Musser, 1984) characterized by 1) large body size; 2) short, soft, and dense fur composed of thin and pliable hairs; 3) short guard hairs barely extending beyond over overfur layer of dorsal coat; 4) gray or dark grayish brown upperparts in most specimens, reddish brown in a few (some examples of *T. celebensis* and all specimens of *T. punicans*); 5) long and slender hind feet with six fleshy plantar pads, which includes a hypothenar; 6) length of tail shorter to slightly longer than length of head and body in all species except *T. celebensis* (much longer), and bicolored (basal one-third to one-half brown or blackish brown, distal segment white) in all species except *T. punicans* (monocolored); 7) smooth, glistening tail surfaces due to non-overlapping rings of thin scales and very short, fine scale hairs; 8) either three pairs of teats (one postaxillary pair and two inguinal pairs; most samples) or two inguinal pairs only; 9) slim and long rostrum (except *T. celebensis*); 10) wide zygomatic plate (but its anterior margin not covering nasolacrimal capsule) and deep zygomatic notch (except *T. celebensis*); 11) low and inconspicuous postorbital ridges, barely evident temporal ridges; 12) short incisive foramina, their posterior margins ending anterior to molar rows in *T. celebensis*, *T. callitrichus*, and *T. arcuatus*, but slightly projecting between the rows in *T. taerae* and *T. hamatus*; 13) posterior margin of bony palate either even with end of third molars or extending slightly beyond them; 14) sphenopalatine vacuities short and narrow; 15) large stapedial foramen and prominent groove in pterygoid plate for infraorbital branch of stapedial artery (which reflects the primitive murine state of cephalic arterial supply); 16) no alisphenoid strut; 17) small auditory bulla relative to cranium (except *T. celebensis*); 18) configuration of upper incisors mainly orthodont, slightly opisthodont in some species; 19) each first upper molar anchored by five roots, the second molar by four, and the third by three, each first lower molar with four roots, the second and third with three roots; 20) large molars relative to size of cranium and mandible, crowns slightly hypsodont; 21) rows of cusps on upper molars more laminar than cuspidate, anterolabial cusps (t3) on first and second upper molars reduced in size or absent, posterior cingulum usually absent from each first and second upper molar, no cusp t7, and no anterocentral cusp on each first lower molar; 22) entepicondylar foramen present; 23) stomach unilocular-hemiglandular; 24) $2n = 42$, $FN = 60$ for an undescribed species, $2n = 39$, $FN = 50$ for males of *T. celebensis* and $2n = 39$ or 40, $FN = 49$ or 50 for females (X1X1X2X2 female/XsX2Y male system); Musser, ms).

Most species of *Taeromys* are difficult to trap (Musser's experience in the forest), which is reflected by their meager representation in museum collections. Current patterns of geographic distributions and character variation revealed by morphometric analyses (Musser, ms) requires testing by larger samples and exposure to analyses of other data sets, both morphological and molecular. An additional species not listed here is found in montane forest formations dominated by species of oak (*Lithocarpus*) and chestnut (*Castanea*) in C Sulawesi, is allied to *T. arcuatus*, and will be described by Musser (ms).

Taeromys arcuatus (Tate and Archbold, 1935). Am. Mus. Novit., 802:9.
> COMMON NAME: Southeastern Mountain Taeromys.
> TYPE LOCALITY: Indonesia, SE Sulawesi, Pegunnungan Mekongga (= Mengkoka), Tanke Salokko (highest spot on the mountain), 03°35′S/121°15′E, 1500 m.
> DISTRIBUTION: Known only from the type locality.
> STATUS: IUCN – Vulnerable.
> COMMENTS: Closest relative is an undescribed species in mountains of C Sulawesi (Musser and Holden, 1991; Musser, ms). This geographic pattern, in which a species in the central part of the island has a close ally on the SE Peninsula, is repeated by *Taeromys callitrichus* (central) and *T. microbullatus* (southeast), an undescribed species of *Maxomys* (central) and *M. dollmani* (southeast), and *Rattus marmosurus* (central) and *R. salocco* (southeast). Sody (1941) proposed *Arcuomys* for *arcuatus*.

Taeromys callitrichus (Jentink, 1879). Notes Leyden Mus., p. 12.
> COMMON NAME: Greater Taeromys.
> TYPE LOCALITY: Indonesia, NE Sulawesi, Kakas.
> DISTRIBUTION: Sulawesi: recorded from a few places on the NE peninsula, 600-1100 m; and in the central core, 760-2260 m (Musser, 1970d; Musser and Holden, 1991).

STATUS: IUCN – Lower Risk (lc).

SYNONYMS: *jentinki* (Laurie and Hill, 1954); *maculipilis* (Laurie and Hill, 1954).

COMMENTS: Reviewed by Musser (1970*d*). The largest in body size among the species. The original description is vague and based upon 12 specimens, all of which Jentink (1879*b*, 1888:65) regarded as types. Subsequently, Tate (1940) selected one of the specimens as a lectotype, and Musser (1970*d*) identified all 12 specimens: the lectotype and two others are examples of *callitrichus*, one is *Paruromys dominator*, another is *Rattus hoffmanni*, three are *Bunomys chrysocomus*, and four are *Bunomys fratrorum*. Musser (1970*d*) also described the historical allocations of *callitrichus* with *Mus*, *Rattus*, *Lenomys*, and *Eropeplus*. *Taeromys callitrichus* is the central and north-eastern peninsular highland and morphological counterpart to the southeastern highland *T. microbullatus*; judged by cranial and dental traits, both species are more closely related to *T. arcuatus* and its undescribed ally than to any other species of *Taeromys* (Musser's research). Because *T. callitrichus* is not easily trapped, it is represented by few specimens, and significance of the appreciable morphological variation present among the small geographic samples will have to be determined by study of larger series from more places.

Taeromys celebensis (Gray, 1867). Proc. Zool. Soc. Lond., 1867:598.

COMMON NAME: Long-tailed Taeromys.

TYPE LOCALITY: Indonesia, NE Peninsula, Manado.

DISTRIBUTION: Sulawesi: recorded from throughout the island in lowland tropical evergreen rainforest, sea level to 1200 m; absent from montane forest formations (Musser and Holden, 1991).

STATUS: IUCN – Lower Risk (lc).

COMMENTS: Documented from SW peninsula only by subfossil fragment (Musser, 1984), but from elsewhere by extant specimens. An arboreal species with a very long tail relative to head and body length, and morphologically unlike any other species in the genus, all of which are terrestrial and have relatively short tails (Breed and Musser, 1991; Musser and Newcomb, 1983; Musser, ms). *Taeromys celebensis* is relatively common in primary forest, not difficult to trap, and represented by many more specimens than the other species of *Taeromys*. It is superficially similar to *Paruromys dominator* and often misidentified as that species in museum collections.

Taeromys hamatus (Miller and Hollister, 1921). Proc. Biol. Soc. Wash., 34:96.

COMMON NAME: Central Mountain Taeromys.

TYPE LOCALITY: Indonesia, C Sulawesi, Gunung Lehio, 6000 ft (1830 m).

DISTRIBUTION: Sulawesi: known only from a few montane localities in the central core of the island, 1280-2287 m; absent from lowland tropical evergreen rainforest formations.

STATUS: IUCN – Vulnerable.

COMMENTS: Closest relative is *T. taerae* from the NE peninsular highlands.

Taeromys microbullatus (Tate and Archbold, 1935). Am. Mus. Novit., 802:8.

COMMON NAME: Small-eared Taeromys.

TYPE LOCALITY: Indonesia, SE Peninsula, Pegunungan Mekongga (= Mengkoka), Tanke Salokko (highest place in Pegunungan Mekongga), 03°35'S, 121°15'E, 1500 m.

DISTRIBUTION: Recorded only from the type locality.

COMMENTS: Still represented only by two adults and a juvenile collected in 1932. Although described as a species of *Rattus*, *microbullatus* has been allied to the *Rattus xanthurus* group (Ellerman, 1941), included in *Paruromys* (Ellerman, 1949*a*; Laurie and Hill, 1954) or synonymized with *T. callitrichus* (Corbet and Hill, 1992; Musser, 1970*d*; Musser and Carleton, 1993), but morphometric analyses indicate the small sample represents a distinctive entity that is more closely related to *T. callitrichus* than to any other species of *Taeromys* (Musser, ms).

Taeromys punicans (Miller and Hollister, 1921). Proc. Biol. Soc. Wash., 34:98.

COMMON NAME: Reddish-furred Taeromys

TYPE LOCALITY: Indonesia, C Sulawesi, Pinedapa, 100 ft (30 m).

DISTRIBUTION: Sulawesi: known only from the central part and the SW peninsula.

STATUS: IUCN – Lower Risk (nt).

COMMENTS: A lowland species represented by two specimens caught in 1918 from the type locality and a few subfossil fragments from the SW peninsula (Musser, 1984). Other than that it inhabits tropical lowland evergreen rainforest, nothing is known about ecology of this species or its actual distribution over the island.

Taeromys taerae (Sody, 1932). Natuurh. Maandbl. Maastricht., 21:158.
COMMON NAME: Northeastern Mountain Taeromys.
TYPE LOCALITY: Indonesia, NE Sulawesi, Lembean, near Tondano.
DISTRIBUTION: Sulawesi: recorded only from highlands on the NE peninsula, 600-800 m.
STATUS: IUCN – Lower Risk (lc).
SYNONYMS: *simpsoni* (Ellerman, 1949) [not Philippi, 1900]; *tatei* (Sody, 1941) [not Ellerman, 1941].
COMMENTS: Reviewed by Musser (1971*d*). *Taeromys taerae* is the NE peninsular highland and morphological relative of *T. hamatus*, which is found only in mountains of central Sulawesi.

Tarsomys Mearns, 1905. Proc. U.S. Natl. Mus., 28:453.
TYPE SPECIES: *Tarsomys apoensis* Mearns, 1905.
COMMENTS: *Rattus* Division. Revised by Musser and Heaney (1992), who recorded history of the incorrect inclusion of *Tarsomys* in *Rattus* and placed the genus among the Philippine New Endemics. Phylogenetic analyses of complete mtDNA cytochrome *b* sequences for 13 of the 16 genera of endemic Philippine murines place *Tarsomys* as sister-taxon to *Limnomys*, and in a clade with *Bullimus* and *Rattus everetti*, which are also New Endemics (Jansa and Heaney, 2001); this alliance and membership in the *Rattus* Group also supported by chromosomal data (Rickart and Heaney, 2002).

Tarsomys apoensis Mearns, 1905. Proc. U.S. Natl. Mus., 28:453.
COMMON NAME: Dusky Tarsomys.
TYPE LOCALITY: Philippines, S Mindanao Isl, Davao City Province, Mt Apo, 6750 ft (2058 m).
DISTRIBUTION: Greater Mindanao Faunal Region. Endemic to the mountains on Mindanao from 1550 to 2400 m in tropical lower and upper montane rainforest (Heaney et al, 1998; Musser and Heaney, 1992).
STATUS: IUCN – Lower Risk (lc).
COMMENTS: This dark-furred, short-tailed mountain rat may be the ecological equivalent of the Sulawesian *Bunomys chrysocomus*, which it superficially resembles in some morphological traits (Musser and Heaney, 1992). Standard karyotype (2n = 42, FN = 61/62) identical to that of *Limnomys* in autosomes and X chromosome but differs by its larger telocentric Y chromosome (Rickart and Heaney, 2002).

Tarsomys echinatus Musser and Heaney, 1992. Bull. Am. Mus. Nat. Hist., 138:33.
COMMON NAME: Spiny Tarsomys.
TYPE LOCALITY: Philippines, S Mindanao, South Cotabato Province, Mt Matutum, Tupi, Balisong, 2700-3700 ft (823-1128 m).
DISTRIBUTION: Greater Mindanao Faunal Region. Endemic to tropical lowland evergreen rainforests on Mindanao and known only from the vicinity of type locality and Mt Katanglad in Bukidnon Province (Heaney et al., 1998; Musser, 1994); probably occurs elsewhere on Mindanao where lowland forest formations persist.
STATUS: IUCN – Vulnerable.
COMMENTS: This distinctive species is replaced at higher elevations by *T. apoensis* in the region of Mt Katangland.

Tateomys Musser, 1969. Am. Mus. Novit., 2384:2.
TYPE SPECIES: *Tateomys rhinogradoides* Musser, 1969.
COMMENTS: *Melasmothrix* Division. Originally, and incorrectly, linked to the "*Rattus chrysocomus* group" by Musser (1969*b*), but morphological, spermatozoal, and ecological characteristics unite *Tateomys* and the other Sulawesian shrew rat *Melasmothrix* in the same monophyletic group to the exclusion of either *Rattus*, *Bunomys*, or any other Sulawesian endemic (Breed and Musser, 1991; Musser, 1982*c*). Listed as part of the

Sulawesi Old Endemics by Musser (1981c) and revised by him (Musser, 1982c); united with *Melasmothrix* by Corbet and Hill (1992). *Tateomys*, along with *Melasmothrix*, possibly are specialized remnants of one of the earliest lineages to have diverged from the ancestral *Progonomys*-like murine stock. See *Melasmothrix* generic account.

Tateomys macrocercus Musser, 1982. Bull. Am. Mus. Nat. Hist., 174:64.

> COMMON NAME: Long-tailed Sulawesian Shrew Rat.
> TYPE LOCALITY: Indonesia, C Sulawesi, Gunung Nokilalaki, 7500 ft (2286 m).
> DISTRIBUTION: Known only between 1980 and 2286 m in tropical upper montane rain forest on Gunung Nokilalaki, but probably occurs in other mountainous regions of C Sulawesi.
> STATUS: IUCN – Vulnerable.
> COMMENTS: Stomach morphology of *T. macrocercus*, which is scansorial, nocturnal, and vermivorous, described and compared with the arboreal and insectivorous *Sommeromys* and other Sulawesian endemics by Musser and Durden (2002).

Tateomys rhinogradoides Musser, 1969. Am. Mus. Novit., 2384:3.

> COMMON NAME: Tate's Sulawesian Shrew Rat.
> TYPE LOCALITY: Indonesia, C Sulawesi, Gunung Latimodjong, 2200 m.
> DISTRIBUTION: Known from the type locality, Gunung Tokala, and Gunung Nokilalaki at high elevations in tropical upper montane rain forest in C core of Sulawesi; most likely will be found on other mountains in C Sulawesi in similar forest formations.
> STATUS: IUCN – Vulnerable.
> COMMENTS: This terrestrial vermivore is the largest-bodied of the species of *Tateomys* and *Melasmothrix*. Stomach morphology closely similar to that of *Tateomys macrocercus* and *Melasmothrix naso* and unlike the large-bodied, Sulawesian shrew rat, *Echiothrix centrosa*, which occurs in lowland tropical evergeen rain forest at lower altitudes in the same region (Musser and Durden, 2002).

Thallomys Thomas, 1920. Ann. Mag. Nat. Hist., ser. 9, 5:141.

> TYPE SPECIES: *Mus nigricauda* Thomas, 1882.
> COMMENTS: *Oenomys* Division. After being proposed as a genus by Thomas (1920a), *Thallomys* was used in checklists (e.g., G. M. Allen, 1939; Ellerman, 1941) until Ellerman et al. (1953) united it with subgenus *Aethomys* of *Rattus*. *Thallomys* was reinstated by Lundholm (1955c), who also suggested it was closely related to *Thamnomys*. Misonne (1969) pointed out that *Thallomys* has nothing to do with *Rattus*, and is most closely related to Tertiary European *Parapodemus*, an evaluation based on molar occlusal patterns. Analysis of microcomplement fixation of albumin groups *Thallomys* with *Grammomys*, *Pelomys*, *Lemniscomys*, and *Rhabdomys* (Watts and Baverstock, 1995a). Although unrelated to *Rattus*, the phylogenetic position of *Thallomys* within the diversity of African murines is still unresolved; our allocation to the *Oenomys* Division must be tested by analyses of gene sequences and suites of morphological traits. Gastric anatomy described by Perrin (1986).
> Early checklists and faunal accounts recognized several species (Ellerman, 1941; Roberts, 1951), but for the last 30 years, *Thallomys* has been either treated as monotypic (D. H. S. Davis, 1965; Meester et al., 1986; Misonne, 1974), or containing two species (F. Petter, 1973a; Skinner and Smithers, 1990). That considerable and significant morphological diversity exists in the genus has been acknowedged. D. H. S. Davis (1965), for example, recognized five groups of subspecies, and F. Petter (1973a) demonstrated that at least two species could be diagnosed. Regional faunal reports of Southern African Subregion mammals discussed presence of two distinct kinds (Skinner and Smithers, 1990; Smithers and Wilson, 1979). Meester et al. (1986) announced the need for revision of *Thallomys*, and that both morphological and chromosomal data suggested the presence of two species (see Gordon, 1987). Recently, Taylor et al. (1995:59; also see review in Taylor, 2000b) used morphometric analyses of southern African samples that "clearly confirms the existence of two biological species, establishes a reliable morphometric criterion for separating species based on cranial characters, and redefines species distribution limits with a reasonable degree of accuracy." They also noted that "A complete revision of the genus. . . awaits further multidisciplinary studies involving karyotypes, morphometrics, external characters and genetic/molecular studies, based on samples from throughout

the range of *Thallomys* in Africa and including an assessment of all available type specimens." We could not agree more. We list four species. Our review is based on original descriptions of taxa, Roberts' (1951) useful monograph and other regional faunal reports
and revisions, large collections of specimens, and holotypes. *Thallomys* is represented by
fossils recovered from late Pliocene and early Pleistocene sediments in East Africa (Denys,
1987a; Jaeger, 1976; Jaeger and Wesselman, 1976; Wesselman, 1984) and Pleistocene formations in South Africa and Namibia (Avery, 1998, 2000; Senut et al., 1992); see review by
Denys (1999).

Thallomys loringi (Heller, 1909). Smithson. Misc. Coll., 52:471.
 COMMON NAME: Loring's Thallomys.
 TYPE LOCALITY: E Kenya, Lake Naivasha.
 DISTRIBUTION: Known only by specimens from E Kenya, and N and E Tanzania; limits
  unknown.
 STATUS: IUCN – Lower Risk (lc).
 COMMENTS: Originally described as a species of *Thamnomys* (Heller, 1909), then listed as a
  subspecies of *Thallomys nigricauda* (G. M. Allen, 1939; Ellerman, 1941), and finally
  allocated to *T. damarensis* (here included in *nigricauda*) as a subspecies (F. Petter,
  1973a). *Thallomys loringi* is morphologically and probably phylogenetically allied to
  T. nigricauda, but no critical documentation of its conspecificity with that species has
  ever been presented. Because *T. loringi* can be diagnosed by pelage and other distinctions, we list it as a species, a hypothesis that can be tested by careful systematic revision of samples from regions outside of southern Africa. Morphometric analyses
  presented by Taylor et al. (1995) identified a skull from Kenya as possibly being *T. nigricauda*, which highlights the need to determine the relationship between that species
  and *T. loringi*.

Thallomys nigricauda (Thomas, 1882). Proc. Zool. Soc. Lond., 1882:266.
 COMMON NAME: Black-tailed Thallomys.
 TYPE LOCALITY: Namibia, Great Namaqualand, Hountop (= Hudup or Hutop) River, west of
  Gibeon (Meester et al., 1986).
 DISTRIBUTION: W and S Angola, Namibia, N South Africa, Zimbabwe, N Botswana, and
  SE Zambia; northern and eastern limits unknown. The range described here is mapped
  by Taylor et al. (1995), who also identified a skull from Kenya as possibly an example
  of *T. nigricauda*.
 STATUS: IUCN – Lower Risk (lc).
 SYNONYMS: *bradfieldi* Roberts, 1933; *damarensis* (de Winton, 1897); *davisi* Lundholm, 1955;
  herero Thomas, 1926; *kalaharicus* (Dollman, 1911); *leuconoe* Thomas, 1926; *molopensis*
  Roberts, 1933; *nitela* Thomas and Hinton, 1923; *quissamae* F. Petter and Beaufort, 1960;
  robertsi (Ellerman, Morrison-Scott, and Hayman, 1953).
 COMMENTS: *Thallomys nigricauda* and *T. paedulcus* are now recognized as occurring in the
  Southern African Subregion by Skinner and Smithers (1990), who also summarized
  some of the chromosomal, morphological, and ecological distinctions between the
  two species. They also suggested that *herero* and *leuconoe* may represent samples of
  T. paedulcus, but we examined the holotypes and they are examples of *T. nigricauda*.
  Musser and Carleton (1993:669) wrote that "Morphological and geographic
  definitions of *T. nigricauda* are unsatisfactory. Appreciable geographic variation in
  body size, length of molar row, pelage coloration, and tail pilosity exists among
  samples and its significance will have to be assessed by critical systematic revision."
  Results of such a study are presented by Taylor et al. (1995), who employed multivariate analyses of southern African samples already identified by chromosomal traits
  as either *T. nigricauda* (2n = 47-50) or *T. paedulcus* (2n = 43-47) and documented
  morphometric traits distinguishing samples of the two species, provided clearer
  resolution of their geographic distributions, and demonstrated broad sympatry in
  ranges. Crawford-Cabral (1998) reviewed and mapped Angolan records, and P. J.
  Taylor (1998) discussed specimens from KwaZulu-Natal.

Thallomys paedulcus (Sundevall, 1846). Ofv. K. Svenska Vet.-Akad. Forhandl. Stockholm, 3:120.
 COMMON NAME: Acacia Thallomys.

TYPE LOCALITY: South Africa, "In Caffraria interiore, prope tropicum," (Ellerman et al., 1953); "type locality was in the Magaliesberg area and has been provisionally fixed as Crocodile Drift, Brits, Transvaal [= Gauteng]" (D. H. S. Davis, 1965:127).

DISTRIBUTION: From NE South Africa (N KwaZulu-Natal, W Mpumalanga, Limpopo, Gauteng, and North West), Swaziland, and Botswana north through Zimbabwe, S Zambia, Mozambique, Malawi, Tanzania, Kenya to S Ethiopia and S Somalia; limits unknown. See Taylor et al. (1995) for distribution in Southern African Subregion.

STATUS: IUCN – Lower Risk (lc).

SYNONYMS: *acaciae* (Roberts, 1915); *lebomboensis* Roberts, 1931; *moggi* (Roberts, 1913); *rhodesiae* (Osgood, 1910); *ruddi* (Thomas and Wroughton, 1908); *scotti* Thomas and Hinton, 1923; *somaliensis* (Roche, 1964); *stevensoni* Roberts, 1933; *zambesiana* Lundholm, 1955.

COMMENTS: In body size, the smallest of all the species. Identification of *paedulcus* as a separate species compared with the larger *T. damarensis* and association of *scotti*, were correctly perceived and documented by F. Petter (1973a). Thomas and Wroughton's (1908) *ruddi*, described as a species of *Thamnomys*, is a *Thallomys* (Lawrence and Loveridge, 1953) and another example of *T. paedulcus*. Identification of the holotype of *paedulcus*, along with critical measurements, recorded by Ellerman et al. (1953) and verified by F. Petter (1973a). The names *acaciae*, *lebomboensis*, and *stevensoni* were listed by Roberts (1951) as subspecies of *T. moggi*; *rhodesiae* was described by Osgood (1910) as a subspecies of *Mus damarensis* (which is here synonymized in *T. nigricauda*); and *zambesiana* was proposed by Lundholm (1955a) as a subspecies of *T. nigricauda*. Roche (1964) described *somaliensis* as a subspecies of *T. paedulcus*. Morphometric contrasts between *T. paedulcus* and *T. nigricauda* provided by Taylor et al. (1995), who also identified a skull from Kenya as *T. paedulcus* (a different skull than the one from Kenya identified by Taylor et al., 1995, as *T. nigricauda*; see that account). Because Taylor et al. (1995) found *T. nigricauda* to range widely through the Transvaal, they questioned the identity of the holotype of *paedulcus*, even though specimens from 3 km NNE Brits had 2n = 46, within the range of *T. paedulcus*. Neither the holotype of *paedulcus* nor that of *nigricauda* was included in the analyses made by Taylor et al. (1995). Denys et al. (1999) found fragments of *T. paedulcus* in owl pellets in N Malawi. Southern African population reviewed by Lovegrove (1997a), and records from KwaZulu-Natal discussed by P. J. Taylor (1998).

Thallomys shortridgei Thomas and Hinton, 1923. Proc. Zool. Soc. Lond., 1923:492.

COMMON NAME: Shortridge's Thallomys.

TYPE LOCALITY: South Africa, Western Cape Province, Louisvale, S bank of Orange River.

DISTRIBUTION: South Africa; known only from south bank of Orange River from about Upington west to Goodhouse, Little Namaqualand, in Western Cape Province; limits unknown.

STATUS: IUCN – Lower Risk (nt).

COMMENTS: Thomas and Hinton's (1923b) acute description points to a distinctive species distinguished by a diagnostic combination of chromatic and cranial traits (especially the small bullae). Ellerman (1941) treated *shortridgei* as a species, its distinction has been recognized by others (Musser and Carleton, 1993; Roberts, 1951; Van Rooyen *in* De Graff, 1978), and despite recent checklists where it is listed as a subspecies of either *T. paedulcus* (Meester et al., 1986) or *T. nigricauda* (Skinner and Smithers, 1990), the name identifies a valid species (our study of the holotype and specimens in MCZ).

Thamnomys Thomas, 1907. Ann. Mag. Nat. Hist., ser. 7, 19:121.

TYPE SPECIES: *Thamnomys venustus* Thomas, 1907.

COMMENTS: *Oenomys* Division. Many have asserted that species of *Thamnomys* and *Grammomys* are in the same monophyletic group and separable only at the subgeneric level (G. M. Allen, 1939; Hatt, 1940a; Hollister, 1919; Misonne, 1974; F. Petter and Tranier, 1975; and others, see references in Meester et al., 1986). *Thamnomys* is a distinct genus, however, as explained by Ellerman (1941), Hutterer and Dieterlen (1984), Misonne (1969), Rosevear (1969), and other workers (see references in Meester et al., 1986). Those systematists also included *rutilans*, which is now regarded to be a synonym of *Grammomys poensis* (see that account). The considerable geographic and individual variation in body size, pelage coloration, and craniodental dimensions in *Thamnomys* needs to be assessed in a careful systematic revision of the genus. Until then we recognize the three species listed below based upon our examination of specimens and preliminary principle components analysis of cranial and dental measurements.

Oenomys is the closest phylogenetic relative, a view presented more than fifty years ago by Hatt (1940*a*:522): "*Oenomys* and *Thamnomys* bear much resemblance to each other, and I am inclined to believe that they represent respectively semi-arboreal and arboreal descendants of a common stock." Hatt's insight has recently been supported by a phylogenetic study incorporating 17 cranial and 23 dental traits (Lopez-Martinez et al., 1998).

Thamnomys kempi Dollman, 1911. Ann. Mag. Nat. Hist., ser. 8, 8:658.

> COMMON NAME: Dollman's Thamnomys.
> TYPE LOCALITY: E Dem. Rep. Congo, Nord-Kivu Province, Buhamba, near Lake Kivu, 6000 ft (1830 m).
> DISTRIBUTION: Apparently endemic to Kivu and Virunga volcanos in E Dem. Rep. Congo (Dollman, 1911; Gyldenstolpe, 1928; Hatt, 1934; Rahm, 1967; specimens in AMNH, MCZ, USNM), SW Uganda (specimens in FMNH, MCZ, USNM), W Rwanda (Elbl et al., 1966; specimens in USNM), and E Burundi (specimens in FMNH); 1670-3900 m.
> STATUS: IUCN – Lower Risk (lc).
> COMMENTS: The more than 30 specimens we examined indicate *T. kempi* should be treated as a species (G. M. Allen, 1939; Ellerman, 1941), not a subspecies of *T. venustus* (Misonne, 1974; Rahm, 1967). *Thamnomys kempi* is larger, with a longer skull and molar row (29.5-38.9 mm, 6.1-6.8 mm, respectively in *T. kempi*; 30.8-35.5 mm, 5.4-5.8 mm in *T. venustus*) and generally occurs at higher altitudes. Both are sympatric at Kibati (specimens in MCZ) and Irangi (R. Hutterer, in litt., 2002) in the Kivu highlands of E Dem. Rep. Congo.

Thamnomys major Hatt, 1934. Am. Mus. Novit., 708:10.

> COMMON NAME: Hatt's Thamnomys.
> TYPE LOCALITY: E Dem. Rep. Congo, Nord-Kivu Province, Lukumi, N slope Mt Karisimbi, Kivu volcanos, 12,000 ft (3658 m).
> DISTRIBUTION: Known only from the type locality.
> COMMENTS: Still represented only by the holotype (AMNH 82693). Originally described as a subspecies of *Thamnomys kempi*, but the holotype is significantly larger than any specimen of *T. kempi* we have examined, has saturated buffy underparts suffused with underlying gray (whitish gray or with buff tinge in *T. kempi*), and conspicuously longer molar rows (7.5 mm for *major*, 6.1-6.8 for 31 *T. kempi* examined). Gyldenstolpe (1928) described a specimen collected at 3900 m on Mt Karisimbi and mentioned that it was larger than the holotype of *T. kempi*, which suggested to Hatt (1934) that it might be another example of *major*. Dimensions of Gyldenstolpe's specimen, however, fall within the range of variation characteristic of *T. kempi*. Hatt (1934:10) designated AMNH 82695 and 82689 "from Kabara, a site at 3354 m on the saddle between Mtns. Mikeno and Karisimbi" as paratypes of *major*, but their cranial and dental dimensions along with venter coloration are also typical of *T. kempi*.

Thamnomys venustus Thomas, 1907. Ann. Mag. Nat. Hist., ser. 7, 19:122.

> COMMON NAME: Thomas's Thamnomys.
> TYPE LOCALITY: Uganda, Mubuku Valley, Ruwenzori East, 7000 ft (2134 m); see G. M. Allen and Loveridge (1942) for details.
> DISTRIBUTION: From lowlands of NE Dem Rep. Congo at Medje (617 m; specimens in AMNH) south through Ruwenzori Mtns of W Uganda (G. M. Allen and Loveridge, 1942; Delany, 1975; Thomas, 1907*a*; specimens in FMNH and MCZ) to the Kivu volcanos in E Dem. Rep. Congo (G. M. Allen and Loveridge, 1942; specimens in MCZ), 600-2100 m; distributional limits unresolved.
> STATUS: IUCN – Lower Risk (lc).
> SYNONYMS: *kivuensis* G. M. Allen and Loveridge, 1942; *schoutedeni* Hatt, 1934.
> COMMENTS: An appreciably smaller-bodied species than *T. major* (see that account) and recorded from a more expansive range and generally at lower altitudes. Both *kivuensis*

(G. M. Allen and Loveridge, 1942) and *schoutedeni* (Hatt, 1934) were described as a subspecies of *T. venustus*. Altitudinal distribution on Ugandan slopes of Ruwenzori Mtns reviewed by Kerbis Peterhans et al. (1998) and compared with distributions of entire small mammal fauna in those highlands.

Tokudaia Kuroda, 1943. Biogeographica, 139:61.
> TYPE SPECIES: *Rattus jerdoni osimensis* Abe, 1933 (see Kaneko and Maeda, 2002).
> SYNONYMS: *Acanthomys* Tokuda, 1941 [not Lesson, 1842, or Gray, 1867]; *Tokudamys* Johnson, 1943.
> COMMENTS: *Apodemus* Division. Kaneko (2001) provided the first detailed morphological study comparing specimens (of all age groups) from S Okinawa Isl (*muenninki*) with those from N Amami-oshima Isl (*osimensis*). The significant morphological differences he documented parallel impressive chromosomal distinctions between the two insular populations (Honda et al., 1978; Tsuchiya, 1981; Tsuchiya et al., 1989), which support the specific status of each as listed by Musser and Carleton (1993). Kaneko also detailed the taxonomic history of the names now associated with *Tokudaia*.
>
> Based on molar morphology, Kawamura (1989) placed *Tokudaia* in a group containing *Apodemus*, Pliocene *Rhagapodemus*, and Quaternary *Rhagamys*; and suggested that *Tokudaia* evolved from Miocene *Parapodemus-Apodemus* ancestral stock; its evolutionary link to primitive species of *Apodemus*, Kawamura suggested, may be found in Miocene or Pliocene sediments of China. Analyses of mtDNA cytochrome *b* and nuclear IRBP gene sequences from samples of *Tokudaia*, *Apodemus*, *Micromys*, *Mus*, and *Rattus* by Suzuki et al. (2000:15) suggested that they "diverged at similar evolutionary times, namely at the time of the radiation of Murinae". This suggestion is not so different from that supposed by Kawamura (1989) since it is in the Miocene that the initial radiation of Murinae occurred, and species of primitive *Apodemus* are among the basal members (see generic account of *Apodemus*). The close phylogenetic association between *Tokudaia* and *Apodemus* has been reinforced by analyses of sequences from mtDNA cytochrome *b* and 12S rRNA genes (Michaux et al., 2002a).
>
> The species of *Tokudaia* join the murine *Diplothrix legata* and leporid *Pentalagus furnessi* as endemics to the Ryukyu Isls (Dobson, 1994; Kaneko, 1994; Suzuki et al., 1999a). For the rodents, Suzuki et al. (2000:23) noted that "the genetic constitution of the rodents in the Okinawa area is distinct from that of the rodents in the surrounding areas, including the Japanese mainland and Taiwan. These findings imply that, in the region of the Central Ryukyus, mammals with a unique and ancient origin have inhabited a small geographic region."

Tokudaia muenninki (Johnson, 1946). Proc. Biol. Soc. Wash., 59:170.
> COMMON NAME: Okinawa Island Spiny Rat.
> TYPE LOCALITY: Japan, Ryukyu Isls (= Nansei Isls), N Okinawa Isl, Hentona (western coast).
> DISTRIBUTION: Known by modern specimens from N Okinawa, and late Pleistocene and Holocene samples from Okinawa and adjacent island of Le-jima (Kawamura, 1989; Kowalski and Hasegawa, 1976).
> STATUS: IUCN – Critically Endangered.
> COMMENTS: Originally described as a subspecies of *T. osimensis* (Johnson, 1946a), but chromosomal evidence (2n = 44 for *muenninki*, 2n = 25 for *osimensis*), as well as external and cranial morphology distinguish *muenninki* as a separate species (Kaneko, 2001; Tsuchiya, 1981, Tsuchiya et al., 1989). Based upon molar measurements, Kaneko (2001) suggested that the late Pleistocene specimens identified as *T. osimensis* by Kowalski and Hasegawa (1976) and Kawamura (1989, 1991, 1994) actually represent *T. muenninki* and *T. osimensis*, indicating that both species occurred on Okinawa during late Pleistocene. Kaneko cautioned, however, that the fossils must be reexamined to confirm these possible identifications.
>
> A population of *Tokudaia* also occurs on Tokuno-shima Isl, south of Amami-oshima and north of Okinawa. Chromosomal distinctions between *osimensis* (2n = 25) from Amami-oshima, *muenninki* (2n = 44) from Okinawa, and the population from Tokuno-shima Isl (2n = 45) were documented by Honda et al. (1978), Tsuchiya (1981) and Tsuchiya et al. (1989). While that data suggested the Tokuno-Shima Isl sample to

represent a third species, Kaneko unfortunately could not include the population in his morphometric study because it is protected and no voucher specimens have been preserved. Significant differences between samples from Tokuno-shima and Amami-oshima in mtDNA cytochrome *b* sequences and restriction fragment length polymorphism in the nuclear ribosomal RNA gene also led Suzuki et al. (1999*a*) to regard the two insular populations as independent species. Molecular data from *T. muenninki* and morphological information from the population on Tokuno-Shima Isl must be analyzed to test the results implied by present chromosomal information.

Tokudaia osimensis (Abe, 1933). Botany and Zoology, 1:942.
COMMON NAME: Amami-oshima Island Spiny Rat.
TYPE LOCALITY: Japan, Ryukyu Isls (= Nansei Isls), Amami-oshima Isl, Mt Kiyago-kan, village of Sumiyo.
DISTRIBUTION: Known only by modern samples from Amami-oshima Isl.
STATUS: IUCN – Endangered.
COMMENTS: Abe, 1934, J. Sci. Hiroshima Univ., ser. B, div. 1, 3:107, is the usual date and citation for *osimensis* (see Corbet and Hill, 1992 and Musser and Carleton, 1993), but the name was proposed a year earlier in a different journal according to Kaneko (2001); also see Kaneko and Maeda (2002).The unique chromosomal complement of this species (2n = 25, with no X in the female or visible Y in the male) first documented by Honda et al. (1977) and corroborated by Kimiyuki et al. (1989). Testes devopment depends upon inheritance of the *Sry* gene encoded on the Y chromosome, but *T. osimensis* lacks this gene (Suzuki et al., 1999*b*), a phenomenon found elsewhere among murids only in species of the arvicoline *Ellobius* (Just et al., 1995). Suzuki et al. (1999*b*:590) noted that to describe the mechanism of sex determination in these species "will require a more precise genetic analysis . . . and will bring us new information on sex determination mechanisms, evolution, and plasticity." Reviewed by Kaneko (1994), who also provided a beautiful color photograph of the live rat.

Tryphomys Miller, 1910. Proc. U.S. Natl. Mus., 38:399.
TYPE SPECIES: *Tryphomys adustus* Miller, 1910.
COMMENTS: *Rattus* Division. Musser and Newcomb (1983) and Musser and Heaney (1992) revised the genus and documented the history of its past association with *Rattus*; identified as a Philippine New Endemic (Musser and Heaney, 1992). *Tryphomys* and *Abditomys*, another Luzon endemic, form a monophyletic cluster; their phylogenetic alliances to other genera of Philippine New Endemics, or to genera native to other regions in the Indo-Australian region, remain unresolved.

Tryphomys adustus Miller, 1910. Proc. U.S. Natl. Mus., 38:399.
COMMON NAME: Luzon Tryphomys.
TYPE LOCALITY: Philippines, Luzon Isl, Benguet, Haights-in-the-Oaks.
DISTRIBUTION: Greater Luzon Faunal Region. Endemic to Luzon; known only from a few localities in highlands of the Central Cordillera and lower portion of Mount Makiling (Barbehenn et al., 1972-1973; Heaney et al., 1998; Sanborn, 1952*a*).
STATUS: IUCN – Vulnerable.
COMMENTS: Nothing is known about ecology of this species, but its morphology indicates *T. adustus* is terrestrial, probably inhabiting grassy or shrubby clearings in forests.

Uromys Peters, 1867. Monatsb. K. Preuss. Akad. Wiss. Berlin, p. 343.
TYPE SPECIES: *Mus macropus* Gray, 1866 (= *Hapalotis caudimaculatus* Krefft, 1867).
SYNONYMS: *Cyromys* Thomas, 1910; *Gymnomys* Gray, 1867; *Melanomys* Winter, 1983 [not Thomas, 1902].
COMMENTS: *Uromys* Division. Member of the New Guinea and Australian Old Endemics (Musser, 1981*c*). Using albumin immunology, Watts and Baverstock (1994*a*) clustered *Uromys* with *Melomys* and *Solomys* within a larger clade that includes members of our *Hydromys*, *Xeromys*, and *Pseudomys* Divisions (the "Australasian clade" of Watts and Baverstock, 1995*b*, 1996), which occur on New Guinea and Australia, and not in our endemic New Guinea *Pogonomys* Division or *Lorentzimys* Division. This cladistic

configuration is consistent with chromosomal traits (Baverstock et al., 1977c) and sperm morphology (Breed, 1997; Breed and Aplin, 1994). Revised by Groves and Flannery (1994), who arranged the species in subgenera *Uromys* and *Cyromys* (type species = *Mus imperator*).

Uromys anak Thomas, 1907. Ann. Mag. Nat. Hist., ser. 7, 20:72.
    COMMON NAME: Black-tailed Uromys.
    TYPE LOCALITY: Papua New Guinea, Central Province, Brown River, Efogi, "not less than" 4000 ft (1220 m).
    DISTRIBUTION: New Guinea; throughout the Central Cordillera from Weyland Range in the west to Mt Dayman in the east as well as the Huon Peninsula; not recorded from any of the N coastal ranges or the Vogelkop region; 850-2950 m (Flannery, 1995b).
    STATUS: IUCN – Lower Risk (nt).
    SYNONYMS: *albiventer* Groves and Flannery, 1994; *rothschildi* Thomas, 1912.
    COMMENTS: Subgenus *Uromys*. Closely related to *U. neobritannicus*. Geographic variation of morphological traits analyzed by Groves and Flannery (1994) in context of a revision of *Uromys*. Sperm morphology documented by Breed and Aplin (1994). Reviewed by Flannery (1995a).

Uromys boeadii Groves and Flannery, 1994. Rec. Aust. Mus., 46:157.
    COMMON NAME: Biak Island Uromys.
    TYPE LOCALITY: Indonesia, Prov. of Papua (= Irian Jaya), Geelvinck Bay, Pulau Biak, 25 km northeast of Biak (town), 65 m (Flannery, 1995b).
    DISTRIBUTION: Known only by the holotype collected on Pulau Biak.
    COMMENTS: Subgenus *Uromys*. Regarded as a primitive member of the subgenus (Flannery, 1995b) and "the plesiomorphic sister-group to all other species of subgenus *Uromys*" (Groves and Flannery (1994:164).

Uromys caudimaculatus (Krefft, 1867). Proc. Zool. Soc. Lond., 1867:316.
    COMMON NAME: Giant White-tailed Uromys.
    TYPE LOCALITY: Australia, Queensland, Cape York (see Mahoney and Richardson, 1988).
    DISTRIBUTION: Australia: NE coastal Queensland in tropical forests from Townsville area north to tip of Cape York, and a few islands off the coast of N Queensland (Moore, 1995:640; Watts and Aslin, 1981:91). New Guinea: widespread throughout lowland and midmontane regions on the mainland, sea level to 1925 m; also on Aru Isls, Kai Isls, Waigeo Isl, Yapen Isl, and Normanby and Fergusson in the D'Entrecasteaux Arch. (Flannery, 1995a, b; Leary and Seri, 1997).
    STATUS: IUCN – Lower Risk (lc).
    SYNONYMS: *aruensis* Gray, 1873; *ductor* Thomas, 1913; *exilis* Troughton and Le Soeuf, 1929; *lamington* Troughton, 1937; *macropus* (Gray, 1866) [not Hodgson, 1845]; *multiplicatus* (Jentink, 1907); *nero* Thomas, 1913; *papuanus* (Ramsay, 1883) [not von Meyer, 1876, a nomen nudum]; *prolixus* Thomas, 1913; *scaphax* Thomas, 1913; *sherrini* Thomas, 1923; *validus* Peters and Doria, 1881; *waigeuensis* Frechkop, 1932.
    COMMENTS: Subgenus *Uromys*. The Australian population has been studied from viewpoints of chromosomal morphology (Baverstock et al., 1977c), heterochromatin variation (Baverstock et al., 1976b, 1982), electrophoretic data (Baverstock et al., 1981), G-banding homologies (Baverstock et al., 1983b), morphology of male reproductive tract (Breed, 1986), and spermatozoal structure (Breed, 1984; Breed and Sarafis, 1978). Donnellan (1987) provided chromosomal information for samples from New Guinea, Breed and Aplin (1994) reported spermatozoal morphology, and Lidicker (1968) described phallic anatomy. Morphology of gastrointestinal tract and its significance covered by Comport and Hume (1998). Mahoney and Richardson (1988:189) cataloged taxonomic, distributional, and biological references covering Australian populations. Two different chromosomal forms of Australian *U. caudimaculatus* exist, one extending from McIlwraith Ranges northward, the other from Cooktown southward; they are separated by a 200 km break in rainforest (Baverstock et al., 1976b, 1977c; Donnellan, 1989). Significance of morphological variation among samples from mainland New Guinea assessed in the context of a systematic revision of *Uromys* (Groves and Flannery, 1994), but their results should be tested by new analyses. The

taxa *nero* and *scaphax*, for example, which were treated as synonyms of *U. caudimaculatus multiplicatus* by Groves and Flannery (1994:153), are in K. Helgen's (in litt., 2004; he has recently studied holotypes and other specimens at the BMNH) ". . . assessment extremely distinctive and represent biological species not immediately allied to *caudimaculatus . . . .*" Australian populations reviewed by Moore (1995), New Guinea by Flannery (1995*a*, *b*). Aplin et al. (1998) reported material of this species from a late Pleistocene archaeological site on the Ayamaru Plateau, central Bird's Head Peninsula of Prov. of Papua (= Irian Jaya).

Uromys emmae Groves and Flannery, 1994. Rec. Aust. Mus., 46:159.
COMMON NAME: Emma's Uromys.
TYPE LOCALITY: Indonesia, Prov. of Papua (= Irian Jaya), Geelvinck Bay, Schouten Isl Group, Pulau Awai (Owi Isl).
DISTRIBUTION: Recorded only from Pulau Awai, but may also occur on adjacent Biak and Supiori Isls (Flannery, 1995*b*).
COMMENTS: Subgenus *Uromys*. Still known only by the holotype. Morphologically and phylogenetically most closely related to the Australian-New Guinea *U. caudimaculatus* and NE Australian *U. hadrourus* (Groves and flannery, 1994).

Uromys hadrourus (Winter, 1984). Mem. Queensland Mus., 21:519.
COMMON NAME: Masked White-tailed Uromys.
TYPE LOCALITY: Australia, NE Queensland, Thornton Peak summit area, 1200 m; see Winter (1984) and Mahoney and Richardson (1988:163) for additional information.
DISTRIBUTION: Originally known by only a few specimens from rainforest on Mareeba Franite of the Thornton Peak massif north of the Daintree River valley (Winter, 1983, 1984), but now also recorded south of the Daintree River in the Mount Carbine Tableland and farther south in the Lamins Hill area of the Atherton Tableland (Winter and Moore, 1995).
STATUS: IUCN – Lower Risk (nt).
COMMENTS: Subgenus *Uromys*. Originally described as a species of *Melomys*, subsequently placed in *Uromys* and allied with *U. caudimaculatus* (Groves and Flannery, 1994). *Uromys hadrourus* is a member of a group of species that are endemic to "the Townsville to Cooktown region, and considered to be relicts of a wet- and cool-adapted fauna which may have originated in Australia from a common pre-Pleistocene stock of Australia and New Guinea" (Winter, 1984:525). Chromosomal morphology reported by Baverstock et al. (1977*c*). McAllan and Bruce (1989) claimed 1983 to be the publication date for *hadrourus* instead of 1984 as is usually accepted, noting that the species was first given the name *Melanomys hadrourus*. That name was in the title of report cited by Winter (1983:379) as "in press" and placed in a sidebar and not in the main account, which was headed by the name *Melomys* sp. *Melanomys* is a *nomen nudum* as entered in Winter (1983) and has nothing to do with *Melanomys* Thomas, 1902, which is a sigmodontine. Mahoney and Richardson (1988:163) maintain 1984 as publication date. Reviewed by Winter and Moore (1995).

Uromys imperator (Thomas, 1888). Ann. Mag. Nat. Hist., ser. 6, 1:157.
COMMON NAME: Emperor Uromys.
TYPE LOCALITY: Solomon Isls, Guadalcanal Isl, Aola.
DISTRIBUTION: Endemic to Guadalcanal Isl.
STATUS: IUCN – Extinct. May be extinct (see Flannery, 1991; Groves and Flannery, 1994).
COMMENTS: Subgenus *Cyromys*. Type species of *Cyromys*; historically assigned to either genus *Uromys* or *Cyromys*, but is now regarded as a member of the former (Flannery and Wickler, 1990; Groves and Flannery, 1994). Known only by a few historical and archaeological specimens (Flannery, 1991; Groves and Flannery, 1994). Closest relative is apparently *U. rex*, also endemic to Guadalcanal Isl. Reviewed by Flannery (1995*b*).

Uromys neobritannicus Tate and Archbold, 1935. Am. Mus. Novit., 803:4.
COMMON NAME: New Britain Island Uromys.
TYPE LOCALITY: Papua New Guinea, Bismarck Arch., New Britain Isl.
DISTRIBUTION: Endemic to New Britain Isl.

STATUS: IUCN – Lower Risk (lc).

COMMENTS: Subgenus *Uromys*. Included in *U. anak* by Ziegler (1982b:882) without explanation, but *neobritannicus* is diagnosed by distinctive external and cranial traits that are outside the range of morphological variation characteristic of *U. anak* (Groves and Flannery, 1994). Chromosomal data reported by Donnellan (1987). Still known only by a few specimens. Extensive survey of living and fossil mammals on nearby New Ireland did not yield the species, and Groves and Flannery (1994) think it unlikely that *U. neobritannicus* inhabits any of the smaller islands adjacent to New Britain.

Uromys porculus Thomas, 1904. Ann. Mag. Nat. Hist., ser. 7, 14:400.

COMMON NAME: Guadalcanal Uromys.

TYPE LOCALITY: Solomon Isls, Guadalcanal Isl, Aola.

DISTRIBUTION: Endemic to Guadalcanal Isl.

STATUS: IUCN – Extinct. Probably extinct (see Flannery, 1995b).

COMMENTS: Subgenus *Cyromys*. Although originally described as a species of *Uromys*, it was transferred to *Melomys* (Ellerman, 1941; Rümmler, 1938), placed once more in *Uromys* (Tate, 1951), again put in *Melomys* (Laurie and Hill, 1954), and currently returned to *Uromys* (Flannery and Wickler, 1990; Groves and Flannery, 1994). Still known only by the holotype. Morphologically closely related to *U. rex* and *U. imperator*. Reviewed by Flannery (1995b).

Uromys rex (Thomas, 1888). Ann. Mag. Nat. Hist., ser. 6, 1:157.

COMMON NAME: King Uromys.

TYPE LOCALITY: Solomon Isls, Guadalcanal Isl, Aola.

DISTRIBUTION: Found only on Guadalcanal Isl.

STATUS: IUCN – Critically Endangered.

COMMENTS: Subgenus *Cyromys*. Another Guadalcanal species that has historically been placed in either genus *Uromys* or *Cyromys*, but is currently allocated to the former (Flannery and Wickler, 1990; Groves and Flannery, 1994). Still represented by less than a dozen specimens (Flannery, 1991), and likely the only species in subgenus *Cyromys* that is not extinct (Flannery, 1995b).

Uromys siebersi Thomas, 1923. Treubia, 3:422.

COMMON NAME: Great Key Island Uromys.

TYPE LOCALITY: Indonesia, Maluku Tengah, Pulau Kai Besar (Great Key Isl), Gunung Daab.

DISTRIBUTION: Recorded only from Pulau Kai Besar in Kepulauan Kai (Ewab), between Seram Isl and the Aru Isls.

COMMENTS: Originally described as a species but subsequently treated as a subspecies of *Uromys caudimaculatus* (Ellerman, 1941; Rümmler, 1938) or synonym of *U. caudimaculatus aruensis* (Laurie and Hill, 1954). In their revision of *Uromys*, Groves and Flannery (1994) identified *siebersi* as a taxon of "uncertain status" because they could not allocate it to any of the subspecies of *U. caudimaculatus* they recognized. Still known only by two skins and a single skull (Groves and Flannery, 1994), Thomas's *siebersi* should be highlighted as a species. Among its diagnostic characteristics are very short tail relative to head and body length, small slit-like incisive foramina, and distinctly bowed skull (K. Helgen, in litt., 2004). The phylogenetic relationship of *U. siebersi* with samples of *U. caudimaculatus* along with the other recognized species of *Uromys* should be critically assessed in the context of a new taxonomic evaluation of geographic variation within samples now referred to *U. caudimaculatus* and *Uromys* itself. *Uromys siebersi* joins *Melomys bannisteri* as endemics of Pulau Kai Besar. The Kai Isls are in deep water and not on the continental shelf connecting Australia and New Guinea beneath the Arafura Sea, and any endemic species such as *Melomys bannisteri* and *Uromys siebersi* are likely the product of a longer evolutionary history in isolation than is true of faunas on continental shelf islands associated with Australia and New Guinea.

Vandeleuria Gray, 1842. Ann. Mag. Nat. Hist., [ser. 1], 10:265

TYPE SPECIES: *Mus oleraceus* Bennett, 1832.

COMMENTS: *Micromys* Division. An Indomalayan endemic allied to *Chiropodomys*, *Vernaya*, and

Micromys, a hypothesis based on dental morphology (Misonne, 1969). Ellerman (1949: 132) also tied *Vandeleuria* to *Chiropodomys* and *Micromys* through their complex molar occlusal patterns. Using albumin immunology, Watts and Baverstock (1995*b*) pointed to *Micromys* as the closest phylogenetic extant relative of *Vandeleuria*. Phylogenetic analysis of molar traits by Chaimanee (1998) indicated *Vandeleuria* formed a clade with *Pithecheir* and *Lenothrix*; the close alliance between *Vandeleuria* and *Pithecheir* was foretold by Ellerman (1949:132). Because of the immunological relationship between *Micromys* and *Vandeleuria*, the early split of *Micromys* from the central murine lineage represented by a *Pogonomys*-like ancestor (see account of *Micromys*), and the close cladistic relationship among the Indomalayan *Vandeleuria*, *Pithecheir*, and *Lenothrix* based upon molar traits, Chaimanee speculated that those three genera along with *Micromys* may be remnants of the "first divergent lineage in the history of Murinae" (1998:208). Fossils extracted from late Pleistocene cave sediments from the Sichuan-Guizhou region of S China were identified as *Vandeleuria* sp. (Zheng, 1993). Evolutionary history in Thailand, based upon isolated molars recovered from cave strata, extends back to late Pliocene (Chaimanee, 1998).

Vandeleuria nilagirica Jerdon, 1867. Mammals of India, p. 203.
   COMMON NAME: Nilgiri Vandeleuria.
   TYPE LOCALITY: SW India, F Tamil Nadu, Nilgiri Hills, Ootacamund.
   DISTRIBUTION: Recorded only from S end of the Western Ghats (= Sahyadris) in the Nilgiri Hills of SW Peninsular India (Corbet and Hill, 1992), but may occur farther north along the Western Ghats wherever tropical evergreen rain forest has not been eliminated.
   COMMENTS: Originally described as a species by Jerdon (1867:203), who wrote of having "on several occasions found this tree-mouse in woods on the summit of the Neelgherries, near Ootacamund." Jerdon was also familiar with *oleracea*, noting on the previous page that "This very pretty little mouse has been found in all parts of India, from the Himalayas to the extreme south." Subsequently retained as a species by Ellerman (1941), thereafter usually included in *V. oleracea* (Agrawal, 2000; Agrawal and Chakraborty, 1980; Ellerman, 1961; Musser and Carleton, 1993), but separated as a separate species by Corbet and Hill (1992) on the basis of the type description (no holotype exists) and a specimen from Kutta, S Coorg in the Nilgiri Hills (much longer tail than any sample of *V. oleracea* and whitish gray rather than pure white under-parts). A typical *V. oleracea* was also collected at Kutta. See Corbet and Hill (1992) for expanded discussion and comparisons. The murines *Mus famulus*, *Vandeleuria nilagirica* and *Rattus satarae* join the lion-tailed macaque (*Macaca silenus*) in being endemic to tropical rainforests along the Western Ghats.

Vandeleuria nolthenii Phillips, 1929. Ceylon J. Sci., Sec. B, 15:165.
   COMMON NAME: Sri Lankan Vandeleuria.
   TYPE LOCALITY: Sri Lanka, Uva, Ohiya, West Haputale, 6000 ft (1830 m).
   DISTRIBUTION: Forested highlands (above 1158 m) of Central and Uva provinces of Sri Lanka.
   STATUS: IUCN – Vulnerable.
   COMMENTS: Described as a subspecies of *V. nilagirica*, kept that way by Ellerman (1941), but usually considered a subspecies of *V. oleracea* (Agrawal and Chakraborty, 1980; Eller-man, 1961; Phillips, 1980). Howver, *nolthenii* is distinct in its montane distribution, pelage (longer, dark reddish brown upperparts, gray venter), and external and cranial traits (longer tail, larger skull with more inflated braincase, larger bullae, and longer and heavier rostrum) from *V. oleracea* occurring at lower elevations, and should be treated as a separate species (Musser, 1979), an arrangement endorsed by Corbet and Hill (1992). Phillips' (1929) original description and subsequent review (Phillips, 1980) encapsule all that is known about the species.

Vandeleuria oleracea (Bennett, 1832). Proc. Zool. Soc. Lond., 1832:121.
   COMMON NAME: Indomalayan Vandeleuria.
   TYPE LOCALITY: India, Madras, Deccan region.
   DISTRIBUTION: Recorded from Sri Lanka (lowlands; Phillips, 1980), peninsular India,

S Nepal, Burma (Ellerman, 1941), SE China (W Yunnan; Wang, 2003), Thailand (except peninsula south of Isthmus of Kra, 10°E,30′N; J. T. Marshall, Jr., 1977a), SW Cambodia (Cardamom Mtns; A. Smith, in litt., 2002), and S Vietnam (Osgood, 1932; Dang et al., 1994); probably occurs in S Laos in suitable habitat. See Corbet and Hill (1992).

STATUS: IUCN – Lower Risk (lc).

SYNONYMS: *badius* (Blyth, 1859); *domecolus* (Hodgson, 1841) [*nomen nudum*]; *dumeticola* (Hodgson, 1845); *marica* Thomas, 1915; *modesta* Thomas, 1914; *povensis* (Hodgson, 1845); *rubida* Thomas, 1914; *sibylla* Thomas, 1914; *scandens* Osgood, 1932; *spadicea* Ryley, 1914; *wroughtoni* Ryley, 1914.

COMMENTS: Musser and Carleton (1993) noted that "*oleracea* is possibly a composite of species, and despite Agrawal and Chakraborty's (1980) review of geographic variation, needs careful systematic revision." Chromosomal features vary geographically: 2n = 26 or 28 for N, NE and E Thailand samples (Gropp et al., 1972; Winking et al., 1979); 2n = 29 for N India sample (Sharma and Raman, 1972), and 2n = 28 for SW India (Prakash and Aswathanarayana, 1973, 1976). Agrawal (2000) reviewed Indian populations, pointing out that samples from N India have rusty brown upperparts while those from S India and Gujarat have a dull brown dorsum; slight chromosomal differences are concordant with the chromatic distribution. Morphological variation in what has been defined as *V. oleracea* excludes *nilagirica*, which Corbet and Hill (1992) treat as a separate species (see that account). Ecology and distribution in the Aravalli Ranges in Rajasthan, India documented by Prakash et al. (1995a, b) and in Gujarat State by Chakraborty and Agrawal (2000). Excellent description of climbing ability, diet, and other aspects for the Sri Lankan population provided by Phillips (1926, 1980). Chinese localities mapped and listed by Zhang et al. (1997), who also recorded a locality in N Sichuan far to the north of any other record, which is probably a misidentification. Although living populations occur in Thailand only north of the Isthmus of Kra (10°30′N), *V. oleracea* once ranged south into peninsular Thailand, and its evolutionary history in that country extends to late Pliocene (Chaimanee, 1998).

Vernaya Anthony, 1941. Field Mus. Nat. Hist. Publ., Zool. Ser., 27:110.

TYPE SPECIES: *Chiropodomys fulvus* G. M. Allen, 1927.

SYNONYMS: *Octopodomys* Sody, 1941.

COMMENTS: *Micromys* Division. Dental morphology interpreted by Misonne (1969) to indicate close relationship with *Chiropodomys*; hypothesis requires testing with other characters. About the same time Anthony described *Vernaya*, Sody (1941) proposed *Octopodomys*. There is a single extant species, the remains of which have been found in late Pleistocene cave sediments in the Sichuan-Guizhou region of S China, and four extinct species have been described based on early to late Pleistocene fossils from the same region (Zheng, 1993).

Vernaya fulva (G. M. Allen, 1927). Am. Mus. Novit., 270:11.

COMMON NAME: Vernay's Climbing Mouse.

TYPE LOCALITY: China, Yunnan, Yinpankai, Mekong River.

DISTRIBUTION: S China (N Sichuan, W Yunnan, S Gansu, and SW Shaanxi; Li and Wang, 1995) and N Burma (Anthony, 1941); recorded only above 2135 m.

STATUS: IUCN – Vulnerable.

SYNONYMS: *foramena* Wang, Hu, and Chen, 1980.

COMMENTS: Originally described by G. M. Allen as a species of *Chiropodomys*, but later reidentified by him as the only example of *Vandeleuria dumeticola* known from Yunnan (G. M. Allen, 1940), which was refuted by Ellerman (1949). Anthony (1941) correctly pointed out the morphological uniqueness of *fulva* by erecting a new genus to contain it. Still known only by few specimens. Wang et al. (1980) described *foramena* as a species of *Vernaya*, but diagnostic traits simply represent individual and geographic variation found in *V. fulva* (Corbet and Hill, 1992; Li and Wang, 1995; and Musser's study of the Chinese material described by Wang et al., 1980); Wang (2003) listed *foramena* as a synonym of *V. fulva* in his checklist of Chinese mammals.

Xenuromys Tate and Archbold, 1941. Am. Mus. Novit., 1101:3.

TYPE SPECIES: *Mus barbatus* Milne-Edwards, 1900.

COMMENTS: *Pogonomys* Division. Member of the New Guinea Old Endemics (Musser, 1981*c*). Generally regarded as morphologically similar to *Uromys* and phylogenetically related to it (Flannery, 1995*a*; Tate, 1951; Watts and Baverstock, 1994*a*), but sperm morphology is unlike most members of our *Uromys* Division and resembles many forms in our *Pogonomys* Division (Breed and Aplin, 1994). Our allocation to the latter cluster is provisional and must be tested by phylogenetic analyses of gene sequences and other kinds of morphological character sets.

Xenuromys barbatus (Milne-Edwards, 1900). Bull. Mus. Hist. Nat. Paris, 6:167.

COMMON NAME: Rock-dwelling Giant Rat.

TYPE LOCALITY: Papua New Guinea, "British New Guinea."

DISTRIBUTION: New Guinea; represented by only a few specimens collected along the Central Cordillera from Mt Dayman in E Papua to the Idenberg River in Prov. of Papua (= Irian Jaya), and in E end of Torricelli Mtns on north coast; 75-1200 m (Flannery, 1990*b*, 1995*a*; Flannery et al., 1985).

STATUS: IUCN – Lower Risk (nt).

SYNONYMS: *guba* Tate and Archbold, 1941.

COMMENTS: Leary and Seri (1997) reported a specimen taken at 600 m in karst limestone habitat in the Kikori River Basin of S Papua New Guinea. Very little is known about the actual distribution of this species on New Guinea and its natural history, other than it seems to reside in rocky habitats, is terrestrial, and may be frugivorous (Flannery et al., 1985). Aplin et al. (1999) reported a specimen of this species from a late Pleistocene archaeological site on the Ayamaru Plateau, central Bird's Head Peninsula of Prov. of Papua (= Irian Jaya).

Xeromys Thomas, 1889. Proc. Zool. Soc. Lond., 1889:248.

TYPE SPECIES: *Xeromys myoides* Thomas, 1889.

COMMENTS: *Xeromys* Division. Member of the Australian Old Endemics (Musser, 1981*c*:166) or Hydromyini of Baverstock (1984) and Lee et al. (1981). *Xeromys* is phylogenetically distantly related to *Hydromys* (Watts et al., 1992), and albumin immunology allies *Xeromys* with *Crossomys* (but see generic account), *Leptomys*, and *Pseudohydromys* (which includes *Neohydromys* and *Mayermys*) in a clade that is part of a larger assemblage, the Hydromyini, which contains genera in our *Hydromys*, *Pseudomys*, and *Uromys* Divisions (the "Australasian clade" of Watts and Baverstock, 1994*b*, 1995*b*, 1996).

Xeromys myoides Thomas, 1889. Proc. Zool. Soc. Lond., 1889:248.

COMMON NAME: False Water Rat.

TYPE LOCALITY: Australia, Queensland, Mackay.

DISTRIBUTION: Australia: C and S Queensland (Dwyer et al., 1979), North Stradbroke Isl off the coast of SEQueensland (Van Dyck, 1996; Van Dyck et al., 1979), Northern Territory, and Melville Isl off the coast of Northern Territory; probably has a wider range (Van Dyck, 1995, 1996; Watts and Aslin, 1981; Woinarski et al., 2000). S Papua New Guinea: Western Province, Bensbach River in lowlands of the Trans-Fly (Hitchcock, 1998).

STATUS: CITES – Appendix I; U.S. ESA – Endangered; IUCN – Vulnerable.

COMMENTS: Closely associated with tidal mangrove swamps and known by few samples (some of them large, however), most collected since 1970 (Dwyer et al., 1979; Van Dyck, 1996; Watts and Aslin, 1981; Woinarski et al., 2000). Morphology of sperm head structure and male reproductive tract studied by Breed (1984, 1986) in context of comparisons of reproductive structures among Australian murines. Chromosomal complement similar to that of *Hydromys chrysogaster* (Baverstock et al., 1977*c*), as is primitive pattern (for Murinae) of cephalic arterial configuration (Musser's unpublished observations). Mahoney and Richardson (1988:190) cataloged references to taxonomy and natural history for Australian records. Reviewed by Van Dyck (1995, 1996) and Woinarski et al. (2000). *Xeromys myoides* is one of 17 species of Australian mammals that are also found only in the Trans-Fly region of S New Guinea (Norris and Musser, 2001).

Zelotomys Osgood, 1910. Field Mus. Nat. Hist. Publ., Zool. Ser., 10:7.

TYPE SPECIES: *Mus hildegardeae* Thomas, 1902.

SYNONYMS: *Ochromys* Thomas, 1920.

COMMENTS: *Colomys* Division. Similar to *Colomys* in molar occlusal patterns (Misonne, 1969, and our observations), but strength of evolutionary link to either that genus or other murines requires assessment by phylogenetic analyses of morphological traits incorporating more characters than those associated with molars. The phyletic link between *Zelotomys* and *Colomys*, however, is reinforced by mtDNA cytochrome *b* (Lecompte et al., 2002*b*) and nuclear IRBP gene sequences (E. Lecompte, in litt., 2002), which place *Zelotomys* as sister to *Colomys* and align both with *Myomyscus verreauxii* in a monophyletic cluster within a larger group composed of *Praomys*, *Mastomys*, *Myomyscus*, *Hylomyscus*, *Heimyscus*, and *Stenocephalemys*. The relationship between *Zelotomys* and the *Stenocephalemys* Division was earlier inferred by D. H. S. Davis (1965) and Jaeger (1976). Earliest records of *Zelotomys* comes from the late Pliocene of Botswana, and Pleistocene fossils have been found in East Africa, South Africa, and Namibia (Avery, 2000; Denys, 1999; Jaeger, 1976; Pickford and Mein, 1988; Senut et al., 1992).

Zelotomys hildegardeae (Thomas, 1902). Ann. Mag. Nat. Hist., ser. 7, 9:219.

COMMON NAME: Hildegarde's Zelotomys.

TYPE LOCALITY: Kenya (Kenya Colony), Machakos.

DISTRIBUTION: W Angola (Crawford-Cabral, 1998), Zambia (Ansell, 1978), N Malawi (Nyika Plateau; Ansell and Dowsett, 1988), Tanzania, Kenya (Hollister, 1919), SW Uganda (Delany, 1975), Rwanda, Burundi, NE Dem. Rep. Congo (G. M. Allen, 1939), S Sudan (Setzer, 1956), and Central African Republic; see Misonne (1974).

STATUS: IUCN – Lower Risk (lc).

SYNONYMS: *instans* Thomas, 1916; *kuvelaiensis* St. Leger, 1936; *lillyana* Bohmann, 1950; *shortridgei* Hinton, 1920; *vinaceus* Heller, 1912.

COMMENTS: Terrestrial, partly diurnal, primarily insectivorous, emitting high-pitched whistles (Ansell, 1960; Delany, 1975), *Z. hildegardeae* inhabits tall grassland biomes and is infrequently encountered by collectors (Ansell, 1978). G. M. Allen (1939) listed *hildegardeae*, *shortridgei* and *instans* as separate species, as did Ellerman (1941:238), who noted that "The species are very closely allied, and may later be regarded all as races of the type," which reflects modern taxonomic treatment.

Zelotomys woosnami (Schwann, 1906). Proc. Zool. Soc. Lond., 1906:108.

COMMON NAME: Woosnam's Zelotomys.

TYPE LOCALITY: S Botswana, Molopo River.

DISTRIBUTION: A Southern African Subregion endemic ranging from N South Africa (Northern Cape Province) through N and W Botswana (Smithers, 1971) to E and N Namibia (de Graaff, 1997*f*; Skinner and Smithers, 1990).

STATUS: IUCN – Lower Risk (lc).

COMMENTS: Past taxonomic allocations (in *Ochromys*) have been with *Aethomys*, *Rattus*, and *Thallomys* (D. H. S. Davis, 1965; Ellerman, 1941; Meester et al., 1986, and references therein). Graminivorous and carnivorous, terrestrial, nocturnal, and inhabiting arid savanna biome in southern Africa (de Graaff, 1981; Mugo et al., 1995). Excellent reviews of distribution, ecology, and morphological traits provided by Skinner and Smithers (1990) and de Graaff (1981, 1997*f*).

Zyzomys Thomas, 1909. Ann. Mag. Nat. Hist., ser. 8, 3:372.

TYPE SPECIES: *Mus argurus* Thomas, 1889.

SYNONYMS: *Laomys* Thomas, 1909.

COMMENTS: *Pseudomys* Division. Member of the Australian Old Endemics (Musser, 1981*c*) or Conilurini of Baverstock (1984) and Lee et al. (1981), or Hydromyini of Baverstock and Watts (1994*a*). Taxonomy of species appraised by Kitchener (1989). Chromosomal morphology distinct from other Australian murines (Baverstock et al., 1977*c*), but not allozymic data (Baverstock et al., 1981), and G-banding homologies suggested karyotype of *Zyzomys* "to be derived from the ancestral karyotype characterizing *Conilurus*, *Mesembriomys*, and *Leggadina*" (Baverstock et al., 1983*b*). Molar traits resemble those characterizing *Conilurus* (Misonne, 1969), but phallic characters link *Zyzomys* to *Pseudomys* and its close relatives (Lidicker and Brylski, 1987). Albumin immunology indicates *Zyzomys* belongs in a clade with *Mesembriomys*, *Conilurus*, *Leporillus*, and *Melomys* (Watts et al., 1992), and Watts and Baverstock (1994*a*, 1995*b*, 1996) included it in their expanded version of Hydromyini (an "Australasian clade"), which incorporates the former Conilurini. External morphology of glans penis and spermatozoa reported by Breed (1984), Breed and Sarafis (1978), and Morrissey and Breed (1982), and phylogenetic significance discussed by Breed (1997). Mahoney and Richardson (1988) cataloged taxonomic, distributional, and biological references to some of the species. *Zyzomys rackhami*, the earliest record of the genus, comes from late Pliocene sediments (Aplin, 2005; Godthelp, 1997).

Zyzomys argurus (Thomas, 1889). Ann. Mag. Nat. Hist., ser. 6, 3:433.
    COMMON NAME: Common Australian Rock Rat.
    TYPE LOCALITY: Australia, probably Northern Territory (see Mahoney and Richardson, 1988:191).
    DISTRIBUTION: Northern Australia; always in rocky outcrops in Pilbara region of Western Australia through Kimberleys to N coastal Queensland between Cooktown and Townsville; also on offshore islands (Fleming, 1995*a*:621; Watts and Aslin, 1981:138).
    STATUS: IUCN – Lower Risk (lc).
    SYNONYMS: *indutus* (Thomas, 1909).
    COMMENTS: Patterns of allozymic comparisons between populations discordant with chromosomal differences (Baverstock et al., 1977*d*). Reviewed by Watts and Aslin (1981) and Fleming (1995*a*); also see Churchill (1996).

Zyzomys maini Kitchener, 1989. Rec. West. Aust. Mus., 14:357.
    COMMON NAME: Arnhem Land Rock Rat.
    TYPE LOCALITY: Australia, Northern Territory, Djawamba Massif, 1.5 km east of Ja Ja Billabong, 150 m (see Kitchener, 1989:357).
    DISTRIBUTION: Australia, Northern Territory in region of East and South Alligator Rivers on outliers of stony Arnhem Land escarpment (Kitchener, 1989).
    STATUS: IUCN – Lower Risk (lc).
    COMMENTS: Reviewed by Fleming (1995*b*).

Zyzomys palatilis Kitchener, 1989. Rec. West. Aust. Mus., 14:361.
    COMMON NAME: Carpentarian Rock Rat.
    TYPE LOCALITY: Australia, Northern Territory, Echo Gorge, Wollogorang Station, 180 m (see Kitchener, 1989:361).
    DISTRIBUTION: Australia; endemic to Gulf region of Northern Territory near the border with Queensland where it appears to be restricted to monsoon rainforest on scree slopes (Churchill, 1996).
    STATUS: IUCN – Critically Endangered.
    COMMENTS: Reviewed by Woinarski et al. (1995*b*). Distribution, habitat, status, and sympatry with *Z. argurus* documented by Churchill (1996).

Zyzomys pedunculatus (Waite, 1896). Rept. Horn Sci. Exped. Cent. Aust., Zool., 2:395.
    COMMON NAME: Central Australian Rock Rat.
    TYPE LOCALITY: Australia, Northern Territory, Alice Springs, but this locality is suspect (Kitchener, 1989; see also Mahoney and Richardson, 1988:191).
    DISTRIBUTION: C Australia; Northern Territory (Watts and Aslin, 1981:143; Wurst, 1995:625).
    STATUS: CITES – Appendix I; U.S. ESA – Endangered; IUCN – Critically Endangered.
    SYNONYMS: *brachyotis* (Waite, 1896).
    COMMENTS: Rare and restricted to rocks. Distribution of fossils indicated species "may once have extended across the rocky ranges of central Western Australia, through the Hamersleys to the coast" (Watts and Aslin, 1981:143). This "is either one of Australia's rarest rodents or its most recently extinct mammal. . . . The last specimen was collected in 1960 by a stockman in the Western MacDonnell Ranges near Mount Leibig, about 300 kilometres west of Alice Springs, while it was attempting to break into the camp

food supplies." (Wurst, 1995:624). Recently the species has been rediscovered living in two national parks in the Northern Territory (Wurst, 1997, who also provided first photograph of the live animal).

Zyzomys woodwardi (Thomas, 1909). Ann. Mag. Nat. Hist., ser. 8, 3:373.

COMMON NAME: Kimberley Rock Rat.

TYPE LOCALITY: Australia, Western Australia, E Kimberly, Parry Creek, 5 miles (8 km) west of Trig Station HJ9, 100 ft (30.5 m); Mahoney and Richardson (1988:191) provided coordinates and other information.

DISTRIBUTION: Australia; known only from the north Kimberleys of Western Australia (Fleming and McKenzie, 1995:626). Watts and Aslin (1981:141) included Arnhem Land in the Northern Territory within the range, but those animals represent *Z. maini* (Kitchener, 1989).

STATUS: IUCN – Lower Risk (lc).

COMMENTS: Restricted to rocky regions, especially boulders at the base of cliffs; limited in distribution but common at some localities (Watts and Aslin, 1981). Reviewed by Fleming and McKenzie (1995).

Subfamily Otomyinae Thomas, 1897. Proc. Zool. Soc. Lond., 1896:1017 [1897].

SYNONYMS: Otomyini Tullberg, 1899; Otomyidae Roberts, 1951.

COMMENTS: Morphologically, a strongly circumscribed group of species indigenous to Subsaharan Africa. Early on ranked as a subfamily of Muridae, whether defined *sensu lato* (Thomas, 1896; Ellerman, 1941; Roberts, 1951) or *sensu stricto* (Tullberg, 1899; Miller and Gidley, 1918; Simpson, 1945; Reig, 1981); or even a separate family (Roberts, 1951); and later considered a subfamily of Cricetidae (Misonne, 1974) or Nesomyidae (Chaline et al., 1977; Lavocat, 1978). Paleontological evidence and anatomical considerations unequivocally affirm their phyletic origin from African murines, especially arvicanthine-like forms (Pocock, 1976; Carleton and Musser, 1984; Bernard et al., 1991; Breed, 1995*d*; Sénégas, 2001). Similarly, evolutionary affinities inferred from DNA-DNA hybridization (Chevret et al., 1993*b*), immunological assays (Contrafatto et al., 1994; Watts and Baverstock, 1995*a*), and mitochondrial and nuclear DNA sequences (Ducroz et al., 2001; Jansa and Weksler, 2004; Michaux and Catzeflis, 2000; Michaux et al., 2001 *b*) uniformly represent *Otomys* as a "murine," either as sister group to *Arvicanthis-Oenomys* (Chevret et al., 1993*b*) or to arvicanthines in the broad sense (Ducroz et al., 2001). Molecular clock estimates place the split from arvicanthine murids about 7-8.5 million years ago (Chevret et al. 1993*b*; Ducroz et al., 2001), a time frame in broad agreement with fossil data insofar as known (Sénégas, 2001).

The above phylogenetic perspective indicates that otomyines should be merged under Murinae, probably as a tribe, as specifically urged by some (Ducroz et al., 2001; Jansa and Weksler, 2004; Michaux et al., 2001*a*; Watts and Baverstock, 1995*a*). Certain of these studies (Ducroz et al., 2001; Watts and Baverstock, 1995*a*), moreover, convey a deep, monophyletic separation of the African radiation (including otomyines) within Murinae as currently constituted. The larger nomenclatural issue involves the stability of this clade and its possible formal recognition, in which case Otomyinae Thomas, 1897, has priority as the applicable family-group name. The rank accorded and genera allocated to otomyines thus seem premature until relationships among African, Asian, and Indo-Australian murines are more confidently rendered; many of the latter are already type genera for variously recognized family-group taxa (see Murinae account). In lieu of this broader picture, we acknowledge otomyines and their distinctive combination of traits, unobserved elsewhere within Murinae, as a subfamily within Muridae.

Aspects of morphology surveyed and discussed by Bernard et al. (1990, 1991), Bohmann (1952), Breed (1995*d*), Carleton and Musser (1984), Jackson and Spinks (1998), Perrin and Curtis (1980), Taylor and Kumirai (2001), Thomas (1918*b*), and Tullberg (1899). Chromosomal information reviewed by Meester et al. (1992) and Taylor (2000*b*). For extensive paleontological coverage, including the annectant fossil genus *Euryotomys* (Mio-Pliocene, South Africa) and probable southern African origin of the subfamily, see Avery (1998), Chevret et al. (1993*b*), Denys (1989*a*), Denys et al. (1987), Pocock (1976, 1987), Sénégas (2001), and Sénégas and Avery (1998). Meester et al. (1986) consolidated nomenclatural information and identified provisional subspecies arrangements for taxa in the Southern African Subregion; taxonomy of East African forms too poorly understood to justify formal infraspecific divisions. Habitat preferences and conservation implications considered by Mugo et al. (1995) for species in South Africa.

Vlei and whistling rats have been variously classified into a single genus (Bohmann, 1952), commonly as the two genera *Otomys* and *Parotomys* (Ellerman, 1941; Ellerman et al., 1953; Misonne, 1974; De Graaff, 1981; Smithers, 1983; Meester et al., 1986; Corbet and Hill, 1991; Musser and Carleton, 1993), or as many as three (Thomas, 1918b; Pocock, 1976) or five (Roberts, 1951). The diverse generic arrangements principally reflect the emphasis on dentition versus bullar development as the pivotal trait of kinship. Recent multispecific surveys and phylogenetic studies of allozymes (Taylor et al., 1989), chromosomes (Meester et al., 1992), immunological data (Contrafatto et al., 1994), sperm structure (Bernard et al., 1991), and mtDNA (Ducroz et al., 2001) reveal that *Otomys*, as its specific contents are usually denoted, is polyphyletic; in particular, the species *sloggetti* and-or *unisulcatus* demonstrate close kinship with members of *Parotomys*. Several morphological traits also support this nearer relationship: absence of lower incisor sulci (present in all *Otomys*); weak sulcation of the upper incisors (deeply creased in *Otomys*); fewer M3 laminae (4-5 versus 6 or more in *Otomys*); gradually tapered nasals (distal nasals abruptly expanded in *Otomys*); vascular foramen basal to zygomatic plate small (large in *Otomys*); zygomatic plate narrow, not reaching the premaxillary-maxillary suture (plate broader, overlapping suture in *Otomys*); ventrolateral maxillary not extended anteriorly (ventrolateral projection, bearing insertion scar of the superficial masseter, conspicuous in *Otomys*). Such plesiomorphic features suggest the earlier cladistic origin of *sloggetti-unisulcatus* and *Parotomys* relative to the appearance and radiation of *Otomys*. Still, *sloggetti* and *unisulcatus* lack the hypertrophied ectotympanic bullae that characterize *Parotomys*, as well as other traits noted by Thomas (1918b). We provisionally return to the classification of Thomas (1918b), who provided useable diagnoses of his three genera (*Myotomys*, *Otomys*, and *Parotomys*) and whose arrangement is so far consistent with the emerging phylogenetic perspective, admitting its very preliminary and incomplete nature.

Myotomys Thomas, 1918. Ann. Mag. Nat. Hist., ser. 9, 2:206.
> TYPE SPECIES: *Otomys unisulcatus* F. Cuvier, 1829.
> SYNONYMS: *Metotomys* Broom, 1937.
> COMMENTS: Described by Thomas (1918b) as a genus, a rank occasionally observed (Pocock, 1976; Roberts, 1951) but not conventionally so (e.g., Bohmann, 1952; Ellerman et al., 1953; Misonne, 1974; Meester et al., 1986). In merging *Myotomys* as junior synonym of *Otomys*, Ellerman (1941) considered its component species to be morphologically linked with those of that genus. However, cladistic assessments of allozymic data disclose closer affinity of *sloggetti* and *unisulcatus* to species of *Parotomys* (Meester et al., 1992; Taylor et al., 1989), as do the morphological traits enumerated above and those mentioned by Pocock (1976). Phylogenetic investigations should focus on whether *sloggetti* and *unisulcatus* are separate branches in a lineage that also includes *brantsii* and *littledalei* (all as *Parotomys*) or whether they are early diverging sister species (*Myotomys*) relative to *brantsii-littledalei* (*Parotomys*) and species of *Otomys*.

Myotomys sloggetti (Thomas, 1902). Ann. Mag. Nat. Hist., ser. 7, 10:311.
> COMMON NAME: Rock Karroo Rat.
> TYPE LOCALITY: South Africa, N Western Cape Province, Deelfontein, north of Richmond.
> DISTRIBUTION: Subalpine and alpine zones, above 2000 m, in Eastern Cape Province, Lesotho, and W KwaZulu-Natal Province, South Africa (Lynch and Watson, 1992: Fig. 1; P. J. Taylor, 1998:Fig. 78).
> STATUS: IUCN – Lower Risk (nt) as *Otomys sloggetti*.
> SYNONYMS: *basuticus* Roberts, 1929; *jeppei* Roberts, 1929; *robertsi* (Hewett, 1927); *turneri* (Wroughton, 1907).
> COMMENTS: Viewed as closely related to *O. unisulcatus* (Bohmann, 1952; Roberts, 1951; Taylor et al., 1989; Thomas, 1918b), but details of sperm morphology question the closeness of their relationship (Bernard et al., 1991). Taxonomy, distributional records,

and ecology reviewed by Lynch and Watson (1992) and Lynch (1994); karyotypic and genetic data and comparisons supplied by Contrafatto et al. (1992a).

Myotomys unisulcatus (F. Cuvier, 1829). *In* E. Geoffroy and F. Cuvier, Hist. Nat. Mammifères, pt. 3, 6(60):1-2 "Otomys cafre."

COMMON NAME: Bush Karroo Rat.

TYPE LOCALITY: South Africa, Western Cape Province, southwestern Karroo, Matjiesfontein, southwest of Laingsburg (as designated by Roberts, 1946:318).

DISTRIBUTION: Little Namaqualand in W Northern Cape Province, through the Great and Little Karroo, to Eastern Cape Province, South Africa (De Graaff, 1981:155).

STATUS: IUCN – Least Concern as *Otomys unisulcatus*.

SYNONYMS: *albaniensis* Roberts 1946; *bergensis* Roberts, 1929; *broomi* (Thomas, 1902); *grantii* (Thomas, 1902).

COMMENTS: A species having a relictual distribution and exhibiting many traits interpreted as plesiomorphic for the subfamily (e.g., Bohmann, 1952). Genetic distance data (Taylor et al., 1989) suggest the inclusion of *unisulcatus* with species of *Parotomys*, but sperm morphology (Bernard et al., 1991) indicates its singular differentiation from both the *Parotomys* and *O. irroratus* groups. All named forms listed, with reservation, as subspecies by Meester et al. (1986), but allozymic variation unappreciable over species range and erodes validity of these divisions (Van Dyk et al., 1991).

Otomys F. Cuvier, 1824. Dents des Mammifères, p. 255.

TYPE SPECIES: *Euryotis irrorata* Brants, 1827.

SYNONYMS: *Anchotomys* Thomas, 1918; *Euryotis* Brants, 1827; *Lamotomys* Thomas, 1918; *Oreinomys* Trouessart, 1880 [replacement name for *Oreomys*]; *Oreomys* Heuglin, 1877; *Palaeotomys* Broom, 1937; *Prototomys* Broom, 1948.

COMMENTS: Whereas Bohmann (1952) included all otomyine species in *Otomys*, most systematists have accorded the large-bullar forms separate generic status as *Parotomys* (see below), and we follow Thomas (1918b) in recognizing *Myotomys* as genus (see subfamily remarks). Roberts (1951) also treated *Lamotomys* as generically distinct, but cladistic interpretation of allozymic and immunologcial data, albeit limited to few species so far, clearly affiliate its type species *laminatus* with other *Otomys* (Contrafatto et al., 1994; Taylor et al., 1989). Oldest known fossil *Otomys* species date from the middle to late Pliocene (2-3.5 million years ago) in South Africa and from early Pleistocene (1-2 million years ago) in East Africa (see Denys, 1989a; Sénégas, 2001; Sénégas and Avery, 1998). Synonymy of the fossil *Prototomys* follows the observations of Avery (1998), who noted the marginal distinction of its type species, *P. campbelli*, from living *O. saundersiae*.

The species richness and biogeographic diversification within *Otomys* were grossly obscured by forcing its exceptional morphological variation into the polytypic application of the biological species concept that prevailed over the middle 1900s (e.g., Bohmann, 1952; Dieterlen, 1968; Ellerman et al., 1953; Misonne, 1974; Petter, 1982). Deconstructing these polyphyletic accretions into diagnosable, genetically homogeneous species is much progressed in southern Africa (see Taylor and Kumirai, 2001, for overview) and selectively in other areas (e.g., Dieterlen and Van der Straeten, 1992). As we noted in 1993, the occurrence of greater specific endemism throughout the isolated East African highlands and volcanoes deserves more serious consideration than it has received to date. Persuant to such revisionary attention, the early syntheses of Wroughton (1906), Dollman (1915), and Hollister (1919) offer a sounder foundation from which to address questions of specific status and distribution in the East African region. Such investigations should emphasize comparisons both among mountain ranges and along their slopes, drawing upon topotypic examples to explicitly address taxonomic and biogeographic problems at issue. In particular, the possibility of multiple independent originations (speciation) of otomyines in afro-alpine environments on East African mountain tops should be the working hypothesis to be disproven.

Otomys anchietae (Bocage, 1882). J. Sci. Acad. Lisbon, 9:26.

COMMON NAME: Angolan Vlei Rat.

TYPE LOCALITY: Angola, Huila, Caconda.

DISTRIBUTION: C and NE Angola (Crawford-Cabral, 1998:Map 23).

STATUS: IUCN – Lower Risk (nt).

COMMENTS: Type species of *Anchotomys*, employed as a subgenus of *Otomys* by Thomas (1918*b*). Cranial form and dental traits strongly differentiated compared with other *Otomys* (Taylor and Kumirai, 2001). Dieterlen and Van der Straeten (1992) reasonably argued the recognition of *barbouri* and *lacustris*, formerly arranged as subspecies (Misonne, 1974), as species (see those accounts).

Otomys angoniensis Wroughton, 1906. Ann. Mag. Nat. Hist., ser. 7, 18:274.

COMMON NAME: Angoni Vlei Rat.

TYPE LOCALITY: Malawi, Misuku Range, Matipa Forest, 7000 ft (2134 m; as amended by Ansell and Dowsett, 1991).

DISTRIBUTION: SE savannah and grasslands, from S Kenya to SE Botswana and NE South Africa (provinces of Limpopo, KwaZulu-Natal, Mpumalanga, Gauteng, Free State, North West; also Lesotho).

STATUS: IUCN – Lower Risk (lc).

SYNONYMS: *canescens* Osgood, 1910; *divinorum* Thomas, 1910; *elassodon* Osgood, 1910; *mashona* Thomas, 1918; *nyikae* Wroughton, 1906; *pretoriae* Roberts, 1929; *rowleyi* Thomas, 1918; *sabiensis* Roberts, 1929; *tugelensis* Roberts, 1929.

COMMENTS: Populations confused under *O. irroratus* by Bohmann (1952) and Ellerman et al. (1953), but morphological, karyotypic, and allozymic differences decidedly support their separate specific status (D. H. S. Davis, 1962; De Graaff, 1981; Matthey, 1964; Misonne, 1974; Taylor et al., 1989). G-banding comparisons with the highly variable chromosomes of *O. irroratus* conducted by Contrafatto et al. (1992*a*). Synonymy of the listed species-group epithets and distributional limits need verification in the context of a full revision, including the form *maximus* (see account below), which many have viewed as another subspecies of *O. angoniensis* (e.g., De Graaff, 1981; Meester et al., 1986; Misonne, 1974). Reported presence in NE Cape Province (=Eastern Cape Province) and Lesotho (e.g., Meester et al., 1986; Skinner and Smithers, 1990) questioned by Lynch (1994). See Bronner and Meester (1988, Mammalian Species, 306).

Otomys barbouri Lawrence and Loveridge, 1953. Bull. Mus. Comp. Zool., Harvard, 110:63.

COMMON NAME: Barbour's Vlei Rat.

TYPE LOCALITY: Uganda, Mount Elgon, Kaburomi, 10,500 ft (3200 m); 01°14'N, 34°31'E.

DISTRIBUTION: Afro-alpine zone of Mount Elgon, ca. 3500-4300 m, Uganda and Kenya.

COMMENTS: Lawrence and Loveridge (1953) emphasized the contrasts between their new species and *O.* (*Anchotomys*) *anchietae*, but Misonne (1974) reassigned it as a synonym of the latter, an opinion erroneously followed by others (Honacki et al., 1982; Corbet and Hill, 1991; Musser and Carleton, 1993). Trenchant differences from *O. anchietae* explicated by Dieterlen and Van der Straeten (1992), who reinstated *barbouri* as species and suggested its closer relationship to *O. occidentalis* and *O. lacustris*; morphological and morphometric discrimination amplified by Taylor and Kimurai (2001). Knowledge of altitudinal distribution, autecology, and natural history vastly improved by the studies of Clausnitzer (2000, 2001; Clausnitzer and Kityo, 2001).

Otomys burtoni Thomas, 1918. Ann. Mag. Nat. Hist., ser. 9, 2:210.

COMMON NAME: Burton's Vlei Rat.

TYPE LOCALITY: Cameroon, Cameroon Mountains, 7000 ft (2134 m).

DISTRIBUTION: Highlands of NW Cameroon; range limits unknown.

COMMENTS: Retained as a species until submerged within the pan-African rassenkreis of *O. irroratus* (Bohmann, 1952; Petter, 1982). As noted by Thomas (1918*b*), among the smallest species of *Otomys*, resembling *nubilus* (here = *O. tropicalis*) in Kenya.

Otomys cuanzensis Hill and Carter, 1937. Am. Mus. Novitates, 913:7.

COMMON NAME: Cuanza Vlei Rat.

TYPE LOCALITY: Angola, Chitau, 4930 ft (1503 m).

DISTRIBUTION: C Angola; limits uncertain.

COMMENTS: Described as a species, but Bohmann (1952) drew it within his polytypic interpretation of *O. irroratus*, its usual allocation thereafter (Crawford-Cabral, 1986*a*; Ellerman et al., 1953; Meester et al., 1986; Misonne, 1974). Musser and Carleton (1993) mistakenly aligned *cuanzensis* as a synonym of *O. maximus*, but its affinity lies with *O. irroratus* proper, as appreciated by Crawford-Cabral (1986*a*). Hill and Carter's (1937) specific description is terse, contrasting *cuanzensis* solely with Angolan populations of the very different form *maximus*, not *irroratus* in the strict sense. The holotype (AMNH 85841) and referred specimens (AMNH series) of *cuanzensis* approximate *O. irroratus* in overall size and uniformly possess a six-laminated M3 and large stapedial foramen; however, they differ from *O. irroratus* in their predominantly brown dorsal pelage and narrower nasals, a contrast also remarked by Bohmann (1952). In view of such differences, the apparently wide geographic hiatus from southern African *O. irroratus*, and the pronounced karyotypic and genetic substructuring known for even contiguous populations of the latter (e.g., Contrafatto et al., 1997; Taylor, 2000*b*), we retain *cuanzensis* as species until its relationships and status can be critically assessed.

Otomys dartmouthi Thomas, 1906. Ann. Mag. Nat. Hist., ser. 7, 18:141.
> COMMON NAME: Ruwenzori Vlei Rat.
> TYPE LOCALITY: Uganda, Ruwenzori East, Mubuku Valley, 12,500 ft (3180 m).
> DISTRIBUTION: Open moorland, 3300-3900 m, in the Ruwenzori Mtns (Kerbis Peterhans et al., 1998, as *O. typus*).
> COMMENTS: Treated as another montane variant of a highly polymorphic *O. irroratus* (Petter, 1982) or *O. typus* (Bohmann, 1952; Misonne, 1974). As noted by Thomas (1906*a*) and Dollman (1915), this form consistently possesses only 6 M3 laminae (confirmed in FMNH 144327, 144328, 144330), unlike the 8-9 in *O. typus* proper, and is much smaller with more darkly colored, somewhat woolly fur. Morphological and genetic conformity with Ethiopian populations should be demonstrated.

Otomys denti Thomas, 1906. Ann. Mag. Nat. Hist., ser. 7, 18:142.
> COMMON NAME: Dent's Vlei Rat.
> TYPE LOCALITY: Uganda, east slope of Mount Ruwenzori, Mubuku Valley, 6000 ft (1829 m; as restricted by Moreau et al., 1946:420).
> DISTRIBUTION: Intermittently found in EC Africa—Ruwenzori Mtns, SW Uganda and contiguous Dem. Rep. Congo; through the Virunga volcanoes, E Dem. Rep. Congo (Kivu), Rwanda, and Burundi; to the Nyika Plateau, N Malawi and NE Zambia, and the Usambara and Uluguru mountains, EC Tanzania.
> STATUS: IUCN – Lower Risk (nt).
> SYNONYMS: *kempi* Dollman, 1915; *sungae* Bohmann, 1943.
> COMMENTS: The form *kempi* was treated as a distinct species (e.g., Ellerman, 1941; Thomas, 1918*b*), as originally described (Dollman, 1915), and later reduced to subspecific rank by Bohmann (1952), as observed currently (Ansell, 1978; Delany, 1975; Misonne, 1974; Musser and Carleton, 1993). Confirmation of its status and that of *sungae*, isolated in the Nyika Plateau and Eastern Arc Mtns, deserves critical evaluation.

Otomys dollmani Heller, 1912. Smithsonian Misc. coll., 59:5
> COMMON NAME: Dollman's Vlei Rat.
> TYPE LOCALITY: Kenya, Matthews Range, Mount Gargues (Urguess), 7000 ft (2134 m).
> DISTRIBUTION: Known only from the type locality.
> COMMENTS: Described as a subspecies of *orestes* and afterwards swept under either *O. irroratus* (Bohmann, 1952; Petter, 1982) or *O. tropicalis* (Misonne, 1974; Musser and Carleton, 1993). Hollister (1919) noted *dollmani*'s differentiation from both *orestes* and *tropicalis* on Mount Kenya (notably its smaller size, lack of postauricular patches, six-laminated M3, flatter skull profile, as verified with USNM type series) and long ago elevated it to species. We agree.

Otomys irroratus (Brants, 1827). Het Geslacht der Muizen, p. 94.
> COMMON NAME: Southern African Vlei Rat.
> TYPE LOCALITY: South Africa, Western Cape Province, Cape Town district, near Constantia (fixed by A. Smith, 1834:149, in supplying replacement name *typicus*; see Meester et al., 1986:250).

DISTRIBUTION: Mesic savannah and grasslands of southern Africa—S Western Cape Province to Limpopo Province, South Africa; disjunct populations in W South Africa and in E Zimbabwe and contiguous Mozambique (De Graaff, 1981:146).

STATUS: IUCN – Lower Risk (lc).

SYNONYMS: *auratus* Wroughton, 1906; *bisulcatus* F. Cuvier, 1829; *capensis* G. Cuvier, 1830; *coenosus* Thomas, 1918; *cupreoides* Roberts, 1946; *cupreus* Wroughton, 1906; *natalensis* Roberts, 1929; *obscura* (Lichtenstein, 1842); *orientalis* Roberts, 1946; *randensis* Roberts, 1929; *saundersiae* Roberts, 1929; *typicus* (A. Smith, 1834).

COMMENTS: Bohmann (1952) established a broad, improbably polymorphic definition of *O. irroratus*, including some 23 subspecies, a concept further enlarged by Dieterlen (1968) and Petter (1982) and collectively enveloping the following forms here (and elsewhere) treated as separate species (see individual accounts): *O. anchietae*, *O. angoniensis*, *O. barbouri*, *O. burtoni*, *O. cuanzensis*, *O. dollmani*, *O. jacksoni*, *O. laminatus*, *O. maximus*, *O. orestes*, *O. tropicalis*, *O. typus*, and *O. uzungwensis*. Although followed to a greater or lesser extent (e.g., Delany, 1975; Honacki et al., 1982; Kingdon, 1974*b*), such an inclusive species construct has been refuted by others who identify *O. irroratus* proper as a species indigenous to southern Africa, south of the Zambezi River (e.g., De Graaff, 1981; Meester et al., 1986; Misonne, 1974; Taylor and Kumirai, 2001). Morphometrically distinguishable from East African forms, such as *O. typus* and *O. tropicalis*, and M3 characteristically possessing 6 laminae compared with 7-8 in those forms (Dollman, 1915; Thomas, 1918*b*; Taylor and Kumirai, 2001).

A subject of intensive investigation of intraspecific patterns of differentiation, *O. irroratus sensu stricto* is becoming a model organism for pursuit of evolutionary questions. Some results suggest "incipient speciation" among the population moieties so far identified (see overview and mapping of chromosomal races by Taylor, 2000*b*:Fig. 1), although the different data sets are not necessarily wholly concordant. G-banded comparisons reveal a karyotype that is highly derived (Robinson and Elder, 1987). Extensive cytogenetic variation (2n = 23-32; FN = 24-50) reported among South African population samples (Contrafatto et al., 1992*a,b*; Taylor, 2000*b*), complemented by examinations of variation in mtDNA (Lamb et al., 1996), protein electrophoresis (Contrafatto et al., 1997; Taylor et al., 1992; Taylor, 2000*b*), and reproductive isolating mechanisms (Pillay et al., 1992; 1995*a,b*). Correlation of cytotype distributions with climatic variables studied by Taylor et al. (1994*b*). Included *saundersiae* according to Taylor et al. (1993) but afterwards reinstated to species by Taylor (in litt.); see that account. See Bronner et al. (1988, Mammalian Species, 308).

Otomys jacksoni Thomas, 1891. Ann. Mag. Nat. Hist., ser. 6, 7:304.

COMMON NAME: Mount Elgon Vlei Rat.

TYPE LOCALITY: Uganda, crater of Mount Elgon, 13,200 ft (4023 m).

DISTRIBUTION: Restricted to Mount Elgon, ca. 3300-4200 m, Uganda and Kenya.

COMMENTS: Typically considered a subspecies of *O. typus* subsequent to Bohmann's (1952) monograph (e.g., Misonne, 1974; Musser and Carleton, 1993). Although their lower incisors do possess two well-defined grooves, as in *O. typus*, the Mount Elgon populations contrast sharply in their smaller size, dark brown pelage, and M3 with only 7 laminae. Clausnitzer and Kityo (2001) recorded the species (as *O. typus*) as common in afro-alpine habitats, where it exists sympatrically with *O. barbouri*.

Otomys lacustris G. M. Allen and Loveridge, 1933. Bull. Mus. Comp. Zool., 75:120.

COMMON NAME: Tanzanian Vlei Rat.

TYPE LOCALITY: Tanzania, Ukinga Mtns, north end of Lake Nyasa, Madehani, 7000 ft (2134 m).

DISTRIBUTION: Isolated populations in mountains, ca. 1400-2300 m, of N Malawi, SC Tanzania (e.g., Stanley et al., 1998), and SW Kenya (Aberdare Range as per Taylor and Kumirai, 2001:Fig. 1).

COMMENTS: A distinctive form diagnosed as a subspecies of *O. anchietae* by its describers and thereafter placed in taxonomic compendia (Bohmann, 1952; Misonne, 1974; Musser and Carleton, 1993). Morphological contrasts noted and specific status reasserted by Dieterlen and Van der Straeten (1992); morphological and morphometric discrimination amplified by Taylor and Kimurai (2001).

Otomys laminatus Thomas and Schwann, 1905. Abst. Proc. Zool. Soc. Lond., 1905(18):23.

COMMON NAME: KwaZulu Vlei Rat.

TYPE LOCALITY: South Africa, KwaZulu-Natal, Zululand, Nkandhla, Sibudeni, 1050 m.

DISTRIBUTION: SE Mpumalanga and Free State provinces through KwaZulu-Natal Province, to Transkei; isolated segment in SW Western Cape Province, South Africa (Taylor et al., 1994a:Fig. 6).

STATUS: IUCN – Least Concern.

SYNONYMS: *fannini* (Roberts, 1951); *mariepsi* Roberts, 1929; *pondoensis* Roberts, 1924; *silberbaueri* Roberts, 1919.

COMMENTS: A species with highly derived M3 and m1 patterns, designated as the type of *Lamotomys*, a taxon employed as a subgenus of *Otomys* by Thomas (1918b) and as a genus by Roberts (1951). Believed to intergrade with *O. irroratus* by Petter (1982) but usually retained as a distinct species in systematic works (Bohmann, 1952; Misonne, 1974) and faunal studies (De Graaff, 1981; Smithers, 1983; Meester et al., 1986). In addition to the many and obvious morphological differences, allozymic assessments demonstrate the genetic separation of *O. laminatus* from *O. irroratus* (Taylor et al., 1989). Karyotype reported by Taylor et al. (1994a), who collected *O. laminatus* in sympatry with *O. angoniensis* and *O. irroratus* in Natal.

Otomys maximus Roberts, 1924. Ann. Transvaal Mus., 10:70.

COMMON NAME: Okavango Vlei Rat.

TYPE LOCALITY: Zambia, Machile River, a northern tributary of the Zambezi (1725Ac quadrant as restricted by Davis, 1974:173).

DISTRIBUTION: Angola (Crawford-Cabral, 1998:Map 22), SW Zambia, Okavango region of Botswana, NE Namibia (Caprivi Strip), and extreme W Zimbabwe.

STATUS: IUCN – Lower Risk (lc).

SYNONYMS: *davisi* Lundholm, 1955.

COMMENTS: Described as a subspecies of *O. irroratus* but Roberts (1951) reconsidered its status as a full species. Thereafter returned to subspecies of *O. irroratus* (Bohmann, 1952; Ellerman et al., 1953); or viewed as a subspecies of *O. angoniensis* by Davis (1974), the commonly observed synonymy in faunal and systematic treatises (e.g., Ansell, 1978; De Graaff, 1981; Meester et al., 1986; Misonne, 1974); or continued as a distinct species in others (Corbet and Hill, 1991; Musser and Carleton, 1993; Smithers, 1983; Swanepoel et al., 1980). Based on examinations of AMNH and USNM series from Angola and Botswana, we still favor the last treatment as the best working hypothesis. Although the two are apparently closely related, sharing a nearly occluded or absent stapedial foramen, *O. maximus* is a larger animal in most external and craniodental measurements (particularly as seen in the robust hindfoot, longer molar row, and deeper mandibular ramus); typically has an M3 with 6 lamina, occasionally 7 (typically 7, occasionally 6 in *O. angoniensis*); and possesses a dorsal pelage dominated by slate and grayish tones (brown to buffy-brown overtones in *O. angoniensis*). Smithers (1983) also emphasized the belt of dry, inimical terrain in NE Botwana and SW Zambia that effectively separates the ranges of *O. maximus* and *O. angoniensis*. Their consanguinity should be demonstrated using other data sources and comprehensive analyses. Formerly included *cuanzensis* (see above account).

Otomys occidentalis Dieterlen and Van der Straeten, 1992. Bonn. Zool. Beitr., 43:386.

COMMON NAME: Western Vlei Rat.

TYPE LOCALITY: Nigeria, Gotel Mountains, Chappal Waddi.

DISTRIBUTION: Recorded only from the Gotel Mtns, E Nigeria, and Mount Oku, W Cameroon.

STATUS: IUCN – Endangered.

COMMENTS: A species with 5 laminae on m1, closely related to *O. barbouri* and *O. lacustris* in mountains of eastern Africa according to Dieterlen and Van der Straeten (1992).

Otomys orestes Thomas, 1900. Proc. Zool. Soc. Lond., 1900:175.

COMMON NAME: Afroalpine Vlei Rat.

TYPE LOCALITY: Kenya, Mount Kenya, Teleki Valley, 13,000 ft (3962 m).

DISTRIBUTION: Discontinuous in alpine settings, ca. 3200-4500 m, of W and C Kenya and NE Tanzania (Grimshaw et al., 1995).

SYNONYMS: *malleus* Dollman, 1915; *percivali* Dollman, 1915; *thomasi* Osgood, 1910; *squalus* Dollman, 1915; *zinki* Bohmann, 1943.

COMMENTS: Populations of *O. orestes* are apparently confined to open habitats above treeline and exhibit a characteristic morphology: medium-sized species with relatively short tail; fur very deep, soft, and dense with creamy-buff post-auricular patches present; cranial arching strongly pronounced, anterior zygomatic arches squared, and distal expansion of nasals less exaggerated; M3 laminae 6-7, lower incisors with deep lateral and shallow medial grooves. Thomas (1900*b*) appreciated such morphological distinctions between *O. orestes* and samples from the middle slopes of Mount Kenya that he later (1902*c*) recognized as *O. tropicalis*, as did other early workers with East African *Otomys* (Wroughton, 1906; Dollman, 1915; Hollister, 1919; Lawrence and Loveridge, 1953). Bohmann (1952) also considered the two Mount Kenyan series as separate species, although he viewed *orestes* as a synonym of a broadly, if patchily, distributed *O. typus* (as followed by Misonne, 1974; Musser and Carleton, 1993; and others). M3 lamination (8-9 in *typus*) and incisor corrugation (two deep grooves in *typus*) contradict this assignment and recommend rigorous evaluation of their level of relationship and taxonomic status. Similar attention should be devoted to the homogeneity of the included taxa, which vary in development of the second lower incisor groove (indistinct to moderate) and modal number of M3 laminae (6 or 7).

Otomys saundersiae Roberts, 1929. Ann. Transvaal Mus., 13:115.

COMMON NAME: Saunders' Vlei Rat.

TYPE LOCALITY: South Africa, Eastern Cape Province, Grahamstown.

DISTRIBUTION: Isolated populations in Western Cape, Eastern Cape, and Free State provinces, South Africa, and in Lesotho.

STATUS: IUCN – Lower Risk (lc).

SYNONYMS: *karoensis* Roberts, 1931.

COMMENTS: Roberts (1929) named *saundersiae* as a subspecies of *tugelensis* (= *O. angoniensis*) but later (1951) raised it to full species, as commonly recognized in the literature (e.g., Corbet and Hill, 1991; De Graaff, 1981; Ellerman and Morrison-Scott, 1953; Misonne, 1974; Meester et al., 1986). The form *karoensis*, in Western Cape Province, has been variously treated as a species (G. M. Allen, 1939; Bohmann, 1952; Roberts, 1931; Taylor et al., 1993) or as a distinctive subspecies of *O. saundersiae* (Roberts, 1951; Taylor in litt.). Taylor et al. (1993, as *karoensis*) provided morphological bases for discrimination of *O. saundersiae* from *O. irroratus*, evaluated geographic and nongeographic variation, and contrasted G-banded karyotypes. The apparent rarity of *O. saundersiae* populations in the region of its type locality, Eastern Cape Province, is discussed by Taylor (in litt.).

Otomys tropicalis Thomas, 1902. Ann. Mag. Nat. Hist., ser. 7, 10:314.

COMMON NAME: East African Vlei Rat.

TYPE LOCALITY: Kenya, west slope of Mount Kenya, 10,000 ft (3048 m).

DISTRIBUTION: S Sudan, S Ethiopia, NE and E Dem. Rep. Congo, Uganda, Rwanda, W Kenya, and NE Tanzania; limits unknown.

STATUS: IUCN – Lower Risk (lc).

SYNONYMS: *elgonis* Wroughton, 1910; *faradjius* Hatt, 1934; *ghigii* de Beaux, 1924; *giloensis* Setzer, 1953; *nubilus* Dollman, 1915; *rubeculus* Dollman, 1915; *vivax* Dollman, 1915; *vulcanis* Lönnberg and Gyldenstolpe 1925.

COMMENTS: Regarded as conspecific with *O. irroratus* by Bohmann (1952) and accordingly recognized in regional treatments (Delany, 1975; Kingdon, 1974*b*); however, others have recognized eastern African *tropicalis* as morphologically divergent and specifically distinct from the southern African *O. irroratus* (De Graaff, 1981; Meester et al., 1986; Misonne, 1974; Musser and Carleton, 1993; Taylor and Kumirai, 2001).

Even removed from *O. irroratus* and divorced of *burtoni* and *dollmani* (see those accounts), at least three assemblages are apparent among the populations embraced by this nominal species. The taxon *tropicalis* in the strict sense inhabits middle to upper slopes, 2300-4000 m (specimens in USNM), of Mt Kenya and Aberdare Mtns. Examples of the *elgonis* complex (also *faradjius, ghigii?, giloensis, nubilus, vivax*) exhibit a dark russet-brown pelage that is somewhat sleek and moderately long; tail relatively longer; skull flatter, nasals narrower and flaring less abruptly; M3 with 7 laminae, upper incisors wider and medial sulcus of lower incisor faint. As noted by Hollister (1919), this dark-brown form occurs on the lower western slopes of Mount Kenya (2150 m; USNM 164277), seemingly distinct from populations of *tropicalis* found on middle to upper slopes. Elsewhere, specimens exhibiting this morphology occur sympatrically with *thomasi* (here = *O. orestes*) at Molo, Kenya (FMNH 16693, 16695), and with *squalus* (here = *O. orestes*) in the Aberdare Mtns (series in USNM), and are altitudinally parapatric with *O. typus* on Mount Albasso (FMNH 28165), with *O. dartmouthi* in the Ruwenzori Mtns (series in AMNH, FMNH), and with *O. jacksoni* on Mount Elgon (FMNH 25379, 25380; Wroughton, 1906). A third moiety (*rubeculus*) consists of populations in Western Rift mtns that are large-bodied and have lower incisors with a moderately deep medial groove. Variation in pelage color, size, and cranial and dental characteristics within and among these assemblages is appreciable, however, and the complex will require careful revision and multiple data sources to assess the number and distributions of species. Dense altitudinal transects, especially on Mt Kenya and the Abedare Mtns, would be helpful.

Otomys typus (Heuglin, 1877). Reise in Nordost-Afrika, 2:77.
   COMMON NAME: Ethiopian Vlei Rat.
   TYPE LOCALITY: Ethiopia, Gonder Province, Simien Mtns.
   DISTRIBUTION: Highlands, ca. 1800-4000 m, of NC (Gonder) to SC (Gamo Gofa, N Sidamo, and Bale) Ethiopia (Yalden et al., 1976:Fig. 32).
   STATUS: IUCN – Lower Risk (nt).
   SYNONYMS: *degeni* Thomas, 1902; *fortior* Thomas, 1906; *helleri* Frick, 1914; *malkensis* Frick, 1914.
   COMMENTS: Dieterlen (1968) and Petter (1982) viewed *typus* as another variant of a highly polymorphic *O. irroratus*, a conclusion that conflicts with their morphological discrimination as presented elsewhere (e.g., Ansell, 1978; Bohmann, 1952; Kingdon, 1974*b*; Misonne, 1974). Nonetheless, the synonyms listed and wide East African distribution as conveyed by the classifications of Bohmann (1952) and Misonne (1974) embrace such immense morphological heterogeneity that diagnosis as a single species is incomprehensible and geographically improbable. In our view, *O. typus* proper corresponds to those populations with grizzled brown pelage, a moderately vaulted skull, M3 with 8-9 laminae, and two strongly creased grooves on the lower incisors, a morphology endemic to the highlands of Ethiopia. Even so narrowly delineated, the conspecific stature of Ethiopian populations is highly suspect, as suggested by the karyotypic and allozymic diversity reported for *Otomys* in the Bale Mountains (Lavrenchenko et al., 1997; 2000). Whether the exceptional variation reported by those authors is partly attributable to the presence of the *tropicalis-elgonis* complex (see above) at lower elevations or yet another species confused under *O. typus* requires further study. Ethiopia's rugged landscape likely harbors at least three species of *Otomys*.

Otomys uzungwensis Lawrence and Loveridge, 1953. Bull. Mus. Comp. Zool., 110:61.
   COMMON NAME: Uzungwe Vlei Rat.
   TYPE LOCALITY: Tanzania, Iringa District, Uzungwe Mtns, Dadaga, 6000 ft (1829 m).
   DISTRIBUTION: Uzungwe Mtns, WC Tanzania, and Nyika Plateau, N Malawi and NE Zambia, as so far known.
   COMMENTS: Lawrence and Loveridge (1953) understood their new species as generally affiliated with the *jacksoni* group, but they presented a combination of traits (especially smaller size, lack of postauricular patches, flatter skull, and broader nasals) that differentiate it from other members of that complex. Relegated to a subspecies of *O. typus* by Misonne (1974), presumably applying Bohmann's (1952) expansive concept of the species. Certainly not a form of either *O. irroratus* or *O. typus* in the strict sense; its relationship and status with regard to *O. orestes* and included taxa invite investigation.

Parotomys Thomas, 1918. Ann. Mag. Nat. Hist., ser. 9, 2:205.
>
> TYPE SPECIES: *Euryotis brantsii* A. Smith, 1834.
>
> SYNONYMS: *Liotomys* Thomas, 1918.
>
> COMMENTS: Diagnosed as genus by Thomas (1918*b*) and maintained at that rank in most systematic and faunal accounts (Ellerman, 1941; Ellerman et al., 1953; Misonne, 1974; De Graaff, 1981; Meester et al., 1986; Musser and Carleton, 1993). Bohmann (1952) allocated *Parotomys* as another synonym of an all- inclusive genus *Otomys*; Roberts (1951) elevated *Liotomys* to generic rank. Generic boundaries may also include *sloggetti* and *unisulcatus*, species typically classified as *Otomys*; see comments and references under subfamily and *Myotomys*.

Parotomys brantsii (A. Smith, 1834). South African Quart. J., Ser. 2, 2: 150.
>
> COMMON NAME: Brants's Whistling Rat.
>
> TYPE LOCALITY: South Africa, Northern Cape Province, Little Namaqualand, "toward the mouth of the Orange River."
>
> DISTRIBUTION: Western, Eastern, and Northern Cape provinces, South Africa, to SW Botswana and SE Namibia (De Graaff, 1981:160).
>
> STATUS: IUCN – Lower Risk (lc).
>
> SYNONYMS: *deserti* Roberts, 1933; *luteolus* (Thomas and Schwann, 1904); *pallida* (Wagner, 1841); *rufifrons* (Rüppell, 1842).
>
> COMMENTS: Although Port Nolloth, purportedly as restricted by Thomas and Schwann (1904:178), is usually cited as the type locality of *P. brantsii* (e.g., Ellerman et al., 1953; Meester et al., 1986), a careful reading of Thomas and Schwann indicates that they had actually associated one of Smith's cotypes with a series collected at Klipfontein, a place some 50 miles inland from Port Nolloth. The notion that the type locality was "restricted" to Port Nolloth apparently stems from an inadvertent indication in Roberts (1951). A future revisor of the species should clarify this matter. Meester et al. (1986) recognized *deserti* and *rufifrons* as subspecies in addition to the nominate form.

Parotomys littledalei Thomas, 1918. Ann. Mag. Nat. Hist., ser. 9, 2:205.
>
> COMMON NAME: Littledale's Whistling Rat.
>
> TYPE LOCALITY: South Africa, Northern Cape Province, Bushmanland, Kenhardt, Tuin.
>
> DISTRIBUTION: N Western Cape and Northern Cape provinces, South Africa, to S and W Namibia.
>
> STATUS: IUCN – Lower Risk (lc).
>
> SYNONYMS: *molopensis* Roberts, 1933; *namibensis* Roberts, 1933.
>
> COMMENTS: Type species of *Liotomys*, named as a subgenus of *Parotomys* by Thomas (1918*b*) and viewed as a genus by Roberts (1951). Sister species to *P. brantsii* according to phylogenetic interpretation of allozymic data (Meester et al., 1992; Taylor et al., 1989). Meester et al. (1986) retained all species-group taxa as subspecies.

SUBORDER ANOMALUROMORPHA Bugge, 1974.

FAMILY ANOMALURIDAE
by Fritz Dieterlen

Family Anomaluridae Gervais, 1849. In D'Orbigny, Dict. Univ. Hist. Nat., 11:203.
COMMENTS: See Misonne (1974, pt. 6:3-5).

Subfamily Anomalurinae Gervais, 1849. In D'Orbigny, Dict. Univ. Hist. Nat., 11:203.

Anomalurus Waterhouse, 1843. Proc. Zool. Soc. Lond., 1842:124 [1843].
TYPE SPECIES: *Anomalurus fraseri* Waterhouse, 1843 (= *Pteromys derbianus* Gray, 1842).
SYNONYMS: *Anomalurella* Matschie, 1914; *Anomalurops* Matschie, 1914; *Aroaethrus* Waterhouse, 1843.
COMMENTS: See Misonne (1974, pt. 6:3–5).

Anomalurus beecrofti Fraser, 1853. Proc. Zool. Soc. Lond., 1852:17 [1853].
COMMON NAME: Beecroft's Scaly-tailed Squirrel.
TYPE LOCALITY: Equatorial Guinea, "Fernando Po" (= Bioko).
DISTRIBUTION: Guinea-Bissau, Senegal, Sierra Leone, Ghana, Togo, Nigeria, Cameroon, Equatorial Guinea, N Angola, Dem. Rep. Congo, W Uganda, NW Zambia.
STATUS: CITES – Appendix III (Ghana); IUCN – Lower Risk (lc). Not mentioned in Schlitter (1989).
SYNONYMS: *argenteus* Schwann, 1904; *chapini* (J. A. Allen, 1922); *citrinus* Thomas, 1916; *fulgens* Gray, 1869; *hervoi* Dekeyser and Villiers, 1951; *laniger* Temminck, 1853; *schoutedeni* Verheyen, 1968.
COMMENTS: Belongs to the distinct genus *Anomalurops* according to Ansell (1978:73), Grubb et al. (1998:185), Rosevear (1969:159), and Verheyen (1968:404). Taxa *argenteus* and *hervoi* were considered "valid as a race" by Rosevear (1969); *chapini* and *citrinus* were considered "well defined subspecies" by Verheyen (1968); *schoutedeni* was considered a distinct species by Verheyen (1968) and by Cabral (1971). There is no proof of the validity of subspecies (A. C. Schunke, in litt.).

Anomalurus derbianus (Gray, 1842). Ann. Mag. Nat. Hist., [ser. 1], 10:262.
COMMON NAME: Lord Derby's Scaly-tailed Squirrel.
TYPE LOCALITY: "Sierra Leone".
DISTRIBUTION: Sierra Leone, Côte d'Ivoire, Ghana, Togo, Nigeria, Cameroon, Equatorial Guinea, Gabon, Angola, Dem. Rep. Congo, Uganda, Kenya, Tanzania, Zambia, N Malawi, Mozambique.
STATUS: CITES – Appendix III (Ghana); IUCN – Lower Risk (lc). Not mentioned in Schlitter (1989).
SYNONYMS: *beldeni* Du Chaillu, 1860; *chrysophaenus* Dubois, 1888; *cinereus* Thomas, 1895; *erythronotus* Milne-Edwards, 1879; *fortior* Lönnberg, 1917; *fraseri* (Waterhouse, 1843); *griselda* Dollman, 1914; *imperator* Dollman, 1911; *jacksoni* de Winton, 1898; *jordani* St. Leger, 1935; *laticeps* d'Aquilar-Amat, 1922; *neavei* Dollman, 1909; *nigrensis* Thomas, 1904; *orientalis* Peters, 1880; *perustus* Thomas, 1914; *squamicaudus* (Schinz, 1845).
COMMENTS: See Misonne (1974, pt. 6:4-5). Distribution in Côte d'Ivoire formerly questionable, but occurrence confirmed by Fischer et al. (2002). Taxa *fraseri, imperator* and *nigrensis* were considered "valid as a race" by Rosevear (1969); *beldeni, neavei* and *perustus* were considered "well defined subspecies" by Verheyen (1968); *laticeps* and *squamicaudus* are synonyms of *fraseri* (Allen 1939); *fortior* is a synonym of *jacksoni* (Verheyen 1968). There is no proof of the validity of subspecies (A. C. Schunke, in litt.).

Anomalurus pelii (Schlegel and S. Müller, 1845) In Temminck, Verh. Nat. Gesch. Nederland. Overz. Bezitt., Zool., Mammalia, 1: pt. 2, p. 109, 1843-1845.
COMMON NAME: Pel's Scaly-tailed Squirrel.

TYPE LOCALITY: Ghana, "Daboeram" ("Daboerom (or Dabocrom), aan de Goudkust" were mis-spellings, see Grubb et al., 1998:184).

DISTRIBUTION: NE Liberia, Côte d'Ivoire to Ghana.

STATUS: CITES – Appendix III (Ghana); IUCN – Lower Risk (nt). Not mentioned in Schlitter (1989).

SYNONYMS: *auzembergeri* Matschie, 1914.

COMMENTS: Kuhn (1966:337) considered *auzembergeri* a distinct subspecies, which was confirmed by A. C. Schunke (in litt.).

Anomalurus pusillus Thomas, 1887. Ann. Mag. Nat. Hist., ser. 5, 20:440.

COMMON NAME: Dwarf Scaly-tailed Squirrel.

TYPE LOCALITY: Dem. Rep. Congo, Monbuttu (Mangbetu) region, Bellima and Tingasi (= Niangara).

DISTRIBUTION: Liberia (Rosevear 1969:159), S Cameroon, Gabon, NE and E Dem. Rep. Congo, W Uganda (?).

STATUS: IUCN – Lower Risk (lc).

SYNONYMS: *batesi* de Winton, 1897.

COMMENTS: Rosevear (1969:158) considered *batesi* a valid race, but not valid according to A. C. Schunke (in litt.).

Subfamily Zenkerellinae Matschie, 1898. Sitzb. Ges. Naturf. Freunde Berlin, 4:26.

SYNONYMS: *Idiurinae* Miller and Gidley, 1918.

Idiurus Matschie, 1894. Sitzb. Ges. Naturf. Freunde Berlin 8:194.

TYPE SPECIES: *Idiurus zenkeri* Matschie, 1894

COMMENTS: Revised by Verheyen (1963).

Idiurus macrotis Miller, 1898. Proc. Biol. Soc. Wash., 12:73.

COMMON NAME: Long-eared Scaly-tailed Flying Squirrel.

TYPE LOCALITY: Cameroon, "Efulen, Cameroon district, West Africa".

DISTRIBUTION: Sierra Leone, Liberia, Ghana, Cameroon, Equatorial Guinea, Gabon (see Julliot et al., 1998). N, NE, and E Dem. Rep. Congo, W Tanzania (see Schunke and Hutterer, 2000).

STATUS: CITES – Appendix III (Ghana); IUCN – Lower Risk (nt). Not mentioned in Schlitter (1989).

SYNONYMS: *cansdalei* Hayman, 1946; *langi* J. A. Allen, 1922; *panga* J. A. Allen, 1922.

COMMENTS: The names *langi* and *panga* were considered synonyms of *I. m. macrotis* by Verheyen (1963:183); *cansdalei* was regarded as a subspecies by Hayman (1946:211) and by Verheyen (1963:183). In the second edition of this text, Dieterlen (1993) included *kivuensis* Lönnberg, 1917, by mistake in *macrotis* instead of including it in *zenkeri*, as was correctly done by Verheyen (1963:182).

Idiurus zenkeri Matschie, 1894, Sitzb. Ges. Naturf. Freunde Berlin p. 197, text-f., 1894.

COMMON NAME: Pygmy Scaly-tailed Flying Squirrel.

TYPE LOCALITY: S Cameroon, Yaounde.

DISTRIBUTION: Cameroon, Equatorial Guinea, NE and E Dem. Rep. Congo, W Uganda. See Schunke and Hutterer (2000).

STATUS: IUCN – Lower Risk (nt); "Insufficiently known" according to Schlitter (1989).

SYNONYMS: *haymani* Verheyen, 1963; *kivuensis* Lönnberg, 1917.

COMMENTS: Includes *kivuensis,* originally considered (by Lönnberg), as a subspecies of *zenkeri*, then considered a valid species by Hayman (1946:211); Verheyen (1963:183) regarded it as a synonym of *I. z. zenkeri*. Dieterlen (1993:758) included *kivuensis* erronously in *I. macrotis,* see comments therein.

Zenkerella Matschie, 1898. Sitzb. Ges. Naturf. Freunde Berlin, 4:23.

TYPE SPECIES: *Zenkerella insignis* Matschie, 1898.

SYNONYMS: *Aethurus* de Winton, 1898.

Zenkerella insignis Matschie, 1898. Sitzb. Ges. Naturf. Freunde Berlin 4:24.
>   COMMON NAME: Cameroon Scaly-tail.
>   TYPE LOCALITY: "Kamerun ((Cameroon)) Afr. occ., Yaunde".
>   DISTRIBUTION: Cameroon, Equatorial Guinea (incl. Bioko), Central African Republic,
>   Republic of Congo. No records from Gabon, see Pérez de Val et al. (1995), Schunke and
>   Hutterer (2000).
>   STATUS: IUCN – Lower Risk (nt); "Insufficiently known" according to Schlitter (1989).
>   SYNONYMS: *glirinus* (de Winton, 1898).

FAMILY PEDETIDAE
by Fritz Dieterlen

Family Pedetidae Gray, 1825. Ann. Philos., n.s., 10:342.
COMMENTS: See Misonne (1974, Pt. 6:8). The phylogenetic position of the Pedetidae (as is also the case with ctenodactylids) has traditionally been uncertain because it shares both hystricognathic and sciurognathic characters. Ellerman (1940) suggested placing them in an independent Superfamily Pedetoidea within the Suborder Sciuromorpha, which was supported by Fischer and Mossman (1969), Lavocat (1974), and Wood (1974). Otianga'a-Owiti et al. (1992) investigated fetal membranes and placental development of the East African Springhare and confirmed a close relationship to Suborder Sciuromorpha, in which they would place springhares in their own Superfamily Pedetoidea. Landry (1999) recognized the pedetids as a group of high phylogenetic position and suggested the creation of a separate Suborder Pedetomorpha, of equal rank with Entodacrya and Sciurognathi. According to Huchon et al. (2000), the pedetids form an independent and early diverging major lineage because of their incisors possessing a multiserial enamel, thus suggesting a convergent evolution with the Hystricognathi. In their opinion, the pedetids could be a sister-clade of the Ctenohystrica and the earliest offshoot among rodents.

Pedetes Illiger, 1811. Prodr. Syst. Mamm. Avium., p. 81.
TYPE SPECIES: *Yerbua capensis* Forster, 1778.
SYNONYMS: *Helamis* G. Cuvier, 1821; *Helamys* G. Cuvier, 1816 [variant]; *Pedestes* Gray, 1843 [*lapsus cal.*]; *Yerbua* Forster, 1778.

Pedetes capensis (Forster, 1778). K. Svenska Vet.-Akad., Handl. Stockholm, (1)39:109.
COMMON NAME: South African Spring Hare.
TYPE LOCALITY: South Africa, Western Cape Prov., Cape of Good Hope.
DISTRIBUTION: South Africa, Namibia, Angola, Botswana, Mozambique, Zimbabwe, Zambia, S Dem. Rep. Congo.
STATUS: IUCN – Vulnerable.
SYNONYMS: *albaniensis* Roberts, 1946; *angolae* Hinton, 1920; *cafer* (Pallas, 1778); *damarensis* Roberts, 1926; *fouriei* Roberts, 1938; *orangiae* Wroughton, 1907; *salinae* Wroughton, 1907; *typicus* A. Smith, 1834.
COMMENTS: See Misonne (1974, pt. 6:8); de Graaff (1981:43). According to Matthee and Robinson (1997), there is no proof of the validity of subspecies.

Pedetes surdaster (Thomas, 1902). Ann. Mag. Nat. Hist., ser. 7, 9:440.
COMMON NAME: East African Spring Hare.
TYPE LOCALITY: Kenya, Naivasha Prov., Mordat, mile 365 of Uganda Railway.
DISTRIBUTION: Tanzania, Kenya.
STATUS: IUCN – Vulnerable (as included in *P. capensis*?).
SYNONYMS: *currax* Hollister, 1918; *dentatus* Miller, 1927; *larvalis* Hollister, 1918; *taborae* G. M. Allen and Loveridge, 1927.
COMMENTS: A distinct species according to Thomas (1902), Hollister (1918), Coe (1969), and Davies (1982). The separation of *surdaster* from *capensis* was strongly supported by Matthee and Robinson (1997) based upon genetic, morphological, and ethological differences between the East African and South African Springhares. There is no proof of the validity of subspecies.

SUBORDER HYSTRICOMORPHA Brandt, 1855.
INFRAORDER CTENODACTYLOMORPHI Chaline and Mein, 1979.

FAMILY CTENODACTYLIDAE
by Fritz Dieterlen

Family Ctenodactylidae Gervais, 1853. Ann. Sci. Nat. (Paris), ser. 3, 20:245.
COMMENTS: Reviewed by W. George (1979*a*). The phylogenetic position of the Ctenodactylidae has traditionally been uncertain, being either grouped with the Hystricognathi or considered a quite separate and early offshoot of the rodent stem (Hartenberger, 1985). According to Luckett (1985), Bugge (1985), and George (1985), ctenodactylids are an early offshoot of the hystricognathous rodents. Beintema et al. (1991), stating that no molecular data were available, investigated tissues, blood, and proteins and concluded that it was not yet possible to determine whether ctenodactylids and hystricognathous rodents "share a common ancestor or are located on separate branches." Huchon et al. (2000), using the von Willebrand factor (vWF) gene and working with two different molecular dating methods, concluded that their analyses strongly supported a sister-clade relationship between Ctenodactylidae and Hystricognathi and rejected the possibility that the Ctenodactylidae alone might be the earliest branching among rodents. The authors considered the new clade as a monophyletic suborder and named it "Ctenohystrica", including all extant Ctenodactylidae and Hystricognathi.

Ctenodactylus Gray, 1830. Spicilegia Zoologica, 2:10.
TYPE SPECIES: *Ctenodactylus massonii* Gray, 1830 (= *Mus gundi* Rothmann, 1776).
COMMENTS: Reviewed by W. George (1979*a*).

Ctenodactylus gundi (Rothmann, 1776). *In* Schlözers Briefwechsel, 1:339.
COMMON NAME: Common Gundi.
TYPE LOCALITY: Libya, Gharian, 80 km S of Tripoli.
DISTRIBUTION: N Morocco, N Algeria, Tunisia, NW Libya.
STATUS: IUCN – Lower Risk (lc).
SYNONYMS: *arabicus* (Shaw, 1801); *massonii* Gray, 1830; *typicus* A. Smith, 1834.
COMMENTS: Reviewed by W. George (1979*a*).

Ctenodactylus vali Thomas, 1902. Proc. Zool. Soc. Lond., 1902(2):11.
COMMON NAME: Val's Gundi.
TYPE LOCALITY: Libya, "Wadi Bey", NW of Bonjem, Tripoli.
DISTRIBUTION: S Morocco, W Algeria, S Tunisia, NW Libya.
STATUS: IUCN – Lower Risk (lc).
SYNONYMS: *joleaudi* Heim de Balsac 1936.
COMMENTS: Ranck (1968:253) and Corbet (1978:160) included *vali* in *gundi*, but George (1982) and Corbet and Hill (1991) listed both as distinct species.

Felovia Lataste, 1886. Le Naturaliste, 7(36):287.
TYPE SPECIES: *Massoutiera (Felovia) vae* Lataste, 1886.
COMMENTS: Proposed as a subgenus of *Massoutiera*; recognized as a valid genus by O. Thomas (1913*a*:31) and St. Leger (1931:978).

Felovia vae (Lataste, 1886). Le Naturaliste, 7(36):287.
COMMON NAME: Felou Gundi.
TYPE LOCALITY: Senegal, upper Senegal River, Medina District, Felou.
DISTRIBUTION: Senegal, Mauritania, Mali.
STATUS: IUCN – Vulnerable; see also Schlitter (1989).

Massoutiera Lataste, 1885. Le Naturaliste, 7(3):21.
TYPE SPECIES: *Ctenodactylus mzabi* Lataste 1881.

Massoutiera mzabi (Lataste 1881). Bull. Soc. Zool. de France, 6:314.
COMMON NAME: Mzab Gundi.
TYPE LOCALITY: Algeria, Mzab, Ghardaia.
DISTRIBUTION: SE Algeria, SW Libya, NE Mali, N Niger, N Chad.
STATUS: IUCN – Lower Risk (lc).
SYNONYMS: *harterti* Thomas, 1913; *rothschildi* Thomas and Hinton, 1921.

Pectinator Blyth, 1856. Journ. Asiatic Soc. Bengal, for 1855, (2)24:294 [1856].
TYPE SPECIES: *Pectinator spekei* Blyth, 1856.
SYNONYMS: *Petrobates* Heuglin, 1860.

Pectinator spekei Blyth, 1856. Journ. Asiatic Soc. Bengal. for 1855, (2)24:294, pl. 2, f. 1 [1856].
COMMON NAME: Speke's Pectinator.
TYPE LOCALITY: Somalia (between Goree Bunder and Nogal, ca. 09°N, 47° E).
DISTRIBUTION: Ethiopia, Eritrea, Djibouti, Somalia.
STATUS: IUCN – Lower Risk (lc).
SYNONYMS: *legerae* de Beaux, 1934; *meridionalis* de Beaux, 1922.

INFRAORDER HYSTRICOGNATHI Brandt, 1855.
by Charles A. Woods and C. William Kilpatrick

INFRAORDER HYSTRICOGNATHI Brandt, 1855.

COMMENTS: Originally used as a tribe (Tribus) by Tullberg under his Order Glires, Suborder Simplicidentati, the first time this natural group was united together without various hystricomorphous forms. Includes: Hystricidae, Thryonomyoidea, Bathyergoidea, "Caviomorpha", and the Eocene-Oligocene Franimorpha. See Tullberg (1899:69-71) and Wood (1985:478-495) for definitions of hystricognath characters and lists of taxa. The Ctenodactylidae have been identified as the sister taxon of the Hystricognathi from analyses of molecular data (Adkins et al., 2003; Huchon et al., 2000) and Huchon et al. (2000) suggested that the two taxa be recognized as the Ctenohystricha. The name Ctenohystricha is predated by Entodacrya Landry, 1999. Woods (1993) suggested that the term "Caviomorpha" was inappropriate since it was unlikely that all New World forms were part of a single radiation, however, molecular data (Huchon and Douzery, 2001; Nedbal et al., 1994) have consistently found that the Caviomorpha form a monophyletic assemblage distinct from the Old World Phiomorpha. Molecular data also support the division of the New World hystricognaths into superfamily groupings (Chinchilloidea, Cavioidea, Erethizontoidea, and Octodontoidea) suggested by Woods (1982). McKenna and Bell (1997) divided the Hystricognathi into two parvorders the Bathyergomorphi containing the Bathyergidae and the Caviida including all of the New World families of Hystricognathi with the exception of the Erethizontidae. The Erethizontidae, Hystrichidae, Thryonomyidae and Petromuridae were not assigned to either parvorder. Although molecular data (Huchon and Douzery, 2001) support the distinctiveness of the Hystricidae, the Erethizontidae are clearly part of a New World clade and the Bathyergidae form a clade with the Thryonomyidae and Petromuridae (Huchon and Douzery, 2001; Nedbal et al., 1994).

Family Bathyergidae Waterhouse, 1841. Ann. Mag. Nat. Hist., [ser.1], 8:81.

SYNONYMS: Orycterideae Lesson, 1842.

COMMENTS: The family traditionally has been divided into two subfamilies (Roberts, 1951): Bathyerginae with grooved upper incisors (*Bathyergus*); and Georychinae with ungrooved upper incisors (*Cryptomys, Georychus, Heliophobius, Heterocephalus*). Ellerman et al. (1953:227) suggested that a third subfamily should be recognized for the aberrant East African *Heterocephalus*. Molecular data (Allard and Honeycutt, 1992; Faulkes et al., 1997; Honeycutt et al., 1991; Janecek et al., 1992; Nevo et al., 1987; Walton et al., 2000) do not support the traditional division into two subfamilies, but *Heterocephalus* is the most basal taxon of this family in nearly all of these studies. McKenna and Bell (1997) recognized two subfamilies, the Bathyerginae and the Heterocephalinae and placed the family in the parvorder Bathyergomorphi. Recent protein electrophoretic (Filippucci et al., 1994; 1997; Janecek et al., 1992; Nevo et al., 1987), karyotypic (Aguilar, 1993; Burda et al., 1999; Macholán et al., 1993; 1998; Nevo et al., 1986; Scharff et al., 2001) and molecular (Allard and Honeycutt, 1992; Faulkes et al., 1997; Honeycutt et al., 1991; Nedbal et al., 1994; Walton et al., 2000) studies have contributed substantially to the resolution of the systematics of mole-rats.

Subfamily Bathyerginae Waterhouse, 1841. Ann. Mag. Nat. Hist., [ser.1], 8:81.

SYNONYMS: Georychinae Roberts, 1951.

COMMENTS: Includes both the Bathyerginae and the Georychinae of Roberts (1951), thus including all living genera of the Bathyergidae except *Heterocephalus*.

Bathyergus Illiger, 1811. Prodr. Syst. Mamm. Avium., p. 86.

TYPE SPECIES: *Mus maritimus* Gmelin, 1788 (= *Mus suillus* Schreber, 1782).

SYNONYMS: *Orycterus* Cuvier, 1829.

Bathyergus janetta Thomas and Schwann, 1904. Abstr. Proc. Zool. Soc. Lond., 1904(2):6.

COMMON NAME: Namaqua Dune Mole-rat.

TYPE LOCALITY: South Africa, Northern Cape Prov., coastal Little Namaqualand, Port Nolloth.
DISTRIBUTION: SW South Africa; S Namibia.
STATUS: IUCN – Lower Risk (nt).
SYNONYMS: *inselbergensis* Shortridge and Carter, 1938; *plowesi* Roberts, 1946.
COMMENTS: Ellerman et al. (1953) included *janetta* as a subspecies of *suillus*, but de Graaff (1975) regarded *janetta* and *suillus* as separate species. The genetic distance (Janeck et al., 1992; Nevo et al., 1987) and sequence divergence (Allard and Honeycutt, 1992) observed between these two taxa are smaller or similar to values reported between subspecies of *Cryptomys hottentotus*. Roberts (1951) recognized three subspecies of *janetta* but de Graaff (1981) considered this taxon monotypic. Karyotype has 2n=54 and FN=104 (Nevo et al., 1986).

Bathyergus suillus (Schreber, 1782). Die Säugethiere, 4:715.
COMMON NAME: Cape Dune Mole-rat.
TYPE LOCALITY: South Africa, Cape of Good Hope.
DISTRIBUTION: S South Africa.
STATUS: IUCN – Lower Risk (lc).
SYNONYMS: *africana* (Lamarck, 1796); *intermedius* Roberts, 1926; *maritimus* (Gmelin, 1788).
COMMENTS: De Graaff (1981) considered *suillus* monotypic. Karyotype has 2n=56 and FN=102 (Nevo et al., 1986).

Cryptomys Gray, 1864. Proc. Zool. Soc. Lond., 1864:124.
TYPE SPECIES: *Georychus holosericeus* Wagner, 1842 (= *Bathyergus hottentotus* Lesson, 1826).
SYNONYMS: *Coetomys* Gray, 1864; *Typhloryctes* Fitzinger, 1867.
COMMENTS: Originally a subgenus of *Georychus*; a relationship generally not supported by molecular data (Allard and Honeycutt, 1992; Honeycutt et al., 1991; Janecek et al., 1992; Nevo et al., 1987; Walton et al., 2000), which suggest a sister taxon relationship between *Georychus* and *Bathyergus*. This genus has long been problematic because of extreme morphological variation and is in need of further revision.

Cryptomys amatus (Wroughton, 1907). Manchester Mem., 51(5):28.
COMMON NAME: Zambian Mole-rat.
TYPE LOCALITY: Zambia, Alala Plateau.
DISTRIBUTION: Zambia and Dem. Rep. Congo.
SYNONYMS: *molyneuxi* (Chubb, 1908).
COMMENTS: Included as a subspecies of *hottentotus* by de Graaff (1975) but considered a distinct species based on chromosomal differentiation by Macholán et al. (1998). Karyotype has 2n=50 and FN=92 (Macholán et al., 1998).

Cryptomys anselli Burda, Zima, Scharff, Macholán, and Kawalika, 1999. Z. Säugetierk., 64(1):37.
COMMON NAME: Ansell's Mole-rat.
TYPE LOCALITY: Zambia, Lusaka Prov., NE part of Lusaka, Chainama Hills Golf Club.
DISTRIBUTION: Vicinity of Lusaka, S Zambia.
COMMENTS: Protein electrophoretic data (Filippucci et al., 1997) suggest a close relationship with *mechowi*. Karyotype has 2n=68 and FN=75-78 (Burda et al., 1999).

Cryptomys bocagei (de Winton, 1897). Ann. Mag. Nat. Hist., ser. 6, 20:323.
COMMON NAME: Bocage's Mole-rat.
TYPE LOCALITY: Angola, Hanha.
DISTRIBUTION: C Angola, NW Zambia, S Dem. Rep. Congo.
STATUS: IUCN – Lower Risk (lc).
SYNONYMS: *kubangensis* (Monard, 1933).
COMMENTS: Included as a subspecies of *hottentotus* by de Graaff (1975, 1981) but considered a distinct species based on characteristics of the infraorbital foramina by Honeycutt et al. (1991).

Cryptomys damarensis (Ogilby, 1838). Proc. Zool. Soc. Lond., 1838:5.
COMMON NAME: Damara Mole-rat.
TYPE LOCALITY: Namibia, Damaraland.

DISTRIBUTION: E Namibia, Botswana, W Zimbabwe, S Zambia, S Angola.

STATUS: IUCN – Lower Risk (lc).

SYNONYMS: *lugardi* (de Winton, 1898); *micklemi* (Chubb, 1909); *ovamboensis* Roberts, 1946.

COMMENTS: Included as a subspecies of *hottentotus* by de Graaff (1975, 1981) but considered a distinct species based on characteristics of the infraorbital foramina by Honeycutt et al. (1991). Protein electrophoretic data (Filippucci et al., 1994; 1997) suggest this taxon is more closely related to *anselli*, *kafuensis*, and *mechowi* than to *hottentotus*. Karyotype has 2n=74 or 78 and FN=92 (Nevo et al., 1986).

Cryptomys darlingi (Thomas, 1895). Ann. Mag. Nat. Hist., ser. 6, 16:239.

COMMON NAME: Darling's Mole-rat.

TYPE LOCALITY: Zimbabwe, Salisbury (= Harare) (1524 m).

DISTRIBUTION: E Zimbabwe and W Mozambique.

SYNONYMS: *beirae* (Thomas and Wroughton, 1907); *nimrodi* (de Winton, 1896); *zimbitiensis* Roberts, 1946.

COMMENTS: Included as a subspecies of *hottentotus* by de Graaff (1975, 1981) and Honeycutt et al. (1991) but considered a distinct species based on chromosomal differentiation by Aguilar (1993). Karyotype has 2n=54 and FN=80 (Aguilar, 1993).

Cryptomys foxi (Thomas, 1911). Ann. Mag. Nat. Hist., ser. 8, 7:462.

COMMON NAME: Nigerian Mole-rat.

TYPE LOCALITY: Nigeria, Panyam (1212 m).

DISTRIBUTION: C Nigeria.

STATUS: IUCN – Lower Risk (nt).

COMMENTS: Honeycutt et al. (1991:50) recognized *foxi* as distinct. Karyotype has 2n=66 and FN=126 or 2n=70 and FN=134 (Williams et al., 1984).

Cryptomys hottentotus (Lesson, 1826). Zool., 1:166.

COMMON NAME: Southern African Mole-rat.

TYPE LOCALITY: South Africa, Western Cape Prov., near Paarl (east of Capetown).

DISTRIBUTION: South Africa to Tanzania, S Dem. Rep. Congo, and Namibia.

STATUS: IUCN – Lower Risk (lc).

SYNONYMS: *albus* (Roberts, 1913); *bigalkei* Roberts, 1924; *caecutiens* (Brants, 1827); *cradockensis* Roberts, 1924; *exenticus* (Trouessart, 1899); *holosericeus* (Wagner, 1842); *jorisseni* (Jameson, 1909); *ludwigii* (Smith, 1829); *nemo* G. Allen, 1939; *orangiae* Roberts, 1926; *pallidus* (Roberts, 1917); *talpoides* (Thomas and Schwann, 1906); *transvaalensis* Roberts, 1924; *valschensis* Roberts, 1946; *vandami* (Roberts, 1917); *vetensis* Roberts, 1926; *vryburgensis* (Roberts, 1917); **natalensis** Roberts, 1913; *aberrans* (Roberts, 1913); *anomalus* (Roberts, 1913); *arenarius* (Roberts, 1913); *jamesoni* (Roberts, 1913); *junodi* Roberts, 1926; *komatiensis* (Roberts, 1917); *langi* Roberts, 1929; *mahali* (Roberts, 1913); *melanoticus* Roberts, 1926; *montanus* Roberts, 1926; *palki* (Roberts, 1917); *pretoriae* (Roberts, 1913); *rufulus* (Roberts, 1917); *stellatus* (Roberts, 1917); *streeteri* Roberts, 1946; *zuluensis* Roberts, 1951; **whytei** Thomas, 1897; *occlusus* Allen and Loveridge, 1933.

COMMENTS: Includes *holosericeus* and *natalensis* (de Graaff, 1975:3-4; Honeycutt et al., 1991:51). Corbet and Hill (1991:208) recognized *natalensis* as a distinct species without comment. De Graaff (1975, 1981) recognized seven subspecies (*amatus*, *bocagei*, *damarensis*, *darlingi*, *hottentatus*, *natalensis*, and *whytei*) but *amatus*, *bocagei*, *damarensis* and *darlingi* have subsequently been considered distinct species. Karyotype of *C. h. hottentotus* has 2n=54 and FN=106 (Nevo et al., 1986) and *C. h. natalensis* has 2n=54 and FN=104 (Nevo et al., 1986).

Cryptomys kafuensis Burda, Zima, Scharff, Macholán, and Kawalika, 1999. Z. Säugetierk., 64(1):39.

COMMON NAME: Kafue Mole-rat.

TYPE LOCALITY: Zambia, Southern Prov., Kafue National Park, "hot springs" in Itezhi-Tezhi.

DISTRIBUTION: Vicinity of Itezhi-Tezhi, S Zambia.

COMMENTS: Protein electrophoretic data (Filippucci et al., 1997) suggest affinity with *anselli* and *mechowi*. Karyotype with a 2n=58 and FN=78 (Burda et al., 1999).

Cryptomys mechowi (Peters, 1881). Sitzb. Ges. Naturf. Fr., Berlin, p. 133.

COMMON NAME: Giant Mole-rat.

TYPE LOCALITY: Angola, Malange.
DISTRIBUTION: Angola, S Dem. Rep. Congo, Malawi, Zambia, Tanzania.
STATUS: IUCN – Lower Risk (lc).
SYNONYMS: *ansorgei* (Thomas and Wroughton, 1905); *blainei* Hinton, 1921; ***mellandi***
(Thomas, 1906).
COMMENTS: Includes *ansorgei, blainei,* and *mellandi* (de Graaff, 1975:3). Corbet and Hill
(1991:207) spelled the name *mechowii*. Karyotype has 2n=40 and FN=76 (Macholán
et al., 1993). Filippucci et al. (1997) suggested affinity with *anselli, kafuensis* and
damarensis.

Cryptomys ochraceocinereus (Heuglin, 1864). Nouv. Acta Acad. Caes. Leop. Dresden, 31:3.
COMMON NAME: Ochre Mole-rat.
TYPE LOCALITY: Sudan, Upper Bahr-el-Ghazal.
DISTRIBUTION: E Nigeria, Central African Republic, N Dem. Rep. Congo, S Sudan,
NW Uganda.
STATUS: IUCN – Lower Risk (lc).
SYNONYMS: *kummi* (Thomas, 1911); *lechei* (Thomas, 1895); ***oweni*** Setzer, 1956.
COMMENTS: Includes *kummi* and *lechei* (de Graaff, 1975:3).

Cryptomys zechi (Matschie, 1900). Sitzb. Ges. Naturf. Fr., Berlin, p. 146.
COMMON NAME: Togo Mole-rat.
TYPE LOCALITY: Togo, near Kete-Kradji.
DISTRIBUTION: EC Ghana, WC Togo.
STATUS: IUCN – Lower Risk (nt).
COMMENTS: Included in *ochraceocinereus* by de Graaff (1975) but considered a distinct species
based on characteristics of the infraorbital foramina by Honeycutt et al. (1991).

Georychus Illiger, 1811. Prodr. Syst. mamm. Avium., p. 87.
TYPE SPECIES: *Mus capensis* Pallas, 1778.
SYNONYMS: *Fossor* Lichtenstein, 1844; *Georhychus* Wagner, 1843; *Georrychus* Minding, 1829.

Georychus capensis (Pallas, 1778). Njova. Spec. Quad. Glir. Ord., 76:172.
COMMON NAME: Cape Mole-rat.
TYPE LOCALITY: South Africa, Cape of Good Hope.
DISTRIBUTION: South Africa.
STATUS: IUCN – KwaZulu Natal population Critically Endangered, otherwise Lower Risk (lc).
SYNONYMS: *buffonii* (Cuvier, 1834); *canescens* (Thomas and Schwann, 1906); *leucops*
(Lichtenstein, 1844); *yatesi* (Roberts, 1913).
COMMENTS: Roberts (1951) recognized three subspecies *capensis, canescens* and *yatesi* but
de Graaff (1975) considered this taxon monotypic. Karyotype has 2n=54 and FN=100
(Nevo et al., 1986). Honeycutt et al. (1991:53) indicated there may be two species in
South Africa, based on differences in mtDNA sequences and allozyme frequencies.

Heliophobius Peters, 1846. Bericht Verhandl. K. Preuss. Akad. Wiss. Berlin, 11:259.
TYPE SPECIES: *Heliophobius argenteocinereus* Peters, 1846.
SYNONYMS: *Myoscalops* Thomas, 1890.

Heliophobius argenteocinereus Peters, 1846. Bericht Verhandl. K. Preuss. Akad. Wiss. Berlin, 11:259.
COMMON NAME: Silvery Mole-rat.
TYPE LOCALITY: Mozambique, Tete (on the Zambezi River).
DISTRIBUTION: Zimbabwe, E Zambia, and N Mozambique to Dem. Rep. Congo, Kenya, and
N Tanzania.
STATUS: IUCN – Lower Risk (nt).
SYNONYMS: ***albifrons*** (Gray, 1864); *pallidus* (Gray, 1864); ***angonicus*** Thomas, 1917; ***emini***
Noack, 1894; ***kapiti*** Heller, 1909; ***marungensis*** Noack, 1887; ***mottoulei*** (Schouteden,
1913); ***robustus*** Thomas, 1906; ***spalax*** Thomas, 1910.
COMMENTS: Honeycutt et al. (1991:54-55) concluded that the characters used to separate
H. spalax from *H. argenteocinereus* are due to age variation, and that the genus is mono-
typic. See also de Graaff (1975, 1981) who considered the species polytypic with nine

subspecies. Karyotype of specimens from Kenya has 2n=60 and FN=114 (W. George, 1979*b*) whereas specimens from Zambia have 2n=62 and FN=114 (Scharff et al., 2001).

Subfamily Heterocephalinae Landry, 1957. Univ. Calif. Publ. Zool., 56:74.
COMMENTS: Molecular data (Allard and Honeycutt, 1992; Faulkes et al., 1997; Nedbal et al., 1994; Walton et al., 2000) support the recognition of the aberrant East African *Heterocephalus* as a distinct subfamily.

Heterocephalus Rüppell, 1842. Mus. Senckenbergianum Abh., 3(2):99.
TYPE SPECIES: *Heterocephalus glaber* Rüppell, 1842.
SYNONYMS: *Fornarina* Thomas, 1903.

Heterocephalus glaber Rüppell, 1842. Mus. Senckenbergianum Abh., 3(2):99.
COMMON NAME: Naked Mole-rat.
TYPE LOCALITY: Ethiopia, Shoa.
DISTRIBUTION: C Somalia, C and E Ethiopia, C and S Kenya.
STATUS: IUCN – Lower Risk (lc). Common.
SYNONYMS: *ansorgei* Thomas, 1903; *dunni* Thomas, 1909; *phillipsi* Thomas, 1885; *progrediens* Lönnberg, 1911; *scortecci* de Beaux, 1934; *stygius* Allen, 1912.
COMMENTS: Allen (1939) recognized two subspecies, however, this taxon is in need of a thorough revision (Honeycutt et al., 1991). According to Honeycutt et al. (1991:58), genetic data indicate there are two geographic groups of this species in Kenya. Karyotype has 2n=60 (W. George, 1979*b*).

Family Hystricidae G. Fischer, 1817. Mem. Soc. Imp. Nat., Moscow, 5:372.
COMMENTS: Reviewed by Mohr (1965) and by Van Weers (1976, 1977, 1978, 1979, 1983). Usually divided into two subfamilies (Atherurinae, Hystricinae) see McKenna and Bell (1997), but Van Weers (1977, 1978, 1979) does not agree. This family appears to be the most divergent of the suborder (Huchon and Douzery, 2001).

Atherurus F. Cuvier, 1829. Dict. Sci. Nat., 59:483.
TYPE SPECIES: *Hystrix macroura* Linnaeus, 1758.
COMMENTS: See Van Weers (1977:213).

Atherurus africanus Gray, 1842. Ann. Mag. Nat Hist., [ser. 1], 10:261.
COMMON NAME: African Brush-tailed Porcupine.
TYPE LOCALITY: Sierra Leone.
DISTRIBUTION: Gambia, Sierra Leone, Liberia, Ghana, Dem. Rep. Congo, Kenya, Uganda, S Sudan.
STATUS: IUCN – Lower Risk (lc).
SYNONYMS: *armata* Gervais, 1854; *burrowsi* Thomas, 1902; *centralis* Thomas, 1895; *turneri* St. Leger, 1932.
COMMENTS: Includes *centralis* and *turneri*; see Misonne (1974:8).

Atherurus macrourus (Linnaeus, 1758). Syst. Nat., 10th ed., 1:57.
COMMON NAME: Asiatic Brush-tailed Porcupine.
TYPE LOCALITY: "Habitat in Asia", restricted to Malaysia, Malacca by Lyon (1907:584).
DISTRIBUTION: E Assam (India), Szeechwas, Yunnan, Hupei, and Hainan (China) to Malaya, Sumatra, and adjacent islands, Thailand, Laos, Vietnam, Burma.
STATUS: IUCN – Endangered as *A. m. assamensis*, otherwise Lower Risk (lc).
SYNONYMS: *angustiramus* Mohr, 1964; *assamensis* Thomas, 1921; *hainanus* Allen, 1906; *pemangilis* Robinson, 1912; *retardatus* Mohr, 1964; *stevensi* Thomas, 1925; *terutaus* Lyon, 1907; *tionis* Thomas, 1908; *zygomatica* Miller, 1903.
COMMENTS: Includes *angustiramus*, *assamensis*, *hainanus*, *retardatus*, *stevensi*, *terutaus*, *tionis*, and *zygomatica* (Van Weers, 1977).

Hystrix Linnaeus, 1758. Syst. Nat., 10th ed., 1:56.
TYPE SPECIES: *Hystrix cristata* Linnaeus, 1758.

SYNONYMS: *Acanthion* Cuvier, 1823; *Acanthochoerus* Gray, 1866; *Oedocephalus* Gray, 1866 [see Allen, 1939]; *Thecurus* Lyon, 1907.

COMMENTS: Divided into three subgenera: *Acanthion, Hystrix,* and *Thecurus* (see Van Weers, 1978; 1979).

Hystrix africaeaustralis Peters, 1852. Reise nach Mossambique, Säugeth., p. 170.
COMMON NAME: Cape Porcupine.
TYPE LOCALITY: Mozambique, Querimba coast and Tette, about 10°30' to 12°S, 40°30'E, sea level.
DISTRIBUTION: Mouth of the Congo River to Rwanda, Uganda, Kenya, W and S Tanzania, Mozambique, and South Africa. Sympatric with *H. cristata.*
STATUS: IUCN – Lower Risk (lc), common.
SYNONYMS: *capensis* Grill, 1858; *prittwitzi* Müller, 1910; *stegmanni* Müller, 1910; ***zuluensis*** Roberts, 1936.
COMMENTS: Subgenus *Hystrix.* Revised by Corbet and Jones (1965); see Van Weers (1983).

Hystrix brachyura Linnaeus, 1758. Syst. Nat., 10th ed., 1:57.
COMMON NAME: Malayan Porcupine.
TYPE LOCALITY: Malaysia, Malacca.
DISTRIBUTION: Nepal, Sikkim, and Assam (India), C and S China, Burma, Thailand, Indochina, Malaya, Sumatra, Borneo, Malay Isls of Pinang, Singapore.
STATUS: IUCN – Vulnerable.
SYNONYMS: *grotei* (Gray, 1866); *longicauda* Marsden, 1811; *mulleri* Marshall, 1871; ***bengalensis*** Blyth, 1851; ***hodgsoni*** (Gray, 1847); *alophus* Hodgson, 1847; ***subcristata*** Swinhoe, 1870; *klossi* (Thomas, 1916); *millsi* (Thomas, 1922); *papae* (Allen, 1927); ***yunnanensis*** Anderson, 1878.
COMMENTS: Subgenus *Acanthion* (Van Weers, 1979:233); also see Lekagul and McNeely (1988:492).

Hystrix crassispinis (Günther, 1877). Proc. Zool. Soc. Lond., 1876:736 [1877].
COMMON NAME: Thick-spined Porcupine.
TYPE LOCALITY: Malaysia, Sabah, opposite Labuan Isl.
DISTRIBUTION: N Borneo.
STATUS: IUCN – Lower Risk (nt).
SYNONYMS: *major* (Schwarz, 1939).
COMMENTS: Subgenus *Thecurus* (Van Weers, 1978:22).

Hystrix cristata Linnaeus, 1758. Syst. Nat., 10th ed., 1:56.
COMMON NAME: Crested Porcupine.
TYPE LOCALITY: "Asia", restricted to near Rome, Italy by Thomas (1911a:141).
DISTRIBUTION: Morocco to Egypt; Senegal to Ethiopia and N Tanzania; Sicily, Italy, Albania, and N Greece (European population possibly introduced). Sympatric with *H. africaeaustralis* in C Africa.
STATUS: CITES – Appendix III (Ghana); IUCN – Lower Risk (nt).
SYNONYMS: *aerula* Thomas, 1925; *ambigua* Lönnberg, 1908; *conradsi* Müller, 1910; *cuvieri* Gray, 1847; *daubentoni* Cuvier, 1822; *europaea* Kerr, 1792; *galeata* Thomas, 1893; *lademanni* Müller, 1910; *lonnebergi* Müller, 1910; *occidanea* Cabrera, 1924; *senegalica* Cuvier, 1822.
COMMENTS: Subgenus *Hystrix.* Includes *galeata;* see Corbet (1978c:159) and Corbet and Jones (1965). Allen (1939:441) commented on the status of *cuvieri* in N Africa. Karyotype has 2n=60-66 and FN=112-116 (George, 1980; George and Weir, 1974).

Hystrix indica Kerr, 1792. In Linnaeus, Anim. Kingdom, p. 213.
COMMON NAME: Indian Crested Porcupine.
TYPE LOCALITY: India.
DISTRIBUTION: Transcaucasus; Asia Minor; Israel; Arabia to S Kazakhstan and India; Sri Lanka; Tibet (China).
STATUS: IUCN – Lower Risk (lc). Locally common.
SYNONYMS: *aharonii* Müller, 1911; *blanfordi* Müller, 1911; *cuneiceps* Wroughton, 1912; *hirsutirostris* Brandt, 1835; *leucurus* Sykes, 1831; *malabarica* Sclater, 1865; *mersinae*

Müller, 1911; *mesopotamica* Müller, 1911; *narynensis* Müller, 1911; *satunini* Müller, 1911; *schmidtzi* Müller, 1911; *zeylonensis* Blyth, 1851.
COMMENTS: Subgenus *Hystrix*. Citation based on Smellie's translation of Buffon (1781:206). Gromov and Baranova (1981:102) employed the name *leucura* for this species without reference to Kerr, 1792.

Hystrix javanica (F. Cuvier, 1823). Mem. Mus. Hist. Nat. Paris, 9:431.
COMMON NAME: Sunda Porcupine.
TYPE LOCALITY: Indonesia, Java.
DISTRIBUTION: Java, Bali, Sumbawa, Flores, Lombok, Madura, Tonahdjampea, and S Sulawesi (Indonesia).
STATUS: IUCN – Lower Risk (lc).
SYNONYMS: *brevispinosa* Wagner, 1844; *ecaudata* van der Hoeven and de Vriese, 1836; *javanicum* (Cuvier, 1823); *sumbawae* (Schwarz, 1911); *torquata* van der Hoeven and de Vriese, 1836.
COMMENTS: Subgenus *Acanthion*. Formerly included in *brachyura* by Chasen (1940), but see Van Weers (1979).

Hystrix pumila (Günther, 1879). Ann Mag. Nat. Hist., ser. 5, 4:106.
COMMON NAME: Indonesian Porcupine.
TYPE LOCALITY: Philippines, Palawan, Paragua (=Perto Princesa).
DISTRIBUTION: Palawan and Busuanga Isls (Philippines).
STATUS: IUCN – Lower Risk (lc).
COMMENTS: Subgenus *Thecurus*.

Hystrix sumatrae (Lyon, 1907). Proc. U.S. Natl. Mus., 32:583.
COMMON NAME: Sumatran Porcupine.
TYPE LOCALITY: Indonesia, E coast of Sumatra, Aru Bay.
DISTRIBUTION: Sumatra.
STATUS: IUCN – Lower Risk (lc).
COMMENTS: Subgenus *Thecurus*. Treated as a subspecies of *crassispinis* by Chasen (1940), but see Van Weers (1978).

Trichys Günther, 1877. Proc. Zool. Soc. Lond., 1876:739 [1877].
TYPE SPECIES: *Trichys lipura* Günther, 1877 (= *Hystrix fasciculata* Shaw, 1801).
COMMENTS: Reviewed by Van Weers (1976).

Trichys fasciculata (Shaw, 1801). Gen. Zool., 2:11.
COMMON NAME: Long-tailed Porcupine.
TYPE LOCALITY: "Malacca", but no holotype. Van Weers (1976) designated a neotype from "Runuk Tanjong," Malaysia.
DISTRIBUTION: Borneo, Sumatra, Malaya.
STATUS: IUCN – Lower Risk (lc).
SYNONYMS: *guentheri* Thomas, 1889; *lipura* Günther, 1877; *macrotis* Miller, 1903.
COMMENTS: Medway (1977:135) considered *lipura* a distinct species, but did not mention Van Weers (1976).

Family Petromuridae Wood, 1955. J. Mammal, 36(2):184.
SYNONYMS: Petromyidae Tullberg, 1899.
COMMENTS: Swanepoel et al. (1980:161) employed Petromuridae instead of Petromyidae for this family (see *Petromus*). The group had an elaborate Tertiary radiation composed of at least two subfamilies and seven known genera. The placement of this family along with the Thryonomyidae into the superfamily Thryonomuroidea is supported by many studies (Huchon and Douzery, 2001; Lavocat, 1973; Luckett and Harteberger, 1985*b*; Nedbal et al., 1994; Patterson and Wood, 1982; Sarich, 1985). The superfamily Thryonomuroidea was not recognized by McKenna and Bell (1997).

Petromus A. Smith, 1831. S. Afr. Quart. J., 1(5):10.
TYPE SPECIES: *Petromus typicus* A. Smith, 1831.

SYNONYMS: *Petromys* A. Smith, 1834.

COMMENTS: Originally named *Petromus*, but unjustifiably emended to *Petromys* by Smith (1834), see Swanepoel et al. (1980:161).

Petromus typicus A. Smith, 1831. S. Afr. Quart. J., 1(5):11.
COMMON NAME: Dassie Rat.
TYPE LOCALITY: South Africa, "Mountains toward mouth of Orange River".
DISTRIBUTION: W South Africa, Namibia, to SW Angola.
STATUS: IUCN – Lower Risk (lc). Common.
SYNONYMS: ***ausensis*** Roberts, 1938; ***barbiensis*** Roberts, 1938; ***cinnamomeus*** Roberts, 1946; ***coetzeei*** (Cabral, 1966; ***cunealis*** Thomas, 1926; ***greeni*** Lundholm, 1955; ***guinasensis*** Roberts, 1938; ***karasensis*** Roberts, 1946; ***kobosensis*** Roberts, 1938; ***majoriae*** Bradfield, 1936; ***namaquensis*** Roberts, 1938; ***pallidior*** Lundholm, 1955; ***tropicalis*** Thomas and Hinton, 1925; ***windhoekensis*** Roberts, 1938.
COMMENTS: Fifteen subspecies have been named but no attempt has been made to assess their validity or geographic limits, but see Roberts (1951). Karyotype has 2n=56 and FN=112 (George, 1980).

Family Thryonomyidae Pocock, 1922. Proc. Zool. Soc. Lond., 1922:423.
SYNONYMS: Thrynomyidae Pocock, 1922; Ulacodidae Brandt, 1855.
COMMENTS: Includes several fossil forms from the Oligocene of North Africa. Placed in the superfamily Thryonomuroidea, see comments under Petromuridae.

Thryonomys Fitzinger, 1867. Sitzb. Akad. Wiss. Wein, 56(1):141.
TYPE SPECIES: *Aulacodus swinderianus* Temminck, 1827.
SYNONYMS: *Aulacodus* Temminck, 1827 [preoccupied by *Aulacodus* Eschscholtz, 1822, a genus of Coleoptera]; *Choeromys* Thomas, 1922; *Triaulacodus* Lydekker, 1896.

Thryonomys gregorianus (Thomas, 1894). Ann. Mag. Nat. Hist., ser. 6, 13:202.
COMMON NAME: Lesser Cane Rat.
TYPE LOCALITY: Kenya, Luiji Reru River (00°35'S, 37°05'E).
DISTRIBUTION: Cameroon, Central African Republic, Dem. Rep. Congo, S Sudan, Ethiopia, Kenya, Uganda, Tanzania, Malawi, Zambia, Zimbabwe, Mozambique.
STATUS: IUCN – Lower Risk (lc). Common.
SYNONYMS: *camerunensis* Monard, 1949; *congicus* Thomas, 1922; *harrisoni* Thomas and Wroughton, 1907; *logonensis* Jeannin, 1936; *pusillus* Heller, 1912; *rutshuricus* Lönnberg, 1917; *sclateri* Thomas, 1897.
COMMENTS: Rosevear (1969:549) recognized *logonensis* and *camerunensis* as valid species; Misonne (1974:7) concluded that the affinities of *logonensis* and *camerunensis* were not clear and de Graaff (1981) indicated that their affinities should be reassessed. Misonne (1974:7) included *congicus, harrisoni, pusillus*, and *rutshuricus* in *gregoriamus* without comment of subspecific status. Meester et al. (1986) recognized two subspecies in southern Africa but did not comment on the subspecific validity of the other forms. Karyotype has 2n=40 and FN=80 (George, 1980).

Thryonomys swinderianus (Temminck, 1827). Monogr. Mamm., 1:248.
COMMON NAME: Greater Cane Rat.
TYPE LOCALITY: Sierra Leone.
DISTRIBUTION: Africa, south of the Sahara.
STATUS: IUCN – Lower Risk (lc).
SYNONYMS: *angolae* Thomas, 1922; *calamophagus* (De Beerst, 1897); *logani* Romer and Nesbitt, 1930; *raptorum* Thomas, 1922; *semipalmatus* (Heuglin, 1864); *variegatus* (Peters, 1852).

Family Erethizontidae Bonaparte, 1845. Cat. Meth. Mamm. Europe, p. 5.
SYNONYMS: Coendidae Trouessart, 1897.
COMMENTS: Spelled Erithizontidae by Corbet and Hill (1980:189); but see comments under *Erethizon*. Extinct forms date to the Oligocene of South America. Not included in the parvorder Caviida with the other New World Hystricognathi by McKenna and Bell (1997)

but placement into a New World clade well supported by molecular data (Huchon and Douzery, 2001; Nedbal et al., 1994).

Subfamily Chaetomyinae Thomas, 1897. Proc. Zool. Soc. Lond., 1896:1026 [1897].
>SYNONYMS: Cercolabidae Cabrera, 1901 [cited to Ameghino 1889 and listed as a synonym of the Erethizontidae by McKenna and Bell (1997)].
>COMMENTS: Patterson and Wood (1982) removed Chaetomyinae from the Erethizontidae and placed in the Echimyidae because this form retains deciduous premolars (an echimyid derived character) unlike all known erethizontids. *Chaetomys* shares characteristics of the enamel ultrastructure with taxa of the Erethizontidae (T. Martin, 1994*b*).

Chaetomys Gray, 1843. List Specimens Mamm. Coll. Brit. Mus., p. 123.
>TYPE SPECIES: *Hystrix subspinosus* Olfers, 1818.
>SYNONYMS: *Plectrochorerus* Pictet, 1843.
>COMMENTS: Included in Erethizontidae by Corbet and Hill (1991), Emmons and Feer (1997), McKenna and Bell (1997) and Voss and Angermann (1997) but some authors (e.g., Miller and Gidley, 1918; Patterson and Wood, 1982; Stehlin and Schaub, 1951; Woods, 1993) have included in the Echimyidae.

Chaetomys subspinosus (Olfers, 1818). Neue Bibl. Reisenb., p. 211.
>COMMON NAME: Bristle-spined Rat.
>TYPE LOCALITY: Brazil, N Bahia, Salvador.
>DISTRIBUTION: Atlantic coastal forest of SE Brazil from S Sergipe state to N Rio de Janeiro, including easternmost Minas Gerais.
>STATUS: U.S. ESA – Endangered; IUCN – Vulnerable. More common in suitable habitats than previously thought (Oliver and Santos, 1991:17), and may be quite common near Urucuca in S Bahia.
>SYNONYMS: *moricandi* (Pictet, 1943); *tortilis* (Olfers, 1820).
>COMMENTS: Voss and Angermann (1997) indicated that Wied-Neuwied (1826) incorrectly gave the type locality as Cametá, in Pará state of Brazil, that Ávila-Pires (1967) erroneously emended the type to Ilhéus in Bahia state, and document that the type material was collected from Salvador. See Oliver and Santos (1991) for discussion of the confusion surrounding the distribution of this species. Hershkovitz (1959*a*) and Cabrera (1961) listed *rutila* (Olfers, 1818) and *volubilis* (Olfers, 1818) in their synonymies for *C. subspinosus* but *H. volubilis* Olfers, 1818 is a *nomen nudum*, and *H. rutila* Olfers, 1818, is unidentifiable, although possibly a *Coendou nycthemera* (Voss and Angermann, 1997).

Subfamily Erethizontinae Bonaparte, 1845. Cat. Meth. Mamm. Europe, p. 5.
>SYNONYMS: Coendinae Pocock, 1922; Sphingurinae Alston, 1876.

Coendou Lacépède, 1799. Tabl. Mamm., p. 11.
>TYPE SPECIES: *Hystrix prehensilis* Linnaeus, 1758.
>SYNONYMS: *Coandu* Fischer, 1814; *Coendu* Lesson, 1827; *Coendus* Geoffroy, 1803; *Cuandu* Liais, 1872.
>COMMENTS: Cabrera (1961), Hall (1981), Corbet and Hill (1991), Handley and Pine (1992), Emmons and Feer (1997), McKenna and Bell (1997), Voss and Angerman (1997), Alberico et al. (1999) and Voss and da Silva (2001) included *Sphiggurus* in *Coendou*; but see Husson (1978:484-490). Woods (1984, 1993) and Nowak (1991) followed Husson (1978) and recognized *Sphiggurus* as distinct from *Coendou*. Handley and Pine (1992), Voss and Angermann (1997), and Alberico et al. (1999) concluded that *Sphiggurus* could not be meaningfully diagnosed as a taxon distinct from *Coendou*. Bonvicino et al. (2000) suggested that *Coendou* and *Sphiggurus* represent distinct lineages that have evolved by different chromosomal mechanisms. *Coendou* appears to be karyotypically conserved with 2n=74 and FN=82 (Bonvicino et al., 2000). Although karyotypes have been reported in only two taxa of *Coendou*, the chromosomal and molecular data (Bonvicino et al., 2002*a*) tentatively support the distinction of *Coendou* and *Sphiggurus*.

Coendou bicolor (Tschudi, 1844). Fauna Peruana, p. 186.
>COMMON NAME: Bicolored-spined Porcupine.
>TYPE LOCALITY: Peru, Junín Debt., between Tulumayo and Chanchamayo Rivers.
>DISTRIBUTION: Bolivia, Peru, Andean and W Ecuador, N Colombia, SW Colombia; up to
>2,500 m.
>STATUS: IUCN – Lower Risk (lc). Locally common.
>SYNONYMS: *quichua* Thomas, 1899; *richardsoni* Allen, 1913; *simonsi* Thomas, 1902.
>COMMENTS: May include *rothschildi* (Hall, 1981:854), which Corbet and Hill (1991:200)
>listed as a subspecies of *bicolor* (see comments under *rothschildi*). Emmons and Feer
>(1997:218) and Alberico et al. (1999) considered *quichua* of the Andes of Ecuador and
>Colombia to be a distinct species based on small size and shorter tail. Emmons and
>Feer (1997:218) suggested that *richardsoni* may be a valid species or allied with
>*rothschildi*, whereas Alberico et al. (1999) recognized it as a distinct species. We
>tentatively retain these three subspecies, awaiting a needed comprehensive review of
>*Coendou*.

Coendou nycthemera (Olfers, 1818). Neue Bibl. Reisenb., p.211.
>COMMON NAME: Black Dwarf Porcupine.
>TYPE LOCALITY: Brazil, E Amazonia, south of the main channel of the Rio Amazonas; as
>restricted by Voss and Angermann (1997).
>DISTRIBUTION: Amazonian lowlands east of Rio Madeira and south of the Rio Amazonas,
>including at least part of Marajó Isl.
>STATUS: IUCN – Lower Risk (lc) as *C. koopmani*.
>SYNONYMS: *koopmani* Handley and Pine, 1992.
>COMMENTS: Sympatric with *C. prehensilis* throughout its range.

Coendou prehensilis (Linnaeus, 1758). Syst. Nat., 10th ed., 1:57.
>COMMON NAME: Brazilian Porcupine.
>TYPE LOCALITY: Brazil, Pernambuco.
>DISTRIBUTION: E Venezuela, Guyanas, C and E Brazil, N Argentina, Uruguay, E Paraguay,
>Bolivia, Trinidad (see Goodwin and Greenhall, 1961:202); up to 1,500 m.
>STATUS: IUCN – Lower Risk (lc).
>SYNONYMS: *boliviensis* Gray, 1850; *brandtii* Jentink, 1879; *centralis* Thomas, 1903; *cuandu*
>Desmarest, 1822; *longicatus* Lacépède, 1799; *platycentrotus* Brandt, 1835; *sanctamartae*
>Allen, 1904; *tricolor* Gray, 1850.
>COMMENTS: Emmons and Feer (1997:217) suggested that *centralis* and *sanctamartae* may be
>valid species and Alberico et al. (1999) considered *sanctamartae* a distinct species. Voss
>and da Silva (2001) suggested *prehensilis* may represent a complex of closely related
>species. Karyotype has 2n=74 and FN=82 (Lima, 1994).

Coendou rothschildi Thomas, 1902. Ann. Mag. Nat. Hist., ser. 7, 10:169.
>COMMON NAME: Rothchild's Porcupine.
>TYPE LOCALITY: Panama, Chiriqui, Sevilla Isl.
>DISTRIBUTION: Panama.
>STATUS: IUCN – Lower Risk (lc).
>COMMENTS: Possibly a subspecies of *bicolor* (Goldman, 1920:135; Hall, 1981:854). Corbet
>and Hill (1991:200) listed *rothschildi* as a subspecies of *bicolor*. Emmons and Feer
>(1997:218) stated that if *C. rothschildi* is a valid species, the *C. "bicolor"* west of the
>Andes are possibly *rothschildi*. Karyotype has 2n=74 and FN=84 (George and Weir,
>1974).

Echinoprocta Gray, 1865. Proc. Zool. Soc. Lond., 1865:321.
>TYPE SPECIES: *Erithizon* (*Echinoprocta*) *rufescens* Gray, 1865.

Echinoprocta rufescens Gray, 1865. Proc. Zool. Soc. Lond., 1865:321.
>COMMON NAME: Stump-tailed Porcupine.
>TYPE LOCALITY: Colombia.
>DISTRIBUTION: Colombia, montane areas of the eastern cordillera of Andes between 800 and
>2,000 m (Eisenberg, 1989).

STATUS: IUCN – Lower Risk (lc).

SYNONYMS: *epixanthus* (Martínez, 1873); *sneiserni* (Lönnberg, 1937) [see comments under *Sphiggurus*].

COMMENTS: Included in *Coendou* (*Sphiggurus*) by McKenna and Bell (1997) and Alberico et al. (1999).

Erethizon F. Cuvier, 1823. Mem. Mus. Hist. Nat. Paris, 9:432.

TYPE SPECIES: *Hystrix dorsata* Linnaeus, 1758.

SYNONYMS: *Erethison* Cuvier, 1829; *Eretison* McMurtrie, 1831; *Eretizon* Cuvier, 1825; *Erithizon* Burnett, 1830; *Erythizon* Alston, 1876.

COMMENTS: *Erithizon* Burnett, 1830, is a later spelling (Corbet and Hill, 1980:189).

Erethizon dorsatum (Linnaeus, 1758). Syst. Nat., 10th ed., 1:57.

COMMON NAME: North American Porcupine.

TYPE LOCALITY: E Canada (= Quebec Prov.).

DISTRIBUTION: C Alaska (USA) to S Hudson Bay and Labrador (Canada), south to E Tennessee, C Iowa, and C Texas (USA), N Coahuila, Chihuahua, and Sonora (Mexico), and S California (USA).

STATUS: IUCN – Lower Risk (lc). Common.

SYNONYMS: *doani* Bailey, 1937; **bruneri** Swenk, 1916; **couesi** Mearns, 1897; **epixanthus** Brandt, 1835; **myops** Merriam, 1900; **nigrescens** Allen, 1903; **picinum** Bangs, 1900.

COMMENTS: Reviewed by Miller and Kellogg (1955:631), who restricted the type locality; also see Anderson and Rand (1943). Includes *couesi* (see review by Woods, 1973). Karyotype has 2n=42 and FN=78 (George and Weir, 1974).

Sphiggurus F. Cuvier, 1825. Dentes des Mammiferes, p. 256.

TYPE SPECIES: *Hystrix spinosa* F. Cuvier, 1823.

SYNONYMS: *Cercolabes* Brandt, 1835; *Sinethere* F. Cuvier, 1822; *Sinotherus* F. Cuvier, 1825; *Synetheres* G. Cuvier, 1829.

COMMENTS: This genus possibly dates from F. Cuvier, 1823. Mem. Mus. Hist. Hat. Paris, 9:427, 433-435, where *Sphiggurus* seems only a French name "Sphiggure" except on pp. 433-434, where it is abbreviated "*S. spinosa*". Formerly included in *Coendou* by Cabrera (1961), Walker et al. (1975), and Hall (1981); see also comments under *Coendou*. Considered a distinct genus by Husson (1978:484-490). Voss and Angermann (1997) agreed with the conclusion of Handley and Pine (1992) that *Sphiggurus* could not be meaningfully diagnosed as a taxon distinct from *Coendou*. Bonvicino et al. (2000) suggested that *Coendou* and *Sphiggurus* represent distinct lineages that have evolved by different chromosomal mechanisms. *Sphiggurus* appears to have evolved by Robertsonian mechanisms and demonstrates variation in 2n ranging from 42 to 72 while retaining a FN=76 (Bonvicino et al., 2000). The Robertsonian pattern of chromosomal differentiation and molecular data (Bonvicino et al., 2002*a*) tentatively supports the distinction of *Sphiggurus* from *Coendou*. Woods (1984, 1993) and Nowak (1991) followed Husson (1978) and recognized *Sphiggurus* as distinct from *Coendou*. Emmons and Feer (1990:201) recognized *Coendou* [*Sphiggurus*] *sneiderni*, the White-fronted Hairy Dwarf Porcupine from Cauca Department of Colombia, which is apparently known from only one specimen, but this species is not recognized here, and is tentatively placed in the synonymy of *Echinoprocta* following Cabrera (1961).

Sphiggurus ichillus (Voss and da Silva, 2001). Am. Mus. Novit., 3351:17.

COMMON NAME: Streaked Dwarf Porcupine.

TYPE LOCALITY: Ecuador, Río Pastaza.

DISTRIBUTION: Amazonian lowlands E Ecuador but see Voss and da Silva (2001).

COMMENTS: Assigned to *vestitus*-group based on presence of bristle-quills (Voss and da Silva, 2001).

Sphiggurus insidiosus (Olfers, 1818). Neue Bibl. Reisenb., p. 211.

COMMON NAME: Bahia Porcupine.

TYPE LOCALITY: Brazil, Bahía, Salvador.

DISTRIBUTION: Atlantic coastal region, SE Brazil.

STATUS: Unknown. IUCN – Lower Risk (lc) as *S. insidiosus*, Extinct as *S. pallidus*.

SYNONYMS: *pallidus* (Waterhouse, 1848).

COMMENTS: Reviewed by Voss and Angermann (1997), who included *pallidus*. Husson (1978) placed in *Sphiggurus* and included *melanurus* but see Voss and Angermann (1997). Karyotype has 2n=62 and FN=76 (Lima, 1994).

Sphiggurus melanurus (Wagner, 1842). Archiv. für Naturg. I:360.

COMMON NAME: Black-tailed Hairy Dwarf Porcupine.

TYPE LOCALITY: Brazil, Rio Negro, Barra.

DISTRIBUTION: North of the Amazon in the Amazon Basin of Brazil, the Guianas, Colombia, and probably Venezuela.

COMMENTS: Considered conspecific with *insidiosus* by Cabrera (1961) and Husson (1978), but see Voss and Angermann (1997). Considered a distinct species by Emmons and Feer (1997), Handley and Pine (1992), and Voss and Angermann (1997), a position supported by molecular and karyotypic data (Bonvicino et al., 2002*a*). Karyotype has 2n=72 and FN=76 (Bonvicino et al., 2002*a*).

Sphiggurus mexicanus (Kerr, 1792). *In* Linnaeus, Anim. Kingdom, 1:214.

COMMON NAME: Mexican Hairy Dwarf Porcupine.

TYPE LOCALITY: Mexico, in the mountains.

DISTRIBUTION: San Luis Potosi and Yucatan (Mexico) to W Panama, to 3,000 m (Emmons and Feer, 1997).

STATUS: CITES – Appendix III (Honduras); IUCN – Lower Risk (lc).

SYNONYMS: *liebmani* (Reinhart, 1844); **laenatus** (Thomas, 1903); **yucataniae** (Thomas, 1902).

COMMENTS: Emmons and Feer (1997:222) indicated that the small forms from Chiriquí, Panama had been recognized as a distinct species (*laenatus*).

Sphiggurus pruinosus (Thomas, 1905). Ann. Mag. Nat. Hist., ser. 7, 16:310.

COMMON NAME: Frosted Hairy Dwarf Porcupine.

TYPE LOCALITY: Venezuela, Montañas de la Pedregosa near Mérida, 2500 m.

DISTRIBUTION: C and N Colombia and W and N Venezuela.

COMMENTS: Included in *vestitus* by Cabrera (1961), Concepcion and Molinari (1991), Woods (1993), and Soriano and Ochoa (1997), but see Voss and da Silva (2001) for a discussion of the morphological distinctiveness of these two taxa. Assigned to *vestitus*-group based on presence of bristle-quills (Voss and da Silva, 2001). Karyotype has 2n=42 and FN=76 (Concepción and Molinari, 1991).

Sphiggurus roosmalenorum (Voss and da Silva, 2001). Am. Mus. Novit., 3351:24.

COMMON NAME: Roosmalen's Dwarf Porcupine.

TYPE LOCALITY: Brazil, Amazonas, Novo Jerusalem.

DISTRIBUTION: Banks of the middle Rio Maderia between 5 and 9° S latitude, Brazil.

COMMENTS: Assigned to *vestitus*-group based on presence of bristle-quills (Voss and da Silva, 2001).

Sphiggurus spinosus (F. Cuvier, 1823). Mem. Mus. Hist. Nat. Paris, 9:433.

COMMON NAME: Paraguaian Hairy Dwarf Porcupine.

TYPE LOCALITY: Paraguay, along the Paraná River.

DISTRIBUTION: Paraguay, S and E Brazil, NE Argentina, Uruguay.

STATUS: CITES – Appendix III (Uruguay); IUCN – Lower Risk (lc).

SYNONYMS: *affinis* (Brandt, 1835); *nigricans* (Brandt, 1835); *paragayensis* (Oken, 1816); *roberti* (Thomas, 1902); *sericeus* (Cope, 1889).

COMMENTS: Possibly restricted to the shores of the Paraná River. Formerly included in *Coendou*; see Husson (1978) and comments under *Sphiggurus*. Emmons and Feer (1997:221) considered *paragayensis* to be a separate species (Paraguay Hairy Dwarf Porcupine). Cabrera (1961:602) noted that Oken's names are not recognized, but if Oken's name is accepted, the type species for *Sphiggurus* becomes *paragayensis* (see Tate, 1935:307). Emmons and Feer (1997) included *villosus* and suggested that *spinosus* may intergrade with *insidosus*.

Sphiggurus vestitus Thomas, 1899. Ann. Mag. Nat. Hist., ser. 7, 4:284.
> COMMON NAME: Brown Hairy Dwarf Porcupine.
> TYPE LOCALITY: Colombia.
> DISTRIBUTION: Colombia, small area of the E Andean cordillera about 60 km WNW of
> > Bogotá at about 1300 m.
> STATUS: IUCN – Vulnerable.
> COMMENTS: Reviewed by Voss and da Silva (2001), who included *Sphiggurus* in *Coendou*.
> > Includes *pruinosus* according to Cabrera (1961), Concepción and Molinari (1991),
> > Woods (1993), and Soriano and Ochoa (1997), but see Voss and da Silva (2001) for a
> > discussion of the morphological distinctiveness of these two taxa. Assigned to *vestitus*-
> > group based on presence of bristle-quills (Voss and da Silva, 2001).

Sphiggurus villosus (F. Cuvier, 1823). Mam. Mus. Hist. Nat. Paris, 9:434.
> COMMON NAME: Orange-spined Hairy Dwarf Porcupine.
> TYPE LOCALITY: Brazil, mountains near Rio de Janeiro, Corcoracto.
> DISTRIBUTION: Minas Gerais to Rio Grande do Sul (SE Brazil).
> STATUS: IUCN – Lower Risk (lc).
> COMMENTS: Considered a distinct species by Husson (1978:489). Cabrera (1961:600-601)
> > included this species in *insidiosus*, whereas Emmons and Feer (1997:221) included this
> > species in *spinosus*. Formerly included in *Coendou* (see comments under *Sphiggurus*).
> > Karyotype has 2n=42 and FN=76 (Bonvicino et al., 2000).

Family Chinchillidae Bennett, 1833. Proc. Zool. Soc. Lond., 1833:58.
> SYNONYMS: Eriomyidae Burmeister, 1854; Lagostomidae Bonaparte, 1838; Viacacidae
> > Ameghino, 1904; Viscacciidae Roverto, 1914.
> COMMENTS: Placed in the superfamily Chinchilloidea by Woods (1982) and McKenna and Bell
> > (1997). McKenna and Bell (1997) recognized two subfamilies, Chinchillinae and
> > Lagostominae.

Chinchilla Bennett, 1829. Gard. Menag. Zool. Soc., 1:1.
> TYPE SPECIES: *Chinchilla lanigera* Bennett, 1829.
> SYNONYMS: *Eriomys* Lichtenstein, 1829.
> COMMENTS: Osgood (1941) critically reviewed the early descriptions of chinchillas and
> > suggested that Molina's (1782) *Mus laniger* was a composite based on no known
> > specimens. Bennett (1829) used *Chinchilla lanigera*, derived from Molina's *Mus laniger*,
> > without stating that the animal described is the same as that described by Molina (1782).
> > Geographic variation in this taxon is poorly understood making it difficult to determine
> > the number of valid species. Osgood (1943) recognized a single species, Cabrera (1961)
> > and Woods (1993) recognized two species, and Bidlingmaier (1937) recognized three
> > species.

Chinchilla chinchilla (Lichtenstein, 1829). Darst, neu o. wenig. Bekannt. Säugeth., 2 unnumbered
> pages.
> COMMON NAME: Short-tailed Chinchilla.
> TYPE LOCALITY: Peru, vicinity of Lima.
> DISTRIBUTION: Andes of S Bolivia, S Peru, NW Argentina, and Chile.
> STATUS: CITES – Appendix I (non-domesticated forms only); U.S. ESA – Endangered as
> > *C. brevicauda boliviana*; IUCN – Critically Endangered as *C. brevicauda*.
> SYNONYMS: *brevicauda* Waterhouse, 1848; *major* (Trouessart, 1896); **boliviana** Brass, 1911;
> > *intermedia* (Dennler, 1939).
> COMMENTS: Recognized as *brevicauda* by Cabrera (1961) and Woods (1993) but see Osgood
> > (1943) and Anderson (1997). Pine et al. (1979) included *brevicaudata* in *lanigera*
> > without comment.

Chinchilla lanigera Bennett, 1829. Gard. Menag. Zool. Soc., 1:1.
> COMMON NAME: Long-tailed Chinchilla.
> TYPE LOCALITY: Chile, Coquimbo Prov., Coquimbo.
> DISTRIBUTION: N Chile, in foothills of the Andes and coastal mountains south to Talca.
> STATUS: CITES – Appendix I; IUCN – Vulnerable.

SYNONYMS: *chincilla* (Fischer, 1814) [a renaming of *Mus laniger* Molina, 1782]; *velligera* Prell, 1934.

COMMENTS: See Jimenez (1996) for current status of wild populations. Karyotype has 2n=64 and FN=126 (George and Weir, 1974).

Lagidium Meyen, 1833. Nouv. Acta Acad. Caes. Leop.-Carol., 16(2):576.

TYPE SPECIES: *Lagidium peruanum* Meyen, 1833.

SYNONYMS: *Lagotis* Bennett, 1833.

COMMENTS: Revised by Osgood (1943). *Viscaccia* Oken, 1816 has precedence over *Lagidium*, but *Lagidium* was adopted in preference to *Viscaccia* by suspension of the rules (Opinion 110, International Commission on Zoological Nomenclature, 1929*a*). However, names from Oken's 1816 "Lehrbuch der Naturgeschichte" are non-Linnaean and not available and neither an opinion nor suspension of the Rules was required for the recognition of *Lagidium* (Hershkovitz, 1949*b*).

Lagidium peruanum Meyen, 1833. Nouv. Acta Acad. Caes. Leop.-Carol., 16(2):578.

COMMON NAME: Northern Mountain Viscacha.

TYPE LOCALITY: Peru Puno Dept., Pisacoma.

DISTRIBUTION: C and S Peru, N Chile.

STATUS: IUCN – Lower Risk (lc). Unknown.

SYNONYMS: *arequipe* Thomas, 1907; *inca* Thomas, 1907; *pallipes* Bennett, 1835; *punensis* Thomas, 1907; *saturata* Thomas, 1907; *subrosea* Thomas, 1907.

COMMENTS: Karyotype has 2n=64 and FN=126 (George and Weir, 1974).

Lagidium viscacia (Molina, 1782). Sagg. Stor. Nat. Chile, p. 307.

COMMON NAME: Southern Moutain Viscacha.

TYPE LOCALITY: Chile, Santiago Prov., Cordillera de Santiago.

DISTRIBUTION: W Argentina, S and W Bolivia, N Chile, S Peru.

STATUS: IUCN – Data Deficient.

SYNONYMS: *aureus* (Geoffroy and D'Orbigny, 1830); *chilensis* (Oken, 1816); *crassidens* Philippi, 1896; *crinigerum* Philippi, 1896; *viscaccica* Brandis, 1786; *boxi* Thomas, 1921; *cuscus* (Thomas, 1907); *cuvieri* (Bennett, 1833); *lutea* (Thomas, 1907); *lutescens* Philippi, 1896; *famatinae* Thomas, 1920; *lockwoodi* Thomas, 1919; *moreni* Thomas, 1897; *perlutea* (Thomas, 1907); *sarae* Thomas and St. Leger, 1926; *tontalis* Thomas, 1921; *tucumana* (Thomas, 1907); *viatorum* Thomas, 1921; *vulcani* Thomas, 1919.

COMMENTS: Karyotype of *boxi* reported to have 2n=64 and FN=126 (George and Weir, 1974).

Lagidium wolffsohni (Thomas, 1907). Ann. Mag. Nat. Hist., ser. 7, 19:440.

COMMON NAME: Wolffsohn's Mountain Viscacha.

TYPE LOCALITY: Argentina, Santa Cruz, Baguales and Vizcachas Mtns (50°50'S, 72°20'W).

DISTRIBUTION: SW Argentina and adjacent Chile.

STATUS: IUCN – Lower Risk (nt).

Lagostomus Brookes, 1828. Trans. Linn. Soc. London, 16:96.

TYPE SPECIES: *Lagostomus trichodactylus* Brookes, 1828 (= *Dipus maximus* Desmarest, 1817).

SYNONYMS: *Lagostomopsis* Kraglievich, 1926; *Viscaccia* Schinz, 1825 [not Oken, 1816].

Lagostomus crassus Thomas, 1910. Ann. Mag. Nat. Hist., ser. 8, 5:246.

COMMON NAME: Peruvian Plains Viscacha.

TYPE LOCALITY: Peru, Dist. Cuzco, Santa Ana.

DISTRIBUTION: Know only from the type locality.

STATUS: Extinct.

COMMENTS: Known only by single skull from S Peru, presumably of an animal that lived in recent times (Nowak, 1999).

Lagostomus maximus (Desmarest, 1817). Nouv. Dict. Hist. Nat., Nouv. Ed., 13:117.

COMMON NAME: Argentine Plains Viscacha.

TYPE LOCALITY: Unknown; possibly from pampas of Buenos Aires, Argentina; see Cabrera (1961).

DISTRIBUTION: N, C, and E Argentina, S and W Paraguay, SE Bolivia.

STATUS: IUCN – Lower Risk (lc).
SYNONYMS: *americana* Schinz, 1825; *criniger* Lesson, 1842; *diana* Griffith, 1827; *pamparum* Schinz, 1825; *trichodactylus* Brooks, 1828; *viscaccia* Geoffroy and D'Orbigny, 1830; **inmollis** Thomas, 1910; **petilidens** Hollister, 1914.
COMMENTS: Reviewed by Jackson et al. (1996). Karyotype has 2n=56 and FN=110 (George and Weir, 1974).

Family Dinomyidae Peters, 1873. Monatsb. K. Preuss. Akad. Wiss. Berlin, p. 552.
COMMENTS: Includes a diversity of very large extinct species. Included in the superfamily Cavioidea by Woods (1982) and McKenna and Bell (1997).

Dinomys Peters, 1873. Monatsb. K. Preuss. Akad. Wiss. Berlin, p. 551.
TYPE SPECIES: *Dinomys branickii* Peters, 1873.

Dinomys branickii Peters, 1873. Monatsb. K. Preuss. Akad. Wiss. Berlin, p. 551.
COMMON NAME: Pacarana.
TYPE LOCALITY: Peru, Junin Dept., Montana de Vitoc, Amable Maria.
DISTRIBUTION: Venezuela, Colombia, Ecuador, Peru, Brazil, and Bolivia.
STATUS: IUCN – Endangered; rare.
SYNONYMS: *gigas* Anthony, 1921; *occidentalis* Lönnberg, 1921; *pacarana* de Miranda-Ribeiro, 1919.
COMMENTS: Reviewed by White and Alberico (1992). The placement within the Chinchilloidea by Mones (1981) and Reig (1986) is supported in the molecular analysis of Huchon and Douzery (2001).

Family Caviidae Fischer de Waldheim, 1817. Mem. Soc. Imp. Nat., Moscow, 5:372.
COMMENTS: Name often accredited to Gray, 1821. The hydrochoerids were included as a subfamily by Tate (1935) and Ellerman (1940) but others (McKenna and Bell, 1997; Miller and Gidley, 1918; Pocock, 1922*a*; Woods, 1993) have recognized them as a distinct family. Molecular data (Rowe and Honeycutt, 2002) clearly support the inclusion of *Hydrochoerus* within the Caviidae. Placed in the superfamily Cavioidea by Woods (1982) and McKenna and Bell (1997).

Subfamily Caviinae Fischer de Waldheim, 1817. Mem. Soc. Imp. Nat., Moscow, 5:372.
COMMENTS: Numerous extinct genera are present in the fossil record.

Cavia Pallas, 1766. Misc. Zool., p. 30.
TYPE SPECIES: *Cavia cobaya* Pallas, 1766 (= *Mus porcellus* Linnaeus, 1758).
SYNONYMS: *Anoema* F. Cuvier, 1809; *Cauia* Storr, 1780; *Cobaia* Aymard, 1854; *Cobaya* G. Cuvier, 1817; *Coiza* Billberg, 1828; *Mamcaviaus* Herrera, 1899.
COMMENTS: Reviewed by Hückinghaus (1961) and Cabrera (1961). See Tate (1935:343) for the reason *porcellus* is "excluded from consideration" as the type because the name was not included under the generic name at the time of its original publication. This genus is in need of revision as the origin of the domesticated form and the number of species that occur in the wild remains unclear.

Cavia aperea Erxleben, 1777. Syst. Regni Anim., 1:348.
COMMON NAME: Brazilian Guinea Pig.
TYPE LOCALITY: Brazil, Pernambuco.
DISTRIBUTION: Colombia, Ecuador, Venezuela, Guianas, Brazil, N Argentina, Uruguay, Paraguay.
STATUS: IUCN – Lower Risk (lc).
SYNONYMS: *azarae* Lichtenstein, 1823; *leucopyga* Brandt, 1815; **guianae** (Thomas, 1901); *caripensis* Ojasti, 1964; *venezuelae* (J. A. Allen, 1911); **hypoleuca** Cabrera, 1953; **pamparum** (Thomas, 1901); **patzelti** Schliemann, 1982; **rosida** (Thomas, 1917).
COMMENTS: Includes *guianae*; see Hückinghaus (1961:58) and Husson (1978:449); but also see Cabrera (1961:578) who placed *guianae* in *porcellus*. Includes *pamparum* according to Massoia and Fornes (1967) and Hückinghaus (1961:57), but Cabrera (1961:577)

listed *pamparum* as a distinct species. Geographic variation in fundamental number has been reported; the karyotype of *C. a. aperea* has 2n=64 and FN=116 or 128 (George et al., 1972; Maia, 1984) and *C. a. pamparum* has 2n=64 and FN=128 (Gava et al., 1998).

Cavia fulgida Wagler, 1831. Isis., 24:512.
COMMON NAME: Shiny Guinea Pig.
TYPE LOCALITY: "Amazonia"; probably an error (Cabrera, 1961:577).
DISTRIBUTION: E Brazil, between Minas Gerais and Santa Catarina.
STATUS: IUCN – Lower Risk (lc).
SYNONYMS: *nigricans* Wagner, 1844; *rufescens* Lund, 1841.
COMMENTS: Karyotype has 2n=64 and FN=128 (Gava et al., 1998).

Cavia intermedia Cherem, Olimpio, and Ximenez, 1999. Biotemas, 12(1): 100.
COMMON NAME: Moleques do Sul Guinea Pig.
TYPE LOCALITY: Brazil, Santa Catarina, Moleques do Sul Archipelago (27°51′S, 48°26′W).
DISTRIBUTION: Known only from the type locality in S Brazil.
COMMENTS: Karyotype has 2n=62 and FN=112 (Gava et al., 1998).

Cavia magna Ximinez, 1980. Rev. Nordest. Biol., 3 (especial):148.
COMMON NAME: Greater Guinea Pig.
TYPE LOCALITY: "en las orillas del arroyo Imbé, municipio de Tramandaí, estado de Rio Grande del Sur, Brasil".
DISTRIBUTION: Dept. of Rocha, Uruguay, to Estados Rio Grande del Sur and Santa Catarina, N Brazil.
STATUS: IUCN – Lower Risk (lc).
COMMENTS: Karyotype has 2n=64 and FN=128 (Gava et al., 1998).

Cavia porcellus (Linnaeus, 1758). Syst. Nat., 10th ed., 1:59.
COMMON NAME: Domesticated Guinea Pig.
TYPE LOCALITY: Brazil, Pernambuco (questionable).
DISTRIBUTION: Domesticated worldwide; possibly feral in N South America.
STATUS: IUCN – Lower Risk (lc).
SYNONYMS: *anolaimae* J. A. Allen, 1916; *cobaya* Pallas, 1766; *cutleri* Bennett, 1836; *leucopyga* Cabanis, 1848; *longipilis* Fitzinger, 1879.
COMMENTS: Husson (1978:451) reserved the use of *porcellus* to denote domesticated guinea pigs, which are probably derived from *tschudii* (Corbet and Hill, 1991:201), but also see Hückinghaus (1961:96), who regarded *porcellus* as a synonym of *aperea*. This species may be a domesticated animal with no established wild populations, K. F. Koopman (pers. comm.) believed that N South American populations may be feral domesticated guinea pigs; but see comments under *aperea*. *Cavia aperea* can be differentiated from *porcellus* by the presence of the processus cupularis (da Silva Neto, 2000). Karyotype has 2n=64 and FN=128 (George and Weir, 1974).

Cavia tschudii Fitzinger, 1857. Sitzb. Akad. Wiss. Wien, p. 154.
COMMON NAME: Montane Guinea Pig.
TYPE LOCALITY: Peru, Ica Dept., Ica.
DISTRIBUTION: Peru, S Bolivia, NW Argentina, N Chile.
STATUS: IUCN – Lower Risk (lc). Common.
SYNONYMS: *atahualpae* Osgood, 1913; *cutleri* Tschudi, 1844; *umbrata* Thomas, 1917; *arequipae* Osgood, 1919; *pallidior* Thomas, 1917; *festina* Thomas, 1927; *osgoodi* Sandborn, 1949; *sodalis* Thomas, 1926; *nana* Thomas, 1917; *stolida* Thomas, 1926.
COMMENTS: Formerly included in *aperea* by Hückinghaus (1961:57); but see Cabrera (1961:579) and Pine et al. (1979:361), who considered *tschudii* a distinct species. Anderson (1997) recognized *nana* as a subspecies but Hückinghaus placed in synonymy with *sodalis*. Includes *stolida* (Cabrera, 1961:579), but also see Hückinghaus (1961:58), who considered *stolida* a distinct species.

Galea Meyen, 1833. Nouv. Acta Acad. Caes. Léop.-Carol., 16:597.
TYPE SPECIES: *Galea musteloides* Meyen, 1833.

COMMENTS: Reviewed by Hückinghaus (1961) and Cabrera (1961).

Galea flavidens (Brandt, 1835). Mem. Acad. Sci. St. Petersbourg, ser. 6, 3:439.
   COMMON NAME: Brazilian Yellow-toothed Cavy.
   TYPE LOCALITY: Unknown; possibly Minas Gerais, Brazil.
   DISTRIBUTION: Brazil.
   STATUS: IUCN – Lower Risk (lc).
   SYNONYMS: *bilobidens* (Lund, 1841).
   COMMENTS: Paula Couto (1950:232) considered *flavidens* synonymous with *spixii*, but see Cabrera (1961:573) who believed both to be distinct species and discussed the type locality.

Galea musteloides Meyen, 1832. Nouv. Acta Acad. Caes. Leop.-Carol., 16:597.
   COMMON NAME: Common Yellow-toothed Cavy.
   TYPE LOCALITY: Peru, Paso de Tacna, on road to Lake Titicaca.
   DISTRIBUTION: S Peru, Bolivia, Argentina, N Chile.
   STATUS: IUCN – Lower Risk (lc). Common.
   SYNONYMS: *boliviensis* (Waterhoues, 1848); *comes* Thomas, 1919; **auceps** (Thomas, 1911); **demissa** (Thomas 1921); **leucoblephara** (Burmeister, 1861); **littoralis** (Thomas, 1901); *negrensis* Thomas, 1919.
   COMMENTS: Reviewed by Mann (1950). Karyotype has 2n=68 and FN=136 (George et al., 1972).

Galea spixii (Wagler, 1831). Isis, 24:512.
   COMMON NAME: Spix's Yellow-toothed Cavy.
   TYPE LOCALITY: Brazil, restricted to Minas Gerais, Lagoa Santa by Cabrera (1961:575).
   DISTRIBUTION: Brazil, Bolivia, east of the Andes.
   STATUS: IUCN – Lower Risk (lc).
   SYNONYMS: *campicola* Doutt, 1938; *saxatilis* (Lund, 1841); **palustris** (Thomas, 1911); **wellsi** (Osgood, 1915).
   COMMENTS: Includes *wellsi* (Corbet and Hill, 1991:201); also see Cabrera (1961:575). Huckinghaus (1961:71) considered *wellsi* a distinct species. Karyotype has 2n=64 and FN=118 (Maia, 1984).

Microcavia H. Gervais and Ameghino, 1880. Mamm. Fos. Am. Sud., p. 50.
   TYPE SPECIES: *Microcavia typus* H. Gervais and Ameghino, 1880 (fossil).
   SYNONYMS: *Caviella* Osgood, 1915; *Monticavia* Thomas, 1916; *Nanocavia* Thomas, 1925.
   COMMENTS: Includes *Monticavia*. Reviewed by Hückinghaus (1961), Cabrera (1961), and Quintana (1996). Three extinct species recognized by Quintana (1996).

Microcavia australis (I. Geoffroy and d'Orbigny, 1833). Mag. Zool. Paris, 3:3.
   COMMON NAME: Southern Mountain Cavy.
   TYPE LOCALITY: Argentina, Pategonia, vicinity of the lower part of the Rio Negro.
   DISTRIBUTION: Argentina between Jujuy and Santa Cruz Provs.; Aisen Prov. (Chile); possibly extreme S Bolivia but Anderson (1997) reported no known specimens.
   STATUS: IUCN – Lower Risk (lc). Common.
   SYNONYMS: *joannia* (Thomas, 1921); *kingii* (Bennett, 1836); *nigricana* (Thomas, 1921); **maenas** (Thomas, 1898); **salinia** (Thomas, 1921); see Cabrera (1954).
   COMMENTS: Reviewed by Thomas (1921*b*:445) and Tognelli et al. (2001).

Microcavia niata (Thomas, 1898). Ann. Mag. Nat. Hist., ser. 7, 1:282.
   COMMON NAME: Andean Mountain Cavy.
   TYPE LOCALITY: Bolivia, La Paz Dept., Esperanza, Monte Sajama, 4,000 m.
   DISTRIBUTION: Altiplano of SW Bolivia and N Chile.
   STATUS: IUCN – Lower Risk (lc).
   SYNONYMS: **pallidior** (Thomas, 1902).

Microcavia shiptoni (Thomas, 1925). Ann. Mag. Nat. Hist., ser. 9, 15:419.
   COMMON NAME: Shipton's Mountain Cavy.
   TYPE LOCALITY: Argentina, Catamarca Prov., Laguna Blanca, 3,400 m.

DISTRIBUTION: NW Argentina in the mountains of Tucuman, Catamarca, and Salta
Provinces between 3,000 to 4,000 m.
STATUS: IUCN – Lower Risk (lc).

Subfamily Dolichotinae Pocock, 1922. Proc. Zool. Soc. Lond., 1922:426.

Dolichotis Desmarest, 1820. Jour. Phys. Chim. Hist. Nat. Arts Paris, 88:205.
TYPE SPECIES: *Cavia patachonica* Shaw, 1801 (= *Cavia patagonum* Zimmerman, 1780).
SYNONYMS: *Chloromys* Desmoulins, 1823; *Lagospedius* Marelli, 1928; *Mara* d'Orbigny, 1829;
Pediolagus Marelli, 1927.
COMMENTS: Includes *Pediolagus* (Starrett, 1967:263); but see Cabrera (1961:580) who
considered it distinct.

Dolichotis patagonum (Zimmermann, 1780). Geogr. Gesch. Mensch. Vierf. Thiere, 2:328.
COMMON NAME: Patagonian Mara.
TYPE LOCALITY: Argentina, Santa Cruz Prov., Puerto Deseado.
DISTRIBUTION: Argentina, approx 28°S (Bolson de Pipanco, Catamarca Prov.) to 50°S.
STATUS: IUCN – Lower Risk (nt).
SYNONYMS: *australis* Yepes, in Cabrera and Yepes, 1940; *magellanica* (Kerr, 1792);
patachonicha (Shaw 1801); *centricola* (Thomas, 1902).
COMMENTS: Reviewed by Campos et al. (2001). Two subspecies (*centricola*, *patagonum*)
recognized by Cabrera (1954). Karyotype has 2n=64 and FN=126 (Wurster et al., 1971).

Dolichotis salinicola Burmeister, 1876. Proc. Zool. Soc. Lond., 1875:634 [1876].
COMMON NAME: Chacoan Mara.
TYPE LOCALITY: Argentina, SW Catamarca Prov., between Totoralejos and Recreo.
DISTRIBUTION: Chaco of Paraguay; NW Argentina as far south as Cordoba Prov.; extreme
S Bolivia.
STATUS: IUCN – Lower Risk (nt).
SYNONYMS: *ballivianensis* Krumbiegel, 1941; *centralis* Weyenbergh, 1877; *cyniclus* (Cabrera,
1953).
COMMENTS: Formerly included in *Pediolagus* (Starrett, 1967:263); also see comments under
Dolichotis.

Subfamily Hydrochoerinae Gray, 1825. Ann. Philos., n.s., 10:341.
COMMENTS: Molecular data (Rowe and Honeycutt, 2002) clearly support the inclusion of
Hydrochoeris within the Caviidae. *Kerodon* is included due to the consistently observed
sister taxon relationship between *Hydrochoerus* and *Kerodon* (dos Reis, 1994; Rowe and
Honeycutt, 2002; Woods, 1984). Several distinct linages of the Hydrochoerinae are
known in the fossil record and some fossil forms extend into North America during the
late Pliocene and Pleistocene (Mones, 1984).

Hydrochoerus Brisson, 1762. Regn. Anim. 2nd ed., p. 12.
TYPE SPECIES: *Sus hydrochaeris* Linnaeus, 1766.
SYNONYMS: *Capibara* Moussy, 1860; *Capiguara* Liais, 1872; *Hydrochaeris* Brunnich, 1772;
Hydrochaerus Erxleben, 1777; *Hydrocheirus* Hollande and Batisse, 1959; *Hydrocherus*
F. Cuvier, 1829; *Hydrochoerus* Wagler, 1830; *Xenohydrochoerus* Rusconi, 1934.
COMMENTS: Husson (1978:456-457) discussed the spelling of the generic name and concluded
that *Hydrochoerus* Brisson, 1762 is not valid because Brisson's (1762) work was not
consistently binominal. The International Commission on Zoological Nomenclature
(1998) ruled that *Hydrochoerus* Brisson, 1762 is valid. Reviewed by Mones (1991).

Hydrochoerus hydrochaeris (Linnaeus, 1766). Syst. Nat., 12th ed., 1:103.
COMMON NAME: Capybara.
TYPE LOCALITY: "Habitat in Surinamo". Cabrera (1961:583) gave Perambuco, Brazil, but see
Husson (1978:451) for restriction to Suriname.
DISTRIBUTION: E Colombia, E Venezuela, the Guyanas and Peru, south through Brazil,
Paraguay, NE Argentina, and Uruguay.
STATUS: IUCN – Lower Risk (lc). Locally common and farmed for meat in some areas.

SYNONYMS: *capybara* (Pallas, 1766); *cobaya* (Buffon, 1802); *dabbeni* Rovereto, 1914; *irroratus* F. Ameghino, 1889; *notalis* Hollister, 1914; *uruguayensis* C. Ameghino and Roverto, 1914.

COMMENTS: Reviewed by Mones and Ojasti (1986). Karyotype has 2n=66 and FN=102 (Wurster et al., 1971).

Hydrochoerus isthmius Goldman, 1912. Smithsonian Misc.Coll., 60(2):11.
COMMON NAME: Lesser Capybara.
TYPE LOCALITY: Panama, Darién, Río Tuyra, Marragantí.
DISTRIBUTION: Panama, W Colombia and W Venezuela.
STATUS: Uncommon in Panama.
COMMENTS: Included in *hydrochaeris* by Handley (1966a:785) but Mones (1991) recognized as a distinct species. Karyotype has 2n=64 and FN=104.

Kerodon F. Cuvier, 1825. Dentes des Mammiferes, p. 151.
TYPE SPECIES: *Kerodon moco* Lesson, 1827 (= *Cavia rupestris* Wied, 1820).
SYNONYMS: *Cerodon* Waterhouse, 1848; *Kerodons* F. Cuvier, 1829.
COMMENTS: Reviewed by Hückinghaus (1961). Placed in the Caviinae by most workers but a sister taxa relationship between *Kerodon* and *Hydrochoerus* has been consistently observed (dos Reis, 1994; Rowe and Honeycutt, 2002; Woods, 1984).

Kerodon acrobata Moojen, Locks, and Langguth, 1997. Bol. Mus. Nac. Rio Janeiro Zool., 377:1.
COMMON NAME: Climbing Cavy.
TYPE LOCALITY: Brazil, Goiás, Fazenda Santa Helena, at Rio São Mateus about 72 km from São Domingos (13°50'W, 46°50'W).
DISTRIBUTION: C Brazil in NE Goiás and probably Tocantins, west of the Espigão Mestre, Serra Geral de Goiás.

Kerodon rupestris (Wied, 1820). Isis, 6:43.
COMMON NAME: Rock Cavy.
TYPE LOCALITY: Brazil, Bahia, Rio Belmonte. Designated by Moojen (1952) as Rio Grande de Belmonte, Rio Pardo, Rio San Francisco; but see Cabrera (1961:580).
DISTRIBUTION: E Brazil.
STATUS: IUCN – Lower Risk (lc).
SYNONYMS: *moco* Lesson, 1827; *sciurens* Geoffroy St. Hilaire, 1826.
COMMENTS: Karyotype has 2n=52 and FN=92 (Maia, 1984).

Family Dasyproctidae Bonaparte, 1838. Syn. Vert. Syst., *in* Nuovi Ann. Sci. Nat., Bologna 2:112.
COMMENTS: The Cuniculidae (Agoutidae) has been included in Dasyproctidae by several authors, but see comments under Cuniculidae. McKenna and Bell (1997) included the dasyproctids as a subfamily (Dasyproctinae) within the Agoutidae. Molecular data (Rowe and Honeycutt, 2002) strongly support designation of two separate families: the Dasyproctidae and the Cunniculidae (Agoutidae). Placed in the superfamily Cavioidea by Woods (1982) and McKenna and Bell (1997).

Dasyprocta Illiger, 1811. Prodr. Syst. Mamm. Avium., p. 93.
TYPE SPECIES: *Mus aguti* Linnaeus, 1766 (= *Mus leporinus* Linnaeus, 1758).
SYNONYMS: *Chloromys* Lesson, 1927; *Cloromis* Cuvier, 1812; *Mamdasyproctaus* Herrera, 1899.
COMMENTS: This group is in need of revision, and many of the species are questionable and based on geographic distribution at this time.

Dasyprocta azarae Lichtenstein, 1823. Verz. Doublet. Zool. Mus. Berlin, p. 3.
COMMON NAME: Azara's Agouti.
TYPE LOCALITY: Brazil, São Paulo.
DISTRIBUTION: EC and S Brazil, Paraguay, NE Argentina.
STATUS: IUCN – Vulnerable.
SYNONYMS: *acuti* (Cuvier, 1812); *aurea* Cope, 1889; *catrinae* Thomas, 1917; *caudata* Lund, 1841; *felicia* Thomas, 1917; *paraguayensis* Liais, 1872.

Dasyprocta coibae Thomas, 1902. Novit. Zool., 9:136.
    COMMON NAME: Coiban Agouti.
    TYPE LOCALITY: Panama, Coiba Isl.
    DISTRIBUTION: Endemic to Coiba Isl.
    STATUS: IUCN – Endangered.
    COMMENTS: Reviewed by Hall (1981:862).

Dasyprocta cristata (E. Geoffroy, 1803). Cat. Mamm. Mus. Nat. d'Hist. Natur, p. 165.
    COMMON NAME: Crested Agouti.
    TYPE LOCALITY: Suriname.
    DISTRIBUTION: Guianas.
    STATUS: IUCN – Data Deficient.
    COMMENTS: May be synonymous with *leporina*; see Husson (1978:464-466) and Hershkovitz
        (1972b:311-341). Not recognized by Emmons and Feer (1997) or Eisenberg (1989).

Dasyprocta fuliginosa Wagler, 1832. Isis, 25:1221.
    COMMON NAME: Black Agouti.
    TYPE LOCALITY: Brazil, Amazonas, Borba, on lower Rio Madeira (="Amazon River").
    DISTRIBUTION: Colombia, S Venezuela, Surinam, N Brazil, Peru.
    STATUS: IUCN – Lower Risk (lc). Common in many areas.
    SYNONYMS: ***candelensis*** J. A. Allen, 1915.
    COMMENTS: Reviewed by J. A. Allen (1915a:625).

Dasyprocta guamara Ojasti, 1972. Mem. Soc. Cienc. Nat. La Salle, 32:176.
    COMMON NAME: Orinoco Agouti.
    TYPE LOCALITY: Venezuela, Delta Amacuro, Araguabisi, 9°13'27"N, 61°0'16"W.
    DISTRIBUTION: Orinoco Delta (Venezuela).
    STATUS: IUCN – Lower Risk (lc).

Dasyprocta kalinowskii Thomas, 1897. Ann. Mag. Nat. Hist., ser. 6, 20:219.
    COMMON NAME: Kalinowski's Agouti.
    TYPE LOCALITY: Peru, Cuzco Dept., Santa Ana Valley, Idma.
    DISTRIBUTION: SE Peru.
    STATUS: IUCN – Data Deficient.

Dasyprocta leporina (Linnaeus, 1758). Syst. Nat., 10th ed., 1:59.
    COMMON NAME: Red-rumped Agouti.
    TYPE LOCALITY: Suriname, Peninka, Peninka Creek and Cennewijne River.
    DISTRIBUTION: Lesser Antilles, Venezuela, Guianas, Amazonian and E Brazil, introduced into
        the Virgin Isls.
    STATUS: IUCN – Lower Risk (lc). Common.
    SYNONYMS: *aguti* (Linnaeus, 1766); ***albida*** Gray, 1842; ***cayana*** Lacépède, 1802; *cayennae*
        Thomas, 1903; *flavescens* Thomas, 1898; *lucifer* Thomas, 1903; *rubrata* Thomas, 1898;
        croconota Wagler, 1831; ***fulvus*** (Kerr, 1792); *antillensis* Sclater, 1874; ***lunaris*** Thomas,
        1917; ***maraxica*** Thomas, 1923; ***noblei*** Allen, 1914.
    COMMENTS: Although the name *aguti* has been traditionally used, Husson (1978:462-463)
        explained why *Mus leporinus* has priority and Husson (1978) and Voss et al. (2001)
        explained why *aguti* should by synonymized. Recognized as *D. aguti cayana* by Cabrera
        (1961:585-586); may also include *cristata*. Includes *albida*, *antillensis*, and *noblei*; see
        Varona (1974:75), who included these forms in *aguti*. Several subspecies have been
        recognized by Cabrera (1961) and Ojasti (1972), but a comprehensive review of
        geographic variation is needed. The West Indian agoutis are descendents of forms
        introduced to the islands. The pattern appears to be *D. leporina* (from Brazil) to the
        Virgin Islands; *D. l. albida* on St. Vincent and Granada; *D. l. fulvus* on Martinique and
        St. Lucia; and *D. l. noblei* on Guadeloupe, St. Kitts, Dominica, and Montserrat.
        Karyotype has 2n=64 and FN=124 (Fredga, 1966).

Dasyprocta mexicana Saussure, 1860. Rev. Mag. Zool. Paris, ser. 2, 12:53.
    COMMON NAME: Mexican Agouti.
    TYPE LOCALITY: Mexico, probably Veracruz (="hot zone of Mexico"); see Hall (1981:859).

DISTRIBUTION: C Veracruz and E Oaxaca (Mexico); introduced into W and E Cuba.

STATUS: IUCN – Lower Risk (nt).

Dasyprocta prymnolopha Wagler, 1831. Isis, 24:619.

COMMON NAME: Black-rumped Agouti.

TYPE LOCALITY: Brazil, Pará. Original designation of Guyana was probably an error (Cabrera, 1961:588).

DISTRIBUTION: NE Brazil.

STATUS: IUCN – Lower Risk (lc).

SYNONYMS: *nigriclunis* Osgood, 1915.

COMMENTS: Ojasti (1972:164) considered the status of this form uncertain.

Dasyprocta punctata Gray, 1842. Ann. Mag. Nat. Hist., [ser. 1], 10:264.

COMMON NAME: Central American Agouti.

TYPE LOCALITY: Nicaragua, Chinadega, El Realejo.

DISTRIBUTION: Chiapas and Yucatan Peninsula (S Mexico) to S Bolivia, N Argentina, and SE Brazil. Introduced into W and E Cuba and the Cayman Isls.

STATUS: CITES – Appendix III (Honduras); IUCN – Lower Risk (lc).

SYNONYMS: **bellula** Kellogg, 1946; **boliviae** Thomas, 1917; **callida** Bangs, 1901; **chiapensis** Goldman, 1913; **chocoensis** Allen, 1915; **columbiana** Bangs, 1898; **dariensis** Goldman, 1913; **isthmica** Alston, 1876; **nuchalis** Goldman, 1917; **pallidiventris** Bole, 1937; **pandora** Thomas, 1917; **richmondi** Goldman, 1917; **underwoodi** Goldman, 1931; **urucuma** Allen, 1915; **variegata** Tschudi, 1845; **yucatanica** Goldman, 1913; **yungarum** Thomas, 1910; **zamorae** Allen, 1915.

COMMENTS: Includes *variegata* (Goldman, 1913:11); but also see Handley (1976:56) and Emmons and Feer (1997:227) who listed *variegata* as a distinct species. Hall (1981) recognized 11 subspecies in Central America and Cabrera (1961) recognized 8 subspecies in South America, however, a comprehensive evaluation of geographic variation is needed. Karyotype of *variegata* reported as 2n=64 and FN=124 (George and Weir, 1974).

Dasyprocta ruatanica Thomas, 1901. Ann. Mag. Nat. Hist., ser. 7, 8:272.

COMMON NAME: Roatán Island Agouti.

TYPE LOCALITY: Honduras, Roatán Isl.

DISTRIBUTION: Endemic to Roatán Isl.

STATUS: IUCN – Endangered.

COMMENTS: Similar to *D. punctata* but much smaller.

Myoprocta Thomas, 1903. Ann. Mag. Nat. Hist., ser 7, 12:464.

TYPE SPECIES: *Cavia acouchy* Erxleben, 1777.

COMMENTS: Voss et al. (2001) contributed substantially to the nomenclatural stability of this genus (see comments under *M. acouchy*) but additional revision, especially for the green acouchi is needed. Although two species have long been recognized, the green and the red acouchies, the nomenclature for these two forms is confusing (see Voss et al., 2001). Emmons and Feer (1997:230) suggested that hybridization may have occurred in Colombia in the headwaters of the Río Uaupés and the hybrid forms have been given the name *M. milleri*. Voss et al. (2001) suggested that these two forms are allopatrically distributed.

Myoprocta acouchy (Erxleben, 1777). Syst. Regn. Anim., 1:354.

COMMON NAME: Red Acouchi.

TYPE LOCALITY: French Guiana, Cayenne.

DISTRIBUTION: Guyana, Surinam. French Guiana, and Brazil north of the Amazon and east of the Rio Branco.

STATUS: IUCN – Lower Risk (lc) as *M. acouchy*, Data Deficient as *exilis*.

SYNONYMS: *acuschy* (Linnaeus, 1788); *demararae* Tate, 1939; *exilis* (Wagler, 1831); *leptura* (Wagner, 1844).

COMMENTS: The name *acouchy* has been applied to the red acouchi by Thomas (1926d:639), Cabrera (1961:591) and Emmons and Feer (1997:230-231) and to the green acouchi by

Tate (1939), Husson (1978) and Woods (1993). Voss et al. (2001) pointed out that the disagreement among these authors is whether geography or color should be given greater importance in applying the name *acouchy*. Voss et al. (2001) favored geography and selected a neotype of *Cavia acouchy* Erxleben (1777) that stabilizes the nomenclature. Includes *exilis* (Voss et al., 2001:151), which has been applied to the red acouchi by Husson (1978:468-472) and Tate (1939:151-229) but Allen (1916) considered *exilis* a green acouchi (see Voss et al., 2001). Karyotype has 2n=62 and FN=120 (Fredga, 1966).

Myoprocta pratti Pocock, 1913. Ann. Mag. Nat. Hist., ser 8, 12:110.
COMMON NAME: Green Acouchi.
TYPE LOCALITY: Peru, Rio Marañon, Pongo de Rentema.
DISTRIBUTION: S Venezuela (headwaters of the Orinoco), E Colombia, E Ecuador, N Peru, W Brazil.
STATUS: Unknown.
SYNONYMS: *archidonae* Lönnberg, 1925; *caymanum* Thomas, 1926; *limanus* Thomas, 1920; *milleri* Allen, 1913; *parva* Lönnberg, 1921; *puralis* Thomas, 1926.
COMMENTS: Voss et al. (2001) concluded the green acouchies exhibit substantial geographic variation and may represent a complex of closely related species. We have followed Voss et al. (2001) in provisionally recognizing the green acouchies as *M. pratti* pending further revision.

Family Cuniculidae Miller and Gidley, 1918. J. Wash. Acad. Sci., 8(13): 446.
SYNONYMS: Agoutidae Gray, 1821; Coelogenyidae Gervais, 1849.
COMMENTS: The familial and subfamilial status of this taxon has been debated; Husson (1978:472), Cabrera (1961:593) and McKenna and Bell, 1997:197) placed in Agoutidae; Hall (1981:858) placed in Agoutinae of the Dasyproctidae: Starrett (1967:269) placed in Cuniculinae of the Dasyproctidae; and Ellerman (1940:221) in the Cuniculidae. With the ruling by the International Commission on Zoological Nomenclature (1998) that *Cuniculus* is to be conserved, Cuniculidae is the proper familial name. Chromosomal (George and Weir, 1974), allozyme (Woods, 1982) and sequence data (Rowe and Honeycutt, 2002) support recognition as a distinct family. Placed in the superfamily Cavioidea by Woods (1982) and McKenna and Bell (1997).

Cuniculus Brisson, 1762. Regn. Anim. 2nd ed., pp. 13, 98.
TYPE SPECIES: *Mus paca* Linnaeus, 1766.
SYNONYMS: *Agouti* Lacépède, 1799; *Caelogenus* Fleming, 1822; *Caelogenys* Agassiz, 1842; *Coelogenus* Cuvier, 1807; *Coelogenys* Illiger, 1811; *Mamcoelogenysus* Herrera, 1899; *Osteopera* Harlan, 1825; *Paca* Fischer, 1814; *Stictomys* Thomas, 1924.
COMMENTS: The generic name of this taxon has been debated (see Pérez, 1992), but the nomenclatural instability of this genus was resolved by the ruling by the International Commission on Zoological Nomenclature (1998) for the conservation of *Cuniculus* Brisson, 1762.

Cuniculus paca (Linnaeus, 1766). Syst. Nat., 12th ed., 1:81.
COMMON NAME: Lowland Paca.
TYPE LOCALITY: French Guiana, Cayenne.
DISTRIBUTION: SE San Luis Potosi (Mexico) to Paraguay, Guianas, S Brazil, and NE Argentina. Introduced into Cuba.
STATUS: CITES – Appendix III (Honduras) as *Agouti paca*; IUCN – Lower Risk (lc) as *Agouti paca*.
SYNONYMS: *alba* (Kerr, 1792); *fulvus* (Cuvier, 1807); *sublaevis* (Gervais, 1854); *subniger* (Cuvier, 1807); *venezuelica* Krumbriegel, 1940; **guanta** (Lönnberg, 1921); **mexicanae** (Hagmann, 1908); **nelsoni** (Goldman, 1913); **virgata** (Bangs, 1902).
COMMENTS: Reviewed by Pérez (1992). Five subspecies are recognized (Cabrera, 1961; Hall, 1981) but their status needs reevaluation. Karyotype has 2n=74 and FN=56 (Fredga, 1966). Preliminary data of the extent of genetic differentiation between geographic separated populations (Rowe and Honeycutt, 2002) suggests that *paca* may represent populations of more than a single species.

Cuniculus taczanowskii (Stolzmann, 1865). Proc. Zool. Soc. Lond., 1865:161.
  COMMON NAME: Mountain Paca.
  TYPE LOCALITY: Ecuador, Andes.
  DISTRIBUTION: Mountains of Peru, Ecuador, Colombia, and NW Venezuela.
  STATUS: IUCN – Lower Risk (nt) as *Agouti taczanowskii*.
  SYNONYMS: *andina* (Lönnberg, 1913); *sierrae* (Thomas, 1905).
  COMMENTS: Placed in *Stictomys* by Cabrera (1961:595) and in *Agouti* by Handley (1976:55)
    and Gardner (1971:1088). Karyotype has 2n=42 and FN=80 (Gardner, 1971).

Family Ctenomyidae Lesson, 1842. Nouv. Tabl. Règne Animal., Mamm., p. 105.
  COMMENTS: This assemblage of approximately 85 named taxa is still in need of revision. The
    species are variable in chromosome number (2n=10-70), but fairly uniform in morphol-
    ogy, suggesting that the major radiation of species was in the Pleistocene (Roig and Reig,
    1969; Reig et al., 1990*b*). Cook and Lessa (1998) and Lessa and Cook (1998) suggested an
    early burst in diversification. *Ctenomys* is most closely allied to *Octodontomys*. Whether
    the group should be recognized as a subfamily (Ctenomyinae, Reig, 1958), or as a family is
    debated. Although recognition as a subfamily within the Octodontidae best reflects the
    evolutionary history of this group, it is more common to treat the group as a distinct
    family specialized for fossorial life. McKenna and Bell (1997) considered it a tribe within
    the Octodontinae. Glanz and Anderson (1990) gave a cladogram and list of synapomor-
    phies for the Ctenomyidae. Reig et al. (1990*b*) suggested that there might be as many as
    55 living species. Here 60 species are recognized, many primarily based on descriptions of
    unique karyotypes in allopatric populations. In addition, several undescribed forms
    appear to represent distinct species (Anderson, 1997; Giménez et al. 1999; Lessa and
    Cook, 1998; Massarini et al., 1991*a*). See Reig et al. (1990*b*) for an overview of ctenomyid
    taxonomy and Reig et al. (1992), Ortells (1995), and Bidau et al. (1996) for a review of
    karyotypic variation. The importance of chromosomal rearrangements in the speciation
    of ctenomyids was suggested by Reig (1989) and Reig et al. (1990); Rossi et al. (1990,
    1995) suggested a possible mechanism for the extensive rearrangements. See also Cook
    et al. (1990) for a discussion of the possible significance of the extensive chromosomal
    variation in *Ctenomys* and octodontids (2n=10-102), which is nearly as great as the known
    variation for all mammals. Sage et al. (1986*a*) characterized this genus as being in a state
    of "taxonomic chaos", though progress has been made in recent years in the study of
    several species groups using banded karyotypes (Gallardo, 1991; Ortells, 1995), protein
    electrophoresis (Cook and Yates, 1994; Gallardo and Kohler, 1992; Gallardo and Palma,
    1992; Moreira et al., 1991; Ortells and Barrantes, 1994), and DNA sequencing (D'Elía
    et al., 1999; Lessa and Cook, 1998).

Ctenomys Blainville, 1826. Bull. Sci. Soc. Philom. Paris, p. 62.
  TYPE SPECIES: *Ctenomys brasiliensis* Blainville, 1826.
  SYNONYMS: *Chacomys* Osgood, 1946; *Haptomys* Thomas, 1916.
  COMMENTS: Cabrera (1961; based on Osgood, 1946) and Nowak (1999) listed two subgenera:
    Chacomys (*C. conoveri* only) and *Ctenomys* (all other species). The validity of subgenus
    Chacomys has not been supported by subsequent studies (Cook and Yates, 1994; Lessa and
    Cook, 1998). Thomas (1916*c*) placed *C. leucodon* in the subgenus *Haptomys*, an arrange-
    ment that is supported by sequence divergence (Lessa and Cook, 1998). Penis morphol-
    ogy appears to be useful in examining relationships among species (Balbontin et al.,
    1996). Sperm morphology (Feito and Gallardo, 1982; Vitullo et al., 1988; Vitullo and
    Cook, 1991) has been used to suggest two major divisions but the sequence analysis of
    D'Elía et al. (1999) does not support those lineages. A *mendocinus* group was designated by
    Massarini et al. (1991*a*) composed of morphologically similar species: *australis*, *azarae*,
    mendocinus, *porteousi*, and *C.* sp. from Chasicó. Freitas (1994) included *flamarioni* and
    rionegrensis within this group, a conclusion supported by sequence analysis (D'Elia et al.,
    1999). In contrast with the predominant mode of speciation within the genus *Ctenomys*
    associated with chromosomal rearrangements (Ortells, 1995; Reig and Kiblisky, 1969),
    members of the *mendocinus* group share relatively uniform karyotypes. Although a
    karyotype of 2n=48 is predominant, limited polymorphism has been reported in a single

pair of chromosomes in populations of *azarae* and *porteousi* (Massarini et al., 1998) resulting in karyotypes with a 2n=46-48.

Ctenomys argentinus Conterras and Berry, 1982. Hist. Nat. (Argentina), 2(20):166.
COMMON NAME: Argentine Tuco-tuco.
TYPE LOCALITY: Argentina, Chaco Prov., Libertador Gen. San Martin Dept., Campo Araos (26°31'S, 59°15'W).
DISTRIBUTION: Formosa, Chaco, Santiago del Estero, and Santa Fe Provinces (NC Argentina).
STATUS: IUCN – Lower Risk (lc).
COMMENTS: See Contreras and Berry (1985) for distribution. Karyotype has 2n=44 and FN=50-52 (Giménez et al., 1997; Ortells et al., 1990).

Ctenomys australis Rusconi, 1934. Rev. Chil. Nat. Hist., 38:108.
COMMON NAME: Southern Tuco-tuco.
TYPE LOCALITY: Argentina, Buenos Aires Prov.
DISTRIBUTION: E Argentina in Buenos Aires Prov.
STATUS: IUCN – Lower Risk (lc).
COMMENTS: Considered distinct from *porteousi* by Contreras and Reig (1965) and Roig and Reig (1969). Placed in the *mendocinus* group by Massarini et al. (1991*a*). Karyotype is 2n=48, FN=76 (Massarini et al., 1991*a*). In addition, to being the only member of the *mendocinus* group to have an invariant karyotype, *australis* exhibits much lower genetic variation than *porteousi* (Apfelbaum et al., 1991).

Ctenomys azarae Thomas, 1903. Ann. Mag. Nat. Hist., ser. 7, 11:228.
COMMON NAME: Azara's Tuco-tuco.
TYPE LOCALITY: Argentina, Central Pampas, 780 km SW Buenos Aires (37°45'S, 65°W).
DISTRIBUTION: Córdoba and La Pampa Provs. (Argentina).
STATUS: IUCN – Lower Risk (lc).
COMMENTS: Type location originally given as Sapucay, Paraguay (Thomas, 1903) but corrected to 780 km SW Buenos Aires by Thomas (1903). The type locality corresponds approximately to General Acha, La Pampa Province (Tate, 1935). Cabrera (1961) synonymized *azarae* under *mendocinus*. Considered distinct from *mendocinus* by Roig and Reig (1969). Placed in the *mendocinus* group by Massarini et al. (1991*a*). Karyotype has 2n=46-48 and FN=68-74 (Massarini et al., 1998). Topotypes of *azarae*, *mendocinus* and *porteousi* have G-band equivalence and these three taxa share a Robertsonian polymorphism leading Massarini et al. (1998) to suggest that *azarae* and *porteousi* may be conspecific with *mendocinus*.

Ctenomys bergi Thomas, 1902. Ann. Mag. Nat. Hist., ser. 7, 9:241.
COMMON NAME: Berg's Tuco-tuco.
TYPE LOCALITY: Argentina, Córdoba Prov., Cruz del Eje (30°44'S, 64°48'W).
DISTRIBUTION: SW Córdoba Prov. (C Argentina).
COMMENTS: Included in *mendocinus* by Cabrera (1961). Karyotype has 2n=48 and FN=90 (Giménez et al., 1999).

Ctenomys boliviensis Waterhouse, 1848. Nat. Hist. Mamm., 2:278.
COMMON NAME: Bolivian Tuco-tuco.
TYPE LOCALITY: Bolivia, Santa Cruz Dept., Santa Cruz de la Sierra.
DISTRIBUTION: Santa Cruz Dept. (C Bolivia), Mato Grosso (SW Brazil), W Paraguay, and Formosa Prov., Argentina.
STATUS: IUCN – Lower Risk (nt) as *C. nattereri*, Lower Risk (lc) as *C. boliviensis*.
SYNONYMS: *braziliensis* Pelzen, 1883; ***nattereri*** Wagner, 1848; *rondoni* Ribeiro, 1914.
COMMENTS: Several different karyotypes have been reported including 2n=36, FN=64; 2n=42, FN=64 and 2n=44-45, FN=68 (Anderson et al., 1987). Anderson et al. (1987) suggested that three or more cryptic species with different karyotypes may be included within this taxon, one of which (*goodfellowi*) was elevated to specific rank (Anderson, 1997). Anderson et al. (1987) considered *nattereri* a subspecies of *boliviensis* but Woods (1993) recognized it as a species without comment.

Ctenomys bonettoi Conterras and Berry, 1982. Hist. Nat. (Argentina), 2(14):123.
    COMMON NAME: Bonetto's Tuco-tuco.
    TYPE LOCALITY: Argentina, Chaco Prov., Dept. Sargento Cabral, 7.5 km SE of Capitán Solari
        (26°48'S, 59°33'W).
    DISTRIBUTION: Argentina, SE Chaco Prov.
    STATUS: IUCN – Lower Risk (lc).
    COMMENTS: Very similar to *mendocinus* and *haigi* based on isozymes (Sage et al., 1986*a*).

Ctenomys brasiliensis Blainville, 1826. Bull. Sci. Soc. Philom. Paris, 3:62.
    COMMON NAME: Brazilian Tuco-tuco.
    TYPE LOCALITY: Brazil, Minas Gerais.
    DISTRIBUTION: E Brazil.
    STATUS: IUCN – Lower Risk (lc).

Ctenomys budini Thomas, 1913. Ann. Mag. Nat. Hist., ser. 8, 11:141.
    COMMON NAME: Budin's Tuco-tuco.
    TYPE LOCALITY: Argentina, Jujuy Prov., Cerro de Lagunita, E Maimara 4500 m.
    DISTRIBUTION: SE Jujuy Prov., Argentina.
    SYNONYMS: **barbarus** (Thomas, 1921).
    COMMENTS: Considered a synonym of *frater* by Redford and Eisenberg (1992) and Woods
        (1993) without comment but listed as a distinct species by Galliari et al. (1996).
        Thomas (1921*b*) recognized four subspecies but Ellerman (1940) treated *sylvanus* as a
        separate species and considered *utibilis* as a subspecies of that taxon. Galliari et al.
        (1996) listed *barbarus* as a distinct species.

Ctenomys colburni J. A. Allen, 1903. Bull. Am. Mus. Nat. Hist., 19:188.
    COMMON NAME: Colburn's Tuco-tuco.
    TYPE LOCALITY: Argentina, Santa Cruz Prov., Arroyo Aiken, 95 km SE Lake Buenos Ayres.
    DISTRIBUTION: Extreme W Santa Cruz Prov. (Argentina).
    STATUS: IUCN – Lower Risk (lc).
    COMMENTS: Karyotype has 2n=34 and FN=64 (Gallardo, 1991).

Ctenomys coludo Thomas, 1920. Ann. Mag. Nat. Hist., ser. 9, 6:119.
    COMMON NAME: Puntilla Tuco-tuco.
    TYPE LOCALITY: Argentina, Catamarca Prov., La Puntilla 1000 m.
    DISTRIBUTION: Catamarca Prov., Argentina.
    COMMENTS: Included in *fulvus* by Cabrera (1961) and Woods (1993), but considered
        provisionally distinct by Contreras et al. (1977) and Galliari et al. (1996) commented
        on the absence of data for placing this taxon in synonomy.

Ctenomys conoveri Osgood, 1946. Fieldiana Zool., 31:47.
    COMMON NAME: Chacoan Tuco-tuco.
    TYPE LOCALITY: Paraguay, Boqueron, 16 km W of Filadelfia (22°15'S, 60°10'W).
    DISTRIBUTION: Gran Chaco of Paraguay, NE Argentina, and SE Boliva.
    STATUS: IUCN – Lower Risk (lc).
    COMMENTS: Cabrera (1961), based on Osgood (1946), placed this form in a separate
        subgenus *Chacomys*, as did Nowak (1999). Recognition of this monotypic subgenus is
        not supported by allozymes (Cook and Yates, 1994) or sequence analysis (Lessa and
        Cook, 1998). Karyotype has 2n=48-50 and FN=70 (Anderson et al., 1987).

Ctenomys coyhaiquensis Kelt and Gallardo, 1994. J. Mammal., 75:344.
    COMMON NAME: Coyhaique Tuco-tuco.
    TYPE LOCALITY: Chile, Prov. General Carrera, about 2 km S Chile Chico and 1 km W Chile
        Chico aerodromo (46°33'S, 71°46'W).
    DISTRIBUTION: Known from type locality and Prov. Coyhaique, 4.5 km SE Coyhaique Alto
        (Chile).
    COMMENTS: Karyotype has 2n=28 and FN=44.

Ctenomys dorbignyi Contreras and Contreras, 1984. Hist. Nat., 4(13):131.
    COMMON NAME: D'Orbigny's Tuco-tuco.

TYPE LOCALITY: Argentina, Prov. Corrientes, Dept. Berón de Astrada, Paraje Mbarigüí (27°33'S, 57°31'W).

DISTRIBUTION: Prov. Corrientes (Argentina).

COMMENTS: See Contreras and Scolaro (1986) for details of geographic distribution. Karyotype has 2n=70 and FN=84 (Ortells et al., 1990) but Garcia et al. (2000) reported a FN=88. The karyotype of *dorbignyi* is indistinguishable from some of the karyotypes assigned to *C. pearsoni* in gross morphology (Ortells et al., 1990) and G-band patterns (Garcia et al., 2000). Garcia et al. (2000) reported differences in the C-bands of the two species but results for *dorbignyi* are different from that reported by Reig et al. (1992). Phylogenetic analysis of chromosomal characters indicates that *dorbignyi* is paraphyletic within *pearsoni* and Garcia et al. (2000) concluded that there are insufficient cytogenetic data to consider *dorbignyi* distinct from *pearsoni*. The problem is that several distinct karyotypes from different localities have been reported or associated with *pearsoni*, none of which can be associated with specimens from the type locality (see comments under *pearsoni*). Although there may be insufficient cytogenetic data to consider *dorbignyi* distinct from populations of *pearsoni* with a 2n=70, Novello and Lessa (1986) have suggested that more than a single species was represented among populations of *pearsoni*. Without knowledge of which karyotype is associated with the specimens from the type locality of *pearsoni* it is not possible to determine whether *dorbignyi* represents one of the chromosomally differentiated species within the *pearsoni* complex or should be placed in synonymy with *pearsoni*. Little or no differentiation of isozymes was observed among *dorbignyi*, *perrensi*, and *roigi* (Ortells and Barrantes, 1994) and all four of these species have symmetric sperm (Vitullo et al., 1988) and a glans penis with pairs of spiny bulbs (Balbontin et al., 1996).

Ctenomys dorsalis Thomas, 1900. Ann. Mag. Nat. Hist., ser. 7, 6:385.

COMMON NAME: Black-backed Tuco-tuco.

TYPE LOCALITY: Paraguay, Boqueron Prov., N Chaco.

DISTRIBUTION: Paraguay, west of the River of Paraguay in the N Chaco.

STATUS: IUCN – Lower Risk (lc).

COMMENTS: The dubious taxonomic history of this taxon was discussed by Contreras and Roig (1991).

Ctenomys emilianus Thomas and St. Leger, 1926. Ann. Mag. Nat. Hist., ser. 9, 18:637.

COMMON NAME: Emilio's Tuco-tuco.

TYPE LOCALITY: Argentina, Neuquén Prov., Chos Malal.

DISTRIBUTION: Neuquén Prov. (Argentina), confined to sand dune habitats at about 800 m.

STATUS: IUCN – Lower Risk (lc).

Ctenomys famosus Thomas, 1920. Ann. Mag. Nat. Hist., ser. 9, 6:420.

COMMON NAME: Famatina Tuco-tuco.

TYPE LOCALITY: Argentina, Rioja Prov., La Invernada.

DISTRIBUTION: Sierra Famatina of NW Rioja Prov., Argentina.

COMMENTS: Recognized as a subspecies of *fulvus* by Cabrera (1961) and Woods (1993), but considered provisionally distinct by Contreras et al. (1977) and Galliari et al. (1996) recognized it as distinct.

Ctenomys flamarioni Travi, 1981. Iheringia, 60:123.

COMMON NAME: Flamarion's Tuco-tuco.

TYPE LOCALITY: Brazil, Rio Grande do Sul, Estação Ecológica do Taim, Fazenda Caçapava.

DISTRIBUTION: Coastal sand dunes in Rio Grande do Sul, S Brazil.

STATUS: Endangered from urbanization (Freitas, 1995).

COMMENTS: Karyotype has 2n=48 and FN=50-78 (Freitas, 1994). Freitas (1994) suggested that this species might belong to the *mendocinus* group.

Ctenomys fochi Thomas, 1919. Ann. Mag. Nat. Hist., ser. 9, 3:117.

COMMON NAME: Foch's Tuco-tuco.

TYPE LOCALITY: Argentina, Catamarca Prov., Chumbicha 600 m.

DISTRIBUTION: SW Catamarca Prov., Argentina.

COMMENTS: Included in *mendocinus* by Cabrera (1961) and Woods (1993), but Galliari et al. (1996) indicated that neither make a case for its synonymy.

Ctenomys fodax Thomas, 1910. Ann. Mag. Nat. Hist., ser. 8, 5:243.
COMMON NAME: Lago Blanco Tuco-tuco.
TYPE LOCALITY: Argentina, Chubut, Lago Blanco (46°S, 71°W).
DISTRIBUTION: Known only from the type locality but see comments.
COMMENTS: Included in *magellanicus* by Osgood (1943). Karyotype has 2n=28 and FN=42 (Reig et al., 1992). Reig et al. (1992) suggested that a specimen with this karyotype reported by Gallardo (1991) as *Ctenomys* sp. (aff. *colburni*) from Chile Chico on the west shore of General Carrera Lake (46°32'S, 71°45'W) was actually *fodax*.

Ctenomys frater Thomas, 1902. Ann. Mag. Nat. Hist., ser. 7, 9:185.
COMMON NAME: Reddish Tuco-tuco.
TYPE LOCALITY: Bolivia, Potosí, Potosí, 4300 m.
DISTRIBUTION: SW Bolivia at 600-4,300 m.
STATUS: IUCN – Lower Risk (lc). Uncommon.
SYNONYMS: *mordosus* (Thomas, 1926); *sylvanus* Thomas 1925.
COMMENTS: Woods (1993) included *barbarus*, *budini*, and *utibilis* but see Galliari et al. (1996). Argentine populations previously recognized as *frater* are here recognized as *budini* or *sylvanus*. Closely related to *lewisi* (Cook et al., 1990; Cook and Yates, 1994; Lessa and Cook, 1998). Karyotype has 2n=52 and FN=78 (Cook et al., 1990).

Ctenomys fulvus Philippi, 1860. Reise Wuste Atacama, p. 157.
COMMON NAME: Tawny Tuco-tuco.
TYPE LOCALITY: Chile, Antofagasta Prov., Pingo-Pingo.
DISTRIBUTION: Mountains and Monte desert of NW Argentina and N Chile.
STATUS: IUCN – Lower Risk (lc).
SYNONYMS: *atracamensis* Philippi, 1860; *chilensis* Philippi, 1896; *pallidus* Philippi, 1896; *pernix* Philippi, 1896; *robustus* Philippi, 1896.
COMMENTS: Includes *robustus*, a Rassenkreis subspecies restricted to the oasis of Pica in Tarapaca Prov., Chile (Mann, 1978:292). Cabrera (1961) and Woods (1993) included *coludo*, *famosus*, *johannis*, and *tulduco*; but Contreras et al. (1977) considered these forms provisionally distinct and Galliari et al. (1996) indicated a lack of data to support change from specific status. Nowak (1999) considered *johannis* and *robustus* as distinct species. Karyotype has 2n=26 and FN=48 (Gallardo, 1991), but see Redford and Eisenberg (1992) who reported karyotypes of 2n=26; FN=52 for *C. f. fulvus* and 2n=25; FN=52 for *C. f. robustus*. Gallardo (1991) noted that the C-band pattern of *robustus* differs from that of *fulvus* but Gallardo and Palma (1992) found little genetic divergence between these two subspecies based upon protein electrophoresis.

Ctenomys goodfellowi Thomas, 1921. Ann. Mag. Nat. Hist., ser. 9, 7:136.
COMMON NAME: Goodfellow's Tuco-tuco.
TYPE LOCALITY: Bolivia, Prov. Nuflo de Chaves, Esperanza, near Concepcion.
DISTRIBUTION: Santa Cruz Prov., SE Bolivia.
COMMENTS: Considered a subspecies of *boliviensis* by Cabrera (1961) and by Woods (1993). Cook and Yates (1994) noted that "this taxon is distinct from *boliviensis* at the species level" and Anderson (1997) recognized it as a distinct species. Karyotype is 2n=46; FN=68 (Anderson et al., 1987).

Ctenomys haigi Thomas, 1917. Ann. Mag. Nat. Hist., ser. 9, 3:210.
COMMON NAME: Haig's Tuco-tuco.
TYPE LOCALITY: Argentina, Chubut Prov., El Maitén, 700 m.
DISTRIBUTION: Chubut and Rio Negro Provs., Argentina.
STATUS: IUCN – Lower Risk (lc).
SYNONYMS: *lentulus* Thomas, 1919.
COMMENTS: Cabrera (1961) synonymized *haigi* under *mendocinus*, but Pearson (1984) elevated it to a species based on the unique karyotype of 2n=50, FN=66 (Gallardo, 1991). Although the isozyme study of Gallardo and Palma (1992) found that *haigi* was paraphyletic, the lack of data from other taxa of the *mendocinus* complex makes it

difficult to evaluate this conclusion. Substantial divergence has been reported between *haigi* and *mendocinus* consistent with their recognition at the species level (Lessa and Cook, 1998; Sage et al., 1986*a*).

Ctenomys johannis Thomas, 1921. Ann. Mag. Nat. Hist., ser. 9, 7:523.
COMMON NAME: San Juan Tuco-tuco.
TYPE LOCALITY: Argentina, San Juan Prov., Cañada Honda, 600 m.
DISTRIBUTION: S San Juan Prov., Argentina.
COMMENTS: Originally described as a subspecies of *coludo* by Thomas (1921) but Cabrera (1961) and Woods (1993) included *coludo* and *johannis* in *fulvus*. Contreras et al. (1977) considered these forms provisionally distinct and Nowak (1999) and Galliari et al. (1996) considered *johannis* a distinct species. Contreras et al. (1977) corrected the elevation of the type locality to 600 m and suggested a close relationship with *validus*.

Ctenomys juris Thomas, 1920. Ann. Mag. Nat. Hist., ser. 9, 5:194.
COMMON NAME: Jujuy Tuco-tuco.
TYPE LOCALITY: Argentina, Jujuy Prov., El Chaguaral 500 m.
DISTRIBUTION: SE Jujuy Prov., Argentina.
COMMENTS: Cabrera (1961) synonymized *juris* under *mendocinus*. Included within *mendocinus* by Redford and Eisemberg (1992) and Woods (1993), but considered specifically distinct by Galliari et al. (1996).

Ctenomys knighti Thomas, 1919. Ann. Mag. Nat. Hist., ser. 9, 3:498.
COMMON NAME: Catamarca Tuco-tuco.
TYPE LOCALITY: Argentina, Catamarca Prov., Otro Cerro.
DISTRIBUTION: Mountains of La Rioia (W Argentina), north to Salta.
STATUS: IUCN – Lower Risk (lc).
COMMENTS: Cabrera (1961) believed *viperinus* synonymous with *knighti*. Karyotype reported for this taxon (2n=36 and FN=64; Reig et al., 1992) is from specimens now recognized as *C. scagliai*.

Ctenomys lami Freitas, 2001. Stud. Neotrop. Fauna Envir., 36(1):2.
COMMON NAME: Lami Tuco-tuco.
TYPE LOCALITY: Brazil, Rio Grande do Sul, Beco dos Cegos (30°51′S, 51°10′W).
DISTRIBUTION: Coxilha das Lombas, from the type locality NE of Guaiba River nearly 80 km to SW banks of Barros Lake, Rio Grande do Sul, S Brazil.
COMMENTS: Karyotype has 2n=54-58 and FN=76-82.

Ctenomys latro Thomas, 1918. Ann. Mag. Nat. Hist., ser. 9, 1:38.
COMMON NAME: Mottled Tuco-tuco.
TYPE LOCALITY: Argentina, Tucumán Prov., Tapia, 600 m.
DISTRIBUTION: NW Argentina in Tucumán and Salta Provs.
STATUS: IUCN – Lower Risk (nt).
COMMENTS: Cabrera (1961) synonymized *latro* under *mendocinus*. Considered distinct from *mendocinus* by Roig and Reig (1969) and Reig and Kiblisky (1969). Karyotype has 2n=40-42 and FN=46-47 (Reig et al., 1992).

Ctenomys leucodon Waterhouse, 1848. Nat. Hist. Mamm., 2:281.
COMMON NAME: White-toothed Tuco-tuco.
TYPE LOCALITY: Bolivia, La Paz Dept., San Andres de Machaca.
DISTRIBUTION: W Bolivia and E Peru around Lake Titicaca.
STATUS: IUCN – Lower Risk (lc).
COMMENTS: Osgood (1946) suggested that this taxon was distinct at the subgeneric level (*Haptomys*) and sequence analysis is consistant with that suggestion (Lessa and Cook, 1998). Karyotype has 2n=36 and FN=68 (Cook et al., 1990).

Ctenomys lewisi Thomas, 1926. Ann. Mag. Nat. Hist., ser. 9, 17:323.
COMMON NAME: Lewis' Tuco-tuco.
TYPE LOCALITY: Bolivia, Tarija Dept., Sama, 4,000 m.
DISTRIBUTION: S Bolivia.
STATUS: IUCN – Lower Risk (lc).

COMMENTS: Closely related to *frater* (Cook et al., 1990; Cook and Yates, 1994; Lessa and Cook, 1998). May be semi-aquatic (Anderson, pers. comm.). Karyotype has 2n=56 and FN=74 (Cook et al., 1990).

Ctenomys magellanicus Bennett, 1836. Proc. Zool. Soc. Lond., 1835:190 [1836].
COMMON NAME: Magellanic Tuco-tuco.
TYPE LOCALITY: Chile, Magallanes, Bahia de San Gregorio.
DISTRIBUTION: Extreme S Chile and S Argentina.
STATUS: IUCN – Vulnerable; reduced in numbers by sheep grazing.
SYNONYMS: *neglectus* Nehring, 1900; *dicki* Osgood, 1943; *fueginus* Philippi, 1880; *osgoodi* Allen, 1905; *robustus* J. A. Allen, 1903.
COMMENTS: Osgood (1943) included *fodax* but see coments under that taxon. Karyotype has 2n=34-36 and FN=64 (Gallardo, 1991).

Ctenomys maulinus Philippi, 1872. Z. Ges. Naturw., N.F., 6:442.
COMMON NAME: Maule Tuco-tuco.
TYPE LOCALITY: Chile, Talca Prov., Laguna de Maule.
DISTRIBUTION: Between Talca and Cautin Provs. (SC Chile), and Neuquén Prov. (Argentina).
STATUS: IUCN – Lower Risk (lc).
SYNONYMS: *brunneus* Osgood, 1943.
COMMENTS: Karyotype has 2n=26 and FN=48 (Gallardo, 1991).

Ctenomys mendocinus Philippi, 1869. Arch. Naturgesch., 1:38.
COMMON NAME: Mendoza Tuco-tuco.
TYPE LOCALITY: Argentina, Mendoza Prov., Mendoza.
DISTRIBUTION: Eastern slopes of the Andes from Santa Cruz north to Mendoza Prov. (Argentina).
STATUS: IUCN – Lower Risk (lc).
COMMENTS: Karyotype has 2n=47-48 and FN=75-76 (Massarini et al., 1991*b*). Cabrera (1961) synonymized *bergi, fochi, haigi, juris, azarae, latro, pundti, occultus*, and *tucumanus* under *mendocinus*. Roig and Reig (1969) considered *azarae, latro*, and *tucumanus* distinct from *mendocinus* based on the results of precipitin tests and Reig and Kiblisky (1969) recognized *occultus, latro*, and *tucumanus* as distinct based on strikingly different karyotypes. Several additional forms have been recognized as distinct species, including *bergi* (Giménez et al., 1999), *haigi* (Pearson, 1984), and *pundti* (Reig et al., 1992), based in part on karyotypic differentiation. Galliari et al. (1996) also recognized *fochi* and *juris* as species based in part on the lack of justification for their synonymy. A *mendocinus* group was designated by Massarini et al. (1991*a*) composed of morphologically similar species: *australis, azarae, mendocinus*, and *porteousi*, which all share similar karyotypes. Topotypes of *azarae, mendocinus* and *porteousi* have G-band equivalence and these three taxa share a Robertsonian polymorphism, leading Massarini et al. (1998) to suggest that *azarae* and *porteousi* may be conspecific with *mendocinus*.

Ctenomys minutus Nehring, 1887. Sitzb. Ges. Naturf. Fr. Berlin, p. 47.
COMMON NAME: Tiny Tuco-tuco.
TYPE LOCALITY: Brazil, Rio Grande do Sul, Campos E of Mondo Novo.
DISTRIBUTION: Rio Grande do Sul, Santa Catarina, and Mato Grosso (S Brazil) and E Bolivia, but see Freitas (1995) who suggested that *minutus* is endemic to the Coastal Plains of S Brazil and that specimens described as *minutus* from Bolivia represent another species.
STATUS: IUCN – Lower Risk (lc).
SYNONYMS: *beltzeri* Contreras and Contreras, 1984; *marthae* Contreras and Contreras, 1984; *monesi* Contreras and Contreras, 1984; *reigi* Contreras and Contreras, 1984; *bicolor* Ribeiro, 1914.
COMMENTS: Reviewed by Langguth and Abella (1970). Six subspecies were recognized by Contreras and Contreras (1984), however, no justification was given for the four new subspecies named in that paper. Altuna and Lessa (1985) and Altuna et al. (1985) regarded *C. m. rionegrensis* as a full species (see comments under *rionegrensis*). Karyotype has 2n=42, 45-50 and FN=76 (Freitas, 1997).

Ctenomys occultus Thomas, 1920. Ann. Mag. Nat. Hist., ser. 9, 6:243.

COMMON NAME: Furtive Tuco-tuco.

TYPE LOCALITY: Argentina, Monteagudo, 80 km SE Tucumán City.

DISTRIBUTION: NW Argentina in Tucumán and adjacent Provs.

STATUS: IUCN – Lower Risk (lc).

COMMENTS: Revised by Reig et al. (1966); also see Cabrera (1961) who included it in *mendocinus*. Karyotype is 2n=22; FN=38 (Reig et al., 1992).

Ctenomys opimus Wagner, 1848. Arch. Naturgesch., 1:75.

COMMON NAME: Highland Tuco-tuco.

TYPE LOCALITY: Bolivia, Oruro Dept., Monte Sajama.

DISTRIBUTION: NW Argentina, SW Bolivia, S Peru, N Chile between 2,000 and 5,000 m on the high Andean steppe (=Puna).

STATUS: IUCN – Lower Risk (lc). Common.

SYNONYMS: *luteolus* Thomas, 1900; *nigriceps* Thomas, 1900.

COMMENTS: May be conspecific with *fulvus*; *opimus* and *fulvus* cannot be separated on the basis of karyotypes (see Gallardo, 1979 and Cook et al., 1990). Karyotype has 2n=26 and FN=48 (Gallardo, 1991).

Ctenomys osvaldoreigi Contreras, 1995. Notulas Faunisticas, 84:1.

COMMON NAME: Reig's Tuco-tuco.

TYPE LOCALITY: Argentina, Córdoba Prov., Estancia San Luis (31°24'S, 64°48'W).

DISTRIBUTION: Known only from the type locality above 2000 m in the Sierras Grandes (Argentina).

COMMENTS: Not recognized by Galliari et al. (1996). Karyotpe has 2n=52, FN=56 (Giménez et al., 1999).

Ctenomys pearsoni Lessa and Langguth, 1983. Com. Jour. Cienc. Nat. Montevideo (Uruguay), 3:86.

COMMON NAME: Pearson's Tuco-tuco.

TYPE LOCALITY: Uruguay, Dept. Colonia, 25 km SE of Carmelo, Arroyo Limetas.

DISTRIBUTION: Soriana, San José, and Colonia Depts. (Uruguay).

STATUS: IUCN – Lower Risk (lc).

COMMENTS: Several karyotypes from different localities have been reported or associated with this taxon, including 2n=68-70, FN=80 (Novello et al., 1990); 2n=70, FN=80 (Kiblisky et al., 1977; associated with *pearsoni* by Lessa and Langguth, 1983); and 2n=56 and FN=77-79 (Novello et al., 1990). Three different fundamental numbers have been reported for the 2n=70 karyotype (80 by Novello et al., 1990; 84 by Ortells et al., 1990; and 88 by Garcia et al., 2000). No karyotype can be associated with specimens from the type locality. The lack of G-band homology between the 2n=70 karyotype and the 2n=56 karyotype led Novello and Lessa (1986) to suggest that more than a single species was represented in *C. pearsoni*. Garcia et al. (2000) concluded that the 2n=70 karyotype of *pearsoni* was identical to the karyotype of *dorbignyi* and that there was no cytogenetic data for the regognition of *dorbignyi*.

Ctenomys perrensi Thomas, 1898. Ann. Mag. Nat. Hist., ser. 6, 18:311.

COMMON NAME: Goya Tuco-tuco.

TYPE LOCALITY: Argentina, Corrientes Prov., Goya.

DISTRIBUTION: Central area of W Corrientes Prov. (NE Argentina).

STATUS: IUCN – Lower Risk (lc).

COMMENTS: See Contreras et al. (1985) for distribution. Specimens from the type locality have karyotypes of 2n=50, FN=84 (Ortells et al., 1990). Specimens taken north of the type locality with FN=84 and 2n=53-56 or 58 were tentatively assigned to this taxon (Ortells et al., 1990; Reig et al., 1992) but specimens from two other localities with FN=84 and 2n=62 were not included. Little or no genetic differentiation and no fixed differences of isozymes were observed among any of the populations with FN=84 from Corrientes Prov., suggesting that these populations are conspecific (Ortells and Barrantes, 1994). Further, no fixed differences of isozymes were observed among *perrensi*, *dorbignyi*, and *roigi* (Ortells and Barrantes, 1994).

Ctenomys peruanus Sanborn and Pearson, 1947. Proc. Biol. Soc. Wash., 60:13.
COMMON NAME: Peruvian Tuco-tuco.
TYPE LOCALITY: Peru, Puno Dept., Pisacoma.
DISTRIBUTION: Altiplano of extreme S Peru.
STATUS: IUCN – Lower Risk (lc).

Ctenomys pilarensis Contreras, 1993. Resúmenes VI Congreso Iberoamericano de Conservación v Zoología de Vertebrados, 44.
COMMON NAME: Pilar Tuco-tuco.
TYPE LOCALITY: Paraguay, Dept. Ñembucú, 4.5 Km E Pilar, Yatayty (26°52'S, 58°18'W).
DISTRIBUTION: E Paraguay, Depts. Ñembucú and Misiones, between the Paraguay and Parana Rivers.
COMMENTS: Karyotype has 2n=48 or 50, FN=50 (Giménez et al., 1997).

Ctenomys pontifex Thomas, 1918. Ann. Mag. Nat. Hist., ser. 9, 1:39.
COMMON NAME: San Luis Tuco-tuco.
TYPE LOCALITY: Argentina, Mendoza Prov., San Rafael.
DISTRIBUTION: W Argentina, east of the Andes in San Luis and Mendoza Provs.
STATUS: IUCN – Lower Risk (lc).
COMMENTS: Pearson and Lagiglia (1992) noted that *pontifex* may be a synonym of *maulinus*.

Ctenomys porteousi Thomas, 1916. Ann. Mag. Nat. Hist., ser. 8, 18:304.
COMMON NAME: Porteous's Tuco-tuco.
TYPE LOCALITY: Argentina, Buenos Aires Prov., Bonifacio.
DISTRIBUTION: Buenos Aires and La Pampa Provs. (E Argentina).
STATUS: IUCN – Lower Risk (lc).
COMMENTS: Cabrera (1961) included *australis* in *porteousi*; but see Contreras and Reig (1965) and Roig and Reig (1969:670) who considered *australis* as a distinct species. Placed in the *mendocinus* group by Massarini et al. (1991*a*). Karyotype has 2n=47-48 and FN=71-73 (Massarini et al., 1991*a*). Although populations of *porteousi* demonstrate both highly polymorphic karyotypes and high allozymic heterozygosity, Massarini et al. (1992) inferred high rates of gene flow to account for the homogeneity observed among local populations. See comments under *azarae*.

Ctenomys pundti Nehring, 1900. Zool. Anz., 23:423.
COMMON NAME: Pundt's Tuco-tuco.
TYPE LOCALITY: Argentina, Prov. Cordoba, Alejo Ledesna.
DISTRIBUTION: S Cordoba and S San Luis Provs. (Argentina).
COMMENTS: Included in *mendocinus* by Cabrera (1961). Redford and Eisenberg (1992) and Woods (1993) included *pundti* in *mendocinus* but Galliari et al. (1996) listed it as a distinct species. Karyotype has 2n=50 and FN=86 (Reig et al., 1992).

Ctenomys rionegrensis Langguth and Abella, 1970. Comn. Zool. Mus. Hist. Nat. Montevideo, 10 (129):13.
COMMON NAME: Rio Negro Tuco-tuco.
TYPE LOCALITY: Uruguay, Dept. Rio Negro, Las Canas, 7 km S Fray Bentos.
DISTRIBUTION: Entre Rios Prov. (Argentina) and Dept. Rio Negro (Uruguay).
COMMENTS: Included in *minutus* by Redford and Eisenberg (1992) but it is distinct based on cranial (Langguth and Abella, 1970), penial, and sperm morphology (Altuna and Lessa, 1985). The relationship with the four subspecies of *minutus* from Entre Rio Province named without describtion by Contreras and Contreras (1984) is unknown. Karyotype has 2n=48, 50, 52, or 56; FN=68-70, 72, 74, or 80 (Reig et al., 1992).

Ctenomys roigi Contreras, 1988. Bol. Inst. Estud. Almerienses, 53.
COMMON NAME: Roig's Tuco-tuco.
TYPE LOCALITY: Argentina, Corrientes Prov., Costa Mansión, 10 km S Empedrado (28°02'S, 58°49'W).
DISTRIBUTION: Rio Paraná in Corrientes Prov. (Argentina).
COMMENTS: Not listed by Redford and Eisenberg (1992) but recognized by Galliari et al. (1996). Karyotype has 2n=48 and FN=80 (Ortells et al., 1990, but see Reig et al. 1992).

Ctenomys saltarius Thomas, 1912. Ann. Mag. Nat. Hist., ser. 8, 10:639.
COMMON NAME: Salta Tuco-tuco.
TYPE LOCALITY: Argentina, Salta Prov., Salta.
DISTRIBUTION: Salta and Jujuy Provs. (N Argentina).
STATUS: IUCN – Lower Risk (lc). Common.

Ctenomys scagliai Contreras, 1999. Ciencia Siglo XXI, 3:10.
COMMON NAME: Scaglia's Tuco-tuco.
TYPE LOCALITY: Argentina, Tucumán Prov., Los Cardones (26°40′S, 65°51′W) 2,500 m.
DISTRIBUTION: Known only from the type locality.
COMMENTS: Contreras (1999) concluded that this taxon is related to a primitive stem of
    the genus including *C. opimus – C. pundit – C. talarum*. The karyotype attributed to
    C. knighti with a 2n=36 and FN=64 (Reig et al., 1992) are from representatives of this
    species.

Ctenomys sericeus J. A. Allen, 1903. Bull. Am. Mus. Nat. Hist., 19:187.
COMMON NAME: Silky Tuco-tuco.
TYPE LOCALITY: Argentina, Santa Cruz Prov., Rio Chico.
DISTRIBUTION: SW Argentina in Santa Cruz, Chubut, and Río Negro Provs.
STATUS: IUCN – Lower Risk (lc).

Ctenomys sociabilis Pearson and Christie, 1985. Hist. Nat. (Argentina), 5:338.
COMMON NAME: Colonial Tuco-tuco.
TYPE LOCALITY: Argentina, Prov. Neuquén, 2 km W Cerro Puntudo, Estancia Fortín
    Chacabuco (1075 m).
DISTRIBUTION: In region of Reserva Nacional del Parque Nacional Nahuel Huapi, Neuquén
    Prov., Argentina.
STATUS: IUCN – Lower Risk (nt).
COMMENTS: Not included in Redford and Eisenberg (1992). Karyotype has 2n=56 and FN=72
    (Gallardo, 1991) but see Reig et al. (1992) for comments concerning FN.

Ctenomys steinbachi Thomas, 1907. Ann. Mag. Nat. Hist., ser. 7, 20:164.
COMMON NAME: Steinbach's Tuco-tuco.
TYPE LOCALITY: Bolivia, Santa Cruz Dept., near Santa Cruz de la Sierra. Anderson et al. (1987)
    noted that the type locality was a large area only partially occupied by *C. steinbachi*
    and therefore restricted the type locality to 6 km N Buen Retiro (17°13′S, 63°38′W).
DISTRIBUTION: Bolivia, W Santa Cruz Dept.
STATUS: IUCN – Lower Risk (lc).
COMMENTS: Karyotype has 2n=10 and FN=16 (Anderson et al., 1987).

Ctenomys sylvanus Thomas, 1919. Ann. Mag. Nat. Hist., ser. 9, 4:155.
COMMON NAME: Forest Tuco-tuco.
TYPE LOCALITY: Argentina, Salta Prov., Tartagal.
DISTRIBUTION: SE Jujuy Prov. and Salta Prov. (NW Argentina).
SYNONYMS: *utibilis* (Thomas, 1920).
COMMENTS: Considered a synonym of *frater* by Woods (1993) without comment but listed as
    a distinct species by Galliari et al. (1996).

Ctenomys talarum Thomas, 1898. Ann. Mag. Nat. Hist., ser. 7, 1:285.
COMMON NAME: Los Talas Tuco-tuco.
TYPE LOCALITY: Argentina, Buenos Aires Prov., Los Talas.
DISTRIBUTION: Along the coast in Buenos Aires Prov. (E Argentina), possibly to Santa Fe Prov.
STATUS: IUCN – Lower Risk (lc).
SYNONYMS: *antonii* Thomas 1910; *occidentalis* Justo, 1992; *recessus* (Thomas, 1912).
COMMENTS: Considered a subspecies of *mendocinus* by Cabrera (1961). Karyotype has
    2n=46-50 and FN =73-86 (see Reig et al., 1992).

Ctenomys torquatus Lichtenstein, 1830. Darst. Säugeth., text of pl. 31.
COMMON NAME: Collared Tuco-tuco.
TYPE LOCALITY: Brazil, S provinces, and banks of Uruguay River. See Moojen (1952:188).
DISTRIBUTION: Uruguay, NE Argentina, extreme S Brazil.

STATUS: IUCN – Lower Risk (lc). Endangered from strip coal mining (Freitas, 1995).
COMMENTS: Considered a distinct species by Ellerman (1940) and Nowak (1999); but see Cabrera (1961:547) who provisionally included it in *brasiliensis*. Although six different karyotypes attributed to *C. torquatus* have been reported (Freitas and Lessa, 1984; Kiblisky et al., 1977; Reig et al., 1966), two of those (2n=56 and 70) were from populations currently recognized as *C. pearsoni* (Novello and Lessa, 1986). Populations of *C. torquatus* from S Brazil and N Uruguay have 2n=44-46 (FN=72) and populations from S Uruguay and NE Argentina have 2n=64 or 68 (FN=96); see Freitas and Lessa (1984).

Ctenomys tuconax Thomas, 1925. Ann. Mag. Nat. Hist., ser. 9, 15:583.
COMMON NAME: Robust Tuco-tuco.
TYPE LOCALITY: Argentina, Tucumán Prov., Concepcion, 500 m.
DISTRIBUTION: East of the mountains to 3,000 m in Tucumán Prov. (NW Argentina).
STATUS: IUCN – Lower Risk (lc).
COMMENTS: Karyotype has 2n=58-61 and FN=80.

Ctenomys tucumanus Thomas, 1900. Ann. Mag. Nat. Hist., ser. 7, 6:301.
COMMON NAME: Tucumán Tuco-tuco.
TYPE LOCALITY: Argentina, Tucumán Prov., 450 m.
DISTRIBUTION: NW Argentina.
STATUS: IUCN – Lower Risk (lc).
COMMENTS: Cabrera (1961) synonymized *tucumanus* under *mendocinus*. Considered distinct from *mendocinus* by Roig and Reig (1969) and Reig and Kiblisky (1969). Karyotype has 2n=28 and FN=52 (Reig et al., 1992).

Ctenomys tulduco Thomas, 1921. Ann. Mag. Nat. Hist., ser. 9, 8:218.
COMMON NAME: Sierra Tontal Tuco-tuco.
TYPE LOCALITY: Argentina, San Juan Prov., Los Sombreros, 2700 m.
DISTRIBUTION: Sierra Tontal, San Juan Prov., Argentina
COMMENTS: Included in *fulvus* by Cabrera (1961), but considered provisionally distinct by Contreras et al. (1977) and listed as a distinct species by Galliari et al. (1996).

Ctenomys validus Contreras, Roig, and Suzarte, 1977. Physis. Sec. C., 36(92):160.
COMMON NAME: Guanacache Tuco-tuco.
TYPE LOCALITY: Argentina, Mendoza Prov., Guaymallén Dept., Médanos del Borbollón, El Algarrobal (near city of Mendoza).
DISTRIBUTION: Mendoza Prov. (Argentina).
STATUS: IUCN – Lower Risk (lc).
COMMENTS: Closely related to *johannis* according to Contreras et al. (1977).

Ctenomys viperinus Thomas, 1926. Ann. Mag. Nat. Hist., ser. 9, 17:605.
COMMON NAME: Vipos Tuco-tuco.
TYPE LOCALITY: Argentina, Tucuman Prov., Vipos, tableland above Ñorco.
DISTRIBUTION: N Tucuman Prov. (Argentina).
COMMENTS: Cabrera (1961) considered *viperinus* synonymous with *knighti* but Galliari et al. (1996) listed it as a distinct species.

Ctenomys yolandae Contreras and Berry, 1984. Resumenes VII Jornadas Argentinas Zoologia, Mar del Plata: 75.
COMMON NAME: Yolanda's Tuco-tuco.
TYPE LOCALITY: Argentina, Santa Fe Prov., Las Palmas (29°25'S, 59°40'W).
DISTRIBUTION: Along the Paraná and San Javier Rivers, Santa Fe Prov.
COMMENTS: Not recognized by Redford and Eisenberg (1992) but listed by Galliari et al. (1996). Karyotype has 2n=50 and FN=78 and shares several chromosomal characteristics with *rionegrensis* (Ortells et al. 1990). Distinct in both penial (Balbontin et al., 1996) and sperm morphology (Vitullo et al., 1988).

Family Octodontidae Waterhouse, 1840. Proc. Zool. Soc. Lond., 1839:172 [1840].
SYNONYMS: Spalacopidae Lilljeborg, 1866.

COMMENTS: Sometimes considered the most primitive group of South American hystricognaths with numerous fossil genera from Oligocene on, but Reig (1986:418) reserved this distinction for the Echimyidae. *Ctenomys* is often placed here as a subfamily, and is closely related to octodontids (see comments under Ctenomyidae and in Cook et al. (1990:22-23). Chromosomal variation is not as conservative as previously thought, 2n=38-102 and includes the only known mammalian tetraploid (Gallardo et al., 1999). Molecular data (Gallardo and Kirsch, 2000; Honeycutt et al., 2003) support the monophyly of the Octodontidae and the sister taxon relationship with the Ctenomyidae.

Aconaemys Ameghino, 1891. Rev. Argent. Hist. Nat., 1:245.
  TYPE SPECIES: *Schizodon fuscus* Waterhouse, 1842.
  COMMENTS: Reig (1986:408) noted that *Aconaemys* is not separable from the Pliocene/Pleistocene form *Pithanotomys* Ameghino, 1887 at the generic level. If indeed the two are not separable, *Pithanotomys* would have priority as a name over *Aconaemys*. McKenna and Bell (1997) included in *Pithanotomys*. *Aconaemys* appears to by paraphyletic to *Spalacopus* (Gallardo and Kirsch, 2000; Honeycutt et al., 2003).

Aconaemys fuscus (Waterhouse, 1842). Proc. Zool. Soc. Lond., 1841:91 [1842].
  COMMON NAME: Chilean Rock Rat.
  TYPE LOCALITY: Argentina, Mendoza Prov., Valle de las Cuevas; see comments.
  DISTRIBUTION: High Andes of Chile and Argentina (between 33° and 41°S).
  STATUS: IUCN – Lower Risk (lc). Locally abundant.
  COMMENTS: Waterhouse never designated a type specimen, which was finally designated by Thomas (1917*a*). Waterhouse's original type locality was "from Chile", but in 1848 he added that the locality was from the Valle de las Cuevas near the Volcano of Peteroa, which would make the locality in Argentina (Pearson, 1984). Karyotype has 2n=56 and FN=108 (Gallardo and Reise, 1992).

Aconaemys porteri Thomas, 1917. Ann. Mag. Nat. Hist., ser. 8., 19:281.
  COMMON NAME: Porter's Rock Rat.
  TYPE LOCALITY: Chile, Osorno Prov., Osorno.
  DISTRIBUTION: NE slope of Villarrica Volcano to Puyehue (Chile) and to Nahuelhuapi Lake district (Argentina); between 900 to 2,000 m.
  COMMENTS: Pearson (1984) synomyized *porteri* with *fuscus*, but Gallardo and Reise (1992) demonstrated that the two are morphologically and chromosomally distinct. Karyotype has 2n=58 and FN=112 (Gallardo and Reise, 1992).

Aconaemys sagei Pearson, 1984. J. Zool. Lond., 202:229.
  COMMON NAME: Sage's Rock Rat.
  TYPE LOCALITY: Argentina, Neuquén Prov., 2 km S Cerro Quillén at Pampa de Hui Hui (1050 m).
  DISTRIBUTION: Known only from near Lago Quillén and Lago Hui Hui at Pampa de Hui Hui in Neuquén Prov. of Argentina. but reports suggest it might also occur in Malleco Prov. of Chile.
  STATUS: IUCN – Lower Risk (lc). Locally abundant, but may have very limited distribution.
  COMMENTS: Much smaller than *A. fuscus*. Karyotype has 2n=54 and FN=104 (Gallardo and Reise, 1992).

Octodon Bennett, 1832. Proc. Zool. Soc. Lond., 1832:46.
  TYPE SPECIES: *Octodon cumingii* Bennett, 1832 (= *Sciurus degus* Molina, 1782).
  SYNONYMS: *Dendrobius* Meyen, 1833; *Dendroleius* Meyen, 1833.

Octodon bridgesi Waterhouse, 1845. Proc. Zool. Soc. Lond., 1844:155 [1845].
  COMMON NAME: Bridges's Degu.
  TYPE LOCALITY: Chile, Curico Prov., Rio Teno.
  DISTRIBUTION: Chilean Andes from 34°15' to at least 40°S (Redford and Eisenberg, 1992).
  STATUS: IUCN – Lower Risk (lc). Once common, but now quite rare (Redford and Eisenberg, 1992).

COMMENTS: Massoia (1979:36) reported *Octodon*, perhaps *bridgesi*, from S Argentina. Karyotype has 2n=58 and FN=112 (Gallardo, 1992).

Octodon degus (Molina, 1782). Sagg. Stor. Nat. Chile, p. 303.
COMMON NAME: Degu.
TYPE LOCALITY: Chile, Santiago Prov., Santiago.
DISTRIBUTION: Chile, west slope of the Andes between Vallenar and Curico, to 1,200 m.
STATUS: IUCN – Lower Risk (lc). Common.
SYNONYMS: *alba* Fitzinger, 1867; *clivorum* Thomas, 1927; *cumingii* Bennett, 1832; *getulus* (Poeppig, 1829); *kummingii* (Schinz, 1845); *pallidus* Wagner, 1845; *peruana* Waterhouse, 1848.
COMMENTS: Reviewed by Woods and Boraker (1975). Karyotype has 2n=58 and FN=112 (Reig et al., 1972).

Octodon lunatus Osgood, 1943. Field Mus. Nat. Hist. Publ., Zool. Ser., 30:110.
COMMON NAME: Moon-toothed Degu.
TYPE LOCALITY: Chile, Valparaiso Prov., Olmue.
DISTRIBUTION: Coastal mountains of Valparaiso, Aconcagua, and Coquimbo Provs. (Chile) between 31°30' and 35°S.
STATUS: IUCN – Lower Risk (lc). Common.
COMMENTS: Included in *Octodon bridgesi* by Tamayo and Frassinetti (1980:354). Karyotype has 2n=78 (Spotorno et al., 1988).

Octodon pacificus Hutterer, 1994. Z. Säugetierk., 59:28.
COMMON NAME: Pacific Degu.
TYPE LOCALITY: Chile, Arauca Prov., Isla Mocha (38°22'S, 73°55'W).
DISTRIBUTION: Known only from type locality.
STATUS: IUCN – Vulnerable.
COMMENTS: Larger and heavier than the three mainland species of this genus.

Octodontomys Palmer, 1903. Science, n.s., 17:873.
TYPE SPECIES: *Neoctodon simonsi* Thomas, 1902 (= *Octodon gliroides* Gervais and d'Orbigny, 1844).
SYNONYMS: *Neoctodon* Thomas, 1902 [preoccupied].

Octodontomys gliroides (Gervais and d'Orbigny, 1844). Bull. Sci. Soc. Philom. Paris, p. 22.
COMMON NAME: Mountain Degu.
TYPE LOCALITY: Bolivia, La Paz Dept., near La Paz.
DISTRIBUTION: Andes of N Chile, SW Bolivia, and NW Argentina; occurs between 2,000 and 5,000 m in xeric habitats.
STATUS: IUCN – Lower Risk (lc). Common.
SYNONYMS: *simonsi* Thomas, 1902.
COMMENTS: Karyotype has 2n=38 and FN=64 (Gallardo, 1992; George and Weir, 1972a).

Octomys Thomas, 1920. Ann. Mag. Nat. Hist., ser. 9, 6:117.
TYPE SPECIES: *Octomys mimax* Thomas, 1920.
COMMENTS: Formerly included *Octomys barrerae*, which is now separated as *Tympanoctomys*.

Octomys mimax Thomas, 1920. Ann. Mag. Nat. Hist., ser. 9, 6:118.
COMMON NAME: Viscacha Rat.
TYPE LOCALITY: Argentina, Catamarca Prov., La Puntilla.
DISTRIBUTION: Foothills and lower montane slopes of the Andes, and portions of the Monte Desert of Catamarca, La Rioja, San Juan, and N Mendoza Provs. (Argentina).
STATUS: IUCN – Lower Risk (lc).
SYNONYMS: *joannius* Thomas, 1921.
COMMENTS: Occurs at higher elevations in Andes, and is similar in habits to *Neotoma*, the North American Woodrat (Redford and Eisenberg, 1992). Karyotype has 2n=56 and FN=108 (Contreras et al., 1994).

Pipanacoctomys Mares, Braun, Barquez, and Diaz, 2000. Occas. Papers, Mus. Texas Tech Univ, 203:3.
TYPE SPECIES: *Pipanacoctomys aureus* Mares, Braun, Barquez, and Diaz, 2000.

Pipanacoctomys aureus Mares, Braun, Barquez, and Diaz, 2000. Occas. Papers, Mus. Texas Tech Univ., 203:3.

COMMON NAME: Golden Viscacha-rat.

TYPE LOCALITY: Argentina, Catamarca Prov., Dept. Pomán, 28 km S, 9.3 km W Andalgalá (27°50'03"S, 66°15'50"W), 680 m.

DISTRIBUTION: Known only from the type locality.

COMMENTS: Analysis of morphological data (Mares et al., 2000) suggests sister taxon relationship with *Salinoctomys* and close relationship with *Tympanoctomys*.

Salinoctomys Mares, Braun, Barquez, and Diaz, 2000. Occas. Papers, Mus. Texas Tech Univ., 203:6.

TYPE SPECIES: *Salinoctomys loschalchalerosorum* Mares, Braun, Barquez, and Diaz, 2000.

Salinoctomys loschalchalerosorum Mares, Braun, Barquez, and Diaz, 2000. Occas. Papers, Mus. Texas Tech Univ., 203:6.

COMMON NAME: Chalchalero Viscacha-rat.

TYPE LOCALITY: Argentina, La Rioja Prov., Dept. Chamical, 26 km SW Quimilo (30°02'43.4"S, 65°31'13.4"W), 581 m.

DISTRIBUTION: Known only from the type locality.

COMMENTS: Analysis of morphological data (Mares et al., 2000) suggests sister taxon relationship with *Pipanacoctomys* and close relationship with *Tympanoctomys*.

Spalacopus Wagler, 1832. Isis, 25:1219.

TYPE SPECIES: *Spalacopus poeppigii* Wagler, 1832 (= *Mus cyanus* Molina, 1782).

SYNONYMS: *Poephagomys* F. Cuvier, 1834; *Psammoryctes* Poeppig, 1835.

COMMENTS: *Aconaemys* appears to by paraphyletic to *Spalacopus* (Gallardo and Kirsch, 2000; Honeycutt et al., 2003).

Spalacopus cyanus (Molina, 1782). Sagg. Stor. Nat. Chile, p. 300.

COMMON NAME: Coruro.

TYPE LOCALITY: Chile, Valparaiso Prov.

DISTRIBUTION: Chile, west of the Andes between 27° and 36°S.

STATUS: IUCN – Lower Risk (lc). Locally common, very fossorial.

SYNONYMS: *ater* (F. Cuvier, 1834); *noctivagus* (Poeppig, 1835); **maulinus** Osgood, 1943; **poeppigii** Wagler, 1832; *tabanus* Thomas, 1925.

COMMENTS: Reviewed by Torres-Mura and Contreras (1998). Includes *tabanus* (Cabrera, 1961:517; Corbet and Hill, 1991:203; Osgood, 1943). Three subspecies (*cyanus*, *poeppigi*, *maulinus*) recognized by Contreras et al. (1987). Karyotype has 2n=58 and FN=112 (Reig et al., 1972).

Tympanoctomys Yepes, 1941. Rev. Inst. Bact., 9, 1940 [1941]:569.

TYPE SPECIES: *Octomys barrerae* Lawrence, 1941.

COMMENTS: Considered a synonym of *Octomys* (see Reig, 1986:408; Corbet and Hill, 1991:204), but Cabrera (1961:516) and Redford and Eisenberg (1992) considered it a separate genus based on very enlarged tympanic bullae. *Octomys* identified as sister taxon by molecular data (Gallardo and Kirsch, 2001; Honeycutt et al., 2003), however, Mares et al. (2000) suggested *Pipanacoctomys* and *Salinoctomys* are the sister taxa. Although demonstrated to be a tetraploid, its chromosomal evolution is more complicated than a simple duplication of the chromosomes of *Octomys* (Gallardo et al., 1999).

Tympanoctomys barrerae (Lawrence, 1941). Proc. New England Zool. Club 18:43.

COMMON NAME: Red Viscacha-rat.

TYPE LOCALITY: Argentina, Mendoza Prov., La Paz.

DISTRIBUTION: Arid plains of Mendoza Province.

STATUS: IUCN – Vulnerable; rare.

COMMENTS: Reviewed by Diaz et al. (2000). Karyotype is 2n=102 (Contreras and Tores-Mura, 1987) and Gallardo et al. (1999) demonstrated that this species is tetraploid.

Family Abrocomidae Miller and Gidley, 1918. Jour. Washington Acad. Sci., 8(13):447.

COMMENTS: The phylogenetic affinity of *Abrocoma* has been problematic; Ellerman (1940) placed within the family Echimyidae, Landry (1957) within the family Octodontidae whereas Patterson and Wood (1982) assigned familial status. Glanz and Anderson (1990) suggested a sister relationship between Abrocomidae and Chinchillidae. Molecular data (Gallardo and Kirsch, 2001; Honeycutt et al., 2003; Huchon and Douzery, 2001) and allozyme data (Köhler et al., 2000) support familial status within the superfamily Octodontoidea and most suggest a basal position for the Abrocomidae.

Abrocoma Waterhouse, 1837. Proc. Zool. Soc. Lond., 1837:30.
TYPE SPECIES: *Abrocoma bennettii* Waterhouse, 1837.
SYNONYMS: *Habrocoma* Wagner, 1842.

Abrocoma bennettii Waterhouse, 1837. Proc. Zool. Soc. Lond., 1837:31.
COMMON NAME: Bennett's Chinchilla Rat.
TYPE LOCALITY: Chile, Aconcagua Prov., vicinity Aconcagua, flanks of Cordillera.
DISTRIBUTION: Chile from Copiapo to the area of Rio Biobio.
STATUS: IUCN – Lower Risk (lc). Common.
SYNONYMS: *cuvieri* Waterhouse, 1837; *helvina* (Wagner, 1842); *laniger* Prell, 1934; **murrayi** Wolffsohn, 1916.
COMMENTS: Karyotype has 2n=64 and FN=110 (Contreras et al., 1990).

Abrocoma boliviensis Glanz and Anderson, 1990. Am. Mus. Novit., 2991:23.
COMMON NAME: Bolivian Chinchilla Rat.
TYPE LOCALITY: Bolivia, Dept. Santa Cruz, Manual M. Caballero Prov., Comarapa.
DISTRIBUTION: Only known from the vicinity of the type locality.
STATUS: IUCN – Vulnerable.

Abrocoma budini Thomas, 1920. Ann. Mag. Nat. Hist., ser. 9, 5:475.
COMMON NAME: Budin's Chinchilla Rat.
TYPE LOCALITY: Argentina, Catamarca Prov., Otro Cerro, 3,000 m.
DISTRIBUTION: Known only from the type locality but Braun and Mares (2002) suggested probable restriction to Sierra de Ambato of Catamarca and La Rioja Provinces.
COMMENTS: Ellerman (1940) considered *budini* a subspecies of *cinerea* but Braun and Mares (2002) recognized it as distinct.

Abrocoma cinerea Thomas, 1919. Ann. Mag. Nat. Hist., ser. 9, 4:132.
COMMON NAME: Ashy Chinchilla Rat.
TYPE LOCALITY: Argentina, Jujuy Prov., Cerro Casabindo.
DISTRIBUTION: High elevations of SE Peru, SW Bolivia, N Chile, and NW Argentina (Jujuy, Salta and Tucumán Provinces).
STATUS: IUCN – Lower Risk (lc). Rare.
COMMENTS: Ellerman (1940) included *budini*, *famatina*, *schistacea*, and *vaccarum* as subspecies but Braun and Mares (2002) recognized each as a distinct species.

Abrocoma famatina Thomas, 1920. Ann. Mag. Nat. Hist., ser. 9, 6:419.
COMMON NAME: Famatina Chinchilla Rat.
TYPE LOCALITY: Argentina, La Rioja Prov., La Invernada, Cadena Famatina, 3,800 m.
DISTRIBUTION: Known from the Sierra de Famatina but Cabrera (1961) suggested possible occurrence in eastern portions of San Juan Prov., Argentina.
COMMENTS: Ellerman (1940) considered *famatina* a subspecies of *cinerea* but Braun and Mares (2002) recognized it as distinct.

Abrocoma shistacea Thomas, 1921. Ann. Mag. Nat. Hist., ser. 9, 8:216.
COMMON NAME: Sierra del Tontal Chinchilla Rat.
TYPE LOCALITY: Argentina, San Juan Prov., Los Sombreros, Sierra Tontal, 3,800 m.
DISTRIBUTION: Known from the Sierra del Tontal of southern San Juan Prov., Argentina.
COMMENTS: Ellerman (1940) considered *shistacea* a subspecies of *cinerea* but Braun and Mares (2002) recognized it as distinct.

Abrocoma uspallata Braun and Mares, 2002. J. Mammal., 83:9.
COMMON NAME: Uspallata Chinchilla Rat.
TYPE LOCALITY: Argentina, Mendoza Prov., Quebrada de la Vena, ca. 7 km SSE Uspallata (32°39.405'S, 69°20.970'W).
DISTRIBUTION: Known only from the type locality.
COMMENTS: Karyotype has 2n=66.

Abrocoma vaccarum Thomas, 1921. Ann. Mag. Nat. Hist., ser. 9, 8:217.
COMMON NAME: Mendozan Chinchilla Rat.
TYPE LOCALITY: Argentina, Mendoza Prov., Punta de Vacas, 3,000 m.
DISTRIBUTION: Known only from the type locality (Braun and Mares, 2002), although Thomas (1921e) and Cabrera (1961) extended the distribution to NW Mendoza Prov. and SW San Juan Prov., respectively.
COMMENTS: Ellerman (1940) considered *vaccarum* a subspecies of *cinerea* but Braun and Mares (2002) recognized it as distinct.

Cuscomys Emmons, 1999. Am. Mus. Novit., 3279:2.
TYPE SPECIES: *Abrocoma oblativa* Eaton, 1916.

Cuscomys ashaninka Emmons, 1999. Am. Mus. Novit., 3279:2.
COMMON NAME: Ashaninka Arboreal Chinchilla Rat.
TYPE LOCALITY: Peru, Dept. de Cusco, N Cordillera de Vilcabamba (11°39'36"S, 73°40'02"W).
DISTRIBUTION: Known only from the type locality.
COMMENTS: An arboreally adapted abrocomid rodent.

Cuscomys oblativa (Eaton, 1916). Mem. Conn. Acad. Arts Sci., 5:87.
COMMON NAME: Machu Picchu Arboreal Chinchilla Rat.
TYPE LOCALITY: Peru, Dept. de Cusco, Machu Picchu.
DISTRIBUTION: Known only from Inca burial sites of Machu Picchu.
STATUS: Considered probably extinct by Thomas (1920b); but see Emmons (1999a:13).

Family Echimyidae Gray, 1825. Ann. Philos., n.s., 10:341.
SYNONYMS: Echinomyidae Ameghino, 1889; Loncheridae Burmeister, 1850.
COMMENTS: This group is complex, and although several important revisions have occurred (Emmons and Vucetich, 1998; Emmons et al., 2002; Laura and Patton, 2000; Patton and Emmons, 1985), additional revisions are needed. The family includes the most primitive fossil New World hystricognaths from the Early Oligocene of Patagonia. Reig (1986) noted that some living taxa in this family with brachyodont and pentalophodont molars (*Mesomys* and *Lonchothrix*) are of the type expected in the ancestral New World Hystricognathi. The family is also the most diverse of all Hystricognathi. Patterson and Wood (1982) included *Chaetomys* in the Echimyidae (subfamily Chaetomyinae) based on retention of deciduous premolars but see T. Martin (1994b) for a persuasive argument for returning this taxon to the Erethizontidae. Molecular data (Lara et al., 1996; Leite and Patton, 2002) suggest that the remaining living subfamilies (Dactylomyinae, Echimyidae, Eumysopinae) cannot be defined by monophyly, because of what appears to be a very rapid initial radiation (Leite and Patton, 2002).

Subfamily Dactylomyinae Tate, 1935. Bull. Am. Nat. Hist., 68:295.
COMMENTS: Reig (1986) questioned placement of this subfamily in Echimyidae, and raised the possibility that dactylomyines are capromyids. Woods (1993) suggested that if West Indian spiny rats were placed in Capromyidae (see Woods, 1982) then dactylomyines might be too, but left the West Indian forms in their own subfamily, Heteropsomyinae, of the Echimyidae (see comments under Heteropsomyinae). Molecular data clearly place dactylomyine genera (*Dactylomys* and *Kannabateomys*) within echimyids (Lara et al., 1996; Leite and Patton, 2002) and to the exclusion of *Capromys* (Leite and Patton, 2002).

Dactylomys I. Geoffroy, 1838. Ann. Sci. Nat. Zool. (Paris), ser. 2, 10:126.
TYPE SPECIES: *Dactylomys typus* I. Geoffroy, 1838 (=*Echimys dactylinus* Desmarest, 1817).
SYNONYMS: *Lachnomys* Thomas, 1916 [see Cabrera, 1961:543].

Dactylomys boliviensis Anthony, 1920. J. Mammal., 1:82.
    COMMON NAME: Bolivian Bamboo Rat.
    TYPE LOCALITY: Bolivia, Cochabamba Dept., Misión San Antonia, Río Chimoré, 390 m.
    DISTRIBUTION: C Bolivia, SE Peru, SW Brazil (Acre).
    STATUS: IUCN – Lower Risk (lc).
    COMMENTS: Emmons and Feer (1990) recognized only a single species, *dactylinus*, in the
        lowland Amazon basin, however, da Silva and Patton (1993) demonstrated the
        presence of two divergent forms of *Dactylomys* and Patton et al. (2000) provided
        morphological diagnoses and map ranges for *boliviensis* and *dactylinus*. Anderson
        (1997) recognized *boliviensis* as the only species of *Dactylomys* occurring in Bolivia.

Dactylomys dactylinus (Desmarest, 1817). Nouv. Dict. Hist., Nouv. ed., 10:57.
    COMMON NAME: Amazon Bamboo Rat.
    TYPE LOCALITY: Upper Amazon area (no locality in original description).
    DISTRIBUTION: N Brazil from near the mouth of the Amazon west to the base of the Andes in
        N Peru, Ecuador, and SW Colombia and south to N Bolivia; see Patton et al. (2000) for
        details.
    STATUS: IUCN – Lower Risk (lc). Locally common.
    SYNONYMS: *typus* I. Geoffroy, 1838; **canescens** Thomas, 1912; **modestus** Lönnberg, 1921.
    COMMENTS: The genetic (da Silva and Patton, 1993) and morphological (Patton et al., 2000)
        divergence clearly support the recognition of *dactylinus*.

Dactylomys peruanus J. A. Allen, 1900. Bull. Am. Mus. Nat. Hist., 13:220.
    COMMON NAME: Montane Bamboo Rat.
    TYPE LOCALITY: Peru, Juliaca; corrected by J. A. Allen (1901) to Inca Mines, "about 200 miles
        [322 km] northeast of Jaliaca" 1,830 m, 13°30'S, 70°W.
    DISTRIBUTION: Cloud forest of SE Peru.
    STATUS: IUCN – Data Deficient. Unknown.
    COMMENTS: Emmon and Feer (1997) indicated that this taxon occurs at 1,000 to 3,000 m in
        SE Peru and Bolivia, but see Anderson (1997), who recorded no specimens from Bolivia.

Kannabateomys Jentink, 1891. Notes Leyden Mus., 13:109.
    TYPE SPECIES: *Dactylomys amblyonyx* Wagner, 1845.
    SYNONYMS: *Cannabateomys* Lydekker, 1892.

Kannabateomys amblyonyx (Wagner, 1845). Arch. Naturgesch., 1:146.
    COMMON NAME: Atlantic Bamboo Rat.
    TYPE LOCALITY: Brazil, São Paulo, Ipanema.
    DISTRIBUTION: E Brazil, Paraguay, NE Argentina; lives in bamboo thickets.
    STATUS: IUCN – Lower Risk (lc). Locally common.
    SYNONYMS: **pallidior** Thomas, 1903.

Olallamys Emmons, 1988. J. Mammal., 69(2):241.
    TYPE SPECIES: *Thrinacodus albicauda* Günther, 1879.
    COMMENTS: Emmons (1988) indicated that *Thrinacodus* Günther, 1879, the original generic
        name, is preoccupied by *Thrinacodus* St. John and Worthen, 1875 (a shark).

Olallamys albicauda (Günther, 1879). Proc. Zool. Soc. Lond., 1879:144.
    COMMON NAME: White-tailed Olalla Rat.
    TYPE LOCALITY: Colombia, Antioquia Dept., Medellin.
    DISTRIBUTION: NW and C Colombia, west of the Cordillera Central; occurs up to 3,000 m,
        often in dense bamboo thickets (Eisenberg, 1989).
    STATUS: IUCN – Lower Risk (nt).
    SYNONYMS: *apolinari* (J.A. Allen, 1914).

Olallamys edax (Thomas, 1916). Ann. Mag. Nat. Hist., ser. 8, 18:299.
    COMMON NAME: Greedy Olalla Rat.
    TYPE LOCALITY: Venezuela, Merida, Sierra de Merida.
    DISTRIBUTION: W Venezuela and adjacent N Colombia.
    STATUS: IUCN – Lower Risk (nt).

Subfamily Echimyinae Gray, 1825. Ann. Philos., n.s., 10:341.
  SYNONYMS: Loncherinae Thomas, 1896.
  COMMENTS: Recent revisions by Emmons (1993*a*), Emmons and Vucetich (1998), and Emmons
    et al. (2002) clarified the taxonomy and systematics of this subfamily. Molecular data
    (Leite and Patton, 2002) do not support the monopyly of this subfamily. L. Emmons
    (pers. comm.) is completing a further revision.

Callistomys Emmons and Vucetich, 1998. Am. Mus. Novit., 3223:3.
  TYPE SPECIES: *Nelomys pictus* Pictet, 1843.
  COMMENTS: Fossil mandible tentatively identified by Winge (1888) as *Lasiuromys villosus* placed
    in *Callistomys* by Emmons and Vucetich (1998).

Callistomys pictus (Pictet, 1841). Notice Anim. Nouv. Mus. Geneve, p. 29.
  COMMON NAME: Painted Tree-rat.
  TYPE LOCALITY: Brazil, Bahia.
  DISTRIBUTION: Known only from the vicinity of Ilhéus, Bahia, Brazil.
  STATUS: IUCN – Data Deficient as *Echimys pictus*.
  COMMENTS: Previously included in the genus *Isothrix* (Cabrera, 1961; Ellerman, 1940; Patton
    and Emmons, 1985; Waterhouse, 1848), *Nelomys* (Goldman, 1916; Pictet, 1843;
    Thomas, 1916*e, f*), or *Echimys* (Moojen, 1952; Tate, 1935), but placed in a new genus,
    Callistomys, by Emmons and Vucetich (1998).

Diplomys Thomas, 1916. Ann. Mag. Nat. Hist., ser. 8, 18:240.
  TYPE SPECIES: *Loncheres caniceps* Günther, 1877.
  COMMENTS: The genus is under review by L. Emmons (pers. comm.), and generic assignment of
    the species listed is likely to change.

Diplomys caniceps (Günther, 1877). Proc. Zool. Soc. Lond., 1876:745[1877].
  COMMON NAME: Colombian Soft-furred Spiny-rat.
  TYPE LOCALITY: Colombia, Antioquia Dept., Medellin.
  DISTRIBUTION: W Colombia, NW Ecuador.
  STATUS: IUCN – Lower Risk (nt).
  COMMENTS: See description in Emmons and Feer (1990:222).

Diplomys labilis (Bangs, 1901). Am. Nat., 35:638.
  COMMON NAME: Rufous Soft-furred Spiny-rat.
  TYPE LOCALITY: Panama, San Miguel Isl.
  DISTRIBUTION: Panama (including San Miguel Isl), W Colombia, and (probably) N Ecuador.
  STATUS: IUCN – Lower Risk (lc). Locally common.
  SYNONYMS: *darlingi* (Goldman, 1913).
  COMMENTS: Includes *darlingi* (Handley, 1966*a*:787; Hall, 1981:874).

Diplomys rufodorsalis (J. A. Allen, 1899). Bull. Am. Mus. Nat. Hist., 12:197.
  COMMON NAME: Red Crested Soft-furred Spiny-rat.
  TYPE LOCALITY: Colombia, Magdalena Dept., Onaca.
  DISTRIBUTION: Known only from the type locality in the Sierra Nevada de Santa Marta of
    NE Colombia.
  STATUS: IUCN – Vulnerable; rare (known only from two specimens).
  COMMENTS: Taxonomy being revised (Emmons, In Press).

Echimys G. Cuvier, 1809. Bull Sci. Soc. Philom. Paris, 24:394.
  TYPE SPECIES: *Myoxus chrysurus* Zimmermann, 1780.
  SYNONYMS: *Echinomys* Wagner, 1840; *Loncheres* Illiger, 1811; *Nelomys* Jourdan, 1837.
  COMMENTS: Formerly included *pictus* by some authors (e.g., Pictet, 1843; Goldman, 1916;
    Thomas, 1916*e, f*; Tate, 1935; Moojen, 1952), now in genus *Callistomys* (see account
    above); *armatus, grandis, macrurus,* and *rhipidurus,* which are now placed in the genus
    Makalata (see that account); and *blainvillei, braziliensis, dasythrix, lamarum, nigrispinus,*
    thomasi, and *unicolor,* which are now placed in the genus *Phyllomys* (see that account and

Emmons et al., 2002). The genus is under review by L. Emmons (pers. comm.), and generic assignment of the species listed is likely to change.

Echimys chrysurus (Zimmermann, 1780). Geogr. Gesch. Mensch. Vierf. Theire, 2:352.
COMMON NAME: White-faced Spiny Tree-rat.
TYPE LOCALITY: Suriname.
DISTRIBUTION: Guianas to lower Amazonian NE Brazil.
STATUS: IUCN – Vulnerable.
SYNONYMS: *cristatus* Desmarest, 1817; *paleaceus* (Lichtenstein, 1820).

Echimys saturnus Thomas, 1928. Ann. Mag. Nat. Hist., ser. 10, 2:409.
COMMON NAME: Dark Spiny Tree-rat.
TYPE LOCALITY: Ecuador, Napo-Pastaza Prov., Río Napo.
DISTRIBUTION: Ecuador and N Peru, east of the Andes to at least 1,000 m.
STATUS: IUCN – Lower Risk (lc). Rare.

Echimys semivillosus (I. Geoffroy, 1838). Ann. Sci. Nat. Zool. (Paris), ser. 2, 10:125.
COMMON NAME: Speckled Spiny Tree-rat.
TYPE LOCALITY: Colombia, Bolivar Dept., Cartagena.
DISTRIBUTION: N Colombia, Venezuela, Margarita Isl.
STATUS: IUCN – Lower Risk (lc). Locally common.
SYNONYMS: *carrikeri* (J. A. Allen, 1911); *flavidus* (Hollister, 1914); *punctatus* (Thomas, 1899).
COMMENTS: See Cabrera (1961:542-543). Taxonomy being revised (Emmons, In Press).

Isothrix Wagner, 1845. Arch. Naturgesch., 1:145.
TYPE SPECIES: *Isothrix bistriata* Wagner, 1845 (selected by Goldman, 1916).
SYNONYMS: *Lasiuromys* Deville, 1852.
COMMENTS: Formerly included *pictus* (Cabrera, 1961; Patton and Emmons, 1985), which Emmons and Vucetich (1998) placed in a new genus, *Callistomys* (see that account).

Isothrix bistriata Wagner, 1845. Arch. Naturgesch., 1:146.
COMMON NAME: Yellow-crowned Brush-tailed Rat.
TYPE LOCALITY: Brazil, Mato Grosso, Río Guaporé.
DISTRIBUTION: E Bolivia and E Peru, SW to NC Brazil, S Venezuela, adjacent Colombia.
STATUS: IUCN – Lower Risk (nt).
SYNONYMS: *boliviensis* Petter and Cuenca Aguirre, 1982; *molliae* Thomas, 1924; *villosus* (Deville, 1852); **orinoci** Thomas, 1899.
COMMENTS: Patton and Emmons (1985) recognized 2 subspecies (*bistriata* and *orinoci*) and synonymized *villosus* in *I. b. bistriata*. Listed as *bistriatus* by Corbet and Hill (1991:206). The sequence analyses of da Silva and Patton (1993) and Patton et al. (2000) identified two geographic and taxonomic forms within *bistriata*. Bonvicino et al. (2003a) recognized *negrensis* as a distinct species, *bistriata* and *negrensis*. The karyotype has 2n=60, FN=116-118 (Leal-Mesquita, 1991; Patton et al., 2000).

Isothrix negrensis Thomas, 1920. Ann. Mag. Nat. Hist., ser. 9, 6:277.
COMMON NAME: Rio Negro Brush-tailed Rat.
TYPE LOCALITY: Brazil, Acajutuba, lower Rio Negro, near its mouth.
DISTRIBUTION: NC Brazil along the Rio Negro and Rio Solimoes.
COMMENTS: Included in *bistriata* until recognized as distinct species based on sequence, chromosomal, and morphological differentiation (Bonvicino et al., 2003a). Karyotype has 2n=60, FN=112 (Bonvicino et al., 2003a).

Isothrix pagurus Wagner, 1845. Arch. Naturgesch., 1:146.
COMMON NAME: Plain Brush-tailed Rat.
TYPE LOCALITY: Brazil, Amazonas, Borba, Lower Rio Madeira.
DISTRIBUTION: Amazon Basin of Central Brazil from Rio Madeira east to Rio Tapajoz and north to lower Rio Negro.
STATUS: IUCN – Lower Risk (nt).
COMMENTS: Karyotype has 2n=22 and FN=38 (Patton and Emmons, 1985).

Isothrix sinnamariensis Vie et al., 1996. Mammalia, 60:395.
>COMMON NAME: Sinnamary Brush-tailed Rat.
>TYPE LOCALITY: French Guiana, along the Sinnamary River, 21 km upstream of the Petit Saut
>>Dam (4°56′80″N, 53°01′90″W).
>DISTRIBUTION: Known only from the vicinity of the type locality.
>STATUS: IUCN – Data Deficient.
>COMMENTS: Karyotype has 2n=28 and FN=42.

Makalata Husson, 1978. The Mammals of Suriname, p. 445.
>TYPE SPECIES: *Nelomys armatus* (I. Geoffroy 1838) (= *Echimys didelphoides* Desmarest, 1817).
>COMMENTS: Not recognized as a separate genus by Eisenberg (1989) or Corbet and Hill (1991).

Makalata didelphoides (Desmarest, 1817). Nouv. Dict. Nat., Nouv. ed., 10:58.
>COMMON NAME: Red-nosed Armored Tree-rat.
>TYPE LOCALITY: Unknown, probably Brazil.
>DISTRIBUTION: Andes of N Ecuador and Colombia, Venezuela, Guyanas, Amazon Basin of
>>Brazil, Trinidad and Tobago; perhaps Martinique (record probably erroneous, see Hall,
>>1981:1180).
>STATUS: IUCN – Data Deficient as *Mesomys didelphoides*, Lower Risk (lc) as *Makalata armatus*.
>SYNONYMS: *armatus* (I. Geoffroy, 1838); *castaneus* (Allen and Chapman, 1893); *guianae*
>>(Thomas, 1888); *hispidus* (Lichtenstein, 1830); *longirostris* (Anthony, 1921).
>COMMENTS: Emmons (1993*a*) concluded that the holotype for *Echimys didelphoides* was a
>>young *Makalata armata* and therefore *E. didelphoides* Desmarest 1817 antedates
>>*Nelomys armatus* I. Geoffroy (1838). Karyotype has 2n=66 and FN=110 (Lima et al.,
>>1998). Sequence data suggest more than one species present in *didelphoides* (da Silva
>>and Patton, 1993; Patton et al., 2000).

Makalata grandis (Wagner, 1845). Arch. Naturgesch., 1:145.
>COMMON NAME: Giant Armored Tree-rat.
>TYPE LOCALITY: Brazil, Amazonas, Manaqueri.
>DISTRIBUTION: Amazonian Brazil along the banks of the Amazon River from Río Negro to
>>Ilha Caviana.
>STATUS: IUCN – Lower Risk (lc) as *Echimys grandis*.
>COMMENTS: Formerly included *rhipidurus* as a valid subspecies (Cabrera, 1961), which was
>>given specific status by Emmons and Feer (1990).

Makalata macrura (Wagner, 1842). Arch. Naturgesch., 1:360.
>COMMON NAME: Long-tailed Armored Tree-rat.
>TYPE LOCALITY: Brazil, Amazonas, Borba.
>DISTRIBUTION: C Brazil west to N Peru and E Ecuador (see Patton et al., 2000).
>STATUS: IUCN – Lower Risk (lc) as *Echimys macrurus*.
>COMMENTS: Considered a large form of *Makalata didelphoides* by Emmons and Feer (1997)
>>that might represent a distinct species, a hypothesis now supported by morphological
>>and molecular evidence (Patton et al., 2000).

Makalata obscura (Wagner, 1840). Abh. Akad. Wiss. Munich, 3:196.
>COMMON NAME: Dark Armored Tree-rat.
>TYPE LOCALITY: Unknown.
>DISTRIBUTION: Brazil.
>STATUS: IUCN – Data Deficient as *Mesomys obscurus*.
>COMMENTS: Known only from the original description. Tate (1935) placed it first in the
>>genus *Mesomys* (p. 413) and then in the genus *Echimys* (p. 432). Cabrera (1960)
>>followed Tate's first allocation and retained it in *Mesomys* with reservation, a
>>placement followed by subsequent authors (Corbet and Hill, 1991; Woods, 1993).
>>Emmons (1993*a*) suggested that the original illustrations precluded identity with
>>*Mesomys* and Emmons (In Press) placed it in the genus *Makalata*.

Makalata occasius (Thomas, 1921). Ann. Mag. Nat. Hist., ser. 9, 7:450.
>COMMON NAME: Bare-tailed Armored Tree-rat.
>TYPE LOCALITY: Ecuador, Mt. Pichincha, Gualea.

DISTRIBUTION: Ecuador and Peru east of the Andes.

STATUS: IUCN – Critically Endangered.

COMMENTS: Recognized as a distinct species by Emmons and Feer (1990:217) and placed in the genus *Makalata* by Emmons and Feer (1997:236). Distinction further documented by Emmons (In Press), who placed this taxon in a new genus.

Makalata rhipidura (Thomas, 1928). Ann. Mag. Nat. Hist., ser. 10, 2:291.

COMMON NAME: Peruvian Armored Tree-rat.

TYPE LOCALITY: Peru, Dept. Loreto, Pebas.

DISTRIBUTION: C and N Amazonian Peru.

STATUS: IUCN – Data Deficient as *Echimys rhipidurus*.

COMMENTS: Elevated to species by Emmons and Feer (1990:216) and placed in the genus *Makalata* by Emmons and Feer (1997:239).

Phyllomys Lund, 1839. Ann. Sci. Nat. (Zool.), Paris, sér. 2, 11:225.

TYPE SPECIES: *Nelomys blainvilii* Jordan, 1837.

SYNONYMS: *Loncheres* Lichtenstein, 1820.

COMMENTS: Species of the Brazilian laminated-toothed echimyid rodents have historically been grouped with the nonlaminate-toothed arboreal echimyids under the generic name *Echimys* (Tate, 1935; Cabrera, 1961; Woods, 1993). These laminated-toothed forms were segregated as *Nelomys* by Thomas (1916*a*, *b*) and by Emmons and Feer (1990; 1997) and as *Phyllomys* by Moojen (1952) and Emmons et al. (2002). Emmons et al. (2002) consider *Nelomys* a junior synonym of *Echimys* and document that *Phyllomys* is the next available generic name. Revised by Leite (2003), who described two new species.

Phyllomys blainvillii (Jordan, 1837). Comptes Rendus Hebdomadaires des Seances de L'Academie des Sciences, 15:522.

COMMON NAME: Golden Atlantic Tree-rat.

TYPE LOCALITY: Brazil, Bahia, Isla de Deos; restricted to Seabra (ca. 12°25'S, 41°46'W) by Emmons et al. (2002).

DISTRIBUTION: NE Brazil, S Ceara to N Minas Gerais.

STATUS: IUCN – Lower Risk (nt) as *Echimys blainvilei* (sic). Most abundant species of *Phyllomys* in museum collections.

COMMENTS: Name incorrectly spelled as *blainvillei* by Wagner (1840). Karyotype has 2n=50 and FN=94 (Leite, 2003).

Phyllomys brasiliensis Lund, 1840. Ann. Sci. Nat. (Zool.), Paris, sér. 2, 12:208.

COMMON NAME: Orange-brown Atlantic Tree-rat.

TYPE LOCALITY: Brazil, Minas Gerais, cave (Lapa das Quatro Bocas) near Lagos Santa.

DISTRIBUTION: Valleys of the Paraopeba and das Velhas Rivers in Minas Gerais state (S Brazil).

STATUS: IUCN – Lower Risk (lc) as *Echimys brasiliensis*.

SYNONYMS: *armatus* Winge, 1887.

COMMENTS: Cabrera (1961:540) considered *Phyllomys brasiliensis* a *nomen nudum*; but see Emmons et al. (2002).

Phyllomys dasythrix Hensel, 1872. Abh. Konigl. Akad. Wiss. Berlin, p. 49.

COMMON NAME: Drab Atlantic Tree-rat.

TYPE LOCALITY: Brazil, Rio Grande do Sul, Porto Alegre (30°04'S, 51°07'W); as restricted by Emmons et al. (2002).

DISTRIBUTION: S Parana to Rio Grande do Sul, usually below 800 m (S Brazil).

STATUS: IUCN – Lower Risk (lc) as *Echimys dasythrix*.

COMMENTS: Karyotype has 2n=72 (Leite, 2003).

Phyllomys kerri (Moojen, 1950). Rev. Brasil. Biol., 10:489.

COMMON NAME: Kerr's Atlantic Tree-rat.

TYPE LOCALITY: Brazil, São Paulo, Ubatuba, Estação Experimental de Ubatuba (23°25'S, 45°07'W); as restricted by Emmons et al. (2002).

DISTRIBUTION: Known only from Ubatuba, on N coast of São Paulo.

Phyllomys lamarum (Thomas, 1916). Ann. Mag. Nat. Hist., ser. 8, 18:297.
   COMMON NAME: Pallid Atlantic Tree-rat.
   TYPE LOCALITY: Brazil, Bahia, about 70 miles (113 km) NW Salvador, Lamarão; as amended
      by Emmons et al. (2002).
   DISTRIBUTION: E Brazil, Paraiba to Minas Gerais.
   STATUS: IUCN – Lower Risk (lc) as *Echimys lamarum*.

Phyllomys lundi Leite, 2003. Univ. Cal. Publ. Zool., 132:19.
   COMMON NAME: Lund's Atlantic Tree-rat.
   TYPE LOCALITY: Brazil, Minas Gerais, 4 km SE Passa Vinte, Fazenda do Bené (22°14'S,
      44°12'W, elev. 680 m).
   DISTRIBUTION: Known only from two localities 200 km apart in southern Minas Gerais and
      Rio de Janeiro (Brazil).
   COMMENTS: One of the smallest species in the genus; possibly related to *P. nigrispinus*.

Phyllomys mantiqueirensis Leite, 2003. Univ. Cal. Publ. Zool., 132:25.
   COMMON NAME: Mantiqueira Atlantic Tree-rat.
   TYPE LOCALITY: Brazil, Minas Gerais, 13 km SW Delfim Moreira, Fazenda da Onça (22°36'S,
      45°20'W, elev. 1850 m).
   DISTRIBUTION: Brazil.

Phyllomys medius (Thomas, 1909). Ann. Mag. Nat. Hist., ser. 8, 4:239.
   COMMON NAME: Long-furred Atlantic Tree-rat.
   TYPE LOCALITY: Brazil, Paraná, Serra do Mar, Roça Nova (25°28'S, 49°01'W, elev. 1000 m); as
      amended by Emmons et al. (2002).
   DISTRIBUTION: Mainly along the coast in the states of Minas Gerais and Rio de Janeiro to
      Rio Grande do Sul, extends west into Araucaria forest in Paraná (S Brazil).
   COMMENTS: Cabrera (1961) considered *medius* a subspecies of *blainvillii* but Emmons et al.
      (2002) and Leite (2003) documented species status. Karyotype has 2n=96 (Leite, 2003).

Phyllomys nigrispinus (Wagner, 1842). Arch. Naturgesch., 1:361.
   COMMON NAME: Black-spined Atlantic Tree-rat.
   TYPE LOCALITY: Brazil, São Paulo, Floresta Nacional de Ipanema, 20 km NW Sorocaba
      (23°26'S, 47°37'W); as amended by Emmons et al. (2002).
   DISTRIBUTION: SE Brazil, from the state of Rio de Janeiro to Paraná, mainly along the coastal
      zone, but extending inland to W São Paulo to at least 850 m.
   STATUS: IUCN – Lower Risk (lc) as *Echimys nigrispinus*.
   COMMENTS: Karyotype has 2n=52 (Leite, 2003).

Phyllomys pattoni Emmons, Leite, Kock, and Costa, 2002. Amer. Mus. Novit., 3380:30.
   COMMON NAME: Rusty-sided Atlantic Tree-rat.
   TYPE LOCALITY: Brazil, Bahia, Caravelas, Mangue do Caritoti (17°43'30"S, 39°15'35"W).
   DISTRIBUTION: E Brazil from the state of Paraibo to NE São Paulo, chiefly along the coast but
      occurring inland to 1000 m in rainforest.
   COMMENTS: Karyotype has 2n=72 or 80; FN=112 or 114.

Phyllomys thomasi (Ihering, 1871). Rev. Mus. Paulista, 2:171.
   COMMON NAME: Giant Atlantic Tree-rat.
   TYPE LOCALITY: Brazil, São Paulo, Isla de São Sabastião (23°46'S, 45°21'W); as amended by
      Emmons et al. (2002).
   DISTRIBUTION: Endemic to the Isla de São Sabastião (Brazil).
   STATUS: IUCN – Vulnerable as *Echimys thomasi*.

Phyllomys unicolor (Wagner, 1842). Arch. Naturgesch., 1:361.
   COMMON NAME: Short-furred Atlantic Tree-rat.
   TYPE LOCALITY: "Brasilia"; restricted by Emmons et al. (2002) to Bahia, 50 km SW Caravelas,
      Colonia Leopoldina (now Helvecia) (17°48'S, 39°39'W, elev. 59 m).
   DISTRIBUTION: Known only from the type locality.
   STATUS: IUCN – Lower Risk (lc) as *Echimys unicolor*.
   COMMENTS: According to Cabrera (1961:543), Thomas believed that *unicolor* was a synonym

of *brasiliensis*. Emmons et al. (2002) suggested that the affinities of *unicolor* may be with *medius*, with which it shares many features.

Subfamily Eumysopinae Rusconi, 1935. Bol. Paleont. Buenos Aires, 5:2.
COMMENTS: Proposed as a subfamily by Patton and Reig (1989:76). This subfamily includes the Oligocene fossil forms, and the most primitive living genera of South American echimyids. Included in the Heteropsomyinae by McKenna and Bell (1997). Molecular data (Leite and Patton, 2002) suggest that the genera *Clyomys* and *Euryzygomatomys* are allied with the myocastorids and capromyids and that this subfamily is not a monophyletic unit.

Carterodon Waterhouse, 1848. Nat. Hist. Mamm., 2:351.
TYPE SPECIES: *Echimys sulcidens* Lund, 1841.

Carterodon sulcidens (Lund, 1841). Afh. Kongl. Danske Vid. Selsk., p. 49.
COMMON NAME: Owl's Spiny Rat.
TYPE LOCALITY: Brazil, Minas Gerais, Lagoa Santa.
DISTRIBUTION: E Brazil.
STATUS: IUCN – Lower Risk (nt).
SYNONYMS: *temmincki* (Lund, 1842).

Clyomys Thomas, 1916. Ann. Mag. Nat. Hist., ser. 8, 18:300.
TYPE SPECIES: *Echimys laticeps* Thomas, 1909.
COMMENTS: Reviewed by Ávila-Pires and Wutke (1981:530). Reig (1986:409, footnote) noted that *Clyomys* might not be distinct from *Euryzygomatomys*.

Clyomys bishopi Ávila-Pires and Wutke, 1981. Revta Bras. Biol., 41:530.
COMMON NAME: Bishop's Fossorial Spiny Rat.
TYPE LOCALITY: Brazil, São Paulo, Itapetininga.
DISTRIBUTION: Known only from the type locality.
STATUS: IUCN – Vulnerable.

Clyomys laticeps (Thomas, 1909). Ann. Mag. Nat. Hist., ser. 8, 4:240.
COMMON NAME: Broad-headed Spiny Rat.
TYPE LOCALITY: Brazil, Santa Catarina, Joinville.
DISTRIBUTION: Between Minas Gerais and Santa Catarina (E Brazil) in savanna habitats.
STATUS: IUCN – Lower Risk (lc). Common.
SYNONYMS: *spinosus* (Winge, 1887).
COMMENTS: Highly fossorial and colonial. Karyotype has 2n=34 and FN=64 (Yonenaga, 1975).

Euryzygomatomys Goeldi, 1901. Bol. Mus. Para., 3:179.
TYPE SPECIES: *Rattus spinosus* G. Fischer, 1814.
COMMENTS: Reig (1986:409, footnote) noted that *Clyomys* might not be distinct from *Euryzygomatomys*.

Euryzygomatomys spinosus (G. Fischer, 1814). Zoognosia, 3:105.
COMMON NAME: Guiara.
TYPE LOCALITY: Paraguay, Cordillera, Atira.
DISTRIBUTION: S and E Brazil, NE Argentina, Paraguay.
STATUS: IUCN – Lower Risk (lc).
SYNONYMS: *brachyura* (Rengger, 1830); *catellus* Thomas, 1916; *guiara* (Brandt, 1835); *rufa* (Lichtenstein, 1820).
COMMENTS: Use of the names derived from Fischer (1814) is provisional pending clarification of the availability of the work. Should Fischer (1814) become unavailable, the name would be *rufa* Lichtenstein, 1820 (Adhandl. Preuss. Akad, Wiss., 1818-1819 [1820], p. 192). Karyotype has 2n=46 and FN=88 (Yonenaga, 1975).

Hoplomys J. A. Allen, 1908. Bull Am. Mus. Nat. Hist., 24:649.
  TYPE SPECIES: *Hoplomys truei* J. A. Allen, 1908 (= *Echimys gymnurus* Thomas, 1897).
  COMMENTS: Revised by Handley (1959*a*). Patton and Reig (1989:90) demonstrated that this
    genus is very close to *Proechimys*, and may be congeneric, a view supported by DNA
    sequence analyses (Lara et al., 1996).

Hoplomys gymnurus (Thomas, 1897). Ann. Mag. Nat. Hist., ser. 6, 20:550.
  COMMON NAME: Armored Rat.
  TYPE LOCALITY: Ecuador, Esmeraldas Prov., Cachavi.
  DISTRIBUTION: EC Honduras to NW Ecuador.
  STATUS: IUCN – Lower Risk (lc). Locally common.
  SYNONYMS: *goethalsi* Goldman, 1912; *truei* J. A. Allen, 1908.
  COMMENTS: Cabrera (1961) placed *hoplomyoides* Tate, 1939 as a subspecies of *gymnurus*, but
    this taxon has been consistently placed in the genus *Proechimys* by other authors
    (Handley, 1976; Patton, 1987; Tate, 1939).

Lonchothrix Thomas, 1920. Ann. Mag. Nat. Hist., ser. 9, 6:113.
  TYPE SPECIES: *Lonchothrix emiliae* Thomas, 1920.

Lonchothrix emiliae Thomas, 1920. Ann. Mag. Nat. Hist., ser. 9, 6:113.
  COMMON NAME: Tuft-tailed Spiny Tree-rat.
  TYPE LOCALITY: Brazil, Río Tapajoz, Villa Braga.
  DISTRIBUTION: C Brazil, south of the Amazon River in area of Río Tapajoz and Río Madeira.
  STATUS: IUCN – Lower Risk (lc). Locally common.

Mesomys Wagner, 1845. Arch. Naturgesch., 1:145.
  TYPE SPECIES: *Mesomys ecaudatus* Wagner, 1845 (= *Echimys hispidus* Desmarest, 1817).
  COMMENTS: Revision of this genus is needed; see Husson (1978:440), Emmons and Feer
    (1997:234), Voss et al. (2001) and Orlando et al. (2003). Formerly included *didelphoides*
    and *obscurus*, both now placed in the genus *Makalata* (Emmons, 1993*a*).

Mesomys hispidus (Desmarest, 1817). Nouv. Dict. Nat., Nouv. ed., 10:58.
  COMMON NAME: Ferreira's Spiny Tree-rat.
  TYPE LOCALITY: Brazil, Amapá (see comments).
  DISTRIBUTION: Amazon Basin of N Bolivia, Peru, E Ecuador, SE Colombia east through
    S Venezuela, the Guianas, and all of Brazil except east of the Rio Tapajos and south of
    the Amazon.
  STATUS: IUCN – Lower Risk (lc).
  SYNONYMS: *ecaudatus* Wagner, 1845; *ferrugineus* (Günther, 1876); *spicatus* Thomas, 1924.
  COMMENTS: Type locality unknown, given as Amérique méridionale by Desmarest (1817),
    restricted to Borba, Rio Madeira, Brazil by Tate (1939). However, sequence data from
    holotype identified close affinities to material from French Guiana and, in
    consideration of the travel route of the original collector, support the correct
    placement of the type locality in the state of Amapá, Brazil (Orlando et al., 2003). More
    than a single species may be represented in this taxon, which demonstrates
    considerable variation in body size, cranial morphology, and molecular characters
    (Orlando et al., 2003; Patton et al., 2000; Voss et al., 2001). Includes *ferrugineus* and
    spicatus, although both have tasseled tails, a characteristic of the recently described
    occultus; neither have other morphological characteristics of *occultus* (Patton et al.,
    2000). Tentatively includes *ecaudatus* but see comments under *occultus*. Formerly
    included *stimulax*, see Husson (1978) and comment under that taxon. Karyotype has
    2n=60 and FN=116 (Leal-Mesquita, 1991).

Mesomys leniceps Thomas, 1926. Ann. Mag. Nat. Hist., ser. 9, 18:348.
  COMMON NAME: Woolly-headed Spiny Tree-rat.
  TYPE LOCALITY: Peru, Dept Amazonas, Yambasbramba.
  DISTRIBUTION: Higher elevations in Peru (over 2,000 m).
  STATUS: IUCN – Lower Risk (lc).

COMMENTS: Recognized as a species by Emmons and Feer (1990) but not by Eisenberg and Redford (1999), who placed it within *hispidus*.

Mesomys occultus Patton, da Silva, and Malcolm, 2000. Bull. Am. Mus. Nat Hist., 244:194.
COMMON NAME: Tufted-tailed Spiny Tree-rat.
TYPE LOCALITY: Brazil, Amazonas, Colocação Vira-Volta, left bank Rio Juruá on Igarape Arabidi, affluent of Paraná Breu.
DISTRIBUTION: Central Amazon of Brazil.
COMMENTS: Although similar to and sympatric with *hispidus* the two can be differentiated by morphology, karyotype and cytochrome *b* sequences (Patton et al., 2000). It is not clear at this time whether *ecaudatus* should be included in *hispidus* or is a senior synonym of *occultus*. Karyotype has 2n=42 and FN=54.

Mesomys stimulax Thomas, 1911. Ann. Mag. Nat. Hist., ser. 8, 7:607.
COMMON NAME: Pará Spiny Tree-rat.
TYPE LOCALITY: Brazil, Pará, Amazon estuary, "Cametá, Lower Tocantins."
DISTRIBUTION: Brazil, east of the Rio Tapajós and south of the Amazon.
STATUS: IUCN – Lower Risk (lc).
COMMENTS: Formerly included in *hispidus* by Cabrera (1961); considered a distinct species by Husson (1978). Material from south of the Amazon and east of the Rio Tapajós in Brazil is karyotypically identical (2n=60, FN=116) with *hispidus* (Patton et al., 2000) but genetically divergent (da Silva and Patton, 1993; Patton et al., 2000). Voss et al. (2001) examined the holotype and grouped it with "small-toothed" forms from Guiana and SE Amazonia. Small-toothed forms from NE South America, however, have affinities with *hispidus*, whereas *stimulax* appears to be restricted to areas east of the Rio Tapajós and south of the Amazon (Orlando et al., 2003; Patton et al., 2000).

Proechimys J. A. Allen, 1899. Bull. Am. Mus. Nat. Hist., 12:257.
TYPE SPECIES: *Echimys trinitatis* J. A. Allen and Chapman, 1893.
COMMENTS: Traditionally divided into 2 subgenera, *Proechimys* and *Trinomys* (Moojen, 1948; Thomas, 1921f; Woods, 1993) but *Trinomys* was elevated to a genus by Lara et al. (1996). *Proechimys* represents one of the most diverse groups of Neotropical rodents with 63 named forms. Reviewed, in part, by Reig et al. (1980), Gardner and Emmons (1984), and Patton (1987). May include *Hoplomys* (see Patton and Reig, 1989; Lara et al., 1996). The species-level taxonomy of the group is controversial, and in need of a major revision. Gardner and Emmons (1984) and Patton (1987) grouped taxa into species groups. Further revision of the genus is in progress by J. Patton (pers. comm.).

Proechimys brevicauda (Gunther, 1877). Proc. Zool. Soc. Lond., 1876:748 [1877].
COMMON NAME: Short-tailed Spiny-rat.
TYPE LOCALITY: Peru, Dept. de Loreto, Chamicuros (Río Huallago).
DISTRIBUTION: Western Amazonia from S Colombia, E Ecuador, E Peru, N Bolivia, and W Brazil.
STATUS: IUCN – Lower Risk (lc) as *P. brevicauda*, *P. bolivianus*, and *P. gularis*.
SYNONYMS: *bolivianus* Thomas, 1901; *elassopus* Osgood, 1944; *gularis* Thomas, 1911; *securus* Thomas, 1902.
COMMENTS: Listed as a subspecies of *longicaudatus* by Cabrera (1961:524) and as a member of the *longicaudatus* group by Patton (1987), but treated as a species by Patton et al. (2000). Karyotype has 2n=28-30 and FN=48-50 (Gardner and Emmons, 1984; Patton et al., 2000). Likely polytypic, as Gardner and Emmons (1984) suggested *gularis* is a separate species based on karyotypic grounds and Patton et al. (2000) suggested that *elassopus* is also distinct, based on morphological and molecular criteria. Anderson (1997) included *bolivianus* in *brevicauda*.

Proechimys canicollis (J. A. Allen, 1899). Bull. Am. Mus. Nat. Hist., 12:200.
COMMON NAME: Colombian Spiny-rat.
TYPE LOCALITY: Colombia, Dept. Magdalena, Bonda near Santa Marta (NW base of Sierra Nevada de Santa Marta).
DISTRIBUTION: NC Colombia, NW Venezuela.

STATUS: IUCN – Lower Risk (lc).

COMMENTS: Placed in the monotypic *canicollis* group by Patton (1987). Karyotype has 2n=24 and FN=44 (Gardner and Emmons, 1984).

Proechimys chrysaeolus (Thomas, 1898). Ann. Mag. Nat. Hist. ser. 7, 1:244.

COMMON NAME: Boyacá Spiny-rat.

TYPE LOCALITY: Colombia, Dept. Boyacá, Muzo (valley of Río Carare).

DISTRIBUTION: E Colombia.

STATUS: IUCN – Lower Risk (lc).

COMMENTS: Included in the *semispinosus* group by Gardner and Emmons (1984) but placed in the *trinitatus* group by Patton (1987). Listed as a subspecies of *guyannensis* by Cabrera (1961).

Proechimys cuvieri Petter, 1978. C. R. Acad. Sci. Paris, ser. D, 287:263.

COMMON NAME: Cuvier's Spiny-rat.

TYPE LOCALITY: French Guiana, Saul.

DISTRIBUTION: French Guiana, Surinam, Guyana, Brazil west along both sides of the Amazon River to N Peru.

STATUS: IUCN – Lower Risk (lc).

COMMENTS: Placed in the monotypic *cuvieri* group by Patton (1987). Considered closely related to *guyannensis* by Gardner and Emmons (1984), but see Voss et al. (2001) for review of morphological differentiation. Karyotype has 2n=28 and FN=46-48 (Maia and Langguth, 1993; Patton et al. 2000; Reig et al., 1979).

Proechimys decumanus (Thomas, 1899). Ann. Mag. Nat. Hist., ser. 7, 4:282.

COMMON NAME: Pacific Spiny-rat.

TYPE LOCALITY: Ecuador, Prov. de Guayas, Changón.

DISTRIBUTION: NW Peru, SW Ecuador, Pacific lowlands.

STATUS: IUCN – Lower Risk (lc).

COMMENTS: Placed in *brevicauda* group by Gardner and Emmons (1984), but Patton (1987) recognized it as the sole representative of a *decumanus* group. Listed as a subspecies of *guyannensis* by Cabrera (1961), but as a species by Emmons and Feer (1990, 1997). Karyotype has 2n=30 and FN=54 (Gardner and Emmons, 1984).

Proechimys echinothrix da Silva, 1998. Proc. Biol. Soc. Wash., 111:441.

COMMON NAME: Stiff-spine Spiny-rat.

TYPE LOCALITY: Brazil, Amazonas, Colocação Vira-volta, left bank Rio Juruá on Igarapé Arabidi (66°14'W, 03°17'S).

DISTRIBUTION: W Brazilian Amazon and, possibly, SE Colombia (see Patton et al., 2000).

COMMENTS: Karyotype has 2n=32 and FN=60.

Proechimys gardneri da Silva, 1998. Proc. Biol. Soc. Wash., 111:460.

COMMON NAME: Gardner's Spiny-rat.

TYPE LOCALITY: Brazil, Amazonas, Altamira, right bank Rio Juruá (68°54'W, 06°35'S).

DISTRIBUTION: W Amazonia of Brazil and N Boliva between the Rio Juruá and the Rio Madeira.

COMMENTS: Weksler et al. (2001) erroneously cited Patton (1987) as placing in *guyannensis* group; molecular analysis suggests affinities with *pattoni* and *kulinae* (Patton et al., 2000). Karyotype has 2n=40 and FN=56.

Proechimys goeldii Thomas, 1905. Ann. Mag. Nat. Hist., ser. 7, 15:587.

COMMON NAME: Goeldi's Spiny-rat.

TYPE LOCALITY: Brazil, Pará, Santarem.

DISTRIBUTION: Amazonian Brazil between Jamunda and Tapajoz Rivers, W Brazil.

STATUS: IUCN – Lower Risk (lc).

SYNONYMS: *hyleae* Moojen, 1948; *leioprymna* Moojen, 1948; *nesiotes* Moojen, 1948.

COMMENTS: Placed in the *goeldii* group by Patton (1987); Patton et al. (2000) included *hyleae*, *leioprymna*, and *nesiotes*. Karyotype has 2n=24 and FN=42 (Patton et al., 2000).

Proechimys guairae Thomas, 1901. Proc. Biol. Soc. Wash., 14:27.

COMMON NAME: Guaira Spiny-rat.

TYPE LOCALITY: Venezuela, Federal Dist., La Guaira.
DISTRIBUTION: NC Venezuela, east of Lake Maracaibo and the Merida Andes.
STATUS: IUCN – Lower Risk (lc).
SYNONYMS: *ochraceus* Osgood, 1912.
COMMENTS: Placed in *trinitatus* group by Patton (1987). Includes *ochraceus*; *guairae* and
    ochraceus were formerly included in *guyannensis* (Reig et al., 1980). Eisenberg (1989)
    included *urichi* within *guairae*. Karyotype is highly variable, 2n=44-52; FN=66-74
    (Aguilera and Corti, 1994; Reig, 1989; Reig and Useche, 1976). Reig et al. (1980)
    mentioned a closely related, undescribed species ("Barina's") from south of the Merida
    Andes.

Proechimys guyannensis (E. Geoffroy, 1803). Cat. Mamm. Mus. Nat. Hist. Nat., p. 194.
    COMMON NAME: Guyenne Spiny-rat.
    TYPE LOCALITY: French Guiana, Cayenne.
    DISTRIBUTION: SC Venezuela, the Guianas, southward to C Brazil.
    STATUS: IUCN – Lower Risk (lc) as *P. warreni*.
    SYNONYMS: *cayennensis* Desmarest, 1817; *warreni* Thomas, 1905; **arabupu** Moojen, 1948;
        arescens Osgood, 1944; **cherriei** Thomas, 1899; **riparum** Moojen, 1948; **vacillator**
        Thomas 1903.
    COMMENTS: Hershkovitz (1948*a*) resurrected *guyannensis* E. Geoffroy, 1803 for this species as
        this name antedates *cayennensis*. Patton (1987) included *cherriei, roberti, vacillator, oris,
        warreni, boimensis, arescens, riparum,* and *arabupu*, but suggested that taxa from south
        of the Amazon were likely a different species from those to the north. Weksler et al.
        (2001) confirmed this opinion on morphological, karyotypic, and molecular grounds,
        and showed that *P. roberti* was the best name to apply to that species south of the
        Amazon. Husson's (1978) recognition of *warreni* as a species separate from *guyannensis*
        was based on his mistaken application of the latter name to the species *P. cuvieri* (see
        Voss et al., 2001). However, it remains likely that more than a single species is included
        in the geographically restricted view of *guyannensis* presented here. Karyotype has
        2n=40 and FN=54 (Reig et al., 1979).

Proechimys hoplomyoides (Tate, 1939). Bull. Am. Mus. Nat. Hist., 76:179.
    COMMON NAME: Guyanan Spiny-rat.
    TYPE LOCALITY: Venezuela, Bolivar Prov., Mt. Roraima (2,000 m).
    DISTRIBUTION: SE Venezuela, adjacent Guyana and Brazil.
    STATUS: IUCN – Lower Risk (lc).
    COMMENTS: Erroneously considered a subspecies of *Hoplomys gymnurus* by Cabrera (1961).
        Placed in the *trinitatus* group by Patton (1987).

Proechimys kulinae da Silva, 1998. Proc. Biol. Soc. Wash., 111:451.
    COMMON NAME: Kulina Spiny-rat.
    TYPE LOCALITY: Brazil, Amazonas, Seringal Condor, left bank Rio Juruá (70°51'W, 06°45'S).
    DISTRIBUTION: W Brazil north of the Rio Juruá to NE Peru, Dept. Loreto.
    COMMENTS: Molecular analysis suggests affinities with *pattoni* and *gardneri* (da Silva, 1995;
        Patton et al., 2000). Karyotype has 2n=34 and FN=52.

Proechimys longicaudatus (Rengger, 1830). Naturgesch. Säugeth Paraguay, p. 236.
    COMMON NAME: Long-tailed Spiny-rat.
    TYPE LOCALITY: N Paraguay.
    DISTRIBUTION: S Bolivia, N Paraguay, C Brazil.
    STATUS: IUCN – Lower Risk (lc).
    SYNONYMS: *leucomystax* Ribeiro, 1914; *ribeiroi* Moojen 1948; *villacauda* Moojen, 1948.
    COMMENTS: Placed in *longicaudatus* group by Patton (1987). Formerly included *brevicauda*
        and *roberti* (Moojen, 1948). Includes *ribeiroi* and *villacauda* (J. Patton in litt.).
        Karyotype has 2n=28 and FN=48 (Leal-Mesquita, 1991).

Proechimys magdalenae (Hershkovitz, 1948). Proc. U. S. Natl. Mus., 97:136.
    COMMON NAME: Magdalena Spiny Rat.
    TYPE LOCALITY: Colombia, Dept. de Bolívar, Río San Pedro near Norosí (178 m) in the
        foothills of the Cordillera Central.

DISTRIBUTION: Colombia west of the Río Magdalena.
STATUS: IUCN – Lower Risk (lc).
COMMENTS: Combined with *guyannensis* (Cabrera, 1961:521), but recognized as a distinct species by Emmons and Feer (1990:274). Included in the *brevicauda* group by Gardner and Emmons (1984) but placed in the *trinitatus* group by Patton (1987).

Proechimys mincae (J. A. Allen, 1899). Bull. Am. Mus. Nat. Hist., 12:198.
COMMON NAME: Minca Spiny-rat.
TYPE LOCALITY: Colombia, Dept. Magdalena, Minca near Santa Marta.
DISTRIBUTION: N Colombia below 500 m in the Sierra Nevada de Santa Marta.
STATUS: IUCN – Lower Risk (lc).
COMMENTS: Placed in the *trinitatis* group (Patton, 1987). Karyotype has 2n=48 and FN=68 (Gardner and Emmons, 1984).

Proechimys oconnelli J. A. Allen, 1913. Bull. Am. Mus. Nat. Hist., 32(24):479.
COMMON NAME: O'Connell's Spiny-rat.
TYPE LOCALITY: Colombia, Villavicencio, 480 m.
DISTRIBUTION: C Colombia east of Cordillera Oriental.
STATUS: IUCN – Lower Risk (lc).
COMMENTS: Placed in the *semispinosus* group by Patton (1987). Karyotype has 2n=32 and FN=52 (Gardner and Emmons, 1984).

Proechimys pattoni da Silva, 1998. Proc. Biol. Soc. Wash., 111:454.
COMMON NAME: Patton's Spiny-rat.
TYPE LOCALITY: Brazil, Acre, Igarapé Porongaba, right bank Rio Juruá (72°47'W, 08°40'S).
DISTRIBUTION: SE Peru, W Amazonia in headwaters of Rio Juruá (Brazil).
COMMENTS: Weksler et al. (2001) erroneously cited Patton (1987) as placing in *guyannensis* group. Molecular analysis suggests affinities with *gardneri* and *kulinae* (Patton et al., 2000). Karyotype has 2n=40 and FN=56.

Proechimys poliopus Osgood, 1914. Field Mus. Nat. Hist. Publ., Zool. Ser., 10:135.
COMMON NAME: Gray-footed Spiny-rat.
TYPE LOCALITY: Venezuela, Táchira State, San Juan de Colón (797 m) on N slope of Sierra de Merida. Listed in Honacki et al. (1982:590) as Venezuela, Zulia Prov., Río Aurare, El Panorama, for unknown reason.
DISTRIBUTION: NW Venezuela, between Lake Maracaibo and the Sierra de Perija and adjacent Colombia.
STATUS: IUCN – Lower Risk (lc).
COMMENTS: Placed in *trinitatus* group (Patton, 1987). Formerly included in *guyannensis*; see Reig et al. (1980). Karyotype has 2n=42 and FN=72-76 (Aguilera and Corti, 1994; Reig and Useche, 1976).

Proechimys quadruplicatus Hershkovitz, 1948. Proc. U. S. Natl. Mus., 97:138.
COMMON NAME: Napo Spiny-rat.
TYPE LOCALITY: Ecuador, Napo-Pastaza Prov., Río Napo, Isla Llunchi (18 kms below mouth of Rio Coca).
DISTRIBUTION: N Peru, E Ecuador and SE Colombia east across S Venezuela and adjacent Brazil to the vicinity of Manaus, west of Rio Negro.
STATUS: IUCN – Lower Risk (lc) as *P. quadruplicatus* and *P. amphichoricus*.
SYNONYMS: *amphichoricus* Moojen, 1948.
COMMENTS: Placed in *goeldii* group by Patton (1987); Patton et al. (2000) included *amphichoricus*. Karyotype has 2n=26-28 and FN=42-44 (Gardner and Emmons, 1984; Patton et al., 2000).

Proechimys roberti Thomas, 1901. Ann. Mag. Nat. Hist., ser. 7, 14:531.
COMMON NAME: Roberto's Spiny-rat.
TYPE LOCALITY: Brazil, Minas Gerais, Araguari, Rio Jordão.
DISTRIBUTION: Cerrado of C Brazil and E Amazon.
STATUS: IUCN – Lower Risk (lc) as *P. oris*.
SYNONYMS: *arescens* Osgood, 1944: *boimensis*saz Allen, 1916; *oris* Thomas, 1904.

COMMENTS: Thomas (1901) considered *roberti* a close relative of *longicaudatus* and some subsequent reports have treated it as a subspecies of *longicaudatus* (Cabrera, 1961; Moojen, 1948; Woods, 1993). Thomas (1904*b*) suggested affinities with *guyannensis* and Ellerman (1940) listed as a subspecies of *guyannensis*. Patton (1987) assigned it to the *guyannensis* group. Considered a distinct species based on morphometrics (Pessôa et al., 1990) and molecular data (Weksler et al., 2001). Eisenberg and Redford (1999) suggested that *oris* is probably a synonym for *roberti*, supported by Weksler et al. (2001). Karyotype has 2n=30 and FN=54-56 (Gardner and Emmons, 1984; Weksler et al., 2001).

Proechimys semispinosus (Tomes, 1860). Proc. Zool. Soc. Lond., 1860:265.
COMMON NAME: Tome's Spiny-rat.
TYPE LOCALITY: Ecuador, Esmeraldas Prov., Esmeraldas.
DISTRIBUTION: SE Honduras to SW Ecuador.
STATUS: IUCN – Lower Risk (nt) as *P. gorgonae*, Lower Risk (lc) as *P. semispinosus*.
SYNONYMS: **burrus** Bangs, 1901; **calidior** Thomas, 1911; **centralis** (Thomas, 1896); **colombianus** Thomas, 1914; *gorgonae* Bangs, 1905; **goldmani** Bole, 1937; **ignotus** Kellogg, 1946; **panamensis** Thomas, 1900; *chiriquinus* Thomas, 1900; **rosa** Thomas, 1900; **rubellus** Hollister, 1914.
COMMENTS: Placed in *semispinosus* group (Patton, 1987). Formerly included in *amphichoricus*; see Reig et al. (1980) who used the name *centralis* for animals assigned to *semispinosus* from N Venezuela. Gardner (1983*b*) discussed the taxonomic history of this species and corrected the type locality. Karyotype has 2n=30 and FN=50-54 (Gardner and Emmons, 1984; Patton and Gardner, 1972). Includes *gorgonae* based on lack of karyotypic difference between populations on Gorgonae Isl and mainland of Colombia (Gómez-Laverde et al., 1990).

Proechimys simonsi Thomas, 1900. Ann. Mag.Nat. Hist., ser. 7, 6:300.
COMMON NAME: Simon's Spiny-rat.
TYPE LOCALITY: Peru, Dept. Junín, Río Perené (240 m).
DISTRIBUTION: Western Amazon Basin from S Colombia to N Bolivia and W Brazil.
STATUS: IUCN – Lower Risk (lc) as *P. simonsi* and *P. hendeei*.
SYNONYMS: *hendeei* Thomas, 1926; *nigrofulvus* Osgood, 1944.
COMMENTS: Placed in *simonsi* group by Patton (1987). Gardner and Emmons (1984) included *hendeei*; considered a separate species by Eisenberg and Redford (1999). Karyotype has 2n=32 and FN=58 (Gardner and Emmons, 1984; Patton and Gardner, 1972).

Proechimys steerei Goldman, 1911. Proc. Biol. Soc. Wash., 24:238.
COMMON NAME: Steere's Spiny-rat.
TYPE LOCALITY: Brazil, Amazonas State, Hyutanahan (above Purús).
DISTRIBUTION: C Peru and N Bolivia eastward into W Brazil.
STATUS: IUCN – Lower Risk (lc).
SYNONYMS: *hilda* Thomas, 1924; *kermiti* Allen, 1915; *liminalis* Moojen, 1948; *pachita* Thomas, 1904; *rattinus* Thomas, 1926.
COMMENTS: Placed in *goeldii* group by Patton (1987). Listed as a subspecies of *goeldii* in Cabrera (1961:519), but as a distinct species by Anderson (1997) and Patton et al. (2000). Karyotype has 2n=24 and FN=40-42 (Gardner and Emmons, 1984; Patton et al., 2000). Patton et al. (2000) suggested synonyms listed above.

Proechimys trinitatus (J. A. Allen and Chapman, 1893). Bull. Am. Mus. Nat. Hist., 5:223.
COMMON NAME: Trinidad Spiny-rat.
TYPE LOCALITY: Trinidad and Tobago, Trinidad.
DISTRIBUTION: Trinidad.
STATUS: IUCN – Lower Risk (lc).
COMMENTS: Placed in *trinitatus* group by Patton (1987). Formerly included in *guyannensis* (Reig et al., 1980). Karyotype has 2n=62 and FN=80 (Aguilera and Corti, 1994).

Proechimys urichi (J. A. Allen, 1899). Bull. Am. Mus. Nat. Hist., 12:199.
COMMON NAME: Sucre Spiny-rat.

TYPE LOCALITY: Venezuela, Sucre, Quebrada Seca.

DISTRIBUTION: N Venezuela.

STATUS: IUCN – Lower Risk (lc).

COMMENTS: Placed in *trinitatus* group by Patton (1987). Formerly included in *guyannensis* (Reig et al., 1980). Included within *guairae* by Eisenberg (1989). Karyotype has 2n=62 and FN=88 (Reig and Useche, 1976).

Thrichomys Trouessart, 1880. Cat. Mamm. Bull. Soc. Etudes Sci. Angers, 1881:179.

TYPE SPECIES: *Nelomys apereoides* Lund, 1839.

COMMENTS: Formerly referred to as *Cercomys*, which was based on *Cercomys cunicularius*, a composite (Petter, 1973*b*).

Thrichomys apereoides (Lund, 1839). Afh. K. Danske Vid. Selsk., p. 38.

COMMON NAME: Common Punaré.

TYPE LOCALITY: Brazil, Minas Gerais, Lagoa Santa.

DISTRIBUTION: S and SE Brazil.

STATUS: IUCN – Lower Risk (lc). Common.

SYNONYMS: *antricola* (Lund, 1841); **laurenteus** (Thomas, 1904).

COMMENTS: Formerly referred to as *Cercomys cunicularius*, a composite (Petter, 1973*b*; Mares et al., 1981*a*:120). Braggio and Bonvicino (2004) have recognized *inermis* and *pachyurus* as distinct species on the basis of karyotypic and sequence divergence. Even with the recognition of these two additional species, considerable geographic variation in karyotypes and sequence data remains among populations in Brazil. Specimens collected along the Tocantins and Paraná Rivers in the states of Tocantins and Goiás have a 2n=30 and a FN=56 (Bonvicino et al., 2002*b*), whereas specimens from the states of Bahia and Pernambuco (*T. a. laurenteus*) have a 2n=30 and FN=54 (Bonvicino et al., 2002; Leal-Mesquita et al., 1993; Souza and Yonenaga-Yassuda, 1982). Specimens collected from near the type locality (*T. a. apereoides*) have a 2n=28 and a FN=50, whereas specimens from Jaborandi have a 2n=28 and a FN=52 (Bonvicino et al., 2002). More than a single species may still be present. Braggio and Bonvicino (2004) concluded based on analysis of sequence data that the karyotypic form with a 2n=30 and a FN=56 represents an undescribed species. Based on morphological differentiation Reis et al. (2002) suggested that more than one species is present within E Brazil, however, it is unclear how inclusion of material now recognized as *inermis* may have affected their analyses.

Thrichomys inermis (Pictet, 1941). Not. Anim. Nouv. Mus. Geneve, 2:23.

COMMON NAME: Highlands Punaré.

TYPE LOCALITY: Brazil, Bahia.

DISTRIBUTION: Highlands of Chapada Diamantia, Bahia, Brazil.

COMMENTS: Included in *apereoides* by Moojen (1952) and Cabrera (1961) but recognized as a distinct species based on chromosomal differentiation (Bonvicino et al., 2002) and sequence divergence (Braggio and Bonvicino, 2004). Karyotype has 2n=26; FN=48 (Bonvicino et al., 2002; Leal-Mesquita et al., 1993).

Thrichomys pachyurus (Wagner, 1845). Archiv. Naturg., 1:146.

COMMON NAME: Paraguayan Punaré.

TYPE LOCALITY: Brazil, Matto Grosso, Cuiabá.

DISTRIBUTION: Mato Grosso (S Brazil), Paraguay.

SYNONYMS: *crassicaudus* (Wagner, 1947); *fosteri* (Thomas, 1903).

COMMENTS: Originally described in the genus *Isothrix* by Wagner (1845) and included in *Isothrix* by Tate (1935) and Ellerman (1940). Included in *Thrichomys apereoides* by Cabrera (1961) but recognized as a distinct species based on chromosomal differentiation (Bonvicino et al., 2002) and sequence divergence (Braggio and Bonvicino, 2004). Karyotype has 2n=34; FN=64 (Bonvicino et al., 2002).

Trinomys Thomas, 1921. Ann. Mag. Nat. Hist., ser. 9, 8:140.

TYPE SPECIES: *Echimys albispinus* I. Geoffroy, 1838.

COMMENTS: Considered a subgenus of *Proechimys* by Moojen (1948) and Woods (1993) but Lara et al. (1996) demonstrated that *Trinomys* is not the sister taxon of *Proechimys* based on mtDNA sequences and should be regarded as a separate genus. Revised by Lara and Patton (2000).

Trinomys albispinus (I. Geoffroy, 1838). Ann. Sci. Nat. Zool. (Paris), 10:125.
    COMMON NAME: White-spined Atlantic Spiny-rat.
    TYPE LOCALITY: Brazil, Bahía, Ilha de Madre Deos, Itaparica (near Salvador).
    DISTRIBUTION: States of Sergipe, Bahía and Minas Gerais (NE and SE Brazil).
    STATUS: IUCN – Lower Risk (nt) as *Proechimys albispinus*.
    SYNONYMS: **minor** (Reis and Pessôa, 1995); **serotinus** (Thomas 1921).
    COMMENTS: Reviewed by Pessôa and Reis (2002). Karyotype has 2n=60 and FN=116 (Leal-Mesquita et al., 1992). Divergent from other forms of *Trinomys* (Lara and Patton, 2000).

Trinomys dimidiatus (Günther, 1877). Proc. Zool. Soc. Lond., 1876:747 [1877].
    COMMON NAME: Soft-spined Atlantic Spiny-rat.
    TYPE LOCALITY: Brazil, Rio de Janeiro, unknown probably SW Rio de Janeiro.
    DISTRIBUTION: E Brazil including the states of Rio de Janeiro and Distrito Federal.
    STATUS: IUCN – Lower Risk (lc) as *Proechimys dimidiatus*.
    COMMENTS: Reviewed by Pessôa and Reis (1993). Lara and Patton (2000) concluded that *dimidiatus* was the sister taxon of *iheringi*. Karyotype has 2n=56 (Pessôa and Reis, 1993).

Trinomys eliasi (Pessôa and Reis, 1993). Z. Säugetierk., 58:183.
    COMMON NAME: Elias' Atlantic Spiny-rat.
    TYPE LOCALITY: Brazil, Rio de Janeiro, Município de Maricá, Reatinga da Barra de Maricá (22°31′S, 47°17′W).
    DISTRIBUTION: Known only from the type locality.
    COMMENTS: Although *eliasi* was described and recognized as a subspecies of *iheringi* (Possôa and Reis, 1993, 1994, 1996), Lara and Patton (2000) elevated it to a species based on analysis of sequence data. The sister taxon of *eliasi* is *paratus* and these two taxa are included in a clade with *setosus* and *yonenagae* (Lara and Patton, 2000). Karyotype has 2n=58 and FN=112 (Pessôa and Reis, 1996).

Trinomys gratiosus (Moojen, 1948). Univ. Kansas Publ., Nus Nat. Hist., 1(19):379.
    COMMON NAME: Gracile Atlantic Spiny-rat.
    TYPE LOCALITY: Brazil, Espirito Santo, Santa Teresa, Floresta da Caixa Dagua (19°55′S, 40°36′W).
    DISTRIBUTION: South bank of River Doce, Espírito Santo southward to Teresopolis, Rio de Janeiro (SE Brazil).
    SYNONYMS: *panema* (Moojen, 1948); **bonafidei** (Moojen, 1948).
    COMMENTS: Recognized as a subspecies of *iheringi* by Moojen (1948) and Pessôa and Reis (1996). Lara and Patton (2000) elevated *gratiosus* to a species, retained *bonafidei* as a subspecies of *gratiosus*, and suggested that *panema* was not morphologically distinct from *gratiosus*. Karyotype has 2n=56 and FN=108 (Pessôa and Reis, 1996).

Trinomys iheringi (Thomas, 1911). Ann. Mag. Nat. Hist., ser. 8, 8:252.
    COMMON NAME: Ihering's Atlantic Spiny-rat.
    TYPE LOCALITY: Brazil, São Paulo, São Sabastião Isl.
    DISTRIBUTION: Mainland and offshore islands of the states of São Paulo and Rio de Janeiro (E Brazil).
    STATUS: IUCN – Lower Risk (lc) as *Proechimys iheringi*.
    COMMENTS: Reviewed by Pessôa and Reis (1996), who recognized the six subspecies listed by Moojen (1948: *bonafidei, denigratus, gratiosus, iheringi, panema, paratus*) plus the newly described *eliasi*. Pessôa and Reis (1994) concluded that *bonafidei, denigratus, eliasi, gratiosus, panema* may be part of a taxon specifically distinct from *iheringi*. Lara and Patton (2000) concluded that *bonafidei* and *gratiosus* were not allied with *iheringi* and elevated *gratiosus* to a species and retained *bonafidei* as a subspecies of *gratiosus*. Lara and Patton (2000) also elevated *eliasi* and *paratus* to species and concluded that *denigratus* was a subspecies of *setosus*. The sister taxon of *iheringi* is *dimidiatus* (Lara and

Patton, 2000). Karyotype has 2n=60-65 depending on the number of B chromosomes (Yonenaga-Yassuda et al., 1985).

Trinomys mirapitanga Lara, Patton, and Hingst-Zaher, 2002. Mamm. Biol., 67(10):236.
COMMON NAME: Dark-caped Atlantic Spiny-rat.
TYPE LOCALITY: Brazil, Bahia, 16 km W Porto Seguro, Estação Ecológica do Pau Brasil (16°22′S, 39°11′W, 40 m).
DISTRIBUTION: Known only from the type locality.
COMMENTS: Listed as *Trinomys* sp. in Lara and Patton (2000) and placed as the basal member of a clade also containing *dimitiatus* and *iheringi* (see also Lara et al., 2002).

Trinomys moojeni (Pessôa, Oliveira, and Reis, 1992). Z. Säugetierk., 57:40.
COMMON NAME: Moojen's Atlantic Spiny-rat.
TYPE LOCALITY: Brazil, Minas Gerais, Conceição do Matro Dentro.
DISTRIBUTION: Known only from the type locality.
COMMENTS: Molecular affinities of this species not yet examined.

Trinomys myosuros (Lichtenstein, 1820). Abh. Konig. Akad. Wiss., Berlin, p. 192.
COMMON NAME: Mouse-tailed Atlantic Spiny-rat.
TYPE LOCALITY: Brazil, Bahia.
DISTRIBUTION: Bahia (Brazil).
STATUS: IUCN – Lower Risk (lc) as *Proechimys myosuros*.
SYNONYMS: *cinnamomeus* (Lichtenstein, 1830); *leptosoma* (Brants, 1827).
COMMENTS: This taxon is only tentatively placed in *Trinomys* because the teeth remain undescribed. Thomas (1921*a*) synonymized it with *setosus,* whereas Moojen (1948) concluded that its affinities were with *albispinus.*

Trinomys paratus (Moojen, 1948). Univ. Kansas Publ., Nus Nat. Hist., 1(19):382.
COMMON NAME: Spiked Atlantic Spiny-rat.
TYPE LOCALITY: Brazil, Espírito Santo, Santa Teresa, Floresta de Capela de São Braz (630 m) (19°50′S, 40°40′W).
DISTRIBUTION: Espírito Santo, from Capela de São Braz southward to Cariacia and Itapemirim (SE Brazil).
COMMENTS: Recognized as a subspecies of *iheringi* by Moojen (1948) and Pessôa and Reis (1996) but elevated to a species based on sequence analysis by Lara and Patton (2000).

Trinomys setosus (Desmarest, 1817). Nouv. Dict. Hist. Nat., Nouv. ed., 10:59.
COMMON NAME: Hairy Atlantic Spiny-rat.
TYPE LOCALITY: Unknown, but probably Brazil, Bahia (Moojen, 1948:386).
DISTRIBUTION: E Brazil (states of Espírito Santo and Minas Gerais).
STATUS: IUCN – Lower Risk (lc) as *Proechimys setosus* and *P. cayennensis.*
SYNONYMS: *cayennensis* (Pictet, 1841); *fuliginosus* (Wagner, 1843); **denigratus** (Moojen, 1948); **elegans** (Lund, 1841).
COMMENTS: Although *denigratus* was described as a subspecies of *iheringi* (Moojen, 1948) and is often recognized as such (Possôa and Reis, 1994; 1996), Lara and Patton (2000) concluded from analysis of sequence data and morphology that it is one of three subspecies of *setosus.*

Trinomys yonenagae Rocha, 1995. Mammalia, 59(4):541.
COMMON NAME: Yonenaga's Atlantic Spiny-rat.
TYPE LOCALITY: Brazil, Bahia, Ibiraba (10°48′S, 42°50′W).
DISTRIBUTION: Left bank of the São Francisco River from Bara to Pilão Arcado (NE Brazil).
COMMENTS: Karyotype has 2n=54 and FN=108 (Leal-Mesquinta et al., 1992).

Subfamily Heteropsomyinae Anthony, 1917. Bull. Am. Mus. Nat. Hist., 37(4):18.
COMMENTS: Named by Anthony to reflect perceived relationship between *Heteropsomys* and *Dasyprocta.* Expanded by Kraglievich (1965) and Patterson and Pascual (1968*a*) to include *Proechimys* and associated genera. Modified by Woods (1982:386) wherein West Indian spiny rats were classified with capromyids. McKenna and Bell (1997) included the genera of the Eumysopinae within the Heteropsomyinae. West Indian spiny rats are transitional

in characters between Echimyidae and Capromyidae, and here are placed in their own subfamily within the Echimyidae until the two families are revised. Molecular data (Leite and Patton, 2002) suggest that the sister taxa of both the myocastorids and capromyids is a group of the Eumysopinae including the genera *Clyomys* and *Euryzygomatomys*, but do not allow for evaluation of the distinctiveness of a West Indian clade of spiny rats (Heteropsomynae) suggested by Woods (1982). Varona (1974:73) placed all West Indian spiny rats in the genus *Heteropsomys*. However, there are clear differences between the forms on each island, so the original generic names are maintained here, except that *Homopsomys* is combined with *Heteropsomys* (Woods, 1989a, b).

Boromys Miller, 1916. Smithson. Misc. Coll., 66(12):7.
TYPE SPECIES: *Boromys offella* Miller, 1916.

Boromys offella Miller, 1916. Smithson Misc. Coll., 66(12):8.
COMMON NAME: Oriente Cave Rat.
TYPE LOCALITY: Cuba, Oriente Prov., Baracoa, Maisí.
DISTRIBUTION: Cuba and Isla Juventud (Isle of Pines).
STATUS: IUCN – Extinct.

Boromys torrei Allen, 1917. Bull. Mus. Comp. Zool., 61:6.
COMMON NAME: Torre's Cave Rat.
TYPE LOCALITY: Cuba, Matanzas Prov., Cave in Sierra de Hato-Nuevo.
DISTRIBUTION: Cuba and Isla Juventud (Isle of Pines).
STATUS: IUCN – Extinct.

Brotomys Miller, 1916. Smithson. Misc. Coll., 66(12):6.
TYPE SPECIES: *Brotomys voratus* Miller, 1916.

Brotomys contractus Miller, 1929. Smithson Misc. Coll., 81(9):13.
COMMON NAME: Haitian Edible Rat.
TYPE LOCALITY: Haiti, Dept. de l'Artibonite, Saint Michel de l'Atalye (small cave) (19°22'N, 72°20'W).
DISTRIBUTION: Hispaniola.
STATUS: Extinct.

Brotomys voratus Miller, 1916. Smithson Misc. Coll., 66(12):7.
COMMON NAME: Hispaniolan Edible Rat.
TYPE LOCALITY: Dominican Republic, San Pedro de Macoris, (kitchen midden).
DISTRIBUTION: Haiti, Dominican Republic, and La Gonave Isl.
STATUS: IUCN – Extinct (within the last 60 years).

Heteropsomys Anthony, 1916. Ann. New York Acad. Sci., 27:203.
TYPE SPECIES: *Heteropsomys insulans* Anthony, 1916.
SYNONYMS: *Homopsomys* Anthony, 1917.
COMMENTS: Includes *Homopsomys*; see comments under the subfamily.

Heteropsomys antillensis (Anthony, 1917). Bull. Am. Mus. Nat. Hist., 37:187.
COMMON NAME: Antillean Cave Rat.
TYPE LOCALITY: Puerto Rico, Cave at Utuado.
DISTRIBUTION: Puerto Rico.
STATUS: Extinct.

Heteropsomys insulans Anthony, 1916. Ann New York Acad. Sci., 27:202.
COMMON NAME: Puerto Rican Cave Rat.
TYPE LOCALITY: Puerto Rico, Cueva de la Ceiba, Hacienda Jobo (near Utuado).
DISTRIBUTION: Puerto Rico in cave deposits.
STATUS: Extinct.
COMMENTS: Nearly as large in body size as the Hispaniolan Hutia *Plagiodontia aedium*.

Family Myocastoridae Ameghino, 1904. Anales Soc. Cient. Argentina, 56-58:103.

COMMENTS: The higher-level classification of *Myocastor* remains unresolved. Myocastorids presumably evolved in the Oligocene of South America from an echimyid of the subfamily Adelophomyinae (Woods et al., 1992). The myocastorids have been included in Capromyidae by Hall (1981), Corbet and Hill (1991), and others, included in the Echimyidae (McKenna and Bell, 1997), and placed in the family Myocastoridae by Ameghino (1904), Woods and Howland (1979), and Woods (1993). Patterson and Pascual (1968b) and Patterson and Wood (1982) considered both myocastorids and capromyids to be subfamilies of the Echimyidae, based on the retention of the deciduous premolar in these taxa. Sequence data support this placement, as Leite and Patton (2002) suggested the inclusion of *Myocastor* and *Capromys* within the Echimyidae. Although Leite and Patton (2002) identified *Capromys* as the sister taxon of *Myocastor*, their placement within the Echimyidae clade is not well supported and is in need of further examination. Though myocastorids are related to echimyids and capromyids, Woods (1972, 1982) and Woods and Howland (1979) concluded that there are too many morphological differences to unite myocastorids with either of these taxa in the same family.

Myocastor Kerr, 1792. *In* Linnaeus, Anim. Kingdom, p. 225.

TYPE SPECIES: *Mus coypus* Molina, 1782.

SYNONYMS: *Mastonotus* Wesmael, 1841; *Myopotamus* Geoffroy, 1805; *Potamys* Desmarest, 1825.

COMMENTS: Patterson and Pascual (1968b:6) included *Myocastor* and several fossil forms as the subfamily Myocastorinae of the Echimyidae. Based on sequence data, the sister taxon of *Myocastor* appears to be *Capromys* (Leite and Patton, 2002). Although this sister taxon relationship is well supported in both parsimony and maximum likelihood analyses (Leite and Patton, 2002), potential problems including taxon bias and long branch attraction have not been fully addressed.

Myocastor coypus (Molina, 1782). Sagg. Stor. Nat. Chile, p. 287.

COMMON NAME: Coypu.

TYPE LOCALITY: Chile, Santiago Prov., Rio Maipo.

DISTRIBUTION: S Brazil, Paraguay, Uruguay, Boliva, Argentina, Chile.

STATUS: IUCN – Lower Risk (lc). Common.

SYNONYMS: *albomaculatus* (Fitzinger, 1867); *chilensis* (Lesson, 1842); *dorsalis* (Fitzinger, 1867); *popelairi* (Wesmael, 1841); **bonariensis** (Geoffroy, 1805[1806]); *castoroides* (Burrow, 1815); **melanops** (Osgood, 1943); **sanctaecruzae** Hollister, 1914.

COMMENTS: Widely introduced into North America, Europe, N Asia, and E Africa. The common name coypu is preferable to nutria, since nutria in Spanish means otter. The species and subspecies were reviewed by Woods et al. (1992).

Family Capromyidae Smith, 1842. The Naturalist's Library (London), 15:308.

COMMENTS: Does not include *Myocastor*, and apparently represents an entirely intra-Caribbean radiation (Woods 1989a, b; Woods and Howland, 1979); but see comments under Myocastoridae. Woods (1982:386-387) included West Indian Heteropsomyinae in this family, but here this group is assigned to the Echimyidae (see comments under Heteropsomyinae). Varona (1974) and Corbet and Hill (1991:203) classified most of the following species in either of two genera (*Capromys* or *Plagiodontia*), whereas Woods (1989a, b) separated them into several distinct subfamilies and genera.

Subfamily Capromyinae Smith, 1842. The Naturalist's Library (London), 15:30.

COMMENTS: Reviewed by Kratochvíl et al. (1978), Varona and Arredondo (1979), and Borroto (2002). The status of *Brachycapromys*, *Mysateles*, *Mesocapromys*, *Pygmaeocapromys*, *Paracapromys*, and *Stenocapromys* as genera or subgenera is unresolved (Hall, 1981; Rodriquez et al., 1979; Woods and Howland, 1979). This group is in need of revision to standardize the taxonomic levels proposed for the Cuban radiation of capromyids with taxonomic categories established for Hispaniola. The living Cuban capromyids are placed in three genera (*Capromys*, *Mesocapromys*, *Mysateles*) as accepted by the majority of systematists (see Kratochvil et al., 1978 and Borroto, 2002). Woods et al. (2001) demonstrated that *Mysateles* is paraphyletic, and placed *melanurus* within the genus

Mesocapromys. In addition to the extant forms described below, 13 extinct forms of *Capromys*-like hutias have been described (Varona and Arredondo, 1979): *Capromys antiquus, C. arredondoi, C. latus, C. pappus, C. robustus, Mesocapromys barbouri, M. beatrizae, M. delicatus, M. gracilis, M. kraglievichi, M. minimus, M. silvai,* and *Mysateles jaumei*. Some of these are described in separate subgenera (i.e. *Pygmaeocapromys, Brachycapromys, Palaeocapromys*).

Capromys Desmarest, 1822. Bull. Sci. Soc. Philom. Paris, p. 185.
    TYPE SPECIES: *Capromys fourniere* Desmarest, 1822 (= *Isodon pilorides* Say, 1822).
    SYNONYMS: *Macrocapromys* Arredondo, 1958.
    COMMENTS: "*Capromys*" *geayi* (Pousarges, 1899) was described from Venezuela, where it was collected in the mountains between La Guayra and Caracas. It was placed in its own genus (*Procapromys*) by Chapman (1901:322). The type specimen is Mus. Nat. Hist. Nat. (Paris) No. 1898-1785 (1834A). No further specimens have ever been collected, and it is possible that the collecting locality was incorrectly recorded. This specimen is likely a juvenile *Capromys* from Cuba erroneously reported from Venezuela.

Capromys gundlachianus (Varona, 1983). Carib. J. Sci., 19(3-4):77.
    COMMON NAME: Archipiélago de Sabana Hutia.
    TYPE LOCALITY: Small cayo west of Cayo Bahía de Cádiz, Bahía de Santa Clara, Archipiélago de Sabana, Cuba.
    DISTRIBUTION: In the region of Cayo Bahía de Cádiz in the Archipiélago de Sabana, as well as Cayo Fragoso, Cayo Santa María, Cayo Guillermo, and Cayo Patabán. Extirpated on Cayo las Brujas. On Cayo Fragoso it is sympatric with *Mesocapromys auritus* north of Caibarién.
    STATUS: Widespread.
    COMMENTS: Originally described as a subspecies of *C. pilorides* but tentatively elevated to species status because of sequence divergence of a specimen from Cayo Ballenato del Medio (Woods et al., 2001), however, this locality is outside the published range of *gundlachianus*.

Capromys pilorides (Say, 1822). Jour. Acad. Nat. Sci. Philadelphia, 2:333.
    COMMON NAME: Desmarest's Hutia.
    TYPE LOCALITY: "South America or one of the West Indian islands."
    DISTRIBUTION: Mainland Cuba, Isle of Youth, Archipiélago de las Doce Lequnas, Archipiélago de Sabana, and many other islands and cays in the Cuban archipelago.
    STATUS: IUCN – Lower Risk (lc). Common (extremely abundant in some areas, including Guantanamo Bay Naval Base).
    SYNONYMS: *acevedo* (Arredondo, 1958); *fourniere* Desmarest, 1822; *intermedius* (Arredondo, 1958); *doceleguas* Varona, 1980; *relictus* (Allen, 1911); *ciprianoi* Borroto, Camacho and Ramos, 1992.
    COMMENTS: The first mention of this species name is as *Mus pilorides* Pallas 1778, however, Tate (1935:309) noted that it is not associated with this genus. Sometimes placed in the subgenus *Capromys*; see Hall (1981:863). This species is very variable in size, coloration, and habits. There are five named subspecies (*ciprianoi, doceleguas, gundlachianus, pilorides,* and *relictus*; see Varona (1983a:77) and Borroto Paez et al., 1992:98). No genetic differentiation of cytochrome *b* was reported for the two subspecies (*ciprianoi* and *relictus*) on the Isle of Youth (Woods et al., 2001). However, *gundlachianus* from Cayo Fragoso was reported to show over 5% sequence divergence and here we are elevating it to species level. *Macrocapromys acevedo* described by Arredondo (1958:10) is not distinct at the generic level, and is probably synonymous with *pilorides*. Arredondo's spelling was *acevedo*, but this was noted as [sic] by Varona (1974) who changed the spelling to *acevedoi*. This spelling was followed by Hall (1981) but it should be noted that *acevado* is the correct usage according to Brodkorb (1961). Karyotype has 2n=40; FN=64 (Hsu and Benirschke, 1971; Milišnikov et al., 1990).

Geocapromys Chapman, 1901. Bull. Am. Mus. Nat. Hist., 14:314.
    TYPE SPECIES: *Capromys* (*Geocapromys*) *brownii* Fischer, 1830.

COMMENTS: Original use of the name was as a subgenus of *Capromys*. Included in *Capromys* by Mohr (1939:75), Varona (1974:67), Hall (1981:865), and Corbet and Hill (1991:203). Considered a distinct genus by Miller (1929), Woods and Howland (1979:112), and Morgan (1985:30). Includes *columbianus*, *pleistocenicus* (see Hall, 1981:866), and *megas* (see Varona and Arredondo, 1979), which are known only from Holocene fossils from Cuba.

Geocapromys brownii (Fischer, 1830). Synopis. Mamm., Addenda, p. 389 (=589).
COMMON NAME: Jamaican Hutia.
TYPE LOCALITY: "Jamaica".
DISTRIBUTION: Jamaica.
STATUS: IUCN – Vulnerable.
SYNONYMS: *brachyurus* (Hill, 1851).
COMMENTS: Sometimes includes *G. thoracatus* from Little Swan Isl (see that account). Reviewed by Anderson et al. (1983). Karyotype has 2n=88; FN=136 (George and Weir, 1972*b*).

Geocapromys ingrahami (J. A. Allen, 1891). Bull Am. Mus. Nat. Hist., 3:329.
COMMON NAME: Bahamian Hutia.
TYPE LOCALITY: "Plana Keys, Bahamas" (East Plana Key).
DISTRIBUTION: Known from the type locality and introduced populations on Little Wax Cay (1973) and Warderick Wells Cay (1981).
STATUS: IUCN – Vulnerable.

Geocapromys thoracatus (True, 1888). Proc. U.S. Natl. Mus., 11:469.
COMMON NAME: Swan Island Hutia.
TYPE LOCALITY: "Little Swan Island, one of the two small islands lying at the entrance of the Gulf of Honduras".
DISTRIBUTION: Little Swan Isl.
STATUS: IUCN – Extinct; Clough (1976) stated that *thoracatus* became extinct in 1950s, possibly as a result of introduced cats on the island.
COMMENTS: Sometimes included as a subspecies of *G. brownii*; see Mohr (1939:77), Hall (1981:866), and Varona (1974:67). Morgan (1985) reconfirmed its status as a distinct species. Reviewed by Morgan (1989*a*).

Mesocapromys Varona, 1970. Poeyana, 73:8.
TYPE SPECIES: *Capromys* (*Mesocapromys*) *auritus* Varona, 1970.
SYNONYMS: *Paracapromys* Kratochvíl, Rodriguez, and Barus, 1978; *Pygmaeocapromys* Varona, 1979; *Stenocapromys* Varona and Arredondo, 1979.
COMMENTS: Original use of name was as a subgenus, which was elevated to a genus by Kratochvíl et al. (1978:15). All members of this genus are small (less than 1 kg), and construct large nests made of sticks, unlike *Mysateles*.

Mesocapromys angelcabrerai (Varona, 1979). Poeyana, 194:6.
COMMON NAME: Cabrera's Hutia.
TYPE LOCALITY: Cuba, Ciego de Avila Prov., Cayos de Ana Maria (21°30′N, 78°40′W).
DISTRIBUTION: Cayos de Ana Maria, Cuba.
STATUS: U.S. ESA – Endangered as *Capromys angelcabrerai*; IUCN – Critically Endangered.
COMMENTS: Placed in the subgenus *Pygmaeocapromys* by Varona (1979:5). Reviewed by Camacho et al. (1994). Unusual among capromyines in being sexually dimorphic. There are reports of the presence of *M. angelcabrerai* on the coast near Jucaró, but no confirmed specimens are known. This species and *M. nanus* are the two smallest hutias in Cuba.

Mesocapromys auritus (Varona, 1970). Poeyana, ser. A, 73:1.
COMMON NAME: Large-eared Hutia.
TYPE LOCALITY: Cuba, Las Villas Prov., Archipiélago de Sabana, Cayo Fragoso (22°41′N, 79°27′W).
DISTRIBUTION: Known only from the type locality.
STATUS: U.S. ESA – Endangered as *Capromys auritus*; IUCN – Critically Endangered.

COMMENTS: This form is known only from mangrove habitats along the edge of canals passing across Cayo Fragoso. This form constructs nests that are very similar to those of *angelcabrerai*. Karyotype has 2n=36 (Hemandez and Sanchez, 1987).

Mesocapromys melanurus (Poey, 1865). In Peters, 1865, Monatsb. K. Preuss. Akad. Wiss. Berlin, P. 384.
  COMMON NAME: Black-tailed Hutia.
  TYPE LOCALITY: Cuba, Manzanillo, Gramma Prov. (Oriente region).
  DISTRIBUTION: Eastern provinces of Cuba.
  STATUS: IUCN – Lower Risk (nt) as *Mysateles melanurus*. Locally abundant in Guisa, Gramma Prov., but uncommon in other regions of the eastern provinces (Borroto et al., 2001).
  SYNONYMS: *arboricolus* Kratochvíl, Rodriguez, and Barus, 1978; *rufescens* Mohr, 1839.
  COMMENTS: The author and date of publication for this species are usually given as Peters, 1864, but Varona (1974:63), established the correct author and date as Poey, 1865. Includes *arboricolus* of Kratochvíl et al. (1978:48), who placed it in the genus *Mysateles*, subgenus *Leptocapromys*. However, this specimen is considered to be a young female *M. melanurus* by Varona (1986:7). We have examined this specimen, and concur with Varona. Woods et al. (2001) concluded that *Mysateles* was paraphyletic and that the affinities of *melanurus* are with taxa of the genus *Mesocapromys* rather than with *Mysateles p. prehensilis* or *M. p. gundlachi*.

Mesocapromys nanus (G. M. Allen, 1917). Proc. New England Zool. Club, 6:54.
  COMMON NAME: Dwarf Hutia.
  TYPE LOCALITY: Cuba, Matanzas Prov., Sierra de Hato Nuevo.
  DISTRIBUTION: Cienaga (swamp) de Zapata (Matanzas Prov., Cuba).
  STATUS: U.S. ESA – Endangered as *Capromys nana*; ICUN – Critically Endangered. Some mammalogists consider this species to be extinct since no specimens have been collected since 1937, but it is still likely to survive in remote areas of the Zapata Swamp (Jorge de la Cruz, pers. comm.).
  COMMENTS: Placed in newly created subgenus *Pygmaeocapromys* by Varona (1979:5). In genus *Mesocapromys*, subgenus *Paracapromys* by Kratocvil et al. (1978:15) and Rodriguez et al. (1979). However, retained in *Capromys* subgenus *Mysateles* by Hall (1981:863) because of its long tail and small body size. Varona (1979:5), however, states that even though this species "automatically" is associated with *Mysateles* because of tail length, there are important cranial differences between *nanus* and other *Mysateles*. Originally based on fossil material, but subsequently found living in the Zapata Swamp.

Mesocapromys sanfelipensis (Varona and Garrido, 1970). Poeyana, ser. A, 75:3.
  COMMON NAME: San Felipe Hutia.
  TYPE LOCALITY: Cuba, Pinar del Rio Prov., Archipiélago de los Canarreos, Cayo Juan Garcia (21°59'N, 83°31'W).
  DISTRIBUTION: Known only from four specimens collected at the type locality.
  STATUS: U.S. ESA – Endangered as *Capromys sanfelipensis*; IUCN – Critically Endangered. Fire destroyed much of its habitat on Cayo Juan Garcia (Frias et al., 1988) and thus may now be extinct.
  COMMENTS: Placed in subgenus *Mesocapromys* by Varona (1974), and in genus *Mesocapromys*, subgenus *Paracapromys* by Kratochvíl et al. (1978:15). The habitat of this species is uncertain, but all known specimens were captured in grasslands (*Salicornia perennis* = "yerba de vidrio") rather than mangroves.

Mysateles Lesson, 1842. Nouv. Tabl. Regne Animal Mammifères, p 124.
  TYPE SPECIES: *Capromys prehensilis* Poeppig, 1824.
  SYNONYMS: *Brachycapromys* Varona and Arredondo, 1979; *Leptocapromys* Kratochvíl et al., 1978.
  COMMENTS: Lesson separated *Capromys prehensilis* as *Mysateles poeppingi*; see Varona (1979:4-5). Kratochvíl et al. (1978:15) treated *Mysateles* as a genus in their new classification. *Mysateles* was included in *Capromys* by Woods (1989a:781), Hall (1981:863), and Corbet and Hill (1991:203), however, there is biochemical and morphological evidence to support *Mysateles* as a valid genus (Borroto, 2003; Camacho et al., 1995).

Woods et al. (2001) found that *Mysateles* was paraphyletic in relationship to *Mesocapromys angelcabrerai* but was distinct from *Capromys*.

Mysateles garridoi (Varona, 1970). Poeyana, ser. A, 74:2.

COMMON NAME: Garrido's Hutia.

TYPE LOCALITY: Cuba, Archipiélago de los Canarreos, "Cayo Majá"; see comments.

DISTRIBUTION: Known only from a single specimen from a small islet northwest of Cayo Largo, Cuba.

STATUS: IUCN – Critically Endangered.

COMMENTS: Placed in the genus *Mysateles*, subgenus *Leptocapromys* by Kratochvíl et al. (1978:15). However, retained in *Capromys* by Varona (in litt.) and by Woods et al. (2001). The type locality has been clarified by Silva-Taboada (pers. comm.), who identified the correct name for the locality as a yet unnamed small cay adjacent to Cayo Largo in the Archipiélago de los Canarreos. Varona (1970) incorrectly applied the name Cayo Majá to this islet. It is known that this hutia does not occur on the adjacent Cayo Majá (three cays to the northwest of Cayo Largo) based on 1990 expedition to this region by R. Borroto (pers. comm.). The single known specimen may have been left on the site where it was collected by fishermen, and not be from the area, so it is necessary to search more widely in the region to determine the status of this species (A. Camacho, pers. comm.).

Mysateles meridionalis (Varona, 1986). Poeyana, 315:4.

COMMON NAME: Southern Hutia.

TYPE LOCALITY: Cuba, north of Caleta Cocodrilos, SW Isla de la Juventud (former Isle of Pines).

DISTRIBUTION: Restricted to lowland forests SW of the central savanna on the Isle of Youth (south of W Cuba).

STATUS: IUCN – Lower Risk (nt), however, Borroto and Ramos (2003) recommend status of critically endangered.

COMMENTS: Closely related to *Mysateles prehensilis*, but differs in the proportion of the tail, which is 62 percent of body length in *meridionalis*, 73 percent in *gundlachi*, and 79 percent in *prehensilis*; see also comments under *prehensilis*.

Mysateles prehensilis (Poeppig, 1824). Jour. Acad. Nat. Sci. Philadelphia, 4:11.

COMMON NAME: Prehensile-tailed Hutia.

TYPE LOCALITY: Cuba, wooded south coast.

DISTRIBUTION: Cuba, mainly west of Camaguey Province. The status of this species in the eastern provinces (old Oriente Prov.) is not clear.

STATUS: IUCN – Vulnerable as *M. gundlachi*, Lower Risk (lc) as *M. prehensilis*. Common.

SYNONYMS: *pallidus* (Poey, 1865); *poeppingi* Lesson, 1842; *poeyi* (Guerin, 1834); **gundlachi** (Chapman, 1901).

COMMENTS: Placed in subgenus *Mesocapromys* by Mohr (1939:54) and Varona (1974), and in the genus and subgenus *Mysateles* by Kratochvíl et al. (1978:15). This is the largest species of *Mysateles*. *Mysateles gundlachi* was described on the basis of the baculum, which is distinctly broad-based unlike that of *M. meridionalis*. Based on levels of sequence divergence, Woods et al. (2001) considered *gundlachi* as an insular subspecies of *M. prehensilis*, which we follow here. Karyotype has 2n=34; FN=54-56 (Milišnokov et al., 1990).

Subfamily Hexolobodontinae Woods, 1989. Los Angeles Co. Mus. Nat. Hist., Sci. Ser., 33:76.

Hexolobodon Miller, 1929. Smithson. Misc. Coll., 81:19.

TYPE SPECIES: *Hexolobodon phenax* Miller, 1929.

Hexolobodon phenax Miller, 1929. Smithson. Misc. Coll., 81:19.

COMMON NAME: Imposter Hutia.

TYPE LOCALITY: Haiti, Dept. de l'Artibonite, Small cave northeast of Saint Michael de l'Atalye.

DISTRIBUTION: Hispaniola and La Gonave Isl.

STATUS: IUCN – Extinct.
SYNONYMS: *poolei* Rímoli, 1976 [1977].
COMMENTS: Includes *H. poolei* (Rímoli, 1976:21), which is known only by the type specimen (USNM No. 255881), because the normal dental variation of *H. phenax* included the diagnostic condition in *H. poolei*.

Subfamily Isolobodontinae Woods, 1989. Los Angeles Co. Mus. Nat. Hist., Sci. Ser., 33:76.
COMMENTS: Early Miocene form, *Zazamys veronicae*, has been reported from Cuba by MacPhee et al. (2003).

Isolobodon J. A. Allen, 1916. Ann. New York Acad. Sci., 27:19.
TYPE SPECIES: *Isolobodon portoricensis* J. A. Allen, 1916.
SYNONYMS: *Aphaetreus* Miller, 1922; *Ithydontia* Miller, 1922.

Isolobodon montanus (Miller, 1922). Smithson. Misc. Coll., 74:3.
COMMON NAME: Montane Hutia.
TYPE LOCALITY: Haiti, Dept. de l'Artibonite, Cave northeast of Saint Michael de l'Atalye.
DISTRIBUTION: Hispaniola and La Gonave Isl.
STATUS: IUCN – Extinct.
COMMENTS: Originally described in a separate genus from *I. portoricensis* because the enamel folds of the cheekteeth connect to form separate laminar plates, but a large series of each species indicates that this character is clinal, and the two forms are very closely related.

Isolobodon portoricensis J. A. Allen, 1916. Ann. N.Y. Acad. Sci., 27:19.
COMMON NAME: Puerto Rican Hutia.
TYPE LOCALITY: Puerto Rico, Jobo Dist., near Utuado.
DISTRIBUTION: Hispaniola (Haiti and Dominican Republic) and offshore islands. Introduced on Puerto Rico, St. Thomas, St. Croix, and Mona Isl. The only known hutia on La Gonave Isl.
STATUS: IUCN – Critically Endangered, Probably extinct, but possibly surviving on La Tortue Isl off the N coast of Haiti (Woods et al., 1985).
SYNONYMS: *levir* (Miller, 1922).
COMMENTS: Even though the type locality is Puerto Rico, the species was apparently introduced there by Amerindians and the natural range is restricted to Hispaniola. Reported as extinct by Hall (1981:868), but this species survived in Hispaniola and Puerto Rico until the last few decades, and may still survive in certain remote areas (Woods et al., 1985). Includes *levir* (Reynolds et al., 1953).

Subfamily Plagiodontinae Ellerman, 1940. The Families and Genera of Living Rodents, 1:25.

Plagiodontia F. Cuvier, 1836. Ann. Sci. Nat. Zool. (Paris), ser 2, 6:347.
TYPE SPECIES: *Plagiodontia aedium* F. Cuvier, 1836.
SYNONYMS: *Hyperplagiodontia* Rímoli, 1977.
COMMENTS: Reviewed by Mohr (1939), Johnson (1948), Anderson (1965), and Woods (1989*a*, *b*).

Plagiodontia aedium F. Cuvier, 1836. Ann. Sci. Nat. Zool. (Paris), ser 2, 6:347.
COMMON NAME: Hispaniolan Hutia.
TYPE LOCALITY: "Saint-Dominguie" (probably Haiti).
DISTRIBUTION: Hispaniola, La Gonave Isl, not recorded from La Tortue Isl.
STATUS: IUCN – Vulnerable.
SYNONYMS: *spelaem* Miller, 1929; **hylaeum** Miller, 1927.
COMMENTS: Includes *hylaeum* as a separate subspecies (Anderson, 1965).

Plagiodontia araeum Ray, 1964. Breviora, Mus. Comp. Zool., 203:2.
COMMON NAME: Wide-toothed Hutia.
TYPE LOCALITY: Dominican Republic, Prov. San Rafael, unnamed cave 2 km SE Rancho de la Guardia (18°43′N, 71°39′W).

DISTRIBUTION: Hispaniola.

STATUS: Extinct.

SYNONYMS: *stenocoronalis* (Rímoli, 1976 [1977]).

COMMENTS: Type description based on only left upper cheek tooth (DP4). Subsequent fossil material collected in Haiti and deposited at the Florida Museum of Natural History includes complete cranial and dentary material, confirming the validity of this taxon. *Hyperplagiodontia stenocoronalis* of Rimoli (1976:34) is not distinct from *P. araeum*. A very large wide-toothed hutia.

Plagiodontia ipnaeum Johnson, 1948. Proc. Biol. Soc. Wash., 61:72.

COMMON NAME: Samana Hutia.

TYPE LOCALITY: Dominican Republic, Prov. de Samana, Anadel (19°12'N, 69°19'W).

DISTRIBUTION: Hispaniola.

STATUS: IUCN – Extinct. Probably extinct (Woods et al., 1985).

SYNONYMS: *caletensis* Rímoli, 1976 [1977]; *velozi* Rímoli, 1976 [1977].

COMMENTS: A large version of *P. aedium*.

Rhizoplagiodontia Woods, 1989. Los Angeles Co. Mus. Nat. Hist., Sci. Ser., 33:62.

TYPE SPECIES: *Rhizoplagiodontia lemkei* Woods, 1989.

Rhizoplagiodontia lemkei Woods, 1989. Los Angeles Co. Mus. Nat. Hist., Sci. Ser., 33:62.

COMMON NAME: Lemke's Hutia.

TYPE LOCALITY: Haiti, Dept. du Sud, 17 km W Camp Perrin (18°20'N, 74°03'W).

DISTRIBUTION: Endemic to SW Haiti in the Massif de la Hotte.

STATUS: IUCN – Extinct.

Family Heptaxodontidae Anthony, 1917. Bull. Am. Mus. Nat. Hist., 37(4):183.

SYNONYMS: Amblyrhizinae Schaub, 1951; Elasmodontomyinae Anthony, 1917.

COMMENTS: Known only from sub-Recent fossils from Greater and N Lesser Antilles (Woods, 1989a). Whether *Amblyrhiza* and *Clidomys* became extinct before or after humans arrived in the West Indies is debatable. This family is often placed near the Chinchillidae based on similar laminar plates of molariform teeth. One genus (*Quemisia*) is very similar in dental morphology to capromyids, however, and it is possible to derive all of the conditions seen in heptaxodontids from dental patterns found within the Capromyidae. This family should be placed adjacent to the Capromyidae.

Subfamily Clidomyinae Woods, 1989. Biogeography of the West Indies, p. 753.

Clidomys Anthony, 1920. Bull Am. Mus. Nat. Hist., 42:469.

TYPE SPECIES: *Clidomys osborni* Anthony, 1920.

SYNONYMS: *Alterodon* Anthony, 1920; *Speoxenus* Anthony, 1920; *Spirodontomys* Anthony, 1920.

Clidomys osborni Anthony, 1920. Bull Am. Mus. Nat. Hist., 42:469.

COMMON NAME: Osborn's Key Mouse.

TYPE LOCALITY: Jamaica, Balaclava, Wallingford Roadside Cave.

DISTRIBUTION: Jamaica.

STATUS: Extinct.

SYNONYMS: *cundalli* (Anthony, 1920); *jamaicensis* (Anthony, 1920); *major* (Anthony, 1920); *parvus* (Anthony, 1920).

COMMENTS: *C. parvus* is here synonymized with *C. osborni* based on Morgan and Wilkins (2003), who noted the wide size range of *C. osborni* specimens from Slue's Cave, Jamaica. They interpreted this size range to indicate that *C. osborni* is the only valid species of *Clidomys* present in Jamaica.

Subfamily Heptaxodontinae Anthony, 1917. Bull Am. Mus. Nat. Hist., 37(4):183.

Amblyrhiza Cope, 1868. Proc. Am. Philos. Soc., p. 313.

TYPE SPECIES: *Amblyrhiza inundata* Cope, 1868.

SYNONYMS: *Loxomylus* Cope, 1869.

Amblyrhiza inundata Cope, 1868. Proc. Am. Philos. Soc., p. 313.
 COMMON NAME: Blunt-toothed Giant Hutia.
 TYPE LOCALITY: West Indies, Anguilla, Bat Cave.
 DISTRIBUTION: Anguilla, St. Martin.
 STATUS: Extinct.
 SYNONYMS: *latidens* (Cope, 1870); *longidens* (Cope, 1871); *quadrans* (Cope, 1871).

Elasmodontomys Anthony, 1916. Ann. N.Y. Acad. Sci., 27:199.
 TYPE SPECIES: *Elasmodontomys obliquus* Anthony, 1916.
 SYNONYMS: *Heptaxodon* Anthony, 1917.

Elasmodontomys obliquus Anthony, 1916. Ann. N.Y. Acad. Sci., 27:199.
 COMMON NAME: Plate-toothed Giant Hutia.
 TYPE LOCALITY: Puerto Rico, Utuado.
 DISTRIBUTION: Puerto Rico.
 STATUS: Extinct.
 SYNONYMS: *bidens* Anthony, 1917.

Quemisia Miller, 1929. Smithson. Misc. Coll., 81(9):22.
 TYPE SPECIES: *Quemisia gravis* Miller, 1929.

Quemisia gravis Miller, 1929. Smithson. Misc. Coll., 81(9):22.
 COMMON NAME: Twisted-toothed Giant Hutia.
 TYPE LOCALITY: Haiti, Dept. de l'Artibonite, 6 km east of Saint Michel de L'Atlaye.
 DISTRIBUTION: Hispaniola.
 STATUS: IUCN – Extinct.

Literature Cited

Abad, P. L. 1987. Biologia y ecologia del liron careto (*Eliomys quercinus*) en Leon. Ecologia, 1:153-159.

Abe, H. 1967. Classification and biology of Japanese Insectivora (Mammalia). I. Studies on variation and classification. Journal of the Faculty of Agriculture, Hokkaido University, Sapporo, Japan, 55:191-265, 2 pls.

Abe, H. 1971. Small mammals of central Nepal. Journal of the Faculty of Agriculture, Hokkaido University, Sapporo, Japan, 56:367-423.

Abe, H. 1973*a*. Growth and development in two forms of *Clethrionomys*. II. Tooth characters, with special reference to phylogenetic relationships. Journal of the Faculty of Agriculture, Hokkaido University, Sapporo, Japan, 57:229-254.

Abe, H. 1973*b*. Growth and development in two forms of *Clethrionomys*. III. Cranial characters, with special reference to phylogenetic relationships. Journal of the Faculty of Agriculture, Hokkaido University, Sapporo, Japan, 57:255-274.

Abe, H. 1977. Variation and taxonomy of some small mammals from central Nepal. Journal of the Mammalogical Society of Japan, 7(2):63-73.

Abe, H. 1982. Age and seasonal variations of molar patterns in a red-backed vole population. Journal of the Mammalogical Society of Japan, 9:9-13.

Abe, H. 1983. Variation and taxonomy of *Niviventer fulvescens* and notes on *Niviventer* group of rats in Thailand. Journal of the Mammalogical Society of Japan, 9:151-161.

Abe, H. 1984. The classification of mice and voles in Hokkaido. Pp. 1-20, *in* Study on Wild Rodents in Hokkaido (K. Ohta, ed.). Hokkaido University Press, Sapporo (in Japanese).

Abe, H. 1988. [The phylogenetic relationships of Japanese moles]. Honyurui Kagaku (Mammalian Science), 28:63-68 (in Japanese).

Abe, H. 1995. Revision of the Asian moles of the genus *Mogera*. Journal of the Mammalogical Society of Japan, 20:51-68.

Abe, H. 1996. Habitat factors affecting the geographic size variation of Japanese moles. Mammal Study, 21:71-87.

Abe, H. 2003. Trapping, habitat, and activity of the Japanese water shrew, *Chimarrogale platycephala*. Honyurui Kagaku (Mammalian Science), 43:51-61 (in Japanese, with English summary).

Abe, H., and N. Ishii. 1987. Mammals of Tsushima Island. A natural resource exploration report of Tsushima Island. Nature of Tsushima Island (Nagasaki Prefecture, 1987):79-109.

Abe, H., S. Shiraishi, and S. Arai. 1991. A new mole from Uotsuri-jima, the Ryukyu Islands. Journal of the Mammalogical Society of Japan, 15:47-60.

Abe, H., S. Ohdachi, and K. Mackawa. 1996. A survey of small terrestrial mammals in southern Sakhalin, conducted in 1994 and 1995. Wildlife Conservation Japan, 2(1):17-21.

Abe, H., N. Ishii, Y. Kaneko, K. Maeda, S. Miura, and M. Yoneda. 1997. [A pictorial guide to the mammals of Japan.] Tokyo: Tokai University Press. Second ed. 195 pp. (in Japanese).

Abramov, A. V. 1996. Comments on the proposed conservation of usage of 15 mammal specific names based on wild species which are antedated by or contemporary with those based on domestic animals. Bulletin of Zoological Nomenclature, 53:287.

Abramov, A. V. 1999. A taxonomic review of the genus *Mustela* (Mammalia, Carnivora). Zoosystematica Rossica, 8(2):357-364.

Abramov, A. V. 2000. [The taxonomic status of the Japanese weasel, *Mustela itatsi* (Carnivora, Mustelidae)]. Zoologicheskii Zhurnal, 79(1):80-88 (in Russian).

Abramov, A. V. 2001. [Notes on the taxonomy of the Siberian badgers (Mustelidae: *Meles*)]. Proceedings of the Zoological Institute Russian Academy of Sciences, Saint-Petersburg, 288:221-233 (In Russian).

Abramov, A. V. 2002. Variation of the baculum structure of the Palaearctic badger (Carnivora, Mustelidae, *Meles*). Russian Journal of Theriology, 1:57-60.

Abramov, A. V., and G. F. Baryshnikov. 1995. [The structure of os malleus in Palaearctic Mustelidae (Carnivora)]. Zoologicheskii Zhurnal, 74:129-142 (In Russian).

Abramov, A. V., and G. F. Baryshnikov. 1999. Geographic variation and intraspecific taxonomy of weasel *Mustela nivalis* (Carnivora, Mustelidae). Zoosystematica Rossica, 8:365-402.

Abramson, N. I. 1993. Evolutionary trends in the dentition of true lemmings (Lemmini, Cricetidae, Rodentia): Functional-adaptive analysis. Journal of Zoology, London, 230:687-699.

Acharya, L. 1992. *Epomophorus wahlbergi*. Mammalian Species, 394:1-4.

Achverdjan, M. R., E. A. Lyapunova, and N. N. Vorontsov. 1992. [Karyology and systematics of shrub voles of the Caucasus and Transcaucasia (*Terricola*, Arvicolinae, Rodentia)]. Zoologicheskii Zhurnal, 71:96-110 (in Russian).

Acosta, C. E., and R. D. Owen. 1993. *Koopmania concolor*. Mammalian Species, 429:1-3.

Adam, F. 1977. Données préliminaires sur l'habitat et la stratification des rongeurs en forêt de Basse Côte-d'Ivoire. Mammalia, 41(3):283-290.

Adam, F., V. Aellen, and M. Tranier. 1993. Nouvelles données sur le genre *Myopterus*. Le statut de *Myopterus daubentonii* Desmarest, 1820 (Chiroptera: Molossidae). Revue Suisse de Zoologie, 100:317-326.

Adams, J. K. 1989. *Pteronotus davyi*. Mammalian Species, 346:1-5.

Adams, M., P. R. Baverstock, C. H. S. Watts, and T. Reardon. 1987. Electrophoretic resolution of species boundaries in Australian Microchiroptera. I. *Eptesicus* (Chiroptera: Vespertilionidae). Australian Journal of Biological Science, 40:143-162.

Adams, M., T. R. Reardon, P. R. Baverstock, and C. H. S. Watts. 1988. Electrophoretic resolution of species boundaries in Australian Microchiroptera. IV. The Molossidae (Chiroptera). Australian Journal of Biological Science, 41:315-326.

Ade, M. 1999. External morphology and evolution of the rhinarium of Lagomorpha with special reference to the Glires hypothesis. Mitteilungen aus dem Museum fur Naturkunde in Berlin, Zoologische Reihe, 75(2):191-216.

Adkins, R. M., E. L. Gelke, D. Rowe, and R. L. Honeycutt. 2001. Molecular phylogeny and divergence time estimates for major rodent groups: Evidence from multiple genes. Molecular Biology and Evolution, 18(5):777-791.

Adkins, R. M., A. H. Walton, and R. L. Honeycutt. 2003. High-level systematics of rodents and divergence time estimates based on two congruent nuclear genes. Molecular Phylogenetics and Evolution, 26(3):409-420.

Adler, G. H. 1995. Habitat relations within lowland grassland rodent communities in Taiwan. Journal of Zoology, London, 237:563-576.

Adler, G. H. 1996. Habitat relations of two endemic species of highland forest rodents in Taiwan. Zoological Studies, 35(2):105-110.

Adler, S. 1948. Origin of the golden hamster *Cricetus auratus* as a laboratory animal. Nature, 162:256-257.

Aellen, V. 1959. Contribution à l'étude de la faune d'Afghanistan. 9. Chiroptères. Revue Suisse de Zoologie, 66:353-386.

Aellen, V. 1965. Les rongeurs de basse Cote-d'Ivoire (Hystricomorpha et Gliridae). Revue Suisse de Zoologie, 72(37):755-767.

Aellen, V. 1966. Notes sur *Tadarida teniotis* (Raf.)(Mammalia, Chiroptera). I. Systématique, paléontologie et peuplement, repartition géographique. Revue Suisse de Zoologie, 73:119-159.

Aellen, V. 1973. Un *Rhinolophus* nouveau d'Afrique centrale. Periodicum Biologorum, 75:101-105.

Agadzhanyan, A. K. 1986. The history of collared lemmings in the Pleistocene. Pp. 379-388, *in* Beringia in the Cenozoic Era (V. L. Kontrimavichus, ed.). Russian translation series, no. 28. Rotterdam, A. A. Balkema, 724 pp.

Agadzhanyan, A. K. 1993. A new volelike rodent (Mammalia, Rodentia) from the Pliocene of the Russian Plain. Paleontological Journal, 27(2):126-140 (Translated from Paleontologicheskii Zhurnal, 2:99-111).

Agadzhanyan, A. K. 2000. [Phylogeny.] Pp. 22-54, *in* [Species of the fauna of Russia and the contiguous countries. The water vole: Mode of the species (P. A. Panteleyev, ed.)]. Nauka Publishers, Moscow, 527 pp. (in Russian).

Agadzhanyan, A. K., and V. N. Yatsenko. 1984. [Phylogenetic interrelationships in voles of northern Eurasia]. Sbornik Trudov Zoologicheskovo Muzeya MGU, 22:135-190 (in Russian).

Aggundey, I. R., and D. A. Schlitter. 1986. Annotated checklist of the mammals of Kenya. 2. Insectivora and Macroscelidea. Annals of Carnegie Museum, 55:325-347.

Agnelli, P., and A. Lazzeretti. 1995. On the distribution of *Micromys minutus* in Italy. Bollettino di Zoologia, 62:395-399.

Agrawal, V. C. 1975. Taxonomic study of the Indo-Burmese subspecies of the common tree-shrew, *Tupaia glis* Diard. Pp. 385-394, *in* Dr. B. S. Chaun Commemoration Volume 1975 (K. K. Tiwari and C. B. Srivastava, eds.). Zoological Society of India, Orissa, 439 pp.

Agrawal, V. C. 2000. Taxonomic studies on Indian Muridae and Hystricidae (Mammalia: Rodentia). Records of the Zoological Survey of India, Occasional Paper No. 180, 186 pp.

Agrawal, V. C., and S. Chakraborty. 1970. Occurrence of the woolly flying squirrel, *Eupetaurus cinereus*

Thomas (Mammalia: Rodentia: Sciuridae) in North Sikkim. Journal of the Bombay Natural History Society, 66:615-616.

Agrawal, V. C., and S. Chakraborty. 1971. Notes on a collection of small mammals from Nepal, with the description of a new mouse-hare (Lagomorpha: Ochotonidae). Proceedings of the Zoological Society of Calcutta, 24:41-46.

Agrawal, V. C., and S. Chakraborty. 1976. Revision of the subspecies of the lesser bandicoot rat *Bandicota bengalensis* (Gray) (Rodentia: Muridae). Records of the Zoological Survey of India, 69:267-274.

Agrawal, V. C., and S. Chakraborty. 1979. Catalogue of mammals in the Zoological Survey of India. Part 1. Sciuridae. Records of the Zoological Survey of India, 74:333-481.

Agrawal, V. C., and S. Chakraborty. 1980. Intraspecific geographical variations in the Indian long-tailed tree mouse, *Vandeleuria oleracea* (Bennett). Bulletin of the Zoological Survey of India, 3:77-85.

Agrawal, V. C., P. K. Das, S. Chakraborty, R. K. Ghose, A. K. Mandal, T. K. Chakraborty, A. K. Poddar, J. P. Lal, T. P. Bhattachryya, and M. K. Ghosh. 1992. Mammalia. State Fauna Series 3: Fauna of West Bengal, Part 1:27-169.

Aguilar, G. H. 1993. The karyotype and taxonomic status of *Cryptomys hottentotus darlingi* (Rodentia: Bathyergidae). South African Journal of Zoology, 28(4):201-204.

Aguilar, J.-P., and J. Michaux. 1989-1990. Un *Lophiomys* (Cricetidae, Rodentia) nouveau dans le Pliocène du Maroc; rapport avec les Lophiomyinae fossiles et actuels. Paleontologia I Evolució, 23:205-211.

Aguilar, J.-P., and L. Thaler. 1987. *Protolophiomys ibericus nov. gen., nov. sp.* (Mammalia, Rodentia) du Miocène supérieur de Salobreña (Sud de l'Espagne). Comptes Rendus des Séances de l'Académie des Sciences (Paris), 304, ser. 2:859-863.

Aguilar, J.-P., L. D. Brandy, and L. Thaler. 1984. Les rongeurs de Salobreña (sud de L'Espagne) et le probleme de la migration Messinienne. Paleobiologie Continentale, 14(2):3-17.

Aguilera, A., and A. Perez-Zapata. 1989. Cariologia de *Holochilus venezuelae* (Rodentia, Cricetidae). Acta Científica Venezolana, Genetica, 40:198-207.

Aguilera, M., and M. Corti. 1994. Craniometric differentiation and chromosomal speciation of the genus *Proechimys* (Rodentia: Echimyidae). Zeitschrift für Säugetierkunde, 59:366-377.

Aguilera, M., A. Pérez-Zapata, N. Sangines, and A. Martino. 1993. Citogenetica evolutiva en dos generos de rodesores Suramericanos: *Holochilus y Proechimys*. Boletin de la Sociedad Zoologica del Uruguay, 8:49-61.

Aguilera, M., A. Pérez-Zapata, A. Martino, M. A. Barros, and J. L. Patton. 1994. Karyosystematics of *Aepeomys* and *Rhipidomys* (Rodentia: Cricetidae). Acta Cientifica Venezolana, 48:247-248.

Aguilera, M., A. Pérez-Zapata, and A. Martino. 1995. Cytogenetics and karyosystematics of *Oryzomys albigularis* (Rodentia, Cricetidae) from Venezuela. Cytogenetics and Cell Genetics, 69:44-49.

Aguilera, M., A. Pérez-Zapata, J. Ochoa G., and P. Soriano. 2000. Karyology of *Aepeomys* and two species of *Thomasomys* (Rodentia: Muridae) in Venezuela. Journal of Mammalogy, 81:52-58.

Agulnik, S. I., and L. M. Silver. 1996. The Cairo spiny mouse *Acomys cahirinus* shows a strong affinity to the Mongolian gerbil *Meriones unguiculatus*. Molecular Biology and Evolution, 13(1):3-6.

Agulnik, S., S. Adolph, H. Winking, and W. Traut. 1993. Zoogeography of the chromosome 1 HSR in natural populations of the house mouse (*Mus musculus*). Hereditas, 119:39-46.

Agusti, J. 1989. The Miocene rodent succession in Eastern Spain: A zoogeographical appraisal. Pp. 375-404, *in* European Neogene Mammal Chronology (E. H. Lindsay, V. Fahlbusch, and P. Mein, eds.). Plenum Press, New York, 658 pp.

Agusti, J. 1991. The *Allophaiomys* complex in southern Europe. Geobios, 25(1):133-144.

Agustí, J., and I. Casanovas-Vilar. 2003. Neogene gerbils from Europe. Pp. 13-21, *in* Distribution and Migration of Tertiary Mammals in Eurasia. A volume in honour of Hans de Bruijn (J. W. F. Reumer and W. Wessels, eds.). DEINSEA, 10:580 pp.

Ahmed, A. A., and M. Klima. 1978. Zur entwicklung und funktion der lendenwirbelsäule bei der panzerspitzmaus *Scutisorex somereni* (Thomas, 1910). Zeitschrift für Säugetierkunde, 43:1-17.

Ahverdyan, M. R., N. N. Vorontsov, and E. A. Lyapunova. 1991*a*. [The species independence of Shidlovskii's vole–*Microtus schidlovskii*, Argyropulo 1933 (Rodentia, Cricetidae) from western Armenia.] [Academy of Sciences of Armenia, Biological Journal of Armenia], 4:260-265 (in Russian, with English summary).

Ahverdyan, M. R., N. N. Vorontsov, and E. A. Lyapunova. 1991*b*. [Shidlovskii's vole *Microtus schidlovskii*, Argyropulo 1933 (Rodentia, Cricetidae) is an independent species in Armenian fauna.] [Academy of Sciences of Armenia, Biological Journal of Armenia], 4:266-271 (in Russian, with English summary).

Aimi, M. 1967. [Similarity between the voles of Kii Peninsula and of northern part of Honshu.] Zoological Magazine (Japan), 76:44-49 (in Japanese, with English summary).

Aimi, M. 1980. A revised classification of the Japanese red-backed voles. Memoirs of the Faculty of Science, Kyoto University, Series of Biology, 8:35-84.

Aimi, M., H. S. Hardjasasmita, A. Sjarmidi, and D. Yuri. 1986. Geographical distribution of *aygula*-group of the genus *Presbytis* in Sumatra. Kyoto University Overseas Research Report of Studies on Asian Non-human Primates, 5:45-58.

Albayrak, I., and N. Asan. 2002. Taxonomic status and karyotype of *Myotis capaccinii* (Bonaparte, 1837) from Turkey (Chiroptera: Vespertilionidae). Mammalia, 66:63-70.

Alberico, M. S. 1990. Systematics and distribution of the genus *Vampyrops* (Chiroptera: Phyllostomidae) in northwestern South America. Pp. 345-354, *in* Vertebrates in the tropics. Proceedings of the international symposium on vertebrate biogeography and systematics in the tropics, June 5-9, 1989 (G. Peters and R. Hutterer, eds.). Alexander Koenig Zoological Research Insititute and Zoological Museum, Bonn, 424 pp.

Alberico, M. S., and E. Velasco. 1991. Description of a new broad-nosed bat from Colombia. Bonner Zoologische Beiträge, 42:237-239.

Alberico, M. S.,V. Rojas-Díaz, and J. G. Moreno. 1999. Aporte sobre la taxonomía y distribución de los puercoespines (Rodentia: Erethizontidae) en Colombia. Revista da la Academia Colombiana de Ciencias Exactas Fisica, 23 (Suplemento especial):595-612.

Alberico, M., A. Cadena, J. Hernández-Camacho, and Y. Muñoz-Saba. 2000. Mamíferos (Synapsida: Theria) de Colombia. Biota Colombiana, 1:43-75.

Albignac, R. 1970. Notes éthologiques sur quelques carnivores Malagaches: Le *Cryptoprocta ferox* (Bennett). La Terre et la Vie, 24(3):395-402.

Albignac, R. 1973. Faune de Madagascar. 36. Mammifères carnivores. O. R. S. T. O. M., Paris, 206 pp.

Albignac, R. 1974. Observations eco-éthologiques sur le genre *Eupleres*, Viverridae de Madagascar. La Terre et la Vie, 28:321-351.

Albuja V., L. 1982. Murcielagos del Ecuador. Escuela Politécnica Nacional, Departamento de Ciéncias Biológicas, Quito, Ecuador, 285 pp.

Alcantara, M. 1991. Geographical variation in body size of the wood mouse *Apodemus sylvaticus* L. Mammal Review, 21:143-150.

Alcantara, M., M. Diaz, and F. J. P. Pulido. 1991. Variabilidad en las relaciones alométricas entre el peso y las medidas craneales en el raton de campo *Apodemus sylvaticus* L. efectos sobre su utilidad en estudios de ecologia trofica de aves rapaces. Doñana, Acta Vertebrata, 18:205-216.

Alcover, J. A. 1986. Troballa de restes osteologiques de *Eliomys quercinus* (Mammalia, Rodentia, Gliridae) a l'illa de Cabrera. Bolletino de la Sociedad de Historia Natural de Baleares, 30:137-139.

Alcover, J. A., and J. Agusti. 1985. *Eliomys (Eivissia) canarreiensis* n. sgen., n. sp., nou glirid del Pleistocene de la Cova Ca Na Reia (Pitiuses). de Endins, 10-11:51-56.

Alcover, J. A., and J. Gosalbez. 1988. Estudio comparado de la fauna de micromamiferos de las Islas Baleares y Pitiusas. Bulletin d'Écologie, 19(2-3):321-328.

Alcover, J. A., J. Gosalbez, and Ph. Orsini. 1985. *Mus spretus parvus* n. ssp. (Rodentia, Muridae): Un ratoli nan de l'Illa d'Eivissa. Bolletino de la Sociedad de Historia Natural de Baleares, 29:5-17.

Alcover, J. A., X. Campillo, M. Macias, and A. Sans. 1998. Mammal Species of the World: Additional data on insular mammals. American Museum Novitates, 3248:1-29.

Alcover, J. A., M. Llabrés, and L. Moragues (coord.). 2000. Les Balears abans dels humans. Monografias de la Societat d'Historia Natural de les Balears, 8:1-78.

Alexander, L. F. 1996. A morphometric analysis of geographic variation within *Sorex monticolus* (Insectivora: Soricidae). University of Kansas Natural History Museum Miscellaneous Publication, 88:1-54.

Alexander, P. S., L.-K. Lin, and B.-M. Huang. 1987. Ecological notes on two sympatric mountain shrews (*Anourosorex squamipes* and *Soriculus fumidus*) in Taiwan. Journal of the Taiwan Museum, 40:1-7.

Alf, R., A. Hille, and G. Kneitz. 1997. Genetische Populationsstruktur von Gelbhalsmäusen, *Apodemus flavicollis*, in einer intensive genutzten Agrarlandschaft im östlichen Westfalen. Abhandlungen aus dem Westälischen Museum für Naturkunde, 59(3):117-134.

Alibert, P., S. Renaud, B. Dod, F. Bonhomme, and J.-C. Auffray. 1994. Fluctuating asymmetry in the *Mus musculus* hybrid zone: A heterotic effect in disrupted co-adapted genomes. Proceedings of the Royal Society of London, 258:53-59.

Al-Jumaily, M. M. 1998. Review of the mammals of the Republic of Yemen. Fauna of Arabia, 17:477-502.

Al-Jumaily, M. 1999. First record of *Otomops martiensseni* (Matschie 1897) for the Republic of Yemen. Senkenbergiana Biologica, 78:241-245.

Allabergenov, K. 1986. [On the ecology of the forest dormouse in the Fergana Valley]. Uzbekskii Biologicheskii Zhurnal, 2:41-42 (in Russian).

Allard, M. W., and I. F. Greenbaum. 1988. Morphological variation and taxonomy of chromosomally differentiated *Peromyscus* from the Pacific Northwest. Canadian Journal of Zoology, 66:2734-2739.

Allard, M. W., and R. L. Honeycutt. 1991. Ribosomal DNA variation within and between species of rodents, with emphasis on the genus *Onychomys*. Molecular Biology and Evolution, 8:71-84.

Allard, M. W., and R. L. Honeycutt. 1992. Nucleotide sequence variation in the mitochondrial 12S rRNA gene and the phylogeny of African mole-rats (Rodentia: Bathyergidae). Molecular Biology and Evolution, 9(1):27-40.

Allard, M. W., S. J. Gunn, and I. F. Greenbaum. 1987. Mensural discrimination of chromosomally characterized *Peromyscus oreas* and *Peromyscus maniculatus*. Journal of Mammalogy, 68:402-406.

Allard, M. W., S. D. Baker, G. L. Emerson, J. A. Ottenwalder, and C. W. Kilpatrick. 2001. Characterization of the mitochondrial control region in *Solenodon paradoxus* from Hispaniola and the implications for biogeography, systematics, and conservation management. Pp. 331-334, *in* Biogeography of the West Indies: Patterns and perspectives (C. A. Woods and F. E. Sergile, eds.). CRC Press, Boca Raton, Florida. 582 pp.

Allen, G. M. 1912. Some Chinese vertebrates. Mammalia. Memoirs of the Museum of Comparative Zoology at Harvard College, 15(4):201-247.

Allen, G. M. 1915. Mammals obtained by the Phillips Palestine Expedition. Bulletin of the Museum of Comparative Zoology at Harvard College, 59:1-14.

Allen, G. M. 1926. Rats (genus *Rattus*) from the Asiatic Expeditions. American Museum Novitates, 217: 16 pp.

Allen, G. M. 1929. Mustelids from Asiatic expeditions. American Museum Novitates, 358:1-12.

Allen, G. M. 1938-1940. The mammals of China and Mongolia [Natural History of Central Asia (W. Granger, ed.)]. Central Asiatic Expeditions of the American Museum of Natural History, New York, 11:pt. 1:1-620[1938]; pt. 2:621-1350[1940].

Allen, G. M. 1939. A checklist of African mammals. Bulletin of the Museum of Comparative Zoology at Harvard College, 83:1-763.

Allen, G. M. 1942. Extinct and vanishing mammals of the western hemisphere. American Committee for International Wildlife Protection, Special Publication No. 11:xv + 619 pp.

Allen, G. M., and B. Lawrence. 1936. Scientific results of an expedition to rainforest regions in eastern Africa. III. Mammals. Bulletin of the Museum of Comparative Zoology at Harvard College, 79:31-126.

Allen, G. M., and A. Loveridge. 1933. Reports on the scientific results of an expedition to the southwestern highlands of Tanganyika Territory. II. Mammals. Bulletin of the Museum of Comparative Zoology at Harvard College, 75:47-140.

Allen, G. M., and A. Loveridge. 1942. Scientific results of a fourth expedition to forested areas in East and Central Africa. I. Mammals. Bulletin of the Museum of Comparative Zoology at Harvard College, 89:147-214.

Allen, J. A. 1879. On the coatis (genus *Nasua*, Storr). Bulletin of the United States Geological and Geographical Survey, 5:153-174.

Allen, J. A. 1880. History of North American pinnipeds: A monograph of the walruses, seal-lions, sea-bears and seals of North America. United States Geological and Geographical Survey of the Territories, Miscellaneous Publication, 12:1-785.

Allen, J. A. 1887. Sciuridae. Pp. 631-939, *in* Monographs of North American Rodentiae (E. Coues and J. A. Allen, co-authors). United States Geological Survey of the Territories, 11:1-1091.

Allen, J. A. 1892. A synopsis of the pinnipeds, or seals and walruses, in relation to their commercial history and products, Pp. 367-391, *in* Fur seal arbitration, appendix to the case of the United States before the tribunal of arbitration to convene at Paris. United States Government Printing Office, Washington, D.C., 1:605 pp.

Allen, J. A. 1895*a*. On the species of the genus *Reithrodontomys*. Bulletin of the American Museum of Natural History, 7:107-143.

Allen, J. A. 1895*b*. Descriptions of new American mammals. Bulletin of the American Museum of Natural History, 7:327-340.

Allen, J. A. 1898. Revision of the chickarees, or North American red squirrels (subgenus *Tamiasciurus*). Bulletin of the American Museum of Natural History, 10:249-298.

Allen, J. A. 1900. Description of new American marsupials. Bulletin of the American Museum of Natural History, 13:191-199.

Allen, J. A. 1901*a*. Note on the names of a few South American mammals. Proceedings of the Biological Society of Washington, 14:183-185.

Allen, J. A. 1901*b*. On a further collection of mammals from southeastern Peru, collected by Mr. H. H. Keays, with descriptions of new species. Bulletin of the American Museum of Natural History, 14:41-46.

Allen, J. A. 1901*c*. A preliminary study of the North American opossums of the genus *Didelphis*. Bulletin of the American Museum of Natural History, 14:149-188, pls. 22-25.

Allen, J. A. 1902. Mammal names proposed by Oken in his "Lehrbuch der Zoologie". Bulletin of the American Museum of Natural History, 16:373-379.

Allen, J. A. 1905. The Mammalia of southern Patagonia. Reports of the Princeton University Expedition to Patagonia, 1896-1899, 3(Zool):1-210.

Allen, J. A. 1908. Mammals from Nicaragua. Bulletin of the American Museum of Natural History, 24:647-670.

Allen, J. A. 1910. Mammals from Palawan Island, Philippine Islands. Bulletin of the American Museum of Natural History, 28(3):13-17.

Allen, J. A. 1912. Historical and nomenclatorial notes on North American sheep. Bulletin of the American Museum of Natural History, 31:1-29.

Allen, J. A. 1913. Revision of the *Melanomys* group of American Muridae. Bulletin of the American Museum of Natural History, 32:535-555.

Allen, J. A. 1915a. New South American mammals. Bulletin of the American Museum of Natural History, 34:625-634.

Allen, J. A. 1915b. Review of the South American Sciuridae. Bulletin of the American Museum of Natural History, 34:147-309.

Allen, J. A. 1916. Mammals collected on the Roosevelt Brazilian Expedition, with field notes by Leo E. Miller. Bulletin of the American Museum of Natural History, 35:559-610, 6 pls.

Allen, J. A. 1919a. Notes on the synonymy and nomenclature of the smaller spotted cats of tropical America. Bulletin of the American Museum of Natural History, 41(7):341-419.

Allen, J. A. 1919b. Preliminary notes on African Carnivora. Journal of Mammalogy, 1(1):23-31.

Allen, J. A. 1920. Note on Gueldenstaedt's names of certain species of Felidae. Journal of Mammalogy, 1:90-91.

Allen, J. A. 1924. Carnivora collected by the American Museum Congo Expedition. Bulletin of the American Museum of Natural History, 47:73-281.

Almaça, C. 1993. On the Portuguese population of *Microtus agrestis* (Linnaeus, 1761). Arquivos do Museu Bocage, n.s., 2(17):319-115

Alonso-Mejia, A., and R. A. Medellín. 1991. *Micronycteris megalotis*. Mammalian Species, 376:1-6.

Al Robaae, K., and H. Felton. 1990. Was ist *Erythronesokia* Khajuria, 1981 (Mammalia: Rodentia: Muridae)? Zeitschrift für Säugetierkunde, 55:253-259.

Al Saleh, A. A. 1988. Cytological studies of certain desert mammals of Saudi Arabia. 6. First report on chromosome number and karyotype of *Acomys dimidiatus*. Genetica, 76:3-5.

Al Saleh, A. A., and M. A. Khan. 1984. Cytological studies of certain desert mammals of Saudi Arabia. 1. The karyotype of *Jaculus jaculus*. Journal of the College of Science, King Saud University, 15(1):163-168.

Al-Saleh, A. A., and M. A. Khan. 1987. Cytological studies of certain desert mammals of Saudi Arabia. 5. The karyotype of *Meriones rex*. Genetica, 73:185-187.

Alston, E. R. 1876. On the classification of the order Glires. Proceedings of the Zoological Society of London, 1876:61-98.

Altuna, C. A., and E. P. Lessa. 1985. Penial morphology in Uruguayan species of *Ctenomys* (Rodentia: Octodontidae). Journal of Mammalogy, 66:483-488.

Altuna, C. A., M. Ubilla, and E. P. Lessa. 1985. Estado actual del conocimiento de *Ctenomys rionegrensis* Langguth y Abella, 1970 (Rodentia, Octodontidae). Actas de las Jornadas de Zoologia del Uruguay, 1:8-9.

Alvarez, J., M. R. Willig, J. K. Jones, Jr., and W. D. Webster. 1991. *Glossophaga soricina*. Mammalian Species, 379:1-7.

Alvarez, T. 1961. Taxonomic status of some mice of the *Peromyscus boylii* group in eastern Mexico, with description of a new subspecies. University of Kansas Publications, Museum of Natural History, 14:111-120.

Alvarez, T. 1963a. The Recent mammals of Tamaulipas. University of Kansas Publications, Museum of Natural History, 14:363-473.

Alvarez, T. 1963b. The type locality for *Nyctomys sumichrasti* Saussure. Journal of Mammalogy, 44:582-583.

Alvarez, T., and T. Alvarez-Castañeda. 1991. Notas sobre el estado taxónomico de *Pteronotus davyi* en Chiapas y de *Hylonycteris* en Mexico (Mammalia: Chiroptera). Anales de la Escuela Nacionál de Ciéncias, 34:223-229.

Alvarez, T., and S. T. Alvarez-Castañeda. 1996. Descripción de una nueva subespecie de tuza *Cratogeomys goldmani* (Rodentia: Geomyidae) de San Luis Potosí, Mexico. Acta Zoologica Mexicana, n.s., 68:37-43.

Alvarez, T., and N. González-Ruíz. 2001. Nuevos registros de *Notiosorex crawfordi* (Insectivora: Soricidae) para México. Acta Zoologica Mexicana, n.s., 84:175-177.

Alvarez, T., and J. J. Hernández Chávez. 1993. Taxonomía del metorito *Microtus mexicanus* en el centro de México con la descripción de una nuvea subespecie. Pp. 137-156, *in* Avances en el estudio de los mamíferos de México (R. A. Medellín and G. Cebassos, eds.). Publicaciones Especiales, Vol. 1. Asociación Mexicana de Mastozoología, México, D. F., 464 pp.

Alvarez, T., and O. Polaco. 1984. Estudio de los mamíferos capturados en la Michilía, sureste de Durango, México. Anales de la Escuela Nacionál de Ciéncias Biológicas (México City), 28:99-148.

Alvarez, T., N. Sánchez-Casas, and A. Ocaña. 1998. La Mastofauna de la región de Ticumán, Morelos, México. Anales de la Escuela Nacionál de Ciencias Biológicas, México, 43:51-66.

Alvarez, Y., J. B. Juste, E. Tabares, A. Garrido-Pertierra, C. Ibanez, and J. M. Bautista. 1999. Molecular phylogeny and morphological homoplasy in fruitbats. Molecular Biology and Evolution, 16:1061-1067.

Alvarez-Castañeda, S. T. 1994. Current status of the rice rat, *Oryzomys couesi peninsularis*. Southwestern Naturalist, 39:99-100.

Alvarez-Castañeda, S. T. 1996. Los mamíferos del estado de Morelos. Centro de Investigaciones Biológicas del Noroeste, La Paz, México, 211 pp.

Alvarez-Castañeda, S. T. 2002. *Peromyscus hooperi*. Mammalian Species, 709:1-3.

Alvarez-Castañeda, S. T., and M. A. Bogan. 1998. *Myotis peninsularis*. Mammalian Species, 573:1-2.

Alvarez-Castañeda, S. T., and P. Cortes-Calva. 1999. Familia Muridae. Pp. 445-568, *in* Mamíferos del Noroeste de México (S. T. Alvarez-Castañeda and J. L. Patton, eds.). Centro de Investigaciones Biológicas del Noroeste, La Paz, México, 583 pp.

Alvarez-Castañeda, S. T., and P. Cortes-Calva. 2002. *Peromyscus slevini*. Mammalian Species, 705:1-2.

Alvarez-Castañeda, S. T., and A. Ortega-Rubio. 2003. Current status of rodents on islands in the Gulf of California. Biological Conservation, 109:157-163.

Alvarez-Casteñeda, S. T., G. Arnaud, and E. Yensen. 1996. *Spermophilus atricapillus*. Mammalian Species, 521:1-3.

Alvarez-Castañeda, S. T., E. Rios, and A. Butierrez-Ramos. 2001. Noteworthy records of the little pocket mouse (Heteromyidae: *Perognathus longimembris*) on the Baja California Peninsula. Southwestern Naturalist, 46:243-245.

Amani, F., and D. Geraads. 1993. Le gisement moustérien du Djebel Irhoud, Maroc: Précisions sur la faune et la biochronologie, et description d'un nouveau reste humain. Comptes Rendus des Séances de l'Académie des Sciences (Paris), 316, ser. 2:847-852.

Amato, G., M. G. Egan, G. B. Schaller, R. H. Baker, H. C. Rosenbaum, W. G. Robichaud, and R. DeSalle. 1999. Rediscovery of Roosevelt's barking deer (*Muntiacus rooseveltorum*). Journal of Mammalogy, 80:639-643.

Amato, G., M. G. Egan, and G. B. Schaller. 2000. Mitochrondrial DNA variation in muntjac: Evidence for discovery, rediscovery, and phylogenetic relationships. Pp. 285-295, *in* Antelopes, deer, and relatives. Fossil record, behavioral ecology, systematics, and conservation (E. S. Vrba and G. B. Schaller, eds.). Yale University Press, New Haven, 341 pp.

Ameghino, F. 1889. Contribucion al conocimiento de los mamiferos fosiles del la Republica Argentina. Actas de la Academia Nacional de Ciencias, 6:1027 pp.

Ameghino, F. 1904. Myocastoridae. Annales de la Sociedad Científica de Argentina, 56-58:103.

Ameur, R. 1984. Découverte de nouveaux rongeurs dans la formation Miocène de Bou Hanifia (Algérie Occidentale). Géobios, 17(2):167-175.

Ameur-Chehbeur, A. 1991. Un nouveau genre de Gerbillidae (Rodentia, Mammalia) du Mio-Pliocene d'el Eulma, Algérie Orientale. Geobios, 24:509-512.

Amori, G. 1993. Italian insectivores and rodents: Extinctions and current status. Supplemento alle Ricerche di biologia della selvaggina, 21:115-134.

Amori, G., and L. Contoli. 1994. Morphotypic, craniometric and genotypic diversification in *Apodemus flavicollis* and *Apodemus sylvaticus*. Bollettino di Zoologia, 61:353-357.

Amori, G., and M. Masseti. 1996. Does the occurrence of predators on central Mediterranean Islands affect the body size of micromammals? Vie Milieu, 46(3/4):205-211.

Amori, G., M. Cantini, and V. Rota. 1995. Distribution and conservation of Italian dormice. Pp. 331-336, *in* Proceedings of II Conference on Dormice (Rodentia, Myoxidae) (M. G. Filippucci, ed.). Hystrix, n.s., 6(1-2):1-340.

Amori, G., F. M. Angelici and L. Boitani. 1999. Mammals of Italy: A revised checklist of species and subspecies (Mammalia). Senckenbergiana biologica, 79(2):271-286.

Amori, G., L. Corsetti, and C. Esposito. 2002*a*. Mammiferi dei Monti Lepini. Quaderni di Conservazione della Natura, 11:1-210.

Amori, G., S. Ciarlantini, and M. G. Filippucci. 2002*b*. Environmental genetic variation in two sympatric rodent species (*Apodemus flavicollis* and *A. sylvaticus*) in central Italy. Folia Zoologica, 50(3):185-192.

Amrine-Madsen, H., K.-P. Koepfli, R. K. Wayne, and M. S. Springer. 2003. A new phylogenetic marker, apolipoprotein B, provides compelling evidence for eutherian relationships. Molecular Phylogenetics and Evolution, 28:225-240.

Amtmann, E. 1975. Family Sciuridae. Part 6. 1. Pp. 1-12, *in* The mammals of Africa: An identification manual (J. Meester and H. W. Setzer, eds.) [issued 10 Dec 1975]. Smithsonian Institution Press, Washington, D.C., not continuously paginated.

Anadu, P. A. 1979. The occurrence of *Steatomys jacksoni* Hayman in south-western Nigeria. Acta Theriologica, 24:513-526.

Andayani, N., J. C. Morales, M. R. J. Forstner, J. Supriatna, and D. J. Melnick. 2001. Genetic variability in mtDNA of the silvery gibbon: Implications for the conservation of a critically endangered species. Conservation Biology, 15:770-775.

Andera, M. 1986. Dormice (Gliridae) in Czechoslovakia. Part 1. *Glis glis, Eliomys quercinus* (Rodentia, Mammalia). Folia Musei Rerum Naturalium Bohemiae Occidentalis, Zoologica, 24:1-47.

Andera, M. 1987. Dormice (Gliridae) in Czechoslovakia. Part 2. *Muscardinus avellanarius, Dryomys nitedula* (Rodentia, Mammalia). Folia Musei Rerum Naturalium Bohemiae Occidentalis, Zoologica, 26:3-78.

Andera, M. 1995. The present status of dormice in the Czech Republic. Pp. 155-159, *in* Proceedings of II Conference on Dormice (Rodentia, Myoxidae) (M. G. Filippucci, ed.). Hystrix, n.s., 6(1-2):1-340.

Andera, M., and J. Èervený. 1994. Atlas of distribution of the mammals of the Šumava Mts. Region (SW-Bohemia). Acta Scientiarum Naturalium Academiae Scientiarum Bohemicae Brno, n.s., 28(2-3):1-111.

Andersen, K. 1907. On *Pterocyon, Rousettus* and *Myonycteris*. Annals and Magazine of Natural History, ser. 7, 19:501-515.

Andersen, K. 1912. Catalogue of the Chiroptera in the collection of the British Museum. Vol. 1. Megachiroptera. Second ed. British Museum (Natural History), London, 854 pp.

Anderson, A. E., and O. C. Wallmo. 1984. *Odocoileus hemionus*. Mammalian Species, 219:1-9.

Anderson, E. 1970. Quaternary evolution of the genus *Martes* (Carnivora, Mustelidae). Acta Zoologica Fennica, 130:1-132.

Anderson, E. 1977. Pleistocene Mustelidae (Mammalia, Carnivora) from Fairbanks, Alaska. Bulletin of the Museum of Comparative Zoology, 148:1-21.

Anderson, H. L., Jr. 1974. Natural history and systematics of the tundra hare (*Lepus othus* Merriam). M.S. Thesis, University of Alaska, Fairbanks. x + 106 pp.

Anderson, J. 1879. Anatomical and zoological researches in western Yunnan. Quaritch, London, 984 pp.

Anderson, J. 1881. Catalogue of Mammalia in the Indian Museum, Indian Museum, Calcutta, 223 pp.

Anderson, R. M. 1942*a*. Canadian voles of the genus *Phenacomys* with description of two new Canadian subspecies. Canadian Field-Naturalist, 56:56-60.

Anderson, R. M. 1942*b*. Six additions to the list of Quebec mammals with descriptions of four new forms. Annual Report, Provancher Society of Natural History of Canada, Quebec, 1941:31-43 (English); 45-57 (French).

Anderson, R. M. 1947. Catalogue of Canadian Recent mammals. National Museum of Canada Bulletin, Biological Series, 102:1-238.

Anderson, R. M., and A. L. Rand. 1943. Variation in the porcupine (Genus *Erethizon*) in Canada. Canadian Journal of Research, section D, 21:292-309.

Anderson, R. P. 1999 [2000]. Preliminary review of the systematics and biogeography of the spiny pocket mice (*Heteromys*) of Colombia. Revista Academia Colombiana de Ciencias Exactas, Fisicas y Naturals, 23 (Suplemento especial):613-630.

Anderson, R. P. 2003*a*. Taxonomy, distribution, and natural history of the genus *Heteromys* (Rodentia: Heteromyidae) in western Venezuela, with the description of a dwarf species from the Península de Paraguaná. American Museum Novitates, 3396:1-43.

Anderson, R. P. 2003*b*. Real vs. artefactual absences in species distributions: Tests for *Oryzomys albigularis* (Rodentia: Muridae) in Venezuela. Journal of Biogeography, 30:591-605.

Anderson, R. P., and P. Jarrín-V. 2002. A new species of spiny pocket mouse (Heteromyidae: *Heteromys*) endemic to western Ecuador. American Museum Novitates, 3382:1-26.

Anderson, R. P., and P. J. Soriano. 1999. The occurrence and biogeographic significance of the southern spiny pocket mouse *Heteromys australis* in Venezuela. Zeitschrift für Säugetierkunde, 64:121-125.

Anderson, S. 1956. Subspeciation in the meadow mouse, *Microtus pennsylvanicus*, in Wyoming, Colorado, and adjacent areas. University of Kansas Publications, Museum of Natural History, 9:85-104.

Anderson, S. 1959. Distribution, variation, and relationships of the montane vole, *Microtus montanus*. University of Kansas Publications, Museum of Natural History, 9:415-511.

Anderson, S. 1960. The baculum in microtine rodents. University of Kansas Publications, Museum of Natural History, 12:181-216.

Anderson, S. 1962. Tree squirrels (*Sciurus colliaei* group) of western Mexico. American Museum Novitates, 2093:1-13.

Anderson, S. 1964. The systematic status of *Perognathus artus* and *Perognathus goldmani* (Rodentia). American Museum Novitates, 2184:1-27.

Anderson, S. 1965. Conspecificity of *Plagiodontia aedium* and *P. hylaeum* (Rodentia). Proceedings of the Biological Society of Washington, 78:95-98.

Anderson, S. 1966. Taxonomy of gophers, especially *Thomomys* in Chihuahua, Mexico. Systematic Zoology, 15:189-198.

Anderson, S. 1967. Primates. Pp. 151-177, *in* Recent mammals of the world: A synopsis of families (S. Anderson and J. K. Jones, Jr., eds.). Ronald Press Co., New York, 453 pp.

Anderson, S. 1969a. *Macrotus waterhousii*. Mammalian Species, 1:1-4.

Anderson, S. 1969b. Taxonomic status of the woodrat, *Neotoma albigula*, in southern Chihuahua, Mexico. Miscellaneous Publications, Museum of Natural History, University of Kansas, 51:25-50.

Anderson, S. 1972. Mammals of Chihuahua: Taxonomy and distribution. Bulletin of the American Museum of Natural History, 148:149-410.

Anderson, S. 1982. *Monodelphis kunsi*. Mammalian Species, 190:1-3.

Anderson, S. 1993. Los mamíferos bolivianos: Notas de distribucion y claves de identificacion. Instituto de Ecologia, Universidad Mayor de San Andrés, La Paz, Bolivia, 159 pp.

Anderson, S. 1996. Notes on Bolivian mammals, 8. Small species of *Platyrrhinus*. Pp. 89-93, *in* Contributions in mammalogy: A memorial volume in honor of J. Knox Jones, Jr. (H. H. Genoways and R. J. Baker, eds.). Museum of Texas Tech University, Lubbock, Texas, 315 pp.

Anderson, S. 1997. Mammals of Bolivia, taxonomy and distribution. Bulletin of the American Museum of Natural History, 231:1-652.

Anderson, S., and A. S. Gaunt. 1962. A classification of white-sided jack rabbits of Mexico. American Museum Novitates, 2088:1-16.

Anderson, S., and J. P. Hubbard. 1971. Notes on geographic variation of *Microtus pennsylvanicus* (Mammalia, Rodentia) in New Mexico and Chihuahua. American Museum Novitates, 2460:1-8.

Anderson, S., and J. K. Jones, Jr. (eds.). 1967. Recent mammals of the world, a synopsis of families. Ronald Press Co., New York, 453 pp.

Anderson, S., and J. K. Jones, Jr. (eds.). 1984. Orders and families of recent mammals of the world. John Wiley and Sons, New York, 686 pp.

Anderson, S., and C. E. Nelson. 1965. A systematic revision of *Macrotus* (Chiroptera). American Museum Novitates, 2212:1-39.

Anderson, S., and N. Olds. 1989. Notes on Bolivian mammals. 5. Taxonomy and distribution of *Bolomys* (Muridae, Rodentia). American Museum Novitates, 2935:1-22.

Anderson, S., and T. L. Yates. 2000. A new genus and species of phyllotine rodent from Bolivia. Journal of Mammalogy, 81:18-36.

Anderson, S., K. F. Koopman, and G. K. Creighton. 1982. Bats of Bolivia: An annotated checklist. American Museum Novitates, 2750:1-24.

Anderson, S., C. A. Woods, G. S. Morgan, and W. L. R. Oliver. 1983. *Geocapromys brownii*. Mammalian Species, 201:1-5.

Anderson, S., T. L. Yates, and J. A. Cook. 1987. Notes on Bolivian Mammals 4: The genus *Ctenomys* (Rodentia, Ctenomyidae) in the eastern lowlands. American Museum Novitates, 2891:1-20.

Anderson, S., B. R. Riddle, T. L. Yates, and J. A. Cook. 1993. Los mamíferos del Parque Nacional Amboró y la región de Santa Cruz de la Sierra, Bolivia. Special Publication, The Museum of Southwestern Biology, 2:1-58.

Ando, A., S. Shiraishi, M. Harada, and T. A. Uchida. 1988. A karyological study of two intraspecific taxa in Japanese *Eothenomys* (Mammalia: Rodenta). Journal of the Mammalogical Society of Japan, 13:93-104.

Ando, A., M. Harada, S. Shiraishi, and T. A. Uchida. 1991. Variation of the X chromosome in the Smith's red-backed vole *Eothenomys smithii*. Journal of the Mammalogical Society of Japan, 15:83-90.

Andrades-Miranda, J., L. F. B. Oliveira, C. A. V. Lima-Rosa, P. Nunes, N. I. T. Zanchin, and M. S. Mattevi. 2001a. Chromosome studies of seven species of *Oligoryzomys* (Rodentia: Sigmodontinae) from Brazil. Journal of Mammalogy, 82:1080-1091.

Andrades-Miranda, J., L. F. B. Oliveira, N. I. T. Zanchin, and M. S. Mattevi. 2001b. Chromosomal description of the rodent genera *Oecomys* and *Nectomys* from Brazil. Acta Theriologica, 46:269-278.

Andrades-Miranda, J., L. F. B. Oliveira, C. A. V. Lima-Rosa, D. A. Sana, A. P. Nunes, and M. S. Mattevi. 2002a. Genetic studies in representatives of genus *Rhipidomys* (Rodentia, Sigmodontinae) from Brazil. Acta Theriologica, 47:125-135.

Andrades-Miranda, J., N. I. T. Zanchin, L. F. B. Oliveira, A. R. Langguth, and M. S. Mattevi. 2002b. $(T_2AG_3)_n$ telomeric sequence hybridization indicating centric fusion rearrangments in the karyotype of the rodent *Oryzomys subflavus*. Genetica, 114:11-16.

Andreae, P., and I. Krumbiegel. 1976. Die Damagazelle, *Nanguer dama* (Pallas, 1767), und ihr gegenwärtiger Status. Säugetierkundliche Mitteilungen, 23:268-278.

Andreotti, A., N. Baccetti, A. Perfetti, M. Besa, P. Genovesi, and V. Guberti. 2001. Mammiferi e uccelli esotici in Italia: Analisi del fenomeno, impatto sulla biodiversità e linee guida gestionali. Quaderni Di Conservazione Della Natura, 2, Ministero dell'Ambiente Servizio Conservazione Natura-Istituto Nazionale per la Fauna Selvatica "A. Ghigi", 189 pp.

Andrews, C. W. 1900. A monograph of Christmas Island (Indian Ocean). Bulletin of the British Museum. Natural History. Geology Series, 13:1-337.

Andrews, C. W. 1909. An account of Andrews' visit to Christmas Island in 1908. Proceedings of the Zoological Society of London, 1909:101-103.

Andrews, R. C. 1932. The new conquest of central Asia: A narrative of the explorations of the Central Asiatic Expeditions in Mongolia and China, 1921-1930. [Natural History of Central Asia (W. Granger, ed.)]. Central Asiatic Expeditions of the American Museum of Natural History, New York, 1:1-678.

Angas, G. F. 1848[1849]. Description of Tragelaphus Angasii Gray, with some account of its habits. Proceedings of the Zoological Society of London, 1848:89-90.

Angelici, F. M., F. Pinchera, and F. Riga. 1992. First records of Crocidura sp. and Mus domesticus and notes on the mammals of Astipálaia Island (Dodecanese, Greece). Mammalia, 156(1):159-161.

Angerbjorn, A., and J. E. C. Flux. 1995. Lepus timidus. Mammalian Species, 495:1-11.

Angermann, R. 1966. Beiträge zur Kenntnis der Gattung Lepus (Lagomorpha, Leporidae). I. Abgrenzung der Gattung Lepus. II. Der taxonomische Status von Lepus brachyurus Temminck und Lepus mandshuricus Radde. Mitteilungen aus dem Zoologische Museum in Berlin, 42:127-335.

Angermann, R. 1967. Beiträge zur Kenntnis der Gattung Lepus (Lagomorpha, Leporidae). III. Zur Variabilitat palaearktischer Schneehasen. IV. Lepus yarkandensis Günther, 1875 und Lepus oiostolus Hodgson, 1840 — zwei endemische Hasenarten Zentralasiens. Mitteilungen aus dem Zoologische Museum in Berlin, 43:64-203.

Angermann, R. 1972. Hares, rabbits, and pikas. Pp. 419-462, in Grzimek's Animal Life Encyclopedia (B. Grzimek, ed.). Van Nostrand Reinhold Company, New York, NY, 12(Mammals III):1-657.

Angermann, R. 1983. The taxonomy of Old World Lepus. Acta Zoologica Fennica, 174:17-21.

Angermann, R. 1984. Intraspezifische Variabilitat der Molarenmuster bei der Nordischen Wühlmaus (Microtus oeconomus [Pallas, 1776]) (Mammalia, Rodentia, Microtinae). Zoologische Abhandlungen Staatliches Museum für Tierkunde in Dresden, 39:115-136.

Angermann, R., and A. Feiler. 1988. Zur Nomenklatur, Artabgrenzung und Variabilität der Hasen (Gattung Lepus) im westlichen Africa (Mammalia, Lagomorpha, Leporidae). Zoologische Abhandlungen Staatliches Museum für Tierkunde in Dresden, 43:149-167.

Angst, R., and P. Mann. 1971. Zur Variabilität von Urogale everetti. Folia Primatologica, 15:148-158.

Aniskin, V. M., L. A. Lavrenchenko, A. A. Varshavskii, and A. N. Milishnikov. 1997. [Karyotypic differentiation of three harsh-furred mouse species of genus Lophuromys (Murinae, Rodentia) from the Bale Mountains National Park, Ethiopia]. Genetika, 33(7):967-973 (in Russian, with English summary).

Anon. 1803-1804. Nouveau dictionnaire d'histoire naturelle, appliquèe aux art, principalement à l'agriculture et à l'economie rurale et domestique; par une société de naturalistes et d'agriculteurs: Avec des figures tirees des trois regnes de la nature. Deterville, Paris, 24 volumes.

Anon. 1829. [Dates and parts of Cretzchmar's Atlas zu der reise im nördlicihen Afrika von Eduard Rüppell, Säugetiere]. Isis von Oken, 1829:1291-1294.

Ansell, W. F. H. 1967. Additional records of Zambian Chiroptera. Arnoldia (Rhodesia), 2(38):1-29.

Ansell, W. F. H. 1972. Order Artiodactyla. Part 15. Pp. 1-84, in The mammals of Africa: An identification manual (J. Meester and H. W. Setzer, eds.) [issued 2 May 1972]. Smithsonian Institution Press, Washington, D.C., not continuously paginated.

Ansell, W. F. H. 1974a. Order Perissodactyla. Part 14. Pp. 1-14, in The mammals of Africa: An identification manual (J. Meester and H. W. Setzer, eds.) [issued 10 Sep 1974]. Smithsonian Institution Press, Washington, D.C., not continuously paginated.

Ansell, W. F. H. 1974b. Some mammals from Zambia and adjacent countries. Puku, supplement, 1:1-49.

Ansell, W. F. H. 1978. The mammals of Zambia. National Parks and Wildlife Service, Chilanga, 126 pp.

Ansell, W. F. H. 1980. Antilope zebra Gray, 1838 (Mammalia) revised proposals for conservation. Z. N. (S.) 1908. Bulletin of Zoological Nomenclature, 37:152-153.

Ansell, W. F. H. 1982. The southeastern range limit of Manis tricuspis Rafinesque, and the type of Manis tridentata Focillon. Mammalia, 46(4):559-560.

Ansell, W. F. H. 1989a. African mammals, 1938-1988. Trendrine Press, Cornwall, 77 pp.

Ansell, W. F. H. 1989b. Mammals from Malawi: Part II. Nyala, 13(1/2):41-65.

Ansell, W. F. H., and P. D. H. Ansell. 1973. Mammals of the north-eastern montane areas of Zambia. Puku, 7:21-69.

Ansell, W. F. H., and C. F. Banfield. 1979. The subspecies of Kobus leche Gray, 1850 (Bovidae). Säugetierkundliche Mitteilungen, 27:168-176.

Ansell, W. F. H., and R. J. Dowsett. 1988. Mammals of Malawi: An annotated check list and atlas. Trendrine Press, Zennor, St. Ives, United Kingdom, 170 pp.

Ansell, W. F. H., and R. J. Dowsett. 1991. The type locality of Otomys angoniensis Wroughton, 1906. Bonner Zoologische Beiträge, 42:17-19.

Anstee, S. D. 1996. Use of external mound structures as indicators of the presence of the pebble-mound mouse, *Pseudomys chapmani*, in mound systems. Wildlife Research, 23:429-434.

Anthony, H. E. 1923. Preliminary report on Ecuadorean mammals. No. 3. American Museum Novitates, 55:1-14.

Anthony, H. E. 1924. Preliminary report on Ecuadorean mammals. No. 6. American Museum Novitates, 139:1-9.

Anthony, H. E. 1929. Two new genera of rodents from South America. American Museum Novitates, 383: 6 pp.

Anthony, H. E. 1941. Mammals collected by the Vernay-Cutting Burma Expedition. Field Museum of Natural History, Zoological Series, 27:37-123.

Apfelbaum, L. I., and O. A. Reig. 1989. Allozyme genetic distances and evolutionary relationships in species of akodontine rodents (Cricetidae: Sigmodontinae). Biological Journal of the Linnean Society, 38:257-280.

Apfelbaum, L. I., A. I. Massarini, L. E. Daleffe, and O. A. Reig. 1991. Genetic variability in the subterranean rodents *Ctenomys australis* and *Ctenomys porteousi* (Rodentia: Octodontidae). Biochemical Systematics and Ecology, 19 (6):467-476.

Aplin, K. P. 2005. Ten million years of rodent evolution in the Australasian region: A review of the phylogenetic evidence and a speculative historical biogeography. Pp. 707-744 *in* Evolution and biogeography of Australasian vertebrates (J. R. Merrick, M. Archer, G. M. Hickey, and M. S. Y. Lee, eds.). Australian Scientific Publishing, Sydney, 800 pp.

Aplin, K. P., and M. Archer. 1987. Recent advances in marsupial systematics with a new syncretic classification. Pp. xv-lxxii, *in* Possums and opossums: Studies in evolution (M. Archer, ed.). Surrey Beatty and Sons Pty. Ltd. and Royal Zoological Society of New South Wales, Sydney, 1:1-400.

Aplin, K. P., P. R. Baverstock, and S. C. Donnellan. 1993. Albumin immunological evidence for the time and mode of origin of the New Guinean terrestrial mammal fauna. Science in New Guinea,19(3):131-145.

Aplin, K. P., J. M. Pasveer, and W. E. Boles. 1999. Late Quaternary vertebrates fom the Bird's Head Peninsula, Irian Jaya, Indonesia, including descriptions of two previously unknown marsupial species. Records of the Western Australian Museum, Supplement No. 57:351-387.

Aplin, K. P., T. Chesser, and J. ten Have. 2003*a*. Evolutionary biology of the genus *Rattus*: Profile of an archetypal rodent pest. Pp. 487-498, *in* Rats, mice and people: Rodent biology and management (G. R. Singleton, L. A. Hinds, C. J. Krebs, and D. M. Spratt, eds.). ACIAR Technical Report 96, ACIAR, Canberra, 564 pp.

Aplin, K. P., A. Frost, N. P. Tuan, N. M. Hung, and L. P. Lan. 2003*b*. Identification of rodents of the genus *Bandicota* in Vietnam and Cambodia. Pp. 531-535, *in* Rats, mice and people: Rodent biology and management (G. R. Singleton, L. A. Hinds, C. J. Krebs, and D. M. Spratt, eds.). ACIAR Technical Report 96, ACIAR, Canberra, 564 pp.

Aplin, K. P., P. R. Brown, J. Jacob, C. J. Krebs, and G. R. Singleton. 2003*c*. Field methods for rodent studies in Asia and the Indo-Pacific. ACIAR Monograph 100, ACIAR, Canberra, 223 pp.

Aprile, G., and D. Chicco. 1999. New exotic mammal species in Argentina: South-east tree squirrel (*Callosciurus erythraeus*). Mastozoologia Neotropical, 6(1):7-14.

Arai, S., T. Mori, H. Yoshida, and S. Shiraishi. 1985. A note on the Japanese water shrew, *Chimarrogale himalayica platycephala*, from Kyushu. Journal of the Mammalogical Society of Japan, 10:193-203.

Arana-Cardo, R., and C. F. Ascorra. 1994. Observaciones sobre la distribucion de algunos sigmodontinos (Rodentia, Muridae) altoandinos del Departamento de Lima, Peru. Publicaciones del Museo de Historia Natural, Universidad Nacional Mayor de San Marcos, Serie A Zoologia, 47:1-7.

Aranda, M., J. E. Escobedo, and C. Pozo. 1997. Registros recientes de *Otonyctomys hatti* (Rodentia: Muridae) en Quintana Roo, Mexico. Acta Zoologica Mexicana, n.s., 72:63-65.

Arata, A. 1964. The anatomy and taxonomic significance of the male accessory reproductive glands of muroid rodents. Bulletin of the Florida State Museum, Biological Sciences, 9:1-42.

Arbogast, B. S. 1999. Mitochondrial DNA phylogeography of the New World flying squirrels (*Glaucomys*): Implications for Pleistocene biogeography. Journal of Mammalogy, 80(1):142-155.

Arbogast, B. S., R. A. Browne, and P. D. Weigl. 2001. Evolutionary genetics and Pleistocene biogeography of North American tree squirrels (*Tamiasciurus*). Journal of Mammalogy, 82(2):302-319.

Archer, M. 1975. *Ningaui*, a new genus of tiny dasyurids (Marsupialia) and two new species, *N. timealeyi* and *N. ridei* from arid Western Australia. Memoirs of the Queensland Museum, 17(2):237-249.

Archer, M. 1976. Revision of the marsupial genus *Planigale* Troughton (Dasyuridae). Memoirs of the Queensland Museum, 17:341-365.

Archer, M. 1977. Revision of the dasyurid marsupial genus *Antechinomys* Krefft. Memoirs of the Queensland Museum, 18:17-29.

Archer, M. 1979. Two new species of *Sminthopsis* Thomas from northern Australia, *S. butleri* and *S. douglasi*. Australian Museum Zoologist, 20:327-345.

Archer, M. 1981. Results of the Archbold Expeditions. No. 104. Systematic revision of the marsupial dasyurid genus *Sminthopsis* Thomas. Bulletin of the American Museum of Natural History, 168:61-224.

Archer, M. 1982. Review of the dasyurid (Marsupialia) fossil record, integration of data bearing on phylogenetic interpretation, and suprageneric classification. Pp. 397-443, *in* Carnivorous Marsupials (M. Archer, ed). Royal Society of New South Wales, Syndey, 2:1-804.

Archer, M. 1984. The Australian Marsupial Radiation. Pp. 633-808, *in* Vertebrate Zoogeography and Evolution in Australasia (M. Archer and G. Clayton, eds.). Hesperian Press, Carlisle, W. A., Australia, 1203 pp.

Archer, M., and A. A. Bartholomai. 1978. Tertiary mammals of Australia: A synoptic review. Alcheringa, 2:1-19.

Archer, M., and J. A. W. Kirsch. 1977. The case for the Thylacomyidae and Myrmecobiidae, Gill, 1872, or why are marsupial families so extended. Proceedings of the Linnean Society of New South Wales, 102:18-25.

Archer, M., G. Clayton, and S. Hand. 1984. A checklist of Australasian fossil mammals. Pp. 1027-1087, *in* Vertebrate Zoogeography and Evolution in Australasia (M. Archer and G. Clayton, eds.). Hesperian Press, Carlisle, W. A., Australia, 1203 pp.

Archer, M., S. J. Hand, and H. Godthelp. 1991. Australia's lost world. Prehistoric animals of Riversleigh. Indiana University Press, Bloomington and Indianapolis, 264 pp.

Archibald, J. D. 2003. Timing and biogeography of the eutherian radiation: Fossils and molecules compared. Molecular Phylogenetics and Evolution, 28:350-359.

Archibald, J. D., A. O. Averlanov, and E. G. Ekdale. 2001. Late Cretaceous relatives of rabbits, rodents, and other extant eutherian mammals. Nature, 414:62-65.

Arctander, P., P. W. Kat, R. A. Aman, and H. R. Siegismund. 1995. Extreme genetic differences among populations of *Gazella granti*, Grant's gazelle, in Kenya. Heredity, 76:465-475.

Arctander, P., C. Johansen, and M.-A. Coutellec-Vreto. 1999. Phylogeography of three closely related African bovids (Tribe Alcelaphini). Molecular Biology and Evolution, 16:1724-1739.

Arellano, E., and D. S. Rogers. 1994. *Reithrodontomys tenuirostris*. Mammalian Species, 477:1-3.

Arellano, E., D. S. Rogers, and F. A. Cervantes. 2003. Genic differentiation and phylogenetic relationships among tropical harvest mice (*Reithrodontomys*: Subgenus *Aporodon*). Journal of Mammalogy, 84:129-143.

Argyropulo, A. I. 1948. Obzor retsentnykh vidov cem. Lagomyidae Lilljeb., 1886 (Lagomorpha, Mammalia) [A review of Recent species of the family . . .]. Trudy Zoologicheskovo Instituta, Akademiya Nauk, Leningrad, 7:124-128 (in Russian).

Aristov, A. A., and G. F. Baryshnikov. 2001. [Mammals of Russia and adjacent territories. Carnivores and pinnipeds.] Saint Petersburg, Russian Academy of Sciences, 560 pp. (in Russian).

Arita, H. T., and S. R. Humphrey. 1988 [1989]. Revisión taxonómica de los murciélagos magueyeros del género *Leptonycteris* (Chiroptera: Phyllostomidae). Acta Zoológica Mexicana, n.s., 29:1-60.

Armitage, P. L. 1993. Commensal rats in the New World, 1492-1992. Biologist, 40(4):174-178.

Armitage, P. L. 1994. Unwelcome companions: Ancient rats reviewed. Antiquity, 68:231-240.

Armstrong, D. M. 1972. Distribution of mammals in Colorado. Monograph of the Museum of Natural History, The University of Kansas, 3:1-415.

Armstrong, D. M., and J. K. Jones, Jr. 1971*a*. Mammals from the Mexican state of Sinaloa, 1. Marsupialia, Insectivora, Edentata, Lagomorpha. Journal of Mammalogy, 52:747-757.

Armstrong, D. M., and J. K. Jones, Jr. 1971*b*. *Sorex merriami*. Mammalian Species, 2:1-2.

Armstrong, D. M., and J. K. Jones, Jr. 1972*a*. *Megasorex gigas*. Mammalian Species, 16:1-2.

Armstrong, D. M., and J. K. Jones, Jr. 1972*b*. *Notiosorex crawfordi*. Mammalian Species, 17:1-5.

Armstrong, K. N. 2002. Morphometric divergence among populations of *Rhinonycteris aurantius* (Chiroptera: Hipposideridae) in northern Australia. Australian Journal of Zoology, 50:649-669.

Armstrong, L. A., C. Krajewski, and M. Westerman. 1998. Phylogeny of the dasyurid marsupial genus *Antechinus* based on cytochrome-*b*, 12S-rRNA, and protamine-P1 genes. Journal of Mammalogy, 79:1379-1389.

Árnason, Ú. 1977. The relationship between the four principal pinniped karyotypes. Hereditas, 87:227-242.

Árnason, U., and A. Gullberg. 1996. Cytochrome *b* nucleotide sequences and the identification of five primary lineages of extant Cetaceans. Molecular Biology and Evolution, 13(2):407-417.

Árnason, Ú., and B. Widegren. 1986. Pinniped phylogeny enlightened by molecular hybridizations using highly repetitive DNA. Molecular Biology and Evolution, 3:356-365.

Árnason, Ú., K. Bodin, A. Gullberg, C. Ledje, and S. Mouchaty. 1995. A molecular view of pinniped relationships with particular emphasis on the true seals. Journal of Molecular Evolution, 40:78-85.

Arnold, M. L., L. W. Robbins, R. K. Chesser, and J. C. Patton. 1983. Phylogenetic relationships among six species of *Reithrodontomys*. Journal of Mammalogy, 64:128-132.

Arnold, P., H. Marsh, and G. Heinsohn. 1987. The occurrence of two forms of minke whales in east Australian waters with a description of external characters and skeleton of the diminutive or dwarf form. Scientific Reports of the Whales Research Institute (Toyko), 38:1-46.

Arredondo, O. 1958. Aves gigantes de nuestra pasado prehistórico. El Cartero Cubano, La Habana, 17(7):10-12.

Arrighi, F. E., M. W. Sorenson, and L. R. Shirley. 1969. Chromosomes of the tree shrews (Tupaiidae). Cytogenetics, 8:199-208.

Arroyo, J., C. Rodriguez-Murcia, M. Delibes, and F. Hiraldo. 1982. Comparative karyotype studies between Spanish and French populations of *Eliomys quercinus* L. Genetica, 59:161-166.

Arroyo-Cabrales, J., and A. L. Gardner. 2003. The type specimen of *Anoura geoffroyi lasiopyga* (Chiroptera: Phyllostomidae). Proceedings of the Biological Society of Washington, 116:737-741.

Arroyo-Cabrales, J., and J. K. Jones, Jr. 1988*a*. *Balantiopteryx plicata*. Mammalian Species, 301:1-4.

Arroyo-Cabrales, J., and J. K. Jones, Jr. 1988*b*. *Balantiopteryx io* and *Balantiopteryx infusca*. Mammalian Species, 313:1-3.

Arroyo-Cabrales, J., R. R. Hollander, and J. K. Jones, Jr. 1987. *Choeronycteris mexicana*. Mammalian Species, 291:1-5.

Arroyo-Cabrales, J., R. A. Van Den Bussche, K. H. Sigler, R. A. Chesser, and R. J. Baker. 1997. Genic variation of mainland and island populations of *Natalus stramineus* (Chiroptera: Natalidae). Occasional Papers, Museum of Texas Tech University, 171:1-9.

Ascorra, C. F., D. E. Wilson, and A. L. Gardner. 1991. Geographic distribution of *Micronycteris schmidtorum* Sanborn (Chiroptera: Phyllostomidae). Proceedings of the Biological Society of Washington, 104:351-355.

Asher, R. J. 1999. A morphological basis for assessing the phylogeny of the "Tenrecoidea" (Mammalia, Lipotyphla). Cladistiscs, 15:231-252.

Asher, R. J. 2000. Phylogenetic history of Tenrecs and other insectivoran mammals. Ph.D. Dissertation, State University of New York, Stony Brook.

Asher, R. J. 2001. Cranial anatomy of tenrecid insectivorans: Character evolution across competing phylogenies. American Museum Novitates, 3352:1-54.

Asher, R. J., M. C. McKenna, R. J. Emry, A. R. Tabrum, and D. G. Kron. 2002. Morphology and relationships of *Apternodus* and other extinct, zalambdodont, placental mammals. Bulletin of the American Museum of Natural History, 273:1-117.

Asher, R J., M. J. Novacek, and J. H. Geisler. 2003. Relationships of endemic African mammals and their fossil relatives based on morphological and molecular evidence. Journal of Mammalian Evolution, 10:131-194.

Ashley, M. V., J. E. Norman, and L. Stross. 1996. Phylogenetic analysis of the perissodactylan family Tapiridae using mitochondrial cytochrome c oxidase (COII) sequences. Journal of Mammalian Evolution, 3:315-326.

Ashley, T., and K. Fredga. 1994. The curious normality of the synaptic association between the sex chromosomes of the two arvicoline rodents: *Microtus oeconomus* and *Clethrionomys glareolu*s. Hereditas, 120:105-111.

Associated Press, 2003. 'Garden of Eden' devastated under Saddam (The New York Times, April 29, 2003) and Iraqis long to restore Eden that Saddam destroyed (The Post and Courier, April 30, 2003).

Aswathanarayana, N. V. 1987. Evolutionary trends in the genus *Funabulus* (Rodentia—Sciuridae). Current Science (Bangalore), 56(24):1298-1301.

Aswathanarayana, N. V. 2000. Chromosomes and karyotype of the Indian pangolin, *Manis crassicaudata* Gray (Pholidota: Mammalia). Cytologia, 65(4):379-382.

Atallah, S. I. 1967. A new species of spiny mouse (*Acomys*) from Jordan. Journal of Mammalogy, 48:258-261.

Atallah, S. I. 1978. Mammals of the eastern Mediterranean region: Their ecology, systematics and zoogeographical relationships. Part 2. Säugetierkundliche Mitteilungen, 26:1-50.

Atallah, S. I., and D. L. Harrison. 1968. On the conspecificity of *Allactaga euphratica* Thomas, 1881 and *Allactaga williamsi* Thomas 1897 (Rodenta: Dipodidae) with a complete list of subspecies. Mammalia, 32:628-638.

Atanassov, N., and Z. Peschev. 1963. Die Säugetiere Bulgariens. Säugetierkundliche Mitteilungen, 12:101-112.

Atkins, D. L., and L. S. Dillon. 1971. Evolution of the cerebellum in the genus *Canis*. Journal of Mammalogy, 52:96-107.

Atkinson, I. A. E., and H. Moller. 1990. Kiore, Polynesian rat. Pp. 175-192, *in* The handbook of New Zealand mammals (C. M. King, ed.). Oxford University Press, Auckland, 600 pp.

Audebert, J. B. 1799. Histoire naturelle des singes et des makis. Desray, Paris, 172 pp.

Audet, A. M., C. B. Robbins, and S. Lariviere. 2002. *Alopex lagopus*. Mammalian Species, 713:1-10.

Audet, D., M. D. Engstrom, and M. B. Fenton. 1993. Morphology, karyology, and echolocation calls of *Rhogeessa* (Chiroptera: Vespertilionidae) from the Yucatán Peninsula. Journal of Mammalogy, 74:498-502.

Audubon, J. J., and J. Bachman. 1846 [1845]-1854. The viviparous quadrupeds of North America. Privately published by J. J. Audubon, New York, 1:1-389[1846], 2:1-334[1851], 3:1-348[1854].

Auffray, J.-C., and J. Britton-Davidian. 1992. When did the house mouse colonize Europe? Biological Journal of the Linnean Society, 45:187-190.

Auffray, J.-C., E. Tchernov, and E. Nevo. 1988. Origine du commensalisme de la souris domestique (*Mus musculus domesticus*) vis-à-vis de l'homme. Comptes Rendus de l'Académie des Sciences (Paris), ser. 3, 307(9):517-522.

Auffray, J.-C., K. Belkhir, J. Cassaing, J. Britton-Davidian, and H. Croset. 1990a. Outdoor occurrence in robertsonian and standard populations of the house mouse. Vie et Milieu, 40:111-118.

Auffray, J.-C., E. Tchernov, F. Bonhomme, G. Heth, S. Simson, and E. Nevo. 1990b. Presence and ecological distribution of *Mus "spretoides"* and *Mus musculus domesticus* in Israel. Circum-Mediterranean vicariance in the genus *Mus*. Zeitschrift für Säugetierkunde, 55:1-10.

Auffray, J.-C., F. Vanlerberghe, and J. Britton-Davidian. 1990c. The house mouse progression in Eurasia: A palaeontological and archaeozoological approach. Biological Journal of the Linnean Society, 41:13-25.

Auffray, J.-C., P. Alibert, and C. Latieule. 1996. Relative warp analysis of skull shape across the hybrid zone of the house mouse (*Mus musculus*) in Denmark. Journal of the Zoological Society of London, 240:441-445.

Auffray, J.-C., A. Orth, J. Catalan, J.-P. Gonzalez, E. Desmarais, and F. Bonhomme. 2003. Phylogenetic position and description of a new species of subgenus *Mus* (Rodentia, Mammalia) from Thailand. Zoologica Scripta, 32:119-127.

Aulagnier, S. 1991. Zoogéographie des rongeurs du Maroc: I. Gerbillidae et Muridae. Pp. 309-321, *in* Le rongeur et l'espace, Actes du Colloque International, Lyon, 1989 (M. Le Berre and L. Le Guelte, eds.). Raymond Chabaud, Paris, 362 pp.

Aulagnier, S., and M. Thévenot. 1986. Catalogue des mammifères sauvages du Maroc. Travaux de l'Institut Scientifique, Série Zoologie, 41:1-164.

Aulagnier, S., D. Barreau, and A. Rocher. 1993. *Dipodillus maghrebi* Schlitter et Setzer, 1972 et *Gerbillus campestris* Levaillant, 1857 (Rodentia, Gerbillidae) dans le nord du Maroc: Morphologie et biométrie crâniennes, éléments de répartition. Mammalia, 57(1):35-42.

Autino, A. G., G. L. Claps, and R. M. Barquez. 1999. Insectos ectoparásitos de murciélagos de las Yungas de la Argentina. Acta Zoológica Mexicana, n.s., 78:119-169.

Averianov, A. O. 1998. Podrodovaya sistematika zaitsev roda *Lepus* (*Lagomorpha, Leporidae*) [Sub-generic systematics of the hares, genus *Lepus* . . .]. Byulleten' Moskovskovo Obshchestva Ispytatelei Prirody, Otdel Biologicheskii, 103(1):3-8 (in Russian with English summary).

Averianov, A. O. 1999. Phylogeny and classification of Leporidae (Mammalia, Lagomorpha). Vestnik Zoologii, 33(1-2):41-48.

Averianov, A. O, A. V. Abramov, and A. N. Tikhonov. 2000. A new species of *Nesolagus* (Lagomorpha, Leporidae) from Vietnam with osteological description. Contributions from the Zoological Institute, St. Petersburg, No. 3, 22 pp.

Avery, D. M. 1977. Past and present distribution of some rodent and insectivore species in the southern Cape Province, South Africa: New information. Annals of the South African Museum, 74:201-209.

Avery, D. M. 1991. Late Quaternary incidence of some micromammalian species in Natal. Durban Museum Novitates, 16:1-11.

Avery, D. M. 1995. A preliminary assessment of the micromammalian remains from Gladysvale Cave, South Africa. Palaeontologia Africana, 32:1-10.

Avery, D. M. 1996. Late Quaternary micromammals from Mumbwa Caves, Zambia. Journal of African Zoology, 110(3):221-234.

Avery, D. M. 1998. An assessment of the lower Pleistocene micromammalian fauna from Swartkrans Members 1-3, Gauteng, South Africa. Geobios, 31(3):393-414.

Avery, D. M. 2000. Notes on the systematics of micromammals from Sterkfontein, Gauteng, South Africa. Palaeontologica Africana, 36:83-90.

Avila-Flores, R., J. J. Flores-Martínez, and J. Ortega. 2002. *Nyctinomops laticaudatus*. Mammalian Species, 697:1-6.

Avila-Pires, F. D., de. 1960a. Sobre *Oryzomys* do grupo *ratticeps*. Actas y Trabajos del Primer Congreso Sudamerican de Zoologia, sec. 5, 4:3-7.

Avila-Pires, F. D., de. 1960b. Um novo genero de roedor Sul-Americano. Boletim do Museu Nacional, Zoologia (Rio de Janeiro), 220:1-6.

Avila-Pires, F. D., de. 1960c. Roedores colecionados na regiao de Lagoa Santa, Minas Gerais Brasil. Arquivos do Museu Nacional, 50:25-46.

Avila-Pires, F. D., de. 1964. Mamíferos colecionados na região do Rio Negro (Amazonas, Brasil). Boletim do Museu Paraense Emílio Goeldi, n.s., Zoologia, 42:1-23.

Avila-Pires, F. D., de. 1965. The type specimens of Brazilian mammals collected by Prince Maximilian zu Wied. American Museum Novitates, 2209:1-21.

Avila-Pires, F. D., de. 1967. The type-locality of "Chaetomys subspinosus" (Olfers, 1818) (Rodentia, Caviomorpha). Revista Brasileira de Biologia, 27(2):177-179.

Avila-Pires, F. D., de. 1972. A new subspecies of Kunsia fronto (Winge, 1888) from Brazil (Rodentia, Cricetidae). Revista Brasileira de Biologia, 32:419-422.

Avila-Pires, F. D., de. 1985. On the validity and geographical distribution of Callithrix argentata emiliae Thomas, 1920 (Primates, Callithrichidae). Pp. 319-322, in A Primatologia no Brasil (M. T. de Mello, ed.). Instituto de Ciências Biológicas, Brasil, 530 pp.

Avila-Pires, F. D., de. 1994. Mamíferos descritos do Estado do Rio Grande do Sul, Brasil. Revista Brasiliera de Biologia, 54:367-384.

Avila-Pires, F. D., de, and M. R. C. Wutke. 1981. Taxonomia e evoluçao de Clyomys Thomas, 1916 (Rodentia, Echimyidae). Revista Brasileira de Biologia, 41(3):529-534.

Avise, J. C., M. H. Smith, and R. K. Selander. 1974a. Biochemical polymorphism and systematics in the genus Peromyscus. VI. The boylii species group. Journal of Mammalogy, 55:751-763.

Avise, J. C., M. H. Smith, R. K. Selander, T. E. Lawlor, and P. R. Ramsey. 1974b. Biochemical polymorphism and systematics in the genus Peromyscus. V. Insular and mainland species of the subgenus Haplomylomys. Systematic Zoology, 23:226-238.

Avise, J. C., M. H. Smith, and R. K. Selander. 1979. Biochemical polymorphism and systematics in the genus Peromyscus. VII. Geographic differentiation in members of the truei and maniculatus species groups. Journal of Mammalogy, 60:177-192.

Azara, F., de. 1801. Essais sur l'histoire naturelle des quadrupèdes de la province du Paraguay. Traduits sur le manuscrit inédit de l'auteur, Pra. M. L. E. Moreau-Saint-Méry. Charles Pougens, Paris, 1:1-366; 2:1-499.

Azara, F., de. 1802. Apuntamientos para la historia natural de los quadrupedos del Paraguay, y Rio de la Plata. En la imprenta de la viuda de Ibarra, Madrid, 1:1-318; 2:1-328.

Azzaroli, A. 1984. On some verteberate remains of middle Pleistocene age from the Upper Valdarno and Val di Chiana, Tuscany. Palaeontographica Italica, 73:104-115.

Azzaroli, L., and A. M. Simonetta. 1966. Carnivori della Somalia ex-Italiana. Monitore Zoologico Italiano Supplemento, 74:102-195.

Azzaroli-Puccetti, M. L. 1987a. On the hares of Ethiopia and Somalia and the systematic position of Lepus whytei Thomas, 1894 (Mammalia, Lagomorpha). Atti della Accademia Nazionale dei Lincei. Memorie, Classe di scienze fisiche, matematiche e naturali, ser. 8, vol. 19, sec. 3a, fasc. 1:1-19+ 5 pls.

Azzaroli-Puccetti, M. L. 1987b. The systematic relationships of hares (genus Lepus) of the horn of Africa. Cimbebasia, ser. A, 9:1-22.

Bâ, K., L. Granjon, R. Hutterer, and J.-M. Duplantier. 2000. Les micromammifères du Djoudj (Delta du Sénégal) par l'analyse du régime alimentaire de la chouette effraie, Tyto alba. Bonner Zoologische Beiträge, 49(1-4):31-38.

Bâ, K., C. Mathiot, M. Diallo, P. Nabeth, L. Lochouarn, Y. Kâne, M. O. Abdalahi, and L. Granjon. 2001. Preliminary study on some rodents of southern Mauritania as reservoir of human pathogenic viruses. Pp. 101-107, in African Small Mammals (C. Denys, L. Granjon, and A. Poulet, eds.). IRD Éditions, Collection Colloques et Séminaires, Paris, 570 pp.

Baagøe, H. J. 2001a. Myotis bechsteinii (Kuhl, 1818)—Bechsteinfledermaus. Pp. 443-447, in Handbuch der Säugetiere Europas. Band 4: Fledertiere. Teil I: Chiroptera I, Rhinolophidae, Vespertilionidae I (J. Niethammer and F. Krapp, eds.). Aula Verlag, Wiebelsheim. 476 pp.

Baagøe, H. J. 2001b. Vespertilio murinus Linnaeus, 1758—Zweifarbfledermaus. Pp. 473-514, in Handbuch der Säugetiere Europas. Band 4: Fledertiere. Teil I: Chiroptera I, Rhinolophidae, Vespertilionidae I (J. Niethammer and F. Krapp, eds.). Aula Verlag, Wiebelsheim., 476 pp.

Baagøe, H. J. 2001c. Eptesicus serotinus (Schreber, 1774)—Breitflügelfledermaus. Pp. 519-559, in Handbuch der Säugetiere Europas. Band 4: Fledertiere. Teil I: Chiroptera I, Rhinolophidae, Vespertilionidae I (J. Niethammer and F. Krapp, eds.). Aula Verlag, Wiebelsheim., 476 pp.

Bachmayer, F., and R. W. Wilson. 1970. Small mammals (Insectivora, Chiroptera, Lagomorpha, Rodentia) from the Kohfidish Fissures of Burgenland, Austria. Annalen des Naturhistorischen Museums in Wien, 74:533-587.

Bacon, P. J., J. F. Dallas, and S. B. Piertney. 1999. Post-glacial re-colonization of European biota. Biological Journal of the Linnean Society, 68:87-112.

Badr, F. M., and R. L. Asker. 1980. Prevalence of non-Robertsonian polymorphism in the gerbil *Gerbillus cheesmani* from Kuwait. Genetica, 52/53:17-22.

Bahadur, M., and T. Sharma. 1995. Heterochromatin and karyotype evolution in the speciating "*Mus terricolor* complex". Proceedings of the Zoological Society of Calcutta, 48(1):45-48.

Bailey, F. M. 1914. Exploration on the Tsangpo or Upper Brahmaputra. Geographical Journal, 44:341-364.

Bailey, F. M. 1915. Notes from southern Tibet. Journal of the Bombay Natural History Society, 24:72-78.

Bailey, V. E. 1900. Revision of American voles of the genus *Microtus*. North American Fauna, 17:1-88.

Bailey, V. E. 1902. Synopsis of the North American species of *Sigmodon*. Proceedings of the Biological Society of Washington, 15:101-116.

Bailey, V. E. 1906. Identity of *Thomomys umbrinus* (Richardson). Proceedings of the Biological Society of Washington, 19:3-6.

Bailey, V. E. 1915. Revision of the pocket gophers of the genus *Thomomys*. North American Fauna, 39:1-136.

Bailey, V. E. 1926. A biological survey of North Dakota. North American Fauna, 49:1-226.

Bailey, V. E. 1936. The mammals and life zones of Oregon. North American Fauna, 55:1-416.

Baird, S. F. 1857. Mammals: General report upon the zoology of the several Pacific railroad routes. Vol. 8, pt. 1, *in* Reports of explorations and surveys to ascertain the most practicable and economical route for a railroad from the Mississippi River to the Pacific Ocean. Senate executive document no. 78, Washington, D.C., 757 pp.

Baker, A. J., J. L. Eger, R. L. Peterson, and T. H. Manning. 1983. Geographic variation and taxonomy of arctic hares. Acta Zoologica Fennica, 174:45-48.

Baker, A. N. 1977. Spectacled porpoise *Phocoena dioptrica* new to the subantarctic Pacific Ocean. New Zealand Journal of Marine and Freshwater Research, 11(2):401-406.

Baker, A. N. 1985. Pygmy right whale—*Caperea marginata*. Pp. 345-354, *in* Handbook of marine mammals: The sirenians and baleen whales, (S. H. Ridgway and R. Harrison, eds.). Academic Press, London, 3:1-362.

Baker, A. N., A. N. H. Smith, and F. B. Pichler. 2002. Geographical variation in Hector's dolphin: Recognition of new subspecies of *Cephalorhynchus hectori*. Journal of The Royal Society of New Zealand, 32(4):713-727 (December 2002).

Baker, C. M. 1992. *Atilax paludinosus*. Mammalian Species, 408:1-6.

Baker, R. H. 1952. Geographic range of *Peromyscus melanophrys*, with description of a new subspecies. University of Kansas Publications, Museum of Natural History, 5:251-258.

Baker, R. H. 1954. The silky pocket mouse (*Perognathus flavus*) of México. University of Kansas Publications, Museum of Natural History, 7:339-347.

Baker, R. H. 1969. Cotton rats of the *Sigmodon fulviventer* group. Miscellaneous Publications, Museum of Natural History, University of Kansas, 51:177-232.

Baker, R. H., and K. A. Shump. 1978*a*. *Sigmodon fulviventer*. Mammalian Species, 94:1-4.

Baker, R. H., and K. A. Shump. 1978*b*. *Sigmodon ochrognathus*. Mammalian Species, 97:1-2.

Baker, R. J. 1984. A sympatric cryptic species of mammal: A new species of *Rhogeessa* (Chiroptera: Vespertilionidae). Systematic Zoology, 33:178-183.

Baker, R. J., and C. L. Clark. 1987. *Uroderma bilobatum*. Mammalian Species, 279:1-4.

Baker, R. J., and S. K. Davis. 1989. Ribosomal-DNA, mitochondrial-DNA, chromosomal, and allozymic studies on a contact zone in the pocket gopher, *Geomys*. Evolution, 43:63-75.

Baker, R. J., and T. C. Hsu. 1970. Chromosomes of the desert shrew, *Notiosorex crawfordi* (Coues). Southwestern Naturalist, 14:448-449.

Baker, R. J., and J. L. Patton. 1967. Karyotypes and karyotypic variation of North American vespertilionid bats. Journal of Mammalogy, 48:270-286.

Baker, R. J., and S. L. Williams. 1974. *Geomys tropicalis*. Mammalian Species, 35:1-4.

Baker, R. J., T. Mollhagen, and G. Lopez. 1971. Notes on *Lasiurus ega*. Journal of Mammalogy, 52:844-852.

Baker, R. J., P. V. August, and A. A. Steuter. 1978. *Erophylla sezekorni*. Mammalian Species, 115:1-5.

Baker, R. J., R. K. Barnett, and I. F. Greenbaum. 1979. Chromosomal evolution in grasshopper mice (*Onychomys*, Cricetidae). Journal of Mammalogy, 60:297-306.

Baker, R. J., B. F. Koop, and M. W. Haiduk. 1983*a*. Resolving systematic relationships with G-bands: A study of five genera of South American cricetine rodents. Systematic Zoology, 32:403-416.

Baker, R. J., L. W. Robbins, F. B. Stangl, Jr., and E. C. Birney. 1983*b*. Chromosomal evidence for a major subdivision in *Peromyscus leucopus*. Journal of Mammalogy, 64:356-359.

Baker, R. J., J. W. Bickham, and M. L. Arnold. 1985. Chromosomal evolution in *Rhogeessa* (Chiroptera: Vespertilionidae): Possible speciation by centric fusions. Evolution, 39:233-243.

Baker, R. J., J. C. Patton, H. H. Genoways, and J. W. Bickham. 1988*a*. Genic studies of *Lasiurus* (Chiroptera: Vespertilionidae). Occasional Papers, The Museum, Texas Tech University, 117:1-15.

Baker, R. J., C. G. Dunn, and K. Nelson. 1988b. Allozymic study of the relationships of *Phylloderma* and four species of *Phyllostomus*. Occasional Papers, The Museum, Texas Tech University, 125:1-14.

Baker, R. J., M. B. Qumsiyeh, and I. L. Rautenbach. 1988c. Evidence for eight tandem and five centric fusions in the evolution of the karyotype of *Aethomys namaquensis* A. Smith (Rodentia: Muridae). Genetica, 76:161-169.

Baker, R. J., C. S. Hood, and R. L. Honeycutt. 1989. Phylogenetic relationships and classification of the higher categories of the New World bat family Phyllostomidae. Systematic Zoology, 38:228-238.

Baker, R. J., V. A. Taddei, J. L. Hudgeons, and R. A. Van Den Bussche. 1994. Systematic relationships within *Chiroderma* (Chiroptera: Phyllostomidae) based on cytochrome-*b* sequence variation. Journal of Mammalogy, 75:321-327.

Baker, R. J., C. A. Porter, J. C. Patton, and R. A. Van Den Bussche. 2000. Systematics of bats of the family Phyllostomidae based on *RAG2* DNA sequences. Occasional Papers, The Museum of Texas Tech University, 201:1-16.

Baker, R. J., S. Solari, and F. G. Hoffmann. 2002. A new Central American species from the *Carollia brevicauda* complex. Occasional Papers, The Museum of Texas Tech University, 217:1-12.

Baker, R. J., M. B. O'Neill, and L. R. McAliley. 2003a. A new species of desert shrew, *Notiosorex*, based on nuclear and mitochondrial sequence data. Occasional Papers, Museum of Texas Tech University, no. 222:i+1-12.

Baker, R. J., L. C. Bradley, R. D. Bradley, J. W. Dragoo, M. D. Engstrom, R. S. Hoffmann, C. A. Jones, F. Reid, D. W. Rice, and C. Jones. 2003b. Revised checklist of North American mammals north of Mexico, 2003. Occasional Papers Museum of Texas Tech University, 229:1-23.

Bakó, B., G. Csorba and L. Berty. 1998. Distribution and ecological requirements of dormouse species occurring in Hungary. Natura Croatica, 7(1):1-9.

Balajee, A. S., and T. Sharma. 1994. The chromosomal distribution of *Mus musculus*-like AT-rich heterochromatin in the *M. dunni* complex as revealed by *AluI* digestion of metaphase chromosomes. Cytogenetics and Cell Genetics, 66:89-92.

Balakrishnan, C., S. L. Monfort, A. Gaur, L. Singh, and M. D. Sorenson. 2003. Phylogeography and conservation genetics of Eld's deer (*Cervus eldi*). Molecular Ecology, 12:1-10.

Balbontin, J., S. Reig, and S. Moreno. 1996. Evolutionary relationships of *Ctenomys* (Rodentia: Octodontidae) from Argentina, based on penis morphology. Acta Theriologica, 41(3):237-253.

Balciauskas, L. 1996. Lithuanian mammal fauna review. Hystrix, n.s., 8(1-2):9-15.

Balcomb, K. C. 1989. Baird's beaked whale, *Berardius bairdii* Stejneger, 1883: Arnoux's beaked whale *Berardius arnouxii* Duvernoy, 1851. Pp. 261-288, *in* Handbook of marine mammals: River dolphins and the larger toothed whales (S. H. Ridgway and R. Harrison, eds.). Academic Press, New York, 4:1-442.

Balete, D. S., and L. R. Heaney. 1997. Density, biomass, and movement estimates for murid rodents in mossy forest on Mount Isarog, southern Luzon, Philippines. Ecotropica, 3:91-100.

Baloutch, M. 1978. Etude d'une collection de lièvres d'Iran et description de quatre formes nouvelles. Mammalia, 42:441-452.

Banack, S. A. 2001. *Pteropus samoensis*. Mammalian Species, 661:1-4.

Banerjee, S., and M. Ghosh. 1981. Prehistoric fauna of Kausarabi, near Allahabad, U. P., India. Records of the Zoological Survey of India, 78:113-119.

Banfield, A. W. F. 1961. A revision of the reindeer and caribou. National Museum of Canada Bulletin, 177:1-137.

Banfield, A. W. F. 1963. The post glacial dispersal of American caribou. Proceedings of the 16th International Congress of Zoology, Washington D.C., 1:206.

Banfield, A. W. F. 1974. The mammals of Canada. The University of Toronto Press, Toronto, 438 pp.

Bangs, O. 1902. Chiriqui Mammalia. Bulletin of the Museum of Comparative Zoology at Harvard College, 39:17-51.

Banks, R. C. 1967. The *Peromyscus guardia-interparietalis* complex. Journal of Mammalogy, 48:210-218.

Banks, S., and A. Rushton. 1998. Eastern range extension of Forrest's mouse (*Leggadina forresti*). Memoirs of the Queensland Museum, 42(2):386.

Bannikov, A. G. 1954. Mlekopitayushchie Mongol'skoi Narodnoi Respubliki. [Mammals of the Mongolian People's Republic]. Akademiya Nauk SSSR, Moscow, 669 pp. (in Russian).

Bannikova, A. A., V. A. Dolgov, L. V. Fedorova, A. N. Fedorov, A. V. Troitsky, A. A. Lomov, and B. M. Mednikov. 1996. Taxonomic relationships among hedgehogs of the subfamily Erinaceinae (Mammalia, Insectivora) determined basing on the data of restriction endonuclease analysis of total DNA. Zoologicheskii Zhurnal, 74:95-106.

Bannikova, A. A., V. Y. Oleinichenko, A. A. Lomov, and V. A. Dolgov. 2001a. On taxonomic relationships between *Crocidura suaveolens* and *Crocidura gueldenstaedtii* (Insectivora, Soricidae). Zoologicheskii Zhurnal, 80:721-730.

Bannikova, A. A., L. A. Lavrenchenko, A. A. Lomov, and B. M. Mednikov. 2001*b*. Molecular diversity of some *Crocidura* species (Insectivora, Soricidae) from Ethiopia. Pp. 55-64, *in* African small mammals—Petits mammifères africains (C. Denys, L. Granjon, and A. Poulet, eds.). Collection Colloques et Séminaires, IRD Editions, Paris, 570 pp.

Bannikova, A. A., V. A. Matveev, and D. A. Kramerov. 2002. Using inter-SINE-PCR to study mammalian phylogeny. Russian Journal of Genetics, 38:714-724.

Banwell, D. B. 1997. The Pannonians—*Cervus elaphus pannoniensis*—a race apart. Deer, 10:275-277.

Banwell, D. B. 1998. Identification of the Pannonian or Danubian, red deer. A maraloid—*Cervus elaphus pannnoniensis*. Deer, 10:495-497.

Banwell, D. B. 1999. The sika. Halcyon Press, Auckland, New Zealand. 137 pp.

Banwell, D. B. 2002. In defence of the Pannonian *Cervus elaphus pannoniensis*. Deer, 12:198-203.

Bao, Y., W. Du, Y. Lin, and W. Jin. 1998. [Energy metabolism and the morphology of digestive tract in *Rattus niviventer confucianus* and *Rattus norvegicus*.] Acta Theriologica Sinica, 18(3):202-207 (in Chinese, with English abstract).

Barabasch-Nikiforof, I. I. 1975. Die Desmane. A. Ziemsen Verlag, Wittenberg-Lutherstadt, 100 pp.

Baranova, G. I., and M. V. Zaitsev. 2003. A new name for the Ussurian subspecies of the tundra shrew, *Sorex tundrensis* Merriam, 1900 (Mammalia: Soricidae). Zoosistematica Rossica, 11[2002]:403-404.

Baranova, G. I., A. A. Gureev, and P. P. Strelkov. 1981. [Catalogue of type specimens in the collection of the Zoological Institute of the USSR, mammals (Mammalia). Part 1. shrews (Insectivora), bats (Chiroptera), hares (Lagomorpha)]. Nauka, Leningrad, 22 pp. (in Russian).

Barbehenn, K. R. 1974. Recent invasions of Micronesia by small mammals. Micronesica, 10(1):41-50.

Barbehenn, K. R., J. P. Sumangil, and J. L. Libay. 1972-1973. Rodents of the Philippine croplands. Philippine Agriculturist, 56:217-242.

Barghorn, S. F. 1977. New material of *Vespertiliavus* Schlosser (Mammalia: Chiroptera) and suggested relationships. American Museum Novitates, 2618:1-29.

Barkley, L. J. 1984. Evolutionary relationships and natural history of *Tomopeas ravus* (Mammalia: Chiroptera). Masters thesis, Louisiana State University, Baton Rouge, Louisiana.

Barko, V. A., and G. A. Feldhammer. 2002. Cotton mice (*Peromyscus gossypinus*) in southern Illinois: Evidence for hybridization with white-footed mice (*Peromyscus leucopus*). American Midland Naturalist, 147:109-115.

Barnaby, D. 2001. Ward's zebra. The Bartlett Society Journal, 12:32-36.

Barnes, L. G. 1978. A review of *Lophocetus* and *Liolithax* and their relationships to the delphinoid family Kentriodontidae (Cetacea: Odontoceti). Bulletin of the Los Angeles County Museum of Natural History, 28:1-35.

Barnes, L. G. 1985. Evolution, taxonomy and antitropical distribution of the porpoises (Phocoenidae, Mammalia). Marine Mammal Science, 1(2):149-163.

Barnes, L. G. 1989. A new Enaliarctine pinniped from the Astoria formation, Oregon, and a classification of the Otariidae (Mammalia: Carnivora). Contributions in Science, Natural History Museum of Los Angeles County, 403:1-26.

Barnes, L. G., and S. A. McLeod. 1984. The fossil record and phyletic relationships of gray whales. Pp. 3-32, *in* The gray whale (M. L. Jones, S. L. Swartz, and S. Leatherwood, eds.). Academic Press, New York, 600 pp.

Barnes, L. G., D. P. Domning, and C. E. Ray. 1985. Status of studies on fossil marine mammals. Marine Mammal Science, 1(1):15-53.

Barnett, A. A. 1997. The ecology and natural history of a fishing mouse *Chibchanomys* spec. nov. (Ichthyomyini: Muridae) from the Andes of southern Ecuador. Zeitschrift für Säugertierkunde, 62:43-52.

Barnett, A. A., M. Prangley, P. V. Hayman, D. Diawara, and J. Koman. 1996. A survey of the mammals of the Kounounkan Massif, south-western Guinea, West Africa. Journal of African Zoology, 110:235-240.

Barnosky, A. D. 1993. Mosaic evolution at the population level in *Microtus pennsylvanicus*. Pp. 24-59, *in* Morphological change in Quaternary mammals of North America (R. A. Martin and A. D. Barnosky, eds.). Cambridge University Press, Cambridge, ix + 415 pp.

Barnosky, E. H. 1994. Ecosystem dynamics through the past 2000 years as revealed by fossil mammals from Lamar Cave in Yellowstone National Park, USA. Historical Biology, 8:71-90.

Barome, P.-O., M. Monnerot, and J.-C. Gautun. 1998. Intrageneric phylogeny of *Acomys* (Rodentia, Muridae) using mitochondrial gene cytochrome *b*. Molecular Phylogenetics and Evolution, 9:560-566.

Barome, P.-O., M. Monnerot, and J.-C. Gautun. 2000. Phylogeny of the genus *Acomys* (Rodentia, Muridae) based on the cytochrome *b* mitochondrial gene: Implications on taxonomy and phylogeography. Mammalia, 64(4):423-438.

Barome, P.-O., P. Lymberakis, M. Monnerot, and J.-C. Gautun. 2001*a*. Cytochrome *b* sequences reveal *Acomys minous* (Rodentia, Muridae) paraphyly and answer the question about the ancestral karyotype of *Acomys dimidiatus*. Molecular Phylogenetics and Evolution, 18(1):37-46.

Barome, P.-O., V. Volobouev, M. Monnerot, J. K. Mfune, W. Chitaukali, J.-C. Gautun, and C. Denys. 2001*b*. Phylogeny of *Acomys spinosissimus* (Rodentia, Muridae) from north Malawi and Tanzania: Evidence from morphological and molecular analysis. Biological Journal of the Linnean Society, 73:321-340.

Bárquez, R. M. 1983. La distribucion de *Neotomys ebriosus* Thomas en la Argentina y su presencia en la provincia de San Juan (Mammalia, Rodentia, Cricetidae). Historia Natural, 3:189-191.

Barquez, R. M. 1987. Los murciélagos de Argentina. Ph.D. dissertation, Facultad de Ciencias Naturales e Instituto Miguel Lillo, Universidad Nacional de Tucumán, Argentina, 525 pp.

Barquez, R. M., and M. M. Diaz. 2001. Bats of the Argentine Yungas: A systematic and distributional analysis. Acta Zoologica Mexicana, n.s., 82:29-81.

Bárquez, R. M., and R. A. Ojeda. 1979. Nueva subespecie de *Phylloderma* (Chiroptera, Phyllostomidae). Neotropica, 25:83-89.

Barquez, R. M., and R. A. Ojeda. 1992. The bats (Mammalia: Chiroptera) of the Argentine Chaco. Annals of the Carnegie Museum, 61(3):239-261.

Bárquez, R. M., D. F. Williams, M. A. Mares, and H. H. Genoways. 1980. Karyology and morphometrics of three species of *Akodon* (Mammalia: Muridae) from northwestern Argentina. Annals of Carnegie Museum, 49:379-403.

Barquez, R. M., N. P. Giannini, and M. A. Mares. 1993. Guide to the bats of Argentina. Oklahoma Museum of Natural History, Norman, 119 pp.

Barquez, R. M., M. A. Mares, and J. K. Braun. 1999. The bats of Argentina. Special Publication of Texas Tech University and the Oklahoma Museum of Natural History, 275 pp.

Barrantes, G. E., M. O. Ortells, and O. A. Reig. 1993. New studies on allozyme genetic distance and variability in akodontine rodents (Cricetidae) and their systematic implications. Biological Journal of the Linnean Society, 48:283-298.

Barratt, E. M., M. W. Bruford, T. M. Burland, G. Jones, P. A. Racey, and R. K. Wayne. 1995. Characterization of mitochondrial DNA variability within the microchiropoteran genus *Pipistrellus*: Approaches and applications. Symposia of the Zoological Society of London, 67:377-386.

Barratt, E. M., R. Deaville, T. M. Burland, M. W. Bruford, G. Jones, P. A. Racey, and R. K. Wayne. 1997. DNA answers the call of pipistrelle bat species. Nature, 387:138-139.

Barrett-Hamilton, G. E. H. 1900. On geographical and individual variation in *Mus sylvaticus* and its allies. Proceedings of the Zoological Society of London, 1900:387-428.

Barrett-Hamilton, G. E. H. 1902. Mammalia. Report on the collections of natural history made in the Antarctic regions during the voyage of the "Southern Cross". British Museum (Natural History), London, 1:1-66.

Barros, M. A., R. C. Liascovich, L. Gonzalez, M. S. Lizarralde, and O. A. Reig. 1990. Banding pattern comparison between *Akodon iniscatus* and *Akodon puer* (Rodentia, Cricetidae). Zeitschrift für Säugetierkunde, 55:115-127.

Barros, M. A., O. A. Reig, and A. Perez-Zapata. 1992. Cytogenetics and karyosystematics of South American oryzomyine rodents (Cricetidae: Sigmodontinae). IV. Karyotypes of Venezuelan, Trinidadian, and Argentinian water rats of the genus *Nectomys*. Cytogenetics and Cell Genetics, 59:34-38.

Barry, R., and J. Shoshani. 2000. *Heterohyrax brucei*. Mammalian Species, 645:1-7.

Bárta, Z., and P. Benda. 1998. [The distribution of the black-striped field mouse (*Apodemus agrarius*) in the border area of North Bohemia (Czech Republic)]. Lynx (Praha), n.s., 29:7-10 (in Czech with English summary).

Bartels, M., Jr. 1937. Zur Kentnis der Verbreitung und der Lebensweise javanischer Säugetiere. Treubia, 16:149-164.

Bartels, M. A., and D. P. Thompson. 1993. *Spermophilus lateralis*. Mammalian Species, 440:1-8.

Bartig, J. L., T. L. Best, and S. L. Burt. 1993. *Tamias bulleri*. Mammalian Species, 438:1-4.

Baryshnikov, G. F. 2000. A new subspecies of the honey badger *Mellivora capensis* from Central Asia. Acta Theriologica, 45:45-55.

Baryshnikov, G. F., and A. V. Abramov. 1992. On systematic position of *Canis ekloni*. Zoologicheskii Zhurnal, 71(11):121-127.

Baryshnikov, G. F., and A. V. Abramov. 1997. [Structure of baculum (*os penis*) in Mustelidae (Mammalia: Carnivora). Communication 1.] Zoologicheskii Zhurnal, 76:1399-1410 (in Russian).

Baryshnikov, G. F., and A. V. Abramov. 1998. [Structure of baculum (*os penis*) in Mustelidae (Mammalia: Carnivora). Communication 2.] Zoologicheskii Zhurnal, 77:231-236 (in Russian).

Baryshnikov, G. F., and O. R. Potapova. 1990. [Variability of the dental system in badgers (*Meles*, Carnivora) in the USSR fauna.] Zoologicheskii Zhurnal, 69:84-97 (in Russian).

Baryshnikov, G. F., and A. Tikhonov. 1994. Notes on skulls of Pleistocene saiga of northern Eurasia. Historical Biology, 8:209-234.

Baskevich, M. I. 1996a. On morphologically similar species in the genus *Sicista* (Rodentia, Dipodoidea). Bonner Zoologische Beiträge, 46(1-4):133-140.

Baskevich, M. I. 1996b. [On the karyological differentiation in Caucasian population of common vole (Rodentia, Cricetidae, *Microtus*).] Zoologicheskii Zhurnal, 75(2):297-308 (in Russian).

Baskevich, M. I. 1997. [A comparative analysis of structural features of spermatozoa and karyotypes in three species of shrub voles: *Terricola majori*, *T. daghestanicus* and *T. subterraneus* (Rodentia, Cricetidae) from the former USSR.] Zoologicheskii Zhurnal, 76(5):597-607 (in Russian with English summary).

Baskevich, M. I., and G. V. Kuznetsov. 1998. [Cytogenetic differentiation of species from the genus *Rattus* (Rodentia, Muridae) from the Dalat Plateau in southern Vietnam.] Zoologicheskii Zhurnal, 77(11):1321-1328 (in Russian with English abstract).

Baskevich, M. I., and G. V. Kuznetsov. 2000. Preliminary chromosomal results of *Niviventer* Marshall, 1976 (Mammalia, Rodentia, Muridae) from the Dalat Plateau in southern Vietnam. Pp. 351-356, *in* Isolated vertebrate communities in the tropics (G. Rheinwald, ed.). Bonner Zoologische Monographien, 46, 400 pp.

Baskevich, M. I., and L. A. Lavrenchenko. 1995. On the morphology of spermatozoa in some African murines (Rodentia, Muridae): The taxonomic and phylogenetic aspects. Journal of Systematic and Evolutionary Research, 33:9-16.

Baskevich, M. I., and L. A. Lavrenchenko. 2000. Review of karyological studies and the problems of systematics of Ethiopian *Arvicanthis* Lesson, 1842 (Rodentia: Muridae). Pp. 209-215, *in* Isolated vertebrate communities in the tropics (G. Rheinwald, ed.). Bonner Zoologische Monographie, 46, 400 pp.

Baskevich, M. I., and V. N. Orlov. 1993. [Karyological differentiation of rats of the genus *Mastomys* (Rodentia, Muridae) from the central part of Ethiopian Rift Valley.] Zoologicheskii Zhurnal, 72(2):112-121 (in Russian with English summary).

Baskevich, M. I., I. V. Lukyanova, and Yu. M. Koval'skaya. 1984. [Distribution of two forms of bush voles (*P. majori* Thom. and *P. dagestanicus* Shidl.) in the Caucasus]. Byulleten' Moskovskovo Obshchestva Ispytatelei Prirody, Otdel Biologicheskii, 89:29-33 (in Russian).

Baskevich, M. I., V. N. Orlov, and A. Mebrat. 1992. [Notes on karyology of *Tachyoryctes splendens* (Rodentia, Rhizomyidae) from Ethiopia.] Zoologicheskii Zhurnal, 71(11):108-114 (in Russian, with English summary).

Baskevich, M. I., V. N. Orlov, A. Bekele, and A. Mebrate. 1993. Notes on the karyotype of *Tachyoryctes splendens* (Rüppell 1836) (Rodentia Rhizomyidae) from Ethiopia. Tropical Zoology, 6:81-88.

Baskin, J. A. 1978. *Bensonomys*, *Calomys*, and the origin of the phyllotine group of neotropical cricetines (Rodentia: Cricetidae). Journal of Mammalogy, 59:125-135.

Baskin, J. A. 1982. Tertiary Procyoninae (Mammalia: Carnivora) of North America. Journal of Vertebrate Paleontology, 2:71-93.

Baskin, J. A. 1986. The late Miocene radiation of neotropical sigmodontine rodents in North America. Contributions to Geology, University of Wyoming, Special Paper, 3:287-303.

Baskin, J. A. 1989. Comments on New World Tertiary Procyonidae (Mammalia: Carnivora). Journal of Vertebrate Paleontology, 9:110-117.

Bate, D. M. A. 1920. Note on a new vole and other remains from the Ghar Dalam Cavern, Malta. Geological Magazine, 57:208-211.

Bate, D. M. A. 1935. Two new mammals from the Pleistocene of Malta, with notes on the associated fauna. Proceedings of the Zoological Society of London, 1935:247-264.

Bate, D. M. A. 1942a. New Pleistocene Murinae from Crete. Annals and Magazine of Natural History, ser. 11, 9:41-49.

Bate, D. M. A. 1942b. Pleistocene Murinae from Palestine. Annals and Magazine of Natural History, ser. 11, 9:465-487.

Bates, P. J. J. 1985. Studies of gerbils of genus *Tatera*: The specific distinction of *Tatera robusta* (Cretzschmar, 1826), *Tatera nigricauda* (Peters, 1878) and *Tatera phillipsi* (De Winton, 1898). Mammalia, 49:37-52.

Bates, P. J. J. 1988. Systematics and zoogeography of *Tatera* (Rodentia: Gerbillinae) of north-east Africa and Asia. Bonner Zoologische Beiträge, 39:265-303.

Bates, P. J. J. 1994. The distribution of *Acomys* (Rodentia: Muridae) in Africa and Asia. Israel Journal of Zoology, 40:199-214.

Bates, P. J. J., and D. L. Harrison. 1997. Bats of the Indian Subcontinent. Harrison Zoological Museum, 258 pp.

Bates, P. J. J., D. L. Harrison, and M. Mundi. 1994. The bats of western India revisited. Part 2. Journal of the Bombay Natural History Society, 91:224-240.

Bates, P. J. J., D. L. Harrison, P. D. Jenkins, and J. L. Walston. 1997. Three rare species of *Pipistrellus*

(Chiroptera: Vespertilionidae) new to Vietnam. Acta Zoologica Academiae Scientiarum Hungaricae, 43:359-374.

Bates, P. J. J., D. K. Hendrichsen, J. L. Walston, and B. Hayes. 1999. A review of the mouse-eared bats (Chiroptera: Vespertilionidae: *Myotis*) from Vietnam with significant new records. Acta Chiropterologica, 1:47-74.

Bates, P. J. J., T. Nwe, K. M. Swe, and Si S. H. Bu. 2001. Further new records of bats from Myanmar (Burma), including *Craseonycteris thonglongyai* Hill 1974 (Chiroptera: Craseonycteridae). Acta Chiropterologica, 3:33-42.

Batzli, G. O. 1999. Brown lemming/*Lemmus sibiricus*. Pp. 653-654, *in* The Smithsonian book of North American mammals (D. E. Wilson and S. Ruff, eds.). Smithsonian Institution Press, Washington, D.C., 750 pp.

Bauchau, V., and J. Chaline. 1987. Variabilite de la troisieme molaire superieure de *Clethrionomys glareolus* (Arvicolidae, Rodentia) et sa signification evolutive. Mammalia, 51:587-598.

Baud, F. J., and H. Menu. 1993. Paraguayan bats of the genus *Myotis*, with a redefinition of *M. simus* (Thomas, 1901). Revue Suisse de Zoologie, 100(3):595-607.

Bauer, K. 1960. Die Säugetiere des Neusiedlersee-Gebietes (Österreich). Bonner Zoologische Beiträge, 11:141-344.

Bauer, K. 2000. Evolution und ausbreitungsgeschichte von *Mus spicilegus* Petényi, 1882 und *Mus musculus* Linnaeus, 1758. Pp. 89-108, *in* Beiträge zur Kenntnis der Ährenmaus *Mus spicilegus* Petényi, 1882 (K. Unterholzner, R. Willenig, and K. Bauer, eds.). Biosystematics and Ecology Series No. 17, Österreichische Akademie der Wissenschaften, Wien, 108 pp.

Bauer, K., and F. Spitzenberger. 1996. The Recent mammal fauna of Austria. Hystrix, n.s., 8(1-2):17-21.

Bauer, K., F. Spitzenberger, and K. Unterholzner. 1998. The mound-building mouse (*Mus spicilegus*) is part of the Slovakian fauna. Folia Zoologica, 47(2):158-160.

Baumgardner, G. D. 1991. *Dipodomys compactus*. Mammalian Species, 369:1-4.

Baumgardner, G. D., and M. L. Kennedy. 1994. Patterns of interspecific morphometric variation in kangaroo rats (genus *Dipodomys*). Journal of Mammalogy, 75:203-211.

Baumgardner, G. D., and D. J. Schmidly. 1981. Systematics of the southern races of two species of kangaroo rats (*Dipodomys compactus* and *D. ordii*). Occasional Papers, The Museum, Texas Tech University, 73:1-27.

Baumstart, A., M. Akhverdyan, A. Schulze, I. Reisert, W. Vogel, and W. Just. 2001. Exclusion of *SOX9* as the testis determining factor in *Ellobius lutescens*: Evidence for another testis determining gene besides *SRY* AND *SOX9*. Molecular Genetics and Metabolism, 72:61-66.

Baverstock, P. R. 1984. Australia's living rodents: A restrained explosion. Pp. 913-919, *in* Vertebrate Zoogeography and Evolution in Australasia (M. Archer and G. Clayton, eds.). Hesperian Press, Carlisle, W. A., Australia, 1203 pp.

Baverstock, P. R., J. T. Hogarth, S. Cole, and J. Covacevich. 1976*a*. Biochemical and karyotypic evidence for the specific status of the rodent *Leggadina lakedownensis* Watts. Transactions of the Royal Society of South Australia, 100:109-112.

Baverstock, P. R., C. H. S. Watts, and J. T. Hogarth. 1976*b*. Heterochromatin variation in the Australian rodent *Uromys caudimaculatus*. Chromosoma, 57:397-403.

Baverstock, P. R., C. H. S. Watts, and S. R. Cole. 1977*a*. Electrophoretic comparisons between allopatric populations of five Australian pseudomyine rodents (Muridae). Australian Journal of Biological Science, 30:471-485.

Baverstock, P. R., C. H. S. Watts, and S. R. Cole. 1977*b*. Inheritance studies of glucose phosphate isomerase, transferrin and esterases in the Australian hopping mice *Notomys alexis*, *N. cervinus*, *N. mitchellii*, and *N. fuscus* (Rodentia: Muridae). Animal Blood Groups and Biochemical Genetics, 8:3-12.

Baverstock, P. R., C. H. S. Watts, and J. T. Hogarth. 1977*c*. Chromosome evolution in Australian rodents. I. The Pseudomyinae, the Hydromyinae and the *Uromys/Melomys* group. Chromosoma, 61:95-125.

Baverstock, P. R., C. H. S. Watts, J. T. Hogarth, A. C. Robinson, and J. F. Robinson. 1977*d*. Chromosome evolution in Australian rodents. II. The *Rattus* group. Chromosoma, 61:227-241.

Baverstock, P. R., C. H. S. Watts, and J. T. Hogarth. 1977*e*. Polymorphic patterns of heterochromatin distribution in Australian hopping mice, *Notomys alexis*, *N. cervinus* and *N. fuscus* (Rodentia, Muridae). Chromosoma, 61:243-256.

Baverstock, P. R., C. H. S. Watts, M. Adams, and M. Gelder. 1980. Chromosomal and electrophoretic studies of Australian *Melomys* (Rodentia: Muridae). Australian Journal of Zoology, 28:553-574.

Baverstock, P. R., C. H. S. Watts, M. Adams, and S. R. Cole. 1981. Genetical relationships among Australian rodents (Muridae). Australian Journal of Zoology, 29:289-303.

Baverstock, P. R., M. Gelder, and A. Jahnke. 1982. Cytogenetic studies of the Australian rodent, *Uromys caudimaculatus*, a species showing extensive heterochromatin variation. Chromosoma, 84:517-533.

Baverstock, P. R., M. Gelder, and A. Jahnke. 1983a. Chromosome evolution in Australian *Rattus*—G-banding and hybrid meiosis. Genetica, 60:93-103.

Baverstock, P. R., C. H. S. Watts, M. Gelder, and A. Jahnke. 1983b. G-banding homologies of some Australian rodents. Genetica, 60:105-117.

Baverstock, P. R., M. Adams, L. R. Maxson, and T. H. Yosida. 1983c. Genetic differentiation among karyotypic forms of the black rat, *Rattus rattus*. Genetics, 105:969-983.

Baverstock, P. R., M. Adams, and C. H. S. Watts. 1986. Biochemical differentiation among karyotypic forms of Australian *Rattus*. Genetica, 71:11-22.

Baynes, A., and B. Jones. 1993. The mammals of Cape Range peninsula, north-western Australia. Records of the Western Australian Museum, Supplement 45:207-225.

Baynes, A., A. Chapman, and A. J. Lynam. 1987. The rediscovery, after 56 years, of the heath rat *Pseudomys shortridgei* (Thomas, 1907)(Rodentia: Muridae) in Western Australia. Records of the Western Australian Museum, 13:319-322.

Bazan-Kubik, I., and Z. Skrzypiec. 1992. Recherches morphohistologiques sur le thymus chez *Apodemus agrarius* Pall. Dans le cycle vital. Annales Universitatis Mariae Curie-Sklodowska Lublin–Polonia, 67(5, Sectio C):55-73.

Bazhanov, V. S. 1944. [Marmot hybrids (On the question of interspecific hybridization in nature)]. Doklady Akademii Nauk SSSR, 42(7):307-308 (in Russian).

Bazhanov, V. S. 1951. [Some peculiariries of rodents belonging to the family Seleveniidae and endemic to Kazakhstan]. Doklady Akademii Nauk SSSR, Moscow, 80(3):469-472 (in Russian).

Bazhanov, V. S., and B. A. Belosludov. 1941. A remarkable family of rodents from Kasakhstan, USSR. Journal of Mammalogy, 22(3):311-315.

Beard, K. C. 1998. East of Eden: Asia as an important center of taxonomic origination in mammalian evolution. Pp. 5-39, *in* Dawn of the Age of Mammals in Asia (K. C. Beard and M. R. Dawson, eds.) Bulletin of Carnegie Museum of Natural History, 34:348 pp.

Bearder, S. K., P. E. Honess, and L. Ambrose. 1995. Species diversity among galagos with special reference to mate recognition. Pp. 331-352, *in* Creatures of the dark: The nocturnal prosimians. (L. Altermann, G. A. Doyle, and M. K. Izard, eds.). Plenum Press, New York, 571 pp.

Beaubrun, P.-Ch. 1990. Un cétacé nouveau pour les côtes sud-marocaines: *Sousa teuzii* (Kükenthal, 1892). Mammalia, 54(1):162-164.

Beaufort, F., de. 1963. Les cricetines de Galapagos. Valeur du genre *Nesoryzomys*. Mammalia, 27:338-340.

Beaufort, F., de. 1965. Répartition et taxinomie de la Poiane (Viverridae). Mammalia, 29(2):275-280.

Beaufort, F. de, H. Maurin, and P. Haffner. 1996. Terrestrial mammal fauna and threatened species in France. Hystrix, n.s., 8(1-2):23-29.

Beaumont, G., de. 1964. Remarques sur la classification des Felidae. Eclogae Geologicae Helvetiae, 57:837-845.

Beavan-Athfield, N., and R. J. Sparks. 2001a. Dating of *Rattus exulans* bone from Pleasant River (Otago, New Zealand): Testing the effect of burial contamination. Journal of The Royal Society of New Zealand, 31(4):795-800.

Beavan-Athfield, N., and R. J. Sparks. 2001b. Dating of *Rattus exulans* and bird bone from Pleasant River (Otago, New Zealand): Radiocarbon anomalies from diet. Journal of The Royal Society of New Zealand, 31(4):801-809.

Bechthold, G. 1939. Die asiatischen Formen der Gattung *Herpestes*. Zeitschrift für Säugetierkunde, 14:113-219.

Becker, K. 1978a. *Rattus rattus* (Linnaeus, 1758)—Hausratte (HR). Pp. 382-400, *in* Handbuch der Säugetiere Europas (J. Niethammer and F. Krapp, eds.). Akademische Verlagsgesellschaft (Wiesbaden), 1:1-476.

Becker, K. 1978b. *Rattus norvegicus* (Berkenhout, 1769)—Wanderratte (WR). Pp. 401-420, *in* Handbuch der Säugetiere Europas (J. Niethammer and F. Krapp, eds.). Akademische Verlagsgesellschaft (Wiesbaden), 1:1-476.

Beckoff, M. 1977. *Canis latrans*. Mammalian Species, 79:1-9.

Beckoff, M. 1999. Coyote | *Canis latrans*. Pp. 139-141, *in*, The Smithsonian Book of North American Mammals (D. E. Wilson and S. Ruff, eds.). Smithsonian Institution Press, Washington, D.C., 750 pp.

Becsy, L. 1982. [Protected animals in Hungary, *Glis glis*]. Buvar, 37(4):169 (in Hungarian).

Bednarczyk, A. 1993. Early Pliocene terrestrial fauna with *Glirulus* (Mammalia) from Panska Góra (Czestochowa Upland, Poland). Acta Zoologica Cracoviensia, 36(2):233-240.

Bee, J. W., and E. R. Hall. 1956. Mammals of northern Alaska on the Arctic slope. Miscellaneous Publications, Museum of Natural History, University of Kansas, 8:1-309.

Behboudi, A., E. Sjöstrand, P. Gómez-Fabre, Å. Sjöling, Z. Taib, K. Klinga-Levan, F. Ståhl, and G. Levan. 2002. Evolutionary aspects of the genomic organization of rat chromosome 10. Cytogenetic and Genome Research, 96:52-59.

Beintema, J. J., K. Rodewald, G. Braunitzer, J. Czelusniak, and M. Goodman. 1991. Studies on the phylogenetic position of the Ctenodactylidae (Rodentia). Molecular Biology and Evolution, 8(1):151-154.

Bekasova, T. S., and O. N. Mezhova. 1983. Karyotypes of rats of the genus *Rattus* from the USSR. Experientia, 39:541-542.

Bekasova, T. S., N. N. Vorontsov, K. V. Korobitsyna, and V. P. Korablev. 1980. B-chromosomes and comparative karyology of the mice of the genus *Apodemus*. Genetica, 52/53:33-43.

Bekele, A. 1986. The status of some mole-rats of the genus *Tachyoryctes* (Rodentia: Rhizomyidae) based upon craniometric studies. Revue de Zoologie Africaine, 99:411-417.

Bekele, A. 1995. Post-natal development and reproduction in captive bred *Praomys albipes* (Mammalia: Rodentia) from Ethiopia. Mammalia, 59(1):109-118.

Bekele, A., and M. Corti. 1994. Multivariate morphometrics of the Ethiopian populations of harsh-furred rat (*Lophuromys*: Mammalia, Rodentia). Journal of Zoology, London, 232:675-689.

Bekele, A., and M. Corti. 1997. Forest blocks and altitude as indicators of *Myomys albipes* (Rüppell 1842) (Mammalia Rodentia) distribution in Ethiopia. Tropical Zoology, 10:287-293.

Bekele, A., and D. A. Schlitter. 1989. Two new records of rodents from Kenya and Ethiopia. Mammalia, 53:113-116.

Bekele, A., E. Capanna, M. Corti, L. F. Marcus, and D. A. Schlitter. 1993. Systematics and geographic variation of Ethiopian *Arvicanthis* (Rodentia, Muridae). Journal of Zoology, London, 230:117-134.

Belcher, R. L., and T. E. Lee. 2002. *Arctocephalus townsendi*. Mammalian Species, 700:1-5.

Belcheva, R., and D. Peshev. 1985. Constitutive heterochromatin in the ground squirrel *Citellus citellus* (Sciuridae, Rodentia) from Bulgaria. Zoologischer Anzeiger, 215:385-390.

Belcheva, R. G., Ts. H. Peshev, and D. Ts. Peshev. 1980. Chromosome C- and G-banding patterns in a Bulgarian population of *Microtus guentheri* Danford and Alston (Microtinae, Rodentia). Genetica, 52/53:45-48.

Belcheva, R. G., M. N. Topashka-Ancheva, and N. I. Atanassov. 1989. Karyological studies of five species of mammals from Bulgaria's fauna. Comptes Rendus de l'Academie Bulgare des Sciences, 42(2):125-138.

Belcheva, R. G., M. N. Topashka-Ancheva, and N. I. Atanassov. 1992. Autosome polymorphism in the Norway rat (*Rattus norvegicus* Berk., 1769) populations in Bulgaria. Acta Zoologica Bulgarica, 44:36-43.

Belk, M. C., and H. D. Smith. 1990. *Ammospermophilus leucurus*. Mammalian Species, 368:1-8.

Bell, D. J., W. L. R. Oliver, and R. K. Ghose. 1990. The hispid hare *Caprolagus hispidus*. Pp. 128-136, *in* Rabbits, hares and pikas (J. A. Chapman and J. E. C. Flux, eds.). I.U.C.N., Gland, Switzerland, 168 pp.

Bell, D. M., M. J. Hamilton, C. W. Edwards, L. E. Wiggins, R. M. Martinez, R. E. Strauss, R. D. Bradley, and R. J. Baker. 2001. Patterns of karyotypic megaevolution in *Reithrodontomys*: Evidence from cytochrome-*b* phylogenetic hypotheses. Journal of Mammalogy, 82:81-91.

Bell, T. 1836. A history of British quadrupeds including the Cetacea. John van Voorst, London, 668 pp.

Bellinvia, E., P. Munclinger, and J. Flegr. 1999. Application of the RAPD technique for a study of the phylogenetic relationships among eight species of the genus *Apodemus*. Folia Zoologica, 48(4):241-248.

Belosludov, B. A., and V. S. Bazhanov. 1939. [A new genus and species of rodent from the central Kazakhstan (USSR)]. Uchenye Zapiski Kazakhskovo Gosudarstvennovo Universiteta, Alma-Ata, 1(1):81-86 (in Russian).

Belyaev, D. K., O. K. Baranov, Yu. G. Ternovskaya, D. V. Ternovsky. 1980. [A comparative immunochemical study of serum proteins in the Mustelidae (Carnivora)]. Zoologicheskii Zhurnal, 59:254-260 (in Russian).

Benazzou, T. 1983. Le caryotype d'*Acomys chudeaui* capture dans la region de Tata (Maroc). Mammalia, 47:588.

Benazzou, T., and F. Zyadi. 1990. Presence d'une variabilite biometrique chez *Gerbillus campestris* au Maroc (rongeurs, gerbillides). Mammalia, 54:271-279.

Benazzou, T., E. Viegas-Péquignot, F. Petter, and B. Dutrillaux. 1982*a*. Phylogenie chromosomique de quatre especies de *Meriones* (Rongeur, Gerbillidae). Annales de Genetique, 25:19-24.

Benazzou, T., E. Viegas-Péquignot, F. Petter, and B. Dutrillaux.1982*b*. Phylogénie chromosomique des Gerbillidae. II. Etude de six *Meriones*, de *Taterillus gracilis* et de *Gerbillurus tytonis*. Annales de Génétique, 25(4):212-217.

Benazzou, T., E. Viegas-Péquignot, M. Prod'Homme, M. Lombard, F. Petter, and B. Dutrillaux. 1984. Phylogenie chromosomique des Gerbillidae. III. Etude d'especes de *Tatera*, *Taterillus*, *Psammomys* et *Pachyuromys*. Annales de Genetique, 27:17-26.

Benda, P. 2000. Record of *Hipposideros doriae* (Mammalia: Chiroptera) from Sabah (east Malaysia) and remarks to species status. Acta Societatis Zoologicae Bohemicae, 64(2):143-152.

Benda, P., and I. Horácek. 1998. Bats (Mammalia: Chiroptera) of the eastern Mediterranean. Part 1. Review of distribution and taxonomy of bats in Turkey. Acta Societatis Zoologicae Bohemicae, 62:255-313.

Benda, P., and J. Obuch. 2001. Notes on the distribution of hedgehogs (Insectivora: Erinaceidae) in Syria. Lynx (Praha) n.s., 32:45-53.

Benda, P., and J. Sádlová. 1999. New records of small mammals (Insectivora, Chiroptera, Rodentia, Hyracoidea) from Jordan. Časopis Národního muzea Řada přírodovedná, 168(1-4):25-36.

Benda, P., and K. A. Tsytsulina. 2000. Taxonomic revision of *Myotis mystacinus* group (Mammalia: Chiroptera) in the western Palearctic. Acta Societatis Zoologicae Bohemoslovenicae, 64:331-398.

Benda, P., M. Andreas, and A. Reiter. 2002. Record of *Hypsugo arabicus* from Baluchistan, Iran, with remarks on its ecology and systematic status. Bat Research News, 43:75-76.

Beneski, J. T., Jr., and D. W. Stinson. 1987. *Sorex palustris*. Mammalian Species, 296:1-6.

Bennett, A. F., and L. F. Lumsdem. 1995. Mitchell's hopping-mouse, *Notomys mitchellii*. Pp. 212-214, *in* Mammals of Victoria, distribution, ecology and conservation (P. W. Menkhorst, ed.). Oxford University Press, Melbourne, 360 pp.

Bennett, D. K. 1980. Stripes do not a zebra make, part 1: A cladistic analysis of *Equus*. Systematic Zoology, 29:272-287.

Bennett, D., and R. S. Hoffman. 1999. *Equus caballus*. Mammalian Species, 628:1-14.

Bennett, E. T. 1829. The chinchilla *Chichilla lanigera*. Gardens and Managerie of the Zoological Society, 1:1-12.

Ben-Shlomo, R., and E. Nevo. 1993. Myosin-heavy-chain DNA polymorphisms of subterranean mole rats of the *Spalax ehrenbergi* superspecies in Israel. Genetics, Selection, and Evolution, 25:211-227.

Ben-Shlomo, R., H.-S. Shin, and E. Nevo. 1993. Period-homologous sequence polymorphisms in subterranean mammals of the *Spalax ehrenbergi* superspecies in Israel. Heredity, 70:111-121.

Ben-Shlomo, R., T. Fahima, and E. Nevo. 1996. Random amplified polymorphic DNA of the *Spalax ehrenbergi* superspecies in Israel. Israel Journal of Zoology, 42:317-326.

Benson, D. L., and F. R. Gehlbach. 1979. Ecological and taxonomic notes on the rice rat (*Oryzomys couesi*) in Texas. Journal of Mammalogy, 60:225-228.

Bentham, T. 1908. An illustrated catalogue of the Asiatic horns and antlers in the collection of the Indian Museum. Calcutta, 97 pp.

Bentz, S., and C. Montgelard. 1999. Systematic position of the African Dormouse *Graphiurus* (Rodentia, Gliridae) assessed from cytochrome b and 12S rRNA mitochondrial genes. Journal of Mammalian Evolution, 6(1):67-83.

Berardino, D., and L. Iannuzzi. 1981. Chromosome banding homologies in swamp and Murrah buffaloes. Journal of Heredity, 72:183-188.

Berend, S. A. 1993. Inherent levels of somatic chromosomal aberrations among three populations of *Sigmodon hispidus* from north-central Texas. Texas Journal of Science, 45:211-214.

Berg, L. 1996. Small-scale changes in the distribution of the dormouse *Muscardinus avellanarius* (Rodentia, Myoxidae) in relation to vegetation changes. Mammalia, 60(2):211-216.

Berg, L. 1997. Spatial distribution and habitat selection of the hazel dormouse *Muscardinus avellanarius* in Sweden. SLU Institutionen for Naturvardsbiologi Rapport, 1997, 1:1-76.

Berger, M., and R. Feldmann. 1997. Die ausbreitung der gelbhalsmaus, *Apodemus flavicollis*, in Münsterland. Abhandlungen aus dem Westfälischen Museum für Naturkunde, 59(3):135-142.

Bergers, P. J. M., and H. Bussink. 1995. Eerste vondst van de rosse woelmuis *Clethrionomys glareolus* op Schouwen-Duiveland. Lutra, 38:60-61.

Bergers, P. J. M., and R. P. B. Foppen. 1992. Grote bosmuis, *Apodemus flavicollis* (Melchior, 1834). Pp. 286-288, *in* Atlas van de Nederlandse Zoogdieren (S. Broekhuizen, B. Hoekstra, V. van Laar, C. Smeenk, and J. B. M. Thissen, eds.). Stichting Uitgeverij Koninklijke Nederlandse Natuurhistorische Vereniging, Utrecht, 336 pp.

Bergmans, W. 1975. A new species of *Dobsonia* Palmer, 1898 (Mammalia, Megachiroptera) from Waigeo, with notes on other members of the genus. Beaufortia, 23(295):1-13.

Bergmans, W. 1976. A revision of the African genus *Myonycteris* Matschie, 1899 (Mammalia, Megachiroptera). Beaufortia, 24(317):189-216.

Bergmans, W. 1977. Notes on new material of *Rousettus madagascariensis* Grandidier, 1829 (Mammalia, Megachiroptera). Mammalia, 41:67-74.

Bergmans, W. 1978. On *Dobsonia* Palmer 1898 from the Lesser Sunda Islands (Mammalia, Megachiroptera). Senckenbergiana Biologica, 59:1-18.

Bergmans, W. 1979. Taxonomy and zoogeography of *Dobsonia* Palmer, 1898, from the Louisiade Archipelago, the D'Entrecasteaux Group, Trobriand Island and Woodlark Island (Mammalia, Megachiroptera). Beaufortia, 29(355):199-214.

Bergmans, W. 1988. Taxonomy and biogeography of African fruit bats (Mammalia, Megachiroptera). I. General introduction; materials and methods; results: The genus *Epomophorus* Bennett, 1836. Beaufortia, 38(5):75-146.

Bergmans, W. 1989. Taxonomy and biogeography of African fruit bats (Mammalia: Megachiroptera). 2. The genera *Micropteropus* Matschie, 1899, *Epomops* Gray, 1870, *Hypsignathus* H. Allen, 1861, *Nanonycteris* Matschie, 1899, and *Plerotes* Andersen, 1910. Beaufortia, 39(4):89-153.

Bergmans, W. 1990. Taxonomy and biogeography of African fruit bats (Mammalia: Megachiroptera). 3. The genera *Scotonycteris* Matschie, 1894, *Casinycteris* Thomas, 1910, *Pteropus* Brisson, 1762, and *Eidolon* Rafinesque, 1815. Beaufortia, 40(7):111-177.

Bergmans, W. 1994. Taxonomy and biogeography of African fruit bats (Mammalia, Megachiroptera). 4. The genus *Rousettus* Gray, 1821. Beaufortia, 44(4):79-126.

Bergmans, W. 1997. Taxonomy and biogeography of African fruit bats (Mammalia, Megachiroptera). 5. The genera *Lissonycteris* Andersen, 1912, *Myonycteris* Matschie, 1899 and general remarks and conclusions; annex: Key to all species. Beaufortia, 47:11-90.

Bergmans, W. 2001. Notes on distribution and taxonomy of Australasian bats. 1. Pteropodinae and Nyctimeninae (Mammalia, Megachiroptera, Pteropodidae). Beaufortia, 51:119-152.

Bergmans, W., and F. G. Rozendaal. 1988. Notes on collections of fruit bats from Sulawesi and some off-lying Islands (Mammalia; Megachiroptera). Zoologische Verhandelingen (Leiden), 248:1-74.

Bergmans, W., and S. Sarbini. 1985. Fruit bats of the genus *Dobsonia* Palmer, 1898, from the islands of Biak, Owii, Numfoor and Yapen, Irian Jaya (Mammalia; Megachiroptera). Beaufortia, 34(6):181-189.

Bergmans, W., and P. J. H. van Bree. 1972. The taxonomy of the African bat *Megaloglossus woermanni* Pagenstecher, 1885 (Megachiroptera, Macroglossinae). Biologia Gabonica, 8:291-299.

Bergmans, W., and P. J. H. van Bree. 1986. On a collection of bats and rats from the Kangean Islands, Indonesia (Mammalia: Chiroptera and Rodentia). Zeitschrift für Säugetierkunde, 51:329-344.

Bergstrom, B. J., and R. S. Hoffmann. 1991. Distribution and diagnosis of three species of chipmunks (*Tamias*) in the Front Range of Colorado. Southwestern Naturalist, 36:14-28.

Bernard, E. 2001. First capture of *Micronycteris homezi* Pirlot (Chiroptera, Phyllostomidae) in Brazil. Revista Brasileira de Zoologia, 18:645-647.

Bernard, E. 2003. *Cormura brevirostris*. Mammalian Species, 737:1-3.

Bernard, R. T. F., A. H. Hodgson, J. Meester, K. Willan, and C. Bojarski. 1990. Sperm structure and taxonomic affinites of five African rodents of the subfamily Otomyinae (Muridae). Electron Microscope Society of Southern Africa, 20:161-162.

Bernard, R. T. F., A. N. Anderson, and G. K. Campbell. 1991. Sperm structure and taxonomic affinities of five African rodents of the subfamily Otomyinae. South African Journal of Science, 87:503-506.

Berry, D. L., and R. J. Baker. 1972. Chromosomes of pocket gophers of the genus *Pappogeomys*, subgenus *Cratogeomys*. Journal of Mammalogy, 53:303-309.

Berry, R. J. 1973. Chance and change in British long-tailed field mice (*Apodemus sylvaticus*). Journal of the Zoological Society of London, 170:351-366.

Berry, R. J. 1984. House mouse. Pp. 273-284, *in* Evolution of Domesticated Animals (I. L. Mason, ed.). Longman Group Limited, London and New York, 452 pp.

Berry, R. J. 1996. Small mammal differentiation on islands. Philosophical Transactions of the Royal Society of London, B, 351:753-764.

Berta, A. 1982. *Cerdocyon thous*. Mammalian Species, 186:1-4.

Berta, A. 1985. The status of *Smilodon* in North and South America. Contributions in Science, Natural History Museum of Los Angeles County, 370:1-15.

Berta, A. 1986. *Atelocynus microtis*. Mammalian Species, 256:1-3.

Berta, A. 1987. Origin, diversification, and zoogeography of the South American Canidae. Pp. 455-471, *in* Studies in Neotropical Mammalogy: Essays in honor of Philip Hershkovitz (B. D. Patterson and R. M. Timm, eds.). Fieldiana: Zoology, n.s., 39(1382):1-506.

Berta, A. 1988. Quaternary evolution and biogeography of the large South American Canidae (Mammalia: Carnivora). University of California Publications, Geological Sciences, 132:1-149.

Berta, A. 1991. New *Enaliarctos* (Pinnipedimorpha) from the Oligocene and Miocene of Oregon and the role of "Enaliarctids" in pinniped phylogeny. Smithsonian Contributions to Paleobiology, 69:1-33.

Berta, A. 1994. A new species of phocoid pinniped *Pinnarctidion* from the early Miocene of Oregon. Journal of Vertebrate Paleontology, 14:405-413.

Berta, A., and T. A. Demere. 1986. *Callorhinus gilmorei* n. sp., (Carnivora: Otariidae) from the San Diego formation (Blancan) and its implications for otariid phylogeny. Transactions of the San Diego Society of Natural History, 21:111-126.

Berta, A., and L. G. Marshall. 1978. Fossilium catalogus. I: South American Carnivora. Dr. W. Junk, The Hague, 125:1-48.

Berta, A., and A. R. Wyss. 1994. Pinniped phylogeny. Proceedings of the San Diego Society of Natural History, 29:33-56.

Bertolino, S., and I. Currado. 2001. Ecology of the Garden Dormouse (*Eliomys quercinus*) in the Alpine habitat. Trakya University Journal of Scientific Research B, 2(2):75-78.

Bertolino, S., I. Currado, R. Azzollini and C. Viano. 1997. The social organization, home range and movement of the Garden Dormouse *Eliomys quercinus*. Natura Croatica, 6(3):303-312.

Bertolino, S., I. Currado, and P. J. Mazzoglio. 2000. Finlayson's (variable) squirrel *Callosciurus finlaysoni* in Italy. Mammalia, 63(4):522-525.

Besenecker, H., F. Spitzenberger, and G. Storch. 1972. Eine holozane kleinsäugerfauna von der Insel Chios, Agais (Mammalia: Insectivora, Rodentia). Senckenbergiana Biologica, 53(3/4):145-177.

Best, R. C., and V. M. F. da Silva. 1989. Amazon river dolphin, Boto—*Inia geoffrensis* (de Blainville, 1817). Pp. 1-24, *in* Handbook of marine mammals: River dolphins and the larger toothed whales (S. H. Ridgway and R. Harrison, eds.). Academic Press, London, 4:1-442.

Best, T. L. 1978. Variation in kangaroo rats (genus *Dipodomys*) of the *heermanni* group in Baja California, Mexico. Journal of Mammalogy, 59:160-175.

Best, T. L. 1983*a*. Intraspecific variation in the agile kangaroo rat (*Dipodomys agilis*). Journal of Mammalogy, 64:426-436.

Best, T. L. 1983*b*. Morphologic variation in the San Quintín kangaroo rat (*Dipodomys gravipes* Huey 1925). American Midland Naturalist, 109:409-413.

Best, T. L. 1986. *Dipodomys elephantinus*. Mammalian Species, 255:1-4.

Best, T. L. 1987. Sexual dimorphism and morphometric variation in the Texas kangaroo rat (*Dipodomys elator* Merriam 1894). Southwestern Naturalist, 32:53-59.

Best, T. L. 1988*a*. *Dipodomys spectabilis*. Mammalian Species, 311:1-10.

Best, T. L. 1988*b*. *Dipodomys nelsoni*. Mammalian Species, 326:1-4.

Best, T. L. 1991. *Dipodomys nitratoides*. Mammalian Species, 381:1-7.

Best, T. L. 1992. *Dipodomys margaritae*. Mammalian Species, 400:1-3.

Best, T. L. 1993*a*. *Tamias palmeri*. Mammalian Species, 443:1-6.

Best, T. L. 1993*b*. *Tamias ruficaudus*. Mammalian Species, 452:1-7.

Best, T. L. 1993*c*. *Tamias sonomae*. Mammalian Species, 444:1-5.

Best, T. L. 1993*d*. Patterns of morphologic and morphometric variation in heteromyid rodents. Pp. 197-235, *in* Biology of the Heteromyidae (H. H. Genoways and J. H. Brown, eds.). Special Publication, The American Society of Mammalogists, 10:1-719.

Best, T. L. 1993*e*. *Perognathus inornatus*. Mammalian Species, 450:1-5.

Best, T. L. 1993*f*. *Chaetodipus lineatus*. Mammalian Species, 451:1-3.

Best, T. L. 1994*a*. *Perognathus alticola*. Mammalian Species, 463:1-4.

Best, T. L. 1994*b*. *Chaetodipus nelsoni*. Mammalian Species, 484:1-6.

Best, T. L. 1995*a*. *Sciurus alleni*. Mammalian Species, 501:1-4.

Best, T. L. 1995*b*. *Sciurus colliaei*. Mammalian Species, 497:1-4.

Best, T. L. 1995*c*. *Sciurus deppei*. Mammalian Species, 505:1-5.

Best, T. L. 1995*d*. *Sciurus nayaritensis*. Mammalian Species, 492:1-5.

Best, T. L. 1995*e*. *Sciurus oculatus*. Mammalian Species, 498:1-3.

Best, T. L. 1995*f*. *Sciurus variegatoides*. Mammalian Species, 500:1-6.

Best, T. L. 1995*g*. *Spermophilus adocetus*. Mammalian Species, 504:1-4.

Best, T. L. 1995*h*. *Spermophilus annulatus*. Mammalian Species, 508:1-4.

Best, T. L. 1995*i*. *Spermophilus mohavensis*. Mammalian Species, 509:1-7.

Best, T. L. 1996. *Lepus californicus*. Mammalian Species, 530:1-10.

Best, T. L., and G. Ceballos. 1995. *Spermophilus perotensis*. Mammalian Species, 507:1-3.

Best, T. L., and N. J. Granai. 1994*a*. *Tamias merriami*. Mammalian Species, 476:1-9.

Best, T. L., and N. J. Granai. 1994*b*. *Tamias obscurus*. Mammalian Species, 472:1-6.

Best, T. L., and T. H. Henry. 1993*a*. *Lepus alleni*. Mammalian Species, 424:1-8.

Best, T. L., and T. H. Henry. 1993*b*. *Lepus callotis*. Mammalian Species, 442:1-6.

Best, T. L., and T. H. Henry. 1994*a*. *Lepus arcticus*. Mammalian Species, 457:1-9.

Best, T. L., and T. H. Henry. 1994*b*. *Lepus othus*. Mammalian Species, 458:1-5.

Best, T. L., and L. L. Janecek. 1992. Allozymic and morphologic variation among *Dipodomys insularis*, *Dipodomys nitratoides*, and two populations of *Dipodomys merriami* (Rodentia; Heteromyidae). The Southwestern Naturalist, 37:1-8.

Best, T. L., and J. A. Lackey. 1985. *Dipodomys gravipes*. Mammalian Species, 236:1-4.

Best, T. L., and J. A. Lackey. 1992*a*. *Chaetodipus artus*. Mammalian Species, 418:1-3.

Best, T. L., and J. A. Lackey. 1992*b*. *Chaetodipus pernix*. Mammalian Species, 420:1-3.

Best, T. L., and S. Riedel. 1995. *Sciurus arizonensis*. Mammalian Species, 496:1-5.

Best, T. L., and G. D. Schnell. 1974. Bacular variation in kangaroo rats (genus *Dipodomys*). American Midland Naturalist, 91:257-270.

Best, T. L., and M. P. Skupski. 1994a. *Perognathus flavus*. Mammalian Species, 471:1-10.

Best, T. L., and M. P. Skupski. 1994b. *Perognathus merriami*. Mammalian Species, 473:1-7.

Best, T. L., and H. H. Thomas. 1991a. *Dipodomys insularis*. Mammalian Species, 374:1-3.

Best, T. L., and H. H. Thomas. 1991b. *Spermophilus madrensis*. Mammalian Species, 378:1-2.

Best, T. L., R. M. Sullivan, J. A. Cook, and T. L. Yates. 1986. Chromosomal, genic, and morphologic variation in the agile kangaroo rat, *Dipodomys agilis* (Rodentia: Heteromyidae). Systematic Zoology, 35:311-324.

Best, T. L., N. J. Hildreth, and C. Jones. 1989. *Dipodomys deserti*. Mammalian Species, 339:1-8.

Best, T. L., K. Caesar, A. S. Titus, and C. L. Lewis. 1990a. *Ammospermophilus insularis*. Mammalian Species, 364:1-4.

Best, T. L., C. L. Lewis, K. Caesar, and A. S. Titus. 1990b. *Ammospermophilus interpres*. Mammalian Species, 365:1-6.

Best, T. L., A. S. Titus, K. Caesar, and C. L. Lewis. 1990c. *Ammospermophilus harrisii*. Mammalian Species, 366:1-7.

Best, T. L., A. S. Titus, C. L. Lewis, and K. Caesar. 1990d. *Ammospermophilus nelsoni*. Mammalian Species, 367:1-7.

Best, T. L., J. L. Bartig, and S. L. Burt. 1992. *Tamias canipes*. Mammalian Species, 411:1-5.

Best, T. L., S. L. Burt, and J. L. Bartig. 1993. *Tamias durangae*. Mammalian Species, 437:1-4.

Best, T. L., S. L. Burt, and J. L. Bartig. 1994a. *Tamias quadrivittatus*. Mammalian Species, 466:1-7.

Best, T. L., R. G. Clawson, and J. A. Clawson. 1994b. *Tamias panamintinus*. Mammalian Species, 468:1-7.

Best, T. L., R. G. Clawson, and J. A. Clawson. 1994c. *Tamias speciosus*. Mammalian Species, 478:1-9.

Best, T. L., H. A. Ruiz-Pina, and L. S. Leon-Paniagua. 1995. *Sciurus yucatanensis*. Mammalian Species, 506:1-4.

Best, T. L., R. K. Chesser, D. A. McCullough, and G. D. Baumgardner. 1996. Genic and morphometric variation in kangaroo rats, genus *Dipodomys*, from coastal California. Journal of Mammalogy, 77:785-800.

Best, T. L., J. L. Hunt, L. A. McWilliams, and K. G. Smith. 2001a. *Eumops maurus*. Mammalian Species, 667:1-3.

Best, T. L., J. L. Hunt, L. A. McWilliams, and K. G. Smith. 2001b. *Eumops hansae*. Mammalian Species, 687:1-3.

Best, T. L., J. L. Hunt, L. A. McWilliams, and K. G. Smith. 2002. *Eumops auripendulus*. Mammalian Species, 708:1-5.

Betts, B. J. 1999. Current status of Washington ground squirrels in Oregon and Washington. Northwestern Naturalist, 80(1):35-38.

Bezrodny, S. V. 1991. Rasprostranenie son' (Rodentia, Gliridae) na Ukraine [The distribution of dormice (Rodentia) in the Ukraine]. Vestnik Zoologii, 1991(3):45-50 (in Russian).

Bhattacharyya, T. P. 2002. Taxanomic status of the genus *Harpiola* Thomas, 1915 (Mammalia: Chiroptera: Vespertilionidae), with a report of the occurrence of *Harpiola grisea* (Peters,1872) in Mizoram, India. Proceedings of the Zoological Society (Calcutta) 55(1):73-76.

Bianchi, N. O., O. A. Reig, J. Molina, and F. N. Dulout. 1971. Cytogenetics of South American akodont rodents (Cricetidae). I. A progress record of Argentinian and Venezuelan forms. Evolution, 21:724-736.

Bianchi, N. O., S. Merani, and M. Lizarralde. 1979. Cytogenetics of South American akodont rodents (Cricetidae). V. Segregation of chromosome Nr. 1 polymorphism in *Akodon molinae*. Experientia, 35:1438-1439.

Bianchini, J. J., and L. H. Delupi. 1978 [1979]. El estado sistematico de los ciervos neotropicales de la tribu Odocoileini Simpson 1945. Physis, Asociacion Argentina de Ciencias Naturales, 38(94), Seccion C:83-89.

Bianchini, J., and L. Delupi. 1994. Consideraciones sobre el estado sistemático de *Deltamys kempi* Thomas, 1917 (Cricetidae, Sigmodontinae). Physis, Sec. C., 49:27-35.

Bibikov, D. I. 1991. The steppe marmot—its past and future. Oryx, 25:45-49.

Bickham, J. W. 1987. Chromosomal variation among seven species of lasiurine bats (Chiroptera: Vespertilionidae). Journal of Mammalogy, 68(4):837-842.

Bidau, C. J., M. D. Gimenez, and J. R. Contreras. 1996. Especiacion cromosomica y la conservacion de la variabilidad genetica: El caso del genero *Ctenomys* (Rodentia, Caviomorpha, Ctenomyidae). Mendeliana, 12:25-37.

Bidlingmaier, T. C. 1937. Notes on the genus *Chinchilla*. Journal of Mammalogy, 18:159-163.

Bielecka, K. 1986. Present record of fat dormouse *Glis glis* (Linnaeus, 1766) from Pomerania. Przeglad Zoologiczny, 30(1):115-117.

Bierman, W. H., and E. J. Slijper. 1947. Remarks upon the species of the genus *Lagenorhynchus*. I. Koninklijke Nederlandsche Akademie van Wetenschappen, Proceedings, L(10):1353-1364.

Bigalke, R. 1948. The type locality of the bontbok, *Damaliscus pygargus* (Pallas). Journal of Mammalogy, 29:421-422.

Bigini, I., and R. Turini. 1995. The common vole, *Microtus arvalis* Pallas, 1779: A relict species in northern Garfagnana (Tuscany, Italy). Atti del Museo Civico di Storia Naturale di Trieste, 46:129-131.

Biltueva, L. S., M. B. Rogatcheva, P. L. Perelman, P. M. Borodin, S.-I. Oda, K. Koyasu, M. Harada, J. Zima, and A. S. Graphodatsky. 2001. Chromosomal phylogeny of certain shrews of the genera *Crocidura* and *Suncus* (Insectivora). Journal of Zoological Systematics and Evolutionary Research, 39:69-76.

Birkenholz, D. E. 1972. *Neofiber alleni*. Mammalian Species, 15:1-4.

Bing, S., C. Zhiping, L. Hong, W. Wen, W. Yingxiang, and Z. Yaping. 1996. [Protein polymorphism and genetic divergence in three species of *Apodemus* from Yunnan]. Zoological Research, 17(3):259-262 (in Chinese with English abstract).

Bininda-Emonds, O. R. P., J. L. Gittleman, and A. Purvis. 1999. Building large trees by combining phylogenetic information: A complete phylogeny of the extant Carnivora (Mammalia). Biological Reviews, 74:143-175.

Birney, E. C. 1973. Systematics of three species of woodrats (genus *Neotoma*) in central North America. Miscellaneous Publications, Museum of Natural History, University of Kansas, 58:1-173.

Birney, E. C. 1976. An assessment of relationships and effects of interbreeding among woodrats of the *Neotoma floridana* species-group. Journal of Mammalogy, 57:103-132.

Birney, E. C., and H. H. Genoways. 1973. Chromosomes of *Spermophilus adocetus* (Mammalia, Sciuridae), with comments on the subgeneric affinities of the species. Experientia, 29:228-229.

Birula, A. 1913 [1914]. [Contribution à la synonymie de l'*Otocolobus manul* (Pallas) (Felidae)]. Annuaire du Musée Zoologique de l'Académie Impériale des Sciences de St. Pétersbourg, 18:LVII-LVIII.

Birula, A. 1916 [1917]. Contributions à la classification et à la distribution géographique des mammifères. VI. Sur la position de l'*Otocolobus manul* (Pallas) dans le système de la fam. Felidae et sur ses rasses. Annuaire du Musée Zoologique de l'Académie Impériale des Sciences de St. Pétersbourg, 21:130-163.

Birungi, J., and P. Arctander. 2001. Molecular systematics and phylogeny of the Reduncini (Artiodactyla: Bovidae) inferred from the analysis of mitochondrial cytochrome b gene sequences. Journal of Mammalian Evolution, 8:125-147.

Bisbal-E., F. J. 1990. Inventario preliminary de la fauna del Cerro Santa Ana, Península de Paraguaná—Estado Falcón, Venezuela. Acta Cientifgica Venezolana, 41:177-185.

Bishop, L. R. 1979. Notes on *Praomys* (*Hylomyscus*) in eastern Africa. Mammalia, 43:521-530.

Biswas, B. 1967. Authorship of the name *Presbytis geei* (Mammalia: Primates). Journal of the Bombay Natural History Society, 63:429-431.

Biswas, B., and R. K. Ghose. 1970. Taxonomic notes on the Indian pale hedgehogs of the genus *Paraechinus* Trouessart, with descriptions of a new species and subspecies. Mammalia, 34:467-477.

Bitz, A. 1991. Schlafmäuse – Gliridae. *In* Wirbeltiere – Beiträge zur Fauna von Rheinland-Pfalz (R. Kinzelbach and M. Niehuis, eds.). Mainzer Naturwissenschaftliches Archiv, 13:269-321.

Blaaderen, H. van. 1992. Zwarte rat, *Rattus rattus* (L., 1758). Pp. 296-301, *in* Atlas van de Nederlandse Zoogdieren (S. Broekhuizen, B. Hoekstra, V. van Laar, C. Smeenk, and J. B. M. Thissen, eds.). Stichting Uitgeverij Koninklijke Nederlandse Natuurhistorische Vereniging, Utrect, 336 pp.

Blaaderen, H. van, and B. T. Bosman. 1992. Bruine rat, *Rattus norvegicus* (Berkenhout, 1769). Pp. 292-295, *in* Atlas van de Nederlandse Zoogdieren (S. Broekhuizen, B. Hoekstra, V. van Laar, C. Smeenk, and J. B. M. Thissen, eds.). Stichting Uitgeverij Koninklijke Nederlandse Natuurhistorische Vereniging, Utrecht, 336 pp.

Black, C. C. 1963. A review of the North American Tertiary Sciuridae. Bulletin of the Museum of Comparative Zoology at Harvard College, 130:109-248.

Black, C. C. 1972. Review of fossil rodents from the Neogene Siwalik beds of India and Pakistan. Palaeontology, 15:238-266.

Black, C. C., and L. Krishtalka. 1986. Rodents, bats, and insectivores from the Plio-Pleistocene sediments to the east of Lake Turkana, Kenya. Contributions in Science, Natural History Museum of Los Angeles County, 372:1-15.

Blacket, M. J., C. Krajewski, A. Labrinidis, B. Cambron, S. Cooper, and M. Westerman. 1999. Systematic relationships within the dasyurid marsupial tribe Sminthopsini-A multigene approach. Molecular Phylogenetics and Evolution, 12:140-155.

Blacket, M. J., M. Adams, S. J. B. Cooper, C. Krajewskiand, and M. Westerman. 2001. Systematics and evolution of the dasyurid marsupial genus *Sminthopsis*: I. The *macroura* species group. Journal of Mammalian Evolution, 8:149-170.

Blackler, W. F. G. 1916. On a new species of *Microtus* from Asia Minor. Annals and Magazine of Natural History, ser. 8, 17:426-427.

Blagojević, J., and M. Vujošević. 1995. The role of *B* chromosomes in the population dynamics of yellow-necked wood mice *Apodemus flavicollis* (Rodentia, Mammalia). Genome, 38:472-478.

Blaine, G. 1913. On the relationship of *Gazella isabellina* to *Gazella dorcas*, with a discription of a new species and subspecies. Annals and Magazine of Natural History, ser. 8, 11:291-296.

Blainville, H. M. D., de. 1837. Rapport sur un mémoire de M. Jourdan concernant deux nouvelles espèces de mammifères de l'Indie. Annales des Sciences Naturelles, Zoologie et Biologie Animale, ser. 2, 8:270-281.

Blair, F. W. 1941. Observations on the life history of *Baiomys taylori subater*. Journal of Mammalogy, 22:378-383.

Blair, W. F. 1942. Systematic relationships of *Peromyscus* and several related genera as shown by the baculum. Journal of Mammalogy, 23:196-204.

Blair, W. F., A. P. Blair, P. Broadkorb, F. R. Cagle, and G. A. Moore. 1968. Vertebrates of the United States, Second edition. McGraw-Hill, NY, 616 pp.

Blancou, L. 1960. Destruction and protection of the fauna of French Equatorial and of French West Africa. African Wildlife, 14:101-108.

Blanford, W. T. 1875a. On the species of marmot inhabiting the Himalaya, Tibet, and the adjoining regions. Journal of the Asiatic Society of Bengal, 44:113-127.

Blanford, W. T. 1875b. On the scientific names of the Sind "ibex", the markhor, and the Indian antelope. Journal of the Asiatic Society of Bengal, 44(2):12-20.

Blanford, W. T. 1879. Scientific results of the second Yarkand mission. Mammalia. Office of the Superintendent of Government Printing, Calcutta, 94 + xvi plates.

Blanford, W. T. 1888. The fauna of British India including Ceylon and Burma. Mammalia, Part I. Taylor and Francis, London,. 250 pp.

Blank, R. D., G. R. Campbell, and P. D'Eustachio. 1986. Possible derivation of the laboratory mouse genome from multiple wild *Mus* species. Genetics, 114:1257-1269.

Blasius, J. H. 1857. Naturgeschichte der Säugethiere Deutschlands. Braunschweig, 549 pp.

Blaustein, S. A., R. C. Liascovich, L. I. Apfelbaum, L. Daleffe, R. M. Barquez, and O. A. Reig. 1992. Correlates of systematic differentiation between two closely related allopatric populations of the *Akodon boliviensis* group from NW Argentina (Rodentia, Cricetidae). Zeitschrift für Säugetierkunde, 57:1-13.

Bleich, V. C. 1977. *Dipodomys stephensi*. Mammalian Species, 73:1-3.

Block, S. B., and E. G. Zimmerman. 1991. Allozymic variation and systematics of plains pocket gophers (*Geomys*) of south-central Texas. Southwestern Naturalist, 36:29-36.

Blood, B. R., and M. K. Clark. 1998. *Myotis vivesi*. Mammalian Species, 588:1-5.

Blood, B. R., and D. A. Mcfarlane. 1988. Notes on some bats from northern Thailand. Zeitschrift für Säugetierkunde, 53:276-280.

Blundell, A. G. 1996. A preliminary checklist of mammals at Cabang Panti Research Station, Gunung Palung Nationaal Park, West Kalimantan. Tropical Biodiversity, 3(3):251-259.

Blyth, E. 1859. Proceedings of the Asiatic Society: Report of the Curator, Zoological Department, for February to May Meetings, 1859. Asiatic Society of Bengal, 3:271-298.

Blyth, E. 1860. On the flat-horned taurine cattle of S. E. Asia with a note on the races of rein deer, and a note on domestic animals in general. Journal of the Asiatic Society of Bengal, 29:282-306, 376-392.

Blyth, E. 1862. Report of curator, zoological department, February, 1862, No. 1. Journal of the Asiatic Society of Bengal, 31:331-333.

Blyth, E. 1871. The suppositious "*Bos* (?) *pegasus*" of the late Colonel Charles Hamilton Smith. Annals and Magazine of Natural History, ser. 4, 8:204-207.

Bobrinskii, N. A., B. A. Kuznetsov, and A. P. Kuzyakin. 1944. Opredelitl' mlekopitayushchikh SSSR [Guide to the mammals of the U.S.S.R.]. Sovetskaya Nauka, Moscow, 439 pp. (in Russian).

Bobrinskii, N. A., B. A. Kuznetsov, and A. P. Kuzyakin. 1965. Opredelitl' mlekopitayushchikh SSSR [Guide to the mammals of the U.S.S.R.] Second ed. Proveshchenie, Moscow, 382 pp. (in Russian).

Bock, W. J. 1997. Comments on the proposed conservation of usage of 15 mammal specific names based on wild species which are antedated by or contemporary with those based on domestic animals. Bulletin of Zoological Nomenclature, 54:124-125.

Boddaert, P. 1785. Elenchus animalium, volumen 1: Sistens quadrupedia huc usque nota, erorumque varietates. C. R. Hake, Rotterdam, 174 pp.

Bodenheimer, F. S. 1958. The present taxonomic status of the terrestrial mammals of Palestine. Bulletin of the Research Council of Israel, 7b:165-190.

Boeadi, B. 1990. On two new specimens of the insectivorous bat species *Otomops formosus* Chasen, 1939. Mammalia, 54:489.

Boeadi, B., and J. E Hill. 1986. A new subspecies of *Aethalops alecto* (Thomas, 1923) (Chrioptera: Pteropodidae) from Java. Mammalia, 50:263-266.

Bogan, M. A. 1978. A new species of *Myotis* from the Islas Tres Marias, Nayarit, Mexico, with comments on variation in *Myotis nigricans*. Journal of Mammalogy, 59:519-530.

Bogan, M. A. 1997. On the status of *Neotoma varia* from Isla Datil, Sonora. Pp. 81-87, *in* Life among the muses: Papers in honor of James S. Findley (T. L. Yates, W. L. Gannon, and D. E. Wilson, eds.). The Museum of Southwestern Biology, The University of New Mexico, Albuquerque, v + 290 pp.

Bogan, M. A., H. W. Setzer, J. S. Findley, and D. E. Wilson. 1978. Phenetics of *Myotis blythi* in Morocco. Proceedings of the Fourth International Bat Research Conference, Nairobi, 4:217-230.

Bogdanowicz, W. 1994. *Myotis daubentonii*. Mammalian Species, 475:1-9.

Bogdanowicz, W., and R. D. Owen. 1992. Phylogenetic analyses of the bat family Rhinolophidae. Zeitschrift für Zoologische Systematik und Evolutionsforschung, 30:142-60.

Bogdanowicz, W., and R. D. Owen. 1998. In the Minotaur's labyrinth: The phylogeny of the bat family Hipposideridae. Pp. 27-42, *in* Bat biology and conservation (T. H. Kunz and P. A. Racey, eds.). Smithsonian Institution Press, Washington, DC, 361 pp.

Bogdanowicz, W., and D. Kock. 1998. Quoting and spelling species names from H. Kuhl's "Die deutschen Fledermäuse". Bat Research News, 39:4-5.

Bogdanowicz, W., S. Kasper, and R. D. Owen. 1998. Phylogeny of plecotine bats: Reevaluation of morphological and chromosomal data. Journal of Mammalogy, 79:78-90.

Bohlin, R. G., and E. G. Zimmerman. 1982. Genic differentiation of two chromosomal races of the *Geomys bursarius* complex. Journal of Mammalogy, 63:218-228.

Bohlken, H. 1958. Vergleichende Untersuchungen an Wildrindern (Tribus Bovini Simpson, 1945). Zoologische Jahrbücher (Physiologie), 68:113-202.

Bohlken, H. 1961. Haustier und zoologische Systematik. Zeitschrift für Tierzüchtung und Züchtungbiologie, 75:107-113.

Bohlken, H. 1967. Beitrag zur Systematik der rezenten Formen der Gattung *Bison* H. Smith, 1827. Zeitschrift für Zoologische Systematik und Evolutionsforschung, 5:54-110.

Bohmann, L., von. 1939. Neue Rassen der Gattung *Dendromus*. Zoologischer Anzeiger, 127:170-172.

Bohmann, L., von. 1942. Die Gattung *Dendromus* A. Smith. Versuch einer natürlichen Gruppierung (Ergebnisse der Ostafrika-Reise 1937 Uthmöller-Bohmann. VIII). Zoologischer Anzeiger, 139:33-53.

Bohmann, L., von. 1952. Die afrikanische Nagergattung *Otomys* F. Cuvier. Zeitschrift für Säugetierkunde, 18:1-80.

Böhme, W. 1978a. *Micromys minutus* (Pallas, 1778)—Zwergmaus. Pp. 290-404, *in* Handbuch der Säugetiere Europas (J. Niethammer and F. Krapp, eds.). Akademische Verlagsgesellschaft (Wiesbaden), 1:1-476.

Böhme, W. 1978b. *Apodemus agrarius* (Pallas, 1771)—Brandmaus. Pp. 368-381, *in* Handbuch der Säugetiere Europas (J. Niethammer and F. Krapp, eds.). Akademische Verlagsgesellschaft (Wiesbaden), 1:1-476.

Böhme, W., and R. Hutterer. 1979. Kommentierte Liste einer Säugetier-Aufsammlung aus dem Senegal. Bonner Zoologische Beiträge, 29:303-323.

Bole, B. J., Jr., and P. N. Moulthrop. 1942. The Ohio Recent mammal collection in the Cleveland Museum of Natural History. Scientific Publication of the Cleveland Museum of Natural History, 5:83-181.

Bolles, K. 1981. Variation and alarm call evolution in antelope squirrels, *Ammospermophilus* (Rodentia: Sciuridae). Dissertation Abstracts International, 41:2857B.

Bolliger, T. 1996. A current understanding about the Anomalomyidae (Rodentia): Reflections on stratigraphy, paleobiogeography, and evolution. Pp. 235-245, *in* The evolution of western Eurasian Neogene mammal faunas (R. L. Bernor, V. Fahlbusch, and H.-W. Mittmann, eds.). Columbia University Press, New York, 487 pp.

Bolliger, T. 1999. Family Anomalomyidae. Pp. 411-420, *in* The Miocene land mammals of Europe (G. E. Rössner and K. Heissig, eds.). Dr. Friedrich Pfeil, München, 515 pp.

Bolshakov, V. N., E. A. Gileva, and G. V. Bykova. 1985. Chromosome variation in the Asian mountain vole, *Alticola macrotis* Radde, 1861 (Rodentia, Cricetidae). The Annals of Zoology, 23:53-69.

Bonaccorso, F. J. 1998. Bats of Papua New Guinea. Conservation International Tropical Field Guide Series, Conservation International, Washington, D.C., 489 pp.

Bonaparte, C. L. J. L. 1838. Synopsis vertebratorum systematis. Nuovi Annali delle Scienze Naturali, Bologna, 1:105-133.

Bonaparte, C.-L. J. L. 1845. Catalogo methodico dei mammiferi Europei. L. di Giacomo Pirola, Milano, 36 pp.

Bonaparte, C.-L. J. L. 1850. Conspectus systematis mastozoologie. Editio altera reformata. E. J. Brill, Lugduni Batavorum, Leiden, 2 pp.

Bonde, R. K., and T. J. O'Shea. 1989. Sowerby's beaked whale (*Mesoplodon bidens*) in the Gulf of Mexico. Journal of Mammalogy, 70(2):447-449.

Bonhomme, F. J. 1986. Evolutionary relationships in the genus *Mus*. Pp. 19-34, *in* Current topics in microbiology and immunology, Vol. 127 (M. Potter, J. H. Nadeau, and M. P. Cancro, eds.). Springer-Verlag, Berlin, 395 pp.

Bonhomme, F., J. Catalan, J. Britton-Davidian, V. M. Chapman, K. Moriwaki, E. Nevo, and L. Thaler. 1984. Biochemical diversity and evolution in the genus *Mus*. Biochemical Genetics, 22:275-303.

Bonhomme, F., D. Iskandar, L. Thaler, and F. Petter. 1985. Electromorphs and phylogeney in muroid rodents. Pp. 671-683, *in* Evolutionary relationships among rodents, a multidisciplinary analysis (W. P. Luckett and J.-L. Hartenberger, eds.). Plenum Press, New York, 721 pp.

Bonhomme, F., J. Fernández, F. Palacios, J. Catalan, and A. Machordom. 1986. Charactérisation biochimique du complexe d'espèces du genre *Lepus* en Espagne. Mammalia, 50:495-506.

Bonhomme, F., J.-L. Guenet, B. Dod, K. Moriwaki, and G. Bulfield. 1987. The polyphyletic origin of laboratory inbred mice and their rate of evolution. Biological Journal of the Linnean Society, 30:51-58.

Bonhomme, F., N. Miyashita, P. Boursot, J. Catalan, and K. Moriwaki. 1989. Genetical variation and polyphyletic origin in Japanese *Mus musculus*. Heredity, 63:299-308.

Bonhomme, F., R. Anand, D. Darviche, W. Din and P. Boursot. 1994. The house mouse as a ring species? Pp. 13-23, *in* Genetics in Wild Mice. Its Application to Biomedical Research (K. Moriwaki, T. Shiroishi, and H. Yonekawa, eds.). Japan Scientific Societies Press, Tokyo, 333 pp.

Bonhote, J. L. 1900 [1901]. On the mammals collected during the 'Skeat Expedition' to the Malay Peninsula, 1899-1900. Proceedings of the Zoological Society of London, 1900:869-883.

Bonhote, J. L. 1901*a*. On the squirrels of the *Sciurus erythraeus* group. Annals and Magazine of Natural History, ser. 7, 7:160-167.

Bonhote, J. L. 1901*b*. On the martens of the *Mustela flavigula* group. Annals and Magazine of Natural History, ser. 7, 7:342-349.

Bonvicino, C. R. 1999. New taxonomic status to the French Guianan *Nectomys parvipes* Petter (Rodentia, Sigmodontinae). Revista Brasileira de Zoologia, 16:253-255.

Bonvicino, C. R. 2003. A new species of *Oryzomys* (Rodentia, Sigmodontinae) of the *subflavus* group from the Cerrado of central Brazil. Mammalian Biology/Zeitschrift für Säugetierkunde, 68:78-90.

Bonvicino, C. R., and F. C. Almeida. 2000. Karyotype, morphology, and taxonomic status of *Calomys expulsus* (Rodentia: Sigmodontinae). Mammalia, 64:339-351.

Bonvicino, C. R., and L. Geise. 1995. Taxonomic status of *Delomys dorsalis collinus* Thomas, 1917 (Rodentia, Cricetidae) and description of a new karyotype. Zeitschrift für Säugetierkunde, 60:124-127.

Bonvicino, C. R., and M. A. Martins Moreira. 2001. Molecular phylogeny of the genus *Oryzomys* (Rodentia, Sigmodontinae) based on cytochrome *b* DNA sequences. Molecular Phylogenetics and Evolution, 18:282-292.

Bonvicino, C. R., and I. Otazu. 1999. The *Wilfredomys pictipes* (Rodentia: Sigmodontinae) karyotype with comments on the karyosystematics of Brazilian *Thomasomyini*. Acta Theriologica, 44:329-332.

Bonvicino, C. R., and M. Weksler. 1998. A new species of *Oligoryzomys* (Rodentia, Sigmodontinae) from northeastern and central Brazil. Zeitschrift für Säugetierkunde, 63:90-103.

Bonvicino, C. R., P. S. D'Andrea, R. Cerqueira, and H. N. Seuanez. 1996. The chromosomes of *Nectomys* (Rodentia, Cricetidae) with 2n = 52 and 2n =56, and interspecific hybrids (2n = 54). Cytogenetics and Cell Genetics, 73:190-193.

Bonvicino, C. R., I. Otazu, and M. Wexler. 1998*a*. *Oryzomys lamia* Thomas, 1901 (Rodentia, Cricetidae): Karyotype, geographic distribution, and conservation status. Mammalia, 62:253-258.

Bonvicino, C. R., V. Penna-Firme, and H. N. Seuanez. 1998*b*. The karyotype of *Brucepattersonius griserufescens* Hershkovitz, 1998 (Rodentia, Sigmodontinae) with comments on distribution and taxonomy. Zeitschrift für Säugetierkunde, 63:329-335.

Bonvicino, C. R., I. Otazu, and P. M. Borodin. 1999. Chromosome variation in *Oryzomys subflavus* species group (Sigmodontinae, Rodentia). Cytologia, 64:327-332.

Bonvicino, C. R., F. C. Almeida, and R. Cerqueira. 2000. The karyotype of *Sphiggurus villosus* (Rodentia: Erethizontidae). Studies on Neotropical Fauna and Environment, 35:81-83.

Bonvicino, C. R., J. A. de Oliveira, P. S. D'Andrea, and R. W. de Carvalho. 2001. The endemic Atlantic forest rodent *Phaenomys ferrugineus* (Thomas, 1894) (Sigmodontinae): New data on its morphology and karyology. Boletim do Museu Nacional, n.s., Zoologia, 467:1-12.

Bonvicino, C. R., V. Penna-Firme, and E. Braggio. 2002*a*. Molecular and karyotypic evidence of the taxonomic status of *Coendou* and *Sphiggurus* (Rodentia: Hystricognathi). Journal of Mammalogy, 83(4):1071-1076.

Bonvicino, C. R., I. B. Otazu, and P. S. D'Andrea. 2002*b*. Karyologic evidence of diversification of the genus *Thrichomys* (Rodentia, Echimyidae). Cytogenetic and Genome Research, 97:200-204.

Bonvicino, C. R., A. R. E. A. N. de Menezes, and J. A. Oliveira. 2003*a*. Molecular and karyologic variation in the genus *Isothrix* (Rodentia, Echimyidae). Hereditas, 139:206-211.

Bonvicino, C. R., J. P. Boubli, I. B. Otazú, F. C. Almeida, F. F. Nscimento, J. R. Coura, and H. N. Seuánez. 2003*b*. Morphologic, karyotypic, and molecular evidence of a new form of *Chiropotes* (Primates, Pitheciinae). American Journal of Primatology, 61:123-133.

Bonvicino, C. R., J. F. S. Lima, and F. C. Almeida. 2003*c*. A new species of *Calomys* Waterhouse (Rodentia, Sigmodontinae) from the Cerrado of central Brazil. Revista Brasileira de Zoologia, 20:301-307.

Bonvicino, C. R., L. S. Maroja, J. A. de Oliveira, and J. R. Coura. 2003d. Karyology and morphology of
    Zygodontomys (Rodentia, Sigmodontinae) from the Brazilian Amazon, with a molecular appraisal of
    phylogenetic relationships of this genus. Mammalia, 67:119-131.
Boone, J. L., J. Laerm, and M. H. Smith. 1993. Taxonomic status of *Peromyscus gossypinus anastasae*
    (Anastasia Island Cotton Mouse). Journal of Mammalogy, 74:363-375.
Boone, J. L., M. H. Smith, and J. Laerm. 1999. Allozyme variation in the cotton mouse (*Peromyscus
    gossypinus*). Journal of Mammalogy, 80:833-844.
Booth, C. P. 1968. Taxonomic studies of *Cercopithecus mitis* Wolf (East Africa). National Geographic Society—
    Research Report. Abstracts and reviews of research and exploration authorized under grants from the
    National Geographic Society during the year 1963, pp. 37-51.
Borg, J. J. 1998. The lesser mouse-eared bat *Myotis blythi punicus* Felten, 1977 in Malta. Notes on status,
    morphometrics, movements, and diet (Chiroptera: Vespertilionidae). Naturalista siciliano. S. IV,
    22:365-374.
Borisov, Yu. M., and V. M. Malygin. 1991. [Clinal variability of the B-chromosome system in *Apodemus
    peninsulae* (Rodentia, Muridae) from the Buryatia and Mongolia]. Tsitologiya, 33:106-111 (in Russian).
Borkenhagen, P. 1996. Zweiter nachweis einer birkenmaus (*Sicista betulina*) in Schleswig-Holstein. Bonner
    Zoologische Beiträge, 46(1-4):141-142.
Borkowska, A. 1995. Seasonal changes in gut morphology of the striped field mouse (*Apodemus agrarius*).
    Canadian Journal of Zoology, 73:1095-1099.
Borkowska, A., and M. Ratkiewicz. 1996. Relationships between allozyme heterozygosity and gut
    morphology in *Apodemus agrarius*. Acta Theriologica, 41(4):367-374.
Borobia, M., S. Siciiliano, L. Lodi, and W. Hoek. 1991. Distribution of the South American dolphin *Sotalia
    fluviatilis*. Canadian Journal of Zoology, 69:1025-1039.
Borodin, L. P. 1963. The Russian desman. Saransk, Russia. 303 pp.
Borodin, P. M., M. B. Rogatcheva, K. Koyasu, K. Fukuta, K. Mekada, and S. I. Oda. 1997. Pattern of X-Y
    chromosome pairing in the Japanese field vole, *Microtus montebelli*. Genome, 40:829-833.
Borowik, O. A., and M. D. Engstrom. 1993. Chromosomal evolution and biogeography of collared
    lemmings (*Dicrostonyx*) in the eastern and High Arctic of Canada. Canadian Journal of Zoology,
    71:1481-1493.
Borroto, R. 2002. Systemática de las jutías de las Antillas (Rodentia: Capromyidae). Tesis en Opción al Grado
    Científico de Doctor en Ciencias Biológicas, Instituto de Ecología y Sistemática, CITMA. C. Habana, 100
    pp + 30 fig. + 16 tables + 6 anex.
Borroto, R., and I. Ramos. 2003. Current status of the carabali hutia from South of Isla de la Juventud,
    Mysateles meridionalis. Orsis, 18:7-12.
Borroto, R., A. Camacho, and I. Ramos. 1992. Variation in three populations of *Capromys pilorides* (Rodentia:
    Capromyidae), and the description of a new subspecies from the south of the Isle of Youth (Cuba).
    Miscelanea Zoologica Hungarica, 7:87-99.
Borroto, R., I. Ramos, A. Rodriguez, R. Alonso, C. Mancina, M. Condis, A. Daniel, G. Begue, R. Estrada, R.
    Fernandez, and A. Gonzalez. 2001. Estudio para la conservacion de la fauna de vertebrados del Parque
    Nacional "Alejandro de Humboldt", Guantanamo. Okologische Hefte, 14:16-21.
Boschma, H. 1950. Maxillary teeth in specimens of *Hyperoodon rostratus* (Müller) and *Mesoplodon grayi* von
    Haast stranded on the Dutch coasts. Koninklijke Nederlandsche Akademie van Wetenschappen,
    Proceedings, 53(6):3-14, 4 pls.
Boshell-M., J., and P. K. Rajagopalan. 1968. Small rodents and shrews in the Sagar-Sorab area, Mysore State,
    India. Population Studies–1961-1964. Indian Journal of Medical Research, 46(4):527-540.
Bothma, J. du P. 1971. Order Hyracoidea. Part 12, Pp. 1-8, *in* The mammals of Africa: An identification
    manual (J. Meester and H. W. Setzer, eds.) [issued 15 Jul 1971]. Smithsonian Institution Press,
    Washington, D.C., not continuously paginated.
Bouchard, S. 1998. *Chaerephon pumilus*. Mammalian Species, 547:1-6.
Bouchard, S. 2001. *Chaerephon ansorgei*. Mammalian Species, 660:1-3.
Boulay, M. C., and C. B. Robbins. 1989. *Epomophorus gambianus*. Mammalian Species, 344:1-5.
Boursot, P., T. Jacquart, F. Bonhomme, J. Britton-Davidian, and L. Thaler. 1985. Différenciation
    géographique du génome mitochondrial chez *Mus spretus* Lataste. Comptes Rendus de l'Académie des
    Sciences (Paris), ser. 3, 301:161-166.
Boursot, P., J.-C. Auffray, J. Britton-Davidian, and F. Bonhomme. 1993. The evolution of house mice. Annual
    Review of Ecology and Systematics, 24:119-152.
Boursot, P., W. Din, R. Anand, D. Darviche, B. Dod, F. Von Deimling, G. P. Talwar, and F. Bonhomme. 1996.
    Origin and radiation of the house mouse: Mitochondrial DNA phylogeny. Journal of Evolutionary
    Biology, 9:391-415.

Bowen, W. W. 1968. Variation and evolution of Gulf coast populations of beach mice, *Peromyscus polionotus*. Bulletin of the Florida State Museum, 12:1-91.

Bowers, J. H. 1974. Genetic compatibility of *Peromyscus maniculatus* and *Peromyscus melanotis*, as indicated by breeding studies and morphometrics. Journal of Mammalogy, 55:720-737.

Bowers, J. H., R. J. Baker, and M. H. Smith. 1973. Chromosomal, electrophoretic, and breeding studies of selected populations of deer mice (*Peromyscus maniculatus*) and black-eared mice (*P. melanotis*). Evolution, 27:378-386.

Bowers, K. L., M. J. Hamilton, S. M. White, and R. J. Baker. 1998. Origins of heterochromatic repatterning in white-footed mice, *Peromyscus leucopus*. Journal of Mammalogy, 79:725-735.

Bowyer, R. T., and D. M. Leslie, Jr. 1992. *Ovis dalli*. Mammalian Species, 393:1-7.

Boye, P. 1991. Notes on the morphology, ecology and geographic origin of the Cyprus long-eared hedgehog (*Hemiechinus auritus dorotheae*). Bonner Zoologische Beiträge, 42:115-123.

Boye, P., R. Hutterer, N. López-Martínez, and J. Michaux. 1992. A reconstruction of the lava mouse (*Malpaisomys insularis*), an extinct rodent of the Canary Islands. Zeitschrift für Säugetierkunde, 57:29-38.

Boyeskorov, G. 1999. New data on moose (*Alces*, Artiodactyla) systematics. Säugetierkundliche Mitteilungen, 44:3-13.

Boyeskorov, G. G., U. Timm, and E. A. Lyapunova. 1992. Karyological study of two *Apodemus* species (Rodentia, Muridae) from the Baltic countries. Pp. 81-87, *in* Proceedings of the First Baltic Theriological Conference (A. Kirk, A. Miljutin, and T. Randveer, eds.). Tartu Ülikooli Toimetised (=Acta et Commentationes Universitatis Tartuensis), no. 955, 254 pp.

Boyeskorov, G. G., N. G. Yegorov, and Y. V. Revin. 1993. [*Microtus hyperboreus* (Mammalia, Rodentia) in south-eastern Yakutia.] Vestnik Zoologi, 1993(2):72-74 (in Russian with English abstract).

Boyeskorov, G. G., E. A. Lyapunova, N. N. Vorontsov, and J. Jadav. 1997. [The karyology of *Mus dunni* (Mammalia, Rodentia, Muridae).] Zoologicheskii Zhurnal, 76(10):1225-1227 (in Russian with English summary).

Boyeskorov, G. G., E. I. Zholnerovskaya, N. N. Vorontsov, and E. A. Lyapunova. 1999. [Intraspecific divergence of the black-capped marmot *Marmota camtschatica* (Sciuridae, Marmotinae)]. Zoologicheskii Zhurnal, 78(7):866-877 (in Russian).

Bradley, R. D., and D. J. Schmidly. 1987. The glans penes and bacula in Latin American taxa of the *Peromyscus boylii* group. Journal of Mammalogy, 68:595-616.

Bradley, R. D., R. A. Bussache, C. A. Porter, and M. J. Hamilton. 1988. The harvest mouse, *Reithrodontomys humulis*, in central Oklahoma, with comments on its karyotype. Texas Journal of Science, 40:449-450.

Bradley, R. D., D. J. Schmidly, and R. D. Owen. 1989. Variation in the glans penes and bacula among Latin American populations of the *Peromyscus boylii* species complex. Journal of Mammalogy, 70:712-725.

Bradley, R. D., S. K. Davis, S. F. Lockwood, J. W. Bickham, and R. J. Baker. 1991. Hybrid breakdown and cellular-DNA content in a contact zone between two species of pocket gophers (*Geomys*). Journal of Mammalogy, 72:697-705.

Bradley, R. D., D. J. Schmidly, and C. W. Kilpatrick. 1996a. The relationships of *Peromyscus sagax* to the *P. boylii* and *P. truei* species groups in Mexico based on morphometric, karyotypic, and allozymic data. Pp. 95-106, *in* Contributions in mammalogy: A memorial volume in honor of J Knox Jones, Jr. (H. H. Genoways and R. J. Baker, eds.). Museum of Texas Tech University, Lubbock, Texas, il + 315 pp.

Bradley, R. D., R. D. Owen, and D. J. Schmidly. 1996b. Morphological variation in *Peromyscus spicilegus*. Occasional Papers, Museum of Texas Tech University, 159:1-23.

Bradley, R. D., D. S. Carroll, M. L. Clary, C. W. Edwards, I. Tiemann-Boege, M. J. Hamilton, R. A. Van den Bussche, and C. Jones. 1999a. Comments on some small mammals from the Big Bend and Trans-Pecos regions of Texas. Occasional Papers, Museum of Texas Tech University, 193:1-6.

Bradley, R. D., D. J. Schmidly, and C. Jones. 1999b. The northern rock mouse, *Peromyscus nasutus*, (Mammalia: Rodentia), from the Davis Mountains, Texas. Occasional Papers, Museum of Texas Tech University, 190:1-3.

Bradley, R. D., I. Tiemann-Boege, C. W. Kilpatrick, and D. J. Schmidly. 2000. Taxonomic status of *Peromyscus boylii sacarensis*: Inferences from DNA sequences of the mitochondrial cytochrome *b* gene. Journal of Mammalogy, 81:875-884.

Bradshaw, W. N. 1968. Progeny from experimental mating tests with mice of the *Peromyscus leucopus* group. Journal of Mammalogy, 49:475-480.

Bradshaw, W. N., and T. C. Hsu. 1972. Chromosomes of *Peromyscus* (Rodentia, Cricetidae). III. Polymorphism in *Peromyscus maniculatus*. Cytogenetics, 11:436-451.

Braestrup, F. W. 1935. Report on the mammals collected by Mr. Harry Madsen during professor O. Ofufsen's expedition to French Sudan and Nigeria in the years 1927-28. Videnskabelige Meddeleser fra Dansk Naturhistoriske Forening i Kobenhaven, 99:73-130.

Braestrup, F. W., and R. Hutterer. 1985. *Grammomys macmillani tuareg* Braestrup, 1935: A junior synonym of *Praomys daltoni* (Thomas, 1892). Zeitschrift für Säugetierkunde, 50:240-241.

Braggio, E., and C. R. Bonvicino. 2004. Molecular divergence in the genus *Thrichomys* (Rodentia, Echimyidae). Journal of Mammalogy, 85(2):7-11.

Braithwaite, R. W., and P. R. Baverstock. 1995. Pale field-rat, *Rattus tunneyi*. Pp. 663-664, *in* Mammals of Australia (R. Strahan, ed.). Smithsonian Institution Press, Washington, D.C., 756 pp.

Braithwaite, R. W., and J. Covacevich. 1995. Delicate mouse, *Pseudomys delicatulus*. Pp. 592-593, *in* Mammals of Australia (R. Strahan, ed.). Smithsonian Institution Press, Washington, D.C., 756 pp.

Brand, L. R., and R. E. Ryckman. 1969. Biosystematics of *Peromyscus eremicus*, *P. guardia*, and *P. interparietalis*. Journal of Mammalogy, 50:501-513.

Brandle, R., and K. E. Moseby. 1999. Comparative ecology of two populations of *Pseudomys australis* in northern South Australia. Wildlife Research, 26:541-564.

Brandle, R., K. E. Moseby, and M. Adams. 1999. The distribution, habitat requirements and conservation status of the plains rat, *Pseudomys australis* (Rodentia: Muridae). Wildlife Research, 26:463-477.

Brandon-Jones, D. 1978. Comment on the proposal to conserve Colobidae Blyth, 1875, as a family-group name for the leaf-eating monkeys. Bulletin of Zoological Nomenclature, 35:69-70.

Brandon-Jones, D. 1984. Colobus and leaf-monkeys. Pp. 398-408, *in* The encyclopedia of mammals (D. Macdonald, ed.). Facts on File, New York, 895 pp.

Brandon-Jones, D. 1995. A revision of the Asian pied leaf-monkeys (Mammalia: Cercopithecidae: Superspecies *Semnopithecus auratus*), with a description of a new subspecies. Raffles Bulletin of Zoology, 43:3-43.

Brandon-Jones, D., and C. P. Groves. 2002. Neotropical Primate family-group names replaced by Groves (2001) in contravention of Article 40 of the International Code of Zoological Nomenclature. Neotropical Primates, 10:113-115.

Brandt, J. F. 1844 [1843]. Observations sur les différentes espèces de sousliks de Russie, suivies de remarques sur l'arrangement et la distribution géographique du genre *Spermophilus*, ansé que sur la classification de la familie des ecureuils (Sciurina) en général. Bulletin Scientifique l'Académie Impériale des Sciences de Saint-Pétersbourg, 1844:col. 357-382.

Brandt, J. F. 1855. Beitrage zur nahern Kenntniss der Säugethiere Russland's. Kaiserlichen Akademie der Wissenschaften, Saint Petersburg, Mémoires Mathématiques, Physiques et Naturelles, 7:1-365.

Brandt, J. H., M. Dioli, A. Hasssanin, R. A. Melville, L. E. Olson, A. Seveau, and R. M. Timm. 2001. Debate on the authenticity of *Pseudonovibos spiralis* as a new species of wild bovid from Vietnam and Cambodia. Journal of Zoology, 255:437-444.

Brandt, R. S., and L. M. Pessóa. 1994. Intrapopulational variability in cranial characters of *Oryzomys subflavus* (Wagner, 1842) (Rodentia, Cricetidae), in northeastern Brazil. Zoologischer Anzeiger, 233:45-55.

Brandt, S. V., and G. Ortí. 2002. Molecular phylogeney of short-tailed shrews, *Blarina* (Insectivora: Soricidae). Molecular Phylogenetics and Evolution, 22:163-173.

Brandy, L. D. 1978. Données nouvelles sur l'évolution du Rongeur endémique fossile corso-sarde *Rhagamys* F. Major (1905) (Mammalia, Rodentia). Bulletin de la Société Géologique de France, 20(6):831-835.

Brandy, L. D. 1981. Rongeurs muroïdés du Néogène supérieur d'Afghanistan. Évolution, biogéographie, corrélations. Palaeovertebrata, 11(4):133-179.

Brandy, L. D., M. Sabatier, and J.-J. Jaeger. 1980. Implications phylogenetiques et biogeographiques des dernieres decouvertes de Muridae en Afghanistan, au Pakistan et en Ethiopie. Geobios, 14(4):639-643.

Brants, A. 1827. Het Geslacht der muizen door Linnaeus opgesteld, volgens de tegenswoordige toestand der wetenschap in familiën, geslachten en soorten verdeeld. Academische Boekdrukkerij, Berlijn. 190 pp.

Brass, L. J. 1959. Results of the Archbold Expeditions. No. 86. Summary of the Sixth Archbold Expedition to New Guinea (1959). Bulletin of the American Museum of Natural History, 127:145-216.

Braun, A., C. P. Groves, P. Grubb, Q. Yang, and L. Xia. 2001. Catalogue of the Musée Heude collection of mammal skulls. Acta Zootaxonomica Sinica, 26:608-660.

Braun, J. K. 1988. Systematics and biogeography of the southern flying squirrel, *Glaucomys volans*. Journal of Mammalogy, 69:422-426.

Braun, J. K. 1993. Systematic relationships of the Tribe Phyllotini (Muridae: Sigmodontinae) of South America. Oklahoma Museum of Natural History Special Publication, 2:50 pp.

Braun, J. K., and M. A. Mares. 1989. *Neotoma micropus*. Mammalian Species, 330:1-9.

Braun, J. K., and M. A. Mares. 1995a. A new genus and species of phyllotine rodent (Rodentia: Muridae: Sigmodontinae: Phyllotini) from South America. Journal of Mammalogy, 76:504-521.

Braun, J. K., and M. A. Mares. 1995b. The mammals of Argentina: An etymology. Mastozoología Neotropical, 2:173-206.

Braun, J. K., and M. A. Mares. 2002. Systematics of the *Abrocoma cinerea* species complex (Rodentia: Abrocomidae), with a description of a new species of *Abrocoma*. Journal of Mammalogy, 83(1):1-19.

Braun, J. K., M. A. Mares, and R. A. Ojeda. 2000. A new species of grass mouse, genus *Akodon* (Muridae: Sigmodontinae), from Mendoza Province, Argentina. Zeitschrift für Säugetierkunde, 65:216-225.

Breda, M. 2002. Morphological and biometrical study on cranial and dental remains of *Sorex araneus, Sorex samniticus* and *Sorex arunchi* (Mammalia, Insectivora, Soricidae). Bolletino del Museo Civico di Storia Naturale di Verona, 26:65-73.

Breed, W. G. 1980. Further observations on spermatozoal morphology and male reproductive tract anatomy of *Pseudomys* and *Notomys* species (Mammalia: Rodentia). Transactions of the Royal Society of South Australia, 104:51-55.

Breed, W. G. 1983. Variation in sperm morphology in the Australian rodent genus *Pseudomys* (Muridae). Cell and Tissue Research, 229(3):611-625.

Breed, W. G. 1984. Sperm head structure in the Hydromyinae (Rodentia: Muridae): A further evolutionary development of the subacrosomal space in mammals. Gamete Research, 10:31-44.

Breed, W. G. 1985. Morphological variation in the female reproductive tract of Australian rodents in the genera *Pseudomys* and *Notomys*. Journal of Reproduction and Fertility, 73:379-384.

Breed, W. G. 1986. Comparative morphology and evolution of the male reproductive tract in the Australian hydromyine rodents (Muridae). Journal of Zoology, London, ser. A, 209:607-629.

Breed, W. G. 1990. Reproductive anatomy and sperm morphology of the long-tailed hopping-mouse, *Notomys longicaudatus* (Rodentia: Muridae). Australian Mammalogy, 13:201-204.

Breed, W. G. 1991. Variation in sperm head morphology of the Murinae (Family Muridae): Some taxonomic implications. Pp. 927-929, *in* Comparative spermatology 20 years after (B. Baccetti, ed.). Raven Press, New York, 1112 pp.

Breed, W. G. 1993. Novel organisation of the spermatozoon in two species of murid rodents from southern Asia. Journal of Reproduction and Fertility, 99:149-158.

Breed, W. G. 1995*a*. Spermatozoa of murid rodents from Africa: Morphological diversity and evolutionary trends. Journal of Zoology, London, 237:625-651.

Breed, W. G. 1995*b*. Spinifex hopping-mouse, *Notomys alexis*. Pp. 568-570, *in* Mammals of Australia (R. Strahan, ed.). Smithsonian Institution Press, Washington, D.C., 756 pp.

Breed, W. G. 1995*c*. Sandy inland mouse, *Pseudomys hermannsburgensis*. Pp. 604-605, *in* Mammals of Australia (R. Strahan, ed.). Smithsonian Institution Press, Washington, D.C., 756 pp.

Breed, W. G. 1995*d*. Variation in sperm head morphology of muroid rodents of Africa: Phylogenetic implications. Pp. 409-420, *in* Advances in spermatozoal phylogeny and taxonomy (B. G. M. Jamieson, J. Ausio, and J.-L. Justine, eds.). Mémoires du Muséum National D'Histoire Naturelle, Éditions du Muséum, Paris, 564 pp.

Breed, W. G. 1997. Evolution of the spermatozoon in Australasian rodents. Australian Journal of Zoology, 45:459-478.

Breed, W. G. 1998. Interspecific variation in structural organisation of the spermatozoon in the Asian bandicoot rats, *Bandicota* species (family Muridae). Acta Zoologica, 79(4):277-285.

Breed, W. G. 2000. Taxonomic implications of variation in sperm head morphology of the Australian delicate mouse, *Pseudomys delicatulus*. Australian Mammalogy, 21:193-199.

Breed, W. G. 2004. The spermatozoon of Eurasian murine rodents: its morphological diversity and evolution. Journal of Morphology, 261:52-69.

Breed, W. G., and K. P. Aplin. 1994. Sperm morphology of murid rodents from New Guinea and the Solomon Islands: Phylogenetic implications. Australian Journal of Science, 43:17-30.

Breed, W. G., and W. Head. 1991. Conservation status of the plains rat *Pseudomys australis* (Rodentia: Muridae). Australian Mammalogy, 14:125-128.

Breed, W. G., and G. G. Musser. 1991. Sulawesi and Philippine rodents (Muridae): A survey of spermatozoal morphology and its significance for phylogenetic inference. American Museum Novitates, 3003:1-15.

Breed, W. G., and N. Pillay. 1999. Morphology of the spermatozoon of a murid rodent from Africa, *Dasymys incomtus*. Acta Zoologica (Stockholm), 80:201-208.

Breed, W. G., and V. Sarafis. 1978. On the phylogenetic significance of spermatozoal morphology and male reproductive tract anatomy in Australian rodents. Transactions of the Royal Society of South Australia, 103:127-135.

Breed, W. G., and V. Sarafis. 1983. Variation in sperm head morphology in the Australian rodent *Notomys alexis*. Australian Journal of Zoology, 31:313-316.

Breed, W. G., and J. Taylor. 2000. Body mass, testes mass, and sperm size in murine rodents. Journal of Mammalogy, 81(3):758-768.

Breed, W. G., and H. S. Yong. 1986. Sperm morphology of murid rodents from Malaysia and its possible phylogenetic significance. American Museum Novitates, 2856:1-12.

Breed, W. G., G. A. Cox, C. M. Leigh, and P. Hawkins. 1988. Sperm head structure of a murid rodent from southern Africa: The red veld rat *Aethomys chrysophilus*. Gamete Research, 19:191-202.

Breed, W. G., D. Idriss, and R. J. Oko. 2000. Protein composition of the ventral processes on the sperm head of Australian hydromyine rodents. Biology of Reproduction, 63:629-634.

Brickell, J. 1737. The natural history of North-Carolina: With an account of the trade, manners, and customs of the Christian and Indian inhabitants. J. Brickell, Dublin, 417 pp.

Bridelance, P. 1987. Émissions sonores rythmées engendrées par des tapements de pattes: Les podophones chez les rongeurs. Comptes Rendus de l'Académie des Sciences (Paris), ser. 3, 305(5):125-128.

Briggs, J. C. 1974. Marine Zoogeography. McGraw Hill, New York, 475 pp.

Briggs, K. T., and V. G. Morejohn. 1976. Dentition, cranial morphology and evolution in elephant seals. Mammalia, 40:199-222.

Bright, P. W., and P. A. Morris. 1992. The dormouse. Mammal Society, London, 22 pp.

Bright, P. W., P. A. Morris and A. J. Mitchell-Jones. 1996. A new survey of the dormouse *Muscardinus avellanarius* in Britain, 1993-4. Mammal Review, 26(4):189-195.

Brink, C., S. Erlinge, M. Sandell. 1983. Anal sac secretion in mustelids. A comparison. Journal of Chemical Ecology, 9:727-745.

Briones-Salas, M. A., V. Sánchez-Cordero, and G. Q. Altamirano. 2001. Lista de mamíferos terrestres del norte del estado de Oaxaca, Mexico. Anales del Instituto de Biología UNAM, Serie Zoología, 72:125-161.

Briscoe, D. A., B. J. Fox, and S. Ingleby. 1981. Genetic differentiation between *Pseudomys pilligaensis* and related *Pseudomys*. Australian Mammalogy, 4:89-92.

Briscoe, D. A., J. H. Calaby, R. L. Close, G. M. Maynes, C. E. Murtagh, and G. B. Sharman. 1982. Isolation, introgression and genetic variation in rock-wallabies. Pp. 73-87, *in* Species at risk: Research in Australia (R. H. Groves and W. D. L. Ride, eds.). Australian Academy of Science, Canberra, 216 pp.

Brisson, M. J. 1762. Le regnum animale in classes IX distributum, sive synopsis methodica sistens generalem animalium distributionem in classes IX, & duarum primarum classium, quadrupedum scilicet & cetaceorum, particularem dibvisionem in ordines, sectiones, genera & species. T. Haak, Paris, 296 pp.

Britton-Davidian, J. 1990. Genic differentiation in *M. m. domesticus* populations from Europe, the Middle East and North Africa: Geographic patterns and colonization events. Biological Journal of the Linnean Society, 41:27-45.

Britton-Davidian, J., M. Vahdat, F. Benmehdi, P. Gros, V. Nance, H. Croset, S. Guerassimov, and C. Triantaphyllidis. 1991. Genetic differentiation in four species of *Apodemus* from southern Europe: *A. sylvaticus*, *A. flavicollis*, *A. agrarius* and *A. mystacinus* (Muridae, Rodentia). Zeitschrift für Säugetierkunde, 56:25-33.

Britton-Davidian, J., J. Catalan, L. Granjon, and J.-M. Duplantier. 1995. Chromosomal phylogeny and evolution in the genus *Mastomys* (Mammalia, Rodentia). Journal of Mammalogy, 76(1):248-262.

Britton-Davidian, J., J. Catalan, M. da G. Ramalhinho, G. Ganem, J.-C. Auffray, R. Capela, M. Biscoits, J. B. Searle, and M. da L. Mathias. 2000. Rapid chromosomal evolution in island mice. Nature, 403:158.

Britton-Davidian, J., J. Catalan, and K. Belkhir. 2002. Chromosomal and allozyme analysis of a hybrid zone between parapatric Robertsonian races of the house mouse: A case of monobrachial homology. Cytogenetic and Genome Research, 96:75-84.

Broadbooks, H. E. 1965. Ecology and distribution of the pikas of Washington and Alaska. American Midland Naturalist, 73:299-335.

Brodie, P. F. 1989. The white whale—*Delphinapterus leucas* (Pallas, 1776). Pp. 119-144, *in* Handbook of marine mammals: River dolphins and the larger toothed whales (S. H. Ridgway and R. Harrison, eds.). Academic Press, London, 4:1-442.

Brodkorb, P. 1961. Recently described birds and mammals from Cuban caves. Journal of Paleontology, 35(3):633-635.

Brongersma, L. D. 1940. Note on *Mustela lutreolina* Rob. & Thom. Temminckia, 5:257-263.

Bronner, G. N. 1990. New distribution records for four mammal species, with notes on their taxonomy and ecology. Koedoe, 33(2):1-7.

Bronner, G. N. 1995*a*. Systematic revision of the golden mole genera *Amblysomus*, *Chlorotalpa* and *Calcochloris* (Insectivora:Chrysochloromorpha; Chrysochloridae). Ph.D. thesis, University of Natal, Durban.

Bronner, G. N. 1995*b*. Cytogenetic properties of nine species of golden moles (Insectivora: Chrysochloridae). Journal of Mammalogy, 76:957-971.

Bronner, G. N. 1996. Geographic patterns of morphometric variation in the Hottentot golden mole, *Amblysomus hottentotus* (Insectivora: Chrysochloridae): A multivariate analysis. Mammalia, 60:729-751.

Bronner, G. N. 2000. New species and subspecies of golden mole (Chrysochloridae: *Amblysomus*) from Mpumalanga, South Africa. Mammalia, 64:41-54.

Bronner, G. N., and J. A. J. Meester. 1988. *Otomys angoniensis*. Mammalian Species, 306:1-6.

Bronner, G. N., S. Gordon, and J. A. J. Meester. 1988. *Otomys irroratus*. Mammalian Species, 308:1-6.

Bronner, G. N., M. Hoffman, P. J. Taylor, C. T. Chimimba, P. B. Best, C. A. Mathee, and T. J. Robinson. 2003. A revised systematic checklist of the extant mammals of the southern African subregion. Durban Museum Novitates, 28:56-106.

Brooks, A. C. 1961. A study of the Thomson's gazelle (*Gazella thomsoni* Günther) in Tanganyika. Her Majesty's Stationary Office, London, 147 pp.

Brooks, D. M., R. E. Bodmer, and S. M. Matola (eds). 1997. Tapirs. Status survey and conservation action plan. I.U.C.N., Gland, 164 pp.

Broom, R. 1916. On the structure of the skull in *Chrysochloris*. Proceedings of the Zoological Society of London, 1916:449-459.

Brosset, A. 1988. Le peuplement de mammifères insectivores des forêts du nord-est du Gabon. Revue de Ecologie (La Terre et la Vie), 43:23-46.

Brosset, A., and P. Charles-Dominique. 1990. The bats from French Guiana: A taxonomic, faunistic and ecological approach. Mammalia, 54:509-559.

Brosset, A., G. Dubost, and H. Heim de Balsac. 1965. Mammifères inedits recoltes au Gabon. Biologia Gabonica, 1:148-174.

Brown, K. M., and J. Dunlop. 1997. *Rhinolophus landeri*. Mammalian Species, 567:1-4.

Brown, L. N. 1997. A guide to the mammals of the southeastern United States. The University of Tennessee Press, Knoxville, 236 pp.

Brown, L. N., and R. J. McGuire. 1975. Field ecology of the exotic Mexican red-bellied squirrel in Florida. Journal of Mammalogy, 56:405-419.

Brown, R. E. 1980. Rodents of the Kavir National Park, Iran. Mammalia, 44:89-96.

Brown, R. J. 1974. A comparative study of the chromosomes of three species of shrews, *Sorex bendirii*, *Sorex trowbridgii*, and *Sorex vagrans*. Wasmann Journal of Biology, 32:303-326.

Brown, R. J., and R. L. Rudd. 1981. Chromosomal comparisons within the *Sorex ornatus-S. vagrans* complex. Wasmann Journal of Biology, 32:303-326.

Browne, R. A. 1977. Genetic variation in island and mainland populations of *Peromyscus leucopus*. American Midland Naturalist, 97:1-9.

Brownell, E. 1983. DNA/DNA hybridization studies of muroid rodents: Symmetry and rates of molecular evolution. Evolution, 37:1034-1051.

Brownell, R. L., Jr. 1974. Small odontocetes of the Antarctic. Pp. 13-19, *in* Antarctic mammals (S. G. Brown, R. L. Brownell, Jr., A. W. Erickson, R. J. Hofman, G. A. Llano, and N. A. Mackintosh, eds.). American Geographical Society, New York, Antarctic Map Folio Series, Folio 18, 19 pp.

Brownell, R. L., Jr. 1975a. *Phocoena dioptrica*. Mammalian Species, 66:1-3.

Brownell, R. L., Jr. 1975b. Taxonomic status of the dolphin *Stenopontistes zambezicus* Miranda-Ribeiro, 1936. Zeitschrift für Säugetierkunde, 40(3):173-176.

Brownell, R. L., Jr. 1983. *Phocoena sinus*. Mammalian Species, 198:1-3.

Brownell, R. L., Jr. 1986. Distribution of the vaquita, *Phocoena sinus*, in Mexican waters. Marine Mammal Science, 2:299-305.

Brownell, R. L., Jr. 1989. Franciscana—*Pontoporia blainvillei* (Gervais and d'Orbigny, 1844). Pp. 45-68, *in* Handbook of marine mammals: River dolphins and the larger toothed whales (S. H. Ridgway and R. Harrison, eds.). Academic Press, London, 4:1-442.

Brownell, R. L., Jr., and E. S. Herald. 1972. *Lipotes vexillifer*. Mammalian Species, 10:1-4.

Brownell, R. L., Jr., and R. Praderi. 1984. *Phocoena spinipinnis*. Mammalian Species, 217:1-4.

Brownell, R. L., Jr., and R. Praderi. 1985. Distribution of Commerson's dolphin, *Cephalorhynchus commersonii*, and the rediscovery of the type of *Lagenorhynchus floweri*. Scientific Reports of the Whales Research Institute (Toyko), 36:153-164.

Brownell, R. L., Jr., J. E. Heyning, and W. F. Perrin. 1989. A porpoise *Australophocoena dioptrica* previously identified as *Phocoena spinipinnis* from Heard Island. Marine Mammal Science, 5(2):193-195.

Brownell, R. L. Jr., P. J. Clapham, T. Miyashita, and T. Kasuya. 2001. Conservation status of North Pacific right whales. Journal of Cetacean Research and Management, (Special Issue) 2:269-286.

Brum, N., N. Lafuente, and P. Kiblisky. 1973. Cytogenetic studies in the cricetid rodent *Scapteromys tumidus* (Rodentia-Cricetidae). Experientia, 28:1373.

Brum Zorilla, N., G. Oliver, T. G. De Fronza, and R. Wainberg. 1986. Karyological studies of South American rodents (Rodentia, Cricetidae). I. Comparative analysis in *Scapteromys* rodents. Caryologia, 39:131.

Brum-Zorilla, N., G. H. Decatalfo, C. Degiovanangelo, R. L. Wainberg, and T. G. Defronza. 1990. *Calomys laucha* chromosomes (Rodentia; Cricetidae) from Uruguay and Argentina. Caryology, 43:65-77.

Brunelli, G., and G. Fasella. 1929. Su di un rarismo cetaceo spiaggiato nel litorale di Nettuna. Atti della Reale Accademia Nazionale dei Lincei Roma, 8:85-87.

Brunet, A. K., R. M. Zink, K. M. Kramer, R. C. Blackwell-Rago, S. L. Farrell, T. V. Line, and E. C. Birney. 2002.

Evidence of introgression between masked shrews (*Sorex cinereus*) and prairie shrews (*S. haydeni*), in Minnesota. American Midland Naturalist, 147:116-122.

Brunet-Lecomte, P. 1991. Répartition géographique des campagnols du genre *Microtus* (Arvicolidae, Rodentia) dans le nord-oest Ibérique. Arquivos do Museu Bocage, n.s., 2 (2):11-29.

Brunet-Lecomte, P. 1995. Le campagnol souterrain des Alps *Microtus* (*Terricola*) *multiplex* (Rodentia, Arvicolidae) dans la partie est de la Vallée du Rhône. Bulletin Mensuel de la Société Linnéenne de Lyon, 64(10):467-473.

Brunet-Lecomte, P., and J. Chaline. 1991. Morphological evolution and phylogenetic relationships of the European ground voles (Arvicolidae, Rodentia). Lethaia, 24:45-53.

Brunet-Lecomte, P., and J. Chaline. 1992. Morphological convergences versus biochemical divergences in the holarctic ground voles: *Terricola* and *Pitymys* (Arvicolidae, Rodentia). Neues Jahrbuch für Geologie und Paläontologie. Monatshefte, 12:721-734.

Brunet-Lecomte, P., and J. Chaline. 1993. Mise au point sur *Microtus* (*Terricola*) *pyrenaicus gerbei* (Gerbe, 1879) (Rodentia, Arvicolidae). Mammalia, 57(1):139-142.

Brunet-Lecomte, P., and B. Kryštufek. 1993. Evolutionary divergence of *Microtus liechtensteini* (Rodentia, Arvicolidae) based on the first lower molar. Acta Theriologica, 38(3):297-304.

Brunet-Lecomte, P., and A. Nadachowski. 1994. Comparative analysis of the characters of the first lower molar in *Microtus* (*Terricola*) *thomasi* (Rodentia, Arvicolidae). Acta Zoologica Cracoviensia, 37(1):157-162.

Brunet-Lecomte, P., and A. Nadachowski. 1998. On the systematic position of the Bavarian vole *Microtus* (*Terricola*) *bavaricus* (Mammalia: Rodentia: Arvicolidae). Zoologische Abhandlungen Staatliches Museum für Tierkunde Dresden, 50(10):143-144.

Brunet-Lecomte, P., and L. Povoas. 1993. Voles (Arvicolidae, Rodentia) from Caldeirão Cave (Tomar, Portugal). Arquivos do Museu Bocage, n.s., 2(24):409-414.

Brunet-Lecomte, P., and V. Volobouev. 1994. Comparative morphometry and cytogenetics of *Microtus* (*Terricola*) *multiplex* (Arvicolidae, Rodentia) of the western French Alps. Zeitschrift für Säugetierkunde, 59:116-125.

Brunet-Lecomte, P. J., A. Nadarowski, and J. Chaline. 1992. *Microtus* (*Terricola*) *grafi* nov sp. du Pléistocène supérieur de la Grotte de Bacho Kiro (Bulgarie). Geobios, 25:505-509.

Brunet-Lecomte, P., J. Chaline, and M. Campy. 1993. *Microtus* (*Terricola*) *multiplex vuillemeyi* nov. ssp., campagnol souterrain du site pléistocène de Gigny (Jura, France). Revue de Paléobiologie, Genève, 12(2):335-344.

Brunet-Lecomte, P. J., P. Thouy, and J. Chaline. 1994. Etude comparee des populations actuelles et fossiles de *Microtus* (*Terricola*) *pyrenaicus* (Rodentia, Arvicolidae). Bulletin de la Société Zoologique de France, 119(1):37-49.

Brunet-Lecomte, P. J., T. Lode, and P. Pailley. 1995. Morphométrie comparée de la première molaire inférieure du campagnol de Gerbe, *Microtus* (*Terricola*) *pyrenaicus gerbei* (Rodentia, Arvicolidae). Mammalia, 59(2):249-254.

Brunet-Lecomte, P. J., A. Nadachowski, D. Sirugue, and N. Indelicato. 1996. À propos de l'observation d'un rhombe pitymyen à la première molaire inférieure chez les campagnols *Microtus arvalis* et *M. agrestis* (Rodentia, Arvicolidae). Mammalia, 60(3):491-495.

Brunet-Lecomte, P. J., S. Montuire, and V. Dimitrijevic. 2001. The Pleistocene subterranean voles *Terricola* (Rodentia) of Serbia and Montenegro. Paläontologische Zeitschrift, 75(2):189-196.

Brünner, H., N. Lugon-Moulin, F. Balloux, L. Fumagalli, and J. Hausser. 2002a. A taxonomical re-evaluation of the Valais chromosome race of the common shrew *Sorex araneus* (Insectivora: Soricidae). Acta Theriologica, 47:245-275.

Brünner, H., N. Lugon-Moulin, and J. Hausser. 2002b. Alps, genes, and chromosomes: Their role in the formation of species in the *Sorex araneus* group (Mammalia, Insectivora), as inferred from two hybrid zones. Cytogenetic and Genome Research, 96:85-96.

Brünner, H., H. Turni, H.-J. Kapischke, M. Stubbe, and P. Vogel. 2002c. New *Sorex araneus* karyotypes from Germany and the postglacial recolonization of central Europe. Acta Theriologica, 47:277-293.

Brünnich, M. T. 1771. Zoologiae fundamenta praelectionibus academicis accomodata. Grunde I dyrelaeren. F. C. Pelt, Hafniae et Lipsiae, 254 pp.

Bryant, H. N., A. P. Russell, and W. D. Fitch. 1993. Phylogenetic relationships within the extant Mustelidae (Carnivora): Appraisal of the cladistic status of the Simpsonian subfamilies. Zoological Journal of the Linnean Society, 108:301-334.

Bryant, J. D., and M. C. McKenna. 1995. Cranial anatomy and phylogenetic position of *Tsaganomys altaicus* (Mammalia:Rodentia) from the Hsanda Gol Formation (Oligocene), Mongolia. American Museum Novitates, 3156:1-42.

Bryant, M. D. 1945. Phylogeny of Nearctic Sciuridae. American Midland Naturalist, 33:257-390.

Bryden, M. M. 1995. Southern Elephant Seal. Pp. 686-687, *In* Mammals of Australia (R. Strahan, ed.). Smithsonian Institution Press, Washington, D.C., 756 pp.

Bublitz, J. 1987. Untersuchungen zur Sytematik der Rezenten Caenolestidae Trouessart, 1898: Unter Verwendung craniometrischer Methoden. Bonner Zoologische Monographien, 23:1-96.

Bucher, J. E., and R. S. Hoffmann. 1980. *Caluromys derbianus*. Mammalian Species, 140:1-4.

Büchner, E. 1890. Wissenschaftliche Resultate der von N. M. Przewalski nach central-Asien unternommenen Reisen auf Kosten einer von seiner Kaiserlichen Hoheit dem Grossfürsten Thronfolger Nikolai Alexandrowitsh gespendeten Summe herausgegeben von der Kaiserlichen Akademie der Wissenschaften St. Petersburg. Zoologischer, Theil Band I. Säugethiere, V:3-232.

Buden, D. W. 1976. A review of the bats of the endemic West Indian genus *Erophylla*. Proceedings of the Biological Society of Washington, 89:1-16.

Buden, D. W. 1977. First records of bats of the genus *Brachyphylla* from the Caicos Islands, with notes on geographic variation. Journal of Mammalogy, 58:221-225.

Buden, D. W. 1996*a*. Reptiles, birds, and mammals of Ant Atoll, Eastern Caroline Islands. Micronesica, 29(1):21-36.

Buden, D. W. 1996*b*. Reptiles, birds, and mammals of Pakin Atoll, Eastern Caroline Islands. Micronesica, 29(1):37-48.

Bueler, L. E. 1973. Wild dogs of the world. Stein & Day, New York, 274 pp.

Buffon, G.-L. L., Comte de. 1781. Natural history, general and particular, by the Count de Buffon, illustrated with above six hundred copper plates, the history of man and quadrupeds, translated into English with notes and observations by William Smellie. First ed. vol. 7(Natural history of animals). T. Cadell and W. Davies, London, 452 pp.

Buffon, G.-L. L., Comte de, and L. J. M. Daubenton. 1763. Histoire Naturelle, Générale et Particulière, avec la Description du Cabinet du Roi. L'Imprimerie Royale, Paris, 10(Quadrupeds):1-6 (unnumbered) + 368 pp., 57 pls.

Buffon, G.-L. L., Comte de, and L. J. M. Daubenton. 1764. La grimme. Description de la grimme. Pp. 307-309, 329-330, pl. 41, *in* Histoire Naturelle, Générale et Particulière, avec la Description du Cabinet du Roi. L'Imprimerie Royale, Paris. 12:xvi+451 pp.

Bugge, J. 1970. The contribution of the stapedial artery to the cephalic arterial supply in muroid rodents. Acta Anatomica, 76:313-134.

Bugge, J. 1971*a*. The cephalic arterial system in mole-rats (Spalacidae) bamboo-rats (Rhizomyidae), jumping mice and jerboas (Dipodoidea) and dormice (Gliroidea) with special reference to the systematic classification of rodents. Acta Anatomica, 79:165-180.

Bugge, J. 1971*b*. The cephalic arterial system in New and Old World hystricomorphs, and in bathyergoids, with special reference to the systematic classification of rodents. Acta Anatomica, 80:516-536.

Bugge, J. 1974. The cephalic arteries of hystricomorph rodents. Symposium of the Zoological Society of London, 34:61-78.

Bugge, J. 1978. The cephalic arterial system in carnivores, with special reference to the systematic classification. Acta Anatomica, 101(1):45-61.

Bugge, J. 1985. Systematic value of the carotid arterial pattern in rodents. Pp. 355-379, *in* Evolutionary relationships among rodents: A multidisciplinary analysis (W. P. Luckett and J. L. Hartenberger, eds.). Plenum Press, New York, xiii + 721 pp.

Bühler, P. 1996. Zum taxonomischen Status der Großkopf-Wasserspitzmaus (*Neomys fodiens niethammeri* Bühler, 1963) aus Spanien nebst Festlegung und Beschreibung eines Neotypus. Bonner Zoologische Beiträge, 46:307-314.

Bulatova, N., and E. Kotenkova. 1990. Variants of the Y-chromosome in sympatric taxa of *Mus* in southern USSR. Bollettino di Zoologia, 57:357-360.

Bulatova, N. S., and V. N. Orlov. 1997. Biological versus chromosomal diversity in Ethiopian mammals: A preliminary outlook. Pp. 68-85, *in* Ecological and faunistic studies in Ethiopia, part 1 (Joint Ethio-Russian Biological Expedition, ed.). Addis Ababa, 106 pp.

Bulatova, N., V. N. Orlov, and K. V. Shung. 1992. [Karyotypes of the rats of Vietnam.] Pp. 55-74, *in* Zoologicheski Issledovaníiâ vo Vétname. "Hayka", Mockba, 280 pp (in Russian).

Bulatova, N., L. Lavrenchenko, V. Orlov, and A. Milishnikov. 2002. Notes on chromosomal identification of rodent species in western Ethiopia. Mammalia, 66(1):128-132.

Bunch, T. D., N. N. Vorontzov, E. A. Lyapunova, and R. S. Hoffmann. 1998. Confirmation of diploid chromosome number in Severtzov's sheep (*Ovis amon severtzovi* Nasonov, 1914): G-band karyotype comparisons within *Ovis*. Journal of Heredity, 89:267-269.

Burbidge, A. A., and B. F. J. Manly. 2002. Mammal extinctions on Australian islands: Causes and conservation implications. Journal of Biogeography, 29:365-473.

Burbidge, A. A., M. R. Williams, and I. Abbott. 1997. Mammals of Australian islands: Factors influencing species richness. Journal of Biogeography, 24:703-715.

Burchell, W. J. 1822-1824 [1822-23]. Travels in the interior of Southern Africa. Longman, Hurst, Rees, Orme, Brown, and Green, London, 2 volumes [vol. 2 published in 1823, dated 1824].

Burda, H., E. Nevo, and V. Bruns. 1990. Adaptive differentiation of ear structures in subterranean mole-rats of the *Spalax ehrenbergi* superspecies in Israel. Zoologische Jahrbücher für Systematik, 117:369-382.

Burda, H., J. Zima, A. Scharff, M. Macholán, and M. Kawalika. 1999. The karyotypes of *Cryptomys anselli* sp. nova and *Cryptomys kafuensis* sp. nova: New species of the common mole-rat from Zambia (Rodentia, Bathyergidae). Zeitschrift für Säugetierkunde, 64(1):36-50.

Burdelov, A. S., and O. B. Rossinskaya. 1959. [On the range of *Selevinia betpakdalaensis* Belos. et Bazh.(1938) and on some peculiarities of its ecology]. Zoologicheskii Zhurnal, 38:942- 944 (in Russian).

Burdelov, A. S., B. V. Rasin, S. V. Dyagilev, Z. A. Bilyalov, V. S. Ageev. 1993. [*Meriones erythrourus* Gray in northern Pribalkhashye.] Byulleten' Moskovskovo Obshchestva Ispytatelei Prirody, Otdel Biologicheskii, 98(1):14-23 (in Russian with English summary).

Burgess, N. D., D. Kock, A. Cockle, C. FitzGibbon, P. Jenkins, and P. Honess. 2000. Mammals. Pp. 173-190 and 401-406, *in* Coastal forests of eastern Africa. (N. D. Burgess and G. P. Clarke, eds.). I.U.C.N., Gland and Cambridge, 443 pp.

Burgio, E., and T. Kotsakis. 1986. Presenza di *Pitymys* (Mammalia, Arvicolidae) nel Pleistocene della Sicilia. Naturalista Sicilia, ser. 4, 10(104):27-34.

Burgos, M., R. Jimenez, and R. Diaz de la Guardia. 1989. Comparative study of G- and C-banded chromosomes of five species of Microtidae. Genetica, 78:3-12.

Burk, A., and M. S. Springer. 2000. Intergeneric relationships among Macropodoidea (Metatheria: Diprotodontia) and the chronicle of kangaroo evolution. Journal of Mammalian Evolution, 7:213-237.

Burk, A., M. Westerman, and M. Springer. 1998. The phylogenetic position of the musky rat-kangaroo and the evolution of bipedal hopping in kangaroos (Macropodidae: Diprotodontia). Systematic Biology, 47:457-474.

Burk, A., M. Westerman, D. J. Kao, J. R. Kavanagh, and M. S. Springer. 1999. An analysis of marsupial interordinal relationships based on 12S rRNA, tRNA valine, 16S rRNA, and cytochrome *b* sequences. Journal of Mammalian Evolution, 6:317-334.

Burmeister, H. 1850. Verzeichniss der im Zoologischen Museum der Universität Halle-Wittenberg aufgestellten Säugetheire, Vögel, und Amphibien. Friedrichs-Universität, Halle, 84 pp.

Burnett, G. T. 1829. Illustrations of the Alipeda. Quarterly Journal of Literature, Science and the Arts, 262-269.

Burnett, S. E., J. B. Jennings, J. C. Rainey, and T. L. Best. 2001. *Molossus bondae*. Mammalian Species, 668:1-3.

Burns, J. C., J. R. Choate, and E. G. Zimmerman. 1985. Systematic relationships of pocket gophers (genus *Geomys*) on the Central Great Plains. Journal of Mammalogy, 66:102-118.

Burns, J. J., and F. H. Fay. 1970. Comparative morphology of the skull of the ribbon seal, *Histriophoca fasciata*, with remarks on systematics of Phocidae. Journal of Zoology, London, 161:363-394.

Burns, J. J., F. H. Fay, and G. A. Fedoseev. 1984. Craniological analysis of harbor and spotted seals of the North Pacific region. NOAA Technical Report NMFS, 12:5-16.

Burt, M. S., and R. C. Dowler. 1999. Biochemical systematics of *Geomys breviceps* and two chromosomal races of *Geomys attwateri* in eastern Texas. Journal of Mammalogy, 80:799-809.

Burt, S. L., and T. L. Best. 1994. *Tamias rufus*. Mammalian Species, 460:1-6.

Burt, W. H. 1932. Descriptions of heretofore unknown mammals from islands in the Gulf of California, Mexico. Transactions of the San Diego Society of Natural History, 7:161-182.

Burt, W. H. 1934. Subgeneric allocation of the white-footed mouse, *Peromyscus slevini*, from the Gulf of California, Mexico. Journal of Mammalogy, 15:159-160.

Burt, W. H., and F. S. Barkalow. 1942. A comparative study of the bacula of wood rats (subfamily Neotominae). Journal of Mammalogy, 23:287-297.

Burt, W. H., and R. A. Stirton. 1961. The mammals of El Salvador. Miscellaneous Publications, Museum of Zoology, University of Michigan, 117:1-69.

Buruldag, E., and C. Kurtonur. 2001. Hibernation and postnatal development of the mouse-tailed dormouse, *Myomimus roachi* reared outdoor's [sic] in a cage. Trakya University Journal of Scientific Research B, 2(2):179-186.

Bush, R. M., and K. Paigen. 1992. Evolution of B-glucuronidase regulation in the genus *Mus*. Evolution, 46:1-15.

Bustos, A. R. 2002. Enamel line molar analysis in arvicolid rodents, and its potential use in biostratigraphy and palaeoecology. Micromamíferos y Bioestratigrafía, 1:24 pp.

Butler, P. M. 1941. A comparison of the skulls and teeth of the two species of *Hemicentetes*. Journal of Mammalogy, 22:65–81.

Butler, P. M. 1956. The skull of *Ictops* and the classification of the Insectivora. Proceedings of the Zoological Society of London, 126:453-481.

Butler, P. M. 1972. The problem of insectivore classification. Pp. 253-265, *in* Studies in vertebrate evolution (K. A. Joysey and T. S. Kemp, eds.). Oliver and Boyd, Edinburgh, 284 pp.

Butler, P. M. 1980. The tupaiid dentition. Pp. 171-204, *in* Comparative biology and evolutionary relationships of tree shrews (W. P. Luckett, ed.). Plenum Press, New York, 314 pp.

Butler, P. M. 1988. Phylogeny of the Insectivores. Pp. 117-141, *in* The phylogeny and classification of the tetrapods: Volume 2. (M. J. Benton, ed.). Clarendon Press, Oxford, 329 pp.

Butler, P. M. 1998. Fossil history of shrews in Africa. Pp. 121-132, *in* Evolution of shrews (J. M.Wójcik and M. Wolsan, eds.). Mammal Research Institute, Polish Academy of Sciences, Bialowieza, 458 pp.

Butler, P. M., and M. Greenwood. 1979. Soricidae (Mammalia) from the Early Pleistocene of Olduvai Gorge, Tanzania. Zoological Journal of the Linnean Society, 67:329-379.

Butler, P. M., R. S. Thorpe, and M. Greenwood. 1989. Interspecific relations of African crocidurine shrews (Mammalia: Soricidae) based on multivariate analysis of mandibular data. Zoological Journal of the Linnean Society, 96:373-412.

Butler, P. M., E. Nevo, A. Beiles, and S. Simson. 1993. Variations of molar morphology in the *Spalax ehrenbergi* superspecies: Adaptive and phylogenetic significance. Journal of the Zoological Society of London, 229:191-216.

Bykov, A. V. 1990. Distribution of groups of long-tailed field mice in stands of the semidesert of the Trans-Volga region. Lesovedenie [Soviet Forest Sciences], 1:54-58.

Bykova, G. V., I. A. Vasilyeva, and E. A. Gileva. 1978. Chromosomal and morphological diversity in 2 populations of Asian mountain vole, *Alticola lemminus* Miller (Rodentia, Cricetidae). Experientia, 34:1146-1148.

Byrne, J. M., E. J. Duke, and J. S. Fairley. 1990. Some mitochrondrial DNA polymorphisms in Irish wood mice (*Apodemus sylvaticus*) and bank voles (*Clethrionomys glareolus*). Journal of Zoology, London, 221:299-302.

Byun, S. A., B. F. Koop, and T. E. Reimchen. 2002. Evolution of the Dawson caribou (*Rangifer tarandus dawsoni*). Canadian Journal of Zoology, 80:956-960.

Cabral, J. C. 1971. Existencia em Angola de *Anomalurops beecrofti* (Fraser). Boletim do Instituto de Investigacão Científica de Angola, 8:55-63.

Cabrera, A. 1911. On the specimens of spotted hyaenas in the British Museum (Natural History). Proceedings of the Zoological Society of London, 1911:93-99.

Cabrera, A. 1914. Fauna Ibérica. Mamíferos. Museo Nacional de Ciencias Naturales, Madrid, 418 pp.

Cabrera, A. 1916. El tipo de *Philander laniger* Desm. en el Museo de Ciencias Naturales de Madrid. Boletín de la Real Sociedad Española de Historia Natural, 16:514-517.

Cabrera, A. 1923. Sobre algunos ratones Marroquies. Boletin de la Real Sociedad Española de Historia Natural, 22:162-170.

Cabrera, A. 1924. Una nueva forma de caguan de la isla de Borneo. Boletín de la Real Sociedad Española de Historia Natural, 24:128-131.

Cabrera, A. 1925. Genera mammalium: Insectivora, Galeopithecia. Museo Nacional de Ciencias Naturales, Madrid, 232 pp.

Cabrera, A. 1929. Catálogo descriptivo de las mamíferos de la Guinea Española. Memorias de la Real Sociedad Española de Historia Natural, 16:31-32.

Cabrera, A. 1931. On some South American canine genera. Journal of Mammalogy, 12:54-67.

Cabrera, A. 1932. Los mamíferos de Marruecos. Trabajos del Museo Nacional de Ciencias Naturales, 57:209-222.

Cabrera, A. 1940. Notas sobre carnívoras sudamericanos. Notas del Museo de la Plata, 5(29):1-22.

Cabrera, A. 1953. Los roedores argentinos de la familia Caviidae. Publicaciones de la Escuela Veterinaria, Universidad de Buenos Aires, 6:1-93.

Cabrera, A. 1956. Una nueva forma del genero *Nasua*. Neotropica, 2:2-4.

Cabrera, A. 1957 [1958]-1961. Catálogo de los mamíferos de América del Sur. Revista del Museo Argentino de Ciencias Naturales "Bernardino Rivadavia". Ciencias Zoológicas, 4(1):iv + 308 pp.[1958]; 4(2):v-xxii+309-732[1961].

Cabrera, A. 1958. Dos felidos argentinos ineditos (Mammalia, Carnivora). Neotropica, 3(12):70-72.

Caceres, M. C., and R. M. R. Barclay. 2000. *Myotis septentrionalis*. Mammalian Species, 634:1-4.

Cadena, A., R. P. Anderson, and P. Rivas-Pava. 1998. Colombian mammals from the Chocoan slopes of Nariño. Occasional Papers, Museum of Texas Tech University, 180:1-15.

Cagnin, M., and G. Aloise. 1995. Current status of Myoxidae (Mammalia: Rodentia) in Calabria (Southern Italy). Pp. 169-180, *in* Proceedings of II Conference on Dormice (Rodentia, Myoxidae) (M. G. Filippucci, ed.). Hystrix, n.s., 6(1-2) (1994):1-340.

Cagnin, M., G. Aloise, G. Garofalo, C. Milazzo, and M. Cristaldi. 1996. Les communautés de petits mammifères terrestres de trois–Fiumare–de la Calabre (Italie du Sud). Vie Milieu, 46(3/4):319-326.

Cai, B., and Z. Qiu. 1993. [Murid rodents from the Late Pliocene of Yangquan and Yuxian, Hebei.] Vertebrata PalAsiatica, 31(4):267-293 (in Chinese with English summary).

Cai Gui-quan, and Feng Zuo-jiang. 1981. [On the occurrence of the Himalayan musk-deer (*Moschus chrysogaster*) in China and an approach to the systematics of the genus *Moschus*]. Acta Zootaxonomica Sinica, 6:106-110 (in Chinese).

Cai Gui-quan, and Feng Zuo-jiang. 1982. [A systematic revision of the subspecies of highland hare (*Lepus oiostolus*)—including two new subspecies]. Acta Theriologica Sinica, 2:7-182 (in Chinese).

Caire, W. 1997. Annotated checklist of the Recent land mammals of Sonora, Mexico. Pp. 69-80, *in* Life among the muses: Papers in honor of James S. Findley (T. L. Yates, W. L. Gannon, and D. E. Wilson, eds.). The Museum of Southwestern Biology, Albuquerque, New Mexico, v + 290 pp.

Caire, W., J. E. Vaughan, and V. E. Diersing. 1978. First record of *Sorex arizonae* (Insectivora: Soricidae) from Mexico. Southwestern Naturalist, 23:532-533.

Caire, W., J. D. Taylor, B. P. Glass, and M. A. Mares. 1989. Mammals of Oklahoma. University of Oklahoma Press, Norman, 567 pp.

Cakenberghe, V. van, and F. De Vree. 1985. Systematics of African *Nycteris* (Mammalia: Chiroptera). Pp. 53-90, *in* Proceedings of the International Symposium on African vertebrates: Systematics, phylogeny and evolutionary ecology (K.-L. Schuchmann, ed.). Zoologisches Forschungsinstitut und Museum Alexander Koenig, Bonn., 585 pp.

Calaby, J. H. 1966. Mammals of the Upper Richmond and Clarence Rivers, New South Wales. Technical Papers of the Division of Wildlife Survey, CSIRO, Australia, 10:1-55.

Calaby, J. H., G. Mack, and W. D. L. Ride. 1963. The generic name *Macropus* Shaw, 1790 (Mammalia). Z.N.(S.) 1584. Bulletin of Zoological Nomenclature, 20(5):376-379.

Caldarini, G., E. Capanna, M. V. Civitelli, M. Corti, and A. Simonetta. 1989. Chromosomal evolution in the subgenus *Rattus* (Rodentia, Muridae): Karyotype analysis of two species from the Indian subregion. Mammalia, 53:77-83.

Caldwell, D. K., and M. C. Caldwell. 1989. Pygmy sperm whale—*Kogia breviceps* (de Blainville, 1838); dwarf sperm whale *Kogia simus* Owen, 1866. Pp. 235-260, *in* Handbook of marine mammals: River dolphins and the larger toothed whales (S. H. Ridgway and R. Harrison, eds.). Academic Press, London, 4:1-442.

Calhoun, S. W., and I. F. Greenbaum. 1991. Evolutionary implications of genic variation among insular populations of *Peromyscus maniculatus* and *Peromyscus oreas*. Journal of Mammalogy, 72:248-262.

Calhoun, S. W., I. F. Greenbaum, and K. P. Fuxa. 1988. Biochemical and karyotypic variation in *Peromyscus maniculatus* from western North America. Journal of Mammalogy, 69:34-45.

Calhoun, S. W., M. D. Engstrom, and I. F. Greenbaum. 1989. Biochemical variation in pygmy mice (*Baiomys*). Journal of Mammalogy, 70:374-381.

Callahan, J. R. 1977. Diagnosis of *Eutamias obscurus* (Rodentia: Sciuridae). Journal of Mammalogy, 58:188-201.

Callahan, J. R. 1980. Taxonomic status of *Eutamias bulleri*. Southwestern Naturalist, 25:1-8.

Callahan, J. R., and R. Davis. 1977. A new subspecies of the cliff chipmunk from coastal Sonora. Southwestern Naturalist, 22:67-75.

Callahan, J. R., and R. Davis. 1982. Reproductive tract and evolutionary relationships of the Chinese rock squirrel, *Sciurotamias davidianus*. Journal of Mammalogy, 63:42-47.

Caloi, L., T. Kotsakis, and M. R. Palombo. 1986. La fauna a vertebrati terrestri del Pleistocene delle isole del Mediterraneo. Geologica Romania, 25(6):235-256.

Camacho, A., R. Borroto, and I. Ramos. 1994. *Mesocapromys angelcabrerai* (Varona, 1979): Pequeña jutía endémica de Cuba. Ciencias Biológicas, 26:1-12.

Camacho, A., R. Borroto, and I. Ramos. 1995. Los capromidos de Cuba: Estado actual y perspectives de las investigaciones sobre su sistemática. Marmosiana, 1:43-56.

Cameron, G. N., and S. R. Spencer. 1981. *Sigmodon hispidus*. Mammalian Species, 158:1-9.

Campbell, C. B. G. 1966. Taxonomic status of tree shrews. Science, 153:436.

Campbell, C. B. G. 1974. On the phyletic relationships of the tree shrews. Mammal Review, 4:125-143.

Campos, C. M., M. F. Tognelli, and R. A. Ojeda. 2001. *Dolichotis patagonum*. Mammalian Species, 652:1-5.

Cano, J., A. Pretel, R. Grandfils, J. M. Vargas, and V. Sans-Coma. 1984. Anzahl und Struktur der Chromosomen von *Mus spretus* Lataste, 1883 (Rodentia: Muridae) von der Iberischen Halbinsel. Säugetierkundliche Mitteilungen, 31:161-169.

Cano Perez, M. 1984. Revision der Systematik von *Gazella (Nanger) dama*. Zeitschrift des Kölner Zoo, 27:103-107.

Cantini, M. 1991. Comunità di piccoli Mammiferi (Mammalia: Insectivora, Rodentia, Carnivora) nell'Alto

Lario Orientale (Lombardia, Italia) e valutazioni della qualità ambientale. Il Naturalista Valtellinese–Atti Mus. Civ. Stor. Nat. Morbegno, 2:71-98.

Cao, V. S. 1984. Inventaire des rongeurs du Vietnam. Mammalia, 48(3):391-395.

Cao, V. S. 1985. Sur les rats du bambou (Rhizomyidae) au Vietnam. Mammalia, 48:606-607.

Cao, V. S., and V. M. Tran. 1984. Karyotypes et systematique des rats (genre *Rattus* Fisher) du Vietnam. Mammalia, 48:557-564.

Cao, Y., N. Okada, and M. Hasegawa. 1997. Phylogenetic position of guinea pigs revisited. Molecular Biology and Evolution, 14(4):461-464.

Cao, Y., M. Fujiwara, M. Nikaido, N. Okada, and M. Hasegawa. 2000. Interordinal relationships and timescale of eutherian evolution as inferred from mitochondrial genome data. Gene, 259:149-158.

Cao, Z., S. Liu, and A. Yang. 1995. Glans penis histology of the clawed jird (*Meriones unguiculatus*) and the scanning electron microscopic studies on the glans of five species of rodents. Acta Theriologica Sinica, 15(2):137-140.

Capanna, E. 1974. A re-statement of the problem of chromosomal polymorphism in *Rattus rattus* (L.). Pp. 223-235, *in* Symposium Theriologicum II, Proceedings of the International Symposium on Species and Zoogeography of European Mammals held in Brno, Czechoslovakia on 22nd to 26th November 1971 (J. Kratochvil and R. Obrtel, eds.). Academia Publishing house of the Czechoslovak Academy of Sciences (Praha), 394 pp.

Capanna, E. 1981. Caryotype et morphologie cranienne de *Talpa romana* Thomas de terra typica. Mammalia, 45:71-82.

Capanna, E. 1982. Robertsonian numerical variation in animal speciation: *Mus musculus*, an emblematic model. Pp. 155-177, *in* Mechanisms of Speciation (C. Barigozzi, ed.). A. Liss, New York, 546 pp.

Capanna, E., and M. V. Civitelli. 1988. A cytotaxonomic approach of the systematics of *Arvicanthis niloticus* (Desmarest 1822) (Mammalia, Rodentia). Tropical Zoology, 1:29-37.

Capanna, E., and M. Corti. 1991. Chromosomes, speciation and phylogeny in mammals. Pp. 355-370, *in* Symposium on the evolution of terrestrial vertebrates (G. Ghiara, F. Angelini, E. Olmo, and L. Varano, eds.). Selected Symposia and Monographs U. Z. I., 4, Mucchi, Modena, 666 pp.

Capanna, E., and M. S. Merani. 1981. Karyotypes of Somalian rodent populations. 2. The chromosomes of *Gerbillus dunni* (Thomas, 1904), *Gerbillus pusillus* Peters, 1878 and *Ammodillus imbellis* (De Winton, 1898) (Cricetidae Gerbillinae). Italian Journal of Zoology, n.s., Suppl. 14, 15:227-240.

Capanna, E., and C. A. Redi. 1994. Chromosomes and microevolutionary processes. Bollettino di Zoologia, 61:285-294.

Capanna, E., M. V. Civitell, and A. Ceraso. 1982. Karyotypes of Somalian rodent populations. 3. *Mastomys huberti* (Wroughton, 1908) (Mammalia Rodentia). Italian Journal of Zoology, 16:141-152.

Capanna, E., C. A. Redi, S. Valeri, and G. Gentili. 1994. Robertsonian chromosomes with new arm combination in a natural hybrid zone between two chromosomal races of *Mus domesticus*. Rendiconti Fisiche Accademia lincei, 5(ser. 9):269-276.

Capanna, E., M. V. Civitelli, D. Bizzocco, M. Corti, and A. Bekele. 1996a. The chromosomes of *Desmomys harringtoni* (Rodentia, Muridae). Italian Journal of Zoology, 63:37-40.

Capanna, E., A. Bekele, M. Capula, R. Castiglia, M. V. Civitelli, J. T. Cl. Codjia, M. Corti, and C. Fadda. 1996b. A multidisciplinary approach to the systematics of the genus *Arvicanthis* Lesson, 1842 (Rodentia, Murinae). Mammalia, 60(40):677-696.

Capasso Barbato, L., and E. Gliozzi. 2001. Late Pleistocene micromammal association from Praia a Mare (Calabria, Southern Italy): Palaeoclimatological and biochronological implications. Bollettino della Societa Paleontologica Italiana, 40(2):159-166.

Capllonch, P., A. Autino, M. Díaz, R. Bárquez, and M. Goytia. 1997. Los mamíferos del Parque Biológico Sierra de San Javier, Tucumán, Argentina: Observaciones sobre su sistemática y distribución. Mastozoología Neotropical, 4:49-71.

Capula, M., M. V. Civitelli, M. Corti, A. Bekele, and E. Capanna. 1997. Genetic divergence in the genus *Arvicanthis* (Rodentia: Murinae). Biochemical Systematics and Ecology, 25(5):403-409.

Cardinal, B. R., and L. Christidis. 2000. Mitochondrial DNA and morphology reveal three geographically distinct lineages of the large bentwing bat (*Miniopterus schreibersii*) in Australia. Australian Journal of Zoology, 48:1-19.

Cardini A. 2003. The geometry of the marmot (Rodentia: Sciuridae) mandible: Phylogeny and patterns of morphological evolution. Systematic Biology, 52:186-205.

Carleton, M. D. 1973. A survey of gross stomach morphology in New World Cricetinae (Rodentia, Muroidea), with comments on functional interpretations. Miscellaneous Publications, Museum of Zoology, University of Michigan, 146:1-43.

Carleton, M. D. 1977. Interrelationships of populations of the *Peromyscus boylii* species group (Rodentia,

Muridae) in western Mexico. Occasional Papers of the Museum of Zoology, University of Michigan, 675:1-47.

Carleton, M. D. 1979. Taxonomic status and relationships of *Peromyscus boylii* from El Salvador. Journal of Mammalogy, 60:280-296.

Carleton, M. D. 1980. Phylogenetic relationships in neotomine-peromyscine rodents (Muroidea) and a reappraisal of the dichotomy within New World Cricetinae. Miscellaneous Publications, Museum of Zoology, University of Michigan, 157:1-146.

Carleton, M. D. 1981. A survey of gross stomach morphology in Microtinae (Rodentia, Muroidea). Zeitschrift für Säugetierkunde, 46:93-108.

Carleton, M. D. 1984. Introduction to rodents. Pp. 255-265, *in* Orders and families of Recent mammals of the world (S. Anderson and J. Knox Jones, Jr., eds.). John Wiley and Sons, New York, 686 pp.

Carleton, M. D. 1989. Systematics and Evolution. Pp. 7-141, *in* Advances in the study of *Peromyscus* (Rodentia) (G. L. Kirkland and J. N. Layne, eds.). Texas Tech University Press, Lubbock, 367 pp.

Carleton, M. D. 1994. Systematic studies of Madagascar's endemic rodents (Muroidea: Nesomyinae): Revision of the genus *Eliurus*. American Museum Novitates, 3087:1-55.

Carleton, M. D. 2003. *Eliurus*, tufted-tailed rats. Pp. 1373-1380, *in* The natural history of Madagascar (S. M. Goodman and J. P. Benstead, eds.). The University of Chicago Press, Chicago, xxi + 1709 pp.

Carleton, M. D., and R. E. Eshelman. 1979. A synopsis of fossil grasshopper mice, genus *Onychomys*, and their relationships to Recent species. University of Michigan, Museum of Paleontology, Papers on Paleontology, 21:1-63.

Carleton, M. D., and S. M. Goodman. 1996. Systematic studies of Madagascar's endemic rodents (Muroidea: Nesomyinae): A new genus and species from the Central Highlands. Fieldiana: Zoology, n.s., 85:231-256.

Carleton, M. D., and S. M. Goodman. 1998. New taxa of nesomyine rodents (Muroidea: Muridae) from Madagascar's northern highlands, with taxonomic comments on previously described forms. Fieldiana: Zoology, n.s., 90:163-200.

Carleton, M. D., and S. M. Goodman. 2000. Rodents of the Parc National de Marojejy, Madagascar. Pp. 231-263, *in* A Floral and faunal inventory of the Parc National de Marojejy, Madagascar: With reference to elevational variation (S. M. Goodman, ed.). Fieldiana: Zoology, n.s., 97:285 pp.

Carleton, M. D., and S. M. Goodman. 2003*a*. Rodentia: *Brachytarsomys*, white-tailed tree rats, *Antsangy*. Pp. 1368-1370, *in* The natural history of Madagascar (S. M. Goodman and J. P. Benstead, eds.). The University of Chicago Press, Chicago, xxi + 1709 pp.

Carleton, M. D., and S. M. Goodman. 2003*b*. *Gymnuromys*, voalavoanala. Pp. 1381-1383, *in* The natural history of Madagascar (S. M. Goodman and J. P. Benstead, eds.). The University of Chicago Press, Chicago, xxi + 1709 pp.

Carleton, M. D., and S. M. Goodman. 2003*c*. *Macrotarsomys*, big-footed mice. Pp. 1386-1388, *in* The natural history of Madagascar (S. M. Goodman and J. P. Benstead, eds.). The University of Chicago Press, Chicago, xxi + 1709 pp.

Carleton, M. D., and D. G. Huckaby. 1975. A new species of *Peromyscus* from Guatemala. Journal of Mammalogy, 56:444-451.

Carleton, M. D., and C. Martinez. 1991. Morphometric differentiation among West African populations of the rodent genus *Dasymys* (Muroidea: Murinae), and its taxonomic implications. Proceedings of the Biological Society of Washington, 104:419-435.

Carleton, M. D., and G. G. Musser. 1984. Muroid rodents. Pp. 289-379, *in* Orders and families of Recent mammals of the world (S. Anderson and J. K. Jones, Jr., eds.). John Wiley and Sons, New York, 686 pp.

Carleton, M. D., and G. G. Musser. 1989. Systematic studies of oryzomyine rodents (Muridae, Sigmodontinae): A synopsis of *Microryzomys*. Bulletin of the American Museum of Natural History, 191:1-83.

Carleton, M D., and G. G. Musser. 1995. Systematic studies of oryzomyine rodents (Muridae Sigmodontinae): Definition and distribution of *Oligoryzomys vegetus* (Bangs, 1902). Proceedings of the Biological Society of Washington, 108:338-369.

Carleton, M. D., and P. Myers. 1979. Karyotypes of some harvest mice, genus *Reithrodontomys*. Journal of Mammalogy, 60:307-313.

Carleton, M. D., and S. L. Olson. 1999. Amerigo Vespucci and the rat of Fernando de Noronha: A new genus and species of Rodentia (Muridae: Sigmodontinae) from a volcanic island off Brazil's continental Shelf. American Museum Novitates, 3256:59 pp.

Carleton, M. D., and C. B. Robbins. 1985. On the status and affinities of *Hybomys planifrons* (Miller, 1900) (Rodentia: Muridae). Proceedings of the Biological Society of Washington, 98:956-1003.

Carleton, M. D., and D. F. Schmidt. 1990. Systematic studies of Madagascar's endemic rodents (Muroidea: Nesomyinae): An annotated gazetteer of collecting localities of known forms. American Museum Novitates, 2987:1-36.

Carleton, M. D., and E. Van der Straeten. 1997. Morphological differentiation among Subsaharan and North African populations of the *Lemniscomys barbarus* complex (Rodentia: Muridae). Proceedings of the Biological Society of Washington, 110(4):640-680.

Carleton, M. D., E. T. Hooper, and J. Honacki. 1975. Karyotypes and accessory reproductive glands in the rodent genus *Scotinomys*. Journal of Mammalogy, 56:916-921.

Carleton, M. D., D. E. Wilson, A. L. Gardner, and M. A. Bogan. 1982. Distribution and systematics of *Peromyscus* (Mammalia: Rodentia) of Nayarit, Mexico. Smithsonian Contributions to Zoology, 352:1-46.

Carleton, M. D., R. D. Fisher, and A. L. Gardner. 1999. Identification and distribution of cotton rats, genus *Sigmodon* (Muridae: Sigmodontinae), of Nayarit, Mexico. Proceedings of the Biological Society of Washington, 112:813-856.

Carleton, M. D., S. M. Goodman, and D. Rakotondravony. 2001. A new species of tufted-tailed rat, genus *Eliurus* (Muridae: Nesomyinae), from western Madagascar, with notes on the distribution of *E. myoxinus*. Proceedings of the Biological Society of Washington, 114:972-987.

Carleton, M. D., O. Sanchez, and G. Urbano Vidales. 2002. A new species of *Habromys* (Muroidea: Neotominae) from México, with generic review of species definitions and remarks on diversity patterns among Mesoamerican small mammals restricted to humid montane forests. Proceedings of the Biological Society of Washington, 115:16-61.

Carnero, A., R. Jimenez, M. Burgos, A. Sanchez, and R. Diaz de la Guardia. 1991. Achiasmatic sex chromosomes in *Pitymys duodecimcostatus*: Mechanisms of association and segregation. Cytogenetics and Cell Genetics, 56:78-81.

Carpaneto, G. M. 1995. Occurrence of black colobus *Colobus satanas* in northwestern Congo. African Primates, 1(2):42-44.

Carpaneto, G. M., and M. Cristaldi. 1995. Dormice and man: A review of past and present relations. Pp. 303-330, *in* Proceedings of II Conference on Dormice (Rodentia, Myoxidae) (M. G. Filippucci, ed.). Hystrix, n.s., 6(1-2):1-340.

Carr, S. M., and S. A. Hicks. 1997. Are there two species of marten in North America? Genetic and evolutionary relationships within *Martes*. Pp. 15-28, in *Martes*: Taxonomy, ecology, techniques, and management (G. Proulx, H. N. Bryant, and P. M. Woodard, eds.). Provincial Museum of Alberta, Edmonton. 474 pp.

Carr, S. M., and E. A. Perry. 1998. Intra- and interfamilial systematic relationships of phocid seals as indicated by mitochondrial DNA sequences. Pp. 277-290, *in* Molecular genetics of marine mammals (A. E. Dizon, S. J. Chivers, and W. F. Perrin, eds.). Society for Marine mammalogy Special Publication, 3:1-388.

Carrasco, M. A. 2000. Species discrimination and morphological relationships of kangaroo rats (*Dipodomys*) based on their dentition. Journal of Mammalogy, 81:107-122.

Carrasco, M. A., and J. H. Wahlert. 1999. The cranial anatomy of *Cricetops dormitor*, an Oligocene fossil rodent from Mongolia. American Museum Novitates, 3275:1-14.

Carraway, L. N. 1985. *Sorex pacificus*. Mammalian Species, 231:1-5.

Carraway, L. N. 1990. A morphologic and morphometric analysis of the "*Sorex vagrans* species complex" in the Pacific coast region. Special Publications, The Museum, Texas Tech University, 32:1-76.

Carraway, L. N. 1995. A key to Recent Soricidae of the western United States and Canada based primarily on dentaries. Occasional Papers of the Natural History Museum, The University of Kansas, 175:1-49.

Carraway, L. N., and P. K. Kennedy. 1994. Genetic variation in *Thomomys bulbivorus*, an endemic to the Willamette Valley, Oregon. Journal of Mammalogy, 74:952-962.

Carraway, L. N., and R. M. Timm. 2000. Revision of the extant taxa of the genus *Notiosorex* (Mammalia: Insectivora: Soricidae). Proceedings of the Biological Society of Washington, 113:302-318.

Carraway, L. N., and B. J. Verts. 1985. *Microtus oregoni*. Mammalian Species, 233:1-6.

Carraway, L. N., and B. J. Verts. 1991*a*. *Neotoma fuscipes*. Mammalian Species, 386:1-10.

Carraway, L. N., and B. J. Verts. 1991*b*. *Neurotrichus gibbsii*. Mammalian Species, 387:1-7.

Carraway, L. N., and B. J. Verts. 1993. *Aplodontia rufa*. Mammalian Species, 431:1-10.

Carraway, L. N., and B. J. Verts. 1994. *Sciurus griseus*. Mammalian Species, 474:1-7.

Carraway, L. N., and B. J. Verts. 2002. Geographic variation in pelage color of Piñon Mice (*Peromyscus truei*) in the northern Great Basin and environs. Western North American Naturalist, 62:458-465.

Carraway, L. N., E. Yensen, B. J. Verts, and L. F. Alexander. 1993. Range extension and habitat of *Peromyscus truei* in eastern Oregon. Northwestern Naturalist, 74:81-84.

Carrick, R., S. E. Csordas, and S. E. Ingham. 1962. Studies on the southern elephant seal *Mirounga leonina* (L.). IV. Breeding and development. Commonwealth Scientific and Industrial Research Organization Wildlife Research, 7:161-197.

Carroll, D. S., L. L. Peppers, C. Jones, and R. D. Bradley. 2002. *Sigmodon ochrognathus* is a monotypic species: Evidence from DNA sequences. Southwestern Naturalist, 47:494-497.

Carroll, L. E., and H. H. Genoways. 1980. *Lagurus curtatus*. Mammalian Species, 124:1-6.

Carstens, B. C., B. L. Lundrigan, and P. Myers. 2002. A phylogeny of the Neotropical nectar-feeding bats (Chiroptera: Phyllostomidae) based on morphological and molecular data. Journal of Mammalian Evolution, 9:23-53.

Carter, C. H., and H. H. Genoways. 1978. *Liomys salvini*. Mammalian Species, 84:1-5.

Carter, D. C., and P. G. Dolan. 1978. Catalog of type specimens of Neotropical bats in selected European museums. Special Publications, The Museum, Texas Tech University Press, 15:1-136.

Carter, D. C., and C. S. Rouk. 1973. Status of recently described species of *Vampyrops* (Chiroptera: Phyllostomatidae). Journal of Mammalogy, 54:975-977.

Carter, D. C., W. D. Webster, J. K. Jones, Jr., C. Jones, and R. D. Suttkus. 1985. *Dipodomys elator*. Mammalian Species, 232:1-3.

Carvalho, C. T., de. 1965. Commentarios sobre os mamiferos descritos e figurados por Alexandre Rodriguez Ferreira em 1790. Arquivos de Zoologia, 12:7-70.

Casavant, N. C., and S. C. Hardies. 1994. Shared sequence variants of *Mus spretus* LINE-1 elements tracing dispersal to within the last 1 million years. Genetics, 137:565-572.

Casinos, A. 1984. A note on the common dolphin of the South American Atlantic coast, with some remarks about the speciation of the genus *Delphinus*. Acta Zoologica Fennica, 172:141-142.

Casinos, A., and S. Filella. 1981. Notes on cetaceans of the Iberian coasts. IV. A specimen of *Mesoplodon densirostris* (Cetacea, Hyperoodontidae) stranded on the Spanish Mediterranean littoral. Säugetierkundliche Mitteilungen, 29:61-67.

Casinos, A., and J. Ocaña. 1979. A craniometric study of the genus *Inia* d'Orbigny, 1834 (Cetacea, Platanistoidea). Säugetierkundliche Mitteilungen, 27:194-206.

Castella, V., M. Ruedi, L. Excoffier, C. Ibáñez, R. Arlettaz, and J. Hausser. 2000. Is Gibraltar Strait a barrier to gene flow for the bat *Myotis myotis* (Chiroptera: Vespertilionidae)? Molecular Ecology, 9:1761-1772.

Castién, E., and J. Gosálbez. 1992. Distribución de micromamíferos (Insectivora y Rodentia) en Navarra. Navarra Miscellània Zoològica, 16:183-195.

Castiglia, R., M. Corti, P. Tesha, A. Scanzani, C. Fadda, E. Capanna, and W. Verheyen. 2003*a*. Cytogenetics of the genus *Arvicanthis* (Rodentia, Muridae). 3. Comparative cytogenetics of *A. neumanni* and *A. nairobae* from Tanzania. Genetica, 118:33-39.

Castiglia, R., M. Corti, P. Colangelo, F. Annesi, E. Capanna, W. Verheyen, A. M. Sichilima, and R. Makundi. 2003*b*. Chromosomal and molecular characterization of *Aethomys kaiseri* from Zambia and *Aethomys chrysophilus* from Tanzania (Rodentia, Muridae). Hereditas, 139:81-89.

Castillo, C., and J. Agustí. 1996. Early Pliocene rodents (Mammalia) from Asta Regia (Jerez basin, Southwestern Spain). Proceedings of the Koninklijke Nederlandse Akademie van Wetenschappen, 99(1-2):25-43.

Castro, E. C., M. S. Mattevi, S. W. Maluf, and L. F. B. Oliveira. 1991. Distinct centric fusions in different populations of *Deltamys kempi* (Rodentia, Cricetidae) from South America. Cytobios, 68:153-159.

Castro-Arellano, I., H. Zarza, and R. A. Medellín. 2000. *Philander opossum*. Mammalian Species, 638:1-8.

Castro-Campillo, A., and J. Ramírez-Pulico. 2000. Systematics of the smooth-toothed pocket gopher, *Thomomys umbrinus*, in the Mexican transvolcanic belt. American Museum Novitates, 3297:1-37.

Castro-Campillo, A., H. R. Roberts, D. J. Schmidly, and R. D. Bradley. 1999. Systematic status of *Peromyscus boylii ambiguus* based on morphologic and molecular data. Journal of Mammalogy, 80:1214-1231.

Catalisano, A., and Sarà, M. 1995. L'*Arvicola terrestris* L. in Sicilia. Atti della Società Italiana di Scienze naturali e del Museo civico di Storia naturale di Milano, 134/1993(1):8-12.

Catzeflis, F. M. 1990. DNA hybridization as a guide to phylogenies: Raw data in muroid rodents. Pp. 317-345, *in* Evolution of subterranean mammals at the organismal and molecular levels. Proceedings of the Fifth International Theriological Congress, held in Rome, Italy, August 22-29, 1989 (E. Nevo and O. A. Reig, eds.). Wiley-Liss, New York, 422 pp.

Catzeflis, F. 1995*a*. *Eliomys quercinus* (Linnaeus, 1766). Pp. 244-248, *in* Säugetiere der Schweiz (J. Hausser, dir., Denkschriftenkommission der Schweizerischen Akademie der Naturwissenschaften, ed.). Denkschriften der Schweizerischen Aklademie der Naturwissenschaften, 103:1-501.

Catzeflis, F. 1995*b*. *Dryomys nitedula* (Pallas, 1779). Pp. 249-252, *in* Säugetiere der Schweiz (J. Hausser, dir., Denkschriftenkommission der Schweizerischen Akademie der Naturwissenschaften, ed.). Denkschriften der Schweizerischen Aklademie der Naturwissenschaften, 103:1-501.

Catzeflis, F. 1995*c*. *Glis glis* (Linnaeus, 1766). Pp. 253-257, *in* Säugetiere der Schweiz (J. Hausser, dir., Denkschriftenkommission der Schweizerischen Akademie der Naturwissenschaften, ed.). Denkschriften der Schweizerischen Aklademie der Naturwissenschaften, 103:1-501.

Catzeflis, F. 1995*d*. *Muscardinus avellanarius* (Linnaeus, 1758). Pp. 258-262, *in* Säugetiere der Schweiz (J. Hausser, dir., Denkschriftenkommission der Schweizerischen Akademie der Naturwissenschaften, ed.). Denkschriften der Schweizerischen Aklademie der Naturwissenschaften, 103:1-501.

Catzeflis, F. M., and C. Denys. 1992. The African *Nannomys* (Muridae): An early offshoot from the *Mus* lineage–evidence from scnDNA hybridization experiments and compared morphology. Israel Journal of Zoology, 38:219-231.

Catzeflis, F. M., T. Maddalena, S. Hellwing, and P. Vogel. 1985. Unexpected findings on the taxonomic status of East Mediterranean *Crocidura russula* auct. (Mammalia, Insectivora). Zeitschrift für Säugetierkunde, 50:185-201.

Catzeflis, F. M., F. H. Sheldon, J. E. Ahlquist, and C. G. Sibley. 1987. DNA-DNA hybridization evidence of the rapid rate of muroid rodent DNA evolution. Molecular Biology and Evolution, 4:242-253.

Catzeflis, F. M., E. Nevo, J. E. Ahlquist, and C. G. Sibley. 1989. Relationships of the chromosomal species in the Eurasian mole rats of the *Spalax ehrenbergi* group as determined by DNA-DNA hybridization, and an estimate of the Spalacid-Murid divergence time. Journal of Molecular Evolution, 29:223-232.

Catzeflis, F. M., C. Hanni, P. Sourrouille, and E. Douzery. 1995. Re: Molecular systematics of hystricognath rodents: The contribution of sciurognath mitochondrial 12S rRNA sequences (Letter to the Editor). Molecular Phylogenetics and Evolution, 4(3):357-360.

Cavallini, P. 1992. *Herpestes pulverulentus*. Mammalian Species, 409:1-4.

Ceballos, G. 1990. Comparative natural history of small mammals from tropical forest in western Mexico. Journal of Mammalogy, 71:263-266.

Ceballos-G., G., and R. A. Medellin. 1988. *Diclidurus albus*. Mammalian Species, 316:1-4.

Ceballos, G., and D. Navarro-L. 1991. Diversity and conservation of Mexican mammals. Pp. 167-198, *in* Latin American mammalogy: History, biodiversity, and conservation (M. A. Mares and D. J. Schmidly, eds.). University of Oklahoma Press, Norman, 468 pp.

Ceballos-G., G., and D. E. Wilson. 1985. *Cynomys mexicanus*. Mammalian Species, 248:1-3.

Ceballos, G., J. Arroyo-Cabrales, and R. A. Medellín. 2002*a*. The mammals of México: Composition, distribution, and conservation status. Occasional Papers, Museum of Texas Tech University, 218:1-27.

Ceballos, G., H. Zarza, and M. A. Steele. 2002*b*. *Xenomys nelsoni*. Mammalian Species, 704:1-3.

Cerone, G., and G. Aloise. 1994. La fauna a micromammiferi del comprensorio di Muro Lucano (Potenza, Italia). Hystrix, n.s., 5(1-2) (1993):110-115.

Cerqueira, R. 1985. The distribution of *Didelphis* in South America (Polyprotodontia, Didelphidae). Journal of Biogeography, 12:135-145.

Cervantes, F. A. 1993. *Lepus flavigularis*. Mammalian Species, 423:1-3.

Cervantes, F. A. 1997. *Sylvilagus graysoni*. Mammalian Species, 559:1-3.

Cervantes, F. A., and C. Lorenzo. 1997. *Sylvilagus insonus*. Mammalian Species, 568:1-4.

Cervantes, F. A., C. Lorenzo, and R. S. Hoffmann. 1990. *Romerolagus diazi*. Mammalian Species, 360:1-7.

Cervantes, F. A., C. Lorenzo, J. Vargas, and T. Holmes. 1992. *Sylvilagus cunicularius*. Mammalian Species, 412:1-4.

Cervantes, F. A., V. J. Sosa, J. Martinez, R. Ma. González, and R. C. Dowler. 1993*a*. *Pappogeomys tylorhinus*. Mammalian Species, 433:1-4.

Cervantes, F. A., M. Martinez Coronel, and Y. Hortelano Moncada. 1993*b*. Variacion morfometrica intrapoblacional de *Peromyscus melanocarpus* (Rodentia: Muridae) de Oaxaca, Mexico. Anales Instituto de Biologia, Universidad Nacional Autonoma de Mexico, Serie Zoologia, 64:153-168.

Cervantes, F. A., J. Martínez, and R. M. González. 1994. Karyotypes of the Mexican tropical voles *Microtus quasiater* and *M. umbrosus* (Arvicolinae: Muridae). Acta Theriologica, 39:373-377.

Cervantes, F. A., S. T. Álvarez-Castañeda, B. Villa-Ramirez, C. Lorenzo, and J. Varga. 1996*a*. Natural history of the black jackrabbit (*Lepus insularis*) from Espiritu Santo island, Baja California Sur, Mexico. Southwestern Naturalist, 41:186-189.

Cervantes, F. A., C. Lorenzo, S. T. Álvarez-Castañeda, A. Rojas-Viloria, and J. Vargas. 1996*b*. Chromosomal study of the insular San Jose brush rabbit (*Sylvilagus mansuetus*) from Mexico. Southwestern Naturalist, 41:455-457.

Cervantes, F. A., J. Martínez, and O. G. Ward. 1997. The karyotype of the Tarabundi vole (*Microtus oaxacensis*: Rodentia), relict tropical arvicolid. Pp. 87-96, *in* Homenaje al professor Ticul Álvarez (J. Arroyo-Cabrales and O. Polaco, eds.). Instituto Nacional de Antropologia e Historia, México, D. F., 391 pp.

Cervantes, F. A., J. P. Ramirez-Silva, A. Marin, and G. L. Portales. 1999*a*. Allozyme variation of cottontail rabbits (*Sylvilagus*) from Mexico. Zeitschrift für Säugetierkunde, 64:356-362.

Cervantes, F. A., C. Lorenzo, and O. G. Ward. 1999*b*. Chromosomal relationships among spiny pocket mice, *Liomys* (Heteromyidae), from Mexico. Journal of Mammalogy, 80:823-832.

Cervantes, F. A., A. Rojas-Viloria, C. Lorenzo, and S. T. Álvarez-Castañeda. 1999-2000. Chromosomal differentiation between the jackrabbits *Lepus insularis* and *Lepus californicus* from Baja California Sur, Mexico. Revista Mexicana de Mastozoologia, 4:41-53.

Chaimanee, Y. 1998. Plio-Pleistocene rodents of Thailand. Thai Studies in Biodiversity No. 3:303 pp.

Chaimanee, Y., and J.-J. Jaeger. 2000*a*. Evolution of *Rattus* (Mammalia, Rodentia) during the Plio-Pleistocene in Thailand. Historical Biology, 15:181-191.

Chaimanee, Y., and J.-J. Jaeger. 2000*b*. Occurrence of *Hadromys humei* (Rodentia: Muridae) during the Pleistocene in Thailand. Journal of Mammalogy, 81(3):659-666.

Chakraborty, S. 1975. On a collection of mammals from Bhutan. Records of the Zoological Survey of India, 68:1-20.

Chakraborty, S. 1978. The rusty-spotted cat, *Felis rubiginosa* I. Geoffroy, in Jammu and Kashmir. Journal of the Bombay Natural History Society, 75:478-479.

Chakraborty, S. 1981. Studies on *Sciuropterus baberi* Blyth. Proceedings of the Zoological Society (Calcutta), 32:57-63.

Chakraborty, S. 1983. Contribution to the knowledge of the mammalian fauna of Jammu and Kashmir, India. Records of the Zoological Survey of India, 38:1-129.

Chakraborty, S. 1985. Studies on the genus *Callosciurus* Gray (Rodentia: Sciuridae). Records of the Zoological Survey of India, Miscellaneous Publications, Occasional Papers, No. 63:1-93.

Chakraborty, S., and V. C. Agrawal. 2000. Mammalia. Pp. 15-83, *in* State Fauna Series No. 8, Fauna of Gujarat (Part I), Vertebrates (J. R. B. Alfred, ed.), Zoological Survey of India, 464 pp.

Chakraborty, R., and S. Chakraborty. 1991. Taxonomic review of the genus *Bandicota* Gray and its species with a note on the intraspecific geographical variation in the large bandicoot rat, *Bandicota indica* (Bechstein) [Mammalia: Rodentia]. Records of the Zoological Survey of India, 88(1):87-99.

Chaline, J. 1972. Les rongeurs du Pleistocene moyen et superieur de France (Systematique-biostratigraphie-paleoclimatologie). Cahiers de Paleontologie, Centre National de la Recherche Scientifique, Paris, 410 pp.

Chaline, J. 1974. Esquisse de l'evolution morphologique, biometrique et chromosomique du genre *Microtus* (Arvicolidae, Rodentia) dans le Pleistocene de l'hemisphere nord. Bulletin de la Societe de Geologie France, ser. 7, 14:440-450.

Chaline, J. 1975. Taxonomie des campagnols (Arvicolidae, Rodentia) de la sous-famille des Dolomyinae nov. dans l'hémisphère Nord. Comptes Rendus de l'Academie des Sciences, Paris, ser. D, 281:115-118.

Chaline, J. 1980. *Mimomys salpetrierensis* n. sp. forme relique datee de 14,000 B. P. dans la grotte de la Salpetriere (Gard.). Géobios, Lyon, 13:645-651.

Chaline, J. 1985. Evolutionary data on steppe lemmings (Arvicolinae, Rodentia). Pp. 331-341, *in* Evolutionary relationships among rodents: A multidisciplinary analysis (P. Luckett and J.-L. Hartenberger, eds.). Plenum Press, New York, 721 pp.

Chaline, J. 1987. Arvicolid data (Arvicolidae, Rodentia) and evolutionary concepts. Evolutionary Biology, 21:237-310.

Chaline, J. 1990. An approach to studies of fossil arvicolids. Pp. 45-84, *in* International Symposium: Evolution, Phylogeny and Biostratigraphy of Arvicolids (Rodentia, Mammalia) (O. Fejfar, and W.-D. Heinrich, eds.). Geological Survey, Prague, 448 pp.

Chaline, J., and J.-D. Graf. 1988. Phylogeny of the Arvicolidae (Rodentia): Biochemical and paleontological evidence. Journal of Mammalogy, 69:22-33.

Chaline, J., and R. Matthey. 1971. Hypotheses relatives a la formule chromosomique d'*Allophaiomys pliocaenicus* (Rodentia, Arvicolidae) et a la diversification de cette espece. Comptes Rendus de l'Academie des Sciences, Paris, ser. D, 272(8):1071-1074.

Chaline, J., and P. Mein. 1979. Les rongeurs et l'evolution. Doin Editeurs, Paris, 235 pp.

Chaline, J., P. Mein, and F. Petter. 1977. Les grandes lignes d'une classification evolutive des Muroidea. Mammalia, 41:245-252.

Chaline, J., P. Brunet-Lecomte, and J.-D. Graf. 1988. Validation de *Terricola* Fatio, 1867 pour les campagnols souterrains (Arvicolidae, Rodentia) paléarctiques actuels et fossiles. Comptes Rendus de l'Académie des Sciences, Paris, ser. 3, 306:475-478.

Chaline, J., P. Brunet-Lecomte, G. Brochet, and F. Martin. 1989. Les lemmings fossiles du genre *Lemmus* (Arvicolidae, Rodentia) dans le Pleistocene de France. Geobios, 22:613-623.

Chaline, J., P. Brunet-Lecomte, S. Montuire, L. Viriot, and F. Courant. 1999. Anatomy of the arvicoline radiation (Rodentia): Palaeogeographical, palaeoecological history and evolutionary data. Annales Zoologici Fennici, 36:239-267.

Chamberlain, J. L. 1954. The Block Island meadow mouse, *Microtus provectus*. Journal of Mammalogy, 35:587-588.

Champion, F. W. 1929. The distribution of the mouse-deer (*Moschiola memmina*). Journal of the Bombay Natural History Society, 33:985-986.

Chan, K. L. 1977. Enzyme polymorphism in Malayan rats of the subgenus *Rattus*. Biochemical Systematics and Ecology, 5:161-168.

Chan, K. L., S. S. Dhaliwal, and H. S. Yong. 1978. Protein variation and systematics in Malayan rats of the subgenus *Lenothrix* (Rodentia: Muridae, genus *Rattus* Fischer). Comparative Biochemistry and Physiology, 59B:345-351.

Chan, K. L., S. S. Dhaliwal, and H. S. Yong. 1979. Protein variation and systematics of three subgenera of Malayan rats (Rodentia: Muridae, genus *Rattus* Fischer). Comparative Biochemistry and Physiology, 64B:329-337.

Chandrasekar-Rao, A., and G. G. Musser. 1993. New distribution record for *Mus caroli*. Mammalia, 57(3):462-463.

Chandrasekar-Rao, A., and M. E. Sunquist. 1996. Ecology of small mammals in tropical forest habitats of southern India. Journal of Tropical Ecology, 12:561-571.

Channing, A. 1984. Ecology of the namtap *Graphiurus ocularis* (Rodentia: Gliridae) in the Cedarberg, South Africa. South African Journal of Zoology, 19(3):144-149.

Channing, A. 1987. The namtap in the Cedarberg. African Wildlife, 41(3):135-136.

Channing, A. 1997. Family Gliridae. Spectacled dormouse. Pp. 164-165, *in* The complete book of southern African mammals (G. Mills and L. Hes, eds.). Struik Winchester, Cape Town, 356 pp.

Chapman, F. M. 1901. A revision of the genus *Capromys*. Bulletin of the American Museum of Natural History, 14:313-323, 2 pls.

Chapman, J. A. 1974. *Sylvilagus bachmani*. Mammalian Species, 34:1-4.

Chapman, J. A. 1975*a*. *Sylvilagus nuttallii*. Mammalian Species, 56:1-3.

Chapman, J. A. 1975*b*. *Sylvilagus transitionalis*. Mammalian Species, 55:1-4.

Chapman, J. A., and G. G. Ceballos. 1990. The cottontails. Pp. 95-110, *in* Rabbits, hares and pikas (J. A. Chapman and J. E. C. Flux, eds.). I.U.C.N., Gland, Switzerland, 168 pp.

Chapman, J. A., and G. A. Feldhamer. 1981. *Sylvilagus aquaticus*. Mammalian Species, 151:1-4.

Chapman, J. A., and G. R. Willner. 1978. *Sylvilagus audubonii*. Mammalian Species, 106:1-4.

Chapman, J. A., and G. R. Willner. 1981. *Sylvilagus palustris*. Mammalian Species, 153:1-3.

Chapman, J. A., J. G. Hockman, and M. M. Ojeda C. 1980. *Sylvilagus floridanus*. Mammalian Species, 136:1-8.

Chapman, J. A., K. R. Dixon, W. Lopez-Forment, and D. E. Wilson. 1983. The New World jackrabbits and hares (genus *Lepus*) 1. Taxonomic history and population status. Acta Zoologica Fennica, 174:49-51.

Chapman, J. A., et al. 1990. Conservation action needed for rabbits, hares and pikas. Pp. 154-168, *in* Rabbits, hares and pikas (J. A. Chapman and J. E. C. Flux, eds.). I.U.C.N., Gland, Switzerland, 168 pp.

Chapman, J. A., K. L. Cramer, N. J. Dippenaar, and T. J. Robinson. 1992. Systematics and biogeography of the New England cottontail, *Sylvilagus transitionalis* (Bangs, 1895), with the description of a new species from the Appalachian mountains. Proceedings of the Biological Society of Washington, 105:841-866.

Chapman, N. G., and D. I. Chapman. 1980. The distribution of fallow deer: A worldwide review. Mammal Review, 10:61-138.

Chapskii, K. K. 1955. Opyt peresmotra sistemy i diagnostiki tyulenei podsemeistva Phocinae [An attempt at revision of the systematics and diagnoses of seals of the subfamily Phocinae]. Trudy Zoologicheskovo Instituta, Akademiya Nauk, Leningrad, 17:160-199 (in Russian).

Charles-Dominique, P., A. Brosset, and S. Jouard. 2001. Les chauves-souris de Guyane. Patrimoines Naturels, 49:1-150.

Chasen, F. N. 1936. A note on Malaysian *Gunomys*. Bulletin of the Raffles Museum, Singapore, Straits Settlements, 12:135-136.

Chasen, F. N. 1939. Two new mammals from North Sumatra. Treubia, 17:207-208.

Chasen, F. N. 1940. A handlist of Malaysian mammals: A systematic list of the mammals of the Malay Peninsula, Sumatra, Borneo, and Java, including the adjacent small islands. Bulletin of the Raffles Museum, Singapore, 15:1-209.

Chasen, F. N., and C. B. Kloss. 1927. Spolia Mentawiensia—mammals. Proceedings of the Zoological Society of London, 1927:797-840.

Chasen, F. N., and C. B. Kloss. 1929*a*. Two new races of *Galeopterus*. Bulletin of the Raffles Museum, 2:11-12.

Chasen, F. N., and C. B. Kloss. 1929*b*. Notes on flying lemurs (*Galeopterus*). Bulletin of the Raffles Museum, 2:12-22.

Chasen, F. N., and C. B. Kloss. 1932. Mammals from the lowlands and islands of North Borneo. Bulletin of the Raffles Museum, Singapore, Straits Settlements, 6(1931):1-82.

Chassovnikarova, T., and G. Markov. 1999. [The yellow-necked mouse (*Apodemus flavicollis*, Melch., 1834) in Bulgaria: Biotopic adherence and epigenetic variability]. Forest Science, 36(3-4):89-96 (in Bulgarian with English abstract).

Chaturvedi, Y. 1965. A new house rat (Mammalia: Rodentia: Muridae) from the Andaman and Nicobar Islands. Proceedings of the Zoological Society of Calcutta, 18:141-144.

Chaturvedi, Y. 1980. Mammals of the Andamans and Nicobars: Their zoogeography and faunal affinity. Records of the Zoological Survey of India, 77:127-139.

Chaworth-Musters, J. L. 1937. On the nomenclature of the Palearctic chipmunk. Annals and Magazine of Natural History, ser. 10, 19:158-159.

Chaworth-Musters, J. L., and J. R. Ellerman. 1947. A revision of the genus *Meriones*. Proceedings of the Zoological Society of London, 117:478-504.

Chebez, J. C., and E. Massoia. 1996. Mamíferos de la provincia de Misiones. Pp. 180-308, *in* Fauna Misionera. Catalogo sistemático y zoogeográfico de los vertebrados de la provincia de Misiones (Argentina) (J. C. Chebez, ed.). Literature of Latin America, Buenos Aires, Argentina, 318 pp.

Cheema, I. U., S. M. Raza, and L. J. Flynn. 1997. Note on Pliocene small mammals from the Mirpur District, Azad Kashmir, Pakistan. Geobios, 30(1):115-119.

Cheema, I. U., L. J. Flynn, and A. R. Rajpar. 2003. Late Pliocene murid rodents from Lehri, Jhelum District, northern Pakistan. Pp. 85-92, *in* Advances in Vertebrate Paleontology: Hen to Panta. A tribute to Constantin Radulescu and Petre Mihai Samson (A. Petulescu and E. Stiuca, eds.). Speleological Institute "Emil Racovita", Bucharest, 132 pp.

Cheke, A. S., and A. F. Dahl. 1981. The status of bats on western Indian Ocean Islands, with special reference to *Pteropus*. Mammalia, 45:205-238.

Chelomina, G. N. 1993*a*. [Divergence of two families of the repeated DNA in wood and field mice of the genus *Apodemus* (Muridae, Rodentia)]. Genetika, 29(7):1163-1171 (in Russian with English abstract).

Chelomina, G. N. 1993*b*. [Differentiation of GC-rich restriction sites in the highly repeated DNA in *Apodemus* (Muridae, Rodentia)]. Genetika, 29(7):1172-1179 (in Russian with English abstract).

Chelomina, G. N. 1998. [Molecular phylogeny of forest and field mice of the genus *Apodemus* (Muridae, Rodentia) based on the data on restriction analysis of total nuclear DNA.] Genetika, 34(9):1286-1292 (in Russian with English abstract).

Chelomina, G. N., E. A. Lyapunova, N. N. Vorontsov, and K. Tsuchiya. 1995. Molecular genetic typing of three forms of wood and field mice belonging to a transpalearctic genus (*Apodemus*, Muridae, Rodentia). Russian Journal of Genetics, 31(6):701-704 (translated from Genetika, 31(6):820-824).

Chelomina, G. N., M. V. Pavlenko, I. V. Kartavtseva, G. G. Boeskorov, E. A. Lyapunova, and N. N. Vorontsov. 1998*a*. [Genetic differentiation of Caucasian wood mice: Comparison of isozymic, chromosomal and molecular divergence.] Genetika, 34(2):213-225 (in Russian with English abstract).

Chelomina, G. N., H. Suzuki, K. Tsuchiya, K. Moriwaki, E. A. Lyapunova, and N. N. Vorontsov. 1998*b*. [Sequencing of the mtDNA cytochrome *b* gene and reconstruction of the maternal relationships of wood and field mice of the genus *Apodemus* (Muridae, Rodentia).] Genetica, 34(5):650-661 (in Russian with English abstract).

Chen, G.-J., W. Wang, J.-Y. Mo, Z.-T. Huang, F. Tian, and W.-W. Huang. 2002. Pleistocene vertebrate fauna from Wuyun Cave of Tiandong County, Guangxi. Vertebrata Palasiatica, 40(1):42-51.

Chen Jun, and Wang Ding-guo. 1985. [A preliminary survey on geographical distribution of the rodents in Hexi corridor, Gansu province]. Acta Theriologica Sinica, 5(3):195-200 (in Chinese).

Chen, P. 1989. Baiji—*Lipotes vexillifer* Miller, 1918. Pp. 25-44, *in* Handbook of marine mammals: River dolphins and the larger toothed whales (S. H. Ridgway and R. Harrison, eds.). Academic Press, London, 4:1-442.

Chen, Z., Y. Wang, R. Liu, L. Shi, and Y. He. 1992. [A comparative study on the silver-stained karyotypes of eight species of rats (*Rattus*) in China]. Acta Theriologica Sinica, 12(4):280-286 (in Chinese with English abstract).

Chen, Z., L. Ruiqing, L. Chongyun, and W. Yingxiang. 1996. [Studies on the chromosomes of three species of wood mice]. Zoological Research, 17(3):347-352 (in Chinese with English abstract).

Cheng, C.-H., and C.-T. Yen. 1993. Scanning electron microscopic study of the tail of the rat. Acta Zoologica Taiwanica, 4(1):43-49.

Chenu, J. C., and E. Desmarest. 1850-1858. Carnassiers (vol. 21-22), in Encyclopédie d'histoire naturelle; ou, Traité complet de cette science d'après les travaux des naturalistes les plus éminents de tous les pays et de toutes les époques: Buffon, Daubenton, Lacépède, G. Cuvier, F. Cuvier, Geoffroy Saint-Hilaire, Latreille, De Jussieu, Brongnart, etc. Marescq et Compagnie, Paris [1]-2:1-312.

Chernyavskii, F. B. 1972. O rasprostranenii i geograficheskoi izmenchivosti Amerikanskovo dlinnokhvostovo suslika (*Citellus parryi* Rich., 1827) severo-vostochnoi Sibiri [On the distribution and geographic variation of the American long-tailed suslik . . . in northeastern Siberia]. Trudy Moskovskovo Obshchestva Ispytatelei Prirody, 48:199-214.

Chernyavskii, F. B., and A. I. Kozlovskii. 1980. [Species status and history of Arctic lemmings (*Dicrostonyx*, Rodentia) on Wrangel Island]. Zoologicheskii Zhurnal, 59:266-273 (in Russian).

Chernyavskii, F. B., V. G. Krivosheev, Yu. V. Revin, L. P. Khvorostyanskaya, and V. N. Orlov. 1980. [On the distribution, taxonomy and biology of the Amur lemming (*Lemmus amurensis*)]. Zoologicheskii Zhurnal, 59:1077-1084 (in Russian).

Chernyavskii, F. B., N. I. Abramson, A. A. Tsvetkova, E. M. Anbinder, L. P. Kurysheva. 1993. [On systematics and zoogeography of true lemming of the genus *Lemmus* (Rodentia, Cricetidae) of Beringia.] Zoologicheskii Zhurnal, 72(8):111-122.

Chesser, R. K. 1983. Cranial variation among populations of the black-tailed prairie dog in New Mexico. Occasional Papers, The Museum, Texas Tech University, 84:1-13.

Chetoui, M., K. Said, M. Rezig, and T. L. Cheniti. 2002. Analyse caryologique de quatre espèces de gerbilles (Rongeurs, Gerbillinae) de Tunisie. Bulletin de la Société zoologique de France, 127(3):211-221.

Chevret, P. 1994. Etude évolutive des Murinae (Rogeurs: Mammifères) africains par hybridation AND/AND. Comparaison avec les approches morphologiques et paleontologiques. These de Doctorat, Université des Sciences et Techniques du Languedoc, Montpellier II, 215 pp.

Chevret, P., and C. Hänna. 1994. Systematics of the spiny mouse (*Acomys*: Muroidea): Molecular and biochemical evidence. Israel Journal of Zoology, 40:247-254.

Chevret, P., C. Denys, J.-J. Jaeger, J. Michaux, and F. M. Catzeflis. 1993*a*. Molecular evidence that the spiny mouse (*Acomys*) is more closely related to gerbils (Gerbillinae) than to true mice (Murinae). Proceedings of the National Academy of Sciences, 90:3433-3436.

Chevret, P., C. Denys, J. J. Jaeger, J. Michaux, and F. Catzeflis. 1993*b*. Molecular and paleontological aspects of the tempos and mode of evolution in *Otomys* (Otomyinae: Muridae: Mammalia). Biochemical Systematics and Ecology, 21:123-131.

Chevret, P., L. Granjon, J.-M. Duplantier, C. Denys, and F. M. Catzeflis. 1994. Molecular phylogeny of the *Praomys* complex (Rodentia: Murinae): A study based on DNA/DNA hybridization experiments. Zoological Journal of the Linnean Society, 112:425-442.

Chevret, P., F. Catzeflis, and J. R. Michaux. 2001. "Acomyinae": New molecular evidences for a muroid taxon (Rodentia: Muridae). Pp. 109-125, *in* African Small Mammals (C. Denys, L. Granjon, and A. Poulet, eds.), IRD Éditions, Collection colloques et séminaires, Paris, 570 pp.

Chevret, P., P. Jenkins, and F. Catzeflis. 2003. Evolutionary systematics of the Indian mouse *Mus famulus* Bonhote, 1898: Molecular (DNA/DNA hybridization and 12S rRNA sequences) and morphological evidence. Zoological Journal of the Linnean Society, 137:385-401.

Cheylan, G. 1990. Endemisme et speciation chez les mammiferes Mediterraneens. Vie et Milieu, 40:137-143.

Cheylan, G. 1991. Patterns of Pleistocene turnover, current distribution and speciation among Mediterranean mammals. Pp. 227-262, *in* Biogeography of Mediterranean Invasions (R. H. Groves and F. Di Castri, eds.). Cambridge University Press, Cambridge, 501 pp.

Cheylan, G. 1999. Evolution rapide de petites populations insulaires mediterraneennes de *Rattus rattus*. Monografies de la Societat D'Historia Natural de las Balears, 6/Monografia de L'Institut D'Estudis Balearics, 66:83-104.

Cheylan, G., L. Granjon, and J. Britton-Davidian. 1998. Distribution of genetic diversity within and between western Mediterranean island populations of the black rat *Rattus rattus* (L. 1758). Biological Journal of the Linnean Society, 63:393-408.

Chiappero, M. B., G. E. Calderón, and C. N. Gardenal. 1997. *Oligoryzomys flavescens* (Rodentia, Muridae): Gene flow among populations from central-eastern Argentina. Genetica, 101:105-113.

Chiappero, M. B., A. Blanco, G. E. Calderón, M. S. Sabattini, and C. N. Gardenal. 2002. Genetic structure of populations of *Calomys laucha* (Muridae, Sigmodontinae) from central Argentina. Biochemical Systematics and Ecology, 30:1023-1036.

Chiarelli, A. B. 1975. The chromosomes of the Canidae. Pp. 40-53, *in* The wild canids: Their systematics, behavioral ecology and evolution (M. W. Fox, ed.). Van Nostrand Reinhold Company, New York, 508 pp.

Chimimba, C. T. 1998. A taxonomic synthesis of southern African *Aethomys* (Rodentia: Muridae) with a key to species. Mammalia, 62(3):427-437.

Chimimba, C. T. 2000. Geographic variation in *Aethomys chrysophilus* (Rodentia: Muridae) from southern Africa. Zeitschrift für Säugetierkunde, 65:157-171.

Chimimba, C. T. 2001*a*. Infraspecific morphometric variation in *Aethomys namaquensis* (Rodentia: Muridae) from southern Africa. Journal of Zoology, London, 253:191-210.

Chimimba, C. T. 2001*b*. Geographic variation in the Tete veld rat *Aethomys ineptus* (Rodentia: Muridae) from southern Africa. Journal of Zoology, London, 254:77-89.

Chimimba, C. T., and N. J. Dippenaar. 1994. Non-geographic variation in *Aethomys chrysophilus* (De Winton, 1897) and *A. namaquensis* (A. Smith, 1834) (Rodentia: Muridae) from southern Africa. South African Journal of Zoology, 29(2):107-117.

Chimimba, C. T., and N. J. Dippenaar. 1995. The selection of taxonomic characters for morphometric analysis: A case study based on Southern African *Aethomys* (Mammalia: Rodentia: Muridae). Annals of Carnegie Museum, 64(3):197-217.

Chimimba, C. T., and D. J. Kitchener. 1991. A systematic revision of Australian Emballonuridae. Records of the Western Australian Museum, 15:203-265.

Chimimba, C. T., N. J. Dippenaar, and T. J. Robinson. 1998. Geographic variation in *Aethomys granti* (Rodentia: Muridae) from southern Africa. Annals of the Transvaal Museum, 36(28):405-412.

Chimimba, C. T., N. J. Dippenaar, and T. J. Robinson. 1999. Morphometric and morphological delineation of southern African species of *Aethomys* (Rodentia: Muridae). Biological Journal of the Linnean Society, 67:501-527.

Chitaukali, W. N., H. Burda, and D. Kock. 2001. On small mammals of the Nyika Plateau, Malawi. Pp. 415-426, *in* African small mammals—Petits mammifères africains (C. Denys, L. Granjon, and A. Poulet, eds.). Collection Colloques et Séminaires, IRD Editions, Paris, 570 pp.

Choate, J. R. 1969. Taxonomic status of the shrew, *Notiosorex* (*Xenosorex*) *phillipsii* Schaldach, 1966 (Mammalia: Insectivora). Proceedings of the Biological Society of Washington, 82:469-476.

Choate, J. R. 1970. Systematics and zoogeography of Middle American shrews of the genus *Cryptotis*. University of Kansas, Publications of the Museum of Natural History, 19:195-317.

Choate, J. R. 1973. *Cryptotis mexicana*. Mammalian Species, 28:1-3.

Choate, J. R., and E. D. Fleharty. 1974. *Cryptotis goodwini*. Mammalian Species, 44:1-3.

Choate, J. R., and S. L. Williams. 1978. Biogeographic interpretation of variation within and among populations of the prairie vole, *Microtus ochrogaster*. Occasional Papers, The Museum, Texas Tech University, 49:1-25.

Choate, L. L. 1997. The mammals of the Llano Estacado. Special Publications, Museum of Texas Tech University, 40:1-240.

Choate, L. L., and C. Jones. 1998. Annotated checklist of Recent land mammals of Oklahoma. Occasional Papers, Museum of Texas Tech University, 181:1-13.

Choate, L. L., and F. C. Killebrew. 1991. Distributional records of the California myotis and the prairie vole in the Texas Panhandle. Texas Journal of Science, 43:214-215.

Choate, L. L., J. K. Jones, Jr., R. W. Manning, and C. Jones. 1990. Westward ho: Continued dispersal of the pygmy mouse, *Baiomys taylori*, on the Llano Estacado and in adjacent areas of Texas. Occasional Papers, Museum of Texas Tech University, 134:8 pp.

Chondropoulos, B. P., G. Markakis, and S. E. Fraguedakis-Tsolis. 1995. Morphometric and immunological relationships among some Greek *Mus* L. populations (Mammalia, Rodentia, Muridae). Zeitschrift für Säugetierkunde, 60:361-372.

Chondropoulos, B. P., S. E. Fraguedakis-Tsolis, G. Markakis, and E. Giagia-Athanasopoulou. 1996. Morphometric variability in karyologically polymorphic populations of the wild *Mus musculus domesticus* in Greece. Acta Theriologica, 41(4):375-382.

Chorn, J., and R. S. Hoffmann. 1978. *Ailuropoda melanoleuca*. Mammalian Species, 110:1-6.

Christiaens, B. 1995. [Sleep, little mouse, sleep]. Wielewaal, 61(5):162-164.

Christoff, A. U., V. Fagundes, I. J. Sbalqueiro, M. S. Mattevi, and Y. Yonenaga-Yassuda. 2000. Description of a new species of *Akodon* (Rodentia: Sigmodontinae) from southern Brazil. Journal of Mammalogy, 81:838-851.

Chruszez, B., and R. M. R. Barclay. 2002. *Chalinolobus gouldii*. Mammalian Species, 690:1-4.

Churcher, C. S. 1993. *Equus grevyi*. Mammalian Species, 453:1-9.

Churcher, C. S. 1994. The vertebrate fauna from the Natufian Level at Jebel Es-Saaïdé (Saaïdé II), Lebanon. Paléorient, 20(2):35-58.

Churchfield, S. 1990. The natural history of shrews. C. Helm, London. 178 pp.

Churchfield, S., P. Barriere, R. Hutterer, and M. Colyn. 2004. First results on the feeding ecology of sympatric shrews (Insectivora: Soricidae) in the Tai National Park, Ivory Coast. Acta Theriologica, 49:1-15.

Churchill, S. K. 1996. Distribution, habitat and status of the Carpentarian rock-rat, *Zyzomys palatalis*. Wildlife Research, 23:77-91.

Churchill, S. 1998. Australian bats. Reed New Holland Publishers, Sydney. 230 pp.

Ciofolo, L., and Y. Le Pendu. 2001. The feeding behaviour of giraffe in Niger. Mammalia, 66:183-194.

Civitelli, M. V., P. Consentino, and E. Capanna. 1989. Inter- and intra-individual chromosome variability in *Thamnomys* (*Grammomys*) *gazellae* (Rodentia, Muridae) B-chromosomes and structural heteromorphisms. Genetica, 79:93-105.

Civitelli, M. V., R. Castiglia, J.-Cl. Codjia, and E. Capanna. 1995*a*. Cytogenetics of the genus *Arvicanthis* (Rodentia, Muridae). 1. *Arvicanthis niloticus* from Republic of Benin (West Africa). Zeitschrift für Säugetierkunde, 60:215-225.

Civitelli, M. V., M. G. Filippucci, C. Kurtonur, and B. Özkan. 1995*b*. Chromosome analysis of three species of Myoxidae. Pp. 117-126, *in* Proceedings of II Conference on Dormice (Rodentia, Myoxidae) (M. G. Filippucci, ed.). Hystrix, n.s., 6(1-2):1-340.

Claessen, C. J., and F. De Vree. 1990. Systematic and distributional notes on the larger species of the genus *Epomophorus* Bennett, 1836 (Chiroptera: Pteropodidae). Pp. 177-186, *in* Vertebrates in the tropics. Proceedings of the international symposium on vertebrate biogeography and systematics in the tropics,

June 5-9, 1989 (G. Peters and R. Hutterer, eds.). Alexander Koenig Zoological Research Insititute and Zoological Museum, Bonn, 424 pp.

Claessen, C. J., and F. de Vree. 1991. Systematic and taxonomic notes on the *Epomophorus-anurus-labiatus-minor* complex with the description of a new species (Mammalia; Chiroptera; Pteropodidae). Senckenbergiana Biologica, 71:209-238.

Clark, J. W. 1873*a*. On the skull of a seal. Proceedings of the Zoological Society of London, 1873:556-557.

Clark, J. W. 1873*b* [1874]. On the eared seals of the Auckland Islands. Proceedings of the Zoological Society of London, 1873:750-760.

Clark, M. K., M. S. Mitchell, and K. S. Karriker. 1993. Notes on the geographical and ecological distribution of relict populations of *Synaptomys cooperi* (Rodentia: Arvicolidae) from eastern North Carolina. Brimleyana, 19:155-167.

Clark, T. W. 1975. *Arctocephalus galapagoensis*. Mammalian Species, 64:1-2.

Clark, T. W., R. S. Hoffmann, and C. F. Nadler. 1971. *Cynomys leucurus*. Mammalian Species, 7:1-4.

Clark, T. W., E. Anderson, C. Douglas, and M. Strickland. 1987. *Martes americana*. Mammalian Species, 289:1-8.

Clause, B. T. 1993. The Wistar Rat as a right choice: Establishing mammalian standards and the ideal of a standardized mammal. Journal of the History of Biology, 26(2):329-349.

Clausnitzer, V. 2000. Ecology of *Otomys barbouri* Lawrence and Loveridge, 1953 (Mammalia, rodentia): An endemic of the Afro-alpine zone of Mt Elgon, East Africa. Bonner Zoologische Monographien, 46:233-244.

Clausnitzer, V. 2001. Rodents of the afro-alpine zone of Mt. Elgon. Pp. 427-443, *in* African Small Mammals (C. Denys, L. Granjon, and A. Poulet, eds.). IRD Éditions, Collection colloques et séminaires, Paris, 570 pp.

Clausnitzer, V., and R. Kityo. 2001. Altitudinal distribution of rodents (Muridae and Gliridae) on Mt. Elgon, Uganda. Tropical Zoology, 14:95-118.

Clausnitzer, V., S. Churchfield, and R. Hutterer. 2003. Habitat occurrence and feeding ecology of *Crocidura montis* and *Lophuromys flavopunctatus* on Mt. Elgon, Uganda. African Journal of Ecology, 41:1-8.

Clawson, R. G., J. A. Clawson, and T. L. Best. 1994*a*. *Tamias alpinus*. Mammalian Species, 461:1-6.

Clawson, R. G., J. A. Clawson, and T. L. Best. 1994*b*. *Tamias quadrimaculatus*. Mammalian Species, 469:1-6.

Cleef-Roders, J. T. van, and L. W. van den Hoek Ostende. 2001. Dental morphology of *Talpa europaea* and *Talpa occidentalis* (Mammalia. Insectivora) with a discussion of fossil *Talpa* in the Pleistocene of Europe. Zoologische Mededelingen Leiden, 75:51-68.

Clouet, M. 1979. Note sur la systématique du bouquetin d'Espagne. Bulletin de la Société d'Histoire Naturelle de Toulouse, 115:269-277.

Clough, G. C. 1976. Current status of two endangered Caribbean rodents. Biological Conservation, 10(1):43-48.

Cloutier, D., and D. W. Thomas. 1992. *Carollia perspicillata*. Mammalian Species, 417:1-9.

Clutton-Brock, J., G. B. Corbet, and M. Hills. 1976. A review of the family Canidae, with classification by numerical methods. Bulletin of the British Museum (Natural History), 29(3):119-199.

Cockburn, A. 1995*a*. Silky Mouse, *Pseudomys apodemoides*. Pp. 585-586, *in* Mammals of Australia (R. Strahan, ed.). Smithsonian Institution Press, Washington, D.C., 756 pp.

Cockburn, A. 1995*b*. Smoky Mouse, *Pseudomys fumeus*. Pp. 598-599, *in* Mammals of Australia (R. Strahan, ed.). Smithsonian Institution Press, Washington, D.C., 756 pp.

Cockburn, A. 1995*c*. Heath Rat, *Pseudomys shortridgei*. Pp. 617-618, *in* Mammals of Australia (R. Strahan, ed.). Smithsonian Institution Press, Washington, D.C., 756 pp.

Cockrill, W. R. (ed.). 1974. The husbandry and health of the domestic buffalo. FAO, UN, Rome, 993 pp.

Cockrum, E. L. 1976*a*. On the status of the hairy-footed gerbil, *Gerbillus hirtipes* Lataste, 1881. Mammalia, 40:523-526.

Cockrum, E. L. 1976*b*. Status of the name of a rhinolophid bat, *Rhinolophus euryale tuneti* Deleuil and Labbe, 1955. Mammalia, 40:685-686.

Cockrum, E. L. 1977. Status of the hairy footed gerbil, *Gerbillus latastei* Thomas and Trouessart. Mammalia, 41:75-80.

Cockrum, E. L., and H. W. Setzer. 1976. Types and type localities of North African Rodents. Mammalia, 40:633-670.

Cockrum, E. L., T. C. Vaughan, and P. J. Vaughan. 1976*a*. A review of North African short-tailed gerbils (*Dipodillus*) with description of a new taxon from Tunisia. Mammalia, 40(2):313-326.

Cockrum, E. L., T. C. Vaughan, and P. J. Vaughan. 1976*b*. *Gerbillus andersoni* de Winton, a species new to Tunisia. Mammalia, 40:467-473.

Cockrum, E. L., P. J. Vaughan, and T. C. Vaughan. 1977. Status of the pale sand rat, *Psammomys vexillaris* Thomas, 1925. Mammalia, 41:321-326.

Codjia, J. T. C., C. Chrysostome, M. V. Civitelli, and E. Capanna. 1994. Les chromosomes des rongeurs de Bénin (Afrique de l'Ouest):1. Cricetidae. Rendiconti Fisiche Accademia Lincei, ser. 9, 5(3):277-287.

Codjia, J. T. C., M. V. Civitelli, D. Bizzoco, and E. Capanna. 1996. Les chromosomes de *Mastomys natalensis* et *Mastomys erythroleucus* (Rongeurs, Muridés) du sud Bénin (Afrique de l'Ouest): Nouvelles précisions sur la variabilité chromosomique. Mammalia, 60(2):299-303.

Coe, M. 1975. Mammalian ecological studies on Mt. Nimba, Liberia. Mammalia, 39(4):523-581.

Coe, M. J. 1969. The anatomy of the reproductive tract and breeding in the springhare *Pedetes surdaster larvalis* Hollister. Journal of Reproduction and Fertility (Suppl.), 6:159-174.

Coetzee, C. G. 1970. The relative tail-length of striped mice *Rhabdomys pumilio* Sparrman 1784 in relation to climate. Zoologica Africana, 5:1-6.

Coetzee, C. G. 1977a. Genus *Steatomys*. Part 6.8. Pp. 1-4, *in* The mammals of Africa: An identification manual (J. Meester and H. W. Setzer, eds.) [issued 22 Aug 1977]. Smithsonian Institution Press, Washington, D.C., not continuously paginated.

Coetzee, C. G. 1977b. Order Carnivora. Part 8. Pp. 1-42, *in* The mammals of Africa: An identification manual (J. Meester and H. W. Setzer, eds.) [issued 22 Aug 1977]. Smithsonian Institution Press, Washington, D.C., not continuously paginated.

Cohen, J. A. 1978. *Cuon alpinus*. Mammalian Species, 100:1-3.

Coimbra-Filho, A. F. 1990. Sistemática, distribução geográfica e situação atual dos simios Brasileiros (Platyrrhini-Primates). Revista Brasileira de Biologia, 50:1063-1079.

Çolak, E., E. Kivanç, and N. Yiğit. 1994. A study on taxonomic status of *Allactaga euphratica* Thomas, 1881 and *Allactaga williamsi* Thomas, 1897 (Rodentia: Dipodidae) in Turkey. Mammalia, 58(4):591-600.

Çolak, E., N. Yiğit, M. Sözen, and Ş. Özkurt. 1997. Distribution and taxonomic status of the genus *Microtus* (Mammalia: Rodentia) in southeastern Turkey. Israel Journal of Zoology, 43:391-396.

Çolak, E., N. Yiğit, M. Sözen, and Ş. Özkurt. 1998. On the karyotype of the long-eared hedgehog, *Hemiechinus auritus* (Gmelin, 1770)(Mammalia: Insectivora), in Turkey. Zeitschrift für Säugetierkunde, 62:372-374.

Çolak, E., N. Yiğit, and R. Verimli. 1999. On the karyotype of the long-clawed mole vole, *Prometheomys schaposchnikovi* Satunin, 1901 (Mammalia: Rodentia), in Turkey. Zeitschrift für Säugetierkunde, 64:126-127.

Colangelo, P., M. V. Civitelli, and E. Capanna. 2001. Morphology and chromosomes of *Tatera* Lataste 1882 (Rodentia Muridae Gerbillinae) in West Africa. Tropical Zoology, 14:243-253.

Cole, J. R., and J. C. Z. Woinarski. 2000. Rodents of the arid Northern Territory: Conservation status and distribution. Wildlife Research, 27:437-449.

Cole, R. E., A. Engilis, Jr., and F. J. Radovsky. 1997. Report on mammals collected during the Bishop Museum Expedition to Mt. Dayman, Milne Bay Province, Papua New Guinea. Bishop Museum Occasional Papers, 51:1-36.

Colgan, D. J., and T. F. Flannery. 1992. Biochemical systematic studies in the genus *Petaurus* (Marsupialia: Petauridae). Australian Journal of Zoology, 40:245-256.

Colket, E., and D. E. Wilson. 1998. *Taphozous hildegardeae*. Mammalian Species, 597:1-3.

Collett, R. 1886 [1887]. On *Phascogale viginiae*, a rare pouched mouse from Northern Queensland. Proceedings of the Zoological Society of London, 1886:548-549.

Collier, G. E., and S. J. O'Brien. 1985. A molecular phylogeny of the Felidae: Immunological distance. Evolution, 39:473-487.

Collins, P. W. 1993. Taxonomic and biogeographic relationships of the island fox (*Urocyon littoralis*) and gray fox (*Urocyon cinereoargenteus*) from western North America. Pp. 351-390, *in* Proceedings of the third Channel Islands symposium: Recent advances in California Islands research (F. G. Hochberg, ed.). Santa Barbara Museum of Natural History, Santa Barbara, California. 661 pp.

Collinson, M. E., and J. J. Hooker. 2000. Gnaw marks on Eocene seeds: Evidence for early Rodent behavior. Palaeogeography, Palaeoclimatology, Palaeoecology, 157 (1-2):127-149.

Colyn, M. 1986. Les mammifères de forêt ombrophile entre les rivières Tshopo et Maiko (Région du Haute-Zaire). Bulletin de la Institut Royal des Sciences Naturelles de Belgique: Biologie, 56:21-26.

Colyn, M., and A. M. Dudu. 1986. Releve systematique des rongeurs (Muridae) des iles forestieres du fleuve Zaire entre Kisangani et Kinshasa. Revue de Zoologie Africaine, 99:353-357.

Colyn, M., and H. Van Rompaey. 1994. A biogeographic study of cusimanses (*Crossarchus*) (Carnivora, Herpestidae) in the Zaire Basin. Journal of Biogeography, 21:479-489.

Colyn, M., A. Gautier-Hion, and D. Thys van den Audenaerde. 1991. *Cercopithecus dryas* Schwarz, 1932 and *C. salongo* Thys van den Audenaerde, 1977 are the same species with an age-related coat pattern. Folia Primatologia, 56:167-170.

Comport, S. S., and I. D. Hume. 1998. Gut morpology and rate of passage of fungal spores through the gut of a tropical rodent, the giant white-tailed rat (*Uromys caudimaculatus*). Australian Journal of Zoology, 46:461-471.

Concepción, J. L., and J. Molinari. 1991. *Sphiggurus vestitus pruinosus* (Mammalia, Rodentia, Erethizontidae): The karyotype and its phylogenetic implications, descriptive notes. Studies on Neotropical Fauna and Environment, 26:237-241.

Conner, M. M., and T. M. Shenk. 2003. Distinguishing *Zapus hudsonius preblei* freom *Zapus princeps princeps* by using repeated cranial measurements. Journal of Mammalogy, 84(4):1456-1463.

Conole, L. E. 2000. Acoustic differentiation of Australian populations of the large bentwing-bat *Miniopterus schreibersii* (Kuhl, 1817). Australian Zoologist, 31:443-447.

Conroy, C. J., and J. A. Cook. 1999. MtDNA evidence for repeated pulses of speciation within arvicoline and murid rodents. Journal of Mammalian Evolution, 6:221-245.

Conroy, C. J., and J. A. Cook. 2000a. Molecular systematics of a Holarctic rodent (*Microtus*: Muridae). Journal of Mammalogy, 81:344-359.

Conroy, C. J., and J. A. Cook. 2000b. Phylogeography of a post-glacial colonizer: *Microtus longicaudus* (Rodentia: Muridae). Molecular Ecology, 9:165-175.

Conroy, C. J., Y. Hortelano, F. A. Cervantes, and J. A. Cook. 2001. The phylogenetic position of southern relictual species of *Microtus* (Muridae: Rodentia) in North America. Mammalian Biology/Zeitschrift für Saugetierkunde, 66:332-344.

Conroy, G. C., M. Pickford, B. Senut, J. Van Couvering, and P. Mein. 1992. *Otavipithecus namibiensis*, first Miocene hominoid from southern Africa. Nature, 356:144-148.

Contoli, L. 1990. Further data about *Crocidura cossyrensis* Contoli, 1989, with respect to other species of the genus in the Mediterranean. Hystrix, n.s., 2:53-58.

Contoli, L. 2003. On subspecific taxonomy of *Microtus savii* (Rodentia, Arvicolidae). Hystrix, Italian Journal of Mammalogy, n.s., 14:107-111.

Contoli, L., and G. Aloise. 2001. On the taxonomy and distribution of *Crocidura cossyrensis* and *Crocidura russula* (Insectivora, Soricidae) in Maghreb. Hystrix, n.s., 12:11-18.

Contoli, L., and G. Amori. 1986. First record of a live *Crocidura* (Mammalia, Insectivora) from Pantelleria island, Italy. Acta Theriologica, 31:343-347.

Contoli, L., B. Benincasa-Stagni, and A. R. Marenzi. 1989. Morphometry and morphology of *Crocidura* Wagler 1832 (Mammalia, Soricidae) in Italy, Sardinia and Sicily, with fourier descriptors approach: First results. Hystrix, n.s., 1:113-130.

Contoli, L., G. Amori, and C. Nazzaro. 1992. Tooth diversity in Arvicolidae (Mammalia, Rodentia): Ecochorological factors and speciation time. Hystrix, n.s., 4(2):1-15.

Contrafatto, G., J. A. Meester, G. Bronner, P. J. Taylor, and K. Willan. 1992a. Genetic variation in the African rodent subfamily Otomyinae (Muridae). IV: Chromosome G-banding analysis of *Otomys irroratus* and *O. angoniensis*. Israel Journal of Zoology, 38:277-291.

Contrafatto, G., J. A. Meester, K. Willan, P. J. Taylor, M. A. Roberts, and C. M. Baker. 1992b. Genetic variation in the African rodent subfamily Otomyinae (Muridae). II: Chromosomal changes in some populations *Otomys irroratus*. Cytogenetics and Cell Genetics, 59:293-299.

Contrafatto, G., D. David, and V. Goossens-Le Clerq. 1994. Genetic variation in the African rodent subfamily Otomyinae. V: Phylogeny inferred by immuno-electrotransfer analysis. Durban Museum Novitates, 19:1-7.

Contrafatto, G., J. R. Van den Berg, and J. H. Grace. 1997. Genetic variation in the African rodent subfamily Otomyinae (Muridae): Immuno-electrotransfer of liver proteins of some *Otomys irroratus* (Brants 1827) populations. Tropical Zoology, 10:157-171.

Contreras, J. R. 1982. Notas acerca de *Bolomys temchuki* (Massoia, 1982) en el noreste argentino, con la descripción de dos nuevas subespecies. Historia Natural, Corrientes, 2:174-176.

Contreras, J. R. 1992. Acerca de la localidad tipica de *Calomys callosus* (Rengger, 1830) (Mammalia, Rodentia, Cricetidae). Notulas Faunisticas, 35:1-5.

Contreras, J. R. 1999. El genero *Ctenomys* en la provincial de Tucuman, Republica Argentina, con la descripcion de una nueva especie (Rodentia, Ctenomyidae). Ciencia Siglo, XXI(3):1-31.

Contreras, J. R., and L. M. Berry. 1983. Notas acerca de los roedores del genero *Oligoryzomys* de la Provincia del Chaco, Republica Argentina (Rodentia, Cricetidae). Historia Natural, 3:145-148.

Contreras, J. R., and L. M. Berry. 1985. Acerca de la distribucion de *Ctenomys argentinus* (Rodentia: Ctenomyidae). Historia Natural, 5(13):104.

Contreras, J. R., and A. N. Ch. de Contreras. 1984. La situacion de *Ctenomys minutus*, Nehring, 1887, en la provincia de Entre Rios, con la descripcion de nuevas subspecies (Rodentia: Ctenomyidae), Resultados VII Jornadas Argentinas Zoologicas, Mar del Plata, p. 76.

Contreras, J. R., and O. A. Reig. 1965. Datos sobre la distribución del género *Ctenomys* (Rodentia,

Octodontidae) en la zona costera de la Provincia Buenos Aires comprendida entre Necocea y Bahía Blanca. Physis, Buenos Aires, 25(69):169-186.

Contreras, J. R., and V. G. Roig. 1991. Las especies del genero *Ctenomys* (Rodentia: Octodontidae). I. *Ctenomys dorsalis* Thomas, 1900. Historia Natural, 8(3):6-9.

Contreras, J. R., and M. I. Rosi. 1980. El raton de campo *Calomys musculinus cordovensis* (Thomas) en la provincia de Mendoza. I. Consideraciones taxonomicas. Historia Natural, 1:17-25.

Contreras, J. R., and J. A. Scolardo. 1986. Distribucion y relaciones taxonomicas entre los cuatro nucleos geograficos disyuntos de *Ctenomys dorbignyi* en la provincia Corrientes, Argentina (Rodentia: Ctenomyidae). Historia Natural, 6(3):21-30.

Contreras, J. R., and W. Silvera Avalos. 1995. Incorporación del pequeño marsupial *Monodelphis scalops* Thomas, 1988 [sic] a la mastofauna del Paraguay (Marsupialia: Didelphidae). Notulas Faunisticas, 70:1-2.

Contreras, J. R., V. G. Roig, and C. M. Suzarte. 1977. *Ctenomys validus*, una nueva especie de "tunduque" de la Provincia de Mendoza (Rodentia; Octodontidae). Physis, Buenos Aires, sec. C, 36(92):159-162.

Contreras, J. R., Y. E. Davis, A. O. Contreras and M. Alvarez. 1985. Acerca de la distribucion de *Ctenomys perrensi* Thomas, 1896 y sus relaciones geograficas con las demas especies del genero (Rodentia: Ctenomyidae). Historia Natural, 5(22):173-178.

Contreras, L. C., and J. C. Torres-Mura. 1987. *Tympanectomys*, un mamífero deserticola con el numero diploide mas alto. Archivos de Biología y Medicina Experimentales (Santiago), 20, pages unnumbered, resumen no. R-190.

Contreras, L. C., J. C. Torres-Mura, and J. L. Yáñez. 1987. Biogeography of octodontid rodents: An eco-evolutionary hypothesis. Pp. 401-411, *in* Studies in Neotropical mammalogy. Essays in honor of P. Hershkovitz (B. D. Patterson and R. M. Timm, eds.). Fieldiana: Zoology, n.s., 39:1-506.

Contreras, L. C., J. C. Torres-Mura, and A. E. Spotorno. 1990. The largest known chromosome number for a mammal, in a South American desert rodent. Experientia, 46:506-508.

Contreras, L. C., J. C. Torres-Mura, A. E. Spotorno, and L. I. Walker. 1994. Chromosomes of *Octomys mimax* and *Octodontomys gliroides* and relationships of octodontoid rodents. Journal of Mammalogy, 75(3):768-774.

Convention on International Trade in Endangered Species. 1991. Appendices I, II, and III to The Convention on International Trade in Endangered Species of Wild Fauna and Flora. September 1, 1991. U.S. Government Printing Office, Washington, D.C., 19 pp.

Conway, M., and C. G. Schmitt. 1978. Record of the Arizona shrew (*Sorex arizonae*) from New Mexico. Journal of Mammalogy, 59:631.

Cook, J. A., and E. P. Lessa. 1998. Are rates of diversification in subterranean South American tuco-tucos (genus *Ctenomys*, Rodentia: Octodontidae) unusually high? Evolution, 52:1521-1527.

Cook, J. A., and T. L. Yates. 1994. Systematic relationships of the tuco-tucos, genus *Ctenomys* (Rodentia: Octodontidae). Journal of Mammalogy, 75:583-599.

Cook, J. A., S. Anderson, and T. L. Yates. 1990. Notes on Bolivian mammals. 6. The genus *Ctenomys* (Rodentia, Ctenomyidae) in the highlands. American Museum Novitates, 2980:1-27.

Cook, J. M., R. Trevelyan, S. S. Walls, M. Hatcher, and F. Rakotondraparany. 1991. The ecology of *Hypogeomys antimena*, an endemic Madagascan rodent. Journal of Zoology, London, 224:191-200.

Cooke, H. B. S., and A. F. Wilkinson. 1978. Suidae and Tayassuidae. Pp. 435-482, *in* Evolution of African mammals (V. J. Maglio and H. B. S. Cooke, eds). Harvard University Press, Cambridge, MA, 641 pp.

Coolidge, H. J. 1940. The Indochinese forest ox or kouprey. Memoir of the Museum of Comparative Zoology at Harvard College, 54:417-531.

Cooper, N. K. 1993. Identification of *Pseudomys chapmani, P. hermannsburgensis, P. delicatulus* and *Mus musculus* using footpad patterns. Western Australian Naturalist, 19(2):69-73.

Cooper, N. K. 1994. Identification of *Pseudomys albocinereus, P. occidentalis, P. shortridgei, Rattus rattus* and *R. fuscipes* using footpad patterns. Western Australian Naturalist, 19(4):279-283.

Cooper, N. K., K. P. Aplin, and M. Adams. 2000. A new species of false antechinus (Marsupialia: Dasyuromorphia: Dasyuridae) from the Pilbara region, Western Australia. Records of the Western Australian Museum, 20:115-136.

Cooper, N. K., M. Adams, C. Anthony, and L. H. Schmitt. 2003*a*. Morphological and genetic variation in *Leggadina* (Thomas, 1910) with special reference to Western Australian populations. Records of the Western Australian Museum, 21:333-351.

Cooper, N. K., T. Bertozzi, A. Baynes, and R. J. Teale. 2003*b*. The relationship between eastern and western populations of the heath rat, *Pseudomys shortridgei* (Rodentia: Muridae). Records of the Western Australian Museum, 21:367-370.

Cooper, S. J. B., T. B. Reardon, and J. Skilins. 1998. Molecular systematics of Australian rhinolophid bats (Chiroptera: Rhinolophidae). Australian Journal of Zoology, 46:203-220.

Cooper, S. J. B., M. Adams, and A. Labrinidis. 2000. Phylogeography of the Australian dunnart *Sminthopsis crassicaudata* (Marsupialia: Dasyuridae). Australian Journal of Zoology, 48:461-473.

Cooper, T. W., R. R. Hollander, R. J. Kinucan, and J. K. Jones, Jr. 1993. Systematic status of the deer mouse, *Peromyscus maniculatus*, on the Llano Estacado and in adjacent areas. Texas Journal of Science, 45:3-18.

Copeland, N. G., N. A. Jenkins, D. J. Gilbert, J. T. Eppig, L. J. Maltais, J. C. Miller, W. F. Dietrich, A. Weaver, S. E. Lincoln, R. G. Steen, L. D. Stein, J. H. Nadeau, and E. S. Lander. 1993. A genetic linkage map of the mouse: Current applications and future prospects. Science, 262:57-66.

Copley, P. 1999. Natural histories of Australia's stick-nest rats, genus *Leporillus* (Rodentia: Muridae). Wildlife Research, 26:513-539.

Corbet, G. B. 1974. Potomogalinae. Part 1.2, Pp. 1-2 and Macroscelididae. Part 1.5, Pp. 1-6, *in* The mammals of Africa: An identification manual (J. Meester and H. W. Setzer, eds.) [issued 10 Sep 1974]. Smithsonian Institution Press, Washington, D.C., not continuously paginated.

Corbet, G. B. 1978*a*. *Erinaceus dauuricus* Sundevall, 1842 (Mammalia, Insectivora): Proposed conservation under the plenary powers. Bulletin of Zoological Nomenclature, 35:123-124.

Corbet, G. B. 1978*b*. *Sorex dsinezumi* Temminck, 1844 (Mammalia, Insectivora): Proposed use of the plenary powers to rule a correct original spelling. Bulletin of Zoological Nomenclature, 35:125-126.

Corbet, G. B. 1978*c*. The mammals of the Palaearctic region: A taxonomic review. British Museum (Natural History), London, 314 pp.

Corbet, G. B. 1979. The taxonomy of *Procavia capensis* in Ethiopia, with special reference to the aberrant tusks of *P. c. capillosa* Brauer (Mammalia, Hyracoidea). Bulletin of the British Museum (Natural History), Zoology Series, 36(4):251-259.

Corbet, G. B. 1983. A review of classification in the family Leporidae. Acta Zoologica Fennica, 174:11-15.

Corbet, G. B. 1984. The mammals of the Palearctic region: A taxonomic review. Supplement. British Museum (Natural History), London, 45 pp.

Corbet, G. B. 1988. The family Erinaceidae: A synthesis of its taxonomy, phylogeny, ecology and zoogeography. Mammal Review, 18:117-172.

Corbet, G. B. 1990. The relevance of metrical, chromosomal and allozyme variation to the systematics of the genus *Mus*. Biological Journal of the Linnean Society, 41:5-12.

Corbet, G. B., and J. Hanks. 1968. A revision of the elephant-shrews, family Macroscelididae. Bulletin of the British Museum (Natural History), Zoology Series, 16:45-111.

Corbet, G. B., and S. Harris. 1990. The handbook of British mammals. Third ed. Blackwell Scientific Publications, Oxford, 588 pp.

Corbet, G. B., and J. E. Hill. 1980. A world list of mammalian species. British Museum (Natural History), London, 226 pp.

Corbet, G. B., and J. E. Hill. 1986. A world list of mammalian species. Second ed. British Museum (Natural History), London, 254 pp.

Corbet, G. B., and J. E. Hill. 1991. A world list of mammalian species. Third ed. British Museum (Natural History) Publications, London, 243 pp.

Corbet, G. B., and J. E. Hill. 1992. Mammals of the Indomalayan region. A systematic review. Oxford University Press, Oxford, 488 pp.

Corbet, G. B., and L. A. Jones. 1965. The specific characters of the crested porcupines, subgenus *Hystrix*. Proceedings of Zoological Society of London, 144:285-300.

Corbet, G. B., and P. A. Morris. 1967. A collection of recent and subfossil mammals from southern Turkey (Asia Minor), including the dormouse *Myomimus personatus*. Journal of Natural History, 4:561-569.

Corbet, G. B., and D. W. Yalden. 1972. Recent records of mammals (other than bats) from Ethiopia. Bulletin of the British Museum (Natural History), Zoology Series, 22(8):213-252.

Corbet, G. B., J. Cummins, S. R. Hedges, and W. Krzanowski. 1970. The taxonomic status of British water voles, genus *Arvicola*. Journal of Zoology, London, 161:301-316.

Corkeron, P. J. 1990. Aspects of the behavioral ecology of inshore dolphins *Tursiops truncatus* and *Sousa chinensis* in Moreton Bay, Australia. Pp. 285-292, *in* The bottlenose dolphin (S. Leatherwood, and R. R. Reeves, eds.). Academic Press, New York, 653 pp.

Corneli, P. S. 2002. Complete mitochondrial genomes and eutherian evolution. Journal of Mammalian Evolution, 9:281-305.

Cornely, J. E., and R. J. Baker. 1986. *Neotoma mexicana*. Mammalian Species, 262:1-7.

Cornely, J. E., and B. J. Verts. 1988. *Microtus townsendii*. Mammalian Species, 325:1-9.

Corti, M., and C. M. Ciabatti. 1990. The structure of a chromosomal hybrid zone of house mice (*Mus*) in central Italy: Cytogenetic analysis. Zeitschrift für Zoologische Systematik und Evolutionsforschung, 28:277-288.

Corti, M., and C. Fadda. 1996. Systematics of *Arvicanthis* (Rodentia, Muridae) from the Horn of Africa: A geometric morphometrics evaluation. Italian Journal of Zoology, 63:185-192.

Corti, M., M. S. Merani, and G. de Villafane. 1987. Multivariate morphometrics of vesper mice (*Calomys*): Preliminary assessment of species, population, and strain divergence. Zeitschrift für Säugetierkunde, 52:236-242.

Corti, M., M. V. Civitelli, A. Bekele, R. Castiglia, and E. Capanna. 1995. The chromosomes of three endemic rodents of the Bale Mountains, South Ethiopia. Rendiconti Fisiche Accademia Lincei, ser. 9, 6(2):157-164.

Corti, M., C. Fadda, S. Simson, and E. Nevo. 1996*a*. Size and shape variation in the mandible of the fossorial rodent *Spalax ehrenbergi*: A Procrustes Analysis of three dimensions. Pp. 303-320, *in* Advances in Morphometrics (L. F. Marcus, M. Corti, A. Loy, G. J. P. Naylor, D. E. Slice, eds.). Plenum Press, New York, 620 pp.

Corti, M., M. V. Civitelli, R. Castiglia, A. Bekele, and E. Capanna. 1996*b*. Cytogenetics of the genus *Arvicanthis* (Rodentia, Muridae). 2. The chromosomes of three species from Ethiopia: *A. abyssinicus, A. dembeensis* and *A. blicki*. Zeitschrift für Säugetierkunde, 61:339-351.

Corti, M., A. Scanzani, A. R. Rossi, M. V. Civitelli, A. Bekele, and E. Capanna. 1999. Karyotypic and genetic divergence in the Ethiopian *Myomys-Stenocephalemys* complex (Mammalia, Rodentia). Italian Journal of Zoology, 66:341-349.

Corti, M., C. Di Giuliomaria, and W. Verheyen. 2000. Three-dimensional geometric morphometrics of the African genus *Lophuromys* (Rodentia Muridae). Hystrix, n.s., 11(1):145-154.

Corti, M., R. Castiglia, F. Annesi, and W. Verheyen. 2004. Mitochondrial sequences and karyotypes reveal hidden diversity in African pouched mice (Subfamily Cricetomyinae, Genus *Saccostomus*). Journal of Zoology, London, 262:1-12.

Coryndon, S. C. 1977. The taxonomy and nomenclature of the Hippopotamidae (Mammalia, Artiodactyla) and a description of two new fossil species. Proceedings—Koninklijke Nederlandse Akademie van Wetenschappen, ser. B(Paleontology, Geology, Physics, and Chemistry), 80:61-88.

CoŌkun, Y. 2001. On distribution, morphology and biology of the mole vole, *Ellobius lutescens* Thomas, 1897 (Mammalia: Rodentia) in eastern Turkey. Zoology in the Middle East, 23:5-12.

Cosson, J. F., M. Pascal, and F. Bioret. 1996. Origine et répartition des musaraignes du genre *Crocidura* dans les Iles Bretonnes. Vie et Milieu, 46:233-244.

Costa, L. 2003. The historical bridge between the Amazon and Atlantic Forest of Brazil: A study of molecular phylogeography with small mammals. Journal of Biogeography, 30:71-86.

Costa, L. P., Y. L. R. Leite, and J. L. Patton. 2003. Phylogeography and systematic notes on two species of gracile mouse opossums, genus *Gracilinanus* (Marsupialia, Didelphidae) from Brazil. Proceedings of the Biological Society of Washington, 116:275–292.

Costello, R. K., C. Dickinson, A. L. Rosenberger, S. Boinski, and F. S. Szalay. 1993. Squirrel monkey (genus *Saimiri*) taxonomy. Pp. 177-210, *in* Species, species concepts, and primate evolution. (W. H. Kimbel and L. B. Martin, eds.). Plenum Press, New York, 560 pp.

Cothran, E. G., and R. L. Honeycutt. 1984. Chromosomal differentiation of hybridizing ground squirrels (*Spermophilus mexicanus* and *S. tridecemlineatus*). Journal of Mammalogy, 65:118-122.

Cothran, E. G., E. G. Zimmerman, and C. F. Nadler. 1977. Genic differentiation and evolution in the ground squirrel subgenus *Ictidomys*. Journal of Mammalogy, 58:610-622.

Cotterill, F. P. D. 1996. New distribution records of bats of the families Nycteridae, Rhinolophidae and Vespertilionidae (Microchiroptera: Mammalia) in Zimbabwe. Arnoldia Zimbabwe, 10:71-89.

Cotterill, F. P. D. 2001*a*. Distribution records of leaf-nosed bats (Microchiroptera: Hipposideridae) in Zimbabwe. Arnoldia Zimbabwe, 10:189-198.

Cotterill, F. P. D. 2001*b*. Further notes on large Afrotropical free-tailed bats of the genus *Tadarida* (Molossidae: Mammalia). Arnoldia Zimbabwe, 10:199-210.

Cotterill, F. P. D. 2001*c*. The first specimen of Thomas's flat-headed bat, *Mimetillus moloneyi thomasi* (Microchiroptera: Mammalia) in southern Africa from Mozambique. Arnoldia Zimbabwe, 10:211-218.

Cotterill, F. P. D. 2001*d*. New specimens of lesser house bats (Vespertilionidae: *Scotoecus*) from Mozambique and Zambia. Arnoldia Zimbabwe, 10:219-224.

Cotterill, F. P. D. 2001*e*. New records for two species of fruit bats (Megachiroptera: Mammalia) in southeast Africa, with taxonomic comments. Durban Museum Novitates, 26:53-56.

Cotterill, F. P. D. 2001*f*. Notes on a mammal collection and biodiversity conservation in the Ikelenge Pedicle, Mwinilunga District, northwest Zambia. Bulawayo: Biodiversity Foundation for Africa, Occasional Publications in Biodiversity No. 10:1-16.

Cotterill, F. P. D. 2002. A new species of horseshoe bat (Microchiroptera: Rhinolophidae) from south-central Africa, with comments on its affinities and evolution, and the characterization of rhinolophid species. Journal of Zoology, 256:165-179.

Cotterill, F. P. D. 2003*a*. Geomorphological influences on vicariant evolution in some African mammals in the Zambezi Basin: Some lessons for conservation. Pp. 11-58, *in* Ecology and conservation of small antelope (A. Plowman, ed.). Filander Verlag, Fürth, 262 pp.

Cotterill, F. P. D. 2003*b*. Species concepts and the real diversity of antelopes. Pp. 59-118, *in* Ecology and conservation of small antelope (A. Plowman, ed.). Filander Verlag, Fürth, 262 pp.

Cotterill, F. P. D. 2003*c*. Insights into the taxonomy of tsessebe antelopes *Damaliscus lunatus* (Bovidae: Alcelaphini), in south-central Africa: With the description of a new evolutionary species. Durban Museum Novitates, 28:45-55.

Coues, E. 1874. Synopsis of the Muridae of North America. Proceedings of the Philadelphia Academy of Natural Sciences, 26:173-196.

Coult, T. 2001. Notes on historical distribution of the dormouse (*Muscardinus avellanarius*) in Northumberland and Durham. Vasculum, 86(2):41-45.

Courant, F., B. David, B. Laurin, and J. Chaline. 1997. Quantification of cranial convergences in arvicolids (Rodentia). Biological Journal of the Linnean Society, 62:505-517.

Courant, F., P. Brunet-Lecomte, V. Volobouev, J. Chaline, J.-P. Quere, A. Nadachowski, S. Montuire, G. Bao, L. Viriot, R. Rausch, M. Erbajeva, D. Shi, and P. Giraudoux. 1999. Karyological and dental identification of *Microtus limnophilus* in a large focus of alveolar echinococcosis (Gansu, China). Compte Rendu Academie des Sciences, Paris, Sciences de la Vie, 322:473-480.

Couturier, J., and B. Dutrillaux. 1986. Evolution chromosomique chez les carnivores. Mammalia (Numero special):124-162.

Couturier, M. A. J. 1954. L'Ours brun (*Ursus arctos* L.). Grenoble, 904 pp.

Couturier, M. A. J. 1962. Le bouquetin des Alpes *Capra aegagrus ibex ibex* L. Published by the author, Grenoble, 1565 pp.

Covacevich, J. 1995. Lakeland Downs Mouse, *Leggadina lakedownensis*. Pp. 556-557, *in* Mammals of Australia (R. Strahan, ed.). Smithsonian Institution Press, Washington, D.C., 756 pp.

Cowan, C. F. 1973. *Proc. Zool. Soc. Lond.*, publication dates. Journal of the Society for the Bibliography of Natural History, 6:293-294.

Cowan, I. McT. 1936. Distribution and variation in deer (genus *Odocoileus*) of the Pacific coastal region of North America. California Fish and Game, 22:155-246.

Cowan, I. McT. 1940. Distribution and variation in the native sheep of North America. American Midland Naturalist, 24:505-580.

Cowan, I. McT., and C. J. Guiguet. 1965. The mammals of British Columbia. Third Ed. British Columbia Provincial Museum, Handbook, 11:1-414.

Cowan, I. McT., and W. McCrory. 1970. Variation in the mountain goat, *Oreamnos americanus* (Blainville). Journal of Mammalogy, 51:60-73.

Cramer, M. J., and G. N. Cameron. 2001. *Geomys texensis*. Mammalian Species, 679:1-3.

Cramer, M. J., M. A. Willig, and C. Jones. 2001. *Trachops cirrhosus*. Mammalian Species, 656:1-6.

Cranford, J. A. 1999. Western jumping mouse | *Zapus princeps*. Pp. 668-669, *in* The Smithsonian Book of North American Mammals (D. W. Wilson and S. Ruff, eds.). Smithsonian Institution Press, Washington, D.C., 750 pp.

Crawford-Cabral, J. 1966*a*. Note on the taxonomy of *Genetta*. Zoologica Africana, 2:25-26.

Crawford-Cabral, J. 1966*b*. Some new data on Angolan Muridae. Zoologica Africana, 2(2):192-103.

Crawford-Cabral, J. 1969. As genetas de Angola. Boletim do Instituto de Investigação Científica de Angola, 6(1):3-33.

Crawford-Cabral, J. 1970. As genetas da Africa central (República do Zaire, Ruanda e Burrindi). Boletim do Instituto de Investigação Científica de Angola, 7(2):3-23.

Crawford-Cabral, J. 1973. As genetas da Guiné Portuguesa e de Moçambique, Pp. 133-155, *in* Livro de Homenagem ao Professor Fernando Frade Viegas da Costa. Separata. Lisbon, 66:1-329.

Crawford-Cabral, J. 1981*a*. A new classification of the genets. African Small Mammal Newsletter, 6:8-10.

Crawford-Cabral, J. 1981*b*. Análise de dados craniométricos no género *Genetta* G. Cuvier. (Carnivora, Viverridae). Memórias da Junta de Investigações Científicas do Ultramar Centro de Zoologica. Lisbon, 66:1-329.

Crawford-Cabral, J. 1983. Patterns of allopatric speciation in some Angolan Muridae. Annales Musée Royal de l'Afrique Centrale, Tervuren-Belgique, ser. 8 (Sciences Zoologiques), 237:153-157.

Crawford-Cabral, J. 1986*a*. A discussion of the taxa to be used in a zoogeographical analysis as illustrated in Angolan Muroidea. Cimbebasia, ser. A, 8:161-166.

Crawford-Cabral, J. 1986*b*. A list of Angolan Chiroptera with notes on their distribution. Garcia de Orta, Série de Zoologia, Lisboa, 13(1/2):7-48.

Crawford-Cabral, J. 1987. The taxonomic status of *Crocidura nigricans* Bocage, 1889 (Mammalia, Insectivora). Garcia de Orta, Serie de Zoologica, 14(1):3-12.

Crawford-Cabral, J. 1988. A craniometric study on Angolan gerbils of the subgenus *Tatera* (Mammalia,

Rodentia, Gerbillidae). Part I: Results from a principal components analysis. Zoologische Abhandlungen Staatliches Museum für Tierkunde in Dresden, 43:169-192.

Crawford-Cabral, J. 1989a. The prior scientific name of the larger red mongoose (Carnivora: Viverridae: Herpestinae). Garcia de Orta, Serie de Zoologica, 14(2):1-2.

Crawford-Cabral, J. 1989b. *Praomys angolensis* (Bocage, 1890) and the identity of *Praomys angolensis* auct. (Rodentia: Muridae), with notes on their systematic positions. Garcia de Orta, Serie de Zoologica, 15:1-10.

Crawford-Cabral, J. 1996. The species of *Galerella* (Mammalia: Carnivora: Herpestinae) occurring in the southwestern corner of Angola. Garcia de Orta, Serie de Zoologica, 21(1):7-17.

Crawford-Cabral, J. 1998. The Angolan rodents of the superfamily Muroidea: An account of their distribution. Instituto de Investigação Científica Tropical, Estudos, Ensaios e Documentos, 161:223 pp.

Crawford-Cabral, J. 1999. The short-tailed *Aethomys* of Angola (Mammalia: Rodentia: Muridae). Garcia de Orta, Série de Zoologica, 23(1):169-191.

Crawford-Cabral, J., and C. Fernandes. 1999. A comment on the nomenclature of the rusty-spotted genet. Small Carnivore Conservation, 21:12.

Crawford-Cabral, J., and A. P. Pacheco. 1989. A craniometrical study on some water rats of the genus *Dasymys* (Mammalia, Rodentia, Muridae). Garcia de Orta, Serie de Zoologica, 15:11-24.

Crawford-Cabral, J., and A. P. Pacheco. 1991. A craniometric study on Angolan gerbils of subgenus *Tatera* (Mammalia, Rodentia, Gerbillidae). Part II: Results from discriminant and cluster analysis. Zoologische Abhandlungen Staatliches Museum für Tierkunde in Dresden, 46:215-224.

Crawford-Cabral, J., and A. P. Pacheco. 1992. Are the large-spotted and the rusty-spotted genets separate species? (Carnivora, Viverridae, genus *Genetta*). Garcia de Orta, Serie de Zoologica, 16(1/2):7-17.

Crégut-Bonnoure, E. 1992. Intérêt biostratigraphique de la morphologie dentaire de *Capra* (Mammalia, Bovidae). Acta Zoologica Fennici, 28:273-290.

Cresti, M., S. Marini, L. Rinetti, and A. Zangirolami. 1994. Indagine sur popolamento de micromammiferi nell'Alto Luinese (Varese). Atti della Società Italiana de Scienze Naturali e del Museo Civico di Storia Naturale di Milano, 133(13):153-183.

Cristaldi, M., and G. Amori. 1988. Perspectives pour une interpretation historique des populations eoliennes de rongeurs. Bulletin d'Ecologie, 19(2-3):171-176.

Cronin, J. E., and W. E. Meikle. 1979. The phylogenetic position of *Theropithecus*: Congruence among molecular, morphological, and palaeontological evidence. Systematic Zoology, 28:259-269.

Cronin, M. A. 1991. Mitochondrial-DNA phylogeny of deer (Cervidae). Journal of Mammalogy, 72:533-566.

Crowe, P. E. 1943. Notes on some mammals of the southern Canadian Rocky Mountains. Bulletin of the American Museum of Natural History, 80:391-410.

Crowther, M. S., C. R. Dickman, and A. J. Lynam. 1999. *Sminthopsis griseoventer boullangerensis* (Marsupialia: Dasyuridae), a new subspecies in the *S. murina* complex from Boullanger Island, Western Australia. Australian Journal of Zoology, 47:215-243.

Crucitti, P. 1976. Biometria di una collazione de *Miniopterus schreibersi* (Nutt.)(Chiroptera) cafturati nel Lagio (Italia). Annali del Museo civico di Storia Naturale di Genova, 81:131-138.

Crusafont-Pairo, M., and F. Petter. 1964. Un murine geant fossile des Iles Canaries *Canariomys bravoi* gen. nov., sp. nov. (rongeurs, murides). Mammalia, Supplement, 28:607-611.

Csaikl, F., M. Habernig, and U. Csaikl. 1990. A comparative restriction endonuclease analysis of nuclear DNA from Austrian *Apodemus* species: Intraspecific variability in middle repetitive DNA families contrasts isoenzyme data. Zeitschrift für Säugetierkunde, 55:55-59.

Csorba, G. 1993. A review of the occurrences of mouse-like dormouse (*Myomimus personatus*) in Turkmenia and an additional record. Mammalia, 57(2):282-284.

Csorba, G. 1997. Description of a new species of *Rhinolophus* (Chiroptera: Rhinolophidae) from Malaysia. Journal of Mammalogy, 78:342-347.

Csorba, G. 1998. The distribution of the great evening bat *Ia io* in the Indomalayan region. Myotis, 36:197-201.

Csorba, G. 2002. Remarks on some types of the genus *Rhinolophus*. Annales Historico-Naturales Musei Nationalis Hungarici, 94:217-226.

Csorba, G., and P. D. Jenkins. 1998. First records and a new subspecies of *Rhinolophus stheno* (Chiroptera, Rhinolophidae) from Vietnam. Bulletin of the Natural History Museum, London, 64:207-211.

Csorba, G., and L. L. Lee. 1999. A new species of vespertilionid bat from Taiwan and a revision of the taxonomic status of *Arielulus* and *Thianycteris* (Chiroptera: Vespertilionidae). Journal of Zoology, 248:361-367.

Csorba, G., P. Ujhelyi, and N. Thomas. 2003. Horseshoe bats of the world. Alana Books, Bishop's Castle, U.K., 160 pp.

Csuti, B. A. 1979. Patterns of adaptation and variation in the Great Basin kangaroo rat (*Dipodomys microps*). University of California Publications in Zoology, 111:1-69.

Cuartas, C. A., J. Muñoz, and M. González. 2001. Una nueva specie de *Carollia* Gray, 1838 (Chiroptera: Phyllostomidae) de Colombia. Actualidades Biológicas, 23:63-73.

Cucchi, T., J.-D. Vigne, J.-C. Auffray, P. Croft, and E. Peltenburg. 2002. Introduction involontaire de la souris domestique (*Mus musculus domesticus*) à Chypres dès le Néolithique précéramique ancien (fin IX et VIII millénaires av. J.-C.). Comptes Rendus Palevol, 1:235-241.

Cudmore, W. W. 1986. Ectoparasites of *Neotoma cinerea* and *N. fuscipes* from western Oregon. Northwest Science, 60:174-178.

Cuenca Bescos, G., J. I. Canudo and C. Laplana. 2001. La sequence des rongeurs (Mammalia) des sites du Pleistocene inferieur et moyen d'Atapuerca (Burgos, Espagne). Anthropologie, 105(1):115-130.

Cuervo-Diaz, A., J. Hernandez Camacho, and A. Cadena G. 1986. Lista actualizada de los mamiferos de Colombia: Anotaciones sobre su distribucion. Caldasia, xv:71-75.

Culver, M., W. E. Johnson, J. Pecon-Slattery, and S. J. O'Brien. 2000. Genomic ancestry of the American puma (*Puma concolor*). Journal of Heredity, 91:186-197.

Cumming, D. H. M., R. F. Du Toit, and S. N. Stuart. 1990. African elephants and rhinos: Status survey and conservation action plan. I.U.C.N., Gland, Switzerland, 72 pp.

Cummings, W. C. 1985a. Bryde's whale—*Balaenoptera edeni*. Pp. 137-154, *in* Handbook of marine mammals: The sirenians and baleen whales (S. H. Ridgway and R. Harrison, eds.). Academic Press, London, 3:1-362.

Cummings, W. C. 1985b. Right whales—*Eubalaena glacialis* and *Eubalaena australis*. Pp. 275-304, *in* Handbook of marine mammals: The sirenians and baleen whales (S. H. Ridgway and R. Harrison, eds.). Academic Press, London, 3:1-362.

Currier, M. J. P. 1983. *Felis concolor*. Mammalian Species, 200:1-7.

Custodio, C. C., M. V. Lepiten, and L. R. Heaney. 1996. *Bubalus mindorensis*. Mammalian Species, 520:1-5.

Cuvier, F. G. 1822 [1823]. Examen des especes formation des genres ou sous-genres Acanthion, Eréthizon, Sinéthère et Sphiggure. Mémoires du Muséum d'Histoire Naturelle (Paris), 9(1822):413-484.

Cuvier, F. G. 1825. Éléphant d' Afrique. Pp. VI, *in* E. Geoffroy St.-Hilaire and F. Cuvier, Histoire Naturelle Mammifères, 3(52):2.

Cuvier, F. G. 1829. Zoologie. Mammalogie. In Dictionnaire des sciences naturelles, dan lequel on traite méthodiquement des différens êtres de la nature, considérés soit en eux-mêmes, d'après l'état actuel de nos connoissances, soit relativement a l'utilité qu'en peuvent retirer la médecine, l'agriculture, le commerce et les arts. F. G. Levrault. Paris, 59:1-520.

Cuvier, G. [Baron]. 1817. Le règne animal distribué d'après son organisation, pour servir de base à l'histoire naturelle des animaux et d'introduction à l'anatomie comparée. vol. 1. Les mammifères. Deterville, Paris, 540 pp.

Cuvier, G. [Baron]. 1829. Le règne animal distribue d'après son organisation, pour servir de base a l'histoire naturelle des animaux et d'introduction a l'anatomie comparée. vol. 1. Les mammiféres. Nouvelle edition, revue et augmentée. Deterville, Paris, 584 pp.

Czaplewski, N. J. 1983. *Idionycteris phyllotis*. Mammalian Species, 208:1-4.

Czaplewski, N. J. 1987. Sigmodont rodents (Mammalia; Muroidea; Sigmodontinae) from the Pliocene (early Blancan) Verde Formation, Arizona. Journal of Vertebrate Paleontology, 7:183-199.

Czernay, S. 1987. Spiesshirsche und Pudus. A. Ziemsen Verlag, Wittenberg Lutherstadt, 84 pp.

Daams, R. 1981. The dental pattern of the dormice *Dryomys*, *Myomimus*, *Microdyromys* and *Peridyromys*. Utrecht Micropaleontological Bulletins, Special Publication, 3:1-115.

Daams, R. 1999. Family Gliridae. Pp. 301-318, *in* The Miocene land mammals of Europe (G. E. Rössner and K. Heissig, eds.). Pfeil, München, 515 pp.

Daams, R., and H. De Bruijn. 1995. A classification of the Gliridae (Rodentia) on the basis of dental morphology. Pp. 3-50, *in* Proceedings of II Conference on Dormice (Rodentia, Myoxidae) (M. G. Filippucci, ed.). Hystrix, n.s., 6(1-2):1-340.

Dagg, A. I. 1971. *Giraffa camelopardalis*. Mammalian Species, 5:1-8.

Dalebout, M. L., J. G. Mead, C. S. Baker, A. L. Van Helden, and A. N. Baker. 2002. A new species of beaked whale *Mesoplodon perrini* sp. n. (Cetacea: Ziphiidae) discovered through phylogenetic analyses of mitochondrial DNA sequences. Marine Mammal Science, 18(3):577-608.

Dalquest, W. W. 1948. Mammals of Washington. University of Kansas Publications, Museum of Natural History, 2:1-444.

Dalquest, W. W. 1950. Records of mammals from the Mexican state of San Luis Potosi. Occasional Papers of the Museum of Zoology, Louisiana State University, 23:1-15.

Dalquest, W. W., and F. B. Stangl. 1984. The taxonomic status of *Myotis magnamolaris*, Choate and Hall. Journal of Mammalogy, 65:485-486.

Dammerman, K. W. 1934. On the occurrence of wild buffaloes in Java and Sumatra. Treubia, 14:487-494.

Dandelot, P. 1974. Order Primates. Part 3. Pp. 1-45, in The mammals of Africa: An identification manual (J. Meester and H. W. Setzer, eds.) [issued 10 Sep 1974]. Smithsonian Institution Press, Washington, D.C., not continuously paginated.

Dandelot, P., and J. Prévost. 1972. Contributions à l'étude des primates de l'Ethiopie (simiens). Mammalia, 36:607-633.

Dang, H. H., D. V. T. Dao, V. S. Cao, T. A. Pham, and M. K. Hoang. 1994. [Checklist of mammals in Vietnam.] Publishing House "Science and Technics", Hanoi, 167 pp. (in Vietnamese).

Daniels, M. J., B. Balharry, D. Hirst, A. C. Kitchener, and R. J. Aspinall. 1998. Morphological and pelage characteristics of wild living cats in Scotland: Implications for defining the 'wildcat'. Journal of Zoology, London, 244(2):231-247.

Danielsen, F., D. S. Balete, T. D. Christensen, M. Heegaard, O. F. Jakobsen, A. Jensen, T. Lund, and M. K. Poulsen. 1994. Conservation of biological diversity in the Sierra Madre Mountains of Isabela and southern Cagayan Province, the Philippines. BirdLife International, Manila and Copenhagen, 146 pp.

Danilkin, A. A. 1995. Capreolus pygargus. Mammalian Species, 512:1-7.

Danko, Š. 1994. [Small mammals of the protected landscape area East Carpathians]. Zborník Východoslovenského Múzea v Košiciah, Prírodné Vedy, 35:63-76 (in Czech with English summary).

Dannelid, E. 1991. The genus Sorex (Mammalia, Soricidae)—distribution and evolutionary aspects of Eurasian species. Mammal Review, 21:1-20.

Dannelid, E. 1994. Chromosome polymorphism in Sorex alpinus (Mammalia, Soricidae) in the western Alps and the Swiss Jura. Zeitschrift für Säugetierkunde, 59:161-168.

Dannelid, E. 1998. Dental adaptations in shrews. Pp. 157-174, in Evolution of shrews (J. M.Wójcik and M. Wolsan, eds.). Mammal Research Institute, Bialowieza, Polish Academy of Sciences, 458 pp.

Dao Van Tien. 1960. Recherches zoologiques dans la region de Vinh-Linh (Province de Quang-tri, Centre Vietnam). Zoologischer Anzeiger, 164:221-239.

Dao Van Tien. 1961. Notes sur une collection de micromammiferes de la region de Hon-Gay. Zoologischer Anzeiger, 166:290-308.

Dao Van Tien. 1965. O formakh krasnobrinkhikh belok (Callosciurus erythraeus, Sciuridae) i ikh rasprotranenii vo Vietname [On the forms of the squirrel Callosciurus erythraeus (Sciuridae) and their distribution in Viet-nam]. Zoologicheskii Zhurnal, 44(8):1238-1244 (in Russian).

Dao Van Tien. 1966. Sur deux rongeurs nouveaux (Muridae, Rodentia) au Nord-Vietnam. Zoologischer Anzeiger, 176:438-439.

Dao Van Tien. 1978. Sur une collection de mammiferes du Plateau de Moc Chau (Province de So'n-la, Nord-Vietnam). Mitteilungen aus dem Zoologische Museum in Berlin, 54:377-391.

Dao Van Tien. 1983. On the north Indochinese gibbons (Hylobates concolor) in North Vietnam. Journal of Human Evolution, 12:367-372.

Dao Van Tien. 1985. Scientific results of some mammals surveys in North Vietnam (1957-1971). Scientific and Technical Publishing House, Ha Noi, 330 pp.

Dao Van Tien, and Cao Van Sung. 1990. Six new Vietnamese rodents. Mammalia, 54:233-238.

Daoud, A. 1989. Dormice (Rodentia, Gliridae) and their evolution. Przeglad Zoologiczny, 33(2):279-89.

Daoud, A. 1993. Evolution of Gliridae (Rodentia, Mammalia) in the Pliocene and Quaternary of Poland. Acta Zoologica Cracoviensia, 36(2):199-231.

Darviche, D., and P. Orsini. 1982. De souris sympatriques: Mus spretus et Mus musculus domesticus. Mammalia, 46:205-217.

Darviche, D., F. Benmehdi, J. Britton-Davidian, and L. Thaler. 1979. Donnees preliminaires sur la systematique biochimique des genres Mus et Apodemus en Iran. Mammalia, 43:427-430.

Das, I. 1999. A noteworthy collection of mammals from Mount Harriet, Andaman Islands, India. Journal of South Asian Natural History, 4:181-185.

Das, P. K. 1986. Studies on the taxonomy of bats obtained by the Silent Valley (Kerala, India) Expedition, 1980. Records of the Zoological Survey of India, 8:259-276.

Das, P. K. 1987. On the number of specimens, repository, and the heterogeneity of the type-series of Vespertilio montivagus. Bulletin of the Zoological Survey of India, 8:39-45.

Das, P. K. 1990. Occurrence of Pipistrellus comortae in the Andaman Islands. Journal of the Bombay Natural History Society, 87:135-137.

Dashzeveg, D., J.-L. Hartenberger, T. Martin, and S. Legendre. 1998. A peculiar minute Glires (Mammalia) from the early Eocene of Mongolia. Pp. 194-209, in Dawn of the Age of Mammals in Asia (K. C. Beard and M. R. Dawson, eds.). Bulletin of Carnegie Museum of Natural History, 34:348 pp.

Davidow-Henry, B. R., J. K. Jones, Jr., and R. R. Hollander. 1989. Cratogeomys castanops. Mammalian Species, 338:1-6.

Davidson, M. E. M. 1929. Notes on the northern elephant seal. Proceedings of the California Academy of Science (Fourth Series), 18:600-606.

Davies, C. 1982. The Recent and Fossil Affinities of the Genus *Pedetes* (Mammalia: Rodentia). Doctor of Philosophy thesis, St. Peter's College, Oxford.

Davies, J. L. 1963. The antitropical factor in cetacean speciation. Evolution, 17(1):107-116.

Davis, B. L., and R. J. Baker. 1974. Morphometrics, evolution, and cytotaxonomy of mainland bats of the genus *Macrotus* (Chiroptera: Phyllostomatidae). Systematic Zoology, 23:26-39.

Davis, D. D. 1962. Mammals of the lowland rainforest of north Borneo. Bulletin of the National Museum of Singapore, 31:1-129.

Davis, D. D. 1964. The giant panda: A morphological study of evolutionary mechanisms. Fieldiana: Zoology, Memoirs, 3:1-339.

Davis, D. H. S. 1949. The affinities of the South African gerbils of the genus *Tatera*. Proceedings of the Zoological Society of London, 118:1002-1018.

Davis, D. H. S. 1962. Distribution patterns of southern African Muridae, with notes on some of their fossil antecedents. Annals of the Cape Provincial Museums, 2:56-76.

Davis, D. H. S. 1965. Classification problems of African Muridae. Zoologica Africana, 1:121-145.

Davis, D. H. S. 1966. Contribution to the revision of the genus *Tatera* in Africa. Pp. 49-65, *in* Proceedings of the Colloquium on African rodents. Brussels-Tervuren, 1964. Annales Musée Royal De L'Afrique Centrale (Sciences Zoologiques), 144:49-65.

Davis, D. H. S. 1974. The distribution of some small southern African mammals (Mammalia: Insectivora, Rodentia). Annals of the Transvaal Museum, 29:135-184.

Davis, D. H. S. 1975*a*. Genera *Tatera* and *Gerbillurus*. Part 6.4. Pp. 1-7, *in* The mammals of Africa: An identification manual (J. Meester, and H. W. Setzer, eds.) [issued 10 Dec 1975]. Smithsonian Institution, Washington, D.C., not continuously paginated.

Davis, D. H. S. 1975*b*. Genus *Aethomys* Thomas, 1915. Part. 6.6. Pp. 1-5, *in* The mammals of Africa: An identification manual (J. Meester and W. H. Setzer, eds.) [issued 10 Dec. 1975]. Smithsonian Institution Press, Washington, D. C, not continuously paginated.

Davis, J. A. 1978. A classification of otters. Pp. 14-33, *in* Otters (N. Duplaix, ed.). I.U.C.N. Publication, New Series.

Davis, J., and W. Z. Lidicker. 1975. The taxonomic status of the southern sea otter. Proceedings of the California Academy of Science, 40(14):429-437.

Davis, R., and D. E. Brown. 1989. Role of post-Pleistocene dispersal in determining the modern distribution of Abert's squirrel. Great Basin Naturalist, 49:425-434.

Davis, R., and J. R. Callahan. 1992. Post-Pleistocene dispersal in the Mexican vole (*Microtus mexicanus*): An example of an apparent trend in the distribution of southwestern mammals. Great Basin Naturalist, 52:262-268.

Davis, W. B. 1937. Variations in Townsend pocket gophers. Journal of Mammalogy, 18:145-158.

Davis, W. B. 1939. The Recent mammals of Idaho. Caldwell, Idaho, The Caxton Printers, Ltd., 400 pp.

Davis, W. B. 1940. Distribution and variation of pocket gophers (genus *Geomys*) in the southwestern United States. Bulletin of the Texas Agricultural Experiment Station, 590:1-38.

Davis, W. B. 1944. Geographic variation in brown lemmings (Genus *Lemmus*). Murrelet, 25:19-25.

Davis, W. B. 1965. Review of the *Eptesicus brasiliensis* complex in Middle America with the description of a new subspecies from Costa Rica. Journal of Mammalogy, 46:229-240.

Davis, W. B. 1966. Review of South American bats of the genus *Eptesicus*. Southwestern Naturalist, 11:245-274.

Davis, W. B. 1968. A review of the genus *Uroderma* (Chiroptera). Journal of Mammalogy, 49:676-698.

Davis, W. B. 1969. A review of the small fruit bats (genus *Artibeus*) of Middle America. Southwestern Naturalist, 14:15-29.

Davis, W. B. 1970. A review of the small fruit bats (genus *Artibeus*) of Middle America. Part II. Southwestern Naturalist, 14:389-402.

Davis, W. B. 1976. Geographic variation in the lesser noctilio, *Noctilio albiventris* (Chiroptera). Journal of Mammalogy, 57:687-707.

Davis, W. B. 1980. New *Sturnira* (Chiroptera: Phyllostomidae) from Central and South American, with key to currently recognized species. Occasional Papers, The Museum, Texas Tech University, 70:1-5.

Davis, W. B. 1984. Review of the large fruit-eating bats of the *Artibeus "lituratus"* complex (Chiroptera: Phyllostomidae) in Middle America. Occasional Papers, The Museum, Texas Tech University, 93:1-16.

Davis, W. B., and D. C. Carter. 1978. A review of the round-eared bats of the *Tonatia silvicola* complex, with descriptions of three new taxa. Occasional Papers, The Museum, Texas Tech University, 53:1-12.

Davis, W. B., and D. J. Schmidly. 1994. The mammals of Texas. Texas Parks and Wildlife Press, Austin, 338 pp.

Davis, W. B., D. C. Carter, and R. H. Pine. 1964. Noteworthy records of Mexican and Central American bats. Journal of Mammalogy, 45:375-387.

Davis, W. H., and C. L. Rippy. 1968. Distribution of *Myotis lucifugus* and *Myotis austroriparius* in the southeastern United States. Journal of Mammalogy, 49:113-117.

Davydov, G. S. 1984. [Distribution and ecology of the forest dormouse (*Dryomys nitedula* Pallas, 1779) in Tadzhikistan]. Izvestiya Akademii Nauk Tadzhikskoi SSR, Seriya Biologicheskikh Nauk, 2:55-60 (in Russian).

Dawkins, W. B. 1868. Fossil animals and geology of Attica, by Albert Gaudry. (Critical summary). Quarterly Journal of the Geological Society of London, 24:pt. 2, translations & notices, pp. 1-7.

Dawson, L., and T. Flannery. 1985. Taxonomic and phylogenetic status of living and fossil kangaroos and wallabies of the genus *Macropus* Shaw (Macropodidae: Marsupialia), with a new subgeneric name for the larger wallabies. Australian Journal of Zoology, 33:473-98.

Dawson, L., J. Muirhead, and S. Wroe. 1998. The Big Sink local fauna: A lower Pliocene mammalian fauna from the Wellington Caves complex, Wellington, New South Wales. Records of the Western Australian Museum, Suppl. 57:265-290.

Dawson, M. R. 1967. Lagomorph history and the stratigraphic record. Pp. 287-316, *in* Essays in paleontology and stratigraphy (C. Teichert and E. L. Yochelson, eds.). University of Kansas, Department of Geology, Special Publication, 2:1-626.

Dawson, M. R. 1969. Osteology of *Prolagus sardus*, a Quaternary ochotonid (Mammalia, Lagomorpha). Palaeovertebrata, 2:157-190.

Dawson, M. R., and L. Krishtalka. 1984. Fossil history of the families of Recent mammals. Pp. 11-57, *in* Orders and families of Recent mammals of the world (S. Anderson and J. K. Jones, Jr., eds.). John Wiley & Sons, New York, 686 pp.

Dawson, M. R., L. Krishtalka, and R. K. Stucky. 1990. Revision of the Wind River faunas, early Eocene of central Wyoming. Part 9. The oldest known hystricomorphous rodent (Mammalia:Rodentia). Annals of Carnegie Museum, 59(2):135-147.

Daxner-Höck, G. 1999. Family Zapodidae. Pp. 337-342, *in* The Miocene land mammals of Europe (G. E. Rössner and K. Heissig, eds.). Dr. Friedrich Pfeil, München, 515 pp.

Daxner-Höck, G. 2001. New zapodids (Rodentia) from Oligocene-Miocene deposits in Mongolia. Part. 1. Senckenbergiana lethaea, 81(2):359-389.

Daxner-Höck, G., and W. Wu. 2003. *Plesiosminthus* (Zapodidae, Mammalia) from China and Mongolia: Migrations to Europe. Pp. 127-151, *in* Distribution and Migration of Tertiary Mammals in Eurasia. A volume in honour of Hans de Bruijn (J. W. F. Reumer and W. Wessels, eds.). Deinsea, 10:580 pp.

Day, Y.-T. 1988. Subspecies and geographic variation of *Petaurista alborufus* and *P. petaurista* (Rodentia: Sciuridae). Journal of the Taiwan Museum, 41(1):75-83.

de Beaux, O. 1924. Beitrag zur Kenntnis der Gattung *Potamochoerus* Gray. Zoologische Jahrbücher, 47:379-504.

de Beaux, O. 1934. Mammiferi raccolti dal Prof. G. Scorteccinella Somalia Italiana Centrale e Settentrionale Nel 1931 con l'aggiunta di aleuni mammiferi della Somalia Italiana meridionale. Atti della Societa Italiana De Scienze Naturali e del Museo Civico Di Storia Naturale in Milano, 73:261-300.

DeBlase, A. F. 1971. New distributional records of bats from Iran. Fieldiana: Zoology, 58:9-14.

DeBlase, A. F. 1972. *Rhinolophus euryale* and *R. mehelyi* (Chiroptera, Rhinolophidae) in Egypt and southwest Asia. Israel Journal of Zoology, 21:1-12.

DeBlase, A. F. 1980. The bats of Iran: Systematics, distribution, ecology. Fieldiana: Zoology, n.s., 4:1-424.

Debrot, S., and C. Mermod. 1977. Chimiotaxonomie du genre *Apodemus* Kaup, 1829 (Rodentia, Muridae). Revue Suisse de Zoologie, 84:521-526.

de Bruijn, H. 1967. Gliridae, Sciuridae y Eomyidae (Rodentia, Mammalia) miocenos de Calatayud (provincia de Zaragoza, España) y su relación con la biostratigrafia del area. Boletin del Instituto Geologico y Minero de España, 78:187-373.

de Bruijn, H. 1984. Remains of the mole-rat *Microspalax odessanus* Topachevski, from Karaburun (Greece, Macedonia) and the family Spalacidae. Proceedings of the Koninklijke Nederlandse Akademie van Wetenschappen, ser. B, 87(4):417-425.

de Bruijn, H. 1999. A Late Miocene insectivore and and rodent fauna from the Baynunah formation, Emirate of Abu Dhabi, United Arab Emirates. Pp. 186-197, *in* Fossil vertebrates of Arabia with emphasis on the Late Miocene faunas, geology, and palaeoenvironments of the Emirate of Abu Dhabi, United Arab Emirates (P. J. Whybrow and A. Hill, eds.). Yale University Press, New Haven and London, 523 pp.

de Bruijn, H., and J. G. Moltzer. 1974. The rodents from Rubielos de Mora; The first evidence of the existence of different biotopes in the Early Miocene of eastern Spain. Proceedings of the Koninklijke Nederlandse Akademie van Wetenschappen, Amsterdam, Proceedings, ser. B, 77:129-145.

de Bruijn, H., and E. Uenay. 1989. Petauristinae (Mammalia, Rodentia) from the Oligocene of Spain, Belgium, and Turkish Thrace. Series of the Natural History Museum, Los Angeles County, 33:139-145.

de Bruijn, H., and P. J. Whybrow. 1994. A Late Miocene rodent fauna from the Baynunah formation, Emirate of Abu Dhabi, United Arab Emirates. Proceedings of the Koninklijke Nederlandse Akademie van Wetenschappen, 97(4):407-422.

de Bruijn, H., M. R. Dawson, and P. Mein. 1970. Upper Pliocene Rodentia, Lagomorpha and Insectivora (Mammalia) from the isles of Rhodes (Greece). I, II, and III. Koninklijke Nederlandse Akademie van Wetenschappen, Amsterdam, Proceedings, ser. B, 73(5):535-584.

de Bruijn, H., T. Hussain, and J. J. M. Leinders. 1981. Fossil rodents from the Murree formation near Banda Daud Shah, Kohat, Pakistan. Proceedings of the Koninklijke Nederlandse Akademie van Wetenschappen, Amsterdam, Proceedings, ser. B, 84(1):71-99.

de Bruijn, H., J. A. Van Dam, G. Daxner-Höck, V. Fahlbusch, and G. Storch. 1996. The genera of the Murinae, endemic insular forms excepted, of Europe and Anatolia during the late Miocene and early Pliocene. Pp. 253-260, in The evolution of Western Eurasian Neogene mammal faunas (R. L. Bernor, V. Fahlbusch, and H.-W. Mittmann, eds.). Columbia University Press, New York, 487 pp.

de Bruijn, H., G. Saraç, L. W. van den Hoek Ostende, and S. Roussiakis. 1999. The status of the genus name Parapodemus Schaub, 1938; new data bearing on an old controversy. Pp. 95-112, in Elephants have a Snorkel! Papers in honour of Paul Y. Sondaae (J. W. F. Reumer and J. de Vos, eds.). Deinsea, 7:1-422.

Debruyne, R. 2003. Différenciation morphologique et moléculaire des Elephantinae (Mammalia, Proboscidea): Statut systématique de l'éléphant d'Afrique de forêt, Loxodonta africana cyclotis (Matschie, 1900). Unpublished Ph.D. dissertation, Museum National d'Histoire Naturelle, Paris, France.

Debry, R. W. 2003. Identifying conflicting signal in a multigene analysis reveals a highly resolved tree: The phylogeny of Rodentia (Mammalia). Systematic Biology, 52:604-617.

Debry, R. W., and R. M. Sagel. 2001. Phylogeny of Rodentia (Mammalia) inferred from the nuclear-encolded gene IRBP. Molecular Phylogenetics and Evolution, 19:290-301.

Decher, J., and L. K. Bahian. 1999. Diversity and structure of terrestrial small mammal communities in different vegetation types on the Accra Plains of Ghana. Journal of Zoology, London, 247:395-408.

Decher, J., D. A. Schlitter, and R. Hutterer. 1997. Noteworthy records of small mammals from Ghana with special emphasis on the Accra Plains. Annals of the Carnegie Museum, 66:207-225.

Decker, D. M. 1991. Systematics of the coatis, genus Nasua (Mammalia: Procyonidae). Proceedings of the Biological Society of Washington, 104(2):370-386.

Decker, D. M., and W. C. Wozencraft. 1991. Phylogenetic analysis of Recent procyonid genera. Journal of Mammalogy, 72(1):42-55.

Defler, T. R., M. L. Bueno, and J. I. Hernández-Camacho. 2001. Taxonomic status of Aotus hershkovitzi: Its relationship to Aotus lemurinus lemurinus. Neotropical Primates, 9:37-52.

DeFrees, S. L., and D. E. Wilson. 1988. Eidolon helvum. Mammalian Species, 312:1-5.

Degerbøl, M. 1935. Report of the mammals collected by the fifth Thule Expedtion to Arctic North America: Zoology. I. Mammals. Report of the Fifth Thule Expedition, 1921-1922, 2(4):1-67.

de Graaff, G. 1961. A preliminary investigation of the mammalian microfauna in Pleistocene deposits of caves in the Transvaal system. Palaeontologia Africana, 7:59-118.

de Graaff, G. 1975. Family Bathyergidae. Part 6.9. Pp. 1-5, in The mammals of Africa: An identification manual (J. Meester and H. W. Setzer, eds.) [issued 10 Dec 1975]. Smithsonian Institution Press, Washington, D.C., not continuously paginated.

de Graaff, G. 1978. Notes on the southern African black-tailed tree rat Thallomys paedulcus (Sundevall, 1846) and its occurrence in the Kalahari Gemsbok National Park. Koedoe, 21:181-190.

de Graaff, G. 1981. The rodents of Southern Africa. Butterworths, Durban, 267 pp.

de Graaff, G. 1997a. Groove-toothed mouse, Pelomys fallax. P. 138, in The complete book of southern African mammals (G. Mills and L. Hes, eds.). Struik Winchester, Capetown, 356 pp.

de Graaff, G. 1997b. Spiny mouse, Acomys spinosissimus. P. 138, in The complete book of southern African mammals (G. Mills and L. Hes, eds.). Struik Winchester, Capetown, 356 pp.

de Graaff, G. 1997c. Cape spiny mouse, Acomys subspinosus. P. 139. in The complete book of southern African mammals (G. Mills and L. Hes, eds.). Struik Winchester, Capetown, 356 pp.

de Graaff, G. 1997d. Single-striped mouse, Lemniscomys rosalia. P. 139, in The complete book of southern African mammals (G. Mills and L. Hes, eds.). Struik Winchester, Capetown, 356 pp.

de Graaff, G. 1997e. Striped mouse, Rhabdomys pumilio. P. 140, in The complete book of southern African mammals (G. Mills and L. Hes, eds.). Struik Winchester, Capetown, 356 pp.

de Graaff, G. 1997f. Woosnam's desert rat, Zelotomys woosnami. P. 141, in The complete book of southern African mammals (G. Mills and L. Hes, eds.). Struik Winchester, Capetown, 356 pp.

de Graaff, G. 1997g. Water rat, Dasymys incomtus. P. 141, in The complete book of southern African mammals (G. Mills and L. Hes, eds.). Struik Winchester, Capetown, 356 pp.

de Graaff, G. 1997h. Mozambique woodland mouse, Grammomys cometes. P. 142, in The complete book of southern African mammals (G. Mills and L. Hes, eds.). Struik Winchester, Capetown, 356 pp.

de Graaff, G. 1997i. Woodland mouse, *Grammomys dolichurus*. P. 142, *in* The complete book of southern African mammals (G. Mills and L. Hes, eds.). Struik Winchester, Capetown, 356 pp.

de Graaff, G. 1997j. Setzer's pygmy mouse, *Mus setzeri*. P. 143, *in* The complete book of southern African mammals (G. Mills and L. Hes, eds.). Struik Winchester, Capetown, 356 pp.

de Graaff, G. 1997l. Grey-bellied pygmy mouse, *Mus triton*. P. 143, *in* The complete book of southern African mammals (G. Mills and L. Hes, eds.). Struik Winchester, Capetown, 356 pp.

de Graaff, G. 1997m. Thomas's pygmy mouse, *Mus sorella*. P. 143, *in* The complete book of southern African mammals (G. Mills and L. Hes, eds.). Struik Winchester, Capetown, 356 pp.

de Graaff, G. 1997n. Desert pygmy mouse, *Mus indutus*. P. 143, *in* The complete book of southern African mammals (G. Mills and L. Hes, eds.). Struik Winchester, Capetown, 356 pp.

de Graaff, G. 1997o. Pygmy mouse, *Mus minutoides*. P. 144, *in* The complete book of southern African mammals (G. Mills and L. Hes, eds.). Struik Winchester, Capetown, 356 pp.

de Graaff, G. 1997p. Rudd's mouse, *Uranomys ruddi*. P. 145, *in* The complete book of southern African mammals (G. Mills and L. Hes, eds.). Struik Winchester, Capetown, 356 pp.

de Graaff, G. 1997q. Natal multimammate mouse, *Mastomys natalensis*. Pp. 145-146, *in* The complete book of southern African mammals (G. Mills and L. Hes, eds.). Struik Winchester, Capetown, 356 pp.

de Graaff, G. 1997r. Multimammate mouse, *Mastomys coucha*. P. 146, *in* The complete book of southern African mammals (G. Mills and L. Hes, eds.). Struik Winchester, Capetown, 356 pp.

de Graaff, G. 1997s. Shortridge's mouse, *Mastomys shortridgei*. P. 146, *in* The complete book of southern African mammals (G. Mills and L. Hes, eds.). Struik Winchester, Capetown, 356 pp.

de Graaff, G. 1997t. Verreaux's mouse, *Myomyscus verreauxii*. P. 147, *in* The complete book of southern African mammals (G. Mills and L. Hes, eds.). Struik Winchester, Capetown, 356 pp.

de Graaff, G. 1997u. Selinda rat, *Aethomys silindensis*. P. 148, *in* The complete book of southern African mammals (G. Mills and L. Hes, eds.). Struik Winchester, Capetown, 356 pp.

de Graaff, G. 1997v. Namaqua rock mouse, *Aethomys namaquensis*. P. 149, *in* The complete book of southern African mammals (G. Mills and L. Hes, eds.). Struik Winchester, Capetown, 356 pp.

de Graaff, G. 1997w. Grant's Rock mouse, *Aethomys granti*. P. 149, *in* The complete book of southern African mammals (G. Mills and L. Hes, eds.). Struik Winchester, Capetown, 356 pp.

de Graaff, G. 1997x. Nyika veld rat, *Aethomys nyikae*. P. 149, *in* The complete book of southern African mammals (G. Mills and L. Hes, eds.). Struik Winchester, Capetown, 356 pp.

de Graaff, G. 1997y. Red veld rat, *Aethomys chrysophilus*. P. 150, *in* The complete book of southern African mammals (G. Mills and L. Hes, eds.). Struik Winchester, Capetown, 356 pp.

de Graaff, G. 1997z. White-tailed rat, *Mystromys albicaudatus*. P. 156, *in* The complete book of southern African mammals (G. Mills and L. Hes, eds.). Struik Winchester, Capetown, 356 pp.

de Graaff, G. 1997aa. Large-eared mouse, *Malacothrix typica*. P. 158, *in* The complete book of southern African mammals (G. Mills and L. Hes, eds.). Struik Winchester, Capetown, 356 pp.

de Graaff, G. 1997bb. Nyika climbing mouse, *Dendromus nyikae*. P. 159, *in* The complete book of southern African mammals (G. Mills and L. Hes, eds.). Struik Winchester, Capetown, 356 pp.

de Graaff, G. 1997cc. Brant's climbing mouse, *Dendromus mesomelas*. P. 159, *in* The complete book of southern African mammals (G. Mills and L. Hes, eds.). Struik Winchester, Capetown, 356 pp.

de Graaff, G. 1997dd. Grey climbing mouse, *Dendromus melanotis*. P. 159, *in* The complete book of southern African mammals (G. Mills and L. Hes, eds.). Struik Winchester, Capetown, 356 pp.

de Graaff, G. 1997ee. Chestnut climbing mouse, *Dendromus mystacalis*. P. 160, *in* The complete book of southern African mammals (G. Mills and L. Hes, eds.). Struik Winchester, Capetown, 356 pp.

de Graaff, G. 1997ff. Fat mouse, *Steatomys pratensis*. P. 160, *in* The complete book of southern African mammals (G. Mills and L. Hes, eds.). Struik Winchester, Capetown, 356 pp.

de Graaff, G. 1997gg. Tiny fat mouse, *Steatomys parvus*. P. 161, *in* The complete book of southern African mammals (G. Mills and L. Hes, eds.). Struik Winchester, Capetown, 356 pp.

de Graaff, G. 1997hh. Kreb's fat mouse, *Steatomys krebsii*. P. 161, *in* The complete book of southern African mammals (G. Mills and L. Hes, eds.). Struik Winchester, Capetown, 356 pp.

de Graaff, G. 1997ii. Barbour's rock mouse, *Petromyscus barbouri*. P. 161, *in* The complete book of southern African mammals (G. Mills and L. Hes, eds.). Struik Winchester, Capetown, 356 pp.

de Graaff, G. 1997jj. Pygmy rock mouse, *Petromyscus collinus*. Pp. 161-162, *in* The complete book of southern African mammals (G. Mills and L. Hes, eds.). Struik Winchester, Capetown, 356 pp.

de Graaff, G. 1997kk. Brukkaros pygmy rock mouse, *Petromyscus monticularis*. P. 162, *in* The complete book of southern African mammals (G. Mills and L. Hes, eds.). Struik Winchester, Capetown, 356 pp.

de Graaff, G. 1997ll. Shortridge's rock mouse, *Petromyscus shortridgei*. P. 162, *in* The complete book of southern African mammals (G. Mills and L. Hes, eds.). Struik Winchester, Capetown, 356 pp.

Dehne, A. 1855. *Musculus* (Rafinesque-Schmalz) *Mollissimus*, Dehne. Weichhaariges Schlafmäuscvhen.– Allgemeine Deutsche Naturhistorische Zeitung, Dresden, 1:443-444.

de Jong, W. W., M. van Dijk, C. Poux, G. Kapp, T. van Rheede, and O. Madsen. 2003. Indels in protein-coding sequences of Euarchontoglires constrain the rooting of the eutherian tree. Molecular Phylogenetics and Evolution, 28:328-340.

Dekeyser, P. L. 1955. Les mammiferes de l'Afrique noire francaise. Second ed. Institut francais d'Afrique noire, Dakar, 426 pp.

Delany, M. J. 1975. The rodents of Uganda. Trustees of the British Museum (Natural History), London, 165 pp.

de la Torre, L. 1958. The status of the bat *Myotis velifer cobanensis* Goodwin. Proceedings of the Biological Society of Washington, 71:167-170.

D'Elía, G. 2000. Comments on recent advances in understanding sigmodontine phylogeny and evolution. Mastozoologiá Neotropical, 7:47-54.

D'Elía, G. 2003. Phylogenetics of Sigmodontinae (Rodentia, Muroidea, Cricetidae), with special reference to the akodont group, and with additional comments on historical biogeography. Cladistics, 19:307-323.

D'Elía, G., E. P. Lessa, and J. A. Cook. 1999. Molecular phylogeny of tuco-tucos, genus *Ctenomys* (Rodentia: Octodontidae): Evaluation of the mendocinus species group and the evolution of asymmetric sperm. Journal of Mammalian Evolution, 6:19-38.

D'Elía, G., E. M. González, and U. F. J. Pardiñas. 2003. Phylogenetic analysis of sigmodontine rodents (Muroidea), with special reference to the akodont genus *Deltamys*. Mammalian Biology/Zeitschrift für Säugetierkunde, 68:351-364.

D'Elía, G., U. F. J. Pardiñas, and P. Myers. In Press. An introduction to the genus *Bibimys* (Rodentia: Sigmodontinae): Phylogenetic position and alpha taxonomy. *in* Mammalian diversification: From population genetics to biogeography (E. Lacey and P. Myers, eds.). University of California Press, Berkeley.

Delibes, M., F. Hiraldo, N. Arroyo, and M. Rodriguez. 1980. Disagreement between morphotypes and karyotypes in *Eliomys* (Rodentia, Gliridae): The chromosomes of the Central Morocco garden dormouse. Säugetierkundliche Mitteilungen, 28:289-292.

De L'Isle, M. A. 1865. De l'existence d'une race nègre chez le rat, ou de l'identité spécifique du *Mus rattus* et du *Mus alexandrinus*. Annales Sciences Naturelles, Zoologie et Paléontologie, 4:173-222.

DelPero, M., J. C. Masters, D. Zuccon, P. Cervella, S. Crovella, and G. Ardito. 2000. Mitochondrial sequences as indicators of generic classification in bush babies. International Journal of Primatology, 21:889-904.

Delson, E. 1975. Evolutionary history of the Cercopithecidae. Pp. 167-217, *in* Approaches to primate paleobiology (F. S. Szalay, ed.). Contributions to Primatology, 5:1-326.

Delson, E. 1976. The family-group name of the leaf-eating monkeys (Mammalia, Primates): A proposal to give Colobidae Blyth, 1875, precedence over Semnopithecidae Owen, 1843, and Presbytina Gray, 1825. Bulletin of Zoological Nomenclature, 33:85-89.

Delson, E., and P. Andrews. 1975. Evolution and interrelationships of the catarrhine primates. Pp. 405-446, *in* Phylogeny of the Primates (W. P. Luckett and F. S. Szalay, eds.). Plenum Press, New York, 483 pp.

Delson, E., and P. H. Napier. 1976. Request for determination of the generic names of the baboon and the mandrill (Mammalia: Primates, Cercopithecidae) Z.N.(S.) 2093. Bulletin of Zoological Nomenclature, 33(1):46-60.

Delsuc F., F. M. Catzeflis, M. J. Stanhope, and E. J. P. Douzery. 2001. The evolution of armadillos, anteaters and sloths depicted by nuclear and mitochondrial phylogenies: Implications for the status of the enigmatic fossil *Eurotamandua*. Proceedings of the Royal Society of London, ser. B, 268:1605-1615.

Delsuc, F., M. Scally, O. Madsen, M. J. Stanhope, W. W. de Jong, F. M. Catzeflis, M. S. Springer, and E. J. P. Douzery. 2002. Molecular phylogeny of living xenarthrans and the impact of character and taxon sampling on the placental tree rooting. Molecular Biology and Evolution, 19:1656-1671.

De Marinis, A. M., M. Masseti, and A. Sforzi. 1996. Note on the non-flying terrestrial mammals of the Tuscan Archipelago, northern Tyrrhenian Sea (Italy). Bollettino del Museo Regionale di Scienze Naturali, Torino, 14(1):275-281.

DeMaster, D. P., and I. Stirling. 1981. *Ursus maritimus*. Mammalian Species, 145:1-7.

Demastes, J. W. 1994. Systematics and zoogeography of the Mer Rouge pocket gopher (*Geomys breviceps breviceps*) based on cytochrome-b sequences. Southwestern Naturalist, 39:276-280.

Demastes, J. W., M. S. Hafner, and D. J. Hafner. 1996. Phylogeographic variation in two Central American pocket gophers (*Orthogeomys*). Journal of Mammalogy, 77:917-927.

Demastes, J. W., T. A. Spradling, M. S. Hafner, D. J. Hafner, and D. L. Reed. 2002. Systematics and phylogeography in pocket gophers in the genera *Cratogeomys* and *Pappogeomys*. Molecular Phylogenetics and Evolution, 22:144-154.

Demboski, J. R., and J. A. Cook. 2001. Phylogeography of the dusky shrew, *Sorex monticolus* (Insectivora, Soricidae): Insight into deep and shallow history in northwestern North America. Molecular Ecology, 10:1277-1240.

Demboski, J. R., and J. A. Cook. 2003. Phylogenetic diversification within the *Sorex cinereus* group (Soricidae). Journal of Mammalogy, 84:144-158.

Demboski, J. R., and J. Sullivan. 2003. Extensive mtDNA variation within the yellow-pine chipmunk, *Tamias amoenus* (Rodentia: Sciuridae), and phylogeographic inferences for northwest North America. Molecular Phylogenetics and Evolution, 26:389-408.

Deméré, T. A. 1986. The fossil whale, *Balaenoptera davidsonii* (Cope, 1872), with a review of the other neogene species of Balaenoptera (Cetacea: Mysticeti). Marine Mammal Science, 2(4):277-298.

Demeter, A. 1982. Prey of the spotted eagle-owl *Bubo africanus* in the Awash National Park, Ethiopia. Bonner Zoologische Beiträge, 33:283-292.

Demeter, A. 1983. Taxonomical notes on *Arvicanthis* (Mammalia: Muridae) from the Ethiopian Rift Valley. Pp. 129-136, *in* Proceedings of the Third International Colloquium on the Ecology and Taxonomy of African Small Mammals (E. Van der Straeten, W. N. Verheyen, and F. De Vree, eds.). Annales Musée Royal de l'Afrique Centrale, Tervuren-Belgique, ser. 8 (Sciences Zoologiques), 237:1-227.

Demeter, A., and G. Topal. 1982. Ethiopian mammals in the Hungarian Natural History Museum. Annales Historico-Naturales Musei Nationalis Hungarici (Budapest), 74:331-349.

Demeter, R., and R. Hutterer. 1986. Small mammals from Mt. Meru and its environs (Northern Tanzania). Cimbebasia, ser. A, 8:199-207.

Dempster, E. R. 1996. Pre- and postmating isolation in two allopatric gerbil species, *Tatera afra* and *T. brantsii*. Mammalia, 60(4):557-566.

Dempster, E. R., and M. R. Perrin. 1990. Interspecific aggression in sympatric *Gerbillurus* species. Zeitschrift für Säugetierkunde, 55:392-398.

Dempster, E. R., and M. R. Perrin. 1991. Ultrasonic vocalizations of six taxa of southern African gerbils (Rodentia: Gerbillinae). Ethology, 88:1-10.

Dempster, E. R., and M. R. Perrin. 1994. Divergence in acoustic repertoire of sympatric and allopatric gerbil species (Rodentia: Gerbillinae). Mammalia, 58(1):93-104.

Dempster, E. R., R. Dempster, and M. R. Perrin. 1991. Behaviour associated with ultrasonic vocalizations in six gerbilline rodent species. Ethology, 88:11-19.

Dempster, E. R., R. Dempster, and M. R. Perrin. 1992. A comparative study of the behaviour of six taxa of male and female gerbils (Rodentia) in intra- and interspecific encounters. Ethology, 91:25-45.

Dempster, E. R., R. Dempster, and M. R. Perrin. 1993. Behavioural divergence in allopatric and sympatric gerbil species (Rodentia: Gerbillinae). Ethology, 93:300-314.

Dene, H., M. Goodman, and W. Prychodko. 1978. An immunological examination of the systematics of Tupaioidea. Journal of Mammalogy, 59:697-706.

Deng, X.-Y., Q. Feng, and Y.-X. Wang. 2000. [Differentiation of subspecies of Chinese white-bellied rat (*Niviventer confucianus*) in southwestern China with descriptions of two new subspecies.] Zoological Research, 5:375-382 (in Chinese with English abstract).

Denisov, V. P. 1963. O gibridizatsii vidov roda *Citellus* Oken [On hybridization of species of the genus *Citellus* Oken]. Zoologicheskii Zhurnal, 42:1887-1899 (in Russian).

Denisov, V. P. 1964. Rasprostranenie malova (*Citellus pygmaeus* Pallas) i rezhevatovo (*Citellus major* Pallas) suslikov v Zavolzhe [Distribution of the little . . . and reddish . . . ground squirrels in the Transvolga]. Nauchnie Doklad Vysshie Shkolu, Biologicheskii Nauki, 2:49-54 (in Russian).

Denisov, V. P., and N. I. Smirnova. 1976. Immunological relationships of sousliks genus *Citellus* in the Povolgje region. Acta Theriologica, 21:267-278.

Dennis, E., and J. I. Menzies. 1978. Systematics and chromosomes of New Guinea *Rattus*. Australian Journal of Zoology, 26:197-206.

Dennis, E., and J. I. Menzies. 1979. A chromosomal and morphometric study of Papuan tree rats *Pogonomys* and *Chiruromys* (Rodentia, Muridae). Journal of the Zoological Society of London, 189:315-332.

Denys, C. 1987*a*. Rodentia and Lagomorpha. 6.1. Fossil rodents (other than Pedetidae) from Laetoli. Pp. 118-170, *in* Laetoli, a Pliocene site in northern Tanzania (M. D. Leakey and J. M. Harris, eds.). Clarendon Press, Oxford, 561 pp.

Denys, C. 1987*b*. Micromammals from the West Natron Pleistocene deposits (Tanzania). Biostratigraphy and Paleoecology, Sciences Géologiques Bulletin, 40(102):185-201.

Denys, C. 1988. Apports de l'analyse morphométrique à la détermination des espèces actuelles et fossiles du genre *Saccostomus* (Cricetomyinae, Rodentia). Mammalia, 5:497-532.

Denys, C. 1989*a*. Phylogenetic affinities of the oldest East African *Otomys* (Rodentia, Mammalia) from Olduvai Bed I (Pleistocene, Tanzania). Neues Jahrbuch für Geologie und Paleontologie, Monatshefte, 44:705-725.

Denys, C. 1989*b*. Two new gerbillids (Rodentia, Mammalia) from Olduvai Bed I (Pleistocene, Tanzania). Neues Jahrbuch für Geologie und Paläontologie, Abhandlungen, 178:243-265.

Denys, C. 1990*a*. Implications paleoecologiques et paleobiogeographiques de l'etude de rongeurs plio-pleistocenes d'Afrique orientale et australe. Memoirs Institut Science et Terre, Univsité Pierre et Marie Curie, Paris VI, 428 pp.

Denys, C. 1990*b*. The oldest *Acomys* (Rodentia, Muridae) from the Lower Pliocene of South Africa and the problem of its murid affinities. Palaeontographica, Abt. A, 210:79-91.

Denys, C. 1990*c*. Deux nouvelles espèces d'*Aethomys* (Rodentia, Muridae) à Langebaanweg (Pliocène, Afrique du Sud): Implications phylogénétiques et paléoécologiques. Annales de Paléontologie (Vert.-Invert.), 76(1):41-69.

Denys, C. 1991. Un nouveau rongeur *Mystromys pocockei* sp. nov. (Cricetinae) du Pliocene inferieur de Langebaanweg (Region du Cap, Afrique du Sud). Comptes Rendus de l'Académie des Sciences (Paris), ser. 2, 313:1335-1341.

Denys, C. 1993. Réexamen de la dentition de *Leimacomys buettneri* (Mammalia, Rodentia). Hypothèses sur sa position systématique. Mammalia, 57(4):613-618.

Denys, C. 1994*a*. Diet and dental morphology of two coexisting *Aethomys* species (Rodentia) in Mozambique. Implications for diet reconstruction in related extinct species from South Africa. Acta Theriologica, 39(4):357-364.

Denys, C. 1994*b*. Nouvelles especes de *Dendromus* (Rongeurs, Muroidea) à Langebaanweg (Pliocene, Afrique du Sud) consequences stratigraphiques et paleoecologiques. Palaeovertebrata, 13(1-4):153-176.

Denys, C. 1994*c*. Affinités systématiques de *Stenodontomys* (Mammalia, Rodentia) rongeur Muroidea du Pliocène de Langebaanweg (Afrique du Sud). Comptes Rendus des Séances de l'Académie des Sciences, Paris, 318, ser. 2:411-416.

Denys, C. 1999. Of mice and men. Evolution in East and South Africa during Plio-Pleistocene times. Pp. 226-252, *in* African biogeography, climate change, & human evolution (T. G. Bromage and F. Schrenk, eds.). Oxford University Press, Oxford, 485 pp.

Denys, C., and J. Michaux. 1992. La troisième molaire supérieure chez les Muridae d'Afrique tropicale et le cas des genres *Acomys*, *Uranomys* et *Lophuromys*. Bonner Zoologische Beiträge, 43(3):367-382.

Denys, C., and M. Tranier. 1992. Présence d'*Aethomys* (Mammalia, Rodentia, Muridae) au Tchad et analyse morphométrique préliminaire du complexe *A. hindei*. Mammalia, 56(4):626-656.

Denys, C., J. Michaux, and Q. B. Hendey. 1987. Les rongeurs *Euryotomys* et *Otomys*: Un exemple d'evolution parallele en Afrique tropicale? Comptes Rendus de l'Académie des Sciences, Paris, ser. 3, 305:1389-1395.

Denys, C., J. Michaux, F. Petter, J. P. Aguilar, and J. J. Jaeger. 1992. Molar morphology as a clue to the phylogenetic relationships of *Acomys* to the Murinae. Israel Journal of Zoology, 38:253-262.

Denys, C., J.-C. Gautun, M. Tranier, and V. Volobouev. 1994. Evolution of the genus *Acomys* (Rodentia, Muridae) from dental and chromosomal patterns. Israel Journal of Zoology, 40:215-246.

Denys, C., J. Michaux, F. Catzeflis, S. Ducrocq, and P. Chevret. 1995. Morphological and molecular data against the monophyly of Dendromurinae (Muridae: Rodentia). Bonner Zoologische Beitrage, 45:173-190.

Denys, C., W. Chitaukali, J. K. Mfune, M. Combrexelle, and F. Cacciani. 1999. Diversity of small mammals in owl pellet assemblages of Karonga district, northern Malawi. Acta Zoologica Cracoviensia, 42(3):393-396.

Denys, C., G. Dobigny, L. Granjon, and E. Lecompte. 2002. Morphometric, cytogenetic, and molecular data on *Mastomys* from Chad. Rodens & Spatium, 8th International Conference, Biodiversity Research Centre and Environmetry and Geomatics Unit, Université Catholique de Louvain: Abstract 22.

de Paz, O. 1994. Systematic position of *Plecotus* (Geoffroy, 1818) from the Iberian Peninsula (Mammalia: Chiroptera). Mammalia, 58:423-432.

Deraniyagala, P. E. P. 1950. The elephant of Asia. (Presidential address at Ceylon Association of Sciences, 1949). Proceedings of the Ceylon Association of Sciences, vol. III:1-18.

Deraniyagala, P. E. P. 1955. Some extinct elephants, their relatives and the two living species. Ceylon National Museums Administration, Colombo, 161 pp.

D'Erchia, A. M., C. Gissi, G. Pesole, C. Saccone, and U. Arnason. 1996. The guinea-pig is not a rodent. Nature, 381:597-600.

Dermitzakis, M. D., and J. de Vos. 1987. Faunal succession and the evolution of mammals in Crete during the Pleistocene. Neues Jahrbuch für Geologie und Paläontologie. Abhandlungen, 173(3):377-408.

de Roguin, L. 1991. Donnees historiques nouvelles sur la presence du rat noir *Rattus rattus* (L.) en Europe Occidentale. Pp. 323-325, *in* Le rongeur et l'espace, Actes du Colloque International, Lyon, 1989 (M. Le Berre and L. Le Guelte, eds.). Raymond Chabaud, Paris, 362 pp.

De Santis, L. J. M., E. R. Justo, and M. Kin. 2001. Nuevos datos acerca de la distribucíon geográfica de *Oligoryzomys longicaudatus* (Rodentia: Sigmodontinae) en regions extra-andinas de la República Argentina. Neotropica, 47:109-111.

Desclaux, E., M. Abbassi, J.-C. Marquet, J. Chaline, and T. van Kolfschoten. 2000. Acta Zoologica Cracoviensia, 43(1-2):107-125.

Deslongchamps, E. 1866. Observations sur quelques dauphins appartenant a la section des zyphides et description de le te, te d'une espéce de cette section nouvelle pour la faune Française. Bulletin de la Société Linnéene de Normandie (Caen), 10:168-180.

Desmarest, A. G. 1816a. Bradype, *Bradypus*, Linn.; Erxleben; Cuv.; Illiger, etc.; *Tardigradus*, Brisson; *Choloepus* et *Prochilus*, Illiger. Pp. 319-328, *in* Nouveau dictionnaire d'histoire naturelle, appliquée aux arts, à l'agriculture, à l'économie rurale et domestique, à la médecine, etc. par une société de naturalistes et d'agriculteurs, Nouv. éd. Ch. Deterville, Paris, 4:1-602.

Desmarest, A. G. 1816-1819b. Nouveau dictionnaire d'histoire naturelle, appliquèe aux art, principalement à l'agriculture et à l'economie rurale et domestique; par une société de naturalistes. Nouvelle edition, presqu' entierement refondue et considerablement augmentee. Deterville, Paris, 36 volumes.

Desmarest, A. G. 1816c. Antilope. Nouveau Dictionnaire d'Histoire Naturelle, 2:178-208.

Desmarest, A. G. 1820, 1822. Mammalogie ou description des espèce de mammifères. Encyclopèdie Méthodique. Agasse, Paris, 555 pp., 66 plates.

DeWalt, T. S., P. D. Sudman, M. S. Hafner, and S. K. Davis. 1993a. Phylogenetic relationships of pocket gophers (*Cratogeomys* and *Pappogeomys*) based on mitochondrial DNA cytochrome b sequences. Molecular Phylogenetics and Evolution, 2:193-204.

DeWalt, T. S., E. G. Zimmerman, and J. V. Planz. 1993b. Mitochondrial-DNA phylogeny of species of the *boylii* and *truei* groups of the genus *Peromyscus*. Journal of Mammalogy, 74:352-362.

d'Huart, J. P., and P. Grubb. 2001. Distribution of the common warthog (*Phacochoerus africanus*) and the desert warthog (*Phacochoerus aethiopicus*) in the Horn of Africa. African Journal of Ecology, 39:156-169.

Diaz, G. B., R. A. Ojeda, M. H. Gallardo, and S. M. Giannoni. 2000. *Tympanoctomys barrerae*. Mammalian Species, 646:1-4.

Díaz, M. M. 2000. Key to the native mammals of Jujuy Province, Argentina. Occasional Papers of the Sam Noble Oklahoma Museum of Natural History, 7:1-29.

Díaz, M. M., and R. M. Barquez. 1999. Contributions to the knowledge of the mammals of Jujuy Province, Argentina. Southwestern Naturalist, 44:324-333.

Díaz, M. M., R. M. Barquez, J. K. Braun, and M. A. Mares. 1999. A new species of *Akodon* (Muridae: Sigmodontinae) from northwestern Argentina. Journal of Mammalogy, 80:786-798.

Dice, L. R. 1929. The phylogeny of the Leporidae, with description of a new genus. Journal of Mammalogy, 10:340-344.

Dickerman, A. W., and T. E. Yates. 1995. Systematics of *Oligoryzomys*: Protein-electrophoretic analyses. Journal of Mammalogy, 76:172-188.

Dickman, C. R. 1994. Mammals of New South Wales: Past, present and future. Australian Zoologist, 29(3-4):158-165.

Dickman, C. R., D. H. King, M. Adams, and P. M. Baverstock. 1988. Electrophoretic identification of a new species of *Antechinus* (Marsupialia: Dasyuridae) in South-eastern Australia. Australian Journal of Zoology, 36:455-463.

Dickman, C. R., R. L. Pressey, L. Lim, and H. E. Parnaby. 1993. Mammals of particular conservation concern in the Western Division of New South Wales. Biological Conservation, 65:219-248.

Dickman, C. R., H. E. Parnaby, M. S. Crowther, and D. H. King. 1998. *Antechinus agilis* (Marsupialia, Dasyuridae), a new species from the *A. stuartii* complex in south-eastern Australia. Australian Journal of Zoology, 46:1-26.

Dickman, C. R., D. Lunney, and A. Matthews. 2000a. Ecological attributes and conservation of native rodents in New South Wales. Wildlife Research, 27:347-355.

Dickman, C. R., L. K.-P. Leung, and S. M. Van Dyck. 2000b. Status, ecological attributes and conservation of native rodents in Queensland. Wildlife Research, 27:333-346.

Didier, R. 1953. Étude systematique de l'os penien des mammifères. Mammalia, 17(4):260-269.

Didier, R., and F. Petter. 1960. L'os penien de *Jaculus blanfordi* (Murray) 1884, étude comparée de *J. blanfordi*, *J. jaculus* et *J. orientalis* (Rongeurs, Dipodides). Mammalia, 24(2):171-176.

Diersing, V. A. 1976. An analysis of *Peromyscus difficilis* from the Mexican-United States boundary area. Proceedings of the Biological Society of Washington, 89:451-466.

Diersing, V. A. 1980a. Systematics of flying squirrels, *Glaucomys volans* (Linnaeus), from Mexico, Guatemala, and Honduras. Southwestern Naturalist, 125:57-172.

Diersing, V. A. 1980b. Systematics and evolution of the pygmy shrews (subgenus *Microsorex*) of North America. Journal of Mammalogy, 61:76-101.

Diersing, V. A. 1981. Systematic status of *Sylvilagus brasiliensis* and *S. insonus* from North America. Journal of Mammalogy, 62:539-556.

Diersing, V. A., and D. F. Hoffmeister. 1977. Revision of the shrew *Sorex merriami* and a description of a new species of the subgenus *Sorex*. Journal of Mammalogy, 58:321-333.

Diersing, V. A., and D. E. Wilson. 1980. Distribution and systematics of the rabbits (*Sylvilagus*) of west-central Mexico. Smithsonian Contributions to Zoology, 297:1-34.

Dieterlen, F. 1961. Beiträge zur Biologie der Stachelmaus, *Acomys cahirinus dimidiatus* Cretzchmar. Zeitschrift für Säugetierkunde, 26:1-13.

Dieterlen, F. 1962. Geburt und Geburtshilfe bei der Stachelmaus, *Acomys cahirinus*. Zeitschrift für Tierpsychologie, 19:191-222.

Dieterlen, F. 1963. Vergleichende Untersuchungen zur Ontogenese von Stachelmaus (*Acomys*) und Wanderratte (*Rattus norvegicus*)- Teil I. Beiträge zum Nesthocker-Nestflüuchter-Problem bei Nagetieren. Zeitschrift für Säugetierkunde, 28:193-227.

Dieterlen, F. 1968. Zur Kenntnis der Gattung *Otomys* (Otomyinae; Muridae; Rodentia). Beiträge zur Systematik, Okologie und Biologie zentralafrikanischer Formen. Zeitschrift für Säugetierkunde, 33:321-352.

Dieterlen, F. 1969*a*. *Dendromus kahuziensis* (Dendromurinae; Cricetidae; Rodentia)—eine neue Art aus Zentralafrika. Zeitschrift für Säugetierkunde, 34:348-353.

Dieterlen, F. 1969*b*. Zur Kenntnis von *Delanymys brooksi* Hayman 1962 (Petromyscinae; Cricetidae; Rodentia). Bonner Zoologische Beiträge, 20:384-395.

Dieterlen, F. 1971. Beiträge zur Systematik, Ökologie und Biologie der Gattung *Dendromus* (Dendromurinae, Cricetidae, Rodentia), insbesondere ihrer zentralafrikanischen Formen. Säugetierkundliche Mitteilungen, 19:97-132.

Dieterlen, F. 1974. Bemerkungen zur Systematik der Gattung *Pelomys* (Muridae; Rodentia) in Äthiopien. Zeitschrift für Säugetierkunde, 39:229-231.

Dieterlen, F. 1975. *Lophuromys medicaudatus* (Muridae; Rodentia) Beschreibung einer neuen Art, aufgrund neuer Ergebnisse zur systematischen Stellung von *Lophuromys luteogaster* Hatt (1934). Bonner Zoologische Beiträge, 26:293-318.

Dieterlen, F. 1976*a*. Bemerkungen über *Leimacomys büttneri* Matschie, 1893 (Dendromurinae, Cricetidae, Rodentia). Säugetierkundliche Mitteilungen, 24:224-228.

Dieterlen, F. 1976*b*. Die afrikanische Muridengattung *Lophuromys* Peters, 1874. Vergleiche an Hand neuer Daten zur Morphologie, Ökologie und Biologie. Stuttgarter Beiträge zur Naturkunde, ser. A (Biologie), 285:1-96.

Dieterlen, F. 1976*c*. Zweiter Fund von *Dendromus kahuziensis* (Dendromurinae; Cricetidae; Rodentia) und weitere *Dendromus*- Fänge im Kivu- Hochland oberhalb 2000 m. Stuttgarter Beiträge zur Naturkunde, ser. A, 286:1-5.

Dieterlen, F. 1978*a*. *Acomys minous* (Bate, 1905)—Kreta- Stachelmaus. Pp. 452-461, *in* Handbuch der Säugetiere Europas (J. Niethammer and F. Krapp, eds.). Akademische Verlagsgesellschaft (Wiesbaden), 1:1-476.

Dieterlen, F. 1978*b*. Beiträge zur Kenntnis der Gattung *Lophuromys* (Muridae: Rodentia) in Kamerun und Gabun. Bonner Zoologische Beiträge, 29:287-299.

Dieterlen, F. 1979. Zur Ausbreitungsgeschichte der Hausratte (*Rattus rattus*) in Ostafrika. Zeitschrift für Angewandte Zoologie, 66:173-184.

Dieterlen, F. 1983. Zur Systematik, Verbreitung und Ökologie von *Colomys goslingi* Thomas & Wroughton, 1907 (Muridae; Rodentia). Bonner Zoologische Beiträge, 34:73-106.

Dieterlen, F. 1987. Neue Erkenntnisse über afrikanische Bürstenhaarmäuse, Gattung *Lophuromys* (Muridae; Rodentia). Bonner Zoologische Beiträge, 38:183-194.

Dieterlen, F. 1991. *Lemniscomys hoogstraali*, an new murid species from Sudan. Bonner Zoologische Beiträge, 42:11-15.

Dieterlen, F., and K. G. Gelmroth. 1974. Eine weitere bürstenhaarmaus aus dem Kivugebiet: *Lophuromys cinereus* spec. nov. (Muridae, Rodentia). Zeitschrift für Säugetierkunde, 39:337-342.

Dieterlen, F., and H. Heim de Balsac. 1979. Zur Ökologie und Taxonomie der Spitzmäuse (Soricidae) des Kivu-Gebietes. Säugetierkundliche Mitteilungen, 27:241-287.

Dieterlen, F., and G. Nikolaus. 1985. Zur Säugetierfauna des Sudan—weitere Erstnachweise und bemerkenswerte Funde. Säugetierkundliche Mitteilungen, 32:205-209.

Dieterlen, F., and H. Rupp. 1976. Die Rotnasenratte *Oenomys hypoxanthus* (Pucheran, 1855) (Muriden, Rodentia)—Erstnachweis für Äthiopien und dritter Fund aus Tansania. Säugetierkundliche Mitteilungen, 24:229-235.

Dieterlen, F., and H. Rupp. 1978. *Megadendromus nikolausi*, gen. nov., sp. nov. (Dendromurinae; Rodentia), ein neuer Nager aus Äthiopien. Zeitschrift für Säugetierkunde, 43:129-143.

Dieterlen, F., and B. Statzner. 1981. The African rodent *Colomys goslingi* Thomas and Wroughton, 1907 (Rodentia: Muridae)—a predator in limnetic ecosystems. Zeitschrift für Säugetierkunde, 46:369-383.

Dieterlen, F., and E. Van der Straeten. 1984. New specimens of *Malacomys verschureni* from eastern Zaire (Mammalia, Muridae). Revue de Zoologie Africaine, 98:861-868.

Dieterlen, F., and E. Van der Straeten. 1988. Deux nouveaux specimens de *Lamottemys okuensis* Petter, 1986 du Cameroun (Muridae: Rodentia). Mammalia, 52:379-385.

Dieterlen, F., and E. Van der Straeten. 1992. Species of the genus *Otomys* from Cameroon and Nigeria and their relationship to East African forms. Bonner Zoologische Beiträge, 43:383-392.

Dietz, J. M. 1985. *Chrysocyon brachyurus*. Mammalian Species, 234:1-4.

Dimaki, M. 1999. First record of the edible dormouse *Glis glis* (Linnaeus, 1766), from the Greek island of Andros. Annales Musei Goulandris, 10:181-183.

Din, W., B. David, B. Laurin, J. Chaline, M. Harada, and F. Catzeflis. 1993. DNA/DNA hybridization study of the Clethrionomyini (Arvicolidae, Rodentia): Comparison with morphological data. Comptes Rendus des Séances de l'Académie des Sciences, Paris, 316, Série II:709-716.

Din, W., R. Anand, P. Boursot, D. Darviche, B. Dod, E. Jouvin-Marche, A. Orth, G. P. Talwar, P.-A. Cazenave, and F. Bonhomme. 1996. Origin and radiation of the house mouse: Clues from nuclear genes. Journal of Evolutionary Biology, 9:519-539.

Dioli, M. 1995. A clarification about the morphology of the horns of the female kouprey. A new unknown bovid species from Cambodia. Mammalia, 59:663-667.

Dioli, M. 1997. Notes on the morphology of the horns of a new artiodactyl mammal from Cambodia: *Pseudonovibos spiralis*. Journal of Zoology, London, 241:527-531.

Dippenaar, N. J. 1977. Variation in *Crocidura mariquensis* (A. Smith, 1844) in southern Africa, Part 1 (Mammalia: Soricidae). Annals of the Transvaal Museum, 30:163-206.

Dippenaar, N. J. 1979. Variation in *Crocidura mariquensis* (A. Smith, 1844) in southern Africa, Part 2 (Mammalia: Soricidae). Annals of the Transvaal Museum, 32:1-34.

Dippenaar, N. J. 1980. New species of *Crocidura* from Ethiopia and northern Tanzania (Mammalia: Soricidae). Annals of the Transvaal Museum, 32:125-154.

Dippenaar, N. J. 1995. Geographic variation in *Myosorex longicaudatus* (Soricidae) in the southern Cape Province, South Africa. Journal of Mammalogy, 76:1071-1087.

Dippenaar, N. J., and J. A. J. Meester. 1989. Revision of the *luna-fumosa* complex of Afrotropical *Crocidura* Wagler, 1832 (Mammalia: Soricidae). Annals of the Transvaal Museum, 35:1-47.

Dippenaar, N. J., and I. L. Rautenbach. 1986. Morphometrics and karyology of the southern African species of the genus *Acomys* I. Geoffroy Saint-Hilaire, 1838 (Rodentia: Muridae). Annals of the Transvaal Museum, 34:129-183.

Dippenaar, N. J., J. Meester, I. L. Rautenbach, and D. A. Wolhuter. 1983 [1984]. The status of Southern African mammal taxonomy. Annales Musée Royal de l'Afrique Centrale, Sciences Zoologiques, 237:103-107.

Dippenaar, N. J., P. Swanepoel, and D. H. Gordon, 1993. Diagnostic morphometrics of two medically important southern African rodents, *Mastomys natalensis* and *M. coucha* (Rodentia: Muridae). South African Journal of Science, 89:300-303.

di Stefano, G. 1996. The Mesopotamian fallow deer (*Dama*, Artiodactyla) in the Middle East Pleistocene. Neues Jahrbuch fuer Geologie und Palaeontologie Abhandlungen, 199:295-322.

Ditchfield, A. D. 2000. The comparative phylogeography of Neotropical mammals: Patterns of intraspecific mitochondrial DNA variation among bats contrasted to nonvolant small mammals. Molecular Ecology, 9:1307-1318.

Dixon, J. M. 1981. Selection of a neotype for the southern short-nosed (brown) bandicoot, *Isodon* [sic] *obesulus* (Shaw and Nodder, 1797). Victorian Naturalist, 98(3):130-135.

Dixon, J. M. 1995a. White-footed tree-rat, *Conilurus albipes*. Pp. 552-553, *in* Mammals of Australia (R. Strahan, ed.). Smithsonian Institution Press, Washington, D.C., 756 pp.

Dixon, J. M. 1995b. Short-tailed hopping-mouse, *Notomys amplus*. P. 551, *in* Mammals of Australia (R. Strahan, ed.). Smithsonian Institution Press, Washington, D.C., 756 pp.

Dixon, J. M. 1995c. Long-tailed hopping Mouse, *Notomys longicaudatus*. Pp. 577-578, *in* Mammals of Australia (R. Strahan, ed.). Smithsonian Institution Press, Washington, D.C., 756 pp.

Dixon, J. M. 1995d. Big-eared hopping-mouse, *Notomys macrotis*. Pp. 578-579, *in* Mammals of Australia (R. Strahan, ed.). Smithsonian Institution Press, Washington, D.C., 756 pp.

Dixon, J. M. 1995e. Gould's Mouse, *Pseudomys gouldii*. Pp. 600-601, *in* Mammals of Australia (R. Strahan, ed.). Smithsonian Institution Press, Washington, D.C., 756 pp.

Dixon, K. R., J. A. Chapman, G. R. Willner, D. E. Wilson and W. Lopez-Forment. 1983. The New World jackrabbits and hares (genus *Lepus*)— 2. Numerical taxonomic analysis. Acta Zoologica Fennica, 174:53-56.

Dobigny, G., S. Moulin, R. Cornette, and J.-C. Gautun. 2001a. Rodents from Adrar des Iforas, Mali. Chromosomal data. Mammalia, 65(2):215-220.

Dobigny, G., R. Cornette, S. Moulin, and E. Ag Sidiyène. 2001b. The mammals of Adrar des Iforas (Mali), with special emphasis on small mammals. Pp. 445-458, *in* African small mammals—Petits mammifères africains (C. Denys, L. Granjon, and A. Poulet, eds.). Collection Colloques et Séminaires, IRD Editions, Paris, 570 pp.

Dobigny, G., V. Aniskin, and V. Volobouev. 2002a. Explosive chromosome evolution and speciation in the gerbil genus *Taterillus* (Rodentia, Gerbillinae): A case of two new cryptic species. Cytogenetic and Genome Research, 96:117-124.

Dobigny, G., A. Nomao, and J.-C. Gautun. 2002b. A cytotaxonomic survey of rodents from Niger: Implications for systematics, biodiversity and biogeography. Mammalia, 66(4):495-523.

Dobigny, G., L. Granjon, V. Aniskin, K. Ba, and V. Volobouev. 2003. A new sibling species of *Taterillus* (Muridae, Gerbillinae) from West Africa. Mammalian Biology, 68:299-316.

Dobler, F. C., and K. R. Dixon. 1990. The pygmy rabbit *Brachylagus idahoensis*. Pp. 111-115, *in* Rabbits, hares and pikas (J. A. Chapman and J. E. C. Flux, eds.). I.U.C.N., Gland, Switzerland, 168 pp.

Dobson, G. E. 1882a. A monograph of the Insectivora, systematic and anatomical, part 1: Including the families Erinaceidae, Centetidae, and Solenodontidae. John Van Voorst, London, 96 pp.

Dobson, G. E. 1882b. On the natural position of the family Dipodidae. Proceedings of the Zoological Society of London, 1882:640-641.

Dobson, M. 1994. Patterns of distribution in Japanese land mammals. Mammal Review, 24(3):91-111.

Dobson, M. 1998. Mammal distributions in the western Mediterranean: The role of human intervention. Mammal Review., 28(2):77-88.

Dobson, M. 2000. Faunal relationships and zoogeographical affinities of mammals in north-west Africa. Journal of Biogeography, 27:417-424.

Doğramaci, S. 1989. Türkiye memeli faunasi icin yeni bir kayit *Talpa europaea velessiensis* (Mammalia: Insectivora). DOGA, TU Zooloji D. C. 13:60-66.

Doğramaci, S., and I. Gündüz. 1993. The taxonomy and distribution of the species of *Erinaceus concolor* (Mammalia: Insectivora) in Turkey. Turkish Journal of Zoology, 17:267-288.

Doğramaci, S., and H. Kefelioglu. 1990. [The karyotype of *Dryomys nitedula* (Mammalia: Rodentia) from Turkey.] Turkish Journal of Zoology, 14(3):316-328.

Doğramaci, S., and H. Kefelioglu. 1992. [The karyotype of *Muscardinus avellanarius* (Mammalia: Rodentia) from Turkey.] Turkish Journal of Zoology, 16(1):43-49.

Doğramaci, S., and C. Tez. 1991. [Geographic variation and keryological characteristics of the species *Glis glis* (Mammalia:Rodentia) in Turkey.] Turkish Journal of Zoology, 15(4):275-288.

Doğramaci, S., H. Kefelioglu, and I. Gündüz. 1994a. Karyological analysis of the genus, *Spermophilus* (Mammalia: *Rodentia*) in Turkey. Turkish Journal of Zoology, 18(3):167-170.

Doğramaci, S., H. Kefelioglu, and I. Gündüz. 1994b. [Anadolu *Mesocricetus* (Mammalia: Rodentia) Türlerinin Karyolojik Analizi.] Turkish Journal of Zoology, 18:41-45 (in Turkish with English abstract).

Dokuchaev, N. E. 1994. Siberian shrew *Sorex minutissimus* found in Alaska. Zoologicheskii Zhurnal, 73:254-256.

Dokuchaev, N. E. 1997. A new species of shrew (Soricidae, Insectivora) from Alaska. Journal of Mammalogy, 78:811-817.

Dokuchaev, N. E., S. Ohdachi, and H. Abe. 1999. Morphometric status of shrews of the Sorex caecutiens/shinto group in Japan. Mammal Study, 24:67-78.

Dolan, J. M. 1963. Beitrag zur systematischen Gliederung des Tribus Rupicaprini, Simpson 1945. Zeitschrift für Zoologische Systematik und Evolutionsforschung, 1:311-407.

Dolan, J. M. 1988. A deer of many lands—a guide to the subspecies of the red deer *Cervus elaphus* L. Zoonooz, 62(10):4-34.

Dolan, P. G. 1989. Systematics of Middle American mastiff bats of the genus *Molossus*. Special Publications, The Museum, Texas Tech University Press, 23:1-71.

Dolan, P. G., and D. C. Carter. 1973. *Glaucomys volans*. Mammalian Species, 78:1-6.

Dolan, P. G., and T. L. Yates. 1981. Interspecific variation in *Apodemus* from the northern Adriatic islands of Yugoslavia. Zeitschrift für Säugetierkunde, 46:151-161.

Dolch, D., R. Labes, and J. Teubner. 1994. Beiträge zur Säugetierfauna der Prignitz. Veröffentlichung Potsdam-Museum 31, Beiträge zur Tierwelt der Mark XII, S.:33-68.

Dolgov, V. A. 1979. [*Crocidura leucodon* Hermann, 1780 in Kopetdag (Insectivora, Mammalia)]. Sbornik Trudov Zoologicheskovo Museya MGU, 18:257-263 (in Russian).

Dolgov, V. A. 1985. Burozubki Starovo Sveta [Brown-toothed shrews of the Old World]. Moscow University, Moscow, 220 pp. (in Russian).

Dolgov, V. A., and R. S. Hoffmann. 1977. Tibetskaya burozubka—*Sorex thibetanus* Kastchenko, 1905 (Soricidae Mammalia). Zoologicheskii Zhurnal, 46:1687-1692 (in Russian).

Dolgov, V. A., and I. V. Lukyanova. 1966. O stroenii genitalii Palearkticheskikh burozubko (Insectivora, Soricidae). Zoologicheskii Zhurnal, 56:1852-1861 (in Russian).

Dolgov, V. A., and B. S. Yudin. 1975. [Progress and problems in the investigation of the insectivorous mammals of the USSR.] Trudy Biologicheskovo Instituta, Novosibirsk, 23:5-40 (in Russian).

Dollman, G. 1909. New mammals from British East Africa. Annals and Magazine of Natural History, ser. 8, 4:39-553.

Dollman, G. 1910. Two new African mammals. Annals and Magazine of Natural History, ser. 8, 6:226-230.

Dollman, G. 1911. On *Arvicanthis abyssinicus* and allied East-African species, with descriptions of four new forms. Annals and Magazine of Natural History, ser. 8, 8:334-353.

Dollman, G. 1915. On the swamp-rats (*Otomys*) of East Africa. Annals and Magazine of Natural History, ser. 8, 15:149-170.

Dollman, G. 1932. Mammals collected by Lord Cranbrook and Captain F. Kingdon Ward in Upper Burma. Proceedings of the Linnean Society of London, 145:9-11.

Domning, D. P. 1978. Sirenian evolution in the North Pacific Ocean. University of California Publications in Geological Science, 118:1-176.

Domning, D. P. 1981. Distribution and status of manatees *Trichechus* spp. near the mouth of the Amazon River, Brazil. Biological Conservation, 19:85-97.

Domning, D. P. 1982. Evolution of manatees: A speculative history. Journal of Paleontology, 56:599-619.

Domning, D. P. 1996. Bibliography and index of the Sirenia and Desmostylia. Smithsonian Institution Press, Washington, D.C., 611 pp.

Domning, D. A., H. N. Hoeck, D. W. Rice, and J. Shoshani. 1982. An annotated checklist on the Paenungulata. Pp. 305-307, 312-314, *in* Mammal species of the world, a taxonomic and geographic reference (J. H. Honacki, K. E. Kinman, and J. W. Koeppl, eds.). Allen Press., Inc. and the Association of Systematic Collections. niversity of Kansas, Lawrence, 694 pp.

Donnellan, S. C. 1987. Phylogenetic relationships of New Guinean rodents (Rodentia, Muridae) based on chromosomes. Australian Mammalogy, 12:61-67.

Donnellan, S. C., T. B. Reardon, and T. F. Flannery. 1995. Electrophoretic resolution of species boundaries in tube-nosed bats (Chiroptera: Pteropodidae) in Australia and Papua New Guinea. Australian Mammalogy, 18:61-70.

Dor, M. 1966. Restes subfossiles de *Lophiomys* trouvés en Israel. Mammalia, 30:199-200.

Dorst, J. 1972. Notes sur quelques rongeurs observes en Ethiopie. Mammalia, 36:182-192.

Douady, C. J., F. Catzeflis, D. J. Kao, M. S. Springer, and M. J. Stanhope. 2002*a*. Molecular evidence for the monophyly of tenrecidae (Mammalia) and the timing of the colonization of Madagascar by Malagasy tenrecs. Molecular Phylogenetics and Evolution, 22:357–363.

Douady, C. J., P. I. Chatelier, O. Madsen, W. W. de Jong, F. Catzeflis, M. S. Springer, and M. J. Stanhope. 2002*b*. Molecular phylogenetic evidence confirming the Eulipotyphla concept and in support of hedgehogs as the sister group to shrews. Molecular Phylogenetics and Evolution, 25:200–209.

Dowler, R. C. 1989. Cytogenetic studies of three chromosomal races of pocket gophers (*Geomys bursarius* complex) at hybrid zones. Journal of Mammalogy, 70:253-266.

Dowler, R. C., and D. S. Carroll. 1996. The endemic rodents of Isla Fernandina: Population status and conservation issues. Noticias de Galapagos, 57:8-13.

Dowler, R. C., and H. H. Genoways. 1978. *Liomys irroratus*. Mammalian Species, 82:1-6.

Dowler, R. C., D. S. Carroll, and C. W. Edwards. 2000. Rediscovery of rodents (Genus *Nesoryzomys*) considered extinct in the Galapagos Islands. Oryx, 34:109-117.

Downing, K. F., G. G. Musser, and L. E. Park. 1998. The first fossil record of small mammals from Sulawesi, Indonesia: The large murid, *Paruromys dominator*, from the Late (?) Pliocene Walanae formation. Pp. 105-121, *in* Advances in vertebrate paleontology and geochronology (Y. Tomida, L. J. Flynn, and L. L. Jacobs, eds.). National Science Museum Monographs, No. 14, Tokyo, 292 pp.

Downs, C. T., and M. R. Perrin. 1989. An investigation of the macro- and micro-environments of four *Gerbillurus* species. Cimbebasia, 11:41-54.

Downs, C. T., and M. R. Perrin. 1990. Thermal parameters of four species of *Gerbillurus*. Journal of Thermal Biology, 15:291-300.

Downs, C. T., and J. O. Wirminghaus. 1997. The terrestrial vertebrates of the Bazaruto Archipelago, Mozambique: A biogeographical perspective. Journal of Biogeography, 24:591-602.

Dowsett, R. J., and L. Granjon. 1991. Ch. 18. Liste preliminaire des mammiferes du Congo. Pp. 297-310, *in* Flore et faune du bassin du Kouilou (Congo) et leur exploitation (R. J. Dowsett and F. Dowsett-Lemaire, eds.). Tauraco Research Report 4, Tauraco Press, Jupille-Liege, Belgium, 340 pp.

Dragoo, J. W., and R. L. Honeycutt. 1997. Systematics of mustelid-like carnivores. Journal of Mammalogy, 78:426-443.

Dragoo, J. W., J. R. Choate, T. L. Yates, and T. P. O'Farrell. 1990. Evolutionary and taxonomic relationships among North American arid-land foxes. Journal of Mammalogy, 71(3):318-332.

Dragoo, J. W., R. D. Bradley, R. L. Honeycutt, and J. W. Templeton. 1993. Phylogenetic relationships among the skunks: A molecular perspective. Journal of Mammalian Evolution, 1:255-267.

Duarte, J. M. B. 1996. Guia de identifação de cervídeos Brasileiros. Fundação de Estudos e Pesquisas em Agronomia, Medicina Veterinária e Zootecnia, Jaboticabal, Brazil, 8 pp.

Duarte, J. M. B, and W. Jorge. 2003. Morphologic and cytogentic description of the small red brocket (*Mazama bororo* Duarte, 1996) in Brazil. Mammalia, 67:403-410.

Duarte, J. M. B., and M. L. Merino. 1997. Taxonomia e evolução. Pp. 2-21, *in* Biologia e conservação de cervídeos Sul-Americanos: *Blastocerus, Ozotoceros* e *Mazama* (J. M. B. Duarte, ed.). Fundação de Estudos e Pesquisas em Agronomia, Medicina Veterinária e Zootecnia, Jaboticabal, Brazil, 238 pp.

Dubey, D. D., and R. Raman. 1992. Mammalian sex chromosomes. IV. Replication heterogeneity in the late replicating facultative-and constitutive-heterochromatic regions in the X chromosomes of the mole rats, *Bandicota bengalensis* and *Nesokia indica*. Hereditas, 115:275-282.

Dubois, J.-Y., D. Rakotondravony, C. Hanni, P. Sourouille, and F. M. Catzeflis. 1996. Molecular evolutionary relationships of three genera of Nesomyinae, endemic rodent taxa from Madagascar. Journal of Mammalian Evolution, 3:239-259.

Dubois, J.-Y. F., F. M. Catzeflis, and J. J. Beintema. 1999. The phylogenetic position of "Acomyinae" (Rodentia, Mammalia) as sister group of a Murinae + Gerbillinae clade: Evidence from the nuclear ribonuclease gene. Molecular Phylogenetics and Evolution, 13(1):181-192.

Duckworth, J. W., D. L. Harrison, and R. J. Timmins. 1993. Notes on a collection of small mammals from the Ethiopian Rift Valley. Mammalia, 57:278-282.

Duckworth, J. W., R. E. Salter, and K. Khounboline (eds.). 1999. Wildlife in Lao PDR. 1999 status report. The World Conservation Union, Vientiane, Lao PDR, 275 pp.

Ducommun, M.-A., F. Jeanmaire-Besancon, and P. Vogel. 1994. Shield morphology of curly overhair in 22 genera of Soricidae (Insectivora, Mammalia). Revue Suisse de Zoologie, 101:623-643.

Ducroz, J.-F. 1998. Contribution des approches cytogenetique et moléculaire al'étude systématique et évolutive de genres de rongeurs Murinae de la 'division' *Arvicanthis*. Ph.D. Dissertation. Paris, France: Muséum National d'Histoire Naturelle, 195 pp.

Ducroz, J.-F., L. Granjon, P. Chevret, J. M. Duplantier, M. Lombard, and V. Volobouev. 1997. Characterization of two distinct species of *Arvicanthis* (Rodentia: Muridae) in West Africa: Cytogenetic, molecular and reproductive evidence. Journal of Zoology, London, 241:709-723.

Ducroz, J.-F., V. Bolobouev, and L. Granjon. 1998. A molecular perspective on the systematics and evolution of the genus *Arvicanthis* (Rodentia, Muridae): Inferences from complete cytochrome *b* sequences. Molecular Phylogenetics and Evolution, 10(1):104-117.

Ducroz, J.-F., V. Volobouev, and L. Granjon. 2001. An assessment of the systematics of Arvicanthine Rodents using mitochondrial DNA sequences: Evolutionary and biogeographical implications. Journal of Mammalian Evolution, 8:173-206.

Dudu, A. M., E. Van der Straeten, and W. N. Verheyen. 1989. Premiere capture de *Hylomyscus parvus* Brosset, Dubost et Heim de Balzac, 1965 au Zaire avec quelques donnees biometriques (Rodentia, Muridae). Journal of African Zoology, 103:179-182.

Dudu, A., R. Verhagen, H. Gevaerts, and W. Verheyen. 1997. Population structure and reproductive cycle of *Praomys jacksoni* (De Winton, 1897) and first data on the reproduction of *P. misonnei* Van der Straeten and Dieterlen, 1987 and *P. mutoni* Van der Straeten and Dudu, 1990 (Muridae) from Masako forest (Kisangani, Zaire). Belgian Journal of Zoology, 127 (suppl.):67-70.

Dudu, A., S. Churchfield, and R. Hutterer. In Press. Community structure and food niche relationships of coexisting rain-forest shrews in the Masako Forest, north-eastern Congo. *in,* Biology of the Soricidae II (J. Merritt, S. Churchfield, R. Hutterer, and B. Sheftel, eds.). Carnegie Museum Special Publication.

Duncan, F. M., F. H. Waterhouse, and H. Peavot. 1937. On the dates of publication of the Society's *Proceedings*, 1830-1858, compiled by the late F. H. Waterhouse, and of the *Transactions*, by the late Henry Peavot, originally published in P.Z.S. 1893, 1913. Proceedings of the Zoological Society of London, 1937:71-81.

Duncan, J. F. 1976. Karyotypes of four rats (Rodentia: Muridae) from Sulawesi (Celebes), Indonesia. Cytologia, 41:481-486.

Duncan, J. F., and P. F. D. Van Peenen. 1971. Karyotypes of ten rats (Rodentia: Muridae) from Southeast Asia. Caryologia, 24:331-346.

Duncan, P., and R. W. Wrangham. 1971. On the ecology and distribution of subterranean insectivores in Kenya. Journal of Zoology, London, 164:149-163.

Dung, V. V., P. M. Giao, N. N. Chinh, D. Tuoc, P. Arctander, and J. MacKinnon. 1993. A new species of living bovid from Vietnam. Nature, 363:443-445.

Dung, V. V., P. M. Giao, N. N. Chinh, D. Tuoc, and J. MacKinnon. 1994. Discovery and conservation of the Vu Quang ox in Vietnam. Oryx, 28:16-21.

Dunlop, J. 1997. *Coleura afra*. Mammalian Species, 566:1-4.

Dunlop, J. 1998. The evolution of behavior and ecology in the Emballonuridae. Ph.D. dissertation, York University, North York, Ontario, Canada, 271 pp.

Dunlop, J. 1999. *Mops midas*. Mammalian Species, 615:1-4.

Dunlop, J. N., and I. R. Pound. 1981. Observations on the pebble-mound mouse *Pseudomys chapmani*. Records of the Western Australian Museum, 9:1-5.

Dupal, T. A. 1998. Evolutionary changes of size of the first lower molar in the lineage from *Microtus* (*Terricola*) *hintoni* to the Recent forms of *M.* (*Stenocranius*) *gregalis* (Rodentia, Cricetidae). Paleontological Journal, 32(4):410-417 (translated from Paleontologicheskii Zhurnal, 1998(4):87-94).

Duplantier, J.-M. 1989. Les rongeurs myomorphes forestiers du nord-est du Gabon: Structure du peuplement, démographie, domaines vitaux. Revue d'Ecologie (La Terre et La Vie), 44:329-346.

Duplantier, J.-M., and K. Bâ. 2001. Swimming ability in six West-African rodent species under laboratory conditions. Evaluation of their potentialities to colonize islands. Pp. 331-352, *in* African small mammals (C. Denys, L. Granjon, and A. Poulet, eds.). IRD Éditions, Collection Colloques et Séminaires, Paris, 570 pp.

Duplantier, J.-M., and J.-B. Duchemin. 2003. Introduced small mammals and their ectoparasites: A description of their colonization and its consequences. Pp. 1191-1194, *in* The natural history of Madagascar (S. M. Goodman and J. P. Benstead, eds.). University of Chicago Press, Chicago and London, xix + 1709 pp.

Duplantier, J.-M., and L. Granjon. 1992. Liste revisee des rongeurs du Senegal. Mammalia, 56(3):425-431.

Duplantier, J.-M., J. Britton-Davidian, and L. Granjon. 1990*a*. Chromosomal characterization of three species of the genus *Mastomys* in Senegal. Zeitschrift für Zoologische Systematik und Evolutionsforschung, 28:289-298.

Duplantier, J.-M., L. Granjon, E. Mathieu, and F. Bonhomme. 1990*b*. Structures genetiques comparees de trois especes de rongeurs africains du genre *Mastomys* au Senegal. Genetica, 81:179-192.

Duplantier, J.-M., L. Granjon, and K. Bâ. 1991*a*. Decouverte de trois especes de rongeurs nouvelles pour le Senegal: Un indicateur supplementaire de la desertification dans le nord du pays. Mammalia, 55:313-315.

Duplantier, J.-M., L. Granjon, F. Adam, and K. Bâ. 1991*b*. Repartition actuelle du rat noir (*Rattus rattus*) au Senegal: Facteurs historiques et ecologiques. Pp. 339-346, *in* Le rongeur et l'espace, Actes du Colloque International, Lyon, 1989 (M. Le Berre and L. Le Guelte, eds.). Raymond Chabaud, Paris, 362 pp.

Duplantier, J.-M., L. Granjon, and H. Bouganaly. 1996. Reproductive characteristics of three sympatric species of *Mastomys* in Senegal, as observed in the field and in captivity. Mammalia, 60(4):629-638.

Duplantier, J.-M., L. Granjon, and K. Bâ. 1997. Répartition biogéographique des petits rongeurs du Sénégal. Journal of African Zoology, 111:17-26.

Duplantier, J.-M., A. Orth, J. Catalan, and F. Bonhomme. 2002. Evidence for a mitochondrial lineage originating from the Arabian Peninsula in the Madagascar house mouse (*Mus musculus*). Heredity, 89:154-158.

Duplantier, J.-M., J. Catalan, A. Orth, B. Grolleau, and J. Britton-Davidian. 2003. Systematics of the black rat in Madagascar: Consequences for the transmission and distribution of plague. Biological Journal of the Linnean Society, 78:335-341.

Durán, A. C., A. V. de Andrés, M. Cardo, R. Muñoz-Chápuli, and V. Sans-Coma. 1998. Anatomia de las arterias coronarias en arvicolinos (Rodentia, Muridae). Boletín de la Real Sociedad Española de Historia Natural, Sección Biológica, 94:113-123.

Durant, P., and M. A. Guevara. 2001. A new rabbit species (*Sylvilagus,* Mammalia: Leporidae) from the lowlands of Venezuela. Revista de Biologia Tropica, 49(1):369-381.

Duthie, A. G., and T. J. Robinson. 1990. The African rabbits. Pp. 121-127, *in* Rabbits, hares and pikas (J. A. Chapman and J. E. C. Flux, eds.). I.U.C.N., Gland, Switzerland, 168 pp.

Dutrillaux, B. 1986. Evolution chromosomique chez les primates, carnivores et les rongeurs. Mammalia, 50 (Special Number):1-203.

Dutton, J., and J. Haft. 1996. Distribution, ecology and status of an endemic shrew, *Crocidura thomensis*, from Sao Tomé. Oryx, 30:195-201.

Dwyer, P. D. 1984. From garden to forest: Small rodents and plant succession in Papua New Guinea. Australian Mammalogy, 7:29-36.

Dwyer, P., M. Hockings, and J. Willmer. 1979. Mammals of Cooloola and Beerwah. Proceedings of the Royal Society of Queensland, 90:65-84.

Dyatlov, A. I., and L. A. Avanyan. 1987. [Substantiation of species rank for two subspecies of gerbils (*Meriones*, Cricetidae, Rodentia).] Zoologicheskii Zhurnal, 66:1069-1074 (in Russian).

Dyzenchauz, F. J., and A. I. Massarini. 1999. First cytogenetic analysis of the genus *Bibymys* (Rodentia, Cricetidae). Zeitschrift für Säugetierkunde, 64:59-62.

Dziurdzik, B. 1978. Histological structure of hair in the Gliridae (Rodentia). Acta Zoologica Cracoviensia, 22(1):1-12.

Eadie, R. 1951. A comparative study of the male accessory genital glands of *Neurotrichus*. Journal of Mammalogy, 32:36-43.

East, R. (ed.). 1988. Antelopes. Global survey and regional action plans. Part 1. East and northeast Africa. I.U.C.N., Gland, Switzerland, 96 pp.

East, R. (ed.). 1989. Antelopes. Global survey and regional action plans. Part 2. Southern and south-central Africa. I.U.C.N., Gland, Switzerland, 96 pp.

East, R. (ed.). 1990. Antelopes. Global survey and regional action plans. Part 3. West and central Africa. I.U.C.N., Gland, Switzerland, 171 pp.

East, R., and IUCN/SSC Antelope Specialist Group (eds.) 1999. African antelope database 1988. I.U.C.N., Gland, Switzerland, 434 pp.

Edelman, A. J. 2003. *Marmota olympus*. Mammalian Species, 736:1-5.

Edwards, C. W., and R. D. Bradley. 2001. Molecular phylogenetics of the *Neotoma floridana* species group. Journal of Mammalogy, 82:791-798.

Edwards, C. W., and R. D. Bradley. 2002*a*. Molecular systematics and historical phylobiogeography of the *Neotoma mexicana* species group. Journal of Mammalogy, 83:20-30.

Edwards, C. W., and R. D. Bradley. 2002*b*. Molecular systematics of the genus *Neotoma*. Molecular Phylogenetics and Evolution, 25:489-500.

Edwards, C. W., C. F. Fulhorst, and R. D. Bradley. 2001. Molecular phylogenetics of the *Neotoma albigula* species group: Further evidence of a paraphyletic assemblage. Journal of Mammalogy, 82:267-279.

Eger, J. L. 1977. Systematics of the genus *Eumops*. Royal Ontario Museum, Life Sciences, Contribution, 110:1-69.

Eger, J. L. 1990. Patterns of geographic variation in the skull of Nearctic ermine (*Mustela erminea*). Canadian Journal of Zoology, 68:1241-1249.

Eger, J. L. 1995. Morphometric variation in the Nearctic collared lemming (*Dicrostonyx*). Journal of the Zoological Society of London, 235:143-161.

Eger, J. L. 2001. Emendation of *Glauconycteris curryi*. Acta Chiropterologica, 3:248.

Eger, J. L., and M. B. Fenton. 2003. *Rhinolophus paradoxolophus*. Mammalian Species, 731:1-4.

Eger, J. L., and R. L. Peterson. 1979. Distribution and systematic relationship of *Tadarida bivittata* and *Tadarida ansorgei* (Chiroptera: Molossidae). Canadian Journal of Zoology, 57:1887-1895.

Eger, J. L., and D. A. Schlitter. 2001. A new species of *Glauconycteris* from West Africa (Chiroptera: Vespertilionidae). Acta Chiropterologica, 3:1-10.

Eger, J. L., and M. M. Theberge. 1999. *Thainycteris aureocollaris* (Chiroptera, Vespertilionidae) in Vietnam. Mammalia, 63:237-240.

Egoscue, H. J. 1979. *Vulpes velox*. Mammalian Species, 122:1-5.

Ehrich, D., V. B. Fedorov, N. C. Stenseth, C. J. Krebs, and A. Kenny. 2000. Phylogeography and mitochrondrial DNA (mtDNA) diversity in North American collared lemmings (*Dicrostonyx groenlandicus*). Molecular Ecology, 9:329-337.

Eibl-Eibesfeldt, I. 1984. The Galapagos seals. Part 1. Natural history of the Galapagos sea lion (*Zalophus californianus wollebaeki,* Sivertsen). Pp. 207-214, *in* Key environments: Galapagos (R. Perry, ed.). Pergamon Press. Oxford, 321 pp.

Eisenberg, J. F. 1989. Mammals of the Neotropics. The northern Neotropics, vol. 1, Panama, Colombia, Venezuela, Guyana, Suriname, French Guiana. University of Chicago Press, Chicago, IL, 449 pp.

Eisenberg, J. F. 1981. The mammalian radiations: An analysis of trends in evolution, adaptation, and behaviour. Athlone Press, London, 610 pp.

Eisenberg, J. F., and E. Gould. 1970. The tenrecs: A study in mammalian behavior and evolution. Smithsonian Contributions to Zoology, 27:1-138.

Eisenberg, J. F., and E. Gould. 1984. The insectivores. Pp. 155-165, *in* Key environments: Madagascar (A. Jolly, P. Oberlé, and R. Albignac, eds.) Pergamon Press, Oxford, 250 pp.

Eisenberg, J. F., and K. H. Redford. 1999. Mammals of the Neotropics, 3. The central Neotropics. University of Chicago Press, Chicago, IL, 609 pp.

Eisenberg, J. F., C. P. Groves, and K. MacKinnon. 1987. Tapire. Grzimeks Enzyklopädie, Säugetiere, 4:598-608.

Eisenmann, V., and J. S. Brink. 2000. Koffiefontein quaggas and true Cape quaggas; the importance of basic skull morphology. South African Journal of Science, 96:529-533.

Eisenmann, V., and J.-C. Turlot. 1978. Sur la taxinomie du genre *Equus* (equides). Les Cahiers de l'Analyse des Données, 3:179-201.

Eisentraut, M. 1963. Die Wirbeltiere des Kamerungebirges. Hamburg and Berlin, 353 pp.

Eisentraut, M. 1965. Die Muriden von Fernando Poo. Zoologisches Jahrbuch Systematik, 92:13-40.

Eisentraut, M. 1969a. Das Gaumenfaltenmuster bei westafrikanischen Muriden. Zoologische Jahrbücher, 96:478-490.

Eisentraut, M. 1969b. Die Verbreitung der Muriden-Gattung *Hylomyscus* auf Fernando Po und in Westkamerun. Zeitschrift für Säugetierkunde, 34:296-307.

Eisentraut, M. 1970. Die Verbreitung der Muriden-Gattung *Praomys* auf Fernando Po und in Westkamerun. Zeitschrift für Säugetierkunde, 35:1-15.

Eisentraut, M. 1973. Die Wirbeltierfauna von Fernando Poo und Westkamerun. Unter besonderer Berücksichtigung der Bedeutung der pleistozänen Klimaschwankungen für die heutige Faunenverteilung. Bonner Zoologische Monographien, 3:1-428.

Eisentraut, M. 1976. Das Gaumenfaltenmuster der Säugetiere und seine Bedeutung für Stammesgeschichtliche und taxonomische Untersuchungen. Bonner Zoologische Monographien, 8:1-214.

Eizirik, E., W. J. Murphy, and S. J. O'Brien. 2001. Molecular dating and biogeography of the early placental mammal radiation. Journal of Heredity, 92:212-219.

El Ashmawy, S. H. 1990. Karyotype of two Egyptian biotypes of spiny mice *Acomys cahirinus* and their hybrids. Egyptian Journal of Cytology, 19:189-194.

Elbl, A., U. Rahm, and G. Mathys. 1966. Les mammiferes et leurs tiques dans La Foret du Ruggege (Republique Rwandaise). Acta Tropica, 23:223-263.

Elder, F. B. 1980. Tandem fusion, centric fusion, and chromosomal evolution in the cotton rats, genus *Sigmodon*. Cytogenetics and Cell Genetics, 26:199-210.

Elder, F. B., and M. R. Lee. 1985. The chromosomes of *Sigmodon ochrognathus* and *S. fulviventer* suggest a realignment of *Sigmodon* species groups. Journal of Mammalogy, 66:511-518.

Eldridge, M. D. B., and R. L. Close. 1992. Taxonomy of Rock Wallabies, *Petrogale* (Marsupialia: Macropodinae). I. A revision of the eastern *Petrogale* with the description of three new species. Australian Journal of Zoology, 40:605-625.

Eldridge, M. D. B., and R. L. Close. 1995. Black-footed rock-wallaby, *Petrogale lateralis*. Pp. 377-381, *in* Mammals of Australia. (R. Strahan, ed.) Smithsonian Institution Press, 756 pp.

Eldridge, M. D. B., P. G. Johnston, R. L. Close, and P. S. Lowry. 1989. Chromosomal rearrangements in rock wallabies, *Petrogale* (Marsupialia, Macropodidae) II. G-banding analysis of *Petrogale godmani*. Genome, 32:935-40.

Eldridge, M. D. B., R. L. Close, and P. G. Johnston. 1991. Chromosomal rearrangements in Rock Wallabies, *Petrogale*. IV. G-banding analysis of the *Petrogale lateralis* complex. Australian Journal of Zoology, 39:621-627.

Ellerman, J. R. 1940. The families and genera of living rodents. Vol. 1. Rodents other than Muridae. Trustees of the British Museum (Natural History), London, 689 pp.

Ellerman, J. R. 1941. The families and genera of living rodents. Vol. II. Family Muridae. British Museum (Natural History), London, 690 pp.

Ellerman, J. R. 1947a. A key to the Rodentia inhabiting India, Ceylon, and Burma, based on collections in the British Museum. Journal of Mammalogy, 28:249-278; 357-387.

Ellerman, J. R. 1947b. Notes on some Asiatic rodents in the British Museum. Proceedings of the Zoological Society of London, 1947-1948, 117:259-271.

Ellerman, J. R. 1948. Key to the rodents of south-west Asia in the British Museum collection. Proceedings of the Zoological Society of London, 118 (Part III):765-816.

Ellerman, J. R. 1949a. The families and genera of living rodents. Vol. III, Appendix II [Notes on the rodents from Madagascar in the British Museum, and on a collection from the island obtained by Mr. C. S. Webb]. British Museum (Natural History), London, 210 pp.

Ellerman, J. R. 1949b. On the prior name for the Siberian lemming and the genotype of *Glis* Erxleben. Annals and Magazine of Natural History, ser. 12, 2:893-894.

Ellerman, J. R. 1961. Rodentia. Volume 3, *in* The fauna of India including Pakistan, Burma and Ceylon. Mammalia. Second ed. Manager of Publications, Zoological Survey of India, Calcutta, vol. 3(in 2 parts), 1:1-482; 2:483-884.

Ellerman, J. R., and T. C. S. Morrison-Scott. 1951. Checklist of Palaearctic and Indian mammals 1758 to 1946. Trustees of the British Museum (Natural History), London, 810 pp.

Ellerman, J. R., and T. C. S. Morrison-Scott. 1953. Checklist of Palaearctic and Indian mammals— amendments. Journal of Mammalogy, 34:516-518.

Ellerman, J. R., and T. C. S. Morrison-Scott. 1955. Supplement to Chasen (1940) A handlist of Malaysian mammals, containing a generic synonomy and a complete index. British Museum (Natural History), London, 66 pp.

Ellerman, J. R., and T. C. S. Morrison-Scott. 1966. Checklist of Palaearctic and Indian Mammals 1758 to 1946. Second ed. British Museum (Natural History), London, 810 pp.

Ellerman, J. R., T. C. S. Morrison-Scott, and R. W. Hayman. 1953. Southern African mammals 1758 to 1951: A reclassification. British Museum (Natural History), London, 363 pp.

Elliot, C. L., and J. T. Flinders. 1991. *Spermophilus columbianus*. Mammalian Species, 372:1-9.

Elliot, D. G. 1901. A synopsis of the mammals of North America and the adjacent seas. Field Columbian Museum, Zoological Series, 2, 45:1-471.

Elliot, D. G. 1905. A checklist of mammals of the North American continent, the West Indies and the neighboring seas. Field Columbian Museum, Publication 105, Zoological Series, 6:1-761.

Elliot, D. G. 1907a. A catalogue of the collection of mammals in the Field Columbian Museum. Field Columbian Museum, Zoology Series, 8:1-694.

Elliot, D. G. 1907b. Descriptions of apparently new species and subspecies of mammals belonging to the families Lemuridae, Cebidae, Callitrichidae, and Cercopithecidae in the collection of the Natural History Museum. Annals and Magazine of Natural History, ser. 7, 20:185-196.

Elliot, O. 1971. Bibliography of the tree shrews 1780-1969. Primates, 12:323-414.

Ellis, L. S., and L. R. Maxson. 1979. Evolution of the chipmunk genera *Eutamias* and *Tamias*. Journal of Mammalogy, 60:331-334.

Ellis, M. 1993. Extension to the known range of the fawn hopping-mouse *Notomys cervinus* in New South Wales. Australian Zoologist, 29(1-2):77-78.

Ellis, M. 1995a. A discussion of the large extinct rodents of Mootwingee National Park, western New South Wales. Australian Zoologist, 30(1):39-42.

Ellis, M. 1995b. Addendum to "The mammals of Kinchega National Park, western New South Wales." Australian Zoologist, 30(1):110.

Ellison, G. T. H. 1993. Evidence of climatic adaptation in spontaneous torpor among pouched mice *Saccostomus campestris* from southern Africa. Acta Theriologica, 38:49-59.

Ellison, G. T. H., P. J. Taylor, H. A. Nix, G. N. Bronner, and J. P. McMahon. 1993. Climatic adaptation of body size among pouched mice (*Saccostomus campestris*: Cricetidae) in the southern African subregion. Global Ecology and Biogeography Letters, 3:41-47.

El-Rayah, M. A. 1981. A new species of bat of the genus *Tadarida* (Family Molodidae) from West Africa. Royal Ontario Museum, Life Sciences Occasional Paper, 36:1-12.

Elrod, D. A., and M. L. Kennedy. 1995. Microgeographic variation in morphometric characters of the white-footed mouse, *Peromyscus leucopus*. Southwestern Naturalist, 40:42-49.

Elrod, D. A., E. G. Zimmerman, P. D. Sudman, and G. A. Heidt. 2000. A new subspecies of pocket gopher (genus *Geomys*) from the Ozark Mountains of Arkansas with comments on its historical biogeography. Journal of Mammalogy, 81:852-864.

Emelyanova, L. G. 1994. [Spatial distribution and abundance of wood lemming populations in the north-eastern portion of their range.] Byulleten' Moskovskovo Obshchestva Ispytatelei Prirody, Otdel Biologicheskii, 99(5):37-43 (in Russian with English summary).

Emerson, B. C., and M. L. Tate. 1993. Genetic analysis of evolutionary relationships among deer (subfamily Cervinae). Journal of Heredity, 84:266-273.

Emerson, G. L., C. W. Kilpatrick, B. E. McNiff, J. Ottenwalder, and M. W. Allard. 1999. Phylogenetic relationships of the order Insectivora based on complete 12S rRNA sequences from mitochondria. Cladistics, 15:221-230.

Emmons, L. H. 1988. Replacement name for a genus of South American rodents (Echimyidae). Journal of Mammalogy, 69(2):421.

Emmons, L. H. 1993a. On the identity of *Echimys didelphoides* Desmarest, 1817 (Mammalia: Rodentia: Echimyidae). Proceeding of the Biological Society of Washington, 106 (1):1-4.

Emmons, L. H. 1993b. A new genus and species of rat from Borneo (Rodentia: Muridae). Proceedings of the Biological Society of Washington, 106(4):752-761.

Emmons, L. H. 1997. Mammals of the Rio Urucuti Basin, south central Chuquisaca, Bolivia. Pp. 30-33, *in* A rapid assessment of the humid forest of south central Chuquisaca, Bolivia (T. S. Schulenberg and K. Awbrey, eds.). Conservation International, RAP Working Papers 8, Washington D.C., 87 pp.

Emmons, L. H. 1999a. A new genus and species of Abrocomid rodent from Peru (Rodentia: Abrocomidae). American Museum Novitates, 3279:1-14.

Emmons, L. H. 1999b. Two new species of *Juscelinomys* (Rodentia: Muridae) from Bolivia. American Museum Novitates, 3280:15 pp.

Emmons, L. H. 2000. Tupai: A field study of Bornean tree shrews. University of California Press, Berkeley, 280 pp.

Emmons, L. H. in press. A revision of the genera of arboreal Echimyidae (Rodentia: Echimyidae: Echimyinae); with descriptions of two new genera. *In* Mammalian diversication: From population genetics to Biogeography (E. A. Lacey and P. Myers, eds.). University of California Press, Berkeley.

Emmons, L. H., and F. Feer. 1990. Neotropical rainforest mammals. A field guide. University of Chicago Press, Chicago, IL, 281 pp., 7 pls.

Emmons, L. H., and F. Feer. 1997. Neotropical rainforest mammals. A field guide. Second edition, University of Chicago Press, Chicago, IL, 307 pp., 36 pls.

Emmons, L. H., and F. Kinbag. 2002. Survey of mammals of southern New Ireland. Pp. 67-69 and 100-101, *in* Southern New Ireland, Papua New Guinea: A biodiversity assessment (B. M. Beehler and L. E. Alonso, eds). Rapid Assessment Program Bulletin 21, Conservation International, Washington, DC, 101 pp.

Emmons, L. H., and M. G. Vucetich. 1998. The identity of Winge's *Lasiuromys villosus* and the description of a new genus of echimyid rodent (Rodentia: Echimyidae). American Museum Novitates, 3223:1-12.

Emmons, L. H., Y. L. R. Leite, D. Kock, and L. P. Costa. 2002. A review of the named forms of *Phyllomys* (Rodentia: Echimyidae) with the description of a new species from coastal Brazil. American Museum Novitates, 3380:1-40.

Emry, R. J. 1970. A North American Oligocene pangolin and other additions to the Pholidota. Bulletin of the American Museum of Natural History, 142:459-510.

Emry, R. J. 1981. New material of the Oligocene muroid rodent *Nonomys*, and its bearing on muroid origins. American Museum Novitates, 2712:14 pp.

Emry, R. J., and W. W. Korth. 1989. Rodents of the Bridgerian (middle Eocene) Elderberry Canyon local fauna of eastern Nevada. Smithsonian Contributions to Paleobiology, 67:14 pp.

Emry, R. J., and R. W. Thorington, Jr. 1982. Descriptive and comparative osteology of the oldest fossil squirrel, *Protosciurus* (Rodentia: Sciuridae). Smithsonian Contributions to Paleobiology, 47:1-35.

Emry, R. J., and R. W. Thorington, Jr. 1984. The tree squirrel *Sciurus* (Sciuridae, Rodentia) as a living fossil. Pp. 23-31, *in* Living fossils (N. Eldredge and S. M. Stanley, eds.). Springer-Verlag, New York, 291 pp.

Emry, R. J., L. A. Tyutkova, S. G. Lucas, and B. Wang. 1998. Rodents of the middle Eocene Shinzhaly Fauna of eastern Kazakstan. Journal of Vertebrate Paleontology, 18(1):218-227.

Endepols, S., H. Dietze, and H. Endepols. 2001. The occurrence of roof rats (*Rattus rattus* L., 1758) in Germany during the late 20th century. Mammalian Biology, 66:301-304.

Enders, R. K. 1939. A new rodent of the genus *Rheomys* from Chiriqui. Proceedings of the Academy of Natural Sciences of Philadelphia, 90:295-296.

Enders, R. K. 1953. The type locality of *Syntheosciurus brochus*. Journal of Mammalogy, 34:509.

Enders, R. K. 1980. Observations on *Syntheosciurus*: Taxonomy and behavior. Journal of Mammalogy, 61:725-727.

Endo, H., and H. Yanagawa. 1994. Cardiac musculature of the pulminary vein and cranial vena cava in the gray red-backed vole (*Clethrionomys rufocanus*). Memoirs of the National Science Museum, Tokyo, 27:175-179.

Endo, H., Y. Hayashi, W. Rerkamnuaychoke, N. Nadee, J. Nabhitabhata, Y. Kawamoto, H. Hirai, J. Kimura, T. Nishida, and J. Yamada. 2000. Sympatric distribution of the two morphological types of the common tree shrew in Hat-Yai Districts (south Thailand). Journal of Veterinary Medical Science, 62:759-761.

Endo, H., K. Satoh, J. Cuisin, B. Stafford, and J. Kimura. 2001. Morphological adaptation of the masticatory muscles and related apparatus in Asian and African Rhizomyinae species. Mammal Study, 26:101-108.

Endo, H., A. Hayashida, and T. Ogoh. 2002. Catalogue of *Apodemus* specimens. National Science Museum, Tokyo, 235 pp.

Engel, S. R., K. M. Hogan, J. F. Taylor, and S. K. Davis. 1998. Molecular systematics and paleobiogeography of the South American sigmodontine rodents. Molecular Biology and Evolution, 15:35-49.

Engels, H. 1980. Zur Biometrie und Taxonomie von Hausmäusen (genus *Mus* L.) aus dem Mittelmeergebiet. Zeitschrift für Säugetierkunde, 45:366-375.

Engels, H. 1983*a*. Elektrophoretische Untersuchungen an Hausmaeusen (*Mus musculus brevirostris* Waterhouse, 1837 und *Mus spretus* Lataste, 1883) aus Sued-und Mittelportugal zur Ueberpruefung vermuteter Hybridisierung. Ciencia Biologica, Ecologia e Sistematica (Coimbra), 5:97-104.

Engels, H. 1983*b*. Zur Phylogenie und Ausbreitungsgeschichte mediterraner Hausmäuse (genus *Mus* L.) mit Hilfe von "Compatibility Analysis". Zeitschrift für Säugetierkunde, 48:9-19.

Engesser, B. 1975. Revision der europäischen Heterosoricinae (Insectivora, Mammalia). Eclogae Geologicae Helvetiae, 68:649-671.

Engstrom, M. D. 1984. Chromosomal, genic, and morphological variation in the *Oryzomys melanotis* species group. Unpubl. PhD. dissertation, Texas A & M University, 171 pp.

Engstrom, M. D., and J. W. Bickham. 1982. Chromosome banding and phylogenetics of the golden mouse, *Ochrotomys nuttalli*. Genetica, 59:119-126.

Engstrom, M. D., and J. W. Bickham. 1983. Karyotype of *Nelsonia neotomodon*, with notes on the primitive karyotype of peromyscine rodents. Journal of Mammalogy, 64:685-688.

Engstrom, M. D., and J. R. Choate. 1979. Systematics of the northern grasshopper mouse (*Onychomys leucogaster*) on the central Great Plains. Journal of Mammalogy, 60:723-739.

Engstrom, M. D., and D. E. Wilson. 1981. Systematics of *Antrozous dubiaquercus* (Chiroptera: Vespertilionidae), with comments on the status of *Bauerus* Van Gelder. Annals of Carnegie Museum, 50:371-383.

Engstrom, M. D., R. C. Dowler, D. S. Rogers, D. J. Schmidly, and J. W. Bickham. 1981. Chromosomal variation within four species of harvest mice (*Reithrodontomys*). Journal of Mammalogy, 62:159-164.

Engstrom, M. D., D. J. Schmidly, and P. K. Fox. 1982. Nongeographic variation and discrimination of species within the *Peromyscus leucopus* species group (Mammalia: Cricetinae) in eastern Texas. Texas Journal of Science, 34:149-162.

Engstrom, M. D., H. H. Genoways, and P. K. Tucker. 1987a. Morphological variation, karyology, and systematic relationships of *Heteromys gaumeri* (Rodentia: Heteromyidae). Pp. 289-303, *in* Studies in neotropical mammalogy. Essays in honor of Philip Hershkovitz (B. D. Patterson and R. M. Timm, eds.). Fieldiana: Zoology, n.s., 39:1-506.

Engstrom, M. D., T. E. Lee, and D. E. Wilson. 1987b. *Bauerus dubiaquercus*. Mammalian Species, 282:1-3.

Engstrom, M. D., O. Sanchez-Herrera, and G. Urbano-Vidales. 1992. Distribution, geographic variation, and systematic relationships within *Nelsonia* (Rodentia: Sigmodontinae). Proceedings of the Biological Society of Washington, 105:867-881.

Engstrom, M. D., A. J. Baker, J. L. Eger, R. Boonstra, and R. J. Brooks. 1993. Chromosomal and mitochrondrial DNA variation in four laboratory populations of collared lemmings (*Dicrostonyx*). Canadian Journal of Zoology, 71:42-48.

Epstein, H. 1971. The origin of the domestic animals of Africa. Africana Publishing Corporation, New York. 2 vol., 573 + 719 pp.

Erbajeva, M. A. 1988. Pishchukhi Kainozoya [Pikas of the Cenozoic]. Nauka, Moscow-Leningrad, 223 pp. (in Russian).

Erbajeva, M. A. 1994. Phylogeny and evolution of Ochotonidae with emphasis on Asian Ochotonids. Pp. 1-13, *in* Rodent and Lagomorph Families of Asian Origins and Diversification. (Y. Tomida, C.-k. Li, and T. Setoguchi, eds.). National Science Museum Monographs, No. 8, Tokyo, 195 pp.

Erbajeva, M. A. 1997. The history and systematics of the genus *Ochotona*. Gibier Faune Sauvage, 14:505.

Erdbrink, D. P. 1953. A review of fossil and Recent bears of the old world. Proefschrift, Utrecht, 2 vol, 597 pp.

Ermakov, O. A. 1996. *Citellus major* and *C. pygmaeus* from Volga Region: Their distribution and interrelation. Cand. Sci. (Biol.) Dissertation, Moscow: Mosk. Gos. Univ.

Ernest, K. A. 1986. *Nectomys squamipes*. Mammalian Species, 265:1-5.

Ernest, K. A., and M. A. Mares. 1987. *Spermophilus tereticaudus*. Mammalian Species, 274:1-9.

Ernits, P. 1991a. The garden dormouse (*Eliomys quercinus*) on the island of Suur Tutarsaar. Eesti Loodus, 3:132-136.

Ernits, P. 1991b. Dormice (Gliridae) in Estonia. Eesti Loodus, 3:137-138.

Erridge, N. 1988. Case 275. *Camelus* Linnaeus, 1758 (Mammalia, Artiodactyla): Proposed designation of *Camelus bactrianus* Linnaeus, 1758 as type species. Bulletin of Zoological Nomenclature, 45:141-142.

Erxleben, J. C. P. 1777. Systema regni animalis per classes, ordines, genera, species, varietates, cum synonymia et historia animalium. Classis I. Mammalia. Weygandianis, Lipsiae, 636 pp.

Eschevarria, A. L. 1995. Determinacion de la edad en *Akodon simulator simulator* (Thomas, 1916), (Rodentia: Cricetidae). Acta Zoologica Lilloana, 43:65-72.

Eschricht, D. F., and J. Reinhardt. 1861. Om nordhvalen (*Balaena mysticetus* L.); navning med Hensyn til diens Udbredning i Fortiden of Nutiden of til dens ydre og indre Saekjender. Konigelige Danske Videnskabernes Selskabs Skrifter, 5te Raekke, Naturvidenskabelig og Mathematisk Afdeling, 5te Bind,, pp. [432-435] + 436-591, 6 pls.

Eshelman, B. D., and G. N. Cameron. 1987. *Baiomys taylori*. Mammalian Species, 285:1-7.

Eshelman, B. D., and C. S. Sonnemann. 2000. *Spermophilus armatus*. Mammalian Species, 637:1-6.

Eshelman, R. E. 1975. Geology and paleontology of the Early Pleistocene (Late Blancan) White Rock Fauna from north-central Kansas. University of Michigan, Museum of Paleontology, Papers on Paleontology, 13:1-60.

Espinosa, M. B., and O. A. Reig. 1991. Cytogenetics and karyosystematics of South American oryzomyine rodents (Cricetidae, Sigmodontinae). III. Banding karyotypes of Argentinian *Oligoryzomys*. Zeitschrift für Säugetierkunde, 56:306-317.

Espinosa, M. B., A. Laserre, M. Piantanida, and A. D. Vitullo. 1997. Cytogenetics of vesper mice *Calomys* (Sigmodontinae): A new karyotype from the Puna region and its implication for chromosomal phylogeny. Cellular and Molecular Life Sciences, 53:583-586.

Espinoza M., E., H. Núñez O., P. González D., R. Luna R., M. A. Altamirano, G. O., E. Cruz A., G. Cartas H., and C. Guichard R. 1999a. Listado preliminar de los vertebrados terrestres de la reserva de la biosfera "El Triunfo," Chiapas. Publicaciones Especiales del Instituto de Historia Natural, Tuxtla Gutiérrez, Chiapas, 1:38 pp.

Espinoza M., E., H. Núñez O., P. González D., R. Luna R., D. Navarette, G. E. Cruz A., and C. Guichard R. 1999b. Lista preliminar de los vertebrados terrestres de la selva "El Ocote," Chiapas. Publicaciones Especiales del Instituto de Historia Natural, Tuxtla Gutiérrez, Chiapas, 1:40 pp.

Essop, M. F., E. H. Harley, and I. Baumgarten. 1997a. A molecular phylogeny of some Bovidae based on restriction-site mapping of mitochondrial DNA. Journal of Mammalogy, 78:377-386.

Essop, M. F., N. Mda, J. Flamand, and E. H. Harley. 1997b. Mitochondrial DNA comparisons between the African wild cat, European wild cat and the domestic cat. South African Journal of Wildlife Research, 27:71-72.

Estes, J. A. 1980. *Enhydra lutris*. Mammalian Species, 133:1-8.

Estes, R. D. 1999. Hirola. Generic status supported by behavioral and physiological evidence. S.S.C. Antelope Specialists Group Gnusletter, 18(2):10-11.

Etheridge, R. 1889. Lord Howe Island (general zoology). Australian Museum Memoirs, 2:2-42.

Evans, E. P. 1981. Karyotype of the house mouse. Symposia of the Zoological Society of London, 47:127-139.

Everts, W. 1968. Beitrag zur Systematik des Sonnendachse. Zeitschrift für Säugetierkunde, 33:1-19.

Ewer, R. F. 1957. A collection of *Phacochoerus aethiopicus* teeth from the Kalkbank Middle Stone Age site, central Transvaal. Palaeontologia Africana, 5:5-20.

Ewer, R. F. 1973. The Carnivores. Cornell University Press, Ithaca, NY, 494 pp.

Fa, J. E. 1989. Conservation-motivated analysis of mammalian biogeography in the trans-Mexican neovolcanic belt. National Geographic Research, 5:296-316.

Fa, J. E., and D. J. Bell. 1990. The volcano rabbit *Romerolagus diazi*. Pp. 143-146, *in* Rabbits, hares and pikas (J. A. Chapman and J. E. C. Flux, eds.). I.U.C.N., Gland, Switzerland, 168 pp.

Fadda, C., and M. Corti. 1998. Geographic variation of *Arvicanthis* (Rodentia, Muridae) in the Nile Valley. Zeitschrift für Säugetierkunde, 63:104-113.

Fadda, C., and M. Corti. 2000. Three dimensional geometric morphometric study of the Ethiopian *Myomys-Stenocephalemys* complex (Murinae, Rodentia). Hystrix, n.s., 10(2):131-143.

Fadda, C., and M. Corti. 2001. Three-dimensional geometric morphometrics of *Arvicanthis*: Implications for systematics and taxonomy. Journal of Zoological Systematics and Evolutionary Research, 39:235-245.

Fadda, C., M. Corti, and E. Verheyen. 2001a. Molecular phylogeny of *Myomys/Stenocephalemys* complex and its relationships with related African genera. Biochemical Systematics and Ecology, 29:585-596.

Fadda, C., R. Castiglia, P. Colangelo, M. Corti, R. Machang'u, A. Scanzani, P. Tesha, W. Verheyen, and E. Capanna. 2001b. The rodent fauna of Tanzania: A cytotaxonomic report from the Maasai Steppe (1999). Rendiconti Fisiche Accademia Lincei, ser. 9:29-49.

Fagerstone, K. A. 1982. Ethology and taxonomy of Richardson's ground squirrel (*Spermophilus richardsonii*). Unpubl. Ph. D. dissertation, University of Colorado, Boulder, 298 pp.

Fagundes, V., A. U. Christoff, and Y. Yonenaga-Yassuda. 1998. Extraordinary chromosomal polymorphism with 28 different karyotypes in the Neotropical species *Akodon cursor* (Muridae, Sigmodontinae), one of the smallest diploid number in rodents (2n = 16, 15, and 14). Hereditas, 129:263-274.

Fahlbusch, V. 1966. Cricetidae (Rodentia, Mamm.) aus der millelmiozänen Spaltenfüllung Erkertshofen bei Eichstätt. Mitteilungen der Bayerischen Staatssammlung für Paläontologie und Historische Geologie, 6:109-131.

Fahlbusch, V. 1969. Pliocene and Pleistocene Cricetinae (Rodentia, Mammalia) from Poland. Acta Zoologica Cracoviensia, 14(5):99-138.

Fahlbusch, V. 1985. Origin and evolutionary relationships among geomyoids. Pp. 617-629, *in* Evolutionary Relationships among Rodents: A Multidisciplinary Analysis (W. P. Luckett and J.-L. Hartenberger, eds.). Plenum Press, New York, 721 pp.

Fahlbusch, V. 1992. The Neogene mammalian faunas of Ertemte and Harr Obo in Inner Mongolia (Nei Mongol), China.–10. *Eozapus* (Rodentia). Senckenbergiana Lethaea, 72:199-217.

Fahlbusch, V. 1996. Middle and Late Miocene common cricetids with prismatic teeth. Pp. 216-219, *in* The evolution of Western Eurasian Neogene mammal faunas (R. L. Bernor, V. Fahlbusch, and H.-W. Mittmann, eds.). Columbia University Press, New York, 487 pp.

Fahlbusch, V., and T. Bolliger. 1996. Eomyids and zapodids (Rodentia, Mammalia) in the middle and upper Miocene of Central and Southeastern Europe and the Eastern Mediterranean. Pp. 208-212, *in* The evolution of Western Eurasian Neogene mammal faunas (R. L. Bernor, V. Fahlbusch, and H.-W. Mittmann, eds.). Columbia University Press, New York, 487 pp.

Fahr, J., H. Vierhaus, R. Hutterer, and D. Kock. 2002. A revision of the *Rhinolophus maclaudi* species group with the description of a new species from West Africa. Myotis, 40:95-126.

Fairon, J. 1985. *Myotis daubentoni* en Belgique et solution du problèm "*nathalinae*." Bulletin du Centre de Recherce Cheiropterologique de Belgique, 8:5-38.

Faltin, I. 1988. Untersuchungen zur verbreitung der schlafmäuse (Gliridae) in Bayern. Schriftenreihe Bayerische Landesamt für Umweltschutz, 81:7-15.

Fan Nai-chang, and Shi Yin-zhu. 1982. [A revision of the zokors of subgenus *Eospalax*.] Acta Theriologica Sinica, 2:83-199 (in Chinese).

Fang, J.-S., and G. M. Jagiello. 1992. The unique G-banded mitotic karyotype of the Turkish hamster (*Mesocricetus brandti*). Cytologia, 57:121-126.

Fang, Y.-P., and L.-L. Lee. 2002. Re-evaluation of the Taiwanese white-toothed shrew, *Crocidura tadae* Tokuda and Kano, 1936 (Insectivora: Soricidae) from Taiwan and two offshore islands. Journal of Zoology, 257:145-154.

Fang, Y.-P., L.-L. Lee, F.-H. Yew, and H.-T. Yu. 1997. Systematics of white-toothed shrews (*Crocidura*) (Mammalia: Insectivora: Soricidae) of Taiwan: Karyological and morphological studies. Journal of Zoology, 242:151-166.

Farjanel, G., and P. Mein. 1984. Une association de mammmifères et de pollens dans la formation continentale des "Marnes de Bresse" d'âge Miocène supérieur, à Ambérieu (Ain). Geologie de la France, 1-2:131-148.

Fashing, N. J. 1973. Implications of karyotypic variation in the kangaroo rat, *Dipodomys heermanni*. Journal of Mammalogy, 54:1018-1020.

Faulkes, C. G., N. C. Bennett, M. W. Bruford, H. P. O'Brien, G. H. Aguilar, and J. U. M. Jarvis. 1997. Ecological constraints driving social evolution in the African mole-rats. Proceedings of the Royal Society of London B, 264:1619-1627.

Fay, F. H. 1985. *Odobenus rosmarus*. Mammalian Species, 238:1-7.

Fedorchenko, A. A., and I. V. Zagorodnyuk. 1994. [Mice of the genus *Sylvaemus* from the Lower Danube Region. Communication 2. Distribution and abundance]. Vestnick Zoologii, 1994(4-5):55-64 (in Russian with English abstract).

Fedorov, V. B. 1990. [Allozyme polymorphism in a natural population of wood lemmings (*Myopus schisticolor* Lill.).] Genetika, 26:1324-1328 (in Russian).

Fedorov, V. B. 1993. [Genetic variability of the wood lemming *Myopus schisticolor* based on the combination of isozyme loci.] Ékologiya, 1:70-82 (in Russian with English abstract).

Fedorov, V. B. 1999. Contrasting mitochondrial DNA diversity estimates in two sympatric genera of Arctic lemmings (*Dicrostonyx*: *Lemmus*) indicate different responses to Quaternary environmental fluctuations. Proceedings of the Royal Society of London, ser. B, 266:621-626.

Fedorov, V. B., K. Fredga, and G. H. Jarrell. 1999*a*. Mitochondrial DNA variation and the evolutionary history of chromosome races of collared lemmings (*Dicrostonyx*) in the Eurasian Arctic. Journal of Evolutionary Biology, 12:134-145.

Fedorov, V. B., A. Goropashnaya, G. J. Jarrell, and K. Fredga. 1999*b*. Phylogeographic structure and mitochondrial DNA variation in true lemmings (*Lemmus*) from the Eurasian Arctic. Biological Journal of the Linnean Society, 66:357-371.

Fedyk, S. 1970. Chromosomes of *Microtus* (*Stenocranius*) *gregalis major* (Ognev, 1923) and phylogenetic connections between sub-arctic representatives of the genus *Microtus* Schrank, 1798. Acta Theriologica, 15:143-152.

Feiler, A. 1977. Ueber die intraspezifische Variation des *Phalanger ursinus*. Zoologische Abhandlungen Dresden, 34:187-197.

Feiler, A. 1978*a*. Ueber artliche Abgrenzung und innerartliche Ausformung bei *Phalanger maculatus*. Zoologische Abhandlungen Staatliches Museum für Tierkunde in Dresden, 35:1-30.

Feiler, A. 1978*b*. Zur morphologischen Charakteristik des *Phalanger celebensis*. Zoologische Abhandlungen Staatliches Museum für Tierkunde in Dresden, 35:161-168.

Feiler, A. 1978*c*. Bemerkungen über *Phalanger* der "*orientalis*-Gruppe" nach Tate (1945). Zoologische Abhandlungen Staatliches Museum für Tierkunde in Dresden, 34:385-395.

Feiler, A. 1990. Über die Säugetiere der Sangihe- und Talaud-Inseln—der Beitrag A. B. Meyers für ihre Erforschung (Mammalia). Zoologische Abhandlungen Staatliches Museum für Tierkunde in Dresden, 46:75-94.

Feiler, A. 1998. Das Philippine-Schuppentier, *Manis culionensis* Elera, 1915, eine fast vergessene Art (Mammalia: Pholidota: Manidae). Zoologische Abhandlungen, 50(1):161-164.

Feiler, A., and T. Nadler. 1997. Erstnachweis der Etruskerspitzmaus, *Suncus etruscus* (Savi, 1822) für Vietnam (Mammalia: Insectivora: Soricidae). Faunistische Abhandlungen Staatliches Museum für Tierkunde Dresden, 27:161-162.

Feiler, A., and B. Tegegn. 1998. Zur innerartlichen variation und artabgrenzung bei *Apodemus flavicollis* (Melchior, 1834), *A. sylvaticus* (Linnaeus, 1758), *A. agrarius* (Pallas, 1771) und *A. uralensis* (Pallas, 1811) (Mammalia: Rodentia: Muridae). Zoologische Abhandlungen (Dresden), 50(1):133-141.

Feiler, A., and T. Ziegler. 1999. Zur Kenntnis zweier *Crocidura*-Arten aus dem nördlichen Vietnam: *Crocidura attenuata* Milne-Edwards, 1872 und *Crocidura* spec. (Mammalia: Insectivora: Soricidae). Faunistische Abhandlungen Staatliches Museum für Tierkunde Dresden, 29:377-384.

Feito, R., and M. Gallardo. 1982. Sperm morphology of the Chilean species of *Ctenomys* (Octodontidae). Journal of Mammalogy, 63:658-661.

Fejfar, O. 1972. Ein neuer vertreter der gattung *Anomalomys* Gaillard, 1900 (Rodentia, Mammalia) aus dem Europäischen Miozän (Karpat). Neues Jahrbuch für Geologie und Paläontologie. Abhandlungen,141(2):168-193.

Fejfar, O. 1999a. Microtoid cricetids. Pp. 365-372, *in* The Miocene land mammals of Europe (G. E. Rössner and K. Heissig, eds.). Dr. Friedrich Pfeil, München, 515 pp.

Fejfar, O. 1999b. Subfamily Platacanthomyinae. Pp. 389-394, *in* The Miocene land mammals of Europe (G. E. Rössner and K. Heissig, eds.). Dr. Friedrich Pfeil, München, 515 pp.

Fejfar, O., and C. A. Repenning. 1992. Holarctic dispersal of the arvicolids (Rodentia, Cricetidae). Courier Forschungsinstitut Senckenberg, 153:205-212.

Fejfar, O., and C. A. Repenning. 1998. The ancestors of lemmings (Lemmini, Arvicolinae, Cricetidae, Rodentia) in the early Pliocene of Wolfersheim near Frankfurt am Main; Germany. Senckenbergiana Lethaea, 77:161-193.

Fejfar, O., A. Blasetti, G. Calderoni, M. Coltorti, G. Ficcarelli, F. Masini, L. Rook, and D. Torre. 1993. New finds of cricetids (Mammalia, Rodentia) from the late Pleistocene-Holocene of northern Ecuador. Documents des Laboratoire Géologique de Lyon, 125:151-167.

Fejfar, O., G. Ficcarelli, G. Mezzabotta, M. Moreno Espinosa, L. Rook, and D. Torre. 1996. First record of a copemyne-peromyscine form in South America. Hypotheses of its ancestry in Palearctic. Acta Zoologica Cracoviensia, 39:137-145.

Feldhamer, G. A. 1980. *Cervus nippon*. Mammalian Species, 128:1-7.

Feldhamer, G. A., K. C. Farris-Renner, and C. M. Barker. 1988. *Dama dama*. Mammalian Species, 317:1-8.

Feldhammer, G. A., J. C. Whittaker, and E. M. Charles. 1998. Recent records of the cotton-mouse (*Peromyscus gossypinus*) in Illinois. American Midland Naturalist, 139:178-180.

Feldman, H. W. 1926. Unit character inheritance of color in the black rat, *Mus rattus* L. Genetics, 11:456-465.

Felten, H. 1962. Bemerkungen zu Fledermäusen der Gattungen *Rhinopoma* und *Taphozous* (Mammalia, Chiroptera). Senckenbergiana Biologica, 43:171-176.

Felten, H. 1964a. Zur Taxonomie indo-australischer Fledermäuse der Gattung *Tadarida* (Mammalia, Chiroptera). Senckenbergiana Biologica, 45:1-13.

Felten, H. 1964b. Flughunde der Gattung *Pteropus* von Neukaledonien und den Loyalty-Inseln. Senckenbergiana Biologica, 45:671-683.

Felten, H., and D. Kock. 1972. Weitere Flughunde der Gattung *Pteropus* von den Neuen Hebriden, sowie den Banks-und Torres-Inseln, Pacifischer Ozean (Mammalia: Chiroptera). Senckenbergiana Biologica, 53:179-188.

Felten, H., and G. Storch. 1968. Eine neue Schläfer-Art, *Dryomys laniger* n. sp. aus Kleinasien (Rodentia: Gliridae). Senckenbergiana Biologica, 49(6):429-435.

Felten, H., F. Spitzenberger, and G. Storch. 1971. Zur Kleinsäugerfauna West-Anatoliens. Teil I. Senckenbergiana Biologica, 52:393-424.

Felten, H., F. Spitzenberger, and G. Storch. 1973. Zur Kleinsäugerfauna West-Anatoliens Teil II. Senckenbergiana Biologica, 54:227-290.

Felten, H., F. Spitzenberger, and G. Storch. 1977. Zur Kleinsäugerfauna West-Anatoliens. Teil IIIa. Senckenbergiana Biologica, 58:1-44.

Feng Zuo-jiang [Feng Tso-chien], and Kao Yüeh-ting. 1974. [Taxonomic notes on the Tibetan pika and allied species—including a new subspecies]. Acta Zoologica Sinica, 20:76-88 (in Chinese).

Feng Zuo-jiang, and Zheng Chang-lin. 1985. [Studies on the pikas (genus *Ochotona*) of China—taxonomic notes and distribution]. Acta Theriologica Sinica, 5:269-290 (in Chinese).

Feng Zuo-jiang, Zheng Chang-lin, and Wu Jia-yan. 1983. [A new subspecies of *Apodemus peninsulae* from Qinghai-Xizang (Tibet) Plateau, China]. Acta Zootaxonomica Sinica, 8:108-112 (in Chinese).

Feng Zuo-jiang, Cai Gui-quan, and Zheng Chang-lin. 1986. [The mammals of Xizang. The comprehensive scientific expedition to the Qinghai-Xizang Plateau]. Science Press, Academia Sinica, Beijing, 423 pp. (in Chinese).

Fenton, M. B., and R. M. R. Barclay. 1980. *Myotis lucifugus*. Mammalian Species, 142:1-8.

Fenton, M. B., and J. L. Eger. 2002. *Chaerephon chapini*. Mammalian Species, 692:1-2.

Ferguson, W. W. 1981. The systematic position of *Gazella dorcas* (Artiodactyla: Bovidae) in Israel and Sinai. Mammalia, 45:453-457.

Ferguson, W. W., Y. Porath, and S. Paley. 1985. Late Bronze Period yields first osteological evidence of *Dama dama* (Artiodactyla: Cervidae) from Israel and Arabia. Mammalia, 49:209-214.

Fernandes, C., H. Engels, A. Abade, and M. Coutinho. 1991. Zur genetischen und morphologischen Variabilitat der Gattung *Apodemus* (Muridae) im Westen der Iberischen Halbinsel. Bonner Zoologische Beiträge, 42:261-269.

Fernández-Salvador, R. 1998. Topillo de Cabrera, *Microtus cabrerae* Thomas, 1906. Galemys, 10(2):5-18.

Fernández-Salvador, R. 2002. *Microtus cabrerae* Thomas, 1906. Pp. 386-389, *in* Atlas de los Mamíferos terrestres de España (L. J. Pamomo and J. Gisbert, eds.). Dirección General de Conservación de la Naturaleza-SECEM-SECEMU, 563 pp.

Fernando, P., T. N. C. Vidya, J. Payne, M. Stuewe, G. Davison, R. J. Alfred, P. Andau, E. Bosi, A. Kilbourn, and D. J. Melnick. 2003. DNA Analysis indicates that Asian elephants are native to Borneo and are therefore a high priority for conservation. PLoS Biology, 1(1):110-115 [open access online publication: http://biology.plosjournals.org, DOI: 10.1371/journal.pbio.0000006].

Ferrell, C. S., and D. E. Wilson. 1991. *Platyrrhinus helleri*. Mammalian Species, 373:1-5.

Feustel, H. 1984. Zur Verbreitung der Schläfer (Gliridae) im Odenwald. Bericht Naturwissenschaftlicher Verein Darmstadt, 8:7-18.

Fieldhouse, D., F. Yazdani, and G. Brian Golding. 1997. Substitution rate variation in closely related rodent species. Heredity, 78:21-31.

Fielding, D. C. 1966. The identification of skulls of the two British species of *Apodemus*. Journal of Zoology, London, 150(part 4):498-500.

Filippucci, M. G. 1986. Nuovo stazione appenninica di *Dryomys nitedula* (Pallas, 1779) (Rodentia, Gliridae). Hystrix, 1(1):83-86.

Filippucci, M. G. 1992. Allozyme variation and divergence among European, Middle Eastern, and North African species of the genus *Apodemus* (Rodentia, Muridae). Israel Journal of Zoology, 38:193-218.

Filippucci, M. G. (ed.). 1995. Proceedings of II conference on dormice (Rodentia, Myoxidae). Hystrix, n.s., 6(1-2):1-340.

Filippucci, M. G. 1999. *Eliomys quercinus* (Linnaeus, 1766). Pp. 298-299, *in* The atlas of European mammals (A. J. Mitchell-Jones, G. Amori, W. Bogdanowicz, B. Kryštufek, P. J. H. Reijnders, F. Spitzenberger, M. Stubbe, J. B. M. Thissen, V. Vohralík, and J. Zima, eds.). T. and A. D. Poyser, Ltd., London, 484 pp.

Filippucci, M. G., and E. Capanna. 1996. Allozyme variation and differentiation among chromosomal races and species in the genus Eliomys (Rodentia, Myoxidae). Pp. 259-270, *in* Eurpoean Mammals. Proceedings of the I European Congress of Mammalogy (M. Da Luz Mathias, M. Santos-Reis, G. Amori, R. Libois, A. Mitchell-Jones and M. C. Saint-Girons, eds.). Museu Bocage, Lisboa, 314 pp.

Filippucci, M. G., and T. Kotsakis. 1995. Biochemical systematics and evolution of Myoxidae. Pp. 77-97, *in* Proceedings of II Conference on Dormice (Rodentia, Myoxidae) (M. G. Filippucci, ed.). Hystrix, n.s., 6(1-2):1-340.

Filippucci, M. G., and D. Peshev. 1999. *Myomimus roachi* (Bate, 1937). Pp. 302-303, *in* The atlas of European mammals (A. J. Mitchell-Jones, G. Amori, W. Bogdanowicz, B. Kryštufek, P. J. H. Reijnders, F. Spitzenberger, M. Stubbe, J. B. M. Thissen, V. Vohralík, and J. Zima, eds.). T. and A. D. Poyser, Ltd., London, 484 pp.

Filippucci, M. G., and S. Simson. 1996. Allozyme variation and divergence in Erinaceidae (Mammalia, Insectivora). Israel Journal of Zoology, 42:335-345.

Filippucci, M. G., M. V. Civitelli, and E. Capanna. 1985. Le caryotype du lerotin *Dryomys nitedula* (Pallas) (Rodentia, Gliridae). Mammalia, 49(3):365-368.

Filippucci, M. G., M. V. Civitelli, and E. Capanna. 1986. The chromosomes of *Lemniscomys barbarus* (Rodentia, Muridae). Bollettino di Zoologia, 53:355-358.

Filippucci, M. G., G. Nascetti, E. Capanna, and L. Bullini. 1987. Allozyme variation and systematics of European moles of the genus *Talpa* (Mammalia, Insectivora). Journal of Mammalogy, 68:487-499.

Filippucci, M. G., M. V. Civitelli and E. Capanna. 1988*a*. Evolutionary genetics and systematics of the garden dormouse, *Eliomys* Wagner, 1840. 1. Karyotype divergence. Bollettino di Zoologia, 55:35-45.

Filippucci, M. G., E. Rodino, E. Nevo and E. Capanna. 1988*b*. Evolutionary genetics and systematics of the garden dormouse, *Eliomys* Wagner, 1840. 2. Allozyme diversity and differentiation of chromosomal races. Bollettino di Zoologia, 55:47-54.

Filippucci, M. G., S. Simson, and E. Nevo. 1989. Evolutionary biology of the genus *Apodemus* Kaup, 1829 in Israel. Allozymic and biometric analyses with description of a new species: *Apodemus hermonensis* (Rodentia, Muridae). Bollettino di Zoologia, 56:361-376.

Filippucci, M. G., F. Catzeflis, and E. Capanna. 1990. Evolutionary genetics and systematics of the garden doormouse, *Eliomys* Wagner, 1840 (Gliridae, Mammalia). 3. Further karyological data. Bollettino di Zoologia, 57:149-152.

Filippucci, M. G., V. Fadda, B. Kryštufek, S. Simson, and G. Amori. 1991. Allozyme variation and differentiation in *Chionomys nivalis* (Martins, 1842). Acta Theriologica, 36:47-62.

Filippucci, M. G., H. Burda, E. Nevo, and J. Kocka. 1994. Allozyme divergence and systematics of common mole-rats (*Cryptomys*, Bathyergidae, Rodentia) from Zambia. Zeitschrift für Säugetierkunde, 59(1):42-51.

Filippucci, M. G., B. Kryštufek, S. Simson, C. Kurtonur and B. Özkan. 1995. Allozymic and biometric variation in *Dryomys nitedula* (Pallas, 1778). Pp. 127-140, *in* Proceedings of II Conference on Dormice (Rodentia, Myoxidae) (M. G. Filippucci, ed.). Hystrix, n.s., 6(1-2):1-340.

Filippucci, M. G., G. Storch, and M. Macholán. 1996. Taxonomy of the genus *Sylvaemus* in western Anatolia–morphological and electrophoretic evidence. Senckenbergiana Biologica, 75:1-14.

Filippucci, M. G., M. Kawalika, M. Macholán, A. Scharff, and H. Burda. 1997. Allozyme divergence and systematic relationship of Zambian giant mole-rats, *Cryptomys mechowi* (Bathyergidae, Rodentia). Zeitschrift für Säugetierkunde, 62(3):172-178.

Filippucci, M. G., M. Macholán, and J. R. Michaux. 2002. Genetic variation and evolution in the genus *Apodemus* (Muridae: Rodentia). Biological Journal of the Linnean Society, 75:395-419.

Findley, J. S. 1955*a*. Taxonomy and distribution of some American shrews. University of Kansas, Museum of Natural History, 7:613-618.

Findley, J. S. 1955*b*. Speciation of the wandering shrew. University of Kansas, Museum of Natural History, 9:1-68.

Findley, J. S. 1972. Phenetic relationships among bats of the genus *Myotis*. Systematic Zoology, 21:31-52.

Findley, J. S., and C. Jones. 1960. Geographic variation in the yellow-nosed cotton rat. Journal of Mammalogy, 41:462-469.

Findley, J. S., and C. Jones. 1967. Taxonomic relationships of bats of the species *Myotis fortidens*, *M. lucifugus*, and *M. occultus*. Journal of Mammalogy, 48:429-444.

Findley, J. S., and G. L. Traut. 1970. Geographic variation in *Pipistrellus hesperus*. Journal of Mammalogy, 51:741-765.

Findley, J. S., and T. L. Yates (eds.). 1991. The biology of the Soricidae. Special Publication, The Museum of Southwestern Biology, 1:1-91.

Findley, J. S., A. H. Harris, D. E. Wilson, and C. Jones. 1975. Mammals of New Mexico. University of New Mexico Press, Albuquerque, 360 pp.

Finley, R. B., Jr. 1958. The wood rats of Colorado: Distribution and ecology. University of Kansas Publications, Museum of Natural History, 10:213-552.

Finley, R. B, Jr., and M. A. Bogan. 1995. New records of terrestrial mammals in northwestern Colorado. Proceedings of the Denver Museum of Natural History, 3(10):1-6.

Firestone K. B., M. S. Elphinstone, W. B. Sherwin, and B. A. Houlden. 1999. Phylogeographical population structure of tiger quolls *Dasyurus maculatus* (Dasyuridae: Marsupialia), an endangered carnivorous marsupial. Molecular Ecology, 8:1613-1625.

Fischer, F., M. Gross, and K.-E. Linsenmair. 2002. Updated list of the larger mammals of the Comoe National Park, Ivory Coast. Mammalia, 66(1):83-92.

Fischer [Von Waldheim], G. 1813-1814. Zoognosia tabulis synopticis illustrata. Nicolai Sergeidis Vsevolozsky, Moscow, 3 vols [3-1814].

Fischer, H., and F. Ulbrich. 1968. Chromosomes of the Murrah buffalo and its crossbreed with the Asiatic swamp buffalo (*Bubalus bubalis*). Zeitschrift für Tierzüchtung und Zuchtungsbiologie, 84:110-114.

Fischer, K. 1977. Quartäre Mikromammalia Cubas, vorwiegend aus der Höhle San José de la Lamas, Sante Fé, Provinz Habana. Zeitschrift für geologische Wissenschaften, Berlin, 5:213-255.

Fischer, T. V., and H. W. Mossman. 1969. The fetal membranes of *Pedetes capensis*, and their taxonomic significance. American Journal of Anatomy, 124:89-116.

Fisler, G. F. 1965. Adaptations and speciation in harvest mice of the marshes of San Francisco Bay. University of California Publications in Zoology, 77:1-108.

Fitch, J. H., and K. A. Shump. 1979. *Myotis keenii*. Mammalian Species, 121:1-3.

Fitch, J. H., K. A. Shump, and A. U. Shump. 1981. *Myotis velifer*. Mammalian Species, 149:1-5.

Fitzgerald, C. S., and P. R. Kausman. 2002. *Helarctos malayanus*. Mammalian Species, 696:1-5.

Fitzgerald, J. P., C. A. Meaney, and D. M. Armstrong. 1994. Mammals of Colorado. Denver Museum of Natural History and University Press of Colorado, Denver and Niwot, 467 pp.

Fitzgibbon, C. D., H. Liers, and W. Verheyen. 1995. Distribution, population dynamics and habitat use of the lesser pouched rat, *Beamys hindei*. Journal of Zoology, London, 236:499-512.

Fitzinger, L. J. 1858, 1859. Versuch über die Abstammung des zahmen Pferdes und seiner Racen. Situngsberichte der Kaiserlichen Akademie der Wissenschaften. Mathematisch-Naturwissenschaftliche Classe, 31:131-212, 391-465, 35:273-344.

Fitzinger, L. J. 1860, 1861. Wissenschaftlich-populäre Naturgeschichte der Säugethiere in ihren sämmtlichen Hauptformen. Nebst einer Einleitung in die Naturgeschichte überhaupt un in die Lehre von den Thieren insbesondere. Vienna, Am der Kaiserlich-Königlichen Hof-und Staatsdruckerei, 5 volumes.

Fivush, B., R. Parker, and R. H. Tamarin. 1975. Karyotype of the beach vole, *Microtus breweri*, an endemic island species. Journal of Mammalogy, 56:272-273.

Flagstad, Ø., P. O. Syvertson, N. C. Stenseth, and K. S. Jakobsen. 2001. Environmental change and rates of evolution: The phylogeographic pattern within the hartebeest complex as related to climatic variation. Proceedings of the Royal Society, London B, 268:667-677.

Flaherty, S. P., and W. G. Breed. 1982. Structure of hooks on sperm head of plains mouse *Pseudomys australis*. Micron, 13:343-344.

Flannery, T. F. 1983. Revision of the macropodid subfamily Sthenurinae (Marsupialia: Macropodidae) and the relationships of the species of *Troposodon* and *Lagostrophus*. Australian Mammalogy, 6:15-28.

Flannery, T. F. 1988. *Pogonomys championi* n. sp., a new murid (Rodentia) from montane western Papua New Guinea. Records of the Australian Museum, 40:333-341.

Flannery, T. F. 1989. *Microhydromys musseri* n. sp., a new murid (Mammalia) from the Torricelli Mountains, Papua New Guinea. Proceedings of the Linnean Society of New South Wales, 111:215-222.

Flannery, T. F. 1990a. *Echymipera davidi*, a new species of Perameliformes (Marsupialia) from Kiriwina Island, Papua New Guinea, with notes on the systematics of the genus *Echymipera*. Pp. 29-35, *in* Bandicoots and bilbies (J. H. Seebeck, P. R. Brown, R. L. Wallis, and C. M. Kemper, eds.). Surrey Beatty and Sons Pty. Ltd., Sydney, 392 pp.

Flannery, T. F. 1990b. Mammals of New Guinea. Robert Brown and Associates, 439 pp.

Flannery, T. F. 1990c. The rats of Christmas past. Australian Natural History, 23:394-400.

Flannery, T. F. 1991a. Emperor, King and little pig: The three rats of Guadalcanal. Australian Natural History, 23:635-641.

Flannery, T. F. 1991b. A new species of *Pteralopex* (Chiroptera: Pteropodidae) from montane Guadalcanal, Solomon Islands. Records of the Australian Museum, 43:123-129.

Flannery, T. F. 1992. Taxonomic revision of the *Thylogale brunii* complex (Macropodidae: Marsupialia) in Melanesia, with description of a new species. Australian Mammalogy, 15:7-23.

Flannery, T. F. 1993a. Taxonomy of *Dendrolagus goodfellowi* (Macropodidae: Marsupialia) with description of a new subspecies. Records of the Australian Museum, 45:33-42.

Flannery, T. F. 1993b. Revision of the genus *Melonycteris* (Pteropodidae: Mammalia). Records of the Australian Museum, 45:59-80.

Flannery, T. F. 1994a. Possums of the world: A monograph of the Phalangeroidea. Geo productions, Sydney, 240 pp.

Flannery, T. F. 1994b. Systematic revision of *Emballonura furax* Thomas, 1911 and *E. dianae* Hill, 1956 (Chiroptera: Emballonuridae), with description of new species and subspecies. Mammalia, 58:601-612.

Flannery, T. F. 1995a. Mammals of New Guinea. Revised and updated edition. Comstock/Cornell, Ithaca, NY, 568 pp.

Flannery, T. F. 1995b. Mammals of the South-west Pacific and Moluccan Islands. Comstock/Cornell, Ithaca, NY, 464 pp.

Flannery, T. F. 1995c. Northern hopping-mouse, *Notomys aquilo*. Pp. 572-573, *in* Mammals of Australia (R. Strahan, ed.). Smithsonian Institution Press, Washington, D.C., 756 pp.

Flannery, T. F. 1995d. Great hopping-mouse, *Notomys* sp. P. 582, *in* Mammals of Australia (R. Strahan, ed.). Smithsonian Institution Press, Washington, D.C., 756 pp.

Flannery, T. F. 1995e. Basalt Plains mouse, *Pseudomys* sp. P. 619, *in* Mammals of Australia (R. Strahan, ed.). Smithsonian Institution Press, Washington, D.C., 756 pp.

Flannery, T. F., and Boeadi. 1995. Systematic revision within the *Phalanger ornatus* complex (Phalangeridae: Marsupialia), with description of a new species and subspecies. Australian Mammalogy, 18:35-44.

Flannery, T. F., and D. J. Colgan. 1993. A new species and two new subspecies of *Hipposideros* (Chiroptera) from Western Papua New Guinea. Records of the Australian Museum, 45:43-57.

Flannery, T. F., and C. P. Groves. 1998. A revision of the genus *Zaglossus* (Monotremata, Tachyglossidae), with description of new species and subspecies. Mammalia, 62:367-396.

Flannery, T. F., and L. Seri. 1990. The mammals of southern West Sepik Province, Papua New Guinea: Their distribution, abundance, human use and zoogeography. Records of the Australian Museum, 42:173-208.

Flannery, T. F., and L. Seri. 1993. Rediscovery of *Aproteles bulmerae* (Chiroptera: Pteropodidae). Morphology, ecology and conservation. Mammalia, 57(1):19-25.

Flannery, T. F., and J. P. White. 1991. Animal translocation. Zoogeography of New Ireland mammals. National Geographic Research and Exploration, 7:96-113.

Flannery, T. F., and S. Wickler. 1990. Quaternary murids (Rodentia: Muridae) from Buka Island, Papua New Guinea, with descriptions of two new species. Australian Mammalogy, 13:127-139.

Flannery, T. F., S. van Dyck, and M. Krogh. 1985. Notes on the distribution, abundance, diet and habitat of the New Guinea murid (Rodentia) *Xenuromys barbatus* (Milne-Edwards, 1900). Australian Mammalogy, 8:111-115.

Flannery, T. F., M. Archer, and G. Maynes. 1987. The phylogenetic relationships of living phalangerids with a suggested new taxonomy. Pp. 477-506, *in* Possums and opossums, studies in evolution (M. Archer, ed.). Royal Zoological Society of New South Wales, 1:1-460; 2:461-788.

Flannery, T. F., P. V. Kirch, J. Specht, and M. Spriggs. 1988. Holocene mammal faunas from archaeological sites in island Melanesia. Archaeology in Oceania, 23:89-94.

Flannery, T. F., K. Aplin, C. P. Groves, and M. Adams. 1989. Revision of the New Guinean genus *Mallomys* (Muridae: Rodentia), with descriptions of two new species from subalpine habitats. Records of the Australian Museum, 41:83-105.

Flannery, T. F., D. Colgan, and J. Trimble. 1994. A new species of *Melomys* from Manus Island, Papua New Guinea, with notes on the systematics of the *M. rufescens* complex (Muridae: Rodentia). Proceedings of the Linnean Society of New South Wales, 114:(1):29-43.

Flannery, T. F., P. Bellwood, P. White, A. Moore, Boeadi, and G. Nitihaminoto. 1995a. Fossil marsupials (Macropodidae, Peroryctidae) and other mammals of Holocene age from Halmahera, North Moluccas, Indonesia. Alcheringa, 19:17-25.

Flannery, T. F., Boeadi, and A. L. Szalay. 1995b. A new tree-kangaroo (*Dendrolagus: Marsupialia*) from Irian Jaya, Indonesia, with notes on ethnography and the evolution of tree-kangaroos. Mammalia, 59:65-84.

Flannery, T. F., P. Bellwood, J. P. White, T. Ennis, G. Irwin, K. Schubert, and S. Balasubramaniam. 1998. Mammals from Holocene archaeological deposits on Gebe and Morotai Islands, northern Moluccas, Indonesia. Australian Mammalogy, 20:391-400.

Flegr, J., J. Lukás, and T. Tsukamoto. 1994. Sequence of mitochrondrial D-loop in laboratory and wild strains of mice of genus *Mus*. Acta Societatis Zoologicae Bohemieae, 58:33-38.

Fleharty, E. D. 1960. The status of the gray-necked chipmunk in New Mexico. Journal of Mammalogy, 41:235-242.

Fleming, M. R. 1995a. Common rock-rat, *Zyzomys argurus*. Pp. 620-621, *in* Mammals of Australia (R. Strahan, ed.). Smithsonian Institution Press, Washington, D.C., 756 pp.

Fleming, M. R. 1995b. Arnhem Land rock-rat, *Zyzomys maini*. Pp. 621-622, *in* Mammals of Australia (R. Strahan, ed.). Smithsonian Institution Press, Washington, D.C., 756 pp.

Fleming, M. R., and N. L. McKenzie. 1995. Kimberley rock-rat, *Zyzomys woodwardi*. Pp. 626-627, *in* Mammals of Australia (R. Strahan, ed.). Smithsonian Institution Press, Washington, D.C., 756 pp.

Flerov, C. C. 1931. On the generic characters of the Fam. *Tragulidae* (*Mamm.*, *Artiodactyla*). Doklady Akademii Nauk SSSR, 1931:75-79.

Flerov, C. C. 1960. Fauna of the USSR. Mammals. Vol. I, no. 2. Musk deer and deer [A translation of C. C. Flerov, 1952, Fauna SSSR. Mlekopitayushchie. tom. 1, vyp. 2. Kabargi i oleni]. Israel Program for Scientific Translations, Washington, D.C., 257 pp.

Flerov, C. C. 1979. Sistematika i evolyutsiya [Systematics and evolution]. Pp. 9-127, *in* Zubr: Morfologiya, sistematika, evolyutsiya, ekologiya [European bison: Morphology, systematics, evolution, ecology] (V. E. Sokolov, ed.). Nauka, Moscow, 495 pp. (in Russian).

Flower, S. S. 1932. Notes on the recent mammals of Egypt, with a list of the species recorded from that kingdom. Proceedings of the Zoological Society of London, 1932:368-450.

Flower, W. H. 1869. On the value of the characters of the base of the cranium in the classification of the order Carnivora, and on the systematic position of *Bassaris* and other disputed forms. Proceedings of the Zoological Society of London, 1869:4-37.

Flower, W. H., and J. G. Garson. 1884. Catalogue of the specimens illustrating the osteology and dentition of vertebrated animals Recent and extinct contained in the museum of the Royal College of Surgeons of England. Part II. Class Mammalia other than man. Printed for the College, London, 779 pp.

Flux, J. E. C. 1983. Introduction to taxonomic problems in hares. Acta Zoologica Fennica, 174:7-10.

Flux, J. E. C. 1990. The Sumatran rabbit *Nesolagus netscheri*. Pp. 137-139, *in* Rabbits, hares and pikas (J. A. Chapman and J. E. C. Flux, eds.). I.U.C.N., Gland, Switzerland, 168 pp.

Flux, J. E. C., and R. Angermann. 1990. The hares and jackrabbits. Pp. 61-94, *in* Rabbits, hares and pikas (J. A. Chapman and J. E. C. Flux, eds.). I.U.C.N., Gland, Switzerland, 168 pp.

Flux, J. E. C., A. G. Duthie, T. J. Robinson, and J. A. Chapman. 1990. Exotic populations. Pp. 147-153, *in* Rabbits, hares and pikas (J. A. Chapman and J. E. C. Flux, eds.). I.U.C.N., Gland, Switzerland, 168 pp.

Flynn, J. J., and M. A. Nedbal. 1998. Phylogeny of the Carnivora (Mammalia): Congruence vs incompatibility among multiple data sets. Molecular Phylogenetics and Evolution, 9:414-426.

Flynn, J. J., M. A. Nedbal, J. W. Dragoo, and R. L. Honeycutt. 2000. Whence the red panda? Molecular Phylogenetics and Evolution, 17:190-199.

Flynn, L. J. 1982a. Systematic revision of Siwalik Rhizomyidae (Rodentia). Geobios, 15(3):327-389.

Flynn, L. J. 1982b. A revision of fossil rhizomyid rodents from northern India and their correlation to a rhizomyid biochronology of Pakistan. Geobios, 14(4):583-588.

Flynn, L. J. 1982c. Variability of incisor enamel microstructure within *Gerbillus*. Journal of Mammalogy, 63:162-165.

Flynn, L. J. 1990. The natural history of rhizomyid rodents. Pp. 155-183, *in* Evolution of subterranean mammals at the organismal and molecular levels (E. Nevo and O. A. Reig, eds.). Alan R. Liss, Inc., New York, 422 pp.

Flynn, L. J. 1993. A new bamboo rat from the late Miocene of Yushe Basin. Vertebrata PalAsiatica, 31(2):95-101 (in Chinese and English).

Flynn, L. J. 2003. Small mammal indicators of forest paleoenvironment in the Siwalik depostis of the Potwar Plateau, Pakistan. Pp. 183-196, *in* Distribution and Migration of Tertiary Mammals in Eurasia. A Volume in Honour of Hans de Bruijn (J. W. F. Reumer and W. Wessels, eds.). Deinsea, 10:580 pp.

Flynn, L. J., and L. L. Jacobs. 1999. Late Miocene small-mammal faunal dynamics: The crossroads of the Arabian Peninsula. Pp. 410-419, *in* Fossil vertebrates of Arabia with emphasis on the Late Miocene faunas, geology, and palaeoenvironments of the Emirate of Abu Dhabi, United Arab Emirates (P. J. Whybrow and A. Hill, eds.). Yale University Press, New Haven and London, 523 pp.

Flynn, L. J., and M. Sabatier. 1984. A muroid rodent of Asian affinity from the Miocene of Kenya. Journal of Vertebrate Paleontology, 3:160-165.

Flynn, L. J., L. L. Jacobs, and E. H. Lindsay. 1985. Problems in muroid phylogeny: Relationship to other rodents and origin of major groups. Pp. 589-616, *in* Evolutionary relationships among rodents: A multidisciplinary analysis (W. P. Luckett and J.-L. Hartenberger, eds.). Plenum Press, New York, 721 pp.

Flynn, L. J., L. L. Jacobs, and I. U. Cheema. 1986. Baluchimyinae, a new ctenodactyloid rodent subfamily from the Miocene of Balucistan. American Museum Novitates, 2841:1-58.

Flynn, L. J., E. Nevo, and G. Heth. 1987. Incisor enamel microstructure in blind mole rats: Adaptive and phylogenetic significance. Journal of Mammalogy, 68(3):500-507.

Flynn, L. J., N. A. Neff, and R. H. Tedford. 1988. Phylogeny of the Carnivora, Pp. 73-115, *in* The phylogeny and classification of the tetrapods (M. J. Benton, ed.). Clarendon Press, Oxford, 2(Mammals):1-329.

Flynn, L. J., A. J. Winkler, L. L. Jacobs, and W. Downs. 2003. Tedford's gerbils from Afghanistan. Pp. 603-624, *in* Vertebrate fossils and their context: Contributions in honor of Richard H. Tedford (L. J. Flynn, ed.). Bulletin of the American Museum of Natural History, 279, 666 pp.

Fokanov, V. A. 1966. Novyi podvid surka-baibaka i zamechaniya o geograficheskoi izmenchivosti Marmota bobac Müll. [A new subspecies of bobak marmot, and notes on geographical variability of *Marmota bobac* Müll.]. Zoologicheskii Zhurnal, 45:1862-1866 (in Russian).

Fokin, I. M. 1971. [Comparative anatomy of the muscles of the pelvic appendage of the genera *Sicista* and *Salpingotus* (on the position of the subfamily Cardiocraniinae in the system of the Dipodidae)]. Trudy Zoologicheskovo Instituta, Akademiya Nauk SSSR, Leningrad, 48:181-197 (in Russian).

Fons, R., S. Mas-Coma, M. T. Galan-Puchades, M. A. Valero, and F. Moutou. 1994. Parasitological suggestions on the evolution and systematics of *Suncus* and other genera of Soricidae (Mammalia: Insectivora). Zoologische Jahrbücher, Abteilung Systematik, 121:335-344.

Fonseca, G. A. B., G. Herrmann, Y. L. R. Leite, R. A. Mittermeier, A. B. Rylands, and J. L. Patton. 1996. Lista anotada dos mamíferos do Brasil. Occasional Papers in Conservation Biology, 4:38 pp.

Fooden, J. 1964. Rhesus and crab-eating macaques: Intergradation in Thailand. Science, 143:363-365.

Fooden, J. 1967. Identification of the stump-tailed monkey, *Macaca speciosa* I. Geoffroy, 1826. Folia Primatologica, 5:153-164.

Fooden, J. 1969. Taxonomy and evolution of the monkeys of Celebes (Primates: Cercopithecidae). Bibliotheca Primatologica, 10:1-148.

Fooden, J. 1971. Report on the primates collected in western Thailand January-April, 1967. Fieldiana: Zoology, 59:1-62.

Fooden, J. 1975. Taxonomy and evolution of liontail and pigtail macaques (Primates: Cercopithecidae). Fieldiana: Zoology, 67:1-169.

Fooden, J. 1976. Provisional classification and key to living species of macaques. Folia Primatologia, 25:225-236.

Fooden, J. 1979. Taxonomy and evolution of the *sinica* group of macaques: 1. Species and subspecies accounts of *Macaca sinica*. Primates, 20(1):109-140.

Fooden, J. 1980. Classification and distribution of living macaques (*Macaca* Lacépède, 1799). Pp. 1-9, *in* The macaques (D. G. Lindburg, ed.). Van Nostrand Reinhold Company, New York, NY, 384 pp.

Fooden, J. 1981. Taxonomy and evolution of the *sinica* group of macaques. 2. Species and subspecies accounts of the Indian Bonnet Macaque, *Macaca radiata*. Fieldiana: Zoology, n.s., 9:1-52.

Fooden, J. 1982. Taxonomy and evolution of the *sinica* group of macaques: 3. Species and subspecies accounts of *Macaca assamensis*. Fieldiana: Zoology, n.s., 10:1-52.

Fooden, J. 1983. Taxonomy and evolution of the *sinica* group of macaques: 4. Species account of *Macaca thibetana*. Fieldiana: Zoology, n.s., 17:1-20.

Fooden, J. 1987. Type locality of *Hylobates concolor lecogenys*. American Journal of Primatology, 12:107-110.

Fooden, J. 1995. Systematic review of Southeast Asian longtail macaques, *Macaca fascicularis* (Raffles, 1821). Fieldiana: Zoology, n.s., 81:i-vi, 1-206.

Fooden, J. 2000. Systematic review of the rhesus macaque, *Macaca mulatta* (Zimmermann, 1780). Fieldiana: Zoology, n.s., 96:I-v, 1-180.

Fooden, J., A. Mahabal, and S. Sekhar Saha. 1981. Redefinition of rhesus macaque-bonnet macaque boundary in peninsular India (Primates: *Macaca mulatta*, *M. radiata*). Journal of the Bombay Natural History Society, 78(3):463-474.

Fooden, J., Quan Guo-qiang, Wang Zong-ren, and Wang Ying-xiang. 1985. The stumptail macaques of China. American Journal of Primatology, 8(1):11-30.

Foppen, R. P. B., and P. J. M. Bergers. 1992. Garden dormouse *Eliomys quercinus* (L., 1766). Koninklijke Nederlandse Natuurhistorische Vereniging, 56:311-314.

Foppen, R. P. B., P. J. M. Bergers, and J. J. van Gelder. 1989. Occurrrence and distribution of the garden dormouse *Eliomys quercinus* in the Netherlands. Lutra, 32(1):42-52.

Ford, L. S., and R. S. Hoffmann. 1988. *Potos flavus*. Mammalian Species, 321:1-9.

Ford, S. M. 1994. Taxonomy and distribution of the owl monkey. Pp. 1-57, *in* Aotus: The Owl Monkey, (J. F. Baer, I. Kakoma, and R. E. Weller, eds.). Academic Press, San Diego. 380 pp.

Fordyce, R. E. 1989. Origins and evolution of Antarctic marine mammals. Pp. 269-281, *in* Origin and evolution of the Antarctic Biota (J. A. Crame, ed.). Geological Society Special Publication, 47:1-322.

Formozov, N. A. 1997. Pikas (*Ochotona*) of the world: Systematics and conservation. Gibier Faune Sauvage, 14:506-507.

Formozov, N. A., and I. Yu. Baklushinskaya. 1999. O vidovom statuse Khenteiskoi pishchukhi (*Ochotona hoffmanni* Formozov et al., 1996) i vnesenii ee v sostav fauny Rossii. [On the species status of the Khentei pika (*Ochotona hoffmanni* Formozov et al., 1996) and its inclusion in the Russian faunal list]. Byulleten' Moskovskovo Obshchestva Ispytalelei Prirodym, Otdel Biologicheskii, 104:68-72. (in Russian with English summary).

Formozov, N. A., E. L. Yakhontov, and P. P. Dmitriev. 1996. Novaya forma Altaiskoi pishchukhi (*Ochotona alpina hoffmanni* ssp. n.) iz yuzhnykh otrogov Khenteya i veroyatnaya istoriya areala etovo vida [A new form of Alpine pika from the southern spur of the Khentei Ridge and probable historical distribution of this species]. Byulleten' Moskovskovo Obshchestva Ispytatelei Prirodym, Otdel Biologicheskii, 101:28-36. (in Russian with English summary).

Formozov, N. A., I. Yu. Baklushinskaya, and Ma Y. 2004. Taxonomic status of the Helan-shan pika (*Ochotona argentata*) from the Helan-shan Ridge (Ningxia, China). Zoologicheskii Zhurnal, 83:995–1007.

Forsten, A. 1988. The small caballoid horse of the upper Pleistocene and Holocene. Journal of Animal Breeding and Genetics, 105:161-176.

Forsten, A., and P. M. Youngman. 1982. *Hydrodamalis gigas*. Mammalian Species, 165:1-3.

Forster, G. 1780. Herrn von Buffon's Naturgeschichte der vierfussigen Thiere. Mit vermehrungen, aus dem französischen übersetzt [vol. 6 in the 23 vol. work of Buffon's 1772-1801]. J. Pauli, Berlin, vol. 6 [publ. in 1780].

Forys, E. A., and R. D. Dueser. 1993. Inter-island movements of rice rats (*Oryzomys palustris*). American Midland Naturalist, 130:408-412.

Foster, H. L., J. D. Small, and J. G. Fox (eds.). 1981. The mouse in biomedical research. Volume I. History, genetics and wild mice. American College of Laboratory Animal Medicine Series, Academic Press, New York, 306 pp.

Foster, J. B., and R. L. Peterson. 1961. Age variation in *Phenacomys*. Journal of Mammalogy, 42:44-53.

Foster-Turley, P., S. Macdonald, and C. Mason (eds.). 1990. Otters: An action plan for their conservation. I.U.C.N., Gland, Switzerland, 126 pp.

Fox, B. J. 1995*a*. Eastern chestnut mouse, *Pseudomys gracilicaudatus*. Pp. 601-603, *in* Mammals of Australia (R. Strahan, ed.). Smithsonian Institution Press, Washington, D.C., 756 pp.

Fox, B. J. 1995*b*. Pilliga mouse, *Pseudomys pilligaensis*. Pp. 616-617, *in* Mammals of Australia (R. Strahan, ed.). Smithsonian Institution Press, Washington, D.C., 756 pp.

Fox, B. J., and D. A. Briscoe. 1980. *Pseudomys pilligaensis*, a new species of murid rodent from the Pilliga Scrub, Northern New South Wales. Australian Mammalogy, 3:109-126.

Fraguedakis-Tsolis, S. E. 1992. Contribution to the study of the wild house mouse, genus *Mus* L. (Mammalia, Rodentia, Muridae) in Greece. Zeitschrift für Säugetierkunde, 57:225-230.

Fraguedakis-Tsolis, S. E., B. P. Chondropoulos, J. J. Lykakis, and J. C. Ondrias. 1983. Taxonomic problems of woodmice, *Apodemus* ssp., of Greece approached by electrophoretic and immunological methods. Mammalia, 47:333-337.

Fraguedakis-Tsolis, S. E., B. P. Chondropoulos, and N. P. Nikoletopoulos. 1993. On the phylogeny of the genus *Acomys* (Mammalia: Rodentia). Zeitschrift für Säugetierkunde, 58:240-243.

Frahnert, S. 1999. Morphology and evolution of the Glires rostral cranium. Mitteilungen aus dem Museum fur Naturkunde in Berlin, Zoologische Reihe, 75:229-246.

Francis, C. M. 1997. First record for peninsular Malaysia of the gilded tube-nosed Bat *Murina rozendaali*. Malayan Nature Journal, 50:359-362.

Francis, C. M., and J. E. Hill. 1986. A review of the Bornean *Pipistrellus*. Mammalia, 50:43-55.

Francis, C. M., and J. E. Hill. 1998. New records and a new species of *Myotis* (Chiroptera, Vespertilionidae) from Malaysia. Mammalia, 62:241-252.

Francis, C. M., D. Kock, and J. Habersetzer. 1999*a*. Sibling species of *Hipposideros ridleyi* (Mammalia, Chiroptera, Hipposideridae). Senkenbergiana Biologica, 79:255-270.

Francis, C. M., A. Guillén, and M. F. Robinson. 1999*b*. Order Chiroptera; bats. Pp. 225-235, *in* Wildlife in Lao PDR; 1999 status report (J. W. Duckorth, R. E. Salter, and K. Khounble, eds.). I.U.C.N.—The World Conservation Union / Wildlife Conservation Society / Centre for Protected Areas and Watershed Management Areas. Vientiane, 275 pp.

Franco, D. 1988. Habit, biometry and taxonomy of the dormouse (*Glis glis*, Linnaeus, 1776) of the Asiago Plateau. Lavori, Societa Veneziana di Scienze Naturali, 13:135-142.

Frank, P. A. 1997. First record of *Molossus molossus tropidorhynchus* Gray (1839) from the United States. Journal of Mammalogy, 78:103-105.

Franzmann, A. W. 1981. *Alces alces*. Mammalian Species, 154:1-7.

Frase, B., and R. S. Hoffmann. 1980. *Marmota flaviventris*. Mammalian Species, 135:1-8.

Fraser, F. C. 1966. Comments on the Delphinoidea. Pp. 7-31, *in* Whales, dolphins and porpoises (K. S. Norris, ed.). University of California Press, Berkeley, 789 pp.

Fraser, F. C., and P. E. Purves. 1960. Hearing in cetaceans. Bulletin of the British Museum (Natural History), 7(1):1-140.

Frazier, M. K. 1977. New records of *Neofiber leonardi* (Rodentia: Cricetidae) and the paleoecology of the genus. Journal of Mammalogy, 58:368-373.

Fredga, G. 1966. Chromosome studies in five species of South American rodents (suborder Hystricomorpha). Mammalian Chromosome Newsletter, 20:45-46.

Fredga, K. 1972. Chromosome studies in mongooses (Carnivora, Viverridae). Thesis summary, faculty of sciences of the University of Lund, 11 pp.

Fredga, K. 1995. Editorial comment on the nomenclature of the East European vole. Hereditas, 123:95.

Fredga, K., A. Gropp, H. Winking, and F. Frank. 1976. Fertile XX- and XY-type females in the wood lemming, *Myopus schisticolor*. Nature, 261:225-227.

Fredga, K., A. Gropp, H. Winking, and F. Frank. 1977. A hypothesis explaining the exceptional sex ratio in the wood lemming (*Myopus schisticolor*). Hereditas, 85:101-104.

Fredga, K., M. Jaarola, R. A. Ims, H. Steen, and N. G. Yoccoz. 1990. The 'common vole' in Svalbard identified as *Microtus epiroticus* by chromosome analysis. Polar Research, 8:283-290.

Fredga, K., R. Fredriksson, S. Bondrup-Nielsen, and R. A. Ims. 1993. Sex ratio, chromosomes and isozymes in natural populations of the wood lemming, *Myopus schisticolor*. Pp. 465-491, *in* The biology of lemmings (N. C. Stenseth and R. A. Ims, eds.). Linnean Society Symposium Series 15, Academic Press, London, 683 pp.

Fredga, K., D. Bilton, and J. B. Searle. 1995. Colonization history of the pygmy shrew, *Sorex minutus*, in Britain and Scandinavia revealed by genetic markers. P. 38, *in* Abstracts of Oral and Poster Papers, 2nd European Congress of Mammalogy, 27 March-1 April 1995 (J. Gurnell, ed.). Southampton University, England.

Freeman, P. W. 1981. A multivariate study of the family Molossidae (Mammalia: Chiroptera): Morphology, ecology, evolution. Fieldiana: Zoology, n.s., 7:1-173.

Freitas, T. R. O. 1994. Geographic variation of heterochromatin in *Ctenomys flamarioni* (Rodentia-Octodontidae) and its cytogenetic relationships with other species of the genus. Cytogenetics and Cell Genetics, 67:193-198.

Freitas, T. R. O. 1995. Geographic distribution and conservation of four species of the genus *Ctenomys* in Southern Brazil. Studies on Neotropical Fauna and Environment, 30(1):53-59.

Freitas, T. R. O. 1997. Chromosome polymorphism in *Ctenomys minutus* (Rodentia-Octodontidae). Brazilian Journal of Genetics, 20(1):1-7.

Freitas, T. R. O. 2001. Tuco-tuco (Rodentia, Octodontidae) in Southern Brazil: *Ctenomys lami* spec. nov. separated from *C. minutus* Nehring 1887. Studies on Neotropical Fauna and Environment, 36(1):1-8.

Freitas, T. R. O., and E. P. Lessa. 1984. Cytogenetics and morphology of *Ctenomys torquatus* (Rodentia: Octodontidae). Journal of Mammalogy, 65:637-542.

Freitas, T. R. O., M. S. Mattevi, L. F. B. Oliveira, M. J. Souza, Y. Yonenaga-Yassuda, and F. M. Salz. 1983*a*. Chromosome relationships in three representatives of the genus *Holochilus* (Rodentia, Cricetidae) from Brazil. Genetica, 61:13-20.

Freitas, T. R. O., M. Mattevi, and L. F. B. Oliveira. 1983*b*. G- and C-banded karyotype of *Reithrodon auritus* from Brazil. Journal of Mammalogy, 64:318-321.

French, T. W. 1980. *Sorex longirostris*. Mammalian Species, 143:1-3.

Freudenthal, M. 1976. Rodent stratigraphy of some Miocene fissure fillings in Gargano (prov. Foggia, Italy). Scripta Geologica, 37:1-23.

Freudenthal, M., and E. Martín Suárez. 1999. Family Muridae. Pp. 401-409, *in* The Miocene land mammals of Europe (G. E. Rössner and K. Heissig, eds.). Dr. Friedrich Pfeil, München, 515 pp.

Freudenthal, M., J. I. Lacomba, and A. Sacristan. 1992. Classification of European Oligocene Cricetids. Revista Española de Paleontología, Numero Extraordinario, 1992:49-57.

Frey, J. K. 1999. Mogollon vole | *Microtus mogollonensis*. Pp. 634-635, *in* The Smithsonian book of North American mammals (D. E. Wilson and S. Ruff, eds.). Smithsonian Institution Press, Washington, D.C., xxv + 750 pp.

Frey, J. K., and C. T. LaRue. 1993. Notes on the distribution of the Mogollon vole (*Microtus mogollonensis*) in New Mexico and Arizona. Southwestern Naturalist, 38:176-178.

Frey, J. K., and D. W. Moore. 1990a. Nongeographic morphologic variation in the Mexican vole (*Microtus mexicanus*). Transactions of the Kansas Academy of Science, 93:97-109.

Frey, J. K., and D. W. Moore. 1990b. Range expansion of the meadow vole in Kansas. Prairie Naturalist, 22:259-263.

Frey, J. K., J. H. Fraga, and F. C. Bermudez. 1995. A new locality of the montane vole (*Microtus montanus arizonensis*) in New Mexico. Southwestern Naturalist, 40:421-422.

Frey, J. K., R. D. Fisher, M. A. Bogan, and C. Jones. 2002. First record of the Arizona cotton rat (*Sigmodon arizonae*) in New Mexico. Southwestern Naturalist, 47:491-493.

Freye, H. 1978. *Castor fiber* Linnaeus, 1758.—Europäischer Biber. Pp. 184-200, *in* Handbuch der Säugetiere Europas (J. Niethammer and F. Krapp, eds.). Akademische Verlagsgesellschaft (Wiesbaden), 1:1-476.

Frías, A. I., V. Berovides, and C. Fernandez. 1988. Current status of jutiita, *Capromys sanfelipensis*. Doñana, Acta Vertebrata, 15(2):252-254.

Frick, C. 1914. A new genus and some new species and subspecies of Abyssinian rodents. Annals of Carnegie Museum, 9:7-28.

Friend, G. 1991. Encounters with Nuunjala. Wildlife Australia, summer:9-11.

Friend, G. R., and J. H. Calaby. 1995. Black-footed Tree-rat, *Mesembriomys gouldii*. Pp. 565-566, *in* Mammals of Australia (R. Strahan, ed.). Smithsonian Institution Press, Washington, D.C., 756 pp.

Friend, J. A., and N. D. Thomas. 1990. The water-rat, *Hydromys chrysogaster* (Muridae) on Dorre Island, W.A. Western Australian Naturalist, 18:92-93.

Fries, C., Jr., C. W. Hibbard, and D. H. Dunkle. 1955. Early Cenozoic vertebrates in the Red Conglomerate at Guanajuato, Mexico. Smithsonian Miscellaneous Collections, 123(7):1-25.

Frisch, J. L. 1774 [1775]. Das Natursystem der vierfüssigen Thiere in Tabellen, zum Nutzen der erwachsenen Schuljugend [ICZN Opinion 258. Rejected for nomenclatural purposes]. Glogau.

Fritzell, E. K., and K. J. Haroldson. 1982. *Urocyon cinereoargenteus*. Mammalian Species, 189:1-8.

Froehlich, J. W., and P. H. Froehlich. 1987. The status of Panama's endemic howling monkeys. Primate Conservation, 8:58-66.

Froehlich, J. W., J. Supriatna, V. Hart, S. Akbar and R. Babo. 1998. The balan of Balantak: A possible new species of macaque in Central Sulawesi. Tropical Biodiversity, 5:167-184.

Frost, D. R., and R. M. Timm. 1992. Phylogeny of plecotine bats (Chiroptera: "Vespertilionidae"): Summary of the evidence and proposal of a logically consistent taxonomy. American Museum Novitates, 3034:1-16.

Frost, D. R., W. C. Wozencraft, and R. S. Hoffmann. 1991. Phylogenetic relationships of hedgehogs and gymnures (Mammalia: Insectivora: Erinaceidae). Smithsonian Contributions to Zoology, 518:1-69.

Fry, J. K., R. D. Fisher, and L. A. Ruedas. 1997. Identification and restriction of the type locality of the Manzano Mountains cottontail, *Sylvilagus cognatus* Nelson, 1907 (Mammalia:Lagomorpha: Leporidae). Proceedings of the Biological Society of Washington, 110:329-331.

Frye, M. S., and S. B. Hedges. 1995. Monophyly of the order Rodentia inferred from mitochondrial DNA sequences of the genes for 12S rRNA, 16S rRNA, and tRNA-Valine. Molecular Biology and Evolution, 12:168-176.

Frynta, D., and J. Sádlová. 1998. Hybridisation experiments in spin-mice from the eastern Mediterranean: Systematic position of *Acomys cilicicus* revisited. Zeitschrift für Säugetierkunde, 63(Sonderheft):18.

Frynta, D., P. Mikulová, E. Suchomelová, and J. Sádlova. 2001. Discriminant analysis of morphometric characters in four species of *Apodemus* (Muridae: Rodentia) from eastern Turkey and Iran. Israel Journal of Zoology, 47:243-258.

Fujita, M. S., and T. H. Kunz. 1984. *Pipistrellus subflavus*. Mammalian Species, 228:1-6.

Fuller, F., M. R. Lee, and L. R. Maxson. 1984. Albumin evolution in *Peromyscus* and *Sigmodon*. Journal of Mammalogy, 65:466-473.

Fülling, O. 1992. Ergähzande angaben über gaumenfaltenmuster vonhagetierun (Mammalia: Rodentia) aus Kamerun. Bonner Zoologische Beiträge, 43:415-421.

Fumagalli, L., J. Hausser, P. Taberlet, and D. T. Stewart. 1996. Phylogenetic structure of the holarctic *Sorex araneus* group and its relationships with *S. samniticus*, as inferred from mtDNA sequences. Hereditas, 125:191-199.

Fumagalli, L., P. Taberlet, D. T. Stewart, L. Gielly, J. Hausser, and P. Vogel. 1999. Molecular phylogeny and evolution of *Sorex* shrews (Soricidae: Insectivora) inferred from mitochondrial DNA sequence data. Molecular Phylogenetics and Evolution, 11:222-235.

Funakoshi, K., and T. Kunisaki. 2000. On the validity of *Tadarida latouchei*, with references to morphological divergence among *T. latouchei*, *T. insignis* and *T. teniotis* (Chiroptera, Molossidae). Mammal Study, 25:115-123.

Furano, A. V., B. E. Hayward, P. Chevret, F. Catzeflis, and K. Usdin. 1994. Amplification of the ancient murine *Lx* family of long interspersed repeated DNA occurred during the murine radiation. Journal of Molecular Evolution, 38:18-27.

Furley, C. W., H. Tichy, and H.-P. Uerpmann. 1988. Systematics and chromosomes of the Indian gazelle, *Gazella bennetti* (Sykes, 1831). Zeitschrift für Säugetierkunde, 53:48-54.

Gadi, I. K., and T. Sharma. 1983. Cytogenetic relationships in *Rattus*, *Cremnomys*, *Millardia*, *Nesokia* and *Bandicota*. Genetica, 61:21-40.

Gahsche, J. 1994. Die Alpenspitzmaus (*Sorex alpinus*) im Harz. Säugetierkundliche Informationen, 18:601-609.

Gaisler, J. 1970. The bats (Chiroptera) collected in Afghanistan by the Czechosloslovak Expeditions of 1965-1967. Acta Scientiarum Naturalium, Academiae Scientarium Bohemoslovacea (Brno), n.s., 4(6):1-56.

Gaisler, J. 2001a. *Rhinolophus ferrumequinum* (Schreber, 1774)—Grosse Hufeisennase. Pp. 15-37, *in*: Handbuch der Säugetiere Europas. Band 4: Fledertiere. Teil I: Chiroptera I, Rhinolophidae, Vespertilionidae 1 (J. Niethammer and F. Krapp, eds.). Aula Verlag, Wiebelsheim, 602 pp.

Gaisler, J. 2001b. *Rhinolophus euryale* Blasius, 1853—Mittelmeerhufeisennase. Pp. 59-74, *in*: Handbuch der Säugetiere Europas. Band 4: Fledertiere. Teil I: Chiroptera I, Rhinolophidae, Vespertilionidae 1 (J. Niethammer and F. Krapp, eds.). Aula Verlag, Wiebelsheim, 602 pp.

Gaisler, J. 2001c. *Rhinolophus mehelyi* Matchie, 1901—Mehely-Hufeisennase. Pp. 91-104, *in*: Handbuch der Säugetiere Europas. Band 4: Fledertiere. Teil I: Chiroptera I, Rhinolophidae, Vespertilionidae 1 (J. Niethammer and F. Krapp, eds.). Aula Verlag, Wiebelsheim, 602 pp.

Galaktionov, Yu. K. 1995. Between-cycle and within-cycle variability of continuous traits of the skull in water voles (*Arvicola terrestris* L.). Doklady Akademii Nauk, 340(2):279-281.

Galaz, J. l., J. C. Torres-Mura, and J. Yañez. 1999. *Platalina genovensium* (Thomas, 1928), un quiróptero nuevo para la fauna de Chile (Phyllostomatidae: Glossophaginae). Noticiario Mensual del Museo Nacional de Historia Natural, 337:6-12.

Galbreath, G. J., and R. A. Melville. 2003. *Pseudonovibos spiralis*: Epitaph. Journal of Zoology, London, 259:169-170.

Galiano, H., and D. Frailey. 1977. *Chasmaporthetes kani*, new species from China, with remarks on phylogenetic relationships of genera within the Hyaenidae (Mammalia, Carnivora). American Museum Novitates, 2632:1-16.

Galkina, L. I., and L. Yu. Epifantseva. 1988. [New species of mountain vole from Transbaical region (Rodentia, Cricetidae)]. Vestnik Zoologii, 2:30-33 (in Russian with English summary).

Gallardo, M. 1979. Las especies chilenas de *Ctenomys* (Rodentia, Octodontidae). I. Estabilidad cariotípica. Archivos de Biología y Medicina Experimentales (Santiago), 12:71-82.

Gallardo, M. H. 1991. Karyotypic evolution in *Ctenomys* (Rodentia, Ctenomyidae). Journal of Mammalogy, 72(1):11-21.

Gallardo, M. H. 1992. Karyotypic evolution in octodontid rodents based on C-band analysis. Journal of Mammalogy, 73(1):89-98.

Gallardo, M. H., and J. A. W. Kirsch. 2001. Molecular relationships among Octodontidae (Mammalia: Rodentia: Caviomorpha). Journal of Mammalian Evolution, 8(1):73-89.

Gallardo, M. H., and N. Köhler. 1992. Genetic divergence in *Ctenomys* (Rodentia, Ctenomyidae) from the Andes of Chile. Journal of Mammalogy, 73:99-105.

Gallardo, M. H., and E. Palma. 1990. Systematics of *Oryzomys longicadatus* (Rodentia: Muridae) in Chile. Journal of Mammalogy, 71:333-342.

Gallardo, M. H., and R. E. Palma. 1992. Intra- and interspecific genetic variability in *Ctenomys* (Rodentia: Ctenomyidae). Biochemical Systematics and Ecology, 20:523-534.

Gallardo, M. H., and B. D. Patterson. 1985. Chromosomal differences between two nominal subspecies of *Oryzomys longicaudatus* Bennett. Mammalian Chromosome Newsletter, 25:49-53.

Gallardo, M. H., and D. Reise. 1992. Systematics of *Aconaemys* (Rodentia, Octodontidae). Journal of Mammalogy, 73(4):779-788.

Gallardo, M. H., G. Aguilar, and O. Goicochea. 1988. Systematics of sympatric cricetid *Akodon* (*Abrothrix*) rodents and their taxonomic implications. Medio Ambiente, 9:65-74.

Gallardo, M. H., J. W. Bickham, R. L. Honeycutt, R. A. Ojeda, and N. Köhler. 1999. Discovery of tetraploidy in a mammal. Nature, 401:341.

Gallego, L. 1974. Alveolos molares en *Micromys minutus* del Cantábrico. Munibe Sociedad de Ciencias Naturales Aranzadi San Sebastian, 26(3-4):167-171.

Galleni, L., R. Stanyon, A. Tellini, G. Giordano, and L. Santini. 1992. Karyology of the Savi pine vole, *Microtus savii* De Salys-Longchamps, 1838 (Rodentia, Arvicolidae): G-, C-, DA/DAPI-, and AluI-bands. Cytogenetics and Cell Genetics, 59:290-292.

Galleni, L., A. Tellini, R. Stanyon, A. Cicalo, and L. Santini. 1994. Taxonomy of *Microtus savii* (Rodentia, Arvicolidae) in Italy: Cytogenetic and hybridization data. Journal of Mammalogy, 75(4):1040-1044.

Galleni, L., R. Stanyon, L. Contadini, and A. Tellini. 1998. Biogeographical and karyological data of the *Microtus savii* group (Rodentia, Arvicolidae) in Italy. Bonner Zoologische Beiträge, 47(3-4):277-282.

Galliari, C. A., and U. F. J. Pardiñas. 1995. La identidad de *Akodon leucolimnaeus* Cabrera (Rodentia, Sigmodontinae). X Jornadas Argentinas de Mastozoologia, Resumenes:28.

Galliari, C. A., and U. F. J. Pardiñas. 1999. *Abrothrix lanosus* (Rodentia: Muridae) en la Patagonia continental Argentina. Neotropica, 45:113-114.

Galliari, C. A., and U. F. J. Pardiñas. 2000. Taxonomy and distribution of the sigmodontine rodents of genus *Necromys* in central Argentina and Uruguay. Acta Theriologica, 45:211-232.

Galliari, C. A., U. F. J. Pardiñas, and F. J. Goin. 1996. Lista comentada de los mamiferos Argentinos. Mastozoología Neotropical, 3(1):39-61.

Gallo-Reynoso, J. P., and J. L. Solorzano-Velasco. 1991. Two new sightings of California sea lions on the southern coast of Mexico. Marine Mammal Science, 7(1):96.

Gamarra de Fox, I., and A. J. Martin. 1996. Lista de mamíferos del Paraguay. Pp. 469-573, *in* Colecciones de fauna y flora del Museo Nacional de Historia Natural del Paraguay (O. Romero, ed.). Dirección de Parques Nacionales y Vida Silvestre, Asunción, Paraguay, 573 pp.

Gambaryan, P. P. 1960. [Adaptive specializations of the organs of movement of burrowing mammals.] Erevan:Akademy Nauk Armenia, 198 pp. (in Russian).

Gambaryan, P. P., and J.-P. Gasc. 1993. Adaptive properties of the musculoskeletal system in the mole-rat *Myospalax myospalax* (Mammalia, Rodentia), cinefluorographical, anatomical, and biomechanical analyses of burrowing. Zoologische Jahrbüchen für Anatomie, 123:363-401.

Gambaryan, P. P., E. G. Potapova, and I. M. Fokin. 1980. [Morphofunctional analysis of the myology of the jerboa head]. Trudy Zoologicheskovo Instituta, Akademiya Nauk SSSR, 91:3-51 (in Russian).

Gambell, R. 1985*a*. Sei whale—*Balaenoptera borealis*. Pp. 155-170, *in* Handbook of marine mammals: The sirenians and baleen whales (S. H. Ridgway and R. Harrison, eds.). Academic Press, London, 3:1-362.

Gambell, R. 1985*b*. Fin whale—*Balaenoptera physalus*. Pp. 171-192, *in* Handbook of marine mammals: The sirenians and baleen whales (S. H. Ridgway and R. Harrison, eds.). Academic Press, London, 3:1-362.

Gamperl, R. 1982. Chromosomal evolution in the genus *Clethrionomys*. Genetica, 57:193-197.

Gamperl, R., and G. Vistorin. 1980. Comparative study of G- and C-banded chromosomes of *Gerbillus campestris* and *Meriones unguiculatus* (Rodentia, Gerbillinae). Genetica, 52/53:93-97.

Ganem, G. 1993. Ecological characteristics of Robertsonian populations of the house mouse. Is their habitat relevant to their evolution? Mammalia, 57(3):349-357.

Ganem, G., P. Alibert, and J. B. Searle. 1996. An ecological comparison between standard and chromosomally divergent house mice in northern Scotland. Zeitschrift für Säugetierkunde, 61:176-188.

Gannon, M. R., M. R. Willig, and J. K. Jones, Jr. 1989. *Sturnira lilium*. Mammalian Species, 333:1-5.

Gannon, W. L. 1988. *Zapus trinotatus*. Mammalian Species, 315:1-5.

Gannon, W. L. 1998. Syntopy between two species of long-eared bats (*Myotis evotis* and *Myotis auriculus*). Southwestern Naturalist, 43:394-396.

Gannon, W. L. 1999. Pacific jumping mouse, *Zapus princeps*. Pp. 669-670, *in* The Smithsonian book of North American mammals (D. E. Wilson and S. Ruff, eds.). Smithsonian Institution Press, Washington, D.C., 750 pp.

Gannon, W. L., and R. B. Forbes. 1995. *Tamias senex*. Mammalian Species, 502:1-6.

Gannon, W. L., and T. E. Lawlor. 1989. Variation of the chip vocalization of three species of Townsend chipmunks (genus *Eutamias*). Journal of Mammalogy, 70:740-753.

Gannon, W. L., R. B. Forbes, and D. E. Kain. 1993. *Tamias ochrogenys*. Mammalian Species, 445:1-4.

Ganslosser, U. 1980. Vergleichende Untersuchungen zur Kletterfähigkeit einiger Baumkänguruh-Arten (Dendrolagus, Marsupialia). Zoologischer Anzeiger, 205:43-66.

Gao Yao-ting [Kao Yüeh-ting]. 1963. [Taxonomic notes on the Chinese musk-deer]. Acta Zoologica Sinica, 15:479-488 (in Chinese).

Gao Yao-ting. 1983. Current studies on the Chinese Yarkand hare. Acta Zoologica Fennica, 174:23-25.

Gao Yao-ting. 1985. Classification and distribution of the musk deer (*Moschus*) in China. Pp. 113-116, *in* Contemporary mammalogy in China and Japan (T. Kawamichi, ed.). Mammalogical Society of Japan, 194 pp.

Gao Yao-ting (ed.). 1987. [Fauna Sinica, Mammalia: Carnivora]. Science Press, Academia Sinica, Beijing, 377 pp. (in Chinese).

Gao Yao-ting [Kao Yüeh-ting], and Feng Tso-chien [Feng Zuo-jiang]. 1964. [On the subspecies of the Chinese grey-tailed hare, *Lepus oiostolus* Hodgson]. Acta Zootaxonomica Sinica, 1:19-30 (in Chinese).

Gaona, S. 1997. Variacion no geografica de *Peromyscus difficilis* (Rodentia: Muridae) en la region noroeste de la Cuenca de Oriental en Puebla y Veracruz, Mexico. Pp. 135-156, *in* Homenaje al profesor Ticul Alvarez (J. A. Cabrales and O. J. Polaco, eds.). Instituto Nacional de Antropologia e Historia, D. F. Mexico, 357:389 pp.

Garagna, S., C. A. Redi, P. Veneroni, and E. Capanna. 1992. Chromosome banding by restriction enzyme digestion distinguishes between *Mus domesticus* and *Mus musculus* karyotypes. Rendiconti Fisiche Accademia Lincei, Ser. 9, 3(3):247-255.

Garagna, S., C. A. Redi, E. Capanna, N. Andayani, R. M. Alfano, P. Doi, and G. Viale. 1993. Genome distribution, chromosomal allocation, and organization of the major and minor satellite DNAs in 11 species and subspecies of the genus *Mus*. Cytogenetics and Cellular Genetics, 54:247-255.

Garagna, S., M. Zuccotti, C. A. Redi, and E. Capanna. 1997. Trapping speciation. Nature, 390:241-242.

Garagna, S., M. V. Civitelli, N. Marziliano, R. Castiglia, M. Zuccotti, C. A. Redi, and E. Capanna. 1999. Genome size variations are related to X-chromosome heterochromatin polymorphism in *Arvicanthis* sp. from Benin (West Africa). Italian Journal of Zoology, 66:27-32.

Garagna, S., M. Zuccotti, E. Capanna, and C. A. Redi. 2002. High-resolution organization of mouse telomeric and pericentriomeric DNA. Cytogenetic and Genome Research, 96:125-129.

Garcia, B. A., A. Martino, M. B. Chiappero, and C. N. Gardenal. 1999. Allozyme variation and taxonomic status of *Calomys hummelincki* (Rodentia, Sigmodontinae). Zeitschrift für Säugetierkunde, 64:30-35.

Garcia, I. S. 1992. Albinism in the Mediterranean vole, *Pitymys duodecimcostatus*. Mammalia, 56(2):299-300.

Garcia, L., M. Ponsa, J. Egozcue, and M. Garcia. 2000. Comparative chromosomal analysis and phylogeny in four *Ctenomys* species (Rodentia, Octodontidae). Biological Journal of the Linnean Society, 69:103-120.

García-Perea, R. 1992. New data on the systematics of lynxes. Cat News, 16:15-16.

Garcia-Perea, R. 1994. The pampas cat group (Genus *Lynchailurus* Severtzov, 1858) (Carnivora: Felidae), a systematic and biogeographic review. American Museum Novitates, 3096:1-35.

García-Perea, R. 2002. Andean mountain cat, *Oreailurus jacobita*; morphological description and comparison with other felines from the altiplano. Journal of Mammalogy, 83:110-124.

Garde, J. M., and M. C. Escala. 1993. Fluctuacion estacional del peso corporal de los machos adultos de *Arvicola sapidus* Miller, 1908 (Rodentia, Arvicolidae). Doñana, Acta Vertebrata, 20(2):251-255.

Garde, J. M., and M. C. Escala. 1996. Morphometric characteristics and relative growth of the water vole, *Arvicola sapidus* (Rodentia, Arvicolidae). Folia Zoologica, 45(3):227-236.

Garde, J. M., and M. C. Escala. 1999. Coats and moults of the water vole *Arvicola sapidus* Miller, 1908 (Rodentia, Arvicolinae) in southern Navarra (Spain). Zeitschrift für Säugetierkunde, 64:332-343.

Garde, J. M., M. C. Escala, and J. Ventura. 1993. Determinacion de la edad relativa en la rata de agua merididional, *Arvicola sapidus* Miller, 1908 (Rodentia, Arvicolidae). Doñana, Acta Vertebrata, 20(2):266-276.

Gardenal, C. N., M. B. Chiappero, G. M. de Luca D'Oro, and J. N. Mills. 2002. Allozymic polymorphism and genetic differentiation among populations of *Calomys musculinus* and *Calomys laucha* (Rodentia: Muridae) from eastern Argentina. Mammalian Biology, 67:294-303.

Gardner, A. L. 1966. A new subspecies of the aztec mastiff bat, *Molossus aztecus* Saussure, from southern Mexico. Los Angeles County Museum Contributons Science, 111:1-5.

Gardner, A. L. 1971. Karyotypes of two rodents from Peru, with a description of the highest diploid number recorded from a mammal. Experientia, 27:1088-1089.

Gardner, A. L. 1973. The systematics of the genus *Didelphis* (Marsupialia: Didelphidae) in North and Middle America. Special Publications, The Museum, Texas Tech University, 4:1-81.

Gardner, A. L. 1976. The distributional status of some Peruvian mammals. Occasional Papers of the Museum of Zoology, Louisiana State University, 48:1-18.

Gardner, A. L. 1977. Taxonomic implications of the karyotypes of *Molossops* and *Cynomops* (Mammalia: Chiroptera). Proceedings of the Biological Society of Washington, 89:545-549.

Gardner, A. L. 1982. Virginia opossum. Pp. 3-36, *in* Wild mammals of North America (J. A. Chapman and G. A. Feldhamer, eds.). Johns Hopkins Press, Baltimore, MD, 1147 pp.

Gardner, A. L. 1983*a*. *Oryzomys caliginosus* (raton pardo, raton arrocero pardo, Costa Rican dusky rice rat).

Pp. 483-485, *in* Costa Rican natural history (D. H. Janzen, ed.). University of Chicago Press, Chicago, IL, 816 pp.

Gardner, A. L. 1983*b*. *Proechimys semispinosus* (Rodentia: Echimyidae): Distribution, type locality, and taxonomic history. Proceedings of the Biological Society of Washington, 96(1):134-144.

Gardner, A. L. 1986. The taxonomic status of *Glossophaga morenoi* Martinez and Villa, 1938 (Mammalia: Chiroptera: Phyllostomidae). Proceedings of the Biological Society of Washington, 99:489-492.

Gardner, A. L. 1989 [1990]. Two new mammals from southern Venezuela and comments on the affinities of the highland fauna of Cerro de la Neblina. Pp. 411-424, *in* Advances in Neotropical mammalogy (K. Redford and J. F. Eisenberg, eds.). Sandhill Crane Press, Gainesville, FL, 614 pp.

Gardner, A. L. 1993. Order Didelphimorphia. Pp. 15-24, *in* Mammal species of the world, a taxonomic and geographic reference, Second ed. (D. E. Wilson and D. M. Reeder, eds.). Smithsonian Institution Press, Washington, D.C., xviii + 1207 pp.

Gardner, A. L. 1995. Comments on the proposed conservation of some mammal generic names first published in Brisson's (1762) Regnum Animale. Bulletin of Zoological Nomenclature, 52:79-82.

Gardner, A. L., and D. C. Carter. 1972. A review of the Peruvian species of *Vampyrops* (Chiroptera: Phyllostomatidae). Journal of Mammalogy, 53:72-82.

Gardner, A. L., and G. K. Creighton. 1989. A new generic name for Tate's *microtarsus* group of South American mouse opossums (Marsupialia: Didelphidae). Proceedings of the Biological Society of Washington, 102:3-7.

Gardner, A. L., and L. H. Emmons. 1984. Species groups in *Proechimys* (Rodentia, Echimyidae) as indicated by karyology and bullar morphology. Journal of Mammalogy, 65:10-25.

Gardner, A. L., and C. S. Ferrell. 1990. Comments on the nomenclature of some Neotropical bats (Mammalia: Chiroptera). Proceedings of the Biological Society of Washington, 103:501-508.

Gardner, A. L., and J. P. O'Neill. 1969. The taxonomic status of *Sturnira bidens* (Chiroptera: Phyllostomidae) with notes on its karyotype and life history. Occasional Papers of the Museum of Zoology, Louisiana State University, 38:1-8.

Gardner, A. L., and J. L. Patton. 1976. Karyotypic variation in oryzomyine rodents (Cricetinae) with comments on chromosomal evolution in the Neotropical cricetine complex. Occasional Papers of the Museum of Zoology, Louisiana State University, 49:1-48.

Gardner, A. L., and C. B. Robbins. 1998. Generic names of northern and southern fur seals (Mammalia: Otariidae). Marine Mammal Science, 14:544-551.

Gardner, A. L., and M. Romo R. 1993. A new *Thomasomys* (Mammalia: Rodentia) from the Peruvian Andes. Proceedings of the Biological Society of Washington, 106:762-774.

Gardner, A. L., R. K. LaVal, and D. E. Wilson. 1970. The distributional status of some Costa Rican bats. Journal of Mammalogy, 51:712-729.

Gardner, M. B., C. A. Kozak, and S. J. O'Brien. 1991. The Lake Casitas wild mouse: Evolving genetic resistance to retroviral disease. Trends in Genetics, 7(1):22-27.

Garin, I., J. L. García-Mudarra, J. R. Aihartza, U. Goiti, and J. Juste. 2003. Presence of *Plecotus macrobullaris* (Chiroptera: Vespertilionidae) in the Pyrenees. Acta Chiropterologica, 5:243-250.

Garrison, T. E., and T. L. Best. 1990. *Dipodomys ordii*. Mammalian Species, 353:1-10.

Gaskin, D. E., P. W. Arnold, and B. A. Blair. 1974. *Phocoena phocoena*. Mammalian Species, 42:1-8.

Gatesy, J., G. Amato, E. Vrba, G. Schaller, and R. De Salle. 1997. A cladistic analysis of mitochondrial DNA from the Bovidae. Molecular Phylogenetics and Evolution, 7:303-319.

Gaubert, P. 2003. Description of a new species of genet (Carnivora; Viverridae; genus *Genetta*) and taxonomic revision of forest forms related to the Large-spotted Genet complex. Mammalia, 67(1):85-108.

Gaubert, P., and G. Veron. 2003. Exchaustive sample set among Viverridae reveals the sister-group of felids: The linsangs as a case of extreme morphological convergence within Feliformia. Proceedings of the Royal Society of London, Ser. B. 270:2523-2530.

Gaubert, P., G. Veron, M. Colyn, A. Dunham, S. Shultz, and M. Tranier. 2002*a*. A reassessment of the distribution of the rare *Genetta johnstoni* (Viverridae, Carnivora) with some newly discovered specimens. Mammal Review, 32:132-144.

Gaubert, P., G. Veron, and M. Tranier. 2002*b*. Genets and "genet-like" taxa (Carnivora, Viverrinae): Phylogenetic analysis, systematics and biogeographic implication. Zoological Journal of the Linnean Society, 134:317-334.

Gaubert, P., M. Tranier, G. Veron, D. Kock, A. E. Dunham, P. J. Taylor, C. Stuart, T. Stuart, and C. W. [sic] Wozencraft. 2003*a*. Nomenclatural comments on the Rusty-spotted genet (Carnivora, Viverridae) and designation of a neotype. Zootaxa, 160:1-14.

Gaubert, P., M. Tranier, G. Veron, D. Kock, A. E. Dunham, P. J. Taylor, C. Stuart, T. Stuart, and W. C.

Wozencraft. 2003*b*. Case 3204. *Viverra maculata* Gray, 1830 (currently *Genetta maculate*; Mammalia, Carnivora): Proposed conservation of the specific name. Bulletin of Zoological Nomenclature, 60:45-47.

Gaubert, P., M. Tranier, A.-S. Delmas, M. Colyn, and G. Veron. 2004. First molecular evidence for reassessing phylogenetic affinities between genets (*Genetta*) and the enigmatic genet-like taxa *Osbornictis*, *Poiana* and *Prionodon* (Carnivora, Viverridae). Zoologica Scripta, 33:117-129.

Gaubert, P., W. C. Wozencraft, P. Cordeiro-Estrela, and G. Veron. In Press. Mosaic of Convergences, Noise and Misleading Morphological Phylogenies: what's in a Viverrid-like Carnivoran? Systematic Biology.

Gaucher, P. 1995. First record of Geoffroy's Bat *Myotis emarginatus* Geoffroy, 1806 (Mammalia: Chiroptera: Vespertilionidae) in Saudi Arabia. Mammalia, 59:149-151.

Gaucher, P., and A. Brosset. 1990. Record of *Hipposideros* (*Syndesmotis*) *megalotis* (Heuglin) in Saudi Arabia. Mammalia, 54:653-654.

Gaucher, P., and D. L. Harrison. 1995. Occurrence of Bodeneimer's pipistrelle *Pipistrellus bodenheimeri* Harrison, 1960 (Mammalia, Chiroptera: Vespertilionidae) in Saudi Arabia. Mammalia, 59:672-673.

Gaudin, T. J., and J. R. Wible. 1999. The entotympanic of pangolins and the phylogeny of the Pholidota (Mammalia). Journal of Mammalian Evolution, 6(1):39-65.

Gaunt, W. A. 1961. The development of the molar pattern of the golden hamster (*Mesocricetus auratus* W.), together with a re-assessment of the molar pattern of the mouse (*Mus musculus*). Acta Anatomica, 45:219-251.

Gaur, R. 1986. First report on a fossil *Rattus* (Murinae, Rodentia) from the Pinjor formation of Upper Siwalik of India. Current Science, 55(11):542-544.

Gautherin, H. 1988. Mammalogie. Bulletin Trimestriel de la Societe d'Histoire Naturelle et des Amis du Museum d'Autun, 126:41.

Gauthier-Pilters, H., and A. Innis Dagg. 1981. The camel: Its evolution, ecology, behavior and relationships to man. University of Chicago Press, Chicago, IL, 208 pp.

Gautun, J.-C., M. Tranier, and B. Sicard. 1985. Liste preliminaire des rongeurs du Burkina Faso (ex Haute-Volta). Mammalia, 49:537-542.

Gautun, J.-C., I. Sankhon, and M. Tranier. 1986. Nouvelle contribution a la connaissance des rongeurs du massif guineen des monts Nimba (Afrique occidentale). Systematique et aperçu quantitatif. Mammalia, 50:205-217.

Gava, A., T. R. O. Freitas, and J. Olimpio. 1998. A new karyotype for the genus *Cavia* from a southern island of Brazil (Rodentia, Caviidae). Genetics and Molecular Biology, 21(1):77-80.

Gavin, T. A., P. W. Sherman, E. Yensen, and B. May. 1999. Population genetic structure of the northern Idaho ground squirrel (*Spermophilus brunneus brunneus*). Journal of Mammalogy, 80(1):156-168.

Gavrila, L., A. Lungeanu, D. Murariu, and C. Stepan. 1986. New contributions to the cytogenetic study of *Microtus epiroticus* (Ondrias, 1966) (Mammalia, Arvicolidae). Travaux du Museum d'Historie Naturalae "Grigore Antipa", 28:271-274.

Gazaryan, M. A. 1985. [On certain questions of the life style of the forest dormouse (*Dryomys nitedula* Pall.)]. Izvestiya Sel'skokhozyaistvennykh Nauk, 1985(7):40-44 (in Russian).

Geffen, E. 1994. *Vulpes cana*. Mammalian Species, 462:1-4.

Geider, M., and D. Kock. 1991. Kleinsäuger im Nebelwaldgebiet des Forêt de Nyungwe, Rwanda. Natur und Museum, 121(7):210-216.

Geise, L., R. Cerqueira, and H. N. Seuánez. 1996. Karyological characterization of a new population of *Akodon lindberghi* (Rodentia, Sigmodontinae) in Minas Gerais state (Brazil). Caryologia, 49:57-63.

Geise, L., F. C. Canavez, and H. N. Seuánez. 1998. Comparative karyology in *Akodon* (Rodentia, Sigmodontinae) from southeastern Brazil. Journal of Heredity, 89:158-163.

Geise, L., M. F. Smith, and J. L. Patton. 2001. Diversification in the genus *Akodon* (Rodentia: Sigmodontinae) in southeastern South America: Mitochondrial DNA sequence analysis. Journal of Mammalogy, 82:92-101.

Geissert, F., and J.-J. Merkel. 1994. Observations fleuristiques et faunistiques dans le nord du Bas-Rhin. Bulletin de l'Association Philomathique d'Alsace et de Lorraine, 29:39-50.

Geist, V. 1990. On the taxonomy of giant sheep (*Ovis ammon* Linnaeus, 1766). Canadian Journal of Zoology, 69:706-723.

Geist, V. 1991. Phantom subspecies: The wood bison *Bison bison* "athabascae" Rhoads, 1897 is not a taxon but an ecotype. Arctic, 44:287-300.

Geist, V. 1998. Deer of the world. Their evolution, behavior and ecology. Stackpole Books, Mechanicsburg, Pensylvania, 421 pp.

Geist, V., and P. Karsten. 1977. The wood bison (*Bison bison athabascae* Rhoads) in relation to hypotheses on the origin of the American bison (*Bison bison* Linnaeus). Zeitschrift für Säugetierkunde, 42:119-127.

Gemmeke, H. 1980. Proteinvariation und Taxonomie in der Gattung *Apodemus* (Mammalia, Rodentia). Zeitschrift für Säugetierkunde, 45:348-365.

Gemmeke, H. 1981*a*. Genetische Unterschiede zwischen rechts-und linksrheinischen Waldmäusen (*Apodemus sylvaticus*). Bonner Zoologische Beiträge, 32:265-269.

Gemmeke, H. 1981*b*. Albuminunterschiede bei Wald- und Gelbhalsmäusen (*Apodemus sylvaticus* und *A. flavicollis*, Mammalia, Rodentia) auch in getrockneten Muskeln und Bälgen elektrophoretisch nachweisbar. Zeitschrift für Säugetierkunde, 46:124-125.

Gemmeke, H. 1983. Proteinvariation bei Zwergwaldmäusen (*Apodemus microps* Kratochvil und Rosicky, 1952). Zeitschrift für Säugetierkunde, 48:155-160.

Gemmeke, H., and J. Niethammer. 1982. Zur Charakterisierung der Waldmäuse (*Apodemus*) Nepals. Zeitschrift für Säugetierkunde, 47:33-38.

Gemmeke, H., and J. Niethammer. 1984. Zur Taxonomie der Gattung *Rattus* (Rodentia, Muridae). Zeitschrift für Säugetierkunde, 49:104-116.

Genest, H., and F. Petter. 1975. Part 1.1, Family Tenrecidae. Pp. 1-7, *in* The mammals of Africa: An identification manual (J. Meester and H. W. Setzer, eds.). [issued in Dec 1975]. Smithsonian Institution, Washington, D.C., not continuously paginated.

Genest-Villard, H. 1967. Revision du genre *Cricetomys* (Rongeurs, Cricetidae). Mammalia, 31:390-455.

Genest-Villard, H. 1978. Revision systematique du genre *Graphiurus* (Rongeurs, Gliridae). Mammalia, 42(4):391-426.

Genest-Villard, H. 1980. Régime alimentaire des rongeurs myomorphes de forêt équatoriale région du M'Baiki, République Centrafricaine. Mammalia, 44:423-484.

Genoud, M., and R. Hutterer. 1990. *Crocidura russula* (Hermann, 1780)—Hausspitzmaus. Pp. 429- 452, *in* Handbuch der Säugetiere Europas (J. Niethammer and F. Krapp, eds.). Aula-Verlag, Wiesbaden, 3/I:1-524.

Genov, P. V. 1999. A review of the cranial characteristics of the wild boar (*Sus scrofa* Linnaeus, 1758), with systematic conclusions. Mammal Review, 29:205-238.

Genoways, H. H. 1973. Systematics and evolutionary relationships of spiny pocket mice, genus *Liomys*. Special Publications, The Museum, Texas Tech University, 5:1-368.

Genoways, H. H. 1998. Two new subspecies of bats of the genus *Sturnira* from the Lesser Antilles, West Indies. Occasional Papers, Museum of Texas Tech University, 176:1-7.

Genoways, H. H., and R. J. Baker. 1972. *Stenoderma rufum*. Mammalian Species, 18:1-4.

Genoways, H. H., and R. J. Baker. 1996. A new species of the genus *Rhogeessa*, with comments on geographic distribution and speciation in the genus. Pp. 83-87, *in* Contributions in mammalogy: A memorial volume in honor of J. Knox Jones, Jr. (H. H. Genoways and R. J. Baker, eds.). Museum of Texas Tech University, Lubbock, Texas, 315 pp.

Genoways, H. H., and E. C. Birney. 1974. *Neotoma alleni*. Mammalian Species, 41:1-4.

Genoways, H. H., and J. H. Brown (eds.). 1993. Biology of the Heteromyidae. Special Publication, The American Society of Mammalogists, 10:1-719.

Genoways, H. H., and J. R. Choate. 1972. A multivariate analysis of systematic relationships among populations of the short-tailed shrew (genus *Blarina*) in Nebraska. Systematic Zoology, 21:106-116.

Genoways, H. H., and J. R. Choate. 1998. Natural history of the southern short-tailed shrew, *Blarina carolinensis*. Occasional Papers, Museum of Southwestern Biology, 8:1-43.

Genoways, H. H., and J. K. Jones, Jr. 1969*a*. Notes on pocket gophers from Jalisco, Mexico, with descrptions of two new subspecies. Journal of Mammalogy, 50:748-755.

Genoways, H. H., and J. K. Jones, Jr. 1969*b*. Taxonomic status of certain long-eared bats (genus *Myotis*) from the southwestern United States and Mexico. Southwestern Naturalist, 14:1-13.

Genoways, H. H., and J. K. Jones, Jr. 1971. Systematics of southern banner-tailed kangaroo rats of the *Dipodomys phillipsii* group. Journal of Mammalogy, 52:265-287.

Genoways, H. H., and J. K. Jones, Jr. 1972*a*. Mammals from southwestern North Dakota. Occasional Papers, The Museum, Texas Tech University, 6:1-36.

Genoways, H. H., and J K. Jones, Jr. 1972*b*. Variation and ecology in a local population of the vesper mouse (*Nyctomys sumichrasti*). Occasional Papers, The Museum, Texas Tech University, 3:22 pp.

Genoways, H. H., and S. L. Williams. 1979*a*. Notes on bats (Mammalia: Chiroptera) from Bonaire and Curaçao, Dutch West Indies. Annals of Carnegie Museum, 48:311-321.

Genoways, H. H., and S. L. Williams. 1979*b*. Records of bats (Mammalia, Chiroptera) from Suriname. Annals of the Carnegie Museum, 48:323-335.

Genoways, H. H., and S. L. Williams. 1984. Results of the Alcoa Foundation—Suriname Expeditions. IX. Bats of the genus *Tonatia* (Mammalia: Chiroptera) in Suriname. Annals of the Carnegie Museum, 53:327-346.

Genoways, H. H., J. C. Patton, III, and J. R. Choate. 1977. Karyotypes of shrews of the genera *Cryptotis* and *Blarina* (Mammalia: Soricidae). Experientia, 33:1294-1295.

Genoways, H. H., R. C. Dowler, and C. H. Carter. 1981. Intraisland and interisland variations in Antillean populations of *Molossus molossus* (Mammalia: Molossidae). Annals of the Carnegie Museum, 50:475-492.

Genoways, H. H., C. J. Phillips, and R. J. Baker. 1998. Bats of the Antillean Island of Grenada: A new zoogeographic perspective. Occasional Papers, Museum of Texas Tech University, 177:1-28.

Gentile de Fronza, T. 1970. Cariotipo de *Scapteromys tumidus aquaticus* (Rodentia, Cricetidae). Physis (Buenos Aires), 30:343.

Gentry, A. 1994. Regnum animale , Ed. 2 (M. J. Brisson, 1762): Proposed rejection, with the conservation of the mammalian generic names *Philander* (Marsupialia), *Pteropus* (Chiroptera), *Glis, Cuniculus* and *Hydrochoerus* (Rodentia), *Meles, Lutra* and *Hyaena* (Carnivora), *Tapirus* (Perissodactyla), *Tragulus* and *Giraffa* (Artiodactyla). Bulletin of Zoological Nomenclature, 51:135-146.

Gentry, A., J. Clutton-Brock, and C. P. Groves. 1996. Case 3010. Proposed conservation of usage of 15 mammal specific names based on wild species which are antedated by or contemporary with those based on domestic animals. Bulletin of Zoological Nomenclature, 53:28-37.

Gentry, A., J. Clutton-Brock, and C. P. Groves. 2001. Comments on the proposed conservation of usage of 15 mammal specific names based on wild species which are antedated by or contemporary with those based on domestic animals. Bulletin of Zoological Nomenclature, 58:233-234.

Gentry, A. W. 1964. Skull characters of African gazelles. Annals and Magazine of Natural History, ser. 13, 7:353-382.

Gentry, A. W. 1972. Genus *Gazella*. Part 15.1. Pp. 85-93, *in* The mammals of Africa: An identification manual (J. Meester and H. W. Setzer, eds.) [issued 2 May 1972]. Smithsonian Institution Press, Washington, D.C., not continuously paginated.

Gentry, A. W. 1975. A quagga, *Equus quagga* (Mammalia, Equidae), at University College, London and a note on a supposed quagga in the City Museum, Bristol. Bulletin of the British Museum (Natural History), Zoology Series, 28:217-226.

Gentry, A. W. 1978. Bovidae. Pp. 540-572, *in* Evolution of African mammals (V. J. Maglio and H. B. S. Cooke, eds.). Harvard University Press, Cambridge, Massachusetts, not continuously paginated.

Gentry, A. W. 1990. Evolution and dispersal of African Bovidae. Pp. 195-227, *in* Horns, pronghorns, and antlers (G. A. Bubenik and A. B. Bubenik, eds). Springer-Verlag, New York, 562 pp.

Gentry, A. W. 1992. The subfamilies and tribes of the family Bovidae. Mammal Review, 22:1-32.

Gentry, A. W. 1995. Comments on the proposed conservation of some mammal generic names first published in Brisson's (1762) *Regnum Animale*. Bulletin of Zoological Nomenclature, 52:188-190.

Gentry, A. W., and A. Gentry. 1978. Fossil Bovidae (Mammalia) of Olduvai Gorge, Tanzania. Part I. Bulletin of the British Museum (Natural History), Geology Series, 29:289-446.

Gentry, A. W., and J. J. Hooker. 1988. The phylogeny of the Artiodactyla. Pp. 235-272, *in* The phylogeny and classification of the tetrapods, Volume 2: Mammals (M. J. Benton, ed.). Systematics Association Special Volume no. 35B, Clarendon Press, Oxford.

Geoffroy Saint-Hilaire, E. 1796. Mammifères. Mémoire sur les rapports naturels des Makis Lemur. Magazin encycl., 1:20-50.

Geoffroy Saint-Hilaire, É. 1803. Catalogue des mammifères du Muséum National d'Histoire Naturelle. Muséum National d'Histoire Naturelle, Paris, 272 pp.

Geoffroy Saint-Hilaire, E. 1826. *Mustela striata*. Dictionnaire Classique d'Histoire Naturelle, 10:214.

Geoffroy Saint-Hilaire, I. 1839. Notice sur deux nouveaux genres de mammifères carnassiers, les ichneumies, du continent African, et les galidies, de Madagascar. Magasin de Zoologie, ser. 2, 1:1-39.

Geoffroy Saint-Hilaire, I. 1847. Vie, travaux et doctrine scientifique d'Étienne Geoffroy Saint-Hilaire. Paris, P. Bertrand, 3+1+479 pp.

Geoffroy Saint-Hilaire, I. 1851. Catalogue méthodique de la collection des mammifères de la collection des oiseaux et des collection annexes. Muséum National d'Histoire Naturelle, Gide et Baudry, Paris, 96 pp.

George, G. G. 1979. The status of endangered Papua New Guinea mammals. Pp. 93-100, *in* The Status of endangered Australasian mammals (M. Tyler, ed.). Royal Zoological Society of New South Wales, Sydney, 210 pp.

George, G. G., and Schürer. 1978. Some notes on macropods commonly misidentified in zoos. International Zoo Yearbook, 18:152-156.

George, S. B. 1986. Evolution and historical biogeography of soricine shrews. Systematic Zoology, 35:153-162.

George, S. B. 1988. Systematics, historical biogeography, and evolution of the genus *Sorex*. Journal of Mammalogy, 69:443-461.

George, S. B. 1989. *Sorex trowbridgii*. Mammalian Species, 337:1-5.

George, S. B., and J. D. Smith. 1991. Inter- and intraspecific variation among coastal and island populations of *Sorex monticolus* and *Sorex vagrans*. Pp. 75-91, *in* The biology of the Soricidae (J. S. Findley and T. L. Yates, eds.). Special Publication, The Museum of Southwestern Biology, 1:1-91.

George, S. B., and R. K. Wayne. 1991. Island foxes. A model for conservation genetics. Terra, 30:18-23.

George, S. B., J. R. Choate, and H. H. Genoways. 1981. Distribution and taxonomic status of *Blarina hylophaga* Elliot (Insectivora: Soricidae). Annals of Carnegie Museum, 50:493-513.

George, S. B., H. H. Genoways, J. R. Choate, and R. J. Baker. 1982. Karyotypic relationships within the short-tailed shrews, genus *Blarina*. Journal of Mammalogy, 63:639-645.

George, S. B., J. R. Choate, and H. H. Genoways. 1986. *Blarina brevicauda*. Mammalian Species, 261:1-9.

George, W. 1979a. The chromosomes of the hystricomorphous family Ctenodactylidae (Rodentia: ? Sciuromorpha) and their bearing on the relationships of the four living genera. Zoological Journal of the Linnaean Society, 65:261-280.

George, W. 1979b. Conservatism in the karyotypes of two African mole rats (Rodentia, Bathyergidae). Zeitschrift für Säugetierkunde, 44:278-285.

George, W. 1980. A study in hystricomorph rodent relationships: The karyotypes of *Thryonomys gregorianus*, *Pedetes capensis* and *Hystrix cristata*. Zoological Journal of the Linnean Society, 68:361-372.

George, W. 1981. Blood vacular patterns in rodents: Contributions to an analysis of rodent family relationships. Zoological Journal of the Linnean Society, 73:287-306.

George, W. 1982. *Ctenodactylus* (Ctenodactylidae, Rodentia): One species or two? Mammalia, 46:375-380.

George, W. 1985. Reproductive and chromosomal characters of ctenodactylids as a key to their evolutionary relationships. Pp. 453-474, *in* Evolutionary Relationships among Rodents: A Multidisciplinary Analysis (W. P. Luckett and J.-L. Hartenberger, eds.). Plenum Press, New York, 721 pp.

George, W., and B. J. Weir. 1972a. The chromosomes of some octodontids with special reference to *Octodontomys* (Rodentia: Hystricomorpha). Chromosoma (Berlin), 37:53-62.

George, W., and B. J. Weir. 1972b. Record chromosome number in a mammal? Nature, 236:205-206.

George, W., and B. J. Weir. 1974. Hystricomorph chromosomes. Symposia of the Zoological Society of London, 34:79-108.

George, W., B. J. Weir, and J. Bedford. 1972. Chromosome studies in some members of the family Caviidae (Mammalia: Rodentia). Journal of Zoology, London, 168:81-89.

Georgiadis, N. J., P. N. Kat, and H. Oketch. 1990. Allozyme divergence within the Bovidae. Evolution, 44:2135-2149.

Geraads, D. 1992. Phylogenetic analysis of the tribe Bovini (Mammalia: Artiodactyla). Zoological Journal of the Linnean Society, 104:103-207.

Geraads, D. 1994. Rongeurs et lagomorphes du Pleistocène moyen de la "Grotte des Rhinocéros", Carrière Oulad Hamida l à Casablanca, Maroc. Neues Jahrbuch für Geologie und Paläontologie. Abhandlungen, 191(2):147-172.

Geraads, D. 1995. Rongeurs et insectivores (Mammalia) du Pliocene final de Ahl Al Oughlam (Casablanca, Maroc). Geobios, 28(1):99-115.

Geraads, D. 1998. Rongeurs du Mio-Pliocène de Lissasfa (Casablanca, Maroc). Geobios, 31(2):229-245.

Geraads, D. 2001. Rongeurs du Miocene Superieur de Chorora, Ethiopie: Murinae, Dendromurinae et conclusions. Palaeovertebrata, 30(1-2):89-109.

Geraads, D., F. Amani, J.-P. Raynal, and F.-Z. Sbihi-Alaoui. 1998. La fauna de mammifères du Pliocène terminal d'Ahl al Oughlam, Casablanca, Maroc. Comptes Rendus Académie des Sciences, Paris, Sciences de la Terre et des Planètes, 326:671-676.

Gerasimov, S., H. Nikolov, V. Mihailova, J.-C. Auffray, and F. Bonhomme. 1990. Morphometric stepwise discriminant analysis of the five genetically determined European taxa of the genus *Mus*. Biological Journal of the Linnean Society, 41:47-64.

Gervais, F. L. P. 1841. Mammifères. Pp. 1-68, *in* Zoologie (J. F. T. Eydoux and L. F. A. Souleyet, eds.). *in* Voyage autour du Monde exécuté pendant les années 1836 et 1837 sur la corvette La Bonite commandee par M. Vaillant. Arthus Bertrand, Paris, 4:1-334, pls. 1-12 (Mammifères).

Gharaibeh, B. M., and C. Jones. 1996. *Myosciurus pumilio*. Mammalian Species, 523:1-3.

Ghose, R. K. 1965. A new species of mongoose (Mammalia: Carnivora: Viverridae) from West Bengal. Proceedings of the Zoological Society of Calcutta, 18:173-178.

Ghose, R. K. 1978. Observations on the ecology and status of the hispid hare in Rajagarh forest, Darrang District, Assam, in 1975 and 1976. Journal of the Bombay Natural History Society, 75:206-209.

Ghose, R. K., and T. P. Bhattacharya. 1995. New distributional record of *Petaurista fulvinus* Wroughton, 1911 (Mammalia: Rodentia: Sciuridae), with comments on its taxonomic status. Journal of the Bombay Natural History Society, 92(2):254-255.

Ghose, R. K., and S. S. Saha. 1981. Taxonomic review of Hodgson's giant flying squirrel, *Petaurista magnificus* (Hodgson)(Sciuridae: Rodentia), with description of a new subspecies from Darjeeling District, West Bengal, India. Journal of the Bombay Natural History Society, 78:93-102.

Giagia, E. B., and J. C. Ondrias. 1980. Karyological analysis of eastern European hedgehog *Erinaceus concolor* (Mammalia, Insectivora) in Greece. Mammalia, 44:59-71.

Giagia, E. B., I. Savic, and B. Soldatovic. 1982. Chromosomal forms of the mole rat *Microspalax* from Greece and Turkey. Zeitschrift für Säugetierkunde, 47:231-236.

Giagia, E. B., S. E. Fraguedakis-Tsolis, and B. P. Chondropoulos. 1987. Contribution to the study of the taxonomy and zoogeography of wild house mouse, genus *Mus* L. (Mammalia, Rodentia, Muridae) in Greece. II. Karyological study of two populations from southern Greece. Mammalia, 51:111-116.

Giagia-Athanasopoulou, E.-B., and G. Markakis. 1996. Multivariate analysis of morphological characters in the Eastern hedgehog *Erinaceus concolor* from Greece and adjacent areas. Zeitschrift für Säugetierkunde, 61:129-139.

Giagia-Athanasopoulou, E. B., B. P. Chondropoulos, and S. E. Fraguedakis-Tsolis. 1995. Robertsonian chromosomal variation in subalpine voles *Microtus* (*Terricola*), (Rodentia, Arvicolidae) from Greece. Acta Theriologica, 40(2):139-143.

Giannini, N. P., and R. M. Barquez. 2003. *Sturnira erythromos*. Mammalian Specis, 729:1-5.

Giannini, N. P., and N. B. Simmons. 2003. A phylogeny of megachiropteran bats (Mammalia: Chiroptera: Pteropodidae) based on direct optimization analysis of one nuclear and four mitochondrial genes. Cladistics, 19:496-511.

Giannoni, S. M., C. E. Borghi, and J. P. Martínez-rica. 1993. Swimming ability of a fossil iberic vole *Microtus* (*Pitymys*) *lusitanicus* (Rodentia: Microtinae). Mammalia, 57(3):337-340.

Giannoni, S. M., C. E. Borghi, and J. P. Martínez-rica. 1994. Swimming ability of the Mediterranean pine vole *Microtus* (*Terricola*) *duodecimcostatus*. Acta Theriologica, 39(3):257-265.

Giao, P. M., D. Tuoc, V. V. Dung, E. D. Wikramanayake, G. Amato, P. Arctander, and J. R. Mackinnon. 1998. Description of *Muntiacus truongsonensis*, a new species of muntjac (Artiodactyla: Muntiacidae) from Central Vietnam, and implications for conservation. Animal Conservation, 1:61-68.

Gibb, J. A. 1990. The European rabbit *Oryctolagus cuniculus*. Pp. 116-120, *in* Rabbits, hares and pikas (J. A. Chapman and J. E. C. Flux, eds.). I.U.C.N., Gland, Switzerland, 168 pp.

Gibson-Hill, C. A. 1947. A note on the mammals of Christmas Island. Bulletin of the Raffles Museum, 18:166-167.

Gidley, J. M. 1912. The Lagomorpha as an independent order. Science, 36:285-286.

Giere, P., C. Freyer, and U. Zeller. 1999. Opening of the mammalian vomeronasal organ with respect to the Glires hypotheses: A cladistic reconstruction of the therian morphotype. Mitteilungen aus dem Museum fur Naturkunde in Berlin, Zoologische Reihe, 75(2):247-255.

Gilbert, D. A., N. Lehman, S. J. O'Brien, and R. K. Wayne. 1990. Genetic fingerprinting reflects population differentiaion in the California Channel Island fox. Nature, 344:764-767.

Gileva, E. A. 1972. [Chromosomal polymorphism in two close forms of subarctic mice (*Microtus hyperboreus* and *M. middendorffi*)]. Doklady Akademii Nauk SSSR, 203:698-692 (in Russian).

Gileva, E. A. 1980. Chromosomal diversity and an aberrant genetic system of sex determination in the Arctic lemming, *Dicrostonyx torquatus* Pallas (1779). Genetica, 52/53:99-103.

Gileva, E. A. 1983. A contrasted pattern of chromosome evolution in two genera of lemmings, *Lemmus* and *Dicrostonyx* (Mammalia, Rodentia). Genetica, 60:173-179.

Gileva, E. A., and V. B. Fedorov. 1991. Sex ratio, XY females and absence of inbreeding in a population of the wood lemming, *Myopus schisticolor* Lilljeborg, 1844. Heredity, 66:351-355.

Gileva, E. A., I. E. Benenson, A. V. Pokrovskii, and N. A. Lobanova. 1980. [Analysis of aberrant sex ratio and postnatal mortality in progeny of the Arctic lemming *Dicrostonyx torquatus*]. Ekologiya, 6:46-52 (in Russian).

Gileva, E. A., V. N. Bolshakov, N. F. Chernousova, and V. P. Mamina. 1982. [Cytogenetical differentiation of forms in the group of Pamir (*Microtus juldaschi*) and Carruther's (*M. carruthersi*) voles (Mammalia, Microtinae)]. Zoologicheskii Zhurnal, 61:912-922 (in Russian).

Gileva, E. A., V. N. Bolshakov, O. F. Sadykov, and T. I. Omariev. 1983. Chromosome variability and deviating sex ratio in two Ural populations of the wood lemming *Myopus schisticolor* Lilljeborg, 1884. Doklady Akademii Nauk, SSSR, 270:453-456.

Gileva, E. A., I. A. Kuznetsova, M. I. Cheprakov. 1984. [Chromosome sets and taxonomy of true lemmings of the genus *Lemmus*]. Zoologicheskii Zhurnal, 63:105-114 (in Russian).

Gileva, E. A., D. E. Rybnikov, and G. P. Miroshnichenko. 1989. [DNA-DNA hybridization and phylogenetic relationships in two genera of voles, *Alticola*, and *Clethrionomys* (Microtinae-Rodentia)]. Doklady Akademii Nauk SSSR, 311:477-480 (in Russian).

Gileva, E. A., M. I. Cheprakov, and D. Yu Nochrin. 1996. [Voles of the group *Microtus arvalis* (Rodentia, Cricetidae) in the Urals.] Zoologicheskii Zhurnal, 75(9):1436-1439 (in Russian with English summary).

Gill, A. E. 1980. Partial reproductive isolation of subspecies of the California vole, *Microtus californicus*. Pp. 105-117, *in* Animal genetics and evolution (N. N. Vorontsov and J. M. Van Brink, eds.). Dr. W. Junk, The Hague, 353 pp.

Gill, A., B. Petrov, S. Zivkovic, and D. Rimsa. 1987. Biochemical comparisons in Yugoslavian rodents of the families Arvicolidae and Muridae. Zeitschrift für Säugetierkunde, 52:247-256.

Gill, T. 1866. Prodrome of a monograph of the pinnipeds. Proceedings of the Essex Institute, Communications, 5:1-13.

Gill, T. 1872. Arrangement of the families of mammals with analytical tables. Smithsonian Miscellaneous Collections, 11:1-98.

Gill, T. 1873. The ribbon seal of Alaska. American Naturalist, 7:178-179.

Gill, T. K., and S. Gupta. 1989. Chromosomal polymorphism in *Rattus rattus rufescens*. Research Bulletin (Science) of the Panjab University, 40(pts. I-II):39-43.

Gilmore, R. M. 1933. Notes on the Unalaska collared lemming (*Dicrostonyx unalascensis unalascensis* Merriam). Journal of Mammalogy, 14:257-258.

Giménez, M. D., and C. J. Bidau. 1994. A first report of HSRs in chromosome 1 of *Mus musculus domesticus* from South America. Hereditas, 121:291-294.

Giménez, M. D., J. R. Contreras, and C. J. Bidau. 1997. Chromosomal variation in *Ctenomys pilarensis*, a recently described species from eastern Paraguay (Rodentia, Ctenomyidae). Mammalia, 61 (3):385-398.

Giménez, M. D., C. J. Bidau, C. F. Arguelles, and J. R. Contreras. 1999. Chromosomal characterization and relationship between two new species of *Ctenomys* (Rodentia, Ctenomyidae) from northern Cordoba province, Argentina. Zeitschrift für Saugetierkunde, 64:91-106.

Ginsberg, J. R., and D. W. Macdonald. 1990. Foxes, wolves, jackals, and dogs: An action plan for the conservation of canids. I.U.C.N., Gland, Switzerland, 116 pp.

Ginsburg, L. 1982. Sur la position systematique du petit panda, *Ailurus fulgens* (Carnivora, Mammalia). Geobios, Mémoire Spécial, 6:259-277.

Ginsburg, L., and J. Morales. 2000. Origin and évolution of the Melinae (Mustelidae, Carnivora, Mammalia). Comptes Rendus de l'Academie des Sciences. Serie II A, Sciences de la Terre et des Planetes, 330:221-225.

Girman, D. J., P. W. Kat, M. G. L. Mills, J. R. Ginsberg, M. Borner, V. Wilson, J. H. Fanshawe, C. Fitzgibbon, L. M. Lau, and R. K. Wayne. 1993. Molecular genetic and morphological analyses of the African wild dog (*Lycaon pictus*). Journal of Heredity, 84:450-459.

Girman, D. J., C. Vila, E. Geffen, S. Creel, M. G. L. Mills, J. W. McNutt, J. Ginsberg, P. W. Kat, K. H. Mamiya, and R. K. Wayne. 2001. Patterns of population subdivision, gene flow and genetic variability in the African wild dog (*Lycaon pictus*). Molecular Ecology, 10:1703-1723.

Glanz, W. E., and S. Anderson. 1990. Notes on Bolivian mammals. 7. A new species of *Abrocoma* (Rodentia) and relationships of the Abrocomidae. American Museum Novitates, 2991:1-32.

Glas, G. H., and A. M. Voûte. 1992a. Grote hoefijzerneus, *Rhinolophus ferrumequinum* (Schreber, 1974). Pp. 60-62, *in* Atlas van de Nederlandse zoogdieren (S. Broekhuizen, B. Hoekstra, V. van Laar, C. Smeenk, and J. B. M. Thissen, eds.). Uitgeverij Koninklijke Nederlandse Natuurhistorische Vereniging, Utrecht, 150 pp.

Glas, G. H., and A. M. Voûte. 1992b. Kleine hoefijzerneus *Rhinolophus hipposideros* (Bechstein, 1800). Pp. 63-65, *in* Atlas van de Nederlandse zoogdieren (S. Broekhuizen, B. Hoekstra, V. van Laar, C. Smeenk, and J. B. M. Thissen, eds.). Uitgeverij Koninklijke Nederlandse Natuurhistorische Vereniging, Utrecht, 150 pp.

Glass, B. P. 1947. Geographic variation in *Perognathus hispidus*. Journal of Mammalogy, 28:174-179.

Glass, B. P., and R. J. Baker. 1968. The status of the name *Myotis subulatus* Say. Proceedings of the Biological Society of Washington, 81:257-260.

Glass, G. E., and N. B. Todd. 1977. Quasi-continuous variation of the second upper premolar in *Felis bengalensis* Kerr 1792 and its significance for some fossil lynxes. Zeitschrift für Säugetierkunde, 42(1):36-44.

Glazier, D. S. 1980. Ecological shifts and the evolution of geographically restricted species of North American *Peromyscus* (mice). Journal of Biogeography, 7:63-83.

Glover, I. 1986. Archaeology in eastern Timor, 1966-67. Terra Australis, 11:1-241.

Godmann, O., and M. El Kasabi. 2001. Schutzmaßnahmen für den Feldhamster (*Cricetus cricetus*) in Hessen. Jahrbücher des Nassauischen Vereins für Naturkunde, 122:161-166.

Godthelp, H. J. 1990. *Pseudomys vandycki*, a Tertiary murid from Australia. Memoirs of the Queensland Museum, 28:171-173.

Godthelp, H. J. 1997. *Zyzomys rackhami* sp. nov. (Rodentia, Muridae) a rockrat from Pliocene Rackham's Roost Site, Riversleigh, northwestern Queensland. Memoirs of the Queensland Museum, 41:329-333.

Goetze, J. R. 1998. The mammals of the Edwards Plateau, Texas. Special Publications, Museum of Texas Tech University, 41:263 pp.

Goetze, J. R., F. D. Yancey, II, and A. M. Wallace. 1993. Additional records of the plains harvest mouse (*Reithrodontomys montanus*) from the Edwards Plateau, Texas. Texas Journal of Science, 45:358-359.

Goldman, C. A. 1984. Systematic revision of the African mongoose genus *Crossarchus* (Mammalia: Viverridae). Canadian Journal of Zoology, 62:1618-1630.

Goldman, C. A. 1987. *Crossarchus obscurus*. Mammalian Species, 290:1-5.

Goldman, C. A., and M. E. Taylor. 1990. *Liberiictis kuhni*. Mammalian Species, 348:1-3.

Goldman, D. P., R. Giri, and S. J. O'Brien. 1989. Molecular genetic-distance estimates among the Ursidae as indicated by one- and two-dimensional protein electrophoresis. Evolution, 43(2):282-295.

Goldman, E. A. 1910. Revision of the woodrats of the genus *Neotoma*. North American Fauna, 31:1-124.

Goldman, E. A. 1911. Revision of spiny pocket mice (genera *Heteromys* and *Liomys*). North American Fauna, 34:1-70.

Goldman, E. A. 1913. Descriptions of new mammals from Panama and Mexico. Smithsonian Miscellaneous Collections, 60(22):1-20.

Goldman, E. A. 1916. Notes on the genera *Isothrix* Wagner and *Phyllomys* Lund. Proceedings of the Biological Society of Washington, 29:125-128.

Goldman, E. A. 1918. The rice rats of North America (Genus *Oryzomys*). North American Fauna, 43:1-100.

Goldman, E. A. 1920. Mammals of Panama. Smithsonian Miscellaneous Collections, 69:1-309.

Goldman, E. A. 1932. Review of woodrats of *Neotoma lepida* group. Journal of Mammalogy, 13:59-67.

Goldman, E. A. 1935. New American mustelids of the genera *Martes, Gulo,* and *Lutra.* Proceedings of the Biological Society of Washington, 48:175-186.

Goldman, E. A. 1937. The wolves of North America. Journal of Mammalogy, 18:37-45.

Goldman, E. A. 1950. Raccoons of North and Middle America. North American Fauna, 60:1-153.

Golenishchev, F. N., and O. V. Sablina. 1991. [On taxonomy of *Microtus* (*Blanfordimys*) *afghanus*]. Zoologicheskii Zhurnal, 70:98-110 (in Russian).

Golenishchev, F. N., M. N. Meyer, and N. Sh. Bulatova. 2001. The hybrid zone between two karyomorphs of *Microtus arvalis* (Rodentia, Arvicolinae). Proceedings of the Zoological Institute, Russian Academy of Sciences, 289:89-94.

Golenishchev, F. N., O. V. Sablina, P. M. Borodin, and S. Gerasimov. 2002. Taxonomy of voles of the subgenus *Sumeriomys* Argyropulo, 1933 (Rodentia, Arvicolinae, *Microtus*). Russian Journal of Theriology, 1(1):43-55.

Golenishchev, F. N., V. G. Malikov, F. Nazari, A. Sh. Vaziri, O. V. Sablina, and A. V. Polyakov. 2003. New species of vole of *"guentheri"* group (Rodentia, Arvicolinae, *Microtus*) from Iran. Russian Journal of Theriology, 1(2):117-123.

Gómez-Laverde, M., M. L. Bueno, and A. Cadena. 1990. Poblacionesde ratas (*Proechimys semispinosus*) (Rodentia: Echimyidae). Pp. 244-251, *in* Biota y ecosistema de Gorgona y Gorgonilla (J. Aguirre and O. Rangel, eds.). Fundo Fen Columbia, Editorial Presencia, Bogotá, Columbia.

Gómez-Laverde, M., O. Montenegro-Diaz, H. López-Arevalo, A. Cadena, and M. L. Bueno. 1997. Karyology, morphology, and ecology of *Thomasomys laniger* and *T. niveipes* (Rodentia) in Colombia. Journal of Mammalogy, 78:1282-1289.

Gómez-Laverde, M., M. L. Bueno, and H. López-Arévalo. 1999. Descripcion cariologica y morfologica de *Nectomys magdalenae* (Rodentia: Muridae: Sigmodontinae). Revista de la Academia Colombiana de Ciencias Exactas, Físicas y Naturales, 23:631-640.

Gómez-Laverde, M., R. P. Anderson, and L. F. Garcia. 2004. Integrated systematic evaluation of the Amazonian genus *Scolomys* (Rodentia: Sigmodontinae). Mammalian Biology/Zeitschrift für Säugetierkunde, 69:119-139.

Gompper, M. E. 1995. *Nasua narica*. Mammalian Species, 487:1-10.

Gompper, M. E., and D. M. Decker. 1998. *Nasua nasua*. Mammalian Species, 580:1-9.

Gonçalves, P. R., J. A. de Oliveira, M. A. Correa, and L. M. Pessoa. In Press. Morphological and cytogenetic analyses of *Bibimys labiosus* (Winge, 1887) (Rodentia, Sigmodontinae): Implications for its affinities with the scapteromyine group. *in* Mammalian diversification: From population genetics to biogeography (E. Lacey and P. Myers, eds.). University of California Press, Berkeley.

Gong, Z.-d., Y.-x. Wang, Z.-h Li, and S.-q. Li. 2000. [A new species of pika: Piyanma black pika, *Ochotona nigritia* (Lagomorpha: Ochotonidae) from Yunnan, China]. Zoological Research, 21(3):204-209 (in Chinese with English summary).

Gonzales, P. C., and R. S. Kennedy. 1996. A new species of *Crateromys* (Rodentia: Muridae) from Panay, Philippines. Journal of Mammalogy, 77(1):25-40.

González, E. M. 2000*a*. Un nuevo genero de roedor sigmodontino de Argentina y Brasil (Mammalia: Rodentia: Sigmodontinae). Comunicaciones Zoologicas del Museo de Historia Natural de Montevideo, 12(196):1-12.

González, E. M. 2000*b*. Lista sistemática, afinidades biogeográficas, hábitos y hábitats de los mamíferos terrestres autóctonos de Uruguay (Mammalia): Una introducción. Jornadas sobre animales silvestres, desarrollo sustentable y medio ambiente, pp. 58-73.

González, E. M. 2001. Guía de campo de los mamíferos de Uruguay. Introducción al estudio de los mamíferos. Vida Silvestre, Montevideo, 339 pp.

González, E. M., and E. Massoia. 1995. Revalidacion del genero *Deltamys* Thomas, 1917, con la descripcion de una nueva subespecie de Uruguay y sur del Brasil (Mammalia: Rodentia: Cricetidae). Comunicacions Zoologicas del Museo de Historia Natural de Monevideo, 182:1-8.

González, E. M., and U. F. J. Pardiñas. 2002. *Deltamys kempi*. Mammalian Species, 711:1-4.

González, E. M., J. González, G. Frequeiro, and A. Saralegui. 1995. Mamíferos encontrados en regurgitados de lechuzas del noreste de Uruguay. Comunicacions Zoologicas del Museu de Historia Natural de Montevideo, 12(181):1-4.

González, E. M., A. Langguth, and L. F. de Oliveira. 1998. A new species of *Akodon* from Uruguay and southern Brazil (Mammalia: Rodentia: Sigmodontinae). Comunicacions Zoologicas del Museu de Historia Natural de Montevideo, 12(191):1-8.

Gonzalez, F. X., and F. A. Cervantes. 1996. Karyotype of the white-sided jackrabbit (*Lepus callotis*). Southwestern Naturalist, 41:93-95.

González, J. 1994. Analisis bioestadistico del genero *Scapteromys* en Uruguay (Mammalia: Rodentia: Cricetidae). Comunicacions Zoologicas del Museo de Historia Natural de Montevideo, 12:1-6.

González-Esteban, J., I. Villate, and J. Gosálbez. 1995. Expansion del area de distribucion de *Microtus arvalis asturianus* Miller, 1908 (Rodentia, Arvicolidae) en la Meseta Norte (España). Doñana, Acta Vertebrata, 22(1-2):106-110.

González-Esteban, J., J. Gosálbez, and E. Castién. 1996. Estudio morfometrico y del crecimiento de *Apodemus sylvaticus* L., 1758 (Rodentia, Muridae) en el norte de la Peninsula Iberica. Doñana, Acta Vertebrata, 23(1):63-73.

Good, A. I. 1947. Les rongeurs du Cameroun. Bulletin de la Société L'Etudes Camerounaises, 17-18:5-20.

Good, J. M., J. R. Dembowski, D. W. Nagorsen, and J. Sullivan. 2003. Phylogeography and introgressive hybridization: Chipmunks (Genus *Tamias*) in the Northern Rocky Mountains. Evolution, 57:1900-1916.

Goodall, R. N. P., A. R. Galeazzi, S. Leatherwood, K. W. Miller, I. S. Cameron, R. K. Kastelein, and A. P. Sobral. 1988. Studies of Commerson's dolphins, *Cephalorhynchus commersonii*, off Tierra del Fuego, 1976-1984, with a review of information on the species in the South Atlantic. Pp. 3-70, *in* Biology of the genus *Cephalorhynchus* (R. L. Brownell, Jr., and G. P. Donovan, eds.). Reports of the International Whaling Commission, Special Issue, 9:1-344.

Goodman, M., C. A. Porter, J. Czelusniak, S. L. Page, H. Schneider, J. Shoshani, G. Gunnell, and C. P. Groves. 1998. Toward a phylogenetic classification of Primates based on DNA evidence complemented by fossil evidence. Molecular Phylogenetics and Evolution, 9:585-598.

Goodman, M., J. Czelusniak, S. Page and C. M. Meireles. 2001. Where DNA sequences place *Homo sapiens* in a phylogenetic classification of Primates. Pp. 279-289, *in* Humanity from African naissance to coming millenia, (P. V. Tobias, M. A. Raath, J. Moggi-Cecchi, and G. A. Doyle, eds.). Firenze University Press and Witwatersrand University Press, Florence and Johannesburg. 416 pp.

Goodman, S. M. 1989. The occurrence of *Crocidura floweri* in Wadi el Natrun, Egypt. Mammalia, 53:134-135.

Goodman, S. M. 1995. *Rattus* on Madagascar and the dilemma of protecting the endemic rodent fauna. Conservation Biology, 9:450-453.

Goodman, S. M., and J. P. Benstead (eds.). 2003. Natural History of Madagascar. The University of Chicago Press, Chicago, xxi + 1709 pp.

Goodman, S. M., and M. D. Carleton. 1996. The rodents of the Réserve Naturelle Intégrale d'Andringitra, Madagascar. Fieldiana: Zoology, n.s., 85:257-283.

Goodman, S. M., and M. D. Carleton. 1998. The rodents of the Réserve Spéciale d'Anjanaharibe-Sud, Madagascar. Fieldiana: Zoology, n.s., 90:201-221.

Goodman, S. M., and N. R. Ingle. 1993. Sibuyan Island in the Philippines—threatened and in need of conservation. Oryx, 27:174-180.

Goodman, S. M., and P. D. Jenkins. 1998. The Insectivores of the Réserve Spéciale d'Anjanaharibe-Sud, Madagascar. Pp. 139-161, *in* A floral and faunal inventory of the eastern slopes of the Réserve Spéciale d'Anjanaharibe-Sud, Madagascar: With reference to elevational variation. (S. M. Goodman, ed.). Fieldiana: Zoology, n.s. (90):246 pp.

Goodman, S. M., and P. D. Jenkins. 2000. Tenrecs (Lipotyphla: Tenrecidae) of the Parc National de Marojejy, Madagascar. Pp. 201-229, *in* A floral and faunal inventory of the Parc National de Marojejy, Madagascar: With reference to elevational variation. (S. M. Goodman, ed.). Fieldiana: Zoology, n.s. (97):286 pp.

Goodman, S. M., and D. Rakotondravony. 1996. The Holocene distribution of *Hypogeomys* (Rodentia: Muridae: Nesomyinae). Biogeographie de Madagascar, 1996:283-293.

Goodman, S. M., and B. P. N. Rasolonandrasana. 2001. Elevational zonation of birds, insectivores, rodents and primates on the slopes of the Andrinitra Massif, Madagascar. Journal of Natural History, 35:285-305.

Goodman, S. M., and V. Soarimalala. 2002. Les petits mammifères de la Réserve Spéciale de Manongarivo, Madagascar. Boissiera, 59:383-401.

Goodman, S. M., A. Andrianarimisa, L. E. Olson, and V. Soarimalala. 1996a. Patterns of elevational distribution of birds and small mammals in the humid forest of Montagne d'Ambre, Madagascar. Ecotropica, 2:87-98.

Goodman, S. M., D. Rakotondravony, G. Schatz, and L. Wilme. 1996b. Species richness of forest-dwelling birds, rodents and insectivores in a planted forest of native trees: A test case from the Ankaratra, Madagascar. Ecotropica, 2:109-120.

Goodman, S. M., P. D. Jenkins, and M. Pidgeon. 1999a. Lipotyphla (Tenrecidae and Soricidae) of the Réserve Naturelle Intégrale d'Andohahela, Madagascar. Pp. 187 – 216, in A floral and faunal inventory of the eastern slopes of the Réserve Naturelle Intégrale d'Andohahela, Madagascar: With reference to elevational variation (S. M. Goodman, ed.). Fieldiana: Zoology, n.s. (94):297 pp.

Goodman, S. M., M. D. Carleton, and M. Pidgeon. 1999b. The rodents of the Réserve Naturelle Intégrale d'Andohahela, Madagascar. Pp. 217-249, in A floral and faunal inventory of the eastern slopes of the Réserve Naturelle Intégrale d'Andohahela, Madagascar: With reference to elevational variation (S. M. Goodman, ed.). Fieldiana: Zoology, n.s. (94):297 pp.

Goodman, S. M., D. Rakotondravony, V. Soarimalala, J. B. Duchemin, and J.-M. Duplantier. 2000. Syntopic occurrence of Hemicentetes semispinosus and H. nigriceps (Lipotyphla: Tenrecidae) on the Central Highlands of Madagascar. Mammalia, 64:113–116.

Goodman, S. M., R. Hutterer, and P. R. Ngnegueu. 2001a. A report on the community of shrews (Mammalia: Soricidae) occurring in the Minkébé Forest, northeastern Gabon. Mammalian Biology, 66:22-34.

Goodman, S. M., V. Soarimalala, and D. Rakotondravony. 2001b. The rediscovery of Brachytarsomys villosa F. Petter, 1962 (Rodentia, Nesomyinae), in the northern highlands of Madagascar. Mammalia, 65:83-86.

Goodwin, G. G. 1934. Mammals collected by A. W. Anthony in Guatemala, 1924-1928. Bulletin of the American Museum of Natural History, 68:1-60.

Goodwin, G. G. 1938. Five new rodents from the eastern Elburz Mountains and a new race of hare from Teheran. American Museum Novitates, 1950:1-5.

Goodwin, G. G. 1940. Mammals collected by the Legendre 1938 Iran expedition. American Museum Novitates, 1082:1-17.

Goodwin, G. G. 1946. Mammals of Costa Rica. Bulletin of the American Museum of Natural History, 87:271-471.

Goodwin, G. G. 1953. Catalogue of the type specimens of Recent mammals in the American Museum of Natural History. Bulletin of the American Museum of Natural History, 102:207-412.

Goodwin, G. G. 1955a. Mammals from Guatemala, with the description of a new little brown bat. American Museum Novitates, 1744:1-5.

Goodwin, G. G. 1955b. Three new cotton rats from Tehuantepec, Mexico. American Museum Novitates, 1705:1-5.

Goodwin, G. G. 1958. Three new bats from Trinidad. American Museum Novitates, 1877:1-6.

Goodwin, G. G. 1959a. Descriptions of some new mammals. American Museum Novitates, 1967:1-8.

Goodwin, G. G. 1959b. Bats of the Subgenus Natalus. American Museum Novitates, 1977:1-22.

Goodwin, G. G. 1966. A new species of vole (genus Microtus) from Oaxaca, Mexico. American Museum Novitates, 2243:1-4.

Goodwin, G. G. 1969. Mammals from the state of Oaxaca, Mexico, in the American Museum of Natural History. Bulletin of the American Museum of Natural History, 141:1-269.

Goodwin, G. G., and A. M. Greenhall. 1961. A review of the bats of Trinidad and Tobago. Bulletin of the American Museum of Natural History, 122:189-301.

Goodwin, G. G., and A. M. Greenhall. 1962. Two new bats from Trinidad, with comments on the status of the genus Mesophylla. American Museum Novitates, 2080:1-18.

Goodwin, H. T. 1995. Pliocene-Pleistocene biogeographic history of prairie dogs, genus Cynomys (Sciuridae). Journal of Mammalogy, 76(1):100-122.

Goodwin, R. E. 1979. The bats of Timor: Systematics and ecology. Bulletin of the American Museum of Natural History, 163:73-122.

Goodyear, N. C. 1991. Taxonomic status of the silver rice rat, Oryzomys argentatus. Journal of Mammalogy, 72:723-730.

Goodyear, N. C., and J. D. Lazell, Jr. 1986. Relationships of the silver rice rat, Oryzomys argentatus (Rodentia, Muridae). Postilla, 198:1-7.

Gordon, D. H. 1986. Extensive chromosomal variation in the pouched mouse, Saccostomus campestris (Rodentia, Cricetidae) from southern Africa: A preliminary investigation of evolutionary status. Cimbebasia, ser. A, 8:37-47.

Gordon, D. H. 1987. Discovery of another species of tree rat. Transvaal Museum Bulletin, 22:30-32.

Gordon, D. H. 1991. Chromosomal variation in the water rat *Dasymys incomtus* (Rodentia: Muridae). Journal of Mammalogy, 72:411-414.

Gordon, D. H., and I. L. Rautenbach. 1980. Species complexes in medically important rodents: Chromosome studies of *Aethomys*, *Tatera*, and *Saccostomus* (Rodentia: Muridae, Cricetidae). South African Journal of Science, 76:559-561.

Gordon, D. H., and C. R. B. Watson. 1986. Identification of cryptic species of rodents (*Mastomys*, *Aethomys*, *Saccostomus*) in the Kruger National Park. South African Journal of Zoology, 21:95-99.

Gorecki, A., R. Meczeva, T. Pis, S. Gerasimov, and W. Walkowa. 1990. Geographical variation of thermoregulation in wild populations of *Mus musculus* and *Mus spretus*. Acta Theriologica, 35:209-214.

Gorman, M. L., and R. D. Stone. 1990. The natural history of moles. Comstock, Ithaca, 160 pp.

Görner, M., and A. Henkel. 1988. Zum Vorkommen und zur Ökologie der Schläfer (Gliridae) in der DDR. Säugetierkundliche Informationen, 2(12):515-535.

Gorog, A. J., M. H. Sinaga, and M. D. Engstrom. 2004. Vicariance or dispersal? Historical biogeography of three Sunda shelf murine rodents (*Maxomys surifer*, *Leopoldamys sabanus* and *Maxomys whiteheadi*). Biological Journal of the Linnean Society, 81:91-109.

Gosalbez, J., and V. Sans-Coma. 1977. Datos sobre *Microtus arvalis* Pallas, 1778, del Pirineo Catalan. Publicationes. Departamento de Zoologia (Barcelona), 11:59-68.

Gosalbez-Noguera, J., V. Sans-Coma, and H. Kahmann. 1989. Der Gartenschläfer *Eliomys q. quercinus* L., 1758, im Bergland Andorra: Morphometrie, Erscheinungsbild, Wachstum und Fortpflanzung (Mammalia: Rodentia). Spixiana, 12(3):323-335.

Gould, G. C. 1995. Hedgehog phylogeny (Mammalia, Erinaceidae)—the reciprocal illumination of the quick and the dead. American Museum Novitates, 3131:1-45.

Graf, J.-D. 1982. Genetique biochimique, zoogeographie et taxonomie des Arvicolidae (Mammalia, Rodentia). Revue Suisse de Zoologie, 89:749-787.

Graf, J.-D., and A. Meylan. 1980. Polymorphisme chromosomique et biochimique chez *Pitymys multiplex* (Mammalia, Rodentia). Zeitschrift für Saügetierkunde, 45:133-148.

Graf, J.-D., J. Hausser, A. Farina, and P. Vogel. 1979. Confirmation du statut spécifique de *Sorex samniticus* Altobello, 1926 (Mammalia, Insectivora). Bonner Zoologische Beiträge, 30:14-21.

Graham, R. W., and M. A. Graham. 1994. Late Quaternary distribution of *Martes* in North America. Pp. 26-58, *in* Martens, sables, and fishers: Biology and conservation (S. W. Buskirk, A. S. Harestad, M. G. Raphael, and R. A. Powell, eds.). Cornell University Press, Ithaca, New York. 484 pp.

Graham, R. W., and E. L. Lundelius, Jr. 1994. FAUNMAP: A database documenting Late Quaternary distributions of mammal species in the United States. Scientific Papers, Illinois State Museum, 25(1and 2):1-690.

Grandidier, G. 1905. Recherches sur les lemuriens disparus et en particuliers sur ceux qui vivaient a Madagascar. Nouvelles Archives de la Museum de l'Histoire Naturelle, ser. 4, 8:1-140.

Granjon, L. 1991. Les rongeurs myomorphes du bassin du Kouilou (Congo). Pp. 265-278, *in* Flore et faune du Bassin du Kouilou (Congo) et leur exploitation (R. J. Dowsett and F. Dowsett-Lemaire, eds.). Tauraco Research Report 4, 340 pp.

Granjon, L., J.-M. Duplantier, J. Catalan, and J. Britton-Davidian. 1992. Karyotypic data on rodents from Senegal. Israel Journal of Zoology, 38:263-276.

Granjon, L., J.-M. Duplantier, J. Catalan, J. Britton-Davidian, and G. N. Bronner. 1996. Conspecificity of *Mastomys natalensis* (Rodentia: Muridae) from Senegal and South Africa: Evidence from experimental crosses, karyology and biometry. Mammalia, 60(4):697-706.

Granjon, L., J. F. Cosson, J. Cuisin, M. Tranier, and F. Colas. 1997a. Les mammifès du littoral Mauritanien: 2. Biogéographie et écologie. Pp. 77-89, *in* Environnement et littoral Mauritanien, Actes du Colloque, 12-13 juin 1995, Nouakchott, Mauritanie (F. Colas, ed.). CIRAD (Collection Colloques), Montpellier, France, 193 pp.

Granjon, L., J.-M. Duplantier, J. Catalan, and J. Britton-Davidian. 1997b. Systematics of the genus *Mastomys* (Thomas, 1915) (Rodentia: Muridae). A review. Belgian Journal of Zoology, 127(suppl. 1):7-18.

Granjon, L., A. Bonnet, W. Hamdine, and V. Volobouev. 1999. Reevaluation of the taxonomic status of North African gerbils usually referred to as *Gerbillus pyramidum* (Gerbillinae, Rodentia): Chromosomal and biometrical data. Zeitschrift für Säugetierkunde, 64:298-307.

Granjon, L., V. M. Aniskin, V. Volobouev, and B. Sicard. 2002a. Sand-dwellers in rocky habitats: A new species of *Gerbillus* (Mammalia: Rodentia) from Mali. Journal of Zoology, London, 256:181-190.

Granjon, L., C. Bruderer, J. E. Cosson, A. T. Dia, and F. Colas. 2002b. The small mammal community of a coastal site of south-west Mauritania. African Journal of Ecology, 40:10-17.

Grant, P. R. 1974. Reproductive compatibility of voles from separate continents (Mammalia: *Clethrionomys*). Journal of Zoology, London, 174:245-254.

Graphodatsky, A. S., and I. M. Fokin. 1993. Comparative cytogenetics of Gliridae. Zoologicheskii Zhurnal, 72(11):104-113.

Graphodatsky, A. S., V. T. Volobuev, D. V. Ternovskii, and S. I. Radzhabli. 1976. [G-banding of chromosomes in seven species of Mustelidae (Carnivora).] Zoologischeskii Zhurnal, 55:1704-1709 (in Russian).

Graphodatsky, A. S., S. I. Radzhabli, A. V. Sharshov, and M. V. Zaitsev. 1988. [Karyotypes of five *Crocidura* species of the USSR fauna]. Tsitologiya, 30:1247-1255 (in Russian).

Graphodatsky, A. S., S. I. Radzhabli, M. V. Zaitsev, and A. V. Sharshov. 1993. The levels of chromosome conservatisms in the different groups of insectivores (Mammalia, Insectivora). Pp. 47-57, *in* Questions of Systematics, Faunistics and Palaeontology of Small Mammals (M. V. Zaitsev, ed.). Trudy Zoologicheskogo Instituta, 243.pp (in Russian with English summaries).

Graphodatsky, A. S., O. V. Sablina, M. N. Meyer, V. G. Malikov, E. A. Isakova, V. A. Trifonov, A. V. Polyakov, T. P. Lushnikova, N. V. Vorobieva, N. A. Serdyukova, P. L. Perelman, P. M. Borodin, P. Benda, D. Frynta, L. Leikepová, P. Munclinger, J. Piálek, J. Sádlová, and J. Zima. 2000. Comparative cytogenetics of hamsters of the genus *Calomyscus*. Cytogenetics and Cell Genetics, 88:296-304.

Grassé, P. P., and P. L. Dekeyser. 1955. Ordre des Rongeurs. Pp. 1321-1525, *in* Traite de Zoologie, Tome 17(2), Mammiferes, les ordres: Anatomie, ethologie, systematique (P. P Grasse, ed.). Masson, Paris, pp. 1173-2300.

Graur, D. 1994. Molecular evidence concerning the phylogenetic integrity of the Murinae. Israel Journal of Zoology, 40:255-264.

Graur, D., W. A. Hide, and W.-H. Li. 1991. Is the guinea-pig a rodent? Nature, 351:649-651.

Graur, D., W. A. Hide, A. Zharkikh, and W.-H. Li. 1992. The biochemical phylogeny of guinea-pigs and gundis, and the paraphyly of the order Rodentia. Comp. Biochem. Physiol., 101B:495-498.

Gray, A. P. 1954. Mammalian hybrids: A checklist with bibliography. Commonwealth Agricultural Bureau, Farnham Royal, United Kingdom, 262 pp.

Gray, A. P. 1972. Mammalian hybrids. A check-list with bibliography. Commonwealth Bureau of Animal Breeding and Genetics, Edinburgh, Technical Communication, 10(revised):1-262.

Gray, G. G., and C. D. Simpson. 1980. *Ammotragus lervia*. Mammalian Species, 144:1-7.

Gray, J. E. 1821. On the natural arrangement of vertebrose animals. London Medical Repository, 15(1):296-310.

Gray, J. E. 1825. Outline of an attempt at the disposition of the Mammalia into tribes and families with a list of the genera apparently appertaining to each tribe. Annals of Philosophy, n.s., ser. 2, 10:337-344.

[Gray, J. E.]. 1827. Synopsis of the species of the Class Mammalia, as arranged with reference to their organization, by Cuvier, and other naturalists, with specific characters, synonyma, &c. &c., vol. 5, *in* The animal kingdom arranged in conformity with its organization, by the Baron Cuvier, with additional descriptions of all the species hitherto named, and of many not before noticed. (E. Griffith, C. H. Smith, and E. Pidgeon, eds.). G. B. Whittaker, London, 391 pp.

Gray, J. E. 1828-1830 [-1924]. Spicilegia zoological; or original figures and short systematic descriptions of new and unfigured animals. London, 1:1-8[1828]; 2:9-12[1830]; 3:1-13[1924].

Gray, J. E. 1830-1834. Illustrations of Indian zoology; chiefly selected from the collection of Major-General Hardwicke. Treuttel, Wurtz, Treuttel, Jun. and Richter, London, [vol 1, 1830-1832; vol 2, 1833-1834].

Gray, J. E. 1832. On the family of Viverridae and its generic sub-divisions, with an enumeration of the species of several new ones. Proceedings of the Committee of Science and Correspondence of the Zoological Society of London, 1832(2):63-68.

Gray, J. E. 1837 [1838]. On a new species of paradoxure (*Paradoxurus derbianus*) with remarks on some Mammalia recently purchased by the British Museum, and characters of the new species. Proceedings of the Zoological Society of London, 1837:67.

Gray, J. E. 1842. Description of two new species of Mammalia discovered in Australia by Captain George Gray (*Tarsipes Spenseré* and *Chéropus castanotis*). Annals and Magazine of Natural History, [ser. 1], 9:39-42.

Gray, J. E. 1843. List of the specimens of Mammalia in the collection of the British Museum. British Museum (Natural History) Publications, London, 216 pp.

Gray, J. E. 1846. On the cetaceous animals. Pp. 13-53, *in* The zoology of the voyage of H. M. S. Erebus and Terror, under the command of Capt. Sir J. C. Ross, R. N., F. R. S., during the years 1839 to 1843 (Sir J. Richardson and J. E. Gray, eds.) [1844-1875]. E. W. Janson, London, 2 vols.

Gray, J. E. 1849 [1850]. Vertebrata, *in* The zoology of the voyage of H. M. S. Samarang; under the command of Captain Sir Edward Belcher, C. B., F. R. A. S., F. G. S., during the years 1843-1846 (A. Adams, ed.). Reeve and Benham, London, 250 pp.

Gray, J. E. 1850. Catalogue of the specimens of mammalia in the collection of the British Museum. Part I. Cetacea. British Museum (Natural History), London, 153 pp.

Gray, J. E. 1853 [1855]. Observations on some rare Indian animals. Proceedings of the Zoological Society of London, 1853:190-192.

Gray, J. E. 1854. The ounces. Annals and Magazine of Natural History, ser. 2, 14:394.

Gray, J. E. 1864 [1865]. A revision of the genera and species of viverrine animals (Viverridae) founded on the collection in the British Museum. Proceedings of the Zoological Society of London, 1864:502-579.

Gray, J. E. 1867a. Notes on the variegated or yellow-tailed rats of Australasia. Proceedings of the Zoological Society of London, 1867(2):299-314.

Gray, J. E. 1867b. Notes on certain species of cats in the collection of the British Museum. Proceedings of the Zoological Society of London, 1867:394-405.

Gray, J. E. 1869. Catalogue of carnivorous, pachydermatous and edentate mammals in the British Museum. British Museum (Natural History) Publications, London, 398 pp.

Gray, J. E. 1870. Catalogue of monkeys, lemurs and fruit eating bats in the collections of the British Museum. British Museum, London, 137 pp.

Gray, J. E. 1874. Description of a new species of cat (*Felis badia*) from Sarawak. Proceedings of the Zoological Society of London, 1874:322-323.

Gray, P. A., M. B. Fenton, and V. Van Cakenberghe. 1999. *Nycteris thebaica*. Mammalian Species, 612:1-8.

Grayson, D. K., S. D. Livingston, E. Rickart, and M. W. Shaver, III. 1996. Biogeographic significance of low-elevation records for *Neotoma cinerea* from the northern Bonneville Basin, Utah. The Great Basin Naturalist, 56:191-196.

Green, C. A., H. Keogh, D. H. Gordon, M. Pinto, and E. K. Hartwig. 1980. The distribution, identification, and naming of the *Mastomys natalensis* species complex in southern Africa (Rodentia: Muridae). Journal of Zoology, London, 192(1):17-23.

Green, J. S., and J. T. Flinders. 1980. *Brachylagus idahoensis*. Mammalian Species, 125:1-4.

Green, R. H. 1995. Long-tailed mouse, *Pseudomys higginsi*. Pp. 605-606, *in* Mammals of Australia (R. Strahan, ed.). Smithsonian Institution Press, Washington D.C., 756 pp.

Greenbaum, I. F., and R. J. Baker. 1976. Evolutionary relationships in *Macrotus* (Mammalia: Chiroptera): Biochemical variation and karyology. Systematic Zoology, 25:15-25.

Greenbaum, I. F., and R. J. Baker. 1978. Determination of the primitive karyotype of *Peromyscus*. Journal of Mammalogy, 59:820-834.

Greenbaum, I. F., S. J. Gunn, S. A. Smith, B. F. McAllister, D. W. Hale, R. J. Baker, M. D. Engstrom, M. H. Hamilton, W. S. Modi, and L. W. Robbins. 1994. Cytogenetic nomenclature of deer mice, *Peromyscus* (Rodentia): Revision and review of the standardized karyotype. Cytogenetics and Cell Genetics, 66:181-195.

Greene, E. C. 1935. Anatomy of the rat. Transactions of the American Philosophical Society, n.s., 27:1-370.

Greenhall, A. M., and W. A. Schutt. 1996. *Diaemus youngi*. Mammalian Species, 533:1-7.

Greenhall, A. M., G. Joermann, U. Schmidt, and M. R. Seidel. 1983. *Desmodus rotundus*. Mammalian Species, 202:1-6.

Greenhall, A. M., U. Schmidt, and G. Joermann. 1984. *Diphylla ecaudata*. Mammalian Species, 227:1-3.

Gregorin, R., and V. A. Taddei. 2002. Chave artificial para a identificação de molossídeos Brasileiros (Mammalia, Chiroptera). Mastozoologia Neotropical, 9:13-32.

Gregory, W. K. 1910. The orders of mammals. Bulletin of the American Museum of Natural History, 37:1-524.

Gregory, W. K. 1936. On the phylogenetic relationships of the giant panda (*Ailuropoda*) to other carnivores. American Museum Novitates, 878:1-29.

Gregory, W. K., and M. Hellman. 1939. On the evolution and major classification of the civets (Viverridae) and allied fossil and recent Carnivora: Phylogenetic study of the skull and dentition. Proceedings of the American Philosophical Society, 81(3):309-392.

Grenyer, R., and A. Purvis. 2003. A composite species-level phylogeny of the 'Insectivora' (Mammalia: Order Lipotyphla Haeckel, 1866). Journal of Zoology, London, 260:245-257.

Greth, A., D. Williamson, C. Groves, G. Schwede, and M. Vassart. 1993. Bilkis gazelle in Yemen—status and taxonomic relationships. Oryx, 27:239-244.

Griffin, M. 1990. A review of taxonomy and ecology of gerbilline rodents of the central Namib Desert, with keys to the species (Rodentia: Muridae). Pp. 83-98, *in* Namib ecology: 25 years of Namib research (M. K. Seely, ed.). Transvaal Museum Monograph, Transvaal Museum, Pretoria, 7:1-230.

Griffiths, M. 1978. The biology of monotremes. Academic Press, New York, 367 pp.

Griffiths, M., and N. G. Simms. 1993. Observations on the anatomy of mammary glands in two species of conilurine rodent (Muridae: Hydromyinae) and in an opossum (Marsupialia: Didelphidae). Australian Mammalogy, 16:9-15.

Griffiths, T. A. 1982. Systematics of the New World nectar-feeding bats (Mammalia, Phyllostomidae), based on the morphology of the hyoid and lingual regions. American Museum Novitates, 2742:1-45.

Griffiths, T. A. 1994. Phylogenetic systematics of slit-faced bats (Chiroptera, Nycteridae), based on hyoid and other morphology. American Museum Novitates, 3090:1-17.

Griffiths, T. A., and A. L. Smith. 1991. Systematics of emballonuroid bats (Chiroptera: Emballonuridae and Rhinopomatidae) based on hyoid morphology. Bulletin of the American Museum of Natural History, 206:62-83.

Griffiths, T. A., K. F. Koopman, and A. Starrett. 1991. The systematic relationship of *Emballonura nigrescens* to other species of *Emballonura* and *Coleura* (Chiroptera: Emballonuridae). American Museum Novitates, 2996:1-16.

Griffiths, T. A., A. Truckenbrod, and P. J. Sponholtz. 1992. Systematics of megadermatid bats (Chiroptera, Megadermatidae), based on hyoid morphology. American Museum Novitates, 3041:1-21.

Grimshaw, J. H. 1998. The giant forest hog *Hylochoerus meinertzhageni* in Tanzania—records rejected. Mammalia, 62:123-125.

Grimshaw, J., N. Cordeiro and C. Foley. 1995. The mammals of Kilimanjaro. Journal of East African Natural History, 84:105-139.

Grinnell, J. 1922. A geographical study of the kangaroo rats of California. University of California Publications in Zoology, 24:1-124.

Grinnell, J. 1933. Review of the Recent mammal fauna of California. University of California Publications in Zoology, 40:71-234.

Grinnell, J., and J. Dixon. 1918. Natural history of the ground squirrels of California. Monthly Bulletin of the California State Commission on Horticulture, 7:597-708.

Gromov, I. M. 1990. [On the probable causes of differences in the character of microevolution in some common species of arvicolines (Arvicolinae, Rodentia) of nontropical Palaearctic]. USSR Academy of Sciences, Proceedings of the Zoological Institute, Lenningrad, 225:3-20 (in Russian).

Gromov, I. M., and G. I. Baranova (eds.). 1981. Katalog mlekopitayushchikh SSSR [Catalog of mammals of the USSR]. Nauka, Leningrad, 456 pp. (in Russian).

Gromov, I. M., and M. A. Erbajeva. 1995. [The mammals of Russia and adjacent territories. Lagomorphs and Rodents.] Russian Academy of Sciences, Zoological Institut, St. Petersburg, 520 pp.

Gromov, I. M., and I. Ya. Polyakov. 1977. Fauna SSSR, Mlekopitayushchie, tom 3, vyp. 8 [Fauna of the USSR, vol. 3, pt. 8, Mammals]. Polevki [Voles (Microtinae)]. Nauka, Moscow-Leningrad, 504 pp. (in Russian).

Gromov, I. M., A. A. Gureev, G. A. Novikov, I. I. Sokolov, P. P. Strelkov, and K. K. Chapskii. 1963. Mlekopitayushchie fauny SSSR, Chast' 1. [Mammals of the fauna of the USSR. [Vol.] 1. Insectivora, Chiroptera, Lagomorpha, and Rodentia]. Akademii Nauk SSSR, Moskva, 1:1-640.

Gromov, I. M., D. I. Bibikov, N. I. Kalabukhov, and M. N. N. Meier. 1965. Fauna SSSR, Mlekopitayushchie, tom. 3, vyp. 2 [Fauna of the U.S.S.R. Mammals. vol. 3, No. 2]. Nazemnye belich'e [Ground Squirrels]. Nauka, Moscow-Leningrad, 467 pp. (in Russian).

Gropp, A., J. T. Marshall, and A. Markvong. 1973. Chromosomal findings in the spiny mice of Thailand (genus *Mus*) and occurence of a complex intraspecific variation in *M. shortridgei*. Zeitschrift für Säugetierkunde, 38:159-168.

Gropp, A., H. Winking, F. Frank, G. Noack, and K. Fredga. 1976. Sex-chromosome aberrations in wood lemmings (*Myopus schisticolor*). Cytogenetics and Cell Genetics, 17:343-358.

Groves, C. P. 1967a. On the gazelles of the genus *Procapra* Hodgson, 1846. Zeitschrift für Säugetierkunde, 32:144-149.

Groves, C. P. 1967b. Geographic variation in the black rhinoceros *Diceros bicornis* (L., 1758). Zeitschrift für Säugetierkunde, 32:267-276.

Groves, C. P. 1967c. On the rhinoceroses of South-east Asia. Säugetierkundliche Mitteilungen, 15:221-237.

Groves, C. P. 1969a. On the smaller gazelles of the genus *Gazella* de Blainville, 1816. Zeitschrift für Säugetierkunde, 34:38-60.

Groves, C. P. 1969b. Systematics of the anoa (Mammalia, Bovidae). Beaufortia, 223:1-12.

Groves, C. P. 1970. The forgotten leaf-eaters, and the phylogeny of the Colobinae. Pp. 555-587, *in* Old World Monkeys: Evolution, systematics and behavior (J. R. Napier and P. H. Napier, eds.). Academic Press, London, 660 pp.

Groves, C. P. 1971a. *Pongo pygmaeus*. Mammalian Species, 4:1-6.

Groves, C. P. 1971b. Request for a declaration modifying Article 1 so as to exclude names proposed for domestic animals from Zoological Nomenclature. Bulletin of Zoological Nomenclature, 7:269-272.

Groves, C. P. 1971c. Systematics of the genus *Nycticebus*. Proceedings of the 3rd International Congress of Primatology, 1:44-53.

Groves, C. P. 1972a. *Ceratotherium simum*. Mammalian Species, 8:1-6.

Groves, C. P. 1972b. Systematics and phylogeny of gibbons. Pp. 1-89, *in* Gibbon and siamang: Evolution, ecology, behavior and captive maintenance (D. M. Rumbaugh, ed.). S. Karger, New York, 1:1-263.

Groves, C. P. 1974a. Horses, asses and zebras in the wild. David and Charles, Newton Abbot and London, 192 pp.

Groves, C. P. 1974*b*. Taxonomy and phylogeny of prosimians. Pp. 449-473, *in* Prosimian biology (R. D. Martin, G. A. Doyle, and A. C. Walker, eds.). Duckworth and Co., Ltd., London, 983 pp.

Groves, C. P. 1975*a*. Notes on the gazelles. 1. *Gazella rufifrons* and the zoogeography of central African Bovidae. Zeitschrift für Säugetierkunde, 40:308-319.

Groves, C. P. 1975*b*. Taxonomic notes on the white rhinoceros *Ceratotherium simum* (Burchell, 1817). Säugetierkundliche Mitteilungen, 23:200-212.

Groves, C. P. 1976. The taxonomy of *Moschus* (Mammalia, Artiodactyla), with particular reference to the Indian Region. Journal of the Bombay Natural History Society, 72:662-676.

Groves, C. P. 1978*a*. A note on nomenclature and taxonomy in the Lemuridae. Mammalia, 42:131-132.

Groves, C. P. 1978*b*. Phylogenetic and population systematics of mangabeys (Primates: Cercopithecoidea). Primates, 19:1-34.

Groves, C. P. 1978*c*. The taxonomic status of the dwarf blue sheep (Artiodactyla: Bovidae). Säugetierkundliche Mitteilungen, 26:177-183.

Groves, C. P. 1980*a*. A further note on *Moschus*. Journal of the Bombay Natural History Society, 77:130-133.

Groves, C. P. 1980*b*. Notes on the systematics of *Babyrousa* (Artiodactyla, Suidae). Zoologische Mededelingen, 55:29-46.

Groves, C. P. 1980*c*. Speciation in *Macaca*: The view from Sulawesi. Pp. 84-124, *in* The macaques (D. G. Lindburg, ed.). Van Nostrand Reinhold Company, New York, NY, 384 pp.

Groves, C. P. 1981*a*. Ancestors for the pigs: Taxonomy and phylogeny of the genus *Sus*. Department of Prehistory, Research School of Pacific Studies, Australian National University, Technical Bulletin, 3:1-96.

Groves, C. P. 1981*b*. Notes on gazelles. 2. Subspecies and clines in the springbok (*Antidorcas*). Zeitschrift für Säugetierkunde, 46:189-197.

Groves, C. P. 1981*c*. Notes on the gazelles. 3. The dorcas gazelles of North Africa. Annali del Museo Civico di Storia Naturale di Genova, 83:455-471.

Groves, C. P. 1981*d*. Systematic relationships in the Bovini (Artiodactyla, Bovidae). Zeitschrift für Zoologische Systematik und Evolutionsforschung, 4:264-278.

Groves, C. P. 1982*a*. Cranial and dental characteristics in the systematics of Old World Felidae. Carnivore, 5(2):28-39.

Groves, C. P. 1982*b*. Geographic variation in the barasingha or swamp deer (*Cervus duvauceli*). Journal of the Bombay Natural History Society, 79:620-629.

Groves, C. P. 1982*c*. A note on geographic variation in the Indian blackbuck (*Antilope cervicapra* Linnaeus, 1758). Records of the Zoological Survey of India, 79:489-503.

Groves, C. P. 1982*d*. The systematics of tree kangaroos (*Dendrolagus*; Marsupialia, Macropodidae). Australian Mammalogy, 5:157-186.

Groves, C. P. 1982*e*. *Antelope depressicornis* H. Smith, 1827, and *Anoa quarlesi* Ouwens, 1910 (Mammalia, Artiodactyla): Proposed conservation. Z.N.(S.)2310. Bulletin of Zoological Nomenclature, 39:281-282.

Groves, C. P. 1983. Notes on the gazelles. 4. The Arabian gazelles collected by Hemprich and Ehrenberg. Zeitschrift für Säugetierkunde, 48:371-381.

Groves, C. P. 1985*a*. An introduction to the gazelles. Chinkara, 1:4-16.

Groves, C. P. 1985*b*. Was the quagga a species or a subspecies? African Wildlife, 39:106-107.

Groves, C. P. 1986. The taxonomy, distribution and adaptations of recent equids. Pp. 11-65, *in* Equids in the ancient world (R. H. Meadow and H-P. Uerpmann, eds.). Ludwig Reichert Verlag, Wiesbaden, 421 pp.

Groves, C. P. 1987. On the cuscuses (Marsupialia: Phalangeridae) of the *Phalanger orientalis* group from Indonesian territory. Pp. 569-579, *in* Possums and opossums: Studies in evolution (M. Archer, ed.). Surrey Beatty and Sons Pty. Ltd., and the Royal Zoological Society of New South Wales, Sydney, 1:1-400; 2:401-788.

Groves, C. P. 1988. A catalogue of the genus *Gazella*. Pp. 193-198, *in* Conservation and biology of desert antelopes (A. Dixon and D. Jones, eds). Christopher Helm, London, 238 pp.

Groves, C. P. 1989. A theory of human and primate evolution. Oxford University Press, New York, 375 pp.

Groves, C. P. 1992. [Review of] Titis, New World Monkeys of the genus *Callicebus* (Cebidae, Platyrrhini): A preliminary taxonomic review, by P. Hershkovitz. International Journal of Primatology, 13:111-112.

Groves, C. P. 1996*a*. Taxonomic diversity in Arabian gazelles: The state of the art. Pp. 8-39, *in* Conservation of Arabian gazelles (A. Greth, C. Magin, and M. Ancrennaz, eds.). The National Commission for Wildlife Conservation and Development, Riyadh, Saudi Arabia, 168 pp.

Groves, C. P. 1996*b*. The taxonomy of the Asian wild buffalo from the Asian mainland. Zeitschrift für Säugetierkunde, 61:327-338.

Groves, C. P. 1997*a*. Taxonomy of wild pigs (*Sus*) of the Philippines. Zoological Journal of the Linnean Society, 120:163-191.

Groves, C. P. 1997*b*. Leopard-cats, *Prionailurus bengalensis* (Carnivora: Felidae) from Indonesia and the Philippines, with a description of two new subspecies. Zeitschrift für Säugetierkunde, 62:330-338.

Groves, C. P. 1997c. The taxonomy of Arabian gazelles. National Commission for Wildlife Conservation and Development, Riyadh. Publications, 29:24-51.

Groves, C. P. 1998. *Pseudopotto martini*: A new potto? African Primates, 31(1):42-43.

Groves, C. P. 2000a. The genus *Cheirogaleus*: Unrecognized biodiversity in dwarf lemurs. International Journal of Primatology, 21:943-962.

Groves, C. P. 2000b. Phylogenetic relationships within recent Antilopini (Bovidae). Pp. 223-233, *in* Antelopes, deer, and relatives. Fossil record, behavioral ecology, systematics, and conservation (E. S. Vrba and G. B. Schaller, eds.). Yale University Press, New Haven, 341 pp.

Groves, C. [P.] 2001a. Taxonomy of wild pigs of Southeast Asia. Asian Wild Pigs News, 1(1):3-4.

Groves, C. P. 2001b. Mammals in Sulawesi. Where did they come from and when, and what happened to them when they got there? Pp. 333-341, *in* Faunal and floral migration and evolution in S.E. Asia-Australasia (I. Metcalfe, J. M. B. Smith, M. Morwood, and I. Davidson, eds.). A. A. Balkema Publications, Lisse, The Netherlands. 361 pp.

Groves, C. P. 2001c. Primate taxonomy. Smithsonian Institution Press, Washington, DC, 350 pp.

Groves, C. P. 2003. Taxonomy of ungulates of the Indian Subcontinent. Journal of the Bombay Natural History Society, 100:341-362.

Groves, C. P., and R. H. Eaglen. 1988. Systematics of the Lemuridae (Primates, Strepsirhini). Journal of Human Evolution, 17:513-538.

Groves, C. P., and Feng Zuo-jiang. 1986. The status of musk-deer from Anhui Province, China. Acta Theriologica Sinica, 6:101-106 (in English with Chinese summary).

Groves, C. P., and T. F. Flannery. 1989. Revision of the genus *Dorcopsis* (Macropodidae: Marsupialia). Pp. 117-128, *in* Kangaroos, wallabies and rat-kangaroos (G. Grigg, P. Jarman, and I. Hume, eds.). Surrey Beatty and Sons Pty. Ltd, Sydney, 1:1-835.

Groves, C. P., and T. F. Flannery. 1990. Revision of the families and genera of bandicoots. Pp. 1-11, *in* Bandicoots and bilbies (J. H. Seebeck, R. L. Wallis, P. R. Brown, and C. M. Kemper, eds.). Surrey Beatty and Sons Pty. Ltd., Sydney, 392 pp.

Groves, C. P., and T. F. Flannery. 1994. A revision of the genus *Uromys* Peters, 1867 (Muridae: Mammalia) with descriptions of two new species. Records of the Australian Museum, 46:145-169.

Groves, C. P., and P. Grubb. 1974. A new duiker from Rwanda (Mammalia, Bovidae). Revue de Zoologie Africaine, 88:189-196.

Groves, C. P., and P. Grubb. 1982. The species of muntjac (genus *Muntiacus*) in Borneo: Unrecognised sympatry in tropical deer. Zoologische Mededelingen, 56:203-216.

Groves, C. P., and P. Grubb. 1985. Reclassification of the serows and gorals (*Nemorhaedus*: Bovidae). Pp. 45-50, *in* The biology and management of mountain ungulates (S. Lovari, ed.). Croom Helm, London, 271 pp.

Groves, C. P., and P. Grubb. 1987. Relationships of living deer. Pp. 21-59, *in* Biology and management of the Cervidae (C. M. Wemmer, ed.). Smithsonian Institution Press, Washington, D.C., 577 pp.

Groves, C. P., and P. Grubb. 1990. Muntiacidae. Pp. 134-168, *in* Horns, pronghorns, and antlers (G. A. Bubenik and A. B. Bubenik, eds). Springer-Verlag, New York, 562 pp.

Groves, C. P., and P. Grubb. 1993. The Eurasian suids: *Sus* and *Babyrousa*. Taxonomy and description. Pp. 107-111, *in* Action plan for the Suiformes (W. L. R. Oliver, ed.). I.U.C.N., Gland, Switzerland, 202 pp.

Groves, C. P., and L. B. Holthuis. 1985. The nomenclature of the Orang Utan. Zoologische Mededelingen, 59:411-417.

Groves, C. P., and F. Kurt. 1972. *Dicerorhinus sumatrensis*. Mammalian Species, 21:1-6.

Groves, C. P., and V. Mazák. 1967. On some taxonomic problems of the Asiatic wild asses: With the description of a new subspecies (Perissodactyla, Equidae). Zeitschrift für Säugetierkunde, 32:321-355.

Groves, C. P., and G. B. Schaller. 2000. The phylogeny and biogeography of the newly discovered Annamite artiodactyls. Pp. 261-282, *in* Antelopes, deer, and relatives. Fossil record, behavioral ecology, systematics, and conservation (E. S. Vrba and G. B. Schaller, eds.). Yale University Press, New Haven, 341 pp.

Groves, C. P., and C. Smeenk. 1978. On the type material of *Cervus nippon* Temminck, 1836; with a revision of sika deer from the main Japanese islands. Zoologische Mededelingen, 53:11-28.

Groves, C. P., and I. Tattersall. 1991. Geographical variation in the fork-marked lemur, *Phaner furcifer* (Primates, Cheirogaleidae). Folia Primatologia, 56:39-49.

Groves, C. P., and Wang Ying-xiang. 1990. The gibbons of the subgenus *Nomascus* (Primates, Mammalia). Zoological Research, 11:147-154.

Groves, C. P., and C. R. Westwood. 1995. Skulls of the Blaauwbok *Hippotragus leucophaeus*. Zeitschrift für Säugetierkunde, 60:314-318.

Groves, C. P., and D. P. Willoughby. 1981. Studies on the taxonomy and phylogeny of the genus *Equus*. 1. Subgeneric classification of the recent species. Mammalia, 45:321-354.

Groves, C. P., F. Ziccardi, and A. Toschi. 1966. Sull'Asino selvatico Aficano. Laboratorio de Zoologia

Applicata alla Caccia, Supplemento alle Ricerche di Zoologia Applicata alla Caccia, Universita di Bologna, 5:1-30.

Groves, C. P., Y. Wang, and P. Grubb. 1995. Taxonomy of musk-deer, genus *Moschus* (Moschidae, Mammalia). Acta Theriologica Sinica, 15:181-197.

Groves, C. P., G. B. Schaller, G. Amato, and K. Khounboline. 1997. Rediscovery of the wild pig *Sus bucculentus*. Nature, 386:335.

Grubb, P. 1972. Variation and incipient speciation in the African buffalo. Zeitschrift für Säugetierkunde, 37:121-144.

Grubb, P. 1973. Distribution, divergence and speciation of the drill and mandrill. Folia Primatologica, 20(2-3):161-177.

Grubb, P. 1977. Notes on a rare deer, *Muntiacus feai*. Annali del Museo Civico di Storia Naturale di Genova, 81:202-207.

Grubb, P. 1978. Patterns of speciation in African mammals. Pp. 152-167, *in* Ecology and taxonomy of African small mammals (D. A. Schlitter, ed.). Bulletin, Carnegie Museum of Natural History, 6:1-214.

Grubb, P. 1981. *Equus burchellii*. Mammalian Species, 157:1-9.

Grubb, P. 1982a. The systematics of Sino-Himalayan musk deer (*Moschus*), with particular reference to the species described by B. H. Hodgson. Säugetierkundliche Mitteilungen, 30:127-135.

Grubb, P. 1982b. Systematics of sun-squirrels (*Heliosciurus*) in eastern Africa. Bonner Zoologische Beiträge, 33(2-4):191-204.

Grubb, P. 1985. Geographical variation in the bushbuck of eastern Africa (*Tragelaphus scriptus*; Bovidae). Pp. 11-27, *in* Proceedings of the International Symposium on African Vertebrates (K.-L. Schuchman, ed.). Zoologisches Forschungsinstitut und Museum Alexander Koenig, Bonn, 585 pp.

Grubb, P. 1989. The systematic status of the suni (*Neotragus moschatus*) in Malawi. Nyala, 13:21-27.

Grubb, P. 1990. List of deer species and subspecies. Deer, Journal of the British Deer Society, 8:153-155.

Grubb, P. 1993. The Afrotropical suids: *Phacochoerus*, *Hylochoerus* and *Potamochoerus*. Taxonomy and Classification, Chapter 4.1, *in* Action plan for the Suiformes (W. L. R. Oliver, ed.). I.U.C.N., Gland, Switzerland, 202 pp.

Grubb, P. 1994. Notes on Grant's gazelle. S.S.C. Antelope Specialists Group Gnusletter, 13 (1 & 2):4-5.

Grubb, P. 1999. Types and type localities of ungulates named from southern Africa. Koedoe, 42:13-45.

Grubb, P. 2000a. Valid and invalid nomenclature of living and fossil deer, Cervidae. Acta Theriologica, 45:289-307.

Grubb, P. 2000b. Morphoclinal evolution in ungulates. Pp. 156-170, *in* Antelopes, deer, and relatives. Fossil record, behavioral ecology, systematics, and conservation (E. S. Vrba and G. B. Schaller, eds.). Yale University Press, New Haven, 341 pp.

Grubb, P. 2001a. Catalogue des mammifères du Muséum National d'Histoire Naturelle by Étienne Geoffroy Saint-Hilaire (1803): Proposed placement on the Offical List of Works Available for Zoological Nomenclature. Bulletin of Zoological Nomenclature, 58:41-52.

Grubb, P. 2001b. Review of family-group names of living bovids. Journal of Mammalogy, 82:374-388.

Grubb, P. 2001c. *Hippotragus* Sundevall, 1845 (Mammalia, Artiodactyla): Proposed conservation. Bulletin of Zoological Nomenclature, 58:126-132.

Grubb, P. 2002. Types, type locality and subspecies of the gerenuk *Litocranius walleri* (Artiodactyla; Bovidae). Journal of Zoology, London, 257:539-543.

Grubb, P., and W. F. H. Ansell. 1996. The name *Graphiurus hueti* de Rochebrune, 1883, and a critique of de Rochebrune's "*Faune de la Senegambie, Mammiferes*". The Nigerian Field, 61:164-171.

Grubb, P., and C. P. Groves. 1983. Notes on the taxonomy of the deer (Mammalia, Cervidae) of the Philippines. Zoologischer Anzeiger, 210:119-144.

Grubb, P., and C. P. Groves. 2002. Revision and classification of the Cephalophinae. Pp. 703-728, *in* Duikers and rainforests of Africa (V. J. Wilson, principle author). 2001 [2002]. Chipangali Wildlife Trust, Ascot, Bulawayo, Zimbabwe, 798 pp.

Grubb, P., T. S. Jones, A. G. Davies, E. Edberg, E. D. Starin, and J. E. Hill. 1998. Mammals of Ghana, Sierra Leone and The Gambia. The Trendrine Press, Cornwall, 265 pp.

Grubb, P., C. P. Groves, , J. P. Dudley, and J. Shoshani. 2000. Living African elephants belong to two species: *Loxodonta africana* (Blumenbach, 1797) and *Loxodonta cyclotis* (Matschie, 1900). Elephant, 2(4):1-4.

Gruber, U. F. 1969. Tiergeographische, ökologische und bionomische Untersuchungen an kleinen Säugetieren in Ost-Nepal. Khumbu Himal, 3:197-312.

Grulich, I. 1971. Zum Bau des Beckens (pelvis), eines systematisch-taxonomischen Merkmales, bei der Unterfamilie Talpinae. Zoologické Listy, 20:15-28.

Grulich, I. 1972. Ein Beitrag zur Kenntnis der ostmediterranen kleinwüchsigen, blinden Maulwurfsformen (Talpinae). Zoologické Listy, 21:3-21.

Grulich, I. 1982. Zur Kenntnis der Gattungen *Scaptochirus* und *Parascaptor* (Talpini, Mammalia). Folia Zoologica, 31:1-20.

Grulich, I. 1987*a*. Variability of *Cricetus cricetus* in Europe. Acta Scientiarum Naturalium Academiae Scientiarum Bohemoslovacae Brno, 21(7):1-53.

Grulich, I. 1987*b*. Contribution to the sexual dimorphism of the hamster (*Cricetus cricetus*, Rodentia, Mammalia). Folia Zoologica, 36:291-306.

Grulich, I. 1991. Variabiliby of the ossa memberi thoracici et pelvini *Cricetus cricetus* (Rodentia, Mammalia). Acta Scientiarum Naturalium, Academiae Scientiarum Bohemoslovacae (Brno), 25:1-63.

Grulich, I., and M. Jurik. 1994. A morphometric study of *Eliomys quercinus* from the Slovakian karst. Folia Zoologica, 43(3):209-218.

Grunwald, H. 1992. Uber zwei ungewohnliche Habitate des Siebenschlafers *Glis glis* (Linne, 1766) im Raum Honnetal. Dortmunder Beiträge zür Landeskunde, 26(1):47-57.

Guabloche, A., M. Arana, and O. E. Ramirez. 2002. Diet and gross gastric morphology of *Oryzomys xantheolus* (Sigmodontinae, Rodentia) in a Peruvian loma. Mammalia, 66:405-411.

Guénet, J.-L., and F. Bonhomme. 2003. Wild mice: An ever-increasing contribution to a popular mammalian model. Trends in Genetics, 19(1):24-31.

Guerin, C., and M. Faure. 1987. A propos du *Zygomaturus*, Diprotodontide de Nouvelle-Caledonie. Comptes Rendus des Séances de l'Académie des Sciences, Paris, Ser. 2, 305:815-817.

Guerin C., J. H. Winslow, M. Piboule, and M. Faure. 1981. Le pretendu rhinoceros de Nouvelle Caldeonie est un marsupial (*Zygomaturus diahotensis* nov. sp.); solution d'une enigme et consequences paleogeographiques. Geobios, 14:201-217.

Guereña-Gandara, L., M. Uribe-Alcocer, and F. Cervantes-Reza. 1983. Chromosomal study of the tropical rabbit (*Sylvilagus brasiliensis*). Mammalian Chromosomes Newsletter, 7.1:1-6.

Guerrero Vázquez, S., J. Téllez Lopez, and R. Amparán Salido. 1995. Los mamíferos de Jalisco: Análsis zoogeográfico. Biotam, 6:13-30.

Guilday, J. E. 1982. Dental variation in *Microtus xanthognathus*, *M. chrotorrhinus*, and *M. pennsylvanicus* (Rodentia: Mammalia). Annals of Carnegie Museum, 51:211-230.

Guiler, E. R. 1985. Thylacine: The tragedy of the Tasmanian tiger. Oxford University Press, 207 pp.

Guillotin, M., and F. Petter. 1984. Un *Rhipidomys* nouveau de Guyane francaise, *R. leucodactylus aratayae* ssp. nov. (Rongeurs, Cricetides). Mammalia, 48:541-544.

Güldenstädt, A. I. 1776. Chaus, animal feli affine descriptiom. Novi Commentari Academiae Scientiarum Imperialis Petropolitanae, 20:483-500.

Gulotta, E. F. 1971. *Meriones unguiculatus*. Mammalian Species, 3:1-5.

Gündüz, I., C. Tez, V. Malikov, A. Vaziri, A. V. Polyakov, and J. B. Searle. 2000. Mitochondrial DNA and chromosomal studies of wild mice (*Mus*) from Turkey and Iran. Heredity, 84:458-467.

Gündüz, I., J.-C. Auffray, J. Britton-Davidian, J. Catalan, G. Ganem, M. G. Ramalhinho, M. L. Mathias, and J. B. Searle. 2001. Molecular studies on the colonization of the Madeiran Archipelago by house mice. Molecular Ecology, 10:2023-2029.

Gunn, S. J., and I. F. Greenbaum. 1986. Systematic implications of karyotypic and morphologic variation in mainland *Peromyscus* from the Pacific northwest. Journal of Mammalogy, 67:294-304.

Günther, A. 1875. Notes on some small mammals from Madagascar. Proceedings of the Zoological Society of London, 1875:78-80.

Gupta, S. S., and G. V. R. Prasad. 2001. Micromammals from the Upper Siwalik subgroup of the Jammu region, Jammu and Kashmir State, India: Some constraints on age. Neues Jahrbuch für Geologie und Paläontologie Abhandlungen, 220(2):153-187.

Gureev, A. A. 1964. Fauna SSSR, Mlekopitayushchie, tom. 3, vyp. 10, Zaitseobraznye (Lagomorpha) [Fauna of the USSR, mammals, vol. 3, pt. 10, Lagomorpha]. Nauka, Moscow-Leningrad, 276 pp. (in Russian).

Gureev, A. A. 1971. Zemleroikii (Soricidae) fauny mira. [Shrews . . . of the world fauna]. Nauka, Leningrad, 253 pp. (in Russian).

Gureev, A. A. 1979. Fauna SSSR, Mlekopitayutschie, tom. 4, vyp. 2. Nasekomoyadnye . . . [Fauna of the USSR, Mammals, vol. 4, pt. 2. Insectivores (Mammalia, Insectivora)]. Nauka, Leningrad, 501 pp. (in Russian).

Guthrie, R. D. 1971. Factors regulating the evolution of microtine tooth complexity. Zeitschrift für Säugetierkunde, 36:37-54.

Guthrie, R. D. 1990. Frozen fauna of the Mammoth Steppe. University of Chicago Press, Chicago, 323 pp.

Gyldenstolpe, N. 1919. On a collection of mammals made in eastern and central Borneo. Kungliga Svenska Vetenskapsakademiens, Handlingar, 60:1-62.

Gyldenstolpe, N. 1928. Zoological results of the Swedish Expedition to Central Africa 1921. Vertebrata 5. Mammals from the Birunga Volcanoes, north of Lake Kivu. Arkiv för Zoologi, 20A(4):1-76.

Gyldenstolpe, N. 1932. A manual of Neotropical sigmodont rodents. Kungliga Svenska Vetenskapsakademiens Handlingar, Tredje Serien, 11:1-164.

Gynther, I. C., and P. S. O'Reilly. 1995. A new locality for the Hastings River mouse, *Pseudomys oralis*, in Southeast Queensland. Memoirs of the Queensland Museum, 38(2):513-518.

Ha, N. A. Q. 1997. Another new animal discovered in Vietnam. Vietnam Economic News, 38:46-47.

Haas, G. 1952. The fauna of layer B of the Abu Usba Cave. Israel Exploration Journal, 2(1):35-47.

Habersetzer, J., and G. Storch. 1987. Klassifikation und funktionelle flügelmorphologie paläogener fledermäuse (Mammalia, Chiroptera). Courier Forschungsinstitut Senckenberg, 91:11-150.

Haffner, M. 1996. A tendon-locking mechanism in two climbing rodents, *Muscardinus avellanarius* and *Micromys minutus* (Mammalia, Rodentia). Journal of Morphology, 229:219-227.

Hafner, D. J. 1981. Evolution and historical zoogeography of antelope ground squirrels, genus *Ammospermophilus* (Rodentia: Sciuridae). Unpubl. Ph. D. dissertation, University of New Mexico, Albuquerque, New Mexico, 225 pp.

Hafner, D. J. 1984. Evolutionary relationships of the Nearctic Sciuridae. Pp. 3-23, *in* The biology of ground-dwelling squirrels (J. O. Murie and G. R. Michener, eds.). University of Nebraska Press, Lincoln, 459 pp.

Hafner, D. J. 1992. Speciation and persistence of a contact zone in Mojave Desert ground squirrels, subgenus *Xerospermophilus*. Journal of Mammalogy, 73(4):770-778.

Hafner, D. J., and K. N. Geluso. 1983. Systematic relationships and historical zoogeography of the desert pocket gopher, *Geomys arenarius*. Journal of Mammalogy, 64:405-413.

Hafner, D. J., and B. R. Riddle. 1997. Biogeography of Baja California peninsular desert mammals. Pp. 39-68, *in* Life among the muses: Papers in honor of James S. Findley (T. L. Yates, W. L. Gannon, and D. E. Wilson, eds.). The Museum of Southwestern Biology, University of New Mexico, Albuquerque, v + 290 pp.

Hafner, D. J., and R. M. Sullivan. 1995. Historical and ecological biogeography of Nearctic pikas (Lagomorpha:Ochotonidae). Journal of Mammalogy, 76:302-321.

Hafner, D. J., and T. L. Yates. 1983. Systematic status of the Mohave ground squirrel, *Spermophilus mohavensis* (subgenus *Xerospermophilus*). Journal of Mammalogy, 64:397-404.

Hafner, D. J., J. C. Hafner, and M. S. Hafner. 1979. Systematic status of kangaroo mice, genus *Microdipodops*: Morphometric, chromosomal, and protein analyses. Journal of Mammalogy, 60:1-10.

Hafner, D. J., K. E. Petersen, and T. L. Yates. 1981. Evolutionary relationships of jumping mice (genus *Zapus*) of the southwestern United States. Journal of Mammalogy, 62(3):501-512.

Hafner, D. J., B. R. Riddle, and S. T. Alvarez-Castañeda. 2001. Evolutionary relationships of white-footed mice (*Peromyscus*) on islands in the Sea of Cortéz, Mexico. Journal of Mammalogy, 82:775-790.

Hafner, J. C. 1978. Evolutionary relationships of kangaroo mice, genus *Microdipodops*. Journal of Mammalogy, 59:354-366.

Hafner, J. C., and M. S. Hafner. 1983. Evolutionary relationships of heteromyid rodents. Great Basin Naturalist, Memoirs, 7:3-29.

Hafner, M. S. 1982. A biochemical investigation of geomyoid systematics (Mammalia: Rodentia). Zeitschrift für Zoologische Systematik und Evolutionsforschung, 20:118-130.

Hafner, M. S. 1991. Evolutionary genetics and zoogeography of Middle American pocket gophers, genus *Orthogeomys*. Journal of Mammalogy, 72:1-10.

Hafner, M. S., and L. J. Barkley. 1984. Genetics and natural history of a relictual pocket gopher, *Zygogeomys* (Rodentia: Geomyidae). Journal of Mammalogy, 65:474-479.

Hafner, M. S., and D. J. Hafner. 1987. Geographic distribution of two Costa Rican species of *Orthogeomys*, with comments on dorsal pelage markings in the Geomyinae. Southwestern Naturalist, 32:5-11.

Hafner, M. S., J. C. Hafner, J. L. Patton, and M. F. Smith. 1987. Macrogeographic patterns of genetic differentiation in the pocket gopher *Thomomys umbrinus*. Systematic Zoology, 36:18-34.

Hafner, M. S., L. J. Barkley, and J. M. Chupasko. 1994*a*. Evolutionary genetics of New World tree squirrels (tribe Sciurini). Journal of Mammalogy, 75(1):102-109.

Hafner, M. S., P. D. Sudman, F. X. Villablanca, T. A. Spradling, J. W. Demastes, and S. A. Nadler. 1994*b*. Disparate rates of molecular evolution in co-speciating hosts and parasites. Science, 265:1087-1090.

Hagmeier, E. M. 1961. Variation and relationships in North American marten. Canadian Field Naturalist, 75:122-138.

Hahn, H. 1934. Die Familie der Procaviidae. Zeitschrift für Säugetierkunde, 9:207-358.

Haiduk, M. W., J. W. Bickham, and D. J. Schmidly. 1979. Karyotypes of six species of *Oryzomys* from Mexico and Central America. Journal of Mammalogy, 60:610-615.

Haiduk, M. W., C. Sanchez-Hernandez, and R. J. Baker. 1988. Phylogenetic relationships of *Nyctomys* and *Xenomys* to other cricetine genera based on data from G-banded chromosomes. Southwestern Naturalist, 33:397-403.

Haim, A., and E. Tchernov. 1974. The distribution of myomorph rodents in the Sinai peninsula. Mammalia, 38(2):201-223.

Hall, E. R. 1941. Revision of the rodent genus *Microdipodops*. Field Museum of Natural History, Zoological Series, 27:233-277.

Hall, E. R. 1951. American weasels. University of Kansas Publications, Museum of Natural History, 4:1-466.

Hall, E. R. 1955. A new subspecies of wood rat from Nayarit, Mexico, with new name combinations for the *Neotoma mexicana* group. Journal of the Washington Academy of Sciences, 45:328-332.

Hall, E. R. 1960. *Oryzomys couesi* only subspecifically different from the marsh rice rat, *Oryzomys palustris*. Southwestern Naturalist, 5:171-173.

Hall, E. R. 1971. Variation in the blackish deer mouse, *Peromyscus furvus*. Anales de Instituto de Biologia, Universidad Nacional Autónomica de México, 1:149-154.

Hall, E. R. 1981. The mammals of North America. Second ed. John Wiley and Sons, New York, 1:1-600 + *90*, 2:601-1181 + *90*.

Hall, E. R. 1984. Geographic variation among brown and grizzly bears (*Ursus arctos*) in North America. Special Publication, Museum of Natural History, University of Kansas, 13:1-16.

Hall, E. R., and E. L. Cockrum. 1953. A synopsis of the North American microtine rodents. University of Kansas Publications, Museum of Natural History, 5:373-498.

Hall, E. R., and F. H. Dale. 1939. Geographic races of the kangaroo rat, *Dipodomys microps*. Occasional Papers of the Museum of Zoology, Louisiana State University, 4:47-62.

Hall, E. R., and W. D. Dalquest. 1950. A synopsis of the American bats of the genus *Pipistrellus*. University of Kansas Publications, Museum of Natural History, 1:591-602.

Hall, E. R., and H. H. Genoways. 1970. Taxonomy of the *Neotoma albigula*-group of woodrats in central Mexico. Journal of Mammalogy, 51:504-516.

Hall, E. R., and R. M. Gilmore. 1932. New mammals from St. Lawrence Island, Bering Sea, Alaska. University of California Publications in Zoology, 38:391-404.

Hall, E. R., and D. R. Hoffmeister. 1942. Geographic variation in the canyon mouse, *Peromyscus crinitus*. Journal of Mammalogy, 23:51-65.

Hall, E. R., and J. K. Jones, Jr. 1961. North American yellow bats, *"Dasypterus"*, and a list of the named kinds of the genus *Lasiurus* Gray. University of Kansas Publications, Museum of Natural History, 14:73-98.

Hall, E. R., and K. R. Kelson. 1951. A new subspecies of *Microtus montanus* from Montana and comments on *Microtus canicaudus* Miller. University of Kansas Publications, Museum of Natural History, 5:73-79.

Hall, E. R., and K. R. Kelson. 1959. The mammals of North America. Ronald Press Co., New York, 1:1-546; 2:547-1083 + *79*.

Hall, L. S., and G. C. Richards. 1979. Bats of Eastern Australia. Queensland Museum Booklet, 12:1-66.

Hallett, J. G. 1978. *Parascalops breweri*. Mammalian Species, 98:1-4.

Hallett, J. M. 1979. Chromosome polymorphism in *Praomys* (*Mastomys*) *natalensis* in southern Africa: Diploid studies. South African Journal of Science, 75:413-415.

Haltenorth, T. 1953. Die Wildkatzen der Alten Welt; eine Übersicht über die Untergattung. Geest and Portig, Leipzig, 117 pp.

Haltenorth, T. 1958. Klassifikation der Säugetiere, 1 (1, Ordnung Kloakentiere, Monotremata Bonaparte, 1838, und 2, Ordnung Beuteltiere, Marsupialia Illiger, 1811 [=Didelphia Blainville, 1816]). Handbuch der Zoologie, 8(16):1-40.

Haltenorth, T. 1959. Beitrag zur Kenntnis des Mesopotamischen Damhirsches—*Cervus* (*Dama*) *mesopotamicus* Brooke, 1875—und zur Stammes-und Verbreitungeschichte der Damhirsche allgemein. Säugetierkundliche Mitteilungen, 7(Sonderheft):1-89.

Haltenorth, T. 1963. Klassifikation der Säugetiere: Artiodactyla I. Handbuch der Zoologie, 8(32):1-167.

Haltenorth, T., and H. Diller. 1977. Säugetiere Afrikas und Madagaskars. BLV, Munchen, 403 pp.

Hamer, M., and M. Schutowa. 1965. Neue daten über die geographische veränderlichkeit und die entwicklung der gattung *Mesocricetus* Nehring 1898 (Glires, Mammalia). Zeitschrift für Säugetierkunde, 31:237-251.

Hamilton, H., S. Caballero, A. G. Collins, and R. J. Brownell, Jr. 2001. Evolution of river dolphins. Royal Society of London, Procedings, Series B, 268:549-556.

Hamilton, J. E. 1934. The southern sea lion, *Otaria byronia* (de Blainville). Discovery Reports, 8:269-318.

Hamilton, J. E. 1940. On the history of the elephant seal, *Mirounga leonina* (Linn.). Proceedings of the Linnean Society of London, 1939-40:33-37.

Hamilton, H., S. Caballero, A. G. Collins, and R. J. Brownell, Jr. 2001. Hamlett, G. W. D. 1939. Identity of *Dasypus septemcinctus* Linnaeus with notes on some related species. Journal of Mammalogy, 20:328-336.

Hammer, M. F., J. W. Schilling, E. M. Prager, and A. C. Wilson. 1987. Recruitment of lysozyme as a major enzyme in the mouse gut: Duplication, divergence, and regulatory evolution. Journal of Molecular Evolution, 24:272-279.

Hammond, D. D., and S. Leatherwood. 1984. Cetaceans captured for Ocean Park, Hong Kong, April 1974—February 1983. Reports of the International Whaling Commission, 34:491-495.

Han, K. H., F. H. Sheldon, and R. B. Stuebing. 2000. Interspecific relationships and biogeography of some Bornean tree shrews (Tupaiidae: Tupaia), based on DNA hybridization and morphometric comparisons. Biological Journal of the Linnean Society, 70:1-14.

Han, S. H., S. Wakana, H. Suzuki, Y. Hirai, and K. Tsuchiya. 1996. Variation of the mitochondrial DNA and the nuclear ribosomal DNA in the striped field mouse *Apodemus agrarius* on the mainland and offshore islands of South Korea. Mammal Study, 21:125-136.

Han, S.-H., S. Ohdachi, and H. Abe. 2000. New records of two *Sorex* species (Soricidae) from South Korea. Mammal Study, 25:141-144.

Han, S.-H., M. A. Iwasa, S. D. Ohdachi, H.-S. Oh, H. Suzuki, K. Tsuchiya, and H. Abe. 2002. Molecular phylogeny of *Crocidura* shrews in northeastern Asia: A special reference to specimens on Cheju Island, South Korea. Acta Theriologica, 47:369-379.

Hanak, V. 1969. Zur Kenntnis von *Rhinolophus bocharicus* Kastchenko et Akimov, 1917 (Mammalia: Chiroptera). Vèstník eskoslovenské Spole nosti Zoologické, 33(4):315-327.

Hanak, V., and J. Gaisler. 1969. Notes on the taxonomy and ecology of *Myotis longipes* (Dobson, 1873). Zoologické Listy, 18:195-206.

Hanak, V., and J. Gaisler. 1971. The status of *Eptesicus ognevi* Bobrinskii, 1918, and remarks on some other species of this genus (Mammalia: Chiroptera). Vèstník eskoslovenské Spole nosti Zoologické, 35(1):11-24.

Hanak, V., and I. Horáèek. 1986. Zur Südgrenze des Areals von *Eptesicus nilssoni* (Chiroptera: Vespertilionidae). Annalen des Naturhistorischen Museums in Wien, ser. B (Botanik und Zoologie), 88/89:377-388.

Hand, S. J. 1985. New Miocene megadermatids (Chiroptera: Megadermatidae) from Australia with comments on megadermatid systematics. Australian Mammalogy, 8:5-43.

Hand, S. J. 1996. New Miocene and Pliocene megadermatids (Mammalia, Microchiroptera) from Australia, with comments on broader aspects of megadermatid evolution. Geobios, 29:365-377.

Hand, S. J., and J. A. W. Kirsch. 1998. A southern origin for the Hipposideridae (Microchiroptera). Evidence from the Australian fossil record. Pp. 72-90, *in* Bat biology and conservation (T. H. Kunz and P. A. Racey, eds.). Smithsonian Institution Press, Washington, DC, 365 pp.

Handley, C. O., Jr. 1959a. A review of the genus *Hoplomys* (thick-spinned rats), with description of a new form from Isla Escudo de Veraguas, Panamá. Smithsonian Miscellaneous Collections, 139(4):1-10, 1 pl.

Handley, C. O., Jr. 1959b. A revision of American bats of the genera *Euderma* and *Plecotus*. Proceedings of the United States National Museum, 110:95-246.

Handley, C. O., Jr. 1960. Descriptions of new bats from Panama. Proceedings of the United States National Museum, 112:459-479.

Handley, C. O., Jr. 1966a. Checklist of the mammals of Panama. Pp. 753-795, *in* Ectoparasites of Panama, (R. L. Wenzel and V. J. Tipton, eds.). Field Museum of Natural History, Chicago, 861 pp.

Handley, C. O., Jr. 1966b. Descriptions of new bats (*Choeroniscus* and *Rhinophylla*) from Colombia. Proceedings of the Biological Society of Washington, 79:83-88.

Handley, C. O., Jr. 1966c. A synopsis of the genus *Kogia* (pygmy sperm whales). Pp. 62-69, *in* Whales, dolphins, and porpoises (K. S. Norris, ed.). University of California Press, Berkeley, 789 pp.

Handley, C. O., Jr. 1976. Mammals of the Smithsonian Venezuelan Project. Brigham Young University Science Bulletin, Biological Series, 20:1-91.

Handley, C. O., Jr. 1980. Inconsistencies in formation of family-group and subfamily-group names in Chiroptera. Pp. 9-13, *in* Proceedings of the Fifth International Bat Research Conference (D. E. Wilson and A. L. Gardner, eds.). Texas Tech Press, Lubbock, 434 pp.

Handley, C. O., Jr. 1984. New species of mammals from northern South America: A long-tongued bat, genus *Anoura* Gray. Proceedings of the Biological Society of Washington, 97:513-521.

Handley, C. O., Jr. 1987. New species of mammals from northern South America: Fruit-eating bats, genus *Artibeus* Leach. Pp. 163-172, *in* Studies in Neotropical mammalogy, essays in honor of Philip Hershkovitz (B. D. Patterson and R. M. Timm, eds.). Fieldiana: Zoology, n.s., 39:frontispiece, viii + 1-506 pp.

Handley, C. O., Jr. 1989 [1990]. The *Artibeus* of Gray 1838. Pp. 443-468, *in* Advances in neotropical mammalogy (K. H. Redford and J. F. Eisenberg, eds.). Sandhill Crane Press, Gainesville, FL, 614 pp.

Handley, C. O., Jr. 1991. The identity of *Phyllostoma planirostre* Spix, 1823 (Chiroptera: Stenodermatinae). Bulletin of the American Museum of Natural History, 206:12-17.

Handley, C. O., Jr. 1996. New species of mammals from northern South America: Bats of the genera *Histiotus* Gervais and *Lasiurus* Gray (Chiroptera: Vespertilionidae). Proceedings of the Biological Society of Washington, 109:1-9.

Handley, C. O., Jr., and K. C. Ferris. 1972. Descriptions of new bats of the genus *Vampyrops*. Proceedings of the Biological Society of Washington, 84:519-524.

Handley, C. O., Jr., and A. L. Gardner. 1990. The holotype of *Natalus stramineus* Gray (Mammalia: Chiroptera: Natalidae). Proceedings of the Biological Society of Washington, 103:966-972.

Handley, C. O., Jr., and J. Ochoa. 1997. New species of mammals from northern South America: A sword-nosed bat, genus *Lonchorhina* Tomes (Chiroptera: Phyllostomidae). Memoria Sociedad de Ciencias Naturales La Salle, 57:71-82.

Handley, C. O., Jr., and R. H. Pine. 1992. A new species of *Coendou* Lacépède, from Brazil. Mammalia, 56(2):237-244.

Handley, C. O., Jr., and M. Varn. 1994. Identification of the Carolinian shrews of Bachman 1837. Pp. 393-406, *in* Advances in the biology of shrews (J. F. Merritt, , G. L. Kirkland, Jr., and R. K. Rose, eds.). Carnegie Museum of Natural History, Special Publication, 18:1-458.

Handley, C. O., Jr., and W. D. Webster. 1987. The supposed occurrence of *Glossophaga longirostris* Miller on Dominica and problems with the type series of *Glossophaga rostrata* Miller. Occasional Papers of the Museum, Texas Tech University, 108:1-10.

Hanney, P. 1965. The Muridae of Malawi (Africa: Nyasaland). Journal of Zoology, London, 146:577-633.

Hanney, P., and B. Morris. 1962. Some observations upon the pouched rat in Nyasaland. Journal of Mammalogy, 43:238-248.

Hänni, C., V. Laudet, V. Barriel, and F. M. Catzeflis. 1995. Evolutionary relationships of *Acomys* and other murids (Rodentia, Mammalia) based on complete 12S rRNA mitochondrial gene sequences. Israel Journal of Zoology, 41:131-146.

Hanski, I., and E. Pankakoski (eds.). 1989. Population biology of Eurasian shrews. Annales zoologici fennici, 26:331-479.

Hansson, L., and K. Fredga. 1996. The mammal fauna of Sweden. Hystrix, n.s., 8(1-2):31-33.

Happold, D. C. D. 1967. Guide to the natural history of Khartoum Province, part III. Mammals. Sudan Notes and Records, 48:111-132.

Happold, D. C. D. 1987. The mammals of Nigeria. Clarendon Press, Oxford, 402 pp.

Happold, D. C. D. 1995. Broad-toothed rat, *Mastacomys fuscus*. Pp. 562-564, *in* Mammals of Australia (R. Strahan, ed.). Smithsonian Institution Press, Washington, D.C., 756 pp.

Happold, D. C. D., and M. Happold. 1989. The bats (Chiroptera) of Malawi, central Africa: Checklist and keys for identification. Nyala, 14:89-112.

Happold, D. C. D., and M. Happold. 1996. The social organization and population dynamics of leaf-roosting banana bats, *Pipistrellus nanus* (Chiroptera: Vespertilionidae) in Malawi, east-central Africa. Mammalia, 60 (4):517—544.

Happold, D. C. D., and M. Happold. 1997a. Conservation of mammals on a tobacco farm on the highlands of Malawi. Biodiversity and Conservation, 6:837-852.

Happold, D. C. D., and M. Happold. 1997b. New records of bats (Chiroptera: Mammalia) from Malawi, east-central Africa, with an assessment of their status and conservation. Journal of Natural History, 31:805-836.

Happold, D. C. D., and M. Happold. 1998. New distribution records for bats and other mammals of Malawi. Nyala, 20:17-24.

Happold, D. C. D., M. Happold, and J. E. Hill. 1987. The bats of Malawi. Mammalia, 51:337-414.

Harada, M., and R. Kobayashi. 1980. Studies on the small mammal fauna of Sabah, East Malaysia. II. Karyological analysis of some Sabahan mammals (Primates, Rodentia, Chiroptera). Contributions of the Biology Laboratory. Kyoto University, 26:83-95.

Harada, M., and S. Takada. 1985. Karyotypes of two species of Insectivora from Taiwan (Insectivora, Soricidae). Experientia, 41:510-511.

Harada, M., A. Ando, L.-K. Lin, and S. Takada. 1991. Karyotypes of the Taiwan vole *Microtus kikuchii* and the Pere David's vole *Eothenomys melanogaster* from Taiwan. Journal of the Mammalogical Society of Japan, 16:41-45.

Harada, M., A. Ando, K. Tsuchiya, and K. Koyasu. 2001. Geographical variations in chromosomes of the Greater Japanese mole-shrew, *Urotrichus talpoides* (Mammalia: Insectivora). Zoological Science,18:433-442.

Hardjasasmita, H. S. 1987. Taxonomy and phylogeny of the Suidae (Mammalia) in Indonesia. Scripta Geologica, 85:1-68.

Hardy, R. 1950. A new tree squirrel from central Utah. Proceedings of the Biological Society of Washington, 63:13-14.

Haring, E., B. Herzig-Straschil, and F. Spitzenberger. 2000. Phylogenetic analysis of Alpine voles of the *Microtus multiplex* complex using the mitochondrial control region. Journal of Zoological and Systematic Evolutionary Research, 38:231-238.

Harmer, S. F. 1922. On Commerson's dolphin and other species of *Cephalorhynchus*. Proceedings of the Zoological Society of London, 1922(3):627-638.

Harper, F. 1940. The nomenclature and type localities of certain Old World mammals. Journal of Mammalogy, 2l:191-203; 322-332.

Harper, F. 1945. Extinct and vanishing mammals of the Old World. Special Publication, American Committee for International Wild Life Protection, New York, 12:1-850.

Harper, F. 1952. History and nomenclature of the pocket gopher (*Geomys*) in Georgia. Proceedings of the Biological Society of Washington, 65:35-38.

Harris, A. H. 1998. Fossil history of shrews in North America. Pp. 133-156, *in* Evolution of shrews (J. M.Wójcik and M. Wolsan, eds.). Mammal Research Institute, Polish Academy of Sciences, Bialowieza, 458 pp.

Harris, C. J. 1968. Otters; a study of the Recent Lutrinae. Weidenfeld and Nicolson, London, 397 pp.

Harris, C. P., and D. A. McCullough. 1988. G-banded karyotype of *Neotoma goldmani*. Southwestern Naturalist, 33:236-239.

Harris, D., and D. S. Rogers. 1999. Species limits and phylogenetic relationships among populations of *Peromyscus furvus*. Journal of Mammalogy, 80:530-544.

Harris, D., D. S. Rogers, and J. Sullivan. 2000. Phylogeography of *Peromyscus furvus* (Rodentia; Muridae) based on cytochrome *b* sequence data. Molecular Ecology, 9:2129-2135.

Harris, J. M. 1991. Family Hippopotamidae. Pp. 31-85, *in* Koobi Fora Research Project. Vol. 3 (J. M. Harris, ed.). Clarendon Press, Oxford, 384 pp.

Harris, W. P., Jr. 1937. Revision of *Sciurus variegatoides*, a species of Central American squirrel. Miscellaneous Publications, Museum of Zoology, University of Michigan, 38:1-39.

Harrison, D. L. 1963. Report on a collection of bats (Microchiroptera) from N. W. Iran. Zeitschrift für Säugetierkunde, 28:301-308.

Harrison, D. L. 1964-1972. The mammals of Arabia. Ernest Benn Limited, London, 1:1-192 [1964]; 2:193-381 [1968]; 3:383-670 [1972].

Harrison, D. L. 1975. *Macrophyllum macrophyllum*. Mammalian Species, 62:1-3.

Harrison, D. L. 1978. A critical examination of alleged sibling species in the lesser three-toed jerboas (subgenus *Jaculus*) of the north African and Arabian deserts. Pp. 77-80, *in* Ecology and taxonomy of African small mammals (D. A. Schlitter, ed.). Bulletin of Carnegie Museum of Natural History, 6:1-214.

Harrison, D. L., and P. J. J. Bates. 1991. The mammals of Arabia, Second ed. Harrison Zoological Museum, Sevenoaks, United Kingdom, 354 pp.

Harrison, D. L., and P. J. J. Bates. 1986. New records of a rare African shrew, *Crocidura butleri percivali* Dollman, 1915 (Insectivora: Soricidae). Annalen des Naturhistorischen Museums in Wien, 88/89:207-211.

Harrison, D. L., and J. D. L. Fleetwood. 1960. A new race of the flat-headed bat *Platymops barbatogularis* Harrison from Kenya colony, with observations on the anatomy of the gular sac and genitalia. Durban Museum Novitates, 5:269-284.

Harrison, J. L. 1956. Records of bandicoot rats (*Bandicota*, Rodentia, Muridae) new to the fauna of Malaya and Thailand. Bulletin of the Raffles Museum (Singapore), 27:27-31.

Harrison, J. L. 1958. *Chimarrogale hantu* a new water shrew from the Malay Peninsula, with a note on the genera *Chimarrogale* and *Crossogale* (Insectivora, Soricidae). Annals and Magazine of Natural History, ser. 13, 1:282-290.

Harrison, J. L. 1974. An introduction to mammals of Singapore and Malaya, Second ed. Malayan Nature Society, Singapore, 340 pp.

Harrison, M. J. S. 1988. A new species of guenon (genus *Cercopithecus*) from Gabon. Journal of Zoology, London, 215:561-575.

Harrison, R. G., S. M. Bogdanowicz, R. S. Hoffmann, E. Yensen, and P. W. Sherman. 2003. Phylogeny and evolutionary history of the ground squirrels (Rodentia, Marmotinae). Journal of Molecular Evolution, 10(3):249-276.

Harrison, T. 1960. South China sea dolphins. Malayan Nature Journal, 14(2):87-89.

Harsch, P. 1993. Untersuchung zum Vorkommen von Siebenschlafer (*Glis glis*) und Haselmaus (*Muscardinus avellanarius*) im Bayerischen Allgau. Mitteilungen des Naturwissenschaftlichen Arbeitskreises Kempten Allgau, 32(1):61-68.

Hart, E. B. 1992. *Tamias dorsalis*. Mammalian Species, 399:1-6.

Hartenberger, J.-L. 1971. Contribution a l'etude des genres *Gliravus* et *Microparamys* (Rodentia) de l''eocene d'Europe. Paleovertebrata, 4:97-135.

Hartenberger, J.-L. 1985. The Order Rodentia: Major questions on their evolutionary origin, relationships and suprafamilial systematics. Pp. 1-34, *in* Evolutionary Relationships among Rodents: A Multidisciplinary Analysis (W. P. Luckett and J.-L. Hartenberger, eds.). Plenum Press, New York, 721 pp.

Hartenberger, J.-L. 1994. The evolution of the Gliroidea. Pp. 19-33, *in* Rodent and Lagomorph Families of

Asian Origins and Diversification (Y. Tomida, C. k. Li and T. Setoguchi, eds.). National Science Museum Monographs, No. 8, Tokyo, 195 pp.

Hartenberger, J.-L. 1996. Les debuts de la radiation adaptative des Rodentia (Mammalia). Comptes Rendus Académie des Sciences Paris, série IIa, 323:631-637.

Hartenberger, J.-L. 1998. Description de la radiation des Rodentia (Mammalia) du Paleocene superieur au Miocene; incidence phylogenetiques. Comptes Rendus Académie des Sciences Paris, Sciences de la terre et des planetes, 326:439-444.

Hartl, G. B., F. Suchentrunk, R. Willing, J. Markowski, and H. Ansorge. 1992. Inconsistency of biochemical evolutionary rates affecting allozyme divergence within the genus *Apodemus* (Muridae: Mammalia). Biochemical Systematics and Ecology, 20(4):363-372.

Hartl, G. B., R. Willing, G. Lang, F. Klein, and J. Köller. 1990. Genetic variability and differentiation in red deer (*Cervus elaphus* L.) of Central Europe. Genetics Selection Evolution, 22:289-306.

Hartman, G. D., and T. L. Yates. 1985. *Scapanus orarius*. Mammalian Species, 253:1-5.

Hasler, M. J., J. F. Hasler, and A. V. Nalbandov. 1977. Comparative breeding biology of musk shrews (*Suncus murinus*) from Guam and Madagascar. Journal of Mammalogy, 58:285-290.

Hassanin, A., and E. J. P. Douzery. 1999*a*. The tribal radiation of the family Bovidae (Artiodactyla) and the evolution of the mitochondrial cytochrome b gene. Molecular Phylogenetics and Evolution, 7:303-319.

Hassanin, A., and E. J. Douzery. 1999*b*. Evolutionary affinities of the enigmatic saola (*Pseudoryx nghetinhensis*) in the context of molecular phylogeny of Bovidae. Proceedings of the Royal Society of London. Series B—Biological Sciences, 266:893-900.

Hassanin, A., E. Pasquet, and J.-D. Vigne. 1998. Molecular systematics of the subfamily Caprinae (Artiodactyhla, Bovidae) as determined from cytochrome b sequences. Journal of Mammalian Evolution, 5:217-236.

Hassanin, A., A. Seveau, H. Thomas, H. Bocherens, D. Billiou, and B. X. Nguyen. 2001. Evidence from DNA that the mysterious 'linh duong' (*Pseudonovibos spiralis*) is not a new bovid. Comptes Rendus de l'Académie des Sciences. Série III, Sciences de la Vie, 324:71-80.

Hassinger, J. 1968. Introduction to the mammal survey of the 1965 Street Expedition to Afghanistan. Fieldiana: Zoology, 55:1-81.

Hassinger, J. D. 1970. Shrews of the *Crocidura zarudnyi-pergrisea* group with descriptions of a new subspecies. Fieldiana: Zoology, 58:5-8.

Hassinger, J. D. 1973. A survey of the mammals of Afghanistan resulting from the 1965 Street Expedition (excluding bats). Fieldiana: Zoology, 60:1-195.

Hatt, R. T. 1934*a*. The American Museum Congo Expedition manatee and other recent manatees. Bulletin of the American Museum of Natural History, 66:533-566.

Hatt, R. T. 1934*b*. Fourteen hitherto unrecognized African rodents. American Museum Novitates, 708:1-15.

Hatt, R. T. 1936. The hyraxes collected by the American Museum Congo expedition. Bulletin of the American Museum of Natural History, 72:117-141.

Hatt, R. T. 1940*a*. Lagomorpha and Rodentia other than Sciuridae, Anomaluridae and Idiuridae, collected by the American Museum Congo Expedition. Bulletin of the American Museum of Natural History, 76(9):457-604.

Hatt, R. T. 1940*b*. Mammals collected by the Rockefeller-Murphy Expedition to Tanganyika Territory and the eastern Belgian Congo. American Museum Novitates, 1070:1-8.

Hatt, R. T. 1959. The mammals of Iraq. Miscellaneous Publications, Museum of Zoology, University of Michigan, 106:1-113.

Hattori, S., M. Yoshiyuki, Y. Noboru, and H. Asaki. 1990. Geographical and ecological distributions and morphological variations of *Crocidura watasei* Kuroda, 1924 from the Amami Islands. Memoirs of the Natural Science Museum Tokyo, 23:167-172 (in Japanese with English summary).

Hauffee, H. C., and J. B. Searle. 1993. Extreme karyotypic variation in a *Mus musculus domesticus* hybrid zone: The tobacco mouse story revisited. Evolution, 47(5):1374-1395.

Hausser, J. 1990. *Sorex coronatus* Millet, 1882—Schabrackenspitzmaus; *Sorex granarius* Miller, 1909—Iberische Waldspitzmaus; *Sorex samniticus* Altobello, 1926—Italienische Waldspitzmaus. Pp. 279-294, *in* Handbuch der Säugetiere Europas (J. Niethammer and F. Krapp, eds.). Aula-Verlag, Wiesbaden, 3/I:1-524.

Hausser, J. (ed.). 1991. The cytogenetics of the *Sorex araneus* group and related topics. Mémoires de la Société Vaudoise des Sciences Naturelles, 19:1-151.

Hausser, J. 1995. Säugetiere der Schweiz. Verbreitung. Biologie. Ökologie. Birkhäuser Verlag: Basel, 501 pp.

Hausser, J., J.-D. Graf, and A. Meylan. 1975. Données nouvelles sur les *Sorex* d'Espagne et des Pyrénées (Mammalia, Insectivora). Bulletin de la Societé Vaudoise des Sciences Naturelles, 72:241-252.

Hausser, J., F. Catzeflis, A. Meylan, and P. Vogel. 1985. Speciation in the *Sorex araneus* complex (Mammalia: Insectivora). Acta Zoologica Fennica, 170:125-130.

Hausser, J., E. Dannelid, and F. Catzeflis. 1986. Distribution of two karyotypic races of *Sorex araneus*

(Insectivora, Soricidae) in Switzerland and the post-glacial recolonization of the Valais: First results. Zeitschrift für Zoologische Systematik und Evolutionsforschung, 24:307-314.

Hausser, J., R. Hutterer, and P. Vogel. 1990. *Sorex araneus* Linnaeus, 1758—Waldspitzmaus. Pp. 237-278, *in* Handbuch der Säugetiere Europas (J. Niethammer and F. Krapp, eds.). Aula-Verlag, Wiesbaden, 3/I:1-524.

Häussler, U., A. Nagel, M. Bruan, and A. Arnold. 2000. External characters discriminating sibling species of European pipistrelles, *Pipistrellus pipstrellus* (Schreber, 1774) and *P. pygmaeus* (Leach, 1825). Myotis, 37:27-40.

Hawkins, A. F. A., C. E. Hawkins, and P. D. Jenkins. 2000. *Mungotictis decemlineata lineata* (Carnivora: Herpestidae), a mysterious Malagasy mongoose. Journal of Natural History, 34:305-310.

Hay, K. A., and A. W. Mansfield. 1989. Narwhal—*Monodon monoceros* Linnaeus, 1758. Pp. 145-177, *in* Handbook of marine mammals: River dolphins and the larger toothed whales (S. H. Ridgway and R. Harrison, eds.). Academic Press, London, 4:1-442.

Hay, O. P. 1902. Bibliography and catalogue of the fossil Vertebrata of North America. Bulletin of the United States Geological Survey, 179:1-868.

Hayden, F. V. 1863. On the geology and natural history of the upper Missouri. Transactions of the American Philosophical Society, 12:1-218.

Hayes, J. P., and R. G. Harrison. 1992. Variation in mitochondrial DNA and the biogeographic history of woodrats (*Neotoma*) in the eastern United States. Systematic Biology, 41:331-344.

Hayes, J. P., and M. E. Richmond. 1993. Clinal variation and the morphology of woodrats (*Neotoma*) of the eastern United States. Journal of Mammalogy, 74:204-216.

Hayman, R. W. 1935. On a collection of mammals from the Gold Coast. Proceedings of the Zoological Society of London, 1935:915-937.

Hayman, R. W. 1936. A note on *Dologale dybowskii*. Annals and Magazine of Natural History, ser. 10, 18:626-630.

Hayman, R. W. 1946. Systematic notes on the genus *Idiurus* (Anomaluridae). Annals and Magazine of Natural History, ser. 11, 13:208-212.

Hayman, R. W. 1962*a*. A new genus and species of African rodent. Revue de Zoologie et de Botanique Africaines, 65:129-138.

Hayman, R. W. 1962*b*. The occurrence of *Delanymys brooksi* (Rodentia, Muridae) in the Congo. Institut Royal des Sciences Naturelles de Belgique, Bulletin, 38(51):1-4.

Hayman, R. W. 1963*a*. Further notes on *Delanymys brooksi* (Rodentia, Muridae) in the Congo. Revue de Zoologie et de Botanique Africaines, 67:388-392.

Hayman, R. W. 1963*b*. Mammals from Angola, mainly from the Lunda District. Publiçaoes Culturais Museu do Dundo, 66:84-139.

Hayman, R. W. 1966. On the affinities of *Nilopegamys plumbeus* Osgood. Annales Musée Royal de l'Afrique Centrale, Tervuren, Belgique, ser. 8 (Sciences Zoologiques), 144:29-38.

Hayman, R. W., and J. E. Hill. 1971. Order Chiroptera. Part 2. Pp. 1-73, *in* The mammals of Africa: An identification manual (J. Meester and H. W. Setzer, eds.) [issued 15 Jul 1971]. Smithsonian Institution Press, Washington, D.C., not continuously paginated.

Hayssen, V. 1991. *Dipodomys microps*. Mammalian Species, 389:1-9.

Hayward, B. 1970. The natural history of the cave bat *Myotis velifer*. Western New Mexico University, Research Science, 1(1):1-74.

He L., H., R. García-Perea, M. Li, and F. Wei. 2004. Distribution and conservation status of the endemic Chinese mountain cat *Felis bieti*. Oryx, 38:55-61.

Heaney, L. R. 1978. Island area and body size of insular mammals: Evidence from the tri-colored squirrel (*Callosciurus prevosti*) of Southeast Asia. Evolution, 32:29-44.

Heaney, L. R. 1979. A new species of tree squirrel (*Sundasciurus*) from Palawan Island, Philippines (Mammalia: Sciuridae). Proceedings of the Biological Society of Washington, 92:280-286.

Heaney, L. R. 1985. Systematics of Oriental Pygmy Squirrels of the Genera *Exilisciurus* and *Nannosciurus* (Mammalia: Sciuridae). Miscellaneous Publications, Museum of Zoology, University of Michigan, 170:1-58.

Heaney, L. R. 1986. Biogeography of mammals in SE Asia: Estimates of rates of colonization, extinction and speciation. Biological Journal of the Linnean Society, 28:127-165.

Heaney, L. R. 2001. Small mammal diversity along elevational gradients in the Philippines: An assessment of patterns and hypotheses. Global Ecology and Biogeography, 19:15-39.

Heaney, L. R., and E. C. Birney. 1977. Distribution and natural history notes on some mammals from Puebla, Mexico. Southwestern Naturalist, 21:543-545.

Heaney, L. R., and R. S. Hoffmann. 1978. A second specimen of the Neotropical montane squirrel, *Syntheosciurus poasensis*. Journal of Mammalogy, 59:854-855.

Heaney, L. R., and G. S. Morgan. 1982. A new species of gymnure, *Podogymnura*, (Mammalia: Erinaceidae) from Dinagat Island, Philippines. Proceedings of the Biological Society of Washington, 95:13-26.

Heaney, L. R., and R. L. Peterson. 1984. A new species of tube-nosed fruit bat (*Nyctimene*) from Negros Island, Philippines. Occasional Papers of the Museum of Zoology of the University of Michigan, 708:1-16.

Heaney, L. R., and D. S. Rabor. 1982. Mammals of Dinagat and Siargao islands, Philippines. Occasional Papers of the Museum of Zoology, University of Michigan, 699:1-30.

Heaney, L. R., and M. Ruedi. 1994. A preliminary analysis of biogeography and phylogeny of *Crocidura* from the Philippines. Special Publication, Carnegie Museum of Natural History, 18:357-377.

Heaney, L. R., and R. M. Timm. 1983*a*. Relationships of pocket gophers of the genus *Geomys* from the central and northern Great Plains. Miscellaneous Publications, Museum of Natural History, University of Kansas, 74:1-59.

Heaney, L. R., and R. M. Timm. 1983*b*. Systematics and distribution of shrews of the genus *Crocidura* (Mammalia: Insectivora) in Vietnam. Proceedings of the Biological Society of Washington, 96:115-120.

Heaney, L. R., and R. M. Timm. 1985. Morphology, genetics, and ecology of pocket gophers (genus *Geomys*) in a narrow hybrid zone. Biological Journal of the Linnean Society, 25:301-317.

Heaney, L. R., P. C. Gonzales, and A. C. Acala. 1987. An annotated checklist of the taxonomic and conservation status of land mammals in the Philippines. Silliman Journal, 34:32-66.

Heaney, L. R., and P. C. Gonzales, R. C. B. Utzurrum, and E. A. Rickart. 1991. The mammals of Catanduanes Island: Implications for the biogeography of small land-bridge islands in the Philippines. Proceedings of the Biological Society of Washington, 104:399-415.

Heaney, L. R., D. S. Balete, M. L. Dolar, A. C. Alcala, A. T. L. Dans, P. C. Gonzales, N. R. Ingle, M. V. Lepiten, W. L. R. Oliver, P. S. Ong, E. A. Rickart, B. R. Tabaranza, Jr., and R. C. B. Utzurrum. 1998. A synopsis of the mammalian fauna of the Philippine Islands. Fieldiana: Zoology, n.s., 88:1-61.

Heaney, L. R., D. S. Balete, E. A. Rickart, R. C. B. Uzurrum, and P. C. Gonzales. 1999. Mammalian diversity on Mount Isarog, a threatened center of endemism on southern Luzon Island, Philippines. Fieldiana: Zoology, n.s., 95:1-62.

Hecht-Markou, P. 1995. Description, geographical distribution, habitat, and site changes of *Sciurus anomalus* Gueldenstaedt, 1785 on the Lesbos Island (Greece). Annales-Musei-Goulandris, 9:429-444.

Hedges, R. E. M. 2000. Appraisal of radiocarbon dating of kiore bones (Pacific rat *Rattus exulans*) in New Zealand. Journal of the Royal Society of New Zealand, 30(4):385-398.

Hedges, S. B. 2001. Afrotheria: Plate tectonics meets genomics. Proceedings of the National Academy of Sciences, 98:1-2.

Hedges, S. R. 1969. Epigenetic polymorphism in populations of *Apodemus sylvaticus* and *Apodemus flavicollis* (Rodentia, Muridae). Journal of the Zoological Society of London, 159:425-442.

Heffelfinger, J. R. 2000. Status of the name *Odocoileus hemionus crooki* (Mammalia: Cervidae). Proceedings of the Biological Society of Washington, 113:319-333.

Heidecke, D. 1986. Taxonomische Aspekte des Artenschutzes am Beispiel der Biber Eurasiens. Hercynia N.F., 23:146-161.

Heim de Balsac, H. 1943. Mission Th. Monod. Genre nouveau de rongeur (Gerbillinae) de Mauritanie. Bulletin du Museum National d'Histoire Naturelle (sér 2), 15:287-288.

Heim de Balsac, H. 1956. Morphologie divergente des Potamogalinae (Mammifères, Insectivores) en milieu aquatique. Comptes Rendus de l'Acadèmie des Sciences, Paris, 242:2257-2258.

Heim de Balsac, H. 1957. Insectivores Soricidae du Mont Cameroun. Zoologische Jahrbücher, Abteilung für Systematik, Ökologie und Geographie der Tiere, 85:501-672.

Heim de Balsac, H. 1959. Nouvelle contribution à l'étude des Insectivores Soricidae du Mont Cameroun. Bonner Zoologische Beiträge, 10:198-217.

Heim de Balsac, H. 1963. L "abrasion préalable" sous-gingivale des molaires de certains rongeurs et insectivores. Le cas remarquable d'*Uranomys*. Comptes Rendus de l'Académie des Sciences, Paris, 256:5257-5261.

Heim de Balsac, H. 1966*a*. Contribution à l'étude des Soricidae de Somalie. Monitore Zoologico Italiano, 74 Supplemento:196-221.

Heim de Balsac, H. 1966*b*. Faits nouveaux concernant l'évolution cranio-dentaire des Soricinés (Mammifères Insectivores). Comptes Rendus des Séances de l'Académie des Sciences, Paris, 263D:889-892.

Heim de Balsac, H. 1967. Faits nouveaux concernant les *Myosorex* (Soricidae) de l'Afrique orientale. Mammalia, 31:610-628.

Heim de Balsac, H. 1968*a*. Considérations préliminaires sur le peuplement des montages Africaines par les Soricidae. Biologia Gabonica, 4:299-323.

Heim de Balsac, H. 1968*b*. Contribution à l'étude des Soricidae de Fernando Po et du Cameroun. Bonner Zoologische Beiträge, 19:15-42.

Heim de Balsac, H. 1968c. Contributions à la faune de la région de Yaoundé I.—Premier aperçu sur la faune des Soricidae (Mammifères Insectivores). Annales de la Faculté des Sciences du Cameroun, 1968(2):49-58.

Heim de Balsac, H. 1968d. Recherches sur la faune des Soricidae de l'ouest africain (du Ghana au Senegal). Mammalia, 32:379-418.

Heim de Balsac, H. 1968e. Les Soricidae dans le milieu désertique saharien. Bonner Zoologische Beiträge, 19:181-188.

Heim de Balsac, H. 1972. Insectivores. Pp. 629-660, in Biogeography and ecology in Madagascar (R. Battistini, and G. Richard-Vindard, eds.). W. Junk, The Hague, 765 pp.

Heim de Balsac, H., and V. Aellen. 1965. Les Muridae de basse Cote-d'Ivoire. Revue Suisse de Zoologie, 71:695-753.

Heim de Balsac, H., and J. J. Barloy. 1966. Révision des Crocidures du groupe flavescens-occidentalis-manni. Mammalia, 30:601-633.

Heim de Balsac, H., and R. Hutterer. 1982. Les Soricidae (Mammifères Insectivores) des îles du Golfe de Guinée: Faits nouveaux et problèmes biogeographiques. Bonner Zoologische Beiträge, 33:133-150.

Heim de Balsac, H., and M. Lamotte. 1956. Evolution et phylogénie des Soricidés africaines. Mammalia, 20:140-167.

Heim de Balsac, H., and M. Lamotte. 1958. La reserve naturelle integrale du mont Nimba, Part 4 (15): Mammiferes rongeurs (Muscardinides et Murides). Memoires de l'Institut Francais d'Afrique Noire, 53:339-357.

Heim de Balsac, H., and J. Meester. 1977. Order Insectivora. Part 1. Pp. 1-29, in The mammals of Africa: An identification manual (J. Meester and H. W. Setzer, eds.). [issued 22 Aug 1977]. Smithsonian Institution Press, Washington, D.C., not continuously paginated.

Heim de Balsac, H., and P. Mein. 1971. Les musaraignes momifiées des hypogées des Thebes. Existence d'un metalophe chez les Crocidurinae (sensu Repenning). Mammalia, 35:220-244.

Heinonen Fortabat, S., and J. C. Chebez. 1997. Los mamiferos de los parques nacionales de la Argentina. Monografía Especial, Literature of Latin America, 14:76 pp.

Heinrich, W.-D. 1990. Some aspects of evolution and biostratigraphy of Arvicola (Mammalia, Rodentia) in the central European Pleistocene. Pp. 165-182, in International symposium evolution, phylogeny and biostratigraphy of arvicolids (Rodentia, Mammalia) (O. Fejfar and W.-D. Heinrich, eds.). Geological Survey, Prague, 448 pp.

Helden, A. L. Van, A. N. Baker, M. L. Dalebout, J. C. Reyes, K. Van Waerebeek, and C. S. Baker. 2002. Resurrection of Mesoplodon traversii (Gray, 1874), senior synonym of M. bahamondi Reyes, Van Waerebeek, Cardenas and Yanez, 1995 (Cetacea: Ziphiidae). Marine Mammal Science, 18(3):609-621.

Helgen, K. M. 2001. First record of Rattus rattus in Botswana. Mammalian Biology, 66:60-62.

Helgen, K. M. 2003a. Major mammalian clades: A review under consideration of molecular and palaeontological evidence. Mammalian Biology, 68:1-15.

Helgen, K. M. 2003b. A review of the rodent fauna of Seram, Moluccas, with the description of a new subspecies of mosaic-tailed rat, Melomys rufescens paveli. Journal of Zoology, London, 261:165-172.

Helgen, K. M., and T. F. Flannery. 2003. Taxonomy and distribution of the wallaby genus Lagostrophus. Australian Journal of Zoology, 51:199-212.

Helgen, K. M., and T. L. McFadden. 2001. Type specimens of Recent mammals in the Museum of Comparative Zoology. Bulletin of the Museum of Comparative Zoology, 157(2):93-181.

Helgen, K. M., and D. E. Wilson. 2001. Additional material of the enigmatic golden mole Cryptochloris zyli, with notes on the genus Cryptochloris (Mammalia: Chrysochloridae). African Zoology, 36:110-112.

Helgen, K. M., and D. E. Wilson. 2002. The bats of Flores, Indonesia, with remarks on Asian Tadarida. Breviora, 511:1-12.

Helgen, K. M., and D. E. Wilson. 2003. Taxonomic status and conservation relevance of the raccoons (Procyon spp.) of the West Indies. Journal of Zoology, London, 259:69-76.

Helgen, K. M., and D. E. Wilson. 2005. A systematic and zoogeographic overview of the raccoons of Mexico and Central America. Instituto de Biologia, UNAM, Mexico, 20:219–234.

Heller, E. 1904. Mammals of the Galapagos Archipelago, exclusive of the Cetacea. Papers of the Hopkins Stanford Galapagos Expedition, 1898-1899. Proceedings of the California Academy of Science, ser. 3, 3:233-250.

Heller, E. 1909. Two new rodents from British East Africa. Smithsonian Miscellaneous Collections, 52:471-472.

Heller, E. 1911. New species of rodents and carnivores from equatorial Africa. Smithsonian Miscellaneous Collections, 56:1-16.

Heller, E. 1912. New rodents from British East Africa. Smithsonian Miscellaneous Collections, 59:1-20.

Heller, F. 1968. Die Wühlmäuse (Mammalia, Rodentia, Arvicolidae) des Ältest-und Altpleistozäns Europas. Quartär, 19:23-53.

Heller, K.-G. 1992. The echolocation calls of *Hipposideros ruber* and *Hipposideros caffer*. Pp. 75-77, *in* Prague studies in mammalogy, (I Horácek and V. Vohralík, eds.). Charles University Press, Prague, i-xxi + 1-245.

Heller, K.-G., and M. Volleth. 1984. Taxonomic position of '*Pipistrellus societatis*' Hill, 1972 and the karyological characteristics of the genus *Eptesicus* (Chiroptera: Vespertilionidae). Zeitschrift für Zoologische Systematik und Evolutionsforschung, 22:65-77.

Heller, K.-G., M. Volleth, and D. Kock. 1994. Notes on some vespertilionid bats from the Kivu region, Central Africa. Senckenbergiana Biologica, 74:1-8.

Hellwig, H. 2001. Artenschutzproject feldhamster (*Cricetus cricetus*) in Rheinland-Pfalz. Jahrbücher des Nassauischen Vereins für Naturkunde, 122:201-206.

Hemandez, N., and A. Sanchez. 1987. Apuntes cromosómicos en el género *Capromys* (Rodentia). Ciencias Biológicas, 17:98-100.

Hemmer, H. 1966. Untersuchungen zur Stammesgeschichte der Pantherkatzen (Pantherinae). Teil I. Veröffentlichungen der Zoologischen Staatssammlung München, 11:1-121.

Hemmer, H. 1968. Untersuchungen zur Stammesgeschichte der Pantherkatzen (Pantherinae). Teil II. Studien zur Ethologie des Nebelparders *Neofelis nebulosa* (Griffith 1821) und des Irbis *Uncia uncia* (Schreber 1775). Veröffentlichungen der Zoologischen Staatssammlung München, 12:155-247.

Hemmer, H. 1972. *Uncia uncia*. Mammalian Species, 20:1-5.

Hemmer, H. 1974. Untersuchungen zur Stammegeschichte der Pantherkatzen (Pantherinae). III. Zur Artgeschichte des Lowen *Panthera* (*Panthera*) *leo* (Linnaeus 1758). Veröffentlichungen der Zoologischen Staatssammlung München, 17:167-280.

Hemmer, H. 1978. The evolutionary systematics of living Felidae: Present status and current problems. Carnivore, 1(1):71-79.

Hemmer, H. 1990. Domestication. The decline of environmental appreciation. Cambridge University Press, Cambridge, England, 208 pp.

Hemmer, H., P. Grubb, and C. P. Groves. 1976. Notes on the sand cat *Felis margarita* Loche, 1858. Zeitschrift für Säugetierkunde, 41:286-303.

Hemming, F. (ed.). 1950. Conclusions of the fourteenth meeting held at the Sorbonne in the Amphitheatre Louis-Liard on Monday, 26th July 1948 at 2030 hours. Bulletin of Zoological Nomenclature, 4:425-648.

Henderson, D. 1990. Gray whales and whalers on the China coast in 1869. Whalewatcher, 24(4):14-16.

Hendey, Q. B. 1972. The evolution and dispersal of the Monachinae (Mammalia, Pinnipedia). Annals of the South African Museum, 59:99-113.

Hendey, Q. B. 1974. The late Cenozoic Carnivora of the southwestern Cape Province. Annals of the South African Museum, 63:1-369.

Hendey, Q. B. 1978. Late Tertiary Hyaenidae form Langebaanweg, South Africa, and their relevance to the phylogeny of the family. Annals of the South African Museum, 76(7):265-297.

Hendey, Q. B. 1980a. *Agriotherium* (Mammalia: Ursidae) from Langebaanweg, South Africa, and relationships of the genus. Annals of the South African Museum, 81(1):1-109.

Hendey, Q. B. 1980b. Origin of the giant panda. South African Journal of Science, 76:179-180.

Hendey, Q. B. 1981. Palaeoecology of the Late Tertiary fossil occurrences in 'E' Quarry, Langebaanweg, South Africa, and a reinterpretation of their geological context. Annals of the South African Museum, 84(1):1-104.

Hendrichsen, D. K., P. J. J. Bates, and B. D. Hayes. 2001a. Recent records of bats (Chiroptera) from Cambodia. Acta Chiropterologica, 3:21-32.

Hendrichsen, D. K., P. J. J. Bates, B. D. Hayes, and J. L. Walston. 2001b. Recent records of bats (Mammalia: Chiroptera) from Vietnam with six species new to the country. Myotis, 39:35-122.

Hennings, D., and R. S. Hoffmann. 1977. A review of the taxonomy of the *Sorex vagrans* species complex from western North America. Occasional Papers of the Museum of Natural History, University of Kansas, 68:1-35.

Hensley, A. P., and K. T. Wilkins. 1988. *Leptonycteris nivalis*. Mammalian Species, 307:1-4.

Henttonen, H., and V. A. Peiponen. 1982. *Clethrionomys rutilus* (Pallas, 1779)—polarrotelmaus. Pp. 165-176, *in* Handbuch der Säugetiere Europas (J. Niethammer and F. Krapp, eds.). Akademische Verlagsgesellschaft (Wiesbaden), 2/I:1-649.

Henttonen, H., and J. Viitala. 1982. *Clethrionomys rufocanus* (Sundevall, 1846)—graurotelmaus. Pp. 146-164, *in* Handbuch der Säugetiere Europas (J. Niethammer and F. Krapp, eds.). Akademische Verlagsgesellschaft (Wiesbaden), 2/I:1-649.

Heptner, V. G. 1934. Systematische und tiergeographische Notizen über einige russiche Säuger. Folia Zoologica et Hydrobiologica, Bd. VI, 1:21-23.

Heptner, V. G. 1939. The Turkestan desert shrew, its biology and adaptive peculiarities. Journal of Mammalogy, 20:139-149.

Heptner, V. G. 1941. On the names of geographical forms of *Ochotona pallasi* Gray. Journal of Mammalogy, 22:327-328.

Heptner, V. G. 1948*a*. K nomenklature nekotorykh mlekopitayushchikh [On nomenclature of several mammals]. Doklady Akademii Nauk, ser. biol., 60(4):709-712 (in Russian).

Heptner, V. G. 1948*b*. On the nomenclature of the wood mice (*Apodemus 'flavicollis'-sylvaticus*). Doklady Akademii Nauk, SSSR, 60:177-178.

Heptner, V. G. 1967. Mlekopitayushchie Sovetskovo Soyuza. Morskie kotovy i khishehnye [Mammals of the Soviet Union. Sea cows and Carnivores.] (V. G. Heptner and N. P. Naumov, eds.). Vysshaya Shkola. Moscow, 2(1):1-1004 (in Russian).

Heptner, V. G. 1971. O sistematicheskom polozhenii Amurskovo lesnovo kota i o nekhotorykh drugikh yuzhnoaziatiskikh koshokh, otnosimykh k *Felis bengalensis* Kerr, 1792 [Systematic status of the Amur wildcat and some other east-Asian cats referred to *Felis bengalensis* Kerr, 1792]. Zoologicheskii Zhurnal, 50:1720-1727 (in Russian).

Heptner, V. G. 1975. [Materials on the morphology and systematics of the tridactyl jerboas (genus *Jaculus* Erxl., 1777) and related forms (Mammalia, Dipodidae)]. Byulleten' Moskovskovo Obshchestva Ispytatalei Prirody, Otdel Biologicheskii, 80(3):5-15 (in Russian).

Heptner, V. G. 1984. [Contributions to phylogeny and classification of jerboas (Dipodidae) of the fauna of the USSR]. Sbornik Trudov Zoologicheskovo Museya MGU, 22:37-60 (in Russian).

Heptner, V. G., and V. A. Dolgov. 1967. [Systematic position of *Sorex mirabilis* Ognev, 1937. Mammalia, Soricidae]. Zoologicheskii Zhurnal, 46:1419-1422 (in Russian).

Heptner, V. G., and N. P. Naumov (eds). 1967. Mlekopitayushchie Sovetskovo Soyuza. Morskie korovy i khishchnye [Mammals of the Soviet Union. Sea cows and carnivores]. Vysshaya Shkola, Moscow, 2(1):1-1004 (in Russian).

Heptner, V. G., and A. A. Sludskii. 1992. Mammals of the Soviet Union. Volume II, Part 2. Carnivora (Hyaenas and Cats). Smithsonian Institution Libraries. [translation of Heptner, V. G., and A. A. Sludskii. 1972. Mlekopitayushchie Sovetskovo Soyuza. Khishchye (gieny i koshki) Vysshaya Shkola, Moscow, 2(2):1-551 (in Russian).]

Heptner, V. G., A. A. Nasimovich, and A. G. Bannikov. 1961. Mlekopitayushchie Sovetskovo Soyuza: Parnokopytnie i Neparnokopytnie. [Mammals of the Soviet Union: Even-toed and odd-toed ungulates]. Vysshaya Shkola, Moscow, 1:1-776 (in Russian).

Heptner, V. G., A. A. Nasimovich, and A. G. Bannikov. 1988. Mammals of the Soviet Union: Artiodactyla and Perissodactyla [A translation of Heptner et al., 1961, Mlekopitayushchie Sovetskovo Soyuza: Parnokopytnye i neparnokopytnye]. Smithsonian Institution Libraries and National Science Foundation, Washington, D.C., 1:1-1147.

Herd, R. M. 1983. *Pteronotus parnellii*. Mammalian Species, 209:1-5.

Herd, R. M., and M. B. Fenton. 1983. An electrophoretic, morphological, and ecological investigation of a putative hybrid zone between *Myotis lucifugus* and *Myotis yumanensis* (Chiroptera: Vespertilionidae). Canadian Journal of Zoology, 61:2029-2050.

Hermann, J. 1783. Tabula Affinitatum Anim. p. 115.

Hermanson, J. W., and T. J. O'Shea. 1983. *Antrozous pallidus*. Mammalian Species, 213:1-8.

Hernández-Camacho, J. 1960. Primitiae mastozoologicae Colombianae 1: Status taxonomical de *Sciurus pucheranii santanderensis*. Caldasia, 8:359-368.

Hernández-Camacho, J. 1977. Notas para una monografia de *Potos flavus* (Mammalia: Carnivora) en Colombia. Caldasia, 11(55):147-181.

Hernández-Camacho, J., and A. Cadena-G. 1978. Notas para la revision del género *Lonchorhina* (Chiroptera, Phyllostomidae). Caldasia, 12:200-251.

Hernández -Camacho, J., and R. W. Cooper. 1976. The nonhuman primates of Colombia. Pp. 35-69, *in* Neotropical Primates (R. W. Thorington, Jr. and P. Heltne, eds.). National Academy of Sciences, Washington, D.C., 135 pp.

Herrington, S. J. 1986. Phylogenetic relationships of the wild cats of the world. PhD. Dissertation, University of Kansas, 421 pp.

Hershkovitz, P. 1940. Four new oryzomyine rodents from Ecuador. Journal of Mammalogy, 21:78-84.

Hershkovitz, P. 1944. Systematic review of the Neotropical water rats of the genus *Nectomys* (Cricetinae). Miscellaneous Publications, Museum of Zoology, University of Michigan, 58:1-101.

Hershkovitz, P. 1948*a*. Mammals of northern Colombia. Preliminary report no. 2: Spiny Rats (Echimyidae), with supplemental notes on related forms. Proceedings of the United States National Museum, 97:125-140.

Hershkovitz, P. 1948*b*. Mammals of northern Colombia. Preliminary report no. 3. Water rats (genus

Nectomys), with supplemental notes on related forms. Proceedings of the United States National Museum, 98:49-56.

Hershkovitz, P. 1948c. The technical name of the Virginia deer with a list of South American forms. Proceedings of the Biological Society of Washington, 61:41-48.

Hershkovitz, P. 1949a. Generic names of the four-eyed pouch opossum and the woolly opossum (Didelphidae). Proceedings of the Biological Society of Washington, 62:11-12.

Hershkovitz, P. 1949b. Status of names credited to Oken, 1816. Journal of Mammalogy, 30:289-301.

Hershkovitz, P. 1950. Mammals of northern Colombia. Preliminary report no. 6: Rabbits (Leporidae), with notes on the classification and distribution of the South American forms. Proceedings of the United States National Museum, 100:327-375.

Hershkovitz, P. 1951. Mammals from British Honduras, Mexico, Jamaica and Haiti. Fieldiana: Zoology, 31(47):547-569.

Hershkovitz, P. 1954. Mammals of northern Colombia. Preliminary report no. 7: Tapirs (genus *Tapirus*), with a systematic revision of American species. Proceedings of the United States National Museum, 103:465-496.

Hershkovitz, P. 1955a. South American marsh rats, genus *Holochilus*, with a summary of sigmodont rodents. Fieldiana: Zoology, 37:639-687.

Hershkovitz, P. 1955b. Status of the generic name *Zorilla* (Mammalia): Nomenclature by rule or by caprice. Proceedings of the Biological Society of Washington, 68:185-192.

Hershkovitz, P. 1957a. A synopsis of the wild dogs of Colombia. Novedades Colombianas. Contribuciones Científicas del Museo de Historia Natural de la Universidad del Cauca, 3(1):157-161.

Hershkovitz, P. 1957b. The type locality of *Bison bison* Linnaeus. Proceedings of the Biological Society of Washington, 70:31-32.

Hershkovitz, P. 1958. Technical names of the South American marsh deer and pampas deer. Proceedings of the Biological Society of Washington, 71:13-16.

Hershkovitz, P. 1959a. Nomenclature and taxonomy of the Neotropical mammals described by Olfers, 1818. Journal of Mammalogy, 40:337-353.

Hershkovitz, P. 1959b. Two new genera of South American rodents. Proceedings of the Biological Society of Washington, 72:5-10.

Hershkovitz, P. 1959c. A new species of South American brocket, genus *Mazama* (Cervidae). Proceedings of the Biological Society of Washington, 72:45-54.

Hershkovitz, P. 1960. Mammals of northern Colombia, preliminary report no. 8: Arboreal rice rats, a systematic revision of the subgenus *Oecomys*, genus *Oryzomys*. Proceedings of the United States National Museum, 110:513-568.

Hershkovitz, P. 1961a. On the South American small-eared zorro *Atelocynus microtis* Sclater (Canidae). Fieldiana: Zoology, 39:505-523.

Hershkovitz, P. 1961b. On the nomenclature of certain whales. Fieldiana: Zoology, 39(49):547-565.

Hershkovitz, P. 1962. Evolution of Neotropical cricetine rodents (Muridae) with special reference to the phyllotine group. Fieldiana: Zoology, 46:1-524.

Hershkovitz, P. 1963. The nomenclature of South American peccaries. Proceedings of the Biological Society of Washington, 76:85-88.

Hershkovitz, P. 1966a. Catalog of living whales. Bulletin of the United States National Museum, 246:1-259.

Hershkovitz, P. 1966b. Mice, land bridges and Latin American faunal interchange. Pp. 725-751, *in* Ectoparasites of Panama (R. L. Wenzel and V. J. Tipton, eds.). Field Museum of Natural History, Chicago, 861 pp.

Hershkovitz, P. 1966c. South American swamp and fossorial rats of the scapteromyine group (Cricetinae, Muridae), with comments on the glans penis in murid taxonomy. Zeitschrift für Säugetierkunde, 31:81-149.

Hershkovitz, P. 1969. The evolution of mammals on southern continents. VI. The Recent mammals of the Neotropical Region: A zoogeographical and ecological review. Quarterly Review of Biology, 44:1-70.

Hershkovitz, P. 1970a. Notes on tertiary Platyrrhine monkeys and description of a new genus from the late Miocene of Colombia. Folia Primatologica, 12:1-37.

Hershkovitz, P. 1970b. Supplementary notes on Neotropical *Oryzomys dimidiatus* and *Oryzomys hammondi* (Cricetinae). Journal of Mammalogy, 51:789-794.

Hershkovitz, P. 1971. A new rice rat of the *Oryzomys palustris* group (Cricetinae, Muridae) from northwestern Colombia, with remarks on distribution. Journal of Mammalogy, 52:700-709.

Hershkovitz, P. 1972a. Notes on New World Monkeys. International Zoo Yearbook, 12:3-12.

Hershkovitz, P. 1972b. The recent mammals of the Neotropical region: A zoogeographic and ecological review. Pp. 311-431, *in* Evolution, mammals and southern continents (A. Keast, F. C. Erk, and B. Glass, eds.). State University of New York Press, Albany, 543 pp.

Hershkovitz, P. 1976. Comments on generic names of four-eyed opossums (family Didelphidae). Proceedings of the Biological Society of Washington, 89:295-304.

Hershkovitz, P. 1977. Living New World Monkeys (Platyrrhini). University of Chicago Press, Chicago, IL, 1:1-1117.

Hershkovitz, P. 1979. The species of sakis, genus *Pithecia*, with notes on sexual dichromatism. Folia Primatologia, 31:1-22.

Hershkovitz, P. 1981. *Philander* and four-eyed opossums once again. Proceedings of the Biological Society of Washington, 93:943-946.

Hershkovitz, P. 1982. Neotropical deer (Cervidae). Part 1. Pudus, genus *Pudu* Gray. Fieldiana: Zoology, n.s., 11:1-86.

Hershkovitz, P. 1983. Two new species of night monkeys, genus *Aotus* (Cebidae, Platyrrhini): A preliminary report on *Aotus* taxonomy. American Journal of Primatology, 4(3):209-243.

Hershkovitz, P. 1984. Taxonomy of squirrel monkeys genus *Saimiri* (Cebidae, Platyrrhini): A preliminary report with description of a hitherto unnamed form. American Journal of Primatology, 7:155-210.

Hershkovitz, P. 1985. A preliminary taxonomic review of the South American bearded saki monkeys, genus *Chiropotes*, with the description of a new subspecies. Fieldiana: Zoology, n.s., 27:1-46.

Hershkovitz, P. 1987a. First South American record of Coues' marsh rice rat, *Oryzomys couesi*. Journal of Mammalogy, 68:152-154.

Hershkovitz, P. 1987b. A history of the recent mammalogy of the neotropical region from 1492 to 1850. Fieldiana: Zoology, n.s., 39:11-98.

Hershkovitz, P. 1987c. Uacaries, New World monkeys of the genus *Cacajao* (Cebidae, Platyrrhini): a preliminary taxonomic review with the description of a new subspecies. American Journal of Primatology, 12:1-53.

Hershkovitz, P. 1987d. The taxonomy of South American sakis, genus *Pithecia*: A preliminary report and critical review with the description of a new species and subspecies. American Journal of Primatology, 12:387-468.

Hershkovitz, P. 1990a. The Brazilian rodent genus *Thalpomys* (Sigmodontinae, Cricetidae) with a description of a new species. Journal of Natural History, 24:763-783.

Hershkovitz, P. 1990b. Titis, New World monkeys of the genus *Callicebus* (Cebidae, Platyrrhini): A preliminary taxonomic review. Fieldiana: Zoology, n.s., 55:1-109.

Hershkovitz, P. 1990c. Mice of the *Akodon boliviensis* size class (Sigmodontinae, Cricetidae), with the description of two new species from Brazil. Fieldiana: Zoology, n.s., 57:1-35.

Hershkovitz, P. 1992a. Ankle bones: The Chilean opossum *Dromiciops gliroides* Thomas, and marsupial phylogeny. Bonner Zoologische Beiträge, 43:181-213.

Hershkovitz, P. 1992b. The South American gracile mouse opossums, genus *Gracilinanus* Gardner and Creighton, 1989 (Marmosidae, Marsupialia): A taxonomic review with notes on general morphology and relationships. Fieldiana: Zoology, n.s., 70:frontispiece, vi + 1-56.

Hershkovitz, P. 1993. A new central Brazilian genus and species of sigmodontine rodent (Sigmodontinae) transitional between akodonts and oryzomyines, with a discussion of muroid molar morphology and evolution. Fieldiana: Zoology, n.s., 75:18 pp.

Hershkovitz, P. 1994. The description of a new species of South American hocicudo, or long-nosed mouse, Genus *Oxymycterus* (Sigmodontinae, Muroidea), with a critical review of the generic content. Fieldiana: Zoology, n.s., 79:43.

Hershkovitz, P. 1998. Report on some sigmodontine rodents collected in southeastern Brazil with descriptions of a new genus and six new species. Bonner Zoologische Beiträge, 47:193-256.

Hershkovitz, P. 1999. *Dromiciops gliroides* Thomas, 1894, last of the Microbiotheria (Marsupialia), with a review of the family Microbiotheriidae. Fieldiana: Zoology, n.s., 93:frontispiece, v + 1-60.

Hewison, A. J. M., and A. Danilkin. 2001. Evidence for separate specific status of European (*Capreolus capreolus*) and Siberian (*C. pygargus*) roe deer. Zeitschrift für Säugetierkunde, 66:13-21.

Hewitt, G. M. 1999. Post-glacial re-colonization of European biota. Biological Journal of the Linnean Society, 68:87-112.

Hewson, R. 1991. Mountain hare/Irish hare. Pp. 161-167, *in* The handbook of British mammals (G. B. Corbet and S. Harris, eds.). Blackwell Scientific Publications, Oxford, 588 pp.

Heyning, J. E. 1989a. Comparative facial anatomy of beaked whales (Ziphiidae) and a systematic revison among the families of extant Odontoceti. Contributions in Science, Natural History Museum of Los Angeles County, 405:1-64.

Heyning, J. E. 1989b. Cuvier's beaked whale—*Ziphius cavirostris* G. Cuvier, 1823. Pp. 289-308, *in* Handbook of marine Mammals: River dolphins and the larger toothed whales (S. H. Ridgway and R. Harrison, eds.). Academic Press, London, 4:1-442.

Heyning, J. E., and M. E. Dahlheim. 1988. *Orcinus orca*. Mammalian Species, 304:1-9.

Heyning, J. E., and W. F. Perrin. 1991. Re-examination of two forms of common dolphins (genus *Delphinus*) from the eastern North Pacific; evidence for two species. National Marine Fisheries Service, Southwest Fisheries Center, Administrative Report LJ-91-28, ii + 37 pp., 16 figs.

Heyning, J. E., and W. F. Perrin. 1994. Evidence for two species of common dolphins (Genus *Delphinus*) from the eastern north Pacific. Contributions in Science, no. 442, 35 pp.

Hibbard, C. W. 1963. The origin of the P/3 pattern of *Sylvilagus*, *Caprolagus*, *Oryctolagus* and *Lepus*. Journal of Mammalogy, 44:1-15.

Hibbard, C. W. 1968. Paleontology. Pp. 6-26, *in* Biology of *Peromyscus* (Rodenta) (J. A. King, ed.). Special Publication, American Society of Mammalogists, 2:xiii + 593 pp.

Hibbard, C. W., and W. W. Dalquest. 1973. *Proneofiber*, a new genus of vole (Cricetidae: Rodentia) from the Pleistocene Seymour Formation of Texas, and its evolutionary and stratigraphic significance. Journal of Quaternary Research, 3:269-274.

Hice, C. L. 2001. Records of a few rare mammals from northeastern Peru. Zeitschrift für Säugetierkunde, 66:317-319.

Hice, C. L., and S. Solari. 2002. First record of *Centronycteris maximiliani* (Fischer, 1829) and two additional records of *C. centralis* Thomas, 1912 from Peru. Acta Chiropterologica, 4:217-220.

Hickey, M. B. C., and J. M. Dunlop. 2000. *Nycteris grandis*. Mammalian Species, 632:1-4.

Hielscher, K., A. Stubbe, K. Zernahle, and R. Samjaa. 1992. Karyotypes and systematics of Asian high-mountain voles, genus *Alticola* (Rodentia, Arvicolidae). Results of Mongolian-German biological expeditions since 1962, no. 211. Cytogenetics and Cell Genetics, 59:307-310.

Higham, T. F. G., and F. J. Petchey. 2000. On the reliability of archaeological rat bone for radiocarbon dating in New Zealand. Journal of the Royal Society of New Zealand, 30(4):399-409.

Hight, M. E., M. Goodman, and W. Prychodko. 1974. Immunological studies of the Sciuridae. Systematic Zoology, 23:12-25.

Hill, J. E. 1958. Some observations on the fauna of the Maldive Islands. Part II. Mammals. Journal of the Bombay Natural History Society, 55:3-10.

Hill, J. E. 1959. A north Bornean pygmy squirrel, *Glyphotes simus* Thomas, and its relationships. Bulletin of the British Museum (Natural History), Zoology Series, 5:257-266.

Hill, J. E. 1960. The Robinson Collection of Malaysian mammals. Bulletin of the Raffles Museum (Singapore), 29:1-112.

Hill, J. E. 1961a. Fruit-bats from the Federation of Malaya. Proceedings of the Zoological Society of London, 136:629-642.

Hill, J. E. 1961b. Indo-Australian bats of the genus *Tadarida*. Mammalia, 25:29-56.

Hill, J. E. 1961c. Notes on flying squirrels of the genera *Pteromyscus*, *Hylopetes* and *Petinomys*. Annals and Magazine of Natural History, ser. 13, 4:721-738.

Hill, J. E. 1962a. A little-known fruit-bat from Rennel Island. The Natural History of Rennel Island, British Solomon Islands, 4:7-9.

Hill, J. E. 1962b. Notes on some insectivores and bats from Upper Burma. Proceedings of the Zoological Society of London, 139:119-137.

Hill, J. E. 1962c. Notes flying squirrels of the genera *Pteromyscus*, *Hylopetes*, and *Petinomys*. Annals and Magazine of Natural History, ser. 13, 4:721-737.

Hill, J. E. 1963a [1964]. Notes on some tubed-nosed bats, genus *Murina*, from southeastern Asia, with descriptions of a new species and a new subspecies. Federation Museums Journal, n.s. (Kuala Lumpur), 8:48-59.

Hill, J. E. 1963b. A revision of the genus *Hipposideros*. Bulletin of the British Museum (Natural History), Zoology Series, 11:1-129.

Hill, J. E. 1965. Asiatic bats of the genera *Kerivoula* and *Phoniscus* (Vespertilionidae), with a note on *Kerivoula aerosa* Tomes. Mammalia, 29:524-556.

Hill, J. E. 1966. A review of the genus *Philetor* (Chiroptera; Vespertilionidae). Bulletin of the British Museum (Natural History), Zoology Series, 14:371-387.

Hill, J. E. 1967. The bats of the Andaman and Nicobar Islands. Journal of the Bombay Natural History Society, 64:1-9.

Hill, J. E. 1971a. Bats from the Solomon Islands. Journal of Natural History, 5:573-581.

Hill, J. E. 1971b. The bats of Aldabra Atoll, western Indian Ocean. Philososophical Transactions of the Royal Society of London, B260:573-576.

Hill, J. E. 1971c. A note of *Pteropus* (Chiroptera: Pteropidae) from the Andaman Islands. Journal of the Bombay Natural History Society, 68:1-8.

Hill, J. E. 1971d. The status of *Vespertilio brachypterus* Temminck, 1840 (Chiroptera: Vespertilionidae). Zoologische Mededelingen, 45:143.

Hill, J. E. 1972a. The Gunong Benom Expedition, 1967. 4. New records of Malayan bats, with taxonomic notes and the description of a new *Pipistrellus*. Bulletin of the British Museum (Natural History), Zoology Series, 23:21-42.

Hill, J. E. 1972b. A note on *Rhinolophus rex* and *Rhinolophus paradoxolophus*. Mammalia, 36:428-434.

Hill, J. E. 1974a. A review of *Laephotis* Thomas, 1901 (Chiroptera: Vespertilionidae). Bulletin of the British Museum (Natural History), Zoology Series, 27:73-82.

Hill, J. E. 1974b. New records of bats from southeastern Asia. Bulletin of the British Museum (Natural History), Zoology Series, 27:127-138.

Hill, J. E. 1974c. A review of *Scotoecus* Thomas, 1901 (Chiroptera: Vespertilionidae). Bulletin of the British Museum (Natural History), Zoology Series, 27:167-188.

Hill, J. E. 1976. Bats referred to *Hesperoptenus* Peters, 1869 (Chiroptera: Vespertilionidae) with the description of a new subgenus. Bulletin of the British Museum (Natural History), Zoology Series, 30:1-28.

Hill, J. E. 1977a. African bats allied to *Kerivoula lanosa* (A. Smith, 1847)(Chiroptera: Vespertilionidae). Revue de Zoologie Africaine, 91:623-633.

Hill, J. E. 1977b. A review of the Rhinopomatidae (Mammalia: Chiroptera). Bulletin of the British Museum (Natural History), Zoology Series, 32:29-43.

Hill, J. E. 1980a. A note on *Lonchophylla* (Chiroptera: Phyllostomatidae) from Ecuador and Peru, with the description of a new species. Bulletin of the British Museum (Natural History), Zoology Series, 38:233-236.

Hill, J. E. 1980b. The status of *Vespertilio borbonicus* E. Geoffroy, 1803 (Chiroptera: Vespertilionidae). Zoologische Mededelingen, 55:287-295.

Hill, J. E. 1982. A review of the leaf-nosed bats *Rhinonycteris*, *Cloeotis* and *Triaenops* (Chiroptera: Hipposideridae). Bonner Zoologische Beiträge, 33:165-186.

Hill, J. E. 1983. Bats (Mammalia: Chiroptera) from Indo-Australia. Bulletin of the British Museum (Natural History), Zoology Series, 43:103-208.

Hill, J. E. 1985. The status of *Lichonycteris degener* Miller, 1931 (Chiroptera: Phyllostomidae). Mammalia, 49:579-582.

Hill, J. E. 1987. A note on *Balantiopteryx infusca* (Thomas, 1897) (Chiroptera: Emballonuridae). Mammalia, 50:558-560.

Hill, J. E. 1991. Bats (Mammalia: Chiroptera) from the Togian Islands, Sulawesi, Indonesia. Bulletin of the American Museum of Natural History, 206:168-175.

Hill, J. E., and W. N. Beckon. 1978. A new species of *Pteralopex* Thomas, 1888 (Chiroptera: Pteropodidae) from the Fiji Islands. Bulletin of the British Museum (Natural History), Zoology Series, 34:65-82.

Hill, J. E., and T. D. Carter. 1937. Ten new rodents from Angola, Africa. American Museum Novitates, 913: 9 pp.

Hill, J. E., and T. D. Carter. 1941. The mammals of Angola, Africa. Bulletin of the American Museum of Natural History, 78(1):1-211.

Hill, J. E., and M. J. Daniel. 1985. Systematics of the New Zealand short-tailed bat *Mystacina* Gray, 1843 (Chiroptera: Mystacinidae). Bulletin of the British Museum (Natural History), Zoology Series, 48:279-300.

Hill, J. E., and C. M. Francis. 1984. New bats (Mammalia: Chiroptera) and new records of bats from Borneo and Malaya. Bulletin of the British Museum (Natural History), Zoology Series, 47:303-329.

Hill, J. E., and D. L. Harrison. 1987. The baculum in the Vespertilioninae (Chiroptera: Vespertilionidae) with a systematic review, a synopsis of *Pipistrellus* and *Eptesicus*, and the description of a new genus and subgenus. Bulletin of the British Museum (Natural History), Zoology Series, 52(7):225-305.

Hill, J. E., and K. F. Koopman. 1981. The status of *Lamingtona lophorhina* McKean and Calaby, 1968 (Chiroptera: Vespertilionidae). Bulletin of the British Museum (Natural History), Zoology Series, 41:275-278.

Hill, J. E., and F. G. Rozendaal. 1989. Records of bats (Microchiroptera) from Wallacea. Zoologische Mededelingen, 63:97-122.

Hill, J. E., and D. A. Schlitter. 1982. A record of *Rhinolophus arcuatus* (Chiroptera: Rhinolophidae) from New Guinea, with the description of a new subspecies. Annals of Carnegie Museum, 51:455-464.

Hill, J. E., and J. D. Smith. 1984. Bats: A natural history. University of Texas Press, Austin, 243 pp.

Hill, J. E., and S. E. Smith. 1981. *Craseonycteris thonglongyai*. Mammalian Species, 160:1-4.

Hill, J. E., and K. Thonglongya. 1972. Bats from Thailand and Cambodia. Bulletin of the British Museum (Natural History), Zoology Series, 22:171-196.

Hill, J. E., and G. Topal. 1973. The affinities of *Pipistrellus ridleyi* Thomas, 1898 and *Glischropus rosseti* Oey, 1951 (Chiroptera: Vespertilionidae). Bulletin of the British Museum (Natural History), Zoology Series, 24:447-454.

Hill, J. E., and G. Topal. 1990. Records of Marshall's horseshoe bat, *Rhinlophus marshalli* Tonglongya, 1973 (Chiroptera: Rhinolophidae) from Vietnam. Mammalia, 54:490-491.

Hill, J. E., and M. Yoshiyuki. 1980. A new species of *Rhinolophus* (Chiroptera: Rhinolophidae) from Iriomote Island, Ryukyu Islands, with notes on the Asiatic members of the *Rhinolophus pusillus* group. Bulletin of the National Science Museum (Tokyo), ser. A (Zoology), 6:179-189.

Hill, J. E., A. Zubaid, and G. W. H. Davison. 1986. The taxonomy of leaf-nosed bats of the *Hipposideros bicolor* group (Chiroptera: Hipposideridae) from southeastern Asia. Mammalia, 50:535-540.

Hill, W. C. O. 1953. Note on the taxonomy of the genus *Tarsius*. Proceedings of the Zoological Society of London, 123:13-16.

Hill, W. C. O. 1969. The nomeclature, taxonomy and distribution of chimpanzees. Pp. 22-49, *in* Chimpanzee: Anatomy, behavior and diseases of chimpanzees (G. A. Bourne, ed.). S. Karger, Basel, 1:1-468.

Hill, W. C. O. 1974. Primates: Comparative anatomy and taxonomy, vol. 7. Cynopithecinae: *Cercocebus*, *Macaca*, *Cynopithecus*. Edinburgh University Press, 954 pp.

Hille, A., and H. Meinig. 1996. The subspecific status of European populations of the striped field mouse *Apodemus agrarius* (Pallas, 1771) based on morphological and biochemical characters. Bonner Zoologische Beiträge, 46(1-4):203-231.

Hille, A., and A. Stubbe. 1996. Biochemical systematics of four taxa of Asian high-mountain voles, *Alticola* (Rodentia, Arvicolinae). Folia Zoologica, 45(4):289-299.

Hille, A., D. Tarkhnishvili, H. Meinig, and R. Hutterer. 2002. Morphometric, biochemical and molecular traits in Caucasian wood mice (*Apodemus/Sylvaemus*), with remarks on species divergence. Acta Theriologica, 47(4):389-416.

Hillman, C. N., and T. W. Clark. 1980. *Mustela nigripes*. Mammalian Species, 126:1-3.

Hillman-Smith, A. K. K., and C. P. Groves. 1994. *Diceros bicornis*. Mammalian Species, 455:1-8.

Hillyard, J. R., C. J. Phillips, E. C. Birney, J. A. Monjeau, and R. S. Sikes. 1997. Mitochondrial DNA analysis and zoogeography of two species of silky desert mice, *Eligmodontia*, in Patagonia. Zeitschrift für Säugetierkunde, 62:281-292.

Hilton, C. D., and T. L. Best. 1993. *Tamias cinereicollis*. Mammalian Species, 436:1-5.

Hinesley, L. L. 1979. Systematics and distribution of two chromosome forms in the southern grasshopper mouse, genus *Onychomys*. Journal of Mammalogy, 60:117-128.

Hingst-Zaher, E., L. F. Marcus, and R. Cerqueira. 2000. Application of geometric morphometrics to the study of postnatal size and shape changes in the skull of *Calomys expulsus*. Hystrix, n.s., 11:99-113.

Hinojosa, F., S. Anderson, and J. L. Patton. 1987. Two new species of *Oxymycterus* (Rodentia) from Peru and Bolivia. American Museum Novitates, 2898:1-17.

Hinton, M. A. C. 1918. Scientific results from the mammal survey. No. XVIII. Report on the house rats of India, Burma, and Ceylon. Part I. Journal of the Bombay Natural History Society, 26:59-88.

Hinton, M. A. C. 1919a. Scientific results from the mammal survey. Pt. 2. Journal of the Bombay Natural History Society, 26:384-416.

Hinton, M. A. C. 1919b. Notes on the genus *Cricetomys*, with descriptions of four new forms. Annals and Magazine of Natural History, ser. 9, 4:282-289.

Hinton, M. A. C. 1921. Some new African mammals. Annals and Magazine of Natural History, ser. 9, 7:368-373.

Hinton, M. A. C. 1922. Scientific results from the mammal survey. No. 34. The house rats of Nepal. Journal of the Bombay Natural History Society, 28:1056-1066.

Hinton, M. A. C. 1923. On the voles collected by Mr. G. Forrest in Yunnan; with remarks upon the genera *Eothenomys* and *Neodon* and upon their allies. Annals and Magazine of Natural History, ser. 9, 11:145-162.

Hinton, M. A. C. 1926a. Monograph of the voles and lemmings (Microtinae) living and extinct. Volume 1. British Museum (Natural History), London, 488 pp.

Hinton, M. A. C. 1926b. Note on the occurrence of a vole in Northern Africa. Annals and Magazine of Natural History, ser. 9, 18:304-306.

Hinton, M. A. C. 1943. Preliminary diagnosis of five new murine rodents from New Guinea. Annals and Magazine of Natural History, ser. 11, 10:552-557.

Hirai, H., Y. Hirai, Y. Kawamoto, H. Endo, J. Kimura, and W. Rerkamnuaychoke. 2002. Cytogenetic differentiation of two sympatric tree shrew taxa found in the southern part of the Isthmus of Kra. Chromosome Research, 10:313-327.

Hirikawa, H., T. Kuwahata, Y. Shibata, and E. Yamada. 1992. Insular variation of the Japanese hare (*Lepus brachyurus*) on the Oki islands, Japan. Journal of Mammalogy, 73:672-679.

Hirning, U., W. A. Schulz, W. Just, S. Adolph, and W. Vogel. 1989. A comparative study of the heterochromatin of *Apodemus sylvaticus* and *Apodemus flavicollis*. Chromosoma, 98:450-455.

Hitchcock, G. 1998. First record of the false water-rat, *Xeromys myoides* (Rodentia: Muridae), in New Guinea. Science in New Guinea, 23(3):141-144.

Hocking, G. J., and M. M. Driessen. 2000. Status and conservation of the rodents of Tasmania. Wildlife Research, 27:371-377.

Hodgson, B. H. 1831. Some account of a new species of *Felis*. Gleanings in Science, 3:177-178.

Hodgson, B. H. 1842*a*. On a new species of *Prionodon*, *P. pardicolor* nobis. Calcutta Journal of Natural History, 2:57-60.

Hodgson, B. H. 1842*b*. Notice of the mammals of Tibet, with descriptions and plates of some new species. Journal of the Asiatic Society of Bengal, 11:275-289.

Hodgson, B. H. 1843. Notice of two marmots inhabiting respectively the plains of Tibet and the Himalayan slopes near to the snows, and also of a *Rhinolophus* of the central region of Nepal. Journal of the Asiatic Society of Bengal, 12:409-414.

Hodgson, B. H. 1844. Classified catalogue of mammals of Nepal. Calcutta Journal of Natural History, 4:284-294.

Hodgson, B. H. 1845. On the rats, mice, and shrews of the central region of Nepal. Annals and Magazine of Natural History, [ser. 1], 15:266-270.

Hoeck, H. N. 1978. Systematics of the Hyracoidea: Toward a clarification. Bulletin of Carnegie Museum of Natural History, 6:146-151.

Hoek Ostende, L. W. van den. 2001. A revised generic classification of the Galericini (Insectivora, Mammalia) with some remarks on their palaeobiogeography and phylogeny. Geobios, 34:681-695.

Hoeve, R., and L. C. Wijlaars. 1992. Muskrat, *Ondatra zibethicus* (L., 1766). Pp. 250-255, *in* Atlas van de Nederlandse Zoogdieren (S. Broekhuizen, B. Hoekstra, V. van Laar, C. Smeenk, and J. B. M. Thissen, eds.). Stichting Uitgeverij Koninklije Nederlandse Natuurhistorische Vereniging, Utrecht, 336 pp.

Hoffmann, A. 1997. Small mammals in the Kibale Forest, Uganda. Pp. 241-244, *in* Tropical biodiversity and systematics. (H. Ulrich, ed.). Zoologisches Forschungsinstitut und Museum Alexander Koenig, Bonn, 357 pp.

Hoffmann, F. G., and R. J. Baker. 2001. Systematics of bats of the genus *Glossophaga* (Chiroptera: Phyllostomidae) and phylogeography in *G. soricina* based on the cytochrome-*b* gene. Journal of Mammalogy, 82:1092-1101.

Hoffmann, F. G., and R. J. Baker. 2003. Comparative phylogeography of short-tailed fruit bats (*Carollia*: Phyllostomidae). Molecular Ecology, 12:3403-3414.

Hoffmann, F. G., E. P. Lessa, and M. F. Smith. 2002. Systematics of *Oxymycterus* with description of a new species from Uruguay. Journal of Mammalogy, 83:408-420.

Hoffmann, F. G., J. G. Owen, and R. J. Baker. 2003. mtDNA perspective of chromosomal diversification and hybridization in Peters' tent-making bat (*Uroderma bilobatum*: Phyllostomidae). Molecular Ecology, 12:2981-2993.

Hoffmann, R. S. 1971. Relationships of certain Holarctic shrews, genus *Sorex*. Zeitschrift für Säugetierkunde, 36:193-200.

Hoffmann, R. S. 1977. The identity of Lewis' marmot, *Arctomys lewisii*. Proceedings of the Biological Society of Washington, 90:291-301.

Hoffmann, R. S. 1984. A review of the shrew-moles (genus *Uropsilus*) of China and Burma. Journal of the Mammalogical Society of Japan, 10:69-80.

Hoffmann, R. S. 1985*a*. The correct name for the Palearctic brown, or flat-skulled, shrew is *Sorex roboratus*. Proceedings of the Biological Society of Washington, 98:17-28.

Hoffmann, R. S. 1985*b* [1986]. A review of the genus *Soriculus* (Mammalia: Insectivora). Journal of the Bombay Natural History Society, 82:459-481.

Hoffmann, R. S. 1986. A new locality record for the kouprey from Viet-Nam, and an archaeological record from China. Mammalia, 50:391-395.

Hoffmann, R. S. 1987. A review of the systematics and distribution of Chinese red-toothed shrews (Mammalia: Soricinae). Acta Theriologica Sinica, 7:100-139.

Hoffmann, R. S. 1996*a*. Noteworthy shrews and voles from the Xizang-Qinghai Plateau. Pp. 155-168, *in* Contributions in Mammalogy: A Memorial Volume Honoring Dr. J. Knox Jones, Jr. (H. H. Genoways and R. J. Baker, eds.). Museum of Texas Tech University, il+315 pp.

Hoffmann, R. S. 1996*b*. A research information system for mammals with Palaearctic examples. Bonner Zoologische Beiträge, 46:15-32.

Hoffmann, R. S. 1998. The trouble with *Lepus capensis*. Abstract, Euro-American Mammal Congress, Santiago de Compostela, Spain, 19-24 July, p. 24.

Hoffmann, R. S., and J. W. Koeppl. 1985. Zoogeography. Pp. 84-115, *in* Biology of New World *Microtus* (R. H. Tamarin, ed.). Special Publication, American Society of Mammalogists, 8:1-893.

Hoffmann, R. S., and J. G. Owen. 1980. *Sorex tenellus* and *Sorex nanus*. Mammalian Species, 131:1-4.

Hoffmann, R. S., and R. S. Peterson. 1967. Systematics and zoogeography of *Sorex* in the Bering Strait area. Systematic Zoology, 16:127-136.

Hoffmann, R. S., J. W. Koeppl, and C. F. Nadler. 1979. The relationships of the amphiberingian marmots (Mammalia: Sciuridae). Occasional Papers of the Museum of Natural History, University of Kansas, 83:1-56.

Hoffmann, R. S., C. G. Anderson, R. W. Thorington, Jr., and L. R. Heaney. 1993. Family Sciuridae. Pp. 419-465, *in* Mammal species of the world, a taxonomic and geographic reference. Second ed. (D. E. Wilson and D. M. Reeder, eds.). Smithsonian Institution Press, Washington, D.C., xviii + 1207 pp.

Hoffmeister, D. F. 1951. A taxonomic and evolutionary study of the piñon mouse, *Peromyscus truei*. Illinois Biological Monographs, 21:1-104.

Hoffmeister, D. F. 1969. The species problem in the *Thomomys bottae-Thomomys umbrinus* complex of gophers in Arizona. Miscellaneous Publications, Museum of Natural History, University of Kansas, 51:75-91.

Hoffmeister, D. F. 1974. The taxonomic status of *Perognathus penicillatus minimus* Burt. Southwestern Naturalist, 19:213-214.

Hoffmeister, D. F. 1981. *Peromyscus truei*. Mammalian Species, 161:1-5.

Hoffmeister, D. F. 1986. Mammals of Arizona. University of Arizona Press, Tucson., 602 pp.

Hoffmeister, D. F., and L. de la Torre. 1960. A revision of the wood rat *Neotoma stephensi*. Journal of Mammalogy, 41:476-491.

Hoffmeister, D. F., and L. de la Torre. 1961. Geographic variation in the mouse *Peromyscus difficilis*. Journal of Mammalogy, 42:1-13.

Hoffmeister, D. F., and V. E. Diersing. 1973. The taxonomic status of *Peromyscus merriami goldmani* Osgood, 1904. Southwestern Naturalist, 18:354-357.

Hoffmeister, D. F., and V. E. Diersing. 1978. Review of the tassel-eared squirrels of the subgenus *Otosciurus*. Journal of Mammalogy, 59:402-413.

Hoffmeister, D. F., and L. S. Ellis. 1979. Geographic variation in *Eutamias quadrivittatus* with comments on the taxonomy of other Arizonan chipmunks. Southwestern Naturalist, 24:655-665.

Hoffmeister, D. F., and M. R. Lee. 1963a. The status of the sibling species *Peromyscus merriami* and *Peromyscus eremicus*. Journal of Mammalogy, 44:201-213.

Hoffmeister, D. F., and M. R. Lee. 1963b. Taxonomic review of cottontails, *Sylvilagus floridanus* and *Sylvilagus nuttallii*, in Arizona. American Midland Naturalist, 70:138-148.

Hoffmeister, D. F., and M. R. Lee. 1967. Revision of the pocket mice, *Perognathus penicillatus*. Journal of Mammalogy, 48:361-380.

Hoffstetter, R. 1969. Remarques sur le phylogenie et la classification des Edentes Xenarthres (Mammiferes) actuels et fossiles. Bulletin du Muséum National d'Histoire Naturelle (Paris), 41:92-103.

Hoffstetter, R. 1975. El origen de los Caviomorpha y el problema de los Hystricognathi (Rodentia). Actas del Primer Congreso Argentino de Paleontologia y Bioestratigrafia, Tucuman, Argentina, 2:505-528.

Hoffstetter, R. 1982. Les Primate Simiiformes (=Anthropoidea) (comprehension, phylogénie, histoire biogéographique). Annales de Paléontologie (Vertébrés-Invertébrés), 68:241-290.

Hoffstetter, R., and R. Lavocat. 1970. Decouverte dans le Deseadien de Bolivie de genres Pentalophodontes appuyant les affinities Africaines des rongeurs caviomorphes. Comptes Rendus des Seances de l'Academie des Sciences, Paris, (D), 271:172-175.

Hofmeijer, G. K., and H. de Bruijn. 1985. The mammals from the lower Miocene of Aliveri (island of Evia, Greece). Part 4: The Spalacidae and Anomalomyidae. Proceedings of the Koninklijke Nederlandse Akademie van Wetenschappen, Series B, 88(2):185-198.

Hogan, K. M., M. C. Hedin, H. S. Koh, S. K. Davis, and I. F. Greenbaum. 1993. Systematic and taxonomic implications of karyotypic, electrophoretic, and mitochondrial-DNA variation in *Peromyscus* from the Pacific Northwest. Journal of Mammalogy, 74:819-831.

Hogan, K. M., S. K. Davis, and I. F. Greenbaum. 1997. Mitochondrial-DNA analysis of the systematic relationships within the *Peromyscus maniculatus* species group. Journal of Mammalogy, 78:733-743.

Holdaway, R. N. 1996. Arrival of rats in New Zealand. Nature, 384:225-226.

Holdaway, R. N., and N. R. Beavan. 1999. Reliable ^{14}C AMS dates on bird and Pacific rat *Rattus exulans* bone gelatin, from a CaCO$_3$-rich deposit. Journal of the Royal Society of New Zealand, 29(3):185-211.

Holden, H. E., and H. S. Eabry. 1970. Chromosomes of *Sylvilagus floridanus* and *S. transitionalis*. Journal of Mammalogy, 51:166-168.

Holden, M. E. 1993. Family Myoxidae. Pp. 763-770, *in* Mammal species of the world, a taxonomic and geographic reference, Second ed. (D. E. Wilson and D. M. Reeder, eds.). Smithsonian Institution Press, Washington, 1206 pp.

Holden, M. E. 1996a. Description of a new species of *Dryomys* (Rodentia, Myoxidae) from Balochistan, Pakistan, including morphological comparisons with *Dryomys laniger* Felten and Storch, 1968, and *D. nitedula* (Pallas, 1778). Bonner Zoologische Beiträge, 46(1-4):111-131.

Holden, M. E. 1996b. Systematic revision of Sub-Saharan African dormice (Rodentia: Myoxidae: *Graphiurus*)

Part 1: An introduction to the generic revision, and a revision of *Graphiurus surdus*. American Museum Novitates, 3157:1-44.

Holden, M. E. In Press. Gliridae. *In* The Mammals of Africa. Volume 3 (D. C. D. Happold, J. Kingdon, T. M. Butynski, eds.). Academic Press, London.

Hollander, R. R. 1990. Biosystematics of the yellow-faced pocket gopher, *Cratogeomys castanops* (Rodentia: Geomyidae) in the United States. Special Publications, The Museum, Texas Tech University, 33:1-62.

Hollander, R. R., and M. R. Willig. 1992. Description of a new subspecies of the southern grasshopper mouse, *Onychomys torridus*, from Western Mexico. Occasional Papers. The Museum Texas Tech University, 148:1-4.

Hollar, L. J., and M. S. Springer. 1997. Old World fruitbat phylogeny: Evidence for convergent evolution and an endemic African clade. Proceedings of the National Academy of Science, 94:5716-5721.

Hollister, N. 1911. A systematic synopsis of the muskrats. North American Fauna, 32:1-47.

Hollister, N. 1912. The names of the Rocky Mountain goat. Proceedings of the Biological Society of Washington, 25:185-186.

Hollister, N. 1913a. A synopsis of the American minks. Proceedings of the United States National Museum, 44:471-480.

Hollister, N. 1913b. A review of the Philippine land mammals in the United States National Museum. Proceedings of the United States National Museum, 46:299-341.

Hollister, N. 1914. A systematic account of the grasshopper mice. Proceedings of the United States National Museum, 47:427-489.

Hollister, N. 1915a. The genera and subgenera of raccoons and their allies. Proceedings of the United States National Museum, 49:143-150.

Hollister, N. 1915b. The specific name of the striped Muishond of South Africa. Proceedings of the Biological Society of Washington, 28:184.

Hollister, N. 1916a. A systematic account of the prairie dogs. North American Fauna, 40:1-37.

Hollister, N. 1916b. The type species of *Rattus*. Proceedings of the Biological Society of Washington, 29:206-207.

Hollister, N. 1918. East African mammals in the United States National Museum. Part I. Insectivora, Chiroptera, and Carnivora. Bulletin of the United States National Museum, 99:1-194, 55 pls.

Hollister, N. 1919. East African mammals in the United States National Museum. Part II. Rodentia, Lagomorpha, and Tubulidentata. Bulletin of the United States National Museum, 99(2):1-184.

Holloway, G. L., and R. M. Barclay. 2001. *Myotis ciliolabrum*. Mammalian Species, 670:1-5.

Holthuis, L. B. 1987. The scientific name of the sperm whale. Marine Mammal Science, 3(1):87-88.

Holthuis, L. B., and T. Sakai. 1970. Ph. F. von Siebold and Fauna Japonica: A history of early Japanese zoology. Academic Press of Japan, Tokyo, 323 pp.

Holz, H., and J. Niethammer. 1990. *Erinaceus europaeus* Linnaeus, 1758 — Braunbrustigel, Westigel. Pp. 26-49, *in* Handbuch der Säugetiere Europas (J. Niethammer and F. Krapp, eds.). Aula-Verlag, Wiesbaden, 3/I:1-524.

Homan, J. A., and J. K. Jones, Jr. 1975a. *Monophyllus redmani*. Mammalian Species, 57:1-3.

Homan, J. A., and J. K. Jones, Jr. 1975b. *Monophyllus plethodon*. Mammalian Species, 58:1-2.

Honacki, J. H., K. E. Kinman and J. W. Koeppl (eds.). 1982. Mammal species of the world, a taxonomic and geographic reference. Allen Press, Inc. and The Association of Systematics Collections, Lawrence, Kansas, 694 pp.

Honda, T., H. Suzuki, and M. Itoh. 1977. An unusual sex chromosome constitution found in the Amami spinous country-rat, *Tokudaia osimensis osimensis*. Japanese Journal of Genetics, 52(3):247-249.

Honda, T., H. Suzuki, M. Itoh, and K. Hayashi. 1978. Karyotypical differences of the Amami spinous country-rat, *Tokudaia osimensis osimensis* obtained from two neighbouring islands. Japanese Journal of Genetics, 53:297-299.

Honess, P. E., and S. Bearder. 1996 [1997]. Descriptions of the dwarf galago species of Tanzania. African Primates, 2(2):75-79.

Honeycutt, R. L., and D. J. Schmidly. 1979. Chromosomal and morphological variation in the plains pocket gopher, *Geomys bursarius* in Texas and adjacent states. Occasional Papers, The Museum, Texas Tech University, 58:1-54.

Honeycutt, R. L., and S. L. Williams. 1982. Genic differentiation in pocket gophers of the genus *Pappogeomys*, with comments on intergeneric relationships in the subfamily Geomyinae. Journal of Mammalogy, 63:208-217.

Honeycutt, R. L., M. W. Allard, S. V. Edwards, and D. A. Schlitter. 1991. Systematics and evolution of the family Bathyergidae. Pp. 45-65, *in* The biology of the naked mole rat (P. W. Sherman, J. U. M. Jarvis, and R. D. Alexander, eds.). Princeton University Press, Princeton, 518 pp.

Honeycutt, R. L., D. L. Rowe, and M. H. Gallardo. 2003. Molecular systematics of the South American caviororph rodents: Relationships among species and genera in the family Octodontidae. Molecular Phylogenetics and Evolution, 26:476-489.

Hood, C. S. 1989. Comparative morphology and evolution of the female reproductive tract in macroglossine bats (Mammalia, Chiroptera). Journal of Morphology, 199, 207-21.

Hood, C. S., and J. K. Jones, Jr. 1984. *Noctilio leporinus*. Mammalian Species, 216:1-7.

Hood, C. S., and J. Pitocchelli. 1983. *Noctilio albiventris*. Mammalian Species, 197:1-5.

Hood, C. S., L. W. Robbins, R. J. Baker, and H. S. Shellhammer. 1984. Chromosomal studies and evolutionary relationships of an endangered species, *Reithrodontomys raviventris*. Journal of Mammalogy, 65:655-667.

Hoofer, S. R., and R. A. Van Den Bussche. 2001. Phylogenetic relationships of plecotine bats and allies based on mitochondrial ribosomal sequences. Journal of Mammalogy, 82:131-137.

Hoofer, S. R., J. R. Choate, and N. E. Mandrak. 1999. Mensural discrimination between *Reithrodontomys megalotis* and *R. montanus* using cranial characters. Journal of Mammalogy, 80:91-101.

Hoofer, S. R., S. A Reeder, E. W. Hansen, and R. A. Van Den Bussche. 2003. Molecular phylogenetics and taxonomic review of noctilionoid and vespertilionoid bats (Chiroptera, Yangochiroptera). Journal of Mammalogy, 84(3):809-821.

Hoogerwerf, A. 1970. Udjung Kulon. The land of the last Javan rhinoceros. E. J. Brill, Leiden, 512 pp.

Hoogland, J. L. 1996. *Cynomys ludovicianus*. Mammalian Species, 535:1-10.

Hooijer, D. A. 1956. The valid name of the banteng: *Bibos javanicus* d'Alton. Zoologische Mededelingen, 34:223-226.

Hooijer, D. A. 1962. Quaternary langurs and macaques from the Malay Archipelago. Zoologische Verhandelingen (Leiden), 55:1-64.

Hooper, E. T. 1946. Two genera of pocket gophers should be congeneric. Journal of Mammalogy, 27:397-399.

Hooper, E. T. 1947. Notes on Mexican mammals. Journal of Mammalogy, 28:40-57.

Hooper, E. T. 1952. A systematic review of harvest mice (genus *Reithrodontomys*) of Latin America. Miscellaneous Publications, Museum of Zoology, University of Michigan, 77:1-255.

Hooper, E. T. 1953. Notes on mammals of Tamaulipas, Mexico. Occasional Papers of the Museum of Zoology, University of Michigan, 544:1-12.

Hooper, E. T. 1954. A synopsis of the cricetine rodent genus *Nelsonia*. Occasional Papers of the Museum of Zoology, University of Michigan, 558:1-12.

Hooper, E. T. 1955. Notes on mammals of western Mexico. Occasional Papers of the Museum of Zoology, University of Michigan, 565:1-26.

Hooper, E. T. 1957. Dental patterns in mice of the genus *Peromyscus*. Miscellaneous Publications, Museum of Zoology, University of Michigan, 99:1-59.

Hooper, E. T. 1958. The male phallus in mice of the genus *Peromyscus*. Miscellaneous Publications, Museum of Zoology, University of Michigan, 105:1-24.

Hooper, E. T. 1959. The glans penis in five genera of cricetid rodents. Occasional Papers of the Museum of Zoology, University of Michigan, 613:1-11.

Hooper, E. T. 1960. The glans penis in *Neotoma* (Rodentia) and allied genera. Occasional Papers of the Museum of Zoology, University of Michigan, 618:1-21.

Hooper, E. T. 1968*a*. Anatomy of middle ear walls and cavities in nine species of microtine rodents. Occasional Papers of the Museum of Zoology, University of Michigan, 657:1-28.

Hooper, E. T. 1968*b*. Classification. Pp. 27-74, *in* Biology of *Peromyscus* (Rodentia) (J. A. King, ed.). Special Publication, American Society of Mammalogists, 2:1-593.

Hooper, E. T. 1972. A synopsis of the rodent genus *Scotinomys*. Occasional Papers of the Museum of Zoology, University of Michigan, 665:1-32.

Hooper, E. T., and M. D. Carleton. 1976. Reproduction, growth and development in two contiguously allopatric rodent species, genus *Scotinomys*. Miscellaneous Publications, Museum of Zoology, University of Michigan, 151:1-52.

Hooper, E. T., and B. S. Hart. 1962. A synopsis of Recent North American microtine rodents. Miscellaneous Publications, Museum of Zoology, University of Michigan, 120:1-68.

Hooper, E. T., and G. G. Musser. 1964*a*. The glans penis in Neotropical cricetines (Family Muridae) with comments on classification of muroid rodents. Miscellaneous Publications, Museum of Zoology, University of Michigan, 123:1-57.

Hooper, E. T., and G. G. Musser. 1964*b*. Notes on classification of the rodent genus *Peromyscus*. Occasional Papers of the Museum of Zoology, University of Michigan, 635:1-13.

Hopp, M. J., and D. S. Rogers. 1994. *Reithrodontomys creper*. Mammalian Species, 482:1-3.

Hopwood, A. T. 1947. The generic names of the mandrill and baboons, with notes on some of the genera of Brisson, 1762. Proceedings of the Zoological Society of London, 117:533-536.

Horácek, I. 1986. Fossil records and chronological status of dormice in Czechoslovakia. Part 1: *Glis glis*, *Eliomys quercinus*. Folia Musei Rerum Naturalium Bohemiae Occidentalis, Zoologica, 24:49-59.

Horácek, I. 1987. Fossil records and chronological status of dormice in Czechoslovakia Part II. *Dryomys* and *Muscardinus*. Folia Musei Rerum Naturalium Bohemiae Occidentalis, Zoologica, 26:80-96.

Horácek, I. 1991. Enigma of *Otonycteris*: Ecology, relationship, classification. Myotis, 29:17-30.

Horácek, I. 1997. The status of *Vesperus sinensis* Peters, 1880 and remarks on the genus *Vespertilio*. Vespertilio, Revúca-Praha, 2:59-72.

Horácek, I., and V. Hanák. 1984. Comments on the systematics and phylogeny of *Myotis nattereri* (Kuhl, 1818). Myotis, 21-22:20-29.

Horácek, I., and V. Hanak. 1985-1986. Generic status of *Pipistrellus savii*. Myotis, 23-24:11-16.

Horácek, I., V. Hanák, and J. Gaisler. 2000. Bats of the Palearctic region: A taxonomic and biogeographic review. Pp. 11-157, *in* Proceedings of the VIIIth European bat research symposium. Vol I. Approaches to biogeography and ecology of bats (B. W. Woloszyn, ed.). Publication of the Chiropterological Information Center, Institute of Systematics and Evolution of Animals PAS in Krakow, Poland, 315 pp.

Horath, A., and D. A. Navarrete-Gutierrez. 1997. Ampliación del área de distribución de *Peromyscus zarhynchus* Merriam, 1898 (Rodentia: Muridae). Revista Mexicana de Mastozoología, 2:122-125.

Horovitz, I., R. Zardoya, and A. Meyer. 1998. Platyrrhine systematics: A simultaneous analysis of molecular and morphological data. American Journal of Physical Anthropology, 106:261-282.

Horsfield, M. D. 1855. Brief notices of several new or little known species of Mammalia, lately discovered and collected in Nepal by Brian Houghton Hodgson. Annals and Magazine of Natural History, ser. 2, 16:101-114.

Horsfield, T. 1825. Description of the RIMAU-DAHAN of the inhabitants of Sumatra, a new species of *Felis*, discovered in the forests of Bencoolen, by Sir T. Stamford Raffles, late Lieutenant Governor of Fort Marlborough, &c., &c., &c. Zoological Journal, 1(4):542-554.

Horwood, J. 1987. The sei whale: Population biology, ecology and management. Croon Helm, London, [xvi] + 375 pp.

Hosey, G. R. 1982. The Bosporus land-bridge and mammal distributions in Asia Minor and the Balkans. Säugetierkundliche Mitteilungen, 30:53-62.

Hossack, W. C. 1907. Miscellanea, mammals. Records of the Indian Museum, Calcutta, 1(3):275-276.

Hou Lan-xin, Zhao Xin-chun, and Jiang Wei. 1986. [Investigation of rodents in Dabancheng-Chaiwopu, Urumqi Country, Xinjiang Yugur Autonomous Region.] Acta Theriologica Sinica, 6(4):315-316 (in Chinese).

Hou, L., S. Xue, L. Ma, X. Wang, and Kamili. 1995. [New record of mammals in China–*Alticola barakshin*.] Acta Theriologica Sinica, 15(2):105 (in Chinese).

Houseal, T. W., I. F. Greenbaum, D. J. Schmidly, S. A. Smith, and K. M. Davis. 1987. Karyotypic variation in *Peromyscus boylii* from Mexico. Journal of Mammalogy, 68:281-296.

How, R. A., N. K. Cooper, and J. L. Bannister. 2001. Checklist of the mammals of Western Australia. Records of the Western Australian Museum, Supplement, 63:91-98.

Howell, A. B. 1926a. Anatomy of the wood rat. Monograph of the American Society of Mammalogists, 1:1-225.

Howell, A. B. 1926b. Voles of the genus *Phenacomys*. North American Fauna, 48:1-66.

Howell, A. B. 1927. Revision of the American lemming mice (genus *Synaptomys*). North American Fauna, 50:1-37.

Howell, A. B. 1929. Mammals from China in the collections of the United States National Museum. Proceedings of the United States National Museum, 75(1):1-82.

Howell, A. H. 1914. Revision of the American harvest mice (genus *Reithrodontomys*). North American Fauna, 36:1-97.

Howell, A. H. 1915. Revision of the American marmots. North American Fauna, 37:1-80.

Howell, A. H. 1918. Revision of the American flying squirrels. North American Fauna, 44:1-64.

Howell, A. H. 1929. Revision of the American chipmunks. North American Fauna, 52:1-157.

Howell, A. H. 1936. Description of three new red squirrels (*Tamiasciurus*) from North America. Proceedings of the Biological Society of Washington, 49:133-136.

Howell, A. H. 1938. Revision of the North American ground squirrels, with a classification of the North American Sciuridae. North American Fauna, 56:1-256.

Hoyt, R. A., and R. J. Baker. 1980. *Natalus major*. Mammalian Species, 130:1-3.

Hrabe, V. 1969. Der Bau des Glans penis bei vier Schläferarten (Gliridae, Rodentia). Zoologické Listy, 18(4):317-334.

Hrabe, V. 1977. Tarsal glands of voles of the genus *Pitymys* (Microtidae, Mammalia) from southern Austria. Folia Zoologica, 27:123-128.

Hrabe, V., and P. Koubek. 1985. Notes on the taxonomy of the Tatra chamois (*Rupicapra rupicapra tatrica*

Blahout). Pp. 51-55, *in* The biology and management of mountain ungulates (S. Lovari, ed.). Croom Helm, London, 271 pp.

Hristov, I. P. 1993. Morphometrical and phenetic analysis of *Apodemus agrarius* (Pallas, 1771) from the region of Plovdiv. Comptes Rendus l'Académie Bulgare des Sciences, 46(10):73-76.

Hsu, T. C., and K. Benirschke. 1971. *Capromys pilorides*. *In* An Atlas of Mammalian Chromosomes. Springer-Verlag, Berlin, 6(Folio 282):1-3.

Hsu, T. C., and M. L. Johnson. 1963. Karyotype of two mammals from Malaya. American Naturalist, 97:127-129.

Hsu, T. C., and M. L. Johnson. 1970. Cytological distinction between *Microtus montanus* and *Microtus canicaudus*. Journal of Mammalogy, 51:824-826.

Huang, S.-W., K. G. Ardlie, and H.-T. Yu. 2001. Frequency and distribution of *t*-haplotypes in the Southeast Asian house mouse (*Mus musculus castaneus*) in Taiwan. Molecular Ecology, 10:2349-2354.

Huang, X.-S. 1992. [Zapodidae (Rodentia, Mammalia) from the middle Oligocene of Ulantatal, Nei Mongol.] Vertebrata PalAsiatica, 30(4):249-286 (in Chinese with English summary).

Hubbard, C. A. 1970. A first record of *Beamys* from Tanzania with observations on its breeding and habits in captivity. Zoologica Africana, 5:229-236.

Hubbs, C. L. 1946. First records of two beaked whales, *Mesoplodon bowdoini* and *Ziphius cavirostris*, from the Pacific coast of the United States. Journal of Mammalogy, 27(3):242-255.

Hubert, B. 1978*a*. Revision of the genus *Saccostomus* (Rodentia, Cricetomyinae), with new morphological and chromosomal data from specimens from the lower Omo Valley, Ethiopia. Bulletin of Carnegie Museum of Natural History, 6:48-52.

Hubert, B. 1978*b*. Modern rodent fauna of the lower Omo Valley, Ethiopia. Bulletin of Carnegie Museum of Natural History, 6:109-112.

Hubert, B., F. Adam, and A. Poulet. 1973. Liste preliminaire des rongeurs du Senegal. Mammalia, 37(1):76-87.

Hubert, B., A. Meylan, F. Petter, A. Poulet, and M. Tranier. 1983. Different species in genus *Mastomys* from Western, Central and Southern Africa (Rodentia, Muridae). Annales Musée Royal De L'Afrique Centrale, Sciences Zoologiques, 237:143-148.

Hubner, R. 1988. Populations robertsoniennes chez la souris "sauvage" (*Mus domesticus* Rutty, 1772) en Belgique. Annales de la Societe Royale Zoologique de Belgique, 118:69-75.

Huchon, D., and E. J. P. Douzery. 2001. From the Old World to the New World: A molecular chronicle of the phylogeny and biogeography of hystricognath rodents. Molecular Phylogenetic and Evolution, 20:238-251.

Huchon, D., F. M. Catzeflis, and E. J. P. Douzery. 1999. Molecular evolution of the nuclear Willebrand factor gene in mammals and the phylogeny of rodents. Molecular Biology and Evolution, 16:577-589.

Huchon, D., F. Catzeflis, and E. J. P. Douzery. 2000. Variance of molecular datings, evolution of rodents and the phylogenetic affinities between Ctenodactylidae and Hystricognathi. Proceedings of the Royal Society of London B, 267:393-402.

Huchon, D., O. Madsen, M. J. J. B. Sibbald, K. Ament, M. J. Stanhope, F. Catzeflis, W. W. de Jong, and E. J. P. Douzery. 2002. Rodent phylogeny and a timescale for the evolution of Glires: Evidence from an extensive taxon sampling using three nuclear genes. Molecular Biology and Evolution, 19(7):1053-1065.

Huckaby, D. G. 1980. Species limits in the *Peromyscus mexicanus* group (Mammalia: Rodentia: Muroidea). Contributions in Science, Natural History Museum of Los Angeles County, 326:1-24.

Hückinghaus, F. 1961. Vergleichende Untersuchungen über die Formenmannigfaltigkeit der Unterfamilie Caviinae Murray 1886. (Ergebnisse der Südamerikaexpedition Herre/Röhrs 1956-1957). Zeitschrift für Wissenschaftliche Zoologie, 166:1-98, 62 pls.

Hudson, W. S., and D. E. Wilson. 1986. *Macroderma gigas*. Mammalian Species, 260:1-4.

Huey, L. M. 1960. Comments on the pocket mouse, *Perognathus fallax*, with descriptions of two new races from Baja California, Mexico. Transactions of the San Diego Society of Natural History, 12:415-419.

Huey, L. M. 1964. The mammals of Baja California, Mexico. Transactions of the San Diego Society of Natural History, 13:85-168.

Hufnagl, E. 1972. Libyan mammals. Oleander Press, Cambridge. 85 pp.

Hugueney, M. 1999. Genera *Eucricetodon* and *Pseudocricetodon*. Pp. 347-358, *in* The Miocene land mammals of Europe (G. E. Rössner and K. Heissig, eds.). Dr. Friedrich Pfeil, München, 515 pp.

Hugueney, M., and P. Mein. 1965. Lagomorphes et rongeurs du néogène de Lissieu (Rhône). Travaux des Laborotoire de Géologie de la Faculte des Sciences de l'Université de Lyon, 12:109-123.

Hugueney, M., and P. Mein. 1993. A comment on the earliest Spalacinae (Rodentia, Muroidea). Journal of Mammalian Evolution, 1(3):215-223.

Hugueney, M., and M. Vianey-Liaud. 1980. Les Dipodidae (Mammalia, Rodentia) d'Europe Occidentale au

Paleogene et au Neogene Inferieur: Origine et evolution. Palaeovertebrata, Mémoire jubilaire en homage á R. Lavocat:303-342.

Hulva, P., and I. Horácek. 2002. *Craseonycteris thonglongyai* (Chiroptera: Craseonycteridae) is a rhinolophoid: Molecular evidence from cytochrome *b*. Acta Chiropterologica, 4:107-120.

Humphrey, S. R., and L. N. Brown. 1986. Report of a new bat (Chiroptera: *Artibeus jamaicensis*) in the United States is erroneous. Florida Scientist, 49:261-263.

Humphrey, S. R., and H. W. Setzer. 1989. Geographic variation and taxonomic revision of rice rats (*Oryzomys palustris* and *O. argentatus*) of the United States. Journal of Mammalogy, 70:557-570.

Hunt, J. L., L. A. McWilliams, T. L. Best, and K. G. Smith. 2003. *Eumops bonariensis*. Mammalian Species, 733:1-5

Hunt, R. M., Jr. 1974. The auditory bulla in Carnivora: An anatomical basis for reappraisal of carnivore evolution. Journal of Morphology, 143:21-76.

Hunt, R. M., Jr. 1987. Evolution of the Aeluroid Carnivora: Significance of auditory structure in the nimravid cat *Dinictis*. American Museum Novitates, 2886:1-74.

Hunt, R. M., Jr. 1989. Evolution of the Aeluroid Carnivora: Significance of the ventral promontorial process of the petrosal, and the origin of basicranial patterns in the living families. American Museum Novitates, 2930:1-32.

Hunt, R. M., Jr. 1998. Evolution of the Aeluroid Carnivora: Diversity of the earliest aeluroids from Eurasia (Quercy, Hsanda-Gol) and the origin of felids. American Museum Novitates, 3252:1-65.

Hunt, R. M., Jr. 2001. Basicranial anatomy of the living linsangs *Prionodon* and *Poiana* (Mammalia, Carnivora, Viverridae), with comments on the early evolution of Aeluroid Carnivorans. American Museum Novitates, 3330:1-24.

Hunt, R. M., Jr., and R. H. Tedford. 1993. Phylogenetic relationships within the Aeluroid Carnivora and implications of their temporal and geographic distribution. Pp. 53-73, *in* Mammal Phylogeny: Placentals (F. S. Szalay, M. J. Novacek, and M. C. McKenna, eds.). Springer-Verlag, New York, 2:1-321.

Hurka, L. 1990. Die Säugetierfauna des westlichen Teils der Tschechischen Republik. III. Die Nagetiere (Rodentia). Folia Musei Rerum Naturalium Bohemiae Occidentalis, Zoologica, 31:1-59.

Hurrell, E., and G. McIntosh. 1984. Mammal society dormouse survey, January 1975-April 1979. Mammalian Review, 14(1):1-18.

Husar, S. L. 1977. *Trichechus inunguis*. Mammalian Species, 72:1-4.

Husar, S. L. 1978*a*. *Dugong dugon*. Mammalian Species, 88:1-7.

Husar, S. L. 1978*b*. *Trichechus senegalensis*. Mammalian Species, 89:1-3.

Husar, S. L. 1978*c*. *Trichechus manatus*. Mammalian Species, 93:1-5.

Hussain, S. T., H. de Bruijn, and J. M. Leinders. 1978. Middle Eocene rodents from the Kala Chitta Range (Punjab, Pakistan) (III). Proceedings of the Koninklijke Nederlandse Akademie van Wetenschappen, ser. B, 81:101-112.

Husson, A. M. 1955. Notes on the mammals collected by the Swedish New Guinea Expedition. Nova Guinea, n.s., 6:283-306.

Husson, A. M. 1962. The bats of Suriname. Zoologische Verhandelingen, 58:1-282, 30 pls.

Husson, A. M. 1963. On *Blarina pyrrhonota* and *Echimys macrourus*: Two mammals incorrectly assigned to the Suriname fauna. Studies on the Fauna of Suriname and other Guyanas, 5:34-41, 2 pls.

Husson, A. M. 1970. Status of *Myotis australis* (Dobson). Australian Bat Research News, 9:4-6.

Husson, A. M. 1978. The mammals of Suriname. E. J. Brill, Leiden, 569 pp., 160 pls.

Husson, A. M., and L. B. Holthuis. 1955. The dates of publication of "Verhandelingen over de Natuurlijke Geschiedenis der Nederlandsche Overzeesche Bezittingen" Edited by C. J. Temminck. Zoologische Mededelingen, 34(2):17-24.

Husson, A. M., and L. B. Holthuis. 1974. *Physeter macrocephalus* Linnaeus, 1758, the valid name for the sperm whale. Zoologische Mededelingen, 48(19):205-217.

Husson, A. M., and W. C. van Neurn. 1959. Kleurverscheidenheden van de mol, *Talpa europaea* L., in Nederland waargenomen. Zoologische Bijdragen, 4:1-16.

Hutcheon, J. M., J. A. W. Kirsch, and J. D. Pettigrew. 1998. Base-compositional biases and the bat problem. III. The question of microchiropteran monophyly. Philosophical Transactions of the Royal Society of London B, 353:607-617.

Hutchison, J. H. 1968. Fossil Talpidae (Insectivora, Mammalia) from the Later Tertiary of Oregon. Bulletin of the Museum of Natural History Oregon, 11:1-117.

Hutchison, J. H. 1974. Notes on type specimens of European Miocene Talpidae and a tentative classification of Old World Tertiary Talpidae (Insectivora: Mammalia). Geobios, 7:211-256, pl. 37-39.

Hutchison, J. H. 1987. Moles of the *Scapanus latimanus* group (Talpidae, Insectivora) from the Pliocene and Pleistocene of California. Contributions in Science, 386:1-15.

Hutchison, J. H., and C. R. Harington. 2002. A peculiar new fossil shrew (Lipotyphla, Soricidae) from the High Arctic of Canada. Canadian Journal of Earth Sciences, 39:439-443.

Hutson, A. M., S. P. Mickleburgh, and P. A. Racey. 2001. Global status survey and conservation action plan: Microchiropteran bats. IUCN/SSC Chiroptera Specialist Group. IUCN, Gland, Switzerland, 258 pp.

Hutterer, R. 1979 [1980]. Verbreitung und Systematik von *Sorex minutus* Linnaeus, 1766 (Insectivora; Soricidae) im Nepal- Himalaya und angrenzenden Gebieten. Zeitschrift für Säugetierkunde, 44:65-80.

Hutterer, R. 1980. A record of Goodwin's shrew, *Cryptotis goodwini*, from Mexico. Mammalia, 44:413.

Hutterer, R. 1981*a*. Nachweis der Spitzmaus *Crocidura roosevelti* für Tanzania. Stuttgarter Beiträge zur Naturkunde, Ser. A, 342:1-9.

Hutterer, R. 1981*b*. Range extension of *Crocidura smithii*, with description of a new subspecies from Senegal. Mammalia, 45:388-391.

Hutterer, R. 1981*c*. Zur Systematik und Verbreitung der Soricidae Äthiopiens (Mammalia; Insectivora). Bonner Zoologische Beiträge, 31:217-247.

Hutterer, R. 1981*d*. Der Status von *Crocidura ariadne* Pieper, 1979 (Mammalia: Soricidae). Bonner Zoologische Beiträge, 32:3-12.

Hutterer, R. 1982*a*. *Crocidura manengubae* n. sp. (Mammalia: Soricidae), eine neue Spitzmaus aus Kamerun. Bonner Zoologische Beiträge, 32:241-248.

Hutterer, R. 1982*b*. Biologische und morphologische Beobachtungen an Alpenspitzmäusen (*Sorex alpinus*). Bonner Zoologische Beiträge, 33:3-18.

Hutterer, R. 1983*a*. *Crocidura grandiceps*, eine neue Spitzmaus aus Westafrika. Revue Suisse de Zoologie, 90:699-707.

Hutterer, R. 1983*b*. Taxonomy and distribution of *Crocidura fuscomurina* (Heuglin, 1865). Mammalia, 47:221-227.

Hutterer, R. 1983*c*. Über den Igel (*Erinaceus algirus*) der Kanarischen Inseln. Zeitschrift für Säugetierkunde, 48:261-264.

Hutterer, R. 1984. Status of some African *Crocidura* described by Isidore Geoffroy Saint-Hilaire, Carl J. Sundevall and Theodor von Heuglin. Annales Musée Royal de l'Afrique Centrale, Sciences Zoologiques, 237:207-217.

Hutterer, R. 1985. Anatomical adaptations of shrews. Mammal Review, 15:43-55.

Hutterer, R. 1986*a*. African shrews allied to *Crocidura fischeri*: Taxonomy, distribution and relationships. Cimbebasia, ser. A, 8(4):199-207.

Hutterer, R. 1986*b*. Diagnosen neuer Spitzmäuse aus Tansania (Mammalia: Soricidae). Bonner Zoologische Beiträge, 37:23-33.

Hutterer, R. 1986*c*. Synopsis der Gattung *Paracrocidura* (Mammalia: Soricidae), mit Beschreibung einer neuen Art. Bonner Zoologische Beiträge, 37:73-90.

Hutterer, R. 1986*d*. Südamerikanische Spitzmäuse: *Cryptotis meridensis* und *C. thomasi* als verschiedene Arten. Zeitschrift für Säugetierkunde, 51, Sonderheft:33-34.

Hutterer, R. 1987. The species of *Crocidura* (Soricidae) in Morocco. Mammalia, 50:521-534.

Hutterer, R. 1990. *Sorex minutus* Linnaeus, 1766—Zwergspitzmaus. Pp. 183-206, *in* Handbuch der Säugetiere Europas (J. Niethammer and F. Krapp, eds.). Aula-Verlag, Wiesbaden, 3/I:1-524.

Hutterer, R. 1991. Variation and evolution of the Sicilian shrew: Taxonomic conclusions and description of a possibly related species from the Pleistocene of Morocco (Mammalia: Soricidae). Bonner Zoologische Beiträge, 42:241-251.

Hutterer, R. 1993*a*. Order Insectivora. Pp. 69-130, in Mammal species of the world, a taxonomic and geographic reference, Second ed. (D. E. Wilson and D. M. Reeder, eds.). Smithsonian Institution Press, Washington, D.C., 1206 pp.

Hutterer, R. 1993*b*. Ein Lebensbild der tibetanischen Wasserspitzmaus (*Nectogale elegans*). Pp. 39-51, *in* Semiaquatische Säugetiere (M. Stubbe ed.). Wissenschaftliche Beiträge der Universität Halle, Wittenberg. 468 pp.

Hutterer, R. 1994*a*. Shrews of ancient Egypt: Biogeographical interpretation of a new species. Pp. 407-413, *in* Advances in the biology of shrews (J. F. Merritt, G. L. Kirkland, Jr., and R. K. Rose, eds.). Carnegie Museum of Natural History, Special Publication, 18:1-458.

Hutterer, R. 1994*b*. Generic limits among neomyine and soriculine shrews (Mammalia: Soricidae). Neogene and Quaternary mammals of the Palaearctic. Conference in honor of Professor Kazimierz Kowalski, Kraków, Poland. Abstracts:32:17-21,

Hutterer, R. 1996. Book review: Filipucci, M. G., ed., Proceedings of the II Conference on Dormice (Rodentia, Myoxidae). Bonner Zoologische Beiträge, 46(1-4):132.

Hutterer, R. 1999*a*. *Sorex minutus* Linnaeus, 1766. Pp. 54-55, *in* The atlas of European mammals (A. J. Mitchell-Jones, G. Amori, W. Bogdanowicz, B. Kryštufek, P. J. H. Reijnders, F. Spitzenberger, M. Stubbe, J. B. M. Thissen, V. Vohralík, and J. Zima, eds.). T. and A.D. Poyser, London, 484 pp.

Hutterer, R. 1999b. *Crocidura osorio* Molina & Hutterer, 1989. Pp. 66-67, *in* The atlas of European mammals
    (A. J. Mitchell-Jones, G. Amori, W. Bogdanowicz, B. Kryštufek, P. J. H. Reijnders, F. Spitzenberger, M.
    Stubbe, J. B. M. Thissen, V. Vohralík, and J. Zima, eds.). T. and A.D. Poyser, London, 484 pp.

Hutterer, R. 1999c. *Crocidura canariensis* Hutterer, López-Jurado & Vogel, 1987. Pp. 62-63, *in* The atlas of
    European mammals (A. J. Mitchell-Jones, G. Amori, W. Bogdanowicz, B. Kryštufek, P. J. H. Reijnders, F.
    Spitzenberger, M. Stubbe, J. B. M. Thissen, V. Vohralík, and J. Zima, eds.). T. and A.D. Poyser, London,
    484 pp.

Hutterer, R. 2003. Two replacement names and a note on the author of the shrew family Soricidae
    (Mammalia). Bonner Zoologische Beiträge, 50:369-370.

Hutterer, R., and F. Dieterlen. 1984. Zwei neue Arten der Gattung *Grammomys* aus Äthiopien und Kenia
    (Mammalia: Muridae). Stuttgarter Beiträge zur Naturkunde, ser. A (Biologie), 347:1-18.

Hutterer, R., and F. Dieterlen. 1986. Zur Verbreitung und Variation von *Desmodilliscus braueri*. Annalen des
    Naturhistorischen Museums in Wien, 88/89:213-221.

Hutterer, R., and N. J. Dippenaar. 1987. A new species of *Crocidura* Wagler, 1832 (Soricidae) from Zambia.
    Bonner Zoologische Beiträge, 38:1-7.

Hutterer, R., and A. Dudu. 1990. Redescription of *Crocidura caliginea*, a rare shrew from northeastern Zaire.
    Journal of African Zoology, 104:305-311.

Hutterer, R., and R. Geiger-Roswora. 1997. Drastischer Bestandrückgang des Feldhamsters, *Cricetus cricetus*,
    in Nordrhein-Westfalen. Abhandlungen aus dem Westfälischen Museum für Naturkunde, 59(3):71-82.

Hutterer, R., and D. C. D. Happold. 1983. The shrews of Nigeria (Mammalia: Soricidae). Bonner Zoologische
    Monographien, 18:1-79.

Hutterer, R., and D. L. Harrison. 1988. A new look at the shrews (Soricidae) of Arabia. Bonner Zoologische
    Beiträge, 39:59-72.

Hutterer, R., and U. Hirsch. 1979. Ein neuer *Nesoryzomys* von der Insel Fernandina, Galapagos (Mammalia,
    Rodentia). Bonner Zoologische Beiträge, 30:276-283.

Hutterer, R., and T. Hürter. 1981. Adaptive Haarstrukturen bei Wasserspitzmäusen (Insectivora, Soricinae).
    Zeitschrift für Säugetierkunde, 46:1-11.

Hutterer, R., and P. D. Jenkins. 1983. Species-limits of *Crocidura somalica* Thomas, 1895 and *Crocidura
    yankariensis* Hutterer and Jenkins, 1980 (Insectivora, Soricidae). Zeitschrift für Säugetierkunde,
    48:193-201.

Hutterer, R., and U. Joger. 1982. Kleinsäuger aus dem Hochland von Adamaoua, Kamerun. Bonner
    Zoologische Beiträge, 33:119-132.

Hutterer, R., and D. Kock. 1983. Spitzmäuse aus den Nuba-Bergen Kordofans, Sudan (Mammalia: Soricidae).
    Senckenbergiana Biologica, 63:17-26.

Hutterer, R., and D. Kock. 2002. The shrews of Syria, with notes on *Crocidura katinka* Bate, 1937 (Mammalia:
    Soricidae). Bonner Zoologische Beiträge, 50:249-258.

Hutterer, R., and W. v. Koenigswald. 1993. Knochenfunde aus einer Karsthöhle bei Berndorf in der
    Hillesheimer Kalkmulde (Eifel). Mainzer Naturwissenschaftliches Archiv, 31:223-238.

Hutterer, R., and P. Oromi. 1993. La rata gigante de la Isla Santa Cruz, Galapagos: Algunos datos y
    problemas. Resultados Cientificos del Proyecto Galapagos, Patrimonio de la Humanidad, Museo de
    Ciencias Naturales, Tenerife, 4:63-76.

Hutterer, R., and G. Peters. 2001. The vocal repertoire of *Graphiurus parvus*, and comparisons with other
    species of dormice. Trakya University Journal of Scientific Research, ser. B, 2(2):69-74.

Hutterer, R., and D. A. Schlitter. 1996. Shrews of Korup National Park, Cameroon, with the description of a
    new *Sylvisorex* (Mammalia: Soricidae). Pp. 57-66, *in* Contributions in Mammalogy: A Memorial Volume
    Honoring Dr. J. Knox Jones, Jr. Museum of Texas Tech University, il+315 pp.

Hutterer, R., and M. Tranier. 1990. The immigration of the Asian house shrew *Suncus murinus* into Africa and
    Madagascar. Pp. 309-319, *in* Vertebrates in the Tropics (G. Peters and R. Hutterer, eds.). Museum
    Alexander Koenig, Bonn, 424 pp.

Hutterer, R., and D. W. Yalden. 1990. Two new species of shrews from a relic forest in the Bale Mountain,
    Ethiopia. Pp. 63-72, *in* Vertebrates in the Tropics (G. Peters, and R. Hutterer, eds.). Museum Alexander
    Koenig, Bonn, 424 pp.

Hutterer, R., and M. V. Zaitsev. 2004. Cases of homonymy in some Palaearctic and Nearctic taxa of the genus
    Sorex L. (Mammalia: Soricidae). Mammal Study (Tokyo), 29(1):89-91.

Hutterer, R., L. F. López-Jurado, and P. Vogel. 1987a. The shrews of the eastern Canary Islands: A new species
    (Mammalia: Soricidae). Journal of Natural History, 21:1347-1357.

Hutterer, R., E. Van der Straeten, and W. N. Verheyen. 1987b. A checklist of the shrews of Rwanda and
    biogeographic considerations on African Soricidae. Bonner Zoologische Beiträge, 38:155-172.

Hutterer, R., N. López-Martínez, and J. Michaux. 1988. A new rodent from Quaternary deposits of the

Canary Islands and its relationships with Neogene and Recent murids of Europe and Africa. Palaeovertebrata, 18:241-262.

Hutterer, R., F. Dieterlen, and G. Nikolaus. 1992*a*. Small mammals from forest islands of eastern Nigeria and adjacent Cameroon, with systematical and biogeographical notes. Bonner Zoologische Beiträge, 43:393-414.

Hutterer, R., T. Maddalena, and O. M. Molina. 1992*b*. Origin and evolution of the endemic Canary Island shrews (Mammalia: Soricidae). Biological Journal of the Linnean Society, 46:49-58.

Hutterer, R., E. Ag Sidiyène, and M. Tranier. 1992*c*. A record of *Crocidura somalica* from the Sahara. Mammalia, 55:621-622.

Hutterer, R., P. Barriere, and M. Colyn. 2002. A new myosoricine shrew from the Congo Basin referable to the forgotten genus *Congosorex* (Mammalia: Soricidae). Bulletin de l'Institut royal des Sciences naturelles de Belgique, Biologie, 71-Suppl.:7-16 [2001].

Huxley, T. H. 1869. An introduction to the classification of animals. Churchill, London, viii + 147 pp.

Hwang, Y. T., and S. Larivière. 2001. *Mephitis macroura*. Mammalian Species, 686:1-3.

Hwang, Y. T., and S. Larivière. 2003. *Mydaus javanensis*. Mammalian Species, 723:1-3.

Hyndman, D., and J. I. Menzies. 1980. *Aproteles bulmerae* (Chiroptera: Pteropodidae) of New Guinea is not extinct. Journal of Mammalogy, 61:159-160.

Ibáñez, C. 1980. Descripción de un nuevo género de quiróptero neotropical de la familia Molossidae. Doñana, Acta Vertebrata, 7:104-111.

Ibáñez, C., and R. Fernández. 1985. Systematic status of the long-eared bat *Plecotus teneriffae* Barret-Hamilton, 1907 (Chiroptera; Vespertilionidae). Säugetierkundliche Mitteilungen, 32:143-149.

Ibáñez, C., and R. Valverde. 1985. Taxonomic status of *Eptesicus platyops* (Thomas, 1901)(Chiroptera, Vespertilionidae). Zeitschrift für Säugetierkunde, 50:241-242.

Iguchi, R., K. Iguchi and Y. Sato. 1996. [Discoveries of Japanese Dormouse, *Glirulus japonicus* (Rodentia, Gliridae), in Tokushima Prefecture, Shikoku, Japan.] Bulletin Tokushima Prefecture Museum, 6:89-96 (in Japanese).

Ilani, G., and B. Shalmon. 1983. Rock dormouse on trees. Israel, Land and Nature, 9(1):39.

Ilchenko, O. G., and I. A. Volodin. 1992. [On catching *Selevinia betpakdalensis, Cardiocranius paradoxus* and *Sicista subtilis* in northern Pribalkhashye.] Byulleten' Moskovskovo Obshchestva Ispytatelei Prirody, Otdel Biologicheskii, 97:28-30 (in Russian with English abstract).

Illiger, C. D. 1811. Prodromus systematis mammalium et avium additis terminis zoographicis uttriusque classis. Salfeld, Berlin, 301 pp.

Imaizumi, Y. 1949. [The natural history of Japanese mammals.] Yoyoshobo, Tokyo, 348 pp. (in Japanese).

Imaizumi, Y. 1961. Taxonomic status of *Crocidura dsinezumi orii.* Journal of the Mammalogical Society of Japan, 2:17-22.

Imaizumi, Y. 1962. On the species formation of the *Apodemus speciosus* group, with special reference to the importance of relative values in classification. Part. I. Bulletin of the National Science Museum (Tokyo), 49:163-259.

Imaizumi, Y. 1967. A new genus and species of cat from Iriomote, Ryukyu Islands. Journal of the Mammalogical Society of Japan, 3(4):75-106.

Imaizumi, Y. 1970*a*. Description of a new species of *Cervus* from the Tsushima Islands, Japan, with a revision of the subgenus *Sika* based on clinal analysis. Bulletin of the National Science Museum (Tokyo), 13:185-193.

Imaizumi, Y. 1970*b*. The handbook of Japanese land mammals. Shin-Shichoa-Sha, Tokyo, 350 pp.

Imaizumi, Y. 1971. A new vole of the *Clethrionomys rufocanus* group from Rishiri Island, Japan. Journal of the Mammalogical Society of Japan, 6:99-103.

Imaizumi, Y. 1972. Land mammals of the Hidaka Mountains, Hokkaido, Japan, with special reference to the origin of an endemic species of the genus *Clethrionomys*. Memoirs of the National Science Museum, Tokyo, 5:131-149.

Imaizumi, Y., and M. Yoshiyuki. 1989. Taxonomic status of the Japanese otter (Carnivora, Mustelidae), with a description of a new species. Bulletin of the National Science Museum, ser. A (Zoology) 15(3):177-188.

Indyk, F., and A. Pawlowska-Indyk. 1994. [New localities of fat dormouse *Glis glis* (Linnaeus, 1766) (Mammalia, Gliridae) in the province of Wroclaw]. Przeglad Zoologiczny, 38(3-4):353-355 (in Polish).

Ingle, N. R., and L. R. Heaney. 1992. A key to the bats of the Philippine Islands. Fieldiana: Zoology, n.s., 69:1-44.

Ingles, J. M. 1965. Zambian mammals collected for the British Museum (Natural History) in 1962. Puku, 3:75-86.

Ingles, J. M., P. N. Newton, M. R. W. Rands, and C. G. R. Bowden. 1980. The first record of a rare murine rodent *Diomys* and further records of three shrew species from Nepal. Bulletin of the British Museum (Natural History), Zoology Series, 39:205-211.

Ingoldby, C. M. 1927. Some notes on the African squirrels of the genus *Heliosciurus*. Proceedings of the Zoological Society of London, 1927:471-487.

Ingoldby, C. M. 1929. On the mammals of the Gold Coast. Annals and Magazine of Natural History, ser. 10, 3(17):511-529.

Innes, D. G. L. 1994. Life histories of the Soricidae: A review. Pp. 111-129, *in* Advances in the biology of shrews (J. F. Merritt, , G. L. Kirkland, Jr., and R. K. Rose, eds.). Carnegie Museum of Natural History, Special Publication, 18:1-458.

Innes, J. G. 1990. Ship rat. Pp. 206-225, *in* The handbook of New Zealand mammals (C. M. King, ed.). Oxford University Press, Auckland:1-600.

International Commission on Zoological Nomenclature. 1926. Opinion 91. Thirty-five generic names of mammals placed in the Official List of Generic Names. Opinions rendered by the International Commission on Zoological Nomenclature. Smithsonian Miscellaneous Collection, 73:337-338.

International Commission on Zoological Nomenclature. 1929*a*. Opinion 110. Suspension of Rules for *Lagidium* 1833. Smithsonian Miscellaneous Collections, 73(6):415.

International Commission on Zoological Nomenclature. 1929*b*. Opinion 111. Suspension of Rules for *Nycteris* 1795. Smithsonian Miscellaneous Collections, 73(6):416.

International Commission on Zoological Nomenclature. 1929*c*. Opinion 114. Under suspension *Simia*, *Simia satyrus* and *Pithecus* are suppressed. Smithsonian Miscellaneous Collections, 73(6):423-424.

International Commission on Zoological Nomenclature. 1954*a*. Opinion 257. Rejection for nomenclatorial purposes of the work by Zimmermann (A. E. W. von) published in 1777 under the title *Specimen Zoologiae Geographicae, Quadrupedum Domicilia et Migrationes sistens*, and acceptance for the same purposes of the work by the same author published in the period 1778-1783 under the title *Geographische Geschichte de Menschen, und der allgemein verbreiteten vierfüssigen Thiere*. Opinions and Declarations Rendered by the International Commission on Zoological Nomenclature, 5(18):231-244.

International Commission on Zoological Nomenclature. 1954*b*. Opinion 258. Rejection for nomenclatorial purposes of the work by Frisch (J.L.) published in 1775 under the title "Das Natur-System der vierfüssigen Thiere". Opinions and Declarations Rendered by the International Commission on Zoological Nomenclature, 5(19):245-252.

International Commission on Zoological Nomenclature. 1954*c*. Opinion 212. Designation of the dates to be accepted as the dates of publication of the several volumes of Pallas (P. S.), *Zoographia rosso-asiatica*. Opinions and Declarations Rendered by the International Commission on Zoological Nomenclature, 4:15-24.

International Commission on Zoological Nomenclature. 1955*a*. Second report on the status of the generic names "*Odobenus*" Brisson, 1762, and "*Rosmarus*" Brünnich, 1771 (Class Mammalia) (A report prepared at the request of the thirteenth International Congress of Zoology, Paris, 1948). Bulletin of Zoological Nomenclature, 11(6):196-198.

International Commission on Zoological Nomenclature. 1955*b*. Direction 24. Completion of the entries relative to the names of certain genera in the Class Mammalia made in the "Official list of Generic Names in Zoology" in the period up to the end of 1936. Opinions and Declarations Rendered by the International Commission on Zoological Nomenclature, 11:221-246.

International Commission on Zoological Nomenclature. 1956*a*. Opinion 384. Addition to the official list of generic names in zoology of the names of fifty-two genera of the Order Carnivora (Class Mammalia) including twenty-nine from which have been reported parasites common to man. Opinions and Declarations Rendered by the International Commission on Zoological Nomenclature, 12(5):71-190.

International Commission on Zoological Nomenclature. 1956*b*. Opinion 417. Rejection for nomenclatorial purposes of volume 3 (Zoologie) of the work by Lorenz Oken entitled *Okens Lehrbuch der Naturgeschichte* published in 1815-1816. Opinions and Declarations Rendered by the International Commission on Zoological Nomenclature, 14(1):1-42.

International Commission on Zoological Nomenclature. 1957*a*. Opinion 447. Rejection for nomenclatorial purposes of the original edition published at Philadelphia in 1791 and of the editions published in London and Dublin respectively in 1792 of the work by William Bartram entitled "Travels through North and South Carolina, Georgia, east and west Florida, the Cherokee country, the extensive territories of the Muscogulges or Creek confederacy, and the country of the Chactaws", as being a work in which the author did not apply the principles of binominal nomenclature. Opinions and Declarations Rendered by the International Commission on Zoological Nomenclature, 15(12):211-224.

International Commission on Zoological Nomenclature. 1957*b*. Opinion 460. Validation under the plenary powers of the generic name "*Muntiacus*" Rafinesque, 1815, and designation for the genus so named of a type species in harmony with accustomed usage (Class Mammalia). Opinions and Declarations Rendered by the International Commission on Zoological Nomenclature, 15:457-474.

International Commission on Zoological Nomenclature. 1957*d*. Opinion 464. Action under the plenary powers to secure (a) That the specific name *Gambianus* Ogilby, 1835, as published in the combination *Sciurus gambianus* shall be the oldest available name for the sun squirrel and (b) That the generic name *Heliosciurus* Trouessart, 1880, shall be the oldest available generic name for that species (Class Mammalia). Opinions and Declarations Rendered by the International Commission on Zoological Nomenclature, 16(3):25-42.

International Commission on Zoological Nomenclature. 1957*e*. Opinion 467. Validation under the plenary powers of the generic name *Odobenus* Brisson, 1762, as the generic name for the walrus (Class Mammalia). Opinions and Declarations Rendered by the International Commission on Zoological Nomenclature, 16(6):73-88.

International Commission on Zoological Nomenclature. 1957*f*. Opinion 465. Validation under the plenary powers of the specific name "*silvestris*" Schreber [1777] as published in the combination "*Felis* (*Catus*) *silvestris*", for the European wild cat (Class Mammalia). Opinions and Declarations Rendered by the International Commission on Zoological Nomenclature, 16:43-52.

International Commission on Zoological Nomenclature. 1957*g*. Opinion 466. Validation under the plenary powers of the generic name "*Phacochoerus*" Cuvier (F.), 1826, as the generic name for the Wart Hog (Class Mammalia). Opinions and Declarations Rendered by the International Commission on Zoological Nomenclature, 16:55-72.

International Commission on Zoological Nomenclature. 1958*a*. Direction 98: Interpretation under the plenary powers of the nominal species *Vespertilio murinus* Linnaeus, 1758, and insertion in the Official List of Generic Names in Zoology of a revised entry relating to the generic name name *Vespertilio* Linnaeus, 1758 (Class Mammalia). Opinions and Declarations Rendered by the International Commission on Zoological Nomenclature, Vol. 1, Section F, Part F9, pp. 127-160.

International Commission on Zoological Nomenclature. 1958*b*. Opinion 462. Addition to the Official List of Generic Names in Zoology of the generic name *Mormoops* Leach, 1820 (Class Mammalia). Opinions and Declarations Rendered by the International Commission on Zoological Nomenclature, 16:1-12.

International Commission on Zoological Nomenclature. 1959. Opinion 544. Validation under the plenary powers of the family-group names Muntiacinae Pocock, 1923, and Odobenidae (correction of Odobaenidae) Allen, 1880 (Class Mammalia). Opinions and Declarations Rendered by the International Commission on Zoological Nomenclature, 20(2):119-128.

International Commission on Zoological Nomenclature. 1960. Opinion 581. Determination of the generic names for the fallow deer of Europe and the Virginian deer of America (Class Mammalia). Bulletin of Zoological Nomenclature, 17:267-275.

International Commission on Zoological Nomenclature. 1962. Opinion 678. The suppression under the plenary powers of the pamphlet published by Meigen, 1800. Bulletin of Zoological Nomenclature, 20(5):339-342.

International Commission on Zoological Nomenclature. 1965. Opinion 730. *Yerbua* Forster, 1778 (Mammalia): Suppressed under the plenary powers. Bulletin of Zoological Nomenclature, 22(2):84-85.

International Commission on Zoological Nomenclature. 1966. Opinion 760. *Macropus* Shaw, 1790 (Mammalia): Addition to the official list together with the validation under the plenary powers of *Macropus giganteus* Shaw, 1790. Bulletin of Zoological Nomenclature, 22(5/6):292-295.

International Commission on Zoological Nomenclature. 1967. Opinion 818. *Zorilla* I. Geoffroy, 1826 (Mammalia): Suppressed under the plenary powers. Bulletin of Zoological Nomenclature, 24(3):153-154.

International Commission on Zoological Nomenclature. 1970. Opinion 920. *Inuus fuscatus* Blyth, 1875 (Mammalia): Validated under the plenary powers. Bulletin of Zoological Nomenclature, 27(2):77-78.

International Commission on Zoological Nomenclature. 1971*a*. Opinion 944. *Giraffa camelopardalis australis* de Winton, 1899 (Mammalia): Validated under the plenary powers. Bulletin of Zoological Nomenclature, 27:222-223.

International Commission on Zoological Nomenclature. 1971*b*. Opinion 945. *Sciurus ebii* Pel, 1851 (Mammalia): Suppressed under the Plenary Powers. Bulletin of Zoological Nomenclature, 27(5/6):224-225.

International Commission on Zoological Nomenclature. 1977*a*. Opinion 1067. Suppression of *Delphinus pernettensis* de Blainville, 1817 and *Delphinus pernettyi* Desmarest, 1820 (Cetacea). Bulletin of Zoological Nomenclature, 33:157-158.

International Commission on Zoological Nomenclature. 1977*b*. Opinion 1080. *Didermocerus* Brookes, 1828 (Mammalia) suppressed under the plenary powers. Bulletin of Zoological Nomenclature, 34:21-24.

International Commission on Zoological Nomenclature. 1977*c*. Opinion 1081. Addition of family-group names based on *Alca* (Aves) and *Alces* (Mammalia) to the Official List of Family-Group Names in Zoology. Bulletin of Zoological Nomenclature, 34:25-26.

International Commission on Zoological Nomenclature. 1979*a*. Opinion 1129. *Vulpes* Frisch, 1775 (Mammalia) conserved under the plenary powers. Bulletin of Zoological Nomenclature, 36(2):76-78.

International Commission on Zoological Nomenclature. 1979*b*. Opinion 1138. *Giraffa camelopardalis australis* Rhoads, 1896 (Mammalia) suppressed. Bulletin of Zoological Nomenclature, 36:111-113.

International Commission on Zoological Nomenclature. 1983. Opinion 1256. *Sorex dsinezumi* Temminck, 1843 (Mammalia, Insectivora): Ruled to be a correct original spelling. Bulletin of Zoological Nomenclature, 40:147-148.

International Commission on Zoological Nomenclature. 1985*a*. Opinion 1291. *Antilope zebra* Gray, 1838 (Mammalia): Conserved. Bulletin of Zoological Nomenclature, 42:24-26.

International Commission on Zoological Nomenclature. 1985*b*. Opinion 1289. *Mesoplodon* Gervais, 1850 (Mammalia, Cetacea): Conserved. Bulletin of Zoological Nomenclature, 42:19-20.

International Commission on Zoological Nomenclature. 1985*c*. Opinion 1368. The generic names *Pan* and *Panthera* (Mammalia, Carnivora): Available as from Oken, 1816. Bulletin of Zoological Nomenclature, 42(4):365-370.

International Commission on Zoological Nomenclature. 1985*d*. International Code of Zoological Nomenclature. Third ed. University of California Press, Berkeley, 338 pp.

International Commission on Zoological Nomenclature. 1986. Opinion 1413. *Delphinus truncatus* Montagu, 1821 (Mammalia, Cetacea): Conserved. Bulletin of Zoological Nomenclature, 43:256-257.

International Commission on Zoological Nomenclature. 1988. Corrigenda. Bulletin of Zoological Nomenclature, 45(4):304.

International Commission on Zoological Nomenclature. 1989. Opinion 1565. *Platanista* Wagler (Mammalia, Cetacea): Conserved. Bulletin of Zoological Nomenclature, 46(3):217-218.

International Commission on Zoological Nomenclature. 1990. Opinion 1607. *Mus musculus domesticus* Schwarz & Schwarz, 1943 (Mammalia, Rodentia): Specific name conserved. Bulletin of Zoological Nomenclature, 47(2), 171-172.

International Commission on Zoological Nomenclature. 1991. Opinion 1660. *Steno attenuatus* Gray, 1846 (currently *Stenella attenuate*; Mammalia, Cetacea): Specific name conserved. Bulletin of Zoological Nomenclature, 48(3):277-278.

International Commission on Zoological Nomenclature. 1998. Opinion 1894. *Regnum Animals* , Ed. 2 (M.J. Brisson, 1762): Rejected for nomenclatural purposes, with the conservation of the mammalian generic names *Philander* (Marsupialia), *Pteropus* (Chiroptera), *Glis, Cuniculus* and *Hydrochoerus* (Rodentia), *Meles, Lutra* and *Hyaena* (Carnivora), *Tapirus* (Perissodactyla), *Tragulus* and *Giraffa* (Artiodactyla). Bulletin of Zoological Nomenclature, 55(1):64-71.

International Commission on Zoological Nomenclature. 1999. International Code of Zoological Nomenclature. Fourth Edition. The International Trust for Zoological Nomenclature, London, 306 pp.

International Commission on Zoological Nomenclature. 2000. *Arctocephalus* F. Cuvier, 1826 and *Callorhinus* Gray, 1859 (Mammalia, Pinnipedia): Conserved by the designation of *Phoca pusilla* Schreber, 1775 as the type species of *Arctocephalus* and *Otaria* Péron, 1816 and *Eumetopias* Gill, 1866: Conserved by the designation of *Phoca leonina* Molina, 1782 as the type species of *Otaria*. Bulletin of Zoological Nomenclature, 57(3):193-195.

International Commission on Zoological Nomenclature. 2001*a*. Opinion 1978. *Myoxus japonicus* Schinz, 1845 (currently *Glirulus japonicus*; Mammalia; Rodentia): Specific name conserved as the correct original spelling. Bulletin of Zoological Nomenclature, 58(2):159-160.

International Commission on Zoological Nomenclature. 2001*b*. Opinion 1985. *Cervus gouazoubira* Fischer, 1814 (currently *Mazama gouazoubira*: Mammalia, Artiodactyla): Specific name conserved as the correct original spelling. Bulletin of Zoological Nomenclature, 58:247.

International Commission on Zoological Nomenclature. 2002*a*. Opinion 1995 (Case 3004). Lorisidae Gray, 1821, Galagidae Gray, 1825 and Indriidae Burnett, 1828 (Mammalia, Primates): Conserved as the correct original spellings. Bulletin of Zological Nomenclature, 59:65-67.

International Commission on Zoological Nomenclature. 2002*b*. Opinion 2005 (Case 3022). Catalogue des mammifères du Muséum National d'Hisoire Naturelle by Étienne Geoffroy Saint-Hilaire (1803): Placed on the Official List of Works Approved as Available for Zoological Nomenclature. Bulletin of Zoological Nomenclature, 59:153-154.

International Commission on Zoological Nomenclature. 2003*a*. Opinion 2027. Usage of 17 specific names based on wild species which are pre-dated by or contemporary with those based on domestic animals (Lepidoptera, Osteichthyes, Mammalia): Conserved. Bulletin of Zoological Nomenclature, 60:81-84.

International Commission on Zoological Nomenclature. 2003*b*. Opinion 2028 (Case 3073). *Vespertilio pipistrellus* Schreber, 1774 and *V. pygmaeus* Leach, 1825 (currently *Pipistrellus pipistrellus* and *Pipistrellus pygmaeus*; Mammalia, Chiroptera): Neotypes designated. Bulletin of Zoological Nomenclature, 60:85-87.

International Commission on Zoological Nomenclature. 2003*c*. Opinion 2030. *Hippotragus* Sundevall, 1845 (Mammalia, Artiodactyla): Conserved. Bulletin of Zoological Nomenclature, 60:90-91.

International Union for the Conservation of Nature and Natural Resources. 1990. 1990 IUCN red list of threatened animals. I.U.C.N., Gland, Switzerland, 228 pp.

Intress, C., and T. L. Best. 1990. *Dipodomys panamintinus*. Mammalian Species, 354:1-7.

Iredale, T., and E. le G. Troughton. 1934. A checklist of the mammals recorded from Australia. Memoirs Australian Museum, Sydney, 6:1-22.

Irwin, D. W., and R. J. Baker. 1967. Additional records of bats from Arizona and Sinaloa. Southwestern Naturalist, 12:195.

Isawa, M. A., S. Ohdachi, S.-H. Han, H.-S. Oh, H. Abe, and H. Suzuki. 2001. Karyotype and RFLP of the nuclear rDNA of *Crocidura* sp. on Cheju Island, South Korea (Mammalia, Insectivora). Mammalia, 65:451-459.

Ishibashi, Y., S. Abe, and M. C. Yoshida. 1992. DNA fingerprinting of grey red-backed voles *Clethrionomys rufocanus bedfordiae*. Journal of the Mammalogical Society of Japan, 17(1):19-26.

Iskandar, D., and F. Bonhomme. 1984. Variabilite electrophoretique totale a 11 loci structuraux chez les rongeurs murides (Muridae, Rodentia). Canadian Journal of Genetics and Cytology, 26:622-627.

Iskandar, D., J. M. Duplantier, F. Bonhomme, F. Petter, and L. Thaler. 1988. Mise en evidence de deux especes jumelles sympatriques du genre *Hylomyscus* dans le nord-est du Gabon. Mammalia, 52:126-130.

Ismagilov, M. I. 1961. [Ecology of the native rodents of the Betpak-Dala and southern Lake Balkhash region]. Akademiya Nauk Kazakhskoi SSR (Alma-Ata), 368 pp. (in Russian).

Istomin, A. V. 1994*a*. [Polydonty in common red backed vole (*Clethrionomys glareolus*) from two points of its range.] Vestnik Zoologii, 91:83-86 (in Russian with Engish summary).

Istomin, A. V. 1994*b*. [Phenotypic diversity in continual and discrete populations of the bank vole in the South Taiga.] Zhurnal Obshchei Biologii, 55:(4-5):477-488 (in Russian with English abstract).

Istrate, P. 1998. Les petits mammifères du plateau Târnava, Transylvanie (Roumanie). Travaux du Muséum national d'Histoire naturelle "Grigore Antipa", 40:449-474.

Itoo, T. 1985. New cranial materials of the Japanese sea lion, *Zalophus californianus japonicus* (Peters, 1866). Journal of the Mammalogical Society of Japan, 10:135-148.

Ivanitskaya, E. Yu. 1985. Taksonomicheskii i tsitogeneticheskii analiz transberingiiskikh svyazei zemleroek-burozubok (*Sorex*: Insectivora) i pishchukh (*Ochotona*:Lagomorpha) [Taxonomic and cytogenetic analysis of transberingian connections of brown-toothed shrews . . . and pikas . . .]. Aftoreferat, Minist. Vyssh. i Spred. Spets. Obraz., Leningrad., (in Russian).

Ivanitskaya, E. Y. 1994. Comparative cytogenetics and systematics of *Sorex*: A cladistic approach. Pp. 313-323, *in* Advances in the biology of shrews (J. F. Merritt, G. L. Kirkland, Jr., and R. K. Rose, eds.). Carnegie Museum of Natural History, Special Publication, 18:1-458.

Ivanitskaya, E. Yu., and A. I. Kozlovskii. 1983. Kariologichicheskie dokazatel'stva otsutstviya v Palearktike arkticheskoi burozubki. [Karyological evidence for the absence of the arctic shrew in the Palearctic.] Zoologicheskii Zhurnal, 62:399-408 (in Russian).

Ivanitskaya, E. Yu., and A. I. Kozlovskii. 1985. [Karyotypes of Palearctic shrews of the subgenus *Otisorex* with comments on taxonomy and phylogeny of the group "*cinereus*".] Zoologicheskii Zhurnal, 64:950-953 (in Russian).

Ivanitskaya, E., and E. Nevo. 1998. Cytogenetics of mole rats of the *Spalax ehrenbergi* superspecies from Jordan (Spalacidae, Rodentia). Zeitschrift für Säugetierkunde, 63:336-346.

Ivanitskaya, E. Yu., S. I. Isakov, and K. V. Korobytsina. 1977. Chromosomne nabor dvuch vidov zemleroek iz Tadzhikistana — *Sorex buchariensis* Ognev, 1921 i *Crocidura suaveolens* Pallas, 1811 (Soricidae, Insectivora) [Chromosomal complements of two species of shrews from Tadzhikistan . . .]. Zoologichskii Zhurnal, 56:1896-1900 (in Russian).

Ivanitskaya, E. Yu., A. I. Kozlovskii, N. V. Orlov, Y. M. Kovalskaya, and M. I. Baskevich. 1986. [New data on karyotypes of common shrews (*Sorex*, Soricidae, Insectivora) in fauna of the USSR.] Zoologicheskii Zhurnal, 65:1228-1236 (in Russian).

Ivanitskaya, E., G. Shenbrot, and E. Nevo. 1996*a*. *Crocidura ramona* sp. nov. (Insectivora, Soricidae): A new species of shrew from the central Negev desert, Israel. Zeitschrift für Säugetierkunde, 61:93-103.

Ivanitskaya, E., I. Gorlov, O. Gorlova, and E. Nevo. 1996*b*. Chromosome markers for *Mus macedonicus* (Rodentia, Muridae) from Israel. Hereditas, 124:145-150.

Ivanitskaya, E., Y. Cokun, and E. Nevo. 1997. Banded karyotypes of mole rats (*Spalax*, Spalacidae, Rodentia) from Turkey: A comparative analysis. Journal of Zoological and Systematic Evolutionary Research, 35:171-177.

Iwasa, M. A. 1998. [Chromosomal and molecular variations in red-backed voles.] Mammalian Science, 38:145-158 (in Japanese with English abstract).

Iwasa, M. A. 2000. Variation of skull characteristics of Anderson's red-backed vole *Eothenomys andersoni*, from Nagano City, central Honshu, Japan. Mammal Study, 25:125-139.

Iwasa, M. A., and K. Tsuchiya. 2000. Karyological analysis of the *Eothenomys* sp. from Nagano City, central Honshu, Japan. Chromosome Science, 4:31-38.

Iwasa, M. A., S. H. Han, and H. Suzuki. 1999*a*. A karyological analysis of the Korean red-backed vole, *Eothenomys regulus* (Rodentia, Muridae), using differential staining methods. Mammal Study, 24:35-41.

Iwasa, M. A., Y. Obara, E. Kitahara, and Y. Kimura. 1999*b*. Synaptonemal complex analyses in the XY chromosomes of six taxa of *Clethrionomys* and *Eothenomys* from Japan. Mammal Study, 24:103-113.

Iwasa, M. A., Y. Utsumi, K. Nakata, I. V. Kartavtseva, I. A. Nevedomskaya, N. Kondoh, and H. Suzuki. 2000. Geographic patterns of cytochrome *b* and *Sry* gene lineages in the gray red-backed vole *Clethrionomys rufocanus* from Far East Asia including Sakhalin and Hokkaido. Zoological Science, 17:477-484.

Iwasa, M. A., I. V. Kartavtseva, A. K. Dobrotvorsky, V. V. Panov, and H. Suzuki. 2002. Local differentiation of *Clethrionomys rutilus* in northeastern Asia inferred from mitochondrial gene sequences. Mammalian Biology, 67:157-166.

Izor, R. J., and L. de la Torre. 1978. A new species of weasel (*Mustela*) from the highlands of Colombia, with comments on the evolution and distribution of South American weasels. Journal of Mammalogy, 59:92-102.

Izor, R. J., and R. H. Pine. 1987. Notes on the black-shouldered opossum, *Caluromysiops irrupta*. Pp. 133-136, *in* Studies in Neotropical mammalogy, essays in honor of Philip Hershkovitz (B. D. Patterson and R. M. Timm, eds.). Fieldiana: Zoology, n.s., 39:frontispiece, vii + 1-506.

Jaarola, M., and J. B. Searle. 2002. Phylogeography of field voles (*Microtus agrestis*) in Eurasia inferred from mitochondrial DNA sequences. Molecular Ecology, 11:2613-2621.

Jablonski, N. G., and Y. Peng. 1993. The phylogenetic relationships and classification of the doucs and snub-nosed langurs of China and Vietnam. Folia primatologica, 60:36-55.

Jackson, H. H. T. 1915. A revision of the American moles. North American Fauna, 38:1-100.

Jackson, H. H. T. 1928. A taxonomic review of the American long-tailed shrews. North American Fauna, 51:1-238.

Jackson, J. E. 1987. *Ozotoceros bezoarticus*. Mammalian Species, 295:1-5.

Jackson, J. E., L. C. Branch, and D. Villarreal. 1996. *Lagostomus maximus*. Mammalian Species, 543:1-6.

Jackson, T. P., and A. C. Spinks. 1998. Gut morphology of the otomyine rodents: An arid-mesic comparison. South African Journal of Zoology, 33:236-240.

Jackson, T. P., and R. J. van Aarde. 2003. Sex- and species-specific growth patterns in cryptic African rodents: *Mastomys natalensis* and *M. coucha*. Journal of Mammalogy, 84(3):851-860.

Jacobi, A. 1931. Das Rentier; eine zoologische Monographie der Gattung *Rangifer*. Zoologischer Anzeiger, Ergänzungsband zu Band 96, 264 pp.

Jacobs, D. S., and M. B. Fenton. 2002. *Mormopterus petrophilus*. Mammalian Species, 703:1-3.

Jacobs, L. L. 1977*a*. A new genus of murid rodent from the Miocene of Pakistan and comments on the origin of the Muridae. Paleobios, 25:1-11.

Jacobs, L. L. 1977*b*. Rodents of the Hemphillian Redington Local Fauna, San Pedro Valley, Arizona. Journal of Paleontology, 51:505-519.

Jacobs, L. L. 1978. Fossil rodents (Rhizomyidae & Muridae) from Neogene Siwalik deposits, Pakistan. Museum of Northern Arizona Press, Bulletin Series, 52:1-103.

Jacobs, L. L. 1985. The beginning of the age of murids in Africa. Acta Zoologica Fennica, 170:149-151.

Jacobs, L. L., and W. R. Downs. 1994. The evolution of murine rodents in Asia. Pp. 149-156, *in* Rodent and lagomorph families of Asian origins and diversification (Y. Tomida, C. k. Li, and T. Setoguchi, eds.). National Science Museum Monographs, No. 8, Tokyo, 195 pp.

Jacobs., L. L., and Li Chuan-KIuei. 1982. A new genus (*Chardinomys*) of murid rodent (Mammalia, Rodentia) from the Neogene of China, and comments on its biogeography. Geobios, 15(2):255-259.

Jacobs, L. L., and E. H. Lindsay. 1984. Holarctic radiation of Neogene muroid rodents and the origin of South American cricetids. Journal of Vertebrate Paleontology, 4:265-272.

Jacobs, L. L., L. J. Flynn, and W. R. Downs. 1989. Neogene rodents of southern Asia. Pp. 156-177, *in* Papers on fossil rodents in honor of Albert Elmer Wood (C. C. Black and M. R. Dawson, eds.). Natural History Museum of Los Angeles County Science Series, 33:192 pp.

Jacobs, L. L., L. J. Flynn, W. R. Downs, and J. C. Barry. 1990. *Quo vadis, Antemus?* The Siwalik muroid record. Pp. 573-586, *in* European Neogene mammal chronology (E. H. Lindsay, V. Fahlbusch, and P. Mein, eds.). Plenum Press, New York and London, 658 pp.

Jaeger, J.-J. 1970. Découverte au Jebel Irhoud des premieres faunes de rongeurs du Pléistocene inferiéur et moyen du Maroc. Comptes Rendus des Séances de l'Académie des Sciences, Paris, Ser. D, 270:920-923.

Jaeger, J.-J. 1975. Les Muridae (Mammalia, Rodentia) du Pliocène et du Pléistocene du Maghreb. Origine; Evolution; Données biogéographiques et paléoclimatiques. Thèse Doct. Ès Sciences, USTL, Montpellier, no. A011538.

Jaeger, J.-J. 1976. Les rongeurs (Mammalia, Rodentia) du Pléistocène Inférieur d'Olduvai Bed I (Tanzanie).

Ière Partie: Les Muridés. Pp. 57-120, *in* Fossil vertebrates of Africa, vol. 4 (R. J. G. Savage and S. C. Coryndon, eds.). Academic Press, London, 338 pp.

Jaeger, J.-J. 1977a. Rongeurs (Mammalia, Rodentia) du Miocene de Beni-Mellal. Palaeovertebrata, 7(4):91-125.

Jaeger, J.-J. 1977b. Les rongeurs du Miocene moyen et superieur du Maghreb. Palaeovertebrata, 8(1):1-166.

Jaeger, J.-J. 1979. Les faunes de rongeurs et de lagomorphes du Pliocène et du Pléistocène d'Afrique orientale. Bulletin de la Société Géologique de France, 21(3):301-308.

Jaeger, J.-J. 1988a. Origine et evolution du genre *Ellobius* (Mammalia, Rodentia) en Afrique Nord-Occidentale. Folia Quaternaria, 57:3-50.

Jaeger, J.-J. 1988b. Rodent phylogeny: New data and old problems. Pp. 177-199, *in* The Phylogeny and Classification of the Tetrapods, Volume 2 (M. J. Benton, ed.). Clarendon Press, Oxford, x + 329 pp.

Jaeger, J.-J., and J.-L. Hartenberger. 1989. Diversification and extinction patterns among Neogene perimediterranean mammals. Philosophical Transactions of the Royal Society of London, B, 325:401-420.

Jaeger, J.-J., and B. Wesselman. 1976. Fossil remains of micromammals from the Omo Group deposits. Pp. 351-360, *in* Earliest Man and Environments in the Lake Rudolf Basin. Stratigraphy, paleoecology, and evolution (Y. Coppens, F. C. Howell, G. L. Isaac, and R. E. F. Leakey, eds.). University of Chicago Press, Chicago and London, 615 pp.

Jaeger, J.-J., H. Tong, E. Buffetaut, and R. Ingavat. 1985. The first fossil rodents from the Miocène of northern Thailand and their bearing on the problem of the origin of the Muridae. Revue de Paléobiologie, 4(1):1-7.

Jaeger, J.-J., H. Tong, and C. Denys. 1986. Age de la divergence *Mus-Rattus*: Comparaison des donnees paleontologiques et moleculaires. Comptes Rendus des Séances de l'Académie des Sciences, Paris, 302, ser. 2:917-922.

Jameson, E. W., Jr. 1999. Host-ectoparasite relationships among North American chipmunks. Acta Theriologica, 44(3):225-231.

Jameson, E. W., and G. S. Jones. 1977. The Soricidae of Taiwan. Proceedings of the Biological Society of Washington, 90:459-482.

Jammot, D. 1983. Evolution des Soricidae. Insectivora, Mammalia. Symbioses, 15:253-273.

Janczewski, D. N., W. S. Modi, J. C. Stephens, and S. J. O'Brien. 1995. Molecular evolution of mitochondrial 12S RNA and cytochrome b sequences in the pantherine lineage of felidae. Molecular Biology and Evolution, 12:690-707.

Janecek, L. L. 1990. Genic variation in the *Peromyscus truei* group (Rodentia: Cricetidae). Journal of Mammalogy, 71:301-308.

Janecek, L. L., D. A. Schlitter, and I. L. Rautenbach. 1991. A genic comparison of spiny mice, genus *Acomys*. Journal of Mammalogy, 72:542-552.

Janecek, L. L., R. L. Honeycutt, I. L. Rautenbach, B. H. Erasmus, S. Reig, and D. A. Schlitter. 1992. Allozyme variation and systematics of African mole-rats (Rodentia: Bathyergidae). Biochemical Systematics and Ecology, 20(5):401-416.

Janis, C. M., and K. M. Scott. 1987. The interrelationships of higher ruminant families with special emphasis on the members of the Cervoidea. American Museum Novitates, 2893:1-85.

Jannett, F. J., Jr. 1997. The uncut m1 and systematic position of the North American water vole (*Microtus richardsoni*). American Midland Naturalist, 137:173-177.

Jánossy, D. 1986. Pleistocene vertebrate faunas of Hungary. Akadémia Kiado, Budapest, 208 pp.

Jansa, S. A., and M. D. Carleton. 2003a. Systematics and phylogenetics of Madagascar's native rodents. Pp. 1257-1265, *in* The natural history of Madagascar (S. M. Goodman and J. P. Benstead, eds.). The University of Chicago Press, Chicago, xxi + 1709 pp.

Jansa, S. A., and M. D. Carleton. 2003b. *Brachyuromys*, short-tailed rats, Malagasy voles. Pp. 1370-1373, *in* The natural history of Madagascar (S. M. Goodman and J. P. Benstead, eds.). The University of Chicago Press, Chicago, xxi + 1709 pp.

Jansa, S. A., and L. R. Heaney. 2001. A molecular phylogeny of Philippine murid rodents and its implications for patterns of diversification. Abstract 207, 81st Annual Meeting of the American Society of Mammalogists, University of Montana, Missoula.

Jansa, S., and M. Weksler. 2004. Phylogeny of muroid rodents: Relationships within and among major lineages as determined by IRBP gene sequences. Molecular Phylogenetics and Evolution, 31(1):256-276.

Jansa, S. A., S. M. Goodman, and P. K. Tucker. 1999. Molecular phylogeny and biogeography of the native rodents of Madagascar (Muridae: Nesomyinae): A test of the single-origin hypothesis. Cladistics, 15:253-270.

Jansen van Vuuren, B., and T. J. Robinson. 2001. Retrieval of four adaptive lineages in duiker antelope: Evidence from mitochondrial DNA sequences and fluorescence in situ hybridization. Molecular Phylogenetics and Ecology, 20:409-425.

Janzekovic, F., and B. Kryštufek. 2004. Geometric morphometry of the upper molars in European wood mice *Apodemus*. Folia Zoologica, 53(1):47-55.

Jarrell, G. H., and K. Fredga. 1993. How many kinds of lemmings? A taxonomic overview. Pp. 46-57, *in* The biology of lemmings (N. C. Stenseth and R. A. Ims, eds.). Linnean Society Symposium Series 15, Academic Press, London, 683 pp.

Jawdat, S. Z. 1977. A new record of forest dormouse *Dryomys nitedula* Pallas (Rodentia: Muscardinidae) in Iraq. Bulletin of the Biological Research Centre, University of Baghdad, 9:115.

Jayson, E. A., and G. Christopher. 1995. Sighting of spiny dormouse *Platacanthomys lasiurus* Blyth, 1859 in Peppara Wildlife Sanctuary, Trivandrum District. Kerala. Journal of the Bombay Natural History Society, 92:258.

Jeannin, A. 1936. Les mammiféres sauvages du Cameroun. Encyclopedie Biologique, 16:116-130.

Jefferson, T. A. 1988. *Phocoenoides dalli*. Mammalian Species, 319:1-7.

Jenkins, P. D. 1976. Variation in Eurasian shrews of the genus *Crocidura* (Insectivora: Soricidae). Bulletin of the British Museum (Natural History), Zoology Series, 30:271-309.

Jenkins, P. D. 1982. A discussion of Malayan and Indonesian shrews of the genus *Crocidura* (Insectivora: Soricidae). Zoologische Mededelingen, 56:267-279.

Jenkins, P. D. 1984. Description of a new species of *Sylvisorex* (Insectivora: Soricidae) from Tanzania. Bulletin of the British Museum (Natural History), Zoology Series, 47:65-76.

Jenkins, P. D. 1987. Catalogue of Primates in the British Museum (Natural History) and elsewhere in the British Isles. Part 4: Suborder Strepsirrhini, including the subfossil Madagasan lemurs and family Tarsiidae. British Museum (Natural History), London, 189 pp.

Jenkins, P. D. 1992. Description of a new species of *Microgale* (Insectivora: Tenrecidae) from eastern Madagascar. Bulletin of the British Museum of Natural History (Zoology), 58:53-59.

Jenkins, P. D., and A. A. Barnett. 1997. A new species of water mouse, of the genus *Chibchanomys* (Rodentia, Muridae, Sigmodontinae) from Ecuador. Bulletin of The Natural History Museum, London (Zoology), 63:123-128.

Jenkins, P. D., and J. E. Hill. 1981. The status of *Hipposideros galeritus* Cantor, 1846 and *Hipposideros cervinus* (Gould, 1854) (Chiroptera: Hipposideridae). Bulletin of the British Museum (Natural History), Zoology Series, 415:279-294.

Jenkins, P. D., and J. E. Hill. 1982. Mammals from Siberut, Mentawei Islands. Mammalia, 46:219-224.

Jenkins, P. D., and M. F. Robinson. 2002. Another variation on the gymnure theme: Description of a new species of *Hylomys* (Lipotyphla, Erinaceidae, Galericinae). Bulletin of the Natural History Museum, London (Zoology), 68:1-11.

Jenkins, P. D., and A. L. Smith. 1995. A new species of *Crocidura* (Insectivora: Soricidae) recovered from owl pellets in Thailand. Bulletin of the Natural History Museum, London (Zoology), 61:103-109.

Jenkins P. D., S. M. Goodman, and C. J. Raxworthy. 1996. The shrew tenrecs (*Microgale*) (Insectivora: Tenrecidae) of the Réserve Naturelle Intégrale d'Andringitra, Madagascar. Pp. 191-217, *in* A floral and faunal inventory of the eastern slopes of the Réserve Naturelle Intégrale d'Andringitra, Madagascar: With reference to elevational variation. (S. M. Goodman, ed.). Fieldiana: Zoology, n.s. (85):319 pp.

Jenkins, P. D., C. J. Raxworthy, and R. A. Nussbaum. 1997. A new species of *Microgale* (Insectivora, Tenrecidae), with comments on the status of four other taxa of shrew tenrecs. Bulletin of the Natural History Museum, Zoology, 63:1-12.

Jenkins, P., M. Ruedi, and M. Catzeflis. 1998. A biochemical and morphological investigation of *Suncus dayi* (Dobson, 1888) and discussion of relationship in *Suncus* Hemprich & Ehrenberg, 1833, *Crocidura* Wagler, 1832, and *Sylvisorex* Thomas, 1904 (Insectivora: Soricidae). Bonner Zoologische Beiträge, 47:257-276.

Jenkins, S. H., and P. E. Busher. 1979. *Castor canadensis*. Mammalian Species, 120:1-8.

Jenkins, S. H., and B. D. Eshelman. 1984. *Spermophilus beldingi*. Mammalian Species, 221:1-8.

Jenks, S. M., and L. Werdelin. 1998. Taxonomy and systematics of living hyaenas (Family Hyaenidae), Pp. 8-17, *in* Hyaenas: Status Survey and Conservation Action Plan (G. Mills, and H. Hofer, eds.) IUCN/SSC Hyaena Specialist Group. I.U.C.N., Gland, Switzerland, 154 pp.

Jennings, J. B., T. L. Best, C. Rainey, and S. E. Burnett. 2000. *Molossus pretiosus*. Mammalian Species, 635:1-3.

Jennings, J. B., T. L. Best, S. E. Burnett, and J. C. Rainey. 2002. *Molossus sinaloae*. Mammalian Species, 691:1-5.

Jennings, M. 1979. A new bat for Saudi Arabia. Journal of the Saudi Arabian Natural History Society, 24:4.

Jensen, J. V. 1980. [The common dormouse—Denmark's only dormouse]. Naturens Verden, 1:55-65.

Jentink, F. A. 1879a. On a new genus and species of *Mus* from Madagascar. Notes of the Leyden Museum, 1879:107-109.

Jentink, F. A. 1879b. On various species of *Mus*, collected by S. C. I. Van Musschenbroek Esq. in Celebes. Notes of the Royal Zoological Museum of the Netherlands, Leyden, 1(note 2):7-13.

Jentink, F. A. 1883. A list of species of mammals from West-Sumatra and North-Celebes, with descriptions of undescribed or rare species. Notes of the Leyden Museum, 5:170-181.

Jentink, F. A. 1887. Catalogue ostèologique des mammiféres. Muséum d'Histoire Naturelle des Pays-Bas, 9:1-360.

Jentink, F. A. 1888. Catalogue systématique des Mammifères. Museum d'Histoire Naturalle Pays-Bas, Leiden, 12:1-280.

Jentink, F. A. 1892. Catalogue systématique des mammifères. Museum d'Histoire Naturelle des Pays-Bas, 11:1-219.

Jentink, F. A. 1905. *Sus*-studies in the Leyden Museum. Notes from the Leyden Museum, 26:155-195.

Jerdon, T. C. 1867. The Mammals of India; a natural history of all the animals known to inhabit continental India. Rorkee, Calcutta, 319 pp. (Reprinted, London, 1874, 335 pp.)

Jernvall, J., S. V. E. Keränen, and I. Thesleff. 2000. Evolutionary modification of development in mammalian teeth: Quantifying gene expression patterns and topography. Proceedings of the National Academy of Sciences, 97(26):14444-14448.

Ji, Q., Z.-X. Luo, C.-X. Yuan, J. R. Wible, J.-P. Zhang, and J. A. Beorgi. 2002. The earliest known eutherian mammal. Nature, 416:816-822.

Jiang, X. L., and R. S. Hoffmann. 2001. A revision of the white-toothed shrews (*Crocidura*) of southern China. Journal of Mammalogy, 82:1059-1079.

Jiang, X.-L., and Y.-X. Wang. 2000. [The field mice (*Apodemus*) in Wuliang Mountain with a discussion of *A. orestes*.] Zoological Research, 21(6):473-478 (in Chinese with English abstract).

Jiang Xuelong, Wang Yingxiang, and Wang Qishan. 1996. Taxonomy and distribution of Tibetan macaque (*Macaca thibetana*). Zoological Research, 17:361-369.

Jiang, X. L., Y. X. Wang, and R. S. Hoffmann. 2003. A review of the systematics and distribution of Asiatic short-tailed shrews, genus *Blarinella* (Mammalia: Soricidae). Mammalian Biology, 68:193-204.

Jimenez, J. E. 1996. The extirpation and current status of wild chinchillas *Chinchilla lanigera* and *C. brevicaudata*. Biological Conservation, 77:1-6.

Jimenez, R., A. Carnero, M. Burgos, A. Sanchez, and R. Diaz de la Guardia. 1991. Achiasmatic giant sex chromosomes in the vole *Microtus cabrerae* (Rodentia, Microtidae). Cytogenetics and Cell Genetics, 57:56-58.

Johnson, D. H. 1946a. The spiny rats of the Riu Kiu islands. Proceedings of the Biological Society of Washington, 59:169-172.

Johnson, D. H. 1946b. The rat population of a newly established military base in the Solomon Islands. United States Naval Medical Bulletin, 46:1628-1632.

Johnson, D. H. 1948. A rediscovered Haitian rodent, *Plagiodontia aedium*, with a synopsis of related species. Proceedings of the Biological Society of Washington, 61:69-76.

Johnson, D. H. 1962a. Rodents and other Micronesian mammals collected. IV. A. Pp. 21-38, *in* Pacific island rat ecology. Report of a study made on Ponape and adjacent islands, 1955-1958 (T. I. Storer, ed.). Bernice P. Bishop Museum Bulletin, 225:1-274.

Johnson, D. H. 1962b. Two new murine rodents. Proceedings of the Biological Society of Washington, 75:317-319.

Johnson, D. W., and D. M. Armstrong. 1987. *Peromyscus crinitus*. Mammalian Species, 287:1-8.

Johnson, K. A., and A. D. Roff. 1980. Discovery of Ningauis (*Ningaui* sp.: Dasyuridae: Marsupialia) in the Northern Territory, Australia. Australian Mammalogy, 3:127-129.

Johnson, M. L. 1968. Application of blood protein electrophoretic studies to problems in mammalian taxonomy. Systematic Zoology, 17:23-30.

Johnson, M. L. 1973. Characters of the heather vole, *Phenacomys*, and the red tree vole, *Arborimus*. Journal of Mammalogy, 54:239-244.

Johnson, M. L., and S. B. Benson. 1960. Relationships of the pocket gophers of the *Thomomys mazama-talpoides* complex in the Pacific Northwest. Murrelet, 41:17-22.

Johnson, M. L., and S. B. George. 1991. Species limits within the *Arborimus longicaudus* species-complex (Mammalia: Rodentia) with a description of a new species from California. Contributions in Science, Natural History Museum of Los Angeles County, 429:1-16.

Johnson, M. L., and C. Maser. 1982. Generic relationships of *Phenacomys albipes*. Northwest Science, 56:17-19.

Johnson, M. L., and B. T. Ostenson. 1959. Comments on the nomenclature of some mammals of the Pacific Northwest. Journal of Mammalogy, 40:571-577.

Johnson, W. E., and S. J. O'Brien. 1997. Phylogenetic reconstruction of the Felidae using 16S rRNA and NADH-5 mitochondrial genes. Journal of Molecular Evolution, 44 (Suppl. 1): S98-S116.

Johnson, W. E., and R. K. Selander. 1971. Protein variation and systematics in kangaroo rats (genus *Dipodomys*). Systematic Zoology, 20:377-405.

Johnson, W. E., W. L. Franklin, and J. A. Iriarte. 1990. The mammalian fauna of the northern Chilean Patagonia: A biogeographical dilemma. Mammalia, 54:457-469.

Johnson, W. E., M. Culver, J. A. Iriarte, E. Eizirik, K. L. Seymour, and S. J. O'Brien. 1998. Tracking the evolution of the elusive Andean mountain cat (*Oreailurus jacobita*) from mitochondrial DNA. Journal of Heredity, 89:227-232.

Johnson, W. E., F. Shinyashiki, M. M. Raymond, C. Driscoll, C. Leh, M. Sunquist, L. Johnston, M. Buch, D. Wildt, N. Yuhki, and S. J. O'Brien. 1999. Molecular genetic characterization of two Asian cat species, Bornean Bay Cat and Irimote Cat. Pp. 223-248, *in* Evolutionary Theory and Processes: Modern Perspectives. Papers in Honour of Eviatar Nevo (S. P. Wasser, ed.). Kulwer Academic Publishers, 466 pp.

Johnson-Murray, J. L. 1977. Myology of the gliding membranes of some petauristine rodents (Genera: *Glaucomys*, *Petaurista*, *Petinomys*, and *Pteromys*). Journal of Mammalogy, 58:374-384.

Jolley, T. W., R. L. Honeycutt, and R. D. Bradley. 2000. Phylogenetic relationships of pocket gophers (genus *Geomys*) based on the mitochondrial 12S rRNA gene. Journal of Mammalogy, 81:1025-1034.

Jolly, C. J., and F. L. Brett. 1973. Genetic markers and baboon biology. Journal of Medical Primatology, 2(2):85-99.

Jones, C. 1977. *Plecotus rafinesquii*. Mammalian Species, 69:1-4.

Jones, C. 1978. *Dendrohyrax dorsalis*. Mammalian Species, 113:1-4.

Jones, C., and S. Anderson. 1978. *Callicebus moloch*. Mammalian Species, 112:1-5.

Jones, C., and N. J. Hildreth. 1989. *Neotoma stephensi*. Mammalian Species, 328:1-3.

Jones, C., and R. W. Manning. 1989. *Myotis austroriparius*. Mammalian Species, 332:1-3.

Jones, C., C. A. Jones, J. K. Jones, Jr., and D. E. Wilson. 1996. *Pan troglodytes*. Mammalian Species, 529:1-9.

Jones, C., R. S. Hoffmann, D. W. Rice, M. D. Engstrom, R. D. Bradley, D. J. Schmidly, C. A. Jones, and R. J. Baker. 1997. Revised checklist of North American mammals north of Mexico, 1997. Occasional Papers, Museum of Texas Tech University, 173:1-19.

Jones, G. 1997. Acoustic signals and speciation: The roles of natural and sexual selection in the evolution of cryptic species. Advances in the Study of Behavior, 26:317-354.

Jones, G., and E. M. Barratt. 1999. *Vespertilio pipistrellus* Schreber, 1774, and *V. pygmaeus* Leach, 1825 (currently *Pipistrellus pipstrellus* and *P. pygmaeus*; Mammalia, Chiroptera): Proposed designation of neotypes. Bulletin of Zoological Nomenclature, 56:182-186.

Jones, G., and S. M. van Parijs. 1993. Bimodal echolocation in pipistrelle bats: Are cryptic species present? Proceedings of the Royal Society of London, 251B:119-125.

Jones, G., M. Morton, P. M. Hughes, and R. M. Budden. 1993. Echolocation, flight morphology and foraging strategies of some West African hipposiderid bats. Journal of Zoology, London, 230:385-400.

Jones, G. S. 1975. Catalogue of the type specimens of mammals of Taiwan. Quarterly Journal of the Taiwan Museum, 28:183-217.

Jones, G. S., and R. E. Mumford. 1971. *Chimarrogale* from Taiwan. Journal of Mammalogy, 52:228-232.

Jones, J. K., Jr. 1964. Distribution and taxonomy of mammals of Nebraska. University of Kansas Publications, Museum of Natural History, 16:1-356.

Jones, J. K., Jr. 1965. Taxonomic status of the molossid bat, *Cynomops malagai* Villa-R, 1955. Proceedings of the Biological Society of Washington, 78:93.

Jones, J. K., Jr. 1977. *Rhogeessa gracilis*. Mammalian Species, 76:1-2.

Jones, J. K., Jr. 1978. A new bat of the genus *Artibeus* from the Lesser Antillean Island of St. Vincent. Occasional Papers of the Museum, Texas Tech University, 51:1-6.

Jones, J. K., Jr. 1982. *Reithrodontomys spectabilis*. Mammalian Species, 193:1.

Jones, J. K., Jr. 1989. Distribution and systematics of bats in the Lesser Antilles. Pp. 645-660, *in* Biogeography of the West Indes: Past, present, and future (C. A. Woods, ed.). Sandhill Crane Press, Gainesville, Florida, 878 pp.

Jones, J. K., Jr. 1990. *Peromyscus stirtoni*. Mammalian Species, 361:1-2.

Jones, J. K., Jr., and J. Arroyo-Cabrales. 1990. *Nyctinomops aurispinosus*. Mammalian Species, 350:1-3.

Jones, J. K., Jr., and R. J. Baker. 1980. *Chiroderma improvisum*. Mammalian Species, 134:1-2.

Jones, J. K., Jr., and G. A. Baldassarre. 1982. *Reithrodontomys brevirostris* and *Reithrodontomys paradoxus*. Mammalian Species, 192:1-3.

Jones, J. K., Jr., and D. C. Carter. 1976. Annotated checklist, with keys to subfamilies and genera. Part I. Pp. 7-38, *in* Biology of bats of the New World family Phyllostomatidae. Part I (R. J. Baker, J. K. Jones, Jr., and D. C. Carter, eds.). Special Publications, The Museum, Texas Tech University Press, 10:1-218.

Jones, J. K., Jr., and D. C. Carter. 1979. Systematic and distributional notes. Pp. 7-11, *in* Biology of bats of the New World family Phyllostomatidae, Part III. (R. J. Baker, J. K. Jones, Jr., and D. C. Carter, eds.). Special Publications, The Museum, Texas Tech University Press, 16:1-441.

Jones, J. K., Jr., and M. D. Engstrom. 1986. Synopsis of the rice rats (genus *Oryzomys*) of Nicaragua. Occasional Papers, The Museum, Texas Tech University, 103:1-23.

Jones, J. K., Jr., and H. H. Genoways. 1967. Notes on the Oaxacan vole, *Microtus oaxacensis* Goodwin, 1966. Journal of Mammalogy, 48:320-321.

Jones, J. K., Jr., and H. H. Genoways. 1970. Harvest mice (genus *Reithrodontomys*) of Nicaragua. Occasional Papers of the Western Foundation of Vertebrate Zoology, 2:1-16.

Jones, J. K., Jr., and H. H. Genoways. 1973. *Ardops nichollsi*. Mammalian Species, 24:1-2.

Jones, J. K., Jr., and H. H. Genoways. 1975a. *Dipodomys phillipsii*. Mammalian Species, 51:1-3.

Jones, J. K., Jr., and H. H. Genoways. 1975b. *Sciurus richmondi*. Mammalian Species, 53:1-2.

Jones, J. K., Jr., and H. H. Genoways. 1975c. *Sturnira thomasi*. Mammalian Species, 68:1-2.

Jones, J. K., Jr., and H. H. Genoways. 1978. *Neotoma phenax*. Mammalian Species, 108:1-3.

Jones, J. K., Jr., and J. A. Homan. 1974. *Hylonycteris underwoodi*. Mammalian Species, 32:1-2.

Jones, J. K., Jr., and C. S. Hood. 1993. Synopsis of South American bats of the family Emballonuridae. Occasional Papers, The Museum, Texas Tech University, 155:1-32.

Jones, J. K., Jr., and D. H. Johnson. 1960. Review of the insectivores of Korea. University of Kansas, Publications of the Museum of Natural History, 9:549-578.

Jones, J. K., Jr., and D. H. Johnson. 1965. Synopsis of the lagomorphs and rodents of Korea. University of Kansas Publications, Museum of Natural History, 16:357-407.

Jones, J. K., Jr., and T. E. Lawlor. 1965. Mammals from Isla Cozumel, Mexico, with description of a new species of harvest mouse. University of Kansas Publications, Museum of Natural History, 16:409-419.

Jones, J. K., Jr., and C. J. Phillips. 1970. Comments on systematics and zoogeography of bats in the Lesser Antilles. Studies on the Fauna of Curaçao and Other Caribbean Islands, 32:131-145.

Jones, J. K., Jr., and C. J. Phillips. 1976. Bats of the genus *Sturnira* in the Lesser Antilles. Occasional Papers of the Museum, Texas Tech University, 40:1-16.

Jones, J. K., Jr., and A. Schwartz. 1967. Bredin-Archbold-Smithsonian Biological Survey of Dominica. 6. Synopsis of bats of the Antillean genus *Ardops*. Proceedings of the United States National Museum, 124(3634):1-13.

Jones, J. K., Jr., and T. L. Yates. 1983. Review of the white-footed mice, genus *Peromyscus*, of Nicaragua. Occasional Papers, The Museum, Texas Tech University, 82:1-15.

Jones, J. K., Jr., D. C. Carter, and H. H. Genoways. 1975. Revised checklist of North American mammals north of Mexico. Occasional Papers, The Museum, Texas Tech University, 28:1-14.

Jones, J. K., Jr., P. Swanepoel, and D. C. Carter. 1977. Annotated checklist of the bats of Mexico and Central America. Occasional Papers, The Museum, Texas Tech University, 47:1-35.

Jones, J. K., Jr., D. C. Carter, H. H. Genoways, R. S. Hoffmann, and D. W. Rice. 1982. Revised checklist of North American mammals north of Mexico, 1982. Occasional Papers, The Museum, Texas Tech University, 80:1-22.

Jones, J. K., Jr., D. C. Carter, H. H. Genoways, R. S. Hoffman, D. W. Rice, and C. Jones. 1986. Revised checklist of North American mammals north of Mexico. Occasional Papers, The Museum, Texas Tech University, 107:1-22.

Jones, J. K., Jr., R. S. Hoffmann, D. W. Rice, C. Jones, R. J. Baker, and M. D. Engstrom. 1992. Revised checklist of North American mammals north of Mexico, 1991. Occasional Papers Museum Texas Tech University, 146:1-23.

Jones, K. E., A. Purvis, A. MacLarnon, O. R. P. Bininda-Emonds, and N. B. Simmons. 2002. A phylogenetic supertree of the bats (Mammalia: Chiroptera). Biological Reviews, 77:223-259.

Jones, M. L., S. L. Swartz, and S. Leatherwood (eds.). 1984. The gray whale. Academic Press, New York, 600 pp.

Jong, N., de, and W. Bergmans. 1981. A revision of the fruit bats of the genus *Dobsonia* Palmer, 1898 from Sulawesi and some nearby islands (Mammalia, Megachiroptera, Pteropodinae). Zoologische Abhandlungen Staatliches Museum für Tierkunde in Dresden, 37(6):209-224.

Jonkers, D. A. 1992. Veldmuis, *Microtus arvalis* (Pallas, 1779). Pp. 269-272, *in* Atlas van de Nederlandse Zoogdieren (S. Broekhuizen, B. Hoekstra, V. van Laar, C. Smeenk, and J. B. M. Thissen, eds.). Stichting Uitgeverij Koninklijke Nederlandse Natuurhistorische Vereniging, Utrecht, 336 pp.

Jordan, D. S., and G. A. Clark. 1898. The history, condition, and needs of the herd of fur seals resorting to the Pribilof Islands, pt. 1, 249 pp., *in* The fur seals and fur-seal islands of the North Pacific Ocean (D. S. Jordan). United States Government Printing Office, Washington, D.C., U. S. Treasury Document no. 2017:4 volumes.

Jordan, M. 2001. Reintroduction and restocking programmes for the common hamster (*Cricetus cricetus*)–issues and protocols. Jahrbücher des Nassauischen Vereins für Naturkunde, 122:167-177.

Joshua, J. 1996. Interbreeding between grizzled giant squirrel, *Ratufa macroura* (Pennant) and Malabar giant squirrel, *R. indica* (Erxleben). Journal of the Bombay Natural History Society, 92(1):82-83.

Jotterand, M. 1972. Le polymorphisme chromosomique des *Mus* (Leggadas) africains. Cytogenetique, zoogeographie, evolution. Revue Suisse de Zoologie, 79(1):287-359.

Jotterand-Bellomo, M. 1981. Le caryotype et la spermatogenese de *Mus setulosus* (bandes Q, C, G, et coloration argentique). Genetica, 56:217-227.

Jotterand-Bellomo, M. 1984. L'analyse cytogenetique de deux especes de Muridae africains, *Mus oubanguii* et *Mus minutoides/musculoides*: Polymorphisme chromosomique et ebauche d'une phylogenie. Cytogenetics and Cell Genetics, 38:182-188.

Jotterand-Bellomo, M. 1986. Le genre *Mus* africain, un exemple d'homogeneite caryotypique: étude cytogenetique de *Mus minutoides/musculoides* (Côte d'Ivoire), de *M. setulosus* (Republique Centrafricaine), et de *M. mattheyi* (Burkina Faso). Cytogenetics and Cell Genetics, 42:99-104.

Jotterand-Bellomo, M., and P. Schauenberg. 1988. Les chromosomes du rat de Cuming, *Phloeomys cumingi* Waterhouse, 1839 (Mammalia: Rodentia). Genetica, 76:181-190.

Jourdan, M. 1837. Mémoire sur quelques mammifères nouveaux. Comptes Rendus Hebdomadaires de Séances de l'Académie des Sciences, 5:521-524.

Jourde, P. 2001. Redecouverte du muscardin Muscardinus avellanarius Linne, 1758 en Charente-Maritime. Annales de la Societe des Sciences Naturelles de la Charente-Maritime, 9(1):65-68.

Juckwer, E.-A. 1990. *Galemys pyrenaicus*—Pyrenäen-Desman. Pp. 79-92, *in* Handbuch der Säugetiere Europas (J. Niethammer and F. Krapp, eds.). Aula-Verlag, Wiesbaden, 3/I:1-524.

Judd, S. R. 1980. Observations on the chromosome variation in *Microtus mexicanus* (Rodentia: Microtinae). Mammalian Chromosome Newsletter, 21:110-113.

Judd, S. R., and S. P. Cross. 1980. Chromosomal variation in *Microtus longicaudus* (Merriam). Murrelet, 61:2-5.

Judd, S. R., S. P. Cross, and S. Pathak. 1980. Non-Robertsonian chromosomal variation in *Microtus montanus*. Journal of Mammalogy, 61:109-113.

Jüdes, U. 1981. G- and C-band karyotypes of the harvest mouse, *Micromys minutus*. Genetica, 54:237-239.

Julliot, C., S. Cajani, and A. Gautier-Hion. 1998. Anomalures (Rodentia, Anomaluridae) in Central Gabon, species composition, population densities and ecology. Mammalia, 62:9-21.

Junge, J. A., and R. S. Hoffmann. 1981. An annotated key to the long-tailed shrews (genus *Sorex*) of the United States and Canada, with notes on Middle American *Sorex*. Occasional Papers of the Museum of Natural History, University of Kansas, 94:1-48.

Junge, J. A., R. S. Hoffmann, and R. W. DeBry. 1983. Relationships within the Holarctic *Sorex arcticus-Sorex tundrensis* species complex. Acta Theriologica, 28:339-350.

Jurczyszyn, M. 2001. Reintroduction of the edible dormouse (*Glis glis*) in Sierakowski Landscape Park (Poland). Preliminary results. Trakya University Journal of Scientific Research B, 2(2):111-114.

Jurczyszyn, M., and K. Wolk. 1998. The present status of dormice (Myoxidae) in Poland. Natura Croatica, 7(1):11-18.

Juškaitis, R. 1995*a*. Distribution, abundance and conservation status of dormice (Myoxidae) in Lithuania. Pp. 181-184, *in* Proceedings of II Conference on Dormice (Rodentia, Myoxidae) (M. G. Filippucci, ed.). Hystrix, n.s., 6(1-2):1-340.

Juškaitis, R. 1995*b*. Relations between Common dormice (*Muscardinus avellanarius*) and other occupants of bird nest-boxes in Lithuania. Folia Zoologica, 44(4):289-296.

Juškaitis, R. 1995*c*. White-tipped *Muscardinus avellanarius* L. in Lithuania. Ekologija, 2:60-63.

Juškaitis, R. 1997. Ranging and movement of the Common Dormouse *Muscardinus avellanarius* in Lithuania. Acta Theriologica, 42(2):113-122.

Juškaitis, R. 2001. Weight changes of the common dormouse (Muscardinus avellanarius L.) during the year in Lithuania. Trakya University Journal of Scientific Research B, 2(2):79-83.

Juškaitis, R., and K. Baranauskas. 2001. Diversity of small mammals in the northwestern Lithuania (Mažeikiai District). Acta Zoologica Lituanica, 11(4):343-348.

Juškaitis, R., K. Baranauskas, R. Mažeikytè, and A. Ulevièius. 2001. New data on the pygmy field mouse (*Apodemus uralensis*) distribution and habitats in Lithuania. Acta Zoologica Lituanica, 11(4):349-353.

Just, W., W. Rau, W. Vogel, M. Akhverdian, and K. Fredga. 1995. Absence of *Sry* in species of the vole *Ellobius*. Nature Genetics, 11:117-118.

Just, W., A. Baumstark, H. Hameister, B. Schreiner, I. Reisert, M. Hakhverdyan, and W. Vogel. 2002. The sex determination in *Ellobius lutescens* remains bizarre. Cytogenetic and Genome Research, 96:146-153.

Juste, J. B., and C. Ibáñez. 1992. Taxonomic review of *Minopterus minor* Peters, 1867 (Mammalia: Chiroptera) from west central Africa. Bonner Zoologische Beiträge, 43:355-365.

Juste, J. B., C. Ibáñez, and A. Machordom. 1997. Evolutionary relationships among the African fruit bats: *Rousettus egyptiacus*, *R. angolensis*, and *Myonycteris*. Journal of Mammalogy, 78:766-774.

Juste, J. B., C. Ibáñez, and A. Machordom. 2000. Morphological and allozyme variation of *Eidolon helvum* (Mammalia: Megachiroptera) in the islands of the Gulf of Guinea. Biological Journal of the Linnean Society, 71:359-378.

Juste, J., C. Ibáñez, D. Trujillo, J. Muñoz, and M. Ruedi. 2003. Phylogeography of barbastelle bats (*Barbastella barbastellus*) in the western Mediterranean and the Canary Islands. Acta Chiropterologica, 5:165-175.

Juyal, R. C., B. K. Thelma, and S. R. V. Rao. 1989. Heterochromatin variation and spermatogenesis in *Nesokia*. Cytogenetics and Cell Genetics, 50:206-210.

Kadwell, M., M. Fernandez, H. F. Stanley, R. Baldi, J. C. Wheeler, R. Rosadio, and M. W. Bruford. 2001. Genetic analysis reveals the wild ancestors of the llama and the alpaca. Proceedings of the Royal Society Biological Sciences Series B, 268:2575-2584.

Kahila Bar-Gal, G., P. Smith, E. Tchernov, C. Greenblatt, P. Ducos, A. Gardeisen, and L. K. Horwitz. 2002. Genetic evidence for the origin of the agrimi goat (*Capra aegagrus cretica*). Journal of Zoology, London, 256:369-377.

Kahlke, R.-D. 1999. The history of the origin, evolution and dispersal of the late Pleistocene *Mammuthus-Coelodonta* Faunal Complex in Eurasia (large mammals). Mammoth Site of Hot Springs, South Dakota, 219 pp.

Kahmann, H. 1981. Zur Naturgeschichte des Löffelbilches, *Eliomys melanurus* Wagner, 1840. Spixiana, 4(1):1-37.

Kahmann, H., and J. A. Alcover. 1974. Sobre la bionomia del liron careto (*Eliomys quercinus* L.) en Mallorca. Note preliminar. Bolletino de la Sociedad de Historia Natural de Baleares, 19:57-74.

Kahmann, H., and G. Thoms. 1973a. Der Gartenschläfer (*Eliomys*) Menorcas. Säugetierkundliche Mitteilungen, 21:65-73.

Kahmann, H., and G. Thoms. 1973b. Zur Binomie des Gartenschläfers *Eliomys quercinus denticulatus* Ranck, 1968 aus Libyen. Zeitschrift für Säugetierkunde, 38:197-208.

Kahmann, H., and G. Thoms. 1981. Über den Gartenschläfer (*Eliomys*) in nordafrikanischen Ländern (Mammalia: Rodentia, Gliridae). Spixiana, 4(2):191-228.

Kahmann, H., and G. Thoms. 1987. *Eliomys quercinus tunetae* Thomas, 1903 (Mammalia, Rodentia, Gliridae). Spixiana, 10(2):97-114.

Kahmann, H., and I. Vesmanis. 1976. Morphometrische Untersuchungen an Wimperspitzmäusen (*Crocidura*) (Mammalia: Soricidae) 2. Zur weiteren Kenntnis von *Crocidura gueldenstaedti* (Pallas, 1811) auf der Insel Kreta. Opuscula Zoologica (Budapest), 136:1-12.

Kain, D. E. 1985. The systematic status of *Eutamias ochrogenys* and *Eutamias senex* (Rodentia: Sciuridae). Unpubl. M. A. thesis, Humboldt State University, Arcata, California, 67 pp.

Kälin, D. 1999. Tribe Cricetini. Pp. 373-387, *in* The Miocene land mammals of Europe (G. E. Rössner and K. Heissig, eds.). Dr. Friedrich Pfeil, München, 515 pp.

Kalinin, A. A. 1993. [Analysis of the biotopic distribution of Norway (*Rattus norvegicus*) and black (*R. Rattus*) rats in different altitudinal zones of Adzharia (the Caucasus).] Zoologicheskii Zhurnal, 72(6):114-123 (in Russian with English summary).

Kalko, E. K. V., and C. O. Handley, Jr. 1994. Evolution, biogeography, and description of a new species of fruit-eating bat, genus *Artibeus* Leach (1821), from Panamá. Zeitschrift für Säugetierkunde, 59:257-273.

Kaluza, T. 1987. Locality of common dormouse *Muscardinus avellanarius* (Linnaeus, 1758) in the Wielkopolska district. Przeglad Zoologiczny, 31(2):215-218.

Kaminski, M., M. Rousseau, and F. Petter. 1984. Electrophoretic studies on blood proteins of *Arvicanthis niloticus*. Biochemical Systematics and Ecology, 12:215-224.

Kaneko, Y. 1985. Examinations of diagnostic characters (mammae and bacula) between *Eothenomys smithi* and *E. kageus*. Journal of the Mammalogical Society of Japan, 10:221-229.

Kaneko, Y. 1987. Skull and dental characters, and skull measurements of *Microtus kikuchii* Kuroda, 1920, from Taiwan. Journal of the Mammalogical Society of Japan, 12:31-39.

Kaneko, Y. 1990. Identification and some morphological characters of *Clethrionomys rufocanus* and *Eothenomys regulus* from USSR, Northeast China, and Korea in comparison with *C. rufocanus* from Finland. Journal of the Mammalogical Society of Japan, 14:129-148.

Kaneko, Y. 1992a. [Topographic distribution of three small rodents in Shikoku, Japan.] Bulletin of the Biogeographical Society of Japan, 47(16):127-139 (in Japanese).

Kaneko, Y. 1992b. [Mammals of Japan. 17. *Eothenomys smithii* (Smith's red-backed vole).] Mammalian Science, 32(1):39-54 (in Japanese).

Kaneko, Y. 1992c. Identification and morphological characteristics of *Clethrionomys rufocanus*, *Eothenomys shanseius*, *E. inez*, and *E. eva* from the USSR, Mongolia, and northern and central China. Journal of the Mammalogical Society of Japan, 16:71-95.

Kaneko, Y. 1994. [Muridae.] Pp. 90-110 and 168-183, *in* A pictorial guide to the mammals of Japan (H. Abe, N. Ishii, Y. Kaneko, K. Maeda, S. Miura, and M. Yoneda, eds.). Tokai University Publication, Tokyo, 195 pp. (In Japanese).

Kaneko, Y. 1996a. Age variation of the third upper molar in *Eothenomys smithii*. Mammal Study, 21(1):1-13.

Kaneko, Y. 1996b. Morphological variation, and latitudinal and altitudinal distribution of *Eothenomys chinensis, E. wardi, E. custos, E. proditor,* and *E. olitor* (Rodentia, Arvicolidae) in China. Mammal Study, 21(2):89-114.

Kaneko, Y. 2001. Morphological discrimination of the Ryuky spiny rat (genus *Tokudaia*) between the islands of Okinawa and Amami Oshima, in the Ryukyu Islands, southern Japan. Mammal Study, 26:17-33.

Kaneko, Y. 2002. Morphological variation and geographical and altitudinal distribution in *Eothenomys melanogaster* and *E. mucronatus* (Rodentia, Arvicollinae) in China, Taiwan, Burma, India, Thailand, and Vietnam. Mammal Study, 27:31-63.

Kaneko, Y., and Y. Hasegawa. 1995. Some fossil arvicolid rodents from the Pinza-Abut Cave, Miyako Island, the Ryukyu Islands, Japan. Bulletin of the Biogeographical Society of Japan, 50(1):23-37.

Kaneko, Y., and K. Maeda. 2002. A list of scientific names and the types of mammals published by Japanese researchers. Mammalian Science, 42(1):1-21.

Kaneko, Y., and O. Murakami. 1996. [The history of taxonomy in Japanese small rodents.] Mammalian Science, 36(1):109-128 (in Japanese with English abstract).

Kaneko, Y., and M. Sato. 1993. [Identification and distribution of red-backed voles from Is. Rishiri, Hokkaido (preliminary study).] Rishiri Town Museum Annual Report, 12:37-47 (in Japanese).

Kaneko, Y., and C. Smeenk. 1996. The author and date of publication of the Sikkim vole *Microtus sikimensis*. Mammal Study, 21(2):161-164.

Kaneko, Y., T. Nakashima, and Y. Kimura. 1992. [Identification and vertical distribution of two species of *Eothenomys* on Ryo-Hakusan Mountains, Central Honshyu, Japan.] Bulletin of the Gifu Prefecture Museum, 11:23-34 (in Japanese with English abstract).

Kaneko, Y., K. Nakata, T. Saitoh, N. C. Stenseth, and O. N. Bjornstad. 1998. The biology of the vole *Clethrionomys rufocanus*: A review. Research in Population Ecology, 40:21-37.

Kang, Y. S., and H. S. Koh. 1976. Karyotype studies on three species of the family Muridae. Korean Journal of Zoology, 19:101-112.

Kapischke, H.-J. 1992. Weiteres zur variabilität der molarenmuster bei erdmäusen (*Microtus agrestis*) (Mammalia, Rodentia: Arvicolidae). Zoologische Abhandlungen Staatliches Museum für Tierkunde Dresden, 47(7):87-94.

Kapischke, H.-J. 1997. Zur variabilität der molarenmuster von feldmäusen (*Microtus arvalis*) aus dem Kreis Meißen (Sachsen) (Mammalia: Rodentia: Muridae). Zoologische Abhandlungen Staatliches Museum für Tierkunde Dresden, 49(18):311-314.

Kapitonov, V. I. 1966. Rasprostranenie surkov v Tsentral'nom Kazakhstane i perspektivy ikh promysla [Distribution of marmots in Central Kazakhstan and perspectives on their utilization]. Trudy Zoologicheskovo Instituta, Akademiya Nauk, Kazakh SSR, 26:94-134 (in Russian).

Kapitonov, V. I. 1978. Chernoshaposhnyi surok [black-capped marmot]. Pp. 178-209, *in* Surki: Rasprostranenie i ekologiya [Marmots: Distribution and ecology]. Nauka, Moscow-Leningrad, 222 pp. (in Russian).

Karamysheva, T. V., O. V. Andreenkova, M. N. Bochkaerev, Y. M. Borissov, N. Bogdanchikova, P. M. Borodin, and N. B. Rubtsov. 2002. *B* chromosomes of Korean field mouse *Apodemus peninsulae* (Rodentia, Murinae) analysed by microdissection and FISH. Cytogenetic and Genome Research, 96:154-160.

Karaseva, E. V., G. N. Tikhonova, and P. L. Bogomolov. 1992. [Distribution of the field mouse (*Apodemus agrarius*) and peculiarities of its ecology in differennt parts of its range.] Zoologischeskii Zhurnal, 71:106-115 (in Russian).

Karn, R. C., and S. R. Dlouhy. 1991. Salivary androgen-binding protein variation in *Mus* and other rodents. Journal of Heredity, 82:453-458.

Karn, R. C., A. Orth, F. Bonhomme, and P. Boursot. 2002. The complex history of a gene proposed to participate in a sexual isolation mechanism in house mice. Molecular Biology and Evolution, 19(4):462-471.

Kartavtseva, I. V., M. V. Pavlenko, V. A. Kostenko, and F. B. Chernyavskii. 1998. Chromosomal variation and abnormal karyotypes in the red-backed mouse *Clethrionomys rufocanus* (Rodentia, Microtinae). Russian Journal of Genetics, 34(8):928-935 (translated from Genetika, 34:1106-1113).

Kartavtseva, I. V., G. V. Roslik, M. V. Pavlenko, E. Y. Amachaeva, S. Sawaguchi, and Y. Obara. 2000. The B-chromosome system of the Korean field mouse *Apodemus peninsulae* in the Russian Far East. Chromosome Science, 4:21-29.

Kasangaki, A., R. Kityo, and J. Kerbis. 2003. Diversity of rodents and shrews along an elevational gradient in Bwindi Impenetrable National Park, south-western Uganda. African Journal of Ecology, 41:115-123.

Kashiwabara, S., and K. Onoyama. 1988. Karyotypes and G-banding patterns of the red-backed voles, *Clethrionomys montanus* and *C. rufocanus bedfordiae* (Rodentia, Microtinae). Journal of the Mammalogical Society of Japan, 13:33-41.

Kasper, S., M. P. Husby, and J. S. Hausbeck. 1993. Discovery of *Zapus hudsonius* in northeastern Oklahoma, with comments on its ecological status. Texas Journal of Science, 45(3):274-276.

Kass, D. H., J. Kim, A. Rao, and P. L. Deininger. 1997. Evolution of B2 repeats: The muroid explosion. Genetica, 99:1-13.

Kastschenko, N. F. 1902 [1901]. [About the sandy badger (*Meles arenarius* Satunin) and about the Siberian races of badger.] Ezhegodnik zoologicheskogo muzeya Imperatorskoi Akademii Nauk, 6:609-613 (in Russian).

Kasuya, T. 1973. Systematic consideration of recent toothed whales based on the morphology of the tympano-periotic bone. Scientific Reports of the Whales Research Institute (Tokyo), 25:1-103.

Kasuya, T. 1975. Past occurrence of *Globicephala melaena* in the western North Pacific. Scientific Reports of the Whales Research Institute (Tokyo), 27:95-110.

Kawada, S., M. Harada, Y. Obara, S. Kobayashi, K. Koyasu, and S. Oda. 2001. Karyosystematic analysis of Japanese talpine moles in the genera *Euroscaptor* and *Mogera* (Insectivora, Talpidae). Zoological Science, 18:1003-1010.

Kawada, S., M. Harada, A. S. Grafodatsky, and S. Oda. 2002*a*. Cytogenetic study of the Siberian mole, *Talpa altaica* (Insectivora: Talpidae) and karyological relationships within the genus *Talpa*. Mammalia, 66:53-62.

Kawada, S., M. Harada, K. Koyasu, and S. Oda. 2002*b*. Karyological note on the short-faced mole, *Scaptochirus moschatus* (Insectivora, Talpidae). Mammal Study, 27:91-94.

Kawada, S., K. Koyasu, E. I. Zholnerovskaya, and S. Oda. 2002*c*. A discussion of the dental formula of *Desmana moschata* in relation to the premaxillary suture. Mammal Study, 27:107-111.

Kawada, S., A. Shinohara, M. Yasuda, S. Oda, and Lim Boo Liat. 2003. The mole of peninsular Malaysia: Notes on its identification and ecology. Mammal Study, 28:73-77.

Kawai, K., M. Nikaido, M. Harada, S. Matsumura, L.-K. Lin, Y. Wu, M. Hasegawa, and N. Okada. 2002. Intra- and interfamilial relationships of Vespertilionidae inferred by various molecular markers including SINE insertion data. Journal of Molecular Evolution, 55:284-301.

Kawai, K., M. Nikaido, M. Harada, S. Matsumura, L.-K. Lin, Y. Wu, M. Hasegawa, and N. Okada. 2003. The status of the Japanese and East Asian bats of the genus *Myotis* (Vespertilionidae) based on mitochondrial sequences. Molecular Phylogenetics and Evolution, 28:297-307.

Kawamichi, T. 1971. Daily activities and social pattern of two Himalayan pikas, *Ochotona macrotis* and *O. roylei*, observed at Mt. Everest. Journal of the Faculty of Hokkaido University, Japan, ser. 6 (Zoology), 17:587-609.

Kawamura, Y. 1988. Quaternary rodent faunas in the Japanese Islands (Part 1). Memoirs of the Faculty of Science, Kyoto University, Series of Geology and Mineralogy, 53:31-348.

Kawamura, Y. 1989. Quaternary rodent faunas in the Japanese Islands (Part 2). Memoirs of the Faculty of Science, Kyoto University, Series of Geology and Mineralogy, 54(1-2):1-235.

Kawamura, Y. 1991. Quaternary mammalian faunas in the Japanese Islands. The Quaternary Research, 30(2):213-220.

Kawamura, Y. 1994. Late Pleistocene to Holocene mammalian faunal succession in the Japanese Islands, with comments on the Late Quaternary extinctions. Archaeozoologia, 6(2):7-22.

Kawashima, T., N. Miyashita, K. Tsuchiya, H. Li, F. Wang, C. H. Wang, X.-L. Wu, C. Wang, M.-L. Jin, X.-Q. He, A. P. Kryukov, L. V. Yakimenko, L. V. Frisman, and K. Moriwaki. 1995. Geographical distribution of the *Hbb* haplotypes in the *Mus musculus* subspecies in Eastern Asia. Japanese Journal of Genetics, 70:17-23.

Kazmierczak, B., and I. Kaliczewski. 1989. Note on the mammal fauna of the Mazovian Lowland: New station of fat dormouse *Glis glis* (Linnaeus, 1766). Przeglad Zoologiczny, 33(4):623-624.

Kearney, T., and P. J. Taylor. 1997. New distribution records of bats in KwaZulu-Natal. Durban Museum Novitates, 22:53-56.

Kearney, T., M. Volleth, G. Contrafatto, and P. J. Taylor. 2002. Systematic implications of chromosome GTC-band and bacula morphology for Southern African *Eptesicus* and *Pipistrellus* and several other species of Vespertilioninae (Chiroptera: Vespertilionidae). Acta Chiropterologica, 4:55-76.

Kefelioğlu, H. 1995. [The taxonomy of the genus of *Microtus* (Mammalia: Rodentia) and its distribution in Turkey.] Turkish Journal of Zoology, 19:35-63 (in Turkish with English abstract).

Kefelioğlu, H. 1998. The karyotype of *Hemiechinus auritus calligoni* from Turkey. Zeitschrift für Säugetierkunde, 62:312-314.

Kefelioğlu, H., and S. Gencoglu. 1996. The taxonomy and distribution of *Talpa* (Mammalia: Insectivora) in the Black Sea region. Turkish Journal of Zoology, 20:57-66.

Kefelioğlu, H., and B. Kryštufek. 1999. The taxonomy of *Microtus socialis* group (Rodentia:Microtinae) in Turkey, with the description of a new species. Journal of Natural History, 33:289-303.

Keith, L. B., and L. A. Windberg. 1978. A demographic analysis of the snowshoe hare cycle. Wildlife Monographs, 58:1-70.

Kellogg, R. 1918. A revision of the *Microtus californicus* group of meadow mice. University of California Publications in Zoology, 21:1-42.

Kellogg, R. 1945. Two rats from Morotai Island. Proceedings of the Biological Society of Washington, 58:65-68.

Kelly, T. S. 1992. New Uintan and Duchesnean (middle and late Eocene) rodents from the Sespe Formation, Simi Valley, California. Bulletin of the Southern California Academy of Sciences, 91(3):97-120.

Kelson, K. R. 1952. Comments on the taxonomy and geographic distribution of some North American woodrats (genus *Neotoma*). University of Kansas Publications, Museum of Natural History, 5:233-242.

Kelt, D. A. 1988a. *Dipodomys heermanni*. Mammalian Species, 323:1-7.

Kelt, D. A. 1988b. *Dipodomys californicus*. Mammalian Species, 324:1-4.

Kelt, D. A., R. E. Palma, M. H. Gallardo, and J. A. Cook. 1991. Chromosomal multiformity in *Eligmodontia* (Muridae, Sigmodontinae), and verification of the status of *E. morgani*. Zeitschrift für Säugetierkunde, 56:352-358.

Kemper, C. M. 1995a. Brush-tailed tree-rat, *Conilurus penicillatus*. Pp. 553-554, *in* Mammals of Australia (R. Strahan, ed.). Smithsonian Institution Press, Washington, D.C., 756 pp.

Kemper, C. M. 1995b. New Holland mouse, *Pseudomys novaehollandiase*. Pp. 611-612, *in* Mammals of Australia (R. Strahan, ed.). Smithsonian Institution Press, Washington, D.C., 756 pp.

Kemper, C. M., and L. H. Schmitt. 1992. Morphological variation between populations of the brush-tailed tree rat (*Conilurus penicillatus*) in northern Australia and New Guinea. Australian Journal of Zoology, 40:437-452.

Kennedy, M., A. M. Paterson, J. C. Morales, S. Parsons, A. M. Winnington, and H. G. Spencer. 1999. The long and the short of it: Branch lengths and the problem of placing the New Zealand short-tailed bat, *Mystacina*. Molecular Phylogenetics and Evolution, 13:405-416.

Kennedy, M. L., G. A. Heidt, and P. K. Kennedy. 1982. First record of the meadow jumping mouse (*Zapus hudsonius*) in Mississippi. Journal Mississippi Academy of Science, 27:149-150.

Kenyon, K. W. 1969. The sea otter in the eastern Pacific ocean. North American Fauna, 68:1-352.

Kenyon, K. W. 1977. Caribbean monk seal extinct. Journal of Mammalogy, 58:97-98.

Keogh, H. J. 1985. A photographic reference system based on the cuticular scale patterns and grooves of the hair of 44 species of southern African Cricetidae and Muridae. South African Journal of Wildlife Research, 15:109-159.

Keogh, H. J., and P. J. Price. 1981. The multimammate mice: A review. South African Journal of Science, 77:484-488.

Kerbis Peterhans, J. C., and B. D. Patterson. 1995. The Ethiopian water mouse *Nilopegamys* Osgood, with comments on semi-aquatic adaptations in African Muridae. Zoological Journal of the Linnean Society, 113:329-349.

Kerbis Peterhans, J. C., R. M. Kityo, W. T. Stanley, and P. K. Austin. 1998. Small mammals along an elevational gradient in Rwenzori Mountains National Park, Uganda. Pp. 149-171, *in* The Rwenzori Mountains National Park, Uganda. Exploration, environment and biology. Conservation, management and community relations (H. Osmaston, J. Tukahirwa, C. Basalirwa, and J. Nyakaana, eds.). Department of Geography, Makerere University, Kampala, Uganda, 395 pp.

Kerle, J. A. 1995a. Desert mouse, *Pseudomys desertor*. Pp. 594-595, *in* Mammals of Australia (R. Strahan, ed.). Smithsonian Institution Press, Washington, D.C., 756 pp.

Kerle, J. A. 1995b. Central pebble-mound mouse, *Pseudomys johnsoni*. Pp. 607-608, *in* Mammals of Australia (R. Strahan, ed.). Smithsonian Institution Press, Washington, D.C., 756 pp.

Kerle, J. A. 1995c. Grassland melomys, *Melomys burtoni*. Pp. 632-634, *in* Mammals of Australia (R. Strahan, ed.). Smithsonian Institution Press, Washington, D.C., 756 pp.

Kerr, R. 1792. The animal kingdom, or zoological system, of the celebrated Sir Charles Linnaeus; class I: Mammalia. London, J. Murray & R. Faulder, 664 pp.

Kerridge, D. C., and R. J. Baker. 1978. *Natalus micropus*. Mammalian Species, 114:1-3.

Kersten, A. M. P. 1992. Rodents and insectivores from the Palaeolithic rock shelter of Ksar'Akil (Lebanon) and their palaeoecological implications. Paléorient, 18:27-45.

Kesner, M. H. 1980. Functional morphology of the masticatory musculature of the rodent subfamily Microtinae. Journal of Morphology, 165:205-222.

Kesner, M. H. 1986. The myology of the manus of microtine rodents. Journal of the Zoological Society of London, A, 210:1-22.

Key, G. E., and E. M. Heredia. 1994. Distribution and current status of rodents in the Galápagos. Noticias de Galápagos, 53:21-25.

Key, G. E., and R. D. Woods. 1996. Spool-and-line studies on the behavioural ecology of rats (*Rattus* spp.) in the Gálapagos Islands. Canadian Journal of Zoology, 74:733-737.

Key, G. E., A. H. Fielding, M. J. Goulding, R. S. Holm, and B. Stevens-Woods. 1998. Ship rats *Rattus rattus* on the Shiant Islands, Hebrides, Scotland. Journal of the Zoological Society of London, 245:228-233.

Keyserling A., and J. H. Blasius. 1840. Die Wirbelthiere Europa's. F. Vieweg & Son, Braunschweig, xcviii + 248 pp.

Khajuria, H. 1981. A new bandicoot rat, *Erythronesokia bunnii* gen. et sp. nov. (Rodentia: Muridae), from Iraq. Bulletin of the Natural History Research Centre, 7:157-164.

Khajuria, H. 1982. External genitalia and bacula of some central Indian Microchiroptera. Säugetierkundliche Mitteilungen, 30:287-295.

Khajuria, H., Y. Chaturvedi, and D. K. Ghoshal. 1977. Catalogue Mammaliana. An annotated catalogue of the type specimens of mammals in the collections of the Zoological Survey of India. Records of the Zoological Survey of India, Miscellaneous Publication, Occasional Paper, 7:1-45.

Khakhin, G. V., and A. A. Ivanov. 1990. Vychuchol [The Russian Desman]. Agropromizdat Publ., Moscow, 190 pp., 16 pls. (in Russian).

Khidas, K. 1993. Distribution des rongeurs en Kabylie du Djurdjura (Algerie). Mammalia, 57(2):207-212.

Khidas, K., N. Khammes, S. Khelloufi, S. Lek, and S. Aulagnier. 2002. Abundance of the wood mouse *Apodemus sylvaticus* and the Algerian mouse *Mus spretus* (Rodentia, Muridae) in different habitats of northern Algeria. Mammalian Biology, 67:34-41.

Kiblisky, P., N. Brum-Zorrilla, G. Perez, and F. A. Saez. 1977. Variabilidad cromosómica entre diversas poblaciones uruguayas del roedor cavador del género *Ctenomys* (Rodentia, Octodontridae). Mendeliana, 2:85-93.

Kiefer, A., and M. Veith. 2001. A new species of long-eared bat from Europe (Chiroptera: Vespertilionidae). Myotis, 39:5-16.

Kiefer, A., F. Mayer, J. Kosuch, O. von Helverson, and M. Veith. 2002. Conflicting molecular phylogenies can be explained by cryptic diversity. Molecular Phylogenetics and Evolution, 25:557-566.

Kiersten, G., M. Vallinoto, A. Silva, M. P. Schneider, L. Iannuzzi, and B. Brenig. 2003. Analysis of mitochondrial D-loop region casts new light on domestic water buffalo (*Bubalus bubalis*) phylogeny. Molecular Phylogenetics and Evolution, 30:308-324.

Kiliaridis, S., A. Bresin, J. Holm, and K.-G. Strid. 1996. Effects of masticatory muscle function on bone mass in the mandible of the growing rat. Acta Anatomica, 155:200-205.

Kilpatrick, C. W. 1984. Molecular evolution of the Texas mouse, *Peromyscus attwateri*. Pp. 87-96, *in* Festschrift for Walter W. Dalquest in honor of his sixty-sixth birthday (N. Horner, ed.). Midwestern State University, Wichita Falls, TX, 163 pp.

Kilpatrick, C. W., and K. L. Crowell. 1985. Genic variation of the rock vole, *Microtus chrotorrhinus*. Journal of Mammalogy, 66:94-101.

Kilpatrick, C. W., and E. G. Zimmerman. 1975. Genetic variation and systematics of four species of mice of the *Peromyscus boylii* species group. Systematic Zoology, 24:143-162.

Kilpatrick, C. W., and E. G. Zimmerman. 1976. Biochemical variation and systematics of *Peromyscus pectoralis*. Journal of Mammalogy, 57:506-522.

Kilpatrick, C. W., S. M. Rich, and K. L. Crowell. 1994. Distribution of the genus *Peromyscus* in coastal and inland southwestern Maine. Maine Naturalist, 2:1-10.

Kim, I., C. J. Phillips, J. A. Monjeau, E. C. Birney, K. Noack, D. E. Pumo, R. S. Sikes, and J. A. Dole. 1998. Habitat islands, genetic diversity, and gene flow in a Patagonian rodent. Molecular Ecology, 7:667-678.

Kim, I., B.-Y. Min, and M.-H. Yoon. 2000. Micro-geographic genetic subdivision of the striped field mouse (*Apodemus agrarius*) detected by mitochondrial control region sequences. Korean Journal of Genetics, 22(2):101-116.

Kim Sang Wook, and Kim Woo Ki. 1974. Avi-mammalian fauna of Korea. Wildlife population census in Korea, Forest Research Institute, Office of Forestry, 5:1-118.

Kimiyuki, T., S. Wakana, H. Suzuki, S. Hattori, and Y. Hayashi. 1989. Taxonomic study of *Tokudaia* (Rodentia: Muridae): I. Genetic differentiation. Memoirs of the National Science Museum, 22:227-234.

Kimura, S., B. A. Schaumann, and K. Shiota. 1996. Fetal and postnatal development of palmar, plantar, and digital pads, and flexion creases of the rat (*Rattus norvegicus*). Journal of Morphology, 228:179-187.

Kimura, Y., Y. Kaneko, and T. Yoshida. 1994. [Small mammalian fauna in Adatara mountain regions with special reference to genus *Eothenomys*.] Fukushima Seibutsu, 37:13-19 (in Japanese).

Kimura, Y., Y. Kaneko, and M. A. Iwasa. 1999. [Identification and vertical distribution of two species of *Eothenomys* in the Oze district, northeastern Honshu, Japan.] Mammalian Science, 39:257-268 (in Japanese with English abstract).

King, C. J. (ed.). 1990. The handbook of New Zealand mammals. Oxford University Press, Melbourne, 600 pp.

King, C. M. 1983. *Mustela erminea*. Mammalian Species, 195:1-8.

King, J. A., (ed.). 1968. Biology of *Peromyscus* (Rodentia). Special Publication, American Society of Mammalogists, 2:1-593.

King, J. E. 1954. The otariid seals of the Pacific coast of America. Bulletin of the British Museum (Natural History), Zoology Series, 2(10):311-337.

King, J. E. 1956. The monk seals genus *Monachus*. Bulletin of the British Museum (Natural History), Zoology Series, 3(5):203-256.

King, J. E. 1959a. The northern and southern populations of *Arctocephalus gazella*. Mammalia, 23:19-40.

King, J. E. 1959b. A note on the specific name of the Kerguelen fur seal. Mammalia, 23:381.

King, J. E. 1960. Sea-lions of the genera *Neophoca* and *Phocarctos*. Mammalia, 24:445-456.

King, J. E. 1966. Relationships of the hooded and elephant seals (genera *Cystophora* and *Mirounga*). Journal of Zoology, London, 148:385-398.

King, J. E. 1978. On the specific name of the southern sea lion (Pinnipedia, Otariidae). Journal of Mammalogy, 59(4):861-863.

King, J. E. 1983. Seals of the World. Second ed. Cornell University Press, Ithaca, NY, 240 pp.

Kingdon, J. 1971. East African mammals: An atlas of evolution in Africa. Academic Press, London, 1:1-446.

Kingdon, J. 1974a. East African mammals: An atlas of evolution in Africa. (Insectivores and Bats). Academic Press, London, 2A:1-341 + l.

Kingdon, J. 1974b. East African mammals: An atlas of evolution in Africa. (Hares and Rodents). Academic Press, London, 2B:343-704 + lvii.

Kingdon, J. 1977. East African mammals: An atlas of evolution in Africa. (Carnivores). Academic Press, New York, 3A:1-476.

Kingdon, J. 1979. East African mammals: An atlas of evolution in Africa. (Large Mammals). Academic Press, London, 3B:1-436.

Kingdon, J. 1980. The role of visual signals and face patterns in African forest monkeys (guenons) of the genus *Cercopithecus*. Transactions of the Zoological Society of London, 35:431-475.

Kingdon, J. 1982. East African mammals: An atlas of evolution in Africa. (Bovids). Academic Press, London, 3C:1-393.

Kingdon, J. 1990. Arabian mammals. A natural history. Al Areen Wildlife Park and Reserve, Bahrain, 279 pp.

Kingdon, J. 1997. The Kingdon field guide to African mammals. Nature World Academic Press, Harcourt Brace & Company, San Diego, 465 pp.

Kingston, T., T. H. Kunz, R. Hodgkison, and Z. Akbar. 1997. *Phoniscus jagorii*, a vespertilionid bat newly recorded from peninsular Malaysia. Malayan Nature Journal, 50:363.

Kingston, T., M. C. Lara, G. Jones, Z. Akbar, T. H. Kunz, and C. J. Schneider. 2001. Acoustic divergence in two cryptic *Hipposideros* species: A role for social selection? Procedings of the Royal Society of London, 268:1381-1386.

Kingswood, S. C., and D. A. Blank. 1996. *Gazella subgutturosa*. Mammalian Species, 518:1-10.

Kingswood, S. C., and A. T. Kumamoto. 1996. *Madoqua guentheri*. Mammalian Species, 539:1-10.

Kingswood, S. C., and A. T. Kumamoto. 1997. *Madoqua kirkii*. Mammalian Species, 569:1-10.

Kingswood, S. C., A. T. Kumamoto, S. J. Charter, and M. L. Jones. 1998a. Cryptic chromosomal variation in suni *Neotragus moschatus* (Artiodactyla, Bovidae). Animal Conservation, 1:95-100.

Kingswood, S. C., A. T. Kumamoto, S. J. Charter, R. A. Aman, and O. A. Ryder. 1998b. Centric fusion polymorphisms in waterbuck. Journal of Heredity, 89:96-110.

Kinlaw, A. 1995. *Spilogale putorius*. Mammalian Species, 511:1-7.

Kinze, C. C. 2000. Rehabilitation of *Platanista gangetica* (Lebeck), scientific name of the Ganges dolphin. Zoologische Mededeelingen, Leiden, 74(11) 15.9.200:193-203.

Kipp, H. 1965. Beitrag zur Kenntnis der Gattung *Conepatus* Molina, 1782. Zeitschrift für Säugetierkunde, 30(4):193-232.

Kirkland, G. L., Jr. 1977. A re-examination of the subspecific status of the Maryland shrew, *Sorex cinereus fontinalis* Hollister. Proceedings of the Pennsylvania Academy of Sciences, 51:43-46.

Kirkland, G. L., Jr. 1981. *Sorex dispar* and *Sorex gaspensis*. Mammalian Species, 155:1-4.

Kirkland, G. L., Jr., and F. J. Jannett, Jr. 1982. *Microtus chrotorrhinus*. Mammalian Species, 180:1-5.

Kirkland, G. L., Jr., and J. N. Layne (eds.). 1989. Advances in the study of *Peromyscus* (Rodentia). Texas Tech University Press, Lubbock, 367 pp.

Kirkland, G. L., Jr., and H. M. Van Deusen. 1979. The shrews of the *Sorex dispar* group: *Sorex dispar* Batchelder and *Sorex gaspensis* Anthony and Goodwin. American Museum Novitates, 2675:1-21.

Kirkpatrick, T. H. 1995. Hastings River mouse, *Pseudomys oralis*. P. 615, *in* Mammals of Australia (R. Strahan, ed.). Smithsonian Institution Press, Washington, D.C., 756 pp.

Kirsch, J. A. W. 1977. The comparative serology of Marsupialia, and a classification of marsupials. Australian Journal of Zoology, Supplementary Series, 52:1-152.

Kirsch, J. A. W., and J. H. Calaby. 1977. The species of living marsupials: An annotated list. Pp. 9-26, *in* The biology of marsupials (B. Stonehouse and D. Gilmore, eds.). University Park Press, Baltimore, 486 pp.

Kirsch, J. A. W., and W. E. Poole. 1972. Taxonomy and distribution of the grey kangaroos, *Macropus giganteus* Shaw and *Macropus fuliginosus* (Desmarest), and their subspecies (Marsupialia: Macropodidae). Australian Journal of Zoology, 20:315-339.

Kirsch, J. A., T. F. Flannery, M. S. Springer, and F. Lapointe. 1995. Phylogeny of the Pteropodidae (Mammalia: Chiroptera) based on DNA hybridization, with evidence for bat monophyly. Australian Journal of Zoology, 43:395-428.

Kirsch, J. A. W., F-J. Lapointe, and M. S. Springer. 1997. DNA-hybridisation studies of marsupials and their implications for Metatherian classification. Australian Journal of Zoology, 45:211-280.

Kirsch, J. A. W., J. M. Hutcheon, D. G. P. Byrnes, and B. D. Lloyd. 1998. Affinities and historical zoogeography of the New Zealand short-tailed bat, *Mystacina tuberculata* Gray 1843, inferred from DNA-hybridization comparison. Journal of Mammalian Evolution, 5:33-64.

Kitahara, E. 1995. Taxonomic status of Anderson's red-backed vole on the Kii Peninsula, Japan, based on skull and dental characters. Journal of the Mammalogical Society of Japan, 20(1):9-28.

Kitahara, E., and Y. Kimura. 1995. Taxonomic reexamination among three local populations of Anderson's red-backed voles from crossbreeding experiments. Journal of the Mammalogical Society of Japan, 20:43-49.

Kitchener, A. C. 1991. The natural history of the wild cats. Cornell University Press. New York, 280 pp.

Kitchener, A. C. 1992. A western mouse (*Pseudomys occidentalis*) from King George Sound, Western Australia. Australian Mammalogy, 15:153-154.

Kitchener, A. C., and C. Groves. 2002. New insights into the taxonomy of *Macaca pagensis* of the Mentawai Islands, Sumatra. Mammalia, 66:533-542.

Kitchener, A. C., T. Clegg, N. M. J. Thompson, H. Wiik, and A. A. Macdonald. 1993. First records of the Malay Civet, *Viverra tangalunga* Gray, 1832, on Seram with notes on the Seram bandicoot *Rhynchomeles prattorum* Thomas, 1920. Zeitschrift für Säugetierkunde, 58:378-380.

Kitchener, D. J. 1980. A new species of *Pseudomys* (Rodentia: Muridae) from Western Australia. Records of the Western Australian Museum, 8:405-414.

Kitchener, D. J. 1985. Description of a new species of *Pseudomys* (Rodentia: Muridae) from Northern Territory. Records of the Western Australian Museum, 12:207-221.

Kitchener, D. J. 1988. A new species of false *Antechinus* (Marsupialia: Dasyuridae) from the Kimberley, Western Australia. Records of the Western Australian Museum, 14:61-71.

Kitchener, D. J. 1989. Taxonomic appraisal of *Zyzomys* (Rodentia, Muridae) with descriptions of two new species from the Northern Territory, Australia. Records of the Western Australian Museum, 14:331-373.

Kitchener, D. J. 1991. *Pseudantechinus mimulus* (Thomas, 1906) (Marsupialia: Dasyuridae): Rediscovery and redescription. Records of the Western Australian Museum, 15:191-202.

Kitchener, D. J. 1995. Kimberley mouse, *Pseudomys laborifex*. Pp. 608-609, *in* Mammals of Australia (R. Strahan, ed.). Smithsonian Institution Press, Washington, D.C., 756 pp.

Kitchener, D. J., and N. Caputi. 1985. Systematic revision of Australian *Scoteanax* and *Scotorepens* (Chiroptera: Vespertilionidae), with remarks on relationships to other Nycticeini. Records of the Western Australian Museum, 12:85-146.

Kitchener, D. J., and N. Caputi. 1988. A new species of false *Antechinus* (Marsupialia, Dasyuridae) from Western Australia, with remarks on the generic classification within the Parantechini. Records of the Western Australian Museum, 14:35-59.

Kitchener, D. J., and W. F. Humphreys. 1986. Description of a new species of *Pseudomys* (Rodentia: Muridae) from the Kimberley Region, Western Australia. Records of the Western Australian Museum, 12:419-434.

Kitchener, D. J., and W. F. Humphreys. 1987. Description of a new subspecies of *Pseudomys* (Rodentia: Muridae) from Northern Territory. Records of the Western Australian Museum, 13:285-295.

Kitchener, D. J., and I. Maryanto. 1993*a*. Taxonomic reappraisal of the *Hipposideros larvatus* species complex (Chiroptera: Hipposideridae) in the Greater and Lesser Suns Islands, Indonesia. Records of the Western Australian Museum, 16:119-173.

Kitchener, D. J., and I. Maryanto. 1993*b*. A new species of *Melomys* (Rodentia: Muridae) from Kai Besar Island, Maluku Tengah, Indonesia. Records of the Western Australian Museum, 16(3):427-436.

Kitchener, D. J., and I. Maryanto. 1995*a*. A new species of *Melomys* (Rodentia, Muridae) from Yamdena Island, Tanimbar group, eastern Indonesia. Records of the Australian Museum, 17:43-50.

Kitchener, D. J., and I. Maryanto. 1995*b*. Small *Pteropus* (Chiroptera: Pteropodidae) from Timor and surrounding islands, Indonesia. Records of the Western Australian Museum, 17:147-152.

Kitchener, D. J., and Maharadatunkamsi. 1991. Description of a new species of *Cynopterus* (Chiroptera: Pteropodidae) from Nusa Tenggara, Indonesia. Records of the Western Australian Museum, 15:307-363.

Kitchener, D. J., and Maharadatunkamsi. 1995. The *Hipposideros bicolor* group (Chiroptera: Hipposideridae) from Sumbawa Island, Nusa Tenggara, Indonesia. Records of the Western Australian Museum, 17:309-314.

Kitchener, D. J., and Maharadatunkamsi. 1996. Geographic variation in morphology of *Cynopterus nusatenggara* (Chiroptera, Pteropodidae) in southeastern Indonesia, and description of two new subspecies. Mammalia, 60:255-276.

Kitchener, D. J., and G. Sanson. 1978. *Petrogale burbidgei* (Marsupialia, Macropodidae), a new rock wallaby from Kimberley, Western Australia. Records of the Western Australian Museum, 6:269-285.

Kitchener, D. J., and A. Suyanto. 1995. Morphological variation in bearded tomb bats (*Taphozous*) in Maluku Tenggara and Nusa Tenggara Timur, Indonesia. Records of the Western Australian Museum, 17:213-220.

Kitchener, D. J., and A. Suyanto. 1996. A new species of *Melomys* (Rodentia: Muridae) from Riama Island, Tanimbar Group, Maluku Tenggara, Indonesia. Records of the Western Australian Museum, 18:113-119.

Kitchener, D. J., and A. Suyanto. 2002. Morphological variation in *Miniopterus pusillus* and *M. australis* (sensu Hill, 1992) in southeastern Asia, New Guinea, and Australia. Records of the Western Australian Museum, 21:9-33.

Kitchener, D. J., and M. Yani. 1998. Small non-volant terrestrial mammal diversity along an altitudinal gradient at Gunung Ranaka, Flores Island, Indonesia. Tropical Biodiversity, 5(2):155-159.

Kitchener, D. J., M. Adams, and P. Baverstock. 1984a. Redescription of *Pseudomys bolami* Troughton, 1932. Australian Mammalogy, 7:149-159.

Kitchener, D. J., J. Stoddart, and J. Henry. 1984b. A taxonomic revision of the *Sminthopsis murina* complex (Marsupialia, Dasyuridae) in Australia, including descriptions of four new species. Records of the Western Australian Museum, 11:201-247.

Kitchener, D. J., N. Caputi, and B. Jones. 1986. Revision of Australo-Papuan *Pipistrellus* and *Falsistrellus* (Microchiroptera: Vespertilionidae). Records of the Western Australian Museum, 12:435-495.

Kitchener, D. J., B. Jones, and N. Caputi. 1987. Revision of Australian *Eptesicus* (Microchiroptera: Vespertilionidae). Records of the Western Australian Museum, 13:427-500.

Kitchener, D. J., Boeadi, L. Charlton, and Maharadatunkamsi. 1990. Wild mammals of Lombok Island. Records of the Western Australian Museum, Supplement No. 33:1-129.

Kitchener, D. J., R. A. How, and Maharadatunkamsi. 1991a. *Paulamys* sp. cf. *P. naso* (Musser, 1981) (Rodentia: Muridae) from Flores Island, Nusa Tenggara, Indonesia—description from a modern specimen and a consideration of its phylogenetic affinities. Records of the Western Australian Museum, 15:171-189.

Kitchener, D. J., K. P. Aplin, and Boeadi. 1991b. A new species of *Rattus* from Gunung Mutis, South West Timor Island, Indonesia. Records of the Western Australian Museum, 15:445-461.

Kitchener, D. J., R. A. How, and Maharadatunkamsi. 1991c. A new species of *Rattus* from the mountains of West Flores, Indonesia. Records of the Western Australian Museum, 15:611-626.

Kitchener, D. J., R. A. How, and Maharadatunkamsi. 1991d. A new species of *Nyctophilus* (Chiroptera: Vespertilionidae) from Lembata Island, Nusa Tenggara, Indonesia. Records of the Western Australian Museum, 15:97-107.

Kitchener, D. J., R. A. How, and I. Maryanto. 1992a. A new species of *Otomops* (Chiroptera: Molossidae) from Alor I., Nusa Tenggara, Indonesia. Records of the Western Australian Museum, 15:729-738.

Kitchener, D. J., R. A. How, N. K. Cooper, and A. Suyanto. 1992b. *Hipposideros diadema* (Chiroptera: Hipposideridae) in the Lesser Sunda Islands, Indonesia: Taxonomy and geographic morphological variation. Records of the Western Australian Museum, 16:1-60.

Kitchener, D. J., S. Hisheh, L. H. Schmitt, and I. Maryanto. 1993a. Morphological and genetic variation in *Aethalops alecto* (Chiroptera, Pteropodidae) from Java, Bali and Lombok Islands, Indonesia. Mammalia, 57(2):255-272.

Kitchener, D. J., L. H. Schmitt, S. Hisheh, R. A. How, N. K. Cooper, and Maharadatunkamsi. 1993b. Morphological and genetic variation in bearded tomb bats (*Taphozous*: Emballonuridae) of Nusa Tenggara, Indonesia. Mammalia, 57:63-68.

Kitchener, D. J., S. Hisheh, L. H. Schmitt, and A. Suyanto. 1994a. Shrews (Soricidae: *Crocidura*) from the Lesser Sunda Islands and southeast Maluku, eastern Indonesia. Australian Mammalogy, 17:7-17.

Kitchener, D. J., L. H. Schmitt, and Maharadatunkamsi. 1994b. Morphological and genetic variation in *Suncus murinus* (Soricidae: Crocidurinae) from Java, Lesser Sunda islands, Maluku and Sulawesi, Indonesia. Mammalia, 58:433-451.

Kitchener, D. J., M. Adams, and Boeadi. 1994c. Morphological and genetic relationships among populations of *Scotorepens sanborni* (Chiroptera: Vespertilionidae) from Papua New Guinea, Australia and Indonesia. Australian Mammalogy, 17:31-42.

Kitchener, D. J., W. C. Packer, and I. Maryanto. 1994d. Morphological variation in Maluku populations of *Syctonycteris australis* (Peters, 1867)(Chiroptera: Pteropodidae). Records of the Western Australian Museum, 16:485-498.

Kitchener, D. J., L. H. Schmitt, P. Strano, A. Wheeler, and A. Suyanto. 1995a. Taxonomy of *Rhinolophus simplex* Andersen, 1905 (Chiroptera: Rhinolophidae) in Nusa Tenggara and Maluku, Indonesia. Records of the Western Australian Museum, 17:1-28.

Kitchener, D. J., N. Cooper, and I. Maryanto. 1995b. The *Myotis adversus* (Chiroptera: Vespertilionidae) species complex in eastern Indonesia, Australia, Papua New Guinea and the Solomon Islands. Records of the Western Australian Museum, 17:191-212.

Kitchener, D. J., W. C. Packer, and A. Suyanto. 1995c. Systematic review of *Nyctimene cephalotes* and *N. albiventer* (Chiroptera: Pteropodidae) in the Maluku and Sulawesi regions, Indonesia. Records of the Western Australian Museum, 17:125-142.

Kitchener, D. J., W. C. Packer, and Maharadatunkamsi. 1995d. Morphological variation in *Pteropus lombocensis* (Chiroptera: Pteropodidae) in Nusa Tenggara, Indonesia. Records of the Western Australian Museum, 17:61-67.

Kitchener, D. J., Y. Konishi, and A. Suyanto. 1996. Morphological variation among eastern Indonesian island populations of *Hipposideros bicolor* (Chiroptera: Hipposideridae), with descriptions of three new subspecies. Records of the Western Australian Museum, 18:179-192.

Kitchener, D. J., S. Hisheh, L. H. Schmitt, and Maharadatunkamsi. 1997a. Morphological and genetic variation among island populations of *Dobsonia peronii* (Chiroptera: Pteropodidae) from the Lesser Sunda Islands, Indonesia. Tropical Biodiversity, 4:35-51.

Kitchener, D. J., W. C. Packer, and I. Maryanto. 1997b. Morphological variation among populations of *Scophilus kuhlii* (sensu lato) Leach, 1821 (Chiroptera: Vespertilionidae) from the Greater and Lesser Sunda Islands, Indonesia. Tropical Biodiversity, 4:53-81.

Kitchener, D. J., M. Yani, and C. McCulloch. 1998. Aspects of the biology of *Bunomys naso* (Musser, 1986), a rare murid rodent from Flores Island, Indonesia. Tropical Biodiversity, 5(1):75-80.

Kivanç, E. 1983. Die Haselmaus, *Muscardinus avellanarius* L., in der Türkei. Bonner Zoologische Beiträge, 34:419-428.

Kivanç, E. 1986. *Microtus* (*Pitymys*) *majori* Thomas, 1906 in der europäischen Türkei. Bonner Zoologische Beiträge, 37:39-42.

Kivanç, E. 1988. [Geographic variations of Turkish *Spalax* species (Spalacidae, Rodentia, Mammalia).] Biological Faculty, Ankara University, 88 pp. (in Turkish).

Kivanç, E., M. Sözen, E. Çolak and N. Yiğit. 1997a. Karyological and phallic characteristics of Dryomys laniger Felten and Storch, 1968 (Rodentia: Gliridae) in Turkey. Israel Journal of Zoology, 43:401-403.

Kivanç, E., M. Sözen, E. Çolak, and N. Yiğit. 1997b. Karyological and phallic aspects of the spiny mouse, *Acomys cilicicus* Spitzenberger, 1978 (Rodentia: Muridae) in Turkey. Turkish Journal of Zoology, 21:167-169.

Kleiman, D. G. 1981. *Leontopithecus rosalia*. Mammalian Species, 148:1-7.

Klein, R. G. 1974. On the taxonomic status, distribution and ecology of the blue antelope, *Hippotragus leucophaeus* (Pallas, 1766). Annals of the South African Museum, 65:99-143.

Klein, R. G., and K. Cruz Uribe. 1999. Craniometry of the genus *Equus* and the taxonomic affinities of the extinct South African quagga. South African Journal of Science, 95:81-86.

Kleinenberg, S. E., A. V. Yablokov, B. M. Bel'kovich, and M. N. Tarasevich. 1964. Belukha. Opyt monograficheskovo issledovaniya vida [Beluga. Test of monographic investigation of the species]. Akademiya Nauk SSSR, Moscow, 455 pp. (in Russian).

Kleinenberg, S. E., A. V. Yablokov, B. M. Bel'kovich, and M. N. Tarasevich. 1969. Beluga (*Delphinapterus leucas*), investigation of the species. [A translation of S. E. Kleinenberg, et al., 1964, Belukha. Opyt monograficheskovo issledovaniya vida]. Israel Program for Scientific Translations, Jerusalem, 376 pp.

Klimkiewicz, M. K. 1970. The taxonomic status of the nominal species *Microtus pennsylvanicus* and *Microtus agrestis* (Rodentia: Cricetidae). Mammalia, 34:640-655.

Klingener, D. 1963. Dental evolution of *Zapus*. Journal of Mammalogy, 44(2):248-260.

Klingener, D. 1964. The comparative myology of four dipodoid rodents (genera *Zapus*, *Napaeozapus*, *Sicista*, and *Jaculus*). Miscellaneous Publications, Museum of Zoology, University of Michigan, 124:1-100.

Klingener, D. 1965. Notes on the range of *Napaeozapus* in Michigan and Indiana. Journal of Mammalogy, 45(4):644-645.

Klingener, D. 1971. The question of cheek pouches in *Zapus*. Journal of Mammalogy, 52(2):463-464.

Klingener, D. 1984. Gliroid and dipodoid rodents. Pp. 381-388, *in* Orders and families of Recent mammals of the world (S. Anderson and J. K. Jones, Jr., eds.). John Wiley and Sons, New York, 686 pp.

Klingener, D., and G. K. Creighton. 1984. On small bats of the genus *Pteropus* from the Philippines. Proceedings of the Biological Society of Washington, 97:395-403.

Klingener, D., H. H. Genoways, and R. J. Baker. 1978. Bats from Southern Haiti. Annals of Carnegie Museum, 47:81-99.

Kloss, C. B. 1917. On the mongooses of the Malay peninsula. Journal of the Federation of Malay States Museum, 7(3):123-125.

Kloss, C. B. 1921. Some rats and mice of the Malay Archipelago. Treubia, 2:115-124.

Kloss, C. B. 1929. Some remarks on the gibbons, with the description of a new subspecies. Proceedings of the Zoological Society of London, 1929:113-127.

Kminiak, M. 1996. Faunistic notes on small mammals fauna in the region of Oravská Magura Mountains (subregion Kubínska hol'a, Northern Slovakia). Lynx (Praha), n.s., 27:13-24 (in Czech with English abstract).

Knight, M. H., and A. K. Knight-Eloff. 1987. Digestive tract of the African giant rat, *Cricetomys gambianus*. Journal of Zoology, London, 213:7-22.

Kobayashi, S., and A. Langguth. 1999. A new species of titi monkey, *Callicebus* Thomas, from north-eastern Brazil. Revista Brasileira de Zoologia, 16:531-551.

Kobayashi, T. 1985. Taxonomical problems in the genus *Apodemus*. Pp. 80-82, *in* Contemporary mammalogy in China and Japan (T. Kawamichi, ed.). Mammalogical Society of Japan, 194 pp.

Kock, D. 1967. Ein Neunachweis von *Myotis welwitschi* und der status von *Myotis venustus* (Mammalia, Chiroptera). Senckenbergiana Biologica, 48:319-325.

Kock, D. 1969*a*. Die Fledermaus- Fauna des Sudan. (Mammalia, Chiroptera). Abhandlungen der Senckenbergischen Naturforschenden Gesellschaft, 521:1-238.

Kock, D. 1969*b*. *Dyacopterus spadiceus* (Thomas, 1890) auf den Philippinen (Mammalia: Chiroptera). Senckenbergiana Biologica, 50:1-7.

Kock, D. 1975. Ein Originalexemplar von *Nyctinomus ventralis* Heuglin 1861 (Mammalia: Chiroptera: Molossidae). Stuttgarter Beiträge zur Naturkunde, ser. A(Biologie), 272:1-9.

Kock, D. 1978. Vergleichende Untersuchung einiger Säugetiere im südlichen Niger. Senckenbergiana Biologica, 58:113-136.

Kock, D. 1985. Die saharischen Vorkommen von Eliomys Wagner, 1840. Zeitschrift für Säugetierkunde, 50:51-54.

Kock, D. 1990. Notes on mammals (Insectivora, Rodentia) taken by the tawny owl, *Strix aluco*, in NW Turkey. Zoology in the Middle East, 4:5-9.

Kock, D. 1996. Fledermäuse aus Nepal (Mammalia: Chiroptera). Senckenbergiana Biologica, 75:15-21.

Kock, D. 1999*a*. *Tadarida* (*Tadarida*) *latouchei*, a separate species recorded from Thailand with remarks on related Asian taxa (Mammalia, Chiroptera, Molossidae). Senckenbergiana Biologica, 78:237-240.

Kock, D. 1999*b*. The Egyptian *Vespertilio pipistrellus aegyptius* Fischer 1829, a nomen dubium. Senckenbergiana Biologica, 79:101-105.

Kock, D. 2000*a*. On some bats (Chiroptera) from southern Cambodia with a preliminary checklist. Zeitschrift für Säugetierkunde, 65:199-208.

Kock, D. 2000*b*. The fallow deer *Dama schaeferi* Hilzheimer, 1926 (Mammalia: Cervidae), enigmatic and forgotten. Zoology in the Middle East, 21:9-11.

Kock, D. 2001*a*. *Rousettus aegyptiacus* (E Geoffroy St. Hilaire, 1810) and *Pipistrellus anchietae* (Seabra, 1900), justified emendations of original spellings. Acta Chiropterologica, 3:245-256.

Kock, D. 2001*b*. Identity of the African *Vespertilio hesperida* Temminck 1840 (Mammalia, Chiroptera, Vespertilionidae). Senckenbergiana Biologica, 81:277-283.

Kock, D. 2002. The publication dates of *Plecotus alpinus* Kiefer and Veith, 2002 and of *Plecotus microdontus* Spitzenberger, 2002. Acta Chiropterologica, 4:219-220.

Kock, D., and C. Ebenau. 1996. The desert hedgehog, *Paraechinus aethiopicus* (Ehrenberg, 1833), new to the fauna of Syria. Zeitschrift für Säugetierkunde, 61:189-191.

Kock, D., and H. Felten. 1980. Typen und Typus-Lokalität von *Apodemus sylvaticus rufescens* Saint Girons und Bree 1963 (Mammalia: Rodentia: Muridae). Senckenbergiana Biologica, 60:277-283.

Kock, D., and K. M. Howell. 2000. The enigma of the giant forest hog *Hylochoerus meinertzhageni* (Mammalia: Suidae) in Tanzania reviewed. Journal of East African Natural History, 88:25-34.

Kock, D., and D. Kovac. 2000. *Eudiscops denticulus* (Osgood 1932) in Thailand with notes on its roost (Chiroptera: Vespertilionidae). Zeitschrift für Säugetierkunde, 65:121-123.

Kock, D., and T. Künzel. 1999. The maned rat, *Lophiomys imhausii* Milne-Edwards, 1867, in Djibouti, NE-Africa (Mammalia: Rodentia: Lophiomyinae). Zeitschrift für Säugetierkunde, 64:371-375.

Kock, D., and I. A. Nader. 1983. Pygmy shrew and rodents from the Near East. Senckenbergiana Biologica, 64:13-23.

Kock, D., and I. A. Nader. 1984. *Tadarida teniotis* (Rafinesque, 1814) in the W-Palearctic and a lectotype for *Dysopes rupelii* Temminck, 1826 (Chiroptera: Molossidae). Zeitschrift für Säugetierkunde, 49:129-135.

Kock, D., and I. A. Nader. 1990. Identity of *Apodemus sylvaticus* (Linnaeus, 1758) recorded from Qatar (Rodentia: Muridae). Zeitschrift für Säugetierkunde, 55:66-67.

Kock, D., and H. Posamentier. 1983. *Cannomys badius* (Hodgson, 1842) in Bangladesh (Rodentia: Rhizomyidae). Zeitschrift für Säugetierkunde, 48:314-316.

Kock, D., and G. Storch. 1996. A new genus and species of Vespertilioninae bats from SE Asia. Senkenbergiana Biologica, 76:1-6.

Kock, D., F. Malec, and G. Storch. 1972. Rezente und subfossile kleinsäuger aus dem Vilayet Elazig, Ostanatolien. Zeitschrift für Säugetierkunde, 37:204-229.

Kock, D., I. A. Nader, and A. A. Banaja. 1990. *Bandicota bengalensis* (Gray and Hardwicke, 1833) new to Saudi Arabia (Mammalia: Rodentia: Muridae). Fauna of Saudi Arabia, 11:323-328.

Kock, D., H. Burda, W. N. Chitaukali, and M. J. Overton. 1998. *Plerotes anchietae* (Seabra, 1900) in Malawi, central Africa (Mammalia: Chiroptera). Zeitschrift für Säugetierkunde, 63:114-116.

Kock, D., G. Csorba, and K. M. Howell. 2000*a*. *Rhinolophus maendeleo* n. sp. from Tanzania, a horseshoe bat noteworthy for its systematics and biogeography (Mammalia, Chiroptera, Rhinolophidae). Senkenbergiana Biologica, 80:233-239.

Kock, D., J. Altmann, and L. Price. 2000*b*. A fruit bat new to West Malaysia: *Rousettus leechenaultii* (Desmarest 1820) in Batu Caves. Malayan Nature Journal, 54:63-67.

Kock, D., T. Kunzel, and H. A. Rayaleh. 2000*c*. The African civet, *Civettictis civetta* (Schreber, 1776), of Djibouti representing a new subspecies. Senckenbergiana Biologica, 80:241-246.

Kock, D., M. Al-Jumaily, and A. K. Nasher. 2001. On the genus *Rhinopoma* E. Geoffroy 1818, and a record of *Rh. muscatellum* Thomas 1903 from Yemen (Mammalia, Chiroptera, Rhinopomatidae). Senkenbergiana Biologica, 81:285-287.

Kock, D., L. Barnett, J. Fahr, and C. Emms. 2002. On a collection of bats from Gambia (Mammalia: Chiroptera). Acta Chiropterologica, 4(1):77-97.

Koehler, C. E., and P. R. K. Richardson. 1990. *Proteles cristatus*. Mammalian Species, 363:1-6.

Koenigswald, W. von. 1980. Schmelzstruktur und Morphologie in den Molaren der Arvicolidae (Rodentia). Abhandlungen der Senckenbergischen Naturforschenden Gesellschaft, 539:1-129.

Koenigswald, W. von. 1982. Stammesgeschichte und Schmelzmuster. Pp. 60-69, *in* Handbuch der Säugetiere Europas (J. Niethammer and F. Krapp, eds.). Akademische Verlagsgesellschaft (Wiesbaden), 2/I:1-649.

Koenigswald, W. von. 1993. Die Schmelzmuster in den Schneidezähnen der Gliroidea (Gliridae und Seleviniidae, Rodentia, Mammalia) und ihre systematische Bedeutung. Zeitschrift für Säugetierkunde, 58:92-115.

Koenigswald, W. von. 1995. Enamel differentiations in myoxid incisors and their systematic significance. Pp. 99-107, *in* Proceedings of II Conference on Dormice (Rodentia, Myoxidae) (M. G. Filippucci, ed.). Hystrix, n.s., 6(1-2):1-340.

Koenigswald, W. von, and L. D. Martin. 1984. Revision of the fossil and Recent Lemminae (Rodentia: Mammalia). Carnegie Museum of Natural History, Special Publication, 9:122-137.

Koenigswald, W. von, and A. Tesakov. 1997. The evolution of the schmelzmuster in Lagurini (Arvicolinae, Rodentia). Palaeontographica. Abteilung A, Paläozoologie-Stratigraphie, 245(1-6):45-61.

Koenigswald, W. von, O. Fejfar, and E. Tchernov. 1992. Revision einiger alt- und mittelpleistozäner Arvicoliden (Rodentia, Mammalia) aus dem östlichen Mittelmeergebiet ('Ubeidiya, Jerusalem und Kalymnos-Xi) [Revision of some Early and Middle Pleistocene arvicolids (Rodentia, Mammalia) from the Eastern Mediterranean ('Ubeidiya, Jerusalem and Kalymnos-Xi)]. Neues Jahrbücher für Geologie und Paläontologie, Abhandlungen, 184(1):1-23.

Koenigswald, W. von, P. M. Sander, F. L. S., M. B. Leite, T. Mörs, and W. Santel. 1994. Functional symmetries in the schmelzmuster and morphology of rootless rodent molars. Zoological Journal of the Linnean Society, 110(2):141-179.

Koepfli, K. P., and R. K. Wayne. 1998. Phylogenetic relationships of otters (Carnivora: Mustelidae) based on mitochondrial cytochrome b sequences. Journal of Zoology, London, 246:401-416.

Koeppl, J. W., R. S. Hoffmann, and C. F. Nadler. 1978. Pattern analysis of acoustical behavior in four species of ground squirrels. Journal of Mammalogy, 59:677-696.

Koffler, B. R. 1972. *Meriones crassus*. Mammalian Species, 9:1-4.

Koh, H. S. 1982. G- and C-banding pattern analyses of Korean rodents. I. Chromosome banding patterns of striped field mice (*Apodemus agrarius coreae*) and black rats (*R. rattus rufescens*). Korean Journal of Zoology, 25(2):81-92.

Koh, H. S. 1983. A study on age variation and secondary sexual dimorphism in morphometric characters of Korean rodents: I. An analysis on striped field mice, *Apodemus agrarius coreae* Thomas, from Cheongju. Korean Journal of Zoology, 26:125-134.

Koh, H. S. 1988. Systematic studies of Korean rodents: IV. Morphometric and chromosomal analyses of two species of the genus *Apodemus*. Korean Journal of Systematic Zoology, 4:103-120.

Koh, H. S. 1991. Morphometric analyses with eight subspecies of striped field mice, *Apodemus agrarius* Pallas

(Rodentia, Mammalia), in Asia: The taxonomic status of subspecies *chejuensis* at Cheju island in Korea. Korean Journal of Systematic Zoology, 7(2):179-188.

Koh, H. S. 1994. Systematic studies on Korean rodents: VIII. Analyses of morphometric characters, chromosomal karyotype, and mitochondrial DNA restriction fragments in Siberian chipmunks from Korea (*Tamias sibiricus barberi* Johnson and Jones), with the comparison of morphometric characters of Siberian chipmunks from Manchuria (*Tamias sibiricus orientalis* Bonhote). Korean Journal of Systematic Zoology, 10(2):231-243.

Koh, H. S., and W. J. Lee. 1994. Geographic variation of morphometric characters in three subspecies of Korean field mice, *Apodemus peninsulae* Thomas, in China and Korea. Korean Journal of Zoology, 37:33-39.

Koh, H. S., and R. L. Peterson. 1983. Systematic studies of deer mice, *Peromyscus maniculatus* Wagner (Cricetidae, Rodentia): Analysis of age and secondary sexual variation in morphometric characters. Canadian Journal of Zoology, 61:2618-2628.

Koh, H. S., and E. Randi. 2001. Genetic distinction of roe deer (*Capreolus pygargus* Pallas) samples in Korea. Zeitschrift für Säugetierkunde, 66:371-375.

Koh, H. S., and G. Tikhonova. 1998. Morphometric analyses with 15 subspecies of striped field mouse, *Apodemus agrarius* Pallas (Mammalia, Rodentia) from Eurasia. The Korean Journal of Systematic Zoology, 14(4):341-355.

Koh, H. S., and B. S. Yoo. 1992. Variation of mitochondrial DNA in two subspecies of striped field mice, *Apodemus agrarius coreae* and *Apodemus agrarius chejuensis*, from Korea. Korean Journal of Zoology, 35:332-338.

Koh, H. S., Y. S. Roh, S. B. Kim, and B. S. Yoo. 1995*a*. Variation of mitochrondrial DNA restriction fragments of common rats, *Rattus norvegicus caraco* Pallas (Mammalia, Rodentia), from Cheongju, Korea. The Korean Journal of Systematic Zoology, 11(4):409-416.

Koh, H. S., T. Y. Chun, S. K. Yoo, Y. K. Kim, and Y. J. Song. 1995*b*. Variation of mitochondrial DNA restriction fragments within one subspecies of Korean field mice, *Apodemus peninsulae peninsulae* Thomas (Mammalia: Rodentia), from Korea. The Korean Journal of Systematic Zoology, 11(2):291-299.

Koh, H. S., W. J. Lee, and Y. Ma. 1996. Morphometric variation in three subspecies of Korean field mice, *Apodemus peninsulae* Thomas (Mammalia, Rodentia), in China and Korea. The Korean Journal of Systematic Zoology, 12(3):203-210.

Koh, H. S., Y. K. Kim, W. J. Lee, J. Wang, and H. Lu. 1997. Morphometric variation of six subspecies of striped field mouse, *Apodemus agrarius* Pallas (Mammalia, Rodentia), from China and Korea. Korean Journal of Biological Science, 1:25-29.

Koh, H. S., G. Csorba, M. P. Tiunov, and G. Tikhonova. 1998*a*. Morphometric analyses of the three subspecies of striped field mouse, *Apodemus agrarius* Pallas (Mammalia: Rodentia) from Far Eastern Asia: Taxonomic status of North Korean striped field mice. The Korean Journal of Systematic Zoology, 14(4):327-334.

Koh, H. S., B. Y. Lee, Y. K. Kim, S. K. Yoo, and B. K. Yang. 1998*b*. Taxonomic status of striped field mice (Mammalia, Rodentia) from Wando Island, Korea. The Korean Journal of Systematic Zoology, 14(1):51-57.

Köhler, N., M. H. Gallardo, L. C. Contreras, and J. C. Torres-Mura. 2000. Allozyme variation and systematic relationships of the Octodontidae and allied taxa (Mammalia, Rodentia). Journal of Zoology, 252:243-250.

Köhler-Rollefson, I. U. 1991. *Camelus dromedarius*. Mammalian Species, 375:1-8.

Kohn, P. H., and R. H. Tamarin. 1978. Selection at electrophoretic loci for reproductive parameters in island and mainland voles. Evolution, 32:15-18.

Köhncke, M., and K. Leonhardt. 1986. *Cryptoprocta ferox*. Mammalian Species, 254:1-5.

Kolfschoten, van T. 1990. The evolution of the mammal fauna in the Netherlands and the middle Rhine area (Western Germany) during the late middle Pleistocene. Mededelingen Rijks Geoloische Dienst, 43(3):1-69.

Kolfschoten, van T. 1992. Aspects of the migration of mammals to northwestern Europe during the Pleistocene, in particular the reimmigration of *Arvicola terrestris*. Courier Forschungsinstitut. Senckenberg, 153:213-220.

Kolomiets, O. L., T. E. Borviev, L. D. Safronova, Yu. M. Borisov, and Y. F. Bogdanov. 1988. Synaptonemal complex analysis of B- chromosome behavior in meiotic prophase I in the East-Asiatic mouse *Apodemus peninsulae* (Muridae, Rodentia). Cytogenetics and Cell Genetics, 48:183-187.

Kolomiets, O. L., N. N. Vorontsov, E. A. Lyapunova, and T. F. Mazurova. 1991. Ultrastructure, meiotic behavior, and evolution of sex chromosomes of the genus *Ellobius*. Genetica, 84:179-189.

Kompanje, E. J. O., and C. W. Moeliker. 2001. Some fruit bats from remote Moluccan and West-Papuan islands, with the description of a new subspecies of *Macroglossus minimus* (Megachiroptera: Pteropodidae). Deinsea, 8:143-167.

König, C. 1982. *Microtus bavaricus* (Konig, 1962)—Bayerische Kurzohrmaus. Pp. 447-451, *in* Handbuch der Säugetiere Europas (J. Niethammer and F. Krapp, eds.). Akademische Verlagsgesellschaft (Wiesbaden), 2/I:1-649.

Kooij, J. van der, M. Gillham, R. J. Post, T. Heijerman, and M. D. Wilson. 1997. Morphometric variation in the mandible of *Microtus arvalis* and *Microtus agrestis* (Rodentia: Cricetidae). Netherlands Journal of Zoology, 47(1):47-59.

Kooij, J. van der, K. Isaksen, and K. M. Olsen. 2001. [Occurrence of the harvest mouse *Micromys minutus* confirmed in Norway.] Fauna, 54(4):110-120 (in Norwegian with English summary).

Koontz, F. W., and N. J. Roeper. 1983. *Elephantulus rufescens*. Mammalian Species, 204:1-5.

Koop, B. F., R. J. Baker, and J. T. Mascarello. 1985. Cladistical analysis of chromosomal evolution within the genus *Neotoma*. Occasional Papers, The Museum, Texas Tech University, 96:1-9.

Koopman, K. F. 1958. Land bridges and ecology in bat distribution on islands off the northern coast of South America. Evolution, 12:429-439.

Koopman, K. F. 1965. Status of forms described or recorded by J. A. Allen in "The American Museum Congo Expedition Collection of Bats". American Museum Novitates, 2219:1-34.

Koopman, K. F. 1968. Taxonomic and distributional notes on Lesser Antillean bats. American Museum Novitates, 2333:1-13.

Koopman, K. F. 1971*a*. Taxonomic notes on *Chalinolobus* and *Glauconycteris*. American Museum Novitates, 2451:1-10.

Koopman, K. F. 1971*b*. The systematic and historical status of the Florida *Eumops* (Chiroptera: Molossidae). American Museum Novitates, 2478:1-6.

Koopman, K. F. 1972. *Eudiscopus denticulus*. Mammalian Species, 19:1-2.

Koopman, K. F. 1973. Systematics of Indo-Australian *Pipistellus*. Periodicum Biologorum Zagreb, 75:113-116.

Koopman, K. F. 1975. Bats of the Sudan. Bulletin of the American Museum of Natural History, 154:353-444.

Koopman, K. F. 1978*a*. The genus *Nycticeius* (Vespertilionidae) with special reference to tropical Australia. Proceedings of the Fourth International Bat Research Conference, Nairobi, 4:165-171.

Koopman, K. F. 1978*b*. Zoogeography of Peruvian bats with special emphasis on the role of the Andes. American Museum Novitates, 2651:1-33.

Koopman, K. F. 1979. Zoogeography of mammals from islands off the northeastern coast of New Guinea. American Museum Novitates, 2690:1-17.

Koopman, K. F. 1982. Results of the Archbold Expeditions. No. 109. Bats from eastern Papua and the east Papuan Islands. American Museum Novitates, 2747:1-34.

Koopman, K. F. 1984*a*. Bats. Pp. 145-186, *in* Orders and families of Recent mammals of the World (S. Anderson and J. K. Jones, Jr., eds.). John Wiley and Sons, New York, 686 pp.

Koopman, K. F. 1984*b*. A progress report on the systematics of African *Scotophilus* (Vespertilionidae). Proceedings of the Sixth International Bat Research Conference, Ile-Ife, Nigeria, 4:102-113.

Koopman, K. F. 1984*c*. Taxonomic and distributional notes on tropical Australian bats. American Museum Novitates, 2778:1-48.

Koopman, K. F. 1986. Sudan bats revisited: An update of "Bats of the Sudan". Cimbebasia, ser. A. 8(2):9-13.

Koopman, K. F. 1989*a*. Distributional patterns of Indo-Malayan bats (Mammalia: Chiroptera). American Museum Novitates, 2942:1-19.

Koopman, K. F. 1989*b*. Systematic notes on Liberian bats. American Museum Novitates, 2946:1-11.

Koopman, K. F. 1989*c*. A review and analysis of the bats of the West Indies. Pp. 635-644, *in* Biogeography of the West Indies: Past, present, and future (C. A. Woods, ed.). Sandhill Crane Press, Gainesville, FL, 878 pp.

Koopman, K. F. 1992. Taxonomic status of *Nycteris vinsoni* Dalquest (Chiroptera: Nycteridae). Journal of Mammalogy, 73(3):649-650.

Koopman, K. F. 1993. Order Chiroptera. Pp. 137-241, *in* Mammal species of the world, a taxonomic and geographic reference, Second ed. (D. E. Wilson and D. M. Reeder, eds.). Smithsonian Institution Press, Washington D.C., 1207 pp.

Koopman, K. F. 1994. Chiroptera: Systematics. Handbook of Zoology, vol. VIII, pt. 60, Mammalia, 1-217.

Koopman, K. F. 1997. The subspecies of *Emballonura semicaudata* (Chiroptera: Emballonuridae). Journal of Mammalogy, 78:358-360.

Koopman, K. F., and T. N. Danforth. 1989. A record of the tube-nosed bat (*Murina florium*) from western New Guinea. American Museum Novitates, 2934:1-5.

Koopman, K. F., and J. K. Jones, Jr. 1970. Classification of bats. Pp. 22-28, *in* About bats (B. H. Slaughter and D. W. Walton, eds.). Southern Methodist University Press, Dallas, 339 pp.

Koopman, K. F., M. K. Hecht, and E. Ledecky-Janecek. 1957. Notes on the mammals of the Bahamas with special reference to the bats. Journal of Mammalogy, 38:164-174.

Koopman, K. F., R. E. Mumford, and J. F. Heisterberg. 1978. Bat records from Upper Volta, west Africa. American Museum Novitates, 2643:1-6.

Koopman, K. F., C. P. Kofron, and A. Chapman. 1995. The bats of Liberia: Systematics, ecology, and distribution. American Museum Novitates, 3148:1-24.

Kooyman, G. L. 1981. Weddell seal: Consummate diver. Cambridge University Press, Cambridge, England, 135 pp.

Koppers, D. 1990. Vergleichend anatomisch-systematische Untersuchungen an Schädeln von Talpa europaea und anderen Insektivoren. Zoologische Jahrbücher, Abteilung Anatomie, 120:109-125.

Koprowski, J. L. 1994a. *Sciurus niger*. Mammalian Species, 479:1-9.

Koprowski, J. L. 1994b. *Sciurus carolinensis*. Mammalian Species, 480:1-9.

Korablev, V. P., V. E. Kiriljuk, and M. I. Golovushkin. 1996. Study of the karyotype of Daurian hedgehog *Mesechinus dauricus* (Mammalia, Erinaceidae) from its terra typica. Zoologicheskii Zhurnal, 75:558-564.

Kormos, T. 1932. Neues whulmause aus dem Oberpliocan von Puspokfurdo. Neues Jahrbuch für Mineralogie, Geologie, und Palaontologie, Abteilung B, 69:323-346.

Korobitsyna, K. V., and I. V. Kartavtseva. 1988. [Variation and evolution of the karyotype of gerbils (Rodentia, Cricetidae, Gerbillinae). I. Karyotypic differentiation of midday gerbils (*Meriones meridianus*) of the USSR fauna.] Zoologicheskii Zhurnal, 67:1889-1899 (in Russian).

Korobitsyna, K. V., and I. V. Kartavtseva. 1992. [Variability and evolution of karyotype in gerbils (Rodentia, Cricetidae, Gerbillinae). 2. Heterochromatin and its intraspecific and intrapopulation variations in red-tailed Libyan jird *Meriones libycus*.] Zoologisheskii Zhurnal, 71(3):83-95.

Korobitsyna, K. V., and V. P. Korablev. 1980. The intraspecific autosome polymorphism of *Meriones tristrami* Thomas, 1892 (Gerbillinae, Cricetidae, Rodentia). Genetica, 52/53:209-221.

Korobitsyna, K. V., C. F. Nadler, N. N. Vorontsov, and R. S. Hoffmann. 1974. Chromosomes of the Siberian snow sheep, *Ovis nivicola*, and implications concerning the origin of amphiberingean wild sheep (subgenus *Pachyceros*). Quaternary Research, 4:235-245.

Korobitsyna, K. V., L. V. Yakimenko, and L. V. Frisman. 1993. Genetic differentiation of house mice in the fauna of the former U.S.S.R.: Results of cytogenetic studies. Biological Journal of the Linnean Society, 48:93-112.

Korth, W. W. 1993. *Miosicista angulus*, a new sicistine rodent (Zapodidae, Rodentia) from the Barstovian (Miocene) of Nebraska. Transactions of the Nebraska Academy of Sciences, 20:97-101

Korth, W. W. 1994. The Tertiary record of rodents in North America. Topics in Geobiology, vol. 12. Plenum Press, New York, 319 pp.

Korth, W. W. 2001. Comments on the systematics and classification of the beavers (Rodentia, Castoridae). Journal of Mammalian Evolution, 8:279-296.

Kortlucke, S. M. 1973. Morphological variation in the kinkajou, *Potos flavus* (Mammalia: Procyonidae), in Middle America. Occasional Papers of the Museum of Natural History, University of Kansas, 17:1-36.

Kosior, A. 1996. Nowe stanowisko koszatki *Dryomys nitedula* na Dolnym Slasku [The new site of the dormouse *Dryomys nitedula* in Lower Silesia]. Chronmy Przyrode Ojczysta, 52(4):107-110.

Kosloski, M. A. 1997. Morfometria craniana de uma populacao de *Akodon serrensis* Thomas, 1902 (Rodentia-Cricetidae) do Municipio de Araucaria, Parana, Brasil. Estudos de Biologia PUC-PR, Curitiba, 41:61-88.

Kostenko, V. A., and T. V. Allenoba. 1978. [Special features of biology and morphology of Shikotan Island's voles.] Pp. 119-129, *in* Ecology and zoogeography of some vertebrate animals of the Far East Land. Far East Branch, Russian Academy of Sciences, Vladivostok (in Russian, re-issued in Japan).

Kostroň, K. 1948. Tchoø stepní èili Eversmanno (*Putorius eversmanni* Lesson 1827), nový a znaène rozsíøený èlen zvíøeny Èeskoslovenska. Práce Moravskoslezské Akademie vìd Poírodních [Acta Academiae Scientairum Naturalium Moravo Silesiacae], 20(3):1-95 (in Czech).

Kotenkova, E. V., and N. Sh. Bulatova (eds.). 1994. [Species of the fauna of Russia and contiguous countries: The house mouse, origin, distribution, systematics, behavior]. Nauka, Moscow, 267 pp. (in Russian with English summary and table of contents).

Kotlia, B. S. 1992. Pliocene murids (Rodentia, Mammalia) from Kashmir Basin, northwestern India. Neues Jahrbuch fur Geologie und Palaontologie Abhandlungen, 184(3):339-357.

Kotlia, B. S. 1994. Evolution of Arvicolidae in South Asia. Pp. 151-171, *in* Rodent and lagomorph families of Asian origins and diversification (Y. Tomida, C. k. Li, and T. Setoguchi, eds.). National Science Museum Monographs, No. 8, Tokyo, 195 pp.

Kotlia, B. S. 1995. Upper Pleistocene Soricidae and Muridae from Bhimtal-Bilaspur deposits, Kumaun Himalaya, India. Journal of the Geological Society of India, 46:177-190.

Kotlia, B. S., and W. v. Koenigswald. 1992. Plio-Pleistocene arvicolids (Rodentia, Mammalia) from Kashmir intermontane basin, northwestern India. Palaeontographica, Abt A, 223:103-135.

Kotsakis, T. 1980. Osservazioni sui vertebrati Quaternari della Sardegna. Bolletino della Societa Geologica Italiana, 99:151-165.

Kovacs, K. M., and D. M. Lavigne. 1986. *Cystophora cristata*. Mammalian Species, 258:1-9.

Kovaleva, V. Yu., V. M. Efimov, and V. I. Faleyev. 1996. [Craniometric variation of non-wintering water voles *Arvicolas terrestris* (Rodentia, Cricetidae) in different environmental conditions.] Zoologisheskii Zhurnal, 75(10):1551-1559 (in Russian with English summary).

Kovaleva, V. Yu, A. A. Pozdnyakov, and V. M. Efimov. 2002. [Investigation of variability structure of root vole (*Microtus oeconomus* Pallas) molar morphotypes using bilateral asymmetry.] Zoologisheskii Zhurnal, 81(1):111-117 (in Russian with English summary).

Kovalskaya, Yu. M. 1977. [Chromosome polymorphism in *Microtus maximowiczii* Schrenk, 1858 (Rodentia, Cricetidae).] Byulleten' Moskovskovo Obshchestva Ispytatelei Prirody, Otdel Biologicheskii, 82:38-48 (in Russian).

Kovalskaya, Yu. M. 1989. [Karyotype variability of narrow-skulled vole, *Microtus* (*Stenocranius*) *gregalis* (Rodentia, Cricetidae) from northern Mongolia.] Zoologicheskii Zhurnal, 68:77-84 (in Russian).

Kovalskaya, Yu. M. 1994. [On the distribution of voles of the group *arvalis* (Rodentia) in Kazakhstan.] Zoologischeskii Zhurnal, 73(3):120-125 (in Russian with English summary).

Kovalskaya, Yu. M., V. M. Malygin, I. V. Kartavtseva. 1988. [Karyotype stability and distribution of reed voles—*Microtus fortis* (Rodentia, Cricetidae).] Zoologicheskii Zhurnal, 67:1253-1259 (in Russian).

Kovalskaya, Yu. M., V. M. Aniskin, and I. V. Kartavtseva. 1991. [Geographic variation of C-heterochromatin in *Microtus fortis* (Rodentia, Cricetidae).] Zoologicheskii Zhurnal, 70:97-104 (in Russian).

Kovalskaya, Yu. M., I. A. Tikhonov, G. N. Tikhonova, A. V. Surov, and P. L. Bogomolov. 2000. [New geographical localities of chromosome forms of southern birch mouse (*subtilis* Group) and description of *Sicista severtzovi cimlanica* subsp. N. (Mammalia, Rodentia) from the Middle Don River Basin.] Zoologisheskii Zhurnal, 79(8):954-964 (in Russian with English abstract).

Kowalski, K. 1963. The Pliocene and Pleistocene Gliridae (Mammalia, Rodentia) from Poland. Acta Zoologica Cracoviensia, 8(14):533-567.

Kowalski, K. 1967. *Lagurus lagurus* (Pallas, 1773) and *Cricetus cricetus* (Linnaeus, 1758) (Rodentia, Mammalia) in the Pleistocene of England. Acta Zoologica Cracoviensia, 12(6):111-122.

Kowalski, K. 1993. *Neocometes* Schaub and Zapfe, 1953 (Rodentia, Mammalia) from the Miocene of Bełchatów (Poland). Acta Zoologica Cracoviensia, 36(2):259-265.

Kowalski, K. 2001. Pleistocene rodents of Europe. Folia Quaternaria, 72:3-389.

Kowalski, K., and Y. Hasegawa. 1976. Quaternary rodents from Japan. Bulletin of the National Science Museum (Tokyo), ser. C (Geology and Paleontology), 2(1):31-66.

Kowalski, K., and B. Rzebik-Kowalska. 1991. Mammals of Algeria. Zaklad Narodowy Imienia Ossolinskich Wydawnictwo Polskiej Akademii Nauk Wroclaw, Poland, 370 pp.

Kozlovskii, A. I. 1973. [Results of karyological study of allopatric forms in *Sorex minutus*.] Zoologicheskii Zhurnal, 52:390-398 (in Russian).

Kozlovsky, A. I. 1974. [Karyotype differentiation in northeastern subspecies of the arctic lemming.] Doklady Akademii Nauk SSSR, 219:981984.

Kozlovskii, A. I. 1976. [Karyotypes and systematics of some populations of shrews usually classed with *Sorex arcticus*.] Zoologicheskii Zhurnal, 55:756-762 (in Russian).

Kozlovsky, A. I. 1985. [Chromosome polymorphism and cytogenetic mechanism of sex ratio regulation in the wood lemming *Myopus schisticolor* Lill.] Genetika, 21:60-68 (in Russian with English summary).

Kozlovskii, A. I. 1986. Chromosome forms and autosomal polymorphism of the wood lemming (*Myopus schisticolor*) from North-East Asia. Folia Zoologica, 35:63-71.

Kraft, R. 1995. Xenarthra. Handbuch der Zoologie, De Gruyter, Berlin, 59:8 (unnumbered) + 1-80.

Kraglievich, J. L. 1965. Speciation phyletique dans les rongeurs fossils de genre *Eumysops* Amegh. (Echimyidae, Heteropsomyinae). Mammalia, 29:258-267.

Krajewski, C., A. C. Driskell, P. R. Baverstock, and M. J. Braun. 1992. Phylogenetic relationships of the thylacine (Mammalia, Thylacinidae) among Dasyuroid Marsupials – evidence from cytochrome-*b* DNA sequences. Proceedings of the Royal Society of London, Series B, Biological Sciences, 250:19-27.

Krajewski, C., L. Buckley, P. A. Woolley, and M. Westerman. 1996. Phylogenetic analysis of cytochrome *b* sequences in the dasyurid marsupial subfamily Phascogalinae: Systematics and the evolution of reproductive strategies. Journal of Mammalian Evolution, 3:81-91.

Krajewski, C., J. Young, L. Buckley, P. A. Woolley, and M. Westerman. 1997*a*. Reconstructing the evolutionary radiation of dasyurine marsupials with cytochrome *b*, 12S rRNA, and protamine P1 gene trees. Journal of Mammalian Evolution, 4:217-236.

Krajewski, C., M. Blacket, L. Buckley and M. Westerman. 1997*b*. A multigene assessment of phylogenetic relationships within the dasyurid marsupial subfamily Sminthopsinae. Molecular Phylogenetics and Evolution, 8:236-248.

Krajewski C., L. Buckley, and M. Westerman. 1997*c*. DNA phylogeny of the marsupial wolf resolved. Proceedings of the Royal Society of London, Series B, Biological Sciences, 264:911-917.

Krajewski, C., M. J. Blacket, and M. Westerman. 2000. DNA sequence analysis of familial relationships among dasyuromorphian marsupials. Journal of Mammalian Evolution, 7:95-108.

Král, B. 1967. Karyological analysis of two European species of the genus *Erinaceus*. Zoologéskii Listy, 16:239-252.

Král, B., and J. Zima. 1980. Karyosystematika celedi Felidae. Gazella (Prague), 2/3:45-53.

Král, B., J. Zima, V. Hrabe, J. Libosvarsky, and M. Sebela. 1981. On the morphology of *Microtus epiroticus*. Folia Zoologica, 30:317-330.

Král, B., S. I. Radjabli, A. S. Grafodatskij, and V. N. Orlov. 1984. Comparison of karyotypes, G-bands and NORS in three *Cricetulus* spp. (Cricetidae, Rodentia). Folia Zoologica, 33:85- 96.

Kramerov, D. A. 1999. Evidence of the phylogenetic relation of Gliridae and Sciuridae families based on studying the B1-dID short retrotransposon. Doklady Biological Sciences, 364:47-50.

Kramerov, D., and N. Vassetzky. 2001. Structure and origin of a novel dimeric retroposon B1-dID. Journal of Molecular Evolution, 52:137-143.

Kramerov, D., N. Vassetzky and I. Serdobova. 1999. The evolutionary position of dormice (Gliridae) in Rodentia determined by a novel short retroposon. Molecular Biology and Evolution, 16(5):715-717.

Krapp, F. 1978. *Tamias sibiricus* (Laxmann, 1769) – Burunduk. Pp. 116-121, *in* Handbuch der Säugetiere Europas (J. Niethammer and F. Krapp, eds.). Akademische Verlagsgesellschaft, Wiesbaden, 476 pp.

Krapp, F. 1982*a*. *Microtus nivalis* (Martins, 1842)—Schneemaus. Pp. 261-283, *in* Handbuch der Säugetiere Europas (J. Niethammer and F. Krapp, eds.). Akademische Verlagsgesellschaft (Wiesbaden), 2/I:1-649.

Krapp, F. 1982*b*. *Microtus multiplex* (Fatio, 1905)—Alpen- Kleinwühlmaus. Pp. 419-428, *in* Handbuch der Säugetiere Europas (J. Niethammer and F. Krapp, eds.). Akademische Verlagsgesellschaft (Wiesbaden), 2/I:1-649.

Krapp, F. 1982*c*. *Microtus savii* (de Selys-Longchamps, 1838)— Italienische Kleinwühlmaus. Pp. 429-437, *in* Handbuch der Säugetiere Europas (J. Niethammer and F. Krapp, eds.). Akademische Verlagsgesellschaft (Wiesbaden), 2/I:1-649.

Krapp, F. 1982*d*. *Microtus pyrenaicus* (de Selys-Longchamps, 1847)—Pyrenaen- Kleinwühlmaus. Pp. 444-446, *in* Handbuch der Säugetiere Europas (J. Niethammer and F. Krapp, eds.). Akademische Verlagsgesellschaft (Wiesbaden), 2/I:1-649.

Krapp, F. 1990. *Crocidura leucodon* (Hermann, 1780)—Feldspitzmaus. Pp. 465-484, *in* Handbuch der Säugetiere Europas (J. Niethammer and F. Krapp, eds.). Aula Verlag, Wiesbaden, 3/I:1-524.

Krapp, F., and J. Niethammer. 1982. *Microtus agrestis* (Linnaeus, 1761)—Erdmaus. Pp. 349-373, *in* Handbuch der Säugetiere Europas (J. Niethammer and F. Krapp, eds.). Akademische Verlagsgesellschaft (Wiesbaden), 2/I:1-649.

Krapp, F., and H. Winking. 1976. Systematic von *Microtus* (*Pitymys*) *subterraneus* (de Sélys-Longchamps, 1836) und *savii* (de Sélys-Longchamps, 1838) auf der Appenninen-Halbinsel und Benachbarten Regionen. Säugetierkundliche Mitteilungen, 3:166-179.

Kratochvíl, J. 1967. Der Baumschläfer, *Dryomys nitedula* und andere Gliridae-Arten in der Tschechoslowakei. Zoologické Listy, 16(2):99-110.

Kratochvíl, J. 1969. Haarkleid und vibrissenfeld bei *Pitymys subterraneus* und *Pitymys tatricus* (Rodentia) aus der Hohen Tatra. Zoologicke Listy, 18(4):205-308.

Kratochvíl, J. 1970. *Pitymys*-Arten aus der hohen Tatra (Mamm., Rodentia). Acta Scientiarum Naturalium, Academiae Scientarium Bohemoslovacea (Brno), 4:1-63.

Kratochvíl, J. 1971. Die Hodengrösse als Kriterium der europäischen arten der gattung *Apodemus* (Rodentia, Muridae). Zoologické Listy, 20:293-306.

Kratochvíl, J. 1973. Männliche Sexualorgane und System der Gliridae (Rodentia). Acta Scientiarum Naturalium, Academiae Scientarium Bohemoslovacea (Brno), 7:1-52.

Kratochvíl, J. 1975. Zur Kenntnis der Igel der Gattung *Erinaceus* in der CSSR (Insectivora, Mamm.). Zoologéskii Listy, 24:297-312.

Kratochvíl, J. 1979. Contribution to our knowledge of the wood lemming, *Myopus schisticolor*. Folia Zoologica, 28:192-207.

Kratochvíl, J. 1980. Zur Phylogenie und Ontogenie bei *Arvicola terrestris* (Rodentia, Arvicolidae). Folia Zoologica, 29:209-224.

Kratochvíl, J. 1981a. *Chionomys nivalis* (Arvicolidae, Rodentia). Acta Scientiarum Naturalium, Academiae Scientarium Bohemoslovacea (Brno), 15:1-62.

Kratochvíl, J. 1981b. Is *Arvicola cantiana* still living? Folia Zoologica, 30(4):289-300.

Kratochvíl, J. 1982a. Cavum neurocranii der Arvicolidae. Acta Scientiarum Naturalium, Academiae Scientarium Bohemoslovacea (Brno), 16:1-57.

Kratochvíl, J. 1982b. Ein morphologisches Unterscheidungskriterium der Arten *Microtus epiroticus* und *M. arvalis* (Arvicolidae, Rodentia). Folia Zoologica, 31:97-111.

Kratochvíl, J. 1982c. Karyotyp und System der Familie Felidae (Carnivora, Mammalia). Folia Zoologica, 31:289-304.

Kratochvíl, J. 1983. Variability of some criteria in *Arvicola terrestris* (Arvicolidae, Rodentia). The effect of altitude on some taxonomical criteria of the Asian population group of *Arvicola terrestris*. Acta Scientiarum Naturalium, Academiae Scientarium Bohemoslovacea (Brno), 17:1-40.

Kratochvíl, J. 1986. *Mus abbotti*–eine kleinasiatisch-Balkanische art (Muridae–Mammalia). Folia Zoologica, 35(1):3-20.

Kratochvíl, J., and B. Král. 1972. Karyotypes and phylogenetic relationships of certain species of the genus *Talpa* (Talpidae, Insectivora). Zoologéskii Listy, 21:199-208.

Kratochvíl, J., and B. Král. 1974. Karyotypes and relationship of Palaearctic "54-chromosome" *Pitymys* species (Microtidae, Rodentia). Zoologické Listy, 23:289-302.

Kratochvíl, J., and V. Simionescu. 1983. Sur la question de *Micromys danubialis* (Mammalia, Muridae). Folia Zoologica, 32(1):1-18.

Kratochvíl, J., L. Rodriguez, and V. Barus. 1978. Capromyinae (Rodentia) of Cuba. I. Acta Scientiarum Naturalium. Academiae Scientarium Bohemoslovacae (Brno), n.s., 12(11):1-60.

Krattli, H., and M. Haffner. 1995. Funktionelle grobmorphologische und histologische Untersuchungen am Integument der Fusse des Siebenschlafer *Glis glis* (Linnaeus, 1766). Säugetierkundliche Mitteilungen, 36(2):62-81.

Kraus, F., and M. M. Miyamoto. 1991. Rapid cladogenesis among the pecoran ruminants: Evidence from mitochondrial DNA synthesis. Systematic Zoology, 40:117-130.

Kretteck, A., A. Gullberg, and U. Arnason. 1995. Sequence analysis of the complete mitochondrial DNA molecule of the Hedgehog, *Erinaceus europaeus*, and the phylogenetic position of the Lipotyphla. Journal of Molecular Evolution, 41:952-957.

Kretzoi, M. 1942. Präokkupierte und durch ältere zu ersetzende Säugetiernamen. Földtini Közlöny, 72:345-349.

Kretzoi, M. 1945. Bemerkungen über das Raubtiersystem. Annales Historico-Naturales Musei Nationalis Hungarici (Budapest), 38:59-83.

Kretzoi, M. 1955. *Dolomys* and *Ondatra*. Acta Geologica, Academiae Scientiarum Hungaricae, 3:347-355.

Kretzoi, M. 1961. Zwei myospalaciden aus Nordchina. Vertebrata Hungarica, 3:123-136.

Kretzoi, M. 1962. Arvicolidae oder Microtidae. Vertebrata Hungarica (Budapest), 4:171-175.

Kretzoi, M. 1964. Uber einige homonyme und synonyme Säugetiernamen. Vertebrata Hungarica, 6(1-2):131-138.

Kretzoi, M. 1965. *Drepanosorex*—neu definiert. Vertebrata Hungarica, 7:117-129.

Kretzoi, M. 1967. *Tyrrhenicola* und *Allophaiomys*. Vertebrata Hungarica, 9:171-175.

Kretzoi, M. 1969. Skizze einer Arvicoliden-Phylogenie. Vertebrata Hungarica (Budapest), 11:155-193.

Kretzoi, M. 1970/71. Bemerkungen zur Spalaciden-phylogenie. Vertebrata Hungarica, 12:111-121.

Kretzoi, M., and L. Vértes. 1965a. Upper Biharian (Intermindel) pebble-industry occupation site in Western Hungary. Current Anthropology, 6:74-87.

Kretzoi, M., and L. Vértes. 1965b. The role of vertebrate faunae and palaeolithic industries of Hungary in Quaternary stratigraphy and chronology. Acta Geologica Hungarica, 9:125-144.

Krohne, D. T. 1982. The karyotype of *Dicrostonyx hudsonius*. Journal of Mammalogy, 63:174-176.

Kronfeld, N., T. Dayan, N. Zisapel, and A. Haim. 1994. Coexisting populations of *Acomys cahirinus* and *A. russatus*: A preliminary report. Israel Journal of Zoology, 40:177-183.

Krška, A. 1996. The occurrence of *Apodemus microps* in the Podyjí National Park (Czech Republic). Folia Zoologica, 45(4):382-384.

Krumbiegel, I. 1942. Die Säugetiere der Südamerika—Expeditionen Prof. Dr. Kriegs. 17. Hyrare und Grisons (*Tayra* und *Grison*). Zoologischer Anzeiger, 139(5/6):81-108.

Krumbiegel, I. 1978. Die Kurzschwanz-Stumpfnase, *Simias concolor* (Miller, 1903), und die übrigen Nasenaffen. Säugetierkundliche Mitteilungen, 26:59-75.

Kruskop, S. V. 2002. Subspecific structure of *Myotis daubentonii* and composition of the "*daubentonii*" species group. Bat Research News, 43:94.

Krutzsch, P. H. 1954. North American jumping mice (genus *Zapus*). University of Kansas Publications, Museum of Natural History, 7(4):349-472.

Kryštufek, B. 1983. New subspecies of *Pitymys liechtensteini* Wettstein, 1927 from Yugoslavia. Biološki Vestnik, 31(1):73-82.

Kryštufek, B. 1985*a*. Forest dormouse *Dryomys nitedula* (Pallas 1778)- Rodentia, Mammalia- in Yugoslavia. Scopolia, 1985(9):1-36.

Kryštufek, B. 1985*b*. Variability of *Apodemus agrarius* (Pallas, 1771) (Rodentia, Mammalia) in Yugoslavia and some data on its distribution in the northwestern part of the country. Bioloski Vestnik, 33:27-40.

Kryštufek, B. 1990. Geographic variation in *Microtus nivalis* (Martins, 1842) from Austria and Yugoslavia. Bonner Zoologische Beiträge, 41:121-139.

Kryštufek, B. 1991. Sesalci Slovenije [Mammals of Slovenia]. Prirodoslovni muzej Slovenije, Ljubljana, 294 pp.

Kryštufek, B. 1994. The taxonomy of blind moles (*Talpa caeca* and *T. stankovici,* Insectivora, Mammalia) from south-eastern Europe. Bonner Zoologische Beiträge, 45:1-16.

Kryštufek, B. 1997. Overlooked names for European mammals. Folia Zoologica, 46(1):91-93.

Kryštufek, B. 1999*a*. *Glis glis* (Linnaeus, 1766). Pp. 294-295, *in* The atlas of European mammals (A. J. Mitchell-Jones, G. Amori, W. Bogdanowicz, B. Kryštufek, P. J. H. Reijnders, F. Spitzenberger, M. Stubbe, J. B. M. Thissen, V. Vohralík and J. Zima, eds.). T. and A. D. Poyser, Ltd., London, 484 pp.

Kryštufek, B. 1999*b*. *Dryomys nitedula* (Pallas, 1778). Pp. 300-301, *in* The atlas of European mammals (A. J. Mitchell-Jones, G. Amori, W. Bogdanowicz, B. Kryštufek, P. J. H. Reijnders, F. Spitzenberger, M. Stubbe, J. B. M. Thissen, V. Vohralík and J. Zima, eds.). T. and A. D. Poyser, Ltd., London, 484 pp.

Kryštufek, B. 1999*c*. Snow voles, genus *Chionomys*, of Turkey. Mammalia, 63:323-339.

Kryštufek, B. 2001*a*. The distribution of the Levante mole *Talpa levantis*. Zoology in the Middle East, 23:17-21.

Kryštufek, B. 2001*b*. Skull analysis of small blind moles from Turkey and Iran. Folia Zoologica, 50:19-25.

Kryštufek, B. 2001*c*. Compartmentalization of body of a Fat Dormouse *Glis glis*. Trakya University Journal of Research B, 2(2):95-106.

Kryštufek, B. 2002*a*. Identity of four *Apodemus* (*Sylvaemus*) types from the eastern Mediterranean and the Middle East. Mammalia, 66(1):43-51.

Kryštufek, B. 2002*b*. Cranial variability in the Eastern hedgehog *Erinaceus concolor* (Mammalia: Insectivora). Journal of Zoology, London, 258:365-373.

Kryštufek, B. 2003. First record of the garden dormouse (*Eliomys quercinus*) in Slovenia. Acta Zoologica Academiae Scientiarum Hungaricae (Supplement 1), 49:77-84.

Kryštufek, B. 2004. Nipples in the edible dormouse *Glis glis*. Folia Zoologica, 53(1):107-111.

Kryštufek, B., and P. Benda. 2002. The Caucasian mole *Talpa caucasica*—a new mammal for Iran. Mammalian Biology, 67:113-116.

Kryštufek, B., and W. Haberl. 2001. Dormouse associations in Slovenia – a new approach to an old tradition. Trakya University Journal of Research B, 2(2):171-177.

Kryštufek, B., and H. Kefelioğlu. 2002. Redescription and species limits of *Microtus irani* Thomas, 1921, and description of a new social vole from Turkey (Mammalia: Arvicolidae). Bonner Zoologische Beiträge, 50(1-2):1-14.

Kryštufek, B., and D. Kovacic. 1984. Distribution, habitat requirements and morphometric characteristics of *Micromys minutus* Pallas, 1771 (Rodentia, Mammalia) in Yugoslavia. Biosistematika, 10:99-112.

Kryštufek, B., and D. Kovacic. 1989. Vertical distribution of the snow vole *Microtus nivalis* (Martins, 1842) in northwestern Yugoslavia. Zeitschrift für Säugetierkunde, 54:153-156.

Kryštufek, B., and R. Kraft. 1997. Cranial variation and taxonomy of garden dormice (*Eliomys* Wagner, 1840) in the circum-Mediterranean realm. Mammalia, 61(3):411-429.

Kryštufek, B., and M. Macholán. 1998. Morphological differentiation in *Mus spicilegus* and the taxonomic status of mound-building mice from the Adriatic coast of Yugoslavia. Journal of the Zoological Society of London, 245:185-196.

Kryštufek, B., and L. Stojanovski. 1996. *Apodemus sylvaticus stankovici* is a synonym of *Apodemus flavicollis*. Folia Zoologica, 45(1):1-7.

Kryštufek, B., and N. Tvrtkovic. 1984. Redescription of *Arvicola terrestris illyrica* (Barrett-Hamiltion, 1899)- Rodentia, Mammalia. Biosistematika, 10:91-97.

Kryštufek, B., and V. Vohralík. 1992. New records of small mammals (Insectivora, Rodentia) from Montenegro and their zoogeographical significance. Acta Universitatis Carolinae Biologica, 36:279-290.

Kryštufek, B., and V. Vohralík. 1994. Distribution of the Forest Dormouse *Dryomys nitedula* (Pallas, 1779) (Rodentia, Myoxidae) in Europe. Mammal Review, 24(4):161-177.

Kryštufek, B., and V. Vohralík. 2001. Mammals of Turkey and Cyprus: Introduction, checklist, Insectivora. Knjiznica Annales Majora, Koper, 140 pp.

Kryštufek, B., and V. Vohralík. 2004. Molar size variation in three species of pine vole in Asia Minor: *Microtus subterraneus, M. majori, and M. daghestanicus* (Rodentia: Arvicolinae). Israel Journal of Zoology, 50:207-232.

Kryštufek, B., and M. Zavodnik. 2003. Autumn population density of the edible dormouse (*Gllis glis*) in the mixed montane forest of central Slovenia over 33 years. Acta Zoologica Academiae Scientiarum Hungaricae (Supplement 1), 49:99-108.

Kryštufek, B., M. G. Filippucci, M. Macholán, J. Zima, M. Vujoševi, and S. Simson. 1994. Does *Microtus majori* occur in Europe? Zeitschrift für Säugetierkunde, 59:349-357.

Kryštufek, B., H. I. Griffiths, and V. Vohralík. 1996. The status and use of *Terricola* Fatio, 1867 in the taxonomy of Palaearctic "pine voles" (*Pitymys*) (Rodentia, Arvicolinae). Bulletin de l'Institute Royal des Sciences Naturelles de Belgique, Biologie, 66:237-240.

Kryštufek, B., V. Vohralík, and C. Kurtonur. 1998. A new look at the identity and distribution of water shrews (*Neomys* spp.) in Turkey. Zeitschrift für Säugetierkunde, 63:129-136.

Kryštufek, B., A. Davison, and H. I. Griffiths. 2000a. Evolutionary biogeography of water shrews (*Neomys* spp.) in the western Palaearctic Region. Canadian Journal of Zoology, 78:1616-1625.

Kryštufek, B., K. Kolariè, and M. Paunović. 2000b. Age determination and age structure in Martino's vole *Dinaromys bogdanovi*. Mammalia, 64:361-370.

Kryštufek, B., F. Spitzenberger, and H. Kefelioglu. 2001. Description, taxonomy and distribution of *Talpa davidiana*. Mammalian Biology, 66:35-143.

Kryštufek, B., A. Hudoklin, and D. Pavlin. 2003. Population biology of the edible dormouse *Glis glis* in a mixed montane forest in central Slovenia over three years. Acta Zoologica Academiae Scientiarum Hungaricae (Supplement 1), 49:85-97.

Kryštufek, B., W. Haberl, R. M. Baxter, and J. Zima. 2004a. Morphology and karyology of two populations of the woodland dormouse *Graphiurus murinus* in the Eastern Cape, South Africa. Folia Zoologica, 53(4):339-350.

Kryštufek, B., B. Özkan, and C. Kurtonur. 2004b. Abnormal skull of Roach's mouse-tailed dormouse (*Myomimus roachi*). Lynx (Praha), 35:49-51.

Kuch, M., N. Rohland, J. L. Betancourt, C. Latorre, S. Steppan, and H. N. Poinar. 2002. Molecular analysis of a 11,700-year-old rodent midden from the Atacama Desert, Chile. Molecular Ecology, 11:913-924.

Kucheruk, V. V. 1990. Areal. [Range]. Pp. 34-84, *in* Seraya krysa: Sistematika, ekologiya, regulyatsiya chislenosti [Norway rat, systematics, ecology, population control] (V. E. Sokolov and E. V. Karasjova, eds.). Nauka, Moscow, 452 pp. (in Russian).

Kucheruk, V. V. 1994. [Spreading of black rat in Russia: Siberia and Dal'Niy East.] Byulleten' Moskovskovo Obschchestva Ispytatelei Prirody, Otdel Biologicheskii, 99(5):33-36 (in Russian with English abstract).

Kucheruk, V. V. 2000. [Distribution and specific features of synanthropy in *Rattus turkestanicus*.] Zoologicheskii Zhurnal, 79(4):495-502 (in Russian with English abstract).

Kuhn, H.-J. 1960. *Genetta* (*Paragenetta*) *lehmanni*, eine neue Schleichkatze aus Liberia. Säugetierkundliche Mitteilungen, 8:154-160.

Kuhn, H.-J. 1964. *Epixerus ebii jonesi* in Liberia. Bonner Zoologische Beiträge, 15:149-158.

Kuhn, H.-J. 1965. A provisional check-list of the mammals of Liberia. Senckenbergiana Biologica, 46(5):321-340.

Kuhn, H.-J. 1966. Der Zebraducker, *Cephalophus doria* (Ogilby, 1837). Zeitschrift für Säugetierkunde, 31:282-293.

Kuhn, H.-J. 1968. Der Jentink-Ducker. Natur und Museum, 98:17-23.

Kühnel, W. 1983. Über die glandula infraorbitalis des Goldhamsters (*Mesocricetus auratus* Waterhouse 1838). Acta Anatomica, 116:245-256.

Kuiper, K. 1926. On a black variety of the Malay tapir (*Tapirus indicus*). Proceedings of the Zoological Society of London, 1926:425-426.

Kulik, I. L. 1980. [Distribution and range types of jerboa (Dipodidae, Rodentia).] Pp. 152-167, *in* [Contemporary problems of zoogeography (A. G. Voronov and N. N. Drozdov, eds.)]. Nauka, Moscow, 323 pp. (in Russian).

Kulikov, V. F. 1988. [Morphology and functional significance of the vibrissae apparatus of certain rodents.] Pp. 69-80, *in* Ekologo-funktsionalnaya morfologiya kozhnogo pokrova mlekopitayushchikh [Ecological and functional morphology of mammalian skin] (V. E. Sokolov and R. P. Zhenevskaya, eds.). Nauka, Moscow, 205 pp.

Kumamoto, A. T., S. C. Kingswood, and W. Hugo. 1994. Chromosomal divergence in allopatric populations of Kirk's dik-dik, *Madoqua kirki* (Artiodactyla, Bovidae). Journal of Mammalogy, 75:357-364.

Kumamoto, A. T., S. C. Kingswood, W. E. R. Rebholz, and M. I. Houck. 1995. The chromosomes of *Gazella bennettii* and *Gazella saudiya*. Zeitschrift für Säugetierkunde, 60:159-169.

Kumamoto, A. T., S. J. Charter, M. L. Houck, and M. Frahm. 1996. Chromosomes of *Damaliscus* (Artiodactyla, Bovidae): Simple and complex centric fusion rearrangements. Chromosome Research, 4:614-621.

Kumirai, A., and J. K. Jones, Jr. 1990. *Nyctinomops femorosaccus*. Mammalian Species, 349:1-5.

Kuntz, R. E., and D. E. Ming. 1970. Vertebrates of Taiwan taken for parasitological and biomedical studies by U.S. Naval Medical Research Unit No. 2 Taipei, Taiwan, Republic of China. Quarterly Journal of the Taiwan Museum, 23:1-37.

Kunz, T. H. 1982. *Lasionycteris noctivagans*. Mammalian Species, 172:1-5.

Kunz, T. H., and D. P. Jones. 2000. *Pteropus vampyrus*. Mammalian Species, 642:1-6.

Kunz, T. H., and R. A. Martin. 1982. *Plecotus townsendii*. Mammalian Species, 175:1-6.

Kunz, T. H., and I. M. Pena. 1992. *Mesophylla macconnelli*. Mammalian Species, 405:1-5.

Kunze, B., W. Traut, S. Garagna, D. Weichenhan, C. A. Redi, and H. Winking. 1999*a*. Pericentric satellite DNA and molecular phylogeny in *Acomys* (Rodentia). Chromosome Research, 7:131-141.

Kunze, B., F. Dieterlen, W. Traut, and H. Winking. 1999*b*. Karyotype relationship among four species of spiny mice (*Acomys*, Rodentia). Zeitschrift für Säugetierkunde, 64:220-229.

Künzel, Th., and S. Künzel. 1998. An overlooked population of the beira antelope *Dorcatragus megalotis* in Djibouti. Oryx, 32:75-80.

Kurbanov, A. K., S. L. Kalabin and N. V. Molodova. 1990. [*Myomimus personatus* (Ognev, 1924) in Central Kopetdagh.] Izvestiya Akademii Nauk Turkmenskoi SSR,1:75.

Kuroda, N. 1924. [On new mammals from the Riu Kiu Islands and the vicinity.] Published by the author, Tokyo, 16 pp.

Kuroda, N. 1933. A revision of the genus *Pteropus* found in the islands of the Riu Kiu Chain, Japan. Journal of Mammalogy, 14:312-316.

Kuroda, N. 1935. Korean mammals preserved in the collection of Marquis Yamashina. Journal of Mammalogy, 15:229-239.

Kuroda, N. 1938. A list of the Japanese mammals. Privately published, Tokyo, 122 pp.

Kuroda, N. 1957. A new name for the lesser Japanese mole. Journal of the Mammalogical Society of Japan, 1:74.

Kuroda, S., T. Kano, and K. Muhindo. 1985. Further information on the new monkey species, *Cercopithecus salongo* Thys van den Audenaerde 1977. Primates, 26(3):325-333.

Kurose, N., A. V. Abramov, and R. Masuda. 2000. Intrageneric diversity of the cytochrome b gene and phylogeny of Eurasian species of the genus *Mustela* (Mustelidae, Carnivora). Zoological Science, Tokyo, 17:673-679.

Kurose, N., Y. Kaneko, A. V. Abramov, B. Siriaroonrat, and R. Masuda. 2001. Low genetic diversity in Japanese populations of the Eurasian badger *Meles meles* (Mustelidae, Carnivora) revealed by mitochondrial cytochrome b gene sequences. Zoological Science, Tokyo, 18:1145-1151.

Kurta, A., and R. H. Baker. 1990. *Eptesicus fuscus*. Mammalian Species, 356:1-10.

Kurtén, B. 1966. Pleistocene bears of North America. I. Genus *Tremarctos*, spectacled bears. Acta Zoologica Fennica, 115:100-120.

Kurtén, B. 1968. Pleistocene mammals of Europe. Aldine, Chicago, 317 pp.

Kurtén, B. 1973. Transberingian relationships of *Ursus arctos* Linné (brown and grizzly bears). Commentationes Biologicae, 65:1-10.

Kurtén, B., and E. Anderson. 1980. Pleistocene mammals of North America. Columbia University Press New York, 442 pp.

Kurtén, B., and R. Rausch. 1959. Biometric comparisons between North American and European mammals. I. A comparison between Alaskan and Fennoscandian wolverine (*Gulo gulo* Linnaeus). Acta Arctica, 11:1-21.

Kurtonur, C. 1992. First specimens of *Glis glis* (Linnaeus, 1776) from Turkish Thrace (Mammalia: Rodentia: Gliridae). Senckenbergiana Biologica, 72(1-3):1-6.

Kurtonur, C., and B. Ozkan. 1990. New records of *Myomimus roachi* (Bate 1937) from Turkish Thrace (Mammalia: Rodentia: Glirdae). Senckenbergiana Biologica, 71(4):239-244.

Kuss, S. E., and X. Misonne. 1968. Pleistozäne Muriden del Insel Kreta. Neues Jahrbücher für Geologie und Paläontologie, Abhandlungen, 132(1):55-69.

Kuss, S. E., and G. Storch. 1978. Eine Säugetierfauna (Mammalia: Artiodactyla, Rodentia) des älteren Pleistozäna von der Insel Kalymnos (Dodekanés; Griechenland). Neues Jahrbuch fuer Geologie und Palaeontologie Monatshefte, 1978:206-227.

Kuwayama, R., and T. Ozawa. 2000. Phylogenetic relationships among European red deer, wapiti, and sika deer inferred from mitochondrial DNA sequences. Molecular Phylogenetics and Evolution, 15:115-123.

Kuzhyakin, A. P. 1965. Myospalacinae. Pp. 343-344, *in* Key to mammals of USSR (N. A. Bobrinskii, B. A. Kuznetsov, and A. P. Kuzhyakin, eds.). Enlightenment Publishing House, Moscow, 382 pp.

Kuz'mina, I. E. 1965. Saiga i stepnaya pishchukha v verkhov'yakh Pechory [Saiga and steppe pika in the upper Pechora river]. Zoologicheskii Zhurnal, 44:307-311 (in Russian).

Kuz'mina, I. E, 1997. Loshadi severnoi evrazii ot pliotsena do sovremennosti. Trudy Zoologicheskogo Instituta, 273:1-223.

Kuznetsov, B. A. 1944. Ordo Rodentia. Pp. 262-362, *in* Opredelitel' mlekopitayushchikh SSSR [Guide to the mammals of the USSR] (Bobrinskii, N. A., Kuznetsov, B. A. and A. P. Kuzyakin, eds.). Sovetskaya Nauka, Moscow, 439 pp. (in Russian).

Kuznetsov, B. A. 1948. Zveri Kirgizii [Animals of Kirgiziya]. Byul. Mosk. O-va. Ispyt. Priroda, Otd. Biol. [Bulletin Moscow Society Naturalists]12:1-27.

Kuznetsov, B. A. 1965. Ordo Rodentia. Pp. 236-346, *in* Opredelitel' mlekopitayushchikh SSSR [Guide to the mammals of the USSR] (Bobrinskii, N. A., Kuznetsov, B. A. and A. P. Kuzyakin, eds.). Second edition. Proveshchenie, Moscow, 382 pp. (in Russian).

Kuznetsov, G. 2000. Mammals of coastal islands of Vietnam: Zoogeographical and ecological aspects. Pp. 357-366, *in* Isolated vertebrate communities in the Tropics (G. Rheinwald, ed.). Bonner Zoologische Monographie, 46:400 pp.

Kuznetsov, G. V., E. E. Kalikov, N. B. Petrov, N. V. Ivanova, A. A. Lomov, M. V. Kholodova, and A. B. Poltaraus, 2002. Mitochondrial 12S and DNA sequence relationships suggest that the enigmatic bovid "linh duong" *Pseudonovibos spiralis* is closely related to buffalo. Molecular Phylogenetics and Evolution, 23:91-94.

Kwa'ioloa, M., and B. Burt. 2001. Our Forest of Kwara'ae: Our life in Solomon Islands and the things growing in our home-Na Masu'u Kia'I Kwara'aw:Tualaka'I Solomon Islands Fa'inia Logo na ru Bulao ki Saena Fanoa Kia. British Museum Press, London, 238 pp.

Kwiecinski, G. G. 1998. *Marmota monax*. Mammalian Species, 591:1-8.

Kwiecinski, G. G., and T. A. Griffiths. 1999. *Rousettus egyptiacus*. Mammalian Species, 611:1-9.

Kyselyuk, A. I. 1993. [*Sylvaemus microps* (Rodentia, Muridae) in the East Carpathians.] Vestnik Zoologii, 1994(4):41-47 (in Russian with English abstract).

Labes, R., R. Brasecke, and D. Andresen. 1987. Zum Vorkommen des Siebenschlafers (*Glis glis*) im Kreis Schwerin. Säugetierkundliche Informationen, 2(11):441-447.

Lacépède, B. G. E. de. 1809. *In*, F. d'Azara, Voyages dans l'Amérique Méridionale depuis 1781 jusqu'en 1801; contenant la description géographique, politique et civile du Paraguay et de la rivière de la Plata; l'histoire de la découverte et de la conquête de contrées; des détails nombreux sur leur historie naturelle, et sur les peuples sauvages qui les habitent. Dentu, Paris 4 volumes & atlas.

Lackey, J. A. 1967. Biosystematics of *heermanni* group kangaroo rats in southern California. Transactions of the San Diego Society of Natural History, 14:313-344.

Lackey, J. A. 1991*a*. *Chaetodipus arenarius*. Mammalian Species, 384:1-4.

Lackey, J. A. 1991*b*. *Chaetodipus spinatus*. Mammalian Species, 385:1-4.

Lackey, J. A. 1996. *Chaetodipus fallax*. Mammalian Species, 517:1-6.

Lackey, J. A., and T. L. Best. 1992. *Chaetodipus goldmani*. Mammalian Species, 419:1-5.

Lackey, J. A., D. G. Huckaby, and B. G. Ormiston. 1985. *Peromyscus leucopus*. Mammalian Species, 247:1-10.

Lacy, R. C., and B. E. Horner. 1996. Effects of inbreeding on skeletal development of *Rattus villosissimus*. Journal of Heredity, 87:277-287.

Laerm, J. 1981. Systematic status of the Cumberland Island pocket gopher, *Geomys cumberlandius*. Brimleyana, 6:141-151.

Laerm, J., E. Brown, M. A. Menzel, A. Wotjalik, W. M. Ford, and M. Strayer. 1995. The masked shrew, *Sorex cinereus* (Insectivora: Soricidae), and red-backed vole, *Clethrionomys gapperi* (Rodentia: Muridae), in the Blue Ridge Province of South Carolina. Brimleyana, 22:15-21.

Lagos, J. A., and F. Bozinovic. 1999. Intra and interspecific allometric scaling of intestinal dimensions in phyllotine rodents. Revista Chilena de Historia Natural, 72:57-61.

Laing Jun-xun, and Zhang Jun. 1985. [On the rodent fauna and regionalization of the northeast regions in loess plateau.] Acta Theriologica Sinica, 5(4):299-309 (in Chinese).

Lakhotia, S. C., S. R. V. Rao and S. C Jhanwar. 1973. Studies on rodent chromosomes VI. Co-existence of *Rattus rattus* with 38 and 42 chromosomes in South Western India. Cytologia, 38:403-410.

Lamb, J. M., A. Rimmer, J. Raubenheimer, and G. Contrafatto. 1996. Variation of mitochondrial-DNA restriction fragment length in the species *Otomys irroratus* (Otomyinae: Muridae). Mammalia, 60:790-792.

Lamb, T., T. R. Jones, and P. J. Wettstein. 1997. Evolutionary genetics and phylogeography of tassel-eared squirrels (*Sciurus aberti*). Journal of Mammalogy, 78(1):117-133.

Lamotte, M., and F. Petter. 1981. Une taupe dorée nouvelle du Cameroun (Mt Oku, 6°15′N, 10°26′E): *Chrysochloris stuhlmanni balsaci* ssp. nov. Mammalia, 45:43-48.

Lancaster, W. C., and E. K. V. Kalko. 1996. *Mormoops blainvillii*. Mammalian Species, 544:1-5.

Lance, E. W., and J. A. Cook. 1998. Biogeography of tundra voles (*Microtus oeconomus*) of Beringia and the southern coast of Alaska. Journal of Mammalogy, 79:53-65.

Landry, S. O. 1957. The interrelationships of the New and Old World hystricognath rodents. University of California Publications in Zoology, 56:1-118.

Landry, S. O., Jr. 1974. The fundamental relationship of the Lagomorpha and Rodentia (abstract). Transactions of the First International Theriological Congress, Moscow, 2:202-203.

Landry, S. O., Jr. 1999. A proposal for a new classification and nomenclature for the Glires (Lagomorpha and Rodentia). Mitteilungen des Museums für Naturkunde, Berlin, Zoologische Reihe, 75:283-316.

Langdon, R. 1995. Some iconoclastic thoughts about those Polynesian rat bones at Anakena. Rapa Nui Journal, 9(3):77-80.

Lange, R. 1992. Aardmuis, *Microtus agrestis* (L., 1761). Pp. 261-268, *in* Atlas van de Nederlandse Zoogdieren (S. Broekhuizen, B. Hoekstra, V. van Laar, C. Smeenk, J. B. M. Thissen, eds.). Stichting Uitgeverij Koninklijke Nederlandse Natuurhistorische Vereniging, Utrecht, 336 pp.

Langevin, P., and R. M. R. Barclay. 1990. *Hypsignathus monstrosus*. Mammalian Species, 357:1-4.

Langguth, A. 1967. Sobre la identidad de *Dusicyon culpaeolus* (Thomas) y de *Dusicyon inca* (Thomas). Neotropica, 13:21-28.

Langguth, A. 1969. Die südamerikanischen Canidae unter besonderer Bercksichtigung des Mohnenwolfes, *Chrysocyon brachyurus* Illiger. Zeitschrift für Wissenschaftliche Zoologie, 179:1-188.

Langguth, A. 1975. Ecology and evolution in the South American canids. Pp. 192-206, *in* The wild canids: Their systematics, behavioral ecology, and evolution (M. W. Fox, ed.). Van Nostrand Reinhold Company, New York, NY, 508 pp.

Langguth, A., and A. Abella. 1970. Las especies uruguayas del género *Ctenomys*. Communiciones Zoológicos Museo Historia Natural de Montevideo, 10(129):1-20.

Langguth, A., and C. B. Bonvicino. 2002. The *Oryzomys subflavus* species group, with description of two new species (Rodentia, Muridae, Sigmodontinae). Arquivos do Museu Nacional, Rio de Janeiro, 60:285-294.

Langguth, A., and E. J. Silva Neto. 1993. Morfologia do penis em *Pseudoryzomys wavrini* and *Wiedomys pyrrhorhinos* (Rodentia – Cricetidae). Revista Nordestina de Biologia, 8:55-59.

Lansman, R. A., J. C. Avise, C. F. Aquadro, J. F. Shapira, and S. W. Daniel. 1983. Extensive genetic variation in mitochondrial DNAs among geographic populations of the deer mouse, *Peromyscus maniculatus*. Evolution, 36:1-16.

Lapini, L., and R. Testone. 1998. A new *Sorex* from north-eastern Italy (Mammalia: Insectivora: Soricidae). Gortania—Atti del Museo Friulano di Storia Naturale, Udine, 20:231-250 (in Italian).

Lapini, L., M. G. Filipucci, and S. Filacorda. 2001. Genetic and morphometric comparison between *Sorex arunchi* Lapini and Testone, 1998, and other shrews from Italy. Acta Theriologica, 46:337-352.

Lapshov, V. A. 1992. [Projection analysis of low mandible shape of the black (*Rattus rattus*), brown (*R. norvegicus*) and Turkestan (*R. Turkestanicus*) rats.] Zoologicheskii Zhurnal, 71(6):125-134.

Lara, M. C., and J. L. Patton. 2000. Evolutionary diversification of spiny rats (genus *Trinomys*, Rodentia: Echimyidae) in the Atlantic forest of Brazil. Zoological Journal of the Linnean Society, 130:661-686.

Lara, M. C., J. L. Patton, and M. N. F. da Silva. 1996. The simultaneous diversification of South American echimyid rodents (Hystricognathi) based on complete cytochrome b sequences. Molecular Phylogenetics and Evolution, 5(2):403-413.

Lara, M. C., J. L. Patton, and E. Hingst-Zaher. 2002. *Trinomys mirapitanga*, a new species of spiny rat (Rodentia: Echimyidae) from the Brazilian Atlantic forest. Mammalian Biology, 67:233-242.

Largen, M. J. 1985. Taxonomically and historically significant specimens of mammals in the Merseyside County Museums, Liverpool. Journal of Mammalogy, 66:412-418.

Larivière, S. 1998. *Lontra felina*. Mammalian Species, 575:1-5.

Larivière, S. 1999a. *Lontra provocax*. Mammalian Species, 610:1-4.

Larivière, S. 1999b. *Lontra longicaudis*. Mammalian Species, 609:1-5.

Larivière, S. 1999c. *Mustela vison*. Mammalian Species, 608:1-9.

Larivière, S. 2001a. *Poecilogale albinucha*. Mammalian Species, 681:1-4.

Larivière, S. 2001b. *Ursus americanus*. Mammalian Species, 647:1-11.

Larivière, S. 2001c. *Aonyx congicus*. Mammalian Species, 650:1-3.

Larivière, S. 2001d. *Aonyx capensis*. Mammalian Species, 671:1-6.

Larivière, S. 2002a. *Vulpes zerda*. Mammalian Species, 714:1-5.

Larivière, S. 2002b. *Ictonyx striatus*. Mammalian Species, 698:1-5.

Larivière, S. 2002c. *Lutra maculicollis*. Mammalian Species, 712:1-6.

Larivière, S. 2003. *Amblonyx cinereus*. Mammalian Species, 720:1-5.

Larivière, S., and J. Calzada. 2001. *Genetta genetta*. Mammalian Species, 680:1-6.

Larivière, S., and A. M. Pasitschniak. 1996. *Vulpes vulpes*. Mammalian Species, 537:1-11.

Larivière, S., and P. J. Seddon. 2001. *Vulpes rueppelli*. Mammalian Species, 678:1-5.

Larivière, S., and L. R. Walton. 1997. *Lynx rufus*. Mammalian Species, 563:1-8.

Larivière, S., and L. R. Walton. 1998. *Lontra canadensis*. Mammalian Species, 587:1-8.

Larizza, A., G. Pesole, A. Reyes, E. Sbisà, and C. Saccone. 2002. Lineage specificity of the evolutionary dynamics of the mtDNA D-Loop region in rodents. Journal of Molecular Evolution, 54:145-155.

Larson, S. E. 1997. Taxonomic re-evaluation of the Jaguar. Zoo Biology, 16:107-120.

Lassieur, S., and D. E. Wilson. 1989. *Lonchorhina aurita*. Mammalian Species, 347:1-4.

Lasso, C., R. Hutterer, and A. Rial. 1996. Records of shrews (Soricidae) from Equatorial Guinea, especially from Monte Alen National Park. Mammalia, 60:69-76.

Lataste, F. 1883*a*. Note sur les souris d'Algérie et description d'une espèce nouvelle (*Mus spretus*). Acta Linnaean Société de Sciences Naturelle de Maroc, 17:1-7.

Lataste, F. 1883*b*. Introduction a l'étude des campagnols de France. Historique de la classification des campagnols. Le Naturaliste, 2(41):323-324.

Lataste, F. 1883*c*. Introduction a l'étude des campagnols de France. Historique de la classification des campagnols. Le Naturaliste, 2(44):347-349.

Lataste, F. 1887. Observations sur quelques espèces du genre campagnol (*Microtus* Schranck, *Arvicola* Lacépède). Annali del Museo Civico di Storia Naturale di Genova, ser. 2ª, 4(24):259-274.

Lau, Y.-F.d., T. L. Yang-Feng, B. Elder, K. Fredga, and U. H. Wiberg. 1992. Unusual distribution of *Zfy* and *Zfx* sequences on the sex chromosomes of the wood lemming, a species exhibiting XY sex reversal. Cytogenetics and Cell Genetics, 60:48-54.

Laurie, E. M. O. 1952. Mammals collected by Mr. Shaw Mayer in New Guinea 1932-1949. Bulletin of the British Museum (Natural History), Zoology Series, 1:269-318.

Laurie, E. M. O., and J. E. Hill. 1954. List of land mammals of New Guinea, Celebes and adjacent islands 1758-1952. British Museum (Natural History) Publications, London, 175 pp.

Laurie, W. A., E. M. Lang, and C. P. Groves. 1983. *Rhinoceros unicornis*. Mammalian Species, 211:1-6.

Laursen, L., and M. Bekoff. 1978. *Loxodonta africana*. Mammalian Species, 92:1-8.

LaVal, R. K. 1970. Intraspecific relationships of bats of the species *Myotis austroriparius*. Journal of Mammalogy, 51:542-552.

LaVal, R. K. 1973*a*. A revision of the Neotropical bats of the genus *Myotis*. Science Bulletin, Natural History Museum of Los Angeles County, 15:1-54.

LaVal, R. K. 1973*b*. Systematics of the genus *Rhogeesa* (Chiroptera: Vespertilionidae). Occasional Papers of the Museum of Natural History, University of Kansas, 19:1-47.

Lavocat, R. 1956. La faune des rongeurs des grottes a australopitheques. Paleontologia Africana, 4:69-75.

Lavocat, R. 1959. Origine et affinités des rongeurs de la sous-famille des Dendromurinés. Comptes Rendus des Séances de l'Académie des Sciences (Paris), 248:1375-1377.

Lavocat, R. 1964. On the systematic affinities of the genus *Delanymys* Hayman. Proceedings of the Linnean Society of London, 175:183-185.

Lavocat, R. 1965. Order: Rodentia and Lagomorpha. Pp. 17-19, *in* Olduvai Gorge, 1951-61. A Preliminary Report on the Geology and Fauna (L. S. B. Leakey, ed.). Vol. I. The University Press, Cambridge, 116 pp.

Lavocat, R. 1969. La systematique des rongeurs hystricomorphes et la derive des continents. Comptes rendus hebdomadaires des Seances de l'Academie des Sciences, Paris D, 269:1496-1497.

Lavocat, R. 1971. Affinities systematiques des caviomorphes et des phiomorphes et l'origine Africaine des caviomorphes. Anais da Academia Brasiliera de Ciencias, 43(suppl.):515-522.

Lavocat, R. 1973. Les rongeurs du Miocene d'Afrique Orientale. Memoires et Travaux de l'Ecole Pratique des Hautes Etudes, Institut de Montpellier, 1:1-284.

Lavocat, R. 1974. What is an hystricomorph? Symposium of the Zoological Society of London, 34:7-20.

Lavocat, R. 1978. Rodentia and Lagomorpha. Pp. 69-89, *in* Evolution of African mammals (V. J. Maglio and H. B. S. Cooke, eds.). Harvard University Press, Cambridge, MA, 641 pp.

Lavocat, R. 1980. The implications of rodent paleontology and biogeography to the geographic sources and origin of platyrrhine primates. Pp. 93-102, *in* Evolutionary Biology of the New World Monkeys and Continental Drift (R. L. Ciochon and A. B. Chiarelli, eds.). Plenum Press, New York, xvi + 528 pp.

Lavocat, R., and J.-P. Parent. 1985. Phylogenetic analysis of middle ear features in fossil and living rodents. Pp. 333-354, *in* Evolutionary Relationships among Rodents: A Multidisciplinary Analysis (W. P. Luckett and J.-L. Hartenberger, eds.). Plenum Press, New York, 721 pp.

Lavrenchenko, L. A. 1993. [*Uranomys ruddi* (Rodentia, Muridae)–a new genus and species in the fauna of Ethiopia.] Zoologicheskii Zhurnal, 72(3):133-137 (in Russian with English abstract)

Lavrenchenko, L. A. 1994. The strong sexual dimorphism of Cricetidae – *Oryzomys xantheolus*. Pp. 132-134, *in* Mammals of Peruvian Amazonia (A. N. Severtsova, ed.). Rossiiskaia Akademiia Nauk, Institut Evoliutsionnoi Morfologii I Ekologii Zhivotnykh, 299 pp. (in Russian with English abstract).

Lavrenchenko, L. A. 1996. Variation of generative system structure in some species of the genus *Mastomys* Thomas, 1915 (Rodentia, Muridae). Mammalia, 60(2):277-288.

Lavrenchenko, L. A. 2000. The mammals of the isolated Harenna Forest (southern Ethiopia): Structure and history of the fauna. Pp. 223-231, *in* Isolated vetebrate communities in the Tropics. Proceedings of the 4th International Symposium, Bonn (G. Rheinwald, ed.). Bonner Zoologische Monographien, 46: 400 pp.

Lavrenchenko, L. A. 2003. A contribution to the systematics of *Desmomys* Thomas, 1910 (Rodentia, Muridae) with the description of a new species. Bonner Zoologische Beiträge, 50(4):313-327.

Lavrenchenko, L. A., and M. I. Baskevich. 1996. Variation of generative system structure in some species of the genus *Mastomys* Thomas, 1915 (Rodentia, Muridae). Mammalia, 60(2):277-288.

Lavrenchenko, L. A., and A. G. Dmitriev. 1994. On systematics, ecology, and distribution of mammals in the western part of Department Piura (north-western Peru). Pp. 122-131, *in* Mammals of Peruvian Amazonia (A. N. Severtsova, ed.). Rossiiskaia Akademiia Nauk, Institut Evoliutsionnoi Morfologii I Ekologii Zhivotnykh, 299 pp. (in Russian with English abstract).

Lavrenchenko, L. A., and O. P. Likhnova. 1995. [Allozymic and morphological variability in three syntopic species of wood mice from the subgenus *Sylvaemus* (Rodentia, Muridae, *Apodemus*) from Daghestan.] Zoologicheskii Zhurnal, 74(5):107-119 (in Russian with English abstract).

Lavrenchenko, L. A., O. P. Likhnova, and V. N. Orlov. 1992. [Hemoglobin patterns: A possible implication in systematics of multimammate rats *Mastomys* (Muridae, Rodentia).] Zoologicheskii Zhurnal, 71(8):85-93 (in Russian with English abstract).

Lavrenchenko, L. A., A. N. Milishnikov, V. M. Aniskin, A. A. Warshavsky, and W. Gebrekidan. 1997. The genetic diversity of small mammals of the Bale Mountains, Ethiopia. SINET: Ethiopian Journal of Science, 20:213-233.

Lavrenchenko, L. A., O. P. Likhnova, M. I. Baskevich, and A. Bekele. 1998*a*. Systematics and distribution of *Mastomys* (Muridae, Rodentia) from Ethiopia, with the description of a new species. Zeitschrift für Säugetierkunde, 63:37-51.

Lavrenchenko, L. A., W. N. Verheyen, and J. Hulselmans. 1998*b*. Systematic and distributional notes on the *Lophuromys flavopunctatus* Thomas, 1888 species-complex in Ethiopia (Muridae-Rodentia). Bulletin de L'Institut Royal des Sciences Naturelles de Belgique. Biologie, 68:199-214.

Lavrenchenko, L. A., A. N. Milishnikov, V. M. Aniskin, and A. A. Warshavsky. 1999. Systematics and phylogeny of the genus *Stenocephalemys* Frick, 1914 (Rodentia, Muridae): A multidisciplinary approach. Mammalia, 63(4):475-494.

Lavrenchenko, L. A., A. N. Milishnikov, and A. A. Warshavsky. 2000. Allozymic phylogeny: Evidence for coherent adaptive patterns of speciation in Ethiopian endemic rodents from an isolated montane massif. Pp. 245-253, *in* Isolated vertebrate communities in the Tropics. Proceedings of the 4th International Symposium, Bonn (G. Rheinwald, ed.). Bonner Zoologische Monographien, 46:400 pp.

Lavrov, L. S. 1979. [Species of beavers (*Castor*) of the Palearctic.] Zoologicheskii Zhurnal, 58:88-96 (in Russian).

Lavrov, L. S. 1983. Evolutionary development of the genus *Castor* and taxonomy of the contemporary beavers of Eurasia. Acta Zoologica Fennica, 174:87-90.

Lavrov, L. S., and V. N. Orlov. 1973. [Karyotypes and taxonomy of modern beavers (*Castor*, Castoridae, Mammalia).] Zoologicheskii Zhurnal, 52:734-742 (in Russian).

Lawlor, T. E. 1965. The Yucatan deer mouse, *Peromyscus yucatanicus*. University of Kansas Publications, Museum of Natural History, 16:421-438.

Lawlor, T. E. 1969. A systematic study of the rodent genus *Ototylomys*. Journal of Mammalogy, 50:28-42.

Lawlor, T. E. 1971*a*. Distribution and relationships of six species of *Peromyscus* in Baja California and Sonora, Mexico. Occasional Papers of the Museum of Zoology, University of Michigan, 661:1-22.

Lawlor, T. E. 1971*b*. Evolution of *Peromyscus* on northern islands in the Gulf of California, Mexico. Transactions of the San Diego Society of Natural History, 16:91-124.

Lawlor, T. E. 1982. *Ototylomys phyllotis*. Mammalian Species, 181:1-3.

Lawlor, T. E. 1983. The mammals. Pp. 265-289, *in* Island biogeography in the Sea of Cortez (T. J. Case and M. L. Cody, eds.). University of California Press, Berkeley, 508 pp.

Lawrence, B. 1933. Howler monkeys of the *Palliata* group. Bulletin of the Museum of Comparative Zoology at Harvard College, 75:314-354.

Lawrence, B. 1939. Mammals. Pp. 28-73, *in* Collections from the Phillippine Islands (T. Barbour, B. Lawrence, and J. L. Peters). Bulletin of the Museum of Comparative Zoology at Harvard College, 86(2):25-128.

Lawrence, B. 1941. *Neacomys* from northwestern South America. Journal of Mammalogy, 22:418-427.

Lawrence, B. 1964. Notes on the horseshoe bats *Hipposideros caffer*, *ruber* and *beatus*. Breviora, 207:1-5.

Lawrence, B., and W. H. Bossert. 1967. Multiple character analysis of *Canis lupus*, *latrans* and *niger*. American Zoologist, 7:223-232.

Lawrence, B., and W. H. Bossert. 1975. Relationships of North American *Canis* shown by a multiple character analysis of selected populations, Pp. 73-86, *in* The wild canids, their systematics, behavioral ecology and evolution (M. W. Fox, ed.). Van Nostrand Reinhold, New York, NY. 508 pp.

Lawrence, B., and A. Loveridge. 1953. Zoological results of a fifth expedition to East Africa. I. Mammals from Nyasaland and Tete. With notes on the genus *Otomys*. Bulletin of the Museum of Comparative Zoology at Harvard College, 110:1-80.

Lawrence, B., and A. Novick. 1963. Behavior as a taxonomic clue: Relationships of *Lissonycteris* (Chiroptera). Breviora, 184:1-16.

Lawrence, M. A. 1982. Western Chinese arvicolines (Rodentia) collected by the Sage Expedition. American Museum Novitates, 2745:1-19.

Lawrence, M. A. 1988. The identity of *Sciurus duida* J. A. Allen (Rodentia: Sciuridae). American Museum Novitates, 2919:1-8.

Lawrence, M. A. 1991. A fossil *Myospalax* cranium (Rodentia: Muridae) from Shanxi, China, with observations on Zokor relationships. Pp. 261-286, *in* Contributions to mammalogy in honor of Karl F. Koopman (T. A. Griffiths and D. Klingener, eds.). Bulletin of the American Museum of Natural History, 206:1-432.

Lay, D. M. 1967. A study of the mammals of Iran resulting from the Street expedition of 1962-63. Fieldiana: Zoology, 54:1-282.

Lay, D. M. 1972. The anatomy, physiology, functional significance and evolution of specialized hearing organs of gerbilline rodents. Journal of Morphology, 138:41-120.

Lay, D. M. 1975. Notes on rodents of the genus *Gerbillus* (Mammalia: Muridae: Gerbillinae) from Morocco. Fieldiana: Zoology, 65:89-101.

Lay, D. M. 1983. Taxonomy of the genus *Gerbillus* (Rodentia: Gerbillinae) with comments on the applications of generic and subgeneric names and an annotated list of species. Zeitschrift für Säugetierkunde, 48:329-354.

Lay, D. M., and C. F. Nadler. 1969. Hybridization in the rodent genus *Meriones*. I. Breeding and cytological analyses of *Meriones shawi* (female) X *Meriones libycus* (male) hybrids. Cytogenetics, 8:35-50.

Lay, D. M., and C. F. Nadler. 1972. Cytogenetics and origin of North African *Spalax* (Rodentia: Spalacidae). Cytogenetics, 11:279-285.

Lay, D. M., and C. F. Nadler. 1975. A study of *Gerbillus* (Rodentia: Muridae) east of the Euphrates River. Mammalia, 39:423-445.

Lay, D. M., J. A. W. Anderson, and J. D. Hassinger. 1970. New records of small mammals from West Pakistan and Iran. Mammalia, 34:97-106.

Lay, D. M., K. Agerson, and C. F. Nadler. 1975. Chromosomes of some species of *Gerbillus* (Mammalia: Rodentia). Zeitschrift für Säugetierkunde, 40:141-150.

Layne, J. N. 1990. The Florida mouse. Pp. 1-21, *in* Burrow associates of the gopher tortoise (C. K. Dodd, Jr., R. E. Ashton, Jr., R. Franz, and E. Wester, eds.). Florida Museum of Natural History, Gainesville, 134 pp.

Lazarová, J. 1999. Epigenetic variation and fluctuating asymmetry of the house mouse (*Mus*) in the Czech Republic. Folia Zoologica, 48(suppl. 1):37-52.

Lazell, J. D., Jr. 1981. Field and taxonomic studies of tropical American raccoons. National Geographic Society Research Report, 13:381-385.

Lazell, J. D, Jr. 1993. Review of rare and endangered biota of Florida, Volume 1 (S. R. Humphrey, ed.). Journal of Mammalogy, 74:1077-1079.

Lazell, J. D., and K. F. Koopman. 1985. Notes on bats of Florida's Lower Keys. Florida Scientist, 48:37-41.

Leal-Mesquita, E. R. 1991. Estudos citgeneticos em dez especies de roedores brasileiros da familia Echimyidae. Revista Brasileira de Genetica, 14:1094.

Leal-Mesquita, E. R., Y. Yonenaga-Yassuda, T. H. Chu, and P. L. B. da Rocha. 1992. Chromosomal characterization and comparative cytogenetic analysis of two species of *Proechimys* (Echimyidae, Rodentia) from the Caatinga domain of the state of Bahia, Brazil. Caryologia, 45:197-212.

Leal-Mesquita, E. R., V. Fagundes, Y. Yonenaga-Yassuda, and P. L. B. da Rocha. 1993. Comparative cytogenetic studies of two karyomorphs of *Trichomys apereoides* (Rodentia, Echimyidae). Revista Brasileira de Genetica, 16(3):639-651.

Leamy, L. J., E. J. Routman, and J. M. Cheverud. 2002. An epistatic genetic basis for fluctuating asymmetry of mandible in size in mice. Evolution, 56(3):642-653.

Leary, T., and L. Seri. 1997. An annotated checklist of mammals recorded in the Kikori River Basin, Papua New Guinea. Science in New Guinea, 23(2):79–100.

Leatherwood, S., and R. R. Reeves (eds). 1990. The bottlenose dolphin. Academic Press, New York, 653 pp.

Leatherwood, S., J. S. Grove, and A. E. Zuckerman. 1991. Dolphins of the genus *Lagenorhynchus* in the tropical South Pacific. Marine Mammal Science, 7(2):194-197.

Lebedev, V. S., I. Ya. Pavlinov, M. N. Meyer, and V. G. Malikov. 1998. [Craniometric analysis of mouse-like hamsters of the genus *Calomyscus* (Cricetidae).] Zoologschiskii Zhurnal, 77(6):721-731 (in Russian with English abstract).

Le Berre, M. 1990. Faune du Sahara. 2. Mammiferes. Lechevalier-R. Chabaud, Paris, 360 pp.

Le Berre, M., and L. Le Guelte. 1990. Les mammiferes actuels dans l'espace saharien. Vie et Milieu, 40:223-228.

Lechleitner, R. R. 1969. Wild mammals of Colorado: Their appearance, habits, distribution and abundance. Pruett Publishing Co., Colorado, 254 pp.

Lecompte, É. 2003. Systématique et evolution du groupe *Praomys* (Rodentia, Murinae). These Docteur du Muséum National d'Histoire Naturelle, Paris, 365 pp.

Lecompte, É., E. Van der Straeten, F. Petter, and C. Denys. 1999. Caractéristiques cranio-dentaires de l'holotype de *Praomys morio* (Rodentia, Muridae). Mammalia, 63(4):530-534.

Lecompte, É., C. Denys, and L. Granjon. 2001. An identification key for the species within the genus *Praomys* (Rodentia: Muridae). Pp. 127-139, *in* African small mammals (C. Denys, L. Granjon, and A. Poulet, eds.). IRD Éditions, Collection Colloques et Séminaires, Paris, 570 pp.

Lecompte, É., L. Granjon, and C. Denys. 2002*a*. The phylogeny of the *Praomys* complex (Rodentia: Muridae) and its phylogeographic implications. Journal of Zoological Systematics and Evolutionary Research, 40:8-25.

Lecompte, É., L. Granjon, J. Kerbis Peterhans, and C. Denys. 2002*b*. Cytochrome *b*-based phylogeny of the *Praomys* group (Rodentia, Murinae): A new African radiation? Comptes Rendus Biologies, 325:827-840.

Ledje, C., and U. Árnason. 1996. Phylogenetic relationships within caniform carnivores based on analyses of the mitochondrial 12S rRNA gene. Journal of Molecular Evolution, 43:641-649.

LeDuc, R. G., and B. E. Curry. 1997. Mitochondrial DNA sequence analysis indicates need for revision of the genus *Tursiops*. International Whaling Commission, Reports, 47:393.

Lee, A. K., P. R. Baverstock, and C. H. S. Watts. 1981. Rodents— the late invaders. Pp. 1523-1553, *in* Ecological biogeography of Australia (A. Keast, ed.) [vol. 3, part 6]. Monographiae Biologicae, 41:1-2142 [in 3 vols.].

Lee, D. S., and M. K. Clark. 1993. A second record of the southern bog lemming, *Synaptomys cooperi*, from the Delmarva Peninsula. The Maryland Naturalist, 37:29-35.

Lee, H. Y., and R. J. Baker. 1987. Cladistical analysis of chromosomal evolution in pocket gophers of the *Cratogeomys castanops* complex (Rodentia: Geomyidae). Occasional Papers, The Museum, Texas Tech University, 114:1-15.

Lee, J.-H., and S.-W. Son. 1997. Morphological characteristics of sperm in the Korean striped field mouse, *Apodemus agrarius coreae*: Possible role of sperm neck in the movement of sperm head. Korean Journal of Biological Science, 1:371-379.

Lee, M. R., and F. B. Elder. 1977. Karyotypes of eight species of Mexican rodents (Muridae). Journal of Mammalogy, 58:479-487.

Lee, M. R., and D. F. Hoffmeister. 1963. Status of certain fox squirrels in Mexico and Arizona. Proceedings of the Biological Society of Washington, 76:181-190.

Lee, M. R., D. J. Schmidly, and C. C. Huheey. 1972. Chromosomal variation in certain populations of *Peromyscus boylii* and its systematic implications. Journal of Mammalogy, 53:697-707.

Lee, T. E., Jr. 1990. Geographic distribution of the cytotypes of *Chaetodipus nelsoni canescens* (Rodentia: Heteromyidae). Southwestern Naturalist, 35:454-455.

Lee, T. E., Jr., and R. D. Bradley. 1992. New distributional records of some mammals from Honduras. Texas Journal of Science, 44:109-111.

Lee, T. E., Jr., and M. D. Engstrom. 1991. Genetic variation in the silky pocket mouse (*Perognathus flavus*) in Texas and New Mexico. Journal of Mammalogy, 72:273-285.

Lee, T. E., B. R. Riddle, and P. L. Lee. 1996. Speciation in the desert pocket mouse (*Chaetodipus penicillatus* Woodhouse). Journal of Mammalogy, 77:58-68.

Lee, T. E., J. B. Scott, and M. M. Marcum. 2001. *Vampyressa bidens*. Mammalian Species, 684:1-3.

Lee, T. E., S. R. Hooker, and R. A. Van Den Bussche. 2002. Molecular phylogenetics and taxonomic revision of the genus *Tonatia* (Chiroptera: Phyllostomidae). Journal of Mammalogy, 83:49-57.

Legendre, S. 1982. Hipposideridae (Mammalia: Chiroptera) from the Mediterranean Middle and Late Neogene, and evolution of the genera *Hipposideros* and *Asellia*. Journal of Vertebrate Paleontology, 2:372-385.

Legendre, S. 1984. Étude odotologíque des représentants actuels du groupe *Tadarida* (Chiroptera, Molossidae). Implications phylogéniques, systématiques et zoogéographiques. Revue Suisse de Zoologie, 91:399-442.

Le Gros Clark, W. 1926. On the anatomy of the pen-tailed tree-shrew (*Ptilocercus lowii*). Proceedings of the Zoological Society of London, 1926:1179-1309.

Lehmann, E. von. 1961. Über die kleinsäuger der La Sila (Kalabrien). Zoologischer Anzeiger, 167(5-6):213-227.

Lehmann, E. von. 1963. Die Säugetiere des Fürstentums Liechtenstein. Jahrbuch des Historischen Vereins für das Fürstentum Liechtenstein, 62:159-362.

Lehmann, E. von. 1965. Über die Säugetiere im Waldgebiet N.W.- Syriens. Sitzungsberichte der Gesellschaft Naturforschender Freunde zu Berlin, 5:22-38.

Lehmann, E. von. 1969. Über die Hautdrüsen der Schneemaus (*Chionomys nivalis nivalis* Martins, 1842). Bonner Zoologische Beiträge, 20:373-377.

Lehmann, E. von. 1983. Eine Kleinsäugeranthologie aus Strassburg (in memoriam Johann Hermann, 1738-1800). Annalen des Naturhistorischen Museums in Wien, ser. B, 84:509-514.

Leirs, H. 1992. Population ecology of *Mastomys natalensis* (Smith, 1834) multimammate rats: Possible implications for rodent control in Africa. Universiteit Antwerpen, Antwerpen, 263 pp.

Leirs, H., W. Verheyen, M. Michiels, R. Verhagen, and J. Stuyck. 1989. The relation between rainfall and the breeding season of *Mastomys natalensis* (Smith, 1834) in Morogoro, Tanzania. Annales de la Societe Royale Zoologique de Belgique, 119:59-64.

Leirs, H., J. Stuyck, R. Verhagen, and W. Verheyen. 1990. Seasonal variation in growth of *Mastomys natalensis* (Rodentia: Muridae) in Morogoro, Tanzania. African Journal of Ecology, 28:298-306.

Leirs, H., R. Verhagen, and W. Verheyen. 1993. Productivity of different generations in a population of *Mastomys natalensis* rats in Tanzania. Oikos, 68:53-60.

Leirs, H., W. Verheyen, and R. Verhagen. 1996. Spatial patterns in *Mastomys natalensis* in Tanzania (Rodentia, Muridae). Mammalia, 60(4):545-555.

Leite, Y. L. R. 2003. Evolution and systematics of the Atlantic tree rats, genus *Phyllomys* (Rodentia, Echimyidae), with description of two new species. University of California Publications in Zoology, 132:1-133.

Leite, Y. L. R., and J. L. Patton. 2002. Evolution of South American spiny rats (Rodentia, Echimyidae): The star-phylogeny hypothesis revisited. Molecular Phylogenetics and Evolution, 25:455-464.

Leitner, M., and G. B. Hartl. 1988. Genetic variation in the bank vole *Clethrionomys glareolus*: Biochemical differentiation among populations over short geographic distances. Acta Theriologica, 33:231-245.

Lekagul, B., and H. Felten. 1989. Remarks on the genus *Bandicota* in Thailand (Rodentia: Muridae). Thai Journal of Agricultural Science, 22:197-211.

Lekagul, B., and J. A. McNeely. 1977. Mammals of Thailand. Association for the Conservation of Wildlife, Sahakarnbhat Co., Bangkok, 758 pp.

Lekagul, B., and J. A. McNeely. 1988. Mammals of Thailand. Second ed. Darnsutha Press. Bangkok. 758 pp.

Lemire, M. 1966. Particularities de l'appareil masticateur d'un rongeur insectivore *Deomys ferrugineus* (Cricetidae, Dendromurinae). Mammalia, 30:454-494.

Lemos, B., and R. Cerqueira. 2002. Morphological differentiation in the white-eared opossum group (Didelphidae, Didelphis). Journal of Mammalogy, 83:354-369.

Lemos, B., M. Weksler, and C. R. Bonvicino. 2000. The taxonomic status of *Monodelphis umbristriata* (Didelphimorphia: Didelphidae). Mammalia, 64:329-337.

Lenders, A., and E. Pelzers. 1982. Het voorkomen van de hamster *Cricetus cricetus* (L) aan de noordgrens vanzun verspreidingsgebied in Nederland. Lutra, 25:69-80.

Lent, P. C. 1988. *Ovibos moschatus*. Mammalian Species, 302:1-9.

Lento, G. M., R. E. Hickson, G. K. Chambers, and D. Penny. 1995. Use of spectral analysis to test hypotheses on the origin of pinnipeds. Molecular Biology and Evolution, 12:28-52.

Leo, L. M., and A. L. Gardner. 1993. A new species of a giant *Thomasomys* (Mammalia: Muridae: Sigmodontinae) from the Andes of northcentral Peru. The Proceedings of the Biological Society of Washington, 106:417-428.

León, L., T. C. Monterrubio, and M. S. Hafner. 2001. *Cratogeomys neglectus*. Mammalian Species, 685:1-4.

Léon-Paniagua, L., and E. Romo-Vazquez. 1993. Mastofauna de la Sierra de Taxco, Guerrero. Pp. 45-64, *in* Avances en el estudio de los mamíferos de México (R. A. Medellín and G. Cebassos, eds.). Publicaciones Especiales, Vol. 1. Asociación Mexicana de Mastozoología, México, D. F., 464 pp.

Leroy, P. 1940. Observations on living Chinese mole-rats. Bulletin of the Fan Memorial Institute of Biology. Zoology, 10:167-193.

Lessa, E. P., and J. A. Cook. 1998. The molecular phylogenetics of tuco-tucos (genus *Ctenomys*, Rodentia: Octodontidae) suggests an early burst of speciation. Molecular Phylogenetics and Evolution, 9:88-99.

Lessa, E. P., and A. Langguth. 1983. *Ctenomys pearsoni*, n. sp. (Rodentia, Octodontidae), del Uruguay. Resultados Comunicados Jornadas Ciencias Naturales, Montevideo, 3:86-88.

Lesson, R. P. 1840. Species es mammifères bimanes et quadrimanes . . . Baillière, London, 292 pp.

Leung, L. K.-P. 1995. Cape York Melomys, *Melomys capensis*. Pp. 634-635, *in* Mammals of Australia (R. Strahan, ed.). Smithsonian Institution Press, Washington, D.C., 756 pp.

Leung, L. K.-P. 1999*a*. Ecology of Australian tropical rainforest mammals. II. The Cape York melomys, *Melomys capensis* (Muridae: Rodentia). Wildlife Research, 26:307-316.

Leung, L. K.-P. 1999*b*. Ecology of Australian tropical rainforest mammals. III. The Cape York rat, *Rattus leucopus* (Muridae: Rodentia). Wildlife Research, 1999, 26:317-328.

Leung, L. K.-P., and Sudarmaji. 1999. Techniques for trapping the rice-field rat, *Rattus argentiventer*. Malayan Nature Journal, 53(4):323-333.

Levan, G., K. Klinga, C. Szpirer, and J. Szpirer. 1990. Gene map of the rat (*Rattus norvegicus*) 2N=42. Isozyme Bulletin, 23:34-42.

Levenson, H., R. S. Hoffmann, C. F. Nadler, L. Deutsch, and S. D. Freeman. 1985. Systematics of the Holarctic chipmunks. Journal of Mammalogy, 66:219-242.

Lever, C. 1985. Naturalized Mammals of the World. Longman, London, 487 pp.

Lewis, R. E., J. H. Lewis, and S. I. Atallah. 1967. A review of Lebanese mammals. Lagomorpha and Rodentia. Journal of Zoology, London, 153:45-70.

Lewis, S. E., and D. E. Wilson. 1987. *Vampyressa pusilla*. Mammalian Species, 292:1-5.

Lewis, T. H. 1983. The anatomy and histology of the rudimentary eye of *Neurotrichus*. Northwest Science, 57:8-15.

Lewis-Oritt, N., C. A. Porter, and R. J. Baker. 2001*a*. Molecular systematics of the family Mormoopidae (Chiroptera) based on cytochrome *b* and recombination activating gene 2 sequences. Molecular Phylogenetics and Evolution, 20:426-436.

Lewis-Oritt, N., R. A. Van Den Bussche, and R. J. Baker. 2001*b*. Molecular evidence for evolution of piscivory in *Noctilio* (Chiroptera: Noctilionidae). Journal of Mammalogy, 82:748-759.

Leyhausen, P., and M. Pfleiderer. 1994. The taxonomic status of the Iriomote cat (*Prionailurus iriomotensis* Imaizumi, 1967). Cat News, 21:18-20.

Leyhausen, P., and M. Pfleiderer. 1999. The systematic status of the Iriomote cat (*Prionailurus iriomotensis* Imaizumi 1967) and the subspecies of the leopard cat (*Prionailurus bengalensis* Kerr 1792). Journal of Zoological Systematics and Evolutionary Research, 37:121-131.

Li, B., and F. Chen. 1987. [A comparative study of the karyotypes and LDH isoenzymes from some zokors of the subgenus *Eospalax*, genus *Myospalax*.] Acta Theriologica Sinica, 7(4):275-282 (in Chinese with English summary).

Li, B., and F. Chen. 1989. [A taxonomic study and new subspecies of the subgenus *Eospalax*, genus *Myospalax*]. Acta Zoologica Sinica, 35:89-95 (in Chinese).

Li, C., S. Ma, Y. Wang, and G. Lu. 1987. [Mammals from Honghe Region. Pp. 26-27, *in* Report on the Biological Resources of Honghe Region, Southern Yunnan. Vol. 1. Land Vertebrates. Yunnan National Press, 102 pp.] (in Chinese).

Li, C.-K., and S.-Y. Ting. 1985. Possible phylogenetic relationships of eurymylids and rodents, with comments on mimotonids. Pp. 35-58, *in* Evolutionary Relationships among Rodents: A Multidisciplinary Analysis (W. P. Luckett and J.-L. Hartenberger, eds.). Plenum Press, New York, 721 pp.

Li, C.-K., and S.-Y. Ting. 1993. New cranial and postcranial evidence for the affinities of the eurymylids (Rodentia) and mimotonids (Lagomorpha). Pp. 151-158, *in* Mammal Phylogeny-Placentals (F. S. Szalay, M. J. Novacek, and M. C. McKenna, eds.). Springer-Verlag, New York, xi + 321 pp.

Li, C.-K., R. W. Wilson, M. R. Dawson, and L. Krishtalka. 1987. The origin of rodents and lagomorphs. Pp. 97-108, *in* Current Mammalogy, Volume 1 (H. H. Genoways, ed.). Plenum Press, New York, 519 pp.

Li, J., S. Wen, G. Qi, Y. Xie, and T. Pen. 1998. Studies on the mitochondrial DNA restriction map and RFLP of *Bandicota indica*. Acta Theriologica Sinica, 18(1):54-59.

Li, W., H. Li, H. Xamaridan, and J. Ma. 1988. First report on ecological study of the Ili pika (*Ochotona iliensis*). Abstracts, Symposium of Asian Pacific Mammalogy, Huirou, Beijing, Peoples Republic of China, pages not numbered, abstracts listed alphabetical by author.

Li, W.-H., W. A. Hide, and D. Graur. 1992. Origin of rodents and guinea-pigs. Nature, 359:277-278.

Li, X., and T. Wang. 1995. [Discussion of taxonomy of Vernaya's climbing mouse.] Zoological Research, 16(4):325-328 (in Chinese with English abstract).

Li, X., and T. Wang. 1999. [Comparisons of structure and function on masticatory apparatus between mandarin voles and common Chinese zokors.] Acta Theriologica Sinica, 19(4):308-314 (in Chinese).

Li, Z., and Z. Luo. 1979. [On a new species of wild hare from China.] Journal of the Northeast Forestry Institute, 12:71-81 (in Chinese).

Liascovich, R. C., R. Bárquez, and O. A. Reig. 1989. A karyological and morphological reassessment of *Akodon* (*Abrothrix*) *illuteus* Thomas. Journal of Mammalogy, 70:386-391.

Libois, R. M. 1996. The current situation of wild mammals in Belgium: An outline. Hystrix, n.s., 8(1-2):35-41.

Libois, R. M., and R. Fons. 1990. Le mulot des Iles d'Hyeres: Un cas de gigantisme insulaire. Vie et Milieu, 40:217-222.

Libois, R., R. Fons, and D. Bordenave. 1993. Mediterranean small mammals and insular syndrome: Biometrical study of the long-tailed field mouse (*Apodemus sylvaticus*) (Rodentia-Muridae) of Corsica. Bonner Zoologische Beiträge, 44(3-4):147-163.

Libois, R., J. Torrico, M. G. Ramalhinho, J. Michaux, R. Fons, and M. L. Mathias. 1996. Les populations de rats noirs insulaires de l'ouest de l'Europe. Essai preliminaire de caracterisation genetique (caryotype et AND mitochondrial). Vie Milieu, 46(3/4):213-218.

Libois, R. M., J. R. Michaux, M. G. Ramalhinho, C. Maurois, and M. Sarà. 2001. On the origin of the northern African wood mouse (*Apodemus sylvaticus*) populations: A comparative study of mtDNA restriction patterns. Canadian Journal of Zoology, 79:1503-1511.

Lichaèev, G. N. 1972. Rasprostranenie son' v evropejskoj casti SSSR [The distribution of dormice in the European part of the USSR]. Fauna i ekologija gryzunov, 11:71-115.

Lichtenstein, H. 1828. Über die Springmäuse oder die Arten der Gattung *Dipus*. Abhandlungen der Königlichen Akademie der Wissenschaften in Berlin, 1825:133-162.

Lidicker, W. Z., Jr. 1960. An analysis of intraspecific variation in the kangaroo rat *Dipodomys merriami*. University of California Publications in Zoology, 67:125-218.

Lidicker, W. Z., Jr. 1968. A phylogeny of New Guinea rodent genera based on phallic morphology. Journal of Mammalogy, 49:609-643.

Lidicker, W. Z., Jr. 1983. Dasyurids and their kin. Science, 219(4590):1316-1317.

Lidicker, W. Z., Jr., and P. V. Brylski. 1987. The conilurine rodent radiation of Australia, analyzed on the basis of phallic morphology. Journal of Mammalogy, 68:617-641.

Lidicker, W. Z., Jr., and W. I. Follett. 1968. *Isoodon* Desmarest, 1817, rather than *Thylacis* Illiger, 1811, as the valid generic name of the short-nosed bandicoots (Marsupialia: Peramelidae). Proceedings of the Biological Society of Washington, 81:251-256.

Lidicker, W. Z., Jr., and A. Yang. 1986. Morphology of the penis in the taiga vole (*Microtus xanthognathus*). Journal of Mammalogy, 67:497-502.

Lidicker, W. Z., Jr., and A. C. Ziegler. 1968. Report on a collection of mammals from eastern New Guinea, including species keys for fourteen genera. University of California Publications in Zoology, 87:1-60.

Ligtvoet, W. 1992. Noordse woelmuis, *Microtus oeconomus* (Pallas, 1776). Pp. 273-280, *in* Atlas van de Nederlandse Zoogdieren (S. Broekhizen, B. Hoekstra, V. van Laar, C. Smeenk, and J. B. M. Thissen, eds.). Stichting Uitgeverij Koninklijke Nederlandse Natuurhistorische Vereniging, Utrecht, 336 pp.

Ligtvoet, W., and E. H. F. M. Straetmans. 1992. Ondergrondse woemuis, *Pitymys subterraneus* (De Sélys-Longchamps, 1836). Pp. 256-259, *in* Atlas van de Nederlandse Zoogdieren (S. Broekhuizen, B. Hoekstra, V. van Laar, C. Smeenk, and J. B. M. Thissen, eds.). Stichting Uitgeverij Koninklijke Nederlandse Natuurhistorische Vereniging, Utrecht, 336 pp.

Ligtvoet, W., and A. van Wijngaarden. 1994. The colonization of the island of Noord-Beveland (the Netherlands) by the common vole *Microtus arvalis*, and its consequences for the root vole *M. oeconomus*. Lutra, 37(1):1-28.

Lillieborg, W. 1866. Systematisk öfversigt af de gnagande däggdjuren, Glires. Uppsala, Kungliga Akademi Boktryck:1-59.

Lim, B. K. 1987. *Lepus townsendii*. Mammalian Species, 288:1-6.

Lim, B. K. 1993. Cladistic reappraisal of Neotropical stenodermatine bat phylogeny. Cladistics, 9:147-165.

Lim, B. K. 1997. Morphometric differentiation and species status of the allopatric fruit-eating bats *Artibeus jamaicensis* and *A. planirostris* in Venezuela. Studies on Neotropical Fauna and Environment, 32:65-71.

Lim, B. K., and M. D. Engstrom. 1998. Phylogeny of Neotropical short-tailed fruit bats, *Carollia* spp.— phylogenetic analysis of restriction site variation in mtDNA. Pp. 43-58, *in* Bat biology and conservation (T. H. Kunz and P. A. Racey, eds.). Smithsonian Institution Press, 365 pp.

Lim, B. K., and M. D. Engstrom. 2001. Species diversity of bats (Mammalia: Chiroptera) in Iwokrama Forest, Guyana, and the Guianan subregion: Implications for conservation. Biodiversity and Conservation, 10:613-657.

Lim, B. K., and P. D. Ross. 1992. Taxonomic status of *Alticola* and new record of *cricetulus* from Nepal. Mammalia, 56:300-302.

Lim, B. K., and D. E. Wilson. 1993. Taxonomic status of *Artibeus amplus* (Chiroptera: Phyllostomidae) in northern South America. Journal of Mammalogy, 74:763-768.

Lim, B. K., W. A. Pedro, and F. C. Passos. 2003. Differentiation and species status of the Neotropical yellow-eared bats *Vampyressa pusilla* and *V. thyone* (Phyllostomidae) with a molecular phylogeny of the genus. Acta Chiropterologica, 5:15-29.

Lima, F. S. 1994. Cariótipos em espécies de Dasyproctidae e Erethizontidae, com discussão da evolução cromossômica (Rodentia, Caviomorpha). Brazilian Journal of Genetics, 17(suppl.):135.

Lima, J. F. de S., and S. Kasahara. 2001. A new karyotype of *Calomys* (Rodentia, Sigmodontinae). Iheringia, Séries Zoologia, Porto Alegre, 91:133-136.

Lima, J. F. D., and A. Langguth. 2002. Karyotypes of Brazilian squirrels: *Sciurus spadiceus* and *Sciurus alphonsei* (Rodentia, Sciuridae). Folia Zoologica, 51:201-204.

Lima, J. F. de S., A. Langguth, and L. C. de Sousa. 1998. The karyotype of *Makalata didelphoides* (Rodentia, Echimyidae). Zeitschrift für Säugetierkunde, 63:315-318.

Limpus, C. J., C. J. Parmenter, and C. H. S. Watts. 1983. *Melomys rubicola*, an endangered murid rodent endemic to the Great Barrier Reef of Queensland. Australian Mammalogy, 6:77-79.

Lin L.-K., P. S. Alexander, and Yu Ming-jenn. 1987. A survey of the small mammals of Mt. Ali recreation area. Tunghai Journal, 28:669-682 (in Chinese).

Lin, L.-K., M. Motokawa, and M. Harada. 2002*a*. Karyotype of *Mogera insularis* (Insectivora, Talpidae). Mammalian Biology, 67:176-178.

Lin, L.-K., M. Motokawa, and M. Harada. 2002*b*. Karyology of ten vespertilionid bats (Chiroptera: Vespertilionidae) from Taiwan. Zoological Studies, 41:347-354.

Lin, Y.-H., P. A. McLenachan, A. R. Gore, M. J. Phillips, R. Ota, M. D. Hendy, and D. Penny. 2002. Four new mitochondrial genomes and the increased stability of evolutionary trees of mammals from improved taxon sampling. Molecular Biology and Evolution, 19:2060-2070.

Lin, Y-S., D. R. Progulske, P.-F. Lee, and Y.-T. Day. 1985. Bibliography of Petauristinae (Rodentia: Sciuridae). Journal of the Taiwan Museum, 38(2):49-57.

Linares, O. J. 1971. A new subspecies of funnel-eared bat (*Natalus stramineus*) from western Venezuela. Bulletin of the Southern California Academy of Science, 70:81-84.

Linares, O. J. 1998. Mamíferos de Venezuela. Sociedad Conservacionista Audubon de Venezuela, Caracas, 691 pp.

Lindemayer, D. B., J. Dubach, and K. L. Viggers. 2002. Geographic dimorphism in the mountain brushtail possum (*Trichosurus caninus*): The case for a new species. Australian Journal of Zoology, 50:1-25.

Lindsay, E. H. 1972. Small mammal fossils from the Bartow formation, California. University of California Publications in Geological Sciences, 93:1-104.

Lindsay, E. H. 1977. *Simimys* and origin of the Cricetidae (Rodentia: Muroidea). Geobios, 10(4):597-623.

Lindsay, E. H. 1988. Cricetid rodents from Siwalik deposits near Chinji Village. Part I: Megacricetodontinae, Myocricetodontinae and Dendromurinae. Palaeovertebrata, 18:95-154.

Lindsay, E. H. 1994. The fossil record of Asian Cricetidae with emphasis on Siwalik Cricetids. Pp. 131-147, *in* Rodent and lagomorph families of Asian origins and diversification (Y. Tomida, C. ki. Li, and T. Setoguchi, eds.). National Science Museum Monographs, No. 8, Tokyo, 195 pp.

Lindsay, E. H., and L. L. Jacobs. 1985. Pliocene small mammal fossils from Chihuahua, Mexico. Paleontología Mexicana, 51:1-53.

Lindsay, H. M. 1928. Note on *Viverra civettina*. Journal of the Bombay Natural History Society, 33(1):146-148.

Lindsay, S. L. 1981. Taxonomic and biogeographic relationships of Baja California chickarees (*Tamiasciurus*). Journal of Mammalogy, 62:673-682.

Lindsay, S. L. 1982. Systematic relationship of parapatric tree squirrel species (*Tamiasciurus*) in the Pacific Northwest. Canadian Journal of Zoology, 60:2149-2156.

Ling, J. K. 1992. *Neophoca cinerea*. Mammalian Species, 392:1-7.

Ling, J. K., and M. M. Bryden. 1992. *Mirounga leonina*. Mammalian Species, 391:1-8.

Linnaeus, C. 1758. Systema Naturae per regna tria naturae, secundum classis, ordines, genera, species cum characteribus, differentiis, synonymis, locis. Tenth ed. Vol. 1. Laurentii Salvii, Stockholm, 824 pp.

Linnaeus, C. 1766-1768. Systema naturae per regna tria naturae, secundum classes, ordines, genera, species, cum characteribus, differentiis synonymis, locis. Vol. 1. Regnum Animale. pt. 1, pp. 1-532 [1766]; vol. 3, Appendix pp. 223-235 [1768]. 12th ed. [Vol. 3, Regnum Lapideum, pp. 1-222, rejected for nomenclatural purposes, ICZN, opinion 296]. Laurentii Salvii, Stockholm, 3 vols.

Linnaeus, C. (revised by J. F. Gmelin). 1788. Systema naturae per regna tria naturae, secundum classes, ordines, genera, species, cum characteribus, differentiis synonymis, locis. Vol. I. Regum Animale. Class 1, Mammalia. Thirteenth ed. (revised by J. F. Gmelin). G. E. Beir, Lipsiae, 232 pp.

Linnaeus, C. (translated and revised by R. Kerr). 1792. The animal kingdom; or, zoological system of the celebrated Sir Charles Linnaeus. Class 1. Mammalia and Class II. Birds. Being a translation of that part of the Systema Naturae, as lately published with great improvements by Professor Gmelin, together with numerous additions from more recent zoological writers and illustrated with copperplates. J. Murray, London, 644 pp.

Lint, D. W., J. W. Clayton, W. R. Lillie, and L. Postuma. 1990. Evolution and systematics of the Beluga whale, *Delphinapterus leucas*, and other odontocetes: A molecular approach. Canadian Bulletin of Fisheries and Aquatic Sciences, 224:7-22.

Linzey, A. V. 1983. *Synaptomys cooperi*. Mammalian Species, 210:1-5.

Linzey, A. V., and J. N. Layne. 1969. Comparative morphology of the male reproductive tract in the rodent genus *Peromyscus* (Muridae). American Museum Novitates, 2355:1-47.

Linzey, A. V., and J. N. Layne. 1974. Comparative morphology of spermatozoa of the rodent genus *Peromyscus* (Muridae). American Museum Novitates, 2532:1-20.

Linzey, A. V., M. H. Kesner, C. T. Chimimba, and C. Newbery. 2003. Distribution of veld rat sibling species *Aethomys chrysophilus* and *Aethomys ineptus* in southern Africa. African Zoology, 38(1):169-174.

Linzey, D. W., and R. L. Packard. 1977. *Ochrotomys nuttalli*. Mammalian Species, 75:1-6.

Lisanti, J. A., E. Pinna-Senn, M. Ortiz, G. Dalmasso, and S. Parisi de Fabro. 2000. Karyotypic relationship between *Akodon azarae* and *A. boliviensis* (Rodentia, Sigmodontinae). Cytologia, 65:253-259.

Lissovsky, A. A. 2002. Thesis Summary. State Darwin Museum and Moscow State University. 17 pp. [in Russian].

Lissovsky, A. A., and E. V. Lissovskaya. 2000. New data on pikas (*Ochotona*, Mammalia) of *alpina-hyperborea* group; systematics and distribution. Pp. 183-185, *in* Biodiversity and Dynamics of Ecosystems in North Eurasia (N. Kolchanov et al., eds.). 317 pp.

Lister, A. M., P. Grubb, and S. R. M. Sumner. 1998. Taxonomy, morphology and evolution of European roe deer. Pp. 23-46, *in* The European roe deer: The biology of success (R. Andersen, P. Duncan, and J. D. C. Linnell, eds.). Scandinavian University Press, Oslo, 376 pp.

Literák, I., and J. Zejda. 1996. Colour mutations of the bank vole (*Clethrionomys glareolus*) in the Czech Republic. Folia Zoologica, 44(1):95-96.

Liu Chun-sheng, Li Chuan-bin, Wu Wan-neng, and Meng Ji-hui. 1985. The faunal distribution and geographical divisions of rodents in Anhui Province. Acta Theriologica Sinica, 5:111-118.

Liu Chun-sheng, Wu Wan-neng, Guo Shi-kun, and Meng Ji-hui. 1991. [A study of the subspecies classification of *Apodemus agrarius* in eastern continental China.] Acta Theriologica Sinica, 11:294-299 (in Chinese).

Liu, F.-G. R., M. M. Miyamoto, N. P. Freire, P. Q. Ong, M. R. Tennant, T. S. Young, and K. F. Gugel. 2001. Molecular and morphological supertrees for eutherian (placental) mammals. Science, 291:1786-1789.

Liu Nai-fa, Fan Hua-wei, Jing Kai, and Ning Rui-dong. 1990. [Study of community diversity of rodents in Anxi desert.] Acta Theriologica Sinica, 10(3):215-220 (in Chinese).

Liu, S. 1994. [Studies on using the base height of proximal bacular process to divide the age groups of male David's vole.] Acta Theriologica Sinica, 14(4):281-285.

Liu, S., J. Ran, and Q. Lin. 2000. [Morphology and taxonomic significance of the glans penis of *Apodemus* from Sichuan and Chongquing, China.] Acta Theriologica Sinica, 20(1):48-57 (in Chinese with English abstract).

Liu, X., F. Wei, M. Li, and Z. Feng. 2002. [A review of the phylogenetic study on the genus *Apodemus* of China.] Acta Theriologica Sinica, 22(1):46-52 (in Chinese with English abstract).

Liu, X., F. Wei, M. Li, X. Jiang, Z. Feng, and J. Hu. 2004. Molecular phylogeny and taxonomy of wood mice (genus *Apodemus* Kaup, 1829) based on complete mtDNA cytochrome *b* sequences, with emphasis on Chinese species. Molecular Phylogenetics and Evolution, 33:1-15.

Lloyd, B. D. 2001. Advances in New Zealand mammalogy 1990-2000: Short-tailed bats. Journal of the Royal Society of New Zealand, 31:59-81.

Lloyd, B. D. 2003. Intraspecific phylogeny of the New Zealand short-tailed bat *Mystacina tuberculata* inferred from multiple mitochondrial gene sequences. Systematic Biology, 52:460-476.

Locatelli, R., and P. Paolucci. 1996*a*. Micromammiferi della foresta demaniale de Cadino. Natura Alpina, 46(4):1-68.

Locatelli, R., and P. Paolucci. 1996*b*. L'*Arvicola* delle nnevi (*Microtus nivalis* Martins, 1842) nell'Italia nord orientale: Biometrie, morfologia dentale e scelte dell'habitat. Bollettino del Museo Civico di Storia Naturale di Venezia, 45:195-209.

Lock, M. L., and B. A. Wilson. 1999. The distribution of the New Holland mouse (*Pseudomys novaehollandiae*) with respect to vegetation near Anglesea, Victoria. Wildlife Research, 26:565-577.

Loftus, R. J., D. E. MacHugh, D. G. Bradley, P. M. Sharp, and P. Cunningham. 1994. Evidence for two independent domestications of cattle. Proceedings of the National Academy of Sciences, U.S.A., 91:2757-2761.

Long, C. A. 1972*a*. Taxonomic revision of the North American badger, *Taxidea taxus*. Journal of Mammalogy, 59(4):725-759.

Long, C. A. 1972*b*. Taxonomic revision of the mammalian genus *Microsorex* Coues. Transactions of the Kansas Academy of Science, 74:181-196.

Long, C. A. 1973. *Taxidea taxus*. Mammalian Species, 26:1-4.

Long, C. A. 1974. *Microsorex hoyi* and *Microsorex thompsoni*. Mammalian Species, 33:1-4.

Long, C. A. 1978. A listing of Recent badgers of the world, with remarks on taxonomic problems in *Mydaus* and *Melogale*. Reports on the Fauna and Flora of Wisconsin, The Museum of Natural History, Stevens Point, WI, 14:1-6.

Long, C. A. 1981. Provisional classification and evolution of the badgers. Pp. 55-85, *in* Worldwide Furbearer Conference Proceedings (J. A. Chapman and D. Pursley, eds.). Frostburg, MD, 1:1-652.

Long, C. A., and R. S. Hoffmann. 1992. *Sorex preblei* from the Black Canyon, first record for Colorado. Southwestern Naturalist, 37:318-319.

Long, C. A., and C. A. Killingley. 1983. The badgers of the world. Charles C. Thomas, Springfield, IL, 404 pp.

Long, C. A., and J. E. Long. 1993. Discriminant analysis of geographic variation in long-tailed deer mice from northern Wisconsin and Upper Michigan. Transactions of the Wisconsin Academy of Sciences, 81:107-121.

Lönnberg, E. 1910. On the variation of the sea-elephants. Proceedings of the Zoological Society of London, 1910:580-588.

Lönnberg, E. 1913. Mammals from Ecuador and related forms. Arkiv för Zoologi, 8(16):1-36.

Lönnberg, E. 1923*a*. Notes on *Arctonyx*. Annals and Magazine of Natural History, ser. 9, 11:322-326.

Lönnberg, E. 1923*b*. Remarks on some Palaearctic bears. Proceedings of the Zoological Society of London, 1923:85-95.

Lönnberg, E., and N. Gyldenstolpe. 1925. Vertebrata. 2. Preliminary diagnoses of seven new mammals. Arkiv för Zoologi, 17B, 5:1-6.

Lopatin, A. V. 2002. The earliest shrew (Soricidae, Mammalia) from the Middle Eocene of Mongolia. Paleontological Journal, 36:650-659.

Lopatin, A. V., and V. S. Zazhigin. 2000. [The History of the Dipodoidea (Rodentia, Mammalia) in the Miocene of Asia: 2. Zapodidae.] Paleontologicheskii Zhurnal, 4:86-91 (in Russian with English summary).

Lopatin, A. V., and V. S. Zazhigin. 2003. New Brachyericinae (Erinaceidae, Insectivora, Mammalia) from the Oligocene and Miocene of Asia. Paleontological Journal, 37:62-75.

López-Forment C., W., and G. Urbano V. 1977. Restos de Pequeños mamíferos recuperados en regurgitaciones de lechuza, *Tyto alba*, en México. Anales del Instituto de Biologia, Sería Zoología, 1:231-242.

López-Furster, M. J., J. Ventura, M. Miralles, and E. Castién. 1990. Craniometric characteristics of *Neomys fodiens* (Pennant, 1771) (Mammalia, Insectivora) from the northeastern Iberian Peninsula. Acta Theriologica, 35(3-4):269-276.

López-González, C. 1998. *Micronycteris minuta*. Mammalian Species, 583:1-4.

López-González, C., and S. J. Presley. 2001. Taxonomic status of *Molossus bondae* J. A. Allen, 1904 (Chiroptera: Molossidae), with description of a new subspecies. Journal of Mammalogy, 82(3):760-774.

López-González, C., S. J. Presley, R. D. Owen, and M. R. Willig. 2001. Taxonomic status of *Myotis* (Chiroptera: Vespertilionidae) in Paraguay. Journal of Mammalogy, 82:138-160.

López-Martínez, N., and L. F. López-Jurado. 1987. Un nuevo murido gigante del Cuaternario de Gran Canaria *Canariomys tamarani* nov. sp. (Rodentia, Mammalia). Doñana, Publicacion Ocasional, 2:1-66.

López-Martínez, N., J. Michaux, and R. Hutterer. 1998. The skull of *Stephanomys* and a review of *Malpaisomys* relationships (Rodentia: Muridae): Taxonomic incongruence in murids. Journal of Mammalian Evolution, 5(3):185-215.

Lorenz, L. v. 1894. Ueber die von Herrn Dr. E. Holub gespendeten Südafrikanischen Säugethiere. Annalen des K. K. Naturhistorischen Hofmuseums, 9:59-67.

Lorenzini, R., S. Lovari, and M. Masseti. 2002. The rediscovery of the Italian roe deer: Genetic differentiation and management implications. Italian Journal of Zoology, 69:367-379.

Lorenzo, C., and F. A. Cervantes. 1995. The G-banded karyotype of the Tapeti rabbit (*Sylvilagus brasiliensis*) from Chiapas, Mexico. Revista Sociedad Mexicana Historia Natural, 46:1-5.

Losinger, I. 2001. First results of the conservation plan for the common hamster (*Cricetus cricetus*) in the Alsace. Jahrbücher des Nassauischen Vereins für Naturkunde, 122:191-200.

Lotze, J.-H., and S. Anderson. 1979. *Procyon lotor*. Mammalian Species, 119:1-8.

Loughlin, T. R., and M. A. Perez. 1985. *Mesoplodon stejnegeri*. Mammalian Species, 250:1-6.

Loughlin, T. R., M. A. Perez, and R. L. Merrick. 1987. *Eumetopias jubatus*. Mammalian Species, 283:1-7.

Lovari, S. 1985. Behavioural repertoire of the Abruzzo chamois, *Rupicapra pyrenaica ornata* Neumann, 1899 (Artiodactyla: Bovidae). Säugetierkundliche Mitteilungen, 32:113-136.

Lovari, S. 1987. Evolutionary aspects of the biology of *Rupicapra* spp. (Bovidae, Caprinae). Pp. 51-61, *in* The biology and management of *Capricornis* and related antelopes (H. Soma, ed.). Croom Helm, London, 391 pp.

Lovari, S., and C. Scala. 1980. Revision of *Rupicapra* genus. I. A statistical re-evaluation of Couturier's data on the morphometry of six chamois subspecies. Bollettino di Zoologia, 47:113-l24.

Lovari, S., and C. Scala. 1984. Revision of *Rupicapra* genus. IV. Horn biometrics of *Rupicapra rupicapra asiatica* and its relevance to the taxonomic position of *Rupicapra rupicapra caucasica*. Zeitschrift für Säugetierkunde, 49:246-253.

Lovegrove, B. 1997*a*. Tree rat, *Thallomys paedulcus*. P. 147, *in* The complete book of southern African mammals (G. Mills and L. Hes, eds.). Struik Winchester, Capetown, 356 pp.

Lovegrove, B. 1997*b*. Black-tailed tree rat, *Thallomys nigricauda*. Pp. 147-148, *in* The complete book of southern African mammals (G. Mills and L. Hes, eds.). Struik Winchester, Capetown, 356 pp.

Lowery, G. H., Jr. 1974. The mammals of Louisiana and its adjacent waters. Louisiana State University Press, Baton Rouge, 565 pp.

Lozada, M., J. A. Monjeau, K. M. Heinemann, N. Guthmann, and E. C. Birney. 1996. *Abrothrix xanthorhinus*. Mammalian Species, 540:1-6.

Lozan, M. N. 1970. Gryzuny Moldavii. Istorija stanovlemija fauny i ekologija recentnich vidov. Tom 1. [Rodents of Moldavia. Vol. 1]. Akademia Nauk MoSSR, Kisinev. 170 pp.

Lu, J., and T. Wang. 1996. [Studies on the glires (Rodentia and Lagomorpha) fauna and its division of Henan Province.] Acta Theriologica Sinica, 16(2):119-128 (in Chinese with English abstract).

Lucas, S. G. 2001. Chinese fossil vertebrates. Columbia University Press, New York, 320 pp.

Luckett, W. P. (ed.). 1980. Comparative biology and evolutionary relationships of tree shrews. Plenum Press, New York, 314 pp.

Luckett, W. P. 1985. Superordinal and intraordinal affinities of rodents: Developmental evidence from the dentition and placentation. Pp. 227-276, *in* Evolutionary Relationships among Rodents: A Multidisciplinary Analysis (W. P. Luckett and J.-L. Hartenberger, eds.). Plenum Press, New York, 721 pp.

Luckett, W. P., and J.-L. Hartenberger (eds.). 1985*a*. Evolutionary Relationships among Rodents: A Multidisciplinary Analysis. Plenum Press, New York, 721 pp.

Luckett, W. P., and J.-L. Hartenberger. 1985*b*. Evolutionary relationships among rodents: Comments and conclusions. Pp. 685-712, *in* Evolutionary Relationships among Rodents: A Multidisciplinary Analysis (W. P. Luckett and J.-L. Hartenberger, eds.). Plenum Press, New York, 721 pp.

Luckett, W. P., and J.-L. Hartenberger. 1993. Monophyly or polyphyly of the Order Rodentia: Possible conflict between morphological and molecular interpretations. Journal of Mammalian Evolution, 1:127-147.

Ludwig, D. R. 1984. *Microtus richardsoni*. Mammalian Species, 223:1-6.

Ludwig, A., and J. Knoll. 1998. Multivariate morphometrische Analysen der Gattung *Ovis* Linnaeus, 1758 (Mammalia, Caprinae). Zeitschrift für Säugetierkunde, 63:210-219.

Lukácová, L., E. Dannelid, J. Hausser, M. Nacholán, and J. Zima. 1996. G-banded karyotype of the alpine shrew, *Sorex alpinus* (Mammalia, Soricidae), from the Sumava Mts. Folia Zoologica, 45:223-226.

Lumpkin, S., and K. R. Kranz. 1984. *Cephalophus sylvicultor*. Mammalian Species, 225:1-7.

Luna, L., and V. Pacheco. 2002. A new species of *Thomasomys* (Muridae: Sigmodontinae) from the Andes of southeastern Peru. Journal of Mammalogy, 83:834-842.

Luna, L., and B. D. Patterson. 2003. A remarkable new mouse (Muridae: Sigmodontinae) from southeastern Peru with comments on the affinities of *Rhagomys rufescens* (Thomas, 1886). Fieldiana: Zoology, n.s., 101:1-24.

Lund, P. W. 1841. Blik paa Brasiliens Dyreverden för sidste Jordomvaeltning. Tredie Afhandling: Fortsaettelse af Pattedyrene. Konigelige Danske Videnskabernes Selskabs Afhandlinger, Kjöbenhavn, 8:219-272, pls. 14-24.

Lunde, D. P., and E. E. Sarmiento. 2002. Rodents collected from Kalinzu Forest, Uganda. Mammalian Biology, 67:250-255.

Lunde, D. P., T. M. Conway, and P. Beresford. 2001. Notes on a collection of bats (Chiroptera) from Dzanga-Sangha, Central African Republic. Mammalia, 65:535-540.

Lunde, D. P., G. G. Musser, and P. D. Tien. 2003*a*. Records of some little known bats (Chiroptera: Vespertilionidae) from Vietnam. Mammalia, 67:459-461.

Lunde, D. P., G. G. Musser, and N. T. Son. 2003*b*. A survey of small mammals from Mt. Tay Con Linh II, Vietnam, with the description of a new species of *Chodsigoa* (Insectivora: Soricidae). Mammal Study, 28:31-46.

Lundholm, B. G. 1955*a*. Descriptions of new mammals. Annals of the Transvaal Museum, 22:279-303.

Lundholm, B. G. 1955*b*. A taxonomic study of *Cynictis penicillata* (G. Cuvier). Annals of the Transvaal Museum, 22:305-319.

Lundholm, B. G. 1955*c*. Remarks on some South African Murinae. Annals of the Transvaal Museum, 22:321-329.

Lundrigan, B. L., S. A. Jansa, and P. K. Tucker. 2002. Phylogenetic relationships in the genus *Mus*, based on paternally, maternally, and biparentally inherited characters. Systematic Biology, 5(3):410-431.

Lungeanu, A., L. Gavrila, C. Stepan, and D. Murariu. 1984. Donnees preliminaires concernant l'etude du caryotype de *Micromys minutus* (Pallas, 1771) (Rodentia, Muridae). Travaux du Museum d'Histoire Naturelle "Grigore Antipa", 26:241-244.

Lungeanu, A., L. Gavrila, D. Murariu, and C. Stepan. 1986. The distribution of the constituent heterochromatin and the G- banding pattern in the genome of *Apodemus agrarius* (Pallas, 1771) (Mammalia, Muridae). Travaux du Museum d'Histoire Naturelle "Grigore Antipa", 28:267-270.

Lunney, D. 1995a. Bush rat, *Rattus fuscipes*. Pp. 651-653, *in* Mammals of Australia (R. Strahan, ed.). Smithsonian Institution Press, Washington, D.C., 756 pp.

Lunney, D. 1995b. Swamp rat, *Rattus lutreolus*. Pp. 655-657, *in* Mammals of Australia (R. Strahan, ed.). Smithsonian Institution Press, Washington, D.C., 756 pp.

Luo, Z-x. 1981. [A systematic review of the Chinese Cape hare, *Lepus capensis*, Linnaeus]. Acta Theriologica Sinica, 1:149-157 (in Chinese with English summary).

Lura, H., G. Langhelle, T. Fredriksen, and I. Byrkjedal. 1995. Distribution of the field mice *Apodemus flavicollis* and *A. sylvaticus* in Norway. Fauna of Norway, Series A, 16:1-10.

Lushnikova, T. P., L. V. Omelyanchuk, A. S. Graphodatsky, S. I. Radjabli, Yu. G. Ternovskaya, and D. V. Ternovsky. 1989. [Phylogenetic relationships of closely related species (Mustelidae). Intraspecific variability of blot-hybridization patterns of BamHI repeats.] Genetika, 256:1089-1094 (in Russian).

Luyts, G. 1986. [Is the garden dormouse extending its area of distribution as far as the coast?] Wielewaal, 52(5) 1986:91.

Lyalyukhina, S., E. Kotenkova, W. Walkowa, and K. Adamszyk. 1991. Comparison of craniological parameters in *Mus musculus musculus* Linnaeus, 1758 and *Mus hortulans* Nordmann, 1840. Acta Theriologica, 36:95-107.

Lyapunova, Ye. A., and I. V. Zagorodnyuk. 1990. [A normal karyotype and chromosomal polymorphism in Afghan vole *Microtus* (*Blanfordimys*) *afghanus* (Rodentia).] Zoologischkii Zhurnal, 69(4):151-154 (in Russian with English summary).

Lyapunova, E. A., N. N. Vorontsov, and L. Ya. Martynova. 1974. Cytological differentiation of burrowing mammals in the Palaearctic. Pp. 203-215, *in* Symposium Theriologicum II (J. Kratochvil and R. Obrtel, eds.). Prague, 394 pp.

Lyapunova, E. A., N. N. Vorontsov, K. V. Korobitsyna, E. Yu. Ivanitskaya, Yu. M. Borisov, L. V. Yakimenko, and V. Ye. Dovgal. 1980. A Robertsonian fan in *Ellobius talpinus*. Genetica, 52/53:239-247.

Lyapunova, E. A., I. Yu. Baklushinskaya, O. L. Kolomiets, and T. F. Mazurova. 1990. [Analysis of fertility of hybrids of multi-chromosomal forms in mole-voles of the super-species *Ellobius tancrei* differing in a single pair of Robertsonian metacentrics.] Doklady Akademii Nauk SSSR, 310:721-723 (in Russian).

Lyapunova, E. A., G. G. Boyeskorov, and N. N. Vorontsov. 1992. *Marmota camtschatica* Pall.—Nearctic element in Palearctic *Marmota* fauna. Pp. 185-191, *in* First International Symposium on Alpine Marmot (*Marmota marmota*) and on genus *Marmota*. (B. Bassano, P. Durio, U. G. Orsi, and E. Macchi, eds.). Proceedings: Saint Vincent, Aosta—Italy, October 28-30, 1991. Dipartimento di Produzioni Animali, Epidemiologia ed Ecologia, Torino, 268 pp.

Lydekker, R. 1889. Palaeozoology–Vertebrata. Pp. 889-1474, *in* A manual of palaeontology (H. Alleyne and R. Lydekker, authors). Third Edition, vol. 2, pt. 3. William Blackwood and Sons, Edinburgh and London, 1624 pp.

Lydekker, R. 1895 [1896]. On the affinities of the so-called extinct giant dormouse of Malta. Proceedings of the Zoological Society of London, 1895:860-863.

Lydekker, R. 1912. The ox and its kindred. Methuen & Co. Ltd., London, 271 pp.

Lydekker, R. 1913-1916. Catalogue of the ungulate mammals in the British Museum (Natural History) [Vol 2 and 3 co-authored with G. Blaine]. Trustees of the British Museum, London, 5 volumes [1:1-249 [1913], 2:1-295 [1914a], 3:1-283 [1914b], 4:1-438 [1915], 5:1-207 pp. [1916].

Lyman, C. P., and R. C. O'Brien. 1977. A laboratory study of the Turkish hamster *Mesocricetus brandti*. Breviora, 442:1-27.

Lynch, C. D. 1981. The status of the Cape Grey Mongoose, *Herpestes pulverulentus* Wagner, 1839. (Mammalia: Viverridae). Navorsinge van die Nasionale Museum Bloemfontein, 4(5):121-168.

Lynch, C. D. 1983. The mammals of the Orange Free State. Memoirs van die Nasionale Museum Bloemfontein,18:1-218.

Lynch, C. D. 1986. The ecology of the Lesser dwarf shrew, *Suncus varilla* with reference to the use of termite mounds of *Trinervitermes trinervoides*. Navorsinge van die Nasionale Museum Bloemfontein, Natural Sciences, 5:277-297.

Lynch, C. D. 1989. The mammals of the North-Eastern Cape Province. Memoirs van die Nasionale Museum Bloemfontein, 25:1-116.

Lynch, C. D. 1994. The mammals of Lesotho. Navorsinge van die Nasionale Museum Bloemfontein, 10(4):177-241.

Lynch, C. D., and J. P. Watson. 1992. The distribution and ecology of *Otomys sloggetti* (Mammalia: Rodentia) with notes on its taxonomy. Navorsinge van die Nasionale Museum Bloemfontein, Natural Sciences, 8:141-158.

Lyne, A. G., and P. A. Mort. 1981. A comparison of skull morphology in the marsupial bandicoot genus *Isoodon*: Its taxonomic implications and notes on a new species, *Isoodon arnhemensis*. Australian Mammalogy, 4:107-133.

Lyon, M. W., Jr. 1901. A comparison of the osteology of the jerboas and jumping mice. Proceedings of the United States National Museum, 23(1228):659-671.

Lyon, M. W., Jr. 1907. Notes on the porcupines of the Malay Peninsula and Archipelago. Proceedings of United States National Museum, 32:575-594.

Lyon, M. W. 1908. On a collection of mammals from the Batu Islands, West Sumatra. Annals and Magazine of Natural History, ser. 8, 1:137-140.

Lyon, M. W. 1909. Additional notes on some mammals of the Rio-Linga Archipelago, with descriptions of new species and a revised list. Proceedings of the United States National Museum, 36:479-491.

Lyon, M. W., Jr. 1911. Mammals collected by Dr. Abbott in Borneo and some of the small adjacent islands. Proceedings of the United States National Museum, 40:53-146.

Lyon, M. W., Jr. 1913. Tree shrews: An account of the mammalian family Tupaiidae. Proceedings of the United States National Museum, 45:1-188.

Lyon, M. W., Jr. 1942. Additions to the "Mammals of Indiana". American Midland Naturalist, 27(3):790-791.

Lyons, N. F., D. H. Gordon, and C. A. Green. 1980. G-banding chromosome analysis of species A of the *Mastomys natalensis* complex (Smith, 1834) (Rodentia Muridae). Genetica, 54:209-212.

Ma Shi-lai, and Wang Ying-xiang. 1986. [The taxonomy and distribution of the gibbons in southern China and its adjacent region—with description of three new subspecies.] Zoological Research, 7:393-410 (in Chinese).

Ma Yong. 1964. [A new species of hedgehog from Shansi Province, *Hemiechinus sylvaticus* sp. n.] Acta Zootaxonomica Sinica, 5:212-214 (in Chinese).

Ma, Y., and J.-Q. Jiang. 1996. [The reinstatement of the status of genus *Caryomys* (Thomas, 1911) (Rodentia: Microtinae).] Acta Zootaxonomica Sinica, 21(4):493-498 (in Chinese with English abstract).

Ma Yong, Wang Feng-gui, Jin Shan-ke, Li Si-hua, Lin Yonglie, and Yie Zong-yiao. 1981. [On the Glires of northern Xinjiang.] Acta Theriologica Sinica, 1:177-188 (in Chinese).

Ma Yong, Wang Feng-gui, Jin Shan-ke, and Li Si-hua. 1987. [Glires (rodents and lagomorphs) of Northern Xinjiang and their zoogeographical distribution.] Science Press, Academia Sinica, Beijing, 274 pp. (in Chinese).

MacDonald, S. O., and J. A. Cook. 1996. The land mammal fauna of southeast Alaska. The Canadian Field Naturalist, 110:571-598.

MacDonald, S. O., and C. Jones. 1987. *Ochotona collaris*. Mammalian Species, 281:1-4.

Macêdo, R. H., and M. A. Mares. 1987. Geographic variation in the South American cricetine rodent *Bolomys lasiurus*. Journal of Mammalogy, 68:578-594.

Macêdo, R. H., and M. A. Mares. 1988. *Neotoma albigula*. Mammalian Species, 310:1-7.

Machado, A., de Barros. 1969. Mammiferos de Angola aindo não çitados ou pouca conhecidos. Publicações Culturais da Campanhia de Diamantes de Angola, 46:93-232.

Macholán, M. 1996a. Key to European house mice (*Mus*). Folia Zoologica, 45(3):209-217.

Macholán, M. 1996b. Morphometric analysis of European house mice. Acta Theriologica, 41(3):255-275.

Macholán, M. 1996c. Multivariate morphometric analysis of European species of the genus *Mus* (Mammalia, Muridae). Zeitschrift für Säugetierkunde, 61:304-319.

Macholán, M. 2001. Multivariate analysis of morphometric variation in Asian *Mus* and Sub-Saharan *Nannomys* (Rodentia: Muridae). Zoologischer Anzeiger, 240:7-14.

Macholán, M., and V. Vohralík, V. 1997. Note on the distribution of *Mus spicilegus* (Mammalia: Rodentia) in the south-western Balkans. Acta Societatis Zoologicae Bohemicae, 61:219-226.

Macholán, M., and J. Zima. 1994. *Mus domesticus* in western Bohemia: A new mammal for the Czech Republic. Folia Zoologica, 43(1):39-41.

Macholán, M., and J. Zima. 1997. Absence of the B *B*? chromosomes in karyotypes of the yellow-necked wood mouse (*Apodemus flavicollis*) from Asia Minor. Folia Zoologica, 46(2):191-192.

Macholán, M., H. Burda, J. Zima, I. Misek, and M. Kawalika. 1993. Karyotype of the giant mole-rat, *Cryptomys mechowi* (Rodentia, Bathyergidae). Cytogenetics and Cell Genetics, 64:261-263.

Macholán, M., J. Zima, A. Cervená, and J. Cerveny. 1995. Karyotype of *Acomys cilicicus* Spitzenberger, 1978 (Rodentia, Muridae). Mammalia, 59(3):397-402.

Macholán, M., A Scharff, H. Burda, J. Zima, and O. Grütjen. 1998. The karyotype and taxonomic status of *Cryptomys amatus* (Wroughton, 1907) from Zambia (Rodentia: Bathyergidae). Zeitschrift für Säugetierkunde, 63(3):186-190.

Macholán, M., M. G. Filippucci, and J. Zima. 2001a. Genetic variation and zoogeography of pine voles of the *Microtus subterraneus/majori* group in Europe and Asia Minor. Journal of Zoology, London, 255:31-42.

Macholán, M., M. G. Filippucci, P. Benda, D. Frynta, and J. Sádlová. 2001b. Allozyme variation and systematics of the genus *Apodemus* (Rodentia: Muridae) in Asia Minor and Iran. Journal of Mammalogy, 82(3):799-813.

Macholán, M., B. Krysufek, and V. Vohralik. 2003. The location of the *Mus musculus/M. domesticus* hybrid zone in the Balkans: Clues from morphology. Acta Theriologica, 48(2):177-178.

Mack, G. 1961. Mammals from South-western Queensland. Memoirs of the Queensland Museum, 13:213-229.

MacKinnon, J. R., and S. N. Stuart (eds.). 1989. The kouprey. An action plan for its conservation. I.U.C.N., Gland, Switzerland, 20 pp.

MacPhee, R. D. E. 1987a. The shrew tenrecs of Madagascar: Systematic revision and Holocene distribution of *Microgale* (Tenrecidae, Insectivora). American Museum Novitates, 2889:1-45.

MacPhee, R. D. E. 1987b. Systematic status of *Dasogale fontoynonti* (Tenrecidae, Insectivora). Journal of Mammalogy, 68:133-135.

MacPhee, R. D. E. 1994. Morphology, adaptations, and relationships of *Plesiorycteropus*, and a diagnosis of a new order of eutherian mammals. Bulletin of the American Museum of Natural History, 220:1-214.

MacPhee, R. D. E., and D. A. Grimaldi. 1996. Mammal bones in Dominican amber. Nature, 380:489-490.

MacPhee, R. D. E., and M. J. Novacek. 1993. Definition and relationships of Lipotyphla. Pp. 13-31, *in* Mammal phylogeny: Placentals (F. S. Szalay, M. J. Novacek, and M. C. McKenna, eds.). Springer Verlag, New York, 321 pp.

MacPhee, R. D. E., C. Flemming, and D. P. Lunde. 1999. "Last occurrence" of the Antillean insectivoran *Nesophontes*: New radiometric dates and their interpretation. American Museum Novitates, 3261:1-20.

MacPhee, R. D. E., M. A. Iturralde-Vinent, and E. S. Gaffney. 2003. Domo de Zaza, an early Miocene vertebrate locality in southcentral Cuba, with notes on the tectonic evolution of Puerto Rico and the Mona Passage. American Museum Novitates, 3394:1-42.

MacPherson, A. H. 1965. The origin of diversity in mammals of the Canadian Arctic tundra. Systematic Zoology, 14:153-173.

Maddalena, T. 1990. Systematics and biogeography of Afrotropical and Palaearctic shrews of the genus *Crocidura* (Insectivora: Soricidae): An electrophoretic approach. Pp. 297-308, *in* Vertebrates in the tropics (G. Peters, and R. Hutterer, eds.). Museum Alexander Koenig, Bonn, 424 pp.

Maddalena, T., and G. Bronner. 1992. Biochemical systematics of the endemic African genus *Myosorex* Gray, 1838 (Mammalia; Soricidae). Israel Journal of Zoology, 38:245-252.

Maddalena, T., and M. Ruedi. 1994. Chromosomal evolution in the genus *Crocidura* (Insectivora: Soricidae). Pp. 335-344, *in* Advances in the biology of shrews (J. F. Merritt, G. L. Kirkland, Jr., and R. K. Rose, eds.). Carnegie Museum of Natural History, Special Publication, 18:1-458.

Maddalena, T., and P. Vogel. 1990. Relations génétiques entre crocidures méditerranéennes: Le cas des musaraignes de Gozo (Malte). Vie et Milieu, 40:119-123.

Maddalena, T., A.-M. Mehmeti, G. Bonner, and P. Vogel. 1987. The karyotype of *Crocidura flavescens* (Mammalia, Insectivora) in South Africa. Zeitschrift für Säugetierkunde, 52:129-132.

Maddalena, T., B. Sicard, M. Tranier, and J. C. Gautun. 1988. Note sur la presence de *Gerbillus henleyi* (De Winton, 1903) au Burkina-Faso. Mammalia, 52:282-284.

Maddalena, T., E. Van der Straeten, L. Ntahuga, and A. Sparti. 1989. Nouvelles donnees et caryotypes des rongeurs du Burundi. Revue Suisse de Zoologie, 96:939-948.

Maddock, A. H., and M. R. Perrin. 1981. A microscopical examination of the gastric morphology of the white-tailed rat *Mystromys albicaudatus* (Smith, 1834). South African Journal of Zoology, 16:237-247.

Maddock, A. H., and M. R. Perrin. 1983. Development of the gastric morphology and fornical bacterial/epithelial association in the white-tailed rat *Mystromys albicaudatus* (Smith 1834). South African Journal of Zoology, 18:115-127.

Madkour, G. 1984. Chondral and osteological structures in the cranial region of common Qatarian gerbils. Zoologischer Anzeiger, 213:247-257.

Madsen, O., M. Scally, C. J. Douady, D. Kao, R. W. Debry, R. Adkins, H. M. Amrine, M. J. Stanhope, W. W. de Jong, and M. S. Springer. 2001. Parallel adaptive radiations in two major clades of placental mammals. Nature, 409:610-614.

Madureira, M. D. L. 1983. On the identification of West-European *Microtus* of the sub-genera *Microtus* and *Chionomys* (Mammalia, Rodentia). A principal component analysis. Revista de Biologia, 12(3-4):615-632.

Maeda, K. 1980. Review on the classification of little tube-nosed bats, *Murina aurata* group. Mammalia, 44:531-551.

Maeda, K. 1982. Studies on the classification of *Miniopterus* in Eurasia, Australia, and Melanesia. Honyurui Kagaku (Mammalian Science), Supplement 1:1-176.

Maeda K., and S. Matsumura. 1998. Two new species of vespertilionid bats, *Myotis* and *Murina* (Vespertilionidae: Chiroptera) from Yanbaru, Okinawa Island, Okinawa Prefecture, Japan. Zoological Science, 15:301-307.

Maempel, G. Z., and H. de Bruijn. 1982. The Plio/Pleistocene Gliridae from the Mediterranean Islands reconsidered. Proceedings of the Koninklijke Nederlandse Akademie van Wetenschappen, 85(1):113-128.

Maglio, V. J. 1973. Origin and evolution of the Elephantidae. Transactions of the American Philosophical Society, 63(3):1-149.

Maharadatunkamsi, and D. J. Kitchener. 1997. Morphological variation in *Eonycteris spelaea* (Chiroptera: Pteropodidae) from the Greater and Lesser Sunda Islands, Indonesia, and description of a new subspecies. Treubia, 31:133-168.

Maharadatunkamsi, and I. Maryanto. 2002. Morphological variation of three species fruit bat genus *Megaerops* from Indonesia with its new distribution record. Treubia, 32(1):63-85.

Maharadatunkamsi, S. Hisheh, D. J. Kitchener, and L. H. Schmitt. 2003. Relationships between morphology, genetics and geography in the cave fruit bat *Eonycteris spelaea* (Dobson, 1871) from Indonesia. Biological Journal of the Linnean Society, 79:511-522.

Mahida, H., G. K. Campbell, and P. J. Taylor. 1999. Genetic variation in *Rhabdomys pumilio* (Sparrman 1784)–an allozyme study. South African Journal of Zoology, 34(3):91-101.

Mahmut, H., R. Masuda, M. Onuma, M. Takahashi, J. Nagata, M. Suzuki, and N. Ohtaishi. 2002. Molecular phylogeography of the red deer (*Cervus elaphus*) populations in Xinjiang of China: Comparisons with other Asian, European, and North American Populations. Zoological Science, 19:485-495.

Mahoney, J. A. 1968. *Baiyankamys* Hinton, 1943 (Muridae, Hydromyinae) a New Guinea rodent genus named for an incorrectly associated skin and skull (Hydromyinae, *Hydromys*) and mandible (Murinae, *Rattus*). Mammalia, 32:64-71.

Mahoney, J. A. 1975. *Notomys macrotis* Thomas, 1921, a poorly known Australian hopping mouse (Rodentia: Muridae). Journal of the Australian Mammal Society, 1:367-374.

Mahoney, J. A. 1977. Skull characters and relationships of *Notomys mordax* Thomas (Rodentia: Muridae), a poorly known Queensland hopping-mouse. Australian Journal of Zoology, 25:749-754.

Mahoney, J. A. 1981. The specific name of the honey possum (Marsupialia: Tarsipedidae: *Tarsipes rostratus* Gervais and Verreaux, 1842). Australian Mammalogy, 4:135-138.

Mahoney, J. A., and H. Posamentier. 1975. The occurrence of the native rodent *Pseudomys gracilicaudatus* (Gould, 1845) (Rodentia: Muridae) in New South Wales. Journal of the Australian Mammal Society, 1:333-346.

Mahoney, J. A., and B. J. Richardson. 1988. Muridae. Pp. 154-192, *in* Zoological catalogue of Australia. Mammalia (J. L. Bannister, et. al.). Australian Government Publishing Service, Canberra, 5:1-274.

Mahoney, J. A., and D. W. Walton. 1988. Molossidae. Zoological Catalogue of Australia, 5:146-150.

Maia, V. 1984. Karyotypes of three species of Caviinae (Rodentia, Caviidae). Experientia, 40(6):564-566.

Maia, V., and A. Langguth. 1981. New karyotypes of Brazilian akodont rodents with notes on taxonomy. Zeitschrift für Säugetierkunde, 46:241-249.

Maia. V., and A. Langguth. 1993. Constitutive heterochromatin polymorphism and NORs in *Proechimys cuvieri* Petter, 1978. Brazilian Journal of Genetics, 16:145-154.

Maia Vaz, S. 2000. Sobre a distribuicao geografica de *Phaenomys ferrugineus* (Thomas) (Rodentia, Muridae). Revista Brasileira de Zoologia, 17:183-186.

Maier, W. 1980. Konstruktionsmorphologische Untersuchungen am Gebiss der rezenten Prosimiae (Primates). Abhandlungen der Senckenbergischen Naturforschenden Gesellschaft, 538:1-158 (in German with English summary).

Maier, W., P. Klingler, and I. Ruf. 2002. Ontogeny of the medial masseter muscle, pseudo-myomorphy, and the systematic position of the Gliridae (Rodentia, Mammalia). Journal of Mammalian Evolution, 9:253-269.

Maillaird, J. 1924. Expedition of the California Academy of Sciences to the Gulf of California in 1921. A new mouse (*Peromyscus slevini*) from the Gulf of California, Mexico. Proceedings of the California Academy of Sciences, Fourth Series, 12:953-956.

Maina, J. N., G. M. O. Maloiy, and A. N. Makanya. 1992. Morphology and morphometry of the lungs of two East African mole rats, *Tachyoryctes splendens* and *Heterocephalus glaber* (Mammalia, Rodentia). Zoomorphology, 112:167-179.

Major, C. I. F[orsyth]. 1896. Diagnoses of new mammals from Madagascar. Annals and Magazine of Natural History, ser. 6, 18:319-325.

Major, C. I. F[orsyth]. 1897. On the Malagasy rodent genus *Brachyuromys*; and on the mutual relations of some groups of the Muridae (Hesperomyinae, Microtinae, Murinae, and "Spalacidae") with each other and with the Malagasy Nesomyinae. Proceedings of the Zoological Society of London, 1897:695-720.

Major, C. I. F[orsyth]. 1899. On fossil dormice. Geological Magazine, Decade 4, 6(425):492-501.

Major, C. I. F[orsyth]. 1901. The musk-rat of Santa Lucia (Antilles). Annals and Magazine of Natural History, ser. 7, 7:204-206.

Makori, N., D. Oduor-okelo, and G. Otianga-owiti. 1991. Morphogenesis of the foetal membranes and placenta of the root-rat (*Tachyoryctes splendens* (Rüppel)). African Journal of Ecology, 29:248-260.

Makova, K. D., A. Nekrutenko, and R. J. Baker. 2000. Evolution of microsatellite alleles in four species of mice (genus *Apodemus*). Journal of Molecular Evolution, 51:166-172.

Malcolm, J. R., and J. C. Ray. 2000. Influence of timber extraction routes on central African small mammal communities, forest structure, and tree diversity. Conservation Biology, 14:1623-1638.

Malec, F., and G. Storch. 1970. Zur kenntnis der jungpleistozän wühlmaus *Pitymys melitensis* (Mammalia, Rodentia). Zeitschrift für Säugetierkunde, 35(2):75-80.

Malia, M. J., R. M. Adkins, and M. W. Allard. 2002. Molecular support for Afrotheria and the polyphyly of Lipotyphla based on analyses of the growth hormone receptor gene. Molecular Phylogenetics and Evolution, 24:91-101

Malia, J. M. Jr., D. L. Lipscomb, and M. W. Allard. 2003. The misleading effects of composite taxa in supermatrices. Molecular Phylogenetics and Evolution, 27:522-527.

Malikov, V. G., and M. N. Meyer. 1990. [The characters of breeding and postnatal ontogeny of mountain and plain voles (Rodentia, Arvicolinae) in connection with their distribution.] USSR Academy of Sciences, Proceedings of the Zoological Institute, Leningrad, 225:21-33 (in Russian with English summary).

Malikov, V. G., M. N. Meyer, A. S. Grafodatsky, A. V. Polyakov, O. V. Sablina, A. Sh. Vaziri, F. Nazari, and J. Zima. 1999. On a taxonomic position of some karyomorphs belonging to genus *Calomyscus* (Rodentia, Cricetidae). Proceedings of the Zoological Institute *RAS*, 281:27-32.

Mallari, N. A. D., and A. Jensen. 1993. Biological diversity in northern Sierra Madre, Philippines: Its implication for conservation and management. Asia Life Sciences, 2:101-112.

Mallinson, J. J. C. 1971. A note on the hispid hare *Caprolagus hispidus* (Pearson, 1839). Journal of the Bombay Natural History Society, 68:443-444.

Mallon, D. P., and S. C. Kingswood. 2001. Antelopes. Global survey and regional action plans. Part 4: North Africa, the Middle East, and Asia. I.U.C.N., Gland, Switzerland, 260 pp.

Malygin, V. M. 1978. [A comparative-morphological analysis of species from the group *Microtus arvalis* (Rodentia, Cricetidae).] Zoologicheskii Zhurnal, 67:1062-1073 (in Russian).

Malygin, V. M., and T. M. Panteleichuk Santos Luis. 1996. Morphological criteria for determination of taxonomic holotypes for simple vole species (*Microtus*, Rodentia, Mammalia). Doklady Biological Sciences, 348:266-270 (translated from Doklady Akademii Nauk, 348:282-286).

Malygin, V. M., and M. Rosmiarek. 1996. Morphological differences between two karyoforms of the tiny rice rat from the genus *Neacomys* (Rodentia, Mammalia) from Peruvian Amazonia. Doklady Biological Sciences, 351:604-606.

Malygin, V. M., and M. Rosmiarek. 1997. Comparative analyses of the male reproductive tract and spermatozoa in species of Cricetidae from Peruvian Amazonia, with a discussion of their taxonomy and relationships. Pp. 129-133, *in* Tropical biodiversity and systematics (H. Ulrich, ed.). Zoologisches Forschungsinstitut und Museum Alexander Koenig, Bonn, 357 pp.

Malygin, V. M., and V. N. Yatsenko. 1986. [Taxonomic nomenclature of sibling species of the common vole (Rodentia, Cricetidae).] Zoologicheskii Zhurnal, 65:579-591 (in Russian with English summary).

Malygin, V. M., V. N. Orlov, and V. N. Yatsenko. 1990. O vidovoi samostoyatel' nosti priozernoi polevki *Microtus limnophilus*, ee rodstvennykh svyazyakh s polevkoi-ekonomkoi *M. oeconomus* i rasprostranenii etikh vidov v Mongolii [Species independence of *Microtus limnophilus*, its relations with *M. oeconomus* and distribution of these species in Mongolia]. Zoologicheskii Zhurnal, 69(4):115-127 (in Russian).

Malygin, V. M., N. V. Startzev, and Ya. Zima. 1992. [Karyotypes and distribution of striped hamsters of the group *barabensis* (Rodentia, Cricetidae).] Vestnik Moskokvskogo Universiteta, Ser. VI, Biologiia, 2:32-39 (in Russian).

Malygin, V. M., V. M. Aniskin, S. I. Isaev, and A. N. Milishnikov. 1994. *Amphinectomys savamis* Malygin gen. et sp. n., a new genus and a new species of water rat (Cricetidae, Rodentia) from Peruvian Amazonia (in Russian). Zoologicheskii Zhurnal, 73:195-208.

Manceau, V., L. Després, J. Bouvet, and P. Taberlet. 1999*a*. Systematics of the genus *Capra* inferred from mitochondrial DNA sequence data. Molecular Phylogenetics and Evolution, 13:504-510.

Manceau, V., J.-P. Crampe, P. Boursot, and P. Taberlet. 1999*b*. Identification of evolutionary significant units in the Spanish wild goat, *Capra pyrenaica* (Mammalia, Artiodactyla). Animal Conservation, 2:33-39.

Mandal, A. K., and P. K. Das. 1969. Taxonomic notes on the Szechwan burrowing shrew, *Anourosorex squamipes* Milne-Edwards, from India. Journal of the Bombay Natural History Society, 66:608-612.

Mandal, A. K., and S. Ghosh. 1981. A new subspecies of the meltad, *Millardia meltada* (Gray, 1837) [Rodentia: Muridae] from West Bengal. Bulletin of the Zoological Survey of India, 4:235-238.

Mandal, A. K., and M. K. Ghosh, 1984. Report on the occurrence of *Mus cervicolor* in the Andaman and Nicobar Islands. Journal of the Bombay Natural History Society, 81:465-466.

Mandal, A. K., A. K. Poddar, and T. P. Bhattacharyya. 1995. Occurrence of the Szechuan burrowing shrew, *Anourosorex squamipes squamipes* Milne-Edwards, 1872 (Mammalia: Insectivora: Soricidae) in Mizoram, India. Records of the Zoological Survey of India, 95:15-16.

Mann, G. 1950. Nuevos mamíferos de Tarapacá. Investigacións Zoológicas Chilenas, 1:2.

Mann, G. F. 1978. Los pequeños mamíferos de Chile (marsupiales, quirópteros, edentados y roedores). Guyana (Zoología), 40:1-342.

Manning, R. W. 1993. Systematics and evolutionary relationships of the long-eared myotis, *Myotis evotis* (Chiroptera: Vespertilionidae). Special Publications of the Museum, Texas Tech University, 37:1-58.

Manning, R. W., and J. K. Jones, Jr. 1988*a*. *Perognathus fasciatus*. Mammalian Species, 303:1-4.

Manning, R. W., and J. K. Jones, Jr. 1988*b*. A new subspecies of fringed *Myotis*, *Myotis thysanodes*, from the northwestern coast of the United States. Occasional Papers of the Museum, Texas Tech University, 123:1-6.

Manning, R. W., and J. K. Jones, Jr. 1989. *Myotis evotis*. Mammalian Species, 329:1-5.

Manning, T. H. 1956. The northern red-backed mouse, *Clethrionomys rutilus* (Pallas), in Canada. National Museum of Canada Bulletin, 144:1-67.

Mantooth, S. J., C. Jones, and R. D. Bradley. 2000. Molecular systematics of *Dipodomys elator* (Rodentia: Heteromyidae) and its phylogeographic implications. Journal of Mammalogy, 81:885-894.

Manville, R. H. 1961. The entepicondylar foramen and *Ochrotomys*. Journal of Mammalogy, 42:103-104.

Manville, R. H. 1966. The extinct sea mink, with taxonomic notes. Proceedings of the United States National Museum, 122(3584):1-12.

Mapatuna, Y., M. B. Gunasekera, W. D. Ratinasooriya, N. C. W. Goonesekere, and P. J. J. Bates. 2002. Unravelling the taxonomic status of the genus *Cynopterus* (Chiroptera, Pteropodidae) in Sri Lanka by multivariate morphometrics and mitochondrial DNA sequence analysis. Mammalian Biology, 67:321-337.

Maples, W. R., and T. W. McKern. 1967. A preliminary report on the classification of the Kenya baboon. Pp. 13-22, *in* The baboon in medical research (H. Vagtborg, ed.). University of Texas Press, Austin, 2:1-908.

Maree, S., and W. S. Grant. 1997. Origins of horseshoe bats (*Rhinolophus*, Rhinolophidae) in Southern Africa: Evidence from allozyme variability. Journal of Mammalian Evolution, 4:195-215.

Mares, M. A., and J. K. Braun. 1996. A new species of phyllotine rodent, genus *Andagalomys*, (Muridae: Sigmodontinae), from Argentina. Journal of Mammalogy, 77:928-941.

Mares, M. A., and J. K. Braun. 2000*a*. Three new species of *Brucepattersonius* (Rodentia: Sigmodontinae) from Misiones Province, Argentina. Occasional Papers of the Sam Noble Oklahoma Museum of Natural History, 9:1-13.

Mares, M. A., and J. K. Braun. 2000*b*. *Graomys*, the genus that ate South America: A reply to Steppan and Sullivan. Journal of Mammalogy, 81:271-276.

Mares, M. A., and R. A. Ojeda. 1982. Patterns of diversity and adaptation in South American hystricognath rodents. Pp. 393-432, *in* Mammalian biology in South America (M. Mares and H. Genoways, eds.). Special Publication Series, Pymatuning Laboratory of Ecology, University of Pittsburgh, Pennsylvania, 6:1-539.

Mares, M. A., M. R. Willig, K. Streilein, and T. E. Lacher, Jr. 1981*a*. The mammals of northeastern Brazil: A preliminary assessment. Annals of Carnegie Museum, 50(4):80-137.

Mares, M. A., R. A. Ojeda, and M. P. Kosco. 1981*b*. Observations on the distribution and ecology of the mammals of Salta Province, Argentina. Annals of Carnegie Museum, 50:151-206.

Mares, M. A., J. K. Braun, and D. Gettinger. 1989*a*. Observations on the distribution and ecology of the mammals of the Cerrado grasslands of central Brazil. Annals of Carnegie Museum, 58:1-60.

Mares, M. A., R. A. Ojeda, and R. M. Barquez. 1989*b*. Guide to the Mammals of Salta Province. University of Oklahoma Press, Norman, 303 pp.

Mares, M. A., R. M. Bárquez, and J. K. Braun. 1995. Distribution and ecology of some Argentine bats (Mammalia). Annals of the Carnegie Museum, 64:219-237.

Mares, M. A., R. A. Ojeda, J. K. Braun, and R. A. Barquez. 1997. Systematics, distribution, and ecology of the mammals of Catamarca Province, Argentina. Pp. 89-141, *in* Life among the muses: Papers in honor of James S. Findley (T. L. Yates, W. L. Gannon, and D. E. Wilson, eds.). The Museum of Southwestern Biology, University of New Mexico, Albuquerque, v + 290 pp.

Mares, M. A., J. K. Braun, R. M. Barquez, and M. M. Diaz. 2000. Two new genera and species of halophytic desert mammals from isolated salt flats in Argentina. Occasional Papers, Museum of Texas Tech University, 203:1-27.

Margry, C. J. P. J. 1996. Record of *Apodemus alpicola* from Switzerland. Lutra, 39:106-110.

Marineros, L., and F. M. Gallegos. 1998. Guía de campo de los mamíferos de Honduras. Instituto Nacional de Ambiente y Desarrollo, Honduras, 374 pp.

Marinho-Filho, J. M., M. Guimarães, M. L. Reis, F. H. G. Rodrigues, O. Torres, and G. de Almeida. 1997. The discovery of the Brazilian three banded armadillo in the Cerrado of central Brazil. Edentata, 3:11-13.

Marinina, L. S., N. L. Orlov, and N. A. Sorokina. 1987. [On *Myomimus personatus* Ognev, 1924, found on Maly Balkhan.] Izvestiya Akademii Nauk Turkmenskoi SSR, Seriya Biologicheskikh Nauk, 1987:72-73 (in Russian).

Marivaux, L., and J.-L. Welcomme. 2003. New diatomyid and baluchimyine rodents from the Oligocene of Pakistan (Bugti Hills, Balochistan): Systematic and paleobiogeographic implications. Journal of Vertebrate Paleontology, 23(2):420-434.

Marivaux, L., M. Vianey-Liaud, J.-L. Welcomme, and J.-J. Jaeger. 2002. The role of Asia in the origin and diversication of hystricognathous rodents. Zoologica Scripta, 31(3):225-239.

Markov, G. 2001a. Microgeographical non-metrical cranial diversity of the Fat Dormouse (*Glis glis* L.). T rakya University Journal of Scientific Research B, 2(2):115-119.

Markov, G. 2001b. Cranial sexual dimorphism and microgeographical variability of the Forest Dormouse (*Dryomys nitedula* Pal.). Trakya University Journal of Scientific Research B, 2(2):125-135.

Markov, G., and T. Chassovnikarova. 1999. Epigenetic characteristics and divergence between populations of *Apodemus sylvaticus* in Bulgaria. Folia Zoologica, 48(suppl. 1):63-67.

Markov, G. G., M. W. Sablin, and A. A. Danilkin. 1994. Sexual dimorphism and geographical variability of reindeer (*Rangifer tarandus* L., 1758) of the Palearctic (craniological characteristics). Izvestiya Rossiiskoi Akademii Nauk Seriy Biologischeskaya, 3:507-508.

Markov, G., B. Chovancová, and G. B. Hartl. 1995. Biochemical-genetic characterization of Gunther's vole, *Microtus guentheri strandzensis*. Folia Zoologica, 44(1):19-22.

Markov, G., K. Tzocheva and D. Dobryanov. 1997. Karyotaxonomic characterization of forest dormouse (*Dryomys nitedula* Pall., 1779) in Bulgaria. Acta Zoologica Bulgarica, 49:13-19.

Markowski, J. 1995. Non-metric traits: Remarks on sex dependence, age dependence, and on intercorrelations among characters. Acta Theriologica, Suppl. 3:65-74.

Markvong, A., J. Marshall, and A. Gropp. 1973. Chromosomes of rats and mice of Thailand. Natural History Bulletin of the Siam Society, 25:23-40.

Marochkina, V. V. 1996. [Materials on distribution of mole-voles (Rodentia, *Ellobius*) in eastern Turkmenistan.] Zoologischeskii Zhurnal, 75(11):1722-1728 (in Russian with English summary).

Marques-Aguiar, S. A. 1994. A systematic review of the large species of *Artibeus* Leach, 1821 (Mammalia: Chiroptera) with some phylogeographic inferences. Boletin Museo Emílio Goeldi, série Zoologia, 10:3-83.

Márquez, E. J., M. Aguilera M., and M. Corti. 2000. Morphometric and chromosomal variation in populations of *Oryzomys albigularis* (Muridae: Sigmodontinae) from Venezuela: Multivariate aspects. Zeitschrift für Säugetierkunde, 65:84-99.

Marsh, H., R. Lloze, G. E. Heinsohn and T. Kasuya. 1989. Irrawady dolphin—*Orcaella brevirostris* (Gray, 1866). Pp. 101-118, *in* Handbook of marine mammals: River dolphins and the larger toothed whales (S. H. Ridgway and R. Harrison, eds.). Academic Press, London, 4:1-442.

Marshall, J. T., Jr. 1962. Norway rat. Pp. 39-44, *in* Pacific island rat ecology (T. I. Storer, ed.). Bernice P. Bishop Museum Bulletin, 225:274 pp.

Marshall, J. T., Jr. 1977a. Family Muridae: Rats and mice. Pp. 396-487, *in* Mammals of Thailand (B. Lekagul and J. A. McNeely, eds.). Association for the Conservation of Wildlife, Sahakarnbhat Co., Bangkok, 758 pp.

Marshall, J. T., Jr. 1977b. A synopsis of Asian species of *Mus* (Rodentia, Muridae). Bulletin of the American Museum of Natural History, 158:173-220.

Marshall, J. T., Jr. 1981. Taxonomy. Pp. 17-26, *in* The mouse in biomedical research, volume I, history, genetics, and wild mice (H. L. Foster, J. D. Small, and J. G. Fox, eds.). American College of Laboratory Animal Medicine Series, Academic Press, New York, 306 pp.

Marshall, J. T., Jr. 1986. Systematics of the genus *Mus*. Pp. 12-18, *in* The wild mouse in immunology (M. Potter, J. H. Nadeau, and M. P. Cancro, eds.). Current Topics in Microbiology and Immunology, 127:1-395.

Marshall, J. T., Jr. 1998. Identification and scientific names of Eurasian house mice and their European allies, subgenus *Mus* (Rodentia: Muridae). Privately printed at Springfield, Virginia, 80 pp.

Marshall, J. T., Jr., and E. R. Marshall. 1976. Gibbons and their territorial songs. Science, 193:235-237.

Marshall, J. T., Jr., and R. D. Sage. 1981. Taxonomy of the house mouse. Symposia of the Zoological Society of London, 47:15-25.

Marshall, J. T., and J. Sugardjito. 1986. Gibbon systematics. Comparative Primate Biology, 1:137-185.

Marshall, L. G. 1977. *Lestodelphys halli*. Mammalian Species, 81:1-3.

Marshall, L. G. 1978a. *Lutreolina crassicaudata*. Mammalian Species, 91:1-4.

Marshall, L. G. 1978b. *Dromiciops australis*. Mammalian Species, 99:1-5.

Marshall, L. G. 1978c. *Glironia venusta*. Mammalian Species, 107:1-3.

Marshall, L. G. 1978d. *Chironectes minimus*. Mammalian Species, 109:1-6.

Marshall, L. G. 1979. A model for paleobiogeography of South American cricetine rodents. Paleobiology, 5:126-132.

Marshall, L. G. 1980. Systematics of the South American marsupial family Caenolestidae. Fieldiana: Geology, n.s., 5:1-145.

Marshall, L. G., J. A. Case, and M. O. Woodburne. 1990. Phylogenetic relationships of the families of marsupials. Pp. 433-506, *in* Current mammalogy (H. H. Genoways, ed.). Plenum Press, New York, 2:1-577.

Martens, J., and J. Niethammer. 1972. Die Waldmäuse (*Apodemus*) Nepals. Zeitschrift für Säugetierkunde, 37:144-154.

Martin, C. O., and D. J. Schmidly. 1982. Taxonomic review of the pallid bat, *Antrozous pallidus* (Le Conte). Special Publications, The Museum, Texas Tech University Press, 18:1-48.

Martin, L. D. 1975. Microtine rodents from the Ogallala Pliocene of Nebraska and the early evolution of the Microtinae in North America. Papers on Paleontology, The Museum of Paleontology University of Michigan, 12:101-110.

Martin, L. D. 1979. The biostratigraphy of arvicoline rodents in North America. Transactions of the Nebraska Academy of Sciences, 7:91-100.

Martin, L. D. 1980. The early evolution of the Cricetidae in North America. University of Kansas Paleontological Contributions, 102:1-42.

Martin, R. A. 1974. Fossil mammals from the Coleman IIA Fauna, Sumter County. Pp. 35-99, *in* Pleistocene mammals of Florida (S. D. Webb, ed.). The University Presses of Florida, Gainesville, 270 pp.

Martin, R. A. 1979. Fossil history of the rodent genus *Sigmodon*. Evolutionary Monographs, 2, 36 pp.

Martin, R. A. 1987. Notes on the classification and evolution of some North American fossil *Microtus* (Mammalia; Rodentia). Journal of Vertebrate Paleontology, 7:270-283.

Martin, R. A. 1989a. Early Pleistocene zapodid rodents from the Java local fauna of north-central South Dakota. Journal of Vertebrate Paleontology, 9(1):101-109.

Martin, R. A. 1989b. Arvicolid rodents of the early Pleistocene Java Local Fauna from north-central South Dakota. Journal of Vertebrate Paleontology, 9:438-450.

Martin, R. A. 1991. Evolutionary relationships and biogeography of late Pleistocene prairie voles from the eastern United States. Pp. 261-270, *in* Beamers, bobwhites, and blue-points. Tributes to the career of Paul W. Parmalee (J. R. Purdue, W. E. Klippel, and B. Styles, eds.). Illinois State Museum, Springfield, Scientific Papers 23, ix + 436 pp.

Martin, R. A. 1994. A preliminary review of dental evolution and paleogeography in the zapodid rodents, with emphasis on Pliocene and Pleistocene taxa. Pp. 1-15, *in* Rodent families of Asian origins and diversification (Y. Tomida, C. k. Li, and T. Setoguchi, eds.). National Science Museum Monograph, Number 8, 195 pp.

Martin, R. A. 1995. A new middle Pleistocene species of *Microtus* (*Pedomys*) from the southern United States, with comments on the taxonomy and early evolution of *Pedomys* and *Pitymys* in North America. Journal of Vertebrate Paleontology, 15:171-186.

Martin, R. A. 1996. Dental evolution and size change in the North American muskrat: Classification and tempo of a presumed phyletic sequence. Pp. 431-457, *in* Palaeoecology and palaeoenvironments of Late Cenozoic mammals: Tributes to the career of C. S. (Rufus) Churcher (K. M. Stewart and K. L. Seymour, eds.). University of Toronto Press, Toronto, xxxi + 675 pp.

Martin, R. A., L. Duobinis-Gray, and C. P. Crockett. 2003. A new species of early Pleistocene *Synaptomys* (Mammalia, Rodentia) from Florida and its relevance to southern bog lemming origins. Journal of Vertebrate Zoology, 23(4):917-936.

Martin, R. D. 1990. Primate origins and evolution: A phylogenetic reconstruction. Chapman and Hall, London, 804 pp.

Martin, R. L. 1966. Redescription of the type locality of *Sorex dispar*. Journal of Mammalogy, 47(1):130-131.

Martin, R. L. 1973. The dentition of *Microtus chrotorrhinus* (Miller) and related forms. University of Connecticut Occasional Papers, Biological Sciences, 2:183-201.

Martin, R. L. 1979. Morphology, development, and adaptive values of the baculum of *Microtus chrotorrhinus* (Miller, 1894) and related forms. Säugetierkundliche Mitteilungen, 27:307-311.

Martin, T. 1993. Early rodent incisor enamel evolution: Phylogenetic implications. Journal of Mammalian Evolution, 1:227-254.

Martin, T. 1994a. African origin of caviomorph rodents is indicated by incisor enamel microstructure. Paleobiology, 20:5-13.

Martin, T. 1994b. On the systematic position of Chaetomys subspinous (Rodentia: Caviomorph) based on evidence from the incisor enamel microstructure. Journal of Mammalian Evolution, 2(2):117-123.

Martin, T. 1995. Incisor enamel microstructure and phylogenetic interrelationships of Pedetidae and Ctenocactyloidea (Rodentia). Berliner Geowissenschaftliche Abhandlungen, 16:693-707.

Martin, T. 1999. Phylogenetic implications of Glires (Eurymylidae, Mimotonidae, Rodentia, Lagomorpha) incisor enamel microstructure. Mitteilungen aus dem Museum fur Naturkunde in Berlin, Zoologische Reihe, 75(2):257-273.

Martin, Y., G. Gerlach, C. Schlötterer, and A. Meyer. 2000. Molecular phylogeny of European muroid rodents based on complete cytochrome b sequences. Molecular Phylogenetics and Evolution, 16(1):37-47.

Martínez-Coronel, M., J. Ramírez-Pulido, and T. Alvarez. 1991. Variacion intrapoblacional e interpoblacional de Peromyscus melanotis (Rodentia: Muridae) en el Eje Volcanico Transverso, Mexico. Acta Zoologica Mexicana, n.s., 47:51 pp.

Martínez-Coronel, M., A. Castro-Campillo, and J. Ramírez-Pulido. 1997. Variacion no geografica de Peromyscus furvus (Rodentia: Muridae). Pp. 183-203, in Homenaje al profesor Ticul Alvarez (J. A. Cabrales and O. J. Polaco, eds.). Instituto Nacional de Antropologia e Historia, D. F. Mexico, 357:389 pp.

Martínková, N., and A. Dudich. 2003. The fragmented distribution range of Microtus tatricus and its evolutionary implications. Folia Zoologica, 52:11-22.

Martino, A. M. G. 2000. Caratterizzazione biologica di Calomys hummelincki (Husson 1960) (Rodentia, Sigmodontinae). Genetica, crescita, morfometria ed ecologia. Ph.D. Dissertation, Universita Degli Studi "La Sapienza," Rome, 211 pp.

Martino, A. M. G., and E. Capanna. 2002. Chromosome characterization of an endemic South American rodent, Calomys hummelincki (Husson, 1960) (Sigmodontinae, Phyllotini). Caryologia, 55:331-339.

Martino, A. M. G., M. G. Filippucci, and E. Capanna. 2002. Evolutive pattern of Calomys hummelincki (Husson 1960; Rodentia, Sigmodontinae) inferred from cytogenetic and allozymic data. Mastozoología Neotropical, 9:187-197.

Martino, E., and V. Martino. 1937. Preliminary note on four new rodents from Korab Mountains. Annals Magazine of Natural History, ser. 10, 19:515-518.

Martino, V., and E. Martino. 1940. Preliminary notes on five new mammals from Yugoslavia. Annals and Magazine of Natural History, ser. 5, 11:493-498.

Martín Suárez, E., and P. Mein. 1998. Revision of the genera Parapodemus, Apodemus, Rhagamys and Rhagapodemus (Rodentia, Mammalia). Geobios, 31(1):87-97.

Martynova, L. Ya. 1975. [Chromosomal differentiation in three species of zokors (Rodentia, Myospalacinae.] Zoologicheskii Zhurnal, 55(8):1265-1267 (in Russian with English summary).

Martynova, L. Ya., I. I. Formicheva, and N. N. Vorontsov. 1977. [An electrophoretic study of blood proteins in three species of zokors (Myospalacinae, Rodentia).] Zoologicheskii Zhurnal, 56(10):1538-1542 (in Russian with English summary).

Maryanto, I. 2003. Taxonomic status of the ricefield rat Rattus argentiventer (Robinson and Kloss, 1916) (Rodentia) from Thailand, Malaysia and Indonesia based on morphological variation. Records of the Western Australian Museum, 222:47-65.

Maryanto, I., and M. Yani. 2003. A new species of Rousettus (Chiroptera: Pteropodidae) from Lore Lindu, Central Sulawesi. Mammal Study, 28:111-120.

Mascarello, J. T. 1978. Chromosomal, biochemical, mensural, penile, and cranial variation in desert woodrats (Neotoma lepida). Journal of Mammalogy, 59:477-495.

Mascarello, J. T., and K. Bolles. 1980. C- and G-banded chromosomes of Ammospermophilus insularis (Rodentia: Sciuridae). Journal of Mammalogy, 61:714-716.

Mascarello, J. T., and T. C. Hsu. 1976. Chromosome evolution in woodrats, genus Neotoma (Rodentia: Cricetidae). Evolution, 30:152-169.

Masing, M. 1999. The skull of Microtus levis (Arvicolinae, Rodentia). Folia Theriologica Estonica (4), 1999:76-90.

Masing, M., and U. Timm. 1988. On mammals of the Viljandi district. Eesti Ulukid, 5:66-82.

Masini, F., and M. Sarà. 1998. Asoriculus burgioi sp. nov. (Soricidae, Mammalia) from the Monte Pellegrino faunal complex (Sicily). Acta zoologica cracoviensis, 41:111-124.

Massarini, A. I., M. A. Barros, M. O. Ortells, and O. A. Reig. 1991a. Chromosomal polymorphism and small karyotypic differences in a group of Ctenomys species from Central Argentina (Rodentia: Octodontidae). Genetica, 83:131-144.

Massarini, A. I., M. A. Barros, V. G. Roig, and O. A. Reig. 1991b. Banded karyotypes of Ctenomys mendocinus (Rodentia: Octodontidae) from Mendoza, Argentina. Journal of Mammalogy, 72(1):194-198.

Massarini, A. I., H. J. Dopazo, J. L. Bouzat, E. Hasson, and O. A. Reig. 1992. The population genetic structure of Ctenomys porteousi (Rodentia: Octodontidae). Biochemical Systematics and Ecology, 20 (8):723-734.

Massarini, A. I., F. J. Dyzenchauz, and S. I. Tiranti. 1998. Geographic variation of chromosomal polymorphism in nine populations of *Ctenomys azarae*, tuco-tucos of the *Ctenomys mendocinus* group (Rodentia: Octodontidae). Hereditas, 128:207-211.

Masseti, M. 1993. Post-Pleistocene variations of the non-flying terrestrial mammals on some Italian islands. Supplemento alle Ricerchedi Biologia della Selvaggina, 31:209-217.

Massoia, E. 1963*a*. Sobre la posición sistemática y distribución geográfica de *Akodon* (*Thaptomys*) *nigrita* (Rodentia, Cricetidae). Physis (Buenos Aires), 24:73-80.

Massoia, E. 1963*b*. *Oxymycterus iheringi* (Rodentia-Cricetidae), nueva especie para la Argentina. Physis (Buenos Aires), 24:129-136.

Massoia, E. 1964. Sistemática, distribución geográfica y ragos etoecológicos de *Akodon* (*Deltamys*) *kempi* (Rodentia, Cricetidae). Physis, 24:299-305.

Massoia, E. 1973. Descripción de *Oryzomys fornesi*, nueva especie y nuevos datos sobre algunas especies y subespecies argentinas del subgénero *Oryzomys* (*Oligoryzomys*) (Mammalia-Rodentia-Cricetidae). Revista de Investigaciones Agropecuarias, INTA, Serie 1, Biologia y Producción Animal, 10:21-37.

Massoia, E. 1975. Datos sobre un cricetido nuevo para la Argentina: *Oryzomys* (*Oryzomys*) *capito intermedius* y sus diferencias con *Oryzomys* (*Oryzomys*) *legatus* (Mammalia-Rodentia). Revista de Investigaciones Agropecuarias, INTA, Serie 5, Patologia Vegetal, 11:1-7.

Massoia, E. 1976. Sobre la identidad del holotipo de *Graomys hypogaeus* Cabrera 1934 (Mammalia-Rodentia-Cricetidae). Revista de Investigaciones Agropecuarias, INTA, Serie 5, Patologiá Vegetal, 13:15-20.

Massoia, E. 1979*a*. El género *Octodon* en la Argentina. Neotropica, 25:36.

Massoia, E. 1979*b*. Descripcion de un genero y especie nuevos: *Bibymys torresi* (Mammalia-Rodentia-Cricetidae-Sigmodontinae-Scapteromyini). Physis, C, 38:1-7.

Massoia, E. 1980*a*. El estado sistemático de cuatro especies de cricetidos sudamericanos y comentarios sobre otras especies congenericas (Mammalia-Rodentia). Ameghiniana, 17:280-287.

Massoia, E. 1980*b*. Nuevos datos sobre *Akodon*, *Deltamys* y *Cabreramys*, con la descripcion de una especie y una subespecie nuevas (Mammalia, Rodentia, Cricetidae). Nota preliminar. Historia Natural, 1:179.

Massoia, E. 1981. El estado sistematico y zoogeografia de *Mus brasiliensis* Desmarest y *Holochilus sciureus* Wagner (Mammalia-Rodentia-Cricetidae). Physis (Buenos Aires), Seccion C, 39:31-34.

Massoia, E. 1982. Diagnosis previa de *Cabreramys temchuki*, nueva especie (Rodentia, Cricetidae). Historia Natural, Corrientes, 2(11):91-92.

Massoia, E. 1985. El estado sistematico de algunos muroideos estudiados por Ameghino en 1889 con la revalidacion del genero *Necromys* (Mammalia, Rodentia, Myomorpha). Circular Informativa, Asociacion de Paleontologia Argentina, 14:4.

Massoia, E. 1988. Presas de *Tyto alba* en Campo Ramón, Departamento Oberá, Provincia de Misiones. Boletín Científico, Asociación para la Protección de la Naturaleza, 7:4-16.

Massoia, E. 1998. Roedores vinculados con virosis humanas en la República Argentina. Pp. 243-246, *in* 2do Congreso Argentino de Zoonosis y 1er Congreso Argentino y Latinamericano de Enfermedades Emergentes, Buenos Aires.

Massoia, E., and O. Donadío. 1990. *Phyllotis amicus* Thomas, 1900 (Rodentia-Cricetidae): Especie nueva para la República Argentina. Boletín Científico, APRONA, 16:2-3.

Massoia, E., and A. Fornes. 1964. Notas sobre el género *Scapteromys* (Rodentia-Cricetidae). I. Sistemática, distribución geográfica y rasgos etoecologicos de *Scapteromys tumidus* (Waterhouse). Physis (Buenos Aires), 24:279-297.

Massoia, E., and A. Fornes. 1965. Notas sobre el género *Scapteromys* (Rodentia-Cricetidae). II. Fundamentos de la identidad específica de *S. principalis* (Lund) y *S. gnambiquarae* (M. Ribeiro). Neotropica, 11:1-7.

Massoia, E., and A. Fornes. 1967. El estado sistemático, distribución geográfica y datos etoecologicos de algunos mamíferos neotropicales (Marsupialia y Rodentia), con la descripción de *Cabreramys*, género nuevo (Cricetidae). Acta Zoologica Lilloana, 23:407-430.

Massoia, E., and A. Fornes. 1969. Characteres comunes y distintivos de *Oxymycterus nasutus* (Waterhouse) y *O. iheringi* Thomas (Rodentia, Cricetidae). Physis (Buenos Aires), 28:315-321.

Massoia, E., and U. F. Pardiñas. 1993. El estado sistematico de algunos muroideos estudiados por Ameghino en 1889. Revalidacion del genero *Necromys* (Mammalia, Rodentia, Cricetidae). Ameghiniana, 30:407-418.

Massoia, E., A. Fornes, R. L. Wainberg, and T. G. de Fronza. 1968. Nuevos aportes al conocimiento de las especies bonaerenses del genero *Calomys* (Rodentia-Cricetidae). Revista de Investigaciones Agropecuarias, INTA, Serie 1, Biologia y Producción Animal, 5:63-92.

Massoia, E., J. C. Chebez, and S. H. Fortabat. 1991. Nuevos o poco conocidos craneos de mamiferos vivienties—3. *Abrawayomys ruschi* de la Provincia de Misiones, Republica Argentina. Aprona, Boletín Científico, 19:39-40.

Masson, D., and M. Breuil. 1992. Un *Myotis* (Chiroptera, Vespertilionidae) en Guadeloupe (Petites Antilles). Mammalia, 56:473-475.

Masuda, R., and M. C. Yoshida. 1994a. Nucleotide sequence variation of cytochrome b genes in three species of weasels *Mustela itatsi, Mustela sibirica*, and *Mustela nivalis*, detected by improved PCR product-direct sequencing technique. Journal of the Mammalogical Society of Japan, 19:33-43.

Masuda, R., and M. C. Yoshida. 1994b. A molecular phylogeny of the family Mustelidae (Mammalia, Carnivora), based on comparison of mitochondrial cytochrome b nucleotide sequences. Zoological Science (Tokyo), 11:605-612.

Masuda, R., M. Yoshida, F. Shinyashiki, and G. Bando. 1994. Molecular phylogenetic status of the Iriomote cat *Felis iriomotensis*, inferred from mitochondrial DNA sequence analysis. Zoological Science, 11:597-604.

Mathias, M. L. 1996. Skull size variability in adaptation and speciation of the semifossorial pine voles *Microtus duodecimcostatus* and *M. lusitanicus* (Arvicolidae, Rodentia). Pp. 271-286, *in* Proceedings of the I European Congress of mammalogy, Museu Bocage, Lisbon (M. L. Mathias, M. Santos-Reis, G. Amori, R. Libois, A. Mitchell-Jones, and M. C. Saint-Girons, eds.). Museu Nacional de História Natural, Lisboa, 314 pp.

Mathias, M. L., and A. Mira. 1992. On the origin and colonization of house mice in the Madeira islands. Biological Journal of the Linnean Society, 46:13-24.

Mathias, M. L., and M. G. Ramalhinho. 1992. A preliminary report in Robertsonian karyotype variation in long-tailed house mice (*Mus musculus domesticus* Rutty 1772) from Madeira Islands. Bocagiana, Museu Municipal do Funchal, 156:1-3.

Mathias, M. L., M. G. Ramalhinho, M. Santos-Reis, F. Petrucci-Fonseca, R. Libois, R. Fons, G. Ferraz de Carvalho, M. M. Oom, and M. Collares-Pereira. 1998. Mammals from the Azores islands (Portugal): An updated overview. Mammalia, 62:397-407.

Matisoo-Smith, E., and J. H. Robins. 2004. Origins and dispersals of Pacific peoples: Evidence from mtDNA phylogenies of the Pacific rat. Proceedings of the National Academy of Sciences, 101:9167-9172.

Matisoo-Smith, E., R. M. Roberts, G. J. Irwin, J. S. Allen, D. Penny, and D. M. Lambert. 1998. Patterns of prehistoric human mobility in Polynesia indicated by mtDNA from the Pacific rat. Proceedings of the National Academy of Sciences, 95:15145-15150.

Matocq, M. D. 2002. Phylogeographical structure and regional history of the dusky-footed woodrat, *Neotoma fuscipes*. Molecular Ecology, 11:229-242.

Matocq, M. D. 2003. Morphological and molecular analysis of a contact zone in the *Neotoma fuscipes* species complex. Journal of Mammalogy, 83:866-883.

Matschie, P. 1893. Ueber anscheinend neue Afrikanische säugethiere (*Leimacomys* g.n.). Sitzungs-berichte der Gesellschaft Naturforschender Freunde zu Berlin, 4:107-109.

Matschie, P. 1895. Die Säugethiere Deutsch-Ostafrikas. D. Reimer, Berlin, 147 pp.

Matschie, P. 1900. Über geographische Albarten des Afrikanischen elephantens. Sitzungsberichte Gesellschaft naturforschunde Freunde Berlin, 8:189-197.

Matschie, P. 1911. Über einige Säugetiere aus Muansa am Victoria-Nyansa. Sitzungsberichte der Gesellschaft Naturforschender Freunde, Berlin, 8:333-343.

Matschie, P. 1912. Über einige Rassen des Steppenluchses. Sitzungsberichte der Gesellschaft Naturforschender Freunde zu Berlin, 25:55-67.

Matschie, P. 1914. Ein neuer *Anomalurus* von der Elfenbeinküste. Sitzungs-berichte der Gesellschaft Naturforschender Freunde zu Berlin, 1914:349-351.

Matschie, P. 1915. Zwei vermutlich neue Mäuse aus Deutsch-Ostafrika. Sitzungsberichte der Gesellschaft Naturforschender Freunde zu Berlin, 1915:98-101.

Matschie, P. 1916. Bemerkungen über die Gattung *Didelphis* L. Sitzungsberichte der Gesellschaft Naturforschender Freunde zu Berlin, 1916(1):259-272.

Matson, C. W., and R. J. Baker. 2001. DNA sequence variation in the mitochondrial control region of red-backed voles (*Clethrionomys*). Molecular Biology and Evolution, 18:1494-1501.

Matson, J. O. 1975. *Myotis planiceps*. Mammalian Species, 60:1-2.

Matson, J. O. 1980. The status of banner-tailed kangaroo rats, genus *Dipodomys*, from central Mexico. Journal of Mammalogy, 61:563-566.

Matson, J. O., and J. P. Abravaya. 1977. *Blarinomys breviceps*. Mammalian Species, 74:1-3.

Matson, J. O., and R. H. Baker. 1986. Mammals of Zacatecas. Special Publications, The Museum, Texas Tech University, 24:1-88.

Matson, J. O., and D. P. Christian. 1996. Patterns of variation in cranial size and shape in two coexisting gerbillinae rodents. Zeitschrift für Säugetierkunde, 61:295-303.

Matsuda, Y., and V. M. Chapman. 1992. Analysis of sex-chromosome aneuploidy in interspecific backcross progeny between the laboratory mouse strain C57BL/6 and *Mus spretus*. Cytogenetics and Cell Genetics, 60:74-78.

Mattern, M. Y., and D. A. McLennan. 2000. Phylogeny and speciation of felids. Cladistics, 16:232-253.

Mattevi, M. S., T. Haag, A. P. Nunes, L. F. B. Oliveira, J. L. P. Cordeiro, and J. Andrades-Miranda. 2002. Karyotypes of Brazilian representatives of genus *Zygodontomys* (Rodentia, Sigmodontinae). Mastozoología Neotropical, 9:33-38.

Matthee, C. A., and S. K. Davis. 2001. Molecular insights into the evolution of the family Bovidae: A nuclear DNA perspective. Molecular Biology and Evolution, 18:1220-1230.

Matthee, C. A., and T. J. Robinson. 1997. Mitochondrial DNA phylogeography and comparative cytogenetics of the springhare, *Pedetes capensis* (Mammalia: Rodentia). Journal of Mammalian Evolution, 4:53-73.

Matthee, C. A., and T. J. Robinson. 1999. Mitochondrial DNA population structure of roan and sable antelope: Implications for the translocation and conservation of the species. Molecular Ecology, 8:227-238.

Matthews, L. H. 1939*a*. The subspecies and variation of the spotted hyaena, *Crocuta crocuta* Erxl. Proceedings of the Zoological Society of London, 1939:237-260.

Matthews, L. H. 1939*b*. Reproduction in the spotted hyaena, *Crocuta croucta* (Erxleben). Philosophical Transactions of the Royal Society of London. B. Biological Science, 230:1-78.

Matthey, R. 1954. Nouvelles recherches sur les chromosomes des Muridae. Caryologia, 6:1-44.

Matthey, R. 1955. Nouveaux documents sur les chromosomes des Muridae. Problèmes de cytologie comparée et de taxonomie chez les Microtinae. Revue Suisse de Zoologie, 62:163-206.

Matthey, R. 1958. Les chromosomes et la position systematique de quelques Murinae africains. Acta Tropica, 15:97-117.

Matthey, R. 1959. Formules chromosomiques de Muridae et de Spalacidae. La question du polymorphisme chromosomique chez les mammifères. Revue Suisse de Zoologie, 66(5):175-209.

Matthey, R. 1963. La formule chromosomique chez sept especes et sous-especes de Murinae Africains. Mammalia, 27:157-176.

Matthey, R. 1964. Analyse caryologique de cinq especes de Muridae Africains (Mammalia, Rodentia). Mammalia, 28:403-418.

Matthey, R. 1965*a*. Etudes de cytogenetique sur des *Murinae* Africains appartenant aux genres *Arvicanthis*, *Praomys*, *Acomys* et *Mastomys* (Rodentia). Mammalia, 29:228-249.

Matthey, R. 1965*b*. Le probleme de la determination du sexe chez *Acomys selousi* de Winton-Cytogenetique du genre *Acomys* (Rodentia-Muridae). Revue Suisse de Zoologie, 72:119-144.

Matthey, R. 1966*a*. Cytogénétique et taxonomie des rats appartenant au sous-genre *Mastomys* Thomas (Rodentia-Muridae). Mammalia, 30:105-119.

Matthey, R. 1966*b*. Une inversion péricentrique à l'origine d'un polymorphisme chromosomique non-Robertsonien dans une population de *Mastomys* (Rodentia-Murinae). Chromosoma, 18:188-200.

Matthey, R. 1967*a*. Note sur la cytogenetique de quelques murides Africains. Mammalia, 31:281-287.

Matthey, R. 1967*b*. Cytogénétique de *Mus* (*Leggada*) *minutoides/musculoides*. Temm, et de formes voisines: étude d'une population de Côte d'Ivoíre. Archiv der Julius Klaus-stiftung fur Vererbungsforschung, Sozialanthropologie und Rassenhygiene, 42:21-30.

Matthey, R. 1968. Cytogenetique et taxonomie du genre *Acomys*. *A. percivali* Dollman et *A. wilson* Thomas, especes D'Abyssinie. Mammalia, 32:621-627.

Matthey, R. 1970. Caryotypes de murides et de dendromurids originaires de Republique Centrafricaine. Mammalia, 34:459-466.

Matthey, R., and F. Petter. 1970. Etude cytogenetique et taxonomique de 40 *Tatera* et *Taterillus* provenant de Haute-Volta et de Republique Centrafricaine (Rongeurs, Gerbillidae). Mammalia, 34:585-597.

Matyushkin, E. N. 1979. Rysi Golarktiki [Lynx of the Holarctic]. Pp. 76-162, *in* Mlekopitayushchie: Issledovaniya po faune Sovetskovo Soyuza [Mammals: Investigations on the fauna of the Soviet Union] (O. L. Rossolimo, ed). Sbornik Trudov Zoologicheskovo Muzeya MGU, 13:1-279 (in Russian).

Mauk, C. L., M. A. Houck, and R. D. Bradley. 1999. Morphometric analysis of seven species of pocket gophers (*Geomys*). Journal of Mammalogy, 80:499-511.

Maul, L. C., L. Rekovets, W.-D. Heinrich, T. Keller, and G. Storch. 2000. *Arvicola mosbachensis* (Schmidtgen 1911) of Mosbach2: A basic sample for the early evolution of the genus and a reference for further biostratigraphical studies. Senckenbergiana Lethaca, 80(1):129-147.

Maurer, F. W., J. Op't Hof, and D. R. Osterhoff. 1976. Biochemical gene markers in the gerbil *Tatera brantsii* from Lesotho. Animal Blood Groups and Biochemical Genetics, 7:231-239.

Maurizio, R. 1994. I piccoli mammiferi (Mammalia: Insectivora, Chiroptera, Rodentia, Carnivora) della Bregaglia (Grigioni, Svizzera). Il Naturalista Valtellinese-Atti del Museo Civico di Storia Naturale di Morbegno, 5:91-138.

Mayer, F., and O. von Helversen. 2001*a*. Cryptic species diversity in European bats. Proceedings of the Royal Society of London, B, 268:1825-1832.

Mayer, F., and O. von Helversen. 2001*b*. Sympatric distribution of two cryptic bat species across Europe. Biological Journal of the Linnean Society, 74:365-374.

Mayer, G. C., and J. A. W. Kirsch. 2000. Comments on the proposed conservation of usage of the names *Mystacina* Gray, 1843, *Chalinolobus* Peters, 1866, *M. tuberculata* Gray, 1843 and *C. tuberculatus* (J. R. Forster, 1844)(Mammalia, Chiroptera). Bulletin of Zoological Nomenclature, 57:172-176.

Mayer, G. C., J. A. W. Kirsch, J. M. Hutcheon, F.-J. LaPointe, and J. Gingras. 1999. On the valid name of the lesser New Zealand short-tailed bat (Mammalia: Chiroptera). Proceedings of the Biological Society of Washington, 112:470-490.

Mayer, J. J., and R. M. Wetzel. 1986. *Catagonus wagneri*. Mammalian Species, 259:1-5.

Mayer, J. J., and R. M. Wetzel. 1987. *Tayassu pecari*. Mammalian Species, 293:1-7.

Mayhew, D. F. 1977. The endemic Pleistocene murids of Crete. I. Proceedings of the Koninklijke Nederlandse Akademie van Wetenschappen, Amsterdam, ser. B, 80(3):182-214.

Mayhew, D. F. 1996. The extinct murids of Crete. Pp. 167-171, *in* Pleistocene and Holocene fauna of Crete and its first settlers (D. S. Reese, ed.). Monographs in World Archaeology No. 28, Prehistory Press, Madison, Wisconsin, 422 pp.

Mayr, E. 1986. Uncertainty in science: Is the giant panda a bear or raccoon? Nature, 323:769-771.

Mazák, V. 1971. On the correct spelling of the name *Mustela eversmanii* Lesson, 1827 (Mammalia, Mustelidae). Lynx (Series Nova), 12:57-59.

Mazák, V. 1979. Der tiger, *Panthera tigris*. Die Neue Brehm-Bücherei, 356:1-228.

Mazák, V. 1981. *Panthera tigris*. Mammalian Species, 152:1-8.

Mazurok, N. A., T. B. Nesterova, and S. M. Zakian. 1995. High-resolution G-banding of chromosomes in *Microtus subarvalis* (Rodentia, Arvicolidae). Hereditas, 123:47-52.

Mazurok, N. A., A. A. Isaenko, T. B. Nesterova, and S. M. Zakian. 1996*a*. High-resolution G-banding of chromosomes in the common vole *Microtus arvalis* (Rodentia, Arvicolidae). Hereditas, 124:229-232.

Mazurok, N. A., N. V. Rubtsova, A. A. Isaenko, T. B. Nesterova, and S. M. Zakian. 1996*b*. Comparative analysis of chromosomes in *Microtus transcaspicus* and *Microtus subarvalis* (Arvicolidae, Rodentia): High-resolution G-banding and localization of NORs. Hereditas, 124:243-250.

McAllan, A. W., and M. D. Bruce. 1989. Some problems in vertebrate nomenclature. I. Mammals. Bollettino, Museo Regionale di Scienze Naturali, Torino, 7:443-460.

McAllister, J. A., and R. S. Hoffmann. 1988. *Phenacomys intermedius*. Mammalian Species, 305:1-8.

McBee, K., and R. J. Baker. 1982. *Dasypus novemcinctus*. Mammalian Species, 162:1-9.

McCarthy, T. J. 1987. Distributional records of bats from the Caribbean lowlands of Belize and adjacent Guatemala and Mexico. Pp. 137-162, *in* Studies in Neotropical mammalogy, essays in honor of Philip Hershkovitz (B. D. Patterson and R. M. Timm, eds.). Fieldiana: Zoology, n.s., 39:506 pp.

McCarthy, T. J., A. L. Gardner, and C. O. Handley. 1992. *Tonatia carrikeri*. Mammalian Species, 407:1-4.

McCarthy, T. J., W. B. Davis, J. E. Hill, J. K. Jones, and G. A. Cruz. 1993. Bat (Mammalia: Chiroptera) records, early collectors, and faunal lists for northern Central America. Annals of the Carnegie Museum, 62:191-228.

McCarthy, T. J., L. Albuja, and I. Manzano. 2000. Rediscovery of the brown sac-wing bat, *Balantiopteryx infusca* (Thomas, 1897). Journal of Mammalogy, 81:958-961.

McCarty, R. 1975. *Onychomys torridus*. Mammalian Species, 59:1-5.

McCarty, R. 1978. *Onychomys leucogaster*. Mammalian Species, 87:1-6.

McCullough, D. R. 1969. The Tule elk: Its history, behavior, and ecology. University of California Publications, Zoology, 88:1-209.

McCracken, G. F., J. P. Haynes, J. Cevallos, G. Z. Guffey, and F. C. Romero. 1997. Observations on the distribution, ecology, and behavior of bats on the Galapagos Islands. Journal of Zoology, London, 243:757-770.

McDermid, E. M., and W. N. Bonner. 1975. Red cell and serum protein systems of gray seals and harbour seals. Comparative Biochemistry and Physiology, 50B:97-101.

McDonald, J. N. 1981. North American bison, their classification and evolution. University of California Press, Berkeley, 316 pp.

McDowell, S. B., 1958. The Greater Antillean insectivores. Bulletin of the American Museum of Natural History, 115:115–214.

McFarlane, D. A. 1999*a*. A note on dimorphism in *Nesophontes edithae* (Mammalia: Insectivora), an extinct island-shrew from Puerto Rico. Carribean Journal of Science, 35:142-143.

McFarlane, D. A. 1999*b*. Late Quaternary fossil mammals and last occurrence dates from caves at Barahona, Puerto Rico. Carribean Journal of Science, 35:238-248.

McFarlane, D. A., and B. R. Blood. 1986. Taxonomic notes on a collection of Rhinolophidae from Northern Thailand. Zeitschrift für Säugetierkunde, 51:219-223.

McFarlane, D. A., and A. O. Debrot. 2001. A new species of extinct oryzomyine rodent from the Quaternary of Curacao, Netherlands Antilles. Carribean Journal of Science, 37:182-184.

McFarlane, D. A., and J. Lundberg. 2002. A middle Pleistocene age and biogeography for the extinct rodent *Megalomys curazensis* from Curacao, Netherlands Antilles. Caribbean Journal of Science, 38:278-281.

McGhee, M. E., and H. H. Genoways. 1978. *Liomys pictus*. Mammalian Species, 83:1-5.

McGrew, J. C. 1979. *Vulpes macrotis*. Mammalian Species, 123:1-6.

McKay, G. M. 1982. Nomenclature of the gliding possum genera *Petaurus* and *Petauroides* (Marsupialia: Petauridae). Australian Mammalogy, 5:37-39.

McKay, G. M. 1988. Petauridae. Pp. 87-97, *in* Zoological catalogue of Australia. Mammalia (D. W. Walton, ed.). Australian Government Publishing Service, Canberra, 5:1-274.

McKean, J. L. 1972. Notes on some collections of bats (Order Chiroptera) from Papua-New Guinea and Bougainville Island. Division of Wildlife, Research Technical Paper, Commonwealth Scientific and Industrial Research Organization, Australia, 26:1-35.

McKean, J. L., and J. H. Calaby. 1968. A new genus and two new species of bats from New Guinea. Mammalia, 32:372-378.

McKean, J. L., and W. J. Price. 1967. Notes on some Chiroptera from Queensland, Australia. Mammalia, 31:101-119.

McKean, J. L., and W. J. Price. 1978. *Pipistrellus* (Chiroptera: Vespertilionidae) in northern Australia with some remarks on its systematics. Mammalia, 42:343-347.

McKean, J. L., G. C. Richards, and W. J. Price. 1978. A taxonomic appraisal of *Eptesicus* (Chiroptera: Mammalia) in Australia. Australian Journal of Zoology, 26:529-537.

McKee, M. 1981. Discovery of the holotype of the Irish rat *Mus hibernicus* Thompson, 1837 in the Ulster Museum. Irish Nature Journal, 20(6):254.

McKenna, M. C. 1962. *Eupetaurus* and the living petauristine sciurids. American Museum Novitates, 2104:1-38.

McKenna, M. C. 1975. Toward a phylogenetic classification of the Mammalia. Pp. 21-46, *in* Phylogeny of the primates—A multidisciplinary approach (W. P. Luckett and F. S. Szalay, eds.). Plenum Press, New York, 483 pp.

McKenna, M. C. 1992. The alpha crystalline A chain of the eye lens and mammalian phylogeny. Annales Zooligici Fennici, 28:349-360.

McKenna, M. C., and S. K. Bell. 1997. Classification of mammals above the species level. Columbia University Press, New York, 631 pp.

McKenzie, N. L., and J. A. Kerle. 1995. Golden-backed tree-rat, *Mesembriomys macrurus*. Pp. 566-568, *in* Mammals of Australia (R. Strahan, ed.). Smithsonian Institution Press, Washington, D.C., 756 pp.

McLachlan, G. R., R. Liversidge, and R. M. Tietz. 1966. A record of *Berardius arnouxi* from the south-east coast of South Africa. Annals of the Cape Provincial Museums, 5:91-100.

McLaughlin, C. A. 1967. Aplodontoid, Sciuroid, Geomyoid, Castoroid, and Anomaluroid Rodents. Pp. 210-225, *in* Recent Mammals of the World. A synopsis of families (S. Anderson and J. K. Jones, Jr., eds.). Ronald Press Company, New York, 453 pp.

McLaughlin, C. A. 1984. Protrogomorph, sciuromorph, castorimorph, myomorph (geomyoid, anomaluroid, pedetoid, and ctenodactyloid) rodents. Pp. 267-288, *in* Orders and families of Recent mammals of the world (S. Anderson and J. K. Jones, Jr. eds.). John Wiley and Sons, New York, 686 pp.

McLellan, L. J. 1984. A morphometric analysis of *Carollia* (Chiroptera, Phyllostomidae). American Museum Novitates, 2791:1-35.

McLellan, L. J. 1994. Evolution and phylogenetic affinities of the African species of *Crocidura*, *Suncus*, and *Sylvisorex* (Insectivora: Soricidae). Pp. 379-391, *in* Advances in the biology of shrews (J. F. Merritt, , G. L. Kirkland, Jr., and R. K. Rose, eds.). Carnegie Museum of Natural History, Special Publication, 18:1-458.

McManus, J. J. 1974. *Didelphis virginiana*. Mammalian Species, 40:1-6.

McNeely, J. A. 1981. Conservation needs of *Nesolagus netscheri* in Sumatra. Pp. 929-932, *in* Proceedings of the World Lagomorph Conference (K. Meyers and C. D. MacInnes, eds). University of Guelph, Guelph, Ontario, 983 pp.

McPherson, A. B. 1985. A biogeographical analysis of factors influencing the distribution of Costa Rican rodents. Brenesia, 23:97-273.

McWilliam, A. N. 1982. Adaptive responses to seasonality in four species of Microchiroptera in coastal Kenya. Unpublished Ph.D Thesis, University of Aberdeen, U.K.

McWilliams, L. A., T. L. Best, J. L. Hunt, and K. G. Smith. 2002. *Eumops dabbenei*. Mammalian Species, 707:1-3.

Md Nor, S. 1996. The mammalian fauna on the islands at the northern tip of Sabah, Borneo. Fieldiana: Zoology, n.s., 83:1-51

Md Nor, S. 2001. Elevational diversity patterns of small mammals on Mount Kinabalu, Sabah, Malaysia. Global Ecology and Biogeography, 10:41-62.

Mead, E. M., and J. I. Mead. 1989. Quaternary zoogeography of the Nearctic *Dicrostonyx* lemmings. Boreas, 18:323-332.

Mead, J. G. 1975. Anatomy of the external nasal passages and facial complex in the Delphinidae (Mammalia: Cetacea). Smithsonian Contributions to Zoology, 207:1-72.

Mead, J. G. 1981. First records of *Mesoplodon hectori* (Ziphiidae) from the northern hemisphere and a description of the adult male. Journal of Mammalogy, 62(2):430-432.

Mead, J. G. 1989a. Shepherd's beaked whale—*Tasmacetus shepherdi* Oliver, 1937. Pp. 309-320, *in* Handbook of marine mammals: River dolphins and the larger toothed whales (S. H. Ridgway and R. Harrison, eds.). Academic Press, London, 4:1-442.

Mead, J. G. 1989b. Bottlenose whales—*Hyperoodon ampullatus* (Forster, 1770) and *Hyperoodon planifrons* Flower, 1882. Pp. 321-348, *in* Handbook of marine mammals: River dolphins and the larger toothed whales (S. H. Ridgway and R. Harrison, eds.). Academic Press, London, 1:1-442.

Mead, J. G. 1989c. Beaked whales of the genus—*Mesoplodon*. Pp. 349-430, *in* Handbook of marine mammals: River dolphins and the larger toothed whales (S. H. Ridgway and R. Harrison, eds.). Academic Press, London, 4:1-442.

Mead, J. G., W. A. Walker, and W. J. Houck. 1982. Biological observations on *Mesoplodon carlhubbsi* (Cetacea: Ziphiidae). Smithsonian Contributions to Zoology, 34:1-25.

Mead, J. I. 1989. *Nemorhaedus goral*. Mammalian Species, 335:1-5.

Mead, J. I., C. J. Bell, and L. K. Murray. 1992. *Mictomys borealis* (Northern Bog Lemming) and the Wisconsin paleoecology of the east-central Great Basin. Quaternary Research, 37:229-238.

Mead, J. I., A. E. Spiess, and K. D. Sobolik. 2000. Skeleton of extinct North American sea mink (*Mustela macrodon*). Quaternary Research (Orlando), 53:247-262.

Mead, R. A. 1968. Reproduction in western forms of the spotted skunk (genus *Spilogale*). Journal of Mammalogy, 49:373-390.

Meade, L. 1992. New distributional records for selected species of Kentucky mammals. Transactions of the Kentucky Academy of Science, 53(3-4):127-132.

Meagher, M. 1986. *Bison bison*. Mammalian Species, 266:1-8.

Mearns, E. A. 1896. Preliminary diagnoses of new mammals from the Mexican border of the United States. Proceedings of the United States National Museum, 18:443-447.

Mech, L. D. 1974. *Canis lupus*. Mammalian Species, 37:1-6.

Medel, R. G., J. E. Jiménez, and F. M. Jaksíc. 1990. Discovery of a continental population of the rare Darwin's fox, *Dusicyon fulvipes* (Martin, 1837) in Chile. Biological Conservation, 51:71-77.

Medellín, R. A. 1989. *Chrotopterus auritus*. Mammalian Species, 343:1-5.

Medellín, R. A., and H. T. Arita. 1989. *Tonatia evotis* and *Tonatia silvicola*. Mammalian Species, 334:1-5.

Medellín, R. A., D. E. Wilson, and D. Navarro L. 1985. *Micronycteris brachyotis*. Mammalian Species, 251:1-4.

Medellín, R. A., G. Ceballos, and H. Zarza. 1998a. *Spilogale pygmaea*. Mammalian Species, 600:1-3.

Medellin, R. A., A. L. Gardner, and J. Marelo Aranda. 1998b. The taxonomic status of the Yucatan brown brocket, *Mazama pandora* (Mammalia: Cervidae). Proceedings of the Biological Society of Washington, 111:1-14.

Medvedev, D. G. 2000. Morfologicheskie otlichiya irbisa iz Yuzhnogo Zabaikalia [Morphological differences of the snow leopard from Southern Transbaikalia]. Vestnik Irkutskoi Gosudarstvennoi sel'skokhozyaistvennoi akademyi [Proc. Irkutsk State Agricult. Acad.], vypusk, 20:20-30 (in Russian).

Medway, L. 1961. The status of *Tupaia splendidula* Gray (Primates: Tupaiidae). Treubia, 25:269-272.

Medway, L. 1964. Comments on the status of *Rattus inas* (Bonhote), with observations on the distribution of this and related rats in the Sunda Subregion. Federation Museums Journal, 9:95-101.

Medway, L. 1965. Mammals of Borneo. Malaysian Branch of the Royal Asiatic Society, Singapore, 193 pp.

Medway, L. 1969. The wild mammals of Malaya and offshore islands including Singapore. Oxford University Press, London, 118 pp.

Medway, L. 1970. The monkeys of Sundaland. Ecology and systematics of the cercopithecids of a humid equatorial environment. Pp. 513-553, *in* Old world monkeys (J. R. Napier and P. H. Napier, eds.). Academic Press, New York, 660 pp.

Medway, L. 1977. Mammals of Borneo: Field keys and an annotated checklist. Second ed. Monograph, Malay Branch of the Royal Asiatic Society, 7:1-172.

Medway, L. 1978. The wild mammals of Malaya (Peninsular Malaysia) and Singapore. Oxford University Press, Kuala Lumpur, 128 pp.

Medway, L., and H. S. Yong. 1976. Problems in the systematics of the rats (Muridae) of peninsular Malaysia. Malaysian Journal of Science, 4(A):43-53.

Meek, P. 2000. The decline and current status of the Christmas Island shrew *Crocidura attenuata trichura* on Christmas Island, Indian Ocean. Australian Mammalogy, 22:43-49.

Meek, P. D., and B. Triggs. 1999. A record of Hastings river mouse (*Pseudomys oralis*) in a fox (*Vulpes vulpes*) scat from New South Wales. Proceedings of the Linnean Society of New South Wales, 121:193-197.

Meester, J. 1953. The genera of African shrews. Annals of the Transvaal Museum, 22:205-214.

Meester, J. 1954. On the status of the shrew, genus *Myosorex*. Annals and Magazine of Natural History, ser. 12, 7:947-950.

Meester, J. 1958. Variation in the shrew genus *Myosorex* in southern Africa. Journal of Mammalogy, 39:325-339.

Meester, J. 1963. A systematic revision of the shrew genus *Crocidura* in Southern Africa. Transvaal Museum Memoir, 13:1-127.

Meester, J. 1964. Revision of the Chrysochloridae I. The desert golden mole *Eremitalpa* Roberts. Scientific Papers of the Namib Desert Research Station, 26:1-7.

Meester, J. 1972*a*. Order Philodota. Part 4. Pp. 1-3, *in* The mammals of Africa: An identification manual (J. Meester and H. W. Setzer, eds.) [issued 2 May 1972]. Smithsonian Institution Press, Washington, D.C., not continuously paginated.

Meester, J. 1972*b*. Order Tubulidentata. Part 10. Pp. 1-2, *in* The mammals of Africa: An identification manual (J. Meester and H. W. Setzer, eds.) [issued 2 May 1972]. Smithsonian Institution Press, Washington, D.C., not continuously paginated.

Meester, J. A. J. 1974. Order Insectivora, Family Chrysochloridae. Part 1.3. Pp. 1-7, *in* The mammals of Africa: An identification manual (J. Meester and H. W. Setzer, eds.) [issued 10 Sep 1974]. Smithsonian Institution Press, Washington, D.C., not continuously paginated.

Meester, J., and N. J. Dippenaar. 1978. A new species of *Myosorex* from Knysna, South Africa (Mammalia: Soricidae). Annals of the Transvaal Museum, 31:29-42.

Meester, J., and A. von W. Lambrechts. 1971. The Southern African species of *Suncus* Ehrenberg (Mammalia: Soricidae). Annals of the Transvaal Museum, 27:1-14.

Meester, J. A. J., I. L. Rautenbach, N. J. Dippenaar, and C. M. Baker. 1986. Classification of southern African mammals. Transvaal Museum Monograph, 5:1-359.

Meester, J., P. J. Taylor, G. C. Contrafatto, G. K. Campbell, K. Willan, J. M. Lamb, and N. Pillay. 1992. Chromosomal speciation in southern African Ototmyinae: A review. Durban Museum Novitates, 17:58-63.

Méhely, L. 1909. Species generis *Spalax*. Die arten der blindmäuse in systematischer und phylogenetischer Beziehung. Mathematische und naturwissenschaftliche Berichte aus Ungarn, 28:1-390; 28:pls. 1-33.

Méhely, L. 1913. Die streifenmäuse (Sicistinae) Europas. Annales Historico-Naturales Musei Nationalis Hungarici, 11:220-256.

Meigen, J. W. 1800. Nouvelle classification des mouches a deux ailes, (Diptera L.) d'après un plan tout nouveau [ICZN Opinion 678 suppressed for nomenclatural purposes.] Paris, J. J. Fuchs. 40 pp.

Meijaard, E., and C. P. Groves. 2002. Proposal for taxonomic changes within the genus *Babyrousa*. Asian Wild Pig News, 2(1):9-10.

Meijaard, E., and C. P. Groves. 2004. A taxonomic revision of the *Tragulus* mouse-deer (Artiodactyla). Zoological Journal of the Linnean Society, 140:63-102.

Mein, P. 1970. Les sciuroptères (Mammalia, Rodentia) néogènes d'Europe occidentale. Geobios, 3:7-77.

Mein, P., and M. Freudenthal. 1971. Une nouvelle classification des Cricetidae (Mammalia, Rodentia) du Tertiaire de l'Europe. Scripta Geologica, 2:1-37.

Mein, P., and L. Ginsburg. 1997. Les mammifères du gisement miocène inférieur de Li Mae Long, Thaïland: Systématique, biostratigraphie et paléoenvironnement. Geodiversitas, 19(4):783-844.

Mein, P., and M. Pickford. 1992. Gisements karstiques pleistocenes au Djebel Ressas, Tunisie. Comptes Rendus de l'Académie des Sciences, Paris, 315(2):247-253.

Mein, P., and J.-P. Romaggi. 1991. Un gliridé (Mammalia, Rodentia) planeur dans le Miocène Supérieur de l'Ardèche: Une adaptation non retrouvée dans la nature actuelle. Geobios, 13:45-50.

Mein, P., L. Ginsburg, and B. Ratanasthien. 1990. Nouveaux rongeurs du Miocène de Li (Thaïlande). Comptes Rendus de l'Académie des Sciences, Paris, ser. 2, 310(6):861-865.

Mein, P., E. Martín Suárez, and J. Agustí. 1993. *Progonomys* Schaub, 1938 and *Huerzelerimys* gen. nov. (Rodentia); their evolution in Western Europe. Scripta Geologica, 103:41-64.

Mein, P., M. Pickford and B. Senut. 2000*a*. Late Miocene micromammals from the Harasib karst deposits, Namibia. Part 1—Large muroids and non-muroid rodents. Communications of the Geological Survey of Namibia, 12:375-390.

Mein, P., M. Pickford, and B. Senut. 2000*b*. Late Miocene micromammals from the Harasib karst deposits, Namibia. Part 2a—Myocricetodontinae, Petromyscinae and Namibimyinae (Rodentia, Gerbillidae). Communications of the Geological Survey of Namibia, 12:391-401.

Mein, P., M. Pickford, and B. Senut. 2004. Late Miocene micromammals from the Harasib karst deposits, Namibia. Part 2b-Cricetomyidae, Dendromuridae and Muridae, with an addendum on the Myocricetodontinae. Communications of the Geological Survey of Namibia, 13:43-61.

Meinertz, T. 1941. Das oberflächliche facialisgebiet der nager. Zoologische Jahrbücher. Abteilung für Anatomie und Ontogenie der Tiere, 67:119-270.

Meinig, H. 2000. Notes on the mammal fauna of the southern part of the Republic of Mali, West Africa. Bonner Zoologische Beiträge, 49(1-4):101-114.

Meinig, H. 2002. New records of bats (Chiroptera) from Indonesian Islands. Myotis, 40:59-79.

Mekada, K., K. Koyasu, M. Harada, Y. Narita, K. C. Shrestha, and S.-I. Oda. 2002. Karyotype and X-Y chromosome pairing in the Sikkim vole (*Microtus* (*Neodon*) *sikimensis*). Journal of Zoology, London, 257:417-423.

Mellink, E. 1992. The status of *Neotoma anthonyi* (Rodentia, Muridae, Cricetinae) of Todos Santos Islands, Baja California, Mexico. Bulletin of the Southern Academy of Sciences, 9:137-140.

Mellink, E., G. Ceballos, and J. Luévano. 2002. Population demise and extinction threat of the Angel de la Guarda deer mouse (*Peromyscus guardia*). Biological Conservation, 108:107-111.

Melton, D. A. 1976. The biology of the aardvark (Tubulidentata-Orycteropodidae). Mammal Review, 6(2):75-88.

Melville, R. V. 1977. Opinion 1077 refusal of request to use the plenary powers to suppress the generic name *Cynocephalus* Boddaert, 1768 (Mammalia). Bulletin of Zoological Nomenclature, 33:182-184.

Melville, R. V., and J. D. D. Smith (eds.). 1987. Official lists and indexes of names and works in zoology. International Commission on Zoological Nomenclature, 366 pp.

Mendelssohn, H., and Y. Yom-Tov. 1999. Fauna Palaestina. mammalia of Israel. The Israel Academy of Sciences and Humanities, Keterpress Enterprises, Jerusalem, 439 pp.

Mendelssohn, H., Y. Yom-Tov, and C. P. Groves. 1995. *Gazella gazella*. Mammalian Species, 490:1-7.

Mendez, E. 1993. Los roedores de Panama. Panama, xii + 372 pp.

Meng, J. 1990. The auditory region of *Reithroparamys delicatissimus* and its systematic implications. American Museum Novitates, 2972:1-35.

Meng, J., and A. R. Wyss. 2001. The morphology of *Tribosphenomys* (Rodentiaformes, Mammalia): Phylogenetic implications for basal Glires. Journal of Mammalian Evolution, 8:1-71.

Meng, J., Y. Hu, and C. Li. 2003. The osteology of *Rhombomylus* (Mammalia, Glires): Implication for phylogeny and evolution of Glires. Bulletin of the American Museum of Natural History, 275:247 pp.

Menkhorst, P., and F. Knight. 2001. A field guide to mammals of Australia. Oxford University Press, South Melbourne, Australia. 269 pp.

Menkhorst, P. W. (ed.). 1995*a*. Mammals of Victoria, distribution, ecology, and conservation. Oxford University Press, Melbourne, 360 pp.

Menkhorst, P. W. 1995*b*. Broad-toothed rat, *Mastacomys fuscus*. Pp. 208-210, *in* Mammals of Victoria, distribution, ecology, and conservation (P. W. Menkhorst, ed.). Oxford University Press, Melbourne, 360 pp.

Menkhorst, P. W. 1995*c*. Silky mouse, *Pseudomys apodemoides*. Pp. 214-215, *in* Mammals of Victoria, distribution, ecology, and conservation (P. W. Menkhorst, ed.). Oxford University Press, Melbourne, 360 pp.

Menkhorst, P. W. 1995*d*. Smoky mouse, *Pseudomys fumeus*. Pp. 219-220, *in* Mammals of Victoria, distribution, ecology and conservation (P. W. Menkhorst, ed.). Oxford University Press, Melbourne, 360 pp.

Menkhorst, P. W. 1995*e*. Heath mouse, *Pseudomys shortridgei*. Pp. 223-224, *in* Mammals of Victoria, distribution, ecology and conservation (P. W. Menkhorst, ed.). Oxford University Press, Melbourne, 360 pp.

Menkhorst, P. W., and L. M. Williams. 1995. Bolam's mouse, *Pseudomys bolami*. P. 217, *in* Mammals of Victoria, distribution, ecology, and conservation (P. W. Menkhorst, ed.). Oxford University Press, Melbourne, 360 pp.

Menu, H. 1984. Révision du statut de *Pipistrellus subflavus* (F. Cuvier, 1832). Proposition d'un taxon générique nouveau: *Perimyotis* nov. gen. Mammalia, 48:409-416.

Menu, H. 1987. Morphotypes dentaires actuels et fossiles des chiroptères. Palaeovertebrata, 17:77-150.

Menu, H., S. Hand, and B. Sigé. 2002. Oldest Australian vespertilionid (Microchiroptera) from the early Miocene of Riversleigh, Queensland. Alcheringa, 26:319-331.

Menzies, J. I. 1970. An eastward extension of the known range of the olive colobus monkey (*Colobus verus*, Van Beneden). Journal of the West African Science Association, 15:83-84.

Menzies, J. I. 1974. The status of *Melomys platyops arfakiensis* (Ruemmler) and the races of *Melomys rubex* Thomas. Mammalia, 38:647-655.

Menzies, J. I. 1989. Designation of a lectotype for *Melomys levipes* (Thomas, 1897) (Rodentia: Muridae) of New Guinea. Science in New Guinea, 15:108-110.

Menzies, J. I. 1990. A systematic revision of *Pogonomelomys* (Rodentia: Muridae) of New Guinea. Science in New Guinea, 16:118-137.

Menzies, J. I. 1996. A systematic revision of *Melomys* (Rodentia: Muridae) of New Guinea. Australian Journal of Zoology, 44:367-426.

Menzies, J. I., and E. Dennis. 1979. Handbook of New Guinea rodents. Handbook, Wau Ecology Institute, 6:1-68.

Menzies, J. I., and J. C. Pernetta. 1986. A taxonomic revision of cuscuses allied to *Phalanger orientalis*. Journal of Zoology, London, (B) 1:551-618.

Mercelis, S. 2001. The common hamster in Flanders' Fields: Status and protection. Jahrbücher des Nassauischen Vereins für Naturkunde, 122:187-189.

Mercer, J. M., and L. Roth. 2003. The effects of Cenozoic global change on squirrel phylogeny. *Science*, 299(5612):1568-1572.

Mercure, A., K. Ralls, K. P. Koepfli, and R. K. Wayne. 1993 [1994]. Genetic subdivisions among small canids: Mitochondrial DNA differentiation of swift, kit, and arctic foxes. Evolution, 47:1313-1328.

Merriam, C. H. 1889. Preliminary revision of the North American pocket mice (genera *Perognathus* et *Cricetodipus* auct.) with descriptions of new species and subspecies and a key to the known forms. North American Fauna, 1:1-36.

Merriam, C. H. 1890. Description of a new prairie dog from Wyoming. North American Fauna, 4:33-35.

Merriam, C. H. 1895a. Monographic revision of the pocket gophers, family Geomyidae (exclusive of the species of *Thomomys*). North American Fauna, 8:1-258.

Merriam, C. H. 1895b. Revision of the shrews of the American genera *Blarina* and *Notiosorex*. North American Fauna, 10:5-34.

Merriam, C. H. 1895c. Brisson's genera of mammals, 1762. Science, n.s., 1(14):375-376.

Merriam, C. H. 1897. The generic names *Ictis*, *Arctogale*, and *Arctogalidia*. Science, 5:302.

Merriam, C. H. 1901a. Synopsis of the rice rats (genus *Oryzomys*) of the United States and Mexico. Proceedings of the Washington Academy of Sciences, 3:273-295.

Merriam, C. H. 1901b. Seven new mammals from Mexico, including a new genus of rodents. Proceedings of the Washington Academy of Sciences, 3:559-563.

Merriam, C. H. 1905. Two new chipmunks from Colorado and Arizona. Proceedings of the Biological Society of Washington, 18:163-166.

Merritt, J. F. 1978. *Peromyscus californicus*. Mammalian Species, 85:1-6.

Merritt, J. F. 1981. *Clethrionomys gapperi*. Mammalian Species, 146:1-9.

Merritt, J. F., G. L. Kirkland, Jr., and R. K. Rose (eds.). 1994. Advances in the biology of shrews. Carnegie Museum of Natural History, Special Publication, 18:1-458.

Mertens, R. 1925. Verzeichnis der Säugetier-Typen des Senckenbergischen Museums. Senckenbergiana Biologica, 7:18-37.

Mess, A. 1999. The evolutionary differentiation of the rostral nasal skeleton within Glires. A review with new data on Lagomorph ontogeny. Mitteilungen aus dem Museum fur Naturkunde in Berlin, Zoologische Reihe, 75(2):217-228.

Messenger, S. L., and J. A. McGuire. 1998. Morphology, molecules, and the phylogenetics of cetaceans. Systematic Biology, 47(1):90-124.

Meulman, E. P., N. I. Klomp, and J. L. Samuel. 1999. Observations of the gross and microscopic anatomy of the gastrointestinal tracts of the heath mouse *Pseudomys shortridgei* and the swamp rat *Rattus lutreolus* (Rodentia: Muridae). Australian Mammalogy, 21:131-134.

Meyer, A. B. 1898. Ueber zwei Eichhörnchenarten von Celebes. Abhandlungen und Berichte des Königlichen Zoologischen und Anthropologisch-Ethnographischen Museums zu Dresden, 4:1-3.

Meyer, G. E. 1978. Hyracoidea. Pp. 284-314, *in* Evolution of African mammals (V. J. Maglio and H. B. S. Cooke, eds.). Howard University Press, Cambridge, 641 pp.

Meyer, M. N. 1983. [Evolution and taxonomic status of common voles of the subgenus *Microtus* in the fauna of the USSR.] Zoologicheskii Zhurnal, 62:90-101 (in Russian).

Meyer, M. N., and V. G. Malikov. 1995. [On the distribution, taxonomic status and biology of mouselike hamsters of the genus *Calomyscus* (Rodentia, Cricetidae).] Zoologiskii Zhurnal, 74(7):96-100 (in Russian with English summary).

Meyer, M. N., and V. G. Malikov. 1996. [Peculiarities of biology and postnatal otogenesis in *Calomyscus* (Cricetidae, *Calomyscus*).] Zoologiskii Zhurnal, 75(12):1852-1862 (in Russian with English summary).

Meyer, M. N., and V. G. Malikov. 2000. [New species and subspecies of mouse-like hamsters of the genus *Calomyscus* (Rodentia, Cricetidae) from southern Turkmenistan.] Zoologiskii Zhurnal, 79(2):219-223 (in Russian with English summary).

Meyer, M. N., V. N. Orlov, and E. D. Scholl. 1972. [On the nomenclature of 46- and 54-chromosome voles of the type *Microtus arvalis* (Pall.) (Rodentia, Cricetidae).] Zoologiskii Zhurnal, 51:157-161 (in Russian with English summary).

Meyer, M. N., T. A. Grishchenko, and E. V. Zybina. 1981. [Experimental hybridization as a method of studying the degree of divergence of closely related species of the genus *Microtus*.] Zoologicheskii Zhurnal, 60:290-300 (in Russian).

Meyer, M. N., F. N. Golenishchev, S. I. Radjably, and O. V. Sablina. 1996. [Voles (subgenus *Microtus* Schrank) of Russia and adjacent territories.] Russian Academy of Sciences, Proceedings of the Zoological Institute, 232:320 pp. (in Russian with English table of contents).

Meyer, M. N., F. N. Golenishchev, N. Sh. Bulatova, G. V. Artobolevsky. 1997. [On distribution of two *Microtus arvalis* chromosomal forms in European Russia.] Zoologiskii Zhurnal, 76(4):487-493 (in Russian with English summary).

Meylan, A. 1964. Le polymorphisme chromosomique de *Sorex araneus* L. (Mamm.-Insectivora). Revue Suisse de Zoologie, 71:903-983.

Meylan, A. 1968. Formules chromosomiques de quelques petits mammifères nord-américains. Revue Suisse de Zoologie, 78:691-696.

Meylan, A. 1975. Formule chromosomique de *Sylvisorex megalura* (Jentink) (Mammalia, Insectrivora). Mammalia, 39:319-320.

Meylan, A., and J. Hausser. 1973. Les chromosomes des *Sorex* du groupe *araneus-arcticus* (Mammalia, Insectivora). Zeitschrift für Säugetierkunde, 38:143-158.

Meylan, A., and J. Hausser. 1978. Le type chromosomique A des *Sorex* du groupe *araneus*: *Sorex coronatus* Millet, 1828 (Mammalia, Insectivora). Mammalia, 42:115-122.

Meylan, A., and J. Hausser. 1991. The karyotype of the North American *Sorex tundrensis* (Mammalia; Insectivora). Pp. 125-129, *in* The cytogenetics of the *Sorex araneus* group and related topics (J. Hausser, ed.). Mémoires de la Société Vaudoise des Sciences Naturelles, 19:1-151.

Meylan, A., and P. Vogel. 1982. Contribution à la cytotaxonomie des Soricidés (Mammalia, Insectivora) de l'Afrique occidentale. Cytogenetics and Cell Genetics, 34:83-92.

Mezhzherin, S. V. 1991. [On specific distinctness of *Apodemus* (*Sylvaemus*) *ponticus* (Rodentia, Muridae).] Vesnik Zoologii, 6:34-40 (in Russian).

Mezhzherin, S. V. 1997a. [Revision of mouse genus *Apodemus* (Rodentia, Muridae) of northern Eurasia.] Vestnik Zoologii, 31(4):29-41 (in Russian with English summary).

Mezhzherin, S. V. 1997b. Biochemical systematics of the wood mouse, *Sylvaemus sylvaticus* (L., 1758) sensu lato (Rodentia, Muridae) from eastern Europe and Asia. Zeitschrift für Säugetierkunde, 62:303-311.

Mezhzherin, S. V. 2001a. [Genetic and taxonomic homogeneity of the eastern Asian mouse, *Alsomys major* (Rodentia, Muridae).] Vestnik Zoologii, 35(2):43-48 (in Russian with English abstract).

Mezhzherin, S. V. 2001b. [Species identity of the grey hamster *Cricetulus migratorius isabellinus* from Kopetdagh: Allozyme variation analysis.] Vestnik Zoologii, 35(6):91-94 (in Russian with English abstract).

Mezhzherin, S. V., and E. V. Kotenkova. 1992. Biochemical systematics of house mice from the central Palearctic region. Zeitschrift für Zoologische Systematik und Evolutionsforschung, 30:180-188.

Mezhzherin, S. V., and E. I. Lashkova. 1992. Two closely related mice species—*Sylvaemus syvaticus* and *S. flavicollis* (Rodentia, Muridae) in an area of their overlapping occurrence. Vestnik Zoologii, 5:33-41.

Mezhzherin, S. V., and A. G. Mikhailenko. 1991. O vidovoi prinadlezhnosti *Apodemus sylvaticus tscherga* (Rodentia, Muridae) Altaya [On specific identity of *Apodemus sylvaticus tscherga* (Rodentia, Muridae) from Altai]. Vestnik Zoologii, 3:35-45 (in Russian with English summary).

Mezhzherin, S. V., and M. A. Serbenyuk. 1992. [Biochemical variability and genetic divergence of Palearctic Arvicolidae. The genus *Clethrionomys* Tilesius, 1850.] Genetika, 28(2):143-153 (in Russian with English abstract).

Mezhzherin, S. V., and I. V. Zagorodnyuk. 1989. Novi vid myshei roda *Apodemus* (Rodentia, Muridae) [New species of mouse of the genus *Apodemus* (Rodentia, Muridae)]. Vestnik Zoologii, 4:55-59 (in Russian with English summary).

Mezhzherin, S. V., and A. E. Zykov. 1991. Genetic divergence and allozyme variability in mice of genus *Apodemus s. lato* (Muridae, Rodentia). Cytology and Genetics, 25:51-59.

Mezhzherin, S. V., G. G. Boyeskorov, and N. N. Vorontsov. 1992. [Genetic relations between European and Transcaucasian mice of the genus *Apodemus* Kaup.] Genetika, 28(11):111-121 (in Russian with English abstract).

Mezhzherin, S. V., A. E. Zykov, and S. Yu. Morozov-Leonov. 1993. [Biochemical variation and genetic divergence of Palearctic voles (Arvicolidae). Meadow voles, *Microtus* Schrank, 1798, snow voles, *Chionomys* Miller, 1908, water voles, *Arvicola* Lacepede, 1799.] Genetika, 29(1):28-41 (in Russian with English summary).

Mezhzherin, S. V., S. Yu. Morozov-Leonov, and I. A. Kuznetsova. 1995. [Biochemical variation and genetic divergence in Palearctic voles (Arvicolidae): Subgenus *Terricola*, true lemmings *Lemmus* Link 1975, pied lemmings *Dicrostonyx* Gloger 1841, steppe lemmings *Lagurus* Gloger 1842, mole voles *Ellobius* Fischeer von Waldheim 1814.] Genetika, 31(6):788-797 (in Russian with English summary).

Mezhzherin, S. V., E. V. Kotenkova, and A. G. Mikhailenko. 1998. The house mice, *Mus musculus* s. l., hybrid zone of Transcaucasus. Zeitschrift für Säugetierkunde, 63:154-168.

Mezzabotta, C., F. Masini, and D. Torre. 1995. *Microtus* (*Tyrrhenicola*) *henseli*, endemic fossil vole from Pleistocene and Holocene localities of Sardinia and Corsica: Evolutionary patterns and biochronological meaning. Bollettino della Società Paleontologica Italiana, 34(1):81-104.

Mi Jing-chuan, Wang Guan, and Wang Cheng-guo. 1990. [Cluster analysis on rodents community in the eastern part of desert-steppe of Nei Mongol.] Acta Theriologica Sinica, 5(4):299-309 (in Chinese).

Michaelis, B. 1972. Die Schleichkatzen (Viverriden) Afrikas. Säugetierkundliche Mitteilungen, 20(1-2):1-110.

Michaux, J., and F. Catzeflis. 2000. The bushlike radiation of muroid rodents is exemplified by the molecular phylogeny of the LCAT nuclear gene. Molecular Phylogenetics and Evolution, 17:280-293.

Michaux, J., R. Hutterer, and N. Lopez-Martinez. 1991. New fossil faunas from Fuerteventura, Canary Islands: Evidence for a Pleistocene age of endemic rodents and shrews. Comptes Rendus de l'Acadèmie des Sciences, Paris, ser. 2, 312(6):801-806.

Michaux, J., M.-G. Filippucci, R. M. Libois, R. Fons, and R. F. Matagnes. 1996a. Biogeography and taxonomy of *Apodemus sylvaticus* (the woodmouse) in the Tyrrhenian region: Enzymatic variations and mitochondrial DNA restriction pattern analysis. Heredity, 76:267-277.

Michaux, J., R. M. Libois, and R. Fons. 1996b. Différenciation génétique et morphologique du mulot, *Apodemus sylvaticus*, dans le bassin Méditerranéen Occidental. E Milieu, 46(3/4):193-203.

Michaux, J., N. López-Martínez, and J. J. Hernández-Pacheco. 1996c. A ^{14}C dating of *Canariomys bravoi* (Mammalia, Rodentia), the extinct giant rat from Tenerife (Canary Islands, Spain), and the recent history of the endemic mammals in the Archipelago. Vie Milieu, 46(3/4):261-266.

Michaux, J., M. Sara, R. Libois, and R. Matagne. 1998. Is the woodmouse (*Apodemus sylvaticus*) of Sicily a distinct species? Belgian Journal of Zoology, 128(2):211-214.

Michaux, J., S. Kinet, M.-G. Filippucci, R. Libois, A. Besnard, and F. Catzeflis. 2001a. Molecular identification of three sympatric species of wood mice (*Apodemus sylvaticus, A. flavicollis, A. alpicola*) in western Europe (Muridae: Rodentia). Molecular Ecology Notes, 4 pp.

Michaux, J., A. Reyes, and F. Catzeflis. 2001b. Evolutionary history of the most speciose mammals: Molecular phylogeny of muroid rodents. Molecular Biology and Evolution, 18(11):2017-2031.

Michaux, J., P. Chevret, M.-G. Filippucci, and M. Macholán. 2002a. Phylogeny of the genus *Apodemus* with a special emphasis on the subgenus *Sylvaemus* using the nuclear IRBP gene and two mitochondrial markers: Cytochrome *b* and 12S rRNA. Molecular Phylogenetics and Evolution, 23:123-136.

Michaux, J. P., de Bellocq J. Goüy, and S. Morand. 2002b. Body size increase in insular rodent populations: A role for predators? Global Ecology and Biogeography, 11(5):427-436.

Michaux, J. P., E. Magnanou, E. Pasradis, C. Nieberding, and R. Libois. 2003. Mitochondrial phyogeography of the woodmouse (*Apodemus sylvaticus*) in the Western Palearctic region. Molecular Ecology, 12:685-697.

Michaux, J. R., R. Libois, and M.-G. Filippucci. 2004a [2005]. So close and so different: Comparative phylogeogeography of two small mammal species, the Yellow-necked fieldmouse (*Apodemus flavicollis*) and the woodmouse (*Apodemus sylvaticus*) in the Western Palearctic region. Heredity, 94 [2005]:52-63 [online publication 25 August 2004; doi:10.1038/sj.hdy.6800561].

Michaux, J. R., R. Libois, E. Paradis, and M.-G. Filippucci. 2004b. Phylogeographic history of the yellow-necked fieldmouse *Apodemus flavicollis*) in Europe and in the Near and Middle East. Molecular Phylogenetics and Evolution, 32:788-798.

Michaux, J. R., E. Bellinvia, and P. Lymberakis. 2005. Taxonomy, evolutionary history and biogeography of the broad-toothed field mouse (*Apodemus mystacinus*) in the Eastern Mediterranean area based on mitochondrial and nuclear genes. Biological Journal of the Linnean Society, 85:53–63.

Michener, G. R., and J. W. Koeppl. 1985. *Spermophilus richardsonii*. Mammalian species, 243:1-8.

Mickleburgh, S. P., A. M. Hutson, and P. A. Racey. 1992. Old World Fruit Bats: An Action Plan for their Conservation. IUCN/SSC Chiroptera Specialist Group. I.U.C.N., Gland, Switzerland, 252 pp.

Mikes, M., I. Savic, and V. Habijan. 1982. Osteometrische eigenschaften des beckengurtels (cingulum extremitatis pelvinae) der art *Spalax leucodon* Nordmann, 1840. Zoologischer Anzeiger, 208:417-427.

Milishnikov, A., N. Bulatova, and A. Camacho. 1990. Peculiarities of molecular and chromosomal evolution in the endemic species of Capromyinae (Rodentia) in Cuba. Folia Zoologica, 39:183-192.

Milishnikov, A. N., O. P. Likhnova, and V. N. Orlov. 1992. Protein variation in four species of Ethiopian rodents. Folia Zoologia Brno, 41(1):1-9.

Miljutin, A. 1983. [On the appearance of the Norway rat into Europe.] Eesti Ulukid, 2:51-62 (in Russian with English summary).

Miljutin, A. 1990. [Systematics.] Pp. 7-33, *in* [Species of the fauna of the USSR and the contiguous countries. Norway rat: Systematics, ecology, population control (V. E. Sokolov and E. V. Karasjova, eds.).] Nauka Publishers, Moscow:1- 452 (in Russian with English summary and table of contents).

Miljutin, A. 1997. Ecomorphology of the Baltic rodents: Body form, ecological strategies and adaptive evolution. Folia Theriologica Estonica, 3:1-111.

Miljutin, A. 1998. Ecological strategies of the Baltic rodents. Proceedings of the Latvian Academy of Sciences, Section B, 52(1/2):20-30.

Miljutin, A. 1999. Trends of specialization in rodents: The birch mice, genus *Sicista* (Dipodoidea, Rodentia). Folia Theriologica Estonica, 4:76-90.

Miller, C. A., and D. E. Wilson. 1997. *Pteropus tonganus*. Mammalian Species, 552:1-6.

Miller, G. S., Jr. 1896. The genera and subgenera of voles and lemmings. North American Fauna, 12:1-84.

Miller, G. S., Jr. 1897. Revision of the North American bats of the family Vespertilionidae. North American Fauna, 13:5-135.

Miller, G. S., Jr. 1902. The mammals of the Andaman and Nicobar islands. Proceedings of the United States National Museum, 24:751-795.

Miller, G. S., Jr. 1903*a*. Seventy new Malayan mammals. Smithsonian Miscellaneous Collections, 45:1-55.

Miller, G. S., Jr. 1903*b*. Mammals collected by Dr. W. L. Abbott on the coast and islands of Northwest Sumatra. Proceedings of the United States National Museum, 26:437-484.

Miller, G. S., Jr. 1906. The nomenclature of the flying lemurs. Proceedings of the Biological Society of Washington, 19:41.

Miller, G. S., Jr. 1912*a*. Catalogue of the mammals of Western Europe (Europe exclusive of Russia) in the collection of the British Museum. British Museum (Natural History), London, 1019 pp.

Miller, G. S., Jr. 1912*b*. List of North American land mammals in the United States National Museum, 1911. Bulletin of the United States National Museum, 79:1-455.

Miller, G. S., Jr. 1913*a*. Revision of the bats of the genus *Glossophaga*. Proceedings of the United States National Museum, 46:413-429.

Miller, G. S., Jr. 1913*b*. Notes on bats of the genus *Molossus*. Proceedings of the United States National Museum, 46:85-92.

Miller, G. S., Jr. 1924. List of North American Recent mammals. Bulletin of the United States National Museum, 128:1-673.

Miller, G. S., Jr. 1929. The characters of the genus *Geocapromys* Chapman. Smithsonian Miscellaneous Collections, 82(4):1-3, 1 pl.

Miller, G. S., Jr. 1940*a*. The status of the genus *Aschizomys* Miller. Journal of Mammalogy, 21:94-95.

Miller, G. S., Jr. 1940*b*. Notes on some moles from southeastern Asia. Journal of Mammalogy, 21:442-444.

Miller, G. S., Jr. 1942. Zoological results of the George Vanderbilt Sumatran Expedition, 1936-1939. Part V. Mammals collected by Frederick A. Ulmer, Jr. on Sumatra and Nias. Proceedings of the Academy of Natural Sciences of Philadelphia, 94:107-165.

Miller, G. S., Jr., and G. M. Allen. 1928. The American bats of the genera *Myotis* and *Pizonyx*. Bulletin of the United States National Museum, 144:1-218.

Miller, G. S., Jr., and J. W. Gidley. 1918. Synopsis of the supergeneric groups of rodents. Journal of the Washington Academy of Sciences, 8:431-448.

Miller, G. S., Jr., and R. Kellogg. 1955. List of North American Recent mammals. Bulletin of United States National Museum, 205:1-954.

Miller, G. S., Jr., and J. A. G. Rehn. 1901. Systematic results of the study of North American land mammals to the close of the year 1900. Proceedings of the Boston Society of Natural History, 30(1):1-352.

Miller, R. A., R. Dysko, C. Chrisp, R. Seguin, L. Linsalata, G. Buehner, J. M. Harper, and S. Austad. 2000. Mouse (*Mus musculus*) stocks derived from tropical islands: New models for genetic analysis of life-history traits. Journal of the Zoological Society of London, 250:95-104.

Miller, S. D., J. Rottmann, K. J. Raedeke, and R. D. Taber. 1983. Endangered mammals of Chile: Status and conservation. Biological Conservation, 25:335-352.

Millet, M. C., J. Britton-Davidian, and P. Orsini. 1982. Genetique biochimique comparee de *Microtus cabrerae* Thomas, 1906 et de trois autres especes d'Arvicolidae mediterraneens. Mammalia, 46:381-388.

Millien-Parra, V. 2000*a*. Species differentiation among muroid rodents on the basis of their lower incisor size and shape: Ecological and taxonomical implications. Mammalia, 64(2):221-239.

Millien-Parra, V. 2000*b*. The evolution of *Microtia* Freudenthal, 1976 (Mammalia, Rodentia), an endemic genus from the Neogene of the Gargano Island, southern Italy. Pp. 381-389, *in* Proceedings of the 4th International Symposium, Bonn (G. Rheinwald, ed.). Bonner Zoologische Monographien, 46:400 pp.

Millien, V., and J.-J. Jaeger. 2001. Size evolution of the lower incisor of *Microtia*, a genus of endemic murine rodents from the late Neogene of Gargano, southern Italy. Paleobiology, 27(2):379-391.

Mills, G., and L. Hes. 1997. The complete book of southern African mammals. Struik Publishers, Kapstad, 356 pp.

Mills, M. G. L. 1982. *Hyaena brunnea*. Mammalian Species, 194:1-5.

Milne-Edwards, A. 1867. Observations sur quelques mammifères du nord de la Chine. Annales des sciences naturelles. Zoologie et biologie animale, 7(5):375-377.

Milne-Edwards, M. A. 1868-1874. Recherches pour servir à l'histoire naturelle des mammifères. 2 vols. Masson, Paris, vol. I, 394 pp.; vol. II, atlas.

Milner, J., C. Jones, and J. K. Jones, Jr. 1990. *Nyctinomops macrotus*. Mammalian Species, 351:1-4.

Miloš, A. 1994. Distribution of the harvest mouse (*Micromys minutus*) in the Czech Republic. Folia Musei Rerum Naturalium Bohemiae Occidentalis, Zoologica, 40:1-28.

Milyutin, A. I. 1990. Sistematika [Systematics]. Pp. 25-33, *in* Seraya krysa: Sistematika, ekologiya, reguliatsiya chislennosti [Norway rat: Systematics, ecology, population control] (V. E. Sokolov and E. V. Karasjova, eds.). Nauka, Moscow, 452 pp. (in Russian).

Minato, S., and H. Doei. 1995. Arboreal activity of *Glirulus japonicus* (Rodentia, Myoxidae) confirmed by use of bryophytes as nest materials. Acta Theriologica, 40(3):309-313.

Minezawa, M., M. Harada, O. C. Jordan, and C. J. Valdivia Borda. 1985. Cytogenetics of the Bolivian endemic red howler monkeys (*Alouatta seniculus sara*): Accessory chromosomes and Y-autosome translocation related numerical variations. Kyoto University Overseas Research Reports of New World Monkeys, 5:7-16.

Mishra, A. C., and V. Dhanda. 1975. Review of the genus *Millardia* (Rodentia: Muridae), with description of a new species. Journal of Mammalogy, 56:76-80.

Misonne, X. 1957. Mammiferes de la Turquie sub-orientale et du nord de la Syrie. Mammalia, 21(1):53-67.

Misonne, X. 1965*a*. Presence de *Leggada callewaerti* Thomas au Katanga. Mammalia, 29:426-429.

Misonne, X. 1965*b*. Rongeurs. Exploration du Parc National de la Kagera, Deuxième Série, 1(3):77-118.

Misonne, X. 1966. The systematic position of *Mystromys longicaudatus* Noack and of *Leimacomys buttneri* Matschie. Annales Musée Royal de l'Afrique Centrale, Tervuren, Belgique, Serie IN-8, Sciences Zoologiques, 144:41-45.

Misonne, X. 1969. African and Indo-Australian Muridae: Evolutionary trends. Annales Musée Royal de l'Afrique Centrale, Tervuren, Belgique, Serie IN-8, Sciences Zoologiques, 172:1-219.

Misonne, X. 1974. Order Rodentia. Part 6. Pp. 1-39, *in* The mammals of Africa: An identification manual (J. Meester and H. W. Setzer, eds.). [issued 10 Sep 1974]. Smithsonian Institution Press, Washington, D.C., not continuously paginated.

Misonne, X. 1979. Muridae collected in Irian Jaya, Indonesia. Bulletin de l'Institut Royal des Sciences Naturelles de Belgique Biologie, 51(8):16 pp.

Misonne, X. 1990. New record for Iran: *Golunda elliotti* (Gray) (Rodentia, Muridae). Mammalia, 54:494.

Misonne, X., and J. Verschuren. 1964. Notes sur *Rattus pernanus* Kershaw, 1921. Mammalia, 28:654-658.

Misonne, X., and J. Verschuren. 1966. Les rongeurs et lagomorphes de la region du Parc National du Serengeti (Tanzanie). Mammalia, 30:517-537.

Mitchell, E. D. 1970. Pigmentation pattern evolution in delphinid cetaceans: An essay in adaptive coloration. Canadian Journal of Zoology, 48:717-740.

Mitchell, E. D. (ed.). 1975*a*. Report of the meeting on smaller cetaceans—Montreal, April 1-11, 1974. Pp. 889-893, *in* Review of biology and fisheries for smaller cetaceans. Journal of the Fisheries Research Board of Canada, 32(7):875-1240.

Mitchell, E. D. 1975*b*. Parallelism and convergence in the evolution of Otariidae and Phocidae. Rapports et Procès-Verbaux des Réunions Conseil International pour l'exploration de la Mer, 169:12-26.

Mitchell, E., and R. H. Tedford. 1973. The Enaliarctinae, a new group of extinct aquatic Carnivora and a consideration of the origin of the Otariidae. Bulletin of the American Museum of Natural History, 151:201-284.

Mitchell, R. M. 1978. The *Ochotona* (Lagomorpha: Ochotonidae) of Nepal. Säugetierkundliche Mitteilungen, 26:208-214.

Mitchell, R. M. 1981. The *Ochotona* (Lagomorpha: Ochotonidae) of Asia. Pp. 1031-1038 [Vol. 2], *in* Geological and ecological studies of Qinghai-Xizang plateau (Liu D., ed.). Science Press, Academia Sinica, Beijing, Vol. 1: 974 pp, Vol. 2: 1163 pp.

Mitchell-Jones, A. J., G. Amori, W. Bogdanowicz, B. Kryštufek, P. J. H. Reijnders, F. Spitzenberger, M. Stubbe, J. B. M. Thissen, V. Vohralík, and J. Zima. 1999. Atlas of European mammals. Academic Press, London, 496 pp.

Mitev, D., and E. Miteva. 1991*a*. [On the *Rattus norvegicus* Berk. (Muridae, Rodentia) taxonomy from Plovdiv's Region.] Travaux Scientifiques, 29(6):117-123 (in Bulgarian with English summary).

Mitev, D., and E. Miteva. 1991b. [Investigations on the taxonomy of the water rat *Arvicola terrestris* L. in Thrace Dep-ression.] Travaux Scientifiques, 29(6):125-133 (in Bulgarian with English summary).

Mitev, D., and E. Miteva. 1991c. [On the karyotype of *Microtus arvalis* (Microtidae, Rodentia) in Thrace Depression.] Travaux Scientifiques, 29(6):135-138 (in Bulgarian with English summary).

Mitev, D., D. Peshev, V. Delov and S. Stoikov. 1994. Over the craniological characteristic and the subspecific status of the Forest Dormouse *Dryomys nitedula* Pall. 1779 (Rodentia, Mammalia) in Bulgaria. Nauchni trudove, Biologiia, 1991, 29 (6):89-98.

Miththapala, S., J. Seidensticker, and S. J. O'Brien. 1996. Phylogeographic subspecies recognition in leopards (*Panthera pardus*): Molecular genetic variation. Conservation Biology, 10:1115-1132.

Mittermeier, R. A., and A. F. Coimbra-Filho. 1981. Systematics: Species and subspecies. Pp. 29-109, *in* Ecology and behavior of neotropical primates (A. F. Coimbra-Filho and R. A. Mittermeier, eds.). Academia Brasileira de Ciencias (Rio de Janeiro), 1:1-496.

Mittermeier, R. A., A. B. Rylands, A. F. Coimbra-Filho, and G. A. B. da Foneseca (eds.). 1988. Ecology and behavior of Neotropical Primates. World Wildlife Fund, Washington D.C., 2:1-610.

Miura, S., and M. Yoshihara. 2002. The fate of Philippine brown deer *Cervus mariannus* on the Ogasawara Islands, Japan. Mammalia, 66:451-452.

Mivart, St. G. 1882. On the classification and distribution of the Aeluroidea. Proceedings of the Zoological Society of London, 1882:135-208.

Miyamoto, M. M., S. M. Tanhauser, and P. J. Laipis. 1989. Systematic relationships in the artiodactyl tribe Bovini (family Bovidae), as determined from mitochondrial DNA sequences. Systematic Zoology, 38:342-349.

Miyashita, N., T. Kawashima, C. H. Wang, M.-L. Jin, F. Wang, H. Gotoh, L. V. Yakimenko, A. P. Kryukov, L. V. Frisman, J. Akabarzadeh, and K. Moriwaki. 1994. Pp. 85-93, *in* Genetics in wild mice: Its application to biomedical research (K. Moriwaki, T. Shiroishi, and H. Yonekawa, eds.). Japan Scientific Societies Press, Tokyo, 333 pp.

Mockel, R. 1986. Zum vorkommen des gartenschlafers (*Eliomys quercinus*) im Westerzgebirge. Säugetierkundliche Informationen, 2(10):311-317.

Mockel, R. 1988. Zur Verbreitung, Haufigkeit und Okologie der Haselmaus (*Muscardinus avellanarius*) im Westerzgebirge. Säugetierkundliche Informationen, 2(12):569-588.

Modi, W. S. 1986. Karyotypic differentiation among two sibling species pairs of New World microtine rodents. Journal of Mammalogy, 67:159-165.

Modi, W. S. 1987. Phylogenetic analyses of chromosomal banding patterns among the Nearctic Arvicolidae (Mammalia: Rodentia). Systematic Zoology, 36:109-136.

Modi, W. S. 1993. Comparative analyses of heterochromatin in *Microtus*: Sequence heterogeneity and localized expansion and contraction of satellite DNA arrays. Cytogenetics and Celluar Genetics, 62:142-148.

Modi, W. S. 1996. Phylogenetic history of LINE-1 among arvicolid rodents. Molecular Biology and Evolution, 13:633-641.

Modi, W. S., and R. Gamperl. 1989. Chromosomal banding comparisons among American and European red- backed mice, genus *Clethrionomys*. Zeitschrift für Säugetierkunde, 54:141-152.

Modi, W. S., and M. R. Lee. 1984. Systematic implication of chromosomal banding analyses of populations of *Peromyscus truei* (Rodentia, Muridae). Proceedings of the Biological Society of Washington, 97:716-723.

Moeller, W. 1968. Allometrische Analyse der Gürteltierschädel ein Beiträg zur Phylogenie der Dasypodidae Bonaparte, 1838. Zoologische Jahrbücher, Abteilung für Anatomie und Ontogenie der Tiere, 85:411-528.

Moeschler, P., I. A. Nader, and P. Gaucher. 1990. First record of *Asellia patrizii* De Beaux, 1931 (Chiroptera: Hipposideridae) in Saudi Arabia. Mammalia, 54:654-655.

Mohr, E. 1939. Die Baum- und Ferkelratten- Gattungen *Capromys* Desmarest (sens. ampl.) und *Plagiodontia* Cuvier. Mitteilungen aus dem Hamburgischen Museum und Institut (Hamburg), 48:48-118.

Mohr, E. 1952. Beiträge zur Kenntnis der Mähnenrobben. Zoologische Garten, 19:98-112.

Mohr, E. 1961. Schuppentiere. Neue Brehm-Bücherei. A. Ziemsen Verlag, Wittenberg Lutherstadt, 159 pp.

Mohr, E. 1965. Altweltliche Stachelschweine. Ziemsen Verlag, Wittenberg Lutherstadt, 164 pp.

Mohr, E. 1967. Der Blaubock *Hippotragus leucophaeus* (Pallas, 1766). Eine Dokumentation. Mammalia Depicta. Paul Parey, Berlin, 81 pp.

Molina, G. I. 1782. Saggio sulla storia naturale del Chili. Stamperia di S. Tommaso d'Aquino, Bologna.

Molina, M., and J. Molinari. 1999. Taxonomy of Venezuelan white-tailed deer (Mammalia, Cervidae, *Odococileus*) based on cranial and mandibular traits, Canadian Journal of Zoology, 77:632-645.

Molinari, J. 1994. A new species of *Anoura* (Mammalia Chiroptera Phyllostomidae) from the Andes of northern South America. Tropical Zoology, 7:73-86.

Molinari, J., and P. J. Soriano. 1987. *Sturnira bidens*. Mammalian Species, 276:1-4.

Monard, A. 1933. Mission Scientifique Suisse dans l'Angola. Résultats scientifiques. Mammifères. Part VI: Rongeurs. Bulletin de la Société Neuchâteloise des Sciences Naturelles, 57:53-63.

Monard, A. 1935. Contribution à la mammologie d'Angola et prodrome d'une faune d'Angola. Archivos do Museum Bocage, 6:1-309.

Mones, A. 1981. Sinopsis sistemática preliminary de la familia Disomyidae (Mammalia, Rodentia, Caviomorpha). Anais II Congreso Latino Americano de Paleontología, Porto Alegre, Brasil, 2:605-619.

Mones, A. 1984. Estudios sobre la familia Hydrochoeridae, XIV. Revisión sistemática (Mammalia: Rodentia). Senckenbergiana Biology, 65:1-17.

Mones, A. 1991. Monografía de la familia Hydrochoeridae (Mammalia: Rodentia). Sistemática, Paleotología, Filogenia, Bibliogrfía. Courier Forschungsinstitut Senckenberg, 134:1-235.

Mones, A. 2001. La mastozoología en el Uruguay: Pasado y presente. Comunicaciones Zoologicas del Museo de Historia Natural de Montevideo, 13(197):1-19.

Mones, A., and J. Ojasti. 1986. *Hydrochoerus hydrochaeris*. Mammalian Species, 264:1-7.

Mones, A., and M. E. Philippi. 1992. Bibliografia mastozoológica anotada del Uruguay. Anales del Museo Nacional de Historia Natural de Montevideo, 8:71-161.

Monfort, A. 1992. Première liste commentée des mammifères du Rwanda. Journal of African Zoology, 106(2):141-151.

Monjeau, J. A., N. Bonino, and S. L. Sara. 1994. Annotated checklist of the living land mammals in Patagonia, Argentina. Mastozoologia Neotropical, 1:143-156.

Monteiro-Filho, E. L. de A., L. R. Monteiro, and S. F. Dos Reis. 2002. Skull shape and size divergence in dolphins of the genus *Sotalia*: A tridimensional morphometric analysis. Journal of Mammalogy, 83(1):125-134.

Monterrubio, T., J. W. Demastes, L. Leon-Paniagua, and M. S. Hafner. 2000. Systematic relationships of the endangered Queretaro pocket gopher (*Cratogeomys neglectus*). Southwestern Naturalist, 45:249-252.

Montgelard, C. 1992. Albumin preservation in fossil bones and systematics of *Malpaisomys insularis* (Muridae, Rodentia), an extinct rodent of the Canary Islands. Historical Biology, 6:293-302.

Montgelard, C., S. Bentz, C. Douady, J. Lauquin, and F. M. Catzeflis. 2001. Molecular phylogeny of the sciurognath rodent families Gliridae, Anomaluridae and Pedetidae. Morphological and paleontological implications. Pp. 293-307, *in* African Small Mammals (C. Denys, L. Granjon, and A. Poulet, eds.). IRD Éditions, Collection colloques et séminaires, Paris, 570 pp.

Montgelard, C., S. Bentz, C. Tirard, O.Verneau, and F. M. Catzeflis. 2002. Molecular systematics of Sciurognathi (Rodentia): The mitochondrial cytochrome *b* and 12S rRNA genes support the Anomaluroidea (Pedetidae and Anomaluridae). Molecular Phylogenetics and Evolution, 22:220-233.

Montgelard, C., C. A. Matthee, and T. J. Robinson. 2003. Molecular systematics of dormice (Rodentia: Gliridae) and the radiation of *Graphiurus* in Africa. Proceedings of the Royal Society of London. B, 270:1947-1955.

Montuire, S., J. Michaux, S. Legendre, and J.-P. Aguilar. 1997. Rodents and climate. 1. A model for estimating past temperatures using arvicolids (Mammalia: Rodentia). Palaeogeography, palaeoclimatology, palaeoecology, 128:187-206.

Monty, A.-M., E. R. Wagle, R. E. Emerson, and G. A. Feldhamer. 1995. Recently discovered populations of eastern woodrats (*Neotoma floridana*) in southern Illinois. Transactions of the Illinois State Academy of Science, 88:43-47.

Monty, A.-M., E. J. Heist, E. R. Wagle, R. E. Emerson, E. H. Nicholson, and G. A. Feldhamer. 2003. Genetic variation and population assessment of eastern woodrats in southern Illinois. Southeastern Naturalist, 2:243-260.

Moojen, J. 1948. Speciation in the Brazilian spiny rats (genus *Proechimys*, family Echimyidae). University of Kansas Publications, Museum of Natural History, 1(19):301-406.

Moojen, J. 1952. Os roedores do Brasil. Biblioteca Científica Brasileira, ser. A, 2:1-214.

Moojen, J. 1965. Novo genero de Cricetidae do Brasil central (Glires, Mammalia). Revista Brasileira de Biologia, 25:281-285.

Moore, C. M., and P. W. Collins. 1995. *Urocyon littoralis*. Mammalian Species, 489:1-7.

Moore, D. W., and L. L. Janacek. 1990. Genic relationships among North American *Microtus* (Mammalia: Rodentia). Annals of Carnegie Museum, 59:249-259.

Moore, J. C. 1958. New genera of East Indian squirrels. American Museum Novitates, 1914:1-5.

Moore, J. C. 1959. Relationships among the living squirrels of the Sciurinae. Bulletin of the American Museum of Natural History, 118(4):157-206.

Moore, J. C. 1960. Squirrel geography of the Indian subregion. Systematic Zoology, 9:1-17.

Moore, J. C. 1968. Relationships among the living genera of beaked whales with classifications, diagnoses and keys. Fieldiana: Zoology, 53(4):209-298.

Moore, J. C., and G. H. H. Tate. 1965. A study of the diurnal squirrels, Sciurinae, of the Indian and Indochinese subregions. Fieldiana: Zoology, 48:1-351.

Moore, L. A. 1995. Giant white-tailed rat, *Uromys caudimaculatus*. Pp. 639-640, *in* Mammals of Australia (R. Strahan, ed.). Smithsonian Institution Press, Washington, D.C., 756 pp.

Moore, L. A., and L. Leung. 1995. Cape York rat, *Rattus leucopus*. Pp. 653-655, *in* Mammals of Australia (R. Strahan, ed.). Smithsonian Institution Press, Washington, D.C., 756 pp.

Moors, P. J. 1990. Norway rat. Pp. 192-206, *in* The handbook of New Zealand mammals (C. M. King, ed.). Oxford University Press, Auckland, 600 pp.

Morales, A., and J. Rodriguez. 1997. Black rats (*Rattus rattus*) from medieval Mertola (Baixo Alentejo, Portugal). Journal of the Zoological Society of London, 241:623-642.

Morales, J. C., and J. W. Bickham. 1995. Molecular systematics of the genus *Lasiurus* (Chiroptera: Vespertilionidae) based on restriction-site maps of the mitochondrial ribosomal genes. Journal of Mammalogy, 76:730-749.

Morales, J. C., and M. D. Engstrom. 1989. Morphological variation in the painted spiny pocket mouse, *Liomys pictus* (Family Heteromyidae), from Colima and southern Jalisco, México. Royal Ontario Museum, Life Sciences, Occasional Papers, 38:1-16.

Moreau, R. E., G. H. E. Hopkins, and R. W. Hayman. 1946. The type-localities of some African mammals. Proceedings of the Zoological Society of London, 1946:387-447.

Moreira, D. M., M. H. L. P. Franco, T. R. O. Freitas, and T. A. Weimer. 1991. Biochemical polymorphisms and phenetic relationships in rodents of the genus *Ctenomys* from southern Brazil. Biochemical Genetics, 29:601-615.

Moreno, S. 1989. Variacion geographica del genero *Eliomys* en la Peninsula Iberica. Doñana, Acta Vertebrata, 16(1):123-141.

Moreno, S., and M. Delibes. 1981. Notes on the garden dormouse (*Eliomys*; Rodentia, Gliridae) of northern Morocco. Säugetierkundliche Mitteilungen, 30:212-215.

Moreno, S., J. Delibes, J. C. Blanco, and A. R. Larramendi. 1986. Sobre la sistemática y biologia de *Eliomys quercinus* en la Cordillera Cantabrica. Doñana, Acta Vertebrata, 13:147-156.

Morgan, G. S. 1985. Taxonomic status and relationships of the Swan Island hutia, *Geocapromys thoracatus* (Mammalia: Rodentia: Capromyidae), and the zoogeography of the Swan Islands vertebrate fauna. Proceedings of the Biological Society of Washington, 98(1):29-46.

Morgan, G. S. 1989*a*. *Geocapromys thoracatus*. Mammalian Species, 341:1-5.

Morgan, G. S. 1989*b*. Fossil Chiroptera and Rodentia from the Bahamas, and the historical biogeography of the Bahamian mammal fauna. Pp. 685-740, *in* Biogeography of the West Indies: Past, present, and future (C. A. Woods, ed.). Sandhill Crane Press, Gainesville, Florida, 878 pp.

Morgan, G. S. 1994. Late Quaternary fossil vertebrates from the Caiman Islands. Pp. 465-508, *in* The Cayman Islands: Natural history and biogeography. (M. A. Brunt and J. E. Davies, eds.). Kluwer Academic Press, Boston, 576 pp.

Morgan, G. S., and N. J. Czaplewski. 2003. A new bat (Chiroptera: Natalidae) from the early Miocene of Florida, with comments on natalid phylogeny. Journal of Mammalogy, 84:729-752.

Morgan, G. S., and J. A. Ottenwalder. 1993. A new extinct species of *Solenodon* (Mammalia: Insectivora: Solenodontidae) from the Late Quaternary of Cuba. Annals of Carnegie Museum, 62:151-164.

Morgan, G. S., and L. Wilkins. 2003. The extinct rodent *Clidomys* (Heptaxodontidae) from a Late Quaternary cave deposit in Jamaica. Caribbean Journal of Science, 39(1):34-41.

Morgan, G. S., and C. A. Woods. 1986. Extinction and the zoogeography of West Indian land mammals. Biological Journal of the Linnean Society, 28:167-203.

Mori, T., S. Arai, S. Shirashi, and T. A. Uchida. 1991. Ultrastructural observations on spermatozoa of the Soricidae, with special attention to a subfamily revision of the Japanese water shrew *Chimarrogale himalayica*. Journal of the Mammalogical Society of Japan, 16:1-12.

Moriwaki, K. 1994. Introductory remarks: Wild mouse from a geneticist's viewpoint. Pp. xiii-xxv, *in* Genetics in wild mice. Its application to biomedical research. (K. Moriwaki, T. Shiroishi, and H. Yanekawa, eds.). Japan Scientific Press, Tokyo, 333 pp.

Moriwaki, K., T. Shiroishi, and H. Yanekawa (eds.). 1994. Genetics in wild mice. Its application to biomedical research. Japan Scientific Societies Press, Tokyo, 333 pp.

Morley, R. J. 2000. Origin and evolution of tropical rain forests. John Wiley and Sons, New York, 362 pp.

Morlok, W. F. 1978. Nagetiere aus der Türkei. Senckenbergiana Biologica, 59:155-162.

Moro, D. 2000. Kidney structure in two species of arid zone rodent: Lakeland Downs short-tailed mouse, *Leggadina lakedownensis*, and the house mouse, *Mus domesticus*. Australian Mammalogy, 21:251-255.

Moro, D., N. J. H. Campbell, M. S. Elphinstone, and P. R. Baverstock. 1998. The Thevenard Island mouse: Historic and conservation implications from mitochondrial DNA sequence-variation. Pacific Conservation Biology, 4:282-288.

Morris, K. D. 1995. Ash-grey mouse, *Pseudomys albocinereus*. Pp. 583-584, *in* Mammals of Australia (R. Strahan, ed.). Smithsonian Institution Press, Washington, D.C., 756 pp.

Morris, K. D. 2000. The status and conservation of native rodents in Western Australia. Wildllife Research, 27:405-419.

Morris, K. D., and A. C. Robinson. 1995. Shark Bay mouse, *Pseudomys fieldi*. Pp. 596-597, *in* Mammals of Australia (R. Strahan, ed.). Smithsonian Institutional Press, Washington, D.C., 756 pp.

Morris, P. A. 1997*a*. The edible dormouse (*Glis glis*). Mammal Society, London, 20 pp.

Morris, P. A. 1997*b*. A review of the fat dormouse (*Glis glis*) in Britain. Natura Croatica, 6(2):163-176.

Morris, P. A. 1999. *Muscardinus avellanarius* (Linnaeus, 1758). Pp. 296-297, *in* The atlas of European mammals (A. J. Mitchell-Jones, G. Amori, W. Bogdanowicz, B. Kryštufek, P. J. H. Reijnders, F. Spitzenberger, M. Stubbe, J. B. M. Thissen, V. Vohralík, and J. Zima, eds.). T. and A. D. Poyser, Ltd., London, 484 pp.

Morrison, J. L. 1992. Persistence of the meadow jumping mouse, *Zapus hudsonius luteus*, in New Mexico. Southwestern Naturalist, 37(3):308-311.

Morrison-Scott, T. C. S. 1939. The identity of *Acomys megalotis* (Lichtenstein) described from Arabia. Annals and Magazine of Natural History, ser. 11, 3:238-240.

Morrison-Scott, T. C. S. 1955. Proposed use of the Plenary Powers to validate the generic name "*Phacochoerus*" as from Cuvier (F.), 1826, as the generic name for the wart hog. Bulletin of Zoological Nomenclature, 2:191-195.

Morrissey, B. L., and W. G. Breed. 1982. Variation in external morphology of the glans penis of Australian native rodents. Australian Journal of Zoology, 30:495-502.

Morse, R. C., and B. P. Glass. 1960. The taxonomic status of *Antrozous bunkeri*. Journal of Mammalogy, 41:10-15.

Morshed, S., and J. L. Patton. 2002. New records of mammals from Iran with systematic comments on hedgehogs (Erinaceidae) and mouse-like hamsters (*Calomyscus*, Muridae). Zoology in the Middle East, 26:49-58.

Mošanský, A. 1994. Teriofauna Východného Slovenska a katalóg mamaliologických zbierok Východoslovenského Múzea [The mammalian fauna of East Slovakia and the Catalogue of Mammaliological collections of the Eastslovakian Museum. Part VI (Rodentia 3)]. Zborník Východoslovenského Múzea v Košiciach, Prírodné, 35:113-150 (in Czech with English summary).

Moscarella, R. A., M. Aguilera, and A. A. Escalante. 2003. Phylogeography, population structure, and implications for conservation of white-tailed deer (*Odocoilelus virginianus*) in Venezuela. Journal of Mammalogy, 84:1300-1315.

Moseby, K. E., R. Brandle, and M. Adams. 1999. Distribution, habitat and conservation status of the rare dusky hopping-mouse, *Notomys fuscus* (Rodentia: Muridae). Wildlife Research, 26:479-494.

Mostert, K. 1992*a*. Rosse woelmuis, *Clethrionomys glareolus* (Schreber, 1780). Pp. 240-245, *in* Atlas van de Nederlandse Zoogdieren (S. Broekhuizen, B. Hoekstra, V. van Laar, C. Smeenk, and J. B. M. Thissen, eds.). Stichting Uitgeverij Koninklijke Nederlandse Natuurhistorische Vereniging, Utrecht, 336 pp.

Mostert, K. 1992*b*. Dwergmuis, *Micromys minutus* (Pallas, 1771). Pp. 281-285, *in* Atlas van de Nederlandse Zoogdieren (S. Broekhuizen, B. Hoekstra, V. van Laar, C. Smeenk, and J. B. M. Thissen, eds.). Stichting Uitgeverij Koninklijke Nederlandse natuurhistorische Vereniging, Utrecht, 336 pp.

Motokawa, M. 1998. Reevaluation of the Orii's shrew, *Crocidura dsinezumi orii* Kuroda, 1924 (Insectivora, Soricidae) in the Ryukyu Archipelago, Japan. Mammalia, 62:259-267.

Motokawa, M. 1999. Taxonomic history of the genus *Crocidura* (Insectivora: Soricidae) from Japan. Pp. 63-71, *in* Recent advances in the biology of Japanese Insectivora (Y. Yokohata and S. Nakamura, eds.). Proceedings of the Symposium on the Biology of Insectivores in Japan and on Wildlife Conservation. Hiba Society of Natural History, Shobara (Japan). 391 pp. (in Japanese).

Motokawa, M. 2000. Biogeography of living mammals in the Ryukyu Islands. Tropics, 10:63-71.

Motokawa, M. 2003*a*. *Soriculus minor* Dobson, 1890, senior synonym of *S. radulus* Thomas, 1922 (Insectivora, Soricidae). Mammalian Biology, 68:178-180.

Motokawa, M. 2003*b*. Geographic variation in the Japanese white-toothed shrew, *Crocidura dsinezumi* (Insectivora, Soricidae). Acta Theriologica, 48:145-156.

Motokawa, M., and H. Abe. 1996. On the specific names of the Japanese moles of the genus *Mogera* (Insectivora, Talpidae). Mammal Study, 21:115-123.

Motokawa, M., and M. Harada. 1998. Karyotype of Hill's shrew *Crocidura hilliana* Jenkins and Smith, 1995 (Mammalia: Insectivora: Soricidae) from central Thailand. Raffles Bulletin of Zoology, 46:151-156.

Motokawa, M., and L.-K. Lin. 2002. Geographic variation in the mole-shrew *Anourosorex squamipes*. Mammal Study, 27:113-120.

Motokawa, M., S. Hattori, H. Ota, and T. Hikida. 1996. Geographic variation in the Watase's shrew *Crocidura watasei* from the Ryukyu Archipelago, Japan. Mammalia, 60:243-254.

Motokawa, M., H.-T. Yu, Y.-P. Fang, H.-C. Cheng, L.-K. Lin, and M. Harada. 1997*a*. Re-evaluation of the status of *Chodsigoa sodalis* Thomas, 1913 (Mammalia: Insectivora: Soricidae). Zoological Studies, 36:42-47.

Motokawa, M., M. Harada, L. K. Lin, K. Koyasu, and S. Hattori. 1997*b*. Karyological study of the gray shrew *Crocidura attenuata* (Mammalia: Insectivora) from Taiwan. Zoological Studies, 36:70-73.

Motokawa, M., M. Harada, L. K. Lin, H. C. Cheng, and K. Koyasu. 1998. Karyological differentiation between two *Soriculus* (Insectivora: Soricidae) from Taiwan. Mammalia, 62:541-547.

Motokawa, M., H. Suzuki, M. Harada, L. K. Lin, K. Koyasu, and S. I. Oda. 2000. Phylogenetic relationships among East Asian *Crocidura* (Mammalia: Insectivora) inferred from mitochondrial cytochrome b gene. Zoological Science, 17:497-504.

Motokawa, M., K.-H. Lu, M. Harada, and L.-K. Lin. 2001*a*. New records of the Polynesian rat *Rattus exulans* (Mammalia: Rodentia) from Taiwan and the Ryukyus. Zoological Studies, 40(4):299-304.

Motokawa, M., L.-K. Lin, H.-C. Cheng, and M. Harada. 2001*b*. Taxonomic status of the Senkaku mole, *Nesoscaptor uchidai*, with special reference to variation in *Mogera insularis* from Taiwan (Mammalia: Insectivora). Zoological Science, 18:733-740.

Motokawa, M., L.-K. Lin, M. Harada, and S. Hattori. 2003. Morphometric geographic variation in the Asian lesser white-toothed shrew *Crocidura shantungensis* (Mammalia, Insectivora) in East Asia. Zoological Science, 20:789-795.

Motokawa, M., M. Harada, L. K. Lin, and Y. Wu. 2004. Geographic differences in karyotypes of the mole-shrew *Anourosorex squamipes* (Insectivora, Soricidae). Mammalian Biology, 69:197-201.

Mouchaty, S., J. A. Cook, and G. F. Shields. 1995. Phylogenetic analysis of northern hair seals based on nucleaotide sequences of the mitochondrial cytochrome b gene. Journal of Mammalogy, 76:1178-1185.

Mouchaty, S. K. 1999. Mammalian molecular systematics with emphasis on the insectivore Order Lipotyphla. Ph.D. dissertation. Lund University.

Mouchaty, S. K., A. Gullberg, A. Janke, and U. Arnason. 2000*a*. The phylogenetic position of the Talpidae within Eutheria based on analysis of complete mitochondrias sequences. Molecular Biology and Evolution, 17:60-67.

Mouchaty, S. K., A. Gullberg, A. Janke, and U. Arnason. 2000*b*. Phylogenetic position of the tenrecs (Mammalia: Tenrecidae) of Madagascar based on analysis of the complete mitochondrial genome sequence of *Echinops telfairi*. Zoologica Scripta, 29:307-317.

Mouchaty, S. K., F. Catzeflis, A. Janke, and U. Arnason. 2001. Molecular evidence of an African Phiomorpha-South American Caviomorpha clade and support for Hystricognathi based on the complete mitochondrial genome of the cane rat (*Thryonomys swinderianus*). Molecular Phylogenetics and Evolution, 18:127-135.

Moulia, C., J. P. Aussel, F. Bonhomme, P. Boursot, J. T. Nielsen, and F. Renaud. 1991. Wormy mice in a hybrid zone: A genetic control of susceptibility to parasite infection. Journal of Evolutionary Biology, 4:679-687.

Mouse Genome Sequencing Consortium. 2002. Initial sequencing and comparative analysis of the mouse genome. Nature, 420:520-562.

Moyer, C. A., G. H. Adler, and R. H. Tamarin. 1988. Systematics of New England *Microtus*, with emphasis on *Microtus breweri*. Journal of Mammalogy, 69:782-794.

Mucedda, M., A. Kiefer, E. Pidinchedda, and M. Veith. 2002. A new species of long-eared bat (Chiroptera, Vespertilionidae) from Sardinia (Italy). Acta Chiropterologica, 4:121-135.

Mugo, D. N., A. T. Lombard, G. N. Bronner, C. M. Gelderblom, and G. A. Benn. 1995. Distribution and protection of endemic or threatened rodents, lagomorphs, and macroscelidids in South Africa. South African Journal of Zoology, 30:115-126.

Muirhead, L. 1819. Mazology. Pp. 393-486 [pls. 353-358], *in* The Edinburgh encyclopaedia, (D. Brewster, ed.). Fourth ed. William Blackwood, Edinburgh, 13:1-744, pls. 347-371, 1830.

Muizon, C., de. 1982*a*. Les relations phylogénétiques des Lutrinae (Mustelidae, Mammalia). Geobios, Mémoire Spécial, 6:259-277.

Muizon, C., de. 1982*b*. Phocid phylogeny and dispersal. Annals of the South African Museum, 89(2):175-213.

Müller, S. 1839-1840. Over de Zoogdieren van den Indischen Archipel. Pp. 1-8 [1839], Pp. 9-57+6 unnumbered pp. [1840], *in* Verhandelingen over de Natuurlijke Geschiedenis der Nederlandsche overzeesche bezittingen, door de Leden der Natuurkundige Commissie in Indiö en andere Schrijvers (C. J. Temminck, ed.). [V. 3] Zoology [1839-1845]. J. Luchtmans en C. C. van der Hoek, [not continuously paginated].

Mullin, S. K., N. Pillay, and P. J. Taylor. 2001. Non-geographic morphometric variation in the water rat *Dasymys incomtus* (Rodentia: Muridae) in southern Africa. Durban Museum Novitates, 26:38-44.

Mullin, S. K., N. Pillay, P. J. Taylor, and G. Campbell. 2002. Genetic and morphometric variation in populations of South African *Dasymys incomtus incomtus* (Rodentia, Murinae). Mammalia, 66(3):381-404.

Mumford, R. E. 1969. Distribution of the mammals of Indiana. Indiana Academy of Sciences, Indianapolis, 114 pp.

Munclinger, P., E. Bozíková, M. Sugerková, J. Piálek, and M. Macholán. 2002. Genetic variation in house mice (*Mus*, Muridae, Rodentia) from the Czech and Slovak Republics. Folia Zoologica, 51:81-92.

Muñoz, J., and C. A. Cuartas. 2001. *Saccopteryx antioquensis* n. sp. (Chiroptera: Emballonuridae) del noroeste de Colombia. Actualidades Biológicas, 23:53-61.

Muñoz-Muñoz, F., M. A. Sans-Fuentes, M. J. López-Fuster, and J. Ventura. 2003. Non-metric morphological divergence in the western house mouse, *Mus musculus domesticus*, from the Barcelona chromosomal hybrid zone. Biological Journal of the Linnean Society, 80:313-322.

Muñoz-Pedreros, A. 2000. Orden Rodentia. Pp. 73-126, *in* Mamíferos de Chile (A. Muñoz-Pedreros and J. Y. Valenzuela, eds.). Ediciones CEA, Valdivia, Chile, 464 pp.

Murariu, D., and S. Torcea. 1984. The occurrence of the species *Spalax istricus* Mehely, 1909 (Rodentia, Spalacidae) in the Romanian Plain. Travaux du Museum d'Histoire Naturelle "Grigore Antipa", 26:245-249.

Murariu, D., A. Lungeanu, L. Gavrila, and C. Stepan. 1985. Preliminary data concerning the study of the karyotype of *Eliomys quercinus* (Linnaeus, 1766) (Mammalia, Gliridae). Travaux Museum d'Histoire Naturelle "Grigore Antipa", 27:325-327.

Murbach, H. 1979. Zur kenntnis von inselpopulationen der waldmaus *Apodemus sylvaticus* (Linnaeus, 1758). Zeitschrift für Zoologische Systematik und Evolutionsforschung, 17:116-139.

Murphy, E. C., and C. R. Pickard. 1990. House mouse. Pp. 225-242, *in* The handbook of New Zealand mammals (C. M. King, ed.). Oxford University Press, Auckland, 600 pp.

Murphy, M. R. 1985. History of the capture and domestication of the Syrian golden hamster (*Mesocricetus auratus* Waterhouse). Pp. 3-20, *in* The hamster, reproduction and behavior (H. I. Siegel, ed.). Plenum Press, New York and London, xviii + 440 pp.

Murphy, W. J., E. Eizirik, W. E. Johnson, Ya Ping Zhang, O. A. Ryder and S. O'Brien. 2001*a*. Molecular phylogenetics and the origins of placental mammals. Nature, 409:614-618.

Murphy, W. J., E. Eizirik, S. J. O'Brien, O. Madsen, M. Scally, C. J. Douady, E. Teeling, O. A. Ryder, M. J. Stanhope, W. W. de Jong, , and M. S. Springer. 2001*b*. Resolution of the early placental mammal radiation using Bayesian phylogenetics. Science, 294:2348-2351.

Murray, A. 1866. The geographical distribution of mammals. Day and Son, Ltd., London, 420 pp.

Murray, B. R., I. D. Hume, and C. R. Dickman. 1994. Digestive tract characteristics of the spinifex hopping-mouse, *Notomys alexis* and the sandy inland mouse, *Pseudomys hermannsburgensis* in relation to diet. Australian Mammalogy, 18:93-97.

Murray, J. J., and, G. L. Gardner. 1997. *Leopardus pardalis*. Mammalian Species, 548:1-10.

Mursaloğlu, B. 1973*a*. Turkiye 'nin yabani memelileri. IV Blim Kongresi 6-8 Kasim 1973, Ankara, pp. 1-10.

Mursaloğlu, B. 1973*b*. New records for Turkish rodents (Mammalia). Communications of the Faculty of Sciences University of Ankara, 17:213-219.

Musser, G. G. 1964. Notes on geographic distribution, habitat, and taxonomy of some Mexican mammals. Occasional Papers of the Museum of Zoology, University of Michigan, 636:1-22.

Musser, G. G. 1968. A systematic study of the Mexican and Guatemalan gray squirrel, *Sciurus aureogaster*, F. Cuvier (Rodentia: Sciuridae). Miscellaneous Publications, Museum of Zoology, University of Michigan, 137:1-112.

Musser, G. G. 1969*a*. Notes on *Peromyscus* (Muridae) of Mexico and Central America. American Museum Novitates, 2357:1-23.

Musser, G. G. 1969*b*. Results of the Archbold Expeditions. No. 91. A new genus and species of murid rodent from Celebes, with a discussion of relationships. American Museum Novitates, 2384:1-41.

Musser, G. G. 1969*c*. Results of the Archbold Expeditions. No. 92. Taxonomic notes on *Rattus dollmani* and *Rattus hellwaldi* (Rodentia, Muridae) of Celebes. American Museum Novitates, 2386:1-24.

Musser, G. G. 1970*a*. *Rattus masaretes*: A synonym of *Rattus rattus moluccarius*. Journal of Mammalogy, 51:606-609.

Musser, G. G. 1970*b*. Species-limits of *Rattus brahma*, a murid rodent of Northeastern India and Northern Burma. American Museum Novitates, 2406:27.

Musser, G. G. 1970*c*. Identity of the type-specimens of *Sciurus aureogaster* F. Cuvier and *Sciurus nigriscens* Bennett (Mammalia, Sciurdae). American Museum Novitates, 2438:1-19.

Musser, G. G. 1970*d*. Results of the Archbold Expeditions. No. 93. Reidentification and reallocation of *Mus callitrichus* and allocations of *Rattus maculipilis*, *R. m. jentinki*, and *R. microbullatus* (Rodentia, Muridae). American Museum Novitates, 2440:1-35.

Musser, G. G. 1970e. The taxonomic identity of *Mus bocourti* A. Milne Edwards (1874) (Mammalia: Muridae). Mammalia, 34:484-490.

Musser, G. G. 1971a. *Peromyscus allophylus* Osgood: A synonym of *Peromyscus gymnotis* Thomas (Rodentia, Muridae). American Museum Novitates, 2453:1-10.

Musser, G. G. 1971b. Results of the Archbold Expeditions. No. 94. Taxonomic status of *Rattus tatei* and *Rattus frosti*, two taxa of murid rodents known from middle Celebes. American Museum Novitates, 2454:1-19.

Musser, G. G. 1971c. The taxonomic association of *Mus faberi* Jentink with *Rattus xanthurus* (Gray), a species known only from Celebes (Rodentia: Muridae). Zoologische Mededelingen uitgegeven door het Rijksmuseum van Natuurlijke Historie te Leiden, 45:107-118.

Musser, G. G. 1971d. The identities and allocations of *Taeromys paraxanthus* and *T. tatei*, two taxa based on compositie holotypes (Rodentia, Muridae). Zoologische Mededelingen uitgegeven door het Rijksmuseum van Natuurlijke Historie te Leiden, 45:127-138.

Musser, G. G. 1971e. The taxonomic status of *Rattus tondanus* Sody and notes on the holotypes of *R. beccarii* (Jentink) and *R. thysanurus* Sody (Rodentia: Muridae). Zoologische Mededelingen uitgegeven door het Rijksmuseum van Natuurlijke Historie te Leiden, 45:147-157.

Musser, G. G. 1971f. The taxonomic status of *Rattus dammermani* Thomas and *Rattus toxi* Sody (Rodentia, Muridae) of Celebes. Beaufortia, 18(242):205-216.

Musser, G. G. 1972. Identities of taxa associated with *Rattus rattus* (Rodentia, Muridae) of Sumba Island, Indonesia. Journal of Mammalogy, 53:861-865.

Musser, G. G. 1973a. Notes on additional specimens of *Rattus brahma*. Journal of Mammalogy, 54:267-270.

Musser, G. G. 1973b. Zoogeographical significance of the ricefield rat, *Rattus argentiventer*, on Celebes and New Guinea and the identity of *Rattus pesticulus*. American Museum Novitates, 2511:1-30.

Musser, G. G. 1973c. Species-limits of *Rattus cremoriventer* and *Rattus langbianis*, murid rodents of Southeast Asia and the Greater Sunda Islands. American Museum Novitates, 2525:1-65.

Musser, G. G. 1977a. *Epimys benguetensis*, a composite, and one zoogeographic view of rat and mouse faunas in the Philippines and Celebes. American Museum Novitates, 2624:1-15.

Musser, G. G. 1977b. Results of the Archbold Expeditions. No. 100. Notes on the Philippine rat, *Limnomys*, and the identity of *Limnomys picinus*, a composite. American Museum Novitates, 2636:1-14.

Musser, G. G. 1979. Results of the Archbold Expeditions. No. 102. The species of *Chiropodomys*, arboreal mice of Indochina and the Malay Archipelago. Bulletin of the American Museum of Natural History, 162:377-445.

Musser, G. G. 1981a. A new genus of arboreal rat from West Java, Indonesia. Zoologische Verhandelingen uitgegeven door het Rijksmuseum van Natuurlijke Historie te Leiden, 189:1-35.

Musser, G. G. 1981b. Results of the Archbold Expeditions. No. 105. Notes on systematics of Indo-Malayan murid rodents, and descriptions of new genera and species from Ceylon, Sulawesi, and the Philippines. Bulletin of the American Museum of Natural History, 168:225-334.

Musser, G. G. 1981c. The giant rat of Flores and its relatives east of Borneo and Bali. Bulletin of the American Museum of Natural History, 169:67-176.

Musser, G. G. 1982a. Results of the Archbold Expeditions. No. 107. A new genus of arboreal rat from Luzon Island in the Philippines. American Museum Novitates, 2730:1-23.

Musser, G. G. 1982b. Results of the Archbold Expeditions. No. 108. The definition of *Apomys*, a native rat of the Philippine Islands. American Museum Novitates, 2746:1-43.

Musser, G. G. 1982c. Results of the Archbold Expeditions. No. 110. *Crunomys* and the small-bodied shrew rats native to the Philippine Islands and Sulawesi (Celebes). Bulletin of the American Museum of Natural History, 174:1-95.

Musser, G. G. 1982d. The Trinil rats. Modern Quaternary Research in Southeast Asia, 7:65-85.

Musser, G. G. 1984. Identities of subfossil rats from caves in southwestern Sulawesi. Modern Quaternary Research in Southeast Asia, 8:61-94.

Musser, G. G. 1986. Sundaic *Rattus*: Definitions of *Rattus baluensis* and *Rattus korinchi*. American Museum Novitates, 2862:1-24.

Musser, G. G. 1987a. The mammals of Sulawesi. Pp. 73-93, *in* Biogeographical evolution of the Malay Archipelago (T. C. Whitmore, ed.). Oxford University Press, Oxford, 147 pp.

Musser, G. G. 1987b. The occurrence of *Hadromys* (Rodentia:Muridae) in early Pleistocene Siwalik strata in northern Pakistan and its bearing on biogeographic affinities between Indian and northeastern African murine faunas. American Museum Novitates, 2883:1-36.

Musser, G. G. 1990. Sulawesi rodents: Species traits and chromosomes of *Haeromys minahassae* and *Echiothrix leucura* (Muridae: Murinae). American Museum Novitates, 2989:1-18.

Musser, G. G. 1991. Sulawesi rodents: Descriptions of new species of *Bunomys* and *Maxomys* (Muridae, Murinae). American Museum Novitates, 3001:1-41.

Musser, G. G. 1994. New records of *Tarsomys echinatus* Musser and Heaney 1992 and *Limnomys sibuanus*

Mearns 1905 from Mindanao in the southern Philippines (Mammalia: Rodentia: Muridae). Senckenbergiana Biologica, 73:33-38.

Musser, G. G., and Boeadi. 1980. A new genus of murid rodent from the Komodo islands in Nusatenggara, Indonesia. Journal of Mammalogy, 61:395-413.

Musser, G. G., and E. M. Brothers. 1994. Identification of bandicoot rats from Thailand (*Bandicota*, Muridae, Rodentia). American Museum Novitates, 3110:1-56.

Musser, G. G., and D. Califia. 1982. Results of the Archbold Expeditions. No. 106. Identities of rats from Pulau Maratua and other islands off East Borneo. American Museum Novitates, 2726:1-30.

Musser, G. G., and M. D. Carleton. 1993. Family Muridae. Pp. 501-755, *in*: Mammal species of the world, a taxonomic and geographic reference, Second ed. (D. E. Wilson and D. M. Reeder, eds.). Smithsonian Institution Press, Washington D.C., xviii + 1206 pp.

Musser, G. G., and S. Chiu. 1979. Notes on taxonomy of *Rattus andersoni* and *R. excelsior*, murids endemic to Western China. Journal of Mammalogy, 60:581-592.

Musser, G. G., and M. Dagosto. 1987. The identity of *Tarsius pumilus*, a pygmy species endemic to the montane mossy forests of central Sulawesi. American Museum Novitates, 2867:1-53.

Musser, G. G., and L. A. Durden. 2002. Sulawesi rodents: Description of a new genus and species of Murinae (Muridae, Rodentia) and its parasitic new species of sucking louse (Insecta, Anoplura). American Museum Novitates, 3368:50 pp.

Musser, G. G., and L. K. Gordon. 1981. A new species of *Crateromys* (Muridae) from the Philippines. Journal of Mammalogy, 62:513-525.

Musser, G. G., and L. R. Heaney. 1985. Philippine *Rattus*: A new species from the Sulu Archipelago. American Museum Novitates, 2818:1-32.

Musser, G. G., and L. R. Heaney. 1992. Philippine rodents: Definitions of *Tarsomys* and *Limnomys* plus a preliminary assessment of phylogenetic patterns among native Philippine murines (Murinae, Muridae). Bulletin of the American Museum of Natural History, 211:1-138.

Musser, G. G., and M. E. Holden. 1991. Sulawesi rodents (Muridae: Murinae): Morphological and geographical boundaries of species in the *Rattus hoffmanni* group and a new species from Pulau Peleng. Pp. 322-413, *in* Contributions to mammalogy in honor of Karl F. Koopman (T. A. Griffiths and D. Klingener, eds.). Bulletin of the American Museum of Natural History, 206:1-432.

Musser, G. G., and C. Newcomb. 1983. Malaysian murids and the giant rat of Sumatra. Bulletin of the American Museum of Natural History, 174:327-598.

Musser, G. G., and C. Newcomb. 1985. Definitions of Indochinese *Rattus losea* and a new species from Vietnam. American Museum Novitates, 1814:1-32.

Musser, G. G., and J. L. Patton. 1989. Systematic studies of oryzomyine rodents (Muridae): The identity of *Oecomys phelpsi* Tate. American Museum Novitates, 2961:6 pp.

Musser, G. G., and F. Piik. 1982. A new species of *Hydromys* (Muridae) from western New Guinea (Irian Jaya). Zoologische Mededelingen uitgegeven door het Rijksmuseum van Natuurlijke Historie te Leiden, 56:153-167.

Musser, G. G., and H. G. Sommer. 1992. Taxonomic notes on specimens of the marsupials *Pseudocheirus schlegelii* and *P. forbesi* (Diprotodontia, Pseudocheiridae) in the American Museum of Natural History. American Museum Novitates, 3044:1-16.

Musser, G. G., and M. M. Williams. 1985. Systematic studies of oryzomyine rodents (Muridae): Definitions of *Oryzomys villosus* and *Oryzomys talamancae*. American Museum Novitates, 2810:1-22.

Musser, G. G., J. T. Marshall, Jr., and Boeadi. 1979. Definition and contents of the Sundaic genus *Maxomys* (Rodentia, Muridae). Journal of Mammalogy, 60:592-606.

Musser, G. G., K. F. Koopman, and D. Califia. 1982*a*. The Sulawesian *Pteropus arquatus* and *P. argentatus* are *Acerodon celebensis*; the Philippine *P. leucotus* is an *Acerodon*. Journal of Mammalogy, 63:319-328.

Musser, G. G., L. K. Gordon, and H. Sommer. 1982*b*. Species-limits in the Philippine murid, *Chrotomys*. Journal of Mammalogy, 63:515-521.

Musser, G. G., L. R. Heaney, and D. S. Rabor. 1985. Philippine rats: A new species of *Crateromys* from Dinagat Island. American Museum Novitates, 2821:1-25.

Musser, G. G., A. van de Weerd, and E. Strasser. 1986. *Paulamys*, a replacement name for *Floresomys* Musser, 1981 (Muridae), and new material of that taxon from Flores, Indonesia. American Museum Novitates, 2850:1-10.

Musser, G. G., E. M. Brothers, M. D. Carleton, and R. Hutterer. 1996. Taxonomy and distributional records of Oriental and European *Apodemus*, with a review of the *Apodemus-Sylvaemus* problem. Bonner Zoologische Beiträge, 46(1-4):143-190.

Musser, G. G., L. R. Heaney, and B. R. Tabaranza, Jr. 1998*a*. Philippine rodents: Redefinitions of known species of *Batomys* (Muridae, Murinae) and description of a new species from Dinagat Island. American Museum Novitates, 3237:1-51.

Musser, G. G., M. D. Carleton, E. M. Brothers, and A. L. Gardner. 1998b. Systematic studies of oryzomyine rodents (Muridae: Sigmodontinae): Diagnoses and distributions of species formerly assigned to *Oryzomys* "*capito.*" Bulletin of the American Museum of Natural History, 236:376 pp.

Mustrangi, M. A., and J. L. Patton. 1997. Phylogeography and systematics of the slender mouse opossum *Marmosops* (Marsupialia, Didelphidae). University of California Publications in Zoology, 130:x + 86 pp.

Muul, I., and B. L. Lim. 1971. New locality records for some mammals of West Malaysia. Journal of Mammalogy, 52:430-437.

Muul, I., and K. Thonglongya. 1971. Taxonomic status of *Petinomys morrisi* (Carter) and its relationship to *Petinomys setosus* (Temminck and Schlegel). Journal of Mammalogy, 52:362-369.

Myers, P. 1977. A new phyllotine rodent (genus *Graomys*) from Paraguay. Occasional Papers of the Museum of Zoology, University of Michigan, 676:1-7.

Myers, P. 1982. Origins and affinities of the mammal fauna of Paraguay. Special Publication Series, Pymatuning Laboratory of Ecology, University of Pittsburgh, Pennsylvania, 6:85-93.

Myers, P. 1989. A preliminary revision of the *varius* group of *Akodon*. Pp. 5-54, *in* Advances in Neotropical mammalogy (K. Redford and J. F. Eisenberg, eds.). Sandhill Crane Press, Gainesville, FL, 614 pp.

Myers, P., and M. D. Carleton. 1981. The species of *Oryzomys* (*Oligoryzomys*) in Paraguay and the identity of Azara's "Rat sixieme ou rat a tarse noir". Miscellaneous Publications, Museum of Zoology, University of Michigan, 161:1-41.

Myers, P., and J. L. Patton. 1989a. A new species of *Akodon* from the cloud forests of eastern Cochabamba Department, Bolivia (Rodentia: Sigmodontinae). Occasional Papers of the Museum of Zoology, University of Michigan, 720:1-28.

Myers, P., and J. L. Patton. 1989b. *Akodon* of Peru—revision of the *fumeus* group (Rodentia: Sigmodontinae). Occasional Papers of the Museum of Zoology, University of Michigan, 721:1-35.

Myers, P., J. L. Patton, and M. F. Smith. 1990. A review of the *boliviensis* group of *Akodon* (Muridae: Sigmodontinae), with emphasis on Peru and Bolivia. Miscellaneous Publications, Museum of Zoology, University of Michigan, 177:1-104.

Myers, P., B. Lundrigan, and P. K. Tucker. 1995. Molecular phylogenetics of oryzomyine rodents: The genus *Oligoryzomys*. Molecular Phylogenetics and Evolution, 4:372-382.

Myers, P., B. Lundrigan, B. W. Gillespie, and M. L. Zelditch. 1996. Phenotypic plasticity in skull and dental morphology in the Prairie deer mouse (*Peromyscus maniculatus bairdii*). Journal of Morphology, 229:229-237.

Myers, P., J. D. Smith, H. Lama, B. Lama, and K. F. Koopman. 2000. A recent collection of bats from Nepal, with notes on *Eptesicus dimissus*. Zeitschrift für Säugetierkunde, 65:149-156.

Myers Uncie, S. M., D. W. Hale, and I. F. Greenbaum. 1998. Karyotypic variation in populations of deer mice (*Peromyscus maniculatus*) from eastern Canada and the northeastern United States. Canadian Journal of Zoology, 76:584-588.

Naber, F. 1982. Eerste vondst van de dwergmuis *Micromys minutus* (Pallas, 1771) op Terschelling. Lutra, 25:95-96.

Nachman, M. W. 1992a. Geographic patterns of chromosomal variation in South American marsh rats, *Holochilus brasiliensis* and *H. vulpinus*. Cytogenetics and Cell Genetics, 61:10-16.

Nachman, M. W. 1992b. Meiotic studies of Robersonian polymorphisms in the South American marsh rat, *Holochilus brasiliensis*. Cytogenetics and Cell Genetics, 61:17-24.

Nachman, M. W., and P. Myers. 1989. Exceptional chromosomal mutations in a rodent population are not strongly underdominant. Proceedings of the National Academy of Sciences, Washington, D.C., 86:6666-6670.

Nachman, M. W., and J. B. Searle. 1995. Why is the house mouse karyotype so variable? Trends in Ecology and Evolution, 10:397-402.

Nachtigall, W. 1996. Siebenschlafer (*Glis glis*: Mammalia: Rodentia: Gliridae) als Beute des Mausebussards (*Buteo buteo*; Aves: Falconiformes: Accipitridae). (Beitrage zur Saugetierfauna Sachsens, Nr 1). Faunistische Abhandlungen (Dresden), 20(2):320.

Nadachowski, A. 1990a. Comments on variation, evolution and phylogeny of *Chionomys* (Arvicolinae). Pp. 353-368, *in* International symposium evolution, phylogeny and biostratigraphy of arvicolids (Rodentia, Mammalia) (O. Fejfar and W.-D. Heinrich, eds.). Geological Survey, Prague, 448 pp.

Nadachowski, A. 1990b. On the taxonomic status of *Chionomys* Miller, 1908 (Rodentia: Mammalia) from southern Anatolia (Turkey). Acta Zoologica Cracoviensia, 33:79-89.

Nadachowski, A. 1991. Systematics, geographic variation, and evolution of snow voles (*Chionomys*) based on dental characters. Acta Theriologica, 36:1-45.

Nadachowski, A. 1992. Short-distance migration of Quaternary and Recent mammals: A case study of *Chionomys* (Arvicolidae). Courier Forschungsinstitut Senckenberg, 153:221-228.

Nadachowski, A., and G. Baryshnikov. 1991. Pleistocene snow voles (*Chionomys* Miller, 1908) (Rodentia, Mammalia) from Northern Caucasus (USSR). Acta Zoologica Cracoviensia, 34(2):437-451.

Nadachowski, A., and A. Daoud. 1995. Patterns of myoxid evolution in the Pliocene and Pleistocene of Europe. Pp. 141-149, *in* Proceedings of II Conference on Dormice (Rodentia, Myoxidae) (M. G. Filippucci, ed.). Hystrix, n.s., 6(1-2):1-340.

Nadachowski, A., and I. Zagorodnyuk. 1996. Recent *Allophaiomys*-like species in the Palaearctic: Pleistocene relicts or a return to an initial type. Acta Zoologica Cracoviensia, 39(1):387-394.

Nadachowski, A., B. Rzebik-Kowalska, and A.-H. Kadhim. 1978. The first record of *Eliomys melanurus* Wagner, 1840 (Gliridae, Mammalia), from Iraq. Säugetierkundliche Mitteilungen, 26:206-207.

Nadeau, J. H. 2002. Tackling complexity. Nature, 420:517-518.

Nader, I. A. 1978. Kangaroo rats: Intraspecific variation in *Dipodomys spectabilis* Merriam and *Dipodomys deserti* Stephens. Illinois Biological Monographs, 49:1-116.

Nader, I. A. 1991. *Paraechimus hypomelas* (Brandt, 1836) in Arabia with notes on the species' zoogeography and biology (Mammalia: Insectivora: Erinaceidae). Fauna of Saudi Arabia, 12:400-410.

Nader, I. A., and M. M. Al-Safadi. 1993. The Ethiopian hedgehog *Paraechinus aethiopicus* (Ehrenberg, 1833) and Brandt's hedgehog *Paraechimus hypomelas* (Brandt, 1836) (Mammalia: Insectivora: Erinaceidae) from northern Yemen. Fauna of Saudi Arabia, 13:397-400.

Nader, I. A., and D. Kock. 1979 [1980]. First record of *Tadarida nigeriae* (Thomas, 1913) from the Arabian peninsula (Mammalia: Chiroptera: Molossidae). Senckenbergiana Biologica, 60:131-135.

Nader, I. A., and D. Kock. 1990. *Eptesicus* (*Eptesicus*) *bottae* (Peters 1869) in Saudi Arabia with notes on its subspecies and distribution (Mammalia: Chiroptera: Vespertilionidae). Senckenbergiana Biologica, 70:1-13.

Nader, I. A., D. Kock, and A.-K. Al-Khalili. 1983. *Eliomys melanurus* (Wagner 1839) and *Praomys fumatus* (Peters 1878) from the Kingdom of Saudi Arabia. Senckenbergiana Biologica, 63(5-6):313-324.

Nadjafova, R. S., N. Sh. Bulatova, Z. Chasovlikarova, and S. Gerassimov. 1993. Karyological differences between two *Apodemus* species in Bulgaria. Zeitschrift für Säugetierkunde, 58:232-239.

Nadler, C. F., and R. S. Hoffmann. 1970. Chromosomes of some Asian and South American squirrels (Rodentia, Sciuridae). Experientia, 26:1383-1386.

Nadler, C. F., and R. S. Hoffmann. 1974. Chromosomes of the African ground squirrel, *Xerus rutilus*. Experientia, 30:889-890.

Nadler, C. F., and R. S. Hoffmann. 1977. Patterns of evolution and migration in the arctic ground squirrel, *Spermophilus parryii* (Richardson). Canadian Journal of Zoology, 55:748-758.

Nadler, C. F., and D. M. Lay. 1967. Chromosomes of some species of *Meriones* (Mammalia: Rodentia). Zeitschrift für Säugetierkunde, 32:285-291.

Nadler, C. F., D. M. Lay, and J. D. Hassinger. 1969. Chromosomes of three asian mammals: *Meriones meridianus* (Rodentia: Gerbillinae), *Spermophilopsis leptodactylus* (Rodentia: Sciuridae), *Ochotona rufescens* (Lagomorpha: Ochotonidae). Experientia, 25:774-775.

Nadler, C. F., R. S. Hoffmann, and K. Greer. 1971*a*. Chromosomal divergence during evolution of ground squirrel populations. Systematic Zoology, 20:298-305.

Nadler, C. F., D. M. Lay, and J. D. Hassinger. 1971*b*. Cytogenetic analyses of wild sheep populations in northern Iraq. Cytogenetics, 10:137-152.

Nadler, C. F., K. V. Korobitsina, R. S. Hoffmann, and N. N. Vorontsov. 1973. Cytogenetic differentiation, geographic distribution and domestication in Palearctic sheep (*Ovis*). Zeitschrift für Säugetierkunde, 38:109-125.

Nadler, C. F., R. I. Sukernik, R. S. Hoffmann, N. N. Vorontsov, C. F. Nadler, Jr., and I. I. Fomichova. 1974. Evolution in ground squirrels. I. Transferrins in Holarctic populations of *Spermophilus*. Comparative Biochemistry and Physiology, 47A:663-681.

Nadler, C. F., E. A. Lyapunova, R. S. Hoffmann, N. N. Vorontsov, and N. A. Malygina. 1975*a*. Chromosomal evolution in Holarctic ground squirrels (*Spermophilus*). 1. Giemsa-band homologies in *Spermophilus columbianus* and *S. undulatus*. Zeitschrift für Säugetierkunde, 40:1-7.

Nadler, C. F., R. S. Hoffmann, and M. E. Hight. 1975*b*. Chromosomes of three species of Asian tree squirrels, *Callosciurus* (Rodentia: Sciuridae). Experientia, 31:166-167.

Nadler, C. F., V. R. Rausch, E. A. Lyapunova, R. S. Hoffmann, and N. N. Vorontsov. 1976. Chromosomal banding patterns of the Holarctic rodents, *Clethrionomys rutilus* and *Microtus oeconomus*. Zeitschrift für Säugetierkunde, 41:137-146.

Nadler, C. F., R. S. Hoffmann, J. H. Honacki, and D. Pozin. 1977. Chromosomal evolution in chipmunks, with special emphasis on A and B karyotypes of the subgenus *Neotamias*. American Midland Naturalist, 98:343-353.

Nadler, C. F., N. M. Zhurkevich, R. S. Hoffmann, A. I. Kozlovskii, L. Deutsch, and C. F. Nadler, Jr. 1978.

Biochemical relationships of the Holarctic vole genera (*Clethrionomys*, *Microtus*, and *Arvicola* (Rodentia: Arvicolinae)). Canadian Journal of Zoology, 56:1564-1575.

Nadler, C. F., R. S. Hoffmann, N. N. Vorontsov, J. W. Koeppl, L. Deutsch, and R. I. Sukernik. 1982. Evolution in ground squirrels. II. Biochemical comparisons in Holarctic populations of *Spermophilus*. Zeitschrift für Säugetierkunde, 47:198-215.

Nadler, C. F., E. A. Lyapunova, R. S. Hoffmann, N. N. Vorontsov, L. L. Shaitarova, and Y. M. Borisov. 1984. Chromosomal evolution in Holarctic ground squirrels (*Spermophilus*). II. Giemsa-band homologies of chromosomes and the tempo of evolution. Zeitschrift für Säugetierkunde, 49:78-90.

Nadler, T. 1997. A new subspecies of Douc langur, *Pygathrix nemaeus cinereus* ssp. nov.Zoologischer Garten N.F., 67:165-176.

Nadler, T. 1998. Black langur rediscovered. Asian Primates, 6:10-12.

Naftel, J. P., L. P. Richards, M. Pan, and J. M. Bernanke. 1999. Course and composition of the nerves that supply the mandibular teeth of the rat. The Anatomical Record, 256:433-447.

Nagamine, C. M., P. Boursot, Y.-F. C. Lau, and K. Moriwaki. 1994. Evolution of the Y-chromosome in the wild mouse. Pp. 41-55, *in* Genetics in wild mice (K. Moriwaki, T. Shiroishi, and H. Yonekawa, eds.). Japan Scientific Societies Press, Tokyo, 333 pp.

Naglov, V. A. 1995. [Distribution and population density of *Sylvaemus sylvaticus* (Rodentia, Muridae) in Kharkov Oblast'.] Vestnik Zoologii, 1995(5-6):87-89 (in Russian with English abstract).

Naglov, V. A. 1998. [*Ixodes apronophorus* Schulze, 1924 (Acarina, Ixodidae) in Kharkov Region.]. Vestnik Zoologii, 32(1-2):118 (in Russian).

Nagorsen, D. 1985. *Kogia simus*. Mammalian Species, 239:1-6.

Nagorsen, D. W. 1987. *Marmota vancouverensis*. Mammalian Species, 270:1-5.

Nagorsen, D., and J. R. Tamsitt. 1981. Systematics of *Anoura cultrata*, *A. brevirostrum*, and *A. werckleae*. Journal of Mammalogy, 62:82-100.

Nakatsu, A. 1982. Notes on the pattern of third upper molar (M3) in *Clethrionomys rutilus mikado* (Thomas). Journal of the Mammalogical Society of Japan, 9:104-106.

Nanda, I., and R. Raman. 1981. Cytological similarity between the heterochromatin of the large X and Y chromosomes of the soft-furred field rat, *Millardia meltada* (family: Muridae). Cytogenetics and Cell Genetics, 30:77-82.

Nanez-Jimenez, S., and M. Martínez-Coronel. 1995. Analisis morfometrico entre las dos subspecies de *Peromyscus maniculatus* (Rodentia: Muridae) del la region central de México. Annales de la Escuela Nacional de Ciencias Biologicas, México, 41:197-210.

Napier, J. R., and P. H. Napier. 1967. A handbook of living Primates. Academic Press, London, 456 pp.

Napier, P. H. 1976. Catologue of primates in the British Museum (Natural History), Part 1: Families Callitrichidae and Cebidae. British Museum (Natural History), London, 121 pp.

Napier, P. H. 1985. Catalogue of Primates in the British Museum (Natural History) and elsewhere in the British Isles. Part 3: Family Cercopithecidae, subfamily Colobinae. Publications of the British Museum of Natural History, 894:1, 3-111.

Napier, P. H., and C. P. Groves. 1983. *Simia fascicularis* Raffles, 1821 (Mammalia, Primates): Request for the suppression under the plenary powers of *Simia aygula* Linnaeus, 1758, a senior synonym. Bulletin of Zoological Nomenclature, 40:117-118.

Nappi, A., and N. Majo. 2000. Etude de l'holotype de *Crocidura hydruntina* Costa, 1844 (Insectivora, Soricidae): Implications taxonomiques. Mammalia, 64:383-386.

Narita, Y., S.-I. Oda, O. Takenaka, and T. Kageyama. 2001. Phylogenetic position of Eulipotyphla inferred from the cDNA sequences of pepsinogens A and C. Molecular Phylogentics and Evolution, 21:32-42.

Naruse, M., M. Tsukada, and A. Koizumi. 1996. A novel endogenous mouse mammary tumor virus locus in Asian wild mice and its evolutionary divergency. Zoological Science, 13:299-302.

Nascetti, G., S. Lovari, P. Lanfranchi, C. Berducou, S. Mattiucci, L. Rossi, and L. Bullini. 1985. Revision of *Rupicapra* genus. III. Electrophoretic studies demonstrating species distinction of chamois populations of the Alps from those of the Apennines and Pyrenees. Pp. 56-62, *in* The biology and management of mountain ungulates (S. Lovari, ed.). Croom Helm, London, 271 pp.

Nash, D. J., and R. N. Seaman. 1977. *Sciurus aberti*. Mammalian Species, 80:1-5.

Nash, L. T., S. K. Bearder, and T. R. Olson. 1989. Synopsis of *Galago* species characteristics. International Journal of Primatology, 10:57-79.

Natori, M. 1988. A cladistic analysis of interspecific relationships of *Saguinus*. Primates, 29:263-276.

Natori, M., and T. Hanihara. 1988. An analysis of interspecific relations of *Saguinus* based on cranial measurements. Primates, 29:255-262.

Naumov, N. P., and V. S. Lobachev. 1975. Ecology of desert rodents of the U.S.S.R. (Jerboas and Gerbils). Pp. 465-598, *in* Rodents in desert environments (I. Prakash and P. K. Gosh, eds.). Monographiae Biologicae, 28:1-624.

Naumova, E. I., R. A. Valensiya-Leon, and L. I. Lenets. 1990. [Differences in morphology of the stomach in species-twins: The common and eastern European voles.] Doklady Akademii Nauk SSSR, 310:1016-1020 (in Russian).

Nauroz, M. K. 1984. Die Waldmous, *Apodemus sylvaticus* (Rodentia, Muridae) auf der Insel Mellum. Säugetierkundliche Mitteilungen, 31:141-159.

Navarro L., D., and D. E. Wilson. 1982. *Vampyrum spectrum*. Mammalian Species, 184:1-4.

Neal, B. R. 1982. Reproductive biology of three species of gerbils (genus *Tatera*) in East Africa. Zeitschrift für Säugetierkunde, 47:287-296.

Neal, B. R. 1983. The breeding pattern of two species of spiny mice, *Acomys percivali* and *A. wilsoni* (Muridae: Rodentia), in central Kenya. Mammalia, 47:311-321.

Neal, E. 1948. The badger. Collins, London. 158 pp.

Neas, J. F., and R. S. Hoffmann. 1987. *Budorcas taxicolor*. Mammalian Species, 277:1-7.

Nedbal, M. A., M. W. Allard, and R. L. Honeycutt. 1994. Molecular systematics of hystricognath rodents: Evidence from the mitochondrial 12S rRNA gene. Molecular Phylogenetics and Evolution, 3(3):206-220.

Nedbal, M. A., R. L. Honeycutt, and D. A. Schlitter. 1996. Higher-level systematics of rodents (Mammalia, Rodentia): Evidence from the mitochondrial 12S rRNA gene. Journal of Mammalian Evolution, 3:201-237.

Nellis, D. W. 1989. *Herpestes auropunctatus*. Mammalian Species, 342:1-6.

Nelson, E. W. 1898. What is *Sciurus variegatus* Erxleben? Science, n.s., 8:897-898.

Nelson, E. W. 1899*a*. Mammals of the Tres Marias Islands. North American Fauna, 14:15-19.

Nelson, E. W. 1899*b*. Revision of the squirrels of Mexico and Central America. Proceedings of the Washington Academy of Sciences, 1:15-106.

Nelson, E. W. 1909. The rabbits of North America. North American Fauna, 29:1-314.

Nelson, E. W. 1929. Description of a new lemming from Alaska. Proceedings of the Biological Society of Washington, 42:143-146.

Nelson, E. W., and E. A. Goldman. 1929. Four new pocket gophers of the genus *Heterogeomys* from Mexico. Proceedings of the Biological Society of Washington, 42:147-152.

Nelson, E. W., and E. A. Goldman. 1930. The status of *Orthogeomys cuniculus* Elliot. Journal of Mammalogy, 11:317.

Nelson, E. W., and E. A. Goldman. 1933*a*. Revision of the jaguars. Journal of Mammalogy, 14(3):221-240.

Nelson, E. W., and E. A. Goldman. 1933*b*. Three new rodents from southern Mexico. Proceedings of the Biological Society of Washington, 46:195-198.

Nelson, E. W., and E. A. Goldman. 1934*a*. Revision of the pocket gophers of the genus *Cratogeomys*. Proceedings of the Biological Society of Washington, 47:135-154.

Nelson, E. W., and E. A. Goldman. 1934*b*. Pocket gophers of the genus *Thomomys* of the Mexican mainland and bordering territory. Journal of Mammalogy, 15:105-124.

Nelson, K., R. J. Baker, H. S. Shellhammer, and R. K. Chesser. 1984. Test of alternative hypotheses concerning the origin of *Reithrodontomys raviventris*: Genetic analysis. Journal of Mammalogy, 65:668-673.

Nelson, K., R. J. Baker, and R. L. Honeycutt. 1987. Mitochondrial DNA and protein differentiation between hybridizing cytotypes of the white-footed mouse, *Peromyscus leucopus*. Evolution, 41:864-872.

Neraudeau, D., L. Viriot, J. Chaline, B. Laurin, and T. van Kolfschoten. 1995. Discontinuity in the Plio-Pleistocene Eurasian water vole lineage. Palaeontology, 38(1):77-85.

Nersting, L. G., and P. Arctander. 2001. Phylogeography and conservation of impala and greater kudu. Molecular Ecology, 10:711-719.

Nesin, V. A., and A. F. Skorik. 1989. [First find of the *Dinaromys* vole (Rodentia, Microtinae) in the USSR.] Vestnik Zoologii, 5:14-17 (in Russian).

Nessov, L. A., and A. A. Gureev. 1981. [A jaw of a most ancient shrew from the Upper Cretaceous of the Kizylkum desert.] Doklady Akademii Nauk SSSR, 257(4):1002-1004 (in Russian).

Nessov, L. A., D. Sigogneau-Russell, and D. E. Russell. 1994. A survey of Cretaceous tribosphenic mammals from Middle Asia (Uzbekistan, Kazakhstan and Tajikistan), of their geological setting, age and faunal environment. Palaeovertebrata, 23:51-92.

Nesterenko, V. A. 1999. Nasekomoyadnye yuga Dal'nego Vostoka i ikh soobshchestva [Insectivores of the southern Far East and their communities]. Vladivostok. 172 pp. (in Russian).

Neuhäuser, G. 1936. Die Muriden von kleinasien. Zeitschrift für Säugetierkunde, 11:161-236.

Nevo, E. 1982. Genetic structure and differentiation during speciation in fossorial gerbil rodents. Mammalia, 46:523-530.

Nevo, E. 1985. Genetic differentiation and speciation in spiny mice, *Acomys*. Acta Zoologica Fennica, 170:131-136.

Nevo, E. 1989. Natural selection of body size differentiation in spiny mice, *Acomys*. Zeitschrift für Säugetierkunde, 54:81-99.

Nevo, E. 1991. Evolutionary theory and processes of active speciation and adaptive radiation in subterranean mole rats, *Spalax ehrenbergi* superspecies, in Israel. Evolutionary Biology, 25:1-125.

Nevo, E. 1995. Mode, tempo and pattern of evolution in subterranean mole rats of the *Spalax ehrenbergi* superspecies in the Quaternary of Israel. Quaternary International, 19:13-19.

Nevo, E., and E. Amir. 1961*a*. Biological observations on the forest dormouse *Dryomys nitedula* Pallas, in Israel (Rodentia, Muscardinidae). The Bulletin of the Research Council of Israel, 9B(4):200-201.

Nevo, E., and E. Amir. 1961*b*. Geographic variation in reproduction and hibernation patterns of the forest dormouse. Journal of Mammalogy, 45(1):69-87.

Nevo, E., and E. Amir. 1964. Geographic variation in reproduction and hibernation patterns of the forest dormouse. Journal of Mammalogy, 45(1):69-87.

Nevo, E., and A. Beiles. 1992. MtDNA polymorphisms: Evolutionary significance in adaptation and speciation of subterranean mole rats. Biological Journal of the Linnean Society, 47:385-405.

Nevo, E., E. Tchernov, and A. Beiles. 1988. Morphometrics of speciating mole rats: Adaptive differentiation in ecological speciation. Zeitschrift für Zoologische Systematik und Evolutionsforschung, 26:286-314.

Nevo, E., E. Capanna, M. Corti, J. U. M. Jarvis, and G. C. Hickman. 1986. Karyotypic differentiation in the endemic subterranean mole rats of South Africa (Rodentia, Bathyergidae). Zeitschrift für Saugetierkunde, 51(1):36-49.

Nevo, E., R. Ben-Shlomo, A. Beiles, J. U. M. Jarvis, and G. C. Hickman. 1987. Allozyme differentiation and systematics of the endemic subterranean mole rats of South Africa. Biochemical Systematics and Ecology, 15:489-502.

Nevo, E., R. Ben-Shlomo, A. Beiles, C. P. Hart, and F. H. Ruddle. 1992*a*. Homeobox DNA polymorphisms (RFLPs) in subterranean mammals of the *Spalax ehrenbergi* superspecies in Israel: Patterns, correlates, and evolutionary significance. Journal of Experimental Zoology, 263:430-441.

Nevo, E., S. Simson, G. Heth, C. Redi, and M. G. Filippucci. 1992*b*. Recent speciation of subterranean mole rats of the *Spalax ehrenbergi* superspecies in the El-Hamam isolate, northern Egypt. 6th International Colloquium on the ecology and taxonomy of small African mammals (August 11-16, 1991). Israel Journal of Zoology, 38:431.

Nevo, E., R. L. Honeycutt, H. Yonekawa, K. Nelson, and N. Hanzawa. 1993. Mitochondrial DNA polymorphisms in subterranean mole-rats of the *Spalax ehrenbergi* superspecies in Israel, and its peripheral isolates. Molecular Biology and Evolution, 10(3):590-604.

Nevo, E., M. G. Filippucci, and A. Beiles. 1994*a*. Genetic polymorphisms in subterranean mammals (*Spalax ehrenbergi* superspecies) in the Near East revisited: Patterns and theory. Heredity, 72:465-487.

Nevo, E., M. G. Filippucci, C. Redi, A. Korol, and A. Beiles. 1994*b*. Chromosomal speciation and adaptive radiation of mole rats in Asia Minor correlated with increased ecological stress. Proceedings of the National Academy of Sciences, USA, 9:8160-8164.

Nevo, E., M. G. Filippucci, C. Redi, S. Simson, G. Heth, and A. Beiles. 1995. Karyotype and genetic evolution in speciation of subterranean mole rats of the genus *Spalax* in Turkey. Biological Journal of the Linnean Society, 54:203-229.

Nevo, E., V. Kirzhner, A. Beiles, and A. Korol. 1997. Selection versus random drift: Long-term polymorphism persistence in small populations (evidence and modelling). Philosophical Transactions of the Royal Society of London, B, 352:381-389.

Nevo, E., A. Beiles, and T. Spradling. 1999. Molecular evolution of cytochrome *b* of subterranean mole rats, *Spalax ehrenbergi* superspecies, in Israel. Journal of Molecular Evolution, 49:215-226.

Nevo, E., E. Ivanitskaya, M. G. Filippucci, and A. Beiles. 2000. Speciation and adaptive radiation of subterranean mole rats, *Spalax ehrenbergi* superspecies, in Jordan. Biological Journal of the Linnean Society, 69:263-281.

Nevo, E., E. Ivanitskaya, and A. Beiles. 2001. Adaptive radiation of blind subterranean mole rats: Naming and revisiting the four sibling species of the *Spalax ehrenbergi* superspecies in Israel: *Spalax galili* (2n = 52), *S. golani* (2n = 54), *S. carmeli* (2n = 58) and *S. judaei* (2n = 60). Backhuys Publishers, Leiden, 195 pp.

Ni, X., and Z. Qiu. 2002. The micromammalian fauna from the Leilao, Yuanmou hominoid locality: Implications for biochronology and paleoecology. Journal of Human Evolution, 42(5):535-546.

Nicolas, V., W. Wendelen, W. Verheyen, and M. Colyn. 2003. Geographical distribution and morphometry of *Heimyscus fumosus* (Brosset et al., 1965). Revue Écologie (Terre Vie), 58:197-208.

Nicoll, M. E., and G. B. Rathbun. 1990. African Insectivora and elephant-shrews. An action plan for their conservation. I.U.C.N., Gland, Switzerland, 53 pp.

Niedenführ, A., and D. Rathke. 1996. Erstnachweis derschabrackenspitzmaus (*Sorex coronatus* Millet, 1828) und dergelbhalsmaus (*Apodemus flavicollis* Melchior, 1834) für das südliche Bremer Unland. Abhandlungen Naturwissenschaftlicher Verein zu Bremen, 43(2):567-575.

Niemitz, C., A. Nietsch, S. Water, and Y. Rumpler. 1991. *Tarsius dianae*: A new primate species from Central Sulawesi (Indonesia). Folia Primatologia, 56:105-116.

Niethammer, J. 1959. Die nordafrikanischen Unterarten des Gartenschläfers (*Eliomys quercinus*). Zeitschrift für Säugetierkunde, 24:35-45.

Niethammer, J. 1960. Über die Säugetiere der Niederen Tauern. Mitteilungen aus dem Zoologische Museum in Berlin, 36(2):407-443.

Niethammer, J. 1964. Contribution a la connaissance des mammiferes terrestres de l'ile Indefatigable (=Santa Cruz), Galapagos. Résultats de l'expedition Allemagne aux Galapagos 1962/63. Mammalia, 28:593-606.

Niethammer, J. 1969. Die waldmaus *Apodemus sylvaticus* (Linné, 1758), in Afghanistan. Säugetierkundliche Mitteilungen, 17(2):121-128.

Niethammer, J. 1970. Die Wühlmause (Microtinae) Afghanistans. Bonner Zoologische Beiträge, 21:1-24.

Niethammer, J. 1972. Die Zahl der Mammae bei *Pitymys* und bei den Microtinen. Bonner Zoologische Beiträge, 23:49-60.

Niethammer, J. 1973. Zur Kenntnis der Igel (Erinaceidae) Afghanistans. Zeitschrift für Säugetierkunde, 38:271-276.

Niethammer, J. 1975. Zur Taxonomie und Ausbreitungsgeschichte der Hausratte (*Rattus rattus*). Zoologischer Anzeiger, 194:405-415.

Niethammer, J. 1977. Versuch der Rekonstruktion der phylogenetischen Beziehungen zwischen einigen zentralasiatischen Muriden. Bonner Zoologische Beiträge, 28:236-247.

Niethammer, J. 1978a. *Apodemus mystacinus* (Danford and Alston, 1877)—Felsenmaus. Pp. 306-324, *in* Handbuch der Säugetiere Europas (J. Niethammer and F. Krapp, eds.). Akademische Verlagsgesellschaft (Wiesbaden), 1:1-476.

Niethammer, J. 1978b. *Apodemus flavicollis* (Melchior, 1834)—Gelbhalsmaus. Pp. 325-336, *in* Handbuch der Säugetiere Europas (J. Niethammer and F. Krapp, eds.). Akademische Verlagsgesellschaft (Wiesbaden), 1:1-476.

Niethammer, J. 1978c. *Apodemus sylvaticus* (Linnaeus, 1758)—Waldmaus. Pp. 337-358, *in* Handbuch der Säugetiere Europas (J. Niethammer and F. Krapp, eds.). Akademische Verlagsgesellschaft (Wiesbaden), 1:1-476.

Niethammer, J. 1978d. Gattung *Apodemus* Kaup, 1826. P. 305, *in* Handbuch der Säugetiere Europas (J. Niethammer and F. Krapp, eds.). Akademische Verlagsgesellschaft (Wiesbaden), 1:1-476.

Niethammer, J. 1981. Über *Microtus* (*Pitymys*) *savii* (de Sélys-Longchamps, 1838) vom Monte Gargano, Italien. Säugetierkundliche Mittelungen, 29:45-48.

Niethammer, J. 1982a. Familie Cricetidae Rochebrune, 1881—Hamster. Pp. 1-6, *in* Handbuch der Säugetiere Europas (J. Niethammer and F. Krapp, eds.). Akademische Verlagsgesellschaft (Wiesbaden), 2/I:1-649.

Niethammer, J. 1982b. *Cricetus cricetus* (Linnaeus, 1758)—Hamster (Feldhamster). Pp. 1-28, *in* Handbuch der Säugetiere Europas (J. Niethammer and F. Krapp, eds.). Akademische Verlagsgesellschaft (Wiesbaden), 2/I:1-649.

Niethammer, J. 1982c. *Mesocricetus newtoni* (Nehring, 1898)—Rumänischer Goldhamster. Pp. 29-38, *in* Handbuch der Säugetiere Europas (J. Niethammer and F. Krapp, eds.). Akademische Verlagsgesellschaft (Wiesbaden), 2/I:1-649.

Niethammer, J. 1982d. *Cricetulus migratorius* (Pallas, 1773)—Zwerghamster. Pp. 39-50, *in* Handbuch der Säugetiere Europas (J. Niethammer and F. Krapp, eds.). Akademische Verlagsgesellschaft (Wiesbaden), 2/I:1-649.

Niethammer, J. 1982e. *Microtus guentheri* Danford et Alston, 1880—Levante-Wühlmaus. Pp. 331-339, *in* Handbuch der Säugetiere Europas (J. Niethammer and F. Krapp, eds.). Akademische Verlagsgesellschaft (Wiesbaden), 2/I:1-649.

Niethammer, J. 1982f. *Microtus cabrerae* Thomas, 1906—Cabreramaus. Pp. 340-348, *in* Handbuch der Säugetiere Europas (J. Niethammer and F. Krapp, eds.). Akademische Verlagsgesellschaft (Wiesbaden), 2/I:1-649.

Niethammer, J. 1982g. *Microtus subterraneus* (de Selys-Longchamps, 1836)—Kurzohrmaus. Pp. 397-418, *in* Handbuch der Säugetiere Europas (J. Niethammer and F. Krapp, eds.). Akademische Verlagsgesellschaft (Wiesbaden), 2/I:1-649.

Niethammer, J. 1982h. *Microtus felteni* (Malec und Storch, 1963). Pp. 438-441, *in* Handbuch der Säugetiere Europas (J. Niethammer and F. Krapp, eds.). Akademische Verlagsgesellschaft (Wiesbaden), 2/I:1-649.

Niethammer, J. 1982i. *Microtus duodecimcostatus* (de Selys-Longchamps, 1839)—Mittelmeer-Kleinwühlmaus. Pp. 463-475, *in* Handbuch der Säugetiere Europas (J. Niethammer and F. Krapp, eds.). Akademische Verlagsgesellschaft (Wiesbaden), 2/I:1-649.

Niethammer, J. 1982j. *Microtus lusitanicus* (Gerbe, 1879)—Iberien-wühlmaus. Pp. 476-484, *in* Handbuch der Säugetiere Europas (J. Niethammer and F. Krapp, eds.). Akademische Verlagsgesellschaft (Wiesbaden), 2/I:1-649.

Niethammer, J. 1982k. *Microtus thomasi* Barrett-Hamilton, 1903—Balkan-Kurzohrmaus. Pp. 485-490, *in*

Handbuch der Säugetiere Europas (J. Niethammer and F. Krapp, eds.). Akademische Verlagsgesellschaft (Wiesbaden), 2/I:1-649.

Niethammer, J. 1982*l*. *Microtus tatricus* (Kratochvil, 1952)—Tatra-Wühlmaus. Pp. 491-496, *in* Handbuch der Säugetiere Europas (J. Niethammer and F. Krapp, eds.). Akademische Verlagsgesellschaft (Wiesbaden), 2/I:1-649.

Niethammer, J. 1987*a*. Das Streifenwiesel (*Poecilictis libyca*) in Sudan und seine Gesamtverbreitung. Bonner Zoologische Beiträge, 38(3):173-182.

Niethammer, J. 1987*b*. Uber griechische Nager im Museum A. Koenig in Bonn. Annalen des Naturhistorischen Museums in Wien, Ser. B, 88/89:245-256.

Niethammer, J. 1987*c*. Rodent distribution in the Middle East. Beihefte zum Tübinger Atlas des Vorderen Orients, A 28:318-329.

Niethammer, J., and H. Henttonen. 1982. *Myopus schisticolor* (Lilljeborg, 1844)—Waldlemming. Pp. 70-86, *in* Handbuch der Säugetiere Europas (J. Niethammer and F. Krapp, eds.), vol. 2/I. Akademische Verlagsgesellschaft (Wiesbaden), 649 pp.

Niethammer, J., and F. Krapp (eds.). 1978. Handbuch der Säugetiere Europas. Vol. 1, Rodentia. Akademische Verlagsgesellschaft (Weisbaden), 476 pp.

Niethammer, J., and F. Krapp, (eds.). 1982*a*. Handbuch der Säugetiere Europas, vol. 2/I. Akademische Verlagsgesellschaft (Wiesbaden), 649 pp.

Niethammer, J., and F. Krapp. 1982*b*. *Microtis arvalis* (Pallas, 1779)—Feldmaus. Pp. 284-318, *in* Handbuch der Säugetiere Europas (J. Niethammer and F. Krapp, eds.). Akademische Verlagsgesellschaft (Wiesbaden), 2/I:1-649.

Niethammer, J., and F. Krapp, (eds.). 1990. Handbuch der Säugetiere Europas, 3/I. Aula-Verlag, Wiesbaden, 524 pp.

Niethammer, J., and J. Martens. 1975. Die Gattungen *Rattus* und *Maxomys* in Afghanistan und Nepal. Zeitschrift für Säugetierkunde, 40:325-355.

Nikaido, M., K. Kawai, Y. Cao, M. Harada, S. Tomita, N. Okada, and M. Hasegawa. 2001*a*. Maximum likelihood analysis of the complete mitochondrial genomes of eutherians and a reevaluation of the phylogeny of bats and insectivores. Journal of Molecular Evolution, 53:508-516.

Nikaido, M., F. Matsuno, H. Hamilton, R. L. Brownell, Y. Cao, W. Ding, Z. Zuoyan, A. M. Shedlock, R. E. Fordyce, M. Hasegawa, and N. Okada. 2001*b*. Retroposon analysis of major cetacean lineages: The monophyly of toothed whales and the paraphyly of river dolphins. Proceedings of the National Academy of Sciences, USA, 98(13):7384-7389.

Nikaido, M., Y. Cao, M. Harada, N. Okada, and M. Hasegawa. 2003. Mitochondrial phylogeny of hedgehogs and monophyly of Eulipotyphla. Molecular Phylogenetics and Evolution, 28:276-284.

Nikanorov, A. P. 2000. [Class Mammalia–Mammals.] Pp. 100-110, *in* [Catalog of Vertebrates of Kamchatka and adjacent waters] (R. W. Moiseev and A. Tokronov, eds.). Petropavlovsk-Kamchatsky, Kamchatskiy Petchatniy Dvor, 166 pp. (in Russian with English summary).

Nikolaeva, A. I. 1982. [Adaptive variation in the masticatory surface of molar teeth in *Arvicola terrestris*.] Zoologicheskii Zhurnal, 61:1565-1575 (in Russian).

Nikolaus, G., and R. J. Dowsett. 1989. Small mammals collected in the Gotel Mts. and on the Mambilla Plateau, eastern Nigeria. Tauraco Research Reports, 1:42-47.

Nikoletopoulos, N. P., B. P. Chondropoulos, and S. E. Fraguedakis-Tsolis. 1992. Albumin evolution and phylogenetic relationships among Greek rodents of the families Arvicolidae and Muridae. Journal of Zoology, London, 228:445-453.

Nikol'skii, A. A. 1974. Geograficheskaya izmenchivost' ritmicheskoi organizatsii zvukovo signala surkov gruppy *bobac* (Rodentia, Sciuridae) [Geographic variation in rhythmic organization of sound signal in marmots of the *bobac* group . . .]. Zoologicheskii Zhurnal, 53:436-444 (in Russian).

Nikol'skii, A. A. 1984. K voprosu o granitse arealov bol'shovo (*Citellus major*) i krasnoshchekovo (*C. erythrogenys*) suslikov v severnom Kazakhstane [On the question of the range boundary between large . . . and red-cheeked . . . susliks in northern Kazakhstan]. Zoologicheskii Zhurnal, 63:256-262 (in Russian).

Nikol'skii, A. A., and V. P. Starikov. 1997. Izmenchivost' zvukovo signal, preduprezhdayushchevo ob opasnosti, u ryzhevatovo (*Spermophilus major*) i krasnoshchekevo (*S. erythrogenys*) suslikov (Rodentia, Sciuridae) v zone kontakta na territorii Kurganskoi oblasti [Variability of alarm call in russet . . . and red-cheeked . . . ground squirrels in a contact zone in Kurgan District]. Zoologicheskii Zhurnal, 76(7):845-85.

Nikol'skii, A. A., and D. Wallschläger. 1982. On the specific status of holarctic long-tailed squirrels; a bioacoustical study. Experientia, 38:808-809.

Nikol'skii, A. A., I. Yu. Yanina, M. V. Rutovskaya, and N. A. Formozov. 1983. Izmenchivost' zvukovovo signala stepnovo i serovo surkov (*Marmota bobac*, *M. baibacina*; Sciuridae, Rodentia) v zone vtorichnovo kontakta [Variation in sound signal of steppe and gray marmots . . . in a zone of secondary contact]. Zoologicheskii Zhurnal, 62:1258-1266 (in Russian).

Nikol'skii, A. A., N. A. Formozov, V. N. Vasil'ev, and G. G. Boeskorov. 1991. Geograficheskaya izmenchivost' zvukovo signala chernoshapochnovo surka, *Marmota camtschatica* (Rodentia, Sciuridae) [Geographic variation in sound signals of the black-capped marmot, *Marmota camtschatica* (Rodentia, Sciuridae)]. Zoologicheskii Zhurnal, 70(2):155-159 (in Russian).

Nishioka, Y. 1987. Y-chromosomal polymorphism in mouse inbred strains. Genetical Research, Cambridge, 50:69-72.

Nishioka, Y., and E. Lamothe. 1987. The *Mus musculus musculus* Y chromosome predominates in Asian house mice. Genetical Research (Cambridge), 50:195-198.

Nishioka, Y., B. M. Dolan, and L. Zahed. 1993. Molecular characterization of a mouse Y chromosomal repetitive sequence amplified in distantly related species in the genus *Mus*. Genome, 36:588-593.

Nitikman, L. Z. 1985. *Sciurus granatensis*. Mammalian Species, 246:1-8.

Niu, Y.-d. 2002. Molecular systematics and evolution of genus *Ochotona* in the world. Ph.D. thesis, Institute of Zoology, Chinese Academy of Sciences, Beijing, 215 pp.

Niu, Y.-d., M. Li, Z.-j. Feng and F.-w. Wei. 2001. [Molecular taxonomy and phylogeny of subgenus *Pika* (Ochotonidae, *Ochotona*) inferred from cytochrome *B* gene]. Acta Zootaxonomica Sinica, 26:394-400 (in Chinese with English summary).

Niwa-Kawakita, M. 1994. Reproductive depression of female mice in intersubspecific F$_2$ hybrids of the species *Mus musculus*. Pp. 121-128, *in* Genetics in wild mice (K. Moriwaki, T. Shiroishi, and H. Yonekawa, eds.). Japan Scientific Societies Press, Tokyo, 333 pp.

Noack, R. 1887. Beiträge zur kenntniss der säugethier-fauna von Ost- und Central-Afrika. Zoologische Jahrbücher, 2:246-248.

Nogueira, M. R., A. L. Peracchi, and A. Pol. 2002. Notes on the lesser white-lined bat, *Saccopteryx leptura* (Schreber)(Chiroptera, Emballonuridae) from southeastern Brazil. Revista Brasileira de Zoologia, 19:1123-1130.

Nolet, B. A., and F. Rosell. 1998. Comeback of the beaver *Castor fiber*: An overview of old and new conservation problems. Biological Conservation, 83:165-173.

Nores, C. 1988. Diferenciacion biometrica de *Apodemus sylvaticus* y *Apodemus flavicollis* en la Cordillera Cantabrica. Primeros resultados. Revista de Biología de la Universidad de Oviedo, 6:109-116.

Norman, S. A., and J. G. Mead. 2002. *Mesoplodon europaeus*. Mammalian Species, 688:1-5.

Norris, C. A. 1994. The periotic bones of possums and cuscuses: Cuscus polyphyly and the division of the marsupial family Phalangeridae. Zoological Journal of the Linnean Society, 111:73-98.

Norris, C. A. 1999. *Phalanger lullulae*. Mammalian Species, 621:1-4.

Norris, C. A., and G. G. Musser. 2001. Systematic revision within the *Phalanger orientalis* complex (Diprotodontia, Phalangeridae): A third species of lowland gray cuscus from New Guinea and Australia. American Museum Novitates, 3356:1-20.

Norris, R. W., K. Zhou, C. Zhou, G. Yang, C. W. Kilpatrick, and R. L. Honeycutt. 2004. The phylogenetic position of the zokors (Myospalacinae) and comments on the families of muroids (Rodentia). Molecular Phylogenetics and Evolution, 31:972-978.

Nová, P., B. A. Reutter, M. Rábová, and J. Zima. 2002. Sex-chromosome heterochromatin variation in the wood mouse, *Apodemus sylvaticus*. Cytogenetic and Genome Research, 96:186-190.

Novacek, M. J. 1985. Cranial evidence for rodent affinities. Pp. 59-81, *in* Evolutionary Relationships among Rodents: A Multidisciplinary Analysis (W. P. Luckett and J.-L. Hartenberger, eds.). Plenum Press, New York, 721 pp.

Novacek, M. J. 1986. The skull of leptictid insectivorans and the higher-level classification of eutherian mammals. Bulletin of the American Museum of Natural History, 183(1):1-111.

Novacek, M. J., and A. R. Wyss. 1986. Higher-level relationships of the Recent eutherian orders: Morphological evidence. Cladistics, 2:257-287.

Novacek, M. J., A. R. Wyss, and M. C. McKenna. 1988. The major groups of eutherian mammals. Pp. 31-71, *in* The Phylogeny and Classification of the Tetrapods, Volume 2 (M. J. Benton, ed.). Clarendon Press, Oxford, x + 329 pp.

Novaro, A. J. 1997. *Pseudalopex culpaeus*. Mammalian Species, 558:1-8.

Novello, A. F., and E. P. Lessa. 1986. G-band homology in two karyomorphs of the *Ctenomys pearsoni* complex (Rodentia: Octodontidae) of neotropical fossorial rodents. Zeitschrift für Säugetierkunde, 51:378-380.

Novello, A. F., E. P. Lessa, C. Sambarino, and S. Monzon. 1990. Chromosomal variation in two populations of the genus *Ctenomys* (Rodentia: Octodontidae) from Uruguay. Zeitschrift für Säugetierkunde, 55:43-48.

Novikov, G. A. 1939. Evropeiskaya norka [European mink]. Leningradskovo Gosudarstvennovo Universiteta, Leningrad, 180 pp. (in Russian).

Novikov, G. A. 1956. Khishchye mlekopitayushchie fauna SSSR [Carnivorous mammals of the fauna of the USSR]. Akademiya Nauk SSSR, Moscow-Leningrad, 293 pp. (in Russian).

Nowak, R. M. 1979. North American Quaternary *Canis*. University of Kansas, Museum of Natural History, Monograph, 6:1-154.

Nowak, R. M. 1991. Walker's Mammals of the World. Fifth ed. Johns Hopkins University Press, Baltimore, 1:1-642; 2:643-1629.

Nowak, R. M. 1992. The red wolf is not a hybrid. Conservation Biology, 6:593-595.

Nowak, R. M. 1999. Walker's Mammals of the World. Sixth ed. Johns Hopkins University Press, Baltimore.

Nowak, R. M. 2002. The original status of wolves in Eastern North America. Southeastern Naturalist, 1:95-130.

Nowak, R. M., and J. L. Paradiso. 1983. Walker's Mammals of the World. Fouth Edition. Johns Hopkins University Press, Baltimore, 1:1-568, 2:569-1362.

Nowakowski, J., and J. Terlecki. 1991. A new locality of fat dormouse *Glis glis* (Linnaeus, 1766) from north eastern Poland. Przeglad Zoologiczny, 35(3-4):383-385.

Nowakowski, W. 1998. 24-Hour activity in the forest dormouse (*Dryomys nitedula*). Natura Croatica, 7(1):19-29.

Nowakowski, W. 2000. Habitat preferences of the forest dormouse (*Dryomys nitedula*) in lowland forests. Polish Ecological Studies, 23(3-4):199-207.

Nowakowski, W. 2001*a*. Daily torpor in the Edible Dormouse *Glis glis*. Trakya University Journal of Scientific Research B, 2(2):109-110.

Nowakowski, W. 2001*b*. Winter activity in the forest dormouse *Dryomys nitedula*. Trakya University Journal of Scientific Research B, 2(2):143-144.

Nowakowski, W., and P. Boratynski. 2001. Spatial distribution of the forest dormouse (*D. Nitedula* Pallas, 1778) population in the Białowieża Forest. Trakya University Journal of Scientific Research B, 2(2):137-142.

Nowakowski, W., and A. Rachwald. 2000. Ultrasound and audible sound emission in dormice family (Gliridae: Rodentia). Biological Bulletin of Poznan, 37(1):153-158.

Nunes, A. 2002. First record of *Neusticomys oyapocki* (Muridae: Sigmodontinae) from the Brazilian Amazon. Mammalia, 66:445-447.

Oaks, E. C., P. J. Young, G. L. Kirkland, Jr., and D. F. Schmidt. 1987. *Spermophilus variegatus*. Mammalian Species, 272:1-8.

Oates, J. F. 1985. The Nigerian guenon, *Cercopithecus erythrogaster*: Ecological, behavioural, systematic and historical observations. Folia primatologia, 45:25-43.

Oates, J. F., and T. F. Trocc. 1983. Taxonomy and phylogeny of black-and-white colobus monkeys: Inferences from an analysis of loud call variation. Folia Primatologia, 40:83-113.

Oates, J. F., M. Abedi-Lartey, W. S. McGraw, T. T. Struhsaker, and G. H. Whitesides. 2000. Extinction of a West African red colobus monkey. Conservation Biology, 14:1526-1532.

Obara, Y. 1986. G-band homology between the Japanese red-backed vole, *Clethrionomys a. andersoni* and the grey red-backed vole, *C. rufocanus*. Chromosome Information Service, 40:7-9.

Obara, Y., and T. Tada. 1985. Karyotypes and chromosome banding patterns of the Japanese water shrew, *Chimarrogale platycephala platycephala*. Proceedings of the Japan Academy, Series B, 61:20-23.

Obara, Y., H. Kusakabe, K. Miyakoshi, and S. Kawada. 1995. Revised karyotypes of the Japanese northern red-backed vole, *Clethrionomys rutilus mikado*. Journal of the Mammalogical Society of Japan, 20:125-133.

O'Brien, S. J., W. G. Nash, D. E. Wildt, M. E. Bush, and R. E. Benveniste. 1985. A molecular solution to the riddle of the gaint panda's phylogeny. Nature, 317:140-144.

Obuch, J. 1994. [On the food of Eagle Owl (*Bubo bubo*) and Tawny Owl (*Strix aluco*) in the eastern part of Turkey.] Tichodroma, Bratislava, 7:7-16.

Obuch, J. 1998. Plchy (Gliridae) v potrave sov(Strigiformes) na Slovensku [Dormice in the diet of owls in Slovakia]. Lynx, 29:31-41.

Obuch, J. 2001. Dormice in the diet of owls in the Middle East. Trakya University Journal of Scientific Research B, 2(2):145-150.

Ochoa G., J., H. Castellanos, and C. Ibanez. 1988. Records of bats and rodents from Venezuela. Mammalia, 52:175-180.

Ochoa G., J., M. Aguilera, V. Pacheco, and P. J. Soriano. 2001. A new species of *Aepeomys* Thomas, 1898 (Rodentia: Muridae) from the Andes of Venezuela. Zeitschrift für Säugetierkunde, 66:228-237.

O'Connell, M. A. 1983. *Marmosa robinsoni*. Mammalian Species, 203:1-6.

O'Conner, T. P. 1986. The garden dormouse *Eliomys quercinus* from Roman York. Journal of Zoology, London, 210A(4):620-622.

Oda, S., J. Kitch, H. Ota, and G. Isomura. 1985. *Suncus murinus*—Biology of the laboratory shrew. Japan Scientific Societies Press, Tokyo, 535 pp.

O'Donnell, C. F. J. 2001. Advances in New Zealand mammalogy 1990-2000: Long-tailed bat. Journal of the Royal Society of New Zealand, 31:43-57.

O'Farrell, M. J., and A. R. Blaustein. 1974a. *Microdipodops megacephalus*. Mammalian Species, 46:1-3.

O'Farrell, M. J., and A. R. Blaustein. 1974b. *Microdipodops pallidus*. Mammalian Species, 47:1-2.

O'Farrell, M. J., and E. H. Studier. 1980. *Myotis thysanodes*. Mammalian Species, 137:1-5.

O'Gara, B. W. 1978. *Antilocapra americana*. Mammalian Species, 90:1-7.

O'Gara, B. W. 2002. Taxonomy. Pp. 3-68, *in* North American elk: Ecology and management (Toweill, D. E. and Thomas, J. W., eds.). Smithsonian Institution Press, Washington, D.C., 962 pp.

O'Gara, B. W., and G. Matson. 1975. Growth and casting of horns by pronghorns and exfoliation of horns by bovids. Journal of Mammalogy, 56:829-846.

Ogilby, W. 1835. Descriptions of Mammalia and birds from the Gambia. Proceedings of the Zoological Society of London, 1835:97-105.

Ognev, S. I. 1921. Materialy dlya sistematiki nasekomoyadnykh mlekopitayushchikh Rossi [Contribution á la classification des mammifères insectivores de Russie]. Ezhegodnikh Zoologicheskovo Muzeya, Akadimii Nauk [Annuaire du Musée de l'Académie des Sciences de St. Petersbourg], 22:311-350 (in Russian and English).

Ognev, S. I. 1927. Zur Frage über die systematische Stellung einiger Vertreter von *Paraechinus* Trouessart. Zoologischer Anzeiger, 69:209-218.

Ognev, S. I. 1928. Zveri vostochnoi Evropy i severnoi Azii: Nasekomoyadnye i letychie myshi [Mammals of eastern Europe and northern Asia: Insectivora and Chiroptera]. Glavnauka, Moscow, 1:1-631 (in Russian).

Ognev, S. I. 1931. Zveri vostochnoi Evropy i severnoi Azii: Khishchnye mlekopitayushchie [Mammals of eastern Europe and northern Asia: Carnivorous mammals]. Glavnauka, Moscow, 2:1-776 (in Russian).

Ognev, S. I. 1935. Zveri SSSR i prilezhashchikh stran. Khishchnyei i lastonogie (Zveri vostochnoi Evropy i severnoi Azii) [Mammals of the USSR and adjacent countries: Carnivora and Pinnipedia (Mammals of eastern Europe and northern Asia)]. Glavpushnina NKVT, Moscow, 3:1-752 (in Russian).

Ognev, S. I. 1940. Zveri SSSR i prilezhashchikh stran: Gryzuny. (Zveri vostochnoi Evropy i severnoi Azii) [Mammals of the USSR and adjacent countries: Rodents (Mammals of eastern Europe and northern Asia)]. Akademiya Nauk SSSR, 4:1-615 (in Russian).

Ognev, S. I. 1947. Zveri SSSR i prilezhashchikh stran: Gryzuny (prodolzhenie). (Zveri vostochnoi Evropy i severnoi Azii) [Mammals of the USSR and adjacent countries: Rodents (continued). (Mammals of eastern Europe and northern Asia)]. Akademiya Nauk SSSR, 5:1-809 (in Russian).

Ognev, S. I. 1948. Zveri SSSR i prilezhashchikh stran: Gryzuny (prodolzhenie). (Zveri vostochnoi Evropy i severnoi Azii) [Mammals of the USSR and adjacent countries: Rodents (continued). (Mammals of eastern Europe and northern Asia)]. Akademiya Nauk SSSR, 6:1-559 (in Russian).

Ognev, S. I. 1950. Zveri SSSR i prilezhashchikh stran: Gryzuny (prodolzhenie). (Zveri vostochnoi Evropy i severnoi Azii) [Mammals of the USSR and adjacent countries: Rodents (continued). (Mammals of eastern Europe and northern Asia)]. Akademiya Nauk SSSR, 7:1-706+15 maps (in Russian).

Ognev, S. I. 1962a. Mammals of eastern Europe and northern Asia: Insectivora and Chiroptera [A translation of S. I. Ognev, 1928, Zveri vostochnoi Evropy i severnoi Azii: Nasekomoyadnye i letychie myshi]. Israel Program for Scientific Translations, Jerusalem, 1:1-487+I-XV.

Ognev, S. I. 1962b. Mammals of eastern Europe and northern Asia: Carnivora (Fissipedia) [A translation of S. I. Ognev, 1931, Zveri vostochnoi Evropy i severnoi Azii: Khishchnye mlekopitayushchie]. Israel Program for Scientific Translations, Jerusalem, 2:1-590+I-XV.

Ognev, S. I. 1962c. Mammals of the USSR and adjacent countries: Fissipedia and Pinnipedia [A translation of S. I. Ognev, 1935, Zveri SSSR i prilezhashchikh stran. Khishchnyei i lastonogie (Zveri vostochnoi Evropy i severnoi Azii)]. Israel Program for Scientific Translations, Jerusalem, 3:1-641+I-XV.

Ognev, S. I. 1963a. Mammals of the USSR and adjacent countries: Rodents (continued). (Mammals of eastern Europe and northern Asia) [A translation of S. I. Ognev, 1947, Zveri SSSR i prilezhashchikh stran: Gryzuny (prodolzhenie). (Zveri vostochnoi Evropy i severnoi Azii)]. Israel Program for Scientific Translations, Jerusalem, 5:1-662.

Ognev, S. I. 1963b. Mammals of the USSR and adjacent countries: Rodents (continued). (Mammals of eastern Europe and northern Asia) [A translation of S. I. Ognev, 1948, Zveri SSSR i prilezhashchikh stran: Gryzuny (prodolzhenie). (Zveri vostochnoi Evropy i severnoi Azii)]. Israel Program for Scientific Translations, Jerusalem, 6:1-508.

Ognev, S. I. 1964. Mammals of the USSR and adjacent countries: Rodents (continued). (Mammals of eastern Europe and northern Asia) [A translation of S. I. Ognev, 1950, Zveri SSSR i prilezhashchikh stran: Gryzuny (prodolzhenie). (Zveri vostochnoi Evropy i severnoi Azii)]. Israel Program for Scientific Translations, Jerusalem, 7:1-626.

Ognev, S. I. 1966. Mammals of the USSR and adjacent countries: Rodents (Mammals of eastern Europe and northern Asia) [A translation of S. I. Ognev, 1940, Zveri SSSR i prilezhashchikh stran: Gryzuny. (Zveri vostochnoi Evropy i severnoi Asii)]. Israel Program for Scientific Translations, Jerusalem, 4:1-429+61 tables.

Oguge, N., R. Hutterer, R. Odhiambo, and W. Verheyen. 2004. Diversity and structure of shrew communities in montane forests of southeast Kenya. Mammalian Biology, 69(5):289-301.

Ohdachi, S., R. Masuda, H. Abe, and N. E. Dokuchaev. 1997a. Biogeographical history of northeastern Asiatic soricine shrews (Insectivora, Mammalia). Researches on Population Ecology, 39:157-162.

Ohdachi, S., R. Masuda, H. Abe, J. Adachi, N. E. Dokuchaev, V. Haukisalmi, and M. C. Yoshida. 1997b. Phylogeny of Eurasian soricine shrews (Insectivora, Mammalia) inferred from the mitochondrial cytochrome b gene sequences. Zoological Science, 14:527-532.

Ohdachi, S., N. E. Dokuchaev, M. Hasegawa, and K. Masuda. 2001. Intraspecific phylogeny and geographical variation of six species of northeastern Asiatic Sorex shrews based on the mitochondrial cytochrome b sequences. Molecular Ecology, 10:2199-2213.

Ohnishi, N., Y. Ishibashi, T. Saitoh, S. Abe, and M. C. Yoshida. 1998. Polymorphic microsatellite DNA markers in the Japanese wood mouse Apodemus argenteus. Molecular Ecology, 7:1431-1432.

Ohno, S., J. Jainchill, and C. Stenius. 1963. The creeping vole (Microtus oregoni) as a gonosomic mosaic I. The OY/XY constitution of the male. Cytogenetics, 2:232-239.

Ohtsuka, H. M. Oyanagi, Y. Mafune, N. Miyashita, T. Shiroishi, K. Moriwaki, R. Kominami, and N. Saitou. 1996. The presence/absence polymorphism and evolution of the p53 pseudogene in the genus Mus. Molecular Phylogenetics and Evolution, 5(3):548-556.

Ohuchi, T., N. Kitamura, J. Yamada, and T. Yamashita. 1992. A morphological study on the muscular architecture of the stomach in the Japanese field vole Microtus montebelli, with special reference to the gastric groove. Zoologische Jahrbücher. Abteilung für Anatomie und Ontogenie der Tiere, 122:361-369.

Ojasti, J. 1972. Revisión preliminar de los picures o agutís de Venezuela (Rodentia: Dasyproctidae). Memorias Sociedad Cíencias Naturales La Salle, 32:159-204.

Ojasti, J., and O. J. Linares. 1971. Adiciones a la fauna de murcielagos de Venezuela con notas sobre las especies del género Diclidurus (Chiroptera). Acta Biologica Venezuelica, 7:421-441.

Ojasti, J., and E. Mondolfi. 1968. Esbozo de la fauna de mamíferos de Caracas. Pp. 441-461, in Estudio de Caracas. Vol. 1. Ecología vegetal y fauna. (A. Tovar, ed.). Universidad Central de Venezuela, Caracas, 467 pp.

Okamoto, M. 1999. Phylogeny of Japanese moles inferred from mitochondrial co1 gene sequences. Pp. 21-27, in Recent advances in the biology of Japanese Insectivora (Y. Yokohata and S. Namakura, eds.). Hiba Society of Natural History, Shobara, 150 pp.

Oken, L. 1815-1816. Lehrbuch der Naturgeschichte. Zoologie. [vol. 3 'Zoologie.' published 1815-1816; ICZN Opinion 417—vol. 3 rejected for nomenclatural purposes]. August Schmid und Comp., Jena, 3:1-1270.

Okhotina, M. V. 1977. Palaearctic shrew of the subgenus Otisorex: Biotopic preference, population number, taxonomic revision and distribution history. Acta Theriologica, 22:191-206.

Okhotina, M. V. 1983. A taxonomic revision of Sorex arcticus Kerr, 1792 (Soricidae, Insectivora). Zoologeskii Zhurnal, 62:409-417.

Okhotina, M. V. 1993. Subspecies taxonomic revision of Far East shrews (Insectivora, Sorex) with the description of new subspecies. Trudy Zoologicheskogo Instituta, 242 [1991]:58-71 (in Russian).

Olds, N. 1988. A revision of the genus Calomys (Rodentia: Muridae). Ph. D. dissertation, City University of New York.

Olds, N., and S. Anderson. 1987. Notes on Bolivian mammals. 2. Taxonomy and distribution of rice rats of the subgenus Oligoryzomys. Fieldiana: Zoology, n.s., 39:261-281.

Olds, N., and S. Anderson. 1989 [1990]. A diagnosis of the tribe Phyllotini (Rodentia, Muridae). Pp. 55-74, in Advances in Neotropical mammalogy (K. Redford and J. F. Eisenberg, eds.). Sandhill Crane Press, Gainesville, FL, 614 pp.

Olds, N., and J. Shoshani. 1982. Procavia capensis. Mammalian Species, 171:1-7.

Olds, N., S. Anderson, and T. L. Yates. 1987. Notes on Bolivian mammals 3: A revised diagnosis of Andalgalomys (Rodentia, Muridae) and the description of a new subspecies. American Museum Novitates, 2890:1-17.

Olert, J., F. Dieterlen, and H. Rupp. 1978. Eine neue Muriden-Art aus Südäthiopien. Zeitschrift für Zoologische Systematik und Evolutionsforschung, 16(4):297-308.

Olfers, I. von. 1818. Bemerkungen zu Illiger's Ueberblick der Säugthiere nach ihrer Vertheilung über die Welttheile, rücksichtich der Südamericanischen Arten (Species). Abhandlung X in W. L. von Eschwege, Journal von Brasilien, odor vermischte Nachrichten aus Brasilien, auf wissenschaftlichen Reisen gesammelt. Heft 2 in F. Bertuch (ed.), Neue Bibliothek der wichtigsten Reisebeschreibungen zur Erweiterung der Erd- und Völkerkunde, Band 15:192-237. Weimar.

Oliva, D. 1988. *Otaria byronia* (de Blainville, 1820), the valid specific name for the southern sea lion (Carnivora: Otariidae). Journal of Natural History, 22:767-772.

Oliveira, J. A., de, and C. R. Bonvicino. 2002. A new species of sigmodontine rodents from the Atlantic forest of eastern Brazil. Acta Theriologica, 47(3):307-322.

Oliveira, T. G. de. 1995. The Brazilian three-banded armadillo *Tolypeutes tricinctus* in Maranhão. Edentata, 2:18-19.

Oliveira, T. G., de. 1998a. *Leopardus wiedii*. Mammalian Species, 579:1-6.

Oliveira, T. G., de. 1998b. *Herpailurus yagouaroundi*. Mammalian Species, 578:1-6.

Oliver, W. L. R., and I. B. Santos. 1991. Threatened endemic mammals of the Atlantic forest region of Southeast Brazil. Special Scientific Report of Jersey Wildlife Preservation Trust, 4:1-125.

Oliver, W. L. R., C. R. Cox, P. C. Gonzales, and L. R. Heaney. 1993. Cloud rats in the Philippines—preliminary report on distribution and status. Oryx, 27(1):41-48.

Olmos, F. 1995. Edentates in the caatinga of Serra da Capivara National Park. Edentata, 2:16-17.

Olrog, C. C., and M. M. Lucero. 1981. Guia de los mamiferos Argentinos. Ministerio de Cultura y Educacion, San Miguel de Tucuman, 151 pp.

Olsen, O. 1913. On the external characters and biology of Bryde's whale (*Balaenoptera brydei*), a new rorqual from the coast of South Africa. Proceedings of the Zoological Society of London, 1913(IV):1073-1090, pls. CIX-CXIII.

Olsen, S. J. 1990. Fossil ancestry of the yak, its cultural significance and domestication in Tibet. Proceedings of the Academy of Natural Sciences of Philadelphia, 142:73-100.

Olsen, S. J. 1991. Confused yak taxonomy and evidence of domestication. Pp. 387-393, *in* Beamers, bobwhites and blue-points. Tributes to the career of Paul W. Parmalee (J. R. Purdue, W. E. Klippel, and B. W. Styles, eds.). Illinois State Museum Scientific Papers, vol. 23, and University of Tennessee, Department of Anthropology Report of Investigations, no. 52. Illinois State Museum, Springfield, Illinois, 436 pp.

Olsen, S. J. 1993. Evidence of early domestication of the water buffalo in China. Pp. 151-156, *in* Skeletons in her cupboard. Festschrift for Juliet Clutton-Brock (A. Clason, S. Payne, and H.-P. Uerpmann, eds.). Oxbow Monograph, 34:259 pp.

Olson, P. D. 1995. Water-rat, *Hydromys chrysogaster*. Pp. 628-629, *in* Mammals of Australia (R. Strahan, ed.). Smithsonian Institution Press, Washington, D.C., 756 pp.

Ondrias, J. C. 1966. The taxonomy and geographical distribution of the rodents of Greece. Säugetierkundliche Mitteilungen, 14, sonderheft, 1:1-136.

Orlando, L., J.-F. Mauffrey, J. Cuisin, J. L. Patton, C. Hänni, and F. Catzeflis. 2003. Napoleon Bonaparte and the fate of an Amazonian rat: New data on the taxonomy of *Mesomys hispidus* (Rodentia: Echimyidae). Molecular Phylogenetics and Evolution, 27(1):113-120.

Orlov, V. N. 1969. Khromosomnye nabory ezhei vostochnoi Evropy [Chromosomal complements of hedgehogs in eastern Europe]. Pp. 6-7, *in* Materialy ko II Vsesoyuznykh Teriologicheskich Soveshchen po mlekopitayushchim [Materials in II All-Union Theriological Conference about mammals]. Novosibirsk.

Orlov, V. N., and N. Sh. Bulatova. 1983. Sravnitel'naya tsitogenetika i karyosistematika mlekopitayushchikh [Comparative cytogenetics and karyosystematics of mammals]. Nauka, Moscow, 405 pp. (in Russian).

Orlov, V. N., and N. Sh. Bulatova. 1997. Endemic rodents of the Ethiopian plateaux: Their karyology and possible relationships. Pp. 125-128, *in* Tropical biodiversity and systematics (H. Ulrich, ed.). Zoologisches Forschungsinstitut und Museum Alexander Koenig, Bonn, 357 pp.

Orlov, V. N., and N. Davaa. 1975. O systematicheskom polozhenii Alashanskovo suslika *Citellus alashanicus* Buch. (Sciuridae, Rodentia) [On the systematic position of the Alashan ground squirrel . . .]. Pp. 8-9, *in* Sistematika i tsitogenetica mlekopitayushchikh [Systematics and cytogenetics of mammals] (V. N. Orlov, ed.). Nauka, Moscow, 60 pp. (in Russian).

Orlov, V. N., and E. N. Ischakova. 1975. [Taxonomy of the superspecies *Cricetulus barabensis* (Rodentia, Cricetidae).] Zoologicheskii Zhurnal, 54:597-604 (in Russian, with English summary).

Orlov, V. N., and Yu. M. Kovalskaya. 1978. [*Microtus mujanensis* sp. n. from the Vitim River Basin.] Zoologicheskii Zhurnal, 57:1224-1232 (in Russian).

Orlov, V. N., and A. I. Kozlovsky. 1971. A synopsis of chromosome complements of shrews of the genus *Sorex*. Vestnik Moskovskogo Universiteta, Biologia i Pocvovedenie, 1971:12-16 (in Russian).

Orlov, V. N., and V. M. Malygin. 1988. [A new species of hamster—*Cricetulus sokolovi* sp. n. (Rodentia, Cricetidae) from the People's Republic of Mongolia.] Zoologicheskii Zhurnal, 67:304-308 (in Russian).

Orlov, V. N., and N. M. Okulova. 2001. [Application of the Hardy-Weinberg's equation for analysis of geographical variation of the yellow-necked mouse *Apodemus flavicollis* (Muridae, Rodentia).] Zoologicheskii Zhurnal, 80(5):607-617 (in Russian with English abstract).

Orlov, V. N., V. N. Yatsenko, and V. M. Malygin. 1983. [Karyotype homology and species phylogeny in a group of field mice (Cricetidae, Rodentia).] Doklady Akademii Nauk SSSR, 269:236-238 (in Russian).

Orlov, V. N., M. I. Baskevich, and N. Sh. Bulatova. 1992a. [Chromosomal sets of rats of the genus *Arvicanthis*

(Rodentia, Muridae) from Ethiopia.] Zoologicheskii Zhurnal, 71(2):103-112 (in Russian with English summary).

Orlov, V. N., R. S. Nadjafova, and N. Sh. Bulatova. 1992*b*. [Taxonomic isolation of *Mus abbotti* (Muridae, Rodentia) from Azerbaijan.] Zoologicheskii Zhurnal, 71(7):116-122 (in Russian with English summary).

Orlov, V. N., N. Sh. Bulatova, R. S. Nadjafova, and A. I. Kozlovsky. 1996*a*. Evolutionary classification of European wood mice of the subgenus *Sylvaemus* based on allozyme and chromosome data. Bonner Zoologische Beiträge, 46(1-4):191-202.

Orlov, V. N., A. I. Kozlovsky, R. S. Nadjafova, and N. Sh. Bulatova. 1996*b*. [Karyological diagnoses, distribution, and evolutionary classification of wood mice of the subgenus *Sylvaemus* (*Apodemus*, Muridae, Rodentia) in Europe.] Zoologicheskii Zhurnal, 75(1):88-102 (in Russian with English summary).

Orr, R. T. 1938. A new rodent of the genus *Nesoryzomys* from the Galapagos Islands. Proceedings of the California Academy of Science, ser. 4, 23:303-306.

Orr, R. T. 1945. A study of the *Clethrionomys dawsoni* group of red-backed mice. Journal of Mammalogy, 26:67-74.

Orsini, P., and G. Cheylan. 1988. Les rongeurs de Corse: Modifications de taille en relation avec l'isolement en milieu isolaire. Bulletin d'Ecologie, 19(2-3):411-416.

Ortega, J., and I. Castro-Arellano. 2001. *Artibeus jamaicensis*. Mammalian Species, 662:1-9.

Ortells, M. O. 1995. Phylogenetic analysis of G-banded karyotypes among the South American subterranean rodents of the genus *Ctenomys* (Caviomorpha: Octodontidae), with special reference to chromosomal evolution and speciation. Biological Journal of the Linnean Society, 54:43-70.

Ortells, M. O., and G. E. Barrantes. 1994. A study of genetic distances and variability in several species of the genus *Ctenomys* (Rodentia: Octodontidae) with special reference to a probable causal role of chromosomal speciation. Biological Journal of the Linnean Society, 53:189-208.

Ortells, M. O., O. A. Reig, N. Brum-Zorrilla, and O. A. Scaglia. 1988. Cytogenetics and karyosystematics of phyllotine rodents (Cricetidae, Sigmodontinae). I. Chromosome multiformity and gonosomal-autosomal translocation in *Reithrodon*. Genetica, 77:53-63.

Ortells, M. O., O. A. Reig, R. L. Wainburg, G. E. Hurtado de Catalfo, and T. M. L. Gentile de Fronza. 1989. Cytogenetics and karyosystematics of phyllotine rodents (Cricetidae, Sigmodontinae). II. Chromosome multiformity and autosomal polymorphism in *Eligmodontia*. Zeitschrift für Säugetierkunde, 54:129-140.

Ortells, M. O., J. R. Contreras, and O. A. Reig. 1990. New *Ctenomys* karyotypes (Rodentia, Octodontidae) from north-eastern Argentina and from Paraguay confirm the extreme chromosomal multiformity of the genus. Genetica, 82:189-201.

Orth, A., E. Lyapunova, A. Kandaurov, S. Boissinot, P. Boursot, N. Vorontsov, and F. Bonhomme. 1996. L'espèce polytypique *Mus musculus* en Transcaucasie. Comptes Rendus des Séances de l'Académie des Sciences (Paris), Sciences de la vie/Life sciences, 319:435-441.

Orth, A., J.-C. Auffray, and F. Bonhomme. 2002. Two deeply divergent clades in the wild mouse *Mus macedonicus* reveal multiple glacial refuges south of Caucasus. Heredity, 89:353-357.

Ortiz, P. E., S. Cirignoli, D. H. Podesta, and U. F. J. Pardiñas. 2000*a*. New records of sigmodontine rodents (Mammalia: Muridae) from high-Andean localities of northwestern Argentina. Biogeographica, 76:133-140.

Ortiz, P. E., U. F. J. Pardiñas, and S. J. Steppan. 2000*b*. A new fossil phyllotine (Rodentia: Muridae) from northwestern Argentina and relationships of the *Reithrodon* group. Journal of Mammalogy, 81:37-51.

Orts, S. G. 1970. Le *Xenogale* de J. A. Allen (Carnivora, Viverridae) au sujet d'une capture effectvée au Kivu. Revue de Zoologie et de Botanique Africaines, 82:174-186.

Osborn, D. J. 1962. Rodents of the subfamily Microtinae from Turkey. Journal of Mammalogy, 43:515-529.

Osborn, D. J. 1965. Rodents of the subfamilies Murinae, Gerbillinae, and Cricetinae from Turkey. Journal of Egyptian Public Health Association, 60:401-121.

Osborn, D. J., and I. Helmy. 1980. The contemporary land mammals of Egypt (including Sinai). Fieldiana: Zoology, 5:1-579.

Osborne, M. J., and L. Christidis. 2001. Molecular phylogenetics of Australo-Papuan possums and gliders (family Petauridae). Molecular Phylogenetics and Evolution, 20:211-224.

Osgood, W. H. 1900. Revision of the pocket mice of the genus *Perognathus*. North American Fauna, 18:1-73.

Osgood, W. H. 1907. Some unrecognised and misapplied names of American mammals. Proceedings of the Biological Society of Washington, 20:43-52.

Osgood, W. H. 1909. Revision of the mice of the American genus *Peromyscus*. North American Fauna, 28:1-285.

Osgood, W. H. 1910. Eight new African rodents. Annals and Magazine of Natural History, ser. 8, 5:276-282.

Osgood, W. H. 1912. Mammals from western Venezuela and eastern Colombia. Field Museum of Natural History, Zoological Series, 10:32-67.

Osgood, W. H. 1913. New Peruvian mammals. Field Museum of Natural History, Zoological Series, 10:93-100.

Osgood, W. H. 1918. The status of *Perognathus longimembris* (Coues). Proceedings of the Biological Society of Washington, 31:95-96.

Osgood, W. H. 1925. The long-clawed South American rodents of the genus *Notiomys*. Field Museum of Natural History, Zoological Series, 12:113-125.

Osgood, W. H. 1932. Mammals of the Kelley-Roosevelts and Delacour Asiatic expeditions. Field Museum of Natural History, Zoological Series, 18:193-339.

Osgood, W. H. 1933*a*. The South American mice referred to *Microryzomys* and *Thallomyscus*. Field Museum of Natural History, Zoological Series, 20:1-8.

Osgood, W. H. 1933*b*. Two new rodents from Argentina. Field Museum of Natural History, Zoological Series, 20:11-14.

Osgood, W. H. 1933*c*. The supposed genera *Aepeomys* and *Inomys*. Journal of Mammalogy, 14:161.

Osgood, W. H. 1933*d*. The generic position of *Mus pyrrhorhinus* Wied. Journal of Mammalogy, 14:370-371.

Osgood, W. H. 1936. New and imperfectly known small mammals from Africa. Field Museum of Natural History, Zoological Series, 20:217-256.

Osgood, W. H. 1941. The technical name of the chinchilla. Journal of Mammalogy, 22:407-411.

Osgood, W. H. 1943. The mammals of Chile. Field Museum of Natural History, Zoological Series, 30:1-268.

Osgood, W. H. 1945. A new rodent from Dutch New Guinea. Fieldiana: Zoology, 31:1-2.

Osgood, W. H. 1946. A new octodont rodent from the Paraguayan Chaco. Fieldiana: Zoology, 31(6):47-49.

Osgood, W. H. 1947. Cricetine rodents allied to *Phyllotis*. Journal of Mammalogy, 28:165-174.

O'Shea, T. J. 1980. Roosting, social organization and annual cycle in a Kenya population of the bat *Pipistrellus nanus*. Zeitschrift für Tierpsychologie, 53:171-195.

O'Shea, T. J. 1991. *Xerus rutilus*. Mammalian Species, 370:1-5.

O'Shea, T. J., and T. A. Vaughan. 1980. Ecological observations on an East African bat community. Mammalia, 44:485-496.

Oshida, T., and Y. Obara. 1991. Karyotypes and chromosome banding patterns of a male Japanese giant flying squirrel, *Petaurista leucogenys* Temminck. CIS (Chromosome Information Service), No. 50:26-28.

Oshida, T., and Y. Obara. 1993. C-band variation in the chromosomes of the Japanese giant flying squirrel, *Petaurista leucogenys*. Journal of the Mammalogical Society of Japan, 18(2):61-67.

Oshida, T., and M. C. Yoshida. 1994. Banded karyotype of Asiatic chipmunk, *Tamias sibiricus lineatus* Siebold. CIS (Chromosome Information Service), 57:27-28.

Oshida, T., and M. C. Yoshida. 1997. Comparison of banded karyotypes between the Eurasian red squirrel *Sciurus vulgaris* and the Japanese squirrel *Sciurus lis*. Chromosome Science, 1(1):17-20.

Oshida, T., and M. C. Yoshida. 1998. A note on the chromosomes of the white-bellied flying squirrel *Petinomys setosus* (Rodentia, Sciuridae). Chromosome Science, 2(2):119-121.

Oshida, T., M. Matsushima, and M. C. Yoshida. 1993. Chromosome banding patterns of the Eurasian squirrel, *Sciurus vulgaris orientis* Thomas. CIS (Chromosome Information Service), No. 55:10-12.

Oshida, T., M. Itoya, and M. C. Yoshida. 1996*a*. Q-banded karyotype of a male Japanese squirrel, *Sciurus lis*. CIS (Chromosome Information Service), No. 61:22-24.

Oshida, T., R. Masuda, and M. C. Yoshida. 1996*b*. Phylogenetic relationships among Japanese species of the family *Sciuridae* (Mammalia, Rodentia), inferred from nucleotide sequences of mitochondrial 12S ribosomal RNA genes. Zoological Science, 13(4):615-620.

Oshida, T., H. Yangawa, and M. C. Yoshida. 1996*c*. Chromosome banding patterns of a male Pallas squirrel, *Callosciurus erythraeus*. CIS (Chromosome Information Service), No. 60:7-9.

Oshida, T., H. Yangawa, and M. C. Yoshida. 1996*d*. Comparison of G-banded karyotypes between two beautiful squirrel species, *Callosciurus prevostii* and *C. erythraeus*. CIS (Chromosome Information Service), No. 61:26-27.

Oshida, T., L.-K. Lin, H. Yanagawa, H. Endo, and R. Masuda. 2000*a*. Phylogenetic relationships among six flying squirrel genera, inferred from mitochondrial cytochrome *b* gene sequences. Zoological Science, 17:485-489.

Oshida, T., H. Yanagawa, M. Tsuda, S. Inoue, and M. C. Yoshida. 2000*b*. Comparisons of the banded karyotypes between the small Japanese flying squirrel, *Pteromys momonga* and the Russian flying squirrel, *P. volans* (Rodentia, Sciuridae). Caryologia, 53(2):133-140.

Oshida, T., M. Yasuda, H. Endo, N. A. Hussein, and R. Masuda. 2001. Molecular phylogeny of five squirrel species of the genus *Callosciurus* (Mammalia, Rodentia) inferred from cytochrome *b* gene sequences. Mammalia, 65:473-482.

Ostroff, A. C., and E. J. Finck. 2003. *Spermophilus franklinii*. Mammalian Species. 724:1-5.

Oswald, C. 2002. A commentary on the systematics of red deer *Cervus elaphus* L. Deer, 12:84-89.

Otianga'a-Owiti, D. Odour-Okelo, and S. G. Gombe, 1992. Foetal membranes and placenta of the springhare (*Pedetes capensis larvalis* Hollister). African Journal of Ecology, 30:74-86.

Ottenwalder, J. A. 1999. Observations on the habitat and ecology of the Hispaniolan Solenodon (*Solenodon paradoxus*) in the Dominican Republic. Pp. 123-167, *in* Ecologia de les Illes (J. A. Alcover, ed.). Institut d'Estudis Baleàrics, Ciutat de Mallorca, 204 pp

Ottenwalder, J. A. 2001. Systematics and biogeography of the West Indian genus *Solenodon*. Pp. 253-329, *in* Biogeography of the West Indies: Patterns and perspectives (C. A. Woods and F. E. Sergile, eds.). CRC Press, Boca Raton, Florida, 582 pp.

Ottenwalder, J. A., and H. H. Genoways. 1982. Systematic review of the Antillean bats of the *Natalus micropus*-complex (Chiroptera: Natalidae). Annals of Carnegie Museum, 51:17-38.

Ouahbi, Y., M. Aberkan, and F. Serre. 2001. Climatic effect on Late Pleistocene mammals from the northern Rif Mountains, Northern Morocco. Paleontological Journal, 35(6):641-646 (translated from Paleontologicheskii Zhurnal, 6, 2001, pp.78-83).

Owen, J. G. 1984. *Sorex fumeus*. Mammalian Species, 215:1-8.

Owen, J. G. 1989. Population and geographic variation of *Peromyscus leucopus* in relation to climatic factors. Journal of Mammalogy, 70:98-109.

Owen, J. G., and R. S. Hoffmann. 1983. *Sorex ornatus*. Mammalian Species, 212:1-5.

Owen, J. G., R. J. Baker, and J. K. Jones, Jr. 1990. First record of *Peromyscus gymnotis* (Muridae) from El Salvador, with second records for *Choeroniscus godmani* and *Diaemus youngii* (Phyllostomidae). Texas Journal of Science, 42:417-418.

Owen, J. G., R. J. Baker, and S. L. Williams. 1996. Karyotypic variation in spotted skunks (Carnivora: Mustelidae: *Spilogale*) from Texas, Mexico and El Salvador. Texas Journal of Science, 48:119-122.

Owen, R. D. 1987. Phylogenetic analyses of the bat subfamily Stenodermatinae (Mammalia: Chiroptera). Special Publications, The Museum, Texas Tech University, 26:1-65.

Owen, R. D. 1991. The systematic status of *Dermanura concolor* (Peters, 1865)(Chiroptera: Phyllostomidae), with description of a new genus. Bulletin of the American Museum of Natural History, 206:18-25.

Owen, R. D., and M. B. Qumsiyeh. 1987. The subspecies problem in the trident leaf-nosed bat, *Asellia tridens*: Homomorphism in widely separated populations. Zeitschrift für Säugetierkunde, 52(6):329-337.

Owen, R. D., R. K. Chesser, and D. C. Carter. 1990. The systematic status of *Tadarida brasiliensis cynocephala* and Antillean members of the *Tadarida brasiliensis* group, with comments on the generic name *Rhizomops* Legendre. Occasional Papers, The Museum, Texas Tech University, 133:1-18.

Owen-Ashley, N. T., and D. E. Wilson. 1998. *Micropteropus pusillus*. Mammalian Species, 577:1-5.

Özkan, B., and B. Kryštufek. 1999. Wood mice, *Apodemus* of two Turkish islands: Gökçeada and Bozcaada. Folia Zoologica, 48(1):17-24.

Pacheco, V., and B. D. Patterson. 1991. Phylogenetic relationships of the New World bat genus *Sturnira* (Chiroptera, Phyllostomidae). Bulletin of the American Museum of Natural History, 206:101-121.

Pacheco, V., and B. D. Patterson. 1992. Systematics and biogeographic analyses of four species of *Sturnira* (Chiroptera: Phyllostomidae), with emphasis on Peruvian forms. Memorías del Museo de Historia Natural, 21:57-81.

Pacheco, V., and J. L. Patton. 1995. A new species of the Puna mouse, genus *Punomys* Osgood, 1943 (Muridae, Sigmodontinae) from the southeastern Andes of Peru. Zeitschrift für Säugetierkunde, 60:85-96.

Pacheco, V., and E. Vivar. 1996. Annotated checklist of the nonflying mammals at Pakitza, Manu Reserve Zone, Manu National Park, Peru. Pp. 577-592, *in* Manu, the biodiversity of southeastern Peru (D. E. Wilson and A. Sandoval, eds.). Smithsonian Institution Press, Washington, D.C., 679 pp.

Pacheco, V., B. D. Patterson, J. L. Patton, L. H. Emmons, S. Solari, and C. F. Ascorra. 1993. List of mammal species known to occur in Manu Biosphere Reserve, Peru. Publicaciones del Museo de Historia Natural, Universidad Nacional Mayor de San Marcos, Serie A Zoologia, 44:1-12.

Pacheco, V., H. de Macedo, E. Vivar, C. Ascorra, R. Arana-Cardó, and S. Solari. 1995. Lista anotada de los mamíferos peruanos. Occasional Papers in Conservation Biology, Conservation International, 2:35 pp.

Packard, R. L. 1960. Speciation and evolution of the pygmy mice, genus *Baiomys*. University of Kansas Publications, Museum of Natural History, 9:579-670.

Packard, R. L. 1969. Taxonomic review of the golden mouse, *Ochrotomys nuttalli*. Miscellaneous Publications, Museum of Natural History, University of Kansas, 51:373-406.

Packard, R. L., and J. H. Bowers. 1970. Distributional notes on some foxes from western Texas and eastern New Mexico. Southwestern Naturalist, 14:450-451.

Packard, R. L., and J. B. Montgomery. 1978. *Baiomys musculus*. Mammalian Species, 102:1-3.

Paddle, R. 2000. The last Tasmanian tiger: The history and extinction of the thylacine. Cambridge University Press, Cambridge, U.K., 273 pp.

Padilla, M., and R. C. Dowler. 1994. *Tapirus terrestris*. Mammalian Species, 481:1-8.

Pagels, J. F., and C. O. Handley, Jr. 1989. Distribution of the southeastern shrew, *Sorex longirostris* Bachmann, in western Virginia. Brimleyana, 15:123-131.

Pagels, J. F., C. S. Jones, and C. O. Handley, Jr. 1982. Northern limits of the southeastern shrew, *Sorex longirostris* Bachmann (Insectivora: Soricidae), on the Atlantic coast of the United States. Brimleyana, 8:51-59.

Paglia, A. P., P. De Marco, F. M. Costa, R. F. Pereira, and G. Lessa. 1995. Heterogeneidade estrutural e diversidade de pequenos mamíferos em um fragmento de mata secundária de Minas Gerais. Revista Brasileira de Zoologia, 12:67-79.

Painter, J., C. Krajewski, and M. Westerman. 1995. Molecular phylogeny of the marsupial genus *Planigale* (Dasyuridae). Journal of Mammalogy, 76:406-413.

Palacios, F. 1983. On the taxonomic status of the genus *Lepus* in Spain. Acta Zoologica Fennica, 174:27-30.

Palacios, F. 1989. Biometric and morphologic features of the species of the genus *Lepus* in Spain. Mammalia, 53:227-264.

Palacios, F., and J. Fernández. 1992. A new subspecies of hare from Majorca (Balearic Islands). Mammalia, 56:71-85.

Palacios, F., J. F. Orueta, and G. G. Tapia. 1989. Taxonomic review of the *Lepus europaeus* group in Italy and Corsica. Abstract of papers and posters, Fifth International Theriological Congress, Rome, 1:189-190.

Pallas, P. S. 1767-1780. Spicilegia zoologica, quibus novae imprimus et obscurae animalium species iconibus, descriptionibus atque commentariis illustrantur cura P.S. Pallas. [fasc. 1 imprint 1767; fasc. 11-12 imprint 1777-78; fasc. 13 imprint 1779; fasc. 14 imprint 1780]. Berolini, prostant apud Gottl. August. Langed, 14 fasc in 2 volumes.

Pallas, P. S. 1771-1776. Reise durch verschiedene Provinzen des Russischen Reichs. St. Petersbourg, 3 vol.

Pallas, P. S. 1783. *Galeopithecus volans*, camelli. descriptus. Acta Academiae Scientarium Imperialis Petropolitanae, 1:208-222.

Pallas, P. S. 1811[1831]. Zoographia Rosso-Asiatica, sistens omnium Animalium in extenso Imperio Rossico et adjacentibus maribus observatorum recensionem, domicillia, mores et descriptiones, anatomen atque icones plurimorum. [ICZN Opinion 212—dates of volumes: 1 & 2:1811; 3:1814]. Petropoli, in officina Caes. acadamiae scientiarum. 3 vol. 1:1-568.

Palma, R. E. 1997. *Thylamys elegans*. Mammalian Species, 572:1-4.

Palmeirim, J. M. 1991. A morphometric assessment of the systematic position of the *Nyctalus* from Azores and Madeira (Mammalia: Chiroptera). Mammalia, 55:381-388.

Palmeirim, J. M., and R. S. Hoffmann. 1983. *Galemys pyrenaicus*. Mammalian Species, 207:1-5.

Palmer, R. A. 2001. Southern range extension for the delicate mouse (*Pseudomys delicatulus*). Memoirs of the Queensland Museum, 46(2):460.

Palmer, T. S. 1904. Index generum mammalium: A list of the genera and families of mammals. North American Fauna, 23:1-984.

Palmer, T. S. 1928. An earlier name for the genus *Evotomys*. Proceedings of the Biological Society of Washington, 41:87-88.

Palomo, L. J. 1988. Etude descriptive des poils de *Mus spretus* Lataste, 1883. Revue Suisse de Zoologie, 95:505-512.

Palomo, L. J., M. Espana, J. Lopez-Fuster, J. Gosalbez, and V. Sans-Coma. 1983. Sobre la variabilidad genetica y morfometrica de *Mus spretus* Lataste, 1883 en la Penninsula Iberica. Miscellania Zoologica, 7:171-192.

Pamukoglu, N., and I. Albayrak. 1996. The rodents of Kastamonu Province (Mammalia: Rodentia). Communications de la Faculté des Sciences de Université d'Ankara, Série C, Biologie, 14:1-22.

Pankow, H. 1989. Neues vom Siebenschläfer in Mecklenburg. Archiv des Vereins der Freunde der Naturgeschichte, Mecklenburg, 29:73-74.

Panteleyev, P. A. 1996. [On intraspecific systematics and taxonomic importance of exterior and craniometric characters in vole rat *Arvicola terrestris* subspecies (Rodentia, Cricetidae).] Vestnik Zoologii, 1996(3):21-25.

Panteleyev, P. A. 1998. [The rodents of the Palaearctic fauna: Composition and areas]. Moscow: Russian Academy of Sciences, A. N. Severtzov Institute of Ecology and Evolution, 116 pp. (in Russian with partial English translations).

Panteleyev, P. A. (ed.). 2000. [Species of the fauna of Russia and contiguous countries. The water vole: Mode of the species.] Nauka Publishers, Moscow, 527 pp (in Russian with English table of contents).

Panzironi, C., G. Cerone, M. Cristaldi, and G. Amori. 1994. A method for the morphometric identification of southern Italian populations of *Apodemus* (*Sylvaemus*). Hystrix, n.s., 5(1-2):1-16.

Paolucci, P., A. Battisti, and R. de Battisti. 1987. The forest dormouse (*Dryomys nitedula* Pallas, 1779) in the eastern Alps (Rodentia, Gliridae). Biogeographia, 13:855-866.

Paolucci, P., M. Bon, R. De Battisti, F. Mezzavilla, E. Vernier. 1993. Distribuzione di alcuni mammiferi in Veneto. Supplemento alle Ricerchedi Biologia della Selvaggina, 31:415-424.

Papillon, Y., A. Butet, G. Paillat and N. Milan Pena. 2000. Insectivore et rongeurs de France: Le muscardin – *Muscardinus avellanarius* (Linne, 1758). Arvicola, 12(2):39-51.

Paradiso, J. L. 1967. A review of the wrinkle-faced bats (*Centurio senex* Gray), with description of a new subspecies. Mammalia, 31:595-604.

Paradiso, J. L. 1968. Canids recently collected in east Texas, with comments on the taxonomy of the red wolf. American Midland Naturalist, 80:529-534.

Paradiso, J. L., and R. H. Manville. 1961. Taxonomic notes on the tundra vole (*Microtus oeconomus*) in Alaska. Proceedings of the Biological Society of Washington, 74:77-92.

Paradiso, J. L., and R. M. Nowak. 1971 [1972]. A report on the taxonomic status and distribution of the red wolf. United States Fish and Wildlife Service, Special Scientific Report—Wildlife, 145:1-36.

Paradiso, J. L., and R. M. Nowak. 1972. *Canis rufus*. Mammalian Species, 22:1-4.

Pardiñas, U. F. J. 1993. El registro mas antiguo (Pleistoceno temprano a medio) de *Akodon azarae* (Fischer, 1829) (Mammalia, Rodentia, Cricetidae) en la Provincia de Buenos Aires, Argentina. Ameghiniana, 30:149-153.

Pardiñas, U. F. J. 1995. Sobre las vicisitudes de los generos *Bothriomys* Ameghino, 1889, *Euneomys* Coues, 1874, y *Graomys* Thomas, 1916 (Mammalia, Rodentia, Cricetidae). Ameghiniana, 32:173-180.

Pardiñas, U. F. J. 1996. El registro fósil de *Bibimys* Massoia, 1979 (Rodentia) en la Argentina. Consideraciones sobre los Scapteromyini (Cricetidae, Sigmodontinae) y su distribuición durante el Plioceno-Holoceno en la region Pampeana. Mastozoología Neotropical, 3:15-38.

Pardiñas, U. F. J. 1997. Un nuevo sigmodontino (Mammalia: Rodentia) del Plioceno de Argentina y consideraciones sobre el registro fósil de los Phyllotini. Revista Chilena de Historia Natural, 70:543-555.

Pardiñas, U. F. J. 1999. Fossil murids: Taxonomy, paleoecology, and paleoenvironments. Quaternary of South America and Antarctic Peninsula, 12:225-254.

Pardiñas, U. F. J. 2000*a*. Los sigmodontinos (Mammalia, Rodentia) de la Coleccion Ameghino (Museo Argentino de Ciencias Naturales "Bernardino Rivadavia"): Revision taxonomica. Revista del Museo de La Plata, n.s., Paleontologia, 61:247-254.

Pardiñas, U. F. J. 2000*b*. Los roedores muroideos del Pleistoceno Tardío-Holocene en la región Pampeana (sector este) y Patagonia (República Argentina): Aspectos taxonómicos, importancia bioestratigráfica y sigificación paleoambiental. Mastozoología Neotropical, 7:57-61.

Pardiñas, U. F. J., and C. A. Galliari. 1998*a*. La distribución del ratón toto *Notiomys edwardsii* (Mammalia: Muridae). Neotropica, 44:123-124.

Pardiñas, U. F. J., and C. A. Galliari. 1998*b*. Comentario sobre el trabajo "Los mamíferos del Parque Biológico Sierra de San Javier, Tucumán, Argentina: Observaciones sobre su sistemática y distribución" Capllonch et al., 1997 (Mastozoología Neotropical, 4:49-71). Mastozoología Neotropical, 5:61-62.

Pardiñas, U. F. J., and C. A. Galliari. 1999. La presencia de *Akodon iniscatus* (Mammalia: Rodentia) en la provincia de Buenos Aires (Argentina). Neotropica, 45:115-117.

Pardiñas, U. F. J., and C. A. Galliari. 2001. *Reithrodon auritus*. Mammalian Species, 664:1-8.

Pardiñas, U. F. J., and P. E. Ortiz. 2001. *Neotomys ebriosus*, an enigmatic South American rodent (Muridae, Sigmodontinae): Its fossil record and present distribution in Argentina. Mammalia, 65:244-250.

Pardiñas, U. F. J., and E. P. Tonni. 1998. Procedencia estratigráfica y edad de los más antiguous muroideos (Mammalia, Rodentia) de América del Sur. Ameghiniana, 35:473-475.

Pardiñas, U. F. J., G. D'Elía, and P. E. Ortiz. 2002. Sigmodontinos fósiles (Rodentia, Muroidea, Sigmodontinae) de América del Sur: Estado actual de su conocimiento y prospectiva. Mastozoologia Neotropical, 9:209-252.

Pardiñas, U. F. J., G. D'Elía, and S. Cirignoli. 2003*a*. The genus *Akodon* (Muroidea: Sigmodontinae) in Misiones, Argentina. Mammalian Biology/Zeitschrift für Säugetierkunde, 68:129-143.

Pardiñas, U. F. J., P. Teta, S. Cirignoli, and D. H. Podesta. 2003*b*. Micromamiferos (Didelphimorphia Y Rodentia) de NorPatagonia extra Andina, Argentina: Taxonomia alfa y biogeografia. Mastozoologia Neotropical/Journal of Neotropical Mammalogy, 10:69-113.

Park, N. S., S. S. Lee, and H. S. Koh. 1990. Systematic studies on Korean rodents: VII. Immunological analyses of serum proteins of seven species. Korean Journal of Systematic Zoology, 6:165-172.

Parkinson, A. 1979. Morphologic variation and hybridization in *Myotis yumanensis sociabilis* and *Myotis lucifugus carissima*. Journal of Mammalogy, 60:489-504.

Parnaby, H. E. 1987. Distribution and taxonomy of the long-eared bats, *Nyctophilus gouldi* Tomes, 1858, and *Nyctophilus bifax* Thomas, 1915 (Chiroptera: Vespertilionidae) in eastern Australia. Proceedings of the Linnean Society of New South Wales, 109:153-174.

Parnaby, H. E. 2002a. A new species of long-eared bat (*Nyctophilus*: Vespertilionidae) from New Caledonia. Australian Mammalogy, 23:12-124.

Parnaby, H. E. 2002b. A taxonomic review of the genus *Pteralopex* (Chiroptera: Pteropodidae), the monkey-faced bats of the south-western Pacific. Australian Mammalogy, 23:145-162.

Parsons, J. 1745. An account of a quadruped brought from Bengal and now to be seen in London. Philosophical Transactions of the Royal Society, 43:465-467.

Parsons, F. G. 1894. On the myology of the sciuromorphine and hystricomorphine rodents. Proceedings of the Zoological Society of London, 1894:251-296.

Pascale, E., E. Valle, and A. V. Furano. 1990. Amplification of an ancestral mammalian L1 family of long interspersed repeated DNA occurred just before the murine radiation. Proceedings of the National Academy of Sciences, Washington, D.C., 87:9481-9485.

Pasha, M. K. S., and I. Suhail. 1997. Range extension of the Kashmir flying squirrel (*Hylopetes fimbriatus* Gray). Journal of the Bombay Natural History Society, 94(2):395-396.

Pasichnyk, S. V. 1992. [Morpho-functional analysis of the maxillary organs in mole rats (Mammalia, Spalacidae).] Vestnik Zoologii, 5(4):68-72 (in Russian with English summary).

Pasitschniak-Arts, M. 1993. *Ursus arctos*. Mammalian Species, 439:1-10.

Pasitschniak-Arts, M., and S. Larivière. 1995. *Gulo gulo*. Mammalian Species, 499:1-10.

Passarge, H. 1984. *Sorex isodon marchicus* ssp. nova in Mitteleuropa. Zeitschrift für Säugetierkunde, 49:278-284.

Pasteur, N., J. Worms, M. Tohari, and D. Iskandar. 1982. Genetic differentiation in Indonesian and French rats of the subgenus *Rattus*. Biochemical Systematics and Ecology, 10:191-196.

Patnaik, R. 1995. Narmada Valley microvertebrates: Systematics, taphonomy and palaeoecology. Man and Environment, 20(2):75-90.

Patnaik, R. 1997. New murids and gerbillids (Rodentia, Mammalia) from Pliocene Siwalik sediments of India. Palaeovertebrata, 26(1-4):129-165.

Patnaik, R. 2001. Late Pliocene micromammals from Tatrot Formation (Upper Siwaliks) exposed near village Saketi, Himachal Pradesh, India. Palaeontographica, Abteilung A, Palaozoologie-Stratigraphie, 261(1-3):55-81.

Patnaik, R. 2002. Enamel microstructure of some fossil and extant murid rodents of India. Paleontological Research, 6(3):239-258.

Patnaik, R., M. Bahadur, T. Sharma, and A. Sahni. 1993. A comparative analysis of the molars of *Mus booduga, Mus dunni* and fossil *Mus* of the Indian subcontinent: Phylogenetic and palaeobiogeographic implications. Current Science, 65(10):782-786.

Patnaik, R., J.-C. Auffray, J.-J. Jaeger, and A. Sahni. 1996. House mouse ancestor from late Pliocene Siwalik sediments of India. Comptes Rendus des Séances de l'Académie des Sciences (Paris), Sciences de la vie/Life sciences, 319:431-424.

Patterson, B. 1962. An extinct solenodontid insectivore from Hispaniola. Breviora, 165:1-11.

Patterson, B. 1978. Pholidota and Tubulidentata. Pp. 268-278, in Evolution of African mammals (V. J. Maglio and H. B. S. Cooke, eds.). Harvard University Press, Cambridge, MA, 641 pp.

Patterson, B., and R. Pascual. 1968a. The fossil mammal fauna of South America, V. Evolution of mammals on southern continents. Quarterly Review of Biology, 43:409-451.

Patterson, B., and R. Pascual. 1968b. New echimyid rodents from the Oligocene of Patagonia, and a synopsis of the family. Breviora, Harvard Museum of Comparative Zoology, 301:1-14.

Patterson, B., and A. E. Wood. 1982. Rodents from the Deseadan Oligocene of Bolivia and the relationships of the Caviomorpha. Bulletin Museum of Comparative Zoology, 149:371-543.

Patterson, B. D. 1980. A new subspecies of *Eutamias quadrivittatus* from the Organ Mountains, New Mexico. Journal of Mammalogy, 61:455-464.

Patterson, B. D. 1984. Geographic variation and taxonomy of Colorado and Hopi chipmunks (genus *Eutamias*). Journal of Mammalogy, 65:442-456.

Patterson, B. D. 1992. Mammals in the Royal Natural History Museum, Stockholm, collected in Brazil and Bolivia by A. M. Olalla during 1934-1938. Fieldiana: Zoology, n.s., 66:1-48.

Patterson, B. D. 1992. A new genus and species of long-clawed mouse (Rodentia: Muridae) from temperate rainforests of Chile. Zoological Journal of the Linnean Society, 106:127-145.

Patterson, B. D., and M. H. Gallardo. 1987. *Rhyncholestes raphanurus*. Mammalian Species, 286:1-5.

Patterson, B. D., and L. R. Heaney. 1987. Preliminary analysis of geographic variation in red-tailed chipmunks (*Eutamias ruficaudus*). Journal of Mammalogy, 68:782-791.

Patterson, B. D., M. H. Gallardo, and K. E. Freas. 1984. Systematics of mice of the subgenus *Akodon* (Rodentia: Cricetidae) in southern South America, with the description of a new species. Fieldiana: Zoology, n.s., 23:1-16.

Pattie, D. 1973. *Sorex bendirii*. Mammalian Species, 27:1-2.

Patton, J. C., R. J. Baker, and J. C. Avise. 1981. Phenetic and cladistic analyses of biochemical evolution in peromyscine rodents. Pp. 288-308, *in* Mammalian population genetics (M. H. Smith and J. Joule, eds.). University of Georgia Press, Athens, 380 pp.

Patton, J. L. 1967*a*. Chromosome studies of certain pocket mice, genus *Perognathus* (Rodentia: Heteromyidae). Journal of Mammalogy, 48:27-37.

Patton, J. L. 1967*b*. Chromosomes and evolutionary trends in the pocket mouse subgenus *Perognathus* (Rodentia: Heteromyidae). Southwestern Naturalist, 12:429-438.

Patton, J. L. 1969*a*. Karyotypic variation in the pocket mouse, *Perognathus penicillatus* Woodhouse (Rodentia-Heteromyidae). Caryologia, 22:351-358.

Patton, J. L. 1969*b*. Chromosome evolution in the pocket mouse, *Perognathus goldmani* Osgood. Evolution, 23:645-662.

Patton, J. L. 1970. Karyotypes of five species of pocket mice, *Perognathus* (Rodentia; Heteromyidae), and a summary of chromosomal data for the genus. Mammalian Chromosome Newsletter, 11:3-8.

Patton, J. L. 1973. An analysis of natural hybridization between the pocket gophers *Thomomys bottae* and *Thomomys umbrinus*, in Arizona. Journal of Mammalogy, 54:561-584.

Patton, J. L. 1984. Systematic status of the large squirrels (subgenus *Urosciurus*) of the western Amazon basin. Studies on Neotropical Fauna and Environment, 19:53-72.

Patton, J. L. 1987. Species groups of spiny rats, genus *Proechimys* (Rodentia: Echimyidae). Fieldiana: Zoology, n.s., 39:305-345.

Patton, J. L. 1990. Geomyoid evolution: The historical, selective, and random basis for divergence patterns within and among species. Pp. 49-69, *in* Evolution of subterranean mammals at the organismal and molecular levels (E. Nevo and O. A. Reig, eds.). Alan R. Liss, Inc., New York, 422 pp.

Patton, J. L. 1993*a*. Family Geomyidae. Pp. 469-476, *in* Mammal species of the world, a taxonomic and geographic reference, Second ed. (D. E. Wilson and D. M. Reeder, eds.). Smithsonian Institution Press, Washington, D.C., xviii + 1207 pp.

Patton, J. L. 1993*b*. Family Heteromyidae. Pp. 477-486, *in* Mammal species of the world, a taxonomic and geographic reference, Second ed. (D. E. Wilson and D. M. Reeder, eds.). Smithsonian Institution Press, Washington, D.C., xviii + 1207 pp.

Patton, J. L., and S. T. Alvarez-Castañeda. 1999. Family Heteromyidae. Pp. 351-442, *in* Mamíferos del noroeste de México (S. T. Alvarez-Castañeda and J. L. Patton, eds.). Centro de Investigaciones Biológical del Noroeste, La Paz, México, 211 pp.

Patton. J. L., and S. T. Alvarez-Castañeda. 2005. Phylogeography of the desert woodrat, *Neotoma lepida*, with comments on systematics and biogeographic history. Pp. 375-388, *in* Contribuciones mastozoologicas en honor de Bernardo Villa (V. Sánchez-Cordero and R Medellín, eds.). Instituto de Biología, Universidad Nacional Autónoma de México, Mexico, D.F., 500 p.

Patton, J. L., and L. P. Costa. 2003. Molecular phylogeography and species limits in rainforest didelphid marsupials of South America. Pp. 63-81, *in* Predators with pouches: The biology of carnivorous marsupials (M. E. Jones, C. R. Dickman, and M. Archer, eds.). CSIRO Press, Melbourne, 486 pp.

Patton, J. L., and R. E. Dingman. 1968. Chromosome studies of pocket gophers, genus *Thomomys*. I. The specific status of *Thomomys umbrinus* (Richardson) in Arizona. Journal of Mammalogy, 49:1-13.

Patton, J. L., and L. Emmons. 1985. A review of the genus *Isothrix* (Rodentia, Echimyidae). American Museum Novitates, 2817:1-14.

Patton, J. L., and A. L. Gardner. 1972. Notes on the systematics of *Proechimys* (Rodentia: Echimyidae), with emphasis on Peruvian forms. Occasional Papers of the Museum of Zoology, Louisiana State University, 44:1-30.

Patton, J. L., and M. S. Hafner. 1983. Biosystematics of the native rodents of the Galapagos Archipelago, Ecuador. Pp. 539-568, *in* Patterns of evolution in Galapagos organisms (R. I. Bowman, M. Benson, and A. E. Leviton, eds.). American Association for the Advancement of Science, Pacific Division, San Francisco, CA, 568 pp.

Patton, J. L., and T. C. Hsu. 1967. Chromosomes of the golden mouse, *Peromyscus* (*Ochrotomys*) *nuttalli* (Harlan). Journal of Mammalogy, 48:637-639.

Patton, J. L., and O. A. Reig. 1989. Genetic differentiation among echimyid rodents, with an emphasis on spiny rats, genus *Proechimys*. Pp. 75-96, *in* Advances in Neotropical mammalogy (K. H. Redford and J. F. Eisenberg, eds.). Sandhill Crane Press, Gainesville, FL, 614 pp.

Patton, J. L., and D. S. Rogers. 1993*a* Cytogenetics. Pp. 236-258, *in* Biology of the Heteromyidae (H. H. Genoways and J. H. Brown, eds.). Special Publication, The American Society of Mammalogists, 10:1-719.

Patton, J. L., and D. S. Rogers. 1993*b*. Biochemical Genetics. Pp. 259-269, *in* Biology of the Heteromyidae (H. H. Genoways and J. H. Brown, eds.). Special Publication, The American Society of Mammalogists, 10:1-719.

Patton, J. L., and M. N. F. da Silva. 1995. A review of the spiny mouse genus *Scolomys* (Rodentia: Muridae:

Sigmodontinae) with the description of a new species from the western Amazon of Brazil. Proceedings of the Biological Society of Washington, 108:319-337.

Patton, J. L., and M. N. F. da Silva, 1997. Definition of species of pouched four-eyed opossums (Didelphidae, *Philander*). Journal of Mammalogy, 78:90-102.

Patton, J. L., and M. F. Smith. 1981. Molecular evolution in *Thomomys*: Phyletic systematics, paraphyly, and rates of evolution. Journal of Mammalogy, 62:493-500.

Patton, J. L., and M. F. Smith. 1989. Genetic structure and the genetic and morphologic divergence among pocket gopher species (genus *Thomomys*). Pp. 284-304, *in* Speciation and Its Consequences (D. Otte and J. A. Endler, eds.). Sinauer Associates Incorporated, Sunderland, MA, 679 pp.

Patton, J. L., and M. F. Smith. 1990. The evolutionary dynamics of the pocket gopher *Thomomys bottae*, with emphasis on California populations. University of California Publications in Zoology, 123:1-161.

Patton, J. L., and M. F. Smith. 1992. Evolution and systematics of akodontine rodents (Muridae: Sigmodontinae) of Peru, with emphasis on the genus *Akodon*. Memorias del Museo de Historia Natural, Universidad Nacional Mayor de San Marcos (Lima), 21:83-103.

Patton, J. L., and M. F. Smith. 1993. Molecular evidence for mating asymmetry and female choice in a pocket gopher (*Thomomys*) hybrid zone. Molecular Ecology, 2:3-8.

Patton, J. L., and M. F. Smith. 1994. Paraphyly, polyphyly, and the nature of species boundaries in pocket gophers (genus *Thomomys*). Systematic Zoology, 43:11-26.

Patton, J. L., S. Y. Yang, and P. Myers. 1975. Genetic and morphologic divergence among introduced rat populations (*Rattus rattus*) of the Galapagos Archipelago, Ecuador. Systematic Zoology, 24:296-310.

Patton, J. L., H. MacArthur, and S. Y. Yang. 1976. Systematic relationships of the four-toed populations of *Dipodomys heermanni*. Journal of Mammalogy, 57:159-163.

Patton, J. L., S. W. Sherwood, and S. Y. Yang. 1981. Biochemical systematics of chaetodipine pocket mice, genus *Perognathus*. Journal of Mammalogy, 62:477-492.

Patton, J. L., M. F. Smith, R. D. Price, and R. A. Hellenthal. 1984. Genetics of hybridization between the pocket gophers *Thomomys bottae* and *Thomomys townsendii* in northeastern California. Great Basin Naturalist, 44:431-440.

Patton, J. L., P. Myers, and M. F. Smith. 1989. Electromorphic variation in selected South American akodontine rodents (Muridae: Sigmodontinae), with comments on systematic implications. Zeitschrift für Säugetierkunde, 54:347-359.

Patton, J. L., P. Myers, and M. F. Smith. 1990. Vicariant versus gradient models of diversification: The small mammal fauna of eastern Andean slopes of Peru. Pp. 355-371, *in* Vertebrates in the tropics: Proceedings of the international symposium on vertebrate biogeography and systematics in the tropics, Bonn, June 5-8, 1989 (G. Peters and R. Hutterer, eds.). Museum Alexander Koenig Zoological Research Institute and Zoological Museum, Bonn, 424 pp.

Patton, J. L., M. N. F. da Silva, and J. R. Malcolm. 2000. Mammals of the Rio Juruá and the evolutionary and ecological diversification of Amazonia. Bulletin of the American Museum of Natural History, 244:1-306.

Paula Couto, C. 1950. Footnote number 249. P. 232, *in* Memórias sobre a Paleontologia Brasileira (P. W. Lund with notes and comments by C. Paula Couto). Instituto Nacional do Livro, Rio de Janeiro, 589 pp., 56 pls.

Paulson, D. D. 1988*a*. *Chaetodipus baileyi*. Mammalian Species, 297:1-5.

Paulson, D. D. 1988*b*. *Chaetodipus hispidus*. Mammalian Species, 320:1-4.

Pavlinin, V. N. 1966. Der Zobel. *Martes zibellina* L. Wittenberg Lutherstadt, Ziemsen, 102 pp.

Pavlinov, I. Ya. 1980*a*. [Evolution and taxonomic significance of the morphology of the osseous middle ear in Gerbillinae (Rodentia: Cricetidae).] Byulleten' Moskovskovo Obshchestva Ispitatelei Prirody, Otdel Biologicheskii, 85:20-33 (in Russian).

Pavlinov, I. Ya. 1980*b*. Nadvidovye gruppirovki v podsemeistve Cardiocraniinae Satunin (Mammalia, Dipodidae) [Superspecies groupings in the subfamily Cardiocraniinae Satunin (Mammalia, Dipodidae)]. Vestnik Zoologii, 2:47-51 (in Russian).

Pavlinov, I. Ya. 1980*c*. [Taxonomic status of *Calomyscus* Thomas (Rodentia, Cricetidae) on the basis of structure of auditory ossicles.] Zoologicheskii Zhurnal, 59:312-316 (in Russian).

Pavlinov, I. Ya. 1981*a*. [Taxonomic status of gerbils of the genus *Ammodillus* Thomas, 1904 (Rodentia, Gerbillinae).] Zoologicheskii Zhurnal, 60:472-474 (in Russian).

Pavlinov, I. Ya. 1981*b*. [Subgeneric taxonomy of gerbils of the genus *Tatera* Lataste, 1882 based on ossean middle ear morphology.] Vestnik Zoologii, 1981(2):10-14 (in Russian with English summary).

Pavlinov, I. Ya. 1982*a*. [Phylogeny and classification of the subfamily Gerbillinae.] Byulleten' Moskovskovo Obshchestva Ispitatelei Prirody, Otdel Biologicheskii, 87:19-31 (in Russian).

Pavlinov, I. Ya. 1982*b*. [Molar morphology in *Rhombomys opimus*, with notes on taxonomy of the *Rhombomys-Pliorhombomys* group (Rodentia, Gerbillinae).] Vestnik Zoologii, 1982(3):53-57 (in Russian with English summary).

Pavlinov, I. Ya. 1982c. [Species names of jirds of the group *libycus-erythrourus-shawi-caudatus* (Rodentia, Gerbillinae, *Meriones*).] Zoologischeskii Zhurnal, 61(11):1766-1768 (in Russian with English summary).

Pavlinov, I. Ya. 1984a. [Evolution of auditory ossicles in voles, subfamily Microtinae.] Sbornik Trudov Zoologicheskovo Muzeya MGU, 22:191-212 (in Russian).

Pavlinov, I. Ya. 1984b. [Evolution of dental crown pattern in Gerbillinae.] Sbornik trudov Zoologicheskogo muzeia, 22:93-134 (in Russian with English summary).

Pavlinov, I. Ya. 1985. [Contributions to dental morphology and phylogeny of gerbils (Rodentia, Gerbillinae).] Zoologicheskii Zhurnal, 64:574-582 (in Russian).

Pavlinov, I. Ya. 1986. [Taxonomic significance of the male genital morphology in the subfamily Gerbillinae (Mammalia: Rodentia).] Byulleten' Moskovskogo Obshchestva Ispytatelei Prirody Otdel Biologicheskii, 91(1):8-16 (in Russian with English summary).

Pavlinov, I. Ya. 1987. [Cladistic analysis of the gerbilline tribe Taterillini (Rodentia, Gerbillinae) and some questions on the method of numerical cladistic analysis.] Zoologicheskii Zhurnal, 66:903-913 (in Russian).

Pavlinov, I. Ya. 1988. [Evolution of mastoid part of the bulla tympani in specialized desert rodents.] Zoologicheskii Zhurnal, 67(5):739-750 (in Russian).

Pavlinov, I. Ya. 1996. [New records of *Rhombomys opimus* (Mammalia, Gerbillidae) with rooted molars from Iran.] Vestnik Zoologii, 1996(6):18.

Pavlinov, I. Ya. 1997. [Systematics and distribution of *Tatera phillipsi* (Mammalia, Gerbillidae).] Zoologischeskii Zhurnal, 76(4):506-508 (in Russian with English summary).

Pavlinov, I. Ya. 2000. [The contribution to craniometric variation and taxonomy of jirds from the group "*shawi-grandis*" of the genus *Meriones* (Gerbilidae).] Zoologischeskii Zhurnal, 79(2):201-209 (in Russian with English abstract).

Pavlinov, I. Ya. 2001a. Current concepts of gerbillid phylogeny and classification. Pp. 141-149, *in* African small mammals (C. Denys, L. Granjon, and A. Poulet, eds.). IRD Éditions, Collection Colloques et Séminaires, Paris, 570 pp.

Pavlinov, I. Ya. 2001b. Geometric morphometrics of glirid dental crown patterns. Trakya University Journal of Scientific Research B, 2(2):151-157.

Pavlinov, I. Ya., and E. G. Potapova. 2003. Cladistic analysis of the dormouse genus *Graphiurus* Smuts, 1832 (Rodentia: Gliridae), with comments on evolution of its zygomasseteric construction and subgeneric taxonomy. Russian Journal of Theriology, 2(1):49-58.

Pavlinov, I. Ya., and K. A. Rogovin. 2000. [Relation between size of pinna and auditory bulla in specialized desert rodents.] Zhurnal Obshchei Biologii, 61(1):87-101 (in Russian with English abstract).

Pavlinov, I. Ya., and O. L. Rossolimo. 1987. Sistematika mlekopitayushchikh SSSR [Systematics of the mammals of the USSR.]. Moscow University Press, Moscow, 282 pp. (in Russian).

Pavlinov, I. Ya., and O. L. Rossolimo. 1998. [Systematics of mammals of the USSR. Addenda. M.] Archives of the Zoological Museum, Moscow State University, 38:190 pp. (in Russian).

Pavlinov, I. Ya., and G. I. Shenbrot. 1983. [Male genital structure and supraspecific taxonomy of Dipodidae.] Trudy Zoologicheskovo Instituta, Akademiya Nauk SSSR, Leningrad, 119:67-88 (in Russian).

Pavlinov, I. Ya., and G. I. Shenbrot. 1985. Materiali po sistematikye tushkanchikov Iranskovo Nagorya [Materials on the taxonomy of jerboas of the Iranian highland]. Pp. 55-57, *in* Tushkanchiki fauni SSSR [Jerboas of the USSR fauna] (V. E. Sokolov and V. V. Kucheruk, eds.). Tyezisi Dokladov Vsesoyuznovo soveshchaniya [Abstracts of the All-Union Conference], Alma-Ata, All-Union Mammal Society, Moscow, 246 pp. (in Russian).

Pavlinov, I. Ya., Yu. A. Dubrovsky, O. L. Rossolimo, and E. G. Potapova. 1990. [Gerbils of the world.] Nauka, Moscow, 368 pp. (in Russian).

Pavlinov, I. Ya., O. V. Voltzit, and O. L. Rossolimo. 1994. [Analysis of variation of the shape by means of "geometrical morphometrics": demonstration of some possibilities exemplified by tick gnathosoma (Acari: *Ixodes*) and vole molar (Mammalia: *Alticola*).] Zhurnal Obshchei Biologii, 55(1):110-118 (in Russian).

Pavlinov, I. Ya, E. L. Yakhontov, and A. K. Agadzhanyan. 1995a. [Mammals of Eurasia. I. Rodentia. Taxonomic and geographic guide.] Archives of the Zoological Museum, Moscow State University, 32:289 pp. (in Russian).

Pavlinov, I. Ya, A. V. Borisenko, S. V. Kruskop, and E. L. Yakhontov. 1995b. Mlekopitayushchie Evrazii [Mammals of Eurasia]. II. Non-Rodentia. Moscow University, Moscow, 336 pp.

Payne, J., C. M. Francis, and K. Phillipps. 1985. A field guide to the mammals of Borneo. The Sabah Society, Kuala Lumpur, 332 pp.

Payne, S. 1968. The origin of domestic sheep and goats: A reconsideration in the light of the fossil evidence. Proceedings of the Prehistoric Society, 34:368-384.

Pearch, M. J., P. J. J. Bates, and C. Magin. 2001. A review of the small mammal fauna of Djibouti and the results of a recent survey. Mammalia, 65:387-409.

Pearson, O. P. 1951. Mammals in the highlands of southern Peru. Bulletin of the Museum of Comparative Zoology, 106:117-174.

Pearson, O. P. 1958. A taxonomic revision of the rodent genus *Phyllotis*. University of California Publications in Zoology, 56:391-496.

Pearson, O. P. 1972. New information on ranges and relationships within the rodent genus *Phyllotis* in Peru and Ecuador. Journal of Mammalogy, 53:677-686.

Pearson, O. P. 1984. Taxonomy and natural history of some fossorial rodents of Patagonia, southern Argentina. Journal of Zoology, London, 202:225-237.

Pearson, O. P. 1995. Annotated keys for identifying small mammals living in or near Nahuel Huapi National Park or Lanin National Park, southern Argentina. Mastozoología Neotropical, 2:99-148.

Pearson, O. P., and M. I. Christie. 1991. Sympatric species of *Euneomys* (Rodentia, Cricetidae). Studies on Neotropical Fauna and Environment, 26:121-127.

Pearson, O. P., and H. A. Lagiglia. 1992. "Fuerta de San Rafael": Una localidad tipo ilusoria. Revista del Museo de Historia Natural de San Rafael (Mendoza), 12:35-43.

Pearson, O. P., and J. L. Patton. 1976. Relationships among South American phyllotine rodents based on chromosome analysis. Journal of Mammalogy, 57:339-350.

Pearson, O. P., and A. K. Pearson. 1989. Reproduction in bats in southern Argentina. Pp. 549-566, *in* Advances in Neotropical Mammalogy (K. H. Redford and J. F. Eisenberg (eds.). Sandhill Crane Press, Gainesville, Florida, 554 pp.

Pearson, O. P., and M. F. Smith. 1999. Genetic similarity between *Akodon olivaceus* and *Akodon xanthorhinus* (Rodentia: Muridae) in Argentina. Journal of Zoology, London, 247:43-52.

Pearson, S. 1999. Late Holocene biological records from the middens of stick-nest rats in the central Australian arid zone. Quaternary International, 59:39-46.

Pearson, S., and J. R. Dodson. 1993. Stick-nest rat middens as sources of paleoecological data in Australian deserts. Quaternary Research, 39:347-354.

Pearson, S. G., A. Baynes, and B. E. Triggs. 2001. The record of fauna, and accumulating agents of hair and bone, found in middens of stick-nest rats (Genus *Leporillus*) (Rodentia: Muridae). Wildlife Research, 28:435-444.

Pearson, T. 1981. Geographic and intraspecific cranial variation in North American arctic ground squirrels. Unpubl. M. A. thesis, University of Kansas, Lawrence, 96 pp.

Pechev, T., V. Anguelova, V., and T. Dinev. 1964. Etudes sur la taxonomie du *Myomimus personatus* (Ognev, 1924) (Rodentia) en Bulgarie. Mammalia, 28(3):419-428.

Pecon Slattery, J., and S. J. O'Brien. 1995. Molecular Phylogeny of the red panda (*Ailurus fulgens*). Journal of Heredity, 86:413-422.

Peddie, D. A. 1975. A taxonomic and autoecological study of the genus *Pronolagus* in southern Africa. Unpubl. M. S. thesis, University of Rhodesia, Salisbury.

Pedersen, S. C., H. H. Genoways, and P. W. Freeman. 1996. Notes on bats from Montserrat (Lesser Antilles) with comments concerning the effects of Hurricane Hugo. Caribbean Journal of Science, 32:206-213.

Pei, W. C. 1936. On the mammalian remains from Locality 3 at Choukoutien. Palaeontologia Sinica, Series C, 7(5):124 pp.

Peirce, E. J., H. D. M. Moore, C. M. Leigh, and W. G. Breed. 2003. Studies on sperm storage in the vas deferens of the spinifex hopping mouse (*Notomys alexis*). Reproduction, 125:233-240.

Peirce, K. N., and J. M. Peirce. 2000. Range extensions for the Alaska tiny shrew and pygmy shrew in southwestern Alaska. Northwestern Naturalist, 81:67-68.

Pelz, H.-J., H. Gemmeke, R. Hutterer, and U. Jüdes. 1996. Jugendentwicklung der brandmaus *Apodemus agrarius* (Mammalia: Muridae), im vergleich zu anderen arten der gattung. Bonner Zoologische Beiträge, 46:233-247.

Pelzers, E. 1992. Woelrat, *Arvicola terrestris* (L., 1758). Pp. 246-249, *in* Atlas van de Nederlandse Zoogdieren (S. Broekhuizen, B. Hoekstra, V. van Laar, C. Smeenk, and J. B. M. Thissen, eds.). Stichting Uitgeverij Koninklijke Nederlandse Natuurhistorische Vereniging, Utrecht, 336 pp.

Pelzers, E., and A. J. W. Lenders. 1992. Hamster, *Cricetus cricetus* (L., 1758). Pp. 235-239, *in* Atlas van de Nederlandse Zoogdieren (S. Broekhuizen, B. Hoekstra, V. van Laar, C. Smeenk, and J. B. M. Thissen, eds.). Stichting Uitgeverij Koninklijke Nederlandse Natuurhistorische Vereniging, Utrecht, 336 pp.

Pemberton, J. M., P. W. King, S. Lovari, and V. Bauchau. 1989. Genetic variation in the Alpine chamois, with special reference to the subspecies *Rupicapra rupicapra cartusiana* Coutourier, 1938. Zeitschrift für Säugetierkunde, 54:285-294.

Pembleton, E. F., and S. L. Williams. 1978. *Geomys pinetis*. Mammalian Species, 86:1-3.

Peng Yan-zhang, Ye Zhi-zhang, Zhang Yao-ping, and Pan Ru-liang. 1988. [The classification of snub-nosed monkey (*Rhinopithecus* spp.) based on gross morphological characters.] Zoological Research, 9:239-248 (in Chinese).

Pengilly, D., G. H. Jarrell, and S. D. MacDonald. 1983. Banded karyotypes of *Peromyscus sitkensis* from Baranof Island, Alaska. Journal of Mammalogy, 64:682-685.

Pennant, T. 1769. British Zoology. Class III. Reptiles. IV. Fish. V. the Beaked whale. Benjamin White, London, Volume III, pp. 43-44.

Pennant, T. 1771. Synopsis of Quadrupeds. J. Monk, Chester, 382 pp.

Pennant, T. 1781. History of quadrupeds. B. & J. White, London, 2 volumes.

Penzhorn, B. L. 1988. *Equus zebra*. Mammalian Species, 314:1-7.

Peppers, J. A., J. G. Owen, and R. D. Bradley. 1999. The karyotype of *Peromyscus stirtoni* (Rodentia: Muridae). Southwestern Naturalist, 44:109-112.

Peppers, L. L., and R. D. Bradley. 2000. Molecular systematics of the genus *Sigmodon*. Journal of Mammalogy, 81:332-343.

Peppers, L. L., D. M. Bell, J. C. Cathey, T. W. Jolley, R. Martinez, C. W. Matson, A. Y. Nekrutenko, and R. D. Bradley. 1998. Distribution records of mammals in Texas. Occasional Papers, Museum of Texas Tech University, 183:1-5.

Peppers, L. L., D. S. Carroll, and R. D. Bradley. 2002. Molecular systematics of the genus *Sigmodon* (Rodentia: Muridae): Evidence from the mitochondrial cytochrome *b* gene. Journal of Mammalogy, 83:396-407.

Pereira, E., M.-M. Ottaviani-Spella, and M. Salotti. 2001. Nouvelle datation (Pléistocène Moyen) du gisement de Punta di Calcina (Conca, Corse du Sud) par la découverte de *Talpa tyrrhenica* Bate, 1945 et d'une forme primitive de *Microtus* (*Tyrrhenicola*) *henseli* Forsyth-Major, 1882. Geobios, 34(6):697-705.

Pereira, L. G., S. E. M. Torres, H. S. Da Sivla, and L. Geise. 2001. Non-volant mammals of Ilha Grande and adjacent areas in southern Rio de Janeiro State, Brazil. Boletim do Museu Nacional, n.s., Zoologia, 459:1-15.

Pérez, E. M. 1992. *Agouti paca*. Mammalian Species, 404:1-7.

Pérez de Val, J., J. Juste, and J. Castroviejo. 1995. A review of *Zenkerella ingignis* Matschie, 1898 (Rodentia, Anomaluridae): First records in Bioko island (Equatorial Guinea). Mammalia, 59(3):441-443.

Pérez-Zapata, A., and M. Aguilera. 1996. The banding karyotypes of *Oyrzomys talamancae* and *Oryzomys capito* (Rodentia, Cricetidae). Caryologia, 49:321-326.

Pérez-Zapata, A., A. D. Vitullo, and O. A. Reig. 1987. Karyotypic and sperm distinction of *Calomys hummelincki* from *Calomys laucha* (Rodentia, Cricetidae). Acta Cientifica Venezolana, 38:90-93.

Pérez-Zapata, A., D. Lew. M. Aguilera, and O. A. Reig. 1992. New data on the systematics and karyology of *Podoxymys roraimae* (Rodentia, Cricetidae). Zeitschrift für Säugetierkunde, 57:216-224.

Perret, J.-L., and V. Aellen. 1956. Mammifères du Cameroun de la collection J.-L. Perret. Revue Suisse de Zoologie, 63(26):395-450.

Perrin, M. 1997*a*. Short-tailed gerbil, *Desmodillus auricularis*. P. 151, *in* The complete book of southern African mammals (G. Mills and L. Hes, eds.). Struik Winchester, Capetown, 356 pp.

Perrin, M. 1997*b*. Dune hairy-footed gerbil, *Gerbillurus tytonis*. P. 151, *in* The complete book of southern African mammals (G. Mills and L. Hes, eds.). Struik Winchester, Capetown, 356 pp.

Perrin, M. 1997*c*. Pygmy hairy-footed gerbil, *Gerbillurus paeba*. P. 152, *in* The complete book of southern African mammals (G. Mills and L. Hes, eds.). Struik Winchester, Capetown, 356 pp.

Perrin, M. 1997*d*. Brush-tailed hairy-footed gerbil, *Gerbillurus vallinus*. P. 153, *in* The complete book of southern African mammals (G. Mills and L. Hes, eds.). Struik Winchester, Capetown, 356 pp.

Perrin, M. 1997*e*. Setzer's hairy-footed gerbil, *Gerbillurus setzeri*. P. 153, *in* The complete book of southern African mammals (G. Mills and L. Hes, eds.). Struik Winchester, Capetown, 356 pp.

Perrin, M. 1997*f*. Bushveld gerbil, *Tatera leucogaster*. P. 154, *in* The complete book of southern African mammals (G. Mills and L. Hes, eds.). Struik Winchester, Capetown, 356 pp.

Perrin, M. 1997*g*. Cape gerbil, *Tatera afra*. P. 155, *in* The complete book of southern African mammals (G. Mills and L. Hes, eds.). Struik Winchester, Capetown, 356 pp.

Perrin, M. 1997*h*. Highveld gerbil, *Tatera brantsii*. P. 155, *in* The complete book of southern African mammals (G. Mills and L. Hes, eds.). Struik Winchester, Capetown, 356 pp.

Perrin, M. 1997*i*. Gorongoza gerbil, *Tatera inclusa*. P. 156, *in* The complete book of southern African mammals (G. Mills and L. hes, eds.). Struik Winchester, Capetown, 356 pp.

Perrin, M. R. 1986. Gastric anatomy and histology of an arboreal, folivorous murid rodent: The black-tailed tree rat *Thallomys paedulcus* (Sundevall, 1846). Zeitschrift für Säugetierkunde, 51:224-236.

Perrin, M. R., and B. A. Curtis. 1980. Comparative morphology of the digestive system of 19 species of southern African myomorph rodents in relation to diet and evolution. South African Journal of Zoology, 15:22-33.

Perrin, M. R., and A. H. Maddock. 1983. Preliminary investigations of the digestive processes of the white-tailed rat *Mystromys albicaudatus* (Smith 1834). South African Journal of Science, 18:128-133.

Perrin, W. F. 1975. Variation of spotted and spinner porpoise (genus *Stenella*) in the eastern tropical Pacific and Hawaii. Bulletin of the Scripps Institution of Oceanography, 21:1-206.

Perrin, W. F. 1990. Subspecies of *Stenella longirostris* (Mammalia: Cetacea: Delphinidae). Proceedings of the Biological Society of Washington, 103(2):453-463.

Perrin, W. F., E. D. Mitchell, J. G. Mead, D. K. Caldwell, and P. J. H. van Bree. 1981. *Stenella clymene* a rediscovered tropical dolphin of the Atlantic. Journal of Mammalogy, 62(3):583-598.

Perrin, W. F., E. D. Mitchell, J. G. Mead, D. K. Caldwell, M. C. Caldwell, P. J. H. van Bree, and W. H. Dawbin. 1987. Revision of the spotted dolphins *Stenella* spp. Marine Mammal Science, 3(2):99-170.

Perrow, M., and A. Jowitt. 1995. What future for the harvest mouse? British Wildlife, 6(6):356-365.

Perry, E. A., S. M. Carr, S. E. Bartlett, and W. S. Davidson. 1995. A phylogenetic perspective on the evolution of reproductive behavior in pagophilic seals of the northwest Atlantic as indicated by mitochondrial DNA sequences. Journal of Mammalogy, 76:22-31.

Perry, H. R., Jr. 1982. Muskrats (*Ondatra zibethicus* and *Neofiber alleni*). Pp. 282-325, *in* Wild mammals of North America (J. A. Chapman and G. A. Feldhammer, eds.). Johns Hopkins University Press, Baltimore, 1147 pp.

Peshev, D. 1983. New karyotype forms of the mole rat, *Nannospalax leucodon* Nordmann (Spalacidae, Rodentia), in Bulgaria. Zoologischer Anzeiger, 211:65-72.

Peshev, D. 1989a. Craniological study of the species of genus *Spalax* (Spalacidae, Mammalia). I. Sex dimorphism. Zoologischer Anzeiger, 222:83-91.

Peshev, D. 1989b. Craniological study of the species of genus *Spalax* (Spalacidae, Mammalia). II. Interspecific differences. Zoologischer Anzeiger, 222:92-98.

Peshev, D. 1991. On the systematic position of the mouse-like hamster *Calomyscus bailwardi* Thomas, 1905 (Cricetidae, Rodentia) from the Near East and Middle Asia. Mammalia, 55:107-112.

Peshev, D. 1996. Review of the status of mammals in Bulgaria. Hystrix, n.s., 8(1-2):55-59.

Peshev, D., and V. Delov. 1995a. Chromosome study of three species of dormice from Bulgaria. Pp. 151-153, *in* Proceedings of II Conference on Dormice (Rodentia, Myoxidae) (M. G. Filippucci, ed.). Hystrix, n.s., 6(1-2):1-340.

Peshev, D., and V. Delov. 1995b. Craniological study and subspecific status of three species of dormice from Bulgaria. Pp. 225-230, *in* Proceedings of II Conference on Dormice (Rodentia, Myoxidae) (M. G. Filippucci, ed.). Hystrix, n.s., 6(1-2):1-340.

Peshev, D., D. Mitev, V. Delov, and S. Stoikov. 1990a. Craniological characteristic of the Fat Dormouse *Glis glis* L. 1766 (Rodentia, Mammalia) in Bulgaria. Nauchni trudove, Biologiia, 28(6):409-415.

Peshev, D., D. Mitev, V. Delov, and S. Stoikov. 1990b. Over the craniological characteristic and the subspecific status of the Hazel Dormouse *Muscardinus avellanarius* L. 1758 (Rodentia, Mammalia) in Bulgaria. Nauchni trudove, Biologiia, 28(6):417-422.

Peshev, T., and B. Belcheva. 1979. Karyological studies on snow vole *Microtus nivalis* Martins (Mammalia, Rodentia) collected in Bulgaria. Zoologischer Anzeiger, 203:65-68.

Peshev, T., V. Anguelova, and T. Dinev. 1964. Etudes sur la taxonomie du *Myomimus personatus* (Ognev, 1924) (Rodentia) en Bulgarie. Mammalia, 28(3):419-428.

Pessôa, L. M., and S. F. dos Reis. 1993. *Proechimys dimidiatus*. Mammalian Species, 441:1-3.

Pessôa, L. M., and S. F. Reis. 1994. Systematic implications of craniometric variation in *Proechimys iheringi* Thomas (Rodentia: Echimyidae). Zoologischer Anzeiger, 232:181-200.

Pessôa, L. M., and S. F. Reis. 1996. *Proechimys iheringi*. Mammalian Species, 536:1-4.

Pessôa, L. M., and S. F. dos Reis. 2002. *Proechimys albispinus*. Mammalian Species, 693:1-3.

Pessôa, L. M., J. A. de Oliveira, and S. F. dos Reis. 1990. Quantitative cranial character variation in selected populations of the *guyannensis*-group of *Proechimys* (Rodentia: Echimyidae) from Brazil. Zoologischer Anzeiger, 225:396-400.

Pestano, J., R. P. Brown, N. M. Suárez, and S. Farjado. 2003. Phylogeography of pipistrelle-like bats within the Canary Islands, based on mtDNA sequences. Molecular Phylogenetics and Evolution, 26:56-63.

Peter, W. P., and A. Feiler. 1994a. Hörner von einer unbekannten Bovidenart aus Vietnam (Mammalia: Ruminantia). Faunistische Abhandlungen. Staatliches Museum für Tierkunde Dresden, 19:247-253.

Peter, W. P., and A. Feiler. 1994b. Eine neue Bovidenart aus Vietnam und Cambodia (Mammalia: Ruminantia). Zoologische Abhandlungen. Staatliches Museum für Tierkunde Dresden, 48:169-176.

Peters, G. 1986. Mixed herds of common and defassa waterbuck, *Kobus ellipsiprymnus* (Artiodactyla: Bovidae) in northern Kenya. Bonner Zoologische Beitrage, 37:183-193.

Peters, S. L., B. K. Lim, and M. D. Engstrom. 2002. Systematics of dog-faced bats (*Cynomops*) based on molecular and morphometric data. Journal of Mammalogy, 83(4):1097-1110.

Peters, W. C. H. 1852. Reise nach Mossambique. I. Säugethiere. Druck & Verlag, Berlin, 202 pp.

Peters, W. 1865. Note on the systematic position of *Platacanthomys lasiurus*. Proceedings of the Zoological Society of London, 1865:397-399.

Peters, W. 1866. Über neue oder ungengend bekannte Flederthiere (*Vampyrops, Uroderma, Chiroderma, Ametrida, Tylostoma, Vespertilio, Vesperugo*) und Nager (*Tylomys, Lasiomys*). Monatsberichte der Königlich Preussischen Akademie der Wissenschaften zu Berlin, 1867:392-411, 2 pls.

Peters, W. C. H. 1875. Über eine neue Art von Seebären, *Arctophoca gazella* von den Kerguelen-Inseln. Monatsberichte der Königlich Preussischen Akademie der Wissenschaften zu Berlin, 1875:393-399.

Peters, W. C. H. 1876. Über die Pelzrobbe von der Inseln St. Paul und Amsterdam. Monatsberichte der Königlich Preussischen Akademie der Wissenschaften zu Berlin, 1876:315-316.

Peterson, A. T., and L. R. Heaney. 1993. Genetic differentiation in Philippine bats of the genera *Cynopterus* and *Haplonycteris*. Biological Journal of the Linnean Society, 49:203-218.

Peterson, K. E., and T. I. Yates. 1980. *Condylura cristata*. Mammalian Species, 129:1-4.

Peterson, R. L. 1952. A review of the living representatives of the genus *Alces*. Contributions from the Royal Ontario Museum in Zoology and Palaeontology, 34:1-30.

Peterson, R. L. 1965*a*. A review of the flat-headed bats of the family Molossidae from South America and Africa. Royal Ontario Museum, Life Sciences, Contribution, 64:1-32.

Peterson, R. L. 1965*b*. A review of the bats of the genus *Ametrida*, Family Phyllostomidae. Royal Ontario Museum, Life Sciences, Contribution, 65:1-13.

Peterson, R. L. 1966*a*. The mammals of eastern Canada. Oxford University Press, Toronto, 465 pp.

Peterson, R. L. 1966*b*. Recent mammal records from the Galapagos Islands. Mammalia, 30:441-445.

Peterson, R. L. 1969. Notes on the Malaysian fruit bats of the genus *Dyacopterus*. Royal Ontario Museum, Life Sciences, Occasional Papers, 13:1-4.

Peterson, R. L., 1971. The systematic status of the African molossid bats *Tadarida bemmeleni* and *Tadarida cistura*. Canadian Journal of Zoology, 49:1347-1354.

Peterson, R. L. 1972. Systematic status of the African molossid bats *Tadarida congica, T. niangarae* and *T. trevori*. Royal Ontario Museum, Life Sciences, Contribution, 85:1-32.

Peterson, R. L. 1981. Systematic variation in the *tristis* group of the bent-winged bats of the genus *Miniopterus* (Chiroptera: Vespertilionidae). Canadian Journal of Zoology, 59:828-843.

Peterson, R. L. 1982. A new species of *Glauconycteris* from the east coast of Kenya (Chiroptera: Vespertilionidae). Canadian Journal of Zoology, 60:2521-2525.

Peterson, R. L. 1987. Notes on systematic variation in the *nanus* group of African *Pipistrellus*. Bat Research News, 28:37.

Peterson, R. L. 1991. Systematic variation in the megachiropteran tube-nosed bats *Nyctimene cyclotis* and *N. certans*. Bulletin of the American Museum of Natural History, 206:26-41.

Peterson, R. L., and M. B. Fenton. 1970. Variation in bats of the genus *Harpionycteris*, with the description of a new race. Royal Ontario Museum, Life Sciences, Occasional Papers, 17:1-15.

Peterson, R. L., and D. A. Smith. 1973. A new species of *Glauconycteris* (Vespertilionidae, Chiroptera). Royal Ontario Museum, Life Sciences, Occasional Papers, 22:1-9.

Peterson, R. L., J. L. Eger, and L. Mitchell. 1995. Chiroptères. Faune de Madagascar, 84:1-204.

Petrov, B. 1979. Some questions of the zoogeographical division of the western Palaearctic in the light of the distribution of mammals in Yugoslavia. Folia Zoologica, 28:13-24.

Petrov, B., and A. Ruzic. 1982. *Microtus epiroticus* Ondrias, 1966—Südfeldmaus. Pp. 319-330, *in* Handbuch der Säugetiere Europas (J. Niethammer and F. Krapp, eds.). Akademische Verlagsgesellschaft (Wiesbaden), 2/I:1-649.

Petrov, B., and A. Ruzic. 1985. Taxonomy and distribution of members of the genus *Mus* (Rodentia, Mammalia) in Yugoslavia. Proceedings on the fauna of SR Serbia, 3:209-243.

Petrov, B., and M. Todorovic. 1982. *Dinaromys bogdanovi* (V. et E. Martino, 1922)—Bergmaus. Pp. 193-208, *in* Handbuch der Säugetiere Europas (J. Niethammer and F. Krapp, eds.). Akademische Verlagsgesellschaft (Wiesbaden), 2/I:1-649.

Petrov, B., and S. Zivkovic. 1971. Zur kenntnis der *Pitymys liechtensteini* Wettstein, 1927 (Rodentia, Mammalia) in Jugoslawien. Arhiv bioloskih nauka, Beograd, 23:31-32.

Petrov, B. M. 1992. Mammals of Yugoslavia. Insectivores and rodents. Natural History Museum in Belgrade, Supplement 37:1-186.

Petrov, B. M. 1993/94. *Apodemus stankovici* V. and E. Martino, 1937, new ideas on the taxonomic status of some rodent species in the former and actual Yugoslavia. Bulletin of Natural History Museum, Belgrade, B48:183-188.

Petrov, V. V. 1953. [The data on the intraspecific variability of badgers (genus *Meles*).] Uchenye Zapiski Leningradskogo Pedagogicheskogo Instituta, 7:149-205 (in Russian).

Petter, F. 1954. Remarques biologiques sur des rats épineux du genre *Acomys* répartition au Sahara. Mammalia, 18:389-396.

Petter, F. 1957. Remarques sur la sysématique des *Rattus* Africains et description d'une forme nouvelle de l'Air. Mammalia, 21(2):125-132.

Petter, F. 1959. Evolution du dessin de la surface d'usure des molaires des Gerbillides. Mammalia, 23:304-315.

Petter, F. 1961a. Eléments d'une révision des Lievres européens et asiatiques du sous-genre *Lepus*. Zeitschrift für Säugetierkunde, 26:30-40.

Petter, F. 1961b. Affinities des genres *Spalax* et *Brachyuromys*. Mammalia, 25:485-498.

Petter, F. 1962a. Un noveau rongeur Malgache: *Brachytarsomys albicauda villosa*. Mammalia, 26:570-572.

Petter, F. 1962b. Note de nomenclature sur le genere *Mylomys* (Rongeurs Murides). Mammalia, 26:575.

Petter, F. 1963a. Nouveaux éléments d'une révision des lièvres Africains. Mammalia, 27:238-255.

Petter, F. 1963b. Un nouvel insectivore du nord de l'Assam: *Anourosorex squamipes schmidi* nov. sbsp. Mammalia, 27:444-445.

Petter, F. 1963c. Contribution a la connaissance des souris Africaines. Mammalia, 27:602-607.

Petter, F. 1964. Un étrange rongeur de "La Maboké", *Prionomys batesi*. Science et Nature, Paris, 62:37-38.

Petter, F. 1965. Les *Praomys* d'Afrique Centrale. Zeitschrift für Säugetierkunde, 30:54-60.

Petter, F. 1966a. Affinites des genres *Beamys*, *Saccostomus* et *Cricetomys* (Rongeurs, Cricetomyinae). Annales Musée Royal de l'Afrique Centrale, ser. 8 (Sciences Zoologiques), 144:13-25.

Petter, F. 1966b. *Dendroprionomys rousseloti* gen. nov., sp. nov., rongeur nouveau du Congo (Cricetidae, Dendromurinae). Mammalia, 30:129-137.

Petter, F. 1966c. L'origine des murides plan cricetin et plans murins. Mammalia, 30(2):205-225.

Petter, F. 1967a. Contribution a la faune du Congo (Brazzaville). Mission A. Villiers et A. Descarpentries. LV. Mammiferes rongeurs (Muscardinidae et Muridae). Bulletin de l'Institut Fondatmental d'Afrique Noire, 29A(2):815-820.

Petter, F. 1967b. Particularities dentaires des Petromyscinae Roberts 1951 (Rongeurs, Cricetides). Mammalia, 31:217-224.

Petter, F. 1969. Une souris nouvelle d'Afrique Occidentale *Mus mattheyi* sp. nov. Mammalia, 33:118-123.

Petter, F. 1972a. Caracteres morphologiques et repartition geographique de *Mus tenellus* (Thomas 1903), souris d'Afrique Orientale. Mammalia, 36:533-535.

Petter, F. 1972b. Order Lagomorpha. Part 5. Pp. 1-7, *in* The mammals of Africa: An identification manual (J. Meester and H. W. Setzer, eds.) [issued 2 May 1972]. Smithsonian Institution Press, Washington, D.C., not continuously paginated.

Petter, F. 1972c. The rodents of Madagascar: The seven genera of Malagasy rodents. Pp. 661-666, *in* Biogeography and ecology in Madagascar (R. Battistini and G. Richard-Vincard, eds.). Monographiae Biologicae, 21:1-765.

Petter, F. 1972d. Deux rongeurs nouveaux d'Etiophie: *Stenocephalemys griseicauda* sp. nov. et *Lophuromys melanonyx* sp nov. Mammalia, 36:171-181.

Petter, F. 1973a. Capture de *Thallomys paedulcus scotti* en Ethiopie. Mammalia, 37:360-361.

Petter, F. 1973b. Les noms de genre *Cercomys*, *Nelomys*, *Trichomys*[sic] et *Proechimys* (Rongeurs, Echimyides). Mammalia, 37(3):422-426.

Petter, F. 1973c. Tendances evolutives dans le genre *Gerbillus* (Rongeurs, Gerbillides). Mammalia, 37:631-636.

Petter, F. 1975a. Family Cricetidae: Subfamily Nesomyinae, Part 6.2. Pp. 1-4, *in* The mammals of Africa: An identification manual (J. Meester and H. W. Setzer, eds.) [issued 10 Dec 1975]. Smithsonian Institution Press, Washington, D.C., not continuously paginated.

Petter, F. 1975b. Subfamily Gerbillinae. Part 6.3. Pp. 7-12, *in* The mammals of Africa: An identification manual (J. Meester and H. W. Setzer, eds.) [issued 10 Dec 1975]. Smithsonian Institution, Washington, D.C., not continuously paginated.

Petter, F. 1975c. Les *Praomys* de Republique Centrafricaine (Rongeurs, Murides). Mammalia, 39:51-56.

Petter, F. 1978. Epidémiologie de la leishmaniose en Guyane francaise en relation avec l'existance d'une espece nouvelle de rongeurs échimyidés, *Proechimys cuvieri*, sp. n. Compte Rendu Hebdomadaire des seances de l'Academie de Sciences, Paris, ser. D, 287:261-264.

Petter, F. 1981a. Remarques sur la systématique des Chrysochloridés. Mammalia, 45:49-53.

Petter, F. 1981b. Les souris africaines du groupe *sorella* (Rongeurs, Murides). Mammalia, 45:313-320.

Petter, F. 1982. Les parentes des *Otomys* du Mont Oku (Cameroun) et des autres formes rapportees a *O. irroratus* (Brants, 1827) (Rodentia, Muridae). Bonner Zoologische Beiträge, 33:215-222.

Petter, F. 1983. Elements d'une revision des *Acomys* Africains. Un sous-genre nouveau, *Peracomys* Petter et Roche, 1981 (Rongeurs, Murides). Annales Musée Royal de l'Afrique Centrale, Tervuren-Belgique, ser. 8 (Sciences Zoologiques), 237:109-119.

Petter, F. 1986. Un rongeur nouveau du Mont Oku (Cameroun) *Lamottemys okuensis*, gen. nov., sp. nov.; (Rodentia, Muridae). Cimbebasia, ser. A, 8:97-105.

Petter, F. 1990. Relations de parenté des rongeurs de Madagascar. Accademie Nazionale dei Lincei, Atti dei Convegni Lincei, 85:829-837.

Petter, F., and F. de Beaufort. 1960. Description d'une forme nouvelle de rongeur d'Angola, *Thallomys damarensis quissamae*. Bulletin du Muséum National d'Histoire Naturelle, Paris, ser. 2, 32:269-271.

Petter, F., and H. Genest. 1970. Liste preliminaire des rongeurs myomorphes de Republique Centrafricaine. Description de deux souris nouvelles: *Mus oubanguii* et *Mus goundae*. Mammalia, 34:451-458.

Petter, F., and R. Matthey. 1975. Genus *Mus*, Part 6.7. Pp. 1-4, *in* The mammals of Africa: An identification manual (J. Meester and H. W. Setzer, eds.). Smithsonian Institution Press, Washington, D.C., not continuously paginated.

Petter, F., and J. Roche. 1981. Remarques preliminaires sur la systematique des *Acomys* (Rongeurs, Muridae) *Peracomys*, sous-genre nouveau. Mammalia, 45:381-383.

Petter, F., and M. Tranier. 1975. Contribution a l'etude des *Thamnomys* du groupe *dolichurus* (Rongeurs, Murides). Systematique et caryologie. Mammalia, 39:405-414.

Petter, F., M. Quilici, Ph. Ranque, and P. Camerlynck. 1969. Croisement d'*Arvicanthis niloticus* (Rongeurs, Murides) du Senegal et d'Ethiopie. Mammalia, 33:540-541.

Petter, F., F. Adam, and B. Hubert. 1971. Presence au Senegal de *Mus mattheyi* F. Petter 1969. Mammalia, 35:346-347.

Petter, F., A. Poulet, B. Hubert, and F. Adam. 1972. Contribution a l'etude des *Taterillus* du Senegal *T. pygargus* (F. Cuvier, 1832) et *T. gracilis* (Thomas, 1892) (Rongeurs, Gerbillides). Mammalia, 36:210-213.

Petter, F., F. Lachiver, and R. Chekir. 1984. Les adaptations des rongeurs gerbillides a la vie dans les regions arides. Bulletin de Societe Botanique de France, Actualites Botaniques, 131:365:373.

Petter, G. 1962. Le peuplement en carnivores de Madagascar. Pp. 331-342, *in* Evolution des vertébrés (J. P. Lehman, ed.). Colloques Internationaux du Centre National de la Recherche Scientifique, 104:1-477.

Petter, G. 1969. Interpretation evolutive des caractéres de la denture des viverrides Africans. Mammalia, 33(4):607-635.

Petter, G. 1971. Origine, phylogenie et systematique des blaireaux. Mammalia, 35:567-597.

Petter, G. 1974. Rapports phyletiques des viverrides (Carnivores Fissipédes). Les formes de Madagascar. Mammalia, 38(4):605-636.

Petter, J.-J. 1962. Recherches sur l'écologie et l'éthologie des Lémuriens malgaches. Mémoires du Muséum National d'Histoire Naturelle, Nouvelle Serie, ser. A (Zoologie), 27:1-146.

Petter, J.-J., and A. Petter-Rousseaux. 1979. Classification of the prosimians. Pp. 1-44, *in* The study of prosimian behavior (G. A. Doyle and R. D. Martin, eds.). Academic Press, London, 696 pp.

Petter, J.-J., R. Albignac, and Y. Rumpler. 1977. Faune de Madagascar. 44. Mammifères lemuriens. O.R.S.T.O.M., Paris, 513 pp.

Petzsch, H. 1970. Kritisches über die neuentdeckte Iriomote-Wildkatze. Der Pelzgewerbe, 20(5):3-7.

Pévet, P., and M. Yadav. 1980. The pineal gland of equatorial mammals. I. The pinealocytes of the Malaysian rat (*Rattus sabanus*). Cell and Tissue Research, 210:417-433.

Peyre, A. 1957. La formule chromosomique du desman des Pyrénées, *Galemys pyrenaicus* G. Bulletin de la Societé Zoologique du France, 82:434-437.

Pfleiderer, M. 1994. The taxonomic status of the Iriomote cat (*Prionailurus irimotensis* Imaizumi, 1967). Cat News, 21:18-20.

Philippe, H. 1997. Rodent monophyly: Pitfalls of molecular phylogenies. Journal of Molecular Evolution, 45:712-715.

Philippi, R. A. 1900. Figuras i descripciones de los murideos de Chile. Anales de Museo Nacional de Chile, Zoologie, 14a:1-70.

Philippi, T. 1994. Zur innerartlichen Variabilität von *Arvicanthis niloticus* Desmarest 1822 in Ägypten und im Sudan (Mammalia: Rodentia: Muridae). Senckenbergiana Biologica, 73(1-2):39-44.

Phillips, C. J. 1968. Systematics of megachiropteran bats in the Solomon Islands. University of Kansas Publications, Museum of Natural History, 16:777-837.

Phillips, C. J. 1969. Review of central Asian voles of the genus *Hyperacrius*, with comments on zoogeography, ecology, and ectoparasites. Journal of Mammalogy, 50:457-474.

Phillips, C. J. 1971. The dentition of glossophagine bats: development, morphological characteristics, variation, pathology, and evolution. Miscellaneous Publications, Museum of Natural History, University of Kansas, 54:1-138.

Phillips, C. J., and E. C. Birney. 1968. Taxonomic status of the vespertilionid genus *Anamygdon* (Mammalia: Chiroptera). Proceedings of the Biological Society of Washington, 81:491-498.

Phillips, C. J., D. E. Pumo, H. H. Genoways, P. E. Ray, and C. A. Briskey. 1991. Mitochondrial DNA evolution and phylogeography in two Neotropical fruit bats, *Artibeus jamaicensis* and *Aritbeus lituratus*. Pp. 97-123, *in* Latin American mammalogy: History, biodiversity, and conservation (M. A. Mares and D. J. Schmidly, eds.). University of Oklahoma Press, Norman, 468 pp.

Phillips, W. W. A. 1926. The long-tailed tree mouse (*Vandeleuria oleracea*?). Spolia Zeylanica, Sec. B., 13:307-313.

Phillips, W. W. A. 1929. Two new rodents from the highlands of Ceylon. Spolia Zeylanica, Sec. B., 15:165168.

Phillips, W. W. A. 1933. Survey of the distribution of mammals in Ceylon. Ceylon Journal of Science (Spolia Zeylonica), 18:133-142.

Phillips, W. W. A. 1980-1984. Manual of the mammals of Sri Lanka. Second revised ed. Wildlife and Nature Protection Society of Sri Lanka, 1:1-116 [1980]; 2:117-267 [1980]; 3:268-389 [1984].

Piaggio, A. J., and G. S. Spicer. 2001. Molecular phylogeny of the chipmunks inferred from mitochondrial cytochrome *b* and cytochrome oxidase II gene sequences. Molecular Phylogenetics and Evolution, 20(3):335-350.

Pickering, J., and C. A. Norris. 1996. New evidence concerning the extinction of the endemic murid *Rattus macleari* from Christmas Island, Indian Ocean. Australian Mammalogy, 19:19-25.

Pickford, M., and P. Mein. 1988. The discovery of fossiliferous Plio-Pleistocene cave fillings in Ngamiland, Botswana. Comptes Rendus des Séances de l'Académie des Sciences, Paris, Sér. II:1681-1686.

Pictet, F.-J. 1843. Seconde notice sur les animaux nouveaux ou peu connus de Musée de Genève. Mémoires Société de Physique et d'Histoire Naturelle de Genève, 10:201-213.

Pieczarka, J. C., and C. Y. Nagamachi. 1988. Cytogenetic studies of *Aotus* from eastern Amazonia: Y/autosome rearrangement. American Journal of Primatology, 14(3)255-263.

Pieper, H. 1981. Ein subfossiles vorkommen der hausmaus (*Mus musculus* s. l.) Auf Madeira. Bocagiana, 59:1-3.

Pieper, H. 1984. Eine neue *Mesocricetus*-Art (Mammalia: Cricetidae) von der griechischen Insel Armathia. Stuttgarter Beiträge zur Naturkunde, ser. B, 107:1-9.

Pieper, H. 1990. *Crocidura zimmermanni* Wettstein, 1953—Kretaspitzmaus. Pp. 453-460, *in* Handbuch der Säugetiere Europas (J. Niethammer and F. Krapp, eds.). Aula-Verlag, Wiesbaden, 3/I:1-524.

Pierpaoli, M., F. Riga, V. Trocchi, and E. Randi. 1999. Species distinction and evolutionary relationships of the Italian hare (*Lepus corsicanus*) as described by mitochondrial DNA sequencing. Molecular Ecology, 8:1805-1817.

Pierson, E. D., V. M. Sarich, J. M. Lowenstein, M. J. Daniel, and W. E. Rainey. 1986. A molecular link between the bats of New Zealand and South America. Nature, 6083:60-63.

Pietsch, M. 1980. Biometrische Analyse an Schädeln von neun Kleinsäuger-Arten aus der Familie Arvicolidae (Rodentia). Zeitschrift für Zoologische Systematik und Evolutionsforschung, 18:196-211.

Pietsch, M. 1982. *Ondatra zibethicus* (Linnaeus, 1766)—Bisamratte, Bisam. Pp. 177-192, *in* Handbuch der Säugetiere Europas (J. Niethammer and F. Krapp, eds.), vol. 2/I. Akademische Verlagsgesellschaft (Wiesbaden), 649 pp.

Piette, E., and A. Lametschwandtner. 1995. The fine vasculature of the rat mandibular joint. Acta Anatomica, 153:64-72.

Pigage, J. C., and H. K. Pigage. 1994. The western harvest mouse (*Reithrodontomys megalotis*) moves into northeastern Illinois. Transactions of the Illinois State Academy of Science, 87:47-50.

Pike, G. C., and I. B. MacAskie. 1969. Marine mammals of British Columbia. Fisheries Research Board of Canada, Bulletin, 171:1-54.

Pilastro, A. 1990. Studio di una popolazione di ghiro (*Glis glis* Linnaeus) in un ambiente forestale dei Colli Berici. Lavori, Societa Veneziana di Scienze Naturali, 15:145-155.

Pilâts, V. 1995. Dormice—their present status in Latvia. Pp. 185-194, *in* Proceedings of II Conference on dormice (Rodentia, Myoxidae) (M. G. Filippucci, ed.). Hystrix, n.s., 6(1-2):1-340.

Pilbeam, D., M. Morgan, J. C. Barry, and L. Flynn. 1996. European MN units and the Siwalik faunal sequence of Pakistan. Pp. 96-105, *in* The evolution of Western Eurasian Neogene mammal faunas (R. L. Bernor, V. Fahlbusch, and H.-W. Mittmann, eds.). Columbia University Press, New York, 487 pp.

Pillay, N. 2000. Reproductive isolation in three populations of the striped mouse *Rhabdomys pumilio* (Rodentia, Muridae): Interpopulation breeding studies. Mammalia, 64(4):461-470.

Pillay, N., K. Willan, and J. Meester. 1992. Post-zygotic reproductive isolation in the African vlei rat *Otomys irroratus* (Muridae: Otomyinae). Israel Journal of Zoology, 38:307-313.

Pillay, N., K. Willan, and J. Cooke. 1995*a*. Evidence of pre-mating reproductive isolation in two populations of the vlei rat *Otomys irroratus*: Experiments of intra- and interpopulation male-female encounters. Zeitschrift für Säugetierkunde, 60:352-360.

Pillay, N., K. Willan, and J. Meester. 1995*b*. Post-zygotic reproductive isolation in two populations of the African vlei *Otomys irroratus*. Acta Theriologica, 40:69-76.

Pilleri, G. 1978. William Roxburgh (1751-1815), Heinrich Julius Lebeck (1801) and the discovery of the Ganges dolphin (*Platanista gangetica* Roxburgh, 1801). Investigations on Cetacea, 9:11-21.

Pilleri, G., and Chen Pei-xun. 1980. *Neophocoena phocaenoides* and *Neophocaena asiaeorientalis*: Taxonomic differences. Investigations on Cetacea, 11:25-32.

Pilleri, G., and M. Gihr. 1971. Differences observed in the skulls of *Platanista gangetica* (Roxburgh, 1801) and *indi* (Blyth, 1859). Investigations on Cetacea, 3:13-21.

Pilleri, G., and M. Gihr. 1972. Contribution to the knowledge of cetaceans of Pakistan with particular reference to the genera *Neomeris*, *Sousa*, *Delphinus* and *Tursiops* and description of a new Chinese porpoise (*Neomeris asiaeorientalis*). Investigations on Cetacea, 4:107-162.

Pilleri, G., and M. Gihr. 1973-74. Contribution to the knowledge of the cetaceans of southwest and monsoon Asia (Persian Gulf, Indus Delta, Malabar, Andaman Sea and Gulf of Siam). Investigations on Cetacea, 5:95-149.

Pilleri, G., and M. Gihr. 1975. On the taxonomy and ecology of the finless black porpoise, *Neophocaena* (Cetacea, Delphinidae). Mammalia, 39:657-673.

Pilleri, G., and M. Gihr. 1976*a*. Osteological differences in the cervical vertebrae of *Platanista indi* and *gangetica*. Investigations on Cetacea, 7:105-108.

Pilleri, G., and M. Gihr. 1976*b*. The function and osteology of the manus of *Platanista gangetica* and *Platanista indi*. Investigations on Cetacea, 7:109-118.

Pilleri, G., and M. Gihr. 1977. Observations on the Bolivian (*Inia boliviensis* d'Orbigny, 1834) and the Amazonian bufeo (*Inia geoffrensis* de Blainville, 1817) with description of a new subspecies (*Inia geoffrensis humboldtiana*). Investigations on Cetacea, 8:11-77.

Pilleri, G., and M. Gihr. 1980*a*. Additional considerations of the taxonomy of the genus *Inia*. Investigations on Cetacea, 11:15-24.

Pilleri, G., and M. Gihr. 1980*b*. Checklist of the cetacean genera *Platanista*, *Inia*, *Lipotes*, *Pontoporia*, *Sousa* and *Neophocaena*. Investigations on Cetacea, 11:33-36.

Pinder, L., and A. P. Grosse. 1991. *Blastoceros dichotomus*. Mammalian Species, 380:1-4.

Pinder, L., and F. Leeuwenberg. 1997. Veado-catingueiro (*Mazama gouazoubira*, Fisher 1814). Pp. 60-68, *in* Biologia e conservação de cervídeos Sul-Americanos: *Blastocerus*, *Ozotoceros* e *Mazama* (J. M. B. Duarte, ed.). Fundação de Estudos e Pesquisas em Agronomia, Medicina Veterinária e Zootecnia, Jaboticabal, Brazil, 238 pp.

Pine, R. H. 1967. *Baedon meyeri* Pine (Chiroptera: Vespertilionidae) referred to the genus *Antrozous* H. Allen. Southwestern Naturalist, 12:484-485.

Pine, R. H. 1971. A review of the long-whiskered rice rat, *Oryzomys bombycinus*, Goldman. Journal of Mammalogy, 52:590-596.

Pine, R. H. 1972. The bats of the genus *Carollia*. Technical Monograph, Texas Agricultural Experiment Station, Texas A and M University, 8:1-125.

Pine, R. H. 1973*a*. Anatomical and nomenclatural notes on opossums. Proceedings of the Biological Society of Washington, 86:391-402.

Pine, R. H. 1973*b*. Mammals (exclusive of bats) of Belém, Pará, Brazil. Acta Amazonica, 3:47-79.

Pine, R. H. 1976*a*. *Monodelphis umbristriata* (A. de Miranda-Ribeiro) is a distinct species of opossum. Journal of Mammalogy, 57:785-787.

Pine, R. H. 1976*b*. A new species of *Akodon* (Mammalia: Rodentia: Muridae: Cricetinae) from Isla de los Estados, Argentina. Mammalia, 40:63-68.

Pine, R. H. 1977. *Monodelphis iheringi* (Thomas) is a recognizable species of Brazilian opossum (Mammalia: Marsupialia: Didelphidae). Mammalia, 41:235-237.

Pine, R. H. 1980*a*. Taxonomic notes on "*Monodelphis dimidiata itatiayae* (Miranda-Ribeiro)", *Monodelphis domestica* (Wagner) and *Monodelphis maraxina* Thomas (Mammalia: Marsupialia: Didelphidae). Mammalia, 43:495-499.

Pine, R. H. 1980*b*. Notes on rodents of the genera *Wiedomys* and *Thomasomys* (including *Wilfredomys*). Mammalia, 44:195-202.

Pine, R. H. 1981. Review of the mouse opossums *Marmosa parvidens* Tate and *Marmosa invicta* Goldman (Mammalia: Marsupialia: Didelphidae) with description of a new species. Mammalia, 45:55-70.

Pine, R. H. 1993. A new species of *Thyroptera* Spix (Mammalia: Chiroptera: Thyropteridae) from the Amazon Basin of northeastern Perú. Mammalia, 57:213-225.

Pine, R. H., and J. P. Abravaya. 1978. Notes on the Brazilian opossum *Monodelphis scalops* (Thomas) (Mammalia: Didelphidae). Mammalia, 42:379-382.

Pine, R. H., and C. O. Handley, Jr. 1984. A review of the Amazonian short-tailed opossum *Monodelphis emiliae* (Thomas). Mammalia, 48:239-245.

Pine, R. H., and A. Ruschi. 1976. Concerning certain bats described and recorded from Espírito Santo, Brazil. Anales de Instituto de Biologia, Universidad Nacional Autónomica de México, 47:183-196.

Pine, R. H., and R. M. Wetzel. 1975. A new subspecies of *Pseudoryzomys wavrini* (Mammalia: Rodentia: Muridae: Cricetinae) from Bolivia. Mammalia, 39:649-655.

Pine, R. H., D. C. Carter, and R. K. LaVal. 1971. Status of *Bauerus* Van Gelder and its relationships to other nyctophiline bats. Journal of Mammalogy, 52:663-669.

Pine, R. H., S. D. Miller, and M. L. Schamberger. 1979. Contributions to the mammalogy of Chile. Mammalia, 43:339-376.

Pine, R. H., P. L. Dalby, and J. O. Matson. 1985. Ecology, postnatal development, morphometrics, and taxonomic status of the short-tailed opossum *Monodelphis dimidiata*, an apparently semelparous annual marsupial. Annals of Carnegie Museum, 54:195-231.

Pine, R. H., R. K. LaVal, D. C. Carter, and W. Y. Mok. 1996. Notes on the graybeard bat, *Micronycteris daviesi* (Hill)(Mammalia: Chiroptera: Phyllstomidae), with the first records from Ecuador and Brazil. Pp. 183-190, *in* Contributions in mammalogy: A memorial volume in honor of J. Knox Jones, Jr. (H. H. Genoways and R. J. Baker, eds.). Museum of Texas Tech University, Lubbock, Texas, 315 pp.

Pine, R. H., N. Woodman, and R. M. Timm. 2002. Rediscovery of Enders's small-eared shrew, *Cryptotis endersi* (Insectivora: Soricidae), with a redescription of the species. Mammalian Biology, 67:372-377.

Pirlot, P. 1967. Nouvelle recolte de chiropteres dans l'ouest du Venezuela. Mammalia, 31:260-274.

Pirlot, P. L. 1955. Variabilite intra-generique chez un rongeur africain. Annales du Musee Royal du Congo Belge, Tervuren (Belgique), Serie in 8, Sciences Zoologiques, 39:1-66.

Pitman, R. L., D. M. Palacios, P. L. R. Brennan, B. J. Brennan, K. C. Balcomb, III, and T. Miyashita. 1999. Sightings and possible identity of a bottlenose whale in the tropical Indo-Pacific: *Indopacetus pacificus*? Marine Mammal Science, 15(2):531-549.

Pitra, C., A. J. Hansen, D. Lieckfeldt, and P. Arctander. 2002. An exceptional case of historical outbreeding in African sable antelope populations. Molecular Ecology, 11:1197-1208.

Pizzimenti, J. J. 1975. Evolution of the prairie dog genus *Cynomys*. Occasional Papers of the Museum of Natural History, University of Kansas, 39:1-73.

Pizzimenti, J. J. 1976. Genetic divergence and morphological convergence in the prairie dogs, *Cynomys gunnisoni* and *Cynomys leucurus*. I. Morphological and ecological analyses. II. Genetic analyses. Evolution, 30:345-366; 367-379.

Pizzimenti, J. J., and G. D. Collier. 1975. *Cynomys parvidens*. Mammalian Species, 52:1-3.

Pizzimenti, J. J., and R. S. Hoffmann. 1973. *Cynomys gunnisoni*. Mammalian Species, 25:1-4.

Pizzimenti, J. J., and C. F. Nadler. 1972. Chromosomes and serum proteins of the Utah prairie dog, *Cynomys parvidens* (Sciuridae). Southwestern Naturalist, 17:279-286.

Plakchotnikova, J., T. Aksenova, and A. Karpov. 1992. [Contribution to investigation of morphotypic variability of cheekteeth of the bank vole (*Clethrionomys glareolus* Schreb.).] Pp. 88-97, *in* Proceedings of the First Baltic Theriological Conference (A. Kirk, A. Miljutin, and T. Randveer eds.), Tartu Ülikool Toimetised (= Acta et Commentationes Universitatis Tartuensis), no. 955:254 pp. (in Russian with English summary).

Planz, J. V. 1999. Arizona woodrat | *Neotoma devia*. Pp. 600-601, *in* The Smithsonian Book of North American Mammals (D. E. Wilson and S. Ruff, eds.). Smithsonian Institution Press, Washington, D.C., xxv + 750 pp.

Planz, J. V., E. G. Zimmerman, T. A. Spradling, and D. R. Akins. 1996. Molecular phylogeny of the *Neotoma floridana* species group. Journal of Mammalogy, 77:519-535.

Plassmann, W., W. Peetz, and M. Schmidt. 1987. The cochlea in gerbilline rodents. Brain, Behavior and Evolution, 30:82-101.

Platt, S. G., T. R. Rainwater, B. W. Miller, and C. M. Miller. 2000. Notes on the mammals of Turneffe Atoll Belize. Caribbean Journal of Science, 36:166-168.

Pledge, N. S. 1990. The upper fossil fauna of the Henschke Fossil Cave, Naracoorte, South Australia. Memoirs of the Queensland Museum, 28:247-262.

Pledge, N. S. 1992. The Curramulka local fauna: A new late Tertiary fossil assemblage from Yorke Peninsula, South Australia. The Beagle, Records of the Northern Territory Museum of Arts and Sciences, 9(1):115-142.

Plumpton, D. L., and J. K. Jones, Jr. 1992. *Rhynchonycteris naso*. Mammalian Species, 413:1-5.

Pocock, R. I. 1907*a*. Exhibition of a photograph and the skull of a specimen of Pallas' cat that had recently died in the Menagerie with general remarks on the species. Proceedings of the Zoological Society of London, 1907:299-306.

Pocock, R. I. 1907*b* [1908]. Report upon a small collection of Mammalia brought from Liberia by Mr. Leonard Leighton. Proceedings of the Zoological Society of London, 1907:1037-1046.

Pocock, R. I. 1909*a*. On the agriotype of domestic asses. Annals and Magazine of Natural History, ser. 8, 4:523-528.

Pocock, R. I. 1909*b*. Ward's zebra. The Field, 114:889, 1099.

Pocock, R. I. 1915*a*. On some of the external characters of the genus *Linsang* with notes upon the genera *Poiana* and *Eupleres*. Annals and Magazine of Natural History, ser. 8, 16:341-351.

Pocock, R. I. 1915*b*. On the feet and glands and other external characters of the Viverrinae, with the description of a new genus. Proceedings of the Zoological Society of London, 1915:131-149.

Pocock, R. I. 1916a. A new genus of African mongooses, with a note on *Galeriscus*. Annals and Magazine of Natural History, ser. 8, 17:176-179.

Pocock, R. I. 1916b. On the tooth-change, cranial characters, and classification of the snow leopard or ounce (*Felis uncia*). Annals and Magazine of Natural History, ser. 8, 18:306-316.

Pocock, R. I. 1916c. On the external characters of the mongooses (Mungotidae). Proceedings of the Zoological Society of London, 1916:349-374.

Pocock, R. I. 1917a. The classification of the existing Felidae. Annals and Magazine of Natural History, ser. 8, 20:329-350.

Pocock, R. I. 1917b. The groups of the small and medium-sized South American Felidae. Annals and Magazine of Natural History, ser. 8, 20:43-47.

Pocock, R. I. 1919. The classification of the mongooses (Mungotidae). Annals and Magazine of Natural History, ser. 9, 3:515-524.

Pocock, R. I. 1920a. On the external characters of the South American monkeys. Proceedings of the Zoological Society of London, 1920:91-113.

Pocock, R. I. 1920b. On the external and cranial characters of the European badger (*Meles*) and of the American badger (*Taxidea*). Proceedings of the Zoological Society of London, 1920:423-436.

Pocock, R. I. 1921a. The external characters and classification of the Procyonidae. Proceedings of the Zoological Society of London, 1921:389-422.

Pocock, R. I. 1921b. The auditory bulla and other cranial characters in the Mustelidae. Proceedings of the Zoological Society of London, 1921:473-486.

Pocock, R. I. 1921c. On the external characters of some species of Lutrinae (otters). Proceedings of the Zoological Society of London, 1921:535-546.

Pocock, R. I. 1921d [1922]. On the external characters and classification of the Mustelidae. Proceedings of the Zoological Society of London, 1921:803-837.

Pocock, R. I. 1922a. On the external characters of some hystricomorph rodents. Proceedings of the Zoological Society, London, 1922:365-427.

Pocock, R. I. 1922b. The external characters of *Scarturus* and other jerboas compared with those of *Zapus* and *Pedetes*. Proceedings of the Zoological Society of London, 1922:659-682.

Pocock, R. I. 1923. The classification of Sciuridae. Proceedings of the Zoological Society of London, 1923(1):209-246.

Pocock, R. I. 1924. Some external characters of *Orycteropus afer*. Proceedings of the Zoological Society of London, 1924:697-706.

Pocock, R. I. 1929. Tigers. Journal of the Bombay Natural History Society, 33(3):505-541.

Pocock, R. I. 1930a. The panthers and ounces of Asia. Journal of the Bombay Natural History Society, 34(1):65-82.

Pocock, R. I. 1930b. The panthers and ounces of Asia. Part II. The panthers of Kashmir, India, and Ceylon. Journal of the Bombay Natural History Society, 34(2):307-336.

Pocock, R. I. 1930c. The lions of Asia. Journal of the Bombay Natural History Society, 34(3):638-665.

Pocock, R. I. 1932a. The black and brown bears of Europe and Asia. Part I. European and Asiatic representatives of the brown bear. Journal of the Bombay Natural History Society, 35(4):771-823.

Pocock, R. I. 1932b. The black and brown bears of Europe and Asia. Part II. The sloth bear (*Melursus*), the Himalayan black bear (*Selenarctos*) and the Malayan bear (*Helarctos*). Journal of the Bombay Natural History Society, 36(1):101-138.

Pocock, R. I. 1932c. The leopards of Africa. Proceedings of the Zoological Society of London, 1932(2):543-541.

Pocock, R. I. 1932d. The marbled cat (*Pardofelis marmorata*) and some other Oriental species, with the definition of a new genus of the Felidae. Proceedings of the Zoological Society of London, 1932:741-766.

Pocock, R. I. 1933a. The civet-cats of Asia. Journal of the Bombay Natural History Society, 36(2):421-449.

Pocock, R. I. 1933b. The civet cats of Asia. Part II. Journal of the Bombay Natural History Society, 36(3):629-656.

Pocock, R. I. 1933c. The palm civets or 'toddy cats' of the genera *Paradoxurus* and *Paguma* inhabiting British India. Part I. Journal of the Bombay Natural History Society, 36(4):855-877.

Pocock, R. I. 1933d. The rarer genera of oriental Viverridae. Proceedings of the Zoological Society of London, 1933:969-1035.

Pocock, R. I. 1934a. The geographical races of *Paradoxurus* and *Paguma* found to the east of the bay of Bengal. Proceedings of the Zoological Society of London, 1934:613-683.

Pocock, R. I. 1934b. The palm civets or 'toddy cats' of the genera *Paradoxurus* and *Paguma* inhabiting British India. Part III. Journal of the Bombay Natural History Society, 37(2):314-346.

Pocock, R. I. 1934c[1935]. The races of the striped and brown hyaenas. Proceedings of the Zoological Society of London, 1934:799-825.

Pocock, R. I. 1936*a*. The oriental yellow-throated marten (*Lamprogale*). Proceedings of the Zoological Society of London, 1936:531-553.

Pocock, R. I. 1936*b*. The polecats of the genera *Putorius* and *Vormela* in the British Museum. Proceedings of the Zoological Society of London, 1936(2):691-723.

Pocock, R. I. 1937. The mongooses of British India, including Ceylon and Burma. Journal of the Bombay Natural History Society, 39(2):211-245.

Pocock, R. I. 1939*a*. The fauna of British India, including Ceylan and Burma. Mammalia. Vol. 1. Primates and Carnivora (in part). Taylor and Francis, Ltd., London, 463 pp.

Pocock, R. I. 1939*b*. The races of jaguar (*Panthera onca*). Novitates Zoologicae, 41:406-422.

Pocock, R. I. 1940*a*. The hog-badgers (*Arctonyx*) of British India. Journal of the Bombay Natural History Society, 41(3):461-469.

Pocock, R. I. 1940*b*. Notes on some British Indian otters. Journal of the Bombay Natural History Society, 41:514-517.

Pocock, R. I. 1940*c*. The races of Geoffroy's cat (*Oncifelis geoffroyi*). Annals and Magazine of Natural History, ser. 11, 6:350-355.

Pocock, R. I. 1941*a*. The fauna of British India, including Ceylon and Burma. Mammalia. Vol. II. Carnivora (suborders Aeluroidae (part) and Arctoidae). Taylor and Francis, Ltd., London, 503 pp.

Pocock, R. I. 1941*b*. Some new geographical races of *Leopardus*, commonly known as Ocelots and Margays. Annals and Magazine of Natural History, ser. 11, 8:234-239.

Pocock, R. I. 1941*c*. The races of the ocelot and the margay. Field Museum of Natural History Zoological Series, 27:319-369.

Pocock, R. I. 1941*d*. The examples of the Colocolo and of the Pampas Cat in the British Museum. Annals and Magazine of Natural History, ser. 11, 7:257-274.

Pocock, R. I. 1951. Catalogue of the genus *Felis*. British Museum (Natural History), London, 190 pp.

Pocock, T. N. 1976. Pliocene mammalian microfauna from Langebaanweg: A new fossil genus linking the Otomyinae with the Murinae. South African Journal of Science, 72:58-60.

Pocock, T. N. 1987. Plio-Pleistocene fossil mammalian microfauna of southern Africa—a preliminary report including description of two new fossil muroid genera (Mammalia: Rodentia). Palaeontologia Africana, 26:69-91.

Poduschka, W. 1990. The enigmatic hedgehogs of Somalia. Biogeographia, 14:485-498.

Poduschka, W., and C. Poduschka. 1982. Die taxonomische Zugehörigkeit von *Dasogale fontoynonti* G. Grandidier, 1928. Sitzungsberichte der Österreichischen Akademie der Wissenschaften, Mathematisch-Naturwissenschaftliche Klasse, Abteilung I, 191:253-264.

Poduschka, W., and C. Poduschka. 1983. The taxonomy of the extant Solenodontidae (Mammalia: Insectivora), a synthesis. Sitzungsberichte der Österreichischen Akademie der Wissenschaften, Mathematisch-naturwissenschaftliche Klasse, Abteilung I, 192:225-238.

Poduschka, W., and C. Poduschka. 1985. Beiträge zur Kenntnis der Gattung *Podogymnura* Mearns, 1905 (Insectivora, Echinosoricinae). Sitzungsberichte der Österreichischen Akademie der Wissenschaften, Mathematisch-naturwissenschaftliche Klasse, Abteilung I, 194:1-22.

Poglayen-Neuwall, I. 1965. Gefangenschaftsbeobachtungen an Makibären (*Bassaricyon* Allen 1876). Zeitschrift für Säugetierkunde, 30:321-366.

Poglayen-Neuwall, I., and D. E. Toweill. 1988. *Bassariscus astutus*. Mammalian Species, 327:1-8.

Pohle, H. 1920*a*. Die Unterfamilie der Lutrinae. Archiv für Naturgeschichte, 85A(9):1-247.

Pohle, H. 1920*b*. Die systematische Stellung von *Amphictis* and *Nandinia*. Gesellschaft Naturforschender Freunde (Berlin), 1920(1):48-62.

Pohle, H. 1926. Notizen über africanische Elephanten. Zeitschrift für Säugetierkunde, 1:58-64.

Pokrovski, A. V., I. A. Kuznetsova, and M. I. Cheprakov. 1984. [Hybridological studies of reproductive isolation of the *Lemmus* Palaearctic species (Rodentia, Cricetidae).] Zoologicheskii Zhurnal, 63:904-911 (in Russian with English summary).

Polak, S. 1997. The use of caves by the edible dormouse (*Myoxus glis*) in the Slovenian karst. Natura Croatica, 6(3):313-321.

Polechová, J., and R. Graciasová. 2000. [Return of *Apodemus agrarius* (Rodentia: Muridae) to Southern Moravia (Czech Republic)?] Lynx (Praha), n.s., 31:153-155 (in Czech with English abstract).

Pons, J.-M., V. Volobouev, J.-F. Ducroz, A. Tillier, and D. Reudet. 1999. Is the Guadeloupean racoon (*Procyon minor*) really an endemic species? New insights from molecular and chromosomal analyses. Journal of Zoological Systematics and Evolutionary Research, 37(2):101-108.

Poole, A. J., and V. S. Schantz. 1942. Catalog of the type specimens of mammals in the United States National Museum, including the Biological Survey's collection. Bulletin of the United States National Museum, 178:1-705.

Poole, M. A. 1994. The Hastings River mouse, *Pseudomys oralis* from Gambubal State Forest, Southeast Queensland. Memoirs of the Queensland Museum, 37(1):280.

Poole, W. E. 1979. The status of Australian Macropodidae. Pp. 13-27, *in* The status of endangered Australasian wildlife (M. J. Tyler, ed.). Royal Society of South Australia, Adelaide, 210 pp.

Poole, W. E. 1982. *Macropus giganteus*. Mammalian Species, 187:1-8.

Pope, L. C., A. Estoup, and C. Moritz. 2000. Phylogeography and population strucute of an ecotonal marsupial, *Bettongia tropica*, determined using mtDNA and microsatellites. Molecular Ecology, 9:2041-2053.

Popov, V. V. 1981. Age-dependent and specific peculiarities of the correlations among the morphological features of sympatric populations of species of the genus *Apodemus* Kaup, 1829 (Rodentia, Mammalia). Acta Zoologica Bulgarica, 17:38-51.

Popov, V. V. 1993. Discriminant criteria and comparative study on morphology and habitat selection of *Apodemus sylvaticus* (Linnaeus, 1758) and *Apodemus flavicollis* (Melchior, 1834) (Mammalia, Rodentia, Muridae) in Bulgaria. Acta Zoologica Bulgarica, 46:100-111.

Posamentier, H. 1989. Rodents in agriculture. A review of findings in Bangladesh. Deuhsche Gesellschatt fur Tehnische Zusammenarbeit (GT 2) bH. Eschborn 1989:107 pp.

Potapova, E. G. 2000. [On the position of the long-eared jerboas of the genus *Euchoreutes* in the system of Dipodoidea (Rodentia).] Pp. 135-137, *in* [Systematics and phylogeny of the rodents and lagomorphs] (A. K. Agadzhanyan and V. N. Orlov, eds.). Theriological Society, Moscow, 196 pp. (in Russian).

Potapova, E. G. 2001. Morphological patterns and evolutionary pathways of the middle ear in dormice (Gliridae, Rodentia). Trakya University Journal of Scientific Research, 2(2):159-170.

Potter, M., J. H. Nadeau, and M. P. Cancro (eds.). 1986. The wild mouse in immunology. Current Topics in Microbiology and Immunology, 127:1-395.

Poulet, A. R. 1984. Quelques observations sur la biologie de *Desmodilliscus braueri* Wettstein (Rodentia, Gerbillidae) dans le Sahel du Sénégal. Mammalia, 48(1):59-64.

Pousargues, F. de. 1894. Description d'une nouvelle espéce de mammifères du genre *Crossarchus*, et considerations sur la répartition geographique des crossarques rayes. Archives du Museum National d'Histoire Naturelle (Paris), ser. 3, 6:121-134.

Pousargues, F. de. 1898. Sur l'identitè specifique du *Felis Bieti* (A.M. Edw.) et du *Felis pallida* (Büchn.). Bulletin du Muséum National d'Historie Naturelle (Paris), 4:357-359.

Powell, C. B., and H. van Rompaey. 1998. Genets of the Niger Delta, Nigeria. Small Carnivore Conservation, 19:1-7

Powell, R. A. 1981. *Martes pennanti*. Mammalian Species, 156:1-6.

Pozdnyakov, A. A. 1993. [Morphotypic variability of molars in meadow voles of the "*maximowiczi*" group (Rodentia, Arvicolidae, *Microtus*): An attempt at quantitative statistical analysis.] Zoologischeskii Zhurnal, 72(11):114-125 (in Russian with English summary).

Pozdnyakov, A. A. 1996. [On the phylogeny of voles of the subgenus *Alexandromys* (Rodentia, Arvicolidae, *Microtus*). Variability and paleontoloical records.] Zoologischeskii Zhurnal, 75(1):133-140 (in Russian with English summary).

Pozdnyakov, A. A., and Yu. N. Litvinov. 1994. [Ecogeographic interpretation of morphotypical variability of the molars in *Microtus oeconomus* (Rodentia, Arvicolidae).] Zoologischeskii Zhurnal, 73(2):151-157 (in Russian with English summary).

Pozo de la Tijera, C., and J. E. Excobedo Cabrera. 1999. Mamíferos terrestres de la Reserva de la Biosfera de Sian Ka'an, Quintana Roo, México. Revista de Biología Tropical, 47:251-262.

Pradhan, M. S., and A. M. Bhagwat. 1990. Polyacrylamide gel electrophoretic analysis of some species specific proteins in six Indian rodent genera (subfam.: Gerbillinae and Murinae: Fam: Muridae). Proceedings of the Indian Academy of Science (Animal Science), 99(1):67-72.

Pradhan, M. S., A. Mondal, and V. C. Agrawal. 1989. Proposal of an additional species in the genus *Bandicota* Gray (order: Rodentia; fam: Muridae) from India. Mammalia, 53(3):369-376.

Pradhan, M. S., A. M. Bhagwat, and S. T. Ingaij. 1991. Haemoglobin polymorphism and genetic identities in five Indian commensal rodent species. October 10, 2002 Journal of the Bombay Natural History Society, 88(2):229-233.

Pradhan, M. S., A. K. Mondal, A. M. Bhagwat, and V. C. Agrawal. 1993. Taxonomic studies of Indian bandicoot rats (Rodentia: Muridae: Murinae) with description of a new species. Records of the Zoological Survey of India, 93(1-2):175-200.

Pradhan, M. S., R. M. Sharma, and K. Shankar. 1997. First record of Kelaart's long-clawed shrew, *Ferocolus ferocolus* (Kelaart)(Insectivora, Soricidae, Crocidurinae) from Peninsular India. Mammalia, 61:448-450.

Prager, E. M., R. D. Sage, U. Gyllensten, W. K. Thomas, R. Hübner, C. S. Jones, L. Noble, J. B. Searle, and A. C. Wilson. 1993. Mitochondrial DNA sequence diversity and the colonization of Scandinavia by house mice from East Holstein. Biological Journal of the Linnean Society, 50:85-122.

Prager, E. M., H. Tichy, and R. D. Sage. 1996. Mitochondrial DNA sequence variation in the Eastern house mouse, *Mus musculus*: Comparison with other house mice and report of a 75-bp tandem repeat. Genetics, 143:427-446.

Prager, E. M., C. Orrego, and R. D. Sage. 1998. Genetic variation and phylgeography of Central Asian and other house mice, including a major new mitochondrial lineage in Yemen. Genetics, 150:835-861.

Prakash, I., P. Singh, and A. Saravanan. 1995*a*. Small mammals of the Abu Hill, Aravalli Ranges in Rajasthan, India. A comprehensive taxonomic and ecological study. Journal of Pure and Applied Zoology, 5(1):55-64.

Prakash, I., P. Singh, and A. Saravanan. 1995*b*. Ecological distribution of small mammals in the Aravalli Ranges. Proceedings of the Indian Natural Science Academy, 61B(2):137-148.

Prakash, I., A. Saravanan, and P. Singh. 1995*c*. Ecology and taxonomy of the field mice in the Aravalli Ranges. Journal, Bombay Natural History Society, 92(3):372-377.

Prakash, K. L. S., and N. V. Aswathanarayana. 1973. Diversity of the karyotype in the Indian long-tailed tree mouse, *Vandeleuria oleracea* (Bennett). Current Science, 42:755-756.

Prakash, K. L. S., and N. V. Aswathanarayana. 1976. Chromosome complexity in the Indian long-tailed tree mouse. Journal of Heredity, 67:249-250.

Prasad, M. R. N. 1957. Male genital tract of the Indian and Ceylonese palm squirrels and its bearing on the systematics of the Sciuridae. Acta Zoologica, 38:1-26.

Prater, S. H. 1980. The book of Indian animals. Third ed., corrected. Bombay Natural History Society, 324 pp.

Preble, E. A. 1899. Revision of the jumping mice of the genus *Zapus*. North American Fauna, 15:1-41.

Preble, E. A. 1902. A biological investigation of the Hudson Bay region. North American Fauna, 22:1-140.

Preble, E. A. 1908. A biological investigation of the Athabaska-Mackenzie region. North American Fauna, 27:1-574.

Presley, S. J. 2000. *Eira barbara*. Mammalian Species, 636:1-6.

Price, P. K., and M. L. Kennedy. 1980. Genic relationships in the white-footed mouse, *Peromyscus leucopus*, and the cotton mouse, *Peromyscus gossypinus*. American Midland Naturalist, 103:73-82.

Prigioni, C. 1996. Distribution of mammals in Albania. Hystrix, n.s., 8(1-2):67-73.

Pringle, J. A. 1977. The distribution of mammals in Natal: 2. Carnivora. Annals of the Natal Museum, 23(1):93-116.

Prins, H. H. T. 1990. Geographic variation in skulls of the nearly extinct small black rhinoceros *Diceros bicornis michaeli* in northern Tanzania. Zeitschrift für Säugetierkunde, 55:260-269.

Pritchard, J. S. 1989. Ilin Island cloud rat extinct? Oryx, 23:126.

Pritchard, P. C. H. 1994. Comment on gender and declension of generic names. Journal of Mammalogy, 75:549-550.

Profus, P. 2000. Nowe stanowiska popielicy *Glis glis* w Polsce oraz uwagi o jej wystepowaniu w niektorzych jaskiniach Europy [New stations of *Glis glis* in Poland and remarks on its occurrence in some European caves]. Chronmy Przyrode Ojczysta, 56(2):44-50.

Provensal, M. C., and J. L. Polop. 1993. Morphometric variation in populations of *Calomys musculinus*. Studies in Neotropical Fauna and Environment, 28:95-105.

Pucek, Z. 1982. Familie Zapodidae Coues, 1875—Hüpfmäuse. Pp. 497-538, *in* Handbuch der Säugetiere Europas (Niethammer, H. J. and F. Krapp, eds.). Akademische Verlagsgesellschaft (Wiesbaden), 2/I:1-649.

Pucek, Z. 1983*a*. *Eliomys quercinus* (Linnaeus, 1766). Pp. 134, *in* Atlas of Polish mammals (Z. Pucek and J. Raczynski, eds.). Polish Academy of Sciences, Mammals Research Institute, Warszawa. 1–188+1–183.

Pucek, Z. 1983*b*. *Dryomys nitedula* (Pallas, 1779). Pp. 134-135, *in* Atlas of Polish mammals (Z. Pucek and J. Raczynski, eds.). Polish Academy of Sciences, Mammals Research Institute, Warszawa. 1–188+1–183.

Pucek, Z. 1983*c*. *Glis glis* (Linnaeus, 1766). Pp. 135-136, *in* Atlas of Polish mammals (Z. Pucek and J. Raczynski, eds.). Polish Academy of Sciences, Mammals Research Institute, Warszawa. 1–188+1–183.

Pucek, Z. 1983*d*. *Muscardinus avellanarius* (Linnaeus, 1758). Pp. 137-138, *in* Atlas of Polish mammals (Z. Pucek and J. Raczynski, eds.). Polish Academy of Sciences, Mammals Research Institute, Warszawa. 1–188+1–183.

Pumo, D. E., E. Z. Goldin, B. Ellicot, C. J. Phillips, and H. H. Genoways. 1988. Mitochondrial DNA polymorphism in three Antillean island populations of the fruit bat, *Artibeus jamaicensis*. Molecular Biology and Evolution, 5:79-89.

Pumo, D. E., I. Kim, J. Remsen, C. J. Phillips, and H. H. Genoways. 1996. Molecular systematics of the fruit bat, *Artibeus jamaicensis*: Origin of an unusual island population. Journal of Mammalogy, 77:491-503.

Puzachenko, A. Yu. 1993. [Geographical variation of the skull of gigantic mole rat *Spalax giganteus* (Spalacidae, Rodentia).] Zoologicheskii Zhurnal, 72(1):112-119 (in Russian with English summary).

Puzachenko, G. Yu., and V. A. Lapshov. 1994. [Analysis of morphometric reflections in morphology: Mandible of the rat (*Rattus*, Rodentia), a study case.] Zhurnal Obschchei Biologii, 55(1):96-109 (in Russian with English summary).

Pye, J. D. 1972. Bimodal distribution of constant frequencies in some hipposiderid bats (Mammalia: Hipposideridae). Journal of Zoology, London, 166:323-335.

Pye, T., R. Swain, and R. D. Seppelt. 1999. Distribution and habitat use of the feral black rat (*Rattus rattus*) on subantarctic Macquarie Island. Journal of the Zoological Society of London, 247:429-438.

Qian Yan-wen, Zhang Jie, Zheng Bao-lie, Wang Song, Guan Guan-xun, Shen Xiao-zhou. 1965. [Mammals and birds of southern Xinjiang]. Science Press, Beijing.

Qin Chang-yu. 1991. [On the faunistics and regionalization of glires in Ningxia Autonomous Region.] Acta Theriologica Sinica, 4(4):320 (in Chinese).

Qiu, Z.-D. 1985. The Neogene mammalian faunas of Ertemte and Harr Obo in Inner Mongolia (Nei Mongol), China.–3. Jumping mice–Rodentia: Lophocricetinae. Senckienbergiana Lethaea, 66(1/2):39-67.

Qiu, Z-D. 1989. [Fossil Platacanthomyids from the hominoid locality of Lufeng, Yunnan.] Vertebrata PalAsiatica, 27(4):268-283 (in Chinese with English abstract).

Qiu, Z.-D. 2001. Glirid and gerbillid rodents from the middle Miocene Quantougou fauna of Lanzhou, Gansu. Vertebrata PalAsiatica, 39(4):299-305.

Qiu, Z.-D., and G. Storch. 2000. The early Pliocene micromammalian fauna of Bilike, Inner Mongolia, China (Mammalia: Lipotyphla, Chiroptera, Rodentia, Lagomorpha). Senckenbergianaa Lethaea, 80(1):173-229.

Qiu, Zhuding, and Li Chuankuei. 2003. Rodents from the Chinese Neogene: Biogeographic relationships with Europe and North America. Pp. 586-602, *in* Vertebrate fossils and their context: Contributions in honor of Richard H. Tedford (L. J. Flynn, ed.). Bulletin of the American Museum of Natural History, 279, 666 pp.

Quay, W. B. 1954*a*. The Meibomian glands of voles and lemmings (Microtinae). Miscellaneous Publications, Museum of Zoology, University of Michigan, 82:1-17.

Quay, W. B. 1954*b*. The anatomy of the diastemal palate in microtine rodents. Miscellaneous Publications, Museum of Zoology, University of Michigan, 86:1-41.

Quay, W. B. 1968. The specialized posterolateral sebaceous glandular regions in microtine rodents. Journal of Mammalogy, 49:427-445.

Queale, L. F. 1997. Field identification of female little brown bats *Vespadelus* spp. (Chiroptera: Vespertilionidae) in South Australia. Records of the South Australian Museum, 30:29-33.

Queiroz, A. I., A. Bertrand, and K. Khakin. 1996. Status and conservation of Desmaninae in Europe. Council of Europe Publication. Nature and Environment, 76:1-79.

Queroil, S., R. Hutterer, P. Barrière, M. Colyn, J. C. Kerbis Peterhans, and E. Verheyen. 2001. Phylogeny and evolution of African shrews (Mammalia: Soricidae) inferred from 16s rRNA sequences. Molecular Phylogenetics and Evolution, 20:185-195.

Queroil, S., E. Verheyen, M. Dillen, and M. Colyn. 2003. Patterns of diversification in two African forest shrews: *Sylvisorex johnstoni* and *Sylvisorex ollula* (Soricidae, Insectivora) in relation to paleo-environmental changes. Molecular Phylogenetics and Evolution, 28:24-37.

Querouil, S., P. Barriere, M. Colyn, R. Hutterer, A. Dudu, M. Dillen, and E. Verheyen. In Press. A molecular insight into the systematics of African *Crocidura* (Crocidurinae, Soricidae) using 16s rRNA sequences. *in*, Biology of the Soricidae II. (J. Merritt, S. Churchfield, R. Hutterer, and B. Sheftel, eds.), Carnegie Museum Special Publication.

Qui Yu-huang. 1989. [A systematic cluster for the Chinese cape hare, *Lepus capensis*.] Acta Theriologica Sinica, 9:168-172 (in Chinese).

Quintana, C. A. 1996. Diversidad del roedor *Microcavia* (Caviomorpha, Caviidae) de America del sur. Mastozoologia Neotropical, 3(1):63-86.

Quintana, C. A. 2002. Roedores cricétidos del Sanandresense (Plioceno Tardío) de la Provincia de Buenos Aires. Mastozoología Neotropical, 9:263-275.

Qumsiyeh, M. B. 1985. The bats of Egypt. Special Publications, The Museum, Texas Tech University Press, 23:1-102.

Qumsiyeh, M. B. 1986. Phylogenetic studies of the rodent family Gerbillidae. I. Chromosomal evolution in the southern African complex. Journal of Mammalogy, 67:680-692.

Qumsiyeh, M. B. 1996. Mammals of the Holy Land. Texas Tech University Press, Lubbock, 389 pp.

Qumsiyeh, M. B., and J. W. Bickham. 1993. Chromosomes and relationships of long-eared bats of the genera *Plecotus* and *Otonycteris*. Journal of Mammalogy, 74:376-382.

Qumsiyeh, M. B., and R. K. Chesser. 1988. Rates of protein, chromosome and morphological evolution in four genera of rhombomyine gerbils. Biochemical Systematics and Ecology, 16:89-103.

Qumsiyeh, M. B., and J. K. Jones, Jr. 1986. *Rhinopoma hardwickii* and *Rhinopoma muscatellum*. Mammalian Species, 263:1-5.

Qumsiyeh, M. B., and D. A. Schlitter. 1991. Cytogenetic data on the rodent family Gerbillidae. Occasional Papers, The Museum, Texas Tech University, 144:1-20.

Qumsiyeh, M. B., D. A. Schlitter, and A. M. Disi. 1986. New records and karyotypes of small mammals from Jordan. Zeitschrift für Säugetierkunde, 51:139-146.

Qumsiyeh, M. B., M. J. Hamilton, and D. A. Schlitter. 1987. Problems in using Robertsonian rearrangements in determining monophyly: Examples from the genera *Tatera* and *Gerbillurus*. Cytogenetics and Cell Genetics, 44:198-208.

Qumsiyeh, M. B., C. Sanchez-Hernandez, S. K. Davis, J. C. Patton, and R. J. Baker. 1988a. Chromsomal evolution in *Geomys* as revealed by G- and C-band analysis. Southwestern Naturalist, 33: 1-13.

Qumsiyeh, M. B., R. D. Owen, and R. K. Chesser. 1988b. Differential rates of genic and chromosomal evolution in bats of the family Rhinolophidae. Genome, 30:326-35.

Qumsiyeh, M. B., S. W. King, J. Arroyo-Cabrales, I. R. Aggundey, D. A. Schlitter, R. J. Baker, and K. J. Morrow, Jr. 1990. Chromosomal and protein evolution in morphologically similar species of *Praomys sensu lato* (Rodentia, Muridae). Journal of Heredity, 81:58-65.

Qumsiyeh, M. B., M. J. Hamilton, E. R. Dempster, and R. J. Baker. 1991. Cytogenetics and systematics of the rodent genus *Gerbillurus*. Journal of Mammalogy, 72:89-96.

Qumsiyeh, M. B., A. M. Disi, and Z. S. Amr. 1992. Systematics and distribution of the bats (Mammalia: Chiroptera) of Jordan. Dirasat Series B, Pure and Applied Sciences, 19B(2):101-118.

Rabeder, G. 1972. Die Insectivoren und Chiropteren (Mammalia) aus dem Altpleistozän von Hundsheim (Niederösterreich). Annalen des Naturhistorischen Museums in Wien, 76:375-474.

Rabinowitz, A. 1996. Discovery on Ice Mountain. Wildlife Conservation, 99(4), 10.

Rabinowitz, A., and Saw Tun Khaing. 1998. Status of selected mammal species in North Myanmar. Oryx, 32:201-208.

Rabinowitz, A., G. Amato and Saw Tun Khaing. 1998. Discovery of the black muntjac, *Muntiacus crinifrons* (Artiodactyla, Cervidae) in north Myanmar. Mammalia, 62;105-108.

Rachowiak, P. 1993. [Dentition of root vole, *Microtus oeconomus* (Pallas, 1776) from the isolated population from Notecka Forests.] Badania Fizjograficzne Nad Polska Zachodnia, 39, Ser. C, Zoologia:77-89 (in Polish with English summary).

Radinsky, L. B. 1973. Are stink badgers skunks? Implications of neuroanatomy for mustelid phylogeny. Journal of Mammalogy, 54:585-593.

Radinsky, L. B. 1975. Viverrid neuroanatomy: Phylogenetic and behavorial implications. Journal of Mammalogy, 56(1):130-150.

Radjabli, S. I., M. N. Meier, F. N. Golenishchev, and A. A. Isaenko. 1984. [Karyological peculiarities of the Mongolian vole and its relations within the subgenus *Microtus* (Rodentia, Cricetidae)]. Zoologicheskii Zhurnal, 63:441-446 (in Russian).

Radosavlievic, J., M. Vujosevic, and S. Zivkovic. 1990. [Chromosome banding of five arvicolid rodent species from Yugoslavia]. Arhiv Bioloskih Nauka (Beograd), 42:183-194 (in Russian).

Radtke, M., and J. Niethammer. 1984 [1985]. Zur Stellung der Pestratte (*Nesokia indica*) im System der Murinae. Säugetierkundliche Mitteilungen, 32:13-16.

Raffles, T. S. 1821. Descriptive catalogue of a zoological collection, made on account of the honorable East Indian Company, in the island of Sumatra and its vicinity, under the direction of Sir Thomas Stamford Raffles, Lieutenant-Governor of Fort Marlborough, with additional notices illurstrative of the natural history. Transactions of the Linnean Society of London, 13:239-274.

Rafinesque, C. S. 1815. Analyse de la nature. Palerme, 244 p.

Raghavan, P. 1989. Fossil rodent assemblages from the basal Pinjor formation (subHimalaya), Haryana, India: Taxonomy and palaeoecology. Research Bulletin (Science) of the Panjab University, 40 (pts. III-IV):245-268.

Rahm, U. 1967. Les Murides des environs du Lac Kivu et des regions voisines (Afrique Centrale) et leur ecologie. Revue Suisse de Zoologie, 74:439-519.

Rahm, U. 1970. Ecology, zoogeography and systematics of some African forest monkeys. Pp. 591-626, *in* Old World Monkeys (J. R. Napier and P. H. Napier, eds.). Academic Press, London, 660 pp.

Rahm, U. 1976. Zur Morphologie des Magens von *Tachyoryctes splendens* Ruppell, 1835 (Rodentia, Rhizomyidae). Säugetierkundliche Mitteilungen, 24:148-150.

Rahm, U. 1980. Die afrikanische Wurzelratte, *Tachyoryctes*. Die Neue Brehm-Bücherei, 528:1-60.

Rahm, U., and A. Christiaensen. 1963. Les mammiferes de la region occidentale du Lac Kivu. Annales Musée Royal de l'Afrique Centrale, Tervuren-Belgique, ser. 8, 118:1-83.

Raicu, P., and S. Bratosin. 1965. Le caryotype chez le *Mesocricetus newtoni* (Nehring, 1898). Zeitschrift für Säugetierkunde, 31:251-255.

Rainey, D. G., and R. H. Baker. 1955. The pigmy woodrat, *Neotoma goldmani*, its distribution and systematic position. University of Kansas Publications, Museum of Natural History, 7:619-624.

Rajagopalan, P. K. 1970. A note on the arboreal nests of some rodents. Indian Journal of Medical Research, 58:1192-1194.

Rakotondravony, D., S. M. Goodman, J.-M. Duplantier, and V. Soarimalala. 1998. Les petit mammifères. Pp. 197-212, *in* Inventaire biologique de la forêt littorale de Tampolo (Fenoarivo Atsinanana) (J. Ratsirarson and S. M. Goodman, eds.). Recherches pour le développement, Série Sciences Biologiques, no. 14, Centre d'Information et de Documentation Scientifique et Technique, Antananarivo, 222 pp.

Ralls, K. 1973. *Cephalophus maxwelli*. Mammalian Species, 31:1-4.

Ralls, K. 1978. *Tragelaphus eurycerus*. Mammalian Species, 111:1-4.

Ramalhinho, M. G. 1985. On the taxonomic position of Portuguese mole. Arquivos do Museu Bocage, ser. A, 3:1-12.

Ramalhinho, M. G., and R. Libois. 2001. The karyotype of the Formentera Island garden dormouse, *Eliomys quercinus ophiusae*. Belgian Journal of Zoology, 131(1):83-85.

Ramalhinho, M. G., and R. Libois. 2002. First report on the presence in France of a *B*-chromosome polymorphism in *Apodemus flavicollis*. Mammalia, 66(2):300-303.

Ramalhinho, M. G., and M. da L. Mathias. 1988. *Arvicola terrestris monticola* de Selys-Longchamps, 1838, new to Portugal (Rodentia, Arvicolidae). Mammalia, 52:429-431.

Ramalhinho, M. G., M. L. Mathias, M. Santos-Reis, R. Libois, R. Fons, F. Petrucci-Fonseca, M. M. Oom, and M. Collares-Pereira. 1996. First approach on the skull morphology of the black rat (*Rattus rattus*) from Terceira and Sao Miguel Islands (Azores Archipelago). Vie Milieu, 46(3/4):245-251.

Raman, R., and T. Sharma. 1974. DNA replication, G- and C-bands and meiotic behaviour of supernumerary chromosomes of *Rattus rattus* (Linn.). Chromosoma, 45:111-119.

Raman, R., and T. Sharma. 1977. Karyotype evolution and speciation in genus *Rattus* Fischer. Journal of Scientific and Industrial Research, 36:385-404.

Ramanamanjato, J.-B., and J. U. Ganzhorn. 2001. Effects of forest fragmentation, introduced *Rattus rattus* and the role of exotic tree plantations and secondary vegetation for the conservation of an endemic rodent and a small lemur in littoral forests of southeastern Madagascar. Animal Conservation, 4:175-183.

Rambau, R. V., T. J. Robinson, and R. Stanyon. 2003. Molecular genetics of *Rhabdomys pumilio* subspecies boundaries: mtDNA phylogeography and karyotypic analysis by fluorescence in situ hybridization. Molecular Phylogenetics and Evolution, 28:564-575.

Ramírez, P. B., J. W. Bickham, J. K. Braun, and M. A. Mares. 2001. Geographic variation in genome size of *Graomys griseoflavus* (Rodentia: Muridae). Journal of Mammalogy, 82:102-108.

Ramirez-Pulido, J., A. Castro-Campillo, and M. Martinez-Coronel. 1991. Variacion no geografica de *Microtus quasiater* (Rodentia: Arvicolidae) con notas sobre su ecologia y reproduccion. Anales Instituto Biologia Universidad Nacional Autonoma México, Series Zoologia, 62:341-364.

Ramírez-Pulido, J., A. Castro-Campillo, and U. Aguilera. 1995. Sinopsis de los mamíferos del Estado de México, México. Revista de la Sociedad Mexicana de Historia Natural, 46:205-246.

Ramírez-Pulido, J., A. Castro-Campillo, J. Arroyo-Cabrales, and F. A. Cervantes. 1996. Lista taxonómica de los mamíferos terrestres de México. Occasional Papers The Museum Texas Tech University, 158:62 pp.

Ramírez-Pulido, J., A. Castro-Campillo, and A. Salame-Méndez. 2001. Los *Peromyscus* (Rodentia: Muridae) en la colección de mamíferos de la Universidad Autónoma Metropolitana-Unidad Iztapalapa (UAMI). Acta Zoologica Mexicana, n.s., 83:83-114.

Rana, B. D. 1985. Ecological distribution of *Rattus meltada* in India. Journal of the Bombay Natural History Society, 82:573-580.

Rana, S. A., M. A. Beg, and A. A. Khan. 1998. The pygmy mice of Central Punjab, Pakistan. Pakistan Journal of Zoology, 30(2):91-97.

Ranck, G. L. 1968. The rodents of Libya: Taxonomy, ecology, and zoogeographical relationships. Bulletin of the United States National Museum, 275:1-264.

Randi, E., and B. Ragni. 1986. Multivariate analysis of craniometric characters in European wild cat, domestic cat, and African wild cat (genus *Felis*). Zeitschrift für Säugetierkunde, 51:243-251.

Randi, E., V. Lucchini, and H. D. Chiong. 1996. Evolutionary genetics of the Suiformes as reconstructed using mtDNA sequencing. Journal of Mammalian Evolution, 3:163-194.

Randi, E., N. Mucci, M. Pierpuoli, and E. Douzery. 1998. New phylogenetic perspectives on the Cervidae (Artiodactyla) are provided by the mitochondrial cytochrome b gene. Proceedings of the Royal Society. Biological Sciences, 265:793-801.

Randi, E., N. Mucci, F. Claro-Hergueta, A. Bonnet, and E. J. P. Douzery. 2001. A mitochondrial DNA control region phylogeny of the Cervinae: Speciation in *Cervus* and implications for conservation. Animal Conservation, 4:1-11.

Randi, E., J.-P. d'Huart, V. Lucchini, and R. Aman. 2002. Evidence of two genetically deeply divergent species of warthog, *Phacochoerus africanus* and *P. aethiopicus* (Artiodactyla: Suiformes) in East Africa. Mammalian Biology, 67:91-96.

Rao, K. S., and N. V. Aswathanarayana. 1978. Karyological analysis of the Horsfield's shrew from peninsular India. Journal of Heredity, 69:202-204.

Rao, S. R. V., and S. C. Lakhotia. 1972. Chromosomes of *Rattus blanfordi*. Journal of Heredity, 63:44-47.

Rao, S. R. V., V. C. Shah, and C. Seshadri. 1968. Studies on rodent chromosomes III. The somatic chromosomes of the Indian gerbil, *Tatera indica cuverii* (Waterhouse). Cytologia, 33(3-4):477-481.

Rao, S. R. V., K. Vasantha, B. K. Thelma, R. C. Juyal, and S. C. Jhanwar. 1983. Heterochromatin variation and sex chromosome polymorphism in *Nesokia indica*: A population study. Cytogenetics and Cell Genetics, 35:233-237.

Rasoloarison, R. M., S. M. Goodman, and J. U. Ganzhorn. 2000. Taxonomic revision of mouse lemurs (*Microcebus*) in the western portions of Madagascar. International Journal of Primatology, 21:963-1919.

Ratcliffe, J. M. 2002. *Myotis welwitschii*. Mammalian Species, 701:1-3.

Rathbun, G. B. 1979. *Rhynchocyon chrysopygus*. Mammalian Species, 117:1-4.

Ratomponirina, C., E. Viegas-Péquignot, B. Dutrillaux, F. Petter, and Y. Rumpler. 1986. Synaptonemal complexes in Gerbillidae: Probable role of intercalated heterochromatin in gonosome-autosome translocations. Cytogenetics and Cell Genetics, 43:161-167.

Ratomponirina, C., E. Viegas-Péquignot, F. Petter, B. Dutrillaux, and Y. Rumpler. 1989. Synaptonemal complex study in some species of Gerbillidae without heterochromatin interposition. Cytogenetics and Cell Genetics, 52:23-27.

Rau, R. E. 1978. Additions to the revised list of preserved material of the extinct Cape Colony quagga and notes on the relationship and distribution of southern plains zebras. Annals of the South African Museum, 77:27-45.

Rausch, R. L. 1953. On the status of some Arctic mammals. Arctic (Journal of the Arctic Institute of North America), 6:91-148.

Rausch, R. L. 1963*a*. Geographic variation in size in North American brown bears, *Ursus arctos* L., as indicated by condylobasal length. Canadian Journal of Zoology, 41:33-45.

Rausch, R. L. 1963*b*. A review of the distribution of Holarctic Recent mammals. Pp. 29-43, *in* Pacific basin biogeography (J. L. Gressitt, ed.). Bishop Museum Press, Honolulu, 563 pp.

Rausch, R. L. 1964. The specific status of the narrow-skulled vole (subgenus *Stenocranius* Kashchenko) in North America. Zeitschrift für Säugetierkunde, 29:343-358.

Rausch, R. L. 1977. On the zoogeography of some Beringian mammals. Pp. 162-177, *in* Uspekhi sovremennoi teriologii [Advances in modern theriology] (V. E. Sokolov, ed.). Nauka, Moscow, 296 pp. (in Russian).

Rausch, R. L., and V. R. Rausch. 1965. Cytogenetic evidence for the specific distinction of an Alaskan marmot, *Marmota broweri* Hall and Gilmore (Mammalia: Sciuridae). Chromosoma (Berlin), 16:618-623.

Rausch, R. L., and V. R. Rausch. 1968. On the biology and systematic position of *Microtus abbreviatus* Miller, a vole endemic to the St. Matthew Islands, Bering Sea. Zeitschrift für Säugetierkunde, 33:65-99.

Rausch, R. L., and V. R. Rausch. 1971. The somatic chromosomes of some North American marmots. Mammalia, 35:85-101.

Rausch, R. L., and V. R. Rausch. 1972. Observations on chromosomes of *Dicrostonyx torquatus stevensoni* Nelson and chromosomal diversity in varying lemmings. Zeitschrift für Säugetierkunde, 37:372-384.

Rausch, R. L., and V. R. Rausch. 1975*a*. Relationships of the red-backed vole, *Clethrionomys rutilus* (Pallas) in North America: Karyotypes of the subspecies *dawsoni* and *albiventer*. Systematic Zoology, 24:163-170.

Rausch, R. L., and V. R. Rausch. 1975*b*. Taxonomy and zoogeography of *Lemmus* spp. (Rodentia, Arvicolinae), with notes on laboratory-reared lemmings. Zeitschrift für Säugetierkunde, 40:8-34.

Rausch, R. L., and V. R. Rausch. 1995. The taxonomic status of the shrew of St. Lawrence Island, Bering Sea (Mammalia: Soricidae). Proceedings of the Biological Society of Washington, 108:717-728.

Rausch, R. L., and V. R. Rausch. 1997. Evidence for specific independence of the shrew (Mammalia: Soricidae) of St. Paul Island (Pribilof Islands, Bering Sea). Zeitschrift für Säugetierkunde, 62:193-202.

Rausch, V. R., and R. L. Rausch. 1974. The chromosome complement of the yellow-cheeked vole, *Microtus xanthognathus* (Leach). Canadian Journal of Genetics and Cytology, 16:267-272.

Rausch, V. R., and R. L. Rausch. 1982. The karyotype of the Eurasian flying squirrel, *Pteromys volans* (L.), with a consideration of karyotypic and other distictions in *Glaucomys* spp. (Rodentia: Sciuridae). Proceedings of the Biological Society of Washington, 95:58-66.

Rausch, V. R., and R. L. Rausch. 1993. Karyotypic characteristics of *Sorex tundrensis* Merriam (Mammalia: Soricidae), a Nearctic species of the *S. araneus*-group. Proceedings of the Biological Society of Washington, 106:410-416.

Rautenbach, I. L. 1982. Mammals of the Transvaal. Ecoplan Monograph 1. Ecoplan, Pretoria, 211 pp.

Rautenbach, I. L., and D. A. Schlitter. 1978. Revision of genus *Malacomys* of Africa (Mammalia: Muridae). Annals of Carnegie Museum, 47:385-422.

Rautenbach, I. L., G. N. Bronner, and D. A Schlitter. 1993. Karyotypic data and attendant systematic implications for the bats of southern Africa. Koedoe, 36:87-104.

Raxworthy, C. J., and F. Rakotondraparany. 1988. Mammals report. Pp. 122-131, *in* Manongarivo Special Reserve (Madagascar). 1987/88, expedition report (N. Quansah, ed.). Madagascar Environmental Research Group, United Kingdom.

Ray, C. E. 1962. Oryzomyine rodents of the Antillean subregion. Unpubl. Ph. D. dissertation, Harvard University, 356 pp.

Ray, J. C. 1995. *Civettictis civetta*. Mammalian Species, 488:1-7.

Ray, J. C., and R. Hutterer. 1996. Structure of a shrew community in the Central African Republic based on the analysis of carnivore scats, with the description of a new *Sylvisorex* (Mammalia: Soricidae). Ecotropica, 1:85-97 [1995].

Ray-Chaudhuri, S. P., and S. Pathak. 1970. Chromosome analysis in the genus *Rattus*. Mammalian Chromosomes Newsletter, 11(4):135-136.

Read, D. G. 1993. Body size in Hastings River mouse *Pseudomys oralis* (Rodentia: Muridae) from new and old locations. Australian Zoologist, 29(1-2):117-123.

Read, J., P. Copley, and P. Bird. 1999. The distribution, ecology and current status of *Pseudomys desertor* in South Australia. Wildlife Research, 26:453-462.

Rebhholz, W., and E. Harley. 1999. Phylogenetic relationships in the bovid subfamily Antilopinae based on mitochondrial DNA sequences. Molecular Phylogeny and Evolution, 12:87-94.

Redford, K. H., and J. Eisenberg. 1992. Mammals of the Neotropics, 2. The southern cone. University of Chicago Press, Chicago, IL, 430 pp.

Redford, K. H., and R. M. Wetzel. 1985. *Euphractus sexcinctus*. Mammalian Species, 252:1-4.

Redhead, T. D. 1995*a*. Fawn-footed melomys, *Melomys cervinipes*. Pp. 636-637, *in* Mammals of Australia (R. Strahan, ed.). Smithsonian Institution Press, Washington, D.C., 756 pp.

Redhead, T. D. 1995*b*. Canefield rat, *Rattus sordidus*. Pp. 661-662, *in* Mammals of Australia (R. Strahan, ed.). Smithsonian Institution Press, Washington, D.C., 756 pp.

Redhead, T. D., G. R. Singleton, K. Myers, and B. J. Coman. 1991. Mammals introduced to southern Australia. Pp. 293-308, *in* Biogeography of Mediterranean invasions (R. H. Groves and F. Di Castri, eds.). Cambridge University Press, Cambridge, xvi + 485 pp.

Reducker, D. W., T. L. Yates, and I. F. Greenbaum. 1983. Evolutionary affinites among southwestern long-eared *Myotis* (Chiroptera: Vespertilionidae). Journal of Mammalogy, 64:666-677.

Reese, C. L., J. M. Waters, J. F. Pagels, and B. L. Brown. 2001. Genetic structuring of relict populations of Gapper's red-backed vole (*Clethrionomys gapperi*). Journal of Mammalogy, 82:289-301.

Reeve, N. 1994. Hedgehogs. T. & A. D. Poyser Ltd., London, 313 pp.

Reeves, R. R., and R. L. Brownell, Jr. 1989. Susu—*Platanista gangetica* (Roxburgh, 1801) and *Platanista minor* Owen, 1853. Pp. 69-100, *in* Handbook of marine mammals: River dolphins and the larger toothed whales (S. H. Ridgway and R. Harrison, eds.). Academic Press, London, 4:1-442.

Reeves, R. R., and S. Leatherwood. 1985. Bowhead whale—*Balaena mysticetus*. Pp. 305-344, *in* Handbook of marine mammals: The sirenians and baleen whales (S. H. Ridgway and R. Harrison, eds.). Academic Press, London, 3:1-362.

Reeves, R. R., and S. Tracey. 1980. *Monodon monoceros*. Mammalian Species, 127:1-7.

Rehage, H.-O. 1984. Die Säugetiere Westfalens. Gartenschläfer—*Eliomys quercinus* (Linnaeus, 1766). Abhandlungen aus dem Westfalischen Provinzial Museum für Naturkunde, 46(4):163-167.

Rehage, H.-O., and K. Preywisch. 1984. Die Säugetiere Westfalens. Sieberschläfer—*Glis glis* (Linnaeus, 1766). Abhandlungen aus dem Westfalischen Provinzial Museum für Naturkunde, 46(4):167-172.

Rehage, H.-O., and G. Steinborn. 1984. Die Säugetiere Westfalens. Haselmaus—*Muscardinus avellanarius* (Linnaeus,1758). Abhandlungen aus dem Westfalischen Provinzial Museum für Naturkunde, 46(4):172-181.

Reich, D. E., R. K. Wayne, and D. B. Goldstein. 1999. Genetic evidence for a recent origin by hybridization of red wolves. Molecular Ecology, 8:139-144.

Reich, L. M. 1981. *Microtus pennsylvanicus*. Mammalian Species, 159:1-8.

Reichstein, H. 1957. Schädelvariabilität europäischer Mauswiesel (*Mustela nivalis* L.) und Hermeline (*Mustela erminea* L.) in Beziehung zu Verbreitung und Geschlecht. Zeitschrift für Säugetierkunde, 22:151-182.

Reichstein, H. 1982*a*. *Arvicola sapidus* Miller, 1908—Südwesteuropaische Schermaus. Pp. 211-216, *in* Handbuch der Säugetiere Europas (J. Niethammer and F. Krapp, eds.). Akademische Verlagsgesellschaft (Wiesbaden), 2/I:1-649.

Reichstein, H. 1982*b*. *Arvicola terrestris* (Linnaeus, 1758)—Schermaus. Pp. 217-252, *in* Handbuch der

Säugetiere Europas (J. Niethammer and F. Krapp, eds.). Akademische Verlagsgesellschaft (Wiesbaden), 2/I:1-649.

Reid, F. A. 1997. A field guide to the mammals of Central America and southeast Mexico. Oxford University Press, New York, 334 pp.

Reid, F. A., and C. A. Langtimm. 1993. Distributional and natural history notes for selected mammals from Costa Rica. Southwestern Naturalist, 38:299-302.

Reid, F. A., M. D. Engstrom, and B. K. Lim. 2000. Noteworthy records of bats from Ecuador. Acta Chiropterologica, 2:37-51.

Reid, J. W. R., and S. R. Morton. 1995. Forrest's mouse, *Leggadina forresti*. Pp. 555-556, *in* Mammals of Australia (R. Strahan, ed.) Smithsonian Institution Press, Washington, D.C., 756 pp.

Reig, O. A. 1958. Notas para una actualización del conocimiento de la fauna de la formación Chapadmalal. I. Lista faunística preliminar. Acta Geologica Lilloana, 2:241-253.

Reig, O. A. 1964. Roedores y marsupials del Partido de General Pueyrredon y regions ajyacentes (Provincia de Buenos Aires, Argentina). Publicaciones del Museo Municipal de Ciencias Naturales de Mar del Plata, 1:203-224.

Reig, O. A. 1965. Datos sobre la comunidad de pequenos mamiferos de la region costera del Partido de General Pueyrredon y de los partidos limitrofes (Prov. de Buenos Aires, Argentina). Physis, 25:205-211.

Reig, O. A. 1978. Roedores cricetidos del Plioceno superior de la Provincia de Buenos Aires. Publicaciones del Museo Municipal de Ciencias Naturales de Mar del Plata, 2:164-190.

Reig, O. A. 1980. A new fossil genus of South American cricetid rodents allied to *Wiedomys*, with an assessment of the Sigmodontinae. Journal of Zoology, London, 192:257-281.

Reig, O. A. 1981. Teorio del origen y desarrollo de la fauna de mamíferos de America del Sur. Publicadas por el Museo Municipal de Ciencias Naturales "Lorenzo Scaglia," Monografie Naturae, 1:182 pp.

Reig, O. A. 1984. Distribução geográfica e história evolutiva dos roedores muroideos sulamericanos (Cricetidae: Sigmodontinae). Revista Brasileira de Genetica, 7:333-365.

Reig, O. A. 1986. Diversity patterns and differentiation of high Andean rodents. Pp. 404-440, *in* High altitude tropical biogeography (F. Vuilleumier and M. Monasterio, eds.). Oxford University Press, New York, 649 pp.

Reig, O. A. 1987. An assessment of the systematics and evolution of the Akodontini, with the description of new fossil species of *Akodon* (Cricetidae: Sigmodontinae). Fieldiana: Zoology, n.s., 39:347-399.

Reig, O. A. 1989. Karyotypic repatterning as one triggering factor in cases of explosive speciation. Pp. 246-289, *in* Evolutionary biology of transient unstable populations (A. Fontevidela, ed.). Springer-Verlag, New York. 293 pp.

Reig, O. A. 1994. New species of akodontine and scapteromyine rodents (Cricetidae) and new records of *Bolomys* (Akodontini) from the upper Pliocene and Middle Pleistocene of Buenos Aires Province, Argentina. Ameghiniana, 31:99-113.

Reig, O. A., and P. Kiblisky. 1969. Chromosome multiformity in the genus *Ctenomys* (Rodentia, Octodontidae), a progress report. Chromosoma, 28:211-244.

Reig, O. A., and M. Useche. 1976. Diversidad cariotipica y sistematica en poblaciones Venezolanas de *Proechimys* (Rodentia, Echimyidae), con datos adicionales sobre poblaciones de Peru y Colombia. Acta Cientifica Venezolana, 27:132-140

Reig, O. A., J. R. Contreras, and M. J. Piantanida. 1966. Contribución a la elucidación de la sistemática de las entidades del género *Ctenomys* (Rodentia, Octodontidae). I. Relaciones de parentesco entre muestras de ocho poblaciones de tuco-tucos inferidas del estudio estadístico de variables del fenotipo y su correlación con las características del cariotipo. Contribuciones Científicas Facultad Ciéncias Exactas Naturales, Universidad Buenos Aires (Zoologia), 2:299-352.

Reig, O. A., A. Spotorno, and R. Fernandez. 1972. A preliminary survey of chromosomes in populations of the Chilean burrowing octodon rodent *Spalacopus cyanus* (Caviomorpha, Octodontidae). Biological Journal of the Linnean Society, 4:29-38.

Reig, O. A., M. Trainer, and M. A. Barros. 1979. Sur l'ídentification chromosomique de *Proechimys guyannensis* (E. Geoffroy, 1803) et de *Proechimys cuvieri* Petter, 1978 (Rodentia, Echimyidae). Mammalia, 43:501-505.

Reig, O. A., M. Aguilera, M. A. Barros, and M. Useche. 1980. Chromosomal speciation in a Rassenkreis of Venezuelan spiny rats (genus *Proechimys*; Rodentia: Echimyidae). Genetica, 52/53:291-312.

Reig, O. A., J. A. W. Kirsch, and L. G. Marshall. 1987. Systematic relationships of the living and Neocenozoic American opossum-like marsupials (suborder Didelphimorphia) with comments on the classification of these and of the Cretaceous and Paleogene New World and European metatherians. Pp. 1-92, *in* Possums and opossums: Studies in evolution (M. Archer, ed.). Surrey Beatty and Sons Pty. Ltd. and Royal Zoological Society of New South Wales, Sydney, 1:1-400, 4 pls.

Reig, O. A., M. Aguilera, and A. Perez-Zapata. 1990*a*. Cytogenetics and karyosystematics of South American oryzomyine rodents (Cricetidae: Sigmodontinae). II. High numbered karyotypes and chromosomal heterogeneity in Venezuelan *Zygodontomys*. Zeitschrift für Säugetierkunde, 55:361-370.

Reig, O. A., C. Busch, M. O. Ortells, and J. R. Contreras. 1990*b*. An overview of evolution, systematics, population biology, cytogenetics, molecular biology, and speciation in *Ctenomys*. Pp. 71-96, *in* Evolution of subterranean mammals at the organismal and molecular levels: Proceedings of the fifth International Theriological Congress held in Rome, Italy, August 22-29, 1989 (E. Nevo and O. A. Reig, eds.). A. R. Liss (Wiley-Liss), New York, 422 pp.

Reig, O. A., A. I. Massarini, M. O. Ortells, M. A. Barros, S. I. Tiranti, and F. J. Dyzenchauz. 1992. New karyotypes and c-banding patterns of the subterranean rodents of the genus *Ctenomys* (Caviomorpha, Octodontidae) from Argentina. Mammalia, 56(4):603-623.

Reig, S. 1997. Biogeographic and evolutionary implications of size variation in North American least weasels (*Mustela nivalis*). Canadian Journal of Zoology, 75:2036-2049.

dos Reis, S. F. 1994. Allometry, convergence of ontogenetic trajectories and cranial evolution in caviine rodents (Rodentia: Caviinae). Journal of Morphology, 220:342.

dos Reis, S. F., L. C. Duarte, L. R. Monteiro, and F. J. Von Zuben. 2002. Geographic variation in cranial morphology in *Thrichomys apereoides* (Rodentia: Echimyidae). II. Geographic units, morphological discontinuities, and sampling gaps. Journal of Mammalogy, 83(2):345-353.

Reise, D., and M. H. Gallardo. 1990. A taxonomic study of the South American genus *Euneomys* (Cricetidae, Rodentia). Revista Chilena de Historia Natural, 63:73-82.

Reiter, A., M. Andreas, P. Benda, M. Lipa, and P. Wolf. 1995. Mammalian fauna of the Svjatoj Nos peninsula and isthmus, the Baikal Lake, Russia. Acta Societatis Zoologicae Bohemicae, 59:209-225.

Rekovets, L. I. 1990. Principal developmental stages of the water vole genus *Arvicola* (Rodentia, Mammalia) from the Eastern European Pleistocene. Pp. 369-384, *in* International symposium evolution, phylogeny and biostratigraphy of arvicolids (Rodentia, Mammalia) (O. Fejfar and W.-D. Heinrich, eds.). Geological Survey, Prague, 448 pp.

Rempe, U. 1970. Morphometrische Untersuchungen zur Klärung der Verwandtschaft von Steppeniltis, Waldiltis und Frettchen. Verhand. Deutschen Zool. Gesellschaft, 3(7):186-367.

Renaud, S. 1999. Size and shape variability in relation to species differences and climatic gradients in the African rodent *Oenomys*. Journal of Biogeography, 26:857-865.

Renaud, S., and J. R. Michaux. 2003. Adaptive latitudinal trends in the mandible shape of *Apodemus* wood mice. Journal of Biogeography, 30:1617-1628.

Renaud, S., and V. Millien. 2001. Intra- and interspecific morphological variation in the field mouse species *Apodemus argenteus* and *A. speciosus* in the Japanese archipelago: The role of insular isolation and biogeographic gradients. Biological Journal of the Linnean Society, 74:557-569.

Rengger, J. R. 1830. Naturgeschichte de Säugethiere von Paraguay. Schweighausersche Buchhandlung, Basel, Switzerland, 394 pp.

Rennert, P. D., and C. W. Kilpatrick. 1986. Biochemical systematics of populations of *Peromyscus boylii*. I. Populations from east-central Mexico with low fundamental numbers. Journal of Mammalogy, 67:481-488.

Rennert, P. D., and C. W. Kilpatrick. 1987. Biochemical systematics of *Peromyscus boylii*. II. Chromosomally variable populations from eastern and southern Mexico. Journal of Mammalogy, 68:799-811.

Repenning, C. A. 1967. Subfamilies and genera of the Soricidae. Geological Survey Professional Paper, 565:1-74.

Repenning, C. A. 1968. Mandibular musculature and the origin of the subfamily Arvicolinae (Rodentia). Acta Zoologica Cracoviensia, 13:1-72.

Repenning, C. A. 1980. Faunal exchanges between Siberia and North America. Canadian Journal of Anthropology, 1:37-44.

Repenning, C. A. 1983. *Pitymys meadensis* Hibbard from the Valley of Mexico and the classification of North American species of *Pitymys* (Rodentia: Cricetidae). Journal of Vertebrate Paleontology, 2:471-482.

Repenning, C. A. 1990. Of mice and ice in the Late Pliocene of North America. Arctic, 43:314-323.

Repenning, C. A. 1992. *Allophaiomys* and the age of the Olyor Suite, Krestovka Sections, Yakutia. U. S. Geological Survey Bulletin 2037, 98 pp.

Repenning, C. A. 1998. North American mammalian dispersal routes: Rapid evolution and dispersal constrain precise biochronology. Pp. 39-78, *in* Advances in vertebrate paleontology and geochronology (Y. Tomida, L. J. Flynn, and L. L. Jacobs, eds.). National Science Museum, Monographs, No. 14, Tokyo, 292 pp.

Repenning, C. A. 2003. *Mimomys* in North America. Pp. 469-508, *in* Vertebrate fossils and their context, contributions in honor of Richard H. Tedford (L. J. Flynn, ed.). Bulletin of the American Museum of Natural History, 279:666 pp.

Repenning, C. A., and F. M. Grady. 1988. The microtine rodents of the Cheetah Room Fauna, Hamilton

Cave, West Virginia, and the spontaneous origin of *Synaptomys*. United States Geological Survey Bulletin, 1853:1-32.

Repenning, C. A., and R. H. Tedford. 1977. Otarioid seals of the Neogene. United States Geological Survey, Professional Paper, 992:1-93.

Repenning, C. A., R. S. Peterson, and C. L. Hubbs. 1971. Contributions to the systematics of the southern fur seals, with particular reference to the Juan Fernandez and Guadalupe species. Antarctic Research Series, 18:1-34.

Repenning, C. A., O. Fejfar, and W.-D. Heinrich. 1990. Arvicolid rodent biochronology of the Northern Hemisphere. Pp. 385-417, *in* International symposium evolution, phylogeny and biostratigraphy of arvicolids (Rodentia, Mammalia) (O. Fejfar and W.-D. Heinrich, eds.). Geological Survey, Prague, 448 pp.

Reumer, J. W. F. 1980*a*. On the Pleistocene shrew *Nesiotites hidalgo* Bate, 1944 from Majorca (Soricidae, Insectivora). Proceedings of the Koninklijke Nederlandse Akademie van Wetenschappen, Series B, 83:39-68.

Reumer, J. W. F. 1980*b*. Micromammals from the Holocene of Canet Cave (Majorca) and their biostratigraphical implication. Proceedings of the Koninklijke Nederlandse Akademie van Wetenschappen, Series B, 83:355-360.

Reumer, J. W. F. 1982. Some remarks on the fossil vertebrates from Menorca, Spain. Proceedings of the Koninklijke Nederlandse Akademie van Wetenschappen, Series B, 85:77-87.

Reumer, J. W. F. 1984. Ruscinian and early Pleistocene Soricidae (Insectivora, Mammalia) from Tegelen (The Netherlands) and Hungary. Scripta Geologica, 73:1-173.

Reumer, J. W. F. 1986. Notes on the Soricidae (Insectivora, Mammalia) from Crete. I. The Pleistocene species *Crocidura zimmermanni*. Bonner Zoologische Beiträge, 37:161-171.

Reumer, J. W. F. 1987. Redefinition of the Soricidae and the Heterosoricidae (Insectivora, Mammalia), with the description of the Crocidosoricinae, a new subfamily of Soricidae. Revue de Paléobiologie, 6:189-192.

Reumer, J. W. F. 1994. De evolutiebiologie van de spitsmuizen (Mammalia, Insectivora, Soricidae). I. Anatomie, evolutie en biogeografie. Cranium, 11:9-35.

Reumer, J. W. F. 1995. De evolutiebiologie van de spitsmuizen (Mammalia, Insectivora, Soricidae). II. De subfamilies Crocidosoeicinae, Allosoricinae, Limnoecinae en Crocidurinae. Cranium, 12:53-63.

Reumer, J. W. F. 1998. A classification of the fossil and recent shrews. Pp. 5-22, *in* Evolution of shrews (J. M.Wójcik and M. Wolsan, eds.). Mammal Research Institute, Polish Academy of Sciences, Bialowieza, 458 pp.

Reumer, J. W. F., and A. Meylan. 1986. New developments in vertebrate cytotaxonomy. 9. Chromosome numbers in the order Insectivora (Mammalia). Genetica, 70:119-151.

Reutter, B. A., J. Hausser, and P. Vogel. 1999. Discriminant analysis of skull morphometric characters in *Apodemus sylvaticus*, *A. flavicollis*, and *A. alpicola*. Acta Theriologica, 44(3):299-308.

Reutter, B. A., P. Nová, P. Vogel, and J. Zima. 2001. Karyotypic variation between wood mouse species: Banded chromosomes of *Apodemus alpicola* and *A. microps*. Acta Theriologica, 46(4):353-362.

Reutter, B. A., E. Petit, and P. Vogel. 2002. Molecular identification of an endemic Alpine mammal, *Apodemus alpicola*, using a PCR-based RFLP method. Revue Suisse de Zoologie, 109(1):9-16.

Reutter, B. A., E. Petit, H. Brünner, and P. Vogel. 2003. Cytochrome *b* haplotype divergences in West European *Apodemus*. Mammalian Biology, 68:153-164.

Reyes, A., G. Pesole, and C. Saccone. 1998. Complete mitochondrial DNA sequence of the Fat Dormouse, *Glis glis*: Further evidence of rodent paraphyly. Molecular Biology and Evolution, 15(5):499-505.

Reyes, A., C. Gissi, G. Pesole, F. M. Catzeflis, and C. Saccone. 2000. Where do rodents fit? Evidence from the complete mitochondrial genome of *Sciurus vulgaris*. Molecular Biology and Evolution, 17(6):979-983.

Reynolds, T. E., K. F. Koopman, and E. E. Williams. 1953. A cave faunule from western Puerto Rico with a discussion of the genus *Isolobodon*. Breviora, Harvard Museum of Comparative Zoology, 12:1-8.

Rezsutek, M., and G. N. Cameron. 1993. *Mormoops megalophylla*. Mammalian Species, 448:1-5.

Rhoads, S. N. 1893. Geographic variation in *Bassariscus astutus* with description of a new subspecies. Proceedings of the Academy of Natural Sciences of Philadelphia, 1893:413-418.

Rhoads, S. N. 1896. Mammals collected by Dr. A. Donaldson Smith during his expedition to Lake Rudolf, Africa. Proceedings of the Academy of Natural Sciences of Philadelphia, 1896:517-546.

Rice, D. W. 1977. A list of the marine mammals of the world. Third ed. NOAA Technical Report, NMFS SSRF-711:1-15.

Rice, D. W. 1989. Sperm whale—*Physeter macrocephalus* Linnaeus, 1758. Pp. 177-234, *in* Handbook of marine mammals: River dolphins and the larger toothed whales (S. H. Ridgway and R. Harrison, eds.). Academic Press, London, 4:1-442.

Rice, D. W. 1990. The scientific name of the pilot whale—a rejoinder to Schevill. Marine Mammal Science, 6(4):359-360.

Rice, D. W. 1998. Marine mammals of the world: Systematics and distribution. Society for Marine Mammalogy, Special Publication, 4:1-230.

Rice, D. W., and W. A. Wolman. 1971. Life history and ecology of the gray whale (*Eschrichtius robustus*). American Society of Mammalogists, Special Publication, 3:1-142.

Rich, S. M., C. W. Kilatrick, J. L. Shippee, and K. L. Crowell. 1996. Morphological differentiation and identification of *Peromyscus leucopus* and *P. maniculatus* in northeastern North America. Journal of Mammalogy, 77:985-991.

Rich, T. H. 1991. Monotremes, placentals, and marsupials: Their record in Australia and its biases. Pp. 893-1004, *in* Vertebrate palaeontology of Australasia (P. Vickers-Rich, J. M. Monaghan, R. F. Baird, and T. H. Rich, eds.). Pioneer Design Studio Pty Ltd, Victoria, 1437 pp.

Rich, T. H. V., M. Fortelius, P. V. Rich, and D. A. Hooijer. 1987. The supposed *Zygomaturus* from New Caledonia is a rhinoceros: A second solution to an enigma and its palaeogeographic consequences. Pp. 769-778, *in* Possums and opossums: Studies in evolution (M. Archer, ed.). Sydney: Surrey Beatty and Sons, 1:1-400, 2:400-788.

Rich, T. H., M. Archer, S. J. Hand, H. Godthelp, J. Muirhead, N. S. Pledge, T. F. Flannery, M. O. Woodburne, J. A. Case, R. H. Tedford, W. D. Turnbull, E. L. Lundelius, Jr., L. S. V. Rich, M. J. Whitelaw, A. Kemp, and P. Vickers-Rich. 1991. Australian Mesozoic and Tertiary terrestrial mammal localities. Pp. 1005-1070, *in* Vertebrate palaeontology of Australasia (P. Vickers-Rich, J. M. Monaghan, R. F. Baird, and T. H. Rich, eds.). Pioneer Design Studio Pty Ltd, Victoria, 1437 pp.

Richards, R. L. 1980. Rice rat (*Oryzomys* cf. *palustris*) remains from southern Indiana caves. Proceedings of the Indiana Academy of Science, 89:425-431.

Richardson, B. J., and G. B. Sharman. 1976. Biochemical and morphological observations on the wallaroos (Macropodidae: Marsupialia) with a suggested new taxonomy. Journal of Zoology, London, 179:499-513.

Richter, H. 1970. Zur Taxonomie und Verbreitung der paläarktischen Crociduren. Zoologische Abhandlungen Staatliches Museum für Tierkunde in Dresden, 31:293-304.

Rickart, E. A. 1987. *Spermophilus townsendii*. Mammalian Species, 268:1-6.

Rickart, E. A., and L. R. Heaney. 1991. A new species of *Chrotomys* (Rodentia: Muridae) from Luzon Island, Philippines. Proceedings of the Biological Society of Washington, 104:387- 398.

Rickart, E. A., and L. R. Heaney. 2002. Further studies on the chromosomes of Philippine rodents (Muridae: Murinae). Proceedings of the Biological Society of Washington, 115(3):473-487.

Rickart, E. A., and G. G. Musser. 1993. Philippine rodents: Chromosomal characteristics and their significance for phylogenetic inference among 13 species (Rodentia: Muridae: Murinae). American Museum Novitates, 3064:1-34.

Rickart, E. A., and P. B. Robertson. 1985. *Peromyscus melanocarpus*. Mammalian Species, 241:1-3.

Rickart, E. A., and E. Yensen. 1991. *Spermophilus washingtoni*. Mammalian Species, 371:1-5.

Rickart, E. A., R. S. Hoffmann and M. Rosenfeld. 1985 [1987]. Karyotype of *Spermophilus townsendii artemesiae* (Rodentia: Sciuridae) and chromosomal variation in the *Spermophilus townsendii* complex. Mammalian Chromosome Newsletter, 26:94-102.

Rickart, E. A., L. R. Heaney, and R. B. Utzurrum. 1991. Distribution and ecology of small mammals along an elevational transect in southeastern Luzon, Philippines. Journal of Mammalogy, 72:458-469.

Rickart, E. A., L. R. Heaney, P. D. Heideman, and R. C. B. Utzurrum. 1993.The distribution and ecology of mammals on Leyte, Biliran, and Maripipi Islands, Philippines. Fieldiana: Zoology, n.s., 72:62 pp.

Rickart, E. A., L. R. Heaney, B. R. Tabaranza, Jr., and D. S. Balete. 1998. A review of the genera *Crunomys* and *Archboldomys* (Rodentia: Muridae: Murinae), with descriptions of two new species from the Philippines. Fieldianna: Zoology, n.s., 89:1-24.

Rickart, E. A., L. R. Heaney, and B. R. Tabaranza, Jr. 2002. Review of the genus *Bullimus* (Rodentia: Muridae: Murinae) and description of a new species from Camiguin Island, Philippines. Journal of Mammalogy, 83(2):421-436.

Rickart, E. A., L. R. Heaney, and B. R. Tabaranza, Jr. 2003. A new species of *Limnomys* (Rodentia: Muridae: Murinae) from Mindanao Island, Philippines. Journal of Mammalogy, 84(4):1443-1455.

Riddle, B. R. 1995. Molecular biogeography in pocket mice (*Perognathus* and *Chaetodipus*) and grasshopper mice (*Onychomys*): The late Cenozoic development of a North American aridlands rodent guild. Journal of Mammalogy, 76:283-301.

Riddle, B. R. 1999. Mearn's grasshopper mouse | *Onychomys arenicola*. Pg. 588, *in* The Smithsonian Book of North American Mammals (D. E. Wilson and S. Ruff, eds.). Smithsonian Institution Press,Washington, D.C., xxv + 750 pp.

Riddle, B. R., and J. R. Choate. 1986. Systematics and biogeography of *Onychomys leucogaster* in western North America. Journal of Mammalogy, 67:233-255.

Riddle, B. R., and R. L. Honeycutt. 1990. Historical biogeography in North American arid regions: An approach using mitochondrial-DNA phylogeny in grasshopper mice (genus *Onychomys*). Evolution, 44:1-15.

Riddle, B. R., R. L. Honeycutt, and P. L. Lee. 1993. Mitochrondrial DNA phylogeography in northern grasshopper mice (*Onychomys leucogaster*) – the influence of Quaternary climatic oscillations on population dispersion and divergence. Molecular Ecology, 2:183-193.

Riddle, B. R., D. J. Hafner, and L. F. Alexander. 2000*a*. Comparative phylogeography of Bailey's pocket mouse (*Chaetodipus baileyi*) and the *Peromyscus eremicus* species group: Historical vicariance of the Baja California peninsular desert. Molecular Phylogenetics and Evolution, 17:161-172.

Riddle, B. R., D. J. Hafner, L. F. Alexander, and J. R. Jaeger. 2000*b*. Cryptic vicariance in the historical assembly of a Baja California Peninsular Desert biota. Proceedings of the National Academy of Sciences, 97:14438-14443.

Riddle, B. R., D. J. Hafner, and L. F. Alexander. 2000*c*. Phylogeography and systematics of the *Peromyscus eremicus* species and the historical biogeography of North American warm regional deserts. Molecular Phylogenetics and Evolution, 17:145-160.

Ride, W. D. L. 1956. A new fossil *Mastacomys* (Muridae) and a revision of the genus. Proceedings of the Zoological Society of London, 127:431-439.

Ride, W. D. L. 1962. On the use of generic names for kangaroos and wallabies. Australian Journal of Science, 24:367-372.

Ride, W. D. L. 1964*a*. *Antechinus rosamondae*, a new species of dasyurid marsupial from the Pilbara district of Western Australia; with remarks on the classification of *Antechinus*. Western Australian Naturalist, 9:58-65.

Ride, W. D. L. 1964*b*. A review of Australian fossil marsupials. Journal Proceedings of the Royal Society of Western Australia, 47:97-131.

Ride, W. D. L. 1970. A guide to the native mammals of Australia. Oxford University Press, Melbourne, 249 pp.

Rideout, C. B., and R. S. Hoffmann. 1975. *Oreamnos americanus*. Mammalian Species, 63:1-6.

Rieger, I. 1981. *Hyaena hyaena*. Mammalian Species, 150:1-5.

Rieger, T., A. Langguth, and T. Weimer. 1995. Allozymic characterization and evolutionary relationships in the Brazilian *Akodon cursor* species group (Rodentia-Cricetidae). Biochemical Genetics, 33:283-295.

Riley, J. 2001. Mammals on the Sangihe and Talaud Islands, Indonesia, and the impact of hunting and habitat loss. Oryx, 36:288-296.

Riley, J. 2002. The rediscovery of Talaud Islands Flying Fox, *Acerodon humilis* Andersen, 1909, and notes on other fruit bats from the Sangihe and Talaud Islands, Indonesia (Mammalia: Chiroptera: Pteropodidae). Faunistische Abhandlungen (Dresden) 22 (2):393-410.

Rimoli, R. O. 1976 [1977]. Roedores fosiles de la Hispaniola. Universidad Central del Este, Serie Científica III, San Pedro de Macorís, Dominican Republic, 54 pp., 19 pls.

Rios, E., and S. T. Alvarez-Castañeda. 2002. Mamíferos de La Reserva del Valle de Los Cirios, Baja California, México. Acta Zoologica Méxicana, n.s., 86:51-85.

Rishi, K. K., and U. Puri. 1978. Mitotic, meiotic and G-banded chromosomes of *Mus saxicola sadhu* (Muridae: Rodentia). Caryologia, 31(1):103-107.

Rishi, K. K., and U. Puri. 1984. Chromosomes of *Rattus cutchicus cutchicus* and its systematic position. Folia Biologia (Krakow), 32:209-212.

Riskin, D. K. 2001. *Pipistrellus bodenheimeri*. Mammalian Species, 651:1-3.

Ritte, U., E. Markman, and E. Neufeld. 1992. Can the variability of mitochondrial DNA distinguish between commensal and feral populations of the house mouse? Biological Journal of the Linnean Society, 46:235-245.

Rivas, B. A, and J. E. Pefaur. 1999*a*. Variacion geografica en poblaciones Venezolanas de *Oryzomys albigularis* (Rodentia: Muridae). Mastozoologia Neotropical, 6:47-59.

Rivas, B. A, and J. E. Pefaur. 1999*b*. Variacion craneana entre sexo y edad en *Oryzomys albigularis* (Rodentia: Muridae). Mastozoologia Neotropical, 6:61-70.

Robbins, C. B. 1971. Dental nomenclature for *Taterillus* (Thomas) (Rodentia: Cricetidae). Mammalia, 35:629-635.

Robbins, C. B. 1973. Nongeographic variation in *Taterillus gracilis* (Thomas) (Rodentia: Cricetidae). Journal of Mammalogy, 54:222-238.

Robbins, C. B. 1974. Comments on the taxonomy of the West African *Taterillus* (Rodentia: Cricetidae) with the description of a new species. Proceedings of the Biological Society of Washington, 87:395-404.

Robbins, C. B. 1977. A review of the taxonomy of the African gerbils, *Taterillus* (Rodentia: Cricetidae). Pp. 178-194, *in* Uspekhi sovremennoi teriologii [Advances in modern theriology] (V. E. Sokolov, ed.). Nauka, Moscow, 296 pp. (in Russian).

Robbins, C. B. 1978. Taxonomic identification and history of *Scotophilus nigrita* (Schreber) (Chiroptera: Vespertilionidae). Journal of Mammalogy, 59:212-213.

Robbins, C. B. 1980. Small mammals of Togo and Benin. I. Chiroptera. Mammalia, 44:83-88.

Robbins, C. B., and H. W. Setzer. 1985. Morphometrics and distinctness of the hedgehog genera (Insectivora, Erinaceidae). Proceedings of the Biological Society of Washington, 98:112-120.

Robbins, C. B., and E. Van der Straeten. 1982. A new specimen of *Malacomys verschureni* from Zaire, Central Africa (Rodentia, Muridae). Revue de Zoologie Africaine, 96:216-220.

Robbins, C. B., and E. Van der Straeten. 1989. Comments on the systematics of *Mastomys* Thomas 1915 with the description of a new West African species. Senckenbergiana Biologica, 69:1-14.

Robbins, C. B., J. W. Krebs, Jr., and K. M. Johnson. 1983. *Mastomys* (Rodentia: Muridae) species distinguished by hemoglobin pattern differences. American Journal of Tropical Medicine and Hygiene, 32:624-630.

Robbins, C. B., F. deVree, and V. van Cakenberghe. 1985. A systematic revision of the African bat genus *Scotophilus* (Vespertilionidae). Annales Musée Royal de l'Afrique Centrale, Tervuren, Belgique, Sciences Zoologiques, 246:51-84.

Robbins, L. W., and R. J. Baker. 1978. Karyotypic data for African mammals, with a description of an in vivo bone marrow technique. Bulletin of the Carnegie Museum of Natural History, 6:188-210.

Robbins, L. W., and R. J. Baker. 1980. G- and C-band studies on the primitive karyotype for *Reithrodontomys*. Journal of Mammalogy, 61:708-714.

Robbins, L. W., and R. J. Baker. 1981. An assessment of the nature of rearrangements in eighteen species of *Peromyscus* (Rodentia: Cricetidae). Cytogenetics and Cell Genetics, 31:194-202.

Robbins, L. W., and V. M. Sarich. 1988. Evolutionary relationships in the family Emballonuridae (Chiroptera). Journal of Mammalogy, 69:1-13.

Robbins, L. W., and D. A. Schlitter. 1981. Systematic status of dormice (Rodentia: Gliridae) from southern Cameroon, Africa. Annals of Carnegie Museum, 50(9):271-288.

Robbins, L. W., and H. W. Setzer. 1979. Additional records of *Hylomyscus baeri* Heim deBalsac and Aellen (Rodentia, Muridae) from western Africa. Mammalia, 60:649-650.

Robbins, L. W., J. R. Choate, and R. L. Robbins. 1980. Nongeographic and interspecific variation in four species of *Hylomyscus* (Rodentia: Muridae) in southern Cameroon. Annals of Carnegie Museum, 49:31-48.

Robbins, L. W., M. P. Moulton, and R. J. Baker. 1983. Extent of geographic range and magnitude of chromosomal evolution. Journal of Biogeography, 10:533-541.

Robbins, L. W., M. H. Smith, M. C. Wooten, and R. K. Selander. 1985. Biochemical polymorphism and its relationship to chromosomal and morphological variation in *Peromyscus leucopus* and *Peromyscus gossypinus*. Journal of Mammalogy, 66:498-510.

Roberts, A. 1913. The collection of mammals in the Transvaal Museum. Annals of the Transvaal Museum, 4:65-107.

Roberts, A. 1929. New forms of African mammals. Annals of the Transvaal Museum, 13:82-121.

Roberts, A. 1931. New forms of South African mammals. Annals of the Transvaal Museum, 14:221-236.

Roberts, A. 1938. Descriptions of new forms of mammals. Annals of the Transvaal Museum, 19(2):231-245.

Roberts, A. 1946. Descriptions of numerous new subspecies of mammals. Annals of the Transvaal Museum, 20:303-328.

Roberts, A. 1951. The mammals of South Africa. Trustees of "The mammals of South Africa" book fund, Johannesburg, 700 pp.

Roberts, H. R., D. J. Schmidly, and R. D. Bradley. 1998. *Peromyscus spicilegus*. Mammalian Species, 596:1-4.

Roberts, K. J., F. D. Yancey, II, and C. Jones. 1997. Distributional records of small mammals from the Texas Panhandle. Texas Journal of Science, 49:57-64.

Roberts, M. 1991. Origin, dispersal routes, and geographic distribution of *Rattus exulans*, with special reference to New Zealand. Pacific Science, 45:123-130.

Roberts, M. S., and J. L. Gittleman. 1984. *Ailurus fulgens*. Mammalian Species, 222:1-8.

Roberts, T. J. 1977. The mammals of Pakistan. Ernest Benn Limited, London, 361 pp.

Roberts, T. J. 1997. The mammals of Pakistan (revised edition). Oxford University Press, Oxford, 525 pp.

Robertson, P. B., and G. G. Musser. 1976. A new species of *Peromyscus* (Rodentia: Cricetidae), and a new specimen of *P. simulatus* from southern Mexico, with comments on their ecology. Occasional Papers of the Museum of Natural History, University of Kansas, 47:1-8.

Robertson, P. B., and E. A. Rickart. 1975. *Cryptotis magna*. Mammalian Species, 61:1-2.

Robichaud, W. G. 1998. Physical and behavioral description of a captive saola, *Pseudoryx nghetinhensis*. Journal of Mammalogy, 79:394-405.

Robineau, D. 1973. Sur deux rostres de *Mesoplodon* (Cetcea, Hyperoodontidae). Mammalia, 37(3):504-513.

Robineau, D. 1989. Les cetaces des Iles Kerguelen. Mammalia, 53:265-278.

Robins, J. H. 1998. A revised bibliography of the Petauristinae. Virginia Museum of Natural History Special Publication, 6:275-307.

Robinson, A. C. 1995*a*. Lesser stick-nest rat, *Leporillus apicalis*. Pp. 558-559, *in* Mammals of Australia (R. Strahan, ed.). Smithsonian Institution Press, Washington, D.C., 756 pp.

Robinson, A. C. 1995*b*. Greater stick-nest rat, *Leporillus conditor*. Pp. 560-561, *in* Mammals of Australia (R. Strahan). Smithsonian Institution Press, Washington, D.C., 756 pp.

Robinson, A. C. 1995*c*. Western chestnut mouse, *Pseudomys nanus*. Pp. 609-610, *in* Mammals of Australia (R. Strahan, ed.). Smithsonian Institution Press, Washington, D.C., 756 pp.

Robinson, A. C., C. M. Kemper, G. C. Medlin, and C. H. S. Watts. 2000. The rodents of South Australia. Wildlife Research, 27:379-404.

Robinson, H. C., and C. B. Kloss. 1916. Preliminary diagnoses of some new species and subspecies of mammals and birds obtained in Korinchi, West Sumatra, Feb.-June 1914. Journal of the Straits Branch of the Royal Asiatic Society, 73:269-278.

Robinson, H. C., and C. B. Kloss. 1918*a*. A nominal list of the Sciuridae of the Oriental Reigon with a list of specimens in the collection of the Zoological Survey of India. Records of the Indian Museum, 15(4), no. 21:171-254.

Robinson, H. C., and C. B. Kloss. 1918*b*. Results of an expedition to Korinchi Peak, Sumatra. I. Mammals. Journal of the Federation of Malay States Museum, 8(2):1-81.

Robinson, H. C., and C. B. Kloss. 1919*a*. On a collection of mammals from the Bencoolen and Palembang residencies, South West Sumatra. Journal of the Federation of Malay States Museum, 7(4):257-291.

Robinson, H. C., and C. B. Kloss. 1919*b*. On mammals, chiefly from the Ophir District, West Sumatra. Journal of the Federation of Malay States Museum, 7(4):299-323.

Robinson, H. C., and C. B. Kloss. 1920. Notes on Viverridae. Records of the Indian Museum, Calcutta, 19:175-179.

Robinson, H. C., and C. B. Kloss. 1922. New mammals from French Indo-China and Siam. Annals and Magazine of Natural History, ser. 9, 9:87-99.

Robinson, J. W., and R. S. Hoffmann. 1975. Geographic and interspecific cranial variation in big-eared ground squirrels (*Spermophilus*): A multivariate study. Systematic Zoology, 24:79-88.

Robinson, M., F. Catzeflis, J. Briolay, and D. Mouchiroud. 1997. Molecular phylogeny of rodents, with special emphasis on murids: Evidence from nuclear gene LCAT. Molecular Phylogenetics and Evolution, 8(3):423-434.

Robinson, M. F., and M. Webber. 2000. Survey of bats (Mammalia: Chiroptera) in the Khammouan Limestone National Biodiversity Conservation Area, Lao P.D.R. Natural History Bulletin of the Siam Society, 48:21-45.

Robinson, M. F., A. L. Smith, and S. Bumrungsri. 1995. Small mammals of Thung Yai Naresuan and Huai Kha Khaeng Wildlife Sanctuaries in western Thailand. Natural History Bulletin of the Siam Society, 43:27-54.

Robinson, M. F., P. D. Jenkins, C. M. Francis, and A. J. C. Fulford. 2003. A new species of the *Hipposideros pratti* group (Chiroptera, Hipposideridae) from Lao PDR and Vietnam. Acta Chiropterologica, 5:31-48.

Robinson, R. 1984. Norway rat. Pp. 284-290, *in* Evolution of domesticated animals (I. L. Mason, ed.). Longman Group Limited, London and New York, 452 pp.

Robinson, T. J. 1982. Key to the South African Leporidae. South African Journal of Zoology, 17:220-222.

Robinson, T. J., and N. J. Dippenaar. 1983*a*. Morphometrics of the South African Leporidae. I. Genus *Pronolagus* Lyon, 1904. Annals of the Museum, Royal African Centre for Science, Zoology, 237:43-61.

Robinson, T. J., and N. J. Dippenaar. 1983*b*. The status of *Lepus saxatilis*, *L. whytei* and *L. crawshayi* in southern Africa. Acta Zoologica Fennica, 174:35-39.

Robinson, T. J., and N. J. Dippenaar. 1987. Morphometrics of the South African Leporidae. II. *Lepus* Linnaeus, 1758, and *Bunolagus* Thomas, 1929. Annals of the Transvaal Museum, 34:379-404.

Robinson, T. J., and F. F. B. Elder. 1987. Extensive genome reorganization in the African rodent genus *Otomys*. Journal of Zoology, London, 211:735-745.

Robinson, T. J., and D. J. Morris. 1991. Interspecific hybridization in the Bovidae: Sterility of *Alcelaphus buselaphus* X *Damaliscus dorcas* F1 progeny. Biological Conservation, 58:345-356.

Robinson, T. J., and J. D. Skinner. 1983. Karyology of the riverine rabbit, *Bunolagus monticularis*, and its taxonomic implications. Journal of Mammalogy, 64:678-681.

Robinson, T. J., F. F. B. Elder, and W. Lopez-Forment. 1981. Banding studies in the volcano rabbit, *Romerolagus diazi* and Crawshay's hare, *Lepus crawshayi*. Evidence of the leporid ancestral karyotype. Canadian Journal of Genetics and Cytology, 23:469-474.

Robinson, T. J., F. F. B. Elder, and J. A. Chapman. 1983. Evolution of chromosomal variation in cottontails, genus *Sylvilagus* (Mammalia: Lagomorpha): *S. aquaticus*, *S. floridanus*, and *S. transitionalis*. Cytogenetics and Cell Genetics, 35:216-222.

Robinson, T. J., F. F. B. Elder, and J. A. Chapman. 1984. Evolution of chromosomal variation in cottontails, genus *Sylvilagus* (Mammalia: Lagomorpha). II. *Sylvilagus audubonii*, *S. idahoensis*, *S. nuttallii* and *S. palustris*. Cytogenetics and Cell Genetics, 38:282-289.

Robinson, T. J., J. D. Skinner, and A. S. Haim. 1986. Close chromosomal congruence in two species of ground squirrel: *Xerus inauris* and *X. princeps* (Rodentia: Sciuridae). South African Journal of Zoology, 21:100-105.

Robinson, T. J., A. D. Bastos, K. M. Halanych, and B. Herzig. 1996. Mitochondrial DNA sequence relationship of the extinct blue antelope *Hippotragus leucophaeus*. Naturwissenschaften, 83:178-182.

Robinson-Rechavi, M., L. Ponger, and D. Mouchiroud. 2000. Nuclear gene LCAT supports rodent monophyly. Molecular Biology and Evolution, 17:1410-1412.

Roca, A. L, N. Georgiadis, J. Pecon-Slattery, and S. J. O'Brien. 2001. Genetic evidence for two species of elephant in Africa. Science, 293:1473-1477.

Roche, J. 1964. Description d'une nouvelle sous-espece de *Thallomys* de l'est Africain *Thallomys paedulcus somaliensis*. Mammalia, 28:94-100.

Roche, J. 1972. Systematique du genre *Procavia* et des damans en general. Mammalia, 36:22-49.

Roche, J. 1975. A propos des petites gerbilles a soles plantaires nues (sous-genre *Hendecapleura*) de L'est African. Italian Journal of Zoology, n.s., Suppl. 6, 13:263-268.

Roche, J., and F. Petter. 1968. Faits nouveaux concernant trois gerbillides mal connus de Somalia: *Ammodillus imbellis* (De Winton), *Microdillus peeli* (De Winton), *Monodia juliani* (Saint Leger). Monitore Zoologico Italiano, n.s., 2(suppl.):181-198.

Roche, J., E. Capanna, M. V. Civitelli, and A. Ceraso. 1984. Caryotypes des rongeurs de Somalie. 4. Premiere capture de rongeurs arboricoles du sous-genre *Grammomys* (genre *Thamnomys*, Murides) en Republique de Somalie. Monitore Zoologico Italiano, 7:259-277.

de Rochebrune, A. T. 1883. Faune de la Senegambie, Mammiferes. Actes de la Societe Linneenne de Bordeaux 37, 4(7):49-203.

Rodriguez, D., and R. Bastida. 1993. The southern sea lion: *Otaria byronia* or *Otaria flavescens*? Marine Mammal Science, 9:372-381.

Rodriguez, L., V. Barus, and J. Kratochvil. 1979. The genus *Mesocapromys*, a link between the families Echimyidae and Capromyidae. Folia Zoologica, 28(2):97-102.

Rodríguez-Durán, A., and T. H. Kunz. 1992. *Pteronotus quadridens*. Mammalian Species, 395:1-4.

Roesler, U., and G. R. Witte. 1969. Chorologische Betrachtungen zur Subspeziesbildung einiger Vertebraten im italienischen und balkanischen Raum. Zoologischer Anzeiger, 182:27-51.

Roest, A. I. 1973. Subspecies of the sea otter, *Enhydra lutris*. Contributions in Science, Natural History Museum of Los Angeles County, 140(252):1-17.

Rogatcheva, M. B., Y. S. Aulchenko, and S.-I. Oda. 2000. Chromosomal and genic mechanisms of reproductive isolation: The case of *Suncus murinus*. Acta Theriologica, 45(Suppl. 1):143-159.

Rogers, D. S. 1983. Phylogenetic affinities of *Peromyscus* (*Megadontomys*) *thomasi*: Evidence from differentially stained chromosomes. Journal of Mammalogy, 64:617-623.

Rogers, D. S. 1989. Evolutionary implications of chromosomal variation among spiny pocket mice, genus *Heteromys* (Order Rodentia). Southwestern Naturalist, 34:85-100.

Rogers, D. S. 1990. Genic evolution, historical biogeography, and systematic relationships among spiny pocket mice (subfamily Heteromyinae). Journal of Mammalogy, 71:668-685.

Rogers, D. S., and M. D. Engstrom. 1992. Evolutionary implications of allozymic variation in tropical *Peromyscus* of the *mexicanus* species group. Journal of Mammalogy, 73:55-69.

Rogers, D. S., and E. J. Heske. 1984. Chromosomal evolution of the brown mice, genus *Scotinomys* (Rodentia: Cricetidae). Genetica, 63:221-228.

Rogers, D. S., and J. E. Rogers. 1992*a*. *Heteromys oresterus*. Mammalian Species, 396:1-3.

Rogers, D. S., and J. E. Rogers. 1992*b*. *Heteromys nelsoni*. Mammalian Species, 397:1-2.

Rogers, D. S., and D. J. Schmidly. 1981. Geographic variation in the white-throated woodrat (*Neotoma albigula*) from Nex Mexico, Texas, and northern Mexico. Southwestern Naturalist, 26:167-181.

Rogers, D. S., and D. J. Schmidly. 1982. Systematics of spiny pocket mice (genus *Heteromys*) of the *desmarestianus* species group from México and northern Central America. Journal of Mammalogy, 63:375-386.

Rogers, D. S., E. J. Heske, and D. A. Good. 1983. Karyotype and a range extension of *Reithrodontomys tenuirostris*. Southwestern Naturalist, 21:372-374.

Rogers, D. S., I. F. Greenbaum, S. J. Gunn, and M. D. Engstrom. 1984. Cytosystematic value of chromosomal inversion data in the genus *Peromyscus*. Journal of Mammalogy, 65:457-465.

Rogers, M. A. 1991*a*. Evolutionary differentiation within the northern Great Basin pocket gopher, *Thomomys townsendii*. I. Morphological variation. Great Basin Naturalist, 51:109-126.

Rogers, M. A. 1991*b*. Evolutionary differentiation within the northern Great Basin pocket gopher,

Thomomys townsendii. II. Genetic variation and biogeographic considerations. Great Basin Naturalist, 51:127-152.

Rogovin, K. A. 1984. [A comparative analysis of behaviour and supergeneric groups of jerboas (Rodentia, Dipodidae).] Zoologicheskii Zhurnal, 64(11):1702-1711 (in Russian).

Roguin, L. de. 1988. Notes sur quelques mammiferes du Baluchistan Iranien. Revue Suisse de Zoologie, 95:595-606.

Rohwer, S. A., and D. L. Kilgore. 1973. Interbreeding in the arid-land foxes, *Vulpes velox* and *V. macrotis*. Systematic Zoology, 22:157-165.

Roig, V. G., and O. A. Reig. 1969. Precipitin test relationships among Argentinian species of the genus *Ctenomys* (Rodentia, Octodontidae). Comparative Biochemistry and Physiology, 30:665-672.

Romagnoli, M. L., and M. S. Springer. 2000. Evolutionary relationships among Old World fruitbats (Megachiroptera: Pteropodidae) based on 12S rRNA, tRNA Valine, and 16S rRNA gene sequences. Journal of Mammalian Evolution, 7:259-284.

Ronnefeld, U. 1969. Verbreitung und Lebensweise afrikanischer Feloidea (Felidae et Hyaenidae). Säugetierkundliche Mitteilungen, 17(4):285-350.

Rookmaaker, L. C. 1983. Historical notes on the taxonomy and nomenclature of the recent Rhinocerotidae (Mammalia, Perissodactyla). Beaufortia, 33:37-51.

Rookmaaker, L. C. 1991. The scientific name of the bontebok. Zeitschrift für Säugetierkunde, 66:190-191.

Rookmaaker, L. C. 1992. Additions and revisions to the list of specimens of the extinct blue antelope (*Hippotragus leucophaeus*). Annals of the South African Museum, 102:131-141.

Rookmaaker, L. C., and W. Bergmans. 1981. Taxonomy and geography of *Rousettus amplexicaudatus* (Geoffroy, 1810) with comparative notes on sympatric congeners (Mammalia: Megachiroptera). Beaufortia, 31:1-29.

Roos, C., and T. Geissmann. 2001. Molecular phylogeny of the major hylobatid divisions. Molecular Phylogenetics and Evolution, 19:486-494.

Roosevelt, T., and E. Heller. 1914. Life histories of African game animals. 2 vols. Charles Scribner's Sons, New York, 798 pp.

Roots, E. M., and R. J. Baker. 1998. *Rhogeesa genowaysi*. Mammalian Species, 589:1-3.

Rose, R. K., and R. D. Dueser. 1999. Marsh rice rat | *Oryzomys palustris*. Pp. 554-555, *in* The Smithsonian Book of North American Mammals (D. E. Wilson and S. Ruff, eds.). Smithsonian Institution Press, Washington, D.C., xxv + 750 pp.

Rosel, P. E., M. G. Haygood, and W. F. Perrin. 1995. Phylogenetic relationships among the true porpoises (Cetacea: Phocoenidae). Molecular Phylogenetics and Evolution, 4(4):463-474.

Rosenbaum, H. C., R. L. Brownell Jr., M. W. Brown, C. Schaeff, V. Portway, B. N. White, S. Malik, L. A. Pastene, N. J. Patenaude, C. S. Baker, M. Goto, P. Best, P. J. Clapham, P. Hamilton, M. Moore, R. Payne, V. Rowntree, C. T. Tynan, J. L. Bannister, and R. Desalle. 2000. World-wide genetic differentiation of Eubalaena: Questioning the number of right whale species. Molecular Ecology, 9:1793-1802.

Rosenberger, A. L. 1977. *Xenothrix* and ceboid phylogeny. Journal of Human Evolution, 6(5):461-481.

Rosenberger, A. L., and A. F. Coimbra-Filho. 1984. Morphology, taxonomic status and affinities of the lion tamarins, *Leontopithecus* (Callitrichinae, Cebidae). Folia Primatologica, 42(3-4):149-179.

Rosevear, D. R. 1963. On the West African forms of *Heliosciurus* Trouessart. Mammalia, 27:177-185.

Rosevear, D. R. 1965. The bats of West Africa. British Museum (Natural History), London, 418 pp.

Rosevear, D. R. 1969. The rodents of West Africa. Trustees of the British Museum (Natural History), London, 677:1-604.

Rosevear, D. R. 1974. The carnivores of West Africa. Trustees of the British Museum (Natural History), London, 723:1-548.

Ross, G. J. B., and V. G. Cockroft. 1990. Comments on Australian bottlenose dolphins and the taxonomic status of *Tursiops aduncus* (Ehrenberg, 1832). Pp. 101-128, *in* The bottlenose dolphin (S. Leatherwood and R. R. Reeves, eds.). Academic Press, New York, 653 pp.

Ross, P. D. 1988. The taxonomic status of *Cansumys canus*. Abstracts, Symposium of Asian Pacific Mammalogy, Huirou, Beijing, Peoples Republic of China.

Rossi, M. S., O. A. Reig, and J. Zorzopulos. 1990. Evidence for rolling-circle replication in a major satellite DNA from South American rodents of the genus *Ctenomys*. Molecular Biology and Evolution, 7:340-350.

Rossi, M. S., O. A. Reig, G. Viale, A. I. Massarini, and E. Capanna. 1995. Chromosomal distribution of the major satellite DNA of South American rodents of the genus *Ctenomys*. Cytogenetics and Cell Genetics, 69:179-184.

Rossolimo, O. L. 1971. [Variability and taxonomy of *Dryomys nitedula* Pallas.] Zoologicheskii Zhurnal, 50(2):247-258 (in Russian).

Rossolimo, O. L. 1976a. Novyi vid myshevidnoi soni — *Myomimus setzeri* (Mammalia, Myoxidae) iz Irana [A new species of mouse-like dormouse . . . from Iran]. Vestnik Zoologii, 1976(4):51-53 (in Russian).

Rossolimo, O. L. 1976*b*. [Taxonomic status of the mouse-like dormouse *Myomimus* (Mammalia, Myoxidae) from Bulgaria.] Zoologicheskii Zhurnal, 55(10):1515-1525 (in Russian).

Rossolimo, O. L. 1989. [Revision of Royle's high-mountain vole *Alticola* (*A.*) *argentatus* (Mammalia: Cricetidae).] Zoologicheskii Zhurnal, 68:104-114 (in Russian).

Rossolimo, O. L., and I. Ya. Pavlinov. 1985. External genital morphology and its taxonomic significance in the dormouse genera *Myomimus* and *Glirulus* (Rodentia, Gliridae). Folia Zoologica, 34(2):121-124.

Rossolimo, O. L., and I. Ya. Pavlinov. 1992. Species and subspecies of *Alticola* s. str. (Rodentia: Arvicolidae). Pp. 149-176, *in* Prague studies in mammalogy (I. Horáček and V. Vohralik, eds.). Charles University Press, Praha, 245 pp.

Rossolimo, O. L., I. Ya. Pavlinov, O. I. Podtyazhkin, V. S. Skulkin. 1988. [Variability and taxonomy of mountain voles (*Alticola* s. str.) from Mongolia, Tuva, Baikal Region and Altai.] Zoologicheskii Zhurnal, 67:426-437 (in Russian).

Rossolimo, O. L., I. Ya. Pavlinov, and R. S. Hoffmann. 1994. Systematics and distribution of the rock voles of the subgenus *Alticola* s. str. in the People's Republic of China (Rodentia: Arvicolinae). Acta Theriologica Sinica, 14(2):86-99.

Rossolimo O. L., E. G. Potapova , I. Ya. Pavlinov, S. V. Kruskop, and O. V. Voltzit. 2001. [Dormice (Myoxidae) of the World.] Sbornik Trudov Zoologicheskogo Muzeya MGU 42: 1-232 (in Russian with English summary).

Roth, V. L., and R. W. Thorington, Jr. 1982. Relative brain size among African squirrels. Journal of Mammalogy, 63(1):168-173.

Rothenburg, S., M. Eiben, F. Koch-Nolte, and F. Haag. 2002. Independent integration of rodent identifier (ID) elements into orthologous sites of some RT6 alleles of *Rattus norvegicus* and *Rattus rattus*. Journal of Molecular Evolution, 55:221-259.

Rounsevell, D. E., and S. J. Smith. 1982. Recent alleged sightings of the thylacine (Marsupialia: Thylacinidae) in Tasmania. Pp. 233-236. *in* Carnivorous Marsupials (M. Archer (ed.). Surrey Beatty and Sons, Sydney, 2 volumes, 804 pp.

Rounsevell, D. E., R. J. Taylor, and G. J. Hocking. 1991. Distribution records of native terrestrial mammals in Tasmania. Wildlife Research, 18:699-717.

Rousseau, M. 1983. Etude des *Arvicanthis* du Museum de Paris par analyses factorielles (Rongeurs, Murides). Mammalia, 47:525-542.

Rowe, D. L., and R. L. Honeycutt. 2002. Phylogenetic relationships, ecological correlates, and molecular evolution within the Cavioidea (Mammalia, Rodentia). Molecular Biology and Evolution, 19:263-277.

Roy, M. S., E. Geffen, D. Smith, E. A. Ostrander, and R. K. Wayne. 1994. Patterns of differentiation and hybridization in North American wolflike canids, revealed by analysis of microsatellite loci. Molecular Biology and Evolution, 11:553-570.

Roy, M. S., E. Geffen, D. Smith, and R. K. Wayne. 1996. Molecular genetics of pre-1940 red wolves. Conservation Biology, 10:1413-1424.

Rozhnov, V. V. 1995. Taxonomic notes on the yellow-throated marten *Martes flavigula*. Zoologicheskii Zhurnal, 74:131-138.

Rudolphi, D. K. A. 1820 [1822]. Einige anatomische Bemerkungen über *Balaena rostrata*. Abhandlungen der Physikalische Klasse der Königlich-Preussischen Akademie der Wissenschaften zu Berlin, 1820-1821(1822):27-40.

Ruedas, L. A. 1986. Chromosomal variability in the New England cottontail, *Sylvilagus transitionalis* (Bangs)[sic], 1895 with evidence of recognition of a new species. Unpubl. M. S. thesis, Fordham University, New York, 37 pp.

Ruedas, L. A. 1995. Description of a new large-bodied species of *Apomys* Mearns, 1905 (Mammalia: Rodentia: Muridae) from Mindoro Islands, Philippines. Proceedings of the Biological Society of Washington, 108(2):302-318.

Ruedas, L. A. 1998. Systematics of *Sylvilagus* Gray, 1867 (Lagomorpha: Leporidae) from southwestern North America. Journal of Mammalogy, 79:1355-1378.

Ruedas, L. A., and J. A. W. Kirsch. 1997. Systematics of *Maxomys* Sody, 1936 (Rodentia: Muridae: Murinae): DNA/DNA hybridization studies of some Borneo-Javan species and allied Sundaic and Australo-Papuan genera. Biological Journal of the Linnean Society, 61:365-408.

Ruedas, L. A., R. C. Dowler, and E. Aita. 1989. Chromosomal variation in the New England cottontail, *Sylvilagus transitionalis*. Journal of Mammalogy, 70:860-864.

Ruedi, M. 1995. Taxonomic revision of shrews of the genus *Crocidura* from the Sunda Shelf and Sulawesi with description of two new species (Mammalia: Soricidae). Zoological Journal of the Linnean Society, 115:211-265.

Ruedi, M. 1996. Phylogenetic evolution and biogeography of southeast Asian shrews (genus *Crocidura*: Soricidae). Biological Journal of the Linnean Society, 58:197-219.

Ruedi, M. 1998. Protein evolution in shrews. Pp. 269-294, *in* Evolution of shrews (J. M.Wójcik and M. Wolsan, eds.). Mammal Research Institute, Polish Academy of Sciences, Bialowieza, 458 pp.

Ruedi, M., and R. Arlettaz. 1991. Biochemical systematics of the Savi's bat (*Hypsugo savii*) (Chiroptera: Vespertilionidae). Zeitschrift für Zoologische Systematik und Evolutionsforschung, 29:115-122.

Ruedi, M., and L. Fumagalli. 1996. Genetic structure of gymnures (genus *Hylomys*; Erinaceidae) on continental islands of southeast Asia: Historic effects of fragmentation. Journal of Zoological Systematics and Evolutionary Research, 34:153-162.

Ruedi, M., and F. Mayer. 2001. Molecular systematics of bats of the genus *Myotis* (Vespertilionidae) suggests deterministic ecomorphological convergences. Molecular Phylogenetics and Evolution, 21:436-448.

Ruedi, M., and Vogel, P. 1995. Chromosomal evolution and zoogeographic origin of southeast Asian shrews (genus *Crocidura*). Experientia, 51:174-178.

Ruedi, M., T. Maddalena, H.-S. Yong, and P. Vogel. 1990. The *Crocidura fuliginosa* species complex (Mammalia: Insectivora) in peninsular Malaysia: Biological, karyological and genetical evidence. Biochemical Systematics and Ecology, 18:573-581.

Ruedi, M., T. Maddalena, P. Vogel, and Y. Obara. 1993. Systematic and biogeographic relationships of the Japanese white-toothed shrew (*Crocidura dsinezumi*). Journal of Mammalogy, 74:535-543.

Ruedi, M., M. Chapuisat, and D. Iskandar. 1994. Taxonomic status of *Hylomys parvus* and *Hylomys suillus* (Insectivora: Erinaceidae): Biochemical and morphological analyses. Journal of Mammalogy, 75:965-978.

Ruedi, M., C. Courvoisier, P. Vogel, and F. M. Catzeflis. 1996. Genetic differentiation and zoogegraphy of Asian *Suncus murinus* (Mammalia: Soricidae). Biological Journal of the Linnean Society, 57:307-316.

Ruedi, M., M. Auberson, and V. Savolainen. 1998. Biogeography of Sulawesian shrews: Testing for their origin with a parametric bootstrap on molecular data. Molecular Phylogenetics and Evolution, 9:567-571.

Ruedi, M., P. Jourde, P. Giosa, M. Bartaud, and S. Y. Roue. 2002. DNA reveals the existence of *Myotis alcathoe* in France (Chiroptera: Vespertilionidae). Revue Suisse de Zoologie, 109:643-652.

Rümke, C. G. 1985. A review of fossil and recent Desmaninae (Talpidae, Insectivora). Utrecht Micropaleontological Bulletins, Special Publication, 4:1-241.

Rummel, M. 1999. Tribe Cricetodontini. Pp. 359-364, *in* The Miocene land mammals of Europe (G. E. Rössner and K. Heissig, eds.). Dr. Friedrich Pfeil, München, 515 pp.

Rümmler, H. 1932. Über die Schwimmratten (Hydromyinae). Das Aquarium, 1932:131-135.

Rümmler, H. 1936. Die formen der papuanischen Muridengattung *Melomys*. Zeitschrift für Säugetierkunde, 11:247-253.

Rümmler, H. 1938. Die Systematik und Verbreitung der Muriden Neuguineas. Mitteilungen aus dem Zoologische Museum in Berlin, 23:1-297.

Rumpler, Y. 1975. The significance of chromosomal studies in the systematics of the Malagasy lemurs. Pp. 25-40, *in* Lemur Biology (I. Tattersall and R. W. Sussman, eds.). Plenum Press, New York, 365 pp.

Rumpler, Y., S. Warter, C. Rabarivola, J.-J. Petter, and B. Dutrillaux. 1990. Chromosomal evolution in Malagasy lemurs. XII. Chromosomal banding study of *Avahi laniger occidentalis* (syn.: *Lichanotus laniger occidentalis*) and cytogenetic data in favour of its classifcation in a species apart—*Avahi occidentatis*. American Journal of Primatology, 21:307-316.

Rumpler, Y., B. Ravaoarimanana, M. Hauwy, and S. Warter. 2001. Cytogenetic arguments in favour of a taxonomic revision of *Lepilemur septentrionalis*. Folia Primatologia, 72:308-315.

Rupp, H. 1980. Beiträge zur Systematik, Verbreitung und Ökologie äthiopischer Nagetiere. Ergebnisse mehrerer Forschungsreisen. Säugetierkundliche Mitteilungen, 28(2):81-123.

Ruprecht, A. L. 1979. Kryteria identyfikacji gatunkowej podrodzaju *Sylvaemus* Ognev and Vorobiev, 1923 (Rodentia: Muridae). Przeglad Zoologiczny, 23:340-349.

Ruprecht, A. L., and A. Szwagrzak. 1986. Fat dormouse in the food of the Ural Owl. Przeglad Zoologiczny, 30(4):431-432.

Russell, D. E., and B. Sigé. 1970. Révision des chiroptères lutétiens de Messel (Hesse, Allemagne). Palaeovertebrata, 3:83-182.

Russell, R. J. 1968*a*. Evolution and classification of the pocket gophers of the subfamily Geomyinae. University of Kansas Publications, Museum of Natural History, 16:473-579.

Russell, R. J. 1968*b*. Revision of pocket gophers of the genus *Pappogeomys*. University of Kansas Publications, Museum of Natural History, 16:581-776.

Russo, D., and G. Jones. 2000. The two cryptic species of *Pipistrellus pipistrellus* (Chiroptera: Vespertilionidae) occur in Italy: Evidence from echolocation and social calls. Mammalia, 64:187-197.

Ruxton, A. E., and E. Schwarz. 1929. On hybrid hartebeests and on the distribution of the *Alcelaphus buselaphus* group. Proceedings of the Zoological Society of London, 1929:567-583.

Ryan, A., E. Duke, and J. S. Fairley. 1996. Mitochondrial DNA in bank voles *Clethrionomys glareolus* in

Ireland: Evidence for a small founder population and localized founder effects. Acta Theriologica, 41(1):45-50.

Ryan, J. M. 1989a. Comparative myology and phylogenetic systematics of the Heteromyidae (Mammalia, Rodentia). Miscellaneous Publications, Museum of Zoology, University of Michigan, 176:1-103.

Ryan, J. M. 1989b. Evolution of cheek pouches in African pouched rats (Rodentia: Cricetomyinae). Journal of Mammalogy, 70:267-274.

Ryan, J. M. 2003. *Nesomys*, red forest rat, Voalavo mena. Pp. 1388-1389, *in* The natural history of Madagascar (S. M. Goodman and J. P. Benstead, eds.). The University of Chicago Press, Chicago, xxi + 1709 pp.

Ryan, J. M., and D. Attuquayefio. 2000. Mammal fauna of the Muni-Pomadze Ramsar site, Ghana. Biodiversity and Conservation, 9:541-560.

Ryan, J. M., G. K. Creighton, and L. H. Emmons. 1993. Activity patterns of two species of *Nesomys* (Muridae: Nesomyinae) in a Madagascar rain forest. Journal of Tropical Ecology, 9:101-107.

Ryan, R. M. 1965. Taxonomic status of the vespertilionid genera *Kerivoula* and *Phoniscus*. Journal of Mammalogy, 46:517-518.

Ryan, R. M. 1966. A new and some imperfectly known Australian *Chalinolobus* and the taxonomic status of African *Glauconycteris*. Journal of Mammalogy, 47:86-91.

Rybnikov, D. E., E. A. Gileva, and G. P. Miroshnichenko. 1986. [Study of DNA of Asian mountain voles.] Theriology, ornithology and environmental control. Abstracts of the 2nd All-Union symposium Biological Problems of the North, Yakutsk, 3:69 (in Russian).

Rychlik, L. 1998. Evolution of social systems in shrews. Pp. 347-406, *in* Evolution of shrews (J. M.Wójcik and M. Wolsan, eds.). Mammal Research Institute, Polish Academy of Sciences, Bialowieza, 458 pp.

Rydell, J. 1993. *Eptesicus nilssonii*. Mammalian Species, 430:1-7.

Ryder, O. A., A. T. Kumamoto, B. S. Durrant, and K. Benirschke. 1989. Chromosomal divergence and reproductive isolation in dik-diks. Pp. 208-225, *in* Speciation and its consequences (D. Otte and J. A. Endler, eds). Sinauer Associates Incorporated, Sunderland, MA, 679 pp.

Rylands, A. B., and D. Brandon-Jones. 1998. Scientific nomenclature of the red howlers from the northeastern Amazon in Brazil, Venezuela, and the Guianas. International journal of Primatology, 19:879-905.

Rylands, A. B., A. F. Coimbra-Filho, and R. A. Mittermeier. 1993. Systematics, geographic distribution, and some notes on the conservation status of the Callitrichidae. Pp. 11-77, *in* Marmosets and tamarins: Systematics, behaviour, and ecology, (A. B. Rylands, ed.). Oxford University Press, 416 pp.

Rylands, A. B., R. A. Mittermeier, and E. R. Luna. 1995. A species list for the new world primates (Platyrrhini): distribution by country, endemism, and conservation status according to the Mace-Lande system. Neotropical Primates, 3(Suppl.):113-160.

Rylands, A. B., H. Schneider, A. Langguth, R. A. Mittermeier, C. P. Groves, and E. Rodriguez-Luna. 2000. An assessment of the diversity of new world primates. Neotropical Primates, 8:61-93.

Rzasnicki, A. 1938. Vorläufige Mitteilung über die Züchtung eines sogenannten Ward-Zebras. Annales Musei Zoologici Polonici, 13:197-204.

Rzasnicki, A. 1951. Zebras and quaggas. Annales Musei Zoologici Polonici, 14:203-252.

Rzebik-Kowalska, B. 1981. The Pliocene and Pleistocene Insectivora (Mammalia) of Poland. IV. Soricidae: *Neomysorex* n. gen. and *Episoriculus* Ellerman et Morrison-Scott, 1951. Acta zoologica cracoviensia, 25:227-250.

Rzebik-Kowalska, B. 1988. Soricidae (Mammalia, Insectivora) from the Plio-Pleistocene and Middle Quaternary of Morocco and Algeria. Folia Quaternaria, 57:51-90.

Rzebik-Kowalska, B. 1989. Pliocene and Pleistocene Insectivora (Mammalia) of Poland. V. Soricidae: *Petenyia* Kormos, 1934 and *Blarinella* Thomas, 1911. Acta zoologica cracoviensia, 32:521-546.

Rzebik-Kowalska, B. 1998. Fossil history of shrews in Europe. Pp. 23-92, *in* Evolution of shrews (J. M.Wójcik and M. Wolsan, eds.). Mammal Research Institute, Polish Academy of Sciences, Bialowieza, 458 pp.

Rzebik-Kowalska, B. 2002. The Pliocene and Early Pleistocene Lipotyphla (Insectivora, Mammalia) from Romania. Acta zoologica cracoviensia, 45:251-281.

Saavedra, B., and J. A. Simonetti. 2000. A northern and threatened population of *Irenomys tarsalis* (Mammali: Rodentia) from central Chile. Zeitschrift für Säugetierkunde, 65:243-245.

Saavedra, B., and J. A. Simonetti. 2001. New records of *Dromiciops gliroides* (Microbiotheria: Microbiotheridae) and *Geoxus valdivianus* (Rodentia: Muridae) in central Chile: Their implications for biogegraphy and conservation. Mammalia, 65:96-100.

Sabatier, M. 1978. Un nouveau *Tachyoryctes* (Mammalia, Rodentia) du bassin Pliocène de Hadar (Ethiopie). Géobios, 11:95-99.

Sabatier, M. 1982. Les rongeurs du site Pliocene a hominides de Hadar (Ethiopie). Palaeovertebrata, 12(1):1-56.

Sablina, O. V., S. I. Radzhabli, V. G. Malikov, M. N. Meyer, and G. N. Kuliev. 1988. [Taxonomy of voles of the genus *Chionomys* (Rodentia, Microtinae) based on karyological data.] Zoologicheskii Zhurnal, 67(3):472-475 (in Russian with English summary).

Sablina, O. V., J. Zima, S. I. Radjabli, B. Kryštufek, and F. N. Goleniscev [Golenishchev]. 1989. [New data on karyotype variation in the pine vole, *Pitymys subterraneus* (Rodentia, Arvicolidae).] Vestnik Ceskoslovenske Spolecnosti Zoologicke, 53:295-299 (in Russian).

Sádlova, J., and D. Frynta. 2001. Species differences and temporal variation of non-metric characters in *Apodemus flavicollis* and *A. sylvaticus* populations (Mammalia: Rodentia). Acta Societatis Zoologicae Bohemicae, 65:97-104.

Safronova, L. D., V. M. Malygin, E. S. Levenkova, and V. N. Orlov. 1992. [Cytogenetic sequelae of hybridization of hamsters *Phodopus sungorus* and *Phodopus campbelli*.] Doklady Akademii Nauk, 327(2):266-271 (in Russian).

Sage, R. D., J. R. Contreras, V. G. Roig, and J. L. Patton. 1986a. Genetic variation in the South American burrowing rodents of the genus *Ctenomys* (Rodentia: Ctenomyidae). Zeitschrift für Säugetierkunde, 51:158-172.

Sage, R. D., D. Heyneman, K.-C. Lim, and A. C. Wilson. 1986b. Wormy mice in a hybrid zone. Nature, 324:60-63.

Sage, R. D., W. R. Atchley, and E. Capanna. 1993. House mice as models in systematic biology. Systematic Biology, 42(4):523-561.

Saha, S. S. 1978. On some mammals recently collected in Bhutan. Journal of the Bombay Natural History Society, 74:350-354.

Saha, S. S. 1981. A new genus and a new species of flying squirrel (Mammalia: Rodentia: Sciuridae) from northwestern India. Bulletin of the Zoological Survey of India, 4(3):331-336.

Said, K., A. Saad, J.-C. Auffray, and J. Britton-Davidian. 1993. Fertility estimates in the Tunisian all-acrocentric and Robertsonian populations of the house mouse and their chromosomal hybrids. Heredity, 71:532-538.

Said, K., J.-C. Auffray, P. Boursot, and J. Britton-Davidian. 1999. Is chromosomal speciation occurring in house mice in Tunisia? Biological Journal of the Linnean Society, 68:387-399.

Saint Girons, M. C. 1972. Rectification a propos des auteurs de la description de *Erinaceus algirus*. Mammalia, 36:166-167.

St. Leger, J. 1930. On two species of *Dendromus*. Annals and Magazine of Natural History, ser. 10, 6:622.

St. Leger, J. 1931. A key to the families and genera of African Rodentia. Proceedings of the Zoological Society of London, 1931:957-997.

Saitoh, M., N. Matsuoka, and Y. Obara. 1989. Biochemical systematics of three species of the Japanese long-tailed field mice; *Apodemus speciosus*, *A. giliacus*, and *A. argenteus*. Zoological Science, 6:1005-1018.

Sakai, E., Y. Uematsu, and T. Miyao. 1997. [Studies on regional variation on the molars of the small mammals in Japanese islands.] Journal of Growth, 36:51-63 (in Japanese with English abstract).

Sakamoto, Y. 1996. Histological features of endomysium, perimysium and epimysium in rat lateral pterygoid muscle. Journal of Morphology, 227:113-119.

Sala, B. 1996. *Dinaromys allegranzii* n. sp. (Mammalia, Rodentia) from Rivoli Veronese (Northeastern Italy) in a Villanyian association. Bollettino della Società Paleontologica Italiana, 33(1):3-11.

Salata-Pilacinska, B. 1990. The southern range of the root vole in Poland. Acta Theriologica, 35:53-67.

Salazar-Bravo, J., J. W. Dragoo, D. S. Tinnin, and T. L. Yates. 2001. Phylogeny and evolution of the Neotropical rodent genus *Calomys*: Inferences from mitochondrial DNA sequence data. Molecular Phylogenetics and Evolution, 20:173-184.

Salazar-Bravo, J., J. W. Dragoo, M. D. Bowen, C. J. Peters, T. G. Ksiazek, and T. L. Yates. 2002a. Natural nidality in Bolivian hemorrhagic fever and the systematics of the reservoir species. Infection, Genetics, and Evolution, 1:191-199.

Salazar-Bravo, J., E. Yensen, T. Tarifa, and T. L. Yates. 2002b. Distributional records of Bolivian mammals. Mastozoologia Neotropical, 9:70-78.

Salazar-Bravo, J., T. Tarifa, L. F. Aguirre, E. Yensen, and T. L. Yates. 2003. Revised checklist of Bolivian mammals. Occasional Papers, Museum of Texas Tech University, 220:1-27.

Saldaña-DeLeon, J. L., and C. A. Jones. 1998. Annotated checklist of the Recent mammals of Colorado. Occasional Papers, Museum of Texas Tech University, 179:1-4.

Sale, J. B., and M. E. Taylor. 1970. A new four-toed mongoose from Kenya, *Bdeogale crassicauda nigrescens* ssp. nov. Journal of the East African Natural History Society and National Museum, 28:10-15.

Saleh, M. A., and M. I. Basuony. 1998. A contribution to the mammalogy of the Sinai Peninsula. Mammalia, 62(4):557-575.

Salles, L. O. 1992. Felid phylogenetics: Extant taxa and skull morphology (Felidae, Aeluroidea). American Museum Novitates, 3047:1-67.

Salotti, M. 1984. A Ghjira, ou le loir en Corse. Courrier de la Nature, 89:31-35.

Sanborn, C. C. 1930. Distribution and habits of the three-banded armadillo (*Tolypeutes*). Journal of Mammalogy, 11:61-68, pl. 4.

Sanborn, C. C. 1931. Bats from Polynesia, Melanesia, and Malaysia. Field Museum of Natural History, Zoological Series, 18:7-29.

Sanborn, C. C. 1933. Bats of the genera *Anoura* and *Lonchoglossa*. Field Museum of Natural History, Zoological Series, 20:23-28.

Sanborn, C. C. 1937. American bats of the subfamily Emballonurinae. Field Museum of Natural History, Zoological Series, 20:321-354.

Sanborn, C. C. 1947a. The South American rodents of the genus *Neotomys*. Fieldiana: Zoology, 31:51-57.

Sanborn, C. C. 1947b. Geographical races of the rodent *Akodon jelskii* Thomas. Fieldiana: Zoology, 31:133-142.

Sanborn, C. C. 1949. Bats of the genus *Micronycteris* and its subgenera. Fieldiana: Zoology, 31:215-233.

Sanborn, C. C. 1950. Bats from New Caledonia, the Solomon Islands, and New Hebrides. Fieldiana: Zoology, 31:313-338.

Sanborn, C. C. 1952a. Philippine Zoological Expedition 1946-1947. Fieldiana: Zoology, 33:89-158.

Sanborn, C. C. 1952b. The status of "*Triaenops wheeleri*" Osgood. Chicago Academy of Sciences, Natural History Miscellanea, 97:1-3.

Sanborn, C. C. 1953. Mammals from Mindanao, Philippine Islands collected by the Danish Philippine Expedition 1951-1952. Videnskabelige Meddeleser fra Dansk Naturhistoriske Forening i Kobenhaven, 115:283-289.

Sanborn, C. C., and J. A. Crespo. 1957. El Murciélago Blanquizco (*Lasiurus cinereus*) y sus subspecies. Boletin del Museo Argentino de Ciencias Naturales "Bernardino Rivadavia", 4:1-13.

Sanborn, C. C., and H. Hoogstraal. 1953. Some mammals of Yemen and their ectoparasites. Fieldiana: Zoology, 34(23):229-252.

Sanborn, C. C., and A. J. Nicholson. 1950. Bats from New Caledonia, the Solomon Islands, and New Hebrides. Fieldiana: Zoology, 31:313-338.

Sánchez, J., J. Ochoa G., and R. S. Voss. 2001. Rediscovery of *Oryzomys gorgasi* (Rodentia: Muridae), with notes on taxonomy and natural history. Mammalia, 65:205-214.

Sánchez, O. 1993. Analisis de algunas tendencies ecogeograficas del genero *Reithrodontomys* (Rodentia: Muridae) en México. Pp. 25-44, *in* Avances en el estudio de los mamíferos de México (R. A. Medellín and G. Cebassos, eds.). Publicaciones Especiales, Vol. 1. Asociación Mexicana de Mastozoología, México, D. F., 464 pp.

Sánchez-Cordero, V. 2001. Elevation gradients of diversity for rodents and bats in Oaxaca, Mexico. Global Ecology and Biogeography, 10:63-76.

Sánchez-Cordero, V., and R. Valadez Azua. 1989. Habitat y distribucion del genero *Oryzomys* (Rodentia: Cricetidae). Anales de Instituto de Biologia, UNAM, Serie Zoologica, 59:99-112.

Sánchez-Cordero, V., and B. Villa-Ramirez. 1988. Variación morphométrica en *Peromyscus spicilegus* (Rodentia: Cricetinae) en a parte nordeste de Jalisco, México. Anales de Instituto de Biologia, Universidad Nacional Autónomica de México, Series Zoologia, 58:819-836.

Sánchez H., C., C. J. Alvarez R., and M. de L. Romero A. 1996. Biological and ecological aspects of *Microtus oaxacensis* and *Microtus mexicanus*. Southwestern Naturalist, 41:95-98.

Sánchez-Hernández, C., M. de L. Roméro-Almaraz, R. D. Owen, A. Nuñez-Garduño, and R. López-Wilchis. 1999. Noteworthy records of mammals from Michoacán, México. Southwestern Naturalist, 44:231-235.

Sanderson, I. T. 1940. The mammals of the north Cameroon forest area. Being the results of the Percy Sladen expedition to the Mamfe Division of the British Cameroons. Transactions of the Zoological Society of London, 24:623-725.

Sans-Coma, V., J M. Arque, A. D. Duran, M. Cardo, B. Fernandez, and D. Franco. 1993. The coronary arteries of the Syrian hamster, *Mesocricetus auratus* (Waterhouse 1839). Annals of Anatomy, 175:53-57.

Sans-Coma, V., A. Durán, M. Cardo, and J. M. Arqué. 1995. The coronary arteries of the garden dormouse (*Eliomys quercinus* L., 1766). Pp. 217-224, *in* Proceedings of II Conference on Dormice (Rodentia, Myoxidae) (M. G. Filippucci, ed.). Hystrix, n.s., 6(1-2):1-340.

Santapau, H., and Humayun Abdulali (eds.). 1960. The hispid hare, *Caprolagus hispidus* (Pearson). Journal of the Bombay Natural History Society, 57:400-402.

Santel, W., and W. von Koenigswald. 1998. Preliminary report on the middle Pleistocene small mammal fauna from Yarimburgaz cave in Turkish Thrace. Eiszeitalter und Gegenwart, 48:162-168.

Santos, I. B., G. A. B. da Fonseca, S. E. Rigueira, and R. B. Machado. 1994. The rediscovery of the Brazilian three banded armadillo and notes on its conservation status. Edentata, 1:11-15.

Santos, P. B. dos. 1997. Estudo craniometrico de *Oryzomys nigripes* (Olfers, 1818) (Rodentia-Cricetidae) em regioes fitogeograficamente distintas no Estado do Parana, Brasil. Estudos de Biologia PUC-PR, Curitaba, 41:89-110.

Santos-Reis, M., and M. L. Mathias. 1996. The historical and recent distribution and status of mammals in Portugal. Hystrix, n.s., 8(1-2):75-89.

Santucci, F., B. C. Emerson, and G. M. Hewitt. 1998. Mitochondrial DNA phylogeography of European hedgehogs. Molecular Ecology, 7:1163-1172.

Sarà, M. 1995. The Sicilian (*Crocidura sicula*) and the Canary (*C. canariensis*) shrew (Mammalia, Soricidae); peripheral isolate formation and geographic variation. Bolletino di Zoologico, 62:173-182.

Sarà, M. 1996. A landmark-based morphometrics approach to the systematics of Crocidurinae. A case study on endemic shrews *Crocidura sicula* and *C. canariensis* (Soricidae, Mammalia). NATO ASI (Advanced Science Institutes) Series A Life Sciences, 284:335-344.

Sarà, M., and G. Casamento. 1995a. Distribution and ecology of dormice (Myoxidae) in Sicily: A preliminary account. Pp. 161-168, *in* Proceedings of II Conference on Dormice (Rodentia, Myoxidae) (M. G. Filippucci, ed.). Hystrix, n.s., 6(1-2):1-340.

Sarà, M., and G. Casamento. 1995b. Morphometrics of the wood mouse (*Apodemus sylvaticus*, Mammalia, Rodentia) in the Mediterranean. Bolletino di Zoologia, 62:313-320.

Sarà, M., and R. Vitturi. 1996. *Crocidura* populations (Mammalia, Soricidae) from the Sicilian-Maltese insular area. Hystrix, n.s., 8:121-132.

Sarà, M., and P. Vogel. 1996. Geographic variation of the greater white-toothed shrew (*Crocidura russula* Hermann, 1780, Mammalia, Soricidae). Zoological Journal of the Linnean Society, 116:377-392.

Sarà, M., and L. Zanca. 1992. Metric discrimination and distribution of the species of *Crocidura* occurring in Tunisia. Acta Theriologica, 37:103-116.

Sarà, M., M. Lo Valvo, and L. Zanca. 1990. Insular variation in central Mediterranean *Crocidura* Wagler, 1832 (Mammalia, Soricidae). Bollettino di Zoologia, 57:283-293.

Sarà, M., G. Casamento, and A. Spinnato. 2001. Density and breeding of *Muscardinus avellanarius* L., 1758 in woodlands of Sicily. Trakya University Journal of Scientific Research B, 2(2):85-93.

Saralegui, A. M. 1996. *Eumops patagonicus* Thomas, 1924, en el Uruguay (Mammalia, Chiroptera: Molossidae). Communicaciones Zoologicas del Museo de Historia Natural de Montivideo, 12:1-4.

Sargis, E. J. 2001. A preliminary qualitative analysis of the axial skeleton of tupaiids (Mammalia, Scandentia): Functional morphology and phylogenetic implications. Journal of Zoology, London, 253:473-483.

Sargis, E. J. 2002a. Functional morphology of the forelimb of tupaiids (Mammalia, Scandentia) and its phylogenetic implications. Journal of Morphology, 253:10-42.

Sargis, E. J. 2002b. Functional morphology of the hindlimb of tupaiids (Mammalia, Scandentia) and its phylogenetic implications. Journal of Morphology, 254:149-185.

Sarich, V. M. 1969a. Pinniped phylogeny. Systematic Zoology, 18:416-422.

Sarich, V. M. 1969b. Pinniped origins and the rate of evolution of carnivore albumins. Systematic Zoology, 18:286-295.

Sarich, V. M. 1973. The giant panda is a bear. Nature, 245:218-220.

Sarich, V. M. 1976. Transferrin. Transactions of the Zoological Society of London, 33:165-171.

Sarich, V. M. 1985. Rodent macromolecular systematics. Pp. 423- 452, *in* Evolutionary relationships among rodents, a multidisciplinary analysis (W. P. Luckett and J.-L. Hartenberger, eds.). Plenum Press, New York, 721 pp.

Sarmiento, E. E. 1998. The validity of "*Pseudopotto martini*". African Primates, 31(1):44-45.

Sarmiento, E. E., and J. F. Oates. 2000. The Cross River gorillas: A distinct subspecies, *Gorilla gorilla diehli* Matschie 1904. American Museum Novitates, 3304:1-55.

Satoh, K. 1997. Comparative functional morphology of mandibular forward movement during mastication of two murid rodents, *Apodemus speciosus* (Murinae) and *Clethrionomys rufocanus* (Arvicolinae). Journal of Morphology, 231:131-142.

Satoh, K. 1998. Balancing function of the masticatory muscles during incisal biting in two murid rodents, *Apodemus speciosus* and *Clethrionomys rufocanus*. Journal of Morphology, 236:49-56.

Satoh, K. 1999. Mechanical advantage of area of origin for the external pterygoid muscle in two murid rodents, *Apodemus speciosus* and *Clethrionomys rufocanus*. Journal of Morphology, 240:1-14.

Satoh, T., and Y. Obara. 1995. Nonrandom distribution of sister chromatid exchanges in the chromosomes of three mammalian species. Zoological Science, 12:749-756.

Satunin, K. A. 1914. Opredelitel' mlekopitayushchikh Rossiiskoi Imperii [Guide to the mammals of Imperial Russia]. Tiflis, 1:1-410.

Sausy, F. 2000. [Features of fossorial water vole.] Pp. 172-174, *in* [Species of the fauna of Russia and contiguous countries. The water vole: Mode of life.] Nauka Publishers, Moscow, 527 pp. (in Russian).

Savage, R. J. G. 1988. The African dimension in European early Miocene mammal faunas. Pp. 587-599, *in* European Neogene mammal chronology (E. H. Lindsay, V. Fahlbusch, and P. Mein, eds.). Nato ASI Series, Series A: Life Sciences, 180, Plenum Press, New York and London, 658 pp.

Savic, I. R. 1982*a*. Familie Spalacidae Gray, 1821—Blindmäuse. Pp. 539-542, *in* Handbuch der Säugetiere Europas (J. Niethammer and F. Krapp, eds.). Akademische Verlagsgesellschaft (Wiesbaden), 2/I:1-649.

Savic, I. R. 1982*b*. *Microspalax leucodon* (Nordmann, 1840)— Westblindmaus. Pp. 543-569, *in* Handbuch der Säugetiere Europas (J. Niethammer and F. Krapp, eds.). Akademische Verlagsgesellschaft (Wiesbaden), 2/I:1-649.

Savic, I. R. 1982*c*. *Spalax polonicus* Mehely, 1909—Bodolische Blindmaus. Pp. 571-576, *in* Handbuch der Säugetiere Europas (J. Niethammer and F. Krapp, eds.). Akademische Verlagsgesellschaft (Wiesbaden), 2/I:1-649.

Savic, I. R. 1982*d*. *Spalax graecus* Nehring, 1898—Bukowinische Blindmaus. Pp. 577-584, *in* Handbuch der Säugetiere Europas (J. Niethammer and F. Krapp, eds.). Akademische Verlagsgesellschaft (Wiesbaden), 2/I:1-649.

Savic, I. R., and E. Nevo. 1990. The Spalacidae: Evolutionary history, speciation and population biology. Progress in Clinical and Biological Research, 335:129-153.

Savic, I. R., and B. Soldatovic. 1977. Prilog poznavanju ekogeografskog rasprostranjenja i evolucije hromozomskih formi spalacidaea Balkan-Skog Poluostrva. Arhiv Bioloskih Nauka (Beograd), 29:141-156.

Savic, I. R., and B. Soldatovic. 1979. Distribution range and evolution of chromosomal forms in the Spalacidae of the Balkan Peninsula and bordering regions. Journal of Biogeography, 6:363-374.

Savinov, P. R. 1970. [Jerboas (Dipodidae, Rodentia) from the Neogene of Kazakhstan.–Material on evolution of terrestrial vertebrates.] Byulleten' Moskoskovo Obshchestva Ispytatelei Prirody, Otdel Biologicheskii, 1970:91-134 (in Russian).

Savolainen, P., Y-P. Zhang, J. Luo, J. Lundeberg, and T. Leitner. 2002. Genetic evidence for an East Asian origin of domestic dogs. Science, 298:1610-1613.

Sayer, J. A., and A. A. Green. 1984. The distribution and status of large mammals in Benin. Mammal Review, 14:37-50.

Sbalqueiro, I. J., and A. P. Nascimento. 1996. Occurrence of *Akodon cursor* (Rodentia, Cricetidae) with 14, 15, and 16 chromosome cytotypes in the same geographic area in southern Brazil. Brazilian Journal of Genetics, 19:565-569.

Sbalqueiro, I. J., M. S. Mattevi, and L. F. B. Oliveira. 1984. An $X_1X_1X_2X_2/X_1X_2Y$ mechanism of sex determination in a South American rodent, *Deltamys kempi* (Rodentia, Cricetidae). Cytogenetics and Cellular Genetics, 38:50-55.

Sbalqueiro, I. J., M. S. Mattevi, L. F. B. Oliveira, and M. J. V. Solano. 1991. B chromosome system in populations of *Oryzomys flavescens* (Rodentia, Cricetidae) from southern Brazil. Acta Theriologica, 36:193-199.

Scala, C., and S. Lovari. 1984. Revision of *Rupicapra* genus. II. A skull and horn statistical comparison of *Rupicapra rupicapra ornata* and *R. rupicapra pyrenaica* chamois. Bollettino di Zoologia, 51:285-294.

Scally, M., O. Madsen, C. J. Douady, W. W. de Jong, M. J. Stanhope, and M. S. Springer. 2001. Molecular evidence for the major clades of placental mammals. Journal of Mammalian Evolution, 8:239-277.

Scaramella, D., L. F. Russo, and F. P. D'Enrico. 1974. I mammiferi della Somalia (a livrello della sottospecie). Bollettino della Società di Naturalisti in Napoli, 83, 1974:269-331.

Scaravelli, D., L. Casini, and C. Matteucci. 1995. Dormice distribution in Romagna region (Italy). Pp. 195-198, *in* Proceedings of II Conference on Dormice (Rodentia, Myoxidae) (M. G. Filippucci, ed.). Hystrix, n.s., 6(1-2):1-340.

Scarff, J. E. 1986. Historic and present distribution of the right whale (*Eubalaena glacialis*) in the eastern North Pacific south of 50°N and east of 180°W. Pp. 43-63, *in* Right Whales: Past and present status (R. L. Brownell, Jr., P. B. Best, and J. H. Prescott, eds.). Reports of the International Whaling Commission, Special Issue, 10:1-289.

Schaefer, H. 1975. Die Spitzmäuse der Hohen Tatra seit 30 000 Jahren (Mandibular-Studie). Zoologischer Anzeiger, 195:89-111.

Schaldach, W. J. 1960. *Xenomys nelsoni* Merriam, sus relaciones y sus habitos. Revista de la Sociedad de Historia Natural, 21:425-434.

Schaldach, W. J. 1965. Notas breves sobre algunos mamíferos del sur de México. Anales de Instituto de Biologia, Universidad Nacional Autónomica de México, 35:129-137.

Schaller, G. B. 1977. Mountain monarchs. Wild sheep and goats of the Himalaya. The University of Chicago Press, Chicago, 425 pp.

Schaller, G. B. 1998. Wildlife of the Tibetan Steppe. The University of Chicago Press, Chicago, 373 pp.

Schaller, G. B., and A. Rabinowitz. 1995. The saola or spindlehorn bovid *Pseudoryx nghetinhensis* in Laos. Oryx, 29:107-113.

Schaller, G. B., and E. S. Vrba. 1996. Description of the giant muntjac (*Megamuntiacus vuquangensis*) in Laos. Journal of Mammalogy, 77:675-683.

Scharff, V. A., M. Macholán, J. Zima, and H. Burda. 2001. A new karyotype of *Heliophobius argenteocinereus* (Bathyergidae, Rodentia) from Zambia with field notes on the species. Mammalian Biology, 66:376-378.

Schaub, S. 1934. Über einige fossile Simplicidentaten aus china und der Mongolei. Abhandlungen der Schweizerischen Päleontologischen Gesellschaft, 54:1-40.

Schaub, S. 1938. Tertiäre und Quartäre Murinae. Abhandlungen der Schweizerischen Päleontologischen Gesellschaft, 61:1-39 pp.

Schaub, S. 1953. Remarks on the distribution and classification of the "Hystricomorpha." Verhandlungen der Naturforschenden Gesellschaft in Basel, 64:389-400.

Schaub, S. 1958. Simplicidentata (Rodentia), pp. 659-818, *in* Traité de Paleontologie, VI(2) (J. Piveteav, ed.). Masson, Paris, 957 pp.

Schaub, S., and H. Zapfe. 1953. Die fauna der miozänen Spaltenfüllung von Neudorf an der Marcch (ESR), Simplicidentata. Sitzungsberichte Österreichische Akademie der Wissenschaften, Mathematisch-Naturwissenschaftliche Klasse, Abteilung I, 162:181-215.

Schauenberg, P. 1974. Données nouvelles sur le chat des sables, "*Felis margarita*" Loche, 1858. Revue Suisse de Zoologie, 81(4):949-969.

Scheffer, V. B. 1958. Seals, sea lions, and walruses, a review of the Pinnipedia. Stanford University Press, Stanford, CA, 179 pp.

Schevill, W. E. 1986. The international code of zoological nomenclature and a paradigm: The name *Physeter catodon* Linnaeus 1758. Marine Mammal Science, 2(2):153-157.

Schevill, W. E. 1987a. Note by William E. Schevill [on *Steno bredanensis* in the Mediterranean Sea, Watkins et al.]. Marine Mammal Science, 3(1):77.

Schevill, W. E. 1987b. Reply to Holthius, 1987. Marine Mammal Science, 3(1):89-90.

Schevill, W. E. 1990a. On stability in zoological nomenclature. Marine Mammal Science, 6(2):168-169.

Schevill, W. E. 1990b. Reply to D. W. Rice's rejoinder. Marine Mammal Science, 6(4):360.

Schinz, H. R. 1844-1845. Systematisches verzeichniss aller bis jetzt bekannten Säugethiere, oder, Synopsis mammalium, nach dem Cuvier' schen system. Solothurn, Jent und Gassmann, 2 vols.

Schlawe, L. 1980a. Zur geographischen Verbreitung der Ginsterkatzen Gattung *Genetta* G. Cuvier, 1816. Faunistische Abhandlungen Staatliches Museum für Tierkunde in Dresden, 7(15):147-161.

Schlawe, L. 1980b. Kritischen zur Nomenclatur und taxonomischen Beurteilung von *Equus africanus* (Fitzinger, 1858). Equus, 2:101-127.

Schlawe, L. 1981. Material, Fundort, Text- und Bildquellen als Grundlagen fur eine Artenliste zur Revision der Gattung *Genetta* G. Cuvier, 1816 (Mammalia, Carnivora, Viverridae). Zoologische Abhandlungen Staatliches Museum für Tierkunde in Dresden, 37(4):85-182.

Schlawe, L. 1986. Seltene Pfleglinge aus Dschungarei und Mongolei: Kulane, *Equus hemionus hemionus* Pallas, 1775. Zoologische Garten, Neue Folge, 56:299-323.

Schlegel, H. 1877. Prospectus for museum publication. Annals of the Royal Zoological Museum of the Netherlands at Leyden, pp. not numbered.

Schlegel, H. 1879. Note XIV. *Paradoxurus musschenbroekii*. Notes of the Leyden Museum, 1879:43.

Schliemann, H., and B. Maas. 1978. *Myzopoda aurita*. Mammalian Species, 116:1-2.

Schlitter, D. A. 1989. African rodents of special concern: A preliminary assessment. Pp. 33-39, *in* Rodents. A world survey of species conservation concern (W. Z. Lidicker, Jr., ed.). Occasional Papers of the IUCN Species Survival Commission (SSC), No. 4. I.U.C.N., The World Conservation Union, 60 pp.

Schlitter, D. A. 1993. Order Hyracoidea. Pp. 373-374, *In* Mammal species of the world, a taxonomic and geographic reference. Second ed. (D. E. Wilson and D. M. Reeder, eds.). Smithsonian Institution Press, Washington, D.C., 1207 pp.

Schlitter, D. A., and I. R. Aggundey. 1986. Systematics of African bats of the genus *Eptesicus* (Mammalia: Vespertilionidae). 1. Taxonomic status of the large serotines of eastern and southern Africa. Cimbebasia, ser. A, 18:167-174.

Schlitter, D. A., and H. W. Setzer. 1972. A new species of short-tailed gerbil (*Dipodillus*) from Morocco (Mammalia: Cricetidae: Gerbillinae). Proceedings of the Biological Society of Washington, 84:385-392.

Schlitter, D. A., and H. W. Setzer. 1973. New rodents (Mammalia: Cricetidae, Muridae) from Iran and Pakistan. Proceedings of the Biological Society of Washington, 86:163-174.

Schlitter, D. A., and K. Thonglongya. 1971. *Rattus turkestanicus* (Satunin, 1903), the valid name for *Rattus rattoides* Hodgson, 1845 (Mammalia: Rodentia). Proceedings of the Biological Society of Washington, 84:171-174.

Schlitter, D. A., I. L. Rautenbach, and D. A. Wolhuter. 1980. Karyotypes and morphometrics of two species of *Scotophilus* in South Africa (Mammalia: Vespertilionidae). Annals of the Transvaal Museum, 32:231-239.

Schlitter, D. A., S. L. Williams, and J. E. Hill. 1983. Taxonomic review of Temminck's trident bat, *Aselliscus tricuspidatus* (Temminck, 1834)(Mammalia: Hipposideridae). Annals of Carnegie Museum, 52:337-358.

Schlitter, D. A., I. L. Rautenbach, and C. G. Coetzee. 1984. Karyotypes of southern African gerbils, genus *Gerbillurus* Shortridge, 1942 (Rodentia: Cricetidae). Annals of Carnegie Museum, 53:549-557.

Schlitter, D. A., L. W. Robbins, and S. L. Williams. 1985. Taxonomic status of dormice (genus *Graphiurus*) from west and central Africa. Annals of Carnegie Museum, 54(1):1-9.

Schlitter, D. A., R. Hutterer, T. Maddalena, and L. W. Robbins. 1999. New karyotypes of shrews (Mammalia: Soricidae) from Cameroon and Somalia. Annals of Carnegie Museum, 68:1-13.

Schlund, W. 1997. Die Tibialänge als Maß für Körpergröße und als Hilfsmittel zur Altersbestimmung bei Siebenschläfern (*Myoxus glis* L.). Zeitschrift für Säugetierkunde, 62:187-190.

Schmid, M., T. Haaf, H. Weis, and W. Schempp. 1986. Chromosomal homoeologies in hamster species of the genus *Phodopus* (Rodentia, Cricetinae). Cytogenetics and Cell Genetics, 43:168-173.

Schmid, M., R. Johannisson, T. Haaf, and H. Neitzel. 1987. The chromosomes of *Micromys minutus* (Rodentia, Murinae). II. Pairing pattern of X and Y chromosomes in meiotic prophase. Cytogenetics and Cell Genetics, 45:121-131.

Schmidly, D. J. 1972. Geographic variation in the white-ankled mouse, *Peromyscus pectoralis*. Southwestern Naturalist, 17:113-138.

Schmidly, D. J. 1973a. Geographic variation and taxonomy of *Peromyscus boylii* from Mexico and southern United States. Journal of Mammalogy, 54:111-130.

Schmidly, D. J. 1973b. The systematic status of *Peromyscus comanche*. Southwestern Naturalist, 18:269-278.

Schmidly, D. J. 1974a. *Peromyscus attwateri*. Mammalian Species, 48:1-3.

Schmidly, D. J. 1974b. *Peromyscus pectoralis*. Mammalian Species, 49:1-3.

Schmidly, D. J., and R. D. Bradley. 1995. Morphological variation in the Sinaloan Mouse *Peromyscus simulus*. Revista Mexicana Mastozoologica, 1:44-58.

Schmidly, D. J., and F. S. Hendricks. 1976. Systematics of the southern races of Ord's kangaroo rat, *Dipodomys ordii*. Bulletin of the Southern California Academy of Science, 75:225-237.

Schmidly, D. J., and F. S. Hendricks. 1984. Mammals of the San Carlos Mountains of Tamaulipas, Mexico. Pp. 15-69, *in* Contributions in honor of Robert L. Packard (R. E. Martin and B. R. Chapman, eds.). Special Publication, The Museum, Texas Tech University, 22:1-234.

Schmidly, D. J., and G. L. Schroeter. 1974. Karyotypic variation of *Peromyscus boylii* (Rodentia, Cricetidae) from Mexico and corresponding taxonomic implications. Systematic Zoology, 23:333-342.

Schmidly, D. J., M. R. Lee, W. S. Modi, and E. G. Zimmerman. 1985. Systematics and notes on the biology of *Peromyscus hooperi*. Occasional Papers, The Museum, Texas Tech University, 97:1-40.

Schmidly, D. J., R. D. Bradley, and P. S. Cato. 1988. Morphometric differentiation and taxonomy of three chromosomally characterized groups of *Peromyscus boylii* from east-central Mexico. Journal of Mammalogy, 69:462-480.

Schmidt, C. A. 1999. Variation and congruence of microsatellite markers for *Peromyscus leucopus*. Journal of Mammalogy, 80:522-529.

Schmidt, C. A., and M. D. Engstrom. 1994. Genic variation and systematics of rice rats (*Oryzomys palustris* species group) in southern Texas and northeastern Tamaulipas, Mexico. Journal of Mammalogy, 75:914-928.

Schmidt, C. A., M. D. Engstrom, and H. H. Genoways. 1989. *Heteromys gaumeri*. Mammalian Species, 345:1-4.

Schmidt-Kittler, N. 1981. Zur Stammesgeschichte der marderverwandten Raubtiergruppen (Musteloidea, Carnivora). Eclogae Geologicae Helvetiae, 74:753-801.

Schmitt, L. H., D. J. Kitchener, and R. A. How. 1995. A genetic perspective of mammalian radiation and evolution in the Indonesian archipelago: Biogeographic correlates in the fruit bat genus *Cynopterus*. Evolution, 49:399-412.

Schnake-Greene, J. E., L. W. Robbins, and D. K. Tolliver. 1990. Comparison of genetic differentiation among populations of two species of mice (*Peromyscus*). Southwestern Naturalist, 35:54-60.

Schneider, E. 1997. Brown hare (*Lepus europaeus*) towards the 21st century. Gibier Faune Sauvage, 14:553-554.

Schnell, G. D., T. L. Best, and M. L. Kennedy. 1978. Interspecific morphologic variation in kangaroo rats (*Dipodomys*): degree of concordance with genic variation. Systematic Zoology, 27:34-48.

Schomber, H.-W. 1963. Beitrage zur Kenntnis der Giraffengazelle (*Litocranius walleri* Brooke, 1878). Säugetierkundliche Mitteilungen, 11(Sonderheft):1-44.

Schomber, H.-W. 1964. Beitrage zur Kenntnis der Lamagazelle, *Ammodorcas clarkei* (Thomas, 1891). Säugetierkundliche Mitteilungen, l2:65-90.

Schonewald, C. 1994. *Cervus canadensis* and *C. elaphus*. North American subspecies and evaluation of clinal extremes. Acta Theriologica, 39:431-452.

Schoppe, R. 1986. Die Schlafmäuse (Gliridae) in Niedersachsen. Lebensraum und Verbreitung von Siebenschläfer, Gartenschläfer und Haselmaus. Naturschutz und Landschaftspflege Niedersachsen Beiheft, 14:1-52.

Schouteden, H. 1944, 1945, 1946. De Zoogdieren van Belgish Congo en van Ruanda-Urundi. Annales du Musée du Congo Belge, Tervuren (Belgique). C.-Zoologie, Ser. 2, 3:1-576.

Schouteden, H. 1948. Faune du Congo Belge et du Ruanda-Urundi. I. Mammifères. Annales du Musée du Congo Belge, Zoologie, 8(1):1-331.

Schreber, J. C. D. 1777. Die Säugthiere in Abbildungen nach der Natur mit Beschreibungen 1776-1778. Wolfgang Walther, Erlangen, 3:377-440, pls. 104B, 107Aa, 109B, 110B, 115B, 125B, 127B, 136, 146A.

Schreiber, A., R. Wirth, M. Riffel, and H. van Rompaey. 1989. Weasels, civets, mongooses and their relatives: An action plan for the conservation of mustelids and viverrids. I.U.C.N., Gland, Switzerland, 100 pp.

Schreiber, R. 2001. Feldhamster in Bayern–bestandstrends und geplantes artenhilfskonzept. Jahrbücher des Nassauischen Vereins für Naturkunde, 122:207-208.

Schroering, G. B. 1995. Swamp deer resurfaces. Wildlife Conservation, 98(6):22.

Schulze, W. 1986. Zum Vorkommen und zur Biologie von Haselmaus (*Muscardinus avellanarius* L.) und Siebenschläfer (*Glis glis* L.) in Volgelkästen im Südharz der DDR. Säugetierkundliche Informationen, 2(10):341-348.

Schulze, W. 1987. Zum Mobilitat der Haselmaus (*Muscardinus avellanarius*) im Südharz. Säugetierkundliche Informationen, 2(11):485-488.

Schunke, A. C., and R. Hutterer. 2000. Patchy versus continuous distribution patterns in the African rain forest: The problem of the Anomaluridae (Mammalia: Rodentia). Pp. 145-152, *in* Isolated vertebrate communities in the Tropics (G. Rheinwald, ed.). Proceedings of the 4th International Symposium, Bonn, Bonner Zoologische Monographen, 46: 400 pp.

Schwabe, H. W. 1979. Vergleichend-allometrische Untersuchungen an den Schädeln europaischer und asiatischer Hausratten (*Rattus rattus* L.). Zeitschrift für Säugetierkunde, 44:354-360.

Schwangart, F. 1936. Der Manul, *Otocolobus manul* (Pallas), im System der Feliden. Zentralblatt für Kleintierkunde und Pelztierkunde, 12(8)(Carnivoren Studien 2):19-67.

Schwangart, F. 1943. Die Sohlenzeichnung von *Felis* und Verwandtes zur Systematik und Oekologie des Genus. Abhandlungen der Bayerischen Akademie der Wissenschaften, Mathematisch-naturwissenschaftliche Abteilung, Neue Folge, 52:1-35.

Schwartz, A. 1953. A systematic study of the water rat (*Neofiber alleni*). Occasional Papers of the Museum of Zoology, University of Michigan, 547:1-27.

Schwartz, A., and J. K. Jones, Jr. 1967. Bredin-Archbold-Smithsonian Biological Survey of Dominica. 7. Review of bats of the endemic Antillean genus *Monophyllus*. Proceedings of the United States National Museum, 124(3635):1-20.

Schwartz, J. H., and I. Tattersall. 1985. Evolutionary relationships of living lemurs and lorises (Mammalia, Primates) and their potential affinities with European Eocene Adapidae. Anthropological Papers of the American Museum of Natural History, 60, 3:1-100.

Schwarz, E. 1930. Die Sammlung afrikanischer Säugetiere im Congo Museum, Ginsterkatzen (Gattung *Genetta* Oken). Revue de Zoologie et de Botanique Africaines, 19:275-286.

Schwarz, E. 1933*a*. *Cercopithecus mitis* Wolf für *Simia leucampyx* Fischer. Zeitschrift für Säugetierkunde, 8:279.

Schwarz, E. 1933*b*. The hyrax of the Central Sahara. Annals and Magazine of Natural History, ser. 10, 12:625-626.

Schwarz, E. 1939. On mountain-voles of the genus *Alticola* Blanford: A taxonomic and genetic analysis. Proceedings of the Zoological Society of London, ser. B, 108:663-668.

Schwarz, E. 1947. Colour mutants of the Malay short-tailed mongoose, *Herpestes brachyurus* Gray. Proceedings of the Zoological Society of London, 117(1):79-80.

Schwarz, E. 1948. Revision of the Old World moles of the genus *Talpa*. Proceedings of the Zoological Society of London, 118:36-48.

Schwarz, E., and H. K. Schwarz. 1943. The wild and commensal stocks of the house mouse, *Mus musculus* Linneaus. Journal of Mammalogy, 24:59-72.

Schwarz, E., and H. K. Schwarz. 1967. A monograph of the *Rattus rattus* group. Anales de la Escuela Naciónal de Ciéncias Biológicas (México City), 14:79-178.

Schweigert, R. 1998. Some notes on *Crateromys heaneyi* Gonzales and Kennedy, 1996 (Rodentia: Muridae) of Panay, Philippines. Zeitschrift für Säugetierkunde, 63:121-123.

Sclater, P. L., and O. Thomas. 1894-1900. The book of antelopes. R. H. Porter, London, 1:1-220 [1894-1895]; 2:1-194 [1896-1897]; 3:1-245[1897-1898]; 4:1-242 [1899-1900].

Sclater, W. L. 1891. Catalogue of Mammalia in the Indian museum, Calcutta. Part II. Rodentia, Ungulata, Proboscidea, Hyracoidea, Carnivora, Cetacea, Sirenia, Marsupialia, Monotremata. Calcutta, 350 pp.

Sclater, W. L. 1900-1901. Mammals of South Africa [Vol. I. Primates, Carnivora and Ungulata]. Porter, London, 2 vols.

Scoazec, J.-Y. 1996. An overlooked name of an African ungulate, *Connochaetes taurinus babaulti* Kollman, 1919. Mammalia, 60:156-158.

Scopoli, G. A. 1777. Introductio ad historiam naturalem, sistens genera lapidum, plantarum et animalium, Pragae, 506 pp.

Scriven, P. N., and V. Bauchau. 1992. The effect of hybridization on mandible morphology in an island population of the house mouse. Journal of the Zoological Society of London, 226:573-583.

Seal, U. S., and D. G. Makey. 1974. ISIS mammalian taxonomic directory international species inventory system. Minnesota Zoological Garden, St. Paul, Minnesota, 645 pp.

Searle, J. B. 1984. Three new karyotypic races of the common shrew *Sorex araneus* (Mammalia, Insectivora) and a phylogeny. Systematic Zoology, 33:184-194.

Searle, J. B. 1991. A hybrid zone comprising staggered chromosomal clines in the house mouse (*Mus musculus domesticus*). Proceedings of the Royal Society of London, ser. B, 246:47-52.

Searle, J. B., and J. M. Wójcik. 1998. Chromosomal evolution: The case of *Sorex araneus*. Pp. 219-258, *in* Evolution of shrews (J. M.Wójcik and M. Wolsan, eds.). Mammal Research Institute, Polish Academy of Sciences, Bialowieza, 458 pp.

Searle, J. B., and J. M. Wójcik (eds.). 2000. Evolution in the *Sorex araneus* group: Cytogenetic and molecular aspects. Acta Theriologica, 45 (suppl. 1):1-190.

Searle, J. B., Y. N. Navarro, and G. Ganem. 1993. Further studies of a staggered hybrid zone in *Mus musculus domesticus* (the house mouse). Heredity, 71:523-531.

Seba, A. 1734. Locupletissimi rerum naturalium thesauri accurata descriptio, eti conibus artificiosissimis expressio, per universam physices historiam opus, cui, in hoc rerum genere, nullum par exstitit, ex toto terrarum orbe collegit, digessit, descripsit, et depingendum curavit. Apud. J. Wetstenium & Gul. Smith, & Janssonio-Waesbergios, Amstelaedami, 1:1-78 + 38 (unnumbered), 111 pls.

Seddon, J. M., and P. R. Baverstock. 2000. Evolutionary lineages of *RT1.Ba* in the Australian *Rattus*. Molecular Biology and Evolution, 17(5):768-772.

Seddon, J. M., F. Santucci, N. Reeve, and G. M. Hewitt. 2002. Caucasus Mountains divide postulated postglacial colonization routes in the white-breasted hedgehog, *Erinaceus concolor*. Journal of Evolutionary Biology, 15:463-467.

Seebeck, J. H. 1995a. Terrestrial mammals in Victoria–a history of discovery. Proceedings of the Royal Society of Victoria, 107(1):11-23.

Seebeck, J. H. 1995b. Water rat, *Hydromys chrysogaster*. Pp. 203-205, *in* Mammals of Victoria, distribution, ecology and conservation (P. W. Menkhorst, ed.). Oxford University Press, Melbourne, 360 pp.

Seebeck, J. H. 1995c. Bush rat, *Rattus fuscipes*. Pp. 224-226, *in* Mammals of Victoria, distribution, ecology and conservation (P. W. Menkhorst, ed.). Oxford University Press, Melbourne, 360.

Seebeck, J. H. 1995d. Swamp rat, *Rattus lutreolus*. Pp. 227-228, *in* Mammals of Victoria, distribution, ecology and conservation (P. W. Menkhorst, ed.). Oxford University Press, Melbourne, 360 pp.

Seebeck, J. H., and P. W. Menkhorst. 1995. New Holland mouse, *Pseudomys novaehollandiae*. Pp. 220-222, *in* Mammals of Victoria, distribution, ecology and conservation (P. W. Menkhorst, ed.). Oxford University Press, Melbourne, 360 pp.

Seebeck, J. H., and P. W. Menkhorst. 2000. Status and conservation of the rodents of Victoria. Wildlife Research, 27:357-369.

Selander, R. K., M. H. Smith, S. Y. Yang, W. E. Johnson, and J. B. Gentry. 1971. Biochemical polymorphism and systematics in the genus *Peromyscus*. I. Variation in the old-field mouse (*Peromyscus polionotus*). Studies in Genetics VI, University of Texas Publications, 7103:49-90.

Sélys-Longchamps, E. de. 1839. Etudes de micromammalogie. Revue des musaraignes, des rats et des campagnols, suivie d'une index methodique des mammiferes d'Europe. Librairie Encyclopedique de Roret, Paris, 165 pp.

Sempéré, A. J., V. E. Sokolov, and A. A. Danilkin. 1996. *Capreolus capreolus*. Mammalian Species, 538:1-9.

Şen, S.. 1977. La faune de rongeurs Pliocènes de Çalta (Ankara, Turquie). Bulletin du Muséum National d'Histoire Naturelle. Sciences de la Terre 61, Third series, 465:1-171.

Şen, S. 1983. Rongeurs et lagomorphes du gisement Pliocène de Pul-e Charkhi, bassin de Kabul, Afghanistan. Bulletin du Muséum National d'Histoire Naturelle. Sciences de la Terre, Fourth series, 5(1):33-74.

Şen, S., and T. Sharma. 1983. Role of constitutive heterochromatin in evolutionary divergence: Results of chromosome banding and condensation inhibition studies in *Mus musculus*, *Mus booduga* and *Mus dunni*. Evolution, 37:628-636.

Şen, S., M. Brunet, and E. Heintz. 1979. Decouverte de rongeurs "Africains" dans le Pliocene d'Afghanistan (basin de Sarobi). Implications paleobiogeographiques et stratigraphiques. Bulletin Museum National d'Histoire Naturelle, Paris, 4(1), seccion C:65-75.

Sendor, T., I. Roedenbeck, S. Hampl, M. Ferreri, and M. Simon. 2002. Revision of morphological identification of pipistrelle bat phonic types (*Pipistrellus pipstrellus* Schreber, 1774). Myotis, 40:11-17.

Sénégas, F. 2001. Interpretation of the dental pattern of the South African fossil *Euryotomys* (Rodentia, Murinae) and origin of otomyine dental morphology. Pp. 151-160, *in* African small mammals (C. Denys, L. Granjon, A. Poulet, eds.). IRD Éditions, Collection Colloques et Séminaires, Paris, 570 pp.

Sénégas, F., and M. Avery. 1998. New evidence for the murine origins of the Otomyinae (Mammalia, Rodentia) and the age of Bolt's Farm (South Africa). South African Journal of Science, 94:503-507.

Senut, B., M. Pickford, P. Mein, G. Conroy, and J. Van Couvering. 1992. Discovery of 12 new Late Cainozoic fossiliferous sites in palaeokarsts of the Otavi Mountains, Namibia. Comptes Rendus de l'Acadèmie des Sciences (Paris), 314(ser. II):727-733.

Sera, W. E., and C. N. Early. 2003. *Microtus montanus*. Mammalian Species, 716:1-10.

Serdyuk, V. A. 1979. Izmenchivost' stroeniya zubov u arkticheskovo suslika (*Citellus parryi*); veroyantnye puti rasseleniya etovo vida na severo-vostoke SSSR [Variation in structure of the teeth in the arctic ground squirrel . . . and a possible way of migration of this species in the northeastern U.S.S.R.]. Zoologicheskii Zhurnal, 58:1692-1702 (in Russian).

Serizawa, K., H. Suzuki, and K. Tsuchiya. 2000. A phylogenetic view on species radiation in *Apodemus* inferred from variation of nuclear and mitochondrial genes. Biochemical Genetics, 38(1/2):27-40.

Serñhanin, I. N. 1961. Mlekopitajušie Beloruskoj SSR. 2nd ed. Akademia Nauk BSSR, Minsk, 318 pp.

Setzer, H. W. 1949. Subspeciation in the kangaroo rat, *Dipodomys ordii*. University of Kansas Publications, Museum of Natural History, 1:473-573.

Setzer, H. W. 1952. Notes on mammals from the Nile Delta region, Egypt. Proceedings of the United States National Museum, 102:343-369.

Setzer, H. W. 1956. Mammals of the Anglo-Egyptian Sudan. Proceedings of the United States National Museum, 106(3377):447-587.

Setzer, H. W. 1957. The hedgehogs and shrews (Insectivora) of Egypt. Journal of the Egyptian Public Health Association, 32:1-17.

Setzer, H. W. 1959. The spiny mice (*Acomys*) of Egypt. Journal of the Egyptian Public Health Association, 34(3):93-101.

Setzer, H. W. 1969. A review of the African mice of the genus *Desmodilliscus* Wettstein, 1916. Miscellaneous Publications of the University of Kansas Museum of Natural History, 51:283-288.

Setzer, H. W. 1975. Genus *Acomys*. Part 6.5. Pp. 1-2, *in* The mammals of Africa: An identification manual (J. Meester and H. W. Setzer, eds.) [issued 10 Dec 1975]. Smithsonian Institution Press, Washington, D.C., not continuously paginated.

Severinghaus, W. D. 1977. Description of a new subspecies of prairie vole, *Microtus ochrogaster*. Proceedings of the Biological Society of Washington, 90:49-54.

Severinghaus, W. D., and D. F. Hoffmeister. 1978. Qualitative cranial characters distinguishing *Sigmodon hispidus* and *Sigmodon arizonae* and the distribution of these two species in northern Mexico. Journal of Mammalogy, 59:868-870.

Severtzov, N. 1858. Notice sur la classification multisériale des carnivores, spécialement des Félidés, et les études de zoologie générale qui s'y rattachent. Revue et Magasin de Zoologie, Pure et Appliquée (2), 241-246, 385-393, 9:387-391, 433-439; 10:3-8, 145-150, 193-196.

Seymour, K. L. 1989. *Panthera onca*. Mammalian Species, 340:1-9.

Shackleton, D. M. 1985. *Ovis canadensis*. Mammalian Species, 230:1-9.

Shackleton, D. M. (ed.) 1997. Wild sheep and goats and their relatives. Status survey and conservation action plan for Caprinae. I.U.C.N., Gland, Switzerland. 390 pp.

Sharma, T. 1996. Chromosomal and molecular divergence in the Indian pygmy field mice *Mus booduga-terricolor* lineages of the subgenus *Mus*. Genetica, 97:331-338.

Sharma, T., and I. K. Gadi. 1977. Constitutive heterochromatin variation in two species of *Rattus* with apparently similar karyotypes. Genetica, 47:77-80.

Sharma, T., and R. Raman. 1971. An XO female in the Indian mole rat. Journal of Heredity, 62:384-387.

Sharma, T., and R. Raman. 1972. Odd diploid number in both sexes and a unique multiple sex-chromosome system of a rodent, *Vandeleuria o. oleracea* (Bennett). Cytogenetics, 11:247-258.

Sharma, T., and R. Raman. 1973. Variation of constitutive heterochromatin in the sex chromosomes of the rodent *Bandicota bengalensis bengalensis* (Gray). Chromosoma, 41:75-84.

Sharma, T., N. Cheong, P. Sen, and S. Sen. 1986. Constitutive heterochromatin and evolutionary divergence of *Mus dunni*, *M. booduga* and *M. musculus*. Pp. 35-44, *in* Current topics in microbioloby and immunology, vol. 127 (M. Potter, J. H. Nadeau, and M. P. Cancro, eds.). Springer-Verlag, Berlin, 395 pp.

Sharma, T., A. Bardhan, and M. Bahadur. 2002. Increases and decreases of whole-arm heterochromatin in specific chromosomes: An extraordinary situation in hybrids of the *Mus terricolor* complex. Cytogenetics and Genome Research, 96:244-249.

Sharman, G. B., C. E. Murtagh, P. M. Johnson, and C. M. Weaver. 1980. The chromosomes of a rat-kangaroo attributable to *Bettongia tropica* (Marsupialia; Macropodidae). Australian Journal of Zoology, 28:59-63.

Shaughnessy, P. D., and F. H. Fay. 1977. A review of the taxonomy and nomenclature of north Pacific harbour seals. Journal of Zoology, London, 182:385-419.

Shaw, G. 1800. General zoology or systematic natural history. G. Kearsley, London, 12:1-330.

Shcherbina, E. I., L. S. Marinina, and N. A. Sorokina. 1988. [Mammals of the Malyi Balkhan mountain ridge and piedmont plain.] Izvestiya Akademii Nauk Turkmenskoi SSR, Seriya Biologicheskikh Nauk, 3:15-22 (in Russian).

She, J. X., F. Bonhomme, P. Boursot, L. Thaler, and F. Catzeflis. 1990. Molecular phylogenies in the genus *Mus*: Comparative analysis of electrophoretic, scnDNA hybridization, and mtDNA RFLP data. Biological Journal of the Linnean Society, 41:83-103.

Sheets-Pyenson, S. 1981. From the north to Red Lion Court: The creation and early years of the *Annals of Natural History*. Journal of the Society for the Bibliography of Natural History, 10:221-249.

Sheffield, S. R., and C. M. King. 1994. *Mustela nivalis*. Mammalian Species, 454:1-10.

Sheffield, S. R., and H. H. Thomas. 1997. *Mustela frenata*. Mammalian Species, 570:1-9.

Shehab, A. H., K. Kowalski, and A. Daoud. 1999. *Apodemus mystacinus* (Danford and Alston, 1877) (Muridae, Rodentia) from Al Hermon and Al Arab Mountains, southern Syria. Acta Zoologica Cracoviensia, 42(3):397-401.

Shekelle, M., S. M. Leksono, L. L. S. Ichwan, and Y. Masala. 1998. The natural history of the tarsiers of north and central Sulawesi. Sulawesi Primate Newsletter, 4(2):4-11.

Sheldon, C. 1919. The wilderness of the Upper Yukon. Second ed. Charles Scribner's Sons, New York, 364 pp.

Shellhammer, H. S. 1967. Cytotaxonomic studies of the harvest mice of the San Francisco Bay region. Journal of Mammalogy, 48:549-556.

Shellhammer, H. S. 1982. *Reithrodontomys raviventris*. Mammalian Species, 169:1-3.

Shenbrot, G. I. 1974. [Systematic status of *Allactodipus bobrinskii* (Rodentia, Dipodidae).] Zoologicheskii Zhurnal, 53(11):1697-1702 (in Russian).

Shenbrot, G. I. 1984. [Dental morphology and phylogeny of five-toed jerboas of the subfamily Allactaginae.] Sbornik Trudov Zoologicheskovo Museya MGU, 22:61-92 (in Russian).

Shenbrot, G. I. 1986. [Supergeric relationships of the jerboas (Rodentia, Dipodoidea).] Chetvertii Sezd Vsesoyuznovo Teriologicheske Obshchestva, Tezisy Dokladov, Moscow, 1:106-107 (in Russian).

Shenbrot, G. I. 1988. Noviye danniye rasprostraneniyu i sistematikye tolstokhvostikh tushkanchikov [New data on the distribution and taxonomy of the fat-tailed jerboas, genus *Pygeretmus*]. Pp. 121-124, *in* Tushkanchiki fauni SSSR [Jerboas of the USSR fauna], vol. 2 (V. E., Sokolov, V. S. Lobachev, E. I. Naumova, R. R. Reimov, G. I. Shenbrot, and T. I. Dmitrieva, eds.). Tyezisi Dokladov Vesesoiuznovo soveshchaniye [Abstracts of the All-Union Conference], Nukus, Fan Press, Moscow, 139 pp. (in Russian).

Shenbrot, G. I. 1990*a*. Geograficheskaya izmenchivost' Turkmenskovo tushkanchika *Jaculus turkmenicus* (Rodentia, Dipodidae), i zadachi evo okhrany [Geographical variation of the Turkmenian jerboa, *Jaculus turkmenicus* (Rodentia, Dipodidae) and the problems of its protection]. Zoologicheskii Zhurnal, 69(2):114-121 (in Russian).

Shenbrot, G. I. 1990*b*. Geograficheskaya izmenchivost' i podvidovaya differentisiya tushkanchika Likhtenshteina, *Eremodipus lichtensteini* (Rodentia, Dipodidae) [Geographical variation and subspecific differentiation of the Lichtenstein's jerboa, *Eremodipus lichtensteini* (Rodentia, Dipodidae)]. Zoologicheskii Zhurnal, 69(10):154-159 (in Russian).

Shenbrot, G. I. 1991*a*. Geograficheskaya izmenchivost' mokhnonogovo tushkanchika *Dipus sagitta* (Rodentia, Dipodidae) [Geographical variation of the three-toed brush-footed jerboa . . .]. 2. [Subspecific differentiation in eastern Kazakhstan, Tuva, and Mongolia]. Zoologicheskii Zhurnal, 70(7):91-97 (in Russian).

Shenbrot, G. I. 1991*b*. [Subspecific taxonomy revision of common thick-tailed three-toed jerboa, *Stylodipus telum* (Rodentia, Dipodidae)]. Zoologicheskii Zhurnal, 70(6):118-127.

Shenbrot, G. I. 1991*c*. Geograficheskaya izmenchivost' mokhnonogovo tushkanchika *Dipus sagitta* (Rodentia, Dipodidae) [Geographical variation of the three-toed brush-footed jerboa *Dipus sagitta* (Rodentia, Dipodidae)] 1. [General patterns of intraspecific variability and subspecific differentiation in the western part of the species range]. Zoologicheskii Zhurnal, 70(5):101-110 (in Russian).

Shenbrot, G. I. 1991*d* [1992]. [Subspecific systematic revision of the five-toed jerboa genus *Allactaga* in the USSR.] Trudy Zoologicheskovo Instituta, Akademiya Nauk SSSR, 243:1-16 (in Russian).

Shenbrot, G. I. 1992. [Cladistic approach to the analysis of phylogenetic relationships among dipodoid rodents (Rodentia, Dipodoidea)]. Sbornik Trudov Zoologicheskovo Muzeya MGU, 29:176-201 (in Russian).

Shenbrot, G. I. 1993 [1991]. [Revision of subspecies systematics of jerboas of the genus *Allactaga* of the USSR fauna.] Trudy Zoologicheskovo Instituta, Akademiya Nauk SSSR, 243:81-109 (in Russian with English abstract).

Shenbrot, G. I., and B. Krasnov. 1997. Additional records of two small mammalian species in the Central Negev. Israel Journal of Zoology, 43:299-300.

Shenbrot, G. I., and B. Krasnov. 2001. Geographic variation in the role of gerbils and jirds (Gerbillinae) in rodent communities across the Great Palearctic desert belt. Pp. 511-529, *in* African small mammals (C. Denys, L. Granjon, and A. Poulet, eds.). IRD Éditions, Collection Colloques et Séminaires, Paris, 570 pp.

Shenbrot, G. I., and V. N. Mazin. 1989. [On the taxonomy of *Salpingotus pallidus* (Rodentia, Dipodidae) from south Balkhash region.] Zoologicheskii Zhurnal, 67(1):155-158 (in Russian).

Shenbrot, G. I., V. E. Sokolov, V. G. Heptner, and Yu. M. Koval'skaya. 1995. [Mammals of the fauna of Russia and contiguous countries. Dipodoid rodents.] Nauka Publishers, Moscow, 576 pp. (in Russian).

Sheppe, W., Jr. 1961. Systematic and ecological relations of *Peromyscus oreas* and *P. maniculatus*. Proceedings of the American Philosophical Society, 105:421-446.

Sherborn, C. D. 1902-1933. Index Animalium; sive, Index nominum quae ab A.D. MDCCLVIII generibus et speciebus animalium imposita sunt, societatibus eruditorum adiuvantibus. C. J. Clay and Sons, Cambridge University Press Warehouse, London, 2 vol. (in 10).

Shields, G. F., and T. D. Kocher. 1991. Phylogenetic relationships of North American ursids based on analysis of mitochrondrial DNA. Evolution, 45(1):218-221.

Shih, C. M. 1930. Preliminary report on the mammals from Yaoshan, Kwangsi, collected by the Yaoshan Expedition, Sun Yatsen University, Canton, China. Bulletin of the Department of Biology, College of Science, Sun Yatsen University, 4:10 pp.

Shinohara, A., K. L. Campbell, and H. Suzuki. 2003. Molecular phylogenetic relationships of moles, shrew moles, and desmans from the new and old worlds. Molecular Phylogenetics and Evolution, 27:247-258.

Shnitnikov, V. N. 1936. Mlekopitayushchie Semirech'ya [Mammals of Semirech'ye]. Nauka, Moscow-Leningrad, 323 pp. (in Russian).

Shore, R. F., and S. D. Garbett. 1991. Notes on the small mammals of the Shira Plateau, Mt. Kilimanjaro. Mammalia, 55:601-607.

Shortridge, G. C. 1934. The mammals of South West Africa. William Heinemann Ltd, London, 779 pp.

Shortridge, G. C. 1942. Field notes on the first and second expeditions to the Cape Museum's mammal survey of the Cape Province; with descriptions of some new subgenera and subspecies. Annals of the South African Museum, 36:27-100.

Shortridge, G. C., and T. D. Carter. 1938. A new genus and new species and subspecies of mammals from Little Namaqualand and the north-west Cape Province; and a new subspecies of *Gerbillus paeba* from the eastern Cape Province. Annals of the South African Museum, 3(2):281-291.

Shoshani, J. 2000. Comparing the living elephants. Pp. 36-51, *in* Elephants (J. Shoshani, ed.). Checkmark Books, Facts On File, Inc., New York, 240 pp.

Shoshani, J., and J. F. Eisenberg. 1982. *Elephas maximus*. Mammalian Species, 182:1-8.

Shoshani, J., and M. C. McKenna. 1998. Higher taxonomic relationships among extant mammals based on morphology, with selected comparisons of results from molecular data. Molecular Phylogenetics and Evolution, 9:572-584.

Shoshani, J., C. A. Goldman, and J. G. M. Thewissen. 1988. *Orycteropus afer*. Mammalian Species, 300:1-8.

Shotwell, J. A. 1967*a*. *Peromyscus* of the Late Tertiary of Oregon. Bulletin of the Museum of Natrual History, University of Oregon, 5:1-35.

Shotwell, J. A. 1967*b*. Late Tertiary geomyoid rodents of Oregon. Bulletin of the Museum of Natural History, University of Oregon, 9:1-51.

Shou Zhen-huang (ed.) [Shaw Tsen-Hwang]. 1962. Chung-kuo ching chi tung wu chih [Chinese economic zoology: Mammal section]. Ko Hsueh chu pan she [Scientific Publications Office. Beijing], 554 pp., 72 plates (in Chinese).

Shump, K. A., and R. H. Baker. 1978*a*. *Sigmodon alleni*. Mammalian Species, 95:1-2.

Shump, K. A., and R. H. Baker. 1978*b*. *Sigmodon leucotis*. Mammalian Species, 96:1-2.

Shump, K. A., and A. U. Shump. 1982*a*. *Lasiurus borealis*. Mammalian Species, 183:1-6.

Shump, K. A., and A. U. Shump. 1982*b*. *Lasiurus cinereus*. Mammalian Species, 185:1-5.

Sicard, B., and M. Tranier. 1996. Caractères répartition de trois phénotypes d'*Acomys* (Rodentia, Muridae) au Burkina Faso. Mammalia, 60(1):53-68.

Sicard, B., M. Tranier, and J.-C. Gautun. 1988. Un rongeur nouveau du Burkina Faso (ex Haute-Volta): *Taterillus petteri*, sp. nov. (Rodentia, Gerbillidae). Mammalia, 52:187-198.

Sicard, B., J. Catalan, S. Ag'Atteynine, D. Abdoulaye, and J. Britton-Davidian. 2004. Effects of climate and local aridity on the latitudinal and habitat distribution of *Arvicanthis niloticus* and *Arvicanthis ansorgei* (Rodentia, Murinae) in Mali. Journal of Biogeography, 31:5-18.

Sidiyene, E. A. 1989. Capture de *Crocidura lusitania* dans l'Adrar de Iforas. Mammalia, 53:467.

Sidorowicz, J. 1971. Subspecific taxonomy of the squirrel (*Sciurus vulgaris* L.) in Palaearctic. Zoologische Anzeiger, 187:123-142.

Sidorwicz, J. 1960. Problems of the morphology and zoogeography of representatives of the genus *Lemmus* Link 1795 from the Palaearctic. Acta Theriologica, 4:53-77.

Sidorwicz, J. 1964. Comparison of the morphology of representatives of the genus *Lemmus* Link from Alaska and the Palaearctic. Acta Theriologica, 8:217-225.

Siivonen, L. 1965. *Sorex isodon* Turov (1924) and *S. unguiculatus* Dobson (1890) as independent shrew species. Aquilo (Zool.), 4:1-34.

Sikes, R. S., J. A. Monjeau, E. C. Birney, C. J. Phillips, and J. R. Hillard. 1997. Morphological versus chromosomal and molecular divergence in two species of *Eligmodontia*. Zeitschrift für Säugetierkunde, 62:265-280.

Sikorski, M. D. 1982. Craniometric variation of *Apodemus agrarius* (Pallas, 1771) in urban green areas. Acta Theriologica, 27:71-81.

Sillero-Zubiri, C., and D. Gottelli. 1994. *Canis simensis*. Mammalian Species, 485:1-6.

Sillero-Zubiri, C., F. H. Tattersall, and D. W. Macdonald. 1995*a*. Morphometrics of endemic rodents from the Bale Mountains, Ethiopia. Journal of African Zoology, 109:387-391.

Sillero-Zubiri, C., F. H. Tattersall, and D. W. Macdonald. 1995*b*. Habitat selection and daily activity of giant mole-rats (*Tachyoryctes macrocephalus*): Significance to the Ethiopian wolf (*Canis simensis*) in the Afroalpine ecosystem. Biological Conservation, 72:77-84.

Sillero-Zubiri, C., F. H. Tattersal, and D. W. Macdonald. 1995*c*. Bale Mountains rodent communities and their relevance to the Ethiopian wolf (*Canis simensis*). African Journal of Ecology, 33:301-320.

Silva, C. R., A. R. Percequillo, G. E. Iack Himenes, and M. de Vivo. 2003. New distributional records of *Blarinomys breviceps* (Winge, 1888) (Sigmodontinae, Rodentia). Mammalia, 67:147-152.

da Silva, J. M. C., and D. C. Oren. 1993. Observations on the habitat and distribution of the Brazilian three-banded armadillo *Tolypeutes tricinctus*, a threatened Caatinga endemic. Mammalia, 57:149-152.

Silva, M. J. J., and Y. Yonenaga-Yassuda. 1998. Karyotype and chromosomal polymorphism of an undescribed *Akodon* from Central Brazil, a species with the lowest diploid chromosome number in rodents. Cytogenetics and Cell Genetics, 81:46-50.

Silva, M. J. J., A. R. Percequillo, and Y. Yonenaga-Yassuda. 2000. Cytogenetics and systematic approach on a new *Oryzomys* species of the *nitidus* group (Sigmodontinae, Rodentia) from northeastern Brazil. Caryologica, 53:219-226.

da Silva, M. N. F. 1998. Four new species of spiny rats of the genus *Proechimys* (Rodentia: Echimyidae) from western Amazon of Brazil. Proceedings of the Biological Society of Washington, 111 (2):436-471.

da Silva, M. N. F., and J. L. Patton. 1993. Amazonian phylogeography: mtDNA sequence variation in arboreal Echimyid rodents (Caviomorpha). Molecular Phylogenetics and Evolution, 2:243-255.

da Silva Neto, E. J. 2000. Morphology of the regions ethmoidalis and orbitotemporalis in *Galea musteloides* Mayen 1832 and *Kerodon rupestris* (Wied-Neuwied 1820) (Rodentia: Caviidae) with comments on the phylogenetic systematics of the Caviidae. Journal of Zoological Systematics and Evolutionary Research, 38(4):219-229.

Silva-Taboada, G. 1974. Fossil Chiroptera from cave deposits in central Cuba, with descriptions of two new species (genera *Pteronotus* and *Mormoops*) and the first West Indian record of *Mormoops megalophylla*. Acta Zoologica Cracoviensia, 19:34-73.

Silva-Taboada, G. 1976. Historia y actualización taxonómica de algunas especies Antillanas de murciélagos de los generos *Pteronotus*, *Brachyphylla*, *Lasiurus*, y *Antrozous* (Mammalia: Chiroptera). Poeyana (Academia de Ciencias de Cuba), 153:1-24.

Silva-Taboada, G. 1979. Los murciélagos de Cuba. Editorial Academia, 423 pp.

Silva-Taboada, G., and O. H. Garrido. In press. Compendio de los vertebrados terrestres de Cuba. Academia de Ciencias de Cuba (Habana).

Silva-Taboada, G., and K. F. Koopman. 1964. Notes on the occurrence and ecology of *Tadarida laticaudata yucatanica* in eastern Cuba. American Museum Novitates, 2174:1-6.

Silver, L. M. 1995. Mouse genetics, concepts and applications. Oxford University Press, New York and Oxford, 362 pp.

Silverstov, V. B., V. S. Lobachev, and M. N. Shilov. 1969. [New data on the distribution and biology of *Pygerethmus platyurus* Licht.] Byulleten' Moskovskovo Obshchestva Ispytatelei Prirody, Otdel Biologicheskii, 74(3):118-133 (in Russian).

Simmons, A. D., J. L. Longmire, T. W. Reeder, H. A. Wichman, and R. J. Baker. 1992. Restriction fragment length polymorphisms in satellite DNA distinguish chromosomal races of the white-footed mouse *Peromyscus leucopus*. Molecular Ecology, 1:251-254.

Simmons, N. B. 1996. A new species of *Micronycteris* (Chiroptera: Phyllostomidae) from northeastern Brazil, with comments on phylogenetic relationships. American Museum Novitates, 3158:1-34.

Simmons, N. B. 1998. A reappraisal of interfamilial relationships of bats. Pp. 1-26, *in* Bat biology and conservation (T. H. Kunz and P. A. Racey, eds.). Smithsonian Institution Press, Washington, D.C.. 365 pp.

Simmons, N. B., and T. Conway. 2001. Phylogenetic relationships of mormoopid bats (Chiroptera: Mormoopidae) based on morphological data. Bulletin of the American Museum of Natural History, 258:1-98.

Simmons, N. B., and J. H. Geisler. 1998. Phylogenetic relationships of *Icaronycteris*, *Archaeonycteris*, *Hassianycteris*, and *Palaeochiropteryx* to extant bat lineages, with comments on the evolution of echolocation and foraging strategies in Microchiroptera. Bulletin of the American Museum of Natural History, 235:1-182.

Simmons, N. B., and C. O. Handley, Jr. 1998. A revision of *Centronycteris* Gray (Chiroptera: Emballonuridae) with notes on natural history. American Museum Novitates, 3239:1-28.

Simmons, N. B., and R. S. Voss. 1998. The mammals of Paracou, French Guiana: A Neotropical lowland rainforest fauna. Part 1. Bats. Bulletin of the American Museum of Natural History, 237:1-219.

Simmons, N. B., and A. L. Wetterer. 2002. Phylogeny and convergence in cactophilic bats. Pp. 87-120, *in* Evolution, ecology, and conservation of columnar cacti and their mutualists (T. Flemming and A. Valiente-Banuet, eds.). University of Arizona Press, 400 pp.

Simmons, N. B., R. S. Voss, and H. C. Peckham. 2000. The bat fauna of Säul, French Guiana. Acta Chiropterologica, 2:23-36.

Simonetta, A. M. 1968. A new golden mole from Somalia with an appendix on the taxonomy of the family Chrysochloridae (Mammalia, Insectivora). Monitore Zoologico Italiano, n.s., 2 (Supplemento):27-55.

Simonetti Z., J., and A. E. Spotorno O. 1980. Posición taxonómica de *Phyllotis micropus* (Rodentia: Cricetidae). Annales del Museo de Historia Natural de Valparaiso, 13:285-297.

Simons, E. L., and Y. Rumpler. 1988. *Eulemur*: New generic name for species of *Lemur* other than *Lemur catta*. Comptes Rendus de l'Académie des Sciences (Paris), ser. 3, 307:547-551.

Simoons, F. J. 1953. Notes on the bush-pig (*Potamochoerus*). Uganda Journal, 17:80-81.

Simoons, F. J. 1984. Gayal or mithan. Pp. 34-39, *in* Evolution of domestic animals (Mason, I. L., ed.), Longman, London, 452 pp.

Simpson, G. G. 1931. A new classification of mammals. Bulletin of the American Museum of Natural History, 59:259-293.

Simpson, G. G. 1945. The principles of classification and a classification of mammals. Bulletin of the American Museum of Natural History, 85:1-350.

Simpson, G. G. 1961. Principles of Animal Taxonomy. Columbia University, New York, 246 pp.

Simson, S., B. Lavie, and E. Nevo. 1993. Penial differentiation in speciation of subterranean mole rats *Spalax ehrenbergi* in Israel. Journal of Zoology, London, 229:493-503.

Simson, S., L. Ferrucci, C. Kurtonur, B. Özkan and M. G. Filippucci. 1995. Phalli and bacula of European dormice: Description and comparison. Pp. 231-244, *in* Proceedings of II Conference on Dormice (Rodentia, Myoxidae) (M. G. Filippucci, ed.). Hystrix, n.s., 6(1-2):1-340.

Sinclair, A. R. E., C. J. Krebs, J. N. M. Smith, and S. Boutin. 1988. Population biology of snowshoe hares. III. Nutrition, plant secondary compounds and food limitation. Journal of Animal Ecology, 57:787-806.

Sinclair, E. A., and M. Westerman. 1997. Phylogenetic relationships within the genus *Potorous* (Marsupialia: Potoroidae) based on allozyme electrophoresis and sequence analysis of the cytochrome *b* gene. Journal of Mammalian Evolution, 4:147-161.

Singleton, G. R. 1995. House mouse, *Mus musculus*. Pp. 646-647, *in* Mammals of Australia (R. Strahan, ed.). Smithsonian Institution Press, Washington, D.C., 756 pp.

Sinha, Y. P. 1970. Taxonomic notes on some Indian bats. Mammalia, 34:81-92.

Sinha, Y. P. 1973. Taxonomic studies on the Indian horseshoe bats of the genus *Rhinolophus* Lacepede. Mammalia, 37:603-630.

Sinha, Y. P. 1980. The bats of Rajasthan: Taxonomy and zoogeography. Records of the Zoological Survey of India, 76:7-63.

Sinha, Y. P. 1999. Contribution to the knowledge of bats (Mammalia; Chiroptera) of North East Hills, India. Records of the Zoological Survey of India, Occasional Papers, 174:1-52.

Sinha, Y. P., and S. Chakroborty. 1971. Taxonomic status of the vespertilionid bat *Nycticeius emarginatus* Dobson. Proceedings of the Zoological Society of Calcutta, 24:53-57.

Sivertsen, E. 1954. A survey of the eared seals (Family Otariidae) with remarks on the antarctic seals collected by M/K "Norvegica" in 1928-1929. Det Norske Videnskaps-Akademi i Oslo, Scientific Results of the Norwegian Antarctic Expedition, 36:1-76.

Skaren, U. 1964. Variation of two shrews, *Sorex unguiculatus* Dobson and *S. a. araneus* L. Annales Zoologici Fennici, 1:94-124.

Skead, C. J. 1973. Zoo-historical gazetteer. Annals of the Cape Provincial Museums, 10:1-259.

Skinner, J. D., and R. H. N. Smithers. 1990. The mammals of the southern African subregion. Second ed. University of Pretoria, Republic of South Africa, 771 pp.

Skoczen, S. 1976. Condylurini, Dobson, 1883 (Insectivora, Mammalia) in the Pliocene of Poland. Acta Zoologica Cracoviensia, 21:291-313.

Skoczen, S. 1993. New records of *Parascalops*, *Neurotrichus* and *Condylura* (Talpidae, Insectivora) from the Pliocene of Poland. Acta Theriologica, 38:125-137.

Slaughter, B. H., and J. E. Ubelaker. 1984. Relationship of South American cricetine rodents to rodents of North America and the Old World. Journal of Vertebrate Paleontology, 4:255-264.

Sludskii, A. A. (ed.). 1977. Mlekopitayushchie Kazakhstana. Gryzuny (krome surkov, suslikov, zemlyanoi belki, peschanok i polevok) [Mammals of Kazakhstan. Rodents (except marmots, susliks, long clawed ground squirrels, gerbils and voles)]. Nauka, Kazakhskoi SSR, Alma-Ata, 1(2):1-536 (in Russian).

Sludskii, A. A., S. N. Varshavskii, M. N. Ismagilov, V. I. Kapitonov, and I. G. Shubin. 1969. Mlekopitayushchie Kazakhstana [Mammals of Kazakhstan]. Vol. 1, Gryzuny (surki i susliki) [Rodents (Marmots and susliks)]. Nauka, Kazakhskoi SSR, Alma-Ata, 455 pp. (in Russian).

Sludskii, A. A., A. D. Bernstein, I. G. Shubin, V. A. Fadeev, G. I. Orlov, A. Bekenov, V. I. Kharitonov, and S. R. Utinov. 1980. Mlekopitayushchie Kazakhstana, tom 2, Zaitseobraznye [Mammals of Kazakhstan. vol. 2, Lagomorpha]. Nauka, Kazakhskoi SSR, Alma-Ata, 236 pp. (in Russian).

Smadja, C., and G. Ganem. 2002. Subspecies recognition in the house mouse: A study of two populations from the border of a hybrid zone. Behavioral Ecology, 13(3):312-320.

Šmaha, J. 1996. [Notes on the mammal fauna of the Køivoklátsko Biosphere Reserve.] Lynx (Praha), n.s., 27:37-56 (in Czech with English abstract).

Smeenk, C., and Y. Kaneko. 2000. *Myoxus japonicus* Schinz, 1845 (currently *Glirulus japonicus*; Mammalia, Rodentia): Proposed conservation as the correct original spelling of the specific name. Bulletin of Zoological Nomenclature, 57(1):36-38.

Smeenk, C., Y. Kaneko, and K. Tsuchiya. 1982. On the type material of *Mus argenteus* Temminck, 1844. Zoologische Mededelingen, 56:121-129.

Smets, S., M. Languy, D. Gerard, and V. Bauchau. 1996. A case of recent chromosomal evolution in the house mouse. Pp. 251-257, *in* Proceedings of the I European Congress of Mammalogy (M. L. Mathias, M. Santos-Reis, G. Amori, R. Libois, A. Mitchell-Jones, and M. C. Saint-Girons, eds.). Museu Bocage, Lisboa, 314 pp.

Smiddy, P., and D. P. Sleeman. 1994. The bank vole in County Cork. Ireland Nature Journal, 24(9):360-364.

Smirin, Yu. M., N. A. Formozov, D. I. Bibikov, and D. Myagmarzhav. 1985. Kharakteristika poselenii dvukh vidov surkov (*Marmota*, Rodentia, Sciuridae) v zone ikh kontakta na Mongol'skom Altae [Characteristics of colonies of two species of marmots . . . in their contact zone in the Mongolian Altai]. Zoologicheskii Zhurnal, 64:1873-1885 (in Russian).

Smit, A., H. van der Bank, T. Falk, and A. de Castro. 2001. Biochemical genetic markers to identify two morphologically similar South African *Mastomys* species (Rodentia: Muridae). Biochemical Systematics and Ecology, 29:21-30.

Smith, A. 1833-1834. An epitome of African Zoology; or, a concise description of the objects of the animal kingdom inhabiting Africa, its islands and seas. South African Quarterly Journal, 2:16-32, 49-64, 81-96, 113-128, 145-160, 169-192, 209-224, 233-248.

Smith, A. L., M. F. Robinson, and M. Webber. 1998. Notes on a collection of shrews (Insectivora: Soricidae) from Lao PDR. Mammalia, 62:585-588.

Smith, A. L., M. F. Robinson, and P. D. Jenkins. 2000. A collection of shrews (Insectivora: Soricidae) from north-east Thailand. Mammalia, 64:250-253.

Smith, A. L., M. F. Robinson, and M. Webber. 2004. Muridae (Rodentia) from the Khammouan Limestone and Xe Piane National Biodiversity Conservation Areas, Lao P.D.R. Mammalia, (2-3):167-173.

Smith, A. P., and D. G. Quin. 1996. Patterns and causes of extinction and decline in Australian conilurine rodents. Biological Conservation, 77:243-267.

Smith, A. T., and J. M. Foggin. 1999. The plateau pika (*Ochotona curzoniae*) is a keystone species for biodiversity on the Tibetan plateau. Animal Conservation, 2:235-240.

Smith, A. T., and M. L. Weston. 1990. *Ochotona princeps*. Mammalian Species, 352:1-8.

Smith, A. T., N. A. Formozov, Chan-lin Zheng, M. Erbajeva, and R. S. Hoffmann. 1990. The pikas. Pp. 14-60, *in* Rabbits, hares and pikas (J. A. Chapman and J. E. C. Flux, eds.). I.U.C.N., Gland, Switzerland, 168 pp.

Smith, C. H. 1827. The seventh order of the Mammalia. The Ruminantia. Pp. 1-428, *in* The animal kingdom arranged in conformity with its organization, by the Baron Cuvier, member of the Institute of France, &c. &c. &c. with additional descriptions of all the species hitherto named, and of many not before noticed (E. Griffith, C. H.[amilton] Smith, and E. Pidgeon, eds.). G. B. Whitaker, London, 4:1-498.

Smith, F. A., B. T. Bestelmeyer, J. Biardi, and M. Strong. 1993. Anthrogenic extinction of the endemic woodrat *Neotoma bunkeri* Burt. Biodiversity Letters, 1:149-155.

Smith, F. A., J. L. Betancourt, and J. H. Brown. 1995. Evolution of body size in the woodrat over the past 25,000 years of climate change. Science, 270:2012-2014.

Smith, J. D. 1970. The systematic status of the black howler monkey, *Alouatta pigra* Lawrence. Journal of Mammalogy, 51:358-369.

Smith, J. D. 1972. Systematics of the chiropteran family Mormoopidae. Miscellaneous Publications, Museum of Natural History, University of Kansas, 56:1-132.

Smith, J. D. 1977. On the nomenclatorial status of *Chilonycteris gymnonotus* Natterer, 1843. Journal of Mammalogy, 58:245-246.

Smith, J. D., and J. E. Hill. 1981. A new species and subspecies of bat of the *Hipposideros bicolor*-group from Papua New Guinea and the systematic status of *Hipposideros calcaratus* and *Hipposideros cupidus* (Mammalia: Chiroptera: Hipposideridae). Contributions in Science, Natural History Museum of Los Angeles County, 331:1-19.

Smith, J. D., and C. S. Hood. 1980. Additional material of *Rhinolophus ruwenzorii* Hill, 1942, with comments on its natural history and taxonomic status. Proceedings of the Fifth International Bat Research Conference, 5:163-171.

Smith, J. D., and C. S. Hood. 1983. A new species of tube-nosed fruit bat (*Nyctimene*) from the Bismark Archipelago, Papua New Guinea. Occasional Papers, The Museum, Texas Tech University, 81:1-14.

Smith, K. S. 1992. The prairie vole, *Microtus ochrogaster*, in Caddo County, Oklahoma. Texas Journal of Science, 44:116-117.

Smith, L. R., D. W. Hale, and I. F. Greenbaum. 2000. Systematic implications of chromosomal data from two insular species of *Peromyscus* from the Gulf of California. Journal of Heredity, 91:162-165.

Smith, M. F. 1979. Geographic variation in genic and morphological characters in *Peromyscus californicus*. Journal of Mammalogy, 60:705-722.

Smith, M. F. 1998. Phylogenetic relationships and geographic structure in pocket gophers in the genus *Thomomys*. Molecular Phylogenetics and Evolution, 9:1-14.

Smith, M. F., and J. L. Patton. 1991. Variation in mitochondrial cytochrome *b* sequence in natural populations of South American akodontine rodents (Muridae: Sigmodontinae). Molecular Biology and Evolution, 8:85-103.

Smith, M. F., and J. L. Patton. 1993. The diversification of South American murid rodents: Evidence from mitochondrial DNA sequence data for the akodontine tribe. Biological Journal of the Linnean Society, 50:149-177.

Smith, M. F., and J. L. Patton. 1999. Phylogenetic relationships and the radiation of sigmodontine rodents in South America: Evidence from cytochrome *b*. Journal of Mammalian Evolution, 6:89-128.

Smith, M. F., D. A. Kelt, and J. L. Patton. 2001. Testing models of diversification in mice in the *Abrothrix olivaceus/xanthorhinus* complex in Chile and Argentina. Molecular Ecology, 10:397-405.

Smith, M. H., R. K. Selander, and W. E. Johnson. 1973. Biochemical polymorphism and systematics in the genus *Peromyscus*. III. Variation in the Florida deermouse (*Peromyscus floridanus*), a Pleistocene relic. Journal of Mammalogy, 54:1-13.

Smith, M. J. 1973. *Petaurus breviceps*. Mammalian Species, 30:1-5.

Smith, M. J. 1998. Establishment of a captive colony of *Bettongia tropica* (Marsupialia: Potoroidae) by cross-fostering; and observations on reproduction. Journal of Zoology, London, 244:43-50.

Smith, R. J., D. M. Lavigne and W. R. Leonard. 1994. Subspecific status of the freshwater harbor seal (*Phoca vitulina mellonae*): A reassessment. Marine Mammal Science, 10:105-110.

Smith, S. A. 1990. Cytosystematic evidence against monophyly of the *Peromyscus boylii* species group (Rodentia: Cricetidae). Journal of Mammalogy, 71:654-667.

Smith, S. A., R. D. Bradley, and I. F. Greenbaum. 1986. Karyotypic conservatism in the *Peromyscus mexicanus* group. Journal of Mammalogy, 67:584-586.

Smith, S. A., I. F. Greenbaum, D. J. Schmidly, K. M. Davis, and T. W. Houseal. 1989. Additional notes on karyotypic variation in the *Peromyscus boylii* group. Journal of Mammalogy, 70:603-608.

Smith, S. J. 1985. The Atlas of Africa's Principal Mammals. Natural History Books, San Antonio. 241+xxx pp.

Smith, T. G., D. J. St. Aubin, and J. R. Geraci. 1990. Advances in research on the beluga whale, *Delphinapterus leucas*. Canadian Bulletin of Fisheries and Aquatic Sciences, 224:1-206.

Smith, W. P. 1991. *Odocoileus virginianus*. Mammalian Species, 388:1-13.

Smithers, R. H. N. 1971. The mammals of Botswana. Museum Memoir, National Museums of Rhodesia, Salisbury, 4:1-340.

Smithers, R. H. N. 1983. The mammals of the Southern African Subregion. University of Pretoria, Republic of South Africa, 736 pp.

Smithers, R. H. N., and J. L. P. Lobão Tello. 1976. Check list and atlas of the mammals of Mozambique. Museum Memoir, National Museums and Monuments of Rhodesia, Salisbury, 8:1-184.

Smithers, R. H. N., and V. J. Wilson. 1979. Checklist and atlas of the mammals of Zimbabwe Rhodesia. Museum Memoir, National Museums and Monuments of Rhodesia, Salisbury, 9:1-193.

Smolen, M. J. 1981. *Microtus pinetorum*. Mammalian Species, 147:1-7.

Smolen, M. J., and J. W. Bickham. 1994. Chromosomal variation in pocket gophers (*Geomys*) detected by sequential G-, R-, and C-band analyses. Chromosome Research, 2:343-353.

Smolen, M. J., and J. W. Bickham. 1995. Phylogenetic implications of chromosome evolution in *Geomys*. Journal of Mammalogy, 76:50-62.

Smolen, M. J., and B. L. Keller. 1987. *Microtus longicaudus*. Mammalian Species, 271:1-7.

Smolen, M. J., R. M. Pitts, and J. W. Bickham. 1993. A new subspecies of pocket gopher (*Geomys*) from Texas (Mammalia: Rodentia: Geomyidae). Proceedings of the Biological Society of Washington, 106:5-23.

Smorkacheva, A. V., T. G. Aksenova, and T. A. Zorenko. 1990. Ekologiya Kitaiskoi polevki *Lasiopodomys mandarinus* (Rodentia, Cricetidae) v Zabaikal'e [The ecology of the Chinese vole *Lasiopodomys mandarinus* (Rodentia, Cricetidae) in Transbaikaliya]. Zoologicheskii Zhurnal, 69(12):115-124 (in Russian).

Snow, J. L., J. K. Jones, Jr., and W. D. Webster. 1980. *Centurio senex*. Mammalian Species, 138:1-3.

Snowden, P., P. J. J. Bates, D. L. Harrison, and M. R. Brown. 2000. Recent records of bats and rodents from Oman including three species new to the country. Fauna of Arabia, 18:397-407.

Snyder, D. P. 1982. *Tamias striatus*. Mammalian Species, 168:1-8.

Soarimalala, V., S. M. Goodman, H. Ramiarinjanahary, L. L. Fenohery, and W. Rakotonirina. 2001. Les micro-mammiferes non-volants du Parc National de Ranomafana et du couloir forestier qui le relie au Parc National d'Andringitra. Pp. 197-229, *in* Inventaire biologique du Parc National de Ranomafana et du couloir forestier qui le relie au Parc National d'Andringitra (S. M. Goodman and V. R. Razafindratsita, eds.). Recherches pour le Developpement, Série Sciences Biologiques, 17:xii + 243 pp.

Sobti, R. C., and S. S. Gill. 1984. Chromosomes and protein polymorphs in certain *Rattus* species from North India. Research Bulletin (Science) of the Punjab University, 35:203-210.

Sody, H. J. V. 1931. Six new mammals from Sumatra, Java, Bali and Borneo. Natuurkundig Tijdschrift voor Nederlandsch Indie, 91:349-360.

Sody, H. J. V. 1940. On the mammals of Enggano. Treubia, 17:391-405.

Sody, H. J. V. 1941. On a collection of rats from the Indo- Malayan and Indo- Australian regions (with descriptions of 43 new genera, species and subspecies). Treubia, 18:255-325.

Sody, H. J. V. 1946. [Review of] "Prehistoric and fossil rhinoceroses of the Malay Archipelago and India', door D. A. Hoojer. Diss., Leiden, 1946, VI + 138 pp., 10 pl. Natuurwetenschappelijk Tijdschrift voor Nederlandsch-Indie, 102(7):151.

Sody, H. J. V. 1949*a*. Sciuridae from the Indo-Malayan and Indo-Australian regions. Treubia, 20:57-120.

Sody, H. J. V. 1949*b*. Notes on some Primates, Carnivora, and the Babirusa from the Indo-Malayan and Indo-Australian regions (with descriptions of 10 new species and subspecies). Treubia, 20:121-190.

Sofianidou, T. S., and V. Vohralík. 1991. Notes on the distribution of small mammals (Insectivora, Rodentia) in Epeirus, Greece. Bonner zoologische Beiträge, 42:125-135.

Sokolov, I. I. 1957. On the Artiodactyl-fauna in the southern part of Yunnan Province (China). Zoologicheskii Zhurnal, 36:1750-1760 (in Russian).

Sokolov, I. I. 1973. Napravleniya evolyutsii i estestvennaya klassifikatsiya podsemeistva vydrovykh (Lutrinae, Mustelidae, Fissipedia) [Evolutionary trends and the classification of the subfamily Lutrinae (Mustelidae, Fissipedia)]. Byulleten' Moskovskovo Obshchestva Ispytatelei Prirody, Otdel Biologicheskii, 78(6):45-52 (in Russian).

Sokolov, V. E. 1973-1979. Sistematika mlekopitayushchikh [Systematics of mammals]. vol. 1 [Monotremes, marsupials, insectivores, dermopterans, chiropterans, primates, edentates, pangolins]. vol. 2 [Lagomorphs, rodents]. vol. 3 [Cetaceans, carnivores, pinnipeds, tubulidentates, proboscideans, hyracoids, sirenians, artiodactyls, tylopods, perissodactyls]. Vysshaya Shkola, Moscow, 1:1-430 [1973]; 2:1-494 [1977]; 3:1-528 [1979] (in Russian).

Sokolov, V. E. 1974. *Saiga tatarica*. Mammalian Species, 38:1-4.

Sokolov, V. E., and N. V. Bashenina (eds.). 1994. [Species of the fauna of Russia and the contiguous countries. Common vole, the sibling species: *Microtus arvalis* Pallis, 1779, *M. rossiaemeridionalis* Ognev, 1924.] Nauka Publishers, Moscow, 432 pp. (in Russian).

Sokolov, V. E., and M. I. Baskevich. 1988. [A new species of birch mouse—*Sicista armenica* sp. n. (Rodentia, Dipodoidea) from the Lesser Caucasus.] Zoologicheskii Zhurnal, 67(2):300-304 (in Russian).

Sokolov, V. E., and M. I. Baskevich. 1992. [A new chromosomal form of unstriped birch mice of the Caucasus from North Ossetia (Rodentia, Dipodoidea, *Sicista*).] Zoologicheskii Zhurnal, 71(8):94-103 (in Russian with English abstract)

Sokolov, V. E., and N. K. Dzhemukhadze. 1991. [Histochemistry of specialized skin glands of the narrow-skulled and Dagestan voles.] Doklady Akademii Nauk SSSR, Moscow-Leningrad, 316:731-734.

Sokolov, V. E., and N. K. Dzhemukhadze. 1995. The taxonomic status of simple and Kirgiz voles determined by histoenzymatic indices of their specific skin glands. Doklady Biological Sciences, 341:216-218 (translated from Doklady Akademii Nauk, 341(6):846-848).

Sokolov, V. E., and V. S. Gromov. 1990. The contemporary ideas on roe deer (*Capreolus* Gray, 1821) systematization: Morphological, ethological and hybridological analysis. Mammalia, 54:431-444.

Sokolov, V. E., and E. V. Karasjova (eds.). 1990. [Species of the fauna of the USSR and the contiguous countries. Norway rat: Systematics, ecology, population control.] Nauka Publishers, Moscow, 452 pp. (in Russian with English summary and table of contents).

Sokolov, V. E., and Y. M. Kovalskaya. 1990. Kariotipy myshovok severnovo Tyan'-Shanya i Sikhote-Alinya (*Sicista*, Dipodoidea, Rodentia) [Karyotypes of birch mice (*Sicista*, Dipodoidea, Rodentia) in the northern Tien-Shan and Sikhote-Alin]. Zoologicheskii Zhurnal, 69(5):152-157 (in Russian).

Sokolov, V. E., and N. P. Lavrov (eds.). 1993. [Species of the fauna of the USSR and the contiguous countries. The muskrat: Morphology, systematics, ecology.] Nauka Publishers, Moscow, 542 pp. (in Russian with English summary and table of contents).

Sokolov, V. E., and A. A. Lushchekina. 1997. *Procapra gutturosa*. Mammalian Species, 571:1-5.

Sokolov, V. E., and V. M. Neronov (eds.). 1993. Peschanki roda *Meriones* Rossii b sopredel'nyh territoriy: Bibliografia i areologia [Jirds of the genus *Meriones* of Russia and adjacent regions: Bibliography and areology]. 3 vols. Ekopros, Moscow, 692 pp. (in Russian).

Sokolov, V. E., and V. N. Orlov. 1980. Opredelitel' mlekopitayushchickh Mongol'skoi Narodnoi Respubliki [Guide to the mammals of the Mongolian People's Republic]. Nauka, Moscow, 351 pp. (in Russian).

Sokolov, V. E., and V. I. Prikhod'ko. 1996, 1997. Taxonomy of the musk deer *Moschus moschiferus* (Artiodactyla, Mammalia). Biology Bulletin, 24:557-566; 25:28-36.

Sokolov, V. E., and G. I. Shenbrot. 1987a. [Review of Wang Sibo, Yang Ganyun. "Rodent fauna of Xinjiang."]. Zoologicheskii Zhurnal, 66(1):157-159.

Sokolov, V. E., and G. I. Shenbrot. 1987b. [A new species of thick- tailed jerboa, *Stylodipus sungorus* sp. n. (Rodentia, Dipodidae), from western Mongolia]. Zoologicheskii Zhurnal, 66(4):579-587 (in Russian).

Sokolov, V. E., and G. I. Shenbrot. 1988. [Data on the geographical variability and taxonomy of pygmy jerboas (Rodentia, Cardiocraniinae) in Mongolia.] Zoologicheskii Zhurnal, 67(10):1561-1569 (in Russian).

Sokolov, V. E., and A. K. Tembotov. 1989. Mlekopitayushchie Kavkaza: Hasekomoyadnye [Mammals of the Caucasus: Insectivores]. Nauka, Moscow, 547 pp. (in Russian).

Sokolov, V. E., and A. K. Tembotov. 1993. Mlekopitayushchie: Kopytnye Pozvonochnye Kavkaza. Nauka, Moscow, 527 pp.

Sokolov, V. E., Y. M. Kovalskaya, and M. I. Baskevich. 1980. Gryzuny [Rodents]. Materiali Pyatovo Vsesoyuznovo Soveshchaniye, Moscow, 1980:38-40 (in Russian).

Sokolov, V. E., O. L. Rossolimo, I. Ya. Pavlinov, and O. I. Podtyazhkin. 1981a. [Comparative characteristics of two species of jerboas from Mongolia— *Allactaga bullata* Allen, 1925 and *A. nataliae* Sokolov, 1981]. Zoologicheskii Zhurnal, 60(6):895-906 (in Russian).

Sokolov, V. E., M. I. Baskevich, and Y. M. Kovalskaya. 1981b. [Revision of birch mice of the Caucasus: Sibling species *Sicista caucasica* Vinogradov, 1925 and *S. kluchorica* sp. n. (Rodentia, Dipodidae)]. Zoologicheskii Zhurnal, 60(9):1386-1393 (in Russian).

Sokolov, V. E., Y. M. Kovalskaya, and M. I. Baskevich. 1982. [Taxonomy and comparative cytogenetics of some species of the genus *Sicista* (Rodentia, Dipodidae)]. Zoologicheskii Zhurnal, 61(1):102-108 (in Russian).

Sokolov, V. E., G. G. Markov, A. A. Danilkin, Kh. M. Nikolov, and S. Gerasimov. 1985. Species status of the European (*Capreolus capreolus*) and Siberian (*C. pygargus*) roe deer. Craniometric investigations. Doklady Biological Sciences. Proceedings of the Academy of Sciences of the USSR, 280:90-94 (tranlated from the Russian).

Sokolov, V. E., M. I. Baskevich, and Y. M. Kovalskaya. 1986a. [The karyotype variability in the southern birch mouse (*Sicista subtilis* Pallas) and substantiation of the species validity of *S. severtzovi*]. Zoologicheskii Zhurnal, 65:1684 (in Russian).

Sokolov, V. E., M. I. Baskevich, and Y. M. Kovalskaya. 1986b. [*Sicista kazbegica* sp. n. (Rodentia, Dipodidae) from the upper reaches of the Terek River basin]. Zoologicheskii Zhurnal, 65(6):949-951 (in Russian).

Sokolov, V. E., A. A. Danilkin, and G. G. Markov. 1986c. [Craniometry study of Siberian subspecies of Siberian roe (*Capreolus pygargus pygargus* Pall.).] Izvestiya Akad Nauk SSSR (Biol), 1986 (3):445-448 (in Russian).

Sokolov, V. E., M. I. Baskevich, I. V. Lukyanova, M. A. Tarasov, N. N. Kuryatnikov, and V. G. Topilina. 1987a.

[Distribution of birch mice (Rodentia, Zapodidae) of the Caucasus.] Zoologicheskii Zhurnal, 66(11):1730-1735 (in Russian).

Sokolov, V. E., Y. M. Kovalskaya, and M. I. Baskevich. 1987b. Review of karyological research and the problems of systematics in the genus Sicista (Zapodidae, Rodentia, Mammalia). Folia Zoologica, 36(1):35-44.

Sokolov, V. E., Y. M. Kovalskaya, and M. I. Baskevich. 1989. [On species status of the northern birch mouse Sicista strandi (Rodentia, Dipodidae).] Zoologicheskii Zhurnal, 68(10):95-106 (in Russian).

Sokolov, V. E., V. M. Aniskin, and M. A. Serbenyuk. 1990. Sravnitel'naya tsitogenetika shesti vidov polevok roda Clethrionomys (Rodentia, Microtinae) [Comparative cytogenetics of six vole species of the genus Clethrionomys (Rodentia, Microtinae)]. Zoologicheskii Zhurnal, 69(11):145-151 (in Russian).

Sokolov, V. E., V. N. Orlov, M. I. Baskevich, and A. Mebrate. 1992. [Chromosomal sets of the spiny mice Acomys (Rodentia, Muridae) along the Ethiopian Rift Valley.] Zoologicheskii Zhurnal, 71:116-124 (in Russian).

Sokolov, V. E., V. N. Orlov, M. I. Baskevich, A. Bekele, and A. Mebrate. 1993. A karyological study of the spiny mouse Acomys Geoffroy 1838 (Rodentia, Muridae) along the Ethiopian Rift Valley. Tropical Zoology, 6:227-235.

Sokolov, V. E., E. Yu. Ivanitskaya, V. V. Gruzdev, and V. G. Heptner. 1994. Mlekopitayushchie Rossii i sopredel'nykh regionov. Zaitseobraznye. [Mammals of Russia. and adjoining regions. Lagomorpha.] Nauka, Moscow. 272 pp.

Sokolov, V. E., V. S. Lobachev, and V. N. Orlov. 1996. The mammals of Mongolia. Family Dipodidae, Euchoreutinae, Cardiocraniinae, Dipodinae. Moscow, Nauka Press, 271 pp.

Sokolov, V. E., V. S. Lobachev, and V. N. Orlov. 1998. The mammals of Mongolia. Family Dipodidae. Dipodinae, Allactaginae. Moscow, Nauka Press, 206 pp.

Solari, S. 2003. Diversity and distribution of Thylamys (Didelphidae) in South America, with emphasis on species from the western side of the Andes. Pp. 82-101, in Predators with pouches: The biology of carnivorous marsupials (M. E. Jones, C. R. Dickman, and M. Archer, eds.). CSIRO Press, Melbourne, 486 pp.

Solari, S., V. Pacheco, and E. Vivar. 1999. New distribution records of Peruvian bats. Revista Peruviana Biologia, 6:152-159.

Soldatovic, B., P. Seth, H. Reichstein, and M. Tolksdorf. 1975. Comparative karyological study of the genus Apodemus (Kaup, 1829). Acta Veterinaria (Beograd), 25:1-10.

Soldatovic, B., D. Zimonjic, I. Savic, and E. Giagia. 1984. Comparative cytogenetic analysis of the populations of European ground squirrel (Citellus citellus L.) on the Balkan peninsula. Bulletin. Academie Serbe des Sciences. Classe des Sciences Mathematiques et Naturelles, 86:47-56.

Solleder, E., M. Schmid, B. Inglin, and T. Haaf. 1984. Cytogenetic studies on the mitotic and meiotic chromosomes of Micromys minutus (Rodentia, Murinae). Zeitschrift für Säugetierkunde, 49:284-289.

Sommer, S. 1997. Monogamy in Hypogeomys antimena, an endemic rodent of the deciduous dry forest in western Madagascar. Journal of Zoology, London, 241:301-314.

Sommer, S. 2001. Reproductive ecology of the endangered monogamous Malagasy giant jumping rat, Hypogeomys antimena. Zeitschrift für Säugetierkunde, 66:111-115.

Sommer, S. 2003. Hypogeomys antimena, Malagasy giant jumping rat, Vositse, Votsotsa. Pp. 1383-1385, in The natural history of Madagascar (S. M. Goodman and J. P. Benstead, eds.). University of Chicago Press, Chicago, xxi + 1709 pp.

Sondaar, P. Y. 2000. Early human exploration and exploitation of islands. Tropics, 10(1):203-230.

Sondaar, P. Y., P. L. de Boer, M. Sanges, T. Kotsakis, and D. Esu. 1984. First report on a Paleolithic culture in Sardinia. Pp. 29-47, in The Deya conference of prehistory. Early settlement in the western Mediterranean Islands and the peripheral areas (W. H. Waldren, R. Chapman, J. Lewthwaite, and R.-C. Kennard, eds.). B. A. R. International Series 229, Oxford, 1399 pp.

Sondaar, P. Y., G. D. van den Bergh, B. Mubroto, F. Aziz, J. de Vos, and U. L. Batu. 1994. Middle Pleistocene faunal turnover and colonization of Flores (Indonesia) by Homo erectus. Comptes Rendus des Séances de l'Académie des Sciences (Paris), 319, ser. 2:1255-1262.

Song, S. 1985. [A new subspecies of Cricetulus triton from Shaanxi, China.] Acta Theriologica Sinica, 5(2):137-139 (in Chinese with English abstract).

Soota, T. D., and Y. Chaturvedi. 1980. New locality record of Pipistrellus camortae Miller from Nicobar and its systematic status. Records of the Zoological Survey of India, 77:83-87.

Sopin, L. V. 1982. [On intraspecies structure of Ovis ammon (Artiodactyla, Bovidae).] Zoologicheskii Zhurnal, 61:1882-1892 (in Russian).

Soriano, P. J., and J. Ochoa G. 1997. Lista actualizada de los mamíferos de Venezuela. Pp. 205-227, in Vertebrados actuales y fósiles de Venezuela (E. La Marca, ed.). Museo de Ciencia y Technología, Mérida.

Soriano, P. S., and J. Molinari. 1987. Sturnira aratathomasi. Mammalian Species, 284:1-4.

Soriano, P. S., M. R. Fariñas, and M. E. Naranjo. 2000. A new subspecies of Miller's long-tongued bat (*Glossophaga longirostris*) from a semiarid enclave of the Venezuelan Andes. Zeitschrift für Säugetierkunde, 65(6):369-374.

Soullier, S., C. Hanni, F. Catzeflis, P. Berta, and V. Laudet. 1998. Male sex determination in the spiny rat *Tokudaia osimensis* (Rodentia: Muridae) is not *Sry* dependent. Mammalian Genome, 9:590-592.

Soulounias, N. 1988. Evidence from horn morphology on the phylogenetic relationships of the pronghorn (*Antilocapra americana*). Journal of Mammalogy, 69:140-143.

Sourrouille, P., C. Hanni, M. Ruedi, and F. M. Catzeflis. 1995. Molecular systematics of *Mus crociduroides*, an endemic mouse of Sumatra (Muridae: Rodentia). Mammalia, 59(1):91-102.

Souza, M. J., and Y. Yonenaga-Yassuda. 1982. Chromosomal variability of sex chromosomes and NOR's in *Trichomys apereoides* (Rodentia, Echimyidae). Cytogenetics and Cell Genetics, 33:197-203.

Sowerby, A. de C. 1923. The naturalist in Manchuria. Tientsin Press Ltd. [China], 2:xxvii + 191.

Sözen, M., E. Çolak, N. Yiğit, Ş. Özkurt, and R. Verimli. 1999. Contributions to the karyology and taxonomy of the genus *Spalax* Güldenstaedt, 1770 (Mammalia:Rodentia) in Turkey. Zeitschrift für Säugetierkunde, 64:210-219.

Sparrmann, A. 1786. A voyage to the Cape of Good Hope, towards the Antarctic Polar Circle, and round the world; but chiefly into the country of the Hottentots and Caffres, from the year 1772, to 1776 (translation of Sparrmann, 1783). Second ed. G. G. J. & J. Robinson, London, 2 vols.

Spencer, H. G., and D. E. Lee. 1999. *Mystacina* Gray, 1843, *Chalinolobus* Peters, 1866, *M. tuberculata* Gray, 1843 and *Vespertilio tuberculatus* J. R. Forster, 1844 (currently *C. tuberculatus*) (Mammalia, Chiroptera): Proposed conservation of usage of the names. Case 3095. Bulletin of Zoological Nomenclature, 56:250-254.

Spencer, H. G., and D. E. Lee. 2000. Comments on the proposed conservation of usage of the names *Mystacina* Gray, 1843, *Chalinolobus* Peters, 1866, *M. tuberculata* Gray, 1843 and *C. tuberculatus* (J. R. Forster, 1844)(Mammalia, Chiroptera). Bulletin of Zoological Nomenclature, 57:176.

Spencer, S. R., and G. N. Cameron. 1982. *Reithrodontomys fulvescens*. Mammalian Species, 174:1-7.

Spennemann, D. H. R., and G. Rapp. 1989. Can rats colonise oceanic islands unaided? An assessment and review of the swimming capabilities of the genus *Rattus* with particular reference to tropical waters. Zoologische Abhandlungen Staatlichen Museum für Tierkunde Dresden, 45(7):81-91.

Spicer, S. S., and B. A. Schulte. 1994. Ultrastructural differentiation of the first hensen cell in the gerbil cochlea as a distinct cell type. The Anatomical Record, 240:149-156.

Spitz, F. 1978. Etude craniometrique du genere *Pitymys*. Mammalia, 42:267-304.

Spitzenberger, F. 1970. Erstnachweise der Wimperspitzmaus (*Suncus etruscus*) für Kreta und Kleinasien und die Verbreitung der Art im südwestasiatischen Raum. Zeitschrift für Säugetierkunde, 35:107-113.

Spitzenberger, F. 1971a. Eine neue, tiergeographisch bemerkenswerte *Crocidura* (Insectivora, Mammalia) aus der Türkei. Annalen des Naturhistorischen Museums in Wien, 75:539-562.

Spitzenberger, F. 1971b. Zur Systematik und Tiergeographie von *Microtus* (*Chionomys*) *nivalis* and *Microtus* (*Chionomys*) *gud* (Microtinae, Mamm.) in S-Anatolien. Zeitschrift für Säugetierkunde, 36:370-380.

Spitzenberger, F. 1972. Der Hamster *Mesocricetus brandti* (Nehring, 1898) in Zentralanatolien. Zeitschrift für Säugetierkunde, 37:229-231.

Spitzenberger, F. 1976. Beiträge zur Kenntnis von *Dryomys laniger* Felten and Storch, 1968 (Gliridae, Mammalia). Zeitschrift für Säugetierkunde, 41:237-249.

Spitzenberger, F. 1978a. Die Stachelmaus von Kleinasien, *Acomys cilicicus* n. sp. (Rodentia, Muridae). Annalen des Naturhistorischen Museums in Wien, 81:443-446.

Spitzenberger, F. 1978b. Die säugetierfauna zyperns. Teil I: Insectivora and Rodentia. Annalen des Naturhistorischen Museums in Wien, 81:401-441.

Spitzenberger, F. 1983. Die Schläfer (Gliridae) Österreichs. Mammalia Austriaca 6. (Mammalia, Rodentia). Mitteilungen der Abteilung für Zoologie am Landesmuseum Joanneum, 30:19-64.

Spitzenberger, F. 1990a. *Sorex alpinus* Schinz, 1837 — Alpenspitzmaus. Pp. 295-312, *in* Handbuch der Säugetiere Europas (J. Niethammer and F. Krapp, eds.). Aula-Verlag, Wiesbaden, 3/I:1-524.

Spitzenberger, F. 1990b. Gattung *Neomys* Kaup, 1829. Pp. 313-374, *in* Handbuch der Säugetiere Europas (J. Niethammer and F. Krapp, eds.). Aula-Verlag, Wiesbaden, 3/I:1-524.

Spitzenberger, F. 1990c. *Suncus etruscus* (Savi, 1822)—Etruskerspitzmaus. Pp. 375-392, *in* Handbuch der Säugetiere Europas (J. Niethammer, and F. Krapp, eds.). Aula-Verlag, Wiesbaden, 3/I:1-524.

Spitzenberger, F. 1996. Zoogeography of Austria's mammal fauna – an interim report on atlas work in progress. Pp. 55-65, *in* Eurpoean Mammals. Proceedings of the I European Congress of Mammalogy (M. Da Luz Mathias, M. Santos-Reis, G. Amori, R. Libois, A. Mitchell-Jones and M. C. Saint-Girons, eds.). Museu Bocage, Lisboa, 314 pp.

Spitzenberger, F. 1997. Erstnachweis der Brandmaus (*Apodemus agrarius*) für Österreich. Mammalia Austriaca 22. Zeitschrift für Säugetierkunde, 62:250-252.

Spitzenberger, F., and G. Eberl-Rothe. 1974. Der Sohlenhaftmechanismus von *Dryomys laniger*. Annalen Des Naturhistorischen Museums in Wien, 78:485-494.

Spitzenberger F., and H. Englisch. 1996. Die Alpenwaldmaus (*Apodemus alpicola* Heinrich, 1952) in Österreich. Mammalia Austriaca 21. Bonner Zoologische Beiträge, 46(1-4):249-260.

Spitzenberger, F., P. Mildner, and M. Preleuthner. 1995. Die Säugetiere Kärntens. Teil 1. Insektenfresser, Fledermäuse, Hasentiere, Hörnchenartige, Schläfer und Hüpfmäuse. Carinthia II, 105(1):247-352.

Spitzenberger, F., H. Englisch, S. Hammer, G. B. Hartl, and F. Suchentrunk. 1999. Morphological and genetic differentiation of bank voles, *Clethrionomys glareolus*, from the Eastern Alps. Folia Zoologica, 48(suppl. 1):69-94.

Spitzenberger, F., P. Brunet-Lecomte, A. Nadachowski, and K. Bauer. 2000. Comparative morphometrics of the first lower molar in *Microtus* (*Terricola*) cf. *liechtensteini* of the Eastern Alps. Acta Theriologica, 45(4):471-483.

Spitzenberger, F., E. Haring, and N. Tvrtkovíc. 2002. *Plecotus microdontus* (Mammalia, Vespertilionidae), a new bat species from Austria. Natura Croatica, 11:1-18.

Spitzer, N. C., and J. D. Lazell. 1978. A new rice rat (genus *Oryzomys*) from Florida's Lower Keys. Journal of Mammalogy, 59:787-792.

Spotorno O., A. E. 1976. Analisis taxonomico de tres especies altiplanicas del genero *Phyllotis* (Rodentia, Cricetidae). Annales del Museo de Historia Natural de Valparaiso, 9:141-161.

Spotorno O., A. E., and L. I. Walker B. 1979. Analisis de similitud cromosomica segun patrones de bandas G en cuatro especies chilenas de *Phyllotis* (Rodentia, Cricetidae). Archivos de Biología y Medicina Experimentales (Santiago), 12:83-90.

Spotorno O., A. E., and L. I. Walker B. 1983. Analisis electroforético y biométrico de dos especies de *Phyllotis* en Chile central y sus hibridos experimentales. Revista Chilena de Historia Natural, 56:51-59.

Spotorno O., A. E., L. Walker, L. Contreras, J. Pincheira, and R. Fernández-Donoso. 1988. Cromosomas ancestrales en Octodontidae y Abrocomyidae. Sociedad Genetica Chilensis, resumen no. R-65, not paginated.

Spotorno, A. E., C. A. Zuleta, and A. Cortes. 1990. Evolutionary systematics and heterochrony in *Abrothrix* species (Rodentia, Cricetidae). Evolución Biológica, 4:37-62.

Spotorno, A. E., J. Sufan-Catalan, and L. Walker. 1994. Cytogenetic diversity and evolution of Andean *Eligmodontia* species (Rodentia, Muridae). Zeitschrift für Säugetierkunde, 59:299-308.

Spotorno A. E., J. C. Marin, M. Yevenes, L. I. Walker, R. Fernandez-Donoso, J. Pincheira, M. S. Berrios, and R. E. Palma. 1997. Chromosome divergences among American marsupials and the Australian affinities of the American *Dromiciops*. Journal of Mammalian Evolution, 4:259-269.

Spotorno, A. E., H. Cofre, G. M. Manriquez, Y. Vilina, P. Marquet, and L. I. Walker. 1998. Nueva especie de mamifero filotino *Loxodontomys* en Chile central. Revista Chilena de Historia Natural, 71:359-374.

Spotorno, A. E., L. I. Walker, S. V. Flores, M. Yevenes, J. C. Marin, and C. Zuleta. 2001. Evolucion de los filotinos (Rodentia, Muridae) en los Andes del Sur. Revista Chilena de Historia Natural, 74:151-166.

Springer, M. S., L. J. Hollar, and J. A. Kirsch. 1995. Phylogeny, molecules versus morphology, and rates of character evolution among fruitbats (Chiroptera: Megachiroptera). Australian Journal of Zoology, 43:557-582.

Springer, M. S., M. Westerman, J. R. Kavanagh, A. Burk, M. O. Woodburne, D. J. Kao, and C. Krajewski. 1998. The origin of the Australasian marsupial fauna and the phylogenetic affinities of the enigmatic monito del monte and marsupial mole. Proceedings of the Royal Society of London, B, Biological Sciences, 265:2381-2386.

Springer, M. S., H. A. Amrine, A. Burk, and M. J. Stanhope. 1999. Additional support for Afrotheria and Paenungulata, the performance of mitochondrial versus nuclear genes, and the impact of data partitions with heterogeneous base composition. Systematic Biology, 48:65-75.

Springer, M. S., E. C. Teeling, O. Madsen, M. J. Stanhope, and W. W. deJong. 2001. Integrated fossil and molecular data reconstruct bat echolocation. Proceedings National Academy of Sciences, 98:6241-6246.

Spyropoulos, B., P. D. Ross, P. B. Moens, and D. M. Cameron. 1982. The synaptonemal complex karyotypes of Palearctic hamsters, *Phodopus roborovskii* Satunin and *P. sungorus* Pallas. Chromosoma, 86:397-408.

Srinivasulu, C., and B. Srinivasulu. 2001. Bats of the Indian subcontinent—An update. Current Science, Madras, 80(11):1378-1380.

Stafford, B. J., and F. S. Szalay. 2000. Craniodental functional morphology and taxonomy of dermopterans. Journal of Mammalogy, 81:360-385.

Stains, H. J. 1967. Carnivores, Pp. 325-354, *in* Orders and families of Recent mammals of the world (S. Anderson and J. K. Jones, Jr., eds.). John Wiley and Sons, New York, 453 pp.

Stains, H. J. 1975. Distribution and taxonomy of the Canidae, Pp. 3-26, *in* The wild canids: Their systematics, behavorial ecology and evolution (M. W. Fox, ed.). Van Nostrand Reinhold Company, New York, NY, 508 pp.

Stains, H. J. 1983. Calcanea of members of the Viverridae. Bulletin of the Southern California Academy of Sciences, 82:17-38.

Stål, C. 1865. A list of the names of the genera and subgenera from Linneaus ed. 10 1758 to 1861. Hemiptera Africana, 2:1-256.

Stalling, D. T. 1990. *Microtus ochrogaster*. Mammalian Species, 355:1-9.

Stalling, D. T. 1997. *Reithrodontomys humulis*. Mammalian Species, 565:1-6.

Stalmakova, V. A. 1957. [On the occurrence of the jerboa *Jaculus turcmenicus* Vinogr. et Bondar in the northern Kara-Kum and on some of its ecological and morphological peculiarities]. Zoologicheskii Zhurnal, 36(2):275-279 (in Russian).

Stangl, F. B., Jr. 1986. Aspects of a contact zone between two chromosomal races of *Peromyscus leucopus* (Rodentia: Cricetidae). Journal of Mammalogy, 67:465-473.

Stangl, F. B., Jr. 1992. First record of *Sigmodon fulviventer* in Texas: Natural history and cytogenetic observations. Southwestern Naturalist, 37:213-214.

Stangl, F. B., Jr., and R. J. Baker. 1984a. A chromosomal subdivision in *Peromyscus leucopus*: Implications for the subspecies concept as applied to mammals. Pp. 139-145, *in* Festschrift for Walter W. Dalquest in honor of his sixty-sixth birthday (N. Horner, ed.). Department of Biology, Midwestern State University, 163 pp.

Stangl, F. B., Jr., and R. J. Baker. 1984b. Evolutionary relationships in *Peromyscus*: Congruence in chromosomal, genic, and classical data sets. Journal of Mammalogy, 65:643-654.

Stangl, F. B., Jr., and W. W. Dalquest. 1991. Historical biogeography of *Sigmodon ochrognathus* in Texas. Texas Journal of Science, 43:127-131.

Stangl, F. B., G. F. Birkenfeld, and S. M. Shoen. 1999. First record of the California kangaroo rat, *Dipodomys californicus* (Rodentia: Heteromyidae), from Nevada. Southwestern Naturalist, 44:240-241.

Stanhope, M. J., V. G. Waddell, O. Madsen, W. de Jong, S. Blair Hedges, G. C. Cleven, D. Kao, and M. Springer. 1998. Molecular evidence for multiple origins of Insectivora and for a new order of endemic African insectivore mammals. Proceedings of the National Academy of Science, 95:9967-9972.

Stanko, M. 1995. [Synusia of small ground mammals (Insectivora, Rodentia) of the Biosphere Reserve East Carpathians.] Natura Carpatica, 36:119-126 (in Czech with English summary).

Stanko, M., and L. Mošanský. 1994. [Small mammals (Insectivora, Rodentia) in the area of the Ondava Downstream (Eastern Slovakian lowlands).] Zborník Východoslovenského Muzea V Košiciach, Prírodné Vedy, 35:77-87 (in Czech with English summary).

Stanko, M., and L. Mošanský. 2000. [Small mammals on the East Slovakia in collections of the scientific workers of the Zoological Institute SAS in Košice (Slovakia).] Lynx (Praha), n.s., 31:113-123 (in Czech with English summary).

Stanko, M., L. Mošanský, and J. Obuch. 1994. [Small mammals (Insectivora, Rodentia) of the south part of Košice Basin.] Zborník Východoslovenského Muzea V Košiciach, Prírodné Vedy, 35:105-112 (in Czech with English summary).

Stanko, M., L. Mošanský, and J. Budayová. 2000. [Contribution to the knowledge of small mammals in National Nature Reserve Sivá Brada (Hornádska Kotlina Basin, Slovakia).] Natura Carpatica, 61:101-106 (in Czech with English summary).

Stanley, W. T., and R. Hutterer. 2000. A new species of *Myosorex* Gray, 1832 (Mammalia: Soricidae) from the Eastern Arc Mountains, Tanzania. Bonner Zoologische Beiträge, 49:19-29.

Stanley, W. T., S. M. Goodman, and R. Hutterer. 1996. Notes on the insectivores and elephant shrews of the Chome Forest, South Pare Mountains, Tanzania. Zoologische Abhandlungen Staatliches Museum für Tierkunde Dresden, 49:131-147.

Stanley, W. T., S. M. Goodman and P. M. Kihaule. 1998. Results of two surveys of rodents in the Chome Forest Reserve, South Pare Mountains, Tanzania (Mammalia, Rodentia). Zoologische Abhandlungen, 50(11):145-160.

Stanley, W. T., P. M. Kihaule, K. M. Howell, and R. Hutterer. 2000. Small mammals of the Eastern Arc Mountains, Tanzania. Journal of East African Natural History, 87(for 1998):91-100.

Stanley, W. T., S. M. Goodman, P. M. Kihaule, and K. M. Howell. 2002. A survey of the small mammals of the Gonja Forest Reserve, Tanzania. Journal of East African Natural History, 89(1-2):73-83.

Stanley, W. T., M. A. Rogers, and R. Hutterer. In Press. A morphological assessment of *Myosorex zinki*, an endemic shrew on Mt Kilimanjaro. Belgian Journal of Zoology.

Starrett, A. 1967. Hystricoid, erethizontoid, cavioid, and chinchilloid rodents. Pp. 254-272, *in* Recent mammals of the world, a synopsis of families (S. Anderson and J. K. Jones, Jr., eds.). Ronald Press Co., New York, 453 pp.

Starrett, A. 1972. *Cyttarops alecto*. Mammalian Species, 13:1-2.

Starrett, A., and R. S. Casebeer. 1968. Records of bats from Costa Rica. Contributions in Science, Los Angeles Country Museum, 148:1-21.

Start, A. N., and D. J. Kitchener. 1995. Western pebble-mound mouse, *Pseudomys chapmani*. Pp. 591-592, *in* Mammals of Australia (R. Strahan, ed.). Smithsonian Institution Press, Washington, D.C., 756 pp.

Steadman, D. W., and C. E. Ray. 1982. The relationships of *Megoryzomys curoi*, an extinct cricetine rodent (Muroidea: Muridae) from the Galapagos Islands, Ecuador. Smithsonian Contributions to Paleontology, 51:1-23.

Steele, M. A. 1998. *Tamiasciurus hudsonicus*. Mammalian Species, 586:1-9.

Steele, M. A. 1999. *Tamiasciurus douglasii*. Mammalian Species, 630:1-8.

Stehlin, H. G., and S. Schaub. 1951. Die Trigonodontie der simplicidentaten Nager. Verlag Birkhäuser AG., Basel, 385 pp.

Stein, B. R. 1986. Comparative limb myology of four arvicolid rodent genera (Mammalia, Rodentia). Journal of Morphology, 187:321-342.

Stein, B. R. 1987. Phylogenetic relationships among four arvicolid genera. Zeitschrift für Säugetierkunde, 52:140-156.

Stein, B. R. 1990. Limb myology and phylogenetic relationships in the superfamily Dipodoidea (birch mice, jumping mice, and jerboas). Zeitschrift für Zoologische Systematik und Evolutionsforschung, 28:299-314.

Stein, G. 1933. Weitere mitteilungen zur systematik papuanischer sauger. Zeitschrift für Säugetierkunde, 8:87-95.

Stein, G. H. W. 1960. Schädelallometrien und Systematik bei altweltlichen Maulwürfen (Talpinae). Mitteilungen aus dem Zoologische Museum in Berlin, 36:1-48.

Steiner, M. 1978. *Apodemus microps* Kratochvil und Rosicky, 1952— Zwergwaldmaus. Pp. 359-367, *in* Handbuch der Säugetiere Europas (J. Niethammer and F. Krapp, eds.). Akademische Verlagsgesellschaft (Wiesbaden), 1:1-476.

Stenseth, N. C., and R. A. Ims (eds.). 1993. The Biology of lemmings. Linnean Society Symposium Series 15. Academic Press, London, 683 pp.

Stephan, H., and F. Dieterlen. 1982. Relative brain size in Muridae with special reference to *Colomys goslingi*. Zeitschrift für Säugetierkunde, 47:38-47.

Stephan, H., G. Baron, and H. D. Frahm. 1991. Comparative brain research in mammals, volume 1: Insectivora, with a stereotaxic atlas of the hedgehog brain. Springer-Verlag, New York, 573 pp.

Stephenson, P. J. 1995. Small mammal microhabitat use in lowland rain forest of north-east Madagascar. Acta Theriologica, 40(4):425-438.

Steppan, S. J. 1993. Phylogenetic relationships among the Phyllotini (Rodentia: Sigmodontinae) using morphological characters. Journal of Mammalian Evolution, 1:187-213.

Steppan, S. J. 1995. Revision of the Tribe Phyllotini (Rodentia: Sigmodontinae), with a phylogenetic hypothesis for the Sigmodontinae. Fieldiana: Zoology, n.s., 80:112 pp.

Steppan, S. J. 1996. A new species of *Holochilus* (Rodentia: Sigmodontinae) from the Middle Pleistocene of Bolivia and its phylogenetic significance. Journal of Vertebrate Paleontology, 16:522-530.

Steppan, S. J. 1998. Phylogenetic relationships and species limits within *Phyllotis* (Rodentia: Sigmodontinae): Concordance between mtDNA sequence and morphology. Journal of Mammalogy, 79:573-593.

Steppan, S. J., and U. F. J. Pardiñas. 1998. Two new fossil muroids (Sigmodontinae: Phyllotini) from the early Pleistocene of Argentina: Phylogeny and paleoecology. Journal of Vertebrate Paleontology, 18:640-649.

Steppan, S. J., and J. K. Sullivan. 2000. The emerging statistical perspective in systematics: A comment on Mares and Braun. Journal of Mammalogy, 81:260-270.

Steppan, S. J., M. R. Akhverdyan, E. A. Lyapunova, D. G. Fraser, N. N. Vorontsov, R. S. Hoffmann, and M. J. Braun. 1999. Molecular phylogeny of the marmots (Rodentia: Sciuridae): Tests of evolutionary and biogeographic hypotheses. Systematic Biology, 48(4):715-734.

Steppan, S. J., C. Zawadzki, and L. R. Heaney. 2003. Molecular phylogeny of the endemic Philippine rodent *Apomys* (Muridae) and the dynamics of diversification in an oceanic archipelago. Biological Journal of the Linnean Society, 80:699-715.

Steppan, S. J., B. L. Storz, and R. S. Hoffmann. 2004. Nuclear DNA phylogeny of the squirrels (Mammalia: Rodentia) and the evolution of arboreality from c-myc and RAG1. Molecular Phylogenetics and Evolution, 30:703-719.

Steven, D. M. 1953. Recent evolution in the genus *Clethrionomys*. Pp. 310-319, *in* Symposia of the society for experimental biology, evolution, number VII (R. Brown and J. F. Danielli, eds.). University Press, Cambridge, 448 pp.

Stewart, B. E., and R. E. A. Stewart. 1989. *Delphinapterus leucas*. Mammalian Species, 336:1-8.

Stewart, B. S., and H. R. Huber. 1993. *Mirounga angustirostris*. Mammalian Species, 449:1-10.

Stewart, B. S., and S. Leatherwood. 1985. Minke whale—*Balaenoptera acutorostrata*. Pp. 91-136, *in* Handbook of marine mammals: The sirenians and baleen whales (S. H. Ridgway and R. Harrison, eds.). Academic Press, London, 3:1-362.

Stewart, D. T., and A. J. Baker. 1994. Evolution of mtDNA D-loop sequences and their use in phylogenetic studies of shrews in the subgenus *Otisorex* (*Sorex*: Soricidae: Insectivora). Molecular Phylogenetics and Evolution, 3:38-46.

Stewart, D. T., and A. J. Baker. 1997. A phylogeny of some taxa of masked shrews (*Sorex cinereus*) based on mitochondrial-DNA, D-loop sequences. Journal of Mammalogy, 78:361-376.

Stewart, D. T., N. D. Perry, and L. Fumagalli. 2002. The maritime shrew, *Sorex maritimensis* (Insectivora: Soricidae): A newly recognized Canadian endemic. Canadian Journal of Zoology, 80:94-99.

Stiles, C. W., and M. B. Orleman. 1926. Retention of *Cercopithecus*, type *diana*, for the guenons. Journal of Mammalogy, 7:48-53.

Stinson, D. W. 1994. Birds and mammals recorded from the Mariana Islands. Natural History Research, Special Issue, 1:333-344.

Stirling, I. 1971. *Leptonychotes weddelli*. Mammalian Species, 6:1-5.

Stock, A. D. 1974. Chromosome evolution in the genus *Dipodomys* and its taxonomic and phylogenetic implications. Journal of Mammalogy, 55:505-526.

Stogov, I. I. 1985. [On two little studied species of white-toothed shrews (Insectivora, Soricidae, *Crocidura*) from the mountain regions in the southern USSR.] Zoologicheskii Zhurnal, 64:264-268 (in Russian).

Stogov, I. I., and E. P. Bondar. 1966. [A survey of *Crocidura* in South Turkmenia and Tajikistan.] Zoologicheskii Zhurnal, 45:414-420 (in Russian).

Stollmann, A. 1998. The mound-building mouse–living memorial of Ján Salamun Petényi. Chránené Uzemia Slovenska, 35:33 (in Slovak).

Stollmann, A., and M. Macholán. 1999. The mound-building mouse is part of the Slovakian fauna–a reply. Folia Zoologica, 48:237-239.

Stone, K. D., and J. A. Cook. 2002. Molecular evolution of Holarctic martens (genus *Martes*, Mammalia: Carnivora: Mustelidae). Molecular Phylogenetics and Evolution, 24(2):169-179.

Stone, R. D. 1995. Status survey and conservation action plan. Eurasian Insectivores and tree shrews. I.U.C.N., Gland, Switzerland, 108 pp.

Stonehouse, B., and D. Gilmore (eds.). 1977. Biology of Marsupials. Macmillan, London, 486 pp.

Storch, G. 1974. Neue Zwerghamster aus dem Holozan von Aserbeidschan, Iran (Rodentia: Cricetinae). Senckenbergiana Biologica, 55:21-28.

Storch, G. 1975. Eine mittelpleistozäne Nager-Fauna von der Insel Chios, Ägäis (Mammalia: Rodentia). Senckenbergiana Biologica, 56:165-189.

Storch, G. 1977. Die Ausbreitung der Felsenmaus (*Apodemus mystacinus*): Zur Problematik der Inselbesiedlung und Tiergeographie in der Agäis. Natur und Museum, 107:174-182.

Storch, G. 1978a. Familie Gliridae Thomas, 1897—Schläfer. Pp. 201-280, *in* Handbuch der Säugetiere Europas (H. J. Niethammer, and F. Krapp, eds.). Akademische Verlagsgesellschaft (Wiesbaden), 1:1-476.

Storch, G. 1978b. *Myomimus roachi* (Bate, 1937) —- Mausschläfer. Pp. 238-242, *in* Handbuch der Säugetiere Europas (H. J. Niethammer and F. Krapp, eds.). Akademische Verlagsgesellschaft (Wiesbaden), 1:1-476.

Storch, G. 1982. *Microtus majori* Thomas, 1906. Pp. 452-462, *in* Handbuch der Säugetiere Europas (J. Niethammer and F. Krapp, eds.). Akademische Verlagsgesellschaft (Wiesbaden), 2/I:1-649.

Storch, G. 1987. The Neogene mammalian faunas of Ertemte and Harr Obo in Inner Mongolia (Nei Mongol), China.—7. Muridae (Rodentia). Senckenbergiana Lethaea, 67:401-431.

Storch, G. 1988. Eine jungpleistozäne/alholozäne Kleinsäuger-Abfolge von Antalya, SW-Anatolien (Mammalia, Rodentia). Zeitschrift für Säugetierkunde, 53:76-82.

Storch, G. 1995a. The Neogene mammalian faunas of Ertemte and Harr Obo in Inner Mongolia (Nei Mongol), China.—II. Soricidae (Insectivora). Senckenbergiana Lethaea, 75:221-251.

Storch, G. 1995b. Affinities among living dormouse genera. Pp. 51-62, *in* Proceedings of II Conference on Dormice (Rodentia, Myoxidae) (M. G. Filippucci, ed.). Hystrix, n.s., 6(1-2):1-340.

Storch, G., and T. Dahlmann. 1995. The vertebrate locality Maramena (Macedonia, Greece) at the Turolian-Ruscinian Boundary (Neogene). 10. Murinae (Rodentia, Mammalia). Münchner geowissenschaftliche Abhandlungen, Reihe A, Geologie und Paläontologie, 28:121-132.

Storch, G., and T. Dahlmann. 2000. *Desmanella rietscheli*, ein neuer Talpide aus dem Obermiozän von Dorn-Dürkheim 1, Rheinhessen (Mammalia, Lipotyphla). Carolinea, 58:65-69.

Storch, G., and O. Lutt. 1989. Artstatus der Alpenwaldmaus, *Apodemus alpicola* Heinrich, 1952. Zeitschrift für Säugetierkunde, 54:337-346.

Storch, G., and Z. Qiu. 1983. The Neogene mammalian faunas of Ertemte and Harr Obo in Inner Mongolia (Nei Mongol), China. 2. Moles—Insectivora: Talpidae. Senckenbergiana Lethaea, 64:89-127.

Storch, G., and Z. Qiu. 1991. Insectivores (Mammalia: Erinaceidae, Soricidae, Talpidae) from the Lufeng hominoid locality, Late Miocene of China. Geobios, 24:601-621.

Storch, G., and C. Seiffert. 2002. An extraordinarily preserved fossil specimen of *Eogliravus*, the oldest known glirid genus. Pp. 20, *in* International Conference on Dormouse (Myoxidae). Abstracts (G. Bakonyi, S. Bosze and P. Morris, eds.). Szent Istvan University Department of Zoology and Ecology, Godollo, Hungary, 56 pp.

Storch, G., and H. Winking. 1977. Zur systematik der *Pitymys multiplex-Pitymys liechtensteini*-Gruppe (Mammalia, Rodentia). Zeitschrift für Säugetierkunde, 42:78-88.

Storch, G., Z. Qiu, and V. S. Zazhigin. 1998. Fossil history of shrews in Europe. Pp. 93-120, *in* Evolution of shrews (J. M.Wójcik and M. Wolsan, eds.). Mammal Research Institute, Polish Academy of Sciences, Bialowieza, 458 pp.

Storch, G., C. Seiffert and G. Escarguel. 2000. Neuer nager aus Messel – Prachtstück des "Urschläfers". Spektrum der Wissenschaft, 8/2000:12-13.

Storer, T. I. 1937. The muskrat as native and alien. Journal of Mammalogy, 18:443-460.

Stormark, J. G. 1998. Phenetic analysis of Old World *Myotis* (Chiroptera: Vespertilionidae) based on dental characters. Acta Theriologica, 43:1-11.

Storr, G. C. C. 1780. Prodromus methodi mammalium. Inaugeralem disputationem propositus praeside G. C. C. Storr. Respondente F. Wolffer Litteris Reissimis, Tubingae, 43 pp.

Storz, J. F., and T. H. Kunz. 1999. *Cynopterus sphinx*. Mammalian Species, 613:1-8.

Storz, J. F., and W. C. Wozencraft. 1999. *Melogale moschata*. Mammalian Species, 631:1-4.

Storz, J. F., J. Balasingh, H. R. Bhat, P. T. Nathan, D. P. S. Doss, A. A. Prakash, and T. H. Kunz. 2001. Clinal variation in body size and sexual dimorphism in an Indian fruit bat, *Cynopterus sphinx* (Chiroptera: Pteropodidae). Biological Journal of the Linnean Society, 72:17-31.

Strahan, R. (ed.). 1995. Mammals of Australia. Smithsonian Institution Press, Washington, D.C., 756 pp.

Straney, D. O., and J. L. Patton. 1981. Phylogenetic and environmental determinants of geographic variation of the pocket mouse *Perognathus goldmani* Osgood. Evolution, 34:888-903.

Strelkov, P. P. 1972. *Myotis blythi* (Tomes, 1857): distribution, geographical variability and differences from *Myotis myotis* (Borkhausen, 1797). Acta Theriologica, 17:355-380.

Strelkov, P. P. 1983. [*Myotis mystacinus* and *Myotis brandti* in the USSR and interrelationships of these species.] Zoologicheskii Zhurnal, 62:259-270 (in Russian).

Strelkov, P. P. 1986. [The Gobi bat (*Eptesicus gobiensis* Bobrinskii 1926), a new species of chiropteran of the Palearctic fauna.] Zoologicheskii Zhurnal, 65:1103-1108 (in Russian).

Strelkov, P. P., and E. G. Buntova. 1982. [*Myotis mystacinus* and *M. brandti* (Chiroptera: Vespertilionidae) and interrelations of these species. Part I.] Zoologicheskii Zhurnal, 61:1227-1241 (in Russian).

Strelkov, P. P., V. P. Sosnovtseva, and K. B. Babaev. 1978. [Bats (Chiroptera) of Turkmenistan.] Trudy Zoologicheskovo Instituta, Akademiya Nauk, Leningrad, 79:3-71 (in Russian).

Streubel, D. P., and J. P. Fitzgerald. 1978*a*. *Spermophilus spilosoma*. Mammalian Species, 101:1-4.

Streubel, D. P., and J. P. Fitzgerald. 1978*b*. *Spermophilus tridecemlineatus*. Mammalian Species, 103:1-5.

Stroganov, S. U. 1957. Zveri Sibiri. Nasekomoyadnye. [Animals of Siberia. Insectivores]. Akademiya Nauk SSSR, Moscow, 267 pp.

Stroganov, S. U. 1958. Review of steppe polecat subspecies (*Putorius eversmanni* Lesson) of Siberian fauna. Izvestiya Sibirskogo otdelenya AN SSSR, 11:149-155.

Stroganov, S. U. 1962. Zveri Sibiri. Khishchye [Animals of Siberia. Carnivores]. Akademiya Nauk SSSR, Moscow, 458 pp. (in Russian).

Struhsaker, T. T. 1970. Phylogenetic implications of some vocalizations of *Cercopithecus* monkeys. Pp. 365-444, *in* Old World Monkeys (J. R. Napier and P. H. Napier, eds.). Academic Press, London, 660 pp.

Stuart, C., and T. Stuart. 1995. Mammals of the UAE mountains. Tribulus, 5.2:19-21.

Stuart, C. T. 1980. The distribution and status of *Manis temmincki* Smuts, 1832 (Pholidota: Manidae). Säugetierkundliche Mitteilungen, 28:123-129.

Stuart, C. T. 1984. The distribution and status of *Felis caracal* Schreber, 1776. Säugetierkundliche Mitteilungen, 31:197-203.

Stuart, J. N., and N. J. Scott, Jr. 1992. Range extension of the northern pygmy mouse, *Baiomys taylori*, in New Mexico. Texas Journal of Science, 44:487-489.

Stuenes, S. 1989. Taxonomy, habits, and relationships of the subfossil Madagascan hippopotami *Hippopotamus lemerlei* and *H. madagascariensis*. Journal of Vertebrate Paleontology, 9:241-268.

Su, B. L, R. Q. Liu, Y. X. Wang, and L. M. Shi. 1994. Genetic diversity in the Chinese pangolin (*Manis pentadactyla*) inferred from protein electrophoresis. Biochemical Genetics, 32(9-10):343-349.

Su, B., Y.-X. Wang, H. Lan, W. Wang, and Y. Zhang. 1999. Phylogenetic study of complete cytochrome b genes in musk deer (Genus *Moschus*) using museum samples. Molecular Phylogenetics and Evolution, 12:241-249.

Su, B., Y.-X. Wang, and Q.-S. Wang. 2001. Mitochondrial DNA sequences imply Anhui musk deer a valid species in genus *Moschus*. Zoological Research, 22:169-173.

Suárez, W., and S. Díaz-Franco. 2003. A new fossil bat (Chiroptera: Phyllostomidae) from a Quaternary cave deposit in Cuba. Caribbean Journal of Science, 39:371-377.

Subramanyam, K., and M. P. Nayer. 974. Vegetation and phytogeography of the western Ghats. Pp. 178-196, in Ecology and biogeography in India (M. S. Mani, ed.). W. Junk, The Hague, The Netherlands, 773 pp.

Suchentrunk, F. 2004. Phylogenetic relationships between Indian and Burmese hares (Lepus nigricolis and L. peguensis) inferred from epigenetic dental characters. Mammalian Biology, 69(1):28-45.

Suchentrunk, F., A. Haiden, and G. B. Hartl. 1998. On biochemical genetic variability and divergence of the two Hedgehog species Erinaceus europaeus and E. concolor in central Europe. Zeitschrift für Säugetierkunde, 63:257-265.

Sudman, P. D., and M. S. Hafner. 1992. Phylogenetic relationships among Middle American pocket gophers (genus Orthogeomys) based on mitochondrial DNA sequences. Molecular Phylogenetics and Evolution, 1:17-25.

Sudman, P. D., J. R. Choate, and E. G. Zimmerman. 1987. Taxonomy of chromosomal races of Geomys bursarius lutescens Merriam. Journal of Mammalogy, 68:526-543.

Sudman, P, D., L. J. Barkley, and M. S. Hafner. 1994. Familial affinity of Tomopeas ravus (Chiroptera) based on protein electophoretic and cytochrome B sequence data. Journal of Mammalogy, 75:365-377.

Sugg, D. W., M. L. Kennedy, and G. A. Heidt. 1990. Genetic variation in the Texas mouse, Peromyscus attwateri. Journal of Mammalogy, 71:309-317.

Sugimura, K. 1990. The Amami rabbit Pentalagus furnessi. Pp. 140-142, in Rabbits, hares and pikas (J. A. Chapman and J. E. C. Flux, eds.). I.U.C.N., Gland, Switzerland, 168 pp.

Sulentich, J. M., L. R. Williams, and G. N. Cameron. 1991. Geomys breviceps. Mammalian Species, 383:1-4.

Sulimski, A. 1962. Two new rodents from Weze 1 (Poland). Acta Palaeontologica Polonica, 7(3-4):503-511.

Sulimski, A. 1964. Pliocene Lagomorpha and Rodentia from Weze 1 (Poland). Acta Palaeontologica Polonica, 9(2):149-261.

Sulkava, S. 1978. Pteromys volans (Linnaeus, 1758)—Flughörnchen. Pp. 71-84, in Handbuch der Säugetiere Europas (J. Niethammer and F. Krapp, eds.). Akademische Verlagsgesellschaft (Wiesbaden), 1:1-476.

Sulkava, S. 1990. Sorex caecutiens Laxmann, 1788—Maskenspitzmaus; Sorex isodon Turov, 1924—Taigaspitzmaus. Pp. 215-236, in Handbuch der Säugetiere Europas (J. Niethammer and F. Krapp, eds.). Aula-Verlag, Wiesbaden, 3/I:1-524.

Sullivan, J., and D. L. Swofford. 1997. Are guinea pigs rodents? The importance of adequate models in molecular phylogenetics. Journal of Mammalian Evolution, 4(2):77-86.

Sullivan, J. K., K. E. Holsinger, and C. Simon. 1995. Among-site rate variation and phylogenetic analysis of 12S rRNA in sigmodontine rodents. Molecular Biology and Evolution, 12:988-1001.

Sullivan, J. K., J. A. Markert, and C. W. Kilpatrick. 1997. Phylogeography and molecular systematics of the Peromyscus aztecus species group (Rodentia: Muridae) inferred using parsimony and likelihood. Systematic Biology, 46:426-440.

Sullivan, J. K., E. Arellano, and D. Rogers. 2000. Comparative phylogeography of Mesoamerican highland rodents: Concerted versus independent response to past climatic fluctuations. American Naturalist, 155:755-768.

Sullivan, J. M., and C. W. Kilpatrick. 1991. Biochemical systematics of the Peromyscus aztecus assemblage. Journal of Mammalogy, 72:681-696.

Sullivan, J. M., C. W. Kilpatrick, and P. D. Rennert. 1991. Biochemical systematics of the Peromyscus boylii species group. Journal of Mammalogy, 72:669-680.

Sullivan, R. M. 1985. Phyletic, biogeographic, and ecological relationships among montane populations of least chipmunks (Eutamias minimus) in the southwest. Systematic Zoology, 34:419-448.

Sullivan, R. M. 1994. Micro-evolutionary differentiation and biogeographic structure among coniferous forest populations of the Mexican woodrat (Neotoma mexicana) in the American Southwest: A test of the vicariance hypothesis. Journal of Biogeography, 21:369-389.

Sullivan, R. M., and T. L. Best. 1997a. Systematics and morphologic variation in two chromosomal forms of the Agile Kangaroo Rat (Dipodomys agilis). Journal of Mammalogy, 78:775-797.

Sullivan, R. M., and T. L. Best. 1997b. Effects of environment on phenotypic variation and sexual dimorphism in Dipodomys simulans (Rodentia: Heteromyidae). Journal of Mammalogy, 78:798-810.

Sullivan, R. M., D. J. Hafner, and T. L. Yates. 1986. Genetics of a contact zone between three chromosomal forms of the grasshopper mouse (genus Onychomys): A reassessment. Journal of Mammalogy, 67:640-659.

Sullivan, R. M., S. W. Calhoun, and I. F. Greenbaum. 1990. Geographic variation in genital morphology among insular and mainland populations of Peromyscus maniculatus and Peromyscus oreas. Journal of Mammalogy, 71:48-58.

Sumner, J., and C. R. Dickman. 1998. Distribution and identity of species in the Antechinus stuartii-A. flavipes group (Marsupialia: Dasyuridae) in southeastern Australia. Australian Journal of Zoology, 46:27-41.

Sundevall, C. J. 1846. Methodisk öfversigt af Idislande djuren Linnés Pecora. Kongliga Vetenskaps-Ajademiens Handlinger för år 1844:121-210.

Sunquist, M., and F. Sunquist. 2002. Wild cats of the world. University of Chicago Press, 462 pp.

Surridge, A. K., R. J. Timmins, G. M. Hewitt, and D. J. Bell. 1999. Striped rabbits in Southeast Asia. Nature, 400:726.

Sutcliffe, A. J., and K. Kowalski. 1976. Pleistocene rodents of the British Isles. Bulletin of the British Museum (Natural History), Geology, 27(2):33-147.

Sutton, D. A. 1987. Analysis of Pacific Coast Townsend chipmunks (Rodentia: Sciuridae). Southwestern Naturalist, 32:371-376.

Sutton, D. A. 1992. *Tamias amoenus*. Mammalian Species, 390:1-8.

Sutton, D. A., and C. F. Nadler. 1974. Systematic revision of three Townsend chipmunks (*Eutamias townsendii*). Southwestern Naturalist, 19:199-211.

Suzuki, H., and Y. Kurihara. 1994. Genetic variation of ribosomal RNA in the house mouse, *Mus musculus*. Pp. 107-119, *in* Genetics in wild mice (K. Moriwaki, T. Shiroishi, and H. Yonekawa, eds.). Japan Scientific Societies Press, Tokyo, 333 pp.

Suzuki, H., K. Tsuchiya, M. Sakaizumi, S. Wakana, O. Gotoh, N. Saitou, K. Moriwaki, and S. Sakurai. 1990. Differentiation of restriction sites in ribosomal DNA in the genus *Apodemus*. Biochemical Genetics, 28(3/4):137-149.

Suzuki, H., T. Hosoda, S. Sakurai, K. Tsuchiya, I. Munechika, and V. Korablev. 1994a. Phylogenetic relationship between the Iriomote cat and the leopard cat, *Felis bengalensis*, based on the ribosomal DNA. Japan Journal of Genetics, 69:397-406.

Suzuki, H., K. Tsuchiya, M. Sakaizumi, S. Wakana, and S. Sakurai. 1994b. Evolution of restriction sites of ribosomal DNA in natural populations of the field mouse, *Apodemus speciosus*. Journal of Molecular Evolution, 38:107-112.

Suzuki, H., S. Wakana, H. Yonekawa, K. Moriwaki, S. Sakurai, and E. Nevo. 1996a. Variations in ribosomal DNA and mitochondrial DNA among chromosomal species of subterranean mole rats. Molecular Biology and Evolution, 13(1):85-92.

Suzuki, T., H. Yuasa, and Y. Machida. 1996b. Phylogenetic position of the Japanese river otter *Lutra nippon* inferred from the nucleotide sequence of 224 bp of the mitochondrial cytochrome b gene. Zoological Science (Tokyo), 13:621-626.

Suzuki, H., S. Minato, S. Sakurai, K. Tsuchiya, and I. M. Fokin. 1997. Phylogenetic position and geographic differentiation of the Japanese Dormouse, *Glirulus japonicus*, revealed by variations among rDNA, mtDNA and the Sry gene. Zoological Science, 14:167-173.

Suzuki, H., M. A. Iwasa, N. Ishii, H. Nagaoka, and K. Tsuchiya. 1999a. The genetic status of two insular populations of the endemic spiny rat *Tokudaia osimensis* (Rodentia, Muridae) of the Ryukyu Islands, Japan. Mammal Study, 24:43-50.

Suzuki, H., M. Iwasa, M. Harada, S. Wakana, M. Sakaizumi, S.-H. Han, E. Kitahara, Y. Kimura, I. Kartavtseva, and K. Tsuchiya. 1999b. Molecular phylogeny of red-backed voles in Far East Asia based on variation in ribosomal and mitochrondrial DNA. Journal of Mammalogy, 80(2):512-521.

Suzuki, H., K. Tsuchiya, and N. Takezaki. 2000. A molecular phylogenetic framework for the Ryukyu endemic rodents *Tokudaia osimensis* and *Diplothrix legata*. Molecular Phylogenetics and Evolution, 15(1):15-24.

Suzuki, H., J. J. Sato, K. Tsuchiya, J. Luo, Y.-P. Zhang, Y.-X. Wang, and X.-L. Jiang. 2003. Molecular phylogeny of wood mice (*Apodemus*, Muridae) in East Asia. Biological Journal of the Linnean Society, 80:469-481.

Svartman, M., and E. J. C. Almeida. 1993. Pericentric inversion and X chromosome polymorphism in *Rhipidomys* sp. (Cricetidae, Rodentia) from Brazil. Caryologia, 46:219-225.

Svartman, M., and E. J. C. Almeida. 1994. The karyotype of *Akodon lindberghi* Hershkovitz, 1990 (Cricetidae, Rodentia). Revista Brasileira Genetica, 17:225-227.

Swanepoel, P., and H. H. Genoways. 1978. Revision of the Antillean bats of the genus *Brachyphylla* (Mammalia: Phyllostomatidae). Bulletin of Carnegie Museum of Natural History, 12:1-53.

Swanepoel, P., and H. H. Genoways. 1983a. *Brachyphylla cavernarum*. Mammalian Species, 205:1-6.

Swanepoel, P., and H. H. Genoways. 1983b. *Brachyphylla nana*. Mammalian Species, 206:1-3.

Swanepoel, P., and D. A. Schlitter. 1978. Taxonomic review of the fat mice (genus *Steatomys*) of West Africa (Mammalia: Rodentia). Bulletin of Carnegie Museum of Natural History, 6:53-76.

Swanepoel, P., D. A. Schlitter, and H. H. Genoways. 1979. A study of nongeographic variation in *Tatera leucogaster* (Mammalia: Rodentia) from Botswana. Annals of Carnegie Museum, 48:7-24.

Swanepoel, P., R. H. N. Smithers, and I. L. Rautenbach. 1980. A checklist and numbering system of the extant mammals of the Southern African Subregion. Annals of the Transvaal Museum, 32:155-196.

Swann, C. A., R. M. Hope, and W. G. Breed. 2002. cDNA nucleotide sequence encoding the ZPC protein of

Australian hydromyine rodents: A novel sequence of the putative sperm-combining site within the family Muridae. Zygote, 10:291-299.

Swenk, M. H. 1939. A study of local size variation in the prairie pocket gopher (*Geomys lutescens*), with description of a new subspecies from Nebraska. Missouri Valley Fauna, 1:1-8.

Swenson, J. E. 1981. Distribution of Richardson's ground squirrel in eastern Montana. Prairie Naturalist, 13:27-30.

Swynnerton, G. H., and R. W. Hayman. 1951. A checklist of the land mammals of the Tanganyika Territory and the Zanzibar Protectorate. Journal of the East African Natural History Society, 20(6):274-392.

Syvertsen, P. O., P. Shimmings, and K. Isaksen. 1996. The Norwegian mammal fauna: Status and atlas mapping. Hystrix, n.s., 8(1-2):91-95.

Szalay, F. 1982. A new appraisal of marsupial phylogeny and classification. Pp. 621-640, *in* Carnivorous marsupials (M. Archer, ed.). Royal Zoological Society of New South Wales, Mosman, 1:1-397; 2:397-804.

Szalay, F. S., and E. Delson. 1979. Evolutionary history of the primates. Academic Press, New York, 580 pp.

Taberlet, P., L. Fumagalli, A.-G. Wust-Saucy, and J.-F. Cossons. 1998. Comparative phylogeography and postglacial colonization routes in Europe. Molecular Ecology, 7:453-464.

Taddei, V. A., and W. Uieda. 2001. Distribution and morphometrics of *Natalus stramineus* from South America (Chiroptera, Natalidae). Iheringia, Série. Zoologia., Porto Alegre, 91:123-132.

Taddei, V. A., L. D. Vizotto, and I. Sazima. 1978. Notas sobre *Lionycteris* e *Lonchophylla* nas coleções do Museu Paraense Emílio Goeldi (Mammalia, Chiroptera, Phyllostomidae). Boletim do Museu Paraense Emílio Goeldi, Nova Série, Zoologia, 92:1-14.

Taddei, V. A., L. D. Vizotto, and I. Sazima. 1983. Uma nova espécie de *Lonchophylla* do Brasil e chave para identificação das especies do género (Chiroptera, Phyllostomidae). Ciência e Cultura (São Paulo), 35:625-629.

Tagle, D. A., M. M. Miyamoto, M. Goodman, O. Hofmann, G. Graunitzer, R. Goltenboth, and H. Jalanka. 1986. Hemoglobin of pandas: Phylogenetic relationships of carnivores as ascertained with protein sequence data. Naturwissenschaften, 73:512-514.

Takada, Y., H. Yamada, and T. Tateishi. 1994. Morphometric variation of Japanese wild mice on islands. Journal of the Mammal Society of Japan, 19(2):113-128.

Takada, Y., E. Sakai, Y. Uematsu, and T. Tateishi. 1999. Morphometric variation of house mice (*Mus musculus*) on the Izu Islands. Mammal Study, 24:51-65.

Takada, Y., T. Tateishi, E. Sakai, and Y. Uematsu. 2002. Morphological variation of mice, *Mus musculus* on the Ogasawara Islands, and their relationship to those on the Izu Islands, Japan. Mammal Study, 27:23-30.

Talbot, L. M., and M. H. Talbot. 1966. The tamarau (*Bubalus mindorensis* (Heude)). Observations and recommendations. Mammalia, 30:1-12.

Tamarin, R. H. (ed.). 1985. Biology of New World *Microtus*. Special Publication, American Society of Mammalogists, 8:1-893.

Tamarin, R. H., and T. H. Kunz. 1974. *Microtus breweri*. Mammalian Species, 45:1-3.

Tamayo H., M., and D. Frassinetti C. 1980. Catálogo de los mamíferos fósiles y vivientes de Chile. Boletín de Museo Nacional de Historia Natural Chile (Santiago), 37:323-399.

Tamsitt, J. R., and C. Häuser. 1985. *Sturnira magna*. Mammalian Species, 240:1-4.

Tamsitt, J. R., and D. Nagorsen. 1982. *Anoura cultrata*. Mammalian Species, 179:1-5.

Tanaka, K., C. D. Solis, J. S. Masangkay, K. Maeda, Y. Kawamato, and T. Namiwaka. 1996. Phylogenetic relationship among all living species of the genus *Bubalus* based on DNA sequence of the Cytochrome b Gene. Biochemical Genetics, 34:443-452.

Tanaka, R. 1971. A research into variation in molar and external features among a population of the Smith's red-backed vole for elucidation of its systematic rank. Japanese Journal of Zoology, 16:163-176.

Taranin, A. V., E. G. Ufimtseva, E. S. Belousov, D. V. Ternovsky, Yu. G. Ternovskaya, and O. K. Baranov. 1991. [Evolution of antigenic structure of immunoglobulin chains in mustelids (Carnivora, Mustelidae).] Zoologicheskii Zhurnal, 70:105-112 (in Russian).

Tas'an, and S. Leatherwood. 1984. Cetaceans live-captured for Jaya Ancol Oceanarium, Djakarta, 1974-1982. International Whaling Commission, Report, 34:485-489.

Tast, J. 1982a. *Lemmus lemmus* (Linnaeus, 1758)—Berglemming. Pp. 87-105, *in* Handbuch der Säugetiere Europas (J. Niethammer and F. Krapp, eds.). Akademische Verlagsgesellschaft (Wiesbaden), vol. 2/I:1-649.

Tast, J. 1982b. *Microtus oeconomus* (Pallas, 1776)—Nordische Wühlmaus, Sumpfmaus. Pp. 374-396, *in* Handbuch der Säugetiere Europas (J. Niethammer and F. Krapp, eds.). Akademische Verlagsgesellschaft (Wiesbaden), 2/I:1-649.

Tate, C. M., J. F. Pagels, and C. O. Handley, Jr. 1980. Distribution and systematic relationship of two kinds of short-tailed shrews (Soricidae: *Blarina*) in south-central Virginia. Proceedings of the Biological Society of Washington, 93:50-60.

Tate, G. H. H. 1932a. The taxonomic history of the genus *Reithrodon* Waterhouse (Cricetidae). American Museum Novitates, 529:1-4.

Tate, G. H. H. 1932b. The taxonomic history of the South American cricetid genera *Euneomys* (subgenera *Euneomys* and *Galenomys*), *Auliscomys*, *Chelemyscus*, *Chinchillula*, *Phyllotis*, *Paralomys*, *Graomys*, *Eligmodontia*, and *Hesperomys*. American Museum Novitates, 541:1-21.

Tate, G. H. H. 1932c. The taxonomic history of the Neotropical cricetid genera *Holochilus*, *Nectomys*, *Scapteromys*, *Megalomys*, *Tylomys*, and *Ototylomys*. American Museum Novitates, 562:1-19.

Tate, G. H. H. 1932d. The taxonomic history of the South and Central American cricetid rodents of the genus *Oryzomys*. Part 1: Subgenus *Oryzomys*. American Museum Novitates, 579:1-18.

Tate, G. H. H. 1932e. The taxonomic history of the South and Central American cricetid rodents of the genus *Oryzomys*. Part 2: Subgenera *Oligoryzomys*, *Thallomyscus*, and *Melanomys*. American Museum Novitates, 580:1-16.

Tate, G. H. H. 1932f. The taxonomic history of the South and Central American oryzomine genera of rodents (excluding *Oryzomys*): *Nesoryzomys*, *Zygodontomys*, *Chilomys*, *Delomys*, *Phaenomys*, *Rhagomys*, *Rhipidomys*, *Nyctomys*, *Oecomys*, *Thomasomys*, *Inomys*, *Aepeomys*, *Neacomys*, and *Scolomys*. American Museum Novitates, 581:1-28.

Tate, G. H. H. 1932g. The taxonomic history of the South and Central American akodont rodent genera: *Thalpomys*, *Deltamys*, *Thaptomys*, *Hypsimys*, *Bolomys*, *Chroeomys*, *Abrothrix*, *Scotinomys*, *Akodon*, (*Chalcomys* and *Akodon*), *Microxus*, *Podoxymys*, *Lenoxus*, *Oxymycterus*, *Notiomys*, and *Blarinomys*. American Museum Novitates, 582:1-32.

Tate, G. H. H. 1932h. The taxonomic history of certain South and Central American cricetid Rodentia: *Neotomys*, with remarks upon its relationships; the cotton rats (*Sigmodon* and *Sigmomys*); and the "fish-eating" rats (*Ichthyomys*, *Anotomys*, *Rheomys*, *Neusticomys*, and *Daptomys*). American Museum Novitates, 583:1-10.

Tate, G. H. H. 1933. A systematic revision of the marsupial genus *Marmosa*. Bulletin of the American Museum of Natural History, 66:1-250, 26 pls, 1 table (9 sections, pocketed).

Tate, G. H. H. 1935. The taxonomy of the genera of Neotropical hystricoid rodents. Bulletin of the American Museum of Natural History, 68:295-447.

Tate, G. H. H. 1936. Results of the Archbold Expeditions. No. 13. Some Muridae of the Indo-Australian region. Bulletin of the American Museum of Natural History, 72:501-728.

Tate, G. H. H. 1938. New or little known marsupials: A new species of Phascogalinae, with notes on *Acrobates pulchellus* Rothschild. Novitates Zoologicae, 41:58-60.

Tate, G. H. H. 1939. The mammals of the Guiana region. Bulletin of the American Museum of Natural History, 76:151-229.

Tate, G. H. H. 1940. Notes on the types of certain early described species of monotremes, marsupials, Muridae and bats from the Indo-Australian region. American Museum Novitates, 1061:1-10.

Tate, G. H. H. 1941. Results of the Archbold Expeditions. No. 35. A review of the genus *Hipposideros* with special reference to Indo-Australian species. Bulletin of the American Museum of Natural History, 78:353-393.

Tate, G. H. H. 1942a. Review of the vespertilionine bats, with special attention to genera and species of the Archbold collections. Bulletin of the American Museum of Natural History, 80:221-297.

Tate, G. H. H. 1942b. Results of the Archbold expeditions. No. 48. Pteropodidae (Chiroptera) of the Archbold Collections. Bulletin of the American Museum of Natural History, 80:331-347.

Tate, G. H. H. 1943. Further notes on the *Rhinolophus philippinensis* group (Chiroptera). American Museum Novitates, 1219:1-5.

Tate, G. H. H. 1945. Results of the Archbold Expeditions, No. 52. The marsupial genus *Phalanger*. American Museum Novitates, 1283:1-41.

Tate, G. H. H. 1947. Results of the Archbold Expeditions. No. 56. On the anatomy and classification of the Dasyuridae (Marsupialia). Bulletin of the American Museum of Natural History, 88:102-155.

Tate, G. H. H. 1948a. Results of the Archbold Expeditions. No. 59. Studies on the anatomy and phylogeny of the Macropodidae (Marsupialia). Bulletin of the American Museum of Natural History, 91:237-351.

Tate, G. H. H. 1948b. Results of the Archbold Expeditions. No 60. Studies in the Peramelidae (Marsupialia). Bulletin of the American Museum of Natural History, 92:317-346.

Tate, G. H. H. 1951. Results of the Archbold Expeditions. No. 65. The rodents of Australia and New Guinea. Bulletin of the American Museum of Natural History, 97:183-430.

Tate, G. H. H. 1952. Results of the Archbold Expeditions. No. 66. Mammals of Cape York Peninsula, with notes on the occurrence of rain forest in Queensland. Bulletin of the American Museum of Natural History, 98:563-616.

Tate, G. H. H., and R. Archbold. 1935. Results of the Archbold Expeditions. No. 3. Twelve apparently new forms of Muridae other than *Rattus* from the Indo-Australian region. American Museum Novitates, 803:1-9.

Tate, G. H. H., and R. Archbold. 1936. Results of the Archbold Expeditions. No. 9. A new race of *Hyosciurus*. American Museum Novitates, 846:1.

Tate, G. H. H., and R. Archbold. 1938. Results of the Archbold Expeditions. No. 18. Two new Muridae from the Western Division of Papua. American Museum Novitates, 982:1-2.

Tate, G. H. H., and R. Archbold. 1941. Results of the Archbold Expeditions. No. 31. New rodents and marsupials from New Guinea. American Museum Novitates, 1101:1-9.

Tattersall, I. 1976. Notes on the status of *Lemur macaco* and *Lemur fulvus* (Primates, Lemuriformes). Anthropological Papers of the American Museum on Natural History, 53, 2:257-261.

Tattersall, I. 1982. Primates of Madagascar. Columbia University Press, New York, 382 pp.

Tawill, S. A., and J. Niethammer. 1989. Über *Gerbillus pyramidum* (Rodentia, Gerbillidae) im Sudan. Zeitschrift für Säugetierkunde, 54:57-59.

Taylor, E. H. 1934. Philippine land mammals. Monograph of the Bureau of Science (Manila), 30:1-548.

Taylor, F. H. C., M. Fujinaga, and F. Wilke. 1955. Distribution and food habits of the fur seals of the North Pacific Ocean; Report of cooperative investigations by the Governments of Canada, Japan, and the United States of America, February-July 1952. United States Fish and Wildlife Service, Washington, D.C., 86 pp.

Taylor, J. M., and J. H. Calaby. 1988*a*. *Rattus fuscipes*. Mammalian Species, 298:1-8.

Taylor, J. M., and J. H. Calaby. 1988*b*. *Rattus lutreolus*. Mammalian Species, 299:1-7.

Taylor, J. M., and B. E. Horner. 1973. Results of the Archbold Expeditions. No. 98. Systematics of native Australian *Rattus* (Rodentia, Muridae). Bulletin of the American Museum of Natural History, 150:1-130.

Taylor, J. M., J. H. Calaby, and H. M. Van Deusen. 1982. A revision of the genus *Rattus* (Rodentia, Muridae) in the New Guinean region. Bulletin of the American Museum of Natural History, 173:177-336.

Taylor, J. M., J. H. Calaby, and S. C. Smith. 1983. Native *Rattus*, land bridges, and the Australian region. Journal of Mammalogy, 64:463-475.

Taylor, J. R. E. 1998. Evolution of energetic strategies in shrews. Pp. 309-346, *in* Evolution of shrews (J. M. Wójcik and M. Wolsan, eds.). Mammal Research Institute, Polish Academy of Sciences, Bialowieza, 458 pp.

Taylor, M. E. 1972. *Ichneumia albicauda*. Mammalian Species, 12:1-4.

Taylor, M. E. 1975. *Herpestes sanguineus*. Mammalian Species, 65:1-5.

Taylor, M. E. 1987. *Bdeogale crassicauda*. Mammalian Species, 294:1-4.

Taylor, M. E. 1989. Craniometric analysis of the African mongooses in the subgenus *Galerella*. Abstracts of papers and posters, Fifth International Theriological Congress, Rome, I:473.

Taylor, M. E., and C. A. Goldman. 1993. The taxonomic status of the African mongooses, *Herpestes sanguineus, H. nigratus, H. pulverulentus* and *H. ochraceus* (Carnivora: Viverridae). Mammalia, 57:375-391.

Taylor, M. E., and J. Matheson. 1999. A craniometric comparison of the African and Asian mongooses in the genus *Herpestes* (Carnivora: Herpestidae). Mammalia, 63:449-464.

Taylor, P. J. 1994. New distribution records for six small mammal species in Natal, with notes on their taxonomy and ecology. Durban Museum Novitates, 19:59-66.

Taylor, P. J. 1998. The smaller mammals of KwaZulu-Natal. University of Natal Press, Pietermaritzburg, 139 pp.

Taylor, P. J. 1999. Problems with the identification of southern African *Chaerephon* (Molossidae), and the possibility of a cryptic species from South Africa and Swaziland. Acta Chiropterologica, 1:191-200.

Taylor, P. J. 2000*a*. Bats of Southern Africa: Guide to biology, identification, and conservation. University of Natal, Pietermaritzburg, 206 pp.

Taylor, P. J. 2000*b*. Patterns of chromosomal variation in southern African rodents. Journal of Mammalogy, 81:317-331.

Taylor, P. J., and G. Contrafatto. 1997. Mandible shape and size in three species of small musk shrews (*Crocidura* Wagler, 1832) from southern Africa. Mammalia, 60:753-765.

Taylor, P. J., and A. Kumirai. 2001. Craniometric relationships between the Southern African vlei rat, *Otomys irroratus* (Rodentia, Muridae, Otomyinae) and allied species from north of the Zambezi River. Pp. 161-181, *in* African small mammals (C. Denys, L. Granjon, and A. Poulet, eds.). IRD Éditions, Collection Colloques et Séminaires, Paris, 570 pp.

Taylor, P. J., and J. Meester. 1993. *Cynictis penicillata*. Mammalian Species, 432:1-7.

Taylor, P. J., and Van der Merwe, M. 1998. Taxonomic notes on dark-winged house bats of the genus *Scotoecus* Thomas 1901, in Malawi. Durban Museum Novitates, 23:64-66.

Taylor, P. J., G. K. Campbell, J. Meester, K. Willan, and D. Van Dyk. 1989. Genetic variation in the African rodent subfamily Otomyinae (Muridae). 1. Allozyme divergence among four species. South African Journal of Science, 85:257-262.

Taylor, P. J., G. K. Campbell, J. A. J. Meester, and D. van Dyk. 1991. A study of allozyme evolution in African mongooses (Viverridae: Herpestinae). Zeitschrift für Säugetierkunde, 56(3):135-145.

Taylor, P. J., G. K. Campbell, D. Van Dyk, J. Meester, and K. Willan. 1992. Genetic variation in the African vlei rat *Otomys irroratus* (Muridae: Otomyinae). Israel Journal of Zoology, 38:293-305.

Taylor, P. J., J. Meester, and T. Kearney. 1993. The taxonomic status of Saunders' vlei rat *Otomys saundersiae* Roberts (Rodentia: Muridae: Otomyinae). Journal of African Zoology, 107:571-576.

Taylor, P. J., E. J. Richardson, J. Meester, and L. Wingate. 1994*a*. New distribution records for six small mammal species in Natal, with notes on their taxonomy and ecology. Durban Museum Novitates, 19:59-66.

Taylor, P. J., G. Contrafatto, and K. Willan. 1994*b*. Climatic correlates of chromosomal variation in the African vlei rat, *Otomys irroratus*. Mammalia, 58:623-634.

Taylor, P. J., I. L. Rautenbach, D. Gordon, K. Sink, and P. Lotter. 1995. Diagnostic morphometrics and southern African distribution of two sibling species of tree rat, *Thallomys paedulcus* and *Thallomys nigricauda* (Rodentia: Muridae). Durban Museum Novitates, 20:49-62.

Taylor, W. P. 1918. Revision of the rodent genus *Aplodontia*. University of California Publications in Zoology, 17:435-504.

Tchernov, E. 1968. Succession of rodent faunas during the Upper Pleistocene of Israel. Mammalia Depicta, 3:152 pp.

Tchernov, E. 1979. Polymorphism, size trends and Pleistocene paleoclimatic response of the subgenus *Sylvaemus* (Mammalia: Rodentia) in Israel. Israel Journal of Zoology, 28:131-159.

Tchernov, E. 1986. The rodents and lagomorphs from 'Ubeidiya Formation: Systematics, paleoecology and biogeography. Pp. 235-350, *in* Les mammifères du Pléistocène Inférieur de la Vallée du Jourdain a Oubeidiyeh (E. Tchernov, ed.). Mémoires et Travaux du Centre de Recherche Francais de Jérusalem, 5:1-405.

Tchernov, E. 1988. The paleobiogeographical history of the Southern Levant. Pp. 159-250, *in* The Zoogeography of Israel (Y. Yom-Tov and E. Tchernov, eds.). Dr. W. Junk, the Hague, 600 pp.

Tchernov, E. 1992. The Afro-Arabian component in the Levantine mammalian fauna–a short biogeographic review. Israel Journal of Zoology, 38:155-192.

Tchernov, E. 1994. New comments on the biostratigraphy of the Middle and Upper Pleistocene of the Southern Levant. Pp. 333-350, *in* Late Quaternary chronology and paleoclimates of the Eastern Mediterranean (O. Bar-Yosef and R. S. Kra, eds.). Radiocarbon, 1994, 371 pp.

Tchernov, E. 1996. Rodent faunas, chronostratigraphy and paleobiogeography of the southern Levant during the Quaternary. Acta Zoologica Cracoviensia, 39(1):513-530.

Tedford, R. H. 1976. Relationship of pinnipeds to other carnivores (Mammalia). Systematic Zoology, 25:363-374.

Tedford, R. H., B. E. Taylor, and X. Wang. 1995. Phylogeny of the Caninae (Carnivora: Canidae): The living taxa. American Museum Novitates, 3146:1-37.

Teeling, E. C., M. Scully, D. J. Kao, M. L. Romagnoli, M. S. Springer, and M. J. Stanhope. 2000. Molecular evidence regarding the origin of echolocation and flight in bats. Nature, 403:188-192.

Teeling, E. C., O. Madsen, R. A. Van Den Bussche, W. W. de Jong, M. J. Stanhope, and M. S. Springer. 2002. Microbat paraphyly and the convergent evolution of a key innovation in Old World rhinolophid microbats. Procedings of the National Academy of Sciences, 99:1431-1436.

Teeling, E. C., O. Madsen, W. J. Murphy, M. S. Springer, and S. J. O'Brien. 2003. Nuclear gene sequences confirm an ancient link between New Zealand's short-tailed bat and South American Noctilionoid bats. Molecular Phylogenetics and Evolution, 28(2):308-319.

Tegelstrom, H., and M. Jaarola. 1989. Genetic divergence in mitochondrial DNA between the wood mouse (*Apodemus sylvaticus*) and the yellow necked mouse (*A. flavicollis*). Hereditas, 111:49-60.

Teilhard de Chardin, P. 1926. Mammifères tertiares de Chine et Mongolie. Annales de Paleontologie, 5:48-51.

Teilhard de Chardin, P. 1942. New rodents of the Pliocene and Lower Pleistocene of North China. Institut de Geo-Biologie de Pekin, 9:101 pp.

Teilhard de Chardin, P., and C. C. Young. 1936. On the mammal remains from the archeological site of Anyang. Palaeontologia Sinica, ser. C, 12(1):1-61.

Tellez, G., and J. Ortega. 1999. *Musonycteris harrisoni*. Mammalian Species, 622:1-3.

Tello, J. 1979. Mamíferos de Venezuela. Caracas, 192 pp.

Temminck, C. J. 1827 [1824]-1841. Monographies de Mammalogie, ou description de quelques genres de mammifères, dont les espèces ont été observées dans les différens musées de l'Europe. C. C. Vander Hoek, Leiden, 392 pp.

Temminck, C. J. 1841. Quatorzime monographie. Sur les genres taphien—queue-en-fourreau—queue-cache—et queue-bivalve. Pp. 273-304, *in* Monographies de mammalogie ou description de quelques genres de mammifres sont les espces ont été observées dans les différens musées de l'Europe. Bertrand. Leiden, 2:1-392.

Temminck, C. J. 1847. Coup-d'oeil général sur les possessions néerlandaises dans l'Inde archipélagique. Vol. 2. A. Arnz and Company, Leiden, 471 pp.

Tener, J. S. 1965. Muskoxen in Canada: A biological and taxonomic review. Canada Wildlife Services Monograph, 2:1-166.

Terashima, M., A. Suyanto, K. Tsuchiya, K. Moriwaki, M.-L. Jin, and H. Suzuki. 2003. Geographic variation of *Mus caroli* from East and Southeast Asia based on mitochondrial cytochrome *b* gene sequences. Mammal Study, 28:67-72.

Teta, P., A. Andrade, and U. F. J. Pardiñas. 2002. Novedosos registros de roedores sigmodontinos (Rodentia: Muridae) en La Patagonia central Argentina. Mastozoología Neotropical, 9:79-84.

Thabah, A., and P. J. J. Bates. 2002. Recent record of *Otomops wroughtoni* (Thomas, 1913) (Chiroptera: Molossidae) from Meghalaya, North-East India. Acta Zoologica Academiae Scientiarum Hungaricae, 48:251-253.

Thaeler, C. S., Jr. 1968a. An analysis of the distribution of pocket gopher species in northeastern California (genus *Thomomys*). University of California Publications in Zoology, 86:1-46.

Thaeler, C. S., Jr. 1968b. An analysis of three hybrid populations of pocket gophers (genus *Thomomys*). Evolution, 22:543-555.

Thaeler, C. S., Jr. 1972. Taxonomic status of the pocket gophers, *Thomomys idahoensis* and *Thomomys pygmaeus* (Rodentia, Geomyidae). Journal of Mammalogy, 53:417-428.

Thaeler, C. S., Jr. 1977. Taxonomic status of *Thomomys talpoides confinus*. Murrelet, 58:49-50.

Thaeler, C. S., Jr. 1980. Chromosome numbers and systematic relations in the genus *Thomomys* (Rodentia: Geomyidae). Journal of Mammalogy, 61:414-422.

Thaeler, C. S., Jr. 1985. Chromosomal variation in the *Thomomys talpoides* complex. Acta Zoologica Fennica, 170:15-18.

Thaeler, C. S., Jr., and L. L. Hinsley. 1979. *Thomomys clusius*, a rediscovered species of pocket gopher. Journal of Mammalogy, 60:480-488.

Thaler, L. 1966. Les rongeurs fossiles du Bas-Languedoc dans leur rapports avec l'histoire des faunes et la stratigraphie du Tertiaire d'Europe. Mémoires du Muséum National d'Histoire Naturelle, Série C, 17:1-295.

Thaler, L. 1986. Origin and evolution of mice: An appraisal of fossil evidence and morphological traits. Current Topics in Microbiology and Immunology, 127:3-11.

Thalmann, U., and T. Geissmann. 2000. Distribution and geographic variation in the western woolly lemur (*Avahi occidentalis*) with description of a new species (*A. unicolor*). International Journal of Primatology, 21:915-941.

Theiler, G. R., and A. Blanco. 1996. Patterns of evolution in *Graomys griseoflavus* (Rodentia, Muridae). II. Reproductive isolation between cytotypes. Journal of Mammalogy, 77:776-784.

Theiler, G. R., and C. N. Gardenal. 1994. Patterns of evolution in *Graomys griseoflavus* (Rodentia, Cricetidae). I. Protein polymorphism in population with different chromoseome numbers. Hereditas, 120:225-229.

Theiler, G. R., C. N. Gardenal, and A. Blanco. 1999. Patterns of evolution in *Graomys griseoflavus* (Rodentia, Muridae). IV. A case of rapid speciation. Journal of Evolutionary Biology, 12:970-979.

Thelma, B. K., and S. R. V. Rao. 1982. In vivo sister chromatid exchanges in the euchromatin and constitutive heterochromatin of the Indian mole rat, *Nesokia indica*. Cytogenetics and Cell Genetics, 33:319-326.

Thenius, E. 1953. Zur Analyse des Gesbisses des Eisbaren, *Ursus* (*Thalarctos*) *maritimus* Phipps, 1774. Säugetierkundliche Mitteilungen, 1:1-7.

Thenius, E. 1972. Grundzüge der Verbreitungsgeschichte der Säugetiere. Gustav Fischer, Stuttgart, 321 pp.

Thenius, E. 1976. Zur stammesgeschichtlichen Herkunft von *Tremarctos* (Ursidae, Mammalia). Zeitschrift für Säugetierkunde, 41:109-114.

Thenius, E. 1979. Zur systematischen und phylogenetischen Stellung des Bambusbären: *Ailuropoda melanoleuca* David (Carnivora, Mammalia). Zeitschrift für Säugetierkunde, 44(5):286-305.

Thenius, E. 1981. Bemerkungen zur taxonomischen und stammesgeschichtlichen Position der Gibbons. Zeitschrift für Säugetierkunde, 46:232-241.

Thissen, J. B. M., and H. Hollander. 1996. Mammals in the Netherlands. Hystrix, n.s., 8(1-2):97-105.

Thomas, B. 1973. Evolutionary implications of karyotypic variation in some insular *Peromyscus* from British Columbia, Canada. Cytologia, 38:485-495.

Thomas, H. 1994. Anatomie crânniene et relations phylogénétiques du noueau bovidé (*Pseudoryx nghetinhensis*) découvert dans la cordillère annamatique au Vietnam. Mammalia, 58:453-481.

Thomas, H. 2001. The enigmatic new Indochinese bovid *Pseudonovibos spiralis*: An extraordinary forgery. Comptes Rendus de l'Académie des Sciences. Série III, Sciences de la Vie, 324:81-86.

Thomas, H. H., and T. L. Best. 1994a. *Sylvilagus mansuetus*. Mammalian Species, 464:1-2.

Thomas, H. H., and T. L. Best. 1994b. *Lepus insularis*. Mammalian Species, 465:1-3.

Thomas, J., V. Pastukhov, R. Elsner, and E. Petrov. 1982. *Phoca sibirica*. Mammalian Species, 188:1-6.

Thomas, N. M. 2000. Morphological and mitochondrial-DNA variation in *Rhinolophus rouxii* (Chiroptera). Bonner Zoologische Beiträge, 49:1-18.

Thomas, N. M., D. L. Harrison, and P. J. J. Bates. 1994. A study of the baculum in the genus *Nycteris* (Mammalia, Chiroptera, Nycteridae) with consideration of its taxonomic importance. Bonner Zoologische Beiträge, 45:17-31.

Thomas, O. 1881. Description of a new species of *Mus* from Southern India. Annals and Magazine of Natural History, ser. 5, 7:24.

Thomas, O. 1882. On African mongooses. Proceedings of the Zoological Society of London, 1882:59-93.

Thomas, O. 1887a. Description of a second species of rabbit-bandicoot (*Peragale*). Annals and Magazine of Natural History, ser. 5, 19:397-399.

Thomas, O. 1887b. On the small mammals collected in Demerara by Mr. W. L. Sclater. Proceedings of the Zoological Society of London, 1887:150-153.

Thomas, O. 1887c. Report on a zoological collection made by the officers of H. M. S. 'Flying Fish' at Christmas Island, Indian Ocean. I. Mammalia. Proceedings of the Zoological Society of London, 1887:511-514.

Thomas, O. 1888a. Catalogue of the Marsupialia and Monotremata in the collection of the British Museum (Natural History). British Museum (Natural History), London, 401 pp., 33 pls.

Thomas, O. 1888b [1889]. On the mammals of Christmas Island. Proceedings of the Zoological Society of London, 1888:532-534.

Thomas, O. 1888c. On a new and interesting annectant genus of Muridae, with remarks on the relations of the Old- and New-World members of the family. Proceedings of the Zoological Society of London, 1888:130-135.

Thomas, O. 1888d. On a collection of mammals obtained by Emin Pasha in equatorial Africa, and presented by him to the Natural History Museum. Proceedings of the Zoological Society of London, 1888:3-17.

Thomas, O. 1890. On a collection of mammals obtained by Dr. Emin Pasha in central and eastern Africa. Proceedings of the Zoological Society of London, 1890:443-450.

Thomas, O. 1892a. On some new Mammalia from the East-Indian Archipelago. Annals and Magazine of Natural History, ser. 6, 9:250-254.

Thomas, O. 1892b. On some mammals from Mount Dulit, North Borneo. Proceedings of the Zoological Society of London, 1892:221-227.

Thomas, O. 1892c. On the probable identity of certain specimens formerly in the Lidth de Jeude collection and now in the British Museum, with those figured by Albert Seba in his 'Thesaurus' of 1734. Proceedings of the Zoological Society of London, 1892:309-318.

Thomas, O. 1893. On some new Bornean mammalia. Annals and Magazine of Natural History, ser. 6, 11:341-347.

Thomas, O. 1894. On the mammals of Nyasaland: Third contribution. Proceedings of the Zoological Society of London, 1894:136-142.

Thomas, O. 1895. Preliminary diagnoses of new mammals from northern Luzon, collected by Mr. John Whitehead. Annals and Magazine of Natural History, ser. 6, 16:160-164.

Thomas, O. 1896 [1897]. On the genera of rodents: An attempt to bring up to date the current arrangement of the order. Proceedings of the Zoological Society of London, 1896:1012-1028.

Thomas, O. 1897a. On the mammals collected in British New Guinea by Dr. Lamberto Loria. Annali del Museo Civico di Storia Naturale di Genova, ser. 2, 18:1-19.

Thomas, O. 1897b [1898]. On the Mammals obtained by Mr. A. Whyte in Nyasaland, and presented to the British Museum by Sir H. H. Johnston, K. C. B.; being a fifth contribution to the mammal-fauna of Nyasaland. Proceedings of the Zoological Society of London, 1897:924-938.

Thomas, O. 1898a. On indigenous Muridae in the West Indies; with the description of a new Mexican *Oryzomys*. Annals and Magazine of Natural History, ser. 7, 1:176-180.

Thomas, O. 1898b. On the mammals collected by Mr. John Whitehead during his recent expedition to the Philippines with field notes by the collector. Transactions of the Zoological Society of London, 14:377-414.

Thomas, O. 1900a. The geographical races of the Tayra (*Galictis barbara*), with notes on abnormally coloured individuals. Annals and Magazine of Natural History, ser. 7, 5:145-148.

Thomas, O. 1900b. List of mammals obtained by Mr. H. J. Mackinder during his recent expedition to Mount Kenya, British East Africa. Proceedings of the Zoological Society of London, 1900:173-180.

Thomas, O. 1901a. The generic names *Myrmecophaga* and *Didelphis*. American Naturalist, 35:143-145.

Thomas, O. 1901*b*. On mammals obtained by Mr. Alphonse Robert on the Rio Jordao, Minas Geraes. Annals and Magazine of Natural History, ser. 7, 8:526-536.

Thomas, O. 1902*a* [1903]. On a collection of mammals from Abyssinia including some from Lake Tsana collected by Mr. Edward Degen. Proceedings of the Zoological Society of London, 1902:308-316.

Thomas, O. 1902*b*. On the geographical races of the kinkajou. Annals and Magazine of Natural History, ser. 7, 9:266-270.

Thomas, O. 1902*c*. On some new forms of *Otomys*. Annals and Magazine of Natural History, ser. 7, 10:311-314.

Thomas, O. 1903. Notes on Neotropical mammals of the genera *Felis*, *Hapale*, *Oryzomys*, *Akodon*, and *Ctenomys*, with descriptions of new species. Annals and Magazine of Natural History, ser. 7, 12:234-243.

Thomas, O. 1904*a* [1905]. On *Hylochoerus*, the forest pig of Central Africa. Proceedings of the Zoological Society of London, 1904(2):193-199.

Thomas, O. 1904*b*. New *Callithrix*, *Midas*, *Felis*, *Rhipidomys*, and *Proechimys* from Brazil. Annals and Magazine of Natural History, ser. 7, 14:188-196.

Thomas, O. 1904c. On mammals from northern Angola collected by Dr. W. J. Ansorge. Annals and Magazine of Natural History, ser. 7, 13:405-421.

Thomas, O. 1905*a* [1906]. The Duke of Bedford's zoological exploration in eastern Asia. — 1. List of mammals obtained by Mr. M. P. Anderson in Japan. Proceedings of the Zoological Society of London, 1905(2):331-363.

Thomas, O. 1905*b*. On new Japanese mammals (Insectivora, Rodentia). Annals and Magazine of Natural History, ser. 7, 15:487-495.

Thomas, O. 1906*a*. Descriptions of new mammals from Mount Ruwenzori. Annals and Magazine of Natural History, ser. 7, 18:136-147.

Thomas, O. 1906*b*. Two new genera of small mammals discovered by Mrs. Holms-Tarn in British East Africa. Annals and Magazine of Natural History, ser. 7, 18:222-226.

Thomas, O. 1906*c*. New mammals collected in north-east Africa by Mr. Zaphiro, and presented to the British Museum by W. N. McMilan, Esq. Annals and Magazine of Natural History, ser. 7, 18:300-306.

Thomas, O. 1906*d*. Notes on South American rodents. II. On the allocation of certain species hitherto referred respectively to *Oryzomys*, *Thomasomys*, and *Rhipidomys*. Annals and Magazine of Natural History, ser. 7, 18:442-448.

Thomas, O. 1907*a*. On further new mammals obtained by the Ruwenzori Expedition. Annals and Magazine of Natural History, ser. 7, 19:118-123.

Thomas, O. 1907*b*. On Neotropical mammals of the genera *Callicebus*, *Reithrodontomys*, *Ctenomys*, *Dasypus*, and *Marmosa*. Annals and Magazine of Natural History, ser. 7, 20:161-168.

Thomas, O. 1907*c*. The Duke of Bedford's Zoological Exploration in Eastern Asia.–III. On mammals obtained by Mr. M. P. Anderson in the Philippine Islands. Proceeding of the Zoological Society of London, 1907:140-142.

Thomas, O. 1908*a*. The nomenclature of the flying lemurs. Annals and Magazine of Natural History, ser. 8, 1:252-255.

Thomas, O. 1908*b*. The Duke of Bedford's zoological exploration in eastern Asia—IX. List of mammals from the Mongolian Plateau. Proceedings of the Zoological Society of London, 1908(1):104-110.

Thomas, O. 1909*a*. Mr. O. Thomas on the generic arrangement of the African squirrels. Annals and Magazine of Natural History, ser. 8, 3:467-475.

Thomas, O. 1909*b*. New African mammals. Annals and Magazine of Natural History, ser. 8, 4:542-549.

Thomas, O. 1910*a*. New African mammals. Annals and Magazine of Natural History, ser. 8, 5:83-92.

Thomas, O. 1910*b*. Further new African Mammalia. Annals and Magazine of Natural History, ser. 8, 5:282-285.

Thomas, O. 1910*c*. Notes on African rodents. Annals and Magazine of Natural History, ser. 8, 6:221-226.

Thomas, O. 1910*d*. New African mammals in the British Museum. Annals and Magazine of Natural History, ser. 8, 6:426-432.

Thomas, O. 1910*e*. New genera of Australasian Muridae. Annals and Magazine of Natural History, ser. 8, 6:506-508.

Thomas, O. 1911*a*. The mammals of the tenth edition of Linnaeus; an attempt to fix the types of the genera and the exact bases and localities of the species. Proceedings of the Zoological Society of London, 1911:120-158.

Thomas, O. 1911*b*. On new African Muridae. Annals and Magazine of Natural History, ser. 8, 7:378-383.

Thomas, O. 1911*c*. Mammals collected in the provinces of Kan-su and Sze-chwan, western China, by Mr. Malcolm Anderson, for the Duke of Bedford's exploration of Eastern Asia. Abstracts of the Proceedings of the Zoological Society of London, 90:3-5.

Thomas, O. 1911*d*. The Duke of Bedford's zoological exploration of Eastern Asia.–XIII. On mammals from the provinces of Kan-su and Sze-chwan, western China. Proceedings of the Zoological Society of London, 1911:158-180.

Thomas, O. 1911*e*. The Duke of Bedford's zoological exploration of Eastern Asia.–XV. On mammals from the provinces of Szechwan and Yunnan, western China. Proceedings of the Zoological Society of London, 1911:127-141.

Thomas, O. 1912*a*. Exhibition of skin and skull of a viverrine carnivore from Tonkin. Proceedings of the Zoological Society of London, 1912:17-18.

Thomas, O. 1912*b*. Revised determinations of two Far-Eastern species of *Myospalax*. Annals and Magazine of Natural History, ser. 8, 9:93-95.

Thomas, O. 1912*c*. On a collection of small mammals from the Tsin-ling Mountains, Central China, presented by Mr. G. Fenwick Owen to the National Museum. Annals and Magazine of Natural History, ser. 8, 10:395-403.

Thomas, O. 1912*d*. *Diplogale*. Abstracts of the Proceedings of the Zoological Society of London, 1912:18.

Thomas, O. 1912*e*. Three small mammals from South America. Annals and Magazine of Natural History, ser. 8, 9:408-410.

Thomas, O. 1913*a*. Ernst Hartert's expedition to the Central Western Sahara. Mammals. Novitates Zoologicae, Triung, 20:28-33.

Thomas, O. 1913*b*. On new mammals obtained by the Utakwa Expedition to Dutch New Guinea. Annals and Magazine of Natural History, ser. 8, 12:205-212.

Thomas, O. 1914*a*. On various South American mammals. Annals and Magazine of Natural History, ser. 8, 13:345-363.

Thomas, O. 1914*b*. New Asiatic and Australasian bats, and a new bandicoot. Annals and Magazine of Natural History, ser. 8, 13:439-444.

Thomas, O. 1914*c*. Second list of small mammals from Western Yunnan collected by Mr. F. Kingdon Ward. Annals and Magazine of Natural History, ser. 8, 14:472-475.

Thomas, O. 1915. List of mammals (exclusive of Ungulata) collected on the Upper Congo by Dr. Christy for the Congo Museum, Tervueren. Annals and Magazine of Natural History, ser. 8, 16:465-481.

Thomas, O. 1916*a*. On the rats usually included in the genus *Arvicanthis*. Annals and Magazine of Natural History, ser. 8, 18:67-70.

Thomas, O. 1916*b*. Three new African mice of the genus *Dendromus*. Annals and Magazine of Natural History, ser. 8, 18:241-143.

Thomas, O. 1916*c*. The grouping of the South American Muridae commonly referred to *Akodon*. Annals and Magazine of Natural History, ser. 8, 18:336-340.

Thomas, O. 1916*d*. The bandicoot of Mount Popa, and its allies. Journal of the Bombay Natural History Society, 24:640-643.

Thomas, O. 1916*e*. On the generic names *Rattus* and *Phyllomys*. Annals and Magazine of Natural History, ser. 8, 18:240.

Thomas, O. 1916*f*. Some notes on the Echimyinae. Annals and Magazine of Natural History, ser. 8, 18:294-301.

Thomas, O. 1916*g*. Two new Argentine rodents, with a new subgenus of *Ctenomys*. Annals and Magazine of Natural History, ser. 8, 18:303-306.

Thomas, O. 1917*a*. A new species of *Aconaemys* from southern Chile. Annals and Magazine of Natural History, ser. 8, 19:281-282.

Thomas, O. 1917*b*. The geographical races of *Galago crassicaudatus*. Annals and Magazine of Natural History, ser. 8, 20:47-50.

Thomas, O. 1917*c*. On the arrangement of the South American rats allied to *Oryzomys* and *Rhipidomys*. Annals and Magazine of Natural History, ser. 8, 20:192-198.

Thomas, O. 1917*d*. A new vole from Palestine. Annals and Magazine of Natural History, ser. 8, 19:450-451.

Thomas, O. 1918*a*. On the arrangement of the small Tenrecidae hitherto referred to *Oryzorictes* and *Microgale*. Annals and Magazine of Natural History, ser. 9, 1:302-307.

Thomas, O. 1918*b*. A revised classification of the Otomyinae, with descriptions of new genera and species. Annals and Magazine of Natural History, ser. 9, 2:203-211.

Thomas, O. 1919. On small mammals from "Otro Cerro" north-eastern Rioja, collected by Sr. E. Budin. Annals and Magazine of Natural History, ser. 9, 3:489-500.

Thomas, O. 1920*a*. The generic positions of "*Mus*" *nigricauda*, Thos., and *woosnami*, Schwann. Annals and Magazine of Natural History, ser. 9, 5:140-142.

Thomas, O. 1920*b*. Report on the Mammalia collected by Mr. Edmund Heller during the Peruvian Expedition of 1915 under the auspices of Yale University and the National Geographic Society. Proceedings of the United States National Museum, 58:217-249.

Thomas, O. 1920c. On small mammals from Ceram. Annals and Magazine of Natural History, ser. 9, 6:422-431.

Thomas, O. 1921a. On a new genus and species of shrew, and some new Muride from the East-Indian Archipelago. Annals and Magazine of Natural History, ser. 9, 7:243-249.

Thomas, O. 1921b. On the cavies of the genus *Caviella*. Annals and Magazine of Natural History, ser. 9, 7:445-448.

Thomas, O. 1921c. Geographical races of *Herpestes brachyurus*, Gray. Annals and Magazine of Natural History, ser. 9, 8:134-136.

Thomas, O. 1921d. New *Rhipidomys*, *Akodon*, *Ctenomys*, and *Marmosa* from the Sierra Santa Barbara, S. E. Jujuy. Annals and Magazine of Natural History, ser. 9, 7:183-187.

Thomas, O. 1921e. On mammals from the province of San Juan, western Argentina. Annals and Magazine of Natural History, ser. 9, 8:214-221.

Thomas, O. 1921f. On the spiny rats of the *Proechimys* group from Southeastern Brazil. Annals and Magazine of Natural History, ser. 9, 8:140-143.

Thomas, O. 1921g. On small mammals from the Kachin Province, Northern Burma. Journal of the Bombay Natural History Society, 27:499-505.

Thomas, O. 1921h. *Microtus irani*. Pp. 580-581, *in* Report on a collection of mammals made by Col. J. E. B. Hotson in Shiraz, Persia (R. E. Cheesman, author). Journal of the Bombay Natural History Society, 27(3):573-581.

Thomas, O. 1921i. On a collection of rats and shrews from the Dutch East Indian Islands. Treubia, 2:109-114.

Thomas, O. 1922a. New mammals from New Guinea and neighboring islands. Annals and Magazine of Natural History, ser. 9, 9:261-265.

Thomas, O. 1922b. On mammals from New Guinea obtained by the Dutch Scientific expeditions of recent years. Nova Guinea (Zoologie), 13:723-740.

Thomas, O. 1923. Scientific results from the mammal survey. No. XXX. The mongooses of the *Herpestes smithii* group. Journal of the Bombay Natural History Society, 28(1):23-26.

Thomas, O. 1924a. On some Ceylon mammals. Annals and Magazine of Natural History, ser. 9, 13:239-242.

Thomas, O. 1924b. The geographic races of *Oryzomys ratticeps*. Annals and Magazine of Natural History, 14:143-144.

Thomas, O. 1925. The Spedan Lewis South American exploration. I. On mammals from southern Bolivia. Annals and Magazine of Natural History, ser. 9, 25:575-582.

Thomas, O. 1926a. The Godman-Thomas expedition to Peru. I. On mammals collected by Mr. R. W. Hendee near Lake Junin. Annals and Magazine of Natural History, ser. 9, 17:313-318.

Thomas, O. 1926b. The Spedan Lewis South American exploration. II. On mammals collected in the Tarija Department, southern Bolivia. Annals and Magazine of Natural History, ser. 9, 17:318-328.

Thomas, O. 1926c. The Godman-Thomas expedition to Peru. II. On mammals collected by Mr. Hendee in north Peru between Pacasmayo and Chachapoyas. Annals and Magazine of Natural History, ser. 9, 17:610-616.

Thomas, O. 1926d. On some mammals from the middle Amazons. Annals and Magazine of Natural History, ser. 9, 17:635-639.

Thomas, O. 1926e. On mammals from Ovamboland and the Cunene River, obtained during Capt. Shortridge's Third Percy Sladen and Kaffrarian Museum Expedition into South-West Africa. Proceedings of the Zoological Society of London, 1926(1):285-312.

Thomas, O. 1927. On a further collection of mammals made by Sr. E. Budin in Neuquen, Patagonia. Annals and Magazine of Natural History, ser. 9, 19:650-658.

Thomas, O. 1928a. Some rarities from Abyssinia, with the description of a new mole-rat (*Tachyoryctes*), and a new *Arvicanthis*. Annals and Magazine of Natural History, ser. 10, 1:302-304.

Thomas, O. 1928b. The Godman-Thomas expedition to Peru.—VIII. On mammals obtained by Mr. Hendee at Pebas and Iquitos, upper Amazons. Annals and Magazine of Natural History, ser. 10, 2:285-294.

Thomas, O. 1929. The mammals of Senor Budin's Patagonian expedition. Annals and Magazine of Natural History, ser. 10, 4:35-45.

Thomas, O., and M. A. C. Hinton. 1920. On the group of African zorils represented by *Ictonyx libyca*. Annals and Magazine of Natural History, ser. 9, 5:367-369.

Thomas, O., and M. A. C. Hinton. 1923a. On the mammals obtained in Darfur by the Lynes-Lowe Expedition. Proceedings of the Zoological Society of London, 1923:247-271.

Thomas, O., and M. A. C. Hinton. 1923b. On mammals collected by Captain Shortridge during the Percy Sladen and Kaffrarian Museum Expeditions to the Orange River. Proceedings of the General Meetings for Scientific Business of the Zoological Society of London, 1923:483-499.

Thomas, O., and M. A. C. Hinton. 1925. On mammals collected in 1923 by Captain G. C. Shortridge during the Percy Sladen and Kaffrarian Museum Expedition to South-West Africa. Proceedings of the Zoological Society of London, 1925:221-246.

Thomas, O., and H. Schwann. 1904. On a collection of mammals from British Namaqualand, presented to the National Museum by Mr. C. D. Rudd. Proceedings of the Zoological Society of London, 1904(1):171-183.

Thomas, O., and H. Schwann. 1905. The Rudd exploration of South Africa.—II. List of mammals from the Wakkerstroom District, Southeastern Transvaal. Proceedings of the Zoological Society of London, 1905:129-138.

Thomas, O., and R. C. Wroughton. 1905. On a second collection of mammals obtained by Dr. W. J. Ansorge in Angola. Annals and Magazine of Natural History, ser. 7, 16:169-178.

Thomas, O., and R. C. Wroughton. 1908. The Rudd exploration of S. Africa.—X. List of mammals collected by Mr. Grant near Tette, Zambesia. Proceedings of the General Meetings for Scientific Business of the Zoological Society of London, 1908:535-552.

Thomson, C. E. 1982. *Myotis sodalis*. Mammalian Species, 163:1-5.

Thonglongya, K. 1973. First record of *Rhinolophus paradoxolophus* (Bourret, 1951) from Thailand, with the description of a new species of the *Rhinolophus philippinensis* group (Chiroptera: Rhinolophidae). Mammalia, 37:587-597.

Thorington, R. W., Jr. 1984. Flying squirrels are monophyletic. Science, 225:1048-1050.

Thorington, R. W., Jr. 1988. Taxonomic status of *Saguinus tripartitus* (Milne-Edwards, 1878). American Journal of Primatology, 15:367-371.

Thorington, R. W., Jr., and C. P. Groves. 1970. An annotated classification of the Cercopithecoidea. Pp. 629-647, *in* Old World Monkeys (J. R. Napier and P. H. Napier, eds.). Academic Press, London, 660 pp.

Thorington, R. W., Jr., A. L. Musante, C. G. Anderson, and K. Darrow. 1996. Validity of three genera of flying squirrels: *Eoglaucomys*, *Glaucomys*, and *Hylopetes*. Journal of Mammalogy, 77(1):69-83.

Thorington, R. W., Jr., K. Darrow, and C. G. Anderson. 1998. Wingtip anatomy and aerodynamics in flying squirrels. Journal of Mammalogy, 79:245-250.

Thorington, R. W., Jr., D. Pitassy, and S. A. Jansa. 2002. Phylogenies of flying squirrels (Pteromyinae). Journal of Mammalian Evolution, 9:99-135.

Thornton, W. A., and G. C. Creel. 1975. The taxonomic status of kit foxes. Texas Journal of Science, 26:127-136.

Thouless, C. R., and K. Al Bassri. 1991. Taxonomic status of the Farasan Island gazelle. Journal of Zoology, London, 223:151-159.

Thys van den Audenaerde, D. F. E. 1977. Description of a new monkey skin from East-Central Zaire as a probably new monkey species (Mammalia, Cercopithecidae). Revue de Zoologie Africaine, 91:1000-1010.

Tidemann, C. R. 1986. Morphological variation in Australian and island populations of Gould's wattled bat *Chalinolobus gouldi* (Gray)(Chiroptera: Vespertilionidae). Australian Journal of Zoology, 34:503-514.

Tiemann-Boege, I., C. W. Kilpatrick, D. J. Schmidly, and R. D. Bradley. 2000. Molecular phylogenetics of the *Peromyscus boylii* species group (Rodentia: Muridae) based on mitochondrial cytochrome *b* sequences. Molecular Phylogenetics and Evolution, 16:366-378.

Tikhonov, I. A., G. N. Tikhonova, and L. V. Polyakova. 1998. [Sibling species of *Microtus arvalis* and *Microtus rossiaemeridionalis* in north-eastern Moscow District.] Zoologischeskii Zhurnal, 77(1):95-100 (in Russian with English summary).

Tilesius, W. G. 1850. Gliriumspecies in Bavaria nonnullae. Isis, 2:27-29.

Timm, R. M. 1982. *Ectophylla alba*. Mammalian Species, 166:1-4.

Timm, R. M. 1985. *Artibeus phaeotis*. Mammalian Species, 235:1-6.

Timm, R. M., and J. H. Brandt. 2001. *Pseudonovibos spiralis* (Artiodactyla: Bovidae): New information on this enigmatic South-east Asian ox. Journal of Zoology, London, 253:157-166.

Timm, R. M., and H. H. Genoways. 2003. West Indian mammals from the Albert Schwartz Collection: Biological and historical information. Scientific Papers, Natural History Museum, University of Kansas, 29:1-47.

Timm, R. M., D. E. Wilson, B. L. Clauson, R. K. LaVal, and C. S. Vaughan. 1989. Mammals of the La Selva-Braulio Carillo complex, Costa Rica. North American Fauna, 75:162 pp.

Timm, U., V. Pilâts, and L. Balèiauskas. 1998. Mammals of the East Baltic. Proceedings of the Latvian Academy of Sciences. Section B, 52(594/595):1-9.

Timmins, R. J., T. D. Evans, K. Khounboline, and C. Sisomphone. 1998. Status and conservation of the giant muntjac *Megmuntiacus vuquangensis*, and notes on other muntjac species in Laos. Oryx, 32:59-67.

Tiranti, S. I. 1997. Cytogenetics of silky desert mice, *Eligmodontia* spp. (Rodentia, Sigmodontinae) in central Argentina. Zeitschrift für Säugetierkunde, 62:37-42.

Tiranti, S. I. 1998a. Cytogenetics of *Graomys griseoflavus* (Rodentia: Sigmodontinae) in central Argentina. Zeitschrift für Säugetierkunde, 63:32-36.

Tiranti, S. I. 1998b. Chromosomal variation in the scrub mouse *Akodon molinae* (Rodentia: Sigmodontinae) in central Argentina. Texas Journal of Science, 50:223-228.

Tiranti, S. I., and M. P. Torres. 1998. Observations on bats of Córdoba and La Pampa Provinces, Argentina. Occasional Papers, The Museum of Texas Tech University, 175:1-12.

Tirira S., D. 1999. Mamíferos del Ecuador. Museo de Zoología, Centro de Biodiversidad y Ambiente, Pontificia Universidad Católica del Ecuador. Publicación Especial 2:xiv + 392 pp.

Tirira S., D. 2000. Listado bibliográfico sobre los mamíferos del Ecuador. Boletines Bibliográficos sobre la Biodiversidad del Ecuador, 2:xii + 340 pp.

Tiunov, M. P. 1986. [External structure of the male accessory gland in chiropterans and utilization of this character in their taxonomy.] Zoologicheskii Zhurnal, 65:1275-1279 (in Russian).

Tiunov, M. P. 1997. Rukokrylye Dalnevo Vastoka (Bats of the Far East). Dalnauka, Vladivostok, 134 pp.

Tiwari, K. K., R. K. Ghose, and S. Chakraborty. 1972. Notes on collection of small mammals from Western Ghats with remarks on the status of *Rattus rufescens* (Gray) and *Bandicota i. Malabarica* (Shaw). Journal of the Bombay Natural History Society, 68(2):378-384.

Tobien, H. 1975. Zur Gebissstruktur, Systematik und Evolution der Genera *Piezodus*, *Prolagus* und *Ptychoprolagus* (Lagomorpha, Mammalia) aus einigen Vorkommen im jüngeren Tertiär Mittel -und Westeuropas. Notizblatt des Hessischen Landesamtes für Bodenforschung zu Wiesbaden, 103:103-186.

Todd, N. B., and S. R. Pressman. 1968. The karyotype of the lesser panda (*Ailurus fulgens*) and general remarks on the phylogeny and affinities of the panda. Carnivore Genetics Newsletter, 5:105-108.

Toder, R., D. Vonholst, and W. Schempp. 1992. Comparative cytogenetic studies in tree shrews (*Tupaia*). Cytogenetics and Cell Genetics, 60:55-59.

Todoroviè, M., M. Mikeš, and I. Saviè. 1974. Biometric analysis of the genus *Apodemus* in Vojvodina (Yugoslavia). Pp. 173-181, *in* Symposium Theriologicum II. Proceedings of the International Symposium on species and zoogeography of European mammals held in Brno, Czechoslovakia on 22nd to 26th November 1971 (J. Kratochvil and R. Obrtel, eds.). Academia Publishing house of the Czechoslovak Academy of Sciences, Praha, 394 pp.

Tognelli, M. F., C. M. Campos, and R. A. Ojeda. 2001. *Microcavia australis*. Mammalian Species, 648:1-4.

Tomasi, E. T., and R. S. Hoffmann. 1984. *Sorex preblei* in Utah and Wyoming. Journal of Mammalogy, 65:708.

Tokmergenov, T. 1992. [Analysis of intraspecific polymorphism of masticatory surface of the molars M^3 and M_1 in Royle's high-mountain vole.] Zooloichskeii Zhurnal, 71(8):104-123 (in Russian with English summary).

Tokuda, M. 1935. *Neoaschizomys*, a new genus of Microtinae from Shikotan, a South Kuril Island. Memoirs College Science Kyoto Imperial University, Series B, 10:241-250.

Tokuda, M. 1941. A revised monograph of the Japanese and Manchou-Korean Muridae. Transaction of the Biogeographical Society of Japan, 4:1-155.

Tokumitsu, M., and K. Ogawa. 1994. Homology of p53 intronic sequences between four laboratory mouse strains and Japanese wild mouse (*Mus musculus molossinus* Mishima). Genome, 37:1022-1026.

Tolliver, D. K., J. R. Choate, D. W. Kaufman, and G. A. Kaufman. 1987. Microgeographic variation of morphologic and electrophoretic characters in *Peromyscus leucopus*. American Midland Naturalist, 71:420-427.

Tomich, P. Q. 1968. Coat color in wild populations of the roof rat in Hawaii. Journal of Mammalogy, 49(1):74-82.

Tomich, P. Q. 1986. Mammals in Hawai'i: A synopsis and notational bibliography. Second ed. Bernice P. Bishop Museum Special Publications, 76:1-375.

Tomich, P. Q., and H. T. Kami. 1966. Coat color inheritance of the roof rat in Hawaii. Journal of Mammalogy, 47(3):423-431.

Tomilin, A. G. 1957. Zveri SSSR i prilezhashchikh stran: Kitoobrazyne. (Zveri vostochnoi Evropy i severnoi Azii) [Mammals of USSR and adjacent countries: Cetacea (Mammals of eastern Europe and northern Asia)]. Akademiya Nauk SSSR, 9:1-756 (in Russian).

Tomilin, A. G. 1967. Mammals of the USSR and adjacent countries: Cetacea (Mammals of eastern Europe and northern Asia) [A translation of A. G. Tomilin, 1957, Zveri SSSR i prilezhashchikh stran: Kitoobrazyne. (Zveri vostochnoi Evropy i severnoi Asii)]. Israel Program for Scientific Translations, Jerusalem, 9:1-717.

Tong, H. 1989. Origine et evolution des Gerbillidae (Mammalia, Rodentia) en Afrique du Nord. Memoires de la Societe Geologique de France, Nouvelle Serie—1989, 155:1-120.

Tong, H., and J.-J. Jaeger. 1993. Muroid rodents from the Middle Miocene Fort Ternan locality (Kenya) and their contribution to the phylogeny of muroids. Palaeontographica. Abteilung A, Palaozoologie-Stratigraphie, 229(1-3):51-73.

Tong, Y., and J. Wang. 1993. A new soricomorph (Mammalia, Insectivora) from the Early Eocene of Wutu Basin, Shandong, China. Vertebrata Palasiatica, 31:19-32 (in Chinese).

Topachevskii, V. A. 1969. Fauna SSSR, Mlekopitayushchie, Tom 3, vyp. 3. Slepyshovye (Spalacidae) [Fauna of the USSR: Mammals. Mole rats, Spalacidae]. Nauka Publishers, Leningrad Section, Leningrad, 248 pp.

Topachevskii, V. A. 1976. Fauna of the USSR: Mammals. vol. 3, pt. 3, mole rats, Spalacidae [a translation of V. A. Topachevskii, 1969, Fauna SSSR, Mlekopitayushchie, Tom 3, vyp. 3. Slepyshovye (Spalacidae)]. Smithsonian Institution Libraries and National Science Foundation, Washington, D.C., 308 pp.

Topachevskii, V. A., and L. I. Rekovets. 1982. Novye materialy k sistematike i evolyutsii slepushonok nominativnovo podrod roda *Ellobius* (Rodentia, Cricetidae) [New material for systematic and evolution of the mole vole, nominative subgenus and genus *Ellobius* (Rodentia, Cricetidae)]. Vestnik Zoologii, 5:47-54 (in Russian).

Topachevskii, V. A., and A. F. Skorik. 1984. Pervaya nakhodka iskopaemykh ostatkov kosmatykh khomyakov—Lophomyinae (Rodentia, Cricetidae) [The first find of fossil remains of a maned hamster; Lophomyinae (Rodentia, Cricetidae)]. Vestnik Zoologii, 2:57-60 (in Russian).

Topachevskii, V. A., V. A. Nesin, and I. V. Topachevskii. 1998. [Biozonal microtheriological schema (stratigraphic distribution of small mammals–Insectivora, Lagomorpha, Rodentia) of the Neogene of the northern part of the East Parathetis.] Vestnik Zoologii, 32(1-2):76-87 (in Russian with English summary).

Topál, G. 1970*a*. The first record of *Ia io* Thomas, 1902 in Vietnam and India, and some remarks on the taxonomic position of *Parascotomanes beaulieui* Bourret, 1942, *Ia longimana* Pen, 1962, and the genus *Ia* Thomas, 1902 (Chiroptera: Vespertilionidae). Opuscula Zoologica (Budapest), 10:341-347.

Topál, G. 1970*b*. On the systematic status of *Pipistrellus annectans* Dobson, 1871 and *Myotis primula* Thomas, 1920 (Chiroptera: Vespertilionidae). Annales Historico-Naturales Musei Nationalis Hungarici (Budapest), 62:373-379.

Topál, G. 1975. Bacula of some Old World leaf-nosed bats (Rhinolophidae and Hipposideridae, Chiroptera: Mammalia). Vertebrata Hungarica, 16:21-53.

Topál, G. 1993. Taxonomic status of *Hipposideros larvatus alongensis* Bourret, 1942 and the occurrence of *H. turpis* Bamgs, 1901 in Vietnam (Mammalia, Chiroptera). Acta Zoologica Hungarica, 39:267-288.

Topál, G. 1997. A new mouse-eared bat species, from Nepal, with statistical analyses of some other species of subgenus *Leuconoe* (Chiroptera, Vespertilionidae). Acta Zoologica Academiae Scientiarum Hungaricae, 43:375-402.

Topál, G., and G. Csorba. 1992. The subspecific division of *Rhinolophus luctus* Temminck, 1835, and the taxonomic status of *R. beddomei* Andersen, 1905 (Mammalia, Chiroptera). Miscellanea Zoologica Hungarica, 7:101-116.

Torre, I., J. L. Tella, and A. Arrizabalaga. 1996. Environmental and geographic factors affecting the distribution of small mammals in an isolated Mediterranean mountain. Zeitschrift für Säugetierkunde, 61:365-375.

Torres-Mura, J. C., and L. C. Contreras. 1998. *Spalacopus cyanus*. Mammalian Species, 594:1-5.

Toschi, A. 1954. Elenco preliminare dei mammiferi della Libia. Supplemento alle Ricerche di Zoologia Applicata alla Caccia, 2(7):241-273.

Tosi, A. J., J. C. Morales, and D. J. Melnick. 2000. Comparison of Y chromosome and mtDNA phylogenies leads to unique inference of macaque evolutionary history. Molecular Phylogeny and Evolution, 17:133-144.

Toškan, B., and B. Kryštufek. In Press. Noteworthy records of rodents from the Upper Pleistocene and the Holocene of Slovenia. Mammalia.

Tovpinets, N. N. 1993. [Distribution and habitat preference of *Ellobius talpinus* in the Crimea.] Vestnik Zoologii, 1993(4):56-58.

Townsend, W. R. 2001. *Callithrix pygmaea* (Spix, 1823): Pygmy marmoset. Mammalian Species, 665:1-6.

Tranier, M. 1975*a*. Originalité du caryotype de *Gerbillus nigeriae* (Rongeurs, Gerbillidés). Mammalia, 39(4):703-704.

Tranier, M. 1975*b*. Etude preliminaire du caryotype de l'*Acomys* de l'Air (Rongeurs, Murides). Mammalia, 39:704-705.

Tranier, M., and H. Dosso. 1979. Recherches caryotypiques sur les rongeurs de Cote d'Ivoire: Resultats preliminaires pour les milieux fermes. Annales de l'Universite d'Abidjan, ser. E (Ecologie), 12:181-183.

Tranier, M., and J. C. Gautun. 1979. Recherches caryotypiques sur les rongeurs de Cote d'Ivoire: Resultats preliminaires pour les milieux ouverts. Le cas d'*Oenomys hypoxanthus ornatus*. Mammalia, 43(2):252-254.

Tranier, M., and D. Julien-Laferroere. 1990. A propos de petites gerbilles du Niger et du Tchad (Rongeurs, Gerbillidae, *Gerbillus*). Mammalia, 54:451-456.

Tranier, M., and F. Petter. 1978. Les relations d'*Eliomys tunetae* et de quelques autres formes de Lerots de la region mediterraneenne (Rongeurs, Muscardinides). Mammalia, 42(3):349-353.

Tranier, M., B. Hubert, and F. Petter. 1974. *Taterillus* de l'ouest de Tchad et du nord du Cameroun (Rongeurs, Gerbillides). Mammalia, 37:637-641.

Tranier, M., Y. Papillon, P.-O. Barome, A. Doukar, V. Volobouev, and B. Sicard. 1999. Un *Acomys airensis* (Rodentia: Muridae) en plein coeur du delta intérieur du Niger au Mali. Mammalia, 63(1):113-116.

Traut, W., M. F. Seldin, and H. Winking. 1992. Genetic mapping and assignment of a long-range repeat cluster to band D of chromosome 1 in *Mus musculus* and *Mus spretus*. Cytogenetics and Cell Genetics, 60:128-130.

Treviño-Villareal, J. 1991. The annual cycle of the Mexican prairie dog (*Cynomys mexicanus*). Occasional Papers of the Museum of Natural History, University of Kansas, 139:1-27.

Tribe, C. J. 1996. The Neotropical rodent genus *Rhipidomys* (Cricetidae: Sigmodontinae)—a taxonomic revision. Ph. D. Dissertation, University College, London, 316 pp.

Tripathy, N. K., H. S. Misra, and C. C. Das. 1985. A case of chromosome polymorphism in *Rattus rattus*. Current Science, 54(20):1079-1080.

Tristiani, H., J. Priyono, and O. Murakami. 1998. Seasonal changes in the population density and reproduction of the ricefield rat, *Rattus argentiventer* (Rodentia: Muridae) in West Java. Mammalia, 62(2):227-239.

Tristiani, H., O. Murakami, and E. Kuno. 2000. Rice plant damage distribution and home range distribution of the ricefield rat *Rattus argentiventer* (Rodentia: Muridae). Belgium Journal of Zoology, 130(2):83-91.

Tristram, H. 1877. Notes on *Eliomys melanurus* and some other rodents of Palestine. Proceedings of the Zoological Society of London, 1877:40-42.

Trombulak, S. C. 1988. *Spermophilus saturatus*. Mammalian Species, 322:1-4.

Trouessart, E. L. 1885. Catalogue des Mammifères vivants et fossiles. Fasc. IV.- Carnivores (Carnivora). Bulletin de la Société d'Études scientifiques d'Angers, Supplément a l'année 1884, 14:1-108.

Trouessart, E. L. 1897-1905. Catalogus mammalium tam viventium quam fossilium. Quinquennale supplementum anno 1904. [Tomus 1-1897; Tomus 2-1898; Quinquennale supplementum, fascic. 1 & 2—1904; fascic. 3 & 4—1905]. R. Friedländer and Sohn, Berlin, 1 & 2:1469 pp.; Quin supp:929 pp.

Trouessart, E. L. 1910. Faune des Mammifères d'Europe. R. Friedländer, Berlin, 219 pp.

Troughton, E. L. G. 1925. A revision of the genera *Taphozous* and *Saccolaimus* (Chiroptera) in Australia and New Guinea, including a new species and a note on two Malayan forms. Records of the Australian Museum, 14:313-341.

Troughton, E. L. G. 1927. Fixation of the habitat and extended description of *Pteropus tuberculatus* Peters. Records of the Australian Museum, 25:355-359.

Troughton, E. L. G. 1930. A new species and subspecies of fruit-bats (*Pteropus*) from the Santa Cruz group. Records of the Australian Museum, 18:1-4.

Troughton, E. L. G. 1932a. On five new rats of the genus *Pseudomys*. Records of the Australian Museum, 18:287-294.

Troughton, E. L. G. 1932b. A revision of the Rabbit-Bandicoots. Family Peramelidae, Genus *Macrotis*. Australian Zoologist, 7:219-236.

Troughton, E. L. G. 1936. A redescription of *Solomys* ("*Mus*") *salamonis* Ramsay. Proceedings of the Linnean Society of New South Wales, 61:3-4.

Troughton, E. Le G. 1937. Descriptions of some New Guinea mammals. Records of the Australian Museum, 20:117-127.

True, F. W. 1894. Notes on mammals of Baltistan and the Vale of Kashmir, presented to the National Museum by Dr. W. L. Abbott. Proceedings of the United States National Museum, 17(976):1-16.

Trujillo, D., C. Ibáñez, and J. Juste. 2002. A new subspecies of *Barbastella barbastellus* (Mammalia: Chiroptera: Vespertilionidae) from the Canary Islands. Revue Suisse de Zoologie, 109:543-550.

Tsekoura, N., S. Fraguedakis-tsolis, B. Chondropoulos, and G. Markakis. 2002. Morphometric and allozyme variation in central and southern Greek populations of *Microtus* (*Terricola*) *thomasi*. Acta Theriologica, 47(2):137-149.

Tsuchiya, K. 1974. Cytological and biochemical studies of *Apodemus speciosus* group in Japan. Journal of the Mammalogical Society of Japan, 6:67-87.

Tsuchiya, K. 1979. A contribution to the chromosome study in Japanese mammals. Proceedings of the Japan Academy, Series B, 55:191-195.

Tsuchiya, K. 1981. [On the chromosome variations in Japanese cricetid and murid rodents]. Honyurui Kagaku (Mammalian Science), 42:51-58 (in Japanese, with English abstract).

Tsuchiya, K. 1988. Cytotaxonomic studies of the family Talpidae from Japan. Honyurui Kagaku, 28:49-61 (in Japanese).

Tsuchiya, K., T. H. Yosida, K. Moriwaki, S. Ohtani, S. Kulta-Uthai, and P. Sudto. 1979. Karyotypes of twelve species of small mammals from Thailand. Report of the Hokaido Institute of Public Health, 29:26-29.

Tsuchiya, K., S. Wakana, H. Suzuki, S. Hattori, and Y. Hayashi. 1989. Taxonomic study of *Tokudaia*

(Rodentia: Muridae): I. Genetic differentiation. Memoirs of the National Science Museum, Tokyo, 22:227-234.

Tsuchiya, K., N. Miyashita, C. H. Wang, X.-L. Wu, X.-Q. He, M.-L. Jin, H. Li, F. Wang, L. Shi, and K. Moriwaki. 1994. Taxonomic study of the genus *Mus* in China, Korea, and Japan–morphologic identification. Pp. 3-23, *in* Genetics in wild mice (K. Moriwaki, T. Shiroishi, and H. Yonekawa, eds.). Japan Scientific Societies Press, Tokyo, 333 pp.

Tsuchiya, K., H. Suzuki, A. Shinohara, M. Harada, S. Wakana, M. Sakaizumi, S.-H. Han, L.-K. Lin, and A. P. Kryukov. 2000. Molecular phylogeny of East Asian moles inferred from the sequence variation of the mitochondrial cytochrome b gene. Genes and Genetic Systems, 75:17-24.

Tsytsulina, K. 2001. *Myotis ikonnikovi* (Chiroptera, Vespertilionidae) and its relationships with similar species. Acta Chiropterologica, 3:11-19.

Tsytsulina, K., and P. P. Strelkov. 2001. Taxonomy of the *Myotis frater* species group (Vespertilionidae, Chiroptera). Bonner Zoologische Beiträge, 50:15-26.

Tucker, P. K., and D. J. Schmidly. 1981. Studies of a contact zone among three chromosomal races of *Geomys bursarius* in east Texas. Journal of Mammalogy, 62:258-272.

Tuen, A. A., A. Osman, and C. Putet. 2000. Distribution and abundance of small mammals and birds at Mt. Santubong, Sarawak. The Sarawak Museum Journal, 55, no. 76 (n.s.):235-254.

Tullberg, T. 1896. Zur anatomie des *Haplodon rufus*. Pp. 233-251, *in* Zoologiska Studier; Festskrift Wilhelm Lilljeborg Tillegnäd pa Understod af Hans Majestt Konung Oscar II. Letterstedtska Foreningen och Enskilda Mecenater, 233-251. Almqvist and Wiksells, Upsala, 360 pp.

Tullberg, T. 1899. Uber das system der Nagetiere. Eine phylogenetische studie. Nova Acta Regiae Societatis Scientiarium Upsaliensis, Ser. 3, 18:1-514.

Tumanov, I. L., and A. V. Abramov. 2002. A study of the hybrids between the European mink *Mustela lutreola* and the polecat *M. putorius*. Small Carnivore Conservation, 27:29-31.

Tumlison, R. 1987. *Felis lynx*. Mammalian Species, 269:1-8.

Tumlison, R. 1992. *Plecotus mexicanus*. Mammalian Species, 401:1-3.

Tumlison, R., and M. E. Douglas. 1992. Parsimony analysis and the phylogeny of plecotine bats (Chiroptera: Vespertilionidae). Journal of Mammalogy, 73:276-285.

Tumlison, R., V. R. McDaniel, and J. G. Duffy. 1993. Further extension of the range of the northern pygmy mouse, *Baiomys taylori*, in southwestern Oklahoma. Southwestern Naturalist, 38:285-286.

Turker, M. S., G. E. Cooper, and P. L. Bishop. 1993. Region-specific rates of molecular evolution: A fourfold reduction in the rate of accumulation of "silent" mutations in transcribed versus nontranscribed regions of homologous DNA fragments derived from two closely related mouse species. Journal of Molecular Evolution, 36:31-40.

Turni, H., H.-J. Kapischke, H. Brünner, and A. Feiler. 2001. Der Status von *Sorex isodon marchicus* Passarge, 1984 (Mammalia: Insectivora: Soricidae). Zoologische Abhandlungen Staatliches Museum für Tierkunde Dresden, 51:205-219.

Turton, J. D. Mongolian gerbil. Pp. 266-268, *in* Evolution of Domesticated Animals (I. L. Mason, ed.). Longman Group Limited, London and New York, 452 pp.

Tuttle, R. H. 1967. Knuckle-walking and the evolution of hominoid hands. American Journal of Physical Anthropology, n.s., 26:171-206.

Tvrtkovic, N. 1976. The variability of the postero-external supplemental tubercle (t12) on the first and second upper molars in the species *Apodemus sylvaticus* (Linne, 1758) and *Apodemus flavicollis* (Melchior, 1834) from western Yugoslavia. Periodicum Biologorum, 78:91-100.

Tvrtkovic, N., and G. Dzukic. 1977. Sisavci lesinog (slanog) kopova s posebnim osvrtom na vrstu *Apodemus microps* Krat. and Ros. 1952. Arhiv Bioloskih Nauka (Beograd), 29:161-173 (in Serbo-Croatian).

Tvrtkovic, N., B. Dulic, and M. Grubešic. 1995. Distribution and habitats of dormice in Croatia. Pp. 199-207, *in* Proceedings of II Conference on Dormice (Rodentia, Myoxidae) (M. G. Filippucci, ed.). Hystrix, n.s., 6(1-2):1-340.

Twigg, G. I. 1992. The black rat *Rattus rattus* in the United Kingdon in 1989. Mammal Review, 22:33-42.

Tyler, M. J. (ed.). 1979. The status of endangered Australasian wildlife. Royal Society of South Australia, Adelaide, 210 pp.

Uerpmann, H.-P. 1987. The ancient distribution of ungulate mammals in the Middle East. Beihefte zum Tübinger Atlas des vorderen Orients, Reihe A (Naturwissenschaften), 27:1-173.

Uerpmann, H.-P. 1980. *Ovis musimon* Schreber, 1782, oder *Ovis musimon* Pallas, 1814. Säugetierkundliche Mitteilungen, 29(4):59-60.

Uhlig, U. 2001. The Gliridae (Mammalia) from the Oligocene (MP 24) of Gröben 3 in the folded molasse of southern Germany. Paleovertebrata, 30 (3-4):151-187.

Uhrin, M., and P. Benda. 2000. [First record of *Apodemus agrarius* in the Hron river basin (Central Slovakia).] Lynx (Praha), n.s., 31:156-158 (in Czech with English abstract).

Ulbrich, K., and A. Kayser. 2001. Abschätzung des ausfsterbe-risikos von feldhamsterpopulationen mit einem simulationsmodell. Jahrbücher des Nassauischen Vereins für Naturkunde, 122:187-189.

Ünay, E. 1989. Rodents from the Middle Oligocene of Turkish Thrace. Utrecht Micropaleontological Bulletin, Special Publication 5:1-119.

Ünay, E. 1994. Early Miocene rodent faunas from the eastern Mediterranean area Part IV. The Gliridae. Proceedings of the Koninklijke Nederlandse Akademie van Wetenschappen, 97(4):445-490.

Ünay, E. 1996. On fossil Spalacidae (Rodentia). Pp. 246-252, *in* The Evolution of western Eurasian Neogene mammal faunas (R. L. Bernor, V. Fahlbusch, and H.-W. Mittmann, eds.). Columbia University Press, New York, 487 pp.

Ünay, E. 1999. Family Spalacidae. Pp. 421-425, *in* The Miocene land mammals of Europe (G. E. Rössner and K. Heissig, eds.). Dr. Friedrich Pfeil, München, 515 pp.

UNEP. 2003. "Garden of Eden" in southern Iraq likely to disappear completely in five years unless urgent action taken. United Nations Environment Programme, March 22, 2003 (www.grid.unep.ch/activities/sustainable/tigris/marshlands/).

Unterholzner, K., and R. Willenig. 2000. Zu ökologie, verhalten und morphologie der ährenmaus *Mus spicilegus* Petényi, 1882. Pp. 7-88, *in* Beiträge zur Kenntnis der ährenmaus *Mus spicilegus* Petényi, 1882 (K. Unterholzner, R. Willenig, and K. Bauer, eds.). Biosystematics and Ecology Series No. 17, Österreichische Akademie der Wissenschaften, Wien, 108 pp.

Urban-Ramirez, J., and D. Aurioles-Gamboa. 1992. First record of the pygmy beaked whale *Mesoplodon peruvianus* in the North Pacific. Marine Mammal Science, 8(4):420-425.

Uribe-Alcocer, M., F. Rodriguez-Romero, R. Ruiz-Carus, and A. Laguarda-Figueras. 1978. The chromosomes of *Spermophilus spilosoma cabrerai*. Mammalian Chromosome Newsletter, 19(3):81-83.

Uribe-Alcocer, M., A. Ahumada-Medina, A. Laguarda-Figuera, and F. Rodriguez-Romero. 1979. The karyotype of *Spermophilus perotensis*. Mammalian Chromosome Newsletter, 20(4):139-141.

Uribe-Alcocer, M, F. A. Cervantes-Reza, C. Lorenzo-Monterrubio, and L. Guereña-Gandara. 1989. Karyotype of the tropical hare (*Lepus flavigularis*, Leporidae). Southwestern Naturalist, 34:304-306.

Usdin, K., P. Chevret, F. M. Catzeflis, R. Verona, and A. V. Furano. 1995. L1 (LINE-1) retrotransposable elements provide a "fossil" record of the phylogenetic history of murid rodents. Molecular Biology and Evolution, 12(1):73-82.

U.S. [ESA] Fish and Wildlife Service. 1991. Endangered and threatened wildlife and plants (those covered by the regulations for the U.S. Endangered Species Act), 50 CRF 17.11 17.12, July 15, 1991. U. S. Government Printing Office, Washington, D.C., 37 pp.

Valdez, R. 1982. The wild sheep of the world. Wild Sheep and Goat International, Mesilla, New Mexico, 186 pp.

Valdez, R., C. F. Nadler, and T. D. Bunch. 1978. Evolution of wild sheep in Iran. Evolution, 32:56-72.

Van Bemmel, A. C. V. 1949a. Revision of the rusine deer in the Indo-Australian Archipelago. Treubia, 20:191-262.

Van Bemmel, A. C. V. 1949b. Notes on Indo-Australian mammals. 1. A note on *Lutrogale perspicillata* (I. Geoffroy) (Mustelidae). 2. On the meaning of the name *Cervus javanicus* Osbeck 1765 (Tragulidae). Treubia, 20:375-380.

Van Bemmel, A. C. V. 1952. Contribution to the knowledge of the genera *Muntiacus* and *Arctogalidia* in the Indo-Australian Archipelago (Mammalia, Cervidae & Viverridae). Beaufortia, 16:1-50.

van Bree, P. J. H. 1961. On the type specimen of *Hipposideros larvatus sumbae* Oei Hong Peng, 1960 (Mammalia, Chiroptera). Verhandlungen der Naturforschenden Gesellschaft in Basel, 72:122-123.

van Bree, P. J. H. 1971. On *Globicephala sieboldii* Gray, 1846, and other species of pilot whale (Notes on Cetacea, Delphinoidea III). Beaufortia, 19(249):79-87.

van Bree, P. J. H. 1973. *Neophocaena phocaenoides asiaeorientalis* (Pilleri & Gihr 1973) a synonym of the preoccupied name *Delphinus melas* Schlegel 1841 (Notes on Cetacea Delphinoidea VII). Beaufortia, 21:17-24.

van Bree, P. J. H. 1975. On the alleged occurrence of *Mesoplodon bidens* (Sowerby, 1804) (Cetacea, Ziphioidae) in the Mediterranean. Annali del Museo Civico di Storia Naturale di Genova, 80:226-228.

van Bree, P. J. H. 1976. On the correct Latin name of the Indus Susu (Cetacea, Platanistoidea). Bulletin Zoologisch Museum, Universiteit van Amsterdam, 5(17):139-140.

van Bree, P. J. H., and M. Sc. Boeadi. 1978. Notes on the Indonesian mountain weasel, *Mustela lutreolina* Robinson and Thomas, 1917. Zeitschrift für Säugetierkunde, 43:166-171.

van Bree, P. J. H., and M. D. Gallagher. 1978. On the taxonomic status of *Delphinus tropicalis* van Bree 1971 (Notes on Cetacea Delphinoidea IX). Beaufortia, 28(342):1-8.

van Bree, P. J. H., and I. Kristensen. 1974. On the intriguing stranding of four Cuvier's beaked whales *Ziphius cavirostris* G. Cuvier 1823 on the Lesser Antillean island of Bonaire. Bijdragen Tot De Dierkunde, 44(2):235-238.

van Bree, P. J. H., and P. E. Purves. 1972. Remarks on the validity of *Delphinus bairdii* (Cetacea, Delphinidae). Journal of Mammalogy, 53:372-374.

Van Cakenberghe, V., and F. De Vree. 1985. Systematics of African *Nycteris* (Mammalia: Chiroptera). Pp. 53-90, *in* Proceedings of the International Symposium on African vertebrates: Systematics, phylogeny and evolutionary ecology (K.-L. Schuchmann, ed.). Zoologisches Forschingsinstitut und Museum Alexander Koenig, Bonn, 585 pp.

Van Cakenberghe, V., and F. De Vree. 1993*a*. The systematic status of southeast Asian *Nycteris* (Chiroptera: Nycteridae). Mammalia, 57:227-244.

Van Cakenberghe, V., and F. De Vree. 1993*b*. Systematics of African *Nycteris* (Mammalia: Chiroptera). Part II. The *Nycteris hispida* group. Bonner Zoologische Beiträge, 44:299-332.

Van Cakenberghe, V., and F. De Vree. 1994. A revision of the Rhinopomatidae Dobson 1872, with the description of a new subspecies. Senkenbergiana Biologica, 73(1-2):1-24.

Van Cakenberghe, V., and F. De Vree. 1998. Systematics of African *Nycteris* (Mammalia: Chiroptera). Part III. The *Nycteris thebaica* group. Bonner Zoologische Beiträge, 48:123-166.

Van Cakenberghe, V., F. De Vree, and H. Leirs. 1999. On a collection of bats (Chiroptera) from Kikwit, Democratic Republic of the Congo. Mammalia, 63:291-322.

Van Cura, N. J., and D. F. Hoffmeister. 1966. A taxonomic review of the grasshopper mice, *Onychomys*, in Arizona. Journal of Mammalogy, 47:613-630.

Van den Bergh, G. D., J. de Vos, and P. Y. Sondaar. 2001. The Late Quaternary palaeogeography of mammal evolution in the Indonesian Archipelago. Palaeogeography, Palaeoclimatology, Palaeoecology, 171:385-408.

Van den Brink, F. H. 1967. A field guide to the mammals of Britain and Europe. London, Collins, 221 pp.

Van Den Bussche, R. A., and R. J. Baker. 1993. Molecular phylogenetics of the New World bat genus *Phyllostomus* based on cytochrome-*b* DNA sequence variation. Journal of Mammalogy, 74:793-802.

Van Den Bussche, R. A., and S. R. Hoofer. 2000. Further evidence for inclusion of the New Zealand short-tailed bat (*Mystacina tuberculata*) within *Noctilionoidea*. Journal of Mammalogy, 81:865-874.

Van Den Bussche, R. A., and S. R. Hoofer. 2001. Evaluating monophyly of Nataloidea (Chiroptera) with mitochondrial DNA sequences. Journal of Mammalogy, 82:320-327.

Van Den Bussche, R. A., and S. E. Weyandt. 2003. Mitochondrial and nuclear DNA sequence data provide resolution to sister-group relationships within *Pteronotus*. Acta Chiropterologica, 5:1-13.

Van Den Bussche, R. A., R. J. Baker, H. A. Wichman, and M. J. Hamilton. 1993*a*. Molecular phylogenetics of stenodermatine bat genera: Congruence of data from nuclear and mitochondrial DNA. Molecular Biology and Evolution, 10:944-959.

Van Den Bussche, R. A., R. K. Chesser, M. J. Hamilton, R. D. Bradley, C. A. Porter, and R. J. Baker. 1993*b*. Maintenance of a narrow hybrid zone in *Peromyscus leucopus*: A test of alternative models. Journal of Mammalogy, 74:832-845.

Van Den Bussche, R. A., J. L. Hudgeons, and R. J. Baker. 1998. Phylogenetic accuracy, stability, and congruence: Relationships within and among the New World bat genera *Artibeus*, *Dermanura*, and *Koopmania*. Pp. 59-71, *in* Bat biology and conservation (T. H. Kunz and P. A. Racey, eds.). Smithsonian Institution Press, Washington, D.C., 365 pp.

Van Den Bussche, R. A., S. R. Hoofer, and N. B. Simmons. 2002*a*. Phylogenetic relationships of mormoopid bats using mitochondrial gene sequences and morphology. Journal of Mammalogy, 83:40-48.

Van Den Bussche, R. A., S. R. Hoofer, and E. W. Hansen. 2002*b*. Characterization and phylogenetic utility of the mammalian protamine P1 gene. Molecular Phylogenetics and Evolution, 22:333-341.

Van Den Bussche, R. A., S. A. Reeder, E. W. Hansen, and S. R. Hoofer. 2003. Utility of the dentin matrix protein 1 (DMP1) gene for resolving mammalian intraordinal relationships. Molecular Phylogenetics and Evolution, 26:89-101.

van der Feen, P. J. 1962. Catalogue of the Marsupialia from New Guinea, the Moluccas and Celebes in the Museo Civico di Storia Naturale "Giacomo Doria" in Genoa. Annali di Museo Civico di Storia Naturale "Giacomo Doria", Genova, 73:19-70.

Vanderhaar, J. M., and Y. T. Hwang. 2003. *Mellivora capensis*. Mammalian Species, 721:1-8.

van der Kuyl, A. C., J. T. Dekker, and J. Goudsmit. 2000. Primate genus *Miopithecus*: Evidence for the existence of species and subspecies of dwarf guenons based on cellular and endogenous viral sequences. Molecular Phylogenetics and Evolution, 14:403-413.

van der Meulen, A. J., 1973. Middle Pleistocene smaller mammals from the Monte Peglia (Orvieto, Italy), with special reference to the phylogeny of *Microtus* (Arvicolidae, Rodentia). Quaternaria, 17:1-44.

van der Meulen, A. J. 1978. *Microtus* and *Pitymys* (Arvicolidae) from Cumberland Cave, Maryland, with a comparison of some New and Old World species. Annals of Carnegie Museum, 47:101-145.

van der Meulen, A. J., and H. de Bruijn. 1982. The mammals from the Lower Miocene of Aliveri (Island of

Evia, Greece). Part 2. The Gliridae. Proceedings of the Koninklijke Nederlandse Akademie van Wetenschappen, ser. B, 85(4):485-524.

van der Meulen, A. J., and G. G. Musser. 1999. New paleontological data from the continental Plio-Pleistocene of Java. Pp. 361-368, *in* Elephants have a snorkel! Papers in honour of Paul Y. Sondaar (J. W. F. Reumer and J. De Vos, eds.). Deinsea, 7:1-422.

Van der Straeten, E. 1975. *Lemniscomys bellieri*, a new species of Muridae from the Ivory Coast (Mammalia, Muridae). Revue de Zoologie Africaine, 89:906-908.

Van der Straeten, E. 1976a. *Lemniscomys striatus dieterleni*, a new subspecies of Muridae from Zaire (Mammalia, Muridae). Revue de Zoologie Africaine, 90(2):431-434.

Van der Straeten, E. 1976b. Maatgegevens van *Apodemus sylvaticus* (linnaeus, 1758) en *Apodemus flavicollis* (Melchior, 1834) in België. Lutra, 18:15-21.

Van der Straeten, E. 1977a. *Apodemus flavicollis* (Melchior, 1834) in Nederland. Lutra, 19:20-21.

Van der Straeten, E. 1977b. Chromosomenonderzoek en systematiek van West-Afrikaanse *Lemniscomys* (Muridae). Lutra, 19:70.

Van der Straeten, E. 1980a. Etude biometrique de *Lemniscomys linulus* (Afrique Occidentale) (Mammalia, Muridae). Revue de Zoologie Africaine, 94:185-201.

Van der Straeten, E. 1980b. A new species of *Lemniscomys* (Muridae) from Zambia. Annals of the Cape Provincial Museums, Natural History, 13:55-62.

Van der Straeten, E. 1981. Note sur *Lemniscomys striatus venustus* (Thomas, 1911). Mammalia, 45(1):125-128.

Van der Straeten, E. 1984. Etude biometrique des genres *Dephomys* et *Stochomys* avec quelques notes taxonomiques (Mammalia, Muridae). Revue de Zoologie Africaine, 98:771-798.

Van der Straeten, E. 1985. Note sur *Hybomys basilii* Eisentraut, 1965. Bonner Zoologische Beiträge, 36:1-8.

Van der Straeten, E. 1994. A species- and subject-coded bibliography of *Acomys* from Africa and the Middle East. Israel Journal of Zoology, 40:265-289.

Van der Straeten, E. 1999. Notes on *Mastomys pernanus* (Kershaw, 1921). Bonner Zoologische Beiträge, 48(3-4):225-230.

Van der Straeten, E., and F. Dieterlen. 1983. Description de *Praomys ruppi*, une nouvelle espece de Muridae d'Ethiopie. Annales Musée Royal de l'Afrique Centrale, Tervuren-Belgique, ser. 8 (Sciences Zoologiques), 237:121-128.

Van der Straeten, E., and F. Dieterlen. 1987. *Praomys misonnei*, a new species of Muridae from Eastern Zaire (Mammalia). Stuttgarter Beiträge zur Naturkunde, ser. A (Biologie), 402:1-40.

Van der Straeten, E., and F. Dieterlen. 1992. Craniometrical comparison of four populations of *Praomys jacksoni* captured at different heights in Eastern Zaire (Kivu). Mammalia, 56:125-131.

Van der Straeten, E., and A. M. Dudu. 1990. Systematics and distribution of *Praomys* from the Masako Forest Reserve (Zaire), with the description of a new species. Pp. 73-83, *in* Vertebrates in the tropics (G. Peters and R. Hutterer, eds.). Museum Alexander Koenig, Bonn, 424 pp.

Van der Straeten, E., and R. Hutterer. 1986. *Hybomys eisentrauti*, une nouvelle espece de Muridae du Cameroun (Mammalia, Rodentia). Mammalia, 50:35-42.

Van der Straeten, E., and J. C. Kerbis Peterhans. 1999. *Praomys degraaffi*, a new species of Muridae (Mammalia) from central Africa. South African Journal of Zoology, 34(2):80-90.

Van der Straeten, E., and C. B. Robbins. 1997. Further studies on *Mastomys* (Rodentia: Muridae) types and generic distinctions among African Muridae. Mittheilungen Aus Dem Zoologischen Museum zu Berlin, 73(1):153-163.

Van der Straeten, E., and B. Van der Straeten-Harrie. 1977. Etude de la biometrie cranienne et de la repartition d'*Apodemus sylvaticus* (Linnaeus, 1758) et d'*Apodemus flavicollis* (Melchior, 1834) en Belgique. Acta Zoologica et Pathologica Antverpiensia, 69:169-182.

Van der Straeten, E., and W. N. Verheyen. 1978a. Karyological and morphological comparisons of *Lemniscomys striatus* (Linnaeus, 1758) and *Lemniscomys bellieri* Van der Straeten, 1975, from Ivory Coast (Mammalia: Muridae). Bulletin of Carnegie Museum of Natural History, 6:41-47.

Van der Straeten, E., and W. N. Verheyen. 1978b. Taxonomical notes on the West African *Myomys* with the description of *Myomys derooi* (Mammalia—Muridae). Zeitschrift für Säugetierkunde, 43:31-41.

Van der Straeten, E., and W. N. Verheyen. 1979a. Note sur la position systematique de *Lemniscomys macculus* (Thomas et Wroughton, 1910) (Mammalia, Muridae). Mammalia, 43:377-389.

Van der Straeten, E., and W. N. Verheyen. 1979b. Notes taxonomiques sur les *Malacomys* de l'Ouest africain avec redescription du patron chromosomique de *Malacomys edwardsi* (Mammalia, Muridae). Revue de Zoologie Africaine, 93:10-35.

Van der Straeten, E., and W. N. Verheyen. 1980. Relations biometriques dans le groupe specifique *Lemniscomys striatus* (Mammalia, Muridae). Mammalia, 44:73-82.

Van der Straeten, E., and W. N. Verheyen. 1981. Etude biometrique du genre *Praomys* en Cote d'Ivoire. Bonner Zoologische Beiträge, 32:249-264.

Van der Straeten, E., and W. N. Verheyen. 1983. Nouvelles captures de *Lophuromys rahmi* et *Delanymys brooksi* en Republique Rwandaise. Mammalia, 47:426-430.

Van der Straeten, E., W. N. Verheyen, and B. Harrie. 1986. The taxonomic status of *Hybomys univittatus lunaris* Thomas, 1906 (Mammalia: Muridae). Cimbebasia, ser. A., 8:209-218.

Van der Straeten, E., E. Lecompte, and C. Denys. 2003. *Praomys petteri*: Une nouvelle espèce des Muridae Africains (Mammalia, Rodentia). Bonner Zoologische Beiträge, 50(4):329-345.

Van Deusen, H. M., and G. G. George. 1969. Results of the Archbold Expeditions. No. 90. Notes on the echidnas (Mammalia: Tachyglossidae) of New Guinea. American Museum Novitates, 2383:1-23.

Van Deusen, H. M., and J. K. Jones, Jr. 1967. Marsupials. Pp. 61-86, *in* Recent mammals of the world (S. Anderson and J. K. Jones, Jr. eds.). Ronald Press Co., New York, 453 pp.

Van Deusen, H. M., and K. F. Koopman. 1971. Results of the Archbold Expeditions. No. 95. The genus *Chalinolobus* (Chiroptera, Vespertilionidae). Taxonomic review of *Chalinolobus picatus*, *C. nigrogriseus*, and *C. rogersi*. American Museum Novitates, 2468:1-30.

van de Weerd, A. 1979. Early Ruscinian rodents and lagomorphs (Mammalia) from the lignites near Ptolemais (Macedonia, Greece). I and II. Proceedings of the Koninklijke Nederlandse Akademie van Wetenschappen, ser. B, 82:127-154.

van Dijk M., O. Madsen, F. Catzeflis, M. J. Stanhope, W. W. de Jong, and M. Pagel. 2001. Protein sequence signatures support the African clade of mammals. Proceedings of the National Academy of Sciences, 98:188-193.

Van Dyck, S. 1982. The relationships of *Antechinus stuartii* and *A. flavipes* (Dasyuridae, Marsupialia) with special reference to Queensland. Pp. 723-766, *in* Carnivorous Marsupials (M. Archer, ed.). Royal Zoological Society of New South Wales, 2:1-804.

Van Dyck, S. 1990. *Belideus gracilis*—soaring problems for an old De Vis glider. Memoirs of the Queensland Museum, 28:329-336.

Van Dyck, S. 1991. Raising an old glider's ghost—a devil of an exorcise! Wildlife Australia, 28(2):10-13.

Van Dyck, S. 1993. The taxonomy and distribution of *Petaurus gracilis* (Marsupialia: Petauridae), with notes on its ecology and conservation status. Memoirs of the Queensland Museum, 33:77-122.

Van Dyck, S. 1995. False water-rat, *Xeromys myoides*. Pp. 630-631, *in* Mammals of Australia (R. Strahan, ed.). Smithsonian Institution Press, Washington, D.C., 756 pp.

Van Dyck, S. 1996. *Xeromys myoides* Thomas, 1889 (Rodentia: Muridae) in mangrove communities of North Stradbroke Island, southeast Queensland. Memoirs of the Queensland Museum, 42:337-366.

Van Dyck, S. 1997. Queensland pebble-mound mice—up from the tailings. Nature Australia, 25(10):40-47.

Van Dyck, S. 2002. Morphology-based revision of *Murexia* and *Antechinus* (Marsupialia: Dasyuridae). Memoirs of the Queensland Museum, 48:239-330.

Van Dyck, S., and M. S. Crowther. 2000. Reassessment of northern representatives of the *Antechinus stuartii* complex (Marsupialia: Dasyuridae): *A. subtropicus* sp. nov. and *A. adustus* new status. Memoirs of the Queensland Museum, 45:611-635.

Van Dyck, S., W. W. Baker, and D. D. Gillette. 1979. The false water rat, *Xeromys myoides* on Stradbroke Island, a new locality in southeastern Queensland. Proceedings of the Royal Society of Queensland, 90:84.

Van Dyck, S., J. C. Z. Woinarski, and A. J. Press. 1994. The Kakadu dunnart, *Sminthopsis bindi* (Marsupialia: Dasyuridae), a new species from the stony woodlands of the Northern Territory. Memoirs of the Queensland Museum, 37:311-323.

Van Dyk, D., G. K. Campbell, P. J. Taylor, and J. Meester. 1991. Genetic variation within the endemic murid species, *Otomys unisulcatus* F. Cuvier, 1829 (bush karroo rat). Durban Museum Novitates, 16:12-21.

Van Gelder, R. G. 1959. A taxonomic revision of the spotted skunks (genus *Spilogale*). Bulletin of the American Museum of Natural History, 117:229-392.

Van Gelder, R. G. 1966. Comments on the proposals concerning *Zorilla* Geoffroy 1826 (Mammalia). Z.N.(S.) 758. Bulletin of Zoological Nomenclature, 22(5/6):278-280.

Van Gelder, R. G. 1977*a*. An eland X kudu hybrid, and the content of the genus *Tragelaphus*. Lammergeyer, 23:1-6.

Van Gelder, R. G. 1977*b*. Mammalian hybrids and generic limits. American Museum Novitates, 2635:1-25.

Van Gelder, R. G. 1978. A review of canid classification. American Museum Novitates, 2646:1-10.

Van Hensbergen, H. J., and A. Channing. 1989. Habitat preference and use of space by the namtap *Graphiurus ocularis* (Rodentia: Gliridae). Mammalia, 53(1):25-33.

Van Hooft, W. F., A. F. Groen, and H. H. T. Prins. 2002. Phylogeography of the African buffalo based on mitochondrial and Y-chromosome loci. Pleistocene origins and population expansion of the Cape buffalo subspecies. Molecular Ecology, 11:267-279.

Vanitharani, J., A Rajendaran, P. J. J. Bates, D. L. Harrison, and M. J. Pearch. 2003. A taxonomic reassessment of *Kerivoula lenis* Thomas, 1916 (Chiroptera, Vespertilionidae) including a first record from peninsular India. Acta Chiropterologica, 5:49-60.

Van Laar, V. 1984. Geographical distribution and habitat selection of the hazel dormouse, *Muscardinus avellanarius* (L. 1758) in the Netherlands. Lutra, 27(3) 1984:229-260.

Van Laar, V. 1992. Dormouse *Muscardinus avellanarius* (L., 1758). Koninklijke Nederlandse Natuurhistorische Vereniging, 56:306-310.

Vanlerberghe, F., F. Bonhomme, C. A. Hutchison III, and M. H. Edgell. 1993. A major difference between the divergence patterns within the Lines-1 families in mice and voles. Molecular and Biological Evolution, 10(4):719-731.

Van Mensch, P. J. A., and P. J. H. van Bree. 1969. On the African Golden Cat, *Profelis aurata* (Temminck, 1827). Biologia Gabonica, 5(4):235-269.

Van Peenen, P. F. D., P. F. Ryan, and R. H. Light. 1969. Preliminary identification manual for mammals of South Vietnam. United States Naional Museum, Smithsonian Institution, Washington, D.C., 310 pp.

Van Peenen, P. F. D., R. H. Light, F. J. Duncan, R. See, J. Sulianti Saroso, Boeadi, and W. P. Carney. 1974. Observation on *Rattus bartelsii* (Rodentia: Muridae). Treubia, 28:83-117.

Van Rompaey, H. 1988. *Osbornictis piscivora*. Mammalian Species, 309:1-4.

Van Rompaey, H., and M. Colyn. 1992. *Crossarchus ansorgei*. Mammalian Species, 402:1-3.

Van Rompaey, H., and M. Colyn. 1998. A new servaline genet (Carnivora, Viverridae) from Zanzibar Island. South African Journal of Zoology, 33:42-46.

Van Rompaey, J., W. Verheyen, and M. Selens. 1984. Genetic differences between three species of pigmy-mice in Rwanda (Africa). Revue de Zoologie et de Botanique Africaines, 98(4):886-894.

Van Roosmalen, M. G. M., and T. van Roosmalen. 2003. The description of a new marmoset genus, *Calibella* (Callitrichinae, Primates), including its molecular phylogenetic status. Neotropical Primates, 11:1-10.

Van Roosmalen, M. G. M., T. van Roosmalen, R. A. Mittermeier and A. B. Rylands. 2000. Two new species of marmoset, genus *Callithrix* Erxleben, 1777 (Callitrichidae, Primates), from the Tapajós/Madeira interfluvium, South Central Amazonia, Brazil. Neotropical Primates, 8:2-18.

Van Roosmalen, M. G. M., T. van Roosmaalen, and R. A. Mittermeier. 2002. A taxonomic review of the titi monkeys, genus *Callicebus* Thomas, 1903, with the description of two new species, *Callicebus bernhardi* and *Callicebus stephennashi*, from Brazilian Amazonia. Neotropical Primates, 10(Suppl.):1-52.

Van Staaden, M. 1994. *Suricata suricatta*. Mammalian Species, 483:1-8.

Van Valen, L. 1967. New Paleocene insectivores and insectivore classification. Bulletin of the American Museum of Natural History, 135:217-284.

Van Valen, L. 1979. The evolution of bats. Evolutionary Theory, 4:103-121.

Van Vendeloo, N. H. 1953. On the correlation between the masticatory muscles and the skull structure in the muskrat, *Ondatra zibethica* (L.). I and II. Koninklijke Nederlandse Akademie van Wetenschappen Proceedings. Series C, Biological and Medical Sciences, 56:116-127 and 265-277.

van Weers, D. J. 1976. Notes on Southeast Asian porcupines (Hystricidae, Rodentia) I. On the taxonomy of the genus *Trichys* Günther, 1877. Beaufortia (University of Amsterdam), 25(319):15-31.

van Weers, D. J. 1977. Notes on southeast Asian porcupines (Hystricidae, Rodentia) II. On the taxonomy of the genus *Atherurus* F. Cuvier. Beaufortia (University of Amsterdam), 26(336):205-230.

van Weers, D. J. 1978. Notes on southeast Asian porcupines (Hystricidae, Rodentia) III. On the taxonomy of the subgenus *Thecurus* Lyon, 1907. Beaufortia (University of Amsterdam), 28(344):17-33.

van Weers, D. J. 1979. Notes on southeast Asian porcupines (Hystricidae, Rodentia) IV. On the taxonomy of the subgenus *Acanthion* F. Cuvier. Beaufortia, 29(356):215-272.

van Weers, D. J. 1983. Specific distinction in Old World porcupines. Zoologische Garten, Jena, 53:226-232.

Van Wynsberghe, N. R., and M. D. Engstrom. 1992. Chromosomal variation in collared lemmings (*Dicrostonyx*) from the western Hudson Bay region. Musk Ox, 39:203-209.

van Zyll de Jong, C. G. 1972. A systematic review of the Nearctic and Neotropical river otters (genus *Lutra*, Mustelidae, Carnivora). Royal Ontario Museum, Life Sciences, Contribution, 80:1-104.

van Zyll de Jong, C. G. 1976. A comparison between woodland and tundra forms of the common shrew (*Sorex cinereus*). Canadian Journal of Zoology, 54:963-973.

van Zyll de Jong, C. G. 1979. Distribution and systematic relationships of long eared *Myotis* in Western Canada. Canadian Journal of Zoology, 57:987-994.

van Zyll de Jong, C. G. 1980. Systematic relationships of prairie and woodland forms of the common shrew, *Sorex cinereus* Kerr and *S. haydeni* Baird, in the northern zone of contact. Journal of Mammalogy, 61:66-75.

van Zyll de Jong, C. G. 1982. Relationships of amphiberingian shrews of the *Sorex cinereus* group. Canadian Journal of Zoology, 60:1580-1587.

van Zyll de Jong, C. G. 1983a. Handbook of Canadian mammals. Part I. Marsupials and insectivores. National Museum of Natural Sciences (Ottawa), 210 pp.

van Zyll de Jong, C. G. 1983b. A morphometric analysis of North American shrews of the *arcticus* group, with special consideration of the taxonomic status of *S. a. maritimensis*. Nature Canada (Quebec), 110:373-378.

van Zyll de Jong, C. G. 1984. Taxonomic relationships of Nearctic small-footed bats of the *Myotis leibii* group (Chiroptera: Vespertilionidae). Canadian Journal of Zoology, 62:2519-2526.

van Zyll de Jong, C. G. 1986. A systematic study of Recent bison, with particular consideration of the wood bison (*Bison bison athabascae* Rhoads 1898). Publications in Natural Sciences, National Museum of Natural Sciences, Canada, 6:1-69.

van Zyll de Jong, C. G. 1987. A phylogenetic study of the Lutrinae (Carnivora; Mustelidae) using morphological data. Canadian Journal of Zoology, 65:2536-2544.

van Zyll de Jong, C. G. 1991a. A brief review of the systematics and a classification of the Lutrine. Pp. 79-83, *in* Proceedings of the V International Otter Colloquium (C. Reuther and R. Röchert, eds.). Habitat 6, Hankensbüttel, 344 pp.

van Zyll de Jong, C. G. 1991b. Speciation of the *Sorex cinereus* group. Pp. 65-73, *in* The biology of the Soricidae (J. S. Findley and T. L. Yates, eds.). Special Publication, Museum of Southwestern Biology, 1:1-91.

van Zyll de Jong, C. G. 1992. A morphometric analysis of cranial variation in Holarctic weasels (*Mustela nivalis*). Zeitschrift für Säugetierkunde, 57:77-93.

van Zyll de Jong, C. G., and G. L. Kirkland, Jr. 1989. A morphometric analysis of the *Sorex cinereus* group in central and eastern North America. Journal of Mammalogy, 70:110-122.

Vargas, J. M., A. Antunez, V. Sans-Cano, and J. Cano. 1984. Albinismus bei *Mus spretus* Lataste, 1883. Säugetierkundliche Mitteilungen, 31:260-262.

Varona, L. S. 1970. Nueva especie y nuevo subgénero de *Capromys* (Rodentia: Caviomorpha) de Cuba. Poeyana (Academia de Ciencias de Cuba), 73:1-18.

Varona, L. S. 1974. Catálogo de los mamíferos vivientes y extinguidos de las Antillas. Instituto de Zoologia, Academia de Ciencias de Cuba (Havana), 139 pp.

Varona, L. S. 1979. Subgénero y especie nuevos de *Capromys* para Cuba (Rodentia: Caviomorpha). Poeyana (Academia de Ciencias de Cuba), 194:1-33.

Varona, L. S. 1983a. Nueva subespecie de jutia conga, *Capromys pilorides* (Rodentia: Capromyidae). Caribbean Journal of Science, 19:3-4.

Varona, L. S. 1983b. Remarks on the biology and zoogeography of *Solenodon* (*Atopogale*) *cubanus* Peters, 1861 (Mammalia, Insectivora). Bijdragen tot de Dierkunde, 53:93-98.

Varona, L. S. 1986. Táxones del subgénero *Mysateles* en Isla de la Juventud, Cuba. Descripción de una nueva especie (Rodentia; Capromyidae, *Capromys*). Poeyana (Academia de Ciencias de Cuba), 315:1-12.

Varona, L. S., and O. Arrendondo. 1979. Nuevos taxones fósiles de Capromyidae (Rodentia: Caviomorpha). Poeyana (Academia de Ciencias de Cuba), 195:1-51.

Varshavskii, A. A. 1991. Geograficheskaya izmenchivost' tushkanchika-prygusta *Allactaga sibirica* (Rodentia, Dipodidae) v Mongolii [Geographical variation of the Mongolian five-toed jerboa *Allactaga sibirica* (Rodentia, Dipodidae) in Mongolia]. Zoologicheskii Zhurnal, 70(1):91-98 (in Russian).

Varshavsky, S. N. , M. N. Shylov, N. V. Popov, A. V. Survillo, M. A. Tarasov, V. P. Kozakevitch, P. S. Denisov, B. S. Varshavsky, Z. S. Sorokina, A. O. Adamyan, P. D. Golubev, and P. N. Korzhov. 1989. [Distribution of *Rattus norvegicus* in the south-eastern European part of the USSR and tasks of the anti-plague service.] Zoologischeskii Zhurnal, 68 (10):85-94 (in Russian with English summary).

Vasil'eva, A. G., and I. A. Vasil'eva. 1995. Non-metric variation in red vole populations within the East-Ural Radioactive Track (EURT) zone. Acta Theriologica, Suppl. 3:55-64.

Vasil'eva, I. 1999. Epigenetic divergence of Asian high-mountain voles of the subgenus *Aschizomys* from southern and north-eastern Siberia. Folia Zoologica, 48(suppl. 1):105-114.

Vasil'eva, M. V. 1961. K voprosu o sistematicheskom polozhenii i rasprostranenii suslika evropeiskovo (*Citellus citellus* L.) and maloasiatskovo (*Citellus xanthoprymnus* Benn.) [Toward the question of the systematic position and distribution of the European . . . and Near East . . . ground squirrels]. Spornik Trudov Zoologicheskovo Muzeu MGU, 8:253-260 (in Russian).

Vasiliu, G. 1961. Verzeichnis der Säugetiere Rumaniens. Säugetierkundliche Mitteilungen, 9:56-68.

Vassart, M., A. Séquéla, and H. Hayes. 1995. Chromosomal evolution in gazelles. Journal of Heredity, 86:216-227.

Vaughan, T. A. 1978. Mammalogy. Second ed. W. B. Saunders, Philadelphia, PA, 522 pp.

Vaurie, C. 1972. Tibet and its birds. H. F. and G. Witherby Ltd., London, 407 pp.

Veal, R., and W. Caire. 1979. *Peromyscus eremicus*. Mammalian Species, 118:1-6.

Vela, H. R. 1999. *Notiosorex crawfordi* (Coues, 1877) in the desert forests of Nuevo Leon, Mexico. Vertebrata Mexicana, 5:5-8.

Velazquez, A., F. A. Cervantes, and C. Galindo-Leal. 1993. The volcano rabbit *Romerolagus diazi*, a peculiar lagomorph. Lutra, 36:62-70.

Ventura, J. 1991. Morphological characteristics of the molars of *Arvicola terrestris* (Rodentia, Arvicolidae) in its southwestern distribution area. Zoologischer Anzeiger, 226:64-70.

Ventura, J. 1992. Coats and moults in *Arvicola terrestris* from the northeast of the Iberian Peninsula (Mammalia, Rodentia: Arvicolidae). Zoologische Abhandlungen Staatliches Museum für Tierkunde Dresden, 47 (8):95-110.

Ventura, J. 1993*a*. A discriminant function for sexual determination in *Arvicola terrestris monticola* (Rodentia, Arvicolidae) based on the morphometry of the innominate bone. Mammalia, 57(3):435-440.

Ventura, J. 1993*b*. Relative growth of sexual organs in males of *Arvicola terrestris* (Rodentia, Arvicolidae) from the Iberian Peninsula. Zeitschrift für Säugetierkunde, 58:124-126.

Ventura, J. 1993-1994. Crecimiento relativo de *Arvicola terrestris monticola* (Rodentia, Arvicolidae). Miscellània Zoològica, 17:237-348

Ventura, J. 2000. [Characteristic of *Arvicola sapidus*.] Pp. 163-172, *in* [Species of the fauna of Russia and the contiguous countries. The water vole: Mode of the species.] Nauka Publishers, Moscow, 527 pp. (in Russian).

Ventura, J., and J. Gosalbez. 1989. Taxonomic review of *Arvicola terrestris* (Linnaeus, 1758) (Rodentia, Arvicolidae) in the Iberian Peninsula. Bonner Zoologische Beiträge, 40:227-242.

Ventura, J., and J. Gosalbez. 1990. Caracteristicas de los pelajes y las mudas en *Arvicola sapidus* (Rodentia, Arvicolidae). Doñana, Acta Vertebrata, 17:3-15.

Ventura, J., and J. Gosálbez. 1992*a*. Datos sobre la estructura poblacional de *Arvicola sapidus* Miller, 1908 (Rodentia: Arvicolidae) del Delta del Ebro. Butlleti de la Institucio Catalana d'Historia Natural, 60:133-137.

Ventura, J., and J. Gosálbez. 1992*b*. Criterios para la determinación de la edad relativa en *Arvicola terrestris monticola* (Rodentia, Arvicolidae). Miscellània. Zoològica, 16:197-206.

Ventura, J., and M. J. Lopez-Fuster. 2000. Morphometric analysis of the black rat, *Rattus rattus*, from Congreso Island (Chafarinas Archipelago, Spain). Orsis, 15:91-102.

Ventura, J., and M. A. Sans-Fuentes. 1997. Geographic variation and divergence in nonmetric cranial traits of *Arvicola* (Mammalia, Rodentia) in southwestern Europe. Zeitschrift für Säugetierkunde, 62:99-107.

Ventura, J., J. Gosalbez, and M. J. Lopez-Fuster. 1989. Trophic ecology of *Arvicola sapidus* Miller, 1908 (Rodentia, Arvicolidae) in the Ebro Delta (Spain). Zoologischer Anzeiger, 223:283-290.

Ventura, J., M. J. López-Fuster, and E. Gispert. 1993. The abdominal arterial pattern of the northern water vole, *Arvicola terrestris* (Mammalia, Rodentia). Zoologische Jahrbücher. Abteilung für Anatomie und Ontogenie der Tiere, 123:97-102.

Ventura, J., J. M. Garde, and M. C. Escala. 1994. A method for sex determination in the southwestern water vole, *Arvicola sapidus*, using the innominate bone. Folia Zoologica, 43(3):225-229.

Ventura, J., M. J. López-Fuster, and M. Cabrera-Millet. 1998. The Cabrera Vole, *Microtus cabrerae*, in Spain: A biological and morphometric approach. Netherlands Journal of Zoology, 48(1):83-100.

Ventura, J., M. J. López-Fuster, M. Salazar, and R. Perez-Hernández. 2000. Morphometric analysis of some Venezuelan akodontine rodents. Netherlands Journal of Zoology, 50:487-501.

Vereshchagin, N. K. 1967. The mammals of the Caucasus, a history of the evolution of the fauna [A translation of N. K. Vereshchagin, 1959, Mlekopitayushchie Kavkaza: Istoriya formirovaniya fauny]. Israel Program for Scientific Translations, Jerusalem, 816 pp.

Verheyen, E., M. Colyn, and W. Verheyen. 1995. The phylogeny of some African muroids (Rodentia) based upon partial mitochondrial cytochrome *b* sequences. Belgium Journal of Zoology, 125(2):403-407.

Verheyen, E., M. Colyn, and W. Verheyen. 1996. A mitochondrial cytochrome *b* phylogeny confirms the paraphyly of the Dendromurinae Alston, 1896 (Muridae, Rodentia). Mammalia, 60(4):780-785.

Verheyen, W. N. 1959. Un genre de Sciuride nouveau pour la Faune du Congo belge: *Epixerus* Thomas 1909. Revue de Zoologie et de Botanique Africaines, 40:301-306.

Verheyen, W. N. 1962. Quelques notes sur la zoogeographie et la craniologie d' *Osbornictis piscivora* Allen 1919. Revue de Zoologie et de Botanique Africaines, 65:121-128.

Verheyen, W. N. 1963. Contribution a la systématique du genre *Idiurus* (Rodentia-Anomaluridae). Revue de Zoologie et de Botanique Africaines, 68:157-197.

Verheyen, W. N. 1964*a*. Description of *Lophuromys rahmi* a new species of Muridae from Central Africa. Revue de Zoologie et de Botanique Africaines, 69:206-213.

Verheyen, W. N. 1964*b*. Contribution a la systematique du genre *Uranomys* Dollman 1909. Revue de Zoologie et de Botanique Africaines, 70:386-400.

Verheyen, W. N. 1964*c*. New data on *Lophuromys luteogaster* Hatt 1934. Revue de Zoologie et de Botanique Africaines, 70(3-4):341-350.

Verheyen, W. N. 1965a. Contribution a l'etude systematique de *Mus sorella* (Thomas, 1909). Revue de Zoologie et de Botanique Africaines, 71:194-212.

Verheyen, W. N. 1965b. Some notes on the morphology of *Delanymys brooksi* Hayman 1962. Bulletins de la Societe Royale de Zoologie D'Anvers, 36:1-12.

Verheyen, W. N., and E. Van der Straeten. 1977. Description of *Malacomys verschureni*, a new murid-species from Central Africa (Mammalia-Muridae). Revue de Zoologie et de Botanique Africaines, 91:747-744.

Verheyen, W. N., and E. Van der Straeten. 1980. The karyotype of *Lophuromys nudicaudus* Heller 1911. Revue de Zoologie Africaine, 94:311-316.

Verheyen, W. N., and E. Van der Straeten. 1985. Karyological comparison of three different species of *Hybomys*. (Mammalia, Muridae). Pp. 29-34, *in* Proceedings of the International Symposium on African vertebrates. Systematics, phylogeny and evolutionary ecology (K.-L. Schuchmann, ed.). Zoologisches Forschungsinstitut und Museum Alexander Koenig, Bonn, 585 pp.

Verheyen, W. N., and J. Verschuren. 1966. Rongeurs et lagomorphes. Exploration du Parc National de la Garamba Institut des Parcs Nationaux du Congo, Bruxelles, 50:1-66.

Verheyen, W. N., M. Michiels, and J. van Rompaey. 1986. Genetic differences between *Lophuromys flavopunctatus* Thomas, 1888 and *Lophuromys woosnami* Thomas, 1906 in Rwanda. (Rodentia: Muridae). Cimbebasia, ser. A, 8:141-145.

Verheyen, W. N., C. Colyn, and J. Hulselmans. 1996a. Re-evaluation of the *Lophuromys nudicaudus* Heller, 1911 species-complex with a description of a new species from Zaire (Muridae-Rodentia). Bulletin de L'Institut Royal des Sciences Naturelles de Belgique, Biologie, 66:241-273.

Verheyen, E., M. Colyn, and W. Verheyen. 1996b. A mitochondrial cytochrome *b* phylogeny confirms the paraphyly of the Dendromurinae Alston, 1896 (Muridae, Rodentia). Mammalia, 60(4):780-785.

Verheyen, W. N., J. Hulselmans, M. Colyn, and R. Hutterer. 1997. Systematics and zoogeography of the small mammal fauna of Cameroun: Description of two new *Lophuromys* (Rodentia: Muridae) endemic to Mount Cameroun and Mount Oku. Bulletin de L'Institut Royal des Sciences Naturelles de Belgique, Biologie, 67:163-186.

Verheyen, W., T. Dierckx, and J. Hulselmans. 2000. The brush-furred rats of Angola and southern Congo: description of a new taxon of the *Lophuromys sikapusi* species complex. Bulletin de L'Institut Royal des Sciences Naturelles de Belgique, Biologie, 70:253-267.

Verheyen, W., J. L. J. Hulselmans, T. Dierckx, and E. Verheyen. 2002. The *Lophuromys flavopunctatus* Thomas 1888 species complex: A craniometric study, with the description and genetic characterization of two new species (Rodentia-Muridae-Africa). Bulletin de L'Institut Royal des Sciences Naturelles de Belgique, Biologie, 72:141-182.

Verheyen, W. N., J. L. J. Hulselmans, T. Dierckx, M. Colyn, H. Leirs, and E. Verheyen. 2003. A craniometric and genetic approach to the systematics of the genus *Dasymys* Peters, 1875, selection of a neotype and description of three new taxa (Rodentia, Muridae, Africa). Bulletin de L'Institut Royal des Sciences Naturelles de Belgique, Biologie, 73:27-71.

Verimli, R., E. Çolak, N. Yiğit, and M. Sözen. 2001. Blood serum proteins of *Apodemus flavicollis* and *Apodemus hermonensis* (Mammalia: Rodentia) in Turkey. Turkish Journal of Biology, 25(1):89-92.

Vermeiren, L., and W. N. Verheyen. 1980. Notes sur les *Leggada* de Lamto, Cote d'Ivoire, avec la description de *Leggada baoulei* sp. n. (Mammalia, Muridae). Revue de Zoologie Africaine, 94:570-590.

Vermeiren, L., and W. N. Verheyen. 1983. Additional data on *Mus setzeri* Petter (Mammalia, Muridae). Annales Musée Royal de l'Afrique Centrale, ser. 8 (Sciences Zoologiques), 237:137-141.

Verneau, O., F. Catzeflis, and A. V. Furano. 1997. Determination of the evolutionary relationships in *Rattus* sensu lato (Rodentia: Muridae) using L1 (LINE-1) amplification events. Journal of Molecular Evolution, 45:424-436.

Verneau, O., F. Catzeflis, and A. V. Furano. 1998. Determining and dating recent rodent speciation events by using L1 (LINE-1) retrotransposons. Proceedings of the National Academy of Sciences, 95:11284-11289.

Veron, G. 1992a. Histoire biogéographique du castor d'Europe, *Castor fiber* (Rodentia, Mammalia). Mammalia, 56:87-108.

Veron, G. 1992b. Etude morphometrique et taxonomique du genre *Castor*. Bulletin du Muséum National d'Histoire Naturelle (Paris), ser. 4 (Zoologie), 14:829-853.

Veron, G. 1995. La position systematique de *Cryptoprocta ferox* (Carnivora). Analyse cladistique des caracteres morphologiques de carnivores Aeluroidea actuels ete fossils. Mammalia, 59:551-582.

Veron, G., and F. M. Catzeflis. 1993. Phylogenetic relationships of the endemic Malagasy carnivore, *Cryptoprocta ferox* (Aeluroideae): DNA/DNA hybridization experiments. Journal of Mammalian Evolution, 1:169-185.

Veron, G., and S. Heard. 2000. Molecular systematics of the Asiatic Viverridae (Carnivora) inferred from mitochondrial cytochrome b sequence analysis. Journal of Zoological Systematics and Evolutionary Research, 38:209-217.

Veron, G., M. Colyn, A. E. Dunham, P. Taylor, and P. Gaubert. 2004. Molecular systematics and origin of sociality in mongooses (Herpestidae, Carnivora). Molecular Phylogenetics and Evolution, 30:582-598.

Verschuren, J. 1987. Liste commentée des Mammifères des Parcs Nationaux du Zaïre, du Rwanda et du Burundi. Bulletin de l'Institut Royal des Sciences Naturelles de Belgique, Biologie, 57:17-39.

Verts, B. J., and L. N. Carraway. 1987a. *Microtus canicaudus*. Mammalian Species, 267:1-4.

Verts, B. J., and L. N. Carraway. 1987b. *Thomomys bulbivorus*. Mammalian Species, 273:1-4.

Verts, B. J., and L. N. Carraway. 1998. Land mammals of Oregon. University of California Press, Berkeley, xvi + 668 pp.

Verts, B. J., and L. N. Carraway. 1999. *Thomomys talpoides*. Mammalian Species, 618:1-11.

Verts, B. J., and L. N. Carraway. 2000. *Thomomys mazama*. Mammalian Species, 641:1-7.

Verts, B. J., and L. N. Carraway. 2001. *Tamias minimus*. Mammalian Species, 653:1-10.

Verts, B. J., and G. L. Kirkland, Jr. 1988. *Perognathus parvus*. Mammalian Species, 318:1-8.

Verts, B. J., L. N. Carraway, and A. Kinlaw. 2001. *Spilogale gracilis*. Mammalian Species, 674:1-10.

Vesey-Fitzgerald, D. 1953. Notes on some rodents from Saudi Arabia and Kuwait. Journal of the Bombay Natural History Society, 51:424-428.

Vesey-Fitzgerald, D. F. 1966. The habits and habitats of small rodents in the Congo River Catchment Region of Zambia and Tanzania. Zoologica Africana, 2(1):111-122.

Vesmanis, I. E. 1984. Zur Verbreitung von *Jaculus orientalis* Erxleben, 1777 und *Jaculus jaculus* (Linnaeus, 1758) in Tunesien (Mammalia, Rodentia, Dipodidae). Zoologische Abhandlungen Staatliches Museum für Tierkunde in Dresden, 40(4):59-65.

Vesmanis, I. E. 1987. Zur Verbreitung der Etruskerspitzmaus *Suncus etruscus* (Savi, 1822) in den Maghreb-Ländern, unter besonderer Berücksichtigung Tunesiens (Mammalia, Insectivora, Soricidae). Faunistische Abhandlungen aus dem Staatlichen Museum für Tierkunde in Dresden, 15:35-39.

Vesmanis, I. E., and H. Kahmann. 1978. Morphometrische Untersuchungen an Wimperspitzmäusen (*Crocidura*). 4. Bemerkungen über die Typusreihe der kretaischen *Crocidura russula zimmermanni* Wettstein, 1953 im Vergleich mit *Crocidura gueldenstaedti caneae* (Miller, 1909). Säugetierkundliche Mitteilungen, 26:214-222.

Vesmanis, I. E., and A. Vesmanis. 1980. Ein Nachweis des Siebenschläfers, *Glis glis* (Linnaeus, 1766) aus Eulengewöllen von der Insel Elba, Italien. Faunistische Abhandlungen, 7(17):167-170.

Vianey-Liaud, M. 1974. Les rongeurs de loligocene inferieur d'Escamps. Paleovertebrata, 6:197-241.

Vianey-Liaud, M. 1985. Possible evolutionary relationships among Eocene and Lower Oligocene rodents of Asia, Europe and North America. Pp. 277-309, *in* Evolutionary Relationships Among Rodents. A multidisciplinary analysis. W. P. Luckett and J.-L. Hartenberger, eds. Plenum Press, New York, 721 pp.

Vianey-Liaud, M. 1989. Parallelism among Gliridae (Rodentia): The genus *Gliravus* Stehlin and Schaub. Historical Biology, 2:213-226.

Vianey-Liaud, M. 1994. La radiation des Gliridae (Rodentia) a l'Eocene superieur en Europe Occidentale, et sa descendance Oligocene. Muncher Geowissenschaftlichen Abhandlungen, A. 26:117-160.

Vianey-Liaud, M., and J. J. Jaeger. 1996. A new hypothesis for the origin of African Anomaluridae and Graphiuridae (Rodentia). Paleovertebrata, 25:349-358.

Vianey-Liaud, M., J.-J. Jaeger, J.-L. Hartenberger and M. Mahboubi. 1994. Les rongeurs de l'Eocene d'Afrique nord-occidentale [Glib Zegdou (Algerie) et Chambi (Tunisie)] et l'origine des Anomaluridae. Paleovertebrata, 23(1-4):93-118.

Vidal, O. 1990. Lista de los mamiferos acuaticos de Columbia. Informe del Museo del Mar (Universidad Jorge Tadeo Lozano, Bogota, Colombia), 34:1-18.

Vidal, O. R., R. Riva, and N. I. Baro. 1976. Los cromosomas del genero *Holochilus*. I. Polimorfismo en *H. chacarius* Thomas (1906). Physis (Buenos Aires), 35:75-85.

Viegas-Péquignot, E., C. Derbin, F. Lemeunier, and E. Taillandier. 1982. Identification of left-handed Z-DNA by indirect immunomethods in metaphasic chromosomes of a mammal, *Gerbillus nigeriae* (Gerbillidae, Rodentia). Annales de Génétique, 25(4):218-222.

Viegas-Péquignot, E., B. Dutrillaux, M. Prod'Homme, and F. Petter. 1983. Chromosomal phylogeny of Muridae: A study of 10 genera. Cytogenetics and Cell Genetics, 35:269-278.

Viegas-Péquignot, E., S. Kasahara, Y. Yassuda, and B. Dutrillaux. 1985. Major chromosome homeologies between Muridae and Cricetidae. Cytogenetics and Cell Genetics, 39:258-261.

Viegas-Péquignot, E., D. Petit, T. Benazzou, M. Prod'Homme, M. Lombard, F. Hoffschir, J. Descailleaux, and B. Dutrillaux. 1986. Phylogenie chromosomique chez les Sciuridae, Gerbillidae et Muridae, et etude d'especes appartenant a d'autres familles de rongeurs. Pp. 164-202, *in* Evolution chromosomique chez les primates, les carnivores et les rongeurs (B. Dutrillaux, ed.). Mammalia, Numero special, 50:1-203.

Vieira, C. O. da C. 1949 [1950]. Xenartros e marsupiais do estado de São Paulo. Arquivos de Zoologia do Estado de São Paulo, 7:325-362.

Viette, P. 1991. Principales localités où des insectes ont été recueillis à Madagascar. Faune de Madagascar, supplément 2, 88 pp.

Vigne, J.-D. 1983. Le remplacement des faunes de petits mammifères en Corse, lors de l'arrivé de l'homme. Comptes Rendus Séances de la Societé Biogéographique, 59:41-51.

Vigne, J.-D. 1987. L'extinction holocène du fond de peuplement mammalien indigène des îles de Méditerranée occidentale. Mémoires du Societé Géologique du France, n.s., 150:167-177.

Vigne, J.-D. 1988. Les Mammifères post-glaciaires de Corse. Étude Archéozoologique (26e supplement a Gallia Préhistoire). Éditions du Centre National de la Recherche Scientifique, Paris, 337 pp.

Vigne, J.-D. 1990. Biogeographical history of the mammals on Corsica (and Sardinia) since the final Pleistocene. Pp. 370-392, in Biogeographical Aspects of Insularity (International Symposium, Roma, 1987) (A. Azzarali, ed.). Atti dei Convegni Lincei, 85:840 pp.

Vigne, J.-D. 1992. Zooarchaeology and biogeographical history of the mammals of Corsica and Sardinia since the last Ice Age. Mammal Review, 22(2):87-96.

Vigne, J.-D., and J. A. Alcover. 1985. Incidence des relations historiques entre l'homme et l'animal dans la composition actuelle du peuplement amphibien, reptilien et mammalien des îles de Méditerranée occidentale. Actes du 110e Congrès National des Sociétés Savantes, Montpellier 1985, section des sciences, fascicule 2:79-91.

Vigne, J.-D., and M.-C. Marinval-Vigne. 1990. Nouvelles données sur l'histoire des musaraignes en Corse (Insectivora, Soricidae). Vie et Milieu, 40:207-212.

Vigne, J.-D., and M.-C. Marinval-Vigne. 1991. Réflexions écologiques sur le renouvellement holocène des micromammifères en Corse: Les données préliminaires des fossiles du Monte di Tuda. Pp. 183-193, in Le rongeur et l'espace, Actes du Colloque International, Lyon, 1989 (M. Le Berre and L. Le Guelte, eds.). Raymond Chabaud, Paris, 362 pp.

Vigne, J.-D., G. Cheylan, L. Granjon, and J.-C. Auffray. 1993. Evolution osteometrique de Rattus rattus et de Mus musculus domesticus sur de petites iles: Comparison de populations medievales et actuelles des iles Lavezzi (Corse) et de Corse. Mammalia, 57(1):85-98.

Vilà, C., J. E. Maldonado, and R. K. Wayne. 1999. Phylogenetic relationships, evolution, and genetic diversity of the domestic dog. Journal of Heredity, 90:71-77.

Viljoen, S. 1989. Taxonomy and historical zoogeography of the red squirrel, Paraxerus palliatus (Peters, 1852) in the southern African subregion (Rodentia: Sciuridae). Annals of the Transvaal Museum, 35(2):49-59.

Villa-R., B. 1966 [1967]. Los murciélagos de Mexico. Anales de Instituto de Biologia, Universidad Nacional Autónomica de México, 491 pp.

Villalobos, F., and A. A. Valerio. 2002. The phylogenetic relationships of the bat genus Sturnira Gray, 1842 (Chiroptera: Phyllostomidae). Mammalian Biology, 67:268-275.

Vincent, T. 2001. Le rat brun Rattus norvegicus (Berkenhout, 1769) et le rat noir Rattus rattus (Linné, 1758) en centre-ville et sur le port du Havre (Seine-Maritime, France). Bulletin Trimestriel de la Societe Geologique de Normandie et des Amis du Muséum du Havre, 87(1):53-57.

Vinogradov, B. S. 1925. On the structure of the external genitalia in Dipodidae and Zapodidae (Rodentia) as a classificatory character. Proceedings of the Zoological Society of London, 1925(1):572-585.

Vinogradov, B. S. 1928. A third species of dwarf jerboa Salpingotus thomasi sp. n. Annals and Magazine of Natural History, ser. 10, 1:372-374.

Vinogradov, B. S. 1930. [On the classification of Dipodidae (Rodentia). I. Cranial and dental characters.] Izvestiya Akademii Nauk SSSR, 1930:331-350 (in Russian).

Vinogradov, B. S. 1937. Fauna SSSR; Mlekopitaiushchie, tom. 3, vyp. 4. Tushkanchiki. [Fauna of the USSR; Mammals, vol. 3, pt. 4. Jerboas.], 196 pp. (in Russian).

Vinogradov, B. S., A. I. Argyropulo, and V. G. Heptner. 1936. [Rodents of the Central Asiatic part of USSR.] Akademiya Nauk SSSR, Moscow, 228 pp. (in Russian).

Violani, C., and B. Zava. 1995. Carolus Linnaeus and the edible dormouse. Pp. 109-115, in Proceedings of II Conference on Dormice (Rodentia, Myoxidae) (M. G. Filippucci, ed.). Hystrix, n.s., 6(1-2):1-340.

Viriot, L., D. Zany, J. Chaline, F. Courant, P. Brunet-Lecomte and S. Simone. 1991. Complements aux faunes de rongeurs des gisements d'Aldene (Cesseras, Herault), de la grotte du Prince (Grimaldi, Ligurie) et de l'Observatoire (Monaco). Bulletin du Musee d'Anthropologie Prehistorique de Monaco, 34:7-16.

Viriot, L., J. Chaline, A. Schaaf, and E. Le Boulenge. 1993. Ontogenetic change of Ondatra zibethicus (Arvicolidae, Rodentia) cheek teeth analyzed by digital image processing. Pp. 373-391, in Morphological change in Quaternary mammals of North America (R. A. Martin and A. D. Barnosky, eds.). Cambridge University Press, London, 415 pp.

Viro, P., and J. Niethammer. 1982. Clethrionomys glareolus (Schreber, 1780)—rotelmaus. Pp. 109-146, in Handbuch der Säugetiere Europas (J. Niethammer and F. Krapp, eds.). Akademische Verlagsgesellschaft (Wiesbaden), 2/I:1-649.

Viroux, M.-C., and V. Bauchau. 1992. Segregation and fertility in *Mus musculus domesticus* (wild mice) heterozygous for the Rb(4.12) translocation. Heredity, 68:131-134.

Visser, D. S., and T. J. Robinson. 1986. Cytosystematics of the South African *Aethomys* (Rodentia: Muridae). South African Journal of Science, 21:264-268.

Visser, D. S., and T. J. Robinson. 1987. Systematic implications of spermatozoan and bacular morphology for the South African *Aethomys*. Mammalia, 51:447-454.

Vitullo, A. D., and J. A. Cook. 1991. The role of sperm morphology in the evolution of tuco-tucos, *Ctenomys* (Rodentia: Ctenomyidae): Conformation of results from Bolivian species. Zeitschrift für Säugetierkunde, 56:359-364.

Vitullo, A. D., M. S. Merani, O. A. Reig, A. E. Kajon, O. Scaglia, M. B. Espinosa, and A. Perez-Zapata. 1986. Cytogenetics of South American akodont rodents (Cricetidae): New karyotypes and chromosomal banding patterns of Argentinian and Uruguayan forms. Journal of Mammalogy, 67:69-80.

Vitullo, A. D., E. R. S. Roldan, and M. S. Merani. 1988. On the morphology of spermatozoa of tuco-tucos *Ctenomys* (Rodentia: Ctenomyidae): New data and its implications for the evolution of the genus. Journal of Zoology, London, 215:675-683.

Vitullo, A. D., M. B. Espinosa, and M. S. Merani. 1990. Cytogenetics of vesper mice, *Calomys* (Rodentia: Cricetidae): Robertsonian variation between *Calomys callidus* and *Calomys venustus*. Zeitschrift für Säugetierkunde, 55:99-105.

Vivar, E., V. Pacheco, and M. Valqui. 1997. A new species of *Cryptotis* (Insectivora: Soricidae) from northern Peru. American Museum Novitates, 3202:1-15.

Vivo, M. de. 1985. On some monkeys from Rondônia, Brasil (Primates: Callitrichidae, Cebidae). Pápeis Avulsos de Zoologia, São Paulo, 36(11):103-110.

Vivo, M. de. 1991. Taxonomia de *Callithrix* Erxleben, 1777 (Callitrichidae, Primates). Fundaçao Biodiversitas, Belo Horizonte, 105 pp.

Vizcaíno, S. F. 1995. Identifacación específica de las "mulitas," gënero *Dasypus* L. (Mammalia, Dasypodidae), del noroeste Argentino. Descripción de una nueva especie. Mastozoologia Neotropical, 2:5-13.

Vizcaíno, S. F., and A. Giallombardo. 2001. Armadillos del noroeste Argentino (Provincias de Jujuy y Salta). Edentata, 4:5-9.

Vlasak, P., and J. Niethammer. 1990. *Crocidura suaveolens* (Pallas, 1811)—Gartenspitzmaus. Pp. 397-428, *in* Handbuch der Säugetiere Europas (J. Niethammer, and F. Krapp, eds.). Aula-Verlag, Wiesbaden, 3/I:1-524.

Vogel, P. 1983. Contribution à l'écologie et à la zoogeographie de *Micropotamogale lamottei* (Mammalia, Tenrecidae). Revue de Ecologie (La Terre et la Vie), 38:37-49.

Vogel, P. 1986. Der Karyotyp der Kretaspitzmaus, *Crocidura zimmermanni* Wettstein, 1953 (Mammalia, Insectivora). Bonner Zoologische Beiträge, 37:35-38.

Vogel, P. 1988. Taxonomical and biogeographical problems in Mediterranean shrews of the genus *Crocidura* (Mammalia, Insectivora) with reference to a new karyotype from Sicily (Italy). Bulletin de la Societé Vaudoise des Sciences Naturelles, 79:39-48.

Vogel, P., and F. Besançon. 1979. A propos de la position systématique des genres *Nectogale* et *Chimarrogale* (Mammalia, Insectivora). Revue Suisse de Zoologie, 86:335-338.

Vogel, P., and H. Frey. 1995. L'hibernation du muscardin *Muscardinus avellanarius* (Gliridae, Rodentia) en nature: Nids, fréquence des réveils et température corporelle. Bulletin de la Societe Vaudoise des Sciences Naturelles, 83(3):217-230.

Vogel, P., and T. S. Sofianidou. 1996. The shrews of the genus *Crocidura* on Lesbos, an eastern Mediterranean island. Bonner Zoologische Beiträge, 46:339-347.

Vogel, P., T. Maddalena, and F. Catzeflis. 1986. A contribution to the taxonomy and ecology of shrews (*Crocidura zimmermanni* and *C. suaveolens*) from Crete and Turkey. Acta Theriologica, 31:537-545.

Vogel, P., R. Hutterer, and M. Sarà. 1989. The correct name, species diagnosis, and distribution of the Sicilian shrew. Bonner Zoologische Beiträge, 40:243-248.

Vogel, P., T. Maddalena, and P. J. Schembri. 1990. Cytotaxonomy of shrews of the genus *Crocidura* from Mediterranean islands. Vie et Milieu, 40:124-129.

Vogel, P., T. Maddalena, A. Mabille, and G. Paquet. 1991. Confirmation biochimique du statut specifique du mulot alpestre *Apodemus alpicola* Heinrich, 1952 (Mammalia, Rodentia). Bulletin Societie Vaudoise des Sciences Naturelles, 80:471-481.

Vogel, P., T. Maddalena, and M. Sarà. 1992. The taxonomic status of *Crocidura cossyrensis* Contoli, 1989 and its relationship to African and European *Crocidura russula* (Mammalia, Insectivora). Israel Journal of Zoology, 38:424.

Vogel, P., M. Lawrence, and A. Aghnaj. 2000. Shrews (Soricidae, Mammalia) of the National Park Souss-Massa, Morocco. Revue Suisse de Zoologie, 107:591-599.

Vogel, P., J.-F. Cosson, and L. F. López-Jurado. 2003. Taxonomic status and origin of the shrews (Soricidae) from the Canary Islands inferred from a mtDNA comparison with the European *Crocidura* species. Molecular Phylogenetics and Evolution, 27:271-282.

Vogel, W., P. Steinbach, M. Djalali, K. Mehnert, S. Ali, and J. T. Epplen. 1988. Chromosome 9 of *Ellobius lutescens* is the X chromosome. Chromosoma, 96:112-118.

Vogel, W., S. Jainta, W. Rau, C. Geerkens, A. Baumstark, L.-S. Correa-Cerro, C. Ebenhoch, and W. Just. 1998. Sex determination in *Ellobius lutescens*: The story of an enigma. Cytogenetics and Cell Genetics, 80:314-221.

Vohralík, V. 1985. Notes on the distribution and the biology of small mammals in Bulgaria (Insectivora, Mammalia) I. Acta Universitatis Carolinae, Biologica, 1981:445-461.

Vohralík, V. 1991. A record of the mole *Talpa levantis* (Mammalia: Insectivora) in Bulgaria and the distribution of the species in the Balkans. Acta Universitatis Carolinae Biologica, 35:119-127.

Vohralík, V. 1992. Small mammals (Insectivora, Rodentia) of Thrace, Greece. Acta Universitatis Carolinae Biologica, 36:341-369.

Vohralík, V. 2002. Distribution, skull morphometrics and systematic status of an isolated population of *Apodemus microps* (Mammalia:Rodentia) in NW Bohemia, Czech Republic. Acta Societatis Zoologicae Bohemicae, 66:67-80.

Vohralík, V., and M. Andera. 2000. [New records of the black rat (*Rattus rattus*) in Prague and Central Bohemia (Czech Republic).] Lynx (Praha), n.s., 31:159-160 (in Czech with English abstract).

Vohralík, V., and T. Sofianidou. 1987. Small mammals (Insectivora, Rodentia) of Macedonia, Greece. Acta Universitatis Carolinae, Biologica, 1985:319-354.

Vohralík, V., and T. Sofianidou. 1992a. Small mammals (Insectivora, Rodentia) of Thrace, Greece. Acta Universitatis Carolinae, Biologica, 36:341-369.

Vohralík, V., and T. S. Sofianidou. 1992b. New records of *Apodemus agrarius* (Mammalia: Rodentia) in Greece and the distribution of the species in the south of the Balkans. Pp. 217-220, *in* Prague studies in mammalogy (I. Horáeek and V. Vohralík, eds.). Charles University Press, Praha, 246 pp.

Vohralík, V., T. S. Sofianidou, and D. Frynta. 1996. Morphological comparison between mainland and insular *Apodemus* (Mammalia: Rodentia) populations from the North Aegean Region. Vie et Milieu, 46(3-4):375 (abstract).

Volleth, M. 1992. Comparative analysis of the banded karyotypes of the European *Nyctalus* species (Vespertilionidae; Chiroptera). Pp. 221-226, *in* Prague studies in mammalogy, (I Horácek and V. Vohralík, eds.). Charles University Press, Prague, 246 pp.

Volleth, M., and K.-G. Heller. 1994. Phylogenetic relationships of vespertilionid genera (Mammalia: Chiroptera) as revealed by karyological analysis. Zeitschrift für Zoologische Systematik und Evolutionsforschung, 32:11-34.

Volleth, M., and C. R. Tidemann. 1991. The origin of the Australian Vespertilioninae bats, as indicated by chromosomal studies. Zeitschrift für Säugetierkunde, 56:321-330.

Volleth, M., G. Bonner, M. C. Göpfert, K.-G. Heller, O. von Helversen, and H.-S. Yang. 2001. Karyotype comparison and phylogenetic relationships of *Pipistrellus*-like bats (Vespertilionidae; Chiroptera; Mammalia). Chromosome Research, 9:25-46.

Voloubouev, V. T. (ed.). 2003. Evolution in the *Sorex araneus* group: Cytogenetics and molecular aspects. Sixth meeting of the International *Sorex araneus* Cytogenetics Committee (ISACC) and associated symposium in honour of Professor Karl Fredga. Mammalia, 67:163-320.

Voloubouev, V. T., and V. M. Aniskin. 2000. Comparative chromosome banding analysis of three South American species of rice rats of the genus *Oryzomys* (Rodentia, Sigmodontinae). Chromosome Research, 8:295-304.

Voloubouev, V. T., and F. M. Catzeflis. 2000. Chromosome banding analysis (G-, R-, and C-bands) of *Rhipidomys nitela* and a review of the cytogenetics of *Rhipidomys*. Mammalia, 64:353-360.

Voloubouev, V. T., and L. Granjon. 1996. A finding of the XX/XY$_1$Y$_2$ sex-chromosome system in *Taterillus arenarius* (Gerbillinae, Rodentia) and its phylogenetic implications. Cytogenetics and Cell Genetics, 75:45-48.

Voloubouev, V. T., and C. G. van Zyll de Jong. 1988. The karyotype of *Sorex arcticus maritimensis* (Insectivora, Soricidae) and its systematic implications. Canadian Journal of Zoology, 66:1968-1972.

Voloubouev, V. T., D. V. Ternovsky, and A. S. Graphodatsky. 1974. [Taxonomic status of the ferret (*Putorius putorius furo*) by karyological data.] Zoologicheski Zhurnal, 53:1738-1740 (in Russian).

Voloubouev, V. T., E. Viegas-Péquignot, F. Petter, and B. Dutrillaux. 1987. Karyotypic diversity and taxonomic problems in the genus *Arvicanthis* (Rodentia, Muridae). Genetica, 72:147-150.

Voloubouev, V. T., E. Viegas-Péquignot, M. Lombard, F. Petter, J.-M. Duplantier, and B. Dutrillaux. 1988a. Chromosomal evidence for a polytypic structure of *Arvicanthis niloticus* (Rodentia, Muridae). Zeitschrift für Zoologische Systematik und Evolution Forschung, 26:276-285.

Voloubouev, V. T., E. Viegas-Péquignot, F. Petter, J.-C. Gautun, B. Sicard, and B. Dutrillaux. 1988b. Complex chromosomal polymorphism in *Gerbillus nigeriae* (Rodentia, Gerbillidae). Journal of Mammalogy, 69(1):131-134.

Volobouev, V. T., M. Tranier, and B. Dutrillaux. 1991. Chromosome evolution in the genus *Acomys*: Chromosome banding analysis of *Acomys* cf. *dimidiatus* (Rodentia, Muridae). Bonner Zoologische Beiträge, 42:253-260.

Volobouev, V. T., M. Lombard, M. Tranier, and B. Dutrillaux. 1995*a*. Chromosome-banding study in Gerbillinae (Rodentia). I. Comparative analysis of *Gerbillus poecilops, G. henleyi* and *G. nanus*. Journal of Zoological, Systematic, and Evolutionary Research, 33:54-61.

Volobouev, V. T., N. Vogt, E. Viegas-Péquignot, B. Malfoy, and B. Dutrillaux. 1995*b*. Characterization and chromosomal location of two repeated DNAs in three *Gerbillus* species. Chromosoma, 104:252-259.

Volobouev, V. T., J.-C. Gautun, and M. Tranier. 1996. Chromosome evolution in the genus *Acomys* (Rodentia, Muridae): Chromosome banding analysis of *Acomys cahirinus*. Mammalia, 60(2):217-222.

Volobouev, V. T., A. Hoffmann, B. Sicard, and L. Granjon. 2001. Polymorphism and polytypy for pericentric inversions in 38-chromosome *Mastomys* (Rodentia, Murinae) and possible taxonomic implications. Cytogenetic and Cell Genetics, 92:237-242.

Volobouev, V. T., J.-F. Ducroz, V. M. Aniskin, J. Britton-Davidian, R. Castiglia, G. Dobigny, L. Granjon, M. Lombard, M. Corti, B. Sicard, and E. Capanna. 2002*a*. Chromosomal characterisation of *Arvicanthis* species (Rodentia, Murinae) from western and central Africa: Implications on taxonomy. Cytogenetics and Genome Research, 96:250-260.

Volobouev, V. T., V. M. Aniskin, E. Lecompte, and J. F. Ducroz. 2002*b*. Patterns of karyotype evolution in complexes of sibling species within three genera of African murid rodents inferred from the comparison of cytogenetic and molecular data. Cytogenetic and Genome Research, 96:261-275.

Volpers, T., and A. Kumirai. 1996. A review of the genus *Epomophorus* Bennet, 1836 (Mammalia, Megachiroptera) in Zimbabwe. Arnoldia Zimbabwe, 10:103-109.

von Helversen, O., F. Mayer, D. Kock, A. M. Hutson, and G. Jones. 2000. Comments on the proposed designation of neotypes for the nominal species *Vespertilio pipistrellus* Schreber, 1774 and *V. pygmaeus* Leach, 1825 (currently *Pipistrellus pipistrellus* and *P. pygmaeus*; Mammalia, Chiroptera) (Case 3073). Bulletin of Zoological Nomenclature, 57(2):113-116.

von Helversen, O., K. G. Heller, F. Mayer, A. Nemeth, M. Volleth, and P. Gombkoto. 2001. Cryptic mammalian species: A new species of whiskered bat (*Myotis alcathoe* n. sp.) in Europe. Naturwissenschaften, 88(5):217-223.

Vonhof, M. J. 2000. *Rhogeessa tumida*. Mammalian Species, 633:1-3.

Vonhof, M. J., and M. C. Kalcounis. 1999. *Lavia frons*. Mammalian Species, 614:1-4.

Von Richter, W. 1974. *Connochaetes gnou*. Mammalian Species, 50:1-6.

Von Vietinghoff-Riesch, A. F. 1960. Der Siebenschläfer (*Glis glis* L.). Monographien der Wildsäugetiere, 14:1-196.

Voronov, G. A. 1992. [*Microtus fortis* (Rodentia, Cricetidae), a new species in the fauna of Sakhalin Island.] Zoologischeskii Zhurnal, 71(4):85-89 (in Russian with English summary).

Vorontsov, N. N. 1959. [The system of hamsters (Cricetinae) in the sphere of the world fauna and their phylogenetic relations.] Byulleten' Moskovskovo Obshchestva Ispytatelei Prirody, Otdel Biologicheskii, 64:134-137 (in Russian).

Vorontsov, N. N. 1966. [Taxonomic position and a survey of the hamsters of the genus *Mystromys* Wagn. (Mammalia, Glires).] Zoologicheskii Zhurnal, 45:436-446 (in Russian).

Vorontsov, N. N. 1967. Evolyutsiya pishchevaritel'noi sistemy gryzunov mysheobraznye [Evolution of the alimentary system of myomorph rodents]. Nauka, Novosibirsk, 235 pp. (in Russian).

Vorontsov, N. N. 1969. [The chromosome numbers and taxonomical relations of the members of the superfamily Dipodoidea (Rodentia).] Pp. 92-93, *in* [The mammals: Evolution, karyology, faunistics, systematics. 2nd All-Union Mammalogy Conference, Moscow] (N. N. Vorontsov, ed.). Academy of Sciences of the USSR (Siberian Branch), Novosibirsk, 167 pp. (in Russian).

Vorontsov, N. N. 1979. Evolution of the alimentary system of myomorph rodents [A translation of N. N. Vorontsov, 1967, Evolyutsiya pishchevaritel'noi sistemy gryzunov mysheobraznye]. Indian National Scientific Documentation Centre, New Delhi, 346 pp.

Vorontsov, N. N. 1982. [The hamsters (Cricetidae) of the world fauna. Part I. Morphology and ecology.] Fauna of the USSR (Ser. Nov. No. 125), Mammals, 3(6):1-451, Nauka, Leningrad (in Russian).

Vorontsov, N. N. 1986. [The Szechuan dormouse—a new genus of mammals.] Priroda, Moscow, 3:94-95 (in Russian).

Vorontsov, N. N., and E. Yu. Ivanitskaya. 1973. Comparative karyology of north Palaearctic pikas (*Ochotona*, Ochotonidae, Lagomorpha). Caryologia, 26:213-223.

Vorontsov, N. N., and E. A. Lyapunova. 1970. Khromosomnye chisla i vidoobrazovanie u nazemnykh belich'ikh (Sciuridae:Xerinae et Marmotinae) Golarktiki [Chromosome number and species formation in ground squirrels . . . of the Holarctic]. Byulleten' Moskovskovo Obshchestva Ispytatelei Prirody, Otdel Biologicheskii, 75:122-136 (in Russian).

Vorontsov, N. N., and E. A. Lyapunova. 1986. Genetics and problems of Trans-Beringian connections of Holarctic mammals. Pp. 441-481, *in* Beringia in the Cenozoic Era (V. L. Kontrimavichus, ed.) [translation of Beringiya v Kainozoe, 1976]. A. A. Balkema, Rotterdam, 724 pp.

Vorontsov, N. N., and L. Ya Martynova. 1976. Population cytogenetics of Altai zokor *Myospalax myospalax* Laxm. (Rodentia, myospalacinae). Doklady Biological Sciences, 230(1-6):401-404 (translated from Doklady Akademii Nauk SSSR, 230(2):447-449).

Vorontsov, N. N., and E. G. Potapova. 1979. [Taxonomy of the genus *Calomyscus* (Cricetidae). 2. Status of *Calomyscus* in the system of Cricetinae.] Zoologicheskii Zhurnal, 58:1391-1397 (in Russian).

Vorontsov, N. N., and G. I. Shenbrot. 1984. [A systematic review of the genus *Salpingotus* (Rodentia, Dipodidae), with a description of *Salpingotus pallidus* sp. n. from Kazakhstan.] Zoologicheskii Zhurnal, 63(5):731-744 (in Russian).

Vorontsov, N. N., O. J. Orlov, and V. M. Smirnov. 1969*a*. [Biology and distribution of three-toed dwarf jerboas *Saplingotus crassicauda* in the Zaisan Basin.] Pp. 69-73, *in* [The mammals: Evolution, karyology, faunistics, systematics. 2nd All- Union Mammalogy Conference, Moscow] (N. N. Vorontsov, ed.). Academy of Sciences of the USSR (Siberian Branch), Novosibirsk, 167 pp. (in Russian).

Vorontsov, N. N., O. J. Orlov, and N. A. Malygina. 1969*b*. [Biology and taxonomy of fat-tailed jerboas (*Pygerethmus*) and comparative karyology of the genera *Pygerethmus* and *Alactagulus*.] Pp. 74-84, *in* [The mammals: Evolution, karyology, faunistics, systematics. 2nd All-Union Mammalogy Conference, Moscow] (N. N. Vorontsov, ed.). Academy of Sciences of the USSR (Siberian Branch), Novosibirsk, 167 pp. (in Russian).

Vorontsov, N. N., S. I. Radjabli, and N. A. Malygina. 1969*c*. [The comparative karyology of the five-toed jerboas of the genus *Allactaga* (Allactaginae, Dipodidae, Rodentia).] Pp. 85-87, *in* [The mammals: Evolution, karyology, faunistics, systematics. 2nd All- Union Mammalogy Conference, Moscow] (N. N. Vorontsov, ed.). Academy of Sciences of the USSR (Siberian Branch), Novosibirsk, 167 pp. (in Russian).

Vorontsov, N. N., N. A. Malygina, and S. I. Radjabli. 1969*d*. [Chromosome compliments of the jerboas of the subfamilies Dipodinae and Cardiocraniinae (Dipodidae, Rodentia).] Pp. 88-91, *in* [The mammals: Evolution, karyology, faunistics, systematics. 2nd All- Union Mammalogy Conference, Moscow] (N. N. Vorontsov, ed.). Academy of Sciences of the USSR (Siberian Branch), Novosibirsk, 167 pp. (in Russian).

Vorontsov, N. N., T. S. Bekasova, B. Král, K. V. Korobitsina, and E. Yu. Ivanitskaya. 1977*a*. [On specific status of Asian wood mice of the genus *Apodemus* (Rodentia, Muridae) from Siberia and the Far East.] Zoologicheskii Zhurnal, 56:437-449 (in Russian).

Vorontsov, N. N., L. Ya. Martynova, and I. I. Fomichova. 1977*b*. [An electrophoretic comparison of the blood proteins in mole rats of the fauna of the USSR (Spalacinae, Rodentia).] Zoologicheskii Zhurnal, 56:1207-1215 (in Russian).

Vorontsov, N. N., E. A. Lyapunova, E. Yu. Ivanitskaya, C. F. Nadler, B. Král, A. I. Kozlovskii, and R. S. Hoffmann. 1978. [Variability of mammalian sex chromosomes. I. Geographic variability of the structure of the Y chromosomes in red-backed voles of the genus *Clethrionomys* (Rodentia, Microtinae).] Genetika, 14:1432-1446.

Vorontsov, N. N., I. V. Kartavtseva, and E. G. Potapova. 1979. [Systematics of the genus *Calomyscus* (Cricetidae). 1. Karyological differentiation of the sibling species from Transcaucasia and Turkmenia and a review of species of the genus *Calomyscus*.] Zoologicheskii Zhurnal, 58:1213-1224 (in Russian).

Vorontsov, N. N., E. A. Lyapunova, Yu. M. Borisov, and V. E. Dovgal. 1980. Variability of sex chromosomes in mammals. Genetica, 52/53:361-372.

Vorontsov, N. N., S. V. Mezhzherin, G. G. Boeskorov, and E. A. Lyapunova. 1989. [Genetic differentiation of sibling species of wood mice (*Apodemus*) in the Caucasia and their diagnostics.] Doklady Akademii Nauk SSSR, 309:1234-1238 (in Russian).

Vorontsov, N. N., G. G. Boyeskorov, S. V. Mezhzherin, E. A. Lyapunova, and A. S. Kandaurov. 1992. [Systematics of the Caucasian wood mice of the subgenus *Sylvaemus* (Mammalia, Rodentia, *Apodemus*).] Zoologicheskii Zhurnal, 71:119-131 (in Russian with English summary).

Vosmaer, A. 1766. Natuurlyke historie van het Africaansche Breedsnuitig Varken, of Bosch-Zwyn. P. Meijer, Amsterdam, 15 pp.

Voss, R. S. 1988. Systematics and ecology of ichthyomyine rodents (Muroidea): Patterns of morphological evolution in a small adaptive radiation. Bulletin of the American Museum of Natural History, 188:259-493.

Voss, R. S. 1991*a*. An introduction to the Neotropical muroid rodent genus *Zygodontomys*. Bulletin of the American Museum of Natural History, 210:1-113.

Voss, R. S. 1991*b*. On the identity of "*Zygodontomys*" *punctulatus* (Rodentia: Muroidea). American Museum Novitates, 3026:1-8.

Voss, R. S. 1992. A revision of the South American species of *Sigmodon* (Mammalia: Muridae) with notes on their natural history and biogeography. American Museum Novitates, 3050:1-56.

Voss, R. S. 1993. A revision of the Brazilian muroid rodent genus *Delomys* with remarks on "Thomasomyine" characters. American Museum Novitates, 3073:44 pp.

Voss, R. S. 2003. A new species of *Thomasomys* (Rodentia: Muridae) from eastern Ecuador, with remarks on mammalian diversity and biogeography in the Cordillera Oriental. American Museum Novitates, 3421:1-47.

Voss, R. S., and N. I. Abramson. 1999. *Holochilus* Brandt, 1835, *Proechimys* J. A. Allen, 1899, and *Trinomys* Thomas, 1921 (Mammalia: Rodentia): Proposed conservation by the designation of *H. sciureus* Wagner, 1842 as the type species of *Holochilus*. Bulletin of Zoological Nomenclature, 56:255-261.

Voss, R. S., and R. Angermann. 1997. Revisionary notes on Neotropical porcupines (Rodentia: Erethizontidae). 1. Type material described by Olfers (1818) and Kuhl (1820) in the Berlin Zoological Museum. American Museum Novitates, 3214:1-42.

Voss, R. S., and M. D. Carleton. 1993. A new genus for *Hesperomys molitor* Winge and *Holochilus magnus* Hershkovitz (Mammalia, Muridae) with an analysis of its phylogenetic relationships. American Museum Novitates, 3085:39 pp.

Voss, R. S., and L. H. Emmons. 1996. Mammalian diversity in Neotropical lowland rainforests: A preliminary assessment. Bulletin of the American Museum of Natural History, 230:1-115.

Voss, R. S., and A. V. Linzey. 1981. Comparative gross morphology of male accessory glands among Neotropical Muridae (Mammalia: Rodentia) with comments on systematic implications. Miscellaneous Publications, Museum of Zoology, University of Michigan, 159:1-41.

Voss, R. S., and P. Myers. 1991. *Pseudoryzomys simplex* (Rodentia: Muridae) and the significance of Lund's collections from the caves of Lagoa Santa, Brazil. Bulletin of the American Museum of Natural History, 206:414-432.

Voss, R. S., and M. N. F. da Silva. 2001. Revisionary notes on Neotropical porcupines (Rodentia: Erethizontidae). 2. A review of the *Coendou vestitus* group with descriptions of two new species from Amazonia. American Museum Novitates, 3351:1-36.

Voss, R. S., D. P. Lunde, and N. B. Simmons. 2001. Mammals of Paracou, French Guiana: A neotropical lowland rainforest fauna. Part 2: Nonvolant species. Bulletin of the American Museum of Natural History, 263:1-236.

Voss, R. S., M. Gómez-Laverde, and V. Pacheco. 2002. A new genus for *Aepeomys fuscatus* Allen, 1912, and *Oryzomys intectus* Thomas, 1921: Enigmatic murid rodents from Andean cloud forests. American Museum Novitates, 3373:42 pp.

Vrana, P. B., M. C. Milinkovitch, J. R. Powell, and W. C. Wheeler. 1994. Higher level relationships of the Arctoid Carnivora based on sequence data and "total evidence." Molecular Phylogenetics and Evolution, 3:47-58.

Vrba, E. S. 1979. Phylogenetic analysis and classification of fossil and recent Alcelaphini. Mammalia: Bovidae. Biological Journal of the Linnean Society, 11:207-228.

Vrba, E. S. 1997. New fossils of Alcelaphini and Caprinae (Bovidae: Mammalia) from Awash, Ethiopia, and phylogenetic analysis of Alcelaphini. Palaeontologia Africana, 34:127-198.

Vrba, E. S., and J. Gatesy. 1994. New antelope fossils from Awash, Ethiopia and phylogenetic analysis of Hippotragini (Bovidae, Mammalia). Palaeontologica Africana, 31:55-72.

Vrba, E. S., and G. B. Schaller. 2000. Phylogeny of Bovidae based on behavior, glands, skulls, and postcrania. Pp. 203-222, *in* Antelopes, deer, and relatives. Fossil record, behavioral ecology, systematics, and conservation (E. S. Vrba and G. B. Schaller, eds.). Yale University Press, New Haven, 341 pp.

Vrba, E. A., J. R. Vaisnys, J. E. Gatesy, R. DeSalle, and K.-Y. Wei. 1994. Analysis of paedomorphosis using allometric characters: The example of Reduncini antelopes (Bovidae, Mammalia). Systematic Biology, 43:92-116.

Vucetich, M. G., D. H. Verzi, and J.-L. Hartenberger. 1999. Review and analysis of the radiation of the South American Hystricognathi (Mammalia, Rodentia). Comptes Rendus Académie des Sciences Paris, Sciences de la terre et des planetes, 329:763-769.

Vuillaume-Randriamanantena, M., L. R. Godfrey, and M. R. Sutherland. 1985. Revision of *Hapalemur* (*Prohapalemur*) *gallieni* (Standing, 1905). Folia Primatologia, 45:89-116.

Vujošević, M., and J. Blagojević. 1995. Seasonal changes of B-chromosome frequencies within the population of *Apodemus flavicollis* (Rodentia) on Cer mountain in Yugoslavia. Acta Theriologica, 40(2):131-137.

Vujošević, M., and J. Blagojević. 1997. Y chromosome polymorphism in the bank vole *Clethrionomys glareolus* (Rodentia, Mammalia). Zeitschrift für Säugetierkunde, 62:53-57.

Vujošević, M., and J. Blagojević. 2000. Does environment affect polymorphism of B chromosomes in the yellow-necked mouse *Apodemus flavicollis*? Zeitschrift für Säugetierkunde, 65:313-317.

Vujošević, M., D. Rimsa, and S. Zivkovic. 1984. Patterns of G- and C-bands distribution on chromosomes of three *Apodemus* species. Zeitschrift für Säugetierkunde, 49:234-238.

Vujošević, M., J. Blagojević, J. Radosavljevic, and D. Bejakovic. 1991. B chromosome polymorphism in populations of *Apodemus flavicollis* in Yugoslavia. Genetica, 83:167-170.

Vujošević, M., M. G. Filippucci, J. Blagojević, and B. Kryštufek. 1993. Evolutionary genetics and systematics of the garden dormouse, *Eliomys* Wagner, 1840. 4—Karyotype and allozyme analyses of *E. quercinus dalmaticus* from Yugoslavia. Bollettino di Zoologia, 60:47-51.

Wada, S., and K. Numachi. 1991. Allozyme analyses of genetic differentiation among the populations and species of Balaenoptera. Pp. 125-154, *in* Genetic ecology of whales and dolphins (A. R. Hoelzel, ed.). Reports of the International Whaling Commission, Special Issue, 13:1-311.

Wada, S., M. Oishi, and T. K. Yamada. 2003. A newly discovered species of living baleen whale. Nature, 426:278-281 (20 November 2003).

Waddell, P. J., and S. Shelley. 2003. Evaluating placental inter-ordinal phylogenies with novel sequences including RAG1, gamma-fibrinogen, ND6, and mt-tRNA, plus MCMC-driven nucleotide, amino acid, and codon models. Molecular Phylogenetics and Evolution, 28:197-224.

Waddell, P. J., N. Okada, and M. Hasegawa. 1999. Towards resolving the interordinal relationships of placental mammals. Systematic Biology, 48:1-5.

Waddell, P. J., H. Kishino, and R. Ota. 2001. Phylogenetic Foundation for Comparative Mammalian Genomics. Genome Informatics, 12:141–154.

Wade, C. M., E. J. Kulbokas III, A. W. Kirby, M. C. Zody, J. C. Mullkin, E. S. Lander, K. Lindblad-Toh, and M. J. Daly. 2002. The mosaic structure of variation in the laboratory mouse genome. Nature, 420:574-578.

Wade-Smith, J., and B. J. Verts. 1982. *Mephitis mephitis*. Mammalian Species, 173:1-7.

Wagner, J. A. 1839. Die Säugethiere in Abbildungen nach der Natur, Supplement 1. Weigel, Leipzig, 551 pp.

Wahlert, J. H. 1977. Cranial foramina and relationships of *Eutypomys* (Rodentia, Eutypomyidae). American Museum Novitates, 2626:1-8.

Wahlert, J. H. 1984. Relationships of the extinct rodent *Cricetops* to *Lophiomys* and the Cricetinae (Rodentia, Cricetidae). American Museum Novitates, 2784:1-15.

Wahlert, J. H. 1985. Skull morphology and relationships of geomyoid rodents. American Museum Novitates, 2812:1-20.

Wahlert, J. H. 1988. Skull morphology of *Gregorymys* and relationships of the Entoptychinae (Rodentia, Geomyidae). American Museum Novitates, 2922:1-13.

Wahlert, J. H. 1993. The fossil record. Pp. 1-37, *in* Biology of the Heteromyidae (H. H. Genoways and J. H. Brown, eds.). Special Publication, The American Society of Mammalogists, 10:1-719.

Wahlert, J. H., Sawitzke, S. L., and M. E. Holden. 1993. Cranial anatomy and relationships of dormice (Rodentia, Myoxidae). American Museum Novitates, 3061:1-32.

Wahrman, J., and R. Goitein. 1972. Hybridisation in nature between two chromosome forms of spiny mice. Chromosomes Today, 3:228-237.

Wahrman, J., C. Richler, E. Neufeld, and A. Friedmann. 1983. The origin of multiple sex chromosomes in the gerbil *Gerbillus gerbillus* (Rodentia: Gerbillinae). Cytogenetics and Cell Genetics, 35:161-180.

Waithman, J. 1979. A report on a collection of mammals from southwest Papua. 1972-1973. Australian Zoologist, 20:313-326.

Waithman, J., and A. Roest. 1977. A taxonomic study of the kit fox, *Vulpes macrotis*. Journal of Mammalogy, 58:157-164.

Wakana, S., M. Sakaizumi, K. Tsuchiya, M. Asakawa, S. H. Han, K. Nakata, and H. Suzuki. 1996. Phylogenetic implications of variations in rDNA and mtDNA in red-backed voles collected in Hokkaido, Japan, and in Korea. Mammal Study, 21:15-25.

Wakefield, N. A. 1972*a*. Palaeoecology of fossil mammal assemblages from some Australian caves. Proceedings of the Royal Society of Victoria, 85:1-26.

Wakefield, N. A. 1972*b*. Studies in Australian Muridae: Review of *Mastacomys fuscus*, and description of a new subspecies of *Pseudomys higginsi*. Memoirs of the National Museum of Victoria, Melbourne, Australia, 33:15-31.

Wakefield, N. A., and R. M. Warneke. 1963. Some revision in *Antechinus* (Marsupialia)—1. Victorian Naturalist, 80:194-219.

Walker, E. P., F. Warnick, S. E. Hamlet, K. L. Lange, M. A. Davis, H. E. Uible, and P. F. Wright. 1964. Mammals of the world. John Hopkins Press, Baltimore, 1:1-646; 2:647-1500; 3:1-769.

Walker, E. P., F. Warnick, S. E. Hamlet, K. L. Lange, M. A. Davis, H. E. Uible, and P. F. Wright. 1968. Mammals of the world, Second ed. Johns Hopkins Press, Baltimore, MD, 1:1-647; 2:648-1500.

Walker, E. P., F. Warnick, S. E. Hamlet, K. I. Lange, M. A. Davis, H. E. Uible, and P. F. Wright. 1975. Mammals of the world. Third ed. Johns Hopkins University Press, Baltimore, 1:1-646; 2:647-1500.

Walker, L. I., and A. E. Spotorno. 1992. Tandem and centric fusions in the chromosomal evolution of the South American phyllotines of the genus *Auliscomys* (Rodentia, Cricetidae). Cytogenetics and Cell Genetics, 61:135-140.

Walker, L. I., A. E. Spotorno, and J. Arrau. 1984. Cytogenetic and reproductive studies of two nominal subspecies of *Phyllotis darwini* and their experimental hybrids. Journal of Mammalogy, 65:220-230.

Wall, D. A., S. K. Davis, and B. M. Read. 1992. Phylogenetic relationships in the subfamily Bovinae (Mammalia: Artiodactyla) based on ribosomal DNA. Journal of Mammalogy, 73:262-275.

Wallace, B. M. N., J. B. Searle, and C. A. Everett. 2002. The effect of multiple simple Robertsonian heterozygosity on chromosome pairing and fertility of wild-stock house mice (*Mus musculus domesticus*). Cytogenetic and Genome Research, 96:276-286.

Wallace, R. B., and R. L. E. Painter. 1999. A new primate record for Bolivia: An apparently isolated population of common woolly monkeys representing a southern range extension for the genus *Lagothrix*. Neotropical Primates, 7:111-113.

Wallin, L. 1969. The Japanese bat fauna. Zoologiska Bidrage Fran Uppsala, 37:223-440.

Walpole, D. K., S. K. Davis, and I. F. Greenbaum. 1997. Variation in mitochondrial DNA in populations of *Peromyscus eremicus* from the Chihuahuan and Sonoran Deserts. Journal of Mammalogy, 78:397-404.

Walsh, S. L. 1997. New specimens of *Metanoiamys, Pauromys*, and *Simimys* (Rodentia: Myomorpha) from the Uintan (middle Eocene) of San Diego County, California, and comments on the relationships of selected Paleogene Myomorpha. Proceedings of the San Diego Society of Natural History, 32:1-20.

Walston, J., and P. Bates. 2001. The discovery of Wroughton's free-tailed bat *Otomops wroughtoni* (Chiroptera: Molossidae) in Cambodia. Acta Chiropterologica, 3:249-252.

Walston, J., and G. Veron. 2001. Questionable status of the "Taynguyen civet", *Viverra tainguensis* Sokolov, Rozhnov and Pham Trong Anh, 1997. (Mammalia: Carnivora: Viverridae). Mammalian Biology, 66:181-184.

Walton, A. H. 1993. A new genus of eutypomyid (Mammalia:Rodentia) from the middle Eocene of the Texas Gulf Coast. Journal of Vertebrate Paleontology, 13(2):262-266.

Walton, A. H., M. A. Nedbal, and R. L. Honeycutt. 2000. Evidence from intron 1 of the nuclear transthyretin (prealbumin) gene for the phylogeny of African mole-rats (Bathyergidae). Molecular Phylogenetics and Evolution, 16(3):467-474.

Walton, D. W. 1963. A collection of the bat *Lonchophylla robusta* Miller from Costa Rica. Tulane Studies in Zoology, 10:87-90.

Walton, D. W. (ed.). 1988. Zoological catalogue of Australia. 5. Mammalia. Australian Government Publishing Service, Canberra, 274 pp.

Walton, L. R., and D. O. Joly. 2003. *Canis mesomelas*. Mammalian Species, 715:1-9.

Wammes, D. F. 1992a. Bosmuis, *Apodemus sylvaticus* (L., 1758). Pp. 289-291, *in* Atlas van de Nederlandse Zoogdieren (S. Broekhuizen, B. Hoekstra, V. van Laar, C. Smeenk, and J. B. M. Thissen, eds.). Stichting Uitgeverij Koninklijke Nederlandse Natuurhistorische Vereniging, Utrecht, 336 pp.

Wammes, D. F. 1992b. Huismuis, *Mus domesticus* Rutty, 1772. Pp. 302-305, *in* Atlas van de Nederlandse Zoogdieren (S. Broekhuizen, B. Hoekstra, V. van Laar, C. Smeenk, and J. B. M. Thissen, eds.). Stichting Uitgeverij Koninklijke Nederlandse Natuurhistorische Vereniging, Utrecht, 336 pp.

Wang B., and Li Chuan-kwei. 1990. First Paleogene mammalian fauna from northeast China. Vertebrata Palasiatica, 28:165-205, pls. i-vi.

Wang, B.-Y. 1985. Zapodidae (Rodentia, Mammalia) from the lower Oligocene of Qujing, Yunnan, China. Mainzer Geowissenschaftliches Mitteilungen, 14:345-367.

Wang, B.-Y., and M. R. Dawson. 1994. A primitive cricetid (Mammalia:Rodentia) from the Middle Eocene of Jiangsu Province, China. Annals of Carnegie Museum, 63(3):239-256.

Wang, B.-Y., and Qiu, Z.-X. 2000. Dipodidae (Rodentia, Mammalia) from the lower member of Xianshuihe Formation in Lanzhou Basin, Gansu, China. Vertebrata PalAsiatica, 38(1):10-35 (in English with Chinese summary).

Wang, D., Z. Wang, and R. Sun. 1995. [Variations in digestive tract morphology in root vole (*Microtus oeconomus*) and its adaptive significance.] Acta Theriologica Sinica, 15(1):53-59 (in Chinese with English summary).

Wang, J.-X., and S. K. Hung. 1997. Analysis of mitochondrial DNA restriction fragment patterns in two subspecies of striped field mice *Apodemus agrarius* from northern China and Korea. Acta Theriologica Sinica, 17(3):208-215 (in Chinese).

Wang, J.-X., X.-F. Zhao, Y.-Z. Wang, and J. Tian. 1993. [Studies of chromosome of striped field mouse *Apodemus agrarius pallidior* (Rodentia).] Acta Theriologica Sinica, 13(4):283-287 (in Chinese).

Wang, J.-X., X.-F. Zhao, H.-Y. Qi, and Y.-Z. Wang. 1997. [Karyotypes of *Niviventer confucianus* (Rodentia: Muridae).] Acta Zoologica Sinica, 43(3):324-327 (in Chinese).

Wang, J.-X., X.-F. Zhao, Y. Deng, H.-Y. Qi, Y. Liu, Z. Zhang, K. H. Sun, and X. Shan. 1999. Karyotypes and B chromosome of greater long-tailed hamster (*Cricetulus triton*). Acta Theriologica Sinica, 19(3):197-204.

Wang, Peilie. 1999. Chinese cetaceans. Ocean Enterprises Ltd. 325 pp.

Wang, Q., W. Zhou, Y. Zhang, and N. Fan. 1994. [The observation on burrowing behavior of plateau zokor (*Myospalax baileyi*).] Acta Theriologica Sinica, 14(3):203-208 (in Chinese with English abstract).

Wang Si-bo and Yang Gan-yun. 1983. [Rodent fauna of Xinjiang.] Xinjiang People's Publishing House, Wulumuqi [Urumchi], China, 223 pp. (in Chinese).

Wang Sung. 1959. [Further report on the mammals of northeastern China.] Acta Zoologica Sinica, 11:344-352 (in Chinese).

Wang Sung. 1964. [New species and subspecies of mammals from Sinkiang, China.] Acta Zootaxonomica Sinica, 1:6-18 (in Chinese).

Wang Sung and Zheng Chang-lin. 1973. [Notes on Chinese hamsters (Cricetinae).] Acta Zoologica Sinica, 19:61-68 (in Chinese).

Wang Sung and Zheng Chang-lin. 1981. [On the subspecies of the Chinese sulphur-bellied rat—*Rattus niviventer* Hodgson.] Sinozoologia, 1:1-8 (in Chinese).

Wang Ting-zheng. 1990. [On the fauna and zoogeographical regionalization of Glires (including rodents and lagomorphs) in Shaanxi Province.] Acta Theriologica Sinica, 10(2):128-136.

Wang, T., and W. Xu (eds.). 1992. Glires (Rodentia and Lagomorpha) fauna of Shaanxi province. Shaanxi Normal University Press, Xi'an., 317 pp.

Wang, X. T. (ed.). 1990. Vertebrate Fauna of Ningxia. Ningxia People's Publishing House, Yinchuan, 223 pp.

Wang, Xiaoming and R. S. Hoffmann. 1987. *Pseudois nayaur* and *Pseudois schaeferi*. Mammalian Species, 278:1-6.

Wang, Xiaoming, R. H. Tedford, and B. E. Taylor. 1999. Phylogenetic systematics of the Borophaginae (Carnivora: Canidae). Bulletin of the American Museum of Natural History, 243:1-391.

Wang, W., and Lan, H.. 2000. Rapid and parallel chromosomal number reductions in muntjac deer inferred from mitochondrial DNA phylogeny. Molecular Biology and Evolution, 17:1326-1333.

Wang, Y. 1987. Taxonomic research in Burma-Chinese tree shrew, *Tupaia belangeri* (Wagner), from southern China. Zoological Research, 8:213-230.

Wang, Y. 2003. A complete checklist of mammal species and subspecies in China. A taxonomic and geographic reference. China Forestry Publishing House, (Beijing), 394 pp.

Wang, Y., S. Ma, and C. Li. 1993. The taxonomy, distibution and status of forest musk deer in China. Pp. 22-30, *in* Deer of China. Biology and management (N. Ohtaishi and H.-I. Sheng, eds.). Elsevier, Amsterdam, 418 pp.

Wang Y.-c., Y.-r. Tu, and Wang Sung. 1966. Notes on some small mammals from Szechuan Province with description of a new subspecies. Acta Zootaxonomica Sinica, 3:89-90 (in English and Chinese).

Wang Ying-xiang and Yang G. 1989. [Insectivores]. Pp. 202-210, *in* [A list of medical animals in Yunnan (Yunnan Office for Endemic Disease Control, and Yunnan Sanitation and Antiepidemic Station, eds.)]. Yunnan Science and Technology, Kunming.

Wang Ying-xiang, Luo Ze-xun, and Feng Zuo-jiang. 1985. [Taxonomic revision of Yunnan hare, *Lepus comus* G. Allen, with description of two new subspecies.] Zoological Research, 6:101-109 (in Chinese).

Wang, Y.-X., Li Chongyun, and Chen Zhiping. 1996. [Taxonomy, distribution and differentiation of *Typhlomys cinereus* (Platacanthomyidae, Mammalia).] Acta Theriologica Sinica, 16(1):54-66 (in Chinese, with English abstract).

Wang You-zhi. 1985*a*. [A new genus and species of Gliridae-*Chaetocauda sichuanensis* gen. et sp. nov.] Acta Theriologica Sinica, 5:67-75 (in Chinese).

Wang You-zhi. 1985*b*. Subspecific classification and distribution of *Apodemus agrarius* in Sichuan, China. Pp. 86-89, *in* Contemporary mammalogy in China and Japan (T. Kawamichi, ed.). Mammalogical Society of Japan, 194 pp.

Wang You-zhi, Hu Jin-chu, and Chen Ke. 1980. [A new species of Murinae—*Vernaya foramena* sp. nov.] Acta Zoologica Sinica, 26:393-397 (in Chinese).

Ward, O. G., and D. H. Wurster-Hill. 1990. *Nyctereutes procyonides*. Mammalian Species, 358:1-5.

Warmerdam, M. 1982. Numeriek-taxonomische studie van de twee vormen van de woelrat *Arvicola terrestris* (Linnaeus, 1758) in Nederland en Belgie. Lutra, 24:33-66.

Warner, R. M. 1982. *Myotis auriculus*. Mammalian Species, 191:1-3.

Warner, R. M., and N. J. Czaplewski. 1984. *Myotis volans*. Mammalian Species, 224:1-4.

Wasser, S. K. 1985. Current conservation status of the Mwanihana Rain Forest, Uzungwa Mountains, Sanje, Tanzania. Primate Conservation, 6:34.

Wassif, K. 1956. Studies on gerbils of the subgenus *Dipodillus* recorded from Egypt. Ain Shams Science Bulletin, 1:173-194.

Wassif, K., and H. Hoogstraal. 1954. The mammals of South Sinai, Egypt. Proceedings of the Egyptian Academy of Science, Cairo, 9:63-79.

Wassif, K. I. 1960. Further observations on *Gerbillus simoni* Lataste from the Egyptian Province (U. A. R.). Bulletin of the Zoological Society of Egypt, 15(29):29-31.

Waterhouse, G. R. 1838 [1839]. On the skull and dentition of the American badger (*Meles labradoria*). Proceedings of the Zoological Society of London, 1838:153-154.

Waterhouse, G. R. 1839. Observations on the Rodentia with a view to point out groups as indicated by the structure of the crania in this order of mammals. Magazine of Natural History, ser. 2, 3:90-96.

Waterhouse, G. R. 1842 [1843]. Descriptions of a new species of quadrupeds collected by Mr. Fraser at Fernando Po. Proceedings of the Zoological Society of London, 1842:124-130.

Waterhouse, G. R. 1846-1848. The natural history of the mammalia. Hippolyte Bailliere, Publisher, London, 1:2 (unnumbered) + 1-553, 22 pls.[1846]; 2:1-500, 21 pls. [1848].

Watkins, L. C. 1972. *Nycticeius humeralis*. Mammalian Species, 23:1-4.

Watkins, L. C. 1977. *Euderma maculatum*. Mammalian Species, 77:1-4.

Watkins, L. C., J. K. Jones, Jr., and H. H. Genoways. 1972. Bats of Jalisco, Mexico. Special Publications, The Museum, Texas Tech University Press, 1:1-44.

Watson, E. E., S. Easteal, and D. Penny. 2001. *Homo* genus: A review of the classification of humans and the great apes. Pp. 307-318, *in* Humanity from African Naissance to Coming Millenia, (P. V. Tobias, M. A. Raath, J. Moggi-Cecchi, and G. A. Doyle, eds.). Firenze University Press and Witwatersrand University Press, Florence and Johannesburg. 416 pp.

Watson, J. P. 1990. The taxonomic status of the slender mongoose, *Galerella sanguinea* (Rüppell, 1836), in southern Africa. Navorsinge van die Nasionale Museum Bloemfontein, 6(10):351-492.

Watson, J. P., and N. H. Dippenaar. 1987. The species limits of *Galerella sanguinea* (Rüppell, 1836), *G. pulverulenta* (Wagner, 1839) and *G. nigrata* (Thomas, 1928) in southern Africa (Carnivora: Viverridae). Navorsinge van die Nasionale Museum Bloemfontein, 5(14):356-414.

Watson, V., and I. Plug. 1995. *Oreotragus major* Wells and *Oreotragus oreotragus* (Zimmerman) (Mammalia: Bovidae): Two species? Annals of the Transvaal Museum, 36:183-191.

Watts, C. H. S. 1975. The neck and chest glands of the Australian hopping-mice, *Notomys*. Australian Journal of Zoology, 23:151-157.

Watts, C. H. S. 1976. *Leggadina lakedownensis*, a new species of murid rodent from North Queensland. Transactions of the Royal Society of South Australia, 100:105-108.

Watts, C. H. S. 1995a. Fawn hopping-mouse, *Notomys cervinus*. Pp. 574-575, *in* Mammals of Australia (R. Strahan, ed.). Smithsonian Institution Press, Washington, D.C. 756 pp.

Watts, C. H. S. 1995b. Dusky hopping-mouse, *Notomys fuscus*. Pp. 575-576, *in* Mammals of Australia (R. Strahan, ed.). Smithsonian Institution Press, Washington, D.C., 756 pp.

Watts, C. H. S. 1995c. Mitchell's hopping-mouse, *Notomys mitchelli*. Pp. 579-580, *in* Mammals of Australia (R. Strahan, ed.). Smithsonian Institution Press, Washington, D.C., 756 pp.

Watts, C. H. S. 1995d. Darling Downs hopping-mouse, *Notomys mordax*. P. 581, *in* Mammals of Australia (R. Strahan, ed.). Smithsonian Institution Press, Washington, D.C., 756 pp.

Watts, C. H. S. 1995e. Plains rat, *Pseudomys australis*. Pp. 586-587, *in* Mammals of Australia (R. Strahan, ed.). Smithsonian Institution Press, Washington, D.C., 756.

Watts, C. H. S. 1995f. Bolam's mouse, *Pseudomys bolami*. P. 588, *in* Mammals of Australia (R. Strahan, ed.). Smithsonian Institution Press, Washington, D.C., 756 pp.

Watts, C. H. S. 1995g. Bramble Cay melomys, *Melomys rubicola*. Pp. 637-638, *in* Mammals of Australia (R. Strahan, ed.). Smithsonian Institution Press, Washington, D.C., 756 pp.

Watts, C. H. S. 1995h. Pacific rat, *Rattus exulans*. P. 650, *in* Mammals of Australia (R. Strahan, ed.). Smithsonian Institution Press, Washington, D.C., 756 pp.

Watts, C. H. S. 1995i. Brown rat, *Rattus norvegicus*. Pp. 658-659, *in* Mammals of Australia (R. Strahan, ed.). Smithsonian Institution Press, Washington, D.C., 756 pp.

Watts, C. H. S. 1995j. Black rat, *Rattus rattus*. Pp. 659-660, *in* Mammals of Australia (R. Strahan, ed.). Smithsonian Institution Press, Washington, D.C., 756 pp.

Watts, C. H. S. 1995k. Long-haired rat, *Rattus villosissimus*. Pp. 664-665, *in* Mammals of Australia (R. Strahan, ed.). Smithsonian Institution Press, Washington, D.C., 756 pp.

Watts, C. H. S., and H. J. Aslin. 1981. The rodents of Australia. Angus and Robertson, Sydney, 321 pp.

Watts, C. H. S., and P. R. Baverstock. 1994a. Evolution in New Guinean Muridae (Rodentia) assessed by microcomplement fixation of albumin. Australian Journal of Zoology, 42:295-306.

Watts, C. H. S., and P. R. Baverstock. 1994b. Evolution in some South-east Asian Murinae (Rodentia), as assessed by microcomplement fixation of albumin, and their relationship to Australian murines. Australian Journal of Zoology, 42:711-722.

Watts, C. H. S., and P. R. Baverstock. 1995a. Evolution in some African Murinae (Rodentia) assessed by microcomplement fixation of albumin. Journal of African Zoology, 109:423-433.

Watts, C. H. S., and P. R. Baverstock. 1995b. Evolution in the Murinae (Rodentia) assessed by microcomplement fixation of albumin. Australian Journal of Zoology. 43:105-118.

Watts, C. H. S., and P. R. Baverstock. 1996. Phylogeny and biogeography of some Indo-Australian murid rodents. Pp. 47-50, in Proceedings of the First International Conference on Eastern Indonesian-Australian vertebrate fauna, Manado, Indonesia, November 22-26, 1994 (D. J. Kitchener and A. Suyanto, eds.). Western Australian Museum for Lembaga Ilmu Pengetahuan Indonesia, Perth, 174 pp.

Watts, C. H. S., and T. D. Tweedie. 1993. Three new localities for Pseudomys gracilicaudatus (Rodentia: Muridae) in New South Wales. Australian Mammalogy, 16(1):55-57.

Watts, C. H. S., P. R. Baverstock, J. Birrell, and M. Krieg. 1992. Phylogeny of the Australian rodents (Muridae): A molecular approach using microcomplement fixation of albumin. Australian Journal of Zoology, 40:81-90.

Wayne, R. K. 1992. On the use of morphologic and molecular genetic characters to investigate species status. Conservation Biology, 6:590-592.

Wayne, R. K. 1993. Molecular evolution of the dog family. Trends in Genetics, 9:214-218.

Wayne, R. K. 1995. Red wolves: To conserve or not to conserve. Canid News, 3:7-12.

Wayne, R. K., and J. L. Gittleman. 1995. The problematic red wolf. Scientific American, 273:36-39.

Wayne, R. K., and D. Gottelli. 1997. Systematics, population genetics and genetic management of the Ethiopian wolf. Pp. 43-50, in The Ethiopian wolf—status survey and conservation action plan (C. Sillero-Zubiri and D. MacDonald, eds.). I.U.C.N., Gland, Switzerland, 116 pp.

Wayne, R. K., and S. M. Jenks. 1991. Mitochondrial DNA analysis implying extensive hybridization of the endangered red wolf Canis rufus. Nature, 351:565-568.

Wayne, R. K., and S. J. O'Brien. 1987. Allozyme divergence within the Canidae. Systematic Zoology, 36:339-355.

Wayne, R. K., and E. A. Ostrander. 1999. Origin, genetic diversity, and genome structure of the domestic dog. Bioessays, 21:247-257.

Wayne, R. K., W. G. Nash, and S. J. O'Brien. 1987a. Chromosomal evolution of the Canidae: I. Species with high diploid numbers. Cytogenetics and Cell Genetics, 44:123-133.

Wayne, R. K., W. G. Nash, and S. J. O'Brien. 1987b. Chromosomal evolution of the Canidae: II. Species with low diploid numbers. Cytogenetics and Cell Genetics, 44:134-141.

Wayne, R. K., R. E. Benveniste, and S. J. O'Brien. 1989. Phylogeny and evolution of the Carnivora and carnivore families, Pp. 465-494, in, Carnivore behavior, ecology and evolution (J. L. Gittleman, ed.). Cornell University Press, Ithaca, NY, 620 pp.

Wayne, R. K., S. B. George, D. Gilbert, and P. W. Collins. 1991a. The channel island fox (Urocyon littoralis) as a model of genetic change in small populations. Pp. 639-649, in The unity of evolutionary biology (E. C. Dudley, ed.). Proceedings of the fourth international congress of systematic and evolutionary biology. Dioscorides Press, Portland, Oregon, 2:1-1048.

Wayne, R. K., S. George, D. Gilbert, P. Collins, and D. Girman. 1991b. A morphologic and genetic study of the island fox, Urocyon littoralis. Evolution, 45:1849-1868.

Wayne, R. K., N. Lehman, M. W. Allard, and R. L. Honeycutt. 1992. Mitochondrial DNA variability of the gray wolf: Genetic consequences of population decline and habitat fragmentation. Conservation Biology, 6:559-569.

Wayne, R. K., E. Geffen, D. J. Girman, K. P. Koepfli, L. M. Lau, and C. R. Marshall. 1997. Molecular systematics of the Canidae. Systematic Biology, 46:622-653.

Wayne, R. K., M. S. Roy, and J. L. Gittleman. 1998. Origin of the red wolf: Response to Nowak and Federoff and Gardener. Conservation Biology, 12:726-729.

Webb, N. J., and C. R. Tidemann. 1995. Hybridization between black (Pteropus alecto) and grey-headed (P. poliocephalus) flying-foxes (Megachiroptera: Pteropodidae). Australian Mammalogy, 18:19-26.

Webb, S. D. 1985. The interrelationships of tree sloths and ground sloths. Pp. 105-112, in The evolution and ecology of armadillos, sloths, and vermilinguas (G. G. Montgomery, ed.). Smithsonian Institution Press, Washington, D.C., 10 (unnumbered) + 451 pp.

Webb, S. D. 1992. A cranium of Navahoceros and its phylogenetic place among New World Cervidae. Annales Zoologici Fennici, 28:401-410.

Webb, S. D. 1999. Isolation and interchange. A deep history of South American mammals. Pp. 13-19, in Mammals of the Neotropics. The central Neotropics, Vol. 3, Ecuador, Peru, Bolivia, Brazil (J. F. Eisenberg and K. H. Redford, eds.). The University of Chicago Press, Chicago, x + 609 pp.

Webb, S. D. 2000. Evolutionary history of New World Cervidae. Pp. 38-64, in Antelopes, deer, and relatives.

Fossil record, behavioral ecology, systematics, and conservation (E. S. Vrba and G. B. Schaller, eds.). Yale University Press, New Haven, 341 pp.

Webb, S. D., and B. E. Taylor. 1980. The phylogeny of hornless ruminants and a description of the cranium of *Archaeomeryx*. Bulletin of the American Museum of Natural History, 167:117-158.

Weber, M. 1904. Die Säugetiere. Einführung in die Anatomie und Systematik der recenten und fossilen Mammalia. G. Fischer, Jena, xi + 866 pp.

Weber, M. 1927-1928. Die Säugetiere. Einführung in die anatomie und systematik der recenten und fossilen Mammalia. G. Fischer, Jena, and Band II, Systematischer Teil, Second ed., 879 pp.

Webster, W. D. 1993. Systematics and evolution of bats of the genus *Glossophaga*. Special Publications, The Museum, Texas Tech University, 36:1-184.

Webster, W. D., and C. O. Handley, Jr. 1986. Systematics of Miller's long-tongued bat *Glossophaga longirostris*, with description of two new subspecies. Occasional Papers, The Museum, Texas Tech University, 100:1-22.

Webster, W. D., and J. K. Jones, Jr. 1980. Taxonomic and nomenclatorial notes on bats of the genus *Glossophaga* in North America, with description of a new species. Occasional Papers, The Museum, Texas Tech University, 71:1-12.

Webster, W. D., and J. K. Jones, Jr. 1982a. *Reithrodontomys megalotis*. Mammalian Species, 167:1-5.

Webster, W. D., and J. K. Jones, Jr. 1982b. *Artibeus aztecus*. Mammalian Species, 177:1-3.

Webster, W. D., and J. K. Jones, Jr. 1982c. *Artibeus toletecus*. Mammalian Species, 178:1-3.

Webster, W. D., and J. K. Jones, Jr. 1983. *Artibeus hirsutus* and *Artibeus inopinatus*. Mammalian Species, 199:1-3.

Webster, W. D., and J. K. Jones, Jr. 1984. *Glossophaga leachii*. Mammalian Species, 226:1-3.

Webster, W. D., and J. K. Jones, Jr. 1985. *Glossophaga mexicana*. Mammalian Species, 245:1-2.

Webster, W. D., and J. K. Jones, Jr. 1993. *Glossophaga commissarisi*. Mammalian Species, 446:1-4.

Webster, W. D., and R. D. Owen. 1984. *Pygoderma bilabiatum*. Mammalian Species, 220:1-3.

Webster, W. D., J. K. Jones, Jr., and R. J. Baker. 1980. *Lasiurus intermedius*. Mammalian Species, 132:1-3.

Webster, W. D., L. W. Robbins, R. L. Robbins, and R. J. Baker. 1982. Comments on the status of *Musonycteris harrisoni* (Chiroptera: Phyllostomidae). Occasional Papers, The Museum, Texas Tech University, 78:1-5.

Webster, W. D., C. O. Handley, Jr., and P. J. Soriano. 1998. *Glossophaga longirostris*. Mammalian Species, 576:1-5.

Weddle, G. K., and J. R. Choate. 1983. Dental evolution of the meadow vole in mainland, peninsular, and insular environments in southern New England. Fort Hays Studies, n.s., 3:1-23.

Weerd, A. van der. 1976. Rodent faunas of the Mio-Pliocene continental sediments of the Teruel-Alfambra Region, Spain. Utrecht Micropaleontological Bulletin. Special Publication 2:1-217.

Wehausen, J. D., and R. R. Ramey II. 2000. Cranial morphometric and evolutionary relationships in the northern range of *Ovis canadensis*. Journal of Mammalogy, 81:145-161.

Wei, Fuwen, Zuojian Feng, Zuwang Wang, and Jinchu Hu. 1999. Current Distirbution, status and conservation of wild red pandas *ailurus fulgens* in China. Biological Conservation, 89:285-291.

Weigel, I. 1961. Das Fellmuster der wildlebenden Katzenarten und der Hauskatze in Vergleichender und Stammesgeschicher Hinsicht. Säugetierkundliche Mitteilungen, 9:1-120.

Weigel, I. 1969. Systematische übersicht über die insektenfresser und nager Nepals nebst bemerkungen zur tiergeographie. Khumbu Himal, 3(2):149-196.

Weitzel, V. M., and C. P. Groves. 1985. The nomenclature and taxonomy of the colobine monkeys of Java. International Journal of Primatology, 6:399-409.

Weksler, M. 2003. Phylogeny of Neotropical oryzomyine rodents (Muridae: Sigmodontinae) based on the nuclear IRBP exon. Molecular Phylogenetics and Evolution, 29:331-349.

Weksler, M., L. Geise, and R. Cerqueira. 1999. A new species of *Oryzomys* (Rodentia, Sigmodontinae) from southeast Brazil, with comments on the classification of the *O. capito* species group. Zoological Journal of the Linnean Society, 125:445-462.

Weksler, M., C. R. Bonvicino, I. B. Otazu, and J. S. Silva Júnior. 2001. Status of *Proechimys roberti* and *P. oris* (Rodentia: Echimyidae) from eastern Amazonia and central Brazil. Journal of Mammalogy, 82(1):109-122.

Wells, D. R. 1989. Notes on the distribution and taxonomy of peninsular Malaysian mongooses (*Herpestes*). Natural History Bulletin of the Siam Society, 37(1):87-97.

Wells, D. R., and C. M. Francis. 1988. Crab-eating mongoose, *Herpestes urva*, a mammal new to peninsular Malaysia. Malayan Nature Journal, 42:37-41.

Wells, N. M., and J. Giacalone. 1985. *Syntheosciurus brochus*. Mammalian Species, 249:1-3.

Wells-Gosling, N., and L. R. Heaney. 1984. *Glaucomys sabrinus*. Mammalian Species, 229:1-8.

Wemmer, C. (ed.). 1998. Deer. Status survey and conservation action plan. Compiled by Andrew McCarthy, Raleigh Blouch and Donald Moore. I.U.C.N., Gland, Switzerland, 106 pp.

Wenzel, E., and T. Haltenorth. 1972. System der Schleichkatzen (Viverridae). Säugetierkundliche Mitteilungen, 20(1-2):110-127.

Werbitsky, D., and C. W. Kilpatrick. 1987. Genetic variation and genetic differentiation among allopatric populations of *Megadontomys*. Journal of Mammalogy, 68:305-312.

Werdelin, L. 1981. The evolution of lynxes. Annales Zoologici Fennici, 18:37-71.

Werdelin, L. 1987. Some observations on *Sarcophilus laniarius* and the evolution of *Sarcophilus*. Records of the Queen Victoria Museum, Launceston, 90:1-27.

Werdelin, L., and N. Solounias. 1991. The Hyaenidae: Taxonomy, systematics and evolution. Fossils and Strata, 30:1-104.

Wesselman, H. B. 1984. The Omo micromammals. Systematics and paleoecology of Early Man sites from Ethiopia. Contributions to Vertebrate Evolution, 7:1-219.

Wessels, W. 1996. Myocricetodontinae from the Miocene of Pakistan. Proceedings of the Koninklijke Nederlandse Akademie van Wetenschappen, 99(3-4):253-312.

Wessels, W. 1998. Gerbillidae from the Miocene and Pliocene of Europe. Mitteilungen—Bayerische Staatssammlung für Palaontolgie und Historische Geologie, 38:187-207.

Wessels, W. 1999. Family Gerbillidae. Pp. 395-400, *in* The Miocene land mammals of Europe (G. E. Rössner and K. Heissig, eds.). Dr. Friedrich Pfeil, München, 515 pp.

West, J. D., W. I. Frels, and V. M. Chapman. 1978. *Mus musculus* x *Mus caroli* hybrids: Mouse mules. Journal of Heredity, 69:321-326.

Westerman, M., M. S. Springer, J. Dixon, and C. Krajewski. 1999. Molecular relationships of the extinct pig-footed bandicoot *Chaeropus ecaudatus* (Marsupialia: Perameloidea) using 12S rRNA sequences. Journal of Mammalian Evolution, 6:271-288.

Westerman, M., M. S. Springer, and C. Karjewski. 2001. Molecular relationships of the New Guinean bandicoot genera *Microperoryctes* and *Echymipera* (Marsupialia: Peramelina). Journal of Mammalian Evolution, 8:93-105.

Weston, M. L. 1981. The *Ochotona alpina* complex: A statistical re-evaluation. Pp. 73-89, *in* Proceedings of the world lagomorph conference (K. K. Myers and C. D. MacInnes, eds.). Guelph University Press, Guelph, Ontario, 983 pp.

Weston, M. L. 1982. A numerical revision of the genus *Ochotona* (Lagomorpha: Mammalia) and an examination of its phylogenetic relationships. Unpubl. Ph. D. dissertation, University of British Columbia, Vancouver, 387 pp.

Wetterer, A. L., M. V. Rockman, and N. B. Simmons. 2000. Phylogeny of phyllostomid bats (Mammalia: Chiroptera): data from diverse morphological systems, sex chromosomes, and restriction sites. Bulletin of the American Museum of Natural History, 248:1-200.

Wetzel, R. M. 1955. Speciation and dispersal of the southern bog lemming, *Synaptomys cooperi* (Baird). Journal of Mammalogy, 36:1-20.

Wetzel, R. M. 1975. The species of *Tamandua* Gray (Edentata, Myrmecophagidae). Proceedings of the Biological Society of Washington, 88:95-112.

Wetzel, R. M. 1977. The Chacoan peccary *Catagonus wagneri* (Rusconi). Bulletin of Carnegie Museum of Natural History, 3:1-36.

Wetzel, R. M. 1980. Revision of the naked-tailed armadillos, genus *Cabassous* McMurtrie. Annals of Carnegie Museum, 49:323-357.

Wetzel, R. M. 1981. The hidden Chacoan peccary. Carnegie Magazine, 55:24-32.

Wetzel, R. M. 1985a. The identification and distribution of Recent Xenarthra (=Edentata). Pp. 5-22, *in* The evolution and ecology of armadillos, sloths, and vermilinguas (G. G. Montgomery, ed.). Smithsonian Institution Press, Washington, D.C., 10 (unnumbered) + 451 pp.

Wetzel, R. M. 1985b. Taxonomy and distribution of armadillos, Dasypodidae. Pp. 23-48, *in* The evolution and ecology of armadillos, sloths, and vermilinguas (G. G. Montgomery, ed.). Smithsonian Institution Press, Washington, D.C., 10 (unnumbered) + 451 pp.

Wetzel, R. M., and F. D. de Avila-Pires. 1980. Identification and distribution of the Recent sloths of Brazil (Edentata). Revista Brasileira de Biologia, 40:831-836.

Wetzel, R. M., and E. Mondolfi. 1979. The subgenera and species of long-nosed armadillos, genus *Dasypus* L. Pp. 43-63, *in* Vertebrate ecology in the northern Neotropics (J. F. Eisenberg, ed.). Smithsonian Institution Press, Washington, D.C., 271 pp.

Wetzel, R. M., R. E. Dubos, R. L. Martin, and P. Myers. 1975. *Catagonus*, an "extinct" peccary, alive in Paraguay. Science, 189:379-381.

Weyhausen, J. D., and R. R. Ramey II. 2000. Cranial morphometric and evolutionary relationships in the northern range of *Ovis canadensis*. Journal of Mammalogy, 81:145-161.

Wheeler, J. C. 1995. Evolution and present situation of the South American Camelidae. Biological Journal of the Linnean Society, 54:271-295.

Whidden, H. P. 2000. Comparative myology of moles and the phylogeny of the Talpidae (Mammalia, Lipotyphla). American Museum Novitates, 3294:1-53.

Whidden, H. P., and R. J. Asher. 2001. The origin of the Greater Antillean insectivorans. Pp. 237-252, *in* Biogeography of the West Indies: Patterns and perspectives (C. A. Woods and F. E. Sergile, eds.). CRC Press, Boca Raton, Florida, 582 pp.

Whisson, L., and D. J. Kitchener. 1995. Hastings River mouse, *Pseudomys oralis*. Pp. 614-615, *in* Mammals of Australia (R. Strahan, ed.). Smithsonian Institution Press, Washington, D.C., 756 pp.

Whitaker, J. O., Jr. 1972. *Zapus hudsonius*. Mammalian Species, 11:1-7.

Whitaker, J. O., Jr. 1974. *Cryptotis parva*. Mammalian Species, 43:1-8.

Whitaker, J. O., Jr. 1999a. Woodland jumping mouse | *Napaeozapus insignis*. Pp. 665-666, *in* The Smithsonian Book of North American Mammals (D. E. Wilson and S. Ruff, eds.). Smithsonian University Press, Washington, D.C., 750 pp.

Whitaker, J. O., Jr. 1999b. Meadow jumping mouse | *Zapus hudsonius*. Pp. 666-667, *in* The Smithsonian Book of North American Mammals (D. E. Wilson and S. Ruff, eds.). Smithsonian University Press, Washington, D.C., 750 pp.

Whitaker, J. O., Jr., and W. J. Hamilton, Jr. 1998. Mammals of the eastern United States. Third Edition. Cornell University Press, Ithaca, 583 pp.

Whitaker, J. O., Jr., and R. E. Wrigley. 1972. *Napaeozapus insignis*. Mammalian Species, 14:1-6.

White, J. W. 1953. Genera and subgenera of chipmunks. University of Kansas Publications, Museum of Natural History, 5:543-561.

White, T. G., and M. S. Alberico. 1992. *Dinomys branickii*. Mammalian Species, 410:1-5.

Whitehead, G. K. 1972. Deer of the world. Constable, London, 194 pp.

Wible, J. R., M. J. Novacek, and G. W. Rougier. 2004. New data on the skull and dentition in the Mongolian late Cretaceous eutherian mammal *Zalambdalestes*. Bulletin of the American Museum of Natural History, 281:1-144.

Wied-Neuwied, M. A. P., zu Prinz. 1826. Beiträge zur Naturgeschichte von Brasilien, von Maximillian, prinzen zu Wied, Brand II (Abtheilung II, Mammalia):1-620[1826], Landes-Industrie-Comptoirs, Weimar.

Wiener, G., H. Jianlin, and L. Ruijun (eds.). 2003. The yak. Second ed. Food and Agricultural Organization of the United Nations, Bangkok, 460 pp.

Wiles, G. J. 1981. Abundance and habitat preferences of small mammals in southwestern Thailand. Natural History Bulletin of the Siam Society, 29:44-54.

Wiley, R. W. 1980. *Neotoma floridana*. Mammalian Species, 139:1-7.

Wilhelm, D. D. 1982. Zoogeographic and evolutionary relationships of selected populations of *Microtus mexicanus*. Occasional Papers, The Museum, Texas Tech University, 75:1-30.

Wilkins, K. T. 1986. *Reithrodontomys montanus*. Mammalian Species, 257:1-5.

Wilkins, K. T. 1987a. *Lasiurus seminolus*. Mammalian Species, 280:1-5.

Wilkins, K. T. 1987b. A zoogeographic analysis of variation in Recent *Geomys pinetis* (Geomyidae) in Florida. Bulletin of the Florida State Museum, Biological Sciences, 30:1-28.

Wilkins, K. T. 1989. *Tadarida brasiliensis*. Mammalian Species, 331:1-10.

Wilkins, K. T., and C. D. Swearingen. 1990. Factors affecting historical distribution and modern geographic variation in the south Texas pocket gopher *Geomys personatus*. American Midland Naturalist, 124:57-72.

Williams, C. K. 1995. Dusky rat, *Rattus colletti*. Pp. 648-649, *in* Mammals of Australia (R. Strahan, ed.). Smithsonian Institution Press, Washington, D.C., 756.

Williams, D. F. 1978a. Karyological affinities of the species groups of silky pocket mice (Rodentia: Heteromyidae). Journal of Mammalogy, 59:599-612.

Williams, D. F. 1978b. Systematics and ecogeographic variation of the Apache pocket mouse (Rodentia: Heteromyidae). Bulletin of Carnegie Museum of Natural History, 10:1-57.

Williams, D. F. 1978c. Taxonomic and karyologic comments on small brown bats, genus *Eptesicus*, from South America. Annals of Carnegie Museum, 47:361-383.

Williams, D. F. 1979. Checklist of California mammals. Annals of Carnegie Museum, 48:425-433.

Williams, D. F., and H. H. Genoways. 1979. A systematic review of the olive-backed pocket mouse, *Perognathus fasciatus* (Rodentia: Heteromyidae). Annals of Carnegie Museum, 48:73-102.

Williams, D. F., and K. S. Kilburn. 1991. *Dipodomys ingens*. Mammalian Species, 377:1-7.

Williams, D. F., and M. A. Mares. 1978. A new genus and species of phyllotine rodent (Mammalia: Muridae) from northwestern Argentina. Annals of Carnegie Museum, 47:193-221.

Williams, D. F., J. D. Druecker, and H. L. Black. 1970. The karyotype of *Euderma maculatum* and comments on the evolution of the plecotine bats. Journal of Mammalogy, 51:602-606.

Williams, D. F., H. H. Genoways, and J. K. Braun. 1993. Taxonomy. Pp. 38-196, *in* Biology of the

Heteromyidae (H. H. Genoways and J. H. Brown, eds.). Special Publication, The American Society of Mammalogists, 10:1-719.

Williams, L. M. 1995*a*. Lesser stick-nest rat, *Leporillus apicalis*. Pp. 205-206, *in* Mammals of Victoria, distribution, ecology and conservation (P. W. Menkhorst, ed.). Oxford University Press, Melbourne, 360 pp.

Williams, L. M. 1995*b*. Greater stick-nest rat, *Leporillus conditor*. Pp. 206-207, *in* Mammals of Victoria, distribution, ecology, and conservation (P. W. Menkhorst, ed.). Oxford University Press, Melbourne, 360 pp.

Williams, L. M., 1995*c*. Desert mouse, *Pseudomys desertor*. P. 218, *in* Mammals of Victoria, distribution, ecology and conservation (P. W. Menkhorst, ed.). Oxford University Press, Melbourne, 360 pp.

Williams, L. M., and P. W. Menkhorst. 1995*a*. White-footed rabbit-rat, *Conilurus albipes*. Pp. 202-203, *in* Mammals of Victoria: distribution, ecology and conservation (P. W. Menkhorst, ed.). Oxford University Press, Melbourne, 360 pp.

Williams, L. M., and P. W. Menkhorst. 1995*b*. Plains mouse, *Pseudomys australis*. P. 216, *in* Mammals of Victoria, distribution, ecology and conservation (P. W. Menkhorst, ed.). Oxford University Press, Melbourne, 360 pp.

Williams, L. R., and G. N. Cameron. 1991. *Geomys attwateri*. Mammalian Species, 382:1-5.

Williams, S. L. 1982*a*. *Geomys personatus*. Mammalian Species, 170:1-5.

Williams, S. L. 1982*b*. The phallus of Recent genera and species of the family Geomyidae (Mammalia: Rodentia). Bulletin of Carnegie Museum of Natural History, 20:1-62.

Williams, S. L., and R. J. Baker. 1974. *Geomys arenarius*. Mammalian Species, 36:1-3.

Williams, S. L., and H. H. Genoways. 1977. Morphometric variation in the tropical pocket gopher (*Geomys tropicalis*). Annals of Carnegie Museum, 46:245-264.

Williams, S. L., and H. H. Genoways. 1978. Review of the desert pocket gopher, *Geomys arenarius* (Mammalia: Rodentia). Annals of Carnegie Museum, 47:541-570.

Williams, S. L., and H. H. Genoways. 1980*a*. Results of the Alcoa Foundation-Suriname expeditions. 2. Additional records of bats (Mammalia: Chiroptera) from Suriname. Annals of Carnegie Museum, 49:213-236.

Williams, S. L., and H. H. Genoways. 1980*b*. Morphological variation in the southeastern pocket gopher, *Geomys pinetis* (Mammalia: Rodentia). Annals of Carnegie Museum, 49:405-453.

Williams, S. L., and H. H. Genoways. 1980*c*. Results of the Alcoa Foundation-Suriname expeditions. 4. A new species of bat of the genus *Molossops* (Mammalia: Molossidae). Annals of Carnegie Museum, 49:487-498.

Williams, S. L., and H. H. Genoways. 1981. Systematic review of the Texas pocket gopher *Geomys personatus* (Mammalia: Rodentia). Annals of Carnegie Museum, 50:435-473.

Williams, S. L., and J. Ramirez-Pulido. 1984. Morphometric variation in the volcano mouse, *Peromyscus* (*Neotomodon*) *alstoni* (Mammalia: Cricetidae). Annals of Carnegie Museum, 53:163-183.

Williams, S. L., J. C. Hafner, and P. G. Dolan. 1980. Glans penes and bacula of five species of *Apodemus* (Rodentia: Muridae) from Croatia, Yugoslavia. Mammalia, 44:245-258.

Williams, S. L., D. A. Schlitter, and L. W. Robbins. 1984. Morphological variation in a natural population of *Cryptomys* (Rodentia: Bathyergidae) from Cameroon. Annals of the Museum of the Royal African Center, Section of Zoology, 237:159-172.

Williams, S. L., J. Ramírez-Pulido, and R. J. Baker. 1985. *Peromyscus alstoni*. Mammalian Species, 242:1-4.

Williams, S. L., M. R. Willig, and F. A. Reid. 1995. Review of the *Tonatia bidens* complex (Mammalia: Chiroptera), with description of two new subspecies. Journal of Mammalogy, 76:612-626.

Willig, M. R., and R. R. Hollander. 1987. *Vampyrops lineatus*. Mammalian Species, 275:1-4.

Willig, M. R., and J. K. Jones, Jr. 1985. *Neoplatymops mattogrossensis*. Mammalian Species, 244:1-3.

Willig, M. R., and M. A. Mares. 1989. Mammals from the Caatinga: An updated list and summary of recent research. Revista Brasileira de Biologia, 49:361-367.

Willis, C. R., J. M. Psyllakis, and D. J. H. Sleep. 2002. *Chaerephon nigeriae*. Mammalian Species, 710:1-3.

Willis, K. B., M. R. Willig, and J. K. Jones, Jr. 1990. *Vampyrodes caraccioli*. Mammalian Species, 359:1-4.

Willner, G. R. 1984. Muskrat. Pp. 268-272, *in* Evolution of domesticated animals (I. L. Mason, ed.). Longman Group Limited, London and New York, 452 pp.

Willner, G. R., G. A. Feldhamer, E. E. Zucker, and J. A. Chapman. 1980. *Ondatra zibethicus*. Mammalian Species, 141:1-8.

Wilson, A. C., H. Ochman, and E. M. Prager. 1987. Perspectives molecular time scale for evolution. TIG, 3, 9:241-247.

Wilson, B. A. 1994. The distribution of the New Holland mouse *Pseudomys novaehollandiae* (Waterhouse 1843) in the Eastern Otways, Victoria. The Victorian Naturalist, 111(2):46-53.

Wilson, B. A. 1996. The distribution and status of the New Holland mouse, *Pseudomys novaehollandiae* (Waterhouse 1843) in Victoria. Australian Mammalogy, 19:31-46.

Wilson, B. A., and R. Roede. 1997. Identification of New Holland mouse, *Pseudomys novaehollandiae*, populations using hair sampling tubes. Australian Mammalogy, 20:135-138.

Wilson, C. C., and W. L. Wilson. 1977. Behavioral and morphological variation among primate populations in Sumatra. Yearbook of Physical Anthropology, 20:207-233.

Wilson, D. E. 1973. The systematic status of *Perognathus merriami* Allen. Proceedings of the Biological Society of Washington, 86:175-191.

Wilson, D. E. 1976. Cranial variation in polar bears. Pp. 447-453, *in* Bears—their biology and management (M. R. Pelton, J. W. Lentfer, and G. E. Folk, eds.). International Union for the Conservation of Nature, n.s., 40:1-467

Wilson, D. E. 1978. *Thyroptera discifera*. Mammalian Species, 104:1-3.

Wilson, D. E. 1991. Mammals of the Tres Marías Islands. Bulletin of the American Museum of Natural History, 206:214-250.

Wilson, D. E., and F. R. Cole. 2000. Common names of mammals of the world. Smithsonian Institution Press, Washington and London, xiv + 204 pp.

Wilson, D. E., and J. S. Findley. 1977. *Thyroptera tricolor*. Mammalian Species, 71:1-3.

Wilson, D. E., and R. K. LaVal. 1974. *Myotis nigricans*. Mammalian Species, 39:1-3.

Wilson, D. E., and D. M. Reeder (eds.). 1993. Mammal species of the world, a taxonomic and geographic reference, Second ed. Smithsonian Institution Press, Washington, D.C., xviii + 1207 pp.

Wilson, D. E., and S. Ruff (eds.). 1999. The Smithsonian Book of North American Mammals. Smithsonian Institution Press, Washington, D.C., 750 pp.

Wilson, D. E., M. A. Bogan, R. L. Brownell, Jr., A. M. Burdin, M. K. Mamin. 1991. Geographic variation in sea otters, *Enhydra lutris*. Journal of Mammalogy, 72(1):22-36.

Wilson, G. M., and J. R. Choate. 1997. Taxonomic status and biogeography of the southern bog lemming, *Synaptomys cooperi*, on the central Great Plains. Journal of Mammalogy, 78:444-458.

Wilson, J. W. 1984. Chromosomal variation in pine voles, *Microtus* (*Pitymys*) *pinetorum*, in the eastern United States. Canadian Journal of Genetics and Cytology, 26:496-498.

Wilson, P. J., S. Grewal, I. D. Lawford, J. N. M. Heal, A. G. Granacki, D. Pennock, J. B. Theberge, M. T. Theberge, D. R. Voigt, W. Waddell, R. E. Chambers, P. C. Paquet, G. Goulet, D. Cluff and B. White. 2000. DNA profiles of the eastern Canadian wolf and the red wolf provide evidence for a common evolutionary history independent of the gray wolf. Canadian Journal of Zoology, 78:2156-2166.

Wilson, R. T. 1984. The camel. Longman, London, 223 pp.

Wilson, R. W. 1949. Early Tertiary rodents of North America. Carnegie Institute of Washington Publications, Contributions to Paleontology, 584 (-IV):67-164.

Wilson, V. J. 2002 [2001]. Duikers and rainforests of Africa. Chipangali Wildlife Trust, Ascot, Bulawayo, Zimbabwe, 798 pp.

Wiltafsky, H. 1978. *Sciurus vulgaris* Linnaeus, 1758 – Eichhörnchen. Pp. 86-105, *in* Handbuch der Säugetiere Europas (J. Niethammer and F. Krapp, eds.). Akademische Verlagsgesellschaft (Wiesbaden), 1:476 pp.

Wineski, L. E. 1983. Movements of the cranial vibrissae in the golden hamster (*Mesocricetus auratus*). Journal of the Zoological Society of London, 200:261-280.

Winge, H. 1887. Jordfunde og nulevende Gnavere (Rodentia) fra Lagoa Santa, Minas Geraes, Brasilien: Med udsigt over gnavernes indbyrdes slagtskab. E Museo Lundii, 1(3):1-178.

Winge, H. 1941. The interrelationships of the mammalian genera. C. A. Reitzels Forlag, Copenhagen, Vol. 1, 418 pp., Vol. 2, 376 pp. (translation from 1923 original in Danish by E. Deichmann and G. M. Allen).

Winking, H. 1976. Karyologie und Biologie der beiden iberischen Wühlmausarten *Pitymys mariae* und *Pitymys duodecimcostatus*. Zeitschrift für Zoologische Systematik und Evolutionsforschung, 4:104-129.

Winking, H., A. Gropp, and J. T. Marshall. 1979. Karyotype and sex chromosomes in *Vandeleuria oleracea*. Zeitschrift für Säugetierkunde, 44:195-201.

Winking, H., B. Dulic, and G. Bulfield. 1988. Robertsonian karyotype variation in the European house mouse, *Mus musculus*. Survey of present knowledge and new observations. Zeitschrift für Säugetierkunde, 53:148-161.

Winkler, A. J. 1994. The middle/upper Miocene dispersal of major rodent groups between southern Asia and Africa. Pp. 173-184, *in* Rodent and lagomorph families of Asian origins and diversification (Y. Tomida, C. k. Li, and T. Setoguchi, eds.). National Science Museum Monographs, No. 8, Tokyo, 195 pp.

Winkler, A. J. 1998. A new dendromurine (Rodentia: Muridae) from the Middle Miocene of western Kenya. Pp. 91-104, *in* Advances in vertebrate paleontology and geochronology (Y. Tomida, L. J. Flynn, and L. L. Jacobs, eds.). National Science Museum Monographs, No. 14, Tokyo, 292 pp.

Winkler, A. J. 2002. Neogene paleobiogeography and East African paleoenvironments: Contributions from the Tugen Hills rodents and lagomorphs. Journal of Human Evolution, 42:237-256.

Winkler, A. J. 2003. Rodents and lagomorphs from the Miocene and Pliocene of Lothagam, northern Kenya. Pp. 169-198, *in* Lothagam: dawn of humanity in eastern Africa (M. G. Leakey and J. M. Harris, eds.). Columbia University Press, New York, 688 pp.

Winn, H. E., and N. E. Reichley. 1985. Humpback whale—*Megaptera novaeangliae*. Pp. 241-274, *in* Handbook of marine mammals: The sirenians and baleen whales (S. H. Ridgway and R. Harrison, eds.). Academic Press, London, 3:1-362.

Winter, J. W. 1983. Thornton Peak *Melomys*. P. 379, *in* The Australian museum complete book of Australian mammals (The national photographic index of Australian wildlife) (R. Strahan, ed.). Angus and Robertson, Sydney, 530 pp.

Winter, J. W. 1984. The Thornton Peak *Melomys, Melomys hadrourus* (Rodentia: Muridae): A new rainforest species from northeastern Queensland, Australia. Memoirs of the Queensland Museum, 21:519-539.

Winter, J. W. 1997. Responses of non-volant mammals to Late Quaternary climatic changes in the wet tropics region of North-eastern Australia. Wildlife Research, 24:493-511.

Winter, J. W., and L. A. Moore. 1995. Masked white-tailed rat, *Uromys hadrourus*. Pp. 641-643, *in* Mammals of Australia (R. Strahan, ed.). Smithsonian Institution Press, Washington, D.C., 756 pp.

Winter, J. W., and D. Whitford. 1995. Prehensile-tailed rat, *Pogonomys mollipilosus*. Pp. 643-645, *in* Mammals of Australia (R. Strahan, ed.). Smithsonian Institution Press, Washington, D.C., 756 pp.

Wirth, R. 1984. [Lumping subspecies of the dama gazelle, *Nanger dama*.] S.S.C. Antelope Specialist Group Gnusletter. May 1984:4-5.

Wiseman, R., C. O'Ryan, and E. H. Harley. 2000. Microsatellite analysis reveals that domestic cat (*Felis catus*) and southern African wild cat (*F. lybica*) are genetically distinct. Animal Conservation, 3(3):221-228.

Witte, G. 1962. Zur Systematik und Verbreitung des Siebenschläfers in Italien. Bonner Zoologische Beiträge, 13:115-127.

Witte, G. R. 1997. Der Maulwurf *Talpa europaea*. Die Neue Brehm-Bücherei, 637. Magdeburg, Germany, 219 pp.

Wittouck, P., E. Pinna Senn, C. A. Sonez, M. C. Provensal, J. J. Polop, and J. A. Lisanti. 1995. Chromosomal and synaptonemal complex analysis of Robertsonian polymorphisms in *Akodon dolores* and *Akodon molinae* (Rodentia, Cricetidae) and their hybrids. Cytologia, 60:93-102.

Wodzicki, K. A. 1950. Introduced mammals of New Zealand. Department of Scientific and Industrial Research Bulletin, 98:1-255.

Wodzicki, K. A., and J. E. C. Flux. 1967. Rediscovery of the white-throated wallaby, *Macropus parma* Waterhouse, 1846, on Kawau Island, New Zealand. Australian Journal of Science, 29:429-430.

Wodzicki, K., and R. H. Taylor. 1984. Distribution and status of the Polynesian rat *Rattus exulans*. Acta Zoologica Fennica, 172:99-101.

Woinarski, J. C. Z. 2000. The conservation status of rodents in the monsoonal tropics of the Northern Territory. Wildlife Research, 27:421-435.

Woinarski, J. C. Z., R. W. Braithwaite, and M. Maguire. 1995*a*. Kakadu pebble-mound mouse, *Pseudomys calabyi*. Pp. 589-590, *in* Mammals of Australia (R. Strahan, ed.). Smithsonian Institution Press, Washington, D.C., 756 pp.

Woinarski, J. C. Z., S. Churchill, and C. Trainor. 1995*b*. Carpentarian rock-rat, *Zyzomys palatalis*. Pp. 623-624, *in* Mammals of Australia (R. Strahan, ed.). Smithsonian Institution Press, Washington, D.C., 756.

Woinarski, J. C. Z., N. Gambold, D. Wurst, T. F. Flannery, A. P. Smith, R. Chatto, and A. Fisher. 1999. Distribution and habitat of the northern hopping-mouse, *Notomys aquilo*. Wildlife Research, 26:495-511.

Woinarski, J. C. Z., K. Brennan, A. Dee, J. Njudumul, P. Guthayguthay, and P. Horner. 2000. Further records of the false water-rat *Xeromys myoides* from coastal Northern Territory. Australian Mammalogy, 21:221-223.

Wójcik, A. M. 1993. Genetic variation in a fluctuating population of the yellow-necked mouse *Apodemus flavicollis*. Acta Theriologica, 38(3):273-290.

Wójcik, J. M., and M. Wolsan (eds.). 1998. Evolution of shrews. Mammal Research Institute, Polish Academy of Sciences, Bialowieza, 458 pp.

Wójcik, J. M., M. Ratkiewicz, and J. B. Searle. 2002. Evolution of the common shrew *Sorex araneus* chromosomal and molecular aspects. Acta Theriologica, 47(suppl. 1):139-167.

Wolfe, J. L. 1982. *Oryzomys palustris*. Mammalian Species, 176:1-5.

Wolfe, J. L., and A. V. Linzey. 1977. *Peromyscus gossypinus*. Mammalian Species, 70:1-5.

Wolffsohn, J. A. 1908. Contribuciones a la mamalojia chilena. Revista Chilena de Historia Natural, 12:165-172.

Wolk, K. 1987. New localities of common dormouse *Muscardinus avellanarius* L. in Pojezierze Mazurskie. Przeglad Zoologiczny, 31(2):219-220.

Wolman, A. A. 1985. Gray whale—*Eschrichtus robustus*. Pp. 67-90, *in* Handbook of marine mammals: The sirenians and baleen whales (S. H. Ridgway and R. Harrison, eds.). Academic Press, London, 3:1-362.

Wolsan, M. 1999. Oldest mephitine cranium and its implications for the origin of skunks. Acta Palaeontologica Polonica, 44:223-230.

Wolsan, M., and R. Hutterer. 1998. A list of the living species of shrews. Pp. 425-448, *in* Evolution of shrews (J. M.Wójcik and M. Wolsan, eds.). Mammal Research Institute, Polish Academy of Sciences, Bialowieza, 458 pp.

Womochel, D. R. 1978. A new species of *Allactaga* (Rodentia: Dipodidae) from Iran. Fieldiana: Zoology, 72(5):65-73.

Won, C., and K. G. Smith. 1999. History and current status of mammals of the Korean Peninsula. Mammal Review, 29(1):3-33.

Wood, A. E. 1935. Evolution and relationships of the heteromyid rodents with new forms from the Tertiary of western North America. Annals of Carnegie Museum, 24:73-262.

Wood, A. E. 1950. Porcupines, paleogeography and parallelism. Evolution, 4:87-98.

Wood, A. E. 1955. A revised classification of the rodents. Journal of Mammalogy, 36:165-187.

Wood, A. E. 1959. Eocene radiation and phylogeny of the rodents. Evolution, 13:354-361.

Wood, A. E. 1962. The early Tertiary rodents of the Family Paramyidae. Transactions of the American Philosophical Society, n.s., 52:1-261.

Wood, A. E. 1965. Grades and clades among rodents. Evolution, 19:115-130.

Wood, A. E. 1974. The evolution of the Old World and New World hystricomorphs. Symposium of the Zoological Society of London, 34:21-60.

Wood, A. E. 1975. The problem of the hystricognathus rodents. University of Michigan Papers in Paleontology, 12:75-80.

Wood, A. E. 1980. The origin of the caviomorph rodents from a source in Middle America: A clue to the area of origin of the platyrrhine primates. Pp. 79-91, *in* Evolutionary Biology of the New World Monkeys and Continental Drift (R. L. Ciochon and A. B. Chiarelli, eds.). Plenum Press, New York, xvi + 528 pp.

Wood, A. E. 1985. The relationships, origin and dispersal of the hystricognathous rodents. Pp. 475-513, *in* Evolutionary Relationships among Rodents: A Multidisciplinary Analysis (W. P. Luckett and J.-L. Hartenberger, eds.). Plenum Press, New York, 721 pp.

Wood, A. E., and B. Patterson. 1959. The rodents of the Deseadan Oligocene of Patagonia and the beginnings of South American rodent evolution. Bulletin of the Museum of Comparative Zoology, 120:281-428.

Woodburne, M. O. 1968. The cranial myology and osteology of *Dicotyles tajacu*, the collared peccary, and its bearing on classification. Memoirs of the Southern California Academy of Sciences, 7:1-48.

Woodman, N. 1988. Subfossil remains of *Peromyscus stirtoni* (Mammalia: Rodentia) from Costa Rica. Revista de Biologia Tropical, 36:247-253.

Woodman, N. 1993. The correct gender of mammalian generic names ending in *-otis*. Journal of Mammalogy, 74:544-546.

Woodman, N. 1995. Morphological variation between Pleistocene and Recent samples of *Cryptotis* (Insectivora: Soricidae) from the Yucatan Peninsula, Mexico. Journal of Mammalogy, 76:223-231.

Woodman, N. 1996. Taxonomic status of the enigmatic *Cryptotis avia* (Mammalia: Insectivora: Soricidae), with comments on the distribution of the Colombian small-eared shrew, *Cryptotis colombiana*. Proceedings of the Biological Society of Washington, 109:409-418.

Woodman, N. 2000. *Cryptotis merriami* Choate in Costa Rica: Syntopy with *Cryptotis nigrescens* (Allen) and possible character displacement (Mammalia: Insectivora). Carribean Journal of Science, 36:289-299.

Woodman, N. 2002. A new species of small-eared shrew from Colombia and Venezuela (Mammalia: Soricomorpha: Soricidae: Genus *Cryptotis*). Proceedings of the Biological Society of Washington, 115:249-272.

Woodman, N. 2003. A new small-eared shrew of the *Cryptotis nigrescens*–group from Colombia (Mammalia: Soricomorpha: Soricidae). Proceedings of the Biological Society of Washington, 116:853-872.

Woodman, N., and R. M. Timm. 1992. A new species of small-eared shrew, genus *Cryptotis*, (Insectivora: Soricidae), from Honduras. Proceedings of the Biological Society of Washington, 105:1-12.

Woodman, N., and R. M. Timm. 1993. Intraspecific and interspecific variation in the *Cryptotis nigrescens* species complex of small-eared shrews (Insectivora: Soricidae), with the description of a new species from Colombia. Fieldiana: Zoology, n.s., 1452:1-30.

Woodman, N., and R. M. Timm. 1999. Geographic variation and evolutionary relationships among broad-clawed shrews of the *Cryptotis goldmani*-group (Mammalia: Insectivora: Soricidae). Fieldiana: Zoology, n.s., 91(1497):1-35.

Woodman, N., and R. M. Timm. 2000. Taxonomy and evolutionary relationships of Phillips' small eared shrew, *Cryptotis phillipsi* (Schaldach, 1966), from Oaxaca, Mexico (Mammalia: Insectivora: Soricidae). Proceedings of the Biological Society of Washington, 113:339-355.

Woodman, N., E. Schneider, P. Grant, D. Same, K. E. Schmall, and J. T. Curtis. 2002. A new southern distributional limit for the Central American rodent *Peromyscus stirtoni*. Caribbean Journal of Science, 38:281-284.

Woods, C. A. 1972. Comparative mycology of jaw, hyoid, and pectoral appendicular regions of New and Old World hystricomorph rodents. Bulletin of the American Museum of Natural History, 147:115-198.

Woods, C. A. 1973. *Erethizon dorsatum*. Mammalian Species, 29-1-6.

Woods, C. A. 1982. The history and classification of South American hystricognath rodents: Reflections on the far away and long ago. Pp. 377-392, *in* Mammalian biology in South America (M. Mares and H. Genoways, eds.). Special Publication Series, Pymatuning Laboratory of Ecology, University of Pittsburgh, Pennsylvania, 6:1-539.

Woods, C. A. 1984. Hystricoganth rodents. Pp. 389-446, *in* Orders and families of Recent mammals of the world (S. Anderson and J. K. Jones, Jr., eds.). Wiley, New York, 686 pp.

Woods, C. A. 1989*a*. The biogeography of West Indian rodents. Pp. 741-798, *in* Biogeography of the West Indies: Past, present, and future (C. A. Woods, ed.). Sandhill Crane Press, Gainesville, FL, 878 pp.

Woods, C. A. 1989*b*. A new capromyid rodent from Haiti; the origin, evolution, and extinction of West Indian rodents and their bearings on the origin of New World hystricognaths. Los Angeles County Museum Natural History, Science Series, 33:59-89.

Woods, C. A. 1993. Suborder Hystricognathi. Pp. 771-806, *in* Mammal species of the world, a taxonomic and geographic reference, Second ed. (D. E. Wilson and D. M. Reeder, eds.). Smithsonian Institution Press, Washington D.C., 1206 pp.

Woods, C. A., and D. Boraker. 1975. *Octodon degus*. Mammalian Species, 67:1-5.

Woods, C. A., and E. B. Howland. 1979. Adaptive radiation of capromyid rodents. Journal of Mammalogy, 60:95-116.

Woods, C. A., J. A. Ottenwalder, and W. L. R. Oliver. 1985. Lost mammals of the Greater Antilles: Summarized findings of a ten week field survey in the Dominican Republic, Haiti and Puerto Rico. Dodo (Jersey Wildlife Preservation Trust), 22:23-42.

Woods, C. A., L. Contreras, G. Willner-Chapman, and H. P. Whidden. 1992. *Myocastor coypus*. Mammalian Species, 398:1-8.

Woods, C. A., R. Borroro Paéz, and C. W. Kilpatrick. 2001. Insular patterns and radiations of West Indian rodents. Pp. 335-353, *in* Biogeography of the West Indies: Patterns and perspectives. (C. A. Woods and F. E. Sergile, eds.). CRC Press, Boca Raton, 582 pp.

Woolley, P. A. 1974. The pouch of *Planigale subtilissima* and other dasyurid marsupials. Journal of the Royal Society of Western Australia, 57:11-15.

Woolley, P. A. 1989. Nest location by spool-and-line tracking of dasyurid marsupials in New Guinea. Journal of Zoology, 218:689-700.

Woolley, P. A. 1992. New records of the Julia Creek Dunnart, *Sminthopsis douglasi* (Marsupialia: Dasyuridae). Wildlife Research, 19:779-783.

Wozencraft, W. C. 1984. A phylogenetic reappraisal of the Viverridae and its relationship to other Carnivora. Unpubl. Ph. D. dissertation, University of Kansas, Lawrence, KS, 1129 pp.

Wozencraft, W. C. 1986. New species of striped mongoose from Madagascar. Journal of Mammalogy, 67:561-571.

Wozencraft, W. C. 1987. Emendation of species name. Journal of Mammalogy, 68(1):198.

Wozencraft, W. C. 1989*a*. The phylogeny of the Recent Carnivora. Pp. 495-535, *in* Carnivore behavior, ecology, and evolution (J. L. Gittleman, ed.). Cornell University Press, Ithaca, NY, 620 pp.

Wozencraft, W. C. 1989*b*. Classification of the Recent Carnivora. Pp. 569-593, *in* Carnivore behavior, ecology and evolution (J. L. Gittleman, ed.). Cornell University Press, Ithaca, NY, 620 pp.

Wozencraft, W. C. 1993. Order Carnivora. Pp. 279-348, *in* Mammal species of the world, a taxonomic and geographic reference, Second ed. (D. E. Wilson and D. M. Reeder, eds.). Smithsonian Institution Press, Washington, D.C. 1207 pp.

Wright, A. J., R. A. Van Den Bussche, B. K. Lim, M. D. Engstrom, and R. J. Baker. 1999. Systematics of the genera *Carollia* and *Rhinophylla* based on the cytochrome-*b* gene. Journal of Mammalogy, 80:1202-1213.

Wright, D. B. 1989. Phylogenetic relationships of *Catagonus wagneri*: Sister taxa from the Tertiary of North America. Pp. 281-308, *in* Advances in neotropical mammalogy (K. H. Redford and J. F. Eisenberg, eds). Sandhill Crane Press, Gainesville, FL, 554 pp.

Wrigley, R. E. 1972. Systematics and biology of the woodland jumping mouse, *Napaeozapus insignis*. Illinois Biological Monographs, 47:1-117.

Wroughton, R. C. 1905*a*. The common striped palm squirrel. Journal of the Bombay Natural History Society, 16:406-413.

Wroughton, R. C. 1905*b*. Notes on the various forms of *Arvicanthis pumilio*, Sparrm. Annals and Magazine of Natural History, ser. 7, 16:629-639.

Wroughton, R. C. 1906. Notes on the genus *Otomys*. Annals and Magazine of Natural History, ser. 7, 18:264-278.

Wroughton, R. C. 1908*a*. Notes on the classification of the bandicoots. Journal of the Bombay Natural History Society, 18:736-752.

Wroughton, R. C. 1908*b*. Three new African species of *Mus*. Annals and Magazine of Natural History, ser. 8, 1:255-257.

Wroughton, R. C. 1910. On the nomenclature of the Indian hedgehogs. Journal of the Bombay Natural History Society, 20:80-82.

Wroughton, R. C. 1920. Summary of the results from the Indian Mammal Survey of the Bombay Natural History Society. Part V. Journal of the Bombay Natural History Society, 26:955-967.

Wu, A., G. Wang, H. Zhao, Z. Yang. 1989. The karyotype of *Eothenomys miletus miletus*. Acta Theriologica Sinica, 9(3):231, 234.

Wu, C. H., Y. P. Zhang, T. D. Bunch, S. Wang, and W. Wang. 2003. Mitochondrial control region sequence variation within the argali wild sheep (*Ovis ammon*): Evolution and conservation relevance. Mammalia, 67:109-118.

Wu De-ling. 1982. [On subspecific differentiation of brown rat (*Rattus norvegicus* Berkenhout) in China.] Acta Theriologica Sinica, 2:107-112 (in Chinese).

Wu De-ling and Deng Xiang-fu. 1984. [A new species of tree mice from Yunnan, China.] Acta Theriologica Sinica, 4:207-212 (in Chinese).

Wu De-ling and Wang Guan-huan. 1984. [A new subspecies of *Typhlomys cinereus* Milne- Edwards from Yunnan, China.] Acta Theriologica Sinica, 4:213-215 (in Chinese).

Wu, D.-L., J. Luo, and B. J. Fox. 1996. A comparison of ground-dwelling small mammal communities in primary and secondary tropical rainforests in China. Journal of Tropical Ecology, 12:215-230.

Wu, W., and L. J. Flynn. 1992. [New murid rodents from the Late Cenozoic of Yushe Basin, Shanxi.] Vertebrata PalAsiatica, 30(1):17-38 (in Chinese with English summary).

Wu, W.-Y., J. Ye, S.-C. Bi, and J. Meng. 2000. The discovery of late Oligocene dormice from China. Vertebrata Palasiatica, 38(1):36-42 (in Chinese with English summary.)

Wuczynski, A., and A. Garbowski. 2000. Fat dormouse *Glis glis* (Linnaeus, 1766) in the Krzyzowe Hills (Sudetic Foreland). Przeglad Zoologiczny, 44(1-2):93-97.

Wurst, D. 1995. Central rock-rat, *Zyzomys pedunculatus*. Pp. 624-625, *in* Mammals of Australia (R. Strahan, ed.). Smithsonian Institution Press, Washington, D.C., 756 pp.

Wurst, D. 1997. Rock-rats alive! Nature Australia, Spring, 1997:12-14.

Wurster, D. H., and K. Benirschke. 1968. Comparative cytogenetic studies in the order Carnivora. Chromosoma, 24:336-382.

Wurster, D. H., J. R. Snapper, and K. Benirschke. 1971. Unusually large sex chromosomes: New methods of measuring and description of karyotypes of six rodents (Myomorpha and Hystricomorpha) and one lagomorph (Ochotonidae). Cytogenetics, 10:153-176.

Wurster-Hill, D. H. 1973. Chromosomes of eight species from five families of Carnivora. Journal of Mammalogy, 54(3):753-760.

Wynen, L. P., S. D. Goldsworthy, S. J. Insley, M. Adams, J. W. Bickham, J. Francis, J. Pablo Gallo, A. R. Hoelzel, P. Majluf, R. W. G. White, and R. Slade. 2001. Phylogenetic relationships within the eared seals (Otariidae: Carnivora): Implications for the Historical Biogeography of the family. Molecular Phylogenetics and Evolution, 21:270-284.

Wyss, A. R. 1987. The walrus auditory region and the monophyly of pinnipeds. American Museum Novitates, 2871:1-31.

Wyss, A. R. 1988. On "Retrogression" in the evolution of the Phocinae and phylogenetic affinities of the monk seals. American Museum Novitates, 2924:1-38.

Wyss, A. R., and J. J. Flynn. 1993. A phylogenetic analysis and definition of the Carnivora. Pp. 32-52, *in* Mammal Phylogeny: Placentals (F. S. Szalay, M. J. Novacek, and M. C. McKenna, eds.). Placentals Springer-Verlag, New York, 2:1-321.

Wyss, A. R., and J. Meng. 1996. Application of phylogenetic taxonomy to poorly resolved crown clades: A stem-modified node-based definition of Rodentia. Systematic Biology, 45:559-568.

Xia Wu-ping. 1984. [A study on Chinese *Apodemus* with a discussion of its relations to Japanese species.] Acta Theriologica Sinica, 4:93-98 (in Chinese).

Xia Wu-ping. 1985. A study on Chinese *Apodemus* and its relation to Japanese species. Pp. 76-79, *in* Contemporary mammalogy in China and Japan (T. Kawamichi, ed.). Mammalogical Society of Japan, 194 pp.

Ximenez, A. 1975. *Leopardus geoffroyi*. Mammalian Species, 54:1-4.

Ximénez, A., and A. Langguth. 1970. *Akodon cursor montensis* en el Uruguay (Mammalia-Cricetinae). Comunicaciones Zoologicas del Museo de Historia Natural de Montevideo, 10(128):1-7.

Ximenez, A., A. Langguth, and R. Praderi. 1972. Lista sistemática de los mamíferos del Uruguay. Anales del Museo de Historia Natural de Montevideo, 7:1-49.

Xu Ke-fen. 1986. [Analysis of the karyotypes of *Lepus yarkandensis*.] Acta Theriologica Sinica, 6:249-253 (in Chinese).

Xu Long-hui and Yu Sim-ian [Yu Sim-yan]. 1985. [A new subspecies of Edward's rat from Hainan Island, China.] Acta Theriologica Sinica, 5:131-135 (in Chinese).

Xu, X., A. J. Winkler, and L. L. Jacobs. 1996. Is the rodent *Acomys* a murine? An evaluation using morphometric techniques. Pp. 661-675, *in* Palaeoecology and palaeoenvironments of Late Cenozoic mammals. Tributes to the Career of C. S. (Rufus) Churcher (K. M. Stewart and K. L. Seymour, eds.). University of Toronto Press, Toronto, 675 pp.

Yabe, T., P. Boonsong, and S. Hongnark. 1998. The structure of the pawpad lamellae of four *Rattus* species. Mammal Study, 23:129-132.

Yachontov, E. L., and E. G. Potapova. 1991. On the position of dormice (Gliroidea) in the system of rodents. Proceedings of the Zoological Institute, 243:127-147.

Yahnke, C. J., W. E. Johnson, E. Geffen, D. Smith, F. Hertel, M. S. Roy, C. F. Bonacic, T. K. Fuller, B. Van Valkenburgh, and R. K. Wayne. 1996. Darwin's fox: A distinct endangered species in a vanishing habitat. Conservation Biology, 10:366-375.

Yakimenko, L. V., and A. P. Kryukov. 1997. [On karyotype variation in common vole *Microtus rossiaemeridionalis* (Rodentia, Cricetidae).] Zoologicheskii Zhurnal, 76(3):375-378 (in Russian with English summary).

Yakimenko, L. V., and E. A. Lyapunova. 1986. [Cytogenetic corroboration of belonging of northern mole vole from Turkmenia to *Ellobius talpinus* s. str.] Zoologicheskii Zhurnal, 65:946-949 (in Russian).

Yakimenko, L. V., K. V. Korobitsyna, L. V. Frisman, and A. I. Muntianu. 1990. Cytogenetic and biochemical comparison of *Mus musculus* and *Mus hortulanus*. Experientia, 46:1075-1077.

Yalden, D. W. 1975. Some observations on the giant mole-rat *Tachyoryctes macrocephalus* (Rüppell, 1842) (Mammalia, Rhizomyidae) of Ethiopia. Monitore Zoologico Italiano, n.s., Supplemento VI, 15:275-303.

Yalden, D. W. 1978. A revision of the dik-diks of the subgenus *Madoqua* (*Madoqua*). Monitore Zoologico Italiano, n.s., Supplemento, 11:245-264.

Yalden, D. W. 1985. *Tachyoryctes macrocephalus*. Mammalian Species, 237:1-3.

Yalden, D. W. 1988. Small mammals of the Bale Mountains, Ethiopia. African Journal of Ecology, 26:281-294.

Yalden, D. W. 1999. The history of British mammals. Academic Press, London, 305 pp.

Yalden, D. W., and M. J. Largen. 1992. The endemic mammals of Ethiopia. Mammal Review, 22:115-150.

Yalden, D. W., M. J. Largen, and D. Kock. 1976. Catalogue of the mammals of Ethiopia. 2. Insectivora and Rodentia. Monitore Zoologico Italiano, Supplemento 8(1):1-118.

Yalden, D. W., M. J. Largen, and D. Kock. 1980. Catalogue of the mammals of Ethiopia. 4. Carnivora. Monitore Zoologico Italiano, n.s., Supplemento, 13(8):168-272.

Yalden, D. W., M. J. Largen, and D. Kock. 1984. Catalogue of the mammals of Ethiopia. 5. Artiodactyla. Monitore Zoologico Italiano, n.s., Supplemento, 19:67-221.

Yalden, D. W., M. J. Largen, and D. Kock. 1986. Catalogue of the mammals of Ethiopia. 6. Perissodactyla, Proboscidea, Hyracoidea, Lagomorpha, Tubulidentata, Sirenia and Cetacea. Monitore Zoologico Italiano, n.s., Supplemento, 21:31-103.

Yalden, D. W., M. J. Largen, D. Kock and J. C. Hillman. 1996. Catalogue of the mammals of Ethiopia and Eritrea. 7. Revised checklist, zoogeography and conservation. Tropical Zoology, 9:73-164.

Yamagata, T., A. Ishikawa, Y. Tsubota, T. Namikawa, and A. Hirai. 1987. Genetic differentiation between laboratory lines of the musk shrew (*Suncus murinus*, Insectivora) based on restriction endonuclease cleavage patterns of mitochondrial DNA. Biochemical Genetics, 25:429-446.

Yancey, F. D., II. 1997. The mammals of Big Bend Ranch State Park, Texas. Special Publications, Museum of Texas Tech University, 39:1-210.

Yancey, F. D., II, and C. Jones. 1997. Dispersal of two species of harvest mice (*Reithrodontomys*) between the High Plains and Rolling Plains of Texas. Occasional Papers, Museum of Texas Tech University, 166:1-5.

Yancey, F. D., II, R. W. Manning, and C. Jones. 1996. Distribution, natural history, and status of the Palo Duro mouse, *Peromyscus truei comanche*, in Texas. Texas Journal of Science, 48:3-12.

Yancey, F. D., II, J. R. Goetze, and C. Jones. 1998*a*. *Saccopteryx bilineata*. Mammalian Species, 581:1-5.

Yancey, F. D., II, J. R. Goetze, and C. Jones. 1998*b*. *Saccopteryx leptura*. Mammalian Species, 582:1-3.

Yañez, J. L., W. Siefeld, J. Valencia, and F. Jaksic. 1978. Relaciones entre la sistemática y la morfometriá del subgénero *Abrothrix* (Rodentia: Cricetidae) en Chile. Anales del Instituto de la Patagonia, 9:185-197.

Yañez, J. L., J. Valencia, and F. Jaksic. 1979. Morfometría y sistemática del subgénero *Akodon* (Rodentia) en Chile. Archivos de Biología y Medicina Experimentales (Santiago), 12:197-202.

Yañez, J. L., J. C. Torres-Mura, J. R. Rau, and L. C. Contreras. 1987. New records and current status of *Euneomys* (Cricetidae) in southern South America. Fieldiana: Zoology, n.s., 39:283-287.

Yang An-feng and Fang Li-xiang. 1988. [Phallic morphology of 13 species of the family Muridae from China, with comments on its taxonomic significance.] Acta Theriologica Sinica, 8:275-287 (in Chinese with English summary).

Yang, A., S. Liu, and L. Fang. 1992. [Phallic morphology of eight species in Gerbillinae and Microtinae from China.] Acta Theriologica Sinica, 12(1):31-38 (in Chinese with English abstract).

Yang Chun-wen, Chen Rong-hai, and Zhang Chung-mei. 1991. [A study of the rodent community division in Huangnihe forest region.] Acta Theriologica Sinica, 11(2):118-125.

Yang, D.-M., R. Liu, Ya-P. Zhang, Z. Chen, and Y. Wang. 1998. Chromosome study of Yulong vole (*Eothenomys proditor*). Cytologia, 63:435-440.

Yang, F., P. C. M. O'Brien, J. Wienberg, and M. A. Ferguson-Smith. 1997. Evolution of the black muntjac (*Muntiacus crinifrons*) karyotype revealed by comparative chromosome painting. Cytogenetics and Cell Genetics, 76:159-163.

Yang, G. F., N. P. Carter, L. Shi, and M. A. Ferguson-Smith. 1995. A comparative study of karyotypes of muntjacs by chromosome painting. Chromosoma, 103:642-652.

Yang, G.-r., and D.-I. Wu. 1979. Two new records of Chinese rodents. Acta Zootaxonomica Sinica, 4:192-193.

Yang, Guang, K. Zhou, R. Wenhua, G. Ji, S. Liu, R. Bastida, and L. Rivero. 2002. Molecular systematics of river dolphins inferred from complete mitochondrial cytochrome-b gene sequences. Marine Mammal Science, 18(1):20-29.

Yang Guang-rong and Wang Ying-xiang. 1987. [A new subspecies of *Hadromys humei* (Muridae, Mammalia) from Yunnan, China]. Acta Theriologica Sinica, 7:46-50 (in Chinese).

Yang, Y., and H. Lu. 1998. [The comparative study on morphological and biochemical indexes of striped field mouse (*Apodemus agrarius*) in eastern China]. Acta Theriologica Sinica, 18(1):50-53 (in Chinese).

Yang, Y., Z. Zeng , M. Luo, Z. Song, and J. Liang. 1994. [Nonlinear model and stepwise regression of population dynamics of *Rattus nitidus*.] Acta Theriologica Sinica, 14(2):130-137 (in Chinese with English abstract).

Yates, T. L. 1984. Insectivores, elephant shrews, tree shrews, and dermopterans. Pp. 117-144, *in* Orders and families of Recent mammals of the world (S. Anderson, and J. K. Jones, Jr., eds.). John Wiley and Sons, New York, 686 pp.

Yates, T. L., and I. F. Greenbaum. 1982. Biochemical systematics of North American moles (Insectivora: Talpidae). Journal of Mammalogy, 63:368-374.

Yates, T. L., and D. W. Moore. 1990. Speciation and evolution in the family Talpidae (Mammalia: Insectivora). Pp. 1-22, *in* Evolution of subterranean mammals at the organismal and molecular levels (E. Nevo and O. A. Reig, eds.). Alan R. Liss, New York, 434 pp.

Yates, T. L., and D. J. Schmidly. 1975. Karyotype of the eastern mole (*Scalopus aquaticus*), with comments on the karyology of the family Talpidae. Journal of Mammalogy, 56:902-905.

Yates, T. L., and D. J. Schmidly. 1977. Systematics of *Scalopus aquaticus* (Linnaeus) in Texas and adjacent states. Occasional Papers, The Museum, Texas Tech University, 45:1-36.

Yates, T. L., and D. J. Schmidly. 1978. *Scalopus aquaticus*. Mammalian Species, 105:1-4.

Yates, T. L., R. J. Baker, and R. K. Barnett. 1979. Phylogenetic analysis of karyological variation in three genera of peromyscine rodents. Systematic Zoology, 28:40-48.

Yatsenko, V. N. 1980. [C-heterochromatin and chromosomal polymorphism in the Gobi-Altai vole (*Alticola stolizkanus barakschin* Bannikov, 1968, Rodentia, Cricetidae).] Doklady Akademii Nauk SSSR, 254:1009-1010.

Yatsenko, V. N., V. M. Malygin, V. N. Orlov, I. Yu. Yanina. 1980. [The chromosome polymorphism in the Mongolian vole *Microtus mongolicus* Radde, 1861.] Tsitologiya, 22:471-474 (in Russian).

Ye, X.-D., Y. Ma, J.-S. Zhang, Z.-L. Wang, and Z.-K. Wang. 2002. [A summary of *Eothenomys* (Rodentia: Cricetidae: Microtinae).] Acta Zootaxonomica Sinica, 27(1):173-182 (in Chinese with English summary).

Yeboah, S. 1982-1988. *Steatomys parvus cuppedius* (Coetzee): A new record from the forest region of Ghana, West Africa. Ghana Journal of Science, 22-28(1-2):75-76.

Yee, D. A. 2000. *Peropteryx macrotis*. Mammalian Species, 643:1-4.

Yensen, E. 1991. Taxonomy and distribution of the Idaho ground squirrel. Journal of Mammalogy, 72:583-600.

Yensen, E., and K. L. Seymour. 2000. *Oreailurus jacobita*. Mammalian Species, 644:1-6.

Yensen, E., and P. W. Sherman. 1997. *Spermophilus brunneus*. Mammalian Species, 560:1-5.

Yensen, E., and T. Tarifa. 2003a. *Galictis vittata*. Mammalian Species, 727:1-8.

Yensen, E., and T. Tarifa. 2003b. *Galictis cuja*. Mammalian Species, 728:1-8.

Yepes, J. 1928. Los "Edentata" argentinos. Sistemática y distribución. Revista de la Universidad de Buenos Aires, ser. 2a, 1:461-515, 6 figs [Reprint separate is independently paginated].

Yiğit, N., and E. Çolak. 1998. A new subspecies of *Meriones tristrami* Thomas, 1892 (Rodentia: Gerbillinae) from Kilis (Southeastern Turkey): *Meriones tristrami kilisensis* subsp. n. Turkish Journal of Zoology, 22:99-103.

Yiğit, N., and E. Çolak. 1999. A study of the taxonomy and karyology of *Meriones persicus* (Blanford, 1875) (Mammalia: Rodentia) in Turkey. Turkish Journal of Science, 23:269-274.

Yiğit, N., and E. Çolak. 2002. On the distribution and taxonomic status of *Microtus guentheri* (Danford and Alston, 1880) and *Microtus lydius* Blackler, 1916 (Mammalia: Rodentia) in Turkey. Turkish Journal of Zoology, 26:197-204.

Yiğit, N., E. Çolak, E. Kivanç, and M. Sözen. 1997a. Gerbil record from Turkey: *Gerbillus* (*Hendecapleura*) *dasyurus* (Wagner, 1842) (Rodentia: Gerbillinae). Israel Journal of Zoology, 43:13-18.

Yiğit, N., E. Kivanç, and E. Çolak. 1997b. [Diagnostic characters and distribution of *Meriones* Illiger, 1811 species (Mammalia: Rodentia) in Turkey.] Turkish Journal of Zoology, 21:361-374 (in Turkish with English abstract).

Yiğit, N., E. Kivanç, and E. Çolak. 1998a. On the taxonomic status of *Meriones tristrami* Thomas, 1892 (Rodentia: Gerbillinae) in Turkey. Zoology in the Middle East, 16:19-30.

Yiğit, N., E. Kivanç, and E. Çolak. 1998b. Contribution to taxonomy and karyology of *Meriones meridianus* (Pallas, 1773) and *Meriones crassus* Sundevall, 1842 (Rodentia: Gerbillinae) in Turkey. Zeitschrift für Säugetierkunde, 63:311-314.

Yiğit, N., E. Çolak, M. Sözen, and Ş. Özkurt. 1998c. The taxonomy and karyology of *Rattus norvegicus* (Berkenhout, 1769) and *Rattus rattus* (Linnaeus, 1758) (Rodentia: Muridae) in Turkey. Turkish Journal of Zoology, 22:203-212.

Yiğit, N., R. Verimli, M. Sözen, E. Çolak, and Ş. Özkurt. 2000a. The karyotype of *Apodemus agrarius* (Pallas, 1771) (Mammalia: Rodentia) in Turkey. Zoology in the Middle East, 20:21-23.

Yiğit, N., E. Çolak, M. Sözen, Ş. Özkurt, and R. Verimli. 2000b. The distribution, morphology, and karyology of the genus *Mesocricetus* (Mammalia: Rodentia) in Turkey. Folia Zoologica, 49(3):167-174.

Yiğit, N., E. Çolak, R. Verimli, Ş. Özkurt, and M. Sözen. 2001. A study on the distribution, morphology and karyology of *Tatera indica* (Hardwicke, 1807) (Mammalia: Rodentia) in Turkey. Turkish Journal of Zoology, 25:67-70.

Yoccoz, N. G., R. A. Ims, and H. Steen. 1993. Growth and reproduction in island and mainland populations of the vole *Microtus epiroticus*. Canadian Journal of Zoology, 71:2518-2527.

Yochem, P. K., and S. Leatherwood. 1985. Blue whale—*Balaenoptera musculus*. Pp. 193-240, *in* Handbook of marine mammals: The sirenians and baleen whales (S. H. Ridgway and R. Harrison, eds.). Academic Press, London, 3:1-362.

Yoder, A., B. Rakotosamimanana, and T. J. Parsons. 1999. Ancient DNA in subfossil lemurs: Methodological challenges and their solutions. Pp. 1-17, *in* New directions in lemur studies (H. Rasaminanana, B. Rakotosamimanana, S. Goodman, and J. Ganzhorn, eds.). Plenum Press, New York, 292 pp.

Yoder, A. D., M. M. Burns, S. Zehr, T. Delefosse, G. Veron, S. M. Goodman, and J. J. Flynn. 2003. Single origin of Malagasy Carnivora from an African ancestor. Nature, 421:734-737.

Yom-Tov, Y. 1993. Size variation in *Rhabdomys pumilio*: A case of character release? Zeitschrift für Säugetierkunde, 58:48-53.

Yom-Tov, Y., H. Mendelssohn, and C. P. Groves. 1995. *Gazella dorcas*. Mammalian Species, 491:1-6.

Yonekawa, H., K. Moriwaki, O. Gotoh, J.-I. Hayashi, J. Watanabe, N. Miyashita, M. L. Petras, and Y. Tagashira. 1981. Evolutionary relationships among five subspecies of *Mus musculus* based on restriction enzyme cleavage patterns of mitochondrial DNA. Genetics, 98:801-816.

Yonekawa, H., K. Moriwaki, O. Gotoh, N. Miyashita, Y. Matsushima, L.-M. Shie, W.-S. Cho, X.-L. Zhen, and Y. Tagashira. 1988. Hybrid origin of Japanese mice "*Mus musculus molossinus*": Evidence from restriction analysis of mitochondrial DNA. Molecular Biology and Evolution, 5:63-78.

Yonekawa, H., S. Takahama, O. Gotoh, N. Miyashita, and K. Moriwaki. 1994. Genetic diversity and geographic distribution of *Mus musculus* subspecies based on the polymorphism of mitochrondrial DNA. Pp. 25-40, *in* Genetics in wild mice (K. Moriwaki, T. Shiroishi, H. Yonekawa, eds.). Japan Scientific Societies Press, Tokyo, 333 pp.

Yonenaga, Y. 1975. Karyotypes and chromosome polymorphism in Brazilian rodents. Caryologia, 28:269-286.

Yonenaga, Y., O. Frota-Pessoa, S. Kashara, and E. J. C. Almeida. 1976. Cytogenetic studies on Brazilian rodents. Ciencia e Cultura, São Paulo, 28:202-211.

Yonenaga-Yassuda, Y., M. J. de Souza, S. Kasahara, M. L'Abbate, and H. T. Chu. 1985. Supernumerary system in *Proechimys iheringi iheringi* (Rodentia, Echimyidae), from the state of São Paulo, Brazil. Caryologia, 38:179-194.

Yong, H. S. 1969. Karyotypes of Malayan rats (Rodentia-Muridae, genus *Rattus* Fischer). Chromosoma, 27:245-267.

Yong, H.-S. 1970. A Malayan view of *Rattus edwardsi* and *R. sabanus* (Rodentia: Muridae). Zoological Journal of the Linnean Society, 49(4):359-369.

Yong, H.-S. 1972. The systematic status of Malayan *Rattus rajah* and *Rattus surifer*. Bulletin of the British Museum (Natural History), Zoology, 23:157-165.

Yong, H. S. 1973. Chromosomes of the pencil-tailed tree-mouse *Chiropodomys gliroides*. Malayan Nature Journal, 26:159-162.

Yong, H. S. 1983. Heterochromatin blocks in the karyotype of the pencil-tailed tree-mouse, *Chiropodomys gliroides* (Rodentia, Muridae). Experientia, 39:1039-1040.

Yong, H. S., and S. S. Dhaliwal. 1976. Variations in the karyotype of the red giant flying squirrel *Petaurista petaurista* (Rodentia, Sciuridae). Malaysian Journal of Science, 4:9-12.

Yong, H. S., S. S. Dhaliwal, and B. L. Lim. 1982. Karyotypes of *Hapalomys* and *Pithecheir* (Rodentia: Muridae) from Peninsular Malaysia. Cytologia, 47:535-538.

Yoon, M. H. 1990. Taxonomical study on four *Myotis* (Vespertilionidae) species in Korea. Korean Journal of Systematic Zoology, 6:173-191.

Yoon, M. H., and S. W. Son. 1989. Studies on taxonomy and phylogeny of bats inhabiting Korea. I. Taxonomical review of one rhinolophid and six vespertilionid bats. Korean Journal of Zoology, 32:375-392.

Yoram, Y.-T., S. Yom-Tov, and H. Moller. 1999. Competition, coexistence, and adaptation amongst rodent invaders to Pacific and New Zealand islands. Journal of Biogeography, 26:947-958.

Yoshida, I., Y. Obara, and N. Matsuoka. 1989. Phylogenetic relationships among seven taxa of the Japanese microtine voles revealed by karyological and biochemical techniques. Zoological Science, 6:409-420.

Yoshiyuki, M. 1986. The phylogenetic status of *Mogera tokudae* Kuroda, 1940 on the basis of body skeletons. Memoirs of the National Science Museum, Tokyo, 19:203-213.

Yoshiyuki, M. 1988a. Taxonomic status of the least red-toothed shrew (Insectivora, Soricidae) from Korea. Bulletin of the National Science Museum (Tokyo), ser. A, 14:151-158.

Yoshiyuki, M. 1988b. Notes on Thai mammals 1. Talpidae (Insectivora). Bulletin of the National Science Museum (Tokyo), ser. A, 14:215-222.

Yoshiyuki, M. 1989. A systematic study of the Japanese Chiroptera. National Science Museum, Tokyo, 242 pp.

Yoshiyuki, M. 1990. Notes on Thai mammals. 2. Bats of the *pusillus* and *philippinensis* groups of the genus *Rhinolophus*. Bulletin of the National Science Museum, Tokyo, A, 16:21-40.

Yoshiyuki, M. 1991a. Taxonomic status of *Hipposideros terasensis* Kishida, 1924 from Taiwan (Chiroptera, Hipposideridae). Journal of the Mammalogical Society of Japan, 16(1):27-35.

Yoshiyuki, M. 1991b. A new species of *Plecotus* (Chiroptera, Vespertilionidae) from Taiwan. Bulletin of the National Science Museum, Tokyo A, 17:189-195.

Yoshiyuki, M., and M. Harada. 1995. Taxonomic status of *Rhinolophus formosae* Sandborn, 1939 (Mammalia, Chiroptera, Rhinolophidae) from Taiwan. Special Bulletin of the Japanese Society of Colepterology, Tokyo, 4:497-504.

Yoshiyuki, M., and Y. Imaizumi. 1986. A new species of *Sorex* (Insectivora, Soricidae) from Sado Island, Japan. Bulletin of the National Science Museum (Tokyo), ser. A, 12:185-193.

Yoshiyuki, M., and Y. Imaizumi. 1991. Taxonomic status of the large mole from the Echigo plain, central Japan, with description of a new species (Mammalia, Insectivora, Talpidae). Bulletin of the National Science Museum (Tokyo), ser. A, 17:101-110.

Yoshiyuki, M., S. Hattori, and K. Tsuchiya. 1989. Taxonomic analysis of two rare bats from the Amami Islands (Chiroptera, Molossidae and Rhinolophidae). Memoirs of the National Science Museum, Tokyo, 22:215-225.

Yosida, T. H. 1977. Karyologic studies on hybrids between Asian, Ceylonese, and Oceanian type black rats, with a note on an XO female occurring in the F_2 generation. Cytogenetics and Cell Genetics, 19:262-272.

Yosida, T. H. 1978a. Experimental breeding and cytogenetics of the soft-furred rat, *Millardia meltada*. Laboratory Animals, 12:73-77.

Yosida, T. H. 1978b. An XXY male appeared in the F_2 hybrids between Oceanian and Ceylonese type black rats. Proceedings of the Japan Academy, 54, Ser. B(3):121-124.

Yosida, T. H. 1979. Sex chromosome anomalies in F_2 hybrids between Oceanian and Ceylonese type black rats. The Japanese Journal of Genetics, 54(1):27-34.

Yosida, T. H. 1980. Cytogenetics of the black rat: Karyotype evolution and species differentiation. University of Tokyo Press, Tokyo, 256 pp.

Yosida, T. H. 1981. Chromosome polymorphism of the large naked-soled gerbil, *Tatera indica* (Rodentia, Muridae). Japanese Journal of Genetics, 56:341-248.

Yosida, T. H. 1985. The evolution and geographic differentiation of the house shrew karyotypes. Acta Zoologica Fennica, 170:31-34.

Yosida, T. H., and M. Harada. 1985. A population survey of the chromosome polymorphism in the black rats (*Rattus rattus*) collected in the Osaka-city, Japan. Proceedings of the Japanese Academy, ser. B, 61:208-211.

Yosida, T. H., K. Tsuchiya, and K. Moriwaki. 1971. Karyotypic differences of black rats, *Rattus rattus*, collected in various localities of East and Southeast Asia and Oceania. Chromosoma, 33:252-267.

Yosida, T. H., K. Moriwaki, H. Kato, K. Tsuchiya, T. Sagai, and T. Sadaie. 1974. Studies on the karyotype and serum transferrin in the Ceylon black rat, *Rattus rattus*, having 40 chromosomes. Cytologia, 39:753-758.

Yosida, T. H., K. Moriwaki, H. Kato, K. Tsuchiya, M. Sabaratnam, K. D. Arulpragasam, and H. E. Fernando. 1979. Notes on the distribution of the Ceylonese type black rats in Sri Lanka. Proceedings of the Japan Academy, 55, Ser. B(7):351-356.

Yosida, T. H., T. Udagawa, M. Ishibashi, K. Moriwaki, T. Yabe, and T. Hamada. 1985. Studies on the karyotypes of the black rats distributed in the Pacific and South Pacific islands, with special regard to the border line of the Asian and Oceanian type black rats on the Pacific Ocean. Proceedings of the Japan Academy, 6l, ser. B:71-74.

Young, C.-C. 1934. On the Insectivora, Chiroptera, Rodentia and Primates other than *Sinanthropus* from Locality 1 at Choukoutien. Palaeontologia Sinica, Ser. C, 8(3):166 pp.

Young, C. J., and J. K. Jones, Jr. 1982. *Spermophilus mexicanus*. Mammalian Species, 164:1-4.

Young, C. J., and J. K. Jones, Jr. 1983. *Peromyscus yucatanicus*. Mammalian Species, 196:1-3.

Young, C. J., and J. K. Jones, Jr. 1984. *Reithrodontomys gracilis*. Mammalian Species, 218:1-3.

Young, S. P. 1951. The clever coyote. University of Nebraska Press, Lincoln, 411 pp.

Young, S. P., and E. A. Goldman. 1946. The puma: Mysterious American cat. American Wildlife Institute, 358 pp.

Youngman, P. M. 1967. A new subspecies of varying lemmings *Dicrostonyx torquatus* (Pallas), from Yukon Territory (Mammalia, Rodentia). Proceedings of the Biological Society of Washington, 80:31-34.

Youngman, P. M. 1975. Mammals of the Yukon Territory. National Museum of Natural Sciences (Ottawa), Publications in Zoology, 10:1-192.

Youngman, P. M. 1982. Distribution and systematics of the European mink, *Mustela lutreola* Linnaeus, 1761. Acta Zoologica Fennica, 166:1-48.

Youngman, P. M. 1990. *Mustela lutreola*. Mammalian Species, 362:1-3.

Yu, H.-T. 1993. Natural history of small mammals of subtropical montane areas in central Taiwan. Journal of Zoology, London, 231:403-422.

Yu, H.-T. 1994. Distribution and abundance of small mammals along a subtropical elevational gradient in central Taiwan. Journal of Zoology, London, 234:577-600.

Yu, H.-T. 1995. Patterns of diversification and genetic population structure of small mammals in Taiwan. Biological Journal of the Linnean Society, 55:69-89.

Yu, H.-T. 2001. Relatedness structure and individual identification in a semi-fossorial shrew (Soricidae: *Anourosorex squamipes*)—an application of microsattelite DNA. Zoological Studies, 40:226-232.

Yu, H.-T., Y.-P. Fang, C.-W. Chou, S.-W. Huang, and F.-H. Yew. 1996. Chromosomal evolution in three species of murid rodents of Taiwan. Zoological Studies, 35(3):195-199.

Yu, M.-J. 1996. Checklist of vertebrates of Taiwan. Biological Bulletin, 72:211 pp.

Yu, N., and C.-l. Zheng. 1992*a*. [A taxonomic revision of Nubra pika (*Ochotona nubrica* Thomas, 1922).] Acta Theriologica Sinica, 12:132-138 (in Chinese with English summary).

Yu, N., and C.-l. Zheng. 1992*b*. [A revision of Huanghe pika–*Ochotona huangensis* (Matschie, 1907).] Acta Theriologica Sinica, 12:175-182 (in Chinese with English summary).

Yu, N., C.-l. Zheng, and Z.-j. Feng. 1992. [The phylogenetic analysis of subgenus *Ochotona* of China.] Acta Theriologica Sinica, 12:255-266 (in Chinese with English summary).

Yu, N., C.-l Zheng, Wang Xingliang, He Guangxin, Zhang Zhihe, Zhang Anju, Lu Wenqi, and Tang Fei. 1996. [A revision of genus *Uncia* Gray, 1854 based on mitrochondrian DNA restriction site maps.] Acta Theriologica Sinica, 16:105-108 (in Chinese).

Yu, N., C.-l. Zheng, and L. Shi. 1997. Mitochondrial DNA variation and phylogeny of six species of pika (genus *Ochotona*). Journal of Mammalogy, 78:387-396.

Yu, N., C.-l. Zheng, Y. P. Zhang, and W.-hs. Li. 2000. Molecular systematics of pikas (Genus *Ochotona*) inferred from mitochondrial DNA sequences. Molecular Biology and Evolution, 16:85-95.

Yudin, B. S. 1969. Taxonomy of some species of shrews (Soricidae) from Palaearctic and Nearctic. Acta Theriologica, 14:21-34.

Yudin, B. S. 1972. [Contribution to the taxonomy of the masked transarctic common shrew (*Sorex cinereus* Kerr, 1792) from USSR fauna.] Teriologiya (Novosibirsk), 1:45-50 (in Russian).

Yudin, B. S. 1989. Nasekomoyadnye mlekopitayushchie Sibiri [Insectivorous mammals of Siberia]. Nauka, Sibirskoe Otdelenie, Novosibirsk, 360 pp. (in Russian).

Zafonte, F., and F. Masini. 1992. Enamel structure evolution in the first lower molar of the endemic murids of the genus *Microtia* (Pliocene, Gargano, Italy). Bollettino della Società Paleontologica Italiana, 31(3):335-349.

Zagorodnyuk, I. V. 1988. *Pitymys tatricus* (Rodentia)—novyi vid v faune SSSR [*Pitymys tatricus* (Rodentia)— new species in fauna USSR]. Vestnik Zoologii, 3:54 (in Russian).

Zagorodnyuk, I. V. 1989. Taksonomiya, rasprostranenie i morfologicheskaya izmenchivost' polevok roda *Terricola* vostochnoi Evropy [Taxonomy, distribution and morphological variation of the *Terricola* voles in east Europe]. Vestnik Zoologii, 5:3-14 (in Russian).

Zagorodnyuk, I. V. 1990. Kariotipicheskaya izmenchivost' i sistematika serykh polevok (Rodentia, Arvicolini). Soobshchenie 1. Vidovoi sostav i khromosomnye chisla [Karyotypic variability and systematics of the gray voles (Rodentia, Arvicolini). Communication 1. Species composition and chromosomal numbers]. Vestnik Zoologii, 2:26-37 (in Russian).

Zagorodnyuk, I. V. 1991*a*. Kariotipicheskaya izmenchivost' 46-khromosomnykh form polevok gruppy *Microtus arvalis* (Rodentia): Taksonomicheskaya otsenka [Karyotypic variation of 46-chromosome forms of voles of the *Microtus arvalis* group (Rodentia): Taxonomic evaluation]. Vestnik Zoologii, 1:36-46 (in Russian).

Zagorodnyuk, I. V. 1991*b*. Sistematicheskoe polozhenie *Microtus brevirostris* (Rodentiformes): Materialy po taksonomii i diagnostike gruppy "*arvalis*" [Systematic position of *Microtus brevirostris* (Rodentiformes): Materials toward the taxonomy and diagnostics of the "*arvalis*" group]. Vestnik Zoologii, 3:26-35 (in Russian).

Zagorodnyuk, I. V. 1991*c*. [Spatial karyotype differentiation of Arvicolini (Rodentia).] Zoologicheskii Zhurnal, 70:99-110 (in Russian).

Zagorodnyuk, I. V. 1992*a*. [Geographic distribution and levels of abundance of *Terricola subterraneus* on the USSR territory.] Zoologicheskii Zhurnal, 71:86-97 (in Russian).

Zagorodnyuk, I. V. 1992*b*. [A review of the Recent Muroidea (Mammalia), described from the territory of the Ukraine (1777-1990).] Vestnik Zoologi, 1992(2):39-48.

Zagorodnyuk, I. V. 1992*c*. [Karyotypic variability and systematics of the Arvicolini (Rodentia). Communication 2. Correlation pattern of chromosomal numbers.] Vestnik Zoologi, 1992(5):36-45 (in Russian with English abstract).

Zagorodnyuk, I. V. 1993. [Identification of East European forms of *Sylvaemus sylvaticus* (Rodentia) and their geographic occurrence.] Vestnik Zoologii, 1993(6):37-47 (in Russian with English abstract).

Zagorodnyuk, I. V. 1996*a*. What is *Apodemus sylvaticus arianus* (Blanford, 1881)? Vestnik Zoologi, 30(3):20.

Zagorodnyuk, I. V. 1996*b*. [Taxonomic revision and diagnostics of the rodent genus *Mus* from Eastern Europe. Communication 1.] Vestnik Zoologi, 1996(1-2):28-45.

Zagorodnyuk, I. 1996*c*. Rare shrew species in the territory of Ukraine: Legends, facts and diagnostics. Vestnik Zoologii, 6/1996:53-69.

Zagorodnyuk, I. V. 1998. Specimens of *Eliomys quercinus* (Mammalia) collected in the Ukraine. Vestnik Zoologii, 32(5-6):32.

Zagorodnyuk, I. V. 1999. Taxonomy, biogeography and abundance of the horseshoe bats in Eastern Europe. Acta Zoologica Cracoviensia, 42:407-421.

Zagorodnyuk, I. V. 2000. [Nomenclature and system of genus *Arvicola*.] Pp. 174-192, *in* [Species of the fauna of Russia and the contiguous countries. The water vole: Mode of the species.] Nauka Publishers, Moscow, 527 pp. (in Russian).

Zagorodnyuk, I. V., and V. I. Berezovsky. 1994. [*Mus spicilegus* (Mammalia) in the fauna of Podolia and the northern border of its range in Eastern Europe.] Zoologischeskii Zhurnal, 73(6):110-119 (in Russian with English abstract).

Zagorodnyuk, I. V., and O. V. Kondratenko. 2000. [*Sicista severtzovi* and closely related species in Ukrania: Cytogenetic and biogeographic analyses.] Vestnik Zoologi, 14:101-106 (in Russian with English summary).

Zagorodnyuk, I. V., and S. V. Mezhzherin. 1992. [Diagnostics and distributions of *Terricola* and *Sylvaemus* in Baltic Region.] Pp. 70-80, *in* Proceedings of the first Baltic Theriological Conference (A. Kirk, A. Miljutin, and T. Randveer eds.). Tartu Ülikool Toimetised (=Acta et Commentationes Universitatis Tartuensis), no. 955:254 pp. (in Russian with English summary).

Zagorodnyuk, I. V., and A. V. Mishta. 1995. On species identity of the *Erinaceus* hedgehogs of Ukraine and adjoining countries. Vestnik Zoologii, 2-3/95:50-57.

Zagorodnyuk, I. V., and J. Zima. 1992. *Microtus tatricus* (Kratochvil, 1952) in the Eastern Carpathians: Cytogenetic evidence. Folia Zoologica, 41(2):123-126.

Zagorodnyuk, I. V., N. N. Vorontsov, and V. N. Peskov. 1992. [Tatra vole (*Terricola tatricus*) in the Eastern Carpathians.] Zoologicheskii Zhurnal, 71:96-105 (in Russian).

Zagorodnyuk, I. V., G. G. Boyeskorov, and A. Ye. Zykov. 1997. [Variation and taxonomic status of the steppe forms of genus *Sylvaemus "sylvaticus"* (*falzfeini-fulvipectus-hermonensis-arianus*).] Vestnik Zoologii, 31 (5-6):37-56 (in Russian with English abstract).

Zahavi, A., and J. Wahrman. 1956. Chromosome races in the genus *Acomys* (Rodentia: Murinae). Bulletin of the Research Council of Israel, 5:316.

Zahler, P. 1996. Rediscovery of the woolly flying squirrel (*Eupetaurus cinereus*). Journal of Mammalogy, 77(1):54-57.

Zaime, A. K., and M. Pascal. 1988. Recherche d'un indice craniometrique discriminant deux especes de meriones (*Meriones shawi* et *M. libycus*) vivant en sympatrie sur le site de Guelmime (Maroc). Mammalia, 52:575-582.

Zaitsev, M. V. 1984. A contribution to the taxonomy and diagnostics of the subgenus *Erinaceus* (Mammalia, Erinaceinae) of the fauna of the USSR. Zoologicheskii Zhurnal, 63:720-730.

Zaitsev, M. V. 1988. [On the nomenclature of red-toothed shrews of the genus *Sorex* in the fauna of the USSR.] Zoologicheskii Zhurnal, 67:1878-1888 (in Russian).

Zaitsev, M. V. 1993. Species composition and questions of systematics of white-toothed shrews (Mammalia, Insectivora) of the fauna of USSR. USSR Academy of Sciences, Proceedings of the Zoological Institute, 243:3-46.

Zaitsev, M. V. 1999. Problems of diagnostics and systematics in moles (Insectivora, Talpidae, *Talpa*) from the Caucasus. Zoologeskii Zhurnal, 78:718-731.

Zaitsev, M. V., and V. A. Osipova. 2003. Taxonomy and probable relations between recent and fossil red-toothed shrews (Mammalia, Soricidae) of the *"minutus"* species group from the northern Caucasus. Zoologeskii Zhurnal, 82:62-69 (in Russian).

Zakrzewski, R. J. 1974. Fossil Ondatrini from western North America. Journal of Mammalogy, 55:284-292.

Zakrzewski, R. J. 1993. Morphological change in woodrat (Rodentia: Cricetidae) molars. Pp. 392-409, *in* Morphological change in Quaternary mammals of North America (R. A. Martin and A. D. Barnosky, eds.). Cambridge University Press, 415 pp.

Zambelli, A., F. Dyzenchauz, A. Ramos, N. de Rosa, R. Wainberg, and O. A. Reig. 1992. Cytogenetics and karyosystematics of phyllotine rodents (Cricetidae, Sigmodontinae) III. New data on the distribution and karyomorphs of the genus *Eligmodontia*. Zeitschrift für Säugetierkunde, 57:155-162.

Zambelli, A., L. Vidal-Rioja, and R. Wainberg. 1994. Cytogenetic analysis of autosomal polymorphism in *Graomys griseoflavus* (Rodentia, Cricetidae). Zeitschrift für Säugetierkunde, 59:14-20.

Zanalda, E. 1994. *Dinaromys bogdanovi* (Mammalia: Rodentia) from the Middle Pleistocene of Western Lombardy (Italy). Rivista italiana di paleontologia e stratigrafia, 100(1):143-148.

Zanchin, N. I. T., A. Langguth, and M. S. Mattevi. 1992*a*. Karyotypes of Brazilian species of *Rhipidomys* (Rodentia, Cricetidae). Journal of Mammalogy, 73:120-122.

Zanchin, N. I. T., I. J. Sbalqueiro, A. Langguth, R. C. Bossle, E. C. Castro, L. F. B. Oliveira, and M. S. Mattevi. 1992*b*. Karyotype and species diversity of the genus *Delomys* (Rodentia, Cricetidae) in Brazil. Acta Theriologica, 37:163-169.

Zazhigin, V. S., and A. V. Lopatin. 2000*a*. [Evolution, phylogeny, and classification of Dipodoidea.] Pp. 50-52, *in* [Systematics and Phylogeny of the Rodents and Lagomorphs] (A. K. Agadzhanyan and V. N. Orlov, eds.). Theriological Society, Moscow, 196 pp. (in Russian).

Zazhigin, V. S., and A. V. Lopatin. 2000*b*. [The History of the Dipodoidea (Rodentia, Mammalia) in the Miocene of Asia: 1. *Heterosminthus* (Lophocricetinae).] Paleontologicheskii Zhurnal, 3:90-102 (in Russian, with English summary).

Zazhigin, V. S., and A. V. Lopatin. 2000*c*. The History of the Dipodoidea (Rodentia, Mammalia) in the Miocene of Asia: 3. Allactaginae. Paleontological Journal, 34(5):553-565 (translated from Paleontologicheskii Zhurnal, 5:82-94).

Zazhigin, V. S., and A. V. Lopatin. 2001. The History of the Dipodoidea (Rodentia, Mammalia) in the Miocene of Asia: 4. Dipodinae at the Miocene-Pliocene transition. Paleontological Journal, 35(1):60-74 (translated from Paleontologicheskii Zhurnal, 1:61-75).

Zazhigin, V. S., and A. V. Lopatin. 2002. The history of the Dipodoidea (Rodentia, Mammalia) in the Miocene of Asia: 6. Lophodont Lophocricetinae. Paleontological Journal, 36(4):385-394 (translated from Paleontologicheskii Zhurnal, 4:62-71).

Zazhigin, V. S., A. V. Lopatin, and A. G. Pokatilov. 2002. The history of the Dipodoidea (Rodentia, Mammalia) in the Miocene of Asia: 5. *Lophocricetus* (Lophocricetinae). Paleontological Journal, 36(2):180-194 (translated from Paleontologicheskii Zhurnal, 2:62-75).

Zegers, D. A. 1984. *Spermophilus elegans*. Mammalian Species, 214:1-7.

Zejda, J., J. Zukal, and H. M. Steiner. 1994. Variation in the shape pattern of molars in populations of the bank vole (*Clethrionomys glareolus* Schreb.). Folia Zoologica, 4392):121-133.

Zelebor, J. 1869. Reise der Österreichischen Fregatte Novara um die Erde in den Jahren 1857, 1858, 1859. Zoologischer Theil. (Wirbelthiere) 1. Sägethiere, 42 pp.

Zeng, Z., W. Ding, Y. Yang, M. Luo, J. Liang, R. Xie, Y. Dai, and Z. Song. 1996*a*. Population ecology of *Rattus nitidus* in the western Sichuan Plain I. Population dynamics and body mass. Acta Theriologica Sinica, 16(3):202-210.

Zeng, Z., W. Ding, Y. Yang, M. Luo, J. liang, R. Xie, Y. Dai, and Z. Song. 1996*b*. Population ecology of *Rattus nitidus* in the western Sichuan Plain II. Survival and movement. Acta Theriologica Sinica, 16(4):278-284.

Zeng, Z., Y. Yang, M. Luo, J. Liang, R. Xie, and Z. Song. 1999. Population ecology of *Rattus nitidus* in the western Sichuan plain III. Reproduction. Acta Theriologica Sinica, 19(3):183-196.

Zhang Chieh and Wang Tsung-yi. 1963. [Faunistic studies of the Chinghai province.] Acta Zoologica Sinica, 15(3):195-200.

Zhang Cizu. 1987. *Nemorhaedus cranbrooki* Hayman. Pp. 213-219, *in* The biology and management of *Capricornis* and related antelopes (H. Soma, ed.). Croom Helm, London, 391 pp.

Zhang, C., H. Sheng, and H. Lu. 1984. On the Fea's muntjak from Xizang (Tibet), China. Acta Theriologica Sinica, 4:88, 106.

Zhang, M.-W., C. Guo, Y. Wang, Z.-J. Hu, and A.-G. Chen. 2000. [The buff-breasted rats (*Rattus flavipectus*) in China.] Zoological Research, 21(6):487-497 (in Chinese with English abstract).

Zhang, W. 1990. [A study of karyotype and C-banding pattern of *Nyctalus velutinus*.] Journal of Anhui Normal University, 4:58-63 (in Chinese with English abstract).

Zhang Yaping and Shi Li-ming. 1991*a*. Riddle of the giant panda. Nature, 352:573.

Zhang, Ya-P., and L. M. Shi. 1991*b*. Genetic diversity in the Chinese pangolin *Manis pentadactyla* inferred from restriction enzyme analysis of mitochondrial DNAs. Biochemical Genetics, 29(9-10):501-508.

Zhang, Y., S. Jin, G. Quan, S. Li, Z. Ye, F. Wang, and M. Zhang. 1997. Distribution of mammalian species in China. China Forestry Publishing House, Beijing, 280 pp.

Zhang Zi-yu and Zhao Ming-shan. 1984. [A new subspecies of the sulphur-bellied rat from Jilin—*Rattus niviventer naoniuensis*.] Acta Zoologica Sinica, 30:99-102 (in Chinese).

Zhao, T.-q, C.-l. Zheng, and C.-l. Zhou. 1983. [Comments on Luo's "a systematic review of the Chinese Cape hare, *Lepus capensis* Linnaeus".] Acta Theriologica Sinica, 3:1-3 (in Chinese with English summary).

Zhao Xiao-fan and Lu Hao-quan. 1986. [Comparative observations of several biochemical indexes of *Apodemus agrarius pallidior* and *Apodemus agrarius ninpoensis* of the striped backed field mice.] Acta Theriologica Sinica, 6:57-62 (in Chinese).

Zheng Chang-lin. 1986. [Recovery of Koslow's pika (*Ochotona koslowi* Buchner) in Kunlun Mountains of Xinjiang Uygur Autonimous [sic] Region, China.] Acta Theriologica Sinica, 6:285 (in Chinese).

Zheng Chang-lin and Wang Sung. 1980. [On the taxonomic status of *Pitymys leucurus* Blyth.] Acta Zootaxonomica Sinica, 5:106-112 (in Chinese).

Zheng S. 1985. Remains of the genus *Anourosorex* (Insectivora, Mammalia) from Pleistocene of Guizhou District. Vertebrata Palasiatica, 23:39-51.

Zheng, S. 1993. [Quaternary rodents of Sichuan-Guizhou area, China.] Science Press, Beijing, 270 pp (in Chinese, with English summary).

Zheng, S. 1994. Classification and evolution of the Siphneidae. Pp. 57-76, *in* Rodent and Lagomorph Families of Asian Origins and Diversification (Y. Tomida, C. k. Li, and T. Setoguchi, eds.). National Science Museum Monographs, No. 8, Tokyo, 195 pp.

Zheng S., and Li Chuan-kwei. 1990. Comments on fossil arvicolids of China. Pp. 431-442, *in* International symposium evolution, phylogeny and biostratigraphy of arvicolids (Rodentia, Mammalia) (O. Fejfar and W.-D. Heinrich, eds.). Geological Survey, Prague, 448 pp.

Zheng, S., Z. Zhang, and L. Liu. 1997. [Pleistocene mammals from fissure-fillings of Sunjiashan Hill, Shandong, China.] Vertebrata Palasiatica, 35(3):201-216 (in Chinese with English summary).

Zheng Tao and Zhang Ying-mei. 1990. [The fauna and geographical division on Glires of Gansu province.] Acta Theriologica Sinica, 10(2):137-144.

Zheng, X. G., B. S. Arbogast, and G. J. Kenagy. 2003. Historical demography and genetic structure of sister species: deermice (*Peromyscus*) in the North American temperate rain forest. Molecular Ecology, 12:711-724.

Zhi, L., W. B. Karesh, D. N. Janczewski, H. Frazier-Taylor, D. Sajuthi, F. Gombek, M. Andau, J. S. Martenson, and S. J. O'Brien. 1996. Genomic differentiation among natural populations of orang-utan (*Pongo pygmaeus*). Current Biology, 6:1326-1336.

Zholnerovskaya, E. I., D. I. Bibikov, and V. I. Ermolaev. 1990. Immunogeneticheskii analiz sistematicheskikh

vzaimootnoshenii surkov [Immunogenetic analysis of systematic relationships of marmots]. Byulleten'
Moskovskovo Obshchestva Ispytatelei Prirody, Otdel Biologicheskii, 195:15-24 (in Russian).

Zhou Jia-di, Li Si-hua, and Gu Jing-he. 1985. [The preliminary observation on the mammals in Kunlun-
Altun Basin.] Acta Theriologica Sinica, 5(2):160.

Zhou Kai-ya, Qian Wei-juan, and Li Yue-min. 1978. [Recent advances in the study of the baiji, *Lipotes
vexiliffer* Miller.] Journal of the Nanjing Teacher's College (Natural Science), 1:8-13 (in Chinese).

Zhou Kai-ya, Li Yue-min, and Qian Wei-juan. 1979. [The stomach of the baiji *Lipotes vexillifer*.] Acta
Zoologica Sinica, 25:95-100 (in Chinese).

Zhou, L.-Z., Y. Ma, and Di-Q. Li. 2000. [Distribution of great gerbil (*Rhombomys opimus*) in China.] Acta
Zoologica Sinica, 46(2):130-137 (in Chinese with English summary).

Zhou Yu-can and Xia Wu-ping. 1981. [An electrophoretic comparison of the serum protein and hemoglobin
in three species of mouse-hares — a discussion on the systematical position of *Ochotona curzoniae*.] Acta
Theriologica Sinica, 1:39-44 (in Chinese).

Zhu, B.-c., Y. Xu, and C.-h. Fan. 1994. [G-banding characteristic of sex chromosomes in Chinese vole
Lasiopodomys mandarinus (Rodentia, Cricetidae) from Henan.] Vestnik Moskovskogo Universiteta Seriya
16 Biologiya 3, Iyul'-Sentyabr', 1994:56-59 (in Russian with English summary).

Zhuge, Y. 1993. *Crocidura* (Insectivora: Soricidae). Pp. 19-23, *in* Fauna of Zhejiang: Mammalia (Y. Zhuge and
H.-Q. Gu, eds.). Zhejiang Science and Technology Publishing House, Hangzhou, China.

Ziegler, A. C. 1971. Dental homologies and possible relationships of recent Talpidae. Journal of Mammalogy,
52:50-68.

Ziegler, A. C. 1982*a*. The Australo-Papuan genus *Syconycteris* (Chiroptera: Pteropodidae) with the description
of a new Papua New Guinea species. Occasional Papers of the Bernice P. Bishop Museum, 25(5):1-22.

Ziegler, A. C. 1982*b*. An ecological check-list of New Guinea Recent mammals. Pp. 863-894, *in* Biogeography
and ecology of New Guinea, vol. 2 (J. L. Gressitt, ed.). Monographiae Biologicae, 42:1-983[in 2 vols].

Ziegler, A. C. 1984. A Papua New Guinea specimen of *Hydromys hussoni* Musser and Piik, 1982 (Rodentia:
Muridae). Australian Mammalogy, 7(12):101-105.

Ziegler, R. 1995. Die untermiozänen kleinsäugerfaunen aus den Süßwasserkalken von Engelswies und
Schellenfeld bei Sigmaringen (Baden-Württemberg). Stuttgarter Beiträge zur Naturkunde, Ser. B
(Geologie und Paläontologie), 228:53 pp.

Ziegler, S., G. Nikolaus, and R. Hutterer. 2002. High mammalian diversity in the newly established National
Park of Upper Niger, Republic of Guinea. Oryx, 36:73-80.

Ziegler, T., A. Feiler, and U. Zöphel. 2001. New data on the genital morphology of the midge bat *Pipistrellus
pygmaeus* (Leach, 1825) from Germany (Mammalia: Chiroptera: Vespertilionidae). Zoologische
Abhandlungen, Staatliches Museum für Tierkunde, Dresden, 51:435-444.

Zima, J. 1983. Chromosomes of the harvest mouse, *Micromys minutus*, from the Danube Delta (Muridae,
Rodentia). Folia Zoologica, 32:19-22.

Zima, J. 1986. Chromosomal and epigenetic variation in a population of the pine vole, *Pitymys subterraneus*.
Folia Zoologica, 35:333-345.

Zima, J. 1987. Karyotypes of certain rodents from Czechoslovakia (Sciuridae, Gliridae, Cricetidae). Folia
Zoologica, 36(4) 1987:337-343.

Zima, J., and M. Andera. 1996. A synopsis of the mammals of the Czech Republic. Hystrix, n.s.,
8(1-2):107-111.

Zima, J., and B. Král. 1984*a*. Karyotypes of European mammals. II. Acta Scientiarum Naturalium, Academiae
Scientarium Bohemoslovacae (Brno), 18(8):1-62.

Zima, J., and B. Král. 1984*b*. Karyotypic variability in *Sorex araneus* in Central Europe (Soricidae, Insectivora).
Folia Zoologica, 34:235-243.

Zima, J., and M. Macholán. 1989. Robertsonian fusion 5.12 in a population of *Mus musculus musculus*. Folia
Zoologica, 38:233-238.

Zima, J., and M. Macholán. 1995. B chromosomes in the wood mice (genus *Apodemus*). Acta Theriologica,
Supplement 3:75-86.

Zima, J., J. Èervený, V. Hrabe, B. Kral, and M. Sebela. 1981. On the occurrence of *Microtus epiroticus* in
Rumania (Arvicolidae, Rodentia). Folia Zoologica, 30(2):139-146.

Zima, J., V. A. Gaichenko, M. Macholán, S. I. Radjabli, O. V. Sablina, and J. M. Wójcik. 1990. Are
Robertsonian variations a frequent phenomenon in mouse populations in Eurasia? Biological Journal of
the Linnean Society, 41:229-233.

Zima, J., I. V. Zagorodnyuk, V. A. Gaichenko, and T. O. Zhezherina. 1991. Polimorfizm i khromosomnaya
izmenchivost' *Microtus rossiaemeridionalis* (Rodentiformes) [Polymorphism and chromosomal variability
in *Microtus rossiaemeridionalis* (Rodentiformes)]. Vestnik Zoologii, 4:48-53 (in Russian).

Zima, J., J. B. Searle, and M. Macholán (eds.). 1994. The cytogenetics of the *Sorex araneus* group and related
topics. Folia Zoologica, 43 (suppl. 1):iv+116 pp.

Zima, J., M. Macholán and M. G. Filippucci. 1995. Chromosomal variation and systematics of myoxids. Pp. 63-76, *in* Proceedings of II Conference on Dormice (Rodentia, Myoxidae) (M. G. Filippucci, ed.). Hystrix, n.s., 6(1-2):1-340.

Zima, J., M. Macholán, B. Kryštufek and S. Petkovski. 1997*a*. Karyotypes of certain small mammals (Insectivora, Rodentia) from Macedonia. Scopolia, 38:1-15.

Zima, J., Macholán, M. Andìra and J. Èervený. 1997*b*. Karyotypic relationships of the Garden Dormouse (*Eliomys quercinus*) from central Europe. Folia Zoologica, 46(2):105-108.

Zima, J., L. Slivková, M. Andreas, P. Benda, and A. Reiter. 1997*c*. Karyotypic status of shrews (*Sorex*) from Thrace, European Turkey. Zeitschrift für Säugetierkunde, 62:315-317.

Zima, J., M. Macholán, and L. Slivkova. 1997*d*. Confirmation of the presence of B chromosomes in the wood mouse (*Apodemus sylvaticus*). Folia Zoologica, 46(3):217-221.

Zima, J., L. Lukácová, and M. Macholán. 1998. Chromosomal evolution in shrews. Pp. 175-218, *in* Evolution of shrews (J. M.Wójcik and M. Wolsan, eds.). Mammal Research Institute, Polish Academy of Sciences, Bialowieza, 458 pp.

Zima, J., L. A. Ieradi, F. Allegra, A. Sartoretti, E. Wlosoková, and M. Cristaldi. 1999*a*. Frequencies of B chromosomes in *Apodemus flavicollis* are not directly related to mutagenetic environmental effects. Folia Zoologica, 48(suppl. 1):115-119.

Zima, J., M. Macholán, J. Piálek, L. Slivková, and E. Suchomelová. 1999*b*. Chromosomal banding pattern in the Cyprus spiny mouse, *Acomys nesiotes*. Folia Zoologica, 48(2):149-152.

Zima, J., J. Piálek, and M. Macholán. 2003. Possible heterotic effects of B chromosomes on body mass in a population of *Apodemus flavicollis*. Canadian Journal of Zoology, 81:1312-1317.

Zimina, R. P. (ed.). 1978. Surki. Rasprostranenie i ekologiya [Marmots. distribution and ecology]. Nauka, Moscow, 222 pp. (in Russian).

Zimmermann, E., P. Ehresmann, V. Zietemann, U. Radespiel, B. Randrianambinina, and N. Rakotoarison. 1997. A new primate species in north-western Madagascar: the golden-brown mouse lemur (*Microcebus ravelobensis*). Primate Eye, 63:26.

Zimmermann, E., S. Cepok, N. Rakotoarison, V. Zietemann,, and U. Radespiel. 1998. Sympatric mouse lemurs in north-west Madagascar: A new rufous mouse lemur species (*Microcebus ravelobensis*). Folia-Primatologica 69(2):106-114.

Zimmermann, E. A. W., von. 1778-1783. Geographische Geschichte des Menschen, und der allgemein verbreiteten vierfussigen Thiere, nebst einer Hieher gehorigen zoologischen Weltcharte. Vol. 2. Geographische Geschichte des Menschen, und der vierfussigen Thiere. Zweiter Band. Enthalt ein vollstandiges. Verzeichniss aller bekannten Quadrupeden. Weygandschen Buchhandlung, Leipzig, 3:1-276.

Zimmerman, E. G. 1970. Karyology, systematics and chromosomal evolution in the rodent genus, *Sigmodon*. Publications of The Museum, Michigan State University, Biological Series, 4:385-454.

Zimmerman, E. G., and N. A. Gayden. 1981. Analysis of genic heterogeneity among local populations of the pocket gopher, *Geomys bursarius*. Pp. 272-287, in Mammalian population genetics (M. H. Smith and J. Joule, eds.). University of Georgia Press, Athens, 380 pp.

Zimmerman, E. G., and M. E. Nejtek. 1977. Genetics and speciation of three semispecies of *Neotoma*. Journal of Mammalogy, 58:391-402.

Zimmerman, E. G., B. J. Hart, and C. W. Kilpatrick. 1975. Biochemical genetics of the *boylii* and *truei* groups of the genus *Peromyscus* (Rodentia). Comparative Biochemistry and Physiology, 52B:541-545.

Zimmerman, E. G., C. W. Kilpatrick, and B. J. Hart. 1978. The genetics of speciation in the rodent genus *Peromyscus*. Evolution, 32:565-579.

Zimmermann, K. 1942. Zur Kenntnis von *Microtus oeconomus* (Pallas). Archiv für Naturgeschichte, Neue Folge, 11:174-197.

Zimmermann, K. 1962. Die Untergattungen der Gattung *Apodemus* Kaup. Bonner Zoologische Beiträge, 13:198-208.

Zittel, K. A. 1893. Handbuch der Palaeontologie. 1 Abt. Palaeozoologie. IV. Mammalia. Munich and Leipzig: R. Oldenbourg, 799 pp.

Zorenko, T. A., V. Smorkacheva, and T. G. Aksenova. 1994. [Reproduction and postnatal ontogenesis of mandarin vole *Lasiopodomys mandarinus* (Rodentia, Arvicolinae).] Zoologischeskii Zhurnal, 73(6):120-129.

Zortéa, M., and V. A. Taddei. 1995. Taxonomic status of *Tadarida espiritosantensis* Ruschi, 1951 (Chiroptera: Molossidae). Boletim do Museu de Biologia Mello Leitão (N. Sér.), 2:15-21.

Zubko, J. P. 1937. [A new subspecies the common shrew (*Sorex araneus averini* subsp. nova).] Kharkov A. Gorky-State University, Proceedings of the Zoological-Biological Institute, 4:299-303 (in Ukranian).

Zuccotti, M., S. Garagna, C. A. Redi, S. Simson, E. Nevo, and E. Capanna. 1995. Spermatogenesis of *Spalax ehrenbergi* natural hybrids (Rodentia, Spalacidae). Mammalia, 59(1):119-125.

Zukowsky, L. 1927. Bemerkungen über die rassenweise Verschiedenheit der Hirschziegenantilope. Carl Hagenbecks Illustrierte Tier- und Menschenwelt, 2:124-127.

Zukowsky, L. 1965. Die Systematic der Gattung *Diceros* Gray, 1821. Zoologische Garten, N. F., 17:1-178.

Zunino, G. E., O. B. Vaccaro, M. Canevari, and A. L. Gardner. 1995. Taxonomy of the genus *Lycalopex* (Carnivora: Canidae) in Argentina. Proceedings of the Biological Society of Washington, 108:729-747.

Zykov, A. E. 1987. Novaya nakhodka zacaspiiskoi myshevidnoi soni (*Myomimus personatus* Ognev) na territorii SSSR [New finding of the Transcaspian mouse-like dormouse *Myomimus personatus* Ognev on the territory of USSR]. Vestnik Zoologii, 1:80 (in Russian).

Zykov, A. E., and I. V. Zagorodnyuk. 1988. [On the systematic position of voles (Mammalia, Rodentia) from Kopet Dagh.] Vestnik Zoologii, 5:46-52 (in Russian).

Index

SCIENTIFIC NAMES

Synonyms are in italic, and page locators for valid names are in boldface.

albipes, Tscherskia, 1046

albipilis, Lagostrophus, 53

albipinnis, Taphozous, 383

albirostratus, Stenella, 733

albirostris, Amblysomus, 80

albirostris, Lagenorhynchus, **729**

albirostris, Przewalskium, **668**

albirostris, Tayassu, 644

albispinus, Trinomys, 1589, **1590**

albiventer, Akodon, **1092**

albiventer, Elephantulus, 82

albiventer, Melomys, 1375

albiventer, Microdipodops, 849

albiventer, Monachus, 598

albiventer, Mormopterus, 445

albiventer, Myodes, 1027

albiventer, Noctilio, 429

albiventer, Nycteris, 393

albiventer, Nycticeinops, 461

albiventer, Nyctimene, **329**

albiventer, Oryzomys, 1147

albiventer, Petaurista, 772

albiventer, Rattus, 1484

albiventer, Sorex, 285

albiventer, Thyroptera, 430

albiventer, Uromys, 1514

albiventris, Atelerix, **212**, 213

albiventris, Cephalorhynchus, 727

albiventris, Crocidura, 248

albiventris, Didelphis, **5**, 6

albiventris, Genetta, 556

albiventris, Marmosops, 10

albiventris, Neomys, 279

albiventris, Noctilio, **428**

albiventris, Oenomys, 1429

albiventris, Rattus, 1484

albiventris, Tamias, 813

albivexilli, Calloscurius, 777

albo-fasciatus, Erythrocebus, 160

albo griseus, Hylobates, 179

albo-maculata, Talpa, 308

albo nigrescens, Hylobates, 179

albocaudata, Stenocephalemys, 924, 1500, **1501**

albocaudatus, Colobus, 168

albocinereus, Pseudomys, **1454**

albocollaris, Eulemur, **114**, 115

albofuscus, Scotoecus, **464**

albogularis, Arctonyx, 606

albogularis, Cercopithecus, **154**, 157

albogularis, Petrogale, 68

albogularis, Tamias, 817

alboguttata, Dasyurus, 25

alboguttata, Glauconycteris, **486**

albojubatus, Connochaetes, 676

albolimbatus, Liomys, 852

albolimbatus, Pipistrellus, 476

albolineata, Graphiurus, 825

albolineatus, Lemniscomys, 1344

albomaculatum, Phyllops, 422

albomaculatus, Myocastor, 1593

alboniger, Hylopetes, **768**

albonota, Eudorcas, 679

albonotata, Eudorcas, 679

albonotatus, Sciurus, 764

albonotatus, Tragelaphus, 699

albopunctatus, Dasyurus, **24**

alborufescens, Microtus, 1012

alborufus, Petaurista, **770**

alborusus, Petaurista, 771

alboscapulatus, Melonycteris, 327

albosignatus, Erythrocebus, 160

albostriatus, Apodemus, 1261

albotorquatus, Cercopithecus, 154, 158

albovirgatus, Tragelaphus, 698

albovittatus, Xerus, 790

albula, Mus, 1399

albula, Ovis, 710

albulus, Cebus, 137

albulus, Chaetodipus, 853

albulus, Hemiechinus, 215

albus, Apodemus, 1275

albus, Arvicola, 966

albus, Bos, 692

albus, Canis, 576

albus, Capreolus, 654

albus, Castor, 842

albus, Cervus, 662

albus, Cricetus, 1043

albus, Cryptomys, 1540

albus, Dama, 665

albus, Diclidurus, **387**

albus, Gulo, 607

albus, Lepus, 204

albus, Lynx, 541

albus, Microtus, 991

albus, Molossus, 441

albus, Mus, 1399

albus, Mustela, 617

albus, Myotis, 505

albus, Neomys, 279

albus, Nyctereutes, 581

albus, Ondatra, 1033

albus, Ozotoceros, 659

albus, Propithecus, 120

albus, Rattus, 1478, 1484

albus, Sciurus, 764

albus, Ursus, 588

alcathoe, Myotis, **501**

Alce, 653

alce, Alces, 653

Alceini, 652

Alcelaphinae, **673**

Alcelaphini, 673

Alcelaphus, 653

Alcelaphus, **673**, 675, 676

Alcephalus, 674

Alces, **653**

alces, Alces, **653**

alces, Canis, 576

alces, Taurotragus, 697

alchemillae, Crocidura, 231

Alcini, 652

alcinous, Caryomys, 968

alco, Canis, 576

alcorni, Microtus, 1012

alcorni, Pappogeomys, **866**

alcorni, Peromyscus, 1066

alcyone, Rhinolophus, **351**

alcythoe, Pipistrellus, 476

aldabrensis, Pteropus, **334**

aldridgeanus, Naemorhedus, 706

aleco, Chionomys, 970

alecto, Aethalops, 314, **315**

alecto, Cyttarops, **386**

alecto, Emballonura, **387**

alecto, Molossus, 441

alecto, Pteropus, **335**

Alectops, 410

aleksandrisi, Crocidura, **224**

Aletesciurus, **786**

alethina, Rhinophylla, **413**

alettensis, Stenocephalemys, 1500

alexandrae, Phalanger, **46**

alexandrae, Thomomys, 866

alexandrae, Ursus, 588

alexandri, Chlorocebus, 160

alexandri, Crossarchus, **563**

alexandri, Elephantulus, 83

alexandri, Paraxerus, **795**

alexandriae, Macropus, 65

alexandrino-rattus, Rattus, 1484

alexandrinus, Rattus, 1484

Alexandromys, **989**

alexis, Notomys, **1426**

alfari, Microscurius, **757**

alfari, Sigmodontomys, 1177, **1178**

alfaroi, Oryzomys, **1145**

alfredi, Dipodomys, 846

alfredi, Rusa, **669**

alfredi, Sigmodon, 1175

alfurus, Babyrousa, 637

algagrus, Capra, 702

algazel, Oryx, 719

algeriensis, Vulpes, 585

alghazal, Aethomys, 1255

algidus, Lepus, 204

algidus, Peromyscus, 1069

algira, Caracal, 533

algira, Sus, 642

algirensis, Canis, 574

amalae, Aethomys, 1256
amalae, Crocidura, 235
amankaragai, Stylodipus, 885
amargosae, Thomomys, 867
amasari, Sorex, 298
amatus, Cryptomys, **1539**
amazari, Sorex, 298
amazonica, Alouatta, 150
amazonica, Conepatus, 622
amazonicus, Conepatus, 622
amazonicus, Eumops, 438
amazonicus, Holochilus, 1119
amazonicus, Inia, 739
amazonicus, Leopardus, 540
amazonicus, Nectomys, 1133
amazonicus, Oxymycterus, **1156**
amazonius, Trichechus, 93
ambarvalis, Spilogale, 624
ambersoni, Rattus, 1492
ambigua, Spilogale, 623
ambiguus, Alcelaphus, 674
ambiguus, Chaetodipus, 853
ambiguus, Dipodomys, 846
ambiguus, Microdipodops, 849
ambiguus, Peromyscus, 1071
Ambliodon, 550
amblodon, Lagenorhynchus, 729
Amblonyx, 601
Amblotis, 44
amblyceps, Ursus, 588
Amblychilus, 92
Amblycoptini, 267
amblyodon, Lycalopex, 579
amblyonyx, Kannabateomys, **1576**
Amblyotis, 452
amblyotis, Lophostoma, 406
Amblyrhiza, **1599**
Amblyrhizinae, 1599
amblyrrhynchus, Oligoryzomys, 1142
Amblysominae, **80**

Amblysomus, 77, 78, **80, 81**
amboellensis, Kobus, 721
amboinensis, Hipposideros, 367
amboinensis, Myotis, 502
amboinensis, Phalanger, 47
amboinensis, Rattus, 1490
amboinensis, Sus, 641
ambrosiana, Rusa, 670
ambrosius, Meriones, 1237
Ambysus, 599
amecensis, Pappogeomys, 866
ameghiniana, Lama, 646
ameghinoi, Puma, 545
ameliae, Sciurus, 764
amer, Genetta, 556
Ameranthropoides, 150
americana, Antilocapra, **671**
americana, Arvicola, 964
americana, Conepatus, 621
americana, Glaucomys, 768
americana, Lagostomus, 1552
americana, Lontra, 603
americana, Martes, **608**
americana, Mazama, **655**
americana, Mephitis, 622
americana, Monodelphis, **13**, 15
americana, Ondatra, 1033
americana, Phocoena, 736
americanus, Alces, **653**
americanus, Bison, 689
americanus, Canis, 576
americanus, Homo, 182
americanus, Lepus, **195**
americanus, Noctilio, 428, 429
americanus, Odocoileus, 658
americanus, Oreamnos, **706**
americanus, Rattus, 1478
americanus, Sorex, 14
americanus, Tamias, 818
americanus, Tapirus, 634
americanus, Taxidea, 620
americanus, Trichechus, 93
americanus, Ursus, **588**
americanus, Zapus, 893

Ameridelphia, 3
Ametrida, **416**
amicta, Callicebus, 146
amicus, Phyllotis, **1161**
amir, Paraechinus, 217
amiri, Rhinolophus, 357
Ammelaphus, 697
ammobates, Peromyscus, 1076
Ammodillini, 901, 1210, 1212
Ammodillus, 901, **1212**
Ammodorcas, **678**, 682
Ammodorcini, 677
ammodytes, Perognathus, 856
ammodytes, Peromyscus, 1070
Ammomys, 989
Ammon, 707
ammon, Capra, 701
ammon, Ovis, **707**
ammonoides, Ovis, 708
ammophilus, Chaetodipus, 853
ammophilus, Geomys, 862
ammophilus, Microdipodops, 849
ammophilus, Spermophilus, 811
ammophilus, Sylvilagus, 209
Ammospermophilini, 797
Ammospermophilus, 755, **797**
Ammotragus, 699, **700**
amoenus, Cephalophus, 713
amoenus, Distoechurus, 56
amoenus, Gerbillus, **1224**
amoenus, Hylopetes, 769
amoenus, Neacomys, 1128
amoenus, Necromys, **1129**
amoenus, Perognathus, 858
amoenus, Reithrodontomys, 1082
amoenus, Sorex, 299
amoenus, Tamias, **813**
amoenus, Tragulus, 650
amoles, Reithrodontomys, 1083
amoles, Sigmodon, 1176
amori, Eliomys, 835
Amorphochilus, **429**
amosus, Microtus, 1008
amotus, Menetes, 784

amotus, Myotis, 518
amoyensis, Panthera, 548
Amperta, 24
amphialus, Spilogale, 623
Amphiaulacomys, 1241
Amphibia, 595
Amphibiae, 595
amphibius, Arvicola, **963**
amphibius, Hippopotamus, **645**
amphibius, Neomys, 279
amphichoricus, Proechimys, 1587
Amphidyromys, 839
Amphinectomys, 900, **1100**
Amphiosorex, 282
Amphisorex, 278, 282
amplexicaudata, Carollia, 412
amplexicaudata, Glossophaga, 400
amplexicaudatus, Eumops, 437
amplexicaudatus, Rousettus, **347**
amplexicaudus, Molossus, 440
amplus, Ammospermophilus, 798
amplus, Artibeus, **417**
amplus, Meriones, 1236
amplus, Notomys, **1426**
amplus, Perognathus, **856**
amplus, Peromyscus, 1066
ampullatus, Hyperoodon, **740**
amsiri, Presbytis, 172
amurensis, Erinaceus, **213**
amurensis, Lemmus, **986**
amurensis, Lutra, 604
amurensis, Meles, 611
amurensis, Mus, 1398
amurensis, Mustela, 614
amurensis, Myodes, 1027
amurensis, Myotis, 503
amurensis, Nyctereutes, 581
amurensis, Panthera, 548
amygdali, Pteromys, 774
ana, Chiropodomys, 1302
anacapae, Peromyscus, 1071
anadyrensis, Pteromys, 774
anadyrensis, Sciurus, 764
anadyrensis, Vulpes, 585
Anahyster, 601
anak, Uromys, **1514**

argenteocinereus, Heliophobius, **1541**

argenteogrisea, Lepus, 198

argentescens, Funambulus, 782

argenteus, Anomalurus, 1532

argenteus, Apodemus, **1263**, 1264

argenteus, Glis, 840

argenteus, Graphiurus, 827

argenteus, Leopardus, 538

argenteus, Lycalopex, 580

argenteus, Mus, 1263

argenteus, Neomys, 279

argenteus, Pseudochirops, 53

argenteus, Sciurus, 764

argenteus, Trachypithecus, 177

argenteus, Ursus, 588

argenteus, Vulpes, 584

argentimembris, Macaca, 162

argentina, Mazama, 656

argentinius, Sciurus, 761

argentinus, Ctenomys, **1561**

argentinus, Eptesicus, 453

argentinus, Lasiurus, 459

argentiventer, Rattus, **1465**

argentoratensis, Arvicola, 966

argillaceus, Acomys, 1200

Argocetus, 735

argunensis, Canis, 576

argurus, Alticola, 958

argurus, Calomys, 1108

argurus, Zyzomys, 1520, **1521**

argusensis, Dipodomys, 848

argusensis, Thomomys, 866

argynnis, Casinycteris, **315**

argyraceus, Rattus, 1490

argyrochaetes, Capricornis, 704

argyrodytes, Chironectes, 4

argyropoli, Microtus, 990

argyropuli, Apodemus, 1272

argyropuli, Microtus, 990

argyropuloi, Apodemus, 1272

argyropuloi, Microtus, 990

argyropuloi, Spermophilus, 808

argyropus, Arvicola, 964

argyropus, Hydropotes, 671

Argyrotona, 185

ariadne, Crocidura, 250

arianus, Apodemus, 1266

arianus, Mus, 1273

aricana, Nasua, 626

aridicola, Neotoma, 1056

aridicola, Thomomys, 867

aridula, Crocidura, 253

aridulus, Grammomys, **1323**

aridulus, Peromyscus, 1070

ariel, Hypsugo, **489**

ariel, Petaurus, 55

ariel, Plecotus, 483

ariel, Pteropus, 337

Ariela, 570

Arielulus, **451**

Aries, 700, 707

aries, Ovis, 707, **708**

arietina, Capra, 701

arietina, Ovis, 708

arietinus, Cervus, 664

arietinus, Sus, 641

arimalius, Meriones, **1234**

arispa, Crocidura, **225**

Aristippe, 498

aristippe, Hypsugo, 491

aristotelis, Rusa, 670

aristovi, Microtus, 1016

Ariteus, **416**

arizonae, Dipodomys, 845

arizonae, Neotoma, 1054

arizonae, Peromyscus, 1070

arizonae, Sigmodon, **1174**

arizonae, Sorex, **284**

arizonae, Spermophilus, 812

arizonae, Spilogale, 623

arizonae, Sylvilagus, 208

arizonae, Ursus, 589

arizonensis, Bassariscus, 625

arizonensis, Cynomys, 799

arizonensis, Microtus, 1008

arizonensis, Mustela, 615

arizonensis, Mustela, 615

arizonensis, Myodes, 1023

arizonensis, Panthera, 547

arizonensis, Perognathus, 858

arizonensis, Reithrodontomys, 1083

arizonensis, Sciurus, **759**

arizonensis, Spermophilus, 807

arizonensis, Tamias, 815

arizonensis, Vulpes, 584

Arizostus, 97

arkal, Ovis, 709

arkar, Ovis, 709

armalis, Callosciurus, 779

armandii, Myospalax, 912

armandvillei, Papagomys, **1430**

armata, Atherurus, 1542

armatus, Chaetodipus, 853

armatus, Makalata, 1579

armatus, Phyllomys, 1580

armatus, Spermophilus, **804**

armatus, Tenrec, 77

armeniacus, Bison, 690

armeniacus, Spalax, 921

armeniana, Ovis, 709

armenica, Crocidura, **225**

armenica, Sicista, **887**

armenius, Arvicola, 964

armenius, Capreolus, 654

armiger, Hipposideros, **367**

armillatus, Leopardus, 539

armillatus, Petauroides, 51

armorica, Mustela, 616

armoricanus, Microtus, 990

arna, Bubalus, 694

arnee, Bubalus, 694

arnei, Bos, 692

arnhemensis, Isoodon, 39

arnhemensis, Nyctophilus, **468**

arnhemensis, Trichosurus, 49

arni, Bubalus, 694

arnouxianus, Naemorhedus, 706

arnuxii, Berardius, **739**

aroaensis, Mallomys, **1356**

Aroaethrus, 1532

arquatus, Acerodon, 314

arquatus, Eptesicus, 454

arredondoi, Solenodon, **222**

arrhenii, Cephalophus, 713

arrhenii, Heliosciurus, 794

arrhenii, Perodicticus, 123

arrhenii, Potamochoerus, 639

arriagensis, Thomomys, 870

arrogans, Rattus, **1466**

arrucana, Lama, 646

arsenjevi, Martes, 610

arsenjevi, Myodes, 1026

arsenjevi, Pteromys, 774

arsinoe, Myotis, 513

arsipus, Vulpes, 584

artata, Myrmecophaga, 102

artemesia, Sylvilagus, 210

artemesiae, Spermophilus, 809

artemisiae, Lemmiscus, 986

artemisiae, Peromyscus, 1071

artemisiae, Synaptomys, 1038

arthuri, Mustela, 615

Artibaeus, 417

Artibeus, **417**, 421

Artibius, 417

artinii, Scotoecus, 464

Artiodactyla, **637**

Artobius, 417

artus, Chaetodipus, **853**

aruensis, Echymipera, 42

aruensis, Hipposideros, 367

aruensis, Pteropus, **335**

aruensis, Rhinolophus, 355

aruensis, Sus, 642

aruensis, Uromys, 1514

arul, Petaurus, 55

aruma, Phyllostomus, 410

arunchi, Sorex, **284**

arundinacea, Redunca, 722

arundinaceus, Micromys, 1384

arundinum, Redunca, **721**

arundivaga, Puma, 545

aruscensis, Paraxerus, 796

arushae, Eudorcas, 679

arusinus, Ratufa, 756

balutus, Pteropus, 343
bambhera, Ovis, 708
bamendae, Procavia, 88
bampensis, Tscherskia, 1046
banacus, Maxomys, 1372
banakrisi, Pteropus, **335**
bancana, Arctogalidia, 549
bancana, Ratufa, 756
bancanus, Muntiacus, 667
bancanus, Nannosciurus, 784
bancanus, Nycticebus, 122
bancanus, Pipistrellus, 475
bancanus, Tarsius, **127**
bancanus, Tragulus, 650
bancarus, Sundasciurus, 788
bancrofti, Sciurus, 759
bandahara, Maxomys, 1372
bandarum, Funisciurus, 791
banderanus, Osgoodomys, **1061**, 1062, 1076
Bandicota, 904, 1252, **1292**, 1346, 1415, 1416, 1494
bandicota, Bandicota, 1294
bandiculus, Rattus, 1482
banfieldi, Melomys, 1376
bangae, Alcelaphus, 675
bangchakensis, Bandicota, 1295
bangkanus, Callosciurus, 779
bangsi, Glaucomys, 768
bangsi, Mustela, 614
bangsi, Pecari, 644
bangsi, Perognathus, 858
bangsi, Puma, 545
bangsi, Sciurus, 764
bangsi, Vulpes, 585, 586
bangsii, Lepus, 195
banguei, Rattus, 1492
banguei, Ratufa, 756
banguei, Tragulus, 650
banguei, Tupaia, 108
bangueyae, Sundasciurus, 787
banksi, Callosciurus, 779
banksi, Petaurista, 771
banksiana, Pteropus, 335
banksianus, Canis, 576
banksianus, Macropus, 65

banksicola, Lepus, 196
banksii, Pseudocheirus, 51
bannisteri, Melomys, **1374**, 1516
bantamensis, Callosciurus, 778
banteng, Bos, 691
bantiger, Bos, 691
banting, Bos, 691
baoulei, Mus, **1389**
baptista, Callicebus **141**
baptistae, Meriones, 1237
baptistae, Viverricula, 559
barabensis, Arvicola, 964
barabensis, Cricetulus, **1041**
barabensis, Sorex, 292
barakshin, Alticola, **959**
baralou, Mazama, 655
baramensis, Callosciurus, 775
baramensis, Ratufa, 756
barandana, Rusa, 670
barang, Aonyx, 602
barang, Lutra, 604
barawensis, Paraxerus, 796
barbacoas, Sigmodontomys, 1178
barbar, Genetta, 555
barbara, Capra, 702
barbara, Eira, **606**
barbara, Galictis, 606
barbara, Genetta, 555
barbara, Hyaena, 573
burbara, Panthera, 546, 547
barbara, Vulpes, 585
barbarabrownae, Callicebus, **141**, 145
barbarensis, Caenolestes, 19
barbarica, Capra, 702
barbarica, Ovis, 708
barbarica, Panthera, 546
barbarus, Canis, 574
barbarus, Cervus, 663
barbarus, Ctenomys, 1562
barbarus, Lemniscomys, 1340, **1341**
barbarus, Rhinolophus, 354
barbarus, Sus, 642
Barbastella, **480**
barbastelle, Barbastella, 480
barbastellus, Barbastella, **480**

barbata, Hippotragus, 717
barbatogularis, Platymops, 448
barbatus, Alouatta, 148
barbatus, Ammotragus, 700
barbatus, Cebus, 137
barbatus, Cynogale, 552
barbatus, Eira, 606
barbatus, Erignathus, **595**
barbatus, Eumops, 437
barbatus, Nyctiellus, 432
barbatus, Phacochoerus, 639
barbatus, Pseudochirulus, 52
barbatus, Sus, **640**
barbatus, Taurotragus, 697
barbatus, Xenuromys, **1519**
barbei, Tamiops, 789
barbei, Trachypithecus, **175**, 177
barbensis, Hipposideros, 373
barberi, Sciurus, 759
barberi, Tamias, 817
barbertonensis, Rhinolophus, 354
barbertoni, Cephalophus, 715
barbiensis, Petromus, 1545
barbouri, Otomys, **1525**
barbouri, Petromyscus, **954**
barcaeus, Lepus, 197
barclayanus, Bandicota, 1293
bardus, Apomys, 1282
bargusinensis, Myodes, 1026
barhal, Pseudois, 711
baringoensis, Eudorcas, 679
baritensis, Paradoxurus, 551
barkeri, Bullimus, 1298
barkeri, Phacochoerus, 639
barkeri, Tragelaphus, 698
barlowi, Elephantulus, 83
barnardi, Lasiorhinus, 43
barnesi, Molossus, **440**
barnesi, Ochotona, 191
baroni, Oryzomys, 1156
barrerae, Tympanoctomys, **1573**

barretti, Pronolagus, 207
barrigo, Lagothrix, 152
barroni, Petaurista, 772
barrowensis, Isoodon, 39
barrowensis, Spermophilus, 809
bartelsi, Aeromys, 766
bartelsi, Hylopetes, **769**
bartelsi, Myotis, 507
bartelsii, Crocidura, 241
bartelsii, Maxomys, 1338, 1367, **1369**
Barticonycteris, 404, 407
bartletti, Caluromys, 3
bartlettii, Ateles, 150
bartoni, Callosciurus, 776
bartoni, Zaglossus, **1**, 2
barussanoides, Rattus, 1490
barussanus, Niviventer, 1420
baryceros, Rusa, 670
basaci, Chrysochloris, 1343
basengae, Alcelaphus, 675
bashkiricus, Sciurus, 764
basilanensis, Rusa, 670
basilanus, Rattus, 1469
basilicae, Thomomys, 866
basilii, Hybomys, **1331**
basiliscus, Papio, 166
basiliscus, Pteropus, 345
baskii, Speothos, 582
bassanii, Miniopterus, 521
Bassaricyon, **624**
Bassaricyonidae, 624
Bassaridae, 624
Bassaris, 624
Bassariscidae, 624, 625
Bassariscus, **625**
bassi, Scalopus, 301
bassianus, Pseudocheirus, 51
bassii, Vombatus, 44
bastardi, Macrotarsomys, **951**
bastiani, Nycteris, 392
basuticus, Dendromus, 938
basuticus, Galerella, 565
basuticus, Myotomys, 1523
batamana, Tupaia, 106
batamanus, Maxomys, 1373
batangtuensis, Pongo, 184
batarovi, Micromys, 1384

brandtii, Coendou, 1547

brandtii, Lasiopodomys, **984**

brandtii, Myotis, **503**

brandvleiensis, Macroscelides, 84

branickii, Dinomys, **1552**

branti, Sus, 640

brantsii, Gerbilliscus, **1218**

brantsii, Parotomys, **1531**

brasiliana, Pteronura, 605

brasiliense, Lophostoma, **406**

brasiliensis, Bradypus, 101

brasiliensis, Cerdocyon, 578

brasiliensis, Choloepus, 101

brasiliensis, Ctenomys, 1560, **1562**

brasiliensis, Didelphis, 5, 6

brasiliensis, Eptesicus, **453**

brasiliensis, Galictis, 606, 607

brasiliensis, Holochilus, **1119**

brasiliensis, Lasiurus, 459

brasiliensis, Leontopithecus, 133

brasiliensis, Leopardus, 539

brasiliensis, Lycalopex, 580

brasiliensis, Monodelphis, 14

brasiliensis, Mustela, 615

brasiliensis, Myotis, 513

brasiliensis, Nectomys, 1133

brasiliensis, Phyllomys, **1580**

brasiliensis, Potos, 627

brasiliensis, Procyon, 627

brasiliensis, Pteronura, **605**

brasiliensis, Sylvilagus, **208**

brasiliensis, Tadarida, **449**

brasiliensis, Tamandua, 103

brasiliensis, Tapirus, 634

brassi, Pogonomelomys, 1440

braueri, Dendrohyrax, 87

braueri, Desmodilliscus, **1212**

brauneri, Apodemus, 1266

brauneri, Cervus, 663

brauneri, Micromys, 1384

brauneri, Microtus, 1017

brauneri, Spermophilus, 810

brauneri, Talpa, 308

brauni, Heliosciurus, 794

bravoi, Canariomys, 1357

brazenori, Mastacomys, 1360

brazenori, Pseudomys, 1458

brazenori, Rattus, 1471

brazensis, Geomys, 862

brazierhowelli, Thomomys, 867

braziliensis, Bradypus, 101

braziliensis, Carollia, 412, 413

braziliensis, Ctenomys, 1561

braziliensis, Megaptera, 725

braziliensis, Sylvilagus, 208

brazzae, Cercopithecus, 157

brazziformis, Cercopithecus, 157

bredanensis, Steno, **734**

bregullae, Chaerephon, **433**

brelichi, Rhinopithecus, **174**

brendae, Oligoryzomys, **1140**

bresslaui, Chironectes, 4

brevetianus, Tapirus, 633

breviaculeata, Tachyglossus, 1

breviauritus, Lepus, 204

breviauritus, Onychomys, 1061

brevicauda, Apodemus, 1272

brevicauda, Balaenoptera, 725

brevicauda, Blarina, **269**

brevicauda, Carollia, **412**

brevicauda, Chinchilla, 1550

brevicauda, Crocidura, 227

brevicauda, Microtus, 998

brevicauda, Neotoma, 1053

brevicauda, Proechimys, **1584**

brevicauda, Spermophilus, **805**, 807

brevicauda, Zygodontomys, **1185**

brevicaudata, Chinchilla, 1550

brevicaudata, Felis, 535

brevicaudata, Microgale, **72**

brevicaudata, Monodelphis, **14**, 15

brevicaudata, Ourebia, 686

brevicaudata, Phascolosorex, 27

brevicaudatum, Cynopterus, 316, 317

brevicaudatum, Rhinopoma, 380

brevicaudatus, Abrothrix, 1090

brevicaudatus, Desmodillus, 1213

brevicaudatus, Indri, 120

brevicaudatus, Macaca, 163

brevicaudatus, Meriones, 1237

brevicaudatus, Rattus, 1465

brevicaudatus, Setonix, 69

brevicaudatus, Spilocuscus, 48

brevicaudus, Dasypus, 94

brevicaudus, Lophuromys, **1203**

brevicaudus, Myodes, 1023

brevicaudus, Onychomys, 1061

brevicaudus, Peromyscus, 1070

brevicaudus, Rattus, 1490

brevicaudus, Taphozous, 383

breviceps, Blarinomys, **1104**

breviceps, Cephalophus, 713

breviceps, Didelphis, 6

breviceps, Echymipera, 41

breviceps, Geomys, **862**

breviceps, Hipposideros, **368**

breviceps, Kobus, 720

breviceps, Kogia, **737**

breviceps, Lagenorhynchus, 730

breviceps, Microgale, 72

breviceps, Petaurus, 55

breviceps, Phalanger, 47

breviceps, Pteropus, 341

breviceps, Rusa, 669

breviceps, Thylacinus, 23

brevicorpus, Microtus, 1008

brevicula, Echiothrix, 1321

brevidens, Thomomys, 867

brevimaculatus, Dendrohyrax, 87

brevimanus, Chilonatalus, 430

brevimanus, Otonycteris, 482

brevimanus, Plecotus, 482

brevimanus, Stenella, 733

brevimolaris, Bunomys, 1299

brevinasus, Dipodomys, 847

brevinasus, Lepus, 204

brevinasus, Perognathus, 858

brevinasus, Phalanger, 46

brevipes, Plecotus, 483

brevipes, Tragulus, 650

brevipes, Zapus, 893

brevipilis, Canis, 576

brevipilosus, Lontra, 603

brevirostris, Chaetophractus, 96

brevirostris, Cormura, **386**

brevirostris, Cratogeomys, 861

brevirostris, Dasypus, 94

brevirostris, Elephantulus, 82

brevirostris, Geomys, 862

brevirostris, Glossophaga, 400

brevirostris, Microtus, 991

brevirostris, Mus, 1399

brevirostris, Myotis, 510

brevirostris, Orcaella, **731**

brevirostris, Ornithorhynchus, 2

brevirostris, Ototylomys, 1188

brevirostris, Reithrodontomys, **1081**

brevirostrum, Anoura, 398

brevis, Myotis, 517

brevispinosa, Hystrix, 1544

elseyi, Pteropus, 344
eltonclarki, Ursus, 589
elucus, Procyon, 627
elusus, Peromyscus, 1071
elvira, Cremnomys, *1311*
elymocetes, Microtus, 1010
emarginatus, Myotis, *506*
emarginatus, Scotomanes, 465
emarginatus, Sorex, *287*
Emballonura, 386, *387*, 389
Emballonuridae, *381*
Emballonurinae, *385*
Emballonurini, 385
emerita, Lontra, 603
emeritus, Leopardus, 539
emeritus, Thomasomys, 1182
emersa, Dobsonia, *318*
emesi, Mus, 1396
emiliae, Callicebus, 144
emiliae, Callithrix, *130*
emiliae, Gracilinanus, 7
emiliae, Leopardus, 539
emiliae, Lonchothrix, *1583*
emiliae, Macropus, 64
emiliae, Monodelphis, 14
emiliae, Petaurillus, *770*
emiliae, Rhipidomys, *1168*
emilianus, Ctenomys, *1563*
emini, Cephalophus, 714
emini, Chaerephon, 434
emini, Cricetomys, *932*
emini, Dendrohyrax, 87
emini, Funisciurus, 792
emini, Heliophobius, 1541
emini, Paraxerus, 795
emini, Taterillus, 1243, *1244*
emissus, Heliosciurus, 794
emmae, Uromys, *1515*
emmonsae, Oryzomys, *1148*
emmonsi, Peromyscus, 1070
emmonsii, Ursus, 588
emotus, Thomomys, 870
empetra, Marmota, 802
empusa, Rhinolophus, 352
Enchisthenes, 417, *421*
enclavae, Mus, 1397
encoubert, Euphractus, 97
Encoubertus, 97
Endecapleura, 1222

endemicus, Spermophilus, 805
endersi, Cryptotis, *270*
endersi, Oecomys, 1137
endersi, Scotinomys, 1085
endoecus, Microtus, 1010
endoi, Pipistrellus, *474*
endorobae, Hylomyscus, 1336
energumenos, Neovison, 619
enez-groezi, Microtus, 990
enezsizunensis, Crocidura, 250
engana, Kerivoula, 526
enganus, Hipposideros, 370
enganus, Paradoxurus, 551
enganus, Pteropus, 338
enganus, Rattus, *1468*
enganus, Sus, 642
Engeco, 183
engelhardi, Orthogeomys, 864
engelhardti, Marmota, 801
engraciae, Oligoryzomys, 1141
enguvi, Cricetomys, 932
engytithia, Chlorocebus, 159
Enhydra, *602*
Enhydria, 602
Enhydris, 602
Enhydrus, 602
enid, Acomys, 1200
eniseensis, Myotis, 507
enisseyensis, Meles, 611
enixus, Microtus, 1012
enixus, Thomomys, 870
enkamer, Cercopithecus, 155
ennisae, Pteropus, 336
ensenadensis, Blastocerus, 654
ensenadensis, Lama, 647
ensicornis, Capra, 702
ensicornis, Oryx, 719
enslenii, Lasiurus, 458
entelloides, Hylobates, 180
entellus, Semnopithecus, *174*
Entocacrya, 1538
entomophagus, Saimiri, 138
entrerianus, Cerdocyon, 578
enuchus, Conepatus, 621
enudris, Lontra, 603
Enydris, 602

Eoglaucomys, 755, *767*, 768
eogroenlandicus, Rangifer, 660
eogroenlandicus, Ursus, 589
Eolagurus, 898, *976*, 983
Eomops, 446
Eomuscardinus, 835
eonis, Sorex, 290
Eonycteris, *321*
Eosaccomys, 933
Eoscalops, 304, 305
Eosciurus, 756
Eospalax, 897, 909, *910*
Eothenomys, 898, 959, 967, *977*, 1021, 1022, 1039
eotinus, Pteropus, 335
Eozapus, *891*
ephippiger, Bradypus, 101
ephippium, Rattus, 1469
ephippium, Ratufa, 756
epieni, Piliocolobus, 170
epimelas, Apodemus, *1265*
Epimys, 1253, 1296, 1334, 1422, 1443
epiratticeps, Microtus, 1010
epiroticus, Microtus, 1002
epiroticus, Spalax, 920
episcopi, Scotinomys, 1085
episcopus, Rhinolophus, 358
Episiphneus, 909, 911
Episoriculus, *277*
epixanthus, Echinoprocta, 1548
epixanthus, Erethizon, 1548
Epixerus, 755, *791*
Epomophorus, *322*, 323, 327, 328, 333
epomophorus, Callosciurus, 776
epomophorus, Epomophorus, 322
Epomops, *323*
epsilanus, Myospalax, 912
Eptesicini, *451*
Eptesicops, 472
Eptesicus, *452*, 473, 484, 493, 496
epularius, Pteropus, 340
epupae, Micaelamys, 1382
equatoris, Cryptotis, *271*
equatoris, Rhipidomys, 1169

equatorius, Tayassu, 644
equestris, Histriophoca, 596
Equidae, *629*
equiferus, Equus, 631
equile, Rattus, 1469
equina, Rusa, 671
equinus, Hippocamelus, 655
equinus, Hippotragus, *717*
equinus, Rhinolophus, 355
equioides, Equus, 632
equuleus, Equus, 630
Equus, *629*, 655, 672
eracinius, Tachyglossus, 1
Eraria, 606
erasmus, Peromyscus, 1068
erebus, Callosciurus, 779
Eremaelurus, 534, 536
eremiana, Perameles, *40*
eremicoides, Peromyscus, 1075
eremicus, Chaetodipus, *853*
eremicus, Lepus, 196
eremicus, Lynx, 542
eremicus, Odocoileus, 658
eremicus, Peromyscus, *1066*
eremicus, Sigmodon, 1175
eremicus, Ursus, 588
Eremiomys, 983
eremita, Neomys, 279
eremita, Neotoma, 1058
Eremitalpa, *79*
Eremodipus, *883*
eremoecus, Dipodomys, 848
Eremomys, 983
eremonomus, Spermophilus, 812
eremus, Peromyscus, 1071
Erethison, 1548
Erethizon, *1548*
Erethizontidae, 752, *1545*
Erethizontinae, *1546*
Eretison, 1548
Eretizon, 1548
ereunetes, Ursus, 589
erica, Crocidura, *230*
erica, Myodes, 1023
Ericius, 76, 215
Ericulus, 76, 77
Ericus, 76
erigens, Hipposideros, 368
Erignathini, 595
Erignathus, *595*

famulus, Callosciurus, 778

famulus, Gerbillus, **1226**

famulus, Mus **1393,** 1488

famulus, Rhinolophus, 353

fannini, Otomys, 1528

fannini, Ovis, 710

fantiensis, Cercopithecus, 157

fantozatianus, Nae- morhedus, 706

far, Mus, 1399

faradjius, Atelerix, 212

faradjius, Leptailurus, 540

faradjius, Mops, 442

faradjius, Neoromicia, 495

faradjius, Orycteropus, 86

faradjius, Otomys, 1529

farasani, Gazella, 681

fardoulisi, Melonycteris, **327**

fargesianus, Capricornis, 704

fargesianus, Naemorhe- dus, 706

faroulti, Pachyuromys, 1240

farsi, Meriones, 1236

farsistani, Ellobius, 975

fasciata, Capra, 703

fasciata, Galidictis, **561**

fasciata, Histriophoca, **596**

fasciata, Hyaena, 573

fasciata, Lynx, 542

fasciata, Perameles, 40

fasciatus, Connochaetes, 676

fasciatus, Lagostrophus, **59**

fasciatus, Lemniscomys, 1344

fasciatus, Lynx, 542

fasciatus, Moschus, 652

fasciatus, Mungos, 571

fasciatus, Myrmecobius, **23**

fasciatus, Perognathus, **856**

fasciatus, Sus, 642

fasciatus, Tragelaphus, 698

fascicularis, Macaca, **161,** 171

fasciculata, Trichys **1544**

fasensis, Hipposideros, 378

fassini, Ammotragus, 700

fatioi, Microtus, 1009

fatuellus, Cebus, 137

fatuus, Synaptomys, 1038

faunulus, Pteropus, 107, **337**

Faunus, 183

faunus, Cercopithecus, 155

faurei, Equus, 632

faustus, Perodicticus, 123

favillus, Gerbillus, 1228

favillus, Jaculus, 884

favonicus, Funambulus, 782

favonicus, Jaculus, 884

favonicus, Myotis, 503

fea, Berylmys, 1297

feae, Muntiacus, **666**

feae, Murina, 523

feae, Rhinolophus, 358

feai, Muntiacus, 666

fearonii, Acinonyx, 533

fearonis, Acinonyx, 533

fecundus, Calomys, 1106

federatus, Myotis, 511

fedjushini, Sciurus, 764

feileri, Strigocuscus, 49

feliceus, Rattus, **1470**

felicia, Dasyprocta, 1556

Felidae, **532**

Feliformia, **532**

felina, Genetta, 555

felina, Lontra, **603**

felina, Trichosurus, 50

Felinae, **532**

Felinoidea, 532

felinus, Aotus, 141

felinus, Paradoxurus, 551

felipei, Mustela, **615**

felipensis, Neotoma, 1056

felipensis, Orthogeomys, 864

felipensis, Peromyscus, 1066

Felis, 532, 533, **534,** 537, 538, 539, 540, 541, 544, 545, 546, 548

felix, Tamias, 813

felli, Ratufa, 756

fellowesgordoni, Suncus, **258**

fellowsi, Protochromys, 1312, **1451**

Feloidae, **532**

Feloidea, 532

Felovia, **1536**

felteni, Microtus, **997**

femoralis, Chaetodipus, 853

femoralis, Presbytis, **171,** 172

femoralis, Ratufa, 756

femorosaccus, Nyctino- mops, **446**

femurvillosum, Artibeus, 419

fenestrae, Thylamys, 18

fenestratus, Dasypus, 95

fenisex, Ochotona, 191

Fennecus, 583, 586

fennecus, Vulpes, 586

fenniae, Micromys, 1384

fennicus, Apodemus, 1266

fennicus, Rangifer, 660

fera, Capra, 702

fera, Lama, 647

fera, Ovis, 708

ferculinus, Pseudomys, 1458

Feresa, **728**

ferganensis, Vulpes, 586

ferghanae, Mustela, 614

ferghanicus, Capreolus, 655

fergusoni, Bubalus, 694

fergussoni, Apodemus, 1264

fergussoniensis, Pogon- omys, 1440, **1441**

fergussoniensis, Rattus, 1476

ferminae, Sciurus, 761

fernandezi, Lonchorhina, **405**

fernandinae, Nesoryz- omys, **1134**

fernandoni, Mus, **1393**

Feroculus, **256**

feroculus, Feroculus, **256**

ferox, Cryptoprocta, **560**

ferox, Eumops, 438

ferox, Felis, 536

ferox, Ichneumia, 570

ferox, Macaca, 164

ferox, Rhinopoma, 380

ferox, Sus, 642

ferox, Ursus, 589

ferrarii, Leptailurus, 540

ferreocanus, Berylmys, 1297

ferreogrisea, Melogale, 612

ferreus, Monophyllus, 402

ferreus, Sciurus, 759

ferriginea, Piliocolobus, 169

ferrilata, Vulpes, **584**

ferruginea, Crocidura, 244

ferruginea, Felis, 535

ferruginea, Herpestes, 568

ferruginea, Lutreolina, 8

ferruginea, Neotoma, 1058

ferruginea, Nyctinomops, 447

ferruginea, Ochotona, 189

ferruginea, Sminthopsis, 34

ferruginea, Tupaia, 105, 106

ferrugineus, Arvicola, 964

ferrugineus, Callosciurus, 777

ferrugineus, Cricetulus, 1042

ferrugineus, Deomys, 1191, **1201,** 1308

ferrugineus, Eptesicus, 453

ferrugineus, Herpestes, 568

ferrugineus, Lycalopex, 579

ferrugineus, Mesomys, 1583

ferrugineus, Myotis, 505

ferrugineus, Nyctalus, 471

ferrugineus, Phaenomys, **1160**

ferrugineus, Presbytis, 171

ferrugineus, Procavia, 88

ferrugineus, Suncus, 260

ferruginifrons, Sminthop- sis, 35

ferruginiventris, Sciurus, 760

ferruginosus, Piliocolobus, 169

ferrum-equinum, Vesper- tilio, 350

ferrumequinum, Rhinolo- phus, **355**

fertilis, Hyperacrius, **983**

ferus, Bison, 690

ferus, Bubalus, 694

ferus, Camelus, 645, 646

ferus, Canis, 576

ferus, Equus, 631, 632

ferus, Felis, 536

ferus, Sus, 642

fervidus, Caluromys, 3

fervidus, Rhipidomys, 1171

fervidus, Sigmodon, 1177

festina, Cavia, 1553

festinus, Lepus, 196

festivus, Equus, 630

fetidissima, Mephitis, 622

fraterculus, Peromyscus, **1066**
fraterculus, Sundasciurus, **786,** 1349
fraterculus, Tarsius, 128
fraterculus, Urocyon, 582
fraterculus, Zygodon-tomys, 1186
fraternus, Epimys, 1422
fraternus, Macroglossus, 326
fraternus, Niviventer, **1421**
fraternus, Thomasomys, 1182
Frateromys, 1298
fratrorum, Bunomys, **1300,** 1506
fredericae, Mus, 1399
fredericae, Niviventer, 1424
fredericae, Presbytis, 171
fredericae, Prionodon, 553
frederici, Equus, 633
frederici, Eulemur, 114
frederici, Herpestes, 568
frederici, Philantomba, 715
Fredgia, 282
fremens, Leopoldamys, 1348
fremonti, Conepatus, 621
fremonti, Dipodomys, 847
fremonti, Tamias, 818
fremonti, Tamiasciurus, 765
frenata, Dendrogale, 105
frenata, Ictonyx, 607
frenata, Mustela, **615**
frenatus, Artibeus, 419
frenatus, Dipodomys, 846
frenatus, Mustela, 615
frenatus, Philander, 16
frenatus, Sus, 641
frerei, Paraxerus, 796
fretalis, Sorex, 287
fretensis, Pteropus, 338
fretensis, Ratufa, 756
fretensis, Taphozous, 384
Fretidelphis, 733
freyreisii, Diclidurus, 387
friburgensis, Bos, 692
fricator, Canis, 576
fricatrix, Canis, 576
frida, Calomys, 1108
fridariensis, Apodemus, 1275
frigicola, Melomys, **1377**
frinianus, Cervus, 664
friseus, Talpa, 308

frisius, Bos, 692
frisius, Equus, 630
frisius, Talpa, 308
frithii, Coelops, **366**
fritschi, Spalax, 918
froenatus, Stenella, 733
froggatti, Melomys, 1375
froggatti, Sminthopsis, 35
frommi, Alcelaphus, 675
frommi, Heterohyrax, 88
frommi, Kobus, 720
frommi, Tragelaphus, 699
frondator, Castor, 842
frons, Lavia, **379**
frontalis, Acrobates, 56
frontalis, Atelerix, **213**
frontalis, Ateles, 151
frontalis, Bos, **690**
frontalis, Cryptotis, 271
frontalis, Hemiechinus, 215
frontalis, Oecomys, 1139
frontalis, Ratufa, 756
frontalis, Rhinoceros, 636
frontalis, Rucervus, 669
frontalis, Stenella, **733**
frontata, Mephitis, 622
frontata, Presbytis, **171**
frontatus, Ateles, 151
frontatus, Bos, 690
frontatus, Cebus, 137
frontatus, Inia, 739
frontatus, Steno, 734
fronto, Cerdocyon, 578
fronto, Kunsia, **1122**
frontosa, Vicugna, 647
frontosus, Bos, 692
frontosus, Sus, 642
frosti, Babyrousa, 637
frosti, Neopteryx, **328**
frosti, Paruromys, 1435
fructivorus, Macroglossus, 325
frugilegus, Tremarctos, 587
frugivorus, Rattus, 1484
fruhstorferi, Emballonura, 388
frumentariae, Apodemus, 1275
frumentarius, Cricetus, 1043
frumentor, Sciurus, 760
frustrator, Zygodontomys, 1186
frustror, Canis, 575
frutectanus, Napaeoza-pus, 892
fruticicolus, Phyllotis, 1161
fruticus, Macropus, 65

fryanus, Callosciurus, 776
fryi, Malacothrix, 942
Fsihego, 183
fuchsi, Lycaon, 581
fucosus, Microtus, 1008
fudisanus, Rhinolophus, 355
fueginus, Ctenomys, 1566
fuillus, Canis, 576
fujianensis, Hipposideros, 367
fujianensis, Microtus, 997
fujiensis, Myotis, 509
fulgens, Ailurus, **628**
fulgens, Anomalurus, 1532
fulgens, Hipposideros, 372
fulgens, Melomys, **1378**
fulgens, Oryzomys, 1147
fulgida, Cavia, **1553**
fuliginatus, Baiomys, 1049
fuliginatus, Odocoileus, 658
fuliginosa, Crocidura, **231**
fuliginosa, Dasyprocta, **1557**
fuliginosa, Emballonura, 388
fuliginosa, Nycteris, 393
fuliginosa, Rousettus, 348
fuliginosa, Trichosurus, 50
fuliginosus, Artibeus, 420
fuliginosus, Arvicola, 964
fuliginosus, Ateles, 150
fuliginosus, Caenolestes, **19**
fuliginosus, Cercocebus, 153
fuliginosus, Crocidura, 231
fuliginosus, Glaucomys, 768
fuliginosus, Hipposideros, **371**
fuliginosus, Hydromys, 1333
fuliginosus, Macropus, **64**
fuliginosus, Miniopterus, 522
fuliginosus, Molossus, 441
fuliginosus, Onychomys, 1061
fuliginosus, Pan, 183
fuliginosus, Paradoxurus, 551

fuliginosus, Perognathus, 857
fuliginosus, Piliocolobus, 169
fuliginosus, Pteronotus, 428
fuliginosus, Rattus, 1484
fuliginosus, Sciurus, 760
fuliginosus, Sminthopsis, **34**
fuliginosus, Tadarida, 450
fuliginosus, Thaptomys, 1179
fuliginosus, Trachops, 411
fuliginosus, Trinomys, 1591
fuliginosus, Vulpes, 584
fulminans, Tadarida, **450**
fulminatus, Sciurus, 762
fulmineus, Rattus, 1492
fulva, Microtus, 991
fulva, Mops, 442
fulva, Mustela, 616, 617
fulva, Nasua, 626
fulva, Otaria, 593
fulva, Vernaya, **1518**
fulvaster, Crocidura, 231
fulvaster, Rattus, 1484
fulvastra, Crocidura, **231**
fulvescens, Akodon, 1096
fulvescens, Cratogeomys, 861
fulvescens, Herpestes, 568
fulvescens, Neotetracus, 219
fulvescens, Niviventer, **1422**
fulvescens, Oligoryzomys, 1139, **1141,** 1141
fulvescens, Phyllotis, 1162
fulvescens, Reithrodon-tomys, **1081**
fulvescens, Sylvilagus, 208
fulvicaudus, Lycalopex, 581
fulvicollis, Tragulus, 650
fulvidina, Felis, 535
fulvidior, Galerella, 565
fulvidiventris, Mus, 1390
fulvidus, Rhinolophus, 362
fulvidus, Taphozous, 383
fulvidus, Tylonycteris, 496
fulvifrons, Kobus, 720
fulvinus, Nectomys, 1132
fulvinus, Petaurista, 772

fulvior, Xerus, 790
fulvipectus, Apodemus, 1279
fulvipes, Lycalopex, **580**
fulvirostris, Microryzomys, 1127
fulviventer, Marmosa, 9
fulviventer, Microtus, 1006
fulviventer, Neotoma, 1058
fulviventer, Oecomys, 1139
fulviventer, Rhipidomys, **1168**
fulviventer, Sigmodon, **1174**
fulviventer, Tragulus, 650
fulviventer, Tylomys, **1189**
fulvo-fusca, Chiropotes, 146
fulvocinerea, Suncus, 260
fulvofasciatus, Delphinus, 728
fulvogaster, Hydromys, 1333
fulvogriseus, Trachypithecus, 178
fulvolavatus, Hydromys, 1333
fulvoochraceus, Tragelaphus, 698
fulvorubescens, Raphicerus, 688
fulvorufula, Redunca, **722**
fulvoventer, Hydromys, 1333
fulvus, Aonyx, 602
fulvus, Bubalus, 694
fulvus, Canis, 576
fulvus, Castor, 843
fulvus, Cebus, 137
fulvus, Cricetulus, 1043
fulvus, Cricetus, 1043
fulvus, Ctenomys, **1564**
fulvus, Cuniculus, 1559
fulvus, Dasyprocta, 1557
fulvus, Dendrolagus, 60
fulvus, Eulemur, **115**, 116
fulvus, Hipposideros, **371**
fulvus, Macaca, 163
fulvus, Meriones, 1237
fulvus, Microtus, 991
fulvus, Myodes, 1023
fulvus, Peromyscus, 1071
fulvus, Pipistrellus, 477
fulvus, Pteronotus, 427
fulvus, Sciurus, 759
fulvus, Scotophilus, 466

fulvus, Spermophilus, **807**
fulvus, Thomomys, 867
fulvus, Vulpes, 585
fumarium, Sturnira, 414
fumarius, Promops, 448
fumatus, Cricetulus, 1042
fumatus, Mastomys, 1365
fumatus, Myomyscus, 1414, 1415
fumeolus, Sorex, 285
fumeus, Akodon, **1094**
fumeus, Chilomys, 1111
fumeus, Ototylomys, 1188
fumeus, Pseudomys, **1456**
fumeus, Sorex, **288**
fumeus, Thomasomys, 1184
fumicolor, Rhombomys, 1241
fumidus, Episoriculus, **278**
fumigatus, Callosciurus, 776
fumigatus, Crocidura, 250
fumigatus, Lepus, 202
fumigatus, Pteropus, 335
fumigatus, Rhinolophus, **356**
fumigatus, Sciurus, 763
fumipes, Loxodontomys, 1123
fumosa, Crocidura, **231**
fumosa, Ochotona, 191
fumosus, Cercocebus, 153
fumosus, Chelemys, 1110
fumosus, Cratogeomys, **860**
fumosus, Cuon, 578
fumosus, Heimyscus, **1329**
fumosus, Mystromys, 946
fumosus, Platyrrhinus, 423
fumosus, Taphozous, 383
fumosus, Thomomys, 867
Funambulini, 791, 795
Funambulus, 755, **781**, 791, 794
fundatus, Microtus, 1006
fundatus, Pteropus, **337**
funebris, Lasiurus, 458
funebris, Microtus, 1012
funereus, Hylobates, 180
funereus, Mus, 1398
funereus, Pteropus, 346
funestus, Herpestes, 568
Funisciurus, 755, **791**
fur, Macaca, 162

furax, Emballonura, **388**
furax, Felis, 535
furax, Galictis, 606
furax, Papio, 166
furcata, Blastocerus, 654
furcata, Conepatus, 621
Furcifer, 655
furcifer, Antilocapra, 671
furcifer, Phaner, **114**
furcifer, Rangifer, 660
furculus, Triaenops, **378**
Furia, 429
furinalis, Eptesicus, **454**
furinea, Triaenops, 378
Furipteridae, **429**
Furipterus, **429**
furnessi, Pentalagus, **206**, 1512
furo, Mustela, 617
furoputorius, Mustela, 617
furunculus, Cricetulus, 1042
furva, Chodsigoa, 277
furvogaster, Nomascus, 180
furvus, Ailurops, 45
furvus, Nyctalus, **471**
furvus, Palawanomys, 1302, 1371, **1429**
furvus, Peromyscus, **1067**
furvus, Sigmodon, 1177
furvus, Urocyon, 582
fusca, Alouatta, 149
fusca, Arctogalidia, 549
fusca, Geomys, 862
fusca, Hyaena, 572
fusca, Kerivoula, 526
fusca, Macaca, 163
fusca, Mastomys, 1365
fusca, Mazama, 656
fusca, Microtus, 1017
fusca, Murina, **523**
fusca, Mustela, 615
fusca, Nasua, 626
fusca, Neotoma, 1054
fusca, Panthera, 547
fusca, Phalanger, 47
fusca, Procyon, 627
fusca, Pudu, 659
fusca, Rousettus, 348
fuscata, Macaca, 161, **162**
fuscata, Mazama, 655
fuscatus, Handleyomys, **1118**
fuscatus, Heteromys, 850
fuscatus, Lasiurus, 459
fuscatus, Marmosops, 10
fuscatus, Pipistrellus, 475
fuscatus, Tragulus, 650

fuscescens, Sylvilagus, 208
fusciceps, Ateles, **150**
fusciceps, Ellobius, 976
fuscicollis, Pteropus, 344
fuscicollis, Saguinus, **134**, 135, 136
fuscicolor, Philantomba, 716
fuscicornis, Bos, 690
fuscidorsis, Cricetus, 1043
fuscifrons, Gazella, 680
fuscinus, Necromys, 1130
fuscior, Tupaia, 108
fuscipes, Elephantulus, **82**
fuscipes, Ellobius, 976
fuscipes, Jaculus, 884
fuscipes, Neotoma, **1055**
fuscipes, Ochotona, 191
fuscipes, Pipistrellus, 478
fuscipes, Procyon, 627
fuscipes, Rattus, **1470**
fuscipes, Suncus, 260
fuscipes, Tscherskia, 1046
fuscirostris, Praomys, 1364
fusciventer, Isoodon, 39
fusciventer, Molossus, 441
fusco-ater, Abrothrix, 1090
fusco-ater, Macaca, 163
fusco-flavescens, Martes, 610
fusco-murina, Presbytis, 171
fuscoater, Sciurus, 764
fuscocanus, Dipus, 882
fuscocapillus, Ellobius, **975**
fuscocapillus, Petinomys, **773**
fuscodorsalis, Myodes, 1023
fuscogriseus, Onychomys, 1061
fuscogriseus, Philander, 17
fuscomanus, Tarsius, 128
fuscomurina, Crocidura, **231**
fuscomurinus, Crocidura, 232
fusconigricans, Sciurus, 764
fuscorubens, Sciurus, 764
fuscosa, Crocidura, 244
fuscovariegatus, Sciurus, 764
fusculus, Microsciurus, 757

gigas, Balaenoptera, 725
gigas, Canis, 576
gigas, Dinomys, 1552
gigas, Elephas, 90
gigas, Eumops, 438
gigas, Gorilla, 182
gigas, Grammomys, *1325*
gigas, Hipposideros, *372*
gigas, Hydrodamalis, *92*
gigas, Leopoldamys, 1347
gigas, Lynx, 542
gigas, Macroderma, *379*
gigas, Megasorex, 280, *281*
gigas, Priodontes, 98
gigas, Scotophilus, 467
gigas, Sus, 642
gigas, Taurotragus, 696
gigliolii, Hylochoerus, 638
giglis, Glis, 840
gigot, Callicebus, 144
gikdnabu, Panthera, 547
gilberti, Hipposideros, 367
gilberti, Peromyscus, 1079
gilberti, Sminthopsis, *34*
gilbertii, Potorous, *58*
gilbiventer, Niviventer, 1420
gilesi, Planigale, *36*
gilgitensis, Capra, 701
gilgitianus, Rattus, 1482
giliacus, Apodemus, 1271
gillespiei, Lasiorhinus, 43
gillespii, Zalophus, 593, 594
gilli, Gerbilliscus, 1217
gilliardi, Pteropus, 337
gilliardorum, Pteropus, *337*
gilliesi, Scaptochirus, 307
gillii, Tursiops, 734
gilmorei, Microtus, 1010
giloensis, Otomys, 1529
giluwensis, Rattus, *1471*
gilva, Neotoma, 1056
gilvigularis, Sciurus, *761*
gilvipes, Euphractus, 97
gilviventris, Sciurus, 761
gilvus, Aceradon, 314
gilvus, Mus, 1398
gilvus, Perognathus, 858
gina, Gorilla, 182
ginginianus, Calloisciurus, 778
ginginianus, Xerus, 790
ginkgodens, Mesoplodon, *741*
Giraffa, *672*
giraffa, Giraffa, 672

Giraffidae, *672*
Giraffini, 672
girbanensis, Atelerix, 212
girensis, Rattus, 1484
girgentana, Capra, 702
glaber, Heterocephalus, *1542*
glacialis, Eubalaena, *723*
glacialis, Lepus, 196
glacialis, Myodes, 1027
glacialis, Orcinus, 731
glacialis, Sorex, 293
glacialis, Thomomys, 868
glacilis, Ursus, 588
Gladiator, 731
gladiator, Orcinus, 731
Gladobates, 105
glama, Lama, *646*
glandifera, Eonycteris, 321
glaphyrus, Oligoryzomys, 1142
glareolus, Myodes, *1023*
Glareomys, 1020
glasselli, Peromyscus, 1064
glassi, Crocidura, *232*
glauca, Hippotragus, 718
glaucillus, Pipistrellus, 479
glaucinus, Akodon, 1099
glaucinus, Eumops, *438*
glaucinus, Sciurillus, 757
Glaucomys, 755, *768*
Glauconycteris, 484, *486*
Glauconyina, 767
glauculus, Leopardus, 540
glaucus, Artibeus, *418*
glaucus, Canis, 576
glaucus, Chlorocebus, 159
glaucus, Eschrichtius, 726
glaucus, Pseudomys, *1457*
glauerti, Rattus, 1471
gleadowi, Gerbillus, *1227*
gleadowi, Millardia, *1386*
glebula, Crocidura, 232
gleimi, Genetta, 556
gleni, Glauconycteris, *487*
gleni, Miniopterus, *519*
Glipora, 105
Glires, 185, 746, 1538
Gliridae, 747, 752, 819, 819, 894, 905
gliriformis, Cercartetus, 45
glirina, Monodelphis, 14
Glirinae, *838*
Glirini, 819
glirinus, Phyllotis, 1163
glirinus, Zenkerella, 1534

Gliriscus, *822*, 828
Glirisorex, 105
Gliroidea, 819
gliroides, Chiropodomys, *1302*, 1372, 1423
gliroides, Dromiciops, *21*
gliroides, Microcebus, 113
gliroides, Octodontomys, *1572*
Glironia, 4
Glironiidae, 3
Glirulinae, 838
Glirulus, 821, 838, *839*
Glis, 799, 819, 829, 838, *840*, 917
glis, Glis, *840*
glis, Tupaia, 105, *106*, 107
Gliscebus, 112
Glischropus, *470*, 473, 484, 516
Glisorex, 105
Glisoricina, 104
Glisoricinae, 104
Glisosorex, 105
Globicephala, *728*
Globicephalidae, 726
Globiceps, 728
globiceps, Globicephala, 728
globiceps, Lepilemur, 118
globulus, Tolypeutes, 99
glogeri, Mustela, 616
Gloionycteris, 367
gloriaensis, Oryzomys, 1145
Glossonycteris, 398
Glossophaga, *399*
Glossophaginae, 396, *397*
Glossophagini, *397*
Glyphidelphis, 734
Glyphonycteris, *404*, 407
Glyphotes, 755, 778, *782*
gmelini, Crocidura, *232*
gmelini, Equus, 631
gmelinii, Ovis, 709
gnambiquarae, Kunsia, 1123
gnomus, Artibeus, *418*
gnou, Connochaetes, *676*, 676
gnu, Connochaetes, 676
gobabis, Lycaon, 581
gobicus, Lepus, 204

gobicus, Salpingotus, 881
gobiensis, Eptesicus, *455*
gobiensis, Ursus, 588
godeffroyi, Otaria, 593
godmani, Antechinus, *29*
godmani, Choeroniscus, *399*, 402
godmani, Martes, 610
godmani, Petrogale, *67*
godmani, Sorex, 296
godonga, Alcelaphus, 675
godowiusi, Alcelaphus, 675
goeldi, Oryzomys, 1151
goeldii, Callimico, *129*
goeldii, Proechimys, *1585*
goethalsi, Hoplomys, 1583
goffi, Geomys, 863
golani, Spalax, *919*
goldfinchi, Equus, 630
goldiei, Spilocuscus, 48
goldmani, Canis, 575
goldmani, Chaetodipus, *854*, 855
goldmani, Cratogeomys, *860*
goldmani, Cryptotis, *271*
goldmani, Dipodomys, 846
goldmani, Glaucomys, 768
goldmani, Heteromys, 850
goldmani, Mustela, 615
goldmani, Nelsonia, *1052*
goldmani, Neotoma, *1056*
goldmani, Ochotona, 191
goldmani, Ondatra, 1034
goldmani, Oryzomys, 1147
goldmani, Panthera, 547
goldmani, Peromyscus, 1074
goldmani, Proechimys, 1588
goldmani, Reithrodon-tomys, 1083
goldmani, Sciurus, 764
goldmani, Sigmodon, 1175
goldmani, Spermophilus, 804
goldmani, Sylvilagus, 208
goldmani, Thomomys, 870
goliath, Cricetomys, 933
goliath, Crocidura, *232*
goliath, Hyomys, 1312, *1338*
goliath, Lissonycteris, 325

Ia, **492**
Iacchus, 129
ibeae, Galerella, 566
ibeana, Crocidura, 234
ibeanus, Grammomys, **1325**
ibeanus, Ichneumia, 570
ibeanus, Lophiomys, 1047
ibeanus, Papio, 166
ibeanus, Paraxerus, 795
ibeanus, Perodicticus, 123
ibeanus, Tachyoryctes, **924**
iberica, Ovis, 708
ibericus, Bos, 692
ibericus, Microtus, 996
ibericus, Mustela, 617
ibericus, Sus, 642
Iberomys, 989, 994
Ibex, 700
ibex, Capra, **702**
ibex, Raphicerus, 688
ibicensis, Crocidura, 235
ibseni, Lagenorhynchus, 729
ica, Oryzomys, 1156
icarus, Otomops, 447
ichangensis, Elaphodus, 665
ichillus, Sphiggurus, **1548**
Ichneumia, **570**
Ichneumon, 567
ichneumon, Herpestes, **568**
ichneumon, Viverra, 567
ichnusae, Crocidura, **235**
ichnusae, Vulpes, 586
Ichthyomyini, 900, 1086, 1102, 1118, 1120, 1135, 1166
Ichthyomys, 900, 1110, **1120**, 1166
iconica, Gerbilliscus, 1220
iconicus, Apodemus, 1279
Ictailurus, 542
ictericus, Callosciurus, 778
Icticyon, 582
Ictides, 549
Ictidomoides, 803
Ictidomys, 798, **803**, 807
Ictidonyx, 607
Ictis, 613
Ictomys, 607
Ictonyx, **607**
iculisma, Crocidura, 250
ida, Cyclopes, 102

idahoensis, Brachylagus, **194**
idahoensis, Dipodomys, 847
idahoensis, Myodes, 1023
idahoensis, Perognathus, 858
idahoensis, Sorex, 286
idahoensis, Spermophilus, 809
idahoensis, Thomomys, **868**
idahoensis, Ursus, 589
idahoensis, Zapus, 893
Idionycteris, 480, **481**
Idiurinae, 1533
Idiurus, **1533**
Idomeneus, 1234
idoneus, Castor, 842
idoneus, Dipodomys, 847
idoneus, Melanomys, 1126
ifniensis, Lemniscomys, 1341
ifranensis, Apodemus, 1275
ighesicus, Chionomys, 969
igmanensis, Microtus, 992
ignava, Marmota, 802
Ignavus, 100
ignavus, Bradypus, 101
ignifer, Cephalophus, 715
ignifer, Rhinolophus, 359
ignita, Cercopithecus, 155
ignita, Presbytis, 172
ignitoides, Galerella, 565
ignitus, Acomys, **1196**
ignitus, Callicebus, 143
ignitus, Galerella, 566
ignitus, Gracilinanus, 7
ignitus, Sciurus, **761**
igniventris, Brucepatter-
    sonius, **1105**
igniventris, Sciurus, **762**
ignotus, Cervus, 664
ignotus, Neomys, 279
ignotus, Proechimys, 1588
ignotus, Sorex, 283
igromovi, Microtus, 1001
iharal, Hemitragus, 705
iheringi, Alouatta, 149
iheringi, Brucepatter-
    sonius, **1105**
iheringi, Monodelphis, 14
iheringi, Trinomys, **1590**
ikhwanius, Pipistrellus, 476
ikonnikovi, Myotis, **509**

ilaeus, Microtus, **1000**
ileile, Hyosciurus, **783**
ilensis, Crocidura, 232
ilensis, Spermophilus, 805
ileos, Microtus, 1001
ilex, Apodemus, 1264
ilici, Spalax, 920
iliensis, Ochotona, 187, **189**
ilimpiensis, Martes, 610
illapelinus, Phyllotis, 1162
illectus, Oecomys, 1138
illigeri, Saguinus, 134
illinensium, Didelphis, 6
illinoensis, Geomys, 862
illinoensis, Neotoma, 1055
illovoensis, Mastomys, 1365
illustris, Taterillus, 1244
illuteus, Abrothrix, **1089**
illuteus, Hydromys, 1333
illuteus, Lepus, 201
illyricus, Arvicola, 964
illyricus, Capreolus, 654
iltis, Mustela, 617
ilvanus, Apodemus, 1275
ilya, Eira, 606
imago, Aethomys, 1254
imaizumii, Mogera, **305**
imaizumii, Myodes, **1024**
imaizumii, Rhinolophus, **357**
imarius, Callosciurus, 779
imatongensis, Mus, 1411
imbaburae, Sciurus, 761
imbellis, Ammodillus, **1212**
imberbis, Capra, 702
imberbis, Muriculus, 1386, **1387**
imberbis, Saguinus, 134
imberbis, Saiga, 689
imberbis, Tragelaphus, **698**
imbil, Rattus, 1474
imbrensis, Scotomanes, 465
imbricatus, Hypsugo, **490**
imbutus, Metachirus, 12
imhausi, Lophiomys, **1047**
imitabilis, Thomomys, 867
imitator, Alticola, 960
imitator, Anisomys, 1257, **1258**
imitator, Callosciurus, 777

imitator, Cebus, 137
immunis, Thomomys, 869
imogene, Pharotis, **470**
imparilis, Cratogeomys, 861
impavidus, Canis, 575
impavidus, Marmosops, **10**, 11
imperator, Anomalurus, 1532
imperator, Saguinus, **134**
imperator, Uromys, **1515**
imperator, Ursus, 589
imperialis, Cervus, 664
imperii, Mustela, 614
impexus, Oxymicterus, 1118
imphalensis, Mus, 1392
impiger, Reithrodontomys, 1082
impiger, Ursus, 589
importunus, Rhinolophus, 352
improcera, Puma, 545
improvisum, Chiroderma, **421**
impudens, Macaca, 162
imus, Dremomys, 780
inambarii, Abrothrix, 1089
inambarus, Histiotus, 489
inaquosus, Dipodomys, 847
inas, Maxomys, **1369**, 1419
inaurata, Chrysochloris, 78
inaurea, Macaca, 164
inauris, Xerus, **790**
inaurita, Mellivora, 612
inca, Choeroniscus, 399
inca, Conepatus, 621
inca, Eptesicus, 454
inca, Lagidium, 1551
inca, Lestoros, **19**
inca, Lycalopex, 579
inca, Oxymycterus, **1158**
inca, Sylvilagus, 208
incae, Pteronotus, 427
incana, Ochotona, 191
incana, Pseudocheirus, 51
incanens, Pseudocheirus, 51
incanescens, Callicebus, 145
incanus, Marmosops, 11

kathleenae, Funambulus, 782
kathleenae, Millardia, **1386**
katinka, Crocidura, **236**
katschemakensis, Gulo, 607
katsurai, Mustela, 615
kattlo, Lynx, 541
kaufmanni, Equus, 630
kaufmanni, Hippotragus, 718
kaufmanni, Taurotragus, 697
kaukasikus, Bison, 690
kazakstanicus, Spermo-philus, 810
kazbegica, Sicista, **889**
kaznakovi, Alticola, 961
keasti, Nyctimene, **330**
keatii, Presbytis, 171
keaysi, Marmosops, 11
keaysi, Myotis, **509**
keaysi, Oryzomys, **1149**
keelingensis, Rattus, 1484
keenani, Mimon, 409
keeni, Peromyscus, **1069**
keenii, Myotis, **509**
keewatinensis, Rangifer, 660
keiensis, Dendrolagus, 60
keiensis, Echymipera, 42
Keitloa, 635
keitloa, Diceros, 635
kekrimus, Berylmys, **1297**
kelaarti, Felis, 535
kelaarti, Funambulus, 782
kelaarti, Pteropus, 337, 346
kelaarti, Rattus, 1484
kelaarti, Suncus, 260
kelaarti, Trachypithecus, 178
kelabit, Crocidura, 230
kelabit, Tupaia, 108
kelleni, Graphiurus, **824**
kelleni, Raphicerus, 688
kelleri, Capra, 702
kelleri, Rattus, 1490
kelloggi, Oryzomys, 1154
kelloggi, Sylvilagus, 208
kelloggi, Thomomys, 869
Kemas, 705
kemas, Pantholops, 711
kematoceros, Cervus, 664
kemmisi, Callosciurus, 776
kempi, Acomys, **1197**
kempi, Crocidura, 243

kempi, Deltamys, **1112,** 1131
kempi, Gerbilliscus, **1219**
kempi, Heterohyrax, 88
kempi, Leptailurus, 540
kempi, Otomys, 1526
kempi, Rousettus, 348
kempi, Thamnomys, **1511**
kenaiensis, Martes, 608
kenaiensis, Ovis, 710
kenaiensis, Tamiasciurus, 765
kenaiensis, Ursus, 588
kenaiensis, Vulpes, 584, 586
keniae, Alcelaphus, 674
keniae, Cephalophus, 713
keniae, Helioscurus, 794
keniae, Potamochoerus, 639
keniensis, Rhinolophus, 353
kennardi, Myodes, 1024
kennedyi, Oreamnos, 707
kennerleyi, Ursus, 589
kennethi, Berylmys, **1297**
kennicottii, Microtus, 1013
kennicottii, Spermophilus, 809
kennioni, Gazella, 680
kentucki, Synaptomys, 1038
Kentuckomys, 1037
kenyacola, Glauconyc-teris, **487**
kenyae, Ourebia, 686
kenyensis, Cricetomys, 932
kenyoni, Enhydra, 602
kephalopterus, Trachy-pithecus, 178
kerabau, Bubalus, 694
keramae, Cervus, 664
keraudren, Pteropus, 340
keraudreni, Pteropus, 339
keraudrenii, Callosciurus, 777
kerensis, Mastomys, 1365
kerguilensis, Mirounga, 598
Kerivoula, 486, **525**
Kerivoulinae, **525**
kerkhoveni, Arctictis, 549
kermanensis, Microtus, 1019
kermiti, Proechimys, 1588
kermodei, Ursus, 588
kernensis, Dipodomys, 846

kernensis, Microtus, 994
kernensis, Tamias, 815
Kerodon, 1555, **1556**
kerri, Phyllomys, **1580**
kerstingi, Erythrocebus, 160
kerstingi, Procavia, 88
kerulenica, Mustela, 617
kessleri, Lepus, 204
kessleri, Sciurus, 764
ketloa, Diceros, 635
kevella, Gazella, 681
keyensis, Pteropus, **338**
keyensis, Rhinolophus, 357
keyensis, Syconycteris, 349
keysseri, Thylogale, 69
khanensis, Galerella, 565
khanensis, Lepus, 202
khankae, Sorex, 298
kharanurensis, Microtus, 1010
khorkoutensis, Microtus, 992
khubsugulensis, Alticola, 962
khumbuensis, Mus, 1406
khumbuensis, Rattus, 1482
khur, Equus, 632
khyensis, Rattus, 1490
kiaboaba, Ceratotherium, 634
kiang, Equus, **632**
kibalensis, Okapia, 673
kiboko, Hippopotamus, 645
kibonotensis, Cercopithe-cus, 154
kibonotensis, Crocuta, 572
kidderi, Ursus, 588
kievensis, Erinaceus, 214
kihaulei, Myosorex, **265**
kijabae, Crocidura, 244
kijabius, Rattus, 1484
kikuchii, Microtus, **1001**
kikuyuensis, Colobus, 168
kikuyuensis, Otolemur, 126
kilangmiutak, Dicrosto-nyx, 971
kilikiae, Apodemus, 1279
kilimandjari, Dendromus, 937
kilimanus, Atelerix, 212
Kilimatalpa, 77, 78
kilisensis, Meriones, 1239
kima, Cercopithecus, 154
kina, Niviventer, 1420

kinabalu, Megaderma, 380
kinabaluensis, Maxomys, 1368
kincaidi, Microtus, 1012
kinczumi, Crocidura, 229
kindae, Papio, 166
kinezumi, Crocidura, 229
kingiana, Crocidura, 225
kingii, Macropus, 65
kingii, Microcavia, 1554
Kinkajou, 626
Kinkaschu, 626
kinlochii, Petaurillus, **770**
kinnamon, Myotis, 516
kinneari, Callosciurus, 776
kinneari, Rhinopoma, 381
kinoensis, Ammospermo-philus, 797
kinoensis, Perognathus, 858
kiodotes, Macroglossus, 325
Kiodotus, 325
kirchenpaueri, Neotragus, 685
kirgisica, Ovis, 708
kirgisorum, Microtus, 1001
kirgisorum, Spalax, 918
kirivoula, Kerivoula, 528
kiriwinae, Phalanger, 46
kirkii, Hippotragus, 718
kirkii, Madoqua, **683**
kirkii, Otolemur, 126
kirkii, Piliocolobus, **169,** 170
kirschbaumii, Plecotus, 483
kirtlandi, Blarina, 269
kishidai, Lasiopodomys, 985
kisnyiresiensis, Rhinolo-phus, 357
kitcheneri, Hypsugo, **491**
kittlitzi, Lemmus, 988
kiusiuana, Mogera, 306
kivanci, Allactaga, 875
kivu, Crocidura, 244
kivu, Dendromus, 940
kivuana, Crocidura, **236**
kivuana, Hydrictis, 602
kivuensis, Cephalophus, 714
kivuensis, Cricetomys, **933**
kivuensis, Idiurus, 1533
kivuensis, Leptailurus, 540

lakkhanae, Rhinolophus,
   361
lalandei, Lycaon, 581
lalandi, Otocyon, 582
lalandia, Redunca, 722
lalandianus, Graphiurus,
   826
lalandii, Chlorocebus,
   159
lalandii, Megaptera, 725
lalandii, Proteles, 573
lalawora, Maxomys,
   1370
lalisio, Equus, 631
lalolis, Rattus, 1490
laloumena, Hippopota-
   mus, 644
Lama, **646,** *646*
lama, Alticola, 961
lama, Cricetulus, 1042
lama, Lama, 647
lama, Ochotona, 191
lamarum, Phyllomys,
   1581
lambertoni, Nesomys,
   952
lambi, Chaetodipus, 856
lambi, Molossus, 440
lambi, Myotis, 518
lambi, Oryzomys, 1147
lami, Ctenomys, **1565**
lamia, Chiruromys, **1304**
lamia, Oryzomys, **1149**
Lamictis, 552
laminatus, Otomys, **1528**
lamington, Uromys, 1514
Lamingtona, 468
Lamini, 645
Laminungula, 87
Lamnungia, 87
Lamnunguia, 87
Lamotomys, 1258, 1524
lamottei, Crocidura, **236**
lamottei, Hipposideros,
   373
lamottei, Micropotamo-
   gale, **75**
Lamottemys, 903, **1339,**
   1429
lampensis, Tragulus, 650
lampo, Lenomys, 1345
Lamprogale, 608, 609
Lampronycteris, **404,** 407
lamucotanus, Callosciu-
   rus, 778
lamucotanus, Rattus,
   1492
lamula, Chodsigoa, **276**
lanaceus, Myotis, 506
lanata, Pelea, 721
lanatus, Avahi, 119

lanatus, Caluromys, **3**
lanatus, Myotis, 505
lanatus, Phyllotis, 1163
lanatus, Thylogale, **70**
lancasteri, Chaerephon,
   433
lancasteri, Galerella, 566
lancasteri, Ictonyx, 608
lancavensis, Callosciurus,
   776
lancavensis, Leopolda-
   mys, 1348
lancavensis, Tragulus,
   650
lanceolatum, Carollia,
   412
lanceolatus, Myotis, 510
landakkensis, Pongo, 184
landbecki, Abrothrix,
   1090
landeri, Rhinolophus, **357**
landiana, Redunca, 722
lanea, Acinonyx, 533
lanei, Mops, 443
lanensis, Pteropus, 346
lanensis, Rattus, 1490
laneus, Paradoxurus, 551
langbianis, Niviventer,
   1301, **1423**
langguthi, Deltamys,
   1112
langheldi, Colobus, 168
langheldi, Erythrocebus,
   160
langheldi, Eudorcas, 679
langheldi, Hippotragus,
   717
langheldi, Lycaon, 581
langheldi, Papio, 166
langi, Atelerix, 212
langi, Chaerephon, 435
langi, Cricetomys, 933
langi, Crocidura, 234
langi, Cryptomys, 1540
langi, Elephantulus, 82
langi, Hipposideros, 370
langi, Idiurus, 1533
langi, Lepus, 197
langi, Macroscelides, 84
langi, Madoqua, 683
langi, Piliocolobus, 169
langi, Rhynchogale, 571
langkatensis, Pongo, 184
langsdorffi, Chironectes, 4
langsdorffi, Sciurus, 763
lania, Semnopithecus,
   175
laniarius, Canis, 576
laniarius, Sarcophilus, 28
laniger, Abrocoma, 1574
laniger, Alouatta, 150

laniger, Anomalurus,
   1532
laniger, Antechinomys, **32**
laniger, Avahi, **119**
laniger, Canis, 576
laniger, Cuon, 578
laniger, Daubentonia,
   121
laniger, Dryomys, **830**
laniger, Macropus, 66
laniger, Mus, 1550
laniger, Myotis, **509**
laniger, Paguma, 550
laniger, Pteropus, 338
laniger, Thomasomys,
   1182
laniger, Ursus, 590
lanigera, Caluromys, 3
lanigera, Capra, 702
lanigera, Chinchilla,
   1550
lanigera, Oreamnos, 707
lanigera, Paguma, 550
lanigerus, Macropus, 66
laniginosa, Pseudochei-
   rus, 51
lanius, Aotus, 140
lanius, Orthogeomys, **865**
lanka, Canis, 574
lanka, Herpestes, 568
lanka, Petaurista, 772
lankadiva, Hipposideros,
   373
lankavensis, Viverra, 559
lanosa, Crocidura, **236**
lanosa, Kerivoula, **527**
lanosa, Murina, 523
lanosus, Abrothrix, **1089**
lanosus, Macropus, 66
lanosus, Mammelomys,
   1358, 1432
lanosus, Rhinolophus,
   358
lanosus, Rousettus, **348**
lanuginose, Mystromys,
   946
lanuginosus, Millardia,
   1386
lanuginosus, Tamiasciu-
   rus, 765
laomache, Dremomys,
   781
Laomys, 1520
laosiensis, Bos, 690
laotum, Hylopetes, 769
laotum, Melogale, 613
laotum, Paradoxurus, 551
laotum, Tamiops, 789
laotum, Trachypithecus,
   176, **177**
laotum, Tupaia, 105

lapidarius, Saccostomus,
   934
laplataensis, Neotoma,
   1053
lapponicus, Sorex, 286
lapponum, Rangifer, 660
lapsus, Macaca, 162
laptevi, Odobenus, 595
lar, Crocidura, 232
lar, Hylobates, **179,** 180
larapinta, Sminthopsis,
   35
Laratus, 179
lardarius, Nyctalus, 472
larensis, Myotis, 513
largha, Phoca, **599**
largus, Dipodomys, 845
Laria, 783
laricorum, Sorex, 284
Lariscus, 755, **783**
laristanica, Ovis, 709
larkeni, Poelagus, 206
larkenii, Tragelaphus,
   699
laronesiotes, Rusa, 670
larseni, Leptailurus, 540
lartetiana, Ovis, 709
larus, Apodemus, 1275
larvalis, Pedetes, 1535
larvarum, Chodsigoa, 276
larvata, Didelphis, 17
larvata, Paguma, **550**
larvatus, Ellobius, 976
larvatus, Hipposideros,
   373
larvatus, Mustela, 614
larvatus, Nasalis, 168,
   169
larvatus, Potamochoerus,
   638, **639**
larvatus, Pseudochirulus,
   52
larvatus, Rhinolophus,
   360
lasalensis, Ochotona, 191
lasallei, Tremarctos, 587
lascivus, Glaucomys, 768
lasia, Crocidura, 237
lasiae, Macaca, 162
lasiae, Megaderma, 380
lasiae, Rattus, 1488
Lasiomys, 1173, 1202,
   1202
Lasionycteris, **499**
Lasiopodomys, 898, 966,
   984, 989, 1035, 1036
lasiopterus, Nyctalus, **471**
Lasiopus, 570
Lasiopyga, 154
lasiopyga, Anoura, 398
Lasiorhinus, **43**

mangivorus, Odocoileus, 658
manguensis, Lycaon, 581
Mangusta, 567
manhanensis, Sciurus, 762
manicalis, Maxomys, 1372
manicatus, Macropus, 65
manicatus, Rattus, 1471
manicorensis, Callithrix, 131
maniculata, Felis, 537
maniculatus, Peromyscus, 1071
maniculatus, Rattus, 1478
Manidae, 530
maniemae, Colobus, 168
manipulus, Berylmys, 1296, 1297
manipurensis, Tamiops, 789
Manis, 530
manitobensis, Blarina, 269
manitobensis, Cervus, 663
manium, Mustela, 617
manium, Nasua, 626
manni, Crocidura, 244
manningi, Canis, 576
manningi, Heterohyrax, 88
manoquarius, Rattus, 1469
mansalaris, Leopoldamys, 1348
mansalaris, Macaca, 162
mansalaris, Sundasciurus, 788
mansoensis, Abrothrix, 1090
mansorius, Rattus, 1490
mansuetus, Potos, 627
mansuetus, Sylvilagus, 208, 210
mansumensis, Crocidura, 231
mantchurica, Ochotona, 189
mantchuricus, Apodemus, 1261
mantchuricus, Cervus, 664
mantchuricus, Sciurus, 764
manteufeli, Aethomys, 1256
manteufeli, Lemniscomys, 1344

manteufeli, Lophuromys, 1203
mantiqueirensis, Phyllomys, 1581
mantschuricus, Capreolus, 655
manul, Felis, 535
manuselae, Rattus, 1477
manyarae, Eudorcas, 679
manyema, Gorilla, 182
maorium, Rattus, 1469
mapiriensis, Micoureus, 13
mapogonensis, Elephantulus, 83
maporensis, Callosciurus, 778
maporensis, Crocidura, 242
mapravis, Callosciurus, 776
mapurita, Spilogale, 623
mapurito, Conepatus, 621, 622
maputa, Gerbilliscus, 1218
maquassiensis, Crocidura, 239
Mara, 1555
mara, Rattus, 1492
marabutus, Xerus, 790
maracaibensis, Sciurus, 761
maracajuensis, Oryzomys, 1150
maracaya, Leopardus, 539
maradius, Gerbillus, 1224
marae, Galerella, 566
maraisiana, Rusa, 670
marakovici, Dinaromys, 974
maral, Cervus, 663
marana, Ratufa, 756
maranii, Marmosa, 9
maranonicus, Oligoryzomys, 1140
maranonis, Cebus, 137
maraxica, Dasyprocta, 1557
maraxina, Monodelphis, 15
marcai, Callithrix, 132
marcanoi, Solenodon, 223
marcarum, Calomys, 1108
marchei, Mydaus, 622, 623
marchei, Sus, 641
marchicus, Sorex, 283

marchio, Petaurista, 772
marcolinus, Capricornis, 704
marcosensis, Neotoma, 1056
marcrurus, Mungos, 571
marculus, Paranyctimene, 332
mareeba, Petrogale, 68
maresi, Tonatia, 411
Maresomys, 1102, 1103
mareyi, Bradypus, 100
margae, Presbytis, 171
Margaretamys, 904, 1321, 1359, 1368, 1369
margarettae, Haeromys, 1327, 1328
margarettae, Lophuromys, 1202
margarita, Felis, 536
margarita, Sorex, 298
margarita, Trachypithecus, 176
margaritae, Cebus, 137
margaritae, Chaetodipus, 856
margaritae, Dipodomys, 846
margaritae, Felis, 536
margaritae, Odocoileus, 658
margaritae, Peromyscus, 1071
margaritae, Sylvilagus, 209
Margay, 537
margay, Leopardus, 539
margianus, Jaculus, 883
marginae, Meriones, 1236
marginata, Caperea, 726
marginata, Felis, 536
marginatus, Ateles, 150, 151
marginatus, Cynopterus, 316, 317, 350
marginatus, Pipistrellus, 476
marginatus, Rousettus, 348
marginatus, Spermophilus, 811
marginatus, Zaedyus, 97
marguerittei, Felis, 536
maria, Kobus, 721
mariae, Equus, 630
mariae, Gerbillus, 1227
mariae, Meriones, 1236
mariae, Microtus, 1005
mariae, Steatomys, 944

mariakanae, Elephantulus, 83
marianensis, Meles, 612
marianna, Rusa, 669
mariannus, Pteropus, 340
marica, Eligmodontia, 1113
marica, Gazella, 682
marica, Gracilinanus, 7
marica, Mus, 1397
marica, Nycteris, 392
marica, Petaurista, 771
marica, Vandeleuria, 1518
maricelae, Glossophaga, 400
mariepsi, Otomys, 1528
mariguanensis, Erophylla, 397
Marikina, 133
marikina, Leontopithecus, 133
marikquensis, Mastomys, 1362
marina, Enhydra, 602
marinensis, Chaetodipus, 853
marinhus, Oryzomys, 1150
marinkellei, Lonchorhina, 405
marinsularis, Callosciurus, 778
marinus, Lutra, 604
marinus, Niviventer, 1422
marinus, Procyon, 628
marinus, Reithrodon, 1165
marinus, Ursus, 589
maripensis, Leopardus, 539
maripensis, Myotis, 513
mariposae, Microtus, 994
mariposae, Peromyscus, 1064
mariposae, Sorex, 298
mariposae, Sylvilagus, 208
mariposae, Tamias, 815
mariquensis, Crocidura, 240
maris, Pteropus, 338
marisae, Hipposideros, 374
marisalbi, Delphinapterus, 735
marita, Crocidura, 232
maritanus, Sus, 641
maritima, Antechinus, 30

nigricauda, Thallomys, 1508, **1509**
nigricaudata, Ourebia, 687
nigricaudatus, Galerella, 566
nigricaudatus, Heteromys, 850
nigricaudatus, Lepus, 196
nigricaudatus, Petaurista, 772
nigricaudatus, Sylvilagus, 208
nigricaudus, Gerbilliscus, **1220**
nigricaudus, Ictonyx, 608
nigricaudus, Petinomys, 773
nigriceps, Aotus, **140**
nigriceps, Callithrix, **132**
nigriceps, Ctenomys, 1567
nigriceps, Hemicentetes, **76**
nigriceps, Paguma, 550
nigriceps, Saimiri, 138
nigriclunis, Dasyprocta, 1558
nigricollis, Lepus, 198, **201**, 202
nigricollis, Saguinus, **135**
nigricollis, Tragulus, 650
nigricolor, Crossarchus, 563
nigriculus, Mesocricetus, 1045
nigriculus, Peromyscus, 1067
nigriculus, Sorex, 286
nigridius, Dasymys, 1315
nigridorsalis, Callosciurus, 776
nigrifons, Paradoxurus, 551
nigrifrons, Callicebus, **144**, 145
nigrifrons, Cephalophus, **714**
nigrifrons, Dendromus, 938
nigrifrons, Eulemur, 115
nigrifrons, Oxymycterus, 1158
nigrifrons, Saguinus, 134
nigrigenis, Cercopithecus, 157
nigrimanus, Piliocolobus, 169
nigrimanus, Presbytis, 172
nigrimontana, Ovis, 708

nigrimontanus, Spermophilus, 807
nigrimontis, Chaetodipus, 855
nigrinotatus, Tragelaphus, 698
nigripecta, Felis, 535
nigripectus, Cebus, 137
nigripes, Bdeogale, **563**
nigripes, Cercopithecus, 158
nigripes, Crocidura, **243**
nigripes, Dologale, 564
nigripes, Felis, **536**
nigripes, Hylopetes, **769**
nigripes, Lemmus, 988
nigripes, Macrotis, 38
nigripes, Muntiacus, 667
nigripes, Mustela, **616**
nigripes, Neomys, 279
nigripes, Oligoryzomys, **1143**
nigripes, Oryctolagus, 206
nigripes, Papio, 167
nigripes, Procyon, 627
nigripes, Pygathrix, **173**
nigripes, Sciurus, 761
nigripes, Sus, 642
nigrirostris, Urocyon, 582
nigriseta, Petrodromus, 84
nigrispinus, Phyllomys, **1581**
nigrita, Gerbilliscus, 1219
nigrita, Scotophilus, **467**
nigrita, Thaptomys, 1098, **1179**
nigritalus, Apodemus, 1271
nigritellus, Scotophilus, 467
nigritia, Ochotona, 188, **190**
nigritus, Cebus, **137**
nigrivittata, Saimiri, 139
nigrivittatus, Cebus, 138
nigro-aculeatus, Zaglossus, 1
nigro-argenteus, Vulpes, 585
nigro-fasciatus, Erythrocebus, 160
nigrocaudatus, Vulpes, 585
nigrocinctus, Tragulus, 650
nigrofulvus, Proechimys, 1588
nigrofusca, Crocidura, **243**
nigrofuscus, Myotis, 512

nigrogriseus, Chalinolobus, **484**
nigrogriseus, Tadarida, 451
nigronuchalis, Sylvilagus, 209
nigroscapulatus, Kobus, 720
nigrotibialis, Gerbilliscus, 1220
nigroviridis, Allenopithecus, **153**
nigrovittatus, Callosciurus, **777**
niigatae, Myodes, 1022
nikkonis, Petaurista, 771
nikolajevi, Microtus, 1016
nikolausi, Megadendromus, **942**
nikolskii, Apodemus, 1261
nikolskii, Mustela, 617
nikolskii, Spermophilus, 810
nilagirica, Vandeleuria, 1393, 1488, **1517**
nilgirica, Suncus, 258
Nilopegamys, 902, **1417**
nilosa, Coleura, 386
nilotica, Crocidura, 243
nilotica, Felis, 535
nilotica, Hydrictis, 602
niloticus, Arvicanthis, 1285, **1289**
niloticus, Papio, 166
niloticus, Vulpes, 586
nilssonii, Eptesicus, **456**
nimbae, Crocidura, **243**
nimbasilvanus, Crocidura, 232
nimbosus, Oryzomys, 1146
nimr, Panthera, 547
nimrodi, Cryptomys, 1540
Ningaui, **32**, 33
ningbing, Pseudantechinus, **27**
ninglangensis, Eothenomys, 979
ningpoensis, Apodemus, 1261
ningpoensis, Callosciurus, 776
ningshaanensis, Cansumys, 1041, 1046
ningshaanensis, Tscherskia, 1046
Niniventer, 1314, 1319, 1345, 1368
ninus, Niviventer, 1421

niobe, Crocidura, **243**
niobe, Lariscus, **783**
niobe, Rattus, 1416, **1476**
nipalensis, Macaca, 163
nipalensis, Melogale, 613
nipalensis, Mus, 1399
nipalensis, Myotis, **513**
nipalensis, Paguma, 550
nipalensis, Prionailurus, 543
nipalensis, Semnopithecus, 175
niphanae, Megaerops, **326**
niphoecus, Ovibos, 707
nippon, Cervus, 655, 662, **663**
nippon, Lutra, **604**
nippon, Mustela, 614
nippon, Rhinolophus, 355
Nipponanthropus, 182
nipponicus, Equus, 631
nipponicus, Sus, 642
nirnai, Aonyx, 602
nisa, Crocidura, 245
nitedula, Dryomys, 830, **831**
nitedulus, Oecomys, 1137
nitela, Eliomys, 835
nitela, Rhipidomys, **1171**
nitela, Sminthopsis, 36
nitela, Thallomys, 1509
nitellinus, Nyctomys, 1187
nitendiensis, Pteropus, **341**
nitens, Eptesicus, 453
nitida, Ochotona, 185
nitida, Petaurista, 772
nitida, Tupaia, 108
nitidofulva, Suncus, 258
nitidula, Petaurista, 772
nitidulus, Mus, 1392
nitidus, Myotis, 504
nitidus, Oryzomys, **1152**
nitidus, Pipistrellus, 479
nitidus, Rattus, **1477**
nitratoides, Dipodomys, **847**
nitratus, Dipodomys, 846
nivalis, Chionomys, **969**, 1000
nivalis, Leptonycteris, 400, **401**
nivalis, Microtus, 970
nivalis, Mustela, **616**
nivaria, Marmota, 800
nivarius, Myodes, 1023
nivatus, Uropsilus, 310

obscurus, Rattus, 1469
obscurus, Reithrodon, 1165
obscurus, Reithrodon-tomys, 1085
obscurus, Rhinolophus, 355
obscurus, Sorex, 293
obscurus, Spermophilus, 804
obscurus, Sylvilagus, **210**
obscurus, Tamias, **815**
obscurus, Theropithecus, 167
obscurus, Trachypithecus, **177**
obscurus, Tragelaphus, 698
observandus, Orycteropus, 86
obsidianus, Spermophilus, 811
obsoletus, Rattus, 1477
obsoletus, Spermophilus, 811
obtusa, Phyllonycteris, 397
obtusata, Cephalorhyn-chus, 727
obtusata, Phocaena, 736
obtusirostris, Calcochlo-ris, **81**
obtusirostris, Oryzomys, 1149
obvelatus, Sigmodon, 1176
occasius, Makalata, **1579**
occidanea, Hystrix, 1543
occidentalis, Arvicanthis, 1292
occidentalis, Avahi, **119**
occidentalis, Canis, 576
occidentalis, Cervus, 663
occidentalis, Chaeropus, 39
occidentalis, Chionomys, 970
occidentalis, Colobus, 168
occidentalis, Crocidura, 244
occidentalis, Ctenomys, 1569
occidentalis, Dactylopsila, 54
occidentalis, Diceros, 636
occidentalis, Dinomys, 1552
occidentalis, Dipodomys, 847
occidentalis, Eliomys, 834

occidentalis, Erinaceus, 214
occidentalis, Galidia, 561
occidentalis, Hapalemur, **116**
occidentalis, Heliosciurus, 794
occidentalis, Lagurus, 984
occidentalis, Lepus, 198
occidentalis, Loncho-rhina, 405
occidentalis, Lophostoma, 406
occidentalis, Macropus, 66
occidentalis, Macrotarso-mys, 951
occidentalis, Mephitis, 622
occidentalis, Microgale, 72
occidentalis, Microtus, 1019
occidentalis, Miniopterus, 521
occidentalis, Mops, **442**
occidentalis, Mustela, 614, 617
occidentalis, Myodes, 1023
occidentalis, Myotis, 512
occidentalis, Nanger, 684
occidentalis, Neotoma, 1055
occidentalis, Oecomys, 1137
occidentalis, Oligoryz-omys, 1141
occidentalis, Otomys, **1528**
occidentalis, Ovis, 709
occidentalis, Papio, 167
occidentalis, Petrodromus, 84
occidentalis, Phalanger, 48
occidentalis, Prosciurillus, 785
occidentalis, Pseudochei-rus, 51
occidentalis, Pseudohy-dromys, **1453**
occidentalis, Pseudomys, **1459**
occidentalis, Redunca, 722
occidentalis, Scotonyc-teris, 349
occidentalis, Sturnira, 415
occidentalis, Talpa, **309**

occidentalis, Tupaia, 106
occidentalis, Rousettus, 347
occidentosardinensis, Ovis, 709
occiduus, Gerbillus, **1230**
occipitales, Thomomys, 866
occipitalis, Mops, 444
occipitalis, Ondatra, 1034
occisor, Mustela, 615
occlusus, Cryptomys, 1540
occulta, Feresa, 728
occultidens, Suncus, 260
occultus, Chaetodipus, 856
occultus, Ctenomys, **1567**
occultus, Mesomys, **1584**
occultus, Myotis, **514**
occultus, Promops, 448
occultus, Scapanus, 302
oceana, Rusa, 671
oceanensis, Miniopterus, 522
oceanica, Pagophilus, 599
oceanicus, Peromyscus, 1069
oceanitis, Hipposideros, 371
ocellatus, Marmosops, 11
ocelot, Leopardus, 539
Ochetodon, 1080
Ochetomys, 963
ochotensis, Phoca, 599
ochotensis, Pusa, 600
ochoterenai, Balantiop-teryx, 385
Ochotona, **185**
Ochotonida, 185
Ochotonidae, **185**
ochracea, Galerella, 564, **564**
ochracea, Marmota, 802
ochracea, Martes, 610
ochracea, Neotoma, 1058
ochracea, Poiana, 558
ochraceiventer, Maxomys, **1370**
ochraceocinereus, Cryp-tomys, **1541**
ochraceus, Ailurus, 628
ochraceus, Capreolus, 655
ochraceus, Grammomys, 1326
ochraceus, Mops, 442
ochraceus, Myodes, 1023
ochraceus, Papio, 166
ochraceus, Paraxerus, **796**
ochraceus, Peromyscus, 1070

ochraceus, Procyon, 628
ochraceus, Proechimys, 1586
ochraceus, Saguinus, 135
ochraceus, Sigmodon-tomys, 1178
ochraceus, Tamias, 814
ochraphaeus, Acerodon, 314
ochraspis, Petaurista, 771
ochraventer, Peromyscus, **1075**
ochreata, Macaca, **163**
ochrescens, Sciurus, 761
ochrinus, Sigmodontomys, 1178
ochrocephala, Pithecia, 148
ochrogaster, Funisciurus, 791
ochrogaster, Microtus, **1009**
ochrogaster, Rhipidomys, **1171**
ochrogenys, Tamias, **815**
ochrognathus, Sigmodon, **1176**
ochroleucus, Pan, 183
ochromelas, Galerella, 566
ochromixtus, Hypsugo, 491
Ochromys, 1520
ochropoides, Lepus, 197
ochropus, Arvicanthis, 1290
ochropus, Caluromys, 3
ochropus, Canis, 575
ochropus, Dendromus, 940
ochropus, Lepus, 197
ochropus, Spilocuscus, 48
ochrotis, Abrothrix, 1089
Ochrotomyini, 899, 1048, 1060
Ochrotomys, 899, **1060**, 1062
ochrourus, Odocoileus, 658
ochroxantha, Vulpes, 586
ochrus, Chaetodipus, 853
ocius, Thomomys, 869
oconnelli, Oryzomys, 1145
oconnelli, Proechimys, **1587**
ocotensis, Microtus, 1006
ocotepequensis, Reithro-dontomys, 1083
ocreata, Felis, 537
octavus, Bassariscus, 625

Paracyon, 23
Paradipodinae, 871, 884
Paradipodini, 882
Paradipus, **884**
paradisus, Brucepatter-
    sonius, **1105**
paradoxa, Rusa, 670
paradoxa, Selevinia, 838
Paradoxodon, 257
paradoxolophus, Rhinolo-
    phus, **360**
paradoxura, Crocidura,
    245
Paradoxurida, 549
Paradoxurinae, 548, **549**
Paradoxurini, 549
Paradoxurus, 549, **550**,
    552
paradoxus, Cardiocra-
    nius, **880**
paradoxus, Chaetodipus,
    854
paradoxus, Microtus,
    1012
paradoxus, Ornithorhyn-
    chus, 2
paradoxus, Pseudochi-
    rops, 53
paradoxus, Reithrodon-
    tomys, **1084**
paradoxus, Solenodon,
    222, **223**
Paraechinus, 215, **216**
paraensis, Mustela, 613
paraensis, Sciurus, 761
Paragalia, 38
paraganus, Oryzomys,
    1145
paragayensis, Sphiggurus,
    1549
Paragenetta, 554
Parageomys, 861
Paraglirulus, 839
paraguae, Leopardus, 538
paraguanensis, Pterono-
    tus, 428
paraguayanus, Cebus,
    137
paraguayanus, Micoureus,
    13
paraguayensis, Dasy-
    procta, 1556
paraguayensis, Didelphis,
    5
paraguensis, Chironectes, 4
paraguensis, Panthera,
    547
paraguensis, Pteronura,
    605
paraguensis, Sylvilagus,
    208

Parahodomys, 1053
Parahyaena, 572
Parahydromys, 903,
    1312, 1333, 1339,
    1430
Paralactaga, **874**
Paralariscus, 783
Paralces, 653
Paraleporillus, 1453
Paraleptomys, 903, **1431**
paralios, Chaetodipus,
    853
paralius, Dipodomys, 848
paralius, Liomys, 852
parallelus, Tragulus, 650
Paralomys, 900, **1160**
Paramanis, **530**
Paramelomys, 905, 1374,
    1432, 1451
paramensis, Oxymyc-
    terus, **1158**
Parameriones, **1234**
Paramicrogale, 72
paramicrus, Nesophontes,
    221
paramorum, Thomas-
    omys, **1183**
Paramurexia, **31**
Paramyotis, 500
paranaensis, Akodon,
    1098
paranalis, Lutreolina, 8
paranensis, Lontra, 603,
    604
paranensis, Pteronura,
    605
Parantechinus, **26**, 27, 28
paranus, Cynomops, **437**
Paranyctimene, 313, **332**
Paraonyx, 601, 601
Paraphenacomys, 962
Parapitymys, 989
Parapodemus, 1248
Parascalopina, 301
Parascalops, **301**
Parascaptor, **307**
Parasciurus, 758
Parascotomanes, 492
parasiticus, Peromyscus,
    1064
Parastrellus, 489
parata, Marmosa, 9
Paratatera, **1221**
paratus, Trinomys, **1591**
paraxanthus, Rattus,
    1494
Paraxerus, 755, **795**
parca, Chodsigoa, **277**
parcus, Rhinolophus, 361
parda, Micoureus, 13
Pardalina, 537

Pardalis, 537
pardalis, Leopardus, 537,
    539, 539
pardella, Lynx, 541
pardicolor, Prionodon,
    553
Pardictis, 553
pardictis, Leopardus, 540
Pardina, 541
pardina, Genetta, **556**
pardinoides, Leopardus,
    539
pardinus, Lynx, **541**
pardochrous, Prionailu-
    rus, 542, 543
pardochrous, Prionodon,
    553
Pardofelis, **542**
pardoides, Leopardus, 538
Pardotigris, 546
Pardus, 546
pardus, Panthera, 546,
    547
Pareptesicus, 452
paricola, Oecomys, **1138**
parienti, Phaner, **114**
parilina, Lontra, 603
parisii, Nycteris, **393**
parkeri, Platymops, 448
parleppi, Leopardus, 539
parma, Macropus, **65**
parmentierorum, Pilioco-
    lobus, 169
parnabyi, Hipposideros,
    378
parnassica, Ovis, 709
parnassius, Lepus, 198
parnellii, Pteronotus, **427**
Parneotoma, 1053
Parotomys, 905, 1523,
    1524, **1531**
parowanensis, Thom-
    omys, 869
parryi, Macropus, **65**
parryii, Spermophilus,
    809
parsonsii, Erignathus,
    596
parthianus, Spermophi-
    lus, 807
particeps, Didelphis, 6
Paruromys, 904, **1435**,
    1506
parva, Arctocephalus, 591
parva, Chodsigoa, **277**
parva, Condylura, 301
parva, Cryptotis, 270,
    273
parva, Myoprocta, 1559
parva, Phascomurexia, 31
parva, Rupicapra, 711

parvabullatus, Dipod-
    omys, 845
parvacauda, Crocidura,
    226
parvicaudatus, Sorex, 298
parviceps, Cratogeomys,
    860
parviceps, Liomys, 852
parviceps, Lontra, 603
parviceps, Perognathus,
    857
parviceps, Thomomys,
    870
parvidens, Alticola, 959
parvidens, Cheiromeles,
    436
parvidens, Cynomys, **799**
parvidens, Euroscaptor,
    305
parvidens, Herpestes, 568
parvidens, Lycalopex, 581
parvidens, Marmosops,
    10, **11**, 12
parvidens, Myodes, 1028
parvidens, Neotoma,
    1058
parvidens, Sorex, 293
parvidens, Spermophilus,
    808
parvidens, Sturnira, 414
parvipes, Artibeus, 419
parvipes, Crocidura, **245**
parvipes, Galerella, 566
parvipes, Miniopterus,
    522
parvipes, Moschus, 652
parvipes, Nectomys, 1133
parvipes, Pipistrellus, 474
parvula, Helogale, **566**
parvula, Marmota, 801
parvula, Microgale, **74**
parvula, Rhogeessa, **463**
parvulus, Apodemus,
    1277
parvulus, Graphiurus,
    822
parvulus, Lophuromys,
    1207
parvulus, Micromys, 1384
parvulus, Microtus, 1013
parvulus, Mus, 1399
parvulus, Myotis, 513
parvulus, Spermophilus,
    804
parvulus, Sylvilagus, 208
parvulus, Thomomys,
    867
parvus, Apodemus, 1272,
    1275
parvus, Burramys, **44**
parvus, Clidomys, 1599

septentrionalis, Spermo-
philus, 810
septentrionalis, Tadarida,
450
septentrionalis, Tamias,
814
septentrionalis, Vulpes,
585
septicus, Rattus, 1490
sequanicus, Equus, 631
sequanius, Equus, 631
sequoiensis, Peromyscus,
1079
sequoiensis, Tamias, 817
sequoiensis, Urocyon, 582
seraiae, Callosciurus, 778
seramensis, Suncus, 260
serasani, Cynopterus, 317
serbicus, Spalax, 920
serengetae, Nanger, 684
serezkyensis, Crocidura,
248
sergii, Mus, 1408
seri, Chaetodipus, 855
seri, Neotoma, 1053
sericatus, Bunomys, 1300
sericatus, Scapanus, 302
sericea, Capra, 702
sericea, Scalopus, 301
sericeus, Callithrix, 130
sericeus, Crocidura, 231
sericeus, Ctenomys, **1569**
sericeus, Oreamnos, 707
sericeus, Phalanger, **47**
sericeus, Propithecus, 120
sericeus, Sphiggurus,
1549
Sericonycteris, 334
sericus, Bunopithecus,
178
serii, Emballonura, **389**
seringetica, Eudorcas, 679
sernaria, Aonyx, 602
serotine, Eptesicus, 457
serotinus, Eptesicus, **456**
serotinus, Sturnira, 414
serotinus, Trinomys, 1590
serpens, Microtus, 1011
serpentarius, Suncus, 260
serratus, Dasypus, 95
serratus, Peromyscus,
1071
serratus, Scotophilus, 467
serratus, Taphozous, 384
serrensis, Akodon, **1098**
serridens, Lobodon, 597
serutus, Callosciurus, 778
serutus, Maxomys, 1372
Serval, 540
serval, Leptailurus, **540**
Servalina, 540

servalina, Felis, 537
servalina, Genetta, **557**
servalinus, Leptailurus,
540
servalinus, Prionailurus,
543
servorum, Cricetomys,
933
sestinensis, Reithrodon-
tomys, 1083
setchuanus, Eozapus,
891, **892**
Setifer, **77**
setifera, Bandicota, 1294
setiger, Leopoldamys,
1346
setiger, Platymops, **448**
setonbrownei, Gerbillus,
1229
setoni, Rangifer, 660
Setonix, **69**
setosa, Bettongia, 57
setosa, Tachyglossus, 1
setosus, Arvicanthis,
1292
setosus, Euphractus, 97
setosus, Liomys, 852
setosus, Paradoxurus, 551
setosus, Petinomys, **774**
setosus, Potorous, 58
setosus, Rattus, 1484
setosus, Setifer, **77**
setosus, Sorex, 293
setosus, Sus, 642
setosus, Trinomys, **1591**
setosus, Vombatus, 44
setosus, Xerus, 790
setulosus, Mus, **1407**
setzeri, Gerbillurus, **1222**
setzeri, Lepus, 195
setzeri, Mus, **1407**
setzeri, Myomimus, **838**
setzeri, Sigmodon, 1177
seuanezi, Oryzomys,
1154
seurati, Acomys, **1199**
severtzovi, Allactaga, **876**
severtzovi, Alticola, 959
severtzovi, Capra, 701
severtzovi, Mus, 1398
severtzovi, Ovis, 708
severtzovi, Sicista, **889**
severzovi, Meles, 612
sevia, Abeomelomys,
1252, 1312
sevieri, Thomomys, 867
sewelanus, Pipistrellus,
479
sexcinctus, Euphractus, **97**
sexplicatus, Melomys,
1380

seychellensis, Coleura,
386
seychellensis, Meso-
plodon, 741
seychellensis, Pteropus,
344
sezekorni, Erophylla, **397**
shaanxiensis, Ochotona,
187
shameli, Rhinolophus,
362
shanicus, Callosciurus,
776
shanicus, Trachypithecus,
177
shanorum, Pipistrellus,
474
shanseius, Myodes, **1028**
shantaricus, Microtus,
1010
shantungensis, Crocidura,
249
sharica, Procavia, 88
sharicus, Lycaon, 581
sharmani, Petrogale, **68**
sharonis, Erinaceus, 213
sharpei, Funisciurus, 792
sharpei, Raphicerus, **688**
sharpie, Colobus, 168
shastensis, Castor, 842
shastensis, Sorex, 299
shattucki, Microtus, 1013
shawi, Dasymys, 1315
shawi, Meriones, **1238**
shawi, Paramelomys,
1434
shawi, Sciurus, 762
shawi, Thomomys, 869
shawiana, Felis, 535
shawii, Arctocephalus,
591
shawmayeri, Chiruromys,
1304
shawmayeri, Coccymys,
1307
shawmayeri, Dendrola-
gus, 59
shawmayeri, Hydromys,
1334
shawmayeri, Mammelo-
mys, 1358
sheila, Glauconycteris,
487
sheila, Mazama, 656
sheldoni, Lepus, 196
sheldoni, Marmota, 800
sheldoni, Neotoma, 1053
sheldoni, Odocoileus, 658
sheldoni, Ovis, 710
sheldoni, Thomomys, 870
sheldoni, Ursus, 589

sheltoni, Ochotona, 191
shenseius, Eospalax, 911
shenshiensis, Micromys,
1384
shepherdi, Tasmacetus,
743
sheppardi, Suncus, 259
sherif, Lepus, 197
shermani, Blarina, 269
shermani, Sciurus, 762
sherrini, Nyctophilus, 470
sherrini, Tatera, 1243
sherrini, Uromys, 1514
shevketi, Microtus, 999
shigarus, Rattus, 1482
shikarii, Gazella, 680
shikokensis, Sorex, 296
shinanensis, Urotrichus,
310
shindi, Heliosciurus, 794
shinto, Sorex, **296**
shiptoni, Galictis, 606
shiptoni, Microcavia,
1554
shirasi, Alces, 653
shirasi, Ursus, 589
shirazensis, Eptesicus,
457
shirensis, Alcelaphus, 675
shirensis, Gerbilliscus,
1220
shirensis, Heliosciurus,
793
shirensis, Rhynchocyon, 84
shirensis, Sylvicapra, 716
shirokuma, Rattus, 1478,
1490
shiroumanus, Sorex, 289
shistacea, Abrocoma,
1574
shitkovi, Meriones, 1237
shitkovi, Pygeretmus, **879**
shnitnikovi, Alticola, 959
shnitnikovi, Mustela, 614
sho, Mustela, 615
shoae, Ictonyx, 608
shoana, Gerbilliscus,
1221
shoana, Procavia, 88
shortridgei, Callosciurus,
776
shortridgei, Chaerephon,
435
shortridgei, Chlorotalpa,
78
shortridgei, Chrysochloris,
78
shortridgei, Crocidura,
240
shortridgei, Dendromus,
938

sumbae, Hipposideros, **377**
sumbae, Macaca, 162
sumbae, Rattus, 1490
sumbanus, Dobsonia, 320
sumbanus, Paradoxurus, 551
sumbavana, Rusa, 670
sumbawae, Hipposideros, 377
sumbawae, Hystrix, 1544
Sumeriomys, 989, 1012, 1014, 1015, 1016
sumichrasti, Bassariscus, **625**
sumichrasti, Nyctomys, **1187**
sumichrasti, Reithrodontomys, **1084**
sumneri, Peromyscus, 1076
sunameri, Neophocaena, 735
suncoides, Crunomys, **1313**
suncoides, Ruwenzorisorex, 256, **257**
Suncus, **257**
sunda, Bubalus, 694
sundaicus, Melogale, 613
Sundamys, 904, 1296, 1346, 1435, 1465, **1503**
Sundasciurus, 107, 755, 775, 784, **786**
sundavensis, Bandicota, 1293
sundevalli, Gazella, 681
sundevalli, Philantomba, 716
sundevallii, Phacochoerus, 639
sungae, Otomys, 1526
sungarae, Mus, 1397
sungorus, Phodopus, **1046**
sungorus, Stylodipus, **885**
sunidica, Ochotona, 191
Sunkus, 257
suntaricus, Microtus, 1010
superans, Eliomys, 835
superans, Oecomys, **1139**
superans, Ratufa, 757
superans, Rhinolophus, 351
superans, Vespertilio, 472, 498
superba, Glauconycteris, **487**

superciliaris, Mazama, 656
superciliaris, Philander, 16
superciliaris, Sylvilagus, 209
superciliatum, Artibeus, 419
supercilliosus, Lagenorhynchus, 730
superiorensis, Lynx, 542
supernus, Thomomys, 870
superstes, Damaliscus, **677**
superstes, Nesophontes, **222**
superus, Microtus, 997
suraktta, Suricata, 571
suramensis, Microtus, 995
Suranomys, 969, 989
surberi, Neotoma, 1058
surculus, Cratogeomys, 860
surda, Tupaia, 107
surdaster, Grammomys, 1324
surdaster, Pedetes, **1535**
surdaster, Sylvilagus, 208
surdaster, Viverra, 559
surdescens, Cervus, 664
Surdisorex, 224, **267**
surdus, Akodon, **1100**
surdus, Graphiurus, **829**
surdus, Rattus, 1469
surdus, Sundasciurus, 788
Suricata, **571**
Suricatidae, 562
Suricatinae, 562
suricatta, Suricata, **571**
surifer, Maxomys, **1371**
surinamensis, Ateles, 151
surinamensis, Monodelphis, 14
surinamensis, Myotis, 505
surinamensis, Sorex, 283
surkha, Mus, 1406
surmulottus, Rattus, 1478
Surricata, 571
surrufus, Onychomys, 1061
surrutilus, Exilisciurus, 781
sururae, Crocidura, 244
Sus, 637, **639**, 643, 644, 1555
suschkini, Allactaga, 877
suschkini, Meriones, 1237
suschkini, Talpa, 308

sushkini, Ochotona, 185, 186
susiana, Crocidura, **251**
suslicus, Spermophilus, **811**
Susu, 738
Susuidae, 738
Sutra, 602
suturosa, Addax, 717
svatoshi, Ochotona, 189
svineval, Globicephala, 728
sviridenkoi, Cricetulus, 1043
swainsonii, Antechinus, **30**
swakopensis, Gerbillurus, 1222
swalius, Galerella, 566
swalius, Gerbillurus, 1222
swalius, Steatomys, 945
swarthi, Dipodomys, 846
swarthi, Nesoryzomys, **1134**
swaynei, Alcelaphus, 674
swaynei, Madoqua, 684
swaythlingi, Gerbilliscus, 1221
swellendamensis, Amblysomus, 80
swerevi, Microtus, 1007
swettenhami, Capricornis, 704
swinderianus, Thryonomys, **1545**
swinhoei, Aonyx, 602
swinhoei, Capricornis, **704**
swinhoei, Cervus, 664
swinhoei, Hipposideros, 367
swinhoei, Lepus, 204
swinhoei, Meriones, 1235
swinhoei, Scotophilus, 466
swinhoei, Suncus, 260
swinhoei, Tamiops, **789**
swinhoii, Rusa, 671
swinnyi, Galerella, 566
swinnyi, Myosorex, 265
swinnyi, Rhinolophus, **364**
swirae, Taphozous, 384
swynnertoni, Paraxerus, 796
swynnertoni, Petrodromus, 84
swynnertoni, Rhynchocyon, 84
Syarchus, 605

sybilla, Mus, 1397
sybilla, Petaurista, 771
Syconycteris, **349**
sydneiensis, Tachyglossus, 1
syenitica, Ovis, 709
sylhetanus, Bos, 690
Sylvaemus, 1259
Sylvanus, 161
sylvanus, Akodon, **1100**
sylvanus, Bos, 690
sylvanus, Cervus, 664
sylvanus, Ctenomys, 1564, **1569**
sylvanus, Macaca, 161, **164**
sylvatica, Martes, 610
sylvatica, Rupicapra, 712
sylvaticus, Apodemus, **1275**
sylvaticus, Lepus, 204
sylvaticus, Mesechinus, 216
sylvaticus, Mus, 1279
sylvaticus, Odocoileus, 658
sylvaticus, Oryzomys, 1155
sylvaticus, Sylvilagus, 207, 209
sylvaticus, Tragelaphus, 697, 698
sylvester, Callosciurus, 779
sylvestris, Bison, 689
sylvestris, Bos, 692
sylvestris, Dendrohyrax, 88
sylvestris, Equus, 631
sylvestris, Glyphonycteris, **404**
sylvestris, Mandrillus, 165
sylvestris, Martes, 610
sylvestris, Ozotoceros, 659
sylvestris, Pogonomys, 1306, **1443**
sylvestris, Rangifer, 660
sylvestris, Rattus, 1485
sylvestris, Syncerus, 696
sylvestris, Tragelaphus, 699
sylvia, Crocidura, 240
sylvia, Molossops, 439
Sylvicapra, 648, 712, **716**
Sylvicaprina, 712
Sylvicaprinae, 712
Sylvicola, 989
sylvicola, Mandrillus, 165
sylvicola, Myotis, 508

vulturnus, Neotomys,
1134
vulturnus, Vespadelus,
497
vumbae, Heliosciurus,
793
vumbae, Pelomys, 1437
vumbaensis, Gramm-
omys, 1326
vumbensis, Grammomys,
1326
vuquangensis, Muntiacus,
668

wachei, Cervus, 663
wachei, Gulo, 607
waddelli, Vulpes, 586
wadjakensis, Homo, 182
wagati, Prionailurus, 543
wagleri, Papio, 166
wagneri, Acinonyx, 533
wagneri, Catagonus, **643**
wagneri, Chionomys, 970
wagneri, Dipodomys, 844
wagneri, Manis, 530
wagneri, Mus, 1398
wagneri, Reithrodon-
tomys, 1083
wagneri, Sciurus, 760
wagneri, Sorex, 290
Wagneria, 624
wahema, Microtus, 1013
wahlbergi, Epomophorus,
323
wahlbergi, Equus, 630
wahwahensis, Thom-
omys, 867
waigeuensis, Uromys,
1514
waitei, Leggadina, 1340
walambae, Aethomys,
1256
waldemarii, Suncus, 260
waldronae, Piliocolobus,
169
walgi, Canis, 577
wali, Capra, 703
walie, Capra, **703**
walkeri, Nyctophilus, **470**
walkeri, Sylvicapra, 716
Wallabia, 63, **70**
wallacei, Styloctenium,
349
wallacii, Myoictis, **26**
wallawalla, Lepus, 196
walleri, Litocranius, **682**
walli, Eptesicus, 456
wallichii, Cervus, 663
wallichii, Pongo, 184
wallowa, Thomomys, 869
walteri, Barbastella, 480

waltoni, Phaiomys, 1035
wamae, Mus, 1408
wambutti, Mus, 1390
wapiti, Cervus, 663
warburtoni, Ursus, 589
wardi, Apodemus, 1270
wardi, Bandicota, 1293
wardi, Blarinella, **269**
wardi, Capra, 703
wardi, Cervus, 663
wardi, Eothenomys, **982**
wardi, Equus, 630
wardi, Giraffa, 672
wardi, Helarctos, 587
wardi, Lepus, 195
wardi, Lynx, 541
wardi, Ochotona, 192
wardi, Orycteropus, 86
wardi, Ovibos, 707
wardi, Plecotus, 483
wardi, Redunca, 722
wardi, Rhizomys, 914
wardi, Sorex, 285
waringensis, Callosciurus,
779
waringensis, Sundamys,
1504
warreni, Marmota, 801
warreni, Neotoma, 1057
warreni, Petrodromus, 84
warreni, Proechimys,
1586
warreni, Suncus, 261
warreni, Sylvilagus, 208
warringtoni, Lasiopod-
omys, 984
warryato, Hemitragus,
705
warthae, Capreolus, 654
warwickii, Leopardus,
538
wasatchensis, Ochotona,
191
wasatchensis, Tamiasciu-
rus, 765
wasatchensis, Thom-
omys, 869
washake, Ursus, 589
washburni, Myodes, 1028
washingtoni, Lepus, 195
washingtoni, Mustela,
615
washingtoni, Sorex, 289
washingtoni, Spermophi-
lus, **813**
washoensis, Spermophi-
lus, 809
wasjuganensis, Myodes,
1024
wassifi, Dipodillus, 1214
wassifi, Paraechinus, 217

watasei, Crocidura, **254**
watasei, Myotis, 507
watasei, Neomys, 279
watasei, Petaurista, 771
waterbergensis, Ger-
billiscus, 1220
waterbergensis, Micae-
lamys, 1382
waterbergensis, Procavia,
88
waterbergensis, Pronola-
gus, 207
waterhousii, Heliosciurus,
794
waterhousii, Macrotus,
407
waterhousii, Marmosa, 9
watersi, Gerbillus, **1233**
watkinsi, Scotophilus,
466
watsoni, Artibeus, **420**
watsoni, Callosciurus,
778
watsoni, Golunda, 1322
watsoni, Myodes, 1028
watsoni, Tylomys, **1189**
wattsi, Maxomys, **1373**
wattsi, Pipistrellus, **479**
wavicus, Myoictis, 26
wavrini, Pseudoryzomys,
1164
wavula, Puma, 545
webbi, Eliurus, **950**
webensis, Heterohyrax, 88
weberi, Crocidura, 226
weberi, Myotis, 507
weberi, Prosciurillus, **785**
weberi, Sus, 641
websteri, Chaerephon,
434
weddelli, Saguinus, 134
weddellii, Leptonychotes,
597
weemsi, Ovis, 710
weidholzi, Chlorocebus,
159
weidholzi, Nanger, 684
weigoldi, Sicista, 888
weileri, Hylomyscus,
1335
weinheimensis, Arvicola,
964
wellsi, Berylmys, 1297
wellsi, Callosciurus, 776
wellsi, Galea, 1554
wellsi, Viverricula, 559
welmani, Gerbilliscus,
1219
welsbyi, Wallabia, 70
welsianus, Echymipera,
42

welwitschii, Myotis, **518**
welwitschii, Procavia, 89
wembaerensis, Alcela-
phus, 674
wembaerensis, Eudorcas,
679
wembarensis, Syncerus,
696
weragami, Mus, 1390
werckleae, Anoura, 398
werneri, Chlorocebus, 159
werneri, Papio, 166
wertheri, Orycteropus, 86
westermanni, Conepatus,
622
westermannii, Conepatus,
621
westrae, Microtus, 992
westralis, Pipistrellus,
479
wetarensis, Cynopterus,
317
wetarensis, Myotis, 500
wetmorei, Eptesicus, 454
wetmorei, Megaerops, **326**
wetmorei, Pudu, 659
wettsteini, Microtus, 990
wettsteini, Sorex, 283
wetzeli, Rhipidomys,
1171
weylandi, Mallomys,
1356
weylandi, Paramelomys,
1433
weynsi, Cephalophus,
715
weynsi, Chlorocebus, 159
weynsi, Colobus, 168
weynsi, Lophocebus, 160
whartoni, Phascolosorex,
27
wheeleri, Aselliscus, 366
whitakeri, Crocidura, **254**
whitakeri, Lepus, 197
whitchurchi, Jaculus, 884
white, Bettongia, 57
whiteheadi, Chrotomys,
1305, **1306**
whiteheadi, Exilisciurus,
781
whiteheadi, Harpyionyc-
teris, **324**
whiteheadi, Kerivoula,
528
whiteheadi, Maxomys,
1471
whiteheadi, Maxomys,
1373
whiteheadi, Mustela, 614
whitei, Acomys, 1196
whitei, Arctictis, 549

COMMON NAMES